2017 IEEE 44th Photovoltaic Specialist Conference (PVSC 2017)

Washington, DC, USA
25-30 June 2017

Pages 1421-2127

IEEE Catalog Number: CFP17PSC-POD
ISBN: 978-1-5090-5606-4

**Copyright © 2017 by the Institute of Electrical and Electronics Engineers, Inc.
All Rights Reserved**

Copyright and Reprint Permissions: Abstracting is permitted with credit to the source. Libraries are permitted to photocopy beyond the limit of U.S. copyright law for private use of patrons those articles in this volume that carry a code at the bottom of the first page, provided the per-copy fee indicated in the code is paid through Copyright Clearance Center, 222 Rosewood Drive, Danvers, MA 01923.

For other copying, reprint or republication permission, write to IEEE Copyrights Manager, IEEE Service Center, 445 Hoes Lane, Piscataway, NJ 08854. All rights reserved.

*** This is a print representation of what appears in the IEEE Digital Library. Some format issues inherent in the e-media version may also appear in this print version.

IEEE Catalog Number:	CFP17PSC-POD
ISBN (Print-On-Demand):	978-1-5090-5606-4
ISBN (Online):	978-1-5090-5605-7
ISSN:	0160-8371

Additional Copies of This Publication Are Available From:

Curran Associates, Inc
57 Morehouse Lane
Red Hook, NY 12571 USA

Phone:	(845) 758-0400
Fax:	(845) 758-2633
E-mail:	curran@proceedings.com
Web:	www.proceedings.com

2017 IEEE 44th Photovoltaic Specialist Conference (PVSC 2017)

Washington, DC, USA
25-30 June 2017

Pages 1421-2127

IEEE Catalog Number: CFP17PSC-POD
ISBN: 978-1-5090-5606-4

TABLE OF CONTENTS

OPEN CIRCUIT VOLTAGE CALCULATION USING TEMPERATURE AND IRRADIANCE 1
Andrew Melvin

EFFECT OF CL-DOPING IN ZNTEO ON PHOTOLUMINESCENCE AND PHOTOVOLTAIC PROPERTIES OF ZNTEO-BASED INTERMEDIATE BAND SOLAR CELLS 3
T. Tanaka ; S. Tsutsumi ; Y. Okano ; K. Matsuo ; K. Saito ; Q. Guo ; M. Nishio ; T. Tayagaki ; K. M. Yu ; W. Walukiewicz

TOWARD LEAD HALIDE PEROVSKITE-BASED INTERMEDIATE BAND ABSORBERS 6
Matthew D. Sampson ; Ji-Sang Park ; Richard D. Schaller ; Maria. Y. Chan ; Alex B. F. Martinson

TYPE-II QUANTUM DOTS FOR APPLICATION TO PHOTON RATCHET INTERMEDIATE BAND SOLAR CELLS 10
Ryo Tamaki ; Yasushi Shoji ; Yoshitaka Okada

AN INVESTIGATION OF THE ROLE OF RECOMBINATION PROCESSES IN THE OPERATION OF INAS/GAASL-XSBX QUANTUM DOT SOLAR CELLS 14
Y. Cheng ; A. J. Meleco ; A. J. Roeth ; V. R. Whiteside ; M. C. Debnath ; M. B. Santos ; T. D. Mishima ; S. Hatch ; H.Y. Liu ; I. R. Sellers

TEMPERATURE AND VOLTAGE-BIAS DEPENDENT TWO-STEP PHOTON ABSORPTION IN INAS/GAASL AL0.3GAAS QUANTUM DOT IN A WELL SOLAR CELLS 18
Yushuai Dai ; Brittany L. Smith ; Michael A. Slocum ; Zachary S. Bittner ; Hyun Kum ; Julia D'Rozario ; Seth M. Hubbard

INCREASING CURRENT GENERATION BY PHOTON UP-CONVERSION IN A SINGLE-JUNCTION SOLAR CELL WITH A HETERO-INTERFACE 23
Shigeo Asahi ; Kazuki Kusaki ; Toshiyuki Kaizu ; Takashi Kita

RTP-ASSISTED EX-SITU ANALYSIS OF (AG,CU)(IN,GA)SE2FORMATION USING SELENIZATION 26
Sina Soltanmohammad ; William N. Shafarman

ROLE OF EV+0.98 EV TRAP IN LIGHT SOAKING-INDUCED SHORT CIRCUIT CURRENT INSTABILITY IN CIGS SOLAR CELLS 30
P. K. Paul ; T. Jarmar ; L. Stolt ; A. Rockett ; A. R. Arehart

STUDY OF DEFECT PROPERTIES IN CUGASE2THIN-FILM SOLAR-CELLS USING ADMITTANCE SPECTROSCOPY 33
Muhammad Monirul Islam ; Shogo Ishizuka ; Hajime Shibata ; Shigeru Niki ; Katsuhiro Akimoto ; Takeaki Sakurai

TRANSMISSIVE SPECTRUM-SPLITTING CONCENTRATOR PHOTOVOLTAIC CELLS AND MODULES 37
Yaping Ji ; Qi Xu ; Brian Riggs ; John Robertson ; Kazi Islam ; Vince Romanin ; Dimitri D. Krut ; Jim H. Ermer ; Matthew D. Escarra

ALGAINP/GAAS TANDEM SOLAR CELLS FOR POWER CONVERSION AT 400 C AND 1000X CONCENTRATION 42
Myles A. Steiner ; Emmett E. Perl ; John Simon ; Daniel J. Friedman ; Nikhil Jain ; Paul Sharps ; Claiborne Mcpheeters ; Minjoo L. Lee

GALNASP SOLAR CELLS GROWN BY HYDRIDE VAPOR PHASE EPITAXY FOR ONE-SUN & LOW-CONCENTRATION III-V/SI PHOTOVOLTAICS 46
Nikhil Jain ; John Simon ; Kevin L. Schulte ; David R. Diercks ; Corinne E. Packard ; David Young ; Aaron J. Ptak

PHOTO-ELECTROCHEMICAL HYDROGEN GENERATION FROM INVERTED METAMORPHIC MULTIJUNCTION III-VS 47
Todd G. Deutsch ; James L. Young ; Myles A. Steiner ; Henning Döscher ; Ryan M. France ; John A. Turner

ADVANCED SILICON THIN FILMS FOR HIGH-EFFICIENCY SILICON HETEROJUNCTION-BASED SOLAR CELLS 50
A. Descoeudres ; C. Allebe ; N. Badel ; L. Barraud ; J. Champliaud ; G. Christmann ; L. Curvat ; F. Debrot ; A. Faes ; J. Geissbühler ; J. Horzel ; A. Lachowicz ; J. Levrat ; S. Martin De Nicolas ; S. Nicolay ; B. Paviet-Salomon ; L.-L. Senaud ; A. Tomasi ; C. Ballif ; M. Despeisse

MOOXAND WOXBASED HOLE-SELECTIVE CONTACTS FOR WAFER-BASED SI SOLAR CELLS 55
Stephanie Essig ; Julie Dréon ; Jérémie Werner ; Philipp Löper ; Stefaan De Wolf ; Mathieu Boccard ; Christophe Ballif

METAL NANOPARTICLE HOLE CONTACTS FOR SILICON SOLAR CELLS 59
James Bullock ; Zhaoran Xu ; Mark Hettick ; Yimao Wan ; Ali Javey

NEAR-FIELD TRANSPORT IMAGING APPLICATION OF PHOTOVOLTAIC MATERIALS 62

Chuanxiao Xiao ; Chun-Sheng Jiang ; John Moseley ; John Simon ; Kevin Schulte ; Aaron J. Ptak ; Steve Johnston ; Brian Gorman ; Mowafak Al-Jassim ; Nancy M. Haegel ; Helio Moutinho

APPLICATIONS OF DMD-BASED INHOMOGENEOUS ILLUMINATION PHOTOLUMINESCENCE IMAGING FOR SILICON WAFERS AND SOLAR CELLS 66

Yan Zhu ; Mattias Klaus Juhl ; Ziv Hameiri ; Thorsten Trupke

NUMERICAL MODEL TO EXTRACT MATERIALS PROPERTIES MAP FROM SPECTRALLY RESOLVED LUMINESCENCE IMAGES 70

Nicolas Paul ; Vincent Le Guen ; Daniel Ory ; Laurent Lombez

NON-DESTRUCTIVE CONTACT RESISTIVITY MEASUREMENTS ON SOLAR CELLS USING THE CIRCULAR TRANSMISSION LINE METHOD 74

Geoffrey Gregory ; Andrew M. Gabor ; Andrew Anselmo ; Rob Janoch ; Zhihao Yang ; Kristopher O. Davis

RADIATION RESISTANCE OF LOW COST HIGH EFFICIENCY TRIPLE JUNCTION SOLAR CELLS 76

Roberta Campesato ; Erminio Greco ; Giuseppe Gabetta ; Mariacristina Casale ; Gabriele Gori ; M. Sankaran ; Suresh E. Puthanveettil ; B. R. Uma ; M. Ravindra ; Sheeja Krishnan

AMORPHOUS SILICON CARBIDE REAR-SIDE PASSIVATION AND REFLECTOR LAYER STACKS FOR MULTI-JUNCTION SPACE SOLAR CELLS BASED ON GERMANIUM SUBSTRATES 83

Stefan Janz ; Charlotte Weiss ; Christian Mohr ; Rufi Kurstjens ; Bruno Boizot ; Bianca Fuhrmann ; Victor Khorenko

HOT CARRIER TRANSPORTATION DYNAMICS IN INAS/GAAS QUANTUM DOT SOLAR CELL 85

Tomah Sogabe ; Kohdai Nii ; Katsuyoshi Sakamoto ; Koichi Yarnaquchi ; Yoshitaka Okada

INTEGRATION OF CRACK-TOLERANT COMPOSITE GRIDLINES ON TRIPLE JUNCTION PHOTOVOLTAIC CELLS 88

Omar K. Abudayyeh ; Geoffrey K. Bradshaw ; Steven Whipple ; David M. Wilt ; Sang M. Han

SUBCELL LIGHT CURRENT- VOLTAGE CHARACTERIZATION OF IRRADIATED MULTIJUNCTION SOLAR CELL 93

Don Walker ; John Nocerino ; Yao Yue ; Colin J. Mann ; Simon H. Liu

ANALYTICAL METHOD FOR PREDICTING SPACECRAFT POWER GENERATION ON PARTIALLY SHADED SOLAR PANELS 96

Gordon Wu ; Bao Hoang

EVALUATING THE EMISSIVITY OF PSEUDOMORPHIC GLASS (PMG) 102

Ryan D. Beauchemin ; David M. Wilt ; Paul E. Hausgen

CHARACTERIZING THE IMPACT OF SOLAR SPECTRAL IRRADIANCE ON PV MODULE OUTPUT 107

M. Schweiger ; W. Herrmann

USE OF MEASURED AEROSOL OPTICAL DEPTH AND PRECIPITABLE WATER TO MODEL CLEAR SKY IRRADIANCE 110

Mark M. Mikofski ; Clifford W. Hansen ; William F. Holmgren ; Gregory M. Kimbal

RECENT ADVANCEMENTS IN THE NUMERICAL SIMULATION OF SURFACE IRRADIANCE FOR SOLAR ENERGY APPLICATIONS 116

Yu Xie ; Manajit Sengupta ; Chris Deline

OPTIMAL IRRADIANCE SENSOR PLACEMENT FOR PHOTOVOLTAIC SYSTEMS USING MUTUAL INFORMATION BASED GREEDY ALGORITHM IN GAUSSIAN PROCESS 120

Lian Lian Jiang ; R. Srivatsan ; Douglas L. Maskell

EVALUATING DIFFERENT UPSCALING APPROACHES TO DERIVE THE ACTUAL POWER OF DISTRIBUTED PV SYSTEMS 126

Sven Killinger ; Björn Müller ; Bernhard Wille-Haussmann ; Russell Mckenna

ADVANCES IN LONG-TERM SOLAR ENERGY PREDICTION AND PROJECT RISK ASSESSMENT METHODOLOGY 132

Alemu Tadesse ; Adam Kankiewicz ; Alex Kubiniec ; Richard Perez ; John Dise ; Thomas Hoff

DECOUPLING THIN FILM CDTE GROWTH FROM PACKAGING: TOWARD RECORD SPECIFIC POWER IN LOW COST POLYCRYSTALLINE PV 138

D. Clayton-Warwick ; M.D. Kempe ; M. S. Dabney ; T. M. Barnes ; C. A. Wolden ; M. O. Reese

JUNCTION ACTIVATION OF CDTE/CDS SOLAR CELL USING MGCL2 142

G. Angeles-Ordóñez ; E. Regalado-Pérez ; M.G. Reyes-Banda ; N. R. Mathews ; X. Mathew

VARIATION OF CU CONTENT OF SPRAYED CU(IN, GA)(S,SE)2SOLAR CELLS BASED ON A THIOL-AMINE SOLVENT MIXTURE 146

Panagiota Arnou ; Sona Ulicná ; Alexander Eeles ; Mustafa Togay ; Lewis D. Wright ; Andrei V. Malkov ; John M. Walls ; Jake W. Bowers

CUINSE2 ABSORBER LAYER GROWN UNDER COPPER EXCESS WITH A COPPER POOR SURFACE FORMED BY A KF POST DEPOSITION TREATMENT 151
Finn Babbe ; Hossam Elanzeery ; Michele Melchiorre ; Susanne Siebentritt

CU2ZNSNSE4SOLAR CELLS ONTO POLYIMIDE SUBSTRATES FABRICATED AT LOW TEMPERATURE 155
Ignacio Becerril-Romero ; Simón Lopez-Marino ; Moisés Espíndola-Rodríguez ; Laura Acebo ; Markus Neuschitzer ; Yudania Sánchez ; Edgardo Saucedo ; Paul Pistor

AN OPTIMIZED PHOTOLITHOGRAPHY RECIPE FOR CU(IN1-X,GAX)(SY,SE1-Y)2(CIGSSE) SOLAR CELLS 160
Xia Hao ; Shenghao Wang ; Katsuhiro Akimoto ; Takuya Kato ; Hiroki Sugimoto ; Takeaki Sakurai

EFFECTS OF CDCL2PASSIVATION ON THIN CDTE ABSORBERS FABRICATED BY CLOSE-SPACE SUBLIMATION 164
Anna Wojtowicz ; Alexandra M. Huss ; Jennifer A. Drayton ; James R. Sites

CDS1-XSEXWINDOW LAYER FOR CDTE PREPARED BY THE EXCHANGE OF S WITH SE IN CDS FILMS 170
Geethika K. Liyanage ; Adam B. Phillips ; Zhaoning Song ; Suneth C. Watthage ; Ramez H. Ahanzhamejhad ; Michael J. Heben

EFFECT OF ILLUMINATION ON THERMAL CDCL2TREATMENT OF CDTE 175
Sudhajit Misra ; Carina E. Hahn ; Vasilios Palekis ; Christos Ferekides ; Michael A. Scarpulla

CHALLENGES IN THE INDUSTRIAL PRODUCTION OF CZTS MONOGRAIN SOLAR CELLS 178
Gerhard Peharz ; Valentin Satzinger ; Sandra Pötz ; Gemot Oreski ; Theodoros Dimopoulos ; Stefan Edinger ; Wolfeanz Hackl ; Hannes Starkl ; Parichehr Esfandiari ; Peter Krabb ; Stefan Gahr ; Lukas Plessing ; Dieter Meissner

UNDERSTANDING INSTABILITIES AND DEGRADATION DUE TO MOISTURE INGRESS IN CU(IN, GA)SE2SOLAR CELLS 182
Grace Rajan ; Shankar Karki ; Isaac Butt ; Krishna Aryal ; Tyler J. Grassman ; Angus Rockett ; Sylvain Marsillac

CONTROL OF MOSE2 FORMATION IN HYDRAZINE-FREE SOLUTION-PROCESSED CIS/CIGS THIN FILM SOLAR CELLS 186
Sona Ulicná ; Panagiota Arnou ; Alexander Eeles ; Mustafa Togay ; Lewis D. Wright ; Ali Abbas ; Andrei V. Malkov ; John M. Walls ; Jake W. Bowers

GROWTH AND PROPERTIES OF EPITAXIAL CU(IN, GA)SE2THIN FILMS DEPOSITED BY THE THREE-STAGE PROCESS FOR SOLAR CELLS 192
Takeru Yamagami ; Yuta Ando ; Ishwor Khatri ; Mutsumi Sugiyama ; Tokio Nakada

IMPROVEMENT OF CIS SOLAR CELLS WITH KF POSTDEPOSITION FOLLOWING A SIMPLE TWO-STEP SELENIZATION PROCESS 195
Yang Zhang ; Robert E. Bartolo ; Sang Jik Kwon ; Mario Dagenais

THE TWINS STRUCTURE, ELECTRICAL PROPERTIES AND CELL PERFORMANCE OF MAGNETRON SPUTTERING DEPOSITED CHLORINE DOPED CDTE 198
Ziyao Zhu ; Fu-Kuo Chiang ; Zhongming Du ; Yufeng Zhang ; Xiangxin Liu

INVESTIGATION AND MITIGATION OF SHUNTS FOR HIGHER EFFICIENCY EPITAXIAL GASB/GASB AND GASB/GAAS SOLAR CELLS 202
George T. Nelson ; Bor-Chau Juang ; Steve Johnston ; Michael A. Slocum ; Zachary S. Bittner ; Ramesh B. Lagumavarapu ; Diana Huffaker ; Seth M. Hubbard

DEVELOPMENT OF GASB SOLAR CELLS ON GAAS BY MOVPE VIA INTERFACE MISFIT TECHNIQUE 206
Michael A. Slocum ; Alessandro Giussani ; Emily Kessler ; Phil Ahrenkiel ; George T. Nelson ; Seth M. Hubbard

FABRICATION OF INGAASP SOLAR CELLS FOR CONCENTRATOR APPLICATIONS 210
Mitchell F. Bennett ; Matthew P. Lumb ; Kenneth J. Schmieder ; Brent Fisher ; Eric A. Armour ; Robert J. Walters

DETAILED CHARACTERIZATION FOR TCAD SIMULATIONS OF GAAS0.76P0.24/SI1-YGEY/SI SINGLE JUNCTION SOLAR CELLS 213
Sabina Abdul Hadi ; Timothy Milakovich ; Eugene A. Fitzgerald ; Ammar Nayfeh

COMPARATIVE STUDY OF >2 EV LATTICE-MATCHED AND METAMORPHIC (AL)GAINP MATERIALS AND SOLAR CELLS GROWN BY MOCVD 215
Daniel J. Chmielewski ; Christine Jackson ; Jacob Boyer ; Daniel Lepkowski ; John A. Carlin ; Aaron R. Arehart ; Tyler J. Grassman ; Steven A. Ringel

PERFORMANCE OF GASB PHOTOVOLTAICS WITH GRAPHENE COATING 219
Benjamin P. Conlon ; Daniel J. Herrera ; Shaimaa A. Abdallah ; Jonathan O. Okafor ; Luke F. Lester

HIGH EFFICIENCY SINGLE-JUNCTION INGAP PHOTOVOLTAIC DEVICES UNDER LOW INTENSITY LIGHT ILLUMINATION 222
Yushuai Dai ; Hyun Kum ; Michael A. Slocum ; George T. Nelson ; Seth M. Hubbard

RADIATION RESISTANT OF UPRIGHT METAMORPHIC GAINP/GAINAS/GE TRIPLE JUNCTION SOLAR CELLS FOR SPACE USE 226

Liang Fang ; Abuduwayiti Aierken ; Zhen Pan ; Qiming Zhang ; Zhanhang Li ; Heini Maliya ; Wei Gao ; Hui Gao ; Ronghua Wan ; Bao Zhang ; He Wang ; Qi Guo

HIGH EFFICIENCY GLASS WAVEGUIDING SOLAR CONCENTRATOR 229

Chehao Hu ; Yusuf Dogan ; Matthew Morrison ; A. Nanda ; D. Ma ; R. Atkins ; C. K. Madsen

GAINASP/GAINAS TANDEM SOLAR CELL WITH 32.6% ONE-SUN EFFICIENCY 232

Nikhil Jain ; Kevin L. Schulte ; John F. Geisz ; Ryan M. France ; Myles A. Steiner

EVALUATION OF TANDEM EFFICIENCIES: DILUTE NITRIDE P-I-N (BULK OR MQWS) IN CONJUNCTION WITH PRACTICAL SI SOLAR CELLS 236

Khim Kharel ; Alexandre Freundlich

GALLIUM PHOSPHIDE NANOSTRUCTURE ON SILICON BY SILICA NANOSPHERES LITHOGRAPHY AND METAL ASSISTED CHEMICAL ETCHING 240

Sangpyeong Kim ; Chaomin Zhang ; Som Dahal ; Stuart Bowden ; Christiana B. Honsberg

EFFICIENCY ENHANCEMENT OF INGAP/INGAAS/GE SOLAR CELLS WITH GRADUALLY DOPED P-N JUNCTION ACTIVE LAYERS 244

Youngjo Kim ; Sang Hyun Jung ; Chang Zoo Kim ; Kangho Kim ; Hyun-Beom Shin ; Kyung Ho Park ; Won-Kyu Park ; Jaejin Lee ; Ho Kwan Kang

ANALYSIS OF INGAP OXIDE GROWTH RATE AT HIGH TEMPERATURES AND AMBIENT CONDITIONS FOR TERRESTRIAL PHOTOVOLTAIC APPLICATIONS 247

Nicole A. Kotulak ; Matthew P. Lumb ; Raymond Hoheisel ; Erin Cleveland ; Mitchell Bennett ; Phillip P. Jenkins ; Robert J. Walters

GRAIN BOUNDARIES IN THIN-FILM POLYCRYSTALLINE GAAS SOLAR CELLS: A SIMULATION STUDY 251

Khushboo Kumari ; Sushobhan Avasthi

TIME-RESOLVED PL MEASUREMENTS IN THE GROWTH OF HIGH VOLTAGE (AL)GAINP/GAAS SOLAR CELLS 255

Xinyi Li ; Wei Zhang ; Hongbo Lu

LOW-RESISTANCE AND HIGHLY-TRANSPARENT GASB-BASED TUNNEL JUNCTIONS 259

Matthew P. Lumb ; Shawn Mack ; Maria Gonzalez ; Kenneth J. Schmieder ; Mitchell F. Bennett ; Chaffra A. Affouda ; James E. Moore ; Robert J. Walters

MODULATED PHOTOCURRENT MEASUREMENTS IN DOUBLE JUNCTION SOLAR CELLS 263

Nicolás Márquez Peraca ; Behrang H. Hamadani

EFFECT OF ATMOSPHERIC ABSORPTION BANDS ON THE OPTIMAL DESIGN OF MULTIJUNCTION SOLAR CELLS 268

William E. Mcmahon ; Daniel J. Friedman ; John F. Geisz

EFFECTS OF CONTACT CONFIGURATION AND PERIMETER RECOMBINATION ON OPTIMAL CELL SIZE FOR HIGH CONCENTRATION PHOTOVOLTAICS 272

James E. Moore ; Matthew P. Lumb ; Kenneth J. Schmieder ; Robert J. Walters ; Brent Fisher ; Matt Meitl ; Scott Burroughs

NUMERICAL SIMULATION OF DEFECTS IN III-V PV CELLS: THE EFFECT OF VOLTAGE BIAS AND DOPING CONCENTRATION 276

Vasiliki Paraskeva ; Constantinos Lazarou ; Andreas Livera ; Venizelos Venizelou ; Maria Hadjipanayi ; George E. Georghiou

IMPROVEMENT OF OPEN-CIRCUIT VOLTAGE IN METAMORPHIC GASB CELLS GROWN ON GAAS SUBSTRATES BY USING AN INTERFACIAL MISFIT ARRAY AND AN ALSB BLOCKING LAYER 281

E. J. Renteria ; S. J. Addamane ; D. M. Shima ; A. Mansoori ; A. L. Soudachanh ; G. Balakrishnan

ENERGY YIELD EVALUATION FOR FIELD OPERATION OF SOLAR CELLS IN SINGAPORE: GAAS/GAAS TANDEM VS. GAAS SINGLE-JUNCTION SOLAR CELLS 284

Maung Thway ; Zekun Ren ; Kevin Nay Yaung ; Haohui Liu ; Zhe Liu ; Samuel Raj ; Soo Jin Chua ; Armin G. Aberle ; Tonio Buonassisi ; Ian Marius Peters ; Fen Lin

SIMULATION OF THE PERFORMANCES OF MULTIJUNCTION SOLAR CELLS WITH IMPROVED VOLTAGE BY TRANSFER AND SCATTERING MATRIX METHODS 290

Gianluca Timò ; Lucio Andreani

OPTIMIZED DESIGN OF BACK-CONTACT THIN-FILM GAAS SOLAR CELLS 294

Jia-Ling Tsai ; Chung-Yu Hong ; Tien-Chien Zhan ; Yuh-Renn Wu ; Albert Lin ; Peichen Yu

DESIGN CONSIDERATIONS ON GAINNAS SOLAR CELLS WITH BACK SURFACE REFLECTORS 297

Antti Tukiainen ; Arto Aho ; Timo Aho ; Ville Polojärvi ; Mircea Guina

QUANTITATIVE ELECTROLUMINESCENCE ANALYSIS OF TRIPLE JUNCTION SOLAR CELLS TO DETERMINE SUBCELL VOLTAGE-TEMPERATURE COEFFICIENTS 301

Kevin Tyler ; Geoffrey K. Bradshaw ; Sam Wilt ; David M. Wilt ; Richard R. King

PROGRESS TOWARDS DOUBLE-JUNCTION INGAN SOLAR CELL 305

Ehsan Vadiee ; Evan A. Clinton ; Heather Mcfavilen ; Alec M. Fischer ; Yi Fang ; Joshua J. Williams ; Christiana B. Honsberg ; William A. Doolittle ; Stephen M. Goodnick

A PHYSICS-BASED SIMULATION TOOL FOR LEAKAGE CURRENTS IN C-SI PV MODULES 309

John M. Waddle ; Saroj Dahal ; Marco Nardone

BROADBAND TA2O5 MOTH-EYE ANTIREFLECTION COATINGS FOR TANDEM SOLAR CELLS ON SI 315

Bo Yuan ; Brian Thibeault ; David Payne ; James Mutitu ; Ivan Perez-Wurfl ; Kevin Dobson ; Brianna Conrad ; Allen Barnett ; Robert L. Opila

CARRIER TRANSPORT IN POLYCRYSTALLINE SILICON AT HIGH OPTICAL INJECTION: TRANSIENT PHOTOCONDUCTANCE VS. NUMERICAL MODELING 319

Uchechi Anyanwu ; Christian Harris ; Andrey Semichaevsky

IMPROVING SILICON SURFACE PASSIVATION WITH A SILICON OXIDE LAYER GROWN VIA OZONATED DEIONIZED WATER 322

Sara Bakhshi ; Ngwe Zin ; Kristopher O. Davis ; Marshall Wilson ; Ismail Kashkoush ; Winston V. Schoenfeld

DEPOSITION OF SIOC BY PLASMA-FREE ULTRA-LOW-TEMPERATURE ALD (ULT-ALD) AND ITS PASSIVATION ON P-TYPE SILICON 326

Meixi Chen ; Naoto Noda ; Raphael Rochat ; Abhishek Iyer ; James H. Hack ; Changhee Ko ; Christian Dussarrat ; Robert L. Opila

A METHOD FOR QUANTITATIVELY INVESTIGATING THE REAR-SIDE PASSIVATION PERFORMANCE OF PERC CELLS 329

Tsung-Cheng Chen ; Yung-Sheng Lin ; Chen-Hao Ku ; Ting-Wei Kuo ; Cheng-Shun Hu ; Ching-Chang Wen

FIELD-EFFECT PASSIVATION BY NEGATIVE CHARGE ON BORON EMITTER AND BORON-DOPED SURFACES BY A NOVEL LOW-COST PLASMA CHARGE INJECTION 333

Eunhwan Cho ; Young-Woo Ok ; James Hwang ; Aditi Jain ; Vijay D. Upadhyaya ; John Keith Tate ; Ajeet Rohatgi

INDUSTRY RELEVANT RIE TEXTURING FOR MC-SI DIAMOND WIRE OR DIRECT WAFER® PRODUCT: OPTIMIZED REFLECTIVITY, UNIFORMITY, AND THROUGHPUT 337

Jose Luis Cruz-Campa ; Ray Fraser ; Rob Steeman ; John Linton

SHORT-CIRCUIT CURRENT-DENSITY ENHANCEMENT OF SILICON SOLAR CELLS USING PLASMONICS ANTIREFLECTIVE COATING AND LUMINESCENT DOWNSHIFTING 343

Sheng-Kai Feng ; Wen-Jeng Ho ; Guan-Yi Li ; Jheng-Jie Liu ; Hao-Yu Yang ; Ta-Wei Chuang

EXTREMELY LOW REFLECTIVITY NANOPOROUS BLACK SILICON SURFACE BY COPPER CATALYZED ETCHING FOR EFFICIENT SOLAR CELLS 346

K A S M Ehteshamul Haque ; Wenqi Duan ; Fatima Toor

IMPACT OF FRONT SIDE PYRAMID SIZE ON THE LIGHT TRAPPING PERFORMANCE OF WAFER BASED SILICON SOLAR CELLS AND MODULES 352

Oliver Höhn ; Nico Tucher ; Benedikt Bläsi

A STUDY OF BLISTER CONTROL OF AL2O3 THIN FILM DEPOSITED BY PLASMA-ASSISTED ATOMIC LAYER DEPOSITION AFTER FIRING PROCESS 356

Min Gu Kang ; Jeong In Lee ; Hee-Eun Song ; Myeong Sangjeong ; Kyung Taekjeong ; Hyo Sikchang

PYPVCELL: AN OPEN-SOURCE SOLAR CELL MODELING LIBRARY IN PYTHON 359

Kan-Hua Lee ; Kenji Araki ; Omar Elleuch ; Nobuaki Kojima ; Masafumi Yamaguchi

IMPROVEMENT IN SURFACE PASSIVATION OF C-SI USING GRADIENT-LAYERED A-SI:H FILM FOR HIGH EFFICIENCY SILICON HETEROJUNCTION SOLAR CELLS 363

Soonil Lee ; Leo Mathew ; Rajesh Rao ; Jae Hyun Kim ; Sanjay K. Banerjee ; Edward T. Yu

PHOTOVOLTAIC PERFORMANCE ENHANCEMENT OF TEXTURED SILICON SOLAR CELLS USING LUMINESCENT DOWN-SHIFTING METHYLAMMONIUM LEAD TRIBROMIDE PEROVSKITE NANOPHOSPHORS 367

Guan-Yi Li ; Wen-Jeng Ho ; Sheng-Kai Feng ; Hao-Yu Yang ; Ta-Wei Chuang ; Bang-Jin You ; Zong-Xian Lin ; Zong-Liang Tseng ; Lung-Chien Chen

SINX THIN FILMS WITH APPROPRIATE ANTIREFLECTION AND SHIFT-CONVERSION PROPERTIES FOR SILICON SOLAR CELLS 370

E. Men-Pérez ; J. Salazar ; A. Dutt ; J. Santoyo-Salazar ; G. Santana

NUMERICAL SIMULATION OF CRYSTALLINE SILICON SOLAR CELLS WITH FULL AREA METAL OXIDE REAR CONTACTS 373

James E. Moore ; Woojun Yoon ; Phillip P. Jenkins ; Robert J. Walters

INTERDIGITATED BACK CONTACT SILICON SOLAR CELL WITH PEROVSKITE LAYER FOR FRONT SURFACE PASSIVATION AND ULTRAVIOLET RADIATION STABILITY 377

Rahul Pandey ; Shivam Gupta ; Trijul Khatri ; Rishu Chaujar

POTENTIAL OF A-SI:H/C-SI HETEROJUNCTION SOLAR CELLS WITH VERY THIN WAFERS 381

Hitoshi Sai ; Hiroshi Umishio ; Takuya Matsui ; Shota Nunomura ; Tomoyuki Kawatsu ; Hidetaka Takato ; Koji Matsubara

MANIPULATING FIXED CHARGES IN ZRO2 BY DOPING FOR PASSIVATION AND ANTIREFLECTION ON WAFER-SI SOLAR CELLS 385

Woo Jung Shin ; Laidong Wang ; Wen-Hsi Huang ; Meng Tao

LOW TEMPERATURE ANTIREFLECTION COATING FOR SILICON SOLAR CELLS 389

O. S. Shinde ; Ej Schneller ; N. Dhere ; S. V. Ghaisas

RELATIONSHIP BETWEEN POWER LOSS AND VOLTAGE APPLIED TO SOLAR CELLS IN PID-AFFECTED SOLAR MODULES 392

Fumei Wang ; Baosong Duan ; Wenshuang He ; He Wang ; Hong Yang ; Chengfeng Su ; Bojie Su ; Xue Zhang ; Yunxue Cao ; Hui Zhao

A NEW LOW-COST AND LOW-TEMPERATURE CHEMICAL PASSIVATION PROCESS FOR LARGE AREA INDUSTRIAL SINGLE CRYSTALLINE SILICON WAFERS 396

Tarun S. Yadav ; K. Sandeep ; Ashok K. Sharma ; B. Spandana ; K.L. Narasimhan ; B.M. Arora ; Anil Kottantharayil ; Prabir K. Basu

EVALUATION OF ALD PASSIVATION LAYERS FOR INDUSTRIAL PERC PROCESS 399

Chang Youn Yoo ; Keunkee Hong ; Jisun Kim ; Eunjoo Lee ; Dong Seop Kim

QUANTITATIVE ANALYSIS OF ELECTROLUMINESCENCE AND INFRARED THERMAL IMAGES FOR AGED MONOCRYSTALLINE SILICON PHOTOVOLTAIC MODULES 402

Irene Berardone ; Juan Lopez Garcia ; Marco Paggi

GAP PASSIVATION STRUCTURE FOR SCALABLE N-TYPE INTERDIGITATED ALL BACK CONTACT SILICON HETERO-JUNCTION SOLAR CELL 408

Lei Zhang ; Ujjwal Das ; Steven Hegedus

PROPOSAL OF THE BANDGAP DESIGN USING THE SUN HEIGHT OF THE CULMINATION ON THE WINTER SOLSTICE 412

Kenji Araki ; Kan-Hua Lee ; Masafumi Yamaguchi

PHOTOEXCITED CARRIERS, PHONONS, AND THEIR SCATTERING MEASURED IN SEMICONDUCTOR JUNCTIONS BY TRANSIENT EXTREME ULTRAVIOLET SPECTROSCOPY 417

Scott K. Cushing ; Brett M. Marsh ; Mihai E. Vaida ; Lucas M. Carneiro ; Ilana J. Porter ; Angela Lee ; Stephen R. Leone

ON THE USE OF VOLTAGE MEASUREMENTS FOR DETERMINING CARRIER LIFETIME AT HIGH ILLUMINATION INTENSITY 420

Robert Dumbrell ; Mattias K. Juhl ; Thorsten Trupke ; Ziv Hameiri

HIGH RESOLUTION 3D CHEMICAL CHARACTERISATION OF A CADMIUM TELLURIDE SOLAR CELL BY DYNAMIC SIMS 424

Thomas Fiducia ; Kexue Li ; Chris Grovenor ; Kurt Barth ; Walajabad Sampath ; Michael Walls

HARSH OUTDOOR EVALUATION SETUP AND FIRST POWER PRODUCTION RESULTS FOR SI MINI-MODULES COVERED BY EU3+-BASED DOWN CONVERTERS 429

Benjamín González-Díaz ; Carlos Montes ; Joaquín Sanchiz ; Luis Ocaña ; Carlos Quinto ; Cecilio Hernández-Rodríguez ; Mari Paz Friend ; Manuel Cendagorta-Galarza ; David Cañadillas ; Ricardo Guerrero-Lemus

STUDY OF MICRO-STRUCTURAL PROPERTIES OF ZNO AND TIO2THIN FILM GROWN BY SPRAY PYROLYSIS 433

G. Gordillo ; J.M. Correa ; A.A. Ramirez ; E. A. Ramírez

NONLINEAR RESPONSE OF SILICON SOLAR CELLS 437

Behrang H. Hamadani ; Andrew Shore ; Howard W. Yoon ; Mark Campanelli

EXTENDED LINEAR INTERPOLATION/EXTRAPOLATION PROCEDURE FOR ACCURATE AND VERSATILE TRANSLATION OF THE I-V CURVES OF PV CELLS AND MODULES 441

Y. Hishikawa ; H. Ohshima ; M. Higa ; K. Yamagoe ; T. Takenouchi ; T. Doi

SEVERITY TEST WITH UNEVEN LOAD DUE TO WIND ACTION ON PHOTOVOLTAIC MODULE 445

Shu-Tsung Hsu

STANDARDIZED DURABILITY TEST FOR ORGANIC PHOTOVOLTAIC AND DYE SENSITIZED SOLAR CELL 448

Shu-Tsung Hsu ; Yean-San Long ; Teng-Chun Wu

SPATIAL THICKNESS UNIFORMITY AND STRUCTURAL EVALUATION OF RF SPUTTERED ZNO THIN FILMS FOR SOLAR CELL 451

Babar Hussain ; Taj M. Khan

LOCAL MEASUREMENTS OF SURFACE CAPACITANCE BY ELECTROSTATIC FORCE MICROSCOPY ON CU(IN, GA)SE2MATERIALS 455

Tomoaki Ishii ; Takashi Minemoto ; Takuji Takahashi

A COMPARISON OF SI-BASED CAMERAS FOR IMAGING LUMINESCENCE FROM PHOTOVOLTAIC MATERIALS AND DEVICES 459

Steve Johnston

BLISTERING OF AL2O3/A-SINX:H STACKS: ANALYSIS OF THE SUBMERGED PART OF THE ICEBERG BY COLORED PICOSECOND ACOUSTIC MICROSCOPY 464

Fabien Lebreton ; Arnaud Devos ; Etienne Drahi ; Patricia De Coux ; François Silva ; Sergej Filonovich ; Pere Roca I Cabarrocas

SELF-REFERENCE PROCEDURE TO REDUCE UNCERTAINTY IN MODULE CALIBRATION 467

D.H. Levi ; C.R. Osterwald ; S. Rummel ; L. Ottoson ; A. Anderberg

UNCERTAINTY EVALUATION OF PRIMARY REFERENCE PHOTOVOLTAIC CELL CALIBRATION UNDER OUTDOOR CONDITION IN TIBET 472

Haitao Liu ; Shiyu Sang ; Guomin Zhou ; Yonghui Zhai

REQUIREMENT OF ARTIFICIAL LIGHTING SIMULATOR FOR EVALUATION EMERGING PV PERFORMANCE RATING UNDER INDOOR ENVIRONMENT 476

Yean-San Long ; Shu-Tsung Hsu ; Teng-Chun Wu

NON-CONTACT VOLTAGE MEASUREMENT OF SOLAR CELL WITH ELECTROSTATIC VOLTMETER 480

Sakutaro Miyajima ; Kensuke Nishioka ; Yoshihiro Hishikawa

NREL'S CELL AND MODULE PERFORMANCE GROUP'S ASYMPTOTIC PMAX PROTOCOL FOR PEROVSKITE DEVICES 483

Tom Moriarty ; Dean Levi

OUTDOOR OPERATING TEMPERATURE MODELING OF PHOTOVOLTAIC MODULES INCLUDING TRANSIENT EFFECT 487

Soo-Young Oh ; Min-Soo Kim ; Won-Shup So ; Woo Kyoung Kim ; Jae Hak Jung ; Chinho Park ; Benazzouz Aboubakr ; Ikken Badr ; Naimi Zakaria ; Benlarabi Ahmed

PRIMARY REFERENCE CELL CALIBRATIONS WITH REDUCED MEASUREMENT UNCERTAINTY 490

C.R. Osterwald ; L. Ottoson ; R. Williams ; C. Mack ; T. Moriarty ; K.A. Emery ; D.H. Levi

IMPLEMENTATION OF NOVEL PIN CONNECTION AND TEST ROUTINE FOR IMPROVED ACCURACY IN I-V MEASUREMENTS 496

Samuel Raj ; Johnson Kai Chi Wong ; Mohan Krishan Bhan ; Evan Palmer ; Jian Wei Ho ; Sumukh Ramprasad ; Wang Junci ; Thomas Mueller ; Armin G. Aberle

A NEW METHOD TO QUANTIFY CONTACT RESISTANCE USING LOCALIZED-ILLUMINATION PHOTOLUMINESCENCE TECHNIQUE IN A SOLAR CELL 499

Amit Singh Rajput ; Samuel Raj ; Johnson Wong ; Armin G. Aberle

IMPROVEMENT OF THE PROPERTIES OF CZTS THIN FILMS PREPARED BY SPRAY PYROLYSIS USING DMSO IN ACETONE AS SOLVENT 503

E. A. Ramírez ; A. Ramírez ; G. Gordillo

ASSESSMENT OF CARRIER LIFETIMES AND SURFACE RECOMBINATION VELOCITY THROUGH SPECTRAL MEASUREMENTS 508

John Roller ; Behrang H. Hamadani

IMPACT OF SPACE RADIATION ENVIRONMENT ON CONCENTRATOR PHOTOVOLTAIC SYSTEMS 512

Pilar Espinet-Gonzalez ; Tatiana Vinogradova ; Michael D. Kelzenberg ; Alexander Messer ; Emily C. Warmann ; Chris Peterson ; Nina Vaidya ; Ali Naqavi ; Jing-Shun Huang ; Samuel P. Loke ; Don Walker ; Colin J. Mann ; Sergio Pellegrino ; Harry A. Atwater

EXTRACTING THE FIXED CHARGE DENSITY IN HFOX FILMS GROWN ON HIGHLY-DOPED P-SI SAMPLES 517

Alexander To ; Jie Cur ; Bram Hoex

NEAR-UNITY ULTRA-WIDEBAND THERMAL INFRARED EMISSION FOR SPACE SOLAR POWER RADIATIVE COOLING 521

Ali Naqavi ; Samuel P. Loke ; Michael D. Kelzenberg ; Emily C. Warmann ; Pilar Espinet-González ; Nina Vaidya ; Jing-Shun Huang ; Tatiana A. Roy ; Alexander J. Messer ; Tatiana G. Vinogradova ; Ali Hajimiri ; Sergio Pellegrino ; Harry A. Atwater

LINE-FOCUS AND POINT-FOCUS SPACE PHOTOVOLTAIC CONCENTRATORS USING ROBUST FRESNEL LENSES, 4-JUNCTION CELLS, & GRAPHENE RADIATORS 525

Mark O'Neill ; A.J. Mcdanal ; Michael Piszczor ; Matt Myers ; Paul Sharps ; Claiborne Mcpheeters ; Jeff Steinfedt

SIMULATION OF LIGHT TRAPPING STRUCTURES FOR ENHANCING RADIATION HARDNESS IN SPACE SOLAR CELLS 531

Nizami Z. Vagidov ; Kyle H. Montgomery ; Geoffrey K. Bradshaw ; David M. Wilt

AN ALTERNATIVE METHOD FOR SOLAR CELL INTEGRATION 537

Jessica Buckner ; Tracy Davis ; Eric Muskovin ; Bernard Carpenter

NIEL DOSE ANALYSIS ON TRIPLE JUNCTION CELLS 30% EFFICIENT AND RELATED SINGLE JUNCTIONS 541

Roberta Campesato ; Erminio Greco ; Mariacristina Casale ; Massimo Gervasi ; P.G. Rancoita ; Davide Rozza ; Mauro Tacconi ; Enos Gombia ; Aldo Kingma ; Carsten Baur

THIN AND FLEXIBLE TRIPLE JUNCTION CELLS 30% EFFICIENT: QUALIFICATION RESULTS AND FUTURE SPACE APPLICATIONS 545

Roberta Campesato ; Mariacristina Casale ; Giuseppe Gabetta ; Emilio Fernandez Lisbona ; Laurent D'Abrigeon

PRINTED ASSEMBLIES OF MICROSCALE TRIPLE-JUNCTION (3J) INVERTED METAMORPHIC (IMM) GAINP/GAAS/INGAAS SOLAR CELLS 549

Boju Gai ; John Geisz ; Daniel Friedman ; Jongseung Yoon

COMPARATIVE STUDY ON NONRADIATIVE RECOMBINATION CENTERS IN PROTON IRRADIATED INAS/GAAS QUANTUM DOT STRUCTURE BY TWO WAVELENGTH EXCITED PHOTOLUMINESCENCE 552

M. D. Haque ; N. Kamata ; S-I. Sato ; S. M. Hubbard

DESIGN AND PROTOTYPING EFFORTS FOR THE SPACE SOLAR POWER INITIATIVE ... 558

Michael D. Kelzenberg ; Pilar Espinct-Gonzalez ; Nina Vaidya ; Tatiana A. Roy ; Emily C. Warmann ; Ali Naqavi ; Samuel P. Loke ; Jing-Shun Huang ; Tatiana G. Vinogradova ; Alexander J. Messer ; Christophe Leclerc ; Eleftherios E. Gdoutos ; Fabien Royer ; Ali Hajimiri ; Sergio Pellegrino ; Harry A. Atwater

DEFECT CHARACTERIZATION OF III-V QUANTUM STRUCTURE SOLAR CELLS USING PHOTO-INDUCED CURRENT TRANSIENT SPECTROSCOPY 562

Shin-Ichiro Sato ; Takeyoshi Sugaya ; Tetsuya Nakamura ; Takeshi Ohshima

EFFECT OF LUMINESCENCE COUPLING BETWEEN INGAP AND GAAS SUBCELLS TO EXTERNAL QUANTUM EFFICIENCY IN TRIPLE-JUNCTION SOLAR CELLS 567

Mitsunobu Suga ; Mitsuru Imaizumi ; Tetsuya Nakamur ; Takeshi Ohshima

LIGHTWEIGHT CARBON FIBER MIRRORS FOR SOLAR CONCENTRATOR APPLICATIONS 572

Nina Vaidya ; Michael D. Kelzenberg ; Pilar Espinet-Gonzalez ; Tatiana G. Vinogradova ; Jing-Shun Huang ; Christophe Leclerc ; Ali Naqavi ; Emily C. Warmann ; Sergio Pellegrino ; Harry A. Atwater

GAAS SOLAR CELLS ON V-GROOVED SILICON VIA SELECTIVE AREA GROWTH 578

Michelle Vaisman ; Nikhil Jain ; Qiang Li ; Kei May Lau ; Adele C. Tamboli ; Emily L. Warren

HIGH TEMPERATURE ANNEALING OF IN1-XGAXN MQW SOLAR CELLS 582

Joshua J. Williams ; Heather Mcfavilen ; Steven Young ; Christiana B. Honsberg ; Stephen M. Goodnick

SOLAR PROBE PLUS ARRAY RELIABILITY 585

Anton Yanchilin ; Edward Gaddy

PHOTOVOLTAIC TEMPERATURE ESTIMATION MODEL FOR RAPID IRRADIANCE CHANGE CONDITIONS IN TROPICAL REGIONS USING HEURISTIC ALGORITHMS 589

R. Srivatsan ; Lian L. Jiang ; Douglas L. Maskell

ACCURACY OF CDTE PV ENERGY PREDICTIONS USING SPECTRAL CORRECTIONS 595

Mitchell Lee ; Kendra Passow ; Paul Wolffersdorff

PLANTPREDICT: SOLAR PERFORMANCE MODELING MADE SIMPLE 600

Kendra Passow ; Lauren Ngan ; Geoffrey Rich ; Mitch Lee ; Stephen Kaplan

INTEGRABILITY COMPARISON BETWEEN BIPV AND BAPV IN TROPICAL CONDITIONS: A BANGALORE CASE-STUDY 604

Gayathri Aaditya ; Roshan R Rao ; Monto Mani

A NEW PHOTOVOLTAIC SYSTEM TOPOLOGY THROUGH LOAD MANAGEMENT 608

Joseph A. Azzolini ; Meng Tao

FIRST STEP FOR POWER GENERATION AMOUNT ESTIMATION OF SOLAR MATCHING SYSTEM 613

Kazuya Hosokawa ; Toshiaki Yachi ; Yoichi Hirata ; Yasuyuki Watanabe

IRRADIANCE AND TEMPERATURE DISTRIBUTIONS AT HIGH LATITUDES: DESIGN IMPLICATIONS FOR PHOTOVOLTAIC SYSTEMS 619

Anne Gerdimenes ; Josefine Sclj

STEP-BY-STEP EVALUATION OF PHOTOVOLTAIC MODULE PERFORMANCE RELATED TO OUTDOOR PARAMETERS: EVALUATION OF THE UNCERTAINTY 626

Anne Migan Dubois ; Jordi Badosa ; Fausto Calderón-Obaldía ; Olivier Atlan ; Vincent Bourdin ; Marko Pavlov ; Dae Young Kim ; Yvan Bonnassieux

PERFORMANCE COMPARISONS OF A PV SYSTEM BY MONITORING SOLAR IRRADIANCE WITH DIFFERENT PYRANOMETERS 632

Yasuhiro Matsumoto ; J. Antonio Urbano ; Ramón Peña ; María De La Luz Olvera ; Nun Pitalúa ; Miguel A. Luna ; René Asomoza

FINANCIAL ANALYSIS OF A GRID-CONNECTED PHOTOVOLTAIC SYSTEM IN SOUTH FLORIDA 638

Hadis Moradi ; Amir Abtahi ; Ali Zilouchian

STUDY OF PHOTOVOLTAIC SYSTEMS MONITORING METHODS 643
E. Ortega ; G. Aranguren ; M.J. Sáenz ; R. Gutiérrez ; J.C. Jimeno

GLOBAL DESIGN ASPECTS OF PERSISTENT AND AUTONOMOUS PV POWERED SYSTEMS 648
I. M. Peters ; S. Watson ; N. Sahraei ; T. Buonassisi

**HOW TO CHOOSE THE BEST EMPIRICAL MODEL FOR OPTIMUM ENERGY YIELD
PREDICTIONS** 652
Steve Ransome ; Juergen Sutterlueti

**MODELING AND ANALYSIS OF PHOTOVOLTAIC ELECTROCHEMICAL SYSTEM USING
MODULE-LEVEL POWER ELECTRONICS** 658
Gowri M. Sriramagiri ; Nuha Ahmed ; Kevin D. Dobson ; Steven S. Hegedus

BETAVOLTAIC GENERATION FUNCTION IN SILICON 663
A.V. Sachenko ; I.O. Sokolovskyi ; M. Evstigneev

**MULTI-OBJECTIVE OPTIMIZATION FOR COLOR-TUNABILITY AND TRANSPARENCY IN
COLLOIDAL QUANTUM DOT SOLAR CELLS** 667
*Ebuka S. Arinze ; Botong Qiu ; Nathan Palmquist ; Yan Cheng ; Yida Lin ; Gabrielle Nyirjesy ; Gary Qian ;
Susanna M. Thon*

**CUBIC PHASE INXGA1-XN/GAN QUANTUM WELLS FOR THEIR APPLICATION TO
TANDEM SOLAR CELLS** 670
*C. A. Hernández-Gutiérrez ; Y. L. Casallas-Moreno ; Dagoberto Cardona ; Yu. Kudriavtsev ; A. Morales-Acevedo
; G. Santana-Rodríguez ; M. López-López*

**MODELING OF P-I-N GAASPN/GAP MQWS SOLAR CELL: TOWARDS LATTICE MATCHED
III-V/SI TANDEM** 673
Khim Kharel ; Alexandre Freundlich

INP QUANTUM DOT INTERMEDIATE BAND SOLAR CELL GROWN VIA MOCVD 677
Hyun Kum ; Yushuai Dai ; Michael Slocum ; Zachary Bittner ; Seth Hubbard

**MODIFIED LIMITING EFFICIENCY FOR MULTIPLE EXCITON GENERATION SOLAR
CELLS** 681
Jongwon Lee ; Christiana B. Honsberg

A SIMPLE MONTE CARLO MODEL OF A HOT CARRIER CELL 685
Tor Oskar Saetre

**OPTIMIZATION OF SEMICONDUCTOR QUANTUM DOTS FOR LUMINESCENT SOLAR
CONCENTRATORS: MINIMIZING REABSORPTION LOSSES** 690
*Anatoli I. Shkrebtii ; Anatoliy V. Sachenko ; Igor O. Sokolovskyi ; Vitaliy P. Kostylyov ; Mykola R. Kulish ; Denis
V. Khomcnko ; Mykhaylo A. Evstigneev*

**DEVELOPMENT OF ABSORBER AND ENERGY SELECTIVE CONTACTS FOR HOT
CARRIER SOLAR CELLS** 696
*Santosh Shrestha ; Simon Chung ; Yuanxun Liao ; Wenkai Cao ; Neeti Gupta ; Yi Zhang ; Xiaoming Wen ; Gavin
Conibeer*

GAASBI DEVICES FOR THERMAL ENERGY CONVERSION 701
Margaret Stevens ; Abigail Licht ; Nicole Pfiester ; Emily Carlson ; Kevin Grossklaus ; Thomas E. Vandervelde

ANALYTIC JV-CHARACTERISTICS OF IDEAL IMPURITY PV-CELLS 706
Rune Strandberg

**PHOTOLUMINESCENCE PROPERTIES OF IN-PLANE ULTRAHIGH-DENSITY INAS
QUANTUM DOTS ON GAASSB/GAAS(001) FOR SOLAR CELL APPLICATIONS** 712
Ryo Sugiyama ; Naoki Akimoto ; Tomah Sogabe ; Koichi Yamaguchi

**CARRIER SELECTIVE BACK CONTACT (CSBC) SOLAR CELL USING TRANSITION METAL
OXIDES** 716
Astha Tyagi ; Kunal Ghosh ; Anil Kottantharayil ; Saurabh Lodha

**ANALYSIS OF OPEN-CIRCUIT VOLTAGE AND CONVERSION EFFICIENCY IN QUANTUM-
DOT SOLAR CELLS VIA DETAILED-BALANCE-LIMIT THEORY** 721
Lin Zhu ; Hidefumi Akiyama ; Yoshihiko Kanemitsu

**ZINC SELENIDE SURFACE PASSIVATION LAYER FOR SINGLE-CRYSTALLINE CZTSE
SOLAR CELLS** 726
Michael A. Lloyd ; Douglas Bishop ; Brian E. Mccandless ; Robert Birkmirc

**USE OF SINGLE WALL CARBON NANOTUBE FILMS DOPED WITH TRIETHYLOXONIUM
HEXACHLORANTIMONATE AS A TRANSPARENT BACK CONTACT FOR CDTE SOLAR
CELLS** 730
*Fadhil K. Alfadhili ; Jacob M. Gibbs ; Geethika K. Liyanage ; Patrick W. Krantz ; Suneth C. Watthage ; Zhaoning
Song ; Adam B. Phillips ; Michael J. Heben*

**GRAIN AND GRAIN BOUNDARY GEOMETRICAL SHAPE CONSIDERATIONS ON SODIUM
AND POTASSIUM DIFFUSION THROUGH MOLYBDENUM FILMS** 735
Orlando Ayala ; Chinedum Akwari ; Tasnuva Ashrafee ; Shankar Karki ; Grace Rajan ; Sylvain Marsillac

SOLUTION-PROCESSED NICKEL-ALLOYED IRON PYRITE THIN FILM AS HOLE TRANSPORT LAYER IN CADMIUM TELLURIDE SOLAR CELLS 738

Ebin Bastola ; Khagendra P. Bhandari ; Randy J. Ellingson

USE OF CDS:O AND CDSE AS WINDOW LAYERS FOR CDTE PHOTOVOLTAICS 742

Tom Baines ; Guillaume. Zoppi ; Ken Durose ; Jonathan D. Major

APPLICATIONS OF HYBRID ORGANIC-INORGANIC METAL HALIDE PEROVSKITE THIN FILM AS A HOLE TRANSPORT LAYER IN CDTE THIN FILM SOLAR CELLS 748

Khagendra P. Bhandari ; Suneth C. Watthage ; Zhaoning Song ; Adam Phillips ; Michael J. Heben ; Randy J. Ellingson

MAGNESIUM-DOPED ZINC OXIDE AS A HIGH RESISTANCE TRANSPARENT LAYER FOR THIN FILM CDS/CDTE SOLAR CELLS 752

Francesco Bittau ; Elisa Artegiani ; Ali Abbas ; Daniele Menossi ; Alessandro Romeo ; Jake W. Bowers ; John M. Walls

INVESTIGATION OF ZNL-XMGXO:A1 FILM BY RATIO FREQUENCY MAGNETRON CO-SPUTTERING AS TRANSPARENT CONDUCTIVE OXIDE LAYER 757

Jakapan Chantana ; Yuya Ishino ; Takashi Minemoto

A NEW TCO/WINDOW-BUFFER FRONT STACK FOR CDTE SOLAR CELLS AND ITS IMPLEMENTATION 761

Alan E. Delahoy ; Xuehai Tan ; Akash Saraf ; Payal Patra ; Surya Manda ; Yunfei Chen ; Krishnakumar Velappan ; Bastian Siepchen ; Shou Peng ; Ken K. Chin

SYNTHESIS OF HIGH-QUALITY AZO POLYCRYSTALLINE FILMS VIA TARGET BIAS RADIO FREQUENCY MAGNETRON SPUTTERING 767

Zhongming Du ; Yufeng Zhang ; Xiangxin Liu

CLOSE-SPACE SUBLIMATED CDTE SOLAR CELLS WITH CO-SPUTTERED CDSXSE1-XALLOY WINDOW LAYERS 771

Corey R. Grice ; Maxwell Junda ; Alexander Archer ; Jian Li ; Yanfa Yan

EFFECTS OF GRAPHENE OXIDE BARRIER ON CU2ZNSNSXSE4-XTHIN FILM SOLAR CELLS 777

Woo-Lim Jeong ; Jung-Hong Min ; In-Young Kim ; Hae-Sun Kim ; Jin-Hyeok Kim ; Dong-Seon Lee

13% CDS/CDTE SOLAR CELL USING A NANOCOMPOSITE (CUS)X(ZNS)1-X THIN FILM HOLE TRANSPORT LAYER 781

Kamala Khanal Subedi ; Khagendra P. Bhandari ; Ebin Bastola ; Randy J. Ellingson

MOLYBDENUM OXIDE AND MOLYBDENUM NITRIDE BACK CONTACTS FOR THIN-FILM CDTE SOLAR CELLS 785

Anna Kindvall ; Jason Kephart ; Walajabad Sampath

INVESTIGATION AND OPTIMIZATION OF CD-FREE BUFFER LAYERS IN2S3 AND ZN(O, S) FOR CU2ZNSN(S, SE)4-BASED SOLAR CELLS 791

Willi Kogler ; Thomas Schnabel ; Andreas Bauer ; Stefanie Spiering ; Erik Ahlswede ; Michael Powalla

REAR CONTACT PASSIVATION FOR HIGH BANDGAP CU(IN, GA)SE2 SOLAR CELLS WITH VARYING ABSORBER THICKNESS AND FLAT GA PROFILE 796

Dorothea Ledinek ; Pedro Salome ; Carl Hägglund ; Marika Edoff

LASER ANNEALED BACK CONTACTS FOR CDTE SOLAR CELLS 802

Vasilios Palekis ; Shamara Collins ; Imran Khan ; Vamsi Evani ; Sudhajit Misra ; Michael A. Scarpulla ; Mark Lonergan ; Don Morel ; Chris Ferekides

ENHANCED ANTI-REFLECTIVE COATING FOR THIN FILM SOLAR CELLS 807

Grace Rajan ; Shankar Karki ; Robert W. Collins ; Sylvain Marsillac

INFLUENCE OF AGS LAYER INSERTION AT ABSORBER/ITO INTERFACE ON STRUCTURAL AND PHOTOVOLTAIC PROPERTIES OF ULTRATHIN CU(IN,GA)SE2 SOLAR CELLS 810

Muhammad Saifullah ; Jihye Gwak ; Kihwan Kim ; Joo Hyung Park ; Junsik Cho ; Jae Ho Yun

NOVEL, FACILE BACK SURFACE TREATMENT FOR CDTE SOLAR CELLS 815

Suneth C. Watthage ; Geethika K. Liyanage ; Zhaoning Song ; Fadhil K. Alfadhili ; Rabee B. Alkhayat ; Khagendra P. Bhandari ; Randy J. Ellingson ; Adam B. Phillips ; Michael J. Heben

OPTIMIZING CDS BUFFER LAYER FOR CIGS BASED THIN FILM SOLAR CELL 820

Weijie Zhang ; Korhan Demirkan ; Geordie Zapalac ; David Spaulding ; Jochen Titus ; Neil Mackie

INVESTIGATION OF INP DEFECT CHARACTERISTICS GROWN USING NOVEL TF-VLS TECHNIQUE 823

Abhinav Chikhalkar ; Alec Fischer ; Mark Hettick ; Ali Javey ; Richard R. King

INVESTIGATION OF FAST GROWTH GAAS-BASED SOLAR CELL ON REUSABLE SUBSTRATE BY METALORGANIC CHEMICAL VAPOR DEPOSITION 827

Chaomin Zhang ; Abhinav Chikhalkar ; Ehsan Vadiee ; Richard King ; Christiana Honsberg ; Eric Armour ; Yeongho Kim

DEVELOPMENT OF ALUMINUM EPILAYERS AS BUFFERS FOR GAINAS 831

Phil Ahrenkiel ; Nathan Smaglik ; Nikhil Pokharel ; Alessandro Giussani ; Michael A. Slocum ; Seth M. Hubbard

LASER CRYSTALLIZATION OF AMORPHOUS GERMANIUM ON TITANIUM NITRIDE-COATED STEEL FOR LOW-COST GAAS SOLAR-CELLS 837

Saloni Chaurasia ; Srinivasan Raghavan ; Sushobhan Avasthi

HIGH QUALITY EPITAXIAL GERMANIUM ON SI (100) FOR LOW -COST III–V SOLAR-CELLS 841

Saloni Chaurasia ; Srinivasan Raghavan ; Sushobhan Avasthi

CRYSTALLINITY CONTROL IN LOW-TEMPERATURE GROWTH OF POLY-CRYSTALLINE GE BY ION BEAM DEPOSITION 845

S. I. Maximenko ; N. A. Mahadik ; P. P. Jenkins ; R. J. Walters ; A. Giussani ; E. L. Mcclure ; S. M. Hubbard ; C. Bailey

HIGH EFFICIENCY GAINP/GAAS DOUBLE JUNCTION SOLAR CELL ON SI SUBSTRATE ASSISTED BY THE ELECTRON BEAM TREATMENT 849

Hyo Jin Kim ; Yong Whan Kim

ANALYSIS OF DEPOSITED RESIDUES AND ITS CLEANING PROCESS ON GAAS SUBSTRATE AFTER EPITAXIAL LIFT-OFF 854

Tatsuya Nakata ; Kentaroh Watanabe ; Hassanet Sodabanlu ; Daiki Kimura ; Naoya Miyashita ; Yoshitaka Okada ; Yoshiaki Nakano ; Masakazu Sugiyama

ULTRATHIN SILICON-AN-INSULATOR (SOI) WAFER FOR COMPLIANT SUBSTRATE 858

Shinyoung Noh ; Anita Ho-Baillie ; Stephen Bremner ; Martin A. Green ; Xiaojing Hao

CHARACTERIZATION OF GAAS SOLAR CELLS GROWN BY HYDRIDE VAPOR PHASE EPITAXY IN HORIZONTAL REACTOR 861

Ryuji Oshima ; Kikuo Makita ; Takeyoshi Sugaya ; Akinori Ubukata

FLEXIBLE GAAS SINGLE-JUNCTION SOLAR CELLS BASED ON SINGLE-CRYSTAL-LIKE THIN-FILM MATERIALS DIRECTLY GROWN ON METAL TAPES 866

Sara Pouladi ; Monika Rathi ; Mojtaba Asadirad ; Pavel Dutta ; Seung Kyu Oh ; Devendra Khatiwada ; Shahab Shervin ; Yao Yao ; Venkat Selvamanickam ; Jae-Hyun Ryou

REDUCED DEFECT DENSITY IN SINGLE-CRYSTALLINE-LIKE GAAS THIN FILM ON FLEXIBLE METAL SUBSTRATES BY USING SUPERLATTICE STRUCTURES 869

M. Rathi ; P. Dutta ; D. Khatiwada ; Y. Yao ; Y. Gao ; Y. Li ; S. Sun ; S. Pouladi ; S. Reed ; A. Khadimallah ; J. Ryou ; V. Selvamanickam ; N. Zheng ; P. Ahrenkiel

ECONOMIC ANALYSIS OF TRANSFER PRINTED III–V VIRTUAL SUBSTRATES 873

Kenneth J. Schmieder ; Matthew P. Lumb ; Michael K. Yakes ; Shawn Mack ; Mitchell F. Bennett ; Sergey I. Maximenko ; Laura B. Ruppalt ; Michael A. Meeker ; Chase T. Ellis ; Matthew Meitl ; Joseph G. Tischler ; Robert J. Walters

THIN FILMS OF ZINC-DOPED GAAS BY RF MAGNETRON SPUTTERING FOR USE IN PHOTOVOLTAIC CELLS 876

Kirby Simon ; Kyle Cepeda ; Nishit Shetty ; Elijah Thimsen

SELF ALIGNED ALUMINUM SELECTIVE EMITTER FOR N-TYPE SI CELLS 881

San Theigi ; Robert C. Reedy ; Vincenzo Lasalvia ; Paul Stradins ; Benjamin G. Lee

HOW TO REALIZE SOLAR CELLS WITH LASER STRUCTURED PLATED NI-CU-CONTACTS WITH EXCELLENT ADHESION AND HIGH FILL-FACTORS WITHOUT PARASITIC PLATING 884

A. Büchler ; S. Kluska ; J. Bartsch ; B. Grübel ; A.A. Brand ; S. Gutscher ; M. Glatthaar

EXPLOITING THE POTENTIALS OF THE FRONT SURFACE FIELD (FSF) INDUSTRIAL SILICON SOLAR CELL 888

Ahrar Ahmed Chowdhury ; Yu -Chen Hsu ; Veysel Unsur ; Abasifreke Ebong

PHOTOVOLTAIC PERFORMANCE OF SILICON SOLAR CELLS ENHANCED BY PLASMONIC SILVER NANOPARTICLES OF VARIOUS DIMENSIONS DEPOSITING THROUGH ANODIC ALUMINUM OXIDE TEMPLATE 893

Ta-Wei Chuang ; Wen-Jeng Ho ; Sheng-Kai Feng ; Jheng-Jie Liu ; Guan-Yi Li ; Hao-Yu Yang ; Yun-Chie Yang ; Cho-Chun Chiang ; Yao-Hui Chen

MITIGATION OF POTENTIAL-INDUCED DEGRADATION 896

Orry Faur ; Maria Faur

ELECTRODEPOSITION OF SI-LAYER THROUGH REDUCTION OF DIATOMACEOUS EARTH FOR THE APPLICATION OF SOLAR-CELLS 900

Muhammad Monirul Islam ; Imane Abdellaoui ; Takeaki Sakurai ; Saad Hamzaoui ; Katsuhiro Akimoto

EFFECT OF SI CONTENT IN A1 PASTE ON LOCAL A1 REAR CONTACTS IN PERC CELL 904

Supawan Joonwichien ; Katsuhiko Shirasawa ; Satoshi Utsunomiya ; Hidetaka Takato

NEW SILVER PASTE METALLIZATION APPROACH ON P+ DIFFUSION ZONES OF SILICON SOLAR CELLS 907

Yunjun Li ; Mohshi Yang ; Igor Pavlovsky ; Guoping Zeng

INFLUENCES OF ANNEALING AND DEFECT LIMITATION ON P-TYPE SILICON SOLAR CELL 911
Yu-Hsuan Lin ; Sung-Yu Chen ; Kuen-Yi Wu ; Chien-Hsun Chen ; Chen-Hsun Du ; Chun-Ming Yeh

REDUCED TEMPERATURE SILVER PASTE WITH LOW CONTACT RESISTANCE FOR ADVANCED SOLAR CELL APPLICATIONS 914
Ryan Mayberry ; Daniel Holzmann ; Gerd Schulz ; Lindsey Karpowich ; Mark Naylor ; Matthias Hoerteis

BSF ISLANDS FOR REDUCED RECOMBINATION IN IBC CELLS 917
Agnes A. Mewe ; Nicolas Guillevin ; Ilkay Cesar ; Antonius R. Burgers

THERMAL STABILITY OF HYDROGENATED BORON EMITTERS 921
Khaja H. Mohammed ; Larry C. Cousar ; Philip A. Mcmeans ; Garrett Z. Evans ; Douglas A. Hutchings ; Hameed A. Naseem ; Sergiu C. Pop

LIGHT INDUCED PLATING OF SILICON SOLAR CELLS USING BORIC ACID-FREE NICKEL CHEMISTRY 925
Krystal Munoz ; Lynne Michaelson ; Joseph Karas ; Tom Tyson ; James Rand ; Stuart Bowden

BAKING TEMPERATURE DEPENDENCE OF CU PASTE ON A1-BSF CELL PROPERTIES 931
Tomohiro Saito ; Tetsuya Fukuda ; Hoang Tri Hai ; Yuji Kurimoto ; Daisuke Ando ; Yuji Sutou ; Katsuhiko Shirasawa ; Junichi Koike

THE SILVER CONTACT AND FORMATION MECHANISM OF THE BORON EMITTER AND THE CURRENT FLOW MECHANISM OF THE SOLAR CELL ELECTRODE 935
Seunghyun Shin ; Soohyun Bae ; Sungeun Park ; Yoonmook Kang ; Hae-Seok Lee ; Donghwan Kim

LASER ANNEALING TO ENHANCE PERFORMANCE OF ALL-LASER-BASED SILICON BACK CONTACT SOLAR CELLS 937
Zeming Sun ; Mool C. Gupta

LARGE AREA N-TYPE SELECTIVE EMITTER CELLS USING LASER DOPING THROUGH BORON DOPED SCREEN PRINTED PASTE 940
Ajay D Upadhyaya ; Vijaykumar D Upadhyaya ; Brian Rounsaville ; Keeya Madani ; Ajeet Rohatgi ; Toru Hanada

METALLIZED BORON-DOPED BLACK SILICON EMITTERS FOR FRONT CONTACT SOLAR CELLS 944
Guillaume Von Gastrow ; Hele Savin ; Eric Calle ; Pablo Ortega ; Ramón Alcubilla ; Andreana Daniil ; Elias Z. Stutz ; Anna Fontcuberta I Morral ; Sebastian Husein ; Tara Nietzold ; Mariana Bertoni

CONTACT RESISTANCE MEASUREMENT FOR THERMALLY DIFFUSED POINT CONTACT BY LOCALIZED DIELECTRIC BREAKDOWN SOLAR CELLS 948
Qilin Ye ; Ned J. Western ; Anqi Liao ; Stephen P. Bremner

LOW TEMPERATURE REAR SURFACE METALLIZATION OF MULTI-CRYSTALLINE SILICON SOLAR CELLS FOR IMPROVED BULK LIFETIME 953
N. J. Western ; S. P. Bremner

INVESTIGATION OF HIGH PERFORMANCE PEROVSKITE-BASED SOLAR CELLS GROWN BY HYBRID CHEMICAL VAPOR DEPOSITION TECHNIQUE 958
Huseyin Cem Gokkaya ; Shen Qian ; Zhiwei Ren ; Annie Ng ; Charles Surya

ENHANCED PEROVSKITE SOLAR CELL PERFORMANCE USING FULL SPACE DEVICE OPTIMIZATION 963
Ahmer A.B. Baloch ; Shahzada P. Aly ; Mohammad I. Hossain ; Raka Jovanovic ; Nouar Tabet ; Fahhad H. Alharbi

MEASURING OPTICAL ABSORPTION IN ORGANIC PHOTOVOLTAICS USING MONOCHROMATED ELECTRON ENERGY-LOSS SPECTROSCOPY 966
Jessica A. Alexander ; Frank J. Scheltens ; David W. Mccomb ; Lawrence F. Drummy ; Michael F. Durstock ; James B. Gilchrist ; Sandrine Hentz

ADVANCED DEPOSITION OF PHOTO-CATALYTIC TIO2 FILM BY ATMOSPHERIC SPPS FOR DYE SENSITIZED SOLAR CELLS 970
Ifeanacho Anyadiegwu ; Dickson Kindole ; Geoffrey Kibiegon Ronoh ; Yoshimasa Noda ; Yasutaka Ando

CH3NH3PBI3-XBRXPEROVSKITE SOLAR CELLS VIA SPRAY ASSISTED TWO-STEP DEPOSITION: INFLUENCE OF BROMIDE ON THE DEVICE PERFORMANCE 976
Gaoda Chai ; Shiqiang Luo ; Shizhen Wang ; Hang Zhou

MODULATED STRUCTURE TO MAXIMIZE THE OPEN-CIRCUIT VOLTAGE WITH MODERATE BAND-GAP OF SMALL MOLECULE ORGANIC SOLAR CELLS-DFT APPROACH 980
Saravanan Chinnusamy ; Amita Munshi ; Sukanya Santhosh Kumar ; W. S. Sampath ; Milind S. Dangate

PEROVSKITE GRAIN SIZE MODULATION BY ANNEALING IN METHYL-AMINE ENVIRONMENT 986
Arun Singh Chouhan ; Naga Prathibha Jasti ; Srinivasan Raghavan ; Sushobhan Avasthi ; Shreyash Hadke

FE2O3AS AN ELECTRON TRANSPORT MATERIAL FOR ORGANO-METAL HALIDE PEROVSKITE SOLAR CELLS 989
Dallas Fisher ; Pravakar P. Rajbhandari ; Tara P. Dhakal

OPTICAL EVALUATION OF PEROVSKITE FILMS IN AND FOR SOLAR CELL DEVICE STRUCTURES 993

Kiran Ghimire ; Dewei Zhao ; Changlei Wang ; Yanfa Yan ; Nikolas J. Printraza

HYBRID ORGANIC-INORGANIC SOLAR CELLS WITH A BENZOQUINONE PASSIVATING LAYER 999

James Hack ; Abhishek Iyer ; Meixi Chen ; Nicole Kotulak ; Akirt Sridharan ; Robert Opila

PRECISE 1-V CURVE MEASUREMENT PROCEDURE FOR PEROVSKITE SOLAR CELLS: APPLICATION TO VARIOUS TYPES OF DEVICES 1003

Y. Hishikawa ; M. Yoshita ; H. Shimura ; A. Sasaki ; T. Ueda

ENHANCING THE CRYSTALLINE OF PLANAR-STRUCTURE CH3NH3PBI3PEROVSKITE SOLAR CELLS VIA SANDWICH EVAPORATION TECHNIQUE 1006

Po-Tsun Kuo ; Shang-Pang Lin ; Cheng-Shian Lin ; Ching-Fuh Lin

TOWARD HIGH PERFORMANCE ORGANIC-SILICON HYBRID SOLAR CELLS 1009

Yi Lai ; Hong-Jhang Syu ; Ching-Fuh Lin

NICKEL OXIDE THIN FILMS BY RADIO FREQUENCY SPUTTER FOR INVERTED PEROVSKITE SOLAR CELLS 1012

Hyeonseok Lee ; Yu-Ting Huang ; Shien-Ping Feng

ANOMALOUS EFFICIENCY SCALING WITH DARK CURRENT IN PEROVSKITE SOLAR CELLS 1015

Vikas Nandal ; Pradeep R. Nair

NUMERICAL SIMULATION AND PERFORMANCE OPTIMIZATION OF PEROVSKITE SOLAR CELL 1018

Sai Naga Raghuram Nanduri ; Mahbube K. Siddiki ; Ghulam M. Chaudhry ; Yahya Z. Alharthi

PERFORMANCE PREDICTION FOR LARGE AREA PEROVSKITE SOLAR CELLS 1022

Yojak Raote ; Hitarth Choubisa ; Pradeep R. Nair

PHOTOCONVERSION EFFICIENCY MODELING IN PEROVSKITE SOLAR CELLS 1025

A.V. Sachenko ; V.P. Kostylyov ; A.V. Bobyl ; V.M. Vlasiuk ; I.O. Sokolovskyi ; E.I. Terukov ; M. Evstigneev

INFLUENCE OF MONO- AND DI-VALENT METAL ADDITIVES ON MORPHOLOGY AND CHARGE CARRIER DYNAMICS OF CH3NH3PBI3PEROVSKITE 1030

Niraj Shrestha ; Suneth C. Watthage ; Zhaoning Song ; Paul J. Roland ; Adam B. Phillips ; Michael J. Heben ; Randall J. Ellingson

EFFECT OF DUAL CATHODE BUFFER LAYER ON TERNARY ORGANIC SOLAR CELL 1034

Ashish Singh ; T. Bhim Raju ; Anamika Dey ; Ritesh Kant Gupta ; Parameswar K. Iyer

COPPER PLATED TOP ELECTRODE FOR AN INVERTED ORGANIC PHOTOVOLTAIC 1037

Malia Steward ; Zhan Shi ; Kyoung- Tae Kim ; Seungkeun Choi

INTERFACE BAND GAP AND CHARGE TRAPPING IN BULK HETEROJUNCTION SOLAR CELLS 1040

Marian Tzolov ; Maxwell Mcintyre

FABRICATION OF EFFICIENT CH3NH3PBI3 SOLAR CELLS IN AMBIENT AIR 1044

Feng Wang ; Ye Zhongbiao ; Hojjatollah Sarvari ; Somin Park ; Kenneth Graham ; Yuetao Zhao ; Zhi David Chen

HIGH EFFICIENCY PEROVSKITE SOLAR CELLS BY A MODIFIED LOW-TEMPERATURE SOLUTION PROCESS INTER-DIFFUSION METHOD 1048

Yangyi Yao ; Wei-Lun Hsu ; Mario Dagenais

INTERFACIAL MODIFICATION OF SOL-GEL ZNO/AZO BILAYER AS HIGHLY EFFICIENT ELECTRON TRANSPORT LAYER FOR PEROVSKITE SOLAR CELLS 1051

Shang-Hsuan Wu ; Ming-Yi Lin ; Sheng-Hao Chang ; Wei-Chen Tu ; Chi-Wei Chu ; Via-Chung Chang

THE POTENTIAL OF BIFACIAL PHOTOVOLTAICS: A GLOBAL PERSPECTIVE 1055

Xingshu Sun ; Mohammad R. Khan ; Amir Hanna ; Muhammad M. Hussain ; Muhammad A. Alam

PERFORMANCE ASSESSMENT OF STAND ALONE BIFACIAL SOLAR PANEL UNDER REAL TIME CONDITIONS 1058

Ahmer A.B. Baloch ; Maher Armoush ; Basel Hindi ; Abdelkader Bousselham ; Nouar Tabet

OPERATION AND PERFORMANCE ASSESSMENT OF GRID-CONNECTED PV SYSTEMS IN OPERATION IN MAUI, HAWAII 1061

Severine Busquet ; Jonathan Kobayashi ; Richard E. Rocheleau

A NOVEL MULTILEVEL SOLAR PANEL SYSTEM: IMPLEMENTATION AND VERIFICATION 1067

Tanmoy Debnath ; Syed N. Imtiaz ; Syed F. Nawaz ; Abdullah Al Mahmud ; Mosaddequr Rahman

PREDICTING POWER LOSS DUE TO MODULE MISMATCH IN UTILITY-SCALE PHOTOVOLTAIC SYSTEMS 1071

Stephen Kaplan ; Kendra Passow

APPLICATION OF SHAPED REFLECTORS TO INCREASE THE ENERGY HARVEST OF BIFACIAL PV SYSTEMS - ANALYZED WITH A MINIATURIZED TEST ARRAY 1077

Hartmut Nussbaumer ; Markus Klenk ; Nico Keller ; Dominic Heller ; Remo Kaslin ; Thomas Baumann ; Franz Baumgartner

TOWARDS NEW MODULE AND SYSTEM CONCEPTS FOR LINEAR SHADING RESPONSE 1081

Kostas Sinapis ; Tom T.H. Rooijakkers ; Lenneke H. Slooff ; Lars A.G. Okel ; Mark J. Jansen ; Anna J. Carr

PARTIAL SHADING ABATEMENT THROUGH CASCADED H-BRIDGE TOPOLOGY 1086

Steven Tidwell ; Joseph Latham ; Michael Mcintyre

DATA ANALYSIS FOR EFFECTIVE MONITORING OF PARTIALLY SHADED PHOTOVOLTAIC SYSTEMS 1090

Odysseas Tsafarakis ; Kostas Sinapis ; Wilfried G.J.H.M. Van Sark

BIFACIAL PHOTOVOLTAIC MODULE ENERGY YIELD CALCULATION AND ANALYSIS 1094

Christopher E. Valdivia ; Chu Tu Li ; Annie Russell ; Joan E. Haysom ; Rui Li ; David Lekx ; Mohsen M. Sepeher ; Dan Henes ; Karin Hinzer ; Henry P. Schriemer

DESIGN AND DEVELOPMENT OF A SOLAR PHOTOVOLTAIC MODULE DETECTION CONTROL SYSTEM BASED ON PLC 1100

Yiwang Wang ; Jili Zhang ; Kanglin Liu ; Houjun Tang ; Hui Pan ; Yan Lin ; Peter Yang ; Rui Wang

DETECTING CALIBRATION DRIFT AT GROUND TRUTH STATIONS A DEMONSTRATION OF SATELLITE IRRADIANCE MODELS' ACCURACY 1104

Richard Perez ; James Schlemmer ; Adam Kankiewicz ; John Dise ; Alemu Tadese ; Thomas Hoff

PERFORMANCE OF SOLAR RESOURCE MONITORING STATIONS IN HOT CLIMATE REGIONS 1110

Yahya Z. Alharthi ; Mahbube K. Siddiki ; Ghulam M. Chaudhry ; Saad Muaddi ; Ahmed Alahmed

FIRST RESULTS OF A LOW COST ALL-SKY IMAGER FOR CLOUD TRACKING AND INTRA-HOUR IRRADIANCE FORECASTING SERVING A PV-BASED SMART GRID IN LA GRACIOSA ISLAND 1116

David Cañadillas ; Walter Richardson ; Benjamín Gonzalez-Díaz ; Les E. Shephard ; Ricardo Guerrero Lemus

STATISTICAL ANALYSIS OF PV INSOLATION DATA 1122

Abdulmunim Guwaeder ; Rama Ramakumar

A COMPARISON OF PV POWER FORECASTS USING PVLIB-PYTHON 1127

William F. Holmgren ; Antonio T. Lorenzo ; Clifford Hansen

COMPARING THE TYPICAL GHI YEAR VS TYPICAL POWER YEAR 1132

Alex Kubiniec ; Adam Kankiewicz ; Alemu Tadesse

THE HOLY GRAIL OF RESOURCE ASSESSMENT: LOW COST GROUND-BASED MEASUREMENTS WITH GOOD ACCURACY 1134

Bill Marion ; Benjamin Smith

GLOBAL COMPARISON OF THE IMPACT OF TEMPERATURE AND PRECIPITABLE WATER ON CDTE AND SILICON SOLAR CELLS 1140

I. M. Peters ; L. Haohui ; T. Reindl ; T. Buonassisi

ESTIMATION OF MEAN MONTHLY GLOBAL SOLAR RADIATION USING MODEL BASED ON SUNSHINE HOURS FOR COLOMBIA 1143

Diego J. Rodríguez ; Johan Hernández ; Adolfo Jaramillo

IMPLEMENTATION OF SOLAR DIFFUSE CIE MODEL IN RAY TRACING PROGRAM FOR IRRADIANCE CALCULATIONS 1147

Liliana Ruiz Diaz ; Pierre-Alexandre Blanche ; Robert A. Norwood

INVESTIGATION OF CITY-LEVEL SITE-PAIR CORRELATIONS OF SOLAR VARIABILITY USING EMPIRICAL SATELLITE DATA 1151

Rhythm Singh ; Rangan Banerje

ULTRA-SHORT-TERM PHOTOVOLTAIC GENERATION FORECASTING MODEL BASED ON WEATHER CLUSTERING AND MARKOV CHAIN 1158

Jin Tan ; Changhong Deng

DAILY SOLAR IRRADIANCE PROFILE CHARACTERIZATION AND RAMP RATE ANALYSIS AT DIFFERENT TIME RESOLUTIONS 1163

Spyros Theocharides ; Venizelos Venizelou ; George Makrides ; George E. Georghiou

COMPARISON AND ANALYSIS OF INSTRUMENTS MEASURING PLANE OF ARRAY IRRADIANCE FOR ONE-AXIS TRACKING PV SYSTEMS 1169

Frank Vignola ; Chun-Yu Chiu ; Josh Peterson ; Michael Dooraghi ; Manajit Sengupta

A SKY IMAGE ANALYSIS SYSTEM FOR SUB-MINUTE PV PREDICTION 1175

Rodrigo Verschac ; Li Li ; Shohei Nobuhara ; Takekazu Kato

LARGE AREA NANOSTRUCTURE INTEGRATION FOR BROAD-SPECTRUM, OMNIDIRECTIONAL ANTIREFLECTION IMPROVEMENTS ON POLYMER PACKAGED, MECHANICALLY FLEXIBLE, EPITAXIAL LIFT-OFF III-V SOLAR CELLS 1181

Gabriel Cossio ; Jihwan Lee ; Gautham Ragunathan ; Andre Wibowo ; Sudersena Rao Tatavarti ; Kimberly Sablon ; Edward T. Yu

DEVELOPMENT OF BACK SURFACE TEXTURE FOR LIGHT MANAGEMENT IN EPITAXIAL LIFT OFF (ELO) QUANTUM DOT SOLAR CELLS 1184

Brittany L. Smith ; George T. Nelson ; Yushuai Dai ; Michael A. Slocum ; Andre Wibowo ; Rao Tatavarti ; Seth M. Hubbard

ENABLING HIGH-EFFICIENCY INAS/GAAS QUANTUM DOT SOLAR CELLS BY EPITAXIAL LIFT-OFF AND LIGHT MANAGEMENT 1189

F. Cappelluti ; A. P. Cédola ; A. Khalili ; Farid Elsehrawy ; G. Bauhuis ; P. Mulder ; J. Schermer ; G. Bissels ; T. Aho ; T. Niemi ; M. Guina ; D. Kim ; J. Wu ; H. Liu

CHARACTERIZATION OF ARSENIC DOPED CDTE LAYERS AND SOLAR CELLS 1193

Sachit Grover ; Xiaoping Li ; Wei Zhang ; Ming Yu ; Gang Xiong ; Markus Gloeckler ; Roger Malik

ENHANCING P-TYPE DOPING IN POLYCRYSTALLINE CDTE FILMS 1196

Brian Mccandless ; Wayne Buchanan ; Gowri Sriramagiri ; Christopher Thompson ; Joel Duenow ; David Albin ; Soren Jensen ; John Moseley ; M. Al-Jassim ; Wyatt K. Metzger

SPECTRAL AND CONCENTRATION SENSITIVITY OF MULTIJUNCTION SOLAR CELLS AT HIGH TEMPERATURE 1201

Daniel J. Friedman ; Myles A. Steiner ; Emmett E. Perl ; John Simon

ON THE USE OF TRANSPARENT CONDUCTIVE OXIDES IN HIGH CONCENTRATOR III-V MULTIJUNCTION SOLAR CELLS 1204

Ignacio Rey-Stolle ; Yeonbae Lee ; Iván Garcia ; Luis Cifuentes ; Kin Man Yu ; Carlos Algora ; Wladek Walukiewicz

COMPONENT INTEGRATION EFFECTS IN 4-JUNCTION SOLAR CELLS WITH DILUTE NITRIDE 1EV SUBCELL 1210

I. García ; M. Ochoa ; I. Lombardero ; L. Cifuentes ; P. Caño ; M. Hinojosa ; I. Rey-Stolle ; C. Algora ; A. D. Johnson ; J. I. Davies ; K.H. Tan ; W.K. Loke ; S. Wicaksono ; S. F. Yoon

BISMUTH SURFACTANT-MEDIATED GROWTH OF GANASSB(BI) SOLAR CELLS 1215

Aymeric Maros ; Chaomin Zhang ; Jongwon Lee ; Hongfeng Wang ; Stephen Bremner ; Nikolai Faleev ; Christiana B. Honsberg ; Richard. R. King

AMORPHOUS SILICON CARBIDE FOR SILICON SURFACE PASSIVATION IN CARRIER-SELECTIVE CONTACT DEVICES 1220

Mathieu Boccard ; Christophe Ballif ; Zachary C. Holman

SURFACE PASSIVATION OF BORON DIFFUSED JUNCTIONS BY BOROSILICATE GLASS AND IN SITU GROWN SILICON DIOXIDE INTERFACE LAYER 1222

Valentin D. Mihailetchi ; Haifeng Chu ; Jan Lossen ; Radovan Kopecek

IMPROVED LIGHT INCOUPLING IN PLANAR SOLAR CELLS VIA IMPROVED TEXTURE MORPHOLOGY OF PDMS SCATTERING LAYER 1228

Salman Manzoor ; Zhengshan J. Yu ; Asad Ali ; Waqar Ali ; Zachary C. Holman

DAMAGE-FREE LASER ABLATION FOR EMITTER PATTERNING OF SILICON HETEROJUNCTION INTERDIGITATED BACK-CONTACT SOLAR CELLS 1233

Menglei Xu ; Twan Bearda ; Miha Filipic ; Hariharsudan Sivaramakrishnan Radhakrishnan ; Maarten Debucquoy ; Ivan Gordon ; Jozef Szlufcik ; Jef Poortmans

BENEFITS OF A THERMAL DRIFT DURING ATOMIC LAYER DEPOSITION OF AL2O3FOR C-SI PASSIVATION 1237

Fabien Lebreton ; Andy Zauner ; Pavel Bulkin ; Francois Silva ; Sergej Filonovich ; Pere Roca I Cabarrocas

GROWTH DIFFERENCE OF AMORPHOUS SILICON BETWEEN PLASMA ENHANCED AND CATALYTIC CVD BASED ON SILICON HETEROJUNCTION SOLAR CELLS 1241

Liping Zhang ; Renfang Chen ; Zhuopeng Wu ; Chenguang Sun ; Fanying Meng ; Zhengxin Liu

DEVELOPING AN UNDERSTANDING-BASED SELECTION OF HYBRID-PEROVSKITE COMPOUNDS AND THE CU-IN HYBRID-PEROVSKITE (CIHP) FAMILY 1245

Alex Zunger ; G. Dalpian ; Qihang Liu ; L.B Abdalla ; L.L. Kazmerski

EFFECTS OF ELECTRON AND PROTON RADIATION ON PEROVSKITE SOLAR CELLS FOR SPACE SOLAR POWER APPLICATION 1248

Jing-Shun Huang ; Michael D. Kelzenberg ; Pilar Espinet-González ; Colin Mann ; Don Walker ; Ali Naqavi ; Nina Vaidya ; Emily Warmann ; Harry A. Atwater

TOWARDS PEROVSKITE SILICON TANDEM SOLAR CELLS WITH OPTIMIZED OPTICAL PROPERTIES 1253

Jan Christoph Goldschmidt ; Alexander J. Bett ; Patricia S.C. Schulze ; Nico Tucher ; Martin Bivour ; Markus Kohlstädt ; Seunghun Lee ; Simone Mastroianni ; Laura Mundt ; Markus Mundus ; Paul Ndione ; Karl Wienands ; Kristina Winkler ; Uli Würfel ; Martin Hermle ; Stefan W. Glunz

FIRST-PRINCIPLES DENSITY FUNCTIONAL THEORY CALCULATION OF METAL-SUBSTITUTED LEAD HALIDE PEROVSKITE 1256
Ji-Sang Park ; Matthew D. Sampson ; Alex B.F. Martinson ; Maria K.Y. Chan

ESTIMATING THE EFFECTS OF MODULE AREA ON THIN-FILM PHOTOVOLTAIC SYSTEM COSTS 1259
Kelsey A. W. Horowitz ; Ran Fu ; Xingshu Sun ; Tim Silverman ; Michael Woodhouse ; Muhammad A. Alam

COST ANALYSIS OF TANDEM MODULES 1264
Sarah E. Sofia ; Jonathan Mailoal ; Dirk Weiss ; Tonio Buonassisi ; Ian Marius Peters

CAUSE OF CURRENT-COLLECTION FAILURE OBSERVED INISC-REDUCTION PHASE OF PV CELLS AND MODULES EXPOSED TO ACETIC ACID 1268
Tadanori Tanahashi ; Norihiko Sakamoto ; Hajime Shibata ; Atsushi Masuda

COMPARISON OF PV MODULE PERFORMANCE BEFORE AND AFTER 11, 20, AND 25.5 YEARS OF FIELD EXPOSURE 1271
Jacob Rada ; Charles Chamberlin ; Peter Lehman ; Arne Jacobson

MARRYING QUALITY ASSURANCE WITH DESIGN ENGINEERING – A WINNING PARTNERSHIP! BUT, A CULTURAL DIVIDE? 1275
Sarah Kurtz ; Govind Ramu ; Robert Cornell ; Sumanth Lokanath ; Edward Hsi ; Tony Sample ; Masaaki Yamamichi ; George Kelly ; Ted Spooner ; Jonathan Previtali ; John Wohlgemuth

UPDATED EVALUATION OF SHOCK HAZARDS TO FIREFIGHTERS WORKING IN PROXIMITY OF PV SYSTEMS 1280
Olga Lavrova ; Jimmy E. Quiroz ; Jack Flicker ; Renee Gooding

GROWTH AND OPTIMIZATION OF GAINP/INP NANOWIRE TUNNEL DIODE 1286
Xulu Zeng ; Gaute Otnes ; Magnus Heurlin ; Magnus T Borgström

CATHODOLUMINESCENCE MAPPING FOR THE DETERMINATION OF N-TYPE DOPING IN SINGLE GAAS NANOWIRES 1289
Hung-Ling Chen ; Chalermchai Himwas ; Andrea Scaccabarozzi ; Pierre Rale ; Fabrice Oehler ; Aristide Lemaître ; Laurent Lombez ; Jean-François Guillemoles ; Maria Tchemycheva ; Jean-Christophe Harmand ; Andrea Cattoni ; Stéphane Collin

OPTICAL OPTIMIZATION OF PASSIVATED GAAS NANOWIRE SOLAR CELLS 1294
Kyle W. Robertson ; Ray R. Lapierre ; Jacob J. Krich

HIGH EFFICIENCY GAN NANOWIRE/SI PHOTOCATHODE FOR PHOTOELECTROCHEMICAL WATER SPLITTING 1299
Srinivas Vanka ; Sheng Chu ; Yichen Wang ; Ishiang Shih ; Hong Guo ; Zetian Mi

ANALYTIC DESCRIPTION OF THE IMPACT OF GRAIN BOUNDARIES ON VOC 1303
Paul Haney ; Benoit Gaury

ROLE OF TELLURIUM BUFFER LAYER ON CDTE SOLAR CELLS' ABSORBER/BACK-CONTACT INTERFACE 1308
Tao Song ; James R. Sites

SIMULTANEOUS EXAMINATION OF GRAIN-BOUNDARY POTENTIAL, RECOMBINATION, AND PHOTOCURRENT IN CDTE SOLAR CELLS USING DIVERSE NANOMETER-SCALE IMAGING 1312
C.S. Jiang ; H.R. Moutinho ; J. Moseley ; A. Kanevce ; J.N. Duenow ; E. Colegrove ; C. Xiao ; W.K. Metzger ; M.M. Al-Jassim

NANOPARTICLE/METAL REAR REFLECTORS FOR LOW- AND HIGH-TEMPERATURE SILICON SOLAR CELLS 1317
Syeda Qudsia ; Farah Qazi ; Mehwish Azher Javed ; Mathieu Boccard ; Zhengshan J. Yu ; Peter Firth ; Jonathan Bryan ; Zachary C. Holman

ABSORPTION IN EACH LAYER OF A SILICON HETEROJUNCTION SOLAR CELL 1322
Keith R. Mcintosh ; Malcolm D. Abbott ; Benjamin A. Sudbury ; Salman Manzoor ; Zhengshan J. Yu ; Mehdi Leilaeioun ; Jiatiwei Shi ; Zachary C. Holman

INVESTIGATIONS ON PLASMONIC COLOR TUNING COATING ON C-SI SOLAR CELLS ... 1329
Gerhard Peharz ; Wolfgang Waldhauser ; Christine Prietl ; Bettina Großschädl ; Martin C. Schubert ; Bernhard Michl

INVESTIGATION OF INTERFACE AND BULK LOCALIZED STATES IN A-SI:H SOLAR CELLS 1333
Adrien Bidiville ; Takuya Matsui ; Hitoshi Sai ; Koji Matsubara

EXPERIMENTAL AND THEORETICAL STUDY OF THE INFRARED EMISSIVITY OF CRYSTALLINE SILICON SOLAR CELLS 1339
Alberto Riverola ; Alexander Mellor ; Diego Alonso Alvarez ; Lourdes Ferre Llin ; Ilaria Guarracino ; Christos N. Markides ; Douglas Paul ; Daniel Chemisana ; Ned Ekins-Daukes

HIGH PERFORMANCE MOLECULAR DONORS FOR ORGANIC SOLAR CELLS, MATERIALS DESIGN AND DEVICE OPTIMIZATION 1342
Paul Geraghty ; Haotian Wang ; Calvin Lee ; Jegadesan Subbiah ; David Jones

ADVANCED OPTICAL MODELLING OF MICRO-TEXTURED SOLUTION-PROCESSED SOLAR CELLS WITH CONSIDERATION OF SMALL-AREA EFFECTS.................1346

Benjamin Lipovšek ; Marko Jošt ; Andrej Campa ; Fei Gu ; Christoph J. Brabec ; Karen Forberich ; Janez Krc ; Marko Tonic

IDENTIFICATION OF DEGRADATION PATHWAYS OF ORGANIC SOLAR CELLS USING INFRARED SPECTROSCOPY.................1350

S. Shah ; R Biswas ; T. Koschny ; V L Dalal

A DEVICE-INDEPENDENT SCREENING TECHNIQUE FOR RAPIDLY IDENTIFYING NEXT GENERATION OPV MATERIALS.................1354

Bryon W. Larson ; Andrew J. Ferguson ; Bertrand J. Tremolet De Villers ; Ross E. Larsen

NOVEL ANTHANTHRONE AND ANTHANTHRENE CO-POLYMERS AS P-TYPE CONJUGATED SEMICONDUCTORS FOR ORGANIC PHOTOVOLTAICS.................1360

Suru Vivian John ; Patrick Denk ; Christoph Ulbricht ; Herwig Heilbrunner ; Jean-Benoit Giguère ; Antoine Lafleur-Lambert ; Jean-Francois Morin ; Emmanuel Iwuoha ; Daniel Ayuk Mbi Egbe

REDUCING UV INDUCED DEGRADATION LOSSES OF SOLAR MODULES WITH C-SI SOLAR CELLS FEATURING DIELECTRIC PASSIVATION LAYERS.................1366

Robert Witteck ; Henning Schulte-Huxel ; Boris Veith-Wolf ; Malte Ruben Vogt ; Fabian Kiefer ; Marc Kontges ; Robby Peibst ; Rolf Brendel

LARGE-AREA JUNCTION DAMAGE IN POTENTIAL-INDUCED DEGRADATION OF C-SI SOLAR MODULES.................1371

Chuanxiao Xiao ; Chun-Sheng Jiang ; Steve Johnston ; Steve P. Harvey ; Peter Hacke ; Brian Gorman ; Mowafak Al-Jassim

SEARCH FOR MICROSTRUCTURAL DEFECTS AS NUCLEI FOR PID-SHUNTS IN SILICON SOLAR CELLS.................1376

Volker Naumann ; Otwin Breitenstein ; Jan Bauer ; Christian Hagendorf

INVESTIGATING PID SHUNTING IN POLYCRYSTALLINE SILICON MODULES VIA MULTI-SCALE, MULTI-TECHNIQUE CHARACTERIZATION.................1381

Steven P. Harvey ; John Moseley ; Adam Stokes ; Andrew Norman ; Brian Gorman ; Peter Hacke ; Steve Johnston ; Mowafak Al-Jassim

POTENTIAL-INDUCED DEGRADATION OF A SI NITRIDE/CRYSTALLINE SI INTERFACE OBSERVED THROUGH MINORITY CARRIER LIFETIME MEASUREMENT.................1385

Naoyuki Nishikawa ; Seira Yamaguchi ; Keisuke Ohdaira

FIELD INSPECTION OF PV MODULES: QUANTIFICATION OF EVA BROWNING LEVEL USING AN IMAGE PROCESSING TOOL.................1389

Sushanth Gudla ; Govindasamy Tamizhmani

PREVENTING POTENTIAL-INDUCED DEGRADATION IN CRYSTALLINE SILICON PV MODULES: RELATIONSHIP BETWEEN DEGRADATION AND BILL OF MATERIAL.................1395

Alessandro Virtuani ; Eleonora Annigoni ; Christophe Ballif

IDENTIFYING REVERSE-BIAS BREAKDOWN SITES IN CUINXGA(1-X)SE2.................1400

Steve Johnston ; Elizabeth Palmiotti ; Andreas Gerber ; Harvey Guthrey ; Lorelle Mansfield ; Timothy J. Silverman ; Mowafak Al-Jassim ; Angus Rockett

HIMAWARI-8 ENABLED REAL-TIME DISTRIBUTED PV SIMULATIONS FOR DISTRIBUTION NETWORKS.................1405

Nicholas A. Engerer ; Jamie M. Bright ; Sven Killinger

REDUCED MEASUREMENT UNCERTAINTY IN PV MODULE BATCH TESTING.................1411

Blagovest Mihaylov ; Bengt Jaeckel ; Juergen Arp ; Ralph Gottschalg

CLOUD MOTION IDENTIFICATION ALGORITHMS BASED ON ALL-SKY IMAGES TO SUPPORT SOLAR IRRADIANCE FORECAST.................1415

Lydie Magnone ; Fabrizio Sossan ; Enrica Scolari ; Mario Paolone

AUTOMATIC DETECTION OF INACTIVE SOLAR CELL CRACKS IN ELECTROLUMINESCENCE IMAGES.................1421

Sergiu Spataru ; Peter Hacke ; Dezso Sera

APPLYING SPATIAL DOWNSCALING AND SMART PERSISTENCE TO PROVIDE AN IMPROVED SOLAR FORECAST TO REDUCE COMMERCIAL DEMAND CHARGES.................1427

Alex Kubiniec ; Ted Belanger ; Adam Kankiewicz ; Skip Dise ; Nate Glasgow ; Alemu Tadesse

THERMAL CHARACTERISTICS OF PID-AFFECTED MONOCRYSTALLINE SILICON SOLAR MODULES UNDER ILLUMINATED AND DARK CONDITIONS.................1430

Pan Zhao ; Shuwen Guo ; He Wang ; Hong Yang ; Dengyuan Song ; Shiyu Sang ; Bojie Su ; Xue Zhang ; Yunxue Cao ; Hui Zhao

TARGETED EVALUATION OF UTILITY-SCALE AND DISTRIBUTED SOLAR FORECASTING.................1435

Matthew Lave ; Robert J. Broderick ; Laurie Burnham

RECORD EFFICIENCIES FOR SELENIUM PHOTOVOLTAICS AND APPLICATION TO INDOOR SOLAR CELLS 1441

Douglas M. Bishop ; Teodor Todorov ; Yun Seog Lee ; Oki Gunawan ; Richard Haight

CLOSE-SPACED SUBLIMATION FOR SB2SE3SOLAR CELLS 1445

Laurie J. Phillips ; Peter Yates ; Oliver S. Hutter ; Tom Baines ; Leon Bowen ; Ken Durose ; Jonathan D. Major

FABRICATION OF COPPER ARSENIC SULFIDE THIN FILMS FROM NANOPARTICLES FOR APPLICATION IN SOLAR CELLS 1449

Scott A. Mcclary ; Joseph Andler ; Carol A. Handwerker ; Rakesh Agrawal

ORIENTATION CONTROLLED GE THIN FILMS ON GLASS BY AL-INDUCED CRYSTALLIZATION 1452

Kaveh Shervin ; Khim Kharel ; Alexandre Freundlich

IN-LINE POTASSIUM FLUORIDE TREATMENT OF CIGS ABSORBERS DEPOSITED ON FLEXIBLE SUBSTRATES IN A PRODUCTION-SCALE PROCESS TOOL 1455

Ryan Kaczynski ; Jinwoo Lee ; Jane Van Alsburg ; Baosheng Sang ; Urs Schoop ; Jeffrey Britt

LIGHT-SOAK AND DARK-HEAT INDUCED CHANGES IN CU(IN, GA)SE2 SOLAR CELLS: A MACROSCOPIC TO MICROSCOPIC STUDY 1459

Rouin Farshchi ; Benjamin Hickey ; Dmitry Poplavskyy

A NEW MODEL TO DETERMINE INSTALLED SYSTEM COST AND LCOE FOR ARPA-E'S MOSAIC MICRO-CONCENTRATOR PV PROGRAM 1463

Ran Fu ; Kelsey A.W. Horowitz ; Daniel W. Cunningham ; James Zahler

FIXED-TILT 660 × CONCENTRATING PHOTOVOLTAIC SYSTEM WITH 30% EFFICIENCY 1469

Alex J. Grede ; Jared S. Price ; Baomin Wang ; Michael V. Lipski ; Brent Fisher ; Kyu-Tae Lee ; Junwen He ; Gregory S. Brulo ; Xiaokun Ma ; Scott Burroughs ; Christopher D. Rahn ; Ralph G. Nuzzo ; John A. Rogers ; Noel C. Giebink

WAFER INTEGRATED MICRO-SCALE CONCENTRATING PHOTOVOLTAICS 1473

Tian Gu ; Duanhui Li ; Lan Li ; Bradley Jared ; Gordon Keeler ; Bill Miller ; William Sweatt ; Scott Paap ; Michael Saavedra ; Ujjwal Das ; Steve Hegedus ; Anna Tanke-Pedretti ; Juejun Hu

TOWARD STATIONARY CONCENTRATOR PHOTOVOLTAIC PANELS 1476

Peter Kozodoy ; Christopher Gladden ; Michael Pavilonis ; Tobias Wheeler ; Christopher Rhodes ; Chadwick Casper ; Kevin Schneider

CPV TECHNOLOGIES NOT RELYING ON PERFECTION OF TRACKERS 1479

Kenji Araki ; Yasuyuki Ota ; Kan-Hua Lee ; Kensuke Nishioka ; Masafumi Yamaguchi

THE GETTERING EFFECT OF DIELECTRIC FILMS FOR SILICON SOLAR CELLS 1485

A. Y. Liu ; C. Sun ; V. P. Markevich ; A. R. Peaker ; J. D. Murphy ; D. Macdonald

TABULA RASA: OXYGEN PRECIPITATE DISSOLUTION THOUGH RAPID HIGH TEMPERATURE PROCESSING IN SILICON 1491

Erin E. Looney ; Hannu S. Laine ; Mallory A. Jensen ; Amanda Youssef ; Vincenzo Lasalvia ; Paul Stradins ; Tonio Buonassisi

TOWARD EFFECTIVE GETTERING IN BORON-IMPLANTED SILICON SOLAR CELLS 1494

Hannu S. Laine ; Ville Vähänissi ; Zhengjun Liu ; Ernesto Magaña ; Ashley E. Morishige ; Jan Krügener ; Kristian Salo ; Hele Savin ; Barry Lai ; David P. Fenning

IMPACT OF THE INITIAL GROWTH INTERFACE ON THE GRAIN STRUCTURE IN HPMC-SI INGOT 1498

Giri Wahyu Alam ; Etienne Pihan ; Benoit Marie ; Nathalie Mangelinck-Noël

EFFECT OF CARBON CONCENTRATION AND GROWTH CONDITIONS ON OXYGEN PRECIPITATION BEHAVIOR IN N-TYPE CZ-SI 1504

Takuto Kojima ; Ryota Suzuki ; Kosuke Kinoshita ; Kyotaro Nakamura ; Atsushi Ogura ; Yoshio Ohshita ; Isao Masada ; Shoji Tachibana

NANO-IMAGING OF PERFORMANCE IN PHOTOVOLTAICS 1508

Elizabeth M. Tennyson ; Marina S. Leite

IMPLICATIONS OF CONDUCTIVE GRAIN BOUNDARIES IN CHLORINE-TREATED CDTE SOLAR CELLS 1511

Mohit Tuteja ; Vasilios Palekis ; Allen Hall ; Scott Maclaren ; Chris S. Ferekides ; Angus A. Rockett

IMAGING THE MULTI-TEMPORAL PHOTO-CARRIER DYNAMICS AT THE NANOMETER SCALE IN ORGANIC AND INORGANIC SOLAR CELLS 1516

Pablo A. Fernández Garrillo ; Lukasz Borowik ; Florent Caffy ; Renaud Demadrille ; Benjamin Grévin

NANOSCALE TOMOGRAPHIC CHARGE TRANSPORT IN POLYCRYSTALLINE CHALCOGENIDE ABSORBERS: CDTE VERSUS CIGS 1522

Justin L. Luria ; Andrew Moore ; Sun Yu ; Mark Aindow ; Bryan D. Huey

IMPROVING THE PV MODULE SINGLE-DIODE MODEL ACCURACY WITH TEMPERATURE DEPENDENCE OF THE SERIES RESISTANCE 1526

Kyumin Lee

CELL-TO-MODULE (CTM) ANALYSIS FOR PHOTOVOLTAIC MODULES WITH SHINGLED SOLAR CELLS 1531

Max Mittag ; Tobias Zech ; Martin Wiese ; David Blasi ; Matthieu Ebert ; Harry Wirth

A PRACTICAL IRRADIANCE MODEL FOR BIFACIAL PV MODULES 1537

Bill Marion ; Sara Macalpine ; Chris Deline ; Amir Asgharzadeh ; Fatima Toor ; Daniel Riley ; Joshua Stein ; Clifford Hansen

A DETAILED MODEL OF REAR-SIDE IRRADIANCE FOR BIFACIAL PV MODULES 1543

Clifford W. Hansen ; Renee Gooding ; Nathan Guay ; Daniel M. Riley ; Johnson Kallickal ; Donald Ellibee ; Amir Asgharzadeh ; Bill Marion ; Fatima Toor ; Joshua S. Stein

VIEW FACTOR MODEL AND VALIDATION FOR BIFACIAL PV AND DIFFUSE SHADE ON SINGLE-AXIS TRACKERS 1549

Marc Abou Anoma ; David Jacob ; Ben C. Bourne ; Jonathan A. Scholl ; Daniel M. Riley ; Clifford W. Hansen

A FAST QUASI-STATIC TIME SERIES (QSTS) SIMULATION METHOD FOR PV IMPACT STUDIES USING VOLTAGE SENSITIVITIES OF CONTROLLABLE ELEMENTS 1555

Xiaochen Zhangl ; Santiago Grijalva ; Matthew J. Reno ; Jeremiah Deboever ; Robert J. Broderick

FAST DETERMINATION OF DISTRIBUTION-CONNECTED PV IMPACTS USING A VARIABLE-TIME-STEP QUASI-STATIC TIME-SERIES APPROACH 1561

Barry Mather

SCALABILITY OF THE VECTOR QUANTIZATION APPROACH FOR FAST QSTS SIMULATION 1567

Jeremiah Deboever ; Santiago Grijalva ; Matthew J. Reno ; Xiaochen Zhang ; Robert J. Broderick

MACHINE LEARNING FOR RAPID QSTS SIMULATIONS USING NEURAL NETWORKS 1573

Matthew J. Reno ; Robert J. Broderick ; Logan Blakely

ALGORITHMIC ASPECTS OF A COMMERCIAL-GRADE DISTRIBUTION SYSTEM LOAD FLOW ENGINE 1579

Francis Therrien ; Marc Belletête ; Jean-Sébastien Lacroix ; Matthew J. Reno

RESONANT AND NON-RESONANT DIELECTRIC COATINGS FOR HIGH EFFICIENCY SOLAR CELLS 1585

Dongheon Ha ; Chen Gong ; Marina S. Leite ; Jeremy N. Munday

ENHANCED LIGHT TRAPPING IN THIN SILICON SOLAR CELLS USING EFFECTIVELY TRANSPARENT CONTACTS (ETCS) 1589

Rebecca Saive ; André Augusto ; Stuart G. Bowden ; Harry A. Atwater

ENHANCED POWER CONVERSION EFFICIENCY IN SINGLE NANOWIRE DEVICES THROUGH SYMMETRY BREAKING DESIGN 1594

Jian Zhou ; Yonggang Wu ; Zihuan Xia ; Xuefei Qin ; Zongyi Zhang

CDSE(TE)/CDS/CDSE RODS VS. CDTE/CDS/CDSE SPHERES: MORPHOLOGY-DEPENDENT CARRIER DYNAMICS FOR PHOTON UPCONVERSION 1598

Eric Y. Chen ; Zhuohui Li ; Christopher C. Milleville ; Kyle R. Lennon ; Matthew F. Doty

DRIFT-DIFFUSION INGAN/GAN SOLAR CELL SIMULATOR WITH OPTICAL MANAGEMENT 1603

Y. Fang ; D. Guo ; A. Fischer ; E. Vadiee ; C. Zhang ; J. Williams ; S. M. Goodnick ; D. Vasileska

PERFORMANCE ENHANCEMENT OF A GAAS SOLAR CELL WITH COLLOIDAL QUANTUM DOTS EMBEDDED IN TRENCHES 1606

Chia-Jhe Shu ; Yu-Ming Huang ; Shun-Chieh Hsu ; Jinn-Kong Shu ; Jia-Lin Tsai ; Pei-Chen Yu ; Yung-Jr Hung ; Chien-Chung Lin

ENHANCED PHOTORESPONSE OF INN DEVICES USING INDIUM-TIN OXIDE NANORODS 1610

Lung-Hsing Hsu ; Yuh-Jen Cheng ; Peichen Yu ; Hao-Chung Kuo ; Chien-Chung Lin

PLASMONIC SILVER STRUCTURES FOR IMPROVED PEROVSKITE PHOTOVOLTAIC PERFORMANCE 1614

Arul Varman Kesavan ; Arun D Rao ; Praveen C Ramamurthy

QUANTUM CUTTING LUMINESCENT PMMA FILMS CONTAINING CE3+ - YB3+ CODOPED YAG PHOSPHOR FOR SI CONCENTRATOR SOLAR CELLS 1619

Lu Li ; Chaogang Lou ; Huihui Cao

NUMERICAL EVALUATION ON THE NANO-ROD ARRAY ON A N-SIDE-UP THIN-FILM GAAS SOLAR CELLS 1623

Po-Ching Wu ; Yan-Zhang Lin ; Shun-Chieh Hsu ; Chia-Jhe Hsu ; Chien-Chung Lin

DOWN SHIFTED CONVERSION FOR ENHANCED HIT SOLAR CELL EFFICIENCY 1627

Albert S. Lin ; Parag Parashar ; Wei-Ming Huang ; Yi-Wen Huang ; Ding-Rung Jian ; Ming-Hsuan Kao ; Shi-Wei Chen ; Chang-Hong Shen ; Jia-Min Shieh ; Tzu-Yu Chen ; Chien-Chung Lin ; Hao-Chung Kuo

THE PLANAR THERMOPHOTOVOLTAIC SELECTIVE NEARLY-PERFECT ABSORBERS/EMITTERS 1631

Parag Parashar ; Ding-Rung Jian ; Weiming Huang ; Vi-Wen Huang ; Albert Lin

HYBRID PEDOT:PSS SILICON SOLAR CELLS WITH PENCIL ROD STRUCTURES 1635
Ruei-Ying Wu ; Liang-Chian You ; Hsin-Fei Meng ; Chun-Chi Chen ; Peichen Yu

PL STUDY OF PHOSPHORUS-DOPED CDTE EVT FILMS 1638
Shamara Collins ; Imran Khan ; Vamsi Evani ; Chih An Hsu ; Vasilios Palekis ; Don Morel ; Chris Ferekides

CHARACTERIZATION OF SINGLE-SOURCE DEPOSITED CLOSE-SPACE SUBLIMATION CDTEXSE1-XTHIN FILM SOLAR CELLS 1643
Corey R. Grice ; Jian Li ; Yanfa Yan

THE INFLUENCE OF THE CU-RICH/CU-POOR SEQUENCE ON THE PROPERTIES OF CU(IN, GA)SE2 FILMS DEPOSITED BY IN-LINE CO-EVAPORATION PROCESS 1648
He Wang ; Fang Fang Liu ; Yi Tong Yang ; Li You Yao ; Peng Gao ; Zhi Bin Xiao ; Qiang Sun

DETERMINATION AND MODELING OF INJECTION DEPENDENT SERIES RESISTANCE IN CIGS SOLAR CELLS 1651
Vito Huhn ; Bart E. Pieters ; Andreas Gerber ; Yael Augarten ; Uwe Rau

LARGE GRAIN GROWTH IN CU2ZNSNS4 THIN FILMS IN THE ABSENCE OF NA USING RAPID THERMAL ANNEALING 1656
J. L. Johnson ; A. Bhatia ; J. G. Bolke ; M. A. Scarpulla

CU2ZNSNS4THIN FILMS SYNTHESIZED BY COSPUTTERING AND RAPID THERMAL ANNEALING: EFFECTS OF COMPOSITION AND TEMPERATURE 1661
J.L. Johnson ; W.M. Hlaing Oo ; M. Karmarkar ; M.A. Scarpulla

EARTH-ABUNDANT CZTSSE THIN FILM SOLAR CELLS ON FLEXIBLE STAINLESS STEEL FOIL SUBSTRATES 1665
Hae-Sun Kim ; Woo-Lim Jeong ; Dong-Seon Lee

COMPARISON OF MGCL2AND CDCL2ACTIVATION TREATMENT FOR CDTE SOLAR CELLS: RECRYSTALLIZATION AND DEFECTS 1669
Daniele Menossi ; Elisa Artegiani ; Ivan Rimmaudo ; Alessia Le Donne ; Simona Binetti ; Juan Luis Pena ; Fabio Piccinelli ; Alessandro Romeo

CHARACTERIZATION OF CDTE PHOTOVOLTAIC DEVICES PASSIVATED USING HYDROGEN PLASMA 1674
Amit Munshi ; Piotr Kaminski ; Ali Abbas ; Shiva Tarun Chenna ; Sreeram Chandralal ; John Walls ; Walajabad Sampath

GROUP-V DOPING IMPACT ON CD-RICH CDTE SINGLE CRYSTALS GROWN BY TRAVELING-HEATER METHOD 1679
Akira Nagaoka ; Kenji Yoshino ; Yoshitaro Nose ; Darius Kuciauskas ; Michael A. Scarpulla

BAND-GAP ENGINEERING IN CU2ZNSN(S,SE)4SOLAR CELLS BY POST-SULPHURIZATION OF SELENIZED ABSORBER LAYERS 1682
Markus Neuwirth ; Elisabeth Seydel ; Heinz Kalt ; Michael Hetterich

IMPACT OF GA/III PROFILE ON VOLTAGE-DEPENDENT COLLECTION LOSSES IN CIGS SOLAR CELLS 1686
Dmitry Poplavskyy ; Jeff Bailey ; Rouin Farshchi ; David Spaulding

CL DIFFUSION IN CDTE SOLAR CELLS ACTIVATED BY GASEOUS CHCLF2ATMOSPHERE 1691
I. Rimmaudo ; R. Mis Fernandez ; V. Rejon ; A. Abbas ; F. Lisco ; J.M. Walls ; J.L. Peña

STABILITY OF CD1-XZNXTE ALLOYS UNDER CDTE PROCESSING CONDITIONS 1697
Yegor Samoilenko ; Colin A. Wolden

CIGSE ABSORBER PREPARATION: AN ALTERNATIVE TO H2SE 1701
O.S. Shinde ; E.J. Schenller ; S.R. Jadkar ; S.V Ghaisas ; N. Dhere

CHARGE CONTROLLED SEQUENTIAL ELECTRODEPOSITION FOR SYNTHESIS OF CU2ZNSNS4ON MO-COATED GLASS SUBSTRATE 1704
Ashish K. Singh ; Rajiv Dubey ; Manoj Neergat ; Kavaipatti R. Balasubramaniam

EFFECT OF DEPOSITED PRESSURE ON THE CDTE THIN FILMS BY CLOSED SPACE SUBLIMATION METHOD 1707
Yufeng Zhang ; Zhongming Du ; Xiangxin Liu

ANALYZING THE COST REDUCTION POTENTIAL OF III-V/SI HYBRID CONCENTRATOR PHOTOVOLTAIC SYSTEMS 1711
Kan-Hua Lee ; Kenji Araki ; Masafumi Yamaguchi

GENERALIZED NUMERICAL DESIGN OF AXIALLY-ASYMMETRICAL AND GRID-ARRANGED STATIC CPV ARRAY FOR MAXIMIZING ANNUAL ENERGY GENERATION 1714
Kenji Araki ; Kan-Hua Lee ; Masafumi Yamaguchi

SPECTRAL TRANSMITTANCE ANALYSIS OF LIQUIDS FOR HIGH CONCENTRATION III-V PHOTOVOLTAIC IMMERSION COOLING APPLICATIONS 1719
Xinyue Han ; Yongjie Guo

OPTICAL DESIGN FOR 2-TERMINAL III-V/SI SMAC MODULE 1724
Masaaki Baba ; Kikuo Makita ; Hidenori Mizuno ; Hidetaka Takato ; Takeyoshi Sugaya ; Noboru Yamada

DESIGN OF OPTICAL ELEMENTS FOR LOW PROFILE CPV PANEL WITH SUN TRACKING FOR ROOFTOP INSTALLATION...................1728

Xinbing Liu ; Zhou Lu ; Riccardo Leto ; Carlton Brule ; Nanu Brates

MICRO CHIPLET PRINTER DEVELOPMENT FOR MOSAIC PROGRAM 1733

P.Y. Maeda ; Y. D. Wang ; S. Raychaudhuri ; J. Kalb ; D. K. Biegelsen ; R. Lujan ; Q. Wang ; Y. Wang ; J. Bert ; B. Rupp ; I. Matei ; L. Crawford ; A. Plochowietz ; E.M. Chow ; J.P. Lu ; V. Gupta

MICRO-OPTICAL TANDEM LUMINESCENT SOLAR CONCENTRATOR 1737

David R. Needell ; Zach Nett ; Ognjen Ilic ; Colton R. Bukowsky ; Junwen He ; Lu Xu ; Ralph G. Nuzzo ; Benjamin G. Lee ; John F. Geisz ; A. Paul Alivisatos ; Harry A. Atwater

INCREASE IN MAXIMUM POWER OF A-SI, C-SI AND GAAS.76P.24 SOLAR CELLS UNDER LOW CONCENTRATION...................1741

Hiba Riaz ; Sabina Abdul Hadi ; Ammar Nayfeh

DESIGN AND EVALUATION OF PARTIAL CONCENTRATION III-V/SI MODULE WITH ENHANCED DIFFUSE SUNLIGHT TRANSMISSION 1743

Daisuke Sato ; Noboru Yamada ; Kan-Hua Lee ; Kenji Araki ; Masafumi Yamaguchi

CONTAMINATION CONTROL CHALLENGES ON SHJ SOLAR CELL PROCESSING 1747

G. Condorelli ; P. Rotoli ; A. Canino ; A. Battaglia ; W. Favre ; A. -S. Ozanne ; A. Moustafa ; A. Danel ; D. Muñoz ; P. -J. Ribeyron ; C. Gerardi

>23% SILICON HETEROJUNCTION SOLAR CELLS IN MEYER BURGER'S DEMO LINE: RESULTS OF PILOT PRODUCTION ON MASS PRODUCTION TOOLS...................1752

J. Zhao ; M. König ; A. Wissen ; V. Breus ; D. Deckerl ; M. Fritzsche ; M. Schorch ; H. J. Nonnenmacher ; M. Leonhardt ; T. Große ; J. Hausmann ; A. Waltmger ; D. Landgraf ; S. Burkhardt ; H. Mehlich ; E. Vetter ; F. Schitthelm ; Y. Yao ; T. Söderström ; A. Richter ; D. Habermann ; S. Leu

EXPERIMENTAL AND SIMULATION STUDIES ON TIO2/SILICON HETEROJUNCTION DIODES...................1755

Swasti Bhatia ; Neha Raorane ; Nimisha Sreekumar ; Pradeep R. Nair ; Aldrin Antony

A STUDY ON BLISTER FORMATION AND ELECTRICAL PROPERTIES UNDER VARIOUS ANNEALING CONDITION FOR TUNNELING OXIDE PASSIVATION LAYER...................1758

Sungjin Choi ; Ka-Hyun Kim ; Min Gu Kang ; Jeong In Lee ; Donghwan Kim ; Hee-Eun Song

PROCESSING APPROACHES AND CHALLENGES OF INTERDIGITATED BACK CONTACT SI SOLAR CELLS...................1761

Ujjwal Das ; Lei Zhang ; Steven Hegedus

FABRICATION OF CUI/A-SI:H/C-SI STRUCTURE FOR APPLICATION TO HOLE-SELECTIVE CONTACTS OF HETEROJUNCTION SI SOLAR CELLS...................1765

Kazuhiro Gotoh ; Min Cui ; Nguyen Cong Thanh ; Koichi Koyama ; Isao Takahashi ; Yasuyoshi Kurokawa ; Hideki Matsumura ; Noritaka Usami

CHARACTERISTICS OF THIN CRYSTALLINE SILICON SOLAR CELLS WITH RIB STRUCTURE 1769

Yukimi Ichikawa ; Shuhei Yoshiba ; Masakazu Hirai ; Makoto Konagai

MEASUREMENT OF TIO2/P-SI SELECTIVE CONTACT PERFORMANCE USING A HETEROJUNCTION BIPOLAR TRANSISTOR WITH A SELECTIVE CONTACT EMITTER...................1773

Janam Jhaveri ; Alexander Berg ; Sigurd Wagner ; James C. Sturm

EFFECT OF GROWTH AND POST-OXIDATION ANNEALING TEMPERATURE OF THERMALLY GROWN TUNNELING SIOX, ON THE IIMPLIED VOCOF PASSIVATED CONTACTS FOR C-SI BASED SOLAR CELLS...................1777

Abhijit S. Kale ; William Nemeth ; Matthew Page ; Sumit Agarwal ; Paul Stradins

PARTIALLY CONTACTED SURFACES WITH CONTACT SIZE IN THE 1 µM RANGE FOR C-SI PERC SOLAR CELLS...................1781

R. Khoury ; I. Martín ; G. López ; C. Jin ; J.M. López-González ; L. Zeyu ; P. Bulkin ; E.V. Johnson ; R. Alcubilla

ENTRANCE OF LOW COST FABRICATION OF BACK-CONTACT HETEROJUNCTION SOLAR CELLS BY USING PLASMA ION IMPLANTATION 1787

Koichi Koyama ; Keisuke Ohdaira ; Hideki Matsumura

TLM MEASUREMENTS VARYING THE INTRINSIC A-SI:H LAYER THICKNESS IN SILICON HETEROJUNCTION SOLAR CELLS 1790

Mehdi Leilaeioun ; William Weigand ; Pradyumna Muralidharan ; Mathieu Boccard ; Dragica Vasileska ; Stephen Goodnick ; Zachary Holman

SOLAR CELLS APPLICATION OF P-TYPE POLY-SI THIN FILM BY ALUMINUM INDUCED CRYSTALLIZATION 1794

Shota Masuda ; Kazuhiro Gotoh ; Isao Takahashi ; Kyotaro Nakamura ; Yoshio Ohshita ; Noritaka Usami

A SELF - CONSISTENTLY COUPLED DRIFT DIFFUSION AND MONTE CARLO SIMULATOR TO MODEL SILICON HETEROJUNCTION SOLAR CELLS...................1797

Pradyumna Muralidharan ; Stuart Bowden ; Stephen M. Goodnick ; Dragica Vasileska

DOPANT PATTERNING BY PECVD AND MECHANICAL MASKING FOR PASSIVATED TUNNELING CONTACT IBC CELL ARCHITECTURES 1801

William Nemeth ; Vincenzo Lasalvia ; Benjamin G. Lee ; Abhijit Kale ; Paul Stradins

ALD ALUMINUM OXIDE AS A HOLE SELECTIVE TUNNELING CONTACT FOR CRYSTALLINE SILICON SOLAR CELLS 1804

Kortan Ögütman ; Kristopher O. Davis ; Winston V. Schoenfeld ; Michael Haslinger ; Sofie Robert ; Emanuele Cornagliotti ; Joachim John

SCREEN PRINTED, LARGE AREA BIFACIAL N-PERT CELLS WITH TUNNEL OXIDE PASSIVATED BACK CONTACT 1807

Young-Woo Ok ; Ajay D Upadhyaya ; Brian Rounsaville ; Ying-Yuan Huang ; Vijaykumar D Upadhyaya ; Ajeet Rohatgi

CORRELATION BETWEEN ELECTROLUMINESCENCE AND PHOTOCONVERSION EFFICIENCY IN A-SI:H/C-SI HETEROJUNCTION SOLAR CELLS 1811

A.V. Sachenko ; A.V. Bobyl ; V.N. Verbitskiy ; V.M. Vlasyuk ; D.M. Zhigunov ; V.P. Kostylyov ; I.O. Sokolovskyi ; E.I. Terukov ; P.A. Forsh ; M. Evstigneev

AN ISOTOPE STUDY OF HYDROGEN PASSIVATION OF POLY-SI/SIOXPASSIVATED CONTACTS FOR SI SOLAR CELLS 1817

Manuel Schnabel ; William Nemeth ; Bas W.H. Van De Loo ; Bart Macco ; Wilhelmus M.M. Kessels ; Paul Stradins ; David L. Young

ALLEVIATING HYDROGEN PLASMA DAMAGE TO AMORPHOUS/CRYSTALLINE SILICON INTERFACE PASSIVATION 1820

Jianwei Shi ; Zachary C. Holman

LARGE-AREA N-TYPE TOPCON CELLS WITH SCREEN-PRINTED CONTACT ON SELECTIVE BORON EMITTER FORMED BY WET CHEMICAL ETCH-BACK 1824

Yuguo Tao ; Felix Book ; Barbara Terheiden ; Viiaykumar Upadhvaya ; Keeya Madani ; Brian Rounsaville ; Eunhwan Cho ; Ajeet Rohatgi

HYDROGEN PLASMA POST-DEPOSITION TREATMENT FOR PASSIVATION OF A-SI/C-SI INTERFACE FOR HETEROJUNCTION SOLAR CELL BY CORRELATING OPTICAL EMISSION SPECTROSCOPY AND MINORITY CARRIER LIFETIME 1828

Anishkumar Soman ; Ugochukwu Nsofor ; Lei Zhang ; Ujjwal Das ; Tingyi Gu ; Steve Hegedus

MEASURING DIODE RESISTIVITY OF PASSIVATED CONTACTS 1832

San Theingi ; William Nemeth ; David L. Young ; Paul Stradins ; Benjamin G. Lee

ULTRA-THIN CRYSTALLINE SILICON SOLAR CELLS WITH NICKEL OXIDE INTERLAYER AS HOLE-SELECTIVE CONTACT 1835

Muyu Xue ; Raisul Islam ; Junyan Chen ; Zheng Lyu ; Yusi Chen ; Daniel Dewitt ; Albert Pleus ; Christian Tae ; Ching-Ying Lu ; Kai Zang ; Jieyang Jia ; Yijie Huo ; Ted Kamins ; Krishna Saraswat ; James Harris

CRYSTALLINE SI SOLAR CELLS WITH PASSIVATING, CARRIER-SELECTIVE NICKEL OXIDE CONTACTS 1838

Woojun Yoon ; James Moore ; David Scheiman ; Eunhwan Cho ; Young-Woo Ok ; Nicole Kotulak ; Phillip P. Jenkins ; Ajeet Rohatgi ; Robert J. Walters

GAP/SI HETEROJUNCTION SOLAR CELLS GROWN BY MOLECULAR BEAM EPITAXY 1841

Chaomin Zhang ; Ehsan Vadiee ; Richard R. King ; Christiana B. Honsberg

SPIN COATED NICKEL OXIDE AND VANADIUM OXIDE LAYERS ON SILICON FOR A CARRIER SELECTIVE CONTACT SOLAR CELL 1845

Jing Zhao ; Fa-Jun Ma, Jae-Yun ; Anita Ho-Baillie ; Stephen Bremner

QUANTIFICATION OF PV MODULE DISCOLORATION USING VISUAL IMAGE ANALYSIS 1850

Shashwata Chattopadhyay ; Chetan Singh Solanki ; Anil Kottantharayil ; K.L. Narasimhan ; Juzer Vasi ; Sai Tatapudi ; Govindasamy Tamizhmani

TEMPERATURE AND POWER STUDY OF ADHERED AND RACKED DOUBLE GLASS PHOTOVOLTAIC MODULES 1855

Volker Beutner ; Rubina Singh ; Cameron Stark

FIELD INSPECTION OF PV MODULES: QUANTITATIVE DETERMINATION OF PERFORMANCE LOSS DUE TO CELL CRACKS USING EL IMAGES 1858

Carlos A. Rodríguez Castañeda ; Shashwata Chattopadhyay ; Jaewon Oh ; Sai Tatapudi ; Govindasamy Tamizhmani ; Hailin Hu

SCALE UP DESIGNS FOR HAND-HELD LIGHT-WEIGHT TPV DC POWER SUPPLY 1863

L. M. Fraas ; J. E. Avery ; L. Minkin ; Hui She ; L. Ferguson

HIGH EFFICIENCY ANTI-REFLECTIVE COATING FOR PV MODULE GLASS 1869

Brennen M. Freiburger ; Corey S. Thompson ; Robert A. Fleming ; Douglas Hutchings ; Sergiu C. Pop

INVESTIGATION OF EFFICIENCY FOR PID-AFFECTED SOLAR MODULE AT NONSTANDARD TEST CONDITIONS 1873

Shuwen Guo ; Pan Zhao ; Weijing Huang ; Jipeng Chang ; He Wang ; Hong Yang ; Chengfeng Su ; Bojie Su ; Xue Zhang ; Yunxue Cao ; Hui Zhao

THERMAL UNIFORMITY MAPPING OF PV MODULES AND PLANTS 1877
Ashwini Pavgi ; Jaewon Oh ; Joseph Kuitche ; Sai Tatapudi ; Govindasamy Tamizhmani

CLIMATE-SPECIFIC THERMAL MODEL COEFFICIENTS FOR C-SI AND THIN-FILM PV MODULES 1883
Ashwini Pavgi ; Joseph Kuitche ; Jaewon Oh ; Govindasamy Tamizhmani

EFFECT OF THE THERMOPHYSICAL PROPERTIES OF A PHASE CHANGE MATERIAL ON THE ELECTRICAL OUTPUT OF A CONCENTRATED PHOTOVOLTAIC SYSTEM 1888
Jawad Sarwar ; Ahmed E. Abbas ; Konstantinos E. Kakosimos

PASSIVE COOLING OF PHOTOVOLTAICS WITH DESICCANTS 1893
Lin J. Simpson ; Jason Woods ; Nicolas Valderrama ; Alex Hill ; Nina Vincent ; Timothy Silverman

MODIFIED MAXIMUM POWER EXTRACTION TECHNIQUE FOR RAPIDLY CHANGING NUI AND DYNAMIC LOADS 1898
U Aswani ; S.P. Duttagupta ; T.I. Eldho ; B.V. Rao

REAL-TIME MONITORING OF PHOTO VOLTAIC RELIABILITY ONLY USING MAXIMUM POWER POINT - THE SUNS-VMP METHOD 1904
Xingshu Sun ; Haejun Chung ; Raghu Vamsi Krishna Chavali ; Peter Bermel ; Muhammad Ashraful Alam

PHOTOVOLTAIC MODULE DURABILITY AND RELIABILITY: ANALYSIS OF A 23-YEAR-OLD ARRAY OPERATING IN QUEBEC, CANADA 1908
Christopher Baldus-Jeursen ; Alexandre Côté ; Naveen Goswamy ; Tanya Deer ; Yves Poissant

ARE E-W TRACKERS A BETTER OPTION FOR FUTURE INVESTMENTS IN PV SECTOR-A DETAILED TECHNO-COMMERCIAL STUDY 1912
Rakesh Bohra ; Ramesh Rame Gowda ; Mani R. Krishnan

EXPERIMENTAL EVALUATION OF THE PERFORMANCE OF CRYSTALLINE SI PV MODULE DEGRADATION AFTER 15-YEARS OF FIELD EXPOSURE 1917
Denio A. Cassini ; Antonia Sônia A. C. Diniz ; Marcelo Machado Viana ; Michele C. C. De Oliveira ; F. C. Lins Vanessa De ; Roberto Zilles ; Lawrence L. Kazmerski

FIELD INVESTIGATIONS OF POTENTIAL-INDUCED DEGRADATION (PID) FOR CRYSTALLINE SILICON PV PANELS IN DIFFERENT CLIMATES 1922
Yifeng Chen ; Peter Hacke ; Yong Sheng Khoo ; Kaitlyn Vansant ; Zigang Wang ; Wei Luo ; Jing Chai ; Chris Deline ; Yan Wang ; Armin G. Aberle ; Pietro P. Altermatt ; Zhiqiang Feng ; Sarah Kurtz ; Pierre J. Verlinden

DETERMINING THE POWER RATE OF CHANGE OF 353 PLANT INVERTERS TIME-SERIES DATA ACROSS MULTIPLE CLIMATE ZONES, USING A MONTH-BY-MONTH DATA SCIENCE ANALYSIS 1927
Alan J. Curran ; Yang Hu ; Rojiar Haddadian ; Jennifer L. Braid ; David Meakin ; Timothy J. Peshek ; Roger H. French

PHOTOVOLTAIC ARRAY DIFFERENTIAL BACKSIDE EXPOSURE CONDITIONS: BACKSHEET DEGRADATION AND SITE DESIGN 1933
Andrew Fairbrother ; Julien Avenet ; Yadong Lyu ; Matthew Boyd ; Scott Julien ; Kai-Tak Wan ; Liang Ji ; Kenneth Boyce ; Sebastien Merzlic ; Amy Lefebvre ; Greg O'Brien ; Yu Wang ; Laura Bruckman ; Roger French ; Michael Kempe ; Brian Dougherty ; Xiaohong Gu

STUDY ON RANDOM FAILURE OF CRYSTALLINE SILICON SOLAR MODULES IN THE FIELD 1937
Xuefang Jiang ; Fumei Wang ; Ao Wang ; Hong Yang ; He Wang ; Jie Ding ; Junjun Zhang ; Jingsheng Huang

POTENTIAL INDUCED DEGRADATION (PID) POWER LOSS CORRELATION TO LEAKAGE AND REVERSE BIAS CURRENTS 1941
Michalis Florides ; Georgios Konstantinou ; Venizelos Venizelou ; George Makrides ; George E. Georghiou

PERFORMANCE STUDY OF VARIOUS PV MODULE TECHNOLOGIES IN DESERT CONDITIONS 1946
Jim J John ; Ammar Elnosh ; Anwar Almheiri ; Wadhah Alzahmi ; Marco Stefancich ; Pedro Banda

HIGH-SPEED MEASUREMENTS OF GENERATED POWER AND ITS RELATIONSHIP TO WEATHER OBSERVATIONS AT YOSHINOGARI MEGA SOLAR POWER PLANT 1950
Makoto Kasu ; Shigeomi Hara ; Takumi Uematsu

IMPACT OF MISSING DATA ON THE ESTIMATION OF PHOTOVOLTAIC SYSTEM DEGRADATION RATE 1954
Andreas Livera ; Alexander Phinikarides ; George Makrides ; George E. Georghiou

FIELD DEGRADATION AND FAILURES OF AGED CRYSTALLINE SILICON PV MODULES IN MEXICO 1959
D. Martínez Escobar ; P. A. Sánchez-Pérez ; Rocío De La Luz Santos Magdaleno ; José Ortega Cruz ; Sai Tatapudi ; Aarón Sánchez Juárez ; Govindasamy Tamizhmani

RAPID SHUTDOWN WITH PANEL LEVEL ELECTRONICS-A SUITABLE SAFETY MEASURE? 1965
Adam Cordova ; Christopher Merz ; Gerd Bettenwort ; Markus Hopf ; Hannes Knopf ; Joachim Laschinski

INVESTIGATING A NEW OPERATING POINT FOR PV PANELS SEEKING MAXIMUM LIFE SPAN...... 1968
Bechara Nehme ; Nacer K. M'sirdi ; Tilda Akiki

POWER GENERATION EVALUATION OF LARGE-SCALE PHOTOVOLTAIC SYSTEMS LOCATED ON INCLINED PLANE...... 1973
Naotaka Oka ; Yasuhito Takahashi ; Koji Fujiwara ; Kazuyuki Hidaka ; Hiroshi Morita

INVESTIGATING THE IMPACT OF SOLAR CELLS PARTIAL SHADING ON PHOTOVOLTAIC MODULES BY THERMOGRAPHY...... 1979
David Pera ; José A. Silva ; Sara Costa ; João M. Serra

ANNUAL DEGRADATION RATE AND ITS LINEARITY ANALYSIS USING METERED KWH DATA...... 1984
Christopher Raupp ; Govindasamy Tamizhmani

ELECTRICAL PERFORMANCE ANALYSIS OF A 27 KW GRID-CONNECTED PV SYSTEM WITH SOILING AND SHADING IN MORELOS MEXICO...... 1990
P. A. Sánchez-Pérez ; D. Martínez Escobar ; E. O. Ángel Ruiz ; R. Santos Magdaleno ; José Ortega Cruz ; A. Sánchez Juárez

MODIFIED STC CORRECTION PROCEDURE FOR ASSESSING PV MODULE DEGRADATION IN FIELD SURVEYS...... 1995
Hemant K. Singh ; R. Dubey ; S. Zachariah ; K. L. Narasimhan ; B. M. Arora ; A. Kottantharayil ; J. Vasi

DEGRADATION MODELS OF PHOTOVOLTAIC MODULE BACKSHEETS EXPOSED TO DIVERSE REAL WORLD CONDITION...... 2000
Yu Wang ; Sebastien Merzlic ; Andrew Fairbrother ; Scott Julien ; Lucas Fridman ; Camille Loyer ; Amy L. Lefebvre ; Gregory O'Brien ; Xiaohong Gu ; Liang Ji ; Ken Boyce ; Michael Kempe ; Kai-Tak Wan ; Roger H. French ; Laura S. Bruckman

ADDRESSING HOTSPOTS IN THE PRODUCT ENVIRONMENTAL FOOTPRINT OF CDTE PHOTOVOLTAICS...... 2005
Parikhit Sinha ; Andreas Wade

PHOTOVOLTAIC SMART HOME SYSTEM - DUBAI CASE STUDY...... 2011
Ammar Natsheh ; Marwa Aljaziri ; Maitha Moosa ; Gharibah Essa ; Hassa Moosa

DIRECT DRIVE PHOTOVOLTAIC MILK CHILLING EXPERIENCE IN KENYA...... 2014
Robert Foster ; Brian Jensen ; Brian Dugdill ; Wendy Hadley ; Bruce Knight ; Abudul Faraj ; Johnson Kyalo Mwove

COST OPTIMIZATION OF DECOMMISSIONING AND RECYCLING CDTE PV POWER PLANTS...... 2019
V. Fthenakis ; Z. Zhang ; J. -K Choi

CHALLENGES FOR DECISION MAKERS WHEN FEED-IN TARIFFS OR NET METERING SCHEMES CHANGE TO INCENTIVES DEPENDENT ON A HIGH SHARE OF SELF-CONSUMED ELECTRICITY...... 2025
Mattias Gustafsson

PROCEDURES TO MAKE PROJECTS ABOUT RENEWABLE ENERGY GENERATION CONNECTED TO THE GRID IN COLOMBIA...... 2031
J. A. Hernandez ; C. A. Arredondo ; D. J. Rodriguez

A CRITICAL ANALYSIS ON THE THIN CRYSTALLINE SILICON PV MODULE OF THE LIGHTWEIGHT PV SYSTEM...... 2035
Meixi Chen ; Abhishek Iyer ; Cheng-Hao Shih ; Lado Kurdgelashvili ; Robert Opila

PHOTOVOLTAIC MODULE MANUFACTURING COSTS, AVERAGE PRICES AND INDUSTRY BALANCE 2006–2016...... 2039
Paula Mints ; Zhengshan J Yu

SOLAR CELL AND WIND ENERGY REPLACEMENT OF POWER PLANTS GLOBALLY...... 2042
Larry Partain ; Shirley Hansen ; Dirk Bennett ; Richard Hansen ; Allan Newlands ; Lewis Fraas

ANALYSIS OF LIGHT ENVIRONMENT UNDER SOLAR PANELS AND CROP LAYOUT...... 2048
Deng Wang ; Yaojie Sun ; Yandan Lin ; Yuan Gao

INTERFACE EFFECTS OF ALKALI TREATMENT ON CU-RICH THIN FILM SOLAR CELLS...... 2054
Hossam Elanzeery ; Finn Babbe ; Anastasiya Zelenina ; Michele Melchiorre ; Susanne Siebentritt

INCREASEDVOCAND FF IN ZNO1-XSX-BUFFERED CUIN1-XGAXSE2SOLAR CELLS BY CADMIUM PARTIAL ELECTROLYTE TREATMENT...... 2058
Andreas Bauer ; Dimitrios Hariskos ; Wiltraud Wischmann

PASSIVATING AND CARRIER-SELECTIVE CONTACTS - BASIC REQUIREMENTS AND IMPLEMENTATION...... 2064
S.W. Glunz ; M. Bivour ; C. Messmer ; F. Feldmann ; R. Müller ; C. Reichel ; A. Richter ; F. Schindler ; J. Benick ; M. Hermle

FIRST-PRINCIPLES MODELING OF ALKALI METAL POST DEPOSITION TREATMENT EFFECTS IN CIGS SOLAR CELLS..2070

Maria Fedina ; Hannu-Pekka Komsa ; Ville Havu ; Martti J. Puska

EXPLORING SILICON CARBIDE- AND SILICON OXIDE-BASED LAYER STACKS FOR PASSIVATING CONTACTS TO SILICON SOLAR CELLS...2073

P. Löper ; G. Nogay ; P. Wyss ; M. Hyvl ; P. Procel ; J. Stuckelberger ; A. Ingenito ; I. Mack ; Q. Jeangros ; M. Ledinsky ; A. Fejfar ; C. Allebé ; J. Horzel ; M. Despeisse ; F. Crupi ; F.-J. Haug ; C. Ballif

EFFICIENT ELECTRON CONTACTS FORN-TYPE SILICON SOLAR CELLS USING MAGNESIUM METAL, OXIDE, AND FLUORIDE...2076

Yimao Wan ; Chris Samundsett ; James Bullock ; Di Yan ; Thomas Allen ; Jun Peng ; Jie Cui ; Mark Hettick ; Ali Javey ; Andres Cuevas

GRADED (ALZGA1-Z)XIN1-XP WINDOW-EMITTER STRUCTURES FOR IMPROVED SHORT-WAVELENGTH RESPONSE..2079

Jacob T. Boyer ; Daniel L. Lepkowski ; Daniel J. Chmielewski ; Steven A. Ringel ; Tyler J. Grassman

INTEGRATION OF QUANTUM DOTS AND QUANTUM WELLS INTO INGAAS METAMORPHIC SUBCELL FOR RADIATION HARD 3-J ELO IMM PHOTOVOLTAICS..............2084

Zachary S. Bittner ; Hyun Kum ; Michael A. Slocum ; George T. Nelson ; Rao Tatavarti ; Andre Wibowo ; Seth M. Hubbard

PROTON IRRADIATION OF 3J SOLAR CELLS AT LOW TEMPERATURE......................2087

Seonyong Park ; Jacques C. Bourgoin ; Olivier Cavani ; Sandrine Picard ; Jérôme Bourcois ; Victor Khorenko ; Carsten Baur ; Bruno Boizot

ULTRA-THIN GAAS SOLAR CELLS: RADIATION TOLERANCE AND SPACE APPLICATIONS................2091

Louise C. Hirstl ; Michael K. Yakes ; Jeffery. H. Warner ; Mitchell F. Bennett ; Kenneth J. Schmieder ; Stephanie Tomasulo ; Erin Cleveland ; Sergey Maximenko ; James Moore ; Robert J. Walters ; Phillip P. Jenkins

LARGE AREA MULTIJUNCTION III-V SPACE SOLAR CELLS OVER 31% EFFICIENCY..........................2094

X.Q. Liu ; C. Fetzer ; P. Chiu ; M. Haddad ; X. Zhang ; R. Cravens ; D. Law ; J. Ermer ; J. Krogen ; S. Sharma ; J. Hanley

ADVANCED-ARCHITECTURE HIGH-EFFICIENCY SOLAR CELLS FOR LOW IRRADIANCE LOW TEMPERATURE (LILT) APPLICATIONS...2099

Andreea Boca ; Jonathan Grandidier ; Claiborne Mcpheeters ; Paul Sharps ; Philip Chiu ; Xing-Quan Liu ; James Ermer

ULTRA-LIGHTWEIGHT PV MODULE DESIGN FOR BUILDING INTEGRATED PHOTOVOLTAICS..2104

Ana C. Martins ; Valentin Chapuis ; Alessandro Virtuani ; Christophe Ballif

DESIGN IT WITH LSCS; AN EXPLORATION OF APPLICATIONS FOR LUMINESCENT SOLAR CONCENTRATOR PV TECHNOLOGIES..2109

Wouter Eggink ; Angèle Reinders

INVESTIGATING PV-BATTERY 3-TERMINAL INTEGRATION CONCEPT AS A SELF-SUSTAINING POWER SOLUTION...2114

Solomon N. Agbo ; Oleksandr Astakhov ; Uwe Rau ; Tsvetelina Merdzhanova

PERFORMANCE ASSESSMENT OF A BIPV ROOFING TILE IN OUTDOOR TESTING 2118

Cristina S. Polo Lopez ; Pierluigi Bonomo ; Francesco Frontini ; Vasco Medici ; Lorenzo Nespoli

LIFE CYCLE ASSESSMENT OF TRANSPARENT ORGANIC PHOTOVOLTAIC FOR WINDOW APPLICATIONS 2124

Annick Anctil ; Eunsang Lee ; Jack Stephan ; Anjali Munasinghe ; Christopher Traverse ; Richard R. Lunt

A REDUCED ORDER MODEL FOR A TOV STUDY IN A SOLAR PV PROJECT....................2128

Ahmad Abdullah ; Billy Yancey

CYBER SECURITY ASSESSMENT OF DISTRIBUTED ENERGY RESOURCES.....................2135

Cedric Carter ; Ifeoma Onunkwo ; Patricia Cordeiro ; Jay Johnson

EVALUATION OF FAST-FREQUENCY SUPPORT FUNCTIONS IN HIGH PENETRATION ISOLATED POWER SYSTEMS 2141

Mohamed Elkhatib ; Jason Neely ; Jay Johnson

LOSS OF UTILITY DETECTION CAPABILITIES FOR TODAY'S UTILITY INTERCONNECTED PHOTOVOLTAIC INVERTERS 2147

Sigifredo Gonzalez ; Gregory Kern ; Michael Ropp

PARAMETRIC PV GRID-SUPPORT FUNCTION CHARACTERIZATION FOR SIMULATION ENVIRONMENTS 2153

Javier Hernandez-Alvidrez ; Jay Johnson

COST ANALYSIS AND COST REDUCTION OPPORTUNITIES OF RESIDENTIAL PV SYSTEM IN THE JAPAN 2159

Izumi Kaizuka ; Haruki Yamaya ; Takashi Ohigashi ; Risa Kurihara ; Osamu Ikki

SUPPLY AND DEMAND CONSTRAINTS ON FUTURE PV POWER IN THE USA 2163

Paul A. Basore ; Wesley J. Cole

RESIDENTIAL PHOTOVOLTAIC ELECTRICITY GENERATION IN THE EUROPEAN UNION 2017-OPPORTUNITIES AND CHALLENGES 2167
Arnulf Jäger-Waldau ; Thomas Huld ; Sandor Szabo

INVESTIGATING NANOSCALE DETERMINANTS OF CHARGE COLLECTION IN QUASI-2D PEROVSKITE SOLAR CELLS 2170
Yanqi Luo ; Xueying Li ; Bat-El Cohen ; Barry Lai ; Lioz Etgar ; David P Penning

RECENT DEVELOPMENTS OF SOLAR PHOTOVOLTAIC SYSTEMS IN INDIA 2172
Saravanan Vasudevan ; Arumugam Murugesan

OPERANDO X-RAY DIFFRACTION FOR CHARACTERIZATION OF PHOTOVOLTAIC MATERIALS 2176
Laura T Schelhasl ; Jeffrey A. Christians ; Joseph J. Berry ; Michael F. T Oney ; Christopher J. Tassone ; Joseph M. Luther ; Kevin H. Stone

X-RAY BEAM INDUCED VOLTAGE: A NOVEL TECHNIQUE FOR ELECTRICAL NANOCHARACTERIZATION OF SOLAR CELLS 2179
Michael E. Stuckelberger ; Tara Nietzold ; Bradley M. West ; Barry Lai ; Jörg M. Maser ; Volker Rose ; Mariana I. Bertoni

ELECTRO-LUMINESCENT REFRIGERATION ENABLED BY HIGHLY EFFICIENT PHOTOVOLTAICS 2185
T. Patrick Xiao ; Kaifeng Chen ; Parthiban Santhanam ; Shanhui Fan ; Eli Yablonovitch

MULTIPLE QUANTUM WELLS AS SLOWED HOT CARRIER COOLING ABSORBERS IN HOT CARRIER CELLS 2186
Gavin Conibeer ; Yi Zhang ; Simon Chung ; Yuaxun Liao ; Stephen Bremner ; Santosh Shrestha

QUANTITATIVE OPTOELECTRONIC MEASUREMENTS OF CARRIER THERMODYNAMICS PROPERTIES IN QUANTUM WELL HOT CARRIER SOLAR CELL 2192
Dac-Trung Nguven ; Laurent Lombez ; François Gibelli ; Soline Boyer-Richard ; Alain Le Corre ; Olivier Durand ; Jean-François Guillemoles

ABSORPTION ENHANCEMENT IN INGAASP/INGAP QUANTUM WELL SOLAR CELLS 2195
Islam E.H. Sayed ; Nikhil Jain ; Myles A. Steiner ; John F. Geisz ; Salah M. Bedair

CARRIER COLLECTION MODEL AND DESIGN RULE FOR QUANTUM WELL SOLAR CELLS 2201
Kasidit Toprasertpong ; Boram Kim ; Yoshiaki Nakano ; Masakazu Sugiyama

INFLUENCE OF CONDUCTION BAND OFFSETS AT WINDOW/BUFFER AND BUFFER/ABSORBER INTERFACES ON THE ROLL-OVER OF J-V CURVES OF CIGS SOLAR CELLS 2205
Giovanna Sozzi ; Simone Di Napoli ; Roberto Menozzi ; Florian Werner ; Susanne Siebentritt ; Philip Jackson ; Wolfram Witte

OVERVIEW OF SURFACE PASSIVATION SCHEMES FOR THIN FILM SOLAR CELLS 2209
Ratan Kotipalli ; Bart Vermang

TOWARDS 10% STATE-OF-THE-ART PURE SULFIDE CU2ZNSNS4 SOLAR CELL BY MODIFYING THE INTERFACE CHEMISTRY 2213
Kaiwen Sun ; Jialiang Huang ; Steve Johnston ; Chang Yan ; Fangyang Liu ; Xiaojing Hao ; Martin Green

BAND GAP CHANGES OF THE CDS BUFFER INDUCED BY POST-ANNEALING OF CU2ZNSN(S,SE)4SOLAR CELLS 2216
Mario Lang ; Nicolas Schäfer ; Christian Huber ; Thomas Schnabe ; Heinz Kalt ; Michael Hetterich

22.61 % EFFICIENT FULLY SCREEN PRINTED PERC SOLAR CELL 2220
Weiwei Deng ; Feng Ye ; Ruimin Liu ; Yunpeng Li ; Haiyan Chen ; Zhen Xiong ; Yang Yang ; Yifeng Chen ; Yongqian Wang ; Pietro P. Altermatt ; Zhiqiang Feng ; Pierre J. Verlinden

HOW TO ACHIEVE 23% EFFICIENT LARGE-AREA CU PLATED N-PERT CELLS? 2227
Monica Aleman ; Angel Uruena ; Emanuele Cornagliotti ; Patrick Choulat ; Joachim John ; Richard Russell ; Sukvhinder Singh ; Loic Tous ; Wen-Cheng Sun ; Filip Duerinckx ; Jozef Szlufcik

MICROSTRUCTURE AND RECOMBINATION ACTIVITY OF GRAIN BOUNDARIES FROM FRONT AND REAR SIDE DURING A LID-CYCLE OF MC-PERC SOLAR CELLS 2232
Tabea Luka ; Marko Turek ; Stephan Großer ; Christian Hagendorf

THERMODYNAMIC EFFICIENCY LIMIT OF BIFACIAL SOLAR CELLS FOR VARIOUS SPECTRAL ALBEDOS 2236
Thomas C.R. Russell ; Rebecca Saive ; Harry A. Atwater

PROCESS-INDUCED DEGRADATION RESISTANT N-CZ WAFERS THROUGH TABULA RASA DEFECT ENGINEERING 2242
Vincenzo Lasalvia ; William Nemeth ; Matthew Page ; Wooseok Nam ; Youngsik Han ; Sungsun Baik ; Amanda Youssef ; Tonio Buonassisi ; Paul Stradins

DETECTION OF A SHIFTING BROMINE CONCENTRATION IN HYBRID PEROVSKITES BY X-RAY FLUORESCENCE MICROSCOPY 2245
Yanqi Luo ; Parisa Khoram ; Sarah Brittman ; Barry Lai ; Erik C. Garnett ; David P. Fenning

INFLUENCE OF GRAIN SIZE AND INTERFACES ON PHOTO-STABILITY OF PEROVSKITE SOLAR CELLS 2247

Istiaque Hossain ; Liang Zhang ; Ranjith Kottokkaran ; Mohamed El-Henawey ; Pranav Joshi ; Max Noack ; Vikram Dalal

COLD THOUGHTS ON PEROVSKITE FEVER 2251

Tao Xu ; Jue Gong

LBIC ANALYSIS OF PEROVSKITE BASED SOLAR CELLS STABILITY 2255

Carmen M. Ruiz ; Javier Ramos ; Richard Garuz ; Damien Barakel ; Jean Reusser ; Judikaël Le Rouzo

ASSESSING JOB GROWTH AND SUSTAINABILITY IN THE US PV INDUSTRY 2258

Brion Bob

ENSURING THE RELIABILITY OF PHOTOVOLTAIC POWER SYSTEMS USING INTERNATIONAL STANDARDS AND THE IECRE CONFORMITY ASSESSMENT SYSTEM 2263

George Kelly ; Adrian Häring ; Ted Spooner ; Greg Ball ; Sarah Kurtz ; Matthias Heinze ; Masaaki Yamamichi ; Govind Ramu

A FRAMEWORK TO CALCULATE UNCERTAINTIES FOR LIFETIME ENERGY YIELD PREDICTIONS OF PV SYSTEMS 2267

Bjorn Muller ; Peter Bostock ; Boris Farnung ; Christian Reise

INTEGRATED PV-RECYCLING-MORE EFFICIENT, MORE EFFECTIVE 2272

Wolfram Palitzsch ; Ulrich Loser

ANALYSIS OF GAINP SOLAR CELLS GROWN BY HYDRIDE VAPOR PHASE EPITAXY 2275

Kevin L. Schulte ; John Simon ; David L. Young ; Aaron J. Ptak

INVESTIGATION OF ADHESION FORCES BETWEEN DUST PARTICLES AND SOLAR GLASS 2280

H.R. Moutinho ; C.-S. Jiang ; B. To ; C. Perkins ; M. Muller ; M.M. Al-Jassim ; L. Simpson

ANTI-REFLECTIVE AND ANTI-SOILING PROPERTIES OF A KLEANBOOST™, A SUPERHYDROPHOBIC NANO-TEXTURED COATING FOR SOLAR GLASS 2285

Illya Nayshevsky ; Qianfeng Xu ; Gil Barahman ; Alan Lyons

MULTILAYER-GROWN ULTRATHIN NANOSTRUCTURED GAAS SOLAR CELLS 2291

Boju Gai ; Yukun Sun ; Minjoo Lee ; Jongseung Yoon

LABORATORY STUDIES OF PARTICLE CEMENTATION AND PV MODULE SOILING 2294

Craig L. Perkins ; Matthew Muller ; Lin Simpson

VIRTUAL SUBSTRATES FOR LOW-COST HIGH EFFICIENCY III-V PHOTOVOLTAICS 2298

Sean J. Babcock ; Marlene L. Lichty ; Shankar Karki ; Grace Rajan ; Sylvain Marsillac ; Elisabeth L. Mcclure ; Seth M. Hubbard ; Christopher G. Bailey

SEASONAL TRENDS OF SOILING ON PHOTOVOLTAIC SYSTEMS 2301

Leonardo Micheli ; Daniel Ruth ; Matthew Muller

INTERRELATIONSHIPS AMONG NON-UNIFORM SOILING DISTRIBUTIONS AND PV MODULE PERFORMANCE PARAMETERS, CLIMATE CONDITIONS, AND SOILING PARTICLE AND MODULE SURFACE PROPERTIES 2307

Lawrence L. Kazmerski ; Antonia Sonia A.C. Diniz ; Daniel Sena Braga ; Cristiana Brasil Maia ; Marcelo Machado Viana ; Suellen C. Costa ; Pedro P. Brito ; Cláudio Dias Campos ; Sergio De Morais Hanriot ; Leila R. De Oliveira Cruz

PV MODULE DURABILITY -CONNECTING FIELD RESULTS, ACCELERATED TESTING, AND MATERIALS 2312

T. John Trout ; W. Gambogi ; T. Felder ; K. R. Choudhury ; L. Garreau-Iles ; Y. Heta ; K. Stika

FEMTOSECOND VS NANOSECOND: AN ANALYSIS ON THE LASER ABLATION PROPERTIES OF DIELECTRIC LAYERS FOR SOLAR CELLS 2318

Jaffar Moideen Yacob Ali ; Vinodh Shanmugam ; Carlos D. Rodríguez-Gallegos ; Bianca Lim ; Armin Aberle ; Thomas Mueller

GROWTH OF MOS2 THIN FILMS WITH MICRODOME TEXTURE AS OMNIDIRECTIONAL LIGHT TRAP FOR SOLAR CELL APPLICATIONS 2324

Hussain M. Abouelkhair ; Nina A. Orlovskaya ; Robert E. Peale

STUDY OF SPATIAL DISTRIBUTION OF ELECTRICAL, OPTICAL AND STRUCTURAL PROPERTIES OF MAGNETRON SPUTTERED AZO THIN FILMS 2330

Mohit Agarwal ; Rajiv O Dusane

MULTIBAND FORMATION IN CR DOPED CUGAS2 THIN FILMS SYNTHESIZED BY CHEMICAL SPRAY PYROLYSIS 2334

Nazmul Ahsan ; Sivaperuman Kalainatharr ; Naoya Miyashita ; Takuya Hoshii ; Yoshitaka Okada

EFFECTS OF ANNEALING AND SUBSTRATE TEMPERATURE FOR SN-S THIN FILMS 2338

Yoji Akaki ; Kazuya Iwasaki ; Shigeyuki Nakamura ; Hideaki Araki

MOLYBDENUM OXIDE THIN FILMS FOR HETEROJUNCTION SOLAR CELLS 2342

A. Dominguez ; Ateet Dutt ; O. De Melo ; G. Santana

DUAL ION BEAM SPUTTERED TCO THIN FILMS: SPUTTER-INSTIGATED PLASMONIC FEATURES FOR ULTRATHIN PHOTOVOLTAICS 2345
Vivek Garg ; Brajendra S. Sengar ; Vishnu Awasthi ; Shailendra Kumar ; Shaibal Mukherjee

COMBINATORIAL STUDY OF SN-TI-W-O TRANSPARENT CONDUCTING OXIDE THIN FILMS FOR PHOTOVOLTAIC APPLICATIONS 2349
Michael N. Gona ; Patrick J. M. Isherwood ; Jake W. Bowers ; John M. Walls

BANDGAP AND ELECTRON AFFINITY OPTIMIZATION OF ZINC OXIDE FOR N-ZNO/P-SI SINGLE HETEROJUNCTION SOLAR CELL 2355
Babar Hussain ; Aasma Aslam

MODELING AND OPTIMIZING THE EFFICIENCY OF A ZNO/ZNTE SOLAR CELL USING SCAPS SOFTWARE 2358
Amal Kabalan ; Sam Roy ; Benjamin Chen

TERNARY PHOSPHIDE SEMICONDUCTOR INMG/ZN3P2SOLAR CELLS 2361
Ryoji Katsube ; Kenji Kazumi ; Yoshitaro Nose

NUMERICAL MODELING OF WSE2SOLAR CELLS 2364
H. Kyureghian ; M. Hilfiker ; E. Ediger ; V. Medic ; N.J. Ianno

BIAXIAL-TEXTURED TITANIUM NITRIDE THIN FILMS ON LOW-COST, FLEXIBLE METAL SUBSTRATE AS A CONDUCTIVE BUFFER LAYER FOR THIN FILM SOLAR CELLS 2368
Yongkuan Li ; Yao Yao ; Ying Gao ; Sicong Sun ; Pavel Dutta ; Monika Rathi ; Jae-Hyun Ryou ; Venkat Selvamanickam

SNS BY IONIZED JET DEPOSITION FOR PHOTOVOLTAIC APPLICATIONS 2372
Daniele Menossi ; Simone Di Mare ; Ivan Rimmaudo ; Elisa Artegiani ; Giampiero Tedeschi ; Juan Luis Pena ; Fabio Piccinelli ; Andrei Salavei ; Alessandro Romeo

EFFECT OF VALENCE BAND SPLITTING ON THE ABSORPTION SPECTRA OF MONOLAYER MOS2 IN PRESENCE OF SULPHUR VACANCIES 2376
Himani Mishra ; Sitangshu Bhattacharya

THE STUDY OF SOME MATERIALS AS BUFFER LAYER IN COPPER ANTIMONY SULPHIDE (CUSBS2) SOLAR CELL USING SCAPS 1-D 2381
Muteeu Olopade ; Adeyinka Adewoyin ; Michael Chendo ; Adewumi Bolaji

INFLUENCE OF HETERO-INTERFACES ON PHOTOVOLTAIC PERFORMANCE IN SOLAR CELLS BASED ON ZNSNP2BULK CRYSTAL 2385
Shigeru Nakatsuka ; Shunsuke Akari ; Jakapan Chantana ; Takashi Minemoto ; Yoshitaro Nose

JUNCTION BY DIFFUSION OF ELEMENTAL SODIUM ALONE INTO BRIDGMAN CU(IN, GA) SE2 2388
S. Park ; C. H. Champness ; S. Vanka ; Z. Mi ; I. Shih

OXYGEN SUBSTITUTION AND SULFUR VACANCIES IN NABIS2: A PB-FREE CANDIDATE FOR SOLUTION PROCESSABLE SOLAR CELLS 2392
Robert J Patterson ; Hongze Xia ; Long Hu ; Zhilong Zhang ; Lin Yuan ; Jianfeng Yang ; Weijian Chen ; Zihan Chen ; Yijun Gao ; Yicong Hu ; Binesh Puthen Veettil ; John A. Stride ; Gavin Conibeer ; Shujuan Huang

EFFECT OF ANNEALING ON PERFORMANCE OF SOLAR CELLS WITH NEW OXIDE ABSORBER MN2V2O7 2395
Pramod Ravindra ; Eashwer Athresh ; Rajeev Ranjan ; Srinivasan Raghavan ; Sushobhan Avasthi

ELECTRO-OPTICAL PROPERTIES OF ZN2MO3O8THIN-FILMS: A NOVEL LOW-BANDGAP SOLAR ABSORBER 2399
Pramod Ravindra ; Eashwer Athresh ; Rajeev Ranjan ; Srinivasan Raghavan ; Sushobhan Avasthi

LOW TEMPERATURE SOLUTION PROCESS FOR RANDOM HIGH ASPECT RATIO SILVER NANOWIRE AS PROMISING TRANSPARENT CONDUCTIVE LAYER 2403
Arastoo Teymouri ; Supriya Pillai ; Zi Ouyang ; Xiaojing Hao ; Martin Green

OXYGEN INCORPORATION INTO SI NANOCRYSTAL/SIC MULTILAYERS 2407
Charlotte Weiss ; Andreas Reichert ; Johannes Hofmann ; Stefan Janz

DESIGN OF CASCADED HETEROSTRUCTURED P-I-I-N CDS/CDSE LOW COST SOLAR CELL 2411
M. Zinaddinov ; S. Mil'shtein

FAST C-V METHOD TO MITIGATE EFFECTS OF DEEP LEVELS IN CIGS DOPING PROFILES 2414
P. K. Paull ; J. Bailey ; G. Zapalac ; A. R. Arehart

CRYSTAL GROWTH PHENOMENA IN POLYCRYSTALLINE (CU)ZNTE/CDTE/CDS VIA MOLECULAR DYNAMICS 2419
Rodolfo Aguirre ; Jose J. Chavez ; Xiao W. Zhou ; David Zubia

USING HIGH-RESOLUTION ANOMALOUS-SCATTERING X-RAY DIFFRACTION TO OBSERVE OFF-STOICHIOMETRIC CU2ZNSNS4CRYSTAL STRUCTURES 2423
Christopher J. Bosson ; Max T. Birch ; Douglas P. Halliday ; Chiu C. Tang ; Peter D. Hatton

SIMULATION OF ZNMGO AS THE WINDOW LAYER FORCDTESOLAR CELLS 2427
Yunfei Chen ; Shou Peng ; Xin Cao ; Alan E. Delahoy ; Ken K. Chin

MODELING EFFECT OF DEFECTS ON EFFICIENCY OF NANOWIRE CDS-CDTE SOLAR CELLS 2432
Hongmei Dang ; Esther Ososanya ; Nian Zhang ; Xiaohui Wang ; Hojjatollah Sarvari ; Vijay P. Singlr

ANALYTICAL DESCRIPTION OF CHARGED GRAIN BOUNDARY RECOMBINATION IN POLYCRYSTALLINE THIN FILM SOLAR CELLS 2438
Benoit Gaury ; Paul M. Haney

IMAGING THE EFFECT OF CDSE WINDOW LAYERS IN CDTE PHOTOVOLTAICS 2443
John M. Howard ; Elizabeth M. Tennyson ; William B. Gunnarsson ; Naba R. Paudel ; Yanfa Yan ; Marina S. Leite

INVESTIGATION OF TRAPS DENSITY AND POSITION IN ALKALI TREATED CU(IN,GA)SE2 THIN FILMS AND SOLAR CELLS 2446
Shankar Karki ; Pran K. Paul ; Grace Rajan ; Chinedum Akwari ; Angus Rockett ; Steven A Ringel ; Aaron R. Arehart ; Sylvain Marsillac

THE EFFECT OF DEPOSITION STOICHIOMETRY AND POST-DEPOSITION TREATMENTS ON DEEP DEFECTS IN CDTE 2449
Imran S. Khan ; Vamsi Evani ; Shamara Collins ; Chih An Hsu ; Vasilis Palekis ; Chris Ferekides

TESTING THE LIMITS OF MECHANICALLY-SCRIBED CIGS MICROCELLS 2453
Ombline Lafont ; Nicolas Vandamme ; Leia Ruffini ; Jia Yu ; Philip Jackson ; Jose Alvarez ; Daniel Lincot

PHOTOLUMINESCENCE IMAGING ANALYSIS OF DOPING IN THIN FILM CDS AND CDS/CDTE DEVICES 2457
C. Potamialis ; F. Lisco ; B. Maniscalco ; M. Togay ; A. Abbas ; M. Biiss ; J.W. Bowers ; J.M. Waiis ; I. Rimmaudo ; R. Mis Fernandez ; V. Rejon ; J.L. Peña

APPLICATION OF MAPPING SPECTROSCOPIC ELLIPSOMETRY FOR CDSE/CDTE SOLAR CELLS: OPTIMIZATION OF LOW-TEMPERATURE PROCESSED DEVICES WITH ALL-SPUTTERED SEMICONDUCTORS 2462
Mohammed A. Razooqi ; Adam B. Phillips ; Geethika K. Liyanage ; Fadhil K. Al-Fadhili ; Maxwell M. Junda ; Nikolas J. Podraza ; Michael J. Heben ; Robert W. Collins ; Prakash Koirala

ASSESSING THE VALIDITY AND ACCURACY OF EFFECTIVE ELECTRONIC MATERIALS: CAN 1D SIMULATIONS PREDICT POLYCRYSTALLINE DEVICE PERFORMANCE? 2467
Yubo Sun ; Allison Perna ; Sudhajit Misra ; Vasilios Palekis ; Chris Ferekides ; Jeffrey Aguiar ; Peter Bermel ; Michael A. Scarpulla

CHARACTERIZING RECOMBINATION IN CDTE-BASED SOLAR CELLS BY THE TEMPERATURE AND EXCITATION DEPENDENCE OF OPEN-CIRCUIT VOLTAGE AND PHOTOLUMINESCENCE 2473
Craig H. Swartz ; Sanjoy Paul ; Corey R. Grice ; Yanfa Yan ; Lorelle Mansfield ; Sachit Grover ; Gang Xiong ; Jian V. Li

EXPERIMENTAL EVIDENCE FOR CDS-RELATED TRANSPORT BARRIER IN THIN FILM SOLAR CELLS AND ITS IMPACT ON ADMITTANCE SPECTROSCOPY 2478
Florian Werner ; Anastasiya Zelenina ; Susanne Siebentritt

TRANSPARENT CONDUCTIVE ADHESIVES FOR TANDEM SOLAR CELLS 2482
Talysa R. Klein ; Benjamin G. Lee ; Manuel Schnabel ; Emily L. Warren ; Pauls Stradins ; Adele C. Tamboli ; Maikel F.A.M. Van Hest

MODELING THREE-TERMINAL III- V LSI TANDEM SOLAR CELLS 2488
Emily L. Warren ; Michael G. Deceglie ; Paul Stradins ; Adele C. Tamboli

WAFER BONDING APPROACHES FOR III-V ON SI MULTI-JUNCTION SOLAR CELLS 2492
Laura Vauche ; Elias Veinberg-Vidal ; Clément Weick ; Christophe Morales ; Vincent Larrey ; Christophe Lecouvey ; Mickaël Martin ; Jérémy Da Fonseca ; Christophe Jany ; Thibaut Desrues ; Céline Brughera ; Philippe Voarino ; Thierry Salvetat ; Frank Fournel ; Mathieu Baudrit ; Cécilia Dupré

DESIGN ARITHMETIC OF THE LATERAL III-V / SI HYBRID MODULE 2498
Kenji Araki ; Kyotaro Nakamura ; Kan-Hua Lee ; Takefumi Kamioka ; Yu-Cian Wang ; Nobuaki Kojima ; Yoshio Ohshita ; Masafumi Yamaguchi

GAASP NANOWIRE SOLAR CELL DEVELOPMENT TOWARDS NANOWIRE/SI TANDEM APPLICATIONS 2502
Enrique Barrigon ; Yang Chen ; Gaute Otnes ; Vilgaile Dagyte ; Nicklas Anttu ; Lars Samuelson ; Magnus Borgström

DEMONSTRATION OF GAINP2/SI VOLTAGE MATCHED TANDEM SOLAR CELLS 2506
David C. Bobela ; Kenneth J. Schmieder ; Matthew P. Lumb ; James E. Moore ; Robert J Walters ; Eric A. Armour ; Leo Matthew ; Rajesh Rao ; Angelo Mascarenhas ; Kirstin Alberi

WAFER BONDED III–V ON SILICON MULTI -JUNCTION CELL WITH EFFICIENCY BEYOND 31% 2511
Romain Cariou ; Jan Benick ; Paul Beutel ; Nico Tucher ; Martin Graf ; David Lackner ; Martin Hermle ; Stefan W. Glunz ; Andreas W. Bett ; Frank Dimroth

INTEGRATION OF THIN AL FILMS ON INO.18GAO.82AS METAMORPHIC GRADE STRUCTURES FOR LOW-COST III- V PHOTOVOLTAICS..2514

Alessandro Giussani ; Michael A. Slocum ; Seth M. Hubbard ; Nathan Smaglik ; Nikhil Pokharel ; S. Phillip Ahrenkiel

TEMPERATURE DEPENDENT CHARACTERISTICS OF GAINP/GAAS/GAINNASSB SOLAR CELL UNDER SIMULATED AM0 SPECTRA..2520

Riku Isoaho ; Arto Aho ; Antti Tukiainen ; Mircea Guina

EFFICIENCY OF GAAS P/SI TWO-JUNCTION SOLAR CELLS WITH MULTI-QUANTUM WELLS: A REALISTIC MODELING WITH CARRIER COLLECTION EFFICIENCY...................2524

Boram Kim ; Kasidit Toprasertpong ; Oliver Supplie ; Agnieszka Paszuk ; Thomas Hannappel ; Yoshiaki Nakano ; Masakazu Sugiyama

INVERSE METAMORPHIC III-V/EPI-SIGE TANDEM SOLAR CELL PERFORMANCE ASSESSED BY OPTICAL AND ELECTRICAL MODELING..2528

Raphaël Lachaurne ; Martin Foldyna ; Gwénaëlle Hamon ; Nicolas Vaissiére ; Jean Decobert ; Romain Cariou ; Pere Roca I Cabarrocas ; José Alvarez ; Jean-Paul Kleider

TOWARDS MONOLITHICALLY INTEGRATED GAAS ON SI TANDEM SOLAR CELL............2532

Zhen Liu ; Zekun Ren ; Haohui Liu ; Tonio Buonassisi ; Ian Marius Peters

ZNSIP2 THIN FILM GROWTH FOR SI-BASED TANDEM PHOTOVOLTAICS 2536

Aaron D. Martinez ; Elisa M. Miller ; Andrew G. Norman ; Paul Stradins ; Eric S. Toberer ; Adele C. Tamboli

IN SITU CONTROL OVER THE SUBLATTICE ORIENTATION OF GAP/SI(100): AS VIRTUAL SUBSTRATES FOR TANDEM ABSORBERS 2538

Aznieszka Paszuk ; Oliver Supplie ; Sebastian Brückner ; Matthias M. May ; Anja Dobrich ; Andreas Nägelein ; Boram Kim ; Yoshiaki Nakano ; Masakazu Sugiyama ; Peter Kleinschmidt ; Thomas Hannappel ; Thomas Hannappel

III-V/SI TANDEM CELL TO MODULE INTERCONNECTION - COMPARISON BETWEEN DIFFERENT OPERATION MODES..2543

Henning Schulte-Huxel ; Emily L. Warren ; Manuel Schnabel ; Paul Stradins ; Daniel Friedman ; Adele C. Tamboli

INGAP/GAAS/ITO/SI HYBRID TRIPLE-JUNCTION CELLS WITH GAAS/ITO BONDING INTERFACES 2548

Naoteru Shigekawa ; Tomoya Hara ; Tomoki Ogawa ; Jianbo Liang ; Takefumi Kamioka ; Kenji Araki ; Masafumi Yamaguchi

MEASUREMENTS OF POTENTIALS AT TAP CONTACTS AND ESTIMATION OF RESISTANCE ACROSS BONDING INTERFACES IN INGAP/GAAS/SI HYBRID TRIPLE-JUNCTION CELLS 2551

Naoteru Shigekawa ; Jianbo Liang

OPTIMIZATION OF A GAASP TOP CELL FOR IMPLEMENTATION IN A III-V/SI TANDEM STRUCTURE 2554

Amber C. Silvaggio ; Daniel L. Lepkowski ; Daniel J. Chmielewski ; Jacob T. Boyer ; Steven A. Ringel ; Tyler J. Grassman

THEORETICAL DESIGN OF PEROVSKITE/CDTE FOUR-TERMINAL TANDEM SOLAR CELLS..2558

Tao Tang ; Huan Zhang ; Xingzhi Du ; Yiming Lnr ; Hang Zhou

WAFER-BONDED ALGAAS///SI DUAL-JUNCTION SOLAR CELLS.....................................2562

Elias Veinberg-Vidal ; Laura Vauche ; Clément Weick ; Jérémy Da Fonseca ; Christophe Jany ; Christophe Morales ; Christophe Lecouvey ; Thibaut Desrues ; Philippe Voarino ; Frank Fournel ; Anne Kaminski-Cachopo ; Alejandro Datas ; Pablo Garcia-Linares ; Mathieu Baudrit ; Pierre Mur ; Cécilia Dupré

ENHANCEMENT OF SI PHOTOVOLTAIC MODULE BY INTRODUCING III-V/SI HYBRID CONFIGURATIONS AND COST EVALUATIONS UNDER VARIOUS COST RATIOS OF III-V/SI PHOTOVOLTAICS...2566

Yu-Cian Wang ; Kenii Araki ; Kyotaro Nakamura ; Kan-Hua Lee ; Takefumi Kamioka ; Nobuaki Kojima ; Yoshio Ohshita ; Masafumi Yamaguchi

NUMERICAL SIMULATION OF P-TYPE FRONT JUNCTION PERL SILICON CELL FOR III-V LSI TANDEM DEVICES 2569

Chuqi Yi ; Fa-Jun Ma ; Anita Ho-Baillie ; Stephen Bremner

EPITAXIAL GAP LAYERS GROWN ON SI SUBSTRATES USING MIGRATION ENHANCED AND MOLECULAR BEAM EPITAXY..2573

Chaomin Zhang ; Allison Boley ; Nikolai Faleev ; David J. Smith ; Christiana B. Honsberg

INVESTIGATION OF CARRIER-INDUCED DEFECT BEHAVIOR IN P-TYPE MULTICRYSTALLINE SILICON 2576

Catherine E. Chan ; Tsun H. Fung ; David N.R. Payne ; Daniel Chen ; Malcolm D. Abbott ; Alison M. Ciesla ; Ran Chen ; Brett J. Hallam ; Stuart R. Wenham

MAGNETRON SPUTTERED HYDROGENATED SILICON THIN FILMS: ASSESSMENT FOR APPLICATION IN PHOTOVOLTAICS 2582

Dipendra Adhikari ; Maxwell M. Junda ; Sylvain X. Marsillac ; Robert W. Collins ; Nikolas J. Podraza

HIGH QUALITY AND THIN SILICON WAFER FOR NEXT GENERATION SOLAR CELLS 2588

Yoshio Ohshita ; Takuto Kojima ; Ryota Suzuki ; Kosuke Kinoshita ; Tomoyuki Kawatsu ; Kyotaro Nakamura ; Atsushi Ogura

FIRST DEMONSTRATION OF RADIAL JUNCTION SILICON NANOWIRE SOLAR MINI-MODULES PREPARED BY PECVD AND LASER SCRIBING 2593

Mutaz Al-Ghzaiwat ; Martin Foldyna ; Takashi Fuyuki ; Wanghua Chen ; Erik V. Johnson ; Jacques Meot ; Pere Roca I Cabarrocas

IMPACT OF INDUCED DEFECTS ON DEVICE PERFORMANCE IN SILICON HETEROJUNCTION SOLAR CELLS 2596

Pradeep Balaji ; André Augusto ; Stuart G. Bowden

LASER HYDROGENATION ON HEAVILY DISLOCATED CAST-MONO SILICON CELLS 2600

Alison M. Ciesla ; Catherine E. Chan ; Sisi Wang ; Malcolm D. Abbott ; Cheemun Chong ; Stuart R. Wenham

PERFORMANCE OPTIMIZATION OF SEMI-TRANSPARENT THIN-FILM AMORPHOUS SILICON SOLAR CELLS 2605

Yuan Gao ; Fai Tong Si ; Olindo Isabella ; Rudi Santbergen ; Guangtao Yang ; Jianfei Dong ; Guoqi Zhang ; Miro Zeman

LOW TEMPERATURE SPALLING OF SILICON: A CRACK PROPAGATION STUDY 2610

Pablo Guimera Coll ; Tine Uberg Nærland ; Nathan Stoddard ; Michael Stuckelberger ; Mariana Bertoni

NEW FINDINGS OF THERMAL EFFECT ON PM-SI:H SOLAR CELLS OPTOELECTRONIC PROPERTIES 2614

L. Hamui ; L. A. Górnez-González ; G. Santana

STUDY OF PV MODULE DEGRADATION RATE PREDICTION THROUGH CORRELATION OF FIELD-AGED AND ACCELERATED-AGED MODULE DEGRADATION DATA 2618

Babak T. Hamzavy ; William J. Grieco ; Brian J. Fields ; Cara S. Libby ; William B. Hobbs ; Olga Lavrova ; C. Birk Jones

ADVANCED ANALYSIS OF MULTI WIRE WAFERING PROCESSES 2622

Ringo Koepgel ; Samuel Brinnig ; Felix Kaule ; Hartmut Schwabe ; Stephan Schoenfelder

CONSIDERATION ON OPEN-CIRCUIT VOLTAGE OF SI HETEROJUNCTION SOLAR CELLS UNDER LOW CONCENTRATION CONDITION 2627

Makoto Konagai

CHARACTERIZATION OF MICROCRYSTALLINE SILICON THIN FILM SOLAR CELLS PREPARED BY HIGH WORKING PRESSURE PLASMA-ENHANCED CHEMICAL VAPOR DEPOSITION 2631

Jung-Dae Kwon ; Dong-Ho Kim ; Ji-Hoon Lee ; Myungkwan Song ; Myunghun Shin

ATOMIC-LAYER-DEPOSITEDV2O5-XFILMS AS A HIGHLY-EFFICIENT P-TYPE LAYER FOR THIN FILM A-SI SOLAR CELLS 2634

Ji-Hoon Lee ; Myungkwan Song ; Dong-Ho Kim ; Jung-Dae Kwon

A NOVEL DEFECT PASSIVATION METHOD FOR MULTICRYSTALLINE SI WAFER BY H2S REACTION 2637

Hsiang-Yu Liu ; Ujjwal K. Das ; Robert W. Birkmire

CARRIER TRANSPORTATION AT NOVEL SILVER PASTE CONTACT 2642

Takefumi Kamioka ; Satoshi Kamevama ; Kazuo Muramatsu ; Aki Tanaka ; Naotaka Iwata ; Kyotaro Nakamura ; Atsushi Ogura ; Yoshio Ohshita

INFLUENCE OF DEPOSITION PARAMETERS ON SILICON THIN FILMS DEPOSITED BY MAGNETRON SPUTTERING 2646

Grace Rajan ; Tejaswini Miryala ; Shankar Karki ; Robert W. Collins ; Nikolas Podraza ; Sylvain Marsillac

MINORITY CARRIER LIFETIME VARIATIONS IN MULTICRYSTALLINE SILICON WAFERS WITH TEMPERATURE AND INGOT POSITION 2651

Sissel Tind Søndergaard ; Jan Ove Odden ; Rune Strandberg

CUO NANOWIRES-BASED RADIAL HETERO-JUNCTION THIN FILM SILICON SOLAR CELLS WITH A HIGH OPEN-CIRCUIT VOLTAGE 2656

Xiaolin Sun ; Jiawen Lu ; Fan Yang ; Linwei Yu ; Jun Xu ; Ling Xu ; Kunji Chen

THE EFFECT OF CHEMICAL COMPOSITION ON POROUS ETCHING FOR EPI AND LIFT-OFF WAFER PROCESS 2660

Teng-Yu Wang ; Peng-Wei Chen ; Han-Wen Liu

ELECTRICAL AND OPTICAL PERFORMANCE OF SILICON SOLAR CELLS USING PLASMONICS INDIUM NANOPARTICLES LAYER EMBEDDED IN SIO2ANTIREFLECTIVE COATING 2664

Hao-Yu Yang ; Wen-Jeng Ho ; Sheng-Kai Feng ; Jheng-Jie Liu ; Ta-Wei Chuang ; Guan-Yi Li ; Yun-Chie Yang ; Cho-Chun Chiang ; Yao- Hui Chen

ELECTROLUMINESCENCE ANALYSIS FOR SEPARATION OF SERIES RESISTANCE FROM RECOMBINATION EFFECTS IN SILICON SOLAR CELLS WITH INTERDIGITATED BACK CONTACT DESIGN 2667

Nuha Ahmed ; Lei Zhang ; Ujjwal Das ; Steven Hegedus

INDOOR MEASUREMENT OF ANGLE RESOLVED LIGHT ABSORPTION BY BLACK SILICON 2672

Mekbib W. Amdemeskel ; Beniamino Iandolo ; Rasmus S. Davidsen ; Ole Hansen ; Gisele A. Dos Reis Benatto ; Nicholas Riedel ; Peter B. Poulsen ; Sune Thorsteinsson ; Anders Thorseth ; Carsten Dam-Hansen

IMPACT OF NON- FLAT PHOTOGENERATION AND CARRIER PROFILES ON THE LUMINESCENT EMISSION AND DETECTION OF SILICON SOLAR CELLS 2677

Nekane Azkona ; Federico Recart ; Pedro Rodríguez ; Vanesa Fano ; Aloña Otaegi ; Juan Carlos Jimeno

DEVELOPMENT OF OUTDOOR LUMINESCENCE IMAGING FOR DRONE-BASED PV ARRAY INSPECTION 2682

Gisele A. Dos Reis Benatto ; Nicholas Riedel ; Sune Thorsteinsson ; Peter B. Poulsen ; Anders Thorseth ; Carsten Dam-Hansen ; Claire Mantel ; Soren Forchhammer ; Kenn H. B. Frederiksen ; Jan Vedde ; Michael Petersen ; Henrik Voss ; Michael Messerschmidt ; Harsh Parikh ; Sergiu Spataru ; Dezso Sera

CLIMBING DRUM PEEL (CDP) TEST METHOD FOR CHARACTERIZING ADHESION IN FLEXIBLE PV MODULES 2688

Venkata Bheemreddy ; Kedar Hardikar

ACCURACY OF SOLAR SIMULATOR SPECTRAL DETERMINATION USING BAND-PASS FILTERING METHOD 2692

Weston Dobson ; Harrison Wilterdink ; Cassidy Sainsbury ; Adrienne Blum ; Justin Dinger ; Ronald A. Sinton ; Karsten Bothe ; David Hinken ; Martin Wolf

CORRELATION OF I-V CURVE PARAMETERS WITH MODULE-LEVEL ELECTROLUMINESCENT IMAGE DATA OVER 3000 HOURS DAMP-HEAT EXPOSURE 2697

Justin S. Fada ; Andrew J. Loach ; Alan J. Curran ; Jennifer L. Braid ; Shuying Yang ; Timothy J. Peshek ; Roger H. French

A NOVEL METHOD TO INVESTIGATE STOICHIOMETRY AND PERFORMANCE OF BURIED PASSIVATED CONTACTS UTILIZING TIME-OF-FLIGHT SIMS 2702

Steven P. Harvey ; William Nemeth ; Jeff Aguiar ; Craig Perkins ; Pauls Stradins

A COMPARISON BETWEEN QUASI-STEADY STATE AND TRANSIENT PHOTOCONDUCTANCE LIFETIMES IN SILICON INGOTS: SIMULATIONS AND MEASUREMENTS 2707

Mohsen Goodarzi ; Ronald Sinton ; Daniel Chung ; Bernhard Mitchell ; Thorsten Trupke ; Daniel Macdonald

NEW DEVELOPMENT IN GLOW DISCHARGE OPTICAL EMISSION SPECTROMETRY FOR THE CHARACTERIZATION AND THE THICKNESS MEASUREMENT OF LAYERS FOR PHOTOVOLTAIC APPLICATIONS 2711

Philippe Hunault ; Matthieu Chausseau ; Patrick Chaporr ; Sofia Gaiaschi ; Anais Loubar ; Muriel Bouttcmy ; Arnaud Etcheberry

DEEP LEVEL TRANSIENT SPECTROSCOPY MEASUREMENTS OF SILICON HETEROJUNCTION CELLS 2716

Sanchit Khatavkar ; C. V. Kannan ; Vijay Kumar ; P. R. Nair ; B. M. Arora

CHARACTERIZATION OF MODULES AND ARRAYS WITH SUNS VOC 2719

Alex Killam ; Stuart Bowden

A STUDY OF PERFORMANCE CHARACTERIZATION WITH REAR LIGHT SOURCE IN CONVENTIONAL BIFACIAL SOLAR CELLS 2723

Soo Min Kim ; Sang Hoon Jung ; Rae-Won Choi ; Yong Bae Kim ; Min Gu Kang ; Hee-Eun Sonp ; Gyu-Seok Choi

ELECTRICAL CHARACTERIZATION OF THE CARRIER TRANSPORT PROPERTIES IN ACU(IN,GA)SE2SOLAR CELL 2728

Roberto Lopez ; Sanjoy Paull ; Ingrid Repins ; Jian V. Li

SYSTEMATIC THERMALPHOTOVOLTAIC SOLAR CELL OPTIMIZATION 2732

Zheng Lyu ; Muyu Xue ; Junyan Chen ; Jieyang Jia ; Shanhui Fan ; James Harris

CHARACTERIZATION OF TELLURIUM AS A BACK CONTACT FOR CDTE SOLAR CELLS 2736

C.E. Moffett ; W.S. Sampath

ON THE DIFFERENT EXPLANATIONS OF THE RECOMBINATION CURRENTS WITH HIGH IDEALITY FACTOR IN SILICON SOLAR CELLS 2740

A. Otaegi ; V. Fano ; N. Azkona ; J. R. Gutiérrez ; J. C. Jimeno

IDENTIFICATION OF SHUNTS IN A MONOLITHIC MULTIJUNCTION GAAS/GAAS DEVICE BY SPECTROMETRIC CHARACTERIZATION 2744

Felipe Oviedo ; Liu Zhe ; Zekun Ren ; Kevin Nay Yaung ; Maung Thway ; Liu Haohui ; Tonio Buonassisi ; Ian Marius Peters

A SIMULATION STUDY ON RADIATIVE RECOMBINATION ANALYSIS IN CIGS SOLAR CELL 2749

Sanjoy Paul ; Roberto Lopez ; Md Dalim Mia ; Craig H. Swartz ; Jian V. Li

SIMULATION AND SPECTROSCOPY OF CARRIER RELAXATION IN GASB AND GAAS 2755

A.C. Scofield ; A.I. Hudson ; B.L. Liang ; B.C. Juang ; D.L. Huffaker ; S.M. Hubbard ; W.T. Lotshaw

COMPUTATIONAL DESIGN OF DOPANTS IN CDTE GRAIN BOUNDARIES FOR EFFICIENT PHOTOVOLTAICS 2759

Fatih G. Sen ; Tadas Paulauskas ; Ce Sun ; Moon Kim ; Robert F. Klie ; Maria K.Y. Chan

ANALYSES OF PHOTOVOLTAIC POWER PLANT PERFORMANCE ESTIMATES BASED ON DETAILED LABORATORY MODULE CHARACTERIZATIONS AND TYPICAL REAL-WORLD INPUT DATA SOURCES 2762

Rajeev Singh ; John L.R. Watts ; Kellen Gillispie

CRITICAL EVALUATION OF THE FOUNDATIONS OF SOLAR SIMULATOR STANDARDS 2765

Ronald A. Sinton ; Harrison Wilterdink ; Justin Dinger ; Adrienne L. Blum ; Weston Dobson ; Cassidy Sainsbury

IMPACT OF INFRARED OPTICAL PROPERTIES ON CRYSTALLINE SI AND THIN FILM CDTE SOLAR CELLS 2771

Indra Subedi ; Timothy J Silverman ; Michael Deceglie ; Nikolas J. Podraza

THE IMPACT OF IMPURITIES ON THE RELATIVE EFFICIENCIES OF SOLAR CELLS FROM DIFFERENT SILICON FEEDSTOCKS 2776

Muhammad Tayyib ; Aleksandr Dobroliubov ; Zekija Ramic ; Muhammad Nadeem Akarm ; Jan Ove Odden

ACCURACY EVALUATION OF ABSOLUTE ELECTROLUMINESCENCE-EFFICIENCY MEASUREMENTS OF SOLAR CELLS USING A SENSITIVITY-CALIBRATED-PHOTODETECTOR CONTACT METHOD 2781

Masahiro Yoshita ; Yoshihiro Hishikawa ; Yoshihiko Kanemitsu ; Hidefumi Akiyama

NANOMETER-SCALE CARRIER IMAGING OF POTENTIAL-INDUCED DEGRADATION IN C-SI SOLAR CELLS 2785

C.-S. Jiang ; C. Xiao ; H.R. Moutinho ; S. Johnston ; M.M. Al-Jassim ; X. Yang ; Y. Chen ; J. Ye

NREL EFFORTS TO ADDRESS SOILING ON PV MODULES 2789

Lin J. Simpson ; Matthew Muller ; Michael Deceglie ; Helio Moutinho ; Craig Perkins ; C. S. Jiang ; David C. Miller ; Leonardo Micheli ; Govindasamy Tamizhmani ; Sai Ravi Vasista Tatapudi ; Mowafak Al-Jassim

MODELING POTENTIAL-INDUCED DEGRADATION (PID) OF FIELD-EXPOSED CRYSTALLINE SILICON SOLAR PV MODULES: FOCUS ON A REGENERATION TERM 2794

Eleonora Annigoni ; Alessandro Virtuani ; Fanny Sculati-Meillaud ; Christophe Ballif

SOILING LOSS ON PV MODULES AT TWO LOCATIONS IN INDIA STUDIED USING A WATER BASED ARTIFICIAL SOILING METHOD 2799

Sonali Bhaduri ; Sachin Zachariah ; Lawrence L. Kazmcrski ; Balasubramaniam Kavaipatti ; Anil Kottantharayil

QUANTIFYING YEAR-TO-YEAR VARIATIONS IN SOLAR PANEL SOILING FROM PV ENERGY-PRODUCTION DATA 2804

Michael G. Deceglie ; Leonardo Micheli ; Matthew Muller

ACCURATELY MEASURING PV SOILING LOSSES WITH SOILING STATION EMPLOYING PV MODULE POWER MEASUREMENTS 2808

Michael Gostein ; Bill Stueve ; Mandy Chan

PERFORMANCE OF MONOCRYSTALLINE SILICON SOLAR CELL- INFLUENCE OF DUST ON ULTRA-VIOLET AND VISIBLE REGION DURING EARLY STAGE OF DEPOSITION 2811

Hemaprabha Elangovan ; Upasna Ranjan ; A K Jagdish ; Praveen C. Ramamurthy ; Kamanio Chattopadhyay

A COMPREHENSIVE STUDY OF LIGHT SOAKING EFFECT IN CDTE SOLAR CELLS 2816

D. Guo ; A. Moore ; D. Krasikov ; I. Sankin ; D. Vasileska

CORRECTION FOR METASTABILITY IN THE QUANTIFICATION OF PID IN THIN-FILM MODULE TESTING 2819

Peter Hacke ; Sergiu Spataru ; Steve Johnston

A FINE MODEL OF POWER DEGRADATION FOR CRYSTALLINE SILICON SOLAR MODULES 2823

Wenshuang Hea ; Baosong Duan ; Fumei Wang ; Ao Wang ; Jipeng Chang ; He Wang ; Hong Yang ; Jie Ding ; Junjun Zhang ; Jingsheng Huang

TEST METHODS FOR HYDROPHOBIC COATINGS ON SOLAR COVER GLASS 2827

Kenan Isbilir ; Biancamaria Maniscalco ; Ralph Gottschalg ; John Michael Walls

IMPACT OF DEGRADATION RATES ON SOLAR PV FINANCING FOR PROJECTS LOCATED IN THE UNITED STATES 2833

Rounak A. Kharait ; Phil Stiles ; Jarrett Carriere ; Larry Mcclung

ANALYSIS OF WIND DIRECTION AND SPEED MEASUREMENTS IN ARID REGION - A SITE EVALUATION USING DATA WITH LOW TEMPORAL RESOLUTION 2836

Elisabeth Klimm ; Felix Guischard ; Karl-Anders Weiss

FORECASTING ENVIRONMENTAL DEGRADATION POWER LOSS IN SOLAR PANELS WITH A PREDICTIVE CRACK OPENING TEST 2839

Jason L. Lincoln ; Andrew M. Gabor ; Eric J. Schneller ; Hubert Seigneur ; Joseph Walters ; Rob Janoch ; Andrew Anselmo ; Victor Huayamave ; Winston Schoenfeld

FLUORESCENCE IMAGING ON THE CROSS-SECTION OF PHOTOVOLTAIC LAMINATES AGED UNDER DIFFERENT UV INTENSITIES 2844

Yadong Lyu ; Jae Hyun Kim ; Xiaohong Gu

STATISTICAL ANALYSIS OF DEGRADATION DATA FOR C-SI MODULES OBSERVED IN INDIA IN 2016 2849

Chiranjibi Mahapatra ; Rajiv Dubey ; Shashwata Chattopadhyay ; Sachin Zachariah ; Sanjeev Sabnis

PROCESS INDUCED DEFLECTION AND STRESS ON ENCAPSULATED SOLAR CELLS 2854

Xiaodong Meng ; Michael Stuckelberger ; Peter Hacke ; Mariana Bertoni

A UNIFIED GLOBAL INVESTIGATION ON THE SPECTRAL EFFECTS OF SOILING LOSSES OF PV GLASS SUBSTRATES: PRELIMINARY RESULTS 2858

Leonardo Micheli ; Eduardo F. Fernández ; Greg P. Smestad ; Hameed Alrashidi ; Nabin Sarmah ; Nazmi Sellami ; Ibrahim A. I. Hassan ; Amal Kasry ; Gustavo Nofuentes ; Neeru Sood ; Bala Pesala ; S. Senthilarasu ; Florencia Almonacid ; K.S. Reddy ; Matthew Muller ; Tapas K. Mallick

REFERENCE: PROCEEDINGS OF THE IEEE PVSC CONF., 2017 THE DEVELOPMENT OF A DC BREAKDOWN VOLTAGE TEST FOR PHOTOVOLTAIC INSULATING MATERIALS 2864

David C. Miller ; Bernt Ake-Sultan ; Axel Borne ; Rene Eugen ; Bradley L. Givot ; Jürgen Jung ; Steven W. Macmaster ; Byron K. Mcdanold ; Ulf H. Nilsson ; Nancy H. Phillips ; Ian A. Tappan ; Nick S. Bosco

FIELD-EVALUATION OF ELECTRODYNAMIC SCREENS FOR MAINTAINING HIGH OPTICAL EFFICIENCY OPERATION OF SOLAR COLLECTORS 2870

Cristian Morales ; Annie Bernard ; Ryan Eriksen ; Julius Yellowhair ; Sean Garner ; Ricci La Centra ; Alecia Griffin ; Alexis Lloyd ; Yujie Gao ; Ramakrishnan Lakshmanan ; Mark Horenstein ; Malay Mazumder

EFFECT OF REVERSE BIAS VOLTAGES ON SMALL SCALE GRIDDED CIGS SOLAR CELLS 2875

Soheyl Mortazavi ; Klaas Bakker ; Jome Carolus ; Michael Daenen ; Gabriela De Amorim Soares ; Henk Steijvers ; Arthur Weeber ; Mirjam Theelen

A METHOD TO EXTRACT SOILING LOSS DATA FROM SOILING STATIONS WITH IMPERFECT CLEANING SCHEDULES 2881

Matthew Muller ; Leonardo Micheli ; Alfredo A. Martinez-Morales

ANALYTICAL (S)TEM STUDIES OF DEFECTS ASSOCIATED WITH PID IN STRESSED SI PV MODULES 2887

Andrew Norman ; Adam Stokes ; John Moseley ; Steven Harvey ; Steve Johnston ; Harvey Guthrey ; Mowafak Al-Jassim

DESIGN, DEVELOPMENT, AND EVALUATION OF ELECTRODYNAMIC SCREENS FOR SELF-CLEANING SOLAR PANELS AND CONCENTRATING MIRRORS 2891

Annie Bernard ; Cristian Morales ; Ryan S. Eriksen ; Alecia C. Griffin ; Yujie Gao ; Ramakrishnan Lakshmanan ; Ricci La Centra ; Arash Sayyah ; Julius E. Yellowhair ; Sean M. Garner ; N Mark Horenstein ; Malay K. Mazumder

EVALUATING SOLAR CELL FRACTURE AS A FUNCTION OF MODULE MECHANICAL LOADING CONDITIONS 2897

Eric J. Schneller ; Andrew M. Gabor ; Jason Lincoln ; Rob Janoch ; Andrew Anselmo ; Joseph Walters ; Hubert Seigneur

COMPUTATIONAL STUDY OF THE EFFECT OF PHOTOVOLTAIC (PV) MODULE PARAMETERS ON STRESS DEVELOPMENT IN SILICON UNDER STATIC LOADING 2902

Saurabh Sethia ; Karan Shishir Yadav ; Sudharm Rathore ; Abhishek Shubhrant ; Aparna Singh

A SIMPLE METHOD FOR MEASURING SOLAR RADIATION INTENSITY BY IMAGE ANALYSES 2906

Akiko Takahashi ; Akinori Moriki ; Nobuyuki Yamada ; Jun Imai ; Shigeyuki Funabiki

DEGRADATION OF SOLDER BONDS IN FIELD AGED PV MODULES: CORRELATION WITH SERIES RESISTANCE INCREASE 2912

Abhishiktha Tummala ; Jaewon Oh ; Sai Tatapudi ; Govindasamy Tamizhmani

PERFORMANCE OF LIGHT AND DARK CURRENT-VOLTAGE CHARACTERISTICS FOR PID-AFFECTED MONOCRYSTALLINE SILICON SOLAR MODULES 2918

He Wang ; Pan Zhao ; Shuwen Guo ; Hong Yang ; Weijing Huang ; Shiyu Sang ; Bojie Su ; Xue Zhang ; Yunxue Cao ; Hui Zhao

SOILING RATES OF PV MODULES VS. THERMOPILE PYRANOMETERS 2923

Martin Waters ; Tejas Tirumalai ; Michael Gostein ; Bill Stueve

GRID INTEGRATION OF BUILDING SYSTEMS AND 1 MW PHOTOVOLTAIC ARRAY USING VOLTTRON 2926

David Raker ; Andrew Sellers ; Roshan Kini ; Michael Green ; Thomas Stuart ; Randall Ellingson ; Raghav Khanna ; Michael Heben

INTERCONNECTION STUDY OF DISTRIBUTED PV SYSTEMS BY INTERFACING MATLAB WITH OPENDSS AND GIS 2931

Joseph A. Ahamioje ; Hariharan Krishnaswami

NOVEL MPPT ALGORITHM FOR ACTIVE POWER CONTROL OF MULTI-LEVEL DUAL-ACTIVE BRIDGE PV CONVERTER IMPLEMENTED IN NI MYRIO 2936

Shilpa Marti ; Hariharan Krishnaswami

MODELING A GRID-CONNECTED PV/BATTERY MICROGRID SYSTEM WITH MPPT CONTROLLER 2941

Genesis Alvarez ; Hadis Moradi ; Mathew Smith ; Ali Zilouchian

>94.5%REDUCTION IN GRID-BUY ELECTRICITY AND ELIMINATION OF AM & PM ENERGY PEAKS/SPIKES BY OPTIMIZING ENERGY USAGE AND INTEGRATION OF CUSTOMER SELF-SUPPLY ROOFTOP SOLAR PV WITH ELECTRICAL & THERMAL (HOT & COLD) STORAGE BATTERIES: A CASE STUDY FOR RESIDENTIAL HAWAII 2947

John Borland ; Jay Moore ; Corpuz Poncho ; Takahiro Tanaka ; Harumi Mcclure

A SINGLE-STAGEC"UK-BASED TRANSFORMERLESS INVERTER FOR 1-Φ GRID-CONNECTED PV SYSTEMS 2952

Phani Kumar Chamarthi ; Amit Kumar Gupta ; Madhuwanti S. Joshi ; Vivek Agarwal

A STATE SPACE AVERAGE MODEL FOR DYNAMIC MICROGRID BASED SPACE STATION SIMULATIONS 2957

Rachid Darbali-Zamora ; Eduardo I. Ortiz-Rivera

BUCK CONVERTER AND SEPIC BASED ELECTRONIC POWER SUPPLY DESIGN WITH MPPT AND VOLTAGE REGULATION FOR SMALL SATELLITE APPLICATIONS 2963

Rachid Darbali-Zamora ; Nicolás Cobo-Yepes ; John E. Salazar-Duque ; Eduardo I. Ortiz-Rivera ; Amilcar A. Rincon-Charris

VIRTUAL POWER PLANT FEEDBACK CONTROL DESIGN FOR FAST AND RELIABLE ENERGY MARKET AND CONTINGENCY RESERVE DISPATCH 2969

Mohamed Elkhatib ; Jay Johnson ; David Schoenwald

INTELLIGENT SAMPLING OF PERIODS FOR REDUCED COMPUTATIONAL TIME OF TIME SERIES ANALYSIS OF PV IMPACTS ON THE DISTRIBUTION SYSTEM 2975

Jason Galtieri ; Matthew J. Reno

A PWM SCHEME TO REALISE TWO TIMES EFFECTIVE SWITCHING FREQUENCY WITH CONSTANT COMMON MODE VOLTAGE AND REACTIVE POWER CAPABILITY IN 1-Φ GRID-TIED TRANSFORMERLESS H6 PV INVERTER 2981

Amit Kumar Gupta ; Madhuwanti S. Joshi ; Vivek Agarwal

A SOLAR PV RETROFIT SOLUTION FOR RESIDENTIAL BATTERY INVERTERS 2986

Amit Kumar Gupta ; Vaibhav Pawar ; Madhuwanti S. Joshi ; Vivek Agarwal ; Deepak Chandran

COST BENEFIT AND ALTERNATIVES ANALYSIS OF DISTRIBUTION SYSTEMS WITH ENERGY STORAGE SYSTEMS 2991

Tom Harris ; Adarsh Nagarajan ; Murali Baggu ; Tom Bialek

EVALUATION OF PV HOSTING CAPACITIES OF DISTRIBUTION GRIDS WITH UTILIZATION OF SOLAR-ROOF-POTENTIAL-ANALYSES 2996

Gerd Heilscher ; Falko Ebe ; Basem Idlbi ; Jeromie Morris ; Florian Meier

EXPERIMENTAL DISTRIBUTION CIRCUIT VOLTAGE REGULATION USING DER POWER FACTOR, VOLT-VAR, AND EXTREMUM SEEKING CONTROL METHODS 3002

Jay Johnson ; Sigifredo Gonzalez ; Daniel B. Arnold

DYNAMIC SETPOINT CONTROL OF ELECTRIC HOT WATER HEATER TANKS FOR INCREASED INTEGRATION OF SOLAR PHOTOVOLTAIC SYSTEMS 3008

C. Birk Jones ; Monte Lunacek ; Matthew Lave ; Jay Johnson ; Robert Broderick

SPATIAL ANALYSIS OF RESIDENTIAL COMBINED PHOTOVOLTAIC AND BATTERY POTENTIAL: CASE STUDY UTRECHT, THE NETHERLANDS 3014

Geert Litjens ; Bala Bhavya Kausika ; Ernst Worrell ; Wilfried Van Sark

POWER BALANCE REQUIREMENTS FOR SUSTAINED ISLANDING OF INVERTER BASED DISTRIBUTED GENERATION 3020

Gregory A. Kern ; Michael Ropp ; Sigifredo Gonzalez

FULL-SCALE DEMONSTRATION OF DISTRIBUTION SYSTEM PARAMETER ESTIMATION TO IMPROVE LOW-VOLTAGE CIRCUIT MODELS 3025

Matthew Lave ; Matthew J. Reno ; Robert J. Broderick ; Jouni Peppanen

CREATION AND VALUE OF SYNTHETIC HIGH-FREQUENCY SOLAR INPUTS FOR DISTRIBUTION SYSTEM QSTS SIMULATIONS 3031

Matthew Lave ; Matthew J. Reno ; Robert J. Broderick

A DIRECT MAXIMUM POWER POINT SEARCH USING CURRENT-VOLTAGE BASED POWER-LAW RELATION FOR PHOTOVOLTAIC SYSTEM UNDER UNIFORM IRRADIANCE 3038
Hitesh K. Mehta ; Ashish K. Panchal

PASSIVITY BASED CONTROLLER FOR PHOTOVOLTAIC MODULES USING ZETA CONVERTER 3044
Daniel A. Merced Cirino ; Rachid Darbali Zamora ; Eduardo I. Ortiz Rivera

SIC SWITCH BASED SINGLE-STAGE BUCK-BOOST TRANSFORMERLESS MINI INVERTER WITH LOW LEAKAGE CURRENT AND NEGLIGIBLE DC INJECTION 3050
Soumya Ranjan Mohapatra ; Amit Kumar Gupta ; Madhuwanti S. Joshi ; Vivek Agarwal

OPEN SOURCE TOOLS FOR HIGH PERFORMANCE QUASI-STATIC-TIME-SERIES SIMULATION USING PARALLEL PROCESSING 3055
Davis Montenegro ; Roger C. Dugan ; Matthew J. Reno

MAXIMUM POWER POINT TRACKING OF PV MODULE BASED ON NEW EXPLICIT I-V RELATION 3061
Tejeswar Nukala ; A. K. Panchal

AN AUTOCORRELATION-BASED COPULA MODEL FOR PRODUCING REALISTIC CLEAR-SKY INDEX AND PHOTOVOLTAIC POWER GENERATION TIME-SERIES 3067
Joakim Munkhammar ; Joakim Widén

DYNAMIC RESPONSE OF MAXIMUM POWER POINT TRACKING USING PERTURB AND OBSERVE ALGORITHM WITH MOMENTUM TERM 3073
Gautam A. Raiker

A FRAMEWORK FOR COMPARING THE ECONOMIC PERFORMANCE AND ASSOCIATED EMISSIONS OF GRID-CONNECTED BATTERY STORAGE SYSTEMS IN EXISTING BUILDING STOCK: A NYISO CASE STUDY 3077
Julian Do Nascimento Ricardo ; Vasilis Fthenakis

IMPROVING ANY ARBITRARY MPPT HILL CLIMBER WITH ANN ESTIMATIONS 3083
Jesse Roberts ; Indranil Bhattacharya

INCREASING SOLAR PHOTOVOLTAIC PENETRATION USING THERMAL ENERGY STORAGE 3088
Alexander F. Routhier ; Christiana Honsberg

MODEL PREDICTIVE CONTROL OF GRID CONNECTED MODULAR MULTILEVEL CONVERTER FOR INTEGRATION OF PHOTOVOLTAIC POWER SYSTEMS 3092
Amir Shahirinia ; Amin Hajizadeh

MAXIMIZATION OF SELF-SUFFICIENCY WITH GRID CONSTRAINTS: PV GENERATORS, WIND TURBINES AND STORAGE TO FEED TERTIARY SECTOR USERS 3096
Filippo Spertino ; Jawad Ahmad ; Alessandro Ciocia ; Paolo Di Leo ; Francesco Giordano

SWITCHES CONTROLLING TO IMPLEMENT ADAPTIVE MULTILEVEL INVERTER ON PV SYSTEM 3102
Hadi Suhana ; Ngapuli I Sinisuka ; Muhammad Nurdin ; Yvon Besanger ; Vincent Debusschere

DEMAND RESPONSE FOR THE PROMOTION OF PHOTOVOLTAIC PENETRATION 3107
Venizelos Venizelou ; Spyros Theocharides ; George Makrides ; Venizelos Efthymiou ; George E. Georghiou

GRIDDLER AI: NEW PARADIGM IN LUMINESCENCE IMAGE ANALYSIS USING AUTOMATED FINITE ELEMENT METHODS 3113
Johnson Wong ; Percis Teena ; Daniel Inns

INTERACTION OF O2IDIMERS WITH GA IN SI AND IMPLICATIONS FOR A COMPREHENSIVE MODEL OF LIGHT- INDUCED DEGRADATION 3119
Yu Jin ; Scott T. Dunham

NUMERICAL SIMULATION OF EBIC FOR ANALYSIS OF EXTENDED DEFECTS 3123
Marco Nardone ; John Moseley ; Saroj Dahal ; Anuja V. Parikh ; John M. Waddle

COLLOIDAL QUANTUM DOT SOLAR CELL ELECTRICAL PARAMETER IMAGING USING CAMERA-BASED HIGH-FREQUENCY HETERODYNE LOCK-IN CARRIEROGRAPHY 3129
Lilei Hu ; Mengxia Liu ; Andreas Mandelis ; Qiming Sun ; Alexander Melnikov ; Edward H. Sargent

A NEW PERSPECTIVE ON POTENTIAL-INDUCED DEGRADATION OF THE SHUNTING TYPE BY MICRO RAMAN-SPECTROSCOPY AND MICRO LIGHT-BEAM-INDUCED CURRENT 3135
A. Büchler ; H. Nagel ; M. Breitwieser ; S. Kluska ; F. D. Heinz ; M. C. Schubert ; M. Glatthaar ; S. Glunz

NANOSCALE DETECTION OF DEEP LEVELS IN CIGS USING ELECTRON ENERGY LOSS SPECTROSCOPY 3139
Julia I. Deitz ; Pran K. Paul ; Shankar Karki ; Sylvain Marsillac ; Aaron R. Arehart ; Tyler J. Grassman ; David W. Mccomb

MEASUREMENT OF CARRIER DYNAMICS IN PHOTOVOLTAIC CZTSE BY TIME-RESOLVED TERAHERTZ SPECTROSCOPY 3143

Siming Li ; Michael A. Lloyd ; Andrew A. Golembeski ; Brian E. Mccndless ; Jason B. Baxter

DECOUPLING GRAIN-BOUNDARY, GRAIN-INTERIOR, AND SURFACE RECOMBINATION WITH CATHODOLUMINESCENCE 3147

John Moseley ; Pierre Rale ; Stéphane Collin ; Ana Kanevce ; Eric Colegrove ; Joel Duenow ; Soren Jensen ; Wyatt K. Metzger ; Mowafak M. Al-Jassim

HIGH RESOLUTION THZ SCANNING FOR OPTIMIZATION OF DIELECTRIC LAYER OPENING PROCESS ON DOPED SI SURFACES 3150

P. Spinelli ; F.J.K. Danzl ; D. Deligiannls ; N. Guillevin ; A.R. Burgers ; S. Sawallich ; M. Nage ; I. Cesar

DEGRADATION ASSESSMENT OF FIELDED CIGS PHOTOVOLTAIC ARRAYS 3155

Bruce H. King ; Joshua S. Stein ; Daniel Riley ; C. Birk Jones ; Charles D. Robinson

APPLICATION OF IEC 61724 STANDARDS TO ANALYZE PV SYSTEM PERFORMANCE IN DIFFERENT CLIMATES 3161

Katherine A. Klise ; Joshua S. Stein ; Joseph Cunningham

EFFECTS OF URBAN ENVIRONMENT ON SOLAR PV PERFORMANCE 3167

Panagiotis Moraitis ; Bala Bhavya Kausika ; Wilfried G.J.H.M. Van Sark

IRRADIANCE MEASUREMENT CONSIDERATIONS FOR SYSTEM PERFORMANCE ASSESSMENT WHEN MANAGING FLEETS OF PHOTOVOLTAIC ASSETS ACROSS ASIA 3172

André M. Nobre ; Shravan Karthik ; Chenxi Liu ; Rohit Jaswal ; Rupesh Baker ; Raghav Malhotra ; Alan Khor

MACHINE LEARNING IN PV FAULT DETECTION, DIAGNOSTICS AND PROGNOSTICS: A REVIEW 3178

Sandy Rodrigues ; Helena Geirinhas Ramos ; F. Morgado-Dias

OUTDOOR FIELD PERFORMANCE FROM BIFACIAL PHOTOVOLTAIC MODULES AND SYSTEMS 3184

Joshua S. Stein ; Daniel Riley ; Matthew Lave ; Clifford Hansen ; Chris Deline ; Fatima Toor

DEFINING THRESHOLD VALUES OF ENCAPSULANT AND BACKSHEET ADHESION FOR PV MODULE RELIABILITY 3190

Nick Bosco ; Joshua Eafanti ; Sarah Kurtz ; Jared Tracy ; Reinhold Dauskardt

CHARACTERIZATIONS OF AGED GLASS/ETHYLENE VINYL ACETATE/GLASS USING FLUORESCENCE SPECTROSCOPY AND INSTRUMENTED INDENTATION 3195

Jae Hyun Kim ; Yadong Lyu ; David C. Miller ; Xiaohong Gu

ENCAPSULANT ADHESION TO SURFACE METALLIZATION ON PHOTOVOLTAIC CELLS 3200

Jared Tracy ; Nick Bosco ; Reinhold Dauskardt

IMPACT OF UV LIGHT INTENSITY ON PHOTODEGRADATION OF PV BACKSHEETS 3204

Xiaohong Gu ; Li-Chieh Yu ; Yadong Lyu ; Jae Hyun Kim ; Andrew Fairbrother ; Tinh Nguyen

SURVEY OF MECHANICAL DURABILITY OF PV BACKSHEETS 3208

Michael D. Kempe ; David C. Miller ; Allen Zielnik ; Daniel Montiel-Chicharro ; Jiang Zhu ; Ralph Gottschalg

SOLAR VARIABILITY REDUCTION USING OFF-MAXIMUM POWER POINT TRACKING AND BATTERY STORAGE 3214

Jason Galtieri ; Philip T. Krein

INTEGRATION OF ELECTROCHEMICAL CAPACITORS ON SILICON PHOTOVOLTAIC MODULES FOR RAPID-RESPONSE POWER BUFFERING 3220

Yu Jiang ; Xuanyi Shi ; Derwin Lau ; Da-Wei Wang ; Zi Ouyang ; Alison Lennon

DESIGN & EVALUATION OF A HYBRID SWITCHED CAPACITOR CIRCUIT WITH WIDE-BANDGAP DEVICES FOR COMPACT MVDC PV POWER CONVERSION 3224

J. Stewart ; J. Delhotal ; J. Richards ; J. Neely ; L. Rashkin ; J. D. Flicker ; R. Kaplar ; S. Gonzalez ; J. Lehr

SOLAR ENERGY FOR CLEAN AND AFFORDABLE WATER DESALINATION 3230

V. M. Fthenakis ; Adam A. Atia

GLOBAL RESIDENTIAL AIR-CONDITIONING SECTOR AS A DRIVER FOR PHOTOVOLTAIC INDUSTRY GROWTH DURING THE 21ST CENTURY 3236

Hannu S. Laine ; Jyri Salpakari ; Marius Peters ; Erin E. Looney ; Ashley E. Morishige ; Hele Savin ; Gregory Wilson ; Tonio Buonassisi

MEASURES TO REMOVE ECONOMIC NON-MARKET FAILURE AND INSTITUTIONAL BARRIERS THAT RESTRICT PHOTOVOLTAICS SELF-CONSUMPTION AND NET-METERING IN SPAIN 3240

Enrique Rosalcs-Ascnsio ; Juan A. Méndez ; Benjamín Gonzálcz-Díaz ; Ricardo Guerrero Lemus

COST COMPETITIVE CONCENTRATOR PHOTOVOLTAICS FOR SOLAR THERMAL APPLICATIONS 3245

Brian C. Riggs ; Richard E. Biedenham ; Chris Dougher ; Yaping Vera Ji ; Qi Xu ; Vince Romanin ; Daniel S. Codd ; James M. Zahler ; Matthew D. Escarra

PREDICTING THE EFFICIENCY OF THE SILICON BOTTOM CELL IN A TWO-TERMINAL TANDEM SOLAR CELL 3250

Zhengshan J. Yu ; Zachary C. Holman

MECHANICALLY STACKED 4-TERMINAL III-V/SI TANDEM SOLAR CELLS 3254

Stephanie Essig ; Christophe Allebe ; John F. Geisz ; Myles A. Steiner ; Loris Barraud ; Antoine Descoeudres ; J. Scott Ward ; Manuel Schnabel ; David L. Young ; Matthieu Despeisse ; Christophe Ballif ; Adele Tamboli

PEROVSKITE/SILICON TANDEM SOLAR CELLS: CHALLENGES TOWARDS HIGH-EFFICIENCY IN 4-TERMINAL AND MONOLITHIC DEVICES 3256

Jérémie Werner ; Florent Sahli ; Brett Kamino ; Davide Sacchetto ; Matthias Bräuninger ; Arnaud Walter ; Christophe Ballif ; Matthieu Despeisse ; Sylvain Nicolay ; Bjoern Niesen ; Raphaël Monnard ; Stefaan De Wolf ; Soo-Jin Moon ; Loris Barraud ; Bertrand Paviet-Salomon ; Jonas Geissbuehler ; Christophe Allebé

THE OUTCOME OF REPLACING SN COMPLETELY BY GE IN KESTERITE CU2ZNSNSE4SOLAR CELLS 3260

S. Sahayaraj ; G. Brammertz ; B. Vermang ; T. Schnabel ; E. Ahlswede ; Z. Huang ; S. Ranjbar ; M. Meuris ; J. Vleugels ; J. Poortmans

TRANSITION METAL OXIDES NANO-LAYERS AS EFFICIENT BACK ELECTRON REFLECTORS FOR CU2ZNSNSE4SOLAR CELLS 3265

Sergio Giraldo ; Moisés Espíndola-Rodríguez ; Florian Oliva ; Víctor Izquierdo-Roca ; Alejandro Pérez-Rodríguez ; Edgardo Saucedo

MIXED SULFUR AND SELENIUM ANNEALING STUDY OF COMPOUND-SPUTTERED BILAYER CU2ZNSNS4/ CU2ZNSNSE4PRECURSORS 3269

N. Ross ; S. Grini ; L. Vines ; C. Platzer-Björkman

REVEALING THE ROLE OF MN INCORPORATION IN CU2ZNSN(S, SE)4PHOTOVOLTAIC ABSORBER LAYER 3275

Stener Lie ; Joel M. R. Tan ; Wenjie Li ; Shin Woei Leow ; Oki Gunawan ; Doug Bishop ; Lydia H. Wong

NON-VACUUM SINGLE STEP SYNTHESIS OF LARGE-GRAIN SIZE CZTS PHOTO ABSORBER FOR THIN FILM SOLAR CELLS BY FLUX ASSISTED CHEMICAL SPRAY 3279

Ratheesh R. Thankalekshmi ; Navjot Kaur Sidhu ; A.C. Rastogi

RAMAN SCATTERING ASSESSMENT OF POINT DEFECTS IN KESTERITE SEMICONDUCTORS: UV RESONANT RAMAN CHARACTERIZATION FOR ADVANCED PHOTOVOLTAICS 3285

Florian Oliva ; Laia Arqués Farré ; Sergio Giraldo ; Mirjana Dimitrievska ; Paul Pistor ; Alejandro Martínez-Pérez ; Lorenzo Calvo-Barrio ; Edgardo Saucedo ; Alejandro Pérez-Rodríguez ; Victor Izquierdo-Roca

ASSESSING THE DEFECT RESPONSIBLE FOR LETID: TEMPERATURE- AND INJECTION-DEPENDENT LIFETIME SPECTROSCOPY 3290

Mallory A. Jensen ; Yan Zhu ; Erin E. Looney ; Ashley E. Morishige ; Carlos Vargas ; Ziv Hameiri ; Tonio Buonassisi

MICROSCOPIC DISTRIBUTION OF LUMINESCENCE FROM DISLOCATION CLUSTERS IN MULTICRYSTALLINE SILICON WAFERS 3295

H. T. Nguyen ; M. A. Jensen ; L. Li ; C. Samundsett ; H. C. Sio ; B. Lai ; T. Buonassisi ; D. Macdonald

DO GRAIN BOUNDARIES MATTER? ELECTRICAL AND ELEMENTAL IDENTIFICATION AT GRAIN BOUNDARIES IN LETID-AFFECTED P-TYPE MULTICRYSTALLINE SILICON 3300

Mallory A. Jensen ; Ashley E. Morishige ; Sagnik Chakraborty ; Romika Sharma ; Hang Cheong Sio ; Chang Sun ; Barry Lai ; Volker Rose ; Amanda Youssef ; Erin E. Looney ; Sarah Wieghold ; Jeremy Poindexter ; Juan-Pablo Correa-Baena ; Daniel Macdonald ; Joel B. Li ; Tonio Buonassisi

PERC SOLAR CELL PERFORMANCE PREDICTIONS FROM MULTICRYSTALLINE SILICON INGOT METROLOGY DATA 3304

Bernhard Mitchell ; Daniel Chung ; Qiuxiang He ; Hua Zhang ; Zhen Xiong ; Pietro P. Altermatt ; Peter Geelan-Small ; Thorsten Trupke

PHOTOLUMINESCENCE-IMAGING-BASED EVALUATION OF NON-UNIFORM CDTE DEGRADATION 3305

Steve Johnston ; David Albin ; Peter Hacke ; Steven P. Harvey ; Helio Moutinho ; Mowafak Al-Jassim ; Wyatt K. Metzger

MACHINE LEARNING AND CORRELATIVE MICROSCOPY: HOW 'BIG DATA' TECHNIQUES CAN BENEFIT THIN FILM SOLAR CELL CHARACTERIZATION 3309

Bradley M. West ; Michael Stuckelberger ; Tara Nietzold ; Barry Lai ; Jörg Maser ; Mariana I. Bertoni

METAL INDUCED CONTACT RECOMBINATION MEASURED BY QUASI-STEADY-STATE PHOTOLUMINESCENCE 3315

Robert Dumbrell ; Mattias K. Juhl ; Mengjie Li ; Thorsten Trupke ; Ziv Hameiri

USING TIME-OF-FLIGHT SIMS TO INVESTIGATE GROUP V DOPANT DISTRIBUTION IN CDTE 3319

Steven P. Harvey ; Eric Colegrove ; Brian Mccandless ; David Albin ; Mowafak Al-Jassim ; Wyatt K. Metzger

QUANTITATIVE ANALYSIS OF ACTIVE DOPANT DISTRIBUTION AND ESTIMATION OF EFFECTIVE DIFFUSIVITY IN PHOSPHORUS- IMPLANTED EMITTER OF SI SOLAR CELL USING SCANNING NONLINEAR DIELECTRIC MICROSCOPY 3323

Kotaro Hirose ; Katsuto Tanahashi ; Hidetaka Takato ; Yasuo Cho

SIMULATION OF DRIVE-LEVEL CAPACITANCE PROFILING TO INTERPRET MEASUREMENTS ON CU(IN, GA)SE2SCHOTTKY DEVICES 3327

Geordie Zapalac ; Jeff Bailey

ANALYSIS OF THE IMPACT OF INSTALLATION PARAMETERS AND SYSTEM SIZE ON BIFACIAL GAIN AND ENERGY YIELD OF PV SYSTEMS 3333

Amir Asgharzadeh ; Tomas Lubenow ; Joseph Sink ; Bill Marion ; Chris Deline ; Clifford Hansen ; Joshua Stein ; Fatima Toor

DEPENDENCE OF STRING POWER ON ITS HEIGHT IN THE ARRAY IN YOSHINOGARI MEGA SOLAR POWER PLANT 3339

Shigeomi Hara ; Makoto Kasu ; Yasuki Masutomi

A BOTTOM-UP ENERGY SIMULATION FRAMEWORK TO ACCURATELY COMPARE PV MODULE TOPOLOGIES UNDER NON-UNIFORM AND DYNAMIC OPERATING CONDITIONS 3343

Patrizio Manganiello ; Maro Baka ; Hans Goverde ; Tom Borgers ; Jonathan Govaerts ; Arvid Van Der Heide ; Eszter Voroshazi ; Francky Catthoor

A PERFORMANCE MODEL FOR BIFACIAL PV MODULES 3348

Daniel Riley ; Clifford Hansen ; Joshua Stein ; Matthew Lave ; Johnson Kallickal ; Bill Marion ; Fatima Toor

ACCURATE MODELING OF PARTIALLY SHADED PV ARRAYS 3354

Bennet Meyers ; Mark Mikofski

EVALUATION OF UNCERTAINTY IN PV PROJECT DESIGN: DEFINITION OF SCENARIOS AND IMPACT ON ENERGY YIELD PREDICTIONS.. 3360

Giorgio Belluardo ; Magnus Herz ; Ulrike Jahn ; Mauricio Richter ; David Moser

MONOCRYSTALLINE 1.7 EV MGCDTE DOUBLE-HETEROSTRUCTURE SOLAR CELL WITH 11.2% EFFICIENCY 3366

Calli M. Campbell ; Xin-Hao Zhao ; Yuan Zhao ; Mathieu Boccard ; Cheng- Ying Tsai ; Jacob J. Becker ; Zachary Holman ; Yong-Hang Zhang

MBE GROWTH OF 1.7EV AL0.2GA0.8AS AND 1.42EV GAAS SOLAR CELLS ON SI USING DISLOCATIONS FILTERS: AN ALTERNATIVE PATHWAY TOWARD III-V/ SI SOLAR CELLS ARCHITECTURES 3370

Arthur Onno ; Mingchu Tang ; Mu Wang ; Yurii Maidaniuk ; Mourad Benamara ; Yuriy I. Mazur ; Gregory J. Salamo ; Lars Oberbeck ; Jiang Wu ; Huiyun Liu

III- V/SI TANDEM CELLS UTILIZING INTERDIGITATED BACK CONTACT SI CELLS AND VARYING TERMINAL CONFIGURATIONS 3371

Manuel Schnabel ; Michael Rienacker ; Agnes Merkle ; Talysa R. Klein ; Nikhil Jain ; Stephanie Essig ; Henning Schulte-Huxel ; Emily Warren ; Maikel F.A.M. Van Hest ; John Geisz ; Jan Schmidt ; Rolf Brendel ; Robby Peibst ; Paul Stradins ; Adele Tamboli

TOWARDS HIGH-EFFICIENCY GAASP/SI TANDEM CELLS 3376

S. Fan ; M. Vaisman ; K. Nay Yaung ; E. Perl ; D. Martín-Martín ; M. Leilaeioun ; Z. C. Holman ; M. L. Lee

CHARACTERIZATION OF HETEROEPITAXIAL GAAS FILMS GROWN ON SI USING SELECTIVE AREA NUCLEATION 3381

Emily L. Warren ; Emily A. Makoutz ; Michelle Vaisman ; Benjamin F. Bachman ; William E. Mcmahon ; Jeramy D. Zimmerman ; Adele C. Tamboli

EFFICIENT PHOTON UPCONVERSION IN SEMICONDUCTOR NANOSTRUCTURES: CONSTRAINTS AND OPPORTUNITIES 3384

Matthew F. Doty ; Eric Y. Chen ; Jing Zhang ; Diane G. Sellers ; Zhuohui Li ; Christopher C. Milleville ; Kyle Lennon ; Joshua M. O. Zide

ENHANCED ULTRA-THIN A-GE:H SOLAR CELLS BY PLASMONIC NANOPARTICLES EMBEDDED IN THE OPTICAL RESONANT CAVITY 3388

Brendan Brady ; Volker Steenhoff ; Benedikt Nickel ; Martin Vehse ; Alexander G. Brolo

NATIVE-METAL-OXIDE-COATED PLASMONIC ELECTRODE METASURFACES FOR NANOPHOTONIC LIGHT TRAPPING AND EFFICIENT CHARGE COLLECTION 3393

Deirdre M. O'Carroll ; Christopher E. Petoukhoff ; Zhongkai Cheng ; Zeqing Shen ; Catrice M. Carter

IN-GA PRECURSOR ISLANDS FOR CU(IN, GA)SE2MICRO-CONCENTRATOR SOLAR CELLS 3396

Katharina Eylers ; Franziska Ringleb ; Berit Heidmann ; Sergiu Levcenco ; Thomas Unold ; Hagen W. Klemm ; Gina Peschel ; Alexander Fuhrich ; Thomas Teubner ; Thomas Schmidt ; Martina Schmid ; Torta Boeck

ADVANCES IN SILICON SURFACE TEXTURIZATION BY METAL ASSISTED CHEMICAL ETCHING FOR PHOTOVOLTAIC APPLICATIONS 3402

Sylvain Le Gall ; Raphaël Lachaume ; Encarnacion Torralba ; Mathieu Halbwax ; Vincent Magnin ; Taha El Assimi ; Marin Fouchier ; Joseph Harari ; Jean-Pierre Vilcot ; Christine Cachet-Vivier ; Stéphane Bastide

SINGLE CRYSTALLINE SUBSTRATES FOR III- V GROWTH VIA EXFOLIATION OF BULK SINGLE CRYSTALS 3406

Celeste L. Melamed ; Brenden R. Ortiz ; Aaron D. Martinez ; William E. Mcmahon ; Adele C. Tamboli ; Andrew G. Norman ; Eric S. Toberer

CUZNS HOLE CONTACTS ON MONOCRYSTALLINE CDTE SOLAR CELLS 3410

Jacob J. Becker ; Xiaojie Xu ; Rachel Woods-Robinson ; Calli M. Campbell ; Maxwell Lassise ; Joel Ager ; Yong-Hang Zhang

THE EFFECT OF THE CDCL2 HEAT TREATMENT ON CDSEXTE1-X SOLAR CELLS 3413

Chih An Hsu ; Vasilios Palekis ; Imran Khan ; Shamara Collins ; Don Morel ; Chris Ferekides

EFFECTS OF CDCL2TREATMENT ON THE LOCAL ELECTRONIC PROPERTIES OF POLYCRYSTALLINE CDTE MEASURED WITH PHOTOEMISSION ELECTRON MICROSCOPY 3417

Morgann Berg ; Jason M. Kephart ; Walajabad S. Sampath ; Taisuke Ohta ; Calvin Chan

POINT DEFECTS IN CDTE BULK SINGLE CRYSTALS GROWN IN CD-RICH CONDITIONS 3422

Tursun Ablekim ; Santosh K. Swain ; Teresa M. Barnes ; Kelvin G. Lynn

OPTICAL PROPERTIES OFCDSE1-XSXANDCDSE1-YTEYALLOYS AND THEIR APPLICATION FOR CDTE PHOTOVOLTAICS 3426

Maxwell M. Junda ; Corey R. Grice ; Prakash Koirala ; Robert W. Collins ; Yanfa Yan ; Nikolas J. Podraza

BLISTERING OF MAGNETRON SPUTTERED THIN FILM CDTE DEVICES 3430

P.M. Kaminski ; S. Yilmaz ; A. Abbas ; F. Bittau ; J.W. Bowers ; R.C. Greenhalgh ; J.M. Walls

ENERGY YIELD IN HOT & SUNNY CLIMATES: IMPACT OF SILICON SOLAR CELL ARCHITECTURE AND CELL INTERCONNECTION 3435

Jan Haschke ; Johannes P. Seif ; Yannick Riesen ; Andrea Tomasi ; Jean Cattin ; Loïc Tous ; Patrick Choulat ; Monica Aleman ; Emanuele Comagliotti ; Angel Uruena ; Richard Russell ; Filip Duerinckx ; Jonathan Champliaud ; Jacques Levrat ; Amir A. Abdallah ; Brahim Aïssa ; Nouar Tabet ; Nicolas Wyrsch ; Matthieu Despeisse ; Jozef Szlufcik ; Stefaan De Wolf ; Christophe Ballif

NOVEL REAR SIDE METALLIZATION ROUTE FOR SI SOLAR CELLS USING A TRANSPARENT CONDUCTING ADHESIVE 3439

Manuel Schnabel ; Talysa R. Klein ; Benjamin G. Lee ; William Nemeth ; Vincenzo Lasalvia ; Maikel F.A.M. Van Hest ; Paul Stradins

MULTILAYER FOIL METALLIZATION FOR ALL BACK CONTACT CELLS 3442

David H. Levy ; David E. Carlson

ELECTROLUMINESCENCE EXCITATION SPECTROSCOPY: A NOVEL APPROACH TO NON-CONTACT QUANTUM EFFICIENCY MEASUREMENTS 3448

Kristopher O. Davis ; Greg S. Horner ; Joshua B. Gallon ; Leonid A. Vasilyev ; Kyle B. Lu ; Antonius B. Dirriwachter ; Terry B. Rigdon ; Eric J. Schneller ; Kortan Ogutman ; Richard K. Ahrenkiel

ILLUMINATED OUTDOOR LUMINESCENCE IMAGING OF PHOTOVOLTAIC MODULES 3452

Timothy J Silverman ; Michael G. Deceglie ; Kaitlyn Vansant ; Steve Johnston ; Ingrid Repins

ELECTROLUMINESCENT IMAGE PROCESSING AND CELL DEGRADATION TYPE CLASSIFICATION VIA COMPUTER VISION AND STATISTICAL LEARNING METHODOLOGIES 3456

Justin S. Fada ; Mohammad A. Hossain ; Jennifer L. Braid ; Shuying Yang ; Timothy J Peshek ; Roger H. French

TOWARDS DEVELOPING A STANDARD FOR TESTING BIFACIAL PV MODULES: SINGLE-SIDE VERSUS DOUBLE-SIDE ILLUMINATION METHOD I-V MEASUREMENTS UNDER DIFFERENT IRRADIANCE AND TEMPERATURE 3462

Stefan Roest ; Witek Nawara ; Bas B. Van Aken ; Elias Garcia Goma

ELECTRICAL TRANSPORT PROPERTIES FROM LONG WAVELENGTH ELLIPSOMETRY 3468

Prakash Uprety ; Maxwell M. Junda ; Indra Subedi ; Michael A. Slocum ; David V. Forbes ; Seth M. Rubbard ; Nikolas J. Podraza

IN SITU RAMAN MONITORING OF KESTERITE CU2ZNSNS4 PHASE FORMATION FROM SULFURIZATION OF SOL-GEL OXIDE PRECURSORS 3473

Osama Awadallah ; Joseph Hernandez ; Andriy Durygin ; Zhe Cheng

PERFORMANCE OF FIELD-AGED PV MODULES IN INDIA: RESULTS FROM 2016 ALL INDIA SURVEY OF PV MODULE RELIABILITY 3478

Rajiv Dubey ; Sachin Zachariah ; Shashwata Chattopadhyay ; Vivek Kuthanazhi ; Sugguna Rambabu ; Sonali Bhaduri ; Hemant K. Singh ; Archana Sinha ; Birinchi Bora ; Rajesh Kumar ; O. S. Sastry ; Chetan S. Solanki ; Anil Kottantharayil ; Brij M. Arora ; K. L. Narasimhan ; Juzer Vasi

INFERRING THE PERFORMANCE RATIO OF PV SYSTEMS DISTRIBUTED IN AN REGION: A REAL-CASE STUDY IN SOUTH TYROL 3482

Marco Pierro ; Giorgio Belluardo ; Philip Ingenhoven ; Cristina Cornaro ; David Moser

QUANTIFY PHOTOVOLTAIC MODULE DEGRADATION USING THE LOSS FACTOR MODEL PARAMETERS 3488

C. Birk Jones ; Bruce H. King ; Joshua S. Stein ; Justin S. Fada ; Alan J. Curran ; Roger H. French ; Erdmut Schnabel ; Michael Koehl ; Olga Lavrova

SIMULATING PV SYSTEM PERFORMANCE WITH COMPONENT RELIABILITY DISTRIBUTIONS 3494

Geoffrey T. Klisel ; Janine M. Freeman ; Olga Lavrova

LIFETIME AND DEGRADATION OF PRE-DAMAGED PV-MODULES – FIELD STUDY AND LAB TESTING 3500

Claudia Buerhop ; Sven Wirsching ; Simon Gehre ; Tobias Pickel ; Thilo Winkler ; Andreas Bemrrr ; Julia Merghcim ; Christian Camus ; Jens Hauch ; Christoph J. Brabec

IMM TRIPLE-JUNCTION SOLAR CELLS AND MODULES OPTIMIZED FOR SPACE AND TERRESTRIAL CONDITIONS 3506

Tatsuya Takamoto ; Hiroyuki Juso ; Kohsuke Ueda ; Hidetoshi Washio ; Hiroshi Yamaguchi ; Mitsuru Imaizumi ; Taishi Sumita ; Tetsuya Nakamura

VERY HIGH SPECIFIC POWER ELO SOLAR CELLS (>3 KW/KG) FOR UAV, SPACE, AND PORTABLE POWER APPLICATIONS 3511

D. Cardwell ; A. Kirk ; C. Stender ; A. Wibowo ; F. Tuminello ; M. Drees ; R. Chan ; M. Osowski ; N. Pan

ENHANCED ENDURANCE OF A UNMANNED AERIAL VEHICLES USING HIGH EFFICIENCY SI AND III-V SOLAR CELLS 3514

David Scheiman ; Raymond Hoheisel ; Daniel J Edwards ; Andrew Paulsen ; Justin Lorentzen ; Steve Carruthers ; Sam Carter ; Matthew Kelly ; Phillip Jenkins ; Robert Walters

HIGH PERFORMANCE, LIGHTWEIGHT GAAS SOLAR CELLS FOR AEROSPACE AND MOBILE APPLICATIONS 3520

Aarohi Vijh ; Lori Washington ; Robert C. Parenti

THROUGH-EPITAXIAL-VIA BACK-CONTACT MULTI-JUNCTION SOLAR CELLS FABRICATED USING EPITAXIAL LIFT-OFF 3524

Rekha Reddy ; Marilyn L. Nowakowski ; David Rowell ; Christopher L. Stender ; Christopher Youtsey

DESIGN OF INGAP/GAAS/LNGAAS MULTI-JUNCTION CELLS WITH REDUCED LAYER THICKNESSES USING LIGHT-TRAPPING REAR TEXTURE 3528

Lin Zhu ; Anurag Reddy ; Kentaroh Watanabe ; Masakazu Sugiyama ; Yoshiaki Nakano ; Hidefumi Akiyama

Author Index

Automatic Detection of Inactive Solar Cell Cracks in Electroluminescence Images

Sergiu Spataru[1], Peter Hacke[2], Dezso Sera[1]

[1]Aalborg University, Aalborg, 9220, Denmark
[2]National Renewable Energy Laboratory, Golden, CO 80401, United States

Abstract — **Inactive solar cell regions resulted from their disconnection from the electrical circuit of the cell are considered to most severe type of solar cell cracks, causing the most power loss. In this work, we propose an algorithm for automatic determination of the electroluminescence (EL) signal threshold level corresponding these inactive solar cell regions. The resulting threshold enables automatic quantification of the cracked region size and estimation of the risk of power loss in the module.**

We tested the algorithm for detecting inactive cell areas in standard mono and mc-Si, showing the influence of current bias level and camera exposure time on the detection. Last, we examined the correlation between the size of the detected solar cell cracks and the power loss of the module.

Index Terms — **crystalline silicon, cell crack, detection, diagnosis, electroluminescence, photovoltaic module.**

I. INTRODUCTION

Solar cell micro-cracks can occur due to mechanical stress during the PV panel manufacturing process [1], transportation [2], or installation [3]. It is estimated that ~6% of PV panels develop at least one crack after transportation [4]. These can further evolve, or new ones can be formed during the service of the PV module due to wind or snow loads [5] and temperature cycling [6]. The most severe cracks can cause significant power loss [7], as well as hot-spots [3], which can further shorten the lifetime of the PV panel.

Currently, the most efficient method of solar cell crack detection is electroluminescence (EL) imaging. Nowadays, EL cameras have become widespread, and are starting to be used as field diagnostic tools as well on fixed [8] or mobile [9] imaging platforms. Consequently, machine analysis methods for detecting and evaluating the severity of solar cell cracks are valuable for analyzing a large volume of EL image data from a PV plant, for example.

Previous research on investigating the severity of solar cell cracks [10] has defined three main types: mode A, B, and C. Amongst these, mode C cracks – corresponding to inactive cell areas – cause the most power loss in PV modules [10] and have the highest likelihood of causing hot spots. Thus, EL image machine analysis methods need to be able to detect and quantify such severe solar cell cracks.

In [11] a method was proposed for quantifying mode B and C cracks from EL images, based on analyzing the EL intensity distribution of individual cells or the entire PV panel. The method makes use of certain EL intensity thresholds in the image, determined by image segmentation algorithms, or

manually from the image. These thresholds determine which areas of the cell correspond to cracks and which are undamaged. This method was included in the draft of the new EL imaging standard currently under development "IEC TS 60904-13 Electroluminescence of photovoltaic modules," which focuses on EL imaging requirements, procedures, and methods for quantification of cell characteristics.

This paper continues that work, and proposes an algorithm for determining the EL intensity threshold corresponding to mode C cracks. This algorithm can be used to automatically detect mode C cracks in low-current bias EL images, as well as for detecting *possible* mode C cracks in high-current bias EL images. Detecting such cracks from high-current bias EL images is relevant for applications that may be constrained by a short imaging exposure time, such as outdoor imaging [12].

In the experimental part of this work we apply the method to detect and quantify cell cracks from mono- and mc-Si PV modules, degraded through accelerated thermo-mechanical stress. In this analysis, we investigate the influence of the forward current bias and camera exposure time used for the PV module EL imaging, on the cell crack detection accuracy of the method. Last, we examine the correlation between the size of the detected solar cell cracks and the power loss of the module. This opens the possibility for estimating the power degradation of a module due to cell cracks from EL images alone, which has potential applications in outdoor EL inspection of PV plants.

II. TYPES OF SOLAR CELL CRACKS

Mode A cracks, shown in Fig. 1a and Fig. 1c, represent an incipient form of solar cell cracks, that usually do not cause much power loss, but can develop over time into more severe type of cell cracks (mode B and C) [10]. The second type of cell cracks, denoted mode B, shown in Fig. 1b and Fig. 1d, correspond to partially disconnected cell areas, that cause increased series resistance and losses [10]. These appear black or gray in the EL images, depending on current-bias of the module and camera exposure time.

The most severe type of cracks is considered mode C, shown in Fig. 1d. These correspond to completely disconnected cell areas [10], effectively reducing the area of the cell and its current generation, and causing the most power loss. Mode C cracks appear black in EL images irrespective of current level and camera exposure time, since no photons are being emitted from the affected regions.

978-1-5090-5606-4/17 $31.00 © 2017 IEEE

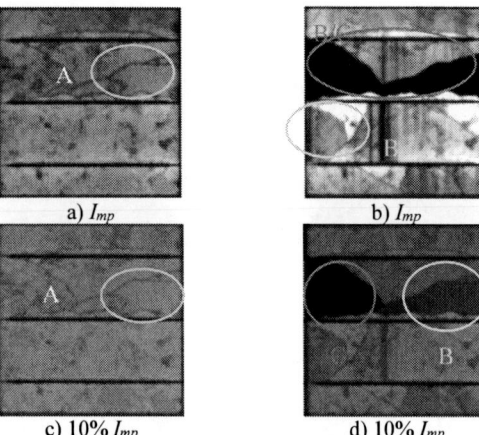

a) I_{mp} b) I_{mp}

c) 10% I_{mp} d) 10% I_{mp}

Fig. 1.   Example of mode A, B and C solar cell cracks as defined in [10]. EL images correspond mc-Si solar cell before and after thermo-mechanical stress testing: a) mode A crack measured at I_{mp} bias b) mixed mode B/C cracks measured at I_{mp} bias, c) mode A crack at 10% I_{mp} current bias, d) mode B, C cracks measured at 10% I_{mp} bias.

Discerning between mode B and C cracks requires imaging at a low-current bias [10], typically ~10% of the PV module short-circuit (I_{sc}) or maximum power point current (I_{mp}). At these lower current levels, voltage losses due to series resistance (R_s) are smaller, thus mode B cracks, which cause increased R_s, appear relatively brighter relative to the surrounding cell area than in the higher current bias images. On the other hand, mode C cracks remain dark irrespective of the current level. In practice, mode C cracked regions usually have a higher than zero EL intensity due to the noise of the camera, ambient, and reflections from adjacent cells [11].

Most often PV modules are imaged at I_{sc} or I_{mp} bias, to shorten camera exposure time and improve the signal-to-noise ratio of the image. Under these conditions, severe mode B cracks appear the same as mode C in the image, as shown in Fig. 1b. This is due to the low EL signal emission of the cracked area and the limited dynamic range of the camera. We denote this type as *mixed mode B/C cracks* for the rest of the paper.

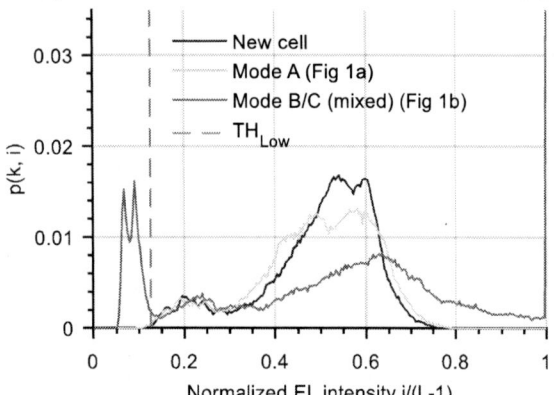

Fig. 2. EL intensity histogram of a solar cell (Fig. 1) imaged at I_{mp} bias, and at different stages of mechanical degradation: blue – new cell; green – affected by mode A cracks (Fig. 1a); red – affected by mode B/C cracks (Fig. 1b).

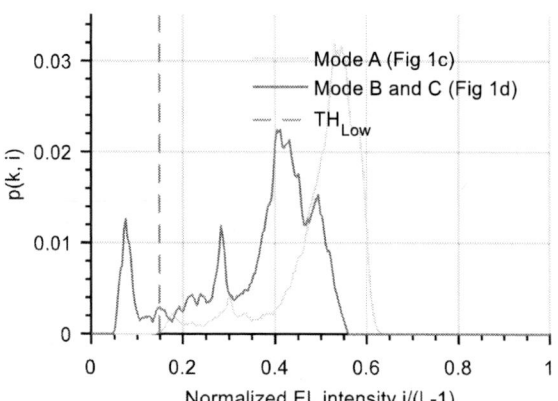

Fig. 3.   EL intensity histogram of the solar cell in Fig. 1 imaged at 10% I_{mp} bias, and two stages of degradation: green – affected by mode A cracks (Fig. 1c); red – affected by mode B and C cracks (Fig. 1d).

III. DETECTION OF MIXED MODE B/C AND MODE C SOLAR CELL CRACKS

Mixed mode B/C and C cracks can be detected and quantified from the *EL intensity (ELI) histogram* of PV module or of individual cells, as proposed in [11]. The method requires the calculation of a normalized ELI histogram $p(k, i)$, as in (1):

$$p(k,i) = \frac{n_i^k}{n^k}, \ 0 \leq i < L, 1 \leq k \leq N_c , \qquad (1)$$

where k is the solar cell number, N_c is the number of cells in the module, n_i^k is the number of pixels of gray level i in cell k, n^k is the total number of pixels in the image of cell k, and L is the total number of gray levels in the image.

By calculating $p(k, i)$ for the cell shown in Fig. 1, through the different stages of degradation, we can quantify the effect of cell cracks on the EL signal of the cell. Fig. 2 shows the ELI histogram $p(k, i)$ of the cell imaged at I_{mp} bias. Here we can observe that mode A (green) cracks influence the higher ELI region of the histogram, as compared to when the cell was new (blue). Whereas mode B/C cracks (red) impact the lower region of $p(k, i)$. By quantifying this increase in the lower ELI region, we can determine the area of new mode B/C cracks in the cell.

The same increase in the lower ELI region of the histogram, can be observed in Fig. 3, determined from low bias current EL images of the cell. In this case the increase in the lower ELI region is mainly due to the mode C cracks.

Quantifying mixed mode B/C and C cracks from the lower ELI region of the $p(k, i)$ histogram, requires the determination of an ELI threshold TH_{Low} – shown in magenta in Fig. 2 and Fig. 3. This threshold must separate the active (EL emitting) regions of the cell from the inactive ones, and its value is influenced by the noise level the EL image. Fig. 4 exemplifies the application of TH_{Low} (determined manually) for segmenting cell EL images in Fig. 1b and Fig. 1d, and determining the location and area of the mixed mode B/C and C cracks, respectively.

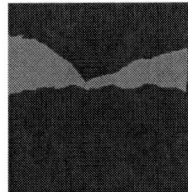

| a) mode B/C (Fig. 1b) | b) mode C (Fig. 1d) |

Fig. 4. Binary cell images showing the location of the mode B/C (a) and C cracks (b) segmented from Figs. 1b and 1d, using a threshold TH_{Low} determined manually from the cell ELI histogram $p(k, i)$.

A. Proposed Method for Automatic Determination of TH_{Low}

One of the main challenges in automating the detection and correct quantification of mode B/C and C cracks using the method described in this paper is precisely determining TH_{Low}. Its value is dependent on the current bias level, camera exposure time, ambient noise level, and can even vary slightly from module to module within the same module type. This variation can be observed also in the cell ELI histograms in Fig. 6, between the undamaged cells within the same module. Thus, we need to determine TH_{Low} from each EL image independently, to minimize false detection errors.

In the following, we propose an algorithm for determining TH_{Low} from EL images, which can be used to detect and quantify mode B/C and C cracks, and that can be automated:

1) Select a representative sample (N) of undamaged cells from the module EL image:

Undamaged cells are defined as solar cells with no cracks, shunting or increased series resistance areas. In this work, N=20 (out of 60) cells in the module were selected automatically, based on the criteria of having the lowest standard deviation in the EL intensity histogram. This parameter has been shown to increase with various types of solar cell degradation [13].

2) Select a WxH area from each undamaged cell image:

To exclude dark areas, close to the cell edges from affecting the analysis, we recommend performing the TH_{Low} determination only on a central cell area, as depicted in Fig. 5a, corresponding to ~70% width and height.

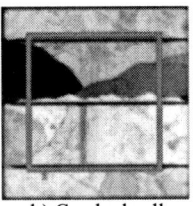

| a) Undamaged cell | b) Cracked cell |

Fig. 5. Example of WxH area of analysis (blue), used for determining the low intensity threshold TH_{Low} of the EL Image, for: a) an undamaged cell; b) a cell with mode B and C cracks.

3) Compute the cumulative EL intensity distribution for each area:

For each selected cell image area k, corresponding to the N undamaged cells, we compute the cumulative EL intensity distribution $cdp(k, i)$, according to (2):

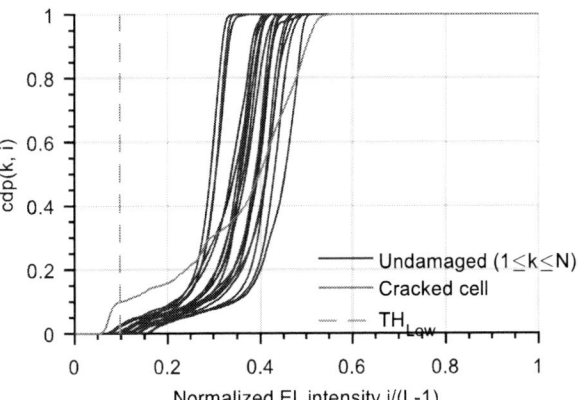

Fig. 6 Cumulative EL intensity profiles for 20 undamaged cells vs. the intensity profile of a cracked cell. The low EL intensity threshold TH_{Low} is calculated from the profile of the undamaged cells.

$$cdp(k,i) = \sum_{j=0}^{i} \frac{n_j^k}{n^k}, 0 \le i < L, 1 \le k \le N \qquad (2)$$

Fig. 6 shows the cdp distribution for $N=20$ undamaged cells (blue) of a mc-Si module, selected based on the lowest EL intensity standard deviation. By comparison, the cdp of a cell with a large mode C crack (Fig. 5b) is shown in magenta. Here we can observe an increase in the number of dark pixels in the image, because of the solar cell crack.

4) Calculate a local threshold for each undamaged cell:

For each $cdp(k, i)$ we calculate a local threshold $TH_{Low}(k)$ as the maximum EL intensity i for which $cdp(k, i)$ is below a fixed threshold A_{IN}:

$$\begin{aligned} TH_{Low}(k) &= max[i] \\ subject\ to\ cdp(k,i) &\le A_{IN} \end{aligned} \qquad (3)$$

where A_{IN} is the average percentage of inactive area in an undamaged cell, and is determined primarily by the number thickness of the cell busbars, size of the WxH area and camera resolution. AIN must be calibrated for the solar cell type and camera setup. In this work, A_{IN} = 0.1% for mc-Si cells and A_{IN} = 0.5% for mono-Si cells, which have thicker busbars.

5) Calculate a global threshold for the entire module:

Given there will likely be some variation between ELI histogram and $cdp(k, i)$ of the N selected undamaged cells, as can be observed in Fig. 6, the cell thresholds $TH_{Low}(k)$ will vary as well. Consequently, we need to calculate an average TH_{Low} for the entire module. However, considering that cells with defects and low ELI standard deviation may be falsely selected as "undamaged", which will skew the distribution of cell thresholds, the module level threshold TH_{Low} should be calculated as the median of the N local threshold values $TH_{Low}(k)$:

$$TH_{Low} = median(TH_{Low}(k)) \qquad (4)$$

IV. RESULTS AND DISCUSSION

To evaluate the mode C crack detection method, we used two sets of standard 60 cell modules (mono- and mc-Si), consisting of four samples each. These were degraded artificially, by several rounds of mechanical loading and humidity freeze cycles, causing the formation of mode A, B, and C cracks. All modules were flash tested under standard test conditions (STC), before and after stress, as well as imaged at 10 % I_{mp} and I_{mp} forward current bias, in a darkroom with a high-resolution Si CCD camera. The mc-Si modules were also imaged at 20% and 50 % I_{mp} bias, as well as two different camera exposures.

A. Influence of Forward Current Bias Level

In the first part of the analysis we investigate the influence of the bias current level during EL imaging on the detection of mode B/C and C cell cracks, in terms of total cracked module area. We applied the algorithm to determine TH_{Low} from each EL image, then the mixed mode B/C and C cracked regions were quantified according to the method described in [11].

Fig. 7 exemplifies the location (in magenta) of the mode C cracks detected from the 10% I_{mp} bias image of one of the mc-Si modules. This solar cell crack map allows for calculating the size of each cracked area isolated from the cell circuit relative to the cell area [11]. Fig. 8 shows the same module, but imaged with I_{mp} bias. Here we can observe a larger number of cell cracks identified as mixed mode B/C, some of which are mode B cracks that show very low EL emission regions, due to the high series resistance, but are not completely disconnected. Nevertheless, they could be considered the most severe mode B cracks in the module based on their low EL emission level.

As can be observed from Fig. 9, the total percent of mode B/C cracks detected per module increases with the module bias-current, which confirms a number of mode B cracked regions confounded as mode C, increases with bias current. This is a limitation of relying on the high-current bias EL images only, where severe mode B cracks will have a similar EL signal level as mode C cracks. Low-bias EL images are necessary to discern between such crack types.

Fig. 7. Low-current bias (10% I_{mp}) EL image of a mc-Si PV module which has sustained thermo-mechanical degradation. The magenta areas represent solar cell cracks that have been identified as Mode C – having an EL intensity below TH_{Low} determined for this bias level.

Fig. 8. High-current bias (100% I_{mp}) EL image of a mc-Si PV module in Fig. 7. The magenta areas represent solar cell cracks that have been identified as mixed mode B/C cracks.

B. Artifacts of Camera Exposure Time

In the previous analysis, one of the mono-Si modules was excluded from the analysis because the cell crack detection method –applied to the 10% I_{mp} bias EL image – yielded a cell crack of 100% for one of the cells, which was clearly erroneous. The cause was underexposure of the 10% I_{mp} bias EL image, which had two important consequences. First, image underexposure causes the ELI histogram to skew towards the low EL intensity region, as shown in Fig. 10 (blue), and the cell crack detection is thus confounded by the camera sensitivity and dynamic range. In this situation, determining a valid TH_{Low} threshold to detect mode C cracks is difficult.

The second consequence of EL image underexposure is seen with cells having excessive mode A cracks, as the cell highlighted in Fig. 11 and Fig. 12, measured under 100% and 10% I_{mp} bias, respectively. Typically, cells with a high series resistance will appear brighter (relative to the other cells in the PV module) in low bias images than in higher bias images. However, if the low bias image is underexposed, cells with excessive micro-cracks, causing additional shunting, can appear darker still, due to the voltage losses associated with recombination currents at the cracks and limited dynamic range of the camera.

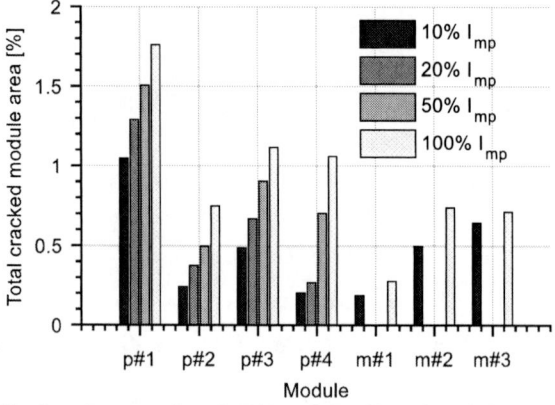

Fig. 9. Percent of mode B/C and C cell cracks relative to the PV module area, for the mc-Si (p#) and mono-Si (m#) modules, determined under different current bias levels.

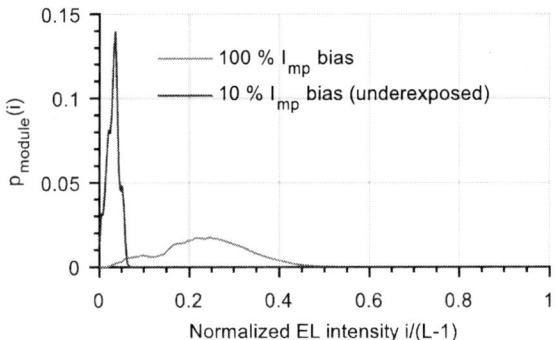

Fig. 10. Normalized ELI histograms of mono-Si module m#4, calculated from EL images measured under I_{mp} (red) and 10% I_{mp} bias (blue) – showing the consequence of image underexposure.

To investigate further the influence of camera exposure time on the cell crack detection method, we analyzed the EL images of the mc-Si modules, measured at I_{mp} bias and two exposure levels (19.2 sec and 25.6 sec). Fig. 13 shows the largest cell crack (relative to cell area) detected in each of the four modules, for the two exposure levels. As can be observed, the differences are negligible – showing that the TH_{Low} calculation method is robust to camera exposure time, if the EL image is not under- or over-exposed.

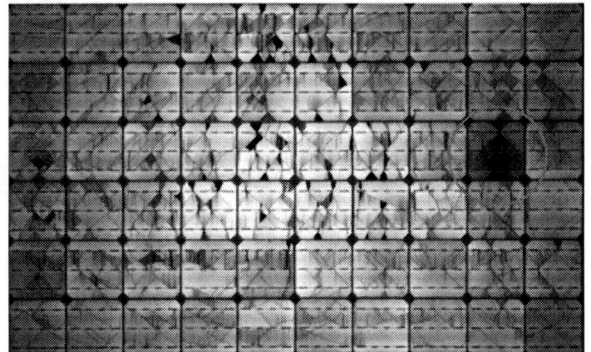

Fig. 11. EL image of module m#4, measured under I_{mp} bias, highlighting a cell with excessive micro-cracks and shunting.

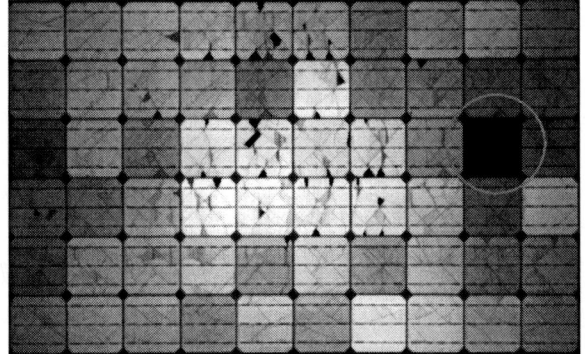

Fig. 12. EL image of module m#4, measured under 10% I_{mp} bias, highlighting a cell with excessive micro-cracks and shunting. The image contrast was adjusted such that the cells are visible.

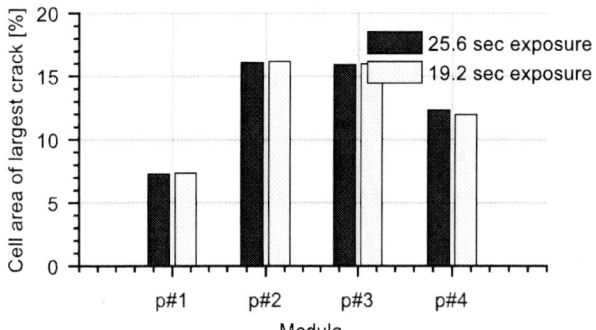

Fig. 13. Largest mode B/C cell cracks relative to the cell area, for the mc-Si modules, measured under two camera exposure and I_{mp} bias.

C. Correlation of Cell Crack Size with Module Power Loss

From a module power loss perspective, mode C cracks are considered severe since they reduce the effective photon collection area of the cell, causing current mismatch in the cell substring. The work in [10] showed that a mode C cracked area lower than ~8% of the total cell area does not cause significant STC power loss. However, for mode C cracks between 8% and 50 % disconnected cell area, the module power loss increases approximately linearly to 33% of module STC power, then saturates due to the bypassing of the cell sub-string.

This mode C crack area vs. STC P_{max} loss characteristic is illustrated in Fig. 14 (dotted red line), which has been obtained through LTSpice simulation of a standard 60-cell 250 Wp mc-Si PV module – where the inactive area of one cell has been varied between 0-25%. This characteristic in Fig. 14 gives us an idea of the lower module power loss boundary, given the size of the largest mode C crack area.

In practice however, modules which have large cracks, often have number of smaller ones, which also cause power loss – thus the total module power loss will be grater. We illustrate this characteristic in Fig. 14, where size of the largest mode B/C crack detected from I_{mp} bias images of the mc- and mono-Si modules, are correlated with the respective module power loss.

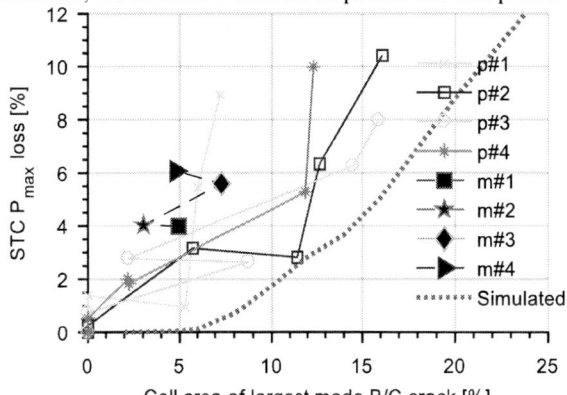

Fig. 14. Largest mode B/C solar cell cracks, measured from I_{mp} bias EL images of the mono- and mc-Si modules, correlated with their respective STC P_{max} degradation due to cracks. Each mc-Si module is imaged at six stages of thermo-mechanical degradation.

978-1-5090-5606-4/17 $31.00 © 2017 IEEE   1425

Since all the modules have sustained multiple mode B and C cracks, no clear dependency can be observed between the largest mode B/C crack and module power loss. But using the characteristic in Fig. 14 we can infer what is the lower limit of power loss. However, if we calculate the total mode B/C cracked area per module, we can observe a better correlation with module power loss, as shown in Fig. 15.

V. SUMMARY AND CONCLUSION

In this work, we proposed a method to automatically determine the EL intensity threshold necessary for quantifying mode C (inactive) solar cell cracks from low current bias EL images, and mixed mode B/C (appearing inactive) cracks in high current bias EL images. The method was primarily developed to support the mode C solar cell crack quantification method proposed in the draft of the new EL imaging standard currently under development "IEC TS 60904-13 Electroluminescence of photovoltaic modules".

Preliminary results showed that 50-60% of the mixed mode B/C cracks detected in high current bias EL images, overlap with the mode C cracks detected in low current bias EL images. However, this percentage may be lower if the EL image is underexposed, since the detection is limited by the sensitivity and dynamic range of the EL camera.

Last, we showed that the area of mode B/C cracks, detected from high current bias EL images, can be used to approximate the lower module power loss boundary due to cracks. This finding can be relevant for outdoor EL inspection applications, where the EL images are usually taken at higher current bias.

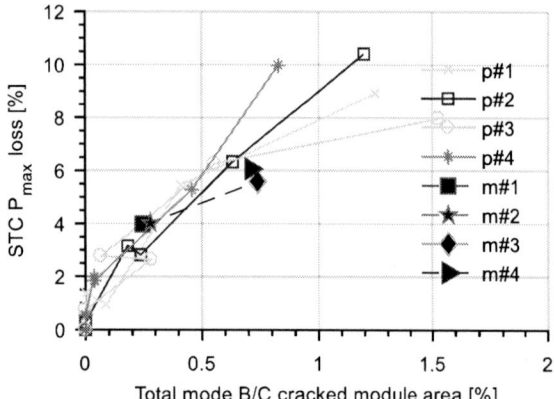

Fig. 15. Total mode B/C cracked module area, measured from I_{mp} bias EL images of the mono- and mc-Si modules, correlated with their respective STC P_{max} degradation due to cracks.

ACKNOWLEDGEMENT

The authors thank Karl Bedrich for their help in performing the EL module measurements. This work was partially supported by the research project "DronEL – Fast and accurate inspection of large photovoltaic plants using aerial drone imaging", project 6154-00012B supported by Innovation Fund Denmark, and Aalborg University. As well as financial support from Otto Mønsteds Fond and the U.S. Department of Energy under Contract No. DE-AC36-08-GO28308 with the National Renewable Energy Laboratory.

REFERENCES

[1] A. M. Gabor, M. Ralli, S. Montminy, L. Alegria, C. Bordonaro, J. Woods, L. Felton, M. Davis, B. Atchley, and T. Williams, "Soldering induced damage to thin Si solar cells and detection of cracked cells in modules," in 21st EUPVSEC, Dresden, Germany, September, 2006, pp. 4-8.

[2] F. Reil, J. Althaus, W. Vaassen, W. Herrmann, and K. Strohkendl, "The Effect of Transportation Impacts and Dynamic Load Tests on the Mechanical and Electrical Behaviour of Crystalline PV Modules," in 25th EUPVSEC, Valencia, Spain, 2010, pp. 3989 - 3992.

[3] M. Köntges, S. Kurtz, C. Packard, U. Jahn, K. A. Berger, K. Kato, T. Friesen, H. Liu, and M. Van Iseghem, *Review of Failures of Photovoltaic Modules*, International Energy Agency, 2014.

[4] M. Köntges, S. Kajari-Schröder, I. Kunze, and U. Jahn, "Crack Statistic of Crystalline Silicon Photovoltaic Modules," in 26th EUPVSEC, Hamburg, Germany, 2011, pp. 3290 - 3294.

[5] S. Kajari-Schröder, I. Kunze, U. Eitner, and M. Köntges, "Spatial and orientational distribution of cracks in crystalline photovoltaic modules generated by mechanical load tests," *Solar Energy Materials and Solar Cells,* vol. 95, no. 11, pp. 3054-3059, 11//, 2011.

[6] M. Sander, S. Dietrich, M. Pander, S. Schweizer, M. Ebert, and J. Bagdahn, "Investigations on crack development and crack growth in embedded solar cells," in Reliability of Photovoltaic Cells, Modules, Components, and Systems Conference, San Diego, California, 2011, pp. 811209-811209-10.

[7] C. Buerhop-Lutz, D. Schlegel, C. Vodermayer, and M. Nieß, "Quality Control of PV-Modules in the Field Using Infrared-Thermography," in 26th EUPVSEC, Hamburg, Germany, 2011, pp. 3894 - 3897.

[8] L. Stoicescu, L. Reuter, and J. Werner, "DaySy: Daylight Luminescence for PV Systems: How to Check 400kWpeak Per Day With Electroluminescence," in 2014 Photovoltaic Module Reliability Workshop, Golden, Colorado, 2014.

[9] S. Koch, T. Weber, T. Sobottka, A. Fladung, P. Clemens, and J. Berghold, "Outdoor Electroluminescence Imaging of Crystalline Photovoltaic Modules: Comparative Study between Manual Ground-Level Inspections and Drone-Based Aerial Surveys," in 32nd EUPVSEC, Munich, Germany, 2016, pp. 1736 - 1740.

[10] M. Köntges, I. Kunze, S. Kajari-Schröder, X. Breitenmoser, and B. Bjørneklett, "The risk of power loss in crystalline silicon based photovoltaic modules due to micro-cracks," *Solar Energy Materials and Solar Cells,* vol. 95, no. 4, pp. 1131-1137, 2011.

[11] S. Spataru, P. Hacke, D. Sera, S. Glick, T. Kerekes, and R. Teodorescu, "Quantifying Solar Cell Cracks in Photovoltaic Modules by Electroluminescence Imaging," in 42nd IEEE Photovoltaic Specialist Conference, New Orleans, 2015, pp. 8.

[12] J. Adams, B. Doll, C. Buerhop-Lutz, T. Pickel, T. Teubner, C. Camus, and C. J. Brabec, "Non-Stationary Outdoor EL-Measurements with a Fast and Highly Sensitive InGaAs Camera" in 32nd EUPVSEC, Munich, Germany, 2016, pp. 1837 - 1841.

[13] S. Spataru, "Characterization and Diagnostics for Photovoltaic Modules and Arrays," Department of Energy Technology, Aalborg university, Aalborg, Denmark, 2015.

Applying Spatial Downscaling and Smart Persistence to Provide an Improved Solar Forecast to Reduce Commercial Demand Charges

Alex Kubiniec[1], Ted Belanger[2], Adam Kankiewicz[1], Skip Dise[1], Nate Glasgow[2], and Alemu Tadesse[1]

[1] Clean Power Research, Kirkland, WA, 98006, USA
[2] EdgePower, Aspen, CO, 81611, USA

Abstract — Commercial PV is an appealing way to reduce electricity costs, however due to the variable nature of PV power and the larger power needs of most commercial applications, demand charges induced by variable load and PV generation can offset savings of generating electricity on site. This paper will explore options to reduce these demand charge costs through forecast improvements by downscaling variability and smart persistence using real time site observations. This improved forecast will be used by a load control system which will employ strategies such as load shifting to attempt to reduce demand charges.

I. INTRODUCTION

On site solar PV provides excellent energy savings to commercial customers but unreliable demand charge reduction benefits. Commercial scale PV installations can incur significant demand charges, especially on partly cloudy days. A reliable forecast of PV power output would help reduce demand charges as a commercial PV installation could be managed along with the buildings load to help avoid demand charges. Solar forecasting technology has typically been optimized for scale utility plants, where plant footprints span significantly larger areas and time resolution requirements are at hourly resolution. Demand charges are usually based on 15-minute time periods, so for a forecast to be useful it needs to be informative at a 15-minute temporal resolution. Interpolating an hourly solar forecast by persisting the clear sky index (the ratio of measured irradiance to clear sky irradiance) would not be effective in imparting any 15-minute variability because it assumes no change in cloud conditions sub-hourly. In practice, interpolation does not add additional forecast accuracy largely because most locations will experience sub-hourly cloud variability. Therefore, an improved solar forecast is needed to optimize for commercial demand charge reduction in buildings with PV. This work was completed in cooperation with the DOE-funded Incubator 10 program, under the project titled, "Predictive Solar-Integrated Commercial Building Load Control". Clean Power Research is partnering with EdgePower to provide a better solar PV forecast to feed into a building load management system that will reduce demand charges.

II. SIMULATION EVALUATION METHODOLOGY AND METRICS

To evaluate the performance of CPR forecasted power against measured power data, we first defined comparison metrics. The error metrics of interest are Mean Absolute Percent Error (MAPE) and Mean Absolute Error (MAE). These error metrics are defined below:

$$MAPE = \frac{\sum_{i=1}^{n}(|S_i - O_i|)}{Average\ Daily\ Power} \qquad \text{(Eqn. 1)}$$

$$MAE = \sum_{i=1}^{n}(|S_i - O_i|) \qquad \text{(Eqn. 2)}$$

where S_i is forecasted power, O_i is observed power. Nighttime values have been excluded from reported error metrics. Furthermore, Average Daily Power from Equation 2 is the average (mean) power from the site for a day over the time period that data is present. Selection of error metrics are based on criteria developed by Hoff et. al. [1] and adapted to the present application. MAPE will be used to determine forecast accuracy at shorter, sub-daily time periods.

III. PROPOSED METHODOLOGY

The goal was to operationally deliver a forecast that is optimized and useful for commercial PV installation in minimizing utility demand charges. To accomplish this, a 15-minute temporal resolution downscaling of variability was applied to an hourly SolarAnywhere® forecast. Hourly SolarAnywhere forecasts are an ensemble of short term forecast values derived from cloud motion vector (CMV) and long term forecast values from numerical weather predictions (NWP) models. Using a year's worth of 15-minute historic ground power measurements from a target site (Figure1)

978-1-5090-5606-4/17 $31.00 © 2017 IEEE

Figure 1: Map of SolarAnywhere 10 km resolution data and site tile used for the forecast, the red dot indicates the location of the ground station. Colorado Mountain College: 39.520 N latitude, -107.734 W.

clear sky power index (Kt power) was calculated as well as delta Kt power. A historic variability polynomial was derived from the relationship between delta power and delta Kt power [2][3]. This polynomial is then used to impart historic 15-minute variability to an interpolated SolarAnywhere forecast. Resulting in a 15-minute forecast that reflects the 15-minute variability of the specific Colorado Mountain College site. Figure 2 shows the data used to derive the polynomial relationship between delta power and delta Kt.

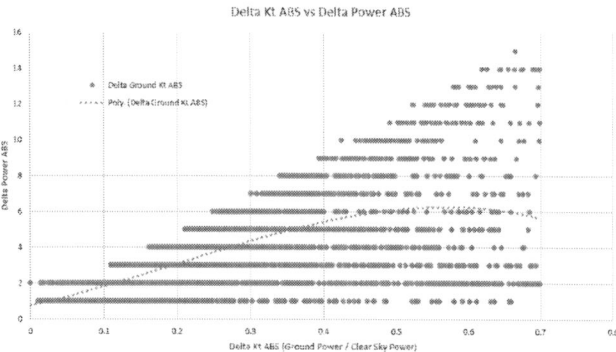

Figure 2: This figure shows the relationship of change in Kt power vs. change in ground power, as well as a polynomial fitted to that relationship. Ground power was reported as integer values.

The second part of improving the forecast is to use smart persistence for the first hour. Smart persistence is applied by using real time ground power values and interpolating them forward in time along the clear sky power profile. After the first hour smart persistence's average mean absolute error (MAE) increases above the average forecast error and so smart persistence values are not included in forecast horizons farther than 1 hour into the future. Smart persistence forecasts are blended with the variability downscaled forecast for forecast values beyond one hour. The combination of both smart persistence and downscaling variability results in a more accurate 15-minute resolution forecast which was used by EdgePower to demonstrate reduced commercial demand charges.

IV. RESULTS AND DISCUSSION

Overall, smart persistence provided substantial improvement in forecast error, whereas downscaling variability provided limited improvements in forecast error at best. Forecast error was binned by forecast time horizon. Both the baseline hourly SolarAnywhere forecasts and the modified forecasts including smart persistence and downscaled variability were compared to historic ground power observations. Hourly MAPE was calculated, normalizing by the daily average power for the system. There is clear forecast improvement from smart persistence in the first 60 minutes with subtle improvement noted after one hour due to the addition of downscaled variability. The downscaled variability does more closely match the ground data's variance, but this does not directly reduce forecast error. Downscaling variability can only help improve the forecast's variability to better match that of the historic ground data. This improved characterization of variability will improve the forecast skill slightly, but given this specific site's characteristic variability the forecast skill improvement is minimal. Additional sites' different variability profiles would need to be tested to determine if a downscaling variability approach would provide increased forecast performance at perhaps a certain type of site or climate.

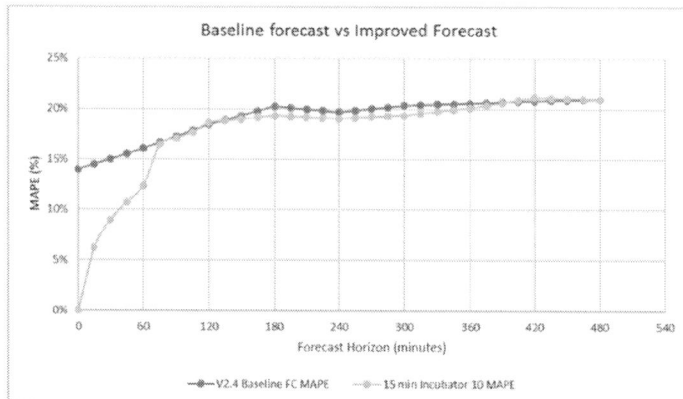

Figure 3: This figure shows the baseline forecast vs the improved Incubator 10 forecast. The v2.4 forecast values on the hour are actual baseline values, the values not on the hour are linearly interpolated between the hours. Values for both the baseline forecast and Incubator are hour ending averages

This improved forecast was then fed over to EdgePower's load control systems. The Solar-Integrated Load Control product went live on June 29th. Between June 29th and September 15th there were 4 billing periods that were evaluated for demand charge savings. During the 4 billing periods, there were 26 control periods forecasted that the Solar-Integrated Load Control took control of the building in an attempt to reduce load. There were no electricity cost savings observed across the billing periods but there were multiple control periods within the billing periods which showed significant load reduction. Load control was shown to reduce load by up to 25kW on a building that typically has a maximum load of 100kW. This load reduction was achieved

978-1-5090-5606-4/17 $31.00 © 2017 IEEE   1428

Sunday	Monday	Tuesday	Wednesday	Thursday	Friday	Saturday
26 June	27 June	28 June	29 June **Control Start**	30 June	1 July	2 July
3 July	4 July	5 July	6 July	7 July 8:15-10:00 & 11:30-16:00 Non-Excluded Max @ 19:45	8 July 9:30-9:30 & 12:00-17:45	9 July
10 July	11 July 8:00 - 16:00	12 July 8:15 - 19:00	13 July 8:15 - 18:15	14 July 8:15 - 19:00	15 July 8:15-9:45 & 13:00-18:00	16 July
17 July	18 July 8:30 - 9:45 Non-Excluded Max @ 6:00	19 July 8:30 - 9:45	20 July 9:45 - 9:45	21 July	22 July	23 July
24 July	25 July	26 July 9:45 - 9:45	27 July	28 July 9:45 - 9:45	29 July	30 July
31 July	1 August	2 August	3 August 9:45 - 9:45	4 August 9:45 - 9:45 - Excluded Raw Max @ 8:00	5 August	6 August
7 August	8 August	9 August	10 August	11 August	12 August	13 August Summer Semester Ends
14 August	15 August 19:30 - 19:30	16 August	17 August Excluded	18 August Excluded	19 August Excluded Raw Max @ 13:45	20 August
21 August	22 August	23 August	24 August	25 August Excluded	26 August	27 August
28 August Fall Semester Starts	29 August 15:30 - 15:30	30 August	31 August 19:30 - 19:45	1 September 19:15 - 19:30	2 September 11:00 - 11:00	3 September
4 September Excluded	5 September Excluded	6 September Excluded Raw Max @ 15:45	7 September 18:30 - 19:45 - Excluded Raw Max @ 20:00	8 September 18:00 - 19:45	9 September	10 September
11 September	12 September 15:30 - 20:30	13 September 18:00 - 20:45 Non-Excluded Max @ 14:00	14 September 16:45 - 20:45	15 September 18:15 - 20:30 Non-Excluded Max @ 19:45		

Figure 4: In this image the green, orange, blue, and gray days represent the 4 different billing periods. The red days are days that were excluded from the analysis because they had irregular building operations which were out of EdgePower's control. The times listed are the individual control periods and the days with Raw Max or Non-Excluded Max are the days and times when the billing period peaks occurred. Since the days with Raw Max were excluded, the Non-Excluded Max becomes the new billing period max. Results courtesy of EdgePower.

without sacrificing occupant comfort. Because of the load reductions observed during the control periods of this pilot, with further refinement, it's believed that significant electricity cost savings could be achieved with the Solar-Integrated Load Control product. Figure 4, shows the days that load control and peak demand occurred. In Figure 4 the green, orange, blue, and gray days represent the 4 different billing periods. The red days are days that were excluded from the analysis because they had irregular building operations which were out of EdgePower's control. The times listed are the individual control periods and the days tagged as Raw Max or Non-Excluded Max are the days and times when the billing period peaks occurred. Since the days with Raw Max were excluded, the Non-Excluded Max becomes the new billing period max.

VII. CONCLUSION

When live feeds of ground data are available, smart persistence can be utilized to improve short term solar forecasts. The addition of downscaled variability provided additional, but limited improvement to the baseline solar forecasts. However, variability is site specific and it may be worth exploring downscaling variability at other sites in regions with different climates. It is important highlight, that although the forecast skill was not directly improved, the

characteristic variability of the forecasts did match the observed ground's variability, and results in a more realistic forecast. The load shifting and load control had difficultly reducing demand charge costs due to the nature of demand charge billing and variable nature in which the building was being used during the performance period. There were some periods within the billing periods where demand was reduced by 25 kW. This was all accomplished with no increase in energy consumption. By refining the forecast with a higher temporal resolution, we hope to improve the short term forecast and reduce costs.

REFERENCES

[1] Hoff, T. E., Perez, R., Kleissl, J., Renne, D. and Stein, J. (2013), Reporting of irradiance modeling relative prediction errors. Prog. Photovolt: Res. Appl., 21: 1514–1519. doi:10.1002/pip.2225

[2] R. Perez, S. Kivalov, J. Schlemmer, K. Hemker Jr., and T. Hoff (2011): Parameterization of site-specific short-term irradiance variability, Solar Energy 85 (2011) 1343–1353

[3] P. Lauret, R. Perez, L. M. Aguiar, E. Tapache`s, H. M. Diagne, and M. David (2015): Characterization of the intraday variability regime of solar irradiation of climatically distinct locations, Solar Energy 125 (2016) 99–11

Thermal Characteristics of PID-Affected Monocrystalline Silicon Solar Modules under Illuminated and Dark Conditions

Pan Zhao[a], Shuwen Guo[a], He Wang[a,*], Hong Yang[a], Dengyuan Song[b], Shiyu Sang[c], Bojie Su[d], Xue Zhang[d], Yunxue Cao[e], Hui Zhao[e]

[a]MOE Key laboratory for Nonequilibrim Synthese and Modulation of Condensed Matter, School of Science, Xi'an Jiaotong University, Xi'an 710049, People's Republic of China

[b]Yingli Group Co., Ltd., Baoding 071051, People's Republic of China

[c]Institute of Electrical Engineering of the Chinese Academy of Sciences, Beijing 100190, People's Republic of China

[d]China Quality Certification Center, Beijing 100070, People's Republic of China

[e]SPIC Power Plant Operation Technology Co., Ltd, Beijing 100190, People's Republic of China

*Corresponding author: He Wang, hw69cn@126.com

Abstract — **This paper focuses on thermal characteristics of PID-affected solar modules under illuminated and dark conditions by thermographic images. It is found that the temperature of PID-affected module becomes higher under negative bias when the sequence of module approaches to negative pole of module string. The thermal distribution under dark conditions shows that substrings neighboring the module frame yield more serious degradation than the middle substring within the PID-affected module. The results obtained under illuminated and dark conditions can reflect the regular power loss for PID-affected solar modules.**

Index Terms — **solar module, thermal characteristics, potential-induced degradation (PID), thermal distribution.**

I. INTRODUCTION

It is clear that potential-induced degradation (PID) has drawn much attention in recent years due to its detrimental effect on solar power plants [1]-[2]. Extensive investigations show that the sodium ions decorated stacking faults is the origin of PID [3]-[8]. Although some measures have been taken to prevent solar cells, modules, and systems from PID [9]-[12], there are still some new grounded photovoltaic modules affected by PID.

The infrared thermography provided a reliable and accurate tool for the diagnosis of optical and electrical degradation defects in PV modules, even the severity of these degradations shown by relative temperature differences [13]. Sometimes, especially the usage of remote controlled drone equipped with a thermographic camera is a fast tool for detecting of PID affected modules under operation [14]-[15].

In this work, we investigated numerous module strings with a handed infrared camera and a I-V tester. It was found that PID-affected modules occurred in grounded module strings. Through the detailed analysis of thermographic images, increasingly approaching to negative pole of module string, the temperature of module under negative bias go up gradually. This demonstrates that the degradation becomes more serious accordingly. Then the PID-affected fourteenth module was dismounted and analyzed in the laboratory. It is detected that substrings neighboring the module frame yield more serious degradation than the middle substring in PID-affected modules.

II. EXPERIMENTAL RESULTS AND DISCUSSION

Our investigation was performed in an integrated PV power plant located in Jingbian County of YuLin in Shaanxi province at 37.61 degrees North latitude and 108.79 degrees East longitude. The field picture of the integrated PV power plant is showed in Fig.1.

In this PV power plant, seventeen solar modules are interconnected into a string and sixteen strings connected in parallel come into being one string box. Ten string boxes connect with an inverter so that direct current is inverted to alternating current. All the modules investigated in this PV power station are monocrystalline silicon modules consisting of 72 solar cells (125mm×125mm) connected in series. For the monitoring PV modules under operation we used an infrared camera with a resolution of 320×240 pixels, a digital thermocouple and a PV 200 tester for taking I-V curves.

978-1-5090-5606-4/17 $31.00 © 2017 IEEE

Fig. 1.   The scene of 50 MW PV power plant.

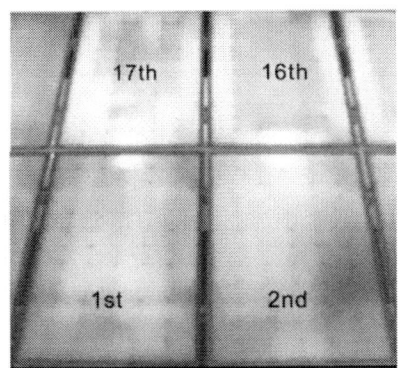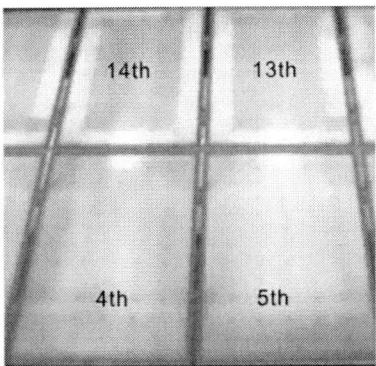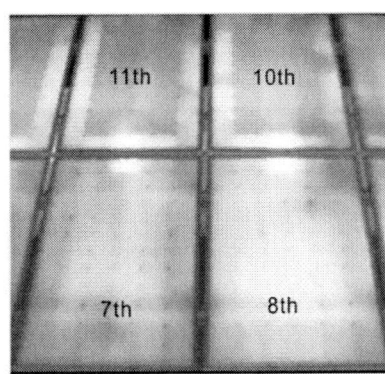

27.6°C   45.8°C

Fig. 2.   Thermographic image of a module string under operation.

A. Thermal distribution of PID-affected modules under operation in a power plant

By employing a PV 200 tester and taking infrared images to investigate plenty of module strings in this power plant, we detected that modules under negative bias underwent PID within grounded module strings. Instead, there is no occurrence of PID within ungrounded module strings.

By the equipment of Solar Survey 100/200R, the ambient temperature is 25.3 °C and the irradiance is ~700 W/m². As seen from Fig. 2, it is found that all the cells within PID-free modules are on the same temperature under operation from the first module to the 9th module. But characteristic thermal distribution occurs to PID-affected modules from the 10th to 17th module. Bright PID-affected cells neighboring the frame in the thermography are at higher temperature than dark PID-free cells in the middle. Increasingly approaching to negative pole of module string, the number of bright cells is larger and the highest temperature of bight cell within PID-affected modules goes up.

It is concluded that modules under negative bias yield different-level PID by the detailed analysis of thermographic images. The degradation becomes more serious in sequence from the 10th to the 17th module. This illustrates that under negative bias, the closer to the negative pole of module string the module is, the more seriously module degrades. The most serious degradation obtained by I-V measurement indicates that the 17th module shows a power degradation of up to 50%.

It can be speculated that due to the augment of dark current through equivalent diode, PID-affected cells exhibit a higher temperature than sound ones.

B. Thermal distribution of PID-affected modules under dark conditions

To study thermal distribution of PID-affected modules under dark conditions, the PID-affected fourteenth module was dismounted and analyzed in the laboratory. Fig. 3 displays thermographic image of this module under positive 30V bias and EL image acquired by EL tester. PID-affected solar cells appear dark in EL images, while PID-free cells appear bright. PID-affected cells neighboring to the frame in the thermography are at lower temperature than PID-free cells in the middle of module, which is a good agreement with the result of EL image.

To study three substrings of this PID-affected module, we removed the bypass diodes and applied the external bias on each substring respectively. Fig. 4 shows thermographic images of three substrings under reverse 140 V bias. It is obvious that the first substring and third substring show a higher temperature that the second substring. It is concluded that substrings neighboring the module frame yield more serious degradation than the middle substring within the PID-affected modules. From Table I, the main electrical parameters of the three substrings were measured at standard test conditions (STC). Both the open-circuit voltage and maximum power of substring 1 and substring 3 exhibit more reductions than those of substring 2. This is compatible with the result of Fig. 4.

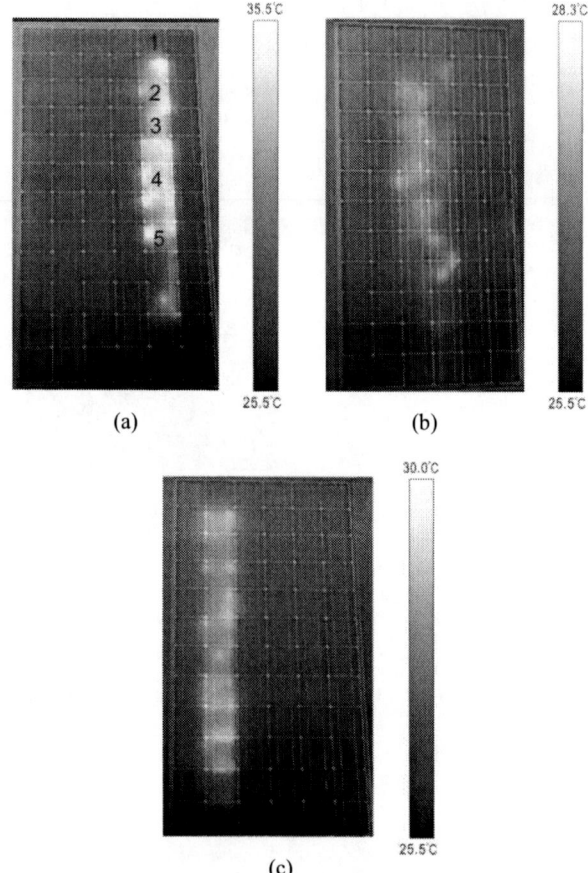

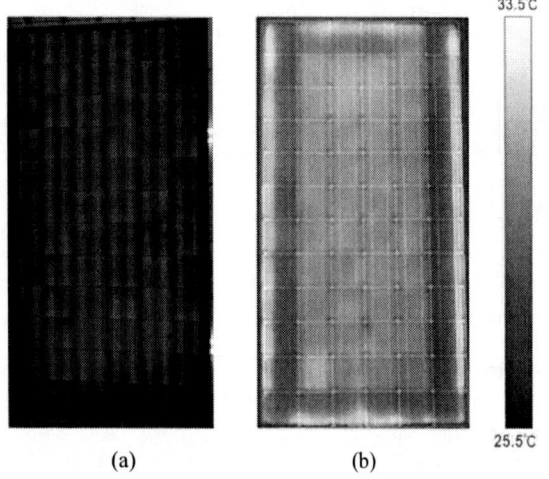

Fig. 3.   (a) Electroluminescence image of the dismounted fourteenth module and (b) thermographic image of the same module under positive bias.

Fig. 4.   Thermographic images of the first substring (a), second substring (b) and third substring (c) of the fourteenth module under reverse 140 V bias.

TABLE I
THE MAIN ELECTRICAL PARAMETERS OF THREE SUBSTRINGS
AT STC.

Substring	$V_{oc}(V)$	$I_{sc}(A)$	$V_m(V)$	$I_m(A)$	FF(%)	$P_{max}(W)$
Substring 1	13.81	5.54	10.35	3.65	50.02	37.76
Substring 2	14.86	5.54	11.23	5.11	69.80	57.50
Substring 3	14.45	5.55	11.78	3.85	57.60	45.44

In order to study the dependence of the average temperature of each thermal cell on the voltage, the procedure was performed by measuring distinguished cell temperatures with thermocouples on the back sheet of the module. Fig. 5 shows that average temperature of several thermal cells in first substring depends on the voltage.

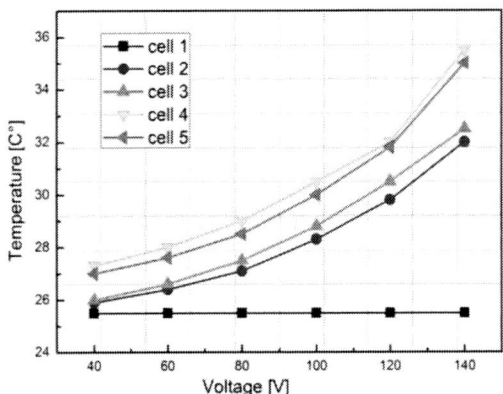

Fig. 5. The dependence of the average temperature of four thermal cells and a dark cell on the voltage applied on the first substring of the fourteenth module.

C. The explanation for thermal distribution of PID-affected modules under dark conditions

To illustrate the thermal distribution of solar modules and substrings, an equivalence principle is proposed. In Fig. 6, cell 1 is a PID-free cell and cell 2 is a PID-affected cell. Under forward bias, because of the significant reduction of parallel resistance of cell 2, the divided voltage on cell 1 is more than that on the cell 2, causing the growth of the current through the parallel resistance of cell 1. Therefore, the PID-free cell will be heated, exhibiting a higher temperature than PID-affected one. Furthermore, due to the larger number of dark cells in module 2, the divided voltage on the PID-free cells in module 2 exceeds that in module 1 so that the heating of module 2 is stronger than module 1. Under reverse bias, the average divided voltage on each cell is less than reverse breakdown voltage of diode so that the equivalent diode can work. The divided voltage on the PID-free cell is more than that on the PID-affected one. As a result of containing more PID-affected cells, the substrings neighboring the frame exhibit a higher temperature than the substring in the middle of the module.

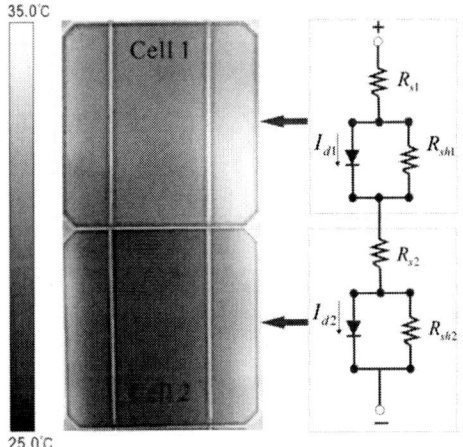

Fig. 6. Test of the equivalence principle used to illustrate the thermal distribution of the interconnection between a PID-free cell and PID-affected one under forward bias.

III. CONCLUSION

Thermal characteristics of PID-affected solar modules under illuminated and dark conditions have been analyzed in this paper. Firstly, the study of thermal distribution under illuminated conditions shows thermal law and degraded law of PID-affected modules in a grounded module string. In addition, the thermal distribution under dark conditions is opposite to illuminated distribution but degraded laws shown by them are identical. Finally, test of the equivalence principle explains thermal distribution of PID-affected modules under dark conditions.

ACKNOWLEDGEMENTS

The authors would like to thank the support of National High Technology Research and Development Program of China (Grant No.2015AA050301). This study was also supported by the Natural Science Foundation of China (Grant No.61274050 and 61376067).

REFERENCES

[1] S. Pingel, O. Frank, M. Winkler, S. Daryan, T. Geipel, H. Hoehne and J. Berghold, "Potential Induced Degradation of solar cells and panels," *in Proceedings of the 35th IEEE Photovoltaic Specialists Conference,* 2010, pp. 2817–2822.

[2] P. Hacke, K. Terwilliger, S. Glick, D. Trudell, N. Bosco, S. Johnston and S. Kurtz, "Test-to-Failure of crystalline silicon modules," *in Proceedings of the 35th IEEE Photovoltaic Specialists Conference*, 2010, pp. 000244–000250.

[3] V. Naumann, D. Lausch, S. Großer, M. Werner, S. Swatek, C. Hagendorf and J. Bagdahn, "Microstructural Analysis of Crystal Defects Leading to Potential-Induced Degradation (PID) of Si Solar Cells," *Energy Procedia*, vol. 33, pp. 76–83, 2013.

[4] V. Naumann, Dominik Lausch, Jan Bauer and Jörg Bagdahn, "The role of stacking faults for the formation of shunts during potential-induced degradation of crystalline Si solar cells," *Phys. Status Solidi RRL*, vol. 7, pp. 315–318, 2013.

[5] V. Naumann, D. Lausch, A. Hahnel, J. Bauer, O. Breitenstein, A. Graff, M. Werner, S. Swatek, S. Großer, J. Bagdahn, and C. Hagendorf, "Explanation of potential-induced degradation of the shunting type by Na decoration of stacking faults in Si solar cells," *Solar Energy Materials & Solar Cells*, vol. 120, pp. 383–389, 2013.

[6] Benedikt Ziebarth, Matous Mrovec, Christian Elsässer, and Peter Gumbsch, "Potential-induced degradation in solar cells: Electronic structure and diffusion mechanism of sodium in stacking faults of silicon," *Journal of Applied Physics*, vol. 116, pp. 093510, 2014.

[7] Volker Naumann, Carlo Brzuska, Martina Werner, Stephan Großer, Christian Hagendorf, "Investigations on the formation of stacking fault-like PID-shunts," *in Proceedings 6th International Conference on Silicon Photovoltaics*, 2016, pp. 569-575.

[8] Atsushi Masuda, Minoru Akitomi, Masanao Inoue, Keizo Okuwaki, Atsuo Okugawa, Kiyoshi Ueno, Toshiharu Yamazaki,

Kohjiro Hara, "Microscopic aspects of potential-induced degradation phenomena and their recovery processes for p-type crystalline Si photovoltaic modules," *Current Applied Physics,* vol. 116, pp. 1659-1665, 2016.

[9] H Nagel，A Metz，K Wangemann, "Crystalline Si Solar Cells and Modules Featuring Excellent Stability Against Potential-Induced Degradation," *in Proceedings of the 29th European Photovoltaic Solar Energy Conference and Exhibition,* 2011, pp. 3107-3112.

[10] X. Gou, X. Li, S. Zhou, S. Wang, W. Fan and Q. Huang, "PID Testing Method Suitable for Process Control of Solar Cells Mass Production," *International Journal of Photoenergy,* vol. 2016, 863248, 2016.

[11] J Kapur, KM Stika, CS Westphal, JL Norwood, "Prevention of Potential-Induced Degradation With Thin Ionomer Film," *IEEE Journal of Photovoltaics,* vol. 5, pp. 219-223, 2015.

[12] Pan Zhao, He Wang, Hong Yang, Fumei Wang, Ao Wang and Dengyuan Song, "Mechanism Analysis of Potential-induced Degradation of P-type Crystalline Si Solar Cells," *in: Proceedings 43rd IEEE Photovoltaic Specialists Conference,* 2016, pp. 2756-2760.

[13] E. Kaplani, "Detection of Degradation Effects in Field-Aged c-Si Solar Cells through IR Thermography and Digital Image Processing," *International Journal of Photoenergy,* vol. 2012, pp. 1-11, 2012.

[14] Thomas Kaden, Katrin Lammers, Stephan Hoffmann, Michael Köhl, Peter Bentz, and Hans Joachim Möller, "Fast Detection of PID Affected Solar Modules Using Flight Thermography," *in 29th European Photovoltaic Solar Energy Conference,* 2014, p. 2994-2996.

[15] Thomas Kaden, Katrin Lammers and Hans Joachim Möller, "Power loss prognosis from thermographic images of PID affected silicon solar modules," *Solar Energy Materials & Solar Cells,* vol. 142, pp. 24-28, 2015.

Targeted Evaluation of Utility-Scale and Distributed Solar Forecasting

Matthew Lave, Robert J. Broderick, Laurie Burnham

Sandia National Laboratories, Livermore, CA and Albuquerque, NM, 94550 and 87185, USA

Abstract — In this paper, we evaluate one-year of measured and forecasted data for PV generation, load and net load for multiple scenarios: a single substation, aggregated substations and a 2.1MW PV farm. By examining systematic deviations between the forecasted and actual power generation, we found we could identify spatial and temporal patterns in forecast errors that can be used to improve the accuracy of the forecasting tools. We also found there are seasonal changes that impact the accuracy of the forecasting results that should be considered.

Index Terms — solar energy, solar power generation, power grids.

I. INTRODUCTION

The variability of solar generation remains a key challenge to its growth, with utilities looking to find better ways to anticipate and manage fluctuations in power flow. As a result, interest in sophisticated forecasting tools that can bring confidence to system operations is growing. At the same time, the exponential buildout of residential solar (as of November 2015, distributed PV accounted for 40% of the solar electric generating capacity in the United States [1]), is spurring utilities to look beyond utility-scale forecasting and to consider forecasting tools that can be applied to distributed PV.

II. BACKGROUND

Forecasting tools that can provide information in near real-time to grid operators have the potential to enable the growth of variable renewables by enabling utilities to anticipate fluctuations in power flow and therefore more effectively balance generation and load. Forecasts can be produced at different timescales, from intra-hour to hour-ahead to day-ahead, or longer, forecasts.

Methodologies vary according to timescale. Intra-hour forecasts, critical for making real-time grid decisions, are often based on sky-imagery or statistical techniques such as extrapolation; hour-ahead forecasts rely on satellite data; and day-ahead or longer projections are based on numerical weather prediction (NWP). Because most utility-scale plants have high-quality monitoring systems and meter their power output, machine learning techniques can also be applied to the forecasting algorithms to further improve their accuracy [2].

Hoff, et al. [3] describe how satellite-derived irradiance and simple irradiance measurements can be fed into forecasting models to simulate the power output of aggregated residential rooftop PV (in this case the authors looked at 130,000 rooftop systems in California.) Another study by Lonij, et al. [4] forecast the power expected from distributed renewables for the state of Vermont, applying machine learning techniques to massive amounts of weather, SCADA, and smart meter data.

Yet rigorous evaluation of these forecasting tools and their underlying methodologies to date has been lacking. To begin with, the metric by which forecast software is commonly evaluated is root-mean-square error (RMSE), which is best suited for stationary statistical processes yet solar power has time-varying statistics.

In addition, most of the solar-forecasting research to date has focused on single utility-scale solar plants, not the aggregate of plants and rooftop installations that typically feed into the grid. Yet forecasts are most valuable when they represent the actual grid, i.e., consider solar generation in aggregate (i.e., more broadly than a single utility plant) and in association with load to understand the net impact to the system.

III. TECHNICAL APPROACH

In this paper, we compare measured and forecasted data (PV power, load, and net load) for a single high-penetration distribution feeder, for an aggregate of multiple substations and for a 2.1MW utility-scale solar plant. Our objectives were threefold; 1) to assess the overall accuracy of solar forecasts by comparing forecast data with measured data; 2) to examine factors that influence a forecast's performance, such as time of day, sky conditions (cloudy/clear) and season; and 3) to pinpoint specific opportunities for improving those forecasts.

The data that formed the basis for our analysis came from commercial forecasts, i.e., forecasts delivered in near real-time, at a resolution of one hour, to system operators. They were provided by a transmission utility that anticipates a significant growth in grid-tied renewables and sees economic value in improving the accuracy of its solar forecasts. This company provided 1-24 hour-ahead and longer forecast data and also capacity, load and generation data needed to augment our analysis.

The forecast data reflects machine learning that includes trends in weather, PV performance and load consumption, with machine-learning algorithms retrained on a regular basis as the historical database expands.

IV. RESULTS

We looked at one-year of measured and forecasted data for three scenarios: a single high-penetration substation, for an aggregate of several tens of substations, and for a single PV farm. For each, we evaluated the forecasted PV power output against measured PV power output. For the single substation and for the aggregate of many substations, we also evaluated forecasted load and net load (the latter measured as the subtraction of PV generation from load) against measured values. We looked at variables that affect forecast performance, including time of day, sky conditions (clear/not clear) and season.

A. High PV-Penetration Substation

The representative high-penetration-PV substation we selected has a maximum load of 5.7MW and has distributed PV, with maximum PV production of 4.7MW (representing a PV penetration rate of 82%). Reverse power flow back through the substation occurs 10% of the time and the maximum reverse power flow was 3.5MW.

Measured and forecasted PV power, load, and net load for a sample week are shown in Figure 1. The biggest discrepancies occur on clear days (those with the high-power output) when the PV power forecast deviates noticeably from the measured power. Negative PV forecast errors (that is, under-prediction of PV production) partially mitigate negative load forecast errors (that is, the under-prediction of load), such that errors in net load are smaller.

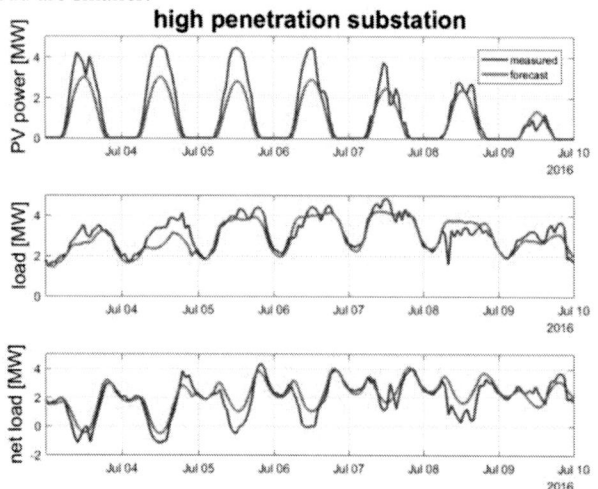

Figure 1: Measured (blue) and forecasted (red) data: [Top] PV power, [Middle] load, and [Bottom] net load for a substation with high PV penetration

Figure 2 shows the measured versus forecast daily PV production for the period from August 2015 to July 2016. Consistent with Figure 1, the forecast is shown to under-predict power generation, especially as the measured PV values increase. The plot is color coded to depict data before and after May 16, 2016, the date when the forecast appears to have adjusted for an increase in PV capacity. Both colors follow roughly linear trends, but the slopes of the two lines are far from the 1:1 line we would expect in a perfect forecast. These linear trends indicate that when the forecast was high, the measured output was also high, and vice-versa, but the mismatched slope shows that the forecast did not accurately predict the magnitude of PV generation, especially at high PV power outputs. The most plausible explanation is that the forecast model under-

calculated the amount of installed PV capacity, as evidenced by the fact that the measured PV production exceeds 40MWh, yet the forecasts have maximum values of around 25MWh (after May 16) or 12MWh (before May 16).

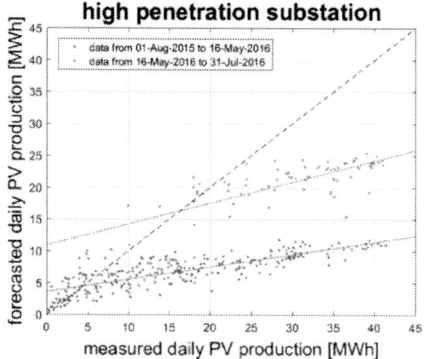

Figure 2: Forecasted versus measured daily PV generation to a single substation before May 16 (blue dots) and after May 16 (red dots). The dashed black line represents a hypothetical perfect forecast.

In contrast, the measured load closely aligns with the forecast load, as seen in Figure 3. The only deviation from the 1:1 line can be seen at the highest loads, where a small amount of under-prediction is apparent, consistent with the missing the midday spikes seen in Figure 1.

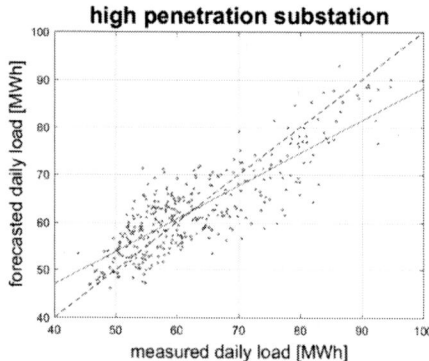

Figure 3: Forecasted versus measured daily load on the high penetration substation. The dashed black 1:1 line is represents a perfect forecast.

We also looked at net load because coupled errors in PV and load forecasts can cause errors in net load forecasts. As shown in Figure 4, we plotted measured to forecasted net load on an hourly basis (rather than the daily comparisons seen in Figures 1 and 2). This hourly comparison allows us to examine the forecast's ability to predict reverse flows on the feeder. Figure 3 is color-coded to show when there was measured reverse power flow and if the forecast correctly predicted. Blue dots indicate normal power flow that was correctly forecasted, and purple dots show reverse power flows that was correctly forecasted. Yellow dots show normal power flows incorrectly forecast as reverse power flow, and red dots show reverse power flow incorrectly forecasted as normal power flow. There are many more red dots (reverse power flow incorrectly forecasted) than there are purple dots (reverse power flow correctly forecasted), showing that the forecast does not predict reverse power flow often enough.

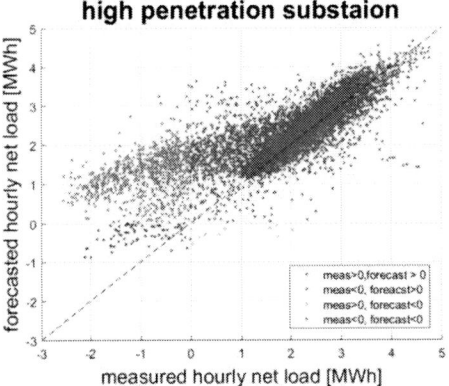

Figure 4: Forecasted versus measured daily load on the high penetration substation. The dots are color coded: based on whether the measured and forecast were positive or negative. The dashed black 1:1 line is what a perfect forecast would follow.

B. Aggregate of Substations

After looking at the forecasting performance for a single substation, we then looked at an aggregate of several substations, which is more representative of a utility's full PV portfolio. As we did for the single substation, we looked at one-year of measured and forecasted data for power output, load, and net load but for the aggregate of substations spread over a large geographic area and varying in both load and PV capacity.

The contrast to the single substation was operationally significant: PV penetration dropped to 12.6% from the 82% at the single substation, and reverse flow never occurred. We found, as a result, that the load forecasts were relatively more accurate, dominating the net load forecast accuracy (see Figure 5).

Similarly, the PV forecast for the aggregated substations out-performed the forecast for the single substation, as seen in Figure 6. We surmise this accuracy can be explained by two factors: (1) the aggregated data represent several PV farms for which forecasting accuracy is generally good (section IV. C.), and (2), forecast errors tend to average out over large spatial areas (e.g., the forecast may be low in one region but high in another region). But, as we saw with the single substation, the forecast for aggregated substations under-predicts production when the measured production is high because the PV capacity values in the model are too low, i.e., not representative of new installations.

As shown in Figure 7, the load forecast for the aggregate matches the measured load even more closely than it does for the single substation., a finding that is expected because aggregate loads are easier to predict.

Because the load for this aggregate is always greater than the PV generation (the latter averages 12.6%), negative net loads never occur. As a result, the net load forecast errors

approximate the load forecast errors. Yet as PV penetrations increase throughout a utility's service territory, PV generation may increasingly outpace the load at certain times, a prediction that underscores the importance of ensuring the accuracy of the PV forecast.

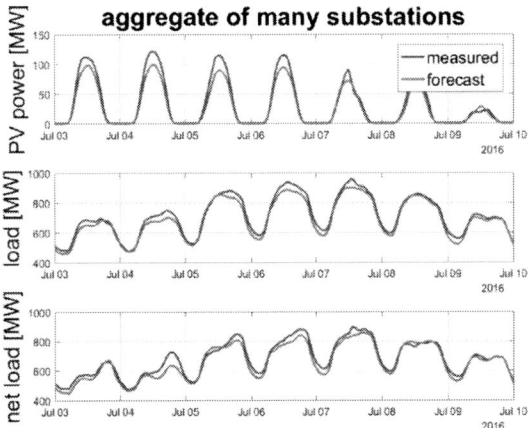

Figure 5: Measured (blue) and forecasted (red) values: [Top] PV power, [Middle] load, and [Bottom] net load for the aggregate of many substations.

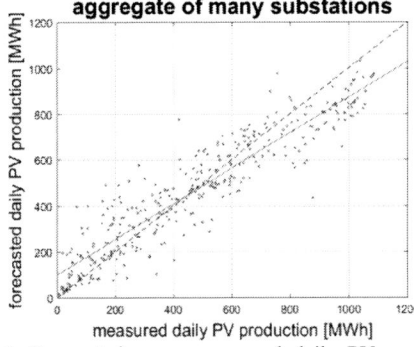

Figure 6: Forecasted versus measured daily PV production for aggregated substations. The dashed black line is the 1:1 line that a perfect forecast would follow.

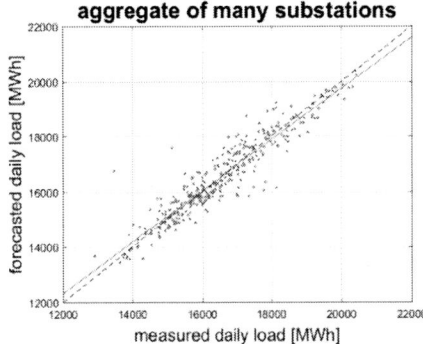

Figure 7: Forecasted versus measured daily load for aggregated substations. The dashed black 1:1 line is what a perfect forecast would follow.

C. Single Utility-Scale PV Farm

We also considered a utility-scale (2.1MW) PV farm, which is sufficiently sized to have a PPA with the local utility. One week of measured and forecasted power output is shown in Figure 8. For this evaluation, we included forecasts made at three different time horizons: 1-24 hours ahead, 25-48 hours ahead, and 49-72 hours ahead and published at midnight coordinated universal time (UTC.)

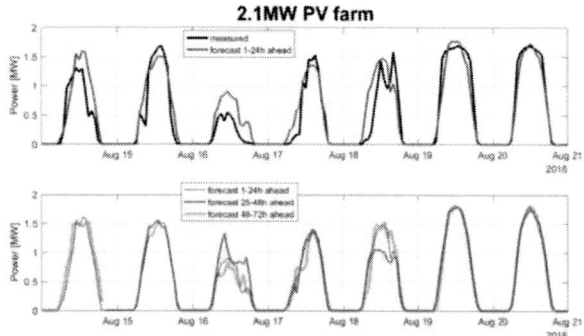

Figure 8: [Top] Measured and most recently forecasted PV farm power output. [Bottom] Comparison of forecasts at different time horizons

The above forecast shows that clear days can be predicted with a reasonably high degree of accuracy (see August 19th and 20th). Cloudy days introduce more uncertainty as can be seen by looking at August 16th. Although the latter is correctly forecasted as cloudy, the forecast over-predicts the power output for that day. Interestingly, we saw no major differences in accuracy between one, two, or three day-ahead forecasts: all three time horizons captured the general trends (clear/cloudy/partly cloudy) of the PV production.

A scatter plot of forecasted versus measured daily PV production (see Figure 9) shows a closer linear alignment to the measured data than was seen for either the single or aggregated substations. The most likely explanation is that machine learning occurs more quickly for an instrumented PV farm, where the collection of high-resolution can better resolve temperature, soiling, shading, etc. trends that are learned and captured into the future. In contrast, the combined substations, which aggregate data from tens or hundreds of PV systems, make it difficult to detect the trends of any one system. Another difference is that new residential systems are constantly being added so the PV capacity at the substation level is ever changing, whereas the PV farm capacity is fixed.

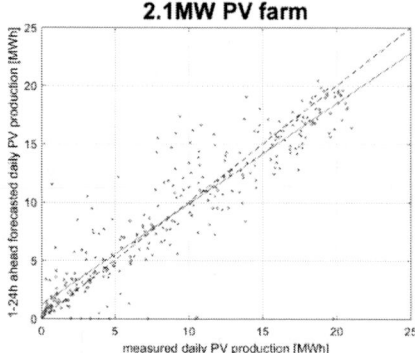

Figure 9: Forecasted versus measured daily PV production for a 2.1MW PV farm. The dashed black line is the 1:1 line that a perfect forecast would follow.

We were also interested in the impact of seasonal changes on the accuracy of the forecasting tools (see Figure 10, which shows seasonal trends for the three time horizons.) We found that the forecasts tend to under-predict in the winter and over-predict in the summer. These seasonal changes are sharp, ranging from a -6% bias in winter to a +7-11% bias in July. The 1-24 hour ahead forecast does not always have the smallest monthly mean bias error, showing that sometimes forecast updates lead to larger errors.

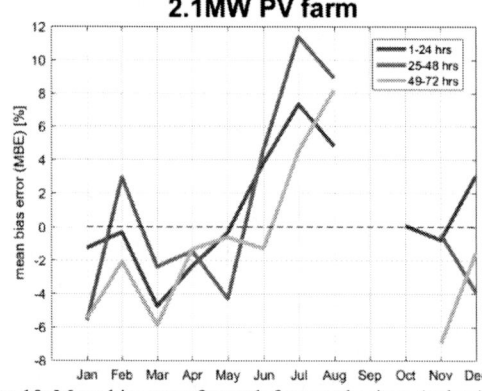

Figure 10: Mean bias error for each forecast horizon (colors), sorted by month of the year. No data was available for September, and only 1-24 hrs ahead forecasts were available in October.

V. SUGGESTED IMPROVEMENTS

As a result of our analysis, we identified three areas where forecast performance can be improved.

1. PV capacity

With installations of distributed solar growing exponentially in many regions of the country, the PV capacity at substations is a moving target. As seen in Figure 1, the distributed forecast sometimes under-predicts PV production on clear days, a finding we attribute largely to newly commissioned (not as yet recognized by the forecasting model) PV installations. Although the forecast model does update PV capacity, the

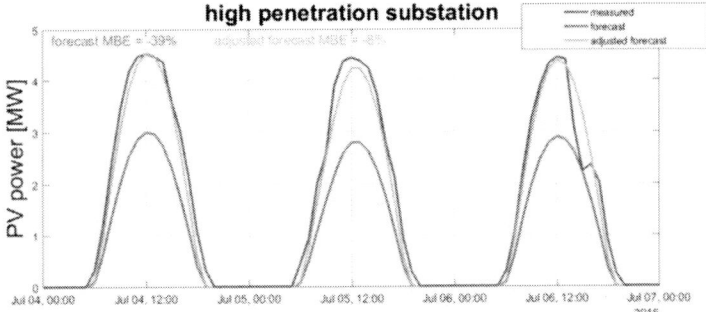

Figure 11: Measured, forecast, and adjusted forecast PV power for three clear days.

update cycle is slow and depends on the timely and accurate updating of the interconnection database, which includes number, size and interconnection dates for all PV systems.

One simple solution to quickly account for this changing capacity could be to scale the forecast by the ratio of maximum-measured to maximum-forecasted power from the previous week:

$$\text{adjusted forecast} = \text{original forecast} \times \left[\frac{\max(\text{measured power})}{\max(\text{forecasted power})} \right]_{\text{previous week}}$$

As shown in Figure 11, by using this adjusted forecast for on three sample clear days, the mean bias error of the forecast was reduced from -39% to -8%. Care must be taken, however, to only apply this correction to clear days, as the forecast does not have the same (low) bias on cloudy days.

2. Tilt Angle and Azimuth

The tilt and azimuth angles of PV modules impact production. Yet, the machine learning in the PV farm forecast only accounts for a possible tilt angle of PV modules, not a possible azimuth deviating from due-south, resulting in inaccuracies in the clear sky model. As can be seen in Figure 12, the clear-sky profile for the 2.1MW solar farm does not follow the same shape as the measured profile. The difference is because the farm has dual-axis trackers for the PV modules, leading to nearly flat, high power output for most of the day on clear days.

The forecast, however, fails to account for solar tracking and therefore predicts the power one might expect from a fixed-tilt

system, with a more distinct peak at solar noon. We therefore would recommend that the underlying clear-sky model be refined to better resemble the actual PV racking system, which could include single- or dual-axis trackers or fixed-tilt systems at not facing due south.

3. Separate Forecast Training for Clear vs. Other Days

To evaluate the difference in forecast performance for clear vs. not-clear days, we used a simple clear-sky detection method: we compared measured PV output to a clear-sky model [5], looking for days that had a similar energy output and shape as the clear-sky model, according to the following criteria:

1. The ratio of measured daily energy to clear-sky daily energy had to be greater than 0.9
2. The midday correlation (10A-2P) had to be greater than 0.95

We applied this filter to a week of PV data from the substation aggregate and were able to demonstrate its effectiveness, as shown in Figure 13. When we segregated the clear days from the not-clear days, we found the forecast under-predicts PV production on clear days, but slightly over-predicts it on other days (see Figure 14). This result is consistent with Figures 2, 6, and 9, where the forecast often under-predicted at high-power outputs (e.g., clear days) and over-predicted at low power outputs.

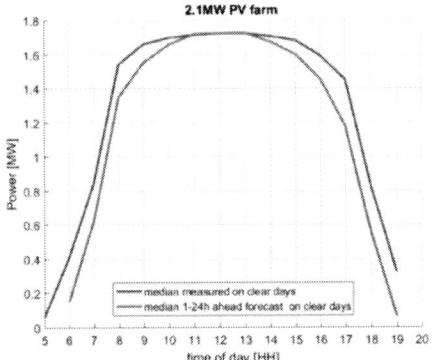

Figure 12: [Left] average measured (blue) and forecasted (red) power output on clear days at 2.1MW PV farm. [Right] Picture from the 2.1MW PV farm, showing that the modules are dual-axis tracking.

978-1-5090-5606-4/17 $31.00 © 2017 IEEE

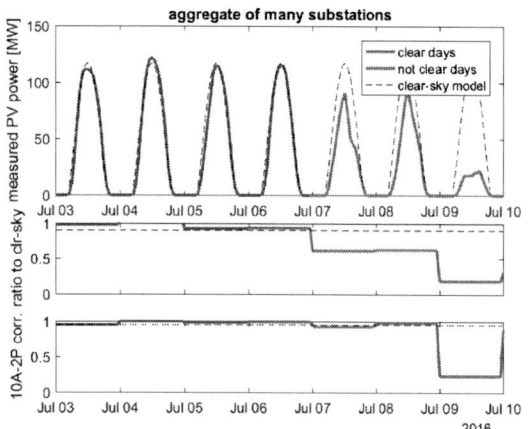

Figure 13: Example of the clear-sky filter, showing days determined to be clear-sky in blue and other days in red. The two bottom plots are the two filters applied to determine if a day was clear.

We attribute this clear vs. not clear inconsistency to forecast training which uses machine learning and considers all days, without segregating based on clear or not clear. On clear days, the actual power output is higher than the average power output over all days. On not clear days, the opposite is true.

Clear days are the special case where we can make a very good prediction using simply a clear-sky model; this simple prediction would eliminate these negative forecast errors on clear days. Thus, a better approach would be to divide the training data into two bins, based on whether a day is forecasted to be clear or cloudy. If a day is forecast to be clear, its forecast should be based solely on previous clear days. Similarly, if a day is forecast to not be clear, it should be trained only on data from other not clear days.

VI. CONCLUSIONS

Our evaluation of how PV forecasts perform relative to measured data has led to a better understanding of the parameters that determine their performance and to the identification of technical areas that need improvement. We

found, for example, that while the load forecasts were generally accurate, the PV forecasts were more errant, depending on the scale of the forecast (single PV plant versus single substation versus multiple substations). As a result, we identified three strategies for increasing the accuracy of future PV forecasts: 1) accurately account for the total amount of distributed solar installed capacity, including fast updates to account for new installations; 2) include data on the azimuth and tilt angles of the PV installations in forecast training; and 3) bring greater specificity to the training data by segmenting clear from not-clear days.

ACKNOWLEDGMENTS

Sandia National Laboratories is a multimission laboratory managed and operated by National Technology and Engineering Solutions of Sandia, LLC, a wholly owned subsidiary of Honeywell International, Inc., for the U.S. Department of Energy's National Nuclear Security Administration under contract DE-NA0003525. Report number SAND2017-5645 C.

REFERENCES

[1]   U.S. Energy Information Administration. *California has nearly half of the nation's solar electricity generating capacity*. Available: http://www.eia.gov/todayinenergy/detail.php?id=24852

[2]   IBM. *Machine Learning Helps IBM Boost Accuracy of U.S. Department of Energy Solar Forecasts by up to 30 Percent*. Available: http://www-03.ibm.com/press/us/en/pressrelease/47342.wss

[3]   T. E. Hoff *et al. Behind-the-Meter PV Fleet Forecasting*. Available: http://cleanpower.com/wp-content/uploads/BehindMeterPVFleetForecast.pdf

[4]   V. P. A. Lonij *et al.*, "A scalable demand and renewable energy forecasting system for distribution grids," in *2016 IEEE Power and Energy Society General Meeting (PESGM)*, 2016, pp. 1-5.

[5]   P. Ineichen and R. Perez, "A new airmass independent formulation for the Linke turbidity coefficient," (in English), *Solar Energy*, vol. 73, no. 3, pp. 151-157, 2002.

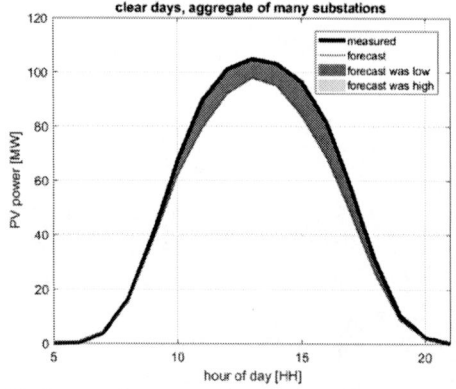
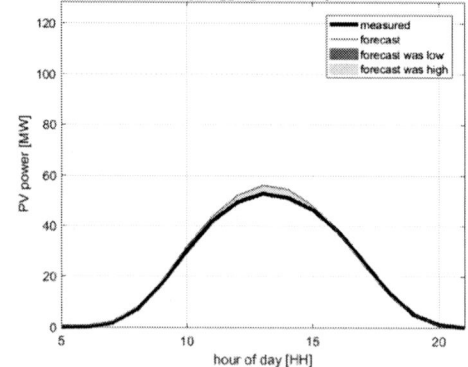

Figure 14: Average forecast performance [Left] on clear days and [Right] on not clear days.

Record Efficiencies for Selenium Photovoltaics and Application to Indoor Solar Cells

Douglas M. Bishop, Teodor Todorov, Yun Seog Lee, Oki Gunawan, Richard Haight

IBM T.J. Watson Research Center, Yorktown Heights, NY, 10598, USA

Abstract — **Until recently, Selenium photovoltaics remained overlooked and stalled with a last efficiency record of 5% established more than 30 years ago. We recently demonstrated a 30% increase in efficiency and a simultaneous reduction in thickness by a factor of 5x. Despite the expectation for continued improvements, selenium, with a bandgap of ~1.9 eV, is not well matched for a single junction outdoor AM1.5 spectrum, but selenium matches the indoor light spectrum of LED's and CFL's perfectly. In indoor lighting conditions we have achieved >13% efficiency which is more efficient than typical commercial a-Si and c-Si in indoor lighting, while fabrication is vastly simpler. Unlike alternative promising technologies such as perovskites, Se is both stable in air and has minimal toxicity risks. In this paper we will highlight the improvements in device structure and annealing that led to the recent record efficiency, and summarize the key variables and needs of solar cells for indoor applications.**

I. Introduction

The element selenium was the basis of the first solid state solar cells in the late 19th century, and it was considered a leading contender for photovoltaics until silicon became the dominant semiconductor with the rise of the microelectronics industry. Selenium photovoltaics research plateaued more than 30 years when the last efficiency record was set at 5% [1]. This can be attributed to both the dominance of silicon, as well as the development of CIS and CdTe which have bandgaps more appropriate for the AM1.5 spectrum. Se, with a bandgap of ~1.9 eV, is better suited for multi-junction cells or the indoor-light spectrum. In fact, Se matches the indoor solar spectrum from either a CFL or LED perfectly [2]. Energy harvesting, and specifically indoor-photovoltaics (IPV) are receiving renewed attention due to the rapid growth of the Internet-of-Things, and to enable wireless and autonomous devices for ubiquitous sensing.

With our recent research on Se PV, we demonstrated 30% increase in efficiency compared to the previous champion and a simultaneous reduction in thickness by a factor of five [3]. Due to the high Eg, the efficiency in AM1.5 is 6.5%, but in the indoor spectrum, Se solar cells typically exhibited ~2x increase in efficiency. Top performing Se cells achieved >13% in the indoor spectrum, which is higher than most competitive technologies such as a-Si, while the fabrication is vastly more simple. Other alternative organic and perovskite materials have similarly been suggested for indoor applications, however Selenium has the benefit that it is both stable in air, and does not contain highly toxic elements.

In this work we highlight the technical aspects that underlie the recent selenium record, including a new device structure and optimized annealing conditions that enabled the efficiency improvements. In the second half of the paper we lay out the key priority differences for indoor PV applications (compared to typical commercial PV). These key qualities include changes in optimal band gap, relative effects of series and shunt resistance, as well as commercial aspects such as toxicity, fabrication temperature, and cost. We examine how selenium solar cells compare to other solar materials in these areas and demonstrate its compelling promise for indoor PV.

III. Selenium Solar Cell Fabrication and Performance

Selenium is a nearly intrinsic p-type material and with hole concentrations ~10^{12} according to our hall measurement. The old champion Se solar cell utilized TiO_2 n-layer with an ITO front contact and Au back contact. Our improved structure is deposited on FTO coated glass in superstrate fashion with the structure: FTO/ZnMgO/Se/MoO$_3$/Au as seen in the cross-section SEM in Figure 1. More details on the fabrication process and the effects of each layer can be found in our recent publication [3]. The achievement of record efficiency for this material was enabled by the change in structure wherein the ZnMgO improves the band alignment and the MoO_3 increases collection and reduces recombination in the thin solar cell due to a back surface field induced by the high work function.

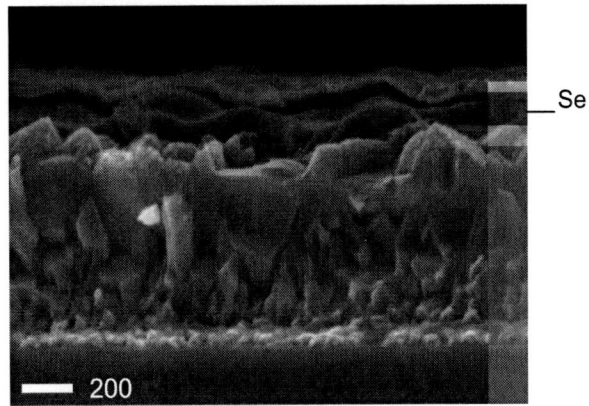

Fig. 1.   SEM cross-section of Se solar cell

978-1-5090-5606-4/17 $31.00 © 2017 IEEE

In our testing, the optimal efficiency was achieved with a Se absorber thickness of just ~100 nm. Although thicker films can improve absorption near the band edge, the field assisted collection from the back contact helps improve carrier separation in thinner films.

Due to the thin Se layer and high vapor pressure (Se = 0.1 torr at 280°C), depositions were completed in less than 20 minutes after pump-down. The substrate remains near room temperature during deposition and therefore the as-deposited Se layer consists of amorphous Se polymeric rings. After the device completion, the devices structure are annealed to form crystalline hexagonal crystal structure as confirmed by Raman and XRD.

The annealing provides two necessary changes for efficient solar cells. First, the bandgap shrinks from ~2.2 eV as deposited, to 1.95 eV as confirmed from UV-VIS measurements. Second, the diffusion length increases with the improved collection, as can be seen in the progression of EQE as a function of annealing temperature in Figure 2.

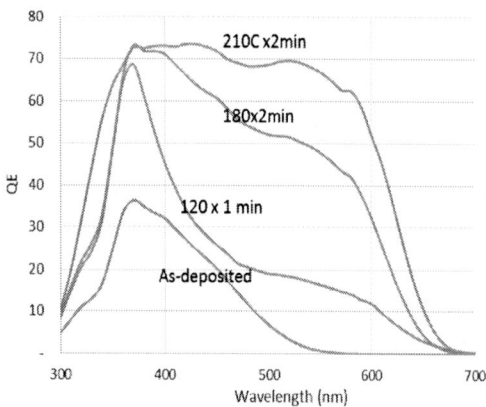

Fig. 2.   Quantum efficiency of selenium solar cells as a function of annealing.

Prior to annealing, the high recombination in the amorphous Se layer led to poor devices with low current and low FF as evidenced in the EQE, however voltages in excess of 1V were achieved. With annealing, current and fill factor increase due to increasing diffusion lengths, however the voltage slightly declines as Se increases in crystallinity resulting in a lower E_g. A champion cell efficiency of 6.5% was achieved under AM1.5 which can be seen in Figure 3. A comparison to the previous champion cell as well as the details of the device analysis and the effect of the high work function back contact and change in buffer layer were recently published [3].

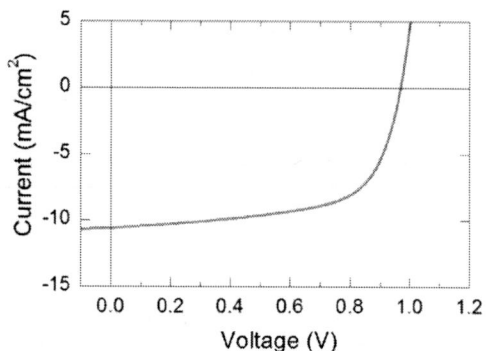

Fig. 3.   IV curve of champion Se device under AM1.5

III. SELENIUM PERFORMANCE IN INDOOR LIGHTING

The QE of the champion Se solar cell is compared to both outdoor AM1.5 and typical office indoor spectrum using measured spectrum from a LED and CFL in Figure 4. The relative match of the QE and indoor spectrum means most of the photons can be absorbed and losses due to thermalization are minimized. The theoretical efficiency potential of different materials can be evaluated using an indoor light spectrum with the well-known Shockley-Quessier detailed balance calculations. Selenium with its bandgap of 1.9 eV has a theoretical maximum efficiency in the AM1.5 spectrum of 23%, while in a typical office environment, with LED or CFL lights, the theoretical maximum efficiency is ~60%. This is almost twice the theoretical max achievable by c-Si [2], [4].

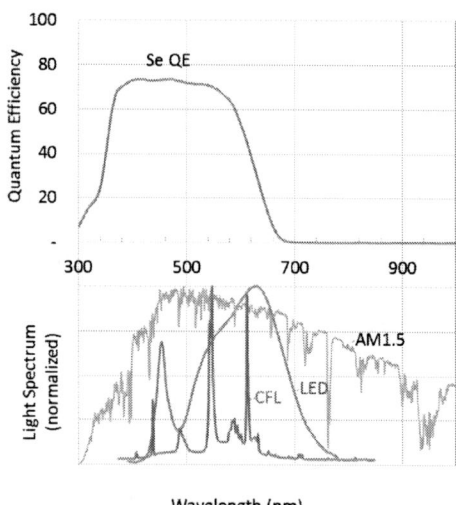

Fig. 4.   QE of Se solar cell and comparison to indoor spectrum and outdoor light spectrum.

The device results as a function of light intensity and light spectrum are shown in Figure 5. Cell efficiency was typically double the AM1.5 performance with best devices exceeding

13%. It is expected that further optimization by reducing non-absorption losses and decreasing shunting will provide further gains for indoor lighting. In Figure 5, for comparison, high-efficiency, high Eg CZTSSe solar cells are shown. The best performing cells were greater than 11% in AM1.5 but decreased to less than 10% under CFL or LED lighting.

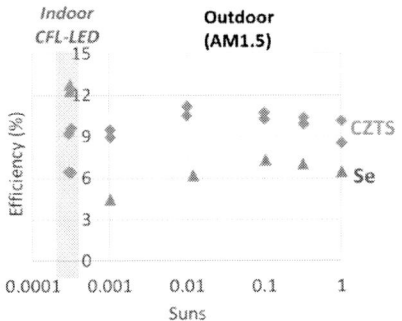

Fig. 5. Selenium solar cell performance as a function of light intensity and using indoor light spectrum.

II. GOALS AND NEEDS FOR INDOOR SOLAR CELLS

The goals and requirements for indoor photovoltaics are very different from outdoor applications. In addition to a shifted spectrum, the power density in typical indoor environments is ~1000x lower than AM1.5 test conditions and due to this low power density it can be clearly concluded that indoor photovoltaics cannot compete on cost with energy produced at the utility scale. Instead, indoor photovoltaics can offer power where it's needed to enable mobility, sensing and increased functionality in the built environment. These applications include mobile devices, or stationary sensors to alleviate the burden of wiring. Unlike an energy utility, the goal is to provide sufficient energy for the application, with power demands measured in μw/hr or mW/hr. Solar-powered calculators are the classic example for indoor-PV, but with the rapid rise of connected devices and mobile sensing, there is a significant potential for further growth. As will be shown, Selenium is well positioned to compete in this growing space.

The largest change in solar cell design for indoor applications stem from the dramatic change in light spectrum. The optimal band-gaps for office lighting is ~1.9 eV which approximately matches Se [2].

A second change is due to 1000x reduction in light intensity which shifts the relative importance of R_{series} and R_{shunt} in solar cell performance. In low light intensities, resistance losses are a relatively smaller parasitic contribution due to lower I^2R power losses. On the other hand, the effect of R_{shunt} is magnified relative to 1-sun, and therefore the balance of series and shunt resistance change in indoor applications [5]. Selenium is photoconductive, but in low light, has quite high resistivity as well as high R_{shunt} resistance which benefits Se in low light.

The potential for IPV to be dispersed widely throughout consumer spaces in the home or a business, powering small sensors or portable devices, suggests that toxicity should be a serious concern. In addition, unlike the lifecycle of a CdTe solar power plant which are easily collected at end of life, it is inevitable that many small consumer sensors will end up in a landfill. For these reasons, solar cells containing lead or highly toxic elements will not be applicable for extensive IPV. Although Selenium is potentially toxic in large doses, the quantities that are ingestible for IPV do not present a hazard. In fact, Se is also required micronutrient, and while ingesting solar cells is not advisable, a 1 cm^2 solar cell (48 μg) would provide just under the daily recommended intake of Se (55 μg)[6]. For reference, a CIGSe or CZTSe solar cell has 10x more Se per cm^2 than a pure Se solar cell due to the greater absorber thickness.

Stability and longevity requirements for IPV may vary from a few years for limited commercial applications to >10 years for other sensor demands. This requirement is less stringent compared to typical utility scale power plants which rely on 25+ years of up time, however the desired stability is still a concern for organic or perovskite solar cells. Selenium solar cells are stable in air over the tested period of 6+ months, and more detailed long-term testing is underway.

Silicon is both stable and non-toxic, however in addition to the bandgap mismatch for IPV applications, other factors such as cell interconnection and small cell size increase the cost and decrease the resulting efficiency. Unlike thin film solar cells which can be connected in series by monolithic integration, silicon must be cut and wired into multiple cells. Monolithic integration provides a particularly large cost benefit as cell size decrease as is required for IOT and mobile IPV applications where total package area may be smaller than a credit card. For silicon with indoor lighting, more than 9 cells would be required in series to achieve >3.7V necessary for charging a Li-ion battery. While it is feasible to cut, package and interconnect many cells in series; it is likely that the additional steps involved as well as the interconnection points would significantly increase cost and add potential failure points. A second challenge for c-Si and to a lesser extent poly-xtal-Si, is that cell efficiency decreases with cell size due to increased recombination at exposed edges, and this detriment is exaggerated at low light levels [7]. The efficiency loss can ameliorated to a limited degree with edge passivation, although this requires patterning the doping profile of each cm^2 sized cell and confirms that cutting up mass produced commercial wafers will likely not be an attractive strategy for small PV cells required for mobility.

Monolithic integration and high voltage in low-light for the Se-technology allow the package to be simplified and the number of series-connections to be reduced by more than 50%. The use of a plastic substrate is also beneficial to allow for flexible conformality, reduced weight and risk of shattering, while also reducing total package costs. The ultimate goal for low cost sensing would likely use monolithic fabrication of both photovoltaic and sensor on a single manufacturing line. For both flexible substrates and integrated sensor manufacturing, a low thermal budget is essential. In the case of Se, we have demonstrated production on flexible plastics and a simple fabrication process and minimal thermal budget of only 200°C for 2 minutes is required [3].

IV. SUMMARY

We have demonstrated new record efficiencies for Se devices after three decades of minimal progress. The improvements can be attributed to a new device stack and improved microstructure. While future improvements are expected, the applications to indoor photovoltaics may be particularly compelling. Due to the high bandgap which matches the typical LEDs and CFLs, selenium outperforms typical commercially available c-Si or a-Si in indoor lighting environments while also having a higher theoretical potential. The fabrication process is both extremely simple and quick, and unlike other alternative solar materials, selenium solar cells were found to be stable in air, and non-toxic in the applied quantities – both of which are critical aspects for indoor PV applications. With the rapid growth of wireless devices and sensors powering the "Internet-of-Things," indoor photovoltaics – and specifically selenium – deserves a second look.

REFERENCES

[1] T. Nakada, A. Kunioka, D. R. S. and G. C. Sinke, J. J. Loferski, R. F. S. and A. K. Ghosh, A. T. and Y. M. H. Ito, M. Oka, T. Ogino, T. N. and A. Kunioka, T. N. and A. Kunioka, H. Hovel, H. G. and A. Tausend, S. M. and M. Schmeits, and D. E. B. and J. H. M. R. Singh, K. Rajkanan, "Polycrystalline Thin-Film TiO2/Se Solar Cells," *Jpn. J. Appl. Phys.*, vol. 536, no. Part 2, No. 7, pp. L536–L538, Jul. 1985.

[2] M. F. Müller, M. Freunek, and L. M. Reindl, "Maximum efficiencies of indoor photovoltaic devices," *IEEE J. Photovoltaics*, vol. 3, no. 1, pp. 59–64, 2013.

[3] T. K. Todorov, S. Singh, Y. S. Lee, T. S. Gershon, D. M. Bishop, P. D. Antunez, and R. A. Haight, "Ultrathin, record-performance solar cells with high band gap from the world's oldest photovoltaic material," *Nat. Commun. (In Rev.*, 2017.

[4] S. Rühle, "Tabulated values of the Shockley-Queisser limit for single junction solar cells," *Sol. Energy*, vol. 130, no. October, pp. 139–147, 2016.

[5] K. Ruhle, S. W. Glunz, and M. Kasemann, "Towards new design rules for indoor photovoltaic cells," *2012 38th IEEE Photovolt. Spec. Conf.*, no. 1, pp. 002588–002591, 2012.

[6] M. P. Rayman, "Selenium and human health Role of selenium : selenoproteins," *Lancet*, vol. 379, no. 9822, pp. 1256–1268, 2012.

[7] S. W. Glunz, J. Dicker, M. Esterle, M. Hermle, J. Isenberg, F. J. Kamerewerd, J. Knobloch, D. Kray, A. Leimenstoll, F. Lutz, D. Osswald, R. Preu, S. Rein, E. Schaffer, C. Schetter, H. Schmidhuber, H. Schmidt, M. Steuder, C. Vorgrimler, and G. Willeke, "High-efficiency silicon solar cells for low-illumination applications," *Conf. Rec. Twenty-Ninth IEEE Photovolt. Spec. Conf. 2002.*, pp. 450–453, 2002.

Close-Spaced Sublimation for Sb₂Se₃ Solar Cells

Laurie J. Phillips,[1]* Peter Yates,[1] Oliver S. Hutter,[1] Tom Baines,[1] Leon Bowen,[2]
Ken Durose,[1] and Jonathan D. Major[1]

1 - Stephenson Institute for Renewable Energy, University of Liverpool, L69 7ZF, UK.
2 - G. J. Russell Microscopy Facility, University of Durham, South Road, DH1 3LE, UK.
*corresponding author – laurie.phillips@liverpool.ac.uk

Abstract — **A CSS technique for depositing large-grained, well-orientated Sb₂Se₃ thin films has been developed, and solar cells based around this material have been fabricated. Adapting a standard CdTe methodology can produce high efficiency devices in a FTO/TiO₂/Sb₂Se₃/P3HT/Au heterostructure. The influence of temperature during deposition is examined to show larger columnar grain growth at increasing source temperature. Optimising the temperature of deposition both increases the proportion of grains in the correct orientation and balances the improvements due to larger grains while limiting void formation. At optimized conditions of 475°C, the peak performing device exhibited a maximum PCE of 4.3%.**

Index Terms — **Antimony Selenide, Sb₂Se₃, CSS**

I. INTRODUCTION

Photovoltaics installations continue to grow around the world. However, if solar is to make a substantial contribution to the switch to sustainable energy sources, new absorber materials are needed that are abundant for terawatt-scale manufacture. Silicon-based technologies dominate the market, as the promise of thin-film photovoltaics is yet to be fully realized. Copper indium gallium di-selenide (CIGS) and cadmium telluride (CdTe) have the most widely accepted thin-film technologies, with efficiencies exceeding 20%. However, the scarcity of indium and tellurium, and the toxicity of cadmium are limiting factors, difficult to overcome for more widespread adoption. The recent emergence of lead-halide-based perovskite absorbers, while rapidly achieving similarly high efficiencies, have stability issues and contain highly toxic, and restricted, lead. The relatively lower toxicity of antinomy selenide (Sb₂Se₃), coupled with its low-cost and higher abundance, means it is a material that shows considerable promise as an alternative.[1],[2] A single phase V-VI chalcogenide, Sb₂Se₃ has a near-ideal band-gap of 1.1-1.3 eV, and high absorption coefficient. It is composed of $(Sb_4Se_6)_n$ 1D ribbon structures that theoretical work has suggested should lead to grain-boundaries that are relatively benign, due to edges terminating without dangling bonds, and therefore should exhibit reduced recombination.[3] It has been proposed that the $6s^2$ electronic configuration of lead is an important factor in the success of the lead-halide perovskites. Other nS^2 materials, such as Sb are likely to offer similar benefits, which is another reason for the increasing interest in Sb₂Se₃.[4] Solar cells using this material have been demonstrated by several methods including spin-coating,

chemical bath deposition, and evaporation with the best PCE currently at 5.9%.[5]-[7] However, closed-spaced sublimation (CSS), a technique related to evaporation and widely used for the fabrication of CdTe, has yet to be investigated. It is a technique whereby material is sublimed from a source onto a substrate separated by a small gap, typically a few millimeters. It is capable of producing films with large grains and high deposition rates, and can be scaled up for industrial manufacturing easily due to modest vacuum and temperature requirements, and high material usage.

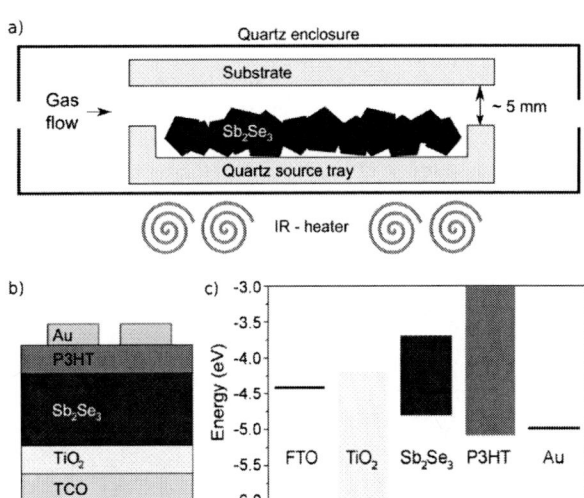

Fig. 1.   a) Schematic of the CSS experimental setup b) Cell layer heterostructure c) Energy level diagram of the cell.

In this work, we investigate Sb₂Se₃ thin films deposited at source temperatures from 400 to 500°C using the CSS method, and solar cells made with layers deposited under the same conditions. We show that Sb₂Se₃ material, deposited by CSS, can be made to form exceptionally large grains by adjusting the temperature of deposition. XRD measurements suggest that grain orientation can be altered by changing the deposition conditions, and optical transmission measurements show a 0.04 eV shift in the bandgap. A power conversion efficiency of 4.3% has been achieved by using a Sb₂Se₃ film deposited at the optimum temperature of 475°C under AM 1.5 illumination.

II. EXPERIMENTAL

Thin-films of Sb_2Se_3 were grown on glass and within cell structures to examine both the material in isolation and its performance in a photovoltaic device. The glass surface was cleaned using sequential ultrasonic baths of DI-water, acetone, and IPA, followed by a 15 minutes UV-Ozone treatment. The Sb_2Se_3 layer was deposited using an entirely quartz-based CSS chamber as shown in figure 1a. The layers were deposited at four source temperatures from 400 to 500°C in 10 mTorr nitrogen for 15 minutes (source material 99.999% Sb_2Se_3 from Alfa Aesar). Cells were grown on FTO coated 2.2 mm glass substrates (TEC7, Sigma Aldrich). The TCO was cleaned in the same way as the glass and a TiO_2 blocking layer was then coated onto the substrates via spin-coating in two sequential steps using 0.15M and 0.30M titanium isopropoxide solutions in ethanol. The solutions were made up under heavy stirring, filtered with a 0.45 um PTFE filter and spin-coated at 3000 rpm for 30s in a nitrogen filled glove-box. Each sub-layer was dried at 120°C for 10 minutes before removing the substrates from the glovebox and sintering the complete layer at 550°C for 30 minutes in air. Thin-films of the Sb_2Se_3 absorber were deposited at the different temperatures in identical conditions to the layers on glass. A thin layer of un-doped P3HT (Mw=34,100 Ossila) was then spin-coated on the back from a 10 mg/ml solution in chlorobenzene at 3000 rpm for 30s. Finally, the cells were completed by depositing 50 nm gold back-contacts through 10 mm^2 apertures in a shadow-mask to create an 8×8 pixel array. The complete cell structure and expected energy level diagram of the resultant heterostructure are shown in figures 1b and 1c, respectively.

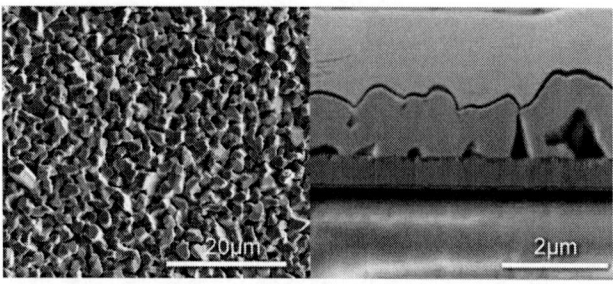

Fig. 2. a) SEM image of the upper surface showing the large micrometer grain structure of Sb_2Se_3 deposited by CSS at 450°C. b) FIB cross-section of the Sb_2Se_3 layer (blue overlay), deposited on TiO_2-coated (green overlay) TEC (red overlay) glass. Platinum coating is also visible (yellow overlay).

XRD analysis was performed using a Rigaku Smartlab X-Ray diffractometer, and the transmission data was obtained using a Shimadzu solid-spec 3700 UV-Vis Nir spectrophotometer and a profilometer. JV characteristics were taken under AM1.5 illumination from a TS Space Systems Unisim 100 AAA solar simulator and EQE measurements were taken with a Bentham PVE300.

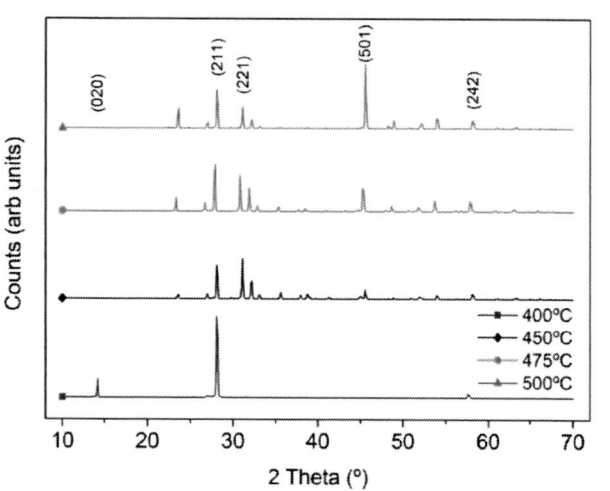

Fig. 3. XRD of Sb_2Se_3 deposited via CSS at source temperatures from 400 °C to 500°C. 400°C is the only temperature with (020) grains, while the presence of (501) is increasingly evident with increasing temperature.

III. RESULTS

Close-spaced sublimation is a technique that generally produces large-grained material. The combination of high-temperature, and low temperature difference between source and substrate, means that island-growth and coalescence is favoured over the growth of many smaller islands and therefore, as can be seen in figure 2a, the grains of material can grow to a diameter of several microns. At lower temperatures, it is possible to form nanorods, which are highly absorbing, but also highly resistive and do not adhere well to the substrate and so are unsuitable for photovoltaic applications. Increasing the temperature of deposition increases the grain size, but this also increases the chance of void formation during coalescence. As can be seen in figure 2b, the grains extend the entire depth of the layer, which should be beneficial to carrier collection as grain boundaries are more likely to be recombination sites. Also visible in figure 2b, is the lack of boundary between the majority of adjacent grains, suggesting either a merging of grains towards the substrate or a grain boundary unable to be distinguished by the SEM. At the TiO_2 surface, there are several voids visible, possibly due to contamination of the surface, strain relief due to lattice mismatch, or simply voids formed during growth. Some of these voids propagate through the entire layer and are likely to increase recombination, and act as shunting pathways between the back-contact and window layers. The main reason for coating the back surface of Sb_2Se_3 with P3HT is to mitigate the deleterious effects of these voids, as the polymer will block these paths and reduce the prevalence of shunting.

Sb_2Se_3 is known to have a 1D ribbon-like crystal structure.[1] Working with this anisotropy will be crucial to the fabrication of high-efficiency devices. Orientating the grains such that charge transport occurs along the covalently-coupled ribbons, rather than relying on hopping between the weakly van-der-waal bound chains is essential for maximal mobility and minimal recombination. XRD was used to examine the crystal structure orientation of the layers deposited at the various temperatures and is shown in figure 3.

At 400°C, there is a peak at ~14.2° which is assigned to (020) orientated ribbons lying parallel to the substrate surface while at higher temperatures, this peak is absent. All the material deposited at higher temperatures exhibits two strong peaks at ~31.1° and ~28.1°, corresponding to (221) and (211) orientations respectively. These are the perpendicular and intermediate orientations conducive to efficient charge extraction in a cell. The peak at 45.5° is absent at 400°C but grows with increasing temperature, and corresponds to (501) orientated material. This is indicative of the higher substrate temperatures inducing reorganization and re-evaporation from the surface.

A UV-Vis spectrophotometer was used to measure the optical transmission of the material, as shown in figure 4, and the band-gap, inset in figure 4, was extracted from a tauc plot. The bandgap increases by 0.04 eV from 1.14 eV at 400°C to 1.18 eV at 475°C. These values are consistent with literature, the former generally matching nanostructured thin films while the latter more consistent with to bulk samples and theoretical calculations.[8] The smaller grains and nanorods of the lower temperature sample are not small enough to be affected by quantum confinement and more work is needed to understand the factors affecting the bandgap of Sb_2Se_3 to determine if, for example, the stoichiometry is different with differing levels of selenization occurring at different temperatures.[9]

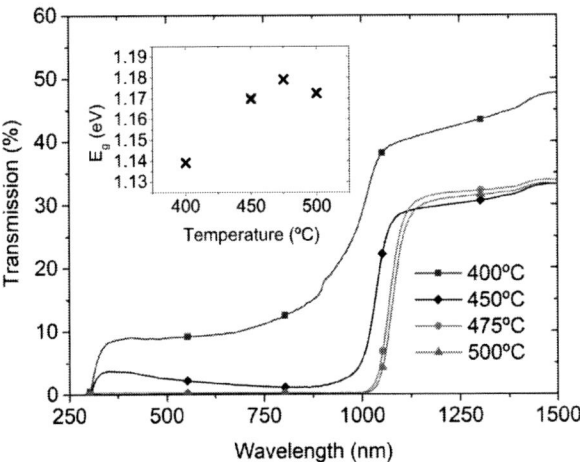

Fig. 4.   Transmission curves of Sb2Se3 layers deposited via CSS at varying source temperatures. The bandgap at each temperature extracted from a tauc plot is also included (inset).

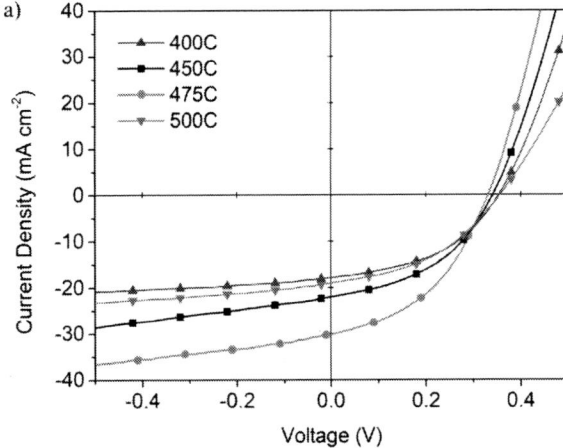

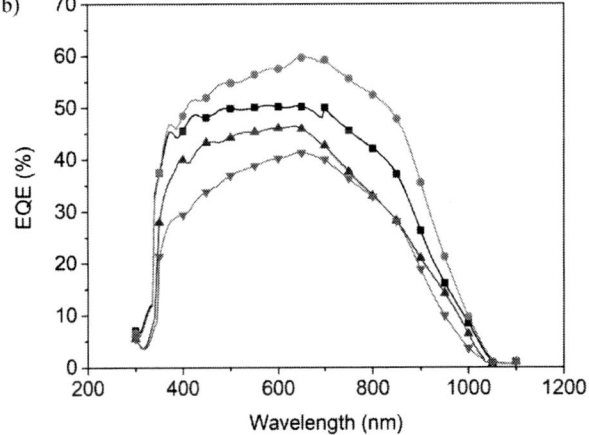

Fig. 5.   a) JV and b) EQE curves for highest efficiency device contacts for cells produced using CSS Sb_2Se_3 at temperatures varying from 400 to 500°C.

Cells were fabricated with Sb_2Se_3 deposited at temperatures of 400, 450, 475, and 500°C. The J-V characteristics of the peak cell pixel from each 5×5 cm sample plate are shown in figure 3a, with the peak and average parameters also shown in table 1. The best cell performance was achieved using a deposition temperature of 475°C, mainly due to the increase in J_{sc}. This increase is likely to be due to the improved charge collection from a thicker film and larger grains. The increased thickness of the film is also the likely cause of the drop in V_{oc} due to higher series resistance. Sb_2Se_3 film thickness increased from ~1.1 μm for 400°C to ~8.5 μm for 475°C with the film at 500°C reducing to ~4.0 μm. The increasing chance of void formation at the highest temperature limits the performance, in addition to material loss due to re-evaporation, and therefore the current is reduced for the 500°C deposition. As the source temperature becomes too high, the substrate reaches a temperature high enough that the rate of re-evaporation from the surface actually reduces the total amount of material deposited. The most consistent limiting factor in all devices is fill-factor, probably caused by a low shunt resistance due to

TABLE I

AVERAGE DEVICE PERFORMANCE PARAMETERS (FROM ~56 DEVICES FOR EACH TEMPERATURE)

	Av. η (%)	Av. V_{OC} (V)	Av. J_{SC} (mA/cm^2)	Av. FF (%)
400°C	1.43 ± 0.77	0.34 ± 0.06	10.76 ± 4.69	36.41 ± 5.88
450°C	1.94 ± 0.86	0.31 ± 0.03	15.62 ± 6.23	39.08 ± 2.48
475°C	3.56 ± 0.49	0.36 ± 0.15	25.74 ± 3.04	40.81 ± 7.69
500°C	1.29 ± 1.12	0.34 ± 0.03	9.79 ± 7.31	33.61 ± 8.37

TABLE II

PARAMETERS FROM THE CONTACTS HAVING HIGHEST EFFICIENCIES

	Peak η (%)	Peak V_{OC} (V)	Peak J_{SC} (mA/cm^2)	Peak FF (%)
400°C	2.84	0.36	17.86	44.19
450°C	3.25	0.34	22.99	41.63
475°C	4.32	0.33	30.19	43.38
500°C	2.86	0.36	19.01	41.79

voids and a high series resistance due to a thick absorber layer. It is notable how high the current is at 475°C with an 8.5 μm thick absorber layer and reinforces our confidence in the potential of this material.

The external quantum efficiencies (EQE) for each peak cell are shown in figure 3b. The wide-bandgap TiO$_2$ window layer allows much of the low wavelength light into the cell and so produces a relatively square EQE curve at high energies. However, more interesting is the shallow gradient of the EQE curve at the band-edge. This suggests that the thicker films could be reducing the charge extraction efficiency towards the back surface of the cell where longer wavelengths are more likely to be collected. Also, while the large grains and 1D ribbon-like structure may minimize grain boundary defects, it appears likely that there are still many defects present within the material which, results in a significant Urbach-tail. This would also be a significant factor in limiting the V_{oc}. Therefore, to achieve higher efficiencies, there is still significant work to do on identifying defects and defect passivation is likely to be required if they cannot be eliminated during growth.

IV. CONCLUSION

We have demonstrated Sb$_2$Se$_3$-based solar cells grown via a CSS technique with very promising performance characteristics. The large, micron-sized, grains produced using this technique have been shown capable of generating and allowing the extraction of high currents, as demonstrated by the maximum J$_{sc}$. The absorber material has grains spanning the entirety of the layer perpendicular to the substrate and which appear to merge towards the substrate, suggesting benign boundaries. Grain orientation has been shown to be tunable by altering deposition conditions and the desirable

(211) and (221) orientations maximised. The peak performance of 4.32% achieved using the CSS method has attained a significant fraction of available current i.e. ~72% of the 42 mA cm^{-2} Shockley-Queisser limit (SQL), whereas the voltage for the best device falls far short (~44% of the possible 0.75 V for the SQL) The slope of the EQE at longer wavelengths at the band-gap suggests significant Urbach-tailing and indicates the presence of defects within the material that require further investigation and passivation before solar cells made with this material can achieve their full potential.

ACKNOWLEDGEMENTS

This work was funded by the UK Engineering and Physical Sciences Research Council Grants number EP/N014057/1, EP/J017361/1, and EP/L01551X/1.

REFERENCES

[1] Y. Zhou, M. Leng, Z. Xia, et al., *Adv. Energy Mater.* **4**, 1301846 (2014).

[2] K. Zeng, D.-J. Xue, and J. Tang., *Semicond. Sci. Technol.* **31** (6), 063001 (2016).

[3] Y. Zhou, L. Wang, S. Chen, et al., *Nat. Photon.*, **9** (6), 409 (2015).

[4] A. Ganose, C. Savory, D. O. Scanlon, *Chem. Commun.*, **53**, 20-44 (2017)

[5] V. Deringer, R. Stoffel, M. Wuttig, and R. Dronskowski, *Chem. Sci.*, **6**, 5255 (2015).

[6] X. Liu, J. Chen, M. Luo, et al., *ACS Appl. Mater. Interfaces* **6**, 10687 (2014).

[7] L. Wang, D.-B. Li, K. Li, *et al. Nat. Energy* **2**, 17046 (2017).

[8] R. Vadapoo, S. Krishnan, H. Yilmaz et al., Phys. *Status Solidi B* **248** 700-705 (2010)

[9] M. Leng, M. Luo, C. Chen, et al., *Phys. Lett.* **105**, 083905 (2014)

978-1-5090-5606-4/17 $31.00 © 2017 IEEE

Fabrication of Copper Arsenic Sulfide Thin Films from Nanoparticles for Application in Solar Cells

Scott A. McClary*, Joseph Andler**, Carol A. Handwerker**, and Rakesh Agrawal*

*Davidson School of Chemical Engineering, Purdue University, West Lafayette, IN 47907, USA
**School of Materials Engineering, Purdue University, West Lafayette, IN 47907, USA

Abstract — **Thin films of copper arsenic sulfide (Cu_3AsS_4) have been fabricated for the first time by annealing nanoparticles in an arsenic (V) sulfide atmosphere. The resulting films contained dense, micron-sized grains when treated at 425 °C, with the crystal structure transforming from the tetragonal luzonite phase to the orthorhombic enargite phase and an amorphous impurity phase appearing in the Raman spectrum. Solar cells were fabricated using CdS as the n-type buffer layer, yielding device efficiencies up to 0.24%.**

Index Terms – nanocrystals, thin-film solar cell, copper arsenic sulfide, novel materials, absorber layer, grain growth

I. INTRODUCTION

Traditional thin-film solar cell materials such as CdTe and $Cu(In,Ga)Se_2$ have achieved efficiencies surpassing 20%, but elemental scarcity makes scale-up to terawatt levels of production unlikely [1]. $Cu_2ZnSn(S,Se)_4$ has emerged as a promising candidate for solar applications, but recent research suggests that the material's ultimate performance is restricted primarily due to band tailing that causes significant V_{oc} deficits in devices [2]-[3]. Researchers are continuously searching for new materials to develop for photovoltaic applications, but finding useful phases that consist of economically viable elements is a difficult task.

Copper antimony sulfides are emerging as promising candidates for thin-film solar cells, with efficiencies exceeding 3% for $CuSbS_2$-based devices [4]. Copper arsenic sulfides are closely related to their antimony-bearing counterparts but have received significantly less attention. Recent computational studies have predicted direct bandgaps (1.2 eV – 1.4 eV) and high absorption coefficients for both the luzonite (LUZ) and enargite (ENG) polymorphs of the Cu_3AsS_4 composition [5]-[6], and both phases have been synthesized in nanocrystalline form [7]-[8]. To the best of our knowledge, however, a high density, contiguous thin film of Cu_3AsS_4 suitable for device applications has not yet been reported.

In this work, we describe a method to fabricate dense films of ENG Cu_3AsS_4 using LUZ nanocrystals (NCs) as the initial material. The thin films display well-rounded micron-sized dense grains. X-ray diffraction (XRD) confirms the presence of ENG as the dominant crystalline phase, though Raman spectroscopy suggests the formation of a secondary phase likely associated with excess sulfur bound to carbon residues in the film. The films were used as absorber layers in solar cells and achieved a champion device efficiency of 0.24%, a promising initial result for this class of material.

II. EXPERIMENTAL METHODS

LUZ NCs were synthesized according to a previous report from our lab [7]. They were dried and suspended in 1-hexanethiol, forming an "ink" which was then cast onto a molybdenum-coated soda-lime glass substrate *via* doctor blading. The substrate was then vacuum-sealed in a borosilicate glass ampoule with either one or a combination of As_2S_5, S, and As powders. The sealed ampoule was heated in a tube furnace; the powders inside vaporized and served as an annealing environment for the NCs. After cooling, the ampule was opened, and the substrate was either characterized or fabricated into a solar cell by chemical bath deposition of ~50 nm CdS, sputtering of ~80 nm ZnO and ~220 nm ITO, and e-beam evaporation of Ni/Al grids. Mechanical scribing isolated individual solar cells with areas of approximately 0.1 cm^2; these were each tested using J-V sweeps under simulated AM1.5G sunlight.

III. RESULTS AND DISCUSSION

In the well-studied $Cu(In,Ga)Se_2$ and $Cu_2ZnSn(S,Se)_4$ systems, a liquid-assisted grain growth mechanism during selenization is typically used to simultaneously transform pure sulfide NCs to their corresponding selenides and to facilitate the growth of large, densely-packed grains [9]. Inspired by this process, we sealed NC-coated substrates in an ampoule with added arsenic (V) sulfide. Upon heating to temperatures greater than 300°C, vapor may condense on the surface and form a liquid phase that aids in grain growth. Such arsenic-sulfur liquids can absorb small amounts of copper, resulting in a ternary liquid that is in equilibrium with Cu_3AsS_4 [10]. This environment also minimizes loss of As and S from the film.

Figs. 1a-1c show SEM micrographs of LUZ nanocrystal films treated with As_2S_5 for 1 h at 375 °C, 425 °C, and 500 °C, respectively. All three conditions resulted in nonuniform crystal growth on the surface. At 375 °C, a porous film composed of small randomly oriented grains was formed, while the 500 °C treatment yielded large, discontinuous patches of grains. Both films are unsuitable for solar cells due to the high likelihood of electrical shunting. In contrast, the

978-1-5090-5606-4/17 $31.00 © 2017 IEEE

425 °C heat treatment resulted in closer-to-equiaxed, unfaceted grains, which form a denser film suitable for further investigation.

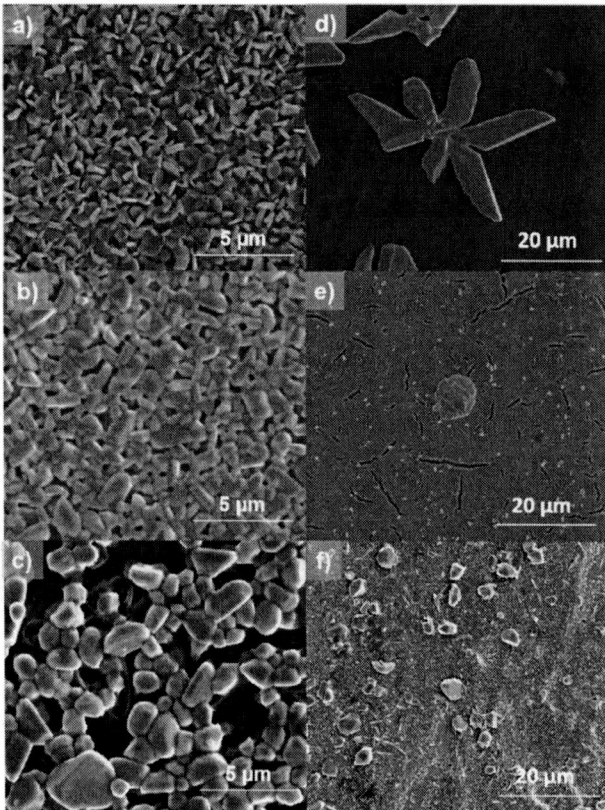

Fig 1. SEM micrographs of LUZ NCs treated for 1 h at the following temperatures and added powders: a) 375 °C, As_2S_5; b) 425 °C, As_2S_5; c) 500 °C, As_2S_5; d) 425 °C, S; e) 425 °C, As; f) 425 °C, elemental As and S in a 2:5 molar ratio.

Elemental sulfur and arsenic were compared with As_2S_5 as vapor sources for treating LUZ NC films. A sulfur atmosphere (Fig. 1d) at 425 °C facilitated the growth of large, heterogeneous grains, while an arsenic atmosphere (Fig. 1e) resulted in a few isolated grains and a highly cracked film. With the elements supplied in a 2:5 As:S molar ratio, the resulting film (Fig. 1f) contained two distinct grain sizes. By comparing the films annealed with individual elements with those for As_2S_5, it is evident that As_2S_5 produces more uniform, higher density films, both necessary for device-quality films from NC precursors.

The film treated at 425 °C for 1 h with added As_2S_5 was characterized using XRD (Fig. 2a). The diffraction peaks match the orthorhombic ENG phase (space group $Pmn2_1$), indicating a transition from the starting tetragonal LUZ crystal structure. Oriented growth along the (002) plane may explain the differences observed between the relative intensities of the obtained XRD pattern and its corresponding standard pattern.

Elucidating the selectivity between ENG and LUZ will be of great interest for future work, as both are promising for solar applications.

The Raman spectrum (Fig. 2b) of the film displays modes consistent with those of a reference spectrum of the mineral ENG. However, the major stretch at 494 cm^{-1} and minor shoulders near 360 cm^{-1} and 560 cm^{-1} do not align with any reported binary or ternary phase in the Cu-As-S system. The identity of this secondary phase is unknown, but a previous report attributes similar stretches to excess sulfur on the surface of carbon-containing NCs [11]. EDX measurements suggest a highly sulfur- and arsenic-rich film, likely due to the condensation of sulfur and arsenic vapor on the surface of the sample during cooling. This phase must be eliminated to minimize recombination centers for high-performance devices.

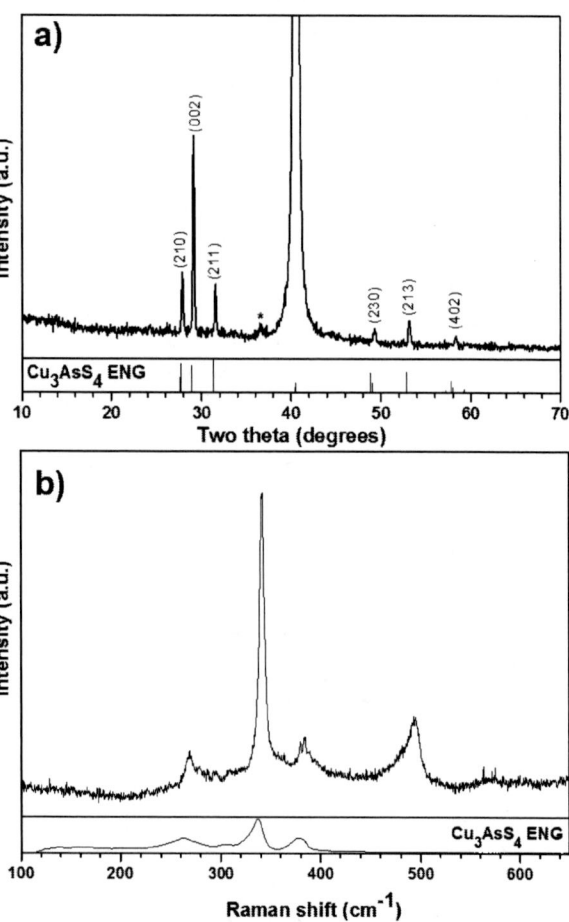

Fig 2. Characterization of film treated at 425 °C for 1 h with added As_2S_5. a) XRD spectrum showing peaks corresponding to an ENG mineral reference (JCPDS 01-082-1464). The peak at 40° is indexed to the Mo substrate and the peak at 36.5° (*) is its associated K beta peak from the Mo. b) Raman spectrum with modes agreeing with reference ENG mineral (RRUFF ID R050373) and additional modes at 360 cm^{-1}, 494 cm^{-1}, and 560 cm^{-1}.

To demonstrate the potential of using our ENG films in solar cells, we fabricated complete devices using them as the p-type absorber layer. Fig. 3 gives the device parameters (including an efficiency of 0.24%) and J-V curve of the champion cell under simulated AM1.5G sunlight. There is clearly significant work that needs to be done to further understand and develop ENG-based solar cells. The removal of the amorphous secondary phase and the growth of more densely-packed films will be critical steps towards improved device performance. Then, a detailed investigation into the electrical and optical properties of the cell will be conducted to identify the factors limiting cell performance and how they relate to the observed structures.

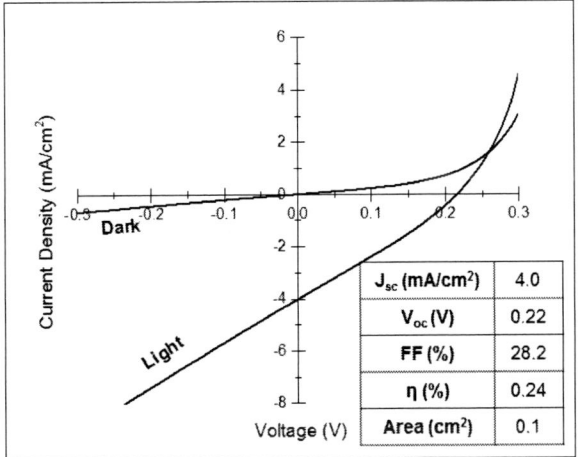

Fig. 3.   J-V curve of a solar cell using an ENG film as the absorber layer. The inset table gives the performance parameters of the device.

IV. CONCLUSIONS

We have fabricated the first thin films and solar cells based on ternary copper arsenic sulfide materials through careful selection of an annealing atmosphere that facilitates grain growth and mitigates elemental losses at high temperatures. Treatment of LUZ nanocrystals in As_2S_5 results in the formation of ENG grains and an amorphous secondary phase on the surface. The device performance (i.e., an efficiency of 0.24%) is encouraging for their further study. Removal of the pores and secondary phase will be critical for improving the quality of ENG-based solar cells, and then detailed characterization will be conducted to determine the parameters that most significantly impact device performance. These preliminary results establish the foundation for Cu-As-S-based solar cells and justify further research into this family of compounds.

ACKNOWLEDGEMENTS

This work was supported by the NSF through the DMREF Program under grant #1534691-DMR and the IGERT Program under grant #1144843-IGERT. The authors would also like to thank Brian Graeser, Robert Boyne, and Xianyi Hu for their expertise in preparing molybdenum-coated soda-lime glass.

REFERENCES

[1]   C. Wadia, a. P. Alivisatos, and D. M. Kammen, "Materials availability expands the opportunity for large-scale photovoltaics deployment," *Environ. Sci. Technol.*, vol. 43, no. 6, pp. 2072–2077, 2009.

[2]   T. Gokmen, O. Gunawan, T. K. Todorov, and D. B. Mitzi, "Band tailing and efficiency limitation in kesterite solar cells," *Appl. Phys. Lett.*, vol. 103, no. 10, p. 103506, 2013.

[3]   C. J. Hages, N. J. Carter, and R. Agrawal, "Generalized quantum efficiency analysis for non-ideal solar cells: Case of Cu2ZnSnSe4," *J. Appl. Phys.*, vol. 119, no. 1, p. 14505, Jan. 2016.

[4]   W. Septina, S. Ikeda, Y. Iga, T. Harada, and M. Matsumura, "Thin film solar cell based on CuSbS2 absorber fabricated from an electrochemically deposited metal stack," *Thin Solid Films*, vol. 550, pp. 700–704, 2014.

[5]   L. Yu, R. S. Kokenyesi, D. A. Keszler, and A. Zunger, "Inverse Design of High Absorption Thin-Film Photovoltaic Materials," *Adv. Energy Mater.*, vol. 3, no. 1, pp. 43–48, Jan. 2013.

[6]   T. Shi, W.-J. Yin, M. Al-Jassim, and Y. Yan, "Structural, electronic, and optical properties of Cu3-V-VI4 compound semiconductors," *Appl. Phys. Lett.*, vol. 103, no. 15, p. 152105, 2013.

[7]   R. B. Balow, E. J. Sheets, M. M. Abu-Omar, and R. Agrawal, "Synthesis and Characterization of Copper Arsenic Sulfide Nanocrystals from Earth Abundant Elements for Solar Energy Conversion," *Chem. Mater.*, vol. 27, no. 7, pp. 2290–2293, Apr. 2015.

[8]   A. Das, A. Shamirian, and P. T. Snee, "Arsenic Silylamide: An Effective Precursor for Arsenide Semiconductor Nanocrystal Synthesis," *Chem. Mater.*, vol. 28, no. 11, pp. 4058–4064, Jun. 2016.

[9]   C. J. Hages, M. J. Koeper, C. K. Miskin, K. W. Brew, and R. Agrawal, "Controlled Grain Growth for High Performance Nanoparticle-Based Kesterite Solar Cells," *Chem. Mater.*, vol. 28, no. 21, pp. 7703–7714, Nov. 2016.

[10]   S. Maske and B. J. Skinner, "Studies of the sulfosalts of copper; I, Phases and phase relations in the system Cu-As-S," *Econ. Geol.*, vol. 66, no. 6, pp. 901–918, Oct. 1971.

[11]   B. D. Chernomordik, A. E. Béland, D. D. Deng, L. F. Francis, and E. S. Aydil, "Microstructure Evolution and Crystal Growth in Cu2ZnSnS4 Thin Films Formed By Annealing Colloidal Nanocrystal Coatings," *Chem. Mater.*, vol. 26, no. 10, pp. 3191–3201, May 2014.

978-1-5090-5606-4/17 $31.00 © 2017 IEEE

Orientation Controlled Ge Thin Films on Glass by Al-Induced Crystallization

Kaveh Shervin, Khim Kharel, and Alexandre Freundlich

Center for Advanced Materials, University of Houston, Houston, Texas, 77204-5004, USA

Abstract — We have empirically investigated a low temperature (<400C) Al induced crystallization of amorphous Ge (a-Ge) layers deposited at room temperature on Al coated glass. The initial Ge nucleation stages were carefully examined by the optical and electron microscope. The X-Ray Diffraction approach was implemented to investigate the critical role of natively formed aluminum oxide (between the a-Ge and the Al) in driving Ge crystallizations and Al/Ge thickness ratio in setting the orientation and crystalline structure of subsequent Ge layers. In addition, the optoelectronic properties of crystalline Ge films were studied by photoluminescence (PL). Here, we show the playing role of layers thickness in crystal orientation. Subsequently, we demonstrate that by fine tuning of annealing conditions, the Al layer thickness and natively formed Aluminum Oxide, it is possible to yield a-Ge crystallization towards (110) orientation. Furthermore, we have shown for the first time possibility of obtaining a novel rhombohedric (100) Ge with a c-plane perpendicular to the interface offering an interesting option toward tuning of the Ge lattice constant to ~0.593 nm (instead of 0.565nm for c-Ge), closely matched to the lattice of InP and related materials.

I. INTRODUCTION

Single crystalline substrates account for more than 70% of the cost for high efficiency solar cells. Therefore, one strategy to reduce the production cost of highly expensive III-V Photovoltaics is to substitute the costly bulk substrates with thin films fabricated on low cost substrates such as glass. Ge as the both substrate and active bottom cell lattice matched to GaAs has derived a huge attention in the fabrication of multi junction III-V solar cells. It has been shown that highly orientation-controlled and large grain Ge polycrystalline (>10 microns) films could serve as templates for development of high efficiency inexpensive III-V Photovoltaics [1]. Al Induced Crystallization (AIC) approach, enabling low temperatures crystallization, has shown promising horizons to comply the critical requirements, large grains and orientation wise controllability, for photovoltaic industry. Therefore, a detailed understanding and highly controlled solid phase transformation is essential for applications in novel solid-state devices. AIC has been initially studied for the fabrication of inexpensive large grain poly-Si on glass [2-3] and subsequently they been investigated for the fabrication of inexpensive solar cells on glass or alternative substrates [4-5]. According these studies, whenever Si atoms are in contact with non-reactive metals such as Al the covalent bands between their atoms are weakened by the screening effect of metal free electrons [6]. The interfacial atoms freely diffuse most favorably into Al grain boundaries (GB) and start to nucleate beyond minimum critical thickness size. Hence, annealing at low temperatures, below their eutectic system at 577 °C [7] could result in Si crystallization. The Al and a-Si layer exchange is widely believed to occur when Al atoms are repelled from the Si crystals while growing out. The outcome of these studies has been relatively successful for the understanding of the exchange layer mechanism, yet not for controlling the crystal orientation and the use in solar cells.

Recently, AIC has derived an increasing interest into the fabrication highly textured large grain poly-Ge on inexpensive substrates in the similar way as Si [8-10].

The Al-Ge lower eutectic temperature, at 424 °C, suits even better for the fabrication of poly-crystalline films on glass based substrates with low melting point. Also, Ge has the eminent benefit of lattice matched to GaAs over Si for photovoltaic devices. In this regard, recently successful efforts have been reported on achieving (111) oriented large grain (>100 microns) Ge on glass. The preferential (111) orientation naturally achieved in Ge, is theoretically justified by its lowest free energy and smallest critical size [11]. However, (111) oriented large grain poly-Ge films do not serve the purpose for stacking lattice matched GaAs multi-junction solar cells on Ge. Therefore, we have investigated the parameters being able to alter the formation of normally achieved (111) preferential orientation.

We have empirically studied the critical role of interfacial oxide layer in severely improving the poly crystalline formation of Ge layer and examined the role of annealing time in improving the preferentially developed orientation.

The Al/a-Ge layer thickness in choosing the preferential crystal orientation has been vastly studied in this work. Characterization tools employed in this experiment include optical microscopy, scanning electron microscopy (SEM), X-ray diffraction (XRD) analysis, X-ray Photoelectron Spectroscopy (XPS), Transmission Electron Microscope (TEM) and Photoluminescence (PL). In the next section, we describe the details of our sample preparation procedures using thermal evaporation machine followed by the characterization and discussion section on the parameters critically controlling crystalline formation and orientation.

II. EXPERIMENTAL

In this experiment, we have used double-side polished fused quartz glass as substrates. A vacuum Thermal

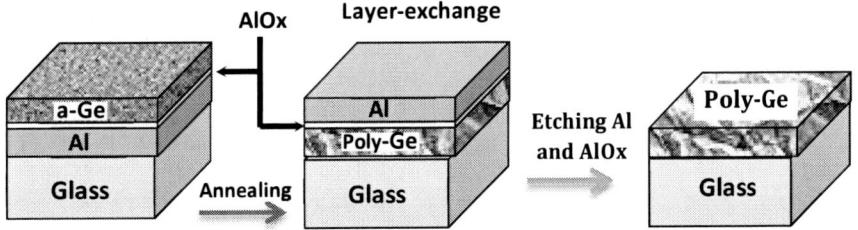

Fig. 1.   The schematic of the preparation of samples on glass substrates.

Evaporation system has been used for Al and Ge deposition at room temperature. Al layer thickness measured by Profilometer varied from 50 nm to 300 nm.

Following Al deposition, majority of samples were exposed in air between 30 minutes and several days for the formation of natively grown AlOx. The AlOx thickness was characterized by XPS from less than 1 nm to 4 nm. After AlOx formation, samples were loaded in the same chamber for Ge layer deposition. Ge was deposited at the same condition as Al with thicknesses varying from 50 nm to 200 nm. In the next stage, samples were annealed in N2 ambient using a quartz-tube furnace. The samples were annealed from 1 hour to 100 hours at temperature ranging from 250C to 400 C. Fig. 1 illustrates the schematic preparation of polycrystalline Ge thin films on glass substrates using AIC process. During the annealing, the Ge atoms initiate to recrystallize towards the formation of grains and consequently a poly-Ge layer, exchanged with Al, on the substrate. The Ge and Al atoms exchange process was visualized by an optical microscopy observing surface morphology and dendrites formation from both back and front surface. Fig. 2 shows the structure of the sample under study (100nm Al and Ge) and the onsite evolution of Ge dendrites. The images were taken from the back side of the sample showing Al/glass interface.

Fig. 2.   The evolution of Ge grains for the given annealing conditions; a-b) Ge dendrites formation stage with the inlet showing the sample structure c) grain boundaries formation d) poly-Ge covering the entire film .

OPTICAL CHARACTERIZATION OF CRYSTALLINE GE FILM

Following Al/Ge exchange layer process and Ge crystallization on glass, samples were studied to remove the Al layer and reveal Ge crystals for optical characterization. The selectively etching of Al layer from Ge is very challenging mostly due to the residual of Ge left in the Al layer and polycrystalline nature of formed Ge.

We managed to lift the Al layer off by removing the interfacial oxide layer underneath using dilute hydrofluoric acid (HF). Fig. 3 shows the poly-Ge gain boundaries on glass taken from sample surface after Al layer removal. The gains average sizes were measure about 25 μm.

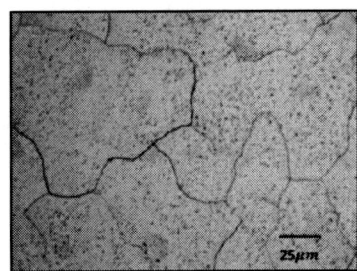

Fig. 3.   Top view of poly-Ge grain boundaries after Al layer removal, taken by optical microscopy.

absorbs the light with wavelengths smaller than its band Ge sintense light at wavelength a relatively emits gap and .(K10 tev a 0.74) coresposing to its bandgapIn order for the optical characterization of the polycrystalline Ge (after Al removal), a PL system equipped with 532 nm laser and InSb detector was operated at 895 mW power. Fig. 4 shows a peak corresponding to the light emitted about Ge ideal bandgap. The peak broadening could be justified by the presence of active recombination stages in Ge bandgap as a result of defects in Ge film. The blue shift in the bandgap may be related to the unintentional p-type doping by Al residuals left in Ge layer.

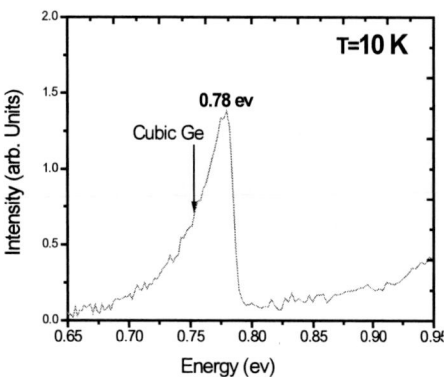

Fig. 4. The Photoluminescence scan of the sample with Al layer removed. The emitted light intensity is showed versus its energy. The ideal cubic Ge peak is marked on the graph.

III. (110) AND (100) CRYSTAL ORIENTATIONS

The a-Ge crystallization at the nucleation sites depends on the physical parameters including surface free energy and the energy of Ge atoms diffused. It is seen that Ge nucleation can take place within the layers of a-Ge, AlOx, bulk Al and at the AlOx/Al and Al/glass interfaces [12]. The crystal orientation in polycrystalline Ge is determined by the cluster critical energy and size where the amount of energy and number of atoms required for a specific crystal orientation is provided. Hence, by fine tuning of annealing conditions and layer thickness it is theoretically and experimentally possible to choose between nucleation sites thereby crystal orientation. In this regard, we have fabricated several samples varying in their Al and Ge layer thickness to investigate the role of layers thickness in altering Ge preferential crystal orientation. The sample with the same thickness of Al as Ge (~100nm) is kept as the reference sample. The approach is to keep the Ge thickness as the reference sample and change the Al layer thickness instead. The samples with 50 nm and 300 nm Al layers were fabricated to compare the role of Al layer thickness in relation with that of Ge. Fig. 5 compares the XRD results taken from the samples varying in their Al thickness compared to the reference sample previously discussed. The sample featuring thinner Al layer than the Ge has registered a outstanding (220) peak despite the very common (111) orientation achieved by the reference sample. However, the energy seems not be enough for the formation (100) orientation dominantly. On the contrary, the sample with thicker Al layer has demonstrated the formation of Ge in the tetragonal system, (200) rhombohedral.

We are preparing to use crystalline Ge films on glass as a template for Ge epitaxial growth using Molecular Beam Epitaxy (MBE) to study the possibility of improving Ge films and increasing their grain sizes. The results will be presented at the conference.

CONCLUSION

We have studied low temperature crystallization of amorphous Ge on glass using Al-induced crystallization. We characterized the optical properties of crystalline Ge films and empirically studied the role of Al/a-Ge layer thickness and showed the possibility of achieving (110) dominated cubic poly-Ge on Glass by fine tuning of annealing condition (specifically temperature) and layers thickness (more importantly Al). The dominated (110) direction as a heteropolar Ge surface could be particularly attractive to mitigate antiphase defect formation during the subsequent III-V heteroepitaxy. Furthermore, we have shown for the first time the possibility of achieving near single crystalline rhomboedric (100) Ge offering an interesting option in tuning of the Ge lattice constant to that of InP and related materials.

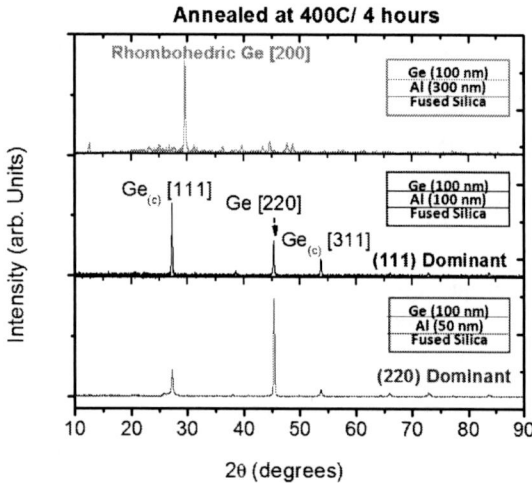

Fig. 5. The XRD results of the samples with the same Ge thickness (100nm) and varying Al thickness (50, 100 and 300 nm). The insert shows the sample structures on the right.

REFERENCES

[1] M. Kurosawa et al. Solid. State. Electron., vol. 60, no. 1, 2011.
[2] O. Nast et al. Appl. Phys. Lett., v. 73, no. 22, 1998.
[3] Z. Wang et al. Nano Lett., vol. 12, no. 12, , 2012.
[4] O. Nast et al. Sol. Energy Mater, vol. 65, no. 1, 2001.
[5] B. Rau et al., PVSEC-15, 2005.
[6] A. Hiraki, no. December, pp. 2662–2665, 1980.
[7] J. Schneider, Technischen Universität Berlin, 2005.
[8] L. R. Muniz et al. J. Phys. Condens. Matter, vol. 19, no. 7, 2007.
[9] Z. M. Wang et al. Acta Mater., vol. 56, no. 18, Oct. 2008.
[10] K. Toko et al. Suemasu CrystEngComm, vol. 16, no. 13, 2014.
[11] R. J. Jaccodine et al. J. Electrochem. Soc., vol. 110, no. 6,1963.
[12] H. Kim et al. Sol. Energy Mater, vol. 74, no. 1–4, 2002.
[13] F. Catalina et al., Mater. Sci. Eng. B, vol. 5, no. 4, 1990.

978-1-5090-5606-4/17 $31.00 © 2017 IEEE

In-line Potassium Fluoride Treatment of CIGS Absorbers Deposited on Flexible Substrates in a Production-Scale Process Tool

Ryan Kaczynski, JinWoo Lee, Jane van Alsburg, Baosheng Sang, Urs Schoop, and Jeffrey Britt

Global Solar Energy Inc., 8500 South Rita Road, Tucson, AZ 85747, USA

Abstract — **Record CIGS solar cell device efficiencies have been achieved recently by the utilization of alkali post-deposition treatments after the growth of the absorber. To realize the continuous goal of higher module efficiencies, these technological improvements must be implemented into manufacturing production lines. Global Solar Energy has investigated the key process variables necessary for the successful integration of a potassium fluoride treatment step into the existing Roll-to-Roll CIGS evaporation manufacturing process. Absolute efficiency improvements of nearly 1% were achieved with potassium-doped samples in conjunction with thinner CdS buffer layers. These gains were the result of elevated open-circuit voltages when an optimal amount of potassium resides near the interface with the buffer. A champion small-area device (1cm²) with an efficiency of 18.7% has been yielded by the production equipment that incorporates an in-line potassium fluoride processing step.**

Index Terms — **cadmium sulfide, CIGS, photovoltaic cells, potassium fluoride.**

I. INTRODUCTION

Record Cu(In,Ga)Se₂ (CIGS) device efficiencies on flexible substrates have been demonstrated by applying a post-deposition potassium treatment process to the absorber layer [1]. This effect has been widely studied on the laboratory-scale, but it must be implemented on manufacturing equipment lines to realize the cost benefits of higher large-area module efficiencies.

Global Solar Energy (GSE) attempted to replicate the concept of a distinct post-deposition potassium treatment process, but after some initial testing it was determined that a separate processing step was incompatible with production goals. The cost in terms of capital expenditures and reduced throughput was much too great. An in-line process that was adaptable to the current CIGS process tool was a necessity. This study encompasses our investigation into the key process variables for the integration of a potassium fluoride (KF) evaporation step into the manufacturing line to make high-efficiency solar cells.

II. EXPERIMENT

All tests were completed on the roll-to-roll production equipment at GSE with typical batch lengths of 1km [2]. The CIGS absorber layer is deposited onto a Mo-coated stainless steel web by a multi-source evaporator (Fig. 1) employing X-ray fluorescence (XRF) for compositional control. Cadmium Sulfide (CdS) is added via solution before the transparent conducting oxide (TCO) layer is sputtered under vacuum.

Small area devices (1cm²) were produced using the standard thin film structure (SS/Mo/CIGS/CdS/TCO) and a proprietary integrated cell interconnection (ICI) grid pattern. Samples used for the analysis of each test condition were removed from the web at pre-determined down web locations during a slitting process.

Fig. 1.   CIGS evaporation chamber at Global Solar Energy.

CIGS deposition follows the standard three-stage co-evaporation process flow with a sodium fluoride (NaF) precursor layer added before metal deposition (Fig. 2). For this battery of tests, evaporation sources were filled with equal amounts of potassium fluoride and then installed into the deposition zone following the third stage of CIGS growth. Each deposition zone allows for the evaporation of metals or alkali compounds, provides a source of selenium if necessary, and includes a means of heating the moving web via a backside substrate heater. A constant amount of selenium was supplied within the potassium treatment zone for all subsequent tests.

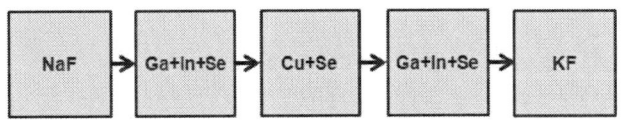

Fig. 2.   Standard CIGS deposition steps with KF treatment.

Potassium fluoride post-deposition treatments of CIGS have allowed for a significant reduction of CdS thickness without device performance degradation [1]. For all tests in this study, the standard buffer thickness was effectively cut in

978-1-5090-5606-4/17 $31.00 © 2017 IEEE

half to ~30nm. The influence of the CdS growth process on performance is an area that will be explored in the future.

There are three main process variables that can be investigated in regards to potassium treatments in CIGS: KF deposition rate, web temperature, and time. Time will not be tested in this experiment because it would require altering a well-established CIGS deposition process. The influence of the potassium deposition rate can be tested by changing the control temperature of the KF sources. The web temperature in the potassium treatment zone, which we will call the annealing temperature, can be controlled by the substrate heater temperature set-point.

III. RESULTS AND DISCUSSION

A. CdS Thickness

Since the main driver for the efficiency improvement of CIGS devices that incorporate a potassium treatment process is the ability to significantly thin the buffer layer, we first needed to find the appropriate CdS processing conditions that enhance performance. The resulting recipe was then used for all of the campaign test runs. The external quantum efficiency (EQE) plot in Fig. 3 shows a gain in the low wavelength spectral response typically associated with the buffer layer for those films made with the thin CdS recipe.

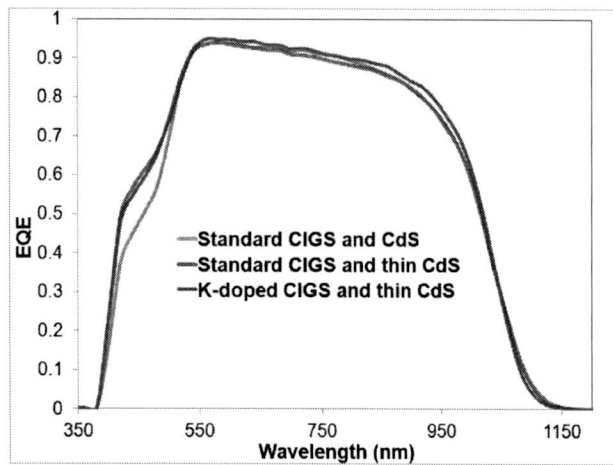

Fig. 3. External Quantum Efficiency plot comparing small-area devices made from various K-doped/thin CdS conditions.

The current-voltage (I-V) parameters were then analyzed to determine the effectiveness of the thinner CdS layer. Devices produced from thin CdS on standard CIGS (i.e. no potassium) showed the commonly observed losses in Fill Factor and open-circuit voltage (V_{OC}) resulting in reduced efficiencies. KF samples, which also include a thin CdS layer, displayed elevated efficiencies when compared to the standard material largely due to an improved V_{OC} as is shown in Fig. 4. Khatri et al. suggest that this improved open-circuit voltage is partially the result of an enhanced minority carrier lifetime [3].

An increase in carrier density is likely not sufficient for the gain in open-circuit voltage that is observed so an interfacial effect is also suspected.

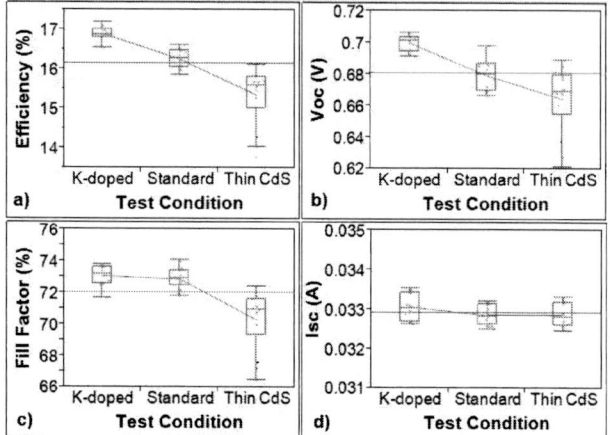

Fig. 4. I-V parameter comparisons of sample groups of small area devices made from the different thin film test conditions.

This inline potassium fluoride treatment process produced $1cm^2$ devices with an average efficiency of nearly 1% absolute greater than the standard CIGS process. Fig. 5 compares the I-V curves of the champion devices produced by the two methods. The electrical characteristics of these cells are listed in Table I along with the 18.7% device that was produced with a MgF_2 anti-reflective coating (ARC) layer. All of the manufacturing and testing procedures occurred in-house at Global Solar Energy, except for the deposition of the anti-reflective coating. This step was kindly provided by Solibro Research AB.

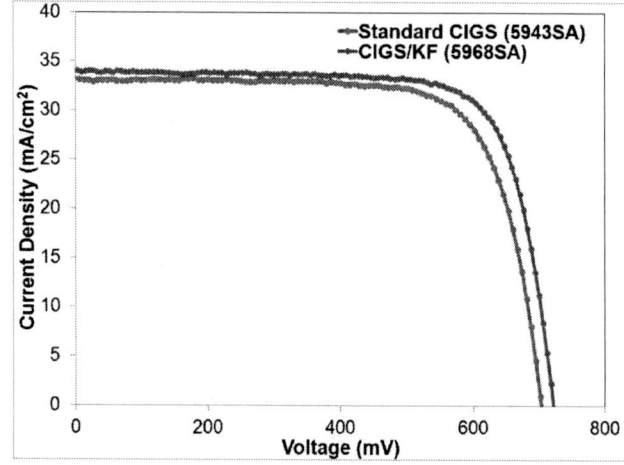

Fig. 5. I-V curves for the GSE champion devices of the standard CIGS/thick CdS process vs. the CIGS/KF/thin CdS process.

TABLE I
BEST IN-HOUSE 1CM² CELL DEVICE PARAMETERS

Cell	Voc (mV)	Isc (mA)	FF (%)	η (%)
Standard CIGS (5943SA)	704	33.0	74.0	17.2
CIGS/KF (5953SA)	721	33.8	75.1	18.3
CIGS/KF (5953SA) w/ARC	720	35.0	74.4	18.7
CIGS/KF (5968SA)	722	33.7	75.8	18.4

B. Potassium Fluoride Deposition Rate

The KF source temperature determines the effusion rate of potassium fluoride. Glow Discharge Optical Emission Spectroscopy (GD-OES) was used to measure the depth profiles of specific elements through the film. Measurements were made on completed solar cells. Fig. 6 shows the elemental depth profile of potassium within the CIGS layer at four different temperature set-points. KF Source Temp 4 represents the highest potassium fluoride deposition rate while KF Source Temp 1 is the lowest (i.e. T1<T2<T3<T4). The potassium accumulates near the CIGS/CdS interface and at the CIGS/Mo interface. The difference within the space charge region is significant in contrast to the back surface when comparing the highest and lowest source temperatures (the red and blue lines in the plot, respectively).

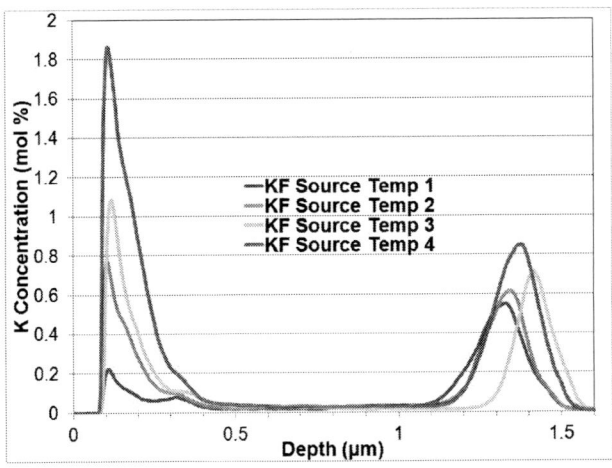

Fig. 6. GD-OES depth profiles of potassium through the CIGS absorber for several KF deposition rates (KF Source Temp).

Some device performance degradation is observed when excess potassium is incorporated into the CIGS film. Significant series and shunt resistance losses (Fig. 7) in these cells resulted in reduced Fill Factors and efficiencies (Fig. 8). This negative impact from excess potassium near the CIGS surface is likely the main culprit for the large variance of efficiencies for some KF/CIGS sample sets. Hence, the uniform distribution of potassium fluoride will likely be required to achieve high-efficiency large-area modules.

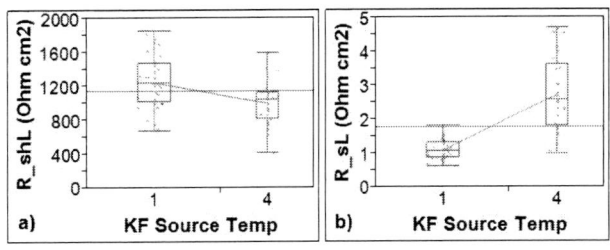

Fig. 7. Shunt (a) and series (b) resistance comparisons of low (KF Source Temp 1) vs. high (KF Source Temp 4) KF deposition rates.

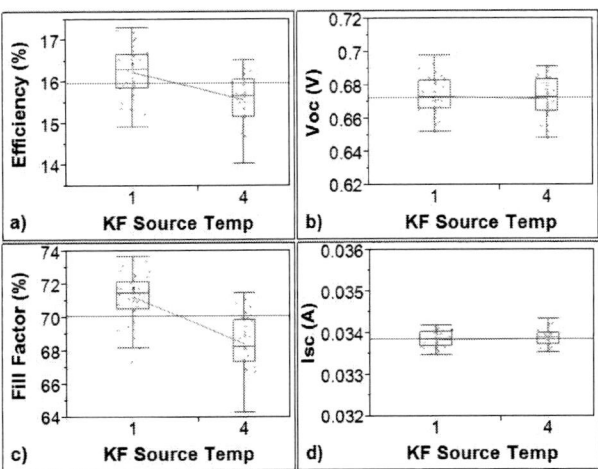

Fig. 8. I-V parameter comparisons of low (KF Source Temp 1) vs. high (KF Source Temp 4) KF deposition rates.

C. Annealing Temperature

The temperature and duration of the potassium treatment process are deemed to be critical components for realizing significant efficiency gains. In this set of experiments, the post-deposition processing time is fixed by an established CIGS deposition speed of 0.5 meters per minute. The annealing temperature (i.e. the web temperature during the potassium treatment process), which is in excess of 500°C, can be controlled by the substrate heater temperature set-point in the final zone. Fig. 9 shows that less potassium accumulates near the CdS interface and more settles at the CIGS/Mo boundary when the annealing temperature is increased under constant KF effusion source temperatures. Potassium build-up near 0.3μm is a process-dependent feature. Its location correlates with the transition from the second stage (copper-rich) to the third stage of CIGS growth.

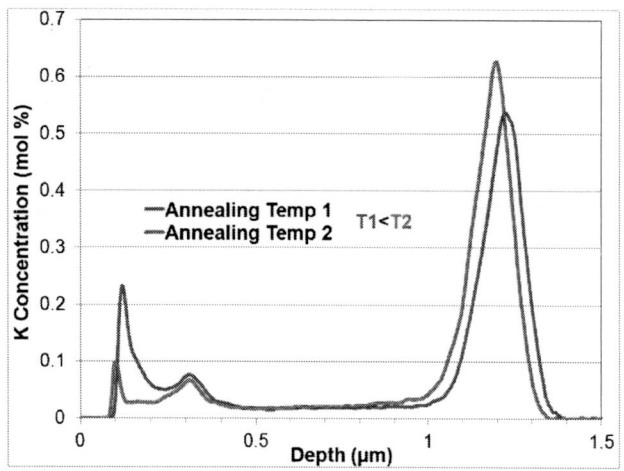

Fig. 9. GD-OES depth profiles of potassium through the CIGS absorber for two annealing temperatures at a constant KF rate.

A minimum amount of potassium near the CIGS/CdS interface appears to be required. Under the Annealing Temp 2 condition (i.e. higher substrate heater temperature), more potassium diffuses towards the Mo surface. Fig. 10 shows reduced device performance for this condition, which is similar to un-doped CIGS (i.e. no potassium treatment) with a thin CdS layer. This efficiency loss was not repeated when the same annealing condition was tested along with a higher KF source effusion rate.

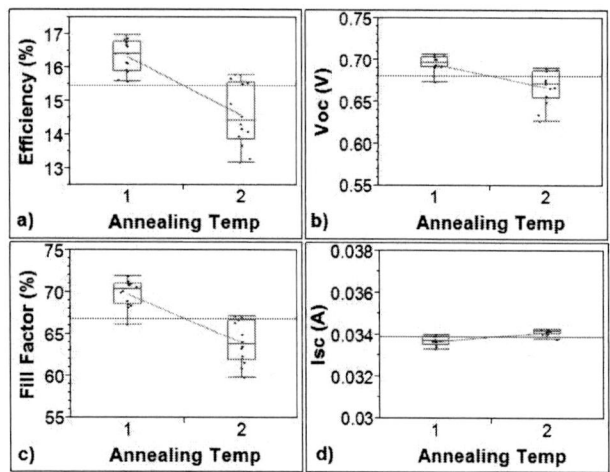

Fig. 10. I-V parameter comparisons of low (Annealing Temp 1) vs. high (Annealing Temp 2) temperature potassium treatment steps.

IV. CONCLUSIONS

This effort, in conjunction with other thin film processing improvements, has yielded a small area (1cm^2) device with an efficiency of 18.7%. An in-line potassium treatment step that produces CIGS devices with elevated efficiencies was shown to be possible, and no extensive scale-up is required since all of the testing was performed on the production equipment.

Many of the same effects on device performance due to the evaporation of KF that have been detailed in the literature were observed in this study of an in-line process. The efficiency improvements of K-doped samples with a thin CdS layer were the result of elevated open-circuit voltages. There is also a minimum potassium requirement for the effective thinning of the buffer layer, and excess potassium leads to device performance degradation.

REFERENCES

[1] A. Chirilă, P. Reinhard, F. Pianezzi, P. Bloesch, A. R. Uhl, C. Fella, L. Kranz, D. Keller, C. Gretener, H. Hagendorfer, D. Jaeger, R. Erni, S. Nishiwaki, S. Buecheler, and A. N. Tiwari, "Potassium-induced surface modification of Cu(In,Ga)Se$_2$ thin films for high efficiency solar cells," *Nature Materials*, vol. 12, pp. 1107-1111, 2013.

[2] J. S. Britt, S. Wiedeman, U. Schoop and D. Verebelyi, "High-volume manufacturing of flexible and lightweight CIGS solar cells," in *33rd IEEE Photovoltaic Specialist Conference*, 2008.

[3] I. Khatri, H. Fukai, H. Yamaguchi, M. Sugiyama and T. Nakada, "Effect of potassium fluoride post-deposition treatment on Cu(In,Ga)Se$_2$ thin films and solar cells fabricated onto sodalime glass substrates," *Solar Energy Materials and Solar Cells*, 155, pp. 280-287, 2016.

Light-Soak and Dark-Heat Induced Changes in Cu(In,Ga)Se$_2$ Solar Cells: A Macroscopic to Microscopic Study

Rouin Farshchi, Benjamin Hickey, and Dmitry Poplavskyy

MiaSole Hi-Tech Corp, Santa Clara, CA, 95051, USA

Abstract — We report the effects of light-soaking and dark-heat on Cu(In,Ga)Se$_2$ solar cells from the macroscopic level of module and cell performance down to the microscopic level of carrier concentration and sodium profile changes in the absorber layer. We find that solar cell performance degradation from each stress is driven by a fundamentally different mechanism. In particular, we find that the Lany-Zunger (L-Z) framework, namely an increase in carrier concentration associated with changes in the charge state of a copper-selenium divacancy complex in the absorber, can adequately explain light-soak induced degradation. Dark-heat induced degradation, on the other hand, is driven by an increase in voltage dependent collection (VDC). Atomic redistribution of sodium at various stages of light-soak and dark-heat is reported, and its potential impact on the observed device performance is discussed.

I. Introduction

While Cu(In,Ga)Se$_2$ solar cell efficiencies continue to steadily improve in recent years [1], challenges still remain in understanding the pathways through which their performance can change during exposure to stresses inherent to operation in the field. In particular, changes in solar cell efficiency during long term light-soaking as well as thermal stresses have been topics of considerable interest for academic and industry researchers [2-6]. While empirical modeling of solar panel performance under such stresses is crucial for predicting long-term field performance, a microscopic understanding of the mechanisms that drive the performance changes is critical for devising new synthesis recipes that mitigate or even eliminate stress-dependent metastabilities.

In this report, we analyze the effects of various degrees of light-soak and dark-heat stresses on Cu(In,Ga)Se$_2$ multi-cell solar modules, single cell modules, and small area devices in the module packaging. The modules provide information on how the J-V performance is affected by each stress and are also analyzed for changes in sodium profiles in the absorber layer. The small area devices are used for capacitance measurements and provide information on carrier concentration changes in the absorber layer resulting from the different stresses. By examining this data in aggregate, we hope to make progress towards identifying the microscopic mechanisms driving performance changes resulting from light-soak and dark-heat stresses. Whether changes in sodium profiles and/or carrier concentration are responsible for the observed performance changes is an ongoing question that we aim to address herein.

II. Experimental Methods

Cu(In,Ga)Se$_2$ solar cells were fabricated in a production scale roll-coater tool consisting of multiple chambers for sputter deposition of each layer of the device stack in a roll-to-cell process. The following layers are deposited on top of a flexible stainless steel substrate: metal back electrode, Cu(In,Ga)Se$_2$ absorber layer, CdS buffer layer, TCO window layer, and a metal grid top electrode. Solar modules were fabricated by laminating the cells in transparent barrier layers, either as strings of 22 cells (N22 modules) or as single cells (N1 modules). Small area capacitance devices were prepared by evaporating circular Ni/Al contacts directly on the TCO layer to serve as the top contact, and isolating the device via scribing to ~2 mm^2 areas. These devices were subsequently laminated into mini-modules using the same materials and processes as the larger modules for protection against ambient moisture degradation during light-soak and dark-heat stresses. Six N22 modules, fourteen N1 modules, and six mini-modules (two devices per mini-module) were used in this study.

Light-soaking (LS) was carried out at a simulated irradiation of 1 sun for 24 and 48 hours, followed by dark-heat (DH) exposure carried out at 85 °C at ambient relative humidity of about 30% for durations of 21, 150, 300, and 1000 hours. Exceptional properties of moisture barriers used in our module packaging ensure that degradation of modules in our study is driven by heat-induced changes in the cell properties rather than moisture ingress [7]. Current-voltage measurements were carried out under 1 sun irradiation at room temperature for N22 and N1 modules, denoted as "light" illumination conditions in the text, and also without illumination, denoted as "dark" conditions, for the N1 modules only. After each stress, one N1 module and one mini-module were removed from the stress sequence. The removed N1 modules were analyzed using secondary ion mass spectrometry (SIMS) to measure sodium profiles through the stack, while the devices in the removed mini-modules were analyzed using room temperature drive-level capacitance profiling (DLCP) [8,9] to measure carrier concentration in the absorber layer.

III. Experimental Results

We will first focus on the J-V performance of the N1 and N22 modules after the applied light-soak and dark-heat

978-1-5090-5606-4/17 $31.00 © 2017 IEEE

stresses. Figure 1 shows changes in open-circuit voltage (Voc) and fill-factor (FF) after each applied stress with respect to the initial (pre-stress) values, where the stresses were carried out sequentially. The N1 modules were measured under both "light" and "dark" illumination conditions. For the "dark" illumination condition, the Voc and FF values were obtained from the Jsc-shifted dark J-V curves [10,11]. In particular, the measured dark J-V curves were fitted using a 2-diode model, corrected for series and shunt resistances (extracted from the fit), and then shifted by Jsc from the "light" measurement into the 4th quadrant. Changes in Jsc with stress were negligible in this experiment, and therefore not discussed here. Comparison of light and Jsc-shifted dark J-V parametrics provides a measure of the voltage-dependent collection effect in the device stack [10,11], as described in more detail below.

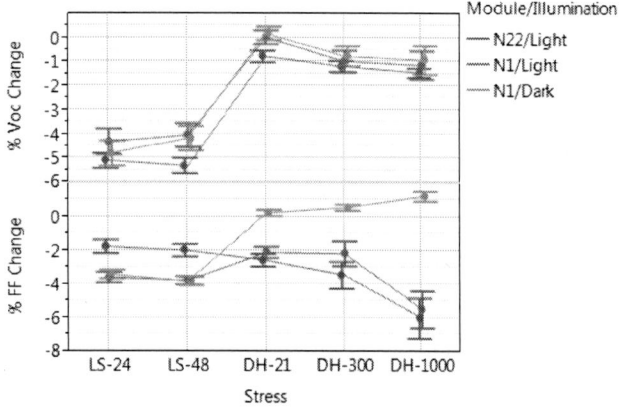

Fig. 1. Mean values of changes in open-circuit voltage (top) and fill-factor (bottom) after various durations of light-soak and dark-heat stresses, measured under various illumination conditions for N22 and N1 modules. The changes are calculated with respect to the initial state.

It can be seen that upon 24 hours of light-soak (LS-24), a strong Voc loss of 4-6% occurs for both N22 and N1 modules, for both illumination conditions of the N1 modules, which remains essentially unchanged after another 24 hours of light-soak (LS-48). This Voc loss is accompanied by a FF loss for both module types and both illuminations. After dark heat exposure for 21 hours (DH-21), Voc almost completely recovers for all three cases. Fill-factor, on the other hand, only fully recovers for the "dark" N1 module measurements, whereas the "light" N1 and N22 measurements still have more than 2% FF loss after DH-21 compared to the initial state. After prolonged dark heat exposure for 300 and 1000 hours, Voc changes only slightly (~1%) for both module types and both illuminations. Remarkably, the FF continues to drop from 2-3% loss at DH-21 to >5% loss at DH-1000 for the N22 and N1 "light" measurements, while the N1 "dark" measurement shows a gradual improvement of FF over the same prolonged dark-heat duration.

Several observations can be made at this point to help distinguish the mechanisms for light-soak and dark-heat degradation. Light-soaking leads to Voc and FF losses for both dark and light illumination conditions, indicating that the diode has fundamentally changed with regards to the extent of recombination currents present at forward bias. This type of behavior is consistent with the Lany-Zunger framework, where light-soaking is theoretically predicted to change the charge state of the copper-selenium divacancy complex (V_{Cu}-V_{Se}) from a compensating shallow donor state to an acceptor state [12,13]. Indeed, DLCP measurements on devices that were light-soaked for 24 hours show a significant increase (5 fold) in carrier concentration that fully relaxes after DH-21, as shown in Fig. 2. According to simulations discussed later, a 5 fold increase in carrier concentration in the absorber layer can indeed lead to reductions in both Voc and FF for both light and dark illumination conditions. Interestingly, the carrier concentration only decreases slightly from DH-21 to DH-300, similarly to the only small Voc changes observed for the modules over the same stress period.

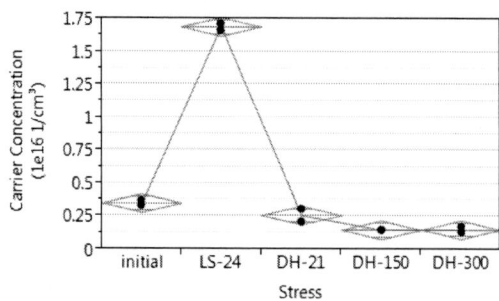

Fig. 2. Carrier concentrations in the absorber layer after various light-soak and dark-heat stresses measured by DLCP.

The other important observation that can be made from the data shown in Fig. 1 is that the fill-factor loss associated with dark heat is only observed under light illumination conditions and not for the dark conditions. As described above, the FF values for the dark conditions are taken from the Jsc-shifted dark J-V curves. This shifting of the J-V curve by Jsc assumes that the photo-generated current collected at short-circuit conditions will also be collected at all forward-bias conditions. In reality, as the electric field in the depletion region is reduced upon going into forward bias, the ability of the junction to collect photo-generated carriers, namely the "collection efficiency", is also reduced, leading to lower Voc and FF values than predicted by shifting the dark J-V curves. This phenomenon is known as voltage-dependent collection, or VDC [14]. An increase in VDC, as observed here with dark-heat exposure, could originate from increased recombination in the depletion region (increased defect concentration) or possibly as result of increased energy barriers that impede current collection under light illumination.

978-1-5090-5606-4/17 $31.00 © 2017 IEEE   1460

Increasing VDC with dark heat has been previously reported for CdTe solar cells [15].

Among the microscopic properties in the space-charge region that can change with heat stresses and can affect the fill-factor through voltage-dependent collection, changes in sodium profiles are a potential candidate. As a positive ion, Na has been hypothesized to passivate native defects in the absorber layer that form during growth [16]. Figure 3 shows sodium profiles throughout the device stack for the initial (pre-stress) state, after 48 hours of light-soak, after 21 hours of dark-heat, and after 1000 hours of dark-heat, where these stresses were applied sequentially. An increase in sodium counts of ~20% relative to the initial state can be observed through most of the CIGS layer after 48 hours of light-soak, which partially relaxes after 21 hours of dark-heat. While this increase in sodium may have contributed to the observed increase in carrier concentration in the absorber layer shown in Fig. 2, previous studies have shown that sodium changes alone cannot account for the total carrier concentration changes with light-soak [3]. After 1000 hours of dark-heat, the sodium profile is ~30% lower in the absorber layer relative to the initial state. The removal of sodium may lead to an increase in VDC due to its potential role in passivating defects in the depletion region. However, direct measurement of the responsible defect and its changes with dark-heat would be necessary to corroborate this hypothesis. An increase in concentration of deep defects in the space-charge region of $Cu(In,Ga)Se_2$ solar cells upon damp heat exposure has been reported previously [6].

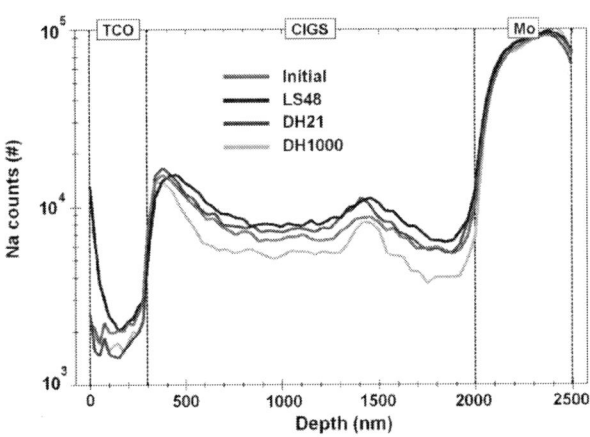

Fig. 3. Sodium concentration profiles at various stages of light-soak and dark-heat stresses measured by SIMS.

Simulations were carried out in SCAPS [17] to model the effects of light-soak and dark-heat on Voc and FF under light and dark illumination conditions. A summary of the basic SCAPS inputs can be found in reference [3]. As discussed above, according to the L-Z model light-soaking leads to the conversion of the V_{Cu}-V_{Se} divacancy complex from a shallow donor to a shallow acceptor, as well as emergence of a deep defect level that appears at ~1eV above the valence band [12,13]. We simulated the effect of a five-fold increase in shallow acceptor concentration (CCx5), consistent with DLCP measurements shown in Fig. 2, as well as the effect of an additional deep defect level 1eV above the valence band (DD), at equal concentrations to the shallow acceptors in the light-soaked state. Voc and FF changes from the initial state (pre-light-soak) for the CCx5 and DD conditions were derived under light (L) and dark (D) illumination conditions and reported in Table 1. For the dark simulations, J-V curves were shifted by Jsc (from the light simulation) into the 4th quadrant, allowing for extraction of Voc and FF. It can be seen from Table 1 that increasing carrier concentration by a factor of 5 in the simulation leads to a reduction of Voc and FF under both light and dark conditions, in qualitative agreement with experimental light-soak data shown in Fig. 1. However, the Voc losses of ~8% are larger than the 4-6% losses with light-soak observed in experiment, and the FF losses with CCx5 according to SCAPS are ~1%, hence smaller than the experimental FF changes of ~3-4% for the N1 modules shown in Fig. 1. By including the effect of the deep defect (DD) in addition to the increase in carrier concentration, Voc is not significantly affected. Fill-factor, on the other hand, drops down to a 2.5% loss for the light illumination case, while having no effect on the dark FF. The fact that differences between light and dark FF, namely the VDC effect, can be induced by the deep defect in the simulation further supports the possibility of additional deep defects forming with dark heat, where a clear VDC effect is observed experimentally in Fig. 1 that increases with dark anneal time.

TABLE I
SIMULATED VOC AND FF CHANGES IN SCAPS

Condition	%Voc change (L)	%Voc change (D)	%FF change (L)	% FF change (D)
CCx5	-7.9%	-7.8%	-0.7%	-1.4%
CCx5 + DD	-8.0%	-7.8%	-2.5%	-1.4%

IV. CONCLUSIONS

We show that light-soak and dark-heat stresses can lead to changes in solar device performance that are driven by fundamentally different mechanisms. Light-soaking leads to performance changes driven by Voc and FF changes that occur for both light and dark illumination conditions of the J-V measurement. This is consistent with an increase in carrier concentration with light-soak, as predicted in the Lany-Zunger framework and measured here using DLCP. The dark-heat stress, on the other hand, while reverting the Voc and carrier concentration changes incurred by light-soaking, leads to FF loss only under light illumination conditions of the J-V measurements. This is characteristic of voltage-dependant collection, or a loss in collection efficiency, as a result of dark-heat that is not observed with light-soaking. Reduction in sodium concentrations in the absorber layer after prolonged dark-heat can potentially explain the VDC effect, given that sodium is expected to passivate defects in the absorber layer. Light-induced interfacial blocking barriers that increase with dark-heat could also result in a VDC effect.

REFERENCES

[1] M.A. Green, K. Emery, Y. Hishikawa, W. Warta, and E.D. Dunlop, "Solar cell efficiency tables (Version 48)", *Progress in Photovoltaics*, vol. 24, pp. 905-913, 2016.

[2] T. Walter, "Reliability issues of CIGS-based thin film solar cells", *Semiconductors and Semimetals*, vol. 92, pp. 111-150, 2015.

[3] R. Farshchi, B. Hickey, G. Zapalac, J. Bailey, D. Spaulding, and D. Poplavskyy, "Mechanisms for light-soaking induced carrier concentration changes in the absorber layer of Cu(In,Ga)Se2 solar cells", *43rd IEEE Photovoltaic Specialist Conference*, pp. 2157-2160, 2016.

[4] S. Chen, T. Jarmar, S. Södergren, U. Malm, E. Wallin, O. Lundberg, S. Jander, R. Hunger, and L. Stolt, "Light soaking induced doping increase and sodium redistribution in Cu(In,Ga)Se$_2$-based thin film solar cells", *Thin Solid Films*, vol. 582, pp. 35-38, 2015.

[5] P. Mack, T. Walter, D. Hariskos, R. Schäffler, and B. Dimmler, "Endurance Testing and Accelerated Ageing of CIGS Thin Film Solar Cells", *24th European Photovoltaic Solar Energy Conference*, pp. 2439-2442, 2009.

[6] M. Schmidt, D. Braunger, R. Schäffler, H. W. Schock, and U. Rau, "Influence of damp heat on the electrical properties of Cu(In,Ga)Se2 solar cells", *Thin Solid Films*, vol. 361, pp. 283-287, 2000.

[7] T. Cao, G. Kimball, and T. Krajewski, "An in-situ method of monitoring flexible CIGS modules moisture induced degradation and lifetime prediction", *43rd IEEE Photovoltaic Specialist Conference*, pp. 854-856, 2016.

[8] G. Zapalac, K. Demirkan, N. Mackie, "Drive-level capacitance profiling of (Cu(In,Ga)Se$_2$ solar cells for different Cu/III ratios", *40th IEEE Photovoltaic Specialist Conference*, pp. 452-455, 2014.

[9] J. T. Heath, J. D. Cohen, W. N. and Shafarman, "Bulk and metastable defects in CuIn$_{1-x}$Ga$_x$Se$_2$ thin films using drive-level capacitance profiling", *Journal of Applied Physics*, vol. 95, p. 1000, 2004.

[10] L. M. Mansfield, R. L. Garris, K. D. Counts, J. R. Sites, C. P. Thompson, W. N. Shafarman, K. Ramanathan, "Comparison of CIGS Solar Cells Made With Different Structures and Fabrication Techniques," IEEE Journal of Photovoltaics, vol. 7(1), pp. 186-193, 2017.

[11] D. Poplavskyy, J. Bailey, R. Farshchi, and David Spaulding, "Impact of Ga/III Profile on Voltage-dependent Collection Losses in CIGS Solar Cells", to be published in Proceedings of the 44th IEEE PVSC (2017).

[12] S. Lany and A. Zunger, "Light- and bias-induced metastabilities in Cu(In, Ga)Se$_2$ based solar cells caused by the V_{Se}–V_{Cu} vacancy complex", *Journal of Applied Physics*, vol. 100, p. 113725, 2006.

[13] S. Lany and A. Zunger, "Limitation of the open-circuit voltage due to metastable intrinsic defects in Cu(In,Ga)Se$_2$ and strategies to avoid these defects", *33rd IEEE Photovoltaic Specialist Conference*, pp. 1-3, 2008.

[14] S. S. Hegedus and W. N. Shafarman, "Thin-film solar cells: device measurement and analysis", *Progress in Photovoltaics: Research and Applications*, vol. 12, pp. 155-176, 2004.

[15] S. Hegedus, D. Desai, and C. Thompson "Voltage dependent photocurrent collection in CdTe/CdS solar cells", *Progress in Photovoltaics: Research and Applications*, vol. 15, pp. 587-602, 2007.

[16] M. A. Contreras, B. Egaas, P. Dippo, J. Webb, J. Granata, K. Ramanathan, S. Asher, A. Swartzlander, and R. Noufi, "On the role of Na and modifications to Cu(In,Ga)Se$_2$ absorber materials using thin-MF (M=Na, K, Cs) precursor layers", *26th IEEE Photovoltaic Specialist Conference*, pp. 359-362, 1997.

[17] M. Burgelman, P. Nollet, and S. Degrave, "Modelling polycrystalline semiconductor solar cells", *Thin Solid Films*, vol. 361, pp. 527-532, 2000.

A New Model to Determine Installed System Cost and LCOE for ARPA-E's MOSAIC Micro-Concentrator PV Program

[1]Ran Fu, [1]Kelsey A.W. Horowitz, [2]Daniel W. Cunningham, [2]James Zahler

[1]National Renewable Energy Laboratory (NREL), Golden, CO 80401, United States

[2]Advanced Research Projects Agency – Energy (ARPA-E), Washington, D.C. 20585, United States

Abstract — **The U.S. Department of Energy's Advanced Research Projects Agency - Energy (ARPA-E) started a program called Microscale Optimized Solar-cell Arrays with Integrated Concentration (MOSAIC). The goal of MOSAIC is to leverage attributes of conventional flat-plate photovoltaic (PV) system configuration and concentrator photovoltaic (CPV) module technology to produce a new class of modules for residential, commercial, and utility-scale markets. This paper describes a collaborative effort between ARPA-E and the National Renewable Energy Laboratory (NREL) to develop a bottom-up system cost model for MOSAIC research teams. The model is a valuable design tool within the program, helping to validate the various technical approaches and compare future deployed system competitiveness versus incumbent technologies in the PV industry.**

Index Terms —**balance of system (BOS), bottom-up cost model, concentrator photovoltaic (CPV), levelized cost of energy (LCOE), solar energy.**

I. INTRODUCTION

Currently, three types of solar power systems exist commercially: solar photovoltaic (PV), concentrated solar power (CSP), and concentrator photovoltaic (CPV). Solar PV, which is the most widely deployed type of solar system, offers a clean, renewable source of electricity at a cost level that has been increasingly competitive. Although it has experienced dramatic growth and cost reductions in recent years, solar PV represents only 1% of U.S. power generation [1]. While today's single-junction, flat-plate modules are making impressive progress, new technologies that achieve higher efficiency and/or lower cost could potentially propel the growth of solar even faster. One way to increase the efficiency of solar PV systems is by using CPV modules, which use optical devices to concentrate sunlight onto a smaller, high efficiency solar PV receiver. This method allows the system to generate more power with a much smaller footprint. However, CPV only converts direct sunlight, not diffuse solar radiation (sunlight scattered by the atmosphere and clouds); therefore, CPV is only viable in a limited geographic range, namely the southwestern United States where direct sunlight predominates. CPV has not been widely adopted because of rapidly declining flat-plate prices, the added material cost of the receiver and solar tracking systems, more limited field experience with these systems [2]-[3] and the surplus overall solar market when CPV was introduced in the field. Led by the Advanced Research Projects Agency – Energy (ARPA-E),

the Microscale Optimized Solar-cell Arrays with Integrated Concentration (MOSAIC) program seeks to overcome these challenges and develop arrays of very small CPV systems (known as microscale CPV technology) that integrate more affordable materials and manufacturing techniques. In addition, MOSAIC seeks solutions that will utilize diffuse sunlight as well as direct sunlight in order to expand the geographic regions in which the benefits of CPV may be exploited cost-effectively [2].

In the MOSAIC program, project teams hope to develop microscale CPV systems that are similar in cost and size to conventional flat-plate solar PV systems but with greatly increased performance levels. To accomplish this goal, MOSAIC's multidisciplinary teams will leverage existing experience and expertise from conventional flat-plate PV, CPV, manufacturing, optical engineering, and material science to produce a new class of PV panels without increasing manufacturing costs. By designing CPV solutions for use in all three primary market sectors (residential, commercial, and utility-scale) and across a wide range of solar resources, the new PV technology will facilitate cost-effective deployment of solar power systems across a wide range of geographical locations, providing consumers with more opportunities to choose their own source of power generation and potentially reducing consumer electricity costs.

Past National Renewable Energy Laboratory (NREL) studies employed in-house bottom-up cost models to estimate PV system costs [4]-[5]. Some prior studies have also published data or top-down estimates of total installed system costs and/or the levelized cost of energy (LCOE) associated with different CPV designs [3]. Paap et al. provide qualitative discussion on how acceptance angle could influence system cost [6], but a bottom-up cost model allowing for quantitative evaluation of system cost drivers for dual-axis trackers with CPV has not previously been published. This paper introduces a bottom-up cost model of a dual-axis tracked ground-mounted system developed by NREL and incorporates our existing models of system costs for fixed-tilt and single-axis tracking systems as a function of efficiency and location. Further, all those cost models are integrated in the MOSAIC model. A simple LCOE calculator is also included. The resulting, integrated MOSAIC model is being used to support techno-economic analysis for MOSAIC awardees for their CPV applications (residential, fixed-tilt, single-axis tracker or dual-axis tracker). In Section II, we describe the cost model structure. In Section III, we review the LCOE results from the MOSAIC model. Finally, in Section IV, we present results using the model for two example systems: a fixed-tilt, residential rooftop system with internal tracking and a ground-mounted, utility-scale system with traditional dual-axis tracking. Throughout this report, we discuss how the model and results can be used to guide MOSAIC program research.

II. SYSTEM COST MODEL FOR MOSAIC

Our MOSAIC cost model includes installed costs for four different system types: fixed-tilt, residential rooftop systems; fixed-tilt, utility-scale ground-mounted systems; utility-scale, ground-mounted single-axis tracked systems; and utility-scale, ground-mounted dual-axis tracked systems. For the fixed-tilt and single-axis tracked cases, we assume that the total non-module costs are the same as associated traditional, flat-plate PV systems because the MOSAIC program targets BOS system designs with similar form factors to incumbent technology. Data from [4] on non-module costs as a function of efficiency and location (state level) are used for these systems in our model. In this paper, the key model functions, such as LCOE calculation and module efficiency impact on system costs are presented.

The dual-axis tracked system cost assumes a dual-axis tracker with a traditional pedestal architecture consisting of a column, tubing, racking, brackets, gear drive for dual-axis tracking, foundation, and a tracker control unit. The functional flow of the model is illustrated in Fig. 1. Users can plug in model inputs, including module characteristics (such as price, weight, dimensions, efficiency, and degradation rate), system configurations (such as acceptance angle, tracker area, first-year energy yield, system life, and operation and maintenance [O&M] costs), location (state level), and financial parameters (capital structure). In this case, efficiency is defined as the rated direct current (DC) module efficiency at Concentrator Standard Test Conditions (CSTC); however, alternative standards for rating efficiency, still under development [2], may eventually be more applicable to MOSAIC-type module designs. Intermediate values, such as wind speed and snow loading, are calculated based on these inputs to further determine engineering design factors including foundation depth, racking component quantity, and dual-axis tracking gear drive capacity. Subsequently, cost factors related to engineering design factors, such as foundation cost, racking cost, labor installation cost, and gear drive cost, are computed. The final outputs include total system cost categories and are grouped into module, tracker, structural balance of systems (BOS), electrical BOS, permitting, and interconnection as well as overhead for Engineering Procurement Construction (EPC) firms and the developer.

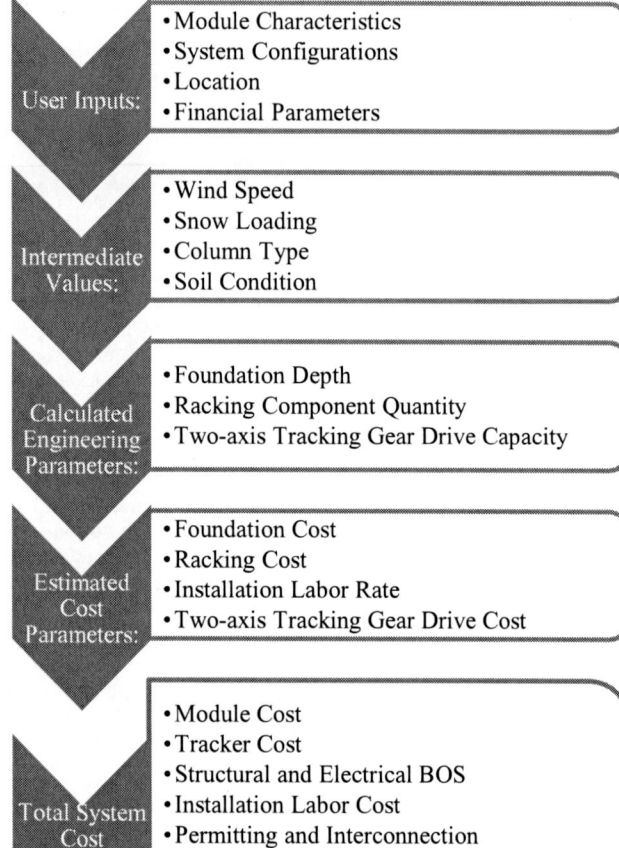

Fig. 1.   Model Process Flow.

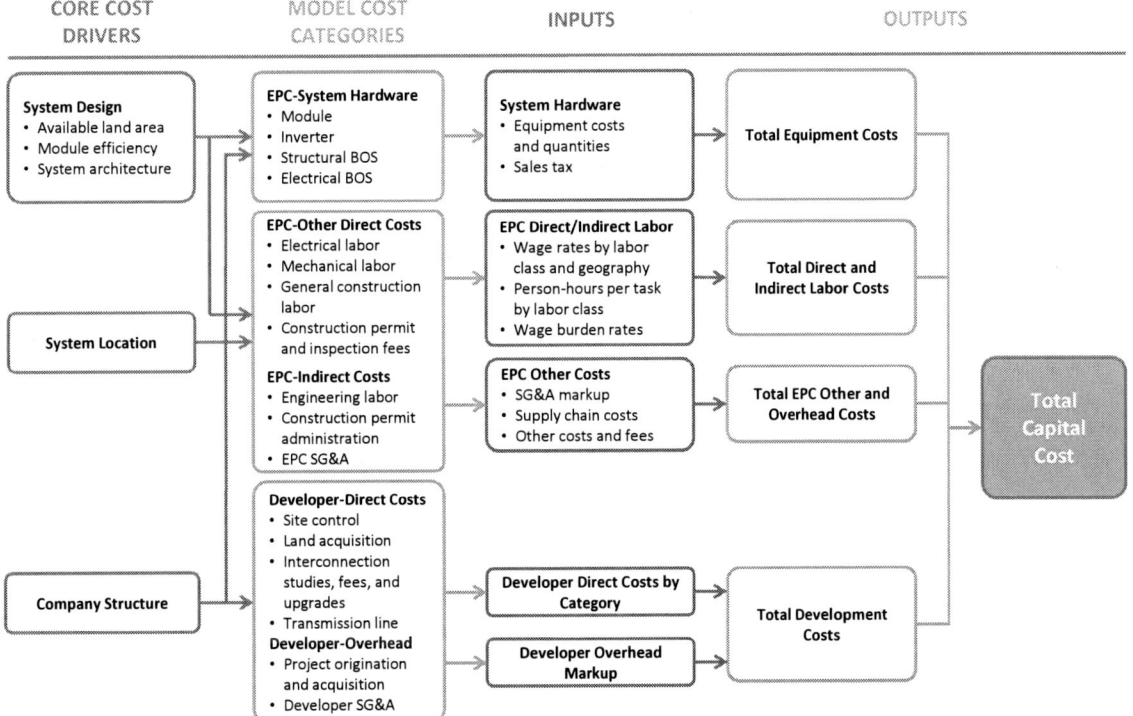

Fig. 2. Structure of the bottom-up cost model.

A detailed bottom-up cost structure of our model is presented in Fig. 2. Overall, total CPV system upfront capital costs are broken into EPC costs and developer costs. An experienced EPC is typically hired by a developer for construction tasks. EPC soft costs are driven by region-specific structural design criteria, such as wind speeds and snow loading; thus, the resulting differences in installed costs can vary significantly across the country. To incorporate these cost drivers, a structural design tool based on the American Society of Civil Engineers (ASCE) design code [7] and a construction cost estimating model are used to determine the EPC hardware costs (including racking, mounting, and foundation) and related EPC soft costs (including related labor and equipment hours required in any given U.S. location).

Developers typically use an internal rate of return (IRR) target or a specific power purchase agreement (PPA) to determine the project's net present value (NPV). Thus, the value of a CPV system, in terms of NPV, is dependent on different corporate strategies, such as capital structures, market competition, and local electricity rates. In this paper, a "cost approach" is used for CPV system installed cost ($/W), and an "income approach" is used for CPV system LCOE [4]. Key model inputs and assumptions are summarized in Table I.

TABLE I
UTILITY-SCALE CPV SYSTEM MODEL INPUTS AND ASSUMPTIONS IN 2016 [4]

Model components:	Model inputs:
Module	User input (for incumbent technologies, commodity average selling prices [ASPs] are used)
Inverter	$0.15/W, residential, fixed-tilt (rooftop)
	$0.09/W, utility-scale, fixed-tilt (ground-mount)
	$0.10/W, utility-scale, single-axis (ground-mount)
	$0.11/W, utility-scale, dual-axis tracker for CPV
	Note: Inverter prices are the same in $/Wac but different in $/Wdc when as DC-to-AC ratios vary.
Racking and foundation	Determined by wind speed, snow loading, and material cost index by state
Balance of system	Determined by the size of the CPV system
Installation labor	Non-union at rates taken from BLS statistics survey average by state
Sales tax (if any)	Determined by location (state level)
EPC overhead and profit	<10 MW, use 10%; >100 MW, use 8% 10~100 MW, use linear interpolation
Transmission line (gen-tie line)	<10 MW, use 0 mile; >200 MW, use 5 miles 10~200 MW, use linear interpolation
Land acquisition	$0.03/W
Interconnection	$0.03/W
Permitting (if any)	$500,000 for California; $250,000 for other states
Contingency	4%
Developer overhead	<10 MW, use 15%; >100 MW, use 10%; 10~100 MW, use linear interpolation

978-1-5090-5606-4/17 $31.00 © 2017 IEEE   1465

III. LCOE MODEL

Although LCOE ($/kWh) cannot capture all power systems or societal costs and benefits, it still provides a transparent metric for understanding costs and differences among technologies. Our MOSAIC cost model links results from the bottom-up system cost model to an LCOE calculator described by Eq. 1. This LCOE calculation involves simplified financial and performance modeling compared with NREL's System Advisory Model (SAM) or other tools used for planning specific installations or projects. This allows researchers to quickly and easily compare their designs to incumbent module types on an "apples-to-apples" technology basis without having to develop their own sets of financial assumptions, learn more complex tools, or, in general, deal with models that include detail beyond what is required at their stage of development and whose use could result in analysis errors.

$$LCOE = \frac{C_{upfront} + \sum_{n=0}^{N} \frac{O\&M_n}{(1+d)^n}}{\sum_{n=0}^{N} \frac{E_1(1-r)^n}{(1+d)^n}}$$

(1)

where:

- $C_{upfront}$ is the total upfront capital cost in $/W_{p(DC)}$
- n is the year index and N is the project life
- E_1 is the first-year energy yield, in kWh/kW_{p(DC)}$
- $O\&M_n$ is the total O&M expenses incurred in year n, in $/W_{p(DC)}$
- r is the degradation rate
- d is the discount rate, which we set equal to the weighted-average cost of capital (WACC).

The first-year energy yield is a key parameter to Eq. 1. This value is calculated by the MOSAIC teams and input into the LCOE model. This energy yield could be increased, by, for example, increasing the acceptance angle of the module in order to collect more diffuse light. Spectral effects and temperature effects must be carefully considered in the calculation of first-year energy yield. While typical multi-junction solar cells used in CPV designs have a lower temperature coefficient than c-Si cells, the temperature response of optics required for concentration and tracking (if internally tracked) for MOSAIC systems must be considered. Some optical systems can be quite sensitive to temperature and/or humidity. In addition, spectral effects at both the cell and module level on energy yield must also be considered. Alignment issues, which may also be coupled with temperature or spectrally dependent effects as well as soiling (which causes proportionally higher losses in concentrated component) should also be considered.

The input assumptions used in the LCOE calculator are shown in Table II. These are based on data for traditional flat-plate systems defined in 2016 [4].

TABLE II
DEFAULT INPUT ASSUMPTIONS USED IN THE LCOE MODEL [4]

Parameter:	Value:
Percent financing from debt	80%
Cost of debt	6.9%
Cost of equity	23.1%
Effective corporate tax rate	41.5%
WACC (calculated using the above)	7.84%
Federal investment tax credit (ITC) level	30%
State and local incentives or tax credits	None
DC-to-AC ratio	1.15 for residential rooftop, 1.2 for utility-scale single-axis tracked, 1.4 for utility-scale ground-mounted fixed-tilt, 1.2 for dual-axis tracked
Project life	30 years
Inverter lifetime	10 years
Inverter replacement cost	$0.15/W for residential rooftop, $0.10/W for utility-scale single-axis tracked, $0.09/W for utility-scale ground-mounted fixed-tilt, $0.11/W for dual-axis tracked

With the model options, we include the ability to compare LCOE for MOSAIC designs to current performance and cost for two incumbent technologies—monocrystalline silicon (c-Si) and cadmium telluride (CdTe)—in three locations—Phoenix, Arizona (high DNI), Kansas City, Missouri (medium DNI), and New York, New York (low DNI). First-year energy yield values for c-Si and CdTe in each of these locations were calculated offline using SAM and input to the MOSAIC LCOE model in the form of a look-up table.

IV. MODEL RESULTS

The MOSAIC cost model may be applied to a range of analyses useful for setting research targets and evaluating the potential of different designs, including assessment of cost-performance tradeoffs, sensitivity analysis, and LCOE comparison of a module design to incumbent technology. In this section, we provide examples of fixed-tilt, residential rooftop systems and ground-mount, dual-axis tracked systems.

A. Fixed-Tilt, Residential Rooftop Systems

CPV designs that incorporate internal tracking mechanisms do not require external tracking. Some MOSAIC projects are developing internally tracked modules with similar form factors as traditional flat-plate modules. These designs would incur the same installation costs as traditional flat-plate systems. For example, say a project team has modeled energy yield (energy produced per rated power of the module) and estimated O&M costs and degradation, and wants to know what module efficiencies and prices to target with their design. This team could calculate the dependence of LCOE on both efficiency and module price, as shown in Figure 3. Efficiency drives reductions in LCOE by decreasing non-module system costs in dollars per watt. Efficiency also reduces the cost of the module in dollars per watt, although the total module cost may be constant or increase depending on if and how area-related module costs were changed in order to achieve increased efficiency; in Figure 3, we vary both module efficiency and module price independently. For this illustrative example, we assume a first-year energy yield of 2,100 kWh/kW$_{p(DC)}$ in Phoenix, Arizona. Using SAM, we calculate the first-year energy yield for a residential rooftop system with traditional monocrystalline silicon (c-Si) panels to be 1,800 kWh/kW$_{p(DC)}$; for a utility-scale, ground-mount single-axis tracked system using c-Si panels in Phoenix, we calculate a first-year energy yield of 2,459 kWh/kW$_{p(DC)}$. Our input assumption lies in the middle of these values and represents a number that cannot be achieved with existing fixed-tilt, flat-plate designs but may be feasible if concentration and an internal tracking mechanism are used. Note that this does not correspond to an energy yield value computed for a particular MOSAIC design, and an understanding of the energy yield for designs in the MOSAIC portfolio is still being developed for each approach within the program. MOSAIC designs may have energy yields above or below this value.

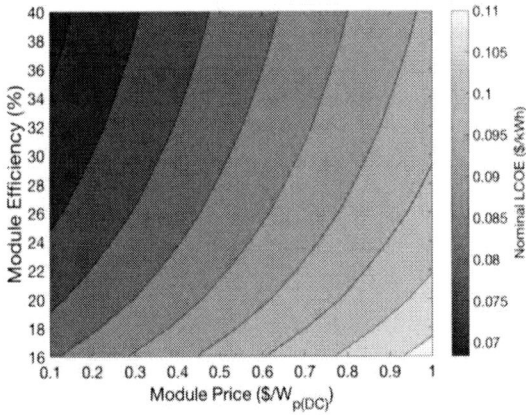

Fig. 3. LCOE versus module cost and efficiency for a fixed-tilt, residential rooftop system in Phoenix, Arizona, assuming a first-year energy yield of 2,100 kWh/kW$_{p(DC)}$, 0.5% degradation rate, 30-year system life, and O&M cost of $20/kW$_{p(DC)}$/year, including a federal ITC at 30%.

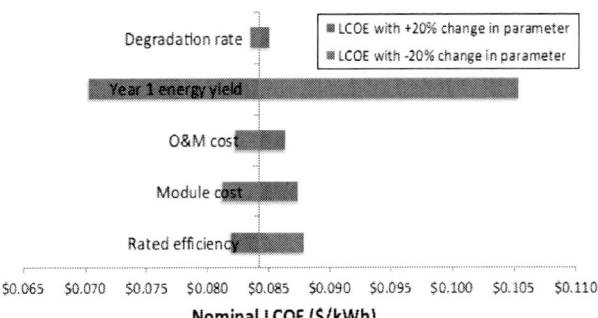

Fig. 4. Sensitivity of LCOE to ±20% changes in key input parameters for fixed-tilt residential rooftop systems. Base case corresponds to a $0.50/W$_{p(DC)}$ module price and 30% module efficiency, including a federal ITC at 30%.

As can be seen from Figure 3, LCOE savings from efficiency are non-linear and start to asymptote as efficiency increases. The curve of efficiency versus LCOE will evolve over time according to how area-dependent, capacity-dependent (rated power), and fixed non-module system costs evolve. Also, our previous Q1 2016 US Solar PV Benchmark [4] demonstrates the US LCOE maps by using the similar financial parameters but conventional flat-plate crystalline silicon PV system configurations.

The results shown in Figure 3 depend on the input values for first-year energy yield, degradation rate, O&M costs, and system life, in addition to financial assumptions. Figure 4 shows the sensitivity of LCOE to these input parameters. The base case—an arbitrarily chosen base case which could correspond to a set of possible values associated with an ultra-high efficiency module—assumes a $0.50/W$_{p(DC)}$ module price and 30% module efficiency. MOSAIC teams can use the model to run sensitivity analysis for their specific set of base case assumptions.

B. Ground-Mount, Dual-Axis Tracked Systems

Our bottom-up cost model for a dual-axis CPV tracker shows that one key driver of tracker cost is total loading on the array. This is determined by the dimensions of the array on the pedestal as well as the wind and snow loading in a location.

Figure 5 compares the sensitivity of non-module total system costs to module efficiency for dual-axis tracked and single-axis tracked system types. Module efficiency improvements have reduced the number of modules required to construct a system of a given size, thus reducing hardware costs as well as soft costs from direct labor and related installation overhead. This figure shows that the cost savings from high module efficiency is more significant for the dual-axis tracker than for the single-axis tracker. This result is caused by higher construction cost per column for dual-axis trackers than for single-axis trackers.

A key question for the MOSAIC program is the value of reducing the required tracking accuracy for a dual-axis tracker. As discussed in Section III, larger acceptance angles allow for the collection of some diffuse light and, in somcases,

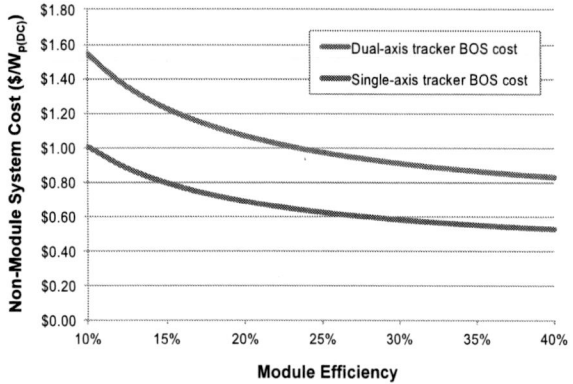

Fig. 5.   Non-module system costs versus module efficiency for dual-axis versus single-axis tracked systems.

additional circumsolar radiation, increasing energy yield and all else equal reducing LCOE. Depending on the location, diffuse light and circumsolar radiation are typically around 10% and 1% to 10% (higher in very hazy locations) of the total irradiance, respectively. The gains associated with increased diffuse light harvesting can be significant if a lower concentration ratios are used, for example by using low concentration onto a silicon bottom cell and higher concentration onto a III-V top cell. These energy yield benefits are calculated by the MOSAIC teams for their specific designs and then input to the MOSAIC cost model to determine the associated LCOE savings. Another potential benefit of increased acceptance angle, which can be estimated using the bottom-up, dual-axis tracker cost model developed here, is reduced tracker costs associated with the relaxation of tracking accuracy requirements. The required tracking accuracy depends on the acceptance angle of the module (the half angle) in addition to any accuracy losses resulting from misalignment. Our interviews with tracker companies and members of the CPV industry indicate that for tracking accuracies ≥0.5°, the structure is typically designed to withstand wind and snow loading in non-operating (stowed) conditions, and thus, the cost of the tracker is not reduced by further relaxations of the required tracking accuracy. For tracking accuracies between 0.2°–0.3° and 0.5°, you may be required to use a higher accuracy slew drive, increasing costs. For even smaller tracking accuracy requirements (≤0.2°–0.3°), the structural components of the tracker may also need to be reinforced to meet deformation criteria. In the latter case, the additional required material is dependent on the specific tracker and module design and requires detailed simulation, such as finite element analysis, to compute. These calculations are beyond the scope of this work but should be considered for

designs anticipated to have tight tracking accuracy requirements. Other possible benefits to larger acceptance angles include greater tolerance to gross misalignment due to installation or operational error of the trackers.

V. SUMMARY

A new bottom-up cost model developed by NREL and ARPA-E can be used for MOSAIC research teams to conduct consistent techno-economic analysis, including CPV installed cost and LCOE. For instance, tradeoffs between module cost and performance—including rated efficiency, energy yield, degradation, and acceptance angle—can be rapidly evaluated to guide research directions. This model is a valuable design tool within the program, helping to validate the various technical approaches and compare future deployed system competitiveness versus incumbent technologies in the PV industry on an apples-to-apples, technology basis.

ACKNOWLEDGMENTS

The authors would like to thank Geoffrey Kinsey, Sarah Kurtz, Michael Haney, and Zigurts Majumdar for insightful discussion and review of this work. The work was funded by ARPA-E in support of the Micro-scale Optimized Solar-Cell Arrays with Integrated Concentration (MOSAIC) program.

REFERENCES

[1] U.S. Energy Information Administration (EIA) Electricity Data Browser, Annual Net Electricity Generation in 2015. Accessed 1/23/2017. https://www.eia.gov/electricity/data/browser/.

[2] Advanced Research Projects Agency – Energy (ARPA-E), MOSAIC website. Accessed 1/23/2017. https://arpa-e.energy.gov/?q=arpa-e-programs/mosaic.

[3] S.P. Philipps, A.W. Bett, K. Horowitz, and S. Kurtz, "Current Status of Concentrator Photovoltaic (CPV) Technology." Golden, CO: National Renewable Energy Laboratory. NREL/TP-6A20-63196,2017. https://www.ise.fraunhofer.de/content/dam/ise/de/documents/publications/studies/cpv-report-ise-nrel.pdf.

[4] R. Fu, D. Chung, T. Lowder, D. Feldman, K. Ardani, and R. Margolis, "U.S. Solar Photovoltaic System Cost Benchmark: Q1 2016." Golden, CO: National Renewable Energy Laboratory. NREL/TP-6A20-66532, 2016.

[5] R. Fu, T. James, D. Chung, D. Gagne, A. Lopez, and A. Dobos, "Economic Competitiveness of U.S. Utility-Scale Photovoltaics Systems in 2015: Regional Cost Modeling of Installed Cost ($/W) and LCOE ($/kWh)." 42nd Photovoltaic Specialist Conference, New Orleans, LA, June 2015.

[6] S. Paap, V. Gupta, J. Luis Cruz-Campa, M. Okandan, W. Sweatt, B. Jared, B. Anderson, G. Nielson, A. Tauke-Pedretti, J. Nelson. "Cost Analysis for Flat-Plate Concentrators Employing Micro-Scale Photovoltaic Cells." 40th Photovoltaic Specialist Conference, Denver, CO, June 2014.

[7] American Society of Civil Engineers (ASCE), *Minimum Design Loads for Buildings and Other Structures (7-05)*, 2006

Fixed-tilt 660× Concentrating Photovoltaic System with 30% Efficiency

Alex J. Grede*, Jared S. Price*, Baomin Wang*, Michael V. Lipski*, Brent Fisher[†], Kyu-Tae Lee[‡],
Junwen He[§], Gregory S. Brulo[¶], Xiaokun Ma[¶], Scott Burroughs[†], Christopher D. Rahn[¶],
Ralph G. Nuzzo[§], John A. Rogers[‡] and Noel C. Giebink*
*Department of Electrical Engineering, The Pennsylvania State University,
University Park, Pennsylvania 16802 USA
[†]Semprius Inc. 4915 Prospectus Drive, Suite C, Durham, North Carolina 27713 USA
[‡]Department of Materials Science and Engineering, University of Illinois at Urbana-Champaign,
Urbana, Illinois 61801 USA
[§]Department of Chemistry, University of Illinois at Urbana-Champaign,
Urbana, Illinois 61801 USA
[¶]Department of Mechanical and Nuclear Engineering, The Pennsylvania State University,
University Park, Pennsylvania 16802 USA

Abstract—**High efficiency concentrating photovoltaics (CPV) can reduce the cost of photovoltaic power in a market dominated by balance of system and soft costs. Planar microtracking aims to provide a CPV solution for rooftops by laterally tracking a dense array of solar cells within a concentrator module having the form factor of standard PV panels. Here, we demonstrate a small-scale proof-of-concept system that reaches 30% power conversion efficiency at over 600× concentration, resulting in 54% more energy generated per unit area than a commercial silicon cell. This represents a milestone for planar microtracking CPV toward a larger scale, fully integrated panel prototype.**

Index Terms—**Photovoltaic systems, optics, optical design.**

I. INTRODUCTION

Commercial multijunction concentrating photovoltaic (CPV) modules are nearly twice as efficient as silicon-based panels. This increased efficiency can reduce the overall cost of an installation by generating more power since balance of system and soft costs increasingly dominate the overall cost of photovoltaic power.

Any high concentration CPV system must continuously orient itself toward the Sun, which is typically accomplished using precision dual-axis rotational trackers that support large module arrays. This is vastly different from the form factor of traditional flat-plate systems used on rooftops, which are compact, lightweight, and operate at fixed tilt. This difference in form factor and tracking mode limits deployment of traditional CPV to large open areas.

Using CPV technology while maintaining a flat-plate like form factor has brought designs with dense optical arrays coupled to individual cells [1], [2] or coupled to photovoltaic cells via a waveguide [3]. Small lateral translations maintain this coupling as the solar incident angle changes (Apostoleris, Stefancich, and Chiesa [4] reviews these and other methods). Planar tracking does suffer from the cosine projection loss of the flat-plate fixed tilt form factor, and the loss of diffuse light from the concentration optics, but in a sunny area where orientational tracking is not practical this method has promise.

Using a catadioptric lens mirror combination sandwiching a sliding middle sheet of microcells (similar to Fig. 1(a)), Price *et al.* [1] experimentally demonstrated a planar microtracking CPV system (μPT) with a wide acceptance angle ($\theta_{acc} = \pm 60°$), while maintaining a high concentration ratio ($C > 100\times$) Further theoretical calculations of μPT designs show even higher concentration ratios are possible while maintaining $>90\%$ optical efficiency (η_{opt}) over a $\pm 70°$ acceptance angle [5]. Figure 1(b) shows that the concentration ratio is strongly dependent on the refractive index the cell is embedded in. Concentration ratios can change by a factor of 2 by using an optic with a refractive index of 1.6 over 1.5.

To both demonstrate the capabilities of μPT and to accentuate design challenges for an eventual prototype tracking-integrated array, the previous optical system in Price *et al.* [1] needed a higher geometric gain, defined as the ratio of areas between the lens and cell, while maintaining a high optical efficiency and wide acceptance angle. Custom turned aspherical optics were used instead of trying to design with common plano-convex lenses and allowed for the use of high-index glass ($n \approx 1.9$). These optics resulted in a fixed-tilt CPV system with a geometric gain of 743×, $\eta_{opt} > 90\%$, and $\theta_{acc} = \pm 70°$.

II. RESULTS

The optical system shown in Fig. 1(a) is a custom lens made from N-BK10N glass ($n \approx 1.5$) forming the top optic while a mirror based on N-LASF31A ($n \approx 1.9$) forms the bottom optic. The high-index glass enables a higher concentration ratio for a simple two-surface optic as shown in Fig. 1(b). Additional broadband anti-reflective coatings limit reflection losses from the air/glass interface with an oblique-angle evaporated Teflon AF™ [6] and an Al_2O_3 coating between the index matching fluid and N-LASF31A glass. The mirror maintains high reflectivity over a broad spectral range with $SiO_2/Al_2O_3/Ag$ layers.

978-1-5090-5606-4/17 $31.00 © 2017 IEEE

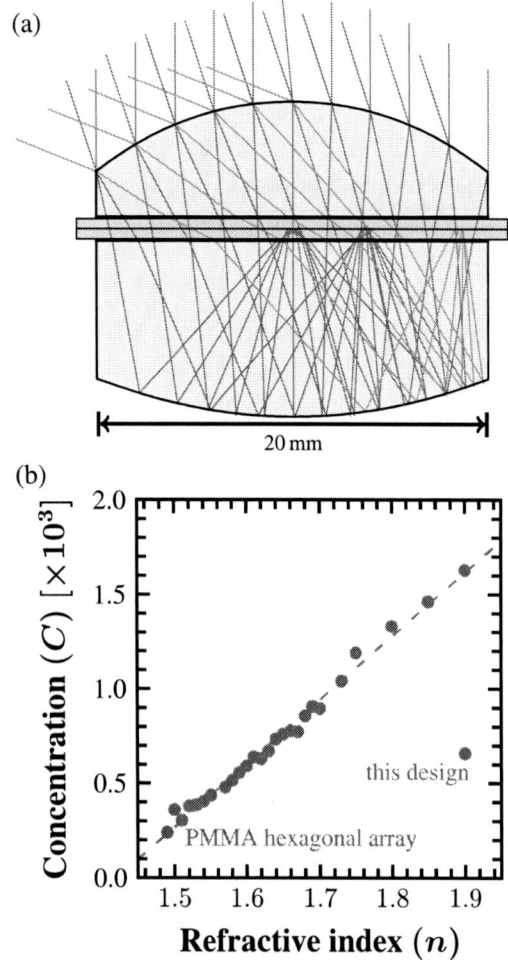

Fig. 1. (a) Planar microtracking (µPT) concept where top lens and bottom mirror sandwich a sliding middle sheet containing the cell. Translating the sheet maintains the cell at the focal point. (b) Concentration ratio vs. refractive index for $\theta_{acc} = \pm 70°$ for reference a PMMA based design and the design from (a) is plotted in red (figure adapted from [5]).

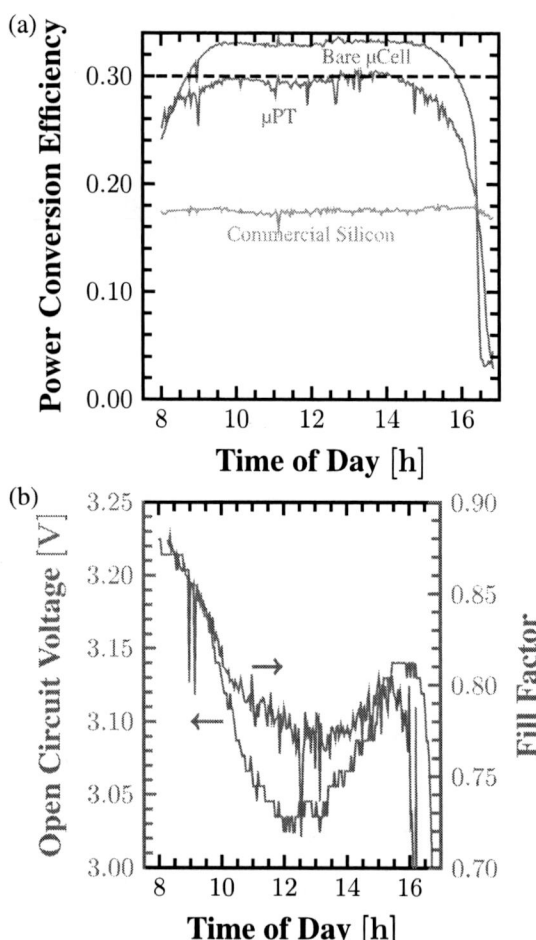

Fig. 2. (a) Power conversion efficiency of outdoor test. CPV reaches 30% and is below the bare cell. (b) Open-circuit voltage and fill factor for the same test. The drop in the open-circuit voltage is due to heating effects and fill factor drops due to series resistance losses. Figure adapted with permission [7]. Copyright 2017, Nature Publishing Group.

Outdoor testing was performed on Nov. 7, 2016 at a National Oceanic and Atmospheric Administration (NOAA) Surface Radiation (SURFRAD) monitoring site near State College, PA. SURFRAD instruments provide direct normal incidence (DNI) and diffuse global horizontal solar irradiance measurements for the power conversion efficiency calculations in Fig. 2(a). The incident DNI power was corrected for the cosine projection loss given the fixed tilt of the test jig (facing south at latitude tilt). The lower efficiency of the CPV compared to an unconcentrated reference bare cell is a result of the drop in open-circuit voltage and fill factor shown in Fig. 2(b). A very simple short-circuit current based closed-loop tracking algorithm sustained automated solar tracking throughout the day.

The cell maintained an efficiency of ~30 % for most of the day. Comparisons of the short-circuit current in the bare cell and CPV along with subsequent Zemax simulations indicate

a ~100 µm misalignment between the top and bottom optics.

The lower efficiency of the CPV cell relative to the bare cell resulted from heating of the CPV cell by 70 K above the ambient as measured by the 190 mV drop in open-circuit voltage, an open-circuit voltage temperature coefficient $-4.7\,mV \cdot K^{-1}$, a cell ideality factor of $n = 4.4$, and assuming minimal heating at the start of the day. Additional losses were from series resistance in the thin contact lines to the cell.

COMSOL simulations of the cell temperature as a function of cell size is shown in Fig. 3(a) using a geometric gain more in line with the gain expected for polymer optics (Fig. 1). The simulation suggests the heating scales as the area to perimeter ratio and a $400 \times 400\,µm^2$ cell would keep the temperature difference with the ambient below 15 K, enough to make a 1 to 2 point improvement in cell efficiency.

Mitigating the drop in fill-factor can be accomplished in a full array by sending most of the power through wider series connection traces, and any mismatch power through

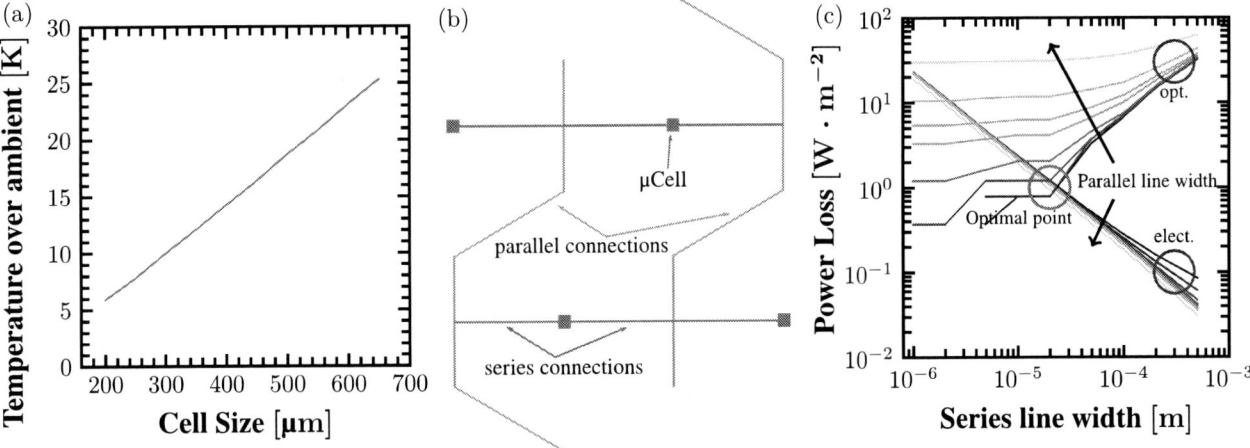

Fig. 3. (a) Simulated cell temperature dependence on cell size with the geometric gain fixed at 400×. Scaling the optical system can help mitigate heating effects. (b) Interconnect strategy to minimize series resistance losses by using thicker series connections for power transfer and thin parallel connections to minimize current mismatch issues between cells. (c) Electrical and optical (shading) loss as a function of the series resistance line width and the parallel line width. Both widths were varied from 1 μm to 500 μm where a 20 μm series line width and a 2 μm parallel resulted in the optimal configuration. Temperature simulations figure adapted with permission [7]. Copyright 2017, Nature Publishing Group.

parallel connections (Fig. 3(b)). Simulations of the optical loss in ZEMAX with varying series and parallel line widths are shown in Fig. 3(c) show the effect on optical loss due to shading. The electrical loss shown is assuming a 10 μm thick copper trace, using the unit cell length of 12 mm, a series current of 13 mA and parallel current of 1.3 mA. The optimal trade-off between shading (where narrower traces are better) and resistive loss (where wider traces are better) is found with a series width of 20 μm and a parallel line width of 2 μm.

Figure 4 shows power and energy density for the CPV and the commercial silicon cell along with a calculation for the same cell in the absence of heating based on the predicted open-circuit voltage with respect to current (Fig. 3(a)) and assuming a constant fill factor of 84%. The μPT design experimentally achieved a 54% higher power generation than a commercial silicon cell over the course of a day, which would increase to 73% (red curve) if heating and series resistive losses are eliminated.

An acrylic based hexagonal array of lenslets is expected to be completed later this year. Designed with a geometric gain of 300×, the distributions of optical efficiency predicted as a function of angle for various different possible fabrication and alignment errors are shown in Fig. 5. The distribution was calculated from a ZEMAX Monte Carlo lens tolerance simulation, using parameters from the mold manufacturer.

III. ANALYSIS

Individual optics are robust to misalignment of the top and bottom lenslets. A 100 μm misalignment between the top and bottom optics did not significantly impact the energy collected. This applies to an array as long as the misalignment does not change between elements. From a the tolerance simulation, the optical efficiency maintains a <5 % full-with at half-maximum spread out to 60° incidence angles.

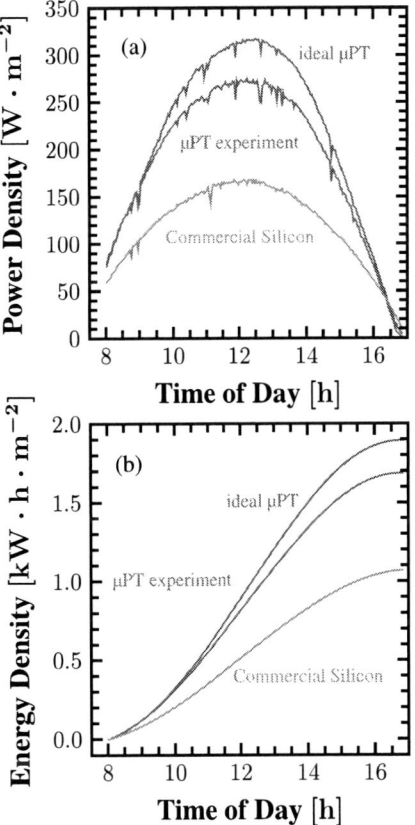

Fig. 4. (a) Output power density comparison between CPV cell with thermal and resistive losses mitigated, CPV cell as measured and 17% efficient commercial silicon cell. (b) Integrated energy density for the day. The CPV system generated 54% more energy than the silicon cell, and with design improvements to mitigate thermal losses could reach 73%. Figure adapted with permission [7]. Copyright 2017, Nature Publishing Group.

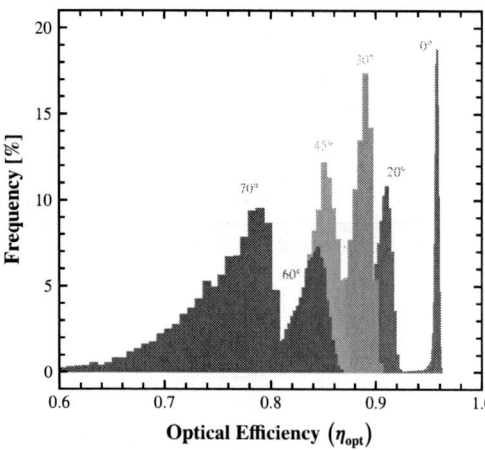

Fig. 5. Expected optical efficiency distribution of a molded acrylic hexagonal array at a geometric gain of 300×. Distribution is from a Monte-Carlo tolerance simulation using manufacturers tolerances. Figure adapted with permission [7] supplementary. Copyright 2017, Nature Publishing Group.

Primary loss mechanisms seem to come from a heat dissipation issue and series resistance losses. Heating related losses can be addressed by scaling the cell to a smaller area since the temperature change scales linearly with the side-length of the cell. Resistive and shading losses from the interconnects can be kept to under 1% by using wider contact traces, to lower the sheet resistance of the trace, and using series connections to transmit a majority of the power, lowering the current through the interconnects.

Scaling carries additional benefits in reducing the mass of the panel with thinner optics and in reducing absorption losses. This comes with some reduction in mechanical tolerances along with some added integration complexity.

IV. DISCUSSION

The measured concentration ratio of 660× results from the high-index glass and shows the capability of μPT based designs. As this design is scaled to an array of cells, reasonable manufacturing constraints suggest an acrylic based optic system at lower concentration (without sacrificing optical efficiency or acceptance angle) is necessary. This prototype highlights several considerations for a larger scale prototype panel.

The observed drop in open-circuit voltage, fill factor, and the subsequent thermal model suggest a 40 % drop in cell size with the 45 % drop in geometric gain is required by the lower refractive index of acrylic and by the passive cooling inherent to the design. This result also suggests that even at high gain and with solar cells surrounded by dielectric, passive cooling is still possible provided the cell size is small enough.

Mechanical tolerances of the test jig used in this experiment showed that the design is tolerant of some misalignment relaxing the assembly tolerances of the module. Further tolerance analysis of the molded optic array confirms this out to 60° incidence angle for an acrylic based lenslet array. Automated tracking using low-cost parts will undoubtedly be more challenging, but given the success of the simple hill-climbing algorithm used here, it remains tenable.

V. CONCLUSION

Planar microtracking based CPV systems can achieve a concentration and optical efficiency similar to orientational based CPV systems and are capable of exceeding the energy generation from silicon cells by 50%. While the fixed-tilt suffers an additional cosine projection loss over orientational tracking, the similarity in form factor to silicon flat-plate systems may allow μPT based CPV systems to operate on rooftops where conventional CPV cannot.

ACKNOWLEDGMENT

This work was funded in part by the Advanced Research Projects Agency-Energy (ARPA-E) MOSAIC program, U.S. Department of Energy, under Award No. DE-AR0000626 and by the National Science Foundation under Grant No. CBET-1508968. The views and opinions of authors expressed herein do not necessarily state or reflect those of the United States Government or any agency thereof.

REFERENCES

[1] J. S. Price, X. Sheng, B. M. Meulblok, J. A. Rogers, and N. C. Giebink, "Wide-angle planar microtracking for quasi-static microcell concentrating photovoltaics," *Nat. Commun.*, vol. 6, p. 6223, Feb. 2015.
[2] F. Duerr, Y. Meuret, and H. Thienpont, "Tailored free-form optics with movement to integrate tracking in concentrating photovoltaics," *Opt. Express*, vol. 21, no. S3, p. A401, May 2013.
[3] J. M. Hallas, K. A. Baker, J. H. Karp, E. J. Tremblay, and J. E. Ford, "Two-axis solar tracking accomplished through small lateral translations," *Appl. Opt.*, vol. 51, no. 25, p. 6117, Sep. 2012.
[4] H. Apostoleris, M. Stefancich, and M. Chiesa, "Tracking-integrated systems for concentrating photovoltaics," *Nat. Energy*, vol. 1, no. 4, p. 16018, Mar. 2016.
[5] A. J. Grede, J. S. Price, and N. C. Giebink, "Fundamental and practical limits of planar tracking solar concentrators," *Opt. Express*, vol. 24, no. 26, pp. A1635, Dec. 2016.
[6] B. Wang, J. S. Price, and N. C. Giebink, "Durable broadband ultralow index fluoropolymer antireflection coatings for plastic optics," *Optica*, vol. 4, no. 2, pp. 239, Feb. 2017.
[7] J. S. Price *et al.*, "High concentration planar microtracking photovoltaic system exceeding 30% efficiency," *Nat. Energy*, in press.

Wafer Integrated Micro-scale Concentrating Photovoltaics

Tian Gu[1,*], Duanhui Li[1], Lan Li[1], Bradley Jared[2], Gordon Keeler[2], Bill Miller[2], William Sweatt[2], Scott Paap[2], Michael Saavedra[2], Ujjwal Das[3], Steve Hegedus[3], Anna Tauke-Pedretti[2], Juejun Hu[1,*]

[1]Massachusetts Institute for Technology, Cambridge, MA, USA
[2]Sandia National Laboratories, Albuquerque, NM, USA
[3]Institute of Energy Conversion, University of Delaware, Newark, DE, USA
*gutian@mit.edu, hujuejun@mit.edu

Abstract — **Development of novel wafer integrated micro-scale photovoltaics is presented. Key notion is a multi-functional silicon platform hybrid-integrated with micro-scale multijunction cells. The IIIV-on-silicon platform simultaneously provides optical micro-concentration, hybrid photovoltaics for both direct and diffuse sunlight, and mechanical micro-assembly functionalities. A >100% improvement on the concentration-acceptance-angle product is demonstrated using the wafer-embedded micro-concentrating elements, leading to dramatically reduced module costs, sufficient angular tolerance to low-cost trackers, and an ultra-compact flat-plate form factor. Leveraging low-cost micro-fabrication and high-level integration techniques, the micro-scale PV approach seamlessly combines the high performance of multijunction cells and the low costs of flat-plate Si PV systems.**

I. INTRODUCTION

Solar energy has been growing fast in recent years with the rapid price falling of Si-wafer based photovoltaic (PV) technology, driven by technological advancements and scaling-up of deployment volume. As the efficiency of Si PV reaches its practical limit, balance-of-system (BOS) costs gradually become the dominant challenges for continued price reduction. High-efficiency, low-cost PV modules beyond Si are critical for further market penetration and can potentially enable new price learning curves of solar technology. By utilizing high performance multijunction cells and concentrator optics, concentrating photovoltaics (CPV) systems can in principle reduce energy production costs. However, while the cell performance has been advancing steadily recently, conventional CPV technologies are plagued by several issues that offset the potential cost effectiveness. Within the module level, trade-offs exist among the cell, optic, and module manufacturing costs; within the system level, trade-offs exist between the module performance and system installation/operation costs. Finally, the inability of conventional CPV to collect diffuse light further limits its geographic and market penetration.

By dramatically scaling down the dimensions of multijunction cells to the 100's microns regime, micro-scale PV integrate arrays of micro-cells and micro-optics within a compact module similar to flat plate Si PV using advanced cell fabrication approaches compatible with large-scale manufacturing [1-4]. Benefits of micro-scale solar cells include enhanced cell performance, reduced semiconductor and optic materials costs, interconnect flexibility, improved heat dissipation and a compact physical profile. However, simply miniaturizing traditional CPV modules and adapting

(a)   (b)

Fig. 1. Prototype micro-scale CPV modules developed in the Microsystem-enabled Photovoltaic program: (a) prototypical module integrating molded two-stage concentrating optics and micro-cell arrays; (b) fabricated micro-cell array and interconnects.

fabrication and assembly techniques optimized for macro-CPV is not a viable route to fulfilling the potential benefits of micro-scale PV. Particularly, high level integration of high performance micro-cells and micro-optics is demanded and the module fabrication and BOS costs must be fully compatible with Si PV technology. Through the Microsystems Enabled Photovoltaics (MEPV) program led by Sandia, we previously demonstrated several prototypical micro-CPV modules (Fig. 1) integrating a variety of advanced micro-cell and micro-optic arrays [1-3] and introduced an integrated hybrid micro-PV/CPV architecture [5].

In this paper, a new micro-scale PV approach developed under ARPA-E's *Micro-scale Optimized Solar-Cell Arrays with Integrated Concentration* (MOSAIC) program is presented, aiming to radically improve PV system's cost effectiveness by exploiting the cell/optic scaling effects. It utilizes III-V micro-cells integrated with a multifunctional Si platform to fully leverage the high performance of multijunction cells and module- and system-level benefits of Si flat-plate PV. The PV system designs are guided by a detailed cost model based on industrial-scale fabrication processes that analyzes and predicts energy production costs [6].

II. Wafer Integrated Micro-scale Photovoltaics (WPV)

A novel *Wafer Integrated Micro-scale Photovoltaic* (WPV) concept is described in this paper. As schematically illustrated in Fig. 2 (a), The key notion is a novel multi-functional silicon platform comprising high concentration multijunction micro-cell arrays hybrid-integrated on a 1X or low-concentration Si solar cell. The Si cell contains etched V-shaped reflective

978-1-5090-5606-4/17 $31.00 © 2017 IEEE

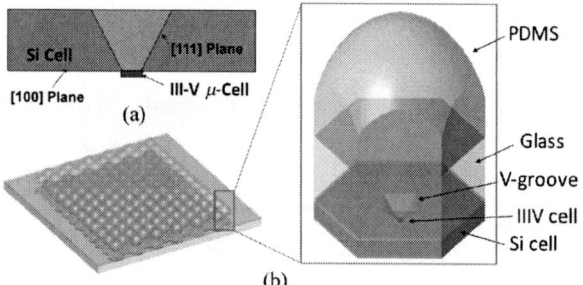

(a)

(b)

Fig. 2. (a) Wafer Integrated Micro-scale Concentrating Photovoltaics concept; (b) a baseline prototype module consists of a molded lens array layer, a middle glass plate, and a multi-functional Si cell integrated with a multijunction micro-cell array.

cavities that serve as efficient non-imaging micro-optical concentrators and/or alignment features embedded at the wafer level. By anisotropic-etching conventional (100) oriented silicon substrates to expose the (111) crystal planes, inverted-pyramid-shaped rectangular cavities are formed with facets of a 35.3° slanting angle and precisely-defined optical apertures matching the micro-scale cells. The Si cell is also designed to capture diffuse and off-alignment sunlight which usually contribute to major optical losses in conventional CPV systems. As depicted in Fig. 2(b), a baseline WPV structure consists of: 1) a Si cell platform embedding the reflective cavities and cell interconnections, 2) a multi-junction micro-scale PV cell array integrated on the Si platform, and 3) a primary concentrating optic array.   The Si platform with wafer level alignment features can be further integrated into a variety single- or multi-stage optical concentrator with a compact form factor.

The WPV approach thus seamlessly integrates hybrid photovoltaics, optical micro-concentration, and mechanical micro-assembly functionalities on an ultra-compact IIIV-on-Si micro-cell platform that benefits from both the high performance of multijunction cells and the low cost of flat plate Si PV infrastructures.

A. Optical micro-concentration

A key figure of merit for evaluating CPV systems is the concentration-acceptance product [7],

$$CAP = \sqrt{C_g} sin\theta_{in} \qquad (1)$$

where C_g and θ_{in} are the geometric concentration ratio and acceptance angle, respectively. Note that CAP is nearly an invariant for a given optical architecture due to the conservation of étendue. Hence Equation (1) reveals the trade-offs between concentration ratio and acceptance angles and accordingly the balance among materials, module, and system level costs. With limited CAP values close or below 0.5, state-of-the-art CPV technologies are typically designed for high concentration to reduce cell costs, at the expenses of complex module designs and tight tolerances to assembly and operation misalignments. For example, for concentration above 1000X, high-precision module assembly and high-accuracy trackers (<1°) are usually

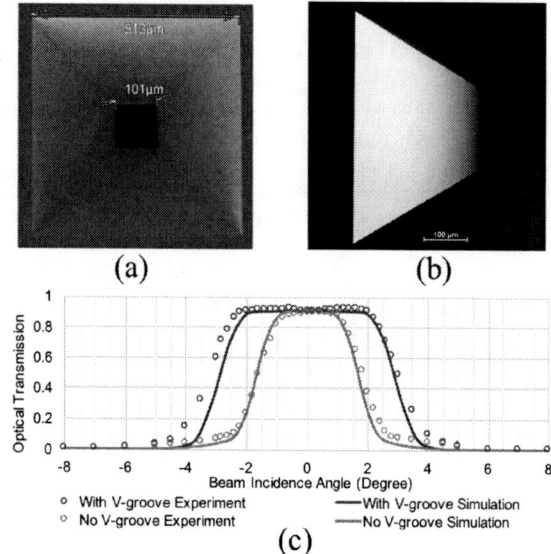

Fig. 3. (a) Top view and (b) side view of an etched Si V-shaped cavity; (c) experimental vs. simulation results indicate >100% improvement on acceptance angle and CAP by incorporating the low-profile Si cavity into the optical system.

necessitated, leading to significantly increased module fabrication and BOS costs that off-set the performance and cost improvements at the cell level.

The etched Si cavity plays a critical role in simultaneously improving the concentration ratio and acceptance angle of a micro-CPV module with minimum induced module complexity and costs. As shown in Fig. 3 (a) and (b), reflective V-groove cavity arrays with 100 μm × 100 μm output apertures designed to match micro-cell arrays are fabricated on a 280-μm thick Si wafer using KOH etching.   The optical performance is characterized by coupling the reflective cavity to an off-the-shelf primary concentrator.   As shown in Fig. 3 (c), experimental results indicate that a >100% improvement on the concentration-acceptance-product is achieved, when compared to the concentrator-only case. Enabled by such wafer-level integrated micro-concentrators, the simple baseline WPV architecture yields prototype designs with concentration ratio ranging between 400X and 2400X while maintaining sufficient angular tolerances (±1°~ ±2°) that are fully compatible with commercial low-cost trackers (1°~1.5° tracking accuracy).

B. Hybrid photovoltaics

The multi-functional Si platform in the WPV approach also enables a highly-integrated hybrid PV/CPV architecture at low cost. The diffuse radiation component of the sunlight (i.e., light scattered by atmospheric aerosols and clouds) constitutes a considerable portion of the total incident power, depending on location, but usually cannot be captured by conventional CPV systems. A hybrid PV/CPV architecture that combines high-performance micro-cells integrated in low-cost flat plate PV would dramatically improve overall power conversion

978-1-5090-5606-4/17 $31.00 © 2017 IEEE

efficiency. Standard solar radiation data for various locations in the USA are used to compare the performance of a hybrid PV/CPV architecture with flat-plate and CPV systems under a range of solar insolations. The contribution from diffuse radiation is approximately 2-2.5 kWh/m^2-day for all locations, but this component represents 20-40% of the global radiation. Analysis indicates that the WPV approach is able to provide 40-50% and 15-40% more energy production per unit area across the USA than conventional flat plate PV and CPV, respectively. Using state-of-the-art 4-junction cells (assuming ~40% DNI efficiency), a hybrid module efficiency of 33% is projected for low DNI regions (i.e., <60% DNI) that were typically considered not suitable for conventional CPV deployment.

III. PROTOTYPING

A first prototypical module is developed based on the baseline design (Fig. 2 (b)). It comprises a 1×1 inch^2 Si platform, a hexagonal array of 100-μm-wide InGaP/GaAs micro-cells hybrid-integrated on the Si platform, a middle glass plate and a 400X-concentration PDMS primary lens array (~2.5 mm sub-lens dia.) directly molded on the glass (Fig. 4 (a) – (c)).

The Si platform and III-V micro-cell arrays are processed in parallel before the hybrid integration. Reflective cavity arrays are first formed on a 280-μm thick Si wafer using KOH etching, followed by metallization. The cavity side walls achieve an RMS surface roughness of (9 ± 1) nm, ensuring high optical reflection. The Si wafer is subsequently processed into an interdigitated back contact (IBC) solar cell with interconnects and contacts formed on its backside for the Si and III-V cells. The III-V multijunction micro-cell arrays are then bonded onto the Si platform matching the apertures. InGaP/GaAs micro-cells bonded on Si have demonstrated cell efficiencies of 29.5% under 200 suns, while other multijunction cells with higher efficiencies can be readily integrated onto the same Si platform.

The Si platform is subsequently assembled onto the backside of a glass plate with the V-groove cavities filled with PDMS. A high-quality PDMS aspherical primary concentrator array is directly cast on the front side of the glass. Including a front cover glass and a backing sheet, the total thickness of the fully-packaged module is less than 1 cm. For advanced multi-stage optic designs, the etched cavity can simultaneously serve as an optical micro-concentrator and a micro-mechanical bench. Fig. 3d shows a 2-mm diameter ball lens concentrator positioned and self-aligned on the Si platform via the V-shaped cavity.

IV. COST MODEL

Cost models are developed for both cell and module production. The cost model is based on an industrial-scale fabrication processes taking into account contributions from raw materials, capital costs, labor, facilities overhead, and consumables. The cell fabrication cost serves as an input to the module cost model, which calculates the expected module cost

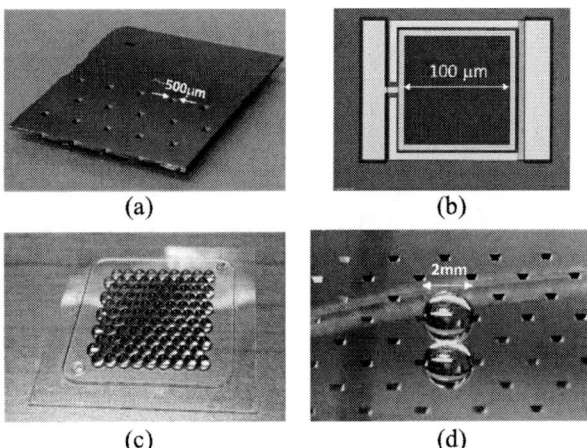

Fig. 4.   (a) Etched cavity arrays on a Si substrate; (b) InGaP/GaAs microcell (c) molded PDMS lens array on glass; (d) self-aligned ball-lens concentrator positioned on a Si cavity.

for a given optical concentration ratio and cell size. Many of the module-level costs are similar to those for conventional non-concentrating silicon-based PV, obtained through direct inquiries to vendors or from [8]. The estimated costs for each of the system elements project that the WPV approach can achieve a module cost of less than $150/m^2.

ACKNOWLEDGMENT

This work was supported by Advanced Research Projects Agency-Energy under the Micro-scale Optimized Solar-Cell Arrays with Integrated Concentration (MOSAIC) program (DE-AR0000632).

REFERENCES

[1] G. N. Nielson, *et al.* "Leveraging scale effects to create next-generation photovoltaic systems through micro- and nanotechnologies," *Proc. SPIE 8373, Micro- and Nanotechnology Sensors, Systems, & Appl. IV*, 837317, 2012.
[2] Jared B, *et al* "Micro-concentrators for a microsystems-enabled photovoltaic system," *Optics Express*. Vol. 22, Issue S2. A521-A527.
[3] T. Gu, *et al.* "Micro-concentrator module for Microsystems-Enabled Photovoltaics: Optical performance characterization, modelling and analysis," in *Proc. PVSC*, 2015.
[4] Sheng, X, *et al.* " Printing-based assembly of quadruple-junction four-terminal microscale solar cells and their use in high-efficiency modules," *Nature Mater.*, 13, 593–598 (2014).
[5] M. Haney, *et al.*, "Hybrid Micro-scale CPV/PV Architecture", in *Proc. PVSC*, 2014.
[6] S. Paap, *et al.* "Cost analysis for flat-plate concentrators employing microscale photovoltaic cells for high energy per unit area applications," in *Proc. PVSC*, 2014.
[7] P. Benitez, *et al.* "High performance Fresnel-based photovoltaic concentrator," *Opt. Express* 18(S1), A25–A40 (2010).
[8] A. Goodrich, *et al.* "A wafer-based monocrystalline silicon photovoltaics roadmap: Utilizing known technology improvement opportunities for further reductions in manufacturing costs," *Solar Energy Materials & Solar Cells*, vol. 114, 2013, pp. 110-135.

Toward Stationary Concentrator Photovoltaic Panels

Peter Kozodoy, Christopher Gladden, Michael Pavilonis, Tobias Wheeler, Christopher Rhodes, Chadwick Casper, Kevin Schneider

Glint Photonics Inc., 1520 Gilbreth Rd., Burlingame CA 94010, USA

Abstract — **Stationary concentrator PV modules offer the potential for high energy production per unit area at low cost. We have explored two optical architectures toward this goal and report on our progress. A novel singlet lens array design incorporating fluidic self-aligned tracking has been demonstrated; this provides an elegant tracking solution but does not provide a sufficiently large effective acceptance angle to be practical for stationary mounting. A catadioptric design is much more promising in this regard. Initial demonstrations show the wide-angle internal tracking compatible with stationary mounting, and detailed modeling predicts high energy output. Practical issues surrounding actuation and backplane assembly must be solved, but these do not represent fundamental obstacles.**

I. INTRODUCTION

Concentrator photovoltaic (CPV) systems offer the promise of high efficiency and low cost by pairing inexpensive focusing optics with small areas of high-efficiency multijunction photovoltaic cells. Unfortunately, the potential cost and performance advantages of such systems have been offset to a considerable degree by the requirement for optical alignment via precision mechanical tracking devices. These trackers add cost and complexity to the CPV systems and limit their installation options. The use of two-axis trackers also limits achievable ground-cover ratio.

These limitations have spurred interest in novel stationary high-concentration CPV architectures. In such designs, the focusing optics remain stationary and small scale internal adjustments in the optical system provide the required sun-tracking. Stationary CPV panels are designed using close-packed arrays of small-scale focusing optics. This configuration can provide thin flat panels that mimic the form factor of conventional flat-plate photovoltaic (PV) modules, and that can be mounted similarly using standard racking. If sufficiently low-cost in their construction, stationary CPV panels can provide a new type of high-efficiency alternative to conventional PV modules in high DNI locations. We have pursued two system architectures for stationary CPV modules and report here on our progress and conclusions.

II. SINGLET DESIGN

As shown in Fig. 1(a), a slab light-guide is covered by an array of *mm*-scale lenses that concentrate incoming sunlight to form an array of focal spots near the bottom face of a slab light-guide. The slab is clad on the top by a passive low-refractive-index material (e.g. a fluoropolymer), and on the bottom by a thin (5 to 30 μm) layer of light-reactive optical

cladding. The entire structure is approximately 2 to 10 mm thick. To form a self-tracking concentrator, coupling of the focused light into the guide must be made to occur automatically at the focal spots, in a way that follows movements of the focal spots with the changing angle of solar incidence. The static elements of this design were first developed by the Ford group at UC San Diego,[1] and approaches to producing self-tracking functionality have been pursued using optical-trapping-induced index shift [2] and various types of phase change materials.[3,4,5] Our design for self-tracking uses thermocapillary forcing in a microfluidic configuration. Compared to prior efforts, it has the advantages of actuating at low incident power and being largely insensitive to environmental temperature.

As shown in Fig. 1(b), a fluid bilayer forms the "smart" or "light-reactive" lower cladding material. This is constructed of a low refractive index cladding fluid that preferentially wets the slab surface, layered above an immiscible liquid of high refractive index, referred to as the coupling fluid. The fluids

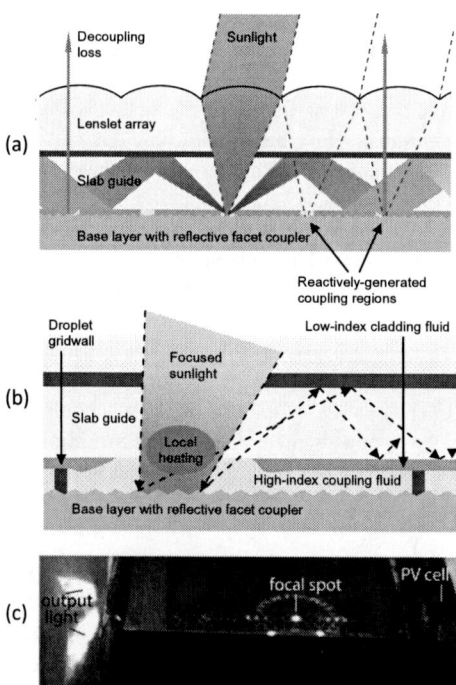

Fig 1 (a) cross-section of singlet design CPV module; (b) close-up of coupling region formation; (c) photograph of prototype under focused sunlight.

are held in place by a network of thin grid walls which separate the coupling fluid into individual droplet volumes so that capillary forces dominate over gravitational forces. Heating generated by the focused light produces a local reduction in interface tension between the two liquids, and the resulting thermocapillary forces deform the fluid interface, locally rupturing the low-index fluid layer and allowing the high-index fluid to provide a continuous high-index optical path between the guide and the underlying faceted reflecting surface. These "coupling regions" are dynamically generated at the focal point and will follow movement of the focal point as the solar position changes over time, thus providing automated tracking.

Focused light striking the facets at these coupling regions is deflected into angles that are trapped by the slab guide, and then travels through the guide by total internal reflection. Light from all the lenses is coupled into a common light-guide, and propagation loss is low because the thermally-generated coupling regions occur over only a very small fraction of the slab surface (typically less than 0.1%), with the remaining area clad by the low-index fluid. Concentrated light is extracted onto PV cells attached to the edge of the slab.

Because the optical properties of the concentrator vary depending upon the incident light characteristics, system design optimization is a complex problem requiring coupled analysis of optical, fluidic, and thermal properties over the anticipated range of illumination conditions. Simulations to determine optimal designs have been undertaken, combining multiphysics finite-element analysis and non-sequential raytrace modeling. These simulations indicate that annualized optical efficiencies above 80% are possible at concentrations above 500x over a wide angular tracking range.

Experimental concentrator devices have been fabricated at sizes up to 4" x 4". Analysis of fluid response dynamics tested with focused laser illumination match predictions of the multiphysics simulation with high accuracy. The device performance of initial prototypes was limited by material selection and geometry. Improved devices have achieved optical efficiencies as high as 72%, and self-tracking at angles up to ± 25° in the axis parallel to the coupling facets. Angular response of one prototype concentrator device is shown in Fig. 2. This wide self-tracking range greatly exceeds the narrow acceptance angle of conventional concentrating optics, and is sufficient to track the seasonal elevation change of the sun.

Angular acceptance in the other axis is lower, due to focal plane curvature of the focusing optics and reduced coupling efficiency into the guide for off-axis reflections from the facets. Thus, the singlet optical design is only practical for implementation on a platform with at least some coarse mechanical tracking in one axis. These results demonstrate the potential of the self-tracking concentrator to eliminate the requirement for precision mechanical tracking in high-concentration photovoltaic systems, but a more complex optical configuration is required to enable truly stationary concentrator panels.

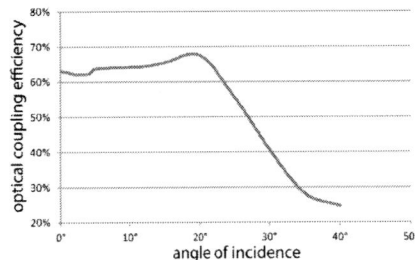

Fig. 2. Optical efficiency of a concentrator prototype as a function of incident light angle

III. CATADIOPTRIC DESIGN

The catadioptric design, as shown in Fig 3, uses a front refractive lens and a back reflective lens to form a focus between the two optics. The use of the combined optics allows for the focal plane to be flattened and eliminates the majority of chromatic aberrations. This optical structure was investigated previously by the Ford group [6] and more recently by the Giebink group at Penn State [7]. Focused light may be aggregated in a central lightguide, but improved efficiency is achieved by placing arrayed solar cells on a transparent substrate suspended at the focal plane, and mechanically translating the sheet inside the cavity. This optical structure enables high efficiency light gathering out to incident angles of ~ 70° in both axes, with only small translational "micro-tracking" movements of the transparent sheet. This design is therefore fundamentally well-suited to realization of stationary CPV modules. However, the optical design introduces a number of practical challenges.

The first of these is achieving the required micro-tracking of the central sheet. Unlike the singlet system, in which movement of fluids alone can provide the required optical changes for tracking, tracking in the catadioptric design requires physical movement of a solid object. This entails larger forces, although still orders of magnitude lower than in conventional two-axis mechanical trackers. It is possible to provide self-tracking orientation of solid objects via thermofluidic effects driven by focused sunlight. We have demonstrated such effects in the laboratory and are currently working to develop high-force versions suitable for micro-tracking within the CPV module. Many other actuation schemes are also possible using embedded low-cost actuators; the slow short-throw low-force requirements of the system are compatible with a variety of electromagnetic, thermoelectric, and piezoelectric actuator types.

A second complication is the requirement to mount the cells on a transparent sheet. This precludes the use of conventional backplane materials. Designs must be developed to prevent registration errors resulting from thermal expansion differences between the glass or polymer sheet and the molded polymer optics. Such errors can result from environmental temperature variation as well as self-heating effects during operation.

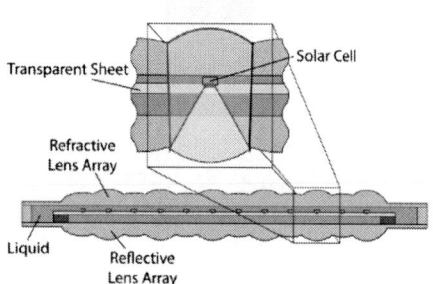

Fig 3 Cross-sectional diagram of catadioptric CPV module

A third complication is in PV cell heat mitigation. Conventional heat-sinking approaches cannot be used for the cells due to the requirement for optical transparency. The transparent polymer sheet and fluid ambient have low thermal conductivity, so heat removal is primarily via the metal traces connecting the cells. These must be designed to optimize performance by balancing heat removal with shadowing impact. Narrow but vertically thick metal traces are optimal. Further improvements are provided by using a network of traces that provide effective heat spreading in the area between PV cells.

We have built initial prototypes of catadioptric-type PV modules as shown in Fig 4. The module in this figure has an aperture area measuring approximately 4.5" on a side. It is a sealed unit filled with fluid and a transparent sheet. For this initial test prototype, the sheet is carrying packaged surface-mount silicon photodiodes as sensor elements, rather than the multi-junction PV cells that will be used in future work. Geometric concentration is only around 30x using the large photodiodes, compared to ~ 400x planned for multijunction cells. Fig 4 also shows the relative optical efficiency of the system, demonstrating that the expected wide-angle response is indeed achieved. The gentle falloff in optical efficiency with angle is a result of shadowing by the large photodiode and not inherent to the design.

Potential performance of an optimized panel has been analyzed using multi-physics models and considering a wide range of site conditions. The models assume stationary mounting facing south, and combine (i) NREL typical meteorological year (TMY) data for environmental temperature and hourly DNI availability, (ii) optical raytrace models of system performance as a function of incident light angle, (iii) finite element models of PV cell temperature rise, and (iv) models of multijunction PV cell performance as a function of temperature and flux, derived from measured data. Comparison to crystalline silicon flat plate modules (of 18% efficiency) shows improved annual energy generation in all three locations evaluated: Tucson, St. Louis, and Syracuse. In Tucson, where DNI is highest, the CPV module produces 80% more energy per year, while the benefit is 37% in St. Louis. These models highlight the potential advantages of such modules for area-constrained locations such as rooftops.

IV. CONCLUSIONS

Stationary CPV modules offer the potential for greatly increased PV energy production per unit area. We have explored two optical architectures toward this goal. A singlet lens array design incorporating fluidic self-aligned tracking has been demonstrated; this provides an elegant solution but does not provide a sufficiently large effective acceptance angle to be practical for stationary mounting. Catadioptric panels are more promising in this regard. Practical issues surrounding actuation and backplane assembly must be solved, but these do not represent fundamental obstacles. Initial prototypes have been built and confirm the wide angular tracking capability. Comprehensive system modeling that includes multiple loss mechanisms indicates the potential for high system efficiency in real world modules. Development of more sophisticated prototypes is ongoing.

V. ACKNOWLEDGMENTS

We gratefully acknowledge funding for this work from ARPA-E through awards DE-AR0000332 (OPEN 2012) and DE-AR0000644 (MOSAIC), and from the California Energy Commission through award EPC14-040.

REFERENCES

[1] J.H. Karp, E.J. Tremblay, J.E. Ford "Planar micro-optic solar concentrator" *Optics Express* **18**(2) pp. 1122-1133 (2010)

[2] K. Baker, J. Karp, E. Tremblay, J. Hallas, and J. Ford, "Reactive self-tracking solar concentrators: concept, design, and initial materials characterization," *Applied Optics* **51**(8), pp. 1086-1094 (2012).

[3] P. H. Schmaelzle and G. L. Whiting, "Lower critical solution temperature (LCST) polymers as a self adaptive alternative to mechanical tracking for solar energy harvesting devices," presented at the MRS Fall Meeting, Boston (29 November–3 December 2010).

[4] V. Zagolla and C. Moser, "Trackfree planar solar concentrator system" *Proc. SPIE 8256, Physics, Simulation, and Photonic Engineering of Photovoltaic Devices*, 825618 (February 9, 2012).

[5] V. Zagolla, D. Dominé, E. Tremblay, and C. Moser, "Self-tracking solar concentrator with an acceptance angle of 32°" *Optics Express*, **22** (S7), pp. A1880-A1894 (2014).

[6] J. E. Ford, J. H. Karp, E. Tremblay and J. M. Hallas, "System and Method for Solar Energy Capture and Related Method of Manufacturing". United States of America Patent US2011/0226332, 22 Sept. 2011.

[7] J. S. Price, X. Sheng, B. M. Meulblok, J. A. Rogers and N. C. Giebink, "Wide-angle planar microtracking for quasi-static microcell concentrating photovoltaics," Nature Communications, vol. 6:6223, 2015.

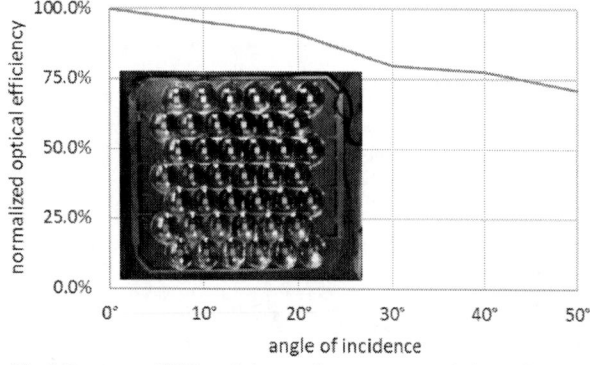

Fig 4 Prototype CPV module angular response and photo (inset)

CPV Technologies Not Relying on Perfection of Trackers

Kenji Araki[1], Yasuyuki Ota[2], Kan-Hua Lee[1], Kensuke Nishioka[2], and Masafumi Yamaguchi[1]

1. Toyota Technological Institute, Nagoya, 468-8611 Japan,

2. University of Miyazaki, Miyazaki, 889-2192 Japan

Abstract — CPV went off the market mainly due to problems of trackers. We are trying to solve the tracking issue.

Step 1. Control of the mating module by imperfection of tackers: We have developed a Monte Carlo model of the behavior of CPV module and system with assemble errors.

Step 2. Advanced CPV module for rough tracking: We have succeeded in demonstrating 100 x and 28.5% CPV module tracking every 30 minutes.

Step 3. No tracking: We have developed a modern design method that enables ultra-wide acceptance by advanced asymmetrical optics.

Index Terms — CPV, tracker, Monte Carlo method, acceptance angle, static concentrator, automobile.

I. INTRODUCTION

CPV went off the market mainly due to problems of trackers (under-performance by poor accuracy, low reliability, high maintenance and warranty cost, and sluggish cost reduction. Despite the withdrawal of the major manufacture, R&D activities of CPV are active, seeking for new path of further development. One of them is trying to solve the tracking issue.

The main interest to CPV has been to improve the cell and the module efficiency. Either CPV cells or CPV modules are highest efficient energy conversion device in the field of photovoltaics. In this atmosphere, and except for a few research institutes and engineers, like ISFOC in Spain, trackers have been regarded as convenient tools to have cells and optics always work in the best conditions. On the other hand, trackers and their installation and maintenance technologies have been tossed about by heavy waves of the cost pressure without good technological guidelines. The over-expectation of the up-stream section (module and cell) and over-promises of trackers quality and accuracy often brought about underperformance and incidents of the CPV system. 85% of the CPV incidents came from trackers [1].

So far, the cost portfolio of the trackers is still high and very few researches have been reported on trackers in these years including IEEE PVSC. It may be an appropriate time to accept that the tracker may not a perfect and convenient tool.

Based on this idea, TTI has started a series of research on a new direction of CPV technology not relying on perfection of trackers with following three steps (Fig. 1).

- Control of the mating module by imperfection of tackers
- Advanced CPV module for rough tracking
- No tracking

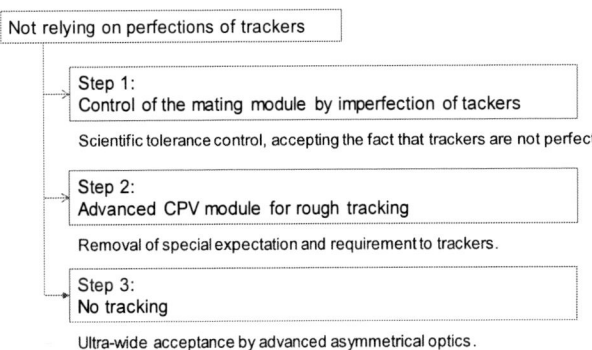

Fig. 1. Summary illustration of the three steps of technology advancement toward CPV system without relying on perfection of trackers.

These three steps do not indicate that the improvement will be done sequentially. The order does not mean degree of the advancement of technology. The merit of the step 1 can be applied to the most of CPV system using trackers, although it still relies on the requirement of trackers of current technology level. The module of the step 2 relaxes requirement of tracker precision, but the module itself has many technological challenges in product development. In the third step (step 3), the CPV module completely releases trackers but the concentration ratio is inevitably low. In this regard, all the three steps are equally important.

The purpose of this paper is to report on the current progress, especially the recent success of demonstrating a 100 x CPV module unit by 30 minutes of intermittent tracking with 28.5 % of outdoor efficiency. Table 1 summarizes the R&D activities done in Toyota Technological Institute for establishment of the technology on CPV without relying perfection of trackers.

TABLE I SUMMARY OF RELATED R&D IN TTI

	R&D Partners	Targets
Control of the mating module by imperfection of tackers	UPM	Design rule
Advanced CPV module for rough tracking	U. Miyazaki NEDO	Intermittent tracking every 30 minutes
No tracking	Toyota U. Miyazaki	> 30% efficiency and competitive to Si

978-1-5090-5606-4/17 $31.00 © 2017 IEEE

II. CONTROL OF THE MATING MODULES

CPV requires accuracy in optical alignment that is relied on accuracy of trackers as well as accurate control of the module assembly. Optical misalignment in the module was measured and analyzed [2]. Such information was useful to identify assembly problems and improved module performance.

1-D and 2-D Monte Carlo simulations were developed to analyze the impact of alignment errors by assembly errors. The 1-D simulation was useful to anticipate the power output of the module [3], whereas 2-D was for acceptance angle [4]. The analysis on that Monte Carlo simulation was applied to the production record of CPV by Daido Steel.

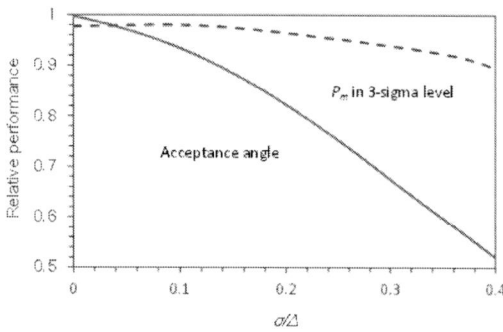

Fig. 2. Impact of the assembly error to the power loss of the module and the acceptance angle

The impact to the CPV module power output and the acceptance angle were analyzed (Fig. 1). In Fig. 1, the power loss is scaled by the relative output (*Pmp*) of the module in the worst performance in the 3-sigma level. The acceptance angle is defined as the 90 % of the power point. The X-axis is the ratio of the standard deviation of assembly errors σ divided by the designed acceptance half angle *Δ* (90% power point). The distribution of assembly errors was assumed as a normal distribution with an average of 0° of assembly error. From Fig. 1, the lack of the inspection of the acceptance angle of individual modules is thought to be the reason of the mismatching between tracker's performance and module's expectation, leading to the unexpectedly low performance in the field. The analyzed result met well to the measured values [5] and it is expected to solve the gap between module performance test result and real performance in the field.

III. ADVANCED CPV MODULE FOR ROUGH TRACKING

The motivation of this research is to remove special expectation and requirements to trackers to CPV applications, including sub-degree of tracking accuracy, and rigidness and flatness of the tracking plate to ensure the same tracking accuracy entire the loading area. To demonstrate a substantial improvement will be possible, our target is to realize the new system consists of new concentrator optics and their tracking strategy while keeping more than 100 x of concentration ratio. The target of acceptance is intermittent tracking in 30 minutes'

interval. Namely, the tracker moves every 30 minutes and it completely stays in the same position by the friction in the gear system in the rest of the time.

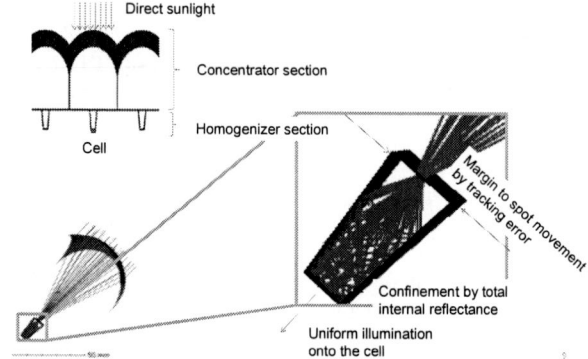

Fig. 3. Overall view of the optical design of the new CPV module.

It was shown that the 30 minutes intermittent tracking would be possible by the combination of the advanced optics with wide acceptance angle and feedforward control [6]. The concentrator optics is the integration of the dielectric concentrator and the homogenizer (Fig. 3).

However, this structure had a fatal problem. The peak flux density reached as high as almost 1 kW/cm² that was beyond the level of safety operation of the glass material that has sufficiently high optical transmittance, high refractive index, and most importantly high endurance to outdoor operation. Based on the experience of 820 x CPV module with the kaleidoscope homogenizer with dome-shaped Fresnel lens [7] and 1,000 x DFK module [8], the highest acceptable level was 500 W/cm². To reduce the peak flux density into half while keeping the acceptance angle, the dielectric lens was re-designed so that the focal points would be distributed around the highest illumination point. The focal distance was adjusted to avoid excessive concentration inside the glass (from 1000 W/cm² to 400 W/cm² < 500 W/cm² as the experienced design rule in the glass homogenizer using the same material). Specifically, the focal distance at the center aperture was modified to shorter and that of the peripheral aperture was to longer so that the focal point became the string, namely the energy at the focal point distributed in the direction of the optical axis. Despite the defocus design by practical compromise, the acceptance angles both along the X and Y axis were sufficiently wide (Fig. 4).

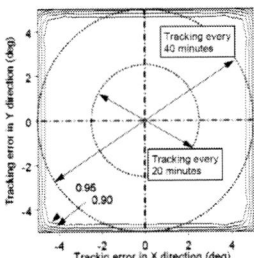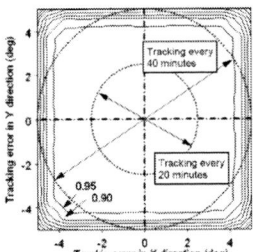

Fig. 4. Calculation of optical efficiency (scaled by photon number) by pointing error in both X and Y directions. The most inner contour corresponds to the 95 % power point. The left graph corresponds to the optically-ideal design and the right graph corresponds to the practical defocused design with the maximum flux is less than 500 W/cm². Note that the tracking error circles (dashed lines) correspond to the ranges of the pointing error by the feed-forward tracking control that will be discussed in the following section (not the conventional feed-back closed-loop control).

The module had of 5.5 × 5.5 mm² InGaP/GaInAs/Ge lattice-matched triple-junction solar cell with 27.4 % efficiency (1 sun and AM1.5G) and the solid lens was made from S-TIM2 glass.

The mono module was prototyped and sent to University of Miyazaki for outdoor evaluation. It was mounted on the same plane with the mono module using the conventional CPV technology (820 X dome-shaped Fresnel lens made from injection-molded PMMA with inverted pyramid kaleidoscope SOE made from S-TIM2 glass). The tracker was made by THK (fig. 5).

Fig. 5. Setup of the outdoor evaluation and verification of 30 minutes' intermittent tracking of the 100 x mono-module in University of Miyazaki site.

The conventional feedback control always tries to correct the tracker pointing vector to the center and does not make full of the acceptance range in both sides. The new method corrects the sun to the edge of the circle so that the range of the sun movement makes double (Fig. 6). This is a kind of the feed-forward control. Unlike the conventional closed-loop feedback control, the tracking direction of the open-loop feed forward system is determined solely by the absolute time. In principle, it does not correct installation and other errors. One practical solution is an automatic calibration to the open-loop system by the help of the sun sensor [9]. The previous work done by Luque-Herrada was model-free error correcting routines [9]. The advantage of the model-free algorithm and automatic calibration is versatility. However, if the tracker structure and mechanical properties are known by the system

designers, it may be wise to simplify the calibration model and to improve the robustness of the error correction.

The error-correction arithmetic we chose was a model-dependent algorithm using only three independent correction parameters [10]. The reduced parameter number has advantages of accomplishment of learning by one-time calibration. However, it is necessary to prove that the error model is precise. The basic structure of the correction model is a linear combination between theoretical direction cosine and an error correction vector [10].

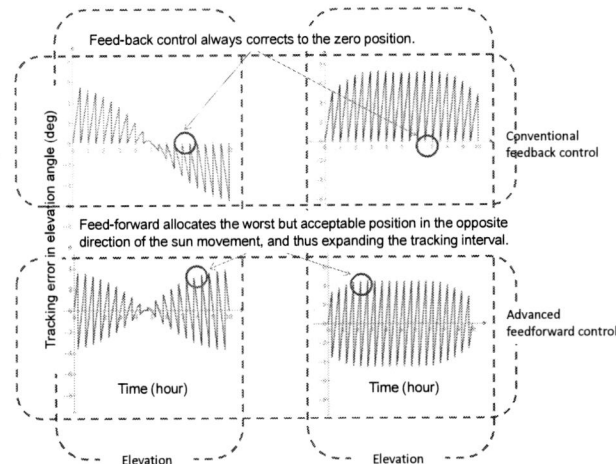

Fig. 6. Advanced feed-forward control for expanding tracking interval with contrast to the conventional feedback control.

The peak efficiency of the mono module was 28.5 % at the closed to the standard operating condition (Fig. 7).

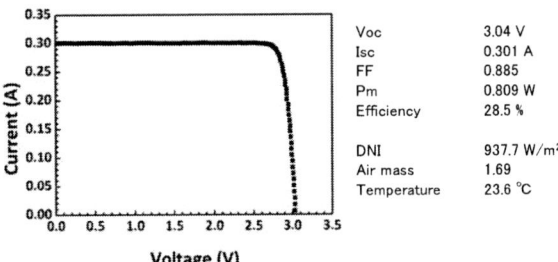

Voc	3.04 V
Isc	0.301 A
FF	0.885
Pm	0.809 W
Efficiency	28.5 %
DNI	937.7 W/m²
Air mass	1.69
Temperature	23.6 ℃

Fig. 7. I-V curve at closed condition of the standard operating condition of CPV module. The outdoor efficiency was 28.5 %. The standard operating condition (CSOC) is 20℃ ambient temperature, 0.97 < SMR < 1.03 of the spectrum matching range and 900 W/m² DNI.

The outdoor efficiency was monitored during 30 minutes intermittent tracking. The efficiency drop by the intermittent tracking was less than 7 % (relative), and the integrated output dropped only 1 %. It was shown the prototyped module continuously generated the power although the tracker only moves at every 30 minutes (Fig. 8). Note that the tracking duration in the conventional CPV system is typically less than

978-1-5090-5606-4/17 $31.00 © 2017 IEEE

15 seconds and 30 minutes intermittent tracking will be two order relaxation of the tracking tolerance.

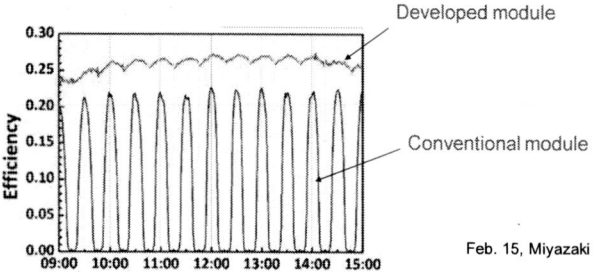

Fig. 8. Efficiency monitoring of the 30 minutes intermittent tracking with reference of the conventional wide-acceptance HCPV module mounted in the same tracking plane.

For the comparison to the optical design, it is important to normalize the output. Especially, influence from spectrum change should be carefully corrected. It was done by the linear correction by Air mass value (Fig. 9).

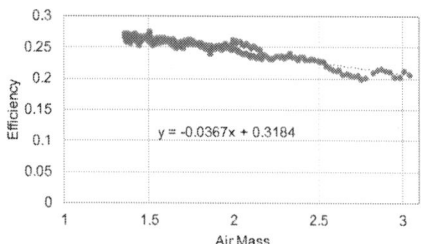

Fig. 9. Correlation between the air mass and corresponding module efficiency fluctuation, showing that correction by the linear regression can be possible due to no abrupt change of atmospheric parameters.

After correction by the air mass, the output at the center position that was equivalent to the continuous tracking was calculated at the average of 5 succeeding points and the relative output was analyzed at the off-axis performance by intermittent tracking and compared with designed off-axis performance calculated by the ray-tracing simulation. Since the off-axis characteristics is not axially symmetrical in this concentrator optics, there might be some error of the optical axis either in the module assembly or setup of the module to the tracking structure (Fig. 10).

If this off-axial error is constant regardless of any positions, namely both the module structure and the tracker frame are rigid enough, the offset vector can be calculated by the least square fit by minimizing the difference of the measured error in the relative efficiency data points from the designed output performance due to the tracking error.

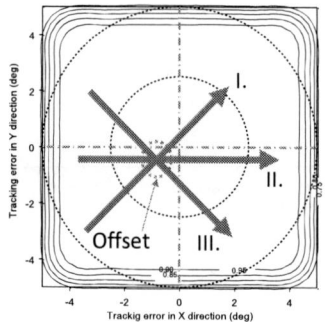

Fig. 10. Movement of the sun in the concentrator aperture considering the offset.

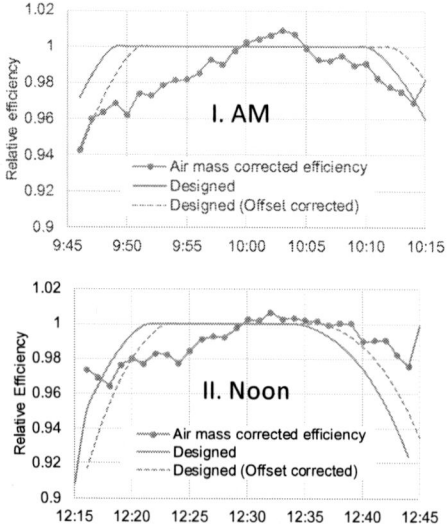

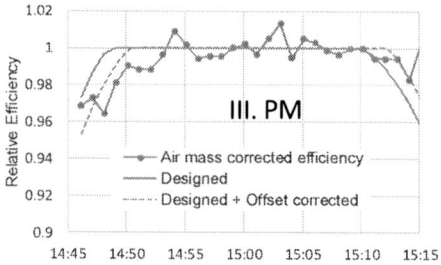

Fig. 11. Comparison to the measured module output to the designed off-axial performance by the concentrator optics in different three time zone of the day. The top is in the morning. The middle is in the noon time. The bottom is in the afternoon. Both the cases of the morning and afternoon were selected when the movement of the sun to the local coordinate of the concentrator aperture is 45 degree and maximizing the tracking acceptance. The movement of the sun becomes 0 degree to the local coordinate of the concentrator aperture at the noon time that is shown in the noon and the acceptance angle becomes minimized.

The comparison to the measured output is shown in Fig. 11. Since, off-axis characteristics is not axially-symmetrical, the acceptance angle differs in the time of the day. The measured response showed a slight asymmetry, but it can be corrected after considering the offset error in installation. It can be concluded that the mono module had wide acceptance angle as it was designed and capable of operating in tracking in every 30 minutes regardless of its 100 x of high concentration ratio.

IV. No TRACKING

The ultimate solution is to remove the trackers. This technology is called static CPV. We did research on two innovative approaches. One is axially-asymmetric lens. Another is partially cascaded static concentrator (PCSC).

A. Axially-asymmetric lens array

Axially-symmetric lens array was developed for car-roof PV [11-12]. Different from the conventional axially-symmetrical lens, it is thin and has much wider acceptance angle (Fig. 12). It was designed by numerically optimization of various asymmetrical profile functions. This lens is effective to grid arrangement rather than honeycomb arrangement. For the design optimized to horizontal direction, the acceptance half angles (90 % power) were 50.5° and 63.4° in 4 x and 3 x concentration.

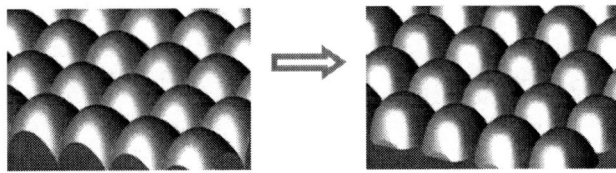

Fig. 12. Conventional symmetrical lens (left) and the new axially-asymmetrical lens (right)

B. Partially cascaded static concentrator to III-V/Si hybrids

To compete to the flat-plate crystalline Si cell module, III-V on Si structure was developed. It is possible to improve the efficiency with more than 30 % without relying on concentration [13]. However, it is likely that the situation of the higher cost of III-V cell relative to the Si cell will be unchanged. Then, it is preferred concentrating III-V cell and further saving the cost. On the other hand, concentration means reduction of the acceptance angle and trackers will be required.

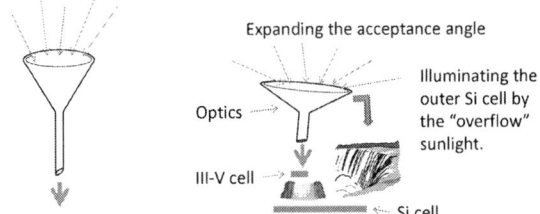

Fig. 13. Conventional CPV (left) vs. PCSC (right).

The concept of the partially cascaded static concentrator (PCSC) is expanding the acceptance angle in spite of the higher concentration ratio with sacrifice to the optical loss (Fig. 13 and Fig. 14). The thermodynamic limit and related constraints are applied to the III-V top cell, but the rays that were not absorbed by the top cell was collected by the bottom Si cell (larger than the top III-V cell). As a result, the III-V on Si region works as higher number of suns in spite of its wide acceptance angle (Fig. 13 and Fig. 14). Note that the aceptance characteristics is completely different from the conventional concentrators.

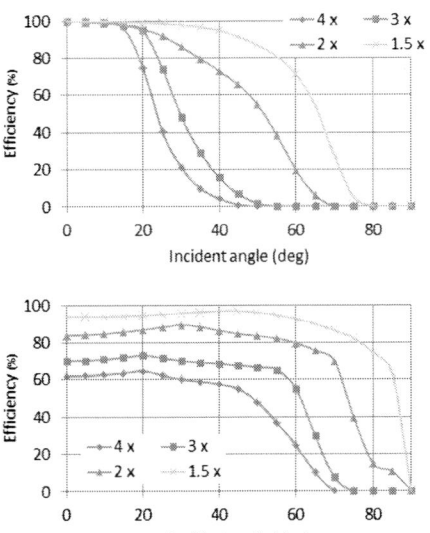

Fig. 14. Acceptance characteristics of conventional concentrator (top) and PCSC (bottom)

There are two types of configuration. One is vertical and another is lateral (Fig. 15).

 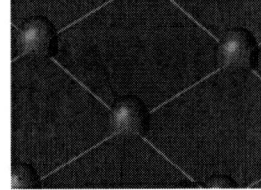

Fig. 15. Vertical PCSC (left) and Lateral PCSC (right)

The vertical configuration is the tandem III-V/Si structure but the top III-V is smaller. The lateral configuration is to place the small III-V cells on four corners of the single crystalline Si cell. Some design example is shown in Table 2 and Table 3. The module efficiency gain was considerably high in both cases.

TABLE II DESIGN EXAMPLES OF **VERTICAL** PCSC

Concentration ratio	III-V cell efficiency	Si cell Effciency	Si Module efficiency	PCSC module efficiency
2 x				30.9 %
3 x	28.3%	20%	19.1%	28.3 %
4 x				27.3 %

TABLE III DESIGN EXAMPLES OF **LATERAL** PCSC

Si cell	III-V cell	Si Module efficiency	PCSC module efficiency
□156	□19	19.1%	20.6%
20%	32%		
□125	□26	22.7%	26.6%
25%	35%		

VI. CONCLUSION

CPV went off the market mainly due to problems of trackers (under-performance by poor accuracy, low reliability, high maintenance and warranty cost, and sluggish cost reduction.

To solve the tracker issue, 3 steps of R&D were carried out.

- Control of the mating module by imperfection of tackers
- Advanced CPV module for rough tracking
- No tracking

We have developed a Monte Carlo model of the behavior of CPV module and system with assemble errors. The simulated characteristics met well to the measurement. Design and production control rules were made so that the CPV system can work properly with presence of field errors.

We have just succeeded in demonstrating 100 x and 28.5% CPV module tracking every 30 minutes. Since the most common tracking interval is 15 seconds, it is equivalent to two-order of relaxation of the tracking tolerance. It was achieved by the combination of the advanced optics with 4.5° of acceptance half angle and the new feed-forward tracking control with linear correction of the installation error.

We have developed a modern design method that enables ultra-wide acceptance by advanced asymmetrical optics. The acceptance half angle of 4 x concentration optimized for horizontal surface reached to 50.5° (90% power point). This static concentrator with ultra-wide acceptance may be suitable to car-roof solar panel. Additionally, a new type of concentrator PV named PCSC, including vertical and horizontal configuration, was proposed so that it achieved wider apparent acceptance angle than the thermodynamic limit.

ACKNOWLEDGMENTS

This work has been partially supported by NEDO in Japan.

REFERENCES

[1] F. Rubio , M. Martinez , R. Coronado, J. L. Pachón, and P. Banda. "Deploying CPV power plants-ISFOC experiences". in *PVSC'08. 33rd IEEE*. 2008, p. 1-4,.

[2] R. Herrero, C. Domínguez, S. Askins, I. Antón, and G. Sala, "Luminescence inverse method for CPV optical characterization", *Optics Express*, vol. 21, no. S6, pp. A1028-A1034, 2013.

[3] K. Araki, H. Nagai, R. Herrero, I. Antón, G. Sala and M. Yamaguchi, "Off-axis Characteristics of CPV Modules Result from Lens-Cell Misalignment − Measurement and Monte Carlo Simulation", *IEEE, J. Photovolt,* VOL. 6, NO. 5, pp. 1353-1359, 2016.

[4] K Araki, R. Herrero, I. Antón, G. Sala, H. Nagai, K. H. Lee, and M. Yamaguchi, (2016, June). "Why are acceptance angle of P_m and I_{sc} different in spite of uniform illumination onto concentrator solar cells?" in *Photovoltaic Specialists Conference (PVSC), 2016 IEEE 43rd* p. 0549-0553.

[5] K. Araki, H. Nagai, R. Herrero, I. Antón, , G. Sala, K. H. Lee, and M. Yamaguchi, "1-D and 2-D Monte Carlo simulations for analysis of CPV module characteristics including the acceptance angle impacted by assembly errors." *Solar Energy*, 147, 448-454. 2017.

[6] K. Araki, Y. Ota, K. H. Lee, K. Nishioka, and M. Yamaguchi, (2016, September). "Intermittent tracking (30 minutes interval) using a wide acceptance CPV module." in *AIP Conference Proceedings*, Vol. 1766, No. 1, 2017. p. 050001.

[7] K. Araki, T. Yano and Y. Kuroda, "30 kW Concentrator Photovoltaic System Using Dome-shaped Fresnel Lenses", *Optics Express*. 04/2010; 18(9):A53-63.

[8] H. Nagai, K. Araki, K. Hobo, P. Zamora, P. Benitez, J.C. Miñano, K. Nishioka, Y. Ota, I. Luque-Heredia, J. Hashimoto, T. Ueda, Y. Hishikawa, R. Herrero, S. Askins, Ignacio Antón, R. Núñez, G. Sala, Marc Steiner, M. Niemeyer, G. Siefer, A.W.Bett, "Dvelopment of the New DFK CPV Module by NGCPV JAPAN-EU Collaboration", *Technical Digest of 6th WCPEC*, 2014.

[9] I. Luque-Heredia, F. Gordillo, and F. A. Rodriguez, "PI based hybrid sun tracking algorithm for photovoltaic concentration." in *Proceedings of the IEEE 19th European Photovoltaic Energy Conversion*, 2004, p. 7-14.

[10] K. Araki "HCPV tracker using a new adjustable open-loop control." In *Proc. 23rd EUPVSEC*, 2008.

[11] K. Araki, H. Nagai, and M. Yamaguchi, "Possibility of solar station to EV." *AIP Conference Proceedings* Vol. 1766, No. 1, p. 080001. 2016.

[12] T. Masuda, K. Araki, K. Okumura, S. Urabe, Y. Kudo, K. Kimura, , ... and M. Yamaguchi, "Static concentrator photovoltaics for automotive applications." *Solar Energy*, 146, 523-531. 2017.

[13] R. Cariou, J. Benick, P. Beutel, N. Razek, C. Flötgen, M. Hermle, , ... and F. Dimroth, "Monolithic Two-Terminal III–V//Si Triple-Junction Solar Cells With 30.2% Efficiency Under 1-Sun AM1. 5g." *IEEE Journal of Photovoltaics*, 7(1), 367-373, 2017.

The gettering effect of dielectric films for silicon solar cells

A. Y. Liu[1], C. Sun[1], V. P. Markevich[2], A. R. Peaker[2], J. D. Murphy[3], and D. Macdonald[1]

[1]Research School of Engineering, Australian National University, Canberra, ACT, 2601, Australia
[2]Photon Science Institute, University of Manchester, Manchester, M13 9PL, United Kingdom
[3]School of Engineering, University of Warwick, Coventry, CV4 7AL, United Kingdom

Abstract — We present our recent findings on the strong gettering effect of dielectric films for removing impurities from the silicon bulk. Iron was used as a model impurity in silicon to study the gettering effects. The iron concentration in silicon is found to be significantly reduced after annealing silicon wafers coated with either plasma-enhanced chemical vapour deposited (PECVD) silicon nitride (SiN_x) films, or atomic layer deposited (ALD) aluminium oxide (Al_2O_3) films. Both PECVD SiN_x coated and ALD Al_2O_3 coated silicon wafers demonstrate largely diffusion-limited iron reduction rates at around 400°C, removing half of the bulk iron concentration within 30 min for a 160-µm thick silicon wafer with dielectric films on both sides. PECVD SiN_x films are shown to achieve diffusion-limited gettering of iron at temperatures below 700°C, meaning that 90% of iron can be gettered by the surface SiN_x films within 5 min for a 260-µm silicon wafer, or within 2 min for a 160-µm wafer at 700°C. Secondary ion mass spectrometry reveals that the gettering effect of PECVD SiN_x films is caused by impurity segregation into the bulk of the SiN_x films, and the gettering effect of ALD Al_2O_3 films comes from impurity accumulation at the Al_2O_3/Si interfaces.

Index Terms — gettering, silicon, iron, silicon nitride, aluminium oxide, plasma enhanced chemical vapour deposition, atomic layer deposition.

I. INTRODUCTION

Improving the silicon bulk material quality for solar cells is a vital aspect for photovoltaic research and industry. Phosphorus diffusion gettering is commonly applied to remove unwanted impurities in silicon, and the gettering process is conveniently coupled with the formation of n+ regions near the silicon wafer surfaces. Effective gettering via phosphorus diffusion, however, relies on the full-area high-temperature heavy diffusion processes, with lighter diffusions or ion implanted phosphorus regions being much less effective at gettering [1-4]. Moreover, novel high efficiency cell architectures, for instance, heterojunction and passivating contact solar cells, no longer require heavy phosphorus or boron diffusion to form p-n junctions. The need for alternative gettering processes is therefore imminent. Solar cells based on *n*-type silicon could also benefit from alternative gettering approaches, as the gettering effectiveness of boron diffusion for the formation of p+ regions is thus far not well integrated into cell processing steps [5].

In this paper, we present our recent findings of the strong gettering effect of the commonly used dielectric films for silicon solar cells, namely plasma enhanced chemical vapour deposited (PECVD) silicon nitride (SiN_x) films, recently published in Ref. [6], and atomic layer deposited (ALD) aluminium oxide (Al_2O_3) films, recently published in Ref. [7]. Experimental evidence for the effective gettering capabilities of these two dielectric films is presented. Iron is used as a model impurity in silicon to demonstrate and assess the gettering effect, as dissolved interstitial iron (Fe_i) in silicon can be easily and accurately quantified using lifetime based techniques [8, 9]. Furthermore, iron is a common efficiency-limiting impurity in silicon solar cells [10]. The gettering effects of PECVD SiN_x films, and ALD Al_2O_3 films, are examined by monitoring the Fe_i concentrations ($[Fe_i]$) in the silicon wafer bulk for samples annealed with the respective film coatings, via photoconductance-based carrier lifetime measurements. The re-distribution of iron in the dielectric films and the near-surface regions of the underlying silicon wafer bulk is studied by secondary ion mass spectrometry (SIMS) depth profiling of the iron concentration. A comparison of the gettering kinetics and mechanisms for the two films is presented and discussed.

II. EXPERIMENTAL DETAILS

High quality boron-doped float-zone (FZ) silicon wafers, of $1 - 2$ Ωcm, with well-controlled doses of implanted Fe, were used in this study. The wafers were $250 - 260$ µm and $160 - 180$ µm thick after saw damage etching in an alkaline solution. The targeted volumetric interstitial Fe concentration ($[Fe_i]$) varied from 2×10^{12} to 1×10^{14} cm^{-3}. After ion implantation of Fe atoms, the wafers were annealed at 1000°C for $2 - 3$ h to uniformly distribute Fe throughout the wafer thicknesses. Dry oxygen was chosen as the ambient gas to enable the growth of silicon dioxide (SiO_2) as surface passivation layers. Some samples were set aside as the SiO_2 controls.

A subset of the samples had the SiO_2 layers removed in dilute HF, and were then coated with either PECVD SiN_x films, or ALD Al_2O_3 films, on both sides. Details of the PECVD deposition can be found in Ref. [6], and ALD in Ref. [7]. It is worth mentioning that the deposition conditions were optimised for the surface passivation and antireflection purposes, not for the gettering effects.

Quasi-steady-state photoconductance (QSSPC) lifetime measurements [11] before and after breaking FeB pairs via illumination were used to determine the Fe_i concentrations in the Si bulk [8, 9, 12].

To study the Fe_i reduction kinetics, wafers with either SiN_x, Al_2O_3, or SiO_2 coatings were co-annealed cumulatively, with

978-1-5090-5606-4/17 $31.00 © 2017 IEEE

[Fe_i] measured after each annealing step using QSSPC. The SiN_x samples, with the SiO_2 controls, were annealed at 250 – 900°C; and the Al_2O_3 samples, including SiO_2 controls, were annealed at 425°C, a typical passivation activation temperature for the ALD Al_2O_3 films.

The re-distribution of Fe after annealing silicon wafers coated with different dielectric films was measured by SIMS depth profiling of the Fe concentration in the annealed SiN_x, Al_2O_3 or SiO_2 layer and the near surface region of the underlying Si substrate. SIMS measurements were conducted by Evans Analytical Group (EAG) Laboratories.

III. RESULTS AND DISCUSSION

A. Reduction kinetics of interstitial iron in silicon wafer bulk

When annealing silicon wafers coated with either PECVD SiN_x films, or ALD Al_2O_3 films, significant reductions of the interstitial Fe concentrations can be observed, as shown in Figs. 1 (400 – 425°C) and 2 (700°C).

Fig. 1(a) compares the Fe_i reduction kinetics in samples coated with different dielectric films – PECVD SiN_x, ALD Al_2O_3, or thermally grown SiO_2. The samples were of the same wafer thickness of 260 μm, and were annealed at a similar temperature of 400°C (SiN_x and SiO_2 samples) or 425°C (Al_2O_3 and SiO_2 samples). As shown in Fig. 1(a), the SiO_2 samples show a much slower rate of Fe_i reduction compared to the SiN_x and Al_2O_3 samples. As evident by the SIMS measurements in Section III B below, the substantial bulk Fe_i losses in the SiN_x and Al_2O_3 coated samples are caused by Fe_i gettering to the surface layers. Therefore, a diffusion-limited surface-loss model from Murphy and Falster [13] is used to fit the measured Fe_i reduction kinetics of the SiN_x and Al_2O_3 samples, with apparent dissolved Fe_i diffusivities being fitting parameters.

The 400°C annealed SiN_x sample yielded an apparent Fe_i diffusivity of 6.5×10^{-9} cm²/s, and the known Fe_i diffusivity from the literature is 9.6×10^{-9} cm²/s [14]. The 425°C annealed Al_2O_3 sample gave rise to an apparent Fe_i diffusivity of 5×10^{-9} cm²/s, and the reported Fe_i diffusivity is 1.5×10^{-8} cm²/s [14]. The Fe_i reduction kinetics in both SiN_x and Al_2O_3 coated samples are therefore close to diffusion-limited surface gettering processes, with the PECVD SiN_x films being a slightly faster gettering agent than the ALD Al_2O_3 films.

Fig. 1(b) illustrates the Fe_i reduction kinetics in thinner silicon wafers of 160 μm, which is more typical of the wafer thicknesses used for silicon solar cells nowadays. The silicon wafers were coated with either thermal SiO_2, or ALD Al_2O_3 of 20 nm or 80 nm. As a typical 20-nm Al_2O_3 film is too thin for SIMS analysis, the Fe_i kinetics of the 80-nm Al_2O_3 coated samples are examined and compared with the kinetics of 20-nm Al_2O_3 films, as shown in Fig. 1(b). This serves to ensure that the SIMS measurements on 80-nm Al_2O_3 films are

representative of the gettering behaviours of typical ALD Al_2O_3 films (commonly less than 30 nm).

As shown in Fig. 1(b), the samples coated with Al_2O_3 films, of either 20 nm or 80 nm, demonstrate a much faster Fe_i decline compared to the co-annealed SiO_2 sample in the initial stages. At this early stage, the Fe_i reduction kinetics can be fitted by the same diffusion-limited surface-gettering model [13] using a fitted diffusivity of 7×10^{-9} cm²/s. This is similar to the fitted diffusivity of the 260-μm Al_2O_3 sample shown in Fig. 1(a), indicating that the slower Fe_i kinetics in the 260-μm samples are largely due to the different wafer thicknesses. The Fe_i kinetics of the Al_2O_3 samples gradually slow down, the cause of which is unclear at this stage, as discussed in Ref. [7].

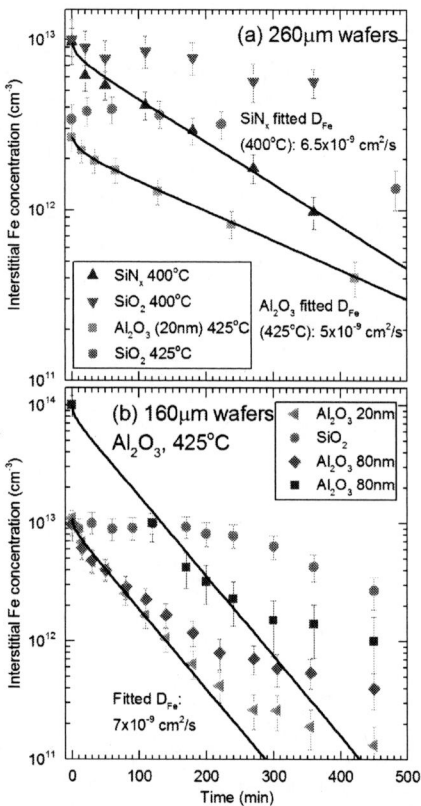

Fig. 1. Measured Fe_i reduction kinetics in the silicon wafer bulk for samples coated with PECVD SiN_x, ALD Al_2O_3, or thermal SiO_2, and cumulatively annealed at 400°C (SiN_x and SiO_2 samples) or 425°C (Al_2O_3 and SiO_2 samples). Fig. 1(a) plots the kinetics for wafers of 260 μm thick, with three different dielectric coatings. Fig. 1(b) shows the kinetics of 160-μm thick wafers, coated with either SiO_2, or Al_2O_3 of different thicknesses and two different initial Fe_i concentrations. The straight black lines are fittings to the experimental data based on a diffusion-limited surface-loss model, yielding fitted apparent Fe_i diffusivities [13].

As evident in Fig. 1(b), the SiO_2 sample again first shows little change in Fe_i concentration. With increasing annealing time, [Fe_i] reduction becomes faster. The initial slow change phase may be related to the nucleation of precipitation sites that later facilitate Fe_i precipitation to take place, and thus a faster $[Fe_i]$ decline at later stages.

The measured Fe_i kinetics and fitted apparent diffusivities in Fig. 1 show that, for a typical 160-μm wafer, a 30-min anneal at $400 - 425^\circ C$ reduces the bulk Fe_i concentration by half. Iron has a moderate diffusivity in silicon. This reported gettering effect should be taken into account in studies that use PECVD SiN_x or ALD Al_2O_3 films as surface passivation coatings, to avoid misinterpretation. PECVD typically operates at $250 - 500^\circ C$, ALD Al_2O_3 films typically require a post-deposition anneal at $300 - 500^\circ C$, and the processing time ranges from minutes to tens of minutes. This means that the fabrication of these dielectric films could result in considerable bulk impurity gettering effects, depending on the processing conditions, as observed in Refs. [15, 16]. The external gettering of Fe_i during the passivation step could also explain lower Fe_i concentrations measured in multicrystalline silicon wafers passivated by PECVD SiN_x compared to temporary liquid passivation schemes [17, 18].

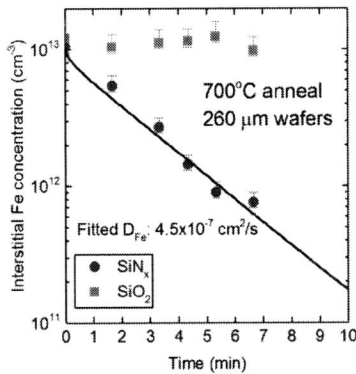

Fig. 2. Measured Fe_i reduction kinetics in 260-μm thick silicon wafers coated with either PECVD SiN_x or thermal SiO_2, cumulatively annealed at $700^\circ C$. The straight black line is a fitting to the experimental data based on a diffusion-limited surface-loss model, yielding a fitted apparent Fe_i diffusivity [13].

Fig. 2 presents the Fe_i reduction kinetics at $700^\circ C$ for the PECVD SiN_x and thermal SiO_2 coated samples. Similar to the $400^\circ C$ results in Fig. 1, the SiN_x sample demonstrates a much faster Fe_i loss than the SiO_2 sample, and the loss kinetics can be fitted by the same diffusion-limited surface-loss model [13], using a fitted diffusivity that is close to the reported Fe_i diffusivity (fitted diffusivity is 4.5×10^{-7} cm^2/s, and the reported Fe_i diffusivity is 3.4×10^{-7} cm^2/s [14]). This indicates a diffusion-limited surface-gettering process at $700^\circ C$ for the PECVD SiN_x films. In fact, for PECVD SiN_x films the Fe_i reduction kinetics over a wide temperature range of $250 - 900^\circ C$ have been investigated in Refs. [6, 15, 19], and the results show that the SiN_x gettering process is largely diffusion-limited for temperatures below $700^\circ C$. This helps to estimate the amount of bulk impurity loss for any given temperature profiles below $700^\circ C$.

As shown by the experimental data in Fig. 2, for a 260-μm silicon wafer coated with SiN_x films on both sides, the bulk Fe_i concentration is reduced by 90% after 5 min at $700^\circ C$. Modelling of the diffusion-limited surface-gettering process [13] predicates that, for a 160-μm wafer with SiN_x films on both sides, 90% of the bulk Fe_i is removed within 2 min at $700^\circ C$.

B. Re-distribution of iron

In order to understand the observed Fe_i reductions in the silicon wafer bulk (Figs. 1 and 2), SIMS analysis was conducted to reveal the re-distribution of Fe after annealing, in samples coated with different types of dielectric films and annealed for extended time to drive large bulk Fe_i reductions. Note that SIMS detects the total Fe concentration regardless of the electrical state, that is, SIMS detects both dissolved and precipitated Fe atoms.

It is observed that, all of the measured SIMS Fe profiles, in Figs. 3 and 4 and in Refs. [6, 7], demonstrate a near-surface Fe increase, for samples with or without previous bulk Fe contamination and with different film coatings. This suggests that this near-surface Fe increase does not come from the gettering of Fe from the silicon bulk, and is likely a measurement artefact in SIMS analysis, possibly caused by unintentional surface contamination during handling and processing.

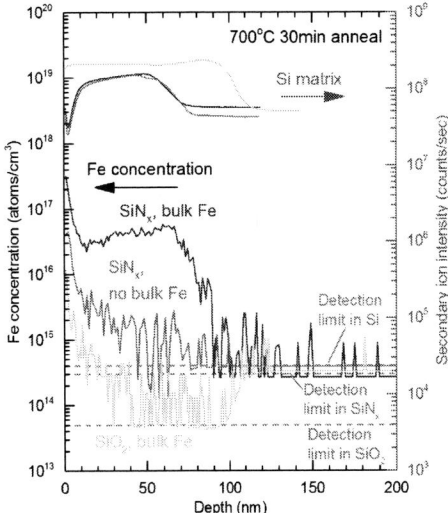

Fig. 3. SIMS measurements of the total Fe distribution in (1) PECVD SiN_x coated silicon wafer with bulk Fe_i contamination of 10^{13} cm^{-3}; (2) PECVD SiN_x coated silicon wafer with no bulk Fe_i contamination; and (3) thermally grown SiO_2 coated silicon wafer with bulk Fe_i contamination of 10^{13} cm^{-3}. All three samples were annealed at $700^\circ C$ for 30min in N_2. The upper curves are a qualitative measure of the Si matrixes (plotted against the y-axis on the right), which are used to indicate the positions of the SiN_x or SiO_2 films and the Si substrate in the depth profiles. The detection limits of total Fe in SiN_x, SiO_2, and Si are marked by dashed lines.

As shown in Fig. 3, a high concentration of Fe is detected in the SiN_x film for the Fe_i-contaminated SiN_x sample, whereas the Fe concentration in the SiO_2 layer in Fe_i-contaminated co-annealed SiO_2 sample is below the detection limit. This is consistent with the lifetime measurements (Fig. 2 and Ref. [6]) that the SiO_2 sample only experienced a small decrease in the bulk Fe_i concentration after a 700°C 30-min anneal, and hence the majority of Fe_i still remains in the silicon wafer bulk, at a concentration well below the detection limit of SIMS for quantifying Fe in Si.

The high Fe concentration in the SiN_x film is shown to arise from the Si bulk, by comparing the SIMS Fe profiles in SiN_x samples with or without previous bulk Fe_i contamination (Fig. 3). The bulk Fe_i-contaminated sample shows a much higher concentration of Fe in the SiN_x film, indicating that this high Fe concentration does not come from the PECVD SiN_x deposition or the subsequent annealing processes, but rather from Fe previously distributed in the Si bulk in the interstitial form (i.e. Fe_i), that is, Fe_i is gettered to the SiN_x film from the Si bulk upon annealing. The quantitative Fe concentration in the SiN_x film from SIMS measurements is also found to be similar to the bulk Fe_i reduction measured from QSSPC, further confirming that Fe_i is gettered from the silicon bulk to the surface SiN_x film.

The uniform Fe distribution in the SiN_x film (shown in Fig. 3), and the effectiveness of SiN_x gettering at temperatures where the Fe_i solubilities are well above the bulk Fe_i concentrations (shown in Refs. [6, 15]), indicate that the SiN_x gettering process is driven by impurity segregation into the PECVD SiN_x films.

The observed [Fe_i] reductions in PECVD SiN_x coated samples were previously conjectured to be due to the hydrogenation of Fe [15, 19] (and the Refs within). This hypothesis, however, was later found to be inconsistent with the results from deep-level transient spectroscopy (DLTS) analysis of the annealed silicon wafers, and thermal stability studies of the Fe_i reductions, as detailed in Ref. [6]. The SIMS measurements of the Fe re-distribution provide conclusive evidence for the gettering effect of PECVD SiN_x films, ruling out the previous hydrogenation hypothesis. Our results do not exclude the possibility of hydrogenation of the silicon from the SiN_x film, but the involvement of hydrogen is not necessary to explain the kinetics of the interstitial iron reduction observed.

Fig. 4 presents the SIMS results of ALD Al_2O_3 coated and thermal SiO_2 coated samples, co-annealed at 425°C for an extended 18 h. After the long cumulative anneals, both Fe_i-contaminated samples, with either Al_2O_3 or SiO_2 coatings, show a significant reduction in the bulk Fe_i concentration (Fig. 1 and Ref. [7]). The SIMS profiles of the two Fe_i-contaminated samples both demonstrate large Fe peaks at the Al_2O_3/Si or SiO_2/Si interface. The quantitative Fe concentrations in these Fe peaks from SIMS measurements show general agreement with the losses of bulk Fe_i concentrations determined from QSSPC lifetime

measurements, indicating that the Fe peaks source from Fe_i originally distributed in the silicon bulk.

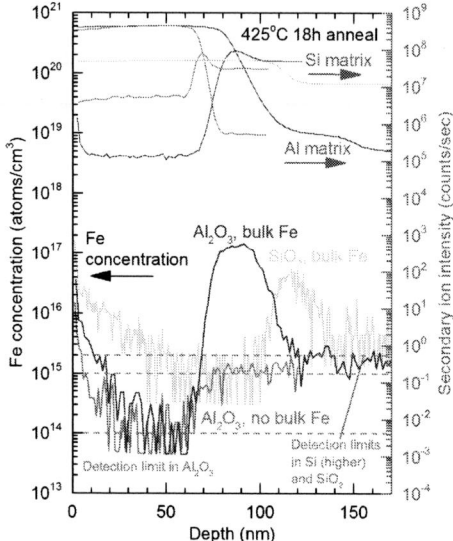

Fig. 4. SIMS measurements of the total Fe distribution in (1) ALD Al_2O_3 coated silicon wafer with bulk Fe_i contamination of 10^{14} cm^{-3}; (2) ALD Al_2O_3 coated silicon wafer with no bulk Fe_i contamination; and (3) thermally grown SiO_2 coated silicon wafer with bulk Fe_i contamination of 10^{13} cm^{-3}. All three samples were cumulatively annealed at 425°C for a total of 18h in air. The upper curves are a qualitative measure of the Al and Si matrixes (plotted against the y-axis on the right), which are used to indicate the positions of the Al_2O_3 or SiO_2 films and the Si substrate in the depth profiles. The detection limits of total Fe in Al_2O_3, SiO_2, and Si are marked by dashed lines.

It is well known in the literature that Fe-silicide precipitates tend to form at the SiO_2/Si interfaces (see, for example, Ref. [10]). The observed Fe peak at the SiO_2/Si interface for the thermal SiO_2 coated sample is therefore likely due to Fe_i precipitation, as the solubility of Fe_i at 425°C is in the range of 10^5 cm^{-3} [14] to 10^8 cm^{-3} (from extrapolation [20]), providing sufficient driving force for relaxation precipitation to take place.

As shown in Fig. 4, an Al_2O_3 coated sample without previous bulk Fe_i contamination illustrates no such Fe peak at the Al_2O_3/Si interface. This further confirms that the Fe peak observed in the Fe_i-contaminated Al_2O_3 sample comes from the gettering of Fe_i from the silicon bulk.

The underlying mechanism for the accumulation of Fe at the Al_2O_3/Si interface remains unknown at this stage. As an ultra-thin silicon oxide (SiO_x) layer is known to reside at the ALD Al_2O_3/Si interface [21], this Fe peak may be caused by Fe_i precipitation at the SiO_x/Si interface, similar to the Fe_i precipitation at the thermal SiO_2/Si interface discussed above. However, the ALD Al_2O_3 and thermal SiO_2 samples show very different Fe_i loss kinetics, as can be seen in Fig. 1. Hence, Fe accumulation at the ALD Al_2O_3/Si interface could be caused by a different mechanism, for example, by impurity segregation at the Al_2O_3/Si interface. Further temperature-dependent gettering studies will be required to assess the hypotheses.

IV. SUMMARY

This paper presents experimental evidence for the strong gettering effects of the commonly used dielectric films for silicon solar cells: PECVD silicon nitride and ALD aluminium oxide films. The gettering process is found to be largely diffusion-limited for both films near 400°C. The PECVD silicon nitride films are shown to result in diffusion-limited gettering effects at temperatures below 700°C. The measured interstitial iron reduction kinetics indicate that significant impurity gettering can be achieved by the dielectric films at moderate temperatures within minutes to tens of minutes, depending on the impurity diffusivity in silicon. International impurity gettering to dielectric films at low temperatures could be a viable alternative to high temperature gettering and may be particularly useful in the development of novel silicon-based solar cells.

Although the two films demonstrate similar diffusion-limited gettering processes at around 400°C, the underlying gettering mechanisms are different: PECVD silicon nitride films relocate bulk impurities into the films via segregation gettering, and ALD aluminium oxide films getter impurities via impurity accumulation at the Al_2O_3/Si interfaces.

ACKNOWLEDGEMENT

This work has been supported by the Australian Renewable Energy Agency (ARENA) through project RND009. Author C Sun acknowledges the financial support from Australian Centre for Advanced Photovoltaics. The work in the UK is supported by the Engineering and Physical Sciences Research Council via the SuperSilicon PV project (EP/M024911/1). We acknowledge access to NCRIS facilities (ANFF and the Heavy Ion Accelerator Capability) at the Australian National University.

REFERENCES

[1] E. Cho, Y.-W. Ok, L. D. Dahal, A. Das, V. Upadhyaya, and A. Rohatgi, "Comparison of POCl3 diffusion and phosphorus ion-implantation induced gettering in crystalline Si solar cells," *Solar Energy Materials and Solar Cells*, vol. 157, pp. 245-249, 2016.

[2] H. S. Laine, V. Vähänissi, Z. Liu, H. Huang, E. Magaña, A. E. Morishige, *et al.*, "Finite- vs. infinite-source emitters in silicon photovoltaics: Effect on transition metal gettering," in *2016 IEEE 43rd Photovoltaic Specialists Conference (PVSC)*, 2016, pp. 0678-0680.

[3] M. Kim, H. Li, D. Payne, S. Wenham, B. Hallam, and M. Abbott, "Application of high efficiency emitters to multi-crystalline silicon," presented at the 32nd European Photovoltaic Solar Energy Conference and Exhibition, Munich, Germany, 2016.

[4] V. Vähänissi, A. Haarahiltunen, M. Yli-Koski, and H. Savin, "Gettering of Iron in Silicon Solar Cells With Implanted Emitters," *IEEE Journal of Photovoltaics*, vol. 4, pp. 142-147, 2014.

[5] S. P. Phang, W. Liang, B. Wolpensinger, M. A. Kessler, and D. Macdonald, "Tradeoffs Between Impurity Gettering, Bulk Degradation, and Surface Passivation of Boron-Rich Layers on Silicon Solar Cells," *IEEE Journal of Photovoltaics*, vol. 3, pp. 261-266, 2013.

[6] A. Y. Liu, C. Sun, V. P. Markevich, A. R. Peaker, J. D. Murphy, and D. Macdonald, "Gettering of interstitial iron in silicon by plasma-enhanced chemical vapour deposited silicon nitride films," *Journal of Applied Physics*, vol. 120, 193103, 2016.

[7] A. Y. Liu and D. Macdonald, "Impurity gettering effect of atomic layer deposited aluminium oxide films on silicon wafers," *Applied Physics Letters*, vol. 110, 191604, 2017.

[8] G. Zoth and W. Bergholz, "A fast, preparation-free method to detect iron in silicon," *Journal of Applied Physics*, vol. 67, pp. 6764-6771, 1990.

[9] D. H. Macdonald, L. J. Geerligs, and A. Azzizi, "Iron detection in crystalline silicon by carrier lifetime measurements for arbitrary injection and doping," *Journal of Applied Physics*, vol. 95, pp. 1021-1028, 2004.

[10] A. A. Istratov, H. Hieslmair, and E. R. Weber, "Iron contamination in silicon technology," *Applied Physics A*, vol. 70, pp. 489-534, 2000.

[11] R. A. Sinton and A. Cuevas, "Contactless determination of current–voltage characteristics and minority-carrier lifetimes in semiconductors from quasi-steady-state photoconductance data," *Applied Physics Letters*, vol. 69, pp. 2510-2512, 1996.

[12] L. Geerligs and D. Macdonald, "Dynamics of light-induced FeB pair dissociation in crystalline silicon," *Applied Physics Letters*, vol. 85, p. 5227, 2004.

[13] J. D. Murphy and R. J. Falster, "The relaxation behaviour of supersaturated iron in single-crystal silicon at 500 to 750 [degree]C," *Journal of Applied Physics*, vol. 112, 113506, 2012.

[14] A. A. Istratov, H. Hieslmair, and E. R. Weber, "Iron and its complexes in silicon," *Applied Physics A*, vol. 69, pp. 13-44, 1999.

[15] A. Liu, C. Sun, and D. Macdonald, "Hydrogen passivation of interstitial iron in boron-doped multicrystalline silicon during annealing," *Journal of Applied Physics*, vol. 116, 194902, 2014.

[16] S. P. Phang and D. Macdonald, "Direct comparison of boron, phosphorus, and aluminum gettering of iron in crystalline silicon," *J. Appl. Phys.*, vol. 109, 073521, 2011.

[17] P. Karzel, A. Frey, S. Fritz, and G. Hahn, "Influence of hydrogen on interstitial iron concentration in multicrystalline silicon during annealing steps," *Journal of Applied Physics*, vol. 113, 114903, 2013.

[18] M. Al-Amin and J. D. Murphy, "Passivation Effects on Low-Temperature Gettering in Multicrystalline Silicon," *IEEE Journal of Photovoltaics*, vol. 7, pp. 68-77, 2017.

[19] C. Sun, A. Liu, S. P. Phang, F. E. Rougieux, and D. Macdonald, "Charge states of the reactants in the hydrogen passivation of interstitial iron in P-type crystalline silicon," *Journal of Applied Physics*, vol. 118, 085709, 2015.

[20]   J. D. Murphy and R. J. Falster, "Contamination of silicon by iron at temperatures below 800 °C," *Phys. Status Solidi RRL,* vol. 5, pp. 370-372, 2011.

[21]   F. Werner, B. Veith, D. Zielke, L. Kühnemund, C. Tegenkamp, M. Seibt, *et al.*, "Electronic and chemical properties of the c-Si/Al2O3 interface," *Journal of Applied Physics,* vol. 109, 113701, 2011.

Tabula Rasa: Oxygen precipitate dissolution though rapid high temperature processing in silicon

Erin E. Looney[1], Hannu S. Laine[1,2], Mallory A. Jensen[1], Amanda Youssef[1], Vincenzo LaSalvia[3], Paul Stradins[3], Tonio Buonassisi[1]

[1]Massachusetts Institute of Technology, Cambridge, MA 02139, USA, email: elooney@mit.edu
[2]Department of Electronics and Nanoengineering, Aalto University, Tietotie 3, 02150 Espoo, Finland
[3]National Renewable Energy Laboratory, Golden, CO 80401, USA

Abstract — Over one fourth of all monocrystalline silicon ingots suffer from a 20% performance degradation due to oxygen precipitates. Tabula Rasa (TR) is a mitigation technique that dissolves these precipitates, making them harmless. This work explores the dependence of oxygen dissolution on annealing time and temperature for the TR process to aid in solar cell process optimization. The dissolution time for oxygen precipitates was found to be more than 10 minutes for total dissolution, longer than normal TR process times in the electronics industry. The activation energy, extracted from the precipitate dissolution curves, is found to be 2.6 ± 0.5 eV. This value when compared to the migration enthalpy of oxygen in silicon can be used to reveal the energy limiting proces in TR.

Index terms – oxygen related defects, monocrystalline silicon, *tabula rasa*, thin wafering, precipitate dissolution

I. INTRODUCTION

The capital expenditure (capex) required to manufacture photovoltaic modules is a central barrier to the rapid adoption of solar technologies needed to meet climate targets [1]. One strategy to reduce the capex is to increase manufacturing yield. The majority of industrial mono-crystalline silicon is made with the Czochralski (Cz-Si) method which comprises 35-40% of the current solar material used today [2]. Between 25-30% of typical Cz-Si ingots are afflicted with a specific defect that accounts for a 20% relative drop or around 4% absolute reduction in conversion efficiency for this material [3]. The portion of the ingot inflicted with these defects , called "ring" or "swirl" defects are sorted out as lower quality material. The "swirl" defects are in reality several types of micro defects that are caused by the interplay of silicon interstitials, vacancies, oxygen precipitates, and other intrinsic point defects during the solidification of the silicon ingot [4].

Oxygen is the most abundant impurity in Cz-Si, usually present at 10-20 ppma in solar cell material. Oxygen in as-grown material takes several forms ranging from interstitial oxygen with low recombination activity to strained oxide precipitates with high recombination activity [5]-[6]. During cell fabrication, high temperature steps such as phosphorus diffusion gettering, thermal oxidation, and firing introduce the wafer to temperatures that allow the oxygen to change between these different states, and therefore reduce the minority carrier lifetime in the material [7].

Process optimization tools have been developed to mitigate detrimental effects of most metal impurities, for example iron defect control during phosphorus diffusion gettering and firing have been optimized using kinetic models [8]. However, currently no such process or modeling tools exist for mitigating the effects of oxygen in the solar industry, partly due to the insufficient quantification of the dissolution kinetics of oxygen in silicon for solar applications.

We work to solve this problem through the study of *tabula rasa* (TR), meaning "blank slate," which is a rapid high temperature process between 1090 – 1150°C maintained for 1-2 minutes aimed at mitigating these defects [9]. In the integrated circuit (IC) industry this process is used to install a certain vacancy profile in the wafer, while erasing the thermal history, leaving silicon wafers "blank slates" as all oxygen is left in interstitial form with no precipitates or nuclei [10]. The solar industry does not incorporate such an anneal before cell processing, leaving as-grown defects that formed during the crystallization process to evolve throughout the high temperature process steps of cell fabrication. By using an optimized TR process on an industrial cell fabrication line, it is hypothesized that these oxygen-related swirl defects can be dissolved, reclaiming the 20% relative drop in efficiency in the affected portion of each Cz-Si ingot and thus improving manufacturing yield [11]. In this work, the dissolution kinetics of the TR process have been quantified so that accurate modeling and optimization can be realized to accomplish this yield gain.

II. METHODS

A. Materials

We use 745 µm thick, *p*-type, 200 mm diameter, double side polished, electronic-grade Cz-Si wafers with a resistivity of 10-12 Ω-cm resistivity. The wafers contain negligible extrinsic impurities other than oxygen content of 13.75 ppma. These wafers are laser cut into 3.5x3.5 cm^2 rectangles for oxygen precipitate growth. After chemical cleaning for metal and organic contaminants, each sample undergoes a series of high temperature steps in an N_2 atmosphere, quartz tube furnace to nucleate and grow large oxygen precipitates in a homogeneously distributed pattern [12]. The time-temperature profile can be seen in Fig. 1, and includes a nucleation step at 550 °C for 6 hours, precipitate growth steps at 800°C for 4 hours, 925°C for 4 hours, and 1000°C for 16 hours, and lastly ambient cooling to 700°C before being pulled out of the furnace. The samples are then cleaved into smaller 1x1 cm^2 pieces, chemically etched with CP4 and HF-dipped to remove any surface damage or oxide left from the high temperature processing.

B. Experiment

A series of TR processes are performed on the samples at five temperatures between 1100°C and 1290°C. For each temperature, TR experiments were performed at six different times between 1 and 30 minutes. We expect the highest temperature, longest time TR process (1290°C for 30 minutes) to completely dissolve the grown-in oxygen precipitates, and for the shortest time, lowest temperature process (1100°C for 1 minutes) to dissolve the precipitates the least. The TR process is done in a horizontal mullite tube furnace (N_2 ambient) with temperatures measured from external thermocouples and a disappearing filament optical pyrometer. The push and pull rate both into and out of the furnace was 10 seconds, and the samples sat at the end of the furnace for 10 seconds before being put on a large silicon heat sink and cooled in ambient air. The samples were then HF dipped to remove any oxide formed on the surface.

C. Characterization

The oxygen distribution in the wafer was characterized using Fourier Transform Infrared Spectroscopy (FTIR) to quantify interstitial oxygen $[O_i]$ concentration and chemical etching to reveal oxygen precipitates [13]. FTIR was used to determine $[O_i]$ concentration before precipitate growth, after precipitate growth, and after the TR step shown in Fig. 1. These measurements were calibrated using ASTM standard 121-83. After the growth process depicted in Fig. 1, the interstitial oxygen content was measured at 5.35 ppma, down from the total oxygen before the anneal at 13.75 ppma, or 6.88×10^{17} atoms/cm^3, indicating that 61% of oxygen atoms precipitated during growth. Through chemical defect etching done for 45 seconds (HF:CH$_3$COOH:HNO$_3$ with a volume ratio of 36:15:2), the precipitate density is determined to be 1.6×10^9 ppt/cm^3 by counting etch pits in a representative volume. Assuming that the oxygen precipitates are all of similar size and spherical, using the radius of an oxygen atom, $r = 6.5 \times 10^{-9}$ cm, the oxygen precipitates are calculated to have a diameter between 70-90 nm or 2-3×10^8 oxygen atoms per precipitate. Finally, after each TR step, FTIR is performed again to determine how many of these precipitates have dissolved back into interstitial oxygen. Using these methods, the macro scale character of the oxygen including total, interstitial, and precipitated density and size is determined for each point in the process.

III. RESULTS

Fig. 2 plots $[O_i]$ concentration as a function of annealing time for each TR temperature. As expected, interstitial oxygen concentration increases steadily at each TR temperature, due to dissolution of oxygen precipitates. To quantify the dependence of the dissolution rate on TR temperature, dissolution curves for the oxygen precipitates are plotted in **Fig. 2**. The dissolution time constant, τ_{diss}, was determined for each temperature by fitting the interstitial oxygen data with the equation:

$$[Oi](t) \propto C_s \left[1 - \exp\left(-\frac{t}{\tau_{diss}} \right) \right] \qquad (1)$$

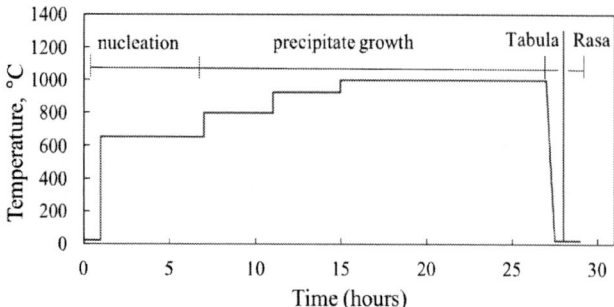

Fig. 1 Time-temperature profile for wafer preparation and experiment.

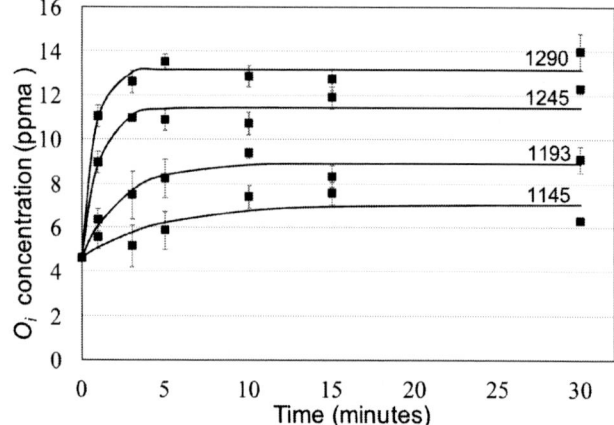

Fig. 2 Dissolution curves are plotted as interstitial oxygen content for four temperatures and six times. The dissolution time constant ranged from around 1 to 6 minutes between the temperatures, and none of the temperatures managed to completely dissolve the precipitates to $[O_i] = 13.75$ ppma which is the total oxygen content of the samples.

where, C_s is the observed saturation concentration of interstitial oxygen at each TR temperature.

The four curves plotted in Fig. 2 represent the four TR process temperatures and demonstrate the dissolution of oxygen precipitates at a steep initial rate with a saturation at a certain concentration. The solid solubility limit of interstitial oxygen in silicon is not reached during the longest 30-minute TR run for any of these temperatures except for the 1145°C curve using the intrinsic solubility described by $9 * 10^{22} \exp\left(\frac{-1.52\,eV}{kT} \right) cm^{-3}$ [14]. The 1100°C curve is not plotted as it shows no change in concentration, because the solid solubility at this temperature (4.74 ppma) is close to the starting concentration of $[O_i]$ for these samples. At 1193°C and 1245°C, the $[O_i]$ concentration reaches a steady dissolution rate below their solid solubility limits.

Our explanation for this behavior is that the process has become kinetically limited and the size of the oxygen precipitates in these samples is too large to dissolve quickly with a TR step at these temperatures. The traditional TR process is around 1050—1200°C for 1—5 minutes and would not be sufficient. It is observed that for TR to be an effective process,

978-1-5090-5606-4/17 $31.00 © 2017 IEEE

the appropriate time and temperature for complete dissolution of oxygen precipitates must be used. This profile must take into consideration the oxygen content and morphologies within the as-grown wafer.

The activation energy for dissolution, E_a, was found using the fitted dissolution time constants, τ_{diss}, and fitting these to an Arrhenius-type temperature-dependent equation.

$$\frac{1}{\tau_{diss}} \propto \exp\left(\frac{E_a}{k_b T}\right) \qquad (2)$$

The activation energy indicates the limiting physical process in dissolution. If the activation energy is larger than the diffusion coefficient for interstitial oxygen in silicon, then there is an additional physical process limiting the dissolution process. This could be, for example, the dissolution reaction in which one oxygen atom breaks away from a precipitate (reaction-limited dissolution). If the activation energy is found to be close to the migration enthalpy, then the diffusion of an oxygen atom away from a precipitate is interpreted to be the limiting step, and no significant extra energy is needed to dissolve precipitates greater than the kinetic limitations of diffusion [15]. If the activation energy is smaller than the migration enthalpy, then the dissolution of oxygen precipitates is enhanced, decreasing the energy needed for an oxygen atom to break from a precipitate and diffuse away. The activation energy was found to be 2.6±0.5 eV which is within the range of the migration enthalpy for $[O_i]$ in silicon which is 2.53±0.3 eV [14]. This information can be used to reveal the limiting energy for the TR process, and dissolution of oxygen precipitates can be modeled based on known kinetic and solubility material properties without any extra energy barrier.

IV. INDUSTRY IMPACT

The TR process is a promising solution to the issue of yield loss for Cz solar cells due to oxygen-related defects. To justify such a high temperature process step, predictive process modeling capabilities are key to implementation. This work enables this modeling as oxygen dissolution during TR is shown to be kinetically limited. To realize this model for industry, efficient methods for characterization of the as-grown oxygen distributions are in development. Work on more

reliable and nondestructive techniques for oxygen precipitate characterization is ongoing [16]. With as-grown oxygen distributions characterized and the physics behind dissolution now quantified, time-temperature profiles can be optimized to put oxygen into the least recombination active form.

V. CONCLUSIONS

The activation energy of oxygen dissolution was determined to be in the same range as the coefficient of diffusion, ~2.6 eV. This information can be used to infer the limiting physical process in the dissolution. The oxygen dissolution saturates before the solid solubility limit. From that saturation point, the precipitates slowly dissolve after an initially quick dissolution rate because the oxygen precipitates are above the critical radius for that temperature. This explains why none of the TR processes fully dissolved the precipitates with diameters averaging 70 nm. For other material, including industrial PV materials, shorter and lesser temperature processes are possible if the material has smaller oxygen precipitates. Knowledge of the size and density of the oxygen precipitates within the wafer is needed to determine which TR process is appropriate.

REFERENCES

[1] D. Berney Needleman, *et al., Ener. & Environ. Sci.*, **9**, 2016.
[2] "ITRPV: Results 2015," VDMA PV Equipment, 2016.
[3] J. Haunschild, *et al., Phys. Status Solidi RRL* **5**, 2011.
[4] V. V. Voronkov, Jour. of Crys. Growth, **59**, pp. 625–643, 1982.
[5] J. Michel, *et al., Semicond. & Semimet.*, **42**, 1994.
[6] R. Falster, *et al., Mat. Sci. Forum,* **573-574**, pp. 45–60, 2008.
[7] J. D. Murphy, *et al., Solid State Phenomen.*, **178–179**, 2011.
[8] J. Hofstetter, et al. Prog. in PV: Res. and App. 2011; **19**
[9] V. LaSalvia *et al.,* 2016 IEEE 43rd PVSC, 2016.
[10] G. Kissinger *et al., Jour. of Electrchem. Soc.,* **154**, 2007.
[11] B. Sopori *et al., IEEE Jour. of PV,* 7, 2017.
[12] R. J. Falster, Montgomery Research Group Europe, **12**, 2002.
[13] B. Sopori, *et al., Jour. of Electrchem. Soc.,* **131**, 1984.
[14] J. C. Mikkelsen, *MRS Proceedings*, **59**, 1985.
[15] F. Shimura, *Appl. Phys. Letters*, **39**, 1981.
[16] A. Youssef, *et al.,* accepted talk, *Silicon PV,* 2017.

Toward Effective Gettering in Boron-Implanted Silicon Solar Cells

Hannu S. Laine[1], Ville Vähänissi[1], Zhengjun Liu[1], Ernesto Magaña[2], Ashley E. Morishige[3], Jan Krügener[4],
Kristian Salo[1], Barry Lai[5], Hele Savin[1], David P. Fenning[2]

[1]Department of Electronics and Nanoengineering, Aalto University, 02150 Espoo, Finland
[2]Department of Nanoengineering, University of California, San Diego, La Jolla, CA 92093, USA
[3]Massachusetts Institute of Technology, Cambridge, MA 02139, USA
[4]Institute of Electronic Materials and Devices, Leibniz Universität Hannover, 30167 Hannover, Germany
[5]Advanced Photon Source, Argonne National Laboratory, Argonne, IL 60439, USA

Abstract — Boron-implantation is a pathway to high-quality, low-cost emitters required for high-efficiency *n*-type silicon solar cells. Compared to the industry standard BBr$_3$ diffusion, B-implant requires no edge isolation or boron rich layer (BRL) removal, which reduces manufacturing complexity and cost. B-implant also offers easy control of the dopant profile, which can translate to lower emitter saturation current density (j_{0e}) and thus higher cell efficiencies. In addition to low emitter saturation current density, an important property of the emitter is its gettering efficiency, or its ability to reduce recombination active bulk defects that degrade bulk minority charge carrier diffusion length. Here, we perform a controlled experiment to map the potential of high-quality ($j_{0e} < 50$ fA/cm^2) B-implanted emitters to reduce bulk iron point defects. We show that the point defect concentration can be reduced by more than 99.9 %. We describe efforts to generalize our results and elucidate the underlying gettering mechanisms via predictive modeling.

Index Terms — silicon, boron, ion-implantation, gettering, iron, modeling.

I. INTRODUCTION

The highest silicon solar cell efficiencies available in commercial production or realized at laboratory scale are based on *n*-type substrates [1,2]. To facilitate broader adoption of *n*-type substrates, alternatives for the traditional BBr$_3$ are under investigation. Boron-implantation avoids some of the inherent problems of the BBr$_3$ diffusion process, such as the need for edge isolation or the formation of a boron rich layer (BRL) [3,4]. Simultaneously, it facilitates easy control of the boron profile, which helps reach a low emitter saturation current density (j_{0e}), a requirement for high solar cell efficiency [5,6].

Besides low j_{0e}, a long minority charge carrier diffusion length is required for high solar cell efficiencies [6]. For >25 % efficiencies, diffusion lengths more than 10 times the wafer thickness are often required [1,2]. For this reason, the ability of both phosphorus and boron emitters to reduce detrimental bulk metal point defects (their gettering efficiency) has been studied in several experiments [7-15]. However, past research on boron emitters is mostly based on diffused emitters [9,10,16,17] or implantations using high (> 100 keV) implant energies [7,8]. For PV emitters, low energies (~10 keV) are desirable to minimize recombination-inducing implantation damage [18,19].

Here, we experimentally study boron emitters implanted at a low (10 keV) energy, which have been demonstrated to offer an excellent j_{0e} (< 50 fA/cm^2) when combined with appropriate surface passivation [20]. In particular, we study the extent to which these emitters can reduce bulk iron point defect concentrations during solar cell processing. Iron deserves attention as it is one, if not the, most common detrimental metal impurity in silicon solar cell processing. Due to iron's simple characterization and the reasonable similarity in diffusivity and solubility amongst the later *3d* transition metals (Cr, Ni, Co, Fe, Cu) [21-23], learnings from previous studies of iron [13, 24] have been successfully applied to other *3d* transition metals [25, 26].

Lastly, we describe an on-going effort in which we generalize our results for arbitrary implantation parameters and iron contamination levels. This work enables computer-assisted optimization of gettering strategies for boron-implanted silicon solar cells, which can help accelerate the development of high-efficiency boron-implanted silicon solar cells.

II. EXPERIMENTAL

Polished, *p*-type, 390±10 µm thick, 3-5 Ωcm resistivity, 100 mm diameter, low oxygen content (< 10 ppma), IC-grade Cz-wafers were used. Cz-Si, as a substrate type, comprises roughly a third of the PV market today, and its share is expected to grow [27]. Low-oxygen Cz-Si has a low density of internal gettering sites, like oxygen precipitates, dislocations or grain boundaries. With the emitter as the only significant impurity sink, the emitter gettering efficiency is easier to characterize.

To ensure reproducibility and to allow accurate modeling of our experimental results, the samples undergo a controlled iron contamination process. The wafers are immersed into a spiked RCA-solution [28], and annealed at i) 845°C for 55 minutes for a "low" contamination level of $[Fe_0] = 3.0 \times 10^{13}$ cm^{-3} or ii) 945°C for 45 minutes for a "high" contamination level of $[Fe_0]$

978-1-5090-5606-4/17 $31.00 © 2017 IEEE

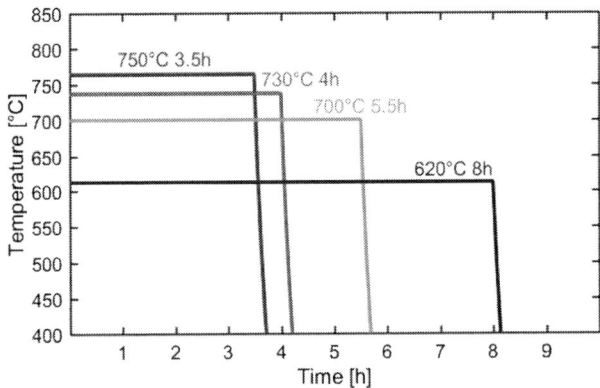

Figure 1. Gettering anneals used in this study.

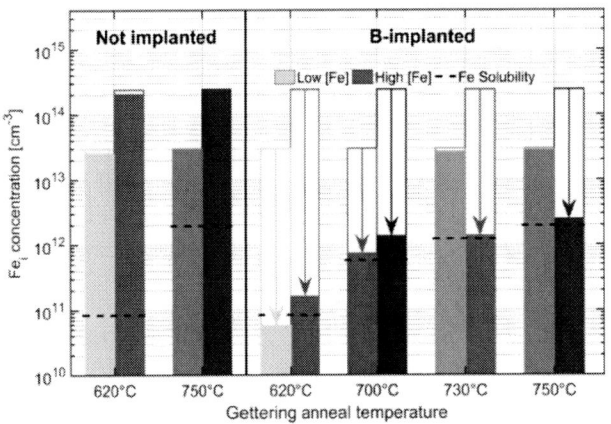

Figure. 2. Observed [Fe$_i$] reductions for the gettering anneals in implanted (right) and non-implanted (left) samples in the lowly-contaminated (light bars) and highly-contaminated (dark bars) samples. The iron solid solubility [32,33] at the gettering temperature is shown as a dashed line.

$= 2.4 \times 10^{14}$ cm^{-3}. Because thermally excited charge carriers make silicon essentially intrinsic at temperatures above 400°C [29], and iron exhibits low diffusivity at temperatures below 400°C [30], our results are relevant for both *n*- and *p*- type substrates.

After the controlled contamination procedure, boron was implanted into the wafers with a dose 3×10^{15} cm^{-2} at an energy of 10 keV. The dopants are driven-in and the implant damage is healed [19] by annealing the wafers at 1050°C for 45 minutes. The electronic properties of these emitters has been studied previously [20], showing that on polished samples, the emitter parameters result in a sheet resistance of approximately 55 Ohms / sq. and a j_{0e} below 50 fA/cm^2.

To explore gettering strategies with different thermal budgets, the implant anneal was directly followed by a 4°C/min ramp down to a gettering anneal at i) 750°C for 3.5 h, ii) 730°C for 4 h, iii) 700°C for 5.5 h or iv) 620°C for 8 h. After the gettering anneals, the furnace ejects the wafers directly to room temperature. The gettering times are chosen to ensure a $\sqrt{Dt}$ interstitial iron diffusion length of at least twice the wafer thickness, facilitating significant iron re-distribution at the gettering temperature. If the wafer thickness was a typical PV wafer thickness, such as 160 μm [27], the dwell times at the gettering temperatures could be shortened by approximately a factor of six while maintaining a similar gettering efficiency. To rule out other gettering sinks besides the boron-implanted emitter, non-implanted contaminated reference samples were included in the 750°C and 620°C gettering anneals. The time-temperature profiles of the gettering anneals are in shown Fig. 1.

The iron point defect reduction during the gettering anneals is characterized with the surface photovoltage technique [30]. To investigate possible iron precipitation and the precipitate size distribution in the implanted layer, synchrotron-based nano-X-ray fluorescence (XRF) is performed at beamline 2-ID-D of the Advanced Photon Source as detailed in [31].

III. RESULTS

Figure 2 illustrates the observed [Fe$_i$] reductions measured after annealing. First, we notice that the bulk interstitial iron concentration is virtually unchanged in the non-implanted samples. This suggests that there are no significant sinks of iron point defects in the non-implanted wafers, such as oxygen precipitates [34,35] or structural defects [36], which was expected from the low-oxygen high-quality Cz-wafers chosen for this study.

The B-emitters, on the other hand, can act as significant sinks for interstitial iron. The highly-contaminated 620°C sample experiences a bulk [Fe$_i$] reduction of over 99.9%, from the high contamination level of 2.4×10^{14} cm^{-3} down to 1.6×10^{11} cm^{-3}. In terms of increased cell efficiency potential on *n*-type substrates, this corresponds to an increase from below 15 % cell efficiency up to a ~21.2 %, which is the limitation of the passivated emitter and rear cell (PERC) architecture on an *n*-type substrate used in a recent simulation study [37]. For *p*-type substrates, this would correspond to an increase from below 10 % up to approximately 20.4 %1 [37]. Other samples exhibit impressive gettering efficiencies as well: for example, in the lowly-contaminated 620°C sample, bulk [Fe$_i$] is reduced from 3.0×10^{13} to 5.7×10^{10} cm^{-3}, a 99.8 % relative reduction. Also the highly- and lowly-contaminated 700°C samples experience impressive relative bulk [Fe$_i$] reductions of 99.4 % and 97.6 %, respectively.

The 730°C and 750°C samples exhibit a different trend than the lower temperature samples. For the lowly-contaminated samples, the bulk [Fe$_i$] remains virtually unchanged during the implant anneal, whereas the highly-contaminated samples exhibit significant (~ 99 %) bulk [Fe$_i$] reductions and ultimately reach bulk [Fe$_i$] values below the original low contamination level. Such "inversion" of contamination level has been observed before for phosphorus-implanted emitters [12,38] and it implies iron precipitation in the implanted layer. The inversion is explained due to the fact that a high initial supersaturation of iron induces a higher density of iron precipitate nuclei, which are then able to getter iron more effectively than a lower density of nuclei induced by a lower initial supersaturation [39]. The fact that all but two samples exhibit final bulk [Fe$_i$] values close to the iron solid solubility values in the gettering temperature is another piece of evidence suggesting that iron precipitation in the implanted layer is the dominant gettering mechanism.

Precipitation was confirmed by synchrotron-based nano-XRF measurements by locating several, large (> 1 μm in length) precipitates in three investigated samples: the highly-contaminated 750°C sample, as well as the lowly- and highly-contaminated 620°C samples. The details of these findings will be discussed in [31].

IV. OUTLOOK: PREDICTIVE MODELING

Ongoing work [31] will generalize our results for arbitrary implantation parameters and contamination levels, which will facilitate computer-assisted optimization of gettering strategies for high-efficiency boron-implanted silicon solar cells. This is achieved by quantitatively modeling the structural defect distribution induced by the implantation process [40], accounting for the changes in the distribution during thermal processing [18, 19], and then using these defects as precipitation sites in a heterogeneous precipitation model for iron [13]. This work will raise the understanding of gettering with implanted emitters to a similar level as has already been reached for traditional emitters, in particular phosphorus-diffused emitters [14, 41-43].

V. SUMMARY

To facilitate cost-effective manufacturing of boron-implanted silicon solar cells, we performed a quantitative test on the gettering induced by PV-typical boron-implants under different gettering anneals. We showed that boron-implantation can reduce bulk iron point defect concentrations by over 99.9 %, and that the dominant gettering mechanism is precipitation in the implanted layer.

ACKNOWLEDGEMENTS

This research used resources of the Advanced Photon Source, a U.S. Department of Energy (DOE) Office of Science User Facility operated for the DOE Office of Science by Argonne National Laboratory under Contract No. DE-AC02-06CH11357. This material is based upon work supported in part by the National Science Foundation (NSF) and the Department of Energy (DOE) under NSF CA No. EEC-1041895. The Aalto University authors acknowledge the financial support from the Finnish Funding Agency for Innovation under project

"BLACK" (project No. 2956/31/2014), Academy of Finland, Okmetic Oyj and Semilab Inc, as well as the provision of facilities and technical support by Aalto University at Micronova Nanofabrication Centre. H. S. L. acknowledges the Fulbright Technology Industries of Finland grant as well as support from Finnish Cultural Foundation and Walter Ahlström Foundation. E.M. and D.P.F. acknowledge the support of start-up funds from the University of California, San Diego.

REFERENCES

[1] D. D. Smith, G. Reich, M. Baldrias, M. Reich, N. Boitnott, and G. Bunea, "Silicon solar cells with total area efficiency above 25%," in *43rd IEEE Photovoltaic Specialist Conference*, 2016, p. 3351.

[2] K. Yoshikawa, H. Kawasaki, W. Yoshida, T. Irie, K. Konishi, K. Nakano, T. Uto, D. Adachi, M. Kanematsu, H. Uzu, and K. Yamamoto, "Silicon heterojunction solar cell with interdigitated back contacts for a photoconversion efficiency over 26%," *Nature Energy*, vol. 2, p. 17032, 2017.

[3] A. Rohatgi, D. L. Meier, B. McPherson, Y.-W. Ok, A. D. Upadhyaya, J.-H. Lai, and F. Zimbardi, "High-throughput ion-implantation for low-cost high-efficiency silicon solar cells," *Energy Procedia*, vol. 15, p. 10, 2012.

[4] V. Prajapati T. Janssens, J. John, J. Poortmans, and R. Mertens, "Diffusion-free high efficiency silicon solar cells," *Progress in Photovoltaics: Research and Applications*, vol 21, p. 980, 2013.

[5] F. Kiefer, R. Peibst, T. Ohrdes, T. Dullweber, J. Krügener, H. J. Osten, C. Schöllhorn, A. Grohe, and R. Brendel, "Influence of the boron emitter profile on V_{OC} and J_{SC} losses in fully ion implanted n-type PERT solar cells," *physica status solidi (a)*, vol. 212, p. 291, 2015.

[6] M. Rudiger, H. Steinkemper, M. Hermle, and S. W. Glunz, "Numerical current density loss analysis of industrially relevant crystalline silicon solar cell concepts," *IEEE Journal of Photovoltaics*, vol. 4, p. 533, 2014.

[7] A. Haarahiltunen, H. Talvitie, H. Savin, O. Anttila, M. Yli-Koski, M. I. Asghar, and J. Sinkkonen "Gettering of iron in silicon by boron implantation," *Journal of Materials Science: Materials in Electronics*, vol. 19, p. 41, 2008.

[8] J. L. Benton, P. A. Stolk, D. J. Eaglesham, D. C. Jacobson, J. Y. Cheng, J. M. Poate, S. M. Myers, and T. E. Haynes, "The mechanisms of iron gettering in silicon by boron ion-implantation," *Journal of The Electrochemical Society*, vol. 143, p. 1406, 1996.

[9] S. P. Phang, and D. Macdonald, "Effect of boron codoping and phosphorus concentration on phosphorus diffusion gettering," *IEEE Journal of Photovoltaics*, vol. 4, p. 64, 2014.

[10] V. Vähänissi, A. Haarahiltunen, H. Talvitie, M. Yli-Koski, J. Lindroos, and H. Savin, "Physical mechanisms of boron diffusion gettering of iron in silicon," *physica status solidi (RRL) - Rapid Research Letters*, vol. 4, p. 136, 2010.

[11] J. Schön, M. C. Schubert, W. Warta, H. Savin, and A. Haarahiltunen, "Analysis of simultaneous boron and phosphorus diffusion gettering in silicon," *physica status solidi (a)*, vol. 207, p. 2589, 2010.

[12] H. S. Laine, V. Vähänissi, A. E. Morishige, J. Hofstetter, A. Haarahiltunen, B. Lai, H. Savin, D. P. Fenning, "Impact of iron precipitation on phosphorus-implanted silicon solar cells," *IEEE Journal of Photovoltaics*, vol. 6, p. 1094, 2016.

[13] A. Haarahiltunen, H. Väinölä, O. Anttila, M. Yli-Koski, and J. Sinkkonen, "Experimental and theoretical study of heterogeneous

iron precipitation in silicon," *Journal of Applied Physics,* vol. 101, p. 043507, 2007.

[14] J. Hofstetter, D. P. Fenning, M. I. Bertoni, J. F. Lelièvre, C. del Cañizo, and T. Buonassisi, "Impurity-to-efficiency simulator: predictive simulation of silicon solar cell performance based on iron content and distribution," *Progress in Photovoltaics: Research and Applications* 19.4 (2011): 487-497.

[15] R. Chen, B. Trzynadlowski, and S. T. Dunham, "Phosphorus vacancy cluster model for phosphorus diffusion gettering of metals in Si," *Journal of Applied Physics*, vol. 115, p. 054906, 2014.

[16] S. P. Phang, and D. Macdonald, "Direct comparison of boron, phosphorus, and aluminum gettering of iron in crystalline silicon," *Journal of Applied Physics*, vol. 109, p. 073521, 2011.

[17] H. Talvitie, V. Vähänissi, A. Haarahiltunen, M. Yli-Koski, and H. Savin, "Phosphorus and boron diffusion gettering of iron in monocrystalline silicon," *Journal of Applied Physics*, vol. 109, p. 093505, 2011.

[18] R. Müller, A. Moldovan, C. Schiller, and J. Benick, "Defect removal after low temperature annealing of boron implantations by emitter etch-back for silicon solar cells," *physica status solidi (RRL) – Rapid Research Letters*, vol. 9, p. 32, 2015.

[19] J. Krügener, R. Peibst, F. A. Wolf, E. Bugiel, T. Ohrdes, F. Kiefer, C. Schöllhorn, A. Grohe, R. Brendel, and H. J. Osten, "Electrical and structural analysis of crystal defects after high-temperature rapid thermal annealing of highly boron ion-implanted emitters," *IEEE Journal of Photovoltaics*, vol. 5, p. 166, 2015.

[20] G. von Gastrow, P. Ortega, R. Alcubilla, S. Husein, T. Nietzold, M. Bertoni, and H. Savin, "Recombination processes in passivated boron-implanted black silicon emitters," *Journal of Applied Physics*, vol. 121, p. 185706, 2017.

[21] E. R. Weber, "Transition metals in silicon," *Applied Physics A: Materials Science & Processing*, vol. 30, p. 1, 1983.

[22] G. Coletti, P. C. P. Bronsveld, G. Hahn, W. Warta, D. Macdonald, B. Ceccaroli, K. Wambach, N. L. Quang, and J. M. Fernandez, "Impact of metal contamination in silicon solar cells," *Advanced Functional Materials*, vol. 21, p. 879, 2011.

[23] S. Riepe, I. E. Reis, W. Kwapil, M. A. Falkenberg, J. Schön, H. Behnken, J. Bauer, D. Kreßner-Kiel, W. Seifert, and W. Koch, "Research on efficiency limiting defects and defect engineering in silicon solar cells - results of the German research cluster SolarFocus," *physica status solidi (c)*, vol. 8, p. 733, 2011.

[24] D. P. Fenning, A. S. Zuschlag, M. I. Bertoni, B. Lai, G. Hahn, and T. Buonassisi, *Journal of Applied Physics,* vol. 113, p. 044521, 2013.

[25] J. Schön, H. Habenicht, W. Warta, and M. C. Schubert, "Chromium distribution in multicrystalline silicon: comparison of simulations and experiments," *Progress in Photovoltaics: Research and Applications,* vol. 21, p. 676, 2013.

[26] M. A. Jensen, J. Hofstetter, A. E. Morishige, G. Coletti, B. Lai, D. P. Fenning, and T. Buonassisi, "Synchrotron-based analysis of chromium distributions in multicrystalline silicon for solar cells," *Applied Physics Letters*, vol. 106, p. 202104, 2015.

[27] SEMI PV Group, "International Technology Roadmap for Photovoltaic," 2017

[28] V. Vähänissi, A. Haarahiltunen, H. Talvitie, M. Yli-Koski, H. Savin, "Impact of phosphorus gettering parameters and initial iron level on silicon solar cell properties," *Progress in Photovoltaics: Research and Applications*, vol. 21, p. 1127, 2013.

[29] A. E. Morishige, M. A. Jensen, J. Hofstetter, P. X. T. Yen, C. Wang, B. Lai, and D. P. Fenning, *Applied Physics Letters*, vol. 108, p. 202104, 2016.

[30] A. A. Istratov, H. Hieslmair, and E. R. Weber, "Iron and its complexes in silicon," *Applied Physics A: Materials Science & Processing* vol. 69, p. 13, 1999.

[31] H. S. Laine, V. Vähänissi, Z. Liu, E. Magaña, J. Krügener, A. E. Morishige, B. Lai, K. Salo, H. Savin, and D. P. Fenning, "Elucidation of iron gettering mechanisms operating in boron-implanted silicon," to be submitted to *Applied Physics Letters*.

[32] M. Aoki, H. Akito Hara, and O. Akira, "Fundamental properties of intrinsic gettering of iron in a silicon wafer," *Journal of Applied Physics*, vol. 72, p. 895, 1992.

[33] J. D. Murphy, and R. J. Falster, "Contamination of silicon by iron at temperatures below 800°C," *physica status solidi (RRL) - Rapid Research Letters,* vol. 5, p. 370, 2011.

[34] J. D. Murphy, R. E. McGuire, K. Bothe, V. V. Voronkov, and R. J. Falster, "Competitive gettering of iron in silicon photovoltaics: oxide precipitates versus phosphorus diffusion," *Journal of Applied Physics*, vol. 116, p. 053514, 2014.

[35] H. Hieslmair, A. A. Istratov, S. A. McHugo, C. Flink, T. Heiser, and E. R. Weber, "Gettering of iron by oxygen precipitates," *Applied Physics Letters*, vol. 72, p. 1460, 1998.

[36] T. Buonassisi, A. A. Istratov, M. Heuer, M. A. Marcus, R. Jonczyk, J. Isenberg, B. Lai, Z. Cai, S. Heald, W. Warta, R. Schindler, G. Willeke, and E. R. Weber, "Synchrotron-based investigations of the nature and impact of iron contamination in multicrystalline silicon solar cells," *Journal of Applied Physics*, vol. 97, p. 074901, 2005.

[37] J. Schmidt, B. Lim, D. Walter, K. Bothe, S. Gatz, T. Dullweber, and P. P. Altermatt, "Impurity-related limitations of next-generation industrial silicon solar cells," *IEEE Journal of Photovoltaics,* vol. 3, p. 114, 2013.

[38] V. Vähänissi, A. Haarahiltunen, M. Yli-Koski, and H. Savin, "Gettering of iron in silicon solar cells with implanted emitters," *IEEE Journal of Photovoltaics*, vol. 4, p. 142, 2014.

[39] A. E. Morishige, H. S. Laine, J. Schön, A. Haarahiltunen, J. Hofstetter, C. del Cañizo, M. C. Schubert, H. Savin, and T. Buonassisi, "Building intuition of iron evolution during solar cell processing through analysis of different process models," *Applied Physics A: Materials Science & Processing*, vol. 120, p. 1357, 2015.

[40] A. F. Wolf, A. Martinez-Limia, D. Stichtenoth, and P. Pichler, "Modeling the annealing of dislocation loops in implanted c-si solar cells," *IEEE Journal of Photovoltaics*, vol. 4, p. 851, 2014.

[41] H. Hieslmair, S. Balasubramanian, A. A. Istratov, and E. R. Weber, "Gettering simulator: physical basis and algorithm," *Semiconductor Science and Technology*, vol. 16, p. 567, 2001.

[42] A. Haarahiltunen, H. Savin, M. Yli-Koski, H. Talvitie, and J. Sinkkonen, "Modeling phosphorus diffusion gettering of iron in single crystal silicon," *Journal of Applied Physics*, vol. 105, p. 023510, 2009.

[43] H. Wagner, A. Dastgheib-Shirazi, B. Min, A. E. Morishige, M. Steyer, G. Hahn, C. del Cañizo, T. Buonassisi, and P. P. Altermatt, "Optimizing phosphorus diffusion for photovoltaic applications: peak doping, inactive phosphorus, gettering, and contact formation," *Journal of Applied Physics*, vol. 119, p. 185704, 2016.

Impact of the initial growth interface on the grain structure in HPMC-Si ingot

Giri Wahyu Alam[1,2,3], Etienne Pihan[1], Benoit Marie[1], Nathalie Mangelinck-Noël[2]

[1] Univ. Grenoble Alpes, INES, CEA, LITEN, Department of Solar Technologies, F-73375 Le Bourget du Lac, France

[2] Aix-Marseille University, CNRS, IM2NP, UMR CNRS 7334, Campus Saint-Jérôme, Case 142, 13397 Marseille Cedex 20, France

[3] Centre of Technology for Materials, Agency for the Assessment and Application of Technology, Bldg. 224 PUSPIPTEK, South Tangerang 15314, Indonesia

Abstract — In High Performance Multi-Crystalline-Silicon (HPMC-Si) ingots, small seed grains generate grain boundaries that can terminate the propagation of dislocation clusters. Here, we focus on the seed template formation and its impact on the initial growth by directional solidification utilizing metallography, photoluminescence, and EBSD analysis. In the seed region, two randomly oriented grain morphologies are found: a genuine non-melted seed from the poly-Si chunks, and a re-solidified infiltrated molten silicon region. All grains grow by epitaxy on the seed grains and grains grown from wider grains in the seed, reach a higher solidification height.

Index Terms — EBSD, grain structure, HPMC-Si, photoluminescence, seeded growth.

I. INTRODUCTION

In both research and industry, seeding is an explored way to control the final structure of Silicon photovoltaic (PV) ingots. In particular, HPMC–Si method makes use of fine grained seeds aiming at creating grain boundaries that can terminate the propagation of dislocation clusters [1-3]. Indeed, dislocation clusters coupled to impurities are very detrimental to PV properties [4] so that it is necessary to limit their propagation. Numerous characterization studies concerning the grain morphology during ingot growth have led to the conclusion that RAGB presence in the growth area and small grain size are of primary importance to provide sufficient termination sites for dislocations clusters [5]. Consequently, the crystalline defect density in the growth ingot can be reduced. However, initial growth on seeds and seed morphology has only been studied in details by few authors. Ronit *et al.* [6] describe the growth from poly-Si seeds as a three stage process. Those authors evidenced an initial growth with spherical grain morphology (stage I), from which grains grow in the solidification direction together with grain selection (stage II) before to reach a steady state in terms of columnar grain size (stage III). From the observation of the grain structure in both FBR (Fluidized Bed Reactor) and poly-Si seeds, Ekstrøm *et al.* [7] observed a coarsening of the microstructure of the two seed materials. In the case of poly-Si chips, the authors also observed a larger grain structure initiated

from silicon infiltrated in-between poly-Si chunks gaps below the main seed interface.

It is the aim of this contribution to deepen the knowledge of initial grain template formation in seeds and to discuss its consequences on initial grain structure at the start of HPMC-Si ingot growth. The ultimate objective is to tailor the seed layer morphology to control the final grain structure. As a consequence, we focused on metallographic observations, photoluminescence imaging, etch pits analysis and EBSD mapping of:

- initial seed layer composed of poly-Si chunks grain structure and inter-chunks gaps,
- seed layer formation from genuine non-melted seed from the poly-Si chunks and from re-solidified infiltrated molten silicon,
- grain structure in the vicinity of the seed / initial growth interface.

II. EXPERIMENTAL TECHNIQUES

A G2 (60kg) silicon ingot was directionally solidified from high purity (> 250 Ω.cm) WACKER poly-Silicon (Size 4) in a silicon nitride coated VESUVIUS silica crucible. Boron doped silicon was added close to the top of the poly-Si chunks to achieve 1 - 2 Ω.cm resistivity target. The melting and growth interface was concave and a dipping rod provided regularly the information on residual solid height in the middle of the ingot. The seed arrangement consisted of 20 mm poly-Si chunks Size 1 (length range 0 – 15 mm chunk size). During the melting segment, the top and bottom zone temperatures were set at 1510°C and 1410°C, respectively while the growth rate was set at 1.2 cm/h. The position of the sample extracted from the ingot and studied in details in the following can be seen in red Fig. 1. This sample is situated at the seed layer- initial growth interface in a region of concavity of the solid/liquid interface. Besides, a reference sample (30 mm in diameter) was created by introducing Epoxy Araldite AY-103 resin in size 1 poly-Si chunks to simulate the initial seed morphology and to observe initial grain size and gaps between chunks.

978-1-5090-5606-4/17 $31.00 © 2017 IEEE

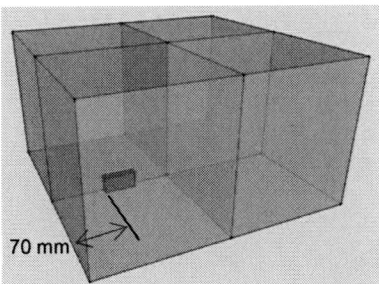

Fig. 1. Position of the sample inside the ingot (red parallelepiped).

After polishing, the selected sample was characterized using photoluminescence (LIS-R2-[8]) to identify seed layer-initial growth interface from doping variation influence on carrier density under illumination. Etch pit studies were performed using Wright etchant [9] to reveal dislocations and grain boundaries. Lastly a SEM – Nova Nanosem 630 was used for EBSD mapping with a 9.9 μm step on a selected 10 x 8.85 mm² area across the seed-initial growth interface to study grain structure and crystallographic orientation.

III. ANALYSIS OF THE RESULTS

A. Initial seed layer

The reference sample was observed during resin pouring. There were many gas bubbles escaping from the gaps simultaneously to the resin filling in the gaps between poly-Si chunks. Gas pockets were trapped and concentrated on the isolated areas between the stacked poly-Si chunks or are due to its irregular shape. After resin hardened, it was horizontally polished and vertically cut in the middle revealing the cross section of the poly-Si chunks. Gap widths between poly-Si chunks are found highly variable in the range of 0.1 mm to 15 mm horizontally and 0.06 mm to 7.3 mm vertically. At the same time, the cross section of poly-Si chunk can be very thin around 0.09 μm in thickness.

Metallography of the initial poly-Si chunks reveals two main microstructure morphologies and are measured by Abrams Three-Circle Procedure (Fig. 2) [10]: isolated concentric columnar grains with grain size over 10 μm, some even reaching 150 μm (red arrows) and abundant small elongated grains, below 10 μm diameter (yellow arrows). Concentric columnar grains are also found in [11] and are called "feather" like structure. This measurement is comparable to the measurement in [11] that used Heyn Intercept Method with a circular line. Additionally, many circle shape defects (yellow rectangle) which, behave as cracks in the chunks are observed on the surface. Their origin is unknown to the authors. Dislocation etch pits are found in all grains of the initial poly-Si under 500x magnification and above. Moreover, Σ3 twins are revealed particularly in the large columnar grains.

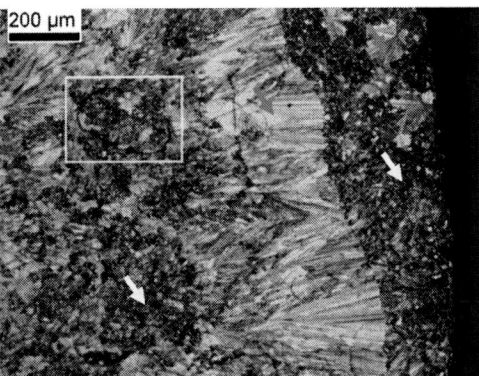

Fig. 2. Reference typical initial poly-Si microstructure (polished and wright etched reference resin infiltrated sample).

B. Seed layer in the grown ingot

PL imaging (Fig. 3) allows identifying precisely the position of the concave initial growth interface separating seed layer (darker bottom domain) and growth (light grey top domain) at the beginning of the process. The unambiguous use of PL is probably related to the different doping concentrations in the seed layer region and in the molten silicon just before the crystallization started. The observation of very bright domains near the growth interface is an indication that dopants are imbedded in the re-solidified areas in the seed layer during the early stages of the melting segment. However, it must be reminded that the PL contrast reveals a complex combination of doping and crystalline quality in the sample. The growth interface is found almost continuous separating spherical and columnar grain structures.

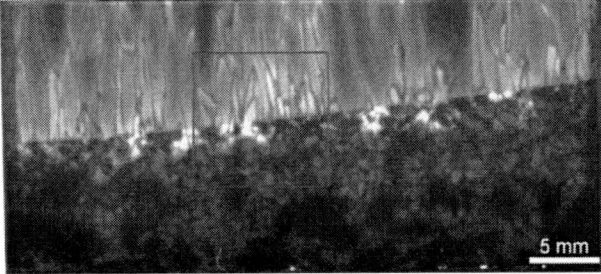

Fig. 3. Photoluminescence imaging using high magnification lens at the bottom of the grown ingot (40x20 mm²) showing both seed layer and growth areas delineated by the initial growth interface.

In Fig. 3, the genuine non-melted seed domains and the re-solidified infiltrated melt domains are revealed by a dark PL intensity and a brighter PL intensity, respectively. Similarly to the reference sample, the gap width between non-melted chunks is around 10 mm at maximum.

Chemical etching of the same area (Fig. 4) allows to identify the associated morphology. One can distinguish two main grain structures and two sublayers in the seed layer. From the interface to a position about 4 mm below it, most of the grains have an aspect ratio close to unity. Below that level, we clearly distinguish genuine non-melted seed domain of poly-Si chunks (size < 200 μm) from re-solidified infiltrated melt domains (size

472 µm in average and Coefficient of Variation (CV) 0.37) [12].

Moreover, we can see in Fig. 4 that genuine non-melted seed from poly-Si chunks undergoes a strong grain coarsening generating an increasing grain size toward the growth interface (from 18 – 72 µm at the seed layer bottom to around 300 µm close to the initial growth interface), everywhere far above the initial grain size.

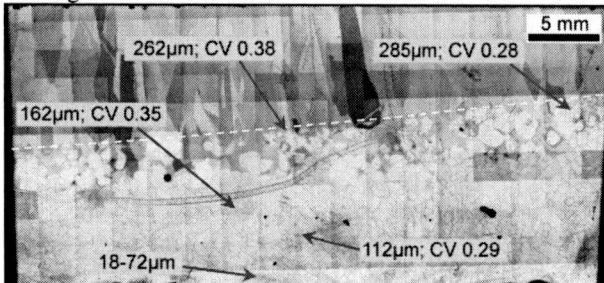

Fig. 4. Build-up of the same area as in Fig. 3 (optical images after chemical attack). The yellow dashed line delineates the initial growth interface determined from PL. Local grain size in genuine non-melted seed are also indicated.

C. Grain structure in the vicinity of the seed layer-growth interface

The 10 x 8.85 mm2 area identified as a red box in Fig. 3 was analyzed in more details using EBSD (Fig. 5). The EBSD technique provides both crystallographic orientation (IPF: Inverse Pole Figure, Fig. 5a) and grain boundary types (Fig. 6a).

At first, EBSD results also confirm that all grains from the seed intersecting the initial interface grow as columnar grains due to the directional solidification process. Second, regarding grain size and morphology measured from Heyn Lineal Intercept Method with straight lines [10] and aspect ratio, it appears that:

- the seed layer up to 4 mm below the initial interface position (Fig. 5a) has an average grain size of 465 µm (CV 0.78) and 453 µm (CV 0.94) in the horizontal and vertical directions respectively. The average aspect ratio is around 1 indicating spherical grains in longitudinal sections.
- columnar grains above the initial interface position (Fig. 5a) grow on seeds with an average grain size of 352 µm (width) near the seed layer-initial growth interface line.
- the average width almost doubles to reach 639 µm after 5 mm of growth on the top side of observed area indicating strong grain selection in the very first 5 millimeters of growth

Both morphologies of seed grains (spherical and columnar) have a random orientation as shown in Fig. 5a. In the growth region just above the interface (Fig. 5c), the crystallographic orientations along the growth direction are mostly in direction near-<115> and nearly none near-<101> and <111>. The texture in the initial growth region (Fig. 5b) and seed layer (Fig. 5c) are not identical due to the fact that the orientations of the grains in the seed region are not all present at the level of the initial growth interface.

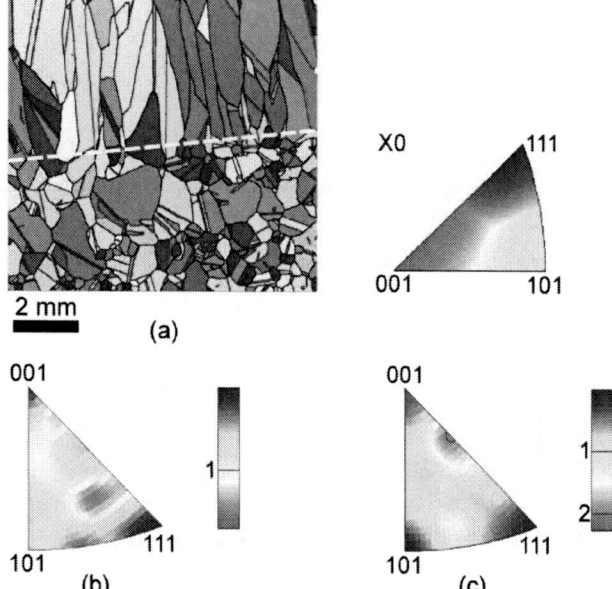

Fig. 5. (a) EBSD IPF (Inverse Pole Figure) map along the growth direction X0. Texture of the seed layer area (b), and of the initial growth area (c).

To understand in deeper details the grain formation and competition from the seed, we measure the growth length of the different grain boundary types. In our process, the length percentage of RAGBs (Fig. 6b) is the highest then followed by Σ3, Σ9, Σ27a, and Σ27b length percentage. The growth lengths of Σ9, Σ27a, and Σ27b are the smallest. These observations confirm that grains start there growth by nucleation on the

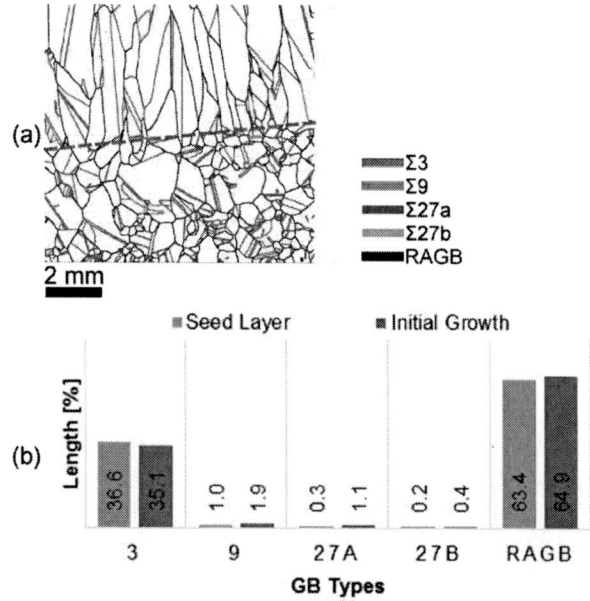

Fig. 6. The grain boundary map showing selected twin grain boundaries (Σ3, Σ9, Σ27a, Σ27b) and RAGB (a) and Length proportion of each grain boundary types in the seed (green) and growth (blue) regions just above the initial interface (b).

randomly oriented seed grains leading to a randomly oriented grain structure at the beginning of growth. This is consistent with the lower occurrence of Σ3 grain boundary type. Another study will be devoted to the evolution of the grain structure upper in the sample where Σ3 grain formation followed by the formation of Σ9, Σ27a, and Σ27b by competition could be expected [13] oppositely to random nucleation.

Measuring the grain width at the level of the initial growth interface shows that the grain width is linked to the columnar grain propagation length. In the EBSD observation area, about 50% of the grains have a width between 200 – 400 μm, and the rest have a width between 400 – 1200 μm.

The growth length of columnar grains issued from a grain width lower than 400 μm is below 1 mm for 70% of them and none reaches a growth length of 3 mm. In the case of grains with an initial width higher than 400 μm, there is no stop of growth before 1 mm, around 20% does not reach a higher growth length than 2 mm and 40% reaches a growth length above 4 mm.

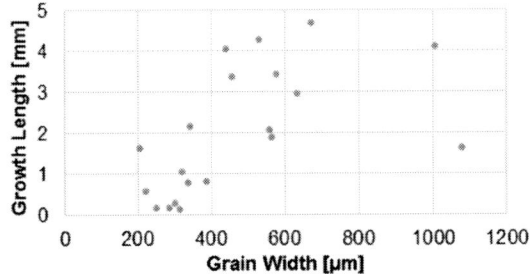

Fig. 7. Growth propagation length perpendicularly to the interface as a function of grain width measured at the initial interface from Fig. 5.

IV. DISCUSSION

A. Formation of the seed layer

During the resin pouring, the resin easily reached the bottom of the mold and filled in the gaps between the chunks. The viscosity of this resin (1800 – 2400 mPa.s at 25°C [14]) is higher than the molten silicon viscosity at the melting point (0.575 mPa.s [15]). As a consequence, we can conclude that the molten silicon can flow down and fill in the seed layer gap without retardation and can finally reach the bottom surface of the crucible. In parallel, there is an intense exchange in the seed layer between the argon gas that previously fills the gap between the poly-Si chunks and the infiltrated molten silicon. However, this exchange cannot completely be carried out in some areas because the escaped gas can be trapped in between the stacked chunks, thus creating gas holes as also observed in metal casting. The cavities are not always caused by the trapped gas. Indeed, due to the heat distribution in the seed layer, a colder area in the corner crucible of a concave furnace type is created. When approaching this area, the temperature of flowing molten silicon reduces rapidly and might freeze before completely filling the seed layer gap.

Our results in Fig. 3 and Fig. 4 show that, on the one hand, some large spherical grains (499 μm (CV 0.68) in average) grow during the melting stage inside the seed layer gap and that on the other hand, poly-Si coarsening is significant (from size lower than 100 μm to grain size larger than 200 μm). In the following, the first and the second regions are denoted respectively re-solidified infiltrated molten silicon and genuine non-melted silicon. The first region is created after the infiltrated silicon is accumulated in the bottom crucible and re-solidified between the poly-Si chunks. The low temperature difference and the higher solidification rate creates re-solidified grains with an aspect ratio close to 1. In average, these grains have a bigger size than the second region of genuine non-melted silicon at the bottom crucible. This phenomenon is also discussed in [16, 17].

The falling molten silicon corresponds to the first silicon melted and can be doped with various amounts of dopants depending on its initial location in the crucible. Indeed, Boron doped silicon was initially situated at the top part of the feedstock arrangement. As a consequence, it is believed that the varying PL contrasts in the seed layer in (Fig. 3) is due to successive melt infiltrations downwards in the gaps even though, the PL contrast can also indicate a defective area. Upon measuring the resistivity, the target value of 1 – 2 Ω.cm in the whole ingot growth is attained only from the middle height of the ingot growth. It implies that some dopant still remains somewhere in the seed layer after the melting segment ends. As a consequence, the molten silicon has a lower than expected concentration of boron in the initial growth.

In the vertical direction, the seed layer structure can be divided in two main regions: re-solidified infiltrated molten silicon and genuine non-melted silicon.

In parallel, the applied thermal gradient also creates conditions for grain coarsening inside the seed. A strong grain coarsening on the genuine non-melted seed increases the seed grain size. The very dark PL intensity as found on the bottom crucible is to be related to a smaller grain sizes and a high etch pits density. When grains reach a specific size (> 300 μm), their emitted PL intensity is similar as the one from re-solidified infiltrated molten silicon. This condition is observed in the seed layer up to 4 mm below the growth interface. Consequently, it is difficult to distinguish between grains originating from solidified infiltrations in-between the seed chunks (~ 500 μm) and large coarsened grains of poly-Si chunks (> 200 μm). However, some genuine non-melted seed is still present close to the interface as indicated from the circular crack in poly-Si chunk also evidenced in the literature [7] and from the dark PL intensity region for different experimental conditions.

Moreover, comparison to the grain structure in the seed layer of another run, with different thermal history during the melting segment, shows that genuine non-melted seed of poly-Si grain coarsening is dependent on thermal history during the melting segment. The grain coarsening might start from the beginning of the melting segment, i.e. earlier than in the case of the infiltrated molten silicon, as long as the poly-Si chunk temperature reaches 1000°C or 1150°C [10, 18]. At this temperature, the grain coarsening could be generated gradually from the poly-Si chunk surface towards the center depending on the heat on its surface. The thickness of the poly-Si chunk might also induce a temperature difference between the surface and the inner area and, finally, influence its grain size as seen in the cross section of genuine non-melted seed near to the crucible bottom.

One of the most important factors in thermal history, that control the grain size, is the temperature setting of heaters in the furnace. The effect of thermal gradient is only found in the vertical cut due to the vertical heat effect whereas the horizontal cut shows similar grain structure

B. Consequences on ingot grain structure

Combined PL and EBSD results prove that initial growth is governed by nucleation on seed so that the seed layer structure is reproduced at the initial growth interface. This simple but important point implies that the initial grain structure is controlled fully by the seed layer properties. As a consequence, the grain size on the seed layer will control the initial size of the columnar grains. Moreover, the wider the grains in the seed layer, the longer the columnar grains whereas a high density of grain boundaries associated to small grains might be preferred in HPMC-Si process. Another important point is that grain selection is strong in the first 5-millimeter of growth, larger grains having a higher survival probability. Further investigation needs to be conducted to find which kind of initial grain size could generate lower dislocated area in the whole ingot.

In the case of the utilization of poly-Si chunks as seed, the combination of the initial grain size variation and of the different thermal gradient might provide a higher variation of grain size in the seed layer. However, lower grain size variation and a bigger grain size might develop in the re-solidified seed. As the larger initial grains have a higher survival probability, the columnar grains grown from the re-solidified seed can develop longer and can be found at the top part of the ingot. We can expect that an appropriate adjustment between the gradient temperature and the poly-Si size can ensure homogenous grain size. Having homogenous grain size might also have an impact on the occurrence of dislocated areas as observed when using FBR as seed in comparison to the poly-Si chunk seed [7]. From this, we can conclude that we should control the formation of specific grain properties from the seeds to be able to tailor the ingot properties.

Additionally, as the growth initiated from the remaining seed grains, the initial crystallographic orientation is mainly random as expected in the growth of HPMC-Si ingot. This orientation is not promoted only by the genuine non-melted seed, but also by the combination with the re-solidified infiltrated silicon melt seed providing a random orientation base for the HPMC-Si initial growth. The random orientation in the genuine non-melted seed is expected from the initial grain orientation in the poly-Si chunks. The random orientation in the re-solidified infiltrated silicon melt seed is a consequence of the small grains development after arrival of the infiltrated melt in the cold region. These small grains are solidified without preferred orientation and without an assistance from the poly-Si chunk. It leads to a random orientation as in the genuine non-melted seed.

At the start of growth, the occurrence of RAGB and Σ3 boundaries in the vertical section are similar to what is also observed in the literature. This initial growth, in the first 5-milimeter, shows similar length of grain boundary type as found in the seed layer near the growth interface. The RAGB length has the highest proportion above 50% followed by Σ3 boundaries. There are several factors that influence the length of RAGB. As the RAGB grow in the same direction together with the columnar grains, they elongate perpendicular to the growth interface. Oppositely, Σ3 boundaries grow in geometrical directions imposed by the {111} facets with a short growth length [13]. Therefore, the growth length of RAGB is much longer in the initial stage.

However, the Σ3 twin grains are easily formed due to their lowest energy of formation and thus they can be found associated to almost all columnar grains. In the columnar grains, the presence of Σ3 boundaries, which is associated with twin growth, can also be initiated by the seed layer. The grains of the seed layer grow jointly with associated Σ3 twin grains. The Σ3 grain boundaries in the seed layer are not only formed by the re-solidified infiltrated melt seed or during the synthesis of poly-Si chunk in the Siemens process. They can also be formed as twin annealing growth during the coarsening of poly-Si chunk [19]. This coarsening is clearly seen by comparing the microstructure of poly-Si chunk before and after the ingot growth.

V. SUMMARY OF THE WORK

The seed layer morphology from HPMC-Si ingots is found to vary from the bottom of the crucible to the initial growth interface. There are two domain morphologies at the bottom of the seeds: a genuine non-melted seed, and a re-solidified seed. The genuine non-melted seed undergoes a strong grain coarsening generating an increasing grain size toward the growth interface due to the thermal history.

The comparison of PL imaging and EBSD shows that at the seed layer-growth interface, growth takes place on all grains intersecting the growth interface. Consequently, a strong correlation between the seed layer grain sizes and crystallographic orientation exists in the first growth area. Grains are randomly oriented, contain twins and RAGB boundaries with equivalent radial grain size in this region. The larger seed grains regrown at the initial growth interface have a longer survival length. It suggests that the largest grains in the seeds are more prone to generate the final grain structure after further growth.

As a consequence, two major parameters control grain size and grain selection: seed layer formation during the melting stage and the first stage of columnar growth for the given processing parameters. In the future, we will also focus on the grain crystallographic orientation in the propagation length analysis and on the analysis of the dislocation density and distribution related to the seeding.

ACKNOWLEDGMENTS

This work was partially funded by the CEA-INES, France and the Ministry of Research, Technology and Higher Education Republic of Indonesia, through its Scholarship Program for Research and Innovation in Science and Technology (RISET-Pro) World Bank Loan No. 8245-ID. The authors would like to acknowledge N. Plassat and D. Ponthenier for ingot growth.

REFERENCES

[1] C. W. Lan, W. C. Lan, T. F. Lee, A. Yub, Y. M. Yang, W. C. Hsu and A. Yang, "Grain control in directional solidification of photovoltaic silicon," *Journal of Crystal Growth*, vol. 360, pp. 68-75, 2012.

[2] C. W. Lan, C. F. Yang, A. Lan, M. Yang, A. Yu, H. P. Hsu, B. Hsu and C. Hsu," Engineering silicon crystals for photovoltaics," *CrystEngComm*, vol.18, pp. 1474–1485, 2016.

[3] Arjan Ciftja and Gaute Stokkan, "Growth of High Performance Multicrystalline Silicon: A Literature Review," 978-82-14-05742-3, 2014.

[4] Supawan Joonwichien, Satoru Matsushima and Noritaka Usami, "Effects of crystal defects and their interactions with impurities on electrical properties of multicrystalline Si," *Journal of Applied Physics*, vol. 113, pp 133503-1 – 6, 2013.

[5] G. Stokkan, Y. Hub, O. Mios and M. Juel, "Study of evolution of dislocation clusters in high performance multicrystalline silicon," *Solar Energy Materials & Solar Cells*, vol. 130, pp. 679-685, 2014.

[6] R. R. Prakash, T. Sekiguchi, K. Jiptner and Y. Miyamura, "Grain growth of cast-multicrystalline silicon grown from small randomly oriented seed crystal," *Journal of Crystal Growth*, vol. 401, pp. 717-719, 2014.

[7] K. Ekstrøm, G. Stokkan, A. Autruffe, R. Søndenå, H. Dalaker, L. Arnberg and M. D. Sabatino, "Microstructure of multicrystalline silicon seeded by polysilicon chips and fluidized bed reactor granules," *Journal of Crystal Growth*, vol. 441, pp. 95-100, 2016.

[8] T. Trupke, B. Mitchell, J. Weber, W. McMillan and R. Bardos, "Photoluminescence Imaging for Photovoltaic Applications," *Energy Procedia*, vol. 15, pp. 135-146, 2012.

[9] P. Walker and W. H, Eds., CRC Handbook of Metal Etchants, Boca Raton, Florida: CRC Press LLC, 1991.

[10] ASTM E112-12, Standard Test Methods for Determining Average Grain Size, West Conshohocken, PA: ASTM International, 2012.

[11] R. W. Fancher, C. M. Watkins, M. G. Norton, D. F. Bahr, and E. W. Osborne, "Grain growth and mechanical properties in bulk polycrystalline silicon," *Journal of Materials Science*, vol. 36, pp. 5441-5446, 2001.

[12] H. Abrams, "Grain Size Measurement by the Intercept Method," *Metallography*, vol. 4, pp. 59-78, 1971.

[13] M. Tsoutsouva, T. Riberi-Beridot, G. Regula, G. Reinhart, J. Baruchel, F. Guittonneau, L. Barrallier and N. Mangelinck-Noël, "In situ investigation of the structural defect generation and evolution during the directional solidification of ⟨110⟩ seeded growth Si," *Acta Materialia*, vol. 115, pp. 210-223, 2016.

[14] Huntsman Advanced Materials, Araldite® AY 103-1/ Hardener HY 991, March 2009.

[15] Yuzuru Sato et al., "Viscosity of molten silicon and the factors affecting measurement," Journal of Crystal Growth, vol. 249, pp. 404-415, 2003.

[16] M. Trempa, C. Reimann, J. Friedrich, G. Müller, A. Krause, L. Sylla and T. Richter, "Defect formation induced by seed-joints during directional solidification of quasi-mono-crystalline silicon ingots," *Journal of Crystal Growth*, vol. 405, pp. 131 – 141, 2014.

[17] Q. Wang, R. P. Liu, Y. Q. Qian, D. C. Lou, Z. B. Su, M. Z. Ma, W. K. Wang, C. Panofen and D.M. Herlach, "Metal-like growth of silicon during rapid solidification by quenching undercooled droplets on a chill plate," *Scripta Materialia*, vol. 54, pp. 37-40, 2006.

[18] C. Daey Ouwens and H. Heijligers, "Recrystallization processes in polycrystalline silicon," *Applied Physics Letters*, vol. 26, pp. 569-571, 1975

[19] R. E. Smallman and R. J. Bishop, Modern Physical Metallurgy and Materials Engineering: Science, Process, Application, 6th ed., Oxford: Butterworth-Heinemann, 1999.

Effect of Carbon Concentration and Growth Conditions on Oxygen Precipitation Behavior in n-type Cz-Si

Takuto Kojima[1]*, Ryota Suzuki[1], Kosuke Kinoshita[1], Kyotaro Nakamura[1], Atsushi Ogura[1],
Yoshio Ohshita[2], Isao Masada[3], and Shoji Tachibana[3]

[1]Meiji University, Kawasaki, Kanagawa, 214-8571, Japan
[2]Toyota Technological Institute, Nagoya, Aichi 468-8511, Japan
[3]Tokuyama Corporation, Shunan, Yamaguchi 745-8648, Japan

Abstract — **The behavior of oxygen precipitates under solar cell fabrication processes and the effect on device performance were investigated using TEM observation. Samples were prepared with different carbon concentration and with two sets of growth conditions. The number of precipitates correlates monotonically with the carbon concentration. When the initial carbon concentration is high, the cell efficiency is improved by decreasing the carbon concentration. When the initial carbon concentration is reduced to smaller than 10^{16} cm^{-3}, the oxygen precipitate grows largely and introduces dislocation. Precipitates grown in a plate form introduce dislocations to the surroundings at a high density, while dislocation density is relatively small around the precipitates polyhedral-grown on the basis of an octahedron. Under the growth conditions for introducing a plate-like precipitate, the cell efficiency was not improved even if the carbon concentration was reduced.**

I. INTRODUCTION

Oxygen in Cz silicon agglomerates into precipitates and the precipitates grow during crystal growth and device processes, and it becomes a major deterioration factor of carrier lifetime [1]. Oxygen precipitates (OPs) change morphology by heat treatment depending on the growth and process conditions. Recombination enhancement at oxygen precipitates has been attributed to the defect levels at the Si/SiO$_x$ interface and/or gettered metals [2-4]. However, the precipitation behavior in the actual solar cell process and the influence on the devices are not sufficiently discussed.

In this study, the dynamics of oxygen precipitation by solar cell processes and the influence on the devices are discussed from the transmission electron microscope (TEM) evaluation. Two sets of growth conditions were used to investigate the influence of the initial state of precipitation.

II. EXPERIMENTAL PROCEDURE

N-type ingots were produced under two sets of growth conditions, conventional (CZ-A) and improved (CZ-B), using polysilicon feedstock with different carbon concentrations. The crystal diameters are 220 mm with 1 m long. The same growth furnace was used under both CZ-A and CZ-B conditions, and the overall growth time was set to the same length. Figure 1 shows the lifetime and resistivity profiles, measured by Sinton

BSL-I bulk lifetime measurement system, along the crystal length for the typical CZ-A and CZ-B. As clearly shown the improved crystal CZ-B exhibits the higher lifetime covering almost complete region.

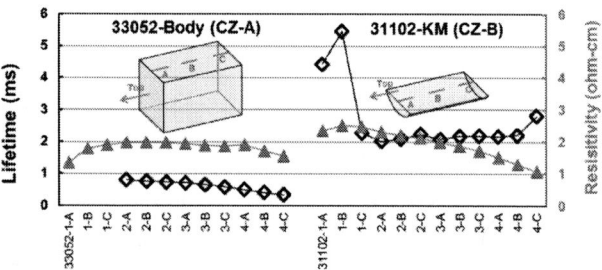

Fig. 1. Lifetime and resistivity profiles along the crystal length for the typical CZ-A and CZ-B.

Substrates sliced from the top of the ingots, containing high concentration of oxygen, were used in the experiment. Concentrations of substitutional carbon at parts of the ingots were measured by FT-IR, and the initial concentrations of carbon in the melt were calculated assuming normal segregation. The used wafers were named A0 to A5, and B0 to B2 for the substrates from CZ-A and CZ-B growth conditions in descending order of carbon concentration, respectively. The wafers were processed into bifacial n-PERT cells. The fabrication process includes thermal processes at approximately 1000 °C for boron diffusion and oxidation. Density of OPs were determined by TEM observation. The morphology of OPs was also observed.

III. RESULTS AND DISCUSSION

Before the cell fabrication process, OPs were not observed by TEM, whereas after the cell fabrication process, OPs of >10 nm were observed in all the samples. The numbers of OPs in a 10,000× field of view by TEM were counted to determine the density of OPs at the center part of the wafers (Fig. 2). The density is calculated assuming thickness the observed ion-milled sample is 30 nm. For both sets of growth conditions, the carbon concentration and OP density showed approximately

978-1-5090-5606-4/17 $31.00 © 2017 IEEE

linear correlation, while the OP density in the CZ-B substrates were nearly one order magnitude smaller than those of CZ-A. The density of OPs could be reduced by modification of thermal history in crystal growth.

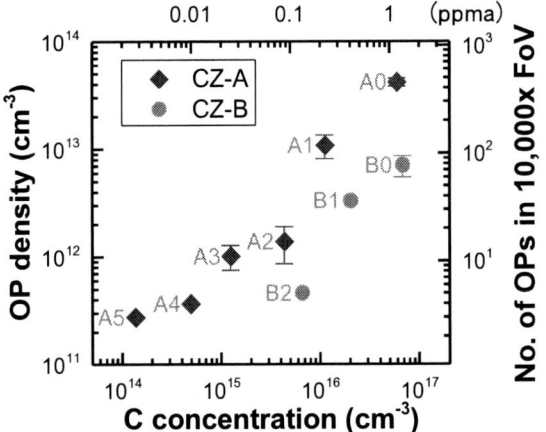

Fig. 2. Oxygen precipitate density for different carbon concentrations and growth conditions. The Error bars represent variations in counting OPs in an identical FoV.

The conversion efficiencies of the fabricated solar cells are shown in Fig. 3. When the initial carbon concentration is the highest (A0 and B0), there is no difference in conversion efficiency between CZ-A and CZ-B growth conditions. The cell efficiency was improved by decreasing the carbon concentration, up to 16.5 % (V_{oc} = 599 mV, J_{sc} = 36.2 mA/cm², and FF = 0.762) in CZ-A cells and 17.8% (V_{oc} = 629 mV, J_{sc} = 37.5 mA/cm², and FF = 0.754) in CZ-B cells. The improvement was greater in CZ-B, whereas in CZ-A, it was saturated under approximately 5×10^{15} cm⁻³.

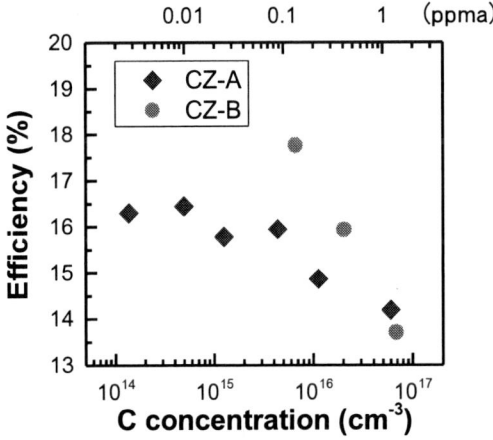

Fig. 3. Cell performance of the n-PERT bifacial cells for different carbon concentrations and the growth conditions.

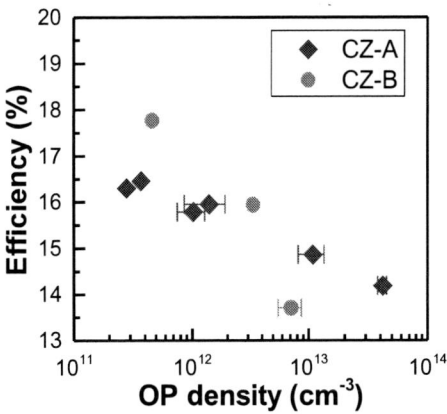

Fig. 4. Cell performance of the n-PERT bifacial cells shown in Fig. 3 against OP density in Fig. 2.

The relation between cell efficiency (in Fig. 3) and OP density (in Fig. 2) is shown in Fig. 4. In CZ-B, the deterioration per one OP is larger than that of CZ-A when the OP density is about 7×10^{12} cm⁻³ at high carbon concentration (B0). On the other hand, at low density, OPs in CZ-B wafers become much more harmless. The degradation per OP may be derived from the size, morphology, and/or OP-induced defects. Therefore, TEM observation for each OP was carried out.

The results of high magnification ⟨110⟩ TEM observation on the precipitates are shown in Fig. 5. The observed dislocations are indicated by yellow arrows. In A0 and B0 with the highest carbon concentration, small octahedral or truncated-octahedral precipitates of approximately 10 nm were observed. The small precipitates do not introduce dislocations. When the initial carbon concentration becomes smaller than 10^{16} cm⁻³, the oxygen precipitate grows largely and produces so-call "punch-out" dislocations. In B2 with a small precipitation density, the precipitate grows larger than B1, and dislocations are generated around. On the other hand, in A2, dislocations were distributed with high density around plates of precipitates. Similar plate-like precipitates with dislocations were also observed in A3 to A5.

978-1-5090-5606-4/17 $31.00 © 2017 IEEE

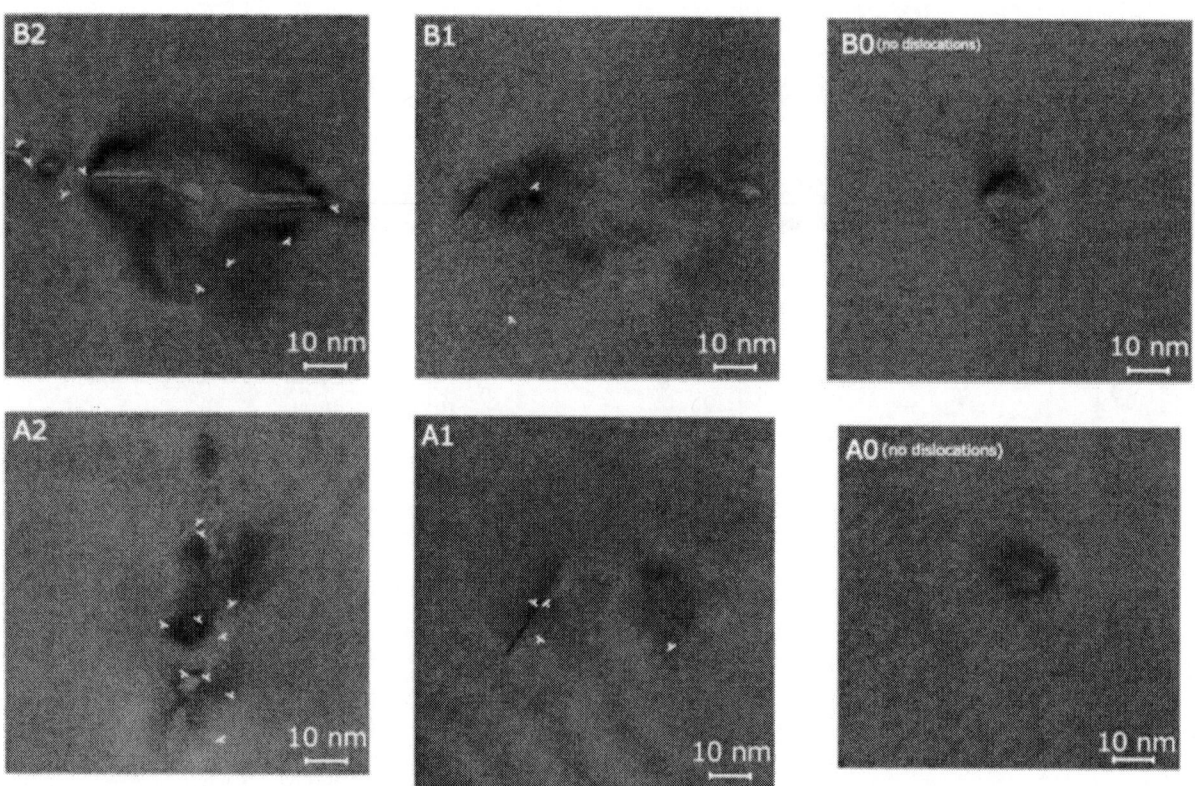

Fig. 5. ⟨110⟩ TEM images at oxygen precipitates in A2, B1, and B2. Yellow arrows point dislocations.

From the results, when the initial carbon concentration is small, the density of the precipitate becomes small and grows largely. The OP density is so high that the distance between OPs is smaller than or comparable to diffusion length of interstitial oxygen (O_i) during thermal processes. Therefore, when OP density is high, mobile oxygen is competed by adjacent OPs for size growth. Precipitates that do not introduce dislocations also degrade cell efficiency, but plate-like precipitates that introduce dislocations at high density suppress efficiency even at low density. The precipitates growing on the basis of the octahedron are considered to have a moderate introduction of dislocation and the influence on cell performance is relatively small.

IV. SUMMARY

The behavior of oxygen precipitates under solar cell fabrication process and the effect on device performance were investigated using TEM observation. Samples were prepared with different carbon concentration and with two sets of growth conditions. The number of precipitates correlates monotonically with the carbon concentration. When the initial carbon concentration is high, the cell efficiency is improved by decreasing the carbon concentration. When the initial carbon concentration becomes smaller than 10^{16} cm^{-3}, the oxygen precipitate grows largely and introduces dislocation. Precipitates grown in a plate form introduce dislocations to the surroundings at a high density, while dislocation density is relatively small in precipitates polyhedral-grown on the basis of an octahedron. Under the growth conditions for introducing a plate-like precipitate, the cell efficiency was not sufficiently improved even if the carbon concentration was reduced. As a result, the CZ-B has been much improved compared to conventional CZ-A not only in their initial life time but also after cell fabrication processes with less degradation.

ACKNOWLEDGMENT

Part of this study was supported by New Energy and Industrial Technology Development Organization (NEDO). We also thank Dr. H. Hashiguchi of JEOL for his support in TEM observation.

REFERENCES

[1] J. Haunschild, I. E. Reis, J. Geilker, and S. Rein, Phys. Status Solidi: Rapid Res. Lett. **5**, 199 (2011).
[2] J. M. Hwang and D. K. Schroder, J. Appl. Phys. **59**, 2476 (1986).
[3] J. D. Murphy, R. E. McGuire, K. Bothe, V. V. Voronkov, and R. J. Falster, Sol. Energy Mater. Sol. Cells **120**, 402 (2014).

[4] J. D. Murphy, M. Al-Amin, K. Bothe, M. Olmo, V. V. Voronkov, and R. J. Falster, J. Appl. Phys. **118**, 215706 (2015).

Nano-imaging of Performance in Photovoltaics

Elizabeth M. Tennyson[1,2] and Marina S. Leite[1,2]

[1]Department of Materials Science and Engineering, [2]Institute for Research in Electronics and Applied Physics, Univ. of Maryland, College Park, MD, 20740, USA

Abstract — The performance of most polycrystalline photovoltaics is still limited by their open-circuit voltage (V_{oc}). Thus, the successful implementation of high-efficiency and low-cost solar cells requires the control and suppression of non-radiative recombination centers within the material. We realize novel high spatial resolution microscopies and image the local V_{oc} of a variety of materials, with spatial resolution <50 nm. In CIGS we map spatial variations in photovoltage >20% between grains, never observed before. Through fast-Kelvin probe force microscopy (16 sec/scan) we measure, in real-time, the transient behavior of perovskites at the nanoscale, attributed to iodine ion migration. In summary, we developed and demonstrated a new functional imaging platform to determine the local electrical response of photovoltaic materials, providing new insights into the loss mechanisms that hinder several types of solar cells.

I. INTRODUCTION

Polycrystalline solar cells, such as CIGS, CdTe and perovskites, are an attractive substitute for Si photovoltaic (PV) devices due to their versatility of substrate, low fabrication costs and high performance. At present, these technologies are all >22% power-conversion efficiency [1]. Typically the photon absorbing films are between 1–3 μm, two orders of magnitude thinner than conventional Si solar cells. Maintaining high-efficiency, flexible substrates broadens PV applications, enabling innovative concepts and designs.

A primary open question in the field of polycrystalline PV is: how does the microstructure influence the local electrical response? It is known that the active layers in these solar cells are composed of grains typically 1 μm in diameter (see Fig. 1(a)). The grain boundaries (GBs) throughout the layers are attributed to sites of charge carrier non-radiative recombination, thus, reducing the open-circuit voltage (V_{oc}) of these solar cells.

Several types of characterization techniques have been used previously to investigate the structural and electrical properties of polycrystalline materials [2]-[3]. In particular, scanning probe microscopy techniques can be applied to help explain the influence of the GBs on the devices' electrical properties [4]-[9]. Despite the remarkable progress achieved in the field, measurements of the *local* electrical properties of devices *in-operando* are still missing. Here, we spatially resolve and quantify the charge recombination and collection characteristics of polycrystalline grains by imaging the local voltage response of solar cells with high spatial resolution. For that, we developed a set of functional imaging methods, as described below.

II. RESULTS AND DISCUSSION

A. Photovoltage Tomography of Charge Carriers in CIGS

We build maps of photovoltage at different depths in the CIGS through a sequence of spectrally dependent voltage scans (Fig. 1). The penetration depth of photons into the CIGS layer is a function of the absorption coefficient, α, as shown in Fig. 1(b). As such, by spectrally resolving the electrical response, we yield 3D information about the charge carrier recombination within the CIGS layer. These measurements are acquired using a confocal optical microscope, providing sub-micron spatial resolution. According to Fig. 1, at short wavelengths (λ) the light penetration into the CIGS material is

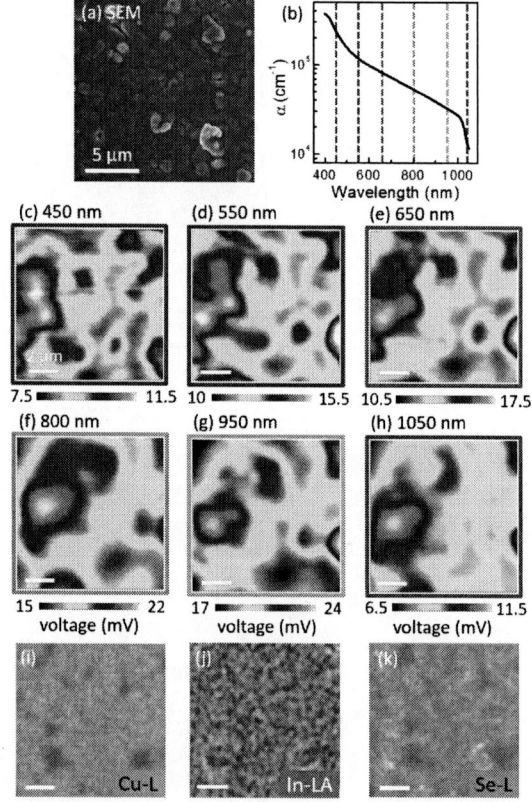

Fig. 1. (a) SEM image of CIGS surface showing grains and interfaces. (b) Absorption coefficient (α). Dashed colored bars highlight the wavelengths (λ) at which photovoltage scans were acquired. (c-h) Scanning photovoltage microscopy for different excitation λ, acquired at the *same* location as shown in (a). (i-k) EDS maps for Cu, In, and Se, respectively.

close to the surface, as expected. However, for incident photons with energy near the material bandgap (Figs. 1(c)-(h)) the distribution of the voltage signal across the scanned area varies substantially. This result demonstrates that the CIGS grains do not uniformly generate voltage in all three dimensions. Further, the mesoscale electrical signal of the material varies at the same length scale as the grains, suggesting that different interfaces in the CIGS layer produce distinct photovoltage responses, dependent upon the incident λ. We determine local electrical enhancements at each λ by dividing the local voltage response value by the average voltage ($\langle V \rangle$) of the entire scanned area, and we find spatial variations >20% [10] (not shown here) in certain locations. To confirm that the non-uniformities in the voltage response are not caused by chemical composition we perform energy dispersion X-ray spectroscopy (EDS), see Figs. 1(i)-(k). The maps show an even distribution of the chemical elements, evidence that the electrical signal is likely due to the structural properties of the CIGS grains.

B. A Universal Method to Map V_{oc} at the Nanoscale

To elucidate the origin of the modest V_{oc} currently observed in all polycrystalline PV devices, we develop and implement a novel imaging method to quantify the local V_{oc} with spatial resolution <50 nm, based on Kelvin-probe force microscopy (KPFM). Briefly, in KPFM the contact potential difference signal measured (V_{CPD}) is proportional to the work function difference between the tip and the surface of the material under investigation. Thus, by subtracting an illuminated-KPFM from a dark-KPFM scan, we acquire a map of the splitting of the quasi-Fermi level ($\Delta\mu$), defined as [11]:

$$V_{oc} = \frac{\Delta\mu}{q} = V_{CPD}^{il} - V_{CPD}^{d} \qquad (1)$$

For each map, the probe passes over the surface twice first to image the topography and then to measure the V_{CPD}. As such, the morphology and electrical response of the polycrystalline grains are acquired simultaneously. The details concerning the instrumentation implemented here is presented in ref [11].

To analyze how specific interfaces of polycrystalline devices influence the overall device performance (and V_{oc}), we

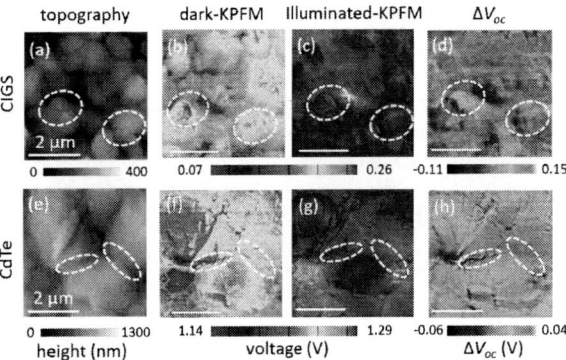

Fig. 2. AFM, dark-KPFM, illuminated-KPFM, and ΔV_{oc} maps for (a-d) CIGS and (e-h) CdTe solar cells, respectively. Illumination conditions: 1-sun illumination, $\lambda = 660$ nm. Adapted from [11].

introduce the term $\Delta V_{oc} = V_{x,y} - \langle V \rangle$, here $V_{x,y}$ is a pixel in the dark- minus illuminated-KPFM map (not shown) and $\langle V \rangle$ is the scan averaged value. Thus, $\Delta V_{oc} < 0$ corresponds to regions that are harmful to the voltage response, and $\Delta V_{oc} > 0$ refers to local electrical enhancements. Note that, in CdTe, different interfaces showed distinct local electrical responses (see white dashed areas in Fig. 2(f)). Our metrology can be applied to any optoelectronic device, including LEDs and photodetectors, does not require full device processing, is nondestructive, and works in ambient environment. Further, it can be applied to identify if specific processing steps are beneficial for the ultimate performance of the device.

C. Assessing the Transient V_{oc} of Perovskites in Real-Time

One of the primary limitations for transforming perovskites into a PV technology is due to the material instabilities when exposed to light [12]-[13]. Thus, to investigate the origin of the light-induced processes that affect device reliability we measure, at the nanoscale, the transient voltage response of perovskites under 1-sun illumination conditions. We leverage our expertise in KPFM to implement fast KPFM scans (16 sec/scan), which enables real-time measurements of the voltage changes. We perform all scans in a low humidity (<15% relative humidity) atmosphere. Fig. 3 shows a sequence of time-dependent scans during and post-

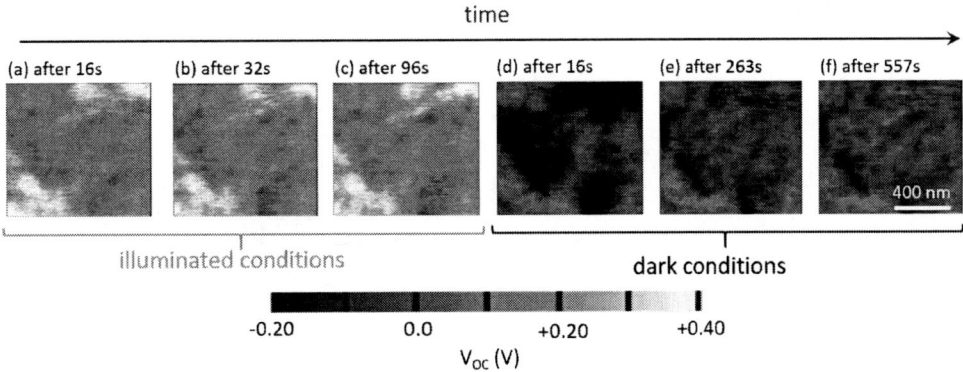

Fig. 3. Time-dependent V_{oc} nanoscale maps of a perovskite solar cell under (a-c) 1-sun illumination λ=500 nm, and (d-f) dark conditions.

978-1-5090-5606-4/17 $31.00 © 2017 IEEE

illumination. While the material is being illuminated, see Figs. 3(a)-(c), we measure spatial variations in voltage >300 mV, which we attribute to the non-uniform distribution of ions during sample preparation. Surprisingly, when the perovskite is under dark conditions (post-illumination), regions out of and in equilibrium co-exist, as shown in Figs. 3(d)-(f). This residual and transient V_{oc} results from iodine migration [14]. It takes ~9 min for the sample to reach its steady state. Next steps include combining the V_{oc} scans with chemical imaging by μ-Raman measurements [15].

IV. CONCLUSIONS

We have developed and implemented a set of functional imaging methods for PV, based on scanning microscopies, to map the local V_{oc} and to address key scientific questions related to a variety of materials. The recombination of charge carriers within polycrystalline grains and interfaces were probed by two high spatial resolution microscopy platforms. By spectrally varying the incident photons on a CIGS material, we can build a 3D tomography of charge carrier interactions. Further, illuminated-KPFM demonstrates at the nanoscale, how the interfaces between grains in both CdTe and CIGS devices can influence the electrical response. Finally, we showed the perovskite's instability as a function of time due to light exposure. For this, we imaged the V_{oc} signal every 16 sec and tracked the location- and time-dependent instabilities. We expect these microscopy methods to become a diagnostic tool for non-epitaxial PV, as it can be expanded to probe a variety of emerging materials, such as high-efficiency Cs-triple cation and lead-free perovskite alternatives.

ACKNOWLEDGMENTS

The authors gratefully acknowledge N. Ballew, R.Y. Bekele, J.A. Frantz, J.L. Garrett, W.B. Gunnarsson, J.M. Howard, M. Hu, J. Huang, J.N. Munday, J.D. Myers, O. Rabin, and J.S. Sanghera, and the technical support of the Maryland NanoCenter and its AIMLab. Financial support from NSF-ECCS (award #16-10833), the 2017-2018 UMERC Hulka Energy Research Fellowship, 2016 UMD Graduate Dean's Dissertation Fellowship, 2015 UMD Graduate School Summer Research Fellowship, the 2015 RASA Award, and the 2014 Minta Martin Award.

REFERENCES

[1] M. A. Green, K. Emery, Y. Hishikawa, W. Warta, E. D. Dunlop, D. H. Levi and A. W. Y. Ho-Baillie "Solar Cell Efficiency Tables (version 49)" *Progress in Photovoltaics: Research and Applications* vol. 25, 1, pp. 3-13, 2017.

[2] T. Kirchartz and U. Rau *Advanced Characterization Techniques for Thin Film Solar Cells* Published Online: 7 APR 2011:Wiley-VCH, 2011.

[3] E. M. Tennyson, J. M. Howard and M. S. Leite "Mesoscale Functional Imaging of Materials for Photovoltaics" *ACS Energy Letters,* **invited perspective**. In Review, 2017.

[4] T. Dittrich, A. Gonzales, T. Rada, T. Rissom, E. Zillner, S. Sadewasser and M. Lux-Steiner "Comparative study of Cu(In,Ga)Se-2/CdS and Cu(In,Ga)Se-2/In2S3 systems by surface photovoltage techniques" *Thin Solid Films* vol. 535, pp. 357-361, 2013.

[5] M. S. Leite, M. Abashin, H. J. Lezec, A. G. Gianfrancesco, A. A. Talin and N. B. Zhitenev "Nanoscale Imaging of Photocurrent and Efficiency in CdTe Solar Cells" *ACS Nano* vol. 8, 11, pp. 11883-11890, 2014.

[6] H. R. Moutinho, R. G. Dhere, C.-S. Jiang, Y. Yan, D. S. Albin and M. M. Al-Jassim "Investigation of potential and electric field profiles in cross sections of CdTe/CdS solar cells using scanning Kelvin probe microscopy" *Journal of Applied Physics* vol. 108, 7, pp. 074503, 2010.

[7] M. Takihara, T. Minemoto, Y. Wakisaka and T. Takahashi "An investigation of band profile around the grain boundary of Cu(InGa)Se2 solar cell material by scanning probe microscopy" *Progress in Photovoltaics: Research and Applications* vol. 21, pp. 595-599, 2013.

[8] I. Visoly-Fisher, S. R. Cohen, K. Gartsman, A. Ruzin and D. Cahen "Understanding the Beneficial Role of Grain Boundaries in Polycrystalline Solar Cells from Single-Grain-Boundary Scanning Probe Microscopy" *Advanced Functional Materials* vol. 16, pp. 649-660, 2006.

[9] W. Li, S. R. Cohen and D. Cahen "Effect of Chemical Treatments on nm-scale Electrical Characteristics of Polycrystalline Thin Film Cu(In,Ga)Se2 Surfaces" *Solar Energy Materials and Solar Cells* vol. 120, pp. 500-505, 2014.

[10] E. M. Tennyson, J. A. Frantz, J. M. Howard, W. B. Gunnarsson, J. D. Myers, R. Y. Bekele, J. S. Sanghera, S.-M. Na and M. S. Leite "Photovoltage Tomography in Polycrystalline Solar Cells" *ACS Energy Letters* vol. 1, pp. 899-905, 2016.

[11] E. M. Tennyson, J. L. Garrett, J. A. Frantz, J. D. Myers, R. Y. Bekele, J. S. Sanghera, J. N. Munday and M. S. Leite "Nanoimaging of Open-Circuit Voltage in Photovoltaic Devices" *Advanced Energy Materials* vol. 5, 23, pp. 1501142, 2015.

[12] J. Berry, T. Buonassisi, D. A. Egger, G. Hodes, L. Kronik, Y.-L. Loo, I. Lubomirsky, S. R. Marder, Y. Mastai, J. S. Miller, D. B. Mitzi, Y. Paz, A. M. Rappe, I. Riess, B. Rybtchinski, O. Stafsudd, V. Stevanovic, M. F. Toney, D. Zitoun, A. Kahn, D. Ginley and D. Cahen "Hybrid Organic-Inorganic Perovskites (HOIPs): Opportunities and Challenges" *Advanced Materials* vol. 27, pp. 5102-5112, 2015.

[13] W.-J. Yin, J.-H. Yang, J. Kang, Y. Yan and S.-H. Wei "Halide perovskite materials for solar cells: a theoretical review" *Journal of Materials Chemistry A* vol. 3, 17, pp. 8926-8942, 2015.

[14] J. L. Garrett, E. M. Tennyson, M. Hu, J. Huang, J. N. Munday and M. S. Leite "Real-Time Nanoscale Open-Circuit Voltage Dynamics of Perovskite Solar Cells" *Nano Letters* vol. 17, 4, pp. 2554-2560, 2017.

[15] W. Tress, N. Marinova, T. Moehl, S. M. Zakeeruddin, M. K. Nazeeruddin and M. Gratzel "Understanding the Rate-Dependent J-V Hysteresis, Slow Time Component, and Aging in $CH_3NH_3PbI_3$ Perovskite Solar Cells: The Role of a Compensated Electric Field" *Energy Environ. Sci.* vol. 8, 3, pp. 995-1004, 2015.

Implications of conductive grain boundaries in Chlorine-treated CdTe solar cells

Mohit Tuteja[1,2], Vasilios Palekis[4], Allen Hall[2], Scott MacLaren[3], Chris S. Ferekides[4], Angus A. Rockett[2]

[1]Institute for Research in Electronics and Applied Physics, University of Maryland at College Park, MD 20742, USA
[2]Department of Materials Science and Engineering and [3]Frederick Seitz Materials Research Lab, University of Illinois at Urbana-Champaign, IL 61801, USA
[4]Department of Electrical Engineering, University of South Florida, Tampa, FL 33620, USA

ABSTRACT -- The rear surfaces of CdTe photovoltaic devices without back contacts, grown by closed-space sublimation (CSS) were analyzed using conductive – atomic force microscopy (C-AFM). As-deposited and $CdCl_2$ treated CdTe samples were compared to clarify the effect of the treatment on charge flow through grains and grain boundaries. The $CdCl_2$ treated samples exhibit a more homogenous and enhanced current flow across the grains as compared to the as-deposited samples. Under high bias, grain boundaries dominate current flow when the main junction is reverse biased and conducting current in reverse breakdown. The results are interpreted as resulting from the improved crystallinity of the CdTe with reduced p-type doping along the grain boundaries. Further, implications of these grain boundaries and guidelines for optimal device microstructure are discussed.

Index terms – conductive atomic force microscopy, CdTe solar cells, grain boundaries

I. INTRODUCTION

$CdCl_2$ annealing is an important step in CdTe solar cell fabrication that leads to improved open circuit voltage, short circuit current, and fill factor of the device [1], [2]. Atomic force microscopy, scanning electron microscopy, x-ray diffraction, and time-resolved photoluminescence measurements [2]–[5] have demonstrated that $CdCl_2$ annealing treatment induces recrystallization of CdTe films and promotes grain growth, which lead to enhanced minority carrier lifetime. CdTe is intrinsically p-type [6] but recently it has been demonstrated that after a $CdCl_2$ anneal, Cl is incorporated in the CdTe grain boundaries [7]–[9]. The presence of Cl, an n-dopant for CdTe, results in local doping, type inversion and carrier depletion, and results in p-doped CdTe grains separated by n-type or depleted p-type grain boundaries [8], [10]–[12].

Investigating individual grain boundary properties is challenging since interaction volumes probed by conventional characterization techniques span multiple grains and thus, the electrical conductivity of the grain boundaries as a function of $CdCl_2$-anneal treatment remains unclear. In order to clarify the effect of Cl incorporation, we compare close-spaced sublimation (CSS) deposited thin-film CdTe, with and without $CdCl_2$ annealing. This was accomplished using conductive atomic-force microscopy (C-AFM), which is a scanning probe technique that enables high-resolution nanoscale mapping of the current from the tip to the sample.

II. EXPERIMENT

The CdTe samples were fabricated on Corning EagleXG glass at the University of South Florida. Details of the growth process can be found in Ref. [13]. Representative device performance parameters for the finished CdTe solar cells fabricated using this method, with and without the $CdCl_2$ annealing, are listed in Table 1. C-AFM measurements were performed on the surfaces of the CdTe films at the University of Illinois using an Asylum Research MFP-3D AFM with commercially available solid-Platinum cantilevers from Rocky Mountain Nanotechnology, LLC. C-AFM utilizes a conductive probe in an AFM instrument. The typical radius of curvature of the tip is 20 nm. A voltage bias was applied between the tip and the sample, and the conductive probe was rastered across the sample in contact mode. During each scan the current flowing between the tip and the sample was measured at each scan pixel, enabling generation of a nanoscale-resolution map of electrical current through the sample. In all of the experiments reported here, the tip was grounded and the voltage bias was applied to the TCO layer underlying the junction.

Sample	V_{OC} (mV)	J_{SC} (mA/cm^2)	FF (%)	η (%)
as-deposited	700	19	53	≈7
$CdCl_2$ annealed	840	23	73	≈14

Table 1: Device performance parameters of CdTe solar cells with- and without $CdCl_2$ anneal treatment.

III. RESULTS AND DISCUSSION

Figures 1(a) and 1(b) show representative C-AFM current vs. sample voltage bias (I-V) plots along the grain bulk and the grain boundary regions for the as-deposited and $CdCl_2$ annealed CdTe samples, respectively. It can be seen that it is only possible to pass relatively high currents through the CdTe samples, both $CdCl_2$ annealed and un-annealed, for relatively high (compared to device operating conditions) sample biases.

The estimates of the valence band energy in CdTe (6 eV, from vacuum level to valence band maximum) [14] and the work function of Pt (5.1-5.5 eV) [15] indicate that we can anticipate creation of a high Schottky barrier between the CdTe film and the Pt-probe. The p-n junction diode (CdS/CdTe) and the Schottky diode (CdTe/Pt) are oriented in opposite directions. Therefore one diode is always reverse biased and would be expected to determine the resistance of the device overall. This is indeed what we observe. With positive voltage applied to the sample, the back Schottky contact is forward biased and the reverse biased junction is the main diode. Thus, the behavior illustrated in Figure 1 for both samples is dominated by reverse breakdown in the main CdS/CdTe junction in the positive sample bias regime. This reverse current flowing through the p-n junction diode increases exponentially with voltage and is referred to as *pre-breakdown reverse current*. This has been a subject of many investigations in the past [16]–[19] and some studies have concluded that such reverse currents primarily arise due to material defects/imperfections in the case of polycrystalline solar cells [18], [19].

Figure 2(a) and 2(c) show a deflection (derivative of topography) images for the as-deposited and the CdCl$_2$ annealed CdTe sample, respectively. We assume here that the grain boundaries are defined primarily by topography such that grains are delimited by the large consistent surface protrusions and the grain boundaries are the regions between these features. This is consistent with electron backscatter diffraction mapping measurements (not shown). The corresponding C-AFM current images acquired at +2V of sample bias on these CdTe samples are shown in Figures 2(b) and 2(d), respectively.

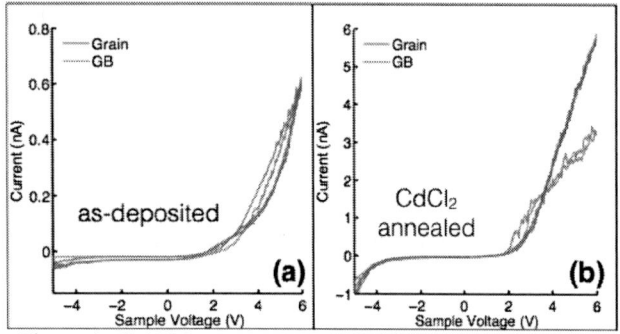

Figure 1: C-AFM current vs. sample bias curves for the (a) as-deposited and (b) CdCl$_2$ annealed CdTe sample.

We have also performed measurements under more extreme bias conditions, which are far from the conventional operation conditions of a CdTe solar cell. Figures 3(a) and 3(b) show the dark C-AFM current maps on the as-deposited and CdCl$_2$ annealed CdTe samples, respectively, along the scan area as shown in Figures 2(a) and 2(c). Figures 3 were obtained at a sample bias of +6V. The C-AFM current for as-deposited CdTe sample is uniformly distributed across the scan area with both the grains and the grain boundaries acting as current

conduction paths. The current map for the CdCl$_2$-annealed sample shows that the largest C-AFM current flows along the grain boundary areas as compared with the grain interiors.

When a Schottky contact is forward biased, it leads to a flow of majority carriers from the semiconductor into the metal [20]. For the p-doped CdTe, this means that the C-AFM current observed in the positive sample bias regime is normally composed of holes flowing from the CdTe layer into the Pt-tip. When the main CdS/CdTe junction is under sufficient reverse bias to be in a breakdown condition, it looses its rectifying characteristics and allows hole current to pass though the p-n junction relatively freely. The C-AFM current variations in this case would normally be expected to represent the variation in microscale electrical conductivity of the CdTe film for the flow of the majority charge carriers (holes) from the p-n junction to the Schottky contact.

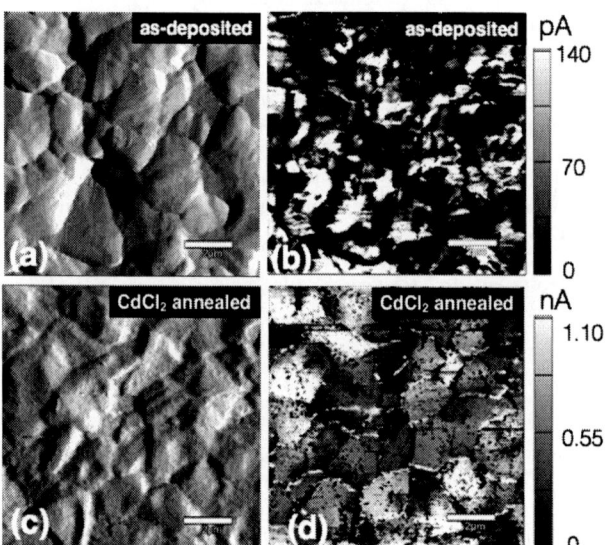

Figure 2: C-AFM analysis of as-deposited and CdCl$_2$ annealed CdTe samples. (a), (c): Deflection (derivative of topography) images for the scan area to highlight the topographical features. (b), (d): C-AFM current maps of the samples at +2V sample bias. Note: RMS roughness of the scan area is 165 nm and 135 nm for (a) and (c), respectively. Scale bars represent 2 μm in length.

Recently, we demonstrated that the CdCl$_2$ annealed CdTe samples exhibit carrier compensation and a reduced doping density along the grain boundaries due to incorporation of Cl. This produces a field in the grains along the grain boundaries that promotes exciton separation [12]. The CdTe layer is p-doped to a carrier concentration of (10^{14}-10^{15} cm^{-3}) [6] and it is likely that an incorporation of Cl along the grain boundaries results in formation of a narrow region of, at least, quasi-intrinsic CdTe. The Schottky diode is primarily a majority carrier device with a very small fraction of minority carrier injection current. However, at large forward voltage biases to the Schottky diode, a low acceptor doping in CdTe (for example in the grain boundaries) and a large Schottky barrier height for majority carrier flow, we could expect electron injection into CdTe (from the Pt-tip) along the grain boundaries to be more favorable than hole injection into the Pt-tip. These electrons would be collected spontaneously at the reverse-biased main junction to the grain boundary (assuming that the boundaries run through the thickness of the CdTe).

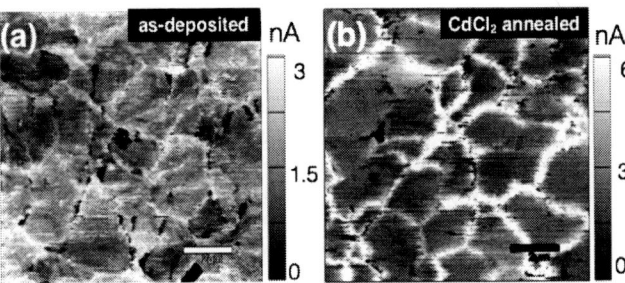

Figure 3: C-AFM current maps acquired at +6V sample bias for the as-deposited [(a)] and CdCl$_2$ annealed [(b)] CdTe samples. Scale bar represents 2 μm in length.

We conclude that when the tip is positioned over the p-doped CdTe grains, the charge motion is holes into the tip from the grains at all positive sample biases, and current is controlled by the main junction breakdown current and the conductivity of the CdTe film. However, along the grain boundaries in the CdCl$_2$ annealed sample, electron injection from the Pt-tip into the CdTe is likely and increases as the positive voltage bias is increased. This current may be collected directly by the reverse-biased main junction. In the negative sample bias regime, the p-n junction is forward biased and current flow should occur by injection of electrons into the p-type CdTe by the p-n junction. This current is limited due to either electron collection by the sample tip or by recombination with the reverse hole current in the Schottky contact.

It can be seen from Figure 2(b) (for the as-deposited sample) that under positive sample bias conditions, the current flows from the TCO layer to the Pt-tip primarily along the grain interiors, presumably as holes. Further, there is large inhomogeneity in electrical conductivity from one grain to the other. Upon comparing Figures 1(a) and 1(b), and Figures 2(b) and 2(d) we can see that, in the +2V sample bias regime, the currents flowing through the CdTe film are up to a factor of ~8 greater after the CdCl$_2$ anneal treatment as compared with the as-deposited samples. This indicates that the electrical conductivity of the CdTe film has improved upon the CdCl$_2$ anneal treatment. By comparison of Figures 2(b) and 3(b) we can also see that after the CdCl$_2$ anneal treatment the current distribution along the grains in the sample becomes much more homogenous as compared to the as-deposited sample. Hence, the CdCl$_2$ treatment makes the electrical properties of the grains in the CdTe film more homogenous and increases their electrical conductivity for the flow of holes.

In Figure 2(d) (low bias) there is considerable variation in grain boundary conductivity relative to the surrounding grains including both significantly higher and significantly lower currents. This is consistent with our previous findings that not all grain boundaries show similar amounts of carrier depletion/compensation following CdCl$_2$ treatment [11], [12]. We observe that at higher +6V bias [Figure 3(b)] there is strong contrast between the grains and grain boundaries with several boundaries conducting much higher current. In this case the current flow along the grain boundaries is due to electron injection from the Pt-tip. A key point here is that the holes injected into the grain interiors by the p-n junction under reverse breakdown do not simply flow through the highly conductive grain boundaries. They are blocked from the grain boundaries by the junction between the more p-type grains and the less p-type or somewhat n-type Cl-doped grain boundaries. In all cases where the grain boundaries are conductive, the adjacent grains do not appear to transfer holes to those conductive channels over significant distances. Finally, we can see that some grain boundaries in Figure 3(b) are at similar current levels to the adjacent grains [one example marked by arrows in Figures 2(c) and 3(b)]. This further proves that the incorporation of Cl along the boundaries and the resultant doping effects are not uniform across all the grain boundaries. There is no major current contrast observed between the grains and the grain boundaries for the as-deposited sample at +6V bias, as shown in Figure 3(a).

IV. OPTIMAL DEVICE STRUCTURE

The results from this work and our previous works combined [11], [12], [21] lead us to conclude that the grain boundaries in Chlorine-treated CdTe solar cells are carrier depleted and conductive for the flow of electrons. The carrier depletion causes downward band bending along the grain-grain boundary interface that is favorable for separation of electron-hole pairs and the high electron conductivity of grain boundaries cause the electrons to get transported to the underlying n-type (CdS) film. Hence, the special charged nature of grain boundaries in CdTe leads to efficient exciton separation and charge transport in the device under operation conditions.

According to the description outlined above, it might seem that an increase in the grain boundary density (through reduced grain size) will increase the charge separation and provide more routes for photo-generated electron transport. However, in polycrystalline CdTe cells, with carrier depleted boundaries, this will imply that holes would have to cross some grain boundary barriers before collection at the back contact, as shown in Figure 4(a). Therefore, decreased grain size, beyond a certain point, will potentially increase the avenues for carrier recombination.

From the advantageous and the harmful effects of grain

boundaries in CdTe solar cells, it seems that the average grain size (diameter of 1–3 μm) in the film, as found in empirically optimized devices, represents a good compromise between the various effects noted above. It is suggested that with columnar rather than isotropic grains, of diameters greater than the depletion width at the grain/grain boundary interface (~100-150 nm) [11], the electron and hole current pathways will be well segregated and run un-interrupted, resulting in improved cell performance. This is, in-fact, observed in the case of empirically optimized polycrystalline CdTe solar cells, where several of the CdTe grains run through the thickness of the CdTe film. It is important to note that the process of deposition of the back contact involves p-doping of the back surface of CdTe, away from the glass side. This prevents shunting of the device.

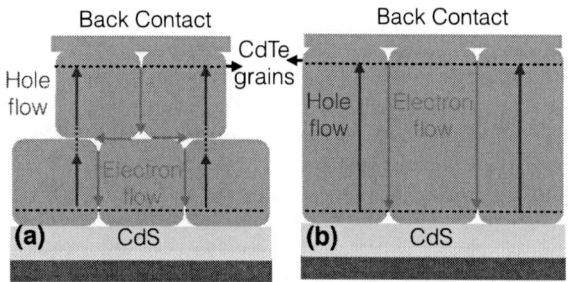

Figure 4: A schematic CdTe grain structure showing (a) small grains such that the grain boundaries cross the hole current paths, and (b) long columnar grains that run through the thickness of the CdTe film, such that the hole and electron current paths run un-interrupted.

V. CONCLUSIONS

We conclude that $CdCl_2$ treatment homogenizes and increases the electrical conductivity of CdTe films. Doping of the grain boundaries promotes charge separation, collection of electrons by the grain boundary, and their conduction to the front contact of the device. This is supported by the observation here of high current in the grain boundary under conditions where electrons are expected to be the dominant charge carrier. In the $CdCl_2$ annealed sample, the lack of conduction of holes injected into the grains through the grain boundaries to the back contact shows why recombination in the device operating as a photovoltaic does not occur at the grain boundaries. In the complete device the p-type doping of the back of the CdTe by the back contact prevents shunting through the conductive grain boundaries.

In order to further enhance the performance of polycrystalline CdTe solar cells, alternate deposition methods/conditions for CdTe should be researched such that the growth of columnar CdTe grains (with axis perpendicular to the CdS/CdTe junction) is promoted, as shown in Figure 4(b). For a more comprehensive discussion on this topic, please refer to Ref. [22].

ACKNOWLEDGEMENTS

The authors are thankful for the financial support from the U.S. Department of Energy, Office of Energy Efficiency and Renewable Energy (Contract DE-EE0005405). This work was carried out in part in the Frederick Seitz Materials Research Laboratory Central Research Facilities, University of Illinois.

REFERENCES

[1] C. S. Ferekides, U. Balasubramanian, R. Mamazza, V. Viswanathan, H. Zhao, and D. L. Morel, "CdTe thin film solar cells: device and technology issues," *Sol. Energy*, vol. 77, no. 6, pp. 823–830, Dec. 2004.

[2] B. E. McCandless, I. Youm, and R. W. Birkmire, "Optimization of vapor post-deposition processing for evaporated CdS/CdTe solar cells," *Prog. Photovolt. Res. Appl.*, vol. 7, no. 1, pp. 21–30, Jan. 1999.

[3] H. R. Moutinho, M. M. Al-Jassim, D. H. Levi, P. C. Dippo, and L. L. Kazmerski, "Effects of CdCl2 treatment on the recrystallization and electro-optical properties of CdTe thin films," *J. Vac. Sci. Technol. A*, vol. 16, no. 3, pp. 1251–1257, May 1998.

[4] H. R. Moutinho, F. S. Hasoon, F. Abulfotuh, and L. L. Kazmerski, "Investigation of polycrystalline CdTe thin films deposited by physical vapor deposition, close-spaced sublimation, and sputtering," *J. Vac. Sci. Technol. A*, vol. 13, no. 6, pp. 2877–2883, Nov. 1995.

[5] M. Tuteja, P. Koirala, J. Soares, R. Collins, and A. Rockett, "Low temperature photoluminescence spectroscopy studies on sputter deposited CdS/CdTe junctions and solar cells," *J. Mater. Res.*, vol. 31, no. 2, pp. 186–194, 2016.

[6] B. E. McCandless and J. R. Sites, "Cadmium Telluride Solar Cells," in *Handbook of Photovoltaic Science and Engineering*, A. Luque and S. Hegedus, Eds. John Wiley & Sons, Ltd, 2011, pp. 600–641.

[7] J. D. Poplawsky *et al.*, "Direct Imaging of Cl- and Cu-Induced Short-Circuit Efficiency Changes in CdTe Solar Cells," *Adv. Energy Mater.*, vol. 4, no. 15, p. 1400454, Oct. 2014.

[8] C. Li *et al.*, "Grain-Boundary-Enhanced Carrier Collection in CdTe Solar Cells," *Phys. Rev. Lett.*, vol. 112, no. 15, p. 156103, Apr. 2014.

[9] S. P. Harvey, G. Teeter, H. Moutinho, and M. M. Al-Jassim, "Direct evidence of enhanced chlorine segregation at grain boundaries in polycrystalline CdTe thin films via three-dimensional TOF-SIMS imaging," *Prog. Photovolt. Res. Appl.*, vol. 23, no. 7, pp. 838–846, Jul. 2015.

[10] I. Visoly-Fisher, S. R. Cohen, A. Ruzin, and D. Cahen, "How Polycrystalline Devices Can Outperform Single-Crystal Ones: Thin Film CdTe/CdS Solar Cells," *Adv. Mater.*, vol. 16, no. 11, pp. 879–883, Jun. 2004.

[11] M. Tuteja *et al.*, "Direct Observation of CdCl2 Treatment Induced Grain Boundary Carrier Depletion in CdTe Solar Cells Using Scanning Probe Microwave

Reflectivity Based Capacitance Measurements," *J. Phys. Chem. C*, vol. 120, no. 13, pp. 7020–7024, Apr. 2016.

[12] M. Tuteja, P. Koirala, S. MacLaren, R. Collins, and A. Rockett, "Direct observation of electrical properties of grain boundaries in sputter-deposited CdTe using scan-probe microwave reflectivity based capacitance measurements," *Appl. Phys. Lett.*, vol. 107, no. 14, p. 142106, Oct. 2015.

[13] C. S. Ferekides *et al.*, "High efficiency CSS CdTe solar cells," *Thin Solid Films*, vol. 361–362, pp. 520–526, Feb. 2000.

[14] S. G. Kumar and K. S. R. K. Rao, "Physics and chemistry of CdTe/CdS thin film heterojunction photovoltaic devices: fundamental and critical aspects," *Energy Environ. Sci.*, vol. 7, no. 1, pp. 45–102, Dec. 2013.

[15] H. B. Michaelson, "The work function of the elements and its periodicity," *J. Appl. Phys.*, vol. 48, no. 11, pp. 4729–4733, Nov. 1977.

[16] M. C. Alonso-García and J. M. Ruíz, "Analysis and modelling the reverse characteristic of photovoltaic cells," *Sol. Energy Mater. Sol. Cells*, vol. 90, no. 7–8, pp. 1105–1120, May 2006.

[17] O. Breitenstein, J. Bauer, J.-M. Wagner, and A. Lotnyk, "Imaging physical parameters of pre-breakdown Sites by lock-in thermography techniques," *Prog. Photovolt. Res. Appl.*, vol. 16, no. 8, pp. 679–685, Dec. 2008.

[18] D. Lausch *et al.*, "Identification of pre-breakdown mechanism of silicon solar cells at low reverse voltages," *Appl. Phys. Lett.*, vol. 97, no. 7, p. 73506, Aug. 2010.

[19] O. Breitenstein, J. Bauer, A. Lotnyk, and J.-M. Wagner, "Defect induced non-ideal dark – characteristics of solar cells," *Superlattices Microstruct.*, vol. 45, no. 4–5, pp. 182–189, Apr. 2009.

[20] S. M. Sze and K. K. Ng, "Metal-Semiconductor Contacts," in *Physics of Semiconductor Devices*, John Wiley & Sons, Inc., 2006, pp. 134–196.

[21] M. Tuteja *et al.*, "CdCl2 Treatment-Induced Enhanced Conductivity in CdTe Solar Cells Observed Using Conductive Atomic Force Microscopy," *J. Phys. Chem. Lett.*, vol. 7, no. 24, pp. 4962–4967, Dec. 2016.

[22] M. Tuteja, "Carrier transport, recombination, and the effects of grain boundaries in polycrystalline cadmium telluride thin films for photovoltaics," University of Illinois at Urbana-Champaign, 2016.

Imaging the Multi-Temporal Photo-Carrier Dynamics at the Nanometer Scale in Organic and Inorganic Solar Cells

Pablo A. Fernández Garrillo, [*,†,‡,§] Łukasz Borowik, [†,‡] Florent Caffy, [§] Renaud Demadrille, [§] and Benjamin Grévin [§]

[†]Université Grenoble Alpes, F-38000 Grenoble, France
[‡]CEA, LETI, MINATEC Campus, F-38054 Grenoble, France
[§]INAC-SYMMES UMR 5819, CEA, CNRS, Université Grenoble Alpes, F-38000 Grenoble, France

Abstract — **Understanding the interplay between the nanostructure and photo-transport mechanisms is central for the development of many emerging photovoltaic technologies. In recent years, there has been an increasing focus on the need to probe the photo-carrier dynamics at the nanometer scale. Recently, we demonstrated that non-contact atomic force microscopy combined with Kelvin probe force microscopy under frequency modulated illumination (FMI-KPFM) can be used to map two-dimensional images of the surface photo-voltage dynamics with a sub-10 nm lateral resolution. Here, a detailed overview of this approach including operating principle, instrumental aspects and data treatment will be given. The capability of FMI-KPFM to map spatial variations of the minority carrier lifetime is demonstrated in the case of silicon nanocrystal-based third generation solar cells. The accuracy of the post-data acquisition processing is discussed by analyzing the evolution of the dynamical contrasts as a function of the illumination intensity in organic bulk heterojunction thin films.**

Index Terms — **carrier recombination, KPFM, photo-carrier dynamics, nanostructured photovoltaics.**

I. Introduction

Photo-carrier recombination is of paramount importance in the overall efficiency of a solar cell as it defines the proportion of photo-generated charges collected at the electrodes. To date, most experimental approaches developed to probe recombination dynamics focus on the use of methods such as transient photo-voltage or charge extraction. Yet, these techniques average out sample properties over macroscopic scales, rendering them unsuitable for directly assessing the influence of the nanostructure and local defects on photo-carrier dynamics. In order to overcome this, few teams started to develop time-resolved scanning probe microscopies focused in addressing photo-carrier dynamics at the nanoscale. [1] In this context, Kelvin Probe Force Microscopy (KPFM) under frequency modulated illumination arises as an alternative in photo-carrier dynamics measurements, as it has been used to probe the recombination dynamics in organic blends [1]-[2] and the minority carrier lifetime in inorganic semiconductors. [3]-[4] However, for the most part, time-resolved KPFM-based methods have only been applied to obtain information coming from single points of the sample.

Recently, we showed that it is possible to image photo-carrier dynamics at different timescales with a sub-10 nm spatial resolution in both organic and inorganic photovoltaic (PV) samples by a two-dimensional (2D) spectroscopy approach based on Kelvin probe force microscopy under modulated illumination (FMI-KPFM). [5] Using this technique we simultaneously recorded time-resolved images of surface photo-voltage dynamics along with the topographic and dissipation channels, paving the way for a proper assessment of the nanostructure influence on the photo-carrier dynamics.

In this paper, detailed insights on this approach's operating principle as well as technical aspects including instrumentation, signals synchronization and data treatment are given. Furthermore, we show how this approach can be applied to image the spatial variation of the minority carrier lifetime in a silicon nanocrystal-based third generation solar cell. Lastly, the impact of the spectroscopic curve fit accuracy on the dynamical imaging resolution is addressed by analyzing the data obtained for different illumination intensities on organic donor-acceptor bulk heterojunction. Thus, this report provides useful guidance for further development of dynamical SPV imaging by FMI-KPFM.

II. Methods

KPFM is a scanning probe microscopy (SPM) method where the electrostatic potential difference, also called contact potential difference (CPD) between an atomic force microscopy (AFM) tip and the surface of a conducting or semi-conducting sample is measured. In essence, KPFM works by detecting electrostatic forces between tip and sample and compensating them by applying a proper DC bias which in turns yields a measure of the CPD. [6] In photovoltaic materials, KPFM can be used to probe the photo-generated charge carriers by analyzing the CPD shift under illumination, *i.e.,* the so-called surface photo-voltage (SPV). [7] Actually, it has been shown that SPV imaging yields a local measurement of the device open circuit voltage, [8] which can be defined as the electron-hole quasi Fermi levels splitting under illumination. [9]

In dynamical surface photo-voltage imaging with KPFM, the SPV is measured as a function of a modulated light source frequency. In this configuration, KPFM measures a temporally averaged surface potential (SP) value for each frequency of the modulated illumination source, as its compensation bias regulation loop cannot follow SPV variations faster than its integration time (typ. a few tens of ms).

Fig. 1 shows a schematic illustration describing the general implementation of dynamical surface photo-voltage imaging

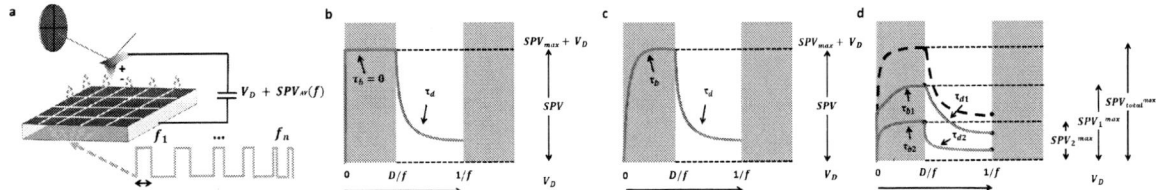

Fig. 1. (a) Schematic illustration of dynamical surface photo-voltage imaging with KPFM. The KPFM surface potential is measured by applying a compensation bias to the oscillating AFM tip. Spectroscopy curves of the average surface potential (Vav) are acquired at each point of a predefined grid over the sample's surface, as a function of the illumination's source modulation frequency. [5] (b),(c),(d) Different SPV time-evolution models used for data treatment and results interpretation. (b) Case of a SPV characterized by an instantaneous build-up time. (c) Case of a SPV characterized by a non-zero build-up time. (d) Case of a SPV defined as the superposition of different SPV contributions, each one with different build-up and decay times. Panels (a),(c) and (d) were adapted with permission under ACS AuthorChoice License from "Photo-Carrier Multi-Dynamical Imaging at the Nanometer Scale in Organic and Inorganic Solar Cells", P. A. Fernández Garrillo, Ł. Borowik, F. Caffy, R. Demadrille, B. Grévin, ACS Appl. Mater. Interfaces, 8 31460–31468 (2016). http://pubs.acs.org/doi/full/10.1021/acsami.6b11423

with KPFM as well as representations of the SPV time-evolution for different scenarios.

In practice, spectroscopic curves of the SP as a function of the illumination modulation frequency f (going from few hundred Hz to few hundred kHz) with a constant duty cycle (D) are acquired at each point of a pre-defined grid area over the sample. The acquisition time of a single spectroscopy curve is typically in the range of 1 to 10 seconds. This time depends on the number of selected frequencies. Typically between five and few tens frequency points are needed, depending on the monotonic or non-monotonic SPV evolution as a function of f (compare Fig. 2h and Fig. 3). Moreover, for each frequency, the data acquisition time is fixed to be at least 3 times the KPFM integration time constant, typically yielding a 100-200 ms data acquisition time. Taking into account that measurements are carried out over a typical 100x100 square grid area, the total duration of a single experiment can overpass a 24-hours' time period. For this reason, to ensure sample stability and to avoid drift artifacts, KPFM measurements are performed in single-pass mode in combination with beam deflection noncontact AFM (nc-AFM) under ultrahigh vacuum (UHV), while the 2D frequency–spectroscopy is operated in close loop mode (*i.e.*, uninterrupted z-regulation).

Furthermore, dynamical surface photo-voltage imaging with KPFM relies on a synchronous acquisition of the 2D frequency–spectroscopy measurements with regards to the illumination modulation frequency sweep. In our setup, we ensure this synchronization by employing logic signals generated by the AFM controller to externally trigger an arbitrary waveform generator. The latter is used to generate the illumination pulse groups at the selected frequencies and to modulate an adjustable output power diode-laser device employed to illuminate the sample.

In terms of data treatment, the temporally averaged surface potential V_{av} measured by KPFM can be defined as the time-domain integral of the SPV. If one considers an instantaneous SPV build-up (Fig. 1b), the SPV decay time-constant can be

extracted by fitting the spectroscopy curves using the following equation:

$$V_{av} = V_D + SPV_{max} D + SPV_{max} f\tau_d \left(1 - e^{\frac{-(1-D)}{f\tau_d}}\right) \quad (1)$$

Here, f and D represent the illumination modulation frequency and duty ratio respectively. V_D is the in-dark surface potential, SPV_{max} is the maximum surface photo-voltage that would be measured under continuous-wave illumination and τ_d represents de SPV decay time-constant.

This approach can be applied in the case of an average SPV signal that give no sign of decrease up to the upper modulation frequency allowed by the illumination chain. Here, we underline that using Eq. 1 follows the approach taken by former FMI-KPFM reports. [3]-[4]

We now consider the case of a SPV displaying a non-instantaneous build-up time (Fig. 1c). In that case a dual set of time-constant (τ_b, τ_d) associated with the SPV build-up (τ_b) and decay (τ_d) processes can be obtained by fitting [5] the spectroscopy curves with the following equation:

$$V_{av} = V_D + SPV_{max} D \left(1 - e^{\frac{-D}{f\tau_b}} e^{\frac{-(1-D)}{f\tau_d}}\right) + SPV_{max} (f\tau_d - f\tau_b) \left(1 - e^{\frac{-D}{f\tau_b}}\right) \left(1 - e^{\frac{-(1-D)}{f\tau_d}}\right) \quad (2)$$

Lastly, if we considered a SPV defined as the superposition of different SPV contributions (SPV_1, SPV_2) as depicted in Fig. 1d, then a set of two independent SPV build-up and decay

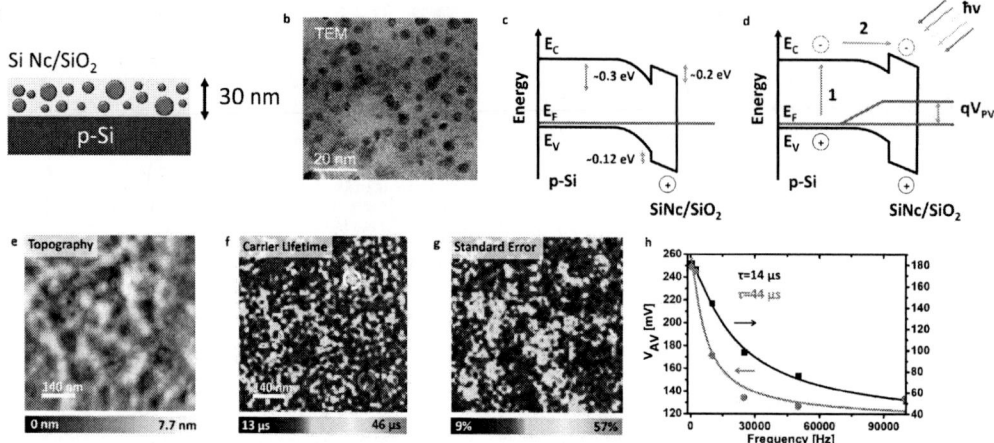

Fig. 2. (a) Cross-sectional schematic of a sample with SiNC embedded in 30 nm of SiO2. The layer of SiO2 is deposited on the p-Si substrate. (b) Top view of the sample made using transmission electron microscopy. Schematic illustration of energy band diagrams of p-Si substrate and SiNC/SiO2 layer in two cases: (c) layers are bounded and the internal electric field permits the Fermi levels alignment of the junction and (d) layers are bounded and under illumination. In this case, photo-generated electrons in the p-Si substrate are injected to the SiNC/SiO2 layer, this injection gives origin to the potential identified as photo-voltage V_{PV}. (e) nc-AFM topographic image (100×100 pixels, 700×700 nm) acquired during the 2D spectroscopy under frequency-modulated illumination (685 nm, 50% duty, 0.61 mW/mm²). (f) Minority carrier lifetime image. (g) Standard error of the minority carrier lifetime image. (h) Spectroscopic curves showing the SPV monotonic evolution with respect to the illumination frequency modulation. Red and black curves were obtained over areas indicated in 2f as white and red dotted circles respectively.

time-constants ($\tau_{b1}, \tau_{b2}, \tau_{d1}, \tau_{d2}$) can be calculated [5] upon spectroscopy curve fitting using the following expression:

$$V_{av} = V_D + SPV_1^{max} D\left(1 - e^{\frac{-D}{f\tau_{b1}}} e^{\frac{-(1-D)}{f\tau_{d1}}}\right) + SPV_1^{max}(f\tau_{d1} - f\tau_{b1})\left(1 - e^{\frac{-D}{f\tau_{b1}}}\right)\left(1 - e^{\frac{-(1-D)}{f\tau_{d1}}}\right) +$$

$$SPV_2^{max} D\left(1 - e^{\frac{-D}{f\tau_{b2}}} e^{\frac{-(1-D)}{f\tau_{d2}}}\right) + SPV_2^{max}(f\tau_{d2} - f\tau_{b2})\left(1 - e^{\frac{-D}{f\tau_{b2}}}\right)\left(1 - e^{\frac{-(1-D)}{f\tau_{d2}}}\right) \quad (3)$$

In order to fit the 2D matrix of spectroscopy curves, an automatic curve fit operation routine based on the least-square method was developed using SCILAB software. This routine allowed us to import the spectroscopic curve files generated by the SPM controller and to recalculate 2-dimensional dynamical images of each curve's characteristic time-constants.

III. RESULTS AND DISCUSSION

In a previous report, to illustrate the application of KPFM under frequency modulated illumination in the field of inorganics photovoltaics, we mapped the spatial variations of the minority carrier lifetime in a small grain polycrystalline silicon thin film grown on n-doped silicon. [5] Here, as a second application example, we imaged the spatial variation of the minority carrier lifetime in a silicon nanocrystal-based third generation solar cell. A schematic structure of the sample is depicted in Fig. 2a. A top view of the sample acquired by transmission electron microscopy (Fig. 2b) shows the nanocrystals embedded in the SiO2 matrix, the nanocrystals

display difference sizes ranging from 3 to 10 nm and are heterogeneously distributed. Moreover, nanocrystals with hydrogenated bonds have been used to reduce defects at the interface with the SiO2 matrix. [10] In this sample, a negative surface photo-voltage is observed (Fig. 2h). Indeed, under illumination, electron-hole pairs are photo-induced in the p-doped substrate. Subsequently, electrons are injected from the p-Si layer to the SiNc/SiO2 matrix due to the band bending at the p-Si/Si-Nc/SiO2 interface, resulting in a negative surface photo-voltage as depicted in Fig. 2c-d. Here, the FMI-KFPM measurements were carried out using a red (685 nm) LuxX (OmicronLaserage GmBH) laser in front side geometry as the illumination source (i.e., sample directly illuminated by the laser). During the 2D spectroscopy the illumination modulation frequencies were swept from 500 Hz to 120 kHz with a 50% duty.

In this case the average photo-voltage displays a monotonic evolution with respect to the frequency-modulation of the illumination source and its magnitude give no sign of decrease up to the upper modulation frequency. Given the above, the spectroscopic curves (Fig. 2h) were fitted using Eq. 1, yielding effective minority carrier lifetimes of a few tens of microseconds. These results are consistent with the lifetime values previously measured on discrete locations of samples from the same series in "single-point" mode [4] and with photoluminescence measurements performed on similar SiNC/SiO2 systems. [11] Moreover, the lifetime values indicate that non-radiative processes are the major pathway for the photo-carrier recombination. Theoretical calculations [12] have

978-1-5090-5606-4/17 $31.00 © 2017 IEEE

indeed shown that radiative recombination should occur on the ms time scale. Despite defects reduction by hydrogen passivation, dangling bonds [13] can still be present and act as local non-radiative recombination centers.

Now, the question that present itself is whether the 2D minority carrier lifetime map (Fig. 2f) displays contrasts that could be related to the sample morphology at the nanoscale. Unfortunately, the nature of the surface morphology does not allow to identify individual nanocrystals (see the topographic image in Fig. 2e). Besides, neither the topographic nor the damping (not shown) contrasts seem to display systematic correlations (or anti-correlations) with the lifetime contrasts. In turn, it is clear that the lifetime image displays local contrasts which cannot be accounted by simple statistical fluctuations. This assumption is supported by the standard error image of the mathematical fit calculated for the minority carrier lifetime (Fig. 2g). Indeed, the absence of systematical correlations between the carrier lifetime image and its associated error image strongly support the hypothesis that local contrasts present in the dynamical image must have a physical origin. Additionally, a weighted image of the minority carrier lifetime and its error (not shown) displays the same general contrast observed in Fig. 2f.

In turn, these contrasts may be accounted by the existence of nanocrystals aggregates (or area displaying a higher density of nanocrystals), which would act as local recombination centers. Transmission electron microscopy images reveal indeed a non-uniform distribution of the nanocrystals within the silicon oxide matrix. Alternatively, surface contaminants could be invoked to account for the existence of localized recombination centers. This latter hypothesis seems however unlikely, as the damping (not shown) and lifetime contrasts are not clearly correlated (or anti-correlated).

Actually, quantifying precisely the lateral and temporal resolution is likely to become a key issue in FMI-KPFM. In the following, we show that the spectroscopy curve fit accuracy is susceptible to have a significant impact on the definition of the dynamical images. For that purpose, we analyze the evolution of the FMI-KPFM data as a function of the illumination intensity in a PDBS-TQx/PC71BM donor-acceptor blend processed to display a heterogeneous morphology at the nanoscale. [15] A schematic structure of the sample and its topography are shown in Fig. 4a and 4b respectively.

For that series of experiments, a green (515 nm) PhoxXplus (OmicronLaserage GmBH) laser was used in backside geometry as the illumination source (*i.e.*, sample indirectly illuminated through a mirror's reflection). The illumination modulation frequencies were swept from 400 Hz to 5 MHz with a 10% duty.

Fig. 3 displays two averaged frequency spectroscopy curves of V_{av} acquired at different illumination intensities, plotted as a function of the light modulation frequency. Here, the optical power is normalized in RMS units and defined per unit of surface taking into account the laser beam diameter. These samples display a negative surface photo-voltage, which is consistent with the open circuit voltage polarity of an organic solar cell in standard configuration (keeping in mind that the

ground is applied to the hole collecting ITO/PEDOT:PSS electrode). For that reason, the surface potential decreases towards more negative values when raising the frequency from 400 Hz to 20 kHz, in other words the (negative) surface photo-voltage increases in absolute value. However, the surface photo-voltage does not saturate when raising the frequency above 20 kHz. Increasing furthermore the frequency results in a photo-voltage diminution observed in the 20 kHz to 200 kHz frequency range. Last, the SPV increases again for frequencies higher than 200 kHz. As above mentioned, this non monotonic behavior can be modeled in a phenomenological approach by summing two photo-voltages characterized by different build up and decay time constants (Eq. 3). The frequency-dependence of the total SPV in the low frequency regime (*i.e.*, in the 300 Hz-200 kHz range) is dominated by the contribution of the "slow" SPV decay (τ_{d1}) and build-up (τ_{b1}) processes. The high frequency part reflects the effect of the "fast" SPV decay (τ_{d2}).

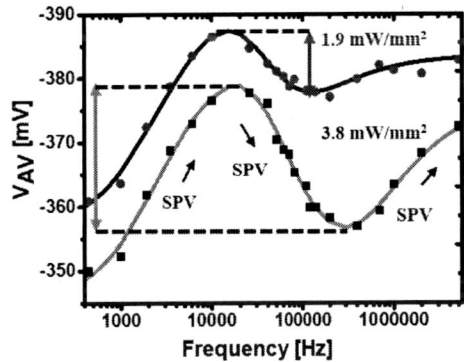

Fig. 3. Spectroscopic curves acquired for different lighting conditions on the nanophase segregated PDBS-TQx/PC71BM blend. Blue and black filled symbols represent the experimental data acquired at 1.9 mW/mm² and 3.8 mW/mm², respectively (all data recorded with a 10% duty). The black and red curves show the numerical fit for 1.9 mW/mm² and 3.8 mW/mm², correspondingly. Colored arrows place above the spectroscopic curves are guidelines to show the magnitude of the spectroscopic curve "rebound" for each illumination condition. Black arrows serves also as guidelines to indicate the SPV evolution trend for each regime.

In a previous report, [5] we tentatively attributed the observed SPV dynamics to (i) non-geminate recombination (τ_{d2}), (ii) carrier trapping (τ_{b1}) and (iii) trap-delayed recombination (τ_{d1}) processes (NB: probing the "fast" SPV build up time constant τ_{b2} remains inaccessible due to the limit of the upper

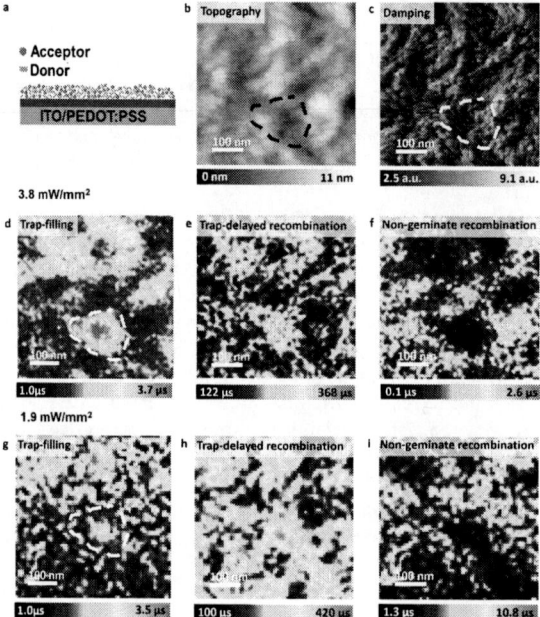

Fig. 4. (a) Model proposed for the nanophase segregated PDBS-TQx/PC$_{71}$BM blend morphology. (b) nc-AFM topographic image (100 × 100 pixels, 500 × 500 nm) acquired during the 2D spectroscopy (c) Dissipation image. (d),(g) Images of the "slow" SPV build-up time constants for 3.8 mW/mm^2 and 1.9 mW/mm^2 illumination intensities respectively, yielding access to the dynamics of trap-filling processes. (e),(h) Images of the "slow" SPV decay time constants for 3.8 mW/mm^2 and 1.9 mW/mm^2 illumination intensities respectively, yielding access to the dynamics of trap-delayed recombination processes. (f),(i) Images of the "fast" SPV decay time constants for 3.8 mW/mm^2 and 1.9 mW/mm^2 illumination intensities respectively, yielding access to the dynamics of nongeminate recombination processes. Panels (a)-(f) were adapted with permission under ACS AuthorChoice License from "Photo-Carrier Multi-Dynamical Imaging at the Nanometer Scale in Organic and Inorganic Solar Cells", P. A. Fernández Garrillo, Ł. Borowik, F. Caffy, R. Demadrille, B. Grévin, ACS Appl. Mater. Interfaces, 8 31460–31468 (2016). http://pubs.acs.org/doi/full/10.1021/acsami.6b11423

modulation frequency allowed by the illumination chain). This interpretation was especially based on the fact that the "fast" decay time is strongly reduced for increasing optical powers, while the slow" decay time (τ_{d1}) is quasi-independent of the illumination intensity. These behaviors are indeed reminiscent of carrier-concentration dependent non-geminate recombination processes and trap-delayed mechanisms.

This interpretation of the SPV dynamics observed for PDBS-TQx/PC$_{71}$BM remains tentative, because it was not possible to definitely conclude about the nature of the defects and the nanoscale morphology in these blends. [5] Here, our goal is not to further discuss the physics of the mechanisms behinds the multiple SPV dynamics. Instead, we will take benefit of the intensity dependence of the "fast decay" time-constant (τ_{d2}) to check to what extent the image contrasts can be affected by the accuracy of the spectroscopy curve fit.

Fig. 4d-i display the results of 2D multi-dynamical SPV imaging for two different illumination intensities. As above mentioned, the timescales for the slow build-up and slow decay time constant images remain nearly unchanged, in opposition

to the fast decay timescale which is strongly reduced under an illumination intensity of 3.8 mW/mm^2. While the images acquired for both intensities display roughly the same contrasts, significant differences can still be seen. For instance, a domain featuring longer τ_{b1} time constants is clearly visible in the image acquired under an illumination output power of 3.8 mW/mm^2 (Fig. 4d), but it becomes poorly resolved under an illumination intensity of 1.9 mW/mm^2 (Fig. 4g).

This effect most likely reflects the impact of the fit accuracy on the global image contrast. In order to prove this, the standard error (i.e., standard deviation) of the mathematical fit was calculated for each pixel of the slow SPV build up time constant image at both illumination intensities. Fig. 5 shows a comparison of the standard error distribution over the images for both optical output powers. It appears evident that an improved fit accuracy was achieved under higher illumination intensities. This can be simply understood by considering the fact that the curves acquired under higher optical powers display a larger decrease in the intermediate frequency range (i.e., frequencies between 20 kHz and 200 kHz), as highlighted by double ended arrow in Fig. 3.

This simple analysis shows that the dynamical image contrast (and by extent the lateral resolution) can be impacted by the accuracy of the post data acquisition numerical fit. This points out the crucial need to optimize the signal to noise ratio of the spectroscopic curves. So far, this was achieved at room temperature by combining single-pass KPFM with beam deflection nc-AFM under UHV. We anticipate that cryogenic nc-AFM/KPFM experiments will help achieving further improvements in terms of lateral imaging resolution.

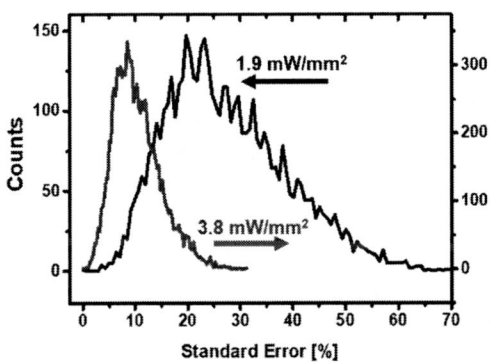

Fig. 5. Mathematical fit standard error distribution for trap filling images under 1.9 mW/mm² (left Y scale) and 3.8 mW/mm² (right Y scale) illumination intensities.

IV. CONCLUSION

In this work, detailed insights on FMI-KPFM dynamical SPV imaging's operating principle, instrumentation, and data treatment were given. We demonstrated how this method can be applied to image the spatial variation of the SPV decay dynamics in a silicon nanocrystal-based third generation solar cell. The findings of this study suggest that local contrasts of the carrier lifetime image obtained over this sample can be accounted by the existence of nanocrystals aggregates acting as local recombination centers. Furthermore, investigations on nano-phase segregated organic donor-acceptor blends at different illumination intensities show that the dynamical images contrast can be impacted by the accuracy of the post data acquisition numerical fit, pointing out the crucial need to optimize the signal to noise ratio of the spectroscopic curves. Taken together, these findings constitute a noteworthy contribution meant to serve as guidance for the further development of dynamical SPV imaging by FMI-KPFM.

REFERENCES

[1] P. Cox, M. Glaz, J. Harrison, S. Peurifoy, D. Coffey, D. Ginger, "Imaging Charge Transfer State Excitations in Polymer/Fullerene Solar Cells with Time-Resolved Electrostatic Force Microscopy," J. Phys. Chem. Lett. Vol. 6, pp. 2852–2858, 2015.

[2] G. Shao, M. Glaz, F. Ma, H. Ju, D. Ginger, "Intensity Modulated Scanning Kelvin Probe Microscopy for Probing Recombination in Organic Photovoltaics," ACS Nano, vol, 8, pp. 10799-10807, 2014.

[3] M. Takihara, T. Takahashi, T. Ujihara, "Minority Carrier Lifetime in Polycrystalline Silicon Solar Cells Studied by Photoassisted Kelvin Probe Force Microscopy," Appl. Phys. Lett., vol. 93, 021902, 2008.

[4] Ł. Borowik,. H. Lepage, N. Chevalier, D. Mariolle, O. Renault, "Measuring the Lifetime of Silicon Nanocrystal Solar Cell PhotoCarriers by Using Kelvin Probe Force Microscopy and X-Ray Photoelectron Spectroscopy," Nanotechnology, vol. 25, 265703, 2014.

[5] P. Fernández Garrillo, Ł. Borowik, F. Caffy, R. Demadrille, B. Grévin, "Photo-Carrier Multi-Dynamical Imaging at the Nanometer Scale in Organic and Inorganic Solar Cells," ACS Appl. Mater. Interfaces, vol. 8, pp 31460–31468, 2016.

[6] W. Melitz, J. Shen, A.C. Kummel, S. Lee, "Kelvin probe force microscopy and its application," Surf. Sci. Rep. vol. 66, pp. 1-27, 2011.

[7] L. Kronik, Y. Shapira, "Surface photovoltage phenomena: theory, experiment, and applications," Surf. Sci. Rep. vol. 37, pp. 1-206, 1999.

[8] E. M. Tennyson, J. L. Garrett, J. A. Frantz, J. D. Myers, R. Y. Bekele, J. S. Sanghera, J. N. Munday, M. S. Leite, "Nanoimaging of Open-Circuit Voltage in Photovoltaic Devices," Adv. Energy Mater. vol. 5, 1501142 (8pp.), 2015.

[9] D. J. Ellison, J. Y. Kim, D. M. Stevens, C. D. Frisbie, J. Am. Chem. Soc. vol. 133, pp 13802–13805, 2011.

[10] K. Surana, H. Lepage, J. Lebrun, B. Doisneau, D. Bellet, L. Vandroux, G. Le Carval, M. Baudrit, P. Thony, P. Mur, "Film-thickness-dependent conduction in ordered Si quantum dot arrays," Nanotechnology, vol 23, 105401, 2012.

[11] L. Van Dao, X. Wen, M. Do, P. Hannaford, E. Cho, Y. Cho, Y. Huang, "Time-resolved and time integrated photoluminescence analysis of state filling and quantum confinement of silicon quantum dots," J. Appl. Phys. vol 97, 013501, 2005.

[12] A. Moskalenko, J. Berakdar, A. Prokofiev, I. Yassievich "Single-particle states in spherical Si/SiO₂ quantum dots," Phys. Rev. B vol. 76, 085427, 2007.

[13] E. Cartier, J. Stathis, D. Buchanan, "Passivation and depassivation of silicon dangling bonds at the Si/SiO₂ interface by atomic-hydrogen," Appl. Phys. Lett. vol. 63, 1510–2, 1993.

[14] I. Horcas, R. Fernández, J. M. Gómez-Rodríguez, J. Colchero, J. Gómez-Herrero, A. M. Baro, "WSXM: A software for scanning probe microscopy and a tool for nanotechnology," Rev. Sci. Inst. vol. 78, 013705, 2015.

[15] F. Caffy, N. Delbosc, P. Chavez, P. Lévêque, J. Faure-Vincent, J-P, Travers, D. Djurado, J. Pecaut, B. Grévin, N. Lemaitre, et al. "Synthesis, Optoelectronic Properties and Photovoltaic Performances of Wide Band-Gap Copolymers Based on Dibenzosilole and Quinoxaline Units, Rivals to P3HT," Polym. Chem. vol. 7, pp. 4160–4175, 2016.

Nanoscale Tomographic Charge Transport in Polycrystalline Chalcogenide Absorbers: CdTe versus CIGS

Justin L. Luria*, Andrew Moore[†], Sun Yu*, Mark Aindow*, Bryan D. Huey*

*Institute of Materials Science, Storrs, CT, 06268 , USA)
[†]Colorado State University, Ft. Collins, CO, 80523 , USA

Abstract—Nanoscale charge transport is uniquely mapped in 3-dimensions for polycrystalline thin-film solar cell absorbers of cadmium telluride (CdTe) and copper indium gallium selenide (CIGS). Using a novel tomographic variation of Atomic Force Microscopy (CTAFM), we site selectively planarize photovoltaic materials while maintaining optoelectronic properties, thereby directly probing interfacial and bulk charge transport at and beneath the surface. This tomography reveals three notable differences and hence recommendations for CdTe versus CIGS. First, CIGS exhibits homogeneous charge transport in the bulk, as compared to CdTe which shows order-of-magnitude grain and grain-boundary dependent performance. Grain boundary engineering approaches, such as passivation and doping, are thus proposed to be more effective for optimizing CdTe as opposed to CIGS. On the other hand, ordered defect compounds and deep acceptor states are detected between CIGS absorber and buffer (CdS) layers, emphasizing the importance of improving this interface for better CIGS efficiency. Lastly, CTAFM is found to enable site-selective comparisons of such local photovoltaic properties with grain orientations via EBSD, to directly explore and ultimately improve nano- and micro- scale connections between microstructure and solar cell performance.

Index Terms—Cadmium Telluride, Copper-Indium-Gallium-Selenide, Scanning Probe Microscopy, Tomography, Imaging

I. INTRODUCTION

Conversion efficiencies and market share continue to improve for 'thin film' photovoltaics. Cadmium Telluride (CdTe) and Copper-Indium-Gallium-Selenide (CIGS) in particular represent a multi-billion dollar annual business and nearly 10% of the global solar cell market. But to optimize these still under-performing polycrystalline devices compared to the Shockley-Queisser limit requires new methods to investigate charge transport, grain boundary passivation, aging, etc. at the nanoscale instead of more common approaches for the bulk or at the panel level. Here a novel scanning probe technique is used to locally planarize and map the local properties of chalcogenide absorber layers, without concomitant ion damage or amorphization as with conventional sectioning techniques. This uniquely enables top-down site-specific investigations of photocurrents, while the resulting smooth surfaces improve topography-sensitive measurements such as electron back-scattering diffraction (EBSD) mapping or near field scanning optical microscopy [1] for directly correlating film

Fig. 1. Rendering of CT-AFM, based on acquiring consecutive images of photocurrents while progressively milling top-down through a polycrystalline solar cell at any location of interest with a conductive AFM probe.

microstructure and performance.

II. EXPERIMENTAL DESIGN AND THEORY

Transport pathways are specifically imaged herein during 1 sun illumination, where a conducting doped-diamond probe simultaneously records topography and pA scale currents between the tip (serving as a positionable top electrode) and the counter-electrode (at the film-substrate interface)1. Planarization is achieved with μN scale forces, such that the specimen is repeatedly sculpted within the scanned region achieving computed tomographic AFM (CTAFM). According to the resolved features in the evolving topography and photocurrent, spatial resolution of 20nm is maintained throughout both in this work and in related recent efforts on CdTe [2]. In this manner, conventional explanations for the reduced performance of polycrystalline thin film solar cells including structural defects [3], [4] or stacking faults and crystalline twins are directly interrogated. In CdTe, our recent work with CTAFM revealed that wurtzite (WZ) laths within nominal zinc-blende grains can act as hole transport layers [2] rather than traps [5].

For CIGS, unlike CdTe, grain boundaries are conventionally understood to be inherently passive, allowing extended intragranular charge transport within the absorber.

The absorber/buffer (CIGS/CdS) interface [6] instead may dominate performance, by controlling the heterojunction band edge energies and limiting majority carrier lifetimes. Reports indicate this can result from an abrupt or compositionally graded p+ layer [7], or grains of distinct phases including a p-type (α phase) chalcopyrite, Cu deficient n-type (β) ordered defect compound (ODC), or Cd-rich (γ) zinc-blende [8]. Each phase possess distinct electronic properties influencing junction performance. Many TEM studies have probed the structure of such near- and sub-surface defects, but locally probing their opto-electronic properties requires novel approaches such as CTAFM.

A. Device Fabrication

CdTe devices were fabricated at CSU using standard close-space sublimation for market compatible performance (0.8 V_{oc}) and efficiency (12.3%). 120 nm of n-type cadmium sulfide was deposited over a fluorinated-tin oxide on a glass substrate. 2.7 μm of p-type CdTe absorber was deposited on the CdS layer. Samples were conventionally treated with $CdCl_2$ and $CuCl_2$. The 'back' electrode, at the top of the stack, consisted of an encapsulating layer of carbon and nickel, which was removed for CTAFM. CIGS samples were fabricated by leading PV company, with a bottom-up architecture of Steel / Molybdenum (Mo) / CIGS / CdS (50nm). Further details are proprietary. The transparent surface electrode (ITO/AZO) was never applied so that AFM measurements could be performed. Nominal macroscopic efficiency for equivalent devices is 11%.

B. Electrical Current Imaging and EBSD

All measurements are performed in air with an Asylum Instruments MFP3D. For ITO/CdS/CdTe, since the underlying substrate is a transparent conducting oxide the illumination is directed from below. This exposes the film to a full visible spectrum light emitting diode (LED) through the substrate via a 40x objective, yielding a controllable intensity of 0.15 to 15 suns. The tip and ITO substrate are electrically grounded, in which case imaging reveals regions of enhanced photocurrent rather than photoconduction. Given the band offsets the current signal then corresponds to free holes (electron vacancies in the valence band) [2]. For the case of CIGS, where the underlying substrate is opaque and the CdS window region is exposed at the free surface, a different experimental method must be used. We thus apply +2V to the substrate and progressively remove the CdS to expose the underlying CIGS. As the Young's moduli for CdS and CIGS are 10GPa and 70GPa, respectively, the 50nm CdS layer can be uniformly removed leaving the CIGS layer intact until more aggressive material removal is initiated. Since the resulting tip-sample architecture is a metal-semiconductor

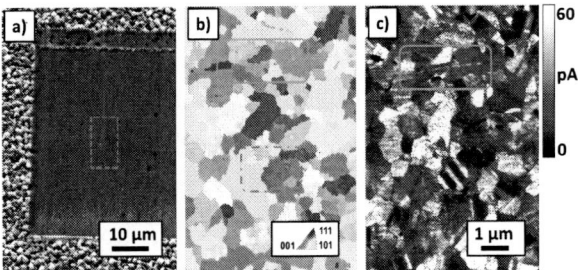

Fig. 2. Site selective surface milling with CTAFM enables direct comparisons between a) SEM, and within the overlain dashed rectangle higher magnification b) EBSD and c) photocurrent imaging. This reveals little apparent correlation between single grain performance and orientation, though inter-grain Wurtzite domains (solid blue) and intra-grain twinning (dot-dashed red) are regularly observed to be photo-active.

junction, for +2V applied to the substrate, regions of enhanced hole conduction are imaged.

EBSD analysis is achieved with a FEI Teneo-LoVac FEG-SEM, operating at an accelerating voltage of 20 kV. As usual the samples are oriented 70 ° from horizontal with a 15 mm working distance.

III. EXPERIMENTAL RESULTS

A. Cadmium Telluride Absorber Layer

To remove topographic effects from EBSD and conductive AFM measurements, early stage CTAFM can planarize a surface leading in this case to a reduction of surface roughness from 200nm RMS to 20nm RMS. Scanning electron microscopy, Figure 2a, shows such a planarized region, with an overlain dashed rectangle indicating the location for higher resolution EBSD and photocurrent mapping. Notably, for the as-provided surface of these cells, little EBSD or meaningful photocurrent images can be acquired due to topographic artifacts. Equivalent top-down polishing with FIB would significantly damage the opto-electronic properties.

Once the CdTe surface is properly prepared, though, photocurrent measurements reveal that the opto-electronic performance for individual grains varies by orders of

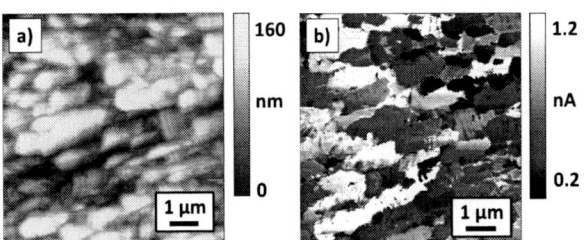

Fig. 3. Surface of Steel/Mo/CIGS(/removed CdS): a) AFM topography revealing grains on the order of one μm. b) Simultaneous conducting AFM with +2V substrate bias, revealing equivalent current flow for grain clusters and hence independent of many grain boundaries.

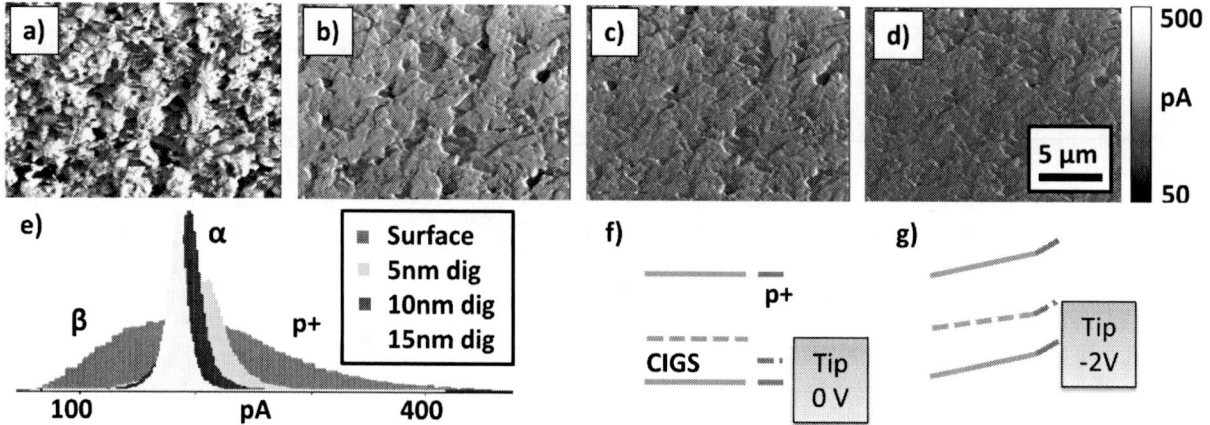

Fig. 4. Conducting AFM for Steel/Mo/CIGS at +2V substrate bias. a) On the surface, strong super-granular contrast variations again exist for the measured current. As b) 5nm, c) 10nm, and d) 15nm of CIGS are removed, weaker contrast variations and a diminishing mean signal are progressively exposed. e) Histograms of the current from (a-d) suggest p+ and β regions at the surface which are gradually removed by planarization. f) Proposed flat-band energy level diagrams of CIGS with p+ before and g) after material contact and tip biasing. P+ domains assist in hole injection.

magnitude. Figures 2(b,c) correlate EBSD grain orientations with local photocurrents based on photoconducting AFM. Such results are able to identify inter-grain planar WZ defects (solid box) [2], [3], as well as the effects of crystal twinning (dot-dashed box). Further work is required to statistically confirm any relationships between the orientation and opto-electronic performance, though early indications suggest that some grain orientations do provide some enhancement in photocurrent. This, however can be obscured by intra-granular effects such as twinning. 3-D EBSD is presently underway to correlate opto-electronic performance, film morphology, and especially grain boundary orientations.

Under different experimental conditions, enhanced electron conduction (conduction band) along CdTe grain boundaries has been observed in 2d by many groups, and by CTAFM in 3-d [2]. The transport of holes through the grain bulk and/or planar defects, with electrons preferentially along grain boundaries, possibly contributes to the exceptionally long minority carrier lifetimes known for high quality CdTe, since such energetically and physically disparate pathways would reduce charge recombination. The grain-boundary dominated electron pathways do not suggest a dependence on orientation, supporting prior observations that they result from dopant diffusion and/or grain boundary passivation. These pathways also exhibit little boundary-to-boundary variation, so they are not likely to be improved through engineering of grain orientation, twinning, or Wurtzite generation [2]. Further analysis of electron transport pathways would be of great use in a characterization-fabrication cycle, but remains beyond the scope of this work.

B. Copper-Indium-Gallium-Selenide Absorber Layer

Since electrical current measurements on half-cells of Steel/Mo/CIGS are acquired at +2V substrate bias, without illumination, the forward bias on the metal-semiconductor junction should image hole-transport. Although cells are initially grown with a CdS top-layer buffer, this 50nm layer was removed as described above. AFM topography of the CdS-free surface (Figure 3a) reveals CIGS grain size on the order of 1 μm. Current maps (b) reveal order of magnitude contrast between, but uniform contrast within, clusters of grains, unlike CdTe.

Changing the milling conditions in order to progressively remove the CIGS absorber, conducting AFM images with +2V substrate bias (Figure 4 allows conduction pathways and heterogeneities to be compared a) at the surface, and b) 5nm, c) 10nm, and d) 15nm beneath the surface. The images become less heterogeneous, with a near uniform contrast 10nm or deeper beneath the CIGS surface that is apparent when comparing image signals (a-d) via overlain histograms (Figure 4e). These results indicate hole conduction in bulk CIGS is not as hindered by grain boundaries and internal interfaces as for CdTe.

According to Figures 3 and 4, large variations in conductivity are still apparent at the CIGS/CdS interface, however. These are hypothesized to result from CIGs/CdS α, β, and/or γ interfacial phases. The bulk phase of CIGS, α, should have little difference in electrical current contrast between the surface and the bulk. On the other hand, a phase with a high density of acceptor-type defect states, effectively a p+ region, should have a lower Fermi energy than α. For example, the ordered defect compound, β phase, is known to be Cu deficient, slightly n-type, with a wider band gap, and a deeper valence band than α [7].

978-1-5090-5606-4/17 $31.00 © 2017 IEEE

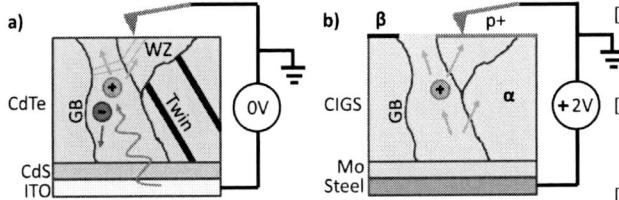

Fig. 5. Proposed charge transport pathways for CdTe and CIGS. a) In CdTe grain boundaries offer pathways for electron conduction, while Wurtzite/zincblende interfaces improve inter-grain hole transport. b) For CIGS, grain boundaries are inherently passivated, while interfacial phases at the CdS/CIGS boundary are critical to device performance. Under biasing conditions, deep acceptor p+ regions, exhibit enhanced electrical current, while β phases inhibit local conduction.

Figure 4f describes a possible resulting energy diagram, theoretically before the layers are brought into contact or any tip biasing. In this scheme, the lower electron density of the β interlayer should inhibit with hole injection into the deep work-function tip. Figure 4g depicts the influence of a strongly p+ after tip bias and material contact, which assist in hole injection.

IV. SUMMARY OF WORK

In conclusion, AFM for tomography and surface planarization opens new avenues for nanoscale characterization of ceramic solar cells. For CdTe, charge transport in the absorption layer is found to be both inhibited and assisted by structural defects, with electrons primarily confined to grain boundaries (Figure 5a). Hole transport is through grain bulk and Wurtzite domains offer intra- and inter-grain transport. We are now able to correlate grain orientation with opto-electronic performance, further guiding grain engineering. In the case of CIGS, milling offers access to the critical interfacial region between the absorber CIGS and buffer CdS (Figure 5b). Conducting probe AFM at the CIGS/CdS interface shows strong contrast between materials phases that develop during processing. Further characterization (Auger, TEM) is required to link our proposed mechanisms to local composition, but this work demonstrates a novel approach for understanding subsurface opto-electronic properties.

REFERENCES

[1] M. S. Leite, M. Abashin, H. J. Lezec, A. Gianfrancesco, A. A. Talin, and N. B. Zhitenev, "Nanoscale imaging of photocurrent and efficiency in cdte solar cells," *ACS Nano*, vol. 8, no. 11, pp. 11 883–11 890, Nov. 2014. [Online]. Available: http://dx.doi.org/10.1021/nn5052585

[2] J. Luria, Y. Kutes, A. Moore, L. Zhang, E. A. Stach, and B. D. Huey, "Charge transport in cdte solar cells revealed by conductive tomographic atomic force microscopy," *Nature Energy*, vol. 1, p. 16150, 2016.

[3] H. Li, X. X. Liu, Y. S. Lin, B. Yang, and Z. M. Du, "Enhanced electrical properties at boundaries including twin boundaries of polycrystalline cdte thin-film solar cells," *Phys. Chem. Chem. Phys.*, vol. 17, p. 11150, 2015. [Online]. Available: http://dx.doi.org/10.1039/c5cp00564g

[4] I. Visoly-Fisher, A. R. Sidney R. Cohen, and D. Cahen, "How polycrystalline devices can outperform single-crystal ones:thin film cdte/cds solar cells," *Advanced Materials*, vol. 16, no. 11, pp. 879–883, 2004. [Online]. Available: 10.1002/adma.200306624

[5] Y. Yan, M. M. Al-Jassim, K. M. Jones, S.-H. Wei, and S. B. Zhang, "Observation and first-principles calculation of buried wurtzite phases in zinc-blende cdte thin films," *Appl. Phys. Lett.*, vol. 77, no. 10, p. 1461, 2000. [Online]. Available: http://dx.doi.org/10.1063/1.1308062

[6] K. Hiepko, J. Bastek, R. Schlesiger, G. Schmitz, R. Wuerz, and N. Stolwijk, "Diffusion and incorporation of cd in solar-grade cu (in, ga) se2 layers," *Applied Physics Letters*, vol. 99, no. 23, p. 234101, 2011.

[7] M. Nichterwitz, R. Caballero, C. A. Kaufmann, H.-W. Schock, and T. Unold, "Generation-dependent charge carrier transport in cigs/cds/zno thin-film solar-cells," *Journal of Applied Physics*, vol. 113, p. 044515, 2013.

[8] Y. Yan, K. Jones, J. Abushama, M. Young, S. Asher, M. Al-Jassim, and R. Noufi, "Microstructure of surface layers in cu (in, ga) se 2 thin films," *Applied physics letters*, vol. 81, no. 6, pp. 1008–1010, 2002.

Improving the PV Module Single-Diode Model Accuracy with Temperature Dependence of the Series Resistance

Kyumin Lee

CFV Solar Test Laboratory, Albuquerque, New Mexico, NM 87106, U.S.A.

Abstract — **PVsyst 6, arguably the most popular PV modeling software in the industry, includes a temperature coefficient for the ideality factor in the single-diode model for PV modules. Multi-temperature and multi-irradiance I-V data from IEC 61853-1 tests on silicon modules, however, reveal no significant temperature dependence of the ideality factor. The series resistance shows a stronger temperature dependence, but this behavior is currently not captured by the PVsyst 6 module performance model. It is shown that, once the temperature dependence of the series resistance is included, the single-diode model can reproduce I-V curve point values Voc, Imp, and Vmp with greater accuracy.**

I. BACKGROUND

PVsyst has become the de-facto industry standard software for the yield simulation of PV systems, by offering a comprehensive set of features that are continuously improved over time. The single-diode model is employed in PVsyst to model the PV module behavior, and the performance model implemented in PVsyst 6 uses the following set of equations for silicon modules [1].

$$I = I_L - I_O \left[\exp\left(\frac{V + I\,R_S}{\gamma\,V_{th}} \right) - 1 \right] - \frac{V + I\,R_S}{R_{SH}} \quad (1)$$

$$I_L = \frac{E}{E_0} [I_{L0} + \alpha_{Isc}(T_C - T_0)] \quad (2)$$

$$I_O = I_{00} \left(\frac{T_C}{T_O} \right)^3 \exp\left(\frac{q\,\varepsilon_G}{k\,\gamma} \left(\frac{1}{T_O} - \frac{1}{T_C} \right) \right) \quad (3)$$

$$R_{SH} = R_{SH,base} + (R_{SH0} - R_{SH,base}) \exp\left(-R_{SHexp} \frac{E}{E_0} \right) (4)$$

$$\gamma = \gamma_0 + \mu_\gamma (T_C - T_0) \quad (5)$$

$$R_S = R_{S0} \quad (6)$$

V and I are the voltage and current measured at the terminal, I_L is the light-generated current, I_O is the dark saturation current, R_S is the series resistance, R_{SH} is the shunt resistance, γ is the diode ideality factor, V_{th} is the thermal voltage, E is the incident irradiance, T_C is the cell temperature, and T_O is the reference temperature (298.15 K). α_{Isc} and μ_γ are the temperature coefficients of the short-circuit current and the diode ideality factor, respectively. The symbols with "0" subscripts refer to the respective values at STC. As for other symbols, I ask the reader to refer to a Sandia publication on the subject [2].

The IEC 61853-1:2011 [3] defines a matrix of temperature and irradiance conditions for the performance measurement of a PV module, and a popular practice in the industry involves deriving an optimum set of PVsyst single-diode model parameters (also known as "PAN files") that can reproduce the IEC 61853-1 test data accurately. As an engineer in a test lab carrying out such "PAN file tests" regularly, it is in my interest to understand to what level of detail the PVsyst single-diode model can describe the actual irradiance- and temperature-dependent behavior of the PV modules.

It will be first shown that the ideality factor of silicon modules, as calculated from the IEC 61853-1 data, has no real dependence on the temperature, contrary to (5). The series resistance, in contrast, shows a significant temperature dependence (~0.3%/°C), but the PVsyst 6 single-diode model assumes a constant Rs value at all temperatures, as seen in (6).

A residual analysis will show that, when an optimum set of PVsyst 6 model parameters are found to minimize the Pmp residuals, the Voc, Imp, and Vmp point values predicted by the PVsyst model show residual errors that are correlated to both temperature and irradiance.

Finally, an improved single-diode model including a linear temperature dependence of the series resistance and excluding the diode ideality factor temperature dependence will be proposed, and the subsequent residual analysis will demonstrate that the "Rs TempCo" model proposed herein can reproduce the Voc, Imp, and Vmp point values in the IEC 61853-1 test data with greater accuracy.

II. EXPERIMENTAL

IEC 61853-1 Data

The IEC 61853-1 performance matrix test data were obtained on three different modules, all manufactured in 2016: (1) a 72-cell polycrystalline-silicon module rated at 315 Wp, (2) a 60-cell monocrystalline-silicon PERC module rated at 295 Wp, and (3) a 72-cell n-type mono-crystalline-silicon PERT module rated at 375 Wp. The performance matrix test was carried out with a tunnel-type Halm module flasher, which is equipped with a glass-window temperature chamber for module I-V measurements at arbitrary temperatures between 15 and 75°C. I-V curves were obtained at 15, 25, 50, and 75°C, at irradiances from 100 to 1100 W/m², as specified in IEC 61853-1:2011. In addition to the 61853-1 test conditions, I-V curves were obtained at low irradiances (100, 200, and 400 W/m²) at high temperatures (50 and 75°C) to determine the ideality factor with a greater accuracy, and 1000 W/m² measurements at all 5-degree steps between 15 and 75°C were further included to improve the accuracy of the temperature coefficients.

Ideality Factor

The ideality factor γ was determined by studying the multi-irradiance data at 15, 25, 50, and 75°C. For each temperature, the differences between Voc at each irradiance E and the open-circuit voltage Voc at the reference-irradiance E_0 (1000 W/m²) were plotted against Vth log(E/E_0). The slope from the linear regression then corresponds to the ideality factor at each temperature. Figure 1 shows the ideality factor regression plot for the 72-cell poly module as an example.

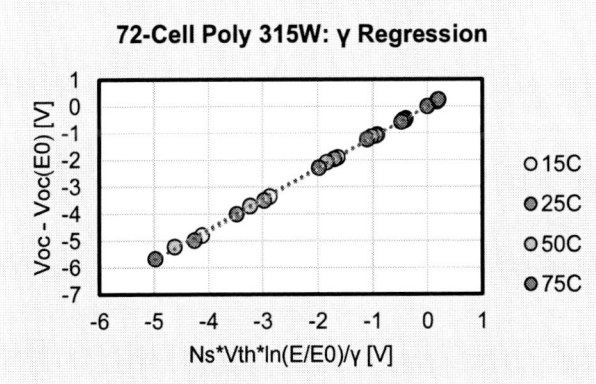

Fig. 1.   Ideality factor regression plot for the 72-cell poly Si module

This method can be applied to silicon modules with high-enough shunt resistance (> 100 Ω) to determine the ideality factor accurately. The method is employed in the Sandia Array Performance Model [4] and in the IEC 60904-5:2011 equivalent-cell-temperature method [5], and is essentially the "Suns-Voc" method [6] used in solar cell analysis.

Series Resistance

The series resistance Rs was determined at each IEC 61853-1 temperature with the Swanson method [7]. This method attributes the difference between the terminal voltage and the diode voltage V_D to the voltage drop across the series resistance (in agreement with the single-diode model), and uses I-V curve data at multiple irradiances to estimate V_D for a given diode current ($Isc_0 - Imp_0$ was used in this work). Rs can then be computed through a linear regression on the terminal current I and the voltage drop ($V_D - V$), both of which varies with irradiance E. An example is shown in Fig. 2, again carried out with the IEC 61853-1 performance matrix I-V curveset data obtained on the 72-cell poly module.

The Swanson method is compatible with the series resistance determination method described in IEC 60891:2009 [8].

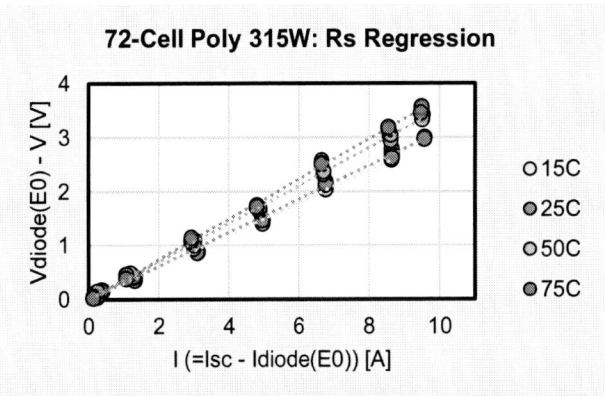

Fig. 2.   Rs regression plot for the 72-cell poly Si module

PVsyst Model Optimization and Residual Analysis

PANOpt®, an in-house developed software at CFV, was employed to find an optimal set of PVsyst single-diode model parameters that gave the lowest root-mean-squared-error (RMSE) of Pmp over the IEC 61853-1 matrix. The STC Isc, Voc, Imp, Vmp, and α_{Isc} values were fixed to the measured values, and Rs was seeded with the average of the per-temperature values determined with the Swanson method. The results were analyzed by studying the Isc, Voc, Imp, and Vmp residuals (PVsyst values – measured values).

Rseries TempCo Model Optimization and Residual Analysis

A separate branch of PANOpt® was prepared for the "Rseries TempCo" model, which replaces (5) and (6) in the PVsyst 6 single-diode model with the following equations.

$$\gamma = \gamma_0 \tag{7}$$

$$R_S = R_{S0} + \mu_{Rs}(T_C - T_O) \tag{8}$$

Simply put, the "Rs TempCo" model being proposed in this paper assumes a constant diode ideality factor, and a series resistance that varies linearly with the cell temperature.

As with the previous case, the modified software found an optimal set of model parameters for the lowest RMSE of Pmp. The residual analysis was carried out in an identical manner.

II. RESULTS

Ideality Factor

Figure 3 shows the ideality factors calculated from the IEC 61853-1 test data, for the three modules, at the four different temperatures (15, 25, 50, and 75°C). The graph in Fig. 3 does not suggest any clear temperature dependence of the ideality factor for the three modules (72-cell poly: -0.041 %/K, 60-cell PERC: +0.006 %/K, 72-cell n-PERT: +0.015 %/K), contrary to (5) of the PVsyst 6 single-diode model.

978-1-5090-5606-4/17 $31.00 © 2017 IEEE

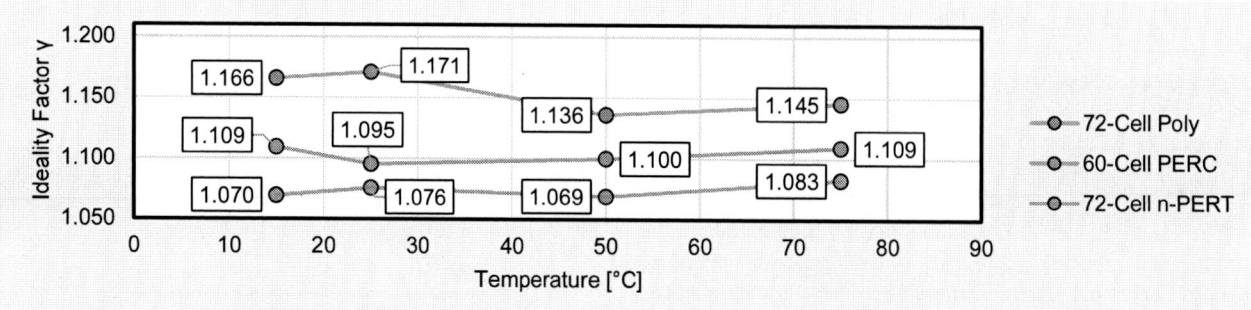

Fig. 3. Ideality factor derived from the IEC 61853-1 data for the three modules, at each test temperature

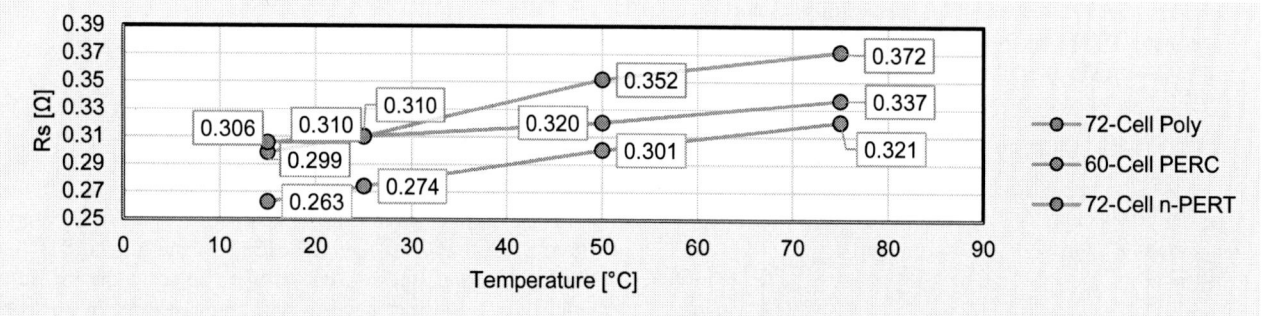

Fig. 4. Series resistance derived from the IEC 61853-1 data for the three modules, at each test temperature

Series Resistance

Figure 4 shows the regression plots for the series resistance, at each of the IEC 61853-1 temperatures. In contrast with the ideality factor, the series resistance shows a clear temperature dependence (72-cell poly: +0.405 %/K, 60-cell PERC: +0.356 %/K, 72-cell n-PERT: +0.164 %/K).

Model Optimization and Residual Analysis

The residual plots for the optimized PVsyst model and the optimized Rs TempCo model are shown one after another for the three modules, in Figs. 5 to 7.

In all figures, the residual errors (model-predicted value – measured value) are shown relative to the STC values, so that the modeling results on the three different module technologies can be compared, without overemphasizing the low-irradiance errors.

For all the modules, the PVsyst model's Voc, Imp and Vmp residual errors show clear correlation to irradiance and temperature. In contrast, the "Rs TempCo" model's Voc, Imp and Vmp residuals are much smaller, and they are less correlated to irradiance and temperature.

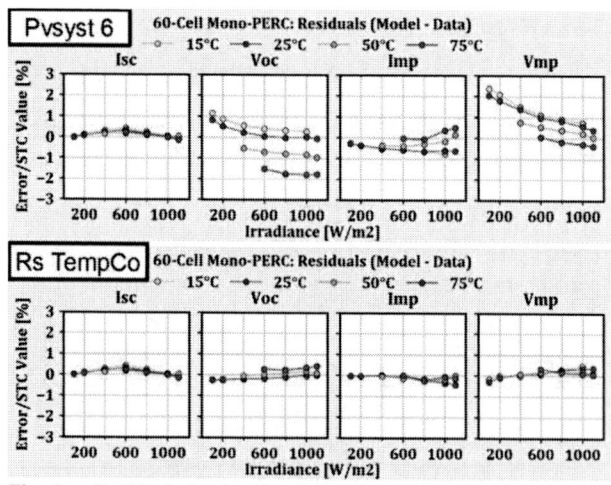

Fig. 5. Residuals for the 72-cell polycrystalline silicon module

Fig. 6. Residuals for the 60-cell PERC module

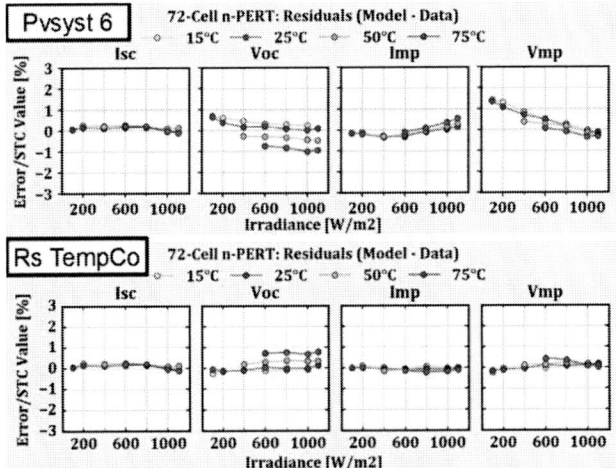

Fig. 7.   Residuals for the 72-cell n-PERT module

III. DISCUSSION

Temperature Dependence of Series Resistance

Metals in general have positive temperature coefficients of resistivity (TCR). Of relevance to the industrial silicon PV modules are the TCRs of copper and silver, whose values are +0.39%/K and +0.38%/K, respectively [9]. In a modern 4-busbar 72-cell polycrystalline silicon module, the copper ribbon wires are responsible for roughly 50% of the module's series resistance, and the front silver grid accounts for roughly 30% [10]. Given how much of the series resistance is due to the metals, it should not be surprising that a PV module also exhibits a positive TCR of similar order.

Furthermore, the temperature coefficient of series resistance is widely used in solar cell modeling, to the point that it is included in the IEC 60891:2009 correction procedures 1 and 2. In PV module modeling, however, I was not able to find a previous work that incorporates this concept. To the best of my knowledge, this is the first work that proposes using a temperature coefficient for the series resistance of a PV module.

PVsyst 6 Model vs "Rs TempCo" Model

The Imp and Vmp residuals are shown as their corresponding Pmp errors in Figs. 8 to 10. The final Pmp residuals are also shown in the figures. The RMS error of Pmp values over the IEC 61853-1 matrix are summarized in Table I, in absolute units and as percentage of the nameplate Pmp.

TABLE I
RMS ERROR OF PMP OVER IEC 61853-1 MATRIX

Module	PVsyst 6 Model	Rs TempCo Model
72-cell Poly	0.338 W (0.11%)	0.287 W (0.09%)
60-cell PERC	0.232 W (0.08%)	0.191 W (0.06%)
72-cell n-PERT	0.256 W (0.07%)	0.326 W (0.09%)

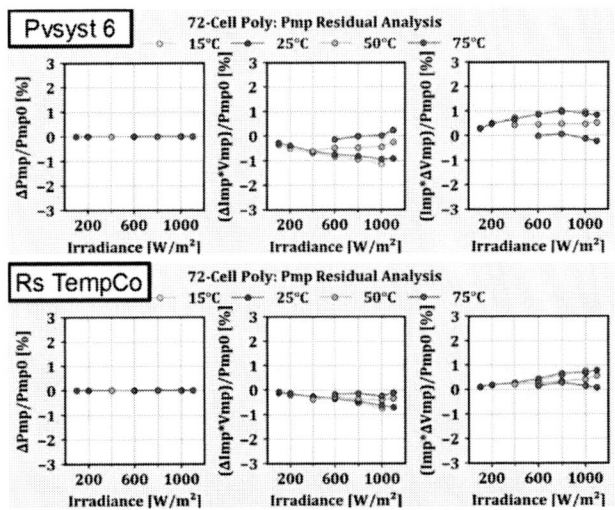

Fig. 8.   Pmp residual analysis for the 72-cell poly module

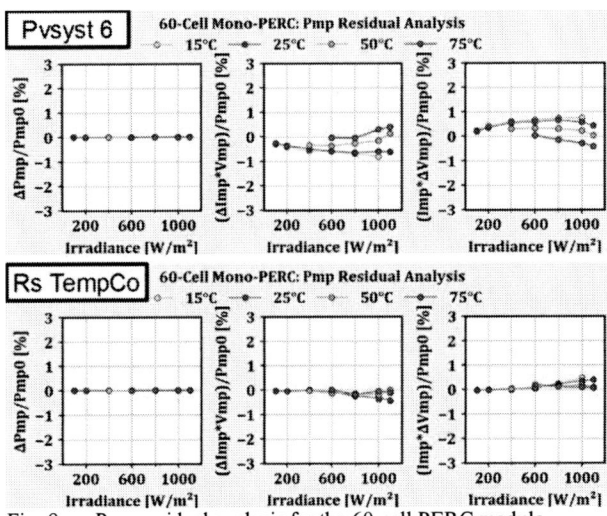

Fig. 9.   Pmp residual analysis for the 60-cell PERC module

Fig. 10.   Pmp residual analysis for the 72-cell n-PERT module

One thing should become clear to the reader: there is no appreciable difference in the Pmp accuracy of the two models. Even though the Imp and Vmp errors are non-negligibly high for the PVsyst 6 model, they cancel each out, resulting in negligible Pmp errors. The lack of any significant improvement in the Pmp accuracy of the Rs TempCo model begs the obvious question: "why should one consider a new model when there is no improvement in the final deliverable?"

In my opinion, the single-diode equivalent circuit model, despite its gross simplification of the complex circuit network that is the PV module, retains popularity over other similarly accurate yet simpler-to-calculate models such as the Sandia Array Performance Model and the Loss Factor model due to one main reason: the single-diode model is described by physical parameters such as the dark saturation current, ideality factor, series resistance, and shunt resistance, which are concepts that the scientists and engineers can easily relate to. There are obvious limitations in representing a complex network of solar cells with the single-diode model, and solar cell scientists would argue that representing a solar cell with one or two diodes is already not accurate enough, but the virtue of the single-diode model is that it still strives to attribute physical meaning to the model parameters. If so, then the parameters should also take on temperature and irradiance dependences that are based on physical observations: Rs should be allowed to vary with temperature, because that is what is observed.

There is also a practical advantage. Errors in the module Imp and Vmp values translate into errors in the estimation of array losses such as module-to-module mismatch and wiring losses. More accurate prediction of the Imp and Vmp values by the "Rs TempCo" model would translate into more accurate estimation of the array losses.

VI. CONCLUSION

A modified single-diode model that incorporates a linear temperature dependence of the series resistance was proposed in this paper. Analyses on 72-cell polycrystalline silicon, 60-cell monocrystalline silicon PERC, and 72-cell n-type silicon PERT modules show that the newly proposed "Rs TempCo" model can predict the I-V point values with greater accuracy than the PVsyst 6 model.

ACKNOWLEDGMENTS

I thank Jim Crimmins and Daniel Zirzow of CFV Solar Test Laboratory for their contributions to the PANOpt® software, which was used for the work presented here.

REFERENCES

[1] A. Mermoud and T. Lejeune, "Performance assessment of a simulation model for PV modules of any available technology", Proc. 25th EU PVSEC, 2010.

[2] C. Hansen, "Estimating parameters for the PVsyst version 6 photovoltaic module performance model", Sandia Report SAND2015-8598, 2015.

[3] International standard "Photovoltaic (PV) module performance testing and energy rating – Part 1: Irradiance and temperature performance measurements and power rating", IEC 61853-1, 2011.

[4] D.L. King, W.E. Boyson, and J.A. Kratochvill, "Photovoltaic Array Performance Model", Sandia Report SAND2004-3535, 2004.

[5] International standard "Photovoltaic devices – Part 5: Determination of the equivalent cell temperature (ECT) of photovoltaic (PV) devices by the open-circuit voltage method", IEC 60904-5, 2011.

[6] R.A. Sinton, "Possibilities for process-control monitoring of electronic material properties during solar cell manufacture", NREL 9th Workshop on Crystalline Silicon Solar Cells and Materials and Processes, 1999.

[7] M. Wolf and H. Rauschenbach, "Series resistance effects on solar cell measurements", Advanced Energy Conversion, vol. 3, pp. 455-479, 1963.

[8] International standard "Photovoltaic devices – Procedures for temperature and irradiance corrections to measured I-V characteristics", IEC 60891, 2009.

[9] http://hyperphysics.phy-astr.gsu.edu/hbase/electric/restmp.html

[10] Based on private calculation, using unit-cell modeling of the solar cell and a SPICE simulation of the PV module.

Cell-to-Module (CTM) Analysis for Photovoltaic Modules with Shingled Solar Cells

Max Mittag, Tobias Zech, Martin Wiese, David Bläsi, Matthieu Ebert, Harry Wirth

Fraunhofer Institute for Solar Energy Systems, Heidenhofstr. 2, 79110 Freiburg, Germany

Abstract — The interconnection of solar cells by shingling increases the active cell area in photovoltaic modules. Cell-to-module (CTM) gains and losses change significantly. We present models to calculate these gains and losses for shingled cells. Module efficiency and power can be increased with the shingle interconnection technology by +33 Wp and +1.86%abs in the analyzed design, when compared to common ribbon-based interconnection. The CTM-ratio for efficiency improves due to shingling and also the CTM-ratio for power increases compared to conventional modules with ribbon or wire cell interconnection.

Index Terms — CTM, cell-to-module, shingle interconnection, efficiency analysis, photovoltaic module, concepts, modelling

I. INTRODUCTION

The integration of solar cells into photovoltaic modules changes power and reference area which define the efficiency of solar devices. Those changes are caused by optical, electrical and geometrical effects and usually lead to a module power different from the sum of the solar cells' initial power [1]-[3].

The cell-to-module power ratio (CTM_{power}) describes the ratio of the power of the module after integration of the solar cells relative to the sum of the initial power of the solar cells before interconnection and module integration.

The individual CTM effects influence each other and render the optimization of the photovoltaic module a non-trivial task. This optimization is nonetheless necessary to further increase the performance of PV modules and to avoid unnecessary losses caused by an unfavorable combination of module components such as encapsulation material, interconnector ribbons or cover materials.

Electrical cell interconnection of common industrial modules relies on ribbons that connect the n-contact of one cell with the p-contact of the next cell. The interconnection concept of cell shingling [4]-[6] omits ribbons and directly connects stripes of solar cells (Figure 1). By doing so, the ribbons as well as the stringing process become redundant. The efficiency of the module increases since the cell spacing area is avoided, resulting in a higher share of active cell area within the module.

Previous work [7] describes a unified methodology to analyze the CTM ratio of photovoltaic modules. Based on this approach we present new models for shingled cells.

Models for shingled solar cells differ from existing models for conventional cell interconnection since the shingling process actually reduces active cell area due to partial cell overlap. The total cell area for conventional modules is always smaller than the module area. For shingled modules, the initial cell area can be larger than the final module area.

We analyze the CTM gain and loss factors for shingled modules and present a detailed model for calculation of power and efficiency based on material properties and the specific module setup. The models are integrated into Fraunhofer ISE's software package SmartCalc.CTM [8], a recently released flexible, precise and user-friendly calculation tool.

II. CELL-TO-MODULE RATIO CALCULATION

A. General model

A model to categorize the single CTM-factors and match them with physical loss mechanisms as well as with module components and layers has been presented in previous work and literature [7][9][10]. We use these models and calculate the module power P_{module} from the CTM-factors k and the sum of the initial solar cell power:

$$P_{module} = \prod_{i=3}^{m} k_i \cdot \sum_{j=1}^{n} P_{cell,j} \qquad (1)$$

$$CTM_{power} = \prod_{i=3}^{m} k_i \qquad (2)$$

We further extended this model and use it to describe the CTM-ratio of shingled modules and discuss relevant factors in the following section.

Figure 2 shows results of a CTM efficiency analysis for a conventional module considering 15 different gain and loss factors.

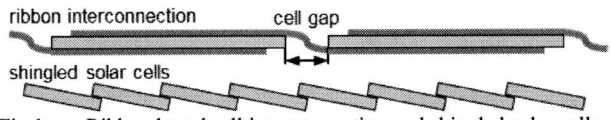

Fig 1:   Ribbon based cell interconnection and shingled solar cells

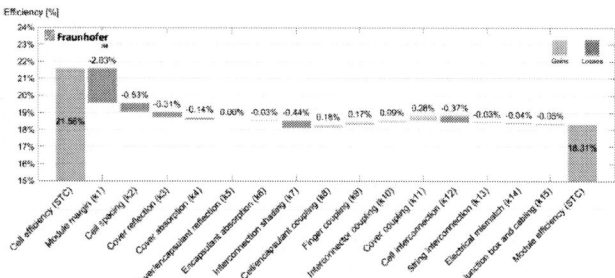

Fig. 2: Cell-to-module (CTM) loss and gain factors for a conventional photovoltaic module

B. Cell-to-Module Gain and Loss Factors for Shingled Modules

Losses by the inactive module margin and the cell and string spacing areas are described by factors k_1 and k_2. They account for geometrical losses of inactive areas that do not contribute to module power but influence efficiency. The latter can be calculated by:

$$\eta_{\text{module}} = \frac{P_{\text{module}}}{E_{STC} \cdot \left(A_{m\,\text{arg}\,in} + A_{cell\,spacing} + A_{cells}\right)} \quad (3)$$

$$\eta_{\text{module}} = \overline{\eta}_{\text{cell}} \cdot \left(k_1 + k_2 - 1\right) \cdot \prod_{i=3}^{m} k_i \quad (4)$$

The efficiency depends on the inactive module area which consists of module margin and cell spacing. The gap between cells in a string and the distance between different strings define the cell spacing area.

Models for factor k_1 (module margin) do not change for shingled modules. We therefore refer to Hädrich [7] who presents a detailed description of this factor.

Changes in k_2 (cell spacing) result from omitting the gaps between cells in a string. String spacing still exists. The factor k_2 describes the geometrical overlap and can be interpreted as the module area that can be saved by shingling.

$$k_2 = 1 + \frac{A_{overlap}}{A_{module}} \quad (5)$$

If shingling not only covers initially inactive (metallized) cell area but also some active cell area, the power loss from this shading is considered in factor k_7 (interconnection shading). In this case the mere geometrical factor k_2 overestimates efficiency gains. A change of cell spacing also affects the gains from reflection on the module rear cover (k_{11}).

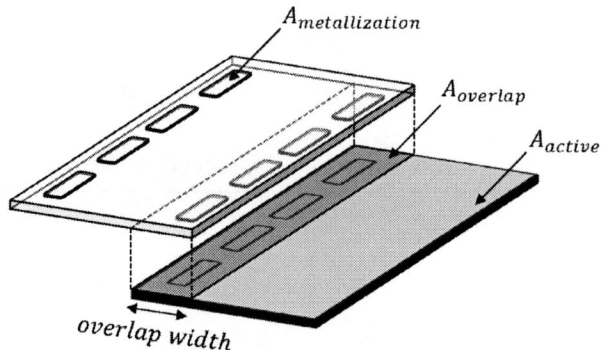

Fig 3: shingle interconnection of two cells with shaded area on the bottom cell ($A_{overlap}$)

Factors k_3, k_4, k_5 and k_6 describe the optical behavior (reflection, absorption) of the encapsulation bulk and do not change for shingled modules.

Changes occur in the modelling of shading by cell interconnection elements (k_7). Shading by interconnector ribbons does no longer occur but instead shading of the lower cell by the upper cell due to interconnection (Figure 3).

The shaded area usually does not only consist of inactive metallization area but also of active area. Therefore, the power output and the current of the lower cell are reduced. We thus use the area shares to calculate k_7:

$$k_7 = 1 - \frac{A_{overlap} - A_{metallizaion}}{A_{cell} - A_{metallizaion}} \quad (6)$$

Strings of shingled solar cells have to be interconnected within a module. Our model assumes that the shading of the string connector ribbon is the same as the cell overlap shading or that the gains of narrower ribbons can be neglected due to the mismatch losses of the shingled cells connected in series. Thus, we do not include the shading effects of string connector ribbons in k_7.

Factors k_8 (cell/encapsulant coupling) and k_9 (finger coupling) remain unchanged compared to other module concepts. The factors describe gains from optical coupling of solar cells after encapsulation.

Reflection gains from interconnector ribbons do not occur due to the absence of these ribbons. The corresponding factor k_{10} becomes unity. Again, we neglect the effects of string interconnection ribbons.

Factor k_{11} describes reflection gains from the rear cover of the module (usually a backsheet). Since a smaller share of backsheet area is visible in shingled modules, this factor is lower compared to conventional modules.

We perform measurements of the rear cover reflection gains using test equipment at Fraunhofer ISE. Results are displayed in figure 4.

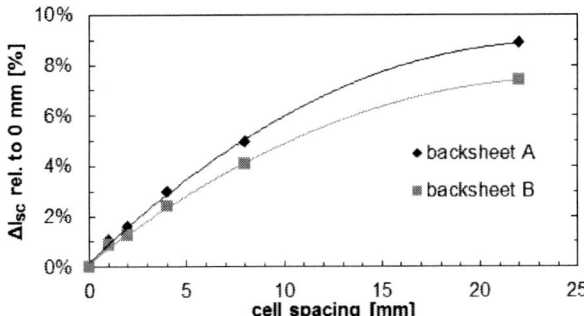

Fig. 4: I_{SC} gains of conventional solar cells totally surrounded by reflective backsheet material for varying cell spacing

An exponential model is used to describe the gains from cell spacing of a cell in a conventional module (7). The parameters a, b and c are fitted to measurement values [7].

$$I_{SC\,gain} = a \cdot \exp\left(\tfrac{cell\,distance}{b}\right) + c. \qquad (7)$$

Measurements are performed using full-size solar cells which feature four active cell edges in a conventional module setup. Cells for shingled modules can be manufactured by splitting full-size cells and cells in a shingled string only feature two active edges that can receive reflected light from the backsheet.

While cover reflection gain data for full-size cells is available, data for shingled cells is rare. We determine the effect of the active edge length on the coupling gain and calculate a corrected rear cover coupling for shingled cells.

Measurements to evaluate the effect of active edges are performed using cells that are partially shaded. This allows the variation of active cell edge. Results are displayed in figure 5.

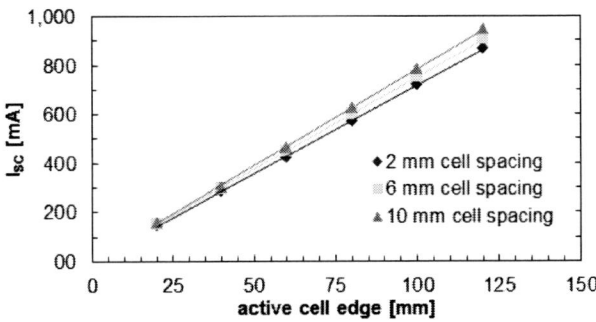

Fig. 5: I_{SC}-measurement results of solar cells with different active edge lengths, cells are shaded to cover parts of the cell edge and cell area, backsheet B.

We find the gains from rear cover reflection to be linear dependent on the active cell edge length. A correction of the I_{SC} gain of a full size cell is performed to fit shingled cells (8). The gain is corrected according to the active cell edge length of a cell in a shingled string.

$$\frac{I_{SC\,gain\,shingle}}{I_{SC\,gain}} = 1 - \frac{2 \cdot (overlap\ width + cell\ width)}{total\ cell\ edge\ length} \qquad (8)$$

The calculation of the cover reflection gain k_{11} is then possible:

$$k_{11} = I_{SC\,gain} \cdot \frac{active\ edge\ length_{shingled\,cell}}{edge\ length_{full-size\,cell}} \qquad (9)$$

The end of each shingled cell string features a cell that has an increased active cell edge and therefore profits from higher irradiation. However, we neglect this effect due to the resulting electrical mismatch.

This method can also be used to correct the rear cover reflection gain for pseudo-square cells or modules with different cell and string distances.

Factor k_{12} describes electrical losses in the cell interconnection. Ohmic losses in ribbons do no longer occur but cells are interconnected using electrical conductive adhesive (ECA). Bulk resistance (ρ_{Bulk}) and contact resistance of ECA ($\rho_{contact}$) have to be considered. Equation (10) shows the electrical resistance R of an ECA-interconnection:

$$R = \frac{\rho_{Bulk} \cdot thickness_{ECA}}{A_{metallization}} + 2 \cdot \frac{\rho_{contact}}{A_{metallization}} \qquad (10)$$

The electrical resistance and the current of the cell string can be used to calculate the power losses of each individual shingled cell. The loss factor k_{12} is the ratio of the loss power and the cell power:

$$k_{12} = 1 - \frac{\sum P_{loss,k12}}{\sum P_{cell}} \qquad (11)$$

While for conventional modules the serial connection of cell strings is common, shingled modules require different module topologies [11]. Strings can be connected in parallel or in networks. Ohmic losses in the string interconnectors can be calculated and (11) can be used accordingly to calculate k_{13}.

The current in a conventional module featuring serial cell and string interconnection is fixed and can be used for all cells. Since cells and strings may deviate regarding electrical properties, for parallel string interconnection the current of every string has to be used and losses have to be calculated separately.

Effects of variations in electrical parameters of cells and strings are considered in k_{14} (electrical mismatch).

Bins of cells feature inherent electrical deviations which lead to electrical mismatch. Cells for shingled modules are usually separated out of larger cells and electrical differences between solar cells may result from the separation of inhomogeneous full-size cells. Since these cells are electrically connected, mismatch occurs.

978-1-5090-5606-4/17 $31.00 © 2017 IEEE

Also, the layup and interconnection process of shingled modules is related to inaccuracies which lead to different shading or variations in electrical cell parameters. This again results in losses due to electrical mismatch.

The precision of the cell layup is crucial for mismatch losses in shingled modules since differences in overlap (fig. 6) result in different shading. The shading results in different currents which lead to increased electrical mismatch of cells in strings.

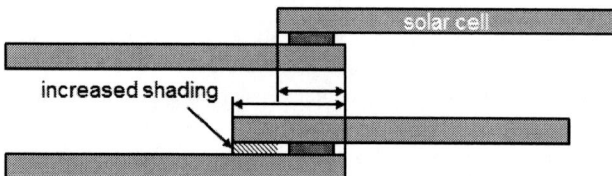

Fig. 6: Increased shading of shingled solar cells due to variations in cell placement during manufacturing.

Again, we use (11) to calculate the k-factor. The loss power for k_{14} is:

$$P_{loss} = \Delta P_{cell\ binning} + \Delta P_{cell\ separation} + \Delta P_{manufacturing} \quad (12)$$

Losses in junction boxes and cabling are considered in k_{15}. Despite the consideration of a possible change in the number of junction boxes and bypass diodes resulting from a different module topology in shingled modules (i.e. parallel string interconnection), no changes in the modelling of k_{15} are necessary for shingled modules.

All models described above are integrated into the SmartCalc.CTM software [8]. We use this software to perform an analysis of shingled modules.

III. CTM-Analysis of Shingled Modules

A. Simulation Setup

Input parameters for the CTM-analysis are determined at Fraunhofer ISE from commercially available module materials by measurements or datasheet analysis. The module we analyze features a low-iron glass with anti-reflective coating and 3.2 mm thickness. The encapsulant foil has a thickness of 0.45 mm and a low UV cut-off. The backsheet is white TPT. For the shingled module we assume a thickness of the ECA of 50 μm with a specific resistance of 0.1 Ωmm²/m.

The monocrystalline shingle cells have a size of 156.75 x 26 mm² and an efficiency of 21.6% (0.88 Wp). We assume a simple busbar metallization on front and rear side with a width of 0.8 mm.

Module dimensions are set to be 1667 x 998 mm (1.66 m²) with margins of 33 mm (top, bottom) and 23.75 mm (left, right).

We assume a cell overlap of 1 mm. Therefore, we receive 64 cells per string using the given module dimensions. String distance is 2 mm and we use 6 strings.

B. Results of the CTM-Analysis for Shingled Modules

The sum of initial cell power is 337.9 Wp. After performing a CTM-analysis with SmartCalc.CTM we find the module efficiency of the shingled module to be 20.2%. Detailed results are displayed in figure 7.

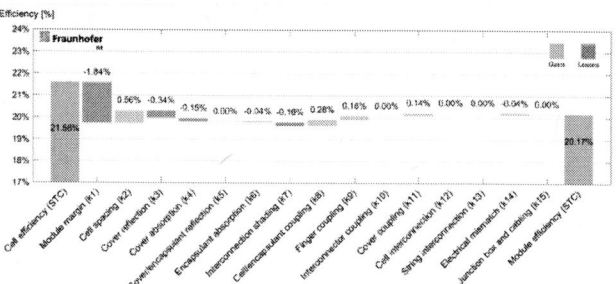

Fig. 7: CTM-analysis (efficiency) of a shingled module

Since the initial cell efficiency was 21.6%, the CTM-ratio for efficiency is $CTM_{efficiency}$ = 93.5%. A major impact factor is the module margin. Other losses are of much lower magnitude or become zero for shingled modules.

The power of the shingled module is 335.8 Wp which results in a CTM_{power} of 99.4%. Detailed results of the CTM-analysis for power are displayed in Figure 8.

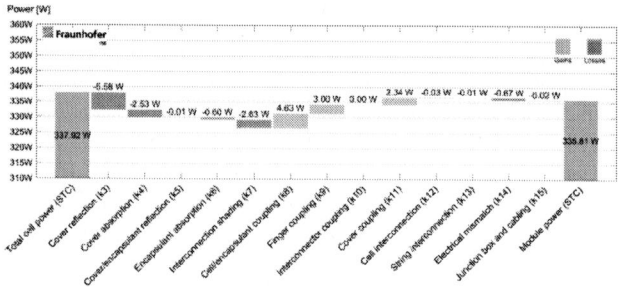

Fig. 8: CTM-analysis (power) of a shingled module

C. Results of the CTM-Analysis for Conventional Modules

We compare the results of the shingled module with a CTM-analysis of a conventional module with ribbon interconnection.

The copper-based ribbons have a cross-section of 1.5 x 0.2 mm². Cell and string distance are set to be 2 mm and margins are adjusted (40.75 mm; 23.75 mm) to have the same module area. The shingled cells are exchanged with pseudo-square cells (diameter 210 mm) of the same efficiency as before (21.6%, 5.24 Wp, 5 busbars).

The CTM-analysis of the conventional module shows a resulting module power of 302.8 Wp (η = 18.3%). The detailed analysis is shown in Figure 2. The CTM ratios for power and efficiency are CTM_{power} = 96.3% and $CTM_{efficiency}$ = 84.8%.

D. Comparison of both Module Concepts

Since both analyzed module setups feature cells of the same efficiency, the same materials and both modules are of the same size, we are able to compare both interconnection concepts. The shingled module has a higher output power, efficiency and CTM-ratios for power and efficiency are higher compared to the conventional module.

The higher output power of the shingled module (+33.0 Wp, +10.9%) is possible due to an improved CTM_{power} but also because more cells can be integrated into the module (+23.5 Wp, +7.5% initial cell power).

An increase in module efficiency can be achieved with shingled modules. The analyzed shingle setup shows an increase of +1.86 %abs in efficiency.

E. Sensitivity Analysis

We perform a sensitivity analysis using SmartCalc.CTM and evaluate the influence of the overlap. We sweep the overlap depth from 1 to 2 mm and keep all other parameters constant (module area changes due to fixed module margins). Results are shown in Figure 9.

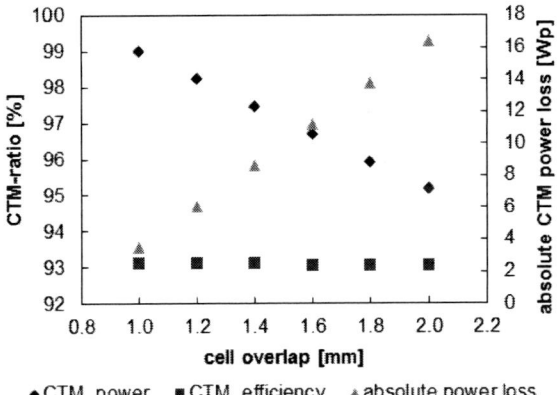

Fig. 9: influence of the overlap depth on CTM-ratios and absolute CTM power loss in shingled modules

We find the shingle overlap to be critical regarding the CTM_{power} ratio. Power loss due to overlap shading becomes a relevant factor for increased overlaps. Module efficiency is only slightly affected since only area shares are affected. The active area changes in size but its efficiency remains unchanged.

A tradeoff in overlap between manufacturing (cell layup precision) and costs (shaded cell parts do not generate power but have to be purchased/manufactured) needs to be achieved.

IV. IMPACT OF SHINGLED CELLS ON MODULES

Shingling solar cells is not only an alternative way of cell interconnection in PV modules, but also influences the design of photovoltaic modules. As mentioned earlier, the CTM ratios

change but also module topology or module size can be affected.

Shingled modules currently feature cells that have approximately the size of 26x156 mm [12][13]. Due to the increased number of (smaller) solar cells in shingled modules, the module voltage increases if a conventional module topology that connects strings of solar cells in series (fig. 10, left) is used.

To be compatible with existing inverters and to not exceed system voltage limitations, electrical properties similar to conventional photovoltaic modules may be desired. Therefore, new module topologies featuring strings connected in parallel or combinations of parallel and serial cell and string interconnection are necessary and can be found in literature [12]-[15]. Shingling requires new solutions for string interconnection, junction boxes, and bypass diode placement. These changes have to be considered in CTM-analyses.

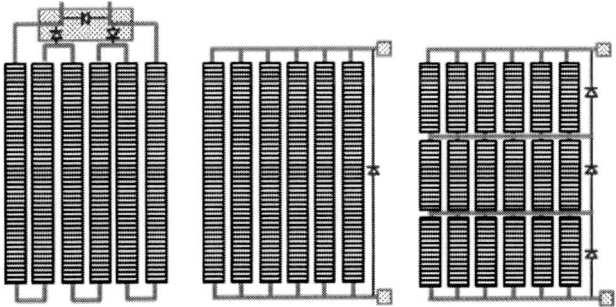

Fig. 10: different module topologies for shingled modules, left: conventional serial interconnection of strings, center: parallel interconnection of strings and two junction boxes, right: network of parallel interconnected strings with two junction boxes

The direct overlap of the cell stripes eliminates the cell gaps and therefore increases the active module area share. Two options are possible to use the resulting gains: A) keeping the module area constant and increasing power and efficiency or B) reducing the module size, keeping the module power constant and saving on module area and materials. We use the CTM-analysis of the conventional and the shingled module to analyze the benefits of both options.

Using the format of the conventional module (option A) and 26 mm shingle cells with 1 mm cell overlap, we are able to fit 64 shingled cells in a string. Module margins for the conventional module are increased (+6.75 mm on short module edges, compared to shingled module) to keep the dimensions of the module constant.

As presented above in the CTM-analysis of the shingled module, the module power is 335.8 Wp compared to 302.8 Wp of the conventional module (+33.0 Wp, +10.9% power in option A).

If we follow option B and reduce the module size, we only need 58 shingled cells per string and module power is calculated to be 304.3 Wp. The size of the module can then be reduced by 9.0%.

V. SUMMARY AND CONCLUSION

We extended an existing methodology to analyze the cell-to-module (CTM) losses and gains by developing new models for shingled cell interconnection. Models are presented and implemented into software. We use this software to perform a CTM-analysis of a shingled setup and compare the shingle concept with a conventional photovoltaic module.

Efficiency and power of the shingled module are higher than for the conventional module (+1.86%$_{abs}$, +33Wp). Also, the cell-to-module ratios for power and efficiency are improved for the shingled module.

We perform a sensitivity analysis of the overlap width and find it to be a crucial factor for CTM power losses in shingled modules.

We analyze a shingled module and a conventional module and find the power output of a shingled module can be increased by 10.9% compared to a conventional module of the same size. Shingled modules allow improved module efficiency.

REFERENCES

[1] Singh, P. J. et al, "Cell-to-module power loss/gain analysis of silicon wafer-based PV modules", *Photovoltaics International, 31st edition*, 2016

[2] Peters, M. et al, "Full loss analysis for a multicrystalline silicon wafer solar cell PV module at short-circuit conditions", *29th PVSEC, DOI: 10.1002/pip.2593*, Amsterdam, 2014

[3] Hanifi, H. et al, "Investigation of cell-to-module (CTM) ratios of PV modules by analysis of loss and gain mechanisms", *Photovoltaics International 33rd edition*, 2016

[4] "Photo-voltaic semiconductor apparatus or the like", *patent US 2938938 A*, 1960

[5] "Shingled array of solar cells", *patent US 3769091 A*, 1973

[6] Zhao, J. et al "Improved Efficiency Silicon Solar Cell Module", *IEEE Electron Device Letters, vol. 18, NO. 2*, 1997

[7] Haedrich, I. et al, "Unified methodology for determining CTM ratios: Systematic prediction of module power", *Solar energy materials and solar cells 131 (2014), pp.14-23, ISSN: 0927-0248*, SiliconPV, 2014

[8] www.cell-to-module.com

[9] Haedrich, I. et al, "PV Module efficiency analysis and optimization", *PV Rollout, ISBN 9783941785-52-6, OTTI*, 2011

[10] H. Wirth et al, "Photovoltaic Modules", *ISBN 978-3-11-034827-9, DeGruyter*, 2016

[11] N. Wörle et al, "Solar Cell Demand for Bifacial and Singulated-Cell Module Architectures", *Photovoltaics International 36th edition*, 2017

[12] Sunpower Corp, "Sunpower Performance Series P17", *technical product datasheet*, 2016

[13] Jiangsu Seraphim Solar System Co., Ltd, "Eclipse", *technical product datasheet*, 2016

[14] R. Morad et al, "Shingled solar cell module", *US 14/605,695*, 2016

[15] J. Zhao, A. Wang, E. Abbaspour-Sani, F. Yun, and M. A. Green, "Improved efficiency silicon solar cell module," *IEEE Electron Device Lett, vol. 18, no. 2, pp. 48–50*, 1997.

A Practical Irradiance Model for Bifacial PV Modules

Bill Marion[1], Sara MacAlpine[1], Chris Deline[1], Amir Asgharzadeh[2], Fatima Toor[2],
Daniel Riley[3], Joshua Stein[3], Clifford Hansen[3]

[1]National Renewable Energy Laboratory, Golden, CO, USA

[2]The University of Iowa, Iowa City, IA, USA

[3]Sandia National Laboratories, Albuquerque, NM

Abstract — **A model, suitable for a row or multiple rows of photovoltaic (PV) modules, is presented for estimating the backside irradiance for bifacial PV modules. The model, which includes the effects of shading by the PV rows, is based on the use of configuration factors to determine the fraction of a source of irradiance that is received by the backside of the PV module. Backside irradiances are modeled along the sloped height of the PV module, but assumed not to vary along the length of the PV row. The backside irradiances are corrected for angle-of-incidence losses and may be added to the front side irradiance to determine the total irradiance resource for the PV cell.**

Model results are compared with the measured backside irradiances for NREL and Sandia PV systems, and with results when using ray tracing software.

Index Terms — **bifacial PV module, irradiance, configuration factor, model, performance.**

I. INTRODUCTION

Bifacial PV modules use radiation received by both front and back surfaces. Unlike the traditional monofacial PV module with an opaque back cover, the bifacial PV module has a transparent back cover to allow the PV cells to receive the backside radiation.

Bifacial PV modules are not new, but there is renewed interest in their deployment because there is presently only a small incremental cost in their manufacture compared to monofacial PV modules.

Guerrero-Lemus et al. [1] recently completed a technical review of nearly 400 papers on bifacial PV modules published since 1979. Their overall recommendation was to make the technology more technically understandable and economically attractive.

To understand the technology and the economics requires the ability to predict the performance of bifacial PV systems. Compared to monofacial PV systems, this requires also modeling the irradiance received by the backside of the PV module.

The beam and diffuse sky irradiance components received on the backside may be modeled with the same model used for the front side, such as the Perez tilted surface model [2], and using the appropriate tilt angle (front tilt angle plus π).

Unless the PV module is mounted vertically, the ground-reflected radiation received by the backside is significantly greater than the beam and diffuse sky radiation received. It is also significantly more difficult to determine because the radiation received by the ground is reduced by shadows from the array and a restricted view of the sky. Additionally, the PV array support structure may prevent ground-reflected radiation from reaching the backside of the PV module.

Ray-tracing software, such as RADIANCE [3], has been used successfully for modeling the backside irradiance [4], but the execution time (hours) is too great an obstacle for routine use for modeling the performance of bifacial PV systems.

To facilitate reasonable execution times, our backside irradiance model uses configuration factors (*CF*s). A *CF* is the fraction of irradiance leaving a surface that is incident on a receiving surface [5]. An annual simulation with an hourly time step may be performed in a few seconds.

As an example of an equation using a *CF*, Eqn. 1 is the familiar equation for the ground-reflected radiation, I_r, incident on the front surface of a PV module:

$$I_r = \rho \cdot \text{GHI} \cdot (1 - \cos\beta) / 2 \qquad (1)$$

where ρ is the ground albedo, GHI is the global horizontal irradiance, and β is the PV module tilt angle from horizontal. The term $\rho \cdot$ GHI is the irradiance leaving the ground surface and the *CF* is equal to $(1 - \cos\beta) / 2$.

The use of *CF*s assumes that the radiation is isotropic, that is, the same intensity for all the angle-of-incidences (AOIs) considered. For ground-reflected radiation for the backside of the PV module, shadows disrupt the isotropic assumption, but the ground area may be divided into areas with equal irradiance distribution and *CF*s applied separately, and then summed to determine the resultant ground-reflected irradiance. A similar technique may be used to determine the diffuse sky irradiance received when the view of the sky is partially obstructed.

II. MODEL

The model is applicable for a row or multiple rows of PV modules. It calculates the backside irradiance for each row of cells to quantify the radiation profile in the PV module slant height direction, but does not distinguish differences in backside irradiance along the row's length. This permits faster program execution because the backside irradiance is not determined for every PV cell in a PV system. Simulations [4] have shown increased backside irradiance for modules on the ends of rows, but this is not thought significant for a PV

978-1-5090-5606-4/17 $31.00 © 2017 IEEE

system with more than a dozen PV modules per row. For rows of shorter length, it may be appropriate to use methods [6]-[9] that can differentiate for positions along the length of the row, but at the expense of complexity and computation time.

The main elements of the model are: (a) identify the ground that is shaded by the PV array, (b) determine irradiance received by the ground by accounting for shading and restricted view of the sky, and (c) determine the irradiance for the backside of the PV module.

A. Ground Shaded by the PV Array

Using the PV array dimensions and orientation, site location, and time, the sun position is calculated and shadows are projected in the row-to-row (rtr) dimension. The rtr is divided into n segments (such as 100) and each segment identified as to whether shaded or unshaded.

B. Irradiance Received by the Ground

The Perez tilted surface model is used with the direct normal irradiance (DNI) and diffuse horizontal irradiance (DHI) to decompose the DHI into its circumsolar (I_{cir}), sky (I_{sky}), and horizon (I_{hor}) components. Using Eqn. 2, the ground irradiance for each of the n segments, GRI_n, is determined.

$$GRI_n = a \cdot (DNI + I_{cir}) + CF_{sky} \cdot I_{sky} \qquad (2)$$

where a is the cosine of the sun zenith angle if the ground segment is unshaded. If the ground segment is shaded, a is the cosine of the sun zenith angle multiplied by the fractional opening of the PV array due to gaps between PV cells of the PV module and gaps between PV modules of the array. CF_{sky} is determined using Eqn. 3 with the view angles of the sky as shown in Fig. 1. For horizontal ground segments, the contribution from I_{hor} is not significant and may be ignored.

$$CF_{sky} = \frac{1}{2} \cdot (\cos \theta_{S1} - \cos \theta_{S2}) \qquad (3)$$

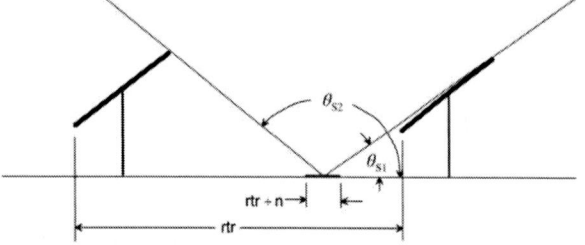

Fig. 1. Field-of-view angles for determining the CFs for the diffuse sky radiation incident a ground segment.

C. Irradiance Received by the Backside of the PV Module

For the location of each row of horizontal PV cells of the PV module or panel, the backside irradiance (BSI) is determined by summing the irradiance from the sky, the irradiance reflected from the ground, the irradiance reflected from the front surface of the PV modules in the row behind, and the irradiance from the sun and circumsolar region of the sky if the AOI is less than 90°. The irradiance reflected from the front surface of the PV modules, I_{refl}, is calculated for only the diffuse radiation incident the front surface. The reflection of the beam and circumsolar radiation from the front surface of the PV module is considered specular and not likely to be reflected to the backside of the PV module in the row to the front for typical PV array configurations.

The diffuse irradiance for the BSI is summed by dividing the field-of-view into 180 one-degree segments, and adding for each segment the product of its CF, AOI correction, and the value of the source's irradiance viewed by the segment (sky, horizon, ground-reflected, or PV module-reflected). The BSI is represented by Eqn. 4:

$$BSI = b \cdot F_b \cdot (DNI + I_{cir}) + \sum_{i=1°}^{180°} CF_i \cdot F_i \cdot I_i \qquad (4)$$

where b = maximum (0, cosine of the AOI of the DNI); F_b is the AOI correction for the DNI using the air-glass model of Sjerps-Koomen et al. [10]; CF_i is the CF for the ith one-degree segment; F_i is the AOI correction for the ith one-degree segment; and I_i is the irradiance viewed by the ith one-degree segment (either I_{sky}, I_{hor}, $\rho \cdot GRI_n$, or I_{refl}). The CF_i is represented by Eqn. 5:

$$CF_i = \frac{1}{2} \cdot [\cos(i-1) - \cos(i)] \qquad (5)$$

where i is in degrees with a range from 1° to 180°. The field-of-view corresponding to a CF_i is shown in Fig. 2.

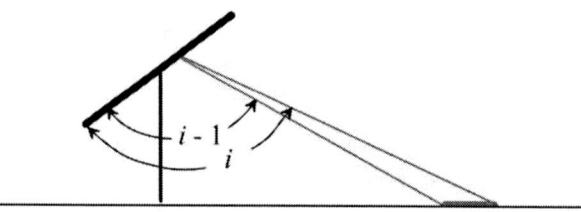

Fig. 2. Field-of-view of the ground for a one-degree segment depicted by the angles i and i-1.

AOI corrections for the one-degree segments of diffuse radiation must consider that the AOI not only varies within the angular i and i-1 limits, but also for radiation originating along the length of row (into or out of the page for Fig. 2). To determine a value of F_i for the one-degree segments, we used a previously developed method [11] where an elemental radiation's AOI correction is weighted by its contribution to the in-plane irradiance. The results are shown in Fig. 3. Note that the F_i is always less than one because the majority of diffuse radiation is always directed other than normal to the surface.

978-1-5090-5606-4/17 $31.00 © 2017 IEEE   1538

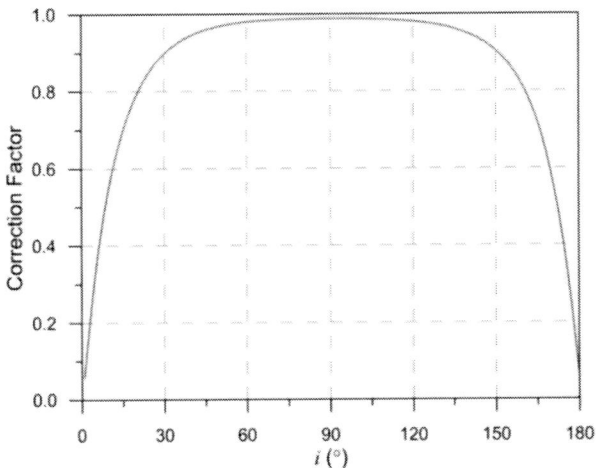

Fig. 3. AOI corrections for the one-degree segments of diffuse radiation as a function of the angle i. For PV modules with an uncoated glass back-surface with a refractive index of 1.526.

Although variations in irradiance for the front side of the PV module are less, the same principles may be applied to account for inter-row shading and variations in field-of- view of the sky due to the presence of rows of other PV modules. For interior rows, the front side irradiance for the bottom of the PV module may be 1-2% less than for the top of the PV module. Backside irradiances have the opposite trend, with the irradiance for the bottom of the PV module being 2 or more times greater than for the top for some circumstances.

III. DATA

For comparison with the model estimates, the irradiances for the backsides of PV modules were measured using reference cells. The measurements were performed on NREL and Sandia National Laboratory PV systems.

A. NREL PV System

The previously installed NREL PV system is shown in Fig. 4. Subarrays are located on two roof levels and measurements were performed for both levels and near the center of the subarrays. The PV modules are monofacial and reference cells were used for short-term measurements of the available BSI.

The reference cells were installed in the center of the subarray, parallel to the PV module back surface, and offset below to represent three locations along the slant-height of the PV module: bottom, middle, and top. A reference cell was also installed in the plane of the PV modules to measure the irradiance for the front side of the PV modules.

The PV modules are oriented with a tilt angle of 10° and an azimuth heading of 165°. Normalized by the PV module slant-height, the horizontal distance between rows is 0.56 and the vertical distance from the roof to the lower edge of the front surface of the PV module is 0.52.

Fig. 4. PV system on the roof of NREL's Science and Technology Facility building which was constructed in 2006.

The white roofing-membrane shows light/medium soiling and the average of the reflectivity measurements over the visible range is 55%. Consequently, an albedo 0.55 was used for modeling purposes. DNI and DHI values from NREL's Solar Radiation Research Laboratory were also used for model input.

B. Sandia PV System

Sandia constructed a facility for testing bifacial PV modules in 2016. As shown in Fig. 5, it consists of four rows of PV modules with monofacial and bifacial PV modules alternating along a row's length. Front to back, the rows are installed south-facing and with tilt angles of 15°, 25°, 35°, and 45°.

Normalized by the PV module slant-height (including the racking), the horizontal distance is 1.07 between the first and second row, 1.42 between the second and third row, and 1.8 between the third and fourth row; the vertical distance from the ground to the lower edge of the front surface of the PV module is 0.58. Compared to the NREL system, the distance between rows will provide more unshaded ground which increases the performance of bifacial PV modules.

For measuring the BSI, reference cells were installed near the middle of each row, parallel to the PV module back surface and offset below to represent two locations along the slant-height of the PV module: bottom and top. A reference cell was also installed in the plane of the PV modules on the east end of the row to measure the irradiance for the front side of the PV modules.

Albedo measurements of the crushed rock ground surface are performed with an inverted pyranometer and DNI and DHI measurements were performed at Sandia's nearby meteorological station.

Data collection began in September 2016 and is ongoing. Other parameters related to the electrical performance of the PV modules are measured, but are not part of this study.

Fig. 5. Bifacial PV module test bed at Sandia. Four rows of PV modules with tilt angles from 15° to 45°. Backside irradiance measured for top and bottom of PV module near the center of each row. Front side irradiances are also measured.

IV. RESULTS

A. NREL PV System

A comparison of the hourly measured and modeled BSI for the top, middle, and bottom reference cells are shown for a cloudy day in Fig. 6 and for a sunny day in Fig. 7 for the subarray on the lower roof of the building. The measured front side reference cell irradiances are included for reference.

While the model results for the cloudy day are favorable, the model underestimated the BSI for all reference cells by a significant amount for the sunny day. During a follow-up site visit on a sunny day, we observed considerable light being reflected to the roof under the subarray from the wall with windows located to the north. (Irradiance enhancements of this type are not addressed by the model.)

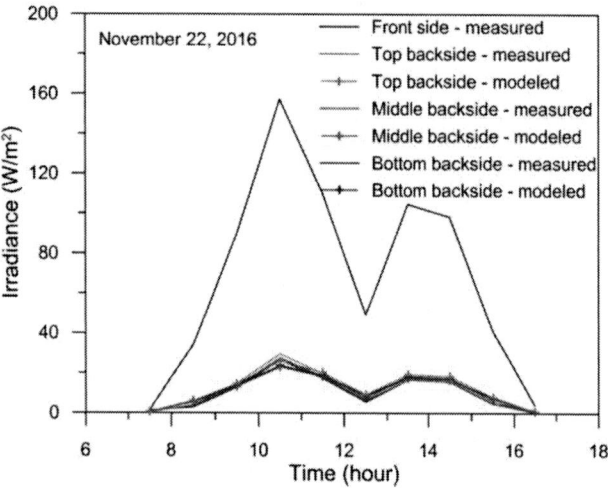

Fig. 6. Cloudy day modeled and measured irradiances for the top, middle, and bottom reference cells located on the backside of a middle row of PV modules of the subarray on the lower roof of the NREL building for November 22, 2016.

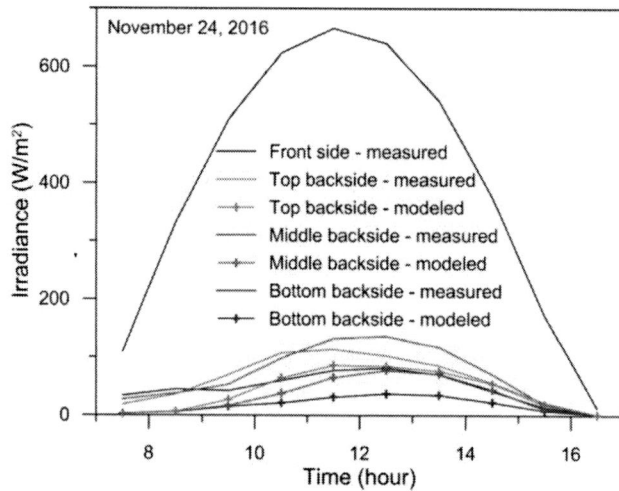

Fig. 7. Sunny day modeled and measured irradiances for the top, middle, and bottom reference cells located on the backside of a middle row of PV modules of the subarray on the lower roof of the NREL building for November 24, 2016.

To confirm our suspicions, the measurement equipment was moved to the middle of the subarray on the top roof to see if the absence of a wall with windows would improve the comparison between modeled and measured values.

For the measurement equipment installation on the top roof, Fig. 8 compares the measured and modeled BSI for the top, middle, and bottom reference cells for a sunny day. Model results are quite good, and they also duplicate the different diurnal profiles measured by the reference cells. Because the azimuth of the subarray is 15° east of south, shadows cast by the PV modules onto the roof have a different pattern in the morning than in the afternoon. This non-symmetry shifts the peak BSI values off-south, with a dependency on the location with respect to the PV module slant-height.

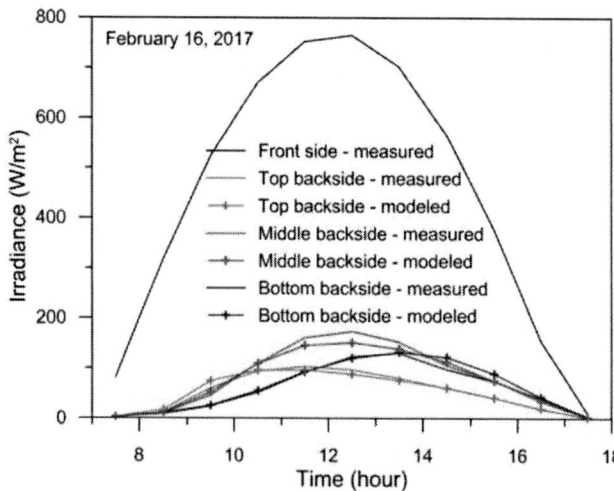

Fig. 8. Sunny day modeled and measured irradiances for the top, middle, and bottom reference cells located on the backside of a middle row of PV modules of the subarray on the upper roof of the NREL building for February 16, 2017.

B. Sandia PV System

The continuous data collection at Sandia permitted statistics comparing modeled and measured BSIs to be determined for the 6-month period from October 1, 2016 through March 31, 2017 using available 15-minute data averages. The statistics used are the mean bias deviation (MBD) and the root-mean-square deviation (RMSD), with the results expressed in both W/m^2 and as a percent of the mean of the measured values. The deviation is the measured value subtracted from the modeled value. For the MBD, a positive value indicates that the model overestimates on average.

The MBD and RMSD statistics for modeling the top and bottom BSIs are provided in Table I. The MBDs were within ±9 W/m^2 and ±16% and the RMSDs were less than 17 W/m^2 and 32%. Because the BSI is added to the front side irradiance to determine the total irradiance for the PV cell, the statistics in units of W/m^2 are more useful for evaluating the error in estimated cell output. The front side irradiances in Table I are generally a factor of 10 greater than the BSI. For additional context, comparing the measured GHI from the meteorological station with the measured GHI from the nearby bifacial module test bed yielded a MBD of 8.6 W/m^2 and a RMSD of 11.7 W/m^2, values not too different than those for the modeled BSIs.

Table II provides the MBDs and RMSDs for the modeled irradiance available to a bifacial PV cell, determined as the sum of the modeled front side irradiance and BSI, compared to the sum of the measured front side irradiance and BSI. The MBDs were within ±2.4% and the RMSDs were less than 6%.

Fig. 9 is a scatterplot of the modeled front side irradiance plus the modeled BSI versus the measured front side irradiance plus the measured BSI for the top reference cell located on the backside of the row of PV modules with a tilt angle of 35°. The diagonal in the Fig. 9 has a slope of one, data points above the diagonal indicate model overestimates, and vice versa for data points below the diagonal. The figure shows good agreement between modeled and measured values.

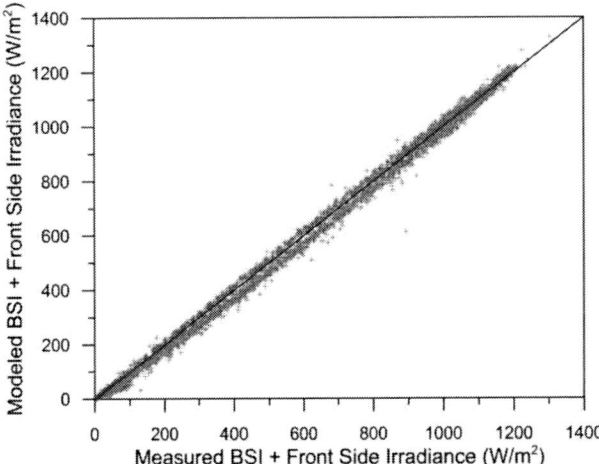

Fig. 9. Scatterplot of the modeled bifacial irradiance (modeled front side irradiance plus the modeled BSI) versus the measured bifacial irradiance (measured front side irradiance plus the measured BSI) for the top reference cell located on the backside of the row of PV modules with a tilt angle of 35°.

TABLE I

MEAN BIAS DEVIATION (MBD) AND ROOT-MEAN-SQUARE DEVIATION (RMSD) FOR MODELED VALUES OF THE BSI FOR TOP AND BOTTOM REFERENCE CELL LOCATIONS AND USING 15-MINUTE DATA MEASURED AT SANDIA FROM OCTOBER 1, 2016 THROUGH MARCH 31, 2017. AVERAGES BASED ON MEASURED VALUES.

PV Row/ Tilt Angle	Front Side Irradiance	Bottom BSI					Top BSI				
	Average (W/m^2)	Average (W/m^2)	MBD (W/m^2)	MBD (%)	RMSD (W/m^2)	RMSD (%)	Average (W/m^2)	MBD (W/m^2)	MBD (%)	RMSD (W/m^2)	RMSD (%)
Row 1 / 15°	512	63.3	8.9	14.1	16.5	26.0	31.6	5.1	16.0	9.9	31.3
Row 2 / 25°	566	49.2	3.9	7.9	14.0	28.4	34.7	-1.0	-3.0	4.8	13.8
Row 3 / 35°	598	55.4	-1.3	-2.3	13.0	23.5	40.9	-3.7	-9.2	6.5	15.9
Row 4 / 45°	596	61.0	8.6	14.1	13.2	21.6	52.8	6.7	12.7	9.5	18.0

TABLE II

MEAN BIAS DEVIATION (MBD) AND ROOT-MEAN-SQUARE DEVIATION (RMSD) FOR THE SUMS OF THE MODELED VALUES OF THE BSI FOR TOP AND BOTTOM REFERENCE CELL LOCATIONS AND THE FRONT SIDE IRRADIANCE AND USING 15-MINUTE DATA MEASURED AT SANDIA FROM OCTOBER 1, 2016 THROUGH MARCH 31, 2017. AVERAGES BASED ON THE SUMS OF THE MEASURED VALUES.

PV Row/ Tilt Angle	Bottom BSI + Front Side Irradiance					Top BSI + Front Side Irradiance				
	Average (W/m^2)	MBD (W/m^2)	MBD (%)	RMSD (W/m^2)	RMSD (%)	Average (W/m^2)	MBD (W/m^2)	MBD (%)	RMSD (W/m^2)	RMSD (%)
Row 1 / 15°	575	13.6	2.4	32.9	5.7	542	9.6	1.8	24.0	4.4
Row 2 / 25°	616	7.6	1.2	35.5	5.8	594	2.7	0.5	25.3	4.3
Row 3 / 35°	653	-8.6	-1.3	33.4	5.1	625	-11.1	-1.8	25.8	4.1
Row 4 / 45°	657	7.1	1.1	27.6	4.2	644	5.2	0.8	28.0	4.3

978-1-5090-5606-4/17 $31.00 © 2017 IEEE

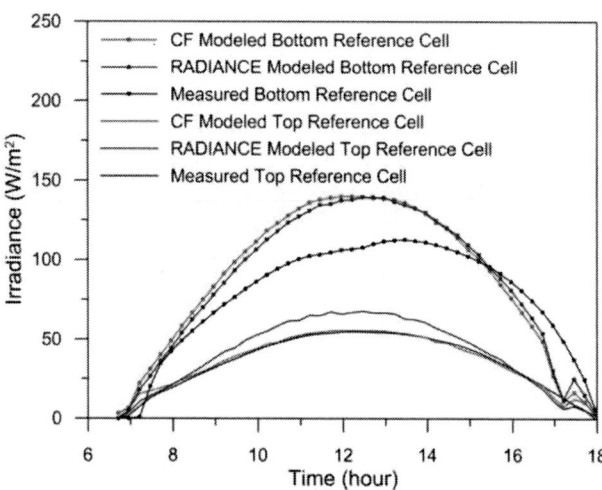

Fig. 10. Modeled and measured irradiances for the top and bottom reference cells located on the backside of the row of PV modules with a tilt angle of 15° for March 1, 2017.

The model did not consider shading by the concrete foundations or their location relative to the reference cells, and this may have adversely impacted the results, particularly for the bottom reference cells. Fig. 10 shows modeled and measured irradiances for the reference cells installed on the row with tilt angle of 15° for a clear day in March. While the model results for the top reference cell are good, the model doesn't duplicate the shift in peak values of the bottom reference cell's measured data toward the afternoon. This is thought to be a consequence of the reference cell being closer to the east concrete foundation than the west and reflections from the concrete foundations

Also shown in Fig. 10 are results when using the RADIANCE ray tracing software to model the irradiance for the top and bottom reference cells that includes the effects of the concrete foundations and the array structure. These results show a slight shift in peak values for the bottom reference cell toward the afternoon, but not to the extent exhibited by the measured data.

V. SUMMARY

A model was presented for estimating the BSI of a bifacial PV module applicable for a row or multiple rows of PV modules. For model efficiency, it calculates the BSI for each row of cells to quantify the radiation profile in the PV module slant height direction, but does not distinguish differences in backside irradiance along the row's length. The model is based on the use of CFs to determine the fraction of a source of irradiance that is incident the PV module, and AOI correction factors are applied to account for both direct and diffuse radiation reflection losses from the front and back PV module surfaces.

For PV systems installed at NREL and Sandia, the model estimates were in agreement with the measured BSIs, with the exception of results influenced by PV system features not addressed by the model. For the NREL system, the subarray on the lower roof received additional radiation reflected from the wall to the north. The concrete foundations at Sandia provided additional reflective surfaces and shadows.

The use of ray-tracing software such as RADIANCE is a useful tool for evaluating how the CF model results might be influenced by PV system features it doesn't directly address.

ACKNOWLEDGEMENT

The U.S. Government retains and the publisher, by accepting the article for publication, acknowledges that the U.S. Government retains a nonexclusive, paid-up, irrevocable, worldwide license to publish or reproduce the published form of this work, or allow others to do so, for U.S. Government purposes.

REFERENCES

[1] R. Guerrero-Lemus, R. Vega, T. Kim, L. Shephard, "Bifacial solar photovoltaics – A technology review", *Renewable and Sustainable Energy Reviews* **60**: 1533-1549, 2016.

[2] R. Perez, P. Ineichen, R. Seals, J. Michalsky, "Modeling daylight availability and irradiance components from direct and global irradiances", *Solar Energy* 44: 271-289, 1990.

[3] G. J. Ward, "The RADIANCE Lighting Simulation and Rendering System," Proceedings of the 21st Annual Conference on Computer Graphics and Interactive Techniques, ACM, 1994. Software at https://github.com/NREL/Radiance/releases

[4] C. Deline, S. MacAlpine, B. Marion, F. Toor, A. Asgharzadeh, J. Stein, "Assessment of Bifacial Photovoltaic Module Power Rating Methodologies – Inside and Out", IEEE Journal of Photovoltaics, in press.

[5] M. Iqbal, *An Introduction to Solar Radiation*. Toronto: Academic Press Canada, 1983.

[6] A. Krenzinger, "Estimation of Radiation Incident on Bifacial Albedo-Collecting Panels", *Int. J. Solar Energy* 4: 297-319, 1986.

[7] A. Lorenzo, G. Sala, S. Lopez-Romero, "Diffusing Reflectors for Bifacial Photovoltaic Panels", *Solar Cells* 13: 277-292, 1985.

[8] U. Yusufoglu, T. Pletzer, L. Koduvelikulathu, C. Comparotto, R. Kopecek, H. Kurz, "Analysis of the Annual Performance of Bifacial Modules and Optimization Methods", IEEE Journal of Photovoltaics 5: 320-328.

[9] J. Appelbaum, "Bifacial photovoltaic panels field", Renewable Energy 85, 338-343, 2016.

[10] E. A. Sjerps-Koomen, E. A. Alsema, W. C. Turkenburg, "A simple model for PV module reflection losses under field conditions", Solar Energy 57: 421–432, 1996.

[11] B. Marion, "Numerical method for angle-of-incidence correction factors for diffuse radiation incident photovoltaic modules", *Solar Energy* 147: 344–348, 2017.

A Detailed Model of Rear-Side Irradiance for Bifacial PV Modules

Clifford W. Hansen[1], Renee Gooding[1], Nathan Guay[1], Daniel M. Riley[1], Johnson Kallickal[1], Donald Ellibee[1], Amir Asgharzadeh[3], Bill Marion[2], Fatima Toor[3], Joshua S. Stein[1]

[1]Sandia National Laboratories, Albuquerque, NM, 87185-1033, USA

[2]National Renewable Energy Laboratory, Golden, CO, 80401, USA

[3]Electrical and Computer Engineering Department, The University of Iowa, Iowa City, IA, 52242, USA

Abstract — We describe and validate a method for modeling irradiance on the back surface of bifacial PV modules at the scale of individual cells using view factors. We compare model results with irradiance measurements on the back of PV modules in various configurations. Our analysis illustrates the relative accuracy of the model as well as the potential variation in back surface irradiance among the cells.

Index Terms — **bifacial PV module, irradiance, view factor.**

I. INTRODUCTION

Bifacial photovoltaic (PV) cells, modules, and systems potentially offer a pathway to significantly lower levelized cost of energy. Bifacial PV arrays are not widely deployed in part because their potential performance advantages are not generally understood. Sandia National Laboratories (Sandia), the National Renewable Energy Laboratory and the University of Iowa are investigating bifacial PV performance and characterization in a joint project funded by the US Department of Energy [1]. The project's main objectives are (1) measure the performance of various bifacial PV technologies using an outdoor test bed, (2) develop and validate models of back surface irradiance, and (3) work with industry to develop rating standards for bifacial PV modules. The outdoor test bed at Sandia in Albuquerque, NM allows investigation of the many factors that influence bifacial PV performance, including ground albedo and array geometry (e.g., height above ground, tilt angle, row position, row-to-row spacing).

Conceptually, total irradiance on the back surface of a rack-mounted module results from the combination of:

- Sky diffuse irradiance. The visible sky depends on the module's tilt and azimuth and is restricted by other nearby structures.
- Ground-reflected irradiance which can vary across the surfaces behind the module due to albedo and the irradiance incident on the ground surfaces.
- Structure-reflected irradiance from nearby objects such as from the front of PV modules in an adjacent row.
- Direct irradiance on the back surface, e.g., when the sun elevation is low and the sun azimuth is to the northeast or northwest of a south-facing array.

Accurate calculation of back surface irradiance remains a challenge. Previously [2] we compared two approaches to modeling back surface irradiance: view factor and ray tracing, with view factor models considering either 3D or a simplified 2D array geometry. At the time, ray tracing models were computationally prohibitive for simulating energy production from bifacial PV arrays. View factor models with simplified 2D geometry cannot represent the full variation of irradiance among the cells in a row of bifacial PV modules, and hence the effect of this variation remains unquantified.

In this paper, we present and validate a computationally efficient approach for modeling back surface irradiance at single-cell resolution using view factors. Our approach represents the spatial non-uniformity in irradiance on the back surface by computing irradiance for each cell in a bifacial PV module.

II. MEASURED BACK SURFACE IRRADIANCE

Sandia National Laboratories is using reference cells to measure rear surface irradiance at high spatial resolution using a sensor array with the form factor of a single PV module (Fig.1) mounted in a variety of tilts and heights. We also measure front and back surface irradiance in conjunction with measuring bifacial PV module I-V curves mounted on several arrays. Fig. 2 illustrates a south-facing rack adjustable in height and tilt with monofacial PV modules on the west half (left in figure) and bifacial PV modules on the right half. Reference cells measure irradiance along the middle of the rack: at the top and bottom of the front, and at the top, middle and bottom of the rear. Reference cells are calibrated outdoors against a primary reference cell (calibrated by NREL) to reduce variation among cells to less than 4 W/m^2 at irradiance of 1000 W/m^2.

III. BACK SURFACE IRRADIANCE MODEL

Our back surface irradiance model uses view factors defined at the resolution of a single cell. Calculation of view factors at a cell level permits an array performance model to directly account for mismatch conditions among cells and modules, and also to represent arrays with subsets of modules in different configurations, e.g., a mix of southward facing, fixed tilt modules and vertical E-W facing modules. Compared to ray tracing simulations, this detailed view factor model is less demanding computationally and require fewer parameters but also represents a PV system with less detail. View factors can

978-1-5090-5606-4/17 $31.00 © 2017 IEEE

be used to efficiently model irradiance for large, regular arrays (e.g., [3]) at the loss of detail regarding the variation in irradiance along the array's rows and at row ends.

Fig. 1. Movable sensor array for high spatial resolution measurements of rear-surface irradiance.

Fig. 2. Adjustable PV array at Sandia's Photovoltaic Systems Evaluation Laboratory.

III.A Detailed View Factor Model

View factors, also termed shape and configuration factors, quantify the fraction of irradiance reflected from one surface that arrives at a receiving surface. View factor models [4], [5] calculate back surface irradiance E_2 (W/m²) by:

$$E_2 = \alpha \times G_1 \times F_{1\to2} \qquad (1)$$

where G_1 is the total irradiance (W/m²) on the reflecting area being considered (e.g., an area of the ground), α is the albedo of the reflecting surface and $VF_{1\to2}$ is the view factor (unitless) from the reflecting area to the receiving surface. The total irradiance on the back surface of a cell is the sum over all contributing reflecting surfaces. A rear surface irradiance model is then assembled by specifying the set of reflecting surfaces, albedos and the irradiance incident on each surface (Fig. 3).

A view factor model assumes that all reflecting surfaces are Lambertian, i.e., irradiance is scattered isotropically. An emitting surface (dA_1) reflects incident irradiance, part of which is incident on the receiving surface: the view factor $F_{1\to2}$ quantifies the fraction of irradiance emitted by A1 that is received by A2. Formally, view factors are calculated by integration (Eq. 2) using terms in the illustration.

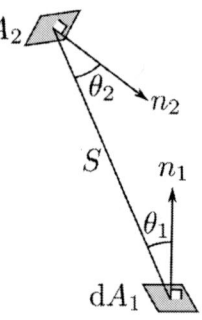

$$F_{1\to2} = \frac{1}{A_1} \int_{A_1} \int_{A_2} \frac{\cos\theta_1 \cos\theta_2}{\pi s^2}\, dA_2\, dA_1 \qquad (2)$$

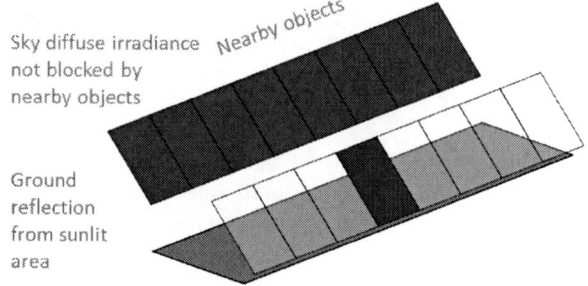

Fig. 3. Features in the detailed view factor model.

Our detailed view factor model calculates irradiance $E_{back}(t)$ at time t on the rear surface of cell k from DHI and DNI by

$$E_{back,k}(t) = E_{ground,k}(t) + E_{sky}(t)VF_{k\to sky} + E_{beam}(t) \quad (3)$$

$$E_{ground,k}(t) = \sum_i \alpha_i G_i(t) VF_{i\to k} \qquad (4)$$

$$G_i(t) = DHI(t)VF_{i\to sky} + \delta_i(t)DNI(t)\cos(Z(t)) \quad (5)$$

$$E_{sky}(t) = \left[(1-A)DHI(t)\left(\frac{1-\cos(\theta_T)}{2} \right) \right] \qquad (6)$$

$$A = DNI(t)/H_o(t)$$

$$E_{beam}(t) = DNI(t)\cos(AOI(t))IAM(AOI(t)) \quad (7)$$

where i enumerates the reflecting surfaces, Z is the solar zenith angle, θ_T is the tilt from horizontal of cell k, $\delta_i(t)=1$ if surface i is sunlit at time t and $\delta_i(t)=0$ otherwise. Eq. 6 is obtained from the Hay and Davies model [6] for sky diffuse irradiance on a tilted surface by omitting its term for circumsolar diffuse irradiance; $H_o(t)$ is the extraterrestrial normal irradiance. In Eq. (7) $IAM(AOI(t))$ reduces any direct irradiance on the rear surface of the cell for specular reflections; we use [7] with $a = 0.18$.

III.B Improving Computational Efficiency

We reported previously [2] on results from a partial implementation in Matlab™ of parts of the detailed view factor model (Algorithm A). Our earlier work assumed $VF_{k \to sky} = 1$, i.e., sky diffuse irradiance was not blocked by nearby objects. Our code was too slow for analysis of bifacial energy production (e.g., 15-minute time steps for a simulation year), nor for optimizing power production when varying bifacial PV module orientation and tilt.

Algorithm A: Inefficient algorithm for view factor model.
1. Compute solar positions and irradiance for all time steps.
2. For each time step:
   a. Use the sun vector to project structures to shadows on the reflecting surfaces.
   b. For sunlit and shaded areas of each surface, compute view factors to each cell using Eq. 2.
   c. Compute irradiance on the rear surface of each cell using Eq. 1.

Analysis of Algorithm A shows two inefficiencies:
- View factors are recomputed at each time step, requiring numerical integration to converge for each cell and each reflecting area. This is highly inefficient because a view factor is a geometric quantity (Eq. 2) which has no dependence on the sun position.
- Numerical integration using the packaged Matlab™ function `integral2` is done separately for each cell, and is not amenable to vectorization for a list of cells.

To reduce computation time, we implement Algorithm B. We grid each reflecting surface in order to approximate the integrand $I(s, n_1, n_2)$ in the calculation of $VF_{i \to k}$ with its value $I(\hat{S}, n_1, n_2)$ at the midpoints of the reflecting and receiving cells. Moreover, calculation of $I(\hat{S}, n_1, n_2)$ can be expressed in terms of matrix operations only:

$$I(s) = \frac{\cos \theta_1 \cos \theta_2}{\pi s^2} \qquad (7)$$

$$I(\hat{S}) = \frac{\dfrac{\langle \hat{S}, n_1 \rangle}{\|\hat{S}\|} \dfrac{\langle -\hat{S}, n_2 \rangle}{\|\hat{S}\|}}{\pi \|\hat{S}\|^2} = \frac{\hat{S}^T \left[-n_1 n_2^T \right] \hat{S}}{\pi \langle \hat{S}, \hat{S} \rangle^2} \qquad (8)$$

Numerical exploration of a wide range of reflecting and receiving cell geometries revealed a rather complex relationship between cell dimensions and the approximation error $\left| I(s, n_1, n_2) - I(\hat{S}, n_1, n_2) \right|$. A rough rule of that sets cell dimension $< 0.25 \|\hat{S}\|$ maintains approximation error $< 1\%$ except when cell normal vectors are nearly parallel. We set the grid boundaries at the intersection with the ground of a ray at an angle of 88° from a vertical ray from the cell's center to the ground. The resulting grid encompasses 97% or more of ground-reflected irradiance which might affect a receiving cell.

Algorithm B: Faster implementation of view factor model.
For each cell k:
1. Grid each reflecting surface.
2. For each combination of grid cell and module cell, compute $VF_{i \to k}$ using Eq. 3.
3. Compute solar positions and $E_{sky}(t)$ (Eq. (6)) for all time steps.
4. Compute $VF_{k \to sky}$.
5. For each time step:
   a. Use the sun vector to project structures to shadows on the reflecting surfaces.
   b. Identify each grid cell as having shaded or unshaded conditions (evaluate $\delta_i(t)$).
   c. Return irradiance on the rear surface of the evaluation of Eq. (3), (4) and (5).

The expression in Eq. 8 involves a set of linear algebra operations which are computed once for each pair of reflecting and receiving cells. The implementation of Algorithm B using Eq. 8 can be done (using CUDA) for GPU rather than CPU processing, further reducing calculation time.

A computational test case using three receiving cells, a grid of 713 × 713 reflecting cells, ten objects which cast shadows and 151 time steps was used for timing analysis. The test case calculated irradiance at the three receiving cells at each time step. CPU processing took approximately 123 seconds to complete while the GPU implementation of the same algorithm completed in 14 seconds. For both implementations the largest amount of computation time is spent in step 5b, where the boundary of each object's shadow is traversed clockwise to determine grid cells whose centers lie to the right of every boundary segment.

IV. ANALYSIS

Fig. 4 compares modeled and measured irradiance at the 6 outer reference cells on a clear day, with the sensor array mounted on open racking (Fig. 1) at 30° tilt and 0.6m from the ground with a relatively clear view of the northern horizon and away from any objects with shadows that might cast shadows near the sensor array. Ground albedo is 0.23 measured with a Kipp and Zonen CMA-11 albedometer. At noon modeled irradiance is within 10 W/m² of measurement. The skewed curves for the middle and bottom rows result from the sensor array's shadow passing underneath the sensor array.

Fig. 5 shows model residuals for each of the 10 cells. Model predictions are generally within 15 W/m^2 (8%) around solar noon, and within 20 W/m^2 at all times. Explanation for the observed negative bias (model < measured) has so far proven elusive. Possible causes include:

1. code errors;
2. error in the sky diffuse irradiance model ([6] in this case) which is used to estimate irradiance incident on shadowed areas on the ground (ground reflections comprise essentially all of irradiance in Fig. 4).
3. spectral changes in irradiance reflected from ground surfaces;
4. deviation from the assumption that reflections are Lambertian.

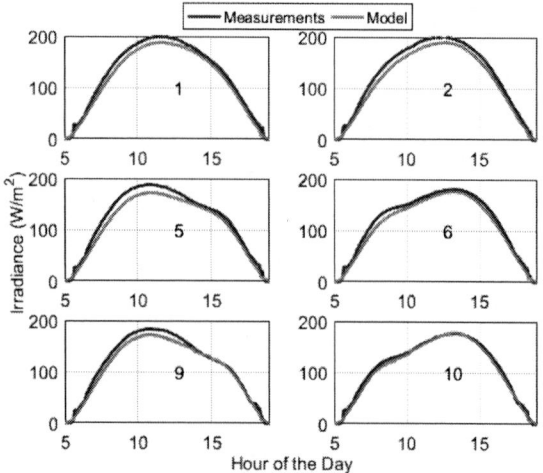

Fig. 4. Modeled and measured rear irradiance for six outside cells in the sensor array: May 5, 2017, a clear day, isolated open rack mount tilted at 30º.

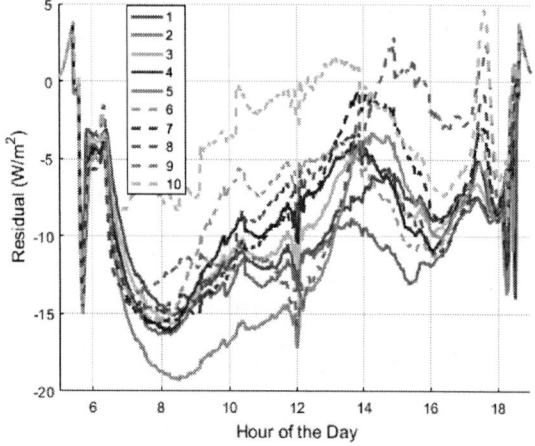

Fig. 5. Model residuals for all 10 cells in the sensor array: May 5, 2017.

Fig. 6 compares measured and modeled irradiance at all times for three days with different conditions (broken clouds persisted through April 30 morning, and May 2 afternoon; otherwise clear sky conditions obtained). For May 2 and May 5, the negative bias is evident. For April 30 morning, a positive bias of approximately 20 W/m^2 is observed (Fig. 7). The variation in prediction bias from –15 W/m^2 (May 5, Fig. 5) to 20 W/m^2 (April 30, Fig. 7) gives a rough envelope of model accuracy for all sky conditions of about ±10%.

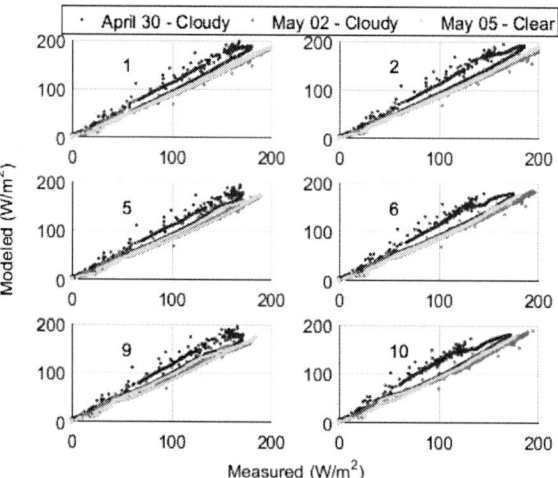

Fig. 6. Modeled and measured rear irradiance for six outside cells in the sensor array: varied sky conditions, isolated open rack mount tilted at 30º.

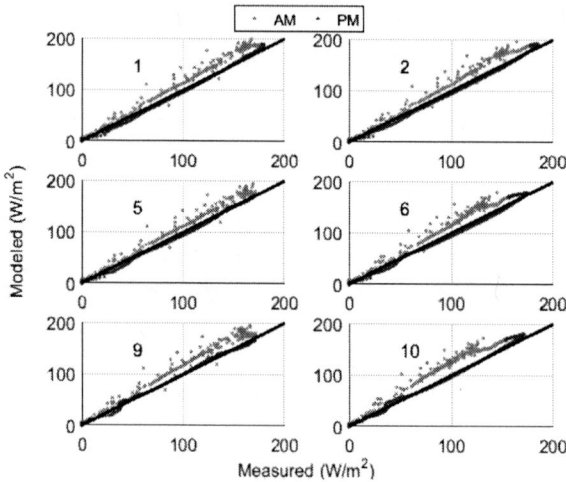

Fig. 7. Modeled and measured rear irradiance for six outside cells in the sensor array: April 30, 2017, isolated open rack mount tilted at 30º. Black line is 1:1.

Fig. 8 compares modeled and measured irradiance at the 6 outer reference cells on a clear day, with the sensor array mounted vertically on isolated racking in landscape orientation 0.63m from the ground. The array is mounted above a concrete block with no gap in between the block and array to keep the shadow shape simple. Direct irradiance on the rear surface of the sensor array cause the 'ears' in each trace; reflection off a nearby row of PV modules appears as a spike just before the afternoon 'ear'. Model results are within 20 W/m² of measurements, with a similar negative bias as is observed for the sensor array on tilted racking.

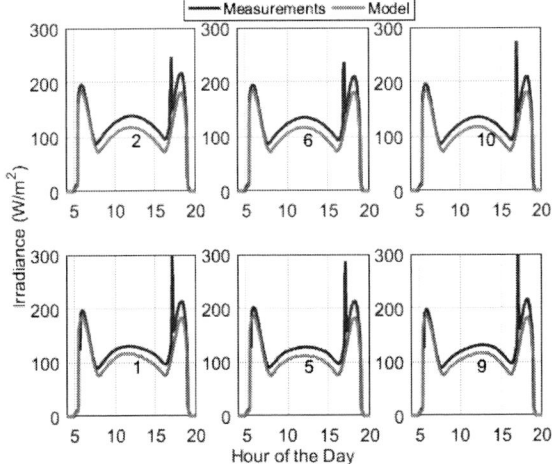

Fig. 8. Modeled and measured rear irradiance for six outside cells in the movable sensor array: May 20, 2017, a clear day, vertical isolated open rack mount.

We simulate irradiance at the three rear-facing reference cells on the adjustable array shown in Fig. 2. These simulations account for shadows cast by the four modules and the array frame structure as depicted in Fig. 9. Fig. 10 compares modeled and measured irradiance for a clear day with the array at a relatively steep tilt. The shape evident for the bottom cell results from the passage of the array's shadows below the cell: the morning and afternoon shoulders correspond to shadows from the modules falling beneath the cell, and the midday peak from the sunlight passing through the gap between the modules.

The model reproduces the shapes evident for each reference cell indicating that the view factor method generally captures the effects of shadows on ground-reflected irradiance. A systematic bias towards underestimating irradiance by approximately 15 W/m² is observed for periods of time (early/late hours) or reference cells (top) for which array shadows do not significantly affect the received irradiance. The consistency of the bias, and the absence of dependence on time of day, suggest the explanation lies with the sky diffuse irradiance model or a reduction in spectral irradiance to which the reference cells respond. If the bias was due to an incorrect value for the ground albedo, we would expect the bias to affect all reference cells in a similar manner, and the discrepancy between model and measurement to shrink at lower irradiance: neither effect is apparent in Fig. 10. The roughness of the model curves results from projecting array shadows onto the grid used for computing view factors, resulting in pixelated shadow boundaries.

Fig. 11 compares modeled and measured irradiance for a clear day with the array at 15° tilt, where nearly all of the received rear-surface irradiance results from ground reflection. Shadows from the array affect all three reference cells. Model bias remains evident although smaller in magnitude than with the steeply tilted configuration.

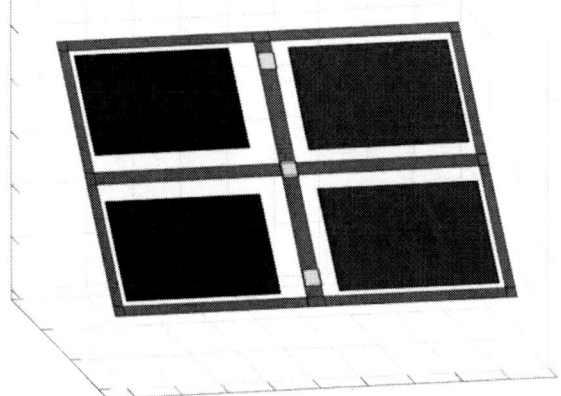

Fig. 9. Schematic of adjustable array features represented in rear irradiance model. Blue: monofacial modules, red: bifacial modules, green: reference cells (facing away), grey: structure.

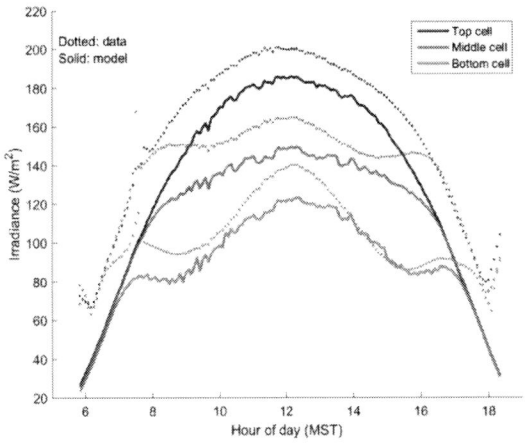

Fig. 10. Modeled and measured irradiance for rear-facing reference cells in the adjustable array: 16 June 2016, a clear day, array titled at 45° with center at 1.63m above ground.

Fig. 12 illustrates the cell-to-cell variation in modeled irradiance for the top-right bifacial module (6 × 10 cells in landscape orientation). The model predictions show a variation as great as 50 W/m², or roughly 5% of total irradiance (front and back) on the bifacial module. The cell-to-cell variation drops nearly to zero during cloudy periods, when the ground surrounding the module's view is all in shadow. The modeled cell-to-cell variation is informative for efforts to develop performance models for bifacial PV modules (e.g. [8]) as this variation contributes to power losses due to mismatched output between cells in series. It is anticipated that cell-to-cell variation widens with increasing albedo, and decreases with increasing height above ground, in conjunction with increased rear-surface irradiance and decreasing view factors, respectively [9].

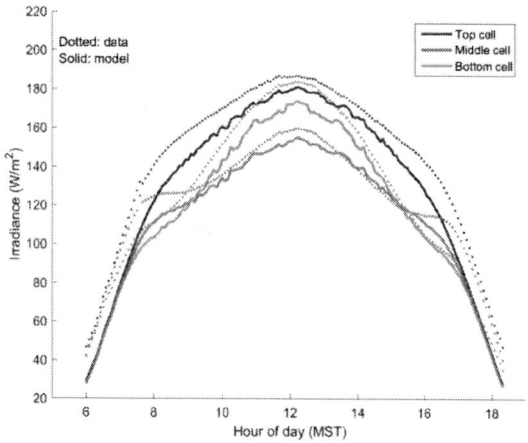

Fig. 11. Modeled and measured irradiance for rear-facing reference cells in the adjustable array: 14 July 2016, a clear day, array titled at 15° with center at 1.63m above ground.

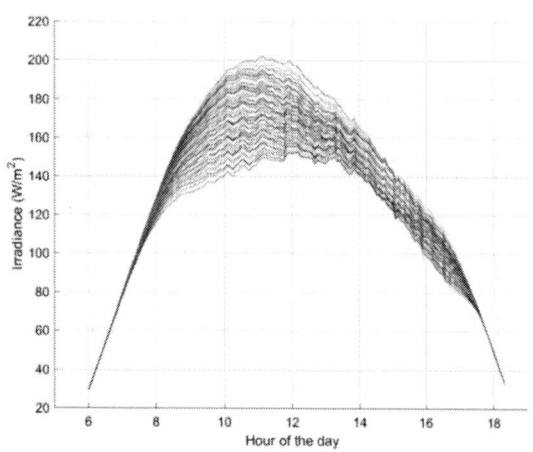

Fig. 12. Modeled rear irradiance by cell on the top right bifacial module in the adjustable array: 6 July 2016, a clear day, array titled at 45° with center at 1.63m above ground.

V. CONCLUSIONS

We present and show validation for a detailed model of rear-surface irradiance at the scale of individual cells. Our model operates on DNI, DHI, albedo and system geometry as input. Irradiance reaching the back surface of a cell is computed as the sum over ground reflections (accounting for shadows on the ground from nearby objects), sky diffuse irradiance and direct irradiance. Within the validation presented, our model shows an accuracy of roughly ±10% when compared to irradiance measured with reference cells.

This detailed rear irradiance model, in its present Matlab™ implementation, is useful for analysis of module- and string-scale performance of bifacial PV systems. Computation time for larger systems remains a challenge: simulating two 6 × 10 cell modules (total of 120 receiving cells) with 10 shadow-casting objects for 100 time steps requires about 5 minutes using CPU processing. Computation time can be dramatically reduced when GPU processing is available, and the algorithm can easily be parallelized.

ACKNOWLEDGEMENTS

This work was supported by the U.S. Department of Energy SunShot Initiative. Sandia National Laboratories is a multi-program laboratory managed and operated by Sandia Corporation, a wholly owned subsidiary of Lockheed Martin Corporation, for the U.S. Department of Energy's National Nuclear Security Administration under contract DE-AC04-94AL85000.

REFERENCES

1. Stein, J., C. Deline, and F. Toor. *Outdoor Field Performance from Bifacial Photovoltaic Modules and Systems*. in *44th IEEE Photovoltaic Specialist Conference*, 2017. Washington, DC.
2. Hansen, C.W., et al. *Analyis of Irradiance Models for Bifacial PV Modules*. in *43rd IEEE Photovoltaic Specialist Conference*, 2016. Portland, OR.
3. Marion, B., et al. *A Practical Irradiance Model for Bifacial PV Modules*. in *44th IEEE Photovoltaic Specialist Conference*, 2017. Washington, DC.
4. Iqbal, M., *An Introduction to Solar Radiation*. 1983, Toronto: Academic Press Canada.
5. Yusufoglu, U.A., et al., *Simulation of Energy Production by Bifacial Modules with Revision of Ground Reflection*. Energy Procedia, 2014. **55**: p. 389-395.
6. Hay, J.E. and J.A. Davies. *Calculations of the solar radiation incident on an inclined surface*. in *First Canadian Solar Radiation Data Workshop*, 59, 1980. Ministry of Supply and Services, Canada.
7. Martin, N. and J.M. Ruiz, *Calculation of the PV modules angular losses under field conditions by means of an analytical model*. Solar Energy, 2001. **70**: p. 25-38.
8. Riley, D., et al. *A Performance Model for Bifacial PV Modules*. in *44th IEEE Photovoltaic Specialist Conference*, 2017. Washington, DC.
9. Kreinin, L., et al. *PV module power gain due to bifacial design. Preliminary experimental and simulation data*. in *2010 35th IEEE Photovoltaic Specialists Conference*, 2010.

View Factor Model and Validation for Bifacial PV and Diffuse Shade on Single-Axis Trackers

Marc Abou Anoma[1], David Jacob[1], Ben C. Bourne[1], Jonathan A. Scholl[1], Daniel M. Riley[2],
Clifford W. Hansen[2]

[1]SunPower Corporation, San Jose, CA, 95134, USA
[2]Sandia National Laboratories, Albuquerque, NM, 87185, USA

Abstract — **In this paper, we use a model based on view factors to estimate the irradiance incident on both surfaces of a single-axis tracker PV array for given direct and diffuse light components of the sky dome. We describe the mathematical formulation of the view factor model that assumes a 2D tracker geometry with Lambertian surfaces while accounting for reflections from all surrounding surfaces. The model allows specifically to calculate the incident irradiance on the back surface of PV modules as well as the diffuse shading effects caused by the presence of neighboring tracker rows in a PV array. We present preliminary results on an experimental validation of the view factor model.**

I. INTRODUCTION

Due to new bifacial technologies and larger utility-scale photovoltaic (PV) arrays, there is a growing need for models that can more accurately account for the multiple diffuse light components and reflections incident on various surfaces of a PV array. The literature shows the use and validation of ray-tracing methods and view factors on 3D geometries to calculate back-surface irradiance on PV modules [1]. View factors have also been used to account more accurately for diffuse light and ground reflections incident on the front surface of PV module 3D geometries [2]-[3]. These methods have been shown to be quite accurate under careful management of the calculation, but they can often be computationally intensive and complex to use.

The method presented here is an application of view factors for a 2D geometry of an array of single-axis trackers, invariant by translation along the tracker axis. It can be used for energy production calculation of large PV arrays thanks to its high computational speed, and also because edge effects occurring in large PV arrays are negligible. We present preliminary work on the experimental validation of this method and show encouraging results.

Calculating view factors for 2D geometries is straightforward and formulas for a large amount of geometries have already been derived and published in the literature [4]. The present work makes use of these formulas, applies them to a geometry in an analytically derived mathematical formulation and enables the calculation of the incident irradiance on all surfaces of the considered 2D geometry while accounting for reflections from all surfaces.

We show that the method is in good agreement with measurements of back-over-front surface irradiance ratio on single-axis trackers installed at a Sandia National Laboratories test site in Albuquerque, New Mexico. We also present preliminary results on how this work can be combined with a diffuse transposition model such as the Perez model [5] to account more accurately for ground reflections as well as shading of diffuse light. Lastly, we compare the model with measurements performed on single-axis trackers installed at the SunPower R&D Ranch located in Davis, California.

II. MODEL DESCRIPTION

In a PV array of modules, neighboring rows have an impact on the irradiance incident on the front and the back surface of PV modules because of the portion of the sky and the ground they obstruct. They also reflect some light and cast shadows on the ground. We illustrate in Fig. 1 the main components of irradiance on a tracker and show the effects of diffuse shading between neighboring trackers and ground reflection.

This work shows how these effects can be captured using view factors on a 2D model of a PV array, when applied in conjunction with a mathematical model for reflections, and optionally with an existing transposition model such as the Perez model [5].

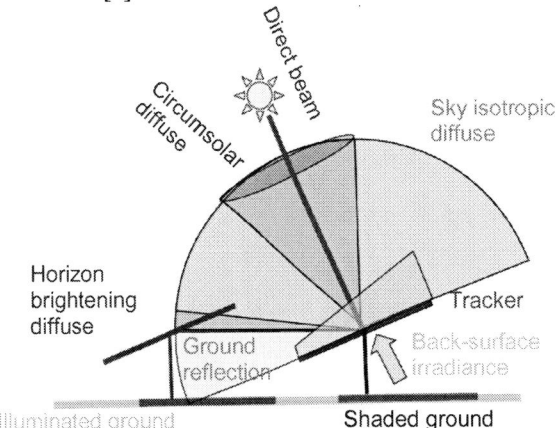

Fig. 1. Schematic showing a 2D representation of a PV array and the effects of neighboring rows on incident irradiance components on the front and back surface of modules. The left tracker obstructs a portion of the sky and casts a shadow on the ground that decreases the reflected irradiance seen by the right tracker.

A. View Factors

View factors, also called configuration factors, are extensively used in thermal radiation heat transfer theory. The view factor from a surface 1 to another surface 2 represents the fraction of the space surrounding and seen by surface 1, and occupied by surface 2.

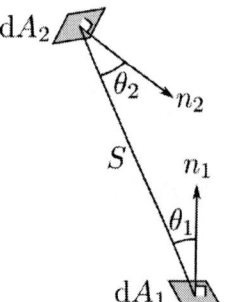

$$F_{1,2} = \frac{1}{A_1} \int_{A_1} \int_{A_2} \frac{cos\theta_2 \ cos\theta_1}{\pi \ S^2} \ dA_2 \ dA_1 \qquad (1)$$

The present method relies on analytical solutions of view factors that have already been calculated and published for 2D geometries [4]. In a 2D-geometry, surfaces are modeled by segments in the same plane. Doing so helps improve the computational speed of the model.

B. Model equations

By making several simplifying assumptions on the modeled PV array surfaces, we can represent the irradiance calculation of a modeled array with a linear system of equations. The dimension of the system is equal to the number of surfaces considered, n. For a surface of index i (integer from 1 to n), we can write that:

$$q_{o,i} = q_{emitted,i} + q_{reflected,i} \qquad (2)$$

where $q_{o,i}$ is the radiosity of surface i, representing the outgoing radiative flux from i. We can write this equation for all n surfaces, and recognize that they are all coupled because surfaces reflect and emit to each other in a PV array. Therefore, we must find all the radiosity terms of all surfaces in order to solve the system and to calculate values of interest like back-surface incident irradiance for instance.

If we assume that the emitted thermal power $q_{emitted,i}$ is negligible compared to $q_{reflected,i}$, we can simplify (2):

$$q_{o,i} \approx q_{reflected,i} \qquad (3)$$

This assumption is reasonable because the temperature of the modules in the field, around 50-60°C, is associated to the emission of much lower energy photons than the photons in the visible spectrum reflected by the surfaces. Assuming Lambertian surfaces we can write that:

$$q_{reflected,i} = \rho_i * q_{incident,i} \qquad (4)$$

where ρ_i is the albedo of surface i. We can further develop (4) into:

$$q_{reflected,i} = \rho_i * (\sum_j q_{o,j} * F_{i,j} + Irradiance_i) \qquad (5)$$

where:

- $\sum_j q_{o,j} * F_{i,j}$ is the contribution of all the surfaces j surrounding i to the incident radiative flux onto surface i. For instance, this could be the irradiance contributions of the sky dome, the ground surfaces and the other trackers to the front surface of a PV module.
- $F_{i,j}$ is the view factor of surface i to surface j.
- $Irradiance_i$ is an irradiance source term specific to surface i, and which includes irradiance contributions from sources not considered to be surfaces in the modeled system. In this work, for instance, it can be equal to the sum of direct, circumsolar, and horizon light components incident on the front surface of the modeled PV modules.

After accounting for the radiosity terms of all the surfaces, we get a system of n coupled equations and can solve it to calculate values of interest such as back-surface incident irradiance. The system of equations indicated by (5) can be written as:

$$(\mathbf{R^{-1} - F}).\mathbf{q_o} = \mathbf{Irr} \qquad (6)$$

or more explicitly:

$$\left[\begin{pmatrix} \rho_1 & 0 & 0 & \cdots & 0 \\ 0 & \rho_2 & 0 & \cdots & 0 \\ \vdots & \vdots & \vdots & \ddots & \vdots \\ 0 & 0 & 0 & \cdots & \rho_n \end{pmatrix}^{-1} - \begin{pmatrix} F_{1,1} & F_{1,2} & F_{1,3} & \cdots & F_{1,n} \\ F_{2,1} & F_{2,2} & F_{2,3} & \cdots & F_{2,n} \\ \vdots & \vdots & \vdots & \ddots & \vdots \\ F_{n,1} & F_{n,2} & F_{n,3} & \cdots & F_{n,n} \end{pmatrix} \right] . \begin{pmatrix} q_{o,1} \\ q_{o,2} \\ \vdots \\ q_{o,n} \end{pmatrix} =$$
$$\begin{pmatrix} Irr_1 \\ Irr_2 \\ \vdots \\ Irr_n \end{pmatrix} \qquad (7)$$

In order to solve this system for $\mathbf{q_o}$ we provide values for the reflectivity and the irradiance term of each surface, and calculate the view factors from each surface to the other surfaces. The vector of incident irradiance on all surfaces $\mathbf{q_{incident}}$ can then be calculated using:

$$\mathbf{q_{incident}} = \mathbf{F}.\mathbf{q_o} + \mathbf{Irr} \qquad (8)$$

C. Application to a 2D PV array of single-axis trackers

In this work, we implemented a geometrical model of a 2D PV array of single-axis trackers using the programming language Python.

After specifying key geometric and environmental inputs such as tracker height, width, and spacing, the algorithm builds a simple 2D geometry representing a PV array with an arbitrary number of rows and their shadows on the ground at a single point in time for which we specify the solar and tracker angles. We built a logic into the algorithm that enables each surface to detect the other surfaces surrounding it, as shown in Fig. 1 and

Fig. 2. The algorithm also allows to divide some surfaces into smaller segments in order to evaluate irradiance distributions.

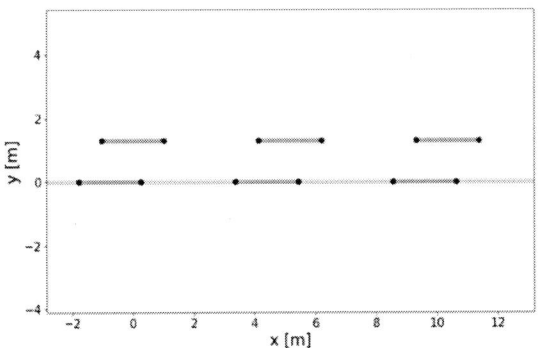

Fig. 2.   Python representation of a 2D PV array of three single-axis trackers in a flat position. The three top lines represent the trackers, and the lines underneath represent the shadows (grey) and the illuminated ground (yellow). In this situation, the trackers are north-south oriented and the sun is in the east with a zenith angle of 30°.

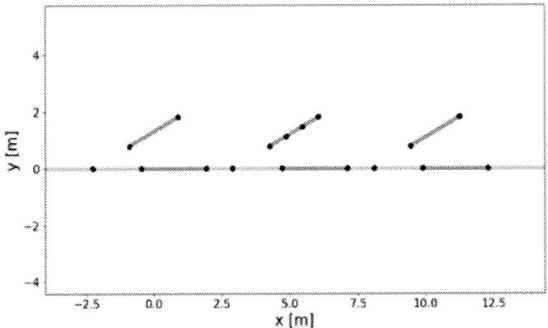

Fig. 3.   Python representation of a 2D PV array of three north-south oriented single-axis trackers at 30° rotation angle, with the sun located in the west at a zenith angle of 30°. The surface of the middle tracker has been discretized into three surfaces in order to calculate the irradiance distribution on it.

The integration of both geometrical and mathematical models provides a fast way to calculate the incident irradiance on all defined surfaces after providing the solar and tracker angles with the direct and diffuse light intensities.

III. VALIDATION AND PRELIMINARY RESULTS

In this section, we use the view factor algorithm described above to calculate the irradiance on north-south oriented single-axis trackers installed at Sandia National Laboratories in New Mexico and at the SunPower R&D Ranch in California.

In addition, we use the pvlib Python [6] implementation of the Perez model [5] to calculate the direct, circumsolar, and horizon plane-of-array (POA) and ground irradiance components, as well as the luminance of the isotropic sky dome. These pre-calculated components allow the view factor

model (VF model) to calculate ground and tracker reflections as well as isotropic diffuse light incident on all surfaces.

A. Bifacial irradiance results

We first compare the modeled front and back surface irradiances to measurements done on two north-south oriented single-axis trackers installed at a Sandia bifacial test site in New Mexico, over a ground with an albedo close to 22% on average. Both east and west trackers have reference cells installed in the plane-of-array (POA) that measure the front and back surface irradiances.

In Fig. 4 we show one clear day dataset, with irradiance levels, tracking angles, and the back-to-front surface irradiance ratios for the two trackers.

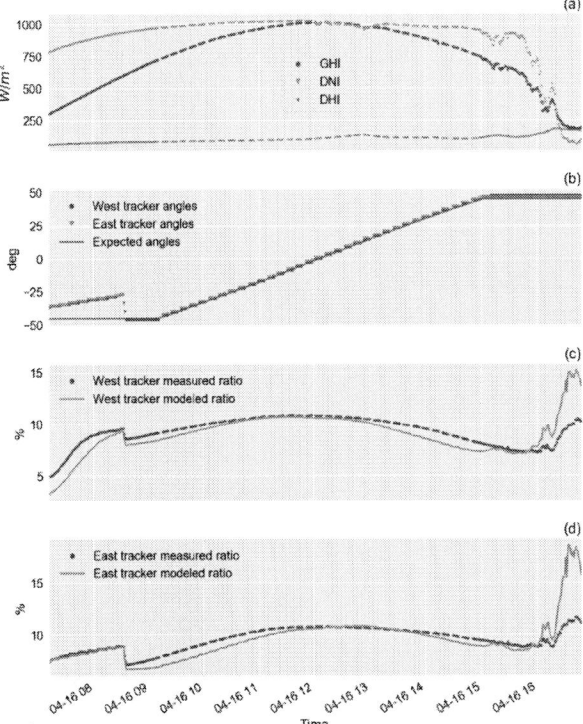

Fig. 4.   Comparison of measured and modeled back to front surface irradiance ratios for two single-axis trackers installed at Sandia (NM) on a clear day with good tracking. (a) Measured irradiance at the site. (b) Measured and expected tracker angles. Angles are positive when trackers are facing west. (c) Measured and modeled back to front surface irradiance ratio of west tracker. (d) Measured and modeled back to front surface irradiance ratio of east tracker.

We observe that there is a very good agreement between the measured and modeled irradiance ratios; they remain within 2% over the course of the day during the tracking period, except at the very end of the day when clouds near the horizon and shading from nearby structures reduce the irradiance on upward facing instruments.

In Fig. 5 we show the back-surface irradiance for the two tracker rows and compare it to the model.

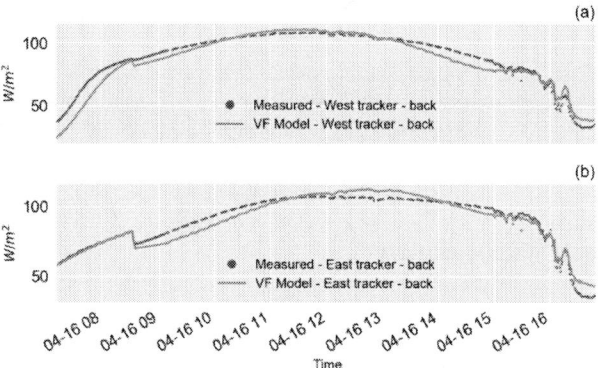

Fig. 5.   Comparison of measured and modeled back-surface irradiance for two single-axis trackers installed at Sandia (NM) on a clear day with good tracking. (a) West tracker. (b) East tracker.

We first note that the irradiance trends predicted by the model are following the measurements within 10 W/m² and that the model is able to distinguish the east and the west trackers. This shows that the model accounts well for the presence of neighboring trackers.

In Fig. 6, we report the modeled contributions of the two main irradiance components on the back-surface irradiance: the irradiance that comes directly from the isotropic sky and the irradiance reflected on both the ground underneath the tracker and the neighboring tracker.

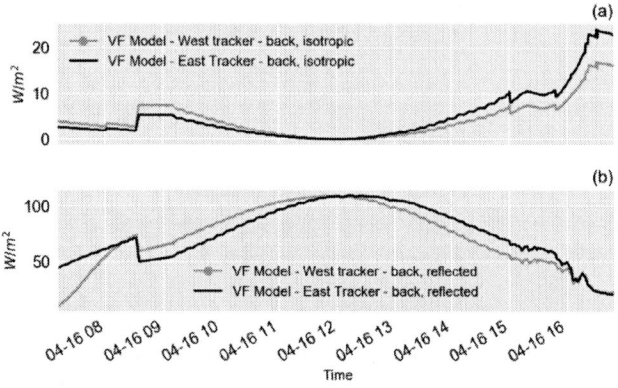

Fig. 6.   Back-surface diffuse light components modeled with the view factor model for two single-axis trackers installed at Sandia (NM) on a clear day with good tracking. (a) Isotropic component. (b) Reflected component (ground and trackers).

This result shows that reflections off the ground and the trackers represent the largest contribution to the back-surface irradiance for that site. Additionally, the model accounts well for the sky shading of the east tracker in the morning by the west tracker (Fig. 6a). It also sees the shadow cast by both

trackers on the ground, which causes lower intensity of incident reflected light compared to the west tracker (Fig. 6b). This trend is opposite between the two trackers in the afternoon hours.

In Fig. 7, we present results for multiple semi-clear days at the site. Looking at the difference between measured and modeled back-to-front irradiance ratios, we can see that the trends are similar.

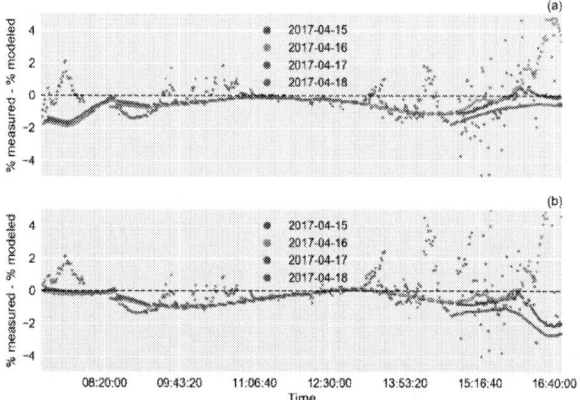

Fig. 7.   Comparison of measured and modeled back-to-front surface irradiance ratios for two single-axis trackers installed at Sandia (NM) on several days. (a) Difference between measured and modeled ratio for the west tracker. (b) Difference between measured and modeled ratio for the east tracker.

These results prove that using view factors with 2D geometries allows to get fast and accurate estimates of bifacial irradiance ratios for single-axis trackers with small edge-effects.

B. Diffuse shading results

We now evaluate the ability of the view factor method to calculate the effect of diffuse shading on surfaces. In the presence of neighboring trackers, the view factor of the front surface to the diffuse sky is not only reduced compared to a free-standing tracker, but there can also be a non-uniform irradiance distribution on the front surface because of differences in view factors of the individual segments of the tracker. In a situation such as the one shown in Fig. 3, the cells on the top of the tracker have a higher view factor to the sky but a lower view factor to the ground than the bottom cells. A non-uniform illumination can be a source of current mismatch within PV strings.

1) Experimental setup

We collect diffuse shade measurements on two north-south oriented single-axis trackers installed with a ground coverage ratio close to 40% at the SunPower R&D Ranch in California. We measure the irradiance in the front plane-of-array using SunPower split-cell reference cells at four locations along the width of the west tracker (Fig. 8). This allows us to measure the distribution of irradiance on the front surface of the PV modules.

Fig. 8.   Picture showing the setup of reference cells installed along the width of the west single-axis tracker at the SunPower R&D Ranch in California.

Due to site obstructions on the west side of the trackers, we are constrained to studying only data in the morning for the irradiance distribution comparisons with our model.

In Fig. 9, we present reference cell irradiance measurements taken during a clear day. They demonstrate the non-uniform irradiance distribution along the width of the tracker: we observe relative differences ranging from 3% to -3% during the course of the day between the outermost reference cells in the east and west. This relative difference changes sign during the day as the reference cells switch positions, suggesting that the top reference cell receives more irradiance than the bottom reference cell at our site.

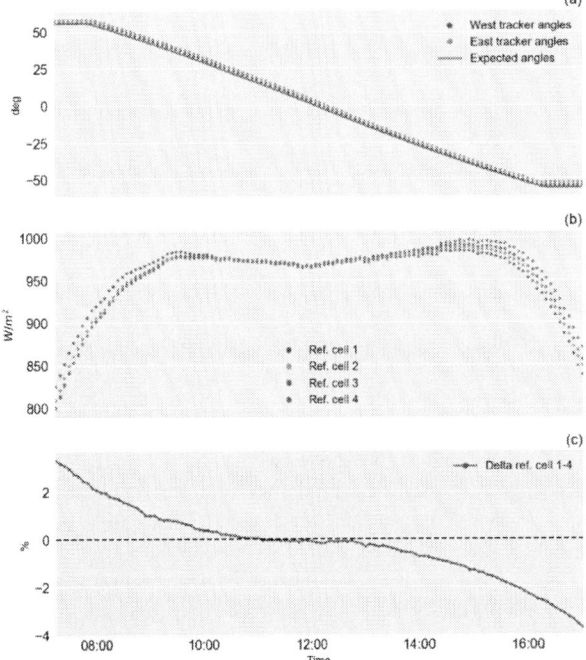

Fig. 9.   Measured data from two north-south oriented single-axis trackers installed at the SunPower R&D Ranch in California, on 5/28/2017. (a) Measured and expected tracker angles. Angles are positive when trackers are facing east. (b) Irradiance measurements of reference cells installed along the width of the west tracker's front surface, where the label 1 is for the westernmost ref. cell, and the label 4 for the easternmost one.   (c) Relative difference between the irradiance measurements of the two edge reference cells (1 and 4).

2)   Average diffuse shading model

In our model, we assumed a site albedo of 15%, and we discretized the front surface of the modeled west tracker in order to calculate the irradiance distribution across the width of the tracker. Note that the average irradiance of the discretized irradiance segments is equal to the irradiance modeled without discretizing of the surface.

In Fig. 10 we plot the average front surface irradiance modeled with the view factor model (combined with Perez pre-calculated components) and the front surface irradiance calculated with the pvlib implementation of the Perez model [6]. The first predicts less front surface POA irradiance than the Perez model in the morning for the west tracker, because the first accounts for isotropic diffuse shading and shadows on the ground. In the afternoon, the view factor model estimates the same irradiance as the Perez model because the west tracker has no more obstructions in front of it. The two models are within 5% of the experimental irradiance curves.

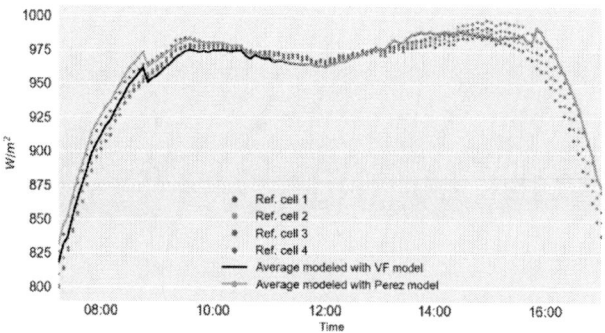

Fig. 10. Comparison of front surface total POA irradiance measured and modeled for the west single-axis tracker installed at the SunPower R&D Ranch, on 5/28/2017.

3) Modeled POA irradiance distribution

We finally present the irradiance non-uniformity modeled with a discretized tracker. In Fig. 11 we show the total irradiance, the relative difference between the two extreme segments, and the two components of irradiance responsible for the non-uniformity: the isotropic sky and the ground reflection.

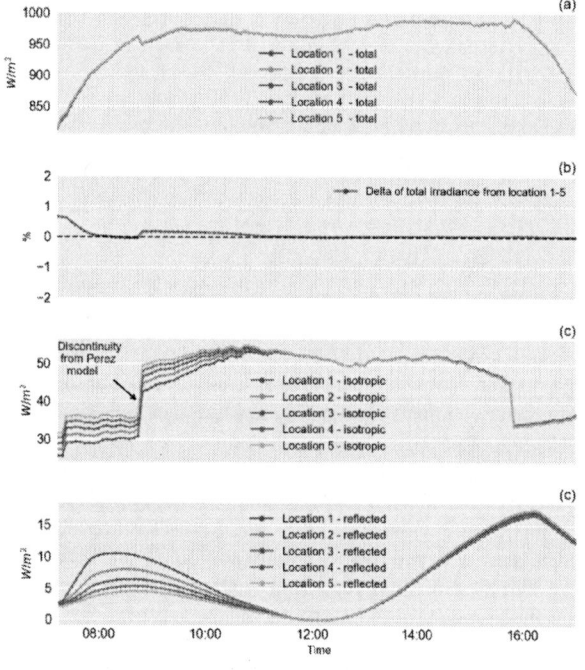

Fig. 11. Modeled diffuse shading effects for the west single-axis tracker installed at the SunPower R&D Ranch. Location 1 is the easternmost segment of the tracker, and location 5 the westernmost segment. (a) Total irradiance modeled for each location. (b) Delta between total irradiance modeled for location 5 and 1. (c) Modeled isotropic light incident on all five locations of the front surface of the tracker. The discontinuity comes from the use of the Perez model in pvlib for the isotropic luminance calculation. (d) Modeled reflected light incident on all five locations of the front side.

The model does not predict the same trend observed experimentally in the morning (Fig. 9c). According to the model, the differences in isotropic and reflected irradiances vary in opposite directions and compensate each other: the bottom segment sees more ground reflected irradiance and less isotropic sky irradiance, and vice versa for the top segment. This discrepancy could have several origins: the ground reflectivity is not broad band [7], resulting in lower reflected ground irradiance measured by the silicon-detector, the reference cells have a different Incidence Angle Modifier [8] compared to the perfect cosine detector assumed by the model, and also, circumsolar and horizon diffuse light shading should play a big role in the measured data, but is not yet implemented in the model. Further refinements of the model could help improve the agreement with the experimental results.

VI. CONCLUSION

A new 2D model relying on view factors and accounting for reflections and diffuse shading is presented as a computationally-efficient method for the calculation of incident irradiance on surfaces of large PV arrays of single-axis trackers. The first results on bifacial back-to-front irradiance ratios are promising, with agreement between model and experiments better than 2% for clear days. Preliminary results on the discretization of surfaces are presented to study the non-uniformities of irradiance. Further development of the model will be performed to improve the agreement between the 2D view factor model and the PV array measured data.

REFERENCES

[1] Hansen, C. W., Stein, J. S., Deline, C., MacAlpine, S., Marion, B., Asgharzadeh, A., & Toor, F. (2016, June). "Analysis of irradiance models for bifacial PV modules". In Photovoltaic Specialists Conference (PVSC), 2016 IEEE 43rd (pp. 0138-0143). IEEE.

[2] Capdevila, H., Herrerías, M., & Marola, A. (2014). Anisotropic Diffuse Shading Model for Sun-tracking Photovoltaic Systems. Energy Procedia, 57, 144-151.

[3] Peled, A., & Appelbaum, J. (2016). Minimizing the current mismatch resulting from different locations of solar cells within a PV module by proposing new interconnections. Solar Energy, 135, 840-847.

[4] Howell, J. R., Menguc, M. P., & Siegel, R. (2010). Thermal radiation heat transfer. CRC press.

[5] Perez, R., Seals, R., Ineichen, P., Stewart, R., & Menicucci, D. (1987). A new simplified version of the Perez diffuse irradiance model for tilted surfaces. Solar energy, 39(3)

[6] Holmgren, W. F., Andrews, R. W., Lorenzo, A. T., & Stein, J. S. (2015, June). Pvlib python 2015. In Photovoltaic Specialist Conference (PVSC), 2015 IEEE 42nd (pp. 1-5). IEEE.

[7] Brennan M.P, A.L.Abramase, R.W.Andrewsc & J.M.Pearce (2014). Effects of spectral albedo on solar photovoltaic devices, Solar Energy Materials and Solar Cells, vol. 124

[8] Winter, S., D. Friedrich, and T. Gerloff. "Effect of the angle dependence of solar cells on the results of indoor and outdoor calibrations." Proceedings of the 25th European Photovoltaic Solar Energy Conference. 2010.

A Fast Quasi-Static Time Series (QSTS) Simulation Method for PV Impact Studies Using Voltage Sensitivities of Controllable Elements

Xiaochen Zhang[1], Santiago Grijalva[1], Matthew J. Reno[2], Jeremiah Deboever[1], and Robert J. Broderick[2]

[1] Georgia Institute of Technology Atlanta, GA, 30318, USA

[2] Sandia National Laboratories Albuquerque, NM, USA

Abstract — **Yearlong quasi-static time series (QSTS) simulations at second-level granularity are required to accurately model controller devices and determine the impact of PV resources on distribution systems. However, the computational time for running such simulations takes 10 to 120 hours for a realistic-sized distribution feeder. This long simulation time is preventing widespread adoption of QSTS simulation for PV impact studies and more generally impact studies needed for all types of distributed energy resources (DERs). This paper proposes a fast QSTS simulation approach by substantially reducing the number of power flow solutions used during the simulation. The proposed method uses voltage sensitivities to model the control logic and behavior of system regulators and capacitors, accurately predicting the control actions of system controllers without having to solve all the power flows through time. The effectiveness and efficiency of the proposed method is demonstrated on the IEEE 13-bus test case with 100 times faster computational time.**

Index Terms — **power system simulation, smart grid, power system planning, photovoltaic systems.**

I. INTRODUCTION

Scenario-based simulation is an approximate approach that has been used in the industry to evaluate the PV integration impacts on distribution circuits. In this method, a few scenarios representing worst-case conditions during a year, are explored through independent power flow analyses. However, scenario-based simulations cannot capture the complex behavior of voltage controllers with thresholds and time delays. Capturing this behavior is critical with variable PV production to determine, for instance, the number of tap changing actions during a year [1].

The state-of-the-art method to evaluate the impact of new distributed resources such as PV systems is quasi-static time series (QSTS) simulation analysis. QSTS simulation takes the time series load and PV temporal profiles as inputs and solves power flow chronologically. Each solution, uses the previous power flow results and takes into account time delays and thresholds of all the control devices [2]-[3].

According to [4], yearlong high-resolution QSTS simulations are required to analyze the impact of PV integrations for seasonal trends and the highly variable PV outputs. This volatile energy output may cause system voltage violations and potential regulator and capacitor status oscillations. Moreover, these voltage violations and system controllable element actions may occur in a few seconds and cannot be identified without higher granularity QSTS simulation. In order to capture these control actions and potential controller oscillations, a 5-second or higher time resolution QSTS simulation is required.

According to [5], a yearlong high- resolution QSTS simulation can take between 10 to 120 hours to run for realistic feeders. Fast QSTS simulations is necessary to ensure distribution system reliability and safety in the face of numerous distributed resources and controllable elements. Therefore, enhanced QSTS approaches that can maintain high accuracy and reduce computational time are highly needed.

This paper describes a voltage sensitivity-based model that can drastically increase the speed of QSTS simulations. The model also provides new insight into the operation of controllable devices in distribution systems. There are many challenges [6] in reducing the computational time of QSTS simulation, including:

a) Presence of multiple valid power flow solutions,
b) Interaction of controllable elements interactions, and
c) Time dependency of the time-series simulation.

In the "brute force" QSTS simulation, the full AC 3-phase unbalanced power flow is solved in chronological order. Using the proposed model with an event-based simulation, we can safely skip the process of solving power flows for many time-points, without missing any controller transition event. The increased in speed is expected to make QSTS simulation a practical approach for high-fidelity PV and DER hosting capacity analysis. In addition, the proposed method provides new insight that will help researchers to understand the state transition process of power system distribution networks due to discrete controller actions.

The remainder of this paper is structured as follows: Section II discusses the sensitivity-based model of distribution system controllable elements including regulators and capacitors. Section III introduces a solution to estimate power system state transitions using the sensitivity-based model and geometric analysis. Section IV provides an iterative method for the sensitivity-based model parameter estimation and proposes a detailed implementation of the fast QSTS simulation approach. Section V tests the proposed method on a test case distribution system with realistic load and PV measurements. Section VI concludes the paper and discusses the potential applications of the proposed sensitivity-based model and fast QSTS simulation method.

II. A DISCRETE SENSITIVITY-BASED STATE-TRANSITION MODEL OF CONTROLLABLE ELEMENTS

A. Bus Voltage Sensitivity Linear Approximation

978-1-5090-5606-4/17 $31.00 © 2017 IEEE

In this section, we introduce a sensitivity-based state-transition model (sensitivity model) for controllable elements on a distribution feeder. The sensitivity model is based on the linear approximation between bus voltage and power injections from load or PV. The model explains various distribution system transition behaviors and can be used for speeding up QSTS simulations.

The control logic of most system controllable elements, such as regulators and capacitors, depends on the system bus voltage. Due to the nonlinear physical property of the distribution network, bus voltages are not strictly linearly-correlated with system loads. However, in most distribution systems and for small changes in the load, we can assume a linear approximation. This linearized assumption is further supported by reference [7], where the authors mathematically derive a tight upper error bound of the linearization assumption.

B. Sensitivity-Based Model for System Regulators

A regulator aims at maintaining the bus voltages within a specific band. Let $V_{regCtrl}$ denote the input control voltage of a regulator. The regulator control keeps $V_{regCtrl}$ within a voltage band (V_{regMin}, V_{regMax}) by changing the tap position. When $V_{regCtrl}$ moves above V_{regMax}, the regulator control will trigger a tap switch event to move the tap to a lower position; similarly, when $V_{regCtrl}$ drops lower than V_{regMin}, regulator will trigger a tap switch event to move the tap to the adjacent higher position. In other words, when $V_{regCtrl}$ moves outside the voltage band, the regulator control will keep adjusting tap position until the $V_{regCtrl}$ falls back in the band, unless the tap is already at extreme positions.

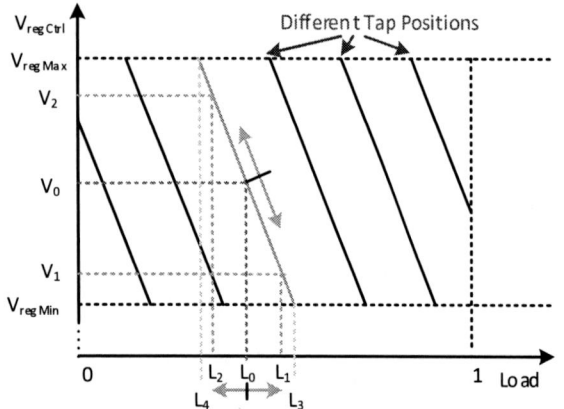

Fig. 1.   Regulator control input voltage vs. system load.

Since $V_{regCtrl}$ increases as the system load increases and decreases as the system load decreases, we introduce a graphical representation of the regulator control, as shown in Fig. 1. When the load increases from L_0 to L_1, $V_{regCtrl}$ will drop from V_0 to V_1, similarly, when the load decreases from L_0 to L_2, $V_{regCtrl}$ will increase from V_0 to V_2. As long as the load remains within L_3 and L_4, no tap event will be triggered. However, when the system load moves beyond the L_3 and L_4,

$V_{regCtrl}$ will moves outside the voltage band (V_{regMin}, V_{regMax}) and triggers a tap action. The regulator tap will move to the adjacent tap position, which corresponds to the adjacent lines in the graphical model.

Let us now also incorporate the PV output profile for PV interconnection studies. The PV production can move much faster compared to the load and can have a larger impact. The PV output profile can be modeled as a fast-varying negative load. In the proposed sensitivity-based model, we add one more dimension of PV output onto the single load profile plot and form a multiple-plane-shaped model as shown in Fig. 2.

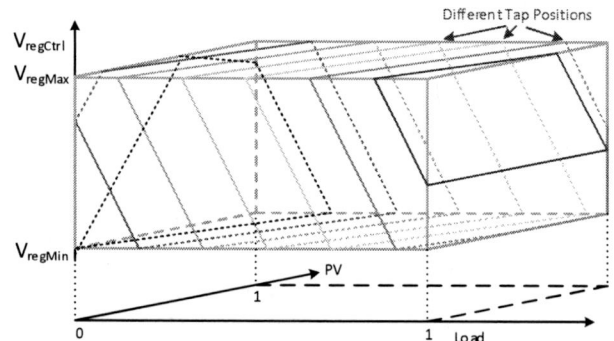

Fig. 2.   Multiple-plane model for different regulator tap positions.

In addition to the abovementioned graphic model, the discrete linear model for a regulator with n load profiles can be mathematically presented as equation (1).

$$V_{regCtrl,i} = \boldsymbol{\beta}_i \boldsymbol{U} \qquad i = 1,2,\dots,m \qquad (1)$$

where $V_{regCtrl,i}$ stands for the regulator control input voltage at tap position i; m stands for the total number of tap positions; $\boldsymbol{U}$ is a $(n+1) \times 1$ vector consists of all input load profiles. For example, in Fig. 1, $\boldsymbol{U}^T = [load, 1]$; similarly, in Fig. 2, $\boldsymbol{U}^T = [load, PV, 1]$. $\boldsymbol{\beta}_i$ is a $1 \times (n+1)$ vector stands for the coefficients of the linear model.

We can easily apply the model to systems with 3 or more load profiles. In those cases, the sensitivity-based model can be represented by a hyper plane. As long as the linear voltage sensitivity assumption holds, the proposed method does not have limitation on the number of load profiles. This property is very appealing, because it allows for multiple PV output profiles in the system and new load measurements, such as smart meter data can be incorporated into the QSTS simulation.

C. Linear Model for Capacitors

Similar to a regulator, a common voltage controlled capacitor maintains the system bus voltages by switching the capacitor banks on and off based on the regulated bus voltage. When the capacitor is on and the voltage rises above the switch-off threshold V_{capOff}, the capacitor will switch off; when the capacitor is off and the voltage falls below the switch-on threshold V_{capOn}, the capacitor will switch on.

978-1-5090-5606-4/17 $31.00 © 2017 IEEE

Similar sensitivity-based model applies to capacitors, as shown in Fig. 3. The red plane represents the operational plane when the capacitor is off, and the blue plane the capacitor is on. The decision boundary for the capacitor to switch on can be derived by the intersection of the plane $ABCD$ and V_{capOn}. Similarly, the other decision boundary for the capacitor to switch off can be derived by the intersection of the plane $EFGH$ and V_{capOff}.

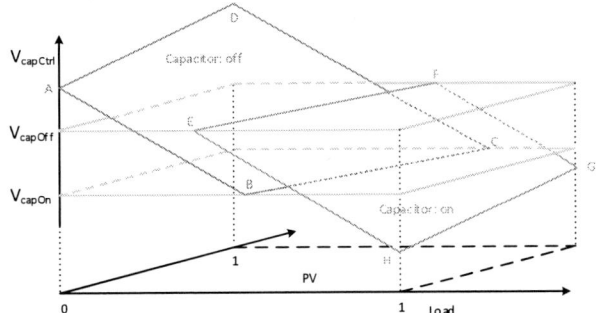

Fig. 3.   Capacitor control input voltage vs. load and PV output.

III. SYSTEM STATE TRANSITION ESTIMATION USING THE SENSITIVITY-BASED MODEL

One of the most important reasons for conducting yearlong high time resolution QSTS simulation is estimating the interactions between the existing system controllable elements and the renewable energy resources. Moreover, if the state of the system controllable elements is given, the relationship between system bus voltage and load becomes a continuous function and is much easier to extrapolate. This section, we show how to predict system controller state transitions without relying on solving power flows, after establishing the proposed model.

A. Use of Sensitivity-based Model to Predict Transitions

According to the proposed sensitivity-based model, for a given regulator tap position, the correlation between $V_{regCtrl}$ and individual power injection (in this case load and PV variables) can be represented as a linearized plane as shown in Fig. 4. Line AB corresponds to $V_{regCtrl} = V_{regMin}$ and line DC corresponds to $V_{regCtrl} = V_{regMax}$. If we project the blue plane $ABCD$ down to the load-PV space, we get a red parallelogram $A'B'C'D'$. $A'B'C'D'$ is also the decision boundary of the current regulator tap position. For example, if the load and PV input combination moves to the right of the red parallelogram, then we have $V_{regCtrl} < V_{regMin}$, which will cause a regulator tap switch-up action. Similarly, if the load and PV combination moves to the left of the red parallelogram, a regulator tap switch-down action will be triggered. In other words, if we get the decision boundary of a tap position on the load-PV plane, we no longer need to solve $V_{regCtrl}$ to predict tap switch actions. Instead, we only need to check whether the load and PV input locates within the decision boundary.

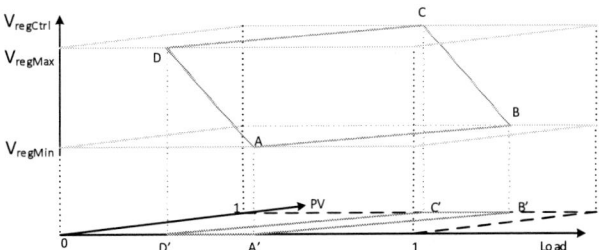

Fig. 4.   Decision boundary for a given regulator tap position.

As shown in Fig. 2, multiple tap positions of a regulator can be represented as multiple planes. Projecting these planes down to the load-PV space will produce a series of overlapping decision boundaries. Fig. 5 shows two adjacent decision boundaries: $A'B'C'D'$ and $E'F'G'H'$. To further illustrate how to use decision boundaries to predict system events, let us assume the load and PV inputs follow the trajectory a-b-c-d-e through time as shown in Figure 5. The load and PV input starts at point a with the regulator tap on the red position. The regulator stays stationary (likely for hundreds or thousands of seconds) until the load and PV inputs move to point b, when the $V_{regCtrl}$ equals to V_{regMin}. Since the load continues to increase after point b, $V_{regCtrl}$ becomes smaller than V_{regMin}. A tap switch action is triggered, which boosts $V_{regCtrl}$ to be above V_{regMin}, and the system now operates on the adjacent green plane. Similarly, when the load moves from c to e, $V_{regCtrl}$ becomes greater than V_{regMax} after point d. This will trigger a tap switch action at point d, and the system jumps from the green plane back on the red plane. In this example, should we know the decision boundary $E'F'$ and $D'C'$, we can predict the system transitions at point b and d without solving time series-power flows for voltages through the entire trajectory.

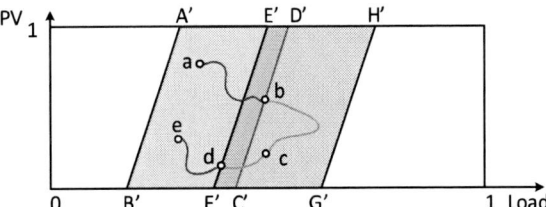

Fig. 5.   Predict regulator actions through decision boundaries.

B. Multiple-Controller Model

In most distribution systems, multiple system controllable elements are presented. The proposed model also applies to systems with multiple regulators and capacitors. Due to the correlation among different controllers, any action of a controller will have impacts on all other controllers. Thus, we need to update the plane model for each controller whenever a controller takes action and changes the system state.

Let us consider an example of a distribution network with three regulators and one capacitor. We first build up the sensitivity-based models for each of the controllers. Then, we

978-1-5090-5606-4/17 $31.00 © 2017 IEEE   1557

combine the decision boundaries of all controllers as shown in Fig. 6. The final decision boundary for the system state is the cut or common area of all decision boundaries, shown as the black dashed lines. If the combination of load and PV moves out of the black decision boundary, a system controller action will be triggered, and the system will move to another state with new decision boundaries.

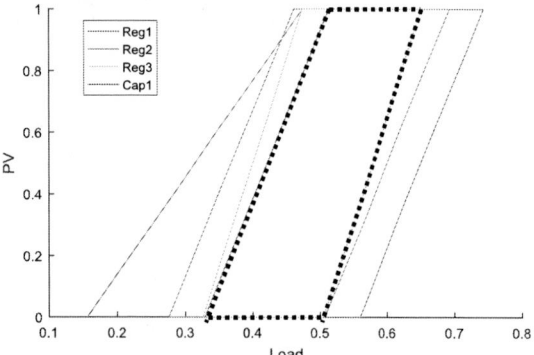

Fig. 6.   Decision boundary for multiple system controllers.

IV. FAST QSTS SIMULATION USING A SENSITIVITY-BASED MODEL

This section discusses the implementation of the proposed sensitivity-based model for fast QSTS simulation, including model parameter estimation.

A. Iterative Method for Model Accuracy

The key for implementing the proposed sensitivity-based model is the estimation of the red decision boundary $A'B'C'D'$ or equivalently the blue plane $ABCD$ shown in Fig. 4. Line AD and line BC are determined by the PV output range (0~1). Line AB and line CD are derived from the regulator settings V_{regMin} and V_{regMax}, which are known. Since the function of the plane $ABCD$ can be found, the decision boundary can also be determined.

In order to uniquely identify a hyper-plane with n-dimensions (n load profiles), we need to solve for β_i in equation (1). Mathematically, we only need n points, which is equivalent to solving $n + 1$ different power flows under the given system controller state. In practice, bus voltage and system load are not strictly linear-correlated. Hence, to increase the accuracy of the estimated plane, we use $2n$ distinct power flow solutions instead of $n + 1$ to estimate the plane. Moreover, an iterative approach is developed to make sure the four power flows are solved close to the edges of the decision boundary. This minimize the error caused by the linearization approximation.

The closer the power flow solutions are to the true decision boundary edges, the more accurate the estimated decision boundary will be. The iterative method keeps updating the $2n$ power flow solutions to make sure the estimated plane is at least accurate at the decision boundary. Fig. 7 shows the flow chart of the iterative method, where a certain iteration number is used as a stopping criteria.

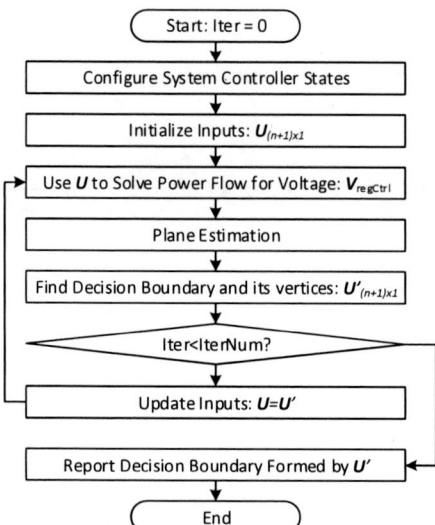

Fig. 7.   Decision boundary estimation using the iterative method.

We further illustrate the iterative method in Fig. 8, where we estimate the decision boundary of a regulator using two iterations. In the initial iteration, we pick four points with two random load levels, combined with two PV scenarios, where the outputs are 0 and 1 pu. These points may not be actual states the system will ever experience, but they define the voltage sensitivity plane. After solving the four power flows, we obtain the four points A1, B1, C1, and D1. From these points we derive the equation of the plane. The boundaries of the voltage sensitivity plane are constrained to the decision boundary by calculating A2-B2-C2-D2 using the voltage thresholds of the voltage regulator and the min and max PV levels as discussed earlier. In the second iteration, we use the load and PV values at A2, B2, C2, and D2 to calculate and update the boundary plane. In the second iteration, the updated plane boundary A3-B3-C3-D3 is drawn using the plane estimated by power flow solutions at A2, B2, C2, and D2.

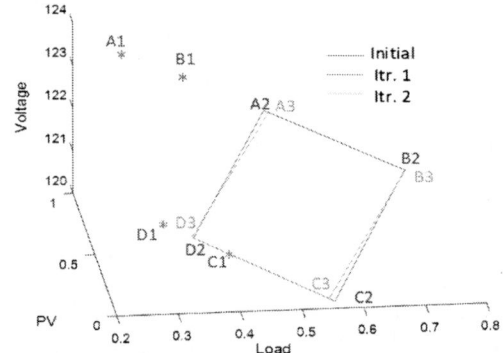

Fig. 8.   Graphic illustration of the iterative estimation method.

B. Flow Chart for Fast QSTS Simulation

Finally, we piece the previous building blocks together and provide the whole flow chart of the proposed sensitivity-based model for fast QSTS simulation. As shown in Fig. 9, the method starts with model initialization where the circuit is compiled. We store the computed plane models in a look-up table. Let *SimLength* stand for the total simulation length. The only building block that requires solving power flow is the green portion of the flow chart, where a system event occurs and the plane model of the new system state has not been solved before. No power flow solve is involved if the simulation stays in the blue block, where no system event occurs or the system state transits to a previously computed plane model.

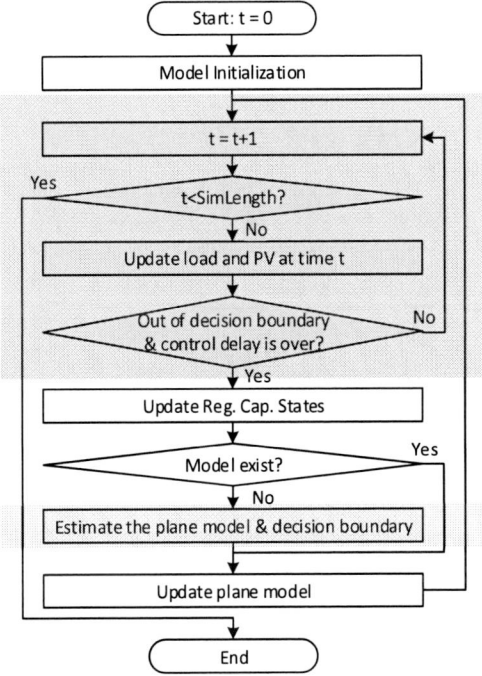

Fig. 9.   Flow chart of the proposed fast QSTS simulation method.

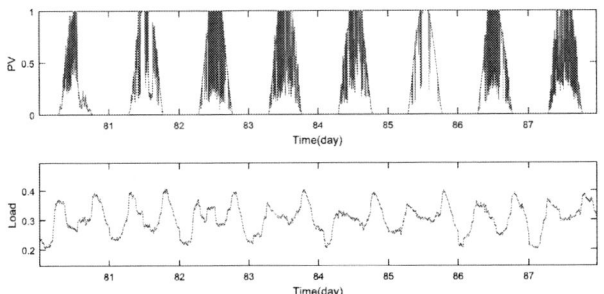

Fig. 10.   Sample PV output and load profiles.

V. TEST RESULT ANALYSIS

The proposed fast QSTS simulation algorithm is tested on an IEEE 13 bus system for hosting capacity analysis. The tested system has one load time-series profile (measured from the substation SCADA) and one PV profile (1-second measured irradiance). Fig. 10 shows sample load and PV output profiles for 8 days. The test system is a modified IEEE 13-bus system,

which has three independent single-phase regulators at the substation and one capacitor at the end of the feeder, where a PV system is installed, as shown in Fig. 11. The PV penetration of the network is set as 40 percent.

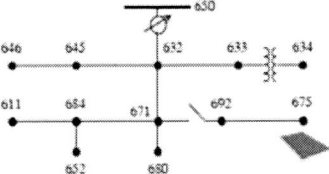

Fig. 11.   IEEE 13-bus system with a PV system installed on bus 675.

In order to acquire the baseline simulation results, we first run a yearlong 1-second QSTS simulation using the brute force method with the abovementioned one load profile and one PV output profile. For the 13-bus small system, the brute force simulation took 13 minutes and 27 seconds. Fig. 12 shows single-phase voltages at bus 675 in per unit with respect to load and PV profiles. Each dot represents a power flow solution for a specific time instance t in the QSTS simulation. We color each dot based on different regulator tap positions. All the dots associated with each tap position lay on separate surfaces which verified our voltage linearization assumption. Since all these surfaces are approximately flat, combined with the previous linear assumption, we refer to them as "planes". As the PV and load change in the system and the solution points forms a trajectory on the given plane. When the controller state changes the operating point "jumps" from one plane to another to continue with a trajectory in the new plane.

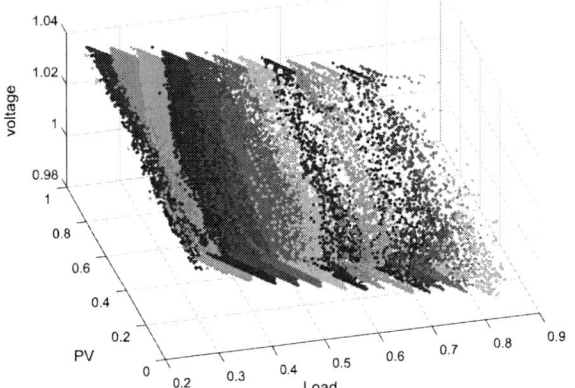

Fig. 12.   Bus 675 voltages for over 31 million power flows in the brute force QSTS simulation.

To test the accuracy of the proposed method, we run the same yearlong 1-second resolution QSTS simulation using the proposed sensitivity-based model, and compare the simulation results. Fig. 13 shows the system controllers' states from both the brute force method and the proposed method for 90 days. Since the states of all system controllable elements are overlapping for both methods, we have demonstrated that the proposed method serves the purpose of predicting system state transitions very well.

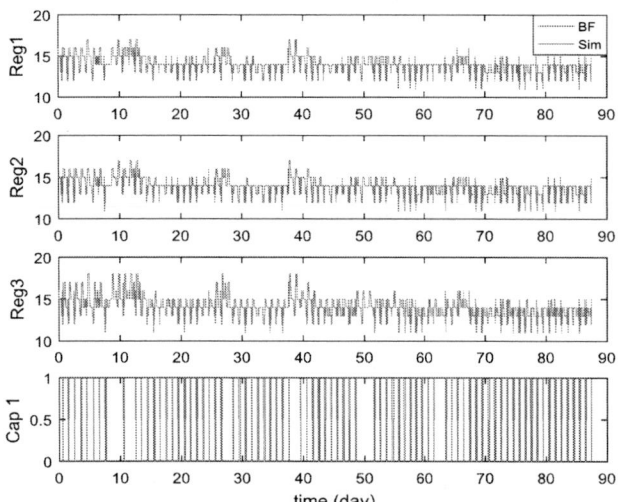

Fig. 13. System controller state comparison for two methods.

TABLE I MODEL ACCURACY AND EFFICIENCY TRADE-OFFS

Num. of Iterations	Reg. Avg. Err (%)	Cap. Avg. Err (%)	Comp. Time (sec)	Comp. Time Reduction (%)
0	3.22	2.35	6.34	99.21
1	2.24	-5.19	6.47	99.20
2	1.91	-4.94	6.57	99.19
3	1.91	-4.94	6.75	99.16
4	1.91	-4.94	6.96	99.14

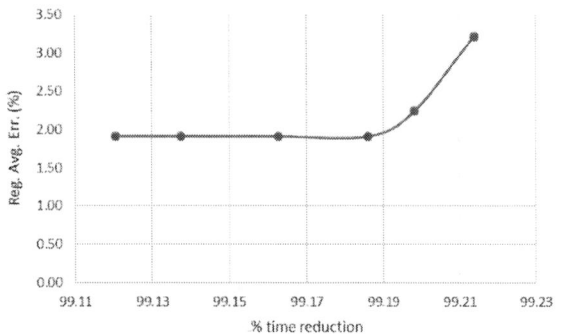

Fig. 14. Model accuracy and computational time trade-off.

The annual QSTS simulation results of the proposed method are shown in TABLE I with percentage error and computational time. TABLE I illustrates how the iterative method helps to improve the accuracy of the algorithm. When we increase the number of iterations in estimating the decision boundary of the plane model, the simulation error decreases but the computational time increases slightly. The simulation error stabilized after just two iterations. This is because the estimated decision boundary converges very fast, which just takes roughly two iterations in this test case. This also provides additional support for the linear voltage sensitivity assumption. Fig. 14 is a more illustrative figure demonstrates the trade-off between model accuracy and efficiency.

VI. CONCLUSION

QSTS simulation is the state-of-the-art distribution system analysis method, which provides a comprehensive and thorough evaluation of possible PV impacts. The major barrier that prevents the pervasive adoption of QSTS simulation is the long computational time for a yearlong high resolution QSTS simulation. In order to speed up QSTS simulation, we proposed a sensitivity-based and discrete transition state model that can capture distribution controller actions with a minimum number of power flow solutions. The proposed fast QSTS simulation method is based on the sensitivity model and has been tested with real load and PV profiles and has demonstrated high simulation accuracy with significant computational time reduction.

The proposed fast QSTS simulation has the potential to make yearlong high resolution QSTS simulation more effective and applicable to PV hosting capacity analysis and distribution system planning. Moreover, the proposed model provides new insight into behaviors of controlling devices and uses voltage sensitivities to vastly speed up the analysis of their operation in distribution networks.

ACKNOWLEDGEMENT

This research was supported by the DOE SunShot Initiative, under agreement 30691. Sandia National Laboratories is a multimission laboratory managed and operated by National Technology and Engineering Solutions of Sandia, LLC., a wholly owned subsidiary of Honeywell International, Inc., for the U.S. Department of Energy's National Nuclear Security Administration under contract DE-NA0003525.

REFERENCES

[1] B. Palmintier, R. Broderick, B. Mather, et al., "On the Path to SunShot: Emerging Issues and Challenges in Integrating Solar with the Distribution System," National Renewable Energy Laboratory, NREL/TP-5D00-65331, 2016.

[2] B. A. Mather, "Quasi-static time-series test feeder for PV integration analysis on distribution systems," IEEE Power Energy Soc. Gen. Meet., pp. 1–8, 2012.

[3] R. J. Broderick, J. E. Quiroz, M. J. Reno, Abraham Ellis, J. Smith, and R. Dugan, "Time Series Power Flow Analysis for Distribution Connected PV Generation," Sandia National Laboratories, SAND2013-0537, 2013.

[4] M. J. Reno, J. Deboever, and B. Mather, "Motivation and requirements for quasi-static time series (QSTS) for distribution system analysis," IEEE PES General Meeting, 2017.

[5] M. J. Reno and R. J. Broderick, "Predetermined time-step solver for rapid quasi-static time series (QSTS) of distribution systems," IEEE Innovative Smart Grid Technologies (ISGT), 2017.

[6] J. Deboever, X. Zhang, M. J. Reno, R. J. Broderick, S. Grijalva, and F. Therrien "Challenges in reducing the computational time of QSTS simulations for distribution system analysis," Sandia National Laboratories, SAND2017-5743, 2017.

[7] S. Volognani and S. Zampieri, "On the existence and linear approximation of the power flow solution in power distribution networks," IEEE Transactions on Power Systems, 2016

Fast Determination of Distribution-Connected PV Impacts Using a Variable-Time-Step Quasi-Static Time-Series Approach

Barry Mather, *Senior Member, IEEE*

National Renewable Energy Laboratory, Golden, CO, 80401, USA

Abstract — The increasing deployment of distribution-connected photovoltaic (DPV) systems requires utilities to complete ever more complex interconnection studies. Relatively simple interconnection study methods worked well for low penetrations of photovoltaic systems, but more complicated quasi-static time-series (QSTS) analysis is required to make better interconnection decisions as DPV penetration levels increase. Tools and methods must be developed to support this type of complex analysis to enable efficient completion of PV impact studies. This paper presents a variable-time-step solver for QSTS analysis that significantly shortens the computational time and effort to complete a detailed analysis of the operation of a distribution circuit with many DPV systems. Specifically, it demonstrates that the proposed variable-time-step solver can reduce the required computational time by as much as 84% without introducing any important errors to metrics, such as the number of voltage regulator tap operations and highest and lowest voltage occurring on the distribution circuit during a 1-yr simulation period. Further improvement in computational speed is possible with the introduction of only modest errors in these metrics, such as a 91% reduction with less than 5% error when predicting voltage regulator operations. Furthermore, an alternate variable time-step method is also shown to perform adequately and shows promise of even greater reductions in simulation runtime.

Index Terms — Distribution system modeling, PV grid integration, QSTS, time-series analysis, variable time-step.

I. INTRODUCTION

Worldwide, many distribution utilities are experiencing increasing deployments of distribution-connected photovoltaic (DPV) systems, which necessitates these utilities to consider and study increasingly complex photovoltaic (PV) interconnection scenarios. When the number of DPV systems interconnected to a single circuit is relatively low, simpler interconnection screening [1] and study processes [2] are adequate to provide the utilities with enough information to guide DPV interconnection approval. These methods employ many worst-case assumptions regarding the operation of the distribution system and the behavior of the DPV systems. These assumptions not only simplify the screening and study processes but also add considerable conservatism to the DPV interconnection outcomes in many circumstances. To provide more accurate, informative, and realistic information regarding the expected impact of DPV on the distribution system, quasi-static time-series (QSTS) analysis methods have been developed and demonstrated for DPV interconnection studies [3]–[4]. These methods have been proven to be very useful,

but they introduce another challenge because QSTS-based DPV interconnection studies require considerable computational effort, making them both costly and potentially untimely for use by utilities on a daily basis or for every requested interconnection. For instance, a yearlong QSTS simulation of a realistic distribution circuit at a temporal resolution of 1 s can take 24–96 hours to complete using business grade computing resources typically available to distribution utility engineers.

To enable regular utility use of QSTS study techniques, the research detailed in this paper was completed with the express goal of significantly shortening the computational time and effort required for DPV interconnection analysis. Specifically, this paper focuses on the temporal nature of the QSTS analysis and presents a method wherein two different time steps are employed to move quickly through the simulation when possible and proceed at the minimum time step during periods when the circuit is undergoing material change. In this paper, this QSTS solution method is referred to as the variable-time-step solver. A number of additional efforts as part of the same project are also under way [5]. Additionally, previous work with this goal included temporal down-sampling and vector quantization (discretization of the system state) [6] and efforts to select better power flow solution initial conditions for each time-step in a time-series analysis to increase simulation speed [7]. Section II provides a description of the two versions of the

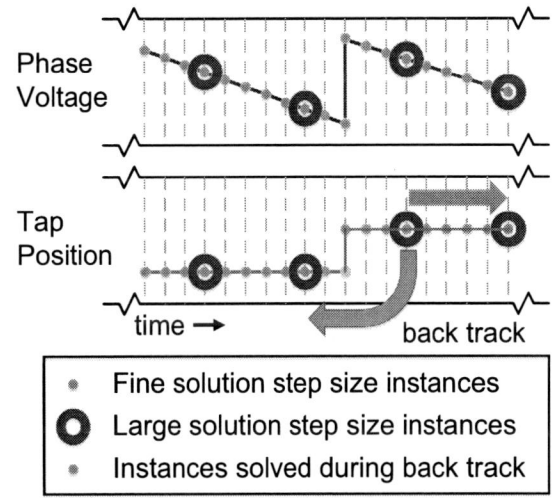

Fig. 1. Conceptual diagram of the operation of the variable-time-step solver.

variable-time-step solver of the QSTS analysis developed. Section III presents the results of the two variable-time-step solvers when applied to two distribution circuits: one simple test circuit and one complex utility circuit. Also included are results on the accuracy of determining the number of automatic voltage regulation (AVR) equipment operations and the minimum and maximum voltages modeled on a circuit over a 1 year period. Section IV concludes the paper.

II. DESCRIPTION OF VARIABLE-TIME-STEP SOLVER

Two variations of variable-time-step solver implementations are described in the following subsections. The first describes the most basic variable-time-step solver, which simply implements a backtrack function to complete higher temporal resolution analysis when needed. The second adds a binary search feature to the backtrack function to further intelligently reduce the number of individual power flow solutions. Additionally, a time-seeking functionality was incorporated wherein once the point at which a pending automatic voltage regulation device is determined using the binary search algorithm, the solution skips ahead to the point in time at which the action will either take place or expire depending on the circuit conditions at that time.

A. Backtrack-Based Variable-Time-Step Solver

The variable-time-step solver was conceptualized from existing methods used in transient simulation solvers. The idea is to vary the temporal time step of the QSTS solution so that periods of interest in circuit operation are analyzed at the highest temporal resolution possible and periods when the circuit is relatively static are analyzed at coarser temporal resolutions to reduce computational time. The variable-time-step solver uses a backtrack method wherein coarse time steps are taken until a system state change is detected. Then the solver backtracks to the previous large time-step instance and proceeds with small time steps for a period of time until large time-stepping is resumed. Fig. 1 shows a conceptual diagram of the variable-time-step solver including the large and small time-step instances as well as the backtrack functionality when a voltage regulator tap position changes.

The primary purpose of DPV interconnection studies is to evaluate the voltage- and equipment-related impacts of the interconnection of additional PV. Thus, the variable-time-step solver uses a change in voltage regulator tap position to detect time periods when a higher resolution analysis would yield higher accuracy in results in terms of determining the number of voltage regulator operations and the minimum and maximum voltages experienced on the circuit being studied.

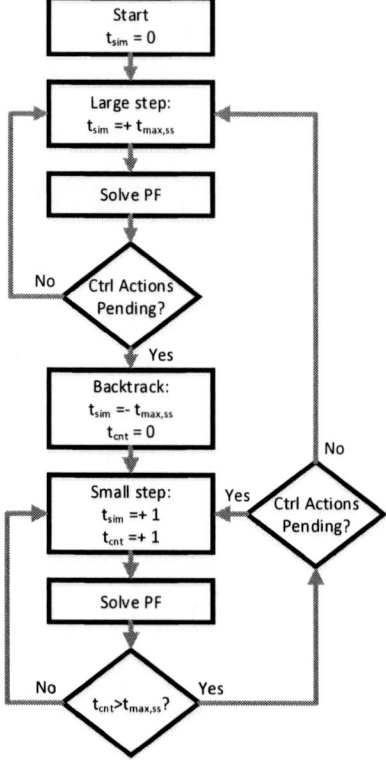

Fig. 2. Flow diagram of the variable-time-step solver.

Fig. 2 shows the programmatic implementation of the variable-time-step solver. It matches the functionality shown in Fig. 1 with the addition of a few features that were added to improve the accuracy of the solver. The number of solutions completed at a small time-step resolution (after a temporal backtrack) is equal to the maximum step size ($t_{max,ss}$) employed in the solver. The existence of other pending control actions is checked again, and additional small time steps are taken until all control actions are resolved. These features were added because multiple, successive changes to the voltage regulator tap position are fairly common, and thus the solver attempts to resolve multiple changes without recursively backtracking.

B. Addition of Binary Search and Time-Seeking to Variable Time-Step Solver

In the variable-time-step solver, as shown in Fig. 2, once a pending control action is detected during a large time-step analysis, no actionable information is available about when a pending control action was instantiated other than the pending control action that occurred between the current solver time and the last time point that showed no pending control actions. The length of time in this window is equal to the large time step. Because this time step can be quite large, a more efficient search method for the possible pending control action time window was implemented. Once the point in time at which a pending control action is determined, a time-seeking function is implemented wherein the time at which the control action will be acted upon—instead of instantiated as pending—is gathered directly from the QSTS power flow solver, and the solution time is advanced directly to that time to avoid unnecessary power flow solutions. Fig. 3 shows the flow diagram of the variable-time-step solver with the binary search and time-seeking implemented. The variables t_l and t_u denote the diminishing time interval being bisected during the binary search. When t_l and t_u are consecutive time points, the binary search is complete. Then the time at which a pending control action is planned to occur (assuming no circuit conditions change and reset the pending control action) is queried directly from the control queue of the QSTS simulation. This time is referred to as t_{act} in Fig. 3. The QSTS simulation time is set to t_{act} and a power flow solution and control iteration are completed to determine if a pending control action took place or expired due to the delay reset functionality of the AVR controllers. This time-seeking functionality eliminates completing a number of power flow solutions that are approximately equal to the length of the delay of the automatic voltage regulation equipment.

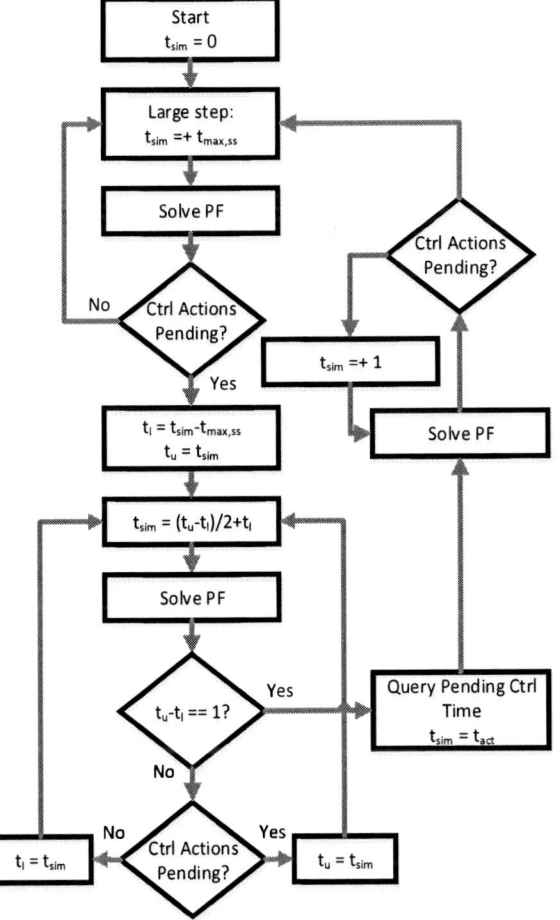

Fig. 3. Flow diagram of the variable-time-step solver with binary search and time-seeking implemented.

III. RESULTS

Both the backtrack-based variable-time-step solver and the binary search and time-seeking solver were evaluated to determine the expected improvement in analysis speed as well as their resulting errors compared to a yearlong 1-s resolution QSTS analysis. Two distribution circuits were used to evaluate the solvers' performance. The first was the IEEE 13-node circuit [8], which contains three single-phase voltage regulators and two fixed shunt capacitors. A 3-MW PV system connected to bus 675 was added to the model for evaluation purposes. Realistic yearlong 1-s resolution load and PV generation profiles were developed to drive the QSTS analysis. The control delay of the voltage regulators was 30 s. This means that the regulators changed tap positons after their control deadband had been exceeded for 30 s or more. Yearlong 1-s resolution analysis of this circuit without the use of a variable-time-step solver was approximately 55 min, making the circuit useful to quickly determine the efficacy of the variable-time-step improvements. The second circuit was a 3,000+ node model of a 15-kV circuit from a utility. The model was modified via the inclusion of 50 PV systems (48 15-kw single-phase residential systems and 2 100-kW three-phase commercial systems). This model includes three single-phase voltage regulators and four three-phase switched capacitors that are controlled based on a measured single-phase voltage. The duration of a yearlong 1-s resolution analysis of this circuit without the use of a variable-time-step solver was approximately 79 h using laptop workstation (2.4 GHz Intel i7-2760QM), making this circuit representative of real-world challenges when completing detailed QSTS analysis on complicated circuits. OpenDSS [9] was used as the base distribution system analysis package but extensive use of the COM via Matlab was used to implement the variable time-step solver routines. Runtime percent reduction, used as a metric to determine the speed increase of the variable time-step methods, is based on actual computer runtimes and thus includes time incurred due to the implementation of the variable time-step solver routines in Matlab and COM communications to OpenDSS. If such routines were added in native formats to OpenDSS or other distribution system analysis packages, runtime reductions should be greater.

A. Voltage Regulator Operations

The only selectable parameter in the presented variable-time-step solvers is the choice of the maximum time step of the simulation, $t_{max,ss}$, (the large time step). The sensitivity of the selection of this parameter was investigated, and the results are presented in this subsection.

The ability to analyze DPV impact on voltage regulation devices is a major benefit of QSTS-based analysis, so the accuracy of such voltage regulation operational impacts is paramount. An acceptable error of ±5% was targeted for this work based on input from practicing distribution engineers.

978-1-5090-5606-4/17 $31.00 © 2017 IEEE

TABLE 3. RUN-TIME REDUCTION AND ERROR/IMPROVEMENT COMPARED TO DOWN-SAMPLED QSTS ANALYSIS OF THE VARIABLE-TIME-STEP SOLVER—IEEE 13-NODE

| max step (s) | Operations (% of base case/% improvement[1]) | | | MAPE | Runtime |
	Reg1-A	Reg1-B	Reg1-C	(Reg. Ops.)	(% reduction)
5	100.00 / 0.74	99.96 / 0.15	99.97 / 1.27	0.02	38.1%
10	100.00 / 2.11	99.96 / 1.11	99.97 / 2.52	0.02	46.8%
15	100.00 / 3.10	100.04 / 1.99	100.03 / 4.28	0.00	73.3%
30	100.00 / 7.25	100.04 / 5.47	99.97 / 7.67	0.02	83.9%
60	95.66 / 23.00	95.83 / 23.09	94.94 / 23.21	4.53	91.3%
120	84.68 / 27.42	84.72 / 27.72	83.29 / 27.23	15.78	95.2%

[1] Improvement in the additional percentage of regulator operations compared to down-sampled QSTS analysis with time step equal to max. time step.

TABLE 2. RUN-TIME REDUCTION AND ERROR/IMPROVEMENT IN THE BACKTRACK VARIABLE-TIME-STEP SOLVER—3,000+ NODE UTILITY CIRCUIT.

| max step (s) | Operations (% of base case) | | | | MAPE | Runtime |
	Reg1-A	Reg1-B	Reg1-C	Cap3	(AVR. Ops.)	(% reduction)
5	100.00	100.00	100.00	100.00	0.00	80.1%
10	100.00	100.00	100.00	100.00	0.00	88.8%
15	100.00	100.00	100.00	100.00	0.00	92.4%
30	100.00	100.00	100.00	100.00	0.00	95.9%
60	100.00	99.55	99.40	100.00	0.26	98.8%
120	100.00	97.30	98.80	100.00	0.98	98.5%
300	93.10	94.14	97.01	100.00	3.94	99.3%

TABLE 1. RUN-TIME REDUCTION AND ERROR/IMPROVEMENT OF THE VARIABLE-TIME-STEP SOLVER WITH BINARY SEARCH AND TIME-SEEKING—IEEE 13-NODE

| max step (s) | Operations (% of base case) | | | MAPE | Runtime |
	Reg1-A	Reg1-B	Reg1-C	(Reg. Ops.)	(% reduction)
5	101.89	104.14	141.10	2.83	22.8%
10	103.27	108.57	141.68	4.91	60.1%
15	102.33	108.71	140.10	4.26	72.5%
30	101.46	108.25	137.88	3.28	85.1%
60	94.05	99.79	127.98	4.41	90.8%
120	80.61	86.92	109.47	17.66	94.7%

Table 1 shows the results of implementing the variable-time-step solver with various maximum step sizes ranging from 5–120 s. The number of voltage regulator operations determined by the analysis is shown for all three phases of the voltage regulator. Additionally, the number of voltage regulator operations resolved by the solver was compared to simply down-sampling the QSTS analysis. Down-sampling simply means that the simulation was run at a fixed step size. This comparison allows for determining the value of the variable-time-step solver in balance with its beneficial reduction in run time because the down-sampled analysis also reduces the run time in a fashion that is similar to that of the variable-time-step solver. The mean absolute percentage error (MAPE) is given for each test run as well as the run-time percentage reduction. As shown, the variable-time-step solver has effectively no error in voltage regulator operations for maximum time steps up to 30 s. For higher maximum time-steps, some error is introduced because the large time steps potentially skip over short-duration tap position changes. However, even at a maximum step size of 60 s, the MAPE is less than the targeted 5%. At this setting, the time required for the simulation is reduced by 91.3%.

When the backtracking variable-time-step solver is used to complete a yearlong QSTS analysis of the 3,000+ node utility circuit, the run-time reduction shown in TABLE 2 is attained. The accuracies of the voltage regulators and the third switched capacitor are also shown. Only the operations for switched capacitor #3 are shown because the other three capacitors were either always off or always on for the duration of the QSTS simulation. The AVR MAPE is less than the estimated 5% accuracy necessary for making good distribution grid planning decisions up to a maximum step size of 300 s. The 3000+ node utility feeder sees less variability than the IEEE 13 node case due to a lower amount of integrated PV and lower impedance construction. Still, the backtrack algorithm provides accurate results with up to 99.3% runtime reduction.

Evaluation of the variable time-step solver with binary event search and time-seeking once a pending event instantiation is determined, using the IEEE 13 node circuit, is shown in TABLE 1. Comparing Table 1 and TABLE 1 reveals that the basic backtrack variable time-step solver has a larger runtime reduction. This is likely due to the specific implementation of the binary event search/time-seeking algorithms as discussed later. For a maximum step-size of 60 s the back track method has a runtime reduction of 91.3% versus 90.8% for the binary search/time-seeking algorithm. The accuracy of determining the regulator operations is comparable between the two methods with the exception of higher error at lower maximum time-steps for the binary search/time-seeking algorithm. The increased error seems to be due to potential for pending regulator control actions to expire between instantiation and the end of the control delay period. Both the binary search and time-seeking part of the algorithm introduce erroneous or mistimed regulator actions when these pending events expire during the delay period. There is also some additional error over the simple backtrack method due to more numerous power flow solutions being solved without previous time-step

initial conditions as happens when the time-series analysis experiences a discontinuity in simulation time.

For the IEEE 13 node test circuit case the amount of time during which no regulator control actions are pending is above 98% - during these period the variable time-step solvers move through the simulation with large time- steps. The remaining time one or more regulator control action(s) are pending the variable time-step algorithms attempt to resolve control action occurrences and timings. The number of power flow solutions required to complete a yearlong analysis of the IEEE 13 node circuit are shown in Figure 5. The figure shows both the minimum number of time-steps, equivalent to the simulation length divided by the maximum time-step and the number of additional solutions incurred to determine voltage regulator operations. The number of additional solutions required by the binary search/time-seeking method is considerably less than the backtrack method. This would indicate that the runtime

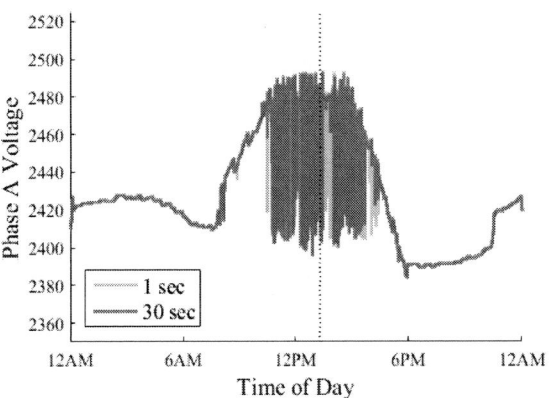

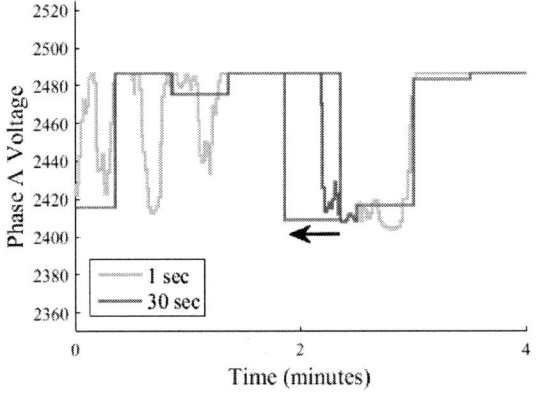

Fig. 4. Example voltage profiles calculated using traditional analysis and the variable-time-step solver with a $t_{max,ss}$ of 30 s. The top plot shows voltage profile during an entire day, and the bottom plot shows the detail of a 4-min. period.

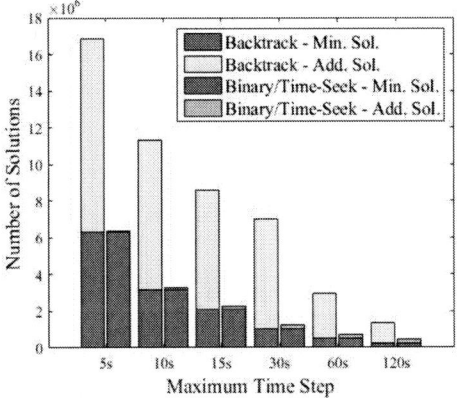

Figure 5. The number of additional solutions completed by the two variable time-step algorithms for a yearlong analysis of the IEEE 13 node test feeder.

reduction of the binary search/time-seeking method could significantly out preform the backtrack method if the algorithm was implemented directly in the distribution system analysis tool. The current implementation which uses Matlab and the COM to manage solutions provided by OpenDSS is too slow to directly demonstrate the runtime reductions possible.

B. Max./Min. Voltages

The maximum and minimum voltages experienced on a circuit is another very important metric for PV interconnection studies. As the maximum time step of the variable-time-step solver increases, the chance of missing a short-duration voltage peak or valley increases. Fig. 4 shows the calculated daily voltage profiles for the point of interconnection of the modeled PV system for standard 1 s resolution analysis and the backtrack variable-time-step solver with a $t_{max,ss}$ of 30 s. The lower plot shows a 4-min period of the solution during which a backtrack event (denoted by an arrow) occurred. During the backtrack period the two voltage time-series overlap. After the back track period the variable time-step algorithm returns to large time steps. Also shown is how short-duration voltage swings are not detected in the voltage profile calculated by the variable-time-step solver with a $t_{max,ss}$ of 30 s. Still, the error of the maximum and minimum voltages for the voltage at the PV system's point of interconnection, calculated by the variable-time-step solver and traditional 1-s analysis, is effectively zero. Any error seems limited to well within the convergence tolerance of the power flow algorithm itself or stems from short-duration voltage excursions. Such excursions certainly occur but their impact in distribution planning activities is not typically of great concern thus not being able to ensure the detection of maximum and minimum voltages occurring from such events is not a great disadvantage of the variable time-step solution methods. Accurately solving for the expected operations of AVR produces accurate maximum and minimum voltages on the timeframe expected by distribution system planners completing PV interconnection impact studies.

IV. CONCLUSIONS

This paper presented variable-time-step solvers for use in advanced, complex DPV interconnection studies. Two versions of the variable-time-step solver are described and evaluated via yearlong 1-s resolution QSTS analysis of PV variability-induced automatic voltage regulation equipment operations on both the IEEE 13-node test feeder and a 3,000+ node utility feeder. The developed backtrack variable-time-step solver reduces the computational time required for such studies by 83.9% with no appreciable error and by 91.3% with acceptable error. The variable time-step solver which uses binary time search and control action time-seeking reduces computational time by 90.8%. The maximum and minimum sustained voltages calculated by the variable time-step solver methods are also accurately determined. The overall number of power flow solutions needed to complete yearlong analysis is significantly lower for the binary time search/time-seeking variable time-step method indicating that the method, if implemented directly in a distribution system analysis package, would outperform the backtrack variable time-step method.

ACKNOWLEDGMENT

This work was supported by the U.S. Department of Energy under Contract No. DE-AC36-08GO28308 with the National Renewable Energy Laboratory. The U.S. Government retains and the publisher, by accepting the article for publication, acknowledges that the U.S. Government retains a nonexclusive, paid-up, irrevocable, worldwide license to publish or reproduce the published form of this work, or allow others to do so, for U.S. Government purposes.

REFERENCES

[1] M. Coddington et al., "Updating Interconnection Screens for PV System Integration," NREL Tech. Report TP-5500-54063, Feb. 2012.

[2] R. Seguin, J. Woyak, D. Costyk, J. Hambrick, and B. Mather, "High-Penetration PV Integration Handbook for Distribution Engineers," NREL Tech. Report TP-5D00-63114, Jan. 2016.

[3] R. J. Broderick et al., "Time Series Power Flow Analysis for Distribution Connected PV Generation," Sandia Report SAND2013-0537, Jan., 2013.

[4] B. Mather, "Quasi-Static Time-Series Test Feeder for PV Integration Analysis on Distribution Systems," in *Proc. IEEE Power and Energy Soc. Gener. Meet.*, San Diego, USA, Jul. 22–26, 2012.

[5] B. Mather, "Presentation: Comparison of Distribution-Connected PV Impact Studies: Steady-State and QSTS," pres. at IEEE Power and Energy Soc. Gener. Meet., Boston, USA, Jul., 17–21, 2016.

[6] C.D. Lopez, B. Idlbi, T. Stetz, and M. Braun, "Shortening Quasi-Static Time-Series Analysis of Low Voltage Network Operation with Photovoltaic Feed-In," in *Proc. Power and Energy Stud. Summit*, Dortmund, DE, Jan. 12–14, 2015.

[7] A. Selim, M. Abdel-Akher, M. M. Aly and S. Kamel, "Efficient Time Series Simulation of Distribution Systems with Voltage Regulation and PV Penetration," in *Proc. Mid. East Pwr. Sys. Conf,* Cairo, Egypt, Dec. 27-29, 2016.

[8] IEEE, PES Distribution System Analysis Subcommittees' Distribution Test Feeder Working Group, www.ewh.ieee.org/soc/pes/dsacom/testfeeders/index.html.

[9] OpenDSS – EPRI Distribution System Simulator, online resource: https://sourceforge.net/projects/electricdss/

Scalability of the Vector Quantization Approach for Fast QSTS Simulation

Jeremiah Deboever[1], Santiago Grijalva[1], Matthew J. Reno[2], Xiaochen Zhang[1], and Robert J. Broderick[2]

[1] Georgia Institute of Technology, Atlanta, GA, 30332, USA
[2] Sandia National Laboratories, Albuquerque, NM, 87123, USA

Abstract — Quasi-static time-series (QSTS) provides the necessary simulation fidelity on the impact that new energy resources would have on a specific distribution system feeder. However, this simulation can often take 10-120 hours for a single study, on traditional computers. The vector quantization approach proposed in [1] demonstrated very attractive computational time reduction for feeders with limited number of input time-series profiles and few distribution system voltage control elements. In this work, we expend on this algorithm and address this issue to model feeders of any complexity while maintaining a high computational time reduction. We demonstrate a 98.7% reduction on a real distribution system feeder with 2969-bus, 3 load/PV profiles, and 8 voltage regulating elements.

Index Terms – power system simulation, smart grid, power system planning, photovoltaic systems.

I. INTRODUCTION

Interconnection studies for distributed energy resources (DERs) on distribution networks are currently performed using scenario-based simulation that includes peak load, back feeding, and high PV output cases. Because this approach solves a limited set of static power flows scenarios, all potential impacts due to DERs will not always be captured. Moreover, scenario-based simulations are not able to capture time-dependent states in the system, e.g. voltage regulating tap changers. A draft of the IEEE standard IEEE P1547.7 D110 recently proposed a time-series simulation meant to capture any temporal effects for interconnection studies [2]. Quasi-static time-series (QSTS) simulation solves static power flows chronologically at a specific granularity, which allows controllable elements to be modeled with their control logic including delays. This is important to model in interconnection impact studies because intermittent resources can create excessive controller actions that will reduce the equipment life.

QSTS simulation has not yet been widely used for impact studies of DERs, and more specifically PV, because of its computational burden. A yearlong study at 1-second granularity is suggested to capture the fast variation of relevant circuit quantities due to PV and the seasonal changes in the power demand [3]. This represents solving 31.5 million chronological power flows. The computation time for large

feeders can be between 10-120 hours on conventional computers. The ability to simulate multiple locations, sizes, advanced inverter settings, or feeder configurations is therefore limited by the speed of the QSTS simulation. This makes fast time-series approximations very desirable.

There are several challenges in reducing the computational time of QSTS simulations due to the discontinuous and nonlinear nature of the simulation [4]. Six of the most significant challenges are: 1) the sheer number of power flows requires significant computational power. 2) some distribution feeders can be very complex when considering unbalanced loads or various controller logics, and their performance cannot be predicted without prior knowledge. 3) the time dependence between time steps requires each time step to be solved chronologically. 4) deadbands in controller logics create multiple feasible power flow solutions within their limits. Models must consider the hysteresis of the controllers. 5) the interaction between controllable elements can be extremely challenging to model, especially when considering the previous two challenges. 6) an accurate analysis for extended time-horizon simulations can require extensive data for post simulation analysis.

Different approaches in speeding up a QSTS simulation have been discussed in the literature: variable time-step, circuit reduction, A-diakoptics method, and vector quantization. Variable time-step increases the granularity of the simulation as a mean of computational time reduction [5]. Circuit reduction reduces the number of buses in the circuit to speed up the power flow solver [6]. A-diakoptics method divides large circuits into sub-areas to solve them separately [7]. Vector quantization takes advantage of similar power flow solutions to alleviate solving the non-linear unbalanced three-phase power flow equations [1], [8], [9]. This paper focuses on the vector quantization method presented in [1]. The scalability aspects discussed in that publication are addressed in this paper. More specifically, the algorithm is adapted to simulate realistic size feeders with more complexity including: size of the feeder, multiple load/PV profile inputs, larger number of controllable elements, and new types of controller logics (e.g. advanced inverter controls). This paper is organized as follows. In Section II, a summary of the vector quantization algorithm proposed in [1] is presented. The scalability of the algorithm in terms of feeder size, number of profiles, number of controllable elements, and types of controller logics are addressed in Section III. In Section IV, a simulation of a realistic test feeder is

---

This research was supported by the DOE SunShot Initiative, under agreement 30691. Sandia National Laboratories is a multimission laboratory managed and operated by National Technology and Engineering Solutions of Sandia, LLC., a wholly owned subsidiary of Honeywell International, Inc., for the U.S. Department of Energy's National Nuclear Security Administration under contract DE-NA0003525.

978-1-5090-5606-4/17 $31.00 © 2017 IEEE

demonstrated. Section V discusses the computational time reduction opportunity. The importance of this work for system planners and operators is presented in the conclusion.

II. Vector Quantization Approach

The vector quantization algorithm proposed in [1] groups similar power flow solutions in clusters to avoid the iterative power flow solver. Each time the power flow equations are solved, the solution is stored in a solution space for consequent time steps. If searching for a previously computed power flow solution is faster than solving the power flow equations, significant computational time reduction can be attained. This section briefly describes the vector quantization algorithm but additional details can be found in [1].

At each time step in the QSTS simulation, the algorithm cycles through two objectives: determining the power flow solution and determining whether an action is taken by the controllable elements on the feeder. The power flow solution depends on two sets of factors: the power injections on the feeder and the previous states of controllable elements. For PV interconnection studies, the power injections at time t is defined as the vector u_t where d_t is a vector of the different load profiles and $p_{pv,t}$ is a vector of the PV output profiles. The previous states of controllable elements at time t defined as l_t includes the states of any elements that affect the voltage on the feeder, for instance any voltage regulating tap changers r_t or capacitor banks c_t. The factors affecting the power flow solution can then be defined as vector h_t.

$$u_t := [d_t, p_{pv,t}] \tag{1}$$

$$l_t := [r_t, c_t] \tag{2}$$

$$h_t := [u_t, l_t] \tag{3}$$

The power flow solution g_t is a vector of the bus voltage magnitudes and angles.

$$g_t := [\,|v_t|, \theta_t\,] \tag{4}$$

From the power flow solution g_t and the vector h_t, any time-series values on the feeder can be analyzed without having to run the time-series simulation. A state vector $x_t := [h_t, g_t]$ can be stored in a state matrix X for post-simulation analysis.

The objective of the vector quantization algorithm is to bypass the iterative AC power flow solver by reassigning a similar solution if it has already been computed. At the beginning of the time step, a quantization logic is used to determine whether a similar (as defined by quantization clustering) power flow solution has been computed. A flow chart of the algorithm is presented in Fig. 1.

The quantization logic used to determine whether a solution exists must be faster than solving the power flow equations to make the algorithm attractive. The logic presented in [1] uses a matrix indexing method as shown in Algorithm 1. Matrix S is the solution space in which solutions are appended each time a new solution is computed. Matrix M is an indexing matrix used to determine whether or not a solution exists. Each value in

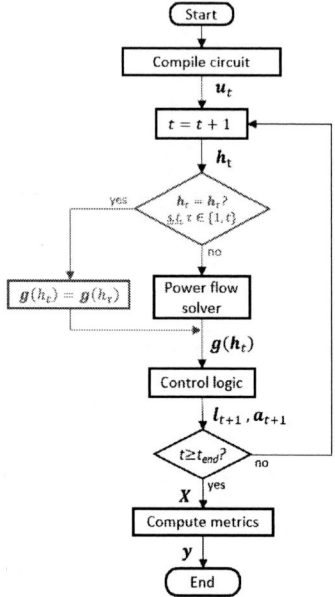

Fig. 1. QSTS flow diagram with proposed quantization algorithm (in red) [1] .

vector h_t is associated to a unique dimension in the matrix M. Thus, for an n-dimensional matrix M, $indx = M([h_t])$ is the unique value at the location $M(h_{t,1}, h_{t,2}, \ldots, h_{t,n})$ that is associated with a power flow solution.

Algorithm 1: Quantization logic using matrix indexing [1]
1:   $indx = M([h_t])$,
2:   $if\ indx \neq 0$:
3:      $g_t = S(indx, :)$
4:   $else$:
5:      $Solve\ power\ flow\ equations\ with\ h_t,$
6:      $indx = size(S, 1) + 1,$
7:      $S(indx, :) = g_t,$
8:      $M([h_t]) = indx,$
9:   end

This matrix indexing logic runs quickly if the two matrices are preallocated in the memory. However, the indexing matrix M can become extremely large considering that it has the same number of dimensions as the length of vector h_t, which is the total number of input time-series profiles and the number of controllable elements. The length of each dimension of M depends on the number of unique combinations of vector h_t, which can be decreased using quantization. However, quantization introduces inherently an error in the data that can affect the accuracy of the simulation. Namely, metrics reported by the QSTS simulation, such as controller actions, under-/over-voltage, power loss, or constraint violations, can be falsely reported when quantization clusters are large and cover a wide range of slightly similar power flow solutions. Because the algorithm requires the indexing matrix M to be preallocated in the memory, a finite number of load and PV profiles and controllable elements can be modeled based on the available memory for this multi-dimensional matrix. The formulation of

978-1-5090-5606-4/17 $31.00 © 2017 IEEE

this algorithm in [1] limited the combined number of profile and controllable elements to 6. These issues are addressed in the following section.

III. SCALABILITY OF THE ALGORITHM

A. Size of the feeder

The test circuit simulated in [1] only has 13 buses and thus, is not representative of a realistic distribution feeder. Although it has been demonstrated in the literature that an accurate model of a large feeder can be created based on a small number of buses [6], the vector quantization algorithm scales well with the size of the feeder. As the number of nodes in the circuit increases, the computational time to solve an individual power flow increases proportionally. The computational time of the proposed algorithm is a function of the number power flows computed ($N_{pf\ solved}$), the controller logic (T_{CL}), and an overhead time associated with the implementation of the algorithm itself (T_{VQ}). The average computational time to solve a power flow can be estimated from the computational time of QSTS simulation with a brute force approach ($T_{brute\ force}$).

$$T_{comp.} = \left(\frac{T_{brute\ force}}{31.5e^6}\right) N_{pf\ solved} + T_{CL} + T_{VQ} \qquad (5)$$

Because the quantization algorithm only computes power flows for a subset of the total number of time steps, the overall computational time of the vector quantization will increase much slower than if the power flow at each time step was computed. Furthermore, the overhead computational time associated with the control logic and the logic required to implement the algorithm is independent of the number of nodes in the circuit (i.e. the size of h_t and M in line 1 in Algorithm 1 do not change based on the number of nodes in the circuit). Therefore, as the feeder size increases, this overhead computational time becomes negligible compared to the computational time of the power flow solver, making the computational time reduction equal to the ratio of the number of power flows computed over the total number of time steps. For instance, a simulation with 31,500 unique computed power flows would reduce the computational time of a QSTS simulation by about a thousand.

B. Number of Load/PV Profiles

In the modified IEEE 13-bus test circuit discussed in [1], the indexing matrix M would be 6-dimensional with the two input time-series profiles, the three voltage regulators, and the switching capacitor bank. If the two profiles are quantized into 100 clusters, the regulators have 33 tap positions, and the capacitor bank has two states, the indexing matrix M would have 719 million entries. Obviously, additional input time-series profiles for each type of customer or different PV locations cannot be done with the current algorithm. Each additional input time-series increased the size of M exponentially. The proposed method to address this issue is not to treat profiles as individual dimensions in the indexing matrix M but as a single dimension representing a 'scenario' of

profiles. A time-series vector $indx_u$ can be created where each entry is the index of the first time that combination of profiles was experienced. The value of each profile at a specific time step t can easily be determined with $u([indx_u])$.

After quantizing the profiles, the time-series vector $indx_u$ can be created representing the 'scenario' of profiles at each time-step. The vector defining unique power flow solutions becomes:

$$h_t = [indx_{u,t}, l_t] \qquad (6)$$

By pre-processing the profiles, the number of dimensions for profiles in the indexing matrix is limited to one no matter how many input time-series profiles are considered. Moreover, the size of the indexing matrix is also reduced since not every unique vector u_t is experienced over the time horizon. This method is only possible because the dataset for the profiles is known prior to the time-series simulation. With this indexing method, the vector quantization algorithm is not affected by the number of profiles considered in the simulation, whether it is individual load profiles based on Advanced Metering Infrastructure (AMI) data or multiple PV profiles based on scale and location.

C. Number of Controllable elements

A similar approach to the one discussed above can be used to reduce the memory requirement for the controllable elements in the indexing matrix. However, unlike the profiles, the states of those elements are not known prior to the time-series simulation. Thus, the entire space must be created for the current indexing method to work. From analyzing the simulation results on the modified IEEE 13-bus test circuit from [1], one can find that this space is extremely sparse as shown in Fig. 2.

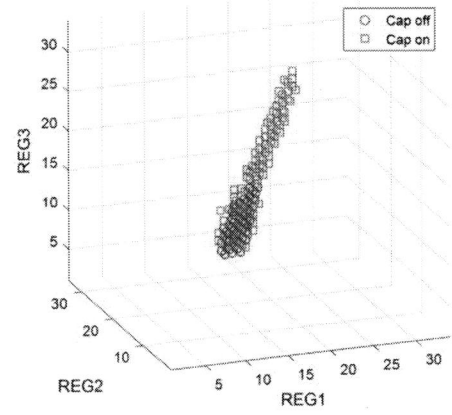

Fig. 2. Regulator tap position and capacitor status combinations that are experienced in a yearlong QSTS simulation for the modified IEEE 13-bus test circuit

To take advantage of this sparsity, a controllable element matrix L can be created with each dimension representing the state of an element. This matrix is initially a zero-matrix that is populated with indices as the simulation progresses. Each time a set of controllable element states are experienced for the first time, a column is appended to the indexing matrix M and an index referring to the column number is stored in matrix L. The following two equalities are then implemented before line 1 of Algorithm 1.

$$indx_L = L([l_t]) \qquad (7)$$
$$h_t = [indx_u, indx_L] \qquad (8)$$

At the end of the yearlong simulation, matrix M is a two-dimensional matrix with a size of the number of profile scenarios by the number of controllable element states experienced. This formulation takes advantage of the sparsity of the space due to the correlation between controllable elements and possibly reduce the pre-allocated space for the controllable elements in the indexing matrix. In the 13-bus test case referenced above, the three phases are relatively balanced, explaining why the voltage regulator tap positions of each phase are closely correlated. Not every entry in matrix L needs to have a value referring to the indexing matrix M.

This matrix decomposition reduces the requirement to preallocate large datasets in memory. This concept can be further expended as the number of controllable elements increases, which increases the size of L but also its sparsity. Furthermore, the computational time of the algorithm will not be affected since Eq. (7) is only performed when a controller action is performed.

D. Type of controllers

Distribution feeders can have a wide variety of devices with control logic: energy storage systems, smart inverters, regulators, capacitor banks, etc. The number of devices implemented on a feeder can rapidly grow as some of them are often installed with other DERs (e.g. advanced inverters with PV systems). In fact, the need to determine the settings for these new types of controls is an important application of QSTS. For example, in order to determine the optimal settings for energy storage controls [10] or advanced inverter controls [11], potentially thousands of QSTS simulations need to be performed with different settings to fully study any potential interactions between controls and potential benefits.

One of the advantages of the proposed quantization algorithm is that it is robust to any type of control logic. Generally speaking, controllers operate based on a specific signal, whether it is time, voltage/current magnitudes, power factor, price, etc. The type of signal will dictate how the control logic is implemented in the quantization algorithm. For this purpose, the controllers can be grouped into three categories.

First, most controllers are dependent on a signal derived from the power flow solution (i.e. voltage/current magnitude, power factor,...) or have an hysteresis that prohibits them to be computed ahead of the time-series simulation. For instance, capacitor banks can operate based on a voltage signal and have delays to reduce excessive operation. To accurately model these types of controllers in a QSTS simulation, their logic is implemented at the end of each time step. States from the power flow solution can be used as control signals and controller states (i.e. equipment states or delay accumulators) can model the hysteresis within the controllers. In the vector quantization algorithm, these equipment states (e.g. tap position) are referred to in the matrix L since they affect the power flow solutions. Because the algorithm goes through each time-step, delay accumulators can easily keep track of any expiring delays within the controllers as the simulation progress through time. As discussed in III.C, the algorithm is not limited by the number of these controllers in the feeder model.

Second, the controllers without delays and no hysteresis will always have the same outcome for a given power flow solution and system state. Thus, their control logic can be implemented separately into the power flow module, with vector quantization only storing the solutions. Because each power flow is defined by a unique vector u, the algorithm does not need to go through that control logic each time a power flow has already been computed. For example, this subtle difference is especially advantageous for feeders with multiple advanced inverters with volt/var capabilities programmed to control the power factor of their output based on the voltage magnitude of the system. In a QSTS simulation, this is simulated with a control logic iteratively changing the output of the inverter and re-computing the power flow until it converges. Under a brute-force approach to the QSTS simulation, this would be done at each time step, which will impact the computational time. However, the proposed algorithm reduces the number of times it goes through this type of controller logic since only the final converged solution is stored by the vector quantization algorithm.

Third, any controllers operating solely on a signal known ahead of the QSTS simulation (i.e. price, time, etc.) and does not depend on any states from the power flow solution could be preprocessed with the other input time-series profiles as a scenario (see subsection III.B). For instance, the power output of an energy storage system or the charging of an electric vehicle can be based on a predetermined schedule. In that case, the power output profile is time-dependent and is treated similar to a PV or load profile in the input vector u.

Not only has the number of controllable elements with hysteresis been addressed in III.C, the proposed vector quantization algorithm is robust to different types of controllers, since most would fit in one of the three previously stated categories.

IV. SIMULATION

The scalability of the algorithm is demonstrated using an actual distribution feeder with 2969 buses (5469 nodes) and 8 controllable elements – 4 three-phase switching capacitor banks, 3 single-phase line voltage regulators, and a three-phase

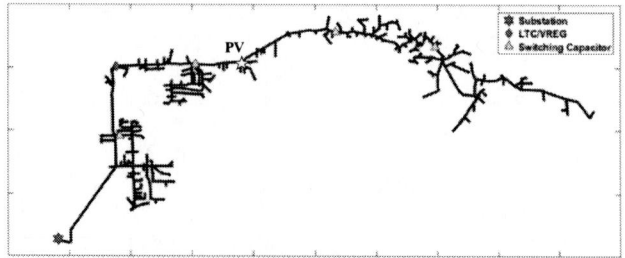

Fig. 3. Topology of the modeled distribution feeder.

substation load tap changer (Fig. 3). This realistic distribution feeder was modeled in OpenDSS through a MATLAB interface and simulated on a Window 10 computer with 32GB of memory and a 3.50GHz processor.

A three-phase centralized PV system is modeled in the middle of the feeder with an irradiance profile similar to [1]. The loads are categorized to simulate two types of customers: residential (single-phase) and commercial customers (three-phase). The 1131 residential customers are mostly on the laterals and account for 4.2 MW of peak power. The 317 commercial customers account for 1.7 MW. The two load profiles are show in boxplots grouped by hour in Fig. 4 and Fig. 5.

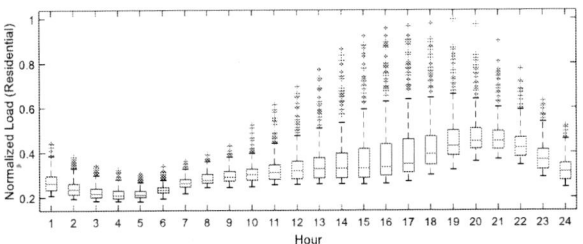

Fig. 4. Boxplot of the normalized power averaged per hour for a yearlong residential load profile.

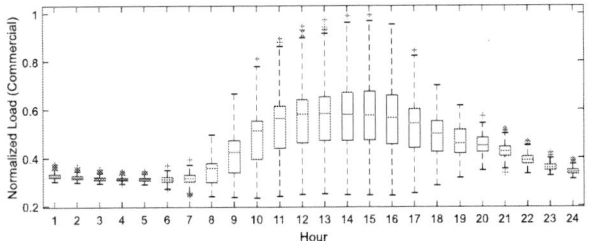

Fig. 5. Boxplot of the normalized power averaged per hour for a yearlong commercial load profile.

As expected, the residential profile has a tendency to peak around 7p-8p while the commercial profile has a higher load during business hours. When both profiles are used to model the load on a distribution feeder, not every combination of load multiplier will be experienced (i.e. commercial load at 1 p.u. and residential load at 0.2 p.u.). Thus, this space is relatively sparse and can be illustrated with a heat map of the reoccurrence of each combination (Fig. 6.).

The white pixels in the heat map illustrate a scenario of profiles that is never experienced in a yearlong simulation. Similar heat maps can be draw between profiles if more are considered. This illustrates the advantage of the formulation in section III.B. Each of the three yearlong profiles have a resolution of 1 second and a precision of 10^{-5} p.u. Without vector quantization, they have a total of 31,448,790 unique scenarios and a power flow is computed for every time step when considering the controllable element states. The number of computed power flows can be reduced through vector quantization, which ultimately reduces the precision of the profiles and clusters values together (TABLE I). Without the updated algorithm discussed in this paper, the indexing matrix

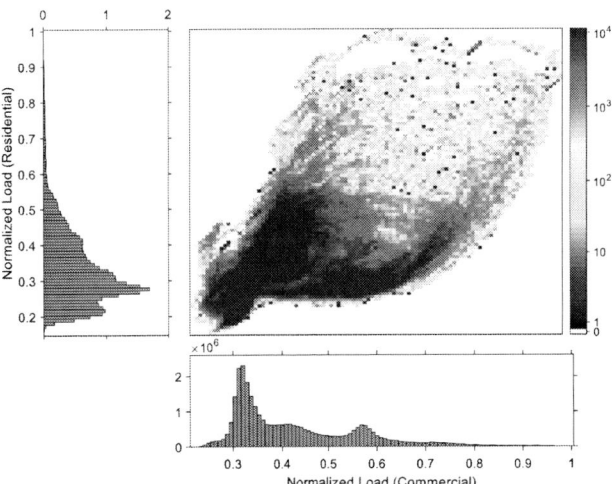

Fig. 6. Heat map of the number of time steps that have the same scenario (combination of multiplier values) for a yearlong profile with 1-second resolution. Note that the profiles were clustered for illustrative purposes.

M used in [1] for the 200-cluster case would have 1.5×10^{14} entries ($200^3 \times 33^4 \times 2^4$) making it impossible to be preallocated on a conventional computer. With the proposed algorithm, the indexing matrix had only 3.3×10^9 entries (863931×3829), which is equivalent to 13.2 GB. The novel implementation of the indexing matrix reduces the memory required to 0.002% of the original size for this larger circuit.

TABLE I. COMPARISON OF RUNNING A QSTS SIMULATION WITH A BRUTE FORCE APPROACH OR WITH THE PROPOSED VECTOR QUANTIZATION (VQ) ALGORITHM.

	# of scenarios	# of computed pf
Brute-force	31,448,790	31,499,600 (100%)
VQ (200 clusters)	863,931	1,722,978 (5.5%)
VQ (100 clusters)	218,187	716,443 (2.3%)
VQ (50 blusters)	43,743	253,870 (0.8%)

By reducing the number of computed power flows through vector quantization, the computational time of the simulation is drastically reduced with minimal impact on the accuracy of the simulation. In TABLE II, the computational time and accuracy is presented for each of the vector quantization cases.

TABLE II. QSTS SIMULATION RESULTS SHOWING THE ACCURACY AND COMPUTATIONAL TIME OF DIFFERENT VECTOR QUANTIZATION (VQ) CASES.

	LTC	REG1	REG2	REG3	Time [hrs]
Brute-force	2072	9031	12656	10783	36.1
VQ (200 clusters)	+0.5%	+2.3%	+2.5%	+2.4%	2.62
VQ (100 clusters)	+1.6%	+7.2%	+5.4%	+5.5%	1.17
VQ (50 blusters)	+2.8%	+5.4%	+5.1%	+5.0%	0.47

978-1-5090-5606-4/17 $31.00 © 2017 IEEE

These results demonstrate the effectiveness of the proposed vector quantization algorithm to run a QSTS simulation of a real distribution feeder and address the robustness of the algorithm. Note that the computational time reported for the vector quantization algorithm includes time-consuming calls through a COM interface that would not exist if implemented directing in OpenDSS or other commercial software such as CYME. (refer to [1] for more information).

V. COMPUTATIONAL TIME REDUCTION OPPORTUNITY

As previously discussed, each power flow solution is defined by a unique vector h based on the time-series profiles and the states of controllable elements with hysteresis. Through vector quantization, the number of unique power flows computed over the time horizon can be reduced to improve the computational time of the simulation. For instance, the precision of load profiles can be reduced to increase the number of repeated values. However, this approximation introduces an error in the simulation that can impact the accuracy of the results. Thus, not all indices should be quantized equally depending on their impact on the power flow solution and the type of study conducted. Each profile may have a different impact on the feeder depending on the location, size, etc. of the element it is associated to. For example, the profile associated with a large centralized PV system may impact the operation of a voltage regulator, while the profile of a smaller system downstream may not. This can also be true for the profiles modeling different types of customers.

Various metrics can be derived from the QSTS simulation results: controller actions, voltage extremes, losses, etc. The accuracy of each is affected differently by the vector quantization. Depending on which metric a simulation should report, the vector quantization might be different and will vary for each specific feeder. Thus, a vector quantization strategy should be used to minimize the introduced error while providing an attractive computational time reduction. This topic is outside the scope of this paper but will be explored in future research.

VI. CONCLUSION

In this paper, the vector quantization algorithm proposed in [1] is adapted to model realistic size, complex feeders. The indexing method proposed can simulate large feeders with multiple profiles and controllable elements, a shortcoming that the previous algorithm could not do. Furthermore, this paper discussed the robustness of the algorithm capable of simulating feeders with various types of controllable elements. The new algorithm is tested with a model of an actual distribution feeder that includes 2969 buses, 3 profiles, and 8 controllable elements. This algorithm provides an opportunity for computational time reduction of QSTS simulations through vector quantization.

The capability to run a realistic distribution system feeder in a 93-99% reduction of the time on a conventional computer eliminates the main limitation to the wide use of QSTS simulation by the industry. Being able to simulate the operation of an actual distribution feeder with controllable elements on them can provide great insights for system planners especially as DER penetration increases. On the other hand, QSTS simulations without their computational burden could become valuable for other applications such as hosting capacity analysis or real time operation.

With an algorithm that can model a wide variety of feeder complexities, future work includes formulating a vector quantization strategy capable of optimally reducing the computational time with minimal losses in accuracy. In addition, this fast time-series approximation algorithm will ultimately be implemented in software such as CYME, OpenDSS, or GridLab-D.

REFERENCES

[1] J. Deboever, S. Grijalva, M. J. Reno, and R. J. Broderick, "Fast quasi-static time-series (QSTS) for yearlong PV impact studies using vector quantization," *Sol. Energy*, p. Under Revision.

[2] IEEE P1547.7 D110, "Draft Guide to Conducting Distribution Impact Studies for Distributed Resource Interconnection," 2013.

[3] M. J. Reno, J. Deboever, and B. A. Mather, "Motivation and Requirements for Quasi-Static Time Series (QSTS) for Distribution System Analysis," in *IEEE PES General Meeting (2017)*.

[4] J. Deboever, X. Zhang, M. J. Reno, R. J. Broderick, and S. Grijalva, "Challenges in reducing computational time of QSTS simulations for distribution system analysis," SAND2017-5743, Albuquerque, NM, 2017.

[5] M. J. Reno and R. J. Broderick, "Predetermined Time-Step Solver for Rapid Quasi-Static Time Series (QSTS) of Distribution Systems," in *IEEE Innovative Smart Grid Technologies (ISGT)*, 2017.

[6] M. J. Reno, K. Coogan, R. Broderick, and S. Grijalva, "Reduction of distribution feeders for simplified PV impact studies," in *IEEE Photovoltaic Specialists Conference*, 2013, pp. 2337–2342.

[7] D. Montenegro, G. A. Ramos, and S. Bacha, "A-Diakoptics for the Multicore Sequential-Time Simulation of Microgrids Within Large Distribution Systems," *IEEE Trans. Smart Grid*, pp. 1–9, 2015.

[8] C. D. López, "Thesis: Shortening time-series power flow simulations for cost-benefit analysis of LV network operation with PV feed-in," Uppsala Universitet, 2015.

[9] A. Pagnetti and G. Delille, "A simple and efficient method for fast analysis of renewable generation connection to active distribution networks," *Electr. Power Syst. Res.*, vol. 125, pp. 133–140, 2015.

[10] M. J. Reno, M. Lave, J. E. Quiroz, and R. J. Broderick, "PV Ramp Rate Smoothing Using Energy Storage to Mititgate Increased Voltage regulator Tapping," *IEEE Photovolt. Spec. Conf.*, 2016.

[11] J. Seuss, M. J. Reno, R. J. Broderick, and S. Grijalva, "Analysis of PV Advanced Inverter Functions and Setpoints under Time Series Simulation," SAND2016-4856, Albuquerque, NM, 2016.

Machine Learning for Rapid QSTS Simulations using Neural Networks

Matthew J. Reno[1], Robert J. Broderick[1], and Logan Blakely[2]

[1] Sandia National Laboratories, Albuquerque, NM, 87185, USA
[2] Portland State University, Portland, OR, 97201, USA

Abstract — Distribution system analysis with high penetrations of distributed PV requires quasi-static time-series (QSTS) analysis to capture the time-varying and time-dependent aspects of the system, but current QSTS algorithms are prohibitively burdensome and computationally intensive. This paper proposes to identify the key time periods throughout the year that need to be run with QSTS simulation and then predict the number of tap changes for the rest of the time periods using an ensemble of neural networks (NN). This ensemble NN approach is a new method of performing time-step simulations based solely on the input data and demonstrates high accuracy for reproducing the full baseline QSTS analysis while performing the simulation up to 4 times faster.

I. INTRODUCTION

Quasi-static time-series (QSTS) simulation is a versatile study method used to understand equipment control operation, power protection coordination, and voltage regulation and reactive power management for distributed PV. QSTS simulations capture the time-varying parameters such as load, and the time-dependent states in the system such as regulator tap positions. Traditional snapshot tools and methods may not be adequate to accurately analyze the fast variability and complex interactions of high levels of distributed energy resources (DER) being interconnected, and snapshot study methods that only analyze peak periods or a peak variability day often lead to over-estimation of normal operating issues [1]. High-resolution yearlong QSTS studies can take from 10 to 120 hours to run using existing methods. New timeseries analysis methods are needed for QSTS to be used by utilities for distribution operation decisions and coordination [2]. In this work, we focus on how to speed up the high resolution quasi-static time-series (QSTS) simulation to make this method viable for both DER distribution impact studies and operational decision making [3].

In this paper, we propose to identify the key time periods throughout the year that need to be run with QSTS simulation and then use neural networks to predict the number of tap changes for the remainder of the time periods to accurately reproduce the full baseline QSTS analysis. Generally, solving the QSTS for half the days using random sampling has very high error compared to running the entire year [4]. This neural network method works by analyzing the simulation input data and selecting a representative sample of inputs throughout the year for QSTS simulation and training data. Then machine learning in the form of neural networks is applied to accurately predict the full year's QSTS simulation results.

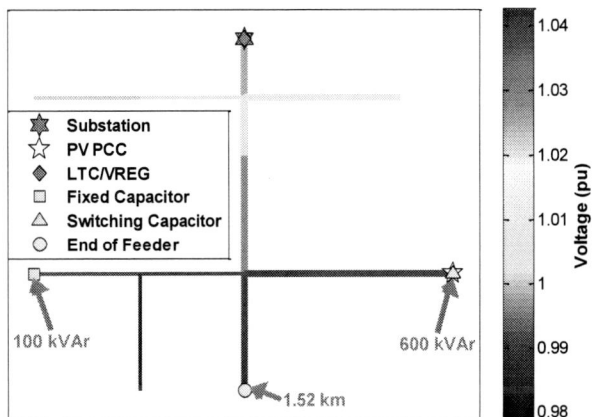

Figure 1. Diagram of the modified IEEE 13-node feeder colored by voltage.

II. QSTS DISTRIBUTION SYSTEM TEST CIRCUIT

The QSTS simulation is on a modified IEEE 13-bus test circuit that incorporates a centralized PV system at the end of the feeder, shown in Figure 1. The circuit has three single-phase voltage regulators at the feeder head, one single-phase capacitor, and a 3-phase capacitor bank. The voltage regulators are modified to provide ±5% regulation, and a voltage switching control is added to the 3-phase capacitor. The phase of some loads are changed to slightly balance the feeder, and all loads were increased by 20% to create some more extreme conditions. The load time-series is a 5-minute resolution normalized profile based on substation SCADA measurements from a feeder in California in 2013. A large 3-phase latitude-tilt 2MW PV system (~40% penetration of peak load) is added at the end of the feeder. The global horizontal irradiance (GHI) time-series measured at 1-second resolution in Oahu is converted to plane-of-array (POA) irradiance using the DIRINT decomposition model and the Hay/Davies transposition model. The Sandia Array Performance model and Sandia Inverter models are used to convert the POA irradiance into PV power output time-series. The circuit is modeled in OpenDSS and the algorithm is coded in MATLAB using the GridPV toolbox to interact with OpenDSS. The simulation is a year at one-second resolution.

III. NEURAL NETWORK METHODOLOGY

Neural Network applications are varied in terms of both fields of study and in goals. The ability of NN's to approximate nonlinear functions has proven invaluable in a multitude of industries. Common goals include forecasting, classifying, pattern recognition, and speed improvements for a piece of software. One example is using NN's to predict inventory demand using an irregular timeseries [5]. Other examples include using NN's to predict the load on a power system using a timeseries [6] and using NN's for financial forecasting [7].

The proposed neural network (NN) methodology performs QSTS simulations for certain parts of the year and then applies machine learning to predict the number of tap changes for the rest of the time periods that were not simulated with the goal of improving the overall time of the simulation. For all time periods in the year, statistics (such as max, min, mean, etc.) are calculated based on the input timeseries profiles. The statistics for each period are used as the inputs to a neural network to learn the correlations between the input data (load and PV) and the number of tap changes.

A. Feature Selection

The input timeseries data is partitioned into time periods, and then a variety of different statistics and potential variables can be calculated for each time period. For example, for the load profile, the maximum load, minimum load, median load, and mean load can be calculated for the period. Since variability has a direct correlation with the number of tap changes [8], several variability statistics are also calculated for the profile, such as the load range (max minus min during the period), standard deviation of the load, and the variability index (VI) of the load during the period. The same statistics are calculated for the PV power output timeseries. Additionally, the hour of the day of the period and the day of the year of the period can be used as training variables. Because the QSTS results are dependent on the events that occurred leading up to the period [3], standard deviation of the load over the previous two weeks is also calculated.

An important part of machine learning is the feature selection, deciding which variables and data attributes should be used to train the model. Selecting a subset of the relevant variables to use in the model construction reduces the model complexity, reduces training times, and improves generalization of the model. The central concept of feature selection is that many features are either redundant or irrelevant and can be removed without loss of information. Redundant features are removed by examining the correlation coefficient matrix shown in Figure 2. For example, the load max, min, median, and mean are all highly correlated, so only one attribute should remain in the final feature selection. The relevance of features is determined by doing a stepwise linear regression using forward selection to iteratively add the most significant variables.

After removing redundant or irrelevant attributes, the final feature list is load mean, load standard deviation, load VI, PV mean, PV standard deviation, PV VI, hour of the day, day of the year, and standard deviation of the load over the previous two weeks.

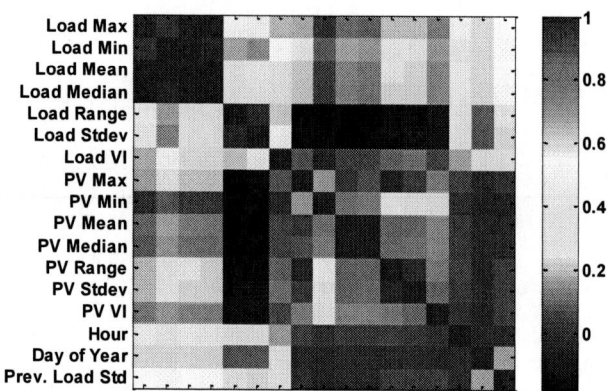

Figure 2. Correlation coefficients between the features calculated from the periods. X-axis is a mirror of the y-axis.

B. Sampling Method to Select Training Data

If periods are randomly selected from the year to perform QSTS simulation, the training data can occasionally have a significant bias in the results because the neural network is trained with only specific types of days (e.g. only clear sky days, or only winter days). An intelligent sample selection method [9] is applied to ensure that a range of time periods has been run with QSTS and is used for training the neural network. The intelligent sampling method is shown in Figure 3, stratifying the periods using the average load and PV variability. The objective is to sample at least one period from each bin to capture training data from the entire solution space. For example, using random sampling, only the low load periods (bottom half of Figure 3) might have biased the results to be too high. The stratification level and grid size are determined by how much of the year is going to be sampled. When sampling less of the year, the larger bins are used to ensure at least one sample from each bin.

Similar clustering methods have been used to group representative distribution systems [10], where a hosting capacity analysis [11] can be performed on subset of feeders to extrapolate the results to the rest of the feeders in the cluster.

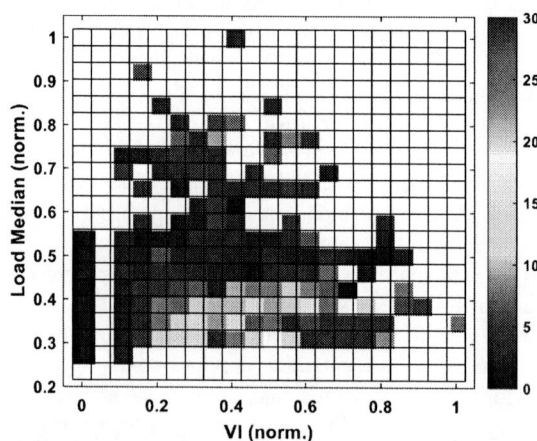

Figure 3. Average number of tap changes in 6 hour periods grouped by the median load and variability index (VI). White pixels have no periods during the year with those values. [9]

C. NN Architecture

Several different neural network architectures were investigated. Specifically, feedforward, cascade-forward, time-delay, and nonlinear autoregressive NN architectures were tested. Initially, it was assumed that the time-delay and autoregressive NN would perform the best because they are designed to solve time-series problems using dynamic neural networks. For this problem, we are sampling periods of the year which means the training data will be randomly spaced. Since time-series data is by definition evenly spaced in time [6], the time-delay and autoregressive models may not perform as well for the irregularly spaced time-series [5]. On the other hand, feedforward neural networks are very good at approximating nonlinear functions. The NN is trying to identify the correlations between load and PV variability and the number of tap changes, which is more similar to a function approximation problem than it is to a timeseries based problem. Since the objective is computation speed, feedforward neural networks also have the advantage that they are fairly fast and learn quickly. Based on literature and initial testing of the various NN architectures, the feedforward NN was selected.

D. NN Hyperparameter Selection

Hyperparameter selection is a key aspect in finalizing the NN architecture after the general type of model has been selected. Choosing specific hyperparameters can be done in many different ways and continues to be an open area of research. Methods for choosing hyperparameters include trial and error, grid search, and various types of optimizing algorithms. The learning rate, momentum, number of layers, number of nodes, the neuron activation function, number of nodes per layer, and the training function are all examples of hyperparameters that needed to be determined for this architecture.

The number of inputs nodes, output nodes, and number of hidden layers were guided by the problem itself. The final result is a single number of the total tap changes that occurred in the period, so there is a single output node. The number of input nodes was driven by the feature selection described in Section III.A. A single hidden layer was chosen since the primary goal of the project was an improvement in speed and multiple intermediate layers can drastically increase the computation time. The number of hidden nodes was determined by extensive trial and error from a third to twice the number of input nodes.

The learning rate, momentum value, and neuron activation function were left at the default values in the Matlab NN toolbox. Optimizing these three hyperparameters is one possible area of future work on this NN architecture.

E. NN Training

There are a wide variety of training algorithms that can be used in a feed-forward NN. This NN is trained using the Levenberg-Marquardt method based on [7]. The mu hyperparameters for the Levenberg-Marquardt method were chosen based on literature review and testing. In this case, it is

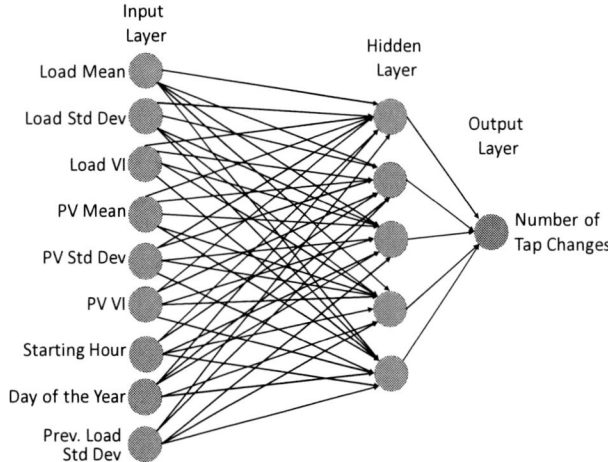

Figure 4. Neural Network Architecture

desirable that the NN not converge too quickly so as to avoid halting training in a local minimum, and the mu values were adjusted accordingly. For the results shown in this paper, the initial mu is set to 1, mu decrease factor to 0.8, and the mu increase factor to 1.5.

Based on using the Levenberg-Marquardt training function the Mean Squared Error method is used as the cost function for the network. The statistics from Section III.A. are normalized prior to being fed into the NN as inputs. In order to keep the neural network from over-fitting to the training data, a portion of the simulated time-periods is used as validation data. The NN training is complete when the error values for the validation data cease decreasing and begin increasing relative to the error on the training data.

F. Ensemble NN

Once a neural network has been trained, the statistics from the time-periods that have not been simulated with QSTS are fed into the neural network to predict the number of tap changes for those periods. Both the training and the prediction with the neural network are very quick, so the vast majority of the computational time goes into running the QSTS simulation for part of the year. Because of this fact, and the fact that the neural network model always returns slightly different predictions based on the initial weights and the split of training/validation data, rapid QSTS is a perfect application of ensemble machine learning using neural networks.

Ensembles are commonly used in machine learning to reduce the prediction variance by training many different models using different partitions and averaging the results over the models. This paper proposes using 50 ensemble neural networks, each with random initial weights, trained separately, and then combining the 50 predictions into a single value. For the ensemble learning, the test dataset will be the same because QSTS was not run for those periods, but the split between the training dataset and validation dataset will vary. Ensemble learning has the advantage of improved performance with low complexity of individual networks.

Three ensemble machine learning algorithms were investigated. First, a k-fold cross validation method, similar to jackknifing, was tried with the total training set being partitioned k times and a single k subsample being left out for the validation data on each ensemble member training. Second, a bootstrap aggregating or bagging algorithm was applied to repeatedly sample randomly and uniformly using a Monte Carlo process to split the dataset into training and validation data, with a unique network trained for each split. The challenge with the k-folds and bagging algorithms is that they split the training and validation data randomly. The dataset was carefully selected using intelligent stratified sampling to ensure the entire state space was sampled, but then randomly sampling to split that data into a training set and a validation set could end up using a small part of the state space for the training data. In order to overcome this, a third method was developed that we are calling weighted bagging. Weighted bagging performs weighted random selection of the training and validation datasets. When selecting the training dataset, the samples from the bins in Section III.B. with fewer periods are weighted higher. By weighting the sampling inversely proportional to the number of samples per bin, the training dataset is slightly weighted to include more of the total state space from the intelligent sampling.

After the 50 NN are trained with different parts of the training and validation dataset using weighted bagging, the NN training Mean Squared Errors are used to remove any outlier NN with extreme training errors. Next, the prediction results from the remaining NN are combined. Because the mean of the NNs can be biased by very high predictions, the median of the ensemble is calculated.

G. NN Length of Time Periods

The neural network is designed to learn the correlation between the input statistics and number of tap changes for given time periods. Any length of time can be used for running the QSTS and training the network. For example, the neural network could learn the correlation to number of tap changes in a day, or number of tap changes in an hour. Figure 5 shows the errors when applying the methodology at different time period resolutions. At the 24-hour day-long resolution, the neural network has trouble distinguishing the correlations between the input data statistics (PV variability) and the number of tap changes, because the statistics lack the details about when during the day the PV variability occurred. Also, as the time periods get longer, the training data for the neural network decreases. For example, when running half of the year, a 24-hour time-period has 182 training points, but the 1-hour time-period has 4380 training points for the same initial effort in performing the QSTS simulation. As the time periods get very short, there is no additional benefit seen, and eventually there are issues with distribution system control time delays overlapping several periods. For the proposed ensemble NN algorithm, a time period of 2-hours was selected based on the lowest MAE in Figure 5.

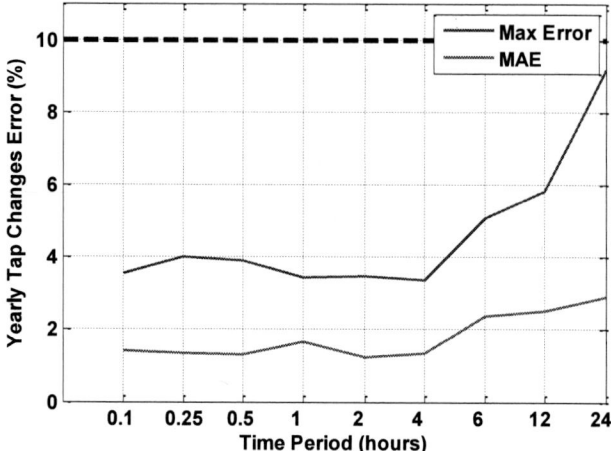

Figure 5. The mean absolute error (MAE) and maximum error for Monte Carlo runs using NN with half the year as training data at different time period resolutions.

IV. NEURAL NETWORK SIMULATION

Section III describes the details of the intelligent sampling, neural network architecture, training process and the ensemble learning algorithm. Each of these pieces is combined into Algorithm 1 to perform rapid QSTS simulations at significantly decreased computational time by fully simulating a portion of the year and using NN's to predict the remainder of the year.

Algorithm 1: Rapid QSTS Algorithm using Neural Networks

1. Separate timeseries into periods
2. Calculate load and PV statistics for each period
3. Intelligently select a percentage of the year for the training data
4. Run QSTS simulation on the selected portion of the year
5. Randomly split the partial-year QSTS results into a training dataset and validation dataset
6. Train NN
7. Use NN to predict total tap changes for the year
8. Go back to 5, repeat 50 times to generate each ensemble member
9. Average NN predictions (without outliers) into a single NN ensemble prediction
10. Calculate error for total number of tap changes in a year

---

Since the NN training data from intelligent sampling in step 3 of Algorithm 1 could come from many different parts of the year, the overall performance and accuracy of the algorithm can be determined by doing a Monte Carlo simulation with different testing datasets. Each Monte Carlo iteration will perform the rapid QSTS algorithm using ensemble NN, giving an approximation of the error introduced by the method. The error distribution expected of the ensemble NN is obtained by performing 1000 Monte Carlo runs.

V. RESULTS

A. Advantage of Ensemble NN

The neural network methodology from Section IV is applied to the test circuit from Section II. Intelligent sampling is used to select 2-hour time periods to train the network. For the 1000 iteration Monte Carlo simulation, the stochastic simulation produces different time periods from the intelligent sampling for the QSTS simulation to be the training data in the neural network. Using this as the input to a single neural network, the distribution of errors is shown in Figure 6a. While the mean absolute error (MAE) in Figure 6a is fairly low with good centering of the error, there are some significant outliers with larger than 40% error. Based on feedback from industry [4], the objective is to ensure all prediction results are within an allowable error threshold of 10% for yearly tap changes. By applying the ensemble NN method described Section III.F, for each Monte Carlo iteration, 50 neural networks are trained, and the results are averaged. The advantage of the overall method proposed in Algorithm 1 is shown in Figure 6b. Not only is the mean absolute error (MAE) less than using the ensemble machine learning, the extreme outliers are now nicely bounded to within 13% error on the number of yearly tap changes.

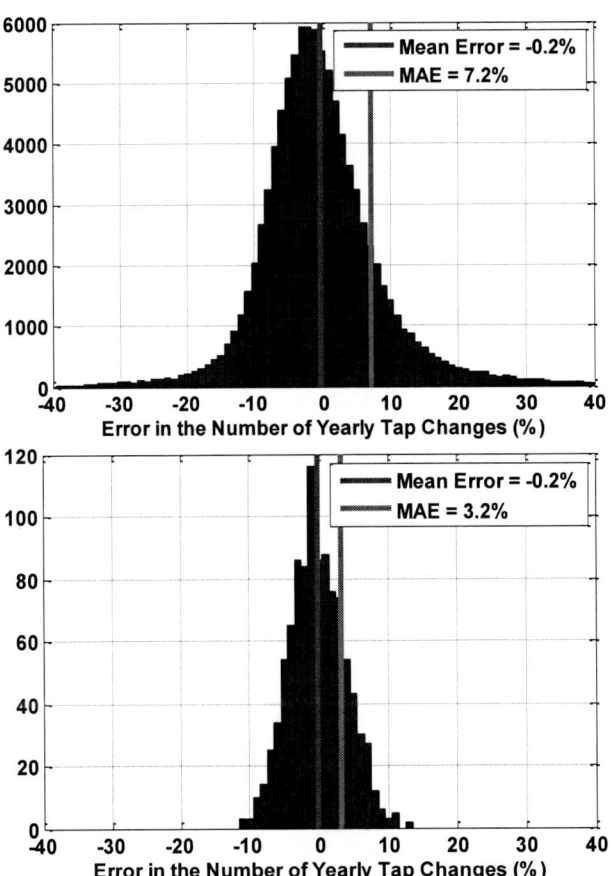

Figure 6. Monte Carlo error distribution for the number of yearly tap changes predicted using 15% of the year as training data by a) a single neural network and b) the ensemble NN.

B. Speed vs. Accuracy Trade off

There is a tradeoff between accuracy and speed based on the length of the QSTS simulation. By simulating more of the year with QSTS, the training data size and accuracy is increased, but additional computational time is required for running the longer QSTS simulation. Figure 7 shows the error decreasing with the longer simulations. Given the 10% error threshold, 21% of the year must be run using QSTS to guarantee that the neural network will appropriately estimate the number of tap changes for the year. This represents a 79% reduction in computational time possible using the algorithm.

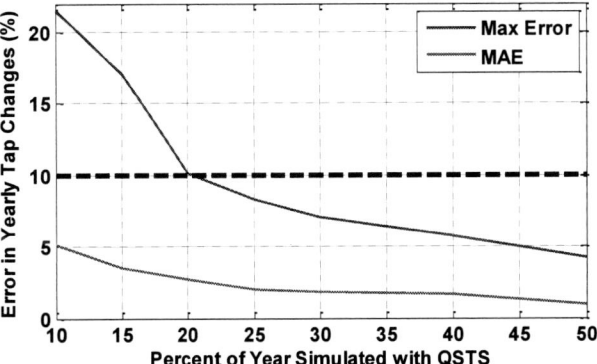

Figure 7. Ensemble NN mean absolute error (MAE) and maximum error for Monte Carlo runs with different lengths of QSTS simulations.

C. Comparison to other Algorithms

The simplest method to reduce the length of a QSTS simulation is to randomly select part of the year to analyze. Random sampling results in very large errors a small percentage of the time [4] as shown in Figure 9.

If an intelligent sampling method is used to select the time periods for running QSTS [9], a model can be fit to the sampled data to predict the number of tap changes during the other periods. Using the same intelligently selected samples, several different models were developed to compare the performance accuracy of the ensemble NN algorithm:

1. Nearest Neighbor – based on the load mean and PV VI of the period, the value for the number of tap changes is determined by finding the nearest neighbor that was analyzed with QSTS
2. Linear Interpolation – based on the load mean and PV VI of the period, performs a local linearization (Delaunay triangulation) to the nearest periods analyzed with QSTS
3. Regression – linear regression model using the full input attribute list discussed in Section III.A.

The results for each method are shown in Figure 8 for the maximum error produced during the Monte Carlo simulations, with the goal being to produce less than 10% error in the number of tap changes in a year. The random sampling approach performs particularly poorly; it requires around 85% of the year to be simulated with QSTS to ensure less than 10%

error. On the other hand, all the models fit to the intelligently sampled data perform much better. The nearest neighbor and linear interpolation both use nearby samples to predict the other periods, so they suffer with errors around outliers. The regression model fit error is due to fitting a linear function to the nonlinear data shown in Figure 3. By fitting the entire state space with a nonlinear model, the ensemble NN is able to both be robust to local outliers and fit various complexities in the distribution system results. As seen in Figure 9, to stay within the acceptable 10% error threshold, the ensemble NN requires 45% less data than the other models fit to the intelligently sampled periods.

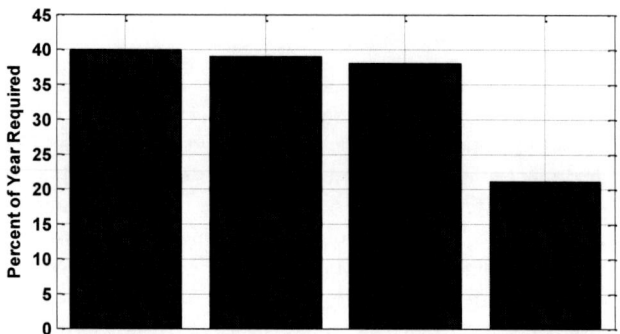

Figure 9. The percent of the year that must be run to guarantee less than 10% error in yearly tap changes for each method fitting to intelligently sampled periods.

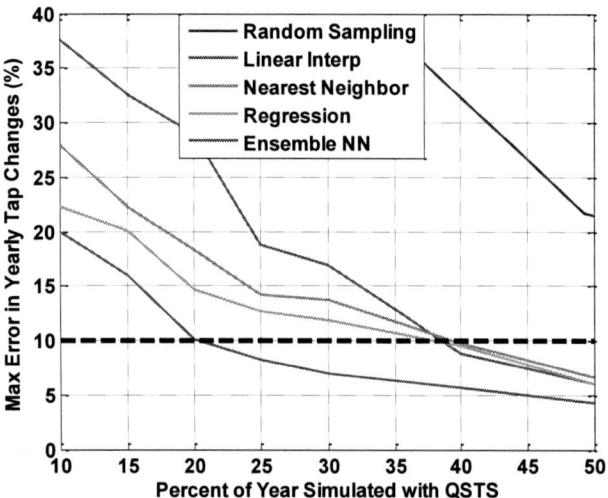

Figure 8. Comparing the maximum error of the Monte Carlo runs for different algorithms and a range of QSTS simulation lengths.

VI. CONCLUSIONS

This paper presents a neural-network-based machine learning algorithm combined with QSTS simulations to quickly and accurately model the impacts of high penetrations of distributed PV on voltage regulation equipment. The NN method is based solely on the input data profiles and does not require any interaction with the power flow engine. Previously QSTS simulations had to be performed on an entire year of data [4], but the contributions from this paper allow subsets of the year to be simulated with QSTS to model the impacts for the whole year. The proposed NN algorithm demonstrates more than a 75% reduction in computational time compared to conventional QSTS analysis, while still maintaining an acceptable level of accuracy. Additionally, the ensemble NN does a better job fitting the intelligently sampled periods than several other algorithms investigated.

This research was supported by the DOE SunShot Initiative, under agreement 30691. Sandia National Laboratories is a multimission laboratory managed and operated by National Technology and Engineering Solutions of Sandia, LLC., a wholly owned subsidiary of Honeywell International, Inc., for the U.S. Department of Energy's National Nuclear Security Administration under contract DE-NA0003525.

REFERENCES

[1] B. Palmintier, R. Broderick, B. Mather, M. Coddington, K. Baker, F. Ding, *et al.*, "On the Path to SunShot: Emerging Issues and Challenges in Integrating Solar with the Distribution System," National Renewable Energy Laboratory NREL/TP-5D00-65331, 2016.

[2] M. J. Reno and R. J. Broderick, "Predetermined Time-Step Solver for Rapid Quasi-Static Time Series (QSTS) of Distribution Systems," in *IEEE Innovative Smart Grid Technologies (ISGT)*, 2017.

[3] J. Deboever, X. Zhang, M. J. Reno, R. J. Broderick, S. Grijalva, and F. Therrien, "Challenges in reducing the computational time of QSTS simulations for distribution system analysis," Sandia National Laboratories SAND2017-5743, 2017.

[4] M. J. Reno, J. Deboever, and B. Mather, "Motivation and Requirements for Quasi-Static Time Series (QSTS) for Distribution System Analysis," in *IEEE PES General Meeting*, 2017.

[5] J. L. Carmo and A. J. Rodrigues, "Adaptive forecasting of irregular demand processes," *Engineering Applications of Artificial Intelligence,* vol. 17, 2004.

[6] C.-M. Lee and C.-N. Ko, "Time series prediction using RBF neural networks with a nonlinear time-varying evolution PSO algorithm," *Neurocomputing,* vol. 73, 2009.

[7] D. Brezak, T. Bacek, D. Majetic, J. Kasac, and B. Novakovic, "A comparison of feed-forward and recurrent neural networks in time series forecasting," in *IEEE Conference on Computational Intelligence for Financial Engineering & Economics (CIFEr)*, 2012.

[8] M. Lave, M. J. Reno, and R. J. Broderick, "Characterizing Local High-Frequency Solar Variability and the Impact to Distribution Studies," *Solar Energy,* 2015.

[9] J. Galtieri and M. J. Reno, "Intelligent Sampling of Periods for Reduced Computational Time of Time Series Analysis of PV Impacts on the Distribution System," in *IEEE Photovoltaic Specialists Conference*, 2017.

[10] R. J. Broderick, K. Munoz-Ramos, and M. J. Reno, "Accuracy of Clustering as a Method to Group Distribution Feeders by PV Hosting Capacity," in *IEEE PES Transmission & Distribution Conference & Exposition*, 2016.

[11] M. J. Reno, K. Coogan, S. Grijalva, R. J. Broderick, and J. E. Quiroz, "PV Interconnection Risk Analysis through Distribution System Impact Signatures and Feeder Zones," in *IEEE PES General Meeting*, 2014.

Algorithmic Aspects of a Commercial-Grade Distribution System Load Flow Engine

Francis Therrien[1], Marc Belletête[1], Jean-Sébastien Lacroix[1], and Matthew J. Reno[2]

[1] CYME International T&D, St-Bruno, QC, J3V 3P8, Canada

[2] Sandia National Laboratories, Albuquerque, NM, 87185, USA

Abstract — **The increasing penetration of PV requires more and more load flow calculations to assess the corresponding impact on planning and operation of distribution systems, e.g., during high-resolution quasi-static time-series calculations. There is therefore a strong incentive to optimize the speed of distribution load flow engines as much as possible without compromising on solution accuracy. This paper details the structure of a commercial-grade topology-independent Newton-Raphson distribution load flow engine, the reasoning behind some of its design choices, and recent improvements made to three rarely discussed aspects: initialization, calculation and discretization of tap changer positions, and solution of linear equations. Computer studies demonstrate the reduction in computational time on various distribution systems, allowing for faster interconnection studies with high PV penetration.**

Index Terms — **distribution systems, initialization, load flow, load tap changer, unbalanced systems.**

I. INTRODUCTION

The large amount of distributed energy resources (DERs) and especially PV now found at the low- and medium-voltage levels require rethinking the planning and operation of distribution grids. Distribution system analysis tools are becoming more important than ever, as they can assess the various impacts of DERs on operating constraints, protection schemes, losses, and so on [1]. While it is now recognized that precise electromagnetic transient (EMT) studies are sometimes needed at the distribution level, load flow calculations remain ubiquitous, whether in "standalone" mode or within quasi-static time-series (QSTS) studies [2]. Improving the speed of load flow engines will therefore facilitate the execution of more comprehensive studies with PV, such as year-long high-resolution QSTS studies [2].

Traditional distribution system load flow engines often use the ladder iterative method (often known as backward-forward sweep), which is tailored for radial and passive systems. While the ladder iterative method can be adapted to handle meshes [3] and voltage-controlled nodes [4], its efficiency decreases quickly as the number of loops and DERs in voltage control mode increases. In fact, ladder iterative methods will rarely converge in urban meshed networks .

Modern commercial-grade load flow engines often use matrix-based methods. Some of these engines are based on a full Newton-Raphson approach [5]–[8], while others use a fixed point scheme [8]–[10]. Full Newton-Raphson

approaches can naturally handle continuous controls by integrating them in the constraint vector, and typically converge in a few iterations; however, they require matrix re-factorizations, which are typically considered as computational bottlenecks. Fixed point schemes keep the same matrix factors, and therefore are considered faster; however, they may have more difficulty converging. The use of a smaller external sensitivity matrix can improve convergence in fixed point techniques [10], but its calculation and solution may require non-negligible computational effort when the number of control devices is large.

The CYME power engineering software has two load flow engines for unbalanced distribution systems: Voltage Drop Unbalanced (VDU, based on the ladder iterative approach), and the more recent Newton-Raphson Unbalanced (NRU, based on a full Newton-Raphson scheme). This paper first gives an overview of NRU, which is built on the work set forth in [8]. We discuss in detail three aspects of NRU that are often overlooked in the literature but that can have a significant impact on computational speed: the initialization method, the calculation and discretization of load tap changer (LTC) positions, and the solution of the resulting system of linear equations. Specifically, we present the advantages and disadvantages of initializing NRU with a fixed point (FP) or flat start (FS) approach, and briefly describe how to generate the FS profile considering NRU's modified nodal formulation. The benefits of integrating the control variables within NRU's set of equations are also discussed. Two LTC position normalization approaches are then presented. Finally, we challenge the often-held belief that constant-matrix engines are necessarily faster than full Newton-Raphson ones unless the former requires many more iterations. A brief discussion on two linear solvers, PARDISO [11] and KLU [12], is also included. Computer studies on various distribution systems are then presented to support the claims and hypotheses made throughout this paper.

II. LOAD FLOW FORMULATION AND ALGORITHMIC ASPECTS

A. Fundamentals

Using the principle of modified nodal analysis (MNA, sometimes referred to as modified *augmented* nodal analysis [8]), the main load flow constraints can be described by

$$\mathbf{i}_n = \mathbf{f}_n\left(\mathbf{v}_n, \mathbf{i}_d, \mathbf{x}_c\right) \qquad (1)$$

$$\mathbf{v}_d = \mathbf{f}_d\left(\mathbf{v}_n, \mathbf{i}_d, \mathbf{x}_c\right) \qquad (2)$$

$$\mathbf{y}_c = \mathbf{f}_c\left(\mathbf{v}_n, \mathbf{i}_d, \mathbf{x}_c\right) \qquad (3)$$

where $\mathbf{f}_n$, $\mathbf{f}_d$, and $\mathbf{f}_c$ are – potentially nonlinear – functions enforcing Kirchhoff's current law at every node, Kirchhoff's voltage law for ideal components (voltage sources, switches, and branch-dependent devices such as transformers and regulators), and additional network/component constraints, respectively; $\mathbf{v}_n$, $\mathbf{i}_d$, and $\mathbf{x}_c$ are state vectors comprising the nodal voltages, the currents of the ideal components, and the additional state variables (e.g., the equivalent admittance of DERs), respectively; and $\mathbf{i}_n$, $\mathbf{v}_d$, and $\mathbf{y}_c$ are nodal injection currents, voltages of ideal voltage sources, and the desired values of the additional network constraints. Complex values are represented using rectangular components. A multiphase framework is considered to model unbalanced conditions and topologies typical of North American distribution systems.

Since some of the constraints are nonlinear, (1)–(3) must be solved iteratively, herein using a full Newton-Raphson approach. Writing the Taylor series of (1)–(3) and truncating it therefore yields the main NRU equation

$$\begin{bmatrix} \mathbf{Y}_n^{"k} & \mathbf{D}_c^k & \mathbf{C}_{c1}^k \\ \mathbf{D}_r^k & \mathbf{S}_d^k & \mathbf{C}_{c2}^k \\ \mathbf{C}_{l1}^k & \mathbf{C}_{l2}^k & \mathbf{C}_d^k \end{bmatrix} \begin{bmatrix} \Delta\mathbf{v}_n^k \\ \Delta\mathbf{i}_d^k \\ \Delta\mathbf{x}_c^k \end{bmatrix} = \begin{bmatrix} \mathbf{i}_n^k \\ \mathbf{v}_d^k \\ \mathbf{y}_c^k \end{bmatrix} - \begin{bmatrix} \mathbf{f}_n^k \\ \mathbf{f}_d^k \\ \mathbf{f}_c^k \end{bmatrix} \qquad (4)$$

or, written in compact form,

$$\mathbf{J}^k \Delta\mathbf{x}^k = \mathbf{b}^k - \mathbf{f}^k . \qquad (5)$$

In the above, the superscript k indicates the iteration count; $\mathbf{Y}_n^{"k}$, $\mathbf{D}_c^k$, $\mathbf{D}_r^k$, $\mathbf{S}_d^k$, $\mathbf{C}_{c1}^k$, $\mathbf{C}_{c2}^k$, $\mathbf{C}_{l1}^k$, $\mathbf{C}_{l2}^k$, and $\mathbf{C}_d^k$ are submatrices of the system Jacobian $\mathbf{J}^k$; and the Δ operator indicates the difference between a given vector at the present and past iterations (e.g., $\Delta\mathbf{v}_n^k = \mathbf{v}_n^k - \mathbf{v}_n^{k-1}$). Additional information regarding the formulation of the Jacobian matrix and the right-hand side vectors $\mathbf{b}^k$ and $\mathbf{f}^k$ can be found in [8], [13]. In particular, $\mathbf{Y}_n^{"k}$ is a modified version of the standard admittance matrix, which accounts for state-variable-dependent equivalent current injections (see Section II-B).

Following the Newton-Raphson approach, (4) or (5) is solved iteratively until the largest normalized voltage mismatch has fallen below a user-specified tolerance, while ensuring that all dependent and independent variables are within their physical boundaries (e.g., minimum and maximum tap positions, reactive power capabilities, ...).

B. Current Injections vs. Additional Network Constraints

Due to the flexibility of MNA, loads can be included either as equivalent current injections in $\mathbf{f}_n^k$ [5], or through additional network constraints in $\mathbf{f}_c^k$ [7], [8]. On one hand, the former approach does not increase the system dimension, but its partial derivatives added in $\mathbf{Y}_n^{"k}$ are fairly involved when voltage-sensitive load models are used [5]. On the other hand,

the latter approach increases the system size (the load currents are added to $\mathbf{x}_c^k$), but only a few straightforward elements must be added to $\mathbf{C}_{l1}^k$ and $\mathbf{C}_d^k$ (e.g., 8 for a single-phase wye-grounded load) [7], [8]. Moreover, $\mathbf{C}_{c1}^k$ acts as an adjacency matrix for loads, and therefore its only non-zero values will be 1s and potentially –1s.

Tests on various networks have shown that no approach is unequivocally faster, owing in part to the high efficiency of modern linear solvers. However, it was noted that the equivalent current injection approach converged more robustly, likely due to its more linear formulation. The robustness of the equivalent current injection approach was further improved by representing loads using a fixed admittance in $\mathbf{Y}_n^{"k}$ and a small current in $\mathbf{i}_n^k$ compensating for the difference between the desired power and the power consumed by the fixed admittance. For a constant impedance load, the load equation becomes fully linear as the current injection is always equal to 0.

Other types of power injections or network constraints, such as DERs in voltage control mode, cannot be formulated linearly or quasi-linearly. In this case, their control equations are added to $\mathbf{f}_c^k$ along with the corresponding states in $\mathbf{x}_c^k$ [7], [8].

C. Initialization

The Newton-Raphson approach is not self-starting: its initial state vector $\mathbf{x}^0$ must be defined to solve (4) or (5). The FS approach remains the initialization method of choice in traditional transmission-level load flow engines, as the state vector is only comprised of nodal voltages in polar coordinates, and the phase shift of transformers is typically ignored. Initialization is more complicated in NRU, as its formulation includes state variables that are not nodal voltages (see (4)), and transformer phase shifts are considered. Moreover, the selected transformer modeling approach [14] involves internal nodes, whose equivalent FS voltage is not always trivial to define.

Due to the above reasons, NRU was originally initialized using a single-iteration FP solution [7], [8]. Therein, all power injection devices are represented using equivalent admittances. The initial reactive power must be guessed for generators in voltage control mode, which can be problematic in some transmission systems. All load tap changers (LTCs, see Section II-D) are fixed. Vector $\mathbf{f}_c^0$ is filled with 0s, $\mathbf{C}_{c1}^0$ and $\mathbf{C}_{c2}^0$ are empty matrices, and $\mathbf{C}_d^0$ is an identity matrix. This creates a linear equivalent to (4), allowing the computation of an initial guess of the state vector. Moreover, if the linearization is accurate, $\mathbf{x}^0$ will be closer to the final solution than a FS, possibly reducing the number of iterations. To increase speed, the FP matrix is padded with "hard" 0s wherever the Jacobian $\mathbf{J}^k$ might have non-zeros. Consequently, $\mathbf{J}^k$ can be built upon the structurally identical FP matrix, thereby significantly reducing CPU time since matrices are stored in a compressed sparse format [15].

The main caveat of this approach is that it requires two successive symbolic factorizations (for the FP solution and the first Newton-Raphson iteration), which are costly operations. More detail on this subject is given in Section II-E. To reduce the number of symbolic factorizations and hopefully the CPU time, an augmented FS approach is proposed. It makes use of a network-model iterator, which computes the complex nominal voltage at each node. For transformers, the internal nodal voltages are initialized based on the corresponding external nodal voltages and winding configuration. Vector $\mathbf{i}_d^0$ can be set to 0, since the corresponding equations are almost always linear; whereas the values of $\mathbf{x}_c^0$ are set using the desired power injection and the corresponding FS voltage.

D. LTC Discretization

There are two main methods to calculate LTC positions in load flows: between iterations (externally, sometimes referred to as the error-feedback approach), or by adding tap positions as state variables (internally) [16]. The former approach usually fits well into constant-matrix solutions [17], [18], but requires the external calculation of a sensitivity factor (empirically or through a more sound analysis [10]). Hunting between different LTCs may also occur [16]. In the latter approach, the LTC voltage set point is added to the constraint vector (e.g., in $\mathbf{f}_c^k$ in NRU). When using a Newton-Raphson approach, the solution usually converges to the desired set point in a few iterations. The main challenge is that the LTC positions are treated as continuous variables, whereas only discrete positions are physically feasible.

Since NRU does not have a constant matrix, and for the sake of simplicity, LTC positions are included in $\mathbf{x}_c^k$ along with the corresponding voltage set-point constraint in $\mathbf{f}_c^k$. Another benefit of this approach is that it meshes naturally with CYME's "infinite taps" mode, where LTC positions are considered as continuous variables. This mode allows planners to overcome the well-known problem of the multiplicity of load flow solutions due to LTC bandwidths, which cannot guarantee a worst-case solution. By using the lower bandwidth value as the voltage set point along with continuous LTC positions, planners can obtain under-voltages that are lower or equal to the worst physically possible under-voltage, and thus plan their network safely and reliably. This approach can be easily tweaked to study over-voltages arising due to high PV penetration and reverse power flow.

Since many distribution engineers require feasible (discrete) LTC positions, a normalization scheme is also needed. A "step discretization" approach was first developed in NRU. Based on the fact that downstream voltage regulators/transformers have little impact on the upstream voltage, after initial convergence, LTCs were discretized one by one following a downstream path. The main disadvantages of this approach are twofold. First, it is numerically costly for networks with many LTCs, since two extra iterations are usually needed per LTC. Second, for meshed networks, the discretization order is not always obvious. To overcome these problems, this paper presents a "joint discretization" approach. After the first convergence, all LTCs are normalized at once. Due to the concurrent discretization, the controlled voltage of some LTCs may end up outside of the specified bandwidth. A detection-correction step has been devised. It checks if all constraints are respected. When it is not the case, the corresponding LTC positions are changed by ±1. This is repeated until all constraints are respected. A break condition is added if hunting is detected, but this is unlikely on real customer networks. Finally, the transformer LTC position (primary or secondary) and the regulator operating mode (e.g., bi-directional, co-generation [19], …) must be considered to choose whether the LTC positions must be incremented or decremented to move towards the bandwidth.

E. Solution of Linear Systems

Roughly speaking, efficient numerical solution of (4) or (5) using sparse methods includes permuting and pre-ordering the unsymmetric indefinite Jacobian matrix to reduce fill-ins (symbolic factorization), calculating the L and U factors (numerical factorization), and solving using backward-forward substitution. When considering matrix-based load flows, it has become almost axiomatic to say that the solution of the linear equations is the "main computational bottleneck [11]". Consequently, since factorization is costlier than backward-forward substitution, it is pretty much accepted that methods with constant matrices (therefore requiring only one factorization) will be more efficient than load flows with changing matrices, unless they require significantly more iterations. For instance, in the 1970s, the Fast Decoupled formulation was shown to be around 5 times faster per iteration than the full Newton-Raphson formulation [16].

However, in view of today's highly efficient linear solvers such as KLU [12] and PARDISO [11], and based on our practical experience, we make the two following claims: 1) in complex commercial-grade packages such as CYME, the percentage of time of a full Newton-Raphson load flow spent on factorization is relatively small (this will be quantified in Section III-D); and 2) numerical factorization is computationally cheap. Since the same symbolic factorization can be reused from iteration to iteration when the matrix non-zero pattern remains constant, based on the 2nd claim, a full Newton-Raphson approach can nowadays be competitive with constant-matrix methods, and sometimes even faster if it requires fewer iterations. This is especially true when care is taken to avoid symbolic factorizations as much as possible, for instance by discretizing all taps jointly as explained in Section II-D. In that approach, the corrective steps do not require additional symbolic factorizations since only the numerical values change.

PARDISO [11], which is included in Intel MKL, was initially used as NRU's solver. It is a general-purpose package that calls different solvers depending on the matrix properties (e.g., real or complex numbers, symmetric or unsymmetric, positive (semi-)definite or indefinite, …). This is an interesting feature for a wide-ranging distribution system analysis program such as CYME, as the matrices generated by different modules often have different properties. KLU [12] was recently tested as it is designed for circuit matrices, which, according to [12], are extremely sparse. This is especially true of the Jacobian of radial systems, but less so for urban meshed networks. The performance of PARDISO and KLU for different test systems are presented in Section III-D.

III. CASE STUDIES

The numerical performance of NRU for each pair of proposed initialization methods, discretization techniques, and solvers are tested on 6 different networks [20]–[21]. The salient features of these test systems are presented in Table I. In particular, the IEEE 342 and Util1 test systems are urban meshed networks, the latter being extremely large for distribution systems. In addition to having several inline voltage regulators, the Util2 and Util3 test systems also have high PV penetration. The number of nodes reported in Table I differ from those of the official IEEE test feeders; this is due to a different (and inconsequential) way of counting nodes. Moreover, to provide additional insight, the detailed CPU timings for one of these systems is provided, and a comparison between the numerical efficiency of NRU and VDU serves as a closing statement.

All tests are executed on a special version of CYME 7.2 used in the course of the Department of Energy-sponsored "Rapid QSTS Project". This version runs on a dedicated benchmark machine (3.4 GHz i7-2600 CPU with 8 GB of RAM). A tolerance of 0.01% on the voltage mismatch is used as the stopping criterion.

A. Initialization

NRU's numerical efficiency using the FP and FS initialization methods is presented in Table II for the subject test systems (all LTCs are operating in "infinite taps" mode and PARDISO is used). The CPU time is presented in normalized form with respect to the FP approach. For two networks, an additional iteration is required with the FS approach; whereas for the other four, the iteration count is identical with both initialization procedures. This is explained by the fact that the FP approach typically yields a state vector that is closer to the final solution. This can be observed in Fig. 1, where the voltage mismatch of four of the test systems is plotted as a function of the iteration. Fig. 1 also demonstrates that both approaches converge almost quadratically. Only the IEEE 342 system converges slower near the end, due to its ungrounded nature and the resulting ill-conditioned matrix. Whether an additional iteration is required or not, the solution time remains smaller with the FS method, ranging from 83% to 92.7% of the CPU time of the FP approach.

TABLE I
MAIN PROPERTIES OF THE TEST SYSTEMS.

Network	Nb. Nodes	Nb. Loops	Nb. LTCs
CKT5 [9]	3003	0	0
IEEE 342 [20]	390	71	0
IEEE 8500 [21]	4875	0	4
Util1	28425	3180	14
Util2	2370	1	10
Util3	4066	0	14

TABLE II
NUMBER OF ITERATIONS AND NORMALIZED CPU TIME FOR A LOAD FLOW SOLUTION USING FIXED POINT AND FLAT START INITIALIZATION.

Network	Fixed Point Init.		Flat Start Init.	
	Nb. Iter.	CPU Time	Nb. Iter.	CPU Time
CKT5	2	100%	3	92.7%
IEEE 342	2	100%	2	83.0%
IEEE 8500	3	100%	4	90.6%
Util1	5	100%	5	90.6%
Util2	3	100%	3	86.8%
Util3	3	100%	3	88.9%

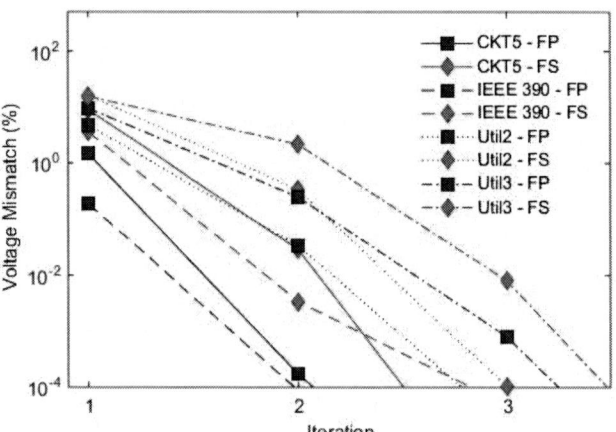

Fig. 1. Convergence pattern of four of the test systems using the fixed point (FP – blue square markers) and flat start (FS – red diamonds markers) initialization approaches.

To verify the generalization of these results, the NRU load flow is executed on all the distribution systems of an entire region of a large utility (657 feeders for a total of 73 independent networks). The majority of these networks are fed by Yg-Delta transformers with a grounding transformer on the secondary side, and the peak load scenario is considered. The combination of these two conditions often causes problems to distribution load flow engines (in fact, 2 of the networks do not converge after 200 iterations when using CYME's robust VDU engine with a relaxed tolerance of 0.1%). With NRU, all networks converge independently of the initialization approach. The FS approach requires an extra iteration for 50 networks; whereas the same number of iterations is needed for the other 23.

TABLE III
NUMBER OF ITERATIONS AND NORMALIZED CPU TIME FOR A LOAD FLOW SOLUTION USING STEP AND JOINT LTC POSITION DISCRETIZATION.

	Step Discret.		Joint Discret.	
Network	Nb. Iter.	CPU Time	Nb. Iter.	CPU Time
IEEE 8500	12	100%	9	87.2%
Util1	28	100%	7	31.5%
Util2	23	100%	5	43.2%
Util3	30	100%	7	42.4%

TABLE IV
NUMBER OF ITERATIONS AND NORMALIZED CPU TIME FOR A POWER FLOW SOLUTION USING THE PARDISO AND KLU SOLVERS.

	PARDISO		KLU	
Network	Nb. Iter.	CPU Time	Nb. Iter.	CPU Time
CKT5	3	100%	3	89.2%
IEEE 342	2	100%	2	92.8%
IEEE 8500	9	100%	9	74.2%
Util1	7	100%	7	225.4%
Util2	5	100%	5	75.8%
Util3	7	100%	7	73.4%

TABLE V
DETAILED CPU TIMINGS OF AN NRU LOAD FLOW SOLUTION.

Total Load Flow	58.61 ms
Create Segment	19.90 ms
Data Exchange (CYME → NRU)	1.57 ms
Initialization (build Jacobian)	14.79 ms
Iteration #1	13.87 ms
Update Jacobian	0.494 ms
Create Vector ($\mathbf{b}^k - \mathbf{f}^k$)	1.162 ms
Symbolic Factorization	9.081 ms
Numerical Factorization	2.879 ms
Solve	0.164 ms
Convergence Check	0.080 ms
Iteration #2	4.93 ms
Update Jacobian	0.527 ms
Create Vector ($\mathbf{b}^k - \mathbf{f}^k$)	1.182 ms
Numerical Factorization	2.990 ms
Solve	0.169 ms
Convergence Check	0.056 ms
Get Results (NRU → CYME)	2.821 ms

B. Discretization

The number of iterations and normalized CPU time for the test systems using the step and joint LTC discretization approaches are summarized in Table III (the FS method is used along with PARDISO). The results for CKT5 and IEEE 342 are not presented as they do not have LTCs. As expected, the iteration count and CPU times are reduced with the joint discretization scheme; this reduction is drastic for networks with several LTCs (e.g., 7 iterations instead of 28 for Util1, requiring only 31.5% of the original CPU time). As each discretization usually results in 2 (sometimes 3) additional iterations, it can be seen by comparing Tables II and III that the first joint discretization sometimes resulted in controlled voltages outside of the specified bandwidth, and that corrective tap changes were needed to find an adequate solution.

C. Solvers

The numerical performance of NRU using the PARDISO and KLU linear solvers are presented in Table IV for the 6 test systems (the FS and joint discretization techniques are used). As it is optimized for extremely sparse circuit matrices, KLU outperforms PARDISO on 5 of the 6 systems; however, for the very large urban meshed network, PARDISO is more than twofold faster than KLU. It is therefore suggested to use KLU for radial and weakly meshed systems, and PARDISO for highly meshed networks. A criterion based on the loop-to-node ratio can be added to the code to automatically select the preferred linear solver for any given system.

D. Detailed Timings

To better assess the numerical performance of NRU, several timers were added to the major code sections. The resulting CPU timings for the IEEE 342 test system with PARDISO are presented in Table V. The "Create Segment" part of the code consists of reading and treating the network file: it accounts for around a third of the CPU time. Creation of the matrix is also non-negligible, as it requires approximately 1/4th of the total CPU time, which is almost the same as for symbolic and numerical factorizations combined. This contradicts the well-known idea that matrix factorization is the main computational bottleneck in load flow calculations. Numerical factorization is also about 3 times faster than symbolic factorization. Due to this, the 2nd iteration accounts for less than 10% of the CPU time. While these exact ratios depend on the test system and the solver, they clearly demonstrate that a full Newton-Raphson approach is not much slower than constant-matrix solutions.

E. Comparison Between NRU and VDU

Finally, the CPU times necessary to solve the load flows of the six test systems with CYME's NRU and VDU engines are presented in Table VI. NRU is initialized using FS, the LTCs are discretized jointly, and either KLU or PARDISO is used, based on the criterion explained in Section III–C. It is observed that VDU does not converge on 3 of the 6 systems. The IEEE 342 and Util1 test systems diverge due to their highly meshed structure; whereas hunting between the 14 LTCs occurs when trying to solve the Util3 network. It is emphasized that Table VI is not meant as the definitive comparison between matrix-based and ladder iterative methods, but simply as a comparison between two optimized

978-1-5090-5606-4/17 $31.00 © 2017 IEEE

TABLE VI
NUMBER OF ITERATIONS AND NORMALIZED CPU TIME USING THE NRU AND VDU LOAD FLOW SOLVERS (THE DASH (–) INDICATES DIVERGENCE).

	NRU		VDU	
Network	Nb. Iter.	CPU Time	Nb. Iter.	CPU Time
CKT5	3	0.174 s	4	0.142 s
IEEE 342	3	0.122 s	–	–
IEEE 8500	9	0.531 s	19	0.733 s
Util1	5	5.400 s	–	–
Util2	5	0.211 s	13	0.205 s
Util3	8	0.298 s	–	–

commercial-grade distribution load flow solvers. As previously demonstrated in [6] in different environments, the cost per iteration is smaller for VDU than NRU; whereas the former typically requires fewer iterations. In particular, NRU converges in much fewer iterations than VDU on highly loaded systems with several controls, such as the IEEE 8500 system. Overall, the computational cost of NRU is absolutely competitive with VDU; in fact, NRU is even faster than VDU on some systems.

IV. CONCLUSION

This paper presents the general structure of a commercial-grade load flow engine suitable for distribution systems of all topologies, called Newton-Raphson Unbalanced (NRU). Three often overlooked aspects of distribution load flow calculations are discussed in detail: initialization, calculation and discretization of tap changer positions, and solution of the resulting linear equations. Computer studies on multiple distribution systems support the design choices made regarding the three aforementioned aspects. They also demonstrate that the versatile NRU engine is very competitive in terms of CPU time with a topology-limited ladder iterative engine. As a consequence of its high numerical efficiency, NRU is well suited to meet the increasing demand of distribution system load flow solutions arising from higher levels of PV penetration.

ACKNOWLEDGEMENT

This research was supported in part by the DOE SunShot Initiative, under agreement 30691. Sandia National Laboratories is a multimission laboratory managed and operated by National Technology and Engineering Solutions of Sandia, LLC., a wholly owned subsidiary of Honeywell International, Inc., for the U.S. Department of Energy's National Nuclear Security Administration under contract DE-NA0003525.

REFERENCES

[1] R. F. Arritt and R. C. Dugan, "Distribution system analysis and the future smart grid," *IEEE Trans. Ind. Appl.*, vol. 47, no. 6, pp. 2343–2350, Nov./Dec. 2011.

[2] M. J Reno, J. Deboever, and B. Mather, "Motivation and requirements for quasi-static time series (QSTS) for distribution system analysis," to appear in *Proc. IEEE Power Energy Soc. General Meeting*, 2017.

[3] D. Shirmohammadi, H. W. Hong, A. Semlyen, and G. X. Luo, "A compensation-based power flow method for weakly meshed distribution and transmission networks," *IEEE Trans. Power Syst.*, vol. 3, no. 2, pp. 753-762, May 1988.

[4] C. S. Cheng and D. Shirmohammadi, "A three-phase power flow method for real-time distribution system analysis," *IEEE Trans. Power Syst.*, vol. 10, no. 2, pp. 671-679, May 1995.

[5] P. A. N. Garcia, "Three-phase power flow calculations using the current injection method," *IEEE Trans. Power Syst.*, vol. 15, no. 2, pp. 508-514, May 2000.

[6] L. R. de Araujo *et al.*, "Comparisons between the three-phase current injection method and the forward/backward sweep method," *Elect. Power Energy Syst.*, vol. 32, pp. 825-833, 2010.

[7] W. Xu, J. R. Martí, and H. W. Dommel, "A multiphase harmonic load flow solution technique," *IEEE Trans. Power Syst.*, vol. 6, no. 1, pp. 174-182, Feb. 1991.

[8] I. Kocar *et al.*, "Multiphase load-flow solution for large-scale distribution systems using MANA," *IEEE Trans. Power Del.*, vol. 29, no. 2, pp. 908-915, Apr. 2014.

[9] R. C. Dugan, T. E. McDermott, "An open source platform for collaborating on smart grid research," in *Proc. IEEE Power Energy Soc. General Meeting*, 2011.

[10] I. Dzafic, R. A. Jabr, E. Halilovic, and B. C. Pal, "A sensitivity approach to model local voltage controllers in distribution networks," *IEEE Trans. Power Syst.*, vol. 29, no. 3, pp. 1419-1428, May 2014.

[11] O. Schenk and K. Gärtner, "Solving unsymmetric sparse systems of linear equations with PARDISO," *Future Generation Comp. Syst.*, vol. 20, no. 3, pp. 475-487, 2004.

[12] T. A. Davis and E. Palamadai Natarajan, "Algorithm 907: KLU, a direct sparse solver for circuit simulation problems," *ACM Trans. Math. Softw*, vol. 37, no. 3, Sep. 2010.

[13] I. Kocar, J.-S. Lacroix, and F. Therrien, "General and simplified computation of fault flow and contribution of distributed sources in unbalanced distribution networks," in *Proc. IEEE Power Energy Soc. General Meeting*, 2012.

[14] I. Kocar and J.-S. Lacroix, "Implementation of a modified augmented nodal analysis based transformer model into the backward forward sweep solver," *IEEE Trans. Power Syst.*, vol. 27, no. 2, pp. 663-670, May 2012.

[15] T. A. Davis, *Direct Methods for Sparse Linear Systems*, Philadelphia, PA: SIAM, 2006.

[16] B. Stott, "Review of load-flow calculations," *Proc. IEEE*, vol. 62, no. 7, pp. 916-929, Jul. 1974.

[17] B. Stott and O. Alsaç, "Fast decoupled load flow," *IEEE Trans. Power App. Syst.*, vol. PAS-93, no. 3, pp. 859-869, 1974.

[18] S.-K. Chang and V. Brandwajn, "Adjusted solutions in fact decoupled load flow," IEEE Trans. Power Syst., vol. 3, no. 2, pp. 726-733, May 1988.

[19] Cooper Power Systems, *S225-10-10: Voltage Regulators*, Oct. 2001.

[20] K. Schneider, P. Phanivong, and J.-S. Lacroix, "IEEE 342-node low voltage networked test system," in *Proc. IEEE Power Energy Soc. General Meeting*, 2014.

[21] R. F. Arritt and R. C. Dugan, "The IEEE 8500-node test feeder," in *Proc. IEEE Power Energy Soc. T&D*, 2010.

Resonant and non-resonant dielectric coatings for high efficiency solar cells

Dongheon Ha[1,2,3], Chen Gong[3,4], Marina S. Leite[3,4], and Jeremy N. Munday[3,5]

[1] Center for Nanoscale Science and Technology, National Institute of Standards and Technology, Gaithersburg, Maryland, 20899, United States

[2] Maryland Nanocenter, University of Maryland, College Park, Maryland, 20742, United States

[3] Institute for Research in Electronics and Applied Physics, University of Maryland, College Park, Maryland, 20742, United States

[4] Department of Materials Science and Engineering, University of Maryland, College Park, Maryland, 20742, United States

[5] Department of Electrical and Computer Engineering, University of Maryland, College Park, Maryland, 20742, United States

Abstract — **Micro-/Nanoscale surface patterning has been extensively applied to solar cells to increase absorptivity and hence photocurrent. However, this approach could also be detrimental to solar cells' performance because of the enhanced surface recombination arising from increased surface area, which can decrease the open-circuit voltage. In this regard, we introduce two new classes of antireflection coatings (ARCs) based on dielectric materials that do not require surface patterning on active materials: a transparent, flexible paper that relies on non-resonant optical scattering and silicon dioxide (SiO_2) nanosphere arrays that rely on collective optical resonances. For the transparent paper-based ARCs, there is reduced reflection on a Si cell for all wavelengths due to a combination of index contrast and surface texturing. It is also shown that power conversion efficiencies are improved for all angles of incident illumination, which reduce the need for expensive tracking of the sun's position. For the SiO_2 nanosphere array-based ARCs, we show that there is a high absorptivity enhancement (> 20%) on a Si substrate. From experiments and finite-difference time-domain (FDTD) calculations, we confirm that the absorptivity enhancement is due to the combinational effects of whispering gallery-like mode excitation within the nanospheres and thin-film interference effects. We also show that the wavelengths for whispering gallery-like resonance excitation can be tuned by either changing the diameter of the nanospheres or using other materials instead of SiO_2 (*e.g.*, silicon nitride (Si_3N_4), titanium dioxide (TiO_2), etc.). Because these dielectric ARCs can be made with an easy and scalable processes, they are excellent candidates to replace conventional thin-film-based antireflection coatings that require complicated high-temperature vacuum deposition processes.**

I. INTRODUCTION

To increase light absorption, many techniques have been applied including single-/multiple-layer thin-film antireflection coatings [1]-[2], plasmonic effects [3]-[5], surface patterning [6]-[8], dielectric micro-/nanostructures [9]-[13], etc.

However, some of these techniques are not ideal options for mass production, as they require expensive and high-temperature fabrication processes. Thus, there is a need to develop more efficient and less expensive antireflection coatings.

Recently, cellulose paper has attracted significant attention for its possible application in optoelectronic devices due to its cost-effectiveness, ease of mass production, flexibility, natural abundance and environmental friendliness. However, opaqueness, resulting from the porous structure of cellulose paper, has delayed its incorporation into industrial optoelectronic applications. To increase transparency, special papermaking techniques (*e.g.*, as a 2,2,6,6-Tetramethylpiperidine-1-oxyl radical (TEMPO) oxidation [14]-[15]) have been applied. The transparent papers have been favorably applied to photovoltaics as ARCs, which increase light absorption and power conversion efficiency [11]-[12].

Due to light trapping and scattering properties of dielectric nanosphere (*e.g.*, silicon dioxide (SiO_2)) arrays, they have been applied on semiconductors to make high efficiency solar cells [9]-[10], [13]. Moreover, a monolayer of dielectric SiO_2 nanosphere arrays is an appealing ARC option as it can be stable with respect to external conditions (*e.g.*, temperature) and is earth abundant. Various techniques including the Langmuir−Blodgett method [16], sedimentation [17], and controlled evaporation [18] have been applied to achieve a monolayer of dielectric nanospheres on top of either flexible or rigid substrates. However, in general, these techniques are not compatible with roll-to-roll processes for mass production. To overcome this problem, we use the Meyer rod rolling technique to deposit a monolayer of SiO_2 nanospheres.

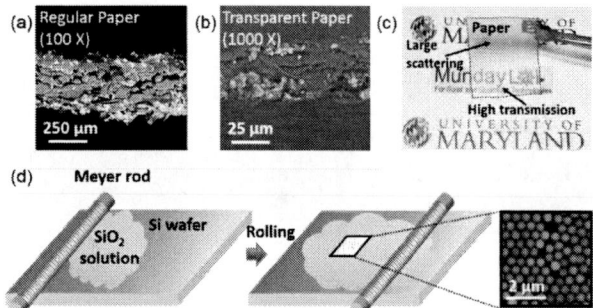

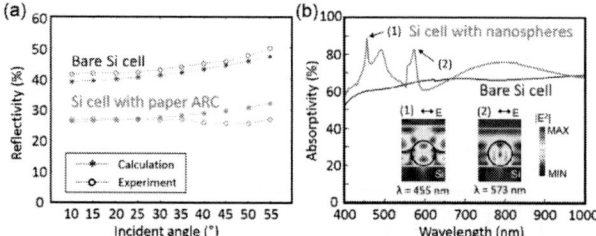

Fig. 1. SEM image of a piece of (a) regular paper and (b) transparent paper used in this study. (c) The transparent paper has good optical characteristics (*i.e.*, high transmission and large scattering) for photovoltaics. (d) Schematic showing nanosphere coating process using a Meyer rod. Inset SEM image shows deposited 500 nm SiO_2 nanosphere arrays on a Si substrate.

In this study, we demonstrate that both paper-based non-resonant ARCs and SiO_2 nanosphere array-based resonant ARCs can lead to increased photocurrent in an underlying solar cell.

II. RESULTS AND DISCUSSION

A. Material properties and fabrication processes

Figure 1(a), (b) show scanning electron microscope (SEM) images of a piece of regular paper and transparent paper, respectively. The TEMPO oxidation undermines the hydrogen bonds between the cellulose fibrils, which results in higher packing density compared to a regular paper and hence high transmittance. Once light travels the transparent paper, it will be scattered broadly due to surface texturing of the transparent paper. Figure 1(c) confirms these two important optical characteristics (*i.e.*, high transmission and large scattering) obtained with this transparent paper. For use as an ARC, this paper is attached to solar cells with a binding material (*e.g.*, polyvinyl acetate) at room temperature.

Figure 1(d) shows the coating process to make a monolayer of SiO_2 nanosphere arrays on a Si wafer. Once a droplet of suspension containing 500 nm SiO_2 nanospheres is placed onto the substrate, a Meyer rod is pulled along the Si surface. For water evaporation, the sample is then annealed at 50 °C for 1 min. A scanning electron microscope (SEM) image in Fig. 1 shows the deposited monolayer of SiO_2 nanospheres on the Si substrate. The mean diameter of the nanospheres is 502 nm with a standard variation of 75 nm.

B. Absorption enhancement

The transparent paper-based non-resonant ARC is applied to a Si substrate. Optical properties are determined from experiments and compared to calculations. Figure 2(a) shows measured and calculated reflection for a bare Si cell and a Si

Fig. 2. (a) Measured (circles with dotted line) and calculated (asterisks with dotted line lines) reflectivity for both the bare Si cell (blue) and the Si cell with the transparent paper ARC (red) for λ = 500 nm. (b) Spectrally-resolved calculated absorptivity for the bare Si cell (blue) and the Si cell with the SiO_2 nanosphere arrays on top (red). Inset (1) and (2) show electric field profiles for resonant peaks (λ = 455 nm and 573 nm, respectively) under transverse magnetic field incident polarization. Black lines in insets indicate SiO_2 nanospheres and white lines show an interface between SiO_2 nanospheres and a Si substrate.

cell with a paper ARC as a function of incident angle at a wavelength of 500 nm. Our optical model considering multiple reflections between layers due to index contrast depicts the experimental results very well for a bare Si cell. Optical constants required for calculations are obtained from Ref. [19]. For a cell with the transparent paper ARC, however, our simple model slightly overestimates experimental results at higher incident angles because the actual sample experiences scattering due to surface texturing of the transparent paper ARC as well as index contrast between layers. This surface texturing enables angle insensitive behavior, which may reduce the need for solar trackers, as confirmed by Fig. 2 (a).

The resonant type SiO_2 nanosphere array-based ARC is applied on a Si substrate and absorptivity is determined (Fig. 2(b)). Based on FDTD calculations, increased light absorption throughout the solar spectrum is observed. The deposited SiO_2 nanospheres effectively constitute a thin-film layer with a thickness of 500 nm, and this explains the broadband absorption enhancement at wavelengths of around 700-900 nm. However, the monolayer of SiO_2 nanospheres also leads to narrowband absorption enhancements, marked with numbers in Fig. 2 (b), in addition to the broadband absorption enhancements. To explain the narrowband enhancements, electric field intensities for transverse magnetic incident polarization are shown in insets of Fig. 2(b) for each marked number. The electric field profile at point 1 (λ = 455 nm at normal incidence) and point 2 (λ = 573 nm at normal incidence) show the excitation of whispering gallery-like modes inside the nanosphere. The close proximity of SiO_2 nanospheres makes excited whispering gallery-like modes couple with each other, which increases the electric field strength within the periodic array of SiO_2 nanospheres. The enhanced fields then couple into the absorbing material (*i.e.*, a Si substrate) that ultimately leads to absorptivity enhancement.

TABLE I

SUMMARY OF AM1.5G SOLAR SPECTRUM-WEIGHTED PHOTOCURRENT ENHANCEMENTS

Antireflective materials on Si cell	Solar spectrum-weighted photocurrent density	Photocurrent density enhancement
Bare Si cell, no antireflection coating	27.7 mA/cm^2	-
500 nm SiO$_2$ nanosphere arrays	29.2 mA/cm^2	5.4%
700 nm SiO$_2$ nanosphere arrays	30.2 mA/cm^2	9.0%
1000 nm SiO$_2$ nanosphere arrays	30.9 mA/cm^2	11.6%
500 nm Si$_3$N$_4$ nanosphere arrays	34.9 mA/cm^2	26.0%
500 nm TiO$_2$ nanosphere arrays	29.5 mA/cm^2	6.5%

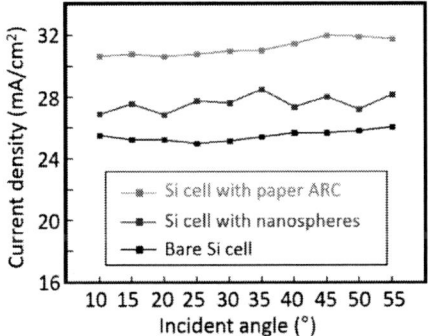

Fig. 3. AM 1.5G solar spectrum-weighted current density for a bare Si cell, a Si cell with transparent paper-based non-resonant ARC (red), and a Si cell with SiO$_2$ nanosphere array-based resonant ARC (blue).

C. Photocurrent enhancement

To investigate whether increased light absorption directly contributes to improved optoelectronic response, we determine the AM1.5G solar spectrum-weighted photocurrent densities. The photocurrent density is determined based on the measured light absorption within the cells. We assume unity internal quantum efficiency and use the measured absorptivity to determine the external quantum efficiency of solar cells. As shown in Fig. 3, both non-resonant and resonant ARCs show enhanced photocurrent density over a large range of incident illumination angles. However, the enhancement is larger with the non-resonant, paper-based ARC for the parameters considered in this manuscript. To confirm these results, we deposit this paper-based ARC on a GaAs solar cell as well, and a ~24% enhancement in the power conversion efficiency is experimentally observed [11].

D. Further improvement of ARCs by resonance tuning

As the solar spectrum has irregular spectral irradiance, absorptivity enhancement peaks from whispering gallery-like mode excitation should correspond to the solar spectrum with high irradiant wavelength ranges to ensure high power conversion efficiency. This necessitates tuning the resonance peaks as needed. This tuning can be achieved by either varying

the diameter or changing the material of the nanospheres [20]-[21]. As the diameter of nanospheres increases, the whispering gallery-like resonances shift to longer wavelengths because the electric field within nanosphere arrays experiences a higher effective index of refraction. This results in a larger periodicity for electromagnetic waves within the layer and a redshift in resonances. Or, alternatively, we can change the refractive index of materials while keeping the same diameter.

To test how the size and material of the nanosphere arrays affect resonances and thus photocurrent, we calculate the AM1.5G solar spectrum-weighted photocurrent density for a Si solar cell with 500 nm, 700 nm, 1000 nm SiO$_2$, 500 nm Si$_3$N$_4$, and 500 nm TiO$_2$ nanosphere array-based ARCs. Table 1 summarizes results for all structures. Resonance tuning can be made by either changing the diameter or material of the nanospheres, but using a close-packed 500 nm Si$_3$N$_4$ nanosphere arrays brings the highest absorptivity enhancement over the entire spectrum. This is because Si$_3$N$_4$ has the optimum refractive index for a single layer of ARC to be placed between air and a Si solar cell.

III. CONCLUSIONS

In conclusion, we introduced two different types of novel dielectric antireflection coatings that improve light absorption and the power conversion efficiency of solar cells. Transparent paper-based non-resonant ARCs showed improved optical response from solar cells independent of incident illumination angle. From experiments and calculations, we attributed the enhancement to the index contrast between layers and surface texturing of the transparent paper ARC. SiO$_2$ dielectric nanosphere array-based resonant ARCs improved light absorption and the power conversion efficiency of a Si solar cell due to combinational effects of whispering gallery-like resonance excitation and thin-film interference. We also showed that whispering-gallery like resonances can be tuned simply by changing either the diameter or material of the nanospheres. Because these dielectric ARCs can be deposited with an easy, scalable process, they may be able to replace conventional thin-film-based ARC technologies that require expensive, high-temperature, and complicated vacuum processes.

978-1-5090-5606-4/17 $31.00 © 2017 IEEE

ACKNOWLEDGEMENT

D. Ha acknowledges support under the Cooperative Research Agreement between the University of Maryland and the Center for Nanoscale Science and Technology at the National Institute of Standards and Technology, Award 70NANB14H209, through the University of Maryland.

REFERENCES

[1] J. Zhao and M. A. Green, "Optimized antireflection coatings for high-efficiency silicon solar cells," *IEEE Trans. Electron Devices*, vol. 38, pp. 1925−1934, 1991.

[2] J. Zhao, A. Wang, P. Altermatt, and M. A. Green, "Twenty-four percent efficient silicon solar cells with double layer antireflection coatings and reduced resistance loss," *Appl. Phys. Lett.*, vol. 66, pp. 3636−3638, 1995.

[3] V. E. Ferry, J. N. Munday, and H. A. Atwater, "Design considerations for plasmonic photovoltaics," *Adv. Mater.*, vol. 22, pp. 4794−4808, 2010.

[4] H. A. Atwater and A. Polman, "Plasmonics for improved photovoltaic devices," *Nat. Mater.*, vol. 9, pp. 205−213, 2010.

[5] J. N. Munday and H. A. Atwater, "Large integrated absorption enhancement in plasmonic solar cells by combining metallic gratings and antireflection coatings," *Nano Lett.*, vol. 11, pp. 2195−2201, 2011.

[6] J. Zhu, Z. Yu, G. F. Burkhard, C. -M. Hsu, S. T. Connor, Y. Xu, Q. Wang, M. McGehee, S. Fan, and Y. Cui, "Optical absorption enhancement in amorphous silicon nanowire and nanocone arrays," *Nano Lett.*, vol. 9, pp. 279-282, 2009.

[7] K. X. Wang, Z. Yu, V. Liu, Y. Cui, and S. Fan, "Absorption enhancement in ultrathin crystalline silicon solar cells with antireflection and light-trapping nanocone gratings," *Nano Lett.*, vol. 12, pp. 1616-1619, 2012.

[8] J. Zhu, C. -M. Hsu, Z. Yu, S. Fan, and Y. Cui, "Nanodome solar cells with efficient light management and self-cleaning," *Nano Lett.*, vol. 10, pp. 1979-1984, 2010.

[9] J. Grandidier, D. M. Callahan, J. N. Munday, and H. A. Atwater, "Light absorption enhancement in thin-film solar cells using whispering gallery modes in dielectric nanospheres," *Adv. Mater.*, vol. 23, pp. 1272−1276, 2011.

[10] J. Grandidier, D. M. Callahan, J. N. Munday, and H. A. Atwater, "Gallium arsenide solar cell absorption enhancement using whispering gallery modes of dielectric nanospheres," *IEEE J. Photovoltaics*, vol. 2, pp. 123−128, 2012.

[11] D. Ha, Z. Fang, L. Hu, and J. N. Munday, "Paper-based anti-reflection coatings for photovoltaics," *Adv. Energy Mater.*, vol. 4, 1301804, 2014.

[12] D. Ha, J. Murray, Z. Fang, L. Hu, J. N. Munday, "Advanced broadband antireflection coatings based on cellulose microfiber paper," *IEEE J. Photovoltaics*, vol. 5, pp. 577−583, 2015.

[13] D. Ha, C. Gong, M. S. Leite, and J. N. Munday, "Demonstration of resonance coupling in scalable dielectric microresonator coatings for photovoltaics," *ACS Appl. Mater. Interfaces*, vol. 8, pp. 24536−24542, 2016.

[14] Z. Fang, H. Zhu, Y. Yuan, D. Ha, S. Zhu, C. Preston, Q. Chen, Y. Li, X. Han, S. Lee, G. Chen, T. Li, J. Munday, J. Huang, and L. Hu, "Novel nanostructured paper with ultrahigh transparency and ultrahigh haze for solar cells," *Nano Lett.*, vol. 14, pp. 765-773, 2014.

[15] H. Zhu, S. Parvinian, C. Preston, O. Vaaland, Z. Ruan, and L. Hu, "Transparent nanopaper with tailored optical properties," *Nanoscale*, vol. 5, pp. 3787-3392, 2013.

[16] C.-M. Hsu, S. T. Connor, M. X. Tang, and Y. Cui, "Wafer-scale silicon nanopillars and nanocones by langmuir−blodgett assembly and etching," *Appl. Phys. Lett.*, vol. 93, 133109, 2008.

[17] P. Jiang, J. F. Bertone, K. S. Hwang, and V. L. Colvin, "Single crystal colloidal multilayers of controlled thickness," *Chem. Mater.*, vol. 11, pp. 2132−2140, 1999.

[18] A. Mihi, C. Zhang, and P. V. Braun, "Transfer of preformed 3D photonic crystals onto dye sensitized solar cells," *Angew. Chem., Int. Ed.*, vol. 50, pp. 5712−5715, 2011.

[19] E. D. Palik, *Handbook of optical constants of solids*, 1st ed.; Academic Press: Cambridge, 1997.

[20] D. Ha, "Microscale dielectric anti-reflection coatings for photovoltaics," Ph.D. Thesis, The University of Maryland, College Park, MD, April 2016.

[21] D. Ha, Y. Yoon, J. N. Munday, and N. B. Zhitenev, "Nanoscale imaging of photocurrent enhancement by resonator array photovoltaic coatings," *in preparation*.

Enhanced Light Trapping in Thin Silicon Solar Cells using Effectively Transparent Contacts (ETCs)

Rebecca Saive[1], André Augusto[2], Stuart G. Bowden[2], and Harry A. Atwater[1]

[1] California Institute of Technology, Pasadena, CA 91125, USA
[2] Arizona State University, Tempe, AZ 85287-5706, USA

Abstract — **We report on the light trapping properties of densely spaced microscale triangular cross-section effectively transparent contacts (ETCs). Optical simulations of thin silicon with closely spaced ETCs were performed and the absorption within the silicon was determined. We found that with a metal coverage of 62.5% the short circuit current density would only be decreased by 0.25%. Such low value of parasitic reflection and transmission is obtained due to the excellent reflection properties of silver microstructures. In the interval 700 nm-1100 nm the short circuit current density is even enhanced by around 1 mA/cm2 compared to a structure without any metal contact. This results from nearly loss free in-coupling in combination with internal reflection at the bottom of the metal lines which increases the path length of light. These results are particularly interesting for 4-terminal silicon tandem devices as the bottom cell only receives the spectrum between around 850 nm and 1100 nm.**

Index Terms — **light trapping, photovoltaic cells, contacts, silicon.**

I. INTRODUCTION

The supply of clean, renewable energy a critical step towards preventing climate change. Photovoltaic (PV) devices can convert the abundant power of the sun into electrical energy and thus, deliver carbon dioxide emission free energy. Increasing the efficiency of solar cells will decrease the cost of electricity generated by photovoltaic power plants and lead to economically profitable large scale application. Silicon solar cells contribute more than 90% of the photovoltaic market. Decreasing the thickness of silicon solar cells introduces novel functionalities such as flexibility and light weight and could decrease the cost for PV even further due to less use of material and ease of installation. Furthermore, in silicon solar cells with excellent surface passivation- such as heterojunction solar cells[1]- bulk recombination plays an important role. Decreasing the thickness of the crystalline silicon can lead to decreased recombination and thus, increased open circuit voltage[2, 3] and efficiency. Silicon is a semiconductor with indirect band gab and hence, the absorption close to the band gap is weak. When decreasing the thickness of the absorber material one has to carefully implement light trapping strategies in order not to lose current from incomplete light absorption. A further optical loss mechanism is the reflection and parasitic absorption of photons at the front contact layers that most solar cells use for the lateral transport of charge carriers. In the case of silicon heterojunction solar cells a combination of a transparent conductive oxide (TCO) and metallic contact fingers provides low resistance for lateral transport. The TCO, usually indium tin oxide (ITO), absorbs a small fraction of the incoming light[4] while the metallic grid fingers reflect around 3-8% of the light equivalent to its surface coverage. Recently, we have demonstrated a new contact design that overcomes shadowing losses and parasitic absorption without decrementing the conductivity [5-9]. By redirecting the light to the active surface of the solar cell microscale triangular cross-section grid lines can perform as effectively transparent and highly conductive top contacts. We found up to 99.9% effective transparency in experiment [6] and demonstrated silicon heterojunction solar cells with 80.7% fill factor and 2 mA/cm^2 short circuit current enhancement [5]. These effectively transparent contacts (ETCs) show record transparency while providing low resistance for lateral charge transport. Furthermore, we showed that it is possible to only use very thin layers of ITO in combination with less absorbing antireflection coating[5] which further increases the current density. Here, we demonstrate that microscale triangular cross-section metal lines not only provide high effective transparency but also improve the light trapping in thin silicon when densely spaced. A schematic of such a thin silicon solar cell with densely spaced ETCs is shown in Fig. 1.

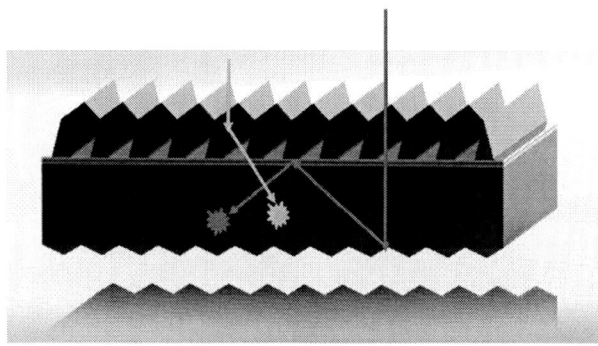

Fig. 1. Schematic description of the investigated device. Densely spaced triangular cross-section effectively transparent contacts provide low sheet resistance and replace transparent conductive oxides while redirecting photons efficiently to the active cell area (green light beam). Furthermore, the contacts provide light trapping through two mechanisms: Previously perpendicular light enters under a more oblique angle, enhancing the path length of light and facilitating total internal reflection. Internal reflection also becomes more likely due to reflection at the bottom of the contact lines.

978-1-5090-5606-4/17 $31.00 © 2017 IEEE

II. COMPUTATIONAL SIMULATIONS OF DENSELY SPACED TRIANGU-LAR CROSS-SECTION CONTACTS

We performed two-dimensional wave optical simulations of solar cells with microscale triangular cross-section metal lines using DifractMOD from RSoft, a solver for rigorous coupled wave analysis (RCWA). To determine the light trapping potential we first simulated 5 μm thin crystalline silicon as absorber material with an optimized TiO_2/SiO_2 antireflection coating and a silver back reflector with 100 nm SiO_2 spacer. Triangular cross-section lines with the optical properties of silver were varied in height, width and period. The overall absorption, absorption within the silicon and all other layers as well as the absorption within the triangular cross-section contacts were computed. Furthermore, the escape and reflection were determined. Note, that RCWA calculates the steady state optical configuration and therefore, reflection and escape are convoluted. For a 5 μm thin crystalline silicon absorber layer with material stack as described above, we simulated the six different geometries shown in Fig. 2. We found 4 μm as an optimal period with a triangle width of 2.5 μm and 5 μm height. The absorption within the silicon for all different geometries in the wavelength interval 850-1100 nm is shown in Fig. 3. A floating average was used to smoothen cavity resonances in the longer wavelength regime. The lowest absorption is obtained for case A with no texture on front and back and no densely spaced effectively transparent contacts (ETCs). Adding densely spaced ETCs increases the absorption as can be seen by case B. Furthermore, in all six cases adding densely spaced ETCs increases the absorption between 850 nm and 1100 nm. In order to quantify this result the absorption can be expressed by the short circuit current density that would be obtained if the solar cell had unity internal quantum efficiency and the absorption is weighted with the solar AM 1.5G spectrum. We obtain the following short circuit current densities

for the six different cases: A) 1.87 mA/cm², B) 2.88 mA/cm², C) 5.12 mA/cm², D) 6.18 mA/cm², E) 5.67 mA/cm², F) 5.99 mA/cm². It can be seen that in all cases adding densely spaced ETCs enhances the short circuit current density. Remarkably, replacing a textured front surface by densely spaced ETCs without texturing increases the short current density. It needs to be noted that the ETCs at the same time also act as lateral charge transport layer and therefore, replace metal grid fingers. If conventional e.g. screen printed metal contacts would be used for the lateral charge transport, the values of case A, C and F would be lowered by 3-6%. Furthermore, it is noteworthy that the short circuit current increases by around 1 mA/cm² for absorber with textured and non-textured rear surface respectively.

In order to demonstrate the light trapping properties of densely spaced ETCs under realistic conditions we investigated the optical properties of silicon heterojunction solar cells. To facilitate the comparison with experimental data we chose a 70 μm crystalline silicon absorber layer and non-textured rear and front surface. The simulated absorption within the crystalline silicon is shown in Fig. 4. For wavelengths between 350 nm and 850 nm no metal contact leads to the highest absorption within silicon. As expected, a state-of-the-art screen printed contact with 50 μm width and a period of 1250 μm performs worse than all other configurations over almost the whole simulated wavelength regime. ETCs with a height of 15 μm, 5 μm width and a period of 18 μm and 20 μm perform well for wavelength shorter than 850 nm and provide some light trapping beyond 850 nm. ETCs with a height of 7.5 μm, 2.5 μm width and a period of 6 μm and 8 μm perform better than a standard contact but not as well as the bigger ETCs for wavelength shorter than 850 nm. They improve light trapping significantly beyond 850 nm.

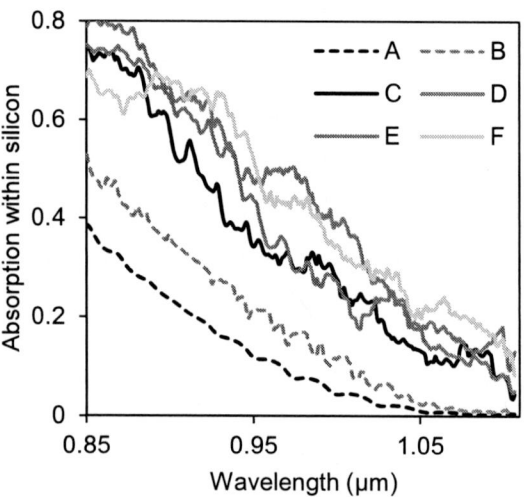

Fig. 3. Computationally simulated absorption within 5 μm silicon in the six different device geometries shown in Fig 2.

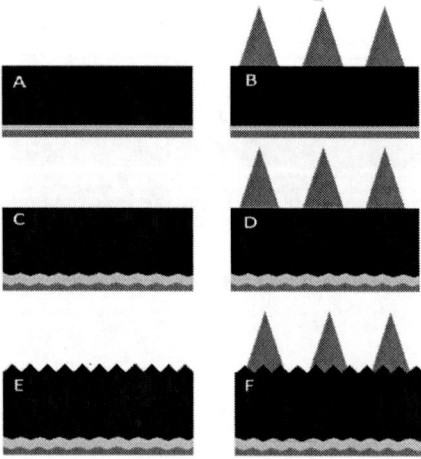

Fig. 2. Different device geometries used for computational optical simulations of 5 μm silicon absorbers with and without densely spaced effectively transparent contacts.

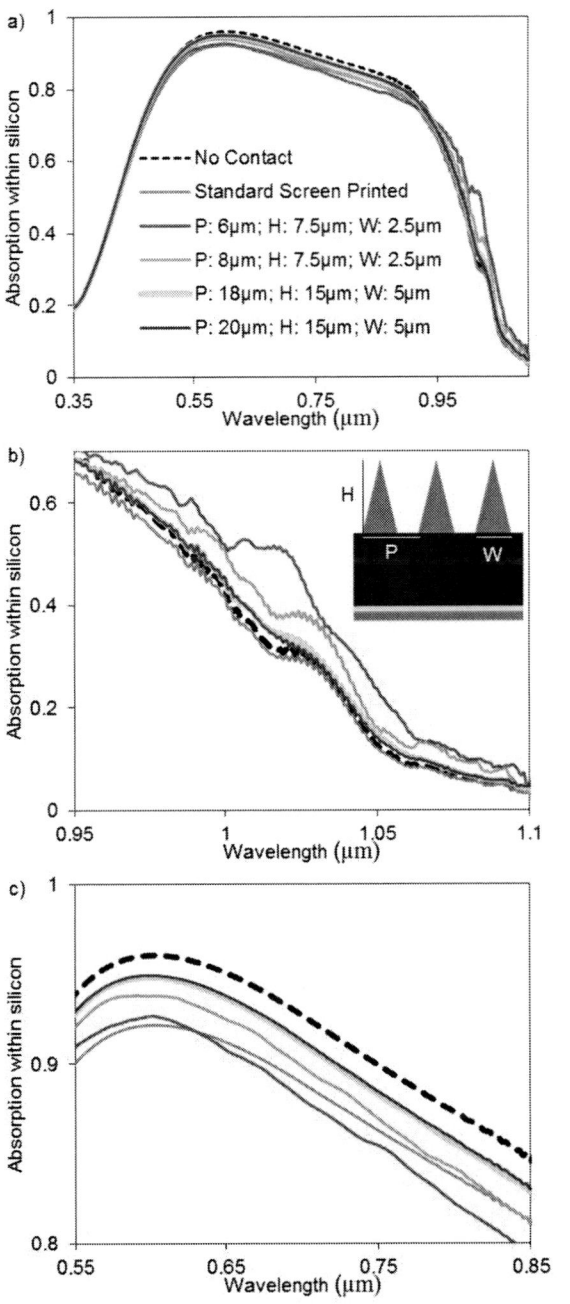

with the AM 1.5G spectrum and the equivalent short circuit current density is reported in mA/cm^2. Values written in bold present the wavelength interval 850-1100 nm. Regular font refers to the interval 350-1100 nm. The first column shows the absorption within the crystalline silicon layer while the third row presents the overall absorption of the full device and includes parasitic absorption within the amorphous silicon, ITO and metals. The table shows that it is important for all light trapping and transparent contact studies to calculate the light absorption within the absorber material. Considering only overall reflection or absorption properties can overestimate the performance of a light management strategy. As an example, the overall absorption of a structure with ETCs with a period of 6 μm, 2.5 μm width and 7.5 μm height even exceeds the overall absorption of a cell without any contact when weighted over the interval 350-1100 nm. Although such a design actually performs significantly better than a state-of-the-art screen printed contact, it does not exceed the absorption within the absorber material of a cell without metal contacts over the wavelength interval 350-1100 nm due to parasitic absorption within the metal. However, it does exceed the absorption in the interval 850-1100 nm as silver inhibits almost no parasitic absorption within this interval. Therefore, between 850 nm and 1100 nm ETCs are not only almost perfectly transparent but additionally increase the absorption within the crystalline silicon by light trapping. This property is of great value as significant effort is made to use silicon solar cells as bottom cells in tandem devices. As the shorter wavelengths are filtered by the top cell, only the wavelength regime around 850-1100 nm reaches the bottom cell. Therefore, the proposed contact scheme is particularly useful for silicon/III-V and silicon/perovskite tandem solar cells.

Fig. 4. Computationally simulated absorption within the crystalline silicon in a 70 μm silicon heterojunction solar cell. No metal contact, a standard screen printed contact and ETCs with different geometries were simulated.

To provide a more quantitative presentation, table 1 gives an overview of the absorption in each single layer as well as the reflection/escape. Again, the optical properties were weighted

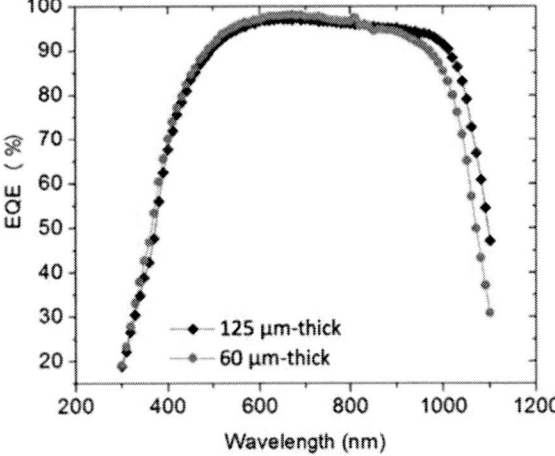

Fig. 5. Measured external quantum efficiency (EQE) of silicon heterojunction solar cells with 60 μm and 125 μm crystalline silicon thickness.

TABLE I

OVERVIEW OF THE ABSORPTION IN EACH SINGLE LAYER AND REFLECTION/ESCAPE OF A 70 μM THIN SILICON HETEROJUNCTION SOLAR CELL WITH NON-TEXTURED FRONT AND REAR SURFACE. THE OPTICAL PROPERTIES WERE WEIGHTED WITH THE AM 1.5G SPECTRUM AND THE EQUIVALENT SHORT CIRCUIT CURRENT DENSITY IS REPORTED IN MA/CM². VALUES WRITTEN IN BOLD PRESENT THE WAVELENGTH INTERVAL 850-1100 NM. REGULAR FONT REFERS TO THE INTERVAL 350-1100 NM. THE FIRST COLUMN SHOWS THE ABSORPTION WITHIN THE CRYSTALLINE SILICON LAYER WHILE THE THIRD ROW PRESENTS THE OVERALL ABSORPTION OF THE FULL DEVICE AND INCLUDES PARASITIC ABSORPTION WITHIN THE AMORPHOUS SILICON, ITO AND METALS.

	Absorption within silicon	Reflection /Escape	Absorption	Metal	a-Si bottom	ITO bottom	Ag bottom	a-Si top	ITO top
No Contact	**5.29**	**5.14**	**5.94**	**0.00**	**0.00**	**0.24**	**0.10**	**0.00**	**0.32**
No Contact	32.50	8.24	35.14	0.00	0.00	0.24	0.10	1.63	0.67
Standard Screen Printed	**5.07**	**5.11**	**5.98**	**0.28**	**0.00**	**0.23**	**0.09**	**0.00**	**0.31**
Standard Screen Printed	31.18	8.44	34.94	1.24	0.00	0.23	0.10	1.56	0.64
P: 6μm; H: 7.5μm; W: 2.5μm	**5.85**	**4.21**	**6.87**	**0.17**	**0.00**	**0.34**	**0.14**	**0.00**	**0.38**
P: 6μm; H: 7.5μm; W: 2.5μm	31.90	8.21	35.17	0.50	0.00	0.34	0.14	1.58	0.72
P: 8μm; H: 7.5μm; W: 2.5μm	**5.56**	**4.66**	**6.42**	**0.10**	**0.00**	**0.30**	**0.12**	**0.00**	**0.34**
P: 8μm; H: 7.5μm; W: 2.5μm	32.05	8.31	35.06	0.30	0.00	0.30	0.12	1.60	0.68
P: 18μm; H: 15μm; W: 5μm	**5.34**	**5.00**	**6.09**	**0.05**	**0.00**	**0.25**	**0.10**	**0.00**	**0.33**
P: 18μm; H: 15μm; W: 5μm	32.14	8.36	35.02	0.22	0.00	0.26	0.11	1.61	0.68
P: 20μm; H: 15μm; W: 5μm	**5.32**	**5.02**	**6.06**	**0.05**	**0.00**	**0.25**	**0.10**	**0.00**	**0.33**
P: 20μm; H: 15μm; W: 5μm	32.17	8.36	35.02	0.22	0.00	0.25	0.11	1.61	0.68

It is important to note that this is only true for the microscale contacts we report here as they interact with the longer wavelength light in a ray optical manner. Nanoscale metal contacts suffer from resonant coupling with the incoming light and therefore, show significantly higher reflection and absorption.

III. FABRICATION OF THIN SILICON HETEROJUNCTION SOLAR CELLS

First, we fabricated silicon heterojunction solar cells with conventional front contact layout with standard (125 μm) and thin (60 μm) crystalline silicon. The cells were prepared on five-inch n-type Czochralski wafers with 3-4 Ω.cm resistivity, <100> orientation, and initial thickness of 145 μm. The wafers were thinned to 65±5 μm and textured using alkaline wet etching, followed by wet chemical cleaning. The heterojunction was formed using plasma enhanced chemical vapor deposition (PECVD) to grow intrinsic and doped a-Si:H layers (7-10 nm) on both sides of the wafer. Indium tin oxide (ITO) was sputtered on both sides of the wafer, and silver on the rear as a mirror and rear contact. The front metallization was screen printed with silver paste[10]. Figure 3 shows the measured external quantum efficiency (EQE) for cells with 125 μm and 60 μm thickness. The EQE shows similar behavior for the major part but drops earlier for the thinner cell as expected from the weak absorption of silicon close to its band gap. The generation current density was found to be 39.7 mA/cm² and 38.7 mA/cm² respectively.

IV. FABRICATION SOLAR CELLS WITH DENSELY SPACED ETCS

We found from our computational study that texturing the front surface does not improve the light trapping when using densely spaced ETCs. Therefore, we also fabricated silicon heterojunction solar cells with the same parameters as de-

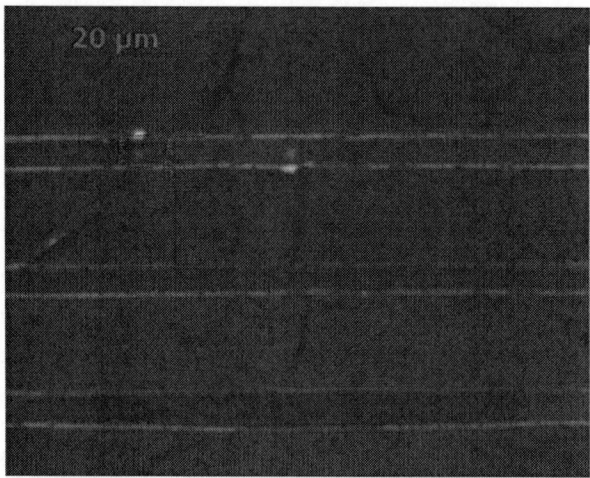

Fig. 6. Optical microscopy image of a 70 μm thin silicon heterojunction solar cell taken under 50 times magnification. ETCs with a period of 20 μm, height 15 μm and width 5 μm are printed on top. The ETCs appear black due to the redirection of incoming light.

978-1-5090-5606-4/17 $31.00 © 2017 IEEE

scribed above but left out the texturing step. We obtained heterojunction solar cells with a crystalline silicon thickness of 70 µm and no intentional texture. However, surfaces showed some texture due to the wafer thinning as can be seen in Fig. 6.

In the next step we fabricated different metal contacts using two-photon lithography and subsequent imprint lithography [5-7, 11]. Residue-free printing with silver ink was achieved using capillary action[7]. Figure 6 shows an optical microscopy image of a 70 µm silicon heterojunction solar cell taken under 50 times magnification. ETCs with a period of 20 µm, height 15 µm and width 5 µm are printed on top. The ETCs appear black due to the redirection of incoming light. Areas around it appear blue as the ITO thickness was optimized for ideal antireflection properties. It can be seen that residue-free printing of silver ink was achieved although the solar cell surface was not perfectly smooth.

V. CONCLUSION

Thin silicon solar cells have the potential to perform better than thicker ones [12] but show decreased EQE for photons close to the band gap. Therefore, we explored densely spaced microscale triangular cross-section effectively transparent contacts (ETCs) as light trapping structures for silicon solar cells. We computationally simulated the absorption within 5 µm silicon and 70 µm silicon heterojunction solar cell with different front contact geometries. We found around $1 mA/cm^2$ absorption enhancement in 5 µm thin silicon using ETCs with a 4.0 µm period, 5 µm height and 2.5 µm width in the wavelength interval 850-1100 nm. For a 70 µm silicon heterojunction solar cells an absorption enhancement of $0.7 mA/cm^2$ is simulated for ETCs with 6 µm period, 2.5 µm width and 7.5 µm height compared to a state-of-the-art screen printed contact between 850 nm and 1100 nm. This wavelength interval is important as it is the approximate spectrum that reaches the bottom cell in a silicon tandem cell (e.g. silicon/perovskite or silicon/III-V materials). Therefore, this concept is extremely useful for 4-terminal tandem devices.

ACKNOWLEDGEMENT

The authors thank Thomas Russell for helpful discussion. This material was based upon work supported by the U.S. Department of Energy through the Bay Area Photovoltaic Consortium under Award Number DE-EE0004946 and by the Engineering Research Center Program of the National Science Foundation and the Office of Energy Efficiency and Renewable Energy of the Department of Energy under NSF Cooperative Agreement No. EEC-1041895.

REFERENCES

[1] M. Tanaka, M. Taguchi, T. Matsuyama, T. Sawada, S. Tsuda, S. Nakano, *et al.*, "Development of new a-Si/c-Si heterojunction solar cells: ACJ-HIT (artificially constructed junction-heterojunction with intrinsic thin-layer)," *Japanese Journal of Applied Physics*, vol. 31, p. 3518, 1992.

[2] H. S. Emmer, "Paths Towards High Efficiency Silicon Photovoltaics," California Institute of Technology, 2016.

[3] S. Y. Herasimenka, W. J. Dauksher, and S. G. Bowden, "> 750 mV open circuit voltage measured on 50 µ m thick silicon heterojunction solar cell," *Applied Physics Letters*, vol. 103, p. 053511, 2013.

[4] Z. C. Holman, A. Descoeudres, L. Barraud, F. Z. Fernandez, J. P. Seif, S. De Wolf, *et al.*, "Current losses at the front of silicon heterojunction solar cells," *Photovoltaics, IEEE Journal of*, vol. 2, pp. 7-15, 2012.

[5] R. Saive, C. R. Bukowsky, S. Yalamanchili, M. Boccard, T. Saenz, A. M. Borsuk, *et al.*, "Effectively transparent contacts (ETCs) for solar cells," in *Photovoltaic Specialists Conference (PVSC), 2016 IEEE 43rd*, 2016, pp. 3612-3615.

[6] R. Saive, A. M. Borsuk, H. S. Emmer, C. R. Bukowsky, J. V. Lloyd, S. Yalamanchili, *et al.*, "Effectively Transparent Front Contacts for Optoelectronic Devices," *Advanced Optical Materials*, vol. 4, pp. 1470–1474, 2016.

[7] R. Saive, M. Boccard, T. Saenz, S. Yalamanchili, C. R. Bukowsky, P. Jahelka, *et al.*, "Silicon heterojunction solar cells with effectively transparent front contacts," *Sustainable Energy & Fuels*, vol. 1, pp. 593-598, 2017.

[8] H. A. Atwater, R. Saive, A. M. Borsuk, H. Emmer, C. Bukowsky, and S. Yalamanchili, "Solar Cells and Methods of Manufacturing Solar Cells Incorporating Effectively Transparent 3D Contacts," ed: Google Patents, 2016.

[9] P. Jahelka, R. Saive, and H. Atwater, "Total Internal Reflection for Effectively Transparent Solar Cell Contacts," *arXiv preprint arXiv:1610.01047*, 2016.

[10] A. Augusto, K. Tyler, S. Y. Herasimenka, and S. G. Bowden, "Flexible Modules Using< 70 µm Thick Silicon Solar Cells," *Energy Procedia*, vol. 92, pp. 493-499, 2016.

[11] R. Saive, C. R. Bukowsky, and H. Atwater, "Three-dimensional nanoimprint lithography using two-photon lithography master samples," *arXiv preprint arXiv:1702.04012v1*, 2017.

[12] A. Richter, M. Hermle, and S. W. Glunz, "Reassessment of the limiting efficiency for crystalline silicon solar cells," *IEEE Journal of Photovoltaics*, vol. 3, pp. 1184-1191, 2013.

Enhanced Power Conversion Efficiency in Single Nanowire Devices through Symmetry Breaking Design

Jian Zhou, Yonggang Wu*, Zihuan Xia, Xuefei Qin, and Zongyi Zhang

School of Physics Science and Engineering, Tongji University, Shanghai 200092, China
*Corresponding author:ygwu@tongji.edu.cn

Abstract — Semiconductor and dielectric nanostructures have highly tunable resonances that have been used to increase light absorption in a variety of photovoltaic and photodetector devices. In this article, we demonstrate that horizontally placed and partially capped nanowires (PCNWs) exhibit strongly enhanced light trapping than conventional bare and core shell nanowires. Based on the temporal coupled-mode theory, we analyze the coupling strength between incident light and leaky mode resonances as a function of the cap angles of the PCNWs. Compared with the conventional bare and core shell nanowires, the PCNWs support more hybrid leaky modes and therefore give rise more coupling channels. Significant increase in short-circuit current and noticeable increase in open-circuit voltage enable the PCNW to possess high power conversion efficiency of 55.31%, increased by 42.7% and 141%, respectively, compared with that of the conventional core-shell nanowire (38.76%) and bare nanowire (22.95%).

Index Terms — PCNW structure, symmetry breaking, leaky modes, light harvesting, photovoltaic cells, conversion efficiency

I. INTRODUCTION

It is well known that single nanowire (NW) has strongly antenna effect that gives it a much larger optical absorption area than its geometrical cross section, which means that much more light surrounding far away from the bare nanowire (BNW) can be harvested. Based on this unique feature that is distinguished from large area nanowire arrays, single nanowire is believed to be the promising candidate for powering ultralow-power electronics and diverse integrated nanosystems, such as photodetectors [1], nanoelectronic power sources [2] and super miniature cells [3]. Recent studies have demonstrated that subwavelength core shell nanowires (CSNWs) and nanocones possess similarly strong, tunable resonances. These structures are combined by metal-core/semiconductor-shell and/or semiconductor-core/dielectric shell, which supports several nearly degenerate leaky mode resonances (LMR)s and the localized surface plasmon (LSP) and/or surface plasmon polariton excitations [4]. The core/shell structure produces a light matter interaction stronger than what would be expected for the bare nanowire. In this article, we show that horizontally placed semiconductor NWs with front partial cap break the centrosymmetry and exhibit higher resonant absorption and power conversion efficiency than the conventional BNWs and CSNWs. The absorption enhancement has origins in the well matched phase and

amplitude of the incident and scattered waves which yields complete destructive interference.

II. PROPOSED STRUCTURE AND SIMULATION STRATEGY

Fig.1 (a) schematically shows the perspective diagram of the front partially capped nanowire (PCNW). As a comparison, cross-sectional configurations of the BNW, CSNW and PCNW are shown in Fig. 1(b). The characterized geometrical parameters shown in Fig. 1(b) are the cap thickness t, the cap angle δ, and nanowire radius r. To investigate the optical properties of the PCNWs, GaAs and ZnO are chosen as examples of semiconductor core and dielectric cap, respectively, and the finite element method based electromagnetic calculations is utilized [5]. The dielectric coefficients of the GaAs is taken from reference [6] and that of the ZnO is set to be a constant 4. The GaAs radius r is predetermined to be 73 nm as the BNW displays the highest efficient at this size [7]. The length of nanowire is assumed to be far beyond the diameter, allowing the use of two-dimensional (2D) simulation. The absorption efficiency (σ_{abs}) is defined as the ratio of the absorption cross section (C_{abs}) to the geometrical cross section (C_{geo}) of the BNW. The unpolarized illumination is represented by averaging TE (electric field normal to the nanowire axis) and TM (magnetic field normal to the nanowire axis) incidences, i.e. $\sigma_{abs} = (\sigma_{abs}^{TE} + \sigma_{abs}^{TM})/2$.

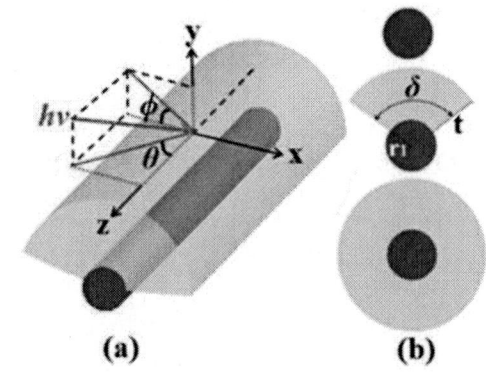

Fig. 1. (a) Schematic diagram of the single PCNW. (b) Cross-sectional views of the single BNW, PCNW and CSNW.

We adopt detailed balance analysis to investigate the relation between the electric current density and voltage in the nanowires [8],

$$J = qF_s + qF_{c0}\left(1 - \exp\left(\frac{qV}{kT_c}\right)\right) \quad (1)$$

where

$$F_s = \int_{E_g}^{\infty} dE I(E)\sigma(E, \theta = 90, \phi = 90) \quad (2)$$

is the radiative generation rate of hole-electron pairs per unit area by the incident sunlight,

$$F_{c0} = \int_0^{2\pi} d\phi \int_0^{\pi} d\theta \int_{E_g}^{\infty} dE \Theta(E)e_m(E, \theta, \phi)|\cos(\theta)|\sin(\theta) \quad (3)$$

is the radiative recombination rate when the cell is in thermal equilibrium with a surrounding blackbody at ambient temperature T_c, and

$$\Theta(E) = \left(\frac{2E^2}{h^3c^2}\right)\left(\exp\left(\frac{E}{kT_c}\right) - 1\right)^{-1} \quad (4)$$

is Planck's law for the incident spectral irradiance at a temperature T_c. $I(E)$ is the incident rate of solar photons per unit area per unit bandwidth at the photon energy E, $\sigma(E, \theta, \phi)$ is the cell's absorption efficiency spectrum summed over both the transverse electric (TE) and transverse magnetic (TM) incident polarizations, $e_m(E, \theta, \phi)$ is the emisson rate, θ is the angle between the propagation vector of the incident light and the nanowire axis, while ϕ is the azimuthal angle of the propagation vector in cylindrical coordinates. E_g is the band gap of the active semiconductor material.

III. RESULTS AND DISCUSSIONS

Fig. 2(a) illustrates the dependence of the optically estimated short circuit current density $J_{sc} = qF_s$, on the shell thickness t, with considering the CSNWs under TE, TM and unpolarized normal incidences. It shows that J_{sc} of the CSNW is greatly improved by introduction of dielectric shell (t > 0), as compared with that of the BNW ($t = 0$), especially when initially increases from 0 to 80 nm. The maximum J_{sc} under CSNW design goes up to 36.49 mA/cm^2 at t=220 nm (unpolarized incidence), i.e., it is increased by 14.57 mA/cm^2 with an enhancement ratio of 66.5% over the BNW (J_{sc} = 21.92 mA/cm^2). However, Fig. 1(a) also shows that keeping increasing t does not lead to higher J_{sc}, which can be attributed to the fact that the maximum electric field intensity of the higher order LMR excited in the core/shell structure is mainly located in shell region.

The partially capped nanowire structure enhances the J_{sc} further, as shown in fig. 2(b). Fig. 2(b) plots J_{sc} versus cap span angle δ of the PCNW with t = 220nm under TE and TM incident polarizations. It is clear that tailoring the δ can dramatically increases J_{sc}. The maximum J_{sc} reaches 50.11, 52.84, and 51.48 mA/cm^2, for the TE, TM and non-polarization incidences, respectively.

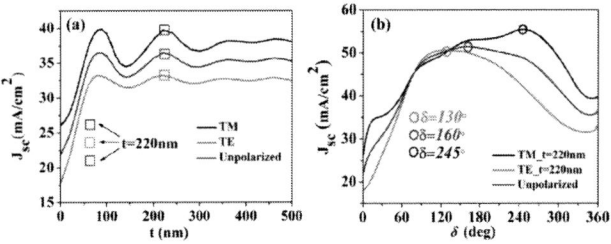

Fig. 2. (a) J_{sc} as a function of the shell thickness t of the CSNWs under TE and TM incidences. (b) J_{sc} as a function of the span angle δ of PCNWs ($t = 220$ nm) under TE and TM and non-polarization incidences, respectively.

To reveal the mechanisms responsible for the improved light-harvesting capability, we examine the absorption cross section spectral of the PCNWs versus the cap span angle at TE incidence, as shown in figure 3(a). From the absorption cross section, it is obvious that PCNW with its span angle between 60 and 240° exhibits better light capture capability than that of the BNWs and CSNWs.

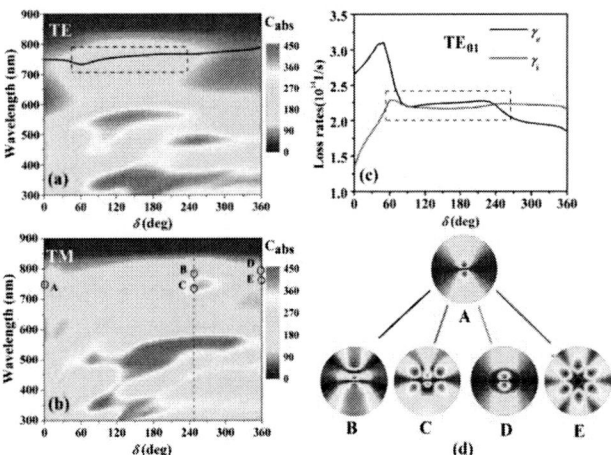

Fig. 3. Light absorption cross section as a function of the incident light wavelength and cap angle of the PCNW for (a) TE and (b) TM light incidence. The color curves in (a) is the dispersion relation of the TE$_{01}$ mode. (c) The radiative loss rates γ_e (black line) and intrinsic loss rate γ_i (red line) of the TE$_{01}$ resonances. (d) The patterns of the eigenmodes denoted in (b).

According to the temporal coupled mode theory [9], the absorption cross section for a single resonance in cylindrical coordinates is:

$$C_{abs} = \frac{2\lambda}{\pi} \frac{\gamma_i \gamma_e}{\left(\omega_i - \omega_0\right)^2 + \left(\gamma_i + \gamma_e\right)^2}. \quad (5)$$

Here λ is the wavelength, ω_i is the frequency of incident light. The absorption cross section reaches a maximum on resonance ($\omega = \omega_0$), when $\gamma_e = \gamma_i$, and $C_{abs,max} = \lambda/2\pi$. It means that the phase and amplitude of the incident and scattered waves are matched to yield complete destructive interference. This leads to the highest energy inside the semiconductor nanowires and hence to the maximum absorption cross section. As a function of cap angles of the PCNW, the real part eigenfrequency ω_0 (wavelengths λ_0) of the TE_{01} (black line) is calculated by FEM, as shown in Fig. 3(a), which agree well with peaks of absorption spectrum. There is an interesting phenomenon that the TE_{01} mode is excited in the PCNWs, but the coupling strength between the incident light and the LMRs at δ = 60-250° is different from that at other angles. We calculate the radiative loss rate γ_e and the intrinsic loss rate γ_i corresponding to the TE_{01} mode, as shown in Fig. 3(c). We find that γ_e and γ_i are very close in the span angle range of 60-250°, while they go far away from each other in the other span angle.

In addition, more eigenmodes are excited in the PCNWs than those in the BNW, e.g. the TM modes shown in Fig. 3(b). Fig. 3(d) shows the patterns of the eigenmodes denoted in (b). As can be seen, only one eigenmode TM_{11} around 749nm exists in the BNW, while two typical eigenmodes are supported in PCNW TM_{21} (790nm) and TM_{31} (731nm). Though the CSNW also support two modes, e.g. TM_{12} (789nm), TM_{31} (771nm), the eigenfeilds distributions in the CSNW are mainly located in the shell rather than in the GaAs core region.

To show the advantage of the PCNWs for the PV application, we calculate the current density–voltage ($J-V$) transport characteristics by utilizing eq. (1-4), as shown in Fig. 4. The non-radiative recombination (including Auger, Shockley-Read-Hall, and surface recombination) has little influence on the radial voltage as GaAs is a direct bandgap semiconductor and has been omitted [7]. The core radius and cap thickness are 73 and 220 nm respectively and the cap angle is 160°. The $J-V$ characteristics of the BNW of the same core radius (r = 73nm), and that of the CSNW (r = 73nm, t = 220nm) are also shown in Fig. 4 for comparison.

It is found that, the short-circuit current of the PCNW, CSNW, and BNW is 51.42, 36.50, and 21.94 mA/cm², and the maximal power density is 55.31, 38.76, and 22.95 mW/cm², respectively. It is worth mentioning that the open-circuit voltage of the PCNW shows noticeable increase in comparison with the CSNW and BNW. The open-circuit voltage of the PCNW, CSNW, and BNW is 1.198, 1.184, and 1.168V respectively. Analysis shows that, F_{c0} drops dramatically in the broad angle range around φ = -π/2 in the PCNW, although both the radiative generation rate F_s and the thermal equilibrium recombination rate F_{c0} in the PCNW are only slightly higher than those in the CSNW and BNW in a relatively narrow angle region around φ = π/2. The high contrast between the intergral F_s and F_{c0} in the PCNW results in a higher V_{oc}. Consequently, the PCNW possess higher power conversion efficiency than that of the CSNW and BNW. If the incident light power is assumed to be sun radiation, the power conversion efficiency is defined as η=FF(($J_{sc}V_{oc}$)/P_{inc}) × 100% where P_{inc} is the total incident sun radiation power per unit cell area about 100mW/cm², and FF is the cell's fill-factor [8], the power conversion efficiency of the three typical configurations of the PCNW, CSNW and BNW structures is 55.31%, 38.76%, and 22.95%, respectively. The former one is increased by 42.7% and 141%, compared with the latter two, respectively.

IV. CONCLUSION

We have shown that the key to increasing absorption in PCNWs structures is to increase the strength and the number of hybrid resonances. The PCNWs can be imagined as a composite miniature version of a highly efficient light absorber device, with the cap playing the role of a light coupling enhancer and the circular nanowire an absorption medium in the structure. Benefited from symmetry breaking design, the PCNW device can achieve maximum absorption from the direction of incident light and suppress radiative recombination in a relatively wide angles. Consequently, significant increase in short-circuit current and noticeable increase in open-circuit voltage enable the PCNWs to possess highly power conversion efficiency. We believe that the PCNWs can be used in super miniature photodetectors, power sources and solar cells applications. The nanowire array composed by the PCNW will be studied in our next research work.

ACKNOWLEDGEMENT

This research was sponsored by the National Natural Science Foundation of China (No.60977028) and Key Project Foundation of Shanghai (No. 09JC1413800).

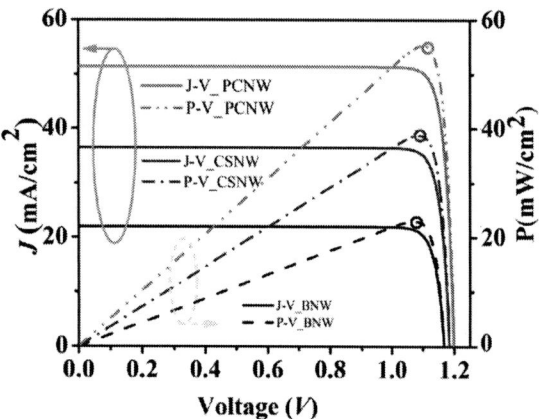

Fig. 4. *J-V* and *P-V* characteristics of the PCNW, CSNW and BNW structures.

REFERENCES

[1] X. Dai, S. Zhang, Z. Wang, G. Adamo, H. Liu, Y. Huang, C. Couteau, and C. Soci, "GaAs/AlGaAs Nanowire Photodetector," *Nano Letters,* vol. 14, pp. 2688-2693, 2014.

[2] B. Tian, X. Zheng, T. J. Kempa, Y. Fang, N. Yu, G. Yu, J. Huang, and C. M. Lieber, "Coaxial silicon nanowires as solar cells and nanoelectronic power sources," *Nature,* vol. 449, pp. 885-889, 2007.

[3] L. Cao, J. S. White, J.-S. Park, J. A. Schuller, B. M. Clemens, and M. L. Brongersma, "Engineering light absorption in semiconductor nanowire devices," *Nature Materials,* vol. 8, pp. 643-647, 2009.

[4] L. Zhou, X. Yu, and J. Zhu, "Metal-Core/Semiconductor-Shell Nanocones for Broadband Solar Absorption Enhancement," *Nano Letters,* vol. 14, pp. 1093-1098, 2014.

[5] J.-M. Jin, "The Finite Element Method in Electromagnetics," ed. New York: Wiley, 2002.

[6] E. D. Palik, "Handbook of Optical Constants of Solids," in *Academic*, ed, 1985.

[7] S. Sandhu, Z. Yu, and S. Fan, "Detailed Balance Analysis and Enhancement of Open-Circuit Voltage in Single-Nanowire Solar Cells," *Nano Letters,* vol. 14, pp. 1011-1015, 2014.

[8] W. Shockley and H. J. Queisser, "Detailed Balance Limit of Efficiency of p-n Junction Solar Cells," *Journal of Applied Physics,* vol. 32, p. 510, 1961.

[9] Z. Ruan and S. Fan, "Superscattering of Light from Subwavelength Nanostructures," *Physical Review Letters,* vol. 105, 2010.

CdSe(Te)/CdS/CdSe Rods vs. CdTe/CdS/CdSe Spheres: Morphology-Dependent Carrier Dynamics for Photon Upconversion

Eric Y. Chen, Zhuohui Li, Christopher C. Milleville, Kyle R. Lennon, Matthew F. Doty

Department of Materials Science and Engineering, University of Delaware, Newark, DE 19716 USA

Abstract—Colloidal quantum dot-hybrid nanostructures have been shown to upconvert two low-energy photons into one high-energy photon and have wide ranging applications, including in photovoltaics, where additional current in the solar cell can be generated from the absorption of upconverted light emitted from the nanostructures. Both double quantum dots (via nanorod formation) and quantum dot-quantum well structures have been studied previously, but there has been no systematic study of the dependence of carrier dynamics on nanostructure morphology. Here, we synthesize both core/multi-shell colloidal QD and core/rod complexes and use steady-state and time-resolved photoluminescence to elucidate carrier separation and transfer dynamics. Understanding these dynamics provides important information for optimizing the nanostructure morphology to create more efficient photon upconversion and thus enhance net solar conversion efficiencies.

Index Terms—core/shell quantum dots, photovoltaics, core/rod nanocrystals, time-resolved photoluminescence, upconversion.

I. INTRODUCTION

Semiconductor nanocrystals, specifically colloidal quantum dots (QDs), have been widely used for sunlight harvesting, e.g. in luminescent solar concentrators with a Si host cell [1] and QD solar cell [2]. Synthesis methods are easily scalable for implementation in automated or roll-to-roll processing and reduce the net cost of solar cells. However, the efficiency of such devices has thus far been limited to 10% net solar conversion efficiency by the challenging balance between high electron and hole mobility and suppression of nonradiative recombination by surface passivation [3]. Recently, semiconductor nanocrystals have been used as an upconversion material for improving spectral utilization beyond what has been demonstrated in existing upconverters [4]. Upconversion is the process by which two incoming low-energy photons are absorbed and a single high-energy photon is emitted. Current upconverter materials include lanthanide-doped nanocrystals and organic emitter-triplet-sensitizer complexes. However, these materials lack the absorption bandwidth necessary to significantly impact the overall solar cell performance since the additional current generated by the solar cell is a function of the increased photon flux above the host cell bandgap. Our previously published model indicates that these existing materials could enable an improvement of at most 2% in overall solar cell efficiency [5]. Semiconductor nanocrystals have a wide absorption bandwidth (>200nm), and we have shown that the overall efficiency of a solar cell backed by a semiconductor photon upconverter can reach 39% [6].

We have previously reported the performance of a solid-state thin film upconversion nanostructure and a solar cell

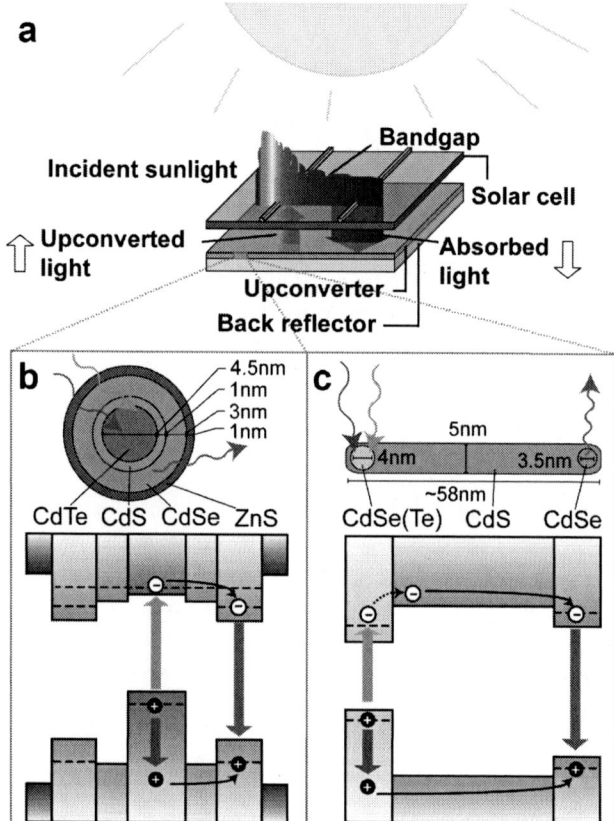

Fig. 1. (a) Schematic of an upconverter-backed solar cell and schematic and band diagrams of (b) Core/multi-shell QDs and c) Core/rod nanocrystals

backed by this upconverter as computed with a kinetic rate model [6]. In this model each carrier energy state is defined, and a series of rate equations describing the carrier dynamics is solved for the equilibrium population in each state. A high degree of "funneling" toward the emitter region of the upconverter leads to higher internal upconversion quantum efficiency and thus higher net solar energy harvesting efficiency despite the energy sacrifice required to achieve the "funneling". In colloidal nanostructures, this funneling could be controlled via the composition and morphology of shells applied around an absorbing core. Here we report synthesis and characterization of spherical and rod-shaped core/shell structures designed for photon upconversion. We first describe the synthesis of the core/multi-shell CdTe/CdS/CdSe

978-1-5090-5606-4/17 $31.00 © 2017 IEEE

QDs and core/rod CdSe(Te)/CdS/CdSe nanocrystals. We then report carrier separation and transfer dynamics measured with time-correlated single photon counting. Finally, we analyze the dependence of the measured dynamics on the particle morphology and discuss methods for improving upconversion and net solar conversion efficiency.

II. SYNTHESIS METHODS

A. Synthesis of CdTe/CdS/CdSe/ZnS Core/multi-shell QDs

The CdTe QDs were synthesized using a hot-injection method based on a previously reported procedure [7]. Typically, the mixture of 0.2 mmol CdO, 0.5 mmol tetradecylphosponic acid (TDPA), and 4mL octadecene (ODE) was heated to 310°C under a slow argon flow. The mixture turned clear and was then cooled to 270°C at which point 2 mL 0.1 M Te precursor (prepared by dissolving 1 mmol tellurium, 5 mL TOP, and 5 mL ODE) was rapidly injected. The as-prepared solution was purified by centrifugation and the supernatant was extracted out. This procedure was repeated at least one more time to remove the vast majority of unreacted precursors. Finally, the CdTe QDs was precipitated by adding excess acetone and re-dissolved in chloroform. The subsequent CdS [7, 8], CdSe [8], and ZnS [9, 10, 11] shell growths were carried out via successive ion layer adsorption and reaction (SILAR) method. The reaction was terminated when the desired size was reached and the solution was allowed to cool down to room temperature. The purification procedure was similar to that applied for the CdTe QDs.

B. Synthesis of CdSe/CdS/CdSe(Te) nanorods

Tellurium-doped cadmium selenide (CdSe(Te)) quantum dot synthesis and the subsequent seeded growth of cadmium sulfide (CdS)/cadmium selenide (CdSe) shells were carried out by adaptation of a previous reported method [4]. The solution of CdSe(Te) nanoparticles was combined with a solution of sulfur (S, 3.7 mmol) in TOP (2 mL). The CdSe(Te)/S solution was quickly injected at 350°C and the nanorods were allowed to grow for 4 minutes before the heating mantle was removed and the nanoparticles were allowed to cool. The flask was then heated to 80°C, the CdSe(Te)/CdS solution was injected, and the temperature was raised to 200°C. CdSe growth progress was monitored by removing reaction mixture aliquots and obtaining absorption and photoluminescence spectra. After 90 minutes, the heating mantle was removed and the nanoparticles were allowed to cool. The CdSe(Te)/CdS/CdSe nanorods were purified using the same precipitation/re-dissolution procedure using acetone/toluene.

III. OPTICAL CHARACTERIZATION

A. Steady-state and Time-Resolved Photoluminescence Spectroscopy Methods

Steady-state (SS) and time-resolved (TR) photoluminescence (PL) measurements were taken of QD samples diluted to OD = 0.5 for minimal re-absorption between QDs (See

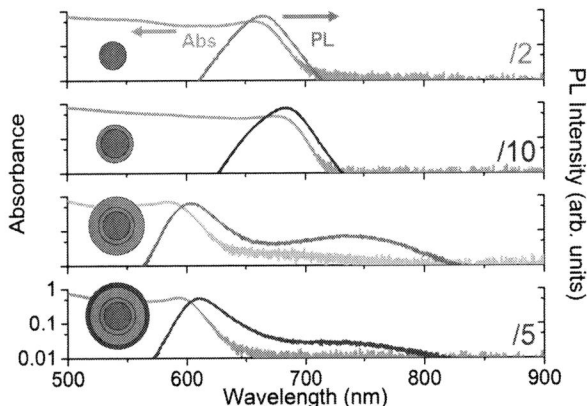

Fig. 2. Absorbance and photoluminescence (log scales) for core/multi-shell QDs under 405nm, 30mW CW excitation

Figures 2 and 3). A 405nm, 30mW continuous wave (CW) laser diode was used for SSPL to excite carriers above the bandgap of all transitions in the QD samples. The emitted light was collected by a microscope objective, which collimates the beam, and was focused down at the 50um entrance slit of a Princeton Instruments Acton SpectraPro 2500i Spectrometer. A grating at 150 grooves/mm directs the beam towards a liquid nitrogen-cooled charge coupled device camera (1340 x 100 pixels). TRPL via time-correlated single photon counting (TCSPC) was measured using a femtosecond pulsed supercontinuum laser (Fianium WhiteLase) with pulse repetition rates of 0.5MHz or 2MHz and wavelength between 450nm and 700nm with average power of 10mW. The emitted light was directed using a flip mirror, through a 3mm exit slit, to a silicon avalanche photodiode (APD) (Micro Photon Devices). A TCSPC module (PicoHarp 300) correlates the electronic pulses from the pulsed laser to the photons detected at the APD to generate a decay curve (See Figures 3(a) and 5(a)). Bi-exponential fits were generated with FluoFit and error bars generated by a bootstrap Markov chain Monte Carlo analysis method. The amplitude fraction and decay constants are plotted below the curves.

B. Optical Characterization of Core/multi-shell QD Complexes

SSPL and absorbance measurements of the core/multi-shell QD complexes are shown in Figure 2. The addition of a single CdS shell (3ML) to the CdTe core results in a redshift of the core emission peak (660nm to 678nm) due to reduced quantum confinement, as seen previously [8]. The SILAR addition of the CdSe shell (9.5ML) results in both a redshift of the core emission peak further (678nm to 740nm) and the generation of PL from carriers recombining across the CdSe bandgap (605nm). Further addition of a ZnS (3ML) passivating layer improves the ratio of PL between the CdSe to CdTe.

Decay constants extracted from TCSPC measurements in Figure 3(a) of the core/multi-shell QDs are shown in Figure 3(b). The center wavelength of the spectrometer grating is adjusted to match the desired peak wavelength emission. In

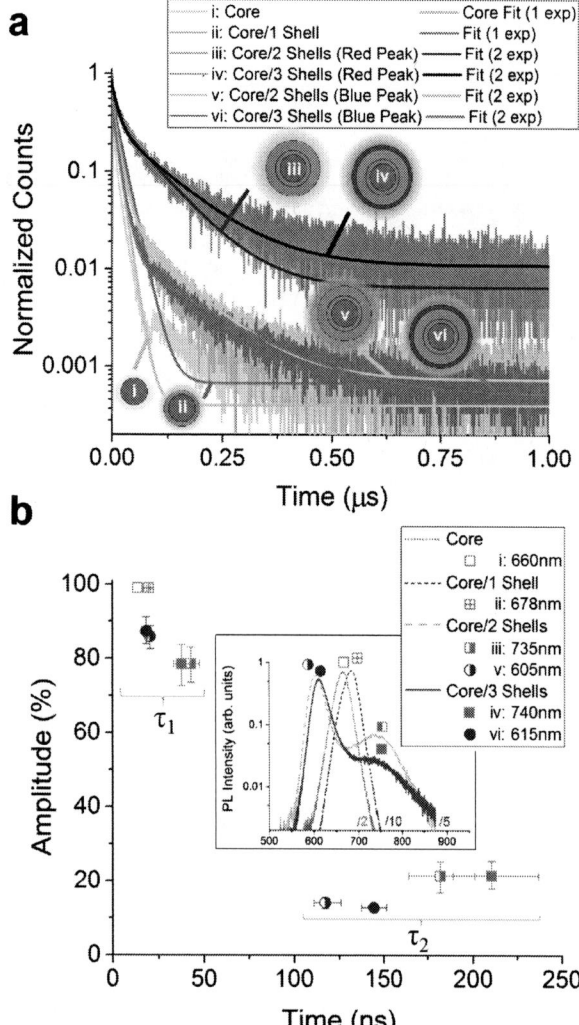

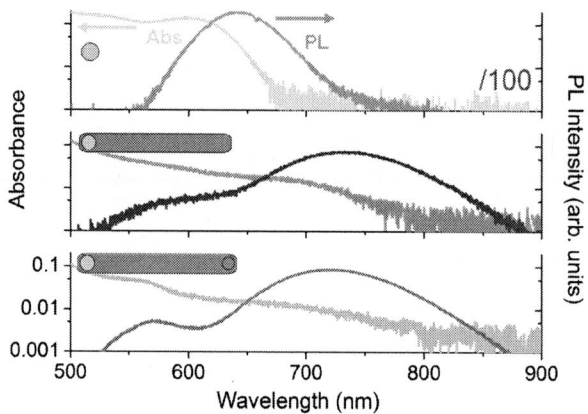

Fig. 4. Absorbance and photoluminescence (log scales) for core/rod QDs under 405nm, 30mW CW excitation

Fig. 3. (a) TRPL decay curves of core/multi-shell complexes with corresponding fits (mono-exponential for core and core/single-shell decays, bi-exponential for all others) and corresponding nanostructure schematic with emission peak indicated by ambient color. (b) Steady-state (inset) and TRPL lifetime vs. amplitude fraction of core/multi-shell QDs. Core emission indicated by red squares, shell emission by blue circles, with emission wavelengths indicated

C. Optical Characterization of Core/rod Nanocrystals

SSPL and absorbance of the core/rod complexes are shown in Figure 4. The growth of an anisotropic CdS rod-like shell around the CdSe(Te) cores results in a large redshift of the core emission peak (635nm to 720nm). The length of the rod (tens of nm) relative to the 1nm CdS shell in the core/multi-shell QD complexes explains the larger redshift as a result of a drastic reduction in quantum confinement. The CdSe QD added to the opposite end of the CdS rod generates a second PL peak due to radiative recombination across the CdSe bandgap (570nm). This blue peak is much lower in intensity relative to the red peak at 725nm. The CdSe(Te) emission wavelength (720nm) does not change significantly because the carrier confinement and overlap is not significantly changed with the addition of the second QD.

Decay constants and amplitude fractions extracted from TCSPC measurements in Figure 5(a) of the core/rod samples are shown in Figure 5(b). Biexponential decays were used to fit all decay curves . The τ_1 lifetime is 37ns for CdSe(Te) core-only QDs, with a τ_2 long lifetime component of 114ns (i). The core emission lifetime decreases with CdS rod (ii) and CdSe QD (iii) additions. The high-energy emission from the CdSe QD (iv) has a τ_1 radiative lifetime of 13ns and a τ_2 lifetime of 73ns.

D. Upconversion Photoluminescence (UCPL) Method and Measurements

A tunable, CW Ti:Sapph laser was used to excite the full structure CdTe/CdS/CdSe/ZnS core/multi-shell QD complex in chloroform and CdSe(Te)/CdS/CdSe core/rod QD complex in toluene. The wavelength of the excitation beam (735nm) was chosen to correspond with the peak of the core absorption in both samples. A 700nm longpass (LP) filter in front of the sample ensured that no short wavelength light would excite the emitter states directly. A 630nm or 650nm shortpass (SP) filter in front of the detector ensured that the laser signal would not be collected. Figure 6 shows the results of the preliminary up-conversion photoluminescence measurements. The solid lines

Figure 3(a), the PL lifetime in the CdTe core (i) is 14ns, much faster than the lifetime of bulk CdTe, due to non-radiative trapping of carriers, presumably in surface-related defect states. Following the addition of CdS (ii), which has a relatively short lifetime as well, the additional CdSe shell (iii) leads to biexponential decay dynamics with longer PL emission lifetimes from the CdTe core. The CdSe shell emitter itself (iv) has a significantly blue-shifted PL emission and shorter lifetimes relative to the core emission lifetimes (iii). The addition of ZnS passivates nonradiative surface trap states, further increasing the τ_2 lifetimes of both core "red" emission (v) and CdSe "blue" emission (vi).

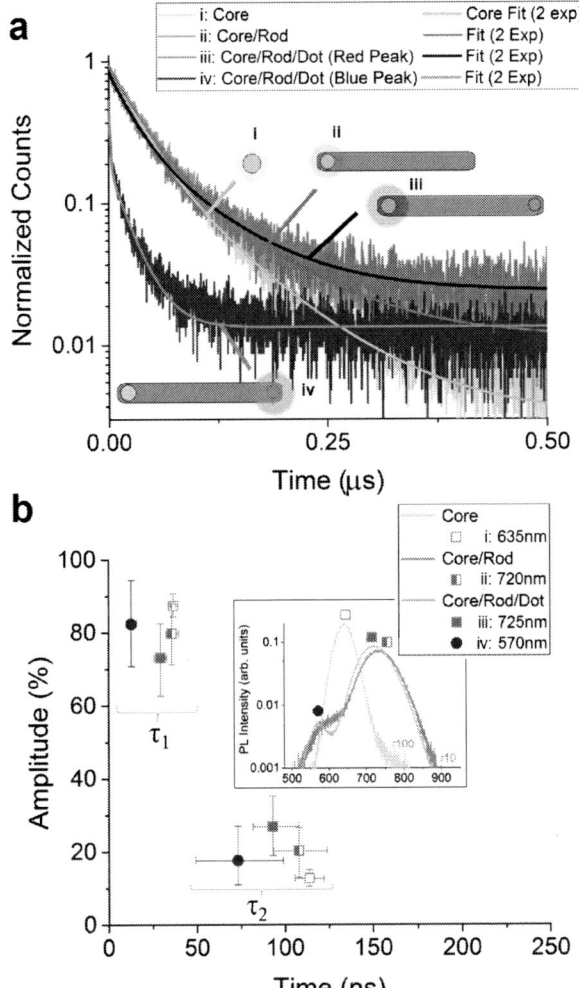

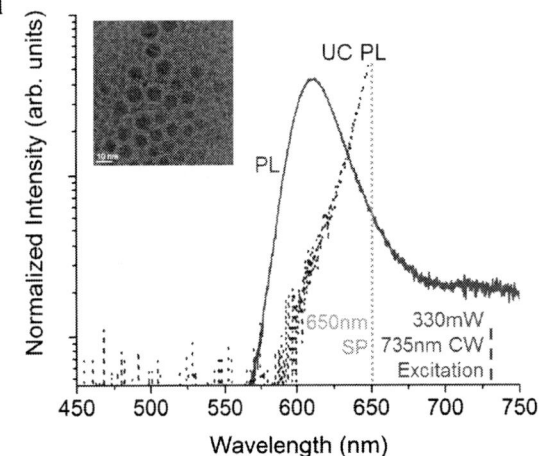

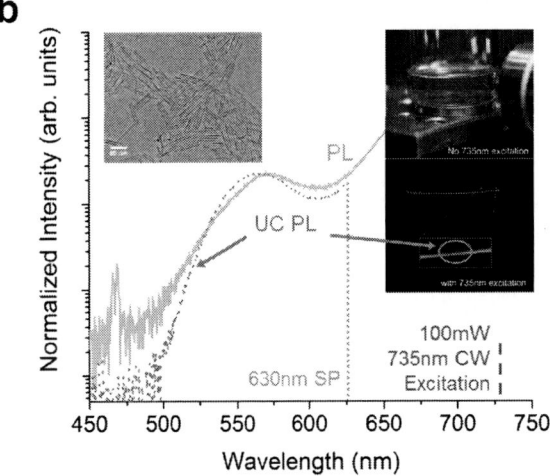

Fig. 5. (a) TRPL decay curves of core/rod nanocrystals with corresponding fits (two exponentials) and corresponding nanostructure schematic with emission peak indicated by ambient color. (b) Steady-state (inset) and time-resolved photoluminescence lifetime vs. amplitude fraction of core/rod nanocrystals. Core emission indicated by red squares, emitter dot emission by blue circles, with emission wavelengths indicated

Fig. 6. Overlay of steady-state photoluminescence emission (solid line) with upconversion photoluminescence (dotted line) of (a) CdTe/CdS/CdSe/ZnS core/multi-shell QDs and (b) CdSe(4% Te)/CdS/CdSe core/rod nanocrystals, with 735nm CW excitation and 630nm/650nm SP filter to eliminate laser peak. Insets show (a) TEM of core/2-shell QDs and (b) TEM of core/rod/dot nanostructure with photograph of vial without and with 735nm excitation, with green circle indicating pale upconversion emission

are the photoluminescence curves of the full structures of the core/multi-shell (purple) and core/rod complex (green), obtained under excitation by a 405nm CW laser at 30mW. This data indicates the expected wavelengths for upconversion emission from the CdSe shell and CdSe QD, respectively. The dotted lines indicate the measured UCPL under 735nm CW excitation at 330mW and 100mW of the spherical nanoparticles and rod-like nanoparticles, respectively. There is clear overlap in Figure 6(b) of the UCPL and PL curves, and no overlap in Figure 6(a), indicating that upconversion via emission from the CdSe state was observed in the rod-like nanoparticles, and not in the spherical nanoparticles. The 'incomplete' upconversion observed in the spherical nanoparticles, by which we mean upconversion to an energy lower than the desired upconversion emission energy, is similar to that observed in single QDs and will be analyzed in future work.

IV. RESULTS AND ANALYSIS

The radiative lifetime of the core in the core/rod nanocrystal is longer than that of the core in the core/multi-shell QD complex. Subsequently, addition of the same CdS/CdSe material to these cores, but in varying morphology (spherical shell vs. rod-QD), leads to changes in the carrier dynamics. We see that in both core/multi-shell QD complexes and core/rod nanocrystals, the addition of shells redshifts the PL due to quantum confinement and creates a longer-lived core emission lifetime component due to delocalization of the electron wavefunction and isolation of the hole from surface-related defect states. The core/rod morphology allows for faster radiative recombination in the CdSe shell relative to the core/shell morphology, potentially indicating higher PL emission efficiency. Upconversion photoluminescence from CdSe emitter states is observed in only the full core/rod/dot nanoparticles and not in the full core/3-shells nor 2-shells QDs. This suggests that the spatial separation of electrons and holes in the rod-like particles is essential to the desired upconversion process and is limited in the spherical morphology. Power-dependent measurements and additional TCSPC data are required to further elucidate the relationship between carrier dynamics and upconversion efficiency, as well as possible upconversion mechanisms, in these different morphologies.

V. Significance to the Field

Determining the appropriate morphology for desired carrier dynamics and charge transfer in semiconducting upconverter nanocrystals is critical to designing materials that can realize improved upconversion quantum yield and thus practical improvements in photovoltaic efficiencies. We have established methods for synthesizing and characterizing upconverter nanostructures of varying morphologies, permitting a systematic analysis of the carrier separation and transfer dynamics.

VI. Conclusion

We have synthesized core/multi-shell QDs and core/rod nanocrystals using a hot-injection method. We compare the energy levels and carrier dynamics of these two morphologies using steady state and time-resolved photoluminescence techniques to determine the rates of various photophysical processes. We analyze the results to identify promising morphologies for improving upconversion performance. We have observed morphology-dependent upconversion, which will facilitate quantification and enhancement of the performance of subsequent upconverter generations and provide the advances necessary for developing practical upconverter-backed solar cell devices

Acknowledgment

The authors acknowledge financial support from the W. M. Keck Foundation.

References

[1] J. Bomm, A. Büchtemann, A. J. Chatten, R. Bose, D. J. Farrell, N. L. A. Chan, Y. Xiao, L. H. Slooff, T. Meyer, A. Meyer, W. G. J. H. M. Van Sark, and R. Koole, "Fabrication and full characterization of state-of-the-art quantum dot luminescent solar concentrators," *Solar Energy Materials and Solar Cells*, vol. 95, no. 8, pp. 2087–2094, 2011.

[2] X. Lan, O. Voznyy, F. P. García De Arquer, M. Liu, J. Xu, A. H. Proppe, G. Walters, F. Fan, H. Tan, M. Liu, Z. Yang, S. Hoogland, and E. H. Sargent, "10.6% Certified Colloidal Quantum Dot Solar Cells Via Solvent-Polarity-Engineered Halide Passivation," *Nano Letters*, vol. 16, no. 7, pp. 4630–4634, 2016.

[3] I. J. Kramer and E. H. Sargent, "Colloidal quantum dot photovoltaics: A path forward," *ACS Nano*, vol. 5, no. 11, pp. 8506–8514, 2011.

[4] Z. Deutsch, L. Neeman, and D. Oron, "Luminescence upconversion in colloidal double quantum dots." *Nature Nanotechnology*, vol. 8, no. 9, pp. 649–653, 2013.

[5] D. G. Sellers, J. Zhang, E. Y. Chen, Y. Zhong, M. F. Doty, and J. M. O. Zide, "Novel nanostructures for efficient photon upconversion and high-efficiency photovoltaics," *Solar Energy Materials and Solar Cells*, vol. 155, pp. 446–453, 2016.

[6] E. Y. Chen, J. Zhang, D. G. Sellers, Y. Zhong, J. M. O. Zide, and M. F. Doty, "A kinetic rate model of novel upconversion nanostructures for high-efficiency photovoltaics," *IEEE Journal of Photovoltaics*, vol. 6, no. 5, pp. 1183–1190, Sept 2016.

[7] J. Wang, Y. Long, Y. Zhang, X. Zhong, and L. Zhu, "Preparation of highly luminescent CdTe/CdS core/shell quantum dots," *ChemPhysChem*, vol. 10, no. 4, pp. 680–685, 2009.

[8] D. G. Sellers and M. F. Doty, "Design, synthesis and photophysical properties of inp/cds/cdse and cdte/cds/cdse (core/shell/shell) quantum dots for photon upconversion," in *Photovoltaic Specialist Conference (PVSC), 2015 IEEE 42nd*, June 2015, pp. 1–5.

[9] H. Shen, H. Wang, Z. Tang, J. Z. Niu, S. Lou, Z. Du, and L. S. Li, "High quality synthesis of monodisperse zinc-blende cdse and cdse/zns nanocrystals with a phosphine-free method," *CrystEngComm*, vol. 11, no. 8, pp. 1733–1738, 2009.

[10] H. Zhu, N. Song, and T. Lian, "Controlling charge separation and recombination rates in CdSe/ZnS type-I core/shell quantum dots by shell thicknesses," *Journal of the American Chemical Society*, vol. 132, no. 42, pp. 15 038–15 045, 2010.

[11] D. Chen, F. Zhao, H. Qi, M. Rutherford, and X. Peng, "Bright and stable purple/blue emitting CdS/ZnS core/shell nanocrystals grown by thermal cycling using a single-source precursor," *Chemistry of Materials*, vol. 22, no. 4, pp. 1437–1444, 2010.

Drift-Diffusion InGaN/GaN Solar Cell Simulator with Optical Management

Y. Fang, D. Guo, A. Fischer, E. Vadiee, C. Zhang, J. Williams, S. M. Goodnick and D. Vasileska

Arizona State University, Tempe, Arizona, 85287, USA

Abstract — **This work reports development of an InGaN/ GaN hetero-structure solar cell simulator by coupling Drift-Diffusion (DD) equations for describing electrical transport and Transfer Matrix Method (TMM) for the optical solver, respectively. A tri-layer Anti-Reflection Coating (ARC) is optimized from 300 to 1200 nm based on TMM to reduce front reflection loss. This broadband ARC could be used for InGaN multi-junction solar cells or hybrid concentrating solar thermal systems in the future. The 2.94 eV InGaN solar cell demonstrates an increase of the conversion efficiency by 23% (relative) under 1 sun AM1.5, room temperature with ARC.**

Index Terms – **High temperature, nitride, multiple quantum wells, photovoltaics, TCAD simulations.**

I. INTRODUCTION

In recent years, there has been increased interest in the Indium Gallium Nitride (InGaN) material system for photovoltaic applications. The InGaN alloy system has demonstrated high performance for high frequency power devices, as well as for optical light emitters. This material system is also promising for photvoltaic applications due to broad range of bandgaps of $In_xGa_{1-x}N$ alloys from 0.65 eV (InN) to 3.42 eV (GaN) [1], which covers most of the electromagnetic spectrum from ultraviolet to infrared wavelengths. InGaN's high absorption coefficient, radiation resistance and thermal stability (operating with temperature > 450 °C) makes it a suitable PV candidate for hybrid concentrating solar thermal systems [2] as well as other high temperature applications. Due to its tunable direct bandgap, InGaN, is a promising candidate for multi-junction solar cells. InGaN/GaN has also demonstrated good growth capability on Si, thus making it suitable for InGaN/Si dual junction tandem cells. In these structures, with an AlN buffer layer, lattice mismatch between GaN and Si is accommodated [3]. A conversion efficiency of 31% [4] has been predicted analytically.

High efficiency solar cell require optimized optical management. An anti-reflection coating should be designed properly to enhance optical absorption in the device. For the potential hybrid thermal system and tandem PV applications, ARC design should be tailored for a broad wavelength range in order to utilize the whole spectrum. In this paper, we report on the ARC design based on the transfer matrix method (TMM) as implemented into our in-house InGaN solar cell Drift-Diffusion solver.

II. MATERIAL PROPERTIES AND COMPUTATIONAL MODEL

Accurate optical and electrical material parameters are needed for reliable solar cell modeling. Material parameters are collected extensively from our measurement and from literature. The bandgaps (E_g), [1] permittivity (ϵ/ϵ_o), affinity χ and effective masses (m_e, m_h) [4] are calculated as follows:

$$Eg(In_xGa_{1-x}N) = 0.65x + 3.42(1-x) - 1.43x(1-x) \tag{1}$$

$$\epsilon(In_xGa_{1-x}N) = 15.3x + 8.9(1-x) \tag{2}$$

$$\chi(GaN) = 4.1\ eV \tag{3}$$

$$m_e(In_xGa_{1-x}N) = 0.12x + 0.2(1-x) \tag{4}$$

$$m_h(In_xGa_{1-x}N) = 0.17x + 1.0(1-x) \tag{5}$$

The InGaN electron affinity χ, can be calculated from the band-offset ratio, which is assumed to be 0.7/0.3. The In mole fraction is assumed as 0.12 with a bandgap 2.94 eV in this simulation. The electron and hole mobilities are calculated using

$$\mu_{e,h} = \mu_{min} + (\mu_{max} - \mu_{min})/(1 + (N/N_{ref})^\gamma) \tag{6}$$

where N is the carrier concentration and $\mu_{max}, \mu_{min}, N_{ref}$, and γ are parameters taken from Ref. [5]. The minority carrier SRH lifetime is assumed to be equal to 10 ns. Auger and band to band recombination models are also included in the theoretical model. A fixed Auger coefficient of $1.5 \times 10^{-30}\ cm^6/s$ is used for electrons and holes. Optical recombination coefficient is assumed to be equal to $1.0 \times 10^{-11} cm^3/s$ [6]. A triple-layer ARC design [7], which consists of SiO_2, SiO_2-TiO_2 and Ta_2O_5 is adopted and optimized for broad wavelength range application. Optical constants of SiO_2, SiO_2-TiO_2 and Ta_2O_5 are obtained from [8]. For GaN and $In_{0.12}Ga_{0.88}N$, optical constants are measured by ellipsometry technique (Fig. 1). SiO_2-TiO_2 can be formed by sol-gel method and has the refractive index between the Ta_2O_5's and SiO_2's [8].

The drift-diffusion model for photo-generated electrons and holes is widely used for semiconductor device simulation. The model solves the Poisson and the electron and hole continuity equations self consistently [9]:

$$\nabla\big(\varepsilon(\nabla\phi)\big) = -q(p - n + N_D^+ - N_A^-) \tag{7}$$

$$\frac{\partial n}{\partial t} = \frac{1}{q}\nabla \cdot J_n - R + G \tag{8}$$

978-1-5090-5606-4/17 $31.00 © 2017 IEEE

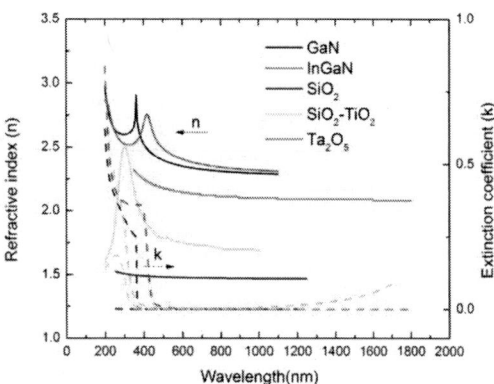

Fig. 1. Optical constants of the materials used in the simulation. Solid lines represent refractive index and dash lines represent extinction coefficient.

$$\frac{\partial p}{\partial t} = -\frac{1}{q}\nabla \cdot J_p - R + G \qquad (9)$$

where ε is the dielectric function, q is the elementary charge, ϕ is the electrostatic potential, n and p are the electron and hole densities, N_D^+ and N_A^- are the concentrations of ionized donors and acceptors, R and G are the recombination and generation rates, J_n and J_p are the current densities of electrons and holes. These equations are solved using the finite difference method. For stable convergence, the linearization of the Poisson equation and Scharfetter-Gummel Discretization of the continuity equations is required.

The TMM [10] is implemented for the ARC design, which determines how much light can be transmitted from the optical source into the InGaN device. The reflectivity and transmissivity of the optical stratified coating films are calculated effectively. Distance dependent generation rate of the solar cell then could be calculated by:

$$G(x) = \int \alpha(\lambda)T(\lambda)\cdot \frac{P(\lambda)}{hc/\lambda}\cdot e^{-\alpha(\lambda)x}\,d\lambda, \qquad (10)$$

where $\alpha(\lambda)$ is the absorption coefficient, $T(\lambda)$ is the transmissivity of the AR coating and $P(\lambda)$ is the incoming spectral power density ($mW/cm^2/nm$) on top of the device. AM1.5 spectrum is used in simulation. The thickness of each layer in the AR coating is optimized in order to minimize the front reflection loss current $\Delta J_{reflection\,loss}$ in the wavelength range from 300 to 1200 nm. The front reflection loss current is calculated by:

$$\Delta J_{reflection\,loss} = \frac{q}{hc}\int \lambda \cdot R(\lambda)P(\lambda)d\lambda \qquad (11)$$

where $R(\lambda)$ is the front reflection of the AR coating.

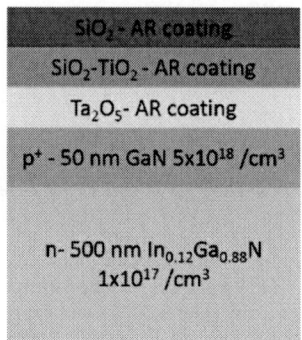

Fig. 2. Schematic diagrams of the InGaN/GaN solar cell with AR coating.

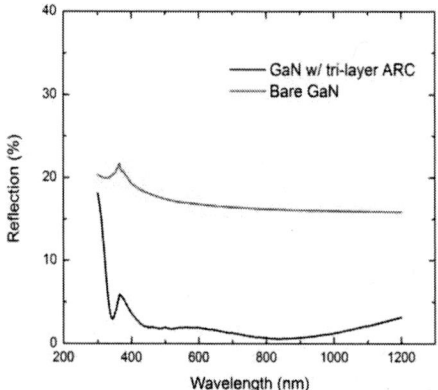

Fig. 3. Comparison of reflectance between a tri-layer ARC with a GaN substrate and a bare GaN substrate.

III RESULTS AND DISCUSSION

Figure 2 shows the simulated device structure which consists of a tri-layer anti-reflection coating stack and a GaN/InGaN bulk solar cell device. The GaN/InGaN heterostructure consists of a 50-nm-thick p-GaN layer and a 500-nm-thick n-InGaN layer. With the ARC, $\Delta J_{reflection\,loss}$ is reduced from 7.74 mA/cm^2 to 0.74 mA/cm^2 for 1sun from 300 to 1200 nm. The thicknesses of the SiO$_2$, SiO$_2$-TiO$_2$ and Ta$_2$O$_5$ layers are 83.33 nm, 50 nm and 55.55 nm, respectively. A reduction of front reflection loss is achieved with ARC as seen in Fig. 3.

The I-V characteristics of the simulated device is shown in Fig. 4. With the ARC, J_{sc} is improved from 1.72 mA/cm^2 to 2.12 mA/cm^2 under 1 sun due solely to a reduction of the front reflection. Although V_{oc} (~2.4 V) almost stays the same, the conversion efficiency is increased from 3.6% to 4.46% with the increase of J_{sc}.

978-1-5090-5606-4/17 $31.00 © 2017 IEEE

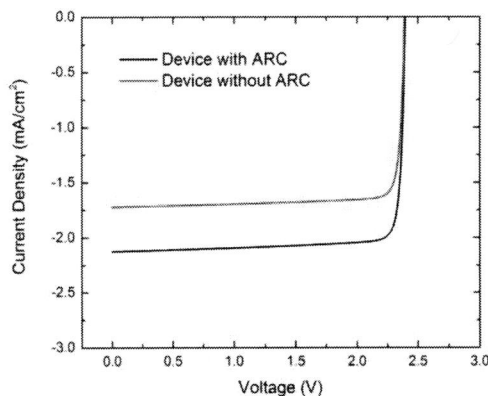

Fig. 4. Comparison of the I-V characteristics between the GaN/InGaN solar cell with and without ARC under 1sun,

IV. CONCLUSION

In summary, a reliable drift-diffusion solver of InGaN/GaN solar cell is developed for device design and optimization. In addition, a broadband optical coating is optimized to reduce the front-reflection loss and improve light absorption. The broadband ARC could potentially be used for multi-junction solar cell or hybrid concentrating solar thermal systems, and here shows a 23% improvement of the conversion efficiency of the device with an ARC.

ACKNOWLEDGEMENT

The work presented here has been funded in part by the Advanced Research Projects Agency-Energy (ARPA-E), U.S. Department of Energy, under Award Number DE-AR0000470.

REFERENCES

[1] C. A. M. Fabien, S. Member, M. Moseley, B. Gunning, W. A. Doolittle, S. Member, A. M. Fischer, Y. O. Wei, and F. A. Ponce, "Simulations, Practical Limitations, and Novel Growth Technology for InGaN-Based Solar Cells," vol. 4, no. 2, pp. 601–606, 2014.

[2] H. M. Branz, W. Regan, K. J. Gerst, J. B. Borak, and E. a. Santori, "Hybrid solar converters for maximum exergy and inexpensive dispatchable electricity," *Energy Environ. Sci.*, vol. 8, no. 11, pp. 3083–3091, 2015.

[3] L. a. Reichertz, I. Gherasoiu, K. M. Yu, V. M. Kao, W. Walukiewicz, and J. W. Ager, "Demonstration of a III–Nitride/Silicon Tandem Solar Cell," *Appl. Phys. Express*, vol. 2, no. 12, p. 122202, 2009.

[4] M. Nawaz and A. Ahmad, "A TCAD-based modeling of GaN/InGaN/Si solar cells," *Semicond. Sci. Technol.*, vol. 27,

no. 3, p. 35019, Mar. 2012.

[5] T. T. Mnatsakanov, M. E. Levinshtein, L. I. Pomortseva, S. N. Yurkov, G. S. Simin, and M. A. Khan, "Carrier mobility model for GaN," *Solid. State. Electron.*, vol. 47, no. 1, pp. 111–115, 2003.

[6] Y. C. Shen, G. O. Mueller, S. Watanabe, N. F. Gardner, a. Munkholm, and M. R. Krames, "Auger recombination in InGaN measured by photoluminescence," *Appl. Phys. Lett.*, vol. 91, no. 14, p. 141101, 2007.

[7] L. J.-H. Lin and Y.-P. Chiou, "Optical design of GaN/In_xGa_1-xN/cSi tandem solar cells with triangular diffraction grating," *Opt. Express*, vol. 23, no. 11, p. A614, Jun. 2015.

[8] S.-Y. Lien, D.-S. Wuu, W.-C. Yeh, and J.-C. Liu, "Tri-layer antireflection coatings (SiO2/SiO2–TiO2/TiO2) for silicon solar cells using a sol–gel technique," *Sol. Energy Mater. Sol. Cells*, vol. 90, no. 16, pp. 2710–2719, 2006.

[9] J. E. Sutherland and J. R. Hauser, "A computer analysis of heterojunction and graded composition solar cells," *IEEE Trans. Electron Devices*, vol. 24, no. 4, pp. 363–372, 1977.

[10] C. C. Katsidis and D. I. Siapkas, "Systems With Coherent, Partially Coherent, and Incoherent Interference," *Appl. Opt.*, vol. 41, no. 19, pp. 3978–3987, 2002.

Performance Enhancement of A GaAs Solar Cell with Colloidal Quantum Dots Embedded in Trenches

Chia-Jhe Shu[1], Yu-Ming Huang[2], Shun-Chieh Hsu[1], Jinn-Kong Shu[3], Jia-Lin Tsai[4], Pei-chen Yu[4],

Yung-Jr Hung[5], Chien-Chung Lin[1]

[1]Institute of Photonic System, National Chiao-Tung University, Tainan, 711, Taiwan
[2]Institute of Imaging and Biomedical Photonic, National Chiao-Tung University, Tainan, 711, Taiwan
[3]Department of Photonics, National Cheng-Kung University, Tainan, 701, Taiwan
[4]Department of Photonics, National Chiao-Tung University, Hsinchu, 300, Taiwan
[5]Department of Photonics, National Sun Yat-sen University Kaohsiung, 804, Taiwan

Abstract — A hybrid colloidal quantum dot (CQD) single junction GaAs solar cell was fabricated and measured. The CQDs are cast into the different sizes of trenches on the device surface. The introduction of CQD layer brings wide-range EQE enhancement, particularly in the ultra-violet band. The analysis reveals possible luminescent down-shifting effect due to our embedded quantum dots. As high as 2 times of EQE enhancement can be observed at specific short wavelength.

I. INTRODUCTION

The cost of fossil fuel raised sharply in the last decade which triggered both the economic downturn and the concerns of human sustainability. To solve this issue, many green technologies have been proposed and demonstrated. One of the promising candidates is the photovoltaic technology, which can change the solar energy into electricity. The photovoltaic devices, or solar cells, can be made of semiconductors such as silicon or GaAs. Currently GaAs still holds the highest efficiency of the single junction solar cell[1] and the silicon solar cell is the most installed one. For these devices, the power conversion efficiency (PCE) can be really high (higher than 25% in most cases), but their external quantum efficiency (EQE) is often suffered at shorter wavelength region due to the shallow penetration depth of the photons and high surface recombination rate. One solution resorts to the luminescent down-shifting (LDS) effect in the material, which converts high energy photons into low energy ones to penetrate deeper into the semiconductors. In the past, people used fluorescence dye doped glass or other materials to transform the high-energy photons (usually in the ultra-violet band) into the visible ones to increase the utilization of the solar UV photons[2]. However, the limited absorption bandwidth and the low Stokes shift between the illumination peak and absorption onset can really deteriorate the overall down-shift efficiency. In our lab, we use colloidal quantum dots (CQDs) for this DS conversion. The CQDs can be made of semiconductor with visible band gaps, and have strong absorption at UV band in general. Most important of all, they usually can re-emit visible photons very efficiently. Different materials, such as Si and GaAs, have been used for single junction solar cell combined with these CQDs for efficiency enhancement [3-5]. Different colors of CQDs have also been tested to find out the best conversion for the GaAs solar cell [5]. On the other hand, most of our previous studies were focused on the planar solar cell, and the effect of surface structure or etch trench was not investigated before. From the previous study for CQD LED, it is known that the proximity of CQDs in the active region (like quantum well) can sometimes activate the non-radiative energy transfer which can in terms improve the energy conversion efficiency[6]. To further exploit this idea, we introduce a GaAs solar cell with trench structure deep to p-n junction such that we hope it can bring strong LDS effect. This conversion capability can be evaluated through EQE and PV I-V measurements.

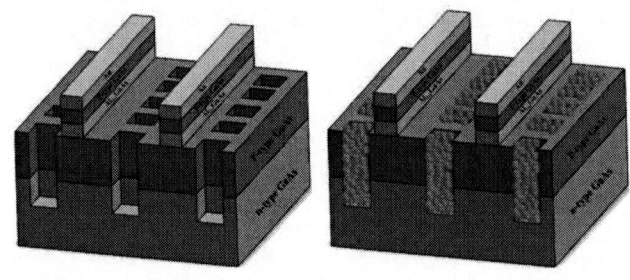

Fig. 1. Schematic diagrams of the trench solar cell with and without QDs.

II. DEVICE DESIGN, GROWTH, AND FABRICATION

The structure of the trenched solar cell with and without QD as shown in Fig. 1, and the top layer is p-type GaAs with AlGaAs as window layer and bottom layer is n-type GaAs. All the single junction solar cells in this study have the same epitaxial layer structure grown in metal organic chemical vapor deposition system. The GaAs wafers then underwent the regular semiconductor processing procedures. First, the wafers were deposited the Cr/Au(50/160 nm) as the p-type contact metal, and the lift-off step can form the grids for current conduction. Then we used the $H_3PO_4:H_2O_2:H_2O$ (1:2:37) etchant to remove AlGaAs and open the window layer. After

978-1-5090-5606-4/17 $31.00 © 2017 IEEE

they were exposed under UV light for 8 seconds and removed the positive photoresist by developer solution, we used the same etchant mentioned before to form the rectangle trench which was etched from the top of the n-type GaAs layer down to about 500nm in depth. These wafers were then lapped down to 200 um to facilitate the subsequent cleaving and dicing. Finally an AuGe/Au layer (50/160 nm) was deposited in the E-beam evaporation system to form the backside n-type contact. The subsequent anneal condition is 400°C for 20 seconds in N_2 environment. When all these steps were finished, the individual cells can be cleaved easily and the isolation between devices is formed by this cutting. The devices under study were cut into 1cm by 1cm square.

After the chip process is finished, the CQDs are casted on the top of the devices. For this experiment, we used the concentration as 5mg/ml and volume as 10μl/cm² of 520nm (CdSe/ZnS) from the UT Dots Company®, and we prepared 4 samples including a normal GaAs solar without trench structure and the other three with different sizes of trenches : 15μm×15 μm, 15 μm×50 μm and 15 μm×100 μm, and a total number of 18081 trenches was dug out uniformly. Each of these four samples was labeled as REF, 15 um, 50 um, and 100 um corresponding to the GaAs solar without trench structure, with sizes of 15μm×15 μm, 15 μm×50 μm and 15 μm×100 μm, respectively.

III. MEASUREMENT AND DISCUSSION

After the wafer processing is done, the finished devices with colloidal QDs were taken to measure their characteristics including photovoltaic J-V curves, external quantum efficiency (EQE) and reflectivity.

Fig. 2 shows the IV characteristics of four samples. All devices (REF, 15um, 50um, and 100um) have Jsc increasing (15.3, 4.9, 5.4, and 4.3 mA/cm² to 16.8, 5.5, 5.8, and 4.7 mA/cm² respectively) and the power conversion efficiencies are also increased (5.9, 1.7, 1.9, and 1.3% to 6.4, 2, 2.1, and 1.5% respectively).

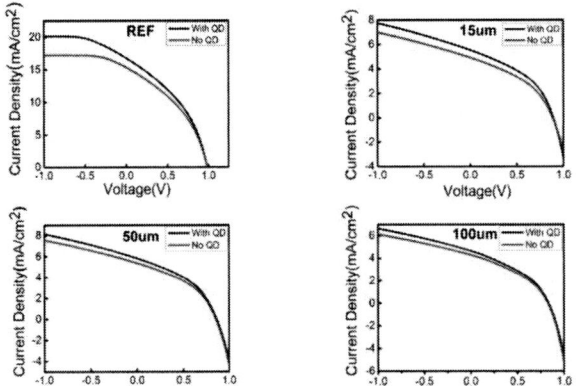

Fig. 2 The photovoltaic J-V characteristics of four samples

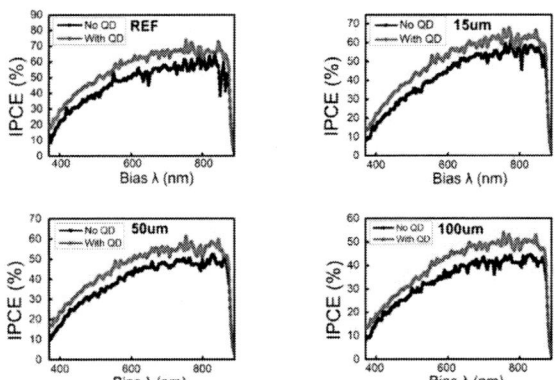

Fig. 3 The photovoltaic EQE of four samples.

From the measurements, the samples with QD have better response in terms of EQE and J-V curves. But to reveal the LDS effect in the enhancement, the surface reflectivity and the EQE spectra must be both measured and analyzed. Fig. 3 shows the EQE spectra from four samples. A wide-range enhancement of the quantum efficiency can be observed in all cases. According to our previous work [5], the EQE enhancement can be attributed to two factors: anti-reflection (AR) and the LDS effects. The AR effect rises from the intermediate refractive index layer introduced by the CQD layers. Their influences on J_{sc} can be expressed by the following formula [5]:

$$\Delta J_{sc} = \int \left(IQE_{QD}\left(1 - R_{QD}\right) - IQE_{ref} \times I(\lambda) \, d\lambda \right.$$

$$= \int \left(\frac{IQE_{QD}}{IQE_{ref}} \frac{1 - R_{QD}}{1 - R_{ref}} - 1 \right) \left(IQE_{ref}\left(1 - R_{ref}\right) \right) \times I(\lambda) d\lambda$$

$$= \int \left(\frac{IQE_{QD}}{IQE_{ref}} \frac{1 - R_{QD}}{1 - R_{ref}} - 1 \right) (EQE_{ref}) \times I(\lambda) d\lambda$$

$$= \int \left((1 + \Delta_{LDS})(1 + \Delta_{AR}) - 1 \right) \times EQE_{ref} \times I(\lambda) d\lambda$$

$$= \int \left(\Delta_{LDS} + \Delta_{AR} + \Delta_{LDS}\Delta_{AR} \right) \times EQE_{ref} \times I(\lambda) d\lambda$$

$$...(1)$$

$$EQE = IQE \times (1\text{-}R)..(2)$$

From equation (1), the factor $(1+\Delta_{LDS})$ is the internal quantum efficiency (IQE) enhancement and the $(1+\Delta_{AR})$ term is the AR enhancement. From equation (2), the IQE can be related to the EQE via the surface reflection (R)[7]. In this modification of surface reflection, the input of photons can be normalized and thus proper comparison can be made. The overall enhancement, which can be obtained by EQE measurement, can be written as[5]:

$$\frac{EQE_{QD}}{EQE_{ref}} = \left((1 + \Delta_{LDS})(1 + \Delta_{AR}) \right) ....(3)$$

By our calculation, the current density of the LDS effect

for the sample of reference sample (the flat case) is approximately 0.77 mA/cm^2, and the current density of the antireflection enhancement effect is approximately 2.08 mA/cm^2. For the sample of 15 μm, the current density of the LDS effect is approximately -0.67 mA/cm^2, and the current density of the antireflection effect is approximately 2.7 mA/cm^2. For the sample of 50 μm, the current density of the LDS effect is approximately -0.23 mA/cm^2, and the current density of the antireflection effect is approximately 2.3 mA/cm^2. For the sample of 100 μm, the current density of the LDS effect is approximately -0.94 mA/cm^2, and the current density of the antireflection effect is approximately 3.43 mA/cm^2. The negative of LDS values indicate that the most of the enhancement comes from the AR effect and very little of LDS enhancement can be detected.

In the above calculation, important notes can be found: If the EQE enhancement in equation (3) is higher than the AR enhancement curve, the value of $(1+\Delta_{LDS})$ would be bigger than one, then the Δ_{LDS} exists. On the other hand, if the EQE enhancement is close to or even lower than the AR enhancement curve, then the LDS probably does not exist and only the refractive index difference of CQD layer works for the efficiency increase. The extra scattering and re-absorption issues in the QDs are not investigated in this case, and could have profound effects on the overall PCE. As Fig. 4 shows, in the range of photon wavelength shorter than 400nm, all the samples show LDS effects because of the above-AR enhancement, and strong peaks can be observed. After 400nm wavelength, EQE enhancement and AR enhancement were very close indicating that there is no LDS effect.

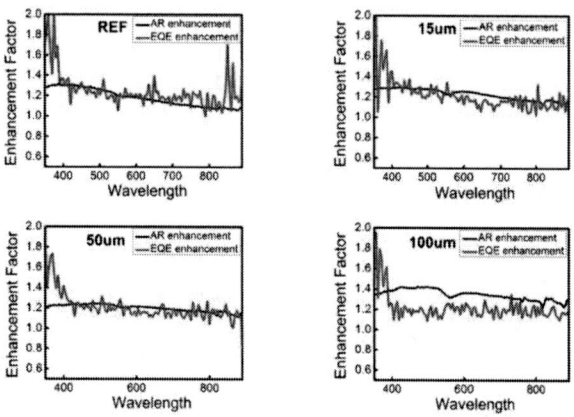

Fig. 4. The photovoltaic enhancement factor of four samples

Currently our devices with trenches have worse PCE when compared to the planar one, which can be attributed to the exposure of the p-n junction during the trench etch. Further optimization on the sidewall protection and less depth in etch can be considered and the PCE can be further increased. The

sidewall protection can also reduce the leakage of the diode which can increase the fill factor of the devices.

Another possible improvement is to better dispense the CQDs into the trenches. We use the manual method to push the CQDs into the trenches in this paper, but the residual nanoparticles on the top surface is inevitable, which might mix our results with the regular planar type. A better dispense method can be developed to fill the trench with CQDs but not on the top surface.

IV. CONCLUSION

In conclusion, we demonstrate the enhancement of hybrid CQD single junction GaAs-based solar cells with different sizes of trenches. The purpose of trench is to bring the LDS effect closer to the p-n junction. The J$_{sc}$ and PCE of the solar cells with QD behave better than those without QD. The EQE and enhancement factor show the dominance of LDS effect in shorter wavelength region while the AR effect prevails in longer wavelength region. The trench structure seems not to be very effective because of possible deep trench depth and sidewall current leakage. In the future work, we can explore different sidewall passivation and trench design to optimize the LDS effect.

V. Acknowledgment

The authors would like to thank the financial support of Ministry of Science and Technology of Taiwan, ROC through the grant number: MOST 104-2628-E-009-013-MY3.

REFERENCES

[1] M. A. Green, K. Emery, Y. Hishikawa, W. Warta, and E. D. Dunlop, "Solar cell efficiency tables (version 48)," *Progress in Photovoltaics: Research and Applications,* vol. 24, pp. 905-913, 2016.

[2] C.-C. Lin, H.-V. Han, H.-C. Chen, K.-J. Chen, Y.-L. Tsai, W.-Y. Lin, *et al.,* "Highly Efficient Multiple-Layer CdS Quantum Dot Sensitized III–V Solar Cells," *Journal of Nanoscience and Nanotechnology,* vol. 14, pp. 1051-1063, // 2014.

[3] C.-C. Lin, H.-C. Chen, Y. L. Tsai, H.-V. Han, H.-S. Shih, Y.-A. Chang, *et al.,* "Highly efficient CdS-quantum-dot-sensitized GaAs solar cells," *Optics Express,* vol. 20, pp. A319-A326, 2012/03/12 2012.

[4] H.-C. Chen, C.-C. Lin, H.-W. Han, Y.-L. Tsai, C.-H. Chang, H.-W. Wang, *et al.,* "Enhanced efficiency for c-Si solar cell with nanopillar array via quantum dots layers," *Optics Express,* vol. 19, pp. A1141-A1147, 2011/09/12 2011.

[5] H.-V. Han, C.-C. Lin, Y.-L. Tsai, H.-C. Chen, K.-J. Chen, Y.-L. Yeh, *et al.,* "A Highly Efficient Hybrid GaAs Solar Cell Based on Colloidal-Quantum-Dot-

Sensitization," *Scientific Reports,* vol. 4, 07/18/online 2014.

[6] S. Chanyawadee, P. G. Lagoudakis, R. T. Harley, M. D. B. Charlton, D. V. Talapin, H. W. Huang, *et al.,* "Increased Color-Conversion Efficiency in Hybrid Light-Emitting Diodes utilizing Non-Radiative Energy Transfer," *Advanced Materials,* vol. 22, pp. 602-606, 2010.

[7] S. M. Sze, *Physics of Semiconductor Devices*, 2nd ed. New York: John Wiley & Sons, 1981.

Enhanced Photoresponse of InN Devices Using Indium-Tin Oxide Nanorods

Lung-Hsing Hsu[1,2], Yuh-Jen Cheng[2], Peichen Yu[3], Hao-Chung Kuo[3], and Chien-Chung Lin[4]*

[1]Institute of Lighting and Energy Photonics, National Chiao Tung University, Tainan 711, Taiwan
[2]Research Center for Applied Sciences, Academia Sinica, Taipei 11529, Taiwan
[3]Institute of Electro-Optical Engineering, National Chiao Tung University, Hsinchu 300, Taiwan
[4]Institute of Photonic System, National Chiao Tung University, Tainan 711, Taiwan

*E-mail: chienchunglin@faculty.nctu.edu.tw

Abstract — **Enhanced infrared photoresponse is observed in InN/ITO rods fabricated by LP-MOCVD and oblique-angle electron beam evaporation. The higher growth temperature and more V/III could trend toward hexagonal InN pillars epitaxy. The peak energy blue-shift phenomenon due to higher Fermi-level to acceptor emission was investigated by temperature dependent photoluminescence (PL) measurements. ITO nanorods enhanced the broadband and angle-independent anti-reflection in the range between 400 nm and 2000 nm. The InN /ITO rods photodetection device was demonstrated with enhanced IR response, and the portion photocurrent (610 nm long pass) of InN detection as high as 24.5% measured via AM1.5G solar simulated spectra.**

Index Terms — **Infrared, Indium compounds, photodetection devices, nanostructured materials.**

I. INTRODUCTION

Recently, nitride-based alloys plays a critical role in the applications of optoelectronic devices such as lasers [1], light emitting diodes [2], [3] solar cell [4], and photodetectors [5], [6]. One of the key features about nitride-based materials is the direct bandgap energy covering from 0.7 eV for InN to 3.4 eV for GaN, and thus provides wide range absorption from ultraviolet to infrared [7]. In order to achieve the concept, these nano-scale structures can either be efficient in light trapping or be embedded inside devices as an extra absorptive layer. Furthermore, indium tin oxide (ITO) emerges as transparent conductive oxides (TCO) which enhances the external efficiency of devices due to the improvement of light extract and anti-reflective mechanism brought by intermediate refractive index [8]–[10]. In this study, we will optically investigate such deployment of nano-structures. High quality InN nanopillars array and ITO nanorods were fabricated into a photodetector to demonstrate the infrared (IR) detection capability in nanoscale materials.

In this study, we employed a low-pressure metal organic chemical vapor deposition (LP-MOCVD) and oblique-angle electron beam evaporation to grow different InN nanostructures/ITO nanorods. The extended InN emission peak and photoluminescence peak shift phenomenon can be observed in cryogenic temperature. Furthermore, we employ a xenon lamp (AM1.5G) with long pass 610 nm / 920 nm spectra filter to evaluate the current-voltage (*I-V*) under dark and different illumination conditions of the InN photodetection device.

II. EXPERIMENT

The nitride-based epitaxy layers with InN pillars structures were grown on c-plane sapphire substrate by a LP-MOCVD. On the bottom sapphire substrate, the epitaxial layers consist of a 25-nm-thick low-temperature GaN nucleation layer, a 1.5-µm-thick un-doped GaN epilayer (u-GaN), a 1-µm-thick Mg doped GaN epilayer (p-GaN), followed by 300-nm-thick InN epilayer that was grown at epitaxy temperature varied from 450 to 550 degrees Celsius. During the growth, trimethylgallium (TMGa), trimethylindium (TMIn), and ammonia (NH_3) were employed as the reactant source element for gallium, indium, and nitrogen sources, respectively. For the growth of InN nanorods and nanopillars, V/III ratio of NH_3 and TMIn was varied from 2×10^4 to 5×10^5 during the InN formation process for several minutes. After the LP-MOCVD growth, a SiO_2 layer was first deposited and the e-beam evaporation was applied to grow a 80-nm-thick ITO thin film and 300-nm-height ITO rods deposited at an oblique-angle method. Then the Ni/Cu (20nm/500nm) is used for p- and n-type contact metals, respectively. The standard mesa process was defined by dry-etch with the size of 18 mil by 18 mil (465 µm by 465 µm). The schematic diagram of the finished standard device (InN film) is depicted in Fig. 1. After the growth of InN layer, the samples were characterized using field-emission scanning electron microscopy (SEM), low temperature photoluminescence (PL) using the 532 nm line of a solid state laser as the excitation source, high resolution X-ray diffraction (HRXRD) performed by a PANalytical X'Pert Pro X-ray diffractometer, UV-Visible spectrophotometer, and bidirectional reflectance distribution function (BRDF). Finally, after normal clean-room processes are finished, a 1000W Class A solar simulator with a xenon lamp equipped with an

978-1-5090-5606-4/17 $31.00 © 2017 IEEE

Air Mass 1.5 Global (AM1.5 G) filter and 610 nm / 910 nm long pass spectra filter was employed to evaluate current-voltage (*I-V*) under dark and illumination conditions.

In this work, different kind of InN nanostructures with varied growth temperature and V/III ratio are grown on c-plane GaN substrate, and different InN formation like film, rods and pillars can be found.

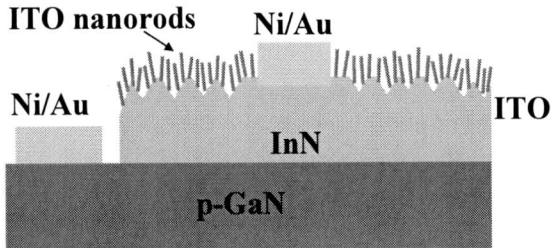

Fig. 1.   Schematic of InN film/ITO rods photodetection devices.

III. RESULTS AND DISCUSSIONS

Fig. 2 shows the SEM images of InN film, InN nanorods, and InN nanopillars on the GaN substrate. Fig. 2(d) also exhibits ITO nanorods grown on InN nanopillars by oblique-angle method. By controlling the more V/III ratio and higher growth temperature, almost perfect (hexagonal) shape of InN nanopillar can be formed. The diameter and height of pillars are about 1.2 μm and 800 nm. The density of InN pillars is about 9.8×10^5 cm^{-2}.

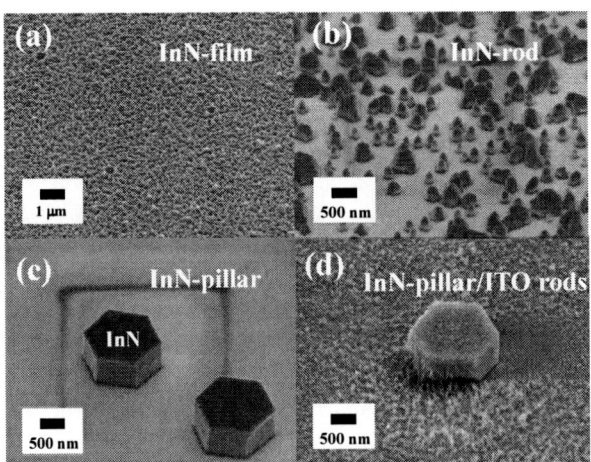

Fig. 2   Bird-view SEM image: (a) InN film and (b) InN nanorods (c) InN nanopillars (d) ITO rods grown on InN nanopillars.

Fig. 3 shows the detailed XRD 2-theta scan of InN samples measured by high resolution X-ray diffractometer. Several pronounced peaks can be observed in this plot: sapphire peak around 42°, GaN (0002) peak at 34.5°, and InN (0002) peak at 31.38°. One particular peak stands out around 33°, and it was

identified as the possible peak for InN(1-101) from previous report [11], which eliminates the possibility of indium existence.

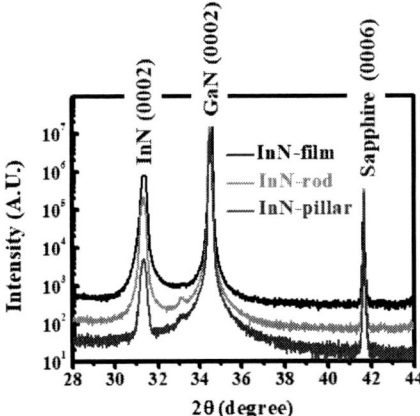

Fig. 3.   The X-ray diffraction pattern of InN samples on GaN substrate.

Fig. 4 shows the PL spectra of InN samples measured under cryogenic and room temperature, which was excited by 10mW of pumping power. The PL spectra of InN samples exhibit the peak energy centered around 0.75 eV (nanorod) and 0.78 eV (nanopillar). From the peak energy position, InN-film reveals a slight blue-shift phenomenon of main peak and a broad band signal at high energy range indicating the higher Fermi-level electron to acceptor state [12]. The room temperature PL intensity is about 13~40 times weaker than the 10K's samples. In the intensity ratio of cryogenic and room temperature, two nanostructure sample are higher than InN-film.

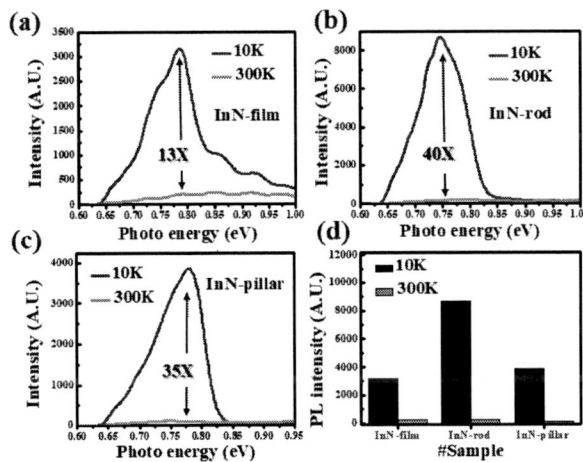

Fig. 4.   Photoluminescence of InN film, rods, pillars under cryogenic and room temperature.

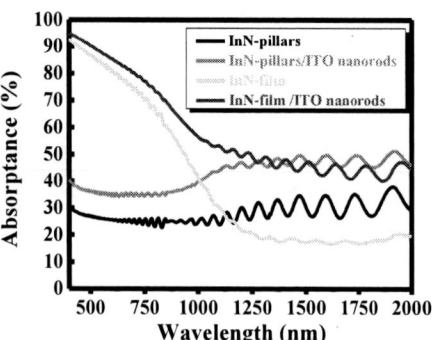

Fig. 5. The absorptance correlated wavelength of InN film/ITO rods and InN nanopillars/ITO rods.

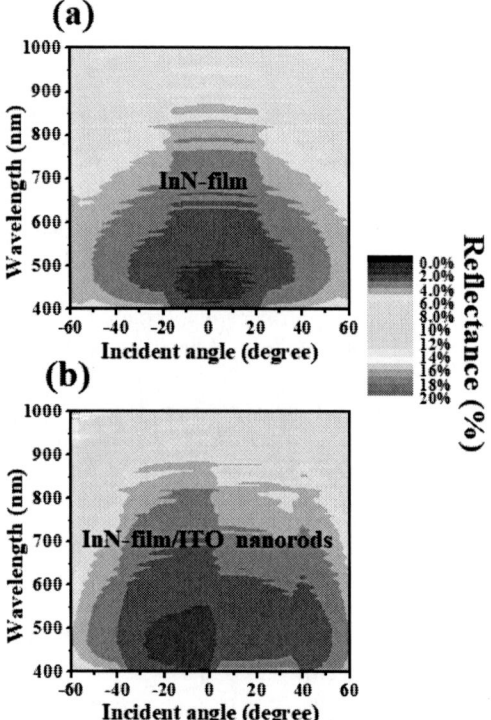

Fig. 6. Measured angle-resolved reflective spectra for (a) InN film and (b) InN film/ITO rods.

As shown in Fig. 5, ITO nanorods effectively enhanced the absorption in the InN templates (InN film and InN pillars) in the range between 400 nm and 2000 nm. We found the deposited ITO nanorod reduce the reflectance compared with initial samples, the wavelength ranging from 400 nm to 2000 nm, related to the previous report [8]–[10]. The absorptance of InN-pillars/ITO nanorods is the highest value correlated wavelength between 1250 nm and 2000 nm. Fig. 6 shows the measured angular reflective spectra for InN film/ITO nanorods that inhibits the broadband reflectivity for normal

incidence and reduced the angular-reflective spectrum. Large-area omnidirectional antireflective structures are very beneficial to photodetection. The nanorods sample shows the inhibited reflectivity over the spectrum from 400 to 1000 nm and the angle-resolved up to 60 degree. In addition to the InN-film/ITO rods sample also exhibits the more inhibited broadband reflectivity at the wavelength up to 800 nm, compared with InN-film sample. However, the previous UV-VIS measurement indicates the stronger absorption owing to the antireflection contributed to ITO nanorods in IR range between 1000 and 2000 nm. Fig. 7 exhibits the *I-V* characteristics under dark and simulated AM1.5G illumination conditions. Enhanced photoresponse rises from antireflection in ITO nanorods. Furthermore, the long pass 610 nm / 920 nm filter are useful to evaluate the photocurrent (I_{photo}) of long wavelength spectra from InN samples. In our series of InN devices, we shows the good results from InN pillar/InN rods photodetection devices. As high as 24.5% (610 nm long pass filter) and 19% (920 nm long pass filter) of the total I_{photo} can be attributed to InN pillars contribution at reverse bias of 4V. It actually increases by 6.3 percent (610 nm long pass filter) and 10 percent (920 nm long pass filter) as compared with InN pillars without ITO nanorods.

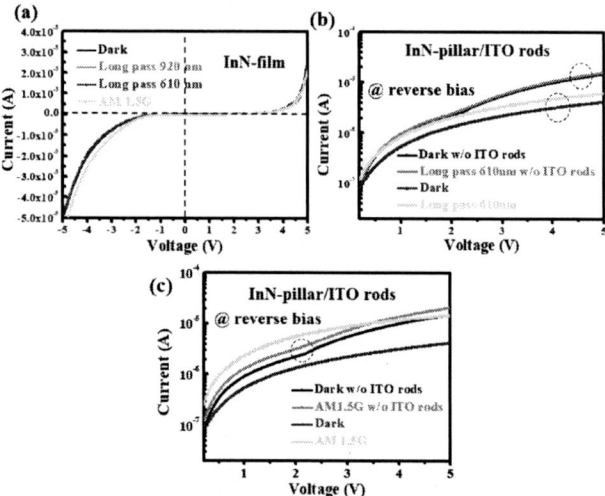

Fig. 7. (a) The generic *I-V* characteristics of InN film photodetector under dark, simulated AM1.5G illumination, and long pass 610 nm / 920 nm of AM 1.5G illumination. (b) InN pillars/ITO rods compared with that w/o ITO rods under long pass 610 nm and (c) simulated AM 1.5G illumination.

IV. CONCLUSION

In summary, we successfully fabricated the novel photodetection device in InN pillars/ITO rods materials. It exhibits the enhanced IR response owing to ITO nanorods contributing to antireflection in wavelength range between 400 and 2000 nm. The high quality InN pillars and ITO nanorods can be integrated into real devices and the visible

/near infrared portion of the photocurrent as high as 24.5% can be measured via filtered AM1.5G simulated spectra. The nanoscale combination InN/ITO could increases by 6.3 percent (610 nm long pass filter) IR photocurrent. It should be the purpose application for the unique long-wavelength operating capability in InN materials and the broadband light-trapping through ITO nanorods.

ACKNOWLEDGEMENT

The work was supported by the National Science Council of Taiwan through contracts: MOST 104-2628-E-009-013-MY3.

REFERENCES

[1] S. Nakamura, M. Senoh, S. Nagahama, N. Iwasa, T. Yamada, T. Matsushita, H. Kiyoku, Y. Sugimoto, T. Kozaki, H. Umemoto, M. Sano, and K. Chocho, "InGaN/GaN/AlGaN-based laser diodes with modulation-doped strained-layer superlattices grown on an epitaxially laterally overgrown GaN substrate," *Applied Physics Letters*, vol. 72, pp. 211-213, 1998.

[2] E. F. Schubert, *Light Emitting Diodes*, 2nd ed. Cambridge, U.K.: Cambridge Univ. Press, 2003, pp. 21–22.

[3] C. H. Chiu, P. M. Tu, C. C. Lin, D. W. Lin, Z. Y. Li, K. L. Chuang, J. R. Chang, T. C. Lu, H. W. Zan, C. Y. Chen, H. C. Kuo, S. C. Wang, and C. Y. Chang, "Highly Efficient and Bright LEDs Overgrown on GaN Nanopillar Substrates," IEEE *Journal of Selected Topics in Quantum Electronics*, vol. 17, pp. 971-978, 2011

[4] N. Hieu Pham Trung, C. Yi-Lu, S. Ishiang, and Z. Mi, "InN p-i-n Nanowire Solar Cells on Si," *IEEE Journal of Selected Topics in Quantum Electronics*, vol. 17, pp. 1062-1069, 2011.

[5] L.-H. Hsu, C.-T. Kuo, J.-K. Huang, S.-C. Hsu, H.-Y. Lee, H.-C. Kuo, C.-C. Lin et al., "InN-based heterojunction photodetector with extended infrared response," *Optics Express*, vol. 23, pp. 31150-31162, 2015.

[6] L.-H. Hsu, C.-C. Lin, H.-V. Han, D.-W. Lin, Y.-H. Lo, Y.-C. Hwang, et al., "Enhanced photocurrent of a nitride–based photodetector with InN dot-like structures," *Optical Materials Express*, vol. 4, pp. 2565-2573, 2014.

[7] J. Wu, "When group-III nitrides go infrared: New properties and perspectives," *Journal of Applied Physics*, vol. 106, pp. 011101, 2009.

[8] P.-C. Tseng, P. Yu, H.-C. Chen, Y.-L. Tsai, H.-W. Han, M.-A. Tsai, et al., "Angle-resolved characteristics of silicon photovoltaics with passivated conical-frustum nanostructures," *Solar Energy Materials and Solar Cells*, vol. 95, pp. 2610-2615, 2011.

[9] Xi, J. Q., et al., "Optical thin-film materials with low refractive index for broadband elimination of Fresnel reflection." *Nat Photon*, vol. 1, pp. 176-179, 2007.

[10] Chia-Hua, C., et al., "Combined micro- and nano-scale surface textures for enhanced near-infrared light harvesting in silicon photovoltaics." *Nanotechnology*, vol. 22, pp. 095201, 2011.

[11] J. Kamimura, K. Kishino, and A. Kikuchi, "Growth of very large InN microcrystals by molecular beam epitaxy using epitaxial lateral overgrowth," *Journal of Applied Physics*, vol. 117, pp. 084314, 2015.

[12] M. Feneberg, J. Däubler, K. Thonke, R. Sauer, P. Schley, and R. Goldhahn, "Mahan excitons in degenerate wurtzite InN: Photoluminescence spectroscopy and reflectivity measurements," *Physical Review B*, vol. 77, pp. 245207, 2008.

Plasmonic Silver Structures for Improved Perovskite Photovoltaic Performance

Arul Varman Kesavan, Arun D Rao and Praveen C Ramamurthy

Department of Materials Engineering, Indian Institute of Science, Bangalore, Karnataka, Bangalore-560012, India

Abstract — In photovoltaics, photon harvesting is one of the important factors for the enhancement of the power conversion efficiency. The prime factor for the device performance is optical absorption by the active layer. Photon harvesting by the active layer can be enhanced by indirect means such as plasmonic light harvesting. In this way, the light is focused towards the active layer to improve the effective light absorption. The mechanism by which light trapping by the metallic nano-structures whose dimension usually less <40nm and generates local surface plasmon resonance (LSPR) which acts like a sub-wavelength antennas as a result plasmonic near field is coupled to the active layer. In the other case, metallic nanostructure having dimension >40 nm behaves like sub-wavelength scattering center and hence traps freely propagating plane wave of incident light and couples to the semiconducting layer. Therefore, it is possible to achieve light trapping by introducing metallic nanostructures at the device interfaces. In this work, poly-dispersed metallic nanostructures (silver nanoparticles AgNP) were embedded at the active layer/cathode interface to study the perovskite photovoltaic device performance. It is observed that device with AgNP at the active layer/cathode interface showed significant enhancement in optical absorption. As a result device with Ag metallic nanostructure showed significant improvement in short circuit current. It is observed that metallic nano-structures at cathode interface aids in light trapping and also reduction in series resistance. These coupled effects of optical and electrical enhancement tend to improved power conversion efficiency in the device.

Index Terms — Plasmonic, metal nanoparticles, perovskite solar cell, carrier concentration, silver.

I. INTRODUCTION

The great attraction of organo-metal halid perovskite photovoltaic material is due to the wide optical absorption range from 380 nm to 800 nm and higher optical absorption coefficient ($10^5 cm^{-1}$) and electrical properties. This type of materials exhibits higher electron and hole mobility as well as high charge carrier diffusion length which is greater than $1 \mu m$. Further, the device fabrication is requires low processing temperature and also easy fabrication steps [1]. The various methodologies have been adopted for the further improvement in the power conversion efficiency. One such method is improving the optical absorption by forming the larger grain size [2]–[4]. The other indirect method is using plasmonic methods. This can be achieved by introducing the plasmonic nanoparticles at the device interface or in the active layer [5]. The plasmonic particle enhances light absorption by mean of increasing the optical path length in the active layer. Increased

light path in the active layer leads to excess exciton formation and enhance more charge carrier generation. Only few works is reported in the literature about plasmonic based perovskite solar cells [6]. In this study, silver metallic nanoparticles were introduced at the perovskite layer/cathode interface. Here, metallic nanoparticle sizes of about 20 to 100 nm were used. Therefore, the resultant light trapping in the device should be combined effect of light absorption and light scattering mechanism. The resultant change in the optical and electrical properties of the device is discussed.

II. EXPERIMENTAL SECTION

A. Materials

Indium tin oxide (ITO) coated glass substrate purchased from Xin-yan Technology Limited Hong Kong, China. poly(3,4-ethylenedioxythiophene)-poly(styrene sulfonate) (PEDOT:PSS(Al4083) obtained from Heraeus Precious Metal GmbH & Co, Germany, bis-[6,6]-Phenyl C61 butyric acid methyl ester (PCBM) obtained from Nano-C Corporation U.S.A. Lead iodide (PbI_2) obtained from Alfa Aesar, Methyl ammonium iodide (MAI) purchased from dysole and Aluminium wire, chlorobenzene and dimethyle sulfoxide (DMSO) were purchased from Sigma Aldrich, γ-butyrolactone (GBL) is obtained from TCI chemical India Pvt.,Limited.

B. Device Fabrication and Characterization

The perovskite photovoltaic devices are fabricated as follows. The patterned ITO was cleaned by soap solution, DI water, acetone and isopropanol for 10 min each and the substrates were dried by nitrogen gas. The dried substrates were UV-Ozone treated to make the surface more hydrophilic. Then PEDOT:PSS was spin coated on ITO surface and annealed at 120°C for 15 min. The perovskite solution was prepared by dissolving 800 mg of PbI_2:MAI::3:1 in the solvent mixture of GBL:DMSO::7:3 ratio. The prepared solution was stirred at 100°C for 7 hours and obtained solution was filtered filter. The filtered solution was spin coated over the PEDOT:PSS layer and annealed at 110°C for 15 min inside the glove box. The PCBM solution was prepared by dissolving 20 mg of PCBM in 1ml of dichlorobenzene and kept for over stirring at room temperature. The obtained solution is filtered and spin coated at 4000 rpm over the active layer. To prepare the composite layer mixture of PCBM solution with various weight ratio of AgNP were mixed. Then the mixture was kept for stirring. After the active layer annealing electron buffer layer (PCBM) was deposited at 4000 rpm. Aluminium cathode was deposited by thermal evaporation. The schematic of the device architecture is shown in Fig 1.

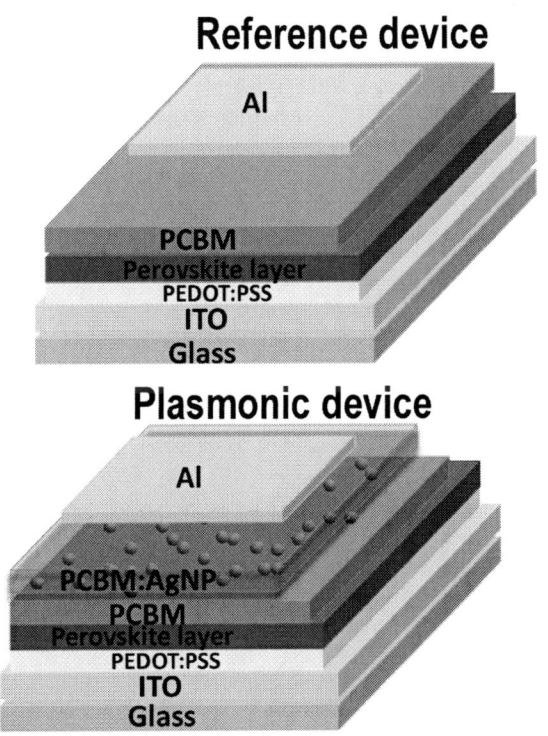

Fig. 1   Schematic of device architecture, with (plasmonic device) and without (reference device) AgNP in ETL.

III. RESULTS AND DISCUSSION

The Ag nanoparticles were characterized for phase conformation by X-ray diffraction method and it is shown in Fig.2 (a). From the obtained XRD spectrum analysis crystallite size of the particle was calculated using Scherrer formula from (111) peak of the spectrum and it was found to be 49±6 nm. Optical absorption properties of the nanoparticles were studied using UV-Visible spectroscopy and obtained spectrum is shown in Fig 2(b). The nanoparticle morphology is studied by scanning electron microcopy (TEM) as shown in Fig 2(b). The Ag particle distribution comprises of the wide size range various from 20 nm to 100 nm. However, Ag nanoparticles are spherical in shape.

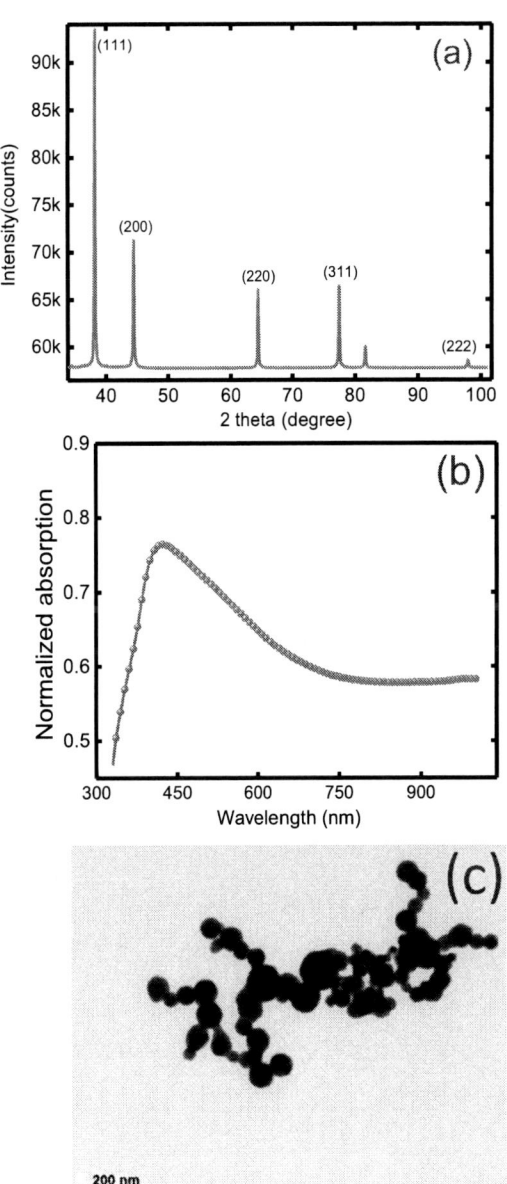

Fig. 2   (a) XRD diffraction pattern, (b) UV-Visible absorption spectrum and (c) TEM micrograph of Ag nanoparticles.

The optical characteristic of the perovskite active layer absorption and emission characteristic was studied by UV-Visible spectrometer and fluorescence spectrometer respectively. The optical characteristics of the device were studied for the optimized $PC_{61}BM$: metallic nanoparticle (MNP) ratio. The absorption spectra were measured for the typical device layer except the cathode which are named as follows reference device: $ITO/PEDOT:PSS/CH_3NH_3PbI_3/PC_{61}BM$,

978-1-5090-5606-4/17 $31.00 © 2017 IEEE   1615

plasmonic device: ITO/PEDOT:PSS/CH₃NH₃PbI₃/PC₆₁BM/PC₆₁BM:AgNP.

plasmonic device: ITO/PEDOT:PSS/CH$_3$NH$_3$PbI$_3$/PC$_{61}$BM/PC$_{61}$BM:AgNP.

Further, the emission property of the perovskite layer was studied to correlate with charge transfer properties. On comparing the emission characteristics, the Reference device emission peak intensity is significantly higher than plasmonic device. This shows that presence of metallic nanoparticle at the cathode interface improves charge transfer to the nanoparticle. This effective charge transport helps in improving device fill factor (FF).

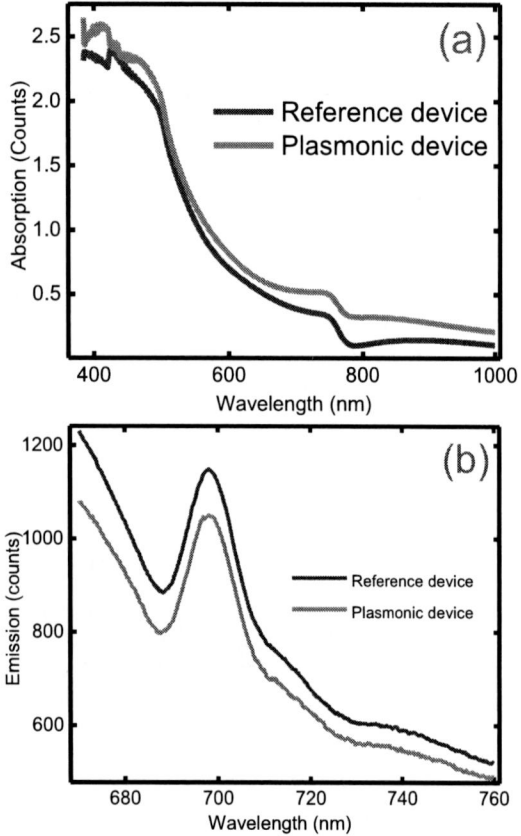

Fig. 3. (a) Optical absorption spectra and (b) photoluminescence spectra of the reference and plasmonic device.

The depletion nature of reference and plasmonic device was analyzed using Mott-Schottky (MS) relation [7]. The MS characteristics were studied using the measured capacitance-voltage characteristics. The C-V spectrum was measured by applying the 1 kHz ac signal. The carrier concentration was studied

$$1/C^2 = \frac{2}{\varepsilon_0 q A^2 N}(V_{bi} - V) \tag{1}$$

Here, C is capacitance, ε_0 is permittivity of free space, ε relative permittivity (17 used for the calculation), q is elementary

charge, N is carrier concentration and A device active area, V is applied voltage. V_{bi} is built-in voltage. The calculated carrier concentration is 7.81×10^{15} cm^{-3} and 2.75×10^{16} cm^{-3} respectively for reference and plasmonic device respectively. The ETL with AgNP showed higher carrier concentration. This sugests that metallic nanoparticles in the ETL at the back contact side helps in improved device performance by means of enhancing the carrier concentration. Hence, the carrier concentration enhancement should be due to optical absorption improvement. Further, the trap density of states was calculated [using equation (3)] in order to evaluate the device performance [8]–[10]. The trap density of states was calculated using C-f spectrum of the device. The depletion width (W) is calculated using equation (2) by substituting the calculated carrier concentration (N) and built-in voltage which is obtained from MS analysis.

$$w^2 = \frac{2\varepsilon_s}{qN}(V_{bi} - V) \tag{2}$$

$$N_t(E_\omega) = -\frac{V_{bi}}{qW} * \frac{dC}{d\omega} * \frac{\omega}{k_B T} \tag{3}$$

Here, V_{bi} is built-in voltage, q is electronic charge, W is depletion width, C is capacitance, ω angular frequency, k_B is Boltzmann constant, T is temperature. The parameter V_{bi} and W is calculated from the using C-V analysis. The calculated trap density of states value is 13.55×10^{17} cm^{-3}eV^{-1} and 3.93×10^{17} cm^{-3}eV^{-1} respectively for reference and plasmonic device. The obtained trap state density value for the plasmonic device is significantly lower than reference device. Obtained results suggest that the presence of AgNP in the PCBM layer reduces the trap density of states. This should be due to effective charge transport due to metal nanoparticles; this is supported by the photoluminescence. That is the intensity of emission peak position at ~ 698 nm for device is significantly lower for plasmonic device. This will occur when there is an effective charge transfer in the device.

The current-voltage characteristic of the photovoltaic is as show in Fig.4. The measured short circuit current (J_{SC}) is 16.71 mAcm^{-2} for reference device and 18.15 mAcm^{-2} for plasmonic device. The plasmonic device showed significantly higher J_{SC} than reference device. A small improvement is observed in fill factor (FF) from 65% to 67.8%. The improved FF is correlated with the reduce intensity of emission at ~ 698 nm in the case of plasmonic device. This also supports the reduced series resistance (8.97 to 6.26 Ωcm^2) and the improved J_{SC} (increment is around 8%) in plasmonic device. Therefore, it can be interpreted that the power conversion efficiency is mainly attributed to the improvement J_{SC}. These arise due to improved optical absorption, effective charge transfer and reduction in the R_s by the incorporation of the AgNP at the ETL/cathode interface.

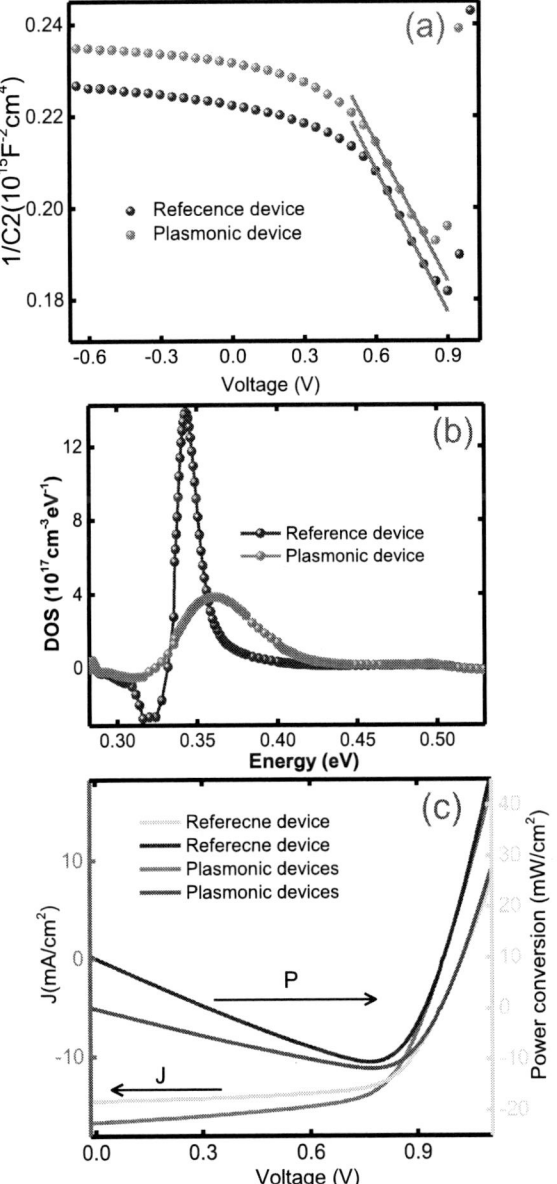

Fig. 4. (a) Mott-Schottky plot the device, (b) trap density of states plot and (c) photovoltaic characteristic of the fabricated device

IV. CONCLUSION

The Ag nanoparticle with random particle size distribution was used embedded with PCBM electron transport layer to study effect of the perovskite photovoltaic. As a result of Ag nanoparticles at the cathode interface improves the optical absorption which leads to enhancement in photocurrent. This

is achieved due to particle size dependent light trapping at the active layer. Further, the effective reduction in the emission peak intensity of the reference device shows that presence of AgNP enhances charge transfer from the perovskite active layer to the NP. This is one of the direct evidence of the charge transfer across semiconductor/Al interface. The obtained result suggests that the resultant improvement in the power conversion efficiency is a combined effect of electrical (reduction in series resistance and increment in current density and reduced trap state density) and improved optical absorption. Since, the light finds longer path way inside the active layer; it increases the light harvesting and hence charges carrier concentration. Moreover, the particle size distributions wide range, the light enhancement should be due to light absorption and the light scattering.

ACKNOWLEDGEMENT

This work is based upon work supported in part under the US– India Partnership to Advance Clean Energy-Research (PACE-R) for the Solar Energy Research Institute for India and the United States (SERIIUS), funded jointly by the U.S. Department of Energy (Office of Science, Office of Basic Energy Sciences, and Energy Efficiency and Renewable Energy, Solar Energy Technology Program, under Subcontract DE-AC36-08GO28308 to the National Renewable Energy Laboratory, Golden, Colorado) and the Government of India, through the Department of Science and Technology under Subcontract IUSSTF/JCERDC-SERIIUS/ 2012 dated 22nd Nov. 2012.

REFERENCES

[1] G. S. Han *et al.*, "Retarding charge recombination in perovskite solar cells using ultrathin MgO-coated TiO2 nanoparticulate films," *J. Mater. Chem. A*, vol. 3, no. 17, pp. 9160–9164, Apr. 2015.

[2] X. Ren *et al.*, "Modulating crystal grain size and optoelectronic properties of perovskite films for solar cells by reaction temperature," *Nanoscale*, vol. 8, no. 6, pp. 3816–3822, 2016.

[3] L. Zheng *et al.*, "Improved light absorption and charge transport for perovskite solar cells with rough interfaces by sequential deposition," *Nanoscale*, vol. 6, no. 14, pp. 8171–8176, 2014.

[4] M. Yin *et al.*, "Annealing-free perovskite films by instant crystallization for efficient solar cells," *J. Mater. Chem. A*, vol. 4, no. 22, pp. 8548–8553, May 2016.

[5] "Light-trapping in perovskite solar cells," *AIP Adv.*, vol. 6, no. 6, p. 065002, Jun. 2016.

[6] Y. Cheng *et al.*, "Considerably enhanced perovskite solar cells via the introduction of metallic nanostructures," *J. Mater. Chem. A*, vol. 5, no. 14, pp. 6515–6521, Apr. 2017.

[7] O. Almora, C. Aranda, E. Mas-Marzá, and G. Garcia-Belmonte, "On Mott-Schottky analysis interpretation of capacitance measurements in organometal perovskite solar cells," *Appl. Phys. Lett.*, vol. 109, no. 17, p. 173903, Oct. 2016.

[8] Q. Wang, Y. Shao, Q. Dong, Z. Xiao, Y. Yuan, and J. Huang, "Large fill-factor bilayer iodine perovskite solar cells fabricated by a low-temperature solution-process," *Energy Environ. Sci.*, vol. 7, no. 7, pp. 2359–2365, 2014.

[9] H.-S. Duan *et al.*, "The identification and characterization of defect states in hybrid organic–inorganic perovskite photovoltaics," *Phys. Chem. Chem. Phys.*, vol. 17, no. 1, pp. 112–116, Dec. 2014.

[10] M. Samiee *et al.*, "Defect density and dielectric constant in perovskite solar cells," *Appl. Phys. Lett.*, vol. 105, no. 15, p. 153502, Oct. 2014.

Quantum Cutting Luminescent PMMA Films Containing Ce^{3+}-Yb^{3+} Co-doped YAG Phosphor for Si Concentrator Solar Cells

Lu Li[1,2], Chaogang Lou[1], and Huihui Cao[1]

[1]School of Electronic Science & Engineering, Southeast University, Nanjing, 210096, Jiangsu Province, People's Republic of China

[2] School of Electronic Information & Engineering, San Jiang University, Nanjing, 210012, Jiangsu Province, People's Republic of China

Abstract — Polymer films placed on solar cells contained quantum cutting phosphor were prepared by spin-coating technique. Ce^{3+}-Yb^{3+} co-doped YAG phosphor was uniformly mixed into polymethyl methacrylate(PMMA) matrix and formed into films on glass or solar cell surface. Transmissivity of glass with PMMA films show that the films has high transparency. Reflectivity of solar cells with PMMA films is lower than the bare solar cells in the whole wavelength range, especially in the range of short wavelength. Compared with bare solar cell, both quantum efficiency (QE) and conversion efficiency of the solar cell with PMMA film are increased.

Index Terms — quantum cutting, phosphor, PMMA film, Si concentrator solar cells, conversion efficiency.

I. INTRODUCTION

The conversion efficiency of Si-based solar cells is lower than the Schockley–Queisser limit (about 30%) [1]. This phenomenon may due to the poor spectral response to short wavelength light [2]. Recently, quantum cutting luminescence of rare earth ions (RE^{3+}) shows potential applications in Si-based solar cells, since such ions has been proposed to reduce the energy loss of short wavelength photons. For instance, Yb^{3+} ion has been widely applied for quantum cutting of visible photons into NIR photons, which has been put into practice to enhance the efficiency of Si-based solar cells [3]-[18]. In order to combine the quantum cutting material with solar cells, polymethyl methacrylate (PMMA) is often used in film form as it has little absorption in the range of 300 to 2500 nm [19]-[21].

Another way to increase the efficiency of the Si solar cells is using the concentrating photovoltaic (CPV) system. CPV systems provide lower energy generation costs by combining concentrator with solar cells with smaller area. Different kinds of concentrator solar cells are used in different CPV systems to attain optimum performance. For the low-medium concentrating system, Si concentrator solar cells are frequently used as its low price [22].

In this paper，PMMA films containing Ce^{3+}-Yb^{3+} co-doped YAG phosphor were fabricated on the Si concentrator solar cells. We test the QE and conversion efficiency in different light intensity to find the effect to the Si concentrator solar cells of the PMMA films.

II. EXPERIMENTAL

The powder sample of $YAG{:}1\%Ce^{3+}15\%Yb^{3+}$ was prepared by solid-state reaction method. The emission and excitation spectra of the sample at different temperature was measured with a Jobin Yvon FLUOROLOG-3-TAU steady state fluorescence spectrometer. Phosphor was uniformly mixed into polymethyl methacrylate(PMMA) matrix and formed into films placed on glass and solar cells (1.5×1.5 cm area) by spin-coating technique. A drop of the above-mentioned mix (0.5 ml approx.) was put on the substrate surface and spinning at 2000 r.p.m. for 60s. After the PMMA films were prepared, transmissivity of glass and reflectivity of solar cells were tested. QE of the solar cell with PMMA film was also tested and compared with the bare solar cell. Furthermore, I-V curves and conversion efficiency of the bare solar cells and solar cells with PMMA film are tested in different light intensity. A 500W xenon lamp was used as solar simulator and the light intensity can be controlled by the drive current.

III. RESULTS AND DISCUSSION

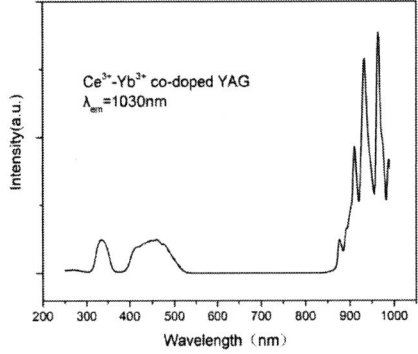

Fig. 1.   PLE spectra of Ce^{3+}-Yb^{3+} co-doped YAG phosphor.

Fig.1 presents the PLE spectra of Ce^{3+}-Yb^{3+} co-doped YAG monitored at 1030 nm. Two excitation bands centered at 330 nm and 450 nm are observed and this is ascribed to the allowed

978-1-5090-5606-4/17 $31.00 © 2017 IEEE

4f-5d transition of Ce^{3+}. Because Yb^{3+} can only absorb the photons around 1000 nm, the excitation bands in Fig.1 indicate that the NIR emission of Yb^{3+} results from the energy transfer from Ce^{3+} to Yb^{3+} and the peaks in the range of 900-1000 nm were assigned to 4f-4f transitions of Yb^{3+}.

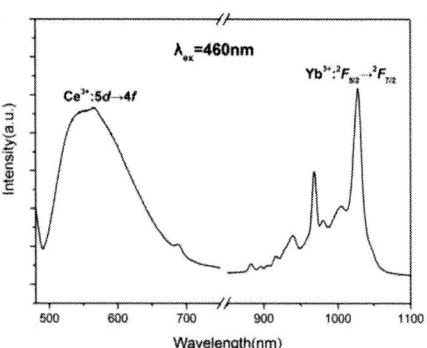

Fig. 2. Emission spectra of Ce^{3+}-Yb^{3+} co-doped YAG phosphor.

Fig. 2 presents the emission spectra of Ce^{3+}-Yb^{3+} co-doped YAG under 460nm excitation. As Yb^{3+} ions cannot absorb the light of 460 nm, it is confirmed again that the NIR emission of Yb^{3+} results from the energy transfer from Ce^{3+} to Yb^{3+}. We can obtain that photons from 350nm to 500nm can be converted to photons around 550nm and 1030nm by these rare earth ions. This conversion process makes the sunlight more suitable to the spectral response of Si-based solar cells. [23]

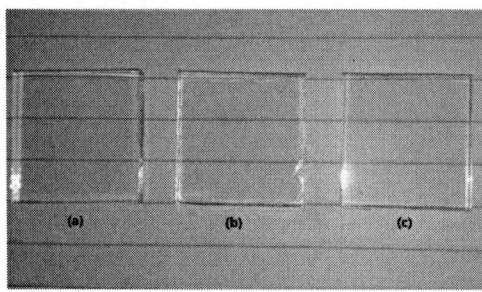

Fig. 3. (a) Original glass; (b) Glass with PMMA film; (c) Glass with PMMA film containing phosphor.

Fig. 3 presents the outward appearances of the original glass(a) and glass with PMMA films(b) and glass with PMMA film containing phosphor(c). Both sample(a) and sample(b) look transparent while sample(c) seems a bit fuzzy and the color shows light yellow. This may due to the light scattering of the YAG phosphor particle in PMMA film. Fig. 4 shows that the transmissivity of these samples is similar in the range of 350nm to 1150nm. So we consider that both the PMMA films and the phosphor have no effect on the transmittance of the glass.

Fig. 5 presents the bare solar cell(a) and solar cell with PMMA film containing phosphor(b). The spacing of grid lines is 1mm on this kind of Si concentrator solar cell while it is 3mm on a regular solar cell. Heavy phosphorus doping at the contact

position of grid lines and solar cell makes the series resistance lower.

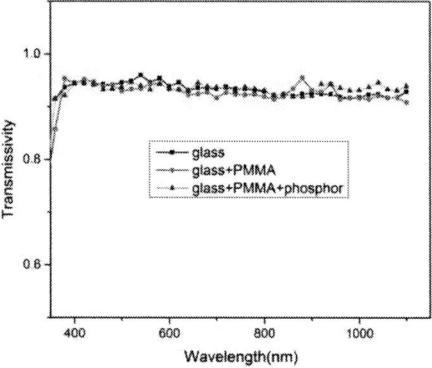

Fig. 4. Transmissivity of the glass samples.

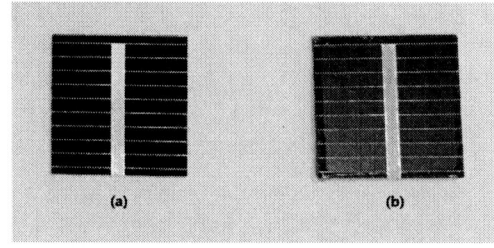

Fig. 5. (a)Bare solar cell; (b)Solar cell with PMMA film containing phosphor.

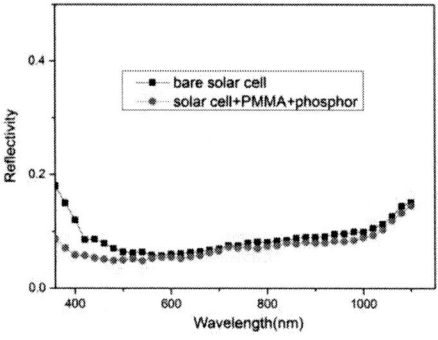

Fig. 6. Reflectivity of the bare solar cell and solar cell with PMMA film containing phosphor.

Sample(b) may has lower reflectivity as it seems darker than sample(a) in appearance. As shown in Fig. 6, reflectivity of solar cell with PMMA film is lower than the bare solar cells in the whole wavelength range, especially in the range of short wavelength. This may due to three reasons: (1) graded refractive index between Si and atmosphere caused by the

978-1-5090-5606-4/17 $31.00 © 2017 IEEE   1620

added PMMA film; (2) absorbing and quantum cutting of the short wavelength photons; (3) light trapping caused by scattering of the phosphor particle in the PMMA film.

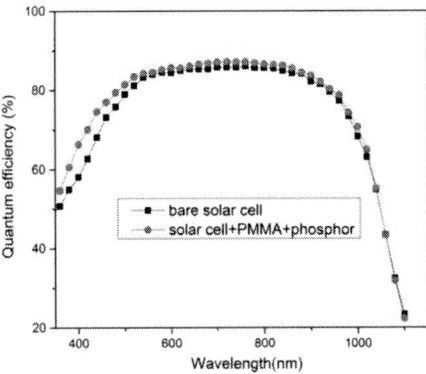

Fig. 7. Quantum efficiency of the bare solar cell and solar cell with PMMA film containing phosphor.

Fig. 7 presents the quantum efficiency of the bare solar cell and solar cell with PMMA film containing phosphor. It is also enhanced in the whole wavelength range, especially in the range of short wavelength. This may due to the quantum cutting effect of the PMMA film with phosphor. Short wavelength photons are transformed into low energy photons, and then generate more carriers inside the solar cells.

TABLE I

IMPROVEMENT IN CONVERSION EFFICIENCY OF THE SOLAR CELL

Sample	Conversion efficiency (%)	Relative growth
Bare solar cell	16.14%	3.66%
Solar cell with PMMA film containing phosphor	16.73%	

In Table I, the conversion efficiency of the solar cell with the PMMA film obtains a relative increase of 3.66%.

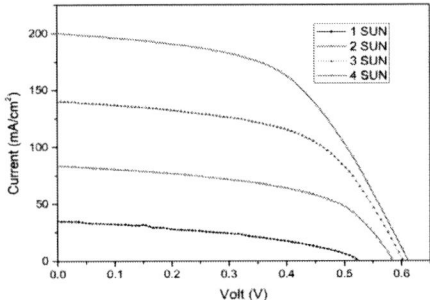

Fig. 8. I-V curves of concentrator solar cell with PMMA film containing phosphor under different light intensity.

Fig. 8 shows the I-V curves of concentrator solar cell with PMMA film containing phosphor under different light intensity. As the light intensity increases linearly, short circuit current increases with the trend of the light intensity while open circuit voltage increases logarithmically.

IV. CONLUCSIONS

In summary, the efficiency enhancement of Si concentrator solar cells by spin-coating PMMA films containing phosphor is demonstrated. Both the quantum efficiency and conversion efficiency were enhanced by the quantum cutting luminescent PMMA films. These films can be potentially used to improve the photovoltaic conversion efficiency of Si concentrator solar cells.

ACKNOWLEDGEMENT

This work is supported by the Natural Science Foundation of Jiangsu colleges and universities (14KJD510009), National Natural Science Foundation of China (61674097).

REFERENCES

[1] W. Shockley and H. J. Queisser, "Detailed Balance Limit of Efficiency of p-n Junction Solar Cells," *J. Appl. Phys.* vol. 32, no. 3, pp. 510–519, 1961.

[2] C. Strümpel, M. McCann, G. Beaucarne, V. Arkhipov, A. Slaoui, V. Švrček, C. Del Canizo, and I. Tobiasd,"Modifying the solar spectrum to enhance silicon solar cell efficiency-An overview of available materials," *Sol.Energy Mater. Sol. Cells,* vol. 91, no. 4, pp. 238–249, 2007.

[3] X. F. Liu, Y. Teng, Y. X. Zhuang, J. H. Xie, Y. B. Qiao, G. P. Dong, D. P. Chen, and J. R. Qiu, "Broadband conversion of visible light to near-infrared emission by Ce^{3+}, Yb^{3+}-codoped yttrium aluminum garnet," *Opt. Lett.* vol. 34, no. 22, pp. 3565–3567, 2009.

[4] D. Q. Chen, Y. S. Wang, Y. L. Yu, P. Huang, and F. Y. Weng, "Quantum cutting down conversion by cooperative energy transfer from Ce^{3+} to Yb^{3+} in borate glasses," *J. Appl. Phys.* vol. 104, no. 11, pp. 116105, 2008.

[5] H. Lin, S. M. Zhou, H. Teng, Y. K. Li, W. J. Li, X. R. Hou, andT. T. Jia, "Near infrared quantum cutting in heavy Yb doped Ce0.03Yb3xY (2.97−3x) Al5O12 transparent ceramics for crystalline silicon solar cells" *J. Appl. Phys.* vol. 107, pp. 043107, 2010.

[6] Y. Li, J. Wang, W.l. Zhou, G. G. Zhang, Y. Chen, and Q. Su, "Ultraviolet-Visible-Near-Infrared Luminescence Properties and Energy Transfer Mechanism of a Novel 5d Broadband Sensitized Sr3SiO5:Ce^{3+}, Yb^{3+} Suitable for Solar Spectral Converter" *Appl. Phys. Express* pp. 082301, 2013.

[7] D. C. Yu, F. T. Rabouw, W. Q. Boon, T. Kieboom, S. Ye, Q. Y. Zhang, and A. Meijerink, "Insights into the energy transfer mechanism in Ce^{3+}-Yb^{3+} codoped YAG phosphors," *Phys. Rev. B.* vol. 90, no. 16, pp. 165126, 2014.

[8] J. Ueda and S. Tanabe, "Visible to near infrared conversion in Ce^{3+}-Yb^{3+} Co-doped YAG ceramics," *J. Appl. Phys.* vol. 106, no. 4, pp. 043101–043105, 2009.

[9] T. Trupke, M. A. Green and P. WOrfel, " Improving solar cell efficiencies by down-conversion of high energy photons", *J Appl. Phys.*, vol. 92, pp. 1668-1674, 2002.

[10] H. Q. Wang , M. Batentschuk, A. Osvet, L. Pinna and C. 1. Brabec, "Rare-earth ion doped up-conversion materials for photovoltaic applications", *Adv. Materials*, vol. 23, pp. 2675-2680,2011.

[11] AShalav, B. S. Richards, T. Trupke, K. W. Kramer, and H. U. Gudel, "Application of NaYF4:Er3+ up-converting phosphors for enhanced near-infrared silicon solar cell response", *App. Phys. Lett.*, vol. 86, pp. 013505, 2005.

[12] M. Bryan, V.D. Ende, A. Linda, and M. Andries, "Lanthanide ions as spectral converters for solar cells," *Phys. Chem. Chem. Phys.*, vol. 11, pp. 11081–11095, 2009.

[13] J.C. Goldschmidt, S. Fischer, P. Löper, K.W. Kramer, D. Biner, M. Hermle, and S.W. Glunz, "Experimental analysis of up conversion with both coherent monochromatic irradiation and broad spectrum illumination," *Sol. Energy Mater. Sol. Cells*, vol. 95, pp. 1960–1963, 2011.

[14] A. Devos, A. Szymanska, and V. Badescu, "Modelling of solar cells with down-conversion of high energy photons, anti-reflection coatings and light trapping," *Energy Conversion and Management*, vol. 50, no. 2, pp. 328-336, Feb. 2009.

[15] A Bednarkiewicz, D. Wawrzynczyk, M. Nyk and M. Samoc, "Tuning red-green-white up-conversion color in nano NaYF4: Er/Yb phosphor", *J. Rare Earths*, vol. 29, pp. 1152-1156, 2011.

[16] F. Wang, Y. Han, C. S. Lim, Y. Lu, J. Wang, 1 Xu, H. Chen, C.Zhang, M.Hong and X. Liu, "Simultaneous phase and size control of up conversion nanocrystals through lanthanide doping", *Nature*, vol. 463, pp. 1061,2010.

[17] AC. Atre and Jennifer A Dionne, "Realistic up converter enhanced solar cells with non-ideal absorption and recombination efficiencies", *J. App. Phys.*, vol. 110, 034505, 2011.

[18] E. Klampaftis, D. Ross, K. R. McIntosh, and B. S. Richards, "Enhancing the performance of solar cells via luminescent down-shifting of the incident spectrum: A review," *Solar Energy Materials and Solar Cells*, vol. 93, no. 8, pp. 1182-1194, 2009.

[19] J. Y. Chen and K. W. Sun, "Enhancement of the light conversion efficiency of silicon solar cells by using nanoimprint anti-reflection layer," *Solar Energy Materials&Solar Cells*, vol. 94, pp. 629, 2010.

[20] S. Ramesh, Koay Hang Leen, K. Kumutha, and A. K. Arof, "FTIR studies of PVC/PMMA blend based polymer electrolytes,"Spectrochimica Acta Part A, vol. 66, pp. 1237, 2007.

[21] Guorong Duan, Chunxiang Zhang, Aimei Li, Xujie Yan, Lude Lu, and Xin Wang, "Preparation and Characterization of Mesoporous Zirconia Made by Using a Poly (methyl methacrylate) Template," *Nanoscale Res Lett*, vol. 3, pp. 118, 2008.

[22] M O'Neill，AJ Mcdanal, D Spears，C Stevenson，D Gelbaum. "Low-Cost 20X Silicon-Cell-Based Linear Fresnel Lens Concentrator Panel," 7th Inter. Conf. on Concentrating Photoviltaic Systems, 2011, p.120.

[23] Lu Li, Chaogang Lou, Xiaolin Sun, Yufei Xie, Lin Hu, and K. Santhosh Kumar. "Temperature dependent energy transfer in Ce^{3+}-Yb^{3+} co-doped YAG phosphor," *ECS J. Solid State Sci. Technol.* vol. 5, no. 9, pp. R146-R149, 2016.

Down Shifted Conversion for Enhanced HIT Solar Cell Efficiency

Albert S. Lin [1], Parag Parashar [1], Wei-Ming Huang [1], Yi-Wen Huang [1], Ding-Rung Jian [1], Ming-Hsuan Kao [2], Shi-Wei Chen [2], Chang-Hong Shen [2], Jia-Min Shieh [2], Tzu-Yu Chen [3], Chien-Chung Lin [4], Hao-Chung Kuo [3]

[1] Department of Electronic Engineering, National Chiao-Tung Univeristy, Hsinchu, Taiwan 30010

[2] National Nano Device Laboratories (NDL), Hsinchu, Taiwan 30010

[3] Department of Photonics, National Chiao-Tung Univeristy, Hsinchu, Taiwan 30010

[4] Institute of Photonic System, College of Photonics, National Chiao-Tung University, Tainan, Taiwan 71150

Abstract — **In this work, HIT solar cell is incorporated with the luminescent down-shifting (LDS) mechanism for the incident solar spectrum. The heterojunction with intrinsic thin-layer(HIT) solar cell is highly efficient, low temperature (200°C), low cost, and highly symmetrical device. Nevertheless, the carrier collection efficiency is low for the short wavelength solar photons due to the surface recombination. Here, a thin phosphor layer is spray coated on the top of HIT solar cells for the luminescent down-shifting of high energy photons and thus enhanced external quantum efficiency (EQE) for the device for λ=300nm-450nm is observed.**

Index Terms — **silicon, thermalization loss, solar cell, luminescent down shifting.**

I. LUMINESCENT DOWN-SHIFTING MECHANISM

Amorphous silicon (a-Si), microcrystalline (μ-Si) and crystalline (c-Si) solar thin-film solar cells have received much attention in research and production for more than two decades [1]–[9]. Currently heterojunction intrinsic thin layer (HIT) solar cells provide the opportunity of higher efficiency by multiple bandgaps and improved stability over amorphous silicon solar cells. Mechanisms based on light trapping using plasmonics, resonances, and quasi guided mode excitations have been used to improve the solar cell absorption [10]–[20]. Yet, losses due to thermalization cannot be avoided using conventional light trapping schemes. In order to eliminate these losses luminescent down-shifting mechanism had been adopted [21]–[23]. Luminescent down-shifting is a mechanism by which one high energy photon, which is inefficiently utilized by the photovoltaic cell due to the surface absorption and surface recombination, is converted into a low energy photon with sufficient energy to be absorbed by the photovoltaic cell. In this process, short-λ photons are absorbed in a thin LDS layer and then re-emitted at more favorable λ before they reach to the photovoltaic cells. LDS mechanism is quite different from photon up-conversion and down-conversion mechanisms where sequential absorption of two or more low energy photons leads to the emission of a high energy photon, or one high energy photon leads to the emission of two longer wavelength photons. Fig. 1 illustrate the down-shifting mechanism.

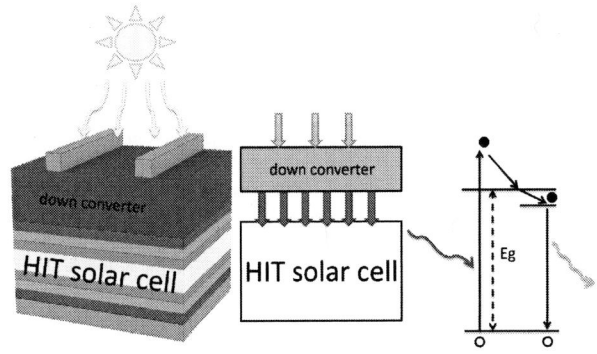

Fig. 1.   Proposed HIT solar cell structure with luminescent down-shifting converter.

LDS in this work is a passive approach that involves applying a luminescent species in a layer, i.e. spray coated phosphor layer in this work, prior to the cells, thus eliminating the need to interfere with the active material of a PV device. The down shifting mechanism shift the high energy solar photons to around λ=600nm where the external quantum efficiency (EQE) is the highest for silicon HIT solar cells, thus enhance the carrier collection efficiency for the short-wavelength photons and the short circuit current density (J_{SC}).

II. GEOMETRY OF THE DEVICE

Fig. 2 demonstrates the geometry configuration of the fabricated HIT structure [24]–[26]. The fabricated HIT solar cell consists of 12 nm thick intrinsic α-Si layer and a 10 nm thick p-type α -Si layer deposited on a randomly textured n-type 180μm thick Czochralski (CZ) crystalline silicon wafer to form a p-i-n heterojunction.

978-1-5090-5606-4/17 $31.00 © 2017 IEEE

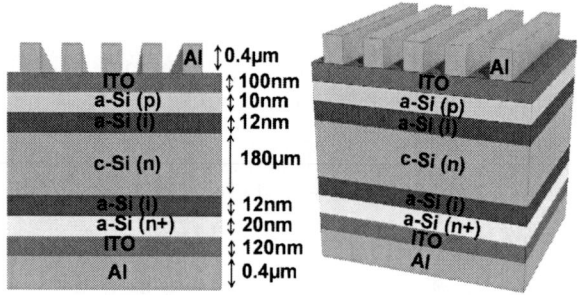

Fig. 2. 2D and 3D view of the HIT solar cell with dimensions labeled.

Then on the opposite side of the wafer, i-type and n-type α-Si layers are deposited with 12 nm and 20 nm thick layers, respectively, to obtain a back surface field structure. On both sides of the doped α-Si layers, indium tin conductive oxide (ITO) layers and Al metal grid electrodes are deposited. All of the processes for HIT solar cell described above are done at low temperature (200°C) to realize low-temperature process. Surface dangling bonds of c-Si are well passivated by inserting the high-quality intrinsic α-Si layer between the c-Si wafer and the doped α-Si layer. On the top of this HIT solar cell, a thin luminescent down-shifting phosphor layer is spray coated for the enhanced performance. For the spray coating, phosphor in the form EY4254 silicate phosphor from Intematix Corporation is mixed at room temperature with vulcanized silicone rubber (KER-2500(A/B)) from Shin-Etsu and n-Heptane in the ratio 1:1:2.5. The phosphor-mixed solution is then stirred for 2 hours. Thereafter, the phosphor-mixed solution is spray-coated on the top of fabricated HIT solar cell structure. Finally, the overall device is put into oven at 150□ for 2 minutes.

III. PL AND PLE SPECTRAL RESPONSE

The photoluminescence excitation (PLE) and photoluminescence (PL) spectral response for our proposed device can be found in [27]. It could be seen that PLE spectral response spans from λ=300 to λ=450 nm wavelength regime. With the inclusion of thin phosphor layer on top of HIT solar cell structure, luminescent down-shifting mechanism of high energy photons to low energy photons (with wavelength near λ=600nm) takes place, and the external quantum efficiency (EQE) is the highest at λ=600nm for silicon HIT solar cells as shown by the EQE data in Fig. 3.

IV. EQE SPECTRAL RESPONSE

Fig. 3 represents the EQE for the baseline HIT solar cells and the HIT solar cells with a phosphor layer. It could be seen that the short wavelength response in the 300-450 nm regime has been enhanced effectively due to the incorporation of

spray-coated down shifting phosphor layer and thus, the down shifting mechanism is indeed present in our proposed device. However, there is a nominal decrease in the long-wavelength EQE probably due to a small parasitic absorption in the Eu doped phosphor silicate in the down shifting layer.

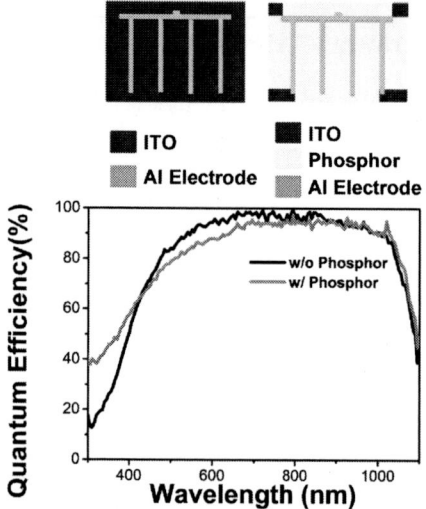

Fig. 3. External quantum efficiency curves for device with and without phosphor.

V. J-V CURVES

Current density vs output voltage curve has been presented in Fig. 4. It could be seen that the J_{SC} has increased to 36.35 mA/cm^2 in our proposed device with a LDS phosphor layer as compared to 31.29 mA/cm^2 structure without a LDS phosphor layer. That amounts to a significant 13.9% enhancement in current density with degradation in open circuit output voltage V_{OC} from 0.66V to 0.61V.

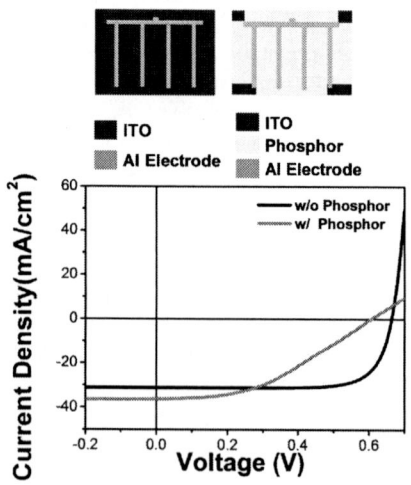

Fig. 4. Current density versus voltage curves.

978-1-5090-5606-4/17 $31.00 © 2017 IEEE

However, overall efficiency has decreased to 9.18% from 15.76%. Fill factor also reduces from 76.3 to 41.4. The reduction in V_{OC} and FF is attributed to the increase in electrode contact series resistance due to the non-conducting phosphor layer on top of the HIT metal electrode. During the spray coating, a small area of Al electrode is covered by tape in order to leave an electrical contact for the measurement purpose afterward. The small and inadequate contact between the probe and the Al metal electrode can be the cause for FF reduction. In addition, the tape contamination on the metal electrode could also be detrimental.

VI. CONCLUSION

The luminescent down-shifting HIT solar cell has been investigated in this work. Phosphor in the form EY4254 silicate phosphor has been spray coated on the top of HIT solar cells. The down shifting mechanism shifted the high energy solar photons to around λ=600nm, where the external quantum efficiency (EQE) is the highest for silicon HIT silicon solar cells, thereby, increasing the short wavelength EQE spectral response for the device. Enhanced effective short wavelength response in the 300-450 nm regime is seen due to the integration of LDS phosphor layer on top of HIT solar cells. Increase in short circuit current density with LDS phosphor layer is reported with 13.9% enhancement with respect to device without LDS phosphor layer. Nevertheless, the reduction in efficiency and fill factor occurs due to the improper electrode contact leading to high series resistance. Thus, future endeavors should be directed into decreasing the electrode contact series resistance by incorporating masked spray coating to enlarge the contact area at Al electrodes for the measurements afterward.

REFERENCES

[1] A. S. Ferlauto, G. M. Ferreira, J. M. Pearce, C. R. Wronski, R. W. Collins, X. Deng, *et al.*, "Analytical model for the optical functions of amorphous semiconductors and its applications for thin film solar cells," *Thin Solid Films* vol. 455 –456 pp. 388–392, 2004.

[2] S. S. Hegedus and R. Kaplan, "Analysis of quantum efficiency and optical enhancement in amorphous Si p–i–n solar cells," *Prog. Photovolt. Res. Appl.* , vol. 10, pp. 257–269 2002.

[3] B.-T. Chou, S.-D. Lin, B.-H. Huang, and T.-C. Lu, "Single-crystalline silver film grown on Si (100) substrate by using electron-gun evaporation and thermal treatment," *J. Vac. Sci. Technol. B*, vol. 32, p. 031209, 2014.

[4] S. A. Mann, M. J. d. Wild-Scholten, V. M. Fthenakis, W. G. J. H. M. v. Sark, and W. C. Sinke, "The energy payback time of advanced crystalline silicon PV modules in 2020: a prospective study," *Prog. Photovolt. Res. Appl.*, vol. 22, pp. 1180–1194, 2014.

[5] S. Hänni, G. Bugnon, G. Parascandolo, M. Boccard, J. Escarré, M. Despeisse, *et al.*, "High-efficiency microcrystalline silicon single-junction solar cells," *Prog. Photovolt: Res. Appl.*, vol. 21, pp. 821–826, 2013.

[6] N. Yamada, T. Ijiro, E. Okamoto, K. Hayashi, and H. Masuda, "Characterization of antireflection moth-eye film on crystalline silicon photovoltaic module," *Opt. Express,* vol. 19, pp. A118–A125, 2012.

[7] U. W. Paetzold, E. Moulin, D. Michaelis, W. Bottler, Wachter, Hagemann, *et al.*, "Plasmonic reflection grating back contacts for microcrystalline silicon solar cells," *Appl. Phys. Lett.,* vol. 99, p. 181105, 2011.

[8] O. Kunz, Z. Ouyang, S. Varlamov, and A. G. Aberlez, "5% efficient evaporated solidphase crystallised Polycrystalline silicon thin-film solar cells," *Prog. Photovolt. Res. Appl.,* vol. 17, pp. 567–573, 2009.

[9] P. Bermel, C. Luo, L. Zeng, L. C. Kimerling, and J. D. Joannopoulos, "Improving thin-film crystalline silicon solar cell efficiencies with photonic crystals," *Opt. Express,* vol. 15, pp. 16986–17000, 2007.

[10] S. M. Fu, Y. K. Zhong, S. L. Yan, N. P. Ju, P. Y. Chen, M.-H. Kao, *et al.*, "Non-reciprocal meta-surfaces for nanophotnic light trapping: optical isolators versus solar cells," in *IEEE Photovoltaics Conference*, New Orleans, LA, USA, 2015.

[11] L. v. Dijk, U. W. Paetzold, G. A. Blab, R. E. I. Schropp, and M. D. Vece, "3D-printed external light trap for solar cells," *Prog. Photovolt. Res. Appl.*, 2015.

[12] L. v. Dijk, E. A. P. Marcus, A. J. Oostra, R. E. I. Schropp, and M. D. Vece, "3D-printed concentrator arrays for external light trapping on thin film solar cells," *Sol. Energ. Mat. Sol.*, vol. 139, pp. 19–26, 2015.

[13] W.-R. Wei, M.-L. Tsai, S.-T. Ho, S.-H. Tai, C.-R. Ho, S.-H. Tsai, *et al.*, "Above-11%-Efficiency Organic–Inorganic Hybrid Solar Cells with Omnidirectional Harvesting Characteristics by Employing Hierarchical Photon-Trapping Structures," *Nano Lett.*, vol. 13, 2013.

[14] F. Pratesi, M. Burresi, F. Riboli, K. Vynck, and D. S. Wiersma, "Disordered photonic structures for light harvesting in solar cells," *Opt. Express*, vol. 21, pp. A460–A468, 2013.

[15] A. Bozzola, M. Liscidini, and L. C. Andreani, "Broadband light trapping with disordered photonic structures in thin-film silicon solar cells," *Prog. Photovolt. Res. Appl.*, 2013.

[16] P. C. Tseng, M. A. Tsai, P. Yu, and H. C. Kuo, "Antireflection and light trapping of subwavelength surface structures formed by colloidal lithography on thin film solar cells," *Prog. Photovolt. Res. Appl.*, vol. 20, pp. 135–142, 2012.

[17] X. Sheng, S. G. Johnson, L. Z. Broderick, J. Michel, and L. C. Kimerling, "Integrated photonic structures for light trapping in thin-film Si solar cells," *Appl. Phys. Lett.*, vol. 100, p. 111110, 2012.

[18] E. R. Martins, J. Li, Y. Liu, J. Zhou, and T. F. Krauss, "Engineering gratings for light trapping in photovoltaics: The supercell concept " *Phys. Rev. B*, vol. 86, p. 041404, 2012.

[19] Z. Yu and S. Fan, "Angular constraint on light-trapping absorption enhancement in solar cells," *Appl. Phys. Lett.*, vol. 98, p. 011106, 2011.

[20] B. G. Lee, P. Stradins, D. L. Young, K. Alberi, T.-K. Chuang, J. G. Couillard, *et al.*, "Light trapping by a dielectric nanoparticle back reflector in film silicon solar cells," *Appl. Phys. Lett.*, vol. 99, p. 064101, 2011.

[21] H.-V. Han, C.-C. Lin, Y.-L. Tsai, H.-C. Chen, K.-J. Chen, Y.-L. Yeh, *et al.*, "A highly efficient hybrid gaas solar cell based on colloidal-quantum-dot-sensitization," *Sci. Rep.*, vol. 4, p. 5734, 2014.

[22] S. Kalytchuk, S. Gupta, O. Zhovtiuk, A. Vaneski, S. V. Kershaw, H. Fu, *et al.*, "Semiconductor nanocrystals as luminescent down-shifting layers to enhance the efficiency of thin-film CdTe/CdS and crystalline Si solar cells," *J. Phys. Chem. C*, vol. 118, pp. 16393–16400, 2014.

[23] E. Klampaftis, D. Ross, K. R. McIntosh, and B. S. Richards, "Enhancing the performance of solar cells via luminescent down-shifting of the incident spectrum: A review," *Sol. Ener. Mat. Sol.,* vol. 93, pp. 1182–1194, 2009.

[24] A. Louwen, W. v. Sark, R. Schropp, and AndréFaaij, "A cost roadmap for silicon heterojunction solar cells," *Sol. Ener. Mat. Sol.,* vol. 147, pp. 295–314, 2016.

[25] M. Taguchi, A. Yano, S. Tohoda, K. Matsuyama, Y. Nakamura, T. Nishiwaki, *et al.*, "24.7% Record Efficiency HIT Solar Cell on Thin Silicon Wafer," *IEEE J. Photovolt.,* vol. 4, p. 96, 2014.

[26] S. Tohoda, D. Fujishima, A. Yano, A. Ogane, K. Matsuyama, Y. Nakamura, *et al.*, "Future directions for higher-efficiency HIT solar cells using a Thin Silicon Wafer," *J. Non-Cryst. Solids,* vol. 358, pp. 2219–2222, 2012.

[27] A. Lin, H.-V. Han, C.-Y. Huang, B.-R. Chen, S. M. Fu, Y. K. Zhong, *et al.*, "The external light trapping using down-conversion polymer and diffuse trench reflector," presented at the IEEE Photovoltaics Specialists Conference, Portland, USA, 2016.

Numerical Evaluation on the Nano-rod Array on a N-side-up Thin-film GaAs Solar Cells

Yan-Zhang Lin, Shun-Chieh Hsu, Po-Ching Wu, Chia-Jhe Hsu, and Chien-Chung Lin

Institute of Photonic System, National Chiao-Tung University, Tainan, 711, Taiwan

Abstract — In this work, we demonstrate the influence of the substrate thickness and the arrangement of nano-rod arrays on the n-side up GaAs thin film solar cell. The J-V characteristics can be altered significantly with different design parameters. The thickness of the substrate can have strong influence on the collected carrier. Also the closeness of the rod to each other can affect the short-circuit current as well due to the limited penetration depth of the photons.

Index Terms — Thin film, solar cells, GaAs, Simulation.

I. INTRODUCTION

Nowadays, people heavily depend on fossil fuels to produce enough energy for our increasing energy demand. A lot of harmful substance was formed in the meantime and cause many severe environmental problems. So, in order to protect our environment and deal with the shortage of fossil fuels, renewable energy alternatives need to be found to replace the fossil fuels. Among these alternatives, solar cell is one of the most favorable choices to achieve this goal because of the abundance of the solar energy. Therefore, many researcher try to promote its power conversion efficiency and reduce the cost in the meantime. One of the top choices is to use the gallium arsenide (GaAs) as the main material for the devices. GaAs has a direct band gap which can boost the photon absorption, and currently sets the world's highest power conversion efficiency record for the single junction design. Many advanced research results point out that the thin film GaAs could be a very versatile choice of the photovoltaic devices due to the possible photon recycling and open-circuit voltage enhancement [1]. Previous results showed very high power conversion efficiencies (PCEs) can be achieved (>28%)[2]. In recent work, researchers also demonstrated highly efficient, ultra-thin and flexible GaAs solar cells with thickness around 1μm and PCE=15% [3]. Most of the works focus on the implementation of the thin film structure, but not the numerical assessment. In addition to the planar solar cell, surface structured photovoltaic devices like nano-rod or nano-pillar also play important role in terms of light trapping or light harvesting [4]. The numerical model of nano-rod solar cells has shown that surface recombination and doping concentration can greatly change the efficiency profile of the devices [5-8]. In this work, a different GaAs thin film solar cell structure is built and numerically evaluated. The GaAs

solar cell is built in an "upside-down" fashion and equipped with nano-rod structure to enhance the light trapping. The devices are set up in a micro-meter to nano-meter scale to evaluate their performances. The influence of the thickness of the substrate will be discussed and the optical field within the nano-rod structures can be revealed. We hope this new design can provide some insights into the design of the next generation of nano-scale solar cells.

II. MODEL DETAILS

In this paper, our model is built in the two-dimension, Frequency Domain and semiconductor module in COMSOL multi-physics ® software. The schematic diagram of our model was shown in Fig.1. The p region is at the bottom and the top layer is n region to form single p-n junction. In both p and n region we add AlGaAs into the layer. Above the regular diode, there is the substrate layer, and in this study we would change the thickness of substrate and also use this substrate as the template for nano-rod formation.

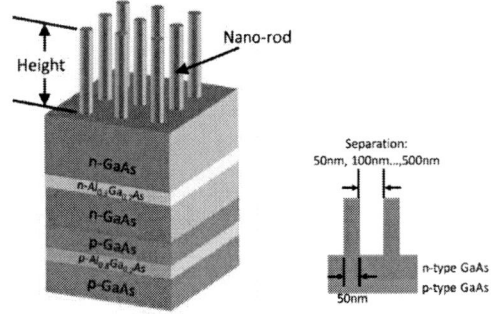

Fig. 1 The model schematic diagram. The inset shows the 2D cross-sectional view and some structural parameters used in our simulation such as separation.

	Doping Concentration	Doping Type	Thickness
substrate	1x10^{18}	n	2~10um
GaAs	1x10^{18}	n	300nm
Al$_{0.3}$GaAs	1x10^{18}	n	120nm
GaAs	1x10^{17}	n	4um
GaAs	2x10^{18}	p	200nm
Al$_{0.8}$GaAs	2x10^{18}	p	30nm
GaAs	2x10^{19}	p	300nm

Table 1 structure information

978-1-5090-5606-4/17 $31.00 © 2017 IEEE

Table 1 shows the basic parameter and doping type of our model structure. This n-side-up design can be practically incorporated into wafer processing because the general GaAs substrate for solar cell is n-type, and our design is just to flip the wafer and thin it down for further fabrication. The thick substrate can be used for nano-rod array without reaching the p-n junction. Also its thickness can be tuned by the lapping and polishing processes.

To setup the basic line of performance, a planar thin film type device was first simulated. Following the structure in Table 1, and varying the top-most n-type GaAs thickness between 0 to 15 μm, we are able to understand how these variations can affect the devices.

In the next step, a nano-rod array can be set up on top of the regular p-n diode in Table. 1. The total thickness of the regular p-n diode is 4.95 μm (not including the substrate layer in Table 1). The length of the rod can be varied between 2μm, 5μm and 8μm, and the separations between the rods are 50nm, 100nm, 150nm, 200nm and 500nm, respectively. The dimension of the rod is 50nm across all the cases. The current density unit was set as A/m^2, which is ten times of the unit mA/cm^2.

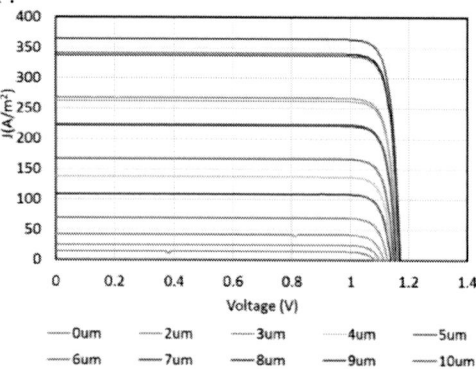

Fig 2 The photovoltaic J-V of a planar thin-film type GaAs solar cells with different substrate thicknesses.

III. SIMULATION RESULT

In electrical simulation, we take a look for the effect of the substrate thickness first. The variation of the thickness was from 0 to 15 μm in our work. In Fig 2, the detailed J-V plot showed the trend of thicker the substrate, the higher short-circuit current density (J_{sc}) and open-circuit voltage (V_{oc}). The general trend of J_{sc} shows that the thicker the substrate, the higher the short-circuit current. However, there is still up and downs if the J_{sc} versus thickness relation is closely watched.

With the planar devices solved, we can move to the nano-rod devices. The general understanding is that the formation of the nano-rod can help the light trapping and thus it is always benefit towards the light harvesting. In our study, we have rods with different lengths and separations to evaluate their effects towards J-V plots. Fig. 3 shows the J-V curves for various separations. Although the photocurrent is greatly enhanced at

larger spacing (like in 500nm), the small spacing such as 50nm or 100nm shows less favorable results with much lower J_{sc}.

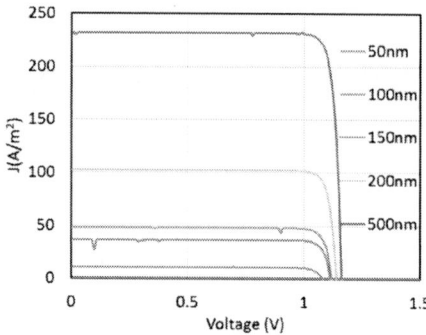

Fig. 3 The J-V characteristics chart with the variation of the nano-rod separations. The rod length is 8μm in this plot.

To investigate this phenomenon, we can extract J_{sc} and V_{oc} of the nano-rod devices under various separations, as shown in Fig. 4. Different rod lengths (2μm, 5μm, and 8μm) were set in the calculation. Meanwhile, one must note that the change of rod length also means the change of effective absorption material (i.e. GaAs volume). This variation can move the baseline of J_{sc} because according to our previous calculation in planar case, the thickness of the device can have positive correlation with the J_{sc}. So it is important for us to mark the upper and lower limit of the planar cell as the reference point. The upper limit comes from the planar thin film result at general diode thickness (4.95μm) + the rod length. The lower limit shall be the general diode result only (no rod structure). Combined with the planar results, we could observe the sharp increase of J_{sc} around 200nm separation. If the separation is too small, the J_{sc} always falls within the upper and lower limits, meaning that there is no optical enhancement from the rod array.

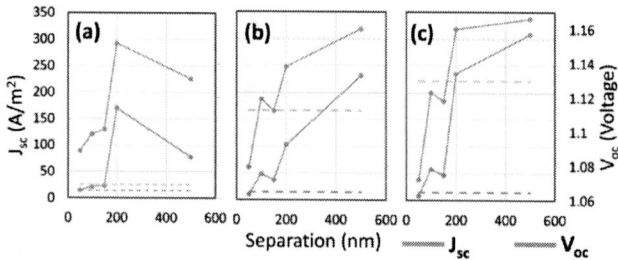

Fig. 4: The J_{sc} and V_{oc} for the different separation varying from 50nm to 500nm. The rod length are (a)2μm, (b)5μm, and (c)8μm, respectively. The dashed line indicates the upper and lower limits of the J_{sc} in planar device.

The reason behind the rather not-so-smooth rise in J_{sc} in Fig. 4 comes from, we believe, the separation of the rods. Our nano-rod is made of GaAs and it can become quite absorptive at visible wavelengths. If the rod is long enough, and dense enough (like separation =50nm or less), the outside photons will not be able to penetrate into bottom p-n junction. Since p-n junction is the highest carrier collection area, this will deteriorate the generation of J_{sc} and thus the overall performances. Once the separation opens up to 200nm or more, (now the gap is no longer sub-wavelength structure), the light can effectively go into the bottom junction (as can be observed in Fig. 5, with stronger optical field ripples in the left plot), the J_{sc} will rise because of the light trapping effect, and exceeds what we got in planar cases. So there is a limit for the density of the nano-rod array, especially for the absorptive semiconductor, when the design exceeds this limit, the photons can not properly propagate through the rods and if the p-n junction is not nearby, the photo-generation and carrier collection will be deteriorated.

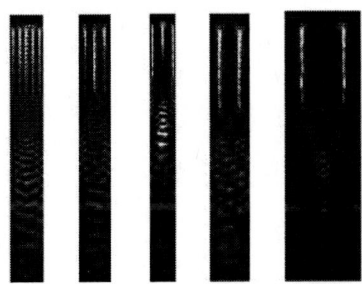

Fig. 5: The optical field at different separations between the nano-rods in our solar cell chips with nano-structures. From the left most to the right : 50nm, 100nm, 150nm, 200nm, and 500nm, respectively. The incident photons are at 800nm band and the rod length is 8μm.

In addition, we calculate the total generation of the thin-film solar cells and nano-rod surface solar cells with the incident light from 300 to 900nm wavelength by using Matlab in Fig. 6. We can see there is just one peak in planar thin-film, and two peaks in nano-rod surface. So if we add nano-rod formation on the top of thin-film surface, the light will be absorb more than planar thin-film. Therefore, the Jsc of the nano-rod top surface would be higher than planar thin-film solar cells.

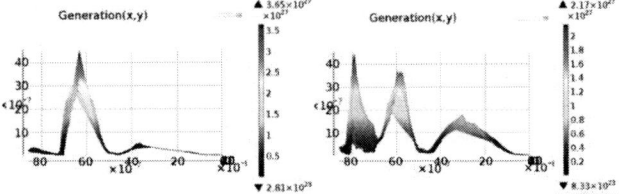

Fig. 6: The total generation of a planar thin-film type GaAs solar cells and nano-rod surface thin-film GaAs solar cells with substrate of 2um.

Otherwise, the surface influence of simple planar thin-film (just two layer, N-side-up or P-side-up) shows in Fig. 7. The curves of N-side-up and P-side-up almost no variation. Due to the structure of them just change their doping type. First, we etch the top region to form the nano-rod formation on the top surface (P-rod & N-rod). Jsc increases with the nano-rod

formation, but Jsc is almost no changes between N-type and P-type rod. As well as the different doping type affect the variation of Voc in the same settings (as can be observed in Fig. 8). And the metal contact position is one of the condition we consider in this case. The top-contact is at the top surface of rod after etching, while the bottom-contact is at the top surface of thin-film after etching. The Jsc of top-contact is higher than bot-contact. If we etch much more until cross the p-n junction of thin-film into the bottom region, Jsc reduces.

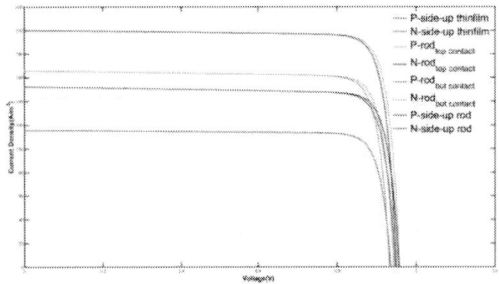

Fig. 7: The photovoltaic J-V of planar thin-film solar cells with different surface and metal contact position.

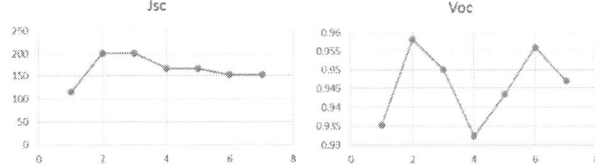

Fig. 8: the Jsc and Voc for the different etching depth and metal contact condition. (1) planar thin-film. (2) etching with top-contact of P-rod. (3) etching with top-contact of N-rod. (4) etching with bot-contact of P-rod. (5) etching with bot-contact of N-rod. (6) etching cross the p-n junction of the P-side-up thin-film. (7) etching cross the p-n junction of the N-side-up thin-film.

IV. CONCLUSION

In conclusion, the simulation set up in COMSOL multi-physics® which can coupled optical and electrical simulation. From the result above, we know that much more substrate thickness leads to higher J_{sc} and V_{oc}. While in top surface design, the nano-rod is not always helpful towards light harvesting due to the self-absorption of the incoming photons, which can be explained by the optical field penetration in the devices. When rod is too close to each other, the solar light can not reach the bottom diode and the high generation on the top of the rod is wasted due to poor carrier collection rate. Until the spacing opens up to 200nm or more, and the photons can be diffracted into the rods, the bottom reception can be improved and thus the short-circuit current can also to dramatically increased.

978-1-5090-5606-4/17 $31.00 © 2017 IEEE

V. ACKNOWLEDGEMENT

The authors would like to thank the Ministry of Science and Technology of Taiwan for financial support through the grant number: MOST104-2628-E-009-013-MY3

REFERENCES

. [1] O. D. Miller, E. Yablonovitch, and S. R. Kurtz, "Strong Internal and External Luminescence as Solar Cells Approach the Shockley–Queisser Limit," *IEEE Journal of Photovoltaics,* vol. 2, pp. 303-311, 2012.

[2] M. A. Green, K. Emery, Y. Hishikawa, W. Warta, and E. D. Dunlop, "Solar cell efficiency tables (version 48)," *Progress in Photovoltaics: Research and Applications,* vol. 24, pp. 905-913, 2016.

[3] J. Kim, J. Hwang, K. Song, N. Kim, J. C. Shin, and J. Lee, "Ultra-thin flexible GaAs photovoltaics in vertical forms printed on metal surfaces without interlayer adhesives," *Applied Physics Letters,* vol. 108, p. 253101, 2016.

[4] H.-C. Chen, C.-C. Lin, H.-W. Han, Y.-L. Tsai, C.-H. Chang, H.-W. Wang, *et al.*, "Enhanced efficiency for c-Si solar cell with nanopillar array via quantum dots layers," *Optics Express,* vol. 19, pp. A1141-A1147, 2011/09/12 2011.

[5] B. M. Kayes, H. A. Atwater, and N. S. Lewis, "Comparison of the device physics principles of planar and radial p-n junction nanorod solar cells," *Journal of Applied Physics,* vol. 97, p. 114302, 2005.

[6] A. C. E. Chia and R. R. LaPierre, "Electrostatic model of radial pn junction nanowires," *Journal of Applied Physics,* vol. 114, p. 074317 2013.

[7] A. C. E. Chia and R. R. LaPierre, "Analytical model of surface depletion in GaAs nanowires," *Journal of Applied Physics,* vol. 112, p. 063705, 2012.

[8] S. L. Lin, H. R. Tseng, S. C. Hsu, Y. H. Chen, and C. C. Lin, "Numerical study of surface-dependent performances of nano-scale solar cells," in *2015 IEEE 42nd Photovoltaic Specialist Conference (PVSC)*, 2015, pp. 1-4.

The Planar Thermophotovoltaic Selective Nearly-Perfect Absorbers/Emitters

Parag Parashar, Ding-Rung Jian, Weiming Huang, Yi-Wen Huang, and Albert Lin

Department of Electronic Engineering, National Chiao-Tung Univeristy, Hsinchu, Taiwan 30010

Abstract — The spectral response of thermophotovoltaic (TPV) absorber is a key factor for TPV system efficiency. We have shown that the planar TPV emitters can be realized by using thin-metal-dielectric stacking (IEEE Photonics J. 8, 1300109, 2016). In this paper, we provide a new structure composed of alternating dielectric and deep-sub-skin-depth (DSSD) metal layers for TPV selective absorbers. The RCWA calculation and the Fourier transform infrared (FTIR) measurement of tantalum (Ta) based planar absorber shows a high absorbance below the cut-off wavelength ($\lambda<\lambda c$) and a suppressed absorbance beyond the cut-off wavelength ($\lambda>\lambda c$).

Index Terms — thermophotovoltaics, spectral absorption, surface roughness, photonics.

I. INTRODUCTION

Theoretically, the efficiency of a single junction solar cell has an upper limit, so-called Shockley-Queisser limit, which is ~40% under fully concentrated solar radiation and ~30% under one sun condition [1]. The thermophotovoltaic (TPV) systems possess the potential to be more efficient than conventional photovoltaics (PV) and the potential to exceed the Shockley-Queisser limit [2]–[4]. In the last two decades, significant progress has been made in order to realize enhanced efficient, close-to-unity absorbance, steep cutoff and wavelength scalable thermophotovoltaic systems [4]–[13].

In our recent publication [14], we show that a planar nearly-ideal TPV emitter can be realized by ultrathin-metal-dielectric stacking with aperiodic dielectric filtering. Here, we further propose an all-planar absorber structure consisting of alternating dielectric and deep-sub-skin-depth (DSSD) metal layers. The adjustment of the dielectric thickness leads to perfect wavelength scalability, which can fit a specific TPV system operation temperature requirement. The spectral absorbance shows this designed absorber has a close-to-unity absorbance in the absorption band, a steep cut-off, and a suppressed absorbance in the long-wavelength regime that can eliminate the thermal re-emission loss and thus achieves an idealized design. Moreover, the angular spectral response and its polarization dependence are also shown to demonstrate its decent omnidirectionality. Finally, we conduct a preliminary experimental effort, and the fully planar design leads to simple device processing and the scalability to large-area TPV absorber arrays at relatively lower cost.

II. METHOD AND PROBLEM SET-UP

The ideal TPV selective absorption spectrum is shown in Fig. 1. For an ideal TPV absorber, the cut-off of absorption is needed to suppress the long wavelength infrared re-emission. The re-emission is detrimental to the TPV operation since the thermal radiation should leave from the emitter side. Fig. 1 demonstrates the basic operation principle behind the proposed deep-sub-skin-depth (DSSD) metal design. The photons with wavelength locating at the absorption band are penetrating the DSSD metal films, leading to gradual power dissipation in the successive metal films and the perfect absorption in the entire absorber structure. The ultrathin metal dimension is the key for the perfect absorption since the DSSD metal films minimize the phase difference between the reflected rays of different orders at a single metallic film. In this work, the structure we proposed consisting of alternating dielectric and ultrathin refractory metal layers, as illustrated in Fig. 1 with physical dimension labeled, ($t_{Dielectric}$, t_{Metal}). The material refractive indices (n) and extinction coefficients (k) are from Rsoft material database [15]. The rigorously coupled wave analysis (RCWA) using Diffractmod^{TM} is used to design the spectral reflectance (R), transmittance (T), and absorbance (A) of the planar ultrathin-metal TPV selective absorber.

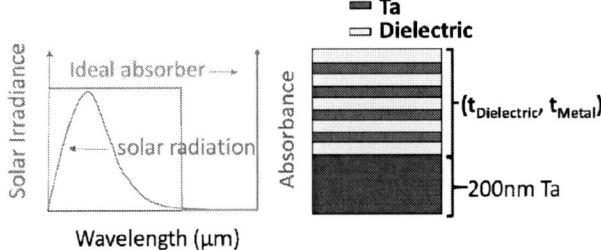

Fig. 1. Illustration of (left) the ideal thermophotovotalic (TPV) selective absorption spectrum with cut-off and (right) The dimension of the proposed deep-sub-skin-depth (DSSD) ultrathin refractory metal TPV absorber. $t_{Dielectric}$ and t_{Metal} are the only two dimension parameters in this simple design. The photons in the absorption band can be dissipated fully by the successively stacked ultrathin metal films. The photons in the suppressed band are reflected back due to the equivalent bulk property of Ta.

III. THEORETICAL RESULT USING COUPLED-WAVE ANALYSIS

The high absorbance value below λ_c, the low absorbance value above λ_c, and a sharp cut-off at λ_c are all observed in our design. It is quite surprising that such a simple planar multi-layer design can achieve this degree of perfection, comparable to many state-of-the-art nanostructured TPV absorbers in literature [4], [16]–[18]. This reflects the versatility and usefulness of employing DSSD ultrathin metal films, which provides a new way of controlling photon interference, dissipation, and penetration. The potential of using field penetration in metallic films may not have been fully explored in photonics and optoelectronics based on our literature review. Fig. 2 shows the spectral responses for the proposed TPV absorbers.

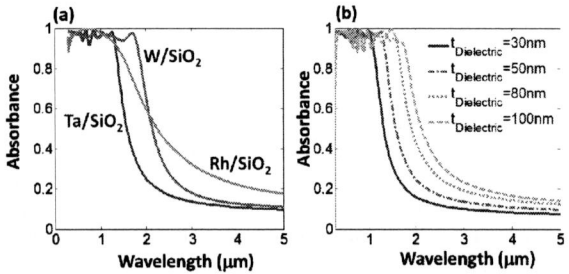

Fig. 2. (a) The spectral absorbance of different DSSD-metal/dielectric combinations, including Ta/SiO₂, W/SiO₂, and Rh/SiO₂. The metallic layer thickness (t_{Metal}) for W, Rh, and Ta layers are 2nm. The dielectric spacer thickness is 50nm for the Ta absorber, 40nm for the W absorber, and 80nm for the Rh absorber. 16 pairs (16X) of alternating dielectric-DSSD metal are stacked. (b) The wavelength scalability of Ta absorbers. t_{Metal} is 2nm and $t_{Dielectric}$ is 40nm. 16 pairs (16X) of Ta-SiO₂ are used.

Fig. 3 shows the angled spectral response for the Ta/SiO₂ configuration. The Ta ultrathin metal thickness is chosen as 2nm. Transverse electric (TE) and transverse magnetic (TM) polarization results are shown. It can be seen that the desired selective spectral absorbance can be sustained until 60° where the close-to-unity absorption still exists at large incident angles, and the cut-off in the spectral absorbance is not deteriorated. Moreover, even at 80° incidence, 70% absorption is still achievable. The broad absorption band exists for both TE and TM polarizations, and thus, a polarization-independent, omnidirectional, broadband selective thermal absorber is proposed.

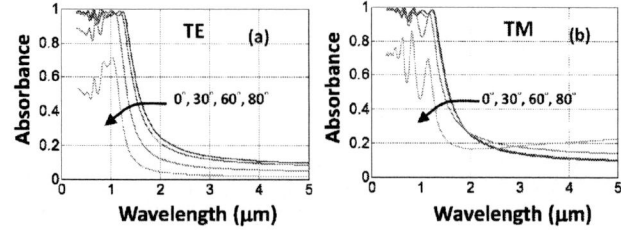

Fig. 3. The spectral response at different incident angles for Ta/SiO₂ configuration for (a) TE polarization (b) TM polarization. Ta thickness (t_{Metal}) is 2nm, SiO₂ ($t_{Dielectric}$) is 50nm. 16 pairs of Ta-SiO₂ are used.

IV. EXPERIMENT

Beside the theoretical calculations, a preliminary experiment is also progressed to demonstrate the concept of ultrathin DSSD metal TPV absorbers. The tungsten oxide (WO₃), hafnium dioxide (HfO₂), silicon dioxide (SiO₂) are selected as the dielectric materials. Tantalum (Ta) is chosen as the ultrathin metal layers. Firstly, we deposited 200nm Ta as the bottom metallic plate. Afterward, we deposited alternating dielectric and ultra-thin metal layers. The deposition of Ta, HfO₂, and WO₃ are done with a ULVAC ENTRON sputter. The deposition of SiO₂ is done with Oxford plasma enhanced chemical vapor deposition (PECVD). The fabrication of this designed planar absorber eliminates lithography or reactive ion etching (RIE) process of any kinds. The surface roughness and thickness uniformity of ultrathin sputtered 2nm and 5nm Ta films are improved over electron-gun (e-gun) evaporated films, but still not comparable to techniques such as atomic layer deposition or molecular beam epitaxy. The most important effect associated with the surface roughness is the photonic light trapping. In fact, as far as the proposed design of a selective thermal absorber is concerned, the light trapping effect is not always adverse. The light trapping due to surface roughness primarily leads to the effect of increased absorbance at all wavelength regimes. In the absorption band below cut-off wavelength (λ_c), the increased absorption is desired, especially in the case where fewer stacking pairs are used.

Fig. 4 shows the UV-VIS-NIR spectral response of Ta/HfO₂, Ta/WO₃, and Ta/SiO₂ configuration. We can see that the experimental results agree with the calculated ones in that the perfect absorption band exists, and a cut-off and suppressed absorbance are also observed at long wavelength.

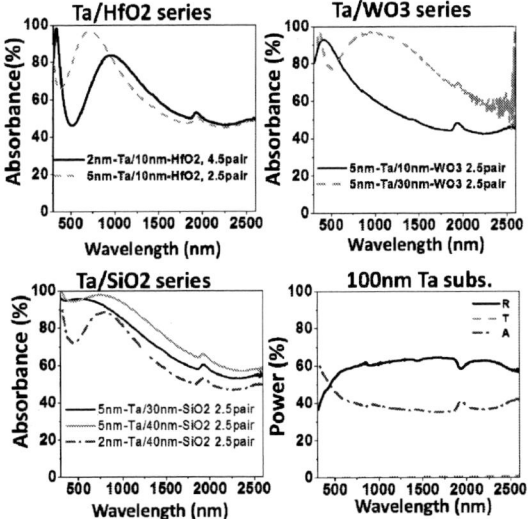

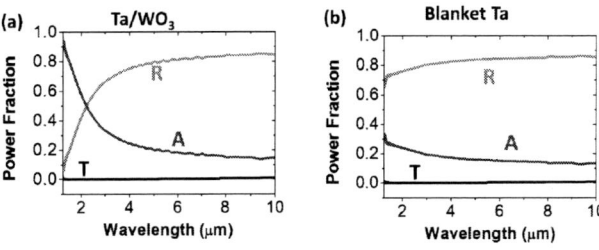

Fig. 4. The measured selective absorption spectrum for the deep-sub-skin-depth (DSSD) metal absorbers. The dielectrics selected include HfO$_2$, SiO$_2$, and WO$_3$. The layer thicknesses of the dielectrics and Ta are specified in the legend. DSSD dimension of 2nm Ta films are deposited. Supplementary 5nm Ta samples are also fabricated for comparison. 5nm Ta films are less preferred due to the fact that in theory, the field penetration and perfect absorption phenomenon can be degraded with increased Ta thickness. Nonetheless, the 5nm Ta film has better thickness uniformity if sputtering is the deposition method used. Four pairs (4X) of 2nm-Ta-dielectric are in theory not sufficient for perfect absorption using DSSD metal films. The light trapping effect due to the surface roughness increases the absorbance at the absorption band but also enhances the absorbance in the rejection band.

Fig. 5 shows the Bruker IFS66V/S Fourier transform infrared spectroscopy (FTIR) measurement from λ=1.26μm to λ=10μm, which supplements the UV-VIS-NIR result at IR regimes. The accuracy of U-4100 at λ>2000nm is not as decent as the FTIR machine, and therefore the absorbance value at the suppression band should be referred to Fig. 5 for a more accurate assessment. In theory, the suppressed absorption at the rejection band can be as low as 0.1 while the measured long wavelength absorbance is ~0.3 evident from Fig. 5. As we mentioned in the previous paragraph, the increased absorbance at the rejection band is mainly due to the surface roughness associated with each deposition layers including the ultrathin metal and the dielectric. To further decrease the absorption at rejection band, a uniform film is necessary by other more elaborate techniques or a polished Ta substrate as in literature [17]–[18]. It is also worth to point out that the spectral characteristics realized here are comparable to many state-of-the-art TPV absorbers in literature [17]–[19], even in terms of the suppressed long wavelength absorption.

Fig. 5. (a) The Fourier transform infrared (FTIR) spectroscopy for the Ta-WO$_3$ sample. Ta thickness is 2nm and WO$_3$ thickness is 30nm. (b) The FTIR measurement of a blanket Ta bottom plate is also shown for comparison.

V. CONCLUSION

Here, we propose a novel one-dimensionally (1D) stacked thermophotovoltaic absorber, composed of alternating dielectric and extremely thin metallic layers. The metal we mainly selected is Ta, and the dielectric chosen is HfO$_2$, SiO$_2$, and WO$_3$ in the initial experiment effort. Through the eigenmode expansion calculation, the proposed thermal absorber indeed has close-to-unity absorbance below the cut-off wavelength (λ<λ_c) and a suppressed absorbance beyond the (λ>λ_c), with decent wavelength scalability covering entire VIS-NIR spectrum. The steep cut-off in its spectral absorbance and a suppressed absorption in long wavelength regime are promising for a reduced re-emission loss in typical TPV systems. Finally, initial experiment effort is carried out to fulfill the novel idea of planar, deep-sub-skin-depth, omnidirectional broadband selective absorbers. The structures we implement are Ta-based 1D stacking with different dielectrics. The spectral measurement of fabricated TPV absorber samples shows agreement with the calculated results and also proves the feasibility of our design. A measured high absorption exists in the absorption band, and a suppressed spectral absorption is observed in the suppression band. The further improvement in the fabrication can be the film thickness uniformity control at ultrathin metal thickness and the surface roughness reduction. We believe the proposed design is not only useful for thermophotovoltaics but also promising for any advanced thermal absorption-emission applications requiring spectral control and shaping.

REFERENCES

[1] Schokley and Queisser, "Detailed balance limit of efficiency of p-n junction solar cell," *J. Appl. Phys.*, vol. 32, pp. 510-519, 1961.

[2] P. A. Davies and A. Luque, "Solar thermophotovoltaics: brief review and a new look," *Sol. Energ. Mat. Sol.*, vol. 33, pp. 11–22, 1994.

[3] Z. Zhou, Q. Chen, and P. Bermel, "Prospects for high-performance thermophotovoltaic conversion efficiencies exceeding the Shockley–Queisser limit," *Energ. convers. manage.*, vol. 97, pp. 63–69, 2015.

[4] P. Bermel, M. Ghebrebrhan, W. Chan, Y. X. Yeng, M. Araghchini, R. Hamam, *et al.*, "Design and global optimization of high-efficiency thermophotovoltaic systems," *Opt. Exp.*, vol. 18, pp. A315-A334, 2010.

[5] J. K. Tong, W.-C. Hsu, Y. Huang, S. V. Boriskina, and G. Chen, "Thin-film 'thermal well' emitters and absorbers for high-efficiency thermophotovoltaics," *Sci. Rep.*, vol. 5, p. 10661, 2015.

[6] H. Daneshvar, R. Prinja, and N. P. Kherani, "Thermophotovoltaics: Fundamentals, challenges and prospects," *Appl. Energy*, vol. 159, pp. 560–575, 2015.

[7] E. S. Sakr, Z. Zhou, and P. Bermel, "High efficiency rare-earth emitter for thermophotovoltaic applications," *Appl. Phys. Lett.*, vol. 105, p. 111107, 2014.

[8] Y. Nam, Y. X. Yeng, A. Lenert, P. Bermel, I. Celanovic, M. Soljačić, *et al.*, "Solar thermophotovoltaic energy conversion systems with two-dimensional tantalum photonic crystal absorbers and emitters," *Sol. Energ. Mat. Sol.*, vol. 122, pp. 287–296, 2014.

[9] A. Lenert, D. M. Bierman, Y. Nam, W. R. Chan, I. Celanovic, M. Soljacic, *et al.*, "A nanophotonic solar thermophotovoltaic device," *Nat. Nanotechnol.*, vol. 9, pp. 126–130, 2014.

[10] C. Ferrari, F. Melino, M. Pinelli, P. R. Spina, and M. Venturini, "Overview and Status of Thermophotovoltaic Systems," *Energy Procedia*, vol. 45, pp. 160–169, 2014.

[11] A. Datas and C.Algora, "Detailed balance analysis of solar thermophotovoltaic systems made up of single junction photovoltaic cells and broadband thermal emitters," *Sol. Energ. Mat. Sol.*, vol. 94, pp. 2137–2147, 2010.

[12] R. E. Nelson, "A brief history of thermophotovoltaic development," *Semicond. Sci. Technol.*, vol. 18, p. S141, 2003.

[13] Y. K. Zhong, "The broadband and selective perfect absorbers for thermophotovoltaics and sensing," Ph.D. Ph.D. Thesis, Department of Electronic Engineering, National Chiao-Tung University, Taiwan, 2017.

[14] S. M. Fu, Y. K. Zhong, M.-H. Tu, B.-R. Chen, and A. Lin, "A Planarized Thermophotovoltaic Emitter with Idealized Selective Emission," *IEEE Photon. J.*, vol. 8, p. 1300109, 2016.

[15] Rsoft, *Rsoft CAD User Manual*, 8.2 ed. New York: Rsoft Design Group, 2010.

[16] C. Wu, B. Neuner, J. John, A. Milder, B. Zollars, S. Savoy, *et al.*, "Metamaterial-based integrated plasmonic absorber/emitter for solar thermo-photovoltaic systems," *J. Opt.*, vol. 14, p. 024005, 2012.

[17] V. Rinnerbauer, A. Lenert, D. M. Bierman, Y. X. Yeng, W. R. Chan, R. D. Geil, *et al.*, "Metallic photonic crystal absorber-emitter for efficient spectral control in high-temperature solar thermophotovoltaics," *Adv. Energy Mater.*, vol. 4, p. 1400334, 2014.

[18] V. Rinnerbauer, Y. X. Yeng, W. R. Chan, J. J. Senkevich, J. D. Joannopoulos, M. Soljačić, *et al.*, "High-temperature stability and selective thermal emission of polycrystalline tantalum photonic crystals," *Opt. Exp.*, vol. 21, pp. 11482–11491, 2013.

[19] J. B. Chou, Y. X. Yeng, Y. E. Lee, A. Lenert, V. Rinnerbauer, I. Celanovic, *et al.*, "Enabling ideal selective solar absorption with 2D metallic dielectric photonic crystals," *Adv. Mats.*, vol. 26, pp. 8041–8045, 2014.

Hybrid PEDOT:PSS Silicon Solar Cells with Pencil Rod Structures

Ruei-Ying Wu, Liang-Chian You, Hsin-Fei Meng, Chun-Chi Chen, Peichen Yu*

Department of Photonics, National Chiao-Tung University, Hsinchu 30010, Taiwan

(*E-mail: peichen.yu@gmail.com)

Abstract — **Pencil-shaped nanorods structure were fabricated on hybrid photovoltaic by nanospheres lithography and metalassisted chemical etching, which showed a low weighted reflectance of 7.4% since the modulation of the tapered tip surface morphology helps to enhance the light harvesting and form radial p-n junction area for efficient carrier separation and collection, thereby effectively increased the J_{SC} of 34 mA/cm^2. Furthermore, by using the atomic layer deposition of a thin dielectric interface passivation layer, Al2O3 , between the Si and PEDOT:PSS layer could improve the V_{OC} of the hybrid solar cells, we successfully achieve a PCE of 13.61% in cells with optimizing the depth of pencil rods.**

Index Terms — heterojunction, hybrid solar cell, metal-assisted chemical etching, nanostructure, Al_2O_3, passivation

I. INTRODUCTION

The cost of the currently dominating silicon-based solar cells remains relatively high due to the high-temperature and high-vacuum processes in making crystalline wafers and cells. Therefore, Organic/Si hybrid solar cells have received considerable attention in recent years because of low-cost and simplified processing of organic materials and the high carrier mobility of Si. The high transparency and conductivity organic layer poly(3,4-ethylene dioxythiophene): poly(styrenesulfonate) (PEDOT:PSS) have been reported not only because of its easy fabrication process and low energy cost but also serves as a passivation layer and antireflection coating.[1] However, the power conversion efficiency (PCE) cannot compete with those of traditional silicon solar cells mostly due to the limited light absorption related to the active c-Si thin substrate and interface recombination losses.[2]

In order to improve cells efficiency, the periodical pencil rod array on the silicon substrate surface which shows the properties of low reflectance, large p-n junction area, and short carrier collection path, is fabricated by combining the polystyrene lithography and the metal-assisted chemical etch (MacEtch) process. The MacEtch having the benefit of less surface damage from wet etching and anisotropic etching from dry etching at the same time. These advantages help us control the etching structure and not produce lots of defects during the process. The MacEtch technique used noble metal as the catalyst. When we immersed the sample into etching solution consisting of HF and some specific oxidant, the oxidation reaction occurred in the area covered by metal. Because of the good electron transportation of noble metal, holes will pass through the metal and oxidize the silicon; the HF will then dissolve the oxidation away. Therefore, the silicon surface morphology is controlled by the pattern of metal catalyst or even different etching solution ratio.[3]

However, the large-area nanostructure will generate defects on the surface which cause carrier recombination.[4] Therefore, controlling the interfacial behavior will be a decisive factor for enhancing the solar cell performance. The advantages of the atomic layer deposition (ALD) technology, including the precisely controlling thickness, fully-covered on high aspect ratio structure, few defects in the large area and stability process have been reported. Therefore, we deposited the ultra-thin aluminum oxide (Al_2O_3) as a passivation layer by ALD technology to enhance the PCE of the hybrid solar cell. [5]

II. FABRICATION

One-side polished N-type Si (100) wafer with the thickness of 625um and resistivity of 2-7Ωcm was used as substrates. The Si wafers were cut into 3 × 3 cm^2 and were pre-cleaned by RCA clean. The fabrication began with the use of nanospheres lithography, as showed in Fig. 1.

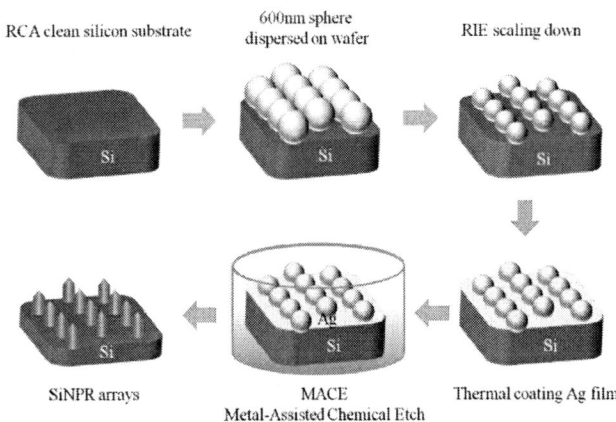

Fig. 1. Flowchart of self-assembly polystyrene nanospheres lithography collaborated with metal-assisted chemical etching

A close-packed monolayer of self-assembled polystyrene (PS) nanospheres with a period of 600 nm was spin-cast and transferred on the polished side of the Si substrate, followed by a reactive ion etching (RIE) process to control the diameter of the nanospheres, such that a non-close-packed monolayer became available for etching. After the RIE process, the Si substrate covered with a monolayer of PS nanospheres as the selective etching mask was deposited with Ag with a thickness of 40 nm by thermal evaporation. Next, the samples were immersed into the mixture solution with hydrogen fluoride (HF), iron(III) nitrate ($Fe(NO_3)_3$), hydrogen peroxide (H_2O_2) and deionized water for fabricating pencil rod array. The PS nanospheres and the Ag film were finally removed from etchant solution with piranha solution and nitric acid (HNO_3), respectively.

Before the hybrid solar cells fabrication, the n-Si substrates with pencil rod were deposited with an ultra thin Al_2O_3 layer on the front and back side of substrates surface by ALD, respectively. Next, we use evaporated aluminum on the rear side of the Si substrate as the back contact. The organic material (PEDOT:PSS) was spin-coated onto the front side of the substrate in order to form the p-n junction. And the sample was annealed. In the end, as for the anode, we use evaporated silver grid, as showed in Fig. 2.

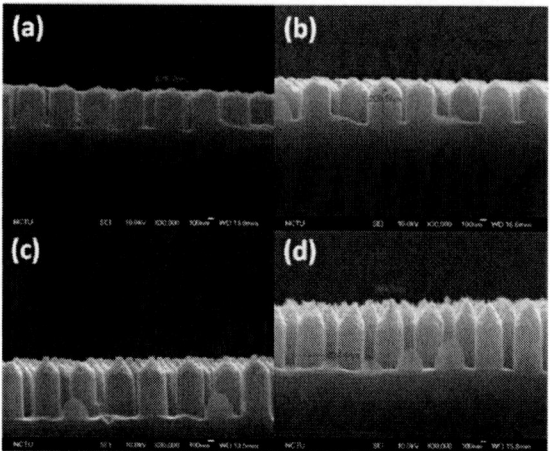

Fig. 3. SEM images of depth nanorods structure (a) 600nm and different depth of pencil rods (b) 700nm (c) 850nm (d) 980nm

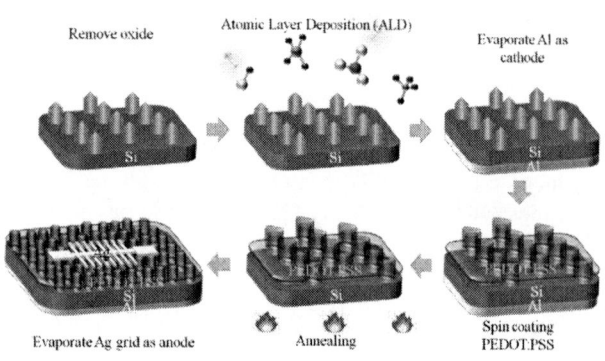

Fig. 2. Flowchart of the fabrication of hybrid organic-silicon solar cells

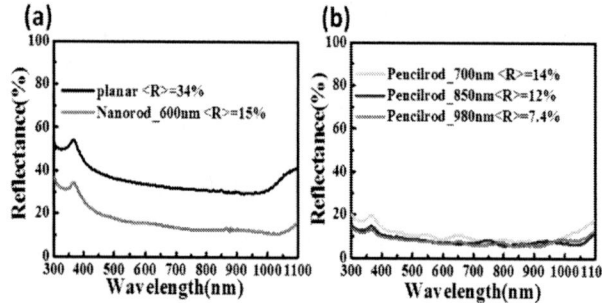

Fig. 4. Reflectance spectrum of (a)planar and nanorods (b)different pencil rod etching time

III. RESULTS AND DISCUSSION

Fig. 3a–d shows the cross-section scanning electron microscopy (SEM) images of periodicity silicon nanorods and silicon pencil rods formed with average diameter 500nm by the etching process. Fig. 3a shows the SEM images of silicon nanorods with the average depth of 600nm formed by the vertical etching step in HF/H_2O_2. Through the mechanism of MacEtch, $Fe(NO_3)$ in the HF/H_2O_2 mixture contributes the excessive holes, which can diffuse to the sidewalls and then oxidize the silicon atoms, whereas HF helps the removal of the oxide layer. Hence, the tapered tip of silicon morphologies can be formed.[6] Fig. 3b–d shows the different depth silicon pencil rod depending on different etching time.

To analyze the light-harvesting properties of the different morphology due to its light-trapping effect, the total reflectance spectra in the wavelength ranging between 300 and 1100 nm were measured, as shown in Fig. 4.

It should be noticed that the nanorods structure is of lower weighted reflectance of 15% compared to planar device 34%, as showed in Fig. 4(a). The reason for the decrease of the reflectance is likely due to the multiple internal reflection of light on the silicon nanostructures that have prolonged optical path before being reflected back to the air, exhibiting a light-trapping effect. And then, the full-band reflectivity decreases because of the tip portion with a smaller pillar diameter for more efficient photon capture. Moreover, as the depth pencil rods increases, the full-band reflectivity following decreases, as showed in Fig. 4(b). The lowest weighted reflectance of 7.4% is obtained by pencil rods with depth of 980nm

However, the surface carrier recombination mostly originates from the increased surface area in the silicon nanostructures. To enhance the performance of nanostructure hybrid silicon solar cells, recombination channels should be suppressed. To resolve this issue further, we deposited the ultrathin Al_2O_3 layer on the front and back side Si substrate respectively reduce the surface defects by ALD.

978-1-5090-5606-4/17 $31.00 © 2017 IEEE

TABLE I

CHARACTERISTICS OF HYBRID SILICON SOLAR CELLS PLOTTED
FOR THE THICKNESS OF AL$_2$O$_3$ LAYER DEPOSITED ON THE BACK
SIDE SI SUBSTRATE

Al$_2$O$_3$ Thickness	Voc (V)	Jsc (mA/cm2)	F.F. (%)	PCE (%)
w/o Al$_2$O$_3$	0.526	27.64	64.75	9.42
Al$_2$O$_3$ 5Å	0.541	31.29	66.47	11.25
Al$_2$O$_3$ 10Å	0.556	30.38	70.37	11.89
Al$_2$O$_3$ 15Å	0.571	31.87	74.44	13.55
Al$_2$O$_3$ 20Å	0.545	32.3	69.16	12.15

TABLE II

CHARACTERISTICS OF HYBRID SILICON SOLAR CELLS PLOTTED
FOR THE THICKNESS OF AL$_2$O$_3$ LAYER DEPOSITED ON THE
INTERFACE BETWEEN PEDOT:PSS AND SI SURFACE
STRUCTURE

Al$_2$O$_3$ Thickness	Voc (V)	Jsc (mA/cm2)	F.F. (%)	PCE (%)
w/o Al$_2$O$_3$	0.576	29.45	72.44	12.29
Al$_2$O$_3$ 3Å	0.572	31.29	74.15	13.28
Al$_2$O$_3$ 5Å	0.556	29.73	67.64	11.20
Al$_2$O$_3$ 10Å	0.522	19.75	40.22	4.15

The photovoltaic performance was measured under AM 1.5G illumination in the standard testing condition. table 1 shows the trend of characteristics of hybrid silicon solar cells depending on the thickness of Al$_2$O$_3$ on the backside of the silicon substrate and indicates that the open circuit voltage (V$_{OC}$) and fill factor (F.F.) are both further enhanced. The thickness of the dielectric barrier over 15Å leads to suppression of carrier transportation, which causes the decrease of the V$_{OC}$ and F.F. In the same way, the short circuit current (J$_{SC}$) is further enhanced when the Al$_2$O$_3$ thickness increases to 3Å which deposited on the interface between PEDOT:PSS and Si surface structure, as showed in table 2. Al$_2$O$_3$ can not only reduce the carrier recombination center but infiltrate the PEDOT:PSS into the gaps of the nanostructure.

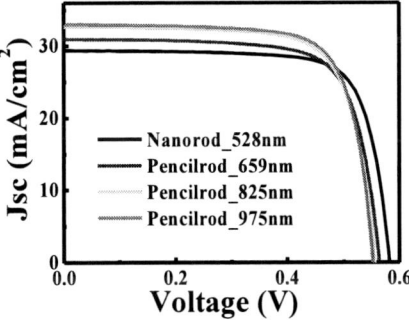

Fig. 5. Current density-voltage characteristics

Fig. 5 and show the properties of hybrid solar cells with different depth of pencil rod. The external quantum efficiency (EQE) improved markedly for the pencil rod compared to the nanorods structure. The significant increase in EQE shows that the broadband light-trapping characteristics of nanostructure could readily contribute to the J$_{SC}$ enhancement. However, we also noted that the V$_{OC}$ and F.F. decrease due to the increasing depth of pencil rods, as well as the high aspect ratio, which limits the coverage of PEDOT:PSS. Even so, we successfully boost the J$_{SC}$ and achieve the PCE of 13.61%.

IV. CONCLUSION

To reduce the costs of silicon-based solar cells, we use the organic material PEDOT:PSS with lower temperature process. However, to improve cells efficiency by using light management, we produced nanostructure on the silicon substrate surface by metal-assisted chemical etching to reduce reflectivity. By this method, we can precisely control the diameter and the length of the structure. The profile control is important to the optical absorption and the organic surface coverage in hybrid organic-silicon solar cells. Besides, by using a thin Al$_2$O$_3$ layer with a thickness of 3Å and 15Å to passivate the pencil-shaped nanorods front surface and the rear surface, we can achieve the hybrid solar cell of high efficiency with 13.61% due to the formation of a favorable internal electric field at the interface for an efficient carrier extraction and optimizing the depth of pencil rod.

REFERENCES

[1] Lining He†‡§, Changyun Jiang‡, Hao Wang†, Donny Lai†§, and *†§, "Si Nanowires Organic Semiconductor Hybrid Heterojunction Solar Cells Toward 10% Efficiency, "*American Chemical Society*, vol. 4, pp. 1704−170, 2012.

[2] Manisha Sharma†, Pushpa Raj Pudasaini‡, Francisco Ruiz-Zepeda‡, David Elam‡, and Arturo A. Ayon*‡, "Ultrathin, Flexible Organic–Inorganic Hybrid Solar Cells Based on Silicon Nanowires and PEDOT:PSS, " *American Chemical Society*, vol. 6, pp. 4356−4363,2014.

[3] Zhipeng Huang, Nadine Geyer, Peter Werner, Johannes de Boor, and Ulrich Gösele, "Metal-Assisted Chemical Etching of Silicon: A Review, "*Adv. Mater*, vol. 23, pp. 285–308, 2011.

[4] Sangmoo Jeong, Michael D. McGehee & Yi Cui, "All-back contact ultra-thin silicon nanocone solar cells with 13.7% power conversion efficiency, " *NATURE COMMUNICATIONS*, 2013.

[5] B. Hoex,aı J. J. H. Gielis, M. C. M. van de Sanden, and W. M. M. Kesselsb, "On the c-Si surface passivation mechanism by the negative-chargedielectric Al2O3, " *JOURNAL OF APPLIED PHYSICS*, 104, 113703, 2008.

[6] Yusaku Asano, Keiichiro Matsuo, Hisashi Ito, Kazuhito Higuchi, Kazuo Shimokawa, and Tsuyoshi Sato, "A Novel Wafer Dicing Method Using Metal-Assisted Chemical Etching, " *IEEE*, 2015.

PL Study of Phosphorus-Doped CdTe EVT Films

Shamara Collins, Imran Khan, Vamsi Evani, Chih An Hsu, Vasilios Palekis, Don Morel and Chris Ferekides

University of South Florida, 4202 East Fowler Avenue, Tampa, Florida, 33620, U.S.A.

Abstract — Low temperature photoluminescence (PL) was used to study the effect of in-situ phosphorus (P) doping on polycrystalline Cadmium Telluride (CdTe) deposited by the Elemental Vapor Transport (EVT) technique. The CdTe films discussed in this paper were deposited under Cd/Te vapor ratios of 1.0, 2.0 and 3.0, and two vapor phase P concentrations of 4,000 and 16,000 ppm. Intensity and temperature dependent measurements revealed PL bands at 1.57, 1.50-1.56, 1.36-1.46, 1.28, 1.00 and 0.70 eV. It was determined that the 1.54 eV was due to transitions from donor to shallow acceptor states, likely P at Te vacancy (P_{Te}). The 1.46 eV band was due to band to level transitions. Deep transitions around 1.00 and 0.70 eV were observed and are believed to be due to native defects.

Index Terms — photoluminescence, CdTe, phosphorus, native defects

I. INTRODUCTION

Thin film polycrystalline Cadmium Telluride (CdTe) is used in the fabrication of thin film photovoltaic devices. Within the past five years, the efficiency of small area CdTe solar cells have increased significantly after years of stagnation. The record efficiency is currently 22% [1], and the material is outcompeting multicrystalline silicon in terms of cost and efficiency [2]. The open-circuit voltage (V_{OC}) is a key performance parameter that historically has limited further advances in efficiency. According to the Shockley-Queisser theoretical limit for single junction photovoltaics, the maximum V_{OC} for CdTe is approximately 1 volt [3]. Reported values for champion polycrystalline CdTe devices are well below this potential maximum, with single crystalline based CdTe cells reaching 1.1 V, indicating that further advancements in this area are possible [4]. V_{OC} can be improved by optimizing many parameters, mainly, the carrier concentration and carrier lifetime, which are influenced by the presence of defects [5]. Defects at mid bandgap create Shockley-Read Hall recombination centers which reduce lifetime. The challenge of maintaining high lifetimes with high doping concentrations is addressed by gaining a thorough understanding of the CdTe native defects and formation energies.

CdTe is a II-VI defect semiconductor that can be intrinsically or extrinsically doped p- or n-type. Intrinsic doping is limited by the formation of compensating native defects [6]. Extrinsic doping of CdTe can be achieved under favorable conditions, where native defects such as vacancies can allow the dopants to be effectively incorporated. The Elemental Vapor Transport (EVT) deposition process allows *in-situ* control over the concentration of native defects. For example, under Te-rich deposition conditions, Group I dopants are favored for substitution at Cd vacancy (V_{Cd}) sites. Alternatively, under Cd-rich deposition conditions Group V dopants are favored for substitution at Te vacancy (V_{Te}) sites. EVT offers the ability for controlled intrinsic doping and the opportunity for systematic extrinsic dopant incorporation through the favorable formation of suitable vacancies. This study investigates the use of a Group V element, phosphorus (P), as the dopant for CdTe thin films grown under Cd-rich conditions, using photoluminescence (PL) measurements.

PL measurements provide information about the presence of defects as they change with deposition conditions through alteration of energetic levels within the CdTe bandgap. The varying EVT deposition conditions include, stoichiometry and P concentration. An understanding of the defect modification will determine which deposition conditions are best suited to effectively incorporate the dopant.

II. EXPERIMENTAL

Polycrystalline CdTe films were deposited by the Elemental Vapor Transport (EVT) technique on glass substrates with RF sputtered indium tin oxide (ITO) as the transparent electrode and CdS as the window layer (via chemical bath deposition process) [7]. The EVT system was designed to control the gas phase Cd/Te stoichiometry and dopant concentration. The term 'Cd/Te ratio' is used to describe the gas phase ratio of the elemental Cd and Te vapors. The CdTe films were grown under stoichiometric and Cd-rich conditions (Cd/Te vapor ratios: 1.0, 2.0, and 3.0). The three gas phase deposition conditions were also carried out at two different P concentrations (4,000 and 16,000 ppm). These concentrations describe the dopant amount in the gas phase, and not what was incorporated into the CdTe films (see discussion on SIMS analysis). Further details on the EVT deposition process are available elsewhere [8], [9].

The PL measurements were carried out using a SPEX 500M monochromator with a 600 groove/mm grating, with energetic resolution of 1 meV. Signal detection was accomplished with a lock-in amplifier (EG&G Model 5209), a chopper at 108Hz (EG&G Model 197), a low noise amplifier (Standard Research Systems Model SR560) and a TE cooled InGaAs photodetector. The photoexcitation source was a Coherent OBIS Fiber Optic laser at 640nm. The incident laser power was varied from 1-25 mW. The PL spectra were collected over a temperature range of 32 to 90K; the temperature was controlled using a CTI-cryogenics closed loop He system (Compressor Model SC) and monitored with a two temperature sensor configuration connected to a Lakeshore Model 325 temperature controller.

III. RESULTS AND DISCUSSION

Secondary Ion Mass Spectroscopy (SIMS) was used as a supportive material characterization technique, to verify the presence of P in the CdTe thin films. SIMS results for the Cd/Te

978-1-5090-5606-4/17 $31.00 © 2017 IEEE

TABLE I

The deconvoluted peaks' energetic position at different ratios present in the photoluminescence spectra at 32K for CdTe doped with Phosphorus (4,000 and 16,000 ppm) and different Cd/Te ratios.

Cd/ Te Ratio	Phosphorus- 4, 000 ppm			Phosphorus- 16, 000 ppm		
	1.0	2.0	3.0	1.0	2.0	3.0
I	1.574	1.571	1.577	1.576	1.571	1.577
II			1.553			
		1.543	1.544		1.543	1.540
			1.527			
		1.516			1.518	1.511
III		1.465	1.468	1.465	1.456	1.467
	1.431					
	1.350	1.380	1.364	1.357		1.369
	1.263	1.200	1.285	1.276	1.296	
IV	1.02	1.056	1.045	1.019	1.064	1.036
		0.775	0.776	0.768	0.774	
		0.762	0.768	0.754	0.765	
		0.734	0.738	0.735	0.757	0.756

1.0 film deposited with 16,000 ppm are presented in Fig. 1. The average P concentration was calculated to be 10^{17} cm^3. This suggests approximately 0.01% of P available in the gas phase during the deposition was incorporated into the film.

The PL measurements and analysis are presented for four spectral regions: (a) Excitonic (1.57-1.59 eV), (b) 1.50-1.55 eV, (c) 1.36-1.46 eV and (d) 0.70-1.20 eV. The complex PL bands in these spectral regions were deconvoluted into elementary components (or sub-bands) using the Lorentzian function. The energy values precision was based on the mathematical accuracy of the deconvolution process as presented elsewhere [8]. It is noteworthy, that elementary bands differing 1-2 meV are considered to have the same origin and ambiguous treatment on the number of bands is possible. The dependence of these elementary components were considered for their dependence on temperature and incident excitation power.

PL analysis of the results led to several conclusions to be drawn. There is a substantial change in PL spectra for the two dopant concentrations and varying Cd/Te ratios. Fig. 2a and 2b

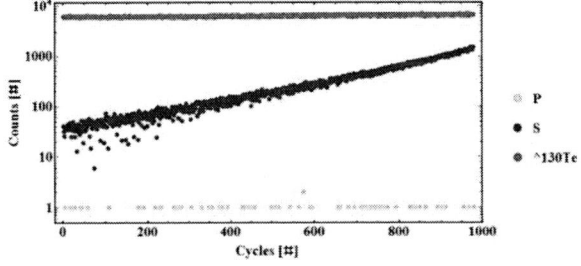

Fig. 1.   Secondary ion mass spectroscopy of the depth profile for the CdTe layer (Cd/Te 1.0) doped with P (16,000 ppm). The phosphorus concentration has been determined to be on the order of 10^{17} cm^3.

are spectra for P at 4,000 and 16,000 ppm, respectively. At the lower doping concentration (4,000 ppm) all films had four PL bands in common: 1.00, 1.36, 1.46 and 1.57 eV. When the dopant concertation increased to 16,000 ppm these common transitions changed. The spectra for the 16,000 ppm films with Cd/Te ratio of 1.0 and 3.0 were comprised of similar transitions. The Cd/Te 2.0 film showed mainly a strong transition at 1.51 eV. Table 1 shows the detailed energetic positions at different ratios for the two phosphorus concentrations. The following sections present a more detailed analysis of the PL transitions separated into their energetic regions.

A. Region I: Excitons (1.57-1.59 eV)

The excitonic transition at 1.57 eV was present in all films. The transition increased in intensity, as the films with P gas phase concentration of 4,000 ppm became more Cd-rich. However, when the P concentration increased to 16,000 ppm, this PL band remained pronounced only for Cd/Te ratios of 1.0 and 3.0. In the film grown under Cd/Te 2.0 condition, the 1.57 eV transition was comparatively weak and a new band emerged at 1.51 eV with high intensity (discussed in the next section).

Considering that the 1.57 eV band changes with dopant and stoichiometry, this suggests a dependence on native defects and compensation between the native defects and phosphorus. In previous photoluminescence study of undoped CdTe EVT films the excitonic transition at 1.585 eV was observed and assigned to the annihilation of electrons bound to an acceptor, while the 1.577 eV transition was considered a phonon replica [9]. Single crystal CdTe films doped via P implantation also exhibited PL bands at 1.591 eV [10] and 1.588 eV [11]. The former was associated with Cd$_i$ and the latter with an acceptor exciton line. The literature partially supports assignment of the 1.57 eV transition in EVT CdTe to an acceptor state effected by excess Cd.

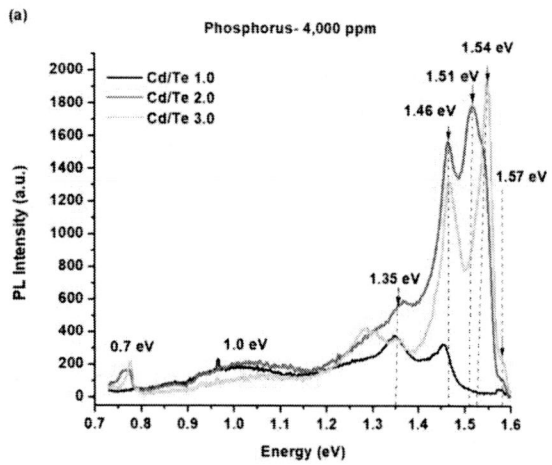
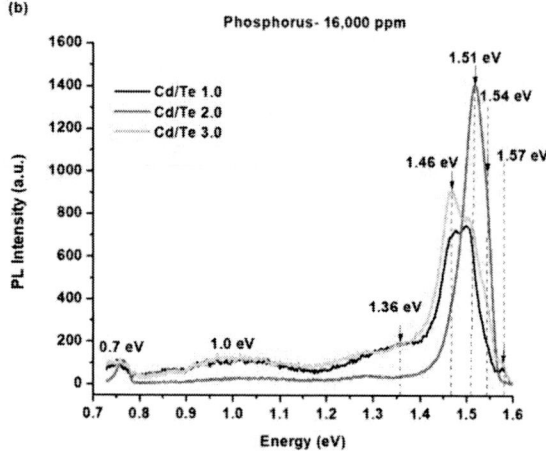

Fig. 2. The photoluminescence spectra at 32K of CdTe films deposited at different Cd/Te vapor ratios and phosphorus concentrations; a) P: 4,000 ppm; b) P: 16,000 ppm. The excitation intensity was 25mW.

B. Region II: 1.50-1.55 eV

Spectra in the range of 1.50-1.55 eV change greatly with increasing Cd/Te ratio and increased doping. As observed in Fig. 2a, the P-doped film of 4,000 ppm has no PL band in this region for films deposited with a Cd/Te ratio of 1.0. A band at 1.51 eV was pronounced for the Cd/Te 2.0 and at 1.54 eV for the Cd/Te 3.0 films. The results for the films with P at 16,000 ppm shows a relatively different spectrum when compared to the lower doping concentration. The trend of no PL band at Cd/Te 1.0 is consistent, as seen in Fig. 2b, but both 1.51 and 1.54 eV bands exist for Cd/Te ratio 2.0 and 3.0.

The 1.54 eV band was predominant in the Cd/Te 2.0 and P-4,000 ppm film. Temperature dependent analysis presented in Fig. 3, suggests that the band is associated with donor-acceptor pairs (DAP) transitions. The acceptor states for the 1.54 eV PL band may be associated with P and/or native defects.

A previous study of intrinsic CdTe films grown by EVT reported a band at 1.547 eV. This band was most pronounced in Te-rich films and decreased in Cd-rich films, indicative of a transition related to the quenching of V_{Cd} [12]. Alternatively, in a study of P-doped CdTe films, the 1.543 eV band was observed and P_{Te} was assigned to be the shallow acceptor involved in the transition [13].

Another study of P-doped CdTe reported an increase in hole concentration after the additional incorporation of Cd. This increase was attributed to increased acceptors (P_{Te}) in a trade for donor states such as P_{Cd} [14]. The defect, P_{Te}, forms a shallow singly ionized acceptor [13]. Reported Hall Effect measurements determine P_{Te} substitutions to form a shallow acceptor state 35-70 meV above the valence band. In support of the above, PL reports indicated that this level is 60-70 meV above the valence band [13]. In addition, the effective mass theory assigned P_{Te} with a transition energy of 68meV [13] - [15].

The above literature results along with the deposition conditions (Cd-rich: Cd/Te 2.0 and 3.0 with high P

concentration) support the assignment of P_{Te} as the shallow acceptor involved in the 1.54 eV transition. The absence of this band in stoichiometric films (Cd/Te 1.0), suggest a screening mechanism involving native defects may occur. The 1.51 and 1.54 eV bands change with dopant and stoichiometry prove there may be a favorable growth condition for efficiently incorporating P into the films.

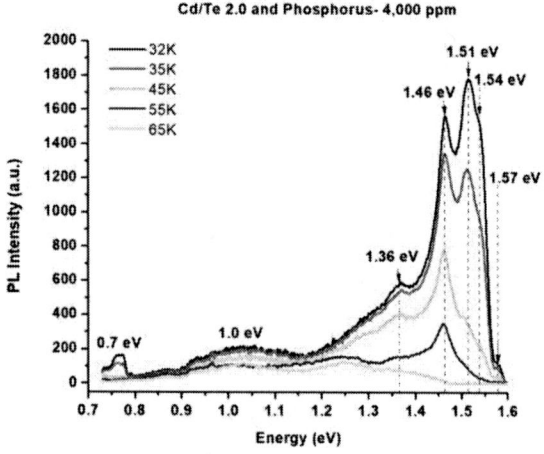

Fig. 3. Temperature dependent photoluminescence spectra for CdTe film with Cd/Te vapor ratio of 2.0 and phosphorus concentration of 4,000 ppm. Selected bands at 1.51 and 1.54 eV show quenching at higher temperatures.

C. Region III: 1.36- 1.46 eV

The nature of the emission bands near 1.4 eV change with Cd/Te ratio and the doping concentrations. Deconvolution of bands in this region resulted in several elementary bands as indicated in Table 1. Most common were transitions at approximately 1.36 and 1.46 eV.

The 1.46 eV emission line was common for all Cd/Te ratios with phosphorus at 16,000 ppm (see Table 1). The films at 4,000 ppm also had this transition. This selected PL band consistently exhibited a linear dependence with excitation power. Fig. 4 shows the PL intensity versus excitation power for this band in films of Cd/Te ratios 2.0 and 3.0 at the lower doping concentration. The linear dependence on excitation intensity, suggests the emission is the result of a band to level transitions (i.e. e-A).

Sub-bands within the 1.4 eV transition have commonly been attributed to native defects with V_{Cd} complexes as the acceptor states [16]. The presence of an impurity band or several acceptors relatively close to each other, such as P_{Te} or P_i, cannot be excluded [14], [16].

The bands near 1.36 eV are present in all films except the Cd/Te 2.0 with highest doping. The transition exhibited an intensity dependence and was associated with band to acceptor transitions [9]. In previous studies, the acceptor states for the 1.36 eV transition were suggested to be present on crystallite boundaries and within the dislocation core [17]. The incorporation of P can cause the bond between Cd-Te atoms to break due to large electronegativity, resultant in Te dangling bonds [16]. The absence of this band in films with higher dopant concentration may be a result of the Te related defects becoming occupied by P.

D. Region IV: 0.7-1.2 eV

The deep radiative transitions in CdTe were observed from 0.70 to 1.20 eV. Through deconvolution, it was determined there was a single wide band transition centered at 1.0 eV and several sub-bands for the 0.7 eV transition.

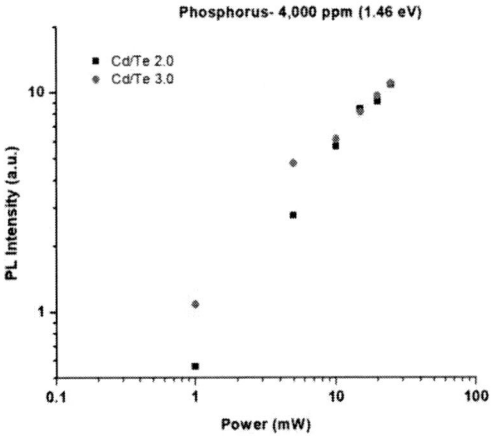

Fig. 4. Photoluminescence intensity of the 1.46eV band versus laser excitation power at 32K, for Cd/Te 3.0 and P concentration of 4,000 ppm.

The 1.0 eV band was observed in all the films, but substantially less intense for the Cd/Te 2.0 at the highest doping concentration, as seen in Fig. 2. In samples with the transition, an increase in the photoexcitation intensity caused the PL emission to be quenched. The intensity dependence, suggest an involvement with transitions from band to defect levels. Ascertaining the origin of the defect in this transition is complicated because the trend with Cd/Te ratio and doping concentration is inconsistent and additional data/measurements will be needed in order to make the appropriate assignments.

In studies of the defect structure of CdTe, a wide luminescence band was reported at 1.0-1.15 eV. Through optical detection and conventional spin resonance, the band observed at 1.1 eV was related to V_{Te} [18]. Also, deep radiative transitions were reported for intrinsic CdTe grown by EVT. Preliminary analysis of this transition showed that the intensity of the band had strong dependence on Cd/Te overpressure ratio and it was comprised of at least two sub-bands (1.042 and 1.126 eV) [9]. The suggested defect assignments were CB to Te_i^{2-}, Cd_i^{2+} to Te_i^{2-} or V_{Te}^{+} to Te_i^{-} [11].

The 0.7 eV band was also present in all films, except for Cd/Te 1.0 at the lowest doping concentration. The deep energy band exhibited a thermal quench, suggesting a donor-acceptor pair transition. Deep energy levels in CdTe were previously studied and widely attributed to an acceptor complex involving native V_{Cd} and impurities [19], [20]. The nature of impurities occupying substitutional sites on the sublattice of the material is interesting and require further investigation.

IV. CONCLUSION

The effects of P doping on the photoluminescence spectra of polycrystalline CdTe films grown by elemental vapor transport (EVT) have been described. The 1.54 eV band was determined to involve a donor-acceptor pair transition. Experimental conditions and theoretical evidence found in literature support the assignment of P_{Te} as the shallow state involved in the transition. Both bands near 1.36 and 1.46 eV were determined to be band to level transitions. The defect in the 1.36 eV was resultant of damage caused by dopant incorporation. Analysis of the deep radiative transitions at 1.0 eV and 0.7 eV determined that they are related to native defects.

Although, the optimum deposition conditions for incorporating phosphorus is yet to be fully understood/determined, the PL results suggest Cd/Te ratios 2.0 and 3.0 are more favorable and there may be an excessive amount for the dopant concentration.

ACKNOWLEDGEMENT

S. Collins is supported by NSF GRFP (1144244); this work was supported by DOE SunShot Initiative under the F-PACE program (DE-EE0005401). The authors acknowledge Dr. Sergiu Vatavu of Moldova State University for fruitful discussions on PL results and Ellie Roe of Dr. Mark Lonergan's research group (University of Oregon) for SIMS measurement and analysis.

REFERENCES

[1] M. Gloeckler, "Realization of the Potential of CdTe Thin-Film PV," in *43rd IEEE Photovoltaic Specialist Conference*, Portland, Oregon, 2016.

[2] E. Wesoff, "First Solar Reaches 16.3% Efficiency in Production PV Modules." *GreenTech Media* <www.greentechmedia.com/articles/read/First-Solar-is-Reaching-16.3-Efficiency-in-Production-PV-modules> Accessed 20 January 2017.

[3] W.S. Queisser and J. Hans, "Detailed Balance Limit of Efficiency of pn Junction Solar Cells," *Journal of Applied Physics*, pp. 510-519, 1961.

[4] Becker et al., "Loss Analysis of Monocrystalline CdTe Solar Cells with 20% Active-Area Efficiency," IEEE Journal of Photovoltaics, vol. 7, no. 3, 2017.

[5] T.A. Gessert et al., "Foundations for Improving thin film CdTe photovoltaic devices beyond 20% conversion efficiency," in *39th IEEE Photovoltaic Specialist Conference*, Tampa, Florida, 2013.

[6] S. Wei and S. B. Zhang. "Chemical trends of defect formation and doping limit in II-VI semiconductors: The case of CdTe", *Physica Review B*, vol. 66.15, pp. 155211, 2002.

[7] C. S. Ferekides, et al. "High efficiency CSS CdTe solar cells," *Thin Solid Films*, vol. 361-362, pp. 520-526, 2000.

[8] M. Khan, et al., "In-Situ Antimony Doping of CdTe," in *42nd IEEE Photovoltaic Specialist Conference*, New Orleans, Louisiana, 2015.

[9] S. Collins, et al., "Radiative recombination mechanisms in CdTe thin films deposited by elemental vapor transport," *Thin Solid Films*, vol. 582, pp. 139-145, 2015.

[10] T. Taguchi, J. Shirafuji and Y. lnnishi, *Physica Status Solidi (b),* vol. 68, pp. 727, 1975.

[11] J. R. Panossian, A.A. Gippius and V. Vavilov, *Physica Status Solidi (b)*, vol. 35, pp. 1069, 1969.

[12] Khan et al., "Stoichiometric Effects in Polycrystalline CdTe," in *41st IEEE Photovoltaic Specialist Conference*, Denver, Colorado, 2014.

[13] J. Burst et al., "Carrier density and lifetime for different dopants in single-crystal and polycrystalline CdTe," *Applied Materials*, vol. 4, 2016.

[14] F. Selim and F. Kroger, "The Defect structure of Phosphorous-doped CdTe," *Journal of Electrochemical Society,* 1977.

[15] M. Said and M.A. Kanehisa, *Journal of Crystal Growth*, vol. 101, 1990.

[16] A. Baldereschi and N.O. Lipari, *Physica Review (b),* vol. 9, 1974.

[17] H.L. Hwang, K. Hsu, and H.Y. Ueng, "Fundamental studies of p-type doping of CdTe*," Journal of Crystal Growth*, vol. 161, pp. 73-81, 1996.

[18] Krustok, Madisson, and Hiie, "Photoluminescence properties of Z-bands in CdTe," *Phys Status Solidi A*, vol. 165, 1998.

[19] P. Fernandez, "Defect structure and luminescence properties of CdTe based compounds," *Journal of Optoelectronics and Advanced Materials*, vol. 5, 2003.

[20] Castaldini, et al., "Deep energy levels in CdTe and CdZnTe", *Journal of Applied Physics*, vol. 83, 1998.

Characterization of Single-Source Deposited Close-Space Sublimation CdTe$_x$Se$_{1-x}$ Thin Film Solar Cells

Corey R. Grice[1], Jian Li[2] and Yanfa Yan[1]

[1]The University of Toledo, Toledo, Ohio, 43607, USA,

[2]Texas State University, San Marcos, Texas, 78666, USA

Abstract — **Thin film CdTe$_x$Se$_{1-x}$ solar cell devices are fabricated using close-space sublimation from homogenized mixtures of CdTe and CdSe powders. Various window layer combinations of CdS and/or CdSe are also explored. The performance of the finished devices, in terms of current-voltage and spectral response, are evaluated while the cross sectional structures and compositional profiles are examined for selected devices. The observations from this study suggest that independent control of the Se alloying content into thin CdTe absorber layers can be independently controlled both near the front interface and within the film bulk while stile using a simple, one-source close-space sublimation film deposition arrangement.**

Index Terms — CdTe solar cells, CdSe, CdTe$_x$Se$_{1-x}$, alloy.

I. INTRODUCTION

Cadmium telluride (CdTe) thin film solar cells have seen significant advances in large scale commercialization efforts over the past few decades, with expectations of up to 5400 MW of deployed global capacity by 2023 [1]. This is, in part, due to their inherent low cost, stability, and high photoconversion efficiency (PCE) [2]. Several research-level advances have led to the high PCEs that have enabled such commercialization, such as increasing the photocurrent generated within the cells upon illumination. One of the methods by which this has been accomplished is with the incorporation of selenium (Se) into the CdTe thin film absorber layer. By incorporating Se as a minor component (typically < 25% atom basis) into the CdTe, the bandgap of the material is reduced from 1.5 eV (for pure CdTe) to values ranging from 1.3-1.45eV, which is a known effect for alloys of CdTe and cadmium selenide (CdSe) [3]. The motivation for doing this is that the reduced bandgap value is more optimized to collect and convert typical 1 sun illumination into usable electrical power, based upon a thermodynamic calculation known as the Shockley-Queisser limit [4].

Initial reports on Se incorporation in CdTe solar cells focused on increasing its concentration near the illuminated, or front, side of the devices. One method of accomplishing this is to introduce a layer of pure CdSe in between the CdTe and a cadmium sulfide (CdS) n-type window layer. Upon high temperature (> 600°C) deposition of the CdTe[5,6], or medium temperature (< 400°C) annealing in the presence of CdCl$_2$ vapor [7], the CdSe diffuses significantly into the CdTe and creates an compositional gradient within the film, hence also creating a gradient in the material bandgap commensurate with the composition. The photon absorption near the p-n device junction is enhanced and thus the operating photocurrents are increased [8]. Later reports improved upon this result by depositing a Te-rich CdTe/CdSe alloy layer between the window layer and the pure CdTe layer [9] using two different deposition sources. While the device performances achieved through this method were exceptional, the added complexity of the deposition process may be difficult for widespread adoption and scalability.

This work explores the deposition of CdTe$_x$Se$_{1-x}$ alloy films (x > 0.5) from a single source of mixed CdTe and CdSe via a close-space sublimation (CSS) technique. Window layers of CdS and CdSe of varying thicknesses are used to identify their effect upon the deposited film properties. The characteristics of the resulting films are also explored as a function of the source material composition and complete solar cells are fabricated from them. The devices are characterized in terms of opto-electronic performance, primarily using current-voltage (I-V) and external quantum efficiency (EQE) measurements, while the cross sectional devices structures and composition gradients are examined using transmission electron microscopy (TEM) and energy dispersive x-ray spectroscopy (EDS).

II. EXPERIMENTAL DETAILS

Commercially available soda-lime glass (TEC-15M, Pilkington North America Inc.) with and without pre-applied SnO$_2$/SnO$_2$:F (FTO) layers were used as substrates for various depositions. These were treated using a detergent-based (Micro-90, International Products Corp.) cleaning procedure assisted by ultra-sonication to ensure chemically clean surfaces.

CdS depositions were performed using an RF magnetron sputtering system (AJA International) using 50W of RF power at 13.56 MHz. The ambient environment in the sputter chamber was maintained at a pressure of 10 mTorr with a constant gas flow rate of 40 sccm with the oxygen content ranging from 0-2% by volume in an argon balance. The substrates were either at ambient temperature or maintained at 270°C. CdSe depositions were performed in a custom built chamber developed in-house. The substrates were heated to

978-1-5090-5606-4/17 $31.00 © 2017 IEEE

approximately 270°C and RF sputtering (50W at 13.56 MHz) was used to deposit CdSe thin films with pure Ar ambient supplied at 5 sccm. All samples were allowed to cool to ambient temperature within the chamber before removal.

Deposition of CSS CdTe$_x$Se$_{1-x}$ films was performed using a custom built fused-silica tube furnace system previously described [10], with a 3mm separation distance between source material and the substrate. The furnace contained an environment of 0-0.5% O$_2$ by volume, balance He, which was maintained at a pressure of approximately 450 Torr during heating. Once the source material had reached a temperature of 660°C, with the substrate at approximately 607°C, the pressure within the furnace was decreased to 50 Torr for the duration of the deposition. Once complete, the pressure was restored to the previous value to end the deposition process and the system was allowed to cool naturally to ambient temperature.

CdCl$_2$ activation treatments were accomplished by exposing the CdTe$_x$Se$_{1-x}$ surface to a saturated CdCl$_2$/methanol solution for activation annealing, which was performed in a similar tube furnace system at 390°C for 30 minutes, with ambient pressure dry air provided at approximately 5 scfh. After cooling to ambient temperature, the sample surfaces were rinsed with methanol to remove residual CdCl$_2$.

Samples used for device fabrication had back contacts of Cu and Au metal bi-layer disks applied via thermal evaporation, with areas of 0.08 cm^2 defined using shadow masks and each sample having between 70 and 140 devices defined in this fashion. A subsequent annealing step at 200°C for 20 minutes in a N$_2$ environment was performed to facilitate Cu diffuse at the device back contact region.

Device characterization was done by measuring I-V responses in near-dark and under AM 1.5G illumination. External quantum efficiencies were also measured at 0mV bias potentials. Scanning electrode microscopy (SEM) images were obtained using a Hitachi-S4800 field-emission gun electron microscope. TEM samples were prepared using Ga focused ion beam (FIB) milling. Cross sectional images were obtained from these samples as well as elemental line profiles using a Hitachi 2300A scanning transmission electron microscope equipped with an energy dispersive x-ray spectroscopy detector. Film structure characterizations were performed using a Rigaku Ultima III x-ray diffractometer equipped with a Cu Kα radiation source, with x-ray diffraction (XRD) measurements taken using a glancing incidence (GIXRD) mode.

III. RESULTS AND DISCUSSIONS

A. Characterization of CSS CdTe$_x$Se$_{1-x}$ films

For the initial studies of this deposition method, powder mixture of 50:50 and 70:30 (molecular ratios) of CdTe and CdSe were prepared by ball milling. The mixed powders were loaded into the source material pocket of the CSS deposition system and films were deposited onto bare glass substrates to evaluate the properties of the resulting films. The x-ray diffraction patterns for the resulting films are shown in Figure 1 and reveal that the films possess a cubic type crystal structure (F-43m, space group 216), which is similar to typical pure CdTe films deposited via the same method. Compositional analysis from EDS measurements reveals that the 70:30 source powder produced a film with a nominal Se concentration of approximately 4.8% (atom basis), corresponding to a CdTe$_x$Se$_{1-x}$ alloy with x = 0.952. Similar analysis for the 50:50 source powder revealed a resulting film Se concentration of 11.5% (atom basis), corresponding to a CdTe$_x$Se$_{1-x}$ alloy with x = 0.730. Top-view SEM images of the film morphologies are shown in Figure 2.

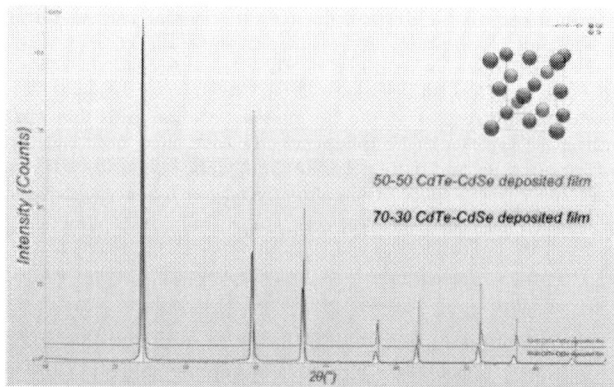

Fig. 1. XRD patterns for CSS deposited CdTexSe1-x films from a 70:30 ratio CdTe:CdSe source powder (dark blue/bottom curve) and 50:50 ratio source powder (green/top curve). Vertical lines (light blue) are for reference pure CdTe peaks.

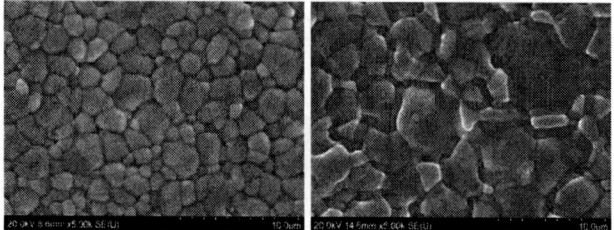

Fig. 2. Top-view SEM images for CSS deposited CdTe$_x$Se$_{1-x}$ films from a 70:30 ratio CdTe:CdSe source powder (left) and 50:50 ratio source powder (right).

B. Characterization of CSS CdTe$_x$Se$_{1-x}$ devices

Thin film solar cell devices were prepared by depositing ∼ 4μm thick CSS CdTe$_x$Se$_{1-x}$ films onto FTO-coated glass substrates with either CdS or CdSe window layers, or bilayer combinations of both. Cross sectional TEM images of the resulting finished solar cell devices made from the 70:30 CdTe:CdSe source powder, along with EDS compositional profiles, were obtained for an 85nm CdS window layer (Figure 3) and a 100nm CdSe window layer (Figure 4). From the cross-sectional images, it can be observed that the $CdTe_xSe_{1-x}$ film deposited on the CdS window layer has smaller grains near the substrate (less than 1μm) compared to those farther from the substrate (greater than 1 μm), whereas the size of the grains deposited on the CdSe layer (above 1μm) appear to be more uniform throughout the thickness of the film. This could be, in part, due to the more facile interdiffusion of CdSe with CdTe as opposed to CdS, which could facilitate resulting grain growth during the deposition process and subsequent $CdCl_2$ activation annealing. The EDS results for both films indicate that the $CdTe_xSe_{1-x}$ composition far from the window layer is nearly constant at ~ 5% Se (atom basis), which is consistent with the results obtained for the films deposited on glass from the same source material. Approaching the window layer, the film deposited on CdSe (Figure 4) shows a gradual increase in Se content to a maximum of ~ 15% (atom basis), which is to be expected. Curiously, the film deposited on CdS shows a decrease in the Se concentration approaching the window layer (to a value of ~2-3%, atom basis) with an increase in S content such that the sum of S + Se in the film remains at ~5% (atom basis). It is somewhat unexpected to see the S diffusion so far into the CdTe-based layer (detectable up to ~0.5μm into the film), when compositional profiles of CSS CdTe devices made on bilayer windows of CdS/CdSe indicated that the S remained at the interface with the FTO, as shown in Figure 5. It may be that nucleation and growth of $CdTe_xSe_{1-x}$ upon the CdS may enhance the transport properties of the S with in alloy and further study of this phenomenon may be warranted.

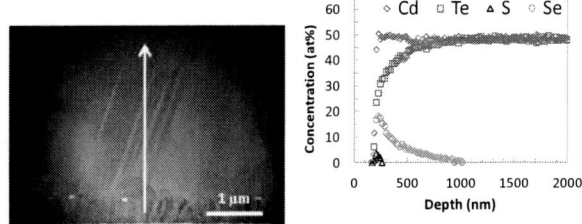

Fig. 5. TEM cross-sectional image (left) and EDS line profile (right) of CSS CdTe device deposited on CdS/CdSe bilayer window.

Similar cross sectional profiling was performed on a device made using 50:50 CdTe:CdSe source material deposited onto a 100 nm CdSe window layer and is shown in Figure 6. This sample displayed an even higher Se elemental composition near the front interface of the device (nearly 20% atom basis) while the Se content in the bulk of the film averaged around 11% at distances of ~ 2 μm from the front interface. For comparison purposes, a similar device was made using pure CdTe deposited onto a 100nm CdSe window layer, with the resulting cross-sectional information shown in Figure 7. Both of these samples show a region of strong Se content within the first ~ 300nm of the front interface and slightly elevated levels in the following ~500nm beyond that – after which the composition becomes relatively uniform and is taken as the bulk value. The primary difference between the two seems to be an offset in the baseline Se content, which can be attributed to the composition of the source material.

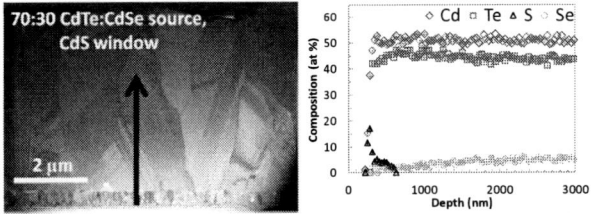

Fig. 3. TEM cross section image (left) and EDS composition profile (right) for $CdTe_xSe_{1-x}$ device made from 70:30 (molecular ratio) CdTe:CdSe source powder onto an 85nm CdS window layer. Black line in TEM image indicates location and direction of EDS linescan.

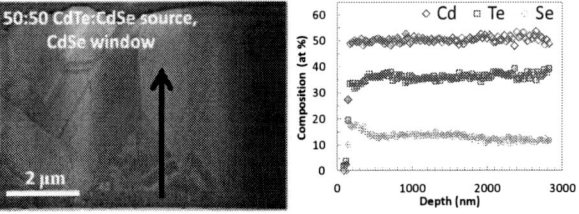

Fig. 6. TEM cross section image (left) and EDS composition profile (right) for $CdTe_xSe_{1-x}$ device made from 50:50 (molecular ratio) CdTe:CdSe source powder onto a 100nm CdSe window layer. Black line in TEM image indicates location and direction of EDS linescan.

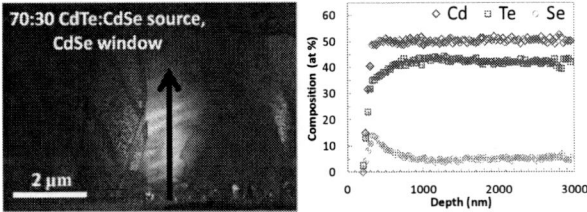

Fig. 4. TEM cross section image (left) and EDS composition profile (right) for $CdTe_xSe_{1-x}$ device made from 70:30 (molecular ratio) CdTe:CdSe source powder onto a 100nm CdSe window layer. Black line in TEM image indicates location and direction of EDS linescan.

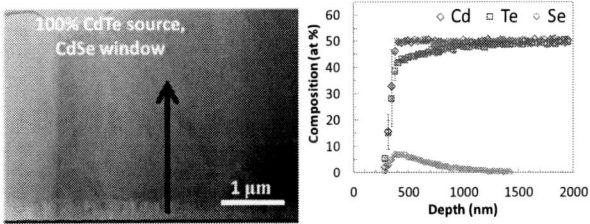

Fig. 7. TEM cross-sectional image (left) and EDS line profile (right) of CSS CdTe device deposited (from pure CdTe source) on 100nm CdSe window layer. EDS data points represent average of 3-

5 line scans at different sample locations, with standard deviation error bars.

Performance characteristics of the devices made from various window layer and CdTe$_x$Se$_{1-x}$ absorber layer combinations were obtained using I-V polarization sweeps performed under AM1.G conditions, as shown in Figures 8 and 9 for the 70:30 and 50:50 CdTe:CdSe source materials, respectively. Baseline performance values are also given in each figure, which represents the performance of typical devices made with pure CSS CdTe deposited on CdS window layer. The results show that, for the 70:30 source material, the device performance improved with increasing CdSe window layer thickness up to 200 nm, with the performances of the devices made with CdSe window layers thicker than 100 nm being better than for pure CdS window layers. However, all devices made with the 70:30 alloy layers showed reduced efficiencies compared to the baseline devices. The 50:50 source material devices showed significantly reduced Voc and efficiency values compared to both the 70:30 devices and the baseline for all combinations of window layers examined. This could be related to the rougher surface morphology for the 50:50 sourced films compared to the 70:30 films, suggesting that the deposition conditions may as yet be un-optimized. Further deposition optimizations for both source materials are expected to yield improved performances.

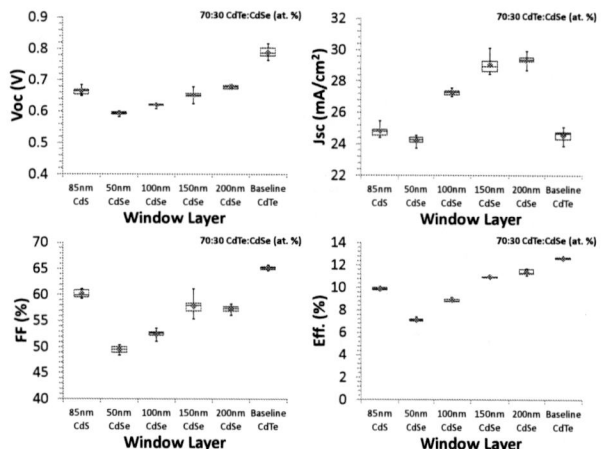

Fig. 8. Performance characteristics obtained from I-V measurements of standard CSS CdTe and CSS CdTexSe1-x devices (deposited from 70:30 CdTe:CdSe source, atom basis) with various window layers.

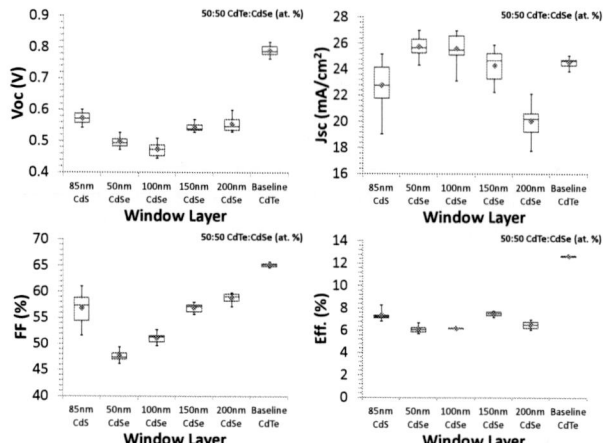

Fig. 9. Performance characteristics obtained from I-V measurements of standard CSS CdTe and CSS CdTexSe1-x devices (deposited from 50:50 CdTe:CdSe source, atom basis) with various window layers.

Comparisons of the EQE spectra for these devices, from both source materials as well as the baseline, are shown in Figures 10 and 11. These spectra clearly display the effect of the Se incorporation with respect to the absorber layer bandgap for the 70:30 sourced devices (Figure 10), with the photocurrent onset red-shifting in the 850 – 950 nm wavelength region with increasing CdSe window layer thickness. This trend is also observed for the 50:50 sourced devices but is less pronounced, likely due to the higher Se content in the bulk absorber film.

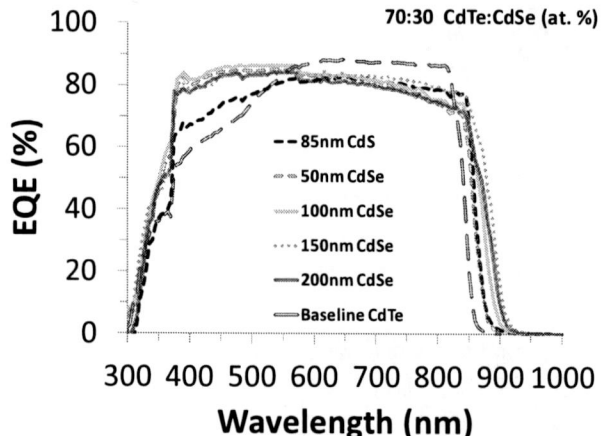

Fig. 10. External quantum efficiency comparison of CdTe solar cells made from 70:30 CdTe:CdSe and pure CdTe source materials with different window layers.

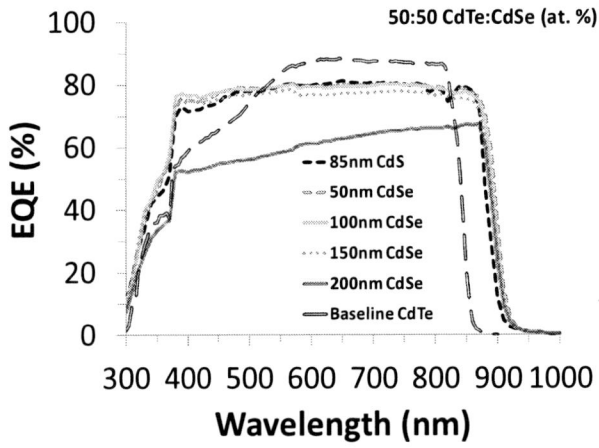

Fig.11. External quantum efficiency comparison of CdTe solar cells made from 50:50 CdTe:CdSe and pure CdTe source materials with different window layers.

IV. SUMMARY AND CONCLUSIONS

The results of this study verify that Se content within thin film CdTe devices can be controlled throughout the thickness of the devices using only a simple, one-source close space sublimation arrangement. The devices made with technique showed reasonable photoconversion performances and it is expected that subsequent optimizations of deposition parameters and associated device fabrication processes will enhance them even further.

ACKNOWLEDGEMENTS

The authors would like to extend our acknowledgement to Dr. David Strickler from NSG, North America for supplying us FTO coated substrates. This work is partially supported by the U.S. Department of Energy F-PACE program and through Sunshot PVRD grant DE-EE-0007541.

REFERENCES

[1] http://www.semiconductor-today.com/news_items/2015/oct/transparency_221015.shtml
[2] investor.firstsolar.com/releasedetail.cfm?ReleaseID=956479
[3] V. A. Kuznetsov, "Analysis of the phase diagrams of the CdS-CdSe-CdTe system," *Semiconductors*, vol. 44, pp. 1117-1120, 2010.
[4] W. Shockley and H. J. Queisser, "Detailed Balance Limit of Efficiency of p-n Junction Solar Cells," *Journal of Applied Physics*, vol. 32, iss. 3, p. 510. 1961.
[5] N. R. Paudel and Y. Yan, "Enhancing the photo-currents of CdTe thin-film solar cells in both short and long wavelength regions," *Applied Physics Letters*, vol. 105, iss. 18, p. 5, 2014.
[6] X. Y. Yang, Z. Bao, R. Luo, B. Liu, P. Tang, B. Li, J. Q. Zhang, W. Li, L. L. Wu and L. H. Feng, "Preparation and characterization of pulsed laser deposited CdS/CdSe bi-layer films for CdTe solar cell application," *Materials Science in Semiconductor Processing*, vol. 48, pp. 27-32, 2016.
[7] C. Grice, A. Archer, S. Basnet, N. R. Paudel, and Y. Yan, "Characterization of CdS/CdSe window layers in CdTe thin film solar cells," in 43rd IEEE Photovoltaic Specialist Conference, 2016, pp. 1459-1463.
[8] N. R. Paudel, J. D. Poplawsky, K. L. Moore and Y. Yan. "Current Enhancement of CdTe-Based Solar Cells," *IEEE Journal of Photovoltaics*, vol. 5, iss. 5, pp. 1492-1496, 2015.
[9] D. E. Swanson, J. R. Sites, and W. S. Sampath, "Co-sublimation of CdSexTe1-x layers for CdTe solar cells," *Solar Energy Materials and Solar Cells*, vol. 159, pp. 389-394, 2017.
[10] N. R. Paudel and Y. F. Yan, "Fabrication and characterization of high-efficiency CdTe-based thin-film solar cells on commercial SnO2:F-coated soda-lime glass substrates," *Thin Solid Films*, vol. 549, pp. 30-35, 2013.

The Influence of the Cu-rich/Cu-poor sequence on the properties of Cu(In,Ga)Se₂ films deposited by in-line co-evaporation process

He Wang[1*], Fang Fang Liu[2], Yi Tong Yang[1], Li You Yao[1], Peng Gao[1], Zhi Bin Xiao[1], and Qiang Sun[1]

1. Tianjin Institute of Power Source, Tianjin 300381, PR China

2. The Institute for Photoelectronic Thin Film Devices and Technology Nankai University, Tianjin 300071, PR China

Abstract — In this study, CIGS films are grown by the in-line evaporation process with Cu-rich /Cu-poor sequence at a low temperature of ~480oC. The influence of various maximum of Cu/(In +Ga) (ym) during film growth on the structural properties of CIGS films is investigated. It is observed that Cu rich growth at low temperature is also important to improve the crystal quality of CIGS film. Both the In/Ga intermixing and the grain size are enhanced as ym increases. However, higher Cu excess (ym>1.5) yields a rougher CIGS morphology, group III‐rich phases near the absorber surface, and deeper crevices extending to the bottom part of CIGS film. It is concluded that a threshold of Cu excess exists for high photovoltaic quality CIGS films, beyond which the inferior structure and undesirable phases of CIGS absorber are obtained, potentially being adverse for the performance of CIGS solar cell.

Index Terms — CIGS films, in-line evaporation process, maximum of Cu/(In +Ga), structural properties, threshold of Cu excess.

I. INTRODUCTION

During the past decades, flexible Cu(In,Ga)Se₂ (CIGS) thin film solar cells on polyimide (PI) substrate have undergone remarkable development. Currently, an efficiency of 18.7 % has been achieved by co-evaporation process in laboratory-scale cells [1]. Moreover, some manufactories in USA and Europe have come true the industrialization of CIGS thin film modules on PI web by in-line co-evaporation process [2]-[3]. As for a successful completion of the in-line production, one critical point is to develop the high photovoltaic quality CIGS films deposited at lower temperature below ~500ºC which is compatible with PI thermal tolerance aspect [4].

A period of Cu rich growth is one of the key elements for improving the crystalline quality of CIGS films deposited by the various co-evaporation processes [5]-[6]. Many previous reports approved this point and put forward several models to explain the mechanism of Cu rich growth. Nishiwaki et al. proposed that CuxSe phases, segregated at surface and grain boundaries of the growing films [7], enhances the crystallization of CIGS layer as well as affects the diffusion of the elements through the layer [8]. Wada et al. introduced the model of "topotactic" reaction illuminating the growth and re-crystallization mechanism of CIGS grains [9]. Seyrling et al. systematically studied the effects of the duration of Cu rich growth on the micro-structure and composition profile of CIGS films as well as the device performance [10].

However, fewer contributions concern for the role of Cu rich compostion on the growth of CIGS films in the in-line co-evaporation process at low temperature. In the present work, we investigate the influence of maximum [Cu]/[In+Ga] ratio (ym) during this process on the structural properties of CIGS films.

II. EXPERIMENTS

PI foils are coated with a molybdenum layer of about 1 μ m by dc sputtering. The process and parameters are optimized to balanced the stress to ensure sufficient adhesion among the PI foils, Mo layer and CIGS film, and to gain high enough conductivity. CIGS layer is then deposited by in-line co-evaporation at constant substrate temperatures of ~480ºC under high vacuum. The Cu-rich/Cu-poor sequence is completed by evaporating the metal element of Ga/In/Cu sequentially onto the substrate, forming over-stoichiometric CIGS with segregated phase of CuxSe, then followed by depositing Ga and In to obtain the under-stoichiometry CIGS films. Se is present in the chamber at a constant overpressure throughout the whole layer growth. During the film growth, ym controll- ed by the evaporating temperature of Copper is measured by X-ray fluorescence (XRF) in the deposition system.

The average composition of final CIGS films and ym are determined by XRF. The structural properties of the CIGS films are measured by a Rigaku TTR-III X-ray diffractometer with Cu K α radiation (XRD) and S-4800 Scanning electron microscopy (SEM) with a cold cathode field emission electron source . The different phases and microstructure of films surface are identified by Raman spectroscopy measurements using a Horiba Jobin-Yvon T64000 spectrometer with an Ar⁺ laser excitation source (514nm) in back-scattering configuration. Penetration depth of the scattered light in the absorber layer is estimated to be below 100 nm.

III. RESULTS AND DISCUSSIONS

The samples of CIGS films with varied y_m of $0.95 \sim 2.5$ are obtained by controlled the evaporation rate of Cu source, as decribed in Table I. Accordingly, either of the evaporation rates of In and Ga elements or the moving velocity of

TABLE I
THE VALUES OF Y_M AND FINAL COMPOSITION OF CIGS SAMPLES

Samples Nr.	y_m (a.u.)	[Cu]/[In+Ga] (a.u.)	[Ga]/[In+Ga] (a.u.)
A	0.95	0.86	0.34
B	1.2	0.85	0.35
C	1.5	0.90	0.35
D	1.8	0.86	0.37
E	2.5	0.88	0.38

substrate webs are changed to ensure the final composition of CIGS films under-stoichiometry uniformly.

A. Morphology results

Figure 1 shows the images of the cross-sections and surfaces of CIGS layers with different ym. The cross-section images are taken at different slantwise angles distorting the actual layer thicknesses. Thus, no conclusion should be drawn about the layer thickness from these images. Actual CIGS layers thickness are approximately 2μm for all samples presented in this paper.

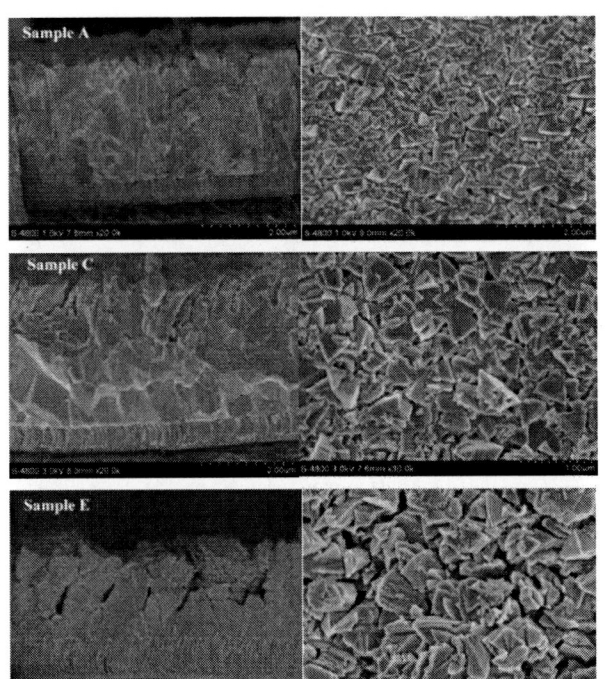

Fig. 1.   SEM cross-sections of CIGS films growth with different y_m

Comparing the images of Sample A with C and E reveals a significantly enhanced grain size, especially at the lower half part of the cross-section of CIGS layers. As $y_m<1$, Sample A without Cu rich growth is composed of slightness grains packing closely. From the surface topography, it can be found that CIGS film is smooth and comprised of many triangle or tiny grains. With the increase of Cu excess , i.e. $y_m>1.2$, larger and compact CIGS grains formed at the lower part of CIGS film firstly, extending to the upper part. Smaller grains with a few of crevices located in the depth of about 200~500nm are visible in the cross-section of Sample C. The above phenomenons accordance with the reports of Shafarman [5] and Kessler [11] indicate that Cu rich growth at low temperature (less than 500ºC) is also important for improving the crystal quality of CIGS film.

For the Sample E with ym=2.5, however, more crevices between the columnar large grains exist and stretch to the deeper regions than those in Sample C. Furthermore, the surface of CIGS films are more rough with some voids which are likely to feed through the crevices. These crevices and voids origin from the CuxSe with lower solubility in stoichiometric CIGS phase [12] segregated in the grain boundaries and surface consumed subsequently during the Cu-poor process. It should be pointed that the crevices are unlikely to reach the interface of CIGS and Mo layer because of the Cu-poor regime at the beginning of CIGS film growth which is determined by the evaporating sequence of metal elements. However, more and deeper crevices resulted from higher Cu excess may be adverse for the performance of CIGS solar cell.

B. X-ray diffraction (XRD) analysis of CIGS films

The XRD patterns acquired on all of the samples reveal a random distribution between (112) and (220)/(204) orientations No correlation between y_m and the preferred crystal orientation is observed. Because the 2θ peak position of (112) reflection shifts toward to higher diffraction angles in direct proportion to the lattice constant determined by Ga content incorporation into CIGS phase. The peak structure can provide information whether CIGS film has a constant and graded composition [13]. A band of peaks correspond to the Ga graded distribution across the layer thickness.

In figure 2 the (112) peaks of Sample A, B, C and E are plotted. The Sample A exhibits shoulder peak toward the higher diffraction angle, indicating the regions with high and low Ga/In concentration ratio existing in CIGS film without Cu excess growth regime. As the increase of y_m, the symmetry of the peaks of Sample B, C and E is improved significantly, which can be interpreted as a sufficient inter-mixing of the group III elements. It exhibits that the extent of Cu excess during film growth has direct influence on the final grading profile of CIGS layer deposited by in-line evaporation process at low temperature.

C. Phase and microstructure on the surface of CIGS fims

Figure 3 presents the information about micro-structure and phases existing in the surface region obtained by Raman spectroscopy with a penetration depth of about 100 nm. For all the samples, the frequency of the main chalcopyrite A1 modes for CIGS phases are around 176-178 cm^{-1}, which corresponding to the vibrations of Se atoms in the chalcopyrite. The modes around 210~250 cm^{-1} correspond to mixed B2/E modes. As ym increasing, the notable distinction is a visible enhanced of the broad shoulder between 150 and 170 cm^{-1}. It indicates the appearance of a group III rich ordered defect compound (ODC) like $Cu(In,Ga)_3Se_5$ or $Cu_2(In,Ga)_4Se_7$ [14] for the absorber layers grown with higher Cu excess, i.e. $y_m > 1.5$. The formation of ODC phase near the surface is attributed to the increase of the evaporation rate of III group elements as an abundance of Cu_xSe phases need to be depleted during the growth of CIGS films.

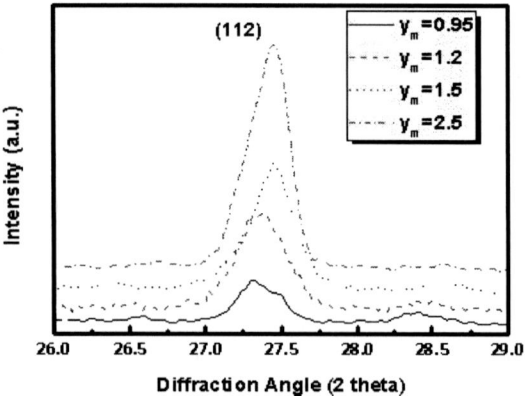

Fig. 2.   SEM cross-sections of CIGS films growth with different y_m

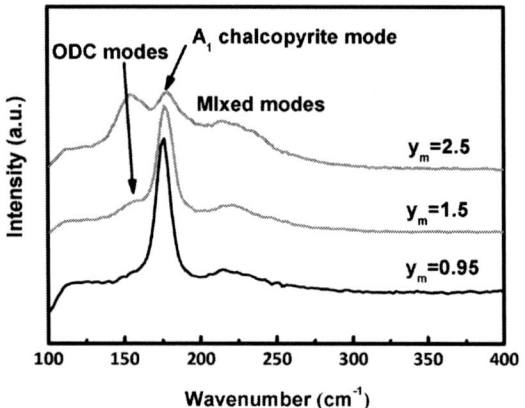

Fig. 3.   Raman spectra of CIGS layers growth with different ym recorded at room temperature. Samples as $y_m > 1.5$ exhibit ODC phases between 150~170cm^{-1}

IV.CONCLUSIONS

CIGS films are grown with various maximum [Cu]/[In+Ga] ratio during film growth by the in-line evaporation process and changes in micro-structures both on the bulk and surface of the films are investigated. It is found that the inter-diffusion of In and Ga determined the final grading profile of CIGS layer is distinctly depended on the ym during film growth. Furthermore, Cu rich growth at low temperature (less than 500ºC) is also important to improve the crystal quality of CIGS film. However, higher value of ym resulting in some crevices extending into CIGS layer and group III‑rich phases at or near the surface of CIGS film is possibly not suitable for the growth of high photovoltaic quality CIGS absorber.

ACKNOWLEDGEMENT

This work was supported by the National Natural Science Foundation of China (61404118), National Natural Science Foundation of China (61504067).

REFERENCES

[1] A. Chirila, S. Buecheler, F. Pianezzi, P. Bloesch, C. Gretener, A. R. Uhl, C. Fella, L. Kranz, J. Perrenoud, S. Seyrling, R. Verma, S. Nishiwaki, Y. E. Romanyuk, G. Bilger and A. N. Tiwari, *NATURE MATERIALS,* vol10, pp857-866, 2011.

[2] J.S. Britt, S. Wiedeman, U. Schoop, D. Verebelyi, *Proceeding of 33th IEEE Photovoltaic Specialist Conference*, San Diego, CA, USA, 2008.

[3] F. Kessler, R. Würz, S. Spiering, D. Bremaud, *Proceeding of 22nd International Photovoltaic Science and Engineering Conference*, Hangzhou China,2012).

[4] D. Brémaud, D. Rudmann, G. Bilger, H. Zogg, A. N. Tiwari, *Proceeding of 31th IEEE Photovoltaic Specialist Conference, Florida.* Piscataway, USA, 2005).

[5] W.N. Shafarman, J Zhu, *Thin Solid Films*, vol361-362, pp 473-477, 2000.

[6] A.M. Gabor, J.R. Tuttle, D.S. Albin, M.A. Contreras, R. Noufi, *Apply Physics Letters*, vol. 65, pp 198‑200, 1994.

[7] S. Nishiwaki, T. Satoh, Y. Hashimoto, T. Negami, T. Wada, *Journal of Materials Research*, vol.16, pp 394‑399, 2001

[8] J. Kessler, J. Lu, J. Schöldström and L. Stolt, Prog. *Photovolt: Res. Ap*pl, vol. 11, pp 319-322, 2003.

[9] T. Wada, N. Kohara, T. Negami, M. Nishitani, *Journal of Materials Research*, vol.12, pp 1456‑1462,1997.

[10] S. Seyrling, A. Chirila, D. Güttler, P. Blösch, F. Pianezzi, R. Verma, S. Bücheler, S. Nishiwaki, Y.E. Romanyuk, P. Rossbach, A.N. Tiwari, *Solar Energy Materials & Solar Cells*, vol.95, pp 1477‑1481, 2011.

[11] J. Kessler , C. Chityuttakan , J. Schöldström , L. Stolt, *Thin Solid Films*, vol 431‑432, pp 1‑5, 2003.

[12] A. Rockett, *Thin Solid Films*, vol. 361‑362, pp 330‑337, 2000.

[13] M. Bodegård , O. Lundberg , J. Lu, L. Stolt, *Thin Solid Films*, vol. 431‑432, pp 46‑52, 2003.

[14] C. Rincón, S.M. Wasim, G. Marin, J.M. Delgado, J.R. Huntzinger, A. Zwick, J. Galibert, *Applied Physics Letters*, vol. 73, pp 441-445, 1998.

Determination and Modeling of Injection Dependent Series Resistance in CIGS Solar Cells

Vito Huhn, Bart E. Pieters, Andreas Gerber, Yael Augarten, and Uwe Rau

Forschungszentrum Jülich IEK-5 Photovoltaics, Jülich, 52425, Germany

Abstract — The total series resistance of industrially produced Cu(In,Ga)Se$_2$ solar cells is investigated using photo- and electro-luminescence imaging at various applied voltages. A strong dependence of the series resistance on the cell operating point was observed. It decreases with increasing illumination intensity and increasing applied voltage. This injection dependent behavior was reproduced using numerical simulations, and was found to be due to the transport properties of the Cu(In,Ga)Se$_2$ bulk. To take the injection dependency of the series resistance into consideration is important when dealing with quantitative luminescence or electrical measurements of CIGS solar cells.

Index Terms — CIGS, series resistance, photovoltaic cells, luminescence, absorber layer.

I. INTRODUCTION

Series Resistance often limits the performance of solar cells and also needs to be considered when interpreting electrical or luminescence measurements [1-4]. It may result from several possible origins, e.g. contact resistance, lateral resistance due to the transport of charge carriers towards the contacts or resistance due to the transport of charge carriers through the bulk of an absorber. The series resistance may change depending on the operating point of the solar cell [5-7]. If such a non-ohmic behavior of the series resistance exists in a solar cell, it may influence its electrical and luminescence behavior to a point where the results cannot be any longer modeled by conventional means (e.g. a one diode model).

This work analyses the injection dependence of the total series resistance in Cu(In,Ga)Se2 (CIGS) solar cells to better understand its influence on luminescence and electrical measurements, but also to find its origin. We compare the junction voltage of a cell determined via luminescence imaging to the external voltage measured at the terminals of a device to calculate the voltage loss across the effective, total series resistance. Dividing this voltage by the current through the sample yields the series resistance value. This procedure has the advantage that a series resistance can be calculated easily with one electrical and optical measurement for each operating point of interest. The method allows for a change of the series resistance with the operating point and does not need any fitting procedures. The experimental results are qualitatively reproduced with numerical simulations using the device simulator SCAPS.

II. THEORY

A. Series Resistance Determination from Luminescence

In this section it will be explained how luminescence imaging can be used to determine the total series resistance of a solar cell. The reciprocity relation [8] relates the local luminescence intensity $\Phi_{\mathrm{EM-PL}}$ to the local junction voltage $V_{\mathrm{j,loc}}$ in the following:

$$\Phi_{\mathrm{EM-PL}}(r, V_{\mathrm{ext}}) = \Phi_{\mathrm{SC}} + C \times \left[\exp\left(\frac{V_{\mathrm{j,loc}}(r,V)}{V_{\mathrm{th}}}\right) - 1\right] \quad (1)$$

The quantity V_{th} is the thermal voltage, V is the external voltage, applied at the cell terminals and Φ_{SC} is a voltage independent part of the luminescence intensity resulting from the illumination. In [9] it is shown how this relation can be used to calculate the total cell junction voltage V_{j} by using an open circuit image to determine the constant C and taking the mean of all the local junction voltages $V_{\mathrm{j,loc}}$. By measuring luminescence images at different applied voltages and simultaneously recording the current through the cell, (1) can be used to determine a characteristic of the current density J through the solar cell vs. the junction voltage V_{j}. Using this J/V_{j} characteristic, the series resistance R_{s} corresponding to a particular cell operating point can then be calculated using the following equation:

$$R_{\mathrm{s}} = (V - V_{\mathrm{j}})/J \quad (2)$$

With this approach, R_{s} can be determined for each operating point using a single calibrated luminescence image, taken at a known voltage and current. This is an advantage to the commonly used short circuit current/ open circuit voltage ($J_{\mathrm{sc}}/V_{\mathrm{oc}}$) method which requires two measurements, a J/V measurement and a $J_{\mathrm{sc}}/V_{\mathrm{oc}}$ measurement, with the illumination of the sample be set so that J_{sc} is equal to the current of the J/V measurement point [5, 10]. The $J_{\mathrm{sc}}/V_{\mathrm{oc}}$ procedure also becomes more complicated when the real generated photocurrent is not externally accessible from the measured J_{sc}, e.g. due to high series resistance.

B. Injection dependent Series Resistance

In [7] the series resistance for pin-type thin film silicon solar cells was analyzed while the sample was in the dark. A strong exponential dependence of R_{s} on the operating point was found. The authors proposed that the origin of that series resistance results from the transport of the charge carriers through the bulk of the absorber. Such a resistance would be given by

$$R = \int \frac{1}{q \, \mu_{e/h} \, n_{e/h}(z)} dz \qquad (3)$$

where the integral has to be calculated over the dominating current path over the depth z of the solar cell. Here, $\mu_{e/h}$ is the mobility of the electrons or holes depending on the band of which the resistance is described and $n_{e/h}$ is the electron or hole density of the conduction or valence band, which depends upon the position z. Thus, the exponential dependence of the series resistance on the applied voltage results from the exponential change of the charge carrier densities with the applied voltage. Additionally, the mobility of the regions, where the most current flows, influences the total series resistance.

III. Experimental Procedure and Results

For the experiments we used, industrially produced CIGS cells. The cells were prepared from larger modules and are 0.4 cm wide and 1.6 cm long. The samples are placed on a temperature controlled stage (25 °C) and are electrically contacted. Imaging and J/V measurements in the dark and under illumination are performed within a luminescence measurement setup, which included a source measurement unit and an 808 nm diode laser. The laser beam is widened to provide homogenous illumination. Images were taken using a Nirvana640 InGaAs camera (640 x 512 pixels) cooled to -80 °C. For illuminated measurements, the laser intensity was set so that the J_{sc} of the device when measured in the luminescence setup is equal to the J_{sc} of the device under a simulated AM1.5 spectrum.

The determination of the J/V_j characteristics was performed with the same procedure as described in [9]. For consistent nomenclature, the J/V_j curves determined from luminescence measurements under constant illumination are referred to as electro-modulated photoluminescence (EM-PL) characteristics while J/V_j curves determined from luminescence measurements in the dark will be called electro-luminescence (EL) characteristics. The only difference to the procedure in [9] was that from the EM-PL images only the dark background was subtracted, before the junction voltage determination and not an image where the sample was at short circuit. This was done as we found that the junction voltage is not zero if the CIGS solar cell is under illumination and at short circuit, due to the large series resistance which was measured under these conditions. Instead, we assumed that the junction voltage independent part of the luminescence signal Φ_{SC} is negligible. Fig. 1 shows the results of the measured EL and EM-PL characteristics for an example cell.

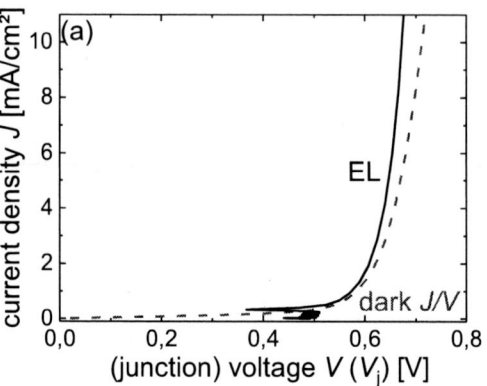

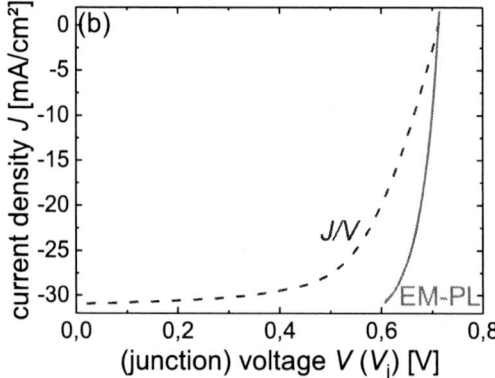

Fig. 1. J/V_j [EL(black)/EMPL(red)] and J/V (blue, dashed) characteristics determined for a CIGS solar cell using electrical measurements and luminescence imaging at various applied voltages: (a) in the dark, (b) under illumination.

The series resistance R_s was determined using (2) and the characteristics shown in Fig. 1. The results are shown in Fig. 2, as a function of the cell voltage for the dark and illuminated case. In both cases, the series resistance decreases with increasing external voltage. In the dark case, this decrease is exponential while under illumination, the series resistance shows a more linear decrease. Additionally, it is possible to determine the series resistance under illumination to much smaller values of the external voltage than in the dark. This is because the larger current flow enhances the influence the series resistance has on the behavior of the junction voltage.

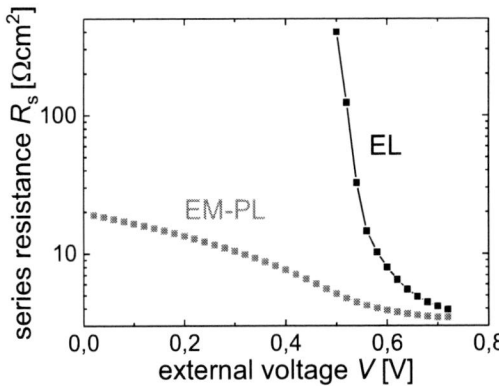

Fig. 2. Total series resistance of a CIGS solar cell plotted versus its operating point.

The results in Fig. 2 show that the series resistance is strongly dependent on the operating point. This behavior of the series resistance could not be expected from the J/V characteristics alone as the high series resistance at low voltages does not influence the behavior of the characteristics much, because the differential diode resistance is even higher than the series resistance. As the resistance changes depend not only on the external voltage but also on the illumination we relate to it as injection dependent. In any case, the behavior of the series resistance in the dark looks very similar to the behavior found for the series resistance of thin film silicon solar cells implying that a similar cause is responsible for it [7].

IV. MODELLING AND DISCUSSION

To understand the origin of the injection dependence of the series resistance, we performed one dimensional device simulation using the program SCAPS [11]. The base parameter set that was used to describe the solar cell is given in [12]. The density of states in the conduction band of the CIGS was adjusted slightly to 9×10^{18} cm^{-3} to obtain a better fit to experimental results. The ZnO layers were removed, leaving only the CIGS and the CdS layer, as it was found that they do not influence the results of the series resistance simulations much and using only a minimum amount of layers ensures to find the origin of the injection dependence of the series resistance more easily. The illumination spectrum was set to peak at 800 nm with a linewidth of 5 nm similar to the illumination in the experiments and the intensity was adjusted to match to the measured short circuit current. The results of the simulated J/V characteristics are compared in Fig. 3 to the measured J/V characteristic. The injection dependent series resistance was calculated from the SCAPS simulations with a procedure similar to experimental measurements. To this end the radiative recombination was calculated by calculating the $n_e n_h$ product at each position in the device. Both can be directly accessed from the simulations. We assumed the

simplified case that the integration over the depth of the device yields the total luminescence signal. Thus, a value for the local junction voltage can be calculated using (1). The proportionality constant was again determined by setting the junction voltage under open circuit conditions equal to the external open circuit voltage. The series resistance was then calculated similarly to the experiment by using (2). The results are shown in Fig. 4. The simulated series resistance behaves very similarly to the experimentally measured values, suggesting that the origin of the injection dependent series resistance needs to originate from either the CIGS or the CdS.

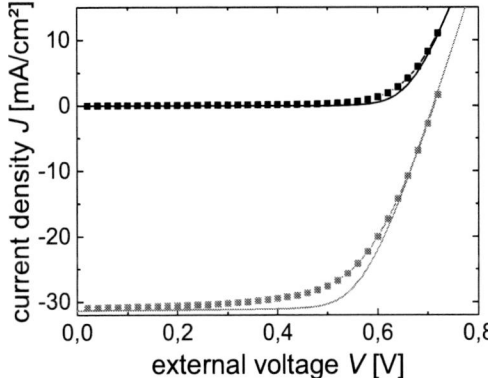

Fig. 3. Measured J/V characteristics (squares) compared to simulated J/V characteristics using SCAPS (lines).

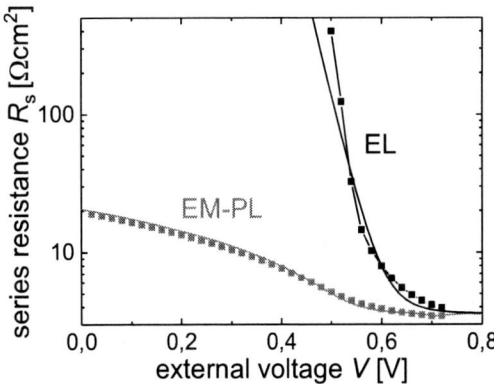

Fig. 4. Measured injection dependence of the series resistance (squares) compared to the simulated injection dependence of the series resistance using SCAPS (lines).

The influence of the CdS was excluded as changes of its parameters had no significant influence on the series resistance. With the help of (3) we can show that the measured injection dependent series resistance mainly results from the transport of the electrons through the CIGS conduction band, as it was found that the mobility of electrons in the bulk mainly influences the determined series resistance, while the hole mobility has much less influence. Additionally, the electron density of states is most dependent on different voltage and illumination conditions. Figure 5 shows the hole and electron charge carrier densities at different voltages and under illumination and in the dark. We can see that the largest changes in the charge carrier densities due to voltage and illumination happen to the electrons in the CIGS bulk. The change with increasing voltage is more significant in the dark explaining the exponential decrease of the series resistance in this case. That the series is much lower under illumination can be also explained as the charge carrier density is already at zero voltage larger than in the dark. Thus, the behavior of the series resistance is explainable with (3).

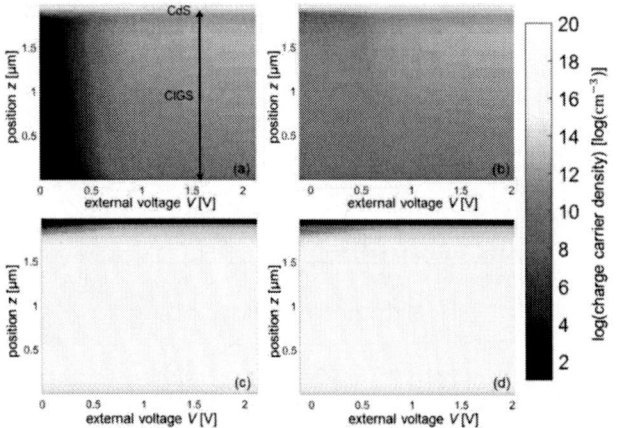

Fig. 5.   Charge carrier density of electrons (a,b) and holes (c,d) in the dark (a,c) and under illumination (b,d) at different voltages and locations in the solar cell.

Using several approximations of the device physics, analytical expressions for the series resistance of a pin thin film silicon device were presented in [7]. Such an approach may also be possible for CIGS devices, too. However, we have so far not been successful in finding appropriate approximations to do so.

V. CONCLUSION

We used luminescence imaging to determine the series resistance of CIGS solar cells at different operating points. It was found that the series resistance is strongly injection dependent. By modeling the experiment using SCAPS we found that the injection dependence of the series resistance results from the minority carrier transport through the bulk of the solar cell, as the behavior of the electron density in the conduction is in alignment with the found behavior of the series resistance. The injection dependence of the series resistance is often neglected when trying to describe CIGS solar cells using equivalent circuit models, as it is difficult to measure and not clearly visible in J/V measurements.

However, the effect is relevant for the interpretation of EL or EM-PL measurements of CIGS devices.

ACKNOWLEDGEMENT

This work has been supported by the "OptiCIGS" Project (FKZ 0325724A). Special thanks go to the project partners from Manz CIGS Technology for providing the samples used in this work.

REFERENCES

[1]   W. Tress, K. Leo, and M. Riede, "Optimum mobility, contact properties, and open-circuit voltage of organic solar cells: A drift-diffusion simulation study," *Physical Review B,* vol. 85, p. 155201, 04/04/ 2012.

[2]   C. Ulzhöfer, P. P. Altermatt, N.-P. Harder, and R. Brendel, "Loss analysis of emitter-wrap-through silicon solar cells by means of experiment and three-dimensional device modeling," *Journal of Applied Physics,* vol. 107, p. 104509, 2010.

[3]   A. Helbig, T. Kirchartz, R. Schaeffler, J. H. Werner, and U. Rau, "Quantitative electroluminescence analysis of resistive losses in Cu(In, Ga)Se2 thin-film modules," *Solar Energy Materials and Solar Cells,* vol. 94, pp. 979-984, 6// 2010.

[4]   T. M. H. Tran, B. E. Pieters, C. Ulbrich, A. Gerber, T. Kirchartz, and U. Rau, "Transient phenomena in Cu(In,Ga)Se2 solar modules investigated by electroluminescence imaging," *Thin Solid Films,* vol. 535, pp. 307-310, 5/15/ 2013.

[5]   M. Turek, "Current and illumination dependent series resistance of solar cells," *Journal of Applied Physics,* vol. 115, p. 144503, 2014.

[6]   J.-M. Wagner, M. Hoppe, A. Schütt, J. Carstensen, and H. Föll, "Injection-level Dependent Series Resistance: Comparison of CELLO and Photoluminescence-based Measurements," *Energy Procedia,* vol. 38, pp. 199-208, 2013/01/01 2013.

[7]   T. C. M. Müller, B. E. Pieters, U. Rau, and T. Kirchartz, "Analysis of the series resistance in pin-type thin-film silicon solar cells," *Journal of Applied Physics,* vol. 113, p. 134503, 2013.

[8]   U. Rau, "Superposition and Reciprocity in the Electroluminescence and Photoluminescence of Solar Cells," *Photovoltaics, IEEE Journal of,* vol. 2, pp. 169-172, 2012.

[9]   V. Huhn, A. Gerber, Y. Augarten, B. E. Pieters, and U. Rau, "Analysis of Cu(In,Ga)Se2 thin-film modules by electro-modulated luminescence," *Journal of Applied Physics,* vol. 119, p. 095704, 2016.

[10]   M. Wolf and H. Rauschenbach, "Series resistance effects on solar cell measurements," *Advanced Energy Conversion,* vol. 3, pp. 455-479, 4// 1963.

[11]   M. Burgelman, P. Nollet, and S. Degrave, "Modelling polycrystalline semiconductor solar cells," *Thin Solid Films,* vol. 361–362, pp. 527-532, 2000.

[12] J. F. L. Salas, S. J. Heise, M. Richter, V. Gerliz, M. S. Hammer, J. Ohland, *et al.*, "Simulation of metastable changes in time resolved photoluminescence of Cu(In,Ga)Se2 thin film solar cells upon light soaking treatment," *Thin Solid Films,* 2016.

Large grain growth in Cu_2ZnSnS_4 thin films in the absence of Na using rapid thermal annealing

J.L. Johnson[1], A. Bhatia[2], J.G. Bolke[2], M.A. Scarpulla[1,2]

[1]Electrical and Computer Engineering, University of Utah, Salt Lake City, UT 84112
[2]Materials Science and Engineering, University of Utah, Salt Lake City, UT 84112

Abstract — **Larger grain sizes and crystallographic texture in the absorber layer are correlated with performance in polycrystalline inorganic thin film solar cells because they lead to reduced densities of recombination centers and traps at grain boundaries. In this work, we compare the effects of rapid thermal annealing (RTA) and furnace annealing on film morphology, grain growth, and crystallographic texture in Cu_2ZnSnS_4 (CZTS) thin films deposited on Mo-coated glass by RF cosputtering. Using x-ray diffraction (XRD), scanning electron microscopy (SEM), and electron backscatter diffraction (EBSD), we show that RTA induces significant grain growth and texture without the degradation of the substrate caused in furnace annealing by the high temperatures required for CZTS phase formation. CZTS grains grown in RTA annealed samples were 2–3 times larger than those in furnace-annealed samples. Significantly, this is demonstrated in RTA on sodium-free glass which implies that RTA and similar processing methods may be capable of producing desirable CZTS grain distributions on alternative substrates such as metal foil, plastic, or thin glass.**

I. INTRODUCTION

$Cu_2ZnSn(S,Se)_4$ (CZTSSe) has attracted the interest of the thin film photovoltaics community as a potential replacements for $Cu(In,Ga)(S,Se)^2$ (CIGSSe) absorber layers because it may allow larger scaling of the industry over 15-20 year timescales. Recently, CZTSSe lab cells have been demonstrated at up to 12.6% power conversion efficiency [1]. The CZTSSe alloy series offers bandgaps in the 1.0-1.5 eV range and high absorption coefficient relevant for thin film photovoltaics. The elements in CZTS are earth- and industrially- abundant and low cost which would allow the scale up to production at levels >100 GWp/year without constraint by raw materials costs [2]. To obtain efficient photovoltaic performance in polycrystalline thin film materials, the minimization of grain boundary recombination and barriers to current flow are desired. Larger grains reduce the grain boundary area to grain volume ratio and are thus a first step towards these goals. Although certainly not the only important factors, the reduction of grain boundary area and misorientation angles through the growth of grain sizes near 1 $\square$m and high crystallographic texture [3] are factors contributing to high efficiency in thin film absorber layers.

While continuous coevaporation is used for some CZTSSe films [4, 5], a two-stage processes involving deposition of a precursor layer followed by annealing in S and/or Sn containing atmospheres is used more widely and is likely more scalable to production. [6-8]. In all of these processes, the high atomic mobility required for grain growth

in thin films of CZTS is provided thermally during the 2nd annealing step. It has been widely observed that large grain growth and the full reaction of precursors to form CZTS requires annealing temperatures above 500-550 °C. These high temperatures will soften and/or warp conventional soda lime glass substrates over long annealing times. The stoichiometry of the CZTS film is also difficult to maintain at high annealing temperatures since the vapor pressures of S and SnS are high above CZTS and will lead to decomposition of the CZTS phase over time unless overpressures are supplied [8-10]. For these reasons, we have compared rapid thermal annealing (RTA) to standard tube furnace annealing.

We show by X-Ray diffraction (XRD), scanning electron microscopy (SEM) and electron backscatter diffraction (EBSD) that significant grain growth and crystallographic texture occur during RTA without degradation of the glass substrate. It is commonly known that the presence of Na during the processing of thin films of CIS, CIGS and CZTS facilitates grain growth (as well as inducing doping and increasing carrier lifetime. Here we show that large CZTS grains can be grown by RTA on boro-aluminosilicate glass (BSG), which is capable of being annealed at higher temperatures and contains much less Na than soda lime glass (SLG). In order to begin to demonstrate applicability to continuous manufacturing (e.g. roll-to-roll), we investigate RTA both with and without relative motion between the lamp and sample. This also provides an additional control parameter for the temperature-time profile without rapidly switching off the lamp.

II. EXPERIMENTAL

Precursor CZTS thin films were RF cosputtered on Mo-coated soda lime and boroaluminosilicate glass substrates using Cu2S, ZnS and SnS2 binary targets. Depositions were carried out in a chamber equipped with three 3" diameter magnetron sputtering sources (Lesker) situated approximately 15 cm from the substrate. The sputtering chamber is pumped by a turbo-pump backed by a dry scroll pump and a liquid nitrogen trap assists in removing water and other condensable species resulting in a base pressure of 2x10-6 Torr. For substrate cleaning as well as sputtering of the sulfides, Ar is provided to the chamber at 20 sccm and the pressure is held at 5 mTorr. Before deposition of the sulfides, the Mo coated substrates were plasma cleaned in the deposition chamber for 10 minutes at 100 W RF by applying power to the sample chuck. During deposition, the powers applied to the sources were: Cu2S = 109 W, ZnS = 92 W and SnS = 44 W. The layers were deposited to a total thickness of 1.2 - 1.4 □m. Additional details are available in Refs. [11, 12].

The standard furnace annealing samples were annealed in a 3" diameter tube furnace with flowing forming gas (4% H2 / 96% N2) at 1 atm and 20 sccm. A boat containing powdered S is placed near the edge of the heated zone of the furnace with additional heating tape and held at 170 □C to 200 □C to generate the S vapor. The other end of the tube is left at ambient so S will flow past the sample and condense on cooler internal surfaces. The samples were placed in the center of the furnace where the specified annealing temperature is set to 550 □C or 600 □C. The tube is purged 3 times to a pressure of 200 mTorr and backfilled with forming gas. Various Sm (m = 1, 2, 3…8) species are present in the elemental S vapor [13] and some H2S will also form via reaction with H2. The temperature in the center of the furnace is ramped up over 2 hours, then held for two hours, and allowed to cool naturally which takes approximately 24 hours because of the highly-insulated furnace. Samples are kept in the tube until the furnace temperature is below 150 □C and then the S source temperature is allowed to cool also. This furnace annealing procedure can be considered an equilibrium process in a

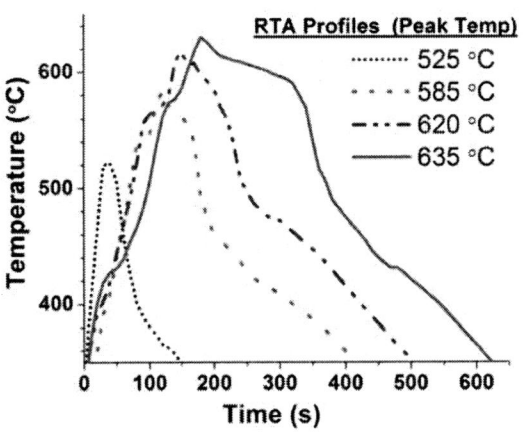

Figure 1 – RTA temperature profile versus sweep time/rate with a base temperature of 300 °C.

closed system.

RTA was carried out in a very small-volume quartz chamber with rectangular cross-section that is connected to a roughing pump, a source of forming gas, a pre-heater and a high-intensity halogen lamp. The chamber is sized such that the light from the heating lamp impinges over the entire chamber to minimize adjacent cold spots that would condense S vapor. The chamber is purged and backfilled with forming gas. A quartz boat containing sulfur is placed next to the

sample such that it is also exposed to the heating lamp at the same time as the sample. The samples are pre-heated over 10 min to 300 or 400 □C by a graphite element outside and below the quartz tube to minimize thermal expansion issues and to ensure S vapor is present when the sample surface is heated. For static RTA, in which the lamp is left stationary, the sample is then heated by the lamp for a specified amount of time and allowed to cool for 30 min after the lamp is turned off. In what we refer to as "dynamic RTA" the lamp begins to the side of the RTA chamber, adjusted to its prescribed intensity, and then swept across the sample and S source simultaneously at a constant rate using a long worm gear. This allowed fast heating and cooling without rapidly changing the power applied to the lamp as well as simulation of rapid annealing in a continuous production process such as roll-to-roll. In both cases, the sample is allowed to cool to below 100 °C before removal from the chamber. For both types of RTA treatments, dozens of samples were annealed

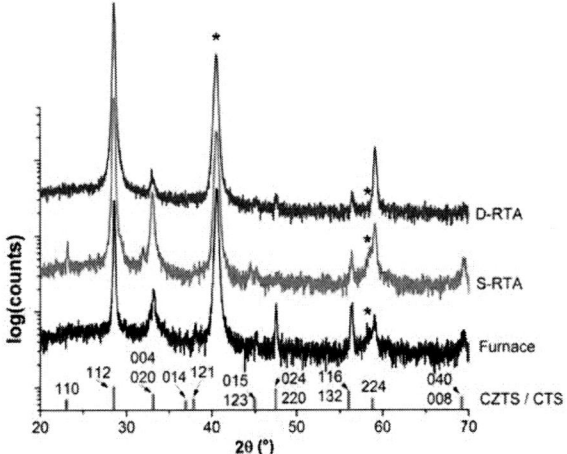

Figure 2 – θ–2θ XRD scan results from CZTS films annealed by tube furnace or by RTA in static (S) or dynamic (D) modes. Peak assignments from the JCPDS card 01-075-4122 for kesterite CZTS are indicated along the 2θ axis and (*) designates Mo peaks.

and characterized for this work.

The actual film temperature during RTA is difficult to measure reliably – optical pyrometry was attempted through the walls of the chamber but failed because of reflections from the lamp. The temperatures during dynamic RTA depicted in Fig. 1 were recorded using a small thermocouple inside a small volume, thin-walled graphite block placed in the same location as the as-sputtered samples. These data represent our best estimate of the film temperature during RTA. While we estimate that the absorptivity/emissivity across the lamp's output spectrum of the rough, grayish as-sputtered films will be similar to that of the grayish graphite jacket around the thermocouple, any such differences would cause errors in the estimated film temperature.

The films for this study were measured by both traditional powder and grazing-incidence x-ray diffraction (XRD). Because of the possibility of the 1D motion of the strip heat source inducing crystallographic texture or grain elongation, the $\theta-2\theta$ XRD scans were taken at two in-plane orientations separated by 90□. Scans from the two directions showed the same results on all samples evaluated with XRD, and no evidence of grain elongation due to the moving RTA lamp was found. As the grain size is ~3 orders of magnitude smaller than the characteristic size of the illuminated spot (and thus thermal gradient regions), this is a consistent finding.

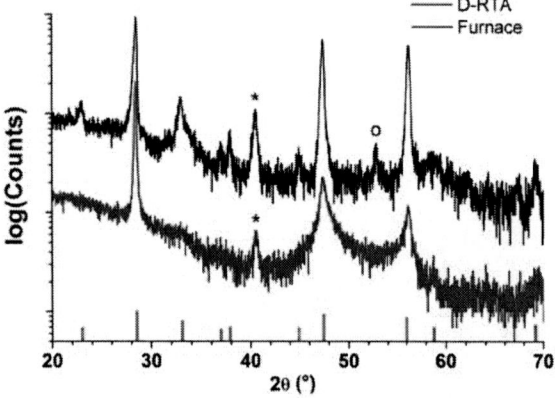

Figure 3 – Comparison of GIXD spectra from samples annealed by furnace and by D-RTA. Peak assignments from CZTS are indicated as for Fig. 2 and (*) designates the (110) Mo peak. One unidentifiable peak appears near $2\theta=53°$ for the furnace annealed sample and is indicated by o.

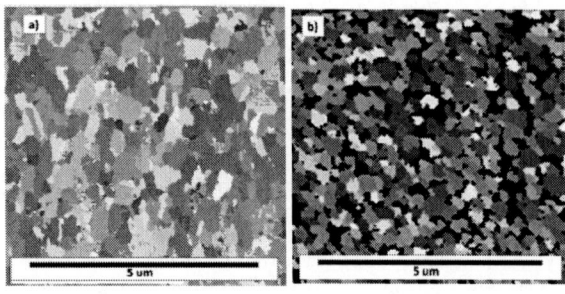

Figure 4 – EBSD inverse pole figure maps of CZTS samples in which a randomly-assigned color identifies the crystallographic direction normal to the film for each pixel. (a) Data from a D-RTA sample showing grains of approximately 1 μm lateral size. (b) Sample furnace annealed at 600 °C for 2 hours. Black denotes pixels which could not be indexed.

Before a sample can be characterized using EBSD, the surface must be polished to remove any surface topography and defects. We accomplished this using a Ga^+ focused ion beam (FIB) with the samples tilted to produce 2° grazing incidence to the CZTS surface. A projected area of approximately 100 μm x 40 μm which exposed a plane extending from the film surface to the Mo back contact was prepared on each EBSD-characterized sample. A finishing

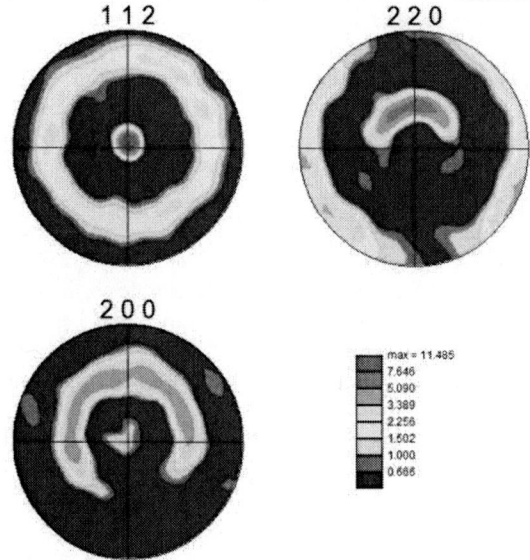

Figure 5 – EBSD pole figures of selected crystallographic orientations of CZTS from the dynamic-RTA process showing the dominant 112 texture with some 220 and 200 also present.

polish using 8 keV Ga^+ at 1.7 nA beam current was then used to remove damage from the rough cut. EBSD data were gathered at 25 keV electron accelerating voltage and 12.5 nm step size.

Figure 2 shows $\theta-2\theta$ diffraction data for a furnace-annealed sample, one annealed by static RTA (S-RTA), and one annealed using dynamic RTA (D-RTA). Since the samples for each batch were all cut from the same precursor film, attenuation of the Mo diffraction peaks should be constant. Thus the data were normalized to the intensity of the Mo (110) peak at 40.4° 2θ to account for varying sample area. Table 1 summarizes the CZTS (112) peak full-width at half-maximum (FWHM) for each type of sample shown in Fig. 1. The higher count rate and smaller FWHM for the D-RTA samples indicate larger crystal coherence lengths as compared to S-RTA or furnace annealed samples.

Annealing	FWHM B_o (°)	FWHM B_i (°)	FWHM B_r (°)
Furnace	0.21	0.063	0.15
S-RTA	0.29	0.063	0.28
D-RTA	0.20	0.063	0.14

Table 1 – CZTS (112) x-ray peak full widths at half-maximum (FWHM) from Fig. 2. B_o is the observed FWHM, B_i is the instrument broadening, and B_r is the remaining peak width caused by finite grain size and inhomogeneous strain. These data indicate that D-RTA produces crystalline quality on par with furnace annealing, but does so in a much shorter time.

During annealing, any elements not formed into CZTS in the bulk may diffuse to the surface to form phases of ZnS, Cu_2S, SnS_2 or other compounds thus we used GIXD to increase sensitivity to surface phases as shown in Fig. 3. A comparison between the GIXD scans of the RTA annealed CZTS sample and the standard tube furnace annealed sample points out that the standard annealed sample has more binary phases on the surface. All samples annealed by standard tube furnace have produced binary phases on the surface.

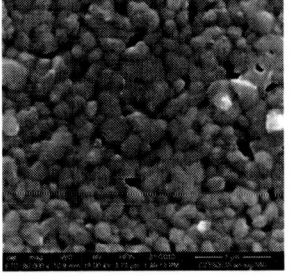

Figure 6 – (Upper left) SEM micrograph of dynamic RTA annealed sample. Significant crystallization is evidenced by the terracing. (upper right) SEM of static RTA annealed sample which is hard to determine grain size or structure. (lower left) Scanning electron micrograph of standard annealed sample at 600 °C. While it appears to show structure similar to grains, this may not be the case.

The sample generated by dynamic RTA shown in Fig. 4a visually shows that the resulting CZTS film has significant (112) texture. The grains do not show any elongation along an in-pane direction and have characteristic size near 1 um2. Similarly, the 2-hour furnace annealed sample shows significant crystallographic texture however the grains are slightly smaller with a significant area of the image not indexable. We believe that this may be caused by degradation of the film preferentially near grain boundaries during the long annealing time. By examining EBSD maps at different positions along the FIB-milled analysis plane, it was confirmed that the grains are roughly uniform in size throughout the film thickness for all types of films. The elevated intensity near the surface normal (center) of the pole figures in Fig. 5 For the D-RTA annealed sample indicate the film has predominant (112) texture but also but also some (002) texture simultaneously. EBSD data from the static RTA samples is not shown as pattern quality was poor. Grain size was on the order of 40-50 nm which is about the size of as deposited CZTS grains.

Many groups have indicated the use of Na to aid in grain growth, but we have accomplished significant grain growth on BSG without the use of Na. CZTS0038B was a sample generated on BSG (corning 1737) and annealed in the RTA with 400 °C base temperature and swept 53 seconds per inch. Since BSG has no Na, the grain growth displayed in figure 4a from EBSD is grown without the use of Na. EBSD calculation indicate that the average grain size is 700 nm.

SEM confirms the results from XRD and EBSD as seen in figure 6 a, c. Figure 6 a. is a SEM of CZTS0038B that was annealed in the RTA using dynamic mode. It can be seen that the CZTS film has been crystallized because one can visually see terracing and sharp angles at the vertex of the terrace points. While crystallization is certain, grain size is not distinguishable. Figure 6 b. is an SEM of a sample that was annealed in the RTA using static mode. From the SEM it's hard to distinguish any crystalline characteristics. It certainly doesn't look anything like the SEM in figure 6 a. The sample annealed by the standard procedure in a static tube furnace shown by SEM in figure 6 c. shows some grain structure and it seems to correlate to the EBSD results shown in figure 4. Grain sizes vary from 200 nm to 700 nm. CZTS samples all show some sign of having semiconducting properties due to charge up in the SEM. We also noticed holes in the CZTS film when taking SEM of the surface. These holes or voids could be from a couple of reasons. First the voids could be the remnants left from the densification of the material during annealing and second the voids could be from elements of the CZTS vaporizing off during annealing. Voids were noticed using all three annealing methods but more pronounced when aggressively using the dynamic RTA.

IV. CONCLUSIONS

In order of preferred crystal characteristics, the samples annealing by dynamic RTA show the best results in that the grain size is larger than static RTA and standard annealing as demonstrated by EBSD and XRD followed by static RTA and standard annealing. Samples annealed by RTA don't have the substrates misshaped or stress as tested by a straight edge and applied force. Samples generated on boro-aluminosilicate and annealed by RTA show great grain growth without the use of Na. Sample annealed by RTA are annealed in less than an hour compared to 30 hours in the standard annealing furnace. In addition to less annealing time this would also mean less power and consumables consumed. On the down side RTA tube furnace annealing has resulted in the decrease of S in the CZTS film. While S concentration is higher than deposited for standard annealing, it is still about 2% low.

V. ACKNOWLEDGEMENTS

This work was supported in full by the U.S. Department of Energy, Office of Basic Energy Sciences, Division of Materials Sciences and Engineering under Award DE-SC0001630.

VI. REFERENCES

[1] W. Wang, M.T. Winkler, O. Gunawan, T. Gokmen, T.K. Todorov, Y. Zhu, and D.B. Mitzi, *Device Characteristics of CZTSSe Thin-Film Solar Cells with 12.6% Efficiency.* Advanced Energy Materials, (2013).

[2] V. Fthenakis, *Sustainability of photovoltaics: The case for thin-film solar cells.* Renewable and Sustainable Energy Reviews, 13(9) 2746-2750 (2009).

[3] W.M. Hlaing Oo, J.L. Johnson, A. Bhatia, E.A. Lund, M.M. Nowell, and M.A. Scarpulla, *Grain Size and Texture of Cu2ZnSnS4 Thin Films Synthesized by Cosputtering Binary Sulfides and Annealing: Effects of Processing Conditions and Sodium.* Journal of Electronic Materials, 40(11) 2214-2221 (2011).

[4] K. Wang, O. Gunawan, T. Todorov, B. Shin, S.J. Chey, N.A. Bojarczuk, D. Mitzi, and S. Guha, *Thermally evaporated Cu2ZnSnS4solar cells.* Applied Physics Letters, 97(14) 143508 (2010).

[5] I. Repins, C. Beall, N. Vora, C. DeHart, DariusKuciauskas, PatDippo, BobbyTo, J. Mann, W.-C. Hsu, A. Goodrich, and RommelNoufi, *Co-evaporated Cu₂ZnSnSe₄ films and devices* Solar Energy Materials and Solar Cells, (2012).

[6] H. Katagiri, K. Jimbo, S. Yamada, T. Kamimura, W.S. Maw, T. Fukano, T. Ito, and T. Motohiro, *Enhanced conversion efficiencies of Cu2ZnSnS4-based thin film solar cells by using preferential etching technique.* Applied Physics Express, 1(4) 041201 (2008).

[7] T.K. Todorov, K.B. Reuter, and D.B. Mitzi, *High-Efficiency Solar Cell with Earth-Abundant Liquid-Processed Absorber.* Advanced Materials, 22(20) E156–E159 (2010).

[8] A. Redinger, D.M. Berg, P.J. Dale, R. Djemour, L. Gutay, T. Eisenbarth, N. Valle, and S. Siebentritt, *Route Toward High-Efficiency Single-Phase Cu 2ZnSn(S,Se) 4 Thin-Film Solar Cells: Model Experiments and Literature Review.* IEEE Journal of Photovoltaics, (2011).

[9] A. Redinger, D.M. Berg, P.J. Dale, and S. Siebentritt, *The consequences of kesterite equilibria for efficient solar cells.* Journal of the American Chemical Society, 133(10) 3320-3323 (2011).

[10] J.J. Scragg, T. Ericson, T. Kubart, M. Edoff, and C. Platzer-Björkman, *Chemical insights into the instability of cu2znsns4 films during annealing.* Chemistry of Materials, 23(20) 4625-4633 (2011).

[11] J.L. Johnson, H. Nukala, E.A. Lund, W.M.H. Oo, A. Bhatia, L.W. Rieth, and M.A. Scarpulla. *Effects of 2nd phases, stress, and Na at the Mo/Cu2ZnSnS4 interface.* in Mat. Res. Soc. Symp. Proc. (2010).

[12] H. Nukala, J.L. Johnson, A. Bhatia, E.A. Lund, W.M.H. Oo, M.M. Nowell, L.W. Rieth, and M.A. Scarpulla. *Synthesis of optimized CZTS thin films for photovoltaic absorber layers by sputtering from sulfide targets and sulfurization.* in Mat. Res. Soc. Symp. Proc. (2010).

[13] B. Mayer, Chem. Rev., 76(382) (1976).

Cu_2ZnSnS_4 thin films synthesized by cosputtering and rapid thermal annealing: effects of composition and temperature

J.L. Johnson[1], W.M. Hlaing Oo[2], M. Karmarkar[2], M.A. Scarpulla[1,2]

[1]Electrical and Computer Engineering, University of Utah, Salt Lake City, UT 84112
[2]Materials Science and Engineering, University of Utah, Salt Lake City, UT 84112

Abstract — **In this paper we study the structural properties of Cu_2ZnSnS_4 (CZTS) synthesized by cosputtering binary sulfide followed rapid thermal annealing (RTA) as functions of temperature and composition. By annealing identical samples at different temperatures we were able to find the optimum temperature for annealing samples in our RTA furnace. With this information, we annealed samples with a variety of discrete compositions at the optimum temperature to study the compositional differences on the optical, electrical and the structural differences on CZTS thin film crystallization. Samples were evaluated by x-ray diffraction, scanning electron microscopy (SEM) and energy dispersive spectroscopy (EDS). Sample rich in Cu tended to form areas of $Cu_{2-x}S$. It was confirmed that Sn loss during annealing increases with increased annealing temperature. Samples rich in tin before annealing grew larger grains indicating that Sn can also play a role in grain growth.**

I. Introduction

Cu_2ZnSnS_4 (CZTS) is an alternative material to $Cu(In,Ga)Se_2$ (CIGSe) for use as thin film photovoltaic absorber layers that is In-free and composed of commodity elements[1,2]. If high efficiencies in the 15-20% range can be realized for CZTS-based solar cells then it may form the basis for a TW-scale terrestrial photovoltaic technology while the large but finite supplies of raw materials for CdTe and CIGSe absorber layers are expected to begin limiting these technologies around the 150 GW/year production level. One large source of photocarrier recombination in polycrystalline inorganic photovoltaic cells is electronic states at grain boundaries. Larger grain sizes and passivation will reduce densities of recombination centers and traps at grain boundaries as well as increasing mobility. In this work, we examine trends in grain growth during synthesis of CZTS by cosputtering all four elements from binary sulfide targets followed by rapid thermal annealing. The knowledge of compositional and temperature effects on grain growth and other properties of the films will be useful for rapid and large-scale synthesis of CZTS.

We have found that processing temperatures of at least 550 °C are required to achieve grain growth on timescales of 2 hours [3, 4]. The composition of CZTS is harder to control at higher annealing temperatures as S and SnS especially are known to evaporate preferentially [5]. Na and excess Cu have been shown to result in more rapid grain growth in CIGSe and similar behavior has been shown for CZTS. The high atomic mobility required for grain growth of semiconductors is typically provided over long times thermally, however materials provide various constraints on temperature or thermal budget. We thus investigate rapid thermal annealing (RTA) which uses higher temperatures and shorter times to achieve similar annealing on the order of seconds to minutes.

II. Experimental

For this study, CZTS films were deposited on three 25 x 75 mm soda lime glass slides per deposition which were cleaned with detergent, solvents, and UV ozone before sputtering. Note, these samples didn't have a Mo back contact. The binary sulfides Cu_2S, ZnS and SnS_2 were magnetron cosputtered uniformly from three RF magnetron heads simultaneously to a thickness of 2 μm [6,7]. Films of different compositions were sputtered in order to assess the effects of composition on the properties of the CZTS films. RTA annealing thermal profiles were investigated on a set of adjacent samples diced from a single film ensuring the same starting composition. The composition of the as-sputtered films was uniform over an approximately 75 mm diameter to within +/-0.5 % and the thickness was uniform to within 20 nm. Discrete temperature and composition samples were generated with temperature and composition as listed in Tables 1 and 2 respectively.

Sample	T (°C)	Cu (at%)	Zn (at%)	Sn (at%)	S (at%)	Description (all Cu-poor)
As-Sputtered		22.8	12.6	14.3	50.3	Zn<Sn
Temp-515	515	24.0	13.5	13.8	48.6	Zn<Sn
Temp-560	560	25.0	12.7	13.8	48.5	Zn<Sn
Temp-610	610	24.0	13.7	13.8	48.6	Zn=Sn

Table 1 – Composition of samples before and after annealing.

The home-built RTA apparatus consists of a high intensity water cooled halogen lamp with collimating reflective enclosure that is translatable across a sample chamber consisting of a low-volume quartz chamber of rectangular cross-section. The samples are held in an open graphite susceptor fitted with a reservoir of powdered sulfur exposed to the lamp and fitted with a thermocouple. Samples of approximately 1 cm^2 were scribed from the deposited films and placed with the film side facing the lamp. The quartz tube was sealed and pumped down to 90 mTorr, then filled with forming gas to a pressure of 1000 Torr. The entire chamber

978-1-5090-5606-4/17 $31.00 © 2017 IEEE

	As-Deposited				RTA				Composition Description	Composition Description
Sample	Cu (at%)	Zn (at%)	Sn (at%)	S (at%)	Cu (at%)	Zn (at%)	Sn (at%)	S (at%)	As-Deposited	After RTA
Comp-1	31.8	13.6	10.3	44.2	27.0	11.0	10.0	52	Cu rich Zn>Sn	Cu rich Zn»Sn
Comp-2	22.8	11.6	16.0	49.6	28.0	12.4	10.1	49	Cu poor Zn<<Sn	Cu rich Zn>>Sn
Comp-3	24.5	13.4	14.8	47.0	24.0	13.7	13.8	49	Cu poor Zn<Sn	Cu poor Zn=Sn
Comp-4	21.5	14.1	13.6	50.8	22.0	12.6	13.2	52	Cu poor Zn≈Sn	Cu poor Zn»Sn
Comp-5	24.9	13.4	13.7	48.0	24.2	13.6	13.1	49	Cu poor Zn≈Sn	Cu poor Zn»Sn

Table 2 – EDS-determined compositions of samples in the composition series before and after annealing.

was then heated from below by an external graphite heater to a base temperature before the lamp was turned on and swept across the sample. The measured RTA temperature profiles plotted in Fig. 1 were controlled by the rate of lamp translation and the power supplied to the lamp. As can be seen, the 615 °C sweep didn't acquire much higher annealing temperature but sustained it for about 3 minutes. Samples were not etched after annealing even if crystals formed on the surface. It was discovered that samples without the Mo back contact lacked sufficient absorber layer adhesion that when exposed to KCN, IPA, acetone or DI water would stay intact.

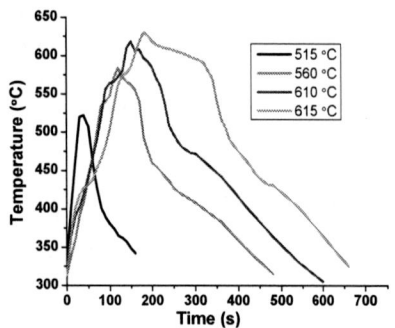

Figure 1 – RTA temperature profiles for the noted maximum temperatures in the different annealing programs.

Samples were characterized by optical transmission, Raman spectroscopy, X-ray diffraction (XRD), energy dispersive spectroscopy (EDS) and scanning electron microscopy (SEM). The θ-2θ XRD patterns were collected on a diffractometer having a divergence slit of 1 °, an anti-scatter slit of ½ °, and a receiver slit of ¼ °. XRD scans contain information about grain growth size, quality, orientation and composition. Compositional information about samples before and after annealing was determined by EDS. EDS scans of areas about 25 um² determined overall material composition and spot mode scans were used to evaluate compositions of

surface crystal. Composition of samples by EDS was confirmed on previous samples by inductively-coupled plasma mass spectroscopy (ICPMS). Compositional variance was on the order of 1%. SEM images help determine surface texture and approximate grain size using 18 kV excitation voltage.

III. RESULTS AND DISCUSSION

Composition as measured by EDS before annealing for discrete temperature samples, indicate slightly higher than desired amounts of Sn. One of the significant observations as

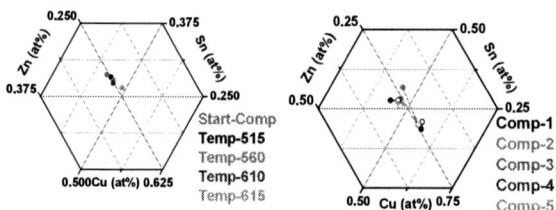

Figure 2 – Sections of CZTS ternary phase diagram showing the samples' metal ratios before and after annealing. (left) Temperature series and (right) composition series. Stoichiometric Cu:Zn:Sn = 2:1:1 is in the center of the diagrams. In all cases, the samples' trajectories are along a line of constant Zn.

shown in Fig. 2a is the loss in Sn during annealing. Figure 2a also indicates that the rate of Sn loss increases with annealing temperature. This has been previously observed and requires an over pressure of Sn to control the loss [5]. Other variations in relation to Cu and Zn after annealing are small enough to be within the variability of the EDS measurement. Table 1 lists the peak annealing temperatures from the RTA temperature profiles in Fig. 1 (515 °C, 560 °C, 610 °C, and 615 °C) along with the pre- and post-RTA sample compositions as measured by EDS. In addition to the annealing temperature used in table 1 a sample was also

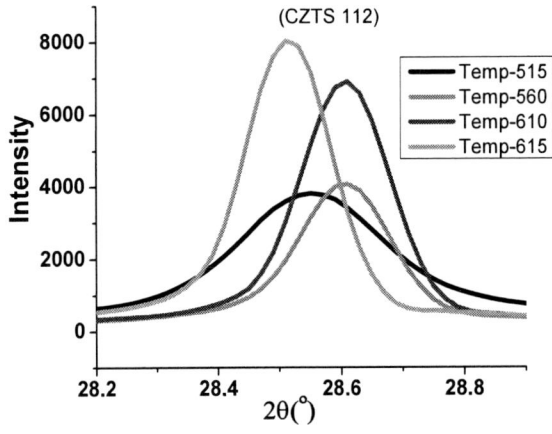

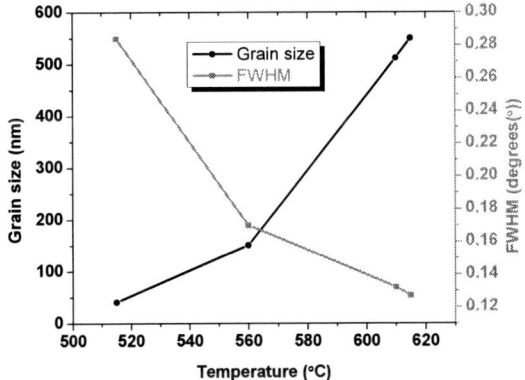

Figure 4 – (top) XRD data from the CZTS (112) peak for temperature series samples. (bottom) Grain size from SEM (black) and (112) peak FWHM from XRD (red) vs. RTA

annealed at 615 °C which is not shown in the some of the graphs and diagrams due to the inability to measure the properties. The higher annealing temperature and annealing duration degraded the deposition by incorporating too many defects causing the sample to be very conductive. SEM's of discrete temperature samples show increased texture with increasing annealing temperature as seen in Fig. 3. XRD scans of discrete temperature samples shown in Fig. 4, which were a focused scan of the CZTS (112) peak indicate for the most part, increased intensity and smaller FWHM with higher annealing temperature. Measurements of the FWHM and grain-size displayed in Fig. 5 conclude that very little improvement is achieved at 615 °C temperature. This most likely was due to the lack of increased temperature of the RTA. At this setting of the RTA, much of the heat from the halogen lamp is radiated rather than being absorbed by the RTA tube and sample holder.

Table 2 shows the compositions before and after annealing for the composition series and these results are plotted in Fig. 2b on a section of the ternary diagram. Again the compositional change for samples Comp-3 and Comp-5

are probably small enough to be considered to be within the margin of error of the EDS measurement, but Comp-1, Comp-2 and Comp-4 could use some explanation. Comp-1 shows a decrease in Cu and Zn to achieve an increase in Sn. Since Comp-1 was significantly Cu rich, the excess Cu diffused to form Cu_2S crystals. XRD of sample Comp-1 indicates the formation of Cu_2S. As is usually noticed, excess Cu results in larger grain growth [10,11]. We observed this effect also during this experiment as displayed in Fig. 5. Comp-1 had accumulations of Cu_2S and Cu_2SnS_3 crystals on the surface, as can be seen in Figure 6, which could account for a reduction in Cu in the overall sample. It is improbable for any sample to have a gain of any metal since there is no source in the annealing chamber. This means that the other metals would have to decrease in at (%). As mentioned earlier, if the deposition and annealing layer is non-uniform, in that a particular element diffuses to the surface, there is reduced electron interaction volume near the surface for EDS measurements. It's also possible that some of the Cu could have also volatized during annealing. Sample Comp-2 shows definite concentrations of Cu_2S as revealed in Raman spectroscopy (not shown).

Figure 5 Scanning electron micrographs of compositional gradient samples which indicate that Cu rich (comp-1) and Sn rich (comp-2) sample grow larger grains. These samples also produced binary compounds (crystals) on the surface.

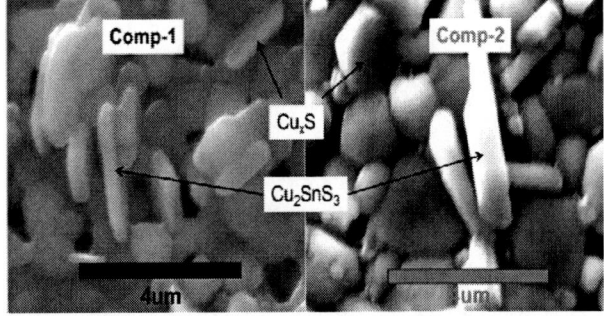

Figure 6 SEM of Comp-1 and Comp-2 samples showing formation of surface crystals after annealing. Crystals include Cu_2S & Cu_2SnS_3.

The Zn loss in sample Comp-4 is a little more difficult to explain since Zn is a harder element to disassociate from the material due to its stronger bonds. Other groups have found that Zn can diffuse to the surface, if this is the case, this could explain the Zn loss. Second, samples high in zinc also can't consume all the zinc resulting in excess ZnS. This is not as big of a problem since ZnS is an insulator with a higher bandgap and doesn't short out the absorber layer.

Figure 5 shows the scanning electron micrographs indicating surface texture and approximate grain size for the samples in table 2. Grain size is large for samples Comp-1 and Comp-2 but surface crystals also formed for these two samples. Discrete compositional samples with significant excess pre-annealing quantities of Cu and Sn result is surface crystals after annealing as seen if Fig. 6. During annealing excess quantities of metals diffused to the top. First, samples with too much Cu can't consume all the copper resulting in residual Cu_2S high conductivity and no bandgap. Third, samples high in tin are not difficult to deal with since the excess Sn will diffuse to the surface and volatizes with the sulfur vapor into a gas and be swept away [16-18]. Too much Sn can be lost as discussed with temperature discrete samples in Fig. 2 and as seen in sample Comp-2. Therefore more attention must be taken to retain the right amount of Sn during the annealing process.

IV. CONCLUSIONS

Many aspects about the deposition and annealing of CZTS were confirmed during this experiment and some new characteristics were discovered. First it was confirmed that the quantity of Sn decreases during annealing and increases with annealing temperature. Second, that grain size increases with annealing temperature. Third, excess metals before annealing diffuse to the surface and form surface crystals or volatize with S vapor and are swept away. Fourth, excess Sn, like Cu, can be used to increase grain size during annealing.

V. ACKNOWLEDGEMENTS

This work was supported in full by the U.S. Department of Energy, Office of Basic Energy Sciences, Division of Materials Sciences and Engineering under Award DE-SC0001630.

VI. REFERENCES

[1] H. Katagiri et al., Applied Physics Express, 1, 041201, (2008)
[2] T. Todorov, et al., Thin Solid Films, (2011)
[3] Weber et al., Mater. Res. Soc. Symp. Proc. Vol. 1012

[4] Schorr & Gonzalez-Aviles, Thin Solid Films 517 2461 (2009)
[5] A. Weber et al., J. of Appl. Phys., 107 (2010)
[6] J. Johnson et al., Mater. Res. Soc. Symp. Proc. EE3.3 (2010).
[7] G. Suresh Baba, et al. Semi. Sci. and Tech. 23 #8 (2008)
[8] B. Schubert, et al., Processes in Photovoltaics 19 #1 (2011)
[9] P.A. Fernandes, et al., Journal of Physics D, 43 (2010)
[10] P.A. Fernandes, et al., J. Alloys and Compd., (2011)
[11] P.A. Fernandes, et al., Phys. Status Solidi C N. 3-4, (2010)
[12] F. Jiang, et al., Applied Physics Express, 4 (2010)
[13] T. Tanaka, et al., Thin Solid Films 518 21 (2010)
[14] WM Hlaing Oo, et al., Journal of Electrical Materials, DOI 10.1007/S 11664-011-1729-3 (2011)
[15] JJ Scragg, et al., Thin Films for Photovoltaics, DOI:1007/978-3-642-22919-0-4 (2011)
[16] A. Weber, et al., Journal of Applied Physics, 107, 013516 (2010)
[17] T.M. Friedmeier, et al., ICTM-11, Salford, 8-12 September 1997, pp 345-348
[18] P.M.P. Salome, et al., Solar Energy Materials & Solar Cells 94 (2010) 2176-2180

Earth-abundant CZTSSe thin film solar cells on flexible stainless steel foil substrates

Hae-Sun Kim[1,2], Woo-Lim Jeong[1,2], and Dong-Seon Lee[1,2,*]

[1]School of Electrical Engineering and Computer Science, Gwangju Institute of Science and Technology (GIST), Gwangju 61005, South Korea

[2]Research Institute for Solar and Sustainable Energies (RISE), Gwangju 61005, South Korea

Abstract — We fabricated $Cu_2ZnSn(S,Se)_4$ (CZTSSe) thin film soar cells on flexible stainless steel foil substrates (SUS). The absorber layers are deposited by magnetron sputtering followed by sulfo-selenization process. The fabricated cell structure was Al/Ni/AZO/i-ZnO/CdS/CZTSSe/Mo/Cr/SUS. The best device represents an open-circuit voltage of 363.4 mV, a short-circuit current density of 29.6 mA/cm^2, a fill factor of 44.7%, and a power conversion efficiency of 4.8%.

Index Terms —$Cu_2ZnSn(S,Se)_4$, Flexible solar cell, Stainless steel foil, C-V measurement.

I. INTRODUCTION

Kesterite structured $Cu_2ZnSnS_xSe_{4-x}$ (CZTSSe) has been a promising photovoltaic material for low-cost and sustainable solar cell devices due to a non-toxic material structure containing only earth-abundant elements [1, 2]. Wang et al. reported device characteristics of CZTSSe thin film solar cells with 12.6% power conversion efficiency (PCE) using a hydrazine pure-solution approach on rigid soda lime glass (SLG) substrate [1].

Most of CZTSSe solar cells have been fabricated on SLG substrates due to its cost-effectiveness and sodium doping effect which enhances grain growth, and cell efficiency. However, CZTSSe thin films grown on flexible substrates are attracting increasing attention for their possible more applications. The flexible CZTSSe solar cells are light weight and can further reduce manufacture cost by roll-to-roll deposition method [3]. Up to now, flexible CZTSSe solar cells have been produced on flexible substrates such as Mo foil [4], Al foil [5], Ti foil [6], stainless steel foil [7, 8], and polyimide (PI) film [9].

In this work, we fabricated CZTSSe solar cells on stainless steel foil (SUS) substrates having a structure of Al/Ni/AZO/i-ZnO/CdS/CZTSSe/Mo/Cr/SUS with various AZO layer thicknesses. As a result, the best cell performed CZTSSe solar cell on flexible stainless steel substrates showed 4.8% conversion efficiency with 300 nm thick Al-doped ZnO (AZO.)

II. EXPERIMENT

0.1-mm-thick stainless steel foils were used as substrates. They were cleaned under an ultrasonic bath with acetone, methanol and isopropyl alcohol during 5 min for each solvent. Then, they were rinsed with deionized water and dried with a nitrogen flux. As a chemical barrier, a thin Cr layer with 200 nm thick was deposited by electron beam evaporation on the cleaned SUS substrate. With the SUS substrates, rigid SLG substrates were also prepared as references. Then, 1 μm in thickness Mo layer was deposited as back electrical contact on the both kinds of substrates.

The base pressure before Cu/Sn/Zn precursor deposition was less than 6.7×10^{-5} Pa. For the deposition of the Zn and Cu layers, DC magnetron sputtering was used. The Sn layers were deposited using a Sn metal-element target by RF magnetron sputtering. Under the 0.4 Pa pressure condition, CZT layers were deposited with 40 W (0.88 W/cm2) sputtering power. Then 10-nm-thick NaF layer was deposited on the precursors to provide Na content in the Na-free SUS substrates. The precursors with/without NaF layer (for SUS/SLG substrate) were sulfo-selenized in a graphite box with 0.5 g of Se powder and 0.02 g of SeS$_2$ powder. The samples were heated from room temperature to 300 °C for ~1000 s and then maintained at 300 °C for 1500 s. Subsequently, they were heated from 300 °C to 550 °C for ~1000 s and then maintained at 550 °C for 1100 s. These absorber films were KCN-etched (5 min in a 0.05 M aqueous solution) to remove Cu-Se binary phases, and then rinsed with deionized water. Right after the KCN-etching, a 70-nm CdS buffer layer was deposited by chemical bath deposition at 80 °C and 50 nm of intrinsic ZnO by RF magnetron sputtering. AZO window-layer with various thicknesses were then sputtered. The AZO thickness was used to name the samples hereafter, i.e. SUS200, SUS300, SUS400, SUS500, SLG200, SLG300, SLG400, SLG500 for the AZO thicknesses of 200 nm, 300 nm, 400 nm, 500 nm for SUS substrates and SLG substrates, respectively. Lastly, 50 nm of Ni and 500 nm of Al top-grid were deposited through a shadow mask by electron beam evaporation.

III. RESULTS AND DISCUSSION

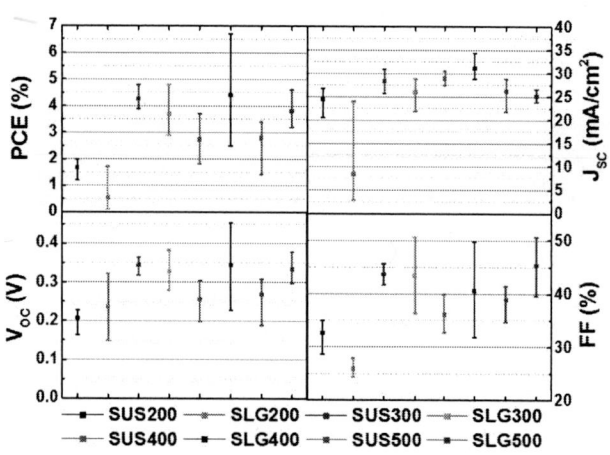

Fig. 1. Solar cell parameters for the CZTSSe solar cells with different substrates and AZO thicknesses

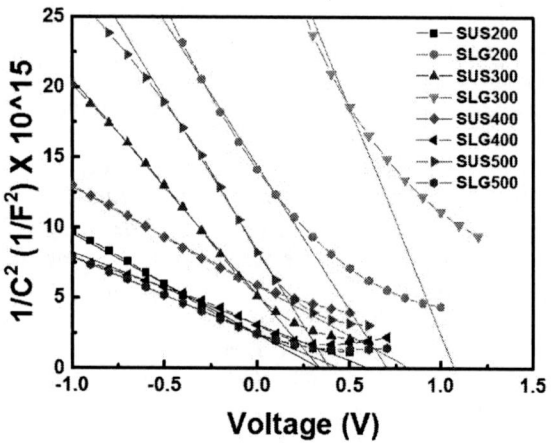

Fig. 2. 1/C² vs. V plot obtained from C-V measurement

The solar cell parameters were measured by solar simulator with an AM 1.5 for the CZTSSe solar cells fabricated with the various AZO thicknesses from 200 to 500 nm on the different substrates. As shown in Fig. 1, the PCE value of flexible cells was firstly increased with AZO thickness, reached the maximum at AZO 300 nm and then decreased. The highest PCE of flexible cell was 4.8% from the SUS300, and that of reference cell was 6.7% from the SLG400. The mean PCE values of each cells were comparable, 4.3%, and 4.4% for the SUS300, and the SLG400, respectively. While the SUS300 cells exhibited a relatively stable PCE from 4% to 5%, the SLG400 cells showed somewhat fluctuated PCE values from 2.5% to 6.7% due mostly to the surface fluctuation of the absorber layer formed during the sulfo-selenization process. The J_{SC} values for flexible cells tended to be higher than SLG

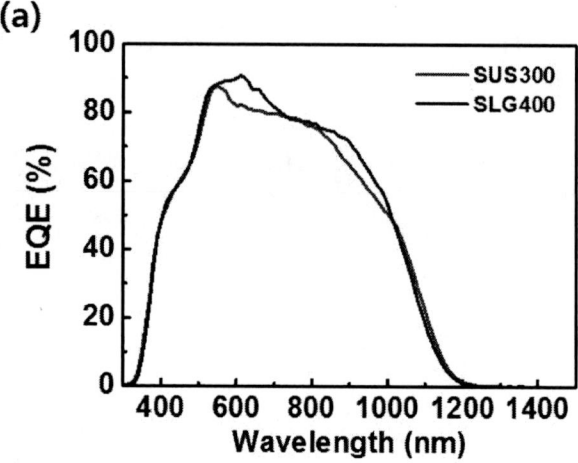

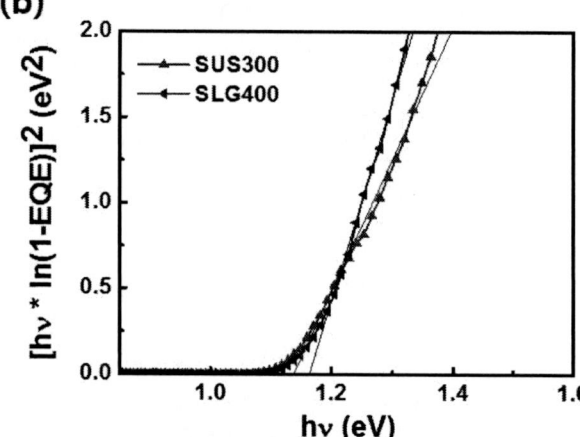

Fig. 3. (a) External quantum efficiency (EQE), and (b) the Tauc plat of the best cells from both substrates.

cells under the same AZO thickness although the V_{OC} and FF values were opposite. The SUS300 sample showed stable J_{SC}, V_{OC}, and FF values between the cells in this sample and high mean values compared with other samples. The difference between SUS and SLG substrate was sodium doping concentration which facilitates CZTSSe film recrystallization.

In order to compare the solar cells from both substrates we performed C-V measurement. Built-in-voltage and doping density were derived from 1/C² vs. V plot of capacitance-voltage data (Fig. 2) as the following expressions [7, 10]:

$$\frac{1}{C^2} = \frac{2}{qN_a\epsilon_0\epsilon_s A^2}(V_{bi} - V) \qquad (1)$$

$$N_a = \frac{2}{q\epsilon_0\epsilon_s A^2\left[\frac{d}{dV}\left(\frac{1}{C^2}\right)\right]} \qquad (2)$$

N_a is doping density, q is the electron charge, ϵ_0 is permittivity of the free space, ϵ_s is dielectric constant, A is area of the cell,

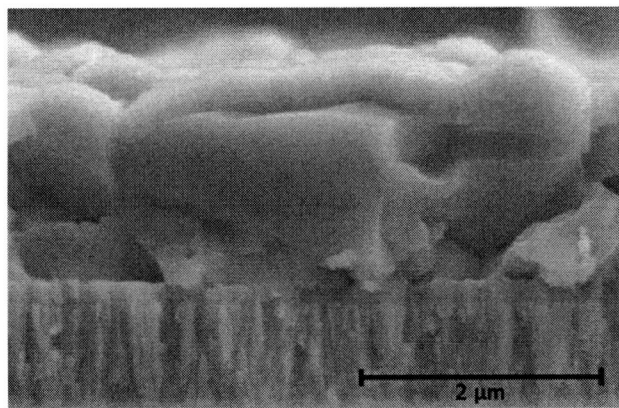

Fig. 4.   Cross-sectional SEM image of the CZTSSe solar cell.

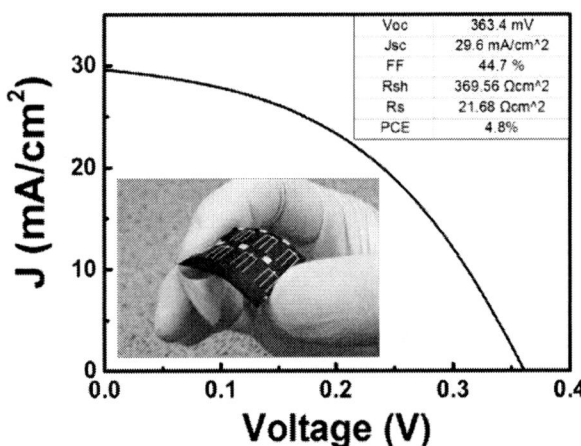

Fig. 5.   J-V curve of the best sample on SUS substrate measured under a solar simulator with AM 1.5. The inset is photograph of a flexible CZTSSe solar cell

C is measured capacitance and V is applied DC voltage. From the above equations, we can obtain the built-in-voltage (V_{bi}) and the doping density. The V_{bi} (V) values from Fig. 1 were 0.34 V, 0.70 V, 0.34 V, 1.08 V, 0.83 V, 0.61 V, 0.39 V, and 0.45 V and the doping densities calculated from Eq. (2) were $6.05 \times 10^{16}/cm^3$, $2.09 \times 10^{16}/cm^3$, $2.8 \times 10^{16}/cm^3$, $1.34 \times 10^{16}/cm^3$, $6.12 \times 10^{16}/cm^3$, $8.39 \times 10^{16}/cm^3$, $2 \times 10^{16}/cm^3$, and $7.89 \times 10^{16}/cm^3$, for SUS200, SLG200, SUS300, SLG300, SUS400, SLG400, SUS500, and SLG500, respectively.

The external quantum efficiency (EQE) curves and the Tauc plot of the two best CZTSSe solar cells are presented in Fig. 3. The SUS300 showed a maximum EQE value of 87.6% at 550 nm and that of SLG400 sample presents 90.8% at 610 nm. The difference came from the thickness of AZO. From the results of Fig. 1, the optimized AZO thickness was considered to be between 300 to 400 nm (Fig. 3. (a)). The band gap calculated by the Tauc plot were 1.14 eV for the SUS300 and 1.16 eV for SLG400. There is no difference between two band gap values because the absorber layers are same.

To get further insights in the device performance, the cross-sectional scanning electron microscopy (SEM) of the CZTSSe solar cells and x-ray diffraction spectroscopy (EDX) of the absorber layer were performed. As shown in Fig. 4, Mo back electrical contact, thin MoSe₂ layer, quite large CZTSSe absorber grains, CdS buffer layer, iZO, and AZO window layer were observed. The thickness of CZTSSe crystal was about 1.6 μm and the atomic percentages of the film measured using EDX showed Cu=20.50%, Zn=16.18%, Sn=12.2%, S=9.28%, and Se=41.85%.

Fig. 5 shows the J-V characteristics of the best SUS300 cell under AM1.5G light illumination. The inset is photograph of the flexible CZTSSe solar cell fabricated on the SUS substrate. The best device represents an open-circuit voltage (V_{OC}) of 363.4 mV, a short-circuit current density (J_{SC}) of 29.6 mA/cm², a fill factor (FF) of 44.7%, and a power conversion efficiency (PCE) of 4.8%.

IV. CONCLUSIONS

Flexible CZTSSe solar cells were fabricated on flexible stainless steel foil substrates with various AZO thicknesses from 200 to 500 nm. Cr layer was deposited as a Fe diffusion barrier and NaF layer was used to provide the sodium content to enhance the grain growth and improve the crystallinity of the absorber layer. From the data of PCE and EQE, the thickness of AZO was optimized between 300 to 400 nm. As a result, 4.8% of power conversion efficiency was achieved by using 300-nm-thick AZO. It is still lower value compared with that of the cell fabricated on SLG substrate (~6.7%). Further optimization of device fabrication will lead to enhanced efficiency for flexible CZTSSe solar cells.

ACKNOWLEDGEMENT

This work was supported by the GIST Research Institute (GRI) grant funded by the GIST in 2017.

REFERENCES

[1] W. Wang, M. T. Winkler, O. Gunawan, T. Gokmen, T. K. Todorov, Y. Zhu, and D. B. Mitzi, "Device Characteristics of CZTSSe Thin-Film Solar Cells with 12.6% Efficiency," *Advanced Energy Materials*, vol. 4, 1301465, 2014.

[2] S. Chen, A. Walsh, J.-H. Yang, X. G. Gong, L. Sun, P.-X. Yang, J.-H. Chu, and S.-H. Wei, "Compositional dependence of structural and electronic properties of Cu₂ZnSn(S,Se)₄ alloys for thin film solar cells," *Physical Review B*, vol. 83, 125201, 2011.

[3] J. Ramanujam, and U. P. Singh, "Copper indium gallium selenide based solar cells – a review," *Energy & Environmental Science*, 2017

[4] Y. Zhang, Q. Ye, J. Liu, H. Chen, X. He, C. Liao, J. Han, H. Wang, J. Mei, and W. Lau, "Earth-abundant and low-cost CZTS solar cell on flexible molybdenum foil," *RSC Advances*, vol. 4, pp. 23666, 2014

[5] Q. Tian, X. Xu, L. Han, M. Tang, R. Zou, Z. Chen, M. Yu, J. Yang, and J. Hu, "Hydrophilic Cu_2ZnSnS_4 nanocrystals for printing flexible, low-cost and environmentally friendly solar cells," *CrystEngComm*, vol. 14, pp. 3847-3850, 2012

[6] J. Li, H. Shen, H. Shang, Y. Li, and W. Wu, "Performance improvement of flexible CZTSSe thin film solar cell by adding a Ge buffer layer," *Materials Letters*, vol. 190, pp. 188-190, 2017

[7] K. Sun, F. Liu, C. Yan, F. Zhou, J. Huang, Y. Shen, R. Liu, and X. Hao, "Influence of sodium incorporation on kesterite Cu_2ZnSnS_4 solar cells fabricated on stainless steel substrates,"

Solar Energy Materials & Solar Cells, vol. 157, pp. 565-571, 2016

[8] X. Zhai, H. Jia, Y. Zhang, Y. Lei, J. Wei, Y. Gao, J. Chu, W. He, J. J. Yin, and Z. Zheng, "In situ fabrication of Cu_2ZnSnS_4 nanoflake thin films on both rigid and flexible substrates," *CrystEngComm*, vol. 16, 6244, 2014

[9] Z. Zhou, Y. Wang, D. Xu, and Y. Zhang, "Fabrication of Cu_2ZnSnS_4 screen printed layers for solar cells," *Solar Energy Material and Solar Cells*, vol. 94, pp. 2042-2045, 2010.

[10] S. K. Saha, A. Guchhait, and A. J. Pal, "Cu_2ZnSnS_4 (CZTS) nanoparticle based nontoxic and earth-abundant hybrid pn-junction solar cells," *Physical Chemistry Chemical Physics*, vol. 14, pp. 8090-8096, 2012

Comparison of MgCl$_2$ and CdCl$_2$ activation treatment for CdTe solar cells: Recrystallization and Defects

Daniele Menossi, Elisa Artegiani, Ivan Rimmaudo°, Alessia Le Donne*, Simona Binetti*, Juan Luis Pena°, Fabio Piccinelli[+] and Alessandro Romeo

Laboratory for Applied Physics, Department of Computer Science, University of Verona, Verona, I-37123, Italy

°Centro de Investigación y de Estudios Avanzados del IPN Unidad Mérida, Depto. Física Aplicada, Km. 6, Antigua Carretera a Progreso, C.P., 97310 Mérida, Yucatán, Mexico.

*MIBSOLAR CENTER and Department of Materials Science, University of Milano-Bicocca, I-20125, Italy

[+]Department of Biotechnology, University of Verona, Verona, I-37123, Italy

Abstract —One of the key factors for the success of CdTe solar cells is the so-called "activation" treatment, consisting in depositing a CdCl$_2$ film on the CdTe layer and subsequent annealing it in air. Recent studies have found MgCl$_2$ as a good alternative. In this work we have studied devices (exceeding 14% efficiency) prepared by low-substrate temperature CdTe deposition and activated with MgCl$_2$ treatment (applied by wet deposition). The CdTe films recrystallized by MgCl$_2$ were analyzed in terms of X-ray diffraction, atomic force microscopy, photoluminescence measurement and compared with the standard CdCl$_2$ treated CdTe cells. Results show that different defects are detected for MgCl$_2$ treated sample. Moreover MgCl$_2$ treated cell does not need a post deposition treatment of the back contact to deliver high current density, showing that different electrical properties of the CdTe are registered.

Index Terms — CdTe, activation treatment, thin films.

I. INTRODUCTION

CdTe has shown superior performance and scalability at the same time delivering high performance exceeding 18% for modules and 22% for solar cells [1][2]. One of the key factors for the success of this technology is the use of a post deposition treatment, typically consisting in a deposition of CdCl$_2$ layer and subsequent annealing in air/argon [3].

However in recent times a different chlorine-containing compound has been introduced by Major et al. [4] to substitute CdCl$_2$ in order to improve the scalability of the production process.

We have applied this innovative recrystallization step for CdTe on our thermal evaporated CdTe based solar cells. Thermally evaporated CdTe were deposited on CdS/ZnO/ITO stack and then treated by MgCl$_2$ by depositing the compound dissolved in methanol solution and annealing the stack in air, in a temperature range between 370 to 400 °C. Previously we have analyzed the electrical properties of the finished cells and compared them with the CdCl$_2$ treated ones [5]. In this work we compare the CdTe layers recrystallized by these two different methods in terms of x-ray diffraction, atomic force microscopy and, most important, photoluminescence. This last

one was made in order to address and define the energy defects that have been identified in our previous work.

Moreover we also have observed a different behavior after Cu/Au deposition before annealing in air, when cells are treated with MgCl$_2$.

II. EXPERIMENTAL DETAILS

CdTe solar cells have been fabricated in superstrate configuration.

The front contact is deposited on top of a 3×3 cm^2, 4 mm thick, soda-lime glass and it is produced as a ITO/ZnO bilayer stack. Indium tin oxide (ITO) and ZnO films are deposited by a RF reactive magnetron sputtering system. ITO with a thickness of 400 nm is fabricated starting from a 90% In$_2$O$_3$:10% SnO$_2$ target in a reactive atmosphere of Ar + 2-3% O$_2$ at a substrate temperature of 400 °C. ZnO with a thickness of 100 nm is fabricated starting from a ZnO target, in a reactive atmosphere of Ar + 10–20% O$_2$ at a substrate temperature of 400 °C.

CdS and CdTe are deposited in the same vacuum chamber with a pressure of 10^{-6} mbar. CdS lumps with 4N purity are evaporated at a temperature higher than 800 °C. CdS layer is grown on the ITO/ZnO stack, at a substrate temperature of 150 °C and with a thickness of 200/300 nm.

CdTe with 5N purity in form of grains is evaporated from an effusion cell at a temperature higher than 700 °C and, the CdTe layer is deposited on the CdS film at a substrate temperature of 340 °C with a thickness of 5 to 6 microns.

CdTe is then activated by two different alternative treatments: both consisting in dissolving the activating agent (CdCl$_2$ or MgCl$_2$) in methanol, applying a specific amount of the resulting solution on the CdTe surface and annealing the stacks in air. Solutions have been prepared with different concentrations and applied in form of drops on the sample surface, corresponding to microliter volumes. Then the stacks are annealed for 30 minutes in air at temperatures between 370 and 400 °C. Successively, bromine-methanol (Br-MeOH) etching is applied on the CdTe surface in order to clean it and to promote the formation of a p⁺-Te rich layer. The back contact is produced by depositing a 2 nm thick Cu and a 50 nm thick Au films, by vacuum evaporation at room temperature (10^{-5} mbar). Finally the device is annealed in air at 190 °C for 20 min.

CdTe layers have been studied in terms of physical properties by: 1) atomic force microscopy (AFM) with a NT-MDT - SMENA A with a gold coated silicon tip from the same company in semi-contact mode and 2) X-ray diffraction (XRD) was processed with a Thermo ARL X'TRA powder diffractometer (in Bragg-Brentano geometry, equipped with a Cu-anode X-ray source (Kα, λ=1.5418 Å) and a Peltier Si (Li) cooled solid state detector). Moreover photoluminescence (PL) spectra were recorded in the 700-1700 nm spectral range at 13 K. All PL measurements were performed using a standard lock-in technique in conjunction with a single grating monochromator and a short wavelength enhanced InGaAs detector with maximum responsivity at 1540 nm. A green laser (λ_{exc}=532 nm) with maximum power of 7 mW was used as excitation source. The samples were illuminated on the CdTe side (i.e. not through the ITO/ZnO/CdS layers). A cooling system consisting of rotary pump, turbo-molecular pump and He closed circuit-cryostat was used to perform PL measurements at low temperature up to 13 K. Finished solar cells were analyzed by current density–voltage (J-V) characteristics performed with a Keithley SourceMeter 2420 at room temperature.

III. ANALYSIS AND DISCUSSION

In Fig. 1 XRD patterns of CdTe layers activated by MgCl₂ (top) and CdCl₂ (bottom) are shown and compared. No extra peaks are registered in case of MgCl₂: all the orientations identified belong to CdTe. The as-deposited CdTe is strongly (111) oriented (not shown here), this preferred orientation is typically lost after recrystallization [6] with CdCl₂ as shown in Fig.1. The XRD analysis shown here confirms this tendency also for the MgCl₂ treatment, after activation the (111) peak is still the main orientation but (311) and (422) peaks are not negligible.

Comparing the CdCl₂ and MgCl₂ patterns we can observe that the loss of orientation is stronger for CdCl₂: more than 7 orientations are shown. When the absorber is annealed at 410 °C, the (311) peak is stronger than the (111) original orientation, as well as in case of 380 °C, with both (311) and (422) orientations larger than the (111). On the other hand this does not happen to the MgCl₂ treated samples, in fact irrespective of the temperature the (111) reflection is always larger than the other ones observed.

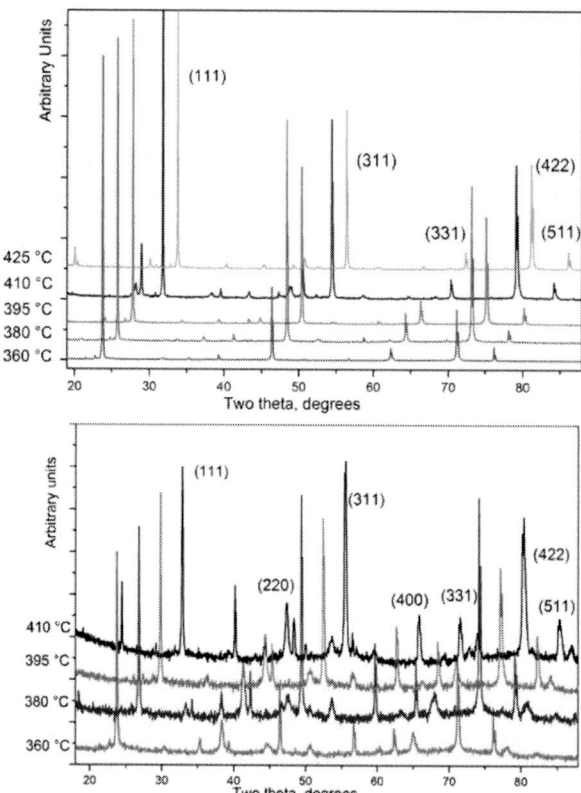

Fig. 1. XRD patterns of MgCl₂ treated (top) and CdCl₂ treated (bottom) CdTe layers.

In Fig. 2 an AFM image of the surface of a MgCl₂-treated CdTe is shown. The morphology demonstrates a clear enhancement of the grain size, as expected and as already reported [4], compared to the as-deposited case (not shown here) [6]: from initial grains of less than 1 micron in diameter, the MgCl₂ treatment enlarges the crystals up to 5 times analogously to the standard treatment [6].

Fig. 2. AFM image of a MgCl₂ treated CdTe layer on CdS/ZnO/ITO/glass stack.

Photoluminescence studies were applied on both CdCl$_2$- and MgCl$_2$- treated samples in low and high resolution and changing the illumination power. In Fig. 3, a first comparison of photoluminescence spectra of differently treated CdTe layers shows a radical different behavior of the CdTe when treated with different Cl-containing compounds. The measurements on MgCl$_2$ treated CdTe were repeated a second time on different position of the absorber, revealing the same peaks but with different intensities.

In particular, for MgCl$_2$ treatment we observe a strong reduction of the 1.59 meV peak, moreover the peak at 1.55 eV (attributed to cadmium vacancy) is not present, while a peak at 1.53 eV is very intense.

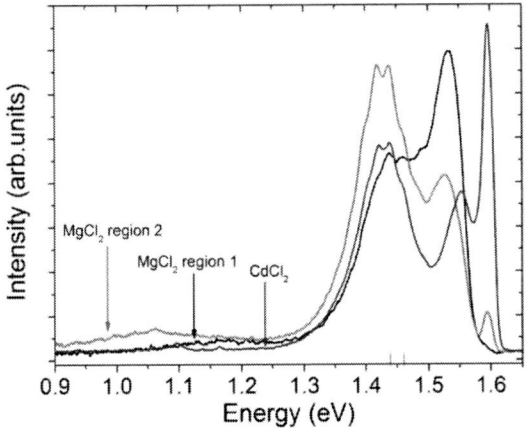

Fig. 3. Photoluminescence spectra of MgCl$_2$ and CdCl$_2$ treated CdTe layers on CdS/ZnO/ITO/glass stack.

A more detailed analysis by high resolution (HR) spectra, (bandpass of 3 nm), reveals the presence of different peaks as shown in figure 4, 5 and 6 respectively for CdCl$_2$ and MgCl$_2$ treated samples.

This HR-analysis reveals the presence of different defects that have been already detected and attributed as reported in different previous works [7][8][9].

The 1.59 eV peak is very near the band gap of CdTe (at 10 K the CdTe band gap is 1.606 eV), the luminescence bands in this energy range can be attributed to free excitons (very near to 1.6 eV), or excitons bound to donors and/or excitons bound to acceptors [10][11][9]. The 1.59 eV peak is present for every sample but if for the CdCl$_2$ treated absorber shows a very high intensity, in the MgCl$_2$ treated CdTe the peak is some cases very low.

In the graph for CdCl$_2$ treated samples, also a 1.55 eV peak is found. This has been reported also in [12] [13] and typically attributed to cadmium vacancies.

Another luminescence band, observed only for MgCl$_2$ treated samples, is the one related to 1.53 eV. According to different papers [14] [15] peaks in this range are attributed to band-related recombination, caused by bound exciton

recombination involving shallow donors, the source of this energy levels are identified with Te vacancies.

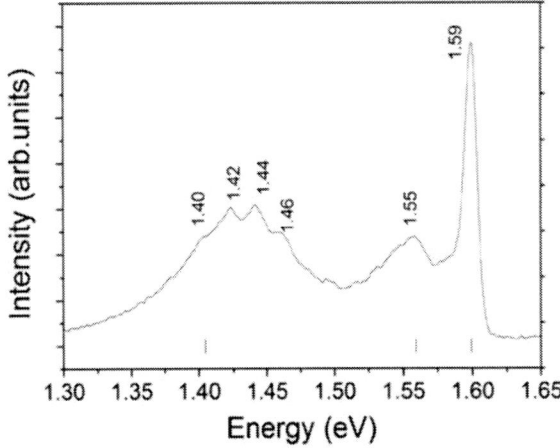

Fig. 4. High-resolution photoluminescence spectra of CdCl$_2$ treated CdTe layers on CdS/ZnO/ITO/glass stack

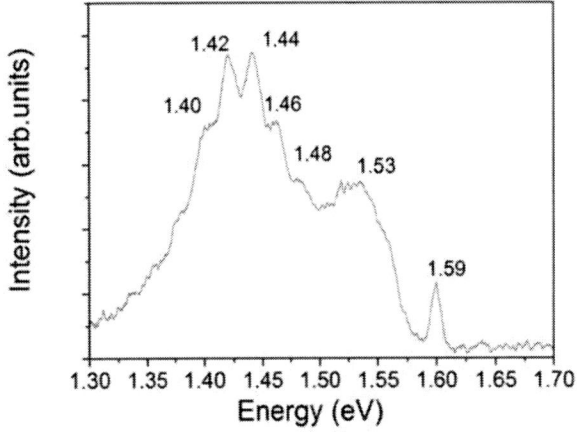

Fig. 5. High-resolution photoluminescence spectra of MgCl$_2$ treated CdTe layers (region 1) on CdS/ZnO/ITO/glass stack.

Finally in the range from 1.4 to 1.48 eV a series of different peaks have been detected for both CdCl$_2$-treated and MgCl$_2$-treated samples. In the range of 1.45 eV the donor acceptor pair (DAP), generally attributed to V_{Cd}-Cl$_{Te}$, has been identified [8] [16], however some of the peaks could also be attributed to CdS$_x$Te$_{1-x}$ intermixing [17].

A similar effect, with most defects observed for both MgCl$_2$ and CdCl$_2$ cases and few different additional energies for the MgCl$_2$, has been observed with admittance spectroscopy, suggesting that some different defects are generated [5]. However what makes a strong difference between CdCl$_2$ and MgCl$_2$ treated CdTe is the absence of the 1.53 eV peak in CdCl$_2$ demonstrating that MgCl$_2$ treatment can provide Te vacancies which act as donors in the CdTe layer. Moreover we

978-1-5090-5606-4/17 $31.00 © 2017 IEEE   1671

can also observe a strong reduction in the 1.59 eV peak especially for region1 which is the one connected with lower performance.

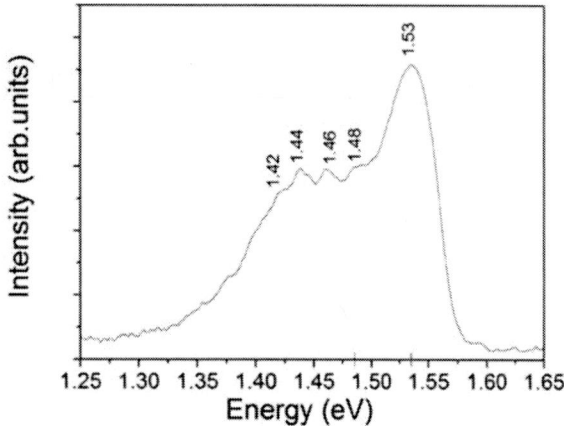

Fig. 6. High-resolution photoluminescence spectra of MgCl$_2$ treated CdTe layers (region 2) on CdS/ZnO/ITO/glass stack.

Finally, finished devices were processed and measured. As already published [5], good efficiencies were performed despite slightly lower than with CdCl$_2$ treatment (~14% vs ~15%).

Table I. List of energy peaks and their attributed defects.

Peak (eV)	CdCl$_2$	MgCl$_2$		Defects
		Area-1	Area-2	
1.6	Yes	No	Yes	Exciton bound to donor/acceptor
1.556	Yes	No	No	V_{Cd}
1.53	No	Yes	Yes	V_{Te}
1.485	Yes	Yes	Yes	V_{Cd}-Cl_{Te}
1.46	Yes	Yes	Yes	V_{Cd}-Cl_{Te}/CdS_xTe_{1-x}
1.44	Yes	Yes	Yes	V_{Cd}-Cl_{Te}
1.42	Yes	Yes	Yes	V_{Cd}-Cl_T/CdS_xTe_{1e}
1.40	Yes	Yes	Yes	V_{Cd}-Cl_{Te}

Most important is the different behavior in performance of devices, before air-annealing of Cu/Au back contact, which is usually applied for enhancing copper diffusion and for forming the CuTe layer, necessary to keep stability and performance [18][19]. What we can observe is that even before annealing, the MgCl$_2$ treated cells show similar high current densities as after annealing, while for the CdCl$_2$ treated case we observe a much lower current density (see Fig. 7). This suggests different electrical properties of the CdTe layer when MgCl$_2$ is applied. However if we look at the different PL peaks we observe 1.55 eV missing peak and 1.53 eV additional peak for the MgCl$_2$ treated case. The first one can

be attributed to cadmium vacancy while the second one can be attributed to tellurium vacancy. This would demonstrate that in case of MgCl$_2$ treatment the CdTe layer has no cadmium vacancies, which is in accordance with the lower carrier concentration measured previously [5]. So one possible explanation of the different behavior of the J-V characteristic could be attributed to the first layers of CdTe where tellurium could have been concentrated in order to enhance Cu-Te compounds, more analysis will be done to investigate this phenomenon.

IV. CONCLUSIONS

We have analyzed 14% CdTe solar cells prepared by MgCl$_2$ treatment and compared them with the typical CdCl$_2$ treated ones. Starting from a previous work on admittance spectroscopy and capacitance voltage analysis, in this work we have studied the physical properties of CdTe treated in the two different methods. We have observed a slightly different recrystallization, with a more preferred orientation for the MgCl$_2$ layer, indicating a less incisive recrystallization effect. On the other way the enlargement of the grain size, typically observed for low-temperature deposited CdTe is confirmed also for MgCl$_2$ with same grain size. Photoluminescence shows the formation of different defects, confirming previous analysis by admittance spectroscopy. The effect of high current density prior to annealing of the back contact suggests the presence of a tellurium rich layer that allows the Cu-Te formation.

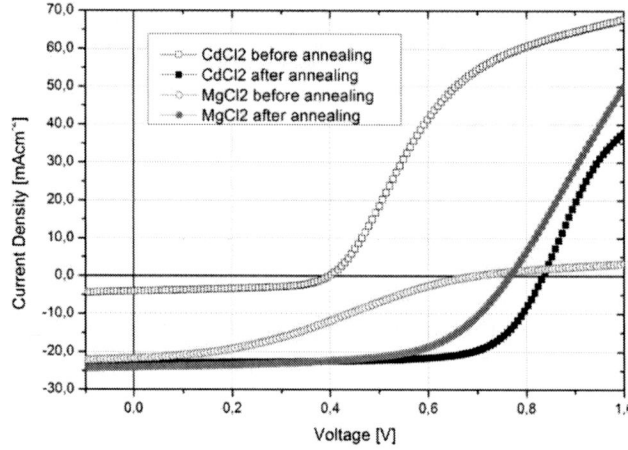

Fig. 7. J-V characteristics of CdTe solar cells before and after back contact annealing for MgCl$_2$ (red) and CdCl$_2$ (black) treatment.

ACKNOWLEDGEMENTS

This work has been supported by CONACYT-SENER (Mexico) and CeMIE-Sol (Grant Nos. 207450/P25).

References

[1] "First Solar Achieves World Record 18.6 % Thin Film Module Conversion Efficiency." [Online]. Available: http://investor.firstsolar.com/releasedetail.cfm?ReleaseID=917926.

[2] S. Krum and S. Haymore, "First Solar Achieves Yet Another Cell Conversion Efficiency World Record," *First Sol.*, 2016.

[3] A. Romeo, M. Terheggen, D. Abou-Ras, D. L. Bätzner, F.-J. Haug, M. Kälin, D. Rudmann, and A. N. Tiwari, "Development of thin-film Cu(In,Ga)Se2 and CdTe solar cells," *Prog. Photovoltaics Res. Appl.*, vol. 12, no. 23, pp. 93–111, 2004.

[4] J. D. Major, R. E. Treharne, L. J. Phillips, and K. Durose, "A low-cost non-toxic post-growth activation step for CdTe solar cells," *Nature*, vol. 511, no. 7509, pp. 334–337, Jul. 2014.

[5] D. Menossi, E. Artegiani, A. Salavei, S. Di Mare, and A. Romeo, "Study of MgCl$_2$ activation treatment on the defects of CdTe solar cells by capacitance-voltage, drive level capacitance profiling and admittance spectroscopy techniques," *Thin Solid Films*.

[6] B. L. Xu, I. Rimmaudo, A. Salavei, F. Piccinelli, S. Di Mare, D. Menossi, A. Bosio, N. Romeo, and A. Romeo, "CdCl2 activation treatment: A comprehensive study by monitoring the annealing temperature," *Thin Solid Films*, vol. 582, pp. 110–114, 2015.

[7] M. D. G. Potter, D. P. Halliday, M. Cousins, and K. Durose, "A study of the effects of varying cadmium chloride treatment on the luminescent properties of CdTe/CdS thin film solar cells," *Thin Solid Films*, vol. 361, pp. 248–252, 2000.

[8] D. . Halliday, J. . Eggleston, and K. Durose, "A photoluminescence study of polycrystalline thin-film CdTe/CdS solar cells," *J. Cryst. Growth*, vol. 186, no. 4, pp. 543–549, 1998.

[9] C. Kraft, H. Metzner, M. Hädrich, U. Reislöhner, P. Schley, G. Gobsch, and R. Goldhahn, "Comprehensive photoluminescence study of chlorine activated polycrystalline cadmium telluride layers," *J. Appl. Phys.*, vol. 108, no. 12, p. 124503, 2010.

[10] V. Consonni, G. Feuillet, and S. Renet, "Spectroscopic analysis of defects in chlorine doped polycrystalline CdTe," *J. Appl. Phys.*, vol. 99, pp. 53502–53507, 2006.

[11] J. M. Francou, K. Saminadayar, and J. L. Pautrat, "Shallow donors in CdTe," *Phys. Rev. B*, vol. 41, no. 17, pp. 12035–12046, 1990.

[12] J. M. Figueroa, F. Sánchez-Sinencio, J. G. Mendoza-Alvarez, O. Zelaya, C. Vázquez-López, and J. S. Helman, "Influence of Cd vacancies on the photoluminescence of CdTe," *J. Appl. Phys.*, vol. 60, no. 1, pp. 452–454, 1986.

[13] R. Furstenberg and J. O. White, "Photoluminescence study of the 1.3-1.55 eV defect band in CdTe," *J. Cryst. Growth*, vol. 305, no. 1, pp. 228–236, 2007.

[14] F. A. Abulfotuh, A. Balcioglu, T. Wangensieen, H. R. Moutinho, F. Hassoon, A. Al-Douri, and A. Alnajjar, "Study of the defect levels, electrooptics, and interface properties of polycrystalline CdTe and CdS thin films and their junction" in *26th IEEE Photovoltaic Specialists Conference*, 1997, pp. 451–454.

[15] J. V Li, J. N. Duenow, D. Kuciauskas, A. Kanevce, R. G. Dhere, M. R. Young, and D. H. Levi, "Electrical Characterization of Cu Composition Effects in CdS / CdTe Thin-Film Solar Cells with a ZnTe : Cu Back Contact Preprint," *IEEE J. Photovoltaics*, vol. 3, no. July, pp. 1095–1099, 2013.

[16] D. . Halliday, J. . Eggleston, and K. Durose, "A study of the depth dependence of photoluminescence from thin film CdS/CdTe solar cells using bevel etched samples," *Thin Solid Films*, vol. 322, pp. 314–318, 1998.

[17] I. Caraman, S. Vatavu, G. Rusu, and P. Gaşin, "The luminescence of Cds and CdTe thin films, components of photovoltaic cells," *Chalcogenide Lett.*, vol. 3, no. 1, pp. 1–7, 2006.

[18] I. Rimmaudo, A. Salavei, B. L. Xu, S. Di Mare, and A. Romeo, "Superior stability of ultra thin CdTe solar cells with simple Cu/Au back contact," *Thin Solid Films*, vol. 582, pp. 105–109, 2015.

[19] I. Rimmaudo, A. Salavei, E. Artegiani, D. Menossi, M. Giarola, G. Mariotto, A. Gasparotto, and A. Romeo, "Improved stability of CdTe solar cells by absorber surface etching," *Sol. Energy Mater. Sol. Cells*, vol. 162, 2017.

Characterization of CdTe Photovoltaic Devices Passivated Using Hydrogen Plasma

Amit Munshi[1], Piotr Kaminski[2], Ali Abbas[2], Shiva Tarun Chenna[1,3], Sreeram Chandralal[1,4], John Walls[2], Walajabad Sampath[1]

[1]NGPV (Next Generation PV Center), Department of Mechanical Engineering, Colorado State University, Fort Collins, CO 80523, United States

[2]CREST (Centre for Renewable Energy Systems Technology), Loughborough University, Leicestershire, LE11 3TU, United Kingdom

[3]Department of Electronics and Communication Engineering, Amrita University, Amritapuri, Kerala-690525, India

[4]Department of Mechanical Engineering, Amrita University, Amritapuri, Kerala-690525, India

Abstract — **Thin-film polycrystalline CdTe photovoltaic devices were studied using electrical and material characterization methods to understand the effects of hydrogen plasma passivation treatment. Devices were fabricated using sublimation and were exposed to hydrogen plasma for 10, 20 and 30 minutes. Current density vs voltage measurements were performed to measure the performance of the devices. Capacitance vs voltage graphs showed that dopants are active and the device behaved like a $CdCl_2$ passivated device. Microscopic characterization was performed using SEM and (S)TEM that showed larger grains and more homogenous film coverage as compared to films without passivation suggesting grain growth during H_2 passivation.**

Index Terms — **thin film devices, II-VI semiconductor materials, cadmium compounds, selenium.**

I. INTRODUCTION

CdTe photovoltaics is an important technology in generation of utility scale electricity generation [1]. It has consistently demonstrated low cost for large scale energy generation is expected expand its share in energy market in near future [2]. With optimization in fabrication process, research scale devices with 22.1% efficiency devices have been reported [3]. Commercial module have achieved efficiencies as high as 18.6% [4] while the average efficiency of production modules has been increased from 13.5% to 16.2% between 2014 [5] and 2016 [6]. Passivation treatment plays a critical role in fabrication of high efficiency CdTe thin-film photovoltaic devices and modules . $CdCl_2$ is the most widely used material for passivation. However, other methods such as passivation using $MgCl_2$, $MnCl_2$, NaCl, etc. have been investigated but with limited success [7], [8]. The study presented here is investigation of hydrogen plasma for passivation of defects in CdTe thin-films. These are amongst the first results reported for use of hydrogen plasma for passivation of CdTe.

Hydrogen plasma passivation has been extensively studied for semiconductor materials such as GaAs, AlGaAs and amorphous as well as crystalline silicon. It is understood to cause surface modification in GaAs [9] wafers and passivation

of deep level defects in molecular beam epitaxial (MBE) GaAs[10]. Hydrogen passivation is also used to passivate point defects such as dangling bonds in crystalline silicon [11] and improve passivation of interface between amorphous and crystalline silicon [12].

These studies suggest that hydrogen introduction using plasma passivation process has a significant impact on semi-conductor properties. Some studies have also been performed to understand shallow donor neutralization in *n*-type indium doped CdTe [13]. Photoluminescence characterization of *p*-type undoped CdTe shows hydrogen introduction within CdTe film suggesting hydrogen actively plays a role in passivation of defects in CdTe as in case of other semi-conductor materials [14]. Since CdTe is now a prominent photovoltaic technology and forms a considerable portion of commercial photovoltaic installations, it is important to understand the effect of hydrogen plasma passivation on CdTe photovoltaic devices. Hydrogen incorporation may cause passivation of certain defects that may not be addressed by $CdCl_2$ passivation alone [15]. CdTe is one of the most economical and commercially viable photovoltaic technologies and further improvement in efficiency would pave way for reduction in manufacturing cost of CdTe photovoltaic modules. Some initial results in another study by the authors have shown that using hydrogen plasma passivation along with $CdCl_2$ passivation gives higher device performance than either one alone. This study will be presented as separate literature with additional characterization.

In this study, CdS/CdTe photovoltaic devices were fabricated by sublimation of CdS and CdTe [16]. Hydrogen plasma was generated in-situ without break of vacuum after CdTe deposition. Device performance and efficiency were measured for devices that were passivated using hydrogen plasma and compared against devices fabricated without passivation. SEM and (S)TEM images of the film surface and film cross-section were collected respectively to understand the effect of hydrogen plasma exposure on grain structure. EDS maps were not collected since EDS would not be able to detect presence of hydrogen. Capacitance vs voltage and

978-1-5090-5606-4/17 $31.00 © 2017 IEEE

capacitance vs frequency measurements were performed to understand the effects of exposure to hydrogen plasma on electrical behavior of the device.

II. FABRICATION AND CHARACTERIZATION METHODS

The CdS and CdTe thin films were deposited on NSG TEC 10 soda lime glass that was coated with fluorine-doped tin oxide (FTO), a transparent conducting oxide (TCO). Deposition of these films was performed in a continuous inline sublimation tool optimized at Colorado State University for fabrication of thin-film CdTe photovoltaic devices [16]. The load-lock section of the sublimation tool had a protruding plasma source that was used to expose the films to hydrogen plasma after deposition of CdTe films. Thickness of CdS film was ~130 nm while CdTe absorber layer thickness was measured to be ~2.4 µm. Flow of hydrogen gas was controlled using a mechanical valve and the pressure in the plasma source was maintained at 800 mTorr. The electrical parameters used to ignite hydrogen plasma were controlled using a Glassman High Voltage power controller and the parameters were maintained at 15 mA current and 450 V voltage. The films were exposed to plasma for 10 and 25 minutes for different substrates.

TABLE I
ELECTRICAL PERFORMANCE OF CDTE DEVICES PASSIVATED USING HYDROGEN PLASMA FOR DIFFERENT AMOUNT OF TIME

Hydrogen plasma exposure time	V_{OC} [mV]	J_{SC} [mA/cm^2]	% Fill-factor	% η
No exposure	460	1.2	40.8	0.22
5 minute	493	2.7	37.0	0.50
10 minute	549	4.8	58.2	1.52
15 minute	511	4.2	36.5	0.78
20 minute	537	10.8	38.7	2.24
25 minute	526	11.7	39.0	2.41

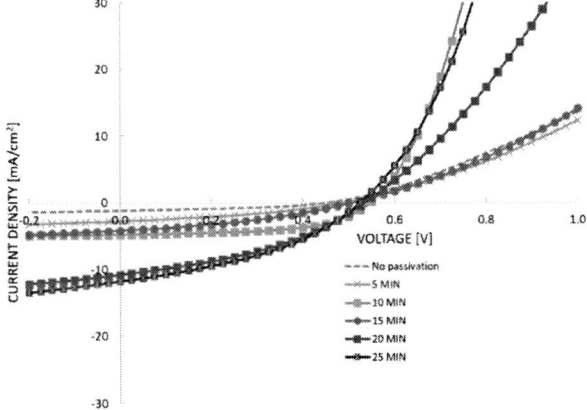

Fig. 1. J-V measurement of CdS/CdTe devices with varying hydrogen plasma exposure time

Thereafter, the films were heated to ~140°C and CuCl was deposited on the film surface for 110 seconds. This was followed by 220 seconds of annealing at 220°C. This process step was performed to form a Cu back contact [17]. Carbon and nickel paint in polymer binder where then sprayed on these films to form the back electrode. The cells were then delineated using a mask and bead blasting to fabricate 25 small scale devices on the substrate. The devices had an area of ~0.65 cm^2.

For electrical characterization, open-circuit voltage [V_{OC}], short-circuit current [J_{SC}], fill-factor and percentage efficiency data were collected using J-V measurements.

In-depth microstructure characterization of the CdTe film was carried out using Transmission Electron Microscopy (TEM) and High Resolution Transmission Electron Microscopy (HRTEM). TEM samples were prepared using Focused Ion Beam (FIB) milling using dual beam FEI Nova 600 Nanolab. A standard in-situ lift out method was used to prepare the cross-section sample through the film stack into the glass substrate. A platinum overlay was deposited on the top of the film to define the area of interest on surface of the sample, homogenize the final thinning of the samples and to avoid damage to the CdTe film surface from the ion beam. STEM bright-field images and high resolution TEM images were collected using a FEI Tecnai F20 (S)TEM operating at 200 kV.

Capacitance-Voltage characteristics were obtained using Kyesight E4990A Impedance Analyzer. The sample bias was swept from -2V to 1V at 20mV step, the capacitance measurements were taken using 500 kHz frequency.

III. ELECTRICAL CHARACTERIZATION

As can be observed from current density vs voltage measurements (figure 1), as the films are exposed to hydrogen plasma for longer periods of time the short-circuit current improves. However, best fill-factor of 58.2% and open-circuit voltage of 549 mV is seen for 10 minutes of passivation time. Longer passivation time for these films show reduction in fill-factor as well as open-circuit voltage. This may be caused by surface damage induced by long exposure to hydrogen plasma. Damage to CdTe surface at long passivation times requires to be verified using SEM.

As deposited CdS/CdTe hetero structures not activated using CdCl$_2$ did not provide response. When the material was H$_2$ passivated the C-V response showed characteristics typical for a working solar cell. Mott-Schotkky characteristics of reference cell and H$_2$ passivated CdTe are shown in figure 2. The V_{bi} of H$_2$ passivated material was higher (~0.67V) than for working solar cell (~0.62V).

Doping profiles were extracted from the C-V curves. The H$_2$ treated device showed ~4.2 10^{14} cm^{-3}, the working cell showed ~7.3 10^{14} cm^{-3}. Calculated doping profiles are plotted in figure 3.

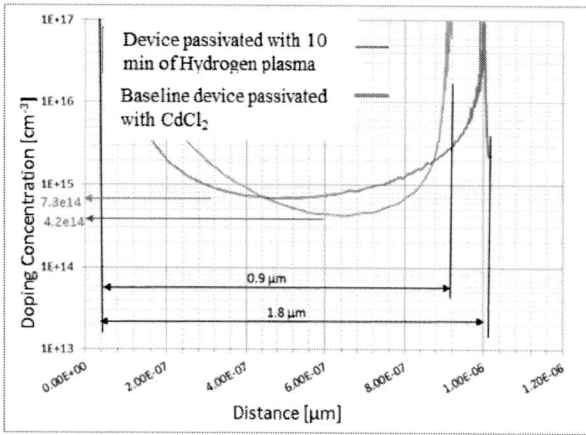

Fig. 2. Doping profiles extracted from Mott-Schottky characteristics for a device with and without hydrogen plasma passivation

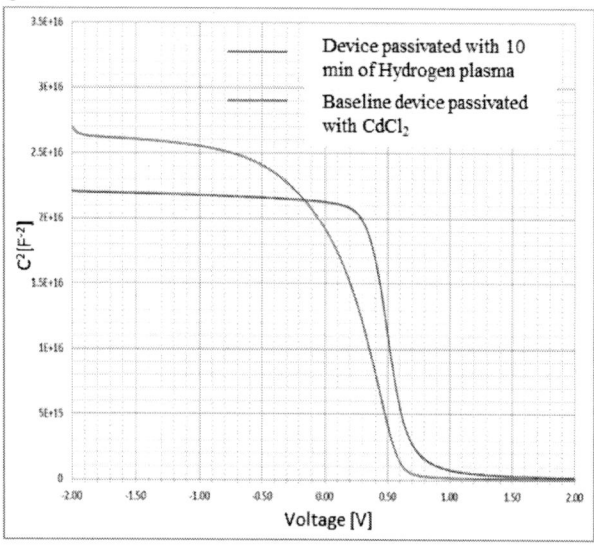

Fig. 3. Mott-Schotkky characteristics of devices with and without hydrogen plasma passivation

IV. MATERIALS CHARACTERIZATION

Scanning electron microscopy (SEM) was used to study the surface of CdTe films after exposure to hydrogen plasma passivation while scanning transmission electron microscopy (STEM) was performed to study the cross-section of these CdS devices. SEM images of the surface showed a more homogenous coverage of CdTe grains with minimal voids after hydrogen plasma passivation (figure 4). These features were similar to the films that are passivated using $CdCl_2$ on similar films.

STEM cross-section images (figure 5) showed stacking faults present in CdTe grains. The grains appeared to small near the interface of CdS and CdTe. These are typical features observed in CdTe grains not passivated using $CdCl_2$. The

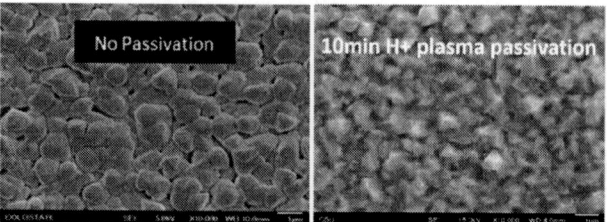

Fig. 4. Surface SEM image of as deposited CdTe film (left) and film surface after exposure to hydrogen plasma for 10 minutes

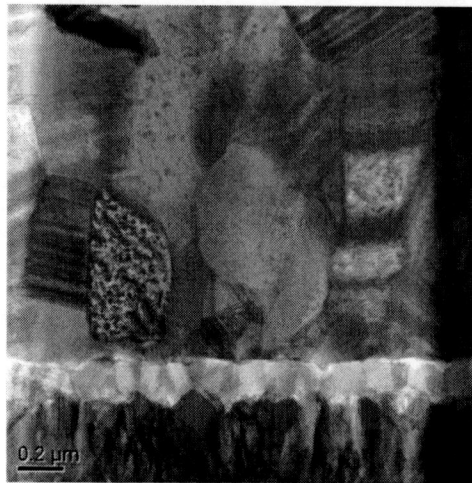

Fig. 5. Cross-section STEM image of CdS/CdTe film after 10 minutes of exposure to hydrogen passivation

grain boundaries appeared to be distinct and there were no voids visible in the film. The growth of CdTe was also conformal on CdS. There were no visible signs of material diffusion and the CdS/CdTe interface also appeared to be distinct and abrupt.

V. DISCUSSION AND FUTURE WORK

Exposing the CdTe films to hydrogen plasma shows improvement in device performance. This improvement in open-circuit voltage as well as short-circuit current is observed with increase in plasma exposure. The fill-factor improves up to 10 minutes of hydrogen plasma exposure and then begins to reduce. This suggests that for given conditions of hydrogen plasma, 10 minutes of hydrogen plasma passivation optimum. Very high device performance is now observed with hydrogen plasma passivation. But hydrogen plasma is playing a critical role in passivation of certain defects in CdTe film cannot be denied.

$CdCl_2$ passivation treatment removed line defects such as stacking faults from CdTe thin-film as it improve device performance [18]–[23]. With hydrogen plasma the stacking faults are not removed but device performance still appears to be improving. This indicates a possibility that hydrogen plasma may be passivating defects within the CdTe thin-film that are not passivated by $CdCl_2$ passivation. It would be important to verify this hypothesis experimentally by fabrication of CdTe devices that undergoes CdTe as well as hydrogen plasma passivation. The proof for this hypothesis would show that a combination of $CdCl_2$ passivation and hydrogen plasma passivation would give device performance better than performance that may be achieved by either one of the treatments alone.

In addition to this, it would also be important to verify the incorporation of hydrogen in the CdTe thin-film. Presence of hydrogen in CdTe film would confirm that hydrogen plasma passivation plays an active role in passivation of certain defects and improvement in device performance. Fundamentally, electron and X-ray based characterization methods cannot identify presence of hydrogen and helium. Extreme ultraviolet laser ablation mass spectroscopy may be able to overcome this limitation and be able to identify and verify the incorporation of hydrogen incorporation in the CdTe thin-film [24], [25].

Thus the logical next step for this study would be to fabricate and characterize CdTe thin-films passivated with $CdCl_2$ as well as hydrogen plasma. In addition, physical incorporation of hydrogen in CdTe thin-films caused by hydrogen plasma passivation must be verified.

VI. CONCLUSIONS

Current density vs voltage measurements show that hydrogen passivation has a positive effect on CdS/CdTe device performance. C-F and C-V measurements also suggest that these devices have characteristics of working devices. However, cross-section electron image shows presence of stacking faults that persist post hydrogen plasma passivation treatment. This suggests hydrogen plasma exposure may be responsible for affecting other point defects such as dangling bonds as well as may be passivating the surface that results in improvement of device performance.

ACKNOWLEDGEMENTS

The CSU authors thank support from NSF's Accelerating Innovation Research, DOE's SunShot and NSF's Industry/University Cooperative Research Center programs. The Loughborough authors are grateful to EPSRC for funding through the EPSRC Supergen SuperSolar Hub.

REFERENCES

[1] A. Munshi and W. Sampath, "CdTe Photovoltaics for Sustainable Electricity Generation," *J. Electron. Mater.*, 2016.

[2] N. M. Haegel, R. Margolis, T. Buonassisi, D. Feldman, A. Froitzheim, R. Garabedian, M. Green, S. Glunz, H.-M. Henning, B. Holder, I. Kaizuka, B. Kroposki, K. Matsubara, S. Niki, K. Sakurai, R. A. Schindler, W. Tumas, E. R. Weber, G. Wilson, M. Woodhouse, and S. Kurtz, "Terawatt-scale photovoltaics: Trajectories and challenges," *Science (80-.).*, vol. 356, no. 6334, 2017.

[3] D. E. D. Green Martin A., Emery Keith, Hishikawa Yoshihiro, Warta Wilhelm, "Solar cell efficiency tables (Version 45)," *Prog. Photovoltaics*, vol. 23, no. 1, pp. 1–9, 2015.

[4] W. Eric, "Exclusive: First Solar's CTO Discusses Record 18.6% Efficient Thin-Film Module," *Greentech Media*, 2015. [Online]. Available: http://www.greentechmedia.com/articles/read/Exclusive-First-Solars-CTO-Discusses-Record-18.6-Efficient-Thin-Film-Mod.

[5] S. K. David Brady, Steve Haymore, "First Solar, INC. Announces First Quarter 2014 Financial Results," Tempe, AZ, 2014.

[6] S. K. Steve Haymore, "First Solar, Inc. 2016 First Quarter Financial Results," Tempe, AZ, 2016.

[7] J. Drayton, Jennifer; Geisthardt, Russell; Raguse, John; Sites, "Metal chloride passivation treatments for CdTe solar cells," in *MRS Proceedings*, 2013, p. Vol. 1538, pp. 269–274.

[8] J. D. Major, R. E. Treharne, L. J. Phillips, and K. Durose, "A low-cost non-toxic post-growth activation step for CdTe solar cells," *Nature*, vol. 511, no. 7509, pp. 334–337, 2014.

[9] E. Yoon, R. A. Gottscho, V. M. Donnelly, and H. S. Luftman, "GaAs surface modification by room-temperature hydrogen plasma passivation," *Appl. Phys. Lett.*, vol. 60, no. 21, pp. 2681–2683, 1992.

[10] W. C. Dautremont-Smith, J. C. Nabity, V. Swaminathan, M. Stavola, J. Chevallier, C. W. Tu, and S. J. Pearton, "Passivation of deep level defects in molecular beam epitaxial GaAs by hydrogen plasma exposure," *Appl. Phys. Lett.*, vol. 49, no. 17, pp. 1098–1100, 1986.

[11] J. L. Benton, C. J. Doherty, S. D. Ferris, D. L. Flamm, L. C. Kimerling, and H. J. Leamy, "Hydrogen passivation of point defects in silicon," *Appl. Phys. Lett.*, vol. 36, no. 8, pp. 670–671, 1980.

[12] A. Descoeudres, L. Barraud, S. De Wolf, B. Strahm, D. Lachenal, C. Guérin, Z. C. Holman, F. Zicarelli, B. Demaurex, J. Seif, J. Holovsky, and C. Ballif, "Improved amorphous/crystalline silicon interface passivation by hydrogen plasma treatment," *Appl. Phys. Lett.*, vol. 99, no. 12, pp. 1–4, 2011.

[13] S. Gurumurthy, H. L. Bhat, B. Sundersheshu, R. K. Bagai, and V. Kumar, "Shallow donor neutralization in CdTe:In by atomic hydrogen," *Appl. Phys. Lett.*, vol. 2424, p. 2424, 1995.

[14] J. Hamann, D. Blass, C. Casimir, T. Filz, V. Ostheimer, C. Schmitz, H. Wolf, T. Wichert, A. Burchard, M. Deicher, and R. Magerle, "Hydrogen-related photoluminescence in CdTe," *Appl. Phys. Lett.*, vol. 72, no. 5, pp. 554–556, 1998.

[15] J. Moseley, W. K. Metzger, H. R. Moutinho, N. Paudel, H. L. Guthrey, Y. Yan, R. K. Ahrenkiel, and M. M. Al-Jassim, "Recombination by grain-boundary type in CdTe," *J. Appl. Phys.*, vol. 118, no. 2, p. 25702, 2015.

[16] D. E. Swanson, J. M. Kephart, P. S. Kobyakov, K. Walters, K. C. Cameron, K. L. Barth, W. S. Sampath, J. Drayton, and J. R. Sites, "Single vacuum chamber with multiple close space sublimation sources to fabricate CdTe solar cells," *J. Vac. Sci. Technol. A Vacuum, Surfaces, Film.*, vol. 34, no. 2, p. 21202, 2016.

[17] R. A. Barth, Kurt L; Sampath, Walajabad S; Enzenroth, "Apparatus and Processes for the Mass Production of Photovoltaic Modules," 2002.

[18] S. Yoo, K. T. Butler, A. Soon, A. Abbas, J. M. Walls, and A. Walsh, "Identification of critical stacking faults in thin-film CdTe solar cells," *Appl. Phys. Lett.*, vol. 105, no. May, p. 62104, 2014.

[19] W. S. S. Ali Abbas, Geoff D. West, Jake W. Bowers, Piotr M. Kaminski, B. Maniscalco, John M. Walls, Kurt L. Barth, "Cadmium Chloride Assisted Re-Crystallization of CdTe: The

Effect of Varying the Annealing Time," *MRS Proc.*, vol. 1638, 2014.

[20] A. Munshi, A. Abbas, J. Raguse, K. Barth, W. S. Sampath, and J. M. Walls, "Effect of varying process parameters on CdTe thin film device performance and its relationship to film microstructure," pp. 1643–1648, 2014.

[21] W. S. S. Ali Abbas, Geoff D. West, Jake W. Bowers, Piotr M. Kaminski, John M. Walls, Kurt L. Barth, "Cadmium Chloride Assisted Re-Crystallisation of CdTe: The Effect on the CdS Window Layer," *MRS Proc.*, vol. 1738, 2015.

[22] C. Li, Y. Wu, J. Poplawsky, T. J. Pennycook, N. Paudel, W. Yin, S. J. Haigh, M. P. Oxley, A. R. Lupini, M. Al-Jassim, S. J. Pennycook, and Y. Yan, "Grain-Boundary-Enhanced Carrier Collection in CdTe Solar Cells," *Phys. Rev. Lett.*, vol. 112, no.

15, p. 156103, 2014.

[23] A. Munshi, "Investigation of Processing, Microstructures and Efficiencies of Polycrystalline CdTe Photovoltaic Films and Devices," Colorado State Univeristy, 2016.

[24] I. Kuznetsov, J. Filevich, F. Dong, M. Woolston, W. Chao, E. H. Anderson, E. R. Bernstein, D. C. Crick, J. J. Rocca, and C. S. Menoni, "Three-dimensional nanoscale molecular imaging by extreme ultraviolet laser ablation mass spectrometry," *Nat. Commun.*, vol. 6, p. 6944, 2015.

[25] T. Green, I. Kuznetsov, D. Willingham, B. E. Naes, G. C. Eiden, Z. Zhu, W. Chao, J. J. Rocca, C. S. Menoni, and A. M. Duffin, "Characterization of extreme ultraviolet laser ablation mass spectrometry for actinide trace analysis and nanoscale isotopic imaging," vol. 32, no. 6, 2017.

Group-V doping impact on Cd-rich CdTe single crystals grown by traveling-heater method

Akira Nagaoka[1,3], Kenji Yoshino[2], Yoshitaro Nose[3], Darius Kuciauskas[4], and Michael A. Scarpulla[1]

[1]Materials Science & Engineering Department, University of Utah, Salt Lake City, UT 84112, USA
[2]Department of Applied Physics and Electronic Engineering, University of Miyazaki, Miyazaki 889-2192, Japan
[3]Department of Materials Science and Engineering, Kyoto University, Kyoto 606-8501, Japan
[4]National Renewable Energy Laboratory, Golden, CO 80401, USA

Abstract — **Cd-rich CdTe single crystals were grown from Cd solvent in the traveling-heater method (THM) rather than the more common Te solvent used for high-resistivity CdTe. The growth process from Cd solution in terms of the solid-liquid interface shape and group-V doping under Cd-rich condition were investigated. Stable, flat solid-liquid interfaces can be obtained by optimizing THM furnace temperature profile. An apparent doping limit in the 10^{16} to 10^{17} cm^{-3} range which scales weakly with increasing group-V element concentrations in the Cd-rich CdTe samples were indicated from capacitance-voltage measurement. The two-photon excitation time-resolved photoluminescence revealed a bulk lifetime of 20-40 ns.**

I. INTRODUCTION

Cadmium telluride (CdTe) based photovoltaic devices have proven to be one of the most important solar energy manufacturing technologies for producing lower cost and higher efficiency. Recently, power conversion efficiency more than 22 % was reported for polycrystalline CdTe cells [1]. However despite several decades of research, this efficiency is still well below the theoretical limits for CdTe, especially in terms of open-circuit voltage (V_{OC}), leaving many opportunities for further optimizing materials properties and device performance.

Generally, most work on CdTe based solar cell devices focused on polycrystalline thin films deposited on substrates using superstrate structure. Polycrystalline thin film devices utilize Cu and Te-rich conditions to reach 10^{14}-10^{15} cm^{-3} doping by Cu substituting on Cd site (Cu_{Cd}), Cd vacancy (V_{Cd}), and Cu_{Cd} + Cl substituting on Te site (Cl_{Te}) (A-center) with minority electron lifetime 1-10 ns – values which have held V_{OC} below 900 meV for decades [2-4]. Considering from this state of the art, the most important variables to optimize in devices with pure CdTe absorber layers for maximizing V_{OC} are the *p*-type doping concentration and minority carrier lifetime. Increasing the doping will lower the Fermi level and thus hole quasi-Fermi level allows V_{OC} to increase by 60 meV/decade of doping. Increasing the minority carrier lifetime will lead to increases in both short-circuit current (J_{SC}) and V_{OC}. It is critical to experimentally understand and manipulate point defect physics in CdTe. Groups-I and -V elements on (respectively) the Cd and Te sites have been reported

experimentally for extrinsic *p*-type doping. The understanding of *p*-type doping in CdTe and its coupling to minority carrier lifetime via recombination has been significantly advanced recently by computational efforts [5, 6]. Recently, group-V doping under Cd-rich condition has attracted much attention in the point of carrier lifetime and *p*-type doping. The combination of experimental and theoretical studies reveals that Cd-rich condition leads long minority carrier lifetimes exceeding 20 ns due to suppression of Te on Cd antisite (Te_{Cd}) recombination centers in Cd-rich conditions [7]. In terms of increasing hole concentration, the phosphorous P-doped CdTe single crystal solar cells annealed in Cd-rich conditions exhibit hole concentrations exceeding 10^{17} cm^{-3} and achieve V_{OC} >1 V, in which P substituting on Te sites (P_{Te}) is dominant [8]. Group-V doping on the Te site should be in general more stable than group-I doping – these smaller cations tend to diffuse readily. Considering their activation energy N, P, and As are good candidates for *p*-type doping under Cd-rich conditions [5].

In this work we investigate the influence of THM growth conditions using Cd solvent on the liquid-solid interface shape and crystal quality. Then we investigate doping with group-V elements substituting on Te sites (P_{Te} and As_{Te}) in Cd-rich condition. These findings will help in the understanding of fundamentals of group-V doping in CdTe to allow eventual use of single crystal wafers as either highly-doped substrates or in transferring group-V doping to polycrystalline layers for thin film photovoltaics.

II. EXPERIMENTAL PROCEDURE

Feed polycrystalline ingots of CdTe were prepared by melting and reacting elemental Cd (99.999%) and Te (99.999%) shot at 1100 °C under high vacuum of ~10^{-6} Torr for 24 h. The feed polycrystalline CdTe ingot, Cd solvent, As (99.999%) or Cd_3P_2 (99.999%) were loaded into a carbon-coated quartz ampoule with 10 mm inner diameter and 2 mm wall thickness. The ampoule was sealed under high vacuum (10^{-6} Torr) and then passed slowly through the hot zone of the THM furnace for single crystal growth. Grown crystals were cut with a diamond blade and polished mechanically with 0.01

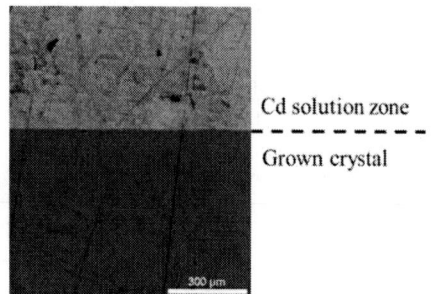

Cd solution zone

Grown crystal

Fig. 1 Cross-sectional view of the flat interface of Cd/CdTe.

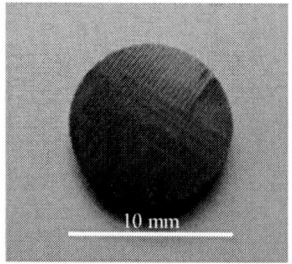

Fig. 2 Cd-rich CdTe single crystal grown from Cd solvent.

μm Al_2O_3 powder and then etched with a 5 % Br_2/Methanol solution for 5 minutes to remove saw and polishing damage before conducting measurements. The cross-sectional morphology and average composition of the samples were analyzed by optical microscopy and a scanning electron microscope equipped with energy dispersive X-ray spectroscopy (EDS). The As dopant and impurity concentration in the grown crystals was determined by inductively coupled plasma atomic emission spectroscopy (ICP-AES; SII Nano Technology SPS3520UV). The electrical properties of all samples were measured using capacitance-voltage (CV; Keithley 4200-SCS) measurement at room temperature. Schottky diode structures of 1 mm diameter Al contacts with Ni-W Ohmic back contact were fabricated by DC sputtering through a shadow mask for CV measurements. The thickness of contacts was 200–300 nm. The bulk lifetime was determined by using two-photon excitation time-resolved photoluminescence (2PE-TRPL), which was performed at National Renewable Energy Laboratory (NREL).

III. RESULT AND DISCUSSIONS

A. Growth of Cd-rich THM

We selected to use 50 mol% Cd + 50 mol% CdTe considering the growth temperature and solubility in Cd solvent. The temperature gradient along the THM furnace axis is controlled by three coil heaters (upper, main, and bottom). The zone temperatures are upper 900 °C, main 950 °C, and bottom 750 °C. The temperature gradient between the main and bottom heaters (which controls the width of the

TABLE I
IMPURITY LEVELS IN NON- AND 10^{17} /cm³ As-DOPED SAMPLES.

Elements	Non-doped	As-doped
As	-	8.6×10^{17}
C	6.2×10^{15}	8.1×10^{15}
O	6.7×10^{15}	6.1×10^{15}
Al	3.6×10^{15}	2.8×10^{15}
Si	4.8×10^{15}	7.0×10^{15}

liquid zone) was about 40 °C/cm which yields a cooling rate of 1.4×10^{-4} °C/s for the 3 mm/day growth rate. Cd solvent/CdTe single crystal interface shape affects the quality of the grown crystals. In case of a concave liquid/solid interface, new grains grow towards the center of the growing crystal resulting in polycrystalline growth. On the other hand, a convex interface causes any wall-nucleated grains to grow outwards preventing the polycrystalline breakdown of the crystal. A nearly flat interface is not as stable with regards to polycrystalline breakdown of the growth front however it can still result in the growth of large single crystals. A nearly flat, smooth interface was obtained and observed from optical microscopy in Fig. 1 which shows a cross section of a boule that was stopped during growth. A large single crystal more than 5 mm long was obtained from the flat smooth interface which is favorable for single crystal growth as seen in Fig. 2.

B. Characterization of samples

The uniformity of the Cd/Te composition along the growth direction can be observed from the growth tip at 5 mm intervals except for the end (tail) region by using EDS measurement. A crystal grown by Bridgeman with the smallest deviation from stoichiometry (as determined by minimal 2nd phase precipitaiton) was obtained and used to calibrate the EDS measurements. Within the errors of the EDS measurement, the CdTe ingots appear to be spatially homogeneous and slightly Cd-rich within +0.5 at.%. The concentrations of As and the impurities that exceed 100 ppb as determined by ICP-AES measurements shown for two exemplary crystals in Table I. The influential impurity such as Cu and Cl in CdTe solar cell were measured at less than 100 ppb [2-4].

Figure 3 shows group-V (As and P elements) doping profiles in CdTe samples obtained from CV measurements. The acceptor concentration N_A data indicate a U-shape characteristic for group-V doped Cd-rich CdTe samples which may due to different factors such as non-Ohmic contacts and the presence of deep level defects [9]. The back diode effect is observed under forward bias due to the absence of a perfect Ohmic contact for CdTe, which means measured capacitance to be artificially reduced leading to an apparent increase in the acceptor concentration under a forward bias [10]. In reverse bias, the apparent increase in acceptor concentration can be due to reasons such as the presence of deep acceptor levels.

978-1-5090-5606-4/17 $31.00 © 2017 IEEE

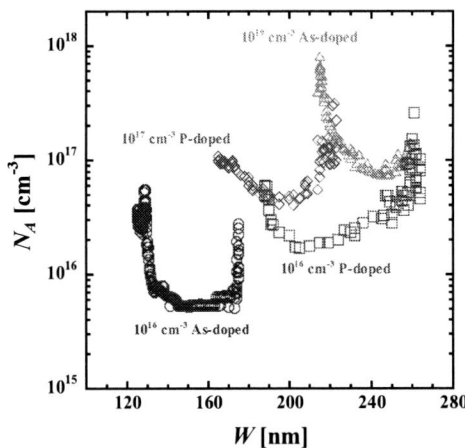

Fig. 3 Depth profiles of the net acceptor concentration in the each As and P-doped CdTe estimated from CV measurement. The data sets are labeled according to the nominal amounts of As and P added to the ampoule.

The acceptor concentration N_A increases with increasing As and P dopant concentrations but remains within the 10^{16}-10^{17} cm^{-3} ranges. The activation efficiences of As and P doping extracted from CV measurement as a function of dopant concentrations are about 5-40 % in the range of less than 10^{18} cm^{-3}. The activation energies of As$_{Te}$ and P$_{Te}$ acceptors indicate that nearly 100% will be ionized at room temperature calculated from thermal activation law, thus we infer that higher doping causes self-compensation by AX centers [5, 6]. This causes the doping efficacy to decrease to less than 1 % above 10^{18} cm^{-3} of As concentration. A strategy of quenching was proposed for increasing the hole concentration further [11].

The bulk lifetime values are 20-40 ns measured by 2PE-TRPL, which is higher than most lifetime values reported on polycrystalline thin-film samples [2, 3]. This result supports Cd-rich condition is better for long lifetime due to suppression of Te$_{Cd}$ recombination centers and increasing of p-type doping by As$_{Te}$ and P$_{Te}$ defects.

IV. CONCLUSION

We report on the growth of non-doped and group-V doped Cd-rich CdTe single crystals using the traveling heater method with metallic Cd solvent. Stable, flat Cd/CdTe interfaces were obtained by optimizing the THM furnace temperature profile for our particular geometry. The composition of the non-doped CdTe single crystals were homogeneous along the growth and radial directions and the crystals were found to be slightly Cd-rich. CV measurement at room temperature showed an apparent doping limit in the 10^{16} to 10^{17} cm^{-3} range which scales weakly with increasing As and P dopants concentration in the Cd-rich CdTe samples, and the activation efficiency

indicates about 5-40 % less than 10^{18} cm^{-3} of dopant concentration. The longer bulk time of 20-40 ns could be obtained by Cd-rich condition and group-V doping.

ACKNOWLEDGMENT

A. N. is supported by JSPS Postdoctoral Fellow for Research Abroad. This work was supported by Department of Energy through the Bay Area Photovoltaic Consortium under Award Number DE-EE0004946. We thank Prof. Kelvin Lynn and Dr. Santosh Swain for supplying the stoichiometric CdTe sample grown by Bridgeman for EDS standard.

REFERENCES

[1] M. A. Green, K. Emery, Y. Hishikawa, W. Warta, and E. D. Dunlop, "Solar cell efficiency tables (version 48)," *Progress in Photovoltaics: Research and Applications*, vol. 24, pp. 905-913, 2016.

[2] J. Sites, and J. Pan, "Strategies to increase CdTe solar-cell voltage," *Thin Solid Films*, vol. 515, pp. 6099-6102, 2007.

[3] T. A. Gessert, S. H. Wei, J. Ma, D. S. Albin, R. G. Dhere, J. N. Duenow, D. Kuciauskas, A. Kanevce, T. M. Barnes, J. M. Burst, W. L. Rance, M. O. Reese, and H. R. Moutinho, "Research strategies toward improving thin-film CdTe photovoltaic devices beyond 20% conversion efficiency," *Solar Energy Materials and Solar Cells*, vol. 119, pp. 149-155, 2013.

[4] J. H. Yang, W. J. Yin, J. S. Park, W. Metzger, and S. H. Wei, "First-principles study of roles of Cu and Cl in polycrystalline CdTe," *Journal of Applied Physics*, vol. 119, 045104 1-7, 2016.

[5] S. H. Wei, and S. B. Zhang, "Chemical trends of defect formation and doping limit in II-VI semiconductors: The case of CdTe," *Physical Review B*, vol. 66, 155211 1-10, 2002.

[6] S. H. Wei, and S. B. Zhang, "First-Principle Study of Doping Limits of CdTe," *Physical Status Solidi (b)*, vol. 229, pp. 305-310, 2002.

[7] J. Ma, D. Kuciauskas, D. Albin, R. Bhattacharya, M. Reese, T. Barnes, J. V. Li, T. Gessert, and S. H. Wei, "Dependence of the Minority-Carrier Lifetime on the Stoichiometry of CdTe Using Time-Resolved Photoluminescence and First-Principles Calculations," *Physical Review Letters*, vol. 111, 067402 1-5, 2013.

[8] J. M. Brust, J. N. Duenow, D. S. Albin, E. Colegrove, M. O. Reese, J. A. Aguiar, C. S. Jiang, M. K. Patel, M. M. Al-Jassim, D. Kuciauskas, S. Swain, T. Ablekim, K. G. Lynn, and W. K. Metzger, "CdTe solar cells with open-circuit voltage breaking the 1 V barrier," *Nature Energy*, vol. 1, 16015 1-7, 2016.

[9] J. V. Li, A. F. Halverson, O. V. Sulima, S. Bansal, J. M. Burst, T. M. Barnes, T. A. Gessert, and D. H. Levi, "Theoretical analysis of effects of deep level, back contact, and absorber thickness on capacitance-voltage profiling of CdTe thin-film solar cells," *Solar Energy Materials and Solar Cells*, vol. 100, pp. 126-131, 2012.

[10] A. Niemegeers, and M. Burgelman,"Effects of the Au/CdTe back contact on *IV* and *CV* characteristics of Au/CdTe/CdS/TCO solar cells," *Journal of Applied Physics*, vol. 81, pp. 2881-2886, 1997.

[11] J. H. Yang, W. J. Yin, J. S. Park, J. Burst, W. K. Metzger, T. Gessert, T. Barnes, and S. H. Wei, "Enhanced p-type dupability of P and As in CdTe using non-equilibrium thermal processing," *Journal of Applied Physics*, vol. 118, 025102 1-6, 2015.

Band-Gap Engineering in $Cu_2ZnSn(S,Se)_4$ Solar Cells by Post-Sulphurization of Selenized Absorber Layers

Markus Neuwirth[1], Elisabeth Seydel[1], Heinz Kalt[1] and Michael Hetterich[2]

[1]Institute of Applied Physics, Karlsruhe Institute of Technology (KIT), 76131 Karlsruhe, Germany
[2]Light Technology Institute, Karlsruhe Institute of Technology (KIT), 76131 Karlsruhe, Germany

Abstract — A post-sulphurization process for kesterite $Cu_2ZnSn(S,Se)_4$ solar cell absorber layers is developed in order to control the sulphur concentration and thus the band gap of the resulting solar cell absorber layers. It is shown that at sulphurization temperatures above 450°C an efficient incorporation of S takes place. Varying temperature between 450°C and 600°C and hold time between 5 min and 1 hour the sulphur content can accurately be controlled, thus allowing band-gap tuning in the complete range between about 1 eV and 1.5 eV.

Index Terms — Kesterite, CZTSSe, Band-gap Engineering

I. INTRODUCTION

Kesterite $Cu_2ZnSn(S,Se)_4$ (CZTSSe) is a promising material system for thin-film photovoltaics due to its non-toxic and earth-abundant constituents. One of its very interesting and useful features is the tunability of the band gap using either the incorporation of Ge as substituent of Sn [1]-[3] or the chalcogenide ratio [4]. In this work we will concentrate on the latter one. Pure selenide $Cu_2ZnSnSe_4$ (CZTSe) has a band gap of $E_g = 1.0$ eV, whereas the substitution of Se with S increases the band gap up to $E_g = 1.5$ eV for the pure sulphide Cu_2ZnSnS_4 (CZTS) [5]. With this property the material system of CZTSSe offers a band-gap range that not only covers the whole favorable spectral range [6] but also offers the possibility of being used in tandem solar cells [7].

There are several ways to control and adjust the Se to S ratio in the CZTSSe absorber. The most commonly used process is to fabricate CZTS precursors and anneal them in a selenium atmosphere [8]-[10]. With this process, the up-to-date record efficiency cell was produced that showed a band gap of 1.13 eV and an efficiency of 12.6% [11]. This process enables to achieve a band gap in-between the pure selenide and pure sulphide case, but depending on process specifics there is almost no sulphur left after selenization [9]. One way to increase the sulphur content is to add S or SnS during the selenization step ending up in a one-step sulfo-selenization [12]. Another approach is to first sulphurize then selenize the sample [13] or the other way around [14]. Sun *et al.* focused on the effect of process parameters of the selenization process as well as the post-sulphurization process on the evolution of phases, the crystallinity and the chalcogenide ratio $[S]/([S]+[Se])$ of $Cu_2ZnSn(S,Se)_4$ films concluding that the process temperature as well as the sulphur addition during the sulphurization are very critical parameters for high quality

thin-films [14]. In this work we take a similar approach of first selenizing our CZTS precursors and post-sulphurizing them at various temperatures for several process hold times to study the possibility of specifically adjusting $[S]/([S]+[Se])$ and thus the band gap of our CZTSSe absorber layers.

II. EXPERIMENTAL DETAILS

A. Sample Fabrication

The samples are fabricated in a two-step process. First, a precursor is fabricated via doctor-blade coating of a metal salt solution onto a molybdenum-covered soda-lime glass. The solution is based on dimethyl sulfoxide (DMSO) and contains copper acetate, zinc chloride, tin chloride and thiourea. For the doctor-blading, a Zehntner ZAA 2300 is used. This process takes place in air. The wet precursor layer is dried at 300°C for 2 min in nitrogen atmosphere. The doctor-blading is repeated six times to reach the desired precursor thickness of around 1 µm. This fabrication step is also described in more detail in Refs. [8],[9]. The dry precursor is then annealed in a selenium atmosphere to form the selenized absorber layer. This process takes place in a home-made oven in a quasi-closed graphite box, which contains the sample and two small ceramic bowls that are filled with selenium pellets. First, the oven is evacuated and then flushed with nitrogen to reach a pressure of 250 mbar. The annealing parameters are as follows: The oven is heated with a heating ramp of 140°C/min up to a temperature of 540°C, which is held for 10 min. Afterwards the oven is naturally cooled down to room temperature. With this process we end up at a sulphur content of our absorber layers of $[S]/([S]+[Se]) \approx 0.1$.

The post-sulphurization of the selenized absorber layers is performed in another but similar oven to ensure consistent process conditions. Therefore, the oven is heated with a heating ramp of 140°C/min to several specific process temperatures. Both the temperature and the time for which the temperature is held are used as process parameters. Afterwards the oven is naturally cooled down to room temperature. Elementary sulphur is used to provide a saturated sulphur atmosphere during the process.

978-1-5090-5606-4/17 $31.00 © 2017 IEEE

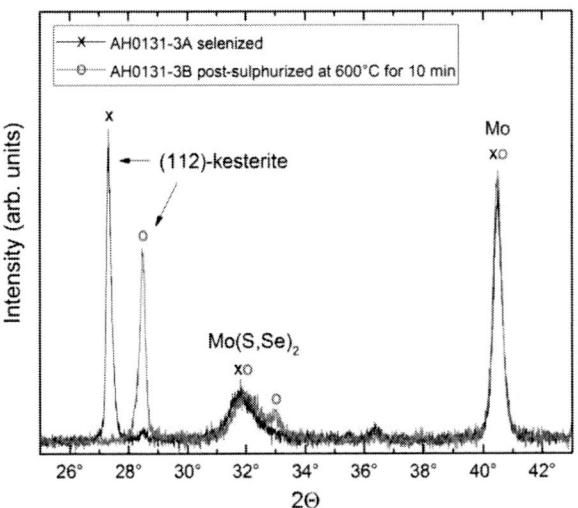

Fig. 1. X-ray diffractograms of a sample before (black, x) and after (red, o) post-sulphurization.

B. Sample Characterization

The sulphur content of our absorber layers is received by utilizing Vegard's law. First a ω-2Θ X-ray diffractogram of the samples before and after post-sulphurization is taken. For this a Bruker AXS D8 DISCOVER with a copper anode ($\lambda_{K\alpha}$ = 0.15406 nm) is used. The peak position of the (112)-kesterite reflex is extracted via a Gaussian fit.

By using the distance of lattice planes d_{hkl} for the tetragonal kesterite structure

$$d_{hkl} = \frac{1}{\sqrt{\frac{h^2+k^2}{a^2}+\frac{l^2}{c^2}}} \quad (1)$$

and Bragg's law we obtain

$$2\theta = 2\arcsin\left(\frac{\lambda}{2}\cdot\sqrt{\frac{h^2+k^2}{a^2}+\frac{l^2}{c^2}}\right) \quad (2)$$

With given lattice constants for pure sulphuric (S) and pure selenide (Se) kesterite a and c (see table 1) the system of equations (3) can be solved numerically regarding x, which is the chalcogenide ratio x = [S]/([S]+[Se]).

$$a = a_{Se}(1-x) + a_S x$$
$$c = c_{Se}(1-x) + c_S x \quad (3)$$

The band gap can be estimated using equation (4), where E_g(CZTS) is set to be 1.5 eV, E_g(CZTSe) is chosen to be 1.0 eV and the bowing parameter b = 0.1 eV [4].

$$E_g = xE_g(CZTS) + (1-x)E_g(CZTSe) - bx(1-x) \quad (4)$$

TABLE I
LATTICE CONSTANTS OF CZTS (FROM ICDD 026-0575) AND CZTSE (FROM ICDD 070-8930)

lattice constants	a (Å)	c (Å)
CZTS	5.427	10.848
CZTSe	5.6772	11.3378

III. RESULTS

In Fig. 1, the X-ray diffractograms of one sample before and after post-sulphurization are shown exemplarily. The black curve shows the typical diffractogram of CZTSSe after selenization with the reflexes being almost at the CZTSe positions as expected due to the small S content after selenization. The (112) reflex at 2Θ = 27.34° corresponds to a sulphur mole fraction of x = 0.14 and an estimated E_g = 1.06 eV. The reflex at 2Θ ≈ 32° can be identified as Mo(S,Se)$_2$ and the reflex at 2Θ = 40.5° matches Mo. These can also be found in the red diffractogram of the sulphurized sample with an additional small peak forming at 2Θ ≈ 33° indicating an increased incorporation of S into Mo(S,Se)$_2$. The kesterite reflex is shifted towards bigger angles indicating a substitution of Se by S. The (112) reflex is shifted to 2Θ = 28.49° corresponding to x = 1.0 and E_g = 1.5 eV. Evaluations like this are done for several samples post-sulphurized at different process temperatures for several hold times.

First, the process temperature is studied. In Fig. 2, x and E_g of samples post-sulphurized for a fixed hold time of t_{hold} = 10 min at temperatures between T = 300°C and T = 600°C are shown. It can clearly be seen that there is no significant incorporation of S into the absorber layer by sulphurizing below temperatures of $T \approx$ 450°C. However, between T = 450°C and T = 600°C the sulphur mole fraction increases from around 0.1 — which is standard for our selenized samples — up to 1.0 and can thus be accurately controlled.

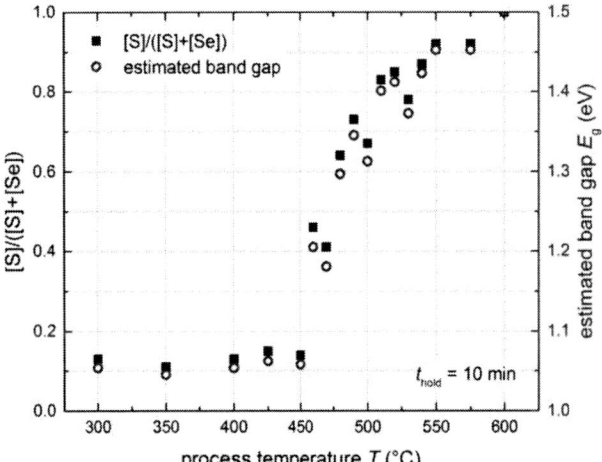

Fig. 2. Chalcogenide ratio [S]/([S]+[Se]) and corresponding estimated band gap of samples post-sulphurized for t_{hold} = 10 min at various temperatures between T = 300°C and T = 600°C.

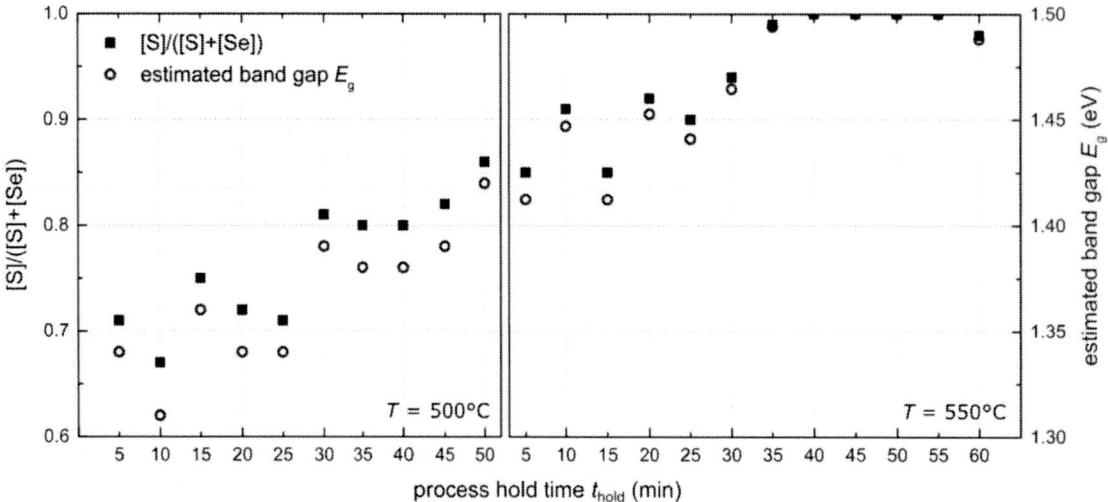

Fig. 3. Band gap E_g and chalcogenide ratio x of several samples post-sulphurized at $T = 500°C$ and $T = 550°C$. The graph is separated into two parts. In the left part the samples are post-sulphurized at $T = 500°C$ and in the right part they are post-sulphurized at $T = 550°C$. The process hold time t_{hold} is varied for both parts of the graph from 5 min to 50 min and 60 min, respectively, in steps of 5 min.

The corresponding estimated band gap of the absorber layers increases from around 1.05 eV to 1.5 eV, respectively.

To take a more detailed look into this, we fixed the process temperature at $T = 500°C$ and $T = 550°C$, respectively, and varied the hold times t_{hold} between 5 min and about one hour in steps of 5 min. The results are shown in Fig. 3. The sulphur mole fraction is plotted as black squares and the corresponding estimated band gap E_g is plotted as blue circles. Using a process temperature of $T = 500°C$ (left part of Fig. 3), the [S]/([S]+[Se]) ratio is increasing almost linearly from 0.67 to 0.86 with increasing process time. For $T = 550°C$ (right part of Fig. 3), it increases further to 1.0. The corresponding E_g is increasing respectively, starting at 1.31 eV and reaching 1.5 eV. The range covered by the $T = 550°C$ process joins perfectly where the $T = 500°C$ process range ends. With the parameters of both experiments of fixed T and fixed t_{hold} we can thus cover nearly the whole band-gap tuning range from x close to zero up to pure sulphuric CZTS and accurately control the chalcogenide mole fraction.

IV. SUMMARY

A post-sulphurization process for CZTSSe is introduced that allows adjusting the sulphur concentration and thus the band gap of solar cell absorber layers. Experiments with fixed process time and varied process temperatures show that for sulphurization temperatures between 450°C and 600°C an efficient S incorporation takes place. The chalcogenide mole fraction can be accurately controlled by the process temperature as well as the used hold time. Varying both parameters, nearly the complete band-gap range of CZTSSe between 1.0 eV and 1.5 eV can be covered.

ACKNOWLEDGMENTS

We acknowledge financial support by the German Federal Ministry of Education and Research (BMBF), FKZ 03SF0530B, and the Karlsruhe School of Optics and Photonics (KSOP) at KIT.

REFERENCES

[1] A. D. Collord, and H. W. Hillhouse, "Germanium Alloyed Kesterite Solar Cells with Low Voltage Deficits," *Chem Mater.*, vol. 28, issue 7, pp. 2067-2073, 2016.

[2] S. Bag, O. Center, T. Gokmen, Y. Zhu, and D. B. Mitzi, "Hydrazine-Processed Ge-Substituted CZTSe Solar Cells," *Chem. Mater.*, vol. 24, pp. 4588-4593, 2012.

[3] G. M. Ford, Q. Guo, R. Agrawal, and H. W. Hillhouse, "Earth Abundant Element $Cu_2Zn(Sn_{1-x}Ge_x)S_4$ Nanocrystals for Tunable Band Gap Solar Cells: 6.8% Efficient Device Fabrication," *Chem. Mater.,* vol. 23, pp. 2626-2629, 2011.

[4] S. Chen, A. Walsh, J.-H. Yang, X. G. Gong, L. Sun, P.-X. Yang, J.-H. Chu, and S.H. Wei, "Compositional dependence of structural and electronic properties of $Cu_2ZnSn(S,Se)_4$ alloys for thin film solar cells," *Phys. Rev. B.*, vol. 83, p. 125201, 2011.

[5] C. Persson, "Electronic and Optical Properties of Cu_2ZnSnS_4 and $Cu_2ZnSnSe_4$," *Appl. Phys.*, vol. 107, issue 5, pp. 053710-053710-8, 2010.

[6] W. Shockley, and H. J. Queisser, "Detailed Balance Limit of Efficiency of p-n Junction Solar Cells," *Appl. Phys.*, vol 32, issue 3, pp. 510-519, 1961.

[7] T. Todorov, T. Gershon, O. Gunawan, C. Sturdevant, and S. Guha, "Perovskite-kesterite monolithic tandem solar cells with high open-circuit voltage," *App. Phys. Lett.,* vol. 105, p. 173902, 2014.

[8] T. Schnabel, M. Löw, and E. Ahlswede, "Vacuum-free preparation of 7.5% efficient $Cu_2ZnSn(S,Se)_4$ solar cells based

on metal salt precursors," *Sol. Energ. Mater. and Sol. Cells*, vol. 117, pp. 324-328, 2013.

[9] M. Neuwirth, H. Zhou, T. Schnabel, E. Ahlswede, H. Kalt, and M. Hetterich, „A multiple-selenization process for enhanced reproducibility of $Cu_2ZnSn(S,Se)_4$ solar cells," *Appl. Physl. Lett.*, vol. 109, p. 233903, 2016.

[10] T. Todorov, H. Sugimoto, O. Gunawan, T. Gokmen, and D. B. Mitzi, "High-Efficiency Devices With Pure Solution-Processed $Cu_2ZnSn(S,Se)_4$ Absorbers," *Proc. of the 39th IEEE Photovoltaic Specialists Conference*, Tampa, FL, USA, June, 2013.

[11] W. Wang, M. T. Winkler, O. Gunawan, T. Gokmen, T. K. Todorov, Y. Zhu, and D. B. Mitzi, „Device Characteristics of CZTSSe Thin-Film Solar Cells with 12.6% Efficiency," *Adv. Energ. Mater.*, vol 4, p. 1301465, 2014.

[12] C.-H. Cai, S.-Y. Wei, W.-C.- Huang, J. Lin, T.-H. Yeh, and C.-H. Lai, „Efficiency enhancement by adding SnS powder during selenization for $Cu_2ZnSn(S,Se)_4$ thin film solar cells," *Sol. Energ. Mater. and Sol. Cells*, vol. 145, pp. 296-302, 2016.

[13] G. Chen, M. Wu, W. Wang, J. Zhang, L. Lin, J. Liu, and F. Lai, „Effects of annealing processing on the oxide derived $Cu_2ZnSn(S_x,Se_{1-x})_4$ thin films," *Superlatt. And Microstruc.*, vol. 100, pp. 209-213, 2016.

[14] R. Sun, M. Zhao, D. Zhuang, Q. Gong, L. Guo, L. Ouyang, Q. Wei, "High-sulfur $Cu_2ZnSn(S,Se)_4$ films by sulfurizing as-deposited CZTSe film: The evolutions of phase, crystallinity and S/(S+Se) ratio," *Journ. of All. And Comp.*, vol. 695, pp. 3139-3145, 2017.

Impact of Ga/III Profile on Voltage-dependent Collection Losses in CIGS Solar Cells

Dmitry Poplavskyy, Jeff Bailey, Rouin Farshchi, and David Spaulding

MiaSolé Hi-Tech Corp, Santa Clara, CA 95051, USA

Abstract — The impact of Ga/III profile changes at the front (junction) side and the back side of the device on *FF* and V_{OC} loss from voltage-dependent collection was studied. Ga/III changes in the back of the device have a minimal impact on voltage-dependent collection, mostly affecting dark diode recombination. In contrast, Ga/III changes at the front side of the absorber could lead to strong *FF* collection losses when the Ga/III gradient creates an opposing effective force field that reduces field strength in the space-charge region. The magnitude of this loss is modulated by the defect recombination strength, as demonstrated by 1D device simulations.

Index Terms — photovoltaic cells. thin film devices

I. INTRODUCTION

Thin film solar cells tend to have *FF* values that are lower than their corresponding crystalline silicon counterparts. This comes from differences in dominant recombination mechanisms in different solar cell architectures as well as from voltage-dependent carrier collection effects that typically exist in thin film solar cells. Voltage-dependent collection leads to the breakdown of the principle of superposition, which states that the dark IV curve, shifted by the magnitude of *J*sc, should reproduce the light IV curve. This superposition principle typically holds true in high efficiency c-Si and III-V solar cells, where recombination losses are independent of voltage and series resistance tends to be low. However thin film solar cells, such as a-Si, CIGS and CdTe, tend to suffer from voltage-dependent collection which reduces fill factor and open circuit voltage [1, 2]. Interface recombination, bulk recombination losses through voltage-dependent depletion width, and depletion region recombination have been discussed as sources of voltage-dependent collection [2]. Voltage dependent collection typically affects *FF* strongly, with comparatively smaller impact on V_{OC} and J_{sc}.

The purpose of this study is to demonstrate how changes in CIGS absorber properties, in particular the Ga/(Ga+In) (referred to as Ga/III) profile, affect *FF* and V_{OC} loss originating from voltage-dependent collection. Understanding how these losses can be controlled via absorber composition is important to reduce losses and enable higher cell efficiencies.

II. EXPERIMENTAL DETAILS

Measurements are done on flexible Cu(In,Ga)Se$_2$ solar cells fabricated at the MiaSolé manufacturing line in Santa Clara, California. The MiaSolé manufacturing process begins with a roll of 50 μm thick 1-meter-wide stainless steel foil. The foil is continuously passed through the interconnected vacuum chambers of the deposition tool where all the layers of the thin film solar cell stack are sequentially deposited by high-rate physical vapor deposition (PVD) without vacuum breaks. Subsequently, the web is cut and then slit into rectangular solar cells, which are interconnected with a low-resistance solder-free wire interconnect (UltraWire™), flash-tested, and binned by an inline IV tester. It takes less than 6 minutes to deposit a 2 μm thick CIGS layer inside the inline deposition tool, and only 60 minutes to convert a bare steel foil into a functional solar cell. After the cells are made, they are typically assembled into a variety of flexible solar module sizes and form factors to meet our customer needs. In this study, however, the cells were not assembled into modules; instead, each cell was additionally subjected to a vacuum lamination process to simulate the module lamination temperature and pressure environment. In this work, CIGS deposition parameters and, in some cases, target compositions were varied in order to modify the CIGS composition profile, in particular the Ga/III ratio. The post-vacuum-laminated cells were subsequently IV-tested in an offline IV tester, where both dark and light IV curves were recorded for each cell. Temperature-dependent V_{OC} measurements are performed on the cells mounted on a temperature controlled stage in the temperature range 10 - 45 °C; activation energies are extracted from the intercept of the linear fit to the V_{OC} vs T dependence. Absorber compositional depth profiles were measured on representative samples using a Physical Electronics 660 Auger Electron Spectrometer (AES) equipped with Zalar rotation, after TCO layers were removed from the CdS layer in an aqueous solution of 30% acetic acid.

III. RESULTS

Typically, IV characteristics of thin film solar cells are analyzed within the framework of a single-diode model, whereby the current J can be expressed as [3]:

$$J(V) = J_0 \exp\left[\frac{q(V - R_s J)}{AkT}\right] + \frac{V}{R_{sh}} - J_L(V) \tag{1}$$

where J_0 is the diode saturation current density, q is electron charge, R_s is the series resistance, R_{sh} is the shunt resistance, A is the diode ideality factor, and $J_L(V)$ is the voltage-dependent photocurrent. As the voltage-dependent collection term is only present under light, its impact on the IV curve can be quantified by comparing dark and light IV curves. Here, we follow the procedure outlined in Ref. [4]:

(1) Each dark IV curve is fitted to the diode equation with $J_L=0$, yielding J_0, A, R_s, and R_{sh};

(2) Using the values of R_s and R_{sh}, extracted from the dark IV fit, shunt- and series-corrected dark and light IV curves are then generated, $J_{corr}(V_{corr})$, where $J_{corr}= J - V/R_{sh}$ and $V_{corr} = V - JR_s$;

(3) The corrected dark IV curve is shifted by the value of J_{sc} to yield the shifted dark IV curve, which represents the (hypothetical) light IV curve not affected by voltage-dependent collection $(J_L(V)=J_{sc})$;

(4) Fill factors can now be extracted from the corrected shifted dark (FF_D) and light (FF_L) IV curves;

(5) Fill factor difference, which represents fill factor losses related to voltage-dependent collection, is calculated next:

$$\Delta FF = FF_D - FF_L. \qquad (2)$$

Voltage-dependent collection losses could originate from a number of different mechanisms, which include interfacial losses, short diffusion length, and space-charge region (SCR) recombination [2]. Changes in the Ga/III profile, which defines the conduction band profile of the CIGS absorber, could have a profound impact on one or more of these mechanisms and could potentially affect collection of minority carriers in the absorber.

In the first experiment the effect of bandgap changes at the back of the device was investigated. Figure 1 shows Ga/III profiles, measured by AES, where the steepness of the Ga/III profile at the bottom part of the film (near the CIGS/Mo interface) was varied, with the rear side ΔGa/III (as defined in Fig. 1) ranging from 0.19 to 0.31. Such changes in the Ga profile could be achieved by process modifications in the beginning of the CIGS growth.

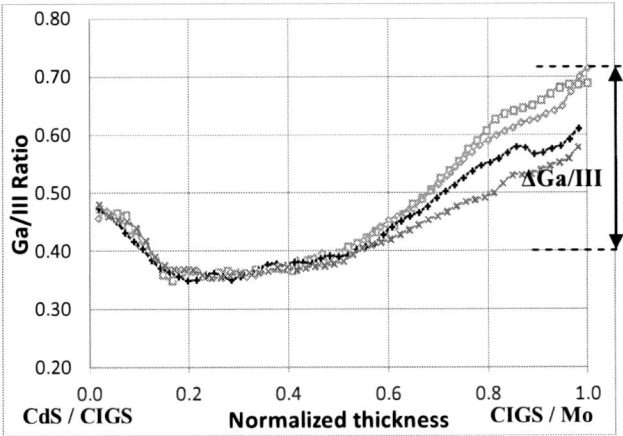

Fig. 1. Ga/III profiles for the back-gradient variation study. ΔGa/III is defined as the difference between the average Ga/III at the bottom 20% of the film (for thicknesses between 0.8-1.0) and the minimum value Ga/III of the profile.

Figure 2 illustrates the impact of such variation on FF collection loss (ΔFF), dark and light fill factors, and V_{OC}. The function of the back Ga/III gradient is to provide a potential gradient pushing electrons away from the highly recombination-active back contact (back surface field), so one may expect that a *reduced* ΔGa/III should result in *increased collection losses* for the photogenerated carriers in the quasi-neutral region (QNR) of the device. Figure 2 shows that indeed there is a slight increase in the value of ΔFF, which represents collection losses, at lower values of ΔGa/III. This increase is accompanied by a rather noticeable reduction of V_{OC} and corresponding increase in the dark FF, which are related to diode recombination. Higher dark FF and lower V_{OC} suggest that the diode ideality factor is reduced, as a result of more dominant QNR recombination and/or back-contact recombination. Overall, however, such variations appear to have a fairly minor impact on collection losses, changing FF loss from collection, ΔFF, by only 0.01 for the range of Ga/III profile changes investigated here.

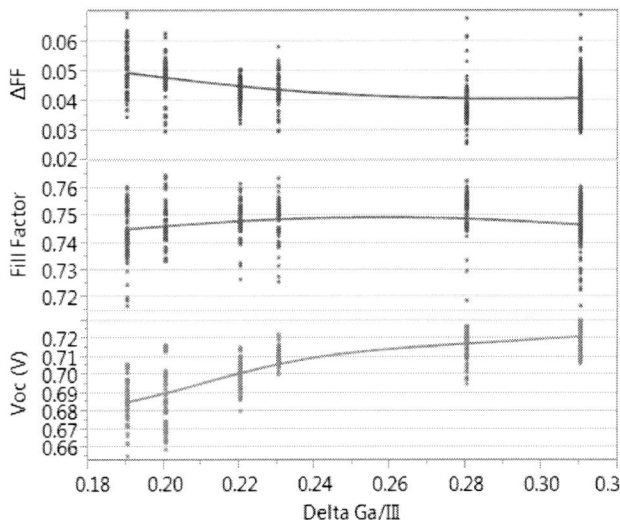

Fig. 2. FF loss, per Eq. (2), series- and shunt-corrected FF, extracted from the dark and light IV curves, and device open circuit voltage V_{OC} as a function of the back-gradient ΔGa/III.

Figure 3 shows Ga/III profiles produced in the second experiment to assess the impact of the absorber front Ga/III gradient on collection losses, where the gradient slope and the surface Ga/III ratio are strongly varied via changes in target composition and deposition parameters. Here the gradient slope changes from positive (profile 1) to negative (profile 7) with the surface Ga/III values at the CdS/CIGS interface changing from 0.1 (profile 1) to 0.59 (profile 7).

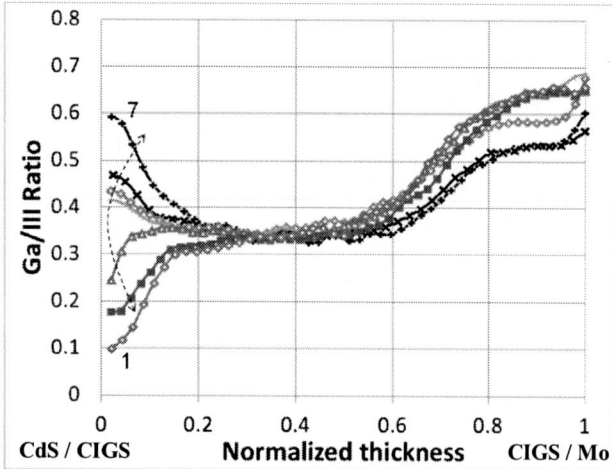

Fig. 3. Ga/III profiles for the front-side variation study.

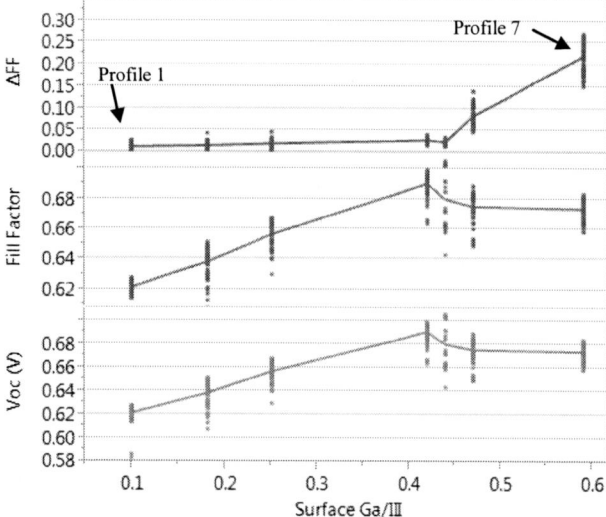

Fig. 4. *FF* loss, per Eq. (2), series- and shunt-corrected *FF*, extracted from the dark and light IV curves, and device open circuit voltage V_{OC} as a function of the surface Ga/III ratio.

Figure 4 illustrates the impact of such variation on *FF* collection loss (ΔFF), dark and light fill factors, and V_{OC}. Clearly, ΔFF stays low (< 0.05) for surface Ga/III ≤ 0.45, above which ΔFF begins to increase steeply, reaching ΔFF ~ 0.2 at a surface Ga/III of 0.59. This can be explained in terms of the additional *opposing* field created by the Ga/III gradient in the depletion region [5], which reduces the collection electric field at the junction and thus increases SCR recombination. Such a strong impact of front Ga/III gradient on collection losses clearly demonstrates that SCR recombination has a strong contribution to the collection losses in the negative Ga/III gradient configuration, where an opposing electric field is created. Note however, that high efficiency CIGS devices typically do have a negative front side gradient, which is required to maintain low interfacial recombination due to increased interface bandgap [6].

In the case of *positive* front gradient, where surface Ga/III < 0.33, V_{OC} and dark *FF* are both reduced when the surface Ga/III is reduced. This can be explained by increasing interfacial recombination as result of reduced surface bandgap; this conclusion is also supported by $V_{OC}(T)$ measurements which yield reduced activation energy for lower surface Ga/III values, as can be seen in Fig. 5.

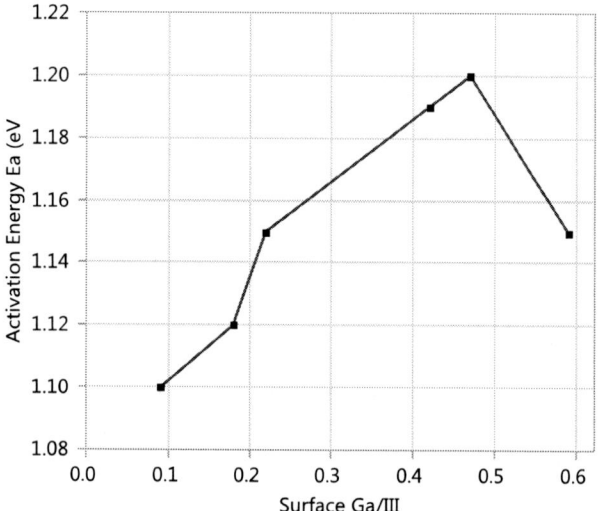

Fig. 5. Activation energy, extracted from $V_{OC}(T)$ measurements, as a function of surface Ga/III.

Clearly, an optimum near-surface Ga/III profile is therefore achieved as a trade-off between the impact on V_{OC} (due to interface recombination) and impact on FF (due to voltage-dependent collection). For the group of devices studied here the optimum performance is achieved at surface Ga/III values between 0.4 and 0.45.

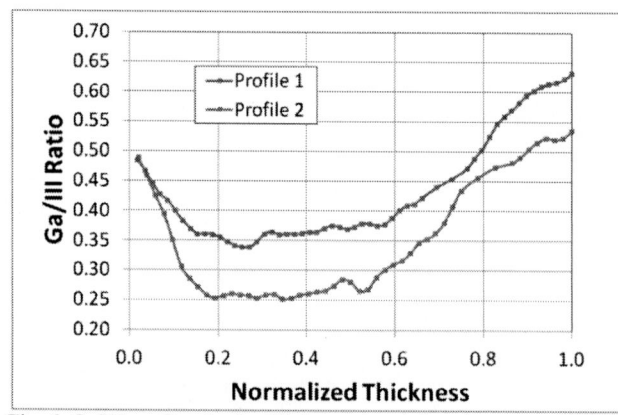

Fig. 6. Ga/III profiles for the third experimental study.

The third experiment was performed to test the impact of the minimum Ga/III while maintaining the same surface Ga/III, as shown by the AES results in Fig. 6. These profiles, while having different back-surface Ga/III values, have similar

978-1-5090-5606-4/17 $31.00 © 2017 IEEE

back-surface gradients thus are not expected to affect the back-surface-field properties of the cell.

Figure 7 shows results of FF analysis per Eq. (2) and V_{OC} for the profiles shown in Fig. 6. As expected from its less negative front gradient, profile 1 exhibits lower FF loss, ΔFF, and a higher V_{OC} due to the higher minimum bandgap.

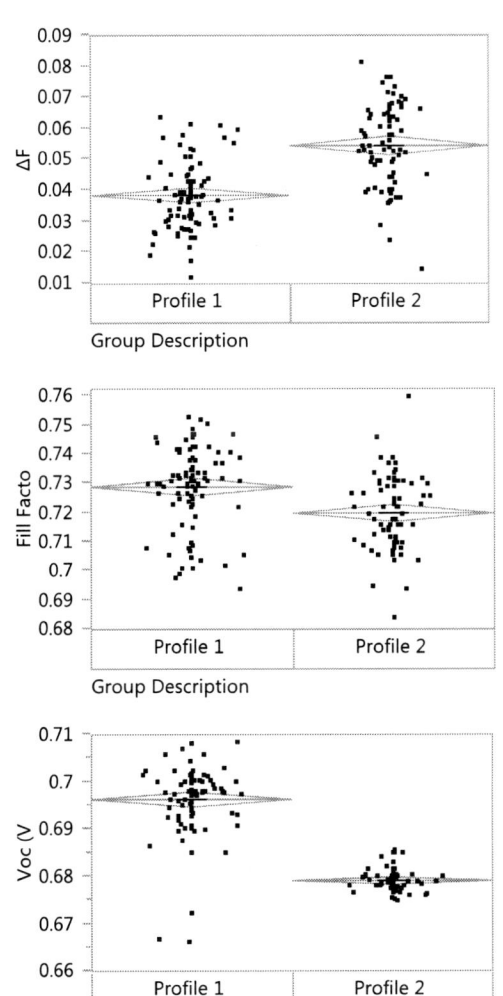

Fig. 7. *FF* loss, per Eq. (2), actual *FF*, and V_{OC} for profiles 1 and 2 shown in Fig. 6.

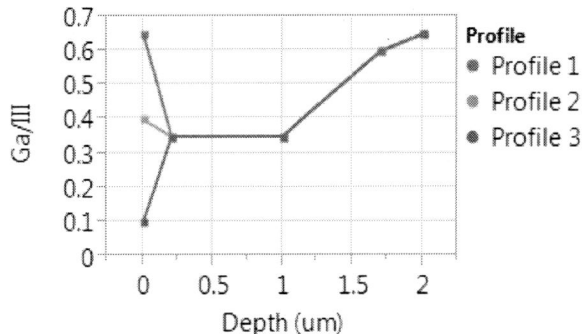

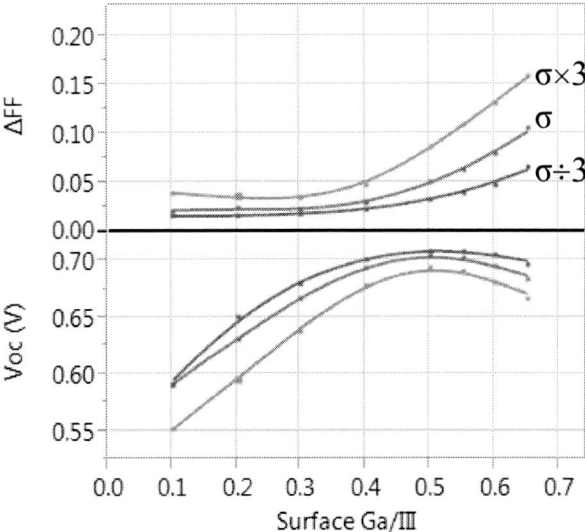

Fig. 8. SCAPS simulation results for ΔFF and V_{OC}, as function of surface Ga/III. Examples of Ga/III depth profiles used for simulations are also shown.

1D device simulations were also performed using SCAPS [7] to model the experimentally observed effects of variations in the front side Ga/III gradient, as shown in Fig. 4. The following values of key parameters were used in our SCAPS model: $N_D = 1.3\times10^{16}$ cm^{-3}, $N_A = 2.4\times10^{15}$ cm^{-3}, $N_t = 2.7\times10^{16}$ cm^{-3}, $\sigma_e / \sigma_h = 3.7\times10^{-17} / 3.7\times10^{-14}$ cm^{-2}, $N_{IF} = 1.2\times10^{11}$ cm^{-2}, where N_D is donor concentration in CdS, N_A is the shallow acceptor concentration in CIGS, N_t is the concentration of mid-gap acceptors, σ_e / σ_h are the electron/hole capture cross-sections of the mid-gap acceptor, and N_{IF} is density of interface mid-gap defects. Light and dark IV curves were simulated for the range of Ga/III profiles, where the front portion of the profile was varied, as shown in the inset to Fig. 8. Series and shunt resistances were not included in this simulation. As the impact of the front Ga/III gradient on voltage dependent collection is expected to depend on minority carrier recombination strength in that region, the recombination strength of the deep acceptor defect was varied by changing the electron and hole capture cross-sections, σ_e and σ_h, as shown in Fig. 8. Finally, dark *FF* was calculated

978-1-5090-5606-4/17 $31.00 © 2017 IEEE

from the J_{sc}-shifted simulated dark IV curve and ΔFF was calculated per Eq. (2). The simulated ΔFF and V_{OC} as a function of surface Ga/III are presented in Fig 8. Clearly, the simulation reproduces the experimentally observed ΔFF and V_{OC} trends shown in Fig. 4. In particular, the simulation also illustrates the effect of recombination strength on FF losses due to voltage-dependent collection in the space-charge region, whereby the collection loss clearly increases with increasing recombination strength of the acceptor defect. Therefore, improvements in the CIGS material quality are expected to not only increase V_{OC} but also reduce voltage-dependent collection FF losses, leading to higher fill factors.

IV. CONCLUSIONS

Changes in the Ga/III profile can have various effects on voltage dependent collection losses, depending on which part of the device is affected. Variations in the Ga/III back-gradient have a relatively small impact on voltage dependent collection, despite a significant change in device V_{OC}, suggesting that diode recombination is the primary reason for performance changes.

In contrast, changes in the Ga/III gradient near the front-side of the absorber result in a significant increase of voltage-dependent FF loss when the Ga/III gradient opposes carrier collection across the junction. This effect can be further modulated by changes in recombination strength in the space-charge region.

REFERENCES

[1] S. Hegedus, W. Shafarman, "Thin-film solar cells: device measurements and analysis", *Prog. Photovolt: Res. Appl.*, vol. 12, pp. 155-176, 2004.

[2] S. Hegedus, D. Desai, and C. Thompson, "Voltage Dependent Photocurrent Collection in CdTe/CdS Solar Cells", *Prog. Photovolt: Res. Appl.*, vol. 15, pp. 587-602, 2007.

[3] J. Nelson, The Physics of Solar Cells, Imperial College Press, 2003.

[4] L. M. Mansfield, R. L. Garris, K. D. Counts, J. R. Sites, C. P. Thompson, W. N. Shafarman, K. Ramanathan, "Comparison of CIGS Solar Cells Made With Different Structures and Fabrication Techniques," *IEEE Journal of Photovoltaics*, vol. 7(1), pp. 186-193, 2017.

[5] Depletion width extends to about 0.2 of the total thickness, based on carrier concentrations of $\sim 5 \times 10^{15}$ cm^{-3}, as measured by DLCP and CV, and absorber thickness of ~ 2 μm.

[6] C. Frisk, C. Platzer-Björkman, J. Olsson, P. Szaniawski, J. T. Wätjen, V. Fjällström, P. Salomé and M. Edoff, "Optimizing Ga-profiles for highly efficient Cu(In,Ga)Se2 thin film solar cells in simple and complex defect models", J. Phys. D: Appl. Phys., vol. 47, pp. 485104, 2014.

[7] M. Burgelman, P. Nollet and S. Degrave, "Modelling polycrystalline semiconductor solar cells", *Thin Solid Films*, vol. 361-362, pp. 527-532, 2000.

Cl Diffusion in CdTe Solar Cells Activated by Gaseous CHClF$_2$ Atmosphere

I. Rimmaudo[a], R. Mis Fernandez[a], V. Rejon[a], A. Abbas[b], F. Lisco[b], J.M. Walls[b] and J.L. Peña[a]

[a]Department of Applied Physics, CINVESTAV- IPN Unidad Merida, Merida, Yucatan, 97310, Mexico

[b]CREST, Wolfson School of Mechanical, Electrical and Manufacturing Engineering, Loughborough University, Loughborough, UK

Abstract — **The activation treatment is consider of dramatic importance for CdTe solar cells. The most effective activation processes are based on chlorine compounds. However it is still not clear how Cl diffusion depends on the chlorination method. In this paper we activated CdTe solar cells by gaseous CHClF$_2$, CdCl$_2$ and MgCl$_2$. We investigated the Cl presence by TEM, EDX and XPS. Cl presence at the interface appears to be responsible of the S and Te inter-diffusion. CdTe re-crystalization has been registered also in absence of evidence of chlorine diffusion.**

I. INTRODUCTION

The photovoltaic (PV) technologies are constantly improving and in the last years new materials and solutions have been presented by the community. In this scenario CdTe thin film solar cells represent one of the most important players of the market.

It is well known that best performance is achievable only when the so called *activation treatment* is applied to the CdTe [1], this is usually performed by means of CdCl$_2$ being applied to the absorber surface in different ways (e.g. vacuum evaporation, dipping, vapor transport, etc..) [1], [2]. Many physical and electrical changes are attributed to the activation of the device: re-crystallization (grain growth and orientation change), grain boundaries passivation [3], doping of the CdTe [4], etc. In the 2010, Romeo et al. [1] demonstrated that the transformation of the absorber is connected to the action of chlorine activating the CdTe by using a Cl containing gas atmosphere, for this reason the activation treatment is often addressed as *chlorination*. More recently, Major et al. [5] presented an effective non-toxic activation treatment replacing the CdCl$_2$ with MgCl$_2$ and obtaining similar results. In [6], it was also claimed that it is possible to control the effects of the activation increasing the amount of chlorine applied to the CdTe [6]. However, there are many open points on the chlorine diffusion and interaction with the lattice atoms during the activation and their dependency on the way how the chlorination is performed.

In this paper we address the effects of chlorination performed in a gaseous atmosphere containing CHClF$_2$, solid state CdCl$_2$ and MgCl$_2$. JVs and EQE confirmed the cells were activated by the three chlorination methods, showing similar performances. TEM was used to reveal the grains transformation and the treatments effects on the CdS/CdTe

interface. Chlorine diffusion was addressed by EDX, XPS depth composition analysis.

II. EXPERIMENTAL

In order to produce comparable solar cells by different chlorination methods a very simple device structure was used. The substrate is commercially available Delta Technologies boro-aluminosilicate glass coated with 200 nm of ITO. Further, 200 nm buffer layer of ZnO was deposited by reactive RF sputtering at 400 °C in Ar/O$_2$ atmosphere, to avoid indium diffusion. CdS was also sputtered at 300 °C with a thickness of 160 nm in argon. The absorber layer was deposited by closed space sublimation (CSS) at 515°C and 0.530 mbar of Ar/O$_2$ dynamic vacuum.

The gaseous CHClF$_2$ activation was performed in a vacuum tube placed in a lamp heating furnace. The base vacuum pressure was $< 10^{-5}$ mbar, and CHClF$_2$ - Air mixture was introduced up to the static pressure of 300 mbar at RT. The temperature was slowly increased up to 400 °C and maintained constant for 15 min, then the sample degassing was performed at 400 °C, 10^{-5} mbar for 10 min, see also [7]. The CdCl$_2$ and MgCl$_2$ were deposited on the CdTe surface by methanol saturated solution, dried and heated at 400°C in air. For brevity in this paper we will refer to the samples prepared as described above, simply as *CHClF$_2$*, *CdCl$_2$* and *MgCl$_2$* with regards to the activation treatment applied to them. Moreover, to present meaningful comparisons, some samples were finished without any activation treatment, in the following sections we will refer to them as *untreated*.

Regardless to the activation treatment, the back contact has been deposited by RF sputtering, 2 nm of Cu and 1.4 μm of Mo and then annealed in argon at 200 °C for 20 min.

Scanning Transmission Electron Microscopy (STEM) images were obtained using a Tecnai F20 (S)TEM equipped with a silicon drift detector allowing high spatial resolution Energy Dispersive X-ray (EDX) measurements and chemical mapping. STEM images were collected using a High Angular Annular Dark Field (HAADF) detector at 200 kV. X-Ray Photoelectron Spectroscopy (XPS) measurements were performed in aThermo Scientific K-Alpha equipment with an Al X-ray source, calibrated by the C1s photoemission emission at 284.6 eV. All spectra were recorded by using a step size of 0.025 eV. The efficiency of the solar cells was evaluated by using a Newport Oriel solar simulator Sol2A under AM 1.5 illumination. The current density vs.voltage (J-V) measurements were obtained using a Source/Meter 2420 from Keithley and home-made Labview^{TM} software. The external quantum efficiency (EQE) was measured by USHIO UXL-150so Xe Lamp into a Newport 67005 lamp housing and H10 Horiba Jobin Yvon monochromator.

III. RESULTS

A. Current-Voltage and External Quantum Efficiency

The typical J-V characterization was performed to check the effectiveness of the different chlorination treatments and the impact on the solar cells parameters. In Fig.1 the comparison between representative curves for $CHClF_2$, $CdCl_2$ and $MgCl_2$ treated samples is made. The J-V curve of solar cells without any chlorination (untreated) is added for comparison.

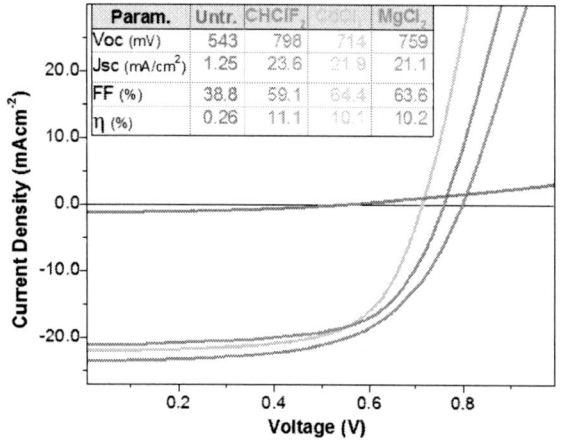

Param.	Untr.	CHClF₂	CdCl₂	MgCl₂
Voc (mV)	543	798	714	759
Jsc (mA/cm²)	1.25	23.6	21.9	21.1
FF (%)	38.8	59.1	64.4	63.6
η (%)	0.26	11.1	10.1	10.2

Fig. 1.   J-V characteristics of the: untreated, $CHClF_2$, $CdCl_2$ and $MgCl_2$ treated solar cells.

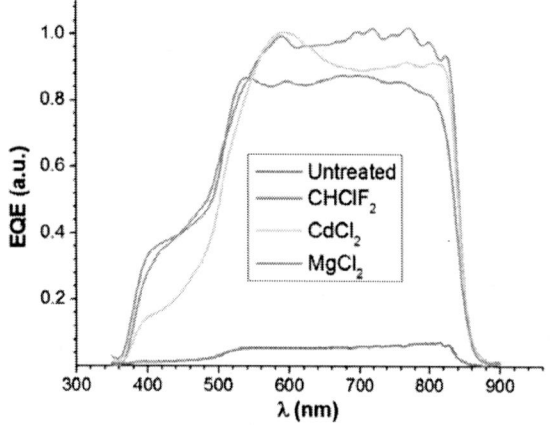

Fig. 2.   Normalized EQE characteristics of the untreated, $CHClF_2$, $CdCl_2$ and $MgCl_2$ treated solar cells.

The dramatic changes in the electrical properties due to the activation treatment involved all the solar cell parameters. In particular, for the untreated samples (red curves) the collected current was very small, indicating a too high resistance of the absorber material. In Fig.1 the increase in Voc after any activation, indicates both an increment of the CdTe doping and a better p-n junction formation. Although efficiency values are comparable, the other parameters changed as different activation treatment was applied on them.

In Fig. 2 the EQE characterization is shown. Since the CdS layer thickness was the same for all the samples, the low UV absorption at short wavelength indicates higher consumption of the CdS layer for $CHClF_2$ and $MgCl_2$ than for $CdCl_2$ samples. More over the EQE intensity in the 500-850 nm range indicates pour current for CdTe device treated with $MgCl_2$. This could be due to high recombination in the CdTe and/or pour junction properties. Devices activated with $CdCl_2$ show a quantum efficiency reduction in the range of 650-800 nm, indicating high recombination in the absorber. Finally, the $CHClF_2$ sample presented more or less constant quantum efficiency in the range of 550-850 nm.

B. Transmission Electron Microscopy and Energy Dispersive X-ray

The TEM analysis (Fig. 3) showed the physical effects of each activation treatment on the films, the untreated sample (Fig. 3a) is shown for comparison. Indeed, in absence of treatment, the shape and size of the CdTe grains varies, small grains are more concentrated at the interface, as also observed in [8]. Very small voids are present at the CdS/CdTe interface and at the grain boundaries. They can be clearly seen in the dark field image in the inset.

After the $CHClF_2$ activation (Fig. 3b), the CdTe grain shape appeared smoother and more homogeneous indicating that re-crystallization occurred. Similarly to the untreated CdTe the grains maintained the columnar shape but they appear more compact, hence no voids are visible at the grain boundaries. However the CdS/CdTe interface shows many voids sensibly bigger than in the untreated device.

When $CdCl_2$ is used to activate the solar cell (Fig. 3c), the grains are big and rounded, with randomized crystalline orientation. Also in this case voids are present at the interface but they are smaller and less concentrated than in the case of $CHClF_2$.

The $MgCl_2$ activation appears much more aggressive and it leads to evident defects at the interface that compromised the CdS film integrity. In some areas CdS is completely consumed and CdTe seems in contact to the ZnO film.

The EDX analysis of the $MgCl_2$ sample is shown in Fig. 4.

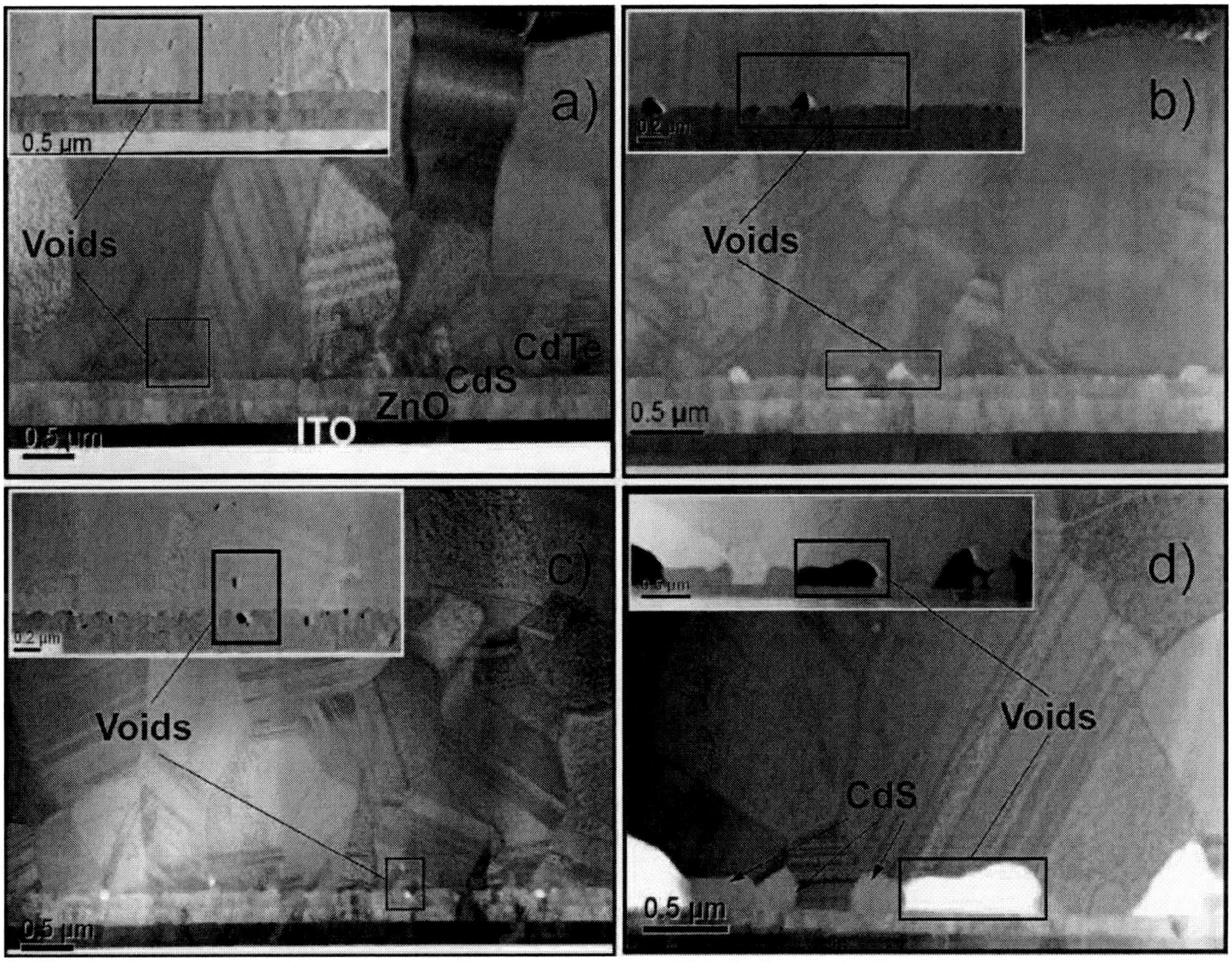

Fig.3. Bright field TEM cross section of (a) untreated sample, (b) CHClF$_2$, (c) CdCl$_2$ and (d) MgCl$_2$. In the insets the dark field TEM showing the voids.

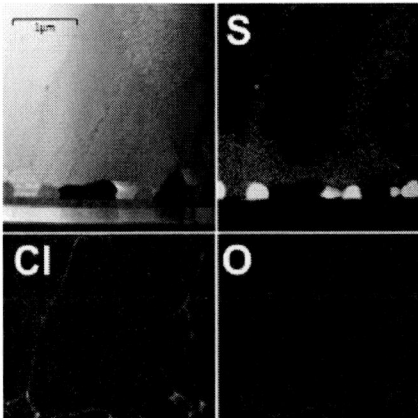

Fig.4. EDX elemental analysis at the interface of the sample treated by MgCl$_2$. Sulphur, chlorine and oxygen images are shown.

The sulphur image yields to an excess of diffusion into the CdTe film, which corroborates the enhanced UV absorption in the EQE. The very feeble oxygen response suggests that also the ZnO film was involved in the reaction leading to oxygen diffusion. Furthermore, the chlorine concentration image is shown. Here, it is clear as chlorine is more concentrated at the grain boundaries and in proximity of the voids formed at the CdS/CdTe interface. Since the source of chlorine is on the back surface of the sample, it can be conclude that, when the absorber was activated by MgCl$_2$, chlorine diffused trough the grain boundaries and accumulated at the interface where probably promoted the sulphur diffusion. The same EDX images in the CHClF$_2$ and CdCl$_2$ treated samples (not shown here for brevity) yielded no significant presence of S, O or Cl.

In Fig. 5 the elemental profiles for the CHClF$_2$ sample are shown. In this case the CdS layer appears in the stoichiometric composition and thickness, sulphur and tellurium inter-

978-1-5090-5606-4/17 $31.00 © 2017 IEEE

diffusion is visible forming $CdTe_xS_{1-x}$, known as *intermixing* layer. Furthermore, a small amount of chlorine (At ~0.35%) is registered corresponding to the inter-diffusion area. Outside of this area the chlorine signal was at noise level.

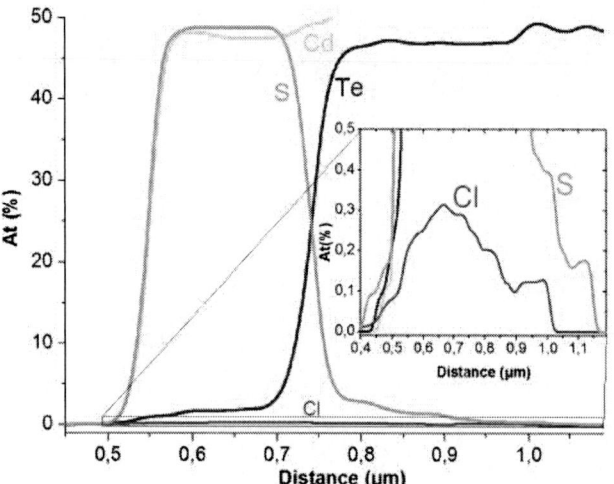

Fig. 5. Elemental analysis in atomic percentage of the CdS/CdTe interface for a $CHClF_2$ treated sample.

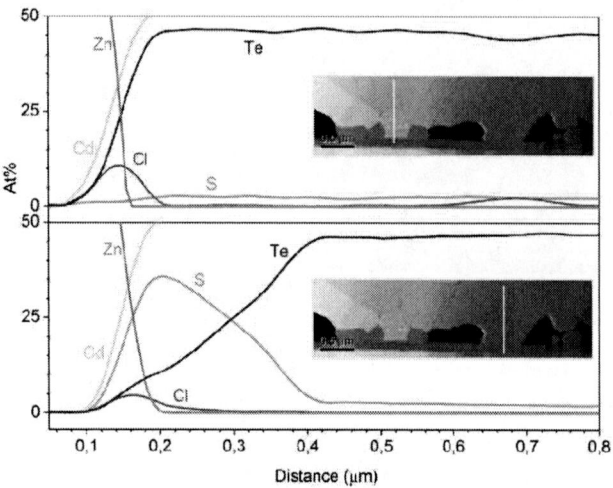

Fig. 6. Elemental analysis of the CdS/CdTe interface for a $MgCl_2$ treated sample in two different points.

In Fig. 6 the same analysis is presented for the $MgCl_2$ sample in two different points of the CdS/CdTe interface. In the first (Fig. 6 bottom) CdS is present but the intermixing layer is much more extended. In this case a discrete amount of chlorine is registered (At= 5%). In the second point (Fig. 6 top) the amount of Cl is sensibly higher (At=10%), and the CdS layer is not present. The same analysis was performed for the $CdCl_2$ activated sample, but in this case the chlorine signal was at noise level everywhere (not shown for brevity).

C. X-ray Photoelectron Spectroscopy

The XPS atomic percentage depth profile was performed to corroborate the different inter-diffusions and the correspondent chlorine presence.

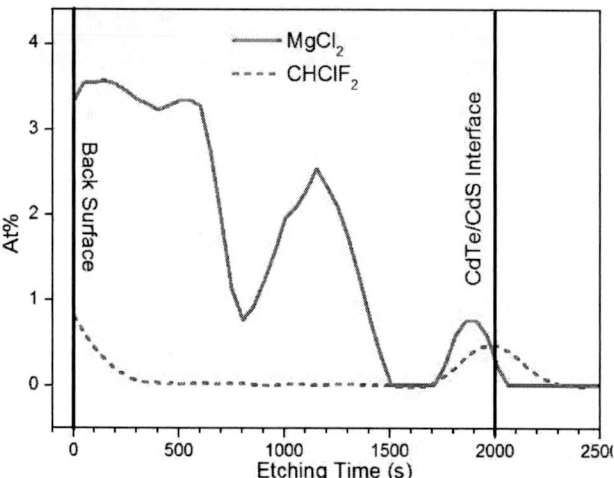

Fig.7. XPS Cl 2p concentration profiles for $MgCl_2$ (solid) and $CHClF_2$ (dashed). $CdCl_2$ profile is not reported because under the noise level.

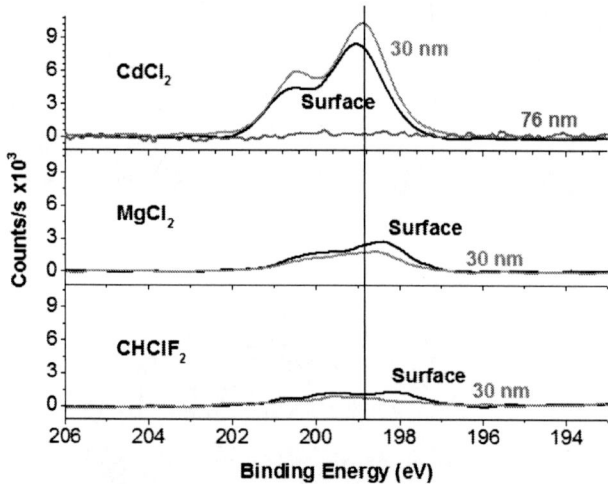

Fig. 8. XPS Cl 2p scan at the back surface (black) for: $CHClF_2$ (bottom), $MgCl_2$ (middle) and $CdCl_2$ (top). Red and green profiles refer to Cl 2p scans at different depth inside the CdTe.

In Fig. 7 the Cl 2p profiles for the $CHClF_2$ and $MgCl_2$ samples. In the first case (dashed line) chlorine is clearly present at the back surface (At ~1%) and it also registered in proximity of the interface (At ~ 0.5%) confirming the EDX results; chlorine seems to be not present in the bulk of the absorber. In the latter case (solid line), the amount of chlorine is remarkably higher at the back surface (At > 3%), in the bulk and also at the interface, At ~ 1%. For XPS the Atomic percentage values should be considered as the average chlorine concentration on the scan length (400 μm). Supposing uniform chlorine diffusion into the CdTe only driven by the concentration gradient, we aspect a regular decay, however profiles in Fig. 7 are very irregular indicating a non-uniform diffusion into the absorber and then supporting the grain boundaries diffusion path.

For the sample activated by $CdCl_2$, the Cl 2p profile is not reported because again at noise level. Considering the XPS sensitivity (At%= 0.1%) we can conclude that chlorine didn't diffuse trough the CdTe as in the other two cases. For this reason an accurate analysis of the samples back surface was performed.

Fig. 8 shows the Cl 2p spectra at the back surface and at few nanometers into the CdTe for $CHClF_2$ (bottom), $MgCl_2$ (middle) and $CdCl_2$ (top).

Fig. 8 leads to the conclusion that for $CdCl_2$ activated cells the chlorine mainly remains at the back surface probably forming stable compounds (doblet) and slightly penetrates into the absorber. On the contrary, when $CHClF_2$ and $MgCl_2$ activations are applied, chlorine is sensibly less concentrated at the back surface.

IV. DISCUSSION

The activation treatment (also known as activation) is critical to high quality CdTe solar cells production. Several methods to perform this treatment have been presented in the past [1][2][4], but it is still not clear the reasons that make one more effective than others. To clarify the chlorine role during the activation will probably help to match the right chlorination method to the absorber deposition process and also reduce the process optimization time. The chlorine diffusion trough the absorber could be the key to understand the mechanism of the CdTe changes. It is in general believed that chlorine diffuses mainly via the grain boundaries [1][4] because there are not direct evidences of its presence within the grains. In other works, chlorine presence after the $CdCl_2$ activation treatment has been registered only at the CdS/CdTe interface [9][10][11]. In this paper we compare the effects on the CdTe film of two solid state activations based on $CdCl_2$ and $MgCl_2$ respectively, with activation performed in gaseous $CHClF_2$ atmosphere addressing the chlorine diffusion.

The cells activated in three different methods present very similar conversion efficiency, much higher than the sample which was not activated (untreated). This implies that all the tested chlorination produce similar changes in the electrical and physical properties of the films. However the EQE analysis indicates that with regard to the different activation method, the UV absorption as well the carriers collection changes [4][12], (e.g. the $MgCl_2$ shows the best UV absorption as well the lowest carriers collection resulting in the lowest Jsc value).

From the TEM analysis it can be concluded that all the activation treatments result in re-crystallization of the CdTe,

but for $CHClF_2$ and $MgCl_2$ it appears columnar while for $CdCl_2$ grains are mainly rounded.

The EDX elemental analysis confirms that voids formation/enlargement at the interface is connected to sulphur diffusion and also to chlorine presence at the interface, since sulphur and tellurium inter-diffusion appears enhanced in the areas where chlorine is more concentrated. This fact is remarkably evident for $MgCl_2$, (where chlorine was registered at the grain boundaries and at the interface), but chlorine was also revealed in the $CHClF_2$ sample, while there is no evidence of its presence for $CdCl_2$ treated sample. XPS yielded that in the case of $CdCl_2$, chlorine remains at the back surface of the CdTe instead of diffusing as in the other two cases.

All these considerations give the idea that when the chlorine stays at the back surface the re-crystallization occurs from the top resulting in more crystal defect in the bulk but less at the interface. On the contrary when chlorine can diffuse trough the grain boundaries reaching the interface it enhances the inter-diffusion of sulphur and tellurium, which should be in the right amount to improve the UV absorption, form the intermixing layer avoiding damage to the junction.

It can be concluded that with regard to the chlorination method it is possible to induce different chlorine diffusion into the CdTe and hence different performance of the solar cells.

V. SUMMARY

The chlorine diffusion trough the CdTe absorber layer due to the gaseous $CHClF_2$, $CdCl_2$ and $MgCl_2$ activation has been addressed, mainly by TEM, EDX and XPS. The solar cell activation is confirmed by the improvement of the electrical and optical properties by J-V and EQE characterization respectively. As expected the chlorine diffusion has a critical role in the solar cell properties. When it reaches the CdS/CdTe it promotes the $CdS_{1-x}Te_x$ intermixing reasonably increasing the UV absorption and the p-n junction properties. However the risk is to exceed and consume the CdS film. On the contrary keeping the chlorine at the back surface anyway produces CdTe re-crystallization, but it reduces the beneficial effects of the intermixing.

ACKNOWLEDGEMENTS

This work was financially supported by CONACYT-SENER (Mexico) and CeMIE-Sol (Grant Nos. 207450/P25).

REFERENCES

[1] N. Romeo, A. Bosio, and A. Romeo, "An innovative process suitable to produce high-efficiency CdTe/CdS thin-film modules," *Sol. EnergyMater.Solar Cells*, vol. 94, pp. 2–7, 2010.

[2] N. Spalatu, J. Hiie, V. Mikli, M. Krunks, V. Valdna, N. Maticiuc, T. Raadik, M. Caraman, "Effect of $CdCl_2$ annealing treatment on structural and optoelectronic properties of close

spaced sublimation CdTe/CdS thin film solar cells vs deposition conditions", *Thin Solid Films* vol. 582, pp. 128-133, 2015.

[4] A.Salavei, I. Rimmaudo, F. Piccinelli, P. Zabierowski, A. Romeo, "Study of difluorochloromethane activation treatment on low substrate temperature deposited CdTe solar cells",*Sol. Energy Mater. Solar Cells,* vol 112 pp.190-195, 2013.

[5] J. D. Major, R. E. Treharne, L. J. Phillips, and K. Durose, "A low-cost non-toxic post-growth activation step for CdTe solar cells," *Nature*, vol.511, pp. 334–337, 2014.

[6] I. Rimmaudo, A. Salavei, A. Romeo, "Effects of activation treatment on the electrical properties of low temperature grown CdTe devices", *Thin Solid Films* vol. 535, pp.253-256, 2013.

[7] A. Rios-Flores O. Are´s, Juan M. Camacho, V. Rejon, J.L. Peña, "Procedure to obtain higher than 14% efficient thin film CdS/CdTe solar cells activated with HCF2Cl gas" *Solar Energy* vol. 86, pp. 780–785, 2012.

[8] B. Maniscalco a,∗, A. Abbas a, J.W. Bowers a, P.M. Kaminski a, K. Bass a, G. West b, J.M.Walls, "The activation of thin film CdTe solar cells using alternative chlorine containing compounds", *Thin Solid Films* vol. 582, pp. 115-119, 2015.

[3] D.L. Bätzner, H. Zogg, A.N. Tiwari, "Recrystallization in CdTe/CdS", *Thin Solid Film,s* vol.361-362, pp. 420–425, 2000.

[9] A. Abbas, G. D. West, J. W. Bowers, P. Isherwood, P. M. Kaminski, B. Maniscalco, P. Rowley, J. M. Walls, K. Barricklow, W. S. Sampath, and K. L. Barth, *IEEE. J. Photovoltaics* vol. 3 (2013), pp.1361-1366.

[10] M. Emziane, K. Durose, N. Romeo, A. Bosio, D.P. Halliday. "Effect of CdCl₂ activation on the impurity distribution in CdTe/CdS solar cell structures". *Thin Solid Films*, vol.480, pp. 377-381, 2005.

[11] K. Durose, M.A. Cousins, D.S. Boyle, J. Beier, D. Bonnet, "Grain boundaries and impurities in CdTeyCdS solar cells". *Thin Solid Films*, vol. 403, pp. 296-404, 2002.

[12] G. Angeles-Ordóñez, E. Regalado-Pérez, M.G. Reyes-Banda, N.R. Mathews, X. Mathew, "CdTe/CdS solar cell junction activation: Study using MgCl₂ as anenvironment friendly substitute to traditional CdCl₂", *Sol. Energy Mater. Solar Cells* vol. 160, pp. 454-462, 2017.

Stability of $Cd_{1-x}Zn_xTe$ alloys under CdTe processing conditions

Yegor Samoilenko and Colin A. Wolden

Colorado School of Mines, Golden, CO 80401, USA

Abstract — Development of II-VI ternary alloys is an important avenue for increasing the efficiency of the CdTe-based solar cells. $Cd_{1-x}Zn_xTe$ ternary alloys offer a wide range flexibility in band gap and are potential candidates for the use both as an absorber and back contact buffer layers. This work investigates the stability of $Cd_{1-x}Zn_xTe$ under CdTe processing conditions. In particular, the effects of chlorine treatment step are studied using both standard $CdCl_2$ and Cl_2 gas methods. X-ray diffraction study shows presence of Te in the Cl_2 gas treated samples, while no Te is detected in $CdCl_2$ treated samples.

Index Terms — cadmium chloride, cadmium telluride, chlorine treatment, II-VI ternary alloys, X-ray diffraction, Zn loss.

I. INTRODUCTION

With a near-ideal band gap of ~1.5eV and high absorption coefficient CdTe has become a recognized leading technology for thin film solar industry with record cell efficiency recently reaching 22.1% [1]. Furthermore, CdTe solar cells are attractive due to their robust manufacturing that leads to low production cost of ~$0.46/Watt [2]. However, there is still a lot to be done to bring the efficiency closer to the theoretical limit of around 30% [3] which will allow to further reduce the cost. While short-circuit current density (J_{sc}) of the CdTe solar cells reached and exceeded its maximum limit [3], open-circuit voltage (V_{oc}) remains low, with best values still below 900 mV for polycrystalline CdTe [3],[4]. Literature suggests several reasons for low open-circuit voltage [4]-[6], such as low carrier concentration, low carrier lifetime, back contact electron reflection, and front contact interface. Development of II-VI ternary alloys may offer solutions to some of those challenges.

Although Cd-S-Te and Cd-Se-Te alloys have been studied and good performance has been demonstrated [7],[8], only limited cases of successfully implementing Cd-Zn-Te alloys into solar cells are present in the literature [9]. One of the reasons is instability of these alloys under CdTe processing conditions, specifically $CdCl_2$ treatment step. $Cd_{1-x}Zn_xTe$ alloys gained increased attention among researchers for use in many types of detectors (x-ray, gamma-ray, etc.) as well as in solar applications in recent years [10],[11]. The popularity of these alloys is attributed to the ability to easily tune the band gap from 1.5 to 2.3 eV by increasing the amount of Zn in the alloy. In particular, $Cd_{1-x}Zn_xTe$ with a band gap of around 1.7 eV has been investigated as a possible absorber layer for tandem cells [10]-[12]. Wide range of deposition techniques such as co-evaporation, MBE, close-space sublimation, and RF magnetron sputtering, etc. have been used to deposit $Cd_{1-x}Zn_xTe$ [12]-[16]. Although many were successful at depositing $Cd_{1-x}Zn_xTe$ alloys with different compositions, the highest photo-conversion efficiency achieved to our knowledge is around 14%, with V_{oc} staying below 900 mV [9]. There are several challenges that arise with the use of $Cd_{1-x}Zn_xTe$. Firstly, Zn loss has been observed to happen by two main mechanisms: reaction of Zn with $CdCl_2$ to form $ZnCl_2$, and diffusion of Zn into CdS layer [9],[15],[17]. Secondly, $CdCl_2$ treatment is carried out in the presence of oxygen and Zn can oxidize to ZnO, which is detrimental to the device efficiency [15]. Shafarman et al. [9] used $ZnCl_2$ instead of $CdCl_2$ treatment to reduce Zn loss from the absorber. However, the efficiencies of devices treated with either $CdCl_2$ or $ZnCl_2$ were identical. Shimpi et al. [17] studying the effect of the $CdCl_2$ treatment on RF sputtered $Cd_{1-x}Zn_xTe$ films explained poor performance of devices by the lack of Cl_2 at the grain boundaries, high density of stacking faults and incomplete recrystallization in the absorber layer. $CdCl_2$ treatment is a critical step in the device processing and is known to result in grain growth, recrystallization, passivation of grain boundaries, and interdiffusion between CdS and CdTe layers [18]. This work presents results of a stability study of $Cd_{1-x}Zn_xTe$ alloys under CdTe processing conditions. Specifically, effects of standard $CdCl_2$ treatment are compared to gaseous Cl_2 treatment.

II. MATERIALS AND METHODS

$Cd_{1-x}Zn_xTe$ alloys were deposited using thermal co-evaporation of CdTe and ZnTe. Substrate temperature was held at 150 °C. The thickness of the films was monitored using quartz monitor crystals (QCMs) and was measured by profilometry. Pure CdTe films were deposited using thermal evaporation as a reference for XRD and UV-Vis. The $CdCl_2$ treatment was performed inside a tube furnace under 50/50 vol% O_2/N_2 atmosphere at 400 °C in a close spaced sublimation geometry. The Cl_2 treatment was performed in the rapid thermal processing (RTP) chamber with 12.7-76.5 ppm Cl_2 in N_2 and 2 min ramp time. X-ray diffraction with Cu source was used to analyze the crystal structure of the as-deposited and treated alloys and to calculate the composition based on Vegard's law. UV-Vis in the 500-1200 nm range was used to study optical properties. Tauc plots were constructed to calculate the band gap of the as deposited alloys.

III. RESULTS AND DISCUSSION

A. As-Deposited Ternary Alloys

978-1-5090-5606-4/17 $31.00 © 2017 IEEE

Fig. 1 shows XRD patterns of (111) peak and a Tauc plot obtained from as deposited alloys and compared to pure CdTe. Films are highly oriented in the (111) orientation and the peak position shifts to higher 2θ values for alloys compared to pure CdTe due to a decrease in the lattice parameter. As deposited alloys are single phase exhibiting cubic zincblende crystal structure. Tauc plot shows band gap increase as expected with increasing Zn content.

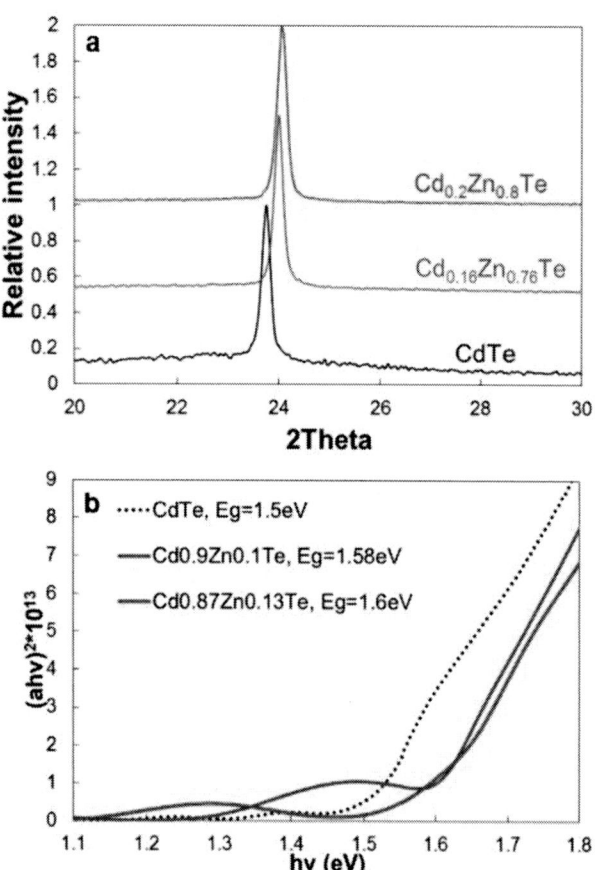

Fig. 1. a) (111) peak XRD and b) Tauc plot for as deposited CdTe and alloys on glass substrates.

B. Stability Study on Glass Substrates

As deposited alloys were subjected to two different treatment methods – standard CdCl$_2$ and gaseous Cl$_2$. Fig. 2 shows X-ray diffraction patterns of the alloys before and after CdCl$_2$ and Cl$_2$ treatment steps. Thickness of the as deposited film on glass substrate was ~450 nm and treatment time was 5 min for both methods. Concentration of Cl$_2$ was set to 76.5 ppm. As can be seen from the figure, CdCl$_2$ and Cl$_2$ treatments produce different results. Alloys after Cl$_2$ treatment exhibit elemental Te peaks. These peaks are not seen for CdCl$_2$ treatment. It is presumed that in the latter case Te freed by Zn removal reacts with Cd vapor, which is not present when using Cl$_2$. For CdCl$_2$ treatment we can see recrystallization

happening in the film, which is common for standard CdCl$_2$ treatment of CdTe [19]. However, the films become (220) oriented rather than random. Cd$_{1-x}$Zn$_x$Te experiences Zn loss as well as formation of ZnO under CdCl$_2$ treatment. Alternative techniques such as X-ray fluorescence and UV-Vis are being investigated at this time to determine whether they can be used to accurately measure the composition of the treated alloys and the amount of Te and ZnO present in the films. Excess Te in the films might be beneficial in small quantities and its effects on the performance of the cell are being investigated. In fact, Te peaks appear after CdCl$_2$ treatment of vapor transport deposited CdTe used for our baseline solar cells.

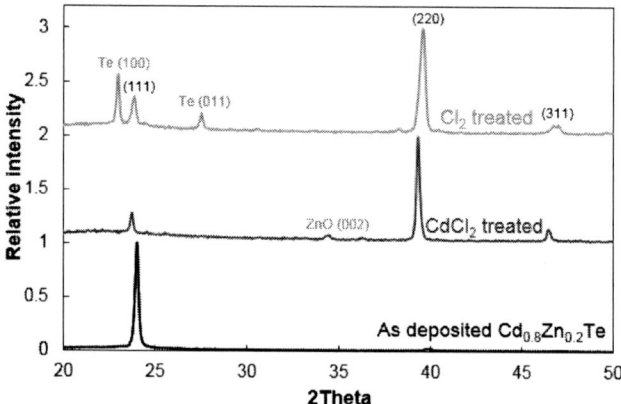

Fig. 2. XRD patterns of as deposited, CdCl$_2$, and Cl$_2$ treated Cd$_{0.8}$Zn$_{0.2}$Te alloy on glass substrates (CdTe peaks are labeled in black).

Fig. 3 illustrates the effects of Cl$_2$ treatment performed at temperatures between 370 and 430 ºC and CdCl$_2$ treatment at 400ºC on the Zn content of the treated alloys. Cl$_2$ concentration was 12.7 ppm in N$_2$ with 10 min dwell time. Alloys had thickness ~850 nm and were deposited on glass substrate. In this figure the Zn content was based on XRD measurements coupled to Vegard's Law. As can be seen from the figure, the treatment temperature plays a role in the final composition of the alloys. CdCl$_2$ treatment seems to lower Zn content of the alloy more than Cl$_2$ treatment at the same temperature. One of the reasons could be the way each treatment is performed – CdCl$_2$ is performed in a tube furnace, while Cl$_2$ treatment is done in the RTP chamber. Heating and cooling rates of these two processes are significantly different. In addition, with Cl$_2$ treatment maximum Zn retention is observed at ~400 ºC. Further experiments and characterization are underway to confirm and refine these trends.

C. Stability Study on TEC15 Substrates

Fig. 4 illustrates the effects of increasing Cl$_2$ concentration on X-ray diffraction pattern of samples deposited on TEC15 substrates. All treatments were performed at 400ºC for 30 min. It can be seen that Zn loss from the alloy is increasing with

978-1-5090-5606-4/17 $31.00 © 2017 IEEE

increasing Cl$_2$ concentration, going from x ~0.18 down to ~0.07 at 12.7 and 76.5 ppm, respectively. CdCl$_2$ treatment at these conditions results in complete Zn loss from the alloy.

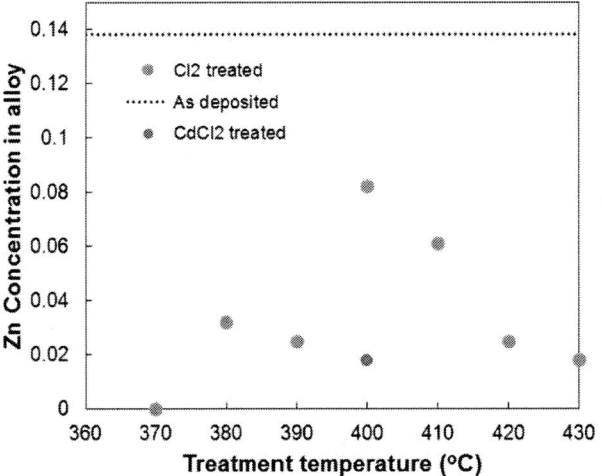

Fig. 3. Effect of Cl$_2$ and CdCl$_2$ treatment temperature on Zn content of Cd$_{1-x}$Zn$_x$Te alloys on glass substrates.

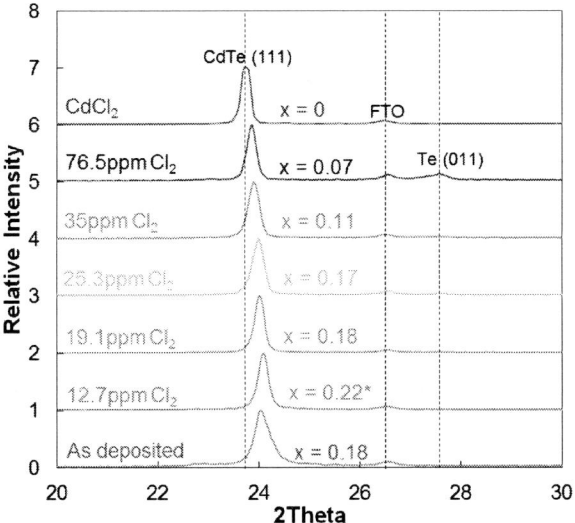

Fig. 4. XRD of alloys treated at different Cl$_2$ concentrations compared to as deposited and CdCl$_2$ treated samples on FTO substrates. *Concentration of Zn is calculated based on peak position. Higher value of remaining Zn may have resulted from initial annealing effect and peak narrowing compared to as deposited sample.

Fig. 5 illustrates the effect of increasing Cl$_2$ concentration on the (111) texture coefficient (TC) of the alloys. TC is calculated using (1) as described by McCandless et al. [18]

$$TC(111) = N \frac{I(111)}{I_o(111)} \left[\sum_{N=1}^{7} \frac{I(hkl)}{I_o(hkl)} \right]^{-1} \qquad (1)$$

where I is the experimental intensity, I_0 is the powder diffraction intensity based on JCPDS no. 15-0770, and N is the number of peaks, which was 7 for this calculation. TC value of one indicates random orientation, while values greater than one reflect preferential orientation in that direction. It can be seen from Fig. 5 that crystal structure becomes more randomly oriented with increasing Cl$_2$ concentration.

Moutinho et al. [19] indicated that the rate of recrystallization during CdCl$_2$ treatment can be different for different substrates. They saw slower rate of recrystallization for CdTe deposited on ITO substrates compared to films deposited on CdS [19]. We saw similar effect when looking at films deposited on glass vs. FTO substrates. Films on glass recrystallized faster than films on FTO. Further work is needed to determine rate of recrystallization and Zn loss for alloys deposited on top of CdS layer. Interestingly, as seen in Fig. 4 Te peaks only start appearing at the highest Cl$_2$ concentration of 76.5 ppm, which also corresponds to completely recrystallized sample.

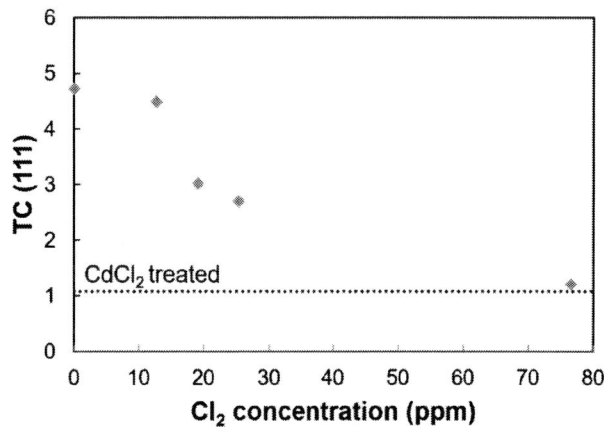

Fig. 5. Effect of Cl$_2$ concentration on texture coefficient of (111) plane of treated alloys on FTO substrates.

IV. SUMMARY AND CONCLUSIONS

In this paper we demonstrate initial results of the stability study of Cd$_{1-x}$Zn$_x$Te alloys under CdTe processing conditions. Alloys were successfully deposited using thermal co-evaporation of CdTe and ZnTe. They have been treated with both standard CdCl$_2$ and Cl$_2$ treatments. CdCl$_2$ and Cl$_2$ treatments of the alloys have different effects on the final composition as seen in the XRD scans. CdCl$_2$ treatment results in formation of ZnO, while Cl$_2$ treatment produces elemental Te. CdCl$_2$ treatment seems to result in greater Zn loss when compared with Cl$_2$ treatment. It was also noted that increasing Cl$_2$ concentration during treatment results in lower Zn content and more random crystal orientation of the alloys as seen by XRD.

978-1-5090-5606-4/17 $31.00 © 2017 IEEE

ACKNOWLEDGEMENT

We would like to thank the Bay Area Photovoltaic Consortium for their support under U.S. Department of Energy Award #DE-EE0004946.

REFERENCES

[1] "Best Research-Cell Efficiencies," *NREL*, 2016. Online at: http://www.nrel.gov/pv/assets/images/efficiency_chart.jpg.

[2] A. Rix, J. D. T Steyl, M. J. Rudman, U. Terblanche, and J. L. van Niekerk, "First Solar's CdTe module technology – performance, life cycle, health and safety impact assessment," *Centre for Renewable and Sustainable Energy Studies*, 2015.

[3] M. A. Green, K. Emery, Y. Hishikawa, W. Warta, and E. D. Dunlop, "Solar cell efficiency tables (version 46)," *Progress in Photovoltaics: Research and Applications*, vol 23, pp. 805-812, 2015.

[4] J. M. Burst, J. N. Duenow, D. S. Albin, E. Colegrove, M. O. Reese, J. A. Aguiar, C. S. Jiang, M. K. Patel, M. M. Al-Jassim, D. Kuciauskas, S. Swain, T. Ablekim, K. G. Lynn, and W. K. Metzger, "CdTe solar cells with open-circuit voltage breaking the 1 V barrier," *Nature Energy*, vol. 1, p. 16015, 2016.

[5] S. H Wei, J. Ma, T. A. Gessert, and K. K. Chin, "Carrier density and compensation in semiconductors with multi dopants and multi transition energy levels: the case of Cu impurity in CdTe," *Presented at 37th IEEE Photovoltaic Specialists Conference (PVSC 37)*, 2011.

[6] J. Sites and J. Pan, "Strategies to increase CdTe solar-cell voltage," *Thin Solid Films*, vol. 515, pp. 6099-6102, 2007.

[7] N. R. Paudel, J. D. Poplawsky, K. L. Moore, and Y. Yan "Current Enhancement of CdTe-Based Solar Cells," *IEEE Journal of Photovoltaics*, vol. 5, pp. 1492-1496, 2015.

[8] J. N. Duenow, R. G. Dhere, H. R. Moutinho, B. To, J. W. Pankow, D. Kuciauskas, and T. A. Gessert, "CdS/CdTe solar cells containing directly deposited CdS$_x$Te$_{1-x}$ alloy layers," *Presented at 37th IEEE Photovoltaic Specialists Conference (PVSC 37)*, 2011.

[9] W. Shafarman and B. McCandless, "Development of a wide bandgap cell for thin film tandem solar cells," *NREL Final Technical Report*, 2007.

[10] C. S. Ferekides, R. Mamazza, U. Balasubramanian, and D. L. Morel, "Cd$_{1-x}$Zn$_x$Te thin films and junctions," *Thin Solid Films*, vol. 480-481, pp. 471-476, 2005.

[11] Y. Eisen, A. Shor, "CdTe and CdZnTe materials for room-temperature X-ray and gamma ray detectors," *Journal of Crystal Growth*, vol. 184/185, pp. 1302-1312, 1998.

[12] R. Dhere, T. Gessert, J. Zhou, J. Pankow, S. Asher, and H. Moutinho, "Development of Cd$_x$Zn$_{1-x}$Te alloy thin films for tandem solar cell application," *Physica status solidi (b)*, vol. 241, pp. 771-774, 2004.

[13] A. Arbaoui, A. Outzourhit, N. Achargui, H. Bellakhder, E. L. Ameziane, and J. C. Bernede, "Effect of the zinc composition on the formation of ternary alloy Cd$_{1-x}$Zn$_x$Te thin films," *Solar Energy Materials and Solar Cells*, vol. 90, pp. 1364-1370, 2006.

[14] O. de Melo, A. Domínguez, K. Gutiérrez Z-B , G. Contreras-Puente, S. Gallardo-Hernández, A. Escobosa, J.C. González, R. Paniago, J. Ferraz Dias, and M. Behar, "Graded composition Cd$_x$Zn$_{1-x}$Te films grown by Isothermal Close Space Sublimation technique," *Solar Energy Materials and Solar Cells*, vol. 138, pp. 17-21, 2015.

[15] A. Rohatgi, R. Sudharsanan, S. A. Ringel and M. H. MacDougal, "Growth and process optimization of CdTe and CdZnTe polycrystalline films for high efficiency solar cells," *Solar Cells*, vol. 30, pp. 109-122, 1990.

[16] S. Hyunlee, A. Gupta, S. Wang, A. Compaan, B. McCandless, "Sputtered CdZnTe films for top junctions in tandem solar cells," *Solar Energy Materials and Solar Cells*, vol. 86, pp. 551-563, 2005.

[17] T. M. Shimpi, J. M. Kephart, D. E. Swanson, A. H. Munshi, W. S. Sampath, A. Abbas, and J. M. Walls, "Effect of the cadmium chloride treatment on RF sputtered Cd0.6Zn0.4Te films for application in multijunction solar cells," *Journal of Vacuum Science & Technology A: Vacuum, Surfaces, and Films*, vol. 34, p. 051202, 2016.

[18] B.E. McCandless, L.V. Moulton, and R.W. Birkmire, "Recrystallization and sulfur diffusion in CdCl$_2$-treated CdTe/CdS thin films," *Progress in Photovoltaics: Research and Applications*, vol. 5, pp. 249-260, 1997.

[19] H. R. Moutinho, M. M. Al-Jassim, D. H. Levi, P. C. Dippo, and L. L. Kazmerski, "Effects of CdCl$_2$ treatment on the recrystallization and electro-optical properties of CdTe thin films," *Journal of Vacuum Science & Technology A: Vacuum, Surfaces, and Films*, vol. 16, pp. 1251-1257, 1998.

CIGSe absorber preparation: an alternative to H_2Se

O.S. Shinde[1], E.J. Schenller[1], S.R.Jadkar[3], S.V Ghaisas[2] and N. Dhere[1]

[1]Florida Solar Energy Center, University Of Central Florida, Fl 32922-5703, USA

[2]IIT Bombay, Mumbai- 400076, India

[3]SP Pune University, Pune-411007

Abstract — Absorber preparation is a crucial process in the CIGS solar cell device fabrication. H_2Se is more commonly used in two stage absorber formation process. It is also important to search for different alternatives for absorber preparation to reduced hazardous and storage risks associated with Hydrogen selenide. Here in this paper, we have used the mixture of hydrogen and Diethyl Selenide for absorber preparation. Raman spectra are recorded for the prepared absorber. Further, Solar cells with device efficiency 9.2% (without antireflection) were obtained from Diethyl selenide processed absorber layer.

Index Terms — CIGSe absorber, two stage process, thin film solar cells.

I. INTRODUCTION

Recently, CIGS has gained recognition due to higher power conversion efficiency among the all other thin film technologies[1]. For higher conversion rate of photon to electron requires a systematic and well planned synthesis of absorber layer. CIGS is acts as p-type material and buffer layer such as CdS or Zn(O,S) acts as n-type material to formed a *p-n* junction by using the chemical bath deposition. At commercial level most of the industries follow the two well-known approaches for absorber preparation. First approach includes, co-evaporation of metals Cu, In, Ga and Se at different temperatures, in the range of 300 to 650^0C [2, 3]. This process of evaporation requires the systematic and critical control over the metal fluxes. In co-evaporation it is very inconvenient to employ a modification and control over the uniformity of a sample [4]. In second approach, absorber preparation follows two step method in which first, stacks of Cu-Ga alloy and In prepared by sputtering and in second step annealing of metal stacks in to the H_2Se environment at temperature range $400-600^0C$[5]. It is also found that this process further extended as SAS (sulfurization after selenization). The H_2Se is used for the selenization and H_2S is used for the sulfurization. The highest efficiency 22.8% is reported with this SAS configuration [6].

However, due to involvement of H_2Se, concerns arise with its safety and storage. It is necessary to seek an alternative for H_2Se. In our group we are developing one such alternative, by using Diethyl-Selenide $((C_2H_5)_2$-Se) based p-CIGS absorber formation at temperature range $400-500^0C$[7]. Here, in this article, we are discussing the Raman characteristics of the absorber prepared by using Diethyl Selenide. Further, in this study we have fabricated the CIGS solar cell device and explore its current voltage characteristics.

II. EXPERIMENTAL

First, a substrate, 2mm thick soda lime glass with SiO_2 barrier diffusion layer thoroughly cleaned by applying the soap solution provided by Alconox. In DC sputtering partial pressure 2.0×10^{-6} Torr was obtained with the help of turbo and cryopump setup. After that substrate went through four-layer Molybdenum back contact deposition. First and third layer deposited at 5mTorr and 200 Watt and Second and fourth layer deposited at 0.1mTorr and 300 Watt. This is due to manage the stress and strain to get adhesion between Mo and CIGS interface[8]. Further, absorber preparation starts with the DC sputtering of five-layer metal stack formation by using the Cu-Ga alloy and In targets. Cu-Ga alloy deposited at 1.5 mTorr and In deposited at 0.7mTorr. Then the sample transfer for selenization. Diethyl selenide with pressure 7.2Torr and H_2 with 384Torr used for selenization reaction in conventional furnace at 500^0C for 30-40 minutes. Prepared absorbers then transferred for the buffer layer junction partner Cadmium Sulfide (CdS) deposition by using chemical bath process. Bath temperature was maintained at 65^0C for 12 minutes. Chemicals used for the CdS depositions are $CdSO_4$(0.15M) ,NH_4OH and CH_4N_2S[9] .After formation of heterojunction the samples subjected for the bilayer (i-ZnO/Al-ZnO) deposition by RF sputtering. A 50nm *i*-ZnO deposited at 200 Watt. And Al:ZnO is deposited at 425 Watt. Final device is formed by evaporating the front contacts Cr/Ag at room temperature.

For the absorber characterization Raman spectroscopy was used and for current-voltage characteristics measurement Oriel Class 2A Solar Simulator from the Newport Corporation was used.

III. RESULT AND DISCUSSION

A. Absorber Raman Spectroscopy

To investigate the quality, of formed C(I,Ga)Se$_2$ we employed the Raman Spectroscopy with incident laser beam of 532nm. We have prepared two absorber samples S1 and S2.

978-1-5090-5606-4/17 $31.00 © 2017 IEEE

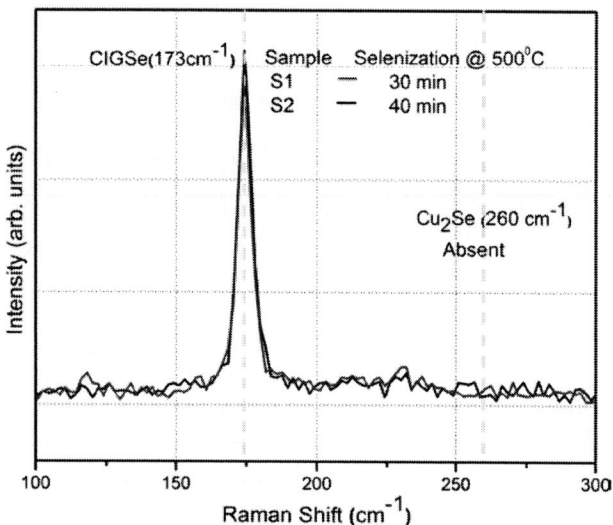

Fig. 1.   Raman spectroscopy of p-type absorber layer.

S1 sample was selenized for 30 minutes (red) and S2 sample 40 minutes (black). The important observation is absence of ODC secondary phases (Cu_2Se) at 260 cm^{-1}. Due to higher decomposition rate of Diethyl selenide prevents the formation of Cu_2Se type of defects during the absorber preparation. Here, both samples shows pure CIGS peaks at 173cm^{-1} and no evidence of Cu_2Se at 260 cm^{-1}[10]. Further, there is no requirement of KCN etching while preparing the devices with this type of absorber preparation method. It is also known that the Cu_2Se formation depends on the concentration of Cu on the surface of the absorber.

B. Absorber FSEM and Cross Sectional SEM :

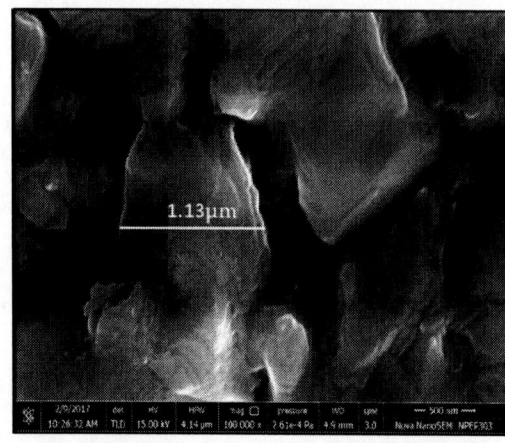

Fig.2.   Top view of the p-type CIGSe absorber Surface

Fig. 2 shows the top view of the FESEM image of CIGSe absorber surface after 40 minutes selenization of metallic stacks of Cu-Ga alloy and In at 500°C in the conventional furnace. It is observed that different 1.13 μm CIGSe crystal

were obtained from the diethyl selenium reaction with hydrogen. Similarly, SEM cross section of Mo/CIGSe interface is shown in Fig.3.

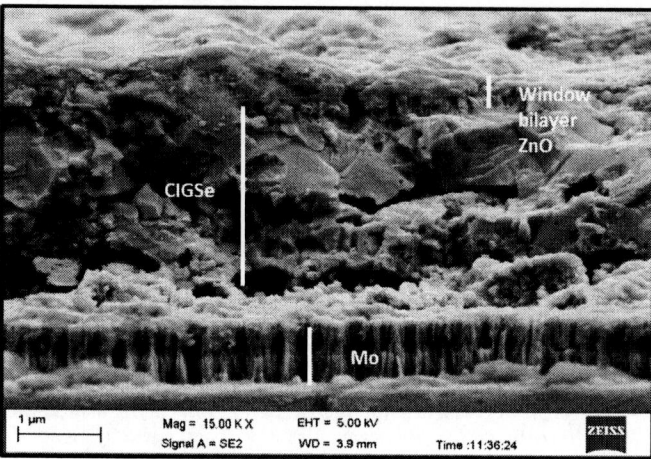

Fig. 3.   Cross Section of a CIGSe device

C. Current Voltage Characteristics

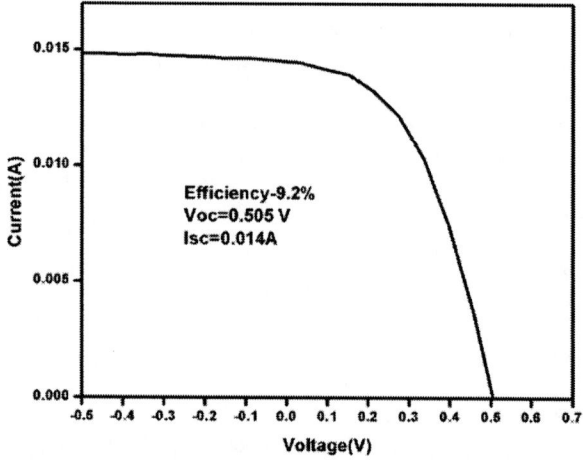

Fig. 4.   Current-voltage characteristics of CIGS/CdS device

In Fig. 4 CIGSe/CdS device IV measurements are shown. The absorber made up of this particular device was selenized at 40 minutes. The 9.2% efficiency measured at standard test conditions with Oriel Class 2A solar simulator with sample area 0.412 cm^2. The measured characteristics are without the antireflection coating. Remarkably, it is possible to fabricate the Solar cell absorber by using diethyl selenide. In order to improve the device performance we need to carefully design and optimized the other layers such as heterojunction partner CdS, ZnO window layers, substitution of the alkali elements and front/ back contacts.

978-1-5090-5606-4/17 $31.00 © 2017 IEEE   1702

IV. CONCLUSION

CIGSe *p*-type absorber layers and solar cells are successfully fabricated by using precursor Diethyl selenide as an alternative to H_2Se. It is expected that after ARC coating and careful optimization of the other layers from the device structure the efficiency would be greater than 12%.

ACKNOWLEDGEMENT

This research is based upon work supported in part by the Solar Energy Research Institute for India and the U.S. (SERIIUS) funded jointly by the U.S. Department of Energy subcontract DE AC36-08G028308 (Office of Science, Office of Basic Energy Sciences, and Energy Efficiency and Renewable Energy, Solar Energy Technology Program, with support from the Office of International Affairs) and the Government of India subcontract IUSSTF/JCERDC-SERIIUS/2012 dated 22nd Nov. 2012.

REFERENCES

[1] P. Jackson, R. Wuerz, D. Hariskos, E. Lotter, W. Witte, and M. Powalla, "Effects of heavy alkali elements in Cu(In,Ga)Se2 solar cells with efficiencies up to 22.6%," *physica status solidi (RRL) - Rapid Research Letters,* vol. 10, no. 8, pp. 583-586, 2016.

[2] A. Chirilă *et al.*, "Potassium-induced surface modification of Cu(In,Ga)Se2 thin films for high-efficiency solar cells," *Nat Mater,* Letter vol. 12, no. 12, pp. 1107-1111, 2013.

[3] I. Repins *et al.*, "19·9%-efficient ZnO/CdS/CuInGaSe2 solar cell with 81·2% fill factor," *Progress in Photovoltaics: Research and Applications,* vol. 16, no. 3, pp. 235-239, 2008.

[4] C. A. Wolden *et al.*, "Photovoltaic manufacturing: Present status, future prospects, and research needs," *Journal of Vacuum Science & Technology A,* vol. 29, no. 3, p. 030801, 2011.

[5] G. Y. Kim *et al.*, "High photoconversion efficiency in doublegraded Cu(In,Ga)(S,Se)2 thin film solar cells with two step sulfurization post treatment," *Progress in Photovoltaics: Research and Applications,* 2016.

[6] T. Feurer *et al.*, "Progress in thin film CIGS photovoltaics – Research and development, manufacturing, and applications," *Progress in Photovoltaics: Research and Applications*, 2016.

[7] A. kadam, "preparation of efficient cuin1-xgaxse2-ysy/cds thin-film solar cells by optimizing the molybdenum back contact and using diethylselenide as selenium precursor," university of central florida.

[8] E. Takahashi, "Correlation Between Preparation Parameters And Properties Of Molybdenum Back Contact Layer For Cigs Thin Film Solar Cells," University of Central Florida, 2010.

[9] A. A. Kadam and N. G. Dhere, "Highly efficient CuIn1−xGaxSe2−ySy/CdS thin-film solar cells by using diethylselenide as selenium precursor," *Solar Energy Materials and Solar Cells,* vol. 94, no. 5, pp. 738-743, 2010.

[10] W. Witte, R. Kniese, and M. Powalla, "Raman investigations of Cu(In,Ga)Se2 thin films with various copper contents," *Thin Solid Films,* vol. 517, no. 2, pp. 867-869, 2008.

Charge controlled sequential electrodeposition for synthesis of Cu_2ZnSnS_4 on Mo–coated glass substrate

Ashish K. Singh [a, *], Rajiv Dubey [b], Manoj Neergat [a], Kavaipatti R. Balasubramaniam [a, †]

[a] Department of Energy Science and Engineering, Indian Institute of Technology Bombay, Mumbai–400076 (INDIA)

[b] Department of Electrical Engineering, Indian Institute of Technology Bombay, Mumbai–400076, (INDIA)

[*]ashishks@iitb.ac.in, [†]bala.ramanathan@iitb.ac.in

Abstract — **Metal precursors are electrodeposited in the sequence of Cu–Sn–Zn on Mo–coated glass substrate to yield CZTS post sulfurization. Precursor deposition is optimized through systematic cyclic voltammograms (CVs) and SEM. All the component layers are deposited *via* controlling the deposition charge of the metals instead of the time–dependent deposition process; the latter method results in the excessive amount of metals and might contribute to the formation of secondary phase. Structural (XRD and Raman) analysis confirms formation of secondary phases of Cu_2S and SnS_2 in Cu and Sn–rich precursors. However, phase–pure CZTS synthesized with Zn–rich precursor.**

Index Terms — **CZTS; thin–film; absorber; Raman mapping; electrodeposition**

I. INTRODUCTION

Cu_2ZnSnS_4 (CZTS) is a p–type semiconductor with desirable band gap of 1.5 eV and high absorption coefficient ($>10^4$ cm^{-1}) [1]. The maximum reported efficiency is ~8.4 % [2] even though it can reach up to 32% (Shockley–Queisser (SQ) analysis) [3]. From the most efficient CZTS solar cell, it is evident that the reduced efficiencies are caused mainly due to losses arising from the open circuit voltage (V_{oc}) and fill factor (FF) [4]. The V_{oc} serves as a measurement of the extent of recombination of the electron–hole pair in the device. The recombination centers arise from impurity phases and the mismatch between the interfaces of different layers. Therefore, synthesis of phase–pure CZTS is essential to maximize the efficiency.

From various methods for CZTS synthesis as reported in the literature, this work involves the use of a sequential electrodeposition as it is cost–effective, environment friendly, and large area deposition is possible even at room temperature. Here, synthesis of phase–pure CZTS is carried out *via* potentiostatic charge controlled sequential electrodeposition wherein, the various layers such as Cu, Sn and Zn are deposited one by one on a Mo–coated substrate. The final component, sulfur is introduced through annealing of metal precursor in a sulfur atmosphere.

II. EXPERIMENTAL DETAILS

About 1 μm thick bi–layer Mo was deposited on soda lime glass *via* DC sputtering to be used as a substrate for absorber layer synthesis. The deposition of the metal precursor was performed by electrodeposition (Bio–logic SP 300) using a three electrode configuration at room temperature. Platinum foil was used as a counter electrode and saturated Ag|AgCl reference electrode. In Fig. 1, these electrodes are indicated as a (W), (C) and (R) respectively. The metal layers were deposited in the sequence of Cu–Sn–Zn (CTZ) as per the scheme shown in Fig. 1. Bath parameters for Cu, Sn, and Zn deposition are given in the Table 1, and their corresponding reduction potential was estimated from cyclic voltammetry (CVs) scan. Samples were labeled using following format; [Cu(R), Cu/Sn+Zn=1.4, Sn/Zn=1], [Zn(R), Cu/Sn+Zn=0.84, Sn/Zn=0.73], [Sn(R), Cu/Sn+Zn=0.84, Sn/Zn=1.36], [Zn(E), Cu/Sn+Zn=1, Sn/Zn=1], [Zn(R1), Cu/Sn+Zn=0.90, Sn/Zn=0.83], [Zn(L), Cu/Sn+Zn=1.06, Sn/Zn=1.13], [Zn(L1), Cu/Sn+Zn=1.11, Sn/Zn=1.25]

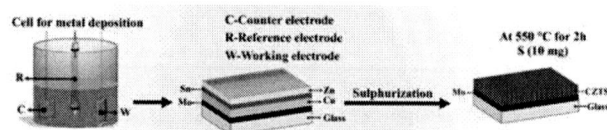

Fig. 1.   Schematic of process flow diagram from solution to CZTS

Sulfurization of electrodeposited substrate was carried out by annealing in a tubular furnace in S and Ar atmosphere at 550 °C for 2 h. A ramp–up rate of 20 °C min^{-1} was fixed for

TABLE 1: DEPOSITION PARAMETERS AND REDUCTION POTENTIAL FOR Cu, Sn, AND Zn

Layer	Bath parameter
Cu	0.2M $CuSO_4.5H_2O$, 0.01M H_2SO_4 and 1mM Additive
Sn	1M methane sulfonic acid, 50 mM Sn(II) methanesulfonate and Empigen BB–0.1% Vol
Zn	1M KCl, 0.05M $ZnCl_2$ and 3 pH Hydrion buffer solution

all experiments, and the furnace was naturally cooled down to ~30 °C to retrieve the samples. Structural properties of films were characterized by X–ray diffraction (XRD) (Rigaku Smartlab), and Laser Raman spectroscopy (Jobin–Yvon, France). Scanning electron microscopy (SEM) (Zeiss Ultra 55 FE) was used to determine the surface uniformity and the transmission electron microscopy (TEM) (Philips CM 200) was used to determine lattice orientation, d spacing and lattice defect.

III. RESULTS AND DISCUSSION

Voltammograms were performed in Cu, Sn and Zn electrolyte to find out the deposition window of Cu, Sn, and Zn, respectively. Fig. 2 (a) shows the voltammogram of Cu electrolyte on Mo substrate and the deposition window was found to be –0.12 to –0.18 V. The Cu film was deposited for 600 s at the different potentials (–0.12, –0.14, –0.16 and –0.18 V) and analyzed using SEM as shown in Fig. 2 (a1), (a2), (a3) and (a4). The optimum quality Cu film (uniform particle size and a lesser number of pinholes) was obtained at the lower negative potential of –0.12 V, as evidenced from the Fig. 2 (a1).

Similarly, Fig. 2 (b) and (c) shows the voltammograms of the analytical electrolyte for Sn and Zn on Cu and Sn films, respectively. From voltammograms, it is observed that the Sn and Zn deposition occurs in the potential range of –0.45 to –0.48 V and –1.13 to -1.16 V, respectively. As in the case of Cu deposition, in a similar way the reduction potential was optimized to obtain a uniform film of Sn and Zn from SEM analysis. The SEM images of Sn and Zn deposition are shown in Fig 2 (b1), (b2), (b3), (b4) and (c1), (c2), (c3), (c4) and the optimum potential for Sn and Zn deposition is –0.48 and –1.15 V respectively.

After optimization of reduction potential, the CTZ precursors were deposited with different proportion ratio and sulfurized. The CZTS phases were identified through XRD analysis (Fig. 3) and the corresponding peaks were compared with the standard data (JCPDS 26–0575 for CZTS and 01–078–2121 for Cu$_2$S). From the Fig. 3 (a), the angles of diffraction at 21.72°, 27.73°, 29.30°, 31.84°, 33.18°, 44.24° and 52.78° correspond to the secondary phases of Cu$_2$S for Cu(R) precursor. All these secondary phases of Cu$_2$S are absent for the film prepared with the other proportions ratio.

Fig. 4 shows the Raman spectra of all proportions of precursors after sulfurization. All the sulfurized samples show a strong mode at 335 to 339 cm^{-1} that belong to the A$_1$ mode of CZTS, and it is related to pure anion modes corresponding to the vibrations of sulfur atoms surrounded by static neighboring atoms. The co–existence of a weak mode around 288 cm^{-1} related to E and B symmetry along with the A$_1$ mode confirms the presence of CZTS in all films. However, for Cu–rich precursor, an impurity phase of Cu$_2$S is shown by the presence of a Raman active mode at 475 cm^{-1} [5]. The appearance of minor phase of Cu$_2$SnS$_3$ (306 cm^{-1}), for Cu(R), Sn(R) and Zn(E) metal precursor. On the contrary, there is shift of 6 cm^{-1} in case of Sn(R) and Zn(E) precursor in Raman spectra. These might have resulted from the change in symmetry of the material, which is not yet clear. The peaks obtained for the ratios of Zn(R) are 288 cm^{-1} and 339 cm^{-1}, that exactly matches with phase–pure CZTS [5]. Therefore, Zn (R) proportion is suitable to obtain a pure–phase CZTS [6].

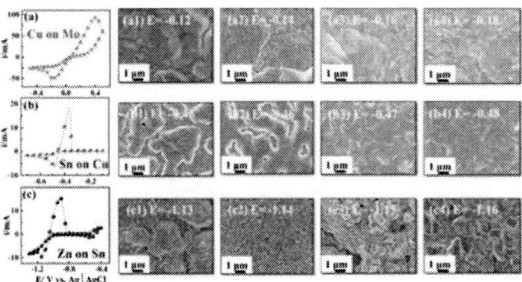

Fig. 2. (a, b and c) CVs of Cu, Sn and Zn recorded in argon–saturated solution at 10 mV s^{-1} scan rate to find out the deposition window. (a1, a2, a3 and a4) SEM images of Cu deposited on Mo; (b1, b2, b3 and b4) Sn deposited on Cu; (c1, c2, c3 and c4) Zn deposited on Sn substrate under different reduction potential.

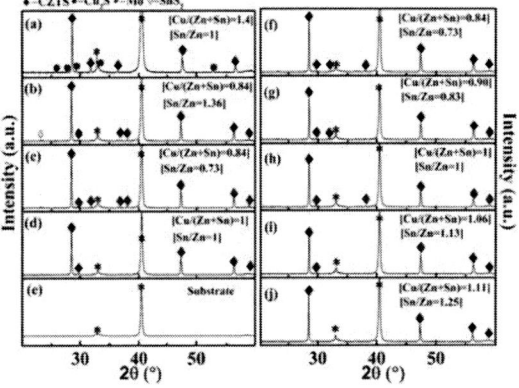

Fig. 3. X–ray diffraction patterns of the thin films synthesized by sulfurization of different proportion of precursor layers. The peaks corresponding to the CZTS phase (♦), Cu$_2$S phase (•), SnS$_2$ phase (◊) and the Mo substrate (∗) are marked.

XRD patterns and Raman spectrum indicates the presence of phase–pure CZTS in the Zn(R) precursor. Further, Raman mapping was carried out to analyze the homogeneity of CZTS films and the extent of Cu$_2$S impurities present in Zn(R) precursor. Fig. 5 shows the integrated Raman scan of CZTS and Cu$_2$S for the films obtained from an area of 10×10 μm^2. In Fig. 5 (a), the color scale of the mapped images varies from black to bright yellow corresponding to the regions where the CZTS phase is present and those devoid of the CZTS phase, respectively.

The uniformity of CZTS with Zn(R) precursor is confirmed by the Fig. 5 (b) where the Cu$_2$S Raman peak (475 cm^{-1}) has been mapped. The bright yellow regions correspond to Cu$_2$S, and the black regions to phases other than Cu$_2$S. Sulfurization of the Zn(R) precursor does not yield Cu_2S impurity phase. Hence, the Raman mapping and Raman spectrum data conclusively prove that phase–pure CZTS is obtained for the Zn(R) precursor.

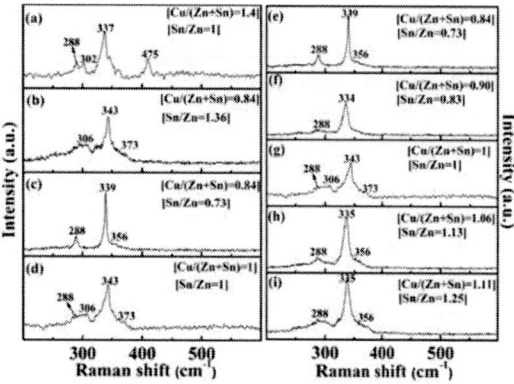

Fig. 4. Raman spectroscopy of thin films synthesized by sulfurization of different proportion of metal precursor.

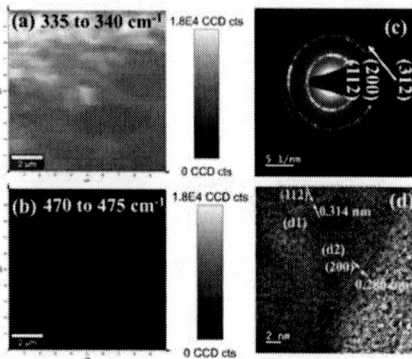

Fig. 5. Raman mapping of the sulfurized thin films obtained over an area of 10 μm×10 μm. The Raman peak corresponding to CZTS at 337 cm^{-1} is mapped (a). Similar mapping of the Raman peak corresponding to the Cu_2S phase at 475 cm^{-1} is shown in (b). The SAED (c) and HRTEM (d) of [Cu/Sn+Zn=0.84] and [Sn/Zn=0.73] precursor.

Selected area electron diffraction patterns (SAED) and high resolution transmission electron microscopy (HRTEM) images are shown in Fig. 5 (c) and (d). Polycrystalline ring pattern shown in Fig. 5 (c) perfectly overlaps the recorded patterns of CZTS [(112), (200), and (312)] and hence corroborates the presence of phase–pure CZTS obtained from the Zn(R) precursor. Fig. 5(d) shows the HRTEM analysis (carried to obtain the lattice spacing and orientations of the films. The lattice spacing of 0.314 nm, obtained from the fringes in the region marked (d1), corresponds well with the d-spacing of the (112) planes of CZTS. Similarly, a lattice spacing of 0.280 nm, obtained from the fringes in the region marked (d2), corresponds well with the d-spacing of the (200) planes of CZTS. This corroborates the presence of polycrystalline CZTS inferred from the SAED analysis.

Therefore, the TEM analysis confirm the conclusions derived from X–Ray diffraction and Raman spectroscopy, establishing that phase–pure CZTS is formed in films synthesized from the Zn(R) precursor, whereas, Cu_2SnS_3, SnS_2 and Cu_2S is the major impurity in the films from the other precursor.

IV. CONCLUSION

Metal precursors Cu–Sn-Zn was deposited *via* cost effective non–vacuum based sequential electrodeposition. Molar ratio of [Cu/Sn+Zn] and [Sn/Zn] plays a crucial role in the formation of CZTS. It is observed that the phase–pure CZTS is formed with the sulfurization of precursor at 550 °C obtained with Zn(R) precursor. Moreover, other proportions formed additional phase of Cu_2SnS_3, SnS_2 and Cu_2S along with the CZTS, which is evidenced through XRD and Raman analysis. Raman mapping clearly reveals the uniformity of phase–pure CZTS.

ACKNOWLEDGEMENTS

This research work is supported by the Solar Energy Research Institute for India and the U.S. (SERIIUS) funded jointly by the U.S. Department of Energy subcontract DE AC36–08G028308 (Office of Science, Office of Basic Energy Sciences, and Energy Efficiency and Renewable Energy, Solar Energy Technology Program, with support from the Office of International Affairs) and the Government of India subcontract IUSSTF/JCERDC-SERIIUS/2012 dated 22nd Nov, 2012

REFERENCES

[1] K. Ito and T. Nakazawa, "Electrical and optical properties of stannite-type quaternary semiconductor thin film," *Jpn. J. Appl. Phys.*, vol. 27, no. 11, pp. 2094–2097, 1988.

[2] S. Ahmed, K. B. Reuter, O. Gunawan, L. Guo, L. T. Romankiw, and H. Deligianni, "A high efficiency electrodeposited Cu_2ZnSnS_4 solar Cell," *Adv. Energy Mater.*, vol. 2, pp. 253–259, 2012.

[3] S. Siebentritt, "Why are kesterite solar cells not 20 % efficient?," *Thin Solid Films*, vol. 535, pp. 1–4, 2013.

[4] I. Repins *et al.*, "Co-evaporated $Cu_2ZnSnSe_4$ films and devices," *Sol. Energy Mater. Sol. Cells*, vol. 101, pp. 154–159, 2012.

[5] A. K. Singh, A. Shrivastava, M. Neergat, and K. R. Balasubramaniam, "Phase evolution during the electrodeposition of Cu–Sn–Zn metal precursor layers and its role on the Cu_2S impurity phase formation in Cu_2ZnSnS_4 thin films," *Sol. Energy*, vol. 155, pp. 627–636, 2017.

[6] P. A. Fernandes, P. M. P. Salomé, and A. F. da Cunha, "Study of polycrystalline Cu_2ZnSnS_4 films by Raman scattering," *J. Alloys Compd.*, vol. 509, no. 28, pp. 7600–7606, 2011.

Effect of deposited pressure on the CdTe thin films by closed space sublimation method

Yufeng Zhang, Zhongming Du, Xiangxin Liu

The Key Laboratory of Solar Thermal and Photovoltaic System,

Institute of Electrical Engineering, Chinese Academy of Sciences, Beijing 100190, China

Abstract — A home-made CSS system in which the CdTe source was above the substrate was used to deposit CdTe thin film. The pressure (3.7~260 Pa) was controlled by nitrogen. The experimental results showed that the deposited rate increased with the decreasing pressure. As the pressure decreased, the center of the film was thicker than that in the edge. Accordingly, the micromorphology obtained by SEM showed that the crystal grain size became larger while the orientation was more random with the decreasing deposition pressure analyzed by XRD. The Non-stoichimometric composition shift of CdTe thin films was characterized by EDX and the Cd/Te atomic ratio was 0.87 ± 0.04, 0.88 ± 0.08 and 0.93 ± 0.13 as the increasing pressure in our experiments. The calculated results showed that Te-5p orbitals of CdTe with Cd vacancy increased comparing with the bulk CdTe at G-point.

I. INTRODUCTION

CdTe is used as an absorbed layer in solar cell for its low cost, high conversion efficiency, band-gap of about 1.5 eV with high absorbed coefficient. To fabricate high performance CdTe thin film solar cell, many methods were used for CdTe thin film deposition, such as close-spaced sublimation (CSS), vapor transport deposition, high vacuum evaporation, electro-deposition, sputtering and so on. In those methods, CSS is one of the most important methods to deposit the high-quality CdTe thin film, which has been used for mass production in China. The growth regimes of CdTe deposition by CSS with source under substrate have been investigated extensively in various labs. However, for large area module production, it is necessary to deposit with substrate under source to prevent deformation of glass plate at high temperature.

It has been shown that the growth rate of CdTe decreased exponentially as the temperature increased, when the temperature was above 450℃ [1]. However, as the thin film was deposited, the high deposition temperature and difference in vapor pressure made that the CdTe usually non-stoichiometry. As shown in phase diagram, the Te-rich and Cd-rich CdTe were formed at high and low temperature in deposition process, respectively [2], which affected the properties of the material. As found in Te-rich film grown by CSS, Te antisite and Te interstitial was considered to be the main deep-level defect in CdTe based on local density approximation, which had a harmful effect on improvement of minority-carrier lifetime [3]. It is also reported that longer minority-carrier lifetimes could be achieved under Cd-rich

condition [3]. Additionally, the effect of deposition temperature and pressure on the defect profile of material via the partial pressure of the constituent elements has been established [4].

In this work, the home-made CSS system in a geometry of CdTe source above glass substrate was used to deposit CdTe films. The non-stoichiometry, morphology and structure of the films were studied at temperature above 450 °C and various pressures.

Experiment:

The home-made CSS system shown in Fig. 1 was used to deposit CdTe thin film on a borosilicate glass of Corning 7059. The 7059 glass which was cleaned with the hot soap and DI water was used as substrate. The ceramic crucible was made of a hole with about 1 mm in diameter along with the central axis of cylinder, so that vapor could diffuse from the CdTe chamber through this hole and reach the glass surface. The entrance of the hole was above the bottom of the CdTe chamber where CdTe powder was contained.

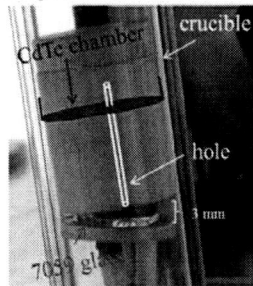

Fig. 1. The schematic of Close-Space Sublimation system.

The source of CdTe powder and 7059 glass substrate were radiatively heated by quartz-halogen lamp under the control of programmable temperature controller. Two K-type thermocouples were used to read the temperature of the substrate and source of CdTe, respectively. The distance between the bottom of the crucible and the substrate was about 1mm.

The chamber was pumped down to a base pressure of approximately 0.3 Pa. The chamber was then purged with high purity N_2 (99.999%) before reaching a designated pressure.

The crystal structures of the CdTe source material and thin films were characterized by X-ray diffraction (XRD). The morphologies were observed in SEM. The compositions of the thin film were investigated by EDX.

Computational method:

The structure and energy of CdTe were calculated with the CASTEP code [5] based on the density function theories (DFT). The calculations were performed with the Local Denstiy Approximate (LDA) in the form of CAPZ. The structural optimizations were conducted using the BFGS minimization [6]. The interactions between the ionic core and valence electrons were described by the ultrasoft pseudo–potential. The valence wave functions were expended in a plane-wave basis set with cut-off energy of 290 eV. A Monkhorst-pack (MP) 4×4×4 sampling was employed to evaluate integrals in the reciprocal space. In order to investigate the effect of Cd Vacancy, A K-point mesh of 2×2×2 was used for a 64-atom supercell.

Results and Discussion:

CdTe deposited at three different pressures but the same source and substrate temperature of 542 °C and 461 °C, respectively. All the three films were deposited for 20 minutes. As shown in Fig. 2, the CdTe film became thicker at lower pressure. In the Fig. 2(a), CdTe under the center of injection hole was much thicker than edge.

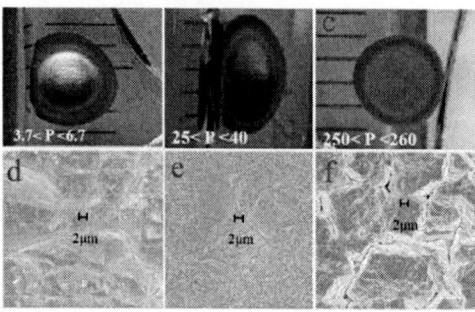

Fig. 2. The morphology of CdTe deposited under the pressure of (a) 3.7~6.7 Pa, (b) 25~40 Pa and (c) 250~260 Pa, which corresponded to the mico-morphology of (d), (c) and (f) obtained by SEM.

Micromorphology of the CdTe films observed in SEM showed that the grain size enlarged as the deposited pressure decreased, which could reach above 100 μm for deposition at 3.7-6.7 Pa. It could also been clearly seen that grains were more closely packed for deposition at lower pressure. Yet the

one deposited at 250-260 Pa shows inter-grain gaps about half micron wide.

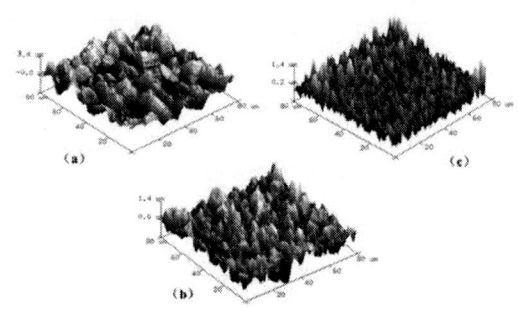

Fig.3. AFM images of the CdTe films deposited at pressure: (a) 3.7~6.7 Pa, (b) 25~40 Pa and (c) 250~260 Pa.

The rough surface of the CdTe films was analyzed over 80 μm × 80 μm area by using AFM. RMS roughness values for films were 1049 nm, 373 nm and 324 nm, respectively, which corresponded to the deposited pressure at 3.7~6.7 Pa, 25~40 Pa and 250~260 Pa (Fig. 3). The surface roughness had a great effect on the back metal contact [7]. If the contacting metal layer thickness than the surface roughness of CdTe film, it could create discontinuities contact between the interface of CdTe and back metal layer. So the appropriate thickness of back metal contacts in CdTe device fabrication would be chose according to the surface roughness of CdTe.

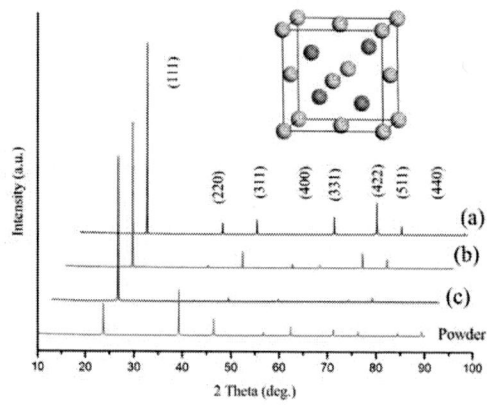

Fig. 4. XRD spectra of the CdTe source powder and films deposited at: (a) 3.7~6.7 Pa, (b) 25~40 Pa and (c) 250~260 Pa.

978-1-5090-5606-4/17 $31.00 © 2017 IEEE

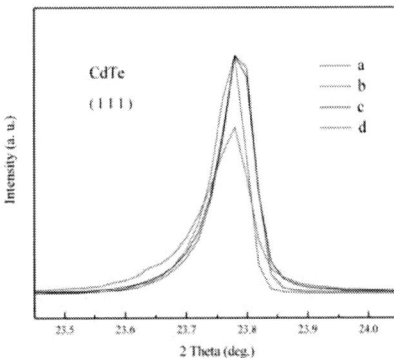

Fig. 5. The (111) peaks of XRD for CdTe films under different pressures: (a) 3.7~6.7 Pa, (b) 25~40 Pa and (c) 250~260 Pa. (d) the (111) peak of CdTe powder.

The crystal structure and orientation of CdTe thin films at different deposition pressure were analyzed by XRD in Fig. 4. The bulk powder CdTe presented a cubic structure with the space group F -4 3 m. As shown in Fig. 4, (111) preferred orientation was enhanced by increasing the deposition pressure from ~3.7 Pa to ~260 Pa. Additionally, it could be seen from Fig. 5, the diffraction angle of (111) peak for CdTe film deposited at about 3.7 GPa was the same as the one of CdTe powder, however, it was a little lower than the values of thin films deposited at about 25 Pa and 250 Pa, which indicated that there was obvious lattice strain in CdTe films at higher deposited pressure.

Non-stoichimometric composition shift of CdTe thin films was characterized by EDX. The Cd/ Te atomic ratio was 1:1 by using CSS method. However, the minor deviation of the composition of thin film CdTe from its perfect stoichiometric crystal has major effect on its electronic properties. The Cd/ Te atomic ratio was 0.87 ± 0.04, 0.88 ± 0.08 and 0.93 ± 0.13 corresponded to the increasing pressure, which could cause vacancies on the Cd-site (V_{cd}). The V_{cd} was caused from the large collision faction of the deposition at atom (ion) with nitrogen atom. Those vacancies which acted as scattering and recombination centers had detrimental effect on the electron mobility and conductivity [7].

It has also been calculated that the V_{cd} was an acceptor and the formation energies was 2.67 eV, which was lower than 3.7 eV of Te_{Cd} (Te on Cd site) [8,9]. Those results inferred that the V_{Cd} was the domain acceptor in CdTe, which was consistent with the experiments [10,11] So in this paper, the Cd vacancies in CdTe were calculated to investigate the electronic properties of CdTe. The calculated lattice parameter was 6.43 nm which was agreed well with the values obtain by Roehl [12]. Theoretical calculation showed that the defect of Cd vacancy was shallow acceptor in CdTe [13]. The band

structures and density of states (DOS) of bulk and Cd vacancy for CdTe were shown in the Fig. 5. The top of the valence band originates from the Te-5p orbitals. The bottom of conduction band was composed of the anti-bonding 5s states of Cd and Te atoms. The Te-5p orbitals of CdTe with Cd vacancy increased comparing to the bulk CdTe at G-point.

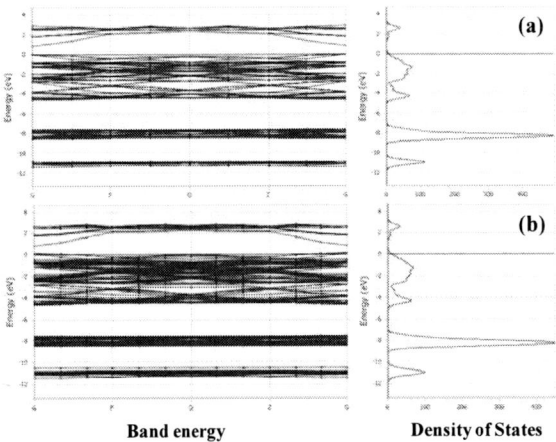

Band energy   **Density of States**

Fig. 6. The Calculated band structures and density of states for (a) the bulk of CdTe and (b) V_{Cd} of CdTe.

In Fig. 6, the valence band maximum (VBM) and conduction band minimum (VBM) occurred at G point, making this compound to be direct band gap material. Using LDA we had obtained a direct band gap for CdTe equal to 0.88 eV. This result was higher than the value, 0.75eV, calculated by Agrawal [14] but it was smaller than the value 1.31 eV, calculated by Raiss [16] and the experimental value 1.45 eV for band gap[17]. It was well known that the calculated values of the energy gap with GGA and LDA methods were strongly underestimated with respect to the experimental value [18]. The Fermi level (E_f) was chosen to locate at 0 eV, which coincided with top of the valence band. Comparing with Fig. 6(a), the valence band in Fig. 6(b) was shifted toward higher energies. It might derive from the fact that the lattice strain caused the band repulsion. Additionally, the defect energy level also had a contribution to the VBM.

Conclusion

Thin films of CdTe have been deposited onto the borosilicate glass by close-spaced sublimation and the effect of the pressures in a range from ~ 3.7 Pa to 250 Pa were investigated. The Cd/ Te atomic ratio of CdTe film increased depending on the increasing pressures while the thickness and the grain size of CdTe film decreased. The first principle calculated showed the Te-5p orbitals of CdTe with Cd vacancy increased comparing with the bulk CdTe at G-point.

REFERENCES

[1] J. Luschitz, K. Lakus-Wollny, A. Klein and W. Jaegermann, *Thin Solid Film*, vol. 515, pp. 5814, 2007.

[2] J. H. Greenberg, *Journal of crystal growth*, vol. 161, pp. 1-11, 1996.

[3] J. Ma, D. Kuciauskas, D. Albin, R. Bhattacharya and M. Reese, *Phys. Rev. Lett.*, vol. 111, pp. 067402, (2013).

[4] E. Ertekin, V. Srinivasan, J. Ravichandran, P. B. Rossen, W. Siemons, A. Majumdar, R. Ramesh, and J. C. Grossman, *Phys. Rev. B* vol. 85, pp.195469 , 2012.

[5] M. D. Segall. J. J. F. Lindan, K. J. Probert, C. J. Pickard, P.J. Hasnip, S. J. Clark, M. C. Payne, *J. Phys.-Condens. Matter* vol. 14, pp. 2717, 2002.

[6] B. G. Pfrommer, M. Cote, S. G. Louie and M. L. Cohen, *J. Comput. Phys.* Vol.131, pp. 233, 1997.

[7] I. M. Dharmadasa, *Coatings*, Vol. 4, pp.282, 2014.

[8] S. H. Wei and S. B. Zhang, *Phys. Rev. B*, vol. 66, pp. 155211, 2002.

[9] H. P. Zhu, M. Q. Gu, L. Huang, J. L. Wang and X. S. Wu, *Mater. Chem. Phys*, vol. 143, pp. 637, 2014.

[10] T. E. Schlesinger, J. E. Tony, H. Yoon, E. Y. Lee, B. A. Brunett, L. Franks and R. B. James, *Mater. Sci. Eng.*, Vol. 23, PP. 103, 2001.

[11] M. Fiederle, V. Babentsov, J. Franc, A. Fauler and J. P. Konrath, *Cryst. Res. Technol.*, Vol.38, PP.588, 2003.

[12] J. L. Roehl and S. V. Khare, *Solar Energy*, vol. 101, pp. 245-253, 2014.

[13] Mao Hua Du, Hiroyuki Takenaka, David J. Singh, *J. Appl. Phys.* Vol. 104 pp. 093521, 2008.

[14] B. K. Agrawal and S. Agrawal, *Phys. Rev. B*, vol. 45, pp. 8322, 1991.

[15] A. A. Raiss, Y. Sbai, L. Bahmad and A. Benyoussef, *J. Magn. Magn. Mater.*, vol. 385, pp. 295, 2015.

[16] K. Durose, et al., *J. Cry. Growth*, vol. 197, pp. 733, 1999.

[17] R. W. Gofby, M. Schluter and L. J. Sham, *Phys. Rev. B*, vol. 37, pp. 10159, 1988.

Analyzing the Cost Reduction Potential of III-V/Si Hybrid Concentrator Photovoltaic Systems

Kan-Hua Lee, Kenji Araki and Masafumi Yamaguchi

Toyota Technological Institute, Nagoya, Aichi, 468-0028, Japan

Abstract—The cost reduction potential of concentrator photovoltaics (CPV) by adding a silicon backplane solar cell is quantified. Using a backplane solar cell module in a CPV system to absorb the diffused sunlight has been proposed and demonstrated as a way to fully utilize the global solar spectrum instead of just direct normal incidence. However, it is not clear that whether the energy price of this system can ultimately be competitive with the state-of-the-art CPV or flat-panel silicon PV system. In this paper, we analyzed the energy yields and the cost of the III-V/Si hybrid CPV system to find its limiting cost reduction potential. Our results suggest that it is unlikely that the energy price of CPV can be dramatically reduced simply by introducing the backplane silicon module into the the state-of-the-art CPV system. One should completely reconsider the designs of CPV components and leverage the unique properties of the III-V/Si hybrid CPV system to make this concept succeed.

Index Terms—concentrator photovoltaics, III-V multi-junction solar cell, silicon solar cell.

I. INTRODUCTION

Concentrator photovoltaics (CPV) system can potentially have the highest energy yields among all the photovoltaic systems. However, its cost has to be further reduced in order to compete with the flat-panel silicon system or even fossil fuel[1]. Apart from reducing the cost of each components in a CPV system, raising the concentration is an obvious way to further reduce the system cost because it can dilute the areal cost of tracker and balance of systems components[2]. With the continuing price drop of flat panel silicon modules, adding silicon modules underneath the concentrator cells to form a hybrid CPV system emerges as another promising pathway to further reduce the cost of CPV system. The idea is to use silicon cell to absorb the diffused sunlight that cannot be focused on the concentrator solar cell. It has been demonstrated that this could increase the overall energy yields of CPV systems[3]. This design also brings other potential benefits. Firstly, it could relax the constraint that CPV has to be installed in locations with constantly high direct normal incidence (DNI) because the hybrid CPV system could utilize the diffused sunlight. Also, it becomes questionable that whether high-precision but costly module and tracking system are still necessary for a hybrid CPV system. With a backplane cell to absorb the misaligned incident light rays, there may be other more cost effective options for the concentrator module and tracking system. However, it has not been quantified yet that how much cost reduction could be achieved by these approaches or what criteria should be met to make the hybrid

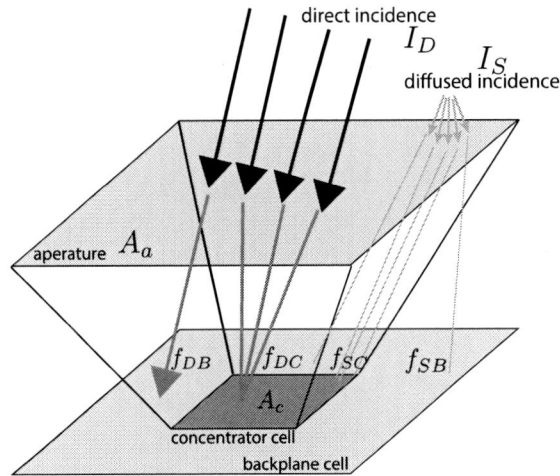

Fig. 1. An illustration of the hybrid CPV system model.

CPV competitive with flat-panel silicon in terms of the energy price.

In this work, we perform theoretical calculations and cost analyses to investigate the cost reduction potential of the hybrid CPV system. By using this result, we will estimate the criteria that a hybrid CPV system should achieve in order to succeed.

II. THE SYSTEM COST MODEL

Our model of the III-V//Si hybrid concentrator module is illustrated in Fig. 1. The concentrator optics concentrates the incident light onto the concentrator solar cell. The concentration ratio is defined as the ratio between the aperture area and the cell area $1/A_c$. This model assumes that all the diffused horizontal incidence (DHI) are absorbed by the backplane cell. It also assumes that a certain portion f_c of the direct normal incidence (DNI) is absorbed by the concentrator cell. The remain DNI is absorbed by the backplane cell. By definition, small f_c thus indicates the error caused by the tracking and concentrator module. However, small f_c also be thought of as the locations with low DNI, although this is not very precise.

We use a simplified model described in [2] to calculate the system price of the conventional III-V and hybrid CPV

978-1-5090-5606-4/17 $31.00 © 2017 IEEE

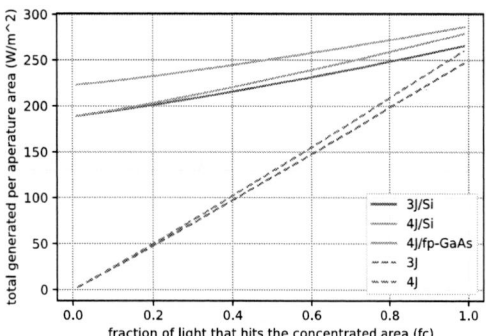

Fig. 2. The total generated power per aperture area of different CPV systems and cells against f_c. The solid lines are hybrid CPV systems, whereas the broken lines are conventional CPV.

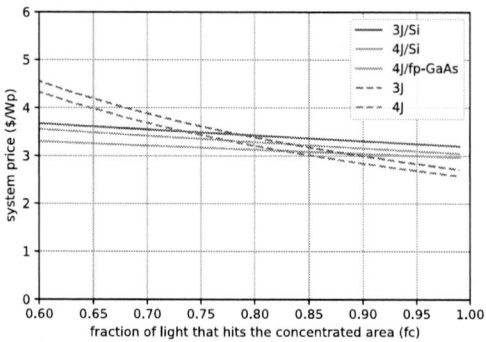

Fig. 3. The system price of different types of CPV systems against f_c. The solid lines are hybrid CPV systems, whereas the broken lines are conventional CPV.

system. The system price ($/W_p$) of the conventional CPV system is described as

$$
\left(\frac{C}{E}\right)_{cpv} = \left(\frac{C_{mod1}}{A_{mod}} + \frac{C_{BOS,area}}{A_{mod}} + \frac{C_{IMF}}{A_{mod}} + \frac{C_{pwr\ cond}}{A_{mod}} + \frac{C_{tracking}}{A_{mod}}\right) / \left(I_c \eta_c\right) \quad (1)
$$

For hybrid CPV system, we simply add the cost of the backplane module and assumes that other parts of the cost remain unchanged, namely,

$$
\left(\frac{C}{E}\right)_{hybrid} = \left(\frac{C_{mod1}}{A_{mod}} + \frac{C_{mod2}}{A_{mod}} + \frac{C_{BOS,area}}{A_{mod}} + \frac{C_{IMF}}{A_{mod}} + \frac{C_{pwr\ cond}}{A_{mod}} + \frac{C_{tracking}}{A_{mod}}\right) / \left(I_c \eta_c + I_b \eta_b\right) \quad (2)
$$

The cost of each component is listed in TABLE I. In the above equation, I_c is the illumination intensity on the concentrator cell, I_b is the illumination intensity on the backplane cell, η_c is the efficiency of the concentrator cell system and η_b is the efficiency of backplane cell system. The system loss of CPV and backplane cell are also included.

The performances of solar cells are modeled by a detailed balance approach with empirical external radiative efficiency values [4]. The parameters of these hypothetical solar cells are listed in TABLE II.

III. RESULTS AND DISCUSSIONS

The projected output power of AM 1.5 Global illumination of CPV and hybrid CPV system against f_c are plotted in Fig. 2. CPV systems have much sharper increase in output power comparing to hybrid CPV because only the photons incident outside the concentrator cells are simply waster. Both of the systems reach their maximum at $f_c = 1$, but hybrid CPV has some extra power gain due to its ability of absorbing DHI.

Fig. 3 compares the system price of CPV and hybrid-CPV with different types of III-V and backplane solar cells. As

expected, the system price of CPV is cheaper than hybrid CPV at high f_c. The break-even point of hybrid CPV occurs between $f_c = 0.8$ and $f_c = 0.9$. This result shows that the system price of hybrid CPV can be cheaper than conventional CPV if the concentrator optics or the tracker cannot fully concentrate the sunlight onto the concentrator cell, or the global solar spectrum contains less portion of DNI, such as cloudy weather.

We further compare the cost of different CPV systems against the price of utility scale flat-panel silicon PV. In this calculation, we choose $2/W_p$ as our price benchmark. Fig. 4 plots the areal cost that needs to be reduced in order to achieve $2/W_p$. At ideal scenario $f_c = 1$, the CPV system still needs to reduce its areal cost by around $100/m^2$ in order to beat the $2/W_p$-benchmark. We can also see that areal cost reduction target of conventional CPV system varies a lot against f_c, indicating that the actual system price strongly depends on the portion of DNI in the input spectrum. For hybrid CPV systems, we found that it requires more than $300/m^2$ of areal cost cut in order to achieve the benchmark. This is nearly the sum of the balance of system cost and the tracker cost, as listed in TABLE I. Thus it is extremely challenging to meet this target with the current cost model of CPV. Therefore, simply inheriting the design of conventional CPV system does not give much room for further cost reduction of hybrid CPV. However, since a hybrid CPV system is less sensitive f_c, which can be utilized to rethink the design of the CPV system for hybrid CPV modules. In other words, a hybrid CPV system can use the module, tracking and other balance of system components that gives in lower f_c but less cost. Also, raising the efficiency of both the III-V and the backplane cell help reduces this price gap between hybrid CPV system and the benchmark.

IV. CONCLUSION

We use a simplified model to analyze how a hybrid CPV module can help further reduce the cost of CPV system. Our calculation shows that a hybrid CPV system can be more

TABLE I
INPUT PARAMETERS OF THE COST MODEL [2]

Cost ($/m²)	CPV	hybrid CPV	Description
C_{mod1}/A_{mod}	282	282	cost of III-V cell and module
C_{mod2}/A_{mod}	0	100	cost of silicon backplane module
$(C_{BOS} + C_{IMF})/A_{mod}$	260	260	cost of balance of system
$C_{tracking}/A_{mod}$	51	51	cost tracking system

TABLE II
INPUT PARAMETERS OF THE SOLAR CELLS.

Solar cell	External radiative efficiency
InGaP/GaAs/Ge (3J)	0.001/0.01/0.0001
InGaP/GaAs/InGaAs/Ge (4J)	0.001/0.01/0.0001/0.0001
Silicon (Si)	0.001
flat-panel GaAs (fp-GaAs)	0.01

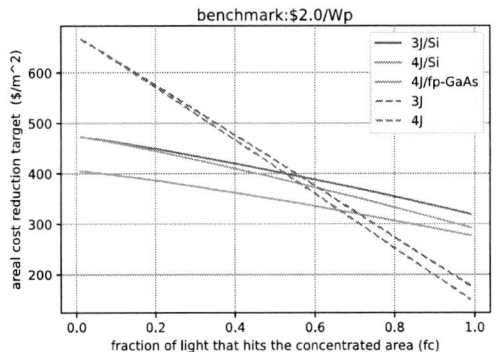

Fig. 4. The areal cost reduction target of different types of CPV system against f_c. The solid lines are hybrid CPV systems, whereas the broken lines are conventional CPV. The target system price for this comparison is $2/Wp$.

cost effective than a conventional CPV system if the fraction of average intensity absorbed by the concentrator cell is less than 80%. The result also shows that it is very difficult for the hybrid CPV system to surpass the $2/W_p$ benchmark based on cost of state-of-the-art CPV components. The cost reduction opportunity lies in leveraging the high tolerance of the optical alignment of hybrid CPV system to redesign and reduce the cost of these components.

ACKNOWLEDGMENT

The authors would like to thank Japan New Energy and Industrial Technology Development Organization (NEDO) for supporting this research (NEDO 15100731-0).

REFERENCES

[1] E. N. J, P. Sandwell, J. Nelson, A. D. Johnson, G. Duggan, and E. Herniak, "What does CPV need to achieve in order to succeed?" *AIP Conference Proceedings*, vol. 1766, p. 020004, 2016.

[2] R. King, D. Bhusari, D. Larrabee, X. Liu, E. Rehder, K. Edmondson, H. Cotal, R. Jones, J. Ermer, C. Fetzer, D. Law, and N. Karam, "Solar cell generations over 40% efficiency," *Prog Photovoltaics Res Appl*, vol. 20, no. 6, pp. 801–815, 2012.

[3] K. T. Lee, Y. Yao, J. He, B. Fisher, X. Sheng, M. Lumb, L. Xu, M. A. Anderson, D. Scheiman, S. Han, Y. Kang, A. Gumus, R. R. Bahabry, J. W. Lee, U. Paik, N. D. Bronstein, A. Alivisatos, M. Meitl, S. Burroughs, M. M. Hussain, J. C. Lee, R. G. Nuzzo, and J. A. Rogers, "Concentrator photovoltaic module architectures with capabilities for capture and conversion of full global solar radiation." *Proc. Natl. Acad. Sci. U.S.A.*, vol. 113, no. 51, pp. E8210–E8218, 2016.

[4] M. Green, "Radiative efficiency of state-of-the-art photovoltaic cells," *Prog Photovoltaics Res Appl*, vol. 20, no. 4, pp. 472–476, 2012.

Generalized Numerical Design of Axially-asymmetrical and Grid-arranged Static CPV Array for Maximizing Annual Energy Generation.

Kenji Araki, Kan-Hua Lee and Masafumi Yamaguchi

Toyota Technological Institute, Nagoya, 468-8611 Japan,

Abstract — **A CAD/CAM friendly design based on a Monte Carlo method was developed to give the optimized axially-asymmetrical lens profile suitable to the grid arrangement with given site conditions such as horizontal plane, and sloped surface in the southern direction. Such CPV panel selectively collects the sunlight from high incident angle and has the substantially wide acceptance half angle with sacrifice of the optical loss in the normal region. The optimization using asymmetrical function successfully adjusted the lens profile for the grid arrangement. With the use of optimized axially-asymmetrical lens, it is not always necessary to rely on honeycomb arrangement.**

Index Terms —**CPV, Optics, Static concentrator, Optimization.**

I. INTRODUCTION

Use of PV to the energy source of the personal cars has been considered and tried by many automobile manufacturers. Since the panel is installed in horizontal position, the sun direction does not have correlation to the panel direction. The car roof is shaded by many surrounding buildings. As a result, the PV panel should collect the sunlight from shallow angles, namely high incident angles [1]. Figure 1 shows an example of the car-roof static low concentrator PV.

Fig. 1. An example of the static concentrator PV for the car-roof.

Static concentrator has a long history [2]. Several types of concentrators were proposed including reflective type [2-3], prism type [4], light-confinement type [5], luminescent concentrator [6] and the lens type [7-8]. More recently, thanks to the progress of bifacial Si solar cells, the static concentrator using bifacial cells are reevaluated [9]. For high latitude area, building-integrated static concentrator for vertical installation was developed [10]. For the lens design algorithm, in addition to the classical geometrical calculation based on Snell's law

[8], freeform design based on the variation method, which is constrained by Fermat's principle, has been developed [11-12]. These freeform optimizations were powerful in searching for optimized lens profile, but they were complicated and difficult to use in flexible product design.

II. TARGET OF THE DEVELOPMENT

Design of the optics is time-consuming and inflexible to practical compromises from non-optical issues like ease of assembly. This difficulty was thought to be enhanced in static concentrator application.

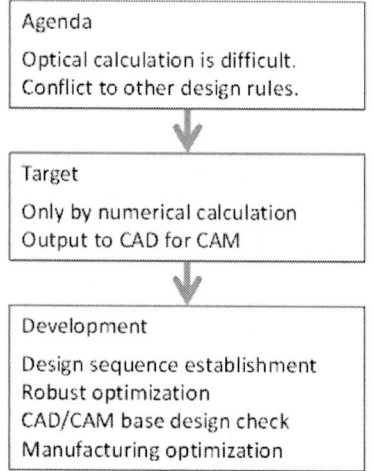

Fig. 2. Diagram of the project of the development on optical design method aiming to CAD/CAM approach

The design method we developed was CAD/CAM friendly so that entire optical design optimization could be done by robust numerical calculation (Fig. 2). We developed it based on a Monte Carlo calculation. The Monte Carlo method is useful to simulate how solar rays are concentrated and lost in the complicated lens structure. This method has two advantages. One is flexibility to the design change. Another is capability of handling complicated lens structure with partial shading and stray rays.

The first target of our research is to develop CADE/CAM based optical design that is friendly mechanical engineers.

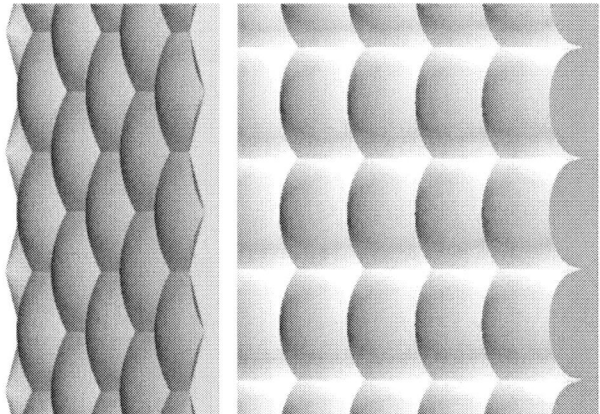

Fig. 3. Honeycomb arrangement (left) vs. grid arrangement (right) The unit lens of both arrangement is conventional axially-symmetrical conic lens.

The honeycomb arrangement has been often used for the CPV lens array especially to high concentration application and small cell/lens application. The honeycomb arrangement has advantages on optical properties, because the aperture shape is closed to the axially-symmetrical disk. On the other hand, the grid arrangement has bigger deflection and higher reflectance loss due to higher incident angle in the four corners of the square lens aperture (Fig. 3).

Despite of the superior optical characteristics, the honeycomb arrangement has practical difficulties. The first one is the dead space at the four sides of the array. Different from the grid arrangement, the edge of the lens array is zigzag that does not match to the car-roof. Secondly, the zigzag array edge has also serious disadvantage of reliability of sealing.

Therefore, our second target is to develop the new grid-arranged optics that has the comparable optical performance of the honeycomb arrangement.

III. METHOD

At first, we need the incident angle distribution. It was generated from 47 representative sites in Japan and 411,720 sets of data consist of direct irradiance, diffused irradiance and the sun position. A continuous distribution function was generated using a linear interpolation. A set of the incident angles was generated at 1,000,000 points using a uniform distribution ranging from 0° to 90°, weighted by the above-mentioned continuous distribution function.

The next step is to define the shape and position. We developed special sub-module structure and the light source model for the ray-tracing calculation. The proposed light source model is 4-folded light source to 9 rectangular lenses [13]. Note that the vertexes of the square light source are placed on the 4 peaks of adjacent lenses and the sides of the square light source are placed on the 8 peaks of adjacent lenses. This is the minimum subset of the lens array considering the shading influence on the single lens-cell pair located in the center position of 9 lenses. The light source was generated by randomly-placed point sources that emit rays in random direction but weighted by the annual incident angle distribution. The position of the collection of the sources in the direction of the light axis was chosen as the highest point of the lens and illumination zone is 4 times of the lens aperture.

The same model can be constructed to honeycomb arrangement for the comparison. In this case, the identifier of the model is 3-7, meaning that 3 times larger light source size to 7 unit lenses. The honeycomb arrangement would be useful, if the lens profile is axially-symmetrical, like the array of the conventional aspheric lens. However, it was found that the optimized axially-asymmetrical lens performed equally well to both grid arrangement and honeycomb arrangement. Therefore, the 3-7 model for honeycomb arrangement was not examined in the following sections.

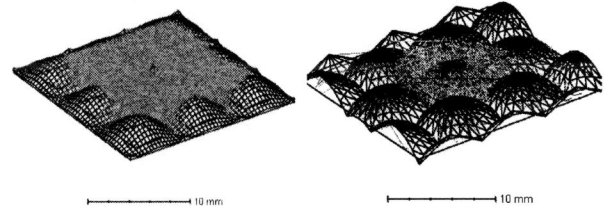

Fig. 4. Models of parquet structure and light sources for numerical optimization (Left: 4-9 model to Rectangular arrangement, right: 3-7 model to Honeycomb arrangement).

The reason why honeycomb arrangement has better optical performance is that the aperture of the lens is closer to the circular disk. This is true if the lens is designed by conventional axially-symmetrical profile like the conic lens. However, if the lens profile is generated by axially-asymmetrical shape, the lens profile may adjust the higher f-number of the corners of the square lens aperture. With this expectation, we tried axially-asymmetrical profile functions.

The single lens unit was given by the solid model and its profile was given by the target equation for optimization.

$$Z(r, \theta) = \sum_{i=1}^{N} c_i f_i(r, \theta) + l \qquad (1)$$

where, $Z(r,\theta)$ is the profile function, N is the maximum number of the index of the orthogonal function system, c_i is the weighting coefficients, $f_i(r,\theta)$ is the orthogonal function system, and l is the working distance from the origin of the lens profile to the cell. It was extracted from Zernike's polynomial [14], considering symmetry of the cell shape and its allocation (grid arrangement).

The maximum number of the function N in Equation (1) is also important in the trade-off between the calculation time and accuracy. We found that azimuthal degree was important and the necessary value was 8 and preferably 12, indicating axially highly-ordered asymmetrical profile was effective to boost the efficiency.

Supposing the ray of the maximum designed incident angle hitting at the center of the lens aperture reaches to the edge of the cell, the initial value of c_1 in Equation (1) is given by Equation (2).

$$c_1 = \frac{2n^2 + \cos 2\theta - 1}{4\sqrt{\frac{2}{3}}\sin\theta \sqrt{C(n-\sin\theta)(n+\sin\theta)}} W \qquad (182)$$

where, C is geometrical concentration ratio, n is refractive index of the medium, and θ is the maximum designed incident angle. Other coefficients, including l were set to zero.

An example of optimized axially-asymmetrical lens array (fixed sloped surface in the south) is shown in Fig. 5.

Fig. 5.   Optimized axially-asymmetrical lens array (fixed sloped surface in the south)

Note that the profile of the optimized lens was no longer axially-asymmetrical. The cross-section perpendicular to the optical axis was closed to square so that the deflection angle along the azimuthal line was kept to closed to constant. It was thought that the optimization routine adjusted

IV. RESULTS

The relationship between the acceptance half angle and the geometrical concentration ratio was investigated among (a) the tracking plane, (b) the sloped plane in the south, and (c) the horizontal plane.

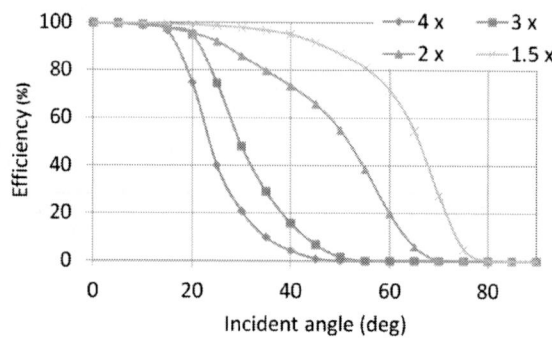

(a)

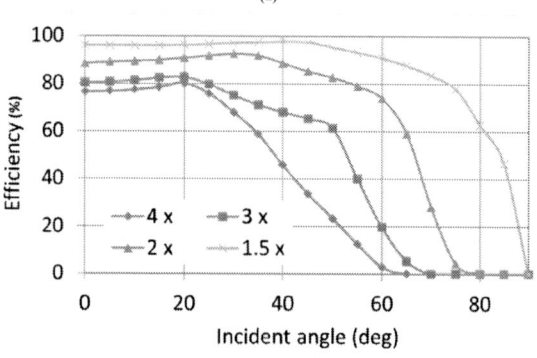

(b)

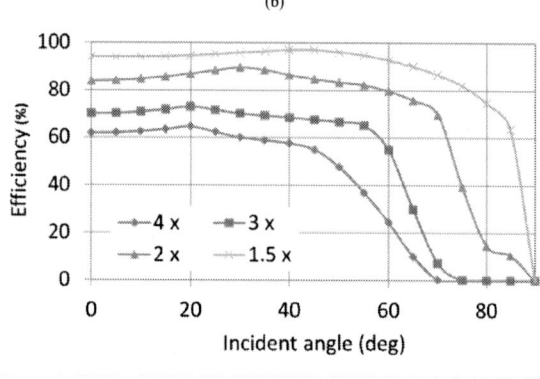

(c)

Fig. 6.   Diagram of the project of the development on optical design method aiming to CAD/CAM approach

The result is shown in Fig. 6. The conventional orientation like the tracking plane (a), the apparent acceptance half angle decreased almost in proportion to the inverse of the square-root of the geometrical concentration ratio like other most of the concentrator optics [16-19] of which on-axis optical efficiency were always 100 %. As a matter of course, the apparent acceptance half angle was always less than the thermodynamic limit and they were about 60 % of the theoretical limits. This trend was not true in other type of the installation, especially (c) the horizontal plane. The on-axis optical efficiency was no longer 100 % and decreased with the increase of the geometrical concentration ratio. This is because the optimized design expanded the acceptance half angle with sacrifices of the performance of the on-axis irradiation that was less important than that of the tracking plane. The apparent acceptance half angle in the optimized lens array to (c) the horizontal plane exceeded the thermodynamic limit of the acceptance angle (48.2° at 4 x and 59.3° at 3 x).

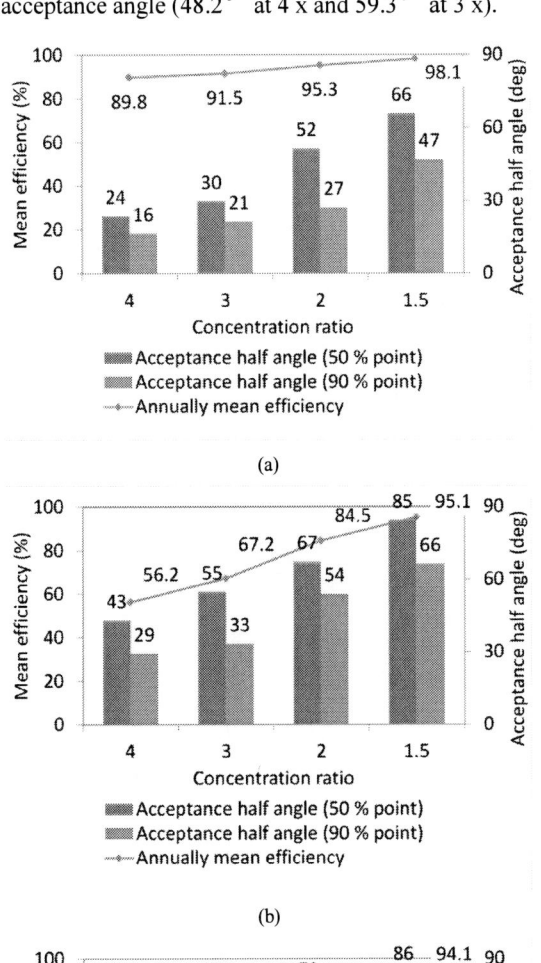

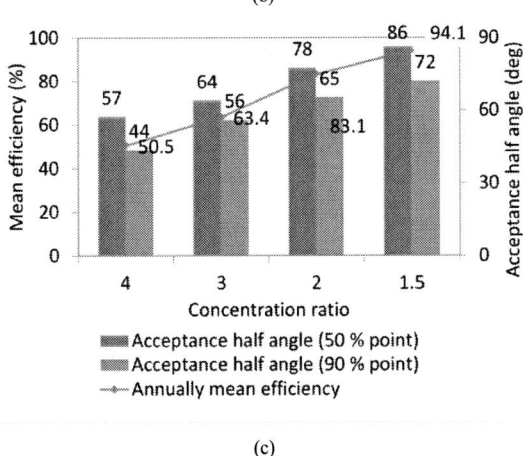

Fig. 7. Summary of the performance of the newly-designed axially-asymmetric and grid-arranged lens array for static CPV.

Fig. 7 summarizes the performance of the optimized lens arrays in this work.

The same kind of the calculation was done to the 3-7 model for honeycomb arrangement (Fig. 4). The optimized profile was closed to the axially-symmetrical and the optimized optical efficiency was also closed to the results in Fig. 7. The optimization using asymmetrical function successfully adjusted the lens profile for the grid arrangement. With the use of optimized axially-asymmetrical lens, it is not always necessary to rely on honeycomb arrangement.

The acceptance angle limit was given by the rigorous thermodynamic acceptance angle that took all the optical loss into consideration. That thermodynamic acceptance half angle *TA* is given by the equation (3), converting the acceptance characteristics is the rectangular window function.

$$TA = \int_0^{\pi/2} \alpha(\phi)d\phi \qquad (3)$$

where, $\alpha(\phi)$ is the optical efficiency as a function of the incident angle ϕ like Fig. 6.

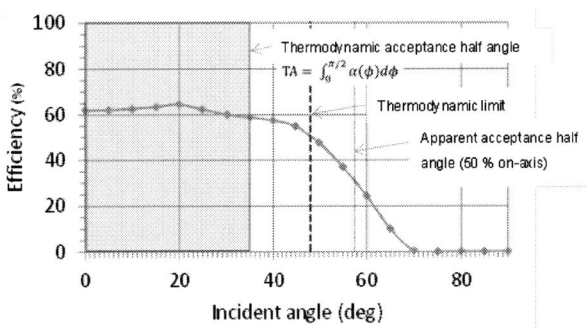

Fig. 8. Illustrated relationship among the apparent and thermodynamic acceptance half angle.

Most of the cases, especially high concentration optics and the low concentration optics that was designed to collect on-axis sunlight, both apparent and thermodynamic acceptance half angle give almost the same value. However, the optics that expanded the apparent acceptance half angle with sacrifice of the on-axis performance, they are not always the same.

The relationship among the apparent acceptance half angle, the thermodynamic acceptance half angle and the thermodynamic limit is illustrated in Fig. 8. Still, the rigorous thermodynamic acceptance half angles of this work were substantially higher than the past works, but they are within the thermodynamic limit.

V. CONCLUSION

For the optimized design of the static concentrator PV, a new and CAD/CAM friendly design approach for the grid-arranged lens array was developed with consideration of the specific incident angle distribution and shading effects from adjacent lenses. The optimized lens design was shown in (a) tracking plane, (b) sloped surface in the south and (c) horizontal plane were presented and compared.

The apparent acceptance angle became substantially wide with sacrifice of the optical performance on-axis. The apparent acceptance half angles (90 % power) were 50.5° and 63.4° in 4 x and 3 x of the geometrical concentration ratio. It looked like that the thermodynamic limit was violated, but after consideration of the rigorous thermodynamic acceptance half angle (34.8° and 44.6° in 4 x and 3 x of the geometrical concentration ratio), they were within the physical limits.

The optimization using asymmetrical function successfully adjusted the lens profile for the grid arrangement. With the use of optimized axially-asymmetrical lens, it is not always necessary to rely on honeycomb arrangement.

ACKNOWLEDGEMENT.

This work has been partially supported by NEDO in Japan.

REFERENCES

[1] T. Masuda, K. Araki, K. Okumura, S. Urabe, Y. Kudo, K. Kimura, , ... and M. Yamaguchi, "Next environment-friendly cars: Application of solar power as automobile energy source." in *Photovoltaic Specialists Conference (PVSC), 2016 IEEE 43rd* pp. 0580-0584. 2016.

[2] A. Rabl, "Comparison of solar concentrators", *Solar Energy*, Vol. 18, pp. 93-111. 1976.

[3] R. Tang and X. Liu, "Optical performance and design optimization of V-trough concentrators for photovoltaic applications", *Soler Energy*, Vol. 85, pp. 2154–2166. 2011.

[4] T. Uematsu, Y. Yazawa, Y. Miyamura, S. Muramatsu, H. Ohtsuka, K. Tsutsui, and T. Warabisako, Static concentrator photovoltaic module with prism array. *Solar energy materials and solar cells*, 67(1), 415-423. 2001.

[5] T. Uematsu, Y. Tazawa, T. Joge, and S. Kokunai, "Fabrication and characterization of a flat-plate static-concentrator photovoltaic module", *Solar Energy Materials and Solar Cells* 67(1-4), 425-434. 2001.

[6] R. Reisfeld, "New developments in luminescence for solar energy utilization." *Optical Materials*, 32(9), 850-856. 2010.

[7] T. Saitoh and K. Yoshioka, "Preparation and properties of photovoltaic static concentrators", *Renewable Energy*, Vol.15, Issues 1–4, pp. 566-571. 1998.

[8] R. Leutz, A. Suzuki, A. Akisawa, and T. Kashiwagi, "Design of a Nonimaging Fresnel Lens for Solar Concentrators", *Solar Energy*, 65, 6, pp. 379-388. 1999.

[9] K. Sopian, P. Ooshaksaraei, S. H. Zaidi, and M. Y. Othman, "Recent Advances in Air-Based Bifacial Photovoltaic Thermal Solar Collectors." in *Photovoltaics for Sustainable Electricity and Buildings* pp. 161-176. Springer International Publishing. 2017

[10] D. Chemisana, J. I. Rosell, A. Riverola, and C. Lamnatou, "Experimental performance of a Fresnel-transmission PVT concentrator for building-façade integration." *Renewable Energy*, 85, 564-572. 2016.

[11] Duerr, F., Benítez, P., Miñano, J. C.,Meuret, Y., and Thienpont, H., 2013. Analytic free-form lens design for imaging applications with high aspect ratio, Opt. Express 21, 31072-31081.

[12] A. Cvetkovic, M. Hernandez, P. Benítez, J. C. Miñano, J. Schwartz, A. Plesniak, R. Jones, and D. Whelan, "The free form XR photovoltaic concentrator: a high performance SMS3D design", *Proc. SPIE 7043, High and Low Concentration for Solar Electric Applications* III, 70430E, 2008.

[13] K. Araki, Y. Ota, K. Ikeda, K-H Lee, K. Nishioka and M. Yamaguchi, "Possibility of Static Low Concentrator PV Optimized for Vehicle Installation", *AIP Proc. CPV-12*, 1766, 020001, 2016.

[14] Noll, RJ. 1976. Zernike polynomials and atmospheric turbulence. J Opt Soc Am 66;207-211. [15] R. Winston *et al.*, 2005. Nonimaging optics, Elsevier-Academic Press.

[16] W. T. Welford, R. Winston, "High Collection Nonimaging Optics", in *ISES Solar World Congress*, San Diego. 1989.

[17] P. Benítez, J. C. Miñano, P. Zamora, R. Mohedano, A. Cvetkovic, M. Buljan, J. Chaves, M. Hernández, "High performance Fresnel-based photovoltaic concentrator", *Optics Express* Volume 18 Issue 9, pp. A25-A40. 2010.

[18] J. Chaves, "Introduction to Nonimaging Optics, Second Edition", CRC Press, pp.33. 2015.

[19] M. Hernandez, A. Cvetkovic, P. Benitez, J. C. Miñano, W. Falicoff, Y. Sun, J. Chaves, R. Mohedano, "CPV and illumination systems based on XR-Köhler devices", *Proc. SPIE 7785, Nonimaging Optics: Efficient Design for Illumination and Solar Concentration* VII, 77850A, 2010.

978-1-5090-5606-4/17 $31.00 © 2017 IEEE

Spectral Transmittance Analysis of Liquids for High Concentration III-V Photovoltaic Immersion Cooling Applications

Xinyue Han, Yongjie Guo

School of Energy and Power Engineering, Jiangsu University, Zhenjiang, Jiangsu 212013, China

Abstract — Direct liquid immersion cooling has been proposed for HCPV systems with densely packed III-V cells. The system output depends on optical properties of immersion liquids. Eight liquids might be adopted for high concentration III-V photovoltaic immersion cooling applications are investigated. Their transmittance is determined using transmittance measurement of two path-length cuvettes via spectrophotometer. J_{np} of bottom subcell of specified 3-junction cells decreased. In relation to bottom subcell generates an excess current, this is a beneficial property. Selected liquids are suitable for immersing 3-junction cells. Therminol VP-1 immersion would cause the smallest power loss of cells, followed by dimethyl silicon oil.

Index Terms — dielectric liquids, direct liquid immersion cooling, high concentration systems, spectral transmittance, triple-junction solar cells.

I. INTRODUCTION

Concentrating photovoltaic (CPV) is of most interest for electricity generation in sun-rich zone. As the increase in efficiency of III-V multi-junction solar cells, more than 90% of the capacity publicly documented to be installed is in the form of high concentration PV (HCPV) systems equipped with III-V cells [1-3]. However, only a fraction of the incident solar energy striking the III-V cells is converted to electric power. The rest of the absorbed solar energy is transformed into heat in the solar cells, causing the reduction of the electrical conversion efficiency [4]. Thus, cooling system is a very important aspect in high concentrating photovoltaic applications with densely packed III-V cells. Several authors have reviewed various cooling approaches applied for CPV solar cells [4-6]. They concluded that the type of cooling depends mainly on the concentration ratio, the size of the cell, and the absorbed concentrated light. For densely packed III-V cells under high concentration, passive cooling is not sufficient and active cooling is used to keep solar cells from over-heating. Han et al. [7] have been proved the high cooling potential of direct liquid immersion cooling for densely packed III-V cells under high concentration.

However, the electrical performance of the densely packed III-V cells with direct liquid immersion cooling is affected by the optical properties of immersion dielectric liquids. Moreover, the spectral transmittance of immersion dielectric liquids is either hardly found in literatures, or is the total transmittance of the layered system window-liquid-window, immersed in air [8,9]. In this paper, the spectral transmittance of several immersion liquid candidates for high concentration

III-V photovoltaic immersion cooling applications is determined using transmittance measurements of two path-length cuvettes via spectrophotometer. The mean transmittance is used as a measure to identify changes in the spectral transmittance with liquid thickness. In addition, the effect of immersion liquid presence on the photocurrent of triple-junction solar cells is analyzed through the introduced normalized photocurrent density (J_{np}).

II. LIQUID CANDIDATES AND METHODOLOGY

A. Liquid Candidates

Using liquids for immersion cooling of densely packed III-V cells, requires different liquid properties to conventional active cooling fluids. Since the immersion liquids perform heat transfer and optical adaptation dual purpose, the requirements for the liquid properties not only include good heat transfer properties, but also include additional optical and electrical properties [10]. Based on their transparency at the visible light wavelengths range, eight liquids were identified as candidates to be the immersion cooling liquid of III-V cells, which were selected in the three categories of liquids, namely synthetic oil, silicone oil and mineral oil. The available properties of the eight liquids are summarized in Table I [11-14]. According to the information listed in Table I, all of them present good heat transfer properties. With respect to the optical properties of the eight liquids, the available data are scarce and should be investigated in depth.

B. Methodology

The spectral transmittance of the eight liquids was measured using a Lambda 950 UV/VIS/NIR spectrophotometer. The measurement resolution of the instrument is ≤0.05 nm in the UV/VIS and ≤0.20 nm in the NIR. All the spectral transmittance measurements were conducted two times and the mean was calculated. The liquids were contained in quartz cuvettes with path-lengths of 5 mm and 10 mm, respectively. The wavelength range has been set from 350 to 1800 nm with an interval of 5 nm.

With air as the reference, the reflections at the interfaces between the air and the quartz window and the quartz window and the liquid sample can not be neglected, as well as the cuvette absorption for the wavelengths of interest. In this paper, a method for determining the spectral transmittance of

978-1-5090-5606-4/17 $31.00 © 2017 IEEE

TABLE I
PROPERTIES OF CANDIDATE DIELECTRIC LIQUIDS

Properties	Therminol VP-1	Dimethyl silicon oil A/B	White oil A/B/C	C14 n-alkane	C16 iso-alkane
Color	Clear	Clear	Clear	Clear	Clear
Density (kg·m^{-3})	1060	818/915	880/822/816	764	767
Specific Heat (J·kg^{-1}·K^{-1})	1570	2000/1758	2130	1578	-
Thermal Conductivity(W·m^{-1}·K^{-1})	0.14	0.10/0.12	0.15	0.14-0.15	-
Viscosity(10^{3}Pa·s)	2.63	0.82/4.58	6.78/2.30/2.45	1.60	1.94
Boiling point(°C)	257	153/170	255-276	254	211
Refractive index	1.65	1.38/1.40	1.48	1.43	1.42
Dielectric constant	3.35	2.28/2.60	2.20	2.03	-

immersion cooling liquids is developed, based on transmittance measurements of two path-length cuvettes via spectrophotometer. The method is described in details as follows.

The spectral transmittance of the liquid $\tau(\lambda)$, with λ wavelength, can be expressed as a function of the spectral absorption coefficient $\alpha(\lambda)$ as:

$$\tau(\lambda) = \exp(-\alpha(\lambda) \cdot x) \qquad (1)$$

where x is the liquid thickness. However, the transmittance $T(\lambda)$ we measure is the total transmittance of the layered system window-liquid-window, immersed in air. At each interface (air-quartz window, quartz window-liquid) the light is partially reflected and partially transmitted. Also, the quartz windows have some absorption for the wavelengths of interest. Under the hypotheses of absence of scattering and negligible coherent effects [15], the total transmittance $T(\lambda)$ can be expressed as a function of $\tau(\lambda)$ and of the reflectances R_1 and R_2 at the interfaces, as well as of the absorption coefficient of cuvette window α_0 and cuvette window thickness d [16].

$$T(\lambda) = \tau(\lambda)(1 - R_1)^2(1 - R_2)^2 \exp(-2\alpha_0 \cdot d) \qquad (2)$$

Therefore, based on the second equation as (2) and the first equation as (1), if we obtain two measured transmittance $T_1(\lambda)$ and $T_2(\lambda)$ at two different optical path-lengths x_1 and x_2, the spectral absorption coefficient of liquid $\alpha(\lambda)$ can be directly obtained from the expression:

$$\alpha(\lambda) = -\frac{1}{x_2 - x_1}\ln\frac{T_2(\lambda)}{T_1(\lambda)} = -\frac{1}{\Delta x}\ln\frac{T_2(\lambda)}{T_1(\lambda)} \qquad (3)$$

According to the spectral absorption coefficient of liquid $\alpha(\lambda)$, the spectral transmittance of the liquid $\tau(\lambda)$ at any thickness x can be calculated through the first equation as (1).

III. RESULTS AND DISCUSSION

Using the transmittance measurements of two path-length cuvettes via spectrophotometer method described in the

previous section, the initial spectral transmittance of the eight immersion liquids for III-V solar cells cooling are presented in Fig. 1, divided into three categories: synthetic oil, silicone oil and mineral oil. The figure also plots the published transmittance of EVA and silicone encapsulants [17,18]. The spectral irradiance of the AM1.5D solar spectrum is provided for reference.

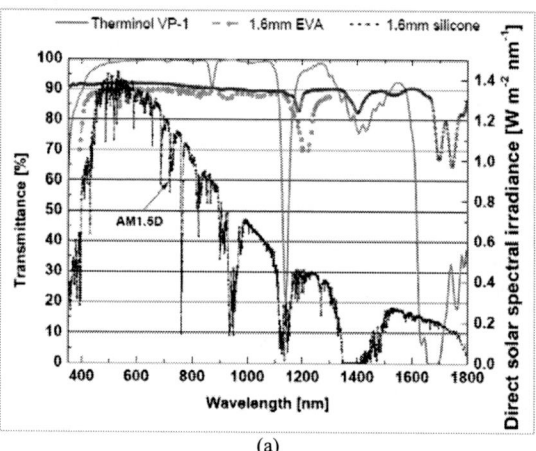

(a)

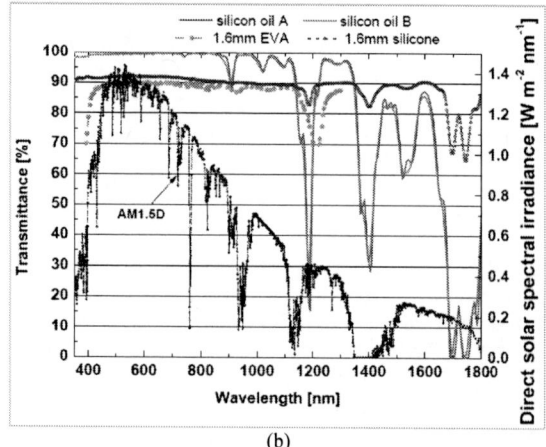

(b)

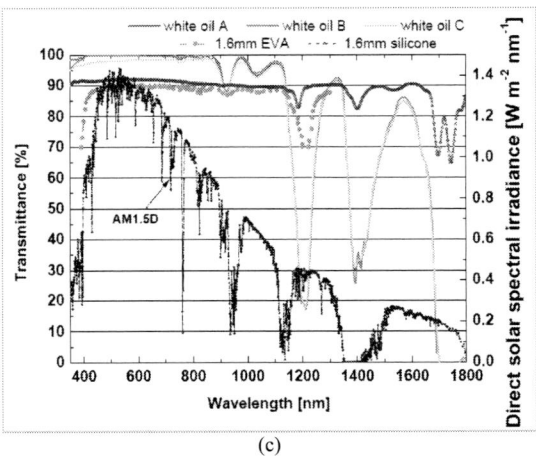

(c)

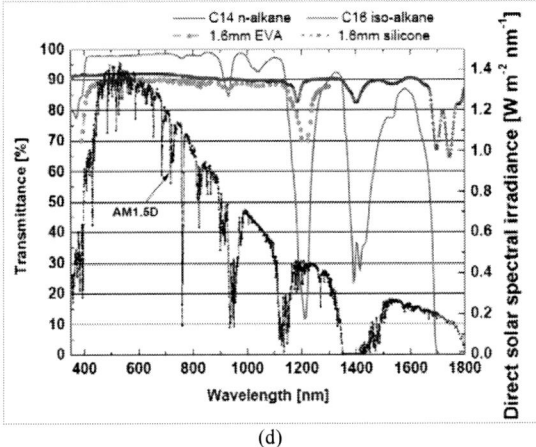

(d)

Fig. 1. Spectral transmittance of (a) Therminol VP-1, (b) dimethyl silicon oil A/B, (c) white oil A/B/C and (d) C14 n-alkane, C16 iso-alkane, as well as the transmittance of EVA and silicone encapsulants, and AM 1.5D irradiance.

As shown in Fig. 1a, the starting point of Therminol VP-1 transmittance plot is at about 350 nm. Therminol VP-1 is highly transparent from 400 nm to 1100 nm and display strong absorption bands at 1150 nm, 1400 nm and again at 1600-1750 nm. Fig. 1b shows the transmittance of two silicone oils with different kinetic viscosities. Their transmittance is very similar to each other. Both of them have almost greater than 98% transmittance in the wavelength range from 300 nm to 1100 nm. From 1100 nm, they display transmittance valleys around 1200 nm, 1400 nm and again 1700 nm. C-H bonds included in both Therminol VP-1 and dimethyl silicone oil can be used to explain the most relative transmittance valleys mentioned above [19]. From Fig. 1b and Fig. 1c, it can be seen that the spectral transmittance of white oil A/B/C, C14 n-alkane and C16 iso-alkane exhibit very similarly to each other. All of them present over 95%

transmittance between 400 and 1100 nm. From 1100 nm, their transmittance fluctuates with significant transmittance troughs at around 1200 nm, 1400 nm and again at 1700-1800 nm. Evidently, compared with C14 n-alkane and C16 iso-alkane, white oil A/B/C perform well in the spectral transmittance of the UV light region. Further, the food grade white oil namely white oil A show higher transmittance in the highly energetic wavelength than the other two white oils.

In addition, when compared to the transmittance of EVA and silicone encapsulants, all of the eight liquid candidates exhibit superior performance at wavelengths of 300-1100 nm. It should be noted that the thickness of both EVA and silicone encapsulants are 1.6 mm, whereas the liquid samples are 10 mm thick.

Fig. 2 shows the mean transmittance of the eight liquid candidates calculated for the wavelength range from 350 nm to 1800 nm with different thicknesses. The results show that the mean transmittance of Therminol VP-1, silicone oil A/B, white oil A/B/C, C14 n-alkane and C16 iso-alkane at liquid thickness of 10 mm are 82.56%, 82.20%, 81.80%, 77.41%, 77.76%, 76.82%, 77.00% and 77.54%, respectively. Also, it can be seen that the mean transmittance at liquid thickness of 5 mm are 87.79%, 87.79%, 88.16%, 83.96%, 83.88%, 83.64%, 83.66% and 83.89%, respectively. Therefore, both synthetic oil and silicone oils exhibit better transmittance than mineral oils. And there is a significant effect of the liquid layer thickness on the spectral transmittance where the transmittance decreases as the liquid layer increases.

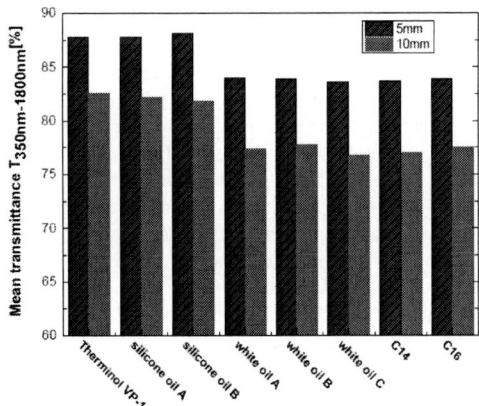

Fig. 2. Mean transmittance of different liquid candidates with different thicknesses.

Further, for densely packed III-V cells with dielectric liquid presence, the transmittance of liquid would affect the power generated by the cells. As proposed before [20,21], to estimate the effect of the selected liquid transmittance, the photocurrent of each subcell when illuminated by the solar spectrum

TABLE II
NORMALIZED PHOTOCURRENT DENSITY (J_{np}) OF TOP, MIDDLE, BOTTOM OF 3-JUNCTION CPV SOLAR CELL WITH 10 MM THICKNESS LIQUID PRESENCE

Liquid Species	Top (350-660nm) GaInP J_{np}	Middle (660-890nm) GaInAs J_{np}	Bottom (890-1800nm) Ge J_{np}	3-junction CPV cell (350-1800nm) GaInP/GaInAs/Ge J_{np}
Therminol VP-1	0.980	0.988	0.904	0.950
Silicon oil A	0.998	0.995	0.861	0.939
Silicon oil B	0.992	0.992	0.857	0.934
White oil A	0.997	0.993	0.785	0.905
White oil B	0.998	0.996	0.792	0.909
White oil C	0.980	0.982	0.783	0.896
C14 n-alkane	0.974	0.982	0.795	0.900
C16 iso-alkane	0.995	0.995	0.796	0.910

filtered by the liquids transmittance is compared to the photocurrent without liquid presence. The normalized photocurrent density (J_{np}) is calculated as:

$$J_{np} = \frac{\int T_{material}(\lambda) QE(\lambda) I_{AM1.5D}(\lambda)(q\lambda/hc)\,d\lambda}{\int QE(\lambda) I_{AM1.5D}(\lambda)(q\lambda/hc)\,d\lambda} \quad (4)$$

where $T_{material}(\lambda)$ is the transmittance of immersion liquid, $QE(\lambda)$ represents the quantum efficiency (QE) of the cell, $I_{AM1.5D}(\lambda)$ is the AM1.5D spectral irradiance, q is a electron charge, h is the Planck's constant and c is light speed in vacuum.

The triple-junction cell considered here is made of $Ga_{1-x}In_xP$ (1.8-1.9 eV), $Ga_{1-y}In_yAs$ (1.3-1.4 eV), and Ge (0.7 eV), which is the commercial standard solar cell for CPV systems. Through the fourth equation as (4), the normalized photocurrent density (J_{np}) of the top, middle, bottom subcells and the complete lattice matched $Ga_{0.49}In_{0.51}P/Ga_{0.99}In_{0.01}As/Ge$ triple-junction cell manufactured by Tianjin Lantian Solar Tech Co., Ltd. are tabulated in Table II for the liquid thickness of 10 mm presence. As seen in Table II, the J_{np} of the bottom subcell for the triple-junction cell is much lower than those of the top and middle subcells. This can be explained by the spectral transmittance of the liquids shown in Fig. 1. All the selected liquids exhibit superior transmittance for the wavelengths of the top and middle subcells interest in the UV and VIS spectrums. However, the spectral transmittance of these liquids show some decrease in wavelength range of 1200-1800 nm and it will result in the bottom subcell Ge limit the cell current when the liquids are used. In fact, this is not a problem for the complete triple-junction solar cells due to much more current can be produced by the bottom subcell Ge.

Further, the J_{np} of the complete triple-junction solar cells shown in Fig. 3 indicates that with each liquid presence, the cells J_{np} is almost more than 0.9 and demonstrate that they are suitable for immersion cooling densely packed III-V cells. The smallest power loss of triple-junction solar cells would be observed with Therminol VP-1 immersion, followed by

dimethyl silicon oil, with the largest loss occurring with white oil C immersion.

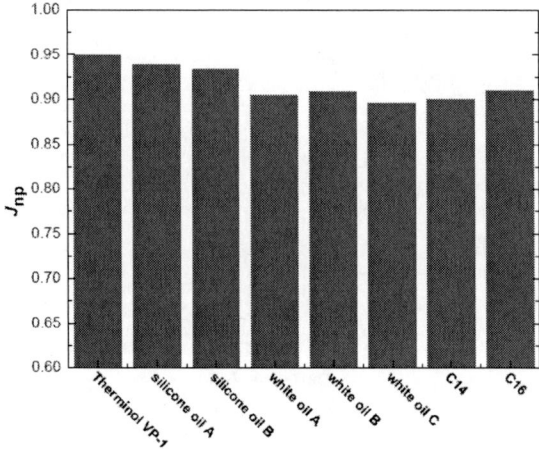

Fig. 3. Normalized photocurrent density (J_{np}) of GaInP/GaInAs/Ge triple-junction solar cell with 10 mm thickness different liquid on.

IV. CONCLUSIONS

Direct liquid immersion cooling of densely packed III-V cells under high concentration is demonstrated as a potential solution for HCPV system thermal management. The spectral transmittance of the immersion cooling liquid is considered to be very important due to it will directly affect the power output of the system. In this paper, the spectral transmittance of several immersion liquid candidates for high concentration III-V photovoltaic immersion cooling application is determined using transmittance measurements of two path-length cuvettes via spectrophotometer. The results show that both synthetic oil and silicone oils exhibit better transmittance than mineral oils. The liquid layer thickness has significant effect on the spectral transmittance where the transmittance decreases as the liquid layer increases. In comparison with the spectral transmittance of the common encapsulants EVA and silicone, the eight candidate immersion liquids perform much higher transmittance. Moreover, the effect of transmittance of liquid candidates on triple-junction solar cells performance is evaluated by calculating the normalized photocurrent density (J_{np}). With the liquids on, the J_{np} of the bottom subcell for the triple-junction cell is reduced. Since the bottom subcell of the specified triple-junction cells generates an excess current, this is a beneficial property. Additionally, results show that the eight candidate liquids are suitable for immersing triple-junction solar cells. Therminol VP-1 immersion would cause the smallest power loss of triple-junction solar cells, followed by dimethyl silicon oil, with the largest loss occurring with white oil C immersion.

ACKNOWLEDGEMENT

The work was supported by the National Natural Science Foundation of China (51306077), Training Project for Young Teachers of Jiangsu University (5521130006) and Oversea Study Scholarship from China Scholarship Council (201608695005).

REFERENCES

[1] G. Zubi, J. L. Bernal-Agustín, and G. V. Fracastoro, "High concentration photovoltaic systems applying III-V cells," *Renewable and Sustainable Energy Reviews*, vol. 13, pp. 2645-2652, 2009.

[2] M. A. Green, K. Emery, Y. Hishikawa, W. Warta, E. D. Dunlop, D. H. Levi, and A. W. Y. Ho-Baillie, "Solar cell efficiency tables (version 49)," *Progress in Photovoltaics: Research and Applications*, vol. 25, pp. 3-13, 2017.

[3] S. P. Philipps, A. W. Bett, K. Horowitz, and S. Kurtz, "Current status of concentrator photovoltaic (CPV) technology," (No. NREL/TP-6A20-63916). *NREL (National Renewable Energy Laboratory (NREL)*, Golden, CO (United States)), 2016.

[4] S. Jakhar, M. S. Soni, and N. Gakkhar, "Historical and recent development of concentrating photovoltaic cooling technologies," *Renewable and Sustainable Energy Reviews*, vol.60, pp. 41-59, 2016.

[5] A. Royne, C. J. Dey, and D. R. Mills, "Cooling of photovoltaic cells under concentrated illumination: a critical review," *Solar Energy Materials and Solar Cells*, vol. 86, pp. 451-483, 2005.

[6] L. Micheli, N. Sarmah, X. Luo, K. S. Reddy, and T. K. Mallick, "Opportunities and challenges in micro-and nano-technologies for concentrating photovoltaic cooling: A review," *Renewable and Sustainable Energy Reviews*, vol. 20, pp. 595-610, 2013.

[7] X. Han, Q. Wang, J. Zheng, and J. Qu, "Thermal analysis of direct liquid-immersed solar receiver for high concentrating photovoltaic system," *International Journal of Photoenergy*, vol. 2015, pp. 1-9, 2015.

[8] R. Looser, M. Vivar, and V. Everett, "Spectral characterisation and long-term performance analysis of various commercial Heat Transfer Fluids (HTF) as Direct-Absorption Filters for CPV-T beam-splitting applications," *Applied Energy*, vol. 113, pp. 1496-1511, 2014.

[9] M. Victoria, S. Askins, C. Domínguez, I. Antón, and G. Sala, "Durability of dielectric fluids for concentrating photovoltaic systems," *Solar Energy Materials and Solar Cells*, vol. 113, pp. 31-36, 2013.

[10] M. Vivar and V. Everett, "A review of optical and thermal transfer fluids used for optical adaptation or beam - splitting in concentrating solar systems," *Progress in Photovoltaics: Research and Applications*, vol. 22, pp. 612-633, 2014.

[11] Solutia Inc. Heat transfer fluids-therminol, product data sheet [DB/OL]. http://www.therminol.com.

[12] T. P. Otanicar, P. E. Phelan, and J. S. Golden, "Optical properties of liquids for direct absorption solar thermal energy systems," *Solar Energy*, vol. 83, pp. 969-977, 2009.

[13] Shin-Etsu Polymer Co. Ltd. silicon-KF-96L, product data sheet [DB/OL]. http://www.shinetsu.co.jp/.

[14] Dirac Delta. Science and engineering encyclopedia [DB/OL]. http://www.diracdelta.co.uk.

[15] C. F. Bohren and D. R. Huffman, *Absorption and Scattering of Light by Small Particles*, Wiley, 2004.

[16] X. Han, Q. Wang, and J. Zheng, "Determination and evaluation of the optical properties of dielectric liquids for concentrating photovoltaic immersion cooling applications," *Solar Energy*, vol. 133, pp. 476-484, 2016.

[17] X. Han, Y. Wang, L. Zhu, H. Xiang, and H. Zhang, "Reliability assessment of silicone coated silicon concentrator solar cells by accelerated aging tests for immersing in de-ionized water," *Solar Energy*, vol. 85, pp. 2781-2788, 2011.

[18] B. Ketola, K. R. McIntosh, A. Norris, and M. K. Tomalia, "Silicones for photovoltaic encapsulation," *in 23rd EU PVSEC*, 2008, pp. 2969-2973.

[19] O. Ziemman, J. Krauser, P. E. Zamzow, W. Daum, and P. O. F. Handbook, *Optical Short Range Transmission Systems*, Springer, 2008.

[20] M. Kempe, "Overview of scientific issues involved in selection of polymers for PV applications," in *37th Photovoltaic Specialist Conference*, 2011, pp. 000085-000090.

[21] X. Han, Y. Wang, and L. Zhu, "Electrical and thermal performance of silicon concentrator solar cells immersed in dielectric liquids," *Applied Energy*, vol. 88, pp. 4481-4489, 2011.

Optical design for 2-terminal III-V/Si SMAC module

Masaaki Baba[1], Kikuo Makita[2], Hidenori Mizuno[3], Hidetaka Takato[3], Takeyoshi Sugaya[2], Noboru Yamada[1]

1 Nagaoka University of Technology, Nagaoka, Niigata, masaaki_baba@stn.nagaokaut.ac.jp, Japan
2 National Institute of Advanced Industrial Science and Technology (AIST), Tsukuba, Japan
3 National Institute of Advanced Industrial Science and Technology (AIST), Koriyama, Japan

Abstract — A static concentrator lens was designed for 2-terminal GaAs/Si SMAC module, in which a GaAs subcell is mechanically stacked on a Si subcell. The Si cell has a larger active area than the GaAs subcell. Energy losses due to a current mismatch and electrodes shadow were considered. The designed lens shows a low peak angular optical efficiency but larger acceptance half angle, achieving annual energy collection efficiency of 75.7% for the module tilt angle of 35°. The prototype module was fabricated to verify the simulation. The measured short-circuit current of prototype SMAC module was close to the simulation results.

I. INTRODUCTION

Concentrator photovoltaic (CPV) system, which consists of a high efficiency solar cell such as III-V multi-junction (MJ) cells and a lens concentrating sunlight to the solar cell, is cost-effective relative to the other photovoltaic systems in the region of rich direct solar radiation. However, III-V MJ cells are generally expensive because of III-V subcells are grown monolithically on a high-cost substrate such as Ge or GaAs. Si is an attractive substitute for the substrate in terms of cost reduction and bandgap suitability [1]–[3]. Mechanical stacking approaches enable to use Si for the subcell of MJ solar cell, although it requires a non-traditional fabrication process. The smart stack technique developed by AIST is an advanced mechanical stacking technique that uses metal nanoparticles, produces excellent optical and electrical properties, and enables flexible device design [4]. In the design of MJ solar cells, current matching among the series-connected subcells is also an important to achieve high efficiency. The areal current matching (ACM) technique has been proposed recently. In this technique, an MJ cell consists of subcells that have different active areas; and the areas of the subcells can be independently varied to reduce the current mismatch [5]. As mentioned above, mechanical stacking, areal current matching, and solar concentration are effective techniques for cost-effective production of III-V MJ photovoltaic modules. Recently, the authors introduced the concept of a 2-terminal (2T) "SMAC" MJ photovoltaic module that integrates a SMart stack, Areal current matching, and solar Concentration [6].

In this study, the optical simulation of concentrator for the SMAC module was performed by ray tracing method to optimize the concentrator design. The concentrator was designed to be a static low-concentration concentrator, which does not need a sun-tracking device, and is suitable for the region of moderate direct solar radiation. The simulation result was verified using the prototype module.

II. CONCENTRATOR DESIGN FOR SMAC MODULE

Fig. 1 shows the conceptual image of SMAC module. The concentrator is a refractive lens encapsulating solar cells. The lens shape is optimized for a 2T GaAs/Si ACM tandem solar cell, in which the bottom cell has two active areas; one is stacked area and the other is the extended area that receives sunlight directly. E_{top} and E_{bottom} represents the sunlight energy incident to the top cell and the bottom cell's extended area, respectively. In this combination of subcell materials (GaAs/Si), the current generated by GaAs subcell and that generated by Si subcell match when the incident energy ratio (E_{bottom} / E_{top}) is 0.38. In the SMAC module, the most of beam sunlight is concentrated to the top cell by the lens, and a part of beam sunlight and diffuse sunlight are captured by the bottom cell's extended area. The lens shape must be designed to minimize the loss due to the current mismatch between the subcells connected in series.

Fig. 2 shows the designed lens exhibiting the highest performance so far for the present SMAC module. The optimization of the lens considered the angular distribution of annual cumulative beam sunlight for various incident angle at Tokyo (due south, tilt angle 35°). The angular distribution of annual cumulative beam sunlight calculated based on the expanded AMeDAS weather data. In the north-south direction, the most of beam sunlight comes from the range within ±30°, and in the east-west direction, the most of beam sunlight

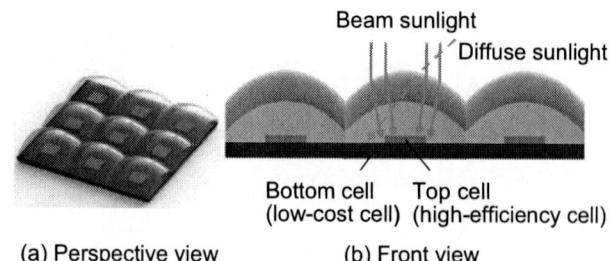

Fig. 1   Conceptual image of the SMAC module [6].

comes from the range within ±60°.

The geometric concentration ratio of the lens to the top cell C_g is 2X. The lens material is silicone.

Fig. 3 shows the simulation model for calculating the optical efficiency at various incident angle α ($0° \leq \alpha \leq 90°$) and azimuth angle β ($0° \leq \beta \leq 45°$). The angular optical efficiency for top cell area and full solar cell area (including top call area and bottom cell's extended area) is simulated by ray-tracing method. To account the effects of the adjacent lens, 3×3 lens array was assumed. Four sidewalls of the lens array are surrounded by flat mirrors with specular reflectance of 1.0. The light source spectrum is AM1.5D and the beam divergence half angle is 0.27°. Full spectrum simulation was carried out by considering absorption loss inside the lens and Fresnel reflection loss. The receivers, i.e., cell surfaces, were assumed to be black body.

Fig. 4 shows the simulated angular optical efficiency for top cell area and full solar cell area for $\beta = 0°$, 22.5°, and 45°. The peak efficiency is not very high, whereas the half acceptance angle is large. The half acceptance of the lens for top cell area and full solar cell area at $\beta = 0$ ° is 59° and 60°, respectively.

III. ANNUAL ENERGY CORRECTION EFFICIENCY

The module efficiency of the SMAC module is defined as

$$\eta_{module} = \eta_{collection} \times \eta_{cell} \qquad (1)$$

where $\eta_{collection}$ is the energy correction efficiency, which is the ratio of the effective sunlight energy, which is supposed to be absorbed by the tandem solar cell and to be converted into electricity, to the sunlight energy incident to the module aperture. The loss associated with modularization is an optical loss E_{opt_loss}, an electrode shadow loss E_{e_loss}, and a current mismatch loss E_{mis_loss} occurring between the top cell and the bottom cell, which are connected in series. The current mismatch loss depends on the irradiance ratio, which primarily depends on the angular optical efficiency and incident solar irradiance. To calculate the E_{mis_loss}, the electrode at the top cell surface must be assumed. The electrode area was assumed to be 13% of the top cell surface, thus the E_{top} becomes 87% of the sunlight energy concentrated on the top cell. For $E_{bottom} / E_{top} = 0.38$, the E_{mis_loss} is zero. For $E_{bottom} / E_{top} < 0.38$, the E_{mis_loss} is calculated as follows:

$$E_{mis\_loss} = (E_{top} + E_{bottom}) - (\frac{E_{bottom}}{0.38} + E_{bottom}) \qquad (2)$$

For $E_{bottom} / E_{top} > 0.38$, the E_{mis_loss} is calculated as follows:

$$E_{mis\_loss} = (E_{top} + E_{bottom}) - ((E_{top} \times 0.38) + E_{top}) \qquad (3)$$

$\eta_{collection}$ is defined by

$$\eta_{collection} = \frac{GTI - E_{opt\_loss} - E_{e\_loss} - E_{mis\_loss}}{GTI} \qquad (4)$$

Where GTI is the global irradiance incident on a tilted surface. The annual energy correction efficiency was calculated based on the annual irradiance data of every hour (expanded AMeDAS weather data) and the simulated angular optical efficiency of the lens. Fig. 5 shows the annual cumulative GTI at Tokyo (due south, tilt angle 35°) and that minus the respective losses. In the calculation, an isotropic angular distribution of diffuse sunlight at $0° \leq \alpha \leq 70°$ was assumed. The resultant $\eta_{collection}$ is 75.7%. E_{opt_loss}, E_{e_loss}, and E_{mis_loss} is 14.1%, 10.3%, and 4.9% of the GTI, respectively.

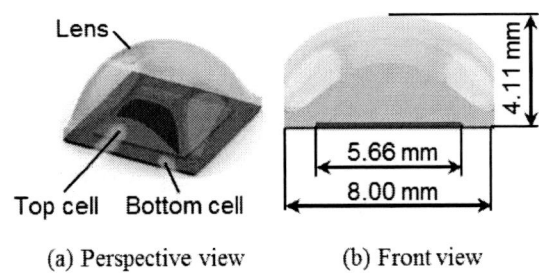

(a) Perspective view   (b) Front view

Fig.2   Schematic diagram of designed SMAC module.

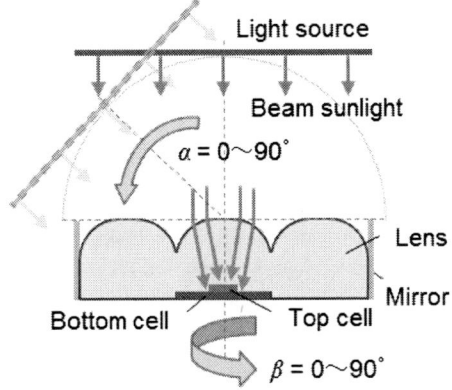

Fig.3   Simulation model for optical efficiency calculation.

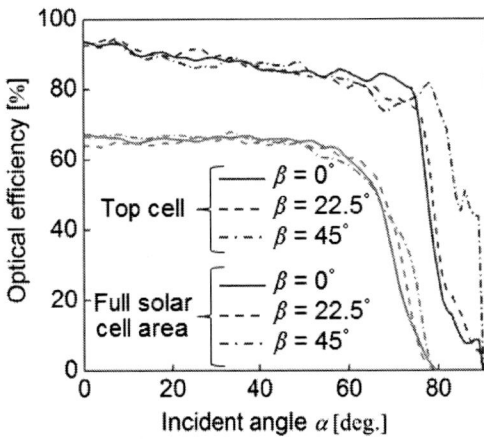

Fig.4   Angular optical efficiency to top cell and full solar cell area.

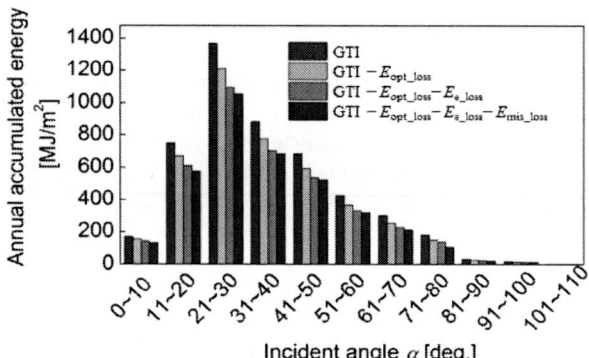

Fig.5   Incident angle dependency of annual accumulated energy (Tokyo, tilt angle 35°, due South).

IV. OUTDOOR CHARACTERIZATION

To verify the simulation results, a prototype 2T GaAs/Si SMAC module (hereafter referred to as SMAC module) having the designed lens was fabricated and tested. Fig. 6 shows the fabricated SMAC module and the 2T GaAs/Si tandem solar cell. The 2T GaAs/Si tandem solar cell was fabricated by smart stack technology [4], and mounted on a substrate by silver paste. The 2X silicone lens was fabricated based on the LED package technology. To obtain the same conditions as the simulation, a 3 × 3 lens array was made and the tandem solar cell was placed in the center of the 3 × 3 lens array. The top cell area A_{top}, the bottom cell area A_{bottom}, and the area ratio of the electrode area to the top cell area R_e are summarized in Fig. 6. Fig. 7 shows measured current-voltage curves of the SMAC module and the tandem solar cell (no lens) under outdoor 1 sun. The measured short-circuit current of the SMAC module J_{sc_m} was 9.28 mA, and that of the tandem solar cell (no lens) J_{sc_0} was 7.28 mA. Those values were close to the expected values in cell design.

The semi-theoretical short-circuit current J_{sc_sim} for the prototype SMAC module is defined by the following equation;

$$J_{sc\_sim}=J_{sc\_0} \times C_g \times \eta_{opt\_top} \times \frac{I}{I_0} \quad (5)$$

where I/I_0 is the ratio of the actual irradiance (GTI) to 1000 W/m². η_{opt_top} is the average value of the simulated angular optical efficiency of the lens to the top cell for $0° \leq \alpha \leq 60°$ ($\eta_{opt\_top} = 0.657$). The ratio J_{sc_m} / J_{sc_sim} is closely related to $\eta_{collection}$ mentioned in the previous section. The estimated J_{sc_sim} by Eq. (5) is 9.57 mA under 1 sun, hence J_{sc_m} / J_{sc_sim} is 0.969, meaning that $\eta_{collection}$ of the module well agreed with that obtained in the simulation.

To observe the incident angle dependency and irradiance dependency, a tentative outdoor experiment was carried out. Fig. 8 shows the experimental setup, which consists of the SMAC module, a pyranometer for GTI measurement, and a current-voltage tester. The SMAC module and the pyranometer were installed on a stand with the tilted angle of 35° facing south.

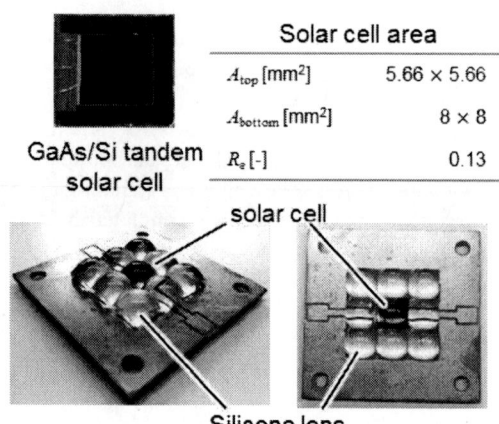

Solar cell area	
A_{top} [mm²]	5.66 × 5.66
A_{bottom} [mm²]	8 × 8
R_e [-]	0.13

Fig.6   Fabricated 2T GaAs/Si tandem solar cell and prototype SMAC module.

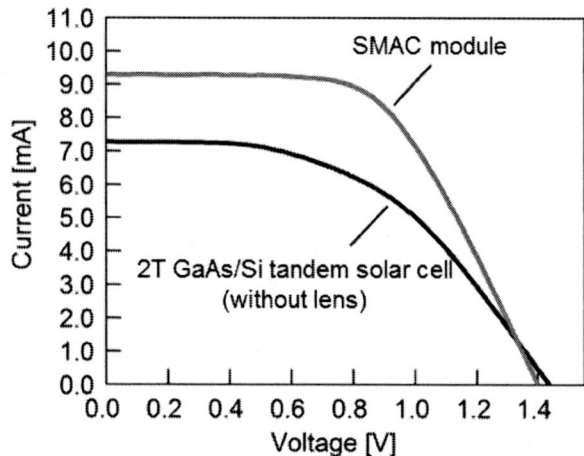

Fig.7   Current−voltage curves of 2T GaAs/Si tandem solar cell and SMAC module under outdoor 1 sun.

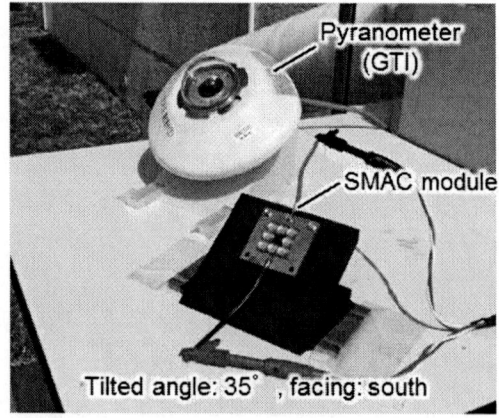

Fig.8   Experimental setup for outdoor characterization of SMAC module.

Fig. 9 shows the time-variation of GTI, the short-circuit current, the short-circuit current ratio J_{sc_m}/J_{sc_sim}, and the GTI-based module conversion efficiency. Fig. 9 (a) shows the GTI varying in the range of 200 ~ 1200 W/m^2 because of partly sunny condition. The measured data was filtered to eliminate the error data due to slow response time of the pyranometer. Fig. 9 (b) shows J_{sc_sim} and J_{sc_m}. $J_{sc\_sim,}$ was calculated by Eq. (5) using measured GTI. J_{sc_sim} and J_{sc_m} were almost the same value without being affected by incident angle and irradiance change. Fig 9 (c) shows J_{sc_m}/J_{sc_sim}, and the average value of J_{sc_m}/J_{sc_sim} during the measurement time was 0.98. This implies that the designed lens concentrates sunlight to the top cell even with a different incident angle at least during 3-hours experiment; and the energy collection performance is unaffected by irradiation fluctuation. On the other hand, Fig 9 (d) shows that the average GTI-based module conversion efficiency during the measurement time was low as 12.7 % probably due to high bonding resistance in the stacking interface of the tandem solar cell.

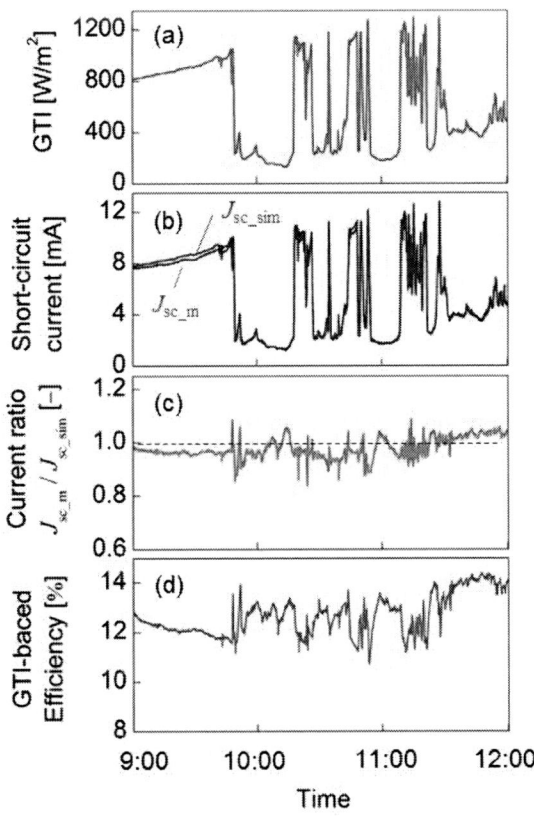

Fig.9   Experimental results of SMAC module. (a) GTI, (b) measured short-circuit current J_{sc_m} and J_{sc_sim}, (c) short-circuit current ratio J_{sc_m}/J_{sc_sim}, (d) GTI-based module conversion efficiency.

V. SUMMARY

The concentrator for 2-terminal GaAs/Si SMAC module with the areal current matching technique was designed to maximize the annual energy collection efficiency. The loss due to the current mismatch between the top cell and the bottom cell was accounted based on the incident energy ratio to each subcells. The designed lens shows a unique characteristic; lower peak angular optical efficiency and larger acceptance half angle as compared to the conventional static concentrator. The result of the tentative outdoor characterization with a prototype SMAC module implies that the energy collection performance of the designed concentrator lens and 2T GaAs/Si tandem solar cell is likely to be consistent with the simulated result. Further outdoor experiments are necessary to confirm the validity of the simulation result. Further improvement of the cell performance is also necessary to demonstrate the advantage of the concept of SMAC module.

ACKNOWLEDGMENT

This work was supported by JSPS KAKENHI Grant Number 26289373 and NICO project.

REFERENCES

[1]   R. Cariou, J. Benick, P. Beutel, N. Razek, C. Flötgen, M. Hermle, D. Lackner, S. W. Glunz, A. W. Bett, M. Wimplinger, and F. Dimroth, "Monolithic Two-Terminal III–V//Si Triple-Junction Solar Cells With 30.2% Efficiency Under 1-Sun AM1. 5G," *IEEE J. Photovoltaics*, vol. 7, no. 1, pp. 367–373, 2017.

[2]   A. C. Tamboli, S. Essig, K. A. W. Horowitz, M. Woodhouse, M. F. A. M. van Hest, A. G. Norman, M. A. Steiner, and P. Stradins, "Indium zinc oxide mediated wafer bonding for III–V/Si tandem solar cells," in *42nd IEEE Photovoltaic Specialist Conference (PVSC)*, 2015, pp. 1–5.

[3]   N. Jain and M. K. Hudait, "III–V multijunction solar cell integration with silicon: Present status, challenges and future outlook," *Energy Harvest. Syst.*, vol. 1, no. 3–4, pp. 121–145, 2014.

[4]   H. Mizuno, K. Makita, and K. Matsubara, "Electrical and optical interconnection for mechanically stacked multi-junction solar cells mediated by metal nanoparticle arrays," *Appl. Phys. Lett.*, vol. 101, p. 191111, 2012.

[5]   J. Yang, Z. Peng, D. Cheong, and R. Kleiman, "Fabrication of high-efficiency III-V on silicon multijunction solar cells by direct metal interconnect," *IEEE J. Photovoltaics*, vol. 4, no. 4, pp. 1149–1155, 2014.

[6]   M. Baba, K. Makita, H. Mizuno, H. Takato, T. Sugaya, and N. Yamada, "Feasibility study of two‐terminal tandem solar cells integrated with smart stack, areal current matching, and low concentration," *Prog. Photovoltaics Res. Appl.*, vol. 25, no. 3, pp. 255–263, 2017.

Design of optical elements for low profile CPV panel with sun tracking for rooftop installation

Xinbing Liu, Zhou Lu, Riccardo Leto, Carlton Brule and Nanu Brates

Panasonic Boston Laboratory, Newton, MA, 02459, USA

Abstract — We are developing a new approach to greatly expand the applicability of concentrator photovoltaic (CPV) from only ground-based, full external 2-axis tracking installations, to fixed-tilt CPV installation with integrated tracking that can be installed on rooftops. We designed and fabricated an optical element for this fixed-tilt CPV installation, and more than 99% of the output energy through the $10 \times 10 mm^2$ optical element is focused on an area with a radius of 180um, which is well within a solar cell with an active area of $0.55 \times 0.55 mm^2$, corresponding to a concentration ratio of 330x. This new strategy ultimately eliminates the costly external 2-axis tracking mechanism, and drives down the all-in system cost of Micro-scale Optimized Solar-cell Arrays with Integrated Concentration (MOSAIC) CPV to achieve cost parity with that of crystalline silicon PV panels.

I. INTRODUCTION

The use of flat-panel solar photovoltaics (FPV) is growing dramatically as costs decrease. By contrast, more efficient CPV systems, which focus direct sunlight onto a single point, have not been widely adopted because of their high cost, large size, and expensive tracking systems. A new approach, micro-scale concentrated photovoltaic systems (micro-CPV), may deliver the cost and size benefits of conventional FPV systems, but with an estimated 50% performance improvement. Micro-CPV modules would use cost-effective trackers and generate more electrical power in a given area. This allows installation on space-constrained residential rooftops and decreased costs for commercial and utility applications. Finally, the MOSAIC systems would have the ability to capture both direct and diffuse sunlight, which could make CPV economical in more geographical regions. These innovations could spur the expanded use of PV to generate clean, renewable energy.

II. PRIOR WORK AND NEW APPROACH

Panasonic AVC Company had carried out prior work with the MOSAIC concept, by working jointly with Solar Junction Corporation [1-4]. The design incorporated high-efficiency triple-junction PV cells [5] scaled down to $0.55 \times 0.55 mm^2$ and a simple polymethylmethacrylate (PMMA) lens array composed of a single refractive optic with a 10-mm array pitch. This yielded a geometric concentration factor, G, of ~ 330x with an array height of only 20 mm (Fig. 1a). In this design, the PV cells were directly bonded to the backside of the optical array and electrically contacted to a printed circuit board (PCB) substrate (Fig. 1b). The cells themselves were

not a back-contact design, but Panasonic carried out substantial packaging work to adapt the cell design so that such back contacting was achieved (Fig. 1c).

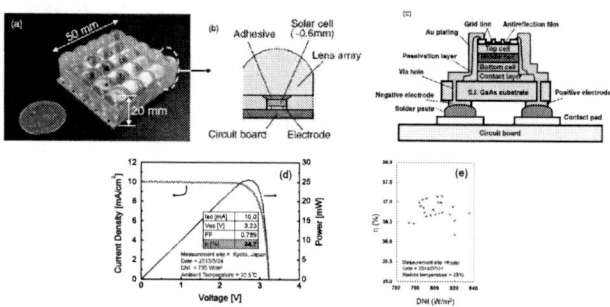

Fig.1. Prior Panasonic / Solar Junction work. (a) 5x5 sub-module. (b) Cross-sectional sketch of single micro-module. (c) PV cell packaging and interconnect detail. (d) IV curve and efficiency of a single cell in the 5x5 sub-module at 735 W/m2 DNI. (e) Single-cell efficiency measurements for one day, ranging between 36.2% and 37.2%, as a function of DNI.

In the course of this work, Solar Junction demonstrated that laboratory cell efficiency > 41% was achievable at up to 500 suns. Panasonic demonstrated that good alignment was achieved for 5 x 5 cm submodules consisting of 25 cells. In outdoor testing of a single micro-module, efficiency of 34.7% was demonstrated under 735 W/m^2 DNI (Fig. 1d) and over 36.5% at 770-830 W/m^2 DNI (Fig. 1e). Module temperature did not depart significantly from ambient temperature.

The concept was subsequently scaled up to larger size (20 x 20 cm) by tiling 16 of the sub-modules, and in outdoor testing of this prototype sub-panel, overall efficiency of 30.6% was achieved (Fig. 2).

Fig.2. Prior Panasonic / Solar Junction work. (a) 20x20 sub-panel, consisting of 16 tiled sub-modules. (b) Sub-panel performance under 887 W/m^2 DNI [3, 4].

However, all these prior work needs an external tracker and thus is not feasible for rooftop installation.

Based on the prior work, we are developing a new approach, called In-Plane Rotation (IPR), to eliminate the external tracker and enable the rooftop installation of the CPV panels.

In this approach, we will maintain on-axis orientation as the sun moves through the sky, enabling us to use a single optical element similar to that used in our prior work. The lens array is separated into rows of lenslets (Fig. 3a) and each row tilts as a rigid body about its long axis (Fig. 3b). A second rotational axis is required to track the sun. This axis is normal to the MOSAIC panel (Fig. 3c). The surface normal of the panel and the sun's rays form a plane, P1. The panel is rotated in-plane around the panel normal so the axis of each lens row is maintained perpendicular to the plane P1. The lens rows are then tilted so that the lens optical axis is parallel to the sun's rays. Fig.4 shows the IPR sun-lens orientation scheme. This micro-tracking subsystem will eliminate the need for bulky trackers, allowing fixed mounting of the panel. As the sun moves throughout the day, the lenses align themselves to the best position to receive sunlight, realizing the efficiency advantages of CPV without the cumbersome tilting of the entire panel.

Please note that the lenses shown in Fig.3 are tapered shaped thick lenses, and for concept illustration purpose only. The actual lens design went through many rounds of iterations and finally a thin lens design (to further reduce material cost and optical absorption loss) was reached, which will be covered in this paper.

This IPR solution could potentially enable rooftop deployment of CPV panels with efficiency much greater than the silicon PV panel, and could encourage greater adoption of solar power in all three primary markets – residential, commercial, and utility.

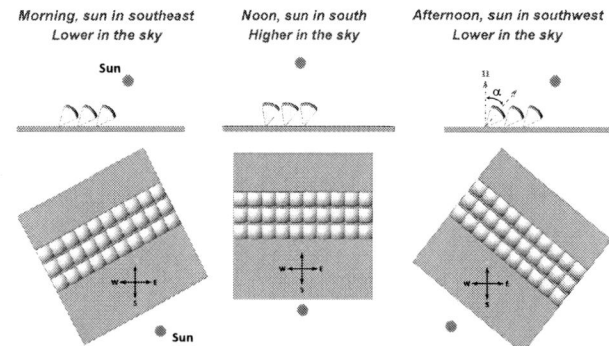

Fig.4. IPR sun-lens orientation scheme. To track the sun, the full array rotates in plane and the lens rows tilt at angle α to maintain alignment between the optical axis and the sun's rays.

III. THE DESIGN AND FABRICATION OF THE OPTICAL ELEMENT

The optical element, or lens, is one of the major corner stones for the IPR approach, and the other will be the fixed-tilt tracking mechanism. In this paper, we will be mainly focusing on the design, fabrication and testing of the IPR lens.

We used ZEMAX, an optical design and simulation software, to optimize our IPR designs with the following considerations: AM1.5D solar spectrum with an angular dispersion angle of 0.54° was used to simulates the sun light; the simulation wavelength was from 400-1600nm considering the working wavelength of a triple junction solar cell; the lens material was chosen to be acrylic (or PMMA) for potentially large scale fabrication (moldable) and cost consideration; the optical property of the actual acrylic was measured and the data is fed back to ZEMAX in the simulation.

By searching through all of the possible solutions in the optical space under sequential mode, ZEMAX reached the optimum optical design resulting in the smallest possible RMS spot radius on the image plane (the solar cell plane), shown in Fig.5. The scale bar is 0.55mm on the spot diagram to simulate the 0.55x0.55mm^2 solar cell active area.

In this process, merit functions in the ZEMAX were setup to satisfy specific requirements of the lens, specifically, the lens curvature constrains, the lens thickness and the edge thickness constrains, to ensure the actual manufacturability of the lens. In addition, the distance from the lens to the image plane was also constrained to ensure a low enough profile comparable to the current commercially available silicon PV panels in the market, for a readily rooftop deployment.

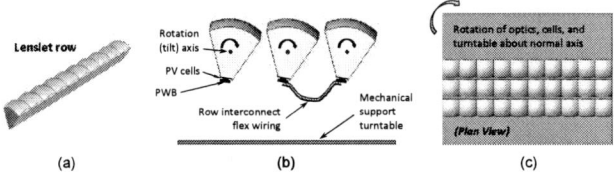

Fig.3. IPR concept. (a) Lenses are grouped in monolithic rows. (b) Rows are tilted about their long axis. (c) The entire array is rotated in plane, about an axis normal to the panel.

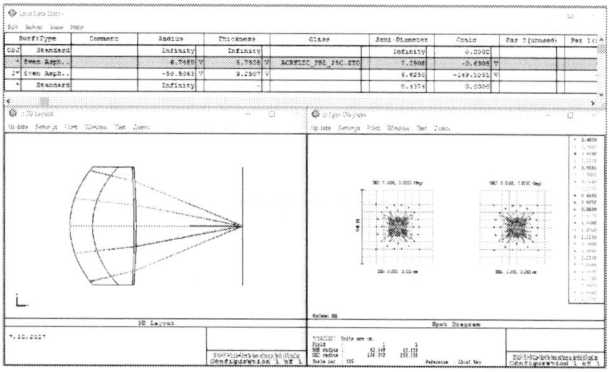

Fig.5. The design of a convex-convex IPR thin lens in ZEMAX.

We used even aspheric surfaces for the lens design with the following lens equation:

$$z = \frac{cr^2}{1+\sqrt{1-(1+k)c^2r^2}} + \alpha_1 r^2 + \alpha_2 r^4 + \alpha_3 r^6 + \alpha_4 r^8 + \alpha_5 r^{10} + \alpha_6 r^{12} + \alpha_7 r^{14} + \alpha_8 r^{16}$$

Where c is the radius of curvature=1/radius, k is the conic constant and α_n is the related n^{th} order term for the even aspheric surfaces.

The specific lens parameters were obtained after ZEMAX reaching the best solution. Specifically, α_n=0, c1=1/6.7489 and c2=-1/50.5063, k1=-0.6305 and k2=-149.1051 for a convex-convex IPR thin lens design.

With these lens parameters available, the actual lens can now be fabricated with the diamond turning machine in house, and the fabrication results are shown in Fig.6. In the fabrication, we used a software called DiffSys to convert lens profile in ZEMAX to coordinates recognizable by the diamond turning machine. Specifically, a ".pgm" tooth path file specifies the x and z coordinates, number of rough and finish passes, and provides compensation for the tool radius.

Convex-convex thin lens (surface 1)

- Size: 10x10mm
- Thickness: 5.7508mm
- Radius of Curvature: 6.7489mm
- Conic: -0.6305

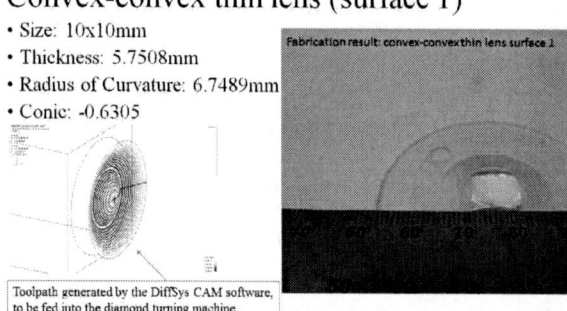

Toolpath generated by the DiffSys CAM software, to be fed into the diamond turning machine.

Convex-convex thin lens (surface 2)

- Size: 10x10mm
- Thickness: 5.7508mm
- Radius of Curvature: -50.5063mm
- Conic: -149.1051

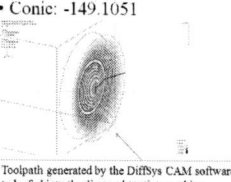

Toolpath generated by the DiffSys CAM software, to be fed into the diamond turning machine.

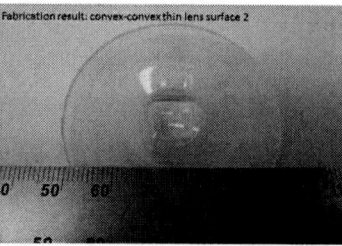

Fig.6. The designed convex-convex IPR thin lens fabricated by the diamond turning machine in house.

IV. EXPERIMENTAL RESULTS

The fabricated lens can now be measured to verify its performance.

For the measurement setup, people generally use solar simulator to simulate the actual sun. However, no matter how good a solar simulator is, it cannot exactly replicate the solar spectrum, which could be an issue for our designs as the designs were optimized based on the actual AM1.5D solar spectrum.

Thus, we use another method to address the solar source issue-we use the actual sun as our solar simulator with the help of an astronomical telescope tracker.

Fig.7 shows the IPR measurement setup with the actual sun tracking by setting up an astronomical telescope tracker at our parking lot in a sunny day.

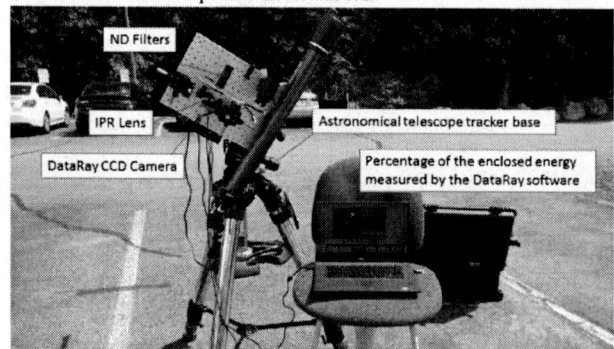

Fig.7. The IPR measurement setup with the actual sun tracking with an astronomical telescope tracker.

We used a DataRay CCD camera to capture the image spot after the sun rays pass the IPR lens. The DataRay software provides information regarding the percentage of the enclosed energy within a specific radius. From this information, we can compare the measurement results (Fig.8) with the simulation results (Fig.9, obtained by ZEMAX through non-sequential simulations) to see if the lens performs according to the design or not.

At the same time, the measured spot radius and the percentage of the focused energy will tell us if the focused energy falls within the 0.55x0.55mm² solar cell area or not, which determines if an IPR design is successful or not.

The measurement results in Fig.10 confirms that our IPR design performs as expected, as the measurement results agrees with the simulation ones very well. The small discrepancy could be from the noise of the DataRay CCD camera and the lens fabrication error.

In addition, we can see that from the actual measurement with the real sun, 99.2% of the output energy is encircled in a circular area with the radius of 180um. This indicates that the majority of the focused sun rays will be captured by the solar cell with an active area of 0.55x0.55mm², with a concentration G factor=330x.

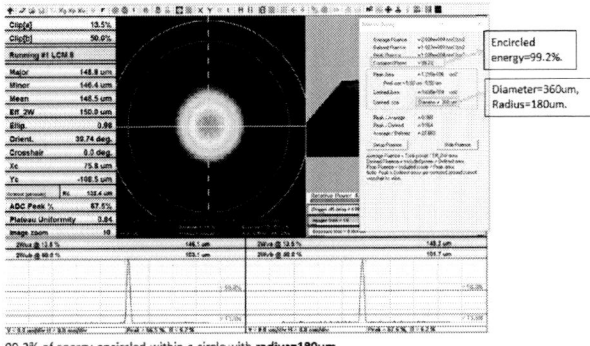

99.2% of energy encircled within a circle with **radius=180um.**

Fig. 8. Measurement result: 99.2% of energy encircled within a circle with radius=180um, which indicates that at least 99.2% of the input energy will be captured on a solar cell with an active area of 0.55x0.55mm², or a concentration ratio of 330x.

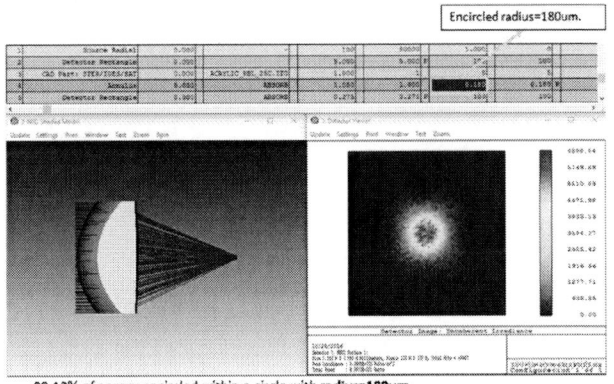

99.12% of energy encircled within a circle with **radius=180um.**

Fig. 9. Simulation result: 99.12% of energy encircled within a circle with radius=180um.

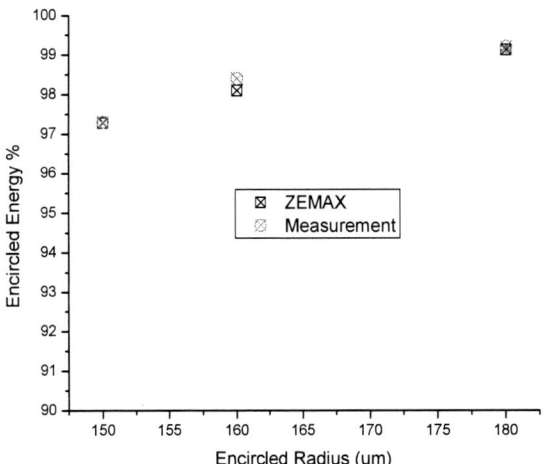

Fig.10. Measurement results with the actual sun, and a comparison with the simulation results.

V. CONCLUSION

A new IPR concept was proposed to enable the fixed-tilt CPV installation on rooftops. We designed, fabricated and tested optical elements for this IPR concept.

More than 99% of the output energy through a 10x10mm² optical element is focused on an area with a radius of 180um, which is well within a solar cell with an active area of 0.55x0.55mm², corresponding to a concentration ratio of 330x. The measurement results agree with the simulation results, indicating a successful design and fabrication of the optical element.

The result is one of the important steps to push forward the idea of deploying CPV panels with much higher efficiency on the rooftops, which ultimately may turn into a serious competitor to the currently dominating rooftop silicon PV panels.

ACKNOWLEDGEMENT

We gratefully acknowledge the financial support for this work provided by ARPA-E MOSAIC Award number DE-AR0000629.

REFERENCES

[1] Fidaner, O, et al, "High efficiency micro solar cells integrated with lens array," Appl. Phys.Lett. 104, 103902, 2014.
[2] Arase, H, et al, "A Novel Thin Concentrator Photovoltaic With Microsolar Cells Directly Attached to a Lens Array," IEEE Journal Of Photovoltaics, 4 (2), 709-712, 2014.
[3] Itou, A, et al, "High-efficiency thin and compact concentrator photovoltaics using micro-solar cells with via-holes sandwiched

between thin lens-array and circuit board," Japanese Journal of Applied Physics 53, 04ER01, 2014.

[4] Nakagawa, T, et al, "High-efficiency Thin and Compact Concentrator Photovoltaics with Micro-solar Cells Directly Attached to Lens Array," in Light, Energy and the Environment, OSA Technical Digest (online), paper RF4B.5., 2014.

[5] Jones-Albertus, R, et al, "Using Dilute Nitrides to Achieve Record Solar Cell Efficiencies," Mater. Res. Soc. Proc. 1538, 161,2013.

Micro Chiplet Printer Development for MOSAIC Program

P.Y. Maeda, Y. D. Wang, S. Raychaudhuri, J. Kalb, D. K. Biegelsen, R. Lujan, Q. Wang, Y. Wang, J. Bert, B. Rupp, I. Matei, L. Crawford, A. Plochowietz, E.M. Chow, J.P. Lu, V. Gupta[†]

PARC, a Xerox Company, 3333 Coyote Hill Rd., Palo Alto, CA 94304
[†]Sandia National Laboratories, Albuquerque, NM 87185 and Livermore, CA 94550 USA

Abstract — The micro-CPV concept being developed under the ARPA-E MOSAIC (Micro-scale Optimized Solar-cell Arrays with Integrated Concentration) program uses arrays of micro unit cells (or elements) such that the material usage, weight, and the required structural strength can all be scaled down favorably. Unfortunately, one of the essential unfavorable scaling factors is the assembly cost due to the many micro scale components that must be deposited, positioned, oriented, and connected over large areas. PARC and Sandia are developing a prototype platform that uses dynamic electric field micro-assembly to demonstrate micro chiplet printing - assembling a desired solar cell chip at a designated location with well controlled orientation. Xerographic printing systems utilizing this method can be extended to provide high-throughput, on-demand heterogeneous assembly of micro-CPV systems.

I. INTRODUCTION

PARC and Sandia are working together to create a capability to fabricate panel-sized micro-scale concentrated photovoltaic (micro-CPV) modules that will require the assembly of tens of thousands to a few hundred thousand components per square meter [1]. A general, massively parallel, high throughput micro assembly tool that is based on generalization of the industrially proven, xerographic process has been developed at PARC. In this envisioned digital manufacturing/printing process, the "inks" are micro scaled PV components that Sandia has developed and the "image" outputs are panel sized micro-CPV substrates with placed and interconnected PV cells. The resulting printing process will deliver the cost-effective and high-yield massively parallel chiplet deposition, alignment, and interconnection necessary to make micro-CPV a reality.

The fundamental problem is clearly shown in Fig.1; the assembly cost will approach or even dominate over the cell price itself at various scaling points. The crossover (assembly cost > component cost) is at mm scale for low cost Si cells, and a few hundred microns for high price, high efficiency multi junction cells, well within the length scales spanned by micro-PV systems. A more cost effective assembly solution is needed.

II. MICRO CHIPLET PRINTER PROTOTYPE DEVELOPMENT

Xerographic digital laser printers are an example of a commercially mature directed assembly manufacturing tool, as they assemble millions of color toner particles onto paper in designated locations. PARC pioneered this technology and

today envisions extending it into the digital assembly of micro-objects and electronics components. As shown in Fig. 1, the massively parallel nature of this assembly technology enables a cost effective manufacturing solution for the micro-CPV vision. While pick-and-place can range from $0.5/component (<500µm precision) to $.01/component (>500µm), the proposed technique could yield <$2x10^{-3}$/cm^2 with <10µm precision.

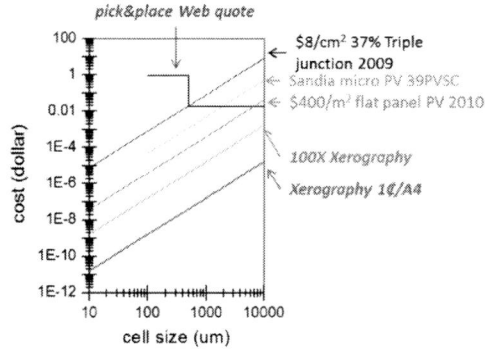

Fig. 1. Scaling calculation for micro CPV cell cost. The price of a solar cell scales with area, whereas the cost of pick-and-place assembly is fixed per cell (with surcharge for small cells due to the higher precision requirement). The cost of assembly [2] with pick-and-place exceeds the materials costs for cell sizes less than 1mm x 1mm, and dominates any cell type for sizes less than 500umX500um. In contrast, the predicted cost of solar cell assembly with xerographic processes is negligible (10^3 times cheaper) compared to material and pick-and-place costs.

Fig. 2 shows a functional block diagram illustrating a prototype micro chiplet printer that resembles a laser printer, but takes micro PV cells in a carrier liquid ("ink"), and places them on a substrate in a predefined pattern with an accuracy of 10µm, such that the high alignment precision necessitated by micro optics and electrical interconnection can be met. The proposed demonstration printer will continuously print a strip of substrate material that is several square centimeters to demonstrate proof-of-concept, which can be later scaled to meter-wide, roll-fed substrates. The proposed micro chiplet printer will have major modules analogous to laser printers: (1) chiplets immersed in dielectric liquid, similar to liquid toner ink, (2) an "image development unit" that sorts and electrophoretically (EP) or dielectrophoretically (DEP) arranges the randomly deposited chiplets, and (3) a transfer mechanism that moves them to the final substrate. A PARC

978-1-5090-5606-4/17 $31.00 © 2017 IEEE

developed Opto Coupler Active Matrix (OptoCAM) device will be used to implement the dynamic electric field image development unit. The OptoCAM technology uses a variation of a-Si:H Thin Film Transistors (TFT) and Photodiode Arrays. Such TFT technology is fully commercialized as the basis for the multi-billion dollar flat panel display industry and digital X-ray imaging in which PARC has previously performed significant research [3]. A customized Digital Light Processing (DLP) projector and projection optics drives the OptoCAM. A high-speed camera provides image data for real-time feedback control of the chip assembly process. To prove the utility of the proposed chiplet printer, we will produce two types of PV cell array backplanes (i.e. electrically connected PV cell array on a substrate) as a proof-of-concept. The first type of demonstrator backplane will use a single type of micro PV cell as the chiplet ink. After successfully demonstrating this first demonstrator, we plan to print, test and characterize a demonstrator backplane that uses at least two types of microscale PV chiplets (i.e. both III-V chips and silicon chips).

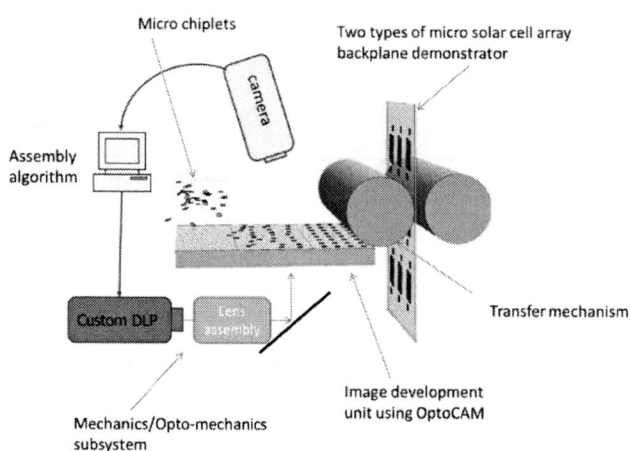

Fig. 2. The PARC SNL prototype micro chiplet printer concept

III. RESULTS

The prototype construction up to this point includes the: (1) imaging system, (2) DLP based projection system, (3) transfer system, and (4) OptoCAM device.

The imaging system consists of a high speed 1696 x 1710, 3MP Optronis CP80-3C-540 camera, and a Leica Z16 APO monocular zoom microscope, with 0.19X to 2.9X magnification that covers a > 25.4 mm field of view (FOV). The DLP projection system uses a 1024 x 768 Digital Light Innovations CEL5500 projector, and 2.31X magnification projection optics with < 25 μm resolution, < 0.06% distortion, that was constructed from off-the-shelf catalog lenses. The imaging and projection system optical paths are combined with a beamsplitter. Sequences of video images from the projector are used to generate a dynamic electric field pattern in the OptoCAM device which is used to move and arrange the chiplets. The imaging system is part of the vision system that provides feedback to the software driven control system that determines the video images sent to the projector. A CAD rendering is shown in Fig. 3.

The transfer process is used to translate assemblies from the OptoCAM development module to a final substrate so that the cell assembly can be deployed for real world applications. The transfer system must be able to transfer assemblies at high fidelity, and form a permanent bond to the substrate such that the assembly can be deployed in the world. The system must achieve these goals without disturbing the development module where the assembly is constructed. The assembly process uses charge directors in the assembly medium (dielectric fluid) for the assembly to work properly. Therefore, it is desirable to prevent possible contaminants from materials on final substrate to come into direct contact with OptoCAM surface. Furthermore, applying pressure to the OptoCAM could damage or soil the electrode array resulting in performance degradation over time.

In order to address all of these challenges we developed a novel continuous transfer concept using a transfer belt mimicking what can be found in some industrial document laser printing systems. The transfer belt is never brought directly into contact with the OptoCAM mitigating the risk of physical damage. Back transfer material on the final substrate can also be minimized to reduce the risk of contaminating the assembly fluid.

The assembled micro chiplets are translated out of the assembly medium, dried and brought in contact with the final substrate to form the final assembly.. The transfer belt can then pass through a cleaning system to ensure no residual material is brought back in contact with the assembly medium. A photograph of the transfer system is shown in Fig. 4.

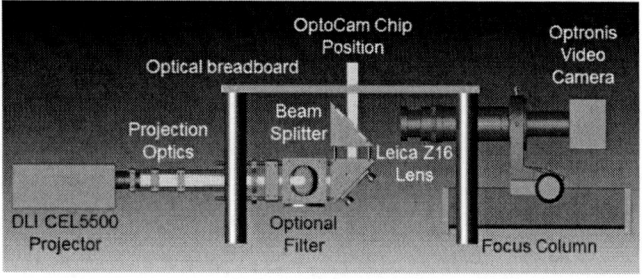

Fig. 3. CAD rendering of projection and imaging systems

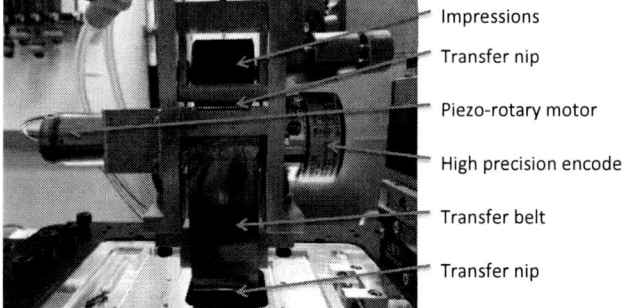

Fig. 4. Transfer system

The goal is to show that the system is capable of transferring 20 chiplets with better than 25 um accuracy. Initial experiments demonstrated the ability to transfer several 10s of chips with an RMS placement error of 13.3 μm.

The working principal of the Image Development Unit of micro chiplet printer, using an Opto Coupler Active Matrix (OptoCAM) device is described as follows. For the prototype micro chiplet printer, the micro solar cells, the chiplets, are randomly dispersed in dielectric fluid as "ink' for the printer. They are sorted and arranged by dynamic electrostatic field, generated by a OptoCAM device. The OptoCAM device can be understood as a massive 2D multiplexer, addressed by projected light from the back. The single high voltage signal, synchronized with the light pattern, is then used to charge the storage capacitor of each pixel. As the storage capacitors are periodically refreshed and reprogramed, arbitrary, dynamic electrostatic potential pattern can be "displayed" at the top electrodes and actuating the chiplets to the desired position and orientation.

Several types of OptoCam designs have been fabricated, where the devices differ in voltage, pitch, and complexity to study the tradeoffs in performance. Fig. 5 shows some fabricated devices. Fig. 6 shows two of the designs and measured I-V characteristics.

Fig. 5. Finished OptoCAM devices

Fig. 7 is a photograph of the prototype micro chiplet printer that has been recently assembled and will be used extensively to demonstrate and validate the technology.

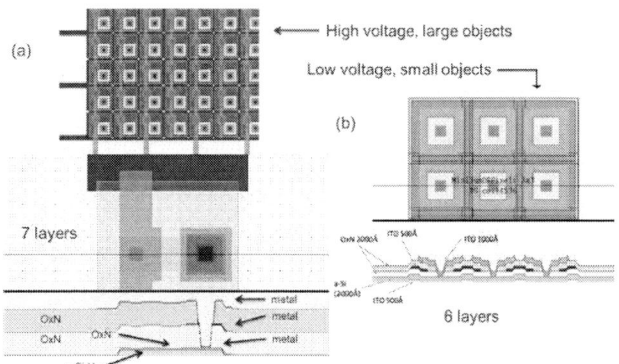

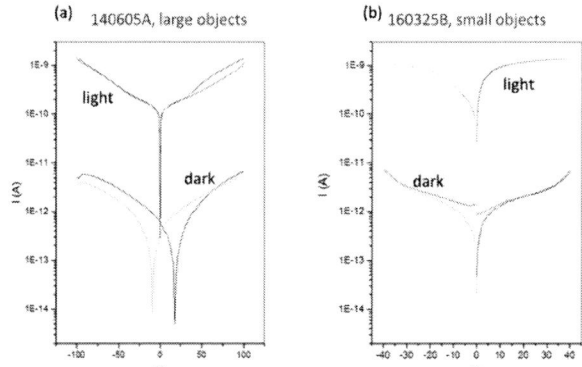

Fig. 6. OptoCAM designs and I-V characteristics

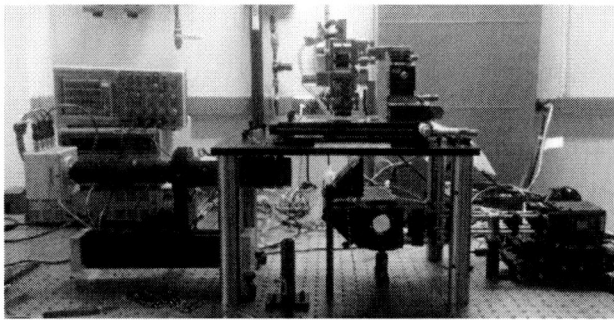

Fig. 7. Prototype micro chiplet printer hardware

Fig. 8. shows the assembly process of a microchiplet ink. A starting solution of 150 x 250μm silicon chiplets in a dielectric fluid was placed on the OptoCAM device, allowing the chiplets to settle in a random orientation. The projector then displays a pattern under the chiplet which is designed to produce electrophoretic and dielectrophoretic motion on the particle. This allows the particle to be dragged to the target

978-1-5090-5606-4/17 $31.00 © 2017 IEEE

location with the movement of the projector image. Once at the target, a different image is designed to orient the particles. This continues until all particles are placed and oriented onto a 1.5mm spaced square lattice. A square lattice was chosen, but any number of configurations is possible, and this can even be changed dynamically. Once assembled, computer vision was used to determine the center of mass of each of the chiplets (blue star) and compared to the distance from the target (red star). The root mean square error (RMSE) in the x- and y-directions was measured to be 15.4µm and 14.4µm, respectively.

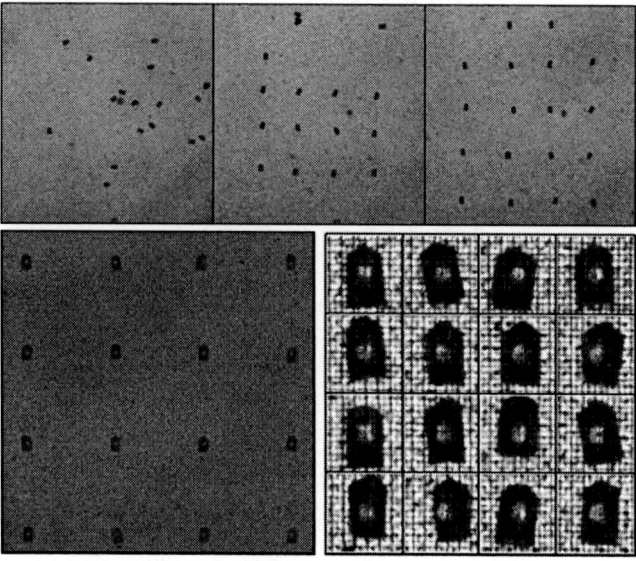

Fig. 8. Chiplet accuracy and assembly from disordered to 1.5mm spaced array

IV. CONCLUSION

We have developed and built the requisite hardware for the prototype micro chiplet printer which is demonstrating the desired imaging and projection system performance, micro chiplet transfer and placement, and OptoCAM device characteristics. Future project work will entail further hardware refinement, software and algorithm development, and device fabrication, complete solar cell backplane assembly, and testing to achieve full technology demonstration.

REFERENCES

[1] Maeda P., Lu J.P., Whiting G., Biegelsen D., Raychaudhuri S., Lujan R., Veres J., Chow E., Gupta V., Nielson G., Paap S., Micro chiplet printer for micro-scale photovoltaic system assembly, 2015 IEEE 42nd PVSC Conference

[2] Deduced from a web quoting tool, assuming 1K boards, 1K components/board, assembled in China, http://www.bittele.com.

[3] Rahn, J. T. ; Lemmi, F. ; Lu, J. P. ; Mei, P. ; Apte, R. B. ; Street, R. A. ; Lujan, R. A. ; Weisfield, R. L.; Heanue, J. A., "High resolution x-ray imaging using amorphous silicon flat panel arrays", IEEE Transactions on Nuclear Science. 1999 June; 46 (3, part 2): 457-461

Micro-optical Tandem Luminescent Solar Concentrator

David R. Needell[*], Zach Nett[†], Ognjen Ilic[*], Colton R. Bukowsky[*], Junwen He[‡], Lu Xu[‡],
Ralph G. Nuzzo[‡], Benjamin G. Lee[§], John F. Geisz[§], A. Paul Alivisatos[†] and Harry A. Atwater[*]

[*]Department of Applied Physics and Materials Science, California Institute of Technology, Pasadena, CA 91125, USA
[†]Department of Chemistry, University of California Berkeley, CA 94720, USA
[‡]Department of Chemistry, University of Illinois at Urbana-Champaign, Urbana, Illinois 61801, USA
[§]National Renewable Energy Laboratory, Golden, CO 80401, USA

Abstract—We propose a design for a concentrated photo-voltaic (CPV) module with features that can capture both direct and diffuse sunlight. This approach uses a luminescent solar concentrator (LSC) sheet that includes quantum dots, and a tailored optical coating that enhances concentration and delivery of sunlight to the micro-PV cells. In addition, the light not captured by the quantum dots impinges on a tandem solar cell beneath the LSC sheet. The design of the LSC focuses on lowering the number of expensive micro-PV cells needed within the concentrator waveguide, thus reducing system costs while maintaining high efficiency. The design also allows the module to be effective without a tracking system, making it attractive to all PV markets.

Index Terms—Luminescent solar concentrator, quantum dots, microoptics, Bragg gratings

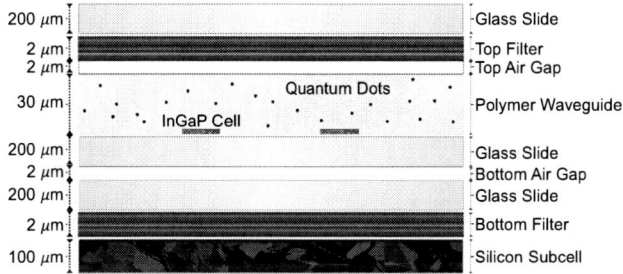

Fig. 1. Two-dimensional schematic of the tandem LSC with top/bottom mirrors included. Note that this figure is not to scale.

I. INTRODUCTION

Much of current photovoltaic (PV) research focuses on decreasing the levelized cost of electricity (LCOE) of the solar module [1]. Concentrated photovoltaic (CPV) devices offer a possible solution to minimizing the overall cost of PV-generated electricity. By concentrating the light that impinges upon a solar cell, the device efficiency can be increased. Additionally, by developing a concentration method that focuses the light to a smaller area, the amount of PV material required decreases. However, two fundamental hurdles for CPV cells exist. First, the complexity and cost of optics involved in concentration can outweigh the decreased area benefit. Second, CPV cells often require tracking hardware in order to efficiently concentrate light. A luminescent solar concentrator (LSC) overcomes both of these shortcomings of traditional geometric concentrators. A LSC device relies upon total internal reflection (TIR) trapping of light by embedding luminophores within a polymer waveguide [2]–[4]. When direct and diffuse light penetrate this waveguide, luminophores absorb and isotropically re-radiate the light at longer wavelengths. Because of the contrast in the index of refraction between the waveguide and the surrounding medium, a significant portion of the emitted light will become trapped [2], [5]–[7]. This trapped light can then be redirected toward solar cells embedded within the waveguide.

While the concept of a LSC device poses an alternative solution to traditional CPV technology, there exist several challenges in developing this type of solar PV concentrator. The two main efficiency-limiting factors include: 1) realizing the necessary optical characteristics (absorption/emission and

quantum efficiency) of the embedded luminophores, and 2) minimizing the amount of light exiting the waveguide via the TIR escape cone [8]–[10]. Here, we propose a LSC device design that mitigates both of these problems as shown in Fig. 1.

We synthesize highly efficient CdSe/CdS core/shell quantum dots (QDs) to be used as the LSC luminophores [11]–[13]. We minimize nonradiative luminophore absorption by optimizing the QD absorption and emission spectra, and photoluminescence quantum yield (PLQY). For our embedded solar cell, we encapsulate an indium gallium phosphide (In-GaP) microcell, planar to the polylaurylmethacrylate (PLMA) waveguide, tuning the bandgap of the cell to match the emission peak of the QD spectra. We additionally fabricate spectrally selective, dielectric stack filters in order to minimize escape cone loss of the waveguide. Finally, we allow longer-wavelength light incident upon our LSC module to propagate unaffected to the underlying Si subcell. By use of a Monte Carlo ray-tracing model, we simulate overall tandem LSC device efficiencies [14]. Fig. 1 shows the complete micro-optical tandem LSC device geometry and architecture as implemented by the Monte Carlo algorithm.

To our best knowledge, previous LSC concentration factors showed a maximum of 22, using a dye with a high PLQY. In 2015, collaborators from this research team achieved a new record of 30 for LSC concentration [12]. The simulation results demonstrated in this research show a concentration factor that approaches 91, and an overall tandem module efficiency approaching 31.1% under low direct normal incidence

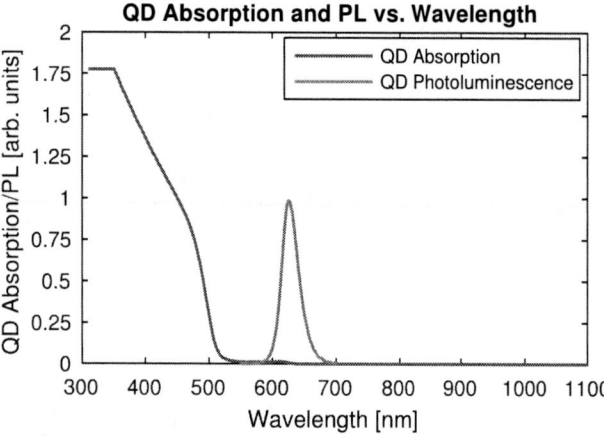

Fig. 2. Absorption and photoluminescence spectra for a CdSe/CdS QD with a 3.45 nm CdSe core, showing a Stokes Ratio over 100.

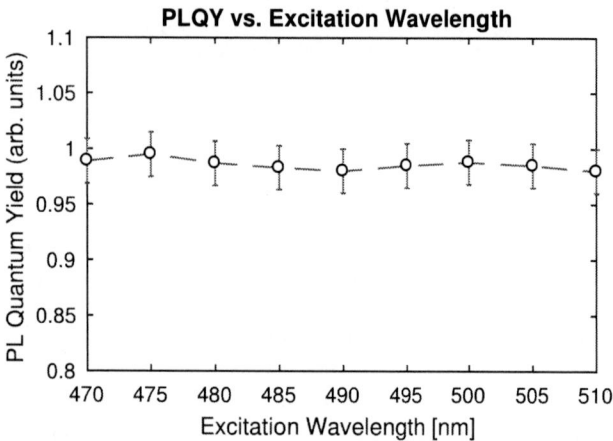

Fig. 3. Measured PLQY for the highest performing QD batch.

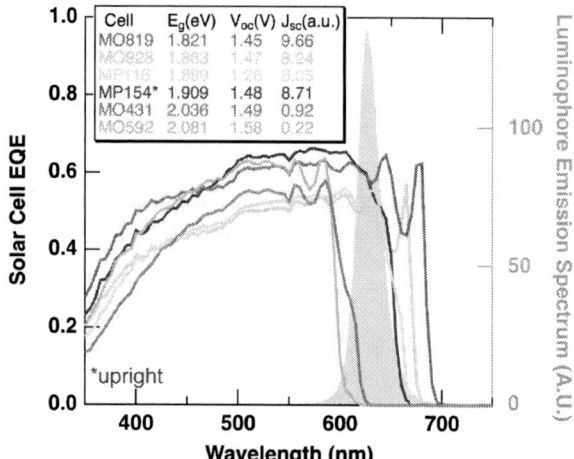

Cell	E_g(eV)	V_{oc}(V)	J_{sc}(a.u.)
MO819	1.821	1.45	9.66
MO928	1.863	1.47	8.24
MP116	1.899	1.26	8.05
MP154*	1.909	1.48	8.71
MO431	2.036	1.49	0.92
MO592	2.081	1.58	0.22

Fig. 4. Measured EQE data for various fabricated InGaP cells. The QD PL spectrum is shown in gray and used to test the performance of the cells under normal incidence.

conditions.

Section II Part A shows the experimental results of synthesized CdSe/CdS QDs and fabricated InGaP microcell characteristics. We additionally detail the process of InGaP microcell array and PLMA waveguide construction. Finally, we show the measured reflectance spectrum for a fabricated wavelength-selective, dielectric stack filter. Section II Part B overviews the Monte Carlo ray-tracing algorithm, and displays its key results. Section III concludes the findings of this research and provides a brief synopsis of future work in this area of tandem LSC technology.

II. RESULTS

A. Experimental Findings

Previous work on CdSe/CdS QDs has shown successful synthesis of high PLQY and large Stokes shifts [12], [15], [16]. Fig. 2 shows the absorption/emission characteristics of synthesized QDs measured while in a cuvette. Fig. 3 displays the PLQY of the QDs with respect to the wavelength of light absorbed by the dots. As seen in Fig. 3, we obtain CdSe/CdS QD samples with greater than 95% efficiency, with a maximum PLQY reaching 98.9 +/- 2%.

We synthesize CdSe/CdS QDs following close literature procedures [12], [15], [16], with slight adjustments in order to tune the spectral characteristics of the dots to our desired luminescence wavelength. We measure the absorption and PL spectra via a Shimadzu UV-3600 double beam spectrometer and a Horiba Jobin-Yvon FluoroLog 2 spectrofluorometer, respectively. We additionally achieve high accuracy measurement of the PLQY via a photothermal detection system.

In addition to tuning the optical characteristics of the QDs, we match the embedded InGaP microcell bandgap to the emission spectrum of the QD luminophores. We grow InGaP cells in an upright orientation, placing a thin emitter layer via metal organic vapor phase epitaxy [17]. We fabricate an array of microcells atop a glass slide via standard microscale transfer printing techniques [18]. Fig. 4 shows the external quantum efficiency (EQE) of various fabricated InGaP cells with corresponding open circuit voltage and short circuit current measurements, performed under the PL emission spectrum as shown in the figure.

We embed this microcell, InGaP array within a 30 μm thick PLMA waveguide with dispersed QD luminophores. For optimal QD absorption, we suspend the CdSe/CdS QDs in hexane solution. We measure the resulting waveguide/QD absorption using a Varian Cary 5G spectrophotometer, achieving an ideal optical density (OD) of 0.30 at 450 nm incident light.

Finally, to achieve increased InGaP cell concentration factors, we fabricate top and bottom spectrally selective filters. We alternate high/low index of refraction dielectric material layers of varying thickness to achieve a well-defined rejection band, as shown in Fig. 5. We measure the filter reflection

978-1-5090-5606-4/17 $31.00 © 2017 IEEE

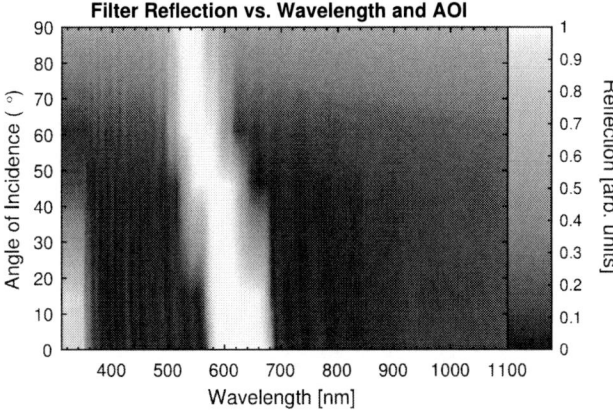

Filter Reflection vs. Wavelength and AOI

Fig. 5. Measured reflection of the dielectric stack filter, with respect to incident photon wavelength and angle of incidence (AOI).

with respect to both incident wavelength and angle. The top and bottom filters must highly reflect light in the rejection band and allow sufficient transmission in the pass bands at a given incident angle in order to enhance overall module LSC efficiencies.

B. Theoretical Findings

To achieve high LSC device efficiency, photoluminesced light must be effectively trapped within the PLMA waveguide. For a polymer waveguide (PLMA polymer matrix with $n \approx 1.44$) bordering air, we find the critical angle for TIR to be $\theta_c \approx 44°$. Light emitted by the embedded QD luminophores at an angle less than $44°$ with respect to normal incidence will not be trapped by the PLMA polymer waveguide.

To mitigate this loss mechanism, we analyze how a top and bottom, omnidirectional, wavelength selective filter boosts the performance of the LSC device when placed directly atop a flat-plate silicon solar cell. For modeling purposes, we assume a standard Si cell with an overall power conversion efficiency (PCE) of $\eta \approx 21.8\%$ under AM 1.5G spectrum at normal incidence. We simulate tandem device performance via a Monte Carlo ray-tracing algorithm, tracking individual photons through the LSC module. We stochastically treat photon trajectory paths, calculating reflection and refraction conditions with Fresnel and Snells laws, respectively [4], [13], [19]. To achieve sufficient statistical distribution, we initialize approximately 10^6 photons, distributed across wavelengths 300 to 1100 nm. To simulate a low direct normal incidence (DNI) environment, we restrict 40% of the initialized photons to normal incidence and the remaining 60% at angles randomly assigned in the incident photon hemisphere, thereby effectively lowering the incident power to approximately 76.75 mW/cm^2. Fig. 6 shows all possible loss mechanisms of the photon trajectory paths, accounted for by this ray-tracing algorithm. The ray-tracing model ends with either absorption of the photon by either the InGaP or Si cell or by photon loss. We integrate the collected photons against the AM 1.5G

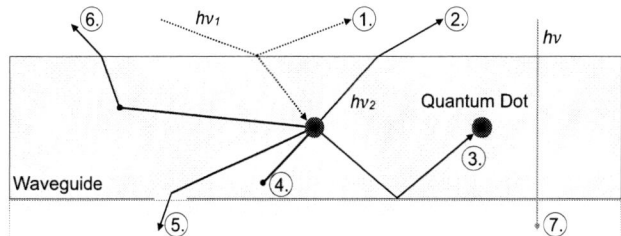

Fig. 6. Loss mechanisms intrinsic to a LSC device and accounted for in the ray-tracing model. 1) Surface reflection of incident light, 2) escape cone loss via QD emission, 3) QD parasitic absorption, 4) polymer parasitic absorption, 5) waveguide surface defect scattering, 6) spontaneous polymer scattering, and 7) transmission of light through the waveguide.

spectrum to obtain the photocurrent for both cells. We then approximate device efficiency by application of the radiative dark current limit with the ideal diode equation to obtain an estimate for the resulting open circuit voltage (Voc) and fill factor (FF) for both the embedded InGaP microcell and the Si subcell [14], [20]–[22]. To properly account for area normalization, we treat the effective area of the embedded, planar InGaP cell as the entire illuminated waveguide top surface.

We additionally compare the resulting efficiencies between the cases with and without ideal top/bottom filters. Tab. 1 summarizes the simulation results. For these simulations, we assume an ideal mirror with pass bands defined as: 300 584 nm and 666 nm 1100 nm, and a rejection band defined as: 585 665 nm, irrespective of incident angle. Given the EQE of the embedded InGaP microcell as described in Fig. 4 and a standard silicon cell with an overall cell efficiency of approximately 21% under full direct illumination, we achieve a maximum of 31.1% device efficiency, simulated under this low DNI environment. This maximum efficiency assumes unity PLQY, a geometric gain (GG) of 91.1, an optical density of the PLMA/QD waveguide of 0.30 measured at 450 nm incident light, ideal top/bottom filters, and QD spectral characteristics shown in Fig. 2. Furthermore, we assume periodic boundary conditions of our model to simulate a larger, InGaP microcell array.

III. Conclusions

We propose and analyze a tandem LSC-Si solar device with estimated module efficiencies approaching 31% under highly diffuse light. We synthesize CdSe/CdS QDs suspended in a PLMA polymer waveguide, embedded InGaP microcell arrays, and spectrally selective filters for use in such a tandem structure. We develop a rigorous Monte Carlo ray-tracing computational tool to serve as a proof of concept in order to demonstrate the possibility of harnessing a LSC device as the top-cell in a tandem LSC-Si module.

Experimental measurements of both the embedded QD luminophores optical characteristics and InGaP microcell

TABLE 1

SUMMARY OF MONTE CARLO RAY-TRACING RESULTS

Mirror Case	PLQY	Jsc (InGaP) [mA/cm^2]	Jsc (Si) [mA/cm^2]	Pout (InGaP) [mW/cm^2]	Pout (Si) [mW/cm^2]	Module Efficiency	GG
Perfect	1.00	7.27	21.87	10.80	13.08	31.1	91.1
None	1.00	2.84	27.20	4.15	16.25	26.6	91.1

EQE demonstrate a possible method of minimizing parasitic photon absorption loss with an LSC. We further minimize the photon escape cone loss by fabricating a dielectric stack filter, centering the rejection band about the QD PL peak. We finally apply our ray-tracing model to these device components to develop a first order approximation to overall module efficiency, simulated under low DNI lighting conditions.

This device design, experimental measurements, and simulation results rely upon the use of this omnidirectional wavelength selective filter to achieve high device efficiency. Our current mirror design efforts focus on continued development and optimization of this aperiodic, one-dimensional dielectric stack filter to selectively reflect the PL wavelengths of interest. Assuming sufficient mirror reflectivity across small angles of incidence, this work shows overall module efficiencies approaching 31%. Future work is directed toward mirror design, simulation, and fabrication as well as overall device design optimization and assembly.

Acknowledgment

This work is made possible by the Advanced Research Projects Agency for Energy (ARPA-E) U.S. Department of Energy grant under the Micro-scale Optimized Solar-cell Arrays with Integrated Concentration (MOSAIC) award, DE-AR0000627. We additionally acknowledge Evaporated Coatings Inc. (ECI) in collaboration for the design and fabrication of the spectrally-selective waveguide filter.

References

[1] B. Vest, "Levelized Cost and Levelized Avoided Cost of New Generation Resources in the Annual Energy Outlook 2016," Us Eia Lcoe, no. August, pp. 1–20, 2016.

[2] J. S. Batchelder, "The Luminescent Solar Concentrator," 1982.

[3] J. Madrid, M. Ropp, D. Galipeau, and S. May, "Investigation of the Efficiency Boost Due to Spectral Concentration in a Quantum-Dot Based Luminescent Concentrator," 2006 IEEE 4th World Conference on Photovoltaic Energy Conference, pp. 154–157, 2006. [Online]. Available: http://ieeexplore.ieee.org/lpdocs/epic03/wrapper.htm?arnumber=4059585

[4] F. Meinardi, S. Ehrenberg, L. Dhamo, F. Carulli, M. Mauri, F. Bruni, R. Simonutti, U. Kortshagen, and S. Brovelli, "Highly Efficient Luminescent Solar Concentrators based on Ultra-Earth-Abundant Indirect Band Gap Silicon Quantum Dots," Nature Photonics, vol. 11, no. 3, pp. 177–185, 2017. [Online]. Available: http://dx.doi.org/10.1038/nphoton.2017.5

[5] a. Goetzberger and W. Greubel, "Applied Physics Solar Energy Conversion with Fluorescent Collectors," Applied Physics, vol. 14, pp. 123–139, 1977.

[6] E. Yablonovitch, "Thermodynamics of the Fluorescent Planar Concentrator," Journal of the Optical Society of America, vol. 70, no. 11, pp. 1362–1363, 1980.

[7] ——, "Statistical ray optics," Journal of the Optical Society of America, vol. 72, no. 7, pp. 899–907, 1982. [Online]. Available: http://www.opticsinfobase.org/abstract.cfm?URI=josa-72-7-899

[8] J. Gutmann, H. Zappe, and J. C. Goldschmidt, "Predicting the performance of photonic luminescent solar concentrators," IEEE Journal of Photovoltaics, pp. 1864–1868, 2013.

[9] A. L. Martínez and D. Gómez, "Design, fabrication, and characterization of a luminescent solar concentrator with optimized optical concentration through minimization of optical losses," Journal of Photonics for Energy, vol. 6, no. 4, p. 045504, 2016. [Online]. Available: http://photonicsforenergy.spiedigitallibrary.org/article.aspx?doi=10.1117/1.JPE.6.045504

[10] W. G. J. H. M. V. Sark, Z. Krumer, C. D. M. Donegá, and R. E. I. Schropp, "Luminescent Solar Concentrators : the route to 10 % efficiency," IEEE Journal of Photovoltaics, pp. 2276–2279, 2014.

[11] N. D. Bronstein, L. Li, L. Xu, Y. Yao, V. E. Ferry, A. P. Alivisatos, and R. G. Nuzzo, "Luminescent solar concentration with semiconductor nanorods and transfer-printed micro-silicon solar cells," ACS Nano, vol. 8, no. 1, pp. 44–53, 2014.

[12] N. D. Bronstein, Y. Yao, L. Xu, E. O'Brien, A. S. Powers, V. E. Ferry, A. P. Alivisatos, and R. G. Nuzzo, "Quantum Dot Luminescent Concentrator Cavity Exhibiting 30-fold Concentration," ACS Photonics, vol. 2, no. 11, pp. 1576–1583, 2015.

[13] F. Meinardi, A. Colombo, K. A. Velizhanin, R. Simonutti, M. Lorenzon, L. Beverina, R. Viswanatha, V. I. Klimov, and S. Brovelli, "Large-area luminescent solar concentrators based on /'Stokes-shift-engineered/' nanocrystals in a mass-polymerized PMMA matrix," Nat Photon, vol. 8, no. 5, pp. 392–399, 2014. [Online]. Available: http://dx.doi.org/10.1038/nphoton.2014.54

[14] U. Rau, U. W. Paetzold, and T. Kirchartz, "Thermodynamics of light management in photovoltaic devices," Physical Review B - Condensed Matter and Materials Physics, vol. 90, 2014.

[15] O. Chen, J. Zhao, V. P. Chauhan, J. Cui, C. Wong, D. K. Harris, H. Wei, H.-S. Han, D. Fukumura, R. K. Jain, and M. G. Bawendi, "Compact high-quality CdSe-CdS core-shell nanocrystals with narrow emission linewidths and suppressed blinking." Nature materials, vol. 12, no. 5, pp. 445–51, 2013. [Online]. Available: http://www.pubmedcentral.nih.gov/articlerender.fcgi?artid=3677691{&}tool=pmcentrez{&}rendertype=abstract

[16] J. Jasieniak, L. Smith, J. van Embden, P. Mulvaney, and M. Califano, "Re-examination of the size dependent absorption properties of CdSe quantum dots," Journal of Physical Chemistry C, vol. 113, pp. 19 468–19 474, 2009.

[17] J. F. Geisz, M. A. Steiner, I. García, S. R. Kurtz, and D. J. Friedman, "Enhanced external radiative efficiency for 20.8% efficient single-junction GaInP solar cells," Applied Physics Letters, vol. 103, no. 4, 2013.

[18] A. Carlson, A. M. Bowen, Y. Huang, R. G. Nuzzo, and J. A. Rogers, "Transfer printing techniques for materials assembly and micro/nanodevice fabrication," Advanced Materials, vol. 24, no. 39, pp. 5284–5318, 2012.

[19] Y. Zhou, D. Benetti, Z. Fan, H. Zhao, D. Ma, A. O. Govorov, A. Vomiero, and F. Rosei, "Near Infrared, Highly Efficient Luminescent Solar Concentrators," Advanced Energy Materials, vol. 6, no. 11, pp. 1–8, 2016.

[20] M. Y. Levy and C. Honsberg, "Rapid and precise calculations of energy and particle flux for detailed-balance photovoltaic applications," Solid-State Electronics, vol. 50, no. 7-8, pp. 1400–1405, 2006.

[21] P. Baruch, A. De Vos, P. Landsberg, and J. Parrott, "On some thermodynamic aspects of photovoltaic solar energy conversion," Solar Energy Materials and Solar Cells, vol. 36, no. 2, pp. 201–222, 1995. [Online]. Available: http://linkinghub.elsevier.com/retrieve/pii/0927024895800042

[22] W. Shockley and H. J. Queisser, "Detailed balance limit of efficiency of p-n junction solar cells," Journal of Applied Physics, vol. 32, no. 3, pp. 510–519, 1961.

978-1-5090-5606-4/17 $31.00 © 2017 IEEE

Increase in Maximum Power of a-Si, c-Si and GaAs$_{.76}$P$_{.24}$ Solar Cells Under Low Concentration

Hiba Riaz, Sabina Abdul Hadi and Ammar Nayfeh

Department of Electrical Engineering and Computer Science (EECS), Masdar Institute of Science and Technology, PO Box. 54224, Abu Dhabi, United Arab Emirates

Abstract — **In this work, three types of solar cells, a-Si PIN, c-Si and GaAs$_{0.76}$P$_{0.24}$, are measured under low solar concentration of up to 52 suns using a plano-convex lens. The effect of solar concentration on the different performance parameters (J$_{sc}$, V$_{oc}$, maximum power, FF and efficiency) of each cell is studied. The J$_{sc}$ increases by ~30-50X for all cells when measured at 52suns. In addition the maximum power increases by a factor of 32x for a-Si cell, 18.4x for c-Si cell and 23x for GaAs$_{.76}$P$_{.24}$ cell.**

I. INTRODUCTION

Concentrator Photovoltaics (CPV) has the potential to reduce costs by replacing expensive cell area with cheaper optical components to concentrate the sunlight. CPV systems can be classified into three types – low concentration from 2-100 suns, medium from 100-300 suns and high from 300-1000 suns [1]. Currently, most research efforts have been directed towards high concentration photovoltaic systems (HCPV) [2]. This has led to research efficiencies of these cells reaching 46% [3]. However, high concentrations can add considerable cost due to the need for active cooling. Another route is to use low concentrator photovoltaic system (LCPV) to improve performance at low cost [4]–[7].

In this work, a-Si PIN, c-Si and GaAs$_{.76}$P$_{.24}$ solar cells tested under low solar concentration. Fabrication details are reported in [8], [9] and [10]. Cross-section images of the cells in Fig, 1.

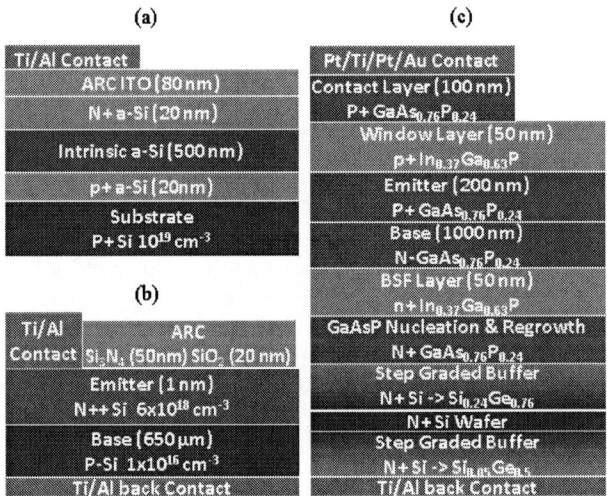

Fig. 1. Cross-section of (a) a-Si PIN, (b) c-Si and (c) GaAs$_{.76}$P$_{.24}$ cell grown on Si substrates.

II. EXPERIMENTAL SETUP

The cells were first characterized under one sun using the under Standard Test Conditions (25°C, AM1.5G [11] and GNI=1000W/m^2) [12]. The performance of the three cells under 1 sun are shown in Table I, while the JV characteristics are shown in Fig. 2.

TABLE I
Cell Performance under One Sun

Cell	Area (cm^2)	J$_{sc}$ (mA/cm^2)	V$_{oc}$ (V)	P$_{max}$ (mW/cm^2)	Fill Factor (%)	Efficiency (%)
a-Si	0.0625	4.31	0.859	2.50	67.5	2.50
c-Si	0.16	32.55	0.615	15.53	77.6	15.53
GaAs$_{.76}$P$_{.24}$	0.16	12.19	1.112	10.31	76.0	10.31

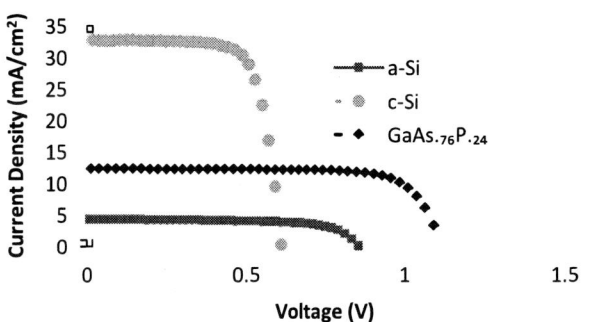

Fig. 2. JV curves of the three cells under one sun.

The lens used to take measurements under low solar concentration is a 75 mm diameter x 75 mm focal length, uncoated, plano-convex lens by Edmund Optics™ with wavelength range 350 – 2200 nm. A lens holder and stand is used to take controlled measurements. The schematics of the final setup is in Fig. 3 (a). In the case of a-Si cells, individual dies with dimensions 2.5 mm × 2.5 mm fabricated on one common sample were measured, while c-Si and GaAs$_{.76}$P$_{.24}$ cells were diced into individual 4 mm × 4 mm dies prior to measurements. The samples are aligned such that the cell being measured is at the center of the concentrated beam. The image of actual measurement setup can be seen in Fig. 3 (b).

978-1-5090-5606-4/17 $31.00 © 2017 IEEE

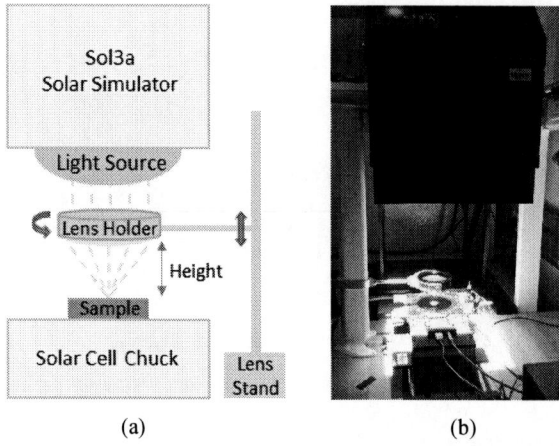

(a)   (b)

Fig. 3. (a) Schematics of lens holder and a stand and (b) image of experimental setup for measuring cells under low solar concentration.

The JV characteristics are measured at different heights resulting in different light concentrations. The measurement was done for heights lower than 13 cm. Repeating the measurements done at 11 cm sometimes resulted in number of suns less and sometimes more than 52 suns. This is caused due to the challenges in setting the exact same height of the lens holder for repeated measurements. This variation of the number of suns around the focal point can be explained using **Error! Reference source not found.**Fig. 4, which shows that the focal point of the lens lies between 11 and 13 cm after which the light diverges again.

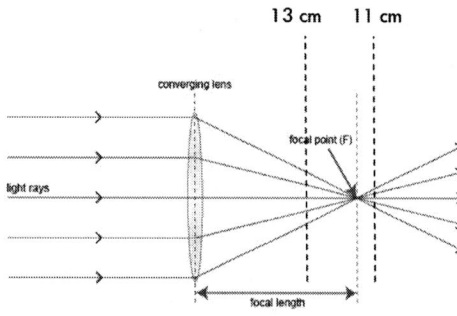

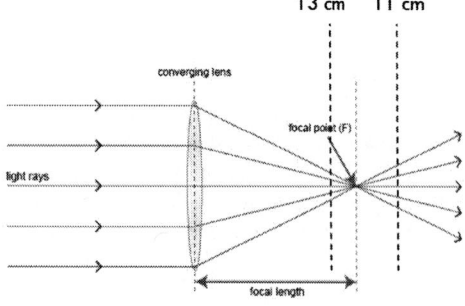

Fig. 4. Focal length of the lens and reason for variation in the number of suns around the focal point.

The JV curves at different lens heights for each cell are shown in Fig. 5.

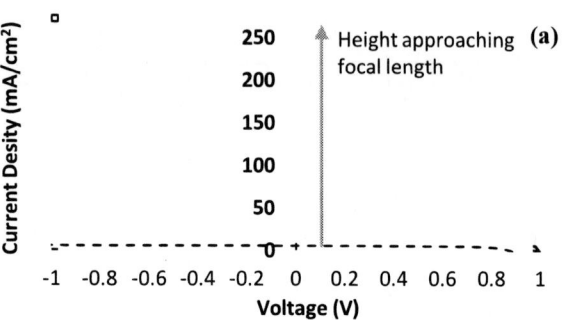

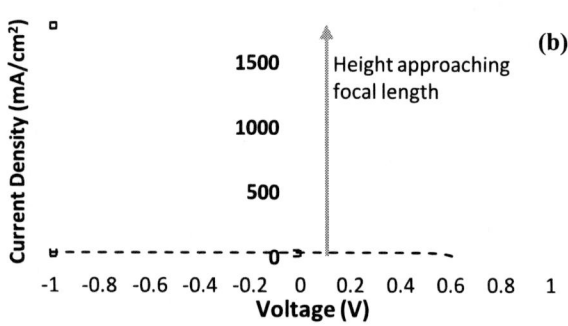

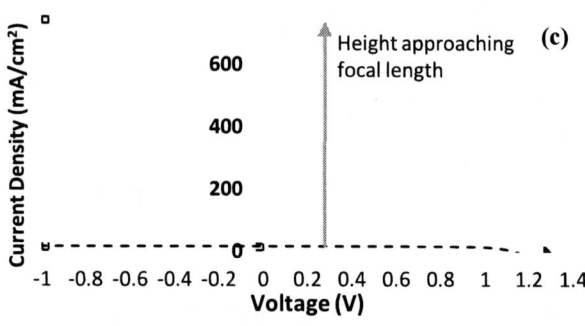

Fig. 5. JV characteristics of (a) a-Si, (b) c-Si and (c) GaAs$_{.76}$P$_{.24}$ cell at different lens height

The number of suns at each lens height is estimated using the linear relationship between J_{sc} and the light intensity. However this approach carries some level of error at higher concentration values, since FF can degrade significantly, resulting in non-linear behavior of J_{sc}. This is most obvious for measurements of c-Si, shown in Fig. 5 (b). In order to minimize errors in concentration estimates, we used maximum measured current density value, J_{max}, instead of J_{sc} (i.e. X = $J_{max@X}$/ $J_{sc@1sun}$). Note that under 1 sun, or low concentration, $J_{max}=J_{sc}$.

978-1-5090-5606-4/17 $31.00 © 2017 IEEE

III. RESULTS AND DISCUSSION

The J_{sc} values for l up to 52 suns are shown in Fig. 6. J_{sc} increases linearly with suns since suns is calculated assuming linear behavior, except for c-Si cells for concentrations above 36 suns, due to significant drop in fill factor (FF), as shown in Fig. 9 (a). For solar concentration from 1 to 52 suns, J_{sc} increases from 4.3 mA/cm^2 to 212 mA/cm^2 for a-Si cell, 32.5 mA/cm^2 to 1043 mA/cm^2 for c-Si cell and 12.2 mA/cm^2 to 582 mA/cm^2 for GaAs$_{.76}$P$_{.24}$ cell.

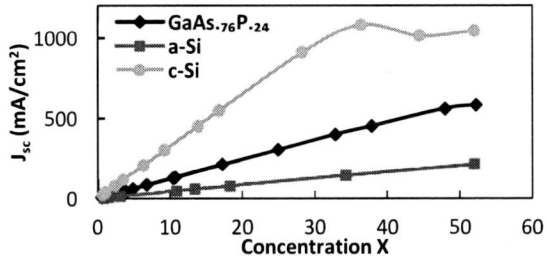

Fig. 6. Short circuit current of a-Si, c-Si and GaAs$_{.76}$P$_{.24}$ cells under low solar concentration

The V_{oc} is plotted in Fig. 7. V_{oc} increases logarthmically with solar concentration. For logarithmic fit, 25°C temperature is used and ideality factor that best fits V_{oc} values at low concentration (1.31 for a-Si, 1.05 for c-Si and 2 for GaAs$_{.76}$P$_{.24}$). Disceprancy between measured and calculated V_{oc} at higher concentration could be due to errors in concentration estimates at higher light intensity and slight changes in cell temperature.

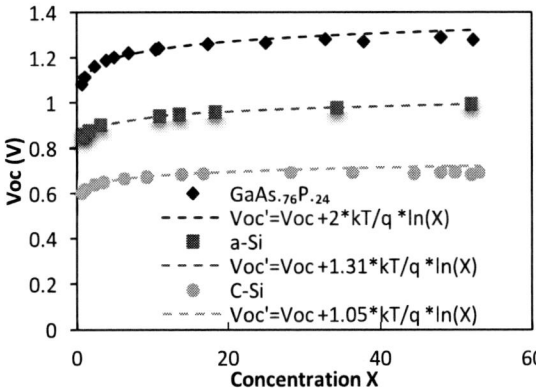

Fig. 7. Open circuit voltage of a-Si, c-Si and GaAs$_{.76}$P$_{.24}$ cells under low solar concentration. Also shown is logarithmic fit for V_{oc} under concentration with best selected ideality factor (dashed lines).

The maximum power as a function of suns is shown in Fig. 8. Maximum power output is continuously increasing with concentration for a-Si and GaAs$_{.76}$P$_{.24}$ cells, while for c-Si cell maximum is achieved at ~ 32 suns, due to FF degradation beyond this point. For solar concentration from 1 to 52 suns,

maximum power increases from 2.5 mW/cm^2 to 80 mW/cm^2 for a-Si cell, 10.2 mW/cm^2 to 188 mW/cm^2 for c-Si cell and 10.3 mW/cm^2 to 237 mW/cm^2 for GaAs$_{.76}$P$_{.24}$ cell.

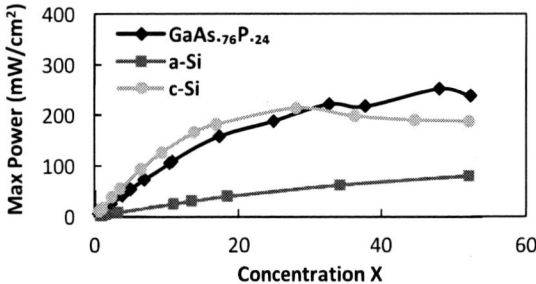

Fig. 8. Maximum Power of a-Si, c-Si and GaAs$_{.76}$P$_{.24}$ cells under low solar concentration

The FF and efficiency from the three cells as a function of solar concentration are shown in Fig. 9 (a) and (b), respectively. The FF drops with concentration which in turn affects maximum power and efficiency. FF is mainly affected by parasitic losses caused by series and shunt resistance due to non-ideal behaviors like contact resistance and manufacturing defects. Under solar concentration the effect of the resistance on the fill factor is magnified due to the higher current [13]. c-Si and GaAs$_{.76}$P$_{.24}$ cells contain fewer defects than a-Si cell and thus start with higher FF at 1 sun. Series resistance is a combination of the resistance of each layer of the cell but is mainly affected by the front grid resistance [14]. Differences in metal coverage and shading of the three cells can be a reason for the difference in FF response of the three cells under varying concentration. It should be noted the a-Si and GaAs$_{.76}$P$_{.24}$ cells retain ther fill factor better than the c-Si cell. Efficiency is improved for all cells for very low level of concentration (< 5 suns).

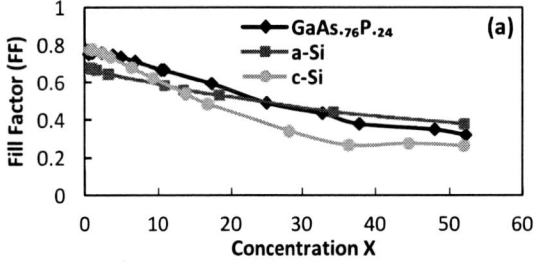

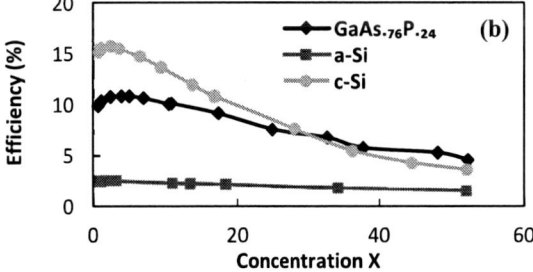

Fig. 9. (a) Fill Factor and (b) efficiency of a-Si, c-Si and GaAs$_{.76}$P$_{.24}$ cells under low solar concentration.

IV. CONCLUSION

In summary, a-Si PIN, c-Si and GaAs$_{.76}$P$_{.24}$ cells were measured under low concentration using plano-convex lens. For solar concentration from 1 to 52 suns the J$_{sc}$ and maximum power is observed to increase for all sun values. Efficiency improvement is observed only for a number of suns less than 5. Major limitation for the efficiency improvement is the decrease in the fill factor with the concentration, mainly due to present series and shunt resistance in the cells. Fill factor degradation with increasing solar concentration could be improved by exploring different metal contact designs that reduce series and shunt resistance in the cells. The results illustrate the potential of low solar concentration for improving the performance of the thin film solar cells at low cost.

ACKNOWLEDGEMENTS

Authors would like to thank professors Dr. Judy L. Hoyt and Dr. Evelina Polyzoeva for fabrication of c-Si solar cells and Dr. Eugene A. Fitzgerald and Dr. Timothy Milakovich for fabrication of GaAs$_{.76}$P$_{.24}$ cells.

REFERENCES

[1] "A Strategic Research Agenda for Photovoltaic Solar Energy Technology", EU Photovoltaic Technology Platform, 2011.

[2] A. Mokri and M. Emziane, "Concentrator Photovoltaic Technologies and Market: A Critical Review," World Renew. Energy Congr. – Sweden, 8–13 May, 2011, Linköping, Sweden, pp. 2738–2742, 2011.

[3] M. Green, K. A. Emery, and E. Dunlop, "Solar cell efficiency tables (version 48): Solar cell efficiency tables (version 48)," no. July, 2016.

[4] M. Castro, I. Anton, and G. Sala, "Pilot production of concentrator silicon solar cells: Approaching industrialization," Sol. Energy Mater. Sol. Cells, vol. 92, no. 12, pp. 1697–1705, 2008.

[5] H. Baig, N. Sarmah, K. C. Heasman, and T. K. Mallick, "Numerical modelling and experimental validation of a low concentrating photovoltaic system," Sol. Energy Mater. Sol. Cells, no. July, 2013.

[6] F. Reis, M. C. Brito, V. Corregidor, J. Wemans, and G. Sorasio, "Modeling the performance of low concentration photovoltaic systems," Sol. Energy Mater. Sol. Cells, vol. 94, no. 7, pp. 1222–1226, 2010.

[7] L. D. P. Lewis M. Fraas, "Solar Cells and Their Applications."

[8] K. Islam, A. Alnuaimi, E. Battal, A. K. Okyay and A. Nayfeh, "Effect of gold nanoparticles size on light scattering for thin film amorphous-silicon solar cells," Solar Energy, vol. 103, pp. 263-268, 2014.

[9] E. Polyzoeva, S. Abdul Hadi, A. Nayfeh, and J. L. Hoyt, "Reducing optical and resistive losses in graded silicon-germanium buffer layers for silicon based tandem cells using step-cell design", AIP Advances, vol. 5, p. 057161, 2015.

[10] T. Milakovich, R. Shah, S. Abdul Hadi, M. Bulsara, A. Nayfeh, E. Fitzgerald, "Growth and characterization of GaAsP top cells for high efficiency III-V/Si tandem PV", 42nd IEEE PVSC, New Orleans, p. 1-4, 2015.

[11] "ASTM G173-03(2012), Standard Tables for Reference Solar Spectral Irradiances: Direct Normal and Hemispherical on 37° Tilted Surface," 2012.

[12] "IEC 60904-1: Photovoltaic devices-Part 1: Measurements of PV current-voltage characteristics," 2006.

[13] H. Cotal, C. Fetzer, J. Boisvert, G. Kinsey, R. King, P. Hebert, H.Yoon and N. Karam, "III–V multijunction solar cells for concentrating photovoltaics", Energy Environ. Sci., 2009, 2, 174–192

[13] D. L. Meier, E. A. Good, R. A. Garcia, B. L. Bingham, S. Yamanaka, V. Chandrasekaran, and C. Bucher, "Determining components of series resistance from measurements on a finished cell," Conf. Rec. 2006 IEEE 4th World Conf. Photovolt. Energy Conversion, WCPEC-4, vol. 2, pp. 1315–1318, 2007.

978-1-5090-5606-4/17 $31.00 © 2017 IEEE

Design and Evaluation of Partial Concentration III–V/Si Module with Enhanced Diffuse Sunlight Transmission

Daisuke Sato, Noboru Yamada, Kan-Hua Lee, Kenji Araki, and Masafumi Yamaguchi

Nagaoka University of Technology, Nagaoka, Niigata, 940-2188, JAPAN
Toyota Technological Institute, Nagoya, Aichi, 468-8511, JAPAN

Abstract — This paper describes the design and evaluation of a partial concentration III-V/Si module that aims to maximize the power generation from global normal irradiation by harvesting not only direct sunlight but also diffuse sunlight. The module structure was designed by optical and thermal analysis, and the fundamental characteristics were investigated by performing outdoor and indoor tests using a prototype mini-module with a geometrical concentration ratio of 100X. Experimental results show that the mini-module achieves enhanced diffuse sunlight transmission of 82% and generates more power from diffuse sunlight than the previous prototype reported by the authors. The tolerance to tracking error was also confirmed, which may contribute to mitigate the requirement of tracking accuracy.

Index Terms — Bi-facial crystalline solar cells, Concentrator photovoltaics, Diffuse sunlight, Direct sunlight, Multi-junction solar cells, Partial concentration.

I. INTRODUCTION

Enhancement of terrestrial photovoltaic (PV) conversion efficiency has been an essential and urgent research goal in recent decades to mitigate energy and climate crisis. Concentrator photovoltaics (CPV) is a promising technology because of high conversion efficiency for direct normal irradiation. However, high-concentration CPV is not able to utilize diffuse sunlight. This feature makes the CPV cost-effective only in high direct normal irradiance (DNI) regions. As a solution to that problem, the partial concentration concept has been experimentally demonstrated [1] and various configurations and the technological potential have been

studied experimentally and/or numerically [2]–[4]. In the partial concentration architecture, a high-efficiency concentrator solar cells on which direct sunlight is concentrated and an additional low-cost solar cells that capture diffuse sunlight and/or a part of direct sunlight are integrated with concentrator optics, as shown in Fig. 1.

In this study, we updated the module design based on the previous study [3] to enhance the diffuse sunlight transmission and conducted indoor and outdoor tests to evaluate the power generation performance as well as the diffuse sunlight transmission.

II. PROTOTYPE MODULE DESIGN AND FABRICATION

Figure 2 illustrates the design of a prototype mini-module that consists of 4×4 lens array, III–V triple-junction (3J) cells, and crystalline Si cell. The 3J cells are soldered to the Ag circuit

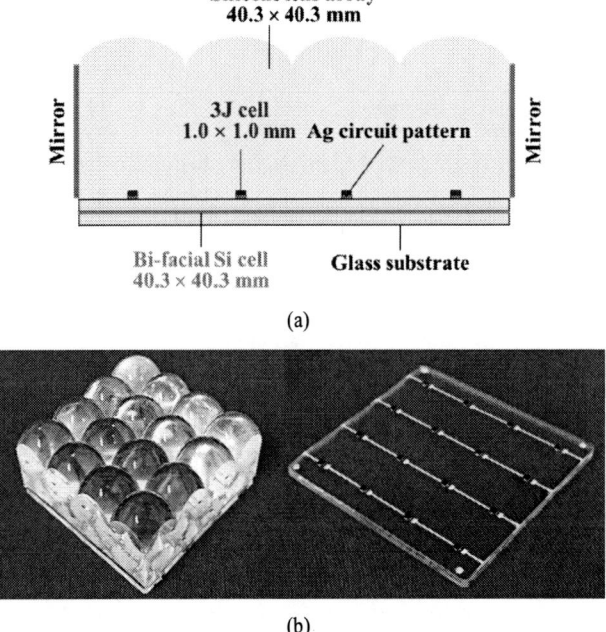

(a)

(b)

Fig. 2. Prototype mini-module design. (a) Schematic of mini-module cross section. (b) Photograph of the fabricated mini-module without Si cell (left) and glass substrate with 3J cells and circuit pattern (right).

Fig. 1. Conceptual image of partial concentration architecture.

978-1-5090-5606-4/17 $31.00 © 2017 IEEE

pattern on the glass substrate that has a minimal non-transparent area to maximize the diffuse sunlight transmission. The ratio of transparent area of glass substrate to its aperture area is 95.7%. Concentrator lens array was made of silicone resin. Geometrical concentration ratio calculated from the ratio between lens aperture area (10×10 mm^2) and 3J cell area (1.0×1.0 mm^2) was 100X. The module size was determined from its performance, as mentioned in section III. The lens shape was designed by optical analysis using a ray-tracing technique. Estimated optical efficiency for direct sunlight to the 3J cell was 87.7%. The CPV module with similar micro-lens design showed the high module conversion efficiency of 37.1% based on DNI [5].

III. MODULE SIZE EFFECT ON SOLAR CELL TEMPERATURE AND CONVERSION EFFICIENCY

To determine the size of the module, the relationship between module size and solar cell temperature and conversion efficiency was examined by three dimensional (3D) heat transfer simulation using the finite element method (FEM). The 3D simulation model comprised silicone lens, 3J cell (Ge), electrode pattern (Ag), glass substrate, and Si cell. The heating power for 3J and Si cells (Q_{3J}, Q_{Si}) were modeled using the following equations:

$$Q_{3J} = I_D \times A_{lens} \times \eta_{opt\_D} \times (1 - \eta_{3J}) , \qquad (1)$$

$$Q_{Si} = I_S \times A_{lens} \times \eta_{opt\_S} \times (1 - \eta_{Si}) , \qquad (2)$$

where I_D and I_S are irradiances of direct sunlight (900 W/m^2) and diffuse sunlight (200 W/m^2), respectively; A_{lens} [m^2] is the aperture area of the lens array (projected area); η_{opt_D} is the transmittance (optical efficiency) for direct sunlight from the lens aperture to the 3J cell (0.877); η_{opt_S} is the transmittance for diffuse sunlight from the lens aperture to the Si cell (0.839); η_{3J} and η_{Si} are the conversion efficiencies of the 3J cell (0.40) and Si cell (0.20). The boundary conditions were applied as follows. The ambient temperature was 35.0 °C. The heat transfer coefficient from the module's outside boundary to the ambient air was 10.0 W/(m^2 K). Periodic boundary conditions were applied around the module. A multiple lens-cell array module [6] was simulated rather than a single lens-cell module. Thermal conductivities of the silicone lens, 3J cell, electrode pattern, glass substrate, and Si cell were 0.2, 60.2, 429, 1.0, and 168 W/(m K), respectively. Emissivity of the silicone lens was 0.95.

Figure 3 shows the simulated steady state temperature and conversion efficiency of each solar cell based on their temperature coefficients for different module sizes under constant geometrical concentration ratio (100X). The simulation results show that in order to maintain the 3J cell temperature under 100 °C and conversion efficiency over 35%, the 3J cell size needs to be less than 1.0 mm. The 3J cell

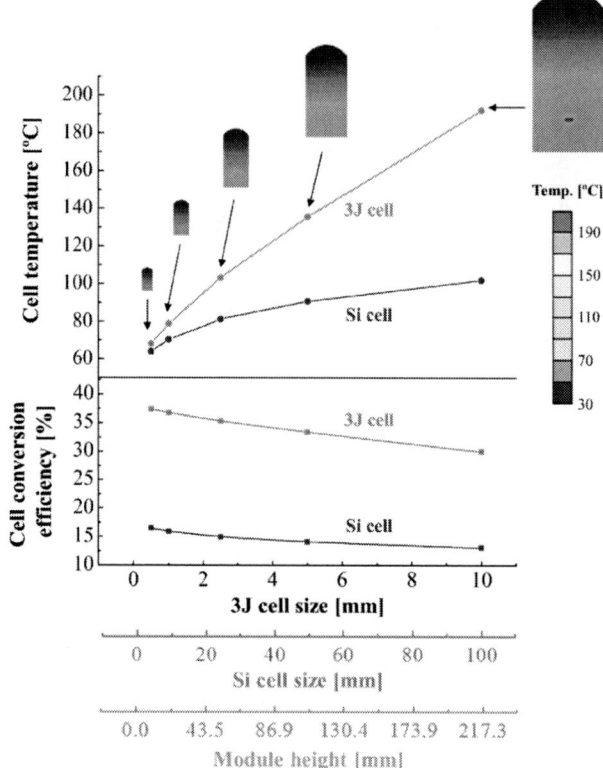

Fig. 3. Simulated temperature and conversion efficiency of cell-lens array. Assumed temperature coefficients of conversion efficiency of 3J cell and Si cell are −0.15%/°C and −0.45%/°C with reference efficiency of 40% (3J) and 20% (Si) at 25 °C.

temperature is expected to be ~80 °C in the present module design (3J cell size: 1.0 mm).

IV. MEASUREMENT OF FUNDAMENTAL CHARACTERISTICS

A. Outdoor Test for Power Generation Characteristics

The prototype mini-module was mounted on a dual-axis sun tracker (EKO, STR-22). DNI and global normal irradiance (GNI) were measured by using a pyrheliometer and pyranometer on the tracked surface. A gray-colored paper with hemispherical solar reflectance for the AM1.5G standard spectrum of ~0.38 was placed on the opposite side of bi-facial Si cell to emulate moderate reflection by ground on the Earth. On-sun current–voltage (I–V) characteristics measurements (4-terminal) were carried out at a university campus in Nagaoka-city in Japan. Figure 4 shows the measured time variation of GNI-based module conversion efficiency, P_{max}, and solar irradiance for clear-sky and partial-cloudy conditions. As shown in Fig. 4 (a), in clear-sky condition, although the 3J cell generated most power, the bi-facial Si cell generated as high as 27% of daily total power, which contributes to boost of the

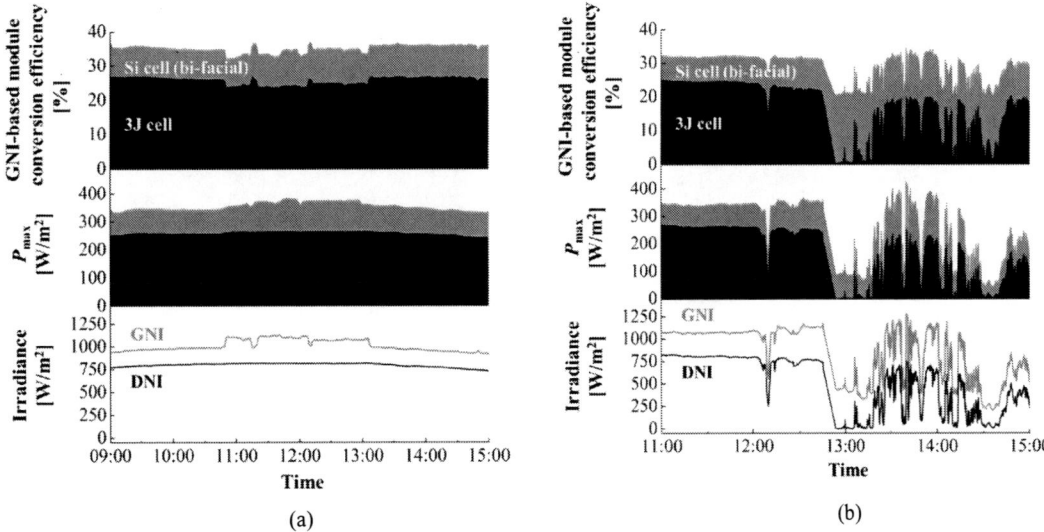

Fig. 4. Measured GNI-based module conversion efficiency of the prototype mini-module, P_{max}, GNI, and DNI (Nagaoka-city, Japan).
(a) Clear-sky condition (2017/4/24). (b) Partial-cloudy condition (2017/5/17).

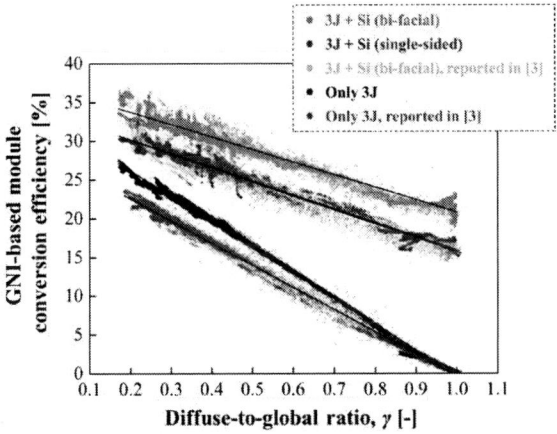

Fig. 5. Relationship between GNI-based module conversion efficiency of the prototype mini-module and diffuse-to-global ratio for newly developed and previous [3] modules.

GNI-based module conversion efficiency. As a result, the module achieved maximum GNI-based conversion efficiency of 37.0%, which is approximately 2 times greater than that of conventional flat PV module. As shown in Fig. 4 (b), in partial-cloudy condition, generated power by the 3J cell diminished with decrease of DNI, in contrast, bi-facial Si cell generated power from diffuse sunlight. The module could maintain the GNI-based conversion efficiency almost as high as that of the conventional flat PV module, i.e. 17~18%, even under less DNI condition.

Figure 5 shows the relationship between the GNI-based module conversion efficiency and diffuse-to-global ratio γ that represents the ratio of diffuse sunlight on the module aperture area to GNI, i.e., $\gamma = (GNI-DNI)/GNI$, obtained by 15 days

measurement during April and June 2017. A larger γ represents larger diffuse component in the GNI, and $\gamma = 1.0$ represents completely cloudy condition, that means the module aperture receives only diffuse sunlight. At the condition of $\gamma = 0.20$, the present module with bifacial Si cell exhibited 5.3%abs greater module conversion efficiency than that previously reported [3], which comes from 3.8%abs and 1.5%abs enhancement of 3J and bi-facial Si cell respectively. Therefore, the enhancement of optical efficiency for both direct sunlight to 3J cell and diffuse sunlight to Si cell contributed to improve the module conversion efficiency. Even in the present module with single-sided Si cell, the module efficiency was almost same level as that of previously reported module with bi-facial Si cell [3] regardless of γ.

B. Indoor Test for Diffuse Sunlight Transmission

The transmittance of the prototype module for diffuse sunlight was measured under the artificial diffuse sunlight emitted from a commercial solar simulator. Figure 6 shows the schematic diagram of experimental setup for the transmittance measurement. Artificial diffuse sunlight has nearly isotropic angular radiance distribution. The amount of light passing through the module was measured by the Si cell placed beneath the glass substrate of the module. The transmittance was determined as a ratio of the short circuit current of the Si cell with the upper module component to that without the upper module component. The active area of the Si cell is the same as that of the lens aperture. Measured transmittance of the present module was 82.0%, which is 12%abs higher than that of the previous module [7].

978-1-5090-5606-4/17 $31.00 © 2017 IEEE

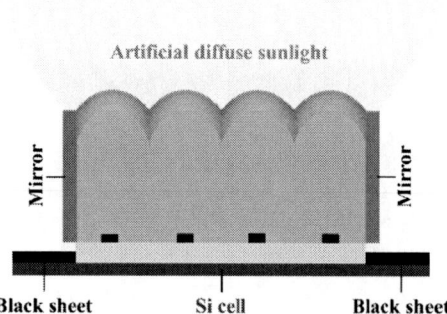

Fig. 6. Schematic diagram of experimental setup for transmittance measurement.

C. Outdoor Test for tolerance to tracking error

The prototype mini-module was mounted on a dual-axis sun tracker with a tracking accuracy sensor (Black Photon Instruments, SE-202-TA) to investigate the tolerance to tracking error. Sun tracking motion of the tracker was intentionally stopped and then variation of the generated power was measured. As a result of the 25-minute continuous measurement, Fig. 7 shows GNI-based module conversion efficiency to tracking error for bi-facial Si cell case under clear-sky condition (γ = 0.24 - 0.25). The tracking error angle was calculated by considering the path of the sun during the measurement. Thus, it is noted that the tracking error dependency of the module conversion efficiency may differ depending on time and season. The prototype mini-module with bi-facial Si cell (3J+Si case) showed greater tolerance to the tracking error than that without bi-facial Si cell (3J case). The 3J+Si case have exhibited higher GNI-based conversion efficiency than the peak efficiency of 3J case for 375 seconds; in the meantime, the tracking error reached up to 1.4°. The result indicates that the partial concentration design remarkably mitigates the requirement of tracking accuracy.

V. SUMMARY

A partial concentration III–V/Si module for high diffuse sunlight transmission was designed and its fundamental characteristics were measured. The results show that the module with a micro-lens array encapsulating both high-efficiency and low-cost solar cells exhibits diffuse sunlight transmission over 80%. This enhancement contributes to the improvement of GNI-based module conversion efficiency and prototype mini-module achieved maximum efficiency of 37.0%. In addition, high diffuse sunlight transmittance and the Si cell mechanically stacked at the bottom of the module contributes the high tolerance to tracking error. The present module design has a potential to enhance the GNI-based conversion efficiency

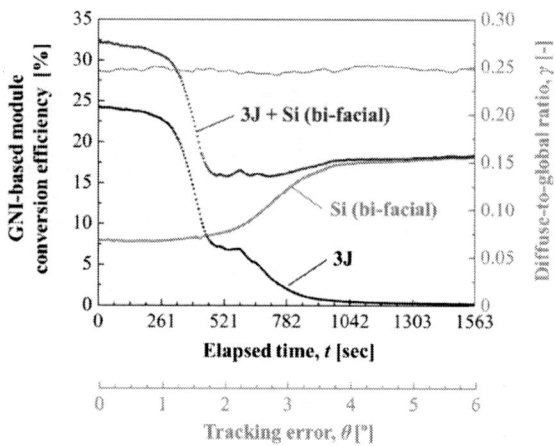

Fig. 7. Tracking error dependency of GNI-based module conversion efficiency of the prototype mini-module. Contributions by each solar cell are also shown. This result was obtained by stopping the dual-axis sun tracker for 25 minutes from 13:16:36 to 13:42:41 on 2017/5/22.

of CPV module and enable to use a low-cost tracker with moderate tracking accuracy. The present module design can be also adapted for a highly transparent CPV module, in which diffuse sunlight passes through and which is utilized in dual land use applications such as agriculture [7].

ACKNOWLEDGMENT

This work was supported by JSPS KAKENHI Grant Number 26289373 and NICO project.

REFERENCES

[1] N. Yamada and K. Okamoto, "Experimental measurements of a prototype high concentration Fresnel lens CPV module for the harvesting of diffuse solar radiation.," *Opt. Express*, vol. 22 Suppl 1, pp. A28–34, 2014.

[2] M. W. Haney, T. Gu, and G. Agrawal, "Hybrid micro-scale CPV/PV architecture," in *2014 IEEE 40th Photovoltaic Spec. Conf. (PVSC)*, 2014, pp. 2122–2126.

[3] N. Yamada and D. Hirai, "Maximization of conversion efficiency based on global normal irradiance using hybrid concentrator photovoltaic architecture," *Prog. Photovoltaics Res. Appl.*, vol. 24, no. 6, pp. 846–854, 2016.

[4] K.-T. Lee *et al.*, "Concentrator photovoltaic module architectures with capabiities for capture and conversion of full global solar radiation.," *Proc. Natl. Acad. Sci. U. S. A.*, vol. 113, no. 51, pp. E8210–E8218, 2016.

[5] N. Hayashi *et al.*, "High-efficiency thin and compact concntrator photovoltaics with micro-solar cells directly attached to a lens array," *Opt. Express*, vol. 23, no. 11, pp. A594–A603, 2015.

[6] N. Ymada *et al.*, "Development of silicone-encapsulated CPV module based on LED package technology," in *2013 IEEE 39th Photovoltaic Spec. Conf. (PVSC)*, 2013, pp. 0493–0496.

[7] D. Hirai, K. Okamoto, and N. Yamada, "Fabrication of highly transparent concentrator photovoltaic module for efficient dual land use in middle DNI region," in *2015 IEEE 42nd Photovoltaic Spec. Conf. (PVSC)*, 2015, pp. 1–4, doi: 10.1109/PVSC.2015.7355759.

Contamination control challenges on SHJ solar cell processing

Condorelli G.[(1)]; Rotoli P.[(1)]; Canino A.[(1)]; Battaglia A.[(1)]; Favre W.[(2)]; A.-S. Ozanne[(2)]; Moustafa A.[(2)]; Danel A.[(2)]; D. Muñoz[(2)]; P.-J. Ribeyron[(2)] and Gerardi C.[(3)]

1. 3SUN, Contrada Blocco Torrazze Zona Industriale 95121, Catania, Italy

2. Univ. Grenoble Alpes, CEA, LITEN, INES, LHET, F-73375 Le Bourget du Lac, France

3. Enel Green Power, Contrada Blocco Torrazze Zona Industriale 95121, Catania, Italy

Abstract — **We have studied the conversion of thin film photovoltaic factory to Silicon Heterojunction (SHJ) technology adapting thin film PECVD reactors for deposition of passivation layers. In the transition from thin film to SHJ cells, it is crucial to study the impact of defectiveness on cell efficiency due to several factors such as transportation and handling of wafers. By performing photoluminescence maps and particle contamination analysis on the surface of wafers and cells different defectiveness sources have been studied. Once the sources of defects have been identified, we elaborated solutions to mitigate effects and we were able to increase the efficiency of SHJ solar cells by an absolute gain of 4.3%.**

Index Terms — **Photovoltaic cells, Heterojunctions, Efficiency**

I. INTRODUCTION

Since the recent expiry of core patents describing the structure of SHJ solar cells with intrinsic thin a-Si layers, research on and development of such cells has regained strong interest.

Silicon heterojunction (SHJ) solar cells consist in crystalline silicon wafers covered by thin amorphous silicon layers deposited on both sides for excellent surface passivation and junction formation. This design enables energy cell level conversion efficiencies above 22% at the industrial production level [1]. Major advantages of the SHJ technology are the low-temperature processing that allows the use of very thin wafers without causing substrate warping, and the small number of process steps to fabricate the device.

Fig. 1.   Schematic view of an SHJ solar cell.

The device is based on n-type silicon wafer on which thin amorphous layers are deposited to form the emitter and the Front Surface Field of the solar cell. These amorphous layers are then covered with Transparent Conductive Oxide (TCO), such as Indium Tin Oxide (ITO). Finally, grids of silver (Ag) electrodes are screen-printed on both sides.

For high-efficiency c-Si based solar cells, high-quality surface passivation is of extreme importance. Intrinsic a-Si:H films have been known for a few decades to yield good c-Si surface passivation [2]-[4].

Industries, wishing to switch to SHJ technology, when called to decide the best tool set among different equipment manufacturers, have to face a dilemma on how to evaluate the technical results, running flying cells at different steps along the process flow of the SHJ cell.

3SUN evaluation on SHJ cells technology involves the possibility to use thin film silicon technology equipment existing in the fab, thus leveraging the know-how on amorphous silicon to develop a-Si:H layers for passivating the c-Si surface.

In collaboration with CEA-INES institute, several process iterations through flying cells have been performed to test and optimize 3SUN a-Si:H layers for passivation on SHJ.

The wafers have been textured and cleaned using in-line automatic wet bench at CEA-INES then shipped to 3SUN for PECVD deposition.

In this paper we focus on the surface defects generated by samples shipping and management for PECVD deposition and propose an improved methodology including wafer carrier adaptation from micro-electronic experience, to minimize the impact of travel on cell efficiency.

II. EXPERIMENTAL DETAIL

As cut 156mm pseudo-square wafers (PSQ) have been textured at CEA-INES using an alkaline based solution, cleaned by industrial ozone-based process step and shipped to

978-1-5090-5606-4/17 $31.00 © 2017 IEEE

3SUN for passivation testing [5]. Prior deposition, HF wet etch has been performed in 3SUN to remove the native oxide.

Doped and undoped a-silicon thin films were prepared at 3SUN using an industrial 13.56MHz multi-chamber plasma-enhanced chemical vapor deposition (PECVD) system, used in 3Sun for thin film modules manufacturing [6]-[7]. The reactor has been developed to ensure high throughput and low cost of ownership. For this reason the process is carried out simultaneously on 8 glasses of 1.4m x 1 m

After extensive investigations we have found that the process conditions to achieve an intrinsic a-Si:H layer with good passivating properties are:

- H_2/SiH_4 ratio = 10÷15
- Substrate temperature =150÷200°C
- RF power density = 50 ÷ 200 mW/cm2

Passivation performances have been measured using a Sinton Consulting WTC-120 quasi-steady-state photoconductance (QSSPC) lifetime tester in the transient mode [8]-[9]. Typically, with 20 nm of intrinsic aSi:H deposited with process parameter here reported on both sides of texturized 156mm pseudo square wafer 2500 µs of lifetime is measured with implied V_{oc} (iV_{oc}) around 744 mV and implied FF (iFF) higher than 84%.

The doped layers (n-type and p-type) were deposited by adding phosphine (PH_3) or diborane (BH_3), respectively, to the gas mixture of silane (SiH_4) and hydrogen (H_2).

After the PECVD deposition cells were shipped to CEA-INES where ITO layers were deposited on both sides of the cell using in-line PVD reactor. Finally, the cells' metallization was performed by screen-printing technique on both front and rear side of the cell using a 4 busbars configuration.

The cells electrical parameters were determined under AM1.5 conditions Chroma solar simulator (dark background).

Photoluminescence (PL) images have been performed using a BT imaging LIS-R2 analysis tool that allows to illuminate uniformly wafers on the full area. In the images presented, the signal intensity is directly related to the radiative recombination activity: defective regions appear darker than the other ones [10].

In order to evaluate the surface wafer particles maps, we processed simultaneously not only the PSQ wafers but also 6 inches microelectronic VLSI silicon wafer. Surface particles have been measured by mean of KLA Tencor Model: 6420 Surfscan, while passivation performance has been measured on the semi-squared wafer processed at the same time.

III. RESULTS

In current 3SUN experimental setup, after the wafers have been textured by CEA-INES first step of the process chain is the wet "HF last" treatment needed to remove the naturally grown thin oxide and promote weak hydrogen passivation.

Being 3SUN fab not under advanced clean room environment, we have tested the cleanliness of the area. To be sure environment where this process is realized is enough clean to avoid wafers surface contamination, TXRF measurements have been performed on wafers placed at various locations in the wet area.

In Fig.2 TXRF maps obtained in sweeping mode are presented. They report in particular the most significant metals (Fe, Zn and Cu) before and after applying a clean concept approach during the tests. It is evident surface metals contamination has been dramatically reduced by one or two order of magnitude improving area sealing, removing all metal and paper source, and settling a better procedure for the access into the area including more adequate clothing.

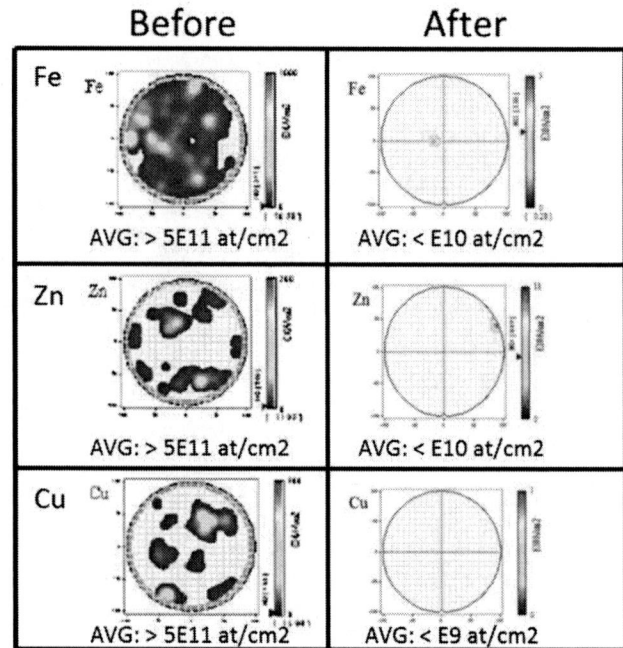

Fig. 2. TXRF sweeping mode map comparison before and after clean concept approach.

The transportation conditions can have severe impact on the cell efficiency especially, as in our case, when the process is interrupted and carried on after several days and after wafer travelling for many kms. We studied the effect of transportation on both on the passivation and cell efficiency.

Different carriers have been evaluated: such as plastic boxes where wafers were horizontally positioned with paper sheet as separation between or carriers with wafer kept on vertical position. Wafers have been processed all together in the same run and 20 nm for passivation of a Si intrinsinc has been deposited on both sides. After the deposition, we split the wafers into different boxes and shipped back to CEA-INES either in horizontal position (inside plastic boxes with interleaf paper) or in vertical position using also an adapted reinforced box with a special feature inside to protect PSQ wafers.

Passivated wafers have been characterized by PL and QSSPC at CEA-INES.

The PL images reported in Fig.3 show a comparison between the different transportation methods. Different level of defectiveness could be found. The wafers carried in horizontal position before the deposition (Fig.3 a and b) are affected by scratch marks, clearly due to wafers rubbing, causing also a significantly increase of surface particles. On the other hand in the case of vertical transportation, see (Fig.3 c and d) rubbing marks are not present and the number of surface defects and black spots is strongly reduced. Lifetime measurement is correlated with the surface defectiveness amount so the wafers less affected by surface damage resulted in lifetime, iVoc and iFF improvements.

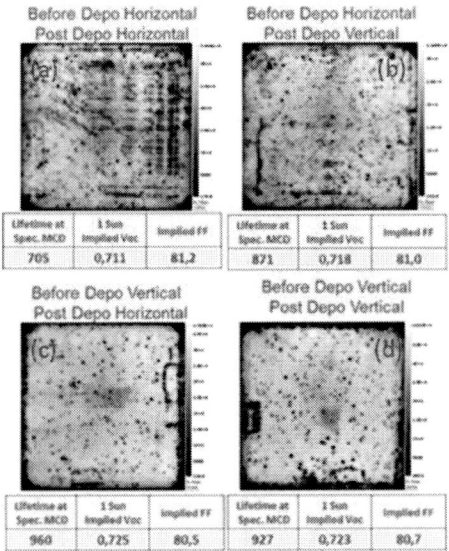

Fig. 3.   PL comparison, lifetime (μs) iVoc (mV) and iFF (%) measurement of wafer passivated after different transportation method.

Focusing our attention on the dark spots, still visible in the best shipping configuration, we have performed a more detailed study by mean of SEM-EDX, as shown in Fig.4. We find that the dark spots are related to the presence of particles with a size of few tens of microns mainly composed of carbon and oxygen. Those particles could not be removed by the HF-dip performed at 3SUN fab affecting the PL emission (and thus minority carriers' effective lifetime) up to several hundreds of microns around since the diffusion length is high in the studied wafers.

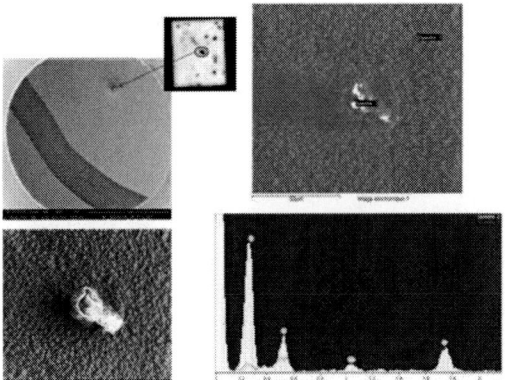

Fig. 4.   SEM analysis and EDX spectrum of in the area of a dark spot.

Similar experiment has been performed to evaluate not only the impact on cell efficiency as well. Wafers texturized at CEA-INES have been shipped to 3SUN and then shipped back to CEA-INES for process up to cells completion. In Fig.5 the impact of the transportation method upon the cell efficiency is presented and compared to samples without transportation (but similar waiting time as shipped wafers, wafers stored after PECVD process) and CEA baseline (typical run without waiting time).

From all the transportation methods used, the one with wafer in vertical position into the reinforced boxes resulted the best in terms of cell efficiency. Transportation with wafer in horizontal position in all the configurations tested (different kind of plastic boxes) produced severe impact on the cell efficiency (up to -2.5% of relative efficiency) mainly due to an increase of series resistance (up to +2.5%) and consequently reduction of FF (up to -2% relative). In addition, we find that even samples without transportation were also affected by efficiency reduction compared to the baseline. It shows how this high efficiency solar cell technology is sensitive to possible surface passivation related effects and thus requires well-controlled fabrication process

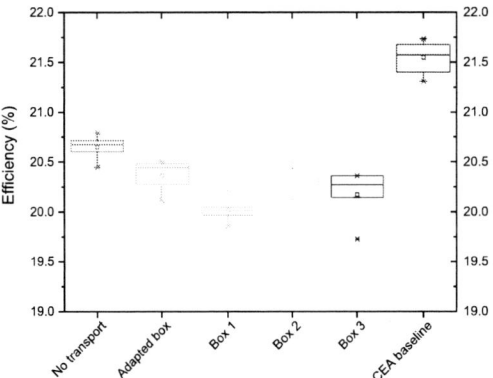

Fig. 5.   Transportation impact on cell efficiency.

978-1-5090-5606-4/17 $31.00 © 2017 IEEE   1749

Impact of wafer tray during PECVD deposition has been also studied. The presence of edge contaminant coming from the tray is clearly shown on PL images performed on cell (Fig 6).

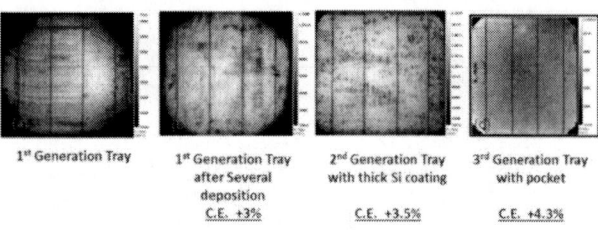

| 1ˢᵗ Generation Tray | 1ˢᵗ Generation Tray after Several deposition | 2ⁿᵈ Generation Tray with thick Si coating | 3ʳᵈ Generation Tray with pocket |
| C.E. +3% | C.E. +3.5% | C.E. +4.3% |

Fig. 6. PL defectiveness maps induced by tray and impact on cell efficiency (C.E.).

The PL images revealed a wide inactive dark region all around the edge of the wafer (Fig.4a). We immediately suspected that this defectiveness originate from the diffusion of contamination induced by the system used to hold the wafer on a non-pocketed tray.

Covering the contaminated tray area with silicon deposition reduced the defective area on the wafer edge (Fig.6 b and c) resulting in a significant increase of cell efficiency up to 3.5% absolute values. However only switching to tray with pocket optimized for this type of wafers the edge defectiveness has been completely removed and it leads to an efficiency increase of 4.3%.

Moreover, a severe impact of the backside defectiveness induced by contact with tray is clearly observed. This defects are collected by the contact between the tray and one of the two active sides of the cell. Fig.7 shows an example were the sliding of the wafer on the tray during loading and unloading can generate a signature defectiveness map.

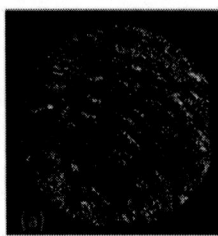 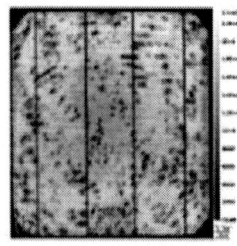

Fig. 7. Particles map and PL images and obtained before the optimization.

Both KLA-Tencor defectiveness map (Fig.7a) and PL mapping (Fig.7b) evidence a clear rubbing signature. Better tray cleaning along with reduction of tray moving speed allowed to strongly reduce the defectiveness level at the wafer backside as shown on Fig.8 (a and b).

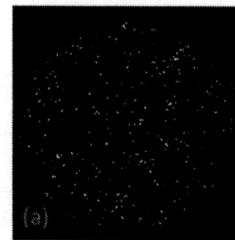 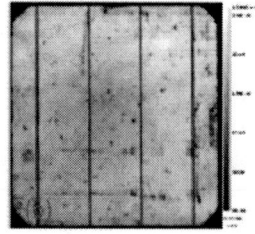

Fig. 8. Particles map and PL images and obtained before the optimization

In Fig.9 the evolution of 3SUN cell efficiency is reported. Many improvements have been achieved since first run in Nov 2015 and a maximum of 19.5% efficiency has been obtained so far.

It is worthwhile to consider that the cell does not undergo to the full process in 3SUN and that cells travel uncovered and with unprotected silicon surface to CEA-INES where they are finished with TCO deposition and metallization processes. We have estimated using witness wafers and cells that, in average, about 0.7% of absolute efficiency loss in the travel due to a-Si contamination and damaging. Therefore we can estimate that the maximum efficiency reached so far in 3SUN by re using thin film designed PECVD is exceeding 20%. More space for efficiency improvement is foreseen as a-Si:H deposition steps are not fully optimized for HJT yet and studies on H_2/SiH_4 ratio or H_2 pre and post treatments are ongoing.

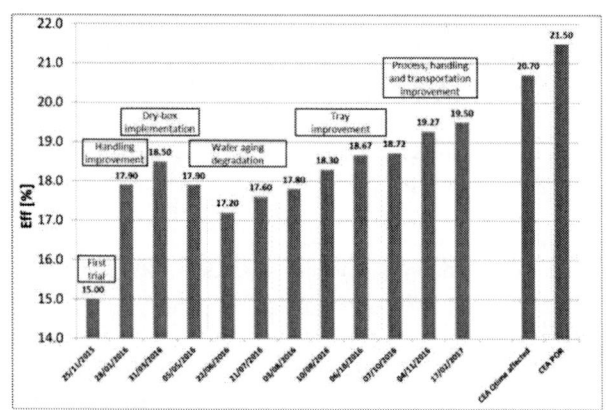

Fig. 9. 3SUN cell efficiency trend over 15 months of activity

IV. CONCLUSION

We focused on the impact of defectiveness induced by transportation and handling of wafers for SHJ solar cell fabrication with industrial thin film PECVD equipment. Several scenarios have been considered. All the process steps of heterojunction solar cell fabrication can be drastically affected by contaminations. Typically, the production environment of a thin film solar cell factory does not require stringent clean-room environment, this is not true for SHJ technology.

We were able to discriminate among all defectiveness sources as carriers, position of wafers inside carriers, trays (including both direct contact and local environment) by using PL maps and particle contamination analysis on the surface of wafers and cells. Once individuated sources of defects we elaborated solutions to mitigate effects and we were able to increase the efficiency with an absolute gain of 4.3%.

This study allows us to define a baseline for the development of a clean methodology for wafer processing in an industrial environment.

ACKNOWLEDGEMENTS

We would like to thank 3SUN technicians D. Nicotra and A. Di Matteo and the CEA-INES LABFAB and CEA-Leti teams for their technical and practical support. We would also like to thank M. Sciuto for his contribution.

REFERENCES

[1] DeWolf, "High-efficiency Silicon Heterojunction Solar Cells: A Review", *Green*, vol 2, pp 7–24, 2012.

[2] J. I. Pankove and M. L. Tarng, "Amorphous silicon as a passivant for crystalline silicon", *Appl. Phys. Lett*, vol 34, pp 156-157, 1979.

[3] M. L. Tarng and J. I. Pankove, "Passivation of p-n junction in crystalline silicon by amorphous silicon", *IEEE Transactions on Electron Devices,* vol 26, pp 1728, 1979.

[4] I. Weitzel, R. Primig and K. Kempter, "Preparation of glow discharge amorphous silicon for passivation layers", *Thin Solid Films*, vol 75, pp 143-150, 1982.

[5] Ribeyron, Gerritsen, Heslinga, Muñoz, Ozanne, Bancel, et al. "Recent Progress on the CEA-INES Heterojunction Solar Cell Pilot Line" *31st Eur Photovolt Sol Energy Conf Exhib*. pp 279–283, 2015.

[6] Gerardi, Foti, Rapisarda, Canino, Battaglia, Condorelli, "Industrial Scale Optimization of SiOx Bottom n-Layer in Tandem Solar Cell", *32nd Eur Photovolt Sol Energy Conf Exhib*, pp 1260-1263, 2016.

[7] Gerardi, Battaglia, Condorelli, Canino. "External Quantum Efficiency as Function of Applied Voltage of Multi-Junction Hydrogenated Amorphous Si Based Cell: Performance Optimization after Stabilization", *32nd Eur Photovolt Sol Energy Conf Exhib*, pp 1097–1099, 2016.

[8] Nagel H, Berge C, Aberle AG, "Generalized analysis of quasi-steady-state and quasi-transient measurements of carrier lifetimes in semiconductors", *J Appl. Phys lett,* vol 86, pp 6218-6221,1999.

[9] Sinton, Cuevas, "Contactless determination of current–voltage characteristics and minority-carrier lifetimes in semiconductors from quasi-steady-state photoconductance data" *Appl Phys Lett.* vol 69, pp 2510-2512, 1996.

[10] Nos O, Favre W, Jay F, Ozanne F, Valla A, Alvarez J, et al. "Quality control method based on photoluminescence imaging for the performance prediction of c-Si/a-Si:H heterojunction solar cells in industrial production lines" *Sol Energy Mater Sol Cells.* vol 144, pp210-220, 2016.

>23% Silicon heterojunction solar cells in Meyer Burger's Demo line: Results of pilot production on mass production tools

J. Zhao[1], M.König[1], A. Wissen[1], V. Breus[1], D. Decker1, M. Fritzsche[1], M. Schorch[1], H.J.Nonnenmacher[1], M. Leonhardt[1], T. Große[1], J. Hausmann[1], A. Waltinger[1], D. Landgraf[1], S. Burkhardt[1], H. Mehlich[1], E. Vetter[1], F. Schitthelm[1], Y. Yao[2], T. Söderström[2], A. Richter[2], D. Habermann[2], S. Leu[2],

[1] Meyer Burger (Germamy) AG; An der Baumschule 6-8, 09337 Hohenstein-Ernstthal, Germany
[2] Meyer Burger AG, Schorenstrasse 39, 3645 Gwatt (Thun), Switzerland

Abstract — after a flat PV market for several years, it recently started to grow strongly again. More and more solar cell manufacturers focus on high efficiency solar cells with low costs. The hydrogenated amorphous (a-Si:H) /crystalline silicon (c-Si) hetero-junction (SHJ) technology provides a well suited application to achieve efficiencies above 23% with process temperatures below 200°C. The whole process flow (Fig.1) is simple and with less process steps compared to the standard solar cell process[1]. The low temperature coefficient (-0.2 %/K) of the SHJ solar cells provides additional benefit in output power for PV systems. The recent world record SHJ cell with an efficiency of 25.6% on a 4 inch n-type solar cell, reached by Panasonic Corporation has shown even more potential for future mass production.

Index Terms — Heterojunctions, Production Control, Optimization

I. INTRODUCTION

As part of Meyer Burger AG group, we have focused since years on equipment and technology development for industrialized PV heterojunction cell manufacture. To realize a fluent technology know-how transfer from a R&D scale line to a mass production line at customer side, the pilot line at Meyer Burger (Germany) AG has been upgraded to a 10~15MW capacity production line for SHJ solar cells with a stable process and high cell efficiency . All the tools, especially the key equipment $HELiA_{PECVD}$ and $HELiA_{PVD}$ (Fig.2) were configured for mass production throughputs of 2400 wafers/hour.

Beside high efficiency, the focus of the process development of this SHJ demonstration line is to achieve reasonable low production costs for cell manufacturing. The rear emitter and bifacial busbar less cell concept has been continually improved and the efficiency has been increased progressively especially by developments for the PECVD, PVD and metallization processes after several R&D activities over years. In Q4 2015, around 41,000 SHJ cells were produced on the demonstration line with the above mentioned cell concept. A high cell performance and process stability of the baseline were achieved in last year. During one year, the cell performance was improved progressively by doing pyramids engineering at wet process, fine line printing and wafer material optimization and so on. At end of 2016 all benefits of process improvements activities in 2016 were integrated in

two runs with nearly 2000 wafers of each for a new record of the cells performance.

II. APPROACH

1) To reduce the reflectivity of the silicon wafer surface, a wet chemical process is needed to form a textured surface of many micro-pyramids. The pyramids size and its distribution uniformity could influence the passivation quality after a-Si:H deposition process and even finger width and paste consumption of printing process. Via the pyramids engineering work, the pyramids size and the uniformity were optimized to improve the cell performance.

2) Thanks the Smart Wire technology, it provides more than 3000 contact points between fingers and wires on one cell and in allows more freedom to print fine line fingers by using 40 μm or even to 30 μm screen opening in production mode. It results less shading, less paste consumption and no fill factor drop in cell and module level at same time.

3) Wafer material plays as well big role for cell efficiency. Especially, for the low temperature process in heterojunction cell technology, the wafers were processed under 200°C in all steps. Therefore the thermal donors and metal impurity in bulk material couldn't be removed by high temperature process like diffusion or firing. Many literature and our previous experiments have showed the content of thermal donors in the seed and tail parts of n-type ingots can influence the bulk resistivity and SHJ cell performance significantly. Beside thermal donors the bulk resistivity of the material itself can also impact the SHJ cell parameters by involving rear emitter cell concept. To obtain the most suitable wafer material for SHJ technology, the impact of bulk resistivity was tested. [2]

4) It's well known, that the wafer price is strongly depended with the thickness, extra for n-type material. By high temperature cell process, the wafer thickness was quite limited by the relative high breakage rate and bow issue. But it's not critical for SHJ technology with low temperature process. The

down limit of the accepted wafer thickness for SHJ was tested and the cell performance was investigated.

5) In these 2 integrative runs, full automatic production mode with 2400 cells/ hour throughput recipes on mass production tools was done and commercial 6 inch n-type production material has a relatively low bulk resistivity and low oxygen concentration to avoid the Oxygen induced Stacking Fault (OSF)[3].

III. RESULTS

1) Via the pyramids engineering the pyramids size and uniformity of pyramids distribution were optimized. As the REM picture (Fig 2. Left) shows much more uniform and smaller pyramids are obtained with new wet process. By involving smaller pyramids size, a significant improvement in I_{sc} was observed and other main cell parameters were similar to the baseline process. The benefit of I_{sc} determined 0.15% abs more cell efficiency for the smaller and more uniform pyramids. (Fig 2. Right)

2) By involving fine line printing process with 40 μm or even 30 μm a clear increasing of current from less shading was demonstrated as we expected to compare with baseline of 50 μm opening. No drop in FF was detected. The higher current results finally the higher cell efficiency and at same time 40% paste consumption. (Fig. 3)

3) In the experiment of different wafer thickness splits (Fig.4), based on same wafer material from 180 (150 μm on cell) to 120 μm (90 μm on cell) thickness in as-cut were tested in cell level. The results show that thinner cells have increased V_{oc} and decreased I_{sc}, no difference in FF in all thickness groups were observed. The ETA in all thickness groups is equal due to compensation of ↑ V_{oc} and ↓ I_{sc}. It shows clearly that wafer thickness 90 μm is not show stopper for SHJ technology! There are even more potential to produce the cells of thickness thinner than 90 μm.

4) In another experiment of impact of 2 different wafer suppliers (Fig.5), one complete ingot from each supplier was qualified by using the baseline process. One n-type commercial ingot from LONGi was specified with Meyer Burger's advanced wafer specification, which was aimed for a low oxygen concentration in bulk and relatively lower bulk resistance, was compared with the ingot from major wafer supplier X for the production. These wafers from LONGi ingot have delivered almost +1.4% FF and +0.45% cell efficiency generally than the wafers from ingot of supplier X in this comparison test.

5) At end of 2016, all the mentioned advantages of the process improvements were integrated in two Best Known Practice (BKP) runs. In these 2 runs the tact time recipes for 2400 cells/hour throughput in each

tool and automation systems were used. The commercial n-type wafers have a low bulk resistivity range and low oxygen concentration. Nearly 2000 wafers for each run were processed. Within each integrative run, slight changes in process parameters were done, to find best process conditions. In the 1st BKP run, a mean value 22,8% of cell efficiency was reached. Based on the results of 1st run, the wet chemical process for the texturing and as well for the CVD tool were optimized in the 2nd BKP run. The V_{oc} and efficiency in 2nd run were improved and the cells in class over 23% cell efficiency were significantly increased. Extremely high cell efficiency with 22,95% was achieved(Fig.6). Additionally in the 2nd run 47,3% cells were measured > 23% and 2 split groups achieved even median over 23,1% (Fig.7). Two cells presented new record cell efficiency > 23,4%. By using the cells of the runs, a 72-cells bifacial module was manufactured and 390 W was measured.

IV. FIGURES AND TABLES

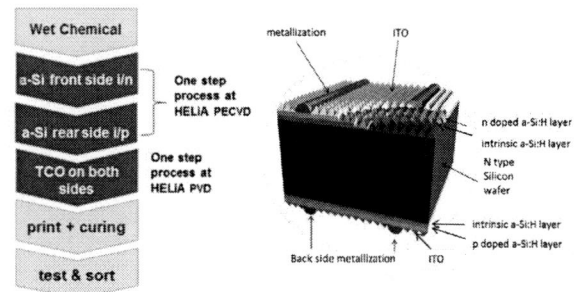

Fig. 1. the process flow (left) of Meyer Burger SHJ production line. For the mentioned production the bifacial and rear emitter cell structure (right) was applied.

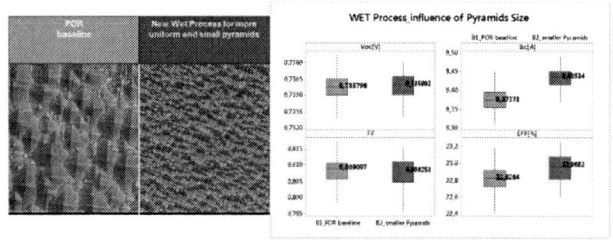

Fig.2. Difference of pyramids size and uniformity in REM (left) and cell performance (right) between baseline and new wet process

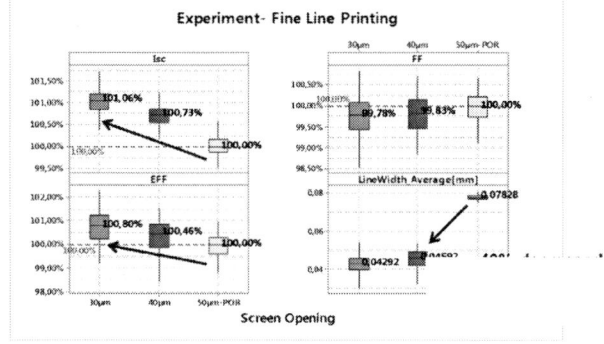

978-1-5090-5606-4/17 $31.00 © 2017 IEEE   1753

Fig.3. cell results in experiment of fine line printing 40 vs 30 vs 50 μm screen opening

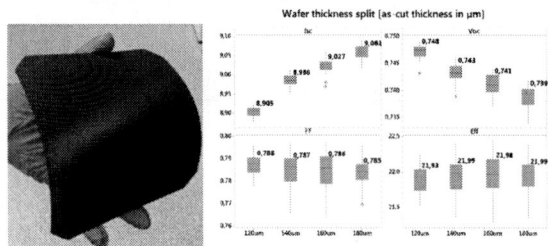

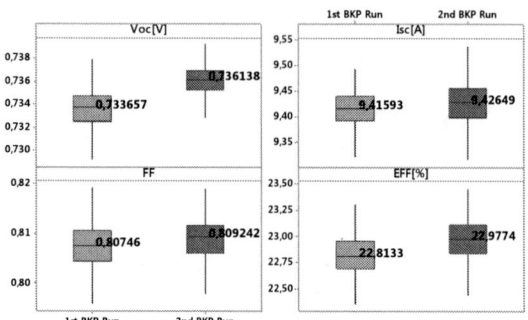

Fig.6. comparison between 1st and 2nd BKP runs

Fig.4. 90 μm cell (left) and results of the thickness splits (right)

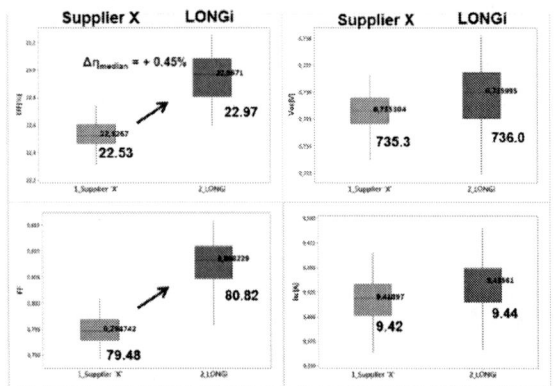

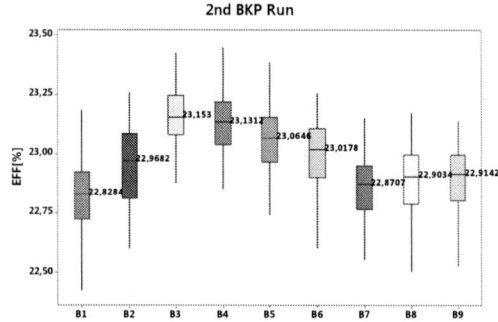

Fig.7. 2 batches reached >23,1% in median cell efficiency

Fig.5. impact of different wafer suppliers and results of Meyer Burger advanced material specification.

REFERENCES

[1] M. Taguchi, A. Yano, S. Tohoda, K. Matsuyama, Y. Nakamura, T. Nishiwaki, K. Fujita, and E. Maruyama, IEEE J. Photovolt., p. 1–4, 2013.

[2] J. Zhao: N Type Mono Crystalline Ingot Characterization for Silicon Heterojunction Technology in Roth&Rau's pilot line (29th European Photovoltaic Solar Energy Conference, 2014)

[3] N. Nakamura: Dependence of Properties for Silicon Heterojunction Solar Cells on Wafer Position in Ingot, 27th European Photovoltaic Solar Energy Conference and Exhibition, 2012

Experimental and Simulation studies on TiO₂/Silicon Heterojunction diodes

Swasti Bhatia[1], Neha Raorane[2], Nimisha Sreekumar[2], Pradeep R. Nair[1], and Aldrin Antony[3]

[1]Dept. of Electrical Engineering; [2]Centre of Excellence in Nanoelectronics (CEN); [3]Dept. of Energy Science and Engineering, Indian Institute of Technology Bombay, Mumbai, Maharashtra, 400076, India
Email: swasti@ee.iitb.ac.in, prnair@ee.iitb.ac.in, aldrin.antony@gmail.com

Abstract — Silicon based heterojunction solar cells with TiO₂ as the electron selective layer is an attractive alternative to diffused junction based solar cells. This study aims to explore the relation between the atomic layer deposition conditions and the electrical performance of Si-TiO₂ heterojunction diodes. Our experimental results indicate that the doping of the TiO₂ film is influenced by the deposition temperature – an inference supported by numerical simulations as well. This optimization allows us to fabricate diodes with comparable performance to that of the best devices reported in literature based on other techniques like low temperature chemical vapor deposition.

Index Terms — Electron selective, titanium dioxide, heterojunction, carrier selective, photovoltaic cells, silicon.

I. Introduction

Titanium oxide thin films have attracted significant interest in the last few years due to its asymmetric band offsets with silicon's valence band and conduction band. The large valence band offset prevents over the barrier transport of holes, while a conduction band offset of less than 0.1 eV enables efficient collection of electrons. Apart from acting as an electron collector, TiO₂ has an added advantage of having a refractive index of around 2.34, which lies between that of air (1) and silicon (3.41), thus reducing the reflection loss. The high bandgap of more than 3.4 eV makes it transparent to visible light which avoids parasitic absorption as compared to the cells that rely on a-Si as carrier selective contacts. These properties make titanium dioxide a viable material to be used as electron extracting layer in carrier selective heterojunction solar cells.

Accordingly, recent reports have experimentally established the electron selectivity of titanium oxide [1] [2]. TiO₂ films deposited by low temperature chemical vapor deposition (CVD) have achieved good passivation after annealing – with reverse saturation current (J_o) of the order of nA/cm² [3]. There have also been reports on enhancement of surface passivation (surface recombination velocity of the order of 10cm/s) for films prepared by atomic layer deposition [4]. However, the dependence of ALD deposition conditions to the eventual device performance is not yet clearly elucidated. In this work, we report TiO₂/Si heterojunction diodes fabricated through ALD method with J_o comparable to that of the best results using low temperature CVD process [3]. Further, through detailed experiments and numerical simulations, we investigate the physical mechanisms that control the performance of such cells

(see Section II & III). Indeed, this work helps to identify the optimum ALD processing conditions for efficient TiO₂/Si diodes.

II. Experiments

The wafers used for experiments were boron doped, CZ, silicon wafers of 4 Ohm-cm resistivity. The wafers underwent standard RCA cleaning process. This was followed by the deposition of TiO₂ films of varying thickness by Atomic layer Deposition (ALD) under different deposition conditions. The precursor used for TiO₂ deposition was TDMAT for Ti and water for oxygen. The deposition was done by thermal ALD without any plasma. Post deposition, Aluminum dots were deposited via shadow mask. The back contact consists of 8nm chromium followed by 80 nm gold layer. Gold was chosen as a back contact since it is a high work function metal and is expected to form an ohmic contact with boron doped silicon.

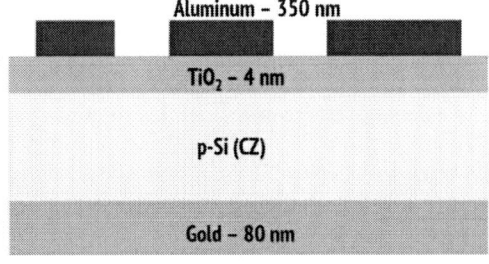

Figure 1. Structure of the Si/TiO₂ diodes fabricated using ALD.

In order to analyze the electrical behavior of electron selective titanium oxide films, diodes with structure as shown in Fig. 1 were fabricated. TiO₂ films of different thickness were prepared at different substrate temperatures. The precursors used were tetrakis (dimethylamino) titanium (TDMAT) and water. TDMAT has a very low boiling point and hence has good vapor pressure at room temperature. Therefore, deposition could be easily done at a temperature as low as 75°C. Films were deposited at 75°C, 100°C, 150°C, 200°C and 250°C, without any change in other ALD process conditions.

978-1-5090-5606-4/17 $31.00 © 2017 IEEE   1755

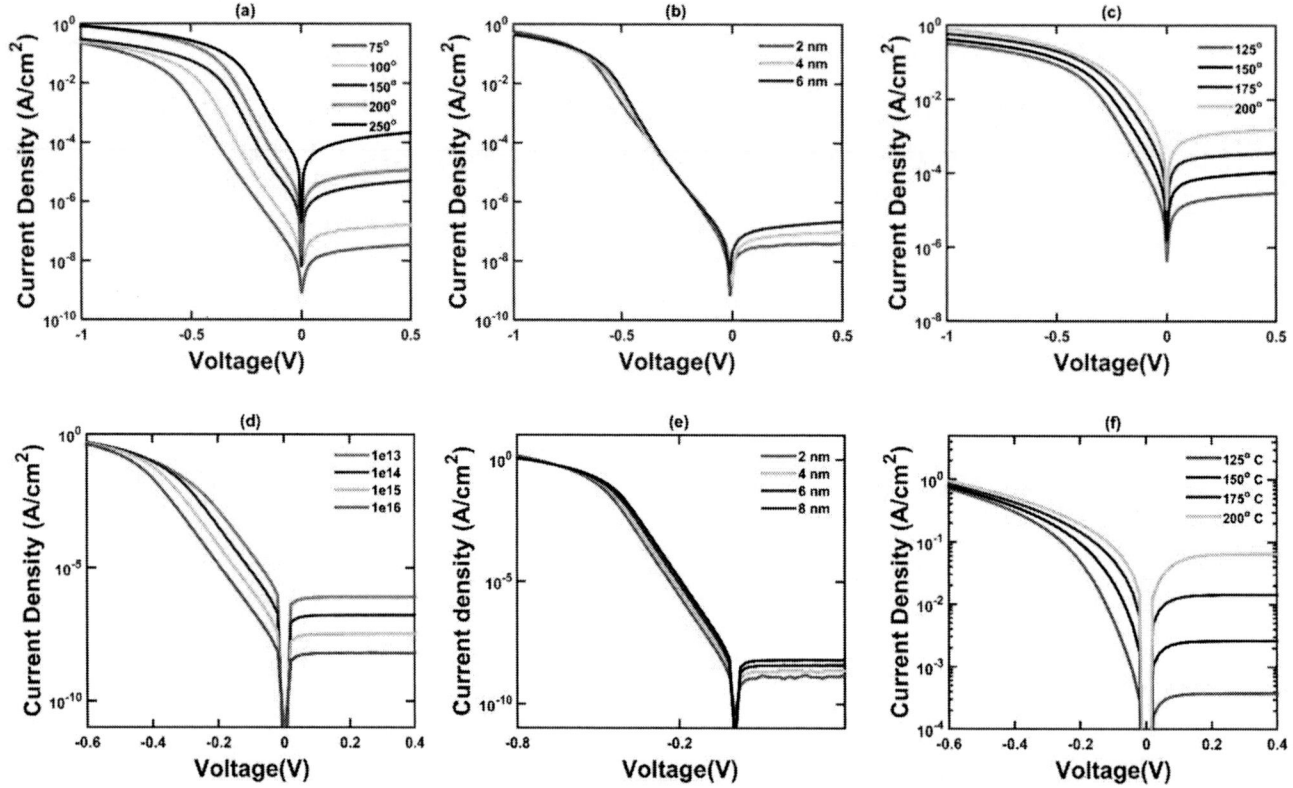

Figure 2. Comparison of experimental (top panel) and numerical simulation results (bottom panel) for Si/TiO_2 heterojunction diodes. Part (a) shows the effect of varying the deposition temperature in fabricated diodes (thickness 8nm), while part (d) shows the simulation results on varying the TiO_2 doping density. Parts (b) and (e) shows the TiO_2 film thickness dependence on current voltage characteristics for the fabricated (deposited at 200°C) and simulated diodes, respectively. The temperature dependent IV characteristics are shown in parts (c) and (f) for fabricated and simulated diodes, respectively.

The current voltage characteristics shown in Fig. 2a clearly indicate a trend of decreasing reverse saturation current with the decrease in deposition temperature (also see Table I), while the ideality factor remains constant. This constant ideality factor suggests that the change in reverse saturation current may not be only due to change in interface state density. The characteristics obtained by varying the thickness as in Fig. 2b show that the reverse bias current increases as the thickness increases.

TABLE I

Substrate temperature (°C)	75	100	150	200	250
J_o (nA/cm^2)	5.6	29	77	120	7700
V_{bi} (eV)	0.55	0.52	0.45	0.40	0.35

To further explore the characteristics, we performed Capacitance voltage measurements (at 10kHz, 30mV ac signal). The Mott-Schottky characteristics obtained showed linear variation with voltage. The extracted built in voltage,

shown in Table I, indicate a decreasing trend with deposition temperature. Further, the temperature dependent IV measurements indicated that the reverse current increases with temperature (Fig. 2c).

To summarize, the experiments indicate the following trends – (a) diode characteristics improve with decrease in deposition temperature, (b) the reverse current increases with increase in TiO_2 thickness, (c) the V_{bi} decreases with increase in deposition temperature, (d) the reverse current increases with temperature with an activation energy of 0.71 eV. We note that a trend opposite to that of (b) has been reported in literature for CVD deposited TiO_2 films [3] – which was attributed to the reduction of tunneling current on increasing thickness. Since we use ALD method, there is uniform coverage even at 2 nm and hence tunneling is not expected to be physical mechanism that results in increase in reverse current as TiO_2 thickness increases.

The variation in deposition conditions, in principle, can affect (a) The effective doping level of TiO_2, (b) band alignment of TiO_2 with Si, and (c) interface trap states at TiO_2/Si interface. Among these, a change in conduction band offset with variation of deposition temperature appears to be highly improbable as the electron affinity of silicon is 4.05 eV while that of TiO_2 is between 4.00 eV to 4.1 eV. Also, a careful analysis of reverse bias current reveals that it is proportional to square root of voltage (see Fig. 2a). In case of significant band offset variation, some of the diodes would show inversion due to collection of electrons at the barrier and thus the depleted region would not expand further into silicon bulk. In such a scenario, the reverse current would not show any square root of voltage dependence. The same observation and the fact that the diode ideality factor is independent of deposition temperature indicates that the interface states are also not significantly influenced by the deposition temperature.

TiO_2 is known to have charge carriers due to the presence of oxygen vacancies. The presence of oxygen vacancies gives rise to Ti^{+3} oxidation state for some Ti atoms. Some of these atoms would lose an electron to achieve the more stable Ti^{+4} oxidation state. The electrons thus lost act as charge carriers. Hence a non-stoichiometric, oxygen deficient film is expected to be more doped than a stoichiometric one. Literature [5] suggests that higher deposition temperatures result in oxygen rich TiO_2. These results also indicate a non-stoichiometric growth of TiO_2 which results in an increase in doping with decreasing deposition temperature. An increase in effective doping concentration will result in decrease in reverse saturation current as well as increase in the built-in voltage – which are consistent with the experimental results. To further test this claim of increase in effective doping density with decrease in deposition temperature, we performed detailed numerical simulations, as described below.

III. SIMULATIONS

The fabricated structure was simulated using Sentaurus. The doping of silicon was kept constant and that of TiO_2 was varied from $10^{13} cm^{-3}$ to $10^{16} cm^{-3}$. The dielectric constant of TiO_2 was taken as 60 and the mobility of electrons in the film as $20 m^2/(Vs)$. Ohmic contacts were assumed for silicon substrate and TiO_2. The simulated current voltage characteristics are shown in Fig. 2 (d), (e) and (f). As the doping decreases, we find that the reverse saturation current increases – in accordance with experimental trends (see Fig. 2a and Fig. 2d). The ideality factor remains the same because the transport is over the built in voltage barrier. Further, our simulations indicate that reverse current increases as the TiO_2 thickness increases (see Fig. 2e). The depletion region width in Si is proportional to the total amount of ions in TiO_2 depletion region. For the thickness range under consideration, TiO_2 is fully depleted. Hence, an increase in TiO_2 thickness results in larger depletion region in Si and consequently a larger reverse bias current. While this trend is

also in accordance with experiments, we find that the temperature dependent IV simulation results (see Fig. 2f) are not in full accordance with experiments. Specifically, the activation energy of reverse saturation current obtained from experiments differ significantly with that of simulations, and further work is required to explain this puzzle. We note that apart from the effective doping density, interface trap density and conduction band offset variations were also explored in numerical simulations. These simulations indicate that the process dependent change in the effective doping density in TiO_2 can explain the observed experimental characteristics.

IV. CONCLUSIONS AND FUTURE WORK

The study suggests that lower ALD deposition temperatures might be desirable for achieving Si/TiO_2 diodes with better J_o. The films deposited via ALD have uniform coverage and diodes with J_o as low as 5.51×10^{-09} A/cm^2 have been achieved. Detailed experiments and numerical simulations indicate that ALD deposition conditions affect the effective doping levels of TiO_2. Further, a first order calculation indicates that it is possible to achieve an efficiency of more than 16% for solar cells with the above mentioned TiO_2 as selective contacts – and further experiments are underway in this direction.

ACKNOWLEDGEMENT

This work was supported by Indo-US Solar Energy Research Institute (SERIIUS), National Centre for Photovoltaic Research and Education (NCPRE), Indian Ministry of New and Renewable Energy (MNRE) through the project AMANSI under the National Solar Science fellowship.

REFERENCES

[1] S. Avasthi *et al.*, "Hole-blocking titanium-oxide/silicon heterojunction and its application to photovoltaics" in *Applied Physics Letters,* 102(20), 203901, 2013.

[2] K. M. Gad *et al.*, "Ultrathin Titanium Dioxide Nanolayers by Atomic Layer Deposition for Surface Passivation of Crystalline Silicon," in *IEEE Journal of Photovoltaics*, vol. 6, no. 3, pp. 649-653, May 2016.

[3] J. Jhaveri *et al.*, "Al/TiO2/p-Si heterojunction as an ideal minority carrier electron injector for silicon photovoltaics," *2016 IEEE 43rd Photovoltaic Specialists Conference (PVSC)*, Portland, OR, 2016, pp. 2444-2447.

[4] B. Liao *et al.*, "Passivation of Boron-Doped Industrial Silicon Emitters by Thermal Atomic Layer Deposited Titanium Oxide," in *IEEE Journal of Photovoltaics*, vol. 5, no. 4, pp. 1062-1066, July 2015.

[5] D. Wei *et al.*, "Influence of Atomic Layer Deposition Temperatures on TiO2/n-Si MOS Capacitor". *ECS Journal of Solid State Science and Technology*, 2(5), N110-N114, 2013.

A Study on Blister Formation and Electrical Properties under Various Annealing Condition for Tunneling Oxide Passivation Layer

Sungjin Choi, Ka-Hyun Kim, Min Gu Kang, Jeong In Lee, Donghwan Kim, and Hee-eun Song

Korea Institute of Energy Research, Daejeon, 34129, Republic of Korea

Reducing the carrier recombination at electrode contact becomes important for high conversion efficiency of the solar cell. In general, rear side contact recombination is reduced by back surface field, it is formed by diffusion process such as POCl3, BBr3 doping. However, lifetime is decreased by crystal defects and inactive dopant according to high temperature during diffusion process. In this study, polycrystalline silicon was formed as back surface field through crystallization of amorphous silicon. Quasi-steady-state photoconductance measurement and spectroscopic ellipsometer were used to analyze electrical properties and crystallinity of a-Si under various annealing condition. Also, we observed the blister after various annealing process at the different surface.

I. INTRODUCTION

Minimizing carrier recombination in the solar cell contacts becomes important for achieving very high conversion efficiency of the solar cell. The rear side contact recombination could be reduced by back surface field. In general, back surface field was formed by diffusion process such as $POCl_3$, BBr_3 doping. But, lifetime was decreased by crystal defects and inactive dopant according to high temperature diffusion. Accordingly, back surface passivation in crystalline silicon solar cell is one of important key technologies to achieve high efficiency. Passivated rear contact suppresses back surface recombination, resulting in high open circuit voltage (Voc). Conventionally, back surface field (BSF) of crystalline silicon is formed by aluminum diffusion. Use of conventional BSF introduces significant recombination at the interface between metal contact and solar cell. Suggested solutions to the back surface passivation have been passivated emitter rear locally diffused (PERL) or passivated emitter rear contact (PERC). Both structures have passivated back surface and small aperture of local openings to the back contact. However, such local contact concepts not only require additional patterning steps, but also suffer with fundamental trade-off between passivation (Voc) and series resistance (Fill Factor). It is due to the limited lateral conductivity of the lightly doped layer. There is another method to make passivating contacts of silicon heterojunction using intrinsic hydrogenated amorphous silicon (a-Si:H), also called heterojunction with intrinsic thin layer (HIT). However, fabrication process of silicon heterojunction solar cells largely relies on the thin film process and is hardly compatible with conventional crystalline silicon solar cell process.

Tunneling oxide passivated contact (TOPCon) is the structure recently spotlighted by achieving high conversion efficiency of 25.3 %. TOPCon structure consists of a thin tunnel oxide and a Phosphorus (P) doped poly-Si layer contact. The P-doped poly-Si layer can be fabricated either crystallization of a-Si:H or directly deposited using LPCVD. Contrary to the rear side passivation using local contacts, one dimensional structure of TOPCon enables implementation into conventional solar cell process without excessive additional production cost. In spite of high efficiency and potential to commercialization of the TOPCon structure, its BSF formation behavior and working mechanism are not yet fully understood and studied.

In this study, polycrystalline silicon was formed as back surface field through crystallization of amorphous silicon. Quasi-steady-state photoconductance measurement and spectroscopic ellipsometer were used to analyze electrical properties and crystallinity of a-Si under various annealing condition. Also, we observed the blister after various annealing process at the different surface.

II. EXPERIMENTAL PROCEDURE

The sample structure was Si substrate/wet chemical oxide/p doped amorphous Si. Thin film BSF were fabricated on both n-type, 200 μm thick ohm·cm solar grade wafers with resistivity of 0.6 ohm·cm. Saw damage removal (SDR) of the solar grade wafers was done using solution of NaOH and deionized water at 80 °C. SDR etches about 10 μm of damaged wafer surfaces on both sides. All wafers were cleaned using following sequence: 10% HF dip ; deionized water+H_2O_2+NH_4OH (RCA1) at 80 °C for 10 min ; deionized water+H_2O_2+HCl (RCA2) at 80 °C for 10min ; 10% HF dip. Then, a thin silicon oxide was grown using the nitric acid (HNO_3) solution. The temperature and time for wet chemical oxidation process were set to be 120 °C and 15 minutes, respectively. Thickness of silicon oxide layer was found to be about 1.5 nm deduced by spectroscopic ellipsometry (SE) measurement and modeling.

On the thin oxide layer, hydrogenated amorphous silicon (a-Si:H) thin films were deposited by the capacitively-coupled-plasma (CCP) radio-frequency (RF, 13.56 MHz) glow discharge PECVD method at substrate temperatures ranging from 175 to 200 °C. P doped a-Si:H films were deposited under carefully controlled plasma conditions using hydrogen-

978-1-5090-5606-4/17 $31.00 © 2017 IEEE

diluted silane gas mixtures. In this work, our a-Si:H layers were deposited at p·d product (pressure×inter-electrode distance) ranging from 1 to 3 Torr·cm and RF power density in a range from 30 to 100 mW/cm2.

After the a-Si:H deposition process, thermal anneal of the wafers was done at the temperature from 600 to 800 °C for one and five minutes, in a quartz tube furnace under nitrogen atmosphere. At elevated temperature above 600 °C, a-Si:H is crystallized in solid phase.1) After annealing, implied Voc of the solar grade wafer was determined by quasi-steady-state photoconductance (QSSPC) method using WCT-120 made by Sinton instruments. And the amount of blister formation was characterized by optical microscope and digital analysis program. As increasing annealing temperature, blister sizes were decreased. However, total area of blister was almost same regardless of annealing temperature.

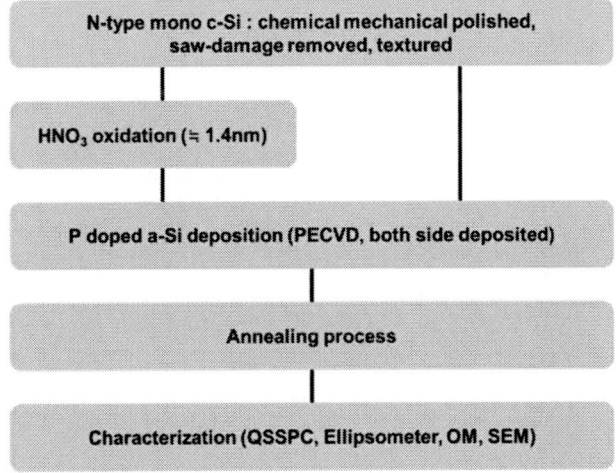

Fig. 1. Experimental procedure for a-Si/wet chemical oxide/c-Si

III. RESULT AND DISCUSSION

Figure 2 shows evolution of implied Voc as functions of annealing conditions of thin film BSF fabricated on 190 μm thick n-type solar grade wafer substrate. Implied Voc of as deposited sample was found to be 556 mV, and showed an increase to 641 mV after annealing at 600 °C for one minute. Implied Voc of the sample decreased to 560 mV after annealing at higher temperature of 700 °C for one minute. It is interesting that for the samples annealed for one minute shows sudden increase of implied Voc after annealing at 600 °C, but shows degradation at higher temperature annealing of 700 and 800 °C for one minute.

There is drastic change in implied Voc of the samples annealed for longer time for five minutes. At 600 °C, implied Voc of the samples showed a large increase to 651 mV after annealing for one minute. However, further annealing for five

minutes leads to drop in implied Voc of the samples to 570 mV.

It should be also pointed out that the samples annealed at higher temperature of 700, 800 °C show continuous increase of their implied Voc upon annealing time. It is necessary to make more detailed study on physical origin of the evolution of implied Voc during annealing for the thin film BSF.

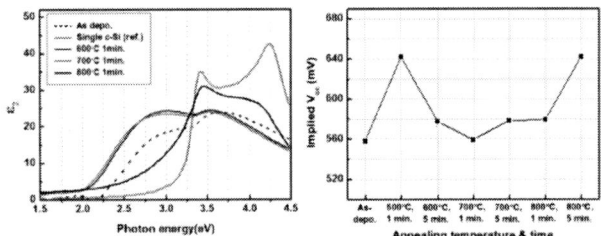

Fig. 2. Crystallinity and implied Voc of thin film back surface field measured by quasi-stead-state photoconductance annealing at various temperature and times

Figure 3 is optical microscope image of blisters according to annealing temperature and time. Blister is formed on the surface of wafer, and it occurs due to releasing of hydrogen in amorphous silicon during the annealing process. Shape and density of blister are influenced by various annealing condition as temperature and time. As the annealing temperature became higher, the average diameter of blister is decreased and total number of blister is increased.

	1 min	3 min	5 min
600°C			
700°C			
800°C			

Fig. 3. Blister formation according to annealing temperature and time

Also, the amount of blister formation was characterized by optical microscope and digital analysis program as Figure 4. As the annealing temperature became higher, the average diameter of blister is decreased and total number of blister is increased. However, total area of blister was almost same regardless of annealing temperature

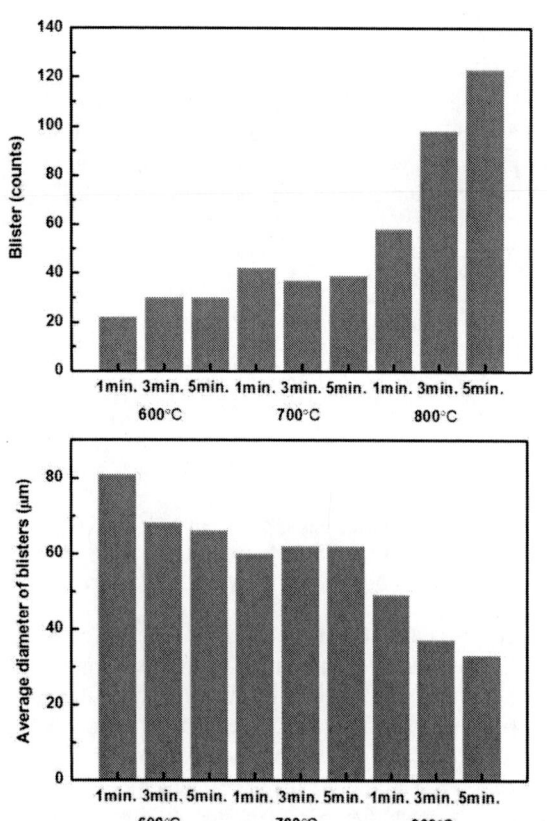

Fig. 4. Total number of blisters (above) and average diameter of blisters (below)

IV. CONCLUSION

In this study, polycrystalline silicon was formed as back surface field through crystallization of amorphous silicon. To crystallize the amorphous silicon, rapid thermal annealing process was used. Annealing temperatures were varied from 600℃ to 800℃. Quasi-steady-state photoconductance measurement and spectroscopic ellipsometer were used to analyze electrical properties and crystallinity of a-Si under various annealing condition. Also, we observed the blister after various annealing process at the different surface. The amount of blister formation was characterized by optical microscope and digital analysis program. As increasing annealing temperature, blister sizes were decreased. However, total area of blister was almost same regardless of annealing temperature.

REFERENCES

[1] S. D. Wolf, S. Olibet and C. Ballif, "Stretched-exponential a−Si:H / c−Sia-Si:H / c-Si interface recombination decay," *Applied Physics Letters*, vol. 93, p. 032101, 2008.

[2] A. F. i Morral, P. R. i Cabarrocas, and C. Clerc, "Structure and hydrogen content of polymorphous silicon thin films studied by spectroscopic ellipsometry and nuclear measurements," *Physical Review B*, vol. 69, p. 125307, 2004.

[3] H. Touir and P. R. i Cabarrocas, "Optical dispersion relations for crystalline and microcrystalline silicon," *Physical Review B*, vol. 65, p. 155330, 2002.

Processing Approaches and Challenges of Interdigitated Back Contact Si Solar Cells

Ujjwal Das, Lei Zhang, and Steven Hegedus

Institute of Energy Conversion, University of Delaware, Newark, DE 19716, USA

Abstract — **Processing approaches and challenges for high efficiency interdigitated back contact (IBC) Si solar cell are discussed in this paper. High efficiency (>25%) solar cells are currently fabricated by a more complex, rate-limiting patterning processes than standard cells contacted on both sides. Therefore, the development of simplified patterning approaches without increasing the carrier recombination rate and degrading performance are of paramount interest for high speed low cost IBC production technology. Several alternative processing techniques are discussed here that show promising early stage results.**

Index Terms — **back surface patterning, high efficiency, interdigitated back contact, silicon heterojunction, simplified manufacturing process, solar cell.**

I. INTRODUCTION

The concept of an interdigitated back contact (IBC) Si solar cell was first reported by Lammert and Schwartz in 1977 [1]. The IBC structure offers several advantages over the standard front contacted solar cell including elimination of front grid shading loss, almost full area metal coverage at the back for low resistance contacts, and coplanar interdigitated contacts for cell interconnect to aid in module fabrication and reliability. To realize these advantages, there have been many process and design modifications to the IBC structures to achieve high efficiency. SunPower Corporation has been at the forefront of developing and manufacturing IBC diffused junction Si solar cell with demonstrated efficiency of 25% in their Gen III cell process [2]. After the first demonstration of an IBC cell using a-Si/c-Si heterojunction (SHJ) structure by Institute of Energy Conversion (IEC) in 2007 [3], several groups have improved processing enabling IBC-SHJ cell efficiency beyond 25% [4 – 6]. Recently Kaneka Corporation in Japan established the record single junction Si solar cell efficiency of 26.6% with the IBC-SHJ structure [6]. However, details of forming the interdigitated strips of p- and n-contacts are not disclosed in most of the reports from industry groups. They likely require multiple photolithography and alignment steps, which increases cost of manufacturing and reduces throughput. Commercialization prospect for high efficiency IBC cells is tempered by lack of simplified IBC design, patterning and fabrication processes.

Si IBC cells can be broadly categorized into two different technologies; a) the junction and base contacts formed by dopant diffusion (either conventional or laser), and b) deposited SHJ for both junction and base contact formation.

This paper provides an overview and presents alternative processing approaches for IBC Si solar cells.

II. IBC CELL DESIGN CRITERIA

Fig. 1 shows a schematic of a simple IBC Si structure on n-type c-Si wafer. The IBC cell requires high bulk quality wafers with excellent surface passivation and low minority carrier recombination velocity (S_f) at the front surface. This is due to the fact that all carriers which are generated near the illuminating front surface need to be diffused to the back IBC collecting junction and contacts. The carrier transport is inherently two dimensional so, as illustrated in Fig.1, the effective minority carrier diffusion length determined by S_f and recombination velocities at the emitter (S_e), base contact (S_b) and gap (S_g) need to be at least 3 times longer than the distance at any point to the nearest collecting junction (L_h) to realize 95% or higher collection efficiency. Moreover, the majority carrier transport length (L_e) needs to be short enough to achieve low series resistance for improved fill factor (FF). This explains why nearly all IBC cells have been made on n-type Si due to its higher diffusion length.

Regarding spatial dimensions, these requirements restrict the emitter (W_e), gap (W_g) and base contact (W_b) widths at the back surface. Poorer bulk and surface passivation quality leading to shorter effective minority carrier diffusion length imposes stricter processing tolerances with need for narrower widths for the rear strip dimensions. Additionally, unless the doped layers and metal are formed simultaneously and are self-aligned, they require multiple alignment steps.

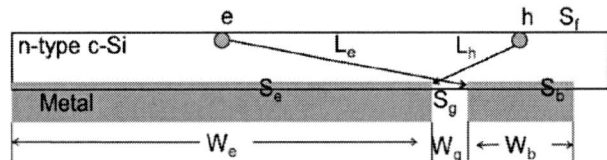

Fig. 1: Schematic of a simple IBC Si solar cell structure on n-type c-Si wafer, where the emitter junction and base contact regions are shown in blue and red respectively. The optimization of their respective widths (W) to achieve high efficiency depends on the relative S values for a given wafer resistivity and bulk lifetime.

III. Processing Approaches and Challenges

A. IBC-Si diffused junction solar cell

A front surface field (FSF) provided by a thin n+ diffused layer can improve the FF but its role is unclear. Some find that the highly doped diffused layer provides a lateral electron transport path with its low sheet resistance on the front surface, while others report that it keeps the cell operating at low injection level at the front surface [7, 8]. However, this FSF need to be optimized for high enough conductivity with low Auger recombination rate.

One of the advantages of the IBC cell is that the back surface can be optimized for a low J_0 junction and contacts independently to the front surface. For a diffused junction cell, the recombination in diffused regions is much higher than the rest of the bulk Si, which limits V_{OC}. Furthermore, emitter J_0 is significantly higher in the region where the diffused emitter contacts to the metal electrode [9]. SunPower has optimized their back surface junction by implementing a "passivated emitter contact" structure where the metal contacts the diffused emitter only through a small localized openings in a passivated dielectric to reduce emitter J_0 [9]. The dielectric layer also provides a better back reflection and light trapping. The base contact is also a point contact through the dielectric. This scheme allows them to achieve 730 mV V_{OC} and 25% efficiency. However, it introduces additional complexity to the patterning process.

B. IBC-Si heterojunction (SHJ) solar cell

The a-Si/c-Si SHJ structure provides excellent surface passivation and through-film carrier transport properties that lead to high V_{OC} with 750 mV reported in a standard front contacted solar cell [10]. The IBC-SHJ cells by Kaneka [6], Panasonic [4] and Sharp [5] exceed the SunPower Gen III cell performance with efficiency = 26.6%, 25.6% and 25.1% respectively, primarily due to higher V_{OC} of 744 mV, 740 mV and 736 mV respectively. One of the critical differences in the IBC-SHJ solar cell compared to the diffused junction IBC cell is that the metal coverage in the emitter junction strip needs to be almost 100% [11]. This is primarily due to 5 − 6 orders of magnitude higher lateral resistance in p-doped a-Si compared to a diffused layer. This large metal coverage increases probability of metal shunts and put restrictions on alignment tolerance with fine metal lines for base contacts.

IV. Future Directions and Alternative Processing

In the quest for high throughput precise interdigitated strip formation of emitter and base contacts, several processing methods that can potentially eliminate photolithography processes have shown promise. These include IBC cells with ion implanted emitter and base contacts, masked deposition of a-Si layers in conjunction with hot melt ink jet printing, laser

assisted ablation of sacrificial layers for IBC-SHJ cell and laser doping of emitter and base contact in IBC structure.

Ion implantation offers several advantages over the standard diffuse junction IBC solar cell including high throughput, high degree of uniformity, controlled depth and profile, and localized selective patterning capabilities [12]. These properties could greatly simplify the complexity in patterning, and number and cost of processing steps for IBC cell. A 21.2% efficient IBC solar cell was demonstrated using ion implanted poly-Si in combination with a tunnel oxide as carrier selective passivating contacts [13]. IBC cells with similar structure to the SunPower Gen III cell have been reported using ion planted junction and contact formation and achieved efficiency = 21.8% using an SiO_x / Al_2O_3 stack as dielectric passivation layer [14]. Thus, the ion implanted IBC cell shows promising efficiency, but above results were obtained using photolithography pattern of mask layers to perform localized ion implantation.

An IBC-SHJ cell with efficiency = 21.5% was fabricated with shadow masking of the PECVD a-Si doped layers followed by an ink jet printing of hot melt resist and chemical etching of interdigitated TCO/metal contacts [15]. The process is photolithography-free, but requires alignment steps to form both mask deposition and inkjet printing. Recently, the same group has further simplified the patterning process by eliminating one of the mask depositions of doped a-Si. A promising efficiency > 22.5% was achieved, while eliminating one of the alignment steps [16]. The base contact was formed by an interband Si tunnel junction, a concept that is widely used as interconnect layer between component cells in monolithic multi-junction solar cells.

Laser patterning approaches offer good precision, flexible pattern design and manufacturable high speed large area processing. A direct laser ablation of a-Si on a c-Si surface was found to degrade passivation quality even for a partial removal of a-Si. A sacrificial laser-absorbing layer on top of a-Si can be patterned by laser ablation without degrading the passivation quality. With a Si-rich a-SiN_X:H acting as a sacrificial laser-absorbing layer and as an etch barrier in subsequent process, a laser patterned IBC-SHJ solar cell using screen printed metal contacts had efficiency = 19% [17]. Using a similar laser patterning process with a bi-layer SiO_2 / p-type a-Si stack as sacrificial laser-absorbing layer, efficiency = 20.5% was reported [18]. However, these techniques also add to the complexity of processing in terms of additional sacrificial layer depositions and wet chemical etching steps. Ortega et al. reported efficiency > 20% using both laser fired emitter and base contact in IBC structure [19]. Both emitter / base contact strips had point emitter / contact formed by Al doping from Al_2O_3 and P doping from n-type a-Si using UV laser firing respectively. However, the interdigitated strips were formed by two photolithography and etching steps that are not compatible with high volume manufacturing.

Laser processes are an enabling technology for IBC because they are programmable, high speed and can drastically reduce number of processing steps. It can be used to create entire IBC pattern including emitter/base strips and their respective metal contacts with high precision due its ability of selective doping, selective metal transfer, dopant diffusion and localized annealing. Laser annealing can mitigate any laser process-induced crystal defects and recovers minority carrier lifetime [20]. An efficiency of 18.9% is reported in all-laser-based IBC solar cells, where a laser annealing step was crucial to mitigate process-induced damage and improve FF [21].

IEC is also developing laser based patterning approaches for IBC-SHJ solar cells. The progress towards simplification of processing steps and their device performance is summarized in Fig. 2. IBC-SHJ cell with an efficiency of 20.2% was achieved using three photolithography and two alignment steps (Fig. 2(a)). It is to be noted that the cell does not have any TCO layer, which is commonly used as a better back reflector and to lower the contact recombination rate at the metal-semiconductor interface between the doped a-Si and the metal contacts. This leads to low J_{SC} and V_{OC} in the completed cell.

The J-V curve shown in Fig. 2(b) shows an IBC-SHJ solar cell fabricated with simplified processing using one photolithography step and one alignment step. In this case, the p- and n-doped a-Si was patterned by a photolithography step without any alignment and subsequently the metal contacts were made by e-beam deposition using shadow mask after alignment. A laser fired contact (LFC) was made to n-type Si using a metal stack of Ti/Sb/Al to form the base contact [22]. An efficiency of 17.3% is achieved with an un-optimized laser firing process that degrades passivation at the base contact strip.

Fig.2 (c) shows the J-V curve of an IBC-SHJ solar cell with the simplest patterning process without any photolithography and alignment step. The emitter structure was formed on the entire back surface followed by a mask deposition of a Ti/Sb/Al stack. The LFC was performed on the narrower strip to form the base contact. A severe degradation of cell performance with an efficiency = 11.9% is observed. The implied V_{OC} estimated from quasi-steady state photo-conductance (QSSPC) decay before metallization was 726 mV. Such drastic degradation in V_{OC} and overall cell performance was identified as due to light and voltage dependent parasitic shunting arising from the induced inversion layer in c-Si. Simulation and experimental proof of this result is presented in another work in this conference [23].

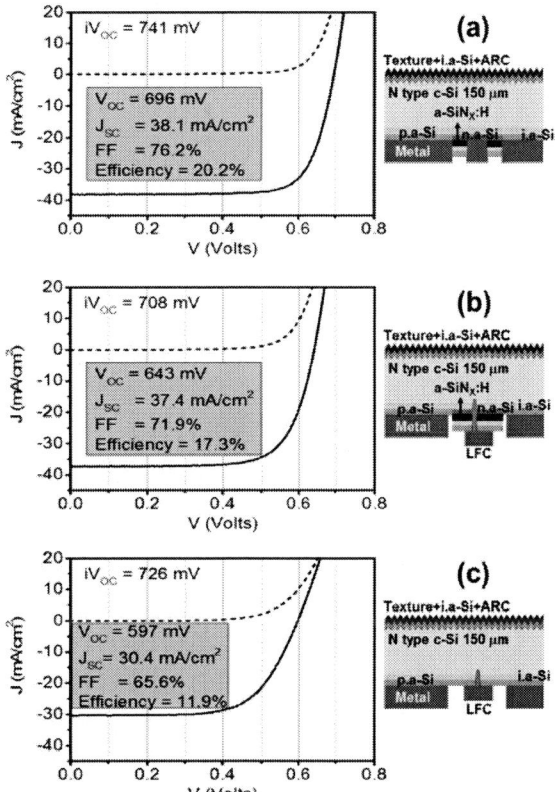

Fig. 2: Dark (dashed) and light (solid) J-V curves of 1.6 cm^2 IBC-SHJ solar cells fabricated using (a) three photolithography and two alignment steps, (b) one photolithography and one alignment step with LFC, (c) no photolithography and no alignment steps with LFC through i/p.a-Si emitter layers. The implied V_{OC} (iV_{OC}) was estimated from QSSPC measurements before metallization. Schematic structures of each processing are shown in the right.

V. CONCLUSION

An overview of current high efficiency IBC solar cell is presented. The IBC-SHJ solar cell performance has advanced significantly in the past few years with efficiency approaching the theoretical limit of single junction Si solar cell. However, the IBC cell processing will require a much simplified high speed formation of interdigitated contacts to increase the production throughput and lower the cost. Such processing would likely need to be photolithography-free with no alignment. Programmable approaches like laser-based or inkjet-based processing seems promising, but have not yet demonstrated high efficiency.

ACKNOWLEDGEMENT

This work is funded by US Department of Energy PVRD program under contract number DE-EE-0007534.

REFERENCES

[1] M. D. Lammert and R. J. Schwartz, "The interdigitated back contact solar cell: A silicon solar cell for use in concentrated sunlight," *IEEE Transactions on Electron Devices*, vol. ED-24, pp. 337-342, 1977.

[2] D. D. Smith, P. Cousins, S. Westerberg, R. D. Jesus-Tabajonda, G. Aniero and Y. Shen, "Towards the practical limits of silicon solar cells," *IEEE J. Photovoltaics*, vol. 4, pp. 1465-1469, 2014.

[3] M. Lu, S. Bowden, U. Das and R. Birkmire, "Interdigitated back contact silicon heterojunction solar cell and the effect of front surface passivation," *Appl. Phys. Lett.*, vol. 91, pp. 063507, 2007.

[4] K. Masuko, M. Shigematsu, T. Hashiguchi, D. Fujishima, M. Kai, N. Yoshimura, T. Yamaguchi, Y. Ichihashi, T. Mishima, N. Matsubara, T. Yamanishi, T. Takahama, M. Taguchi, E. Maruyama and S. Okamoto, "Achievement of more than 25% conversion efficiency with crystalline silicon heterojunction solar cell," *IEEE J. Photovoltaics*, vol. 4, pp. 1433-1435, 2014.

[5] J. Nakamura, N. Asano, T. Hieda, C. Okamoto, H. Katayama and K. Nakamura, "Development of heterojunction back contact Si solar cell," *IEEE J. Photovoltaics*, vol. 4, pp. 1491-1495, 2014.

[6] K. Yoshikawa, H. Kawasaki, W. Yoshida, T. Irie, K. Konishi, K. Nakano, T. Uto, D. Adachi, M. Kanematsu, H. Uzu and K. Yamamoto, "Silicon heterojunction solar cell with interdigitated back contacts for a photoconversion efficiency over 26%," *Nature Energy*, vol. 2, pp. 17032, 2017.

[7] F. Granek, M. Hermle, D. M. Huljic, O. Schultz-Wittmann and S. W. Glunz, "Enhanced lateral current transport via the front surface n$^+$ diffused layer of n-type high efficiency back junction back contact silicon solar cells," *Prog. Photovolt: Res. Appl.*, vol. 17, pp. 47 – 56, 2000.

[8] T. Ohrdes, U. Romer, Y. Larionova, R. Peibst, P. P.Altermatt and N. P. Harder, "High fill factors of back-junction solar cells without front surface field diffusion," *Proc. Of 27th Euro PVSEC*, pp. 866 – 869, 2012.

[9] P. J. Cousins, D. D. Smith, H. C. Luan, J. Manning, T. D. Dennis, A. Waldhauer, K. E. Wilson, G. Harley and W. P. Mulligan, "Generation 3: Improved performance at lower cost," *Presented at 35th IEEE PVSC*, 2010.

[10] M. Taguchi, A. Yano, S. Tohoda, K. Matsuyama, Y. Nakamura, T. Nishiwaki, K. Fujita and E. Maruyama, "24.7% record efficiency HIT solar cell on thin silicon wafer," *IEEE J. Photovoltaics*,vol. 4, pp. 96-99, 2014.

[11] J. Haschke, Y. Y. Chen, R. Gogolin, M. Mews, N. Mingirulli, L. Korte and B. Rech, "Approach for a simplified fabrication process for IBC-SHJ solar cells with high fill factors," *Energy Procedia*, vol. 38, pp. 732 – 736, 2013.

[12] A. Rohatgi, D. L. Meier, B. McPherson, Y. W. Ok, A. D. Upadhyaya, J. H. Lai and F. Zimbardi, "High throughput ion-implantation for low-cost high-efficiency silicon solar cells," *Energy Procedia*, vol. 15, pp. 10 – 19, 2012.

[13] G. Yang, A. Ingenito, O. Isabella and M. Zeman, "IBC c-Si solar cells based on ion-implanted poly-silicon passivating contacts," *Sol. Ener. Mat, and Sol. cells*, vol. 158, pp. 84 – 90, 2016.

[14] R. Muller, C. Reichel, J. Benick and M. Hermle, "Ion implantation for all-alumina IBC solar cells with floating emitter," *Energy Procedia*, vol. 55, pp. 265 – 271, 2014.

[15] A. Tomasi, B. Paviet-Salomon, D. Lachenal, S. M. de Nicolas, A. Descoeudres, J. Geissbuhlar, S. De Wolf and C. Ballif, "Back-contacted silicon heterojunction solar cells with efficiency >21%," *IEEE J. of Photovoltaics*, vol. 4, pp. 1046 – 1054, 2014.

[16] A. Tomasi, B. Paviet-Salomon, K. Jeangros, J. Haschke, G. Christmann, L. Barraud, A. Descoeudres, J. P. Seif, S. Nicolay, M. Despeisse, S. De Wolf and C. Ballif, "Simple processing of back-contacted silicon heterojunction solar cells using selective-area crystalline growth," *Nature Energy*, vol. 2, pp. 17062, 2017.

[17] T. Desrues, S. De Vechhi, F. Souche, D. Munoz and P. J. Ribeyron, "SLASH concept: A novel approach for simplified interdigitated back contact solar cells fabrication," *Proc. Of 38th IEEE PVSC*, pp. 1602 – 1605, 2012.

[18] S. Harrison, O. Nos, G. D'Alonzo, C. Denis, A. Coll and D. Munoz, "Back contact heterojunction solar cells patterned by laser ablation," *Energy Procedia*, vol. 92, pp. 730 – 737, 2016.

[19] P. Ortega, G. Lopez, D. Munoz, I. Martin, C. Voz, C. Molpeceres and R. Alcubilla, "Fully low temperature interdigitated back-contacted c-Si(n) solar cells based on laser-doping from dielectric stacks," *Sol. Ener. Mat, and Sol. cells*, vol. 169, pp. 107 – 112, 2017.

[20] Z. Sun and M. C. Gupta, "Laser annealing of silicon surface defects for photovoltaic applications," *Surf. Science*, vol. 652, pp. 344 – 349, 2016.

[21] Z. Sun and M. C. Gupta, "Laser annealing to enhance performance of all-laser-based silicon back contact solar cells," *in this conference*, 2017.

[22] J. He, S. Hegedus, U. Das, Z. Shu, M. Bennett, L. Zhang and R. Birkmire, "Laser-fired contact for n-type crystalline Si solar cells," *Prog. Photovolt: Res. Appl.*, 2014; DOI: 10.1002/pip.2520.

[23] L. Zhang, U. Das and S. Hegedus, "Gap passivation structure for scalable n-type interdigitated all back contact silicon hetero-junction solar cell," *in this conference*, 2017.

Fabrication of CuI/a-Si:H/c-Si Structure for Application to Hole-selective Contacts of Heterojunction Si Solar Cells

Kazuhiro Gotoh[1], Min Cui[1], Nguyen Cong Thanh[2], Koichi Koyama[2], Isao Takahashi[1], Yasuyoshi Kurokawa[1], Hideki Matsumura[2], Noritaka Usami[1]

[1] Department of Materials Process Engineering, Graduate School of Engineering,

Nagoya University, Furocho, Chikusa, Nagoya 464-8603, Japan

[2] Center for Nano Materials and Technology, Japan Advanced Institute of Science and Technology,

1-1 Asahidai, Nomi, Ishikawa 923-1292, Japan

Abstract — We report on structural and optical properties of copper iodide (CuI) on crystalline Si (c-Si) and properties of the diode with intrinsic type (i-type) hydrogenated amorphous Si (a-Si:H) at CuI/c-Si interface. The i-type a-Si:H layer was deposited on n-type c-Si by catalytic chemical vapor deposition (Cat-CVD) prior to deposition of CuI by spin-coating method. Higher rotational speed of 4000 rpm was found to be useful to suppress generation of large particles and absorption by CuI. The effective carrier lifetime was characterized for CuI/a-Si:H/c-Si/a-Si:H/CuI structures and it strongly depended on a-Si:H. The CuI/a-Si:H/c-Si diodes showed rectification behavior.

Index Terms — copper iodide, hole-selective contacts, catalytic chemical vapor deposition, hydrogenated amorphous silicon, charge carrier lifetime, photovoltaic cells.

I. INTRODUCTION

Recently, heterojunction crystalline silicon (c-Si) solar cells employing transition metal oxides (TMO) such as MoO_x [1-3] have been much attractive. The TMO are used as carrier-selective contacts instead of conventional doped a-Si:H layer due to their wide bandgap larger than 3 eV, leading to increase in short-circuit current density. The MoO_x works as hole-selective contacts, since they possess high work function than 5 eV, which leads to hole selection at the interface of MoO_x/c-Si. These state-of-arts Si solar cells widen material candidates for heterojunction Si solar cells.

Copper iodide (CuI) is p-type transparent conductive material with bandgap energy of 3.1 eV [4,5]. CuI/Si heterojunction shows almost ideal band lineup for extraction of holes in virtue of large conduction band offset of about 2 eV at the CuI/c-Si interface, since ionization energy and electron affinity are 5.2 and 2.1 eV for CuI [4] and 5.1 and 4.0 eV for Si, respectively. Additionally, CuI can be deposited by using liquid phase method such as spin-coating and doctor-blading, which has possibility to reduce process cost of solar cells.

Surface passivation is the key issue for high conversion efficiency of c-Si solar cells [6-8]. So far, conversion efficiency of 19.4% is obtained by employing intrinsic type (i-type) hydrogenated amorphous Si (a-Si:H) at the TMO/c-Si interface, i.e. MoO_x/a-Si:H(i)/c-Si [1]. It is believed that the i-type a-Si:H shows high performance of passivation by hydrogenation of dangling bonds [9,10]. In case of CuI prepared by spin-coating, the interface of CuI/c-Si would be not well passivated. Passivation layer should be employed between CuI and c-Si. In this study, we employed a-Si:H as a passivation layer and characterized effective lifetime of CuI/a-Si:H/c-Si structure. Furthermore, the p-CuI/n-Si diodes utilizing a-Si:H interlayer were fabricated.

II. EXPERIMENTS

Double-side polished, floating zone (Fz) grown n-type c-Si was used as a substrate. The resistivity of 2.8 Ω·cm was measured by four terminal method. Prior to deposition of a-Si:H, RCA cleaning was carried out and then the substrates were dipped into 2.5% HF to remove oxide layer. The a-Si:H was deposited by using a catalytic chemical vapor deposition (Cat-CVD) [11,12]. For characterization of effective lifetime, 7-nm-thick i-type a-Si:H was deposited on both sides of substrates. The growth temperature, flow rate of SiH_4 and total pressure were 90 °C, 20 sccm and 0.85 Pa, respectively. For fabricating diodes, n-type a-Si:H were deposited on rear side of a-Si:H (i)/c-Si/a-Si:H (i) samples. The growth temperature, total pressure and layer thickness were 250 °C, 1.88 Pa and 30 nm, respectively. The flow rate of SiH_4, PH_3 and H_2 were 10, 0.099 and 50 sccm, respectively. Following deposition of a-Si:H, CuI is deposited by spin-coating method. The rotational speed was changed from 1000 to 4000 rotation per minutes (rpm) in order to vary layer thickness. After coating CuI, the substrates were heated at 80°C to evaporate solvent completely. The CuI solution of 0.1 mol/L was prepared by melting CuI powder into the solvent which di-n-propyl sulfide and chlorobenzene were mixed in the ratio of 1 to 39. Indium tin oxide (ITO) and silver electrode was deposited on the sample to fabricate diode after depositing CuI. The schematic structure is depicted in Fig. 1.

Surface morphology and optical properties of samples were characterized by scanning electron microscopy (SEM) and spectrophotometer, respectively. The effective lifetime was measured by micro photoconductivity decay (μ-PCD) method.

978-1-5090-5606-4/17 $31.00 © 2017 IEEE

The wavelength of pulsed light and frequency of microwave were 904 nm and 10 GHz, respectively. For characterizing electric properties of diode, current-voltage measurement was carried out.

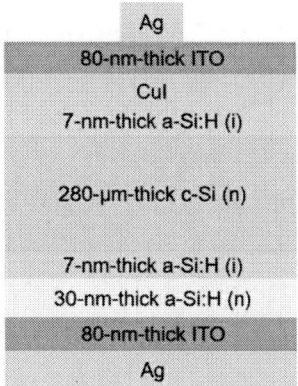

Fig. 1. Schematic structure of CuI/a-Si:H(i)/c-Si/a-Si:H(i)/a-Si:H(n) diode.

III. RESULTS AND DISCUSSION

Figure 2 shows surface SEM images of CuI deposited c-Si with the rotational speed of (a) 1000 rpm and (b) 4000 rpm. The insets show magnified images for (a) and (b). Grains composed of fine particles with size of few tens of nm were observed in insets of (a) and (b). Additionally, particles of few μm and hollows around the particles were observed in (a), while there were no such particles in (b). The particles would be generated by undissolved CuI particles in the CuI solution. Almost all CuI solution including undissolved particles can be flicked off Si substrates by employing higher rotational speed, while lower rotational speed preserves much solution including the undissolved particles. The undissolved particles would be nuclei of the large particles after heating process at 80°C. The hollows in (a) suggest that the undissolved particles grew larger by using surrounding CuI solution on Si.

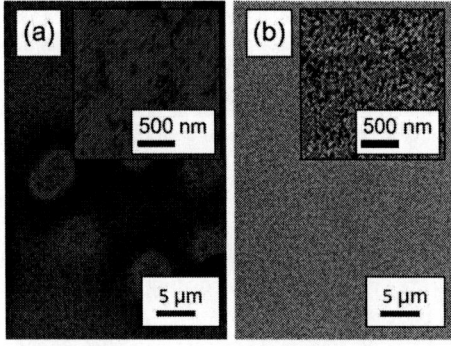

Fig. 2 Surface SEM images for CuI coated c-Si(100) substrates. The rotational speed is 1000 rpm for (a) and 4000 rpm for (b).

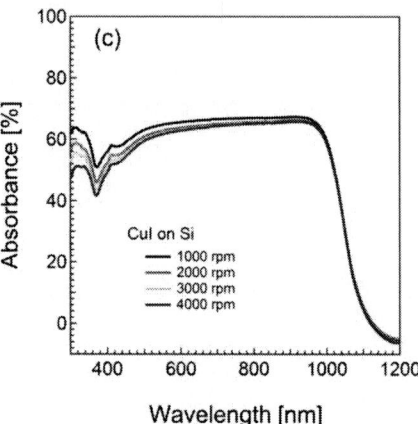

Fig. 3. Absorbance spectra of CuI deposited c-Si. The rotational speed is varied from 1000 to 4000 rpm.

Figure 3 shows absorbance spectra of CuI coated c-Si by spin-coating with the rotational speed of 1000, 2000, 3000 and 4000 rpm. With increasing rotational speed, absorbance decreases possibly due to thinner layer thickness. Generally, layer thickness (t) formed by the spin-coating method follows rotational speed (ω) inversely, i.e. $t \propto \omega^{-1}$. According to separate experiments using glass substrates, absorption edge of CuI was found to be ~420 nm. Thus, the decrease in absorbance from 300 to 420 nm was caused by reduced absorption in CuI owing to thinner layer thickness, while the decrease in absorbance from 420 to 1000 nm would be related to anti-reflection effect owing to lower refractive index of 2.3-2.6 for CuI [5] in comparison with that of 3.5 for c-Si.

Figure 4 shows effective lifetime (τ_{eff}) of CuI/c-Si/CuI, a-Si:H/c-Si/a-Si:H and CuI/a-Si:H/c-Si/a-Si:H/CuI structures as a function of rotational speed. Poor passivation performance of around 10 μs was observed for CuI/c-Si/CuI, indicating there are many recombination sites at the CuI/c-Si interface. The τ_{eff} was about 300-400 μs before CuI deposition, whereas it decreased to 100-200 μs after CuI deposition. Although further study is necessary, we speculate that the reduction of τ_{eff} was caused by relatively high concentration of free carriers in CuI. It is noted that the τ_{eff} after CuI deposition depends on the τ_{eff} before CuI deposition, suggesting that passivation performance of CuI/c-Si interface relies on passivation layer strongly.

Figure 5 shows current density-voltage (J-V) curves of CuI/a-Si:H(i)/c-Si(n)/a-Si:H(i)/a-Si:H(n) diodes in dark. CuI was coated at 4000 rpm. Rectification was observed for, the samples, indicating CuI worked as p-layer. The rectification, however, was not high because of large saturation current of the order of 10^{-1} mA/cm². Photocurrent was observed at reverse bias voltage under 1 sun condition. These results

978-1-5090-5606-4/17 $31.00 © 2017 IEEE

indicate majority carriers are somehow collectable to external circuit, while minority carrier cannot be collected possibly due to large recombination current and/or high resistivity of CuI films. The external quantum efficiency (EQE) response at bias-voltage of -0.25 V is shown in Fig. 6. Though the EQE value is still not enough for high efficiency solar cells, the response was obtained from 300 to 1200 nm. The reduction of EQE at shorter wavelength indicates difficulty of collecting photo-generated carriers to external circuit. It is presumably caused by partially uncovered CuI on Si as shown in Fig. 2 or parasitic absorption by CuI due to its nature of direct bandgap [5]. Although further improvement of CuI films should be needed, these results suggests the CuI formed by spin-coating is a promising candidate as a hole-selective contact.

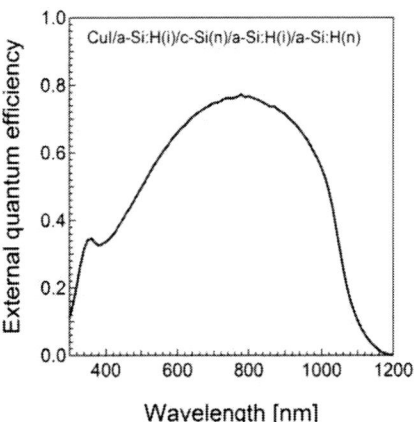

Fig. 6   EQE response of CuI/a-Si:H(i)/c-Si(n)/a-Si:H(i)/a-Si:H(n) diodes at bias voltage of -0.25 V.

IV. SUMMARY

Structural and optical properties of CuI deposited on c-Si by spin-coating method were investigated for application to hole-selective contacts of heterojunction c-Si solar cells. The i-type a-Si:H layer by Cat-CVD was utilized as the passivation layer at the CuI/c-Si interface. The rotational speed was varied in order to change layer thickness of CuI. There were no large particles by using higher rotational speed of 4000 rpm possibly due to less amount of CuI solution and absence of undissolved particles in the solution on Si. With increasing rotational speed, absorption by CuI decreased owing to thinner layer thickness. Furthermore, CuI were deposited on both sides of a-Si:H/c-Si/a-Si:H and effective lifetime were characterized. The effective lifetime relied on that before CuI deposition. Additionally, the diodes using CuI as hole-selective contacts were fabricated and it indicated rectification behavior and EQE response.

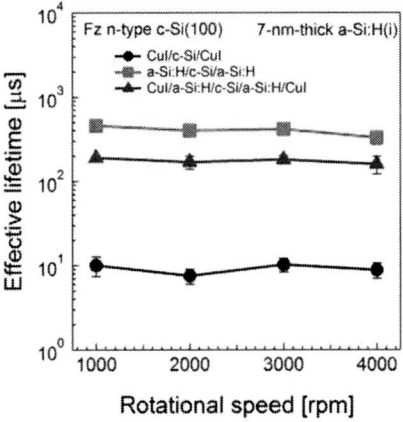

Fig. 4.   Effective lifetime of CuI/c-Si/CuI, a-Si:H/c-Si/a-Si:H and CuI/a-Si:H/c-Si/a-Si:H/CuI samples as a function of rotational speed.

ACKNOWLEDGEMENT

This work was financially supported by The New Energy and Industrial Technology Development Organization (NEDO) of Japan.

REFERENCES

[1] J. Bullock, M. Hettick, J. Geissbühler, A. J. Ong, T. Allen, C. M. Sutter-Fella, T. Chen, H. Ota, E. W. Schaler, S. De Wolf, C. Ballif, A. Cuevas, and A. Javey, "Efficient silicon solar cells with dopant-free asymmetric heterocontacts" Nat. Energy **15031** (2015) 1-7.

[2] L. G. Gerling, S. Mahato, A. Morales-Vilches, G. Masmitja, P. Ortega, C. Voz, R. Alcubilla, and J. Puigdollers, "Transition

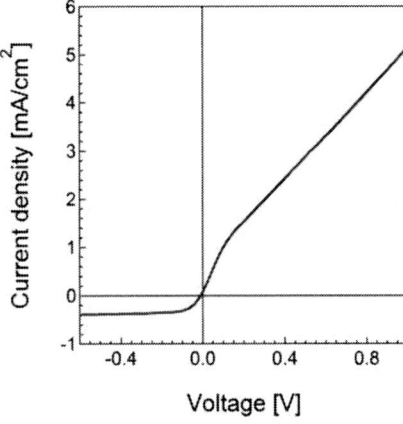

Fig. 5.   Current density-voltage curves of CuI/a-Si:H(i)/c-Si(n)/a-Si:H(i)/a-Si:H(n) diodes in dark. CuI was coated at 4000 rpm.

metal oxides as hole-selective contacts in silicon heterojunction solar cells" Sol. Energy Mater. Sol. Cells **145** (2016) 109-115.

[3] X. Yang, Q. Bi, H. Ali, K. Davis, W. V. Schoenfeld, and K. Weber "Hihg-performance TiO2-Based Electron-Selective Contacts for Crystalline Silicon Solar Cells" Adv. Mater. **28** (2016) 5891-5897.

[4] F-L. Shein, H. Wenckstern, and M. Grundmann. "Transparent p-CuI/ZnO heterojunction diodes" Appl. Phys. Lett. **102**, 092109 (2013) 1-4.

[5] M. Grundmann, F-L. Schein, M. Lorenz, T. Böntgen, J. Lenzner, and H. Wenckstern, "Cuprous iodide – a p-type transparent semiconductor: history and novel application" Phys. Status. Solidi. A **210** (2013) 1671-1703.

[6] M. Bivour, B. Macco, J. Temmler, W. M. M. (Erwin) Kessels, and M. Hermle, "Atomic layer deposited molybdenum oxide for hole-selective contact of silicon solar cells" Energy Procedia **92** (2016) 443-449.

[7] A. Bozzola, P. Kowalczewske, and L.C. Andreani, "Towards high efficiency thin-film crystalline silicon solar cells: The roles of light trapping and non-radiative recombinations" J. Appl. Phys. **115**, 094501 (2014).

[8] A. G. Aberle, "Surface Passivation of Crystalline Silicon Solar Cellls: A Revies" Prog. Photovolt. Res. Appl. **8** (2000) 473-487.

[9] S. De Wolf, S. Olibet, and C. Ballif "Stretched-exponential a-Si:H/c-Si interface recombination decay" Appl. Phys. Lett. **93**, 032101 (2008).

[10] S. De Wolf, and M. Kondo "Nature of doped a-Si:H/c-Si interface recombination" J. Appl. Phys. **105**, 103707 (2009).

[11] H. Matsumura, T. Hasegawa, S.Nishizaki, and K. Ohdaira "Advantages of plasma-less deposition in Cat-CVD to the performance of electric devices" Thin Solid Films **519** (2011) 4568-4570.

[12] H. Matsumura and K. Ohdaira, "Cat-CVD (Caralytic-CVD); Its Fundamentals and Application" ECS Trans. **25** (2009) 53-63.

Characteristics of Thin Crystalline Silicon Solar Cells with Rib Structure

Yukimi Ichikawa[1,2], Shuhei Yoshiba[1], Masakazu Hirai[1], and Makoto Konagai[2]

[1] FUTURE-PV Innovation, JST, 2-2-9 Machiike-dai, Koriyama, Fukushima 963-0215, Japan
[2] Tokyo City University, 8-15-1 Todoroki Setagaya-ku, Tokyo 158-0082, Japan

Abstract — To improve the conversion efficiency of Si solar cells, we have developed a thin Si wafer cell that uses a rib structure. We fabricated a lattice-shaped rib structure on the rear side of a thin Si wafer to improve the wafer's strength. A wafer with thickness of 30 μm was prepared easily using this structure. Fabricating a Si heterojunction on the rib wafer, we demonstrated that a high open-circuit voltage (Voc) could be obtained by thinning the wafer without sacrificing its strength. Experimental and simulation results showed that the rib structure thin wafer could be a candidate for further improvement in conversion efficiency.

Index Terms — crystalline silicon, hetero-junction, open-circuit voltage, thin wafers.

I. INTRODUCTION

To improve the conversion efficiency of single crystalline silicon solar cells, several technologies, including heterojunctions, interdigitated back contacts, and optical confinement structures, have been developed. In addition to these technologies, the wafer thickness remains an important factor for increased efficiency. The open-circuit voltage (V_{OC}) is well known to increase with decreasing wafer thickness [1,2]. Stanislau et al. reported that the excess carrier density increases in thin solar cells with adequate passivation and optical confinement, and this leads to increased V_{OC} [2]. In contrast, the short-circuit current density (J_{SC}) decreases for thinner solar cells, but this effect can be minimized using good optical confinement structures. As a result, there is an optimum wafer thickness that leads to high cell conversion efficiency; this thickness is predicted to be 100 μm or less.

It is, however, very difficult to handle such thin wafers during the cell fabrication processes, and the production yield can be very low because of wafer breakage and warpage. To prevent this, it is necessary to attach some type of support substrate to the thin wafer, but this leads to high process costs. In addition, the wafer cost rises with decreasing thickness in very thin wafers because of their low production yields. As a result, the use of very thin wafers (<100 μm) increases the cost of the final solar cell.

To overcome these difficulties, we developed thin silicon solar cells that use a rib structure. We applied a lattice-shaped rib to the back side of the wafer. Since strict control of these configurations is not required in solar cells, we used laser scribing and alkali etching techniques to form the lattice-shaped rib structure used in this work. Subsequently, amorphous silicon layers were deposited on the wafer to form a heterojunction solar cell.

In this paper, we present the process to fabricate the solar cells with rib structures and the fundamental photovoltaic characteristics including the results of device simulation.

II. FABRICATION OF RIB WAFER

A scanning electron microscope (SEM) image of the fabricated rib structure wafer is shown in Fig. 1. The details of the fabrication processes for the rib structure were described in Ref. [3]. Here, we describe them briefly. To fabricate the thin wafers, we provided 4-in n-type silicon wafers with thickness of 280 μm; the typical wafer resistivity was 2–3 Ωcm and the planar orientation was (100). Silicon nitride films were first deposited on both surfaces of the wafers by sputtering; the nitride film thickness was 100 nm and these films were used as etching masks.

To form the lattice-shaped rib pattern on the back side of the wafer, we used a laser scriber to scribe the silicon nitride film. The laser beam's output power and wavelength were 6 W and 532 nm, respectively and the scribing line width was approximately 80 μm. The rib was aligned in a direction parallel to the (110) orientation and the rib width was varied from 100 μm to 200 μm. The size of the rectangular segment that was surrounded by the rib structure was 1 mm × 2 mm. To etch the silicon within this area, we formed a pattern that consisted of multiple parallel scribing lines; the line pitch was 200 μm and the line direction was at 45° to the rib, i.e., a (100) orientation. Of course, it is easy to remove the whole silicon nitride layer from this area by laser scribing or

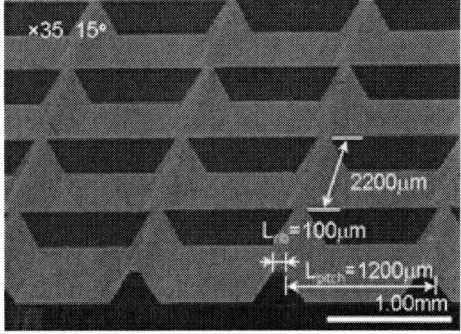

Fig. 1. SEM image of back-side Si wafer with rib structure.

photolithographic processes, but when this multi-line pattern was used, a flat etching surface was obtained across this area because of side-etching. Because the rib pattern is formed parallel to the (110) orientation, the side-etching effect is negligibly small because the (111) plane appears; the etching rate of the (111) plane is two orders of magnitude smaller than that of the (100) orientation plane or that of the (110) orientation plane [4]. The etching was performed using 30% KOH diluted with water at 80°C.

Using this structure, a wafer thickness of 30 μm was easily fabricated. Even in such thin rib wafers, no breakage and no warpage occur, and these wafers had sufficient strength to handle for cell fabrication.

III. RESULTS AND DISCUSSION

A. Simulation

To understand the device properties of the thin solar cells with the rib structure, we carried out a two-dimensional device simulation using SILVACO software. In the two-dimensional simulation modeling, we used a hetero-junction structure and defined the device configuration as shown in Fig. 2. Fig. 3 shows the photovoltaic parameters, V_{OC}, J_{SC}, Fill factor (F.F.), and Efficiency (Eff.), as functions of wafer thickness, W_{wafer}, for the rib structure and the flat wafers. In this simulation, the pitch, L_{pitch}, the initial wafer thickness, $W_{initial}$ and the width of the rib, L_{rib}, were 1200 μm, 280 μm and 100 μm, respectively, and no optical confinement produced by the surface textured structure was applied. Additionally, it was assumed that the surface recombination velocity is zero and the carrier lifetime required to calculate the SRH recombination rate for both electrons and holes is 3 ms in this simulation

When the wafer thickness is 100 μm or larger, the photovoltaic parameters of the rib wafer is in good agreement with those of the flat wafer. However, discrepancy in those parameters appears between the two structures for thinner wafers because the carrier recombination in the rib region becomes dominant, and as a result, the V_{OC} is low as compared with the corresponding value of the flat wafer. For the J_{SC}, however, photoabsorption in the rib region is added to that which occurs in the thin wafer region. As a result, the current in the rib structure increases.

To suppress the influence of the rib region on the

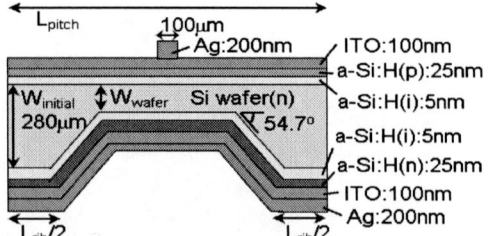

Fig. 2. Cross-sectional schematic diagram of hetero-junction solar cell with rib structure used for simulation.

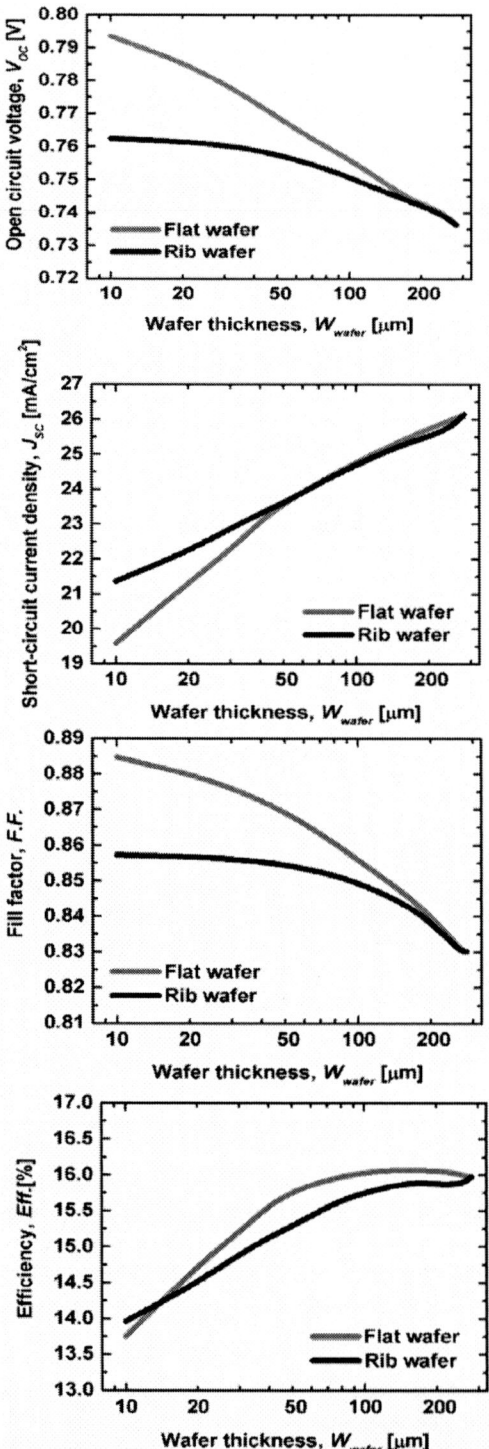

Fig. 3. Photovoltaic parameters obtained from two-dimensional simulation by SILVACO are plotted as a function of wafer thickness, W_{wafer}, for rib wafer cell and flat wafer cell. The dimension of rib structure was L_{pitch}=1200 μm and L_{rib}=100 μm, and it was assumed that the carrier lifetime of Si wafer is 1 ms and the surface recombination velocity, S, is zero.

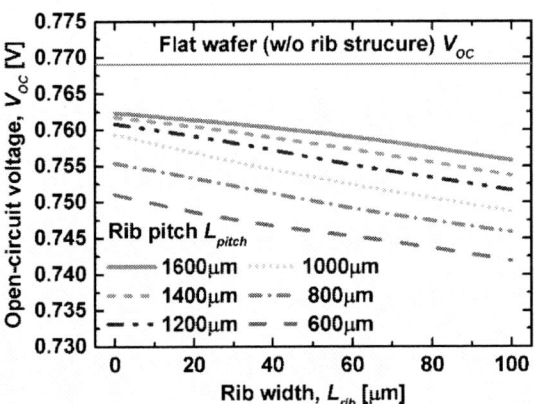

Fig. 4. Simulation results for open-circuit voltage. V_{OC}, as a function of rib width, L_{rib} for Wwafer=50 μm and Winitial=280 μm.

photovoltaic parameters, the volume ratio of the rib region to the thin wafer region must be reduced. In Fig. 4, the V_{OC} obtained from the simulation is plotted as a function of rib width for several L_{pitch}; the thickness of the thin wafer region was set at 50 μm. The results show that a higher V_{OC} value is obtained when the volume fraction of the rib region is smaller, and it approaches that of the flat wafer structure (0.769 V in this case). Therefore, it is very important to find an optimum configuration of the rib wafer from the viewpoints of wafer strength and photovoltaic performance.

B. Fabrication of Solar Cells

To demonstrate the effectiveness of rib wafer solar cells, we fabricated heterojunction solar cells on the rib wafers and measured their photovoltaic performances. The details of the rib wafer provided are listed in TABLE 1. The cell structure fabricated is schematically shown in Fig. 5. Microcrystalline silicon oxide (μc-SiO:H) was used as the p-layer and

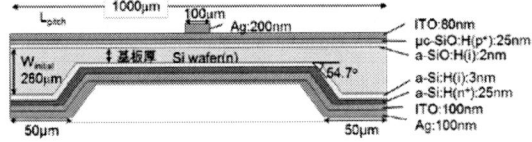

Fig. 5. Schematic diagram of fabricated rib wafer solar cell.

TABLE 1
Properties of Si wafers used for cell fabrication.

Wafer Diameter	4 inches
Wafer Thickness	280 μm
Resistivity of Wafer	2-3 Ωcm (n-type)
Lifetime of Carriers	1-2 ms
Size of One Segment Surrounded by Ribs	1 mm x 2 mm
Rib Width	100 μm

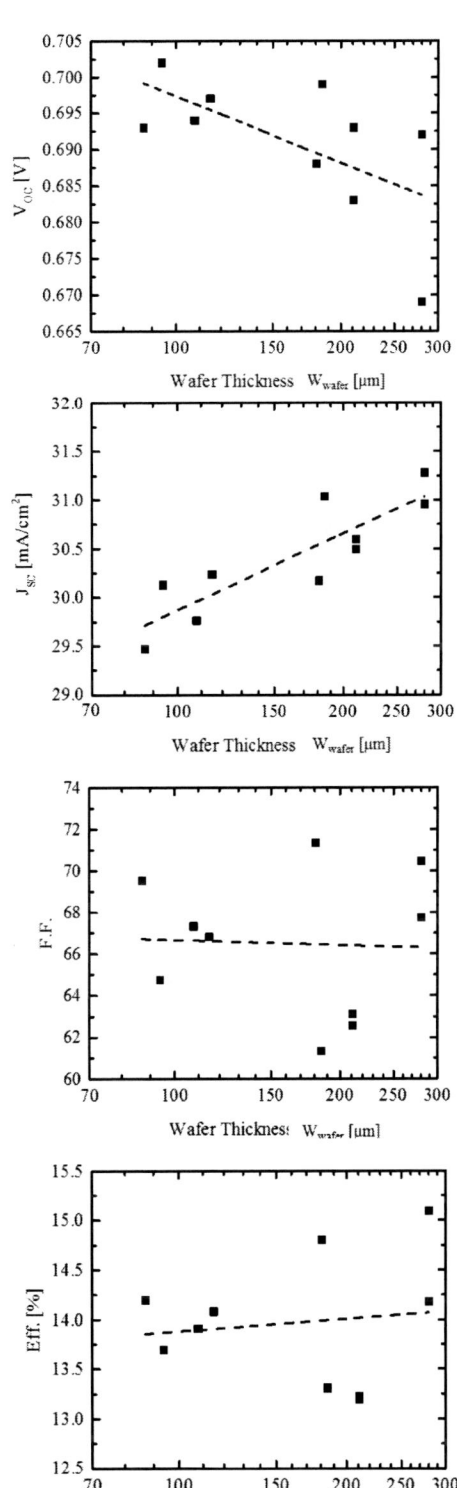

Fig. 6. Photovoltaic parameters of fabricated rib wafer heterojunction solar cells as a function of wafer thickness, W_{wafer}.

TABLE 2
Film deposition conditions for plasma CVD

Film	μc-SiO:H	a-SiO:H	a-Si:H	a-Si:H
Type	p	intrinsic	intrinsic	n
RF Power [W]	30	30	30	20
H_2 [sccm]	340	85	50	90
SiH_4 [sccm]	2	15	10	10
PH_3 or B_2H_6 [sccm]	1	-	-	10
CO_2 [sccm]	0.5	7	-	-
Susceptor Temp. [°C]	260	260	260	260
Pressure [Pa]	90	60	60	60
Thickness [nm]	25	2	3	25

conventional a-Si:H was used as the n-layer. These layers ware deposited by plasma CVD. Deposition conditions for these layers are listed in TABLE 2.

The cell size was 1 cm x 1 cm and the wafer thickness, W_{wafer}, was varied from 280 μm to 87 μm. In Figs. 6, the photovoltaic parameters measured under AM1.5 (100mW/cm²) are shown as a function of Wwafer. The open-circuit voltage, V_{OC}, increases with decreasing W_{wafer} as is expected. In contrast, the short-circuit current density, J_{SC}, decreases as the W_{wafer} reduces. The fill factor, F.F., is independent of the W_{wafer}. As a result, the conversion efficiency does not vary very much for W_{wafer}.

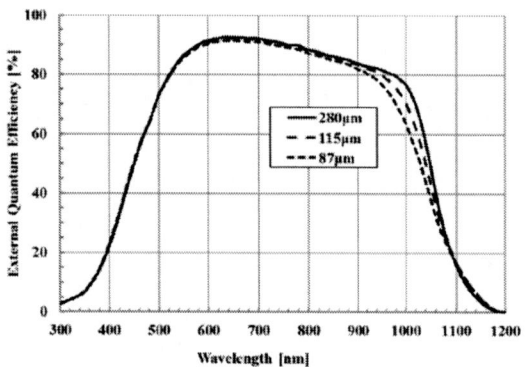

Fig. 7. External quantum efficiency of fabricated rib wafer solar cells.

C. Discussion

Experimental results demonstrated that the rib wafer cell can achieve both high V_{OC} and high wafer strength by reducing the wafer thickness. The reason why the efficiency does not improve in thinner cells is decrease in J_{SC}. We have not applied texture structures to fabricate cells Thus the external quantum efficiency in longer wavelength range reduces with decreasing W_{wafer} in as shown in Fig. 7. If the texture structures were applied, the J_{SC} lowering could be avoided, and the efficiency is expected to increase with decreasing W_{wafer}. Since the rib structure solar cells use the (100) wafer, conventional optical confinement based on surface texturization can be used.

Another point is design optimization of the rib structure. From the viewpoint of thin wafer solar cells, it is preferable that the volume of ribs is negligibly small as compared with the etched thin segments. In contrast, the strength of the wafer significantly depends on the dimension of ribs. Hence there is some optimal dimension for the rib pattern to obtain higher efficiencies.

We have optimized the dimensions of the rib structure and the fabrication process to produce a high-performance rib structured solar cell, and the results will be published in near future.

ACKNOWLEDGEMENT

This work was supported by "FUTURE-PV Innovation" Project of The Ministry of Education, Culture, Sports, Science and Technology.

REFERENCES

[1] S. Hargreavesa, L. E. Blacka, D. Yana and A. Cuevasa, Energy Procedia *38*, 66 (2013).
[2] S. Y. Herasimenka, W. J. Dauksher, and S. G. Bowden, Appl. Phys. Lett. *103*, 053511 (2013).
[3] S. Yoshiba, M. Hirai, Y. Abe, M. Konagai and Y. Ichikawa, AIP ADVANCES *7*, 025104 (2017).
[4] M. Shikida, K. Sato, K. Tokoro and D. Uchikawa, Sensors and Actuators *80*, 179 (2000).

Measurement of TiO₂/p-Si Selective Contact Performance using a Heterojunction Bipolar Transistor with a Selective Contact Emitter

Janam Jhaveri, Alexander Berg, Sigurd Wagner and James C. Sturm

Department of Electrical Engineering and Princeton Institute for the Science and Technology of Materials, Princeton University, Princeton, NJ 08544, USA

Abstract — **In a photovoltaic device, the evaluation and understanding of carrier-selective contacts are difficult, because in both terminals the current measured represents the sum of both electron and hole currents at that contact. We seek to independently know the electron current and hole current across a carrier-selective contact. In this work, we demonstrate a heterojunction bipolar transistor structure using the selective contact as the emitter-base junction, allowing one to separately measure the electron and hole currents at that contact. The method is then used to evaluate the properties of a TiO₂/p-type crystalline Si electron-selective heterojunction contact, and the information learned is used to understand a TiO₂/p-Si heterojunction PV cell, where the TiO₂/p-Si replaces the n⁺-p junction.**

I. Introduction

Dopant-free carrier-selective contacts, utilized to replace p-n or n-n⁺ junctions, are of high interest for silicon photovoltaics [1]. Recently, materials such as poly(3,4- ethylene-dioxythiophene):polystyrenesulfonate (PEDOT:PSS) [2], titanium oxide (TiO₂) [3,4] , nickel oxide (NiOx) [5], and molybdenum oxide (MoOx) [6] have been investigated for selective contact purposes. Carrier-selective contacts work by blocking one type of carrier in one direction while allowing the other carrier to pass through in the other direction (thus selecting one carrier). As such, quantifying the electron and hole current components separately across a terminal could be important to optimize selective contacts. Yet understanding the fundamental performance of such contacts in photovoltaic devices is difficult, because the device current at either contact represents the sum of the electron and hole current at that contact. Thus determining which current is dominant, and how small the smaller current is, is not possible.

In this paper we introduce a general method using the selective contact in a heterojunction bipolar transistor, which allows the independent and direct measurement of the electron and hole current components at the contact. The method is used to study TiO₂/p-type crystalline Si contacts. We show that in a Al/TiO₂/p-type Si single junction cell, the electron and hole currents are of similar magnitude.

II. Current Processes at TiO2/p-Si Heterojunctions

Because of its large valence band offset ($\sim$ 2 eV), one can consider using the TiO₂/p-Si heterojunction as an electron-selective contact, to replace the n-p junction [3]. This would allow Si-based solar cell fabrication using a low-temperature ($\sim$100 °C) CVD step to replace the high-temperature phosphorus diffusion process.

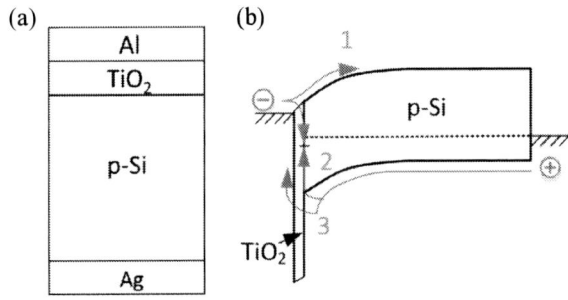

Fig. 1. (a) Device structure and (b) Fundamental current processes at an Al/TiO₂/p-Si selective contact [3]. Process 1 is the ideal electron injection current, process 2 is recombination at the Si/TiO₂ interface and process 3 is hole current through the TiO₂ through tunneling or defect states.

While the high offset should ideally block all hole current from the Si towards the cathode, second-order effects such as recombination at TiO₂/Si interface states and tunneling through the TiO₂ or to TiO₂ defect states (processes 2 and 3 respectively in Fig. 1) would lead to hole current and could negate the blocking effects of the valence band barrier.

Fig. 2 shows the dark I-V curve of an Al/TiO₂/p-Si diode on a high-lifetime FZ substrate. The titanium oxide is deposited using a low temperature chemical vapor deposition method (maximum substrate temperature of 100°C) [7]. Also shown is the ideal injected minority carrier current of electrons, calculated from the following equation:

$$J_{elec,inj} = \frac{qn_i^2 D_n}{N_A W}\left(e^{\frac{qV}{kT}} - 1\right) = J_{0,e} * \left(e^{\frac{qV}{kT}} - 1\right) \qquad (1)$$

978-1-5090-5606-4/17 $31.00 © 2017 IEEE

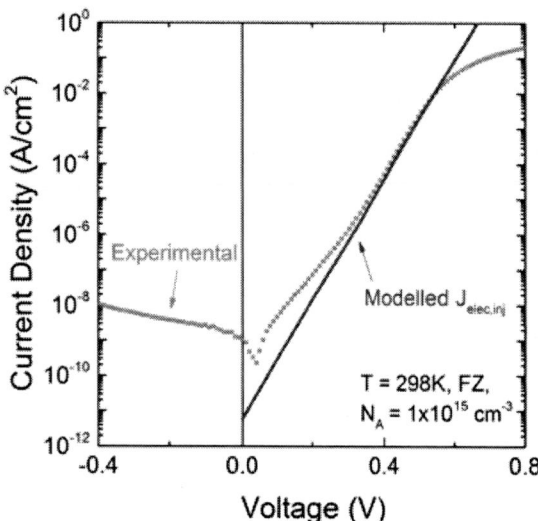

Fig. 2. Measured J-V characteristics of Al/TiO2/FZ p-Si (red) and modelled ideal J-V characteristics for injected electron current (blue) using a diffusion width L_n of 675 um). TiO$_2$ thickness is 4nm. Positive voltage is on the substrate back contact [3]

We assume carriers can diffuse towards the back of the substrate. The ideal injected minority carrier current is comparable to the experimental current. However, from this we can't learn the magnitude of the hole current. The hole current could be nearly as large as the electron current, or orders of magnitude smaller. This is important for PV, since the addition of an minority carrier electron blocker (a second selective contact) at the back interface would have a major impact on lowering the total current (and raising V_{OC}) only if the hole current at the TiO$_2$/p-Si contact were very small. The goal of the work is to measure this hole current.

III. TiO$_2$/P-Si/N-Si Heterojunction Bipolar Transistor

Consider a heterojunction bipolar transistor where the n+ emitter of a conventional n$^+$/p/n device (Fig. 3a) is replaced with a Al/TiO$_2$ selective contact (Fig. 3b). As in the conventional device, under forward bias on the emitter-base and reverse/zero bias on the base-collector, electrons are injected into the p-type base as minority carriers, diffuse across the base, and are collected to become collector current (I$_C$). Hole current from base to emitter (I$_B$) originates from the base contact.

Thus, the hole current, representing processes 2 and 3 in Fig. 1 can be measured as base current I$_B$, independent of the electron current on the selective contact. This independent measurement of electron and hole currents was not possible with the diode device of Fig. 2.

The HBT device was fabricated by starting with an epitaxial p-type base (doping of 5x10^{14} cm^{-3}, width of 6μm) on an n-type

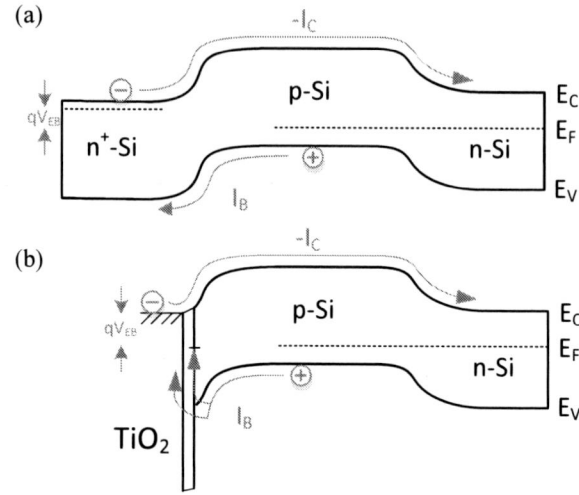

Fig. 3. (a) N/P/N BJT Band diagram and current processes, and (b) TiO$_2$/P/N HBT Band diagram and current processes

substrate (Fig. 4). The area of the base-collector junction was isolated by mesa etching. Shadow masks were used to define the areas of the silver base contact, the TiO$_2$ deposition, and the Al emitter contact. The TiO$_2$ was deposited as with the diode to a thickness of 4 nm. The emitter area was 0.1 cm x 0.1 cm.

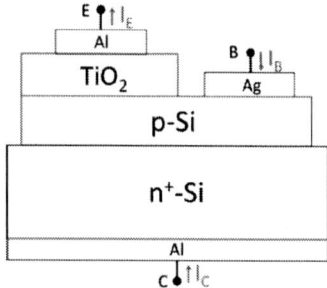

Fig. 4 Cross section of HBT device with n-type Si substrate, p-type Si base and Al/TiO$_2$ electron-selective contact to the base as the minority carrier emitter.

HBT I-V Gummel plots in forward active mode are shown in Fig. 5. The current gain (collector current /base current ratio) is about 180. Our focus is not to make a practical transistor, but rather to isolate different current mechanisms. The large collector current represents electrons injected from emitter to base and diffusing across the base. A dotted line is shown modeling this current with the classical equation for collector current:

$$I_C = \frac{q n_i^2 D_n A_E}{N_A W_D}\left(e^{\frac{qV}{kT}} - 1\right) = I_{0,e} * \left(e^{\frac{qV}{kT}} - 1\right) \qquad (2)$$

where N_A is the base doping (5×10^{14} cm^{-3}), and W_B is the neutral base width (6 µm). Note that replacing N_A and W_B with substrate doping and substrate thickness would give the diode electron current of (1).

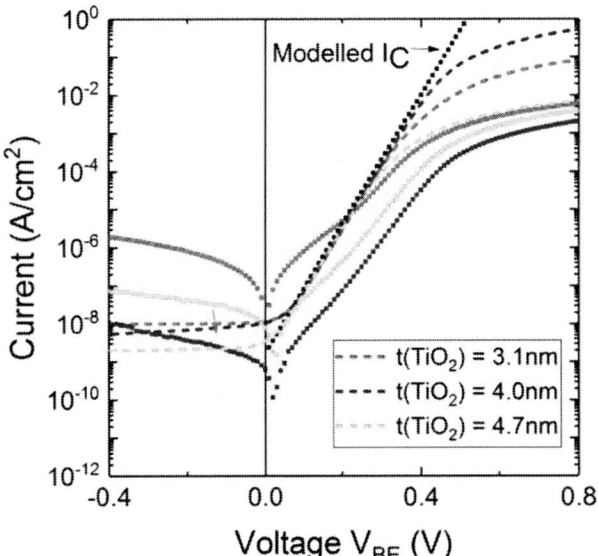

Fig. 6. Comparison of HBT I_C (dotted) and I_B (scatter) for different TiO₂ thicknesses.

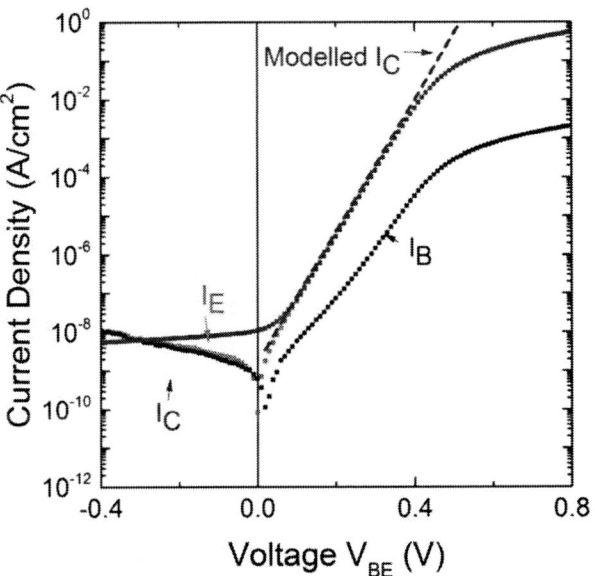

Fig. 5. Gummel plot of I_C and I_B vs base-emitter bias for the HBT of Fig. 4, for $V_{BC} = 0$. I_E is show in red, I_C in blue and I_B in black.

The base current (I_B) of the device is far smaller than I_C, and represents the current of holes from the p-type base to the emitter contact. (Electron recombination in the base can be shown to be negligible). Because the base doping of the HBT and the FZ p-type substrate doping of the diode of Fig. 2 are comparable, the base current of the HBT should thus represent the hole current from substrate to the cathode of the diode device of Fig. 2, specifically recombination at the TiO₂/p-Si interface and/or tunneling into or through the TiO₂ layer (processes 2 and 3 of Fig. 1).

IV. RELEVANCE FOR PHOTOVOLTAICS AND MINORITY CARRIER INJECTION RATIO

For devices where the minority carrier current dominates and the majority carrier current component is hard to detect (such as the device shown in Fig. 2), the HBT can be used as a measurement method to see the effect of processing on the majority carrier component. Fig. 6 shows HBT base current (which corresponds to the hole current) for different thicknesses of TiO₂ – demonstrating the utility of the HBT as a majority carrier probe. Note the lack of variation in I_C for different TiO₂

thicknesses. This is as expected since I_C is given by (2) and thus is independent from the TiO₂/Si heterojunction characteristics, except at higher voltages when series resistance effects become prevalent.

Furthermore, since we now know the magnitude of the hole, current (I_B of the HBT) corresponding to the diode shown in Fig. 2, we can compare it to the total current of Fig. 2. This is done in Fig 7. It shows the hole current (I_B of the HBT) and the total current of the diode of Fig. 2, both versus bias on the Al/TiO₂/p-Si junction. The roll-off of the HBT I_B above V_{BE} = 0.5V is due to excessive lateral base resistance. One can extrapolate the exponential region of I_B to relevant voltages for PV ($\geq$ 0.6V). In this range, the two current levels are similar. The hole current would be expected to vary somewhat from device to device, because it depends on parasitic mechanisms (such as defects at the TiO₂ interface). Also shown is the modelled line for the electron injection current from Fig. 2. This line could vary depending on the exact substrate doping. However, our results show that for the deposition and fabrication processes in this work, the hole and electron currents are of similar magnitude for 0.4V. At higher voltages, utilizing the extrapolated HBT base current and the modelled electron injection current, we obtain better ratios. For example, at the PV-relevant voltage of 0.6V, the electron current component is six times larger than the hole current component.

In crystalline Si PV, it is well known that for high lifetime substrates that injected minority carrier can easily diffuse across the wafer and recombine at the rear contact. Thus, a barrier for the minority carriers (such as a back-side field) is often

978-1-5090-5606-4/17 $31.00 © 2017 IEEE

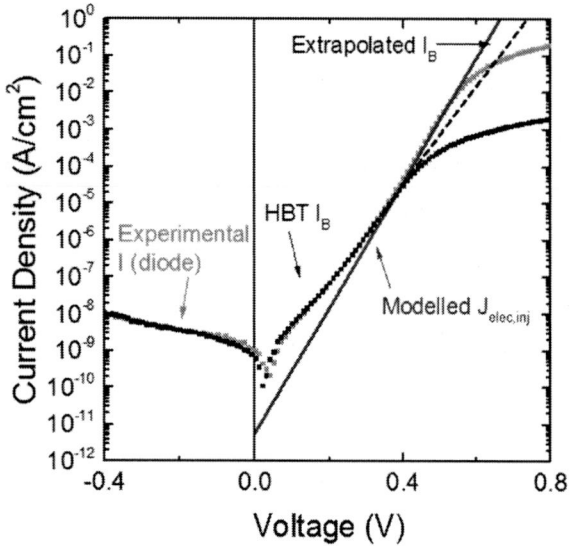

Fig. 7. Comparison of hole current from p-Si to Al/TiO₂ emitter contact (from the HBT) under forward bias with the current in the Al/p-Si diode of Fig's. 1-2.

added at the rear contact to reduce this dark current and raise V_{OC}. This will be effective if the minority carrier current is dominant and other current sources are not significant. This highlights the importance of identifying the magnitude of the hole current at the TiO₂/p-Si interface. Our work suggests that without a back-side field, the electron and hole currents in the structure of Fig. 2 are similar, i.e. the minority carrier injection ratio (ratio of minority carrier current to total current) is on the order of 0.5 for an applied voltage of 0.4V.

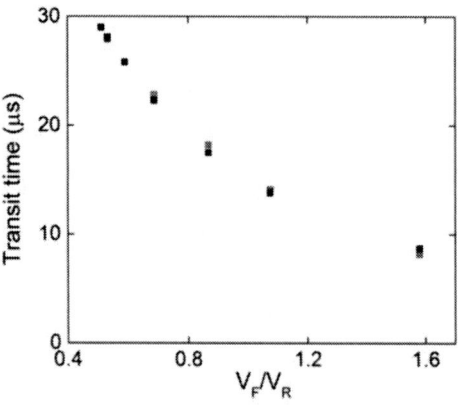

Fig. 8. Experimental reverse recovery data (red) and fit (black) with γ = 0.55.

Reverse recovery experiments further probed this [8], which confirmed that the hole and electron currents are similar. Fig. 8 shows the transit time versus the ratio between forward and reverse current. The fit gives a value of 0.55 for the minority carrier injection ratio γ.

V. CONCLUSIONS

In this work we have introduced a new method to distinguish the hole and electron current across selective contacts: the hybrid heterojunction bipolar junction device (HBT). Using the HBT for TiO₂/Si electron-selective contact, we have shown that the hole and electron currents across the interface are similar. Silicon-based heterojunction cells based on p-type substrates have potential. Further work, possible related to reducing TiO₂/Si interface states, is needed to reduce the majority carrier (hole) current and fulfill the desired open-circuit voltage.

ACKNOWLEDGEMENT

This work was supported by the National Science Foundation Princeton MRSEC Grant No. DMR-0819860 and the Princeton Institute for the Science and Technology of Materials MRSEC Grant No. DMR-0819860 and the Princeton Institute for the Science and Technology of Materials.

REFERENCES

[1] S. De Wolf, A. Descoeudres, Z.C. Holman, and C. Ballif, "High-efficiency silicon heterojunction solar cells: A review," green, vol. 2,1, pp. 7-24, 2012.

[2] K. A. Nagamatsu, S. Avasthi, J. Jhaveri, and J. C. Sturm, "A 12% Efficient Silicon/PEDOT:PSS Heterojunction Solar Cell Fabricated at < 100C," IEEE Journal of Photovoltaics, vol. 4, no. 1, pp. 260–264, Jan. 2014.

[3] S. Avasthi, W. E. McClain, G. Man, A. Kahn, J. Schwartz, and J. C. Sturm, "Hole-blocking titanium-oxide/silicon heterojunction and its application to photovoltaics," *Applied Physics Letters*, vol. 102, pp. 203901, 2013.

[4] K. A. Nagamatsu, S. Avasthi, G. Sahasrabuddhe, G. Man, J. Jhaveri, A. H. Berg, J. Schwartz, A. Kahn, S. Wagner, and J. C. Sturm, "Titanium dioxide/silicon hole-blocking selective contact to enable double-heterojunction crystalline silicon-based solar cell," Applied Physics Letters, vol. 106, no. 12, p. 123906, 2015.

[5] C. Battaglia, S.M. De Nicolas, S. De Wolf, X. Yin, M. Zheng, C. Ballif, and A. Javey, "Silicon heterojunction solar cell with passivated hole selective MoOx contact," Applied Physics Letters, vol. 104, no. 11, p 113902, 2014.

[6] R. Islam, G. Shine, K.C. Saraswat, "Schottky barrier height reduction for holes by Fermi level depinning using metal/nickel oxide/silicon contacts," Applied Physics Letters, vol. 105, no. 18, p. 182103, 2014.

[7] J. Jhaveri, A.H. Berg, S. Wagner and J. C. Sturm, "Al/TiO₂/p-Si heterojunction as an ideal minority carrier electron injector for silicon photovoltaics," in 43rd IEEE Photovoltaic Specialist Conference, 2016, pp. 2444-2447.

[8] B. Lax and S. F. Neustadter, "Transient Response of p-n Junction.," Journal of Applied Physics, vol. 25, no. 9, pp. 1148-1154, 1954.

978-1-5090-5606-4/17 $31.00 © 2017 IEEE

Effect of growth and post-oxidation annealing temperature of thermally grown tunneling SiO$_x$ on the implied V_{oc} of passivated contacts for c-Si based solar cells

Abhijit S. Kale,[1] William Nemeth,[2] Matthew Page,[2] Sumit Agarwal,[1] and Paul Stradins[2]

[1]Colorado School of Mines, Golden, CO, 80401, USA, [2]National Renewable Energy Laboratory, Golden, CO, 80401, USA

Abstract — **Record high efficiencies of monocrystalline Si solar cells have been achieved using passivated contacts based on the ultra-thin tunneling SiO$_x$ and doped pc-Si layers. We have investigated how modifications in the growth and thermal processing of the tunneling SiO$_x$ affect i-V_{oc} and electrical transport properties of p- and n-type passivated contacts. We have shown that either post-oxidation annealing to, or direct growth of the tunneling SiO$_x$ at temperatures >850 °C deteriorates both the contact passivation and the bulk c-Si wafer quality. The resistivity of the contact increases significantly on annealing of the SiO$_x$, likely due to inadvertent increase in the thickness of the SiO$_x$ layer.**

Index Terms — **contact resistivity, passivation, passivated contact, polysilicon, silicon solar cell, tunneling oxide.**

I. INTRODUCTION AND MOTIVATION

Exceptionally high efficiencies for crystalline Si (c-Si) solar cell (25.1%) have been achieved using the tunneling oxide passivated contact approach [1]. These contacts consist of a 1-2 nm tunneling SiO$_x$ on c-Si, and doped polycrystalline Si (pc-Si) on SiO$_x$, to create a c-Si/SiO$_x$/pc-Si structure shown in the inset in Fig. 1. SiO$_x$ is a very good surface passivation layer for c-Si due to the low c-Si/SiO$_x$ interfacial defect densities and still allows electrical transport through tunneling when the SiO$_x$ layer thickness is ~1-2 nm [2]. In addition, heavily doped pc-Si layers on tunneling SiO$_x$ provide good field-effect passivation [3]. The separation of the as-deposited hydrogenated amorphous Si (a-Si:H) layer from c-Si through SiO$_x$ prevents surface recombination and epitaxial growth in the pc-Si layer during the high temperature annealing step. Different methods have been reported for the growth of the tunneling SiO$_x$ using either UV/O$_3$ [4], HNO$_3$ [1] or dry thermal processes [5]. Similarly, the deposition of doped a-Si:H layers has been performed by plasma enhanced chemical vapor deposition (PECVD) [1, 4, 5] or low pressure CVD [6]. Depending on the growth and post-processing conditions, the final doped Si layers can be either polycrystalline [5, 6] or consist of Si nanocrystals embedded in an a-Si matrix [7]. Regardless of the growth method for the tunneling SiO$_x$ layer and the doped Si film, the highest passivation quality of the contact structure (see inset in Fig. 1) as measured by the implied open circuit voltage (i-V_{oc}) is obtained only after the contacts are annealed to a temperature of 800 – 900 °C. However, the reason for this fairly narrow annealing temperature range that is required to improve the

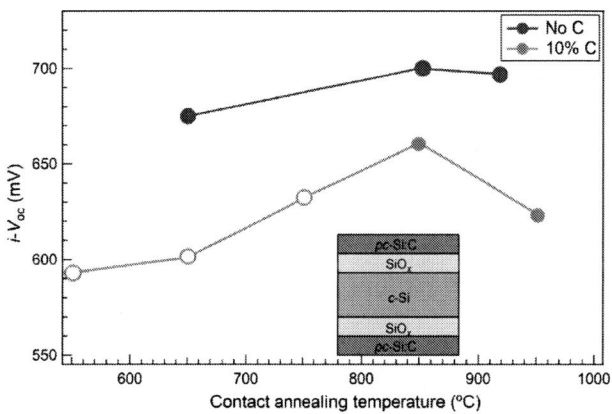

Fig. 1. (inset) Schematic of the passivated contact test structure used to evaluate the passivation quality through the i-V_{oc} measurements. Effect of annealing temperature of the entire contact structure on i-V_{oc} for p-type passivated contacts with varying carbon concentrations in the plasma-deposited a-Si:H film. Solid circles indicate the crystalline phase, and the open circles indicate the amorphous phase of the p-type Si layer.

passivation quality of the contact structures is not completely understood. While the degradation in performance after annealing of the contacts to temperatures of 950 °C may be related to the breakdown of the tunneling SiO$_x$ layer resulting in local loss of passivation, the reason for lower i-V_{oc} below ~800 °C is less clear [8].

Figure 1 shows the i-V_{oc} for symmetric p-type contact test structures (see inset), developed at NREL, that were annealed after a-Si:H deposition over a temperature range of 550 – 950 °C. Similar to previous observations in the literature, we have shown that the optimum annealing temperature is over the range of 800 – 900 °C, which is apparent from the peak in the i-V_{oc} seen in Fig. 1 at 850 °C. Our experiments also show that it is unlikely that the poorer passivation at lower temperatures is related to the phase of doped Si film since we have verified using Raman spectroscopy that the crystallization of a-Si:H films to pc-Si is complete already at 650 °C. Furthermore, increasing the crystallization temperature through carbon incorporation (see Fig. 1) in the as-deposited a-Si:H film does not influence the optimum annealing temperature for the highest i-V_{oc}.

In the passivated contact structures, dopant activation in the

pc-Si film, diffusion of the dopant into the SiO$_x$ layer and/or into the c-Si substrate, and SiO$_x$ structural properties are all strongly dependent on the annealing temperature, and ultimately determine the quality of the contact structure. Given that the high-temperature thermal annealing step is for a complete contact structure, it is difficult to isolate the main reason for the peak in performance at 850 °C. In this study, we have decoupled the effect of the thermal treatments cited above by comparing the performance of the contact structures that undergo high-temperature annealing before a-Si:H deposition with those that do not undergo this intermediate high-temperature treatment.

II. EXPERIMENTAL DETAILS

Textured n-Cz phosphorous doped wafers from NorSun (Oslo, Norway) were subjected to a KOH based etch for planarization. The wafers were then cleaned using standard RCA cleaning procedure and a dry thermal tunneling SiO$_x$ was grown in a tube furnace at 6:1 N$_2$:O$_2$ ratio without stripping off the RCA oxide formed after the RCA cleaning process. The tunneling SiO$_x$ was modified using two approaches: 1) post-oxidation annealing (POA) in a N$_2$ environment of thermal dry SiO$_x$ grown at 700 °C, and 2) a modification of the thermal oxidation temperature of c-Si substrates for SiO$_x$ growth in a dry O$_2$ environment. Both these temperature treatment will hereafter be referred to as 'oxide treatment temperature'. Spectroscopic ellipsometry on polished sister n-Cz wafers confirmed that all SiO$_x$ films were made and remained within the tunneling regime (<2 nm thick) after processing. Doped a-Si:H films were then deposited by PECVD on both sides of the c-Si wafer with thermal SiO$_x$ using SiH$_4$, H$_2$, B$_2$H$_6$ and PH$_3$ to form symmetric test structures of Fig. 1. The resulting samples were then annealed in a quartz tube furnace between 700 and 850 °C in N$_2$ environment to solid-phase crystallize the a-Si:H to pc-Si. This temperature is referred to as 'contact annealing temperature'. A hydrogen induced passivation step followed which involved deposition of Al$_2$O$_3$ via atomic layer deposition and annealing in forming gas (1:9 H$_2$:N$_2$ mixture). Lifetime measurements were performed using a Sinton WCT-120 Lifetime instrument in the generalized (1/1) mode so as to extract the i-V_{oc} values on test structures similar to as shown in inset of Fig. 1. To evaluate the electrical transport properties of the contacts, 1 μm thick rectangular Al pads with varying spacing were deposited on the pc-Si layer via electron beam metal deposition process. The resulting structures were then subject to reactive ion etching using SF$_6$ to etch through the pc-Si and SiO$_x$ layers while using the Al pads as masks. The other side was also etched similarly to remove the pc-Si and SiO$_x$ layers completely. The cross-section of the resulting structure has been shown in the inset of Fig. 4.

III. RESULTS AND DISCUSSION

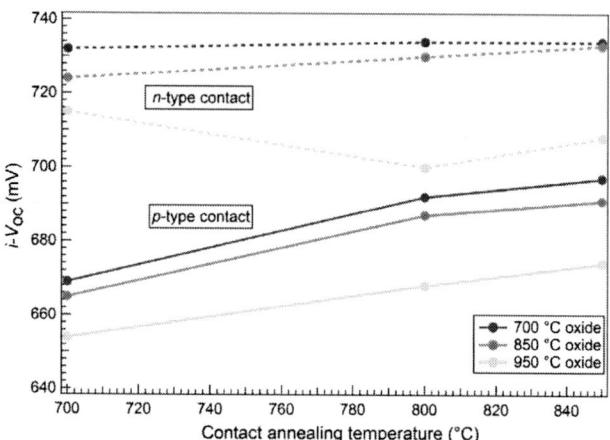

Fig. 2. Effect of annealing temperature of the symmetric contact structure on i-V_{oc} for n-type (shown by dotted lines) and p-type (shown by solid lines) contacts with SiO$_x$ treated to different post-oxidation annealing temperatures prior to a-Si:H deposition.

Figure 2 shows the effect of contact annealing temperature on i-V_{oc} for both p- and n-type contacts. Here, the tunneling SiO$_x$ layer was grown at 700 °C, followed by annealing two sample sets at higher temperatures – 850 and 950 °C – *prior to* a-Si:H deposition. It can be seen that for both n- and p-type of passivated contacts, the higher the SiO$_x$ treatment temperature, the lower is the passivation quality based on the i-V_{oc} values. In fact, for the tunneling SiO$_x$ annealed at 950 °C (green curve), the i-V_{oc} is significantly lower, ~20–50 mV, than for the as-grown SiO$_x$ at 700 °C or the as-grown SiO$_x$ annealed at 850 °C. The 2nd approach involving the growth of SiO$_x$ in dry O$_2$ at different temperatures also gave similar results and has not been shown here.

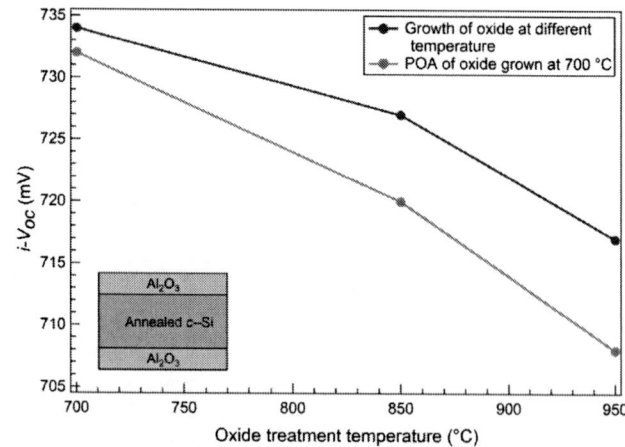

Fig. 3. (inset) Schematic of the test structure used to evaluate the c-Si wafer quality through i-V_{oc} measurements. Effect of thermal treatments during and after SiO$_x$ growth on i-V_{oc} of c-Si wafer passivated with Al$_2$O$_3$.

To investigate the reason for passivated contacts formed using higher temperature SiO_x resulting in lower $i\text{-}V_{oc}$, we monitored the c-Si wafer quality during processing. For this, we etched the thermally treated SiO_x/c-Si sister samples using dilute HF, and then passivated the c-Si wafer using Al_2O_3, followed by annealing in forming gas, resulting in symmetric test structures as shown in inset of Fig. 3. It was confirmed separately that our Al_2O_3 provides excellent passivation ($i\text{-}V_{oc}$ ~ 740 mV) on as-received identical wafers. The measured $i\text{-}V_{oc}$ on such structures has been plotted in Fig. 3. We can conclude that annealing SiO_x/c-Si to high temperatures for a short time (few minutes) results in degradation in c-Si quality as noted by the drop in the measured $i\text{-}V_{oc}$. The growth of oxidation induced stacking faults [9] near the SiO_x-Si interface or structural defects due to rapid thermal processing of the wafers, or residual contamination can be the likely cause of lowered $i\text{-}V_{oc}$ at higher temperatures. Studies in the literature have shown that the extent of such faults can be reduced by extended high temperature annealing in N_2 environment [10].

The next important observation from Fig. 2 is the similar trend for all curves except for the 950 °C annealed SiO_x n-type, i.e. the curves are just shifted along the y-axis as a result of the thermal treatments of the SiO_x prior to a-Si:H deposition. We can infer from this, that the observed 850 − 900 ° C optimum crystallization temperature for passivated contacts is not related to changes in the SiO_x as a result of thermal treatment alone. It is likely that the improvement in the field effect passivation due to dopant activation overshadows possible improvements in chemical passivation due to structural changes in the SiO_x and pc-Si layers over the studied temperature range.

on n-type c-Si wafer (test structure shown in inset) determined using the transmission line method [11] on a logarithmic scale. An important observation here is that the contact resistivity of the passivated contact increases as the SiO_x treatment temperature increases. Analysis of contact resistance for the p-type contact cannot be performed in a similar way using TLM analysis due to the diode behavior of the p^+n structure shown in inset of Fig.4. To compare and quantify the different transport behavior for the p-type contact with different thermally treated SiO_x, we evaluated the total diode resistance of the p^+n diode with tunneling SiO_x by taking the inverse slope of the diode $I\text{-}V$ curves at a fixed voltage of 0.59 V. This has been plotted on the right axis in Fig. 4 and similar to the n-type contact, shows that the resistance of the structure increases as oxide treatment temperature increases. It is interesting to note that for both the n- and p-type contacts the resistance increases significantly when the SiO_x is annealed to 950 °C (thickness ~1.95 nm) as compared to annealing the SiO_x to 850 °C (thickness ~1.7 nm) prior to a-Si:H deposition. It is unlikely that thermally induced changes in the structure of SiO_x can cause such a large variation in electrical transport properties. Studies on the effect of oxide thickness on gate leakage current for NMOSFETs have shown that the leakage current increases exponentially with decrease in SiO_x thickness [12]. It is more likely that the observed increase in resistivity shown in Fig. 4 is due to increase in the SiO_x thickness which occurs inadvertently during the annealing of the SiO_x in a N_2 atmosphere.

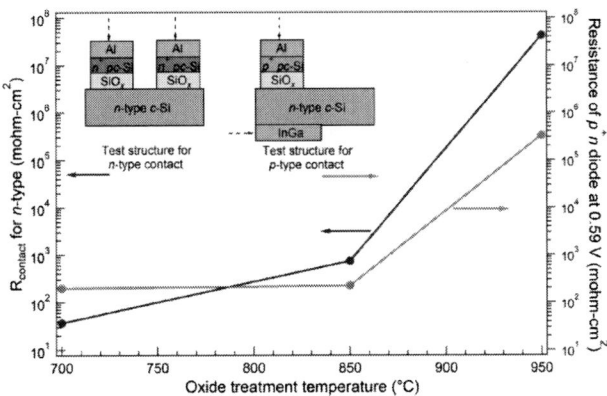

Fig. 4. Effect of SiO_x treatment temperature on contact resistivity of n-type (on left axis) and total diode resistance of p-type (on right axis) passivated contact structure annealed at 850 °C. The inset shows the test structure used for the transport measurement with the black colored dashed arrows showing the contacting points for the probes.

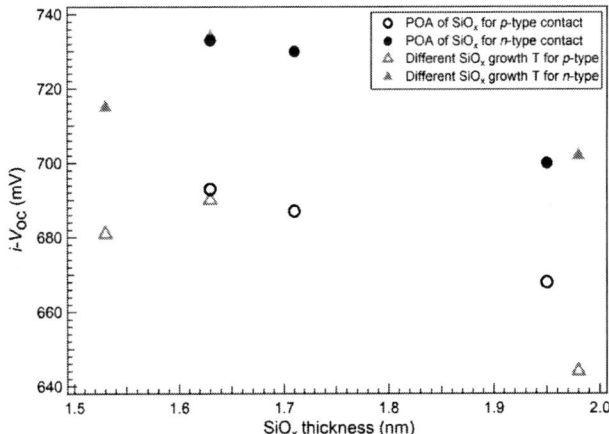

Fig. 5. Effect of SiO_x thickness on $i\text{-}V_{oc}$ of p- and n-type symmetric passivated contact structure annealed at 800 °C. SiO_x treated through both approaches: POA of SiO_x grown at 700 °C (shown in black circles), and varying the growth temperature of SiO_x (shown in red triangles).

Figure 4 shows the effect of post-oxidation annealing of the low temperature tunneling SiO_x on contact resistivity for the n- and p-type passivated contacts annealed at 850 °C. The left axis of Fig. 4 shows the contact resistivity for the n-type pc-Si

Since, the effect of SiO_x thickness is so prominent in the electrical transport measurements shown in Fig. 4, we re-plotted the $i\text{-}V_{oc}$ data for n- and p-type passivated contacts from Fig. 2 and also data from our experiments where the SiO_x was directly grown at different temperatures prior to a-Si:H deposition, as a function of SiO_x thickness which has been shown in Fig. 5. We can conclude that as SiO_x thickness increases, $i\text{-}V_{oc}$ of the symmetric contact structure drops. The thickest oxides, which are the 950 °C annealed and grown oxides, perform worst. Since there was variation in both SiO_x treatment temperature and SiO_x thickness in the current experiment, we are currently studying the isolated effect of SiO_x thickness on passivated contacts by varying oxidation duration while keeping the growth temperature constant.

IV. CONCLUSIONS

We have studied how modifications in the growth and thermal processing of the tunneling SiO_x affect $i\text{-}V_{oc}$ and electrical transport of p- and n-type passivated contacts. Our experiments suggest that for passivated contacts the peak in $i\text{-}V_{oc}$ associated with thermal anneal of the contact between the 800-900 °C, is not related to structural changes in the tunneling SiO_x due to the temperature effects alone. Electrical transport properties of both the contacts show that the significantly higher resistance of the contacts with the SiO_x annealed at 950 °C is likely due to an inadvertent increase in SiO_x thickness. We speculate that the higher temperature SiO_x resulting in poorer n- and p-type passivated contacts is due to the degradation of c-Si bulk, dopant diffusion, interface degradation as a result of rapid thermal treatment or due to inadvertent increase in SiO_x thickness during sample processing.

ACKNOWLEDGEMENTS

Funding for this work was provided by US DOE EERE contract SETP DE-EE00030301 (SuNLaMP) and under Contract No. DE-AC36-08GO28308. The U.S. Government retains and the publisher, by accepting the article for publication, acknowledges that the U.S. Government retains a nonexclusive, paid up, irrevocable, worldwide license to publish or reproduce the published form of this work, or allow others to do so, for U.S. Government purposes.

REFERENCES

[1]   S. W. Glunz *et al.*, "The irresistible charm of a simple current flow pattern—25% with a solar cell featuring a full-area back contact," in *Proceedings of the 31st European Photovoltaic Solar Energy Conference and Exhibition,* 2015, pp. 259-263.

[2]   M. L. Green, E. P. Gusev, R. Degraeve, and E. L. Garfunkel, "Ultrathin (<4 nm) SiO_2 and Si–O–N gate dielectric layers for silicon microelectronics: Understanding the processing, structure, and physical and electrical limits," *Journal of Applied Physics,* vol. 90, no. 5, pp. 2057-2121, 2001.

[3]   J. Gan, "Polysilicon emitters for silicon concentrator solar cells," Stanford University, 1990.

[4]   A. Moldovan, F. Feldmann, M. Zimmer, J. Rentsch, J. Benick, and M. Hermle, "Tunnel oxide passivated carrier-selective contacts based on ultra-thin SiO_2 layers," *Solar Energy Materials and Solar Cells,* vol. 142, pp. 123-127, 2015.

[5]   B. Nemeth *et al.*, "Polycrystalline silicon passivated tunneling contacts for high efficiency silicon solar cells," *Journal of Materials Research,* vol. 31, no. 06, pp. 671-681, 2016.

[6]   G. Yang, A. Ingenito, O. Isabella, and M. Zeman, "IBC c-Si solar cells based on ion-implanted poly-silicon passivating contacts," *Solar Energy Materials and Solar Cells,* vol. 158, Part 1, pp. 84-90, 2016.

[7]   J. Stuckelberger *et al.*, "Passivating electron contact based on highly crystalline nanostructured silicon oxide layers for silicon solar cells," *Solar Energy Materials and Solar Cells,* vol. 158, Part 1, pp. 2-10, 2016.

[8]   G. R. Wolstenholme, N. Jorgensen, P. Ashburn, and G. R. Booker, "An investigation of the thermal stability of the interfacial oxide in polycrystalline silicon emitter bipolar transistors by comparing device results with high - resolution electron microscopy observations," *Journal of Applied Physics,* vol. 61, no. 1, pp. 225-233, 1987.

[9]   S. P. Murarka, "Role of point defects in the growth of the oxidation-induced stacking faults in silicon," *Physical Review B,* vol. 16, no. 6, pp. 2849-2857, 1977.

[10]   Y. Sugita, H. Shimizu, A. Yoshinaka, and T. Aoshima, "Shrinkage and annihilation of stacking faults in silicon," *Journal of Vacuum Science and Technology,* vol. 14, no. 1, pp. 44-46, 1977.

[11]   H. Berger, "Contact resistance on diffused resistors," in *1969 IEEE International Solid-State Circuits Conference. Digest of Technical Papers,* 1969, vol. XII, pp. 160-161.

[12]   M. L. Green *et al.*, "Understanding the limits of ultrathin SiO_2 and Si-O-N gate dielectrics for sub-50 nm CMOS," *Microelectronic Engineering,* vol. 48, no. 1, pp. 25-30, 1999.

Partially contacted surfaces with contact size in the 1 μm range for c-Si PERC solar cells

R. Khoury[1], I. Martín[2], G. López[2], C. Jin[2], J.M. López-González[2], L. Zeyu[1], P. Bulkin[1], E.V. Johnson[1], R. Alcubilla[2]

[1] Laboratoire de Physique des Interfaces et des Couches Minces (UMR 7647) LPICM-CNRS. Ecole Polytechnique, Université Paris Saclay, 91128 Palaiseau, France.

[2] Departament d'Enginyeria Electrònica, Universitat Politècnica de Catalunya, Gran Capità s/n, Mòdul C4, 08034 Barcelona, Spain. Pho: +34 93 405 4193, e-mail: isidro.martin@upc.edu

Abstract — **We examine the electrical benefits of creating contacts with sizes in the 1 μm range on partially contacted surfaces for c-Si PERC/L solar cells. In such a design, a dielectric layer that provides surface passivation is periodically opened to create contacts. Analytical models demonstrate that small contacts in the 1 μm range and separated by a few microns result in lower ohmic losses when compared to a typical contact configuration. In order to obtain a good surface passivation quality, surface recombination velocities below 10^3 cm/s are needed for the contacted area. From PC-1D simulations of a reference structure, a clear advantage for small contacts is deduced for surface recombination at the contacts in the 10^2-10^5 cm/s range. Additionally, a fabrication process to achieve the desired contact configuration is described using Al_2O_3 and thermally grown SiO_2 films. This process is based on colloidal lithography using polystyrene nanoparticles. For Al_2O_3 films, an incomplete contact opening is observed, while much more repeatable results are obtained for SiO_2 films.**

I. Introduction

The passivated emitter and rear cell (PERC) is an attractive structure for cost-effective high-efficiency devices. The main feature of such solar cell is that the rear surface is partially contacted by periodical openings in a dielectric film that provides surface passivation (see fig.1 where a sketch of the cross section of these devices on p-type substrates is shown). As a result, lower effective surface recombination velocity ($S_{eff,rear}$) values compared to the classical Al-BSF solar cell together with a higher rear internal reflectance are obtained and, consequently, short-circuit current (J_{sc}) and open-circuit voltage (V_{oc}) are improved. This type of solar cells was developed by Blakers et al. in 1988 and produced record efficiency for silicon cells at that time of 22.8 % using thermally grown silicon oxide as a passivating layer [1]. In order to passivate the contacts, a local BSF at the contacted regions was introduced leading to passivated emitter rear locally diffused cell (PERL) by the University of New South Wales in Australia in 1999 with a record efficiency of 25 % [2]. Although this knowledge was acquired in the 90's, it is only in recent years that PERC/L solar cells have been introduced into the PV market for p-type substrates due to the combination of aluminum oxide passivation and laser ablation leading to a considerable reduction of the fabrication costs [3].

When the rear contact is defined in a point-like pattern, a trade-off between ohmic losses and surface recombination, i.e. between Fill Factor (*FF*) and V_{oc} is found; the more contacted area fraction (f_c), the lower the ohmic losses, but the higher the surface recombination rate. An optimization is required, trading off recombination losses and resistive losses to maximize the conversion efficiency. Examples of such optimization can be found in the literature [4-5]. Due to the characteristics of the procedure to locally open the contacts in the passivation layer, typical dimensions for the contacts are in the range of tens of microns while optimized *fc* values are well below 10 %. As a consequence, the distance between neighbor contacts or pitch (*p*) is in the order of hundreds of microns introducing a significant series resistance.

In this work, we explore the electrical benefits of using contact opening in the range of 1 μm or below and, consequently, with a pitch shorter than 10 μm. The idea is that shorter pitches would result in negligible ohmic losses while keeping $S_{eff,rear}$ under control. In the first part of the paper, we present theoretical calculations of the trade-off by examining contact geometry and recombination parameters. In the second part, we propose a technological procedure to achieve the desired surface configuration.

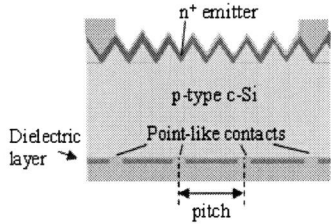

Fig. 1. PERC structure on p-type substrates.

II. Rear Surface Modeling

The c-Si substrate characteristics used in the theoretical calculations are a p-type wafer of 1.8 Ω.cm with a thickness of 160 μm, which is representative of typical substrates in current commercial devices. For the rear surface modelling, a 3D approach is necessary to define point-like patterns. However, due to the high computational cost of such simulations, we reduce the 3D problem to 1D by applying the equation proposed by Fischer [6], improved by Plagwitz and Brendel [7] and generalized by Saint-Cast et al. [8] that allows us to concentrate the ohmic losses in a condensed parameter R_{base} and an equivalent surface recombination velocity, $S_{eff,rear}$. Additionally, 3D simulations with ATLAS/Silvaco TCAD are performed for certain surface configurations in order to check the analytical results. In particular, R_{base} can be calculated using the following expression [6]:

$$R_{base} = \rho \cdot \frac{p^2}{2\pi r} \cdot \tan^{-1}\left(\frac{2w}{r}\right) + \rho \cdot w \left(1 - e^{-\left(\frac{w}{p}\right)}\right) \quad (1)$$

where r is the contact radius, p is the rear pitch, w is the wafer thickness, and ρ is the substrate resistivity. In fig.2, we show calculated R_{base} values as a function of f_c for contact radii of 25 and 0.5 μm together with the results of 3D simulations that perfectly agree with the analytical model. In addition, the solid red line indicates the minimum R_{base} value, corresponding to the case where all the current is flowing vertically through the base: $R_{base} = \rho \cdot w$. As it can be seen, resistance for the case of r = 0.5 μm almost matches this minimum value except for f_c below 2 %. Importantly, these R_{base} values are about one order of magnitude lower than for the case of $r = 25$ μm, introducing a negligible increase in ohmic losses, as expected.

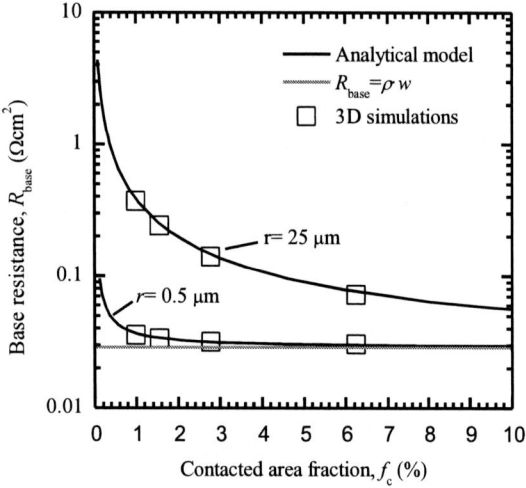

Fig. 2. R_{base} as a function of the contacted area fraction for $r = 25$ and 0.5 μm.

Concerning the effective rear surface recombination velocity $S_{eff,rear}$, it is calculated using [6]:

$$S_{eff,rear} = \left(\frac{R_{base} - \rho \cdot w}{\rho \cdot D_n} + \frac{1}{f_c \cdot S_{cont}}\right)^{-1} + \frac{S_{pass}}{(1 - f_c)} \quad (2)$$

where D_n is the diffusion constant of the minority charge carriers (electrons), S_{cont} is the surface recombination velocity at the contact, S_{pass} is the surface recombination velocity at the dielectrically passivated surface and R_{base} is the base resistance

represented by equation (1). Since we are interested in high-efficiency devices, we keep S_{pass} constant at 1 cm/s while S_{cont} is varied from 5×10^6 cm/s (no passivation at the contacts) to 10^2 cm/s for the two contact radii under study. As it was mentioned, a more sophisticated model is reported in reference [8]. However, under these assumptions, the proposed equation leads to accurate results [8].

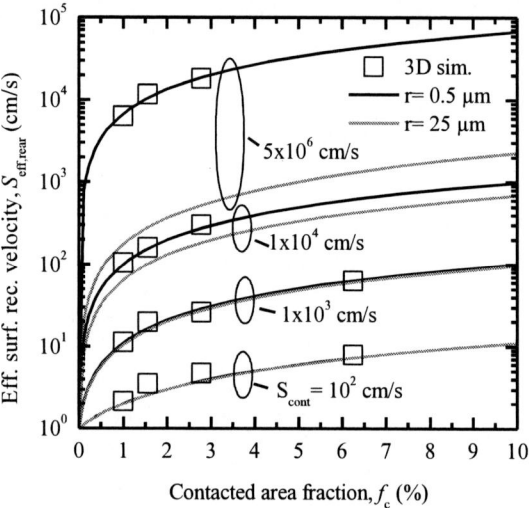

Fig. 3. $S_{eff,rear}$ as a function of contacted area fraction for $r = 0.5$ and 25 μm and S_{cont} values.

Fig.3 shows that for small contacts without passivation, $S_{eff,rear}$ sharply increases with f_c, while reasonable values can be obtained with $r = 25$ μm. This difference comes from the fact that, despite the percentage of non-passivated surface is the same, for small radius the highly-recombining areas are much closer, less than 10 μm compared to 100-500 μm for $r = 25$ μm. As a result, the recombination rate increases, since minority carriers could more easily find contact regions.

The difference in $S_{eff,rear}$ between both contact sizes decreases when S_{cont} is reduced, being negligible for $S_{cont} < 1 \times 10^3$ cm/s. We can therefore conclude that from a recombination point of view contact openings in the 1 μm range are beneficial for technologies that can achieve passivated contacts. For that cases, similar $S_{eff,rear}$ than for big contacts can be obtained without significantly increasing the ohmic losses. For example, in our calculations $S_{eff,rear}$ of 20 cm/s can be obtained for $f_c \approx 2$ % with a contact technology with $S_{cont} = 10^3$ cm/s without significantly increasing R_{base}. Obviously, for passivated contacts with very low S_{cont}, a full contacted area is the best solution, since no additional series resistance is introduced. This is the case of polysilicon contacts [9], tunnel oxide contacts [10] or silicon heterojunction technology [11]. Interestingly, the combination of silicon heterojunction with point-like contacts has already been explored, resulting in an optical improvement in the internal rear reflection, as reported by S. De Vecchi et al. in reference [12] for large point contacts.

III. IMPACT ON SOLAR CELL EFFICIENCY

In this section, we would like to explore the impact of the rear surface configuration on the solar cell efficiency. Since the 3D problem can be accurately described by $S_{\text{eff,rear}}$ and R_{base} values, as demonstrated in the previous section, we use PC-1D simulations with a reference PERC/L structure shown in the inset of fig. 4. From the optical point of view, as external reflectance (R_{ext}) we use experimental data shown in fig. 4 measured for solar cells developed in our group with random pyramid texturized surface and 100 nm SiO_2 passivation [13]. In this measurement, the photons that escape from the semiconductor absorber are also included. Thus, in the simulations we must consider that no photons are coming out from the solar cell, i.e. a front internal reflection of $R_{\text{in,front}}=$ 100 %. At the rear surface, we consider an internal reflectance of $R_{\text{in,rear}}=$ 94 % which is also a realistic value for a partially contacted surface covered with a transparent dielectric, for example SiO_2 or SiN_x, covered with aluminum. Finally, we add a 2.7 % of shadowing due to the metal fingers and busbar that introduce $R_{\text{s,front}}=$ 0.3 $\Omega\cdot\text{cm}^2$ in the series resistance [13].

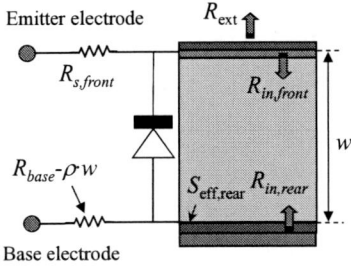

Fig. 4. Experimental reflectance introduced in PC-1D simulations; (inset) solar cell structure and parameter definition used in PC-1D.

Regarding the front emitter, we consider a gaussian profile with peak doping density of 10^{19} cm^{-3} and a junction depth of 1 μm and a surface recombination velocity of 200 cm/s. This doping profile corresponds to a sheet resistance of 123 Ω/sq with a saturation current density of 20 fA/cm^2. This low doped emitter models the n$^+$ regions that could be found in-between contacts in a selective emitter configuration. For the contacted regions, a shunt diode is introduced in PC-1D with a saturation current density of 80 fA/cm^2 (see inset of fig. 4) resulting in a total saturation current density of the emitter (J_{0e}) of 100 fA/cm^2. It should be mentioned that different emitter configurations have been tried with different J_{0e} values. Despite of the strong impact of this parameter on the maximum V_{oc} values reachable by the device, the main conclusions of the study presented hereby are still valid.

The effect of R_{base} is introduced as a lumped resistor at the rear contact without the ρw term (see inset fig. 4). Notice that this term corresponding to the geometrical resistance of the base is already included in the calculations done by PC-1D. Finally, $S_{\text{eff,rear}}$ is defined at the rear surface as the Low Level Illumination value in PC-1D.

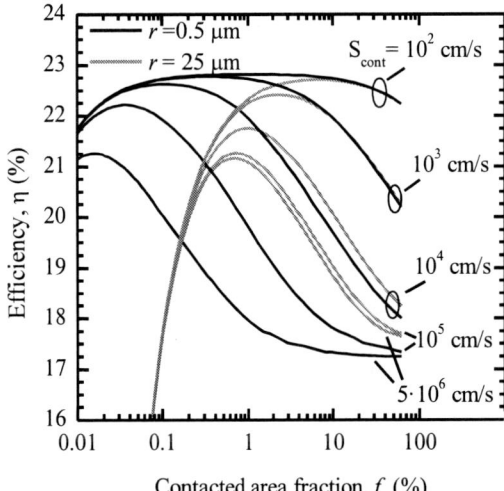

Fig. 5. Solar cell efficiency vs. contacted area fraction curves obtained by PC-1D simulations.

Simulations with PC-1D were carried out varying the contacted area fraction for the two radii under study. The obtained curves of solar cell efficiency vs. f_c are shown in fig. 5. As it can be seen, due to the trade-off between surface passivation and ohmic losses, an optimum f_c value can be determined in all cases. Additionally, we can see that the better the surface passivation at the contacts, the higher the optimum f_c. If the passivation quality at the contacts improves, more area can be contacted without penalizing in surface recombination. Finally, lower S_{cont} values allow higher solar cell efficiencies, as expected.

Focusing on the comparison of small and big contacts, for $S_{\text{cont}}=$ 5·10^6 cm/s we can conclude that similar maximum efficiencies can be obtained. In the previous section, we demonstrate that for this case, surface recombination is much higher for small contacts. However, the reduction in ohmic losses due to the short pitches used for such small contacts compensates the poor surface passivation. For S_{cont} values between 10^5 and 10^3 cm/s, the maximum reachable efficiencies are much better for the case of contacts with $r =$ 0.5 μm. In this case, the reduction in ohmic losses is a clear advantage, since surface recombination is comparable. Finally, for $S_{\text{cont}}=$ 10^2 cm/s similar maximum efficiencies are obtained again. Now, due to the low recombination at the contacts, short pitches, i.e. high f_c values, are not so detrimental for big contacts allowing a compensation of the ohmic losses.

As a summary of this analysis, we can conclude that the definition of contacts whose size is in the 1 μm range is beneficial for contacts with S_{cont} in the range of 10^5-10^3 cm/s, while comparable maximum efficiencies are achievable for small and big contact sizes with both non-passivated and well-passivated contacts. This result demonstrates that small contacts are an attractive solution from the point of view of cell efficiency. In the next section, we describe different methods to obtain the desired contact pattern.

IV. EXPERIMENTAL APPROACH

Two different approaches are considered in order to create a partially contacted rear surface as described in the previous section. One such approach (Approach A) to consider is the definition of passivated contacts for holes by a low temperature technique, such as the use of transition metal oxides like MoO_x [14] that have demonstrated good carrier selectivity with a low contact resistance, or boron-doped amorphous silicon films as used in silicon heterojunction devices [15]. For this study, we choose aluminum oxide (Al_2O_3) deposited by ALD as the dielectric passivation layer, since this layer has been applied to PERC devices by our research group with excellent results [16].

A second approach (Approach B) to consider is passivation at the contacts using a p^+ region created either by thermal diffusion or ion implantation and annealing. In this case, a thick thermal SiO_2 simultaneously functions as a passivating film and as a mask for the doping process. In fig. 6, a sketch of the cross section of the two approaches as applied to PERC/L solar cells can be seen.

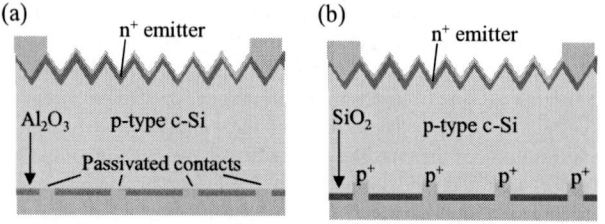

Fig. 6. Rear contact configurations to define partially contacted surfaces with contacts in the 1 μm range; a) Approach A: low temperature passivated contacts and Al_2O_3 passivation; b) Approach B: p^+ regions fabricated by either thermal diffusion or ion implantation and thermally grown SiO_2 passivation.

We herein focus on how to create said contacts through the passivating layer for both approaches using colloidal lithography based on polystyrene (PS) nanoparticles (NPs). For Al_2O_3 films, we use a thin silicon carbide (SiC_x) film deposited by plasma enhanced chemical vapor deposition (PECVD) as a mask for the subsequent wet etching, and for the thermally grown SiO_2 layer, a thin Aluminum (Al) film is evaporated and a (CHF_3,Ar) plasma etching is used for the definition of the holes in the passivating layer. All the steps of approaches A and B are illustrated in the schematic diagrams of fig. 7 and 8, respectively.

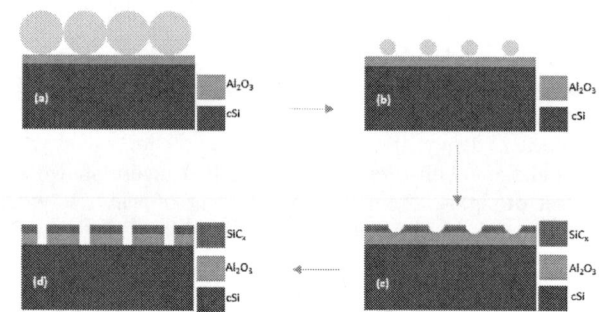

Fig. 7. Schematic diagram of nanohole fabrication process in Al_2O_3 passivation layer (Approach A). (a) NP monolayer deposition, (b) NP etching by O_2 plasma, (c) SiC_x deposition on the NPs followed by NP removal, and (d) Al_2O_3 etching through the holes.

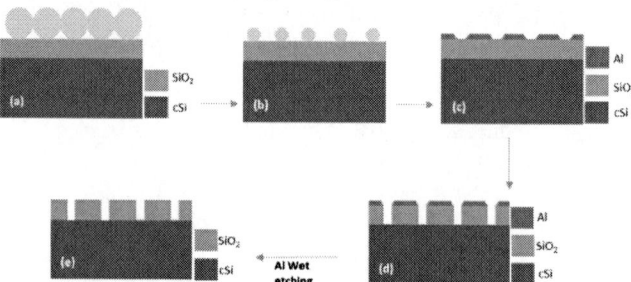

Fig. 8. Schematic diagram of nanohole fabrication process in SiO_2 passivation layer (Approach B). (a) NP monolayer deposition, (b) NP etching by O_2 plasma, (c) 30 nm Al evaporation on NPs followed by NP removal, (d) (CHF_3, Ar) plasma etching of the dielectric layer through aluminum holes, and Wet etching, by (H_3PO_4, IPA) of the aluminum layer (e).

To create holes in both passivation layers with the small sizes desired, we firstly create a monolayer of PS-NPs. The PS-NPs used in this work had nominal diameters of 607±15 nm, and 784±23 nm, obtained as an aqueous suspension of 5 % (Microparticles GmbH). The sphere solution was diluted by mixing it with an equal amount of ethanol, a low-surface tension solvent, which is used as a spreading agent. The floating-transfer technique was used to deposit a compact monolayer of PS NPs on the substrate. The NP monolayer was first formed on water surface in a petri dish and then transferred onto the substrate [17].

The second step consists of etching the NPs from 784 nm to ~240 nm and from 607 nm to ~125 nm using an O_2 plasma in a Matrix Distributed Electron Cyclotron Resonance (MD-ECR) high density plasma reactor with the following settings: 500 W microwave power, cycled with a duty cycle of 33 percent and with a period of 45 seconds to lower the thermal load.

A 30 nm SiC_x film is then deposited on the samples passivated with Al_2O_3, while a 30 nm of Al film is evaporated on the samples passivated with SiO_2. The NPs are then removed, leaving nanoholes in the SiC_x and the Al films. SEM images of the surface after these steps are shown in fig. 9. The initial 607 nm NPs were reduced to around 125±17 nm and

978-1-5090-5606-4/17 $31.00 © 2017 IEEE

create hole diameters around 124 ± 35 nm and 200 ± 18 nm after NP removal for the samples covered with SiC_x and Al, respectively. As for the 784 nm NPs, they were reduced in size to 241 ± 13 nm and create hole diameters of around 134 ± 25 nm and 255 ± 28 nm for the samples covered with SiC_x and Al, respectively. From this data we can conclude that smaller holes are created when SiC_x is used than with Al, probably due to the more conformal PECVD process of the SiC_x.

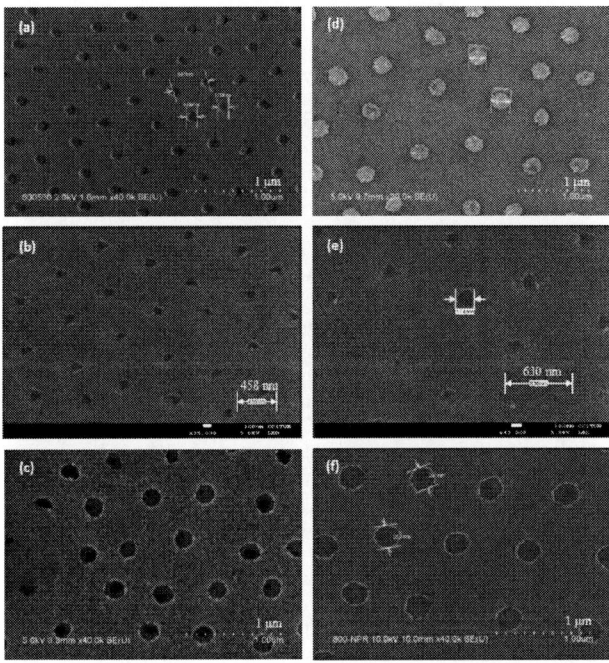

Fig. 9. SEM images at various points during NP processing for 607 nm (a,b,c) and 784 nm (d,e,f) NPs: (a,d) after dry etching, (b,e) after SiC_x coverage and NP removal, and (c,f) after Al coverage and NP removal.

For the Al_2O_3 films, after opening the holes in the SiC_x different wet etching conditions of the passivating film have been explored, all based on the well-known RCA1 cleaning, a mixture of NH_4OH, H_2O_2 and H_2O (1:1:8) [18]. For these samples, we have varied the ramp up time (3.5 minutes for RCA1-3 and RCA1-5 and 9 minutes for the rest) and the final temperature (see fig. 10). The total etching time for all samples was 20 minutes, except for RCA1-7 with 23 minutes and RCA1-8 with 30 minutes. After SEM analysis, we present the percentage of opened holes (fig. 10 (a)), and the size of the opened holes (fig. 10 (b)) as a function of the final temperature.

It is seen from figure 10 (a) that when the temperature is kept lower than 50 °C, less than 38% of holes are opened, independent of all other parameters. However, by increasing the temperature to around 55-58°C, this percentage increases to between 34 % and 70 %. As well, higher temperatures lead to larger holes, but with more dispersion in their sizes.

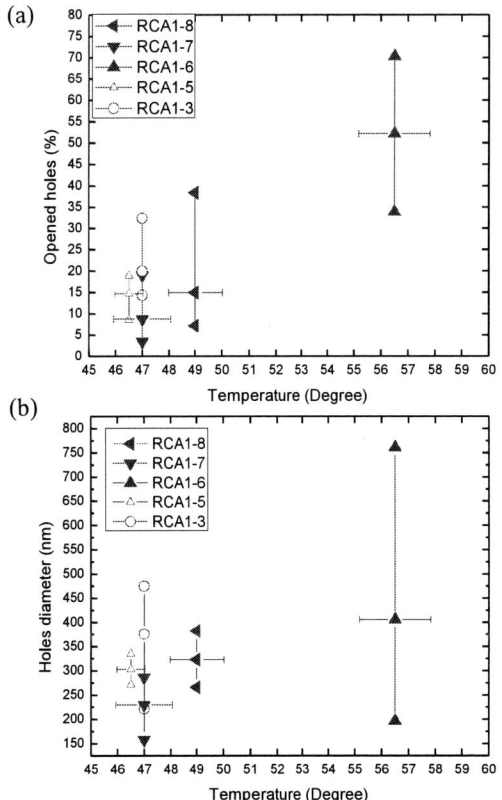

Fig. 10. (a) Fraction of opened holes and (b) opened hole diameter as a function of the temperature used for all RCA1 trials.

Finally, topography measurements done by AFM (figure 11) also show that the Al_2O_3 film is inhomogeneously etched over the surface of the sample.

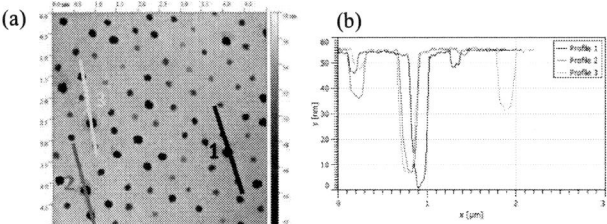

Fig. 11. (a) AFM topography image; (b) depth measurement along the indicated line cuts.

For Approach B, we grew a 240 nm thick thermal SiO_2 layer and followed the contact definition procedure of fig. 8. After (CHF_3,Ar) plasma etching, SEM images show that the holes in the SiO_2 layer are homogenously and completely created (fig. 12). However, their size has increased to 339-370 nm compared to the 200-255 nm initially measured after NP removal. This hole enlargement is attributed to SiO_2 under-etching during the plasma treatment.

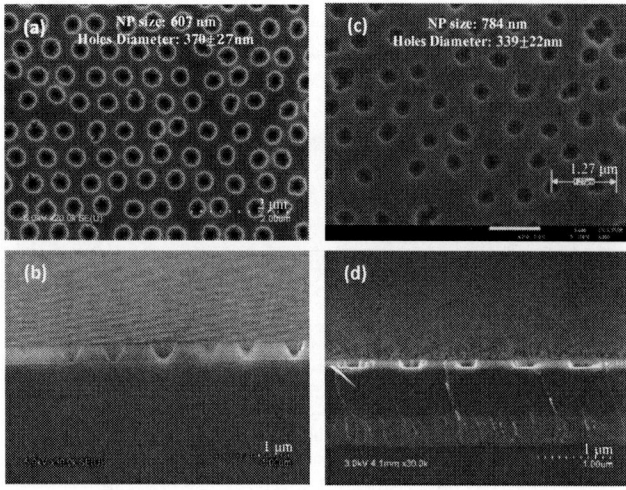

Fig. 12. SEM top view and cross section images of the sample where holes are formed using (a), (b) 607 nm NPs, and (c), (d) 784nm NPs.

This second technique for contact definition on SiO_2 layers has demonstrated much better results. Formation of p^+ regions by thermal diffusion and ion implantation are currently underway, and more complete results will be reported elsewhere.

V. CONCLUSIONS

In the first part of this work, we focused on the advantages of partially rear contacted surfaces on PERC/L cells with contacts in the 1 μm range. We demonstrated that lower ohmic losses with small contacts can be obtained due to the shorter pitches compared to big contacts. However, for similar $S_{eff,rear}$ values passivated contacts with S_{cont} lower than 10^3 cm/s are necessary. Applying these results in PC-1D simulated solar cells, similar maximum efficiencies can be obtained for non- and well-passivated contacts, while a clear advantage is seen for S_{cont} in the 10^3-10^5 cm/s range. In the second part, we proposed two procedures based on polystyrene NPs to define the desired contact pattern on Al_2O_3 and thermally grown SiO_2 films. Through SEM imaging, we show that the holes formed in the Al_2O_3 are not homogeneously defined nor completely etched. On the other hand, much better results are obtained for SiO_2 films and plasma etching. This contact pattern could be used for p^+ region formation.

ACKNOWLEDGEMENT

Our acknowledgement to the InnoEnergy PhD School Programme and the European Institute of Technology (EIT) for supporting part of this research. This work has been supported by MINECO through project TEC2014-59736-R.

REFERENCES

[1] A. W. Blakers, A. Wang, A. M. Milne, J. Zhao, and M. A. Green, "22.8% efficient silicon solar cell," *Applied Physics Letters*, vol. 55, no. 13, pp. 1363–1365, 1989.

[2] M. A. Green, "The path to 25% silicon solar cell efficiency: History of silicon cell evolution", *Progress in Photovoltaics* 17, pp. 183-189, 2009.

[3] "Photovoltaic Report". [Online] Fraunhofer-ISE, 2016. https://www.ise.fraunhofer.de/de/downloads/pdf-files/aktuelles/photovoltaics-report-in-englischer-sprache.pdf.

[4] J.Zhao, A. Wang, M. A. Green, "Series resistance caused by the localized rear contact in high efficiency silicon solar cells", *Solar Energy Materials and Solar Cells*, vol. 32, no.1, pp. 89-94, 1994.

[5] K.R. Catchpole, A.W. Blakers, "Modelling the PERC structure for industrial quality silicon", *Solar Energy Materials and Solar Cells*, vol. 73, no.2, pp. 189-202, 2002.

[6] B. Fischer, "Loss analysis of crystalline silicon cells using photoconductance and quantum efficiency measurements" (PhD Thesis), University of Konstanz, 2003.

[7] H. Plagwitz, R. Brendel, "Analytical model for the diode saturation current of point-contacted solar cells", *Progress in Photovoltaics: Research and Applications*, 14, 1-12, 2006.

[8] P. Saint-Cast, M. Rüdiger, A. Wolf, M. Hofmann, J. Rentsch, R. Preu, "Advanced analytical model for the effective recombination velocity of locally contacted surfaces", *Journal of Applied Physics*, 108, 013705, 2010.

[9] R. Peibst et al., "Implementation of n+ and p+ Polo junctions on front and rear side of double-side contacted industrial silicon solar cells", *Proceedings of 32th EUPVSEC*, 323-327, 2016.

[10] S. W. Glunz, F. Feldmann, A. Richter, M. Bivour, C. Reichel, H. Steinkemper, J. Benick, M. Hermle, "The irresistible charm of a simple current flow pattern - 25% with a solar cell featuring a full-area back contact", *Proceedings of the 31st EU-PVSEC*, pp. 259-263, 2015.

[11] A. Descoeudres, Z. Holman, L. Barraud, S. Morel, S. De Wolf and C. Ballif, ">21% Efficient silicon heterojunction solar cells on n- and p-Type wafers compared", *IEEE J. Photovoltaics*, vol. 3, no.1, pp. 83-89, 2013.

[12] S. De Vecchi, T. Desrues, F. Souche, D. Munoz, P.–J. Ribeyron and M. Lemiti, "Point contact technology for silicon heterojunction solar cells", *Energy Proc.* 27, pp. 549-554, 2012.

[13] J.M. López-González, I. Martín, P. Ortega, R. Alcubilla, "Numerical simulations of rear point-contacted solar cells on 2.2Ωcm p-type c-Si substrates", *Progress in Photovoltaics: Research and Applications* 23, pp- 69-77, 2013.

[14] C. Battaglia, X. Yin, M. Zheng, I.D. Shar, T. Chen, S. McDonnell, A. Azcatl, C. Carraro, B. Ma, R. Maboudian, R.M. Wallace and A. Javey, "Hole selective MoO_x contact for silicon solar cells", *Nano Letters* 14, pp. 967–971 2014.

[15] A. Descoeudres, Z.C. Holman, L. Barraud, ">21% Efficient Silicon Heterojunction Solar Cells on n- and p-Type Wafers Compared", *IEEE J. Photovoltaics* 3, pp. 83-89, 2012.

[16] P. Ortega *et al.*, "p-type c-Si solar cells based on rear side laser processing of Al_2O_3/SiC_x stacks," *Sol. Energy Mater. Sol. Cells*, vol. 106, pp. 80–83, Nov. 2012.

[17] J. R. Oh, J. H. Moon, S. Yoon, C. R. Park, and Y. R. Do, "Fabrication of wafer-scale polystyrene photonic crystal multilayers via the layer-by-layer scooping transfer technique," *J. Mater. Chem.*, vol. 21, no. 37, p. 14167, 2011.

[18] W. Kern, "The Evolution of Silicon Wafer Cleaning Technology", Journal of The Electrochemical Society, vol. 137, no. 6, pp. 1887-1892, 1990.

978-1-5090-5606-4/17 $31.00 © 2017 IEEE

Entrance of Low Cost Fabrication of Back-Contact Heterojunction Solar Cells by Using Plasma Ion Implantation

Koichi Koyama, Keisuke Ohdaira and Hideki Matsumura

Japan Advanced Institute of Science and Technology (JAIST),

Asahidai, Nomi-shi, Ishikawa-ken, k-koyama@jaist.ac.jp, 923-1292, JAPAN

Abstract — Back-contact amorphous-silicon (a-Si) /crystalline-silicon (c-Si) heterojunction appears the best structure to realize solar cells with the world top-class efficiency. However, the fabrication cost is a draw-back due to complicated patterning process for making n- and p- back-electrodes. To overcome it, we attempted to make n-a-Si islands in p-a-Si layers by plasma ion-implantation of phosphorus (P) atoms. We discovered that high temperature annealing was not necessary after plasma ion-implantation, contrary to ion implantation into c-Si, and we succeeded to convert p-type a-Si (p-a-Si)/intrinsic a-Si (i-a-Si)/c-Si hetero-structure to n-type a-Si (n-a-Si)/i-a-Si/c-Si without degradation of carrier lifetimes or heterojunction quality.

Index Terms — Ion implantation, Catalytic chemical vapor deposition, Carrier lifetime, Heterojunction interdigitated back contact solar cells, Secondary ion mass spectrometry.

I. INTRODUCTION

The heterojunction back-contact (HBC) solar cells have achieved the world top and second records of efficiency, 26.3% and 25.6%, owing to low surface recombination velocity and no shadowing loss of front electrodes [1]-[3]. However, the cost for fabrication of HBC solar cells is not so low, since the back electrodes are usually formed by complicated patterning process such as photolithography. Thus, if a new simple technology is developed for conversion of conduction type of heterojunction electrodes at certain patterned areas, it will be the first entrance gate toward low-cost fabrication of HBC solar cells. For converting the conduction type, the ion implantation appears the best method, since the patterning process would be simplified by using a hard mask [4]. Among various methods of ion implantation, plasma ion implantation has advantages of low system cost and short process time for high dose ion-implantation. For instance, the ion implantation of P atoms with a dose of 10^{16} cm^{-2} can be completed within a few second [4]. Thus, here, we attempted to apply plasma ion implantation to make patterned p-a-Si/i-a-Si/c-Si and n-a-Si/i-a-Si/c-Si areas by converting the conduction type of some parts of p-a-Si/i-a-Si/c-Si to n-a-Si/(i-a-Si)/c-Si.

There have been some reports on ion implantation into a-Si; however, the report dealing with ion implantation into HBC solar cells appears only few. Recently, T. Carrere et al. reported on the plasma emersion ion implantation onto HBC

solar cells and claimed that the conductivity of a-Si could be changed [5]. On the other hand, if the p-a-Si conductive type can be change to n-type using phosphine (PH$_3$) ion implantation, so-called *counter doping*, it becomes more cost-effective fabrication method of HBC solar cells.

We have not known what happens after ion implantation of counter doping, particularly, after plasma ion implantation onto a-Si/c-Si heterojunction structures. To make clear this important issues, the present work is to study the feasibility of low cost fabrication of HBC solar cells by using plasma ion implantation technology with various conditions. It is found that p-a-Si layers are converted to n-a-Si layers without degradation of passivation quality of heterojunction, or, without degradation of effective minority carrier lifetimes (τ_{eff}), when the proper requirements for annealing process and thickness of a-Si layers are satisfied.

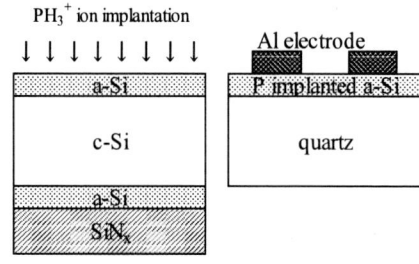

Fig. 1. Schematic view of PH$_3^+$ ion implantation into the passivated c-Si sample for investigation of the interface quality (left) and implanted a-Si on the quartz for conductivity measurements (right).

II. EXPERIMENTAL METHODS

The passivation quality of a-Si/c-Si heterojunction is evaluated by the τ_{eff} measured by microwave photoconductive decay method (μ-PCD). n-type 300 μm-thick (100) 1-5 Ωcm FZ c-Si was used as sample substrates.

The c-Si wafers were cleaned by 5% diluted hydrofluoric acid solution to remove native oxide. Figure 1 shows the passivated structure for ion implanted samples. The back side of the c-Si substrates was coated with Cat-CVD (Hot-Wire

CVD) silicon-nitride (SiN$_x$)/a-Si stacked passivation layers, since the surface recombination velocity (SRV) of such stacked passivation layers can be reduced to as low as 0.18 cm/s [6-8]. i-a-Si layers with various thicknesses were deposited on the front side of c-Si substrate by Cat-CVD. P atoms were implanted with various doses onto them. Ion implantation was carried out with ULVAC PVI-3000 which has no mass spectrometry. The energy of ion implantation was fixed at 5 keV.

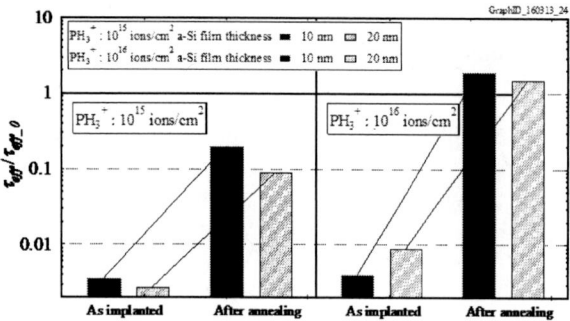

Fig. 2. Normalized τ_{eff} of P-ion-implanted samples with a-Si passivation layers, taking a-Si film thickness as a parameter.

P atoms were provided by plasma decomposition of PH$_3$ gas. Thus, hydrogen (H) atoms may have also been implanted with the energy of 5 keV together with P atoms. After ion implantation, the samples were annealed at various temperatures in air on a hot-plate. For counter doping, the samples having p-a-Si/a-Si stack layers in front side were used.

Profiles of implanted atoms were measured by Substrate Side Depth Profile - Secondary Ion Mass Spectrometry (SSDP -SIMS), which is SIMS measurement from back-side of samples.

The conductivity of samples before and after ion-implantation was measured by forming coplanar electrodes on a-Si layer deposited on quartz substrates.

III. RESULTS AND DISCUSSION

Figure 2 shows the τ_{eff} of samples after P implantation, which is normalized by the values before ion-implantation. The samples with i-a-Si layers with various thicknesses were annealed at 250 °C for 210 min. The τ_{eff} of as-deposited samples are around 1-3 ms. In any samples, τ_{eff} after ion implantation is about only 10 μs. Since a lot of defects are created in samples by ion bombardment, a-Si passivation quality is degraded. However, the situation appears a little bit complicated, when we compare the results of 250 °C annealed samples with 10^{15} and 10^{16} cm^{-2} ion doses and 10 and 20 nm thickness of a-Si layers.

In the case of samples implanted with a dose of 10^{15} cm^{-2}, τ_{eff} is not recovered completely by annealing at 250 °C

although τ_{eff} is increased about 10 times larger than that before annealing. On the other hand, in the samples with a dose of 10^{16} cm^{-2} the τ_{eff} can be completely recovered to original values, which is 2 to 3 ms in this case. It is also demonstrated that thinner a-Si samples tends to recover the passivation quality more. In the conventional ion implantation with a mass-separator, higher dose may introduce more damages and causes to lower τ_{eff}. However, the results of heterojunction structure is different.

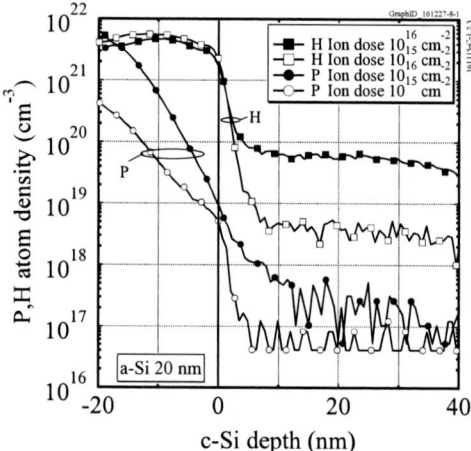

Fig. 3. SIMS profiles of P and H atoms.

Figure 3 shows the profiles of P and H atoms. Amount of P atoms penetrating into c-Si for the sample with a dose of 10^{16} cm^{-2} is 5 times larger than that of 10^{15} cm^{-2} dose sample. Since P is incorporated in the c-Si and induces the change of Fermi level of c-Si at the interface, the band bending effect might be created. This may be the reason why the τ_{eff} for the samples of dose of 10^{16} cm^{-2} is recovered.

In addition, as shown in Fig. 3, H atoms of the order of 10^{18} to 10^{19} cm^{-3} are incorporated inside c-Si, probably due to the implantation of H atoms with the energy of 5 keV. This may eliminate defects inside c-Si and cause the band bending in addition to the effect of the change of Fermi level of c-Si surface.

Figure 4 shows the τ_{eff} after P ion implantation into p-a-Si/i-a-Si/c-Si structure as a function of implanted dose. The τ_{eff} is again recovered to initial values (in this case, about 1 ms) when the dose is set at optimum values. Here, we have to notice that the τ_{eff} of 10^{17} cm^{-2} samples is not recovered completely. This may be due to the residual damage created by ion-implantation.

Figure 5 shows the I-V characteristics for two types of samples shown in the inset of the figure. P atoms were implanted into p-a-Si/i-a-Si/c-Si samples and on them indium tin oxide (ITO) films were deposited by sputtering. The insertion of ITO between a-Si layer and metal layers which were deposited on ITO by vacuum evaporation is to avoid the

978-1-5090-5606-4/17 $31.00 © 2017 IEEE

penetration of metal elements into a-Si layers. After then, the samples were annealed at 200 °C for 30 minutes. The similar process was done for p-a-Si/i-a-Si/c-Si without implanted samples. Both samples were patterned to make coplanar electrodes. By this procedure, a Type-A of samples has the structure of metal/ITO/P-implanted p-a-Si/n-type c-Si and the other of Type-B has metal/ITO/p-a-Si/i-a-Si/n-type c-Si. Since the Type-B includes two diodes in the current path, it becomes very resistive. In the Type-A only the a-Si resistors should be included if p-a-Si is converted to n-a-Si. The I-V characteristics clearly demonstrate the conductance of the Type-A samples, showing the conversion of conduction type from p-a-Si to n-a-Si by plasma ion implantation.

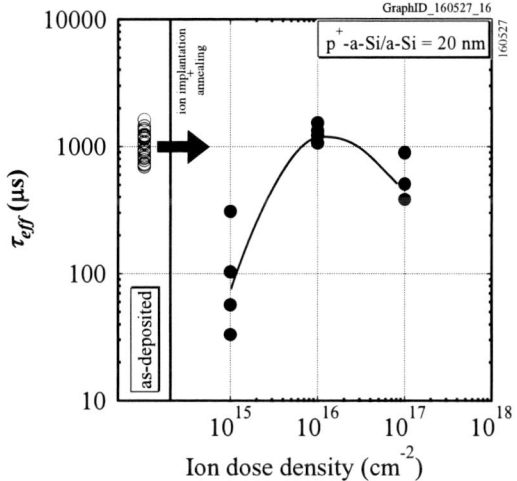

Fig. 4. τ_{eff} of counter-doping samples as a function of ion dose density.

According to the conduction of measurement of after P implantation into p-a-Si/i-a-Si/n-type c-Si samples, the conversion of conduction type from p-a-Si to n-a-Si is clearly verified. Based on the results and discussions above, we may say that the dose of P ion implantation of 10^{16} cm^{-2} is a unique value for the sample with a 20 nm-thick a-Si layer on n-type c-Si. The conductivity can be increased without any residual negative effects on τ_{eff} for such samples.

IV. CONCLUSIONS

As described above, plasma ion implantation into a-Si/c-Si hetero-structures is completely different from the conventional ion-implantation into c-Si using mass-separation machine. By plasma ion-implantation of P atoms, heterojunction p-a-Si/i-a-Si/c-Si layers can be converted to n-a-Si/(i-a-Si)/c-Si layers by the process of only a few second and annealing at only 250 °C. The τ_{eff} after ion implantation and annealing can be kept at the

values before implantation. The results appear to show the promising feasibility for simplifying the fabrication process of HBC solar cells or developing low cost fabrication processes.

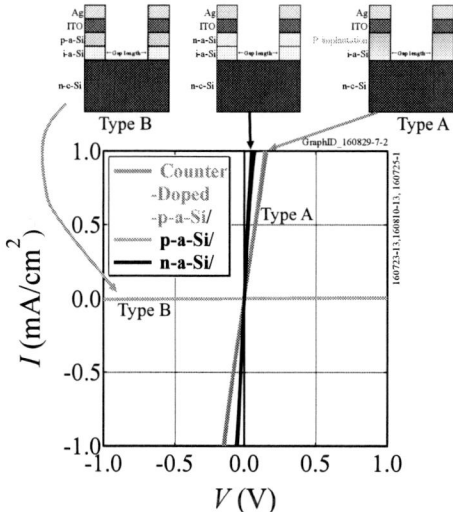

Fig. 5. I-V characteristics of metal/ITO/P-implanted p-a-Si/n-type c-Si (Type-A) and metal/ITO/p-a-Si/i-a-Si/n-type c-Si (Type-B).

ACKNOWLEDGEMENT

This work is supported by The New Energy and Industrial Technology Development Organization (NEDO). We furthermore would like to thank ULVAC, Inc. for the ion implantation.

REFERENCES

[1] K. Masuko, M. Shigematsu, T. Hashiguchi, D. Fujishima, M. Kai, N. Yoshimura, T. Yamaguchi, Y. Ichihashi, T. Mishima, N. Matsubara, T. Yamanishi, T. Takahama, M. Taguchi, E. Maruyama, and S. Okamoto, *IEEE J. Photovoltaics*, 4, 1433 (2014).

[2] H J. Nakamura, N. Asano, T. Hieda , C. Okamoto, H. Katayama, and K. Nakamura, *IEEE J. Photovoltaics*, 4, 1491 (2014).

[3] http://www.kaneka.co.jp/kaneka-e/images/topics/1473811995/1473811995_101.pdf

[4] https://www.ulvac.co.jp/wiki/en/crystalline-sillicon-solar-cells/

[5] T. Carrere, D. Munoz, M. Coig, C. Longeaud and J. P. Kleider, *Sol. RRL* 2017, 1, 1600007

[6] K. Koyama, K. Ohdaira, and H. Matsumura, *Appl. Phys. Lett.*, 97, 082108, (2010).

[7] K. Koyama, K, Ohdaira, and H. Matsumura, *Thin Solid Films*, 519, 4473, (2011).

[8] C. T. Nguyen, K. Koyama, S. Terashima, C. Okamoto, S. Sugiyama, K. Ohdaira, and H. Matsumura, *in Photovoltaic Specialists Conference (PVSC), 2016 IEEE 42th*, #321, in press.

TLM measurements varying the intrinsic a-Si:H layer thickness in silicon heterojunction solar cells

Mehdi Leilaeioun, William Weigand, Pradyumna Muralidharan, Mathieu Boccard, Dragica Vasileska, Stephen Goodnick and Zachary Holman

Arizona State University, Tempe, Arizona, 85287, USA

Abstract — Silicon heterojunction solar cells are a promising device architecture due to their high efficiencies, yet these cells tend to suffer from high series resistances. Until recently, little has been done to understand the main factors contributing to the high resistance. Here we begin a systematic analysis to determine the important interactions between the different layers in the hole-collecting contact consisting of a stack of a-Si:H(i)/a-Si:H(p))/ITO/Ag. We attempt to address how the stack performs when the intrinsic amorphous silicon (a-Si:H(i)) layer thickness is varied. Specifically, we determine how the thickness affects the fill factor of the cell and assess how much of the detriment is due to the contact resistivity. For increasing a-Si:H(i) thicknesses between 4 and 16 nm, we find contact resistivities increasing from 0.48 to 1.9 Ωcm^2 and fill factors decreasing from 76.9% to 71.1%. Additionally, to understand the physics behind these effects, we simulate the contact resistivity variation by modeling the ITO as a Schottky contact and a semiconductor.

Index Terms — TLM, contact resistance, silicon, heterojunction, solar cell, simulation.

I. INTRODUCTION

A complete account of the resistive losses in amorphous silicon/crystalline silicon heterojunction (SHJ) solar cells has been elusive, largely because the resistance of the complete electron- and hole-collecting stacks have not yet been quantified. These resistive losses are detrimental to the overall fill factor (FF) of the cell, which in turn limits the overall efficiency of the device. As is well discussed by Cuevas *et al.*, a good contact material should either have a high or low work function, depending on whether it is a hole or electron contact, and it must have high carrier density or mobility for the carrier type to be collected [1].

The traditional hole-collecting contact stack in SHJ solar cells consists of intrinsic amorphous silicon (a-Si:H(i)), doped amorphous silicon (a-Si:H(p)), indium tin oxide (ITO) and metal on top of a crystalline silicon wafer. This contact structure has been partially, but not fully, characterized: carrier transport through the a-Si:H(i) layer [2,3], as well as through the ITO layer [4], have been simulated and indirectly measured, but the lumped effect of the whole contact stack on device performance has not been described.

Recently, novel materials such as molybdenum oxide (MoO$_x$) and nanocrystalline silicon have been used in place of a-Si:H(p), producing promising results. As part of the development of these materials, techniques from the semiconductor industry for measuring contact resistance have been adapted to entire contact stacks in solar cell. For example, utilizing the measurement method developed by Cox and Strack [4], MoO$_x$ contacts on p and p$^+$ substrates were shown to have minimum contact resistivites of 1 and 0.2 mΩcm^2, respectively [5]. Similarly, transfer length method (TLM) measurements on nanocrystalline silicon revealed contact resistivities on the order of 0.3 Ωcm^2 [6]. These studies provide significant insight into easily measured device parameters (contact resistivity) of SHJ solar cells and their effect on final device performance, yet analogous work describing the traditional a-Si:H-based hole contact has not been reported.

In this paper, we vary the a-Si:H(i) layer thickness between 4 and 16 nm in the hole contact and measure both the FF of complete cells and the contact resistivity of the contact stack using TLM. In order to physically explain the contact resistivity trends, we simulate the TLM structures and focus on the transport between the ITO and a-Si:H(p) layers.

II. MATERIALS AND METHODS

Boron- and phosphorous-doped monocrystalline CZ silicon wafers were double-side textured in KOH and subsequently cleaned in RCA-B and Piranha to remove metal and organic contaminants. The wafers were then dipped in 5% HF solution to remove the native oxide layer and immediately put under vacuum for a-Si:H deposition. a-Si:H stacks were deposited using PECVD (Applied Materials P-5000). Indium tin oxide (ITO) films were reactively sputtered at room temperature using an MRC 944 tool with a DC power supply using a target with a 90/10 In$_2$O$_3$/SnO$_2$ ratio. The thicknesses of a:Si:H and ITO films were measured using a Woollam M-2000 spectroscopic ellipsometer.

To determine the hole contact resistivity and the effect it has on the FF of SHJ cells, varying thicknesses of a-Si:H(i) were deposited on two different sets of samples: p-doped c-Si wafers and n-doped c-Si wafers. In particular, on the back of the cells, the a-Si:H(i) thickness was varied between 4 and 16 nm, while it remained constant on the front at 6 nm. The p-type wafers were used for TLM measurements since an n-type wafer would result in back-to-back diodes after depositing the hole contact stack, whereas the n-type wafers were used to make complete cells to measure the output cell parameters (specifically FF and series resistance). The TLM and cell structures are shown in Figure 1a and b, respectively.

978-1-5090-5606-4/17 $31.00 © 2017 IEEE

TLM measurements were performed at room temperature using pads with lengths of 2 mm, larger than the transfer length of $L_t < 300$ μm. The pad spacing varied between 250 μm and 8 mm along the TLM pattern. Finished cells underwent Suns-Voc measurements to determine the I-V characteristic without series resistance, and they underwent one-sun I-V measurements to determine the actual I-V characteristic. These two curves provide sufficient information to determine the total series resistivity as described by Pysch et al. [7].

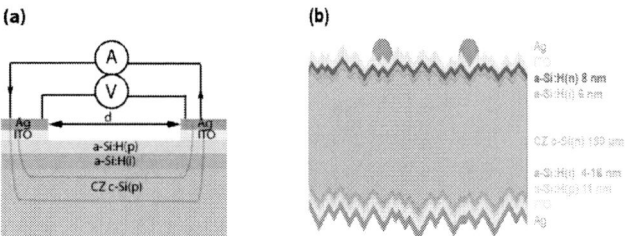

Figure 1: Schematic of (a) TLM structure and (b) cell structure

III. RESULTS AND DISCUSSION

A. Experiment

Figure 2 shows the pseudo-fill factor (pFF) of finished cells extracted from Suns-Voc measurements alongside the FF measured from I-V measurements. As expected, since the pFF is not affected by series resistance, the pFF remains nearly constant at approximately 83%, increasing slightly for thicker a-Si:H(i) layers because of improved surface passivation. The FF, which tends to be strongly affected by the series resistance, drops from 76.1% for the thinnest layer to 71.1% for the thickest layer.

In order to qualitatively understand the effect of series resistivity on the FF, we show series resistivity and contact resistivity in Figure 3 as a function of a-Si:H(i) layer thickness. By increasing the a-Si:H(i) layer thickness from 4

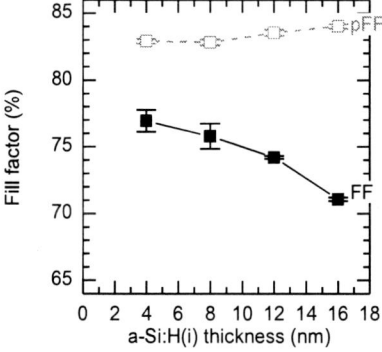

Figure 2: pFF and FF as a function of the a-Si:H(i) layer thickness in the (rear) hole contact. Error bars represent the standard deviation of three cells measured on a single wafer.

to 16 nm, the series resistivity increases from 1.2 to 2.7 Ωcm². The contact resistivity of the full hole contact stack, extracted from TLM measurements, follows a similar trend increasing from 0.48 to 1.9 Ωcm². Within statistical error, the difference between the series and contact resistivity is constant, indicating that all other contributions to the series resistivity are the same and only the changing contact resistivity with increased a-Si:H(i) layer thickness accounts for the increase in series resistivity and drop in FF for this set of experiments.

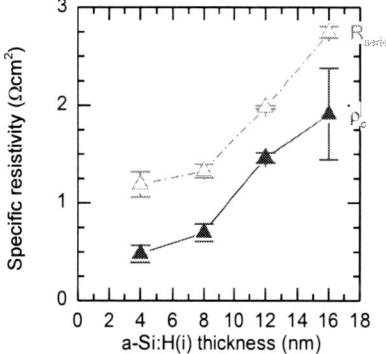

Figure 3: Series resistivity of the cells in Figure 2 and contact resistivity of the hole contact stack as a function of the a-Si:H(i) layer thickness in the (rear) hole contact. Error bars represent the standard deviation of three TLM structures made on the same wafer.

B. Simulation

The Atlas module of the commercial device simulator SILVACO was used to recreate the TLM measurements. The device-level structure contains all of the contact pads along with the appropriate contact spacing to model the electrical characteristics based on various device parameters (a-Si:H(p) layer thickness, a-Si:H(i) layer thickness, doping, etc.). Each pair of contacts was probed separately to generate plots of R_T (total resistance) vs. pad spacing. R_T is calculated by taking the slope of an I-V curve which can be generated by either conducting a voltage sweep or by injecting current into the contacts.

Earlier models designed to study SHJ cells treated the ITO as a metallic contact by assigning it a work function or effective work function [8]. Thus the Schottky barrier formed between the ITO and the a-Si:H(p) layer limited the collection of carriers. However, the ITO is more accurately a degenerately doped n-type semiconductor which forms a conductive layer. Thus the p+/n/n+ cell structure shown in Figure 1 can be interpreted as an n++/p+/n/n+ structure. The major implication of this interpretation is that forward biasing the cell junction results in a reverse-biased ITO(n+)/a-Si:H(p) junction [9]. Using simulations, we have studied the role of tunneling at the Schottky contact and the reverse biased junction, and its effect on contact resistance.

978-1-5090-5606-4/17 $31.00 © 2017 IEEE   1791

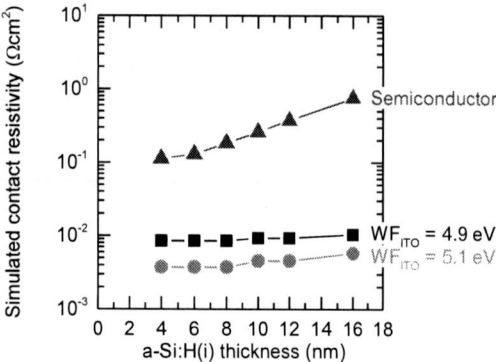

Figure 4: Simulated contact resistivity as a function of a-Si:H(i) layer thickness in the (rear) hole contact with the ITO treated as a Schottky contact with different work functions and as an n-type semiconductor.

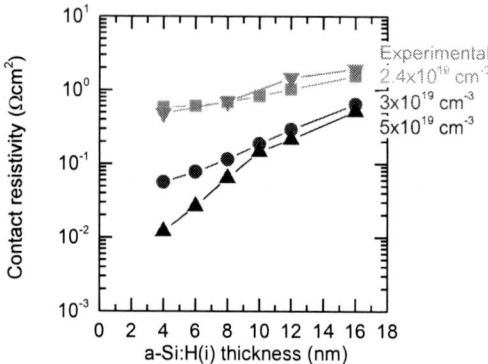

Figure 5: Simulated and experimental contact resistivity as a function of a-Si:H(i) layer thickness in the (rear) hole contact for three a-Si:H(p) doping densities. Also shown is the experimental data from Figure 3.

Figure 4 shows the simulated contact resistance as a function of the a-Si:H(i) layer thickness. The gray and black lines show the variation of contact resistance when the ITO layer is treated as a Schottky contact. The results presented in Figure 4 assumed an a-Si:H(p) layer thickness of 15 nm and a doping density of 3 x 10^{19} cm^{-3}. Due the treatment of the ITO layer as a Schottky contact, the current-voltage characteristic is limited by the barrier ($\Delta E = WF_{ITO} - WF_{a\text{-}Si:H}$) formed at the ITO/a-Si:H(p) interface due the work function mismatch. The work function of a conductive ITO layer is approximately 4.8 eV [10] and the work function of the a-Si:H(p) layer is approximately 5.2 eV [11]. In order to obtain any current through the device, it is imperative to include tunneling through the Schottky barrier. The gray (WF_{ITO} = 5.1 eV) and black (WF_{ITO} = 4.9 eV) lines in Figure 4 show the contact resistivity for relatively high ITO work functions. Simulations indicate that a high ITO work function (>4.9 eV) results in nearly perfect Ohmic behavior and thus very low contact resistances; however, for low ITO work functions (<4.9 eV) the contact resistivity increases by orders of magnitude. This indicates that, with this model, the transport is heavily dependent on the ITO parameters and the band offset created due to the work function mismatch, which is not observed in experiments (several transparent conductive oxides have been used to make SHJ cells with high FF).

As mentioned earlier, the ITO is a semiconductor with a measurable bandgap of approximately 3.7 eV. Thus, in reality, the ITO(n$^+$)/a-Si:H(p) interface forms a p-n junction. The blue line in Figure 4 is calculated by treating the ITO as a semiconductor. As both of these layers are heavily doped, they form a tunnel junction where carriers tunnel from the valence band of the a-Si:H(p) to the conduction band of the ITO. A non-local band-to-band tunneling model is used to describe carrier transport across the junction [12]. The magnitude of the tunneling current is dictated by the doping and thickness of the a-Si:H(p) and ITO layers. This makes the I-V characteristics less sensitive to the properties of the ITO and more dependent on the properties of the p-n junction. The blue line in Figure 4

shows a more realistic description of contact resistance as compared to the Schottky contact treatment of the ITO.

Figure 5 shows a family of simulated contact resistivities (blue) and the experimentally measured (red) contact resistivity. The simulations were fit to the experimentally measured contact resistivity by varying the doping of the ITO and a-Si:H(p) layers. It is evident from Figure 5 that the contact resistivity increases with decreasing a-Si:H(p) doping, as the low doping causes that layer to become depleted: a thin a-Si:H(p) layer with low doping would result in a depleted hole contact giving rise high contact resistances. Thus it is imperative to have an optimum combination of doping and thickness for the a-Si:H(p) layer. The closest fit to the experimental contact resistivity was achieved by using an a-Si:H(p) doping of 2.4 x 10^{19} cm^{-3} and an ITO doping of 10^{20} cm^{-3}. As discussed previously, the inclusion of non-local band-to-band tunneling is required to obtain meaningful current-voltage characteristics. The exclusion of the tunneling model leads to incredibly high (MΩ) contact resistance.

IV. CONCLUSIONS

This paper demonstrates that the a-Si:H(i) layer of the hole contact plays a key role in determining the overall FF of silicon heterojunction cells. TLM measurements with varying a-Si:H(i) layer thickness indicate that this is due to the contact resistivity of the hole contact stack, which can exceed 1 Ωcm^2 for thicknesses greater than 10 nm. The TLM experiments were complemented by simulations that treat carrier transport at the ITO/a-Si:H(p) interface by modeling the ITO as a semiconductor (as opposed to a metallic Schottky contact) and including band-to-band tunneling models to explain trends in contact resistivity variation. This theoretical approach supports the experimental observations of increasing contact resistance with the thicker a-Si:H(i) layers by providing good fits to measured data. The same approach employed here—in which solar cells and complementary TLM structures are simultaneously investigated—can be employed to reveal other sources of FF loss in SHJ cells.

978-1-5090-5606-4/17 $31.00 © 2017 IEEE

REFERENCES

[1] U. Wurfel, A. Cuevas, Peter Wurfel, "Charge Carrier Separation in Solar Cells," *IEEE Journal of Photovoltaics,* vol 5, pp. 461-169, 2015.

[2] R. Crandall, E. Iwaniczko, J. Vi, M. Page "A comprehensive study of hole collection in heterojunction solar cells," *Journal of Applied Physics*, vol 12, 2012.

[3] M. Taguchi, E. Maruyama, M. Tanaka, "Temperature Dependence of Amorphous/Crysatalline Silicon Heterojunction Solar Cells," *Japanese Journal of Applied Physics*, vol 47, 2008.

[4] R.H. Cox, H. Strack, "Ohmic Contacts for GaAs Devices", *Solid-State Electronics*, vol. 10. pp. 1213-1218, 1967

[5] J. Bullock, A. Cuevas, T. Allen, C. Battaglia, "Molybdenum oxide MoOx: A versatile hole contact for silicon solar cells," *Applied Physics Letters,* vol 105, 2014.

[6] G. Nogay, J. Seif, Y. Riesen, A. Tomasi, Q. Jeangros, N. Wyrsch, F. Haug, S. De Wolff, C. Ballif, "Nanocrystalline silicon carrier collectors for silicon heterojunction solar cells and impact on low-temperature device characteristics," *IEEE Journal of Photovoltaics*, vol 6, pp 1652-1662, 2016.

[7] D. Pysch, A. Mette, and S.W. Glunz, "A review and comparison of different methods to determine the series resistance of solar cells," *Solar Energy Materials and Solar Cells*, vol. 91, pp. 1698-1706, 2007.

[8] R. Lachaume, W. Favre, P. Scheiblin, and X. Garros, "Influence of a-Si : H / ITO interface properties on performance of heterojunction solar cells," *Energy Procedia*, vol. 38, pp. 770–776, 2013.

[9] A. Kanevce and W. K. Metzger, "The role of amorphous silicon and tunneling in heterojunction with intrinsic thin layer (HIT) solar cells," *J. Appl. Phys.*, vol. 105, no. 9, 2009.

[10] A. Klein *et al.*, "Transparent conducting oxides for photovoltaics: Manipulation of fermi level,work function and energy band alignment," *Materials (Basel).*, vol. 3, no. 11, pp. 4892–4914, 2010.

[11] M. Bivour, C. Reichel, M. Hermle, and S. W. Glunz, "Improving the a-Si:H(p) rear emitter contact of n-type silicon solar cells," *Sol. Energy Mater. Sol. Cells*, vol. 106, pp. 11–16, 2012.

[12] "SILVACO, Atlas User's Manual," 2015.

Solar Cells Application of p-type poly-Si Thin Film by Aluminum Induced Crystallization

Shota Masuda[1], Kazuhiro Gotoh[1], Isao Takahashi[1], Kyotaro Nakamura[2], Yoshio Ohshita[3] and Noritaka Usami[1]

[1]Department of Materials Process Engineering, Graduate School of Engineering, Nagoya University, Furocho, Chikusa, Nagoya, 464-8603, Japan

[2]Meiji University, 1-1-1 Higashimita, Tama, Kawasaki, 214-8571, Japan

[3]Toyota Technological Institute, 2-12-1 Hisakata, Tempaku, Nagoya, 468-8511, Japan

Abstract — We applied aluminum-induced crystallization (AIC) as a technique to realize heavily doped p-type polycrystalline Si layer at low temperature for the rear side of tunnel oxide passivated contact concept solar cell. Systematic temperature variation of AIC revealed that annealing at a temperature slightly lower than the eutectic temperature (577°C) is effective to improve solar cell performance. This is explained by reduced resistivity for low contact resistance while avoiding melting of Al-Si.

Index Terms — aluminum-induced crystallization, passivation, solar cell, TOPCon, silicon.

I. INTRODUCTION

In 2013, Feldmann *et al.* demonstrated that tunnel oxide passivated contact (TOPCon) concept solar cell achieved conversion efficiency of 23.7 % [1]. One of the advantages of TOPCon concept is to use full coverage of ultra-thin oxide as a passivation layer so that one does not need partial contact schemes like passivated emitter and rear locally diffused (PERL) cell. This concept attracts attention as a high efficiency solar cell with good thermal stability to avoid complex structuring steps [2].

As a fabrication process of p-base TOPCon, thermal treatment for 60 min at a plateau temperature of 800 °C was reported for crystallization of B-doped amorphous Si (a-Si) at the rear side [3]. It would be ideal if we could obtain heavily doped p-type polycrystalline Si (poly-Si) at lower temperature to minimize the thermal budget.

Aluminum-induced crystallization (AIC) is known as a technique to permit fabrication of high quality, heavily doped p-type crystalline Si (c-Si) layer at a temperature below eutectic temperature (577°C) of Al-Si [4,5]. In this method, we deposit stacked layers of a-Si and Al on substrate followed by annealing at about 500 °C. Si atoms diffuse into Al layer and crystal growth takes place driven by supersaturation. After a certain time of annealing, the process was ended by layer exchange to result in poly-Si layer. The AIC grown poly-Si layers have advantageous features for doped layers of solar cells to introduce a back surface field (BSF) since Al atoms

dissolve up to the solubility limit in c-Si. The solubility limit of Al in c-Si increases with increasing temperature [6,7]. As a result, it is assumed that carrier density is also increased and passivation quality is improved. Particularly, the open-circuit voltage (V_{oc}) and filling factor (FF) would be improved by increasing heat treatment temperature for AIC process.

In this work, we investigated the effect of variation of AIC process annealing temperature on solar cell performance.

II. EXPERIMENTAL DETAILS

A. Sample Preparation for Hall Measurement

We prepared glass substrate about 1.7cm square and cleaned with acetone and Semicoclean No.56. Then, we deposited Al and a-Si which thicknesses were 70nm and 100nm, respectively. The samples were annealed at 400-600 °C for AIC growth. To remove the Al layer on AIC-grown poly-Si, the samples were dipped in 5% HF followed by deposition of AuGa as a point electrode by thermal evaporation.

B. Process of Cell Fabrication

Fig. 1 shows the p-based TOPCon cell structure with p^+-Si grown by AIC at the rear side. The substrate was FZ (float zone) p-type (1.0Ωcm) c-Si planar wafer (d=300μm). After cleaning the substrate, n-type emitter was formed by POCl₃ diffusion. SiNx film (80nm) was deposited on the emitter side of the wafer by plasma enhanced chemical vapor deposition. After cutting wafer to about 1.7cm square, we cleaned the sample by Semicoclean No.23 and 5% HF treatment. Then the SiO₂ layer as a chemical passivation layer was formed by ozonized water for 10 minutes. The thickness of the SiO₂ layer was measured as ~1.1 nm by spectroscopic ellipsometry. In order to improve the passivation quality, the samples was annealed at 400 °C in a forming gas (Ar + 3% H₂) for 30 minutes. Next, we fabricated surface electrode with screen printing of Ag paste and fired at 810 °C in a tube furnace for 40 seconds. Al (70nm) and Si (100nm) were deposited on the back side of samples by RF magnetron sputtering. Then, they

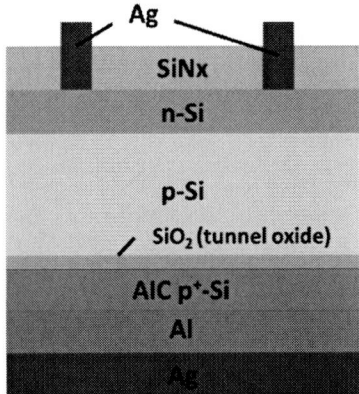

Fig. 1   Cell structure of TOPCon with p⁺-Si grown by AIC

were annealed at 400-600°C for AIC growth. Last, we formed rear Ag electrode with screen printing and dried at 200 °C for 30 minutes.

III. RESULTS AND DISCUSSION

A.   Results of Hall Measurement

Fig.2 and Fig.3 show carrier density and resistivity obtained by Hall measurement, respectively. We confirmed that AIC poly-Si layer is p-type and have sufficient carrier density to induce BSF. The resistivity was found to show minimum at 550-570 °C, which corresponds to a temperature slightly lower than the eutectic temperature. It is considered that Si does not homogeneously dissolve in the Al layer above the eutectic temperature, which explains increased resistivity at 600°C.

B.   J-V Measurement

Fig.4 shows a result of J-V measurements for various temperatures for AIC. At 400-570°C, we confirmed that cell performance improved by increasing temperature for AIC.

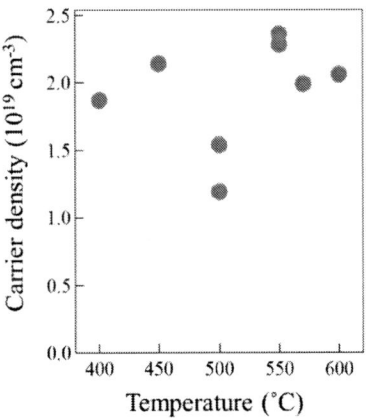

Fig. 2   Carrier density in p⁺-Si layer as a function of the temperature for AIC.

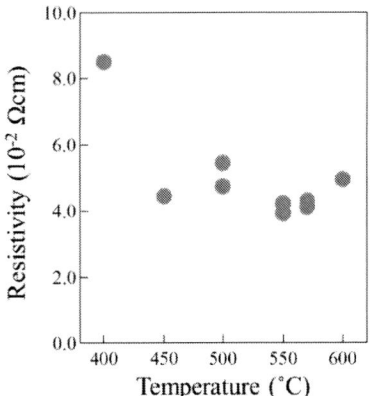

Fig. 3   Resistivity of p⁺-Si layer as a function of the temperature for AIC.

Particularly, FF was drastically increased at 550°C. The maximum efficiency was 12.26% at 570°C near eutectic temperature (see TABLE I). From the result of Hall measurement, the resistivity of AIC poly-Si layer shows minimum at 550-570°C. It seems that this factor improved cell parameter, especially FF.

However, performance of the sample with heat treatment above eutectic temperature was deteriorated. It can be seen from J-V curve of 600°C that the FF decreased due to the influence of the series resistance component. Doped layer with high resistivity formed at 600°C would explain the deterioration of FF. Therefore, we can conclude that the heat treatment temperature of AIC should be chosen at slightly lower than the eutectic temperature.

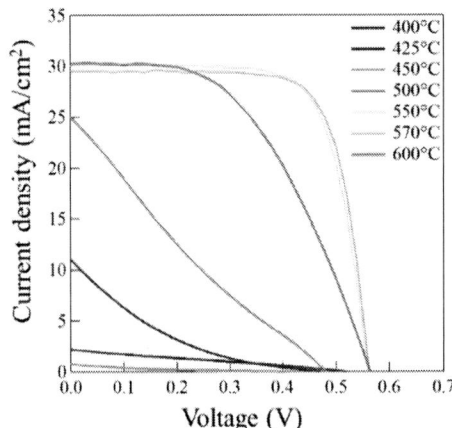

Fig. 4   J-V curves of solar cells with AIC Si grown under different annealing temperatures.

978-1-5090-5606-4/17 $31.00 © 2017 IEEE   1795

TABLE I

SOLAR CELL PARAMETERS OF CELLS WITH AIC UNDER DIFFERENT ANNEALING TEMPERATURE.

T(°C)	J_{sc}(mA/cm^2)	V_{oc}(V)	FF	η(%)
400	2.17	0.520	0.250	0.29
425	10.97	0.471	0.131	0.68
450	0.74	0.405	0.138	0.04
500	24.98	0.481	0.208	2.49
550	30.11	0.560	0.722	12.19
570	29.50	0.563	0.738	12.26
600	30.22	0.565	0.498	8.51

IV. SUMMARY

We attempted to use AIC poly-Si layer for the rear side of p-TOPCon to lower the crystallization temperature. We confirmed that AIC poly-Si layer is p-type by Hall measurement, and the resistivity shows minimum at 550-570°C. J-V characteristics revealed that the annealing slightly lower than the eutectic temperature is the key to realize solar cells with high conversion efficiency by improving FF due to the reduce resistivity of AIC poly-Si.

ACKNOWLEDGEMENT

This work was financially supported by The New Energy and Industrial Technology Development Organization (NEDO) of Japan.

REFERENCES

[1] F. Feldmann, M. Bivour, C. Reichel, M. Hermle, and S. W. Glunz, "A passivated rear contact for high-efficiency n-type silicon solar cells enabling high Vocs and FF > 82%," *in 28th European PV Solar Energy Conference and Exhibition*, 2013, pp. 988 - 992.

[2] F. Feldmann, M. Bivour, C. Reichel, M. Hermle, and S. W. Glunz, "Passivated rear contacts for high-efficiency n-type Si solar cells providing high interface passivation quality and excellent transport characteristics," *Solar Energy Materials & Solar Cells*, vol.120, pp. 270-274, 2014.

[3] F. Feldmann, M. Simon, M. Bivour, C. Reichel, M. Hermle, and S. W. Glunz, "Carrier-selective contacts for Si solar cells," *Applied Physics Letters*, vol.104, pp. 181105, 2014.

[4] O. Nast, T. Puzzer, L. M. Koschier, A. B. Sproul, and S. R. Wenham, "Aluminum-induced crystallization of amorphous silicon on glass substrates above and below the eutectic temperature." *Applied Physics Letters*, vol.73, pp.3214, 1998.

[5] O. Nast, S. Brehme, S. Pritchard, A. G. Aberle, S. R. Wenham, "Aluminium-induced crystallisation of silicon on glass for thin-film solar cells." *Solar Energy Materials & Solar Cells*, vol 65, pp. 385-392, 2001.

[6] I. Takahashi, Y. Sujihara, H. Yating, J. A. Wibowo, S. Tutashkonko, Y. Kurokawa, amd N. Usami. "Application of new doping techniques to solar cells for low temperature fabrication," *in 43rd IEEE Photovoltaic Specialist Conference*, 2016, pp. 2522-2524.

[7] A. Sarikov, J. Schneider, J. Berghold, M. Muske, I. Sieber, and S. Gall. "A kinetic simulation study of the mechanisms of aluminum induced layer exchange process" *Journal of Applied Physics*, vol.107, pp. 114318, 2010.

A Self - Consistently Coupled Drift Diffusion and Monte Carlo Simulator to Model Silicon Heterojunction Solar Cells

Pradyumna Muralidharan, Stuart Bowden, Stephen M. Goodnick and Dragica Vasileska

School of Electrical, Computer and Energy Engineering, Arizona State University, Tempe, Arizona 85287, United States of America

Abstract — Silicon heterojunction (SHJ) solar cell technology has continued to mature in recent years. A tremendous amount of research has been directed towards the optimization of the structure of the cell to obtain higher device performance. Any further improvement of the design requires a comprehensive understanding of all the factors that affect charge transport. Most theoretical attempts to describe charge transport have been based on traditional drift-diffusion (DD) methods, which are unable to capture the physics in crucial regions of the device. In this work we describe a method to self consistently couple a Monte Carlo (MC) solver to a DD solver. The MC solver is implemented to capture the high field behavior of the photogenerated carriers (holes) at the heterointerface. The self-consistent coupling provides a quantitative link between device parameters (doping, thickness, defect densities, etc.) and device performance. This paper describes the first step in our effort to establish a full self-consistent multiscale solver that utilizes a DD (quasi-neutral regions), MC (high field regions) and kinetic Monte Carlo (defect transport) to rigorously study transport.

Index Terms — amorphous silicon, heterojunction, device modeling, multiscale, solar cells.

I. INTRODUCTION

A significant amount of research has been reported on optimizing the design of the SHJ solar cell consisting of hydrogenated amorphous silicon (a-Si:H) and crystalline silicon (c-Si). These cells have consistently displayed higher efficiencies (>20%) and have lead the way towards achieving the Shockley-Queisser limit for single junction solar cell technologies. Panasonic's single junction record of 25.6% [1] was recently broken by Japan's Kaneka Corp. and NEDO by achieving a SHJ solar cell efficiency of 26.33% [2].

The drift-diffusion model has often been used to simulate, study and analyze transport behavior in solar cells, and is the basis of most commercial simulators. It is valid as long as transport is near equilibrium and quantum effects may be neglected. However, as we move towards 2nd and 3rd generation solar cells [3], the novel device structures and new materials often lead to scenarios where the physics is no longer truly semi-classical. Our work focuses on studying interfacial transport across the heterointerface in a SHJ solar cell. Due to the relatively high electric field at the heterointerface, the energy distribution function of the photogenerated carriers is non-Maxwellian [4]. The DD model assumes that all carriers are always in thermal equilibrium. This provides an incomplete picture of transport at higher electric fields.

The Monte Carlo Method is a perfect tool to study high field transport behavior. However, it can be computationally inefficient as a global solver, especially in areas with low electric fields where generation-recombination processes dominate. By coupling Monte Carlo and drift-diffusion solvers we can combine the advantages of each to achieve an efficient solver. By self-consistently coupling the solvers to a global Poisson equation solver, we can quantitatively study the effect of the energy distribution function on performance parameters such as J_{SC}, V_{OC}, FF and efficiency. Such a scheme has previously been applied only to the study of hot carrier effects in MOSFET's [5].

The coupling of the drift-diffusion solver and the Monte Carlo solver described in this paper is a significant step forward in creating a complete multiscale description to rigorously study transport in different regions of the device: (1) the quasi neutral region (drift-diffusion), (2) the heterointerface (Monte Carlo) and (3) the intrinsic amorphous silicon (kinetic Monte Carlo). We have previously conducted decoupled simulations using the solvers mentioned above to study transport in SHJ solar cells [6].

II. THEORETICAL MODEL

In this section, we briefly describe the different components of the multiscale solver and the coupling between the MC and the DD modules. Figure 1 is a schematic diagram of the HIT cell highlighting the different simulation domains. It also contains a schematic representation of the MC domain showing its shared boundary with the DD solver.

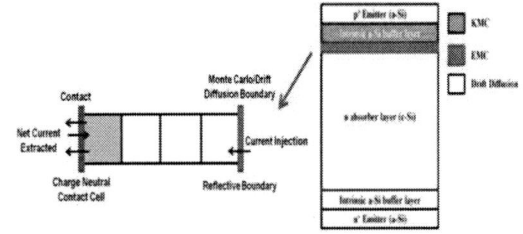

Fig. 1. (Right) Schematic diagram of the HIT cell indicating different simulation domains. (Left) Schematic representation of boundary conditions for drift-diffusion Monte Carlo (DDMC) coupling.

978-1-5090-5606-4/17 $31.00 © 2017 IEEE

A. Drift-Diffusion (DD) Model

First, a complete DD solution is applied to the device to determine potentials, fields and currents in the device. This is done by solving Poisson's equation (Eq. 1) in conjunction with the current continuity equations (Eq. 2).

$$\frac{\nabla . \varepsilon \nabla \phi(x)}{q} = -\left(p(x) - n(x) + N_D(x) - N_A(x) + \sum_{trap} \rho_{trap}(x) \right) \quad (1)$$

$$\frac{1}{q} \nabla . J_n = -(G_n - R_n); \quad \frac{1}{q} \nabla . J_p = (G_p - R_p) \quad (2)$$

$$J_n = q\mu_n n\xi + qD_n \frac{dn}{dx}; \quad J_p = q\mu_p n\xi - qD_p \frac{dp}{dx} \quad (3)$$

Eq. 3 is describe the current equations which consist of drift term (dependent of the electric field 'ξ') and a diffusion term (dependent on the spatial gradient of charge).

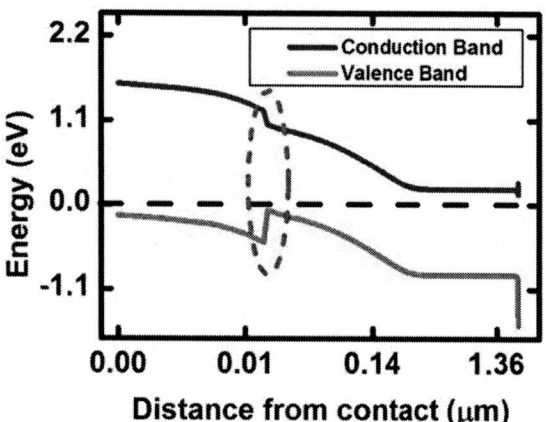

Fig. 2. Energy Band diagram of the HIT cell as calculated by the drift-diffusion solver.

Figure 2 shows the band diagram calculated by the drift-diffusion model. To apply the MC solution, a high field region has to be identified where the energy distribution function is no longer a function of the local electric field. For silicon, an electric field ≈ 10 kV/cm can be considered sufficiency low. Once this region is identified the injection conditions for the MC are setup. The DD solver provides the initial injection current which is used as a flux boundary condition for the MC.

B. Monte Carlo(MC) Domain

The MC domain has two boundaries. One is an injecting boundary (right), wherein current is injected based on the current density calculated by the drift-diffusion (this is a flux boundary condition). The injecting boundary must be placed in a region where the transport can still be described by drift-diffusion (where the electric fields < 10 kV/cm). Great care must be taken while setting up the DD-MC injecting boundary

so that the input carrier distribution for the MC is strongly correlated to the low field assumptions that are applicable in the quasi-neutral region [7]. A second boundary, called the collecting boundary is placed at the edge of the intrinsic amorphous silicon a-Si:H(i)/crystalline silicon (c-Si) heterointerface (see figure 1). Carriers are extracted at this boundary; however carriers also need to be re-injected so that the MC domain is not depleted [8]. We assume that there is no generation-recombination in the MC domain, and thus the flux boundary condition applied at the injecting boundary is used at the collecting boundary. In other words, carriers are re-injected at the collecting boundary to maintain a net flux (Flux = Extracted - Injected). Figure 3 is a schematic diagram that shows the placement of the injecting/collecting boundaries for the MC domain. It is important to note that the collecting boundary for the MC is placed away from the contact, thus standard charge neutrality boundary conditions cannot be applied at this contact.

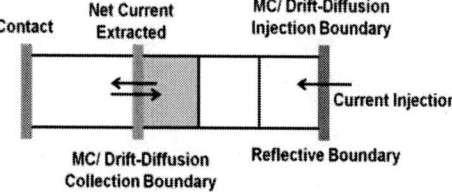

Fig. 3. Schematic diagram representing the injecting/collecting boundary of the MC domain.

We have previously developed the MC module to study the non-Maxwellian behavior of photogenerated carriers at the a-Si:H(i)/c-Si heterointerface [9]. As can be seen from figure 4, the average energy of the photogenerated carriers at the heterointerface is approximately 150 meV, which violates the Maxwellian assumption (≈ 39 meV).

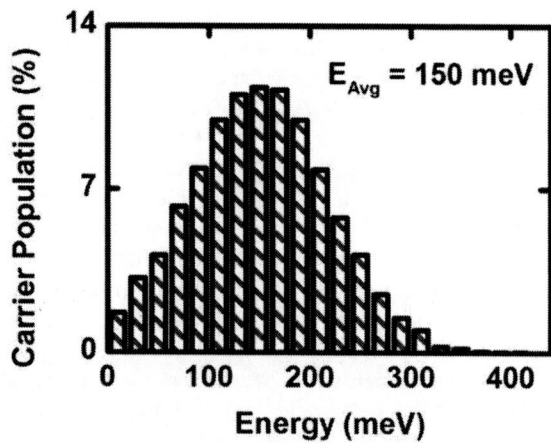

Fig. 4. Energy distribution function of the photogenerated (holes) carriers at the a-Si:H(i)/c-Si heterointerface as calculated by the MC.

C. Kinetic Monte Carlo (KMC)

The KMC solver is an ideal tool to study defect assisted transport. In previous works we have developed KMC solver to study transport in the intrinsic a-Si:H [10]. This is a very important module in terms of developing the multiscale methodology. The multiscale method works on various length and time scales; the KMC works as a buffer between steady state (DD) and picosecond (MC) transport. However, in this work we will limit ourselves to the coupling of the DD and MC solvers.

D. Current Matching

To test the coupling of the DD to the MC solver, we first applied this method to a simple silicon homojunction test structure with a 20 nm p-type emitter to study the transport of holes in the high field depletion region. As mentioned in section B, the MC region operates in the high field region outside the emitter (i.e. depletion region) and simulates hole transport. The test structure was subjected to multiple suns (≈1000) to create a high injection scenario and to provide sufficient injection statistics so that MC solver is not affected by statistical noise. The hole current calculated by the DD solver is injected into the MC domain; the solvers are said to be current matched when the extracted current from the MC is equal to the injected current ($J_{Ext} = J_{Inj}$). Figure 5 (left panel) shows a plot of cumulative charge extracted from the collecting boundary of the MC domain vs. time. As we can see the extracted current (which is given by the slope, $J=qdy/dx$) leads to a current that closely matches the injection current. Also, figure 5 (right panel) shows that the MC domain charge stabilizes after some transient behavior.

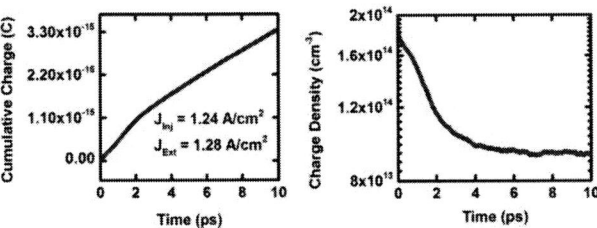

Fig. 5. (Left panel) Cumulative charge vs. time extracted at the collecting boundary of the MC domain for an injection of 1.24 A/cm². (Right Panel) Total charge density in the MC domain vs. time.

Once the MC domain is current matched and in steady state, it provides the continuity equation with the correct carrier density which is utilized to calculate a new injection current for the MC. The MC also resets the mobility for the low field region which is used by DD model to calculate current. This is done till the potential and current both converge. Thus self-consistent coupling is achieved.

E. Gummel loop

To achieve self-consistency, there has to be a continuous flow of information between the MC and DD solvers. Figure 6 shows a flow-chart of the Gummel cycle which is used to couple the MC to the DD. At first, the drift-diffusion model computes an initial set of solutions (potential, electric field, charge density, current density etc.). The MC domain is then determined using the considerations described in section B. As described in section D, the MC solver is current matched to the DD if the extracted current matches the injection. Once current matching and steady state are achieved by the MC, the DD solution is computed in the entire device excluding the MC region. In other words, the charge density calculated by the MC is "frozen", and the continuity equations are solved to recalculate the charge density outside the MC region. We refer to these solutions as the reduced boundary continuity solution. However, the Poisson solution is calculated for the entire device. This results in a new injection current which will be injected into the MC. This process is repeated till the injection current stabilizes. Once stability is achieved, the two solvers are said to be self-consistently coupled.

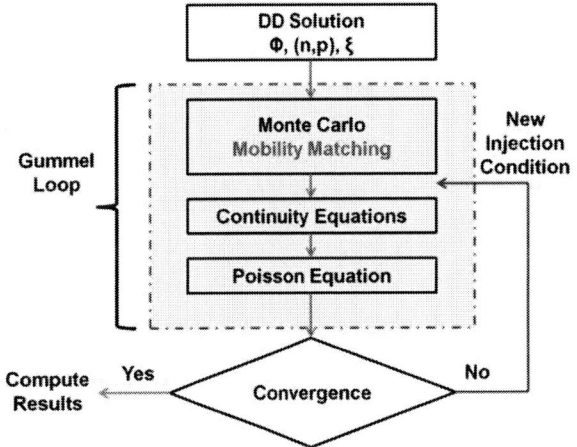

Fig. 6. Flowchart of the Gummel loop which is applied to achieve self consistency.

III. RESULTS AND DISCUSSION

Figure 7 shows the change in hole injection current with every successive Gummel iterations as calculated by the drift-diffusion-Monte Carlo (DD-MC) solver. As can be seen, the current varies during the initial Gummel cycles, but reaches some stability after 4 cycles. An important criteria to establish self-consistency is the continuity of charge density across both the MC and DD domains. Figure 8 shows the carrier density across both the domains. The MC solution is applied to holes in the depletion region. As can be seen the charge is continuous across both the boundaries. From figure 7 and figure 8 we can infer that the DD-MC solver converges and is self-consistently coupled.

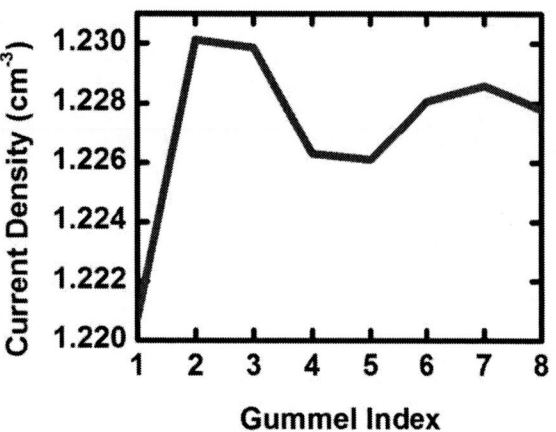

Fig. 7. Hole injection current calculated by the coupled DDMC solver vs. i^{th} Gummel iteration (x-axis).

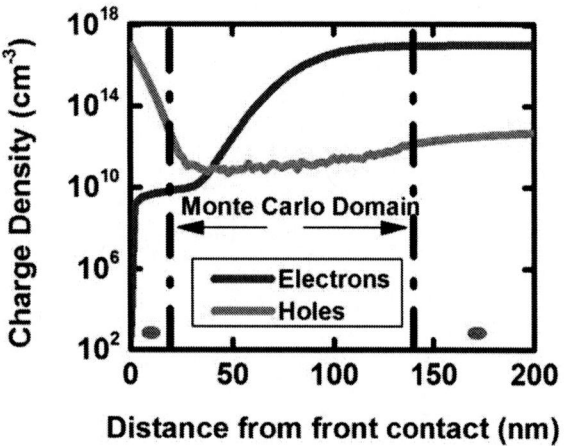

Fig. 8. Electron and hole density across the DD and MC domains. The MC domain is indicated by the dashed lines whilst the DD domain is placed on either side of the dashed lines (indicated by the green dots).

IV. CONCLUSION

In this paper we have shown a self-consistently coupled DDMC solver for a Si homojunction solar cell (test structure). This method uses a MC solver to capture high field physics whilst low field physics can be described by the DD solver. Our main goal now is to utilize this methodology to study SHJ solar cell. This work represents part of a larger theoretical framework (multiscale modeling) which will provide a quantitative correlation between microscopic device behavior and device performance. The microscopic understanding of transport will allow for better design optimization of the SHJ solar cell.

ACKNOWLEDGEMENT

This material is based upon work primarily supported by the Engineering Research Center Program of the National Science Foundation and the Office of Energy Efficiency and Renewable Energy of the Department of Energy under NSF Cooperative Agreement No. EEC-1041895. Any opinions, findings and conclusions or recommendations expressed in this material are those of the author(s) and do not necessarily reflect those of the National Science Foundation or Department of Energy.

REFERENCES

[1] K. Masuko *et al.*,"Achievement of more than 25% conversion efficiency with crystalline silicon heterojunction solar cell", *IEEE Journal of Photovoltaics*, Vol 4., pp. 1433-1435, 2014.

[2] https://www.pv-magazine.com/

[3] Martin A. Green, "Third generation photovoltaics: solar cells for 2020 and beyond", *Physica E: Low-dimensional Systems and Nanostructures*, Volume 14, Issues 1–2, Pages 65-70, 2002.

[4] K. Ghosh *et al.*, "Role of hot carriers in the interfacial transport in amorphous silicon/crystalline silicon heterojunction solar cells", *Phys. Status Solidi A,* 210, No.2, pp 413-319, 2013.

[5] J. M. Higman *et al.*, "Coupled Monte Carlo-drift diffusion analysis of hot-electron effects in MOSFETs," in *IEEE Trans. on Elec. Devices*, vol. 36, no. 5, pp. 930-937,1989.

[6] P. Muralidharan, S. Bowden, S. M. Goodnick and D. Vasileska, "Multiscale modeling of silicon heterojunction solar cells," *IEEE 43rd Photovoltaic Specialists Conference (PVSC)*, 2016.

[7] D. Y. Cheng *et al.*, "PISCES-MC: a multiwindow, multimethod 2-D device simulator," in *IEEE Trans. on Computer-Aided Design of Integrated Circuits and Systems*, vol. 7, no. 9, pp. 1017-1026, 1988.

[8] P.T. Nguyen, D.H. Navon and T.W. Tang, "Boundary conditions in regional Monte Carlo device analysis", in *IEEE Trans. on Elec. Devices*, vol. 32, no. 4, pp. 783-787,1985.

[9] P. Muralidharan, K. Ghosh, D. Vasileska and S. M. Goodnick, "Hot hole transport in a-Si/c-Si heterojunction solar cells", 40th *IEEE Photovoltaics Specialists Conference*, pp. 2519-2523, 2014.

[10] P. Muralidharan, D. Vasileska, Stephen M. Goodnick, and Stuart Bowden, "A kinetic Monte Carlo approach to study transport in amorphous silicon/crystalline silicon HIT cells", *42nd IEEE Photovoltaic Specialist Conference (PVSC)*,pp. 1-4, 2015.

Dopant Patterning by PECVD and Mechanical Masking for Passivated Tunneling Contact IBC Cell Architectures

William Nemeth*, Vincenzo LaSalvia**, Benjamin G. Lee, Abhijit Kale, Paul Stradins

National Renewable Energy Laboratory, Golden, Colorado, CO 80401

* First authors V. LaSalvia and W. Nemeth contributed equally to this work.

Abstract — We present a robust, simple technique for dopant patterning onto poly-Si/SiO$_2$ passivated tunneling contacts on n-Cz wafers. Heavily doped, thin n- and p- type a-Si:H overlayers serve as solid-state dopant sources for blanket intrinsic a-Si:H layer on tunneling SiO$_2$. Subsequent processing results in uniformly doped, functional n- and p-type passivated contacts. Furthermore, physical masking during the PECVD of the dopant source overlayers enables an interdigitated back contact (IBC) cell structure with sufficient doped finger edge fidelity to preserve an intrinsic electrode gap. Lithography-free, passivated contact IBC cell structures are made using a set of n-, p-, and metallization masks from patterned Si wafers with mechanical alignment. Alternatively, doped a-Si:H overlayers are patterned by dielectric layer deposition and lithography.

Index Terms — silicon, polysilicon, passivated contact.

I. INTRODUCTION

The key to Si PV competitiveness is high cell efficiency at low cost. While the current Si PV manufacturing focuses on potentially highly efficient, LID-immune p-PERC on both p-mono and p-multi wafers, n-type Cz technology is regarded as the long-term choice for the highest performance.

Interdigitated back contact (IBC) cell architecture with passivated contacts (IBPC) realizes the high efficiency potential of n-Cz to the fullest extent, as demonstrated by the record cells from SunPower, Panasonic, Sharp, and Kaneka. Moreover, SunPower has proven that the IBC cell despite its complexity can be mass-produced at efficiencies of ~ 24%.

As to the academic studies of dopant pattering that enables these IBC cells, several groups [1]-[3] have demonstrated excellent a-Si:H heterojunction cells by shadow-masked deposition of doped a-Si:H fingers without involving any dopant diffusion. Other groups [4]-[5] have used masked ion implantation including PIII with subsequent dopant thermal drive-in to make patterned passivated contact IBC cells.

In this work, we show that the poly-Si/SiO$_2$ passivated contact IBC structures can be produced by a simple Plasma Enhanced Chemical Vapor Deposition (PECVD) technique without beamline of PIII ion implantation. Namely, a-Si:H dopant source overlayers are deposited by PECVD on instrinsic a-Si:H/SiO$_2$ to form p- and n- passivated contact fingers after thermal crystallization/drive-in. Furthermore, we show that a simple physical masking during PECVD enables high spatial fidelity dopant patterning within 10 - 20 µm. This preserves the intrinsic poly-Si/SiO$_2$ gap between n- and p-type fingers at high sheet resistance and high iV$_{oc}$ over 700 mV. The doped fingers exhibit high passivation and low contact resistivity.

II. EXPERIMENTAL DETAILS

Cells and symmetric test structures are fabricated on n-Cz wafers with resistivity of ~ 2-7 Ω-cm. A ~ 1.5 nm thick tunneling SiO$_2$ is grown thermally at 700 °C [6]. A 20 nm thick intrinsic a-Si:H layer is PECVD-deposited onto the tunneling SiO$_2$ on the planarized back surface. Subsequently, 5- 10 nm thick a-Si:H overlayers doped with B and P to ~ 10^{21} cm^{-3} are deposited onto the intrinsic a-Si:H using either two mechanical masks, or a photolithography / PECVD process sequence involving patterning of SiN$_x$ or Al$_2$O$_3$ dielectric overlayers. Notably, the patterning never extends to the sensitive SiO$_2$ tunnel oxide, being confined to the top of the i-poly, which is much more immune to ambient / handling / lithography damage.

The mechanical masks are made of Si wafers by laser-initiated etching that results in an edge roughness and pattern fidelity of about 10 µm. The masks are mechanically aligned over the substrate with precision and repeatability of about 20 µm. After the masked PECVD and processing, the end result is doped n- and p-type a-Si:H finger IBC pattern over the blanket intrinsic a-Si:H, separated by 50-100 µm wide intrinsic gap. Annealing at 850 °C crystallizes the a-Si:H into poly-Si and drives in the dopants.

The wafer then receives an atomic layer deposited (ALD) Al$_2$O$_3$ layer with additional SiN$_x$ (PECVD) on the textured front surface. A subsequent FGA at 400 °C further passivates the structure by hydrogenation and field effects [6]. After removal of the sacrificial Al$_2$O$_3$ layer from the back, the IBC structure is metallized by thermal evaporated Al through a Si wafer shadow mask. The emitter and back surface field (BSF) contacts as well as the intrinsic poly-Si gap layers are optimized separately using symmetric test structures. These are deposited on planarized and tunnel-oxidized n-Cz wafers. Subjected to the same thermal and passivation treatments as cells, they are measured with a Sinton WCT-120 Lifetime tester for iV$_{oc}$ and J$_o$, and with either Transmission Line Method or Diode Method (see S. Theingi, et al., this conference) for contact resistivity ρ$_c$.

978-1-5090-5606-4/17 $31.00 © 2017 IEEE

III. EXPERIMENTAL RESULTS AND DISCUSSION

A. Passivated contacts from deposited a-Si:H dopant source layers over intrinsic a-Si:H.

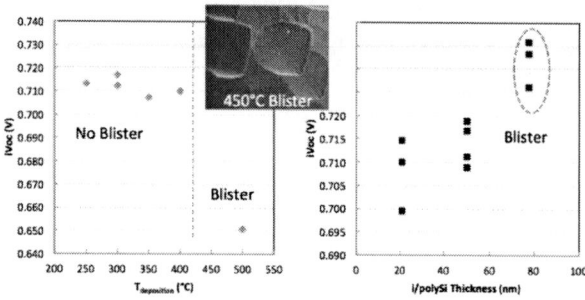

Fig. 1. Passivation quality of intrinsic poly-Si/SiO$_2$ structures on KOH-planarized n-Cz wafers. Left: intrinsic a-Si:H deposition temperature series. Right: a-Si:H thickness series. The pasivation is characterized by the iVoc after 850C crystallization anneal and H-passivation bt Al$_2$O$_3$.

To test our solid-source dopant layer concept, we grow and measure symmetric lifetime structures. Before depositing doped source layers for n- and p-type passivated contacts, we develop intrinsic poly-Si/SiO$_2$ structures with high surface passivation. These undergo the same thermal treatments and passivation by ALD Al$_2$O$_3$ as the sandwich-doped layers. The best degree of passivation of i/poly-Si/SiO$_2$ is obtained using high hydrogen dilution H$_2$/SiH$_4$ ratio during PECVD deposition of i/a-Si:H at low rate < 0.5 Å/s. Other important factors are the deposition temperature and i/a-Si:H film thickness, summarized in Fig. 1. Thicker a-Si:H produce better surface passivation but develop blisters during 850 °C annealing. These do not affect the iV$_{oc}$ initially but cause performance loss upon further cell processing. In contrast, depositing a-Si:H at T > 400 °C already produces films with blisters (see inset of Fig. 1), resulting in low iV$_{oc}$.

After selecting the i/poly-Si process with 20 nm thick film deposited at 300 °C under optimal PECVD conditions, we proceed with deposition of heavily doped n- or p- a-Si:H over it. The dopants are then driven into the rest of the poly-Si by thermal diffusion, similarly as for ion implantation, with a goal of producing the average dopant concentrations within poly-Si of ~ 2x10^{20} cm^{-3}. As seen from the Fig. 2, high performance p- and n- type passivated contacts can be produced by this method. Notably, the typically more problematic p-type passivated contacts exhibit iV$_{oc}$ over 700 mV after this processing. This might be associated with less accumulation of B into the tunneling oxide after diffusion through the poly-Si layer. The n-type contact has remarkably high iVoc after annealing at 650 °C but deteriorates with increasing the annealing temperature. This is somewhat unexpected since uniformly doped n-type contacts after annealing at 850 °C and passivation typically show iV$_{oc}$ > 730 mV peaking near 850 °C anneal temperature. The above effect might be due to too low overall amount of P, leading to a reduced field effect passivation.

B. Patterning of solid-source dopant layers

Having confirmed the high-performance blanket layer passivated contacts produced by the "sandwich" method, we now explore in-plane 2D spatial patterning of B and P dopants. For this purpose, dopant-source layers of Fig. 2 are deposited by spatially-masked PECVD using (i) mechanical mask and (ii) photolithographically patterned, sacrificial dielectric layers on top of the blanket intrinsic a-Si:H.

To ensure the absence of edge effects during PECVD, we use patterned Si wafers as deposition masks for the dopant source overlayers. Figure 3, left shows the resulting pattern of ~ 10 nm thick a-Si:H layer deposited on quartz through the n-type finger (BSF) mask of the 2x2 cm^2 IBC cell.

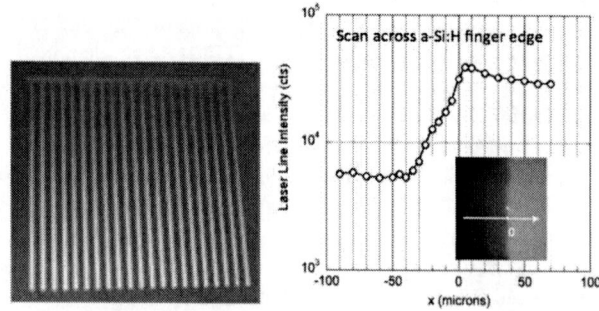

Fig. 3. Left: a-Si:H film deposited on quartz through the BSF finger mask of the IBC cell. Right: Spatial edge fidelity of the BSF finger resolved by optical microreflectance scan.

The a-Si:H finger edge fidelity is ~ 17 µm measured as exponential decay length of the optical microreflectance scan (Fig. 3, right). Taking into account that the poly-Si resistivity increases highly nonlinearly with doping below 10^{19} cm^{-3}, this suggests that the 100 µm wide intrinsic poly-Si gap will

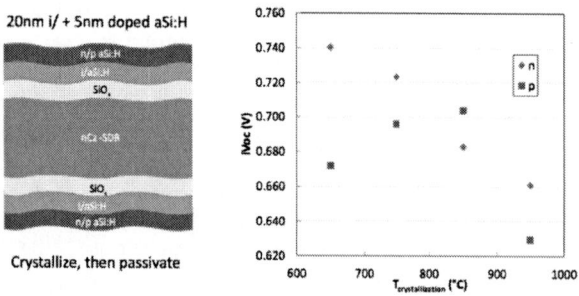

Fig. 2. Left: symmetric sandwich structures as precursors to passivated contacts. Right: iV$_{oc}$ of these structures after crystallization/drive-in anneal at various temparatures, followed by passivation with Al$_2$O$_3$ as H-source layer.

remain insulating between n- and p- fingers. This is very important since the weak shuntage criterion for an IBC cell ($R_{shunt}/R_{char} > 100$) holds only if the sheet resistivity of the i/poly-Si gap is on the order of $M\Omega/\square$ as the total gap length even in 2×2 cm^2 cell is about 80 cm. Note that unlike the diffused IBC cells, the poly-Si contact based IBC cell is easily shunted due to the ohmic behavior of junctions between heavily doped and defective i-, n-, and p- poly-Si layers.

Patterning dopant source layers is also done using dielectric layers to protect intrinsic a-Si:H during doped a-Si:H PECVD, with lithography steps in between n- and p- depositions. This technique provides excellent edge fidelity (Fig. 4). However, several lithography steps and two depositions of dielectric layers makes this process significantly more complex than the above high-throughput mechanical masking.

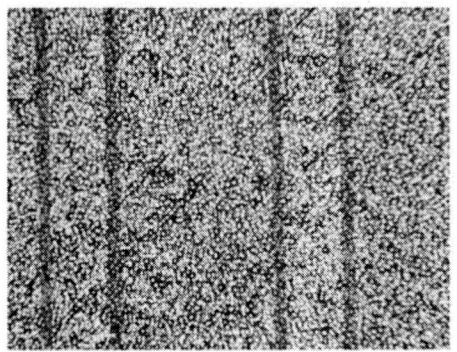

Fig. 4. Image of premetallized, lithography-defined IBC cell with BSF (blue), emitter (yellow), and gap (pink) clearly defined.

C. Solar cell structures

Section A and B results suggest excellent efficiency potential for the IBC cell structure produced with the present method. Our solar cells are produced using both mechanical masking and photolithography described in Section B. The mechanically masked cell process flow shown in Fig. 5. is realized by aligning the same-size 4×4 cm^2 Si substrates with 2×2 cm^2 cells and Si masks with respect to three common points in 2D plane. The alignment precision is about 20 µm confirmed by inspecting the sample/mask common corners.

We are currently metallizing the cells and will report cell performance at the conference.

IV. SUMMARY AND ACKNOWLEDGEMENTS

We have demonstrated that doped a-Si:H PECVD-deposited over intrinsic a-Si:H can serve as solid dopant sources to produce high performance poly-Si/SiO$_2$ passivated contacts. Furthermore, these doped layers can be patterned into an IBC configuration with high fidelity, preserving a highly passivated intrinsic poly-Si gap between n- and p-type fingers. As a result, high performance passivated contact IBC cells can be produced without ion implantation, by a robust masked PECVD deposition process. Funding for this work was provided by the United Sates Department of Energy EERE contract SETP DE-EE00030301 (SuNLaMP) and under Contract No. DE-AC36-08GO28308.

REFERENCES

[1] A. Tomasi, B.Paviet-Salomon, D. Lachenal, S. Martin de Nicolas, M. Ledinsky, A. Descoeudres, S. Nicolay, S. De Wolf, and C. Ballif. "Photolithography-free interdigitated back-contacted silicon heterojunction solar cells with efficiency> 21%." *40th IEEE Photovoltaic Specialist Conference (PVSC),* 2014, p. 3644.

[2] M. Scherff, "Novel method for preparation of interdigitated back contacted a-Si: H/c-Si heterojunction solar cells." *Proc. 26th Eur. Photovoltaic Sol. Energy Conf. Exhib,* 2011, p. 2129.

[3] S. Y. Herasimenka, C. J. Tracy, W. J. Dauksher, C. B. Honsberg, and S. Bowden. "A simplified process flow for silicon heterojunction interdigitated back contact solar cells: Using shadow masks and tunnel junctions." *40th IEEE Photovoltaic Specialist Conference (PVSC),* 2014, p. 2486.

[4] G. Yang, A.Ingenito, O. Isabella, and M. Zeman. "IBC c-Si solar cells based on ion-implanted poly-silicon passivating contacts." *Sol.En. Matls and Solar Cells,* p.84, 158, 2016.

[5] D. L. Young, W. Nemeth, V. LaSalvia, R. Reedy, S. Essig, N. Bateman, and P.Stradins. "Interdigitated back passivated contact (IBPC) solar cells formed by ion implantation." *IEEE J. of Phot,.* 6, no. 1, p.41, 2016.

[6] B. Nemeth, D.L. Young, M.R. Page, V. LaSalvia,, S. Johnston, R. Reedy. and P. Stradins, "Polycrystalline silicon passivated tunneling contacts for high efficiency silicon solar cells." *J. of Matls Res.,* 31(06), p.671, 2016.

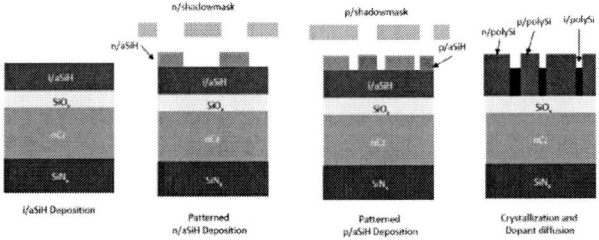

Fig. 5. Schematic of process flow for shadow masked IBC cell.

ALD Aluminum Oxide as a Hole Selective Tunneling Contact for Crystalline Silicon Solar Cells

Kortan Öğütman[1,3], Kristopher O. Davis[1,3],

Winston V. Schoenfeld[1,2,3], Michael Haslinger[4], Sofie Robert[4], Emanuele Cornagliotti[4], Joachim John[4,5].

[1]Florida Solar Energy Center, University of Central Florida, Cocoa, FL 32922, USA

[2]CREOL, the College of Optics and Photonics, University of Central Florida, Orlando, FL 32826, USA

[3]Dept. of Electrical Engineering and Computer Science, University of Central Florida, Orlando, FL 32826, USA

[4]IMEC, 3001 Leuven, Belgium

[5]EnergieVille, 3600 Genk, Belgium

Abstract — Tunneling aluminum oxide (Al_2O_3) layers are proposed as a candidate for hole collecting passivated contacts. These ultra-thin Al_2O_3 films ($\approx$1.5nm) are deposited on boron diffused (113Ω/$\square$) hydrophilic surfaces to act as metal-insulator-semiconductor (MIS) contacts. Record values of emitter recombination current densities (35-40 fA/cm^2) and contact resistivity ($\approx$15 $m\Omega \cdot cm^2$) were measured for this type of contact. Most importantly, all of these processes are carried out using industrially relevant tools on Cz wafers with non-polished surfaces.

Keywords — tunneling layers, aluminum oxide, crystalline silicon passivation, passivated contacts.

I. INTRODUCTION

As the efficiency of crystalline silicon solar cells approaches its intrinsic limit, some sort of passivation scheme within the contact where the extraction of carriers occur becomes essential. This stems from the fact that there exists high a density of recombination centers where the metal touches the silicon. Historically, this problem was dealt in two ways. First introducing a heavily doped region right underneath the contact where the metal-silicon interface takes place (e.g., selective emitter, selective back surface field). However, this introduces complexity in manufacturing due to patterning and alignment requirements, it increases Auger recombination, and ultimately does not achieve the saturation current densities needed to push cells beyond 25%. This limit is closer to around 22% using industrial processes (e.g., screen-printing) [1]. The second approach is limiting these metal-semiconductor interfaces to a smaller fraction of the interface. Therefore, carriers will have to travel laterally to the contacts where the extraction takes places (i.e., three-dimensional carrier transport). This brings the trade-off between the fill-factor and the open-circuit voltage of the cell. Moreover, this scenario also necessitates the use of highly-doped substrates because of resistance problems, which will increase the bulk recombination. A transition is currently taking place in industry from aluminum back surface field (Al-BSF) cells to passivated emitter rear contact (PERC) cells is a prime example of such a phenomenon.

Another approach to the problem is the use of heterostructures to provide contact passivation. This can be done by using a homojunction and tunneling passivational layer, as in a metal-insulator-semiconductor (MIS) structure [2], or by using carrier selective heterojunctions, like amorphous or polycrystalline silicon [3]. Recently there has been a tremendous interest in metal oxide materials for the passivation of the metal-silicon interface. Molybdenum oxide [4], and titanium oxide [5] have been demonstrated in hole and electron extraction layers, respectively. While these somewhat exotic materials have shown promise, being able to achieve the same passivation and carrier extraction properties with a material already in use by industry would likely lead to more rapid adoption by the industry.

Aluminum oxide (Al_2O_3) is an established passivation layer in the solar industry, offering exceptionally high negative built-in charge for field effect passivation of *p*-type surfaces while offering a fairly low interface defect density [6]. Because of these properties, it has been combined with silicon nitride by many research groups and industry to form an insulating passivation stack for *p* and p^+ surfaces (e.g., rear side of PERC cells, front side of *n*-type cells). Therefore, adoption of Al_2O_3 in a metal-insulator-semiconductor contact may bring many advantages. However, this insulating layer has to be thin enough such that there is a high probability for tunneling. This requirement also brings the challenge of growing thin films

978-1-5090-5606-4/17 $31.00 © 2017 IEEE

conformally with low thickness variation such that high passivation quality is maintained over the entire surface of the wafer. Traditionally, the removal of any possible silicon oxide layer is done prior to the deposition of Al_2O_3, typically with hydrofluric (HF) acid. However, as it is shown by previous research groups [7], if these films are deposited on non-OH-terminated surfaces (i.e., HF-last), formation of island-like shapes take place on the surface, prohibiting the conformal growth of the film. This is exteremely crucial in a structure like crystalline-silicon solar cell, which is a fairly large area device.

In this work, we investigate the use of an SiO_x/Al_2O_3 stack as a hole selective, tunneling contact on p^+ surfaces with a ≈ 113 $\Omega/\square$ sheet resistance formed using a BBr_3 diffusion. This structure can be applied to both p-type cells, as a BSF, or n-type cells, for emitter passivation (Figure 1).

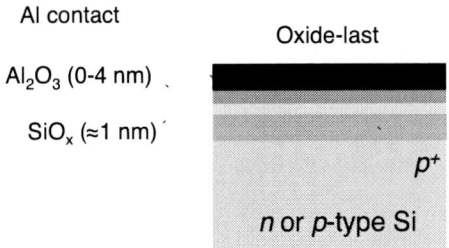

Fig. 1.   Proposed structure for the conformal growth of tunneling layers.

II.   RESULTS AND DISCUSSIONS

To circumvent this phenomenon of non-conformal growth, our group studied the deposition of very thin layers of Al_2O_3 (1-3nm) on hydrophilic surfaces. For this purpose, each building block of this passivation stack is throughly examined. As an initial step, a chemical oxide deposition study took place. Mirror-polished CMOS wafers were used for optimized ellipsometry measurements. All samples were initially dipped in a mixture of $\approx 1\%$ HF, stripping any surface oxide and ensuring that the initial condition of the surface was hydrophobic. Following this, samples were immersed in a chemical bath where ozonated dionized water mixture exists. The variation of the chemical oxide thickness was studied as a function of immersion time. Importantly, since this is a large-area device, uniformity of this chemical oxide was quite crucial. This was verified by 49-point ellipsometry measurements.

TABLE I
Chemical oxide thickness from 49-point Ellipsometry measurements for the solar cell line process.

	0 sec.	10 sec.	25 sec.	45 sec.	60 sec.	90 sec.
Mean (nm)	0.120	0.819	0.882	0.946	0.951	0.950
St. Dev. (nm)	0.016	0.014	0.009	0.01	0.011	0.014

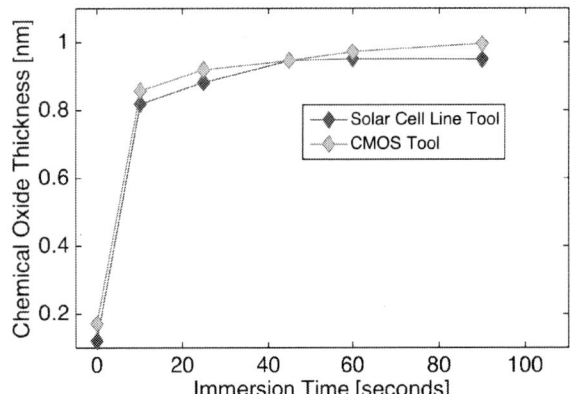

Fig. 2.   Oxide evolution as a function of immersion time in the ozonated water bath for a CMOS tool and a solar cell R&D tool.

What is depicted in Fig. 2. is the extreme similarity in oxide growth process between a CMOS tool and a solar cell R&D tool. The CMOS tool follows the following process: (HF(2%)→DIW/O_3→DIW→Drying in N_2). The solar cell tool, on the other hand, includes the following steps: (HF(2%)→DIW→DIW/O_3→Drying in N_2). As one can see from Table I, the standard deviation in 49-point ellipsometry measurements is extremely low.

After confirming that confromal growth of this chemical oxide was highly stable, our group went on the investigation of the second dielectric layer in this stack, Al_2O_3. Symmetrical lifetime structures were created as a part of this effort. Any problems with the growth mechanism will cause a decrease in lifetime. Below graphs illustrate the evolution of emitter recombination current density as a function of Al_2O_3 thickness deposited by spatial atomic layer deposition (ALD).

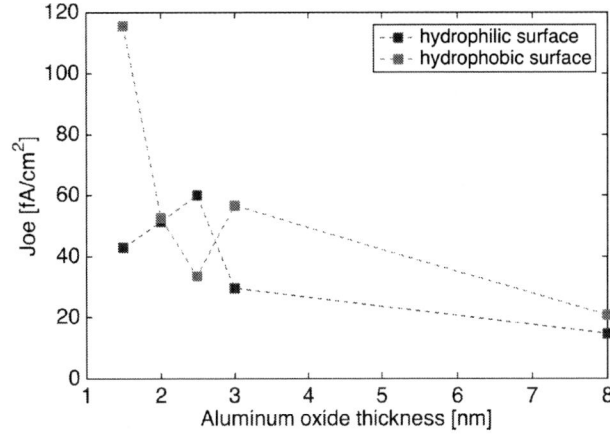

Fig.3.   Evolution of emitter saturation current density (J_{0e}) vs. Al_2O_3 thickness.

The spatial ALD tool from SoLayTec is utilized for this study with a deposition temperature of 200 °C with precursors trimethylaluminum (TMA) and water vapor. The emitter recombination current density (J_{0e}) values are extracted via using the well-known Kane and Swanson method from carrier lifetime mesurements with the aid of Sinton coil [8]. Intrinsic carrier density is assumed to be $9.70 \cdot 10^9 \, cm^{-3}$ at 300°K.

As can be seen, for the first time, J_{0e} values on the order of $40 \, fA/cm^2$ were demonstrated with an Al_2O_3 layer that is 1.5nm thick. This is, to our knowledge, the best passivation and carrier extraction quality obtained with tunnelling Al_2O_3 layer. The key message, though, is the three-fold difference, in recombination factor, between the hydrophobic and hydrophilic case for 1.5nm of Al_2O_3. This is solely due to non-conformal growth of the film for extremely thin layers, which is verified by other experiments.

Given that this layer must also serve as a carrier extraction path while passivating the region of interest, low contact resistivity values are also quite essential. To investigate this, transfer length method (TLM) test structures were created by using thermal evaporation of aluminum. These aluminum stripes were 200nm thick, deposited using a shadow mask. In our experience, these TLM measurements were quite repetable perhaps due to the high uniformity of the process developed.

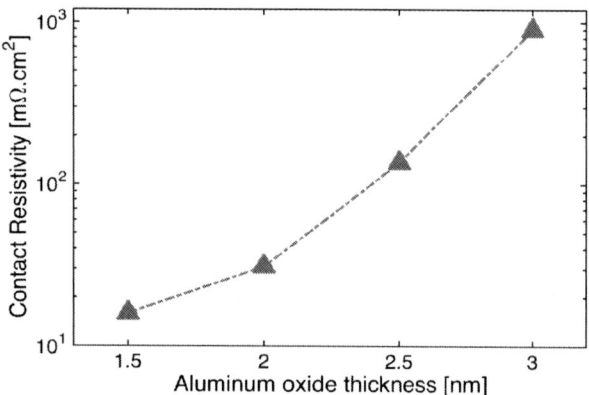

Fig.4.   Contact resistivity obtained from TLM measurements.

This exponential-like growth in contact resistivity for increased Al_2O_3 thickness is expected due to tunneling behavior. If 1D carrier transport is available, one could allow contact resistivities on the order of $100 \, m\Omega \cdot cm^2$, which was achieved with a 2.5nm of Al_2O_3 in combination with a 0.95nm thick chemical oxide. Quite interestingly, the best passivation quality was achieved with the group consisting of 1.5nm Al_2O_3 and the same chemical oxide. At this thickness contact resistivity is around $15 \, m\Omega \cdot cm^2$, which means that this stack could also be a candidate for local contacts in event that the J_{0e} is not low enough.

In future work, we will investigate the trend in lifetime values with the aid of TEM and XPS measurements to gain a better understanding of the interface and how it evolves for thicker Al_2O_3 layers. We are also planning to create test structures to characterize what happens to the passivation quality after contact formation [2].

III.   Conclusion

As a conclusion, our group was able to show that growth of very thin layers of Al_2O_3 on hydrophilic surfaces can enable the use of this very well-known passivation material as a hole collecting layer in crystalline-silicon solar cells. One of the foremost outcomes of this study was that all these depositions were done with industrially relevant processes on $5\Omega \cdot cm$ n-type Cz wafers, resulting in contact resistivity values as low as 15 $m\Omega \cdot cm^2$ while achieving excellent passivation.

References

[1] P. P. Altermatt and K. R. McIntosh, "A Roadmap for PERC Cell Efficiency towards 22%, Focused on Technology-related Constraints," *Energy Procedia,* vol. 55, pp. 17-21, // 2014.

[2] J. Bullock, D. Yan, A. Cuevas, B. Demaurex, A. Hessler-Wyser, and S. D. Wolf, "Passivated contacts to n+ and p+ silicon based on amorphous silicon and thin dielectrics," in *2014 IEEE 40th Photovoltaic Specialist Conference (PVSC),* 2014, pp. 3442-3447.

[3] F. Feldmann, M. Bivour, C. Reichel, M. Hermle, and S. W. Glunz, "Passivated rear contacts for high-efficiency n-type Si solar cells providing high interface passivation quality and excellent transport characteristics," *Solar Energy Materials and Solar Cells,* vol. 120, Part A, pp. 270-274, 1// 2014.

[4] J. Bullock, A. Cuevas, T. Allen, and C. Battaglia, "Molybdenum oxide MoOx: A versatile hole contact for silicon solar cells," *Applied Physics Letters,* vol. 105, p. 232109, 2014.

[5] X. Yang, Q. Bi, H. Ali, K. Davis, W. V. Schoenfeld, and K. Weber, "High-Performance TiO2-Based Electron-Selective Contacts for Crystalline Silicon Solar Cells," *Advanced Materials,* vol. 28, pp. 5891-5897, 2016.

[6] G. Dingemans and W. M. M. Kessels, "Status and prospects of Al2O3-based surface passivation schemes for silicon solar cells," *Journal of Vacuum Science & Technology A: Vacuum, Surfaces, and Films,* vol. 30, p. 040802, 2012.

[7] N. P. Kobayashi, C. L. Donley, S.-Y. Wang, and R. S. Williams, "Atomic layer deposition of aluminum oxide on hydrophobic and hydrophilic surfaces," *Journal of Crystal Growth,* vol. 299, pp. 218-222, 2/1/ 2007.

[8] D. E. Kane and R. M. Swanson, "Measurement of the emitter saturation current by a contactless photoconductivity decay method," in *IEEE Photovoltaic Specialists Conference,* 1985.

Screen Printed, Large Area Bifacial N-PERT cells with Tunnel Oxide Passivated Back Contact

Young-Woo Ok[1], Ajay D Upadhyaya[1], Brian Rounsaville[1], Ying-Yuan Huang[1], Vijaykumar D Upadhyaya[1], Ajeet Rohatgi[1, 2]

[1]Georgia Institute of Technology, 777 Atlantic Drive, Atlanta GA 30332-0250, USA, [2] Suniva Inc., 5765 Peachtree Industrial Blvd., Norcross, GA 30092

Abstract — This paper reports on the effect of screen printed metallization on passivation quality of TOPCon structure. A metallized TOPCon using fire-through screen printing process showed excellent saturation current density (metallized $J_{0,TOPCon} \sim$ 15 fA/cm^2) with very low contact resistance (~6 mohm-cm^2). However, screen printed metallization increased the recombination at low level injection level (≤ 0.1 sun) causing some degradation in iFF. We successfully fabricated 20.1% bifacial n-type PERT cells with ~180 ohm/□ implanted B emitter and TOPCon rear using screen printing and fire through contact technology. Detailed cell analysis showed that high series resistance and saturation current density of metallized B emitter limited the cell performance to ~20%.

Index Terms- passivated contact, TOPCon, screen Printing, bifacial n-type cell

I. INTRODUCTION

Tunnel oxide passivated contact (TOPCon) has become an active area of investigation for next generation high-efficiency Si solar cells. TOPCon provides excellent carrier selectivity with help of an ultra-thin (~15Å) tunnel Si oxide, which also displaces the diffused and metallized regions out of the absorber to minimize the recombination. This tunnel oxide is then capped with a doped poly Si layer and metal. There are several reports showing very low saturation current density (J_0) below 5 fA/cm^2 from TOPCon structures [1]. It is well known that evaporated Ag metallization on top of TOPCon doesn't degrade the passivation quality. Fraunhofer ISE reported >25.0% cell efficiency on 4cm^2 Fz silicon with photolithography contacts on the front and TOPCon back with evaporated Ag [2]. Our group recently reported ~21% efficient, large area n-type Si solar cells with screen printed contact on the front and TOPCon on the rear with evaporated full Ag on the back [3]. However, in order to transfer TOPCon cell technology to industry, evaporated Ag needs to be replaced by simple screen printed technology. Recently, ECN/Tempress presented high performance bifacial n-type solar cells (~20.7%) on 6 inch Cz wafers with TOPCon and fire-through screen printed back metallization employing high throughput LPCVD for the polysilicon [4]. In this paper, we show the effect of fire through screen printed contact on the passivation properties of TOPCon structures. We will also show screen printed, large area >>20% bifacial n-type cells with implanted

B emitter and TOPCon back with detailed analysis for further improvement.

II. EFFECT OF METAL CONTACT ON J_0 OF TOPCON STRUCTURES

In order to explore the effect of screen printed metallization on saturation current density J_0 of the TOPCon structure, SiN_x/n^+-poly/SiO_x/n/SiO_x/n^+-poly/SiN_x symmetric structures were prepared on planar high resistivity Cz-Si wafers. A ~15 Å chemical oxide was grown in nitric acid solution and ~200 nm thickness n^+ poly silicon was grown by LPCVD tube furnace followed by furnace annealing at ~875 °C in N_2 ambient for dopant activation. After that, TOPCon symmetric structures were passivated with PECVD SiN_x on both sides. Metal dots (commercial Ag paste) were screen printed on both sides using different screens with same dots size but different pitch, which give different metal contact coverage. All samples including a control sample without metal dots, were then fired in a belt furnace to simulate contact formation. Finally, excess screen

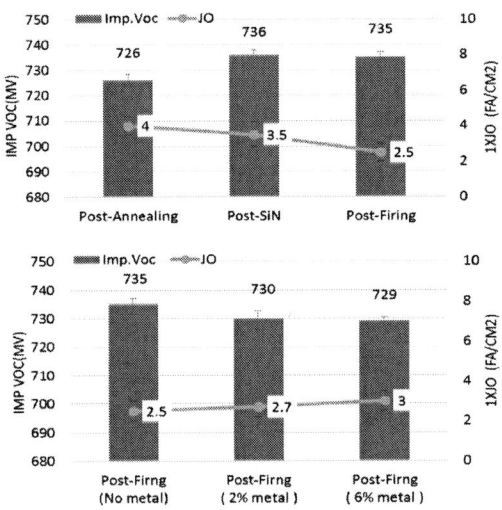

Fig. 1. Measured iVoc and J_0 (a) post annealing, SiN$_x$ deposition and firing (b) firing with different metal contact coverage

printed metal was etched away using HCl cleaning process leaving only the thin glass layer and embedded Ag point contacts onto ploy silicon surface. We measured effective lifetime, iV$_{oc}$ and J_0 using QSSPC tool after each processing step. Figure 1 shows the measured J_0 value from symmetric TOPCon structures ($J_{0, TOPCon}$) after each processing step (Fig. 1a) without metal as well as with different metal contact coverage (Fig. 1b). After activation of n^+ poly-Si without any metal, implied Voc of ~726 mV and $J_{0,TOPCon}$ of ~4 fA/cm^2 were achieved, indicating that the excellent passivation quality without SiN$_x$ and metal contact. After SiN$_x$ deposition on both sides and firing without any metal dots, the passivation quality improved possibly due to hydrogen passivation of interface. As a result, iVoc increased from 726 to 735 mV and $J_{0,TOPCon}$ decreased from 4 to 2.5 fA/cm^2. Hydrogen released from SiN$_x$ layer can passivate some defects at the tunnel oxide/Si bulk interface. It is important to note that $J_{0, TOPCon}$ increased only slightly from 2.5 to 3 fA/cm^2 with 6% screen printed contact coverage (Fig 1b). Following simple calculation, based on measured $J_{0,TOPCON}$ of 3 fA/cm^2 for 6% metal coverage and 2.5fA/cm^2 without metal, gave $J_{0,TOPCon}$ of 11 fA/cm^2 for metallized poly-Si.

$$\text{Measured } J_{0,TOPCon} = (J_{0,TOPCon})*(1\text{-}f)+J_{0, \text{ metallized TOPCon}}*f$$

For 10% metal coverage on the back, metal contribution to J_0 will be only 1.1 fA/cm^2. This value is almost two order magnitude lower than the metallized J_0 of a phosphorus diffused 100 ohm/sq n^+ region [5].

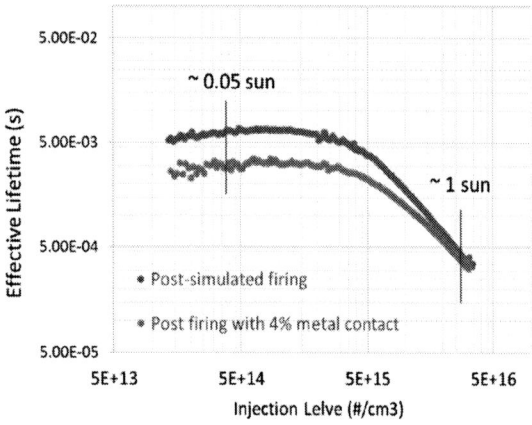

Fig. 2. Measured effective lifetime vs injection level for fired samples with and without metal contact

Although screen printed metallization on TOPCon didn't degrade appreciably $J_{0,n+poly}$ and iV$_{oc}$, it showed higher recombination under low level injection condition. Figure 2 shows the measured effective lifetime as functions of injection level for fired samples with and without metal contact. Very similar effective lifetime (~500 us) was observed for both samples at 1 sun but a huge difference in effective lifetime (1ms vs 5 ms) was observed near V$_{max}$ (~0.05 sun). Due to increased recombination at low level injection, calculated iFF decreased

from 0.855 (without metal) to 0.830 for 4% metal coverage. The reason for this is not clear at this time, but it is important to understand in order to achieve high FF in the cells.

III. FABRICATION OF SCREEN PRINTED, N-TYPE BIFACAIL CELLS WITH TOPCON REAR

With the very low full area metallized $J_{0,TOPCon}$ of 11 fA/cm^2, we expected that screen printing should not degrade cell V$_{oc}$. Figure 3 shows the cell structure and process sequence for the n-type TOPCon bifacial solar cell fabricated in this study. After standard saw damage etching and texturing, a light boron dose was implanted on the front followed by annealing/oxidation step at high temperature. This resulted in ~180 ohm/□ emitter with oxide on both sides. After that, we employed reactive ion etching (RIE) to remove back oxide followed by KOH etching process to planarize and remove RIE damage. After a short HCl/HF cleaning, the samples were placed in a HNO$_3$ solution for 10 minutes to grow tunnel oxide and then immediately loaded in LPCVD tube furnace to deposit phosphorus doped Si layers. The wafers were then annealed at 875 °C for 1 hour under N$_2$ ambient and then capped with PECVD SiN$_x$ layer on top of back poly-Si. Front poly-Si and oxide layer were removed by short KOH and HF dip and then passivated with ALD Al$_2$O$_3$ (~10nm) and PECVD SiN$_x$ on front. Finally, Ag/Al paste was screen printed on front using 89 lines screen and Ag paste was screen printed on the back using 150 lines screen followed by co-firing for contact formation. Cell efficiency was measured using a reference cell measured at Fruahofer ISE. Table 1 shows the light IV data of the cell which was 20.1% with V$_{oc}$ of 672 mV, J$_{sc}$ of 38.7% and FF of 77.4%.

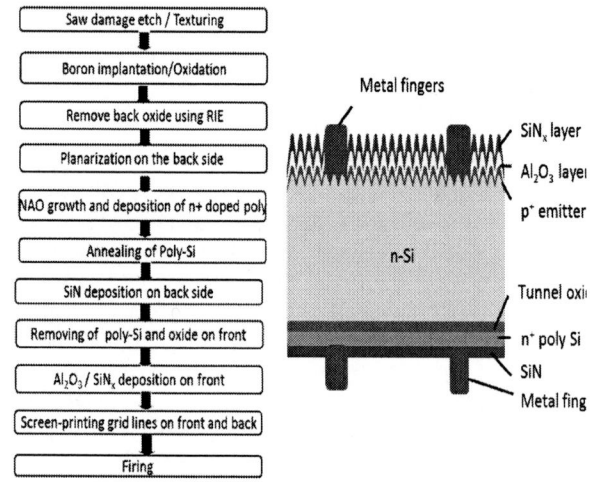

Fig. 3. Cell Processing and Structure

IV. ANALYSIS OF SCREEN PRINTED N-TYPE TOPCON BIFACIAL CELL

Detailed LIV analysis showed that 20.1% TOPCon bifacial n-type Si solar cell is limited by low FF (~77.4%) due to high R_s (>1.1 ohm-cm^2) as shown in table I. To gain a better understanding, each component of the total resistance was determined using the methodology outlined by D.L. Meier. Figure 4 shows the measured specific contact resistance from TLM grid pattern analysis. It shows very high contact resistance of ~16 mohm-cm^2 for B emitter whereas very low contact resistance of ~6 mohm-cm^2 for TOPCon rear.

TABLE I. BEST CELL EFFICIENCY AND LIV PROPERTIES

Bifacial TOPCon	Voc (V)	Jsc (mA/cm^2)	FF (%)	Eff (%)	Rs (ohm-cm^2)	Rshunt (ohm-cm^2)	n	PFF (%)
Best Cell efficiency	672	38.7	77.4	20.1	1.1	1800	1.13	82.8

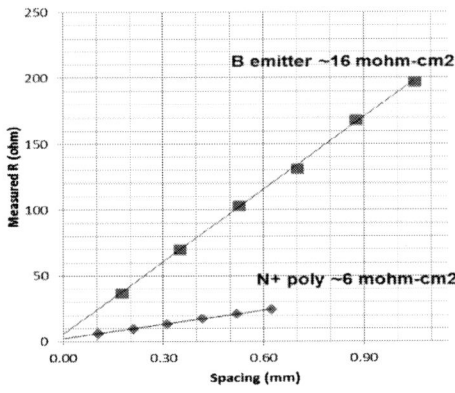

Fig. 4. Measured contact resistance from B emitter and n+ poly using TLM method

Table II shows three main series resistance components including R_{grid}, $R_{contact}$ and R_{sheet} for screen printed B emitter and TOPCon. It shows very high R_s (~0.9 ohm-cm^2) for emitter, but low R_s (~0.24 ohm-cm^2) for TOPCon. This is due to high sheet and contact resistance for B emitter (~180 ohm/□). It is important to note that there is no issue in making good ohmic contact to TOPCon by screen printing technology. In addition, one can reduce R_{grid} further by adding more grid lines without any appreciable increase in metal induced recombination as discussed in section II. Besides high Rs from B emitter, high n factor (~1.13) is also responsible for lower FF. Calculated FF

TABLE II. R_s COMPONENT FOR B EMITTER AND N$^+$ POLY-SI

Front/Back	Rgrid (ohm-cm2)	Rcontact (ohm-cm2)	Rsheet (ohm-cm2)	Rtotal (ohm-cm2)
B emitter	0.2	0.22	0.48	0.9
TOPCon (n+ ploy Si)	0.14	0.06	0.04	0.24

using LIV cell properties including R_s, n factor and R_{shunt} was 0.772 which is consistence with the measured FF.

Detailed Jo analysis for the 20.1% TOPCon bifacial cells is summarized in table III. B emitter had very low saturation current density due to light doping (>150 ohm/□) and Al_2O_3/SiN_x passivation. However, metallized J_{0e} ($J_{0e, metal}$) was ~1600 fA/cm^2 which was calculated from the generated SRV vs J_0 curves and S~10^7 cm/s. Using the metal contact area fraction of 6.5%, J_0 contribution from the emitter are ~113 fA/cm^2 compared to ~3.3 fA/cm^2 from TOPCon rear. $J_{0,bulk}$ was estimated to be ~50 fA/cm^2 for 1 ms the bulk lifetime. This gives total J_0 of 167 fA/cm^2 (113.4+3.3+50) which translates into a V_{oc} of 672 mV for a J_{sc} of 38.7 mA/cm^2 which is in excellent agreement with the measured V_{oc} (Table I). Thus, the main limitation for cell V_{oc} is the metallized B emitter. This can be overcome by formation of selective B emitter with heavy doping under the grid lines. In addition, we are investigating the optimum thickness of poly-Si to minimize light absorption in poly-Si without hurting cell V_{oc} and iFF.

TABLE III. J$_0$ COMPONENTS FOR EACH REGION

Jo (fA/cm2)	B emitter	n+ poly-Si	bulk
J$_{0,passivated}$	10	2.5	~50 (for 1ms)
J$_{0, metallized}$ (100%)	1600	11	
Contact fraction (f)	0.065	0.09	
J$_{0e,cell}$	**113.4**	**3.27**	
J$_{0,total}$		**167**	

V. CONCLUSION

This paper showed the development of a screen printed, large area bifacial n-type Si solar cells with tunnel oxide passivated back contact. We demonstrated that screen printed metal contact on the TOPCon structure does not appreciably degrade the saturation current density, but it increases recombination at low injection level and lowers iFF. Cell efficiency of 20.1% was achieved from screen printed bifacial structure with 180 ohm/□ implanted B emitter on front and TOPcon on back. Detailed analysis showed that high Rs and metallized J_{0e} from lightly doped B emitter on front limited the performance of the cells. Screen printed TOPCon showed very good contact resistance (6 mohm-cm^2) and excellent saturation current density (11 fA/cm^2). We expect that combination of B selective emitter and TOPCon can raise the cell efficiency over 22%.

REFERENCES

[1] Y. Tao, E. L. Chang, A. Upadhyaya, B. Roundaville, Y.-W. Ok, K. Madani, *et al.*, "730 mV implied Voc enabled by tunnel oxide passivated contact with PECVD grown and crystallized n+ polycrystalline Si,"

978-1-5090-5606-4/17 $31.00 © 2017 IEEE

in *Photovoltaic Specialist Conference (PVSC), 2015 IEEE 42nd*, 2015, pp. 1-5.

[2] S. Glunz, F. Feldmann, A. Richter, M. Bivour, C. Reichel, H. Steinkemper*, et al.*, "The irresistible charm of a simple current flow pattern—25% with a solar cell featuring a full-area back contact," in *31st European Photovoltaic Solar Energy Conference and Exhibition*, 2015, pp. 259-263.

[3] Y. Tao, V. Upadhyaya, Y.-Y. Huang, C.-W. Chen, K. Jones, and A. Rohatgi, "Carrier selective tunnel oxide passivated contact enabling 21.4% efficient large-area N-type silicon solar cells," in *Photovoltaic Specialists Conference (PVSC), 2016 IEEE 43rd*, 2016, pp. 2531-2535.

[4] M. Stodolny, M. Lenes, Y. Wu, G. Janssen, I. Romijn, J. Luchies*, et al.*, "n-Type polysilicon passivating contact for industrial bifacial n-type solar cells," *Solar Energy Materials and Solar Cells*, vol. 158, pp. 24-28, 2016.

[5] Y.-W. Ok, A. D. Upadhyaya, Y. Tao, F. Zimbardi, S. Ning, and A. Rohatgi, "Ion-implanted and screen-printed large area 19.6% efficient n-type bifacial Si solar cell," in *Photovoltaic Specialists Conference (PVSC), 2012 38th IEEE*, 2012, pp. 002240-002243.

Correlation between Electroluminescence and Photoconversion Efficiency in a-Si:H/c-Si Heterojunction Solar Cells

A.V. Sachenko[1], A.V. Bobyl[2], V.N. Verbitskiy[2], V.M. Vlasyuk[1], D.M. Zhigunov[3], V.P. Kostylyov[1], I.O. Sokolovskyi[1,4], E.I. Terukov[2,5], P.A. Forsh[3], and M. Evstigneev[4]

[1]V. Lashkaryov Institute of Semiconductor Physics, NAS of Ukraine, 03028 Kiev, Ukraine
[2]Ioffe Institute RAS, 194021 St.-Petersburg, Russia
[3] M. Lomonosov Moscow State University, 119991 Moscow, Russia
[4] Department of Physics and Physical Oceanography, Memorial University of Newfoundland, St. John's, NL, A1B 3X7 Canada
[5]TFTC Ioffe R&D Center, 194021 St.-Petersburg, Russia

Abstract — Photoconversion and electroluminescence measurements in a set of a-Si:H/c-Si heterojunction solar cells (HJSCs) are carried out in a broad temperature range. The samples differed only in surface recombination velocity, *S*, but otherwise they were fabricated identical. Electroluminescence efficiency, q_{el}, is found to be affected by *S* much more strongly than photoconversion efficiency, η: namely, the reduction of η from 20.5% to 18% due to an increase of *S* is accompanied by a decrease of q_{el} by more than an order of magnitude. Electroluminescence intensity as a function of temperature develops a maximum at $T = 223$ K. A theoretical model is developed that quantitatively explains these findings.

Index Terms — electroluminescence, heterojunction solar cells, photoconversion efficiency, surface recombination

I. INTRODUCTION

Among its numerous applications, electroluminescence can be used as an efficient tool to characterize the quality of solar cells (SCs) [1]-[3]. It has been established in a number of publications on the edge luminescence in silicon diode structures that at room temperatures, photoluminescence quantum efficiency can be as high as about 10% [4], and electroluminescence quantum efficiency can reach 1% [5], [6]. The development of highly efficient heterojunction solar cells (HJSCs) based on a-Si:H/c-Si structures motivates investigation of electroluminescence in such SCs as well. In order to explore the interrelation between electroluminescence and recombination processes in HJSCs, this work focuses on the electroluminescence intensity as a function of the applied bias, its spectral characteristics, dark current-voltage curves, and photoconversion efficiency. Electroluminescence and dark I–V curves were measured at different temperatures, whereas photoconversion efficiency measurements were performed under the AM1.5 conditions at room temperature. The experimental results are quantitatively described within a theoretical model developed earlier to analyse current formation in HJSCs [7].

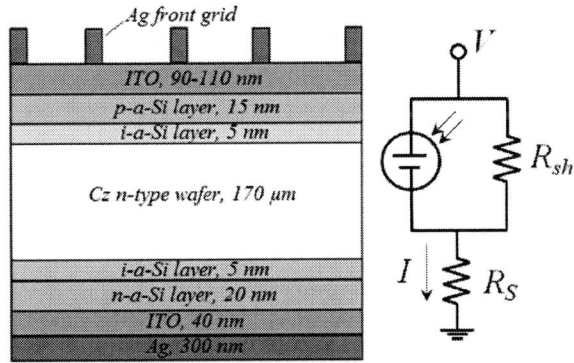

Fig. 1. Schematics of an HJSC element used in this work.

II. SAMPLE PREPARATION AND MEASUREMENTS

The structure of our HJSCs is shown in Fig. 1. It is based on an n-type c-Si wafer of (100) orientation fabricated using Czochralski method. On its front surface are deposited the layers of undoped amorphous a-Si, an p-a-Si:H layer, a conducting transparent ITO layer, and a silver grid for current collection. On the rear side of the wafer are the layer of amorphous i-Si, an n-type a-Si:H layer, an ITO layer, and a silver layer.

The wafers, cut out of the same part of silicon ingot, had the thickness $d = 170$ μm. The doping concentration of the wafers was $N_d = 10^{15}$ cm^{-3}.The HJSC area was $A_{SC} = 239$ cm^2. Because all wafers were cut from the same silicon crystal, they were characterized by the same Shockley-Read-Hall lifetime, $\tau_{SRH} = 2$ ms, and bulk recombination velocity $d/\tau_{SRH} = 8.5$ cm/s. Their radiative recombination velocity at 300 K was not higher than 1 cm/s.

The resulting HJSCs had short-circuit current density of $J_{SC} = 36\text{-}37$ mA/cm^2 and open-circuit voltage of $V_{OC} = 0.69\text{-}0.73$ V. Their output power under AM1.5 conditions was in the 4.2-4.9 W, and photoconversion efficiency varied between 18 and 21 %.

The wafers were identical in all respects except for surface recombination velocity. Variation of surface recombination velocity among the samples was achieved as follows. The texturing process took place in the same chemical solution of volume 10 liters, which gradually accumulated the reaction products of KOH with silicon, such as potassium silicate: $Si + 2KOH + H_2O = K_2SiO_3 + 2H_2\uparrow$. The process proceeded at the temperature of 70-80 °C, at which intensive hydrolysis of K_2SiO_3 takes place, resulting in the formation of silicon dioxide gel. At larger potassium silicate concentrations, the SiO_2 film makes transport of the reagents to silicon surface more difficult, leading to a notable dispersion of etching rates among different microscopic regions of silicon surface, and hence to the sample texturing. Apparently, the probability of defect and step formation is higher on the pyramid faces, and those defects act as recombination centers, which have a negative effect on the final electroluminescence and photoconversion efficiency. Because surface processing of our samples occurred under different conditions, the samples differed in their surface recombination velocity.

The measurements of our solar cells' characteristics at 300 K were performed using cetisPV-Celltest3 setup (HALM, Germany) equipped with a cetisPV-cell-EL-Lab system and a 1.4 MP CCD camera and light filters that cut off the wavelengths below 880 nm. More than a hundred HJSCs were tested. Electroluminescence intensity measurements were performed at constant dark current density $J = 37.65$ mA/cm² maintained by an external bias.

III. THEORETICAL MODEL

The theoretical model describing the I-V curves, electroluminescence internal quantum efficiency, and photoconversion efficiency was introduced in the earlier works [7]-[9] and is briefly reiterated here.

A. Dark I-V curves

The current through the structure equals to the sum of the current through the shunt resistance, $(V - IR_S)/R_{sh}$, and through the HJSC (see Fig. 1). Because the diffusion length, L_d, in the central crystalline n-Si layer is much greater than its thickness, $L_d \gg d$, excess carrier concentration, Δn, in crystalline Si is practically uniform. The current through the HJSC is due to the recombination mechanism and is given by $qA_{SC}d \, \Delta n/\tau_{eff}$, where q is the elementary charge and τ_{eff} is the effective lifetime in the crystalline Si layer. Dividing both contributions – the shunt resistance current and the recombination current in the HJSC – by the SC area, A_{SC}, we obtain the net current density

$$J(V) = \frac{qd\Delta n}{\tau_{eff}} + \frac{VA_{SC}^{-1} - J(V)R_s}{R_{sh}} . \quad (1)$$

Voltage drop across the HJSC is $V - A_{SC}JR_S$. On the other hand, it equals to the sum of potential differences between the central n-Si slab and the upper and lower amorphous layers of

n-a-Si and p-a-Si. These two voltages can be expressed in terms of the equilibrium, n_0 and p_0, and the excess charge carrier concentrations as

$$V - A_{SC}JR_S = \frac{kT}{q}\left(\ln\frac{n_0 + \Delta n}{n_0} + \ln\frac{p_0 + \Delta n}{p_0} \right) . \quad (2)$$

Multiplying both sides by q/kT and performing exponentiation, we obtain a quadratic equation for the excess carrier density. The solution of this equation reads

$$\Delta n(V) = -\frac{n_0 + p_0}{2}$$
$$+ \sqrt{\frac{(n_0 + p_0)^2}{4} + n_i^2\left(e^{\frac{q(V-JA_{SC}R_s)}{kT}} - 1 \right)} \quad (3)$$
$$\cong -\frac{N_d}{2} + \sqrt{\frac{N_d^2}{4} + n_i^2\left(e^{\frac{q(V-JA_{SC}R_s)}{kT}} - 1 \right)} .$$

Here, we used the relation $n_0 p_0 = n_i^2$, where n_i is intrinsic carrier density. For the experimentally relevant doping levels such that $N_d \gg n_i$, we have set $n_0 = N_d \gg p_0$ in the last line.

The inverse effective lifetime in crystalline Si consists of the bulk and the surface contributions:

$$\frac{1}{\tau_{eff}} = \frac{1}{\tau_b} + \frac{S + U_{SC}}{d} . \quad (4)$$

Here, S is the net surface recombination velocity in the $x = 0$ and $x = d$ planes, i.e. on the boundaries between crystalline and amorphous Si layers. The recombination velocity in the space-charge region (SCR) is given by [10]

$$U_{SC}(V) = \frac{1}{\tau_R}\int_0^w dx \exp\left[y_{pn}\left(1 - \frac{x}{w}\right)^2 \right] \times$$
$$\left(1 + b\frac{\Delta n(V)}{(N_d + \Delta n(V))}\exp\left[2y_{pn}\left(1 - \frac{x}{w}\right)^2 \right] \right)^{-1} . \quad (5)$$

Here, τ_R is the Shockley-Read-Hall lifetime in the SCR, b is the ratio of hole and electron capture cross section by a deep recombination centre, $w = 2\sqrt{\ln(N_d/n_i)\varepsilon_0\varepsilon_s kT/(q^2 N_d)}$ is the SCR thickness, $\varepsilon_S = 11.7$ is the relative permittivity of Si, and $y_{pn} = 2\ln(N_d/n_i(T)) - qV/kT$ is the dimensionless inversion potential.

The bulk lifetime is determined by several recombination mechanisms, namely

$$\frac{1}{\tau_b} = \frac{1}{\tau_{SRH}} + \frac{1}{\tau_r} + \frac{1}{\tau_{nr}} + \frac{1}{\tau_{Auger}} ,$$

$$\tau_r = \left(A(N_d + \Delta n)\right)^{-1} , \qquad (6)$$

$$\tau_{nr} = \tau_{SRH} \cdot n_x / (N_d + \Delta n) ,$$

where the first term, τ_{SRH}, is the bulk Shockley-Read-Hall lifetime in c-Si; the second and the third terms, are, respectively, the radiative recombination lifetime characterized by the parameter A [11], and the exciton non-radiative recombination lifetime [12]. The latter depends on the parameter $n_x = 8.2 \cdot 10^{15}$ cm^{-3} at room temperature in Si [12]. The fourth term, τ_{Auger} is Auger band-to-band recombination lifetime, whose somewhat lengthy expression can be found in the work [13] and is not reproduced here.

In order to obtain the dark I-V curve, equation (1) supplemented by the expressions (3)-(6) need to be solved numerically.

B. I-V curves in the presence of illumination

Illumination of the HJSC results in the onset of the light-generated current in the reverse direction. The expression for the excess carrier concentration (3) remains the same, but the current expression becomes:

$$J(V) = \frac{qd\Delta n}{\tau_{eff}} + \frac{VA_{SC}^{-1} - J(V)R_s}{R_{sh}} - \Delta J , \qquad (7)$$

where the light-generated current is chosen so as to have the correct short-circuit current density limit, $J(V=0) = -J_{SC}$:

$$\Delta J = qd\left(\frac{\Delta n}{\tau_{eff}}\right)_{SC} + J_{SC}\left(1 + \frac{R_s}{R_{sh}}\right) \cong J_{SC} , \qquad (8)$$

where $(\Delta n/\tau_{eff})_{SC}$ is given by (3), (4) with $V = 0$, $J = -J_{SC}$. The short-circuit current needs to be determined experimentally. Its substitution into (7) and numerical solution of this equation allows one to describe the I-V curve at arbitrary bias values. In particular, the open-circuit voltage is [cf. (2)]

$$V_{OC} = \frac{kT}{q} \ln\left(\frac{(N_d + \Delta n_{OC})\Delta n_{OC}}{n_i(T)^2}\right) , \qquad (9)$$

where excess charge carrier density in the open-circuit regime is related to the short-circuit current density by the balance equation

$$J_{SC} = q\left[S + U_{SC} + (d/\tau_b)\right]\Delta n_{OC} . \qquad (10)$$

Using the maximal-power condition, $d(VJ(V))/dV = 0$, the voltage V_m at the maximal power is determined. Substitution of V_m into Eq. (8) allows determining the respective photocurrent value J_m. As a result, the photoconversion efficiency and the fill factor are given by

$$\eta = \frac{J_m V_m}{P_S} , \quad FF = \frac{J_m V_m}{J_{SC} V_{OC}} , \qquad (11)$$

where P_S is the incident radiation energy flux.

C. Electroluminescence intensity

Experimental electroluminescence intensity

$$El = \int d\lambda \, I_{el}(\lambda) , \qquad (12)$$

is proportional to the internal electroluminescence quantum efficiency is given by

$$q_{el} = \tau_{eff} / \tau_r \qquad (13)$$

where all relaxation times are given by (5)-(7).

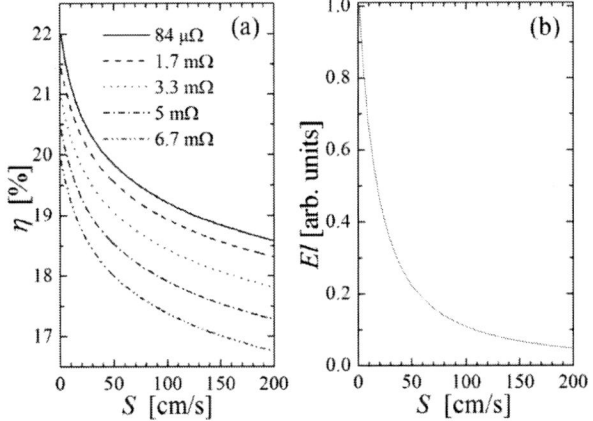

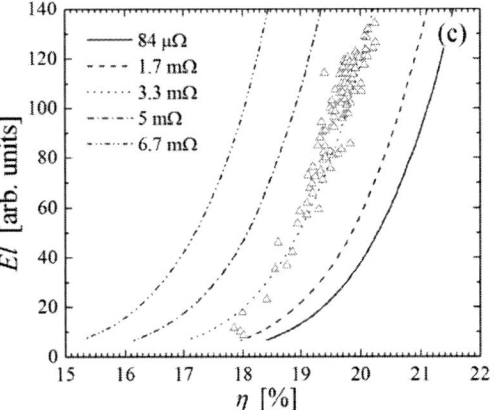

Fig. 2. (a) Theoretical photoconversion efficiency and (b) electroluminescence efficiency vs. surface recombination velocity. In panel (a), the curves are built for different values of the series resistance (from top to bottom): $R_s = 84$ µΩ, 1.7 mΩ, 3.3 mΩ, 5 mΩ, and 6.7 mΩ. In panel (b), the El curves are practically the same for all these values of R_S.

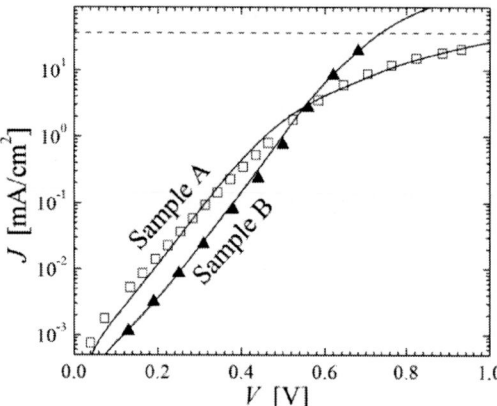

Fig. 3. Dark I-V curves under AM1.5 conditions at room temperature for two representative small-area samples with A_{SC} = 1 cm^2. The dashed horizontal line is at the 40 mA level. The theoretical curve for sample A is obtained for the series resistance R_S = 1.5 Ω, and for sample B, R_S = 12 Ω. The shunt resistance was R_{sh} = 10 kΩ and the surface recombination velocity S = 10 cm/s for both samples. The Schockley-Read-Hall recombination lifetime in the space-charge region τ_R = 1.2 μs, and the ratio of hole and electron capture cross section by a deep recombination level b = 0.01 for both samples.

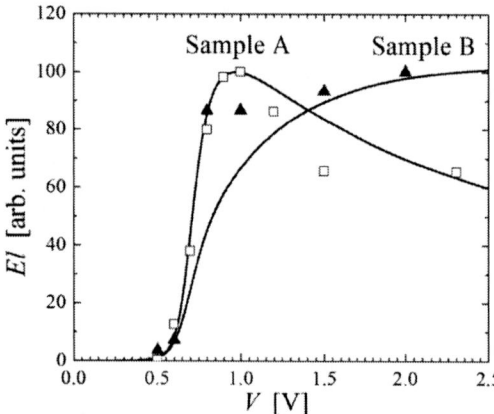

Fig. 4. Integral electroluminescence intensity vs. applied voltage for the two samples (symbols), and the theoretical fit curves (black solid and dashed lines).

IV. EXPERIMENTAL RESULTS AND DISCUSSION

A. Electroluminescence and photoconversion efficiency at room temperature

Figs. 2 (a) and (b) show the theoretical electroluminescence intensity, El, and photoconversion efficiency, η, vs. surface recombination velocity. As seen from Fig. 2, electroluminescence intensity decreases with S much faster than photoconversion efficiency does. Indeed, if the latter

decreases by about 10%, the former decreases by more than an order of magnitude.

Plotted in Fig. 2(c) are the experimental curves of electroluminescence intensity on photoconversion efficiency, measured for all samples, as well as the respective theoretical curves. The latter agrees well with the experiment for the series resistance of R_s = 3.3 mΩ.

One of the large-area HJSCs was cut into small pieces with the area of A_{SC} = 1 cm^2 for temperature-dependent measurements of electroluminescence and photoconversion efficiency. This was necessary because the cryostat used in this study could only accommodate samples of relatively small area. Here, we present the results obtained on two representative samples characterized by relatively strong (sample A) and weak (sample B) electroluminescence intensity. Shown in Fig. 3 are the experimental dark I-V curves for these two samples, measured at T = 300 K. The theoretical I-V curves can fit the experimental data, if one takes the series resistance of the first sample to be 1.5 Ω and 12 Ω for the second sample. The difference in the series resistance values is due to the fact that, after the samples were cut off, each of them was treated individually to produce electrical contacts on its surfaces.

Fig. 4 shows the experimental integral electroluminescence intensity, vs. direct bias for these two samples. The experimental electroluminescence external quantum efficiency value was equal 2.1%. Also shown in this figure are the theoretical curves of electroluminescence internal quantum efficiency $q_{el}(V)$. For sample A, $q_{el}(V)$ behaves non-monotonically, developing a maximum near 1 V, whereas for sample B, $El(V)$ curve is monotonic. The experimental curves agree with the theory for surface recombination velocity of S = 10 cm/s for both samples.

It should be noted that the theoretical value of the luminescence internal quantum efficiency, q_{el}, is strongly influenced by the non-radiative exciton recombination. In particular, for the first sample, neglecting this effect leads to the increase of q_{el} at maximum to 4.08%, as compared to the value of 2.6% obtained with taking this recombination channel into account, i.e. the difference is by a factor 1.5. At the same time, photoconversion efficiency at the parameter values considered here is influenced by this mechanism by only about 2% .

B. Temperature dependence of electroluminescence

Fig. 5 shows the experimental temperature dependence of the integral electroluminescence intensity measured at the recombination current density of 37.65 mA/cm^2. This curve develops a maximum at 223 K. Note that also photoconversion efficiency of HJSCs as a function of temperature has a maximum at about the same temperature, as was experimentally established in [8]. At higher temperatures, electroluminescence intensity decreases due to the decrease of the radiative recombination coefficient with temperature. The reasons for the behaviour at lower temperatures, T < 223 K, are as follows.

978-1-5090-5606-4/17 $31.00 © 2017 IEEE

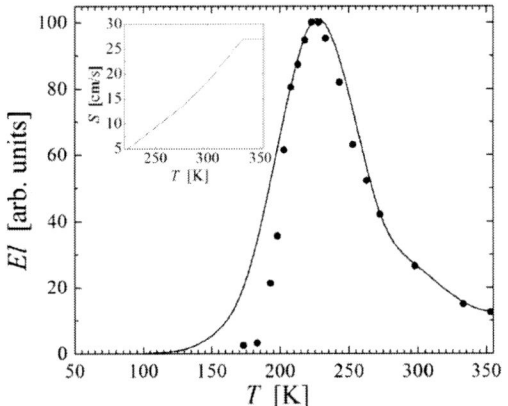

Fig. 5. Experimental (symbols) and theoretical (curve) electroluminescence intensity vs. temperature. The inset shows the assumed temperature dependence of surface recombination velocity.

According to [11], at sufficiently low temperatures ($T <$ 200 K), radiation recombination parameter A is determined mostly by exciton radiative recombination, and can be written as

$$A \approx \frac{N_x e^{E_x/kT}}{N_c N_v \tau_{rx}} , \qquad (14)$$

where N_c, N_v, and N_x are effective densities of states in the conduction, valence, and exciton bands of silicon, E_x is the exciton ground state energy (0.0147 eV in Si), and τ_{rx} is exciton radiative recombination lifetime. The electroluminescence intensity in n-type silicon is proportional to $A(n_0 + \Delta n(V))$. According to (14), the radiative coefficient increases monotonically on cooling, and therefore its temperature dependence cannot be responsible for the experimentally observed decrease of electroluminescence and photoluminescence on cooling (see [4]). To explain this effect, one needs to account for the electron freeze-out at low temperatures. Equilibrium electron concentration can be found from the following neutrality equation

$$n_0(T) = \frac{N_d}{1 + 2e^{(E_f - E_d)/kT}} =$$
$$\frac{2}{\sqrt{\pi}} N_c(T) \int_0^\infty \frac{x^{0.5}}{1 + e^{x + (E_c - E_f)/kT}} dx , \qquad (15)$$

where E_f is Fermi energy, E_d is the energy of a donor level, and N_{c0} the effective density of states in the silicon conduction band at 300 K. Eq. (15) shows that the lower the temperature, the fewer donors are ionized, which leads to the decrease of n_0.

Apart from the experimental curve, Fig. 5 also shows the normalized theoretical internal quantum efficiency of electroluminescence vs. temperature obtained with the help of Eq. (13), and also taking into account the electron freeze-out

in silicon. In addition to exciton radiative recombination, the band-to-band radiative recombination was also accounted for using the expressions from [11]. It was assumed that $E_d = -$ 0.044 eV (corresponding to phosphorus level in Si), and the parameters n_x, τ_{SRH}, and τ_{Auger} had the same values as at $T =$ 300 K. At the same time, temperature dependence of surface recombination velocity, S, had to be assumed, see inset in Fig. 5. Its increase with temperature is due to the shifting of Fermi level to the middle of the bandgap, leading to a variation in the population of surface states, and a corresponding variation in S.

As seen in Fig. 5, the locations of the theoretical and experimental electroluminescence intensity maxima agree well with each other. Below 223 K, the intensity decreases on cooling due to the electron freeze-out, while above 223 K, it decreases with temperature due to the decrease of the radiative recombination coefficient on heating.

Because the binding energy of an exciton in silicon at room temperature is about 14.7 meV, which is about half the thermal energy value, electroluminescence effect has edge character. Then, the temperature dependence of the wavelength at which electroluminescence has maximal intensity is determined mainly by the temperature dependence of bandgap in silicon (see Fig. 6). This maximal wavelength can be related to the bandgap $E_g(T)$ by

$$\lambda_{\max}(T) = \frac{1240\,\text{nm}}{E_g(T)/1\,\text{eV}} + 38\,\text{nm} , \qquad (16)$$

with reasonable accuracy. It should be noted that in crystalline silicon, radiative edge luminescence is accompanied by a TO phonon of energy 58 meV, which alone cannot explain the value of 38 nm in this expression. It can be assumed that several different radiative recombination processes involving different phonon types may be operative.

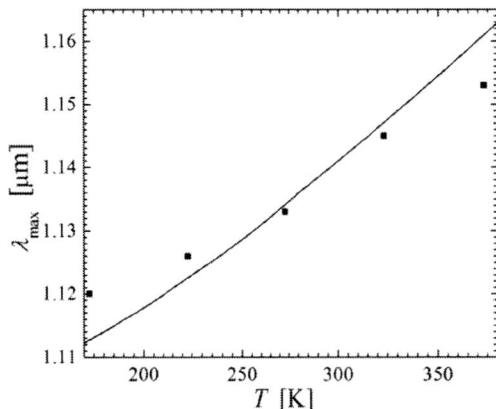

Fig. 6. Experimental temperature dependence of the maximal electroluminescence wavelength (symbols), and the same according to (16) (solid curve).

978-1-5090-5606-4/17 $31.00 © 2017 IEEE

V. CONCLUSIONS

The main results of this work are as follows. First, in order to maximize electroluminescence intensity, it is necessary to minimize both surface and bulk recombination. Second, the sensitivity of electroluminescence in HJSCs to surface recombination velocity is much higher than that of photoconversion efficiency. Thus, measurements of the electroluminescence intensity of HJSCs can serve as a quality indicator of the underlying semiconductor structure, in particular, of the textured wafer surface and the heterojunctions. As a function of temperature, electroluminescence exhibits a sharp maximum at about 223 K, resulting from an interplay between electron freeze-out on cooling and the decrease of the radiative recombination coefficient on heating.

The results obtained here can serve as a methodological basis for the use of electroluminescence efficiency in HJSCs to characterize the degree of surface passivation of monocrystalline silicon.

ACKNOWLEDGEMENT

M.E. is grateful to the Natural Sciences and Engineering Research Council of Canada (NSERC) and to the Research and Development Corporation of Newfoundland and Labrador (RDC) for financial support.

REFERENCES

[1] T. Kirchartz, J. Nelson, and U. Rau, Reciprocity between charge injection and extraction and its influence on the interpretation of electroluminescence spectra in organic solar cells, *Physical Review Applied*, vol. 5, pp. 054003:1-10, 2016.

[2] U. Rau, Reciprocity relation between photovoltaic quantum efficiency and electroluminescent emission of solar cells, *Physical Review B*, vol. 76, pp. 085303:1-8, 2007.

[3] J.M. Raguse and J.R. Sites, Correlation of electroluminescence with open-circuit voltage from thin-film CdTe solar cells, *IEEE Journal of Photovoltaics*, vol. 5, pp. 1175-1178, 2015.

[4] Th. Trupke, J. Zhao, A. Wang, R. Corkish, and M.A. Green, "Very efficient light emission from bulk crystalline silicons," *Applied Physics Letters*, vol. 82, pp. 2996-2998, 2003.

[5] M.A. Green, J. Zhao, S. Wang, P.J. Reece, and M. Gal, "Efficient silicon light-emitting diodes," *Nature*, vol. 412, pp. 805-808, 2001.

[6] W.L. Ng, M. A. Lourenço, R.M. Gwilliam, S. Ledain, G. Shao, and K. P. Homewood, "An efficient room-temperature silicon-based light-emitting diode," *Nature*, vol. 410, 192-194, 2001.

[7] A.V Sachenko, V. P. Kostylev, I.O. Sokolovskii, A.V. Bobyl', V.N. Verbitskii, E.I. Terukov, and M.Z. Shvarts, "Specific Features of Current Flow in α-Si:H/Si Heterojunction Solar Cells", *Technical Physics Letters*, vol. 43, pp. 152-155, 2017.

[8] A.V. Sachenko, Yu.V. Kryuchenko, V. P. Kostylyov, A.V. Bobyl, E.I. Terukov, S. N. Abolmasov, A.S. Abramov, D.A. Andronikov, M.Z. Shvarts, I.O. Sokolovskyi, and M. Evstigneev, "Temperature dependence of photoconversion efficiency in silicon heterojunction solar cells: theory vs experiment," *Journal of Applied Physics*, vol. 119, pp. 225702:1-10, 2016.

[9] A.V. Sachenko, V.P. Kostylyov, V.M. Vlasyuk, I.O. Sokolovskyi, and M. Evstigneev, "The influence of the exciton nonradiative recombination in silicon on the photoconversion, " in *Proceedings of the 32nd European Photovoltaic Solar Energy Conference and Exhibition*, 2016, pp.141-147.

[10] Ch.-T. Sah, R.N. Noyce, and W. Shockley, "Carrier Generation and Recombination in P-NVJunctions and P-N Junction Characteristics", *Proceedings of the IRE*, vol. 45, pp. 1228-1243, 1957.

[11] A.V. Sachenko, A.P. Gorban, V.P. Kostylyov, and I.O. Sokolovsky, "The radiative recombination coefficient and the internal quantum yield of electroluminescence in silicon", *Semiconductors*, vol. 40, pp. 884- 889, 2006.

[12] A.V.Sachenko, V.P.Kostylyov, V.M.Vlasiuk, I.O.Sokolovskyi, and M.Evstigneev, Influence of excitonic effects on luminescence quantum yield in silicon, *Journal of Luminescence*, vol. 183, pp. 299-302, 2016.

[13] A. Richter, S. Glunz, F. Werner, J. Schmidt, and A. Cuevas, "Improved quantitative description of Auger recombination in crystalline silicon", *Physical Review B*, vol. 86, 165202: 1-14, 2012.

An Isotope Study of Hydrogen Passivation of poly-Si/SiO$_x$ Passivated Contacts for Si Solar Cells

Manuel Schnabel[1], William Nemeth[1], Bas W.H. van de Loo[2], Bart Macco[2], Wilhelmus M.M. Kessels[2], Paul Stradins[1], David L. Young[1]

[1]National Renewable Energy Laboratory, Golden CO 80401
[2]Department of Applied Physics, Eindhoven University of Technology, 5600 MB Eindhoven, Netherlands

Abstract — **Recent improvements in Si solar cell efficiency above 25% have used doped poly-Si/SiO$_x$ or a-Si:H passivated contacts. Common to these designs is the need to passivate with hydrogen. In this contribution, we perform a systematic study of *p*-type poly-Si/SiO$_x$ passivation by isotopic hydrogen using atomic layer-deposited (ALD) Al$_2$O$_3$ followed by an activating anneal.**

We observe that annealing *p*-type poly-Si/SiO$_x$ in forming gas, or even nitrogen, is sufficient to provide some passivation. State-of-the-art implied open-circuit voltages exceeding 710 mV for *p*-type passivated contacts, however, were only achieved using Al$_2$O$_3$ and an activating anneal. This suggests that Al$_2$O$_3$ provides a superior source of hydrogen and diminishes the role of ambient hydrogen during the activating anneal.

I. INTRODUCTION

Crystalline silicon (c-Si) dominates the solar cell market [1], with diffused-junction solar cells holding the largest market share. On the other hand, all of the cells that have, in the past two years, improved on the previous Si solar cell efficiency record of 25.0% have used so-called passivated contacts as carrier-selective contacts [2], rather than diffusions, and it is to be expected that a future generation of high-efficiency c-Si modules will utilize this technology. In particular, passivated contacts based on polysilicon (poly-Si) / silicon oxide (SiO$_x$) stacks, have gained a lot of traction [3-6], with both Fraunhofer ISE and ISFH achieving 25% efficient Si solar cells using this approach [7, 8].

Hydrogen plays a pivotal role in achieving good chemical passivation of such passivated contacts, and is added after poly-Si deposition by many groups working on poly-Si/SiO$_x$ passivated contacts to achieve the desired implied open-circuit voltage (iV_{oc}) values [9-11]. Interfaces between c-Si and SiO$_2$ can be hydrogenated using Al$_2$O$_3$ activated by an anneal [12], and Nemeth et al. have shown that this approach can also be successfully applied to poly-Si/SiO$_x$ passivated contacts, and that the passivation persists even if the Al$_2$O$_3$ is etched off after the activating anneal [6]. In the following, we present a more systematic study of this process in order to understand the role played by the different process steps, using symmetric *p*-type poly-Si/SiO$_x$ passivated contact lifetime samples and passivation by atomic layer-deposited (ALD) Al$_2$O$_3$:D and/or annealing in nitrogen or forming gas (FGA:D). Both Al$_2$O$_3$:D and FGA:D are deuterated, rather than hydrogenated, to permit follow-up secondary-ion mass spectrometry studies

(SIMS) which will be the subject of a more detailed, upcoming publication.

II. EXPERIMENTAL DETAILS

We prepared symmetric passivated contact structures as shown schematically in Fig. 1. As-sawn, 160 µm thick 3 Ωcm *n*-type Cz-Si wafers planarized in 25% KOH(aq) were RCA-cleaned [13] and had a 1.5 nm thick thermal oxide grown on them at 700°C in an N$_2$:O$_2$ mixture. Then, 20 nm boron-doped a-Si:H was deposited on both sides of the wafers, and subsequently crystallized by annealing in N$_2$ at 850°C for 30 min. Due to the consensus in the literature that *p*-type passivated contacts require more improvement than *n*-type passivated contacts [5, 6, 14, 15], this study focuses on the former.

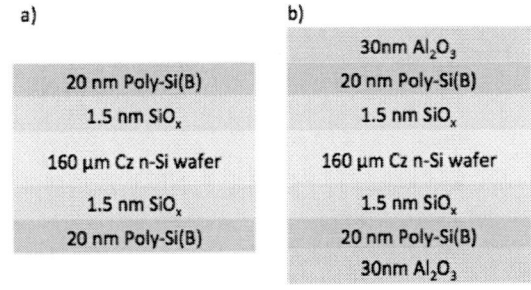

Figure 1. Schematics of samples prepared for this study.

Some samples were maintained as-crystallized (Fig. 1(a)), while others had 30 nm Al$_2$O$_3$:D deposited on both sides by thermal ALD using deuterated trimethylaluminum (Al(CD$_3$)$_3$) and heavy water as precursors (Fig. 1(b)). As-crystallized and Al$_2$O$_3$-coated samples were then exposed to 400°C anneals of different durations in nitrogen and/or deuterated forming gas (FGA:D). However, for simplicity, and because the final isotopic purity of these deuterated hydrogenation treatments is not accurately known at this stage, we will refer to hydrogen rather than deuterium throughout this manuscript. Photoconductance decay measurements performed with a Sinton WCT-120 lifetime tester in generalized (1/64) mode were used to derive iV_{oc} values and injection-level dependent minority carrier lifetimes.

978-1-5090-5606-4/17 $31.00 © 2017 IEEE

III. RESULTS AND DISCUSSION

The iV_{oc} values obtained using lifetime samples prepared for this study are shown in Fig. 2(a),(b). We consider these to be sufficient to describe passivation because the injection-level dependent lifetime curves (shown for three representative samples in Fig. 2(c)) are all of the same shape, indicating that there is no fundamental change in recombination mechanisms upon passivation, merely a decrease in overall recombination rate by decreasing defect-mediated Shockley-Read-Hall recombination.

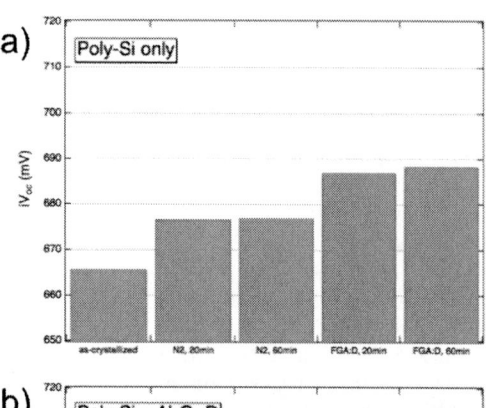

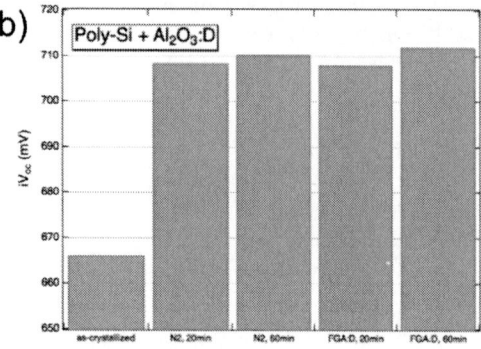

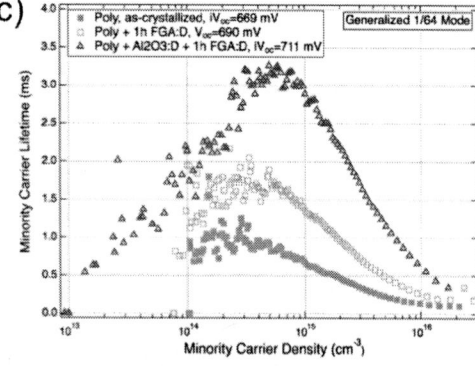

Figure 2: Implied open-circuit voltages of symmetric poly-Si(B)/SiO$_x$ lifetime samples as a function of annealing treatment only ((a), samples as in Fig. 1(a)), and as a function of annealing treatment after Al$_2$O$_3$:D deposition ((b), samples as in Fig. 1(b)). Each data point represents the mean of three identically processed samples. Representative injection-level dependent lifetime curves from which these values are derived are shown in (c).

Figure 2(b) shows that state-of-the-art p-type passivated contacts have been prepared, exceeding 710 mV iV_{oc} after a 1 h activation anneal of the Al$_2$O$_3$:D at 400°C. More importantly, Figure 2(a) provides some clues as to what precisely happens within the material as symmetric poly-Si/SiO$_x$ samples undergo passivation treatments; we observe an initial iV_{oc} of 667 mV that increases by ~10 mV upon 400°C annealing in nitrogen. This is rather unexpected: since the samples had previously been annealed at 850°C, and cooled slowly to room temperature thereby passing through the 400°C temperature regime, no additional changes are expected in a subsequent 400°C anneal. That said, we know from prior studies that as-crystallized samples do still contain some hydrogen. We therefore see two possibilities as to why passivation might improve upon nitrogen annealing: either hydrogen that had not effused from the a-Si:H upon crystallization becomes sufficiently mobile to passivate defects, or additional hydrogen originating from an unintended hydrogen background in the furnace tube diffuses into the sample and to the SiO$_x$/c-Si interface. Subsequently annealing the same samples in FGA:D increases iV_{oc} by another ~10 mV, which we attribute to the passivating effect of the additional hydrogen supplied by that annealing ambient.

Turning to samples on which Al$_2$O$_3$:D was deposited (Fig. 2(b)), we note that simply depositing it has virtually no effect: comparing the first bar in Fig. 2(a) and (b) shows that the iV_{oc} is only ~2 mV higher than on as-crystallized samples without Al$_2$O$_3$:D. This implies that the temperature of 200°C at which Al$_2$O$_3$:D is deposited is insufficient for hydrogen to diffuse through the poly-Si to the SiO$_x$/c-Si interface. 400°C, however, is sufficient, as is evidenced by the fact that post-annealing Al$_2$O$_3$:D at that temperature leads to a leap in iV_{oc}, from below 670 mV to above 710 mV after 1h (Fig. 2(b)).

Interestingly, annealing in N$_2$ and FGA:D leads to similar results. This implies not only that FGA:D supplies negligible hydrogen compared to the Al$_2$O$_3$:D whose hydrogen it mobilizes, but also seems to indicate that effusion of hydrogen from Al$_2$O$_3$:D is not sufficiently detrimental to passivation of the SiO$_x$/c-Si interface for the difference in ambient partial pressure of hydrogen to have an effect. However, more in-depth studies involving compositional analysis will be required to validate this finding. In all cases, we observe some leveling off of the annealing effect after 20 mins, although small additional improvements in iV_{oc} are observed upon extending the treatment to one hour.

IV. CONCLUSION

We have performed a systematic study of p-type poly-Si/SiO$_x$ passivation by isotopic hydrogen using atomic layer-deposited Al$_2$O$_3$ and nitrogen and forming gas anneals. We observed that annealing in forming gas, or even nitrogen, is sufficient to provide some passivation, which we attribute to hydrogen in-diffusion and/or activation of residual hydrogen within as-crystallized samples. However, state-of-the-art implied open-circuit voltages exceeding 710 mV were only achieved on the p-type passivated contacts studied here using Al_2O_3 in combination with an activating anneal. This suggests that Al_2O_3 provides a superior source of hydrogen and diminishes the role of ambient hydrogen during the activating anneal.

ACKNOWLEDGMENTS

Funding for this work was provided by the United Sates Department of Energy EERE contract SETP DE-EE00030301 (SuNLaMP) and under Contract No. DE-AC36-08GO28308, as well as by TKI Urban Energy from the "Toeslag voor Topconsortia voor Kennis en Innovatie" of the Dutch Ministry of Economic Affairs (AAA Project). The United States Government retains and the publisher, by accepting the article for publication, acknowledges that the United States Government retains a non-exclusive, paid-up, irrevocable, world-wide license to publish or reproduce the published form of this manuscript, or allow others to do so, for United States Government purposes.

REFERENCES

[1] Fraunhofer-ISE, "Photovoltaics Report," 17.11.2016 2016.
[2] M. A. Green, K. Emery, Y. Hishikawa, et al., "Solar cell efficiency tables (version 46)," Progress in Photovoltaics: Research and Applications, vol. 23, pp. 805-812, 2015.
[3] F. Feldmann, M. Bivour, C. Reichel, et al., "Tunnel oxide passivated contacts as an alternative to partial rear contacts," Solar Energy Materials and Solar Cells, vol. 131, pp. 46-50, 2014.
[4] M. Rienaecker, M. Bossmeyer, A. Merkle, et al., "Junction Resistivity of Carrier-Selective Polysilicon on Oxide Junctions and Its Impact on Solar Cell Performance," IEEE Journal of Photovoltaics, vol. 7, pp. 11-18, 2016.
[5] J. Stuckelberger, G. Nogay, P. Wyss, et al., "Passivating contacts for silicon solar cells with 800°C stability based on tunnel-oxide and highly crystalline thin silicon layer," in 2016

IEEE 43rd Photovoltaic Specialists Conference (PVSC), 2016, pp. 2518-2521.
[6] B. Nemeth, D. L. Young, M. R. Page, et al., "Polycrystalline silicon passivated tunneling contacts for high efficiency silicon solar cells," Journal of Materials Research, vol. 31, pp. 671-681, 2016.
[7] ISFH, "Silicon solar cell of ISFH yields 25% efficiency with passivating POLO contacts," ed, 2016.
[8] A. Richter, J. Benick, F. Feldmann, et al., "n-Type Si solar cells with passivating electron contact: Identifying sources for efficiency limitations by wafer thickness and resistivity variation," Solar Energy Materials and Solar Cells, in press.
[9] F. Feldmann, M. Bivour, C. Reichel, et al., "Passivated rear contacts for high-efficiency n-type Si solar cells providing high interface passivation quality and excellent transport characteristics," Solar Energy Materials and Solar Cells, vol. 120, pp. 270-274, 2014.
[10] B. Nemeth, D. L. Young, H. C. Yuan, et al., "Low temperature $Si/SiO_x/pc$-Si passivated contacts to n-type Si solar cells," in 2014 IEEE 40th Photovoltaic Specialist Conference (PVSC), 2014, pp. 3448-3452.
[11] J. Stuckelberger, G. Nogay, P. Wyss, et al., "Passivating electron contact based on highly crystalline nanostructured silicon oxide layers for silicon solar cells," Solar Energy Materials and Solar Cells, vol. 158, 2-10, 2016.
[12] G. Dingemans, M. C. M. van de Sanden, and W. M. M. Kessels, "Excellent Si surface passivation by low temperature SiO_2 using an ultrathin Al_2O_3 capping film," physica status solidi (RRL) – Rapid Research Letters, vol. 5, pp. 22-24, 2011.
[13] W. Kern and D. Puotinen, "Cleaning solutions based on hydrogen peroxide for use in silicon semiconductor technology," RCA Review, vol. 31, pp. 187-205, 1970.
[14] F. Feldmann, C. Reichel, R. Müller, et al., "The application of poly-Si/SiO_x contacts as passivated top/rear contacts in Si solar cells," Solar Energy Materials and Solar Cells, vol. 159, pp. 265-271, 2017.
[15] U. Römer, R. Peibst, T. Ohrdes, et al., "Ion Implantation for Poly-Si Passivated Back-Junction Back-Contacted Solar Cells," IEEE Journal of Photovoltaics, vol. 5, pp. 507-514, 2015.

Alleviating hydrogen plasma damage to amorphous/crystalline silicon interface passivation

Jianwei Shi and Zachary C. Holman

School of Electrical, Computer and Energy Engineering, Ira A. Fulton Schools of Engineering, Arizona State University, Tempe, Arizona, 85287, USA

Abstract — **Hydrogen plasma treatment is used to enhance the interface passivation of amorphous/crystalline silicon (a-Si:H/c-Si) heterojunctions, but prolonged hydrogen plasma treatment can cause adverse passivation damage. Although hydrogen plasma etching of the a-Si:H layer has been proposed as a plausible explanation, we find that hydrogen plasma damage to the c-Si substrate could be more important. We cap a-Si:H passivation layers with a ~2.5-nm-thick silicon oxide layer to preserve the a-Si:H/c-Si interface and find that severe hydrogen plasma damage is greatly alleviated (slowed down by a factor of 7-10) while the hydrogen still penetrates into the underlying a-Si:H layer for interface passivation.**

Index Terms — **hydrogen, silicon, heterojunctions, interface, passivation.**

I. INTRODUCTION

Amorphous/crystalline silicon heterojunction solar cells have very high open-circuit voltages that are up to 750 mV [1]. This is due to the excellent passivation properties of thin intrinsic hydrogenated amorphous silicon (a-Si:H) layers on high-quality crystalline silicon (c-Si) wafers. Recently, it was found that performing a hydrogen plasma post-treatment can further passivate the as-deposited a-Si:H/c-Si interface and improve the minority carrier effective lifetime [2-3]. Furthermore, the hydrogen plasma treatment can also be used to recover the surface passivation of thin dehydrogenated intrinsic a-Si:H layers on c-Si substrates after the a-Si:H is annealed at temperatures of up to 450 °C [4].

Although hydrogen plasma treatment can enhance a-Si:H/c-Si interface passivation, prolonged hydrogen plasma treatment is detrimental to the interface passivation [4-5]. This passivation degradation may be linked to an adverse hydrogen etching effect (insufficient a-Si:H layer thickness) [6] and/or undesirable hydrogen plasma damage (defect formation in the near-surface c-Si structure) [5].

In this work, we use a thin silicon oxide (SiO$_x$) capping layer to alleviate the passivation degradation caused by prolonged hydrogen plasma treatment. With such a layer deposited on top of the a-Si:H/c-Si interface before the final hydrogen plasma treatment, we find that the hydrogen etching rate and hydrogen plasma damage are greatly suppressed while hydrogen can still penetrate into the underlying a-Si:H layers for interface passivation. This experiment reveals that hydrogen plasma damage of the c-Si substrates, rather than hydrogen plasma etching of the a-Si:H layer, is the essential reason for interface passivation degradation by prolonged hydrogen plasma treatment.

II. EXPERIMENTS

200-μm-thick, n-type monocrystalline Czochralski silicon wafers were used as substrates. The wafers were textured, cleaned, and dipped in buffered oxide etch (BOE) solution prior to any thin-film deposition and/or hydrogen plasma treatment in a plasma-enhanced chemical vapor deposition (PECVD) tool. The processing sequence of thin-film deposition and hydrogen plasma treatment was varied (see the following Section III), but the identical intrinsic a-Si:H layers, SiO$_x$ layers and hydrogen plasma treatments were performed on both sides of the wafers. Note that the SiO$_x$ layers were deposited at a low temperature so this processing will not dehydrogenate hydrogen from the a-Si:H layers. Effective minority carrier lifetimes were measured by a Sinton photoconductance-decay lifetime tester. The thickness and bandgap of the a-Si:H thin films were determined from fitting a Tauc–Lorentz model to the spectroscopic ellipsometry data, which was taken on polished wafers that were used as witnesses in selected experiments. More experimental details on the sample processing and characterization can be found in [4].

III. RESULTS AND DISCUSSION

Figure 1a shows effective lifetimes of textured c-Si wafers with 10-nm-thick intrinsic a-Si:H on both sides before and after hydrogen plasma treatment. Performing 1-min hydrogen plasma treatment on an a-Si:H/c-Si heterojunction results in passivation enhancement, which is attributed to atomic hydrogen diffusion to the a-Si:H/c-Si interface that reduces dangling bond density [3]. However, negative effects of hydrogen plasma treatment quickly occur—if such treatment is 2 min or longer, the a-Si:H/c-Si interface passivation plummets.

978-1-5090-5606-4/17 $31.00 © 2017 IEEE

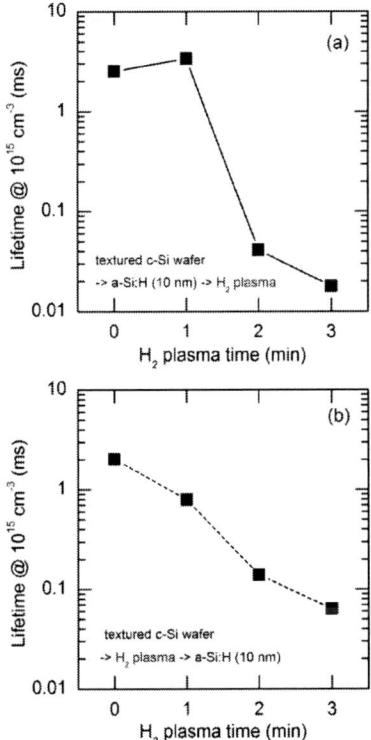

Fig. 1. (a) Effective lifetimes of a textured c-Si wafer with 10-nm-thick a-Si:H before and after H₂ plasma treatment; (b) Effective lifetimes of four textured c-Si wafers encountering time-varied H₂ plasma treatment first and then being passivated with 10-nm-thick a-Si:H. All of the a-Si:H depositions and H₂ plasma treatments were performed on both sides of the samples.

One possibility to explain this lifetime degradation is that the a-Si:H layer after hydrogen plasma treatment is close to the critical minimum thickness (4-5 nm [6]) needed to provide sufficient passivation, because the hydrogen plasma treatment also etches a-Si:H films quite fast. However, in this particular case, we find that the remaining a-Si:H thickness after 2-min hydrogen plasma treatment is above 6 nm. With this film thickness, the lifetime for as-deposited a-Si:H/c-Si/a-Si:H samples is around 1.5 ms [4], but a far lower lifetime (below 0.1 ms) is achieved in Figure 1a. Besides, other samples exhibit low lifetimes (<0.15 ms) even when >10 nm a-Si:H is still retained after prolonged hydrogen plasma treatment (reference samples without any treatment have lifetimes of 1-3 ms). Thus, we suspect that over-etching of the a-Si:H passivation layer is not the essential reason for this lifetime degradation.

Another possibility is that the underlying c-Si substrate is damaged by prolonged hydrogen plasma treatment [5]. To validate this hypothesis, we performed hydrogen plasma treatment directly on c-Si wafers, and subsequently deposited 10-nm-thick a-Si:H layers. We found that hydrogen plasma treatment causes severe damage on c-Si wafers that *could not*

be recovered by subsequent a-Si:H passivation (data in Figure 1b), which agrees with other reports [7-8]. Thus, we think that the hydrogen plasma damage to the c-Si substrate is the essential cause of the interface passivation degradation for prolonged hydrogenation. Of course, as the remaining a-Si:H layer becomes thinner (because of the hydrogen plasma etching effect), the lifetime degradation is more prominent because the hydrogen plasma damage to the underlying c-Si substrates is more severe.

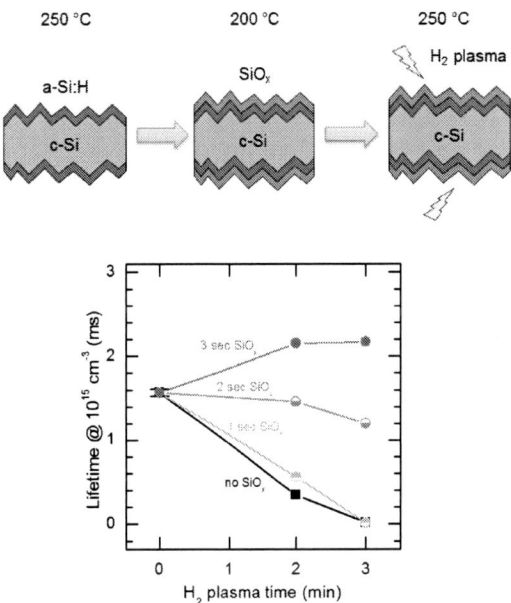

Fig. 2. Schematic structure of sample processing and the effective lifetimes of textured c-Si wafers with SiOₓ/a-Si:H (0-2.5 nm/10 nm) stacks on both sides, before and after H₂ plasma treatment. Before hydrogenation, the symbol represents the average value from different positions within a five-inch wafer while the error bars indicate the maximum and minimum values; this wafer was then cut into four pieces for further SiOₓ deposition and the final hydrogenation.

To alleviate the hydrogen plasma damage to the a-Si:H/c-Si interface, we inserted an additional thin silicon oxide (SiOₓ) capping layer after a-Si:H deposition and before hydrogen plasma treatment. Figure 2 shows effective lifetimes of a-Si:H/c-Si heterojunctions with SiOₓ capping layers of varied thickness, denoted here by the SiOₓ deposition time. The 3-sec SiOₓ layer is around 2.5 nm thick on the textured c-Si substrate. Clearly, as the SiOₓ layer becomes thicker, the lifetime is higher under each hydrogen-plasma-treatment condition. In particular, the sample with 3-sec SiOₓ capping layer actually experiences lifetime enhancement upon 2-3 min of hydrogen plasma treatment, compared to the reference sample without SiOₓ capping layer that incurs lifetime degradation. Therefore, adding a SiOₓ capping layer helps alleviate the lifetime degradation of a-Si:H/c-Si heterojunctions upon prolonged hydrogen plasma treatment, especially for the sample with ~2.5-nm-thick oxide layer.

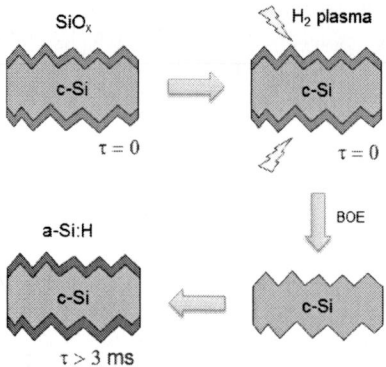

Fig. 3. Schematic processing steps of a reference c-Si sample to show the non-passivation and H₂-plasma-damage-protection properties of SiOₓ layers.

To make sure that the thin SiOₓ is just a capping layer rather than a supporting passivating layer, we processed a reference sample as shown in Figure 3. We first deposited a 3-sec SiOₓ layer on a c-Si substrate and measured its effective lifetime, then performed a 2-min hydrogen plasma directly on this sample and re-measured its effective lifetime again. In both cases, the measured lifetimes are near zero, meaning that the SiOₓ layer alone does not passivate the c-Si surface, even after attempted hydrogenation. Interestingly, upon removing the above-mentioned plasma-treated SiOₓ layers with BOE and then depositing a standard 10-nm-thick a-Si:H layers on both sides of the c-Si substrate, the lifetime increased to several milliseconds. In contrast with the corresponding result without a SiOₓ layer in Figure 1b (lifetime degrades to 0.15 ms after 2-min hydrogen plasma treatment), this indicates that a 3-sec SiOₓ capping layer successfully prevents severe damage to the c-Si wafer, which *can* be passivated by subsequent a-Si:H deposition. Therefore, we confirm that SiOₓ acts as a capping layer that protect the a-Si:H/c-Si heterojunction from hydrogen plasma damage.

To further investigate whether the ~2.5-nm-thick SiOₓ layer can completely protect the a-Si:H/c-Si heterojunction, a series of prolonged hydrogen plasma treatments were carried out until the measured lifetime decreased to near zero. Figure 4a shows the effective lifetimes of textured c-Si wafers with SiOₓ/a-Si:H (2.5 nm/10 nm) stacks or with a-Si:H (10 nm) layers on both sides, before and after hydrogen plasma treatment. Without SiOₓ, the lifetime quickly decreases to 83-195 μs with hydrogen plasma treatment of 2 min; however, a SiOₓ capping layer enables good lifetimes with up-to-12min hydrogen plasma treatment, and the lifetime slowly drops to 123 μs with 21-min hydrogen plasma treatment.

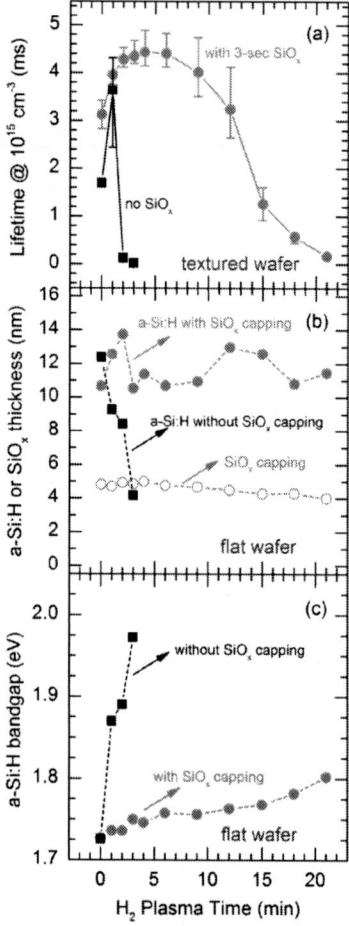

Fig. 4. (a) Effective lifetimes of textured c-Si wafers with SiOₓ/a-Si:H (2.5 nm/10 nm) stacks and with a-Si:H (10 nm) layers on both sides, before and after continuous H₂ plasma treatment. Each symbol represents the average value of five different positions within a five-inch wafer, and the error bars indicate the maximum and minimum values; (b-c) SiOₓ thickness, a-Si:H thickness, and a-Si:H bandgap on separate polished silicon wafers before and after H₂ plasma treatment (each data point corresponds to one separate sample within the same run), fit by a SiOₓ/a-Si:H stack (two-layer fitting) with ellipsometry.

As it is challenging to do ellipsometry on textured wafers, we did a similar set of experiments on polished wafers to study the thin film thickness and bandgap. The results are shown in Figure 4b and Figure 4c. Note that these data may not perfectly correspond with those in Figure 4a because of data variance (each data point corresponds to one separate sample) and different c-Si wafer geometry (flat instead of textured), but the preliminary results are still worth investigating.

First, compared to that of a-Si:H, the hydrogen etching rate of SiOₓ [9] is confirmed to be slow in the SiOₓ/a-Si:H stacks of Figure 3b; thus, the thickness of the underlying a-Si:H layer is protected by the top SiO_x layer. As the a-Si:H layer thickness stays almost the same while plasma etching, we suspect hydrogen plasma damage to the c-Si substrate should be greatly reduced. Meanwhile, Figure 4c shows the top SiO_x layer hinders some atomic hydrogen from going into the a-Si:H/c-Si interface, as the a-Si:H bandgap widening is much slower. Nevertheless, the slow hydrogen incorporation into a-Si:H still enhances the a-Si:H/c-Si interface passivation during the first few minutes of hydrogen plasma treatment. Note that the measured effective lifetime requires a high-quality c-Si substrate without plasma damage, a sharp a-Si:H/c-Si interface during a-Si:H deposition and sufficient hydrogen in the a-Si:H to passivate the interface. In this case, it is beneficial to have such a SiO_x capping layer.

Unfortunately, SiO_x capping cannot prevent the ultimate lifetime degradation after 21-min hydrogen plasma treatment, even though the a-Si:H layer still has sufficient thickness (~10 nm) and reasonable bandgap (~1.8 eV) for what should be good surface passivation. In this case, we suspect that the low lifetime may be due to the slowly accumulated hydrogen damage in the near-surface c-Si [4], which requires a set of high-quality transmission electron microscopy images or other characterization methods to confirm.

IV. CONCLUSION

Prolonged hydrogen plasma treatment causes damage to the a-Si:H/c-Si interface passivation, but we find adding a ~2.5 nm SiO_x capping layer can alleviate such degradation: hydrogen plasma damage to the a-Si:H/c-Si interface is suppressed while hydrogen can still penetrate into the underlying a-Si:H layers for interface passivation. This technique gives a wide plasma-processing-time window for enhancing the a-Si:H/c-Si interface passivation. In addition to the SiO_x layer, other materials are also likely be used as caps to alleviate the prolonged hydrogen plasma damage while allowing for hydrogen running through. Also, it would be scientifically interesting to discover if the hydrogen plasma damage can be fully prevented in the future.

REFERENCES

[1] S. Y. Herasimenka, W. J. Dauksher, and S. G. Bowden. ">750 mV open circuit voltage measured on 50 μm thick silicon heterojunction solar cell," *Appl. Phys. Lett.* vol.103, 053511, 2013.

[2] A. Descoeudres, L. Barraud, S. De Wolf, B. Strahm, D. Lachenal, C. Guérin, Z. C. Holman, F. Zicarelli, B. Demaurex, J. Seif, J. Holovsky, and C. Ballif. "Improved amorphous/crystalline silicon interface passivation by hydrogen plasma treatment," *Appl. Phys. Lett.* vol. 99, 123506, 2011.

[3] M. Mews, T. F. Schulze, N. Mingirulli, and L. Korte. "Hydrogen plasma treatments for passivation of amorphous-crystalline silicon-heterojunctions on surfaces promoting epitaxy," *Appl. Phys. Lett.* vol. 102, 122106, 2013.

[4] J. Shi, M. Boccard, and Z. Holman. "Plasma-initiated rehydrogenation of amorphous silicon to increase the temperature processing window of silicon heterojunction solar cells," *Appl. Phys. Lett.* vol. 109, 031601, 2016.

[5] J. Geissbühler, S. De Wolf, B. Demaurex, J. P. Seif, D. T. Alexander, L. Barraud, and C. Ballif. "Amorphous/crystalline silicon interface defects induced by hydrogen plasma treatments." *Appl. Phys. Lett.* vol. 102, 231604, 2013.

[6] Z. C. Holman, A. Descoeudres, L. Barraud, F. Z. Fernandez, J. P. Seif, S. De Wolf, and C. Ballif. "Current losses at the front of silicon heterojunction solar cells," *IEEE J. Photovoltaics* vol. 2, pp. 7–15, 2012.

[7] J. W. A. Schüttauf, C. H. M. van der Werf, W. G. J. H. M. van Sark, J. K. Rath, R. E. I. Schropp. "Comparison of surface passivation of crystalline silicon by a-Si:H with and without atomic hydrogen treatment using hot-wire chemical vapor deposition," *Thin Solid Film* vol. 519, pp. 4476–4478, 2011.

[8] S. N. Granata, T. Bearda, F. Dross, I.Gordon, J. Poortmans, R. Mertens. "Effect of an in-situ H_2 plasma pretreatment on the minority carrier lifetime of a-Si:H(i) passivated crystalline silicon," *Energy Procedia* vol. 27, pp. 412–418, 2012.

[9] R. P. H. Chang, C. C. Chang, and S. Darack. "Hydrogen plasma etching of semiconductors and their oxides," *J. Vac. Sci. Technol.*, vol. 20, pp. 45-50, 1982.

Large-area n-type TOPCon Cells with Screen-printed Contact on Selective Boron Emitter Formed by Wet Chemical Etch-back

Yuguo Tao[1], Felix Book[2], Barbara Terheiden[2], Vijaykumar Upadhyaya[1], Keeya Madani[1], Brian Rounsaville[1], Eunhwan Cho[1], Ajeet Rohatgi[1, 3]

[1] Georgia Institute of Technology, Atlanta, GA 30332, USA
[2] University of Konstanz, Konstanz D-78464, Germany
[3] Suniva Inc., Norcross, GA 30092, USA

Abstract — This paper presents large-area TOPCon (tunnel oxide passivated contact) cells with a selective boron emitter formed by a screen-printed resist masking and wet-chemical etch-back process. The wet chemical solution with $NaNO_2$ catalyst is developed to uniformly etch-back an APCVD deposited and thermally diffused boron emitter from ~47 to ~95 $\Omega/\square$ by growing a ~250 nm porous Si layer within 3 min and subsequently removing it. After etch-back followed by a thermal oxidation, J_{0e} was reduced from 85 to 21 fA/cm^2 and sheet resistance increased to 65 and 120 $\Omega/\square$ in the p++ and p+ regions, respectively. A solar cell with screen-printed Al/Ag contact on the etched-back selective emitter (65/120 $\Omega/\square$) on the front and TOPCon on the back achieved a V_{oc} of 681.9 mV and efficiency of 21.13% on large-area n-type Cz Si wafer, demonstrating the potential of this technology for high-efficiency industrial solar cells.

Keywords – selective boron emitter, etch-back, screen-printed, Si solar cells, TOPCon

I. INTRODUCTION

In order to achieve high-efficiency front-junction Si solar cells, one of the key challenges is to tackle the minority carrier recombination at the metal/Si interface. In addition to a-Si based heterojunction (HIT) structure [1] that requires a dedicated and expensive back-end metallization process, both selective emitter [2] and tunnel oxide passivated contacts (TOPCon) [3] are promising candidates for allowing efficient majority carrier transport while suppressing minority carrier recombination. However, applying TOPCon scheme on the front side of a solar cell can cause a significant loss in short-circuit current density (J_{sc}) due to parasitic absorption in the *poly*-Si layer. Therefore, in this paper we present large-area n-type front-junction Si solar cells with screen-printed contact on selective boron emitter on the front and TOPCon on the rear, as shown in Fig. 1. The selective emitter is formed on the front by a screen-printed acid resist masking and wet chemical etch-back process [4], and the TOPCon is formed on the back with an ultra-thin oxide capped with a P doped (n^+) *poly*-Si [5].

Our wet chemical etch-back process involves growth of a porous Si layer in a chemical solution composed of DI water, HF/HNO$_3$ acids, and sodium nitrite ($NaNO_2$) catalyst, according to the following reactions [6]:

$$\text{Anode:} \quad Si + 2H_2O + mh^+ \rightarrow SiO_2 + 4H^+ + (4\text{-}m)e^- \quad (1)$$

$$SiO_2 + 6HF \rightarrow H_2SiF_6 + 2H_2O \quad (2)$$

$$\text{Cathode:} \quad HNO_3 + 3H^+ \rightarrow NO + 2H_2O + 3h^+ \quad (3)$$

where m is the average number of holes required to dissociate one Si atom. This is often referred to as "stain etching". When Si dissolves into the etching solution at anodic sites, HNO_3 oxidant is reduced at cathodic sites and its key role is to inject holes (h$^+$) into the valence band. When some sites are anodic much more than they are cathodic, etch pits will form. Then, the quantum confinement effects shield the pore walls, so etching proceeds only at the bottoms of pores and towards the unconfined bulk Si [7]. This leads to porous Si formation. In addition, the $NaNO_2$ can catalyze these reactions by eliminating the induction period and accelerating the etching rate. This is because NO_2^- ions from $NaNO_2$ react with H$^+$ to form nitrous acid (HNO$_2$), then the formed HNO$_2$ reacts with HNO$_3$ to produce NO$_2$ that oxidizes the Si surface by injecting a hole (h$^+$) [8]:

$$NO_2^- + H^+ \rightarrow HNO_2 \quad (4)$$

$$HNO_2 + HNO_3 \rightarrow 2NO_2 + H_2O \quad (5)$$

$$NO_2 \rightarrow NO_2^- + h^+ \quad (6)$$

Furthermore, porous Si is not stable in OH$^-$ aqueous solution [7]. Hence, we used a dilute KOH solution to remove it in this work.

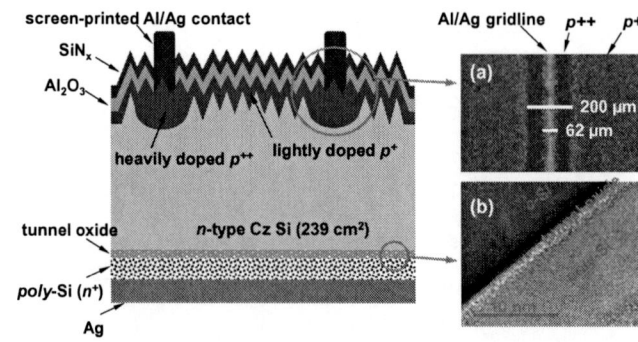

Fig. 1. Schematic of large-area *n*-type TOPCon Si solar cell with etched-back selective boron emitter.

II. EXPERIMENTAL RESULTS & DISCUSSION

In this study, large-area (239 cm^2) *n*-type ~2 Ωm Cz Si wafers were used. After the standard alkaline texturing, a borosilicate glass (BSG) layer was deposited in an APCVD tool, followed by a thermal anneal at 1000°C for 1 hour in N$_2$ and O$_2$ ambient, which resulted in 45~52 Ω/□ boron emitter layer. Then, these wafers were immersed into the HF/HNO$_3$/H$_2$O (volume ratio of 3:2:15) etching solution with NaNO$_2$ concentration of 1.3 g/L to grow porous Si layer of various thickness by changing the etching time. Fig. 2 shows that the emitter sheet resistance increases with etching time, because as Si surface is consumed the thickness of porous Si gradually increases with etching time [4], and the sheet resistance reaches 90~100 Ω/□ after 150s etching, which grows ~250 nm thick porous Si (SEM image in Fig. 3). Thus, this etch-back recipe is suitable for manufacturing because of short process time and etching homogeneity (≤10% variation on the resulting sheet resistance).

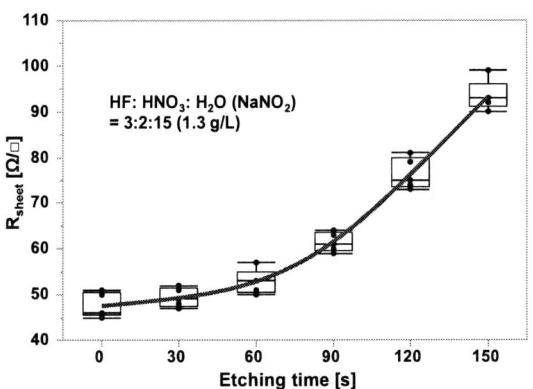

Fig. 2. Sheet resistance of boron emitter as a function of etching time.

Next, 1% KOH solution was applied for 120s to remove the grown porous Si layer at room temperature. After removing porous Si, the measured reflectance slightly increased (Fig. 3), compared to the one without etch-back ("before etch-back"). This is because the angle at the top of pyramid becomes marginally wider since thicker porous Si is grown at the top [4], as well as the pyramid valley is slightly rounded (SEM in Fig. 3) since porous Si grows into the bulk simultaneously on (111) lattice plane surface.

After removing porous Si layer, wafers were subjected to a high temperature thermal oxidation above 1000°C to reduce the boron surface concentration while inducing a deeper junction. The ECV boron profile measurement in Fig. 4 shows that the boron surface concentration is about 6.2×10^{19} cm^{-3} with junction depth of ~0.80 μm (~47 Ω/□) after initial BSG

diffusion anneal, and is reduced to 4.6×10^{19} cm^{-3} with junction depth of ~0.55 μm (~100 Ω/□) after etch-back or porous Si formation for 150s, and further down to 1.0×10^{19} cm^{-3} with a deeper junction of ~0.85 μm (~120 Ω/□) after thermal oxidation. Fig. 4 also shows that the boron emitter is etched-back by about 250 nm during the etching time of 150s. This is in agreement with the SEM image shown in Fig. 3. In addition, the emitter saturation current density (J_{0e}) measured by QSSPC technique on the symmetric structure with ALD Al$_2$O$_3$/SiN$_x$ passivation was decreased from ~85 fA/cm^2 (after BSG anneal) to ~21 fA/cm^2 after etch-back plus oxidation for the p+ emitter.

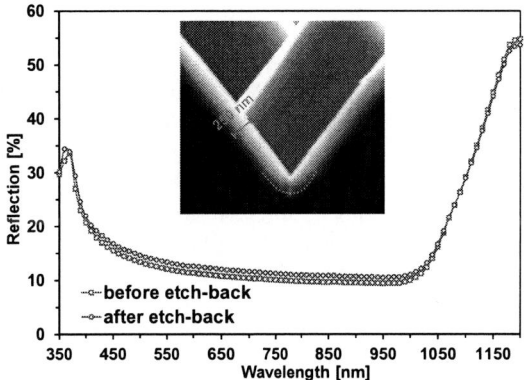

Fig. 3. Reflection comparison between before and after etch-back, and the cross-section SEM image of pyramid with porous Si layer.

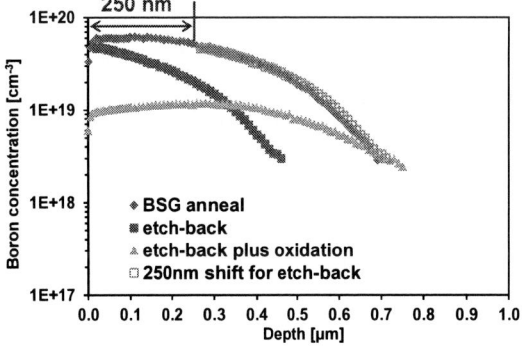

Fig. 4. Boron emitter ECV profiles after each process.

To study the feasibility of this etch-back technology and its potential enhancement on cell performance, front-junction TOPCon Si solar cells were fabricated on large-area (239 cm^2) *n*-type Cz Si wafers (~2 Ωm), as shown in Fig. 1. After texturing, BSG deposition by APCVD, and thermal diffusion, an acid resist mask was screen-printed with the desired gridline pattern on the top of boron emitter. Next, the wafers were etched-back, followed by removal of porous Si and resist mask and finally a thermal oxidation. This formed a selective

978-1-5090-5606-4/17 $31.00 © 2017 IEEE

emitter p^{++}/p^+ (65/120 Ω/□) on the front side, with the p^{++} region width of ~200 μm. Then, TOPCon structure was formed on the planarized rear side, which involved chemical growth of an ultra-thin tunnel oxide layer (~15 Å) capped with a phosphorous doped, PECVD deposited, and thermally crystallized *poly*-Si layer (<20 nm) [5], as shown by TEM image in Fig. 1. Next, Al₂O₃/SiNₓ passivation stack was deposited on the top of emitter. A ~62 μm wide Al/Ag grid pattern was aligned and screen-printed inside the 200 μm p^{++} region (optical microscope image in Fig. 1), followed by a high temperature firing in a belt furnace. Finally, a 200 nm thick Ag layer was evaporated on the entire rear side to form the back contact.

Table I lists the cell parameters from light *I-V* results measured in-house by using the Fraunhofer ISE certificated reference cell [9]. For comparison purpose, a group of cells with ~65 Ω/□ homogeneous emitter (without etch-back) were also fabricated together. The cells with selective emitter yielded an average open-circuit voltage (V_{oc}) of >680 mV, which is about 4 mV higher than the homogeneous emitter cells. This is mainly attributed to the lower Auger recombination in the etched-back field region (p^+). In addition, an increase of ~0.3 mA/cm² in J_{sc} is observed due to better blue response of selective emitter [4]. However, the selective emitter resulted in lower fill factor (*FF*), mainly due to higher series resistance (R_s) that results from higher sheet resistance (~120 Ω/□), compared to the homogeneous emitter (~65 Ω/□). Consequently, the selective emitter gave an average gain of ~0.2% absolute efficiency, with the best efficiency of 21.13% and V_{oc} of ~682 mV.

TABLE I. SUMMARY OF LIGHT *I-V* RESULTS OF FOR LARGE-AREA TOPCON CELLS.

Emitter structure		V_{oc} [mV]	J_{sc} [mA/cm²]	FF [%]	Eff [%]	R_s [Ω-cm²]	R_{sh} [Ω-cm²]
Selective p^{++}/p^+ (65/120 Ω/□)	average	680.5 ±2.5	39.5 ±0.2	78.1 ±0.4	21.0 ±0.2	0.63 ±0.05	7580 ±2567
	best	681.9	39.6	78.2	21.13	0.62	5532
Homogeneous p^{++} (65 Ω/□)	average	676.3 ±1.8	39.2 ±0.2	78.6 ±0.4	20.7 ±0.2	0.58 ±0.05	6785 ±1969
	best	678.1	39.3	78.4	20.90	0.57	8754

TABLE II. DETAILED J_0 COMPONENT DISTRIBUTIONN OF LARGE-AREA TOPCON CELL WITH SELECTIVE BORON EMITTTER.

	$J_{0, total}$	$J_{0e,p+}'$	$J_{0e,p++}'$	$J_{0e,contact}'$	$J_{0b,bulk}$	J_{0b}'
effective value [fA/cm²]	120	18.3	2.8	44	42	13
percentage	100%	15.2%	2.3%	36.6%	35.0%	10.8%

According to our 2D Sentaurus device model [10], this selective emitter should have given higher efficiency gain. Therefore, a detailed J_0 components analysis was performed to

understand and quantify the recombination loss in the best cell with etched-back selective emitter (Table II).

Considering the area fraction of p^+ (87.2%) and p^{++} (8.8%) regions, the effective contribution from these two passivated regions are $J_{0e,p+}'$ of ~18 fA/cm² and $J_{0e,p++}'$ of ~3 fA/cm² ($J_{0e,p++}$ = 32 fA/cm²). The passivated rear contact contributes J_{0b}' of ~13 fA/cm² (after high temperature firing) [5], while the Auger lifetime in bulk region introduces $J_{0b,bulk}$ of ~42 fA/cm² according to [3]:

$$J_0 = J_{0e} + J_{0b} = J_{0e} + J_{0b}' + q n_i^2 \frac{W}{N_D \tau_p} \qquad (7)$$

The saturation current density due to metal contact ($J_{0e,contact}$) was extracted according to the method proposed in [11], in which metal contact with various fraction were screen-printed and fired-through on one side of the wafers with symmetric structure (homogeneous p++ 65 Ω/□ emitter with passivation stack), and then quickly stripped off in a dilute HF solution. A $J_{0e,contact}$ of ~1100 fA/cm² was extracted from the slope and intercept of the linear plot resulting from QSSPC measurements, shown in Fig. 5, according to:

$$2J_0 = (J_{0e,contact} - J_{0e})f + 2J_{0e} \qquad (8)$$

where f is the metal contact fraction and J_0 is the total measured J_0 of the full structure. Hence, the area-weighted $J_{0e,contact}'$ is about 44 fA/cm² for f of 4.0%. Therefore, about 37% of the total recombination is due to the screen-printed contact, while 35% comes from somewhat lower bulk lifetime (~1 ms) after three high temperature anneals. So, in order to fully exploit the advantages of selective emitter structure, heavier doped p^{++} region (i.e., ≤20 Ω/□) needs to be implemented to reduce $J_{0e,contact}$ in combination with a higher quality substrate. However, thicker porous Si layer is required to etch-back the field region to ≥120 Ω/□, which may negatively impact emitter profile uniformity and reflectance. Hence, this trade-off between heavier doping in p^{++} and reflectance needs to be investigated and optimized. High bulk lifetime can be achieved by higher quality wafers or by implementing the Tabrasa treatment on the bare Si wafers prior to cell fabrication process [12].

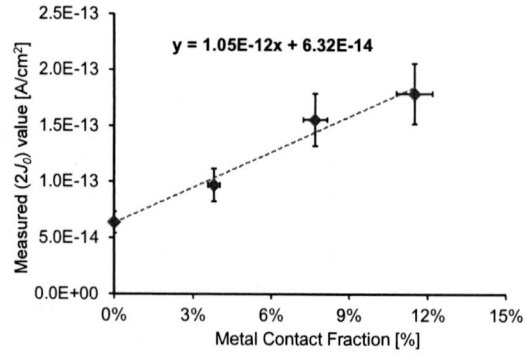

Fig. 5. Measured J_0 value by QSSPC as a function of metal fraction.

III. SUMMARY

In summary, a technology to form selective boron emitter by using wet chemical etch-back was demonstrated and implemented to fabricate large-area n-type TOPCon cells with screen-printed contact on the front side. A cell efficiency of >21% with V_{oc} of >680 mV was achieved. Further optimizing the boron doping profiles of passivated p^+/p^{++} regions and using a more endurable Si substrate with higher quality could increase the efficiency of this cell structure to >22%.

IV. ACKNOWLEDGEMENT

The authors would like to thank all other UCEP group members in Georgia Tech. Also thank Dr. Andrew Norman from National Renewable Energy Laboratory (NREL) for helping TEM image. This work was supported by the U.S. Department of Energy PVRD contract DE-EE-0007554.

REFERENCES

[1] K. Masuko, *et al.*, *IEEE J. Photovoltaics*, vol. 4, p. 1433, 2014.
[2] J. Zhao, A. *et al.*, *Prog. Photovoltaics: Res. Appl.*, vol. 7, p. 471, 1999.
[3] F. Feldmann, *et al.*, *Sol. Energy Mater. Sol. Cells*, vol. 120, p. 270, 2014.
[4] Y. Tao, *et al.*, *Appl. Phys. Lett.*, vol. 110, p. 021101, 2017.
[5] Y. Tao, *et al.*, *Prog. Photovoltaics: Res. Appl.* 24, p. 830, 2016.
[6] D.R. Turner, *J. Electrochem. Soc.*, vol. 107, p. 810, 1960.
[7] K.W. Kolasinski, *Curr. Opin. Solid State and Mater. Sci.*, vol. 9, p. 73, 2005.
[8] L.A. Jones, *et al.*, *Prog. Surf. Sci.*, vol. 50, p. 283, 1995.
[9] Y. Tao, *et al.*, *IEEE J. Photovoltaics*, vol. 4, p. 58, 2014.
[10] C-W. Chen, *et al.*, *Progress in Photovoltaics: Research and Applications*, vol. 25, p. 49, 2017.
[11] J. Deckers, *et al.*, in *28th European PVSEC*, p. 806, 2013.
[12] B. Sopori, *et al.*, *IEEE J. Photovoltaics*, vol. 7, p. 97, 2017.

Hydrogen Plasma Post-Deposition Treatment for Passivation of a-Si/c-Si Interface for Heterojunction Solar Cell by Correlating Optical Emission Spectroscopy and Minority Carrier Lifetime

Anishkumar Soman[1], Ugochukwu Nsofor[1,2], Lei Zhang[1,2], Ujjwal Das[2], Tingyi Gu[1], Steve Hegedus[1,2]

[1] Department of Electrical and Computer Engineering, University of Delaware, Newark, DE, USA 19716

[2] Institute of Energy Conversion, University of Delaware, Newark, DE, USA 19716

Abstract — **Excellent passivation at the a-Si:H/ c-Si interface is required for high efficiency heterojunction solar cells. Hydrogen plasma treatment after the a-Si deposition has been considered as an effective and manufacturable method to saturate the surface dangling bonds. In this paper the correlation between H_2 plasma composition and passivation quality of the a-Si:H films has been investigated using optical emission spectroscopy and minority carrier lifetime. Increasing the H_2 pressure to 2750 mT lead to an implied V_{oc} of 739 mV and effective lifetime >1900 µs, which were 13 mV and 300 µs, respectively, greater than the baseline. Cells with efficiency >18% were fabricated. Comparison of implied JV curves indicates significant loss in V_{oc} during deposition of the doped layers and metal contacts.**

Index Terms — **surface passivation, hydrogen treatment, optical emission spectroscopy, silicon heterojunction solar cell.**

I. INTRODUCTION

With silicon heterojunction solar cells reaching record efficiency of 26.6% [1] it has gained interest as a commercially viable technology which requires lower thermal budget as compared to conventional diffused junction silicon solar cells. The efficiency of heterojunction solar cells depends strongly on achieving superior passivation of the deep level trap states at the crystalline silicon (c-Si)/ amorphous silicon (a-Si:H) interface and thereby reducing the dark current [2]. Hydrogen plasma treatment has been used as an effective method to saturate the dangling bonds at a-Si/ c-Si interface[3][4] and thereby reducing the recombination losses leading to higher V_{oc}. These treatment are very short on the order of ~30 seconds[5] since increased exposure of the ultra-thin (<10 nm) intrinsic a-Si layer to H plasma can result in etching and degrade passivation qualities. Such short duration H_2 plasma treatment happens at the non-equilibrium transient phase of the plasma hence it is critical to monitor the plasma kinetics using time resolved optical emission spectroscopy (OES). In this work, relation between the H_2 plasma process parameters to the H* concentration has been investigated by monitoring the Hα spectral line at 656 nm. The minority carrier lifetime of c-Si wafers after H_2 plasma treatment on surface passivating a-Si:H layers has been co-related with the Hα to explain the passivation quality of the films. The passivation of the films has been explained with reference to ion bombardment energy and H* concentration. The effect of annealing in restructuring the film has also been highlighted.

II. EXPERIMENTAL DETAIL

The a-Si:H films were deposited using dc plasma enhanced chemical vapor deposition (dc-PECVD) with Silane (SiH_4) and hydrogen (H_2) as the precursor gases. The SiH_4 and H_2 flow rates are 10 and 25 sccm respectively. The i-layer was deposited with a dc plasma current of 60 mA and pressure of 1750 mT. The H_2 plasma treatment in this work has also been done using dc plasma at a substrate temperature of 200°C for 20 seconds. The a-Si:H films are deposited on semi-polished n-type CZ wafers having <100> orientation with resistivity of 1-5 Ω-cm and thickness of 140 µm. The wafers undergo standard cleaning procedure before loading into the load lock chamber. The minority carrier lifetime (τ_{eff}) and implied V_{oc} (iV_{oc}) are estimated using Sinton Quasi-steady state photoconductance (QSSPC) measurement [6]. They are reported at an excess carrier density of 10^{15} cm^{-3} and illumination of 1 sun respectively. A 10 nm thick a-Si:H film is deposited on both sides of the wafer and H_2 treatment is done post deposition. In order to understand the dependency of H* species on different process conditions 3 series of experiments were performed by varying dc plasma current, pressure and H_2 flow. In every series the other 2 parameters are kept constant so that we can find the relation between Hα and the varied parameter. The induced plasma voltage relates to the acceleration voltage of the ions and hence the ion bombardment energy. Based on the OES data we performed a second set of experiments to understand the effect of changing plasma voltage (V_{pl}) and Hα on passivation. The process conditions for this set are given below in Table I and have been obtained from the results of the series above. In the table, HV and LV represents high and low voltage. Similarly HH and LH represent high and low Hα in OES. V_{pl} is controlled by the pressure and Hα concentration is controlled by the current. The annealing experiments are done in vacuum at a temperature of 300°C for 25 minutes in the PECVD chamber.

978-1-5090-5606-4/17 $31.00 © 2017 IEEE

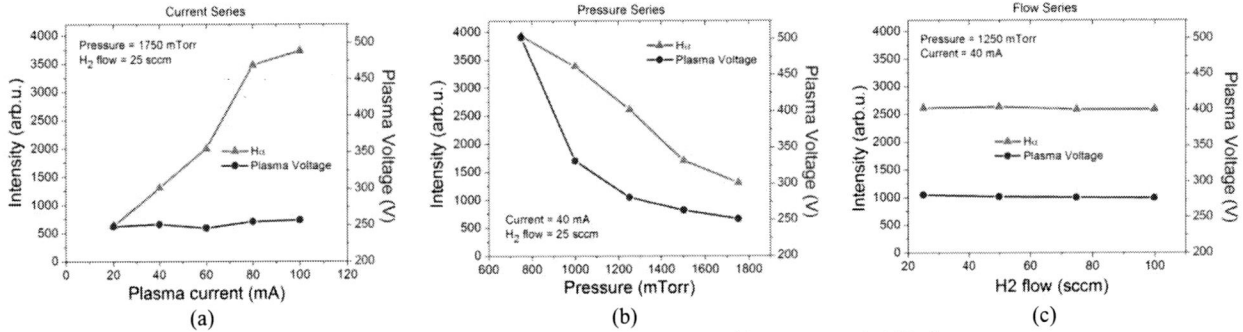

Fig. 1. Variation of Hα and dc plasma voltage with respect to a) plasma current b) pressure and c) H₂ flow

TABLE I. PROCESS CONDITIONS FOR VARYING VOLTAGE AND Hα

	LV	HV
LH	1750 mTorr, 20 mA	750 mTorr, 20 mA
HH	1750 mTorr, 100 mA	750 mTorr, 100 mA

III. RESULTS AND DISCUSSION

A. Dependence of Hα on Process parameters

The Hα, and V_{pl} was monitored with respect to 3 parameters namely – plasma current (I), pressure (P) and H₂ flow (F) and are plotted in Fig.1. The Hα increases steadily with I, while the change in V_{pl} is insignificant with varying I. In the pressure series, Hα and V_{pl} both decrease with increasing pressure. In H₂ flow series, it has been found that varying the flow has no effect on Hα or V_{pl}. For current series the power density increases from 3 to 18 mW/cm² when I increases from 20 to 100 mA. Similarly for pressure series power density decreases from 14 to 7 mW/cm² when P increases from 750 to 1750 mTorr. From these results the two key conclusions are that: 1) increasing plasma current increases the number of charged species in the H₂ plasma, hence the increase in Hα species; and 2) increasing the pressure decreases V_{pl} which in turn reduces the ion bombardment energy and acceleration of the particles. Thus by varying current and pressure as shown in Table I we can understand the effect of ion bombardment energy and total H* available for saturation of dangling bonds, on the passivation quality of a-Si:H films.

B. Co-relation between varying V_{pl} and Hα on passivation of a-Si:H films as deposited

In order to understand the effect of V_{pl} and Hα on passivation we have taken 4 conditions where their values are altered between two values: High (H) and Low (L). Thus the 4 cases are LVLH, LVHH, HVLH and HVHH respectively. Fig.

2 shows τ_{eff} and iV_{oc} of 2 samples for each condition. It is apparent that the passivation is poor for both high voltage (HV) conditions with low τ_{eff} and iV_{oc} values whereas at low voltage (LV) conditions the samples have good passivation with iV_{oc} ~700 mV. This could be explained by the fact that V_{pl} relates to the ion bombardment energy where HV conditions results in surface damage due to high ion bombardment energy. For LV conditions the Hα is changed by varying the current from 20 mA to 100 mA. The optimum condition for H concentration was found to be P = 2750 mTorr and I = 20 mA. It is interesting to note that for LVHH condition the power density is 16 mW/cm² compared to 5.9 mW/cm² for HVLH; ie. the LV condition has a higher lifetime showing the strong contribution of V_{pl} on passivation quality of the film. The power density ranged from 3 to 34 mW/cm², respectively for LVLH to HVHH.

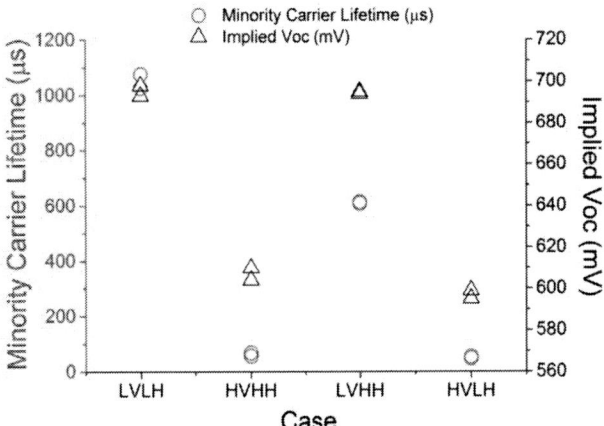

Fig. 2. Minority carrier lifetime and implied Voc of both side a-Si:H coated samples treated with H₂ plasma under different V_{pl} and Hα conditions before thermal annealing.

C. Effect of Annealing

To understand if annealing can recover the damage caused from ion bombardment by restructuring the film, we have annealed all the samples at 300°C in vacuum. In order to compare the effect of H₂ treatment and to understand the importance of degradation due to high V_{pl} we have compared the high and low V_{pl} samples with reference sample in Table II. The reference is 10 nm thick intrinsic a-Si:H layer with no H_2 plasma treatment. We find that annealing considerably improves τ_{eff} and iV_{oc}. The effect of annealing shown in Table II highlights the improvement in iV_{oc} for different V_{pl} conditions. The improvement is more significant in HV samples which initially had very low τ_{eff} and iV_{oc} values. The surface damaged samples had an increase in iV_{oc} by 35 to 50 mV while for LV samples it was considerably less in the range of 10-30 mV as can be seen in Fig. 3. Thus, annealing HV samples show greater improvement in iV_{oc} but cannot completely recover the damage due to ion bombardment hence these samples have significantly lower iV_{oc} even after

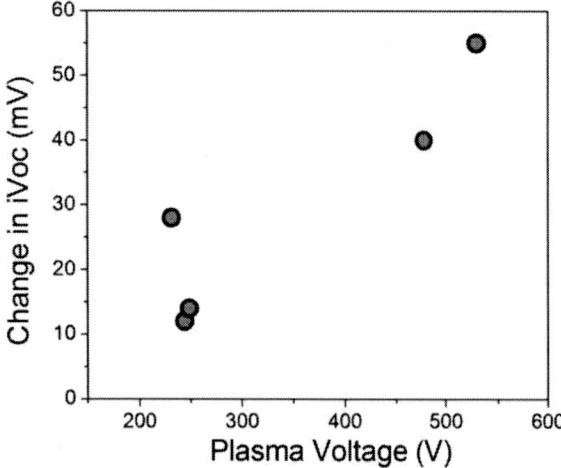

Fig. 3. Change in iV_{oc} of both side a-Si:H coated samples treated with H_2 plasma under different conditions before and after thermal annealing with respect to plasma voltage (V_{pl})

annealing. The maximum τ_{eff} and iV_{oc} obtained after annealing are 1917 μs and 739 mV from the LVLH sample.

TABLE II
BEST IMPLIED V_{OC} FOR VARYING VOLTAGE CONDITION WITH AND WITHOUT ANNEALING. 'REFERENCE' HAS I-LAYER WITHOUT H_2 TREATMENT

	Plasma voltage (V)	Hydrogen pressure (mTorr)	iV_{oc} before annealing (mV)	iV_{oc} after annealing (mV)
Low V_{pl}	231	2750	706	739
Low V_{pl}	244	1750	699	711
High V_{pl}	530	750	600	655
Reference	N.A.	N.A.	699	726

D. Solar Cell Results

A series of front heterojunction solar cells were fabricated as described elsewhere [2]. The doped a-Si p and n layers had been previously optimized at higher power (i.e. plasma current) than the i-layer developed in this work with hydrogen plasma treatment. The cell was completed with sputtered ITO and electron beam deposited Al grids on front and electron beam deposited Al on the back.

Table III shows the iV_{oc} after initial i-layer deposition followed by i-layer with H_2 plasma treatment at P = 2750 mTorr and I = 20 mA, subsequently deposition of doped p and n layers, and then final cell V_{oc} after contacting. The iV_{oc} increases 15 mV with H_2 plasma then decreases 28 mV after doped layer deposition. This is significantly more than what we experience with other i-layers which are deposited at higher power similar to the doped layers, and without H_2 plasma treatment. The considerable drop in V_{oc} could also be associated with hydrogen effusion caused during the doped layer deposition and needs further investigation. This suggests the need to reoptimize the doped layers for the new i-layer even considering a lower temperature p layer to reduce effusion. Another ~30 mV is lost during contacting. We have confirmed this is primarily due to damage caused by electron beam metallization.

TABLE III
IMPLIED V_{OC} AND CELL V_{OC} AFTER I-LAYER DEPOSITION, H_2 PLASMA TREATMENT, DOPED LAYER DEPOSITION AND AFTER CONTACTING. AVERAGE OF 3 CELLS.

	iV_{oc} (mV)	Cell V_{oc} (mV)
i-layer	724	-
H_2 plasma	739	-
Doped layers	711	-
Complete cell	-	679

Implied current voltage curves (iJV) can be constructed from the QSSPC data by assuming $J = J_{sc}$ (1-suns) and V is measured V_{oc} at each intensity (1-suns). The J_{sc} value is obtained from the final completed device. Figure 4 shows the iJV curves extracted from QSSPC measurements after H_2 plasma treatment and after doped layers and JV from the conventional solar cell measurement under a solar simulator, all corresponding to conditions in Table III. The completed cell had V_{oc} = 679 mV, J_{sc} = 35.0 mA/cm^2, FF = 76.5% and efficiency of 18.2%. It clearly shows that the completed cell suffers significant performance loss due to process induced degradation in V_{oc} and series resistance. The V_{oc} loss due to doped layers can be minimized by identifying lower power and lower temperature conditions giving equivalent doping and defects. However, to realize the full potential of excellent i-layer passivation with H_2 plasma treatment, the metallization induced damage needs to be eliminated. If it were eliminated and the device V_{oc} was 711 mV, the cell above would have had an efficiency of 19%.

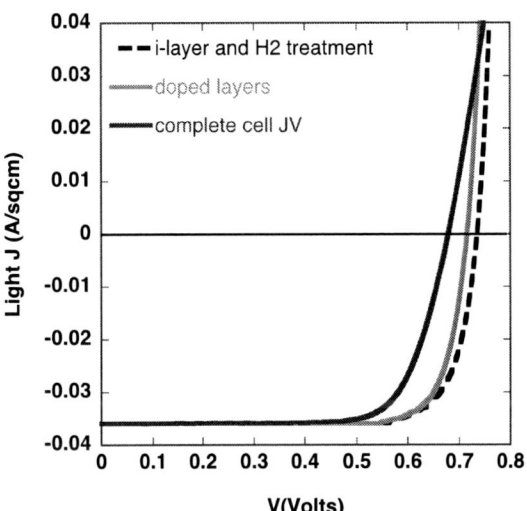

Fig. 4. Implied JV from QSSPC after i-layer plasma treatment, then after doped layer deposition and finally measured JV from completed cell as described in text.

IV. SUMMARY

Hydrogen plasma treatment can be an effective approach to saturate the surface dangling bonds and reduces the recombination losses at the interface. In this paper we have investigated the relation between the H_2 plasma properties to the passivation quality of c-Si surface. The relation between H* species and the process parameters are discussed. Plasma voltage controlled by the pressure and Hα concentration controlled by the plasma current have been found to be two crucial factors which contribute to the passivation quality of the films. Low voltage treatment reduces the damage due to ion bombardment resulting in higher lifetime and implied V_{oc}. Further optimization of the time and pressure will be investigated. The concept has been successfully applied to i-layers deposited with different conditions. The plasma treatment increases iV_{oc} by ~15 mV to ~740 mV and effective lifetime by 500 µs to ~1700 µs with an 8 nm i-layer. The concept has been validated by fabrication and characterization of heterojunction solar cells having V_{oc} of 680 mV and Efficiency> 18%.

ACKNOWLEDGEMENT

The authors would like to acknowledge support from the School of Engineering and Department of Electrical and Computer Engineering at University of Delaware and the US Department of Energy Sunshot Program PVRD Award DE-0007534.

REFERENCES

[1] K. Yoshikawa, H. Kawasaki, W. Yoshida, T. Irie, K. Konishi, K. Nakano, T. Uto, D. Adachi, M. Kanematsu, H. Uzu and K. Yamamoto, *Nature Energy*, vol. 2, pp. 17032, 2017.

[2] Z. Shu, U. Das, J. Allen, R. Birkmire, S. Hegedus "Experimental and simulated analysis of front vs all-back-contact silicon heterojunction solar cells: effect of interface and doped layer defects" Progress in Photovoltaics Vol. 23, pp78-93, 2015

[3] L. Serenelli, R. Chierchia, M. Izzi, M. Tucci, L. Martini, D. Caputo, R. Asquini, and G. de Cesare, "Hydrogen Plasma and Thermal Annealing Treatments on a-Si:H Thin Film for c-Si Surface Passivation," *Energy Procedia*, vol. 60, no. May, pp. 102–108, 2014.

[4] A. Antony and A. Soman, "Hydrogen Plasma Treatment to Enhance a-Si/c-Si Interface Passivation," in *32nd European Photovoltaic Solar Energy Conference and Exhibition*, 2016, pp. 738–741.

[5] A. Descoeudres, L. Barraud, S. De Wolf, B. Strahm, D. Lachenal, C. Guérin, Z. C. Holman, F. Zicarelli, B. Demaurex, J. Seif, J. Holovsky, and C. Ballif, "Improved amorphous/crystalline silicon interface passivation by hydrogen plasma treatment," *Appl. Phys. Lett.*, vol. 99, no. 12, p. 123506, 2011.

[6] R. A. Sinton, A. Cuevas, and M. Stuckings, "Quasi-steady-state photoconductance, a new method for solar cell material and device characterization," in *Conference Record of the Twenty Fifth IEEE Photovoltaic Specialists Conference - 1996*, 1996, pp. 457–460.

Measuring Diode Resistivity of Passivated Contacts

San Theingi[1,2], William Nemeth[1], David L. Young[1], Paul Stradins[1], Benjamin G. Lee[1]

[1]National Renewable Energy Laboratory, Golden CO, 80401 USA

[2]Colorado School of Mines, Golden CO, 80401 USA

Abstract — a technique of examining carrier-selective, passivated contacts is presented, where the contact resistivity is found from the measured diode characteristic. Importantly, this extends the understanding of a passivated contact's resistivity to the case where it is also used to form an emitter or p-n junction. For a p-type poly-Si passivated contact/emitter on n-Cz wafer, we find a low contact resistivity $\rho_c < 5$ mΩ cm^2 using this method. This low ρ_c value is promising for high-efficiency n-type Si cells having p-poly passivated contacts as the emitter. In particular, for interdigitated back contact (IBC) cells using a p-type passivated contact as the partial-area rear emitter.

Index Terms — carrier-selective passivated contacts, contact resistivity, silicon solar cells.

I. INTRODUCTION

Poly-Si/SiO$_x$ passivated contacts [1,2] are a promising pathway for high efficiency Si solar cells, reaching 1-sun efficiencies >25% [3]. A key parameter for these passivated contacts and generally for carrier-selective contacts, is the contact resistivity. The resistivity for the majority carriers in the contact, together with the recombination level for minority carriers, fundamentally determines the selective contact's ultimate performance limit in a cell [4]. While the transmission line method (TLM) is usually employed to find contact resistivity, it is not applicable when the selective contact forms a p-n diode to a differently doped base. Since passivated contact devices have been primarily developed for Si cells with n-type base, then it is the analysis of p-type passivated contacts that is at issue.

Here, we propose a technique for analyzing contact resistivity that is applicable to situations where there is a p-n diode in the device geometry. This enables the full analysis of resistivity for p-type passivated contacts, and could also be applied to the contacts of heterojunction (i.e. HIT) Si cells. We show results demonstrating the applicability of the method and also highlighting the low contact resistivity values achievable with p-type passivated contacts.

II. ANALYSIS PROCEDURE

We outline a method where it is possible to analyze the diode characteristics of a passivated contact – in particular, to extract its resistivity. This technique is applicable to p-type passivated contacts on n-type Si base, or conversely n-type passivated contacts on p-type Si; generally we consider a passivated contact structure where there is a p-n diode.

Since the resistivity of the passivated contact is expected to be low, to measure it we need a test structure where the area of the contact and p-n junction is small. An example test structure for a p-type passivated contact on n-type Si is shown in Fig. 1. The resistance of such a structure has 2 components: the diode resistivity of the passivated contact and the spreading resistance as the current flows into the bulk Si wafer. The resistance of the bottom contact is found to be negligible.

p-poly passivated contact on n-Cz test structures are prepared. Al circular front metal pads are e-beam evaporated and samples are etched with SF$_6$ RIE to isolate individual diode dot mesas with diameters from 300 to 600 μm. InGa paste is used as the back contact.

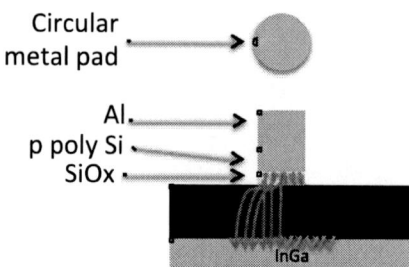

Fig. 1. Cross section of the test structure used to evaluate diode resistivity of passivated contacts. In this case we give the example of a p-type passivated contact on n-type Si wafer.

For the analysis of the diode resistivity, we make the simplifying assumption that the I-V characteristic can be represented by a 1-diode model including a series resistance term R:

$$I = I_0 \exp[\frac{q}{nkT}(V - IR)] \tag{1}$$

Following the example of Hegedus and Sharfarman [5] we take the derivative dV/dI as follows:

$$\frac{dV}{dI} = R_{total} + \frac{nkT}{q}\left(\frac{1}{I}\right) \tag{2}$$

It is clear from this equation that the total resistance of the test structure is found from the intercept of the linear fit to dV/dI vs. 1/I. In addition, the ideality factor of the diode can be obtained from the slope of the fit. The spreading resistance contribution to the total resistance R_{total} can be calculated according to Denhoff [6]. The diode resistivity of the passivated contact is then obtained by subtracting the spreading resistance term from R_{total}.

In Fig. 2 we show an example of the analysis; Fig. 2a shows the I-V curve obtained from the passivated contact diode and 2b shows the fit to dV/dJ vs. 1/J. The calculated contact resistivity $\rho_c = 1$ mΩ cm^2 and the ideality factor $n = 1.1$.

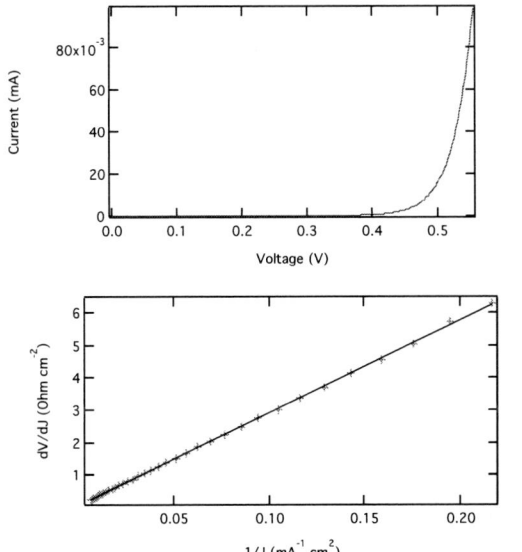

Fig. 2. a) I-V curve of a diode resistivity test structure. b) dV/dJ vs. 1/J plot and calculation of the ideality factor and resistivity from its slope and intercept.

III. RESULTS & DISCUSSIONS

The diode resistivity analysis is performed for a p-poly passivated contact on n-Cz Si. The analysis is done with multiple dot mesas, with diameters ranging from 300 to 600 μm – the results are shown in Fig. 3.

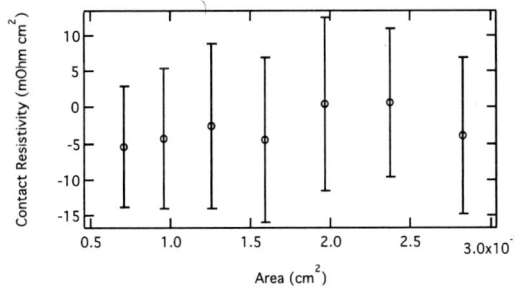

Fig. 3. Contact resistivity of p-type passivated contact on n-type Si wafer, from diode test structures of different diameters.

We note that it is possible to obtain a negative value when the contact resistivity is very low. This is consistent with the low values of $\rho_c < 10$ mΩ cm^2 for p-poly passivated contacts found in the literature [4]. The error bars on individual measurements are large primarily due to variability in

determining the spreading resistance and the fit intercept (i.e. Fig. 2b). After multiple measurements, the error of the mean result is fortunately much lower. Thus, we determine that the averaged, calculated contact resistivity is -2.9 ± 4 mΩ cm^2. We also find that the ideality factor of the diode is ~1.1.

Taken together with the findings of previous literature, we are encouraged by the result that $\rho_c < 5$ mΩ cm^2. The low resistivity of the p-type passivated contact on n-Si means that devices made using it will not be limited by the ρ_c. Moreover, cells employing p-poly passivated contacts can straightforwardly use them on just a small fraction of the area, such as for interdigitated back contact cells [7-9]. This is promising for the application of p-poly passivated contacts in high efficiency cell designs.

IV. CONCLUSION

We have described and demonstrated a technique for calculating the resistivity of a passivated contact, specifically for the case where it forms a p-n junction with a differently doped base and thus also functions as an emitter. Using the method, we find that the contact resistivity of a p-type poly-Si passivated contact on n-Cz wafer is quite low, <5 mΩ cm^2. This is very encouraging for application of this type of p-type passivated contact with high efficiency n-type Si cells; in particular for IBC cells where it forms an emitter over a portion of the rear area. The analysis technique is also applicable to n-type passivated contacts on p-type Si base.

ACKNOWLEDGEMENT

Funding for this work was provided by the United States Department of Energy EERE contract SETP DE-EE00030301 (SuNLaMP) and under Contract No. DE-AC36-08GO28308.

The U.S. Government retains and the publisher, by accepting the article for publication, acknowledges that the U.S. Government retains a nonexclusive, paid-up, irrevocable, worldwide license to publish or reproduce the published form of this work, or allow others to do so, for U.S. Government purposes.

REFERENCES

[1] F. Feldmann et al., "Carrier-selective contacts for Si solar cells," Appl. Phys. Lett. 104(18), 181105 (2014).

[2] U. Römer et al., "Ion implantation for poly-Si passivated back-junction back-contacted solar cells," IEEE Journal of Photovoltaics 5(2), 507-514 (2015).

[3] S.W. Glunz et al., "The irresistible charm of a simple current flow pattern – 25% with a solar cell featuring a full-area back contact," Proceedings of the 31st European Photovoltaic Solar Energy Conference and Exhibition, 2015.

[4] R. Brendel and R. Peibst, "Contact selectivity and efficiency in crystalline silicon photovoltaics," IEEE J. Photovoltaics 6, 1413-1420 (2016).

[5] S.S. Hegedus and W.N. Shafarman, "Thin-film solar cells: device measurements and analysis," Progress in Photovoltaics: Research and Applications 12, 155-176 (2004).

[6] M.W. Denhoff, "An accurate calculation of spreading resistance," Journal of Physics D: Applied Physics 39, 1761 (2006).

[7] G. Yang et al., "Design and application of ion-implanted polySi passivating contacts for interdigitated back contact c-Si solar cells," Applied Physics Letters 108(3), 033903 (2016).

[8] D.L. Young et al., "Low-cost plasma immersion ion implantation doping for interdigitated back passivated contact (IBPC) solar cells," Solar Energy Materials and Solar Cells (2016).

[9] C. Reichel et al., "Tunnel oxide passivated contacts formed by ion implantation for applications in silicon solar cells," J. Appl. Phys. 118 (2015).

Ultra-Thin Crystalline Silicon Solar Cells with Nickel Oxide Interlayer as Hole-selective Contact

Muyu Xue[1], Raisul Islam[2], Junyan Chen[3], Zheng Lyu[2], Yusi Chen[2], Daniel DeWitt[2], Albert Pleus[2],
Christian Tae[2], Ching-Ying Lu[2], Kai Zang[2], Jieyang Jia[2], Yijie Huo[2],
Ted Kamins[2], Krishna Saraswat[1,2], James Harris[1,2]

[1]Department of Material Science and Engineering, Stanford University, Stanford, CA, 94305, US
[2]Department of Electrical Engineering, Stanford University, Stanford, CA, 94305, US
[3]School of Physics, Peking University, Beijng, 100871, China

Abstract — Thin film crystalline silicon (c-Si) solar cells have been a potential candidate to reduce the capital expenditure associated with traditional silicon photovoltaic market. However the contact recombination becomes major concern preventing the thin-film cell performance reaching the theoretical efficiency. To reduce the effect of this problem, low-cost and silicon-compatible process is needed to suppress the contact recombination. In this work, thin film nickel oxide (NiO_x) is proposed as hole-selective interlayer for c-Si solar cells, which could potentially reduce the contact recombination between p-Si/metal interfaces. The thin film NiO_x layer is integrated with 2 μm thick ultra-thin c-Si solar cells by e-beam evaporation. Cell performance including open circuit voltage and efficiency are presented and discussed under different NiO_x thickness and annealing condition. Experimental results showed 6mV Voc enhancement when 8nm thick NiO_x is inserted when annealing condition is optimized. However the interface quality becomes another concern due to the increased series resistance associated with the extra NiO_x layer.

Index Terms — photovoltaic cells, contact passivation, ultra-thin c-Si solar cells, carrier-selective contact

I. INTRODUCTION

Silicon solar cells have been the dominant material for photovoltaic market, however the trend for reducing the capital expenditure associated with silicon cell production to reduce the cost is more important in recent decades. Thin-film crystalline silicon (c-Si) solar cells are proposed and demonstrated to be a good candidate for this requirement [1]. Thin film c-Si solar cells are designed and fabricated not only for reducing the material cost, but also improve the cell efficiency [1-6]. For thin-film c-Si solar cells, bulk recombination is decreased due to the shrinking of cell active region. Surface recombination can be effectively reduced by introducing traditional silicon surface passivation material, such as Al_2O_3 [7-10], silicon nitride [11-13] and amorphous silicon (a-Si) [14]. However, the contact recombination becomes important concern which limits the cell performance [15]. Carrier-selective contact structure is thus proposed to reduce the recombination between silicon/metal interface.

Many metal-semiconductor-silicon (MIS) structures have been investigated as potential carrier-selective contacts for c-Si solar cells. Titanium oxide (TiO_x) has been demonstrated as electron-selective contact material [16-18]. On the other side,

Nickel oxide (NiO_x) has been proposed to reduce the Schottky barrier for holes [19]. Due to the favorable bandgap of 3.6-4 eV [20] and electron affinity of 1.5 eV [21] for NiO_x, the NiO_x/Si band alignment has both small valence band offset ($\Delta E_V < 0.2$ eV) and large conduction band offset ($\Delta E_C > 2$ eV), making the junction hole-selective and electron-blocking. Thus the contact recombination can be reduced.

In this work, ultra-thin c-Si solar cells are fabricated and integrated with NiO_x as hole-selective contact interlayer materials between the p-Si and metal contacts. NiO_x layer with thickness of 2nm, 5nm and 8nm are inserted by e-beam evaporation between p-Si/metal respectively. The corresponding cell performance is shown and discussed. The effect of forming gas annealing temperature on cell performance is also demonstrated. Experimental results showed open circuit voltage (V_{OC}) enhancement of 6mV is observed when 8nm thick NiO_x is inserted compared with the cell without NiO_x as hole-selective interlayer. However the series resistance is also increased which limits the cell efficiency due to the interface quality associated with the NiO_x layer. This study reveals the sensitivity of solar cell performance on the metal contact interface quality especially with extra carrier-selective dielectric layer introduced.

II. SOLAR CELL FABRICATION PROCESS

The fabrication process of the ultra-thin c-Si solar cell is shown in figure 1. The c-Si cell layer is epitaxially grown by chemical vapor deposition on (100) silicon substrate. The total thickness of cell active region is 2 μm. The cell mesa region is defined by standard photolithography followed by reactive ion etching process. Thermal oxidation is done to form SiO_2 layer for surface passivation and single SiO_x layer is deposited by plasma-enhanced chemical vapor deposition (PECVD) for anti-reflection coating. Both p-contact and n-contact region is defined by photolithography, followed by 20:1 buffered oxide etch (BOE) to remove the oxide and form the contact trench. Aluminum is used as n-contact metals. For the p-contact, the NiO_x was deposited on p-Si by e-beam evaporation from NiO_x source directly, and platinum was used as metal contacts. NiO_x layer with different thickness of 0nm (control), 2nm, 5nm and

978-1-5090-5606-4/17 $31.00 © 2017 IEEE

8nm NiO_x is deposited on different wafers. Forming gas (5% H_2, 95% N_2) anneal at different temperature (250°C, 350°C and 450°C) is done after the cell fabrication. The cell results under different annealing condition will be discussed in the next part.

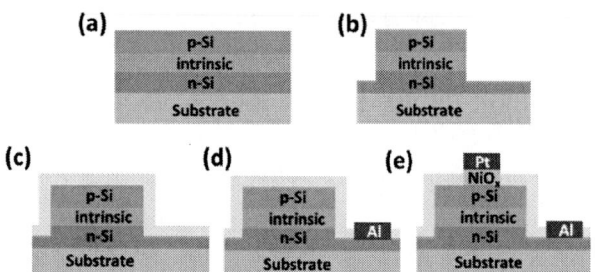

Fig. 1. Schematic of solar cell fabrication process. (a) Epitaxial growth of cell active layer by CVD. (b) Standard dry etching forming the mesa region. (c) Thermal oxidation and PECVD SiO_x anti-reflection coating. (d) n-contact formation after oxide removal by BOE. (e) p-contact formation by combination of e-beam evaporated NiO_x layer followed by platinum deposition.

III. RESULTS AND DISCUSSION

The J-V characteristics of cells are tested under 1 sun illumination (AM 1.5G spectrum) at room temperature. The J-V results for c-Si solar cell with different NiO_x thickness: 0nm (control), 2nm, 5nm and 8nm are summarized in figure 2.

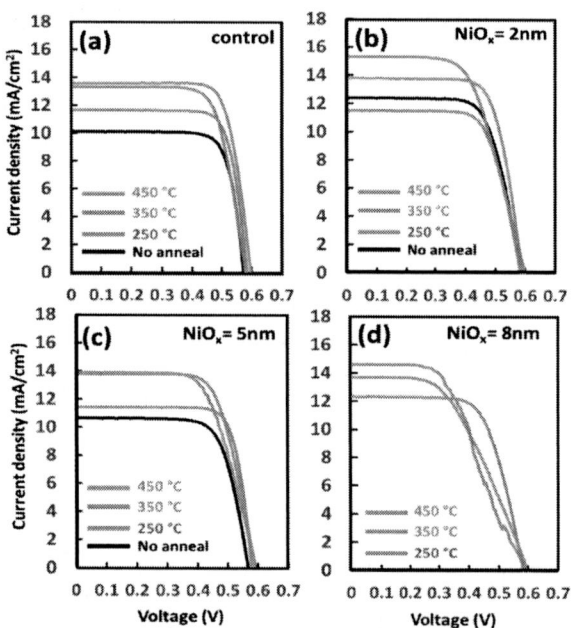

Fig. 2. J-V characteristics under 1 sun for c-Si cells with different thickness of NiO_x between p-Si/metal under multiple forming gas annealing temperature (no anneal, 250°C, 350°C and 450°C). Results for 0nm NiO_x (control), 2nm, 5nm and 8nm thick NiO_x are summarized in subfigure (a), (b), (c) and (d) respectively.

Based on the J-V characteristics above, for each of the groups with NiO_x inserted between p-Si/metal, the open circuit voltage (V_{OC}) and short circuit current (J_{SC}) show different degree of enhancement as annealing temperature increases from 250°C to 450°C, which is due to the better interface quality between Si/NiO_x when forming gas annealing is introduced. Experimental V_{OC} and J_{SC} values are summarized in table I (the best cell result for each group). For the cells with 8nm NiO_x, as annealing temperature increased from 250°C to 450°C, the highest V_{OC} of 603 mV is obtained. However the efficiency degrades and the undesirable "S" shape of the J-V curve is observed. This is due to the huge increase of series resistance from 9.3 Ωcm^2 to 33.6 Ωcm^2. On the other hand, because of the increased series resistance associated with NiO_x layer, the best cell efficiency obtained from each group with NiO_x integrated is still lower that of control group, which has highest efficiency of 6.2% and V_{OC} of 597mV. The reason is still under investigation which may be due to the large tunneling resistance of the NiO_x layer, or the extra energy barrier for majority carriers associated with interface quality of p-Si/NiO_x. These two factors could result in high series resistance on the p-contact of the solar cell thus limits the energy conversion efficiency.

TABLE I
SOLAR CELL PERFORMANCE AFTER 250°C /450°C ANNEAL

	V_{OC} (mV)	J_{SC} (mA/cm^2)	Efficiency (%)
2nm NiO_x	584/595	11.5/15.3	4.8/5.6
5nm NiO_x	580/591	11.4/13.7	5.2/5.7
8nm NiO_x	582/603	12.3/14.6	4.9/4.3

IV. CONCLUSION

In summary, NiO_x is demonstrated as potential hole-selective interlayer material for c-Si solar cells and the NiO_x layer is integrated with 2 μm-thick ultra-thin c-Si solar cell. The solar cell performance dependence on NiO_x thickness and annealing condition is investigated. Voc enhancement of 6 mV is obtained when 8nm evaporated NiO_x is introduced. On the other hand the series resistance is addressed which needs to be minimized by carefully controlling of the interface quality and interlayer thickness. Overall, this work presents a novel design integrating NiO_x with ultra-thin c-Si solar cells, which can be potentially improved towards fabricating low-cost and high-efficiency thin-film c-Si solar cells.

ACKNOWLEDGEMENT

The authors would like to acknowledge the GCEP and BAPVC for financial support.

REFERENCES

[1] SEMI., "International Technology Roadmap for Photovoltaic 2014 Results," Itrpv(March), 1–38 (2015).9.

[2] Jeong, S., McGehee, M. D.., Cui, Y., "All-back-contact ultra-thin silicon nanocone solar cells with 13.7% power conversion efficiency.," Nat. Commun. 4(May), 2950 (2013)

[3] Wang, S., et. al., "Large-area free-standing ultrathin single-crystal silicon as processable materials," Nano Lett. 13(9), 4393–4398 (2013).

[4] Yoon, Jongseung, et al. "Ultrathin silicon solar microcells for semitransparent, mechanically flexible and microconcentrator module designs." Nature materials 7.11 (2008): 907-915.

[5] Branham, Matthew S., et al. "15.7% Efficient 10‐μm‐Thick Crystalline Silicon Solar Cells Using Periodic Nanostructures." Advanced Materials 27.13 (2015): 2182-2188.

[6] Tiedje, T., Yablonovitch, E., Cody, G. D.., Brooks, B. G., "Limiting efficiency of silicon solar cells," IEEE Trans. Electron Devices 31(5), 711–716 (1984).

[7] Hoex, Bram, et al. "Silicon surface passivation by atomic layer deposited Al_2O_3." *Journal of Applied Physics* 104.4 (2008): 044903.

[8] Dingemans, G., et al. "Silicon surface passivation by ultrathin Al_2O_3 films synthesized by thermal and plasma atomic layer deposition." *physica status solidi (RRL)-Rapid Research Letters* 4.1-2 (2010): 10-12.

[9] Agostinelli, G., et al. "Very low surface recombination velocities on p-type silicon wafers passivated with a dielectric with fixed negative charge." *Solar Energy Materials and Solar Cells* 90.18 (2006): 3438-3443.

[10] Schmidt, J., et al. "Surface passivation of high-efficiency silicon solar cells by atomic-layer-deposited Al_2O_3." *Progress in photovoltaics: research and applications* 16.6 (2008): 461-466.

[11] Aberle, Armin G. "Overview on SiN surface passivation of crystalline silicon solar cells." *Solar Energy Materials and Solar Cells* 65.1 (2001): 239-248.

[12] Mackel, H., and R. Ludemann. "Detailed study of the composition of hydrogenated SiN_x layers for high-quality silicon surface passivation." *Journal of Applied Physics* 92.5 (2002): 2602-2609.

[13] Aberle, Armin G., and Rudolf Hezel. "Progress in Low-temperature Surface Passivation of Silicon Solar Cells using Remote-plasma Silicon Nitride." *Progress in photovoltaics: research and applications* 5.1 (1997): 29-50.

[14] Schaper, Martin, et al. "20.1%‐efficient crystalline silicon solar cell with amorphous silicon rear-surface passivation." *Progress in Photovoltaics: Research and Applications* 13.5 (2005): 381-386.

[15] Tiedje, T., Yablonovitch, E., Cody, G. D.., Brooks, B. G., "Limiting efficiency of silicon solar cells," IEEE Trans. Electron Devices 31(5), 711–716 (1984).

[16] Nagamatsu, Ken A., et al. "Titanium dioxide/silicon hole-blocking selective contact to enable double-heterojunction crystalline silicon-based solar cell." *Applied Physics Letters* 106.12 (2015): 123906.

[17] Avasthi, Sushobhan, et al. "Hole-blocking titanium-oxide/silicon heterojunction and its application to photovoltaics." *Applied Physics Letters* 102.20 (2013): 203901.

[18] Xue, M., et al. "Titanium oxide contact passivation layer for thin film crystalline silicon solar cells." Photovoltaic Specialists Conference (PVSC), 2016 IEEE 43rd. IEEE, 2016.

[19] Islam, Raisul, Gautam Shine, and Krishna C. Saraswat. "Schottky barrier height reduction for holes by Fermi level depinning using metal/nickel oxide/silicon contacts." *Applied Physics Letters* 105.18 (2014): 182103.

[20] Sasi, B., et al. "Preparation of transparent and semiconducting NiO films." *Vacuum* 68.2 (2002): 149-154.

[21] Wu, Hongbin, and Lai-Sheng Wang. "A study of nickel monoxide (NiO), nickel dioxide (ONiO), and Ni (O2) complex by anion photoelectron spectroscopy." *Journal of Chemical Physics* 107.1 (1997): 16-21.

Crystalline Si Solar Cells with Passivating, Carrier-selective Nickel Oxide Contacts

Woojun Yoon[1], James Moore[2], David Scheiman[1], Eunhwan Cho[3], Young-Woo Ok[3], Nicole Kotulak[4], Phillip P. Jenkins[1], Ajeet Rohatgi[3] and Robert J. Walters[1]

[1]U.S. Naval Research Laboratory, Washington, DC 20375, USA
[2]The George Washington University, Washington, DC 20037, USA
[3]Georgia Institute of Technology, Atlanta, GA 30332, USA
[4]NRC Postdoctoral Research Associate residing at the U.S. Naval Research Laboratory

Abstract — We examine the potential to enhance cell performance using wide bandgap metal oxide films as full-area rear contacts to *p*-type crystalline Si (*c*-Si) solar cells. We aim to introduce a band offset through wide bandgap nickel oxide rather than introducing a band bending through transition metal oxides (e.g. MoO_x, V_2Ox, WO_x). Our numerical simulation shows it is possible to achieve one-sun efficiency of 21.6% with the open-circuit voltage (V_{oc}) of 652 mV, the short-circuit current density (J_{sc}) of 39.9 mA/cm^2 and the fill factor (FF) of 82.5% when the back surface recombination velocities is 100 cm/s at *p*-Si/NiO_x.

Index Terms— passivating contact, metal oxide, nickel oxide, crystalline silicon, band offset

I. INTRODUCTION

The concept of passivating, carrier-selective contacts based on wide bandgap metal oxides has a potential to reduce the gap between cell efficiencies and the theoretical efficiency limits.

In silicon heterojunction (SHJ) solar cells, metal oxides have been utilized mostly to replace a relatively narrower bandgap (E_g) of amorphous *p*-type Si layers (E_g ~1.8 eV), aiming to minimize parasitic light absorption in the ultraviolet and visible range of the solar spectrum. Recent work by Geissbühler *et al.* has successfully demonstrated this goal using MoO_x layers by improving the short-circuit current density (J_{sc}) while maintaining high fill factor (FF) and open-circuit voltage (V_{oc}) compared to standard SHJ solar cells.

For crystalline Si (*c*-Si) cells with diffused junctions that are without hydrogenated intrinsic amorphous Si passivation layers, thin metal oxides layers have been implemented largely as a full-area rear contact to the Si base.[1-5] Very recently, Yang *et al.* showed over 22% efficient *n*-type Si solar cells with an electron-selective TiO_x contact, demonstrating the potential to achieve very high efficiency with low-cost over conventional Si cell technologies, such as passivated emitter rear contact (PERC) and passivated emitter rear totally diffused (PERT) cells.[4, 6] Other materials, such as MgO_x or MgF have been successfully implemented to *n*-type *c*-Si substrate as an electron-selective contact, exhibiting high efficiency beyond 20%.[7]

Unlike electron-selective contact materials, the cell performance of *c*-Si cells with hole-selective contacts is still limited. Only *n*-type transition metal oxides, such as MoO_x, V_2O_x, and WO_x have been used as hole-selective contacts due to deep workfunction of those oxides. [8, 9] For example, the V_{oc} of *p*-type Si solar cells with full-area hole-selective MoO_x contacts was limited to 616 mV although the saturation current density at *p*-Si/MoO_x was significantly lowered compared to that of conventional Al and boron-doped Si contacts.[2, 10] Recently we found that the values in electron affinity (χ) of MoO_x need to be very high (>5.5 eV) in order to create large enough conduction band bending to effectively block minority carriers when *c*-Si is contacted by MoO_x. [11] However, it is well known that it is difficult to obtain high χ in transition metal oxides because their values in χ are known to be highly sensitive to processing conditions, post-deposition treatments, oxygen, moisture and contaminants.

In this work, we evaluate the potential to enhance cell performance using wide bandgap nickel oxides as full-area rear contacts to *p*-Si solar cells, aiming to introduce a band offset at *p*-Si/NiOx. Our numerical simulation indicates that one-sun efficiency of 21.6% can be achieved with the open-circuit voltage (V_{oc}) of 652 mV, the short-circuit current density (J_{sc}) of 39.9 mA/cm^2 and the fill factor (FF) of 82.5% for the back surface recombination velocities of 100 cm/s at *p*-Si/NiO_x.

II. NUMERICAL SIMULATION

All simulations were performed using the Crosslight 2D modeling tool *Advanced Physical Models of Semiconductor Devices* (APSYS). Crosslight is capable of modeling quantum tunneling, and the models included the intraband tunneling through the metal oxide barrier. [12]. Standard material parameters were used for Si and the results were compared to experimental devices with similar structures, which have been previously reported in literature. The model devices consist of a 180 μm thick, *p*-type Si base (1.5 ohm-cm) with a bulk lifetime of 300 μs and the n^+-emitter (~85 ohm./sq) with a textured surface. The front surface recombination velocity is assumed to be 10000 cm/s. The base and emitter characteristics were held constant throughout all simulations and only the back surface parameters were changed. This was done to better focus on the effects of the surface recombination at *p*-Si/metal oxide contacts. Front surface reflection was taken from measured data for the *p*-type Si cell with standard random pyramid textured front surface coated with SiO_2/SiN_x stacks. [13-15] We assume a front grid shadowing of 7%.

978-1-5090-5606-4/17 $31.00 © 2017 IEEE

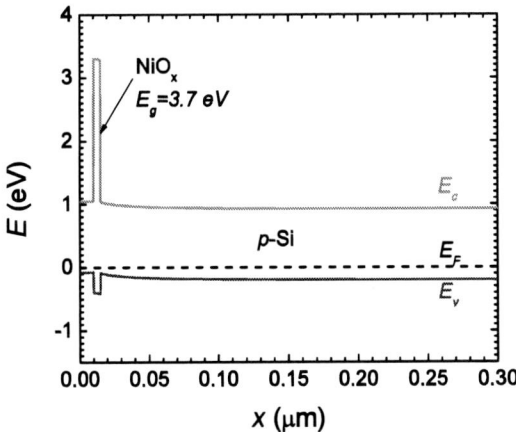

Fig. 1. The band diagrams between NiO$_x$ (E_g=3.7 eV) and p-Si were simulated using CrossLight.

III. RESULTS AND DISCUSSION

In an attempt to find the device and material parameters that appear to most affect the cell performance of the device with NiO$_x$ contacts, we perform numerical simulation using the Crosslight 2D modeling tool APSYS.[16] The cell structure used for the modeling consists of a 180 μm p-Si base with a n^+-

emitter. The thicknesses of NiO$_x$ films were swept from 1 nm to 5 nm and the back surface recombination velocity (BSRV) from 1 cm/s to 10^7 cm/s at p-Si/NiO$_x$ rear contacts. Fig. 1 shows the simulated band diagram between p-Si and NiO$_x$ layer. The large conduction band offset can effectively block minority carriers. However, the valence band offset could act as a barrier to holes tunneling.

Fig. 2 shows the simulated cell performance of NiO$_x$ full-area rear contacted Si solar cells as a function of NiO$_x$ thickness and the BSRV at p-Si/NiO$_x$ under one sun condition. The cell performance of the device with NiO$_x$ contacts was found to be most sensitive to the BSRV at p-Si/NiO$_x$. The cell with high level of rear surface passivation (i.e. BSRV~100 cm/s) exhibits a high efficiency of 21.7% with V_{oc} of 656 mV when a 4 nm thick NiOx was used. For an improved surface passivation with BSRV of 10 cm/s, the efficiency can be further increased to 22.7% with V_{oc} of 685 mV. Simulated cell efficiency of the control device with conventional Al contacts was ~17%, which is mainly limited by low V_{oc} of 580 mV. Our results clearly show that hole-selective NiO$_x$ contacts are very effective to improve the cell performance while maintaining high FF when it is inserted at the rear side between p-Si and metal contact layers.

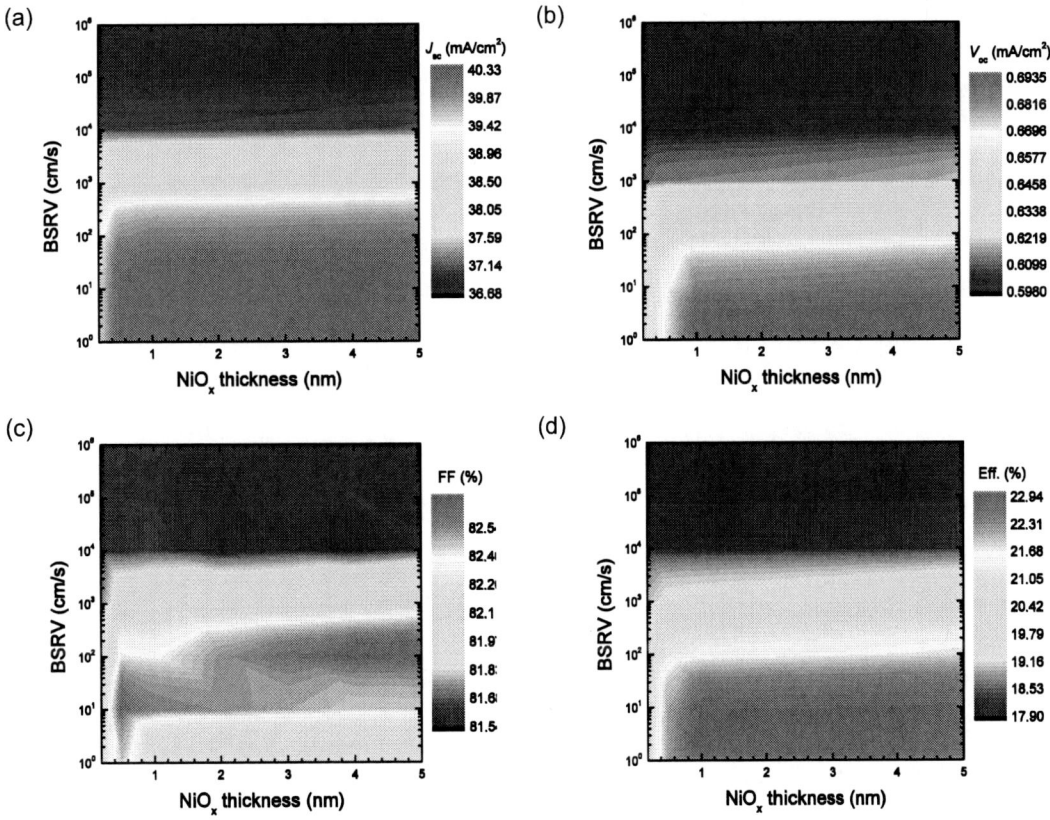

Fig. 2. Simulated cell performance of NiO$_x$ full-area rear contacted p-Si solar cells as a function of the thicknesses of NiO$_x$ film and the back surface recombination velocity (BSRV) at p-Si/NiO$_x$ under one sun condition (AM1.5G, 100 mW/cm^2); (a) J_{sc}, (b) V_{oc}, (c) FF, and (d) efficiency.

978-1-5090-5606-4/17 $31.00 © 2017 IEEE

V. Conclusions

In this paper we have performed numerical simulation for p-Si solar cell structures using NiO$_x$ full-area rear contacts. Several simulation parameters were investigated, but the final analysis focused on the material properties of each structure that appeared to most affect the device performance as measured by the current-voltage characteristics under one sun conditions. Numerical modeling showed that the performance of p-type cells with the NiO$_x$ was most sensitive to the BSRV at the p-Si/NiO$_x$ interface, while the optimum thicknesses of the NiO$_x$ film are in the range of 1 nm to 5 nm.

References

[1] T. G. Allen, M. Ernst, C. Samundsett, and A. Cuevas, "Demonstration of c-Si Solar Cells With Gallium Oxide Surface Passivation and Laser-Doped Gallium p+ Regions," *IEEE J. Photovolt,* vol. 5, pp. 1586-1590, 2015.

[2] J. Bullock, D. Yan, A. Cuevas, Y. Wan, and C. Samundsett, "n- and p-typesilicon Solar Cells with Molybdenum Oxide Hole Contacts," *Energy Procedia,* vol. 77, pp. 446-450, 2015.

[3] L. G. Gerling, S. Mahato, A. Morales-Vilches, G. Masmitja, P. Ortega, C. Voz, R. Alcubilla, and J. Puigdollers, "Transition metal oxides as hole-selective contacts in silicon heterojunctions solar cells," *Sol. Energ. Mat. Sol. Cells,* vol. 145, Part 2, pp. 109-115, 2016.

[4] X. Yang, Q. Bi, H. Ali, K. Davis, W. V. Schoenfeld, and K. Weber, "High-Performance TiO2-Based Electron-Selective Contacts for Crystalline Silicon Solar Cells," *Adv. Mater.,* vol. 28, pp. 5891–5897, 2016.

[5] W. Yoon, E. Cho, J. D. Myers, Y.-W. Ok, M. P. Lumb, J. A. Frantz, N. A. Kotulak, D. Scheiman, P. P. Jenkins, A. Rohatgi, and R. J. Walters, "Transparent conducting oxide-based, passivated contacts for high efficiency crystalline Si solar cells," in *42nd IEEE Photovoltaic Specialists Conference (PVSC),* 2015, pp. 1-4.

[6] X. Yang, "Over 22% Efficient N-type Silicon Solar Cells Featuring A Full-area Electron Selective TiO2 Contact," presented at the the 26th International Photovoltaic Science and Engineering Conference (PVSEC-26), Singapore, 2016.

[7] Y. Wan, C. Samundsett, J. Bullock, M. Hettick, T. Allen, D. Yan, J. Peng, Y. Wu, J. Cui, A. Javey, and A. Cuevas, "Conductive and Stable Magnesium Oxide Electron-Selective Contacts for Efficient Silicon Solar Cells," *Adv. Energy Mater.,* p. 1601863, 2016.

[8] J. Bullock, C. Samundsett, A. Cuevas, D. Yan, Y. Wan, and T. Allen, "Proof-of-Concept p-Type Silicon Solar Cells With Molybdenum Oxide Local Rear Contacts," *IEEE J. Photovolt,* vol. 5, pp. 1591-1594, 2015.

[9] J. Bullock, A. Cuevas, T. Allen, and C. Battaglia, "Molybdenum oxide MoOx: A versatile hole contact for silicon solar cells," *Appl. Phys. Lett.,* vol. 105, p. 232109, 2014.

[10] A. Rohatgi, D. L. Meier, B. McPherson, Y.-W. Ok, A. D. Upadhyaya, J.-H. Lai, and F. Zimbardi, "High-Throughput Ion-Implantation for Low-Cost High-Efficiency Silicon Solar Cells," *Energy Procedia,* vol. 15, pp. 10-19, 2012.

[11] W. Yoon, J. E. Moore, E. Cho, D. Scheiman, N. A. Kotulak, E. Cleveland, Y.-W. Ok, P. P. Jenkins, A. Rohatgi, and R. J. Walters, "Hole-selective Molybdenum Oxide as a Full-area Rear Contact to Crystalline p-type Si Solar Cells," *Japanese Journal of Applied Physics,* vol. 56, 2017.

[12] Y. G. Xiao, M. Lestrade, Z. Q. Li, and Z. M. S. Li, "Modeling of Si-based solar cells with V-grooved surface texture by Crosslight APSYS," 2007, pp. 66510F-66510F-8.

[13] Y.-W. Ok, A. D. Upadhyaya, Y. Tao, F. Zimbardi, K. Ryu, M.-H. Kang, and A. Rohatgi, "Ion-implanted and screen-printed large area 20% efficient N-type front junction Si solar cells," *Solar Energy Materials and Solar Cells,* vol. 123, pp. 92-96, 2014.

[14] M. P. Lumb, W. Yoon, C. G. Bailey, D. Scheiman, J. G. Tischler, and R. J. Walters, "Modeling and analysis of high-performance, multicolored anti-reflection coatings for solar cells," *Opt. Express,* vol. 21, pp. A585-A594, 2013.

[15] W. Yoon, A. Lochtefeld, N. A. Kotulak, D. Scheiman, A. Barnett, P. P. Jenkins, and R. J. Walters, "Enhanced Surface Passivation of Epitaxially Grown Emitters for High-efficiency Ultrathin Crystalline Si Solar Cells " in *43rd IEEE Photovoltaic Specialist Conference (PVSC),* 2016, pp. 1-3.

[16] Y. G. Xiao, M. Lestrade, Z. Q. Li, and Z. M. S. Li, "Modeling of Si-based solar cells with V-grooved surface texture by Crosslight APSYS," presented at the Proc. SPIE, 2007.

GaP/Si Heterojunction Solar Cells Grown by Molecular Beam Epitaxy

Chaomin Zhang, Ehsan Vadiee, Richard R. King, Christiana B. Honsberg

School of Electrical, Computer and Energy Engineering, Arizona State University, Tempe, AZ, 85287, USA

Abstract — **Maintaining Si bulk lifetime during epitaxial growth of III-V materials on Si is critical for achieving high performance Si-based solar cell devices. In this work, we demonstrate two different approaches in which were able to preserve the Si bulk lifetime during the epitaxial growth by a molecular beam epitaxy (MBE) system, by using SiN$_x$ as a coating layer and a phosphorous-rich region as a gettering sink. In order to investigate the solar cell performance, using these two approaches, different MBE grown GaP/Si heterojunction solar cells with different structures were designed, grown, and fabricated. We show that by introducing a phosphorous (P)-rich region (n+Si) between the GaP and Si layers, the efficiency (η) can be significantly increased. The solar cell, fabricated using the phosphorous-rich region, with p-a-Si/a-Si/Si/n+Si/GaP structure shows an open-circuit voltage (V_{OC}) of 617 mV, FF of 64%, and η of 13.1%.**

Index Terms — **Gallium phosphide, minority-carrier lifetime, molecular beam epitaxy, silicon heterojunction solar cells, carrier-selective contact**

I. INTRODUCTION

GaP has a small lattice mismatch (*f=0.4 %*) with Si at room temperature, which allows a high quality epitaxial growth of GaP on Si substrates. This offers a path towards the heterogeneous integration of GaP-based devices with Si ($E_G \approx$ 1.1 eV). GaP as a hetero-emitter integrated with a Si-based solar cell has been theoretically shown to enable higher open-circuit voltages and higher conversion efficiencies compared to the standard PSG solar cells [1], [2]. This device structure provides multiple advantages, such as a lower recombination velocity at the Si interface and a higher carrier mobility compared to Si-based solar cells with a-Si used as hetero-emitter. In addition, it was shown that a high carrier collection efficiency can be achieved by formation of carrier selective interfaces [1]. Thus, the realization of high performance GaP/Si solar cells can be a path for achieving high performance silicon-based multi-junction solar cells. Despite the mentioned advantages, Si bulk lifetime degradation during the epitaxial growth of GaP on Si is a major challenge that needs to be addressed in order to achieve a high efficiency GaP/Si solar cell.

Recent results reported by multiple research groups, show a dramatic deterioration of GaP crystal quality and consequently degradation of Si minority-carrier lifetime during the MBE growth of III-P materials on Si [3]-[6]. The Si lifetime degradation is mainly attributed to the formation of nonradiative recombination centers (known as SRH centers) in Si. The recombination centers are believed to be originated from fast-diffusing species [7]. To improve the Si lifetime, different methods have been proposed, such as using SiN$_x$, as diffusion barrier [3], [7], and phosphorous diffusion, as a post-gettering sink [5].

In this work, we demonstrate that the Si minority-carrier lifetime can be well maintained with two different approaches during the MBE epitaxial growth of GaP on Si. Motivated by those, different GaP/Si structures and their characteristics are demonstrated to investigate the effectiveness of these two life-saving approaches. Furthermore, the n+ region was introduced between the GaP and Si to preserve the Si lifetime and increase the carrier transport through the interface. The crystal quality of the GaP grown on the n+ region was then investigated. Finally, a solar cell with a structure of a-Si/Si/n+ region/GaP is demonstrated.

II. Experimental Details

All the n-type Si (Si) wafers used in this study were 4-inch-diameter, float-zone (FZ), double-side-polished wafers, with (001) orientation. The Si wafers were chemically cleaned with an RCA (Radio Corporation of America) solution and a diluted hydrofluoric acid (HF) solution (5% HF) for 30 seconds before loading into the MBE chamber. Samples were grown using a Veeco GEN III MBE system equipped with P, Sb, and As valved crackers and Ga, In, Al, and Si effusion cells. The 150 nm thick amorphous SiN$_x$ films were deposited at 350 °C by plasma-enhanced chemical vapor deposition (PECVD) system on one side of the Si wafers. After the thermal treatment at 800 °C in the MBE chamber, the SiN$_x$ layer was removed by a concentrated HF (49%) solution. For the P-diffusion, wafers were exposed to a flow of oxygen and POCl$_3$, at 850 °C in a diffusion furnace. After MBE thermal treatment, a mixture of hydrofluoric, nitric and acetic acids (HNA) with a volume ratio of 10:73:17 at room temperature was used to isotropically etch ~10 μm of Si layers from both sides of the wafer to ensure the complete removal of the n+ region. For the minority-carrier lifetime measurement, 50 nm thick intrinsic hydrogenated amorphous silicon (a-Si:H) films prepared at 250 °C were deposited by PECVD on the cleaned Si surfaces. The Si effective minority-carrier lifetimes were measured by quasi-steady state photoconductance decay (QSSPC) using a Sinton lifetime tester.

Unintentionally doped (UID) GaP layers were deposited by the MBE on the Si wafers backside coated with the SiN$_x$, as shown in Fig. 1(a). Approximately 15 nm thick GaP was formed on the n-Si substrate as an electron selective contact. After the GaP deposition and SiN$_x$ removal, ~10 nm thick p-a-Si and ~6 nm thick i-a-Si layers were deposited on top of the bare Si side of the wafer to form a p-a-Si/i-a-Si/n-Si/GaP structure. To make these structures into the solar cells, 75 nm thick ITO and 200 nm thick Ag as current spreading and contact layers, respectively, were deposited on the both sides of the structures using RF-sputtering. In addition to the n-Si/GaP structure, different set of GaP/p-Si solar cells were prepared with 15, 30, and 50 nm thick UID-GaP layers as hetero-emitters with the similar procedure as the n-Si/n-GaP cells. Reference wafers were also prepared with p-a-Si/i-a-Si/n-Si/i-a-Si/n-a-Si structure for an n-type substrate and n-a-Si/i-a-Si/p-Si/i-a-Si/p-a-Si structure for a p-type substrate. External quantum efficiency (EQE) of the fabricated cells were measured by the solar cell spectral response measurement system (QEX10), manufactured by PV Measurements. The IQE curves were obtained by correcting the EQE for reflection.

As shown in Fig. 1(b), the n+ regions were formed on the both sides of the Si wafer with P-diffusion, followed by deposition of a 150 nm SiN$_x$ on the backside by PECVD. The SiN$_x$ layer acts as a diffusion barrier and the P-rich regions act as gettering sinks for impurities. They both can effectively suppress the Si lifetime degradation. To evaluate the GaP crystal quality, X-ray diffraction (XRD) measurement was conducted using a high-resolution XRD diffractometer with CuK$_{\alpha1}$ radiation ($\lambda \approx 1.54$ Å). In addition, an atomic force microscopy (AFM) was performed by a multimode scanning probe microscope (SPM) to study the surface morphology of the GaP. After the removal of SiN$_x$ layer and ~10 μm of the Si wafer, a 16 nm p-a-Si and 9 nm i-a-Si layers were deposited on top of the bare silicon side to form a p-a-Si/i-a-Si/Si/GaP structure. ITO (75 nm) and Ag (200 nm) were also sputtered as contact layers on the both sides.

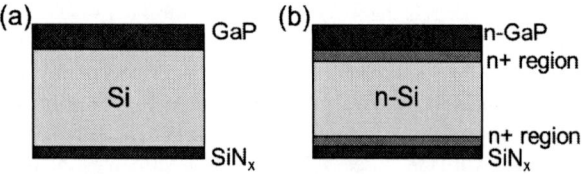

Fig. 1. Schematic structure of (a) GaP deposited on the n-Si or p-Si substrates with the SiN$_x$ coating layer. (b) GaP deposited on the n-Si substrate coated with the SiN$_x$ layer and n+ regions.

III. RESULT AND DISCUSSION

A. Maintaining High Si Bulk Lifetime

Using the a-SiN$_x$:H coating layer on the backside of the Si substrate enables blocking the external contaminants and a gettering effect at the Si/SiN$_x$ interface [7]. After annealing the sample, coated with the a-SiN$_x$:H, at 800 °C, the Si bulk lifetime of ~1.5 ms was measured (the SiN$_x$ layer was etched off before the lifetime measurement). Moreover, by forming the P-rich regions on the both sides of the Si wafer, we were able to maintain the Si bulk lifetime. The annealed Si wafer with the P-rich regions has the lifetime of ~400 μs, which is still comparable to the lifetime of the reference Si wafer. The lifetimes of the both samples are higher than that of the annealed reference bare Si wafer (~0.3 μs). All the annealing processes were performed at 800 °C in the MBE chamber. The lifetime measurement results are depicted in the Fig. 2. As evident, the lifetime of the sample with the P-rich regions is lower than that of the sample with the a-SiN$_x$:H coating layer. This can be related to the rough surface of Si caused by the etching process, as it was revealed by the AFM results (not shown here). These results clearly show that both the SiN$_x$ coating layer and the P-rich regions can maintain the Si bulk lifetime during the annealing process of the Si wafer in the MBE chamber.

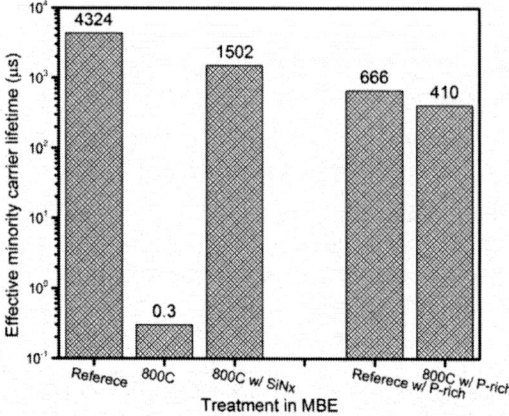

Fig. 2. Effective minority-carrier lifetime of the bare Si wafer compared with the Si wafers coated with SiN$_x$ and the P-rich regions. The annealing process was performed at 800 °C in the MBE chamber.

B. Unintentionally Doped GaP

A 15 nm unintentionally doped GaP layer was deposited on the n-Si wafer and processed into a solar cell, as explained previously, inset of Fig. 3. The solar cell performance with a short-current density of 31.3 mA/cm^2, an open-circuit voltage of 565 mV, a fill factor of 52.1%, and an efficiency of 9.2% was achieved. The external and internal quantum efficiencies are plotted in the Fig. 3. The GaP layer successfully acts as an electron selective contact on the n-emitter side. The GaP/Si cell IQE performance is comparable to that of the reference cell at the wavelength range of 300 – 800 nm. However, it is clear that the IQE of GaP/Si cell is much lower from 800 to 1150 nm, which indicates that the Si bulk lifetime is degraded and/or the GaP/Si interface has a poor passivation quality. Our previous work [7] has revealed that the SiN$_x$ could maintain the Si bulk lifetime at milli-second level after the GaP growth, also verified in this work. Therefore, the recombination at the GaP/Si interface is the primary factor in the IQE loss and the low open-circuit voltage.

In addition to the n-Si/GaP solar cells, the p-Si/GaP cells with different GaP thicknesses were characterized. Figure 4 shows the EQE characteristics of the p-Si/GaP samples. The spectral responses of the p-Si/GaP cells at 300 nm are clearly lower than that of the p-Si reference cell and decrease as the GaP thickness increases, which indicates that the incident photons are lost by a high carrier recombination and parasitic absorption in the GaP layer. For wavelengths in the range of 370 to 600 nm, a significant increase in the EQE is observed. This mainly originates from a better front carrier collection of the GaP emitter. At 450 nm, we have observed more than 16% EQE enhancement for the 15 nm GaP/Si sample. Moreover, more than 96% of the EQE has obtained ~500 nm wavelength for the 15 nm GaP/Si sample, which is in the range of GaP direct (~446 nm) and indirect (~553 nm) bandgaps. At the long wavelengths (900 - 1200 nm), the EQE values of the GaP/Si samples are comparable to that of the reference cell, which indicates that they have similar rear passivation qualities and Si bulk lifetimes. Although the blue response enhancement was observed in the GaP/Si samples, the low open-circuit voltages can be due to the recombination loss at the frontsides of the cells, most probably from the GaP/Si interface and the GaP bulk.

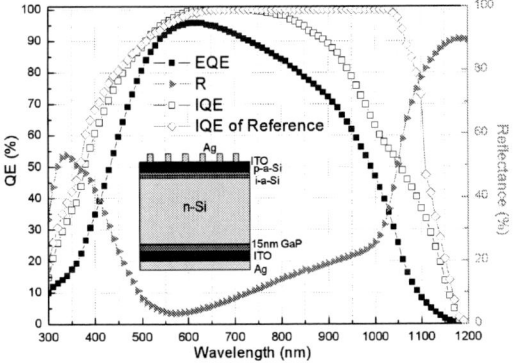

Fig. 3. EQE, IQE and surface reflectance of the n-Si/GaP solar cell and the n-Si reference cell (Inset: the n-Si/GaP schematic structure).

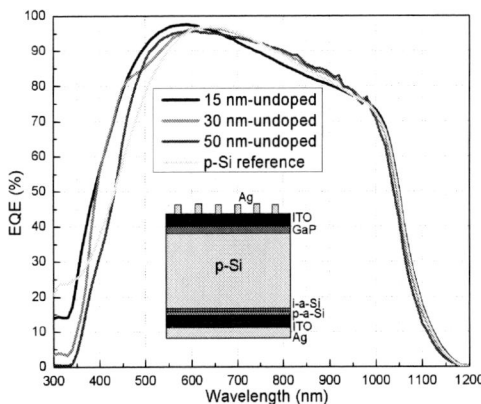

Fig. 4. External quantum efficiency of GaP/p-Si solar cells with various GaP thickness compared to the reference Si cell (Inset: the GaP/p-Si schematic structure).

C. a-Si/Si/n+/GaP Structure

All the GaP/Si samples discussed in section B are suffering from the recombination at the GaP/Si interfaces. One approach to eliminate this recombination loss is using a n+ layer between n-GaP and n-Si layers, as briefly explained previously. To investigate the impact of the P-rich region (n+ layer) on the quality of the MBE grown GaP layer, the XRD measurements were performed. The double-crystal ω-2θ rocking curves are shown in Fig. 5(a), including before and after the GaP growth (GaP/P-rich) and after etching off the GaP and n+ layers (including the P-rich region). As observed in Fig. 5(a), a shoulder appears on the right side of the Si substrate peak for the GaP/n+Si/Si and n+Si/Si samples, which is due to the lattice strain induced by the P-rich region [8]. Furthermore, reciprocal space map (RSM) at the vicinity of the (224) diffraction spots of Gap and So was measured and plotted in Fig. 5(b), which reveals that the GaP is fully strained to the Si. The triple crystal (TC) ω full width at half maximum (FWHM) of the GaP peak is ~14 arcsec, and the threading dislocation density is calculated to be ~2 × 10^6 cm^{-2}. The RMS of the surface roughness was measured to be ~0.49 nm. As a result, a high quality of the GaP layer was achieved on the P-rich surface.

After the SiN$_x$ and n+Si removal, the Si bulk lifetime was measured to be more than ~2 ms (measured without the GaP layer). The structure of p-a-Si/i-a-Si/Si/n+/GaP solar cell was characterized and compared to the sample without the n+ layers. The structure with the n+ region has a higher EQE value from 900 to 1150 nm, as shown in Fig. 6, which reveals that the recombination loss at the rear side of the device is reduced [9]. The n+ region acts as a back surface field (BSF) layer and repels the minority-carriers from the defective GaP/Si interface. A V$_{OC}$ of 617 mV and efficiency of 13.1% were measured for the solar cell with the n+ region. It is evident that deposition of the n+ region on the rear side can improve the GaP/Si solar cell performance.

978-1-5090-5606-4/17 $31.00 © 2017 IEEE

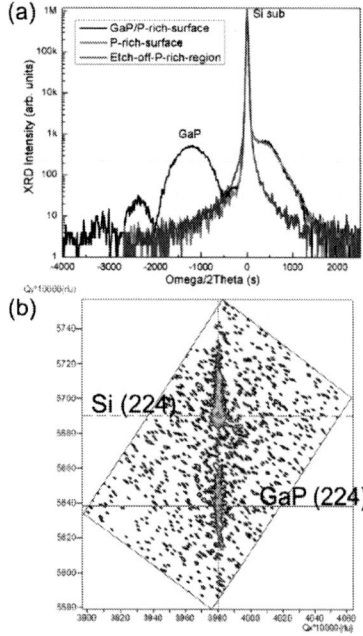

Fig. 5. (a) Double crystal ω-2θ rocking curves scanned in the vicinity of (004) reflection (under symmetric geometry). (b) reciprocal space map of the GaP/n+/Si sample at (224) reflection.

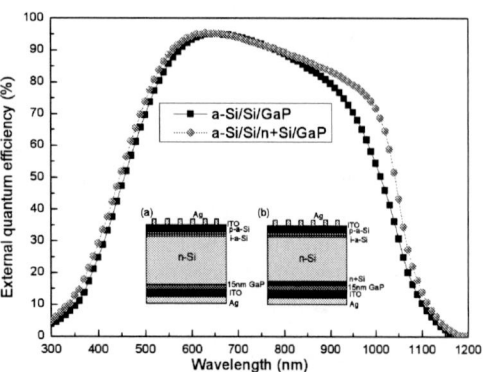

Fig. 6. External quantum efficiencies of a-Si/Si/GaP solar cells. (inset: the schematic structures of (a) a-Si/Si/GaP with no P-rich region and (b) a-Si/Si/GaP with P-rich region).

IV. CONCLUSION

We demonstrated that SiN$_x$ coating layer and P-rich regions, can effectively maintain the Si bulk lifetime for the purpose of GaP growth on Si by a MBE. The GaP was applied as a rear electron contact layer and/or a hetero-emitter layer for Si solar cells. However, the performance of the GaP/Si is limited by the recombination at the GaP/Si interface. It was shown that a high quality GaP layer can be achieved from the epitaxial growth GaP on a P-rich Si surface. The GaP/Si samples with the P-rich regions show device performance improvement and 13.1% efficiency was achieved.

ACKNOWLEDGMENTS

The authors would like to thank L. Ding for her contribution in the wafer preparation. The authors acknowledge funding from the U.S. Department of Energy under contract DE-EE0006335 and the Engineering Research Center Program of the National Science Foundation and the Office of Energy Efficiency and Renewable Energy of the Department of Energy under NSF Cooperative Agreement No. EEC-1041895. We gratefully acknowledge the use of facilities within the LeRoy Eyring Center for Solid State Science at Arizona State University.

REFERENCES

[1] H. Wagner, T. Ohrdes, A. Dastgheib-Shirazi, B. Puthen-Veettil, D. König, and P. P. Altermatt, "A numerical simulation study of gallium-phosphide/silicon heterojunction passivated emitter and rear solar cells," J. Appl. Phys., vol. 115, no. 4, p. 44508, Jan. 2014.

[2] S. Limpert, K. Ghosh, H. Wagner, S. Bowden, C. Honsberg, S. Goodnick, S. Bremner, A. Ho-Baillie, and M. Green, "Results from coupled optical and electrical sentaurus TCAD models of a gallium phosphide on silicon electron carrier selective contact solar cell," *2014 IEEE 40th Photovoltaic Specialist Conference (PVSC)*. pp. 836–840, 2014.

[3] C. Zhang, N. N. Faleev, L. Ding, M. Boccard, M. Bertoni, Z. Holman, R. R. King, and C. B. Honsberg, "Hetero-emitter GaP/Si solar cells with high Si bulk lifetime," in *2016 IEEE 43rd Photovoltaic Specialists Conference (PVSC)*, 2016, pp. 1950–1953.

[4] L. Ding, C. Zhang, T. U. Nœrland, N. Faleev, C. Honsberg, and M. I. Bertoni, "On the source of silicon minority-carrier lifetime degradation during molecular beam heteroepitaxial growth of III-V materials," *2016 IEEE 43rd Photovoltaic Specialists Conference (PVSC)*. pp. 2048–2051, 2016.

[5] L. Ding, C. Zhang, T. U. Nœrland, N. Faleev, C. Honsberg, and M. I. Bertoni, "Silicon Minority-carrier Lifetime Degradation During Molecular Beam Heteroepitaxial III-V Material Growth," *Energy Procedia*, vol. 92, pp. 617–623, 2016.

[6] E. García-Tabarés, J. A. Carlin, T. J. Grassman, D. Martín, I. Rey-Stolle, and S. A. Ringel, "Evolution of silicon bulk lifetime during III–V-on-Si multijunction solar cell epitaxial growth," *Prog. Photovoltaics Res. Appl.*, vol. 24, no. 5, pp. 634–644, May 2016.

[7] C. Zhang, Y. Kim, N. N. Faleev, and C. B. Honsberg, "Improvement of GaP crystal quality and silicon bulk lifetime in GaP/Si heteroepitaxy," *J. Cryst. Growth*, vol. 475, pp. 83–87, 2017.

[8] K. Yagi, N. Miyamoto, and J. Nishizawa, "Anomalous Diffusion of Phosphorus into Silicon," *Jpn. J. Appl. Phys.*, vol. 9, no. 3, pp. 246–254, Mar. 1970.

[9] C. Zhang, E. Vadiee, R. R. King, C. B. Honsberg, Carrier Selective Contact GaP/Si Solar Cells Grown by Molecular Beam Epitaxy, submitted to *Journal of Materials Research*.

Spin Coated Nickel Oxide and Vanadium Oxide Layers on Silicon for a Carrier Selective Contact Solar Cell

Jing Zhao, Fa-Jun Ma, Jae-Yun, Anita Ho-Baillie and Stephen Bremner

Australia Centre for Advanced Photovoltaics, School of Photovoltaic and Renewable Energy Engineering, University of New South Wales, Sydney, NSW 2052, Australia

Abstract — The suitability of two transition metal oxides, nickel oxide and vanadium oxide as carrier selective contacts to silicon were examined. Nickel oxide and vanadium oxide layers on silicon were fabricated by spin-coating and characterized to provide materials parameters as inputs for computer simulations. According to a model we developed in Sentaurus TCAD, the band diagrams of NiOx-Si and VOx-Si solar cells were plotted, followed by a detailed discussion of the transport mechanism. The results demonstrated the potential for fabricating high efficiency carrier selective contact solar cells based on this low-cost and simple method.

Index Terms — Nickel Oxide, Vanadium Oxide, amorphous silicon, silicon photovoltaic cells, spin coating.

I. INTRODUCTION

Following recent excellent device results obtained using amorphous silicon as carrier selective contacts [1], the heterojunction structure is being widely studied. The complex fabrication process and optical losses associated with the parasitic absorption of the amorphous silicon layers are limitations that are sought to be overcome [1]. One alternative is the use of transition metal oxide (TMO) layers instead of a-Si. Transition metal oxide like molybdenum oxide, vanadium oxide, tungsten oxide and nickel oxide are excellent candidates as carrier-selective contacts [2]. TMOs have been widely used in dye-sensitized/organic photovoltaics as hole transporting layer (HTL) or electron transporting layer (ETL) [2] . Such materials have also recently been reported in conjunction with p- and n- type c-Si [3].

As compared with MoOx, less work has been done on NiOx and VOx as carrier selective contacts for silicon solar cells. Stoichiometric NiO and V2O5 are ideal insulators, but their sub-stoichiometric NiOx and VOx show the semi-metallic properties. The electrical properties of NiOx and VOx can be tuned when their stoichiometry is altered by the oxygen defects. Oxygen vacancies act as n-type dopants and raise the Fermi level, thus reducing the work function. Conversely, oxygen interstitials can increase the oxide's electron chemical potential [2]. This principle can be used to tune the energy alignment of an oxide. NiOx is known to form oxygen interstitial defects, whereas VOx is known to form oxygen

vacancy defects [2]. In this article, the performances of NiOx-Si or VOx-Si carrier selective contact (CSC) heterojunction solar cells was modeled to gain insight into carrier transport and the general functioning of CSC solar cells. The p-type a-Si:H of the standard heterojunction with intrinsic thin layer (HIT) solar cell (p-aSi:H/ i-aSi:H/ Si/ i-aSi:H/n-aSi:H) was replaced by the sol-gel processed NiOx and VOx to fabricate the CSC solar cells. The two-dimensional numerical modelling program Sentaurus was used for this purpose. To obtain input parameters for the model, material characterizations were carried out.

II. SIMULATION AND EXPERIMENT

A. Characterization of Spin-Coated NiOx and VOx

In this work, NiOx and VOx films were prepared by sol-gel dip coating method. Spin coating offers a low-cost, low-temperature and easy method to achieve thin and uniform NiOx and VOx coatings. Nickel acetate tetrahydrate was dissolved in ethanolamine ($0.1 \ mol \cdot L-1$) to make the 4wt% NiO precursor solution for characterization. The solution was stirred at 60° for 12h to obtain the sol-gel. Finally the film was annealed for 1h at 500°. Vanadium Oxide was spin-coated from isopropanol solution of vanadium oxitriisopropoxide at 1:10 volume ratio. The film was annealed for 20mins under 100°. Both films were coated on silicon substrate at a withdrawal speed of 3000rpm/s for 30s.

The Omega-Two Theta scan by X-Ray Diffraction (XRD, PANalytical Empyrean Thin-Film XRD) was used to verify the presence and crystal structure of NiOx and VOx. Structural analysis of the NiOx films was carried out and the diffraction patterns of films were recorded by varying diffraction angle (2θ) in the range of 30-50°. From Fig. 1(a), two peaks were observed at 2 θ =37.290° , 43.270° in XRD pattern. The observed pattern matched the standard XRD pattern reported for Rhombohedral NiO very well (ICDD: 04-011-2340). These peaks were attributed to cubic crystalline structure with preferred orientation along (111) and (200) plane. The absence of impurity peaks in all of the XRD patterns suggested the high purity of the nickel oxide. The XRD pattern of VOx showed no well-defined sharp peaks, indicating VOx annealed under 100° was amorphous in this work.1

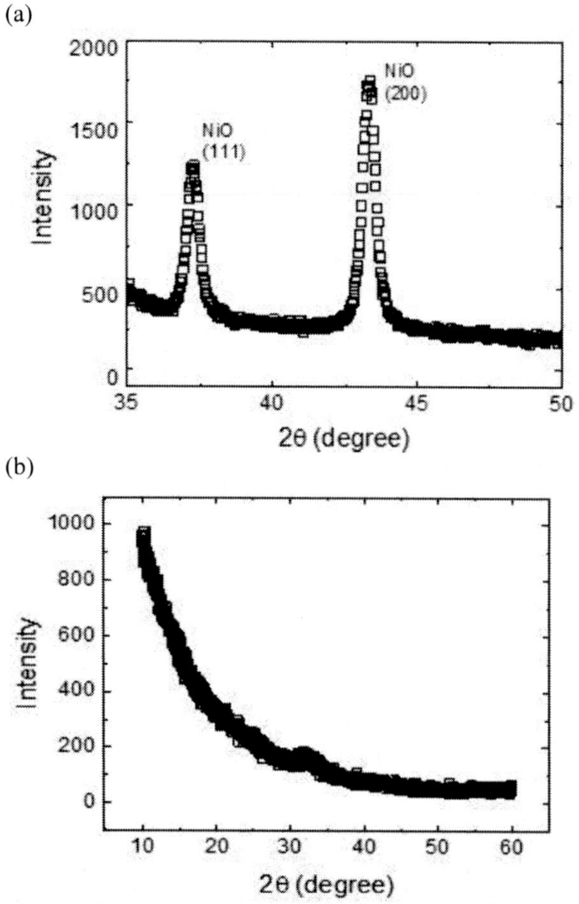

(a)

(b)

Fig.1. X-Ray Diffraction patterns of spin-coated (a) NiOx and (b) VOx

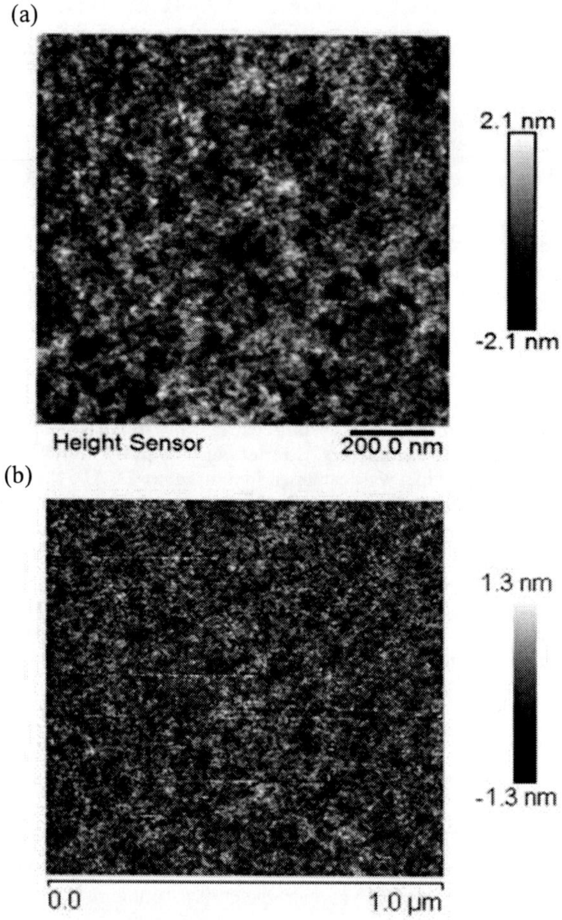

(a)

Height Sensor   200.0 nm

2.1 nm

−2.1 nm

(b)

1.3 nm

−1.3 nm

0.0   1.0 μm

Fig.2. AFM images of NiOx (a) and VOx (b)

In this experiment, optical constants (complex refractive index n&k) in the wavelengths range of 380-1680nm measured by Spectra Ellipsometer were extracted with double Tauc-Lorentz oscillators model for NiOx and triple Lorentz oscillators model for VOx. The optical bandgap were determined from the energy intercept by extrapolating the linear portion of the plot of the square of (absorption coefficient multiply photon energy) versus photon energy hv to absorption coefficient equals to zero. In our case, the optical bandgap for NiOx was estimated to be 3.6 eV and for VOx was 2.7 eV. Compared with p-aSi:H (1.75eV bandgap [1]), the large bandgap of NiOx and VOx promise the great potential to achieving a low optical loss associated with the parasitic absorption of the front side of solar cells.

We used atomic force microscopy (AFM, Bruker Dimension ICON SPM) and scanning electron microscope (SEM, FEI Nova Nano SEM 230) to characterize the morphology of the CSC films. The AFM root mean square roughness (RMS) values obtained for NiOx and VOx were around 0.518 and 0.469 nm respectively. Fig. 2 and Fig. 3 showed that these films were uniformly deposited and smooth after spin coating and annealing of the sol–gel precursor.

(a)

(b)

Fig.3.   SEM images of NiOx (a) and VOx (b)

Hall measurement was taken to measure the electrical parameters of NiOx and VOx with results summarized in Table I. The sign of bulk concentration (+: p type; —: n type) confirmed that non-stoichiometry NiOx is a p type semiconductor and VOx is an n type semiconductor.

TABLE I
ELECTRICAL PARAMETERS OF NiOX AND VOX FILMS

	Bulk concentration (cm^{-3})	Resistivity (Ω.cm)	Carrier mobility (cm^2/Vs)
NiOx	+5.01×10^{16}	99.8	1.21
VOx	-2.07×10^{14}	63.7	0.18

The work function of TMO was also measured using Kelvin Prove Force Microscope (KPFM). KPFM measures a contact potential difference (CPD) between the sample surface and the tip. The CPD (V_{CPD}) between the tip and sample is defined as:

$$V_{CPD} = \frac{\Phi_{tip} - \Phi_{sample}}{e} \qquad (1)$$

, where Φ_{tip} and Φ_{sample} are the work functions of the sample and tip. By measuring the CPD of the silicon substrate and NiOx film separately, the work function of NiOx can be extracted by:

$$V_{CPDSi} - V_{CPDNiO} = \frac{\Phi_{tip} - \Phi_{Si}}{e} - \frac{\Phi_{tip} - \Phi_{NiO}}{e} = \frac{\Phi_{NiO} - \Phi_{Si}}{e} \qquad (2)$$

, where the work function of the silicon substrate used in this study was ~4.95eV. The above equations are also applied to VOx sample. This procedure resulted in an estimated work function of 5.15 eV for NiOx and 6.14 eV for VOx.

B. Advanced Models for CSC-Si Solar Cells

Predictive models for NiOx-Si and VOx-Si CSC solar cell that accepted electrical and optical properties of NiOx and VOx was developed using Sentaurus technology computer-aided design (TCAD). For reduced computation time, optical modelling was decoupled and its solution was fed to the subsequent electrical simulation. The state-of-the-art models highlighted by Altermatt were applied in simulation to accurately predict silicon characteristics [4]. The model had the structure: ITO/ p-aSi:H, NiOx or VOx/i-aSi:H /textured n-cSi/ i-aSi:H/n++-aSi:H. The thickness of ITO as the anti-reflection coating was 130nm. For CSC solar cell, thermionic emission model was used to compute currents and energy fluxes. And a comparison with experimental and simulated HIT solar cell (p-aSi:H/ i-aSi:H/ Si/ i-aSi:H/n-aSi:H) performance was used to check the validity of the model. The input parameters of the HIT solar cell (p-aSi:H/ i-aSi:H/ Si/ i-aSi:H/n-aSi:H) were taken from the previous report [1].

The key optical and electrical input parameters of NiOx and VOx such as bandgap were taken from our measurement results as shown in Table II. Additional parameters like conduction and valence band density of states were taken from the literature values.

TABLE II
TCAD HETEROJUNCTION MODEL PARAMETERS AND THEIR VALUES (FOR THEIR SOURCES, EXP. DENOTES EXPERIMENT; REF. REFERENCE)

Description	Symbol	Unit	Nickel Oxide	Vanadium Oxide
Band gap	Eg	eV	3.6 (Exp.)	2.7 (Exp.)
Work function	ϕ	eV	5.15 (Exp.)	6.14 (Exp.)
Electron affinity	χ	eV	2.10 (Ref.) [5]	5.00 (Ref.) [6]
Conduction band density of states	N_C	cm^{-3}	2.50×10^{19} (Ref.) [7]	1.30×10^{20} (Ref.) [6]
Valence band density of states	N_V	cm^{-3}	2.14×10^{19} (Ref.) [7]	3.68×10^{20} (Ref.) [6]
Carrier mobility	μ	cm^{2}V^{-1}s^{-1}	1.21 (Exp.)	0.18 (Exp.)
Complex refractive index	$n \& k$	NA	(Exp.)	(Exp.)

With the above model parameters, the band diagrams of p-aSi:H-Si, NiOx-Si, VOx-Si solar cells were plotted.

Although the VOx film is n-type material, its high work function builds up enough electric field to transport the photo-generated holes into the VO film. Holes transit through the oxygen vacancy- derived defect states in VOx and be extracted through the metal contact.

As shown in Fig. 4(a-c), a salient feature is the asymmetry of conduction and valence band offsets at the p-aSi:H/n-Si, NiOx/n-Si and VOx/n-Si interface, which makes it possible to achieve high conversion efficiency without the use of heavy

978-1-5090-5606-4/17 $31.00 © 2017 IEEE

doping in the silicon. Transition metal oxide (NiOx and VOx) / p-aSi:H and silicon forms a straddling gap to establish a flat Fermi level at thermal equilibrium condition. Notable energy discontinuities are present for both the conduction band and valence band, which can be quantified by band offset ΔE_C and ΔE_V, respectively. A large ΔE_C leads to a strong upward band bending beneath the silicon surface. The high built in potential leads to the formation of a strong inversion layer in the n-Si substrate next to the NiOx/VOx/p-aSi. The large barrier for electrons resulting from the band bending in n-Si and the conduction band offset between n Si and p graphene blocks the transport of electrons. The comparatively small valence band offset facilitates holes to transit through p-aSi:H/NiOx or the oxygen vacancy- derived defect states in VOx to be collected by the front contact, which shows the good hole-transporting (also electron-blocking) properties of both NiOx and VOx .

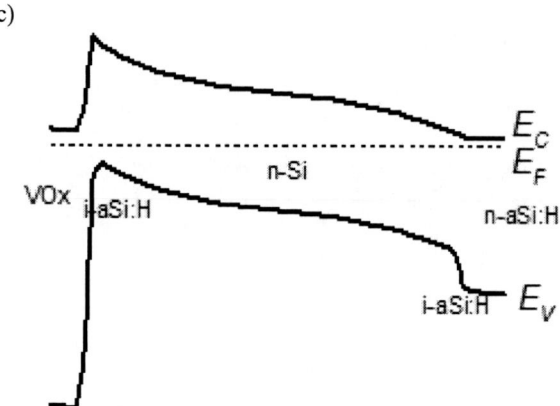

(a)

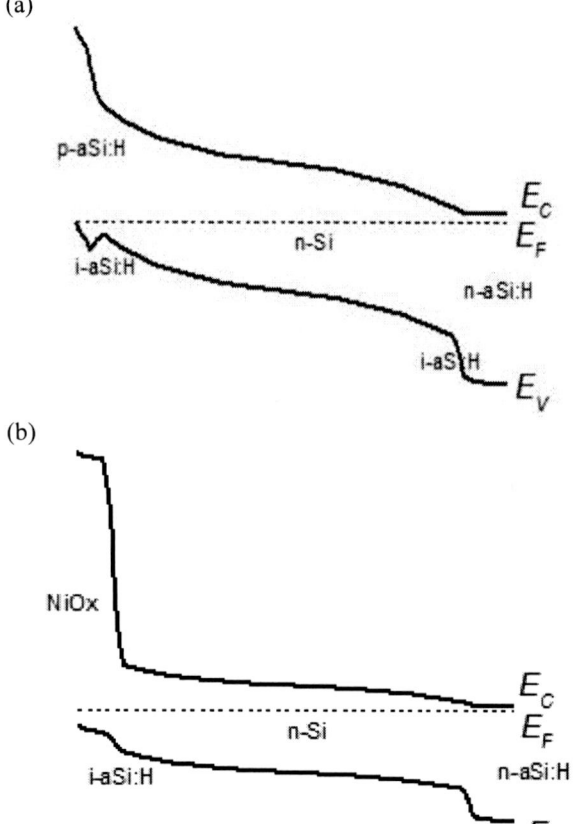

(b)

Fig.4. Schematic band diagram of the heterojunction between (a) p-aSi:H (b) NiOx (c) VOx and silicon under thermal equilibrium condition (d) simulated light J-V characteristics of HIT, NiOx and VOx-Si solar cells

III. CONCLUSION

To conclude, we examined the feasibility of nickel oxide and vanadium oxide as carrier selective contacts to silicon solar cells. Nickel oxide and vanadium oxide layers on silicon were fabricated by spin-coating and characterized to provide materials parameters as inputs for computer simulations. By modelling the band diagram and illuminated current-voltage of Transition Metal Oxide-CSC solar cells, we corroborated that nickel oxide and vanadium oxide have great potential of achieving high carrier selectivity.

ACKNOWLEDGEMENT

The authors acknowledge support from US Department of Energy, FPACEII program contract DE-EEOOO6335 and UNSW Faculty of Engineering.

REFERENCES

[1] M. Rahmouni, A. Datta, P. Chatterjee, J. Damon-Lacoste, C. Ballif, and P. R. i. Cabarrocas, "Carrier transport and sensitivity issues in heterojunction with intrinsic thin layer solar cells on N-type crystalline silicon: A computer simulation study," *Journal of Applied Physics,* vol. 107, p. 054521, 2010.

[2] S. Chen, J. R. Manders, S.-W. Tsang, and F. So, "Metal oxides for interface engineering in polymer solar cells," *Journal of Materials Chemistry,* vol. 22, pp. 24202-24212, 2012.

[3] C. Battaglia, S. M. d. Nicolás, S. D. Wolf, X. Yin, M. Zheng, C. Ballif, *et al.*, "Silicon heterojunction solar cell with passivated hole selective MoOx contact," *Applied Physics Letters,* vol. 104, p. 113902, 2014.

978-1-5090-5606-4/17 $31.00 © 2017 IEEE

[4] P. P. Altermatt, "Models for numerical device simulations of crystalline silicon solar cells—a review," *Journal of Computational Electronics,* vol. 10, p. 314, 2011.

[5] H. Imran, T. M. Abdolkader, and N. Z. Butt, "Carrier-Selective NiO/Si and TiO_2/Si Contacts for Silicon Heterojunction Solar Cells," *IEEE Transactions on Electron Devices,* vol. 63, pp. 3584-3590, 2016.

[6] J. Meyer, K. Zilberberg, T. Riedl, and A. Kahn, "Electronic structure of Vanadium pentoxide: An efficient hole injector for organic electronic materials," *Journal of Applied Physics,* vol. 110, p. 033710, 2011.

[7] M. D. Irwin, J. D. Servaites, D. B. Buchholz, B. J. Leever, J. Liu, J. D. Emery*, et al.*, "Structural and Electrical Functionality of NiO Interfacial Films in Bulk Heterojunction Organic Solar Cells," *Chemistry of Materials,* vol. 23, pp. 2218-2226, 2011/04/26 2011.

Quantification of PV Module Discoloration using Visual Image Analysis

Shashwata Chattopadhyay[1], Chetan Singh Solanki[1], Anil Kottantharayil[1], K.L. Narasimhan[1], Juzer Vasi[1],
Sai Tatapudi[2], Govindasamy TamizhMani[2]

[1]National Centre for Photovoltaic Research and Education (NCPRE), Indian Institute of Technology
Bombay, Mumbai, 400076, India

[2]Arizona State University, Photovoltaics Reliability Lab (ASU-PRL), Mesa, AZ 85212, USA

Abstract — Discoloration of encapsulant is one of the primary causes behind the decrease in power output of field aged PV modules. It would be beneficial to develop a method for quantifying the discoloration based on visual photography and establish a correlation with the power degradation. We hereby describe techniques for analyzing visual images to quantify the discoloration, with the ultimate aim of estimating the short circuit current degradation. Our analysis shows that it is possible to estimate the Yellowness Index of discolored areas in PV modules from the visual images, and it correlates well with the reduction in short circuit current density (calculated from quantum efficiency measurements).

Index Terms — Degradation, encapsulation, photovoltaic modules, silicon, reliability.

I. INTRODUCTION

Encapsulant discoloration is one of the major causes for degradation in PV module performance. It reduces the optical transmittance of solar radiation reaching the solar cells encapsulated inside the PV module thereby reducing the short circuit current and consequently the power output from the module [1]. Ethylene Vinyl Acetate (EVA) is the most commonly used encapsulant in PV modules, though it is well known that it degrades and discolors upon long term exposure to UV radiation (naturally present in sunlight) and high module operating temperatures [2]. Manufacturers add various additives like UV absorbers and anti-oxidants to mitigate these effects. The encapsulants can still degrade within a short period of field exposure, often due to improper manufacturing practices (like insufficient lamination time, or use of expired raw materials beyond their shelf life), or even due to improper mix of additives [2]. Field surveys in India have shown many such cases of early discoloration of the encapsulant within first 5 years of outdoor exposure [3][4]. The pressure for module price reduction from the marketplace is forcing many manufacturers to experiment with low-priced materials which may be of lower grade, and this can lead to faster module discoloration as well. Hence, it becomes important to analyze the encapsulant discoloration in the field-aged PV modules, and understand its impact on the module performance. Visual images can serve as an important tool to quantify the discoloration in the modules, as evident from the works of Mayank Maloo [5] and Sushanth Gudla [6]. This paper gives details of our work in this direction, with the aim to establish an easy method to estimate the short circuit current degradation in discolored modules from the visual images, without interrupting the operation of PV power plant.

II. METHODOLOGY

Digital cameras capture color images by using red, green and blue filters in front of their sensors [7]. The popular JPEG image format also stores images in 3 separate channels (red, green and blue), and it is possible to read the pixel by pixel colour information from the image file as intensities of red, green and blue light. The crystalline silicon solar cells are predominantly dark blue in colour during manufacturing (the actual blue colour depends on the thickness of the anti-reflective coating deposited on the silicon solar cell). As the PV module ages in the field, the discoloration of the encapsulant leads to gradual yellowing, but usually the edges of the solar cells still remain blue in colour (in modules with breathable back-sheets) as the degraded encapsulant undergoes photo-bleaching upon exposure to oxygen diffusing from the edges of the cells [2]. Hence, comparing the color of the encapsulant over the cell, at the centre and the edge, can serve as a useful tool to quantify the discoloration in the module. Table I shows a set of colors starting from blue, and progressively going to full brown. The RGB information of these colors is also given, along with the percentage blue content in the color (which can be called the Blueness Index as defined below).

$$\text{Blueness Index (BI)} = \frac{B}{R+G+B} * 100 \qquad (1)$$

where, R, G and B are the intensities of the red, green and blue channels respectively, in the image.

Since our goal is to determine the extent of discoloration, we define a discoloration index as follows:

$$\text{Discoloration Index (DI}_{BI}) = 1 - \frac{BI_{present}}{BI_{initial}} \qquad (2)$$

where,

$BI_{present}$ = present value of Blueness Index

$BI_{initial}$ = Initial value of Blueness Index

TABLE I
DISCOLORATION INDEX BASED ON BLUE CONTENT

R,G,B	79, 129, 188	86, 122, 170	93, 117, 153	101, 112, 134	107, 105, 116	115, 100, 97	123, 95, 81	129, 90, 61	137, 84, 44	144, 77, 24	151, 72, 6
BI	57.5	51.5	45.6	39.5	33.9	28.4	22.6	17.4	12.9	7.9	2.6
DI$_{BI}$	0	0.11	0.21	0.31	0.41	0.51	0.61	0.70	0.77	0.86	0.95

978-1-5090-5606-4/17 $31.00 © 2017 IEEE

There is an alternative technique to quantify the extent of browning, on similar lines as the Yellowness Index which is widely used to quantify the yellowing of textiles, paints and plastics [8]. It is possible to calculate a Yellowness Index from RGB values, under certain assumptions. One of the important assumptions is regarding the illuminant, and the formula given below assumes the illuminant to be D65. The CIE standard D65 illuminant is meant to simulate the noon-time light from the northern sky, with colour temperature of 6500 K [9] and it is comparable to (though not exactly same as) the AM1.5G spectrum [10]. Since assumptions are involved and it is not a direct measurement using a colorimeter (having a standard D65 illuminant), we refer to this Yellowness Index calculated from the visual image as the *Pseudo Yellowness Index* (PYI).

PYI is calculated by converting the RGB values to the CIE XYZ colour space (formulae given in [11]) and then using the ASTM E313-15 formula for Yellowness Index [12], as given below in Eq. (3). A pre-factor equal to 0.25 is added to this ASTM equation based on the observation that the Yellowness Index obtained from digital image analysis (using the ASTM formula) is about 4 times of the value measured using the colorimeter instrument.

$$PYI = 0.25 * \frac{1.3013\,X - 1.1498\,Z}{Y} * 100 \qquad (3)$$

A discoloration index can again be calculated based on the change in the Yellowness Index of the colour:

$$DI_{PYI} = \frac{PYI_{present} - PYI_{initial}}{PYI_{range}} \qquad (4)$$

where,

$PYI_{present}$ = present value of Pseudo Yellowness Index

$PYI_{initial}$ = initial value of Pseudo Yellowness Index

PYI_{range} = difference in PYI between the worst possible browning and the initial (blue) colour.

PYI_{range} is the normalizing factor which ensures that the discoloration index lies between 0 and 1 (with the worst possible browning denoted by 1). Table II gives the PYI and the discoloration index (DI_{PYI}) values for the same set of colors as in Table I. It can be noted that for PYI values less than -15, the colour is perceived by our eyes as blue and significant browning appears in images with PYI values greater than zero.

TABLE II
DISCOLORATION INDEX BASED ON PSEUDO YELLOWNESS INDEX

R,G,B	79, 129, 188	86, 122, 170	93, 117, 153	101, 112, 134	107, 105, 116	115, 100, 97	123, 95, 81	129, 90, 61	137, 84, 44	144, 77, 24	151, 72, 6
PYI	-57.0	-43.0	-29.3	-14.3	-1.6	9.8	19.7	26.5	32.0	36.8	39.8
DI_{PYI}	0	0.15	0.29	0.44	0.57	0.69	0.79	0.86	0.92	0.97	1

Three PV modules are chosen for this study, such that there is a progressive increase in the extent of discoloration. These modules had been exposed to outdoor environment for more than 10 years. A 12 Mega Pixel digital camera (SONY Cybershot) and a 13 Mega Pixel mobile phone camera (ASUS Zenfone Max) have been used for taking digital photographs of these PV modules. Software has been developed in Octave (which is an open source alternative to Matlab) to compute the Blueness Index and the Pseudo Yellowness Index for a JPEG image, pixel by pixel, and store as a 2 dimensional matrix. These values are shown for points along the centre-line for a solar cell in Fig. 1(a) & 1(b) (overlaid on the image of the cell). This solar cell is located above the name plate in the PV module (which seems to be the cause for the asymmetrical discoloration). Fig. 1(c) shows the false color image formed from the Yellowness Index values, and it is clear that this false color image mimics the discoloration in the actual visual image of the solar cell. In the false colour image, green is used for PYI values less than -15 (so corresponds to the blue areas of the solar cell), and the colour progresses to yellow and then red for higher values of PYI.

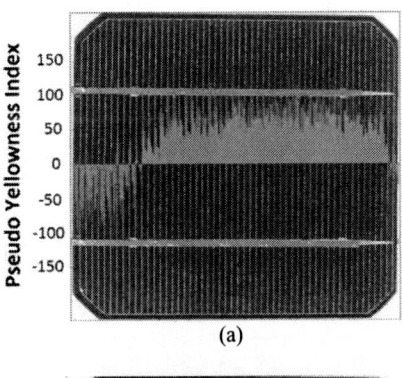

(a)

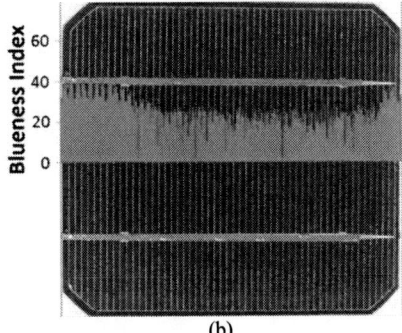

(b)

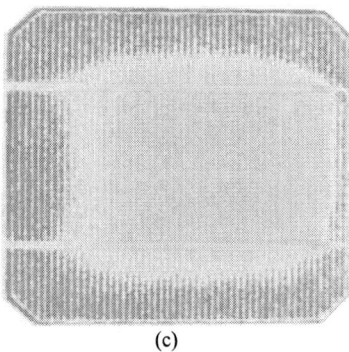

(c)

Fig. 1. (a) Pseudo Yellowness Index of a discolored solar cell (for pixels along the centre line)
(b) Blueness Index of the discolored solar cell (along centre-line)
(c) False color map based on the Pseudo Yellowness Index

III. CORRELATIONS

The Pseudo Yellowness Index is compared with the actual Yellowness Index measured by a colorimeter (Xrite Ci64) and also with the quantum efficiency at different points of the solar cell (obtained by using Module QE machine QEX12M from PV Measurements). The PYI values were computed from digital images of the selected modules, which were taken between 12 noon to 1 PM at Arizona State University on a cloudless day. Fig. 2 shows the comparison with the Yellowness Index measurements, and it can be seen that there is a positive correlation (with correlation coefficient of 0.96). However, the correlation is better at higher degree of browning (refer 1st quadrant of Fig. 2) but not so good if there is no discoloration (refer 3rd quadrant). The Blueness index is also computed for these points. The correlation between the Blueness index and the measured Yellowness Index is also found to be linear, as shown in Fig. 3, but the scatter in the plot is much higher than in the case of Pseudo Yellowness Index. Hence we have used the Pseudo Yellowness Index for further analysis.

Quantum Efficiency (QE) was measured at different points along the centre-line of the discolored solar cell to understand the effect of the discoloration on the short circuit current generation. The various points are marked in Fig. 4(a), and the corresponding short circuit current density (computed from QE) are plotted in Fig. 4(b). The QE curves for these points are shown in Fig. 5. It can be seen that the QE in the wavelength range 400–700 nm reduced significantly in the discolored areas (with a negligible improvement for wavelengths greater than 700 nm). Consequently, the short circuit current density reduces towards the centre of the cell where the discoloration is higher than at the edges.

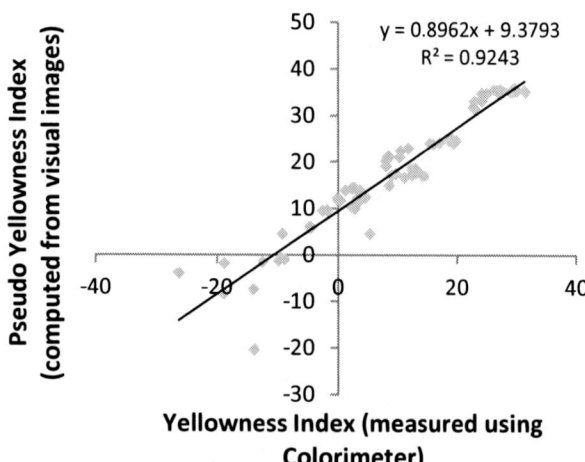

Fig. 2. Correlation of the computed Pseudo Yellowness Index with the measured Yellowness index at different points of solar cells from the three selected modules. (Number of sample points = 70)

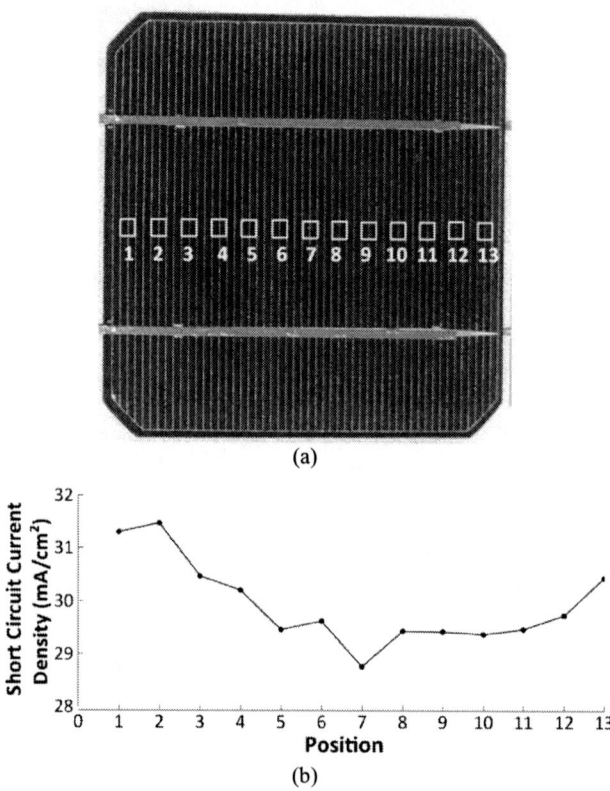

(a)

(b)

Fig. 4. (a) A solar cell in one of the selected PV modules, and (b) Short circuit current density (computed from QE) at various points of the solar cell (Number of sample points = 13)

Fig. 6(a) shows the plot of QE versus the PYI, for 500 nm and 600 nm wavelengths. The relation is linear, with different slopes at different wavelengths. Based on the PYI, the Discoloration Index is calculated using Eq. 4, in which $PYI_{range} = 40 - PYI_{initial}$. The initial PYI value ($PYI_{initial}$) is assumed equal to the present PYI value at the edges of the cell (where photo-bleaching action has prevented the discoloration effect). The maximum possible PYI value seen in any solar cell in this study is ca. 40, so it is considered as the upper limit in this calculation to determine the maximum PYI range. The Quantum Efficiency also follows a linear relation with the Discoloration Index, as shown in Fig. 6(b).

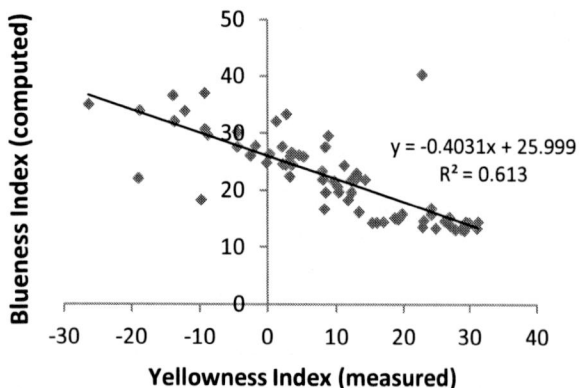

Fig. 3. Correlation of the computed Blueness Index with the measured Yellowness index at different points of solar cells from the three selected modules. (Number of sample points = 70)

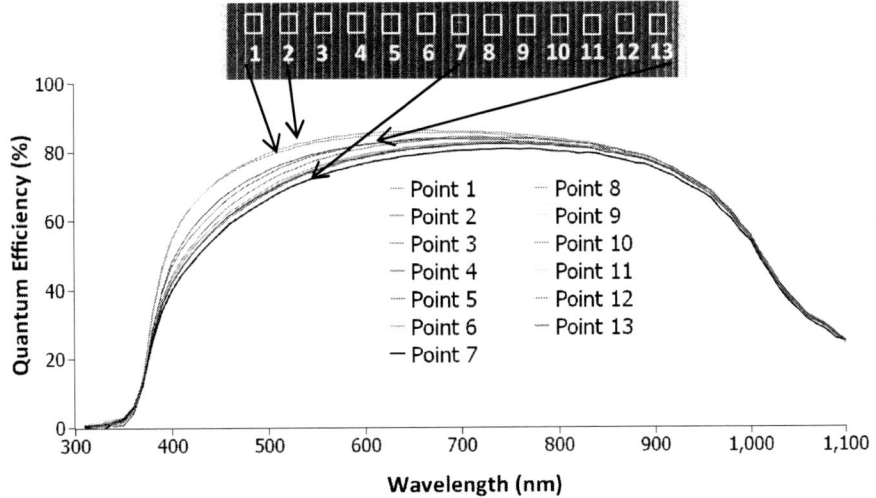

Fig. 5. Quantum Efficiency at different points of the solar cell

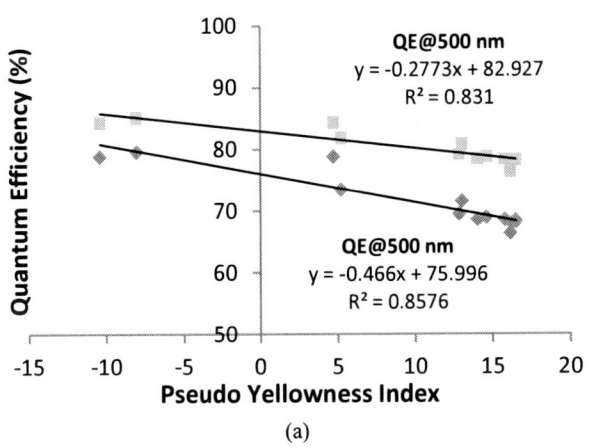

(a)

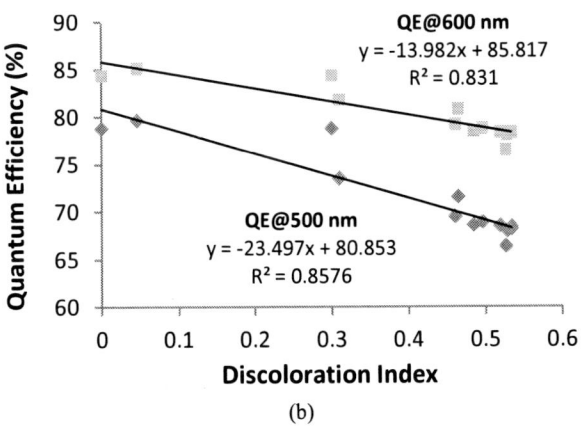

(b)

Fig. 6. (a) Quantum efficiency versus Pseudo Yellowness Index for different points on cell shown in Fig. 4, and (b) Quantum efficiency versus Discoloration Index for different points on the cell (Number of sample points = 13).

The short circuit current loss (computed from the QE data) is plotted against the Discoloration Index, in Fig. 7 and shows a strong correlation. Thus, it is possible to estimate the loss in short circuit current density based on the discoloration index.

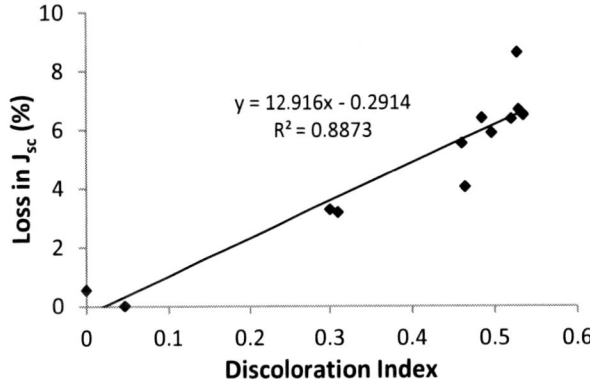

Fig. 7. Short circuit current density loss (computed from QE) versus Discoloration Index at various points of the solar cell.

IV. CONCLUSIONS

Analysis of visual images of PV modules can provide a quantitative indication of the discoloration extent. Our studies indicate that the Pseudo Yellowness Index (PYI) computed from visual image analysis can be used to estimate the actual Yellowness Index for discolored modules. The discoloration Index based on the PYI value has strong correlation with the quantum efficiency. The short circuit current reduction of a discolored cell is also correlated with the discoloration index. This will serve as a useful tool to estimate the decrease in the electrical performance through visual inspection of the modules.

ACKNOWLEDGEMENT

This research is based upon work supported in part by (a) the Solar Energy Research Institute for India and the U.S. (SERIIUS) funded jointly by the U.S. Department of Energy subcontract DE AC36-08G028308 (Office of Science, Office of Basic Energy Sciences, and Energy Efficiency and

Renewable Energy, Solar Energy Technology Program, with support from the Office of International Affairs) and the Government of India subcontract IUSSTF/JCERDC-SERIIUS/2012 dated 22nd Nov. 2012, and (b) the National Centre for Photovoltaic Research and Education funded by Ministry of New and Renewable Energy of the Government of India through the Project No. 31/17/2009-10/PVSC dated 29th September 2010.

REFERENCES

[1] M. Kontges, S. Kurtz, U. Jahn, K.A. Berger, K. Kato, T. Friesen, H. Liu, M.V. Iseghem, "Review of Failure of Photovoltaic Modules", IEA Report No. IEA PVPS T13-01: 2014.[Online] Availabe: http://www.iea-pvps.org/index.php?id=275

[2] A.W. Czanderna and F.J. Pern, "Encapsulation of PV Modules using Ethylene Vinyl Acetate Copolymer as a Pottant: A Critical Review," *Solar Energy Materials and Solar Cells*, 43(2), pp. 101-181, 1996.

[3] R. Dubey, S. Chattopadhyay, V. Kuthanazhi, J. J. John, C. S. Solanki, B. M. Arora, K. L. Narasimhan, A. Kottantharayil, J. Vasi, B. Bora, Y. K. Singh, O. S. Sastry, "All India Survey of Photovoltaic Module Reliability 2014," National Centre For Photovoltaic Research and Education, Mumbai, 2015. [online] Available: http://www.ncpre.iitb.ac.in/research/reports.html

[4] R. Dubey, S. Chattopadhyay, V. Kuthanazhi, A. Kottantharayil, C. S. Solanki, B. M. Arora, K. L. Narasimhan, J. Vasi, B. Bora, Y. K. Singh, O. S. Sastry, "Comprehensive study of performance degradation of field-mounted photovoltaic modules in India," Energy Science and Engineering, 5(1), pp. 51-64, 2017.

[5] M. Maloo, "Encapsulant Discoloration Modelling," M.Tech Thesis, Indian Institute of Technology Bombay, Mumbai, 2015.

[6] S. Gudla, "Quantification of Solar Photovoltaic Encapsulation Browning Level Using Image Processing Tool," M.S. thesis, Arizona State University, Tempe, 2016 [Online]. Available: https://pvreliability.asu.edu/theses

[7] "Digital Color Sensors", [Online]. Available: www.cambridgeincolor.com/tutorials/camera-sensors.htm

[8] Hunterlab, "Application Note – Yellowness Index," [Online] Available: www.hunterlab.se/wp-content/uploads/2012/11/Yellowness-Indices.pdf

[9] CIE Standard Illuminants for Colorimetry, ISO 10526:1999/CIE S005/ E-1998.

[10] M. P. Lamb, W. Yoon, C. G. Bailey, D. Scheiman, J.G.Tischler, and R.J.Walters, " Modelling and Analysis of High Performance, multi-colored, anti-reflective coatings for solar cells," in Optics Express, vol. 21, issue S4, pp. A585-A594, 2013.

[11] "Color Conversion Algorithms", [Online]. Available: https://www.cs.rit.edu/~ncs/color/t_convert.html

[12] Standard Practice for calculating Yellowness and Whiteness indices from Instrumentally Measured Color Co-ordinates, ASTM E313-15, 2015.

Temperature and Power Study of Adhered and Racked Double Glass Photovoltaic Modules

Volker Beutner and Rubina Singh, Cameron Stark

Fraunhofer Center for Sustainable Energy Systems (CSE), Albuquerque, NM, 87106, USA

Abstract— **Frameless, glass-glass photovoltaic (PV) modules have demonstrated superior durability over conventional framed modules. However, their deployment in the residential market has been hampered by limited mounting options. An efficient and relatively unexplored alternative is the use of adhesives to attach the modules to sloped shingled residential roofs. One concern with adhesive mounting is the impact of temperature on module performance due to a reduction in the module/roof gap. This study compares the temperature and performance of three mounting configurations including adhesive mounting of a glass-glass module on a shingled roof. Results indicate an increase of 10.0-15.6°C and a reduction in power of approximately 15 W for the adhesively mounted (no gap) glass-glass module compared with the same module mounted at a gap of 4 and 7 inches.**

Index Terms— **photovoltaic, temperature, outdoor performance, double glass, adhesive, data analysis.**

I. INTRODUCTION

Innovations continue to drive the growth of solar power. One example is the frameless glass-glass module design which has been gaining attention for several reasons. Without the frame, the modules have no exposed metal and thus, do not need grounding. In addition, the risk of potential induced degradation (PID) is eliminated. By reducing moisture ingress [1], the glass-glass construction enables warranties to be extended to 30 years compared with the typical 25-year warranty associated with conventional framed modules [2]. One barrier to the penetration of glass-glass modules into the residential market is the lack of compatible mounting systems. Compared with conventional systems, there are far fewer mounting options for glass-glass modules, though the options are growing. A potential alternative mounting approach is the use of adhesives to attach the module to the roof. Recently, Honeker et al examined the direct adhesive mounting of a frameless, glassless lightweight c-Si module to a shingled roof [3]. One concern with this approach is the effect of an increase in temperature on module performance due to a reduced roof-to-module gap. To address this concern, we decided to study the effect of gap on performance using the glass-glass modules.

This paper presents a performance study of a glass-glass module mounted at three different gap spacings: no gap (adhesively mounted), a four-inch gap and a seven-inch gap. These systems were installed and monitored in Albuquerque, NM for a period of fourteen months. Since cell temperature impacts system performance [4][5], the reduced ventilation of the adhesively mounted system is expected to result in a yield loss compared to the 4-inch and 7-inch systems.

II. EXPERIMENTAL SETUP

Temperature, power and weather data used in this analysis were collected at the Fraunhofer Outdoor Testing Center in Albuquerque, New Mexico. Monocrystalline double glass modules were mounted on test hut roofs at an angle of 26.6 degrees facing south (6/12 pitch). Two modules were mounted at four and seven inch from a shingled roof using a Orion Solar racking system and a third module was mounted using HelioBond PVA 900HM adhesive pads from Royal Adhesives and Sealants. Care was taken to achieve full contact of the adhesive pads with the test hut roof shingles. The test huts were designed to simulate residential housing. The interior space was kept at 24.5°C by a heater/AC unit and an insulated attic separated the interior space from the roof. Current and voltage measurements were collected at a one minute frequency using individual ET Instrumente GmbH ESL-500 IV curve tracers. A Kipp and Zonen pyranometer was placed near the modules and plane of array (POA) irradiance data was collected at one minute frequency. Last, a nearby weather station collected ambient temperatures at 30 second intervals. Figure 1 shows the two test huts with the mounted modules as well as the pyranometer.

III. RESULTS

A. ECT Calculation

The equivalent cell temperature (ECT) was calculated following IEC 60904-5 with two modifications. First, the effect of shading at sunset from the mounting clips is removed by filtering out all values below 500 W/m^2 irradiance. Second, the standard equation to calculate ECT from Voc and irradiance (equation 1):

$$ECT = T_{stc} + \frac{1}{\beta}\left[\Delta V_{OC} + nV_T N_s ln(\frac{G_{stc}}{G})\right] \quad (1)$$

is modified. Here, T_{stc} and G_{stc} are the standard operating condition (STC) module temperature and irradiance, respectively. β and ΔV_{OC} are the open circuit voltage temperature coefficient and change from V_{OC}, respectively. Finally, n is the ideality factor, N_s is the number of cells in series, V_T is the thermal voltage, and G is the measured irradiance. We simplify (1) to:

$$ECT = T_{stc} + \frac{1}{\beta}\left[\Delta V_{OC}\right], \quad (2)$$

by removing the second term of (1). The second term makes only a small contribution to ECT near irradiance values of 1000 W/m^2. Furthermore, recent studies [6][7] question the

978-1-5090-5606-4/17 $31.00 © 2017 IEEE

validity of this as the ideality factor has been found to be a function of temperature.

Figure 2 presents ECT values over the entire measurement period. The amplitude variation is attributed to diurnal and seasonal weather variations. Dropouts are due to equipment maintenance/troubleshooting as well as filtering. Clearly, the adhered glass-glass module experiences a higher temperature than either the 4-inch or 7-inch mounted modules. The racked modules experience quite similar temperatures.

Fig. 1. The adhesively mounted glass-glass module is attached to the left test hut. The 4-inch and 7-inch mounted glass-glass modules are attached to the right test hut. The plane of array (POA) pyranometer is fixed on the left side of the left test hut.

B. Temperature analysis

Figure 3 plots the hourly averaged ECT vs. the ambient temperature as well as the corresponding linear fits. By comparing the fits, we draw several conclusions. First, the adhered module yields a slope over 10% larger than the racked modules indicating a greater sensitivity to ambient temperature. The intercept of the adhered module is also approximately 10°C higher. The increase in average temperature due to adhesive mounting is thus found to range from 10°C at 0°C ambient to 15.6°C at 35°C ambient. Comparison of the fits for the 4-inch and 7-inch modules shows only a slightly lower temperature for the 7-inch modules at lower ambient temperature. The average difference in temperatures between the 4-inch and 7-inch modules ranges from 1°C at 0°C ambient to approximately 0°C at 35°C ambient. Thus, this analysis suggests that the ECT values at higher ambient temperatures are essentially identical. Note that these comparisons are based on linear fits to the data. Although the fit quality is reasonable, (R^2 values of 0.58, 0.67 and 0.74) there may be subtleties, such as thermal lag, that distinguish these mounting configurations, but are averaged out in the fitting.

C. Power Analysis

A second approach to examining the effect of mounting on module performance is to analyze the power output directly. To eliminate the effect of irradiance we scale the power values to STC conditions:

$$Power_G = Power \times \frac{G_{stc}}{G}. \tag{3}$$

Figure 4 plots the corrected power as a function of ambient temperature. The figure also includes the data frequency distribution to indicate that power values at high ambient temperature have greater statistical significance. In all cases the power values decrease with increasing ambient temper-

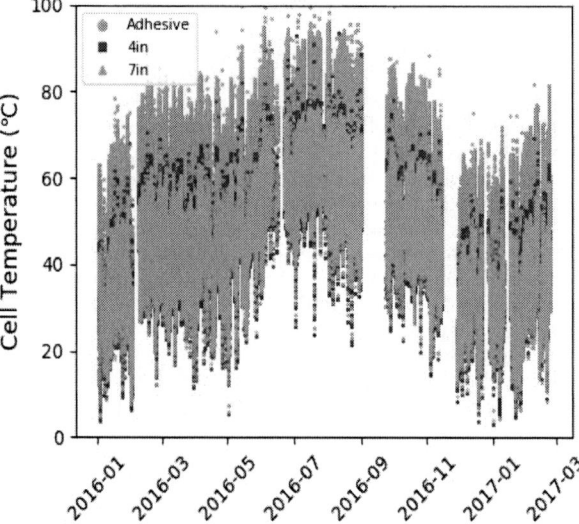

Fig. 2. Cell temperature time series spanning from January 2016 into Feburary 2017. Cell temperature calculated using equation 2.

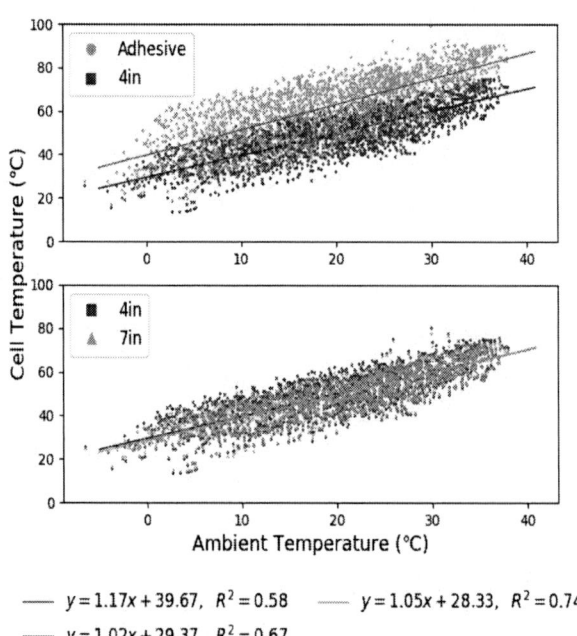

— $y = 1.17x + 39.67$, $R^2 = 0.58$ —— $y = 1.05x + 28.33$, $R^2 = 0.74$
—— $y = 1.02x + 29.37$, $R^2 = 0.67$

Fig. 3. ECT vs. ambient temperature with corresponding linear fits. Top: Comparison of the adhesively mounted module with the 4-inch racked module. Bottom: Comparison of the 4-inch and 7-inch racked modules.

ature with approximately the same slope. The power drop is consistent with the increase in ECT shown in Figure 3. The power output from the adhesively mounted module is approximately 15 W (5% of nameplate power) lower than from the racked modules at all ambient temperatures.

Figure 4 also shows that the 7-inch module produces slightly more power than the 4-inch module at all ambient temperatures. The average increase in power in the 7-inch module compared with the 4-inch module is 1.3%.

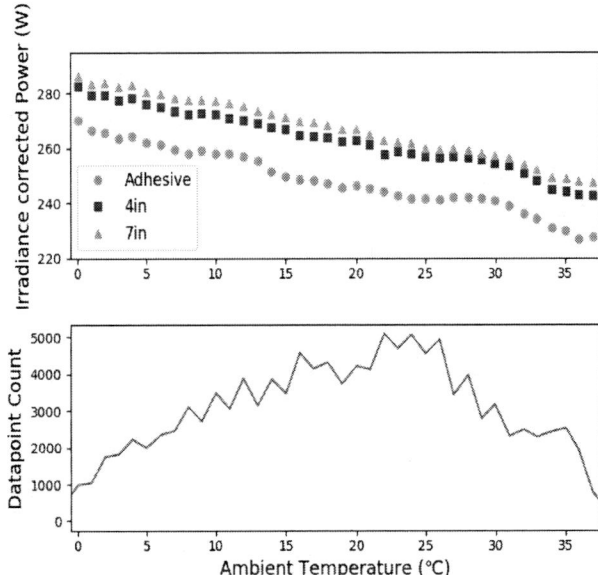

Fig. 4. Top: Irradiance corrected power (ICP) vs ambient temperature for the three mounting configurations. The irradiance correction was performed using equation 3. The data was binned into integer temperature steps. Bottom: Frequency distribution of all the power data

IV. CONCLUSION

The superior performance, durability and aesthetics of glass-glass modules have increased their market appeal. The relatively limited mounting options have lead us to investigate the possibility of adhesive mounting as an alternative. Flush mounting of PV modules to residential roofs is known to increase module temperature reducing module output, however. Here we investigate the magnitude of the temperature increase and corresponding power loss for three gap spacings, no gap (adhesive mounting), 4-inch gap and 7-inch gap for the same glass-glass module in Albuquerque, New Mexico. The effective cell temperature (ECT) was estimated from the open circuit voltage. The adhered module was found to be 10.0-15.6°C higher in temperature than the racked modules. The temperature difference increases with increasing ambient temperature. The 7-inch module was found to be only slightly cooler (1°C) than the 4-inch module at low ambient temperature. A second analysis, in which the irradiance corrected power (ICP) was compared, showed similar results. The

adhesively mounted module showed a power loss of 5% of nameplate power compared with the racked modules. In contrast to the temperature analysis this loss is independent of ambient temperature. The power output of the 7-inch module is slightly higher than the 4-inch module.

One previous numerical study found a highly non-linear relationship between module temperature and gap [8]. As gap decreases module temperature was found to increase asymptotically. In light of these results, the 4-inch and 7-inch gaps may reside in the plateau region showing only a weak dependence of temperature on gap. The adhered module would represent the worst-case condition where no significant ventilation is possible.

The increase in temperature and corresponding decrease in power of the adhesively mounted glass-glass module in this study is significant. Note, however, that special effort was made to directly adhere the glass-glass module to the roof resulting in a negligible gap. To enable this, a portion of the roof was cut-away to provide space for the junction box. For practical purposes the adhesive mounting of a glass-glass module would require sufficient space (at least 1.5 inches) for the junction box. In this case, a gap of 1.5 inches (or more) would enable ventilation to substantially reduce module temperature compared to the directly adhered (no gap) case.

ACKNOWLEDGMENT

The authors would like to acknowledge and thank the following people who all had an important part in writing this paper: Mark Hill, for his outstanding work and feedback on the experimental setup. Christian Honeker, for his keen insights and invaluable input. Cordula Schmidt for her continuous optimism, feedback and support.

REFERENCES

[1] S. Wang, et al., "Double-glass PV modules with silicone encapsulation", *in Photovoltaics International*, pp. 2-7, 2016.

[2] P. Peng. *Dual-Glass Modules to See Rising Opportunities in 2016*, [online]. Available: http://pv.energytrend.com/node/print/10265

[3] C. Honeker, et al., "Reducing Installed Costs of Residential Solar by the Use of Adhesive Mounted Lightweight Solar Modules", *in 43rd IEEE Photovoltaic Specialist Conference*, pp. 95-100, 2016.

[4] N. Umachandran, G. TamizhMani, "Effect of Spatial Temperature Uniformity on Outdoor Photovoltaic Module Performance Characterization", *in 43rd IEEE Photovoltaic Specialist Conferece*, pp. 2731-2737, 2016.

[5] F. Softić, Z. Bundalo, A. Stjepanović, "Temperature Characteristics and Energy Efficiency of Solar Cells and Solar Modules", *in 2012 Mediterranean Conference on Embedded Computing*, pp. 288-291, 2012.

[6] G. Yordanov, O. Midtgard, T. Saetre, "Equivalent Cell Temperature Calculation for PV Modules with Variable Ideality Factors", *in 38th IEEE Photolvoltaic Specialist Conference*, pp. 505-508, 2012.

[7] G. Yordanov, O. Midtgard, T. Saetre, "PV Modules with Variable Ideality Factors", *in 38th IEEE Photolvoltaic Specialist Conference*, pp. 2362-2367, 2012.

[8] G. Gan, "Effect of air gap on the performance of building-integrated photovoltaics", *in Energy*, pp. 913-921, 2009.

978-1-5090-5606-4/17 $31.00 © 2017 IEEE

Field Inspection of PV Modules: Quantitative Determination of Performance Loss due to Cell Cracks using EL Images

Carlos A. Rodríguez Castañeda[1], Shashwata Chattopadhyay[2], Jaewon Oh[3], Sai Tatapudi[3], Govindasamy TamizhMani[3], and Hailin Hu[1]

[1]Universidad Nacional Autónoma de México, México
[2] Indian Institute of Technology Bombay, India
[3] Arizona State University Photovoltaic Reliability Lab (ASU-PRL), Mesa, AZ, United States

Abstract — **Electroluminescence (EL) imaging has been used by the operation and maintenance companies for a qualitative detection of cells cracks in the fielded photovoltaic (PV) modules. This paper presents our attempts to statistically determine the inactive area of the cells in a PV module using processed EL images and to quantitatively correlate the remaining active area with the performance data including quantum efficiency (QE) and short-circuit current (I_{SC}). Commercially available image processing tools have been used to extract the statistics of the defective (inactive) areas of the cells in a module. The power parameter against EL-determined active area can have linear, independent or non-linear correlation depending on whether the cells are having microcracks, part cell isolation or shunting which affect V_{OC}, I_{SC} and FF linearly or non-linearly.**

Index Terms — **electroluminescence, photovoltaic cells, short-circuit currents, silicon.**

I. INTRODUCTION

Under forward bias condition, the radiative recombination of charge carriers causes light emission in the solar cells. The emitted light is captured by an electroluminescence (EL) camera and the emission intensity serves as an indicator of the health of the solar cell [1]. The dark areas in the EL images can be due to various reasons like cracks, metallization problems, shunting and/or other defects in the bulk semiconductor. The issues related to design and manufacturing processes are often attributed to be responsible for the reliability issues seen during field operation [2]. EL imaging has emerged as an important tool for inspection of material and manufacturing quality of photovoltaic (PV) modules, both in manufacturing and field [3-4].

Since the current commercial cells are very thin and large, and the module are also large (300 watts) with glass/polymer construction (instead of glass/glass construction), the cells packaged inside the modules often tend to break during manufacturing, handling, shipping and installation. The operations and maintenance (O&M) companies have traditionally been using the EL images for a qualitative assessment of cell cracks in the field. In this paper, we attempt an approach to quantitatively correlate the cell cracks with the performance loss through a correlation between statistically processed EL images and short-circuit current (I_{SC}) and quantum efficiency (QE) data. This correlative study was initiated with a hope to provide a quantitative measure of performance loss just by using statistically processed EL images without the need of current-voltage (I-V) measurements in the field which is intrusive, expensive and labor intensive.

II. CHARACTERIZATION TECHNIQUES

The following characterization techniques were utilized in this work:

Electroluminescence (EL) Imaging: It was performed in order to identify inactive areas in the module (which show up as black areas in the EL images). This test was performed indoor using an EL imaging setup in a darkroom, with the module under forward bias. The EL images were analyzed using Image J software in order to calculate the inactive areas of the solar cells (through pixel intensity statistics).

Cell Quantum Efficiency (QE) Measurement: The QE was measured at various spots on three different solar cells. For this measurement, the backsheet of the module was cut and contact was made to the cell interconnect ribbons. The QE was measured in a cell-module QE measurement system, which is also capable of taking QE of solar cells in a module non-intrusively [5].

Current-Voltage (I-V) Measurement: The I-V measurements of individual cells of the entire modules were carried out by accessing the cell terminals as that of the QE measurement. These I-V measurements were carried out in a cell-module I-V measurement system.

III. IMAGE PROCESSING

A 220 watt module was chosen for the study. Its specifications are given in Table I. The module was forward biased for current flow equal to its I_{SC}, and its EL image was taken using a Sensovation camera (HR-830). The EL image is shown in Fig. 1. The dark lines and patches in this image correspond to the inactive areas. The busbars also show up as

TABLE I
SPECIFICATIONS OF THE PV MODULE USED IN THIS WORK

Electrical Data:	P_{max}: 220 W, V_{OC}: 35.4 V, I_{SC}: 8.36 A
Size:	1.65 m × 0.992 m (1.63 m² area)

978-1-5090-5606-4/17 $31.00 © 2017 IEEE

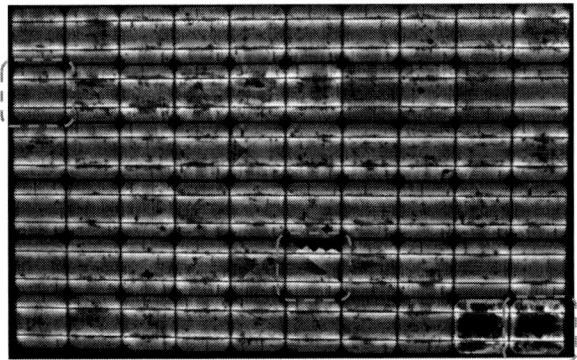

Fig. 1. Original EL image. Grey areas represent active areas on the module and black areas represent inactive parts of the module

dark lines in the EL image, but these busbar areas should not be considered as defective areas. Three solar cells have been highlighted in red in Fig.1, a cell with significant dark (inactive) area in EL image (1st row from bottom, designated as cell #1, with significant shunting due to PID but no cracks), a cell with cracks (2nd row from bottom, designated as cell #16, with cracks and part cell isolation from the circuit but no shunting) and a cell with reasonably good EL image (in 5th row from bottom, designated as cell #50, with no cracks or shunting).

All pixels with intensity higher than the mode of the original EL image histogram were converted to white, while all intensities below it were converted to black. Fig. 2 shows the histogram along with the threshold limit marked in red. It is important to note that if one varies this threshold limit the resultant processed image can turn out to be very different. The processed (binary) EL image is shown in Fig. 3. Based on the dark area in the binary image, the percentage of inactive area is statistically calculated. The total inactive area of the module (based on only black areas inside the solar cells excluding all other areas and busbar areas) is 10.42%.

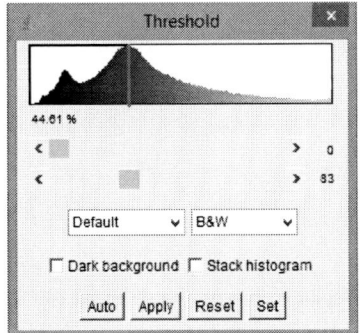

Fig. 2. Threshold function in Image-J, showing the histogram of the original EL image and its modal intensity (highlighted using red line), which was used as the threshold limit.

For the purpose of analysis, each solar cell was divided into 3 zones – A, B and C – with busbars as boundary lines and they are shown in Fig. 4. The percentage of active area is calculated

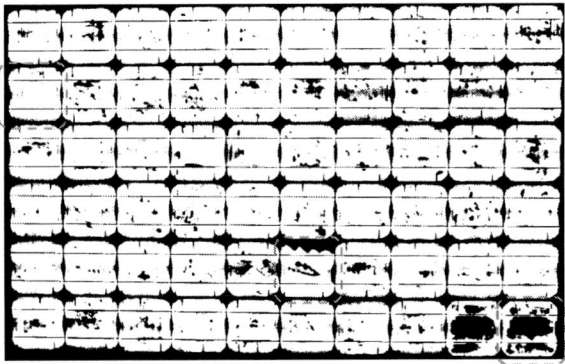

Fig. 3. Processed EL binary image with black and white areas. White areas represent active areas on the module. Black spots in cell areas represent inactive parts of the module.

for each zone of the individual cells. QE was measured at multiple spots in each zone and then averaged to obtain the representative QE for that zone. From this average QE we can calculate an estimated I_{sc} for the respective zone's area:

$$I_{sc, zone} = J_{sc, zone} \times \text{Area of the zone} \qquad (1)$$

Also, we can estimate an ideal I_{sc} for this zone by considering the name plate I_{sc}, as follows:

$$\text{Ideal } I_{sc, zone} = \text{Module } I_{sc} \times \text{Zone area / Solar cell area} \qquad (2)$$

Then the I_{sc} reduction factor can be calculated as follows:

$$\%I_{sc} = I_{sc, zone} / \text{Ideal } I_{sc, zone} \times 100\% \qquad (3)$$

IV. RESULTS AND DISCUSSION

The inactive area for different zones of a solar cell is computed from its processed binary image as shown in Fig. 4. The quantum efficiency is measured in multiple spots in each zone (horizontal line QE scanning in each zone), and the average QE for the three zones of each solar cell is given in Fig. 5. The short circuit current density (J_{sc}) estimated from the average QE curve is used to calculate the effective $\%I_{sc}$. It is plotted against the active area of the respective cell zone in Fig. 6, which shows an expected trend but no good linear correlation for this heavily shunted cell without any cracks.

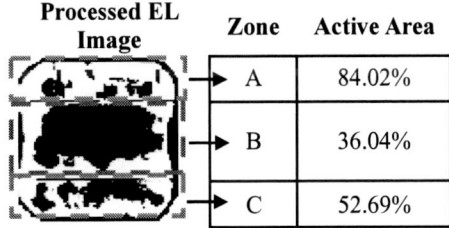

Processed EL Image	Zone	Active Area
	A	84.02%
	B	36.04%
	C	52.69%

Fig. 4. Processed EL Image, showing the active area in each zone of cell #1 (bottom right highlighted cell in Figs. 1 and 3)

978-1-5090-5606-4/17 $31.00 © 2017 IEEE   1859

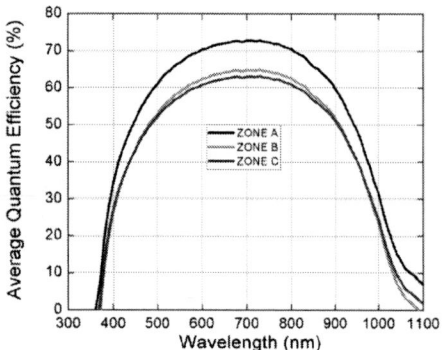

Fig. 5.   Average QE curves for various zones of cell #1.

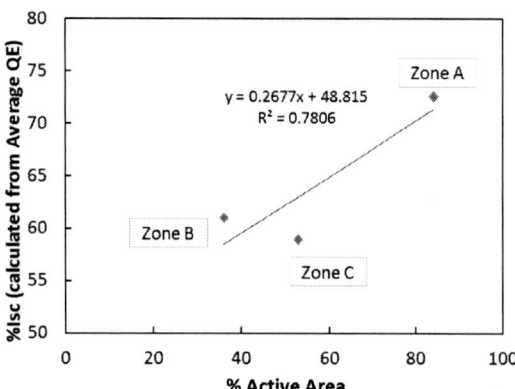

Fig. 6.   Plot of estimated %I_{sc} versus active area of cell #1 (with shunts but without cracks)

This procedure was repeated for a large number of cells in the module. The active area data obtained for the three zones of cell #16 of the same module is shown in Fig. 7. The plot of the estimated %I_{sc} to the percentage active area is shown in Fig. 9. A very good linear correlation between I_{sc} and cell active area can be clearly seen from this figure for this with cracks and part cell isolation from the circuit but without shunts.

Processed EL Image	Zone	Active Area
	A	39.23%
	B	88.4%
	C	94.99%

Fig. 7.   Processed EL Image, showing the active area in each zone of cell #16 (see Figs. 1 & 3: highlighted cell in the second row from the bottom).

As shown in Fig. 10, the influence of the inactive area on the short circuit current has also been investigated using direct I_{sc} measurement or I-V measurement as opposed to integrated QE measurements. As shown in this figure, the short circuit current

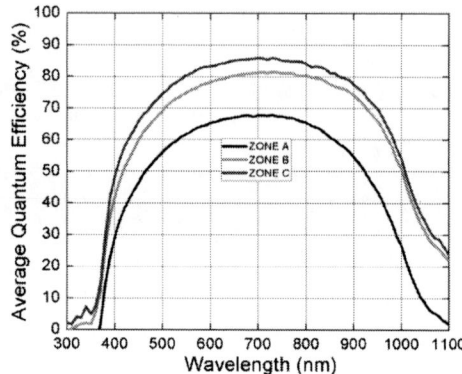

Fig. 8.   Average QE curves for various zones of cell #16.

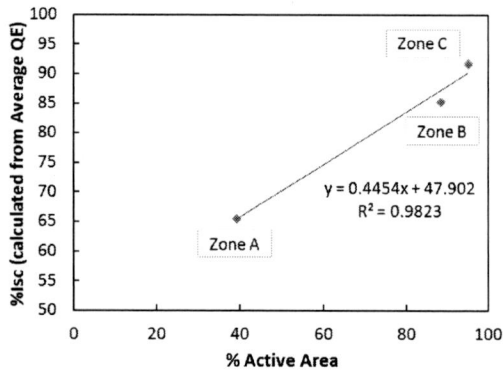

Fig. 9.   Plot of estimated %I_{sc} versus active area of cell #16 (no shunts but with cracks and part cell isolation but not shunts)

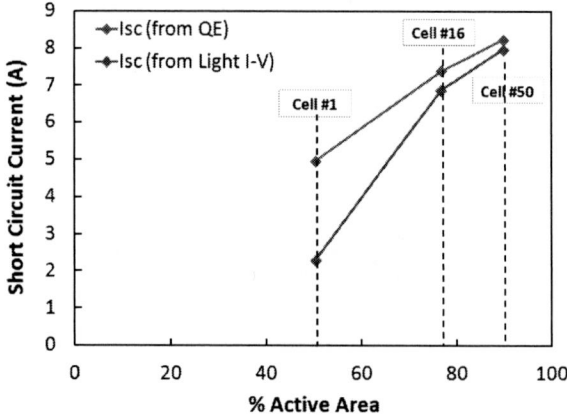

Fig. 10.   Short circuit current (I_{sc}) based on integrated QE and direct measurements, plotted against the percentage active area of cell

derived from the QE measurements is quite close to the actual measured short circuit current for two cells (cells #16 and #50), but there is a significant difference for the 3rd cell (cell #1). This difference is attributed to severe shunting in cell #1. The cell shunting is observed in this cell (a PID susceptible cell located closest to the frame) as it was previously subjected to a short PID (potential induced degradation) stress test. The cell shunting affects the I_{sc} value more than the QE values because

978-1-5090-5606-4/17 $31.00 © 2017 IEEE   1860

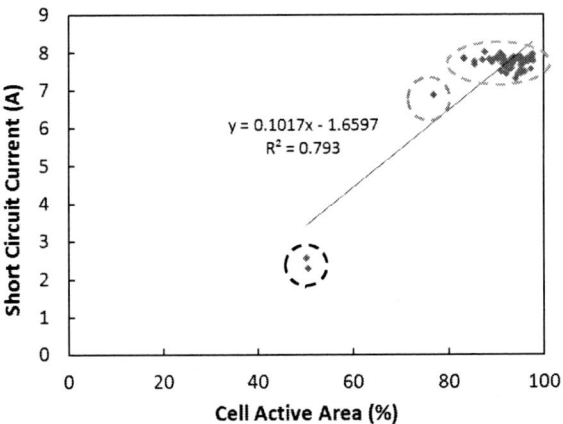

Fig. 11. Short circuit current based on actual measurements versus the percentage active area of the solar cell (black circled cells are shunted cells but no cell cracks; red circled cell is a cracked cell with part cell separation from the circuit but no shunting; green circled cells contain microcracks but absence of part cell isolation or severe shunting)

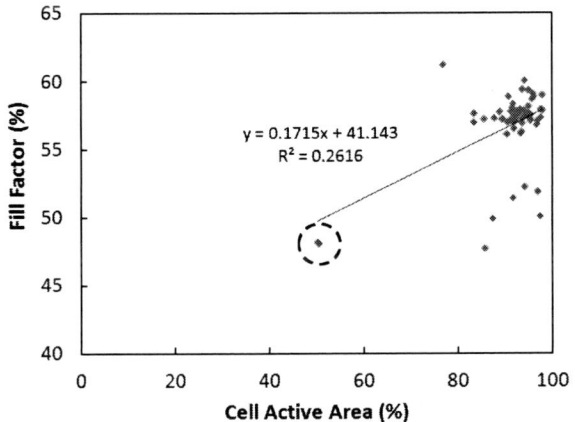

Fig. 12. Fill Factor versus the percentage active area of the solar cell (black circled cells are shunted cells but no cell cracks)

QE measurements were carried out only in three wide zones (more zones needed) in which the top zone is a good quality zone representing ca. 25% of the cell area. Also, the actual measured I_{SC} values of all the cells in the module is plotted against the %Active Area as shown in Fig. 11. Cells with minor finger degradations and benign microcracks do not show much decrease in I_{SC} but cells affected with major cracks with part cell isolation (like cell #16) and heavily shunted cells (like cells #1 and #2) show significant decrease in I_{SC}, which leads to a non-linear correlation. A linear fit has been shown in the figure for reference purpose. This figure indicates that the microcracks do not cause any I_{SC} drop but part cell isolation and severe shunting can cause a significant decrease in I_{SC} as expected.

Fig. 12 shows the plot of the fill factor (FF) versus the %Active area of the cells. The 6 cells showing low FF but high cell Active Area greater than 80% are affected by heavy shunts. Fig. 13 shows that the Open Circuit Voltage is slightly affected

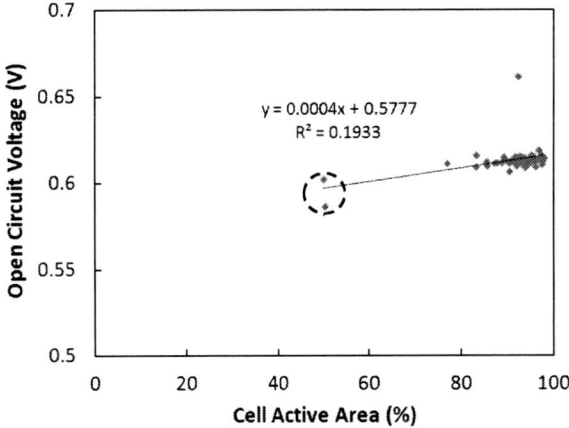

Fig. 13. Open Circuit Voltage versus the percentage active area of the solar cell (black circled cells are shunted cells but no cell cracks)

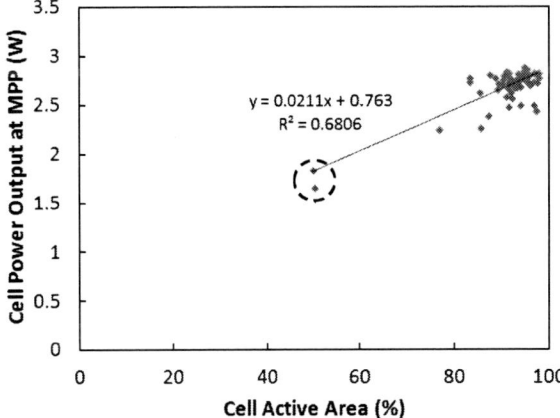

Fig. 14. Cell output power versus the percentage active area of the solar cell cell (black circled cells are shunted cells but no cell cracks)

by the slight loss in cell Active Area but is significantly affected for two cells which are heavily shunted. Finally, Fig. 14 shows that the output power (combined effect of I_{SC}, V_{OC} and FF) is influenced by the cell Active Area, though the relation is not strongly linear due to non-linearity reasons explained above for I_{SC}, V_{OC} and FF. Thus, inactive areas in shunted cells (as detected by EL images) affect the power output by reducing the short circuit current, open circuit voltage and the fill factor. However, the inactive areas in non-shunted cells do not seem to be affected by the microcracks unless the crack is major or part of the cell is completed isolated from the circuit.

V. CONCLUSION

Inactive areas in solar cells due to microcracks, major cracks, part cell isolation or shunts could lead to performance degradation in PV modules. Electroluminescence imaging is an important tool for identification of such inactive areas, but it is often considered as only a qualitative tool. We have attempted to quantify the defective areas provided by the EL images. The cells with microcracks but without part cell isolation or severe shunts, do not seem to affect I_{SC}. The cells with minor shunts do not seem affect FF but the moderate and severe shunts drastically affect FF though the active areas are not significantly affected. The cells with minor and moderate shunts do not seem affect V_{OC} but the severe shunts significantly affect V_{OC}. Since the power output is a combined effect of V_{OC}, I_{SC} and FF, the power parameter against EL-determined active area can have linear, independent or non-linear correlation depending on whether the cells are having microcracks, part cell isolation or shunting. Thus, equipped with the proper image processing tool, it is possible to use EL imaging for semi-quantification of defective area in PV modules, and determination of its impact on module's performance parameters.

ACKNOWLEDGEMENT

This research is based upon work supported by the Solar Energy Research Institute for India and the U.S. (SERIIUS) funded jointly by the U.S. Department of Energy subcontract DE AC36-08G028308 (Office of Science, Office of Basic Energy Sciences, and Energy Efficiency and Renewable Energy, Solar Energy Technology Program, with support from the Office of International Affairs) and the Government of India subcontract IUSSTF/JCERDC-SERIIUS/2012 dated 22nd Nov. 2012.

REFERENCES

[1] Amaury Delamarre, Laurent Lombez, Jean-François Guillemoles, "Characterization of solar cells using electroluminescence and photoluminescence hyperspectral images," J. Photon. Energy. 2(1), 027004, July 2012.

[2] D-C Tseng, Y-S Liu, and C-M Chou "Automatic Finger Interruption Detection in Electroluminescence Images of Multicrystalline Solar Cells," Mathematical Problems in Engineering, February 2015.

[3] R. Ebner, S. Zamini, and G. Újvári "Defect Analysis in Different Photovoltaic Modules Using Electroluminescence (EL) and Infrared (IR)-Thermography," 25th EU-PVSEC Conference, January 2010.

[4] M. Topic, J. Raguse, K. Zaunbrecher, M. Bokalic, and J.R. Sites, "Electroluminescence of Thin Film Solar Cells and PV Modules - Camera Calibration," 26th EU-PVSEC Conference, August 2011.

[5] B. Knisely, J. Kuitche, G. TamizhMani, A. Korostyshevsky and H. Field "Non-Intrusive Cell Quantum Efficiency Measurements of Accelerated Stress Tested Photovoltaic Modules" 40th IEEE Photovoltaic Specialists Conference, June 2014.

Scale Up Designs for Hand-Held Light-Weight TPV DC Power Supply

L. M. Fraas[1], J. E. Avery[1], L. Minkin[1], Hui She[1], L. Ferguson[2]

[1]JX Crystals Inc, Issaquah, WA 98027, USA
[2]C12 Advanced Technologies LLC, Everett, WA, USA

Abstract — Both solar cells and batteries generate quiet DC electric power. Infrared (IR) sensitive GaSb photovoltaic cells can convert energy from a combustion heated glowing ceramic IR emitter into electricity. The result is a lightweight thermophotovoltaic (TPV) battery replacement that operates day and night. We present here the design and operation of a first stand alone TPV generator. Test results on this unit have allowed us to design a scale-up hand-held 10% efficient 50 W TPV generator weighing 1.5 kg. The 50 W scale-up TPV power supply along with 1.5 kg of fuel has a projected weight specific energy density of 645 Wh/kg. This is 4 times larger than for a Li ion battery.

I. INTRODUCTION

Figure 1 shows the design of our portable TPV power supply [1] with the key components labeled. In a TPV unit, a fuel such as propane or butane is burned in a heater or furnace and a ceramic element is located in the flame. The ceramic element emits intense infrared radiation and a photovoltaic array surrounding this emitter converts this infrared energy into electric power. Figures 2, 3, 4, and 5 show the first actual fabrication of an operating portable light weight TPV power supply.

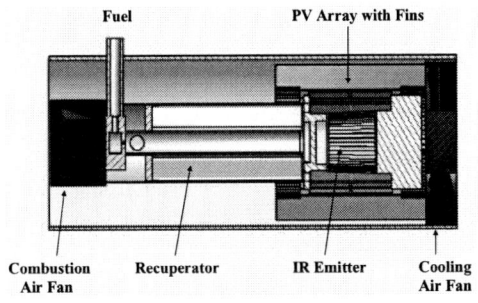

Fig. 1. Design of TPV power supply.

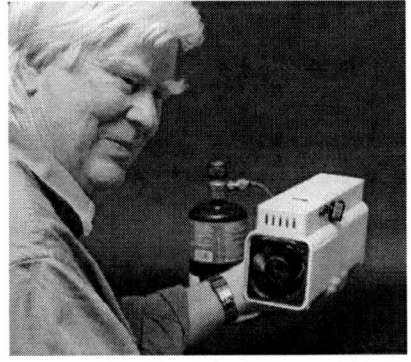

Fig. 2a. Functional hand held TPV power supply.

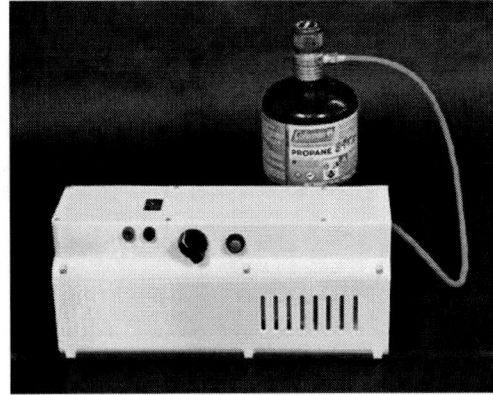

Fig. 2b. Functional stand alone TPV power supply.

Fig. 3. Photo of NiO/MgO IR emitter operating at 1290 C along with test cell. This is part of the Burner-Emitter-Recuperator (BER) assembly [2, 3].

Fig. 4. Power converter array (PCA) with 36 GaSb cells with cooling fins.

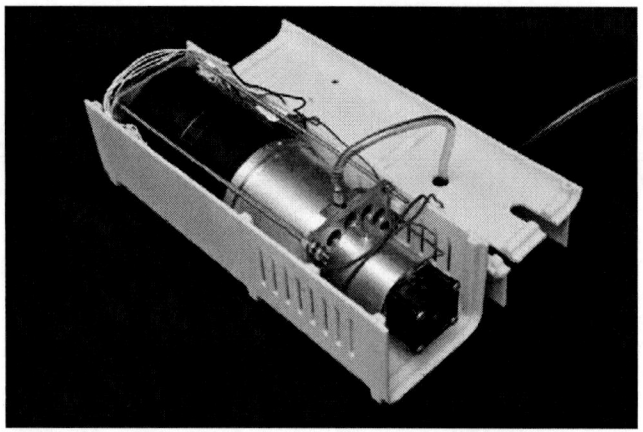

Fig. 5a: Photo of PCA & BER in Housing Cradle

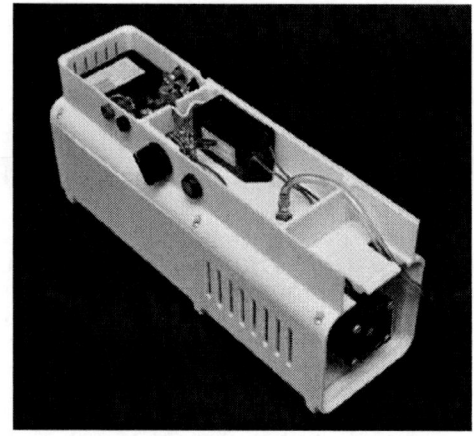

Fig. 5b: TPV with Housing & Controls

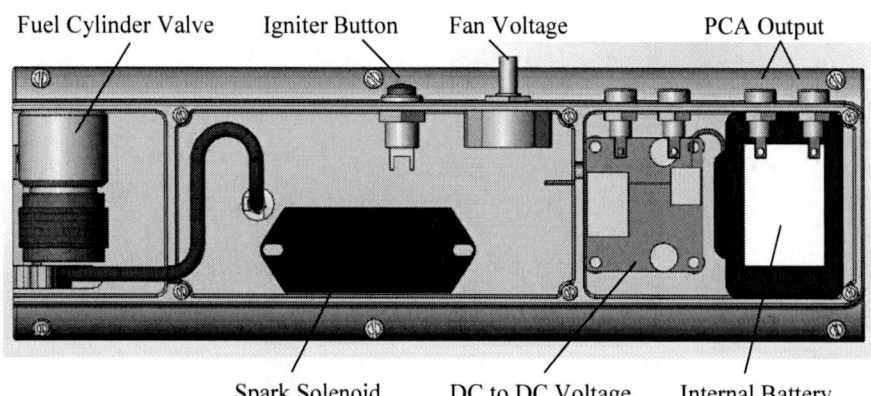

Figure 5c: Control compartment description.

II. RESULTS AND DISCUSSION

To first order, the conversion efficiency of a TPV system is given by the product of four terms: the chemical to radiation conversion efficiency, η_{CR}, the percent of radiation in the cell convertible band known as spectral efficiency, η_{SP}, the cell conversion efficiency, η_{PV}, and the cell to emitter view factor efficiency, VF. While the TPV generator shown in figure 2 is functional, it has limitations. Analysis of these limitations now allows design improvements. As is shown in figures 6 and 7 and Table I, increasing the emitter diameter and height along with the additional improvements described in the following can lead to 50 W and 100 W 10% efficient portable TPV power supplies.

The significant improvements are most evident by comparing figure 6b with 6a. The IR emitter configurations are different. The IR emitter in figure 6a is a flow through emitter. The hot combustion gases flow though the emitter. The figure 6b emitter is a radiant tube emitter where all of the combustion gases are contained inside the emitter assembly. The holes in the emitter base that allow the hot combustion gases to exit are moved from outside the IR emitter in figure 6a to inside the emitter in figure 6b. Note that the IR emitter is now also longer or taller and that there is now a cone at the combustion entrance.

TABLE I: SUMMARY OF SCALE UP TPV PERFORMANCE

TPV System	Present	Projections	
Emitter diameter	25 mm	30 mm	50 mm
Emitter Height	30 mm	60 mm	60 mm
Power	19.2 W	50 W	100 W
Rad / Chem Effic	0.53	0.7	0.7
Spectral Effic	0.4	0.6	0.6
Cell Effic	0.28	0.28	0.28
View Factor	0.58	0.83	0.83
System Effic	3.4%	10%	10%
Dimensions	3.6"x4.6"x12"	3.6"x4.6"x14"	4.6"x5.6"x14
Weight	2 kg	1.5 kg	2 kg
Wh/kg of fuel	439	1290	1290
Wh/kg with 1.5 kg of fuel +TPV	439x1.5/3.5= 188 Wh/kg	1290x1.5/3= 645 Wh/kg	1290x1.5/3.5 553 Wh/kg

Fig. 6. Alternate IR emitter sizes.

   (a) Present 25 mm diameter x 30 mm tall.
   (b) Scale-up 30 mm diameter x 60 mm tall.
   (c) Scale-up 50 mm diameter x 60 mm tall.

Fig. 7. The emitter size has a small impact on the system size but a big impact on the power output and the system efficiency.

   (a) Present 25 mm diameter emitter. TPV unit is 91 mm (3.6') wide.
   (b) Increased emitter diameter to 50 mm. TPV unit is 117 mm (4.6') wide.

III. OPPORTUNITIES FOR PERFORMANCE IMPROVEMENTS

As noted in figure 6, the performance improvements come mainly from increases in the IR emitter height and then diameter but as noted in figure 7, changes in the emitter size has a small impact on the system size but a big impact on the power output and the system efficiency The significance of these design improvements are summarized in Table I and in the following.

A. Spectral Efficiency

Figure 8 shows the measured spectra [2, 3] from the NiO/MgO IR emitter assembly shown in figure 3. The peak at 1.6 microns is the desired peak from the Ni ion matching the spectral response of the GaSb IR cells. However, the undesired longer wavelength radiant energy at 2.8 microns is from the combustion OH and CH gases and the radiant energy beyond 3.5 microns is from the heated fused silica window. The spectral efficiency with the emitter in figure 6a is limited by spurious radiation from the hot window and the combustion process [3]. With the radiant tube configuration emitter in figures 6b and 6c, the exit holes are moved inside the emitter assembly so that the combustion is not seen by the cells and the glass window is not heated by

the exhaust gases. So the spectral efficiency in table I can potentially be improved from 0.4 to 0.6.

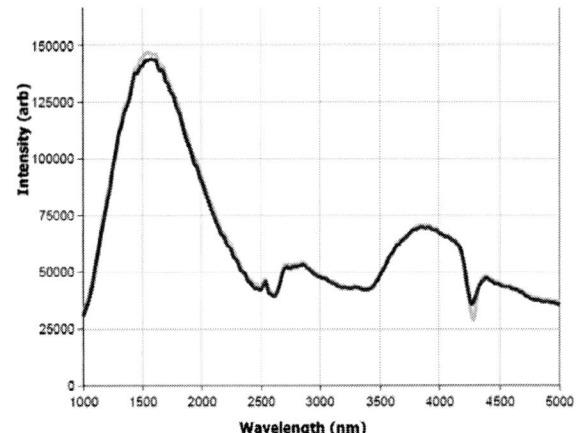

Fig. 8: Measured spectra from the emitter shown in fig. 3.

B. Chemical to Radiation Efficiency

High chemical to radiation efficiency requires complete combustion inside the emitter assembly, high combustion gas coupling efficiency to the IR emitter, minimal heat coupling to the glass window, and good exhaust gas recuperation. The 6b and 6c radiant tube configurations do not allow the hot combustion gases to impinge on the glass window avoiding that chemical energy loss. These configurations also have larger combustion volumes. The insulation base in the figure 6a design is porous and by-passes some fuel and air. A solid conical ceramic cone is inserted in the emitter designs in figs. 6b and 6c to avoid this fuel loss. Combustion is initiated inside this cone. Referring to figure 6b, there are three concentric ceramic cylinders. The inner cylinder is a pillar and it is surrounded by the radiant tube and then that is surrounded by the emitter cylinder. The combustion gases pass upward through the long narrow channel between the pillar and the radiant tube and then turn around and pass downward along the long narrow channel between the radiant tube and the emitter and then exit into the recuperator. This path allows for excellent chemical to radiation energy coupling efficiency. So the chemical to radiation efficiency is increased from 0.53 in 6a to 0.7 in 6b.

C. View Factor Efficiency

Not all of the radiation from the IR emitter reaches the PV cells. Some radiation is lost at the emitter cylinder ends. This is a view factor loss. The view factor efficiency for two coaxial equal length cylinders depends on the length of the cylinders and the spacing between them [4]. As is shown in figure 9 and Table II, increasing the emitter and PCA height from 30 mm to 60 mm increases the view factor from 0.58 to 0.83.

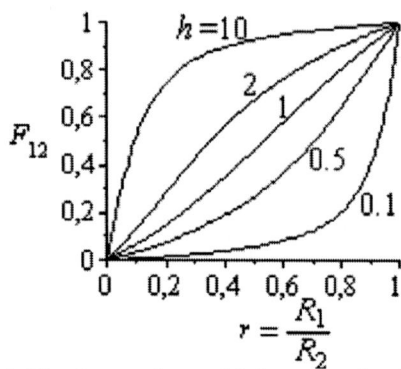

Fig. 9: View Factors for equal finite concentric cylinders

TABLE II: SUMMARY OF VIEW FACTORS

Re	Rc	r=Re/Rc	H	h=H/Re	Fec
No White Reflector					
12	26	0.46	30	2.5	0.58
15	26	0.58	60	4	0.83
25	40	0.625	60	2.4	0.75
White reflectors (add 6 mm to Re)					
31	40	0.78	60	1.94	0.83

D. IR Cell Efficiency

The cells used in the PCA in figure 4 were cells developed for solar applications with high quantum efficiencies (QE) at 1 micron [5]. As shown in figure 10, they can be optimized for TPV with higher QE at 1.6 microns [6]. So the optimized cell efficiency in table I is assumed to be 28%.

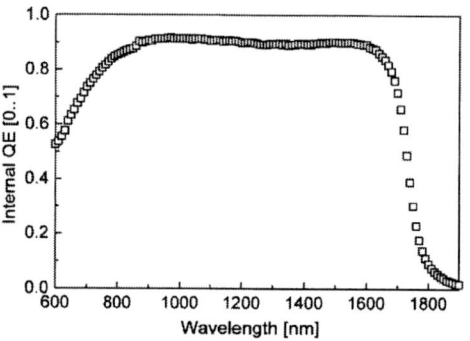

Fig 10a: Internal quantum efficiency of a 4 cm² GaSb PV cell derived from the external QE and reflectance measurements [6].

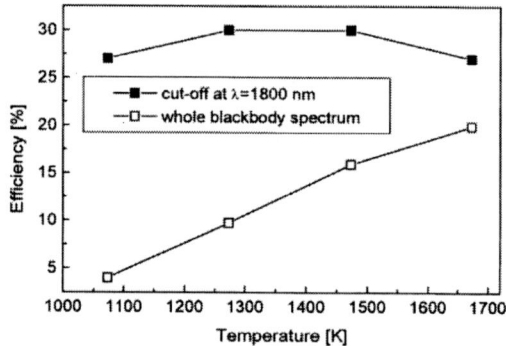

Fig. 10b: Calculated efficiency of a GaSb PV cell with the optimized structure as a function of the IR emitter temperature [6]. The calculations are based on measured parameters (Isc, Voc, FF).

E. Circuit Design

The Power Converter Array shown in figure 4 contains top and bottom rows of 18 series connected cells. There are actually two 9 facet circuits shown in flat and folded forms in figures 11a & 11b.

There are potential but solvable problems with emitter height increase. The first problem is the emitter axial uniformity. Axial non uniformity can be acceptable when the TPV circuit uses two rows of cells with one row at top and the other row at the bottom. In that case, the circuit position can be adjusted along the emitter axis to balance the top and bottom row currents. For the 30 mm tall emitter case of fig 6a, there will be 18 cells per row with the cell length along the emitter axis equal to 15 mm. The height of the two cell rows is then 30 mm equal to the emitter height.

978-1-5090-5606-4/17 $31.00 © 2017 IEEE   1866

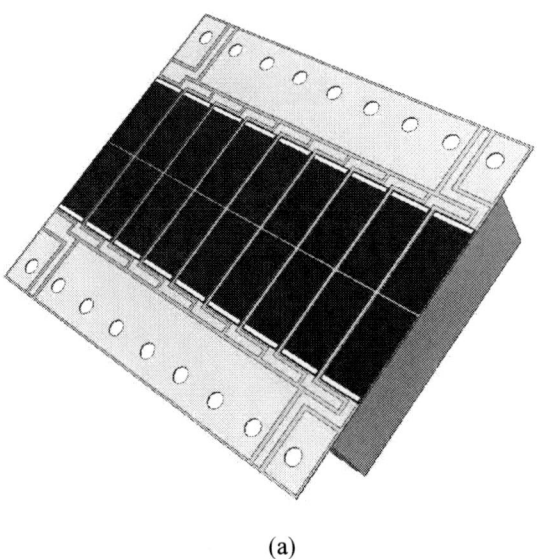

(a)

(b)

Fig. 11: The present TPV generator has two 9 facet circuits with 2 rows of 9 series connected GaSb cells.

For a 60 mm tall emitter, why can't the cell lengths in each row just be 30 mm? There are two problems with this. The first is that the cell current for 50 W and 12 V will be 4.2 A. This is too large for the standard 3 micron thick cell grid and bus bar to handle without a very large series resistance and resistance loss. The second problem is that there is a thermal expansion mismatch between the cell and the copper circuit substrate. Long cells can break. Figure 12 shows the solution to these two problems.

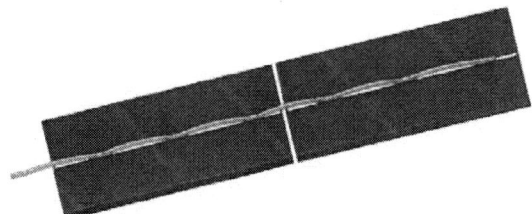

Fig. 12: Two 15 mm long cells with central bus bars with cell currents in parallel and with a wire stitched along the bus bars carrying the current to the external circuit.

Figure 13 shows a 9 facet circuit with 4 cells per facet in top and bottom rows of cell pairs. Two of these 9 facet circuits can be combined in a 50 mm PCA for the 50 W scale-up TPV power supply or three of these 9 facet circuits can be combined to make a 75 mm PCA as shown in figure 14 for the 100 W scale up power supply.

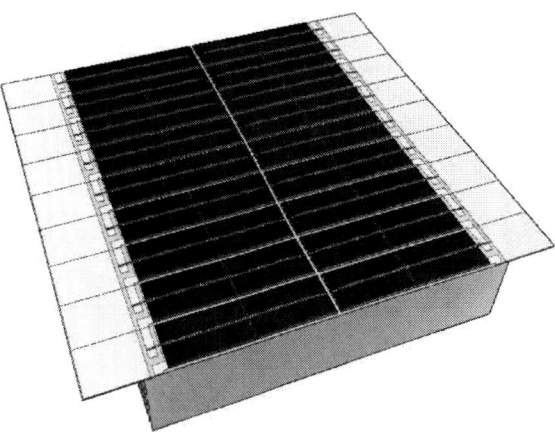

Fig. 13: Nine facet circuit with 4 cells per facet for use with 60 mm tall emitter.

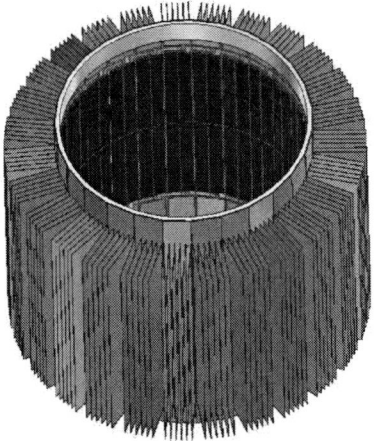

Figure 14: PCA with three 9 facet circuits for 100 W scale-up TPV power supply.

978-1-5090-5606-4/17 $31.00 © 2017 IEEE

IV. Mission Statement Advantage

The improvements just enumerated can increase the TPV conversion efficiency to 10%. Since a hydrocarbon fuel such as Butane or Propane has a specific energy of 12,900 Wh/kg, converting this energy at 10% efficiency offers a potential specific energy of 1290 Wh/kg. However, the weight of the converter hardware also has to be taken into account. Now that we have built the TPV hardware shown in figure 2, we can estimate the weight of the 50 W scale-up unit as 1.5 kg as shown in Table III. With 1.5 kg of fuel, the potential specific energy for a TPV system is then 1.5x1290/3= 645 Wh/kg.

Power systems relying on rechargeable battery operations limit Soldier and platform mobility by the low weight specific energy density of rechargeable batteries (145-160 W·h/kg). A typical squad leader can carry more than 14 pounds (6.35 kg) of batteries for a 72-hour mission. A lightweight standalone propane fired thermophotovoltaic (TPV) battery recharger has been described here [7, 8]. We have described scale-up design for a 1.5 kg (dry), 10% efficient TPV 50 W power source. We calculate that this unit will require 1.5 kg of fuel for a 72 hour (1.8 kWh) mission. It can then achieve a specific energy of over 600 Wh/kg which is 4 times the energy density compared with the rechargeable batteries.

TABLE III: Stand Alone TPV Generator Weights

Component	Present	50 W	100 W
PCA	259 g	259 g	388 g
BER	450 g	430 g	645 g
Housing	560 g	200 g	267 g
Comb Fan	50 g	50 g	45 g
Comb Fan Spacer	125 g	30 g	40 g
Cooling Fan	68 g	68 g	100 g
Threaded Rods	56 g	20 g	20 g
Fuel Valve	170 g	170 g	170 g
Electric Controls	273 g	273 g	273 g
TOTAL	2 kg	1.49 kg	1.95 kg

V. Conclusions

Figure 15 shows the 50 W scale-up design described here. It has a 30 mm diameter 60 mm high IR emitter surrounded by a cell array for a projected output of 50 W. This scale up TPV generator along with 1.5 kg of fuel has a projected weight specific energy density of 645 Wh/kg. A portable light weight TPV generator has applications both for soldiers as well as for unmanned aerial vehicles (UAVs).

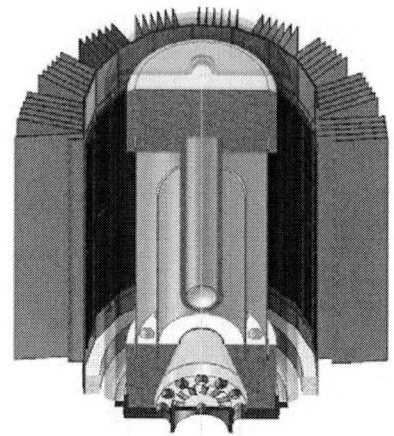

Fig.15. Scale-up IR emitter and cell array design.

Acknowledgement

This TPV development work was funded under an Army Research Lab cooperative agreement. Chris Mike Waits is the ARL Technical Lead.

References

[1] L.M. Fraas, *Low Cost Solar Electric Power*, Chapter 11, "Thermophotovoltaics Using Infrared Sensitive Cells", Springer International Publishing, Switzerland (2014).

[2] L.M. Fraas, L. Minkin, J. Avery, Hui She, F. Dogan, L. Ferguson, "Burner, emitter, and recuperator development for lightweight thermophotovoltaic power supply", in 42nd IEEE Photovoltaic Specialist Conference, New Orleans, LA (14 – 19 June 2015).

[3] L.M. Fraas, L. Minkin, J. Avery, L. Ferguson, J. Samaras, "Spectral Control Development for ThermoPhotoVoltaics", 43rd IEEE Photovoltaic Specialist Conference, Portland, OR, (2016).

[4] Radiation View factors - UPM
webserver.dmt.upm.es/~isidoro/tc3/Radiation%20View%20factors.pdf p. 16.

[5] LM Fraas, JE Avery, PE Gruenbaum, VS Sundaram, K Emery, R Matson, "Fundamental characterization studies of GaSb solar cells", Photovoltaic Specialists Conference, 1991., Conference Record of the Twenty Second IEEE.

[6] O.V. Sulima, A.W. Bett, "Fabrication and Simulation of GaSb Thermophotovoltaic Cells", Solar Energy Materials & Solar Cells, 66 (2001) 533}540

[7] LM Fraas, "Fuel fired ThermoPhotoVoltaic (TPV) cylindrical power supply and battery replacement with catalytic matched emitter or post IR emitter array", US Patent 8,581,090, 2013

[8] LM Fraas, US Patent, "TPV cylindrical generator using radiant tube burner", US 7196263 B2 March 2007.

High Efficiency Anti-Reflective Coating for PV Module Glass

Brennen M. Freiburger[1], Corey S. Thompson[1], Robert A. Fleming[1], Douglas Hutchings[1],
Sergiu C. Pop[2]

[1]WattGlass, Fayetteville, AR 72701
[2]Yingli Green Energy Americas, San Francisco, CA 94108

Abstract — Without antireflective coating, more than 4% of incident light is reflected from the standard front cover glass of photovoltaic (PV) modules. Module efficiency is one of the largest levers to impact the cost-per-watt of solar and recovering some of this reflected light with a simple anti-reflective coating (ARC) has become widespread. The types of ARC can vary in deposition method (roll coating, spray coating, sputtering, etc.) as well as composition and performance. The most widely adopted coatings today are based on a porous silica film with a thickness optimized for the solar spectrum. Current coatings, however, have room for improvement in both the performance and cost which means that manufacturers are actively looking for new solutions that drive down the levelized cost of electricity (LCOE). In this work, we report the test results for a new AR coating from WattGlass showing significantly improved optical performance compared to the traditional AR coatings. The new coating takes advantage of water-based chemistry that is more environmentally friendly than the sol-gel processes used in standard production coatings. This chemistry allows a high performance and durable coating to be deposited and cured at room temperature and is compatible with industry standard glass tempering conditions. The samples under test in this work were subjected to extensive optical performance testing at material and mini-module level. Our results show increased optical performance for the new coating, with solar weighted transmittance improvements as high as 3.1%. This increased optical performance directly translates to increased energy yield, lower LCOE and reduced warranty costs.

Index Terms — antireflective films, glass, efficiency, photovoltaic cells.

I. INTRODUCTION

Due to the mismatch in index of refraction between air and glass, 4% of normally incident light is lost due to reflection at the interface of air and PV module glass. As the PV industry is driving towards higher efficiencies at even lower costs it has widely adopted the anti-reflective coating (ARC) for PV module front cover glass. It has been estimated that greater than 92% of silicon PV modules are now made with ARC glass [1].

Most ARC glass manufacturers create the ARC on the solar cover glass by depositing a coating solution using spray or roll coating methods. The coating solution is typically a sol-gel consisting of silica precursors (e.g., tetraethoxysilane), acids, solvents, and water. The reaction of these components creates a thin porous silica layer that is deposited on the glass surface [2]. These coatings derive their optical and structural properties from their chemical composition and many require

a temperature-controlled process step that occurs when the PV cover glass is tempered in order to reach a cured state.

Porous silica coatings dominate the glass antireflective coating market due to their relative ease of deposition and the well understood sol-gel chemistry. The porosity of these coatings results in an index of refraction less than that of solid silica glass which allows them to be used as quarter-wave thickness antireflective coatings [3]. However, as the index of refraction is further reduced to increase performance, a decrease in the mechanical strength of the coating due to the increased porosity is often seen [4]. This results in an inherent tradeoff between performance and durability of the coating. WattGlass has overcome this tradeoff with a coating that provides best-in-class performance, exceptional durability, easy adaptation to industry standard application methods, water based chemistry, and is shelf-stable at room temperature.

The efficiency of a cell/module is what usually gets headlines, however module manufactures and system owners put emphasis on the levelized cost of electricity (LCOE) which determines the financial return of a given project. As warranty periods extend, module manufactures are exploring coatings that couple long term reliability with improved performance.

II. EXPERIMENTAL DESIGN

In this work, we evaluated side-by-side three sets of samples: bare glass, WattGlass HT coating, and the WattGlass LT coating utilizing improved chemistry. All coatings were evaluated on low iron float glass. The WattGlass coatings were applied to pre-cleaned 200 mm x 200 mm substrates using a custom lab scale dip coater, dried, and then subjected to thermal processing to cure the coating. The WattGlass HT coatings were cured by physical sintering of the constituent nanoparticles during a standard temper. The WattGlass LT coatings includes additional chemistries that are cured at near room temperature, with only slight heating to drive off moisture from the coating solution. The resulting thin-films were homogenous, transparent, and defect-free with thickness in the range of 120 nm with excellent anti-reflection properties.

978-1-5090-5606-4/17 $31.00 © 2017 IEEE

The samples were characterized in terms of optical transmittance and coating abrasion resistance. In addition, the effects of the AR coating on the electrical characteristics of a PERC solar cell were measured with a one-sun solar simulator.

Optical performance was measured using a simultaneous transmittance and reflectance spectrophotometer (aRTie, Filmetrics). The associated analysis software uses advanced curve fitting tools to calculate index of refraction, absorption coefficient, and coating thickness from the measured optical spectra. The abrasion resistance of the coating was characterized with a custom linear reciprocating abrader, using a felt counter face and 400g applied load. After abrasion, the optical transmittance of the sample was measured to determine degradation of the optical performance.

The one-sun simulator experiments were conducted using an Oriel Sol3A Class AAA Solar Simulator (Newport Instruments) with a reference cell for comparison of relative performance between substrates.

III. RESULTS AND DISCUSSIONS

A. Optical performance gains WattGlass HT coating vs. WattGlass LT coating

The transmittance spectra of the coated samples were measured in several locations, averaged, and the solar weighted transmittance calculated. The standard AM 1.5 spectrum [5] was used to calculate solar weighted transmittance (SWT) and solar weighted reflectance (SWR). The solar weighted transmittance was calculated by integrating the product of the transmittance and solar spectra then dividing by the integrated solar spectrum over the measurement range as shown in equation 1, where T_{avg} is the average of the measured transmittance spectrum, $E_{p\lambda}$ is the photon irradiance defined by the AM 1.5 spectrum, and λ is wavelength.

$$T_{solar\ weighted} = \frac{\int_{350\,nm}^{1050\,nm} T_{avg} \cdot E_{p\lambda}(\lambda) \cdot d\lambda}{\int_{350\,nm}^{1050\,nm} E_{p\lambda}(\lambda) \cdot d\lambda} \qquad \text{Eq. 1}$$

The WattGlass LT coated samples showed an increase in maximum transmittance of 3.61% at 550 nm wavelength, while the WattGlass HT coated samples showed a gain of 2.82%. It was found that the average solar weighted transmittance of the WattGlass LT coated samples was 94.8%. This represents a 3.09% increase in solar weighted transmittance over bare glass. In this test, the WattGlass HT coating was found to increase the SWT to 94.11%, representing a 2.39% gain over bare glass. Solar weighted transmittance gains for the best performing "traditional" AR coatings are in the range of 2.9%; however, the majority are at 2.7% and below. These "traditional" coatings almost exclusively require the use of sol-gel based chemistry, sputtering, or etching. In this work, all coatings tested were

water solution based, free from use of acids, solvents, or organics.

Optical performance metrics for the three substrate types tested are reported in Table 1, average spectra for the three samples are shown in Figure 1.

TABLE 1
OPTICAL PROPERTIES

	Bare	WG HT	WG LT
Max T (%)	92.46	95.28	96.07
Peak WL (nm)	548.88	543.11	535.18
% SWT	91.72	94.11	94.81
% SWR	8.84	4.09	5.18
Δ SWT (%)	---	2.39	3.09

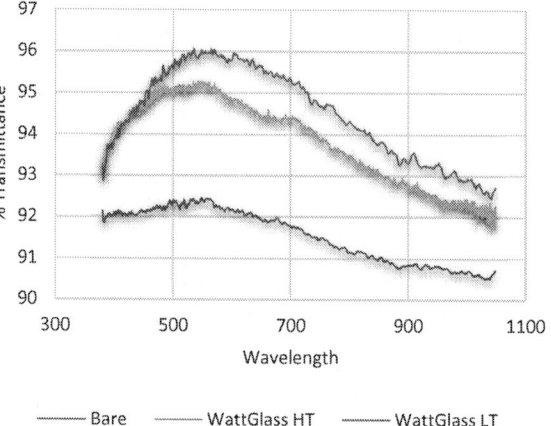

Fig. 1. Average transmittance spectra of the three sample types tested

B. Abrasion testing of WattGlass HT coating vs. WattGlass LT coating

The abrasion resistance of the WattGlass LT and WattGlass HT were characterized after 20, 1000, and 2000 abrasion cycles. A plot in reduction in SWT for both samples is shown in Figure 2. The WattGlass LT coating showed substantially enhanced durability compared to WattGlass HT. After 2000 cycles, the SWT of the WattGlass LT decreased by only 0.34%. In comparision, the SWT of WattGlass HT decreased by 0.28% after only 20 cycles, and degraded by over 1.2% after 2000 cycles.

978-1-5090-5606-4/17 $31.00 © 2017 IEEE

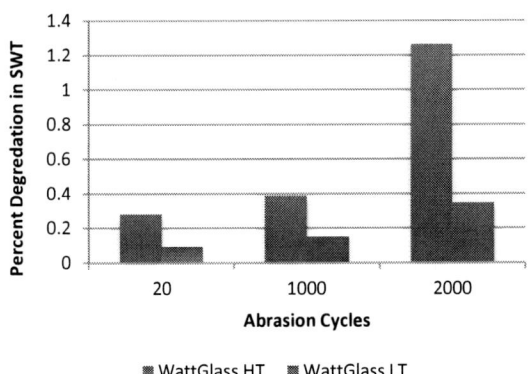

Fig. 2. Reduction in SWT vs. number of abrasion test cycles.

C. Solar Simulator Performance Evaluation

Each of the three sample types were evaluated using the Newport Instruments solar simulator. Only slight variations were seen for V_{oc} voltages and fill factors with values of 0.645 +/- 0.005 volts and 79.37 +/- 0.05, respectively, over all measurements. Current-voltage traces for WattGlass LT coating and bare glass is shown in Figure 3. Based on these measurements, the WattGlass LT coating resulted in about a 3.4% increase in measured output current compared to bare glass. The WattGlass HT coating performed nearly identical to the WattGlass LT coating.

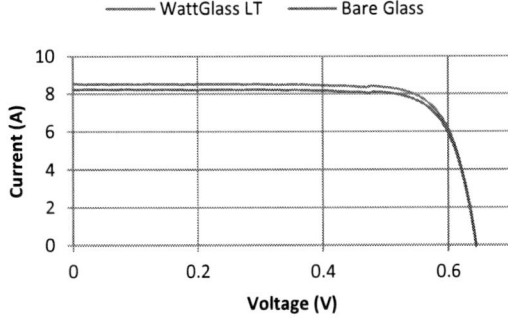

Fig. 3. Efficiency and Current Density values for the three sample types obtained from the solar simulator evaluation.

Enhancements in performance were evident in the short-circuit current density and efficiency of the cell for the WattGlass LT coating over that of bare glass, as shown in Figure 4. The efficiency of the cell covered with WattGlass HT coated substrate was 18.13%, a 0.5% absolute increase over using bare glass alone. Use of WattGlass LT improved

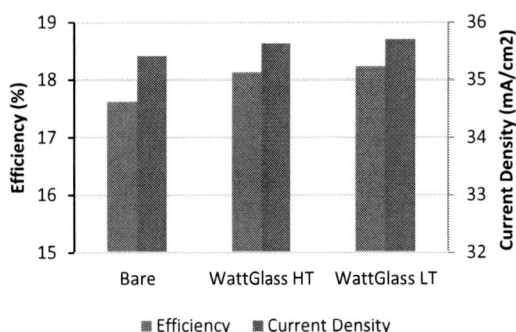

Fig. 4. Efficiency and Current Density values for the three sample types obtained from the solar simulator evaluation.

the efficiency of the cell further to 18.23%, a 0.6% absolute increase in overall efficiency for the cell. The WattGlass LT coating demonstrates a 0.55% relative increase in efficiency over WattGlass HT and a 3.4% relative increase in efficiency over that of bare glass.

The short-circuit current densities showed a 2.73% increase for WattGlass HT and a 3.37% increase for WattGlass LT coating, translating to markedly improved power output gains for the cell/module.

TABLE 2
SOLAR SIMULATOR MEASUREMENTS

	Efficiency (%)	Jsc (mA/cm2)	Power (W)
Bare Glass	17.63	34.47	5.00
WattGlass HT	18.13	35.41	5.15
WattGlass LT	18.23	35.63	5.18

TABLE 3
RELATIVE IMPROVEMENTS FOR SOLAR SIMULATIONS

	Efficiency	Jsc	Power
WG HT	+2.88%	+2.73%	+2.99%
WG LT	+3.43%	+3.36%	+3.56%

IV. CONCLUSION

In this evaluation, bare glass substrates were compared to two generations of coatings developed by WattGlass. Due to the fact that they are economical, easily implemented in traditional manufacturing processes, are water based, and environmentally friendly, both of these coatings are game changers in the world of solar AR coatings. One of these coatings requires curing in parallel with tempering, the other cures readily at near room temperature.

The WattGlass coatings exhibit excellent increases in SWT transmittance up to 3.1% over that of bare glass. This broadband transmission gain translates to efficiency and short-circuit current density increases of 3.43% and 3.37%, respectively, for a commercial PERC cell. The WattGlass LT formulation has superior abrasion resistance compared to WattGlass HT.

Considering that 96% of the solar ARC market is occupied by sol-gel coatings which require harsh chemistries of organics, solvents, and acids, while all the while using expensive precursors. This all impacts the cost to the manufacturer, the system owners, and the financial return of a given project. The combination of performance increase, resiliency, and overall environmental conscientiousness of WattGlass coatings make it a truly exceptional solution.

REFERENCES

[1] *International Technology Roadmap for Photovoltaics (ITRPV) 2015 Results*, March 2016.

[2] D. Chen, "Anti-reflection (AR) coatings made by sol-gel processes: a review," *Solar Energy Materials and Solar Cells,* vol. 68, pp. 313-336, 2001.

[3] R. Prado, G. Beobide, A. Marcaide, J. Goikoetxea, A. Aranzabe, "Development of multifunctional sol–gel coatings: Anti- reflection coatings with enhanced self-cleaning capacity', *Solar Energy Materials & Solar Cells*, vol. 94, pp. 1084-1088, 2010.

[4] S. C. Pop, et al., "A highly abrasive-resistant, long-lasting anti-reflective coating for PV module glass," in the *40th IEEE Photovoltaic Specialists Conference*, 2014.

[5] ASTM G173-03(2012), Standard Tables for Reference Solar Spectral Irradiances: Direct Normal and Hemispherical on 37° Tilted Surface, ASTM International, West Conshohocken, PA, 2012

978-1-5090-5606-4/17 $31.00 © 2017 IEEE

Investigation of Efficiency for PID-affected Solar Module at Non-standard Test Conditions

Shuwen Guo[a], Pan Zhao[a], Weijing Huang[b] Jipeng Chang[a], He Wang[a]*, Hong Yang[a], Chengfeng Su[c], Bojie Su[d], Xue Zhang[d], Yunxue Cao[e], Hui Zhao[e]

[a]MOE Key laboratory for Nonequilibrim Synthese and Modulation of Condensed Matter, School of Science, Xi'an Jaotong University, Xi'an 710049, People's Republic of China

[b]Xi'an Huanghe Photovoltaic Technology Co, Ltd, Xi'an, No.21 North Xingfu Road, Xi'an 710049, People's Republic of China

[c]Taizhou Chisolar Co., Ltd., Taizhou 318020, People's Republic of China

[c]Institute of Electrical Engineering of the Chinese Academy of Sciences, Beijing 100190, People's Republic of China

[d]China Quality Certification Center, Beijing 100070, People's Republic of China

[e]SPIC Power Plant Operation Technology Co., Ltd, Beijing 100190, People's Republic of China

*Corresponding author: He Wang, hw69cn@126.com

Abstract—**Potential-induced degradation (PID) causes a sharp decrease in the efficiency and shunt resistance of the solar module. In this paper, PID-affected modules were studied detailedly under non-standard light conditions for the first time. The trend of efficiency degradation was obtained by measuring irradiance-efficiency (I-η) and the ΔI-$\Delta \eta$. The result shows that the efficiency (η) of PID-affected module reduced by 1.28% which is rather higher than the 0.2% of PID-free module when the light intensity changes from 400 W/m^2 to 200 W/m^2. This means that efficiency of PID-affected module decreases more seriously in low light conditions.**

Index Terms — **photovoltaic power plant, PID, efficiency, shunt resistance, the light intensity**

I. INTRODUCTION

The photovoltaic energy will provide a substantial contribution to future energy demand as one of the promising clean and renewable energy resources. The efficiency of the PV modules is extremely critical for the total power output of the system, the cost effectiveness of the PV(photovoltaics) system and the commercial success of photovoltaic energy [1-4]. In grid-connected photovoltaic systems the individual modules are usually connected serially to strings so as to increase system voltage. Partial solar cells incorporated into PV modules in this architecture can be exposed to a voltage bias of several hundreds of volts. A degradation mechanism named as potential-induced degradation (PID) significantly leads to remarkable efficiency losses [5]. These losses are so significant that they can not be ignored and have fatal impacts on the long-term system performance [6-7]. PID of p-type silicon solar cells is mainly characterized by a significant reduction of the parallel resistance R_{sh} which causes a drop of the efficiency output [8-15]. In this paper, the thermography imaging of the PID-affected modules was achieved in the field, and the electrical properties of PID affected solar module dismounted from power plant were measured by using solar simulator (PASAN Sunsim 3C) in the laboratory. We focus on the PID-affected module's efficiency (η) and the relative change of the efficiency ($\Delta \eta$) at non-standard insolation conditions. The degradation regularity of efficiency were presented by (I-η) and the (ΔI-$\Delta \eta$) curves characteristic. The PID-free module's efficiency (η) increases almost at the rate of 0.2% per 200 W/m^2 at non-standard insolation less than 1000 W/m^2. However, PID-affected module's efficiency is more sensitive to the weak light intensity less than 400 W/m^2. The lower the light intensity is, the faster the efficiency decreases. The maximum degradation of $\Delta \eta$ is about 1.28% which is rather higher than the 0.2% of PID-free module. Besides , the efficiency loss of the last module with PID in module string is up to about 8% at the light intensity of 200 W/m^2, and the yield losses of per string affected by PID is approximately up to 45%. The important and profound efficiency regularity of

978-1-5090-5606-4/17 $31.00 © 2017 IEEE

TABLE I
THE V_{oc} OF SOLAR MODULES

Irradiance(W/m2)	V_{oc1}(V)	V_{oc2}(V)	V_{oc3}(V)	V_{oc4}(V)	V_{oc5}(V)	V_{oc6}(V)	V_{oc7}(V)	V_{oc8}(V)	V_{oc9}(V)
200	41.949	41.948	41.949	41.947	41.945	41.949	41.948	41.949	41.792
400	43.378	43.376	43.377	43.376	43.374	43.376	43.377	43.378	43.232
600	43.567	43.565	43.566	43.564	43.565	43.568	43.565	43.567	43.456
800	44.882	44.881	44.882	44.884	44.883	44.881	44.883	44.882	44.532
1000	45.261	45.262	45.263	45.263	45.261	45.262	45.261	45.261	45.095

TABLE II
THE V_{oc} OF SOLAR MODULES

Irradiance(W/m2))	V_{oc10}(V)	V_{oc11}(V)	V_{oc12}(V)	V_{oc13}(V)	V_{oc14}(V)	V_{oc15}(V)	V_{oc16}(V)	V_{oc17}(V)
200	41.298	40.902	34.263	32.021	31.927	30.148	29.254	25.582
400	43.099	43.066	38.687	37.465	36.158	33.747	33.487	28.223
600	43.126	43.098	39.156	38.321	37.465	35.564	34.256	30.121
800	44.421	44.411	42.213	41.692	39.654	38.565	36.557	32.545
1000	45.088	45.083	43.555	42.645	40.671	39.125	37.288	33.409

PID-affected modules at non-standard insolation has been detected.

II. EXPERIMENTAL RESULTS AND DISCUSSION

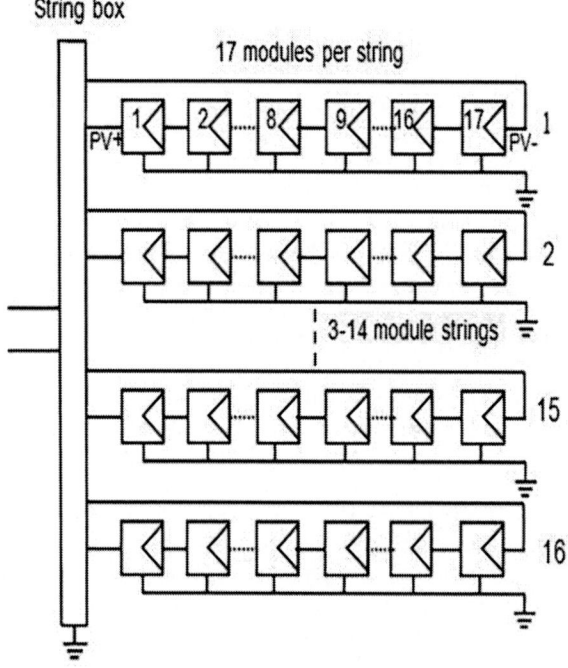

Fig.1. The location of the experimental modules in this PV power plant

The practical survey of PID-free and PID-affected module conducted by our team was in a PV power plant located in Dingbian County of YuLin, a small city in the northwest china in Shaanxi province at 37.6 degrees north latitude and 107.59 degrees east longitude. The location of the experimental modules is shown in Fig.1.

The circuit diagram of sub-array in the power plant is showed by Fig.1. The PV modules have been mounted on the flat field towards south with a 36° inclination. All the modules installed in this PV power station are made of monocrystalline silicon manufactured by some common commercial PV manufactures. Every module (1580 mm × 808 mm) is comprised of 72 solar cells connected in series. The test conditions of the experiment in this PV power plant are average ambient temperature of ~25 °C, solar irradiation of (200 W/m²~1000 W/m²) and relative humid of ~60%. The open circuit voltage (V_{oc}) of the PV modules tested in PASAN3C is summarized in table I and table II.

Table I and table II show the statistical data of modules V_{oc}. In these table, the V_{oc} of modules at different positions in module 1 and module 8 almost unchanged under the same light intensity, however, the V_{oc} in other module, from module 9 to module 17, significantly degrades when the modules position is more close to the end of strings. The thermography imaging to identify the localization and degradation level of PID affected silicon solar modules is displayed in Fig.2. The PID-free module (left) and PID affected module (right) have different temperature distribution pattern. The bottom-middle areas in thermography images of all seventeen modules have the highest temperature spot, because the junction box is built there. So it is a common phenomenon. From Fig. 2 the former half modules 1-8 exhibit a homogeneous temperature distribution at module level. There is no other hotter cell in former half modules 1-8 except the bottom-middle cells. It is suggested that the modules in the former half of grounded module string are free of PID-affected cells.

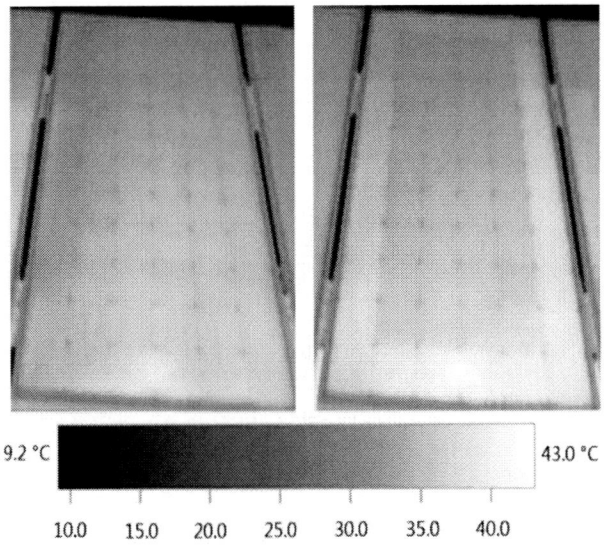

Fig. 2.   The PID-free module (left) and PID affected module (right) have different temperature distribution pattern.

By contrast, different temperature distributions occur in the latter half modules. From Fig.2, more cells with high temperature occur in modules 9-17 and the number of hotter cells is increasing with the module close to the negative pole of grounded module string. An interesting finding is that the hotter cells mainly locate at the edge areas of modules which are adjacent to the Al frame. That is to say the temperature of edge cells is higher than the middle cells. Owing to PID effect, the performance of the edge cells in latter half modules lags for the other cells'. A smaller portion of the irradiated power is converted to electrical energy. Thus, PID-affected cells exhibit a higher temperature than sound ones. This result agreed with other authors' work.[16-18] With PID influence, the shunt resistance (R_{sh}) of cells decreases dramatically. Based on the above analysis, with the enhancement of light, the conversion efficiency of all the modules is increased, and increases the fastest under weak light (200 W/m^2~400 W/m^2). We select the 8 and 9-17 modules as a sample to further calculate the conversion efficiency of the module under weak light.

From the Fig.3, Fig.4, we find that the efficiency of modules is decreasing continuously from module 9 to module 17 in grounded string. In addition, the efficiency of module 8 is hardly changed. However, the efficiency of module 9 to module 17 fluctuates violently, especially in the weak light intensity less than 400 W/m^2. So it is obvious that the more serious the PID of modules is, the greater the rate of efficiency decreases. The reason is that PID effect damaging the depletion region of p-n junction causes parallel resistance decreases. In the standard light intensity, due to the light-

generated current is so large that the influence of the shunt resistance is neglected. But in the weak light, due to the small light-generated current, the current through parallel resistance is not be neglected as well as the influence of shunt resistance should be considered. So the efficiency of PID-affected modules decreases.

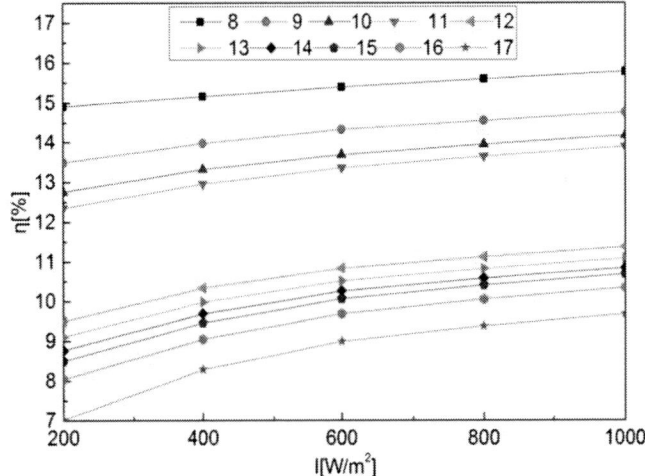

Fig. 3. I-η characteristics of all modules

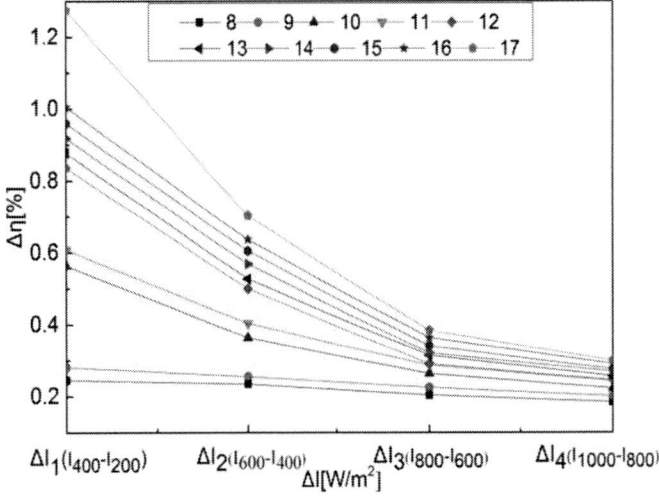

Fig. 4. ΔI-$\Delta\eta$ characteristics of all modules

III. CONCLUSIONS

The efficiency of PID-affected modules is becoming more serious while the module is more close to the end of grounded string under weak light. In the weak light, due to the small light-generated current, the current of parallel resistance is not negligible, and the decreases of shunt resistance should be considered. So the efficiency of PID-affected modules decreases. The end module in grounded string has the worst impact by PID up to about 8% efficiency loss in the weak light intensity of 200 W/m^2, under the same condition, the yield losses of per string by PID are approximately up to 45%. According to this degradation pattern, we propose that the reduction of efficiency which resulting from PID effect is more serious at non-standard insolation especially in the weak light intensity less than 400 W/m^2. Low light intensity has the greatest influence on power generation of PID affected module.

ACKNOWLEDGEMENT

The authors would like to thank the support of Natural Science Foundation of China (Grant No. 61376067 and 61274050). This study was also supported by the National High Technology Research and Development Program of China (Grant No.2015AA050301). This study was also supported by the Bureau of Science and Technology of Taizhou City (2016023).

REFERENCES

[1] A. Ndiaye, et al., "Degradations of Silicon Photovoltaic Modules:A Literature Review," Sol. Energy 96 (2013) 140-151.
[2] H. Yang and H. Wang, "The materials characteristic and the efficiency degradation of solar cells from solar grade silicon from a metallurgical process route," Journal of materials science 46 (2011) 1044-1048.
[3] E. Dunlop and D. Halton, "The Performance of Crystalline Silicon Photovoltaic Solar Modules after 22 Years of Continuous Outdoor Exposure," Prog. Photovolt: Res. Appl 14 (2006) 53-64.
[4] Wang X, Kurdgelashvili L, Byrne J, et al. The value of module efficiency in lowering the levelized cost of energy of photovoltaic systems[J]. Renewable & Sustainable Energy Reviews, 2011, 15(9):4248-4254.
[5] J.A.del Cueto and T. J. McMahon, "Analysis of leakage currents in photovoltaic modules under high-voltage bias in the field," Prog. Photovolt: Res. Appl. 2002; 10:15-18.
[6] Jipeng Chang, He Wang, Hong Yang, Junjun Zhang, Jingsheng Huang. The Real Situation of Potential Induced Degradation in Crystalline Silicon Photovoltaic Power Plant[C].photovoltaic Specialists Conference(PVSC),2016 IEEE 43nd. IEEE,2016:1-3.

[7] R. Swanson, et al., "The surface polarization effect in high efficiency silicon solar cells," in: Proceedings of the 15th IEEE PVSC Kissimmee, FL, USA, 2005, pp.410-411
[8] Dincer F, Meral M E. Critical Factors that Affecting Efficiency of Solar Cells[J]. Smart Grid & Renewable Energy, 2010, 01(1):47-50..
[9] Barbato M, Meneghini M, Giliberto V, et al. Effect of shunt resistance on the performance of mc-Silicon solar cells: a combined electro-optical and thermal investigation[C]// IEEE, 2012:1241-1245.
[10] Mcmahon T J, Basso T S, Rummel S R. Cell shunt resistance and photovoltaic module performance[C]// Photovoltaic Specialists Conference, 1996. Conference Record of the Twenty Fifth IEEE. IEEE Xplore, 1996:1291-1294.
[11] Yunaz I A, Kasashima S, Inthisang S, et al. Effect of light intensity on performance of silicon-based thin film solar cells[C]// Photovoltaic Specialists Conference. IEEE, 2009:000153-000157.
[12] Barbato M, Meneghini M, Giliberto V, et al. Effect of shunt resistance on the performance of mc-Silicon solar cells: a combined electro-optical and thermal investigation[C]// Conference Record of the IEEE Photovoltaic Specialists Conference. IEEE, 2012:1241-1245.
[13] Koster L J A, Mihailetchi V D, Ramaker R, et al. Light intensity dependence of open-circuit voltage and short-circuit current of polymer/fullerene solar cells[C]// Photonics Europe. International Society for Optics and Photonics, 2006:1175-1199.
[14]Yunaz I A, Kasashima S, Inthisang S, et al. Effect of light intensity on performance of silicon-based thin film solar cells[C]// Photovoltaic Specialists Conference. IEEE, 2009:000153-000157.
[15] Barbato M, Meneghini M, Giliberto V, et al. Effect of shunt resistance on the performance of mc-Silicon solar cells: a combined electro-optical and thermal investigation[C]// Photovoltaic Specialists Conference. IEEE, 2012:1241-1245.
[16] J. A. del Cueto and T.J.McMahon, "Analysis of leakage currents in photovoltaic modules under high-voltage bias in the field," Prog. Photovolt: Res. Appl. 2002; 10:15-18.
[17] R. Swanson, et al., "The surface polarization effect in high efficiency silicon solar cells," in: Proceedings of the 15th IEEE PVSC Kissimmee, FL, USA, 2005, pp.410–411
[18] M. Schütze, et al., "Laboratory study of potential induced degradation of silicon photovoltaic modules," in: Proceedings of the 37th IEEE PVSC, Seattle, 2011, pp.821– 826.

Thermal Uniformity Mapping of PV Modules and Plants

Ashwini Pavgi, Jaewon Oh, Joseph Kuitche, Sai Tatapudi and GovindaSamy TamizhMani

Arizona State University Photovoltaic Reliability Laboratory (ASU-PRL), Mesa, Arizona, 85212, USA

Abstract — The conventional assumption of all the cells in a module and all the modules in a plant is operating at a single temperature. This work indicates that it is not the case. The behavior of temperature distribution of PV cells within a module and of PV modules within a plant is presented. ANOVA, a statistical tool, was used to study the influence of various ambient and design factors on temperature variation. In this study, the effect of thermal non-uniformity on I-V parameters of three different PV technologies (crystalline silicon, CdTe, CIGS) is investigated. The temperature mapping data for two power plants (fixed horizontal-tilt and one-axis) located in a desert climate of Arizona showed that the modules placed in the center of one-axis power plant had higher temperatures, whereas in the fixed-tilt power plant, the modules located in the north-west direction had higher operating temperatures. Higher average operating temperature of modules was observed for the one-axis tracker based plant as compared to the fixed-tilt based plant, thereby a higher degradation rate and a lower lifetime are expected for the 1-axis tracker based modules as compared to the fixed-tilt based modules.

Index Terms — temperature variation, thermal uniformity, thermal mapping, performance.

I. INTRODUCTION

The maintenance of temperature uniformity in a photovoltaic (PV) module is critical to accurately measure the performance parameters and temperature coefficients of the module. In the indoor solar simulator based test setups, these parameters and coefficients are obtained by maintaining all the cells within a module at a single uniform temperature. However, in the outdoor test setups, the module experiences varying ambient conditions and hence the cells within a module may encounter non-uniform temperatures; for example, the edge cells may be cooler than the center cells due to cooler frame temperature. This thermal non-uniformity issue is further complicated due to variations in the module technology, module size, module design, materials and mounting methods. This module-level study is a continuation of our previous study performed on the crystalline silicon modules [1]. In the current module-level study, the test results and analysis obtained on the thin-film technologies (CIGS, CdTe and a-Si) are also included and presented along with the c-Si technology. The module-level thermal uniformity study with c-Si and thin-film technologies was performed at the outdoor site of ASU-PRL, Mesa, Arizona.

Traditionally, it is assumed that all the modules in a PV plant are operating at a single temperature. In our previous short-term investigation, we briefly indicated that the modules placed at various locations in a PV plant tend to experience different temperatures [2]. In the current long-term investigation, the dependence of module temperature on the module location of the plant and on the type of module mounting (1-axis vs. fixed-tilt) is extensively analyzed and presented. The plant-level thermal uniformity study was performed at two power plants (1-axis and fixed horizontal-tilt) located in Tempe/Phoenix, Arizona.

II. EXPERIMENTAL METHODS

2.1 Module-level thermal investigation

The test setup used for the module-level thermal uniformity investigation is shown in Fig. 1. The modules were mounted in two rows on the fixed tilt rack at 33°N. The module technology and thermal configuration of each module (corresponding to the code shown in Fig.1) are provided in Table I. The a-Si module and all the c-Si modules are glass/polymer modules whereas the CdTe and CIGS are glass/glass modules.

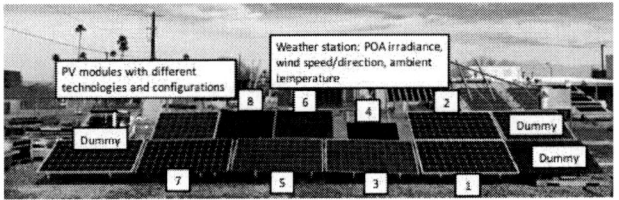

Fig. 1. Module-level thermal mapping setup, Mesa, Arizona

TABLE I
MODULE-LEVEL THERMAL MAPPING FOR DIFFERENT THIN-FILM AND C-SI MODULES WITH DIFFERENT INSULATION CONFIGURATIONS

Code	Insulation Type	Technology
1	Non-insulated	Mono-Si
2	Aluminum tape covered backsheet	Mono-Si
3	Frame insulated	Poly-Si
4	Non-insulated (frameless)	CdTe
5	Frame and backsheet insulated	Poly-Si
6	Non-insulated	a-Si
7	Non-insulated (black frame)	Mono-Si
8	Non-insulated	CIGS

The module temperatures were measured using multiple T-type thermocouples attached to the backsheet at locations defined in the IEC 61853-2 standard [3]. To collect module

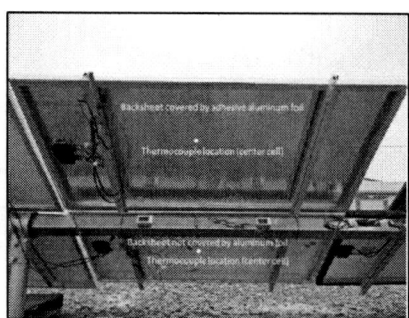

Fig. 2. Two identical modules: Top– Covered by aluminum foil; Bottom–Not covered by aluminum foil

temperature and the module voltage data, HOBO 4-channel data loggers were used. For the module performance monitoring, the performance and temperatures of four cells of each module under maximum power point tracking (MPPT) condition was continuously monitored for several clear sunny days.

The multi-curve tracer was programmed to periodically sweep the I-V curves of all the modules shown in Fig. 1 for 2 consecutive clear sunny days (during the solar window from 10am to 2pm) and the performance parameters were correlated with the module temperatures at four locations. These I-V curves were then translated to STC based on the measured module temperature at each of four locations on the module and the technology-specific temperature coefficients. The technology-specific temperature coefficient measurements were carried out for all the modules at around noon when the angle of incidence was close to zero. To perform these measurements, the modules were first placed in the cold chamber, to bring operating temperatures for modules around 10°C. The IV parameters were recorded for temperature coefficient measurements on each module at four locations on a manual dual-axis tracker on a clear sunny day for a specific range of module operating temperatures (20-30°C). As the modules warm up, multiple curves were taken for the individual modules at different operating temperatures with the thermocouples placed at four locations in each module.

To investigate if the thermal uniformity of the module can be improved by attaching a thermally conductive sheet on the backside of the back sheet, the modules shown by codes 1 and 2 in Fig. 1 were investigated and a photograph of back sides of these two modules is shown in Fig. 2. The back sheet (TPT) of the top module (code 2) was covered with a highly reflective conductive aluminum cover and the bottom module was a control module (code 1) without any aluminum foil.

2.2 Plant-level thermal investigation

Two PV plants with mono-Si modules were investigated in this study. One plant (AZ3) is based on fixed horizontal-tilt arrays and the other plant (AZ5) is based on 1-axis arrays. The fixed-tilt plant has no wind barriers around it. But on the other

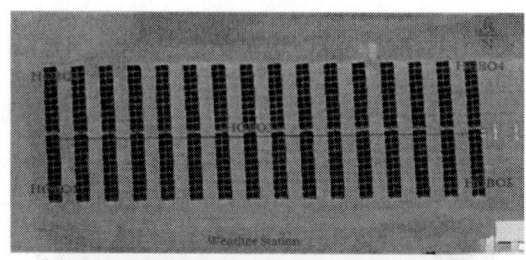

Fig. 3. Thermal mapping at five locations (AZ3 and AZ5 power plant)

hand, 1-axis plant located south of the fixed-tilt plant is about 4 feet lower ground level having some wind obstruction. There is also a 15-foot-high wall on the south side of the 1-axis plant and this wall is about 30 feet away from the array leading to some wind obstructions as well. For the thermal uniformity mapping of power plant, five data loggers were installed at the northwest (NW), northeast (NE), southwest (SW), southeast (SE) and center locations for each of the two power plants to record the temperatures of five modules located in these five locations as shown in Fig. 3.

Each data logger was recording four temperatures per IEC 61853-2 standard [3] for each of the five modules in the plant. The temperature data recorded by HOBO data loggers was retrieved by using HOBO software and converted into an Excel file type. MATLAB was used to interpolate and map the data values on a grid representative of PV module and a power plant.

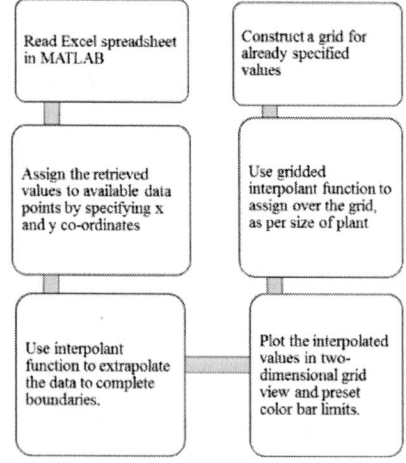

Fig. 4. MATLAB program flowchart

The flowchart representing various steps involved in MATLAB program code is as shown in Fig. 4.

For the analysis of variance (ANOVA) of the five individual modules in each plant, a fixed effect model was performed to study the significance of various factors and their interactions on the module temperature. The three factors (PV technology, electrical condition and thermocouple locations) with different levels were studied through ANOVA design. For the ANOVA of the two individual power plants, the three factors (power plant type, module and thermocouple locations) with different levels were studied on a clear sunny and a cloudy day. The average irradiance recorded from 9 am to 5 pm on a clear sunny day was 940 W/m² and that on the cloudy day was 329 W/m². The average wind speed recorded on clear sunny day was 2 m/s while that on a cloudy day was 4 m/s.

III. RESULTS AND DISCUSSION

3.1. Module-level thermal uniformity

3.1.1 Temperature uniformity: Dependence on thermocouple location and thermal insulation type

Fig. 5 shows the maximum, minimum and median temperature difference between the four thermocouple locations for each insulation type of c-Si modules. This figure also shows the dependence of temperature difference on the operating condition of the module. During these measurements, the average irradiance was in the range of 1007-1015 W/m², the average wind speed was in the range of 0.7-0.8 m/s and the ambient temperature was in the range of 22-24°C. The module was initially maintained at P_{max}, V_{oc} and I_{sc} conditions (for a 30-minute duration each) on a clear sunny day around noon time.

Fig. 5 indicates that there is a least variability in median temperatures, for the four thermocouples, in the black framed module, followed by the frame insulated module for both P_{max}

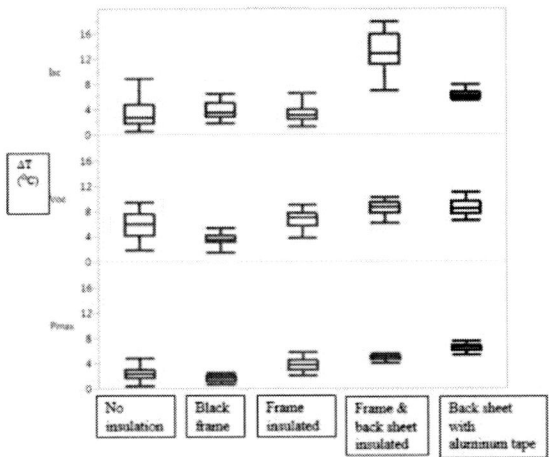

Fig. 5. Temperature difference between four thermocouples for each insulation configuration and each operating condition

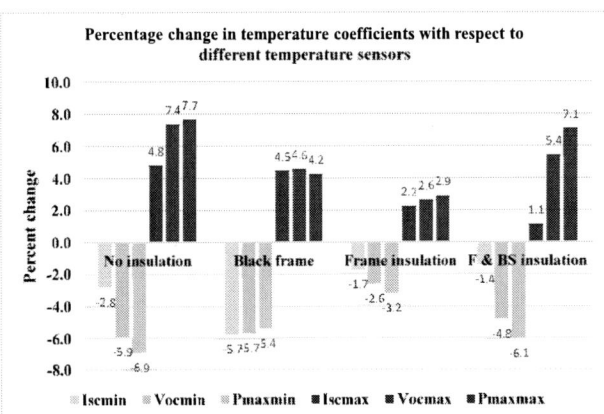

Fig. 6. Percentage change in temperature coefficients with respect to different temperature sensors

and V_{oc} operating conditions. In addition, the least variability in maximum temperatures values within a module is observed at maximum power condition as compared to V_{oc} and I_{sc} conditions. At P_{max}, the general operating conditions of PV modules in the field, the maximum variability was observed in the non-insulated PV modules. Even though the ΔT values and operating temperatures are higher for the module with aluminum cover on back sheet, the standard deviation is found to be lower. Aluminum cover could be a good solution to improve thermal uniformity but the operating temperature shoots up very high due to blockage of the cell radiation through the backsheet. The module with aluminum cover on back sheet was exclusively compared with the conventional white back sheet module in Section 3.1.2.

Fig. 6 represents the percentage change in temperature coefficients with respect to different thermocouples. The least deviation of about ±3 percent is observed in frame insulated modules and the maximum deviation of about ±8 percent is observed in non-insulated modules. When the frame is insulated, no longer are the edge and corner cells exposed to wind directly. This leads to less variation in the temperature across the modules, thereby resulting to least variation in temperature coefficients.

3.1.2 Effect of aluminum cover: Aluminum cover back sheet versus white back sheet module

A temperature difference as high as 15°C was observed on a clear sunny day around solar noon as shown in Fig. 7. This temperature difference was inversely proportional to the open-circuit voltage value. A temperature difference as high as 15°C was observed on a clear sunny day around solar noon as shown in Fig. 7. This temperature difference was inversely proportional to the open-circuit voltage value. The IR images were taken on a clear sunny day with irradiance= 1019 W/m², wind speed= 1.213 m/s and ambient temperature= 24.9°C.

In Fig. 8, the module in the bottom row is the conventional polymer white back sheet PV module and in the top row has aluminum cover on its back sheet. Analyzing the temperature

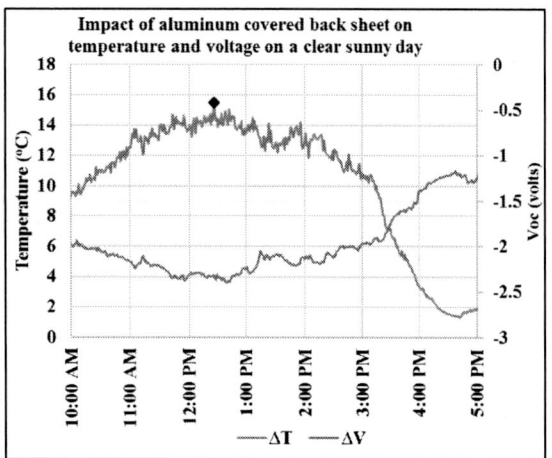

Fig. 7. Impact of aluminum covered back sheet on temperature and voltage on a clear sunny day 10am to 5pm

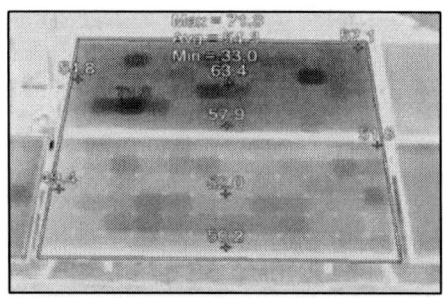

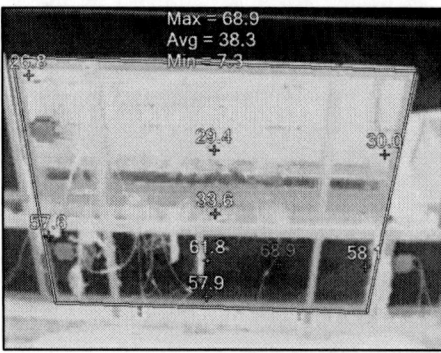

Fig. 8. Front and back side of aluminum cover back sheet and conventional polymer white back sheet PV module

values recorded and IR images, it can be said that the temperature corresponding to the blockage of radiative loss is = 61.9 - 29.4 = 32.5°C. This blocking of radiative loss causes higher operating temperatures. Since IR imaging captures only the surface radiation (not the cell temperature) of the substrate, IR image shows that the white back sheet with aluminum cover is about 32°C cooler but the cell is in fact hotter as can be seen by the attached thermocouples. Therefore, it is cautioned that the temperature determination of the aluminum covered backsheet using IR images would be misleading and cannot be correlated to the measured values.

3.1.3 Temperature coefficients: Dependence on thermocouple location

As shown in Fig. 9, there is a significant dependence of P_{max} temperature coefficient on the location of module back surface thermocouple. The temperature coefficients shown in this figure were obtained on the as-received modules before the installation of thermal insulation or adhesion of aluminum. As noted earlier, the a-Si (double junction) framed and c-Si framed modules are glass/polymer modules whereas the CdTe frameless and CIGS framed modules are glass/glass modules. It is observed that the temperature coefficient of frameless module (CdTe) experiences the least dependence on the thermocouple location of the module.

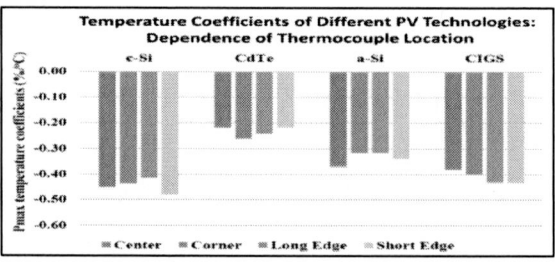

Fig. 9. Dependence of P_{max} temperature coefficient on the thermocouple location of the module

3.1.4 Performance parameters: Dependence on thermocouple location

Temperature variation within a module can have significant impact on the accuracy of module performance parameters' data. For this study, the performance parameters and temperatures of all the modules (while maintaining under MPPT thermal condition at the times when I-V curves are not taken) were continuously monitored for minimum two clear sunny days using the multi-curve tracer. As shown in Fig. 10, about 8% temperature variation between the thermocouples in c-Si and CIGS modules has respectively caused about 2% and 1.5% variations in P_{max}. On the other hand, about 14% variations in temperature in CdTe module seem to cause about 4% variation in P_{max} values.

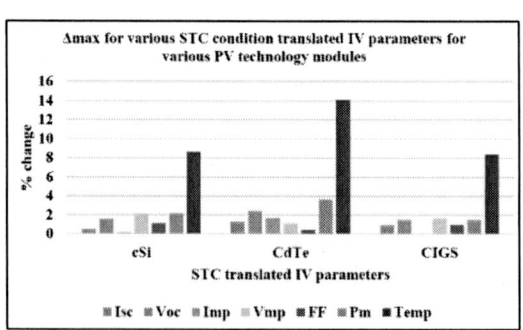

Fig. 10. Dependence of performance parameters (% change) on the thermocouple temperature (% change)

3.2. Plant-level thermal uniformity

3.2.1. The plant and module level temperature distribution

The data for AZ3 and AZ5 PV power plants, which was recorded at five-minute interval, was averaged and analyzed from 9am to 5pm daily from 04/17/2015 to 09/30/2015. Based on the temperature data obtained between 04/17/2015 to 09/30/2015, the following analysis is presented. As shown in Fig. 11, the 1-axis plant experienced higher temperature than the fixed-tilt plant.

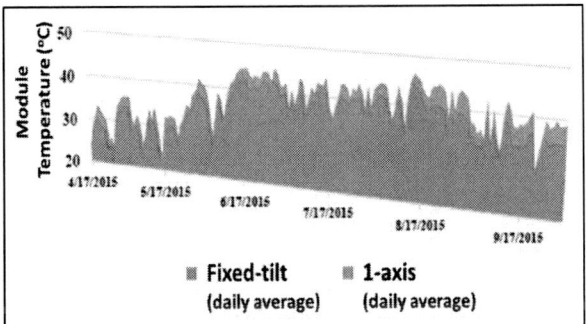

Fig. 11. Average of all five modules in two power plants

In order to study the trend further, thermal mapping was also performed on the individual PV modules (4 thermocouples per module) as well as the complete power plants (5 modules per plant) on a clear sunny day around solar noon time period from 12 to 1 pm. This is shown in Fig. 12.

In the fixed-tilt plant, it is typically observed that the modules

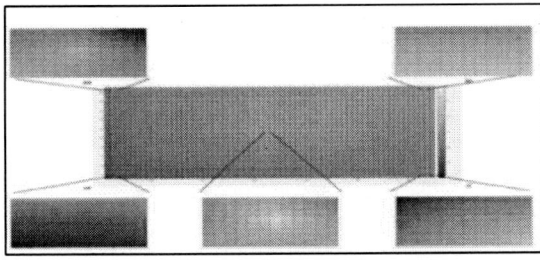

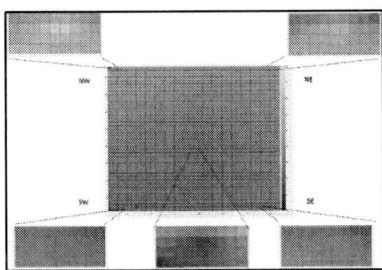

Fig. 12. Thermal mapping of AZ3 (fixed-tilt) and AZ5 (1-axis) PV plant at around noon on a clear sunny day of July 15, 2015.

in NW direction are the hottest while the module in SW direction are the coolest. In the 1-axis plant, it is typically observed that the modules in the center are the hottest, while

TABLE II
ANALYSIS OF VARIANCE (ANOVA) DESIGN SUMMARY FOR AZ3 AND AZ5 PV PLANTS ON A CLEAR SUNNY DAY

Factor	Type	Levels	Values
Plant	fixed	2	1, 3
Module locations	random	5	1, 2, 3, 4, 5
Thermocouple	fixed	4	1, 2, 3, 4

Source	DF	SS	MS	F	P
Plant type	1	0.7317	0.7317	3.67	0.073
Module locations	4	2.7835	0.6959	3.49	0.031
Thermocouple	3	0.8957	0.2986	1.38	0.296
Plant*Thermocouple	3	0.6015	0.2005	1.01	0.416
Module locations*Thermocouple	12	2.5975	0.2165	1.09	0.43
Error	16	3.1907	0.1994		
Total	39	10.8007			

TABLE III
ANALYSIS OF VARIANCE (ANOVA) DESIGN SUMMARY FOR AZ3 AND AZ5 PV PLANTS ON A CLOUDY DAY

Factor	Type	Levels	Values
Plant	Fixed	2	1, 3
Module locations	Random	5	1, 2, 3, 4, 5
Thermocouple	Fixed	4	1, 2, 3, 4

Source	DF	SS	MS	F	P
Plant type	1	8.29	8.29	5.37	0.034
Module locations	4	18.216	4.554	2.95	0.05
Thermocouple	3	17.629	5.876	3.09	0.068
Plant*Thermocouple	3	1.686	0.562	0.36	0.78
Module locations*Thermocouple	12	22.827	1.902	1.23	0.342
Error	16	24.708	1.544		
Total	39	93.358			

the modules in SW direction are the coolest. This trend appears to be mainly dominated by the wind direction.

3.2.2 ANOVA design for AZ3 and AZ5 PV plants

The ANOVA of effect model is performed further to study the effect of various factors on the temperature of AZ3 and AZ5 power plants. The three factors (power plant type, module and thermocouple locations) with different levels are studied on a clear sunny and cloudy day through ANOVA. The average irradiance recorded from 9am-5pm on a clear sunny day was 940 W/m^2 and that on the cloudy day was 329 W/m^2. The average wind speed recorded on clear sunny day was 2 m/s while that on a cloudy day was 4 m/s.

The response values are normally distributed and residual values fitted had satisfactory pattern. Table II and III represent the ANOVA design summary for AZ3 and AZ5 power plants on clear sunny and cloudy days respectively. The p value for plant type and module locations is less than 0.05 on a clear sunny day but on a cloudy day p-value only for module locations is less than 0.05. Therefore, plant type and module location has a significant effect on temperature variation on a sunny day. On the other hand, only module location has a significant effect on temperature variation on a cloudy day.

IV. CONCLUSIONS

Thermal uniformity mapping of both modules and plants have been performed through long-term monitoring studies. At the module level, center cells tend to operate at the highest temperatures and the frame-insulated modules tend to experience more uniform temperatures than the framed modules. The temperature coefficients are significantly dependent on the location of the thermocouple on the module. This dependence can be decreased by the use of black frame or insulating the frame. Based on this study, it is determined that all the cells in a module can be assumed to be operating at a single temperature if a single thermocouple is used after thermally insulating the conventional gray frame (best thermal uniformity) or if a single thermocouple is used with black frame (second best thermal uniformity) or if the average temperature of four thermocouples is used with the conventional gray frame without any thermal insulation (third best thermal uniformity). At the plant level, the center modules of the 1-axis plant and the NW modules of the fixed-horizontal plant tend to operate at the highest temperatures. Overall, the 1-axis modules experience higher operating temperatures than the fixed-tilt modules. Therefore, the degradation rate for the 1-axis modules is expected to be higher than the fixed-tilt modules. For the thermal uniformity mapping study of PV modules in powerplants, we can conclude that position (four corners or center) of the hottest modules depends on the time of the day, wind speed, wind direction and height of array. A detailed analysis of this work is presented elsewhere [4].

ACKNOWLEDGEMENT

This research is based upon work supported by the Solar Energy Research Institute for India and the U.S. (SERIIUS) funded jointly by the U.S. Department of Energy subcontract DE AC36-08G028308 (Office of Science, Office of Basic Energy Sciences, and Energy Efficiency and Renewable Energy, Solar Energy Technology Program, with support from the Office of International Affairs) and the Government of India subcontract IUSSTF/JCERDC-SERIIUS/2012 dated 22nd Nov. 2012. The seed funding to initiate this project was provided by Salt River Project (SRP), Arizona.

REFERENCES

[1] N. Umachandran and G. TamizhMani, "Effect of spatial temperature uniformity on outdoor photovoltaic module performance characterization," *2016 IEEE 43rd Photovoltaic Specialists Conference (PVSC)*, Portland, OR, 2016, pp. 2731-2737.

[2] J. Belmont, K. Olakonu, J. Kuitche and G. TamizhMani, "Degradation rate evaluation of 26-year-old 200 kW power plant in a hot-dry desert climate," *2014 IEEE 40th Photovoltaic Specialist Conference (PVSC)*, Denver, CO, 2014, pp. 3162-3166.

[3] IEC 61853-2:2016, "Photovoltaic (PV) module performance testing and energy rating - Part 2: Spectral responsivity, incidence angle and module operating temperature measurements", 2016

[4] Ashwini Pavgi "Temperature Coefficients and Thermal Uniformity Mapping of PV Modules and Plants," M.S. Thesis, Arizona State University, July 2016 (free downloading: repository.asu.edu).

Climate-specific Thermal Model Coefficients for c-Si and Thin-Film PV Modules

Ashwini Pavgi, Joseph Kuitche, Jaewon Oh and GovindaSamy TamizhMani

Arizona State University Photovoltaic Reliability Laboratory (ASU-PRL), Mesa, Arizona, 85212, USA

Abstract — The thermal model coefficients (U_c and U_v), similar to the thermal loss factors used in PVsyst for various PV technologies specific to a hot climate are presented in this paper. Minitab, a statistics package was used for the analysis of the data. These coefficients were determined by statistically correlating a year-long data for the modules mounted on free-standing arrays of four PV technologies (crystalline silicon, CdTe, CIGS, a-Si) which experienced hot-desert climate conditions. Various statistical approaches were used to analyze the data. Limiting the wind speed in the one hour interval data showed statistically satisfactory patterns in the model adequacy plots. The U_c and U_v values determined for a-Si module with Tefzel superstrate was higher than the a-Si module with glass superstrate. The lowest U_c and U_v values were observed for CdTe PV technology. For crystalline PV technology, U_c value of 25.46 W/m²·K and U_v value of 4.31 W/m³·K·s were obtained.

Index Terms —thermal model, thermal loss factors, PVsyst, hot-desert climate

I. INTRODUCTION

Module temperature is influenced by various design, installation and weather factors. The design factors include module technology type, encapsulant type, superstrated type and substrate type. The installation factors include fixed-open rack, rooftop and 1-axis tracking. The weather factors include ambient temperature, irradiance and wind speed. Therefore, predicting the operating temperature of a module in the field is a complex undertaking due to the influence of these interactive factors. Thermal models help to effectively quantify these factors and estimate the module operating temperature by considering their influences. These models help in reducing inherent uncertainty associated with module temperature determination which in turn improve the accuracy of performance models. These accurately determined performance models play an important role to project annual energy production while designing and operating a photovoltaic system.

Various thermal models are being put forward in the photovoltaics (PV) industry based on either theoretical heat transfer approach or the empirical equations using real time field data. And PVsyst is one of the most used commercially available models. PVsyst is a widely-used PC software package for simulation and data analysis of complete PV systems. It defines the thermal loss (for modules) by using thermal model coefficients of U_c and U_v which is further used in predicting the energy output. PVsyst states that thermal behavior is characterized by a thermal loss factor, U which is split into two components: constant U_c component and wind proportional U_v component. PVsyst proposes U_c and U_v values for three different configurations: wind-dependent and wind-independent weather data for modules on free-standing arrays as well as for modules on fully insulated arrays [1]. Based on a year-long field measured data, this study has statistically determined the thermal model coefficients (U_c and U_v) for modules mounted on free-standing arrays of various thin-film and crystalline silicon (c-Si) PV technologies experiencing hot-desert climate conditions. The primary goal of this study is to provide technology-specific and climate-specific thermal coefficients, U_c and U_v, especially for PVsyst.

II. EXPERIMENTAL METHODS

In this study, the data obtained by ASU-PTL in 2000-2002 has been analyzed and presented. The data was collected at Mesa, Arizona on the field installed modules from different manufacturers covering multiple technologies for a long-term field monitoring [2]. Fourteen (14) PV modules that were tested in this test program is shown in Fig. 1.

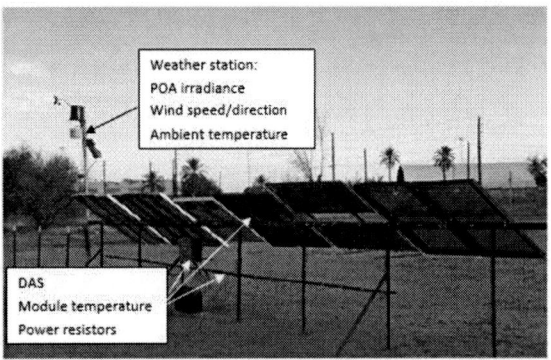

Fig. 1. Modules installed at ASU-PTL site during 2000-2002. Top: front view, bottom: back view [2]

The technologies are: monocrystalline Si (mono-Si), polycrystalline Si (poly-Si), EFG-polycrystalline Si (EPG-Si), amorphous Si (a-Si), Copper Indium Gallium Diselenide (CIGS) and Cadmium Telluride (CdTe). A set of two modules with same electrical specifications and manufacturers for each PV technology was installed. The modules were installed on an open rack system at the site, which experienced hot desert climate conditions. The modules were maintained at near to their P_{max} (maximum power) operating conditions with the help of power resistors. Weather station installed near the test setup monitored the wind speed and direction, ambient temperature, and latitude-tilt global irradiance. All the modules were installed on south facing, latitude-tilt racks with thermocouples attached on the substrate of each module. Table I provides the information about the various PV modules installed on this system along with respective cell technology, front and back sheet material specifications and their manufacturers.

TABLE I
VARIOUS PV MODULES INSTALLED ON THE RACK (2001) [2]

Item	Cell Technology	Model Number	Manufacturer
1	a-Si	US32	USSC
2	a-Si	US32	USSC
3	Mono c-Si	SM55	Siemens
4	Mono c-Si	SM55	Siemens
5	CIGS	ST40	Siemens
6	CIGS	ST40	Siemens
7	EFG-Poly c-Si	50ATF	ASEA
8	EFG-Poly c-Si	50ATF	ASEA
9	Poly c-Si	MSX60	Solarex
10	Poly c-Si	MSX60	Solarex
11	CdTe	N/A	SCI
12	CdTe	N/A	SCI
13	a-Si	Millennium	Solarex
14	a-Si	Millennium	Solarex

The data was stored at every 5-minute interval in the data acquisition system and retrieved periodically. The quality of the collected data was verified periodically by normalized module temperature rise from ambient at 800 W/m² irradiance. Average wind speed measured throughout the year was 1.8 m/s and average ambient temperature was 23.6°C. Average measured plane of array (POA) (annual) during the solar window time was 837 W/m².

III. VALIDATION METHODS

3.1 Thermal models

Various thermal models have been proposed in the industry based on theoretical approach or the empirical equations based on real time data. This paper presents two of those models, which were referred as part of this study.

3.1.1 Faiman module temperature model

David Faiman model uses the simple heat transfer phenomenon to determine module temperature. This approach measured POA irradiance, ambient temperature, wind speed and module temperature for seven module types in order to fit data for heat loss coefficients U_o and U_1 [3]. The equation to determine module temperature is as follows:

$$T_m = T_a + \frac{H}{U_0 + U_1 \times v} \qquad (1)$$

where
T_m: module temperature (°C)
T_a: ambient air temperature (°C)
H: irradiance incident on the plane of the module or array (W/m²)
U_o: constant heat transfer component (W/m²·K)
U_1: convective heat transfer component (W/m³·s·K)
v: wind speed (m/s)

3.1.2 PVsyst thermal model

PVsyst, a PV performance modelling software, have implemented a cell temperature model based on the Faiman module temperature model. The equation to determine the cell temperature is as follows:

$$T_c = T_a + \frac{\alpha \times EPOA \times (1 - \eta_m)}{U_0 + U_1 \times WS} \qquad (2)$$

where
T_c: cell temperature (°C)
T_a: ambient air temperature (°C)
α: adsorption coefficient of PV module (PVsyst default value = 0.9)
EPOA: irradiance incident on the plane of the module or array (W/m²)
η_m: PV module efficiency (PVsyst default value = 0.1)
U_o: constant heat transfer component (W/m²·K)
U_1: convective heat transfer component (W/m³·s·K)
WS: wind speed (m/s)

3.2 Statistical correlation flowchart

Based on the principles similar to these models, Excel and Minitab were used extensively to statistically correlate the year-long data on a monthly and seasonal basis. The methodology as shown in Fig. 2 was used to correlate the data and determine U_c and U_v coefficients.

IV. RESULTS AND DISCUSSION

4.1. Determination of U_c and U_v values using five-minute interval data for year-long

The five-minute interval data available for one year was used to fit the line and obtain U_c and U_v parameters. Even after limiting the y-axis co-ordinate to 120 W/m²·K, the R-square

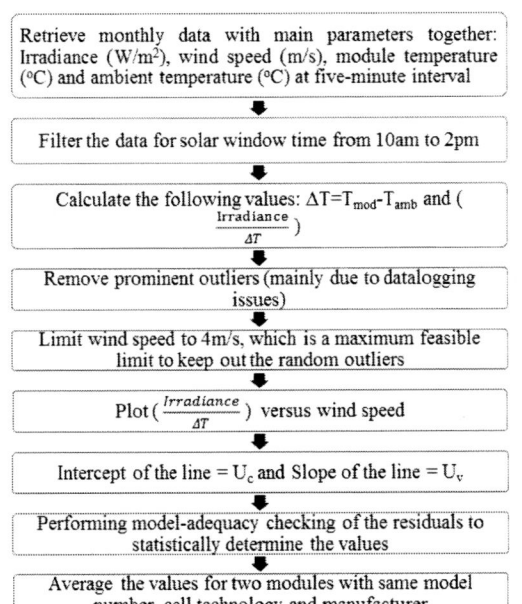

Fig. 2. Flowchart to determine U_c and U_v coefficients

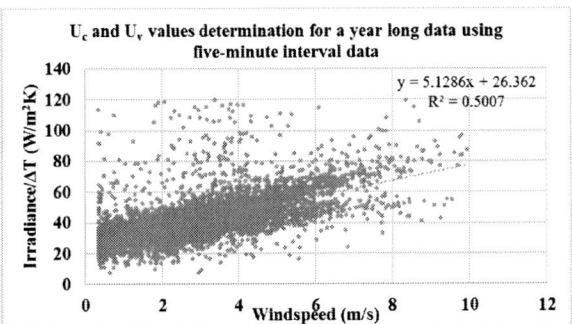

Fig. 3. Determination of U_c and U_v values for a year-long data (2001) at five-minute interval for polycrystalline silicon PV technology

accuracy of the trend was 0.50 as shown in Fig. 3. Additionally, due to presence of more than 10,000 data points, the residual plot obtained in for a year-long data was difficult to analyze. Therefore, the data for each month for each technology was separately analyzed. The model adequacy plot is shown in Fig. 4. This paper only presents the plots for one module but exact same statistical analysis was performed for all 14 modules of four PV technologies and presented elsewhere [4].

4.2 Determination of U_c and U_v Values using Five Minute Interval Data for Each Month

The five-minute interval data available for about 1 year period was analyzed separately for each month to obtain U_c and U_v values technology-specific for each month and understand the trend. The summer and spring season tend to have higher U_c values as compared to winter and fall seasons. On the other

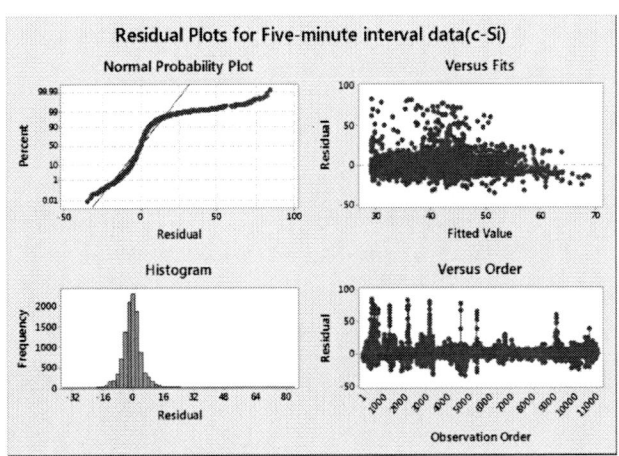

Fig. 4. Residual plots for five-minute interval data for monocrystalline silicon PV technology

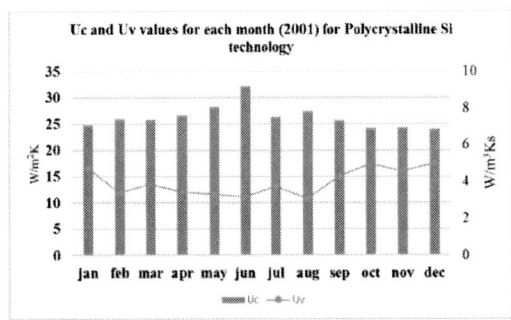

Fig. 5. U_c and U_v values for each month (2001) for polycrystalline PV technology

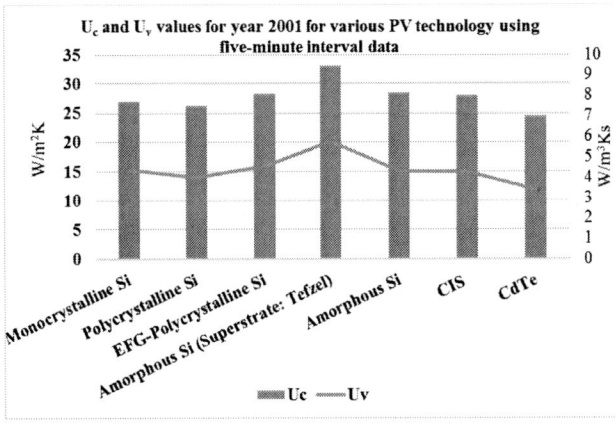

Fig. 6. U_c and U_v values of each month averaged for year-2001 for various PV technology

hand, during the summer-spring season, U_v values are lower. Fig. 5 represents U_c and U_v values based on data for each month separately for polycrystalline PV technology.

An average value of the data for all the months was calculated and plotted as per specific technology. Fig. 6. shows the U_c and

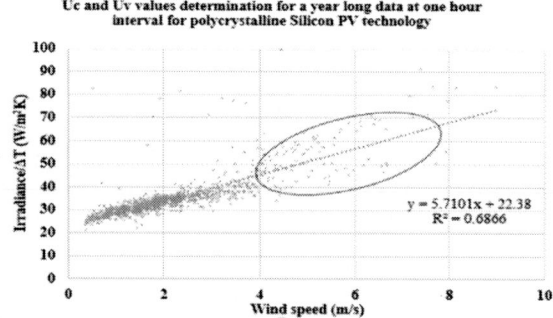

Fig. 7. Determination of U_c and U_v values for a year-long data at one hour interval for polycrystalline silicon technology

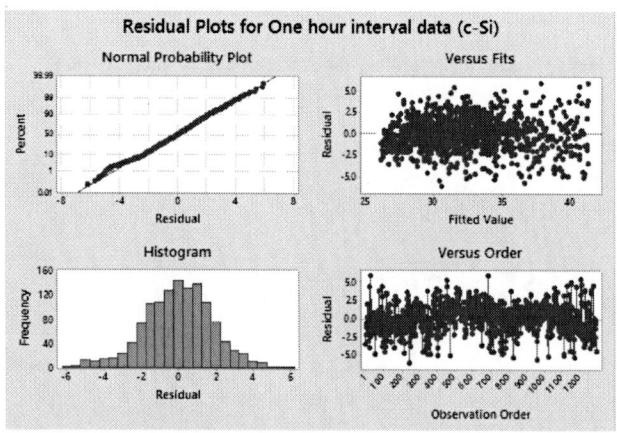

Fig. 8. Residual plots for a year-long data (2001) at one hour interval for polycrystalline silicon PV technology

U_v values for all PV technologies based on one year data. Model adequacy check was performed to statistically correlate the U_c and U_v values and determine the residuals. In most of the plots, the residuals followed satisfactory pattern and the data had normal distribution. But the normal probability plot was lightly tailed, or in other words, did not have normal distribution about the mean. Therefore, another approach was followed.

4.3. Determination of U_c and U_v Values using Hourly Interval Data for Year-long

In order to statistically determine the residuals for model-adequacy checking and to remove the outliers causing tailed distribution, the five-minute interval data was converted to hourly interval data for full one year. The regression fit was obtained for a year-long data at one hour interval for polycrystalline Silicon PV technology with an R-square value of 0.69 as shown in Fig. 7.

A random pattern of data points was observed in Fig. 7 for wind speed values of 4 m/s and above. From feasibility point of view, wind speed values greater than 4 m/s affects the energy yield of PV modules and might affect its performance. Therefore, limiting the wind speed to 4 m/s also improved the R-square value and the model adequacy plots followed

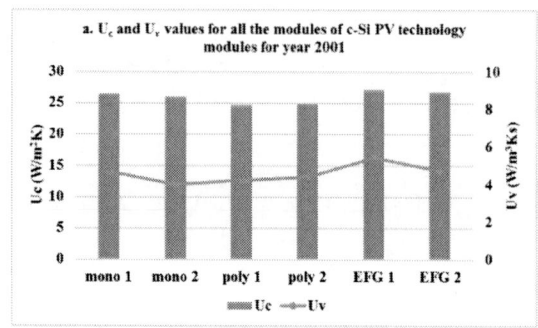

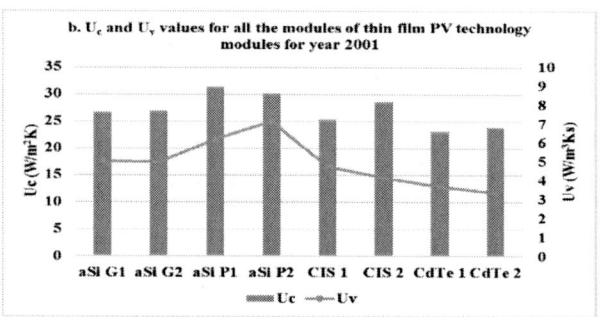

Fig. 9. U_c and U_v values for all the modules of c-Si and thin film PV technologies for year 2001.

satisfactory patterns. Fig. 8 represents a sample model-adequacy check plots for PV technology module.

It can be seen that the fitted values follow satisfactory pattern, the mean is normally distributed about zero and follows normal distribution. Therefore, the plots satisfy model adequacy check and the determination of U_c and U_v is statistically correlated. Moreover, the 95% confidence interval was obtained for each of the parameters. Fig. 9 (a) and (b) represents the U_c and U_v values for all replicates of c-Si and thin film PV technology modules respectively.

There is a stark difference between the values for amorphous silicon technology with glass and Tefzel superstrate, because the modules with polymer superstrate operate at lower temperatures than those with glass superstrate. The following trend is observed in amorphous silicon PV modules:

U_c value for a-Si with Tefzel superstrate > U_c value for a-Si with Glass superstrate.

Two ANOVA designs were performed to determine significance of module replicates and PV technology on U_c and U_v values, if any. The p-value for all the cases was obtained to be greater than 0.05 signifying no dominant effect as shown in Table II.

It can be observed from Fig. 10 that the U_c and U_v values for CdTe technology are the lowest. The lowest values for the CdTe modules are attributed to the glass/glass construction type. Considering all the PV technologies, the following trend is observed for the U_c and U_v values:

Polymer-Polymer > Glass-Polymer > Glass-Glass

TABLE II
ANOVA DESIGN TO DETERMINE SIGNIFICANCE OF MODULE REPLICATES (U_v VALUES)

Source	DF	Adj SS	Adj MS	F-Value	P-Value
Module	1	0.1015	0.1015	0.1	0.762
Error	12	12.7337	1.0611		
Total	13	12.8352			

ANOVA DESIGN TO DETERMINE SIGNIFICANCE OF MODULE REPLICATES (U_c VALUES)

Source	DF	Adj SS	Adj MS	F-Value	P-Value
Module	1	0.3376	0.3376	0.06	0.809
Error	12	66.3226	5.5269		
Total	13	66.6602			

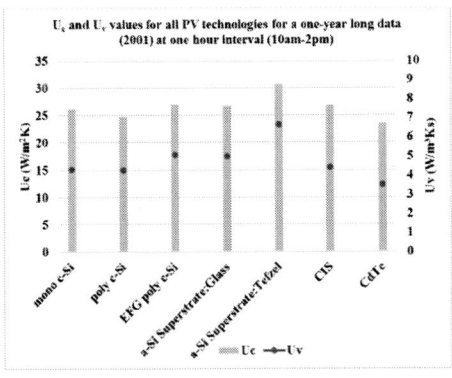

Fig. 10. U_c and U_v values for all PV technologies based on one year data at one hour interval (10am-2pm) for desert climate (Phoenix, Arizona)

IV. SUMMARY

This investigation provides the thermal coefficients (U_c and U_v) for thermal models including PVsyst. These coefficients have been developed for both c-Si and thin-film technologies specifically for the hot-desert climatic conditions. It is observed that the values of these coefficients are heavily influenced by the type of construction (glass/glass, glass/polymer, polymer/polymer).

ACKNOWLEDGEMENT

This research is based upon work supported by the Solar Energy Research Institute for India and the U.S. (SERIIUS) funded jointly by the U.S. Department of Energy subcontract DE AC36-08G028308 (Office of Science, Office of Basic Energy Sciences, and Energy Efficiency and Renewable Energy, Solar Energy Technology Program, with support from the Office of International Affairs) and the Government of India subcontract IUSSTF/JCERDC-SERIIUS/2012 dated 22nd Nov. 2012. The seed funding to initiate this project was provided by Salt River Project (SRP), Arizona.

REFERENCES

[1] "PVSyst Help page," [Online]. Available: http://files.pvsyst.com/help/thermal_loss.htm, May 2016.
[2] Tang Y. "Outdoor Energy Measurements of Photovoltaic modules," M.S. Thesis, Arizona State University, May 2005.
[3] Faiman, David, "Assessing the Outdoor Operating Temperature of Photovoltaic Modules," Photovoltaics Research and Applications, pp. 307-315, 2008
[4] Ashwini Pavgi "Temperature Coefficients and Thermal Uniformity Mapping of PV Modules and Plants," M.S. Thesis, Arizona State University, July 2016 (free downloading: repository.asu.edu).

Effect of the thermophysical properties of a phase change material on the electrical output of a concentrated photovoltaic system

Jawad Sarwar, Ahmed E. Abbas, Konstantinos E. Kakosimos

Sustainable Energy and Clean Air Research Laboratory, Department of Chemical Engineering, Texas A&M University at Qatar, 23874, Doha, Qatar

Abstract — **In this work, we developed a coupled thermal-optical-electrical model for a concentrator photovoltaic system integrated with a phase change material. We investigated the effect of the thermophysical properties of a PCM and ambient temperature on the electrical output. We found that the increase in ambient temperature from 25 °C to 50 °C decreases the electrical output up to 10%. The increase in thermal conductivity of the PCM beyond 12 $Wm^{-1}K^{-1}$ had a negligible effect in low ambient temperature conditions. The selection of the optimum thermophysical properties of a PCM can yield up to 13% gain in the electrical output.**

I. INTRODUCTION

Concentrator Photovoltaic (CPV) systems use optics (concentrators) to focus sunlight onto a photovoltaic (PV) cell, which is converted directly into an electrical DC output. Such systems, due to the concentrating effect, inevitably require cooling of the PV cell, since the increase in temperature of the cell leads to a decreased electrical output [1]. Many schemes exist for cooling, ranging from passive cooling using air as a cooling medium without expending additional energy, or active cooling, commonly using a water pump to circulate water for extended contact time. The thermal energy recovered can then be used for other applications, such as domestic hot water in case of a water heating/cooling system.

One particular method of passive cooling is the use of phase-change materials (PCM) to absorb waste heat and cool a PV cell [2]. Several experimental and numerical studies have been carried out to predict the thermal and the electrical performance of the PV using a PCM as a cooling medium [2-5]. It is reported that the cooling of a PV cell using a PCM leads to an improvement in the operational electrical conversion efficiency of a PV façade [2]. The efficacy of a PCM to improve the efficiency of a PV depends on the climatic conditions. Such an experimental and numerical analysis for the semi-arid climate of Vehari, Pakistan and maritime climate of Dublin, Ireland has shown different temperature regulation potential of the same PCMs [3]. A yearly performance analysis for a hyper-arid climate of Al-Ain, UAE has also demonstrated promising improvement of the electrical conversion efficiency of a PV with the integration of a PCM [4]. An experimental investigation for thermal regulation of a CPV using a PCM and heat spreading cooling method has been carried out. It is reported that combination of heat spreading and a PCM reduces PV temperature by up to 20 °C [6]. An analysis of such a system was done by Kibria et al. (2016) [7]; however, the scope of this analysis focuses on the thermal part of the PVPCM system and reports only temperature reductions and thermal efficiency. The power output changes are investigated by using thermal coefficients for maximum power, which report the power drop per temperature unit, on the assumption that the effect is linear throughout the operating temperature range.

In this work, an electrical model, which uses easily obtained information from a PV cell manufacturer datasheets, is coupled with a previously developed thermal and optical model for a CPV system having a PCM as a cooling medium [8]. Such a system is referred as a CPV-PCM system in the subsequent text. The developed coupled thermal-optical-electrical model is used to analyze the impact of thermophysical properties of the PCM and ambient temperature on the power output of the CPV-PCM system.

II. METHODOLOGY

A. Physical Model

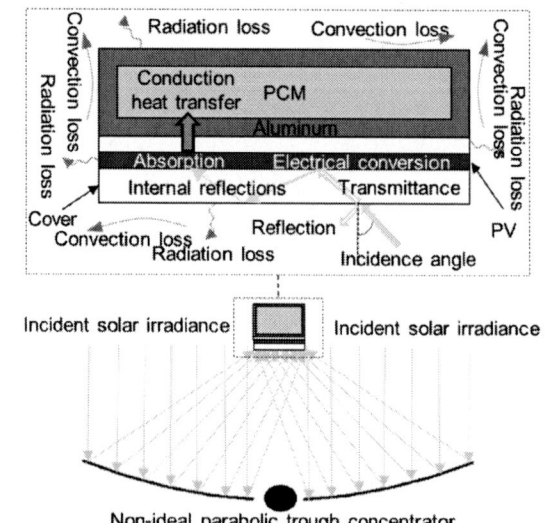

Fig. 1: Physical schematic of CPV-PCM system

The CPV-PCM system consists of a non-ideal parabolic trough concentrator, which focuses incoming sunlight onto a

TABLE I: PARAMETERS OF SYSTEM USED IN THIS WORK

Electrical Parameters – Hyundai S275RG Solar Module										
Datasheet Values			**Single Diode Model Values**							
P_{max}	275 W		I_{ov}	9.301709 A						
I_{mp}	8.8 A		I_0	$4.013 * 10^{-10}$ A						
V_{mp}	31.3 V		R_s	0.075 Ω						
I_{sc}	9.3 A		R_p	955.373 Ω						
V_{oc}	38.7 V		a	1.05						
I_{sc} temperature coefficient	0.032 %/K		No. of series cells	15						
V_{oc} temperature coefficient	-0.32 %/K									
Input thermal parameters for the model										
Property	**Steps**									
	1	2	3	4	5	6	7	8	9	10
Thermal conductivity, k (Wm^{-1}K^{-1})	2	4	6	8	10	12	14	16	18	20
Melting temperature, T_m (°C)	45	47.5	50	52.5	55	57.5	60	62.5	65	67.5
Heat of fusion, H (kJkg^{-1})	150	160	170	180	190	200	210	220	230	240
Ambient temperature, T_a (°C)	25	30	32.5	35	37.5	40	42.5	45	47.5	50
Density, ρ (kgm^{-3})	880									
Heat capcity, c (kJkg^{-1}K^{-1})	1.3 (solid phase), 1.8 (liquid phase)									

PV cell. The design methodology of a non-ideal parabolic concentrator is reported by author elsewhere [9]. A 6" PV cell having solar to electrical conversion efficiency of 16.8% [10] is attached to an aluminum container that contains a PCM. The dimensions of the PCM behind each PV cell are selected as $156.4 \times 156.4 \times 38$ mm^3. A glass cover having a thickness of 5 mm is considered in front of the PV cell while the thickness of aluminum 3 mm. The part of the incident solar energy is converted into electricity while the energy that is not used to generate electricity is conducted as heat through the aluminum and into the PCM, where the PCM absorb/release this heat during phase transformation. The overall thermal, electrical and optical behavior of the CPV-PCM is system is simulated using a developed simulation model.

B. Simulation Model

For electrical modeling, a five parameters model is used. The five parameters are the photocurrent I_L, the reverse saturation diode current I_o, the shunt and series resistances R_P and R_s and finally the diode ideality or quality factor n. The equation that describes the relation between current/voltage and the five parameters is:

$$I = I_{pv} - I_0\left[\exp\left(\frac{V + R_s I}{V_t a}\right)\right] - \frac{V + R_s I}{R_p} \qquad (1)$$

The thermal voltage of the array is $V_t = N_s kT/q$. The algorithm used is based on the work of Villalva et al. (2009)[11]. The electrical model logic coupled with the optical-thermal model is shown in Fig. 2. This algorithm was chosen since it only relies on data that is provided by PV cell manufacturers, simple to implement and accurate compared to other five-parameter model algorithms. [12].

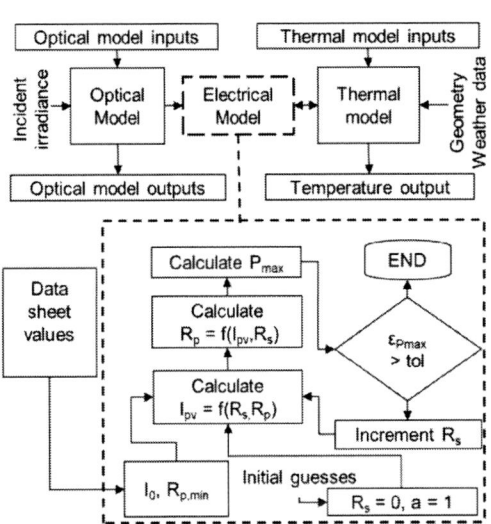

Fig. 2: Flowchart for simulation model, with exploded view of electrical model logic

For the optical-thermal model, we use a finite element based heat transfer model, developed via a 2D differential heat diffusion equation considering heat losses via convection and radiation [8]. The convective and radiative heat transfer coefficients are calculated using following equations:

$$h_c = \frac{Nu\,k}{L} \qquad (2)$$

$$h_r = \sigma\xi F\left(\frac{T^4 - T_a^4}{T - T_a}\right) \qquad (3)$$

Nu_k represents the Nusselt number and calculated using a correlation for an arbitrary shape [13]. The form factor, F, is considered as 1 while emissivity, ξ, is considered as 0.91. The optical behavior is simulated using a Fresnel equations method. The reflectance of the concentrator is assumed as 0.81 while the distribution of the incident flux is considered uniform. The details of the optical-thermal model, relevant equations and assumptions are presented by authors' elsewhere [8].

The input electrical and thermal parameters of the model are shown in Table 1. The thermal and electrical behavior of the CPV-PCM system is investigated by iterating thermophysical properties at a fixed incident irradiation of 1000 Wm^{-2} but variable concentration ratio between 2× - 4× i.e. a total of 30000 simulations. The electrical and thermal output is obtained, compared and analyzed.

III. RESULTS AND DISCUSSION

A. Electrical Model Validation

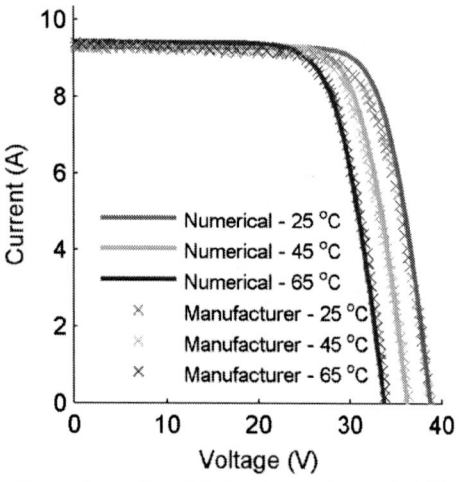

Fig. 3: Comparison of modeled and manufacturer's I-V curves for Hyundai HiS-S275RG at 1000 W/m^2

The electrical model was validated by comparing the I-V curves generated by the model, with those provided by the manufacturer [10] as shown in Fig. 3. The result shows that the I-V curves generated from the model are consistent with manufacturer's data.

B. Results

We examined the effect of the thermal conductivity, latent heat of fusion, the melting point of the PCM, and the ambient temperature on the power generation, the PV cell temperature and their temporal variations. It is found that the temperature of the PV cell varies with the change of these properties at any fixed incident irradiation and concentration ratio. The temperature variations of the PV cell at a concentration ratio

of 3× is shown in Fig. 4. It is found that the increase in the ambient temperature increases the PV cell temperature. The increase in the melting point of the PCM at a fixed ambient temperature also increases the temperature of the PV cell. The increase in the conductivity and the latent heat of fusion under constant other parameters decreases the temperature of the PV cell.

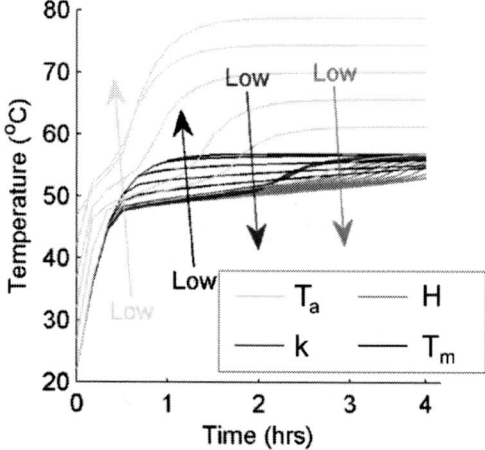

Fig 4: Temperature of the photovoltaic cell at different ambient temperature, thermal conductivity, heat of fusion and melting point at concentration ratio of 3×

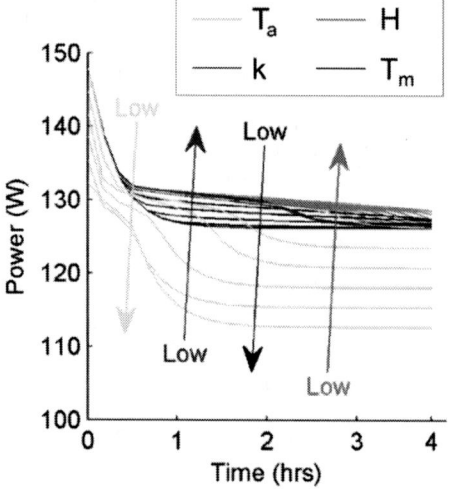

Fig 5: Power of photovoltaic at different ambient temperature, thermal conductivity, heat of fusion and melting point at concentration ratio of 3×

The power generation of the PV cell is also affected by the change of the thermophysical properties of the PCM and the ambient temperature as shown in Fig. 5. It is found that The increase in the conductivity and the latent heat of fusion increases the power output of the PV cell. The increase in the ambient temperature decreases the power output of the PV cell. The increase in the melting temperature of the PCM at a fixed ambient temperature decreases the power output.

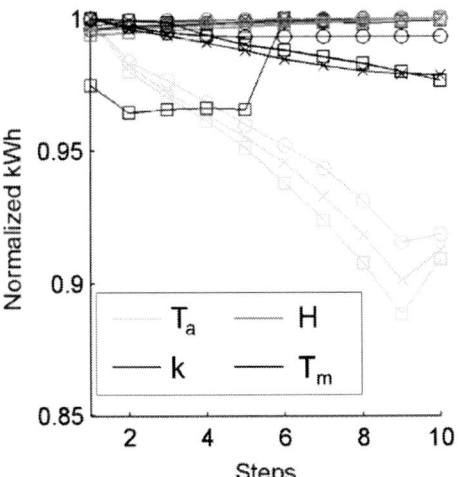

Fig. 6: Normalized energy (kWh) against selected parameters after 1 hour of operation. ○, ×, □ points denote 2×, 3× and 4× solar concentration respectively

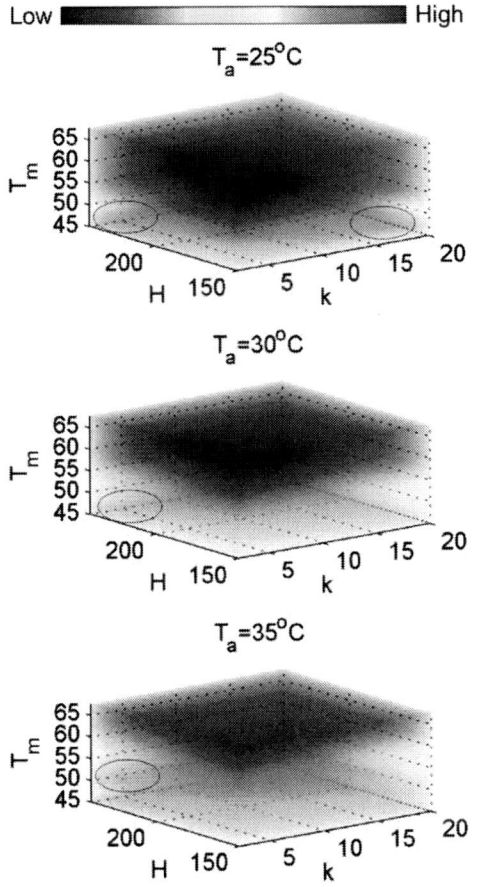

Fig. 7: Power generation at varying thermophysical properties of a PCM at ambient temperatures of 25 °C, 30 °C and 35 °C

The energy produced for one hour (kWh) was calculated and normalized for each set of simulation to compare their output (Fig. 6). We can see that the ambient temperature has the biggest effect on power generation; the power loss is almost 10% with an increase from 25 °C to 50 °C at all flux levels. The change in the heat of fusion, melting temperature, and ambient temperature affects the energy production linearly, but there is a non-linear change in case of thermal conductivity. There is a negligible increase in the energy output of the CPV system is found with an increase in thermal conductivity of the PCM beyond 12 $Wm^{-1}K^{-1}$ for all other fixed thermophysical properties. An overall 3.5% increase in energy output is found for thermal conductivity while the corresponding increase is found as 2% for change in melting temperature and the latent heat of fusion for all other fixed thermophysical properties.

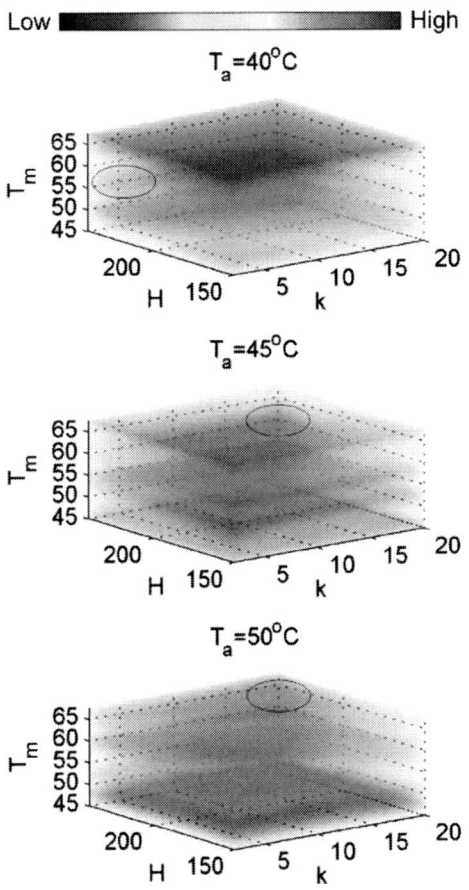

Fig. 8: Power generation at varying thermophysical properties of a PCM at ambient temperatures of 40 °C, 45 °C and 50 °C

We have also investigated the combined effect of the selected thermophysical properties of the PCM on the power generation of the PV cell. We have found thermophysical properties appropriate to generate a maximum power at a given ambient temperature. The result is shown in Fig. 7 and

Fig. 8 for different ambient temperatures and suitable thermophysical properties are encircled. It is found that maximum power at an ambient temperature of 25 °C can be generated by using a PCM having a high heat of fusion (210 – 240 kJkg^{-1}) and a low thermal conductivity (2 – 5 Wm^{-1}K^{-1}) as shown in Fig. 7. Alternatively, maximum power at an ambient temperature of 25 °C can also be generated by using a PCM having a low heat of fusion (150 – 170 kJkg^{-1}) but a high thermal conductivity (15 – 20 Wm^{-1}K^{-1}). A low melting temperature (45 – 50 °C) is found suitable for an ambient temperature of 25 °C. For the ambient temperatures of 30 °C, 35 °C, and 45 °C, the maximum power generation from a PV cell can be obtained by using a PCM having a high heat of fusion and a low thermal conductivity. For high ambient temperatures in the range of 45 – 50 °C, a high heat of fusion, and a high thermal conductivity and a high melting temperature yields maximum power. It is also found that the selection of the optimum thermophysical properties of a PCM can yield up to 13% gain in the electricity generation of a PV cell.

IV. CONCLUSION

In this work, an impact of the thermal conductivity, latent heat of fusion, ambient temperature and a melting point of the PCM on the power output of the CPV system is investigated. It is found that increase in thermal conductivity increases the electrical output by 3.5%. The melting point and latent heat of fusion affect the electrical performance up to 2%, and their selection depends upon utilization of the stored energy. The increase in ambient temperature decreases the output by 10%. The selection of a PCM with the optimum thermophysical properties produces significant gains in the power generation of a PV cell and power gain up to 13% is found.

REFERENCES

[1] E. Radziemska, "The effect of temperature on the power drop in crystalline silicon solar cells," *Renewable Energy,* vol. 28, pp. 1-12, 2003.

[2] M. J. Huang, P. C. Eames, and B. Norton, "Thermal regulation of building-integrated photovoltaics using phase change materials," *International Journal of Heat and Mass Transfer,* vol. 47, pp. 2715-2733, 2004.

[3] A. Hasan, S. J. McCormack, M. J. Huang, J. Sarwar, and B. Norton, "Increased photovoltaic performance through temperature regulation by phase change materials: Materials comparison in different climates," *Solar Energy,* vol. 115, pp. 264-276, 2015.

[4] A. Hasan, J. Sarwar, H. Alnoman, and S. Abdelbaqi, "Yearly energy performance of a photovoltaic-phase change material (PV-PCM) system in hot climate," *Solar Energy,* vol. 146, pp. 417-429, 2017.

[5] J. Sarwar, S. McCormack, M. Huang, and B. Norton, "Experimental validation of CFD modelling for thermal regulation of photovoltaic panels using phase change material," in *Proceedings of the international conference on solar heating, cooling and buildings EuroSun,* 2010.

[6] J. Sarwar, M. Browne, S. McCormack, M. J. Huang, and B. Norton, "Experimental investigation of temperature regulation of concentrated photovoltaic using heat spreading and phase change material cooling method," in *The 2nd International Conference on Sustainable Energy Storage (IC-SES), Dublin, Ireland,* 2013.

[7] M. A. Kibria, R. Saidur, F. A. Al-Sulaiman, and M. M. A. Aziz, "Development of a thermal model for a hybrid photovoltaic module and phase change materials storage integrated in buildings," *Solar Energy,* vol. 124, pp. 114-123, 2016.

[8] J. Sarwar, B. Norton, and K. E. Kakosimos, "Effect of the phase change material's melting point on the thermal behaviour of a concentrated photovoltaic system in a tropical dry climate," in *EuroSun 2016,* Palma de Mallorca, 2016.

[9] J. Sarwar, P. Lemarchand, S. McCormack, M. Huang, and B. Norton, "Novel Method to Design Non-Ideal Parabolic Dish Concentrator for Concentrated Photovoltaic Application; Its Theoratical Analysis and Experimental Characterisation for Design Validation," in *28th European Photovoltaic Solar Energy Conference and Exhibition,* 2013, pp. 632 - 635.

[10] H. I. Hyundai. (2014). *Hyundai Solar Module.*

[11] M. G. Villalva, J. R. Gazoli, and E. R. Filho, "Comprehensive approach to modeling and simulation of photovoltaic arrays," *IEEE Transactions on Power Electronics,* vol. 24, pp. 1198-1208, 2009.

[12] G. Ciulla, V. Lo Brano, V. Di Dio, and G. Cipriani, "A comparison of different one-diode models for the representation of I–V characteristic of a PV cell," *Renewable and Sustainable Energy Reviews,* vol. 32, pp. 684-696, 2014.

[13] A. F. Mills, *Heat transfer,* 2nd ed. ed.: Upper Saddle River, N.J. : Prentice Hall, [1999], 1999.

Passive Cooling of Photovoltaics with Desiccants

Lin J. Simpson,[1] Jason Woods,[1] Nicolas Valderrama,[2] Alex Hill,[3] Nina Vincent,[4] and Timothy Silverman[1]

[1] National Renewable Energy Laboratory, Golden, CO 80401, USA
[2] University of Florida, Gainesville FL 32611, USA
[3] University of Denver, Denver CO 80208, USA
[4] Scripps College, Claremont CA 91711, USA

Abstract — As part of efforts to reduce photovoltaic (PV) costs and improve reliability/durability, temperature control needs to be closely examined in a number of ways including novel passive cooling technology. Modest PV module operating temperature reductions can increase PV power output by 10% or more. The higher the operating temperature of a typical PV device, the lower the overall PV conversion efficiency. Because some incident light is converted to heat in PV cells, the PV module heats up (typically 25°C above ambient temperature in full sun), and the overall energy output of the PV system is reduced (typically a 10% power loss for every 20°C to 30°C increase in temperature). We are integrating material on the PV module that adsorbs water from the air at night when the PV module temperature is cool and the relative humidity in the air is typically high. This sorbed water evaporates as the PV module heats up during the day, taking the excess heat with it and thus effectively cooling the PV module (see Fig. 1). Initial experiments with this novel passive cooling have demonstrated that PV module operating temperatures may be reduced nearly 30°C with evaporative cooling. In this paper, we describe 1) novel materials development, 2) initial modeling that predicts how the cooling will work in different environments, and 3) results demonstrating the PV operating temperature reductions achieved to date.

Index Terms — photovoltaic cells, electronics cooling, electronic packaging thermal management.

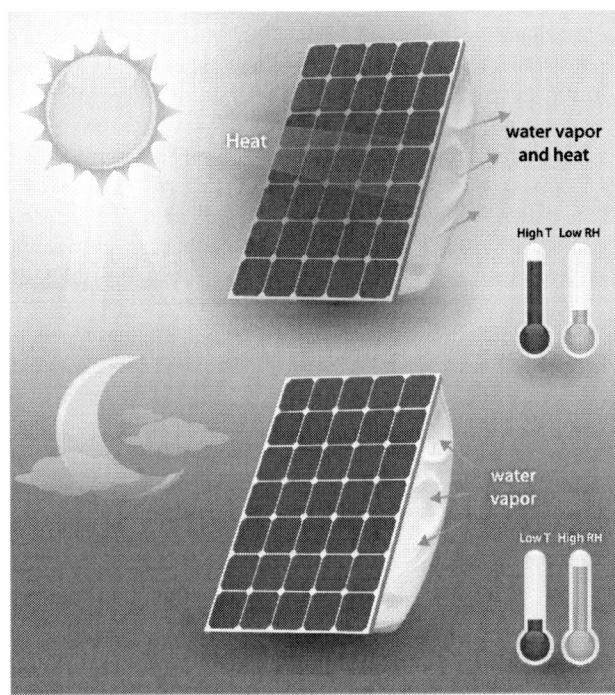

Fig. 1. Illustration demonstrating the use of desiccants to adsorb water from the higher-humidity air at night when the PV modules are cool, and the passive evaporation during the day when the relative humidity (RH) is lower and the PV module temperature is higher.

I. INTRODUCTION/OVERVIEW

For typical single-junction photovoltaics (PV), about 80% of the solar energy collected is ultimately converted to low-grade heat; thus, PV devices can heat up substantially during the day. For example, depending on the specific environmental conditions (e.g., air temperature, humidity, wind), PV device temperatures can reach over 100°C. Furthermore, PV cell performance is known to decrease with increasing temperature (0.3% to 0.5% per degree Celsius) and fully illuminated free-standing modules have operating temperatures typically 25°C above the ambient temperature.[1] This temperature difference can vary twofold depending on the mounting conditions.[2] In these conditions, the PV modules may be producing less than 70% of their rated power.

Ultimately, the higher PV operating temperatures decrease yearly energy output (e.g., perhaps ~10% for climatic conditions similar to those in Phoenix, Arizona), and thus increase the levelized cost of electricity (LCOE) by a commensurate amount. The increased device temperatures may increase the recombination rate of photogenerated carriers within the device, contributing to a substantial decrease in conversion efficiency, and thus, power collected by the PV device. Therefore, reducing PV device temperatures by even 20°C on average may improve the absolute power output by ~10%. This would be equivalent to about a 10% overall reduction in plant costs, regardless of the specific PV technology being used.

978-1-5090-5606-4/17 $31.00 © 2017 IEEE

II. METHODS

One novel [3] way to achieve substantial PV module temperature reductions involves integration of desiccants to provide passive cooling (see Fig. 1).

Passive Cooling: The use of water to cool PV modules has been well established. However, most of this type of cooling balances the operating temperature of the PV with the water temperature that can be used for other low-heat applications (e.g., domestic hot water). This type of balance often settles on relatively high operating temperatures around 40°C to 45°C.[4] Thus, this type of cooling may not provide substantial amounts of PV power increase. Furthermore, the use of water evaporation to cool PV modules has been shown to increase PV power output by 19% through a decrease in module temperature. [5] However, although this approach may be considered "passive" in terms of the cooling, it still requires "active" efforts to bring liquid water to the modules and integrate it onto the back.

The underlying principle of the passive cooling concept is based on the properties of a typical desiccant[6] where the water content is lower at higher temperatures and lower ambient relative humidities (i.e., when the temperature increases farther from the dew point, "dp"). As the desiccant transitions to lower moisture content, the evaporating water takes the heat of vaporization with it. The PV module temperature is thus kept lower. Even in places like Miami, Florida, with a dew point of 15°C, as the PV module heats up from the nighttime temperature of 20°C to ~60°C when illuminated during the day, the effective relative humidity next to the panel will be close to 10%. The desiccant in these conditions will become nearly saturated overnight and dry out substantially during the day, providing cooling.

Due to the large heat capacity associated with the latent heat of vaporization/condensation of water, only a small amount of water is needed to provide the cooling. Equilibrium estimates indicate that only 2 mm to 8 mm of desiccant may be needed to remove the heat load[7] from solar insolation. This correlates to less than 2 ¢/W worth of desiccant (less than 2% of the DOE 2020 1 $/W system cost goal), but it may increase power production by 10%.

The novel passive cooling technology has other benefits: it may be relatively lightweight compared to heat sinks or other related technology such as phase-change materials; it is agnostic to PV technology: and it requires no connections to other resources (e.g., thermal loads, pumps).

III. RESULTS

Novel Desiccant Development: Several materials work well for this application, including liquid (e.g., ionic liquids, deliquescent salts, see Fig. 2) such as lithium chloride (LiCl) or potassium acetate (K-Ac)) or solid-based desiccants. Liquid desiccants require the integration of a container that allows water vapor to pass in and out while retaining the liquid inside. Solid desiccants need to be constructed in a way to enable integration with the PV module while still providing easy access to the water vapor in the air. As shown in Figs. 2 and 3, we have investigated the use of several desiccants that have substantial water vapor sorption at ambient temperature in the nominal time frame of a single night. This includes liquid desiccants such as LiCl and novel compound solid desiccants being developed at NREL. One example is LiCl-SAP-ENG, a compound desiccant developed at NREL involving LiCl salt intercalated into a solid mixture of super absorbent polymer (SAP) and expanded natural graphite (ENG). The ENG was used primarily to increase the thermal conductivity, but with appropriate integration also improves water sorption. LiCl-SAP-ENG adsorbs 10 times more than silica gel). Although not having the water sorption capacities of the liquid desiccants, the solid desiccants remove the constraints of a liquid water-tight container.

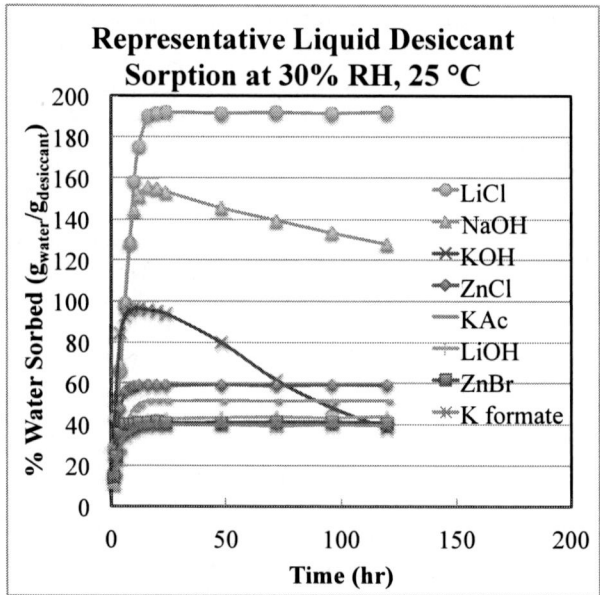

Fig. 2.   Comparison of different desiccant sorption as a function of time at 30% RH and 25 °C.

Desiccant Cooling Model Predictions: A hygrothermal model was developed to validate the concept. We used a one-dimensional heat and moisture transport model typically used for multi-layer building components exposed to weather[8]. The hygrothermal modeling showed (Fig. 4) that in Miami, the desiccant backing on the module could provide up to 8°C peak cooling with LiCl. In Golden, Colorado, the same system may provide up to 5°C peak cooling. The important difference between these climates is the amount of moisture available at night. Thus, these models demonstrate that passive desiccant cooling is a fundamentally sound approach. However, although the hygrothermal model is useful for 1) concept validation, 2) comparison of performance in different climates, and 3) assessment of different desiccants, experimental data are needed to provide the appropriate parameters for our novel desiccants and to help validate the model results.

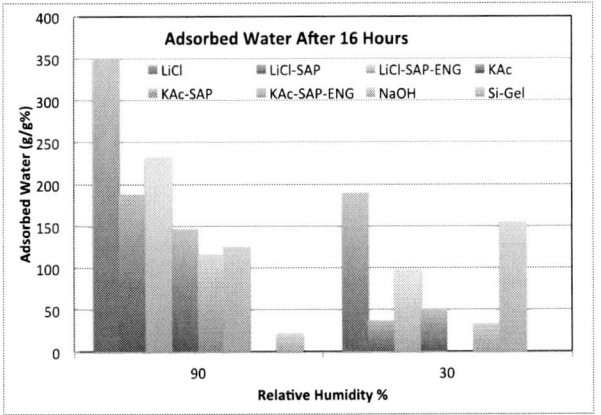

Fig. 4. Comparison of the sorption capacities of different dry desiccants after 16 hours in a 30% and 90% RH environmental chamber.

Novel Desiccant Cooling of PV Modules: In addition to developing novel desiccants and predictive models, NREL also demonstrated that desiccants provide substantial passive cooling. As shown in Fig. 5, two identical PV module configurations were constructed and used to measure the side-by-side temperature performance of one module with desiccant cooling and the second module with an amount of extra glass having equal heat capacity. For calibration, these identical measurement systems demonstrated virtually the same increase in temperature with the heat-capacity-equivalent glass applied to the back. However, when one of the modules contained liquid water or a desiccant, a substantial reduction in the PV module temperature was observed when the PV module was placed in a solar simulator that applied 1000 W/m^2 on the module. Numerous materials were used (e.g., LiCl, K-acetate, KCl, and compound solid desiccants) to demonstrate that with high enough nighttime relative humidity, the desiccant adsorbs water from the air, and then when the relative humidity is lowered and the PV temperature increases with illumination, the water sorbed by the desiccant keeps the PV module cooler while it is being desorbed. Similar experiments demonstrated that LiCl-SAP-ENG kept the PV module ~6°C cooler and KAc was almost as good as pure water.

Desiccant Optimization and Integration for Outdoor Deployment: Synthetic process improvements are needed to increase desiccant performance and reduce the sorption time

and the amount of materials needed for the compound solid desiccants. Furthermore, additional R&D is needed to design and construct ways to appropriately integrate the desiccants with the PV modules. For liquid desiccants, this requires that a liquid tight container be constructed that enables a high rate of water vapor exchange with the ambient air. One approach is to use commercially available water vapor membranes (often used for clothing) on the exposed air-side to allow the water vapor from the air to be adsorbed and desorbed by the desiccant. Figure 6 shows preliminary results where liquid desiccants were integrated on the back of PV modules using a water vapor permeable membrane. The results indicate an initial cooling of the desiccants, and in these preliminary results that LiCl appears to work better than K-acetate.

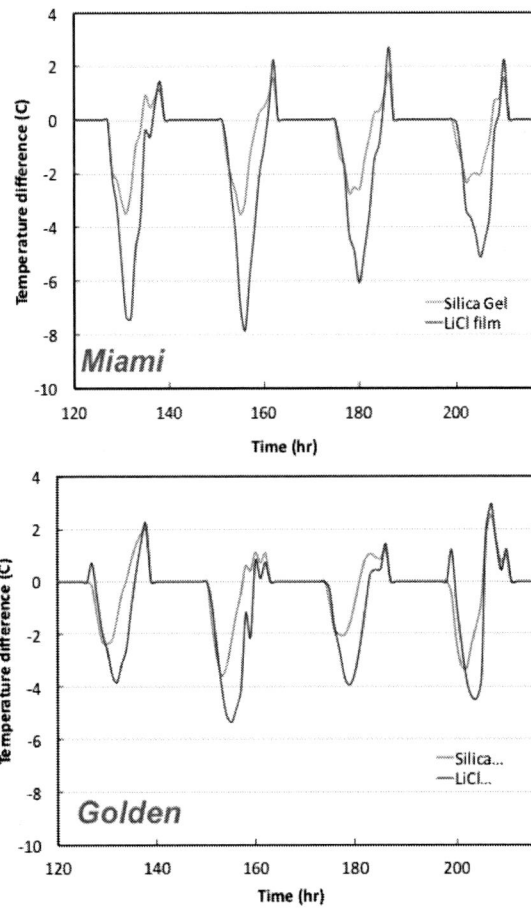

Fig. 3. Calculated temperature difference between desiccant-backed module and standard module, for two climates and two types of desiccant.

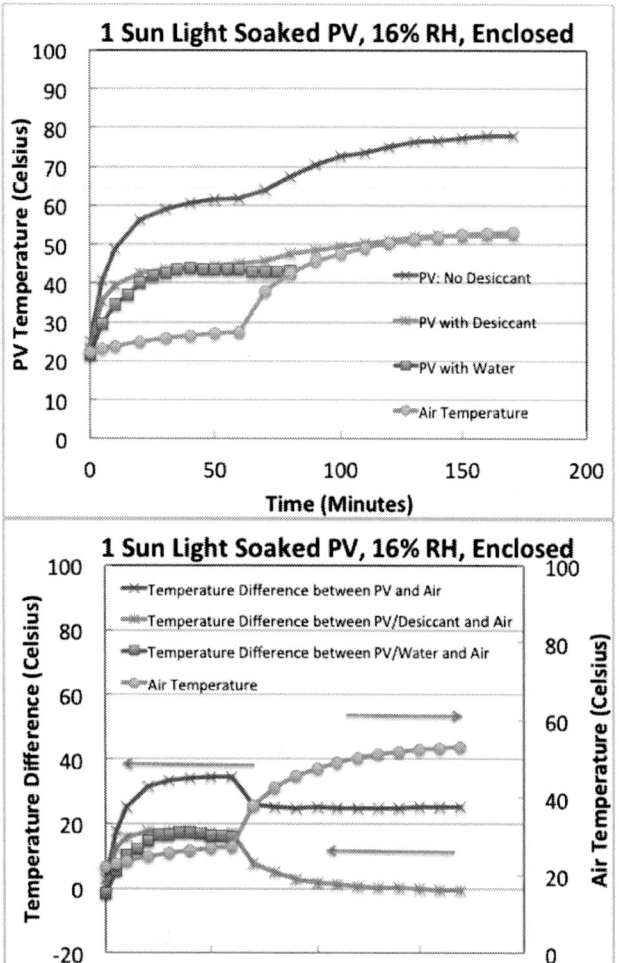

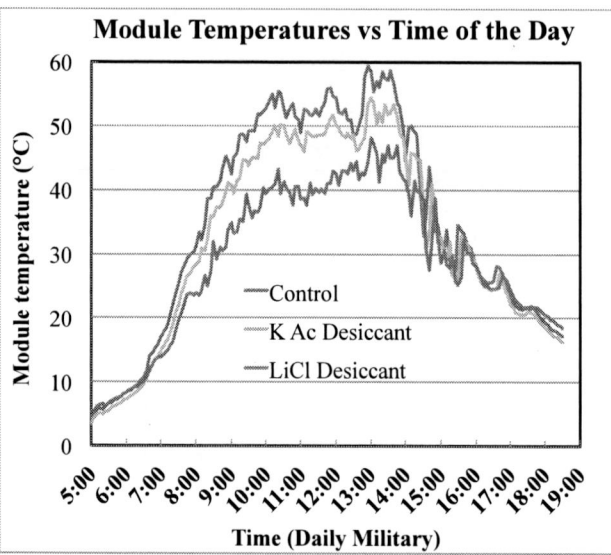

Fig. 6. Preliminary data demonstrating the cooling effect of lithium chloride (LiCl) and potassium acetate (K Ac) desiccants integrated with PV modules. The desiccants sorbed water from the air during the night and then cool the PV during the day.

Fig. 5. Top: PV temperature as a function of time under illumination for modules with no desiccant, desiccant, and water providing cooling. Bottom: Temperature difference between different modules with and without evaporative cooling. The data indicate that as the air temperature increases, the desiccant cooling improves and may keep the module cooler than the ambient air temperature.

ACKNOWLEDGMENTS

The authors would like to acknowledge and thank all those who contributed to this project including Ingrid Repins and Sarah Kurtz. This work was supported by the U.S. Department of Energy under Contract No. DE-AC36-08GO28308 to the National Renewable Energy Laboratory.

The U.S. Government retains and the publisher, by accepting the article for publication, acknowledges that the U.S. Government retains a nonexclusive, paid up, irrevocable, worldwide license to publish or reproduce the published form of this work, or allow others to do so, for U.S. Government purposes.

IV. CONCLUSIONS

PV modules are typically 25°C hotter than the ambient air in sunny conditions. This causes the PV modules to produce less power. NREL's novel approach of using a desiccant that sorbs water at night and then provides evaporative water cooling with that water during the day when the sun is shining has been demonstrated to reduce PV module operating temperatures. Preliminary results indicate that the PV modules may cool to below ambient air temperatures and that the PV modules produce more power (perhaps 10% more in sunny conditions). This initial work provides a good baseline set of results to continue designing and developing improved desiccants for PV module cooling.

REFERENCES

[1] B. Garcia, "Estimation of PV module yearly temperature and performance based on Nominal Operation Cell Temperature calculations," *Renewable Energy* **29**, 2004.

[2] J. Bloem, "Evaluation of a PV-integrated building application in a well-controlled outdoor test environment," Building and Environment 43, 2008.

[3] U.S. Patent Application No. 15/457,178. "Desiccant-Based Cooling of Photovoltaic Modules"

[4] Moharram et al., Ain Shams Engineering Journal, 4, 869, 2013.

[5] A.H. Alami, Energy Conversion and Management 77 (2014) 668–679.

[6] http://www.drytechinc.com/docs/SilicaGel-vs-ZEOLITES-Desiccant-Adsorption-at-Temperature-and-Dewpoint.pdf

[7] C. Ng, et al., "Experimental investigation of the Si gel–water adsorption isotherm characteristics," App. Thermal Eng. 21, 2001, pg 1631.

[8] Kuenzel, H.M., A. Karagiozis, A. Holm, A Hygrothermal Design Tool for Architects and Engineers (WUFI ORNL/IBP), in Moisture Analysis and Condensation Control in Building Envelopes, H.R. Trechsel, Editor. 2001, ASTM International.

Modified Maximum Power Extraction technique for rapidly changing NUI and dynamic loads

Aswani U[1], S.P. Duttagupta[1], T.I. Eldho[2], B.V.Rao[3]

[1]Electrical Engineering, IIT Bombay, Mumbai, Maharashtra, 400076, India

[2]Civil Engineering, IIT Bombay, Mumbai, Maharashtra, 400076, India

[3]Scientist, NSTL, Visakhapatnam, Andhra Pradesh, India

Abstract — In this paper, a modified algorithm for global maximum power extraction is demonstrated using Solar PhotoVoltaic array with distributed sensor network. The SPV array exhibits multiple power peaks during Non-Uniform Illumination (NUI) conditions. The algorithm demonstrated in this paper traces the global power peak by instantaneously predicting the power using multiple sensors data while also offering flexibility for varying load characteristics. Different technology SPV modules are studied to validate the importance of technology in predicting global maximum power point. Simulation studies are carried out and an analysis is presented to validate the effectiveness of the algorithm. The proposed algorithm is also verified by simulation models at different NUI scenarios with different load conditions.

Index Terms —Non-Uniform Illumination, Solar photo voltaic, Distributed sensor network

I. INTRODUCTION

In Solar PV installations, the optimization of power extraction when subjected to Non-Uniform Illumination (NUI) is a major concern. Few algorithms provided in literature are effective in tracking Global Maximum Power Point (GMPP) but need array scanning [1]-[4]. Artificial intelligence based techniques are effective in tracking GMPP but the algorithms are complex to implement [5]-[6]. Array reconfiguration and Distributed Maximum Power Point Tracking (DMPPT) technique helps in improving power output but requires more hardware [7]-[10]. A PWM duty based Maximum Power Extraction (MPE) technique was proposed in which GMPP is tracked instantaneously with power prediction algorithm using technology dependent parameters under dynamic NUI conditions. The total array scanning is avoided and a voltage range is defined in which GMPP exists [11]. A modified algorithm is proposed in this paper which predicts voltage and current at GMPP more accurately and in turn traces GMPP more effectively. Also, this new modified algorithm offers more flexibility for load characteristics there by making it suitable for resistive, battery and dynamically varying load conditions. In this paper, simulation study is carried out with the power circuit which works in closed loop with predicted GMPP voltage as reference input. An error analysis with respect to the predicted GMPP and the actual GMPP is also carried out to validate the algorithm. This modified Maximum

Power Extraction (MPE) technique works effectively for rapidly changing NUI conditions and even for dynamic loads.

Section-II describes the methodology of the proposed modified algorithm. Section-III demonstrates the study of technology parameters K_1 and K_2. Section-IV explains and discusses the results and discussions. Section-V presents the conclusions of the work.

II. METHODOLOGY

For an M×N array, the number of modules connected in series in each string is 'M' whereas 'N' indicates the total number of strings in the array. Fig. 1 shows the block diagram of the proposed MPE technique. The sensor network consists of illumination and temperature sensors. Illumination sensors are arranged across the four corners of each Solar Photovoltaic (SPV) module and a temperature sensor is placed at the bottom of each module. The sensor network provides data of dynamic NUI from all the four sides of module. The DSP based controller takes instantaneous sensor information through ADCs and predicts the voltage at GMPP. This predicted voltage is given as V_{REF} for closed loop control of the boost converter. The minute error in the predicted GMPP voltage by the proposed technique will be further tracked and compensated by Perturb & Observe (P&O) algorithm.

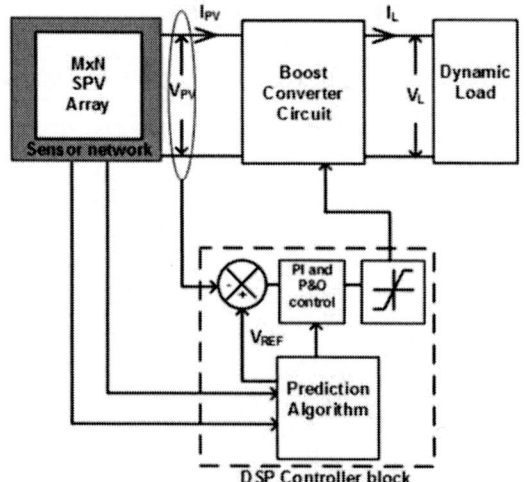

Fig. 1.Block diagram of proposed MPE technique

978-1-5090-5606-4/17 $31.00 © 2017 IEEE

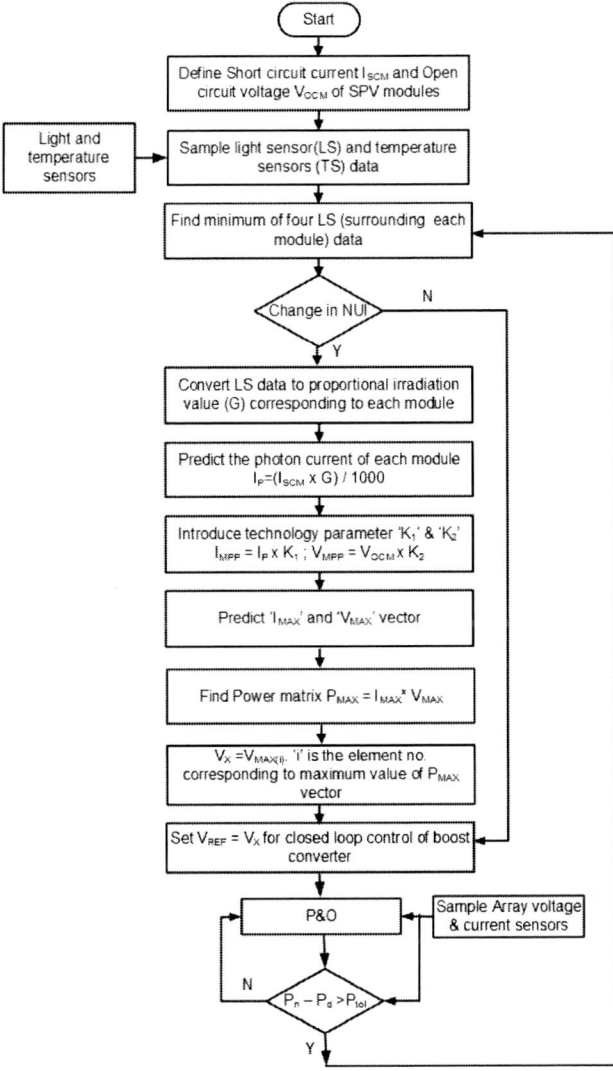

Fig. 2. Flowchart of the algorithm

Fig. 2 describes the flowchart of the proposed algorithm for M×N SPV array. The I_{MPP} and V_{MPP} of each module are calculated based on equation (1) and (2) using the real time irradiation and temperature data from the respective sensors.

$$I_{MPP} = K_1 * I_P \qquad (1)$$
$$K_1 = f(G, TF, D)$$

Where, I_P is the photon generated current in the module. K_1 depends on the physical parameters of the solar module such as irradiation (G), technology factor (TF) and degradation (D) of SPV module.

$$V_{MPP} = K_2 * V_{OCM} \qquad (2)$$
$$K_2 = f(T, TF, D)$$

Where, V_{OCM} is the open-circuit voltage of the module. K_2 depends on temperature (T), technology factor (TF) and degradation (D) of SPV module. These technology dependent factors K_1 and K_2 are to be derived experimentally or by simulation for the modules used in the SPV array. After calculating I_{MPP} and V_{MPP} of each module, the current and voltage boundaries (I_{MAX}, V_{MAX}) of the I-V curve for each string in the M×N array are predicted by the proposed algorithm. I_{MAX} and V_{MAX} vectors are determined by performing matrix sorting operations on I_{MPP} and V_{MPP} vectors. After predicting the desired values as described in the flowchart, the proposed algorithm calculates the power vector (P_{MAX}) at all possible power peaks of array P-V curve for the given NUI case. V_X is the voltage at GMPP taken from V_{MAX} vector which is given as V_{REF} for closed loop control of boost converter. With this approach, a very narrow zone of the SPV array's P-V characteristics is required to be scanned.

The V_{REF} predicted from algorithm is close to the actual V_{GMPP} of the array at the particular irradiation. The error or the difference between the actual V_{GMPP} and predicted V_{GMPP} analysis is presented in next section. This error is further minimised by Perturb & Observe (P&O) algorithm which shifts the array operating point further to GMPP. It is the most commonly used algorithm due to its ease of implementation. The duty cycle of the boost converter connected to the array is perturbed and the power is measured. If the power is increasing, the array is perturbed in the same direction until the array reaches its maximum power point.

Further, any variation or change of load in turn changes the operating point of the array and thereby pushes the array to operate at nearby local maximum in the absence of monitoring the load changes. The proposed algorithm monitors this and changes operating point to GMPP again by observing change in array power output. Along with load fluctuations, sudden changes in NUI are also taken care by the proposed algorithm. The change in power (P_n-P_d) is compared with P_{tol}, which is the predefined tolerable difference in overall array power.

III. STUDY OF K_1 AND K_2

In this section, the procedure for determining K_1 and K_2 for the purpose of the proposed algorithm is presented. It is experimentally found that K_1 varies with the type of SPV technology. Flexible mono-crystalline silicon (F_MCS) SPV module, robust mono-crystalline silicon (MCS) module, poly-crystalline silicon (PCS) module and flexible CIGS (F_CIGS) module are experimentally studied with a 'solar module simulator' to characterize them for K_1 at different irradiations. Dependency of K_1 on temperature is not considered as it is observed to be negligible. Fig. 3 shows the irradiation (G) versus K_1 of the four different technology modules. The dependence of K_1 on irradiation can be expressed by performing curve fitting on obtained data of each module. Table I gives the summary of coefficients for K_1 obtained experimentally for various SPV technology modules.

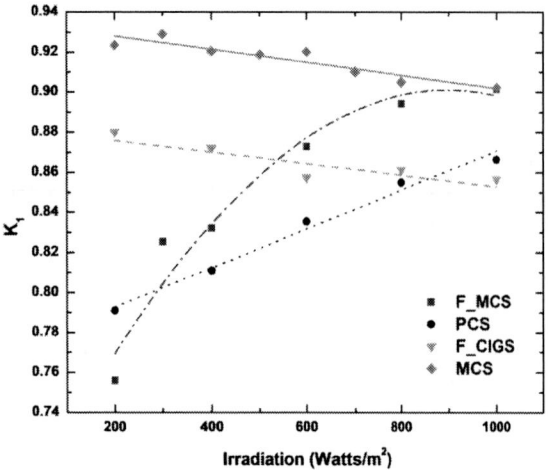

Fig. 3. Irradiation versus K_1 for different technology modules

$$K_1 \text{ (MCS)} = a_1 + b_1 G \quad (3)$$
$$K_1 \text{ (PCS)} = a_1 + b_1 G \quad (4)$$
$$K_1 \text{ (F\_MCS)} = a_1 + b_1 G + c_1 G^2 \quad (5)$$
$$K_1 \text{ (F\_CIGS)} = a_1 + b_1 G \quad (6)$$

TABLE I
COEFFICIENTS OF K_1 FOR DIFFERENT TECHNOLOGY MODULES

SPV module	a_1	b_1	c_1	R^2
MCS	0.93	-3.28e-05	-	0.84
F_MCS	0.68	4.87e-04	-2.72e-07	0.92
PCS	0.77	9.75e-05	-	0.98
F_CIGS	0.88	-2.88e-05		0.90

Where a1, b1 and c1 are regression coefficients. Similarly, K_2 can be derived by observing module P-V characteristics at different temperatures. K_2 is a technology dependent factor which is related to Open-Circuit Voltage (V_{OCM}) of the SPV module. The dependency or variation of V_{OCM} with respect to temperature for three crystalline silicon technology modules is shown in Fig.4. It is observed that the V_{OCM} of all the modules is linearly dependent on temperature.

$$V_{OCM} \text{ (MCS)} = p_1 + q_1 T \quad (7)$$
$$V_{OCM} \text{ (PCS)} = p_1 + q_1 T \quad (8)$$
$$V_{OCM} \text{ (F\_MCS)} = p_1 + q_1 T \quad (9)$$

Where p_1, q_1 are the regression coefficients. Table II gives the summary of coefficients. Apart from experimental approach, the K_1 and K_2 can also be obtained by observing P-V characteristics at different irradiations and temperatures using accurate simulation models of the SPV modules. In this paper, for validating the proposed modified MPE technique, the K_1 and K_2 factors are determined using accurate simulation models of SPV module. Initially, the SPV module is modelled with the available data of that particular module. The modelled module is simulated at different irradiations and temperatures and the P-V curves are plotted. By using equations (1) & (2) K_1 and K_2 of the modules are derived and regression analysis is carried out.

TABLE II
COEFFICIENTS OF K_1 FOR DIFFERENT TECHNOLOGY MODULES

SPV module	p_1	q_1	R^2
MCS	50.25	-8.99e-02	0.98
F_MCS	33.76	-5.09e-02	0.92
PCS	49.75	- 9.61e-02	0.87

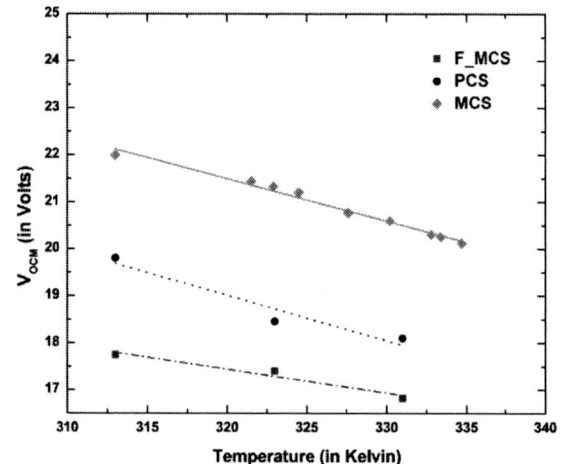

Fig. 4. Temperature versus V_{OCM} of different technology modules

IV. RESULTS & DISCUSSIONS

The proposed modified MPE technique with algorithm is validated by means of simulation studies using MATLAB SIMULINK®. A mono-crystalline SPV module with short-circuit current of 0.9A and open-circuit voltage of 24V is considered for this study. A 2x2 array with Uniform Illumination (UI) and two NUI combinations is considered for the validation as shown in Fig. 5. The actual P-V characteristics for these three different illumination cases are shown in Fig. 6. The actual power peaks and corresponding voltage points are tabulated in table III. To validate the algorithm, the SPV module is simulated at different irradiations and temperatures and technology dependent parameters K_1 and K_2 are determined.

TABLE III
POWER-VOLTAGE VALUES OF A 2X2 ARRAY (FIG. 5)

S.No	Voltage peaks		Power peaks	
UI (a)	-	34.6	-	47.6
NUI-1 (b)	17.3	37.9	23.1	19.7
NUI-2 (c)	16.4	36.4	22.6	32.3

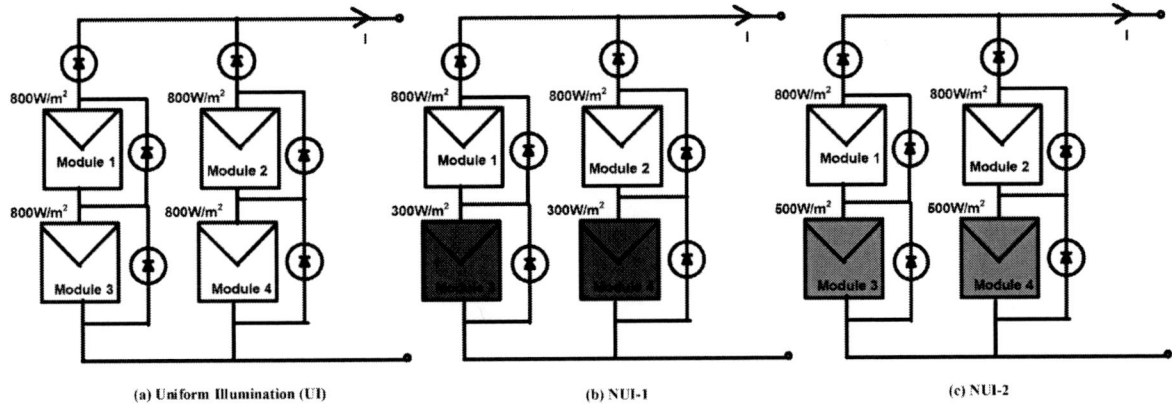

(a) Uniform Illumination (UI)   (b) NUI-1   (c) NUI-2

Fig. 5. Three different illumination conditions of a 2x2 array

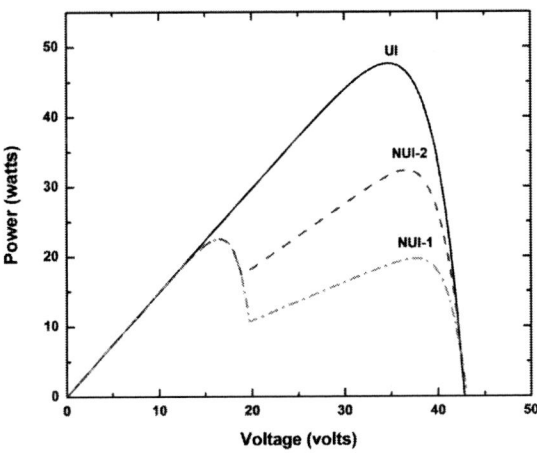

Fig. 6. P-V characteristics for cases given in Fig. 5

Fig. 7 shows the variation of K_1 with irradiation and K_2 with temperature for the selected mono-crystalline module. The equations for these technology dependent parameters to be used in the proposed algorithm are derived by curve fitting approach.

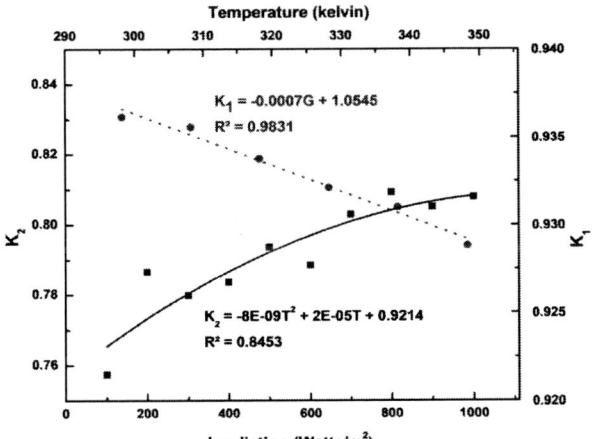

Fig. 7. K_1 vs. Irradiation and K_2 vs. temperature

These equations vary for different technology modules as substantiated by experimental studies presented in earlier sections. The equations shown in Fig. 7 are incorporated in proposed algorithm and simulations are carried out for the case studies shown in Fig.5. The predicted voltage and power at the local and global maximum are listed in Table IV.

TABLE IV
PREDICTED READINGS OF A 2X2 ARRAY (FIG. 5)

S.No	Voltage peaks		Power peaks	
	V_{MPP}(local)	V_{GMPP}	P_{MPP}(local)	P_{GMPP}
UI (a)	-	**34.8**	-	**46.2**
NUI-1 (b)	36.3	**17.3**	18.0	**23.1**
NUI-2 (c)	17.3	**35.6**	23.1	**29.5**

Further, the proposed algorithm is also simulated for these different illumination cases of 2x2 array to validate the tracking of GMPP. Fig. 8(a) shows simulation result of the proposed algorithm for voltage output. At 0.1 seconds, the illumination input to the array changes to NUI-1 scenario and at 0.2 seconds, the illumination input to the array changes to NUI-2 case. It is observed that, with the proposed technique the array is maintained at GMPP in all the cases.

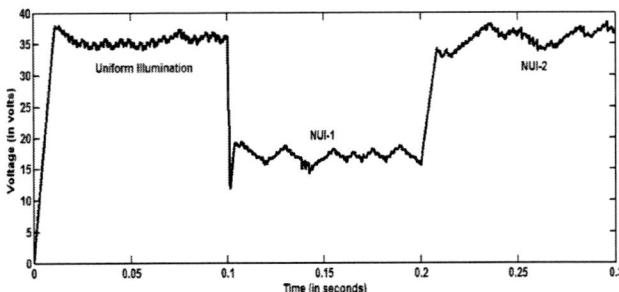

Fig.8(a). Voltage output of 2x2 array with simulated algorithm for test cases (Fig. 5)

978-1-5090-5606-4/17 $31.00 © 2017 IEEE

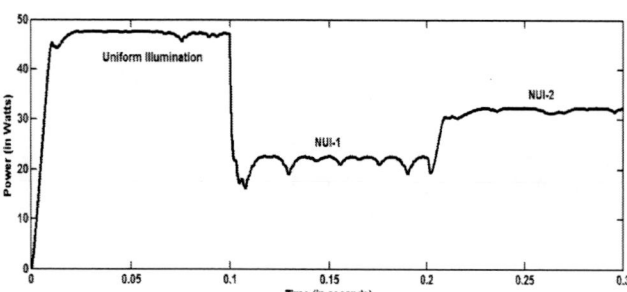

Fig.8(b). Power output of 2x2 array with simulated algorithm for test cases (Fig. 5)

Fig. 8(b) shows simulation result for power output. Table V compares the values of predicted GMPP voltage and power (V_P and P_P) from algorithm with the actual GMPP voltage and power (V_A and P_A) obtained from P-V curve at different NUI scenarios. The % error in predicted voltage (%E(V)) and predicted power (%E(P)) is also shown. The predicted voltage (V_P) is given as V_{REF} to the closed loop controller for tracking the GMPP. After the closed loop control, the remaining error in prediction will be compensated by P&O. The proposed algorithm is also simulated for different MXN arrays with different load combinations. It is observed that GMPP is being tracked in all the cases using the proposed modified maximum power extraction technique followed by the P&O.

The proposed technique also works effectively for dynamically changing loads. In order to validate the effectiveness of proposed technique for dynamically changing loads, a 2x2 array with NUI and an initial resistive load of 100Ω is considered as shown in Fig. 9(a). In general, the load impedance seen by the SPV array is varied using the boost converter by changing its PWM duty cycle (d). The relation between impedance seen by the array (R_{IN}) and load impedance (Ro) in case SPV arrays connected with boost converter is given by equation (10).

$$R_{IN} = R_o(1-d)^2 \qquad (10)$$

Based on the NUI information, the proposed algorithm takes the SPV array's operating point in terms of voltage and current to GMPP condition of the array as shown in Fig 9(b).

TABLE V
COMPARISON OF ACTUAL AND PREDICTED READINGS OF A
2X2 ARRAY FOR DIFFERENT CASE STUDIES (FIG. 5)

S.No	P_A	P_P	V_A	V_P	%E(P)	%E(V)
UI (a)	47.6	46.2	34.6	34.8	2.9	0.57
NUI-1 (b)	22.5	23.1	16.4	17.3	2.7	5.5
NUI-2 (c)	32.3	29.5	36.4	35.6	8.6	2.2

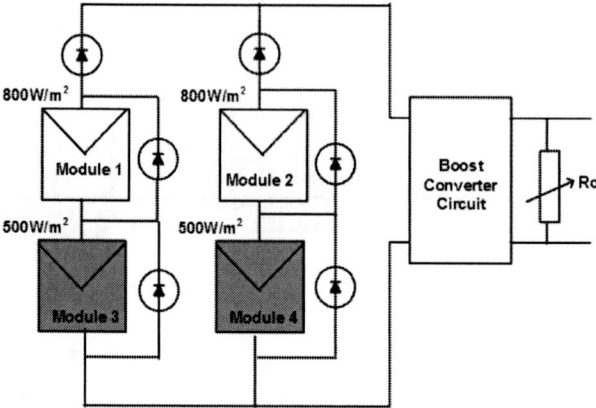

Fig. 9(a). 2x2 SPV array with variable load

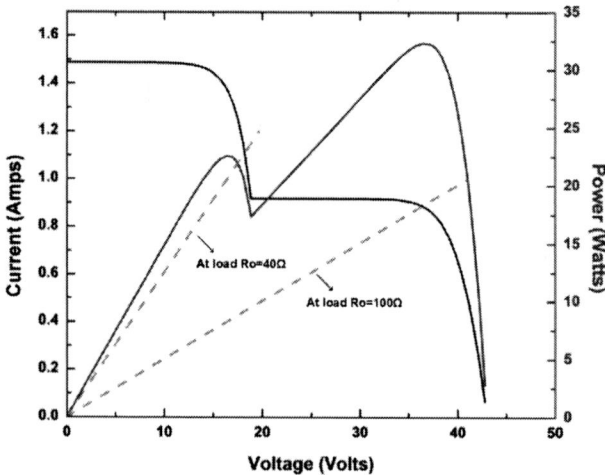

Fig. 9(b).Superimposition of I-V, P-V and load lines of 2x2 array operating at different resistive loads

If the load is suddenly changed to 40Ω (Ro= 40Ω) instantaneously, the impedance R_{IN} seen by array will be 16Ω and the SPV array operating point changes. The P&O further takes this instantaneous change in SPV array operating point (18.23, 1.11) which is not a GMPP. This causes a change in the array power and the proposed algorithm compares the change with respect to a tolerance value. Since the change in array power is considerable (9.3Watts), the proposed algorithm initiates the GMPP tracking and t akes the SPV array operating point to (36.5, 0.89) which is a GMPP. This array power change monitoring based GMPP tracking initiation is an elegant approach in the proposed algorithm and ensures the robust performance of the proposed technique both for rapidly changing NUI conditions and dynamically changing loads.

V. CONCLUSION

A modified MPE technique is proposed which works effectively for rapidly changing NUI and dynamic loads. The proposed MPE technique closely predicts the GMPP value

978-1-5090-5606-4/17 $31.00 © 2017 IEEE

thereby requiring a very less array scanning. The proposed technique has been studied for different NUI cases and also for different loads like resistive, battery and also dynamically varying loads. The algorithm converges fast to the GMPP for illuminations changes and also for load fluctuations in real time unlike other MPE techniques which requires large array scanning. The proposed MPE technique requires minimum hardware and is simple for real time implementation.

REFERENCES

[1] B. Subudhi, R. Pradhan, "A Comparative Study on Maximum Power Point Tracking Techniques for Photovoltaic Power Systems", IEEE Transactions on Sustainable Energy, Vol.4, Issue.1, Pages.89-98, 2013.

[2] Kashif Ishaque, Zainal Salam, "A review of maximum power point tracking techniques of PV system for uniform insolation and partial shading condition", Renewable and Sustainable Energy Reviews, Volume 19, Pages 475-488, March 2013

[3] H. Patel and V. Agarwal, "MATLAB-Based Modeling to Study the Effects of Partial Shading on PV Array Characteristics," in IEEE Transactions on Energy Conversion, vol. 23, no. 1, pp. 302-310, March 2008. doi: 10.1109/TEC.2007.914308

[4] H. Patel, V. Agarwal, "Maximum Power Point Tracking Scheme for PV Systems Operating Under Partially Shaded Conditions", IEEE Transactions on Industrial Electronics, Vol.55, Issue.4, Pages.1689-1698, 2008.

[5] K. Ishaque and Z. Salam, "A Deterministic Particle Swarm Optimization Maximum Power Point Tracker for Photovoltaic System Under Partial Shading Condition," in IEEE Transactions on Industrial Electronics, vol. 60, no. 8, pp. 3195-3206, Aug. 2013. doi: 10.1109/TIE.2012.2200223

[6] S. V. Dhople, R. Bell, J. Ehlmann, A. Davoudi, P. L. Chapman and A. D. Domínguez-García, "A global maximum power point tracking method for PV module integrated converters," 2012 IEEE Energy Conversion Congress and Exposition (ECCE), Raleigh, pp.4762-4767, 2012, doi:10.1109/ECCE.2012.6342172

[7] B. Patnaik, P. Sharma, E. Trimurthulu, S. P. Duttagupta, and V. Agarwal, "Reconfiguration strategy for optimization of solar photovoltaic array under non-uniform illumination conditions," in Proceedings of the 37th IEEE Photovoltaic Specialists Conference (PVSC), pp. 1859–1864, June 2011.

[8] B. Patnaik, Aswani U., G. Sarkar and S. P. Duttagupta, "Image aided dynamic reconfiguration of SPV array under non uniform illumination," 2014 IEEE 40th Photovoltaic Specialist Conference (PVSC), Denver, CO, 2014, pp. 0797-0802. doi: 10.1109/PVSC.2014.6925037.

[9] R. Bell and R. C. N. Pilawa-Podgurski, "Decoupled and Distributed Maximum Power Point Tracking of Series-Connected Photovoltaic Submodules Using Differential Power Processing," in IEEE Journal of Emerging and Selected Topics in Power Electronics, vol. 3, no. 4, pp. 881-891, Dec. 2015. doi: 10.1109/JESTPE.2015.2475607

[10] C. Olalla, C. Deline, D. Clement, Y. Levron, M. Rodriguez and D. Maksimovic, "Performance of Power-Limited Differential Power Processing Architectures in Mismatched PV Systems," in IEEE Transactions on Power Electronics, vol. 30, no. 2, pp. 618-631, Feb. 2015. doi: 10.1109/TPEL.2014.2312980

[11] Aswani U, S. P. Duttagupta and T. I. Eldho, "A novel maximum power extraction technique for rapidly changing NUI conditions," in Proceedings of 42nd IEEE Photovoltaic Specialist Conference (PVSC), pp. 1-6, June 2015.

Real-time Monitoring of Photovoltaic Reliability Only Using Maximum Power Point – the Suns-Vmp Method

Xingshu Sun, Haejun Chung, Raghu Vamsi Krishna Chavali, Peter Bermel, and Muhammad Ashraful Alam

Network of Photovoltaic Technology, Purdue University, West Lafayette, IN, 47907, USA

Abstract — Millions of solar panels have been deployed around the world for electricity production. Meanwhile, these solar panels are routinely generating a massive amount of streaming data of current and voltage as they operate at the maximum power point (MPP). The existing characterization methods, however, cannot effectively mine and decode these datasets to provide useful insights of the degradation for solar panels. In this paper, we propose the new Suns-Vmp methodology, which offers a degradation-agnostic approach to monitoring and diagnosing the reliability of solar panel in real time by physically interpreting the MPP data. The physics-based method reconstructs "IV" curves based on MPP characteristics under varying illumination and temperature, from which time-series circuit parameters can be extracted using equivalent circuits. The proposed method has been applied to analyze the data obtained from an NREL test facility. Our Suns-Vmp based analysis indicates that the solar panels have failed within nine years with a rate of ~3 %/year, mostly likely due to yellowing/delamination of EVA and contact corrosion. Integrated with degradation models or machine learning, the method can also serve to predict the lifetime of PV systems.

I. INTRODUCTION

The levelized cost of electricity (LCOE) of photovoltaics (PV) has gone down dramatically in the past decades, yet it is still higher than fossil fuels. The reliability of solar panels must be significantly improved to drive down the cost further. Therefore, it is crucial to diagnose and understand the geography- and climate-specific degradation of PV to make future solar panels more reliable.

There have been many studies on PV reliability in literature. For example, Ref. [1] has performed statistical investigations on the degradation of installed solar panels aging from 1 year to 30 years under different climates. In these studies, the efficiencies of solar panels were rated either by indoor or outdoor IV measurements. However, the dominant failure modes of PV, which is crucial to understand degradation pathways, were not explicitly reflected by the IV data. More importantly, only the average degradation rate over time was reported, the temporal behavior of the degradation which impacts the LCOE was not presented. Indeed, a method that does not require full IV measurement but still can quantitatively monitor and diagnose the degradation of solar panels in real time can be a powerful and transformative tool to assess PV reliability.

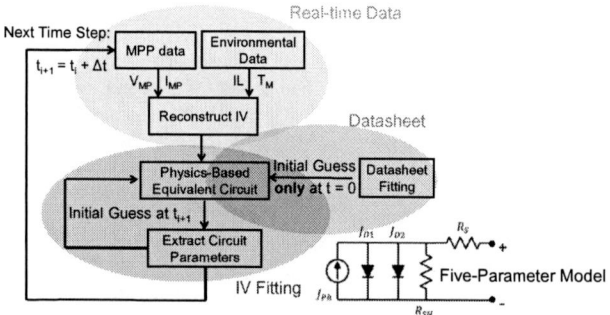

Fig. 1   The flowchart of the Suns-Vmp method. The analytical formulation of the five-parameter model is from [6].

On the other hand, photovoltaic systems across the world routinely generate an enormous amount of time-series data during their operation. In this paper, we demonstrate that effective mining of this data can yield rich knowledge on the performance and reliability of solar panels. The stored electrical data typically consists of current and voltage at maximum power point (MPP). This single-point MPP data in real time is difficult to interpret by existing methods, including the ones that do not require full IV curves [2], [3].

In this paper, we propose the Suns-Vmp method that can deduce circuit parameters as a function time by fitting the reconstructed "IV" based hourly varying MPP data under different illumination and temperature, see Fig. 1. The method can monitor the reliability of solar panels in real time by systematically and physically mining the streaming MPP data. We have applied the proposed method to field data from a National Renewable Energy Laboratory (NREL) test facility. *Our analysis suggests that the solar panels have failed within nine years, and diagnoses yellowing/delamination and corrosion as the primary degradation pathways.* The method is degradation-agnostic and allows comprehensive studies of geographic- and technology-dependent degradation of installed solar panels. The findings will facilitate reliability-aware design for the next-generation PV. The results from the Suns-Vmp method can also serve to calibrate physics-based degradation model [4] and train machine learning algorithms to predict the performance and reliability of PV systems.

978-1-5090-5606-4/17 $31.00 © 2017 IEEE

II. THE SUNS-VMP METHOD

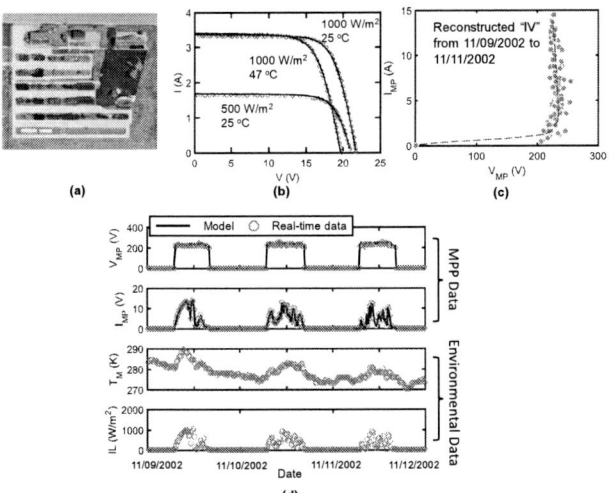

(a)   (b)   (c)

(d)

Fig. 2   (a) Satellite Image of the test facility from Google Map. (b) Initial fitting to the datasheet (Siemens M55 [7]) for time-zero circuit parameters. (c) An illustration of reconstructing "IV" from the MPP data dating from 11/09/2002 to 11/11/2002 (the last three days of this analysis). (d) Three-day fitting result of MPP data using the environmental data by the Suns-Vmp method.

Similar to illumination intensity and ambient temperature, Imp and Vmp (operating current and voltage at maximum power point) of solar panels increase from morning to noon then decrease from noon to evening, see Fig. 2(d). By aggregating the MPP (Vmp, Imp) data at different times of the day, one can create an MPP "IV" as shown in Fig. 2(c). Notably, Vmp in Fig. 2(c) decreases slightly as Imp reaches the maximum because of the elevated temperature of solar panels under high illumination intensity at noon. In contrast to a standard full IV sweep where the temperature and illumination are maintained constant, each data point of the MPP "IV" are measured at varying illumination and temperature. In general, illumination and module temperature are commonly measured and preserved along with the MPP data. Hence, we can fit the equivalent circuit (Fig. 1) to the MPP "IV" by continuously tracking the Vmp and Imp values and matching them to the recorded MPP data; ultimately, calculate the time-dependent circuit parameters. *Note that, unlike existing methods [2], [3], the Suns-Vmp method characterizes degradation through the MPP data, thus does not disturb the normal operation of PV systems.*

Before applying the Suns-Vmp method, there are still, however, two important notices: *1) initial parameterization of equivalent circuits based on datasheets, and 2) measurement window of MPP data for a single time step.* **First**, the Suns-Vmp method can diminish multiple solutions during fitting MPP data by obtaining the time-zero circuit parameters based on the datasheet as robust initial guesses, see Fig. 2 (b).

Subsequently, the circuit parameters extracted at each time step can be used as the initial guess for the following one. Hence, unlike directly fitting MPP data, multiple solutions of circuit parameters can be avoided. **Second**, we assume that the time scale for degradation is slow. Thus the circuit parameters can be considered constant over the course of a day or several days. Hence, the recommended measurement window of MPP data length can be one to several days, during which there is also sufficient variation in illumination and temperature to reconstruct the "Suns-Vmp IV." In the case of catastrophic degradation (such as partial shading degradation that can quickly erode the power in seconds), the extracted circuit parameters become the average of pre- and post-degradation values over time.

III. APPLICATION TO FIELD DATA

We apply the Sun-Vmp method to a test facility (NREL x-Si #4) at Denver, see Fig 2(a). All the on-site MPP and environmental data (illumination and module temperature) were obtained from the publicly accessible NREL PV Data Acquisition (PVDAQ) database [5]. The analysis of this site only applies to 05/13/1994 to 11/11/2002 because of the incompleteness of date from 2003 to 2012.

Figure 3 summarizes the extracted circuit parameters by fitting the five-parameter model (see Fig. 1) in [6] to the 3-day

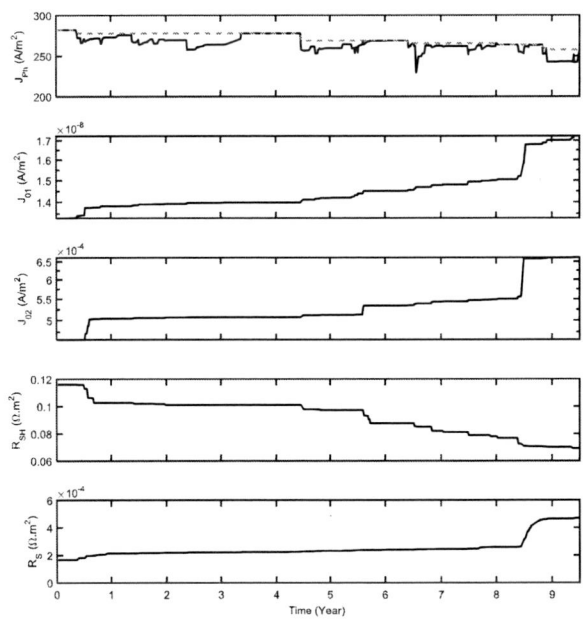

Fig. 3   The extracted circuit parameters of the five-parameter model as a function of time. Notations: J_{Ph}: maximum photocurrent density; J_{01} reverse saturation current density with ideality factor of 1; J_{02} reverse saturation current density with ideality factor of 2; R_{SH}: shunt resistance; R_S: series resistance.

978-1-5090-5606-4/17 $31.00 © 2017 IEEE   1905

Suns-Vmp data over a span of ~10 years. The maximum photocurrent (J_{PH}) fluctuates probably due to accumulation of dust or snow, which may be useful to calculate the soiling/snowing rate (not discussed here). In this paper, we simply adjust J_{PH} to be decreasing monotonically with time. Remarkably, it appears that all the circuit parameters in Fig. 3 are degrading (e.g., shunt resistance (R_{SH}) reduces, and series resistance (R_S) increases). To quantify the degradation rate, we calculate the efficiency at standard test condition at each time step, see Fig. 4(a). The efficiency decreases at a rate of ~3%/year (relative percent) to 80% of the rating within nine years. Hence, it suggests that the lifetime of the panels is more than 3 times shorter than that of warranty (25 years). Next, we will investigate the possible root causes for the rapid degradation.

We deconvolve the power losses associated with each parameter as shown in Fig. 4(b). At the failure time of the panel (~8.5 years), Fig. 4(b) elucidates that yellowing/delamination and contact corrosion are the dominant contributors to efficiency reduction for this specific test site, reflected by the decrease in J_{PH} and increase in R_S, respectively. Yellowing of the EVA encapsulation can be expected because of the relatively high ultraviolet light concentration at Denver (altitude of 1600 m). Given the moderate humidity level at Denver, pronounced corrosion of contact may be caused by moisture ingress through the imperfectly sealed edges of the solar panels. The operating voltage of the panel array is only around 200 V; therefore, the efficiency degradation by potential-induced degradation (PID) is expected to be insignificant. Indeed, our result confirms this conjecture by showing that only 5% power loss is to shunting (R_{SH}) and increased recombination current (J_{02}), which are effective indicators for PID. As demonstrated here, the Suns-Vmp allows us to quantitatively diagnose the reliability of existing installed solar panels by analyzing the time signature of individual circuit parameter.

IV. DISCUSSION AND CONCLUSION

In the previous section, we have illustrated the Suns-Vmp method as a means to analyze photovoltaic degradation in real time. Next, we will explain how this method can also enable prediction of PV reliability.

Recently, several physics-based degradation models that can directly reflect PV degradation to the temporal behavior of circuit parameters have been developed [4]. The extracted circuit parameters by the Suns-Vmp method can be used to calibrate these degradation models (e.g., moisture diffusion coefficient for corrosion). Integrated with the weather forecast, the calibrated degradation models can physically predict the lifespan of solar panels. Alternatively, the time-dependent circuit parameters can train machine learning algorithms; the trained machine learning algorithms may allow prediction of PV lifetime. The validity of these predictive approaches,

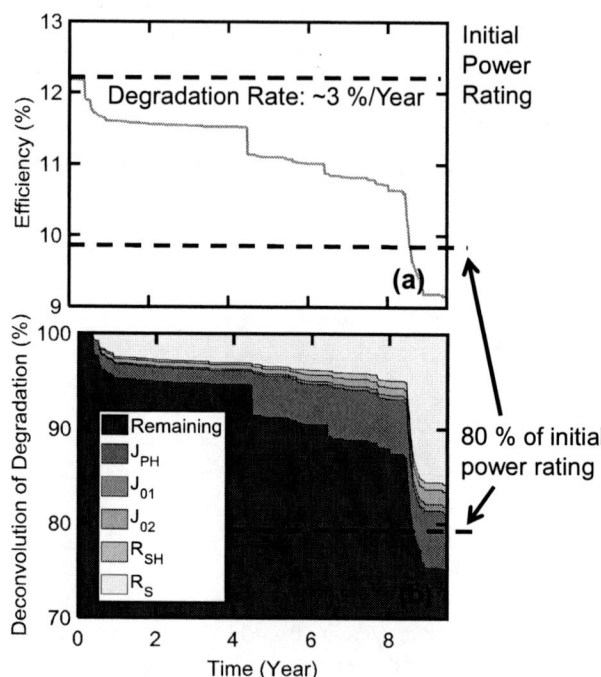

Fig. 4   (a) The calculated efficiency at 1000 W/m² and 25 °C and (b) deconvolution of power loss related to each circuit parameter as a function time.

however, remains an interesting open question and requires more rigorous research efforts.

In this paper, we have proposed the Suns-Vmp method that allows real-time monitoring and diagnosis of solar panels by systemically and physically mining the field data produced by PV systems, which cannot be fully interpreted by the previous methods. By applying the Suns-Vmp method to an NREL test facility, we have extracted the circuit parameters and deduced the efficiency of solar panels as a function of time without any IV measurement. *Our analysis suggests that the solar panel failed within nine years (80% of the nameplate power rating), mainly caused by contact corrosion and yellowing/delamination of EVA.* The breakdown analysis of degradation pathways can shed light on technology- and geographic-specific degradation (e.g., more yellowing with high UV intensity), and can promote environment-specific design for the next-generation reliability-aware solar panels. Also, the results from the Suns-Vmp method can calibrate physics-based degradation models as well as train machine learning algorithms, both of which can then predict future electricity yield and power degradation of PV.

REFERENCES

[1]   D. C. Jordan, S. R. Kurtz, K. VanSant, and J. Newmiller, "Compendium of photovoltaic degradation rates," *Prog. Photovoltaics Res. Appl.*, vol. 24, no. 7, pp. 978–989, Jul. 2016.

[2] M. G. G. Deceglie, T. J. J. Silverman, B. Marion, and S. R. R. Kurtz, "Real-Time Series Resistance Monitoring in PV Systems Without the Need for I – V Curves," *IEEE J. Photovoltaics*, vol. 5, no. 6, pp. 1706–1709, 2015.

[3] M. J. J. Kerr, A. Cuevas, and R. A. A. Sinton, "Generalized analysis of quasi-steady-state and transient decay open circuit voltage measurements," *J. Appl. Phys.*, vol. 91, no. 1, p. 399, 2002.

[4] R. Asadpour, R. V. K. V. K. Chavali, and M. A. A. Alam, "Physics-Based computational modeling of moisture ingress in solar modules: Location-specific corrosion and delamination," in *2016 IEEE 43rd Photovoltaic Specialists Conference (PVSC)*, 2016, pp. 0840–0843.

[5] "PVDAQ (PV Data Acquisition)." [Online]. Available: http://developer.nrel.gov/docs/solar/pvdaq-v3/.

[6] M. Hejri, H. Mokhtari, M. R. Azizian, M. Ghandhari, and L. Soder, "On the Parameter Extraction of a Five-Parameter Double-Diode Model of Photovoltaic Cells and Modules," *IEEE J. Photovoltaics*, vol. 4, no. 3, pp. 915–923, May 2014.

[7] "Siemans Solar Panels Installation Guide." [Online]. Available: http://iodlabs.ucsd.edu/dja/codered/engineering/procedures/solarPower/siemens solar panels.pdf.

Photovoltaic Module Durability and Reliability: Analysis of a 23-Year-Old Array Operating in Quebec, Canada

Christopher Baldus-Jeursen[1], Alexandre Côté[1], Naveen Goswamy[1], Tanya Deer[2], Yves Poissant[1]

[1]CanmetENERGY, Natural Resources Canada, Varennes, Quebec, J3X 1S6, Canada
[2]Relsol Inc., Vaughan, Ontario, L4H 0R6, Canada

Abstract — **Photovoltaic module reliability and performance data during field operation is essential to understanding the mechanisms that cause module degradation. In this work, degradation mechanisms were analyzed for 357 modules in a rooftop array operating for 23 years in Varennes, Quebec. Performance was quantified by visual inspection, solar simulator I-V curves, and comparison of array output to solar irradiance using a responsivity methodology. Array deterioration mechanisms such as potential induced degradation and moisture damage were analyzed.**

Index Terms — **Degradation Mechanisms, Photovoltaic Arrays, Responsivity, Solar Simulator, Yield.**

I. INTRODUCTION

As the installed worldwide photovoltaic (PV) power capacity continues to grow, module performance studies in the field are important for providing data on the physical and chemical processes of module degradation. Modules and arrays may show a variety of independent or interrelated failure modes which can depend on factors such as mounting structure, climate conditions, and cell and module manufacturing methods.

Installed in 1992, the 23.5 kW PV array in Varennes, Quebec, is possibly the oldest continually monitored PV system in Canada. The climate is temperate, with low temperatures, significant snow falls, freeze-thaw events in winter, and high temperatures in summer. Each module, manufactured by Astropower Canada, consists of 36 cells in series with one bypass diode and is either single-crystal or multicrystalline silicon with 47 W nameplate power.

Fig. 1.   Canmet Energy Research Centre in Varennes.

There were 540 modules on open racks divided into four arrays with a fixed tilt angle of 45 degrees. The 357 modules analyzed in this study were divided into arrays A, B, and C. In October 2014, the system was disconnected from the grid. Module inspection was started in January 2016.

II. MAIN EXPERIMENTAL RESULTS

Here, we report performance measurements and causes of module degradation. Further analysis of the modules is anticipated including defect characterization and analysis of monitored system/array performance data.

A. Power Degradation

Power degradation was determined by pulsed solar simulator measurements using a 4 point connection. Rather than using only the nameplate power rating for the baseline performance, all 357 modules were measured in 1992 before installation. The comparison of the module nameplate power to solar simulator measurements made in 1992 is shown in Fig. 2. Module parameters that were measured were: short circuit current and open circuit voltage, maximum power point current and voltage, and power produced.

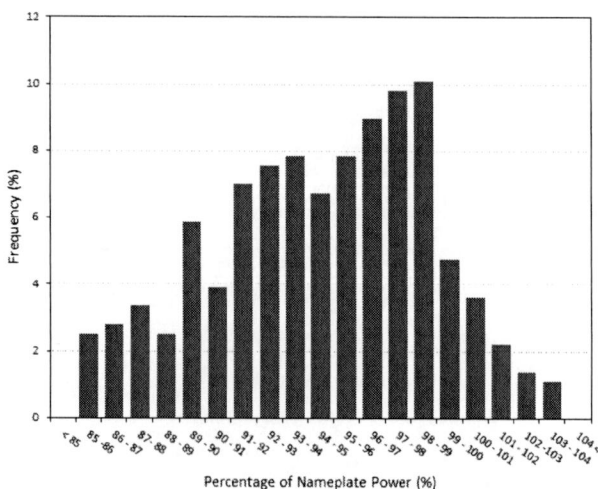

Fig. 2.   Comparison of measured module output power to 47 W nameplate power.

978-1-5090-5606-4/17 $31.00 © 2017 IEEE

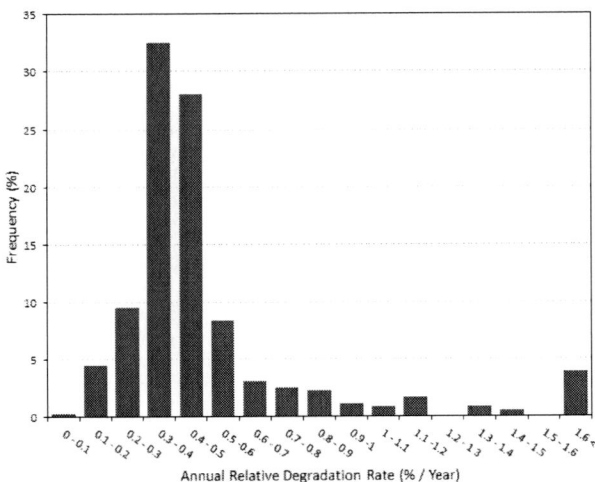

Fig. 3. Histogram of the relative efficiency decline for all modules during the period under load.

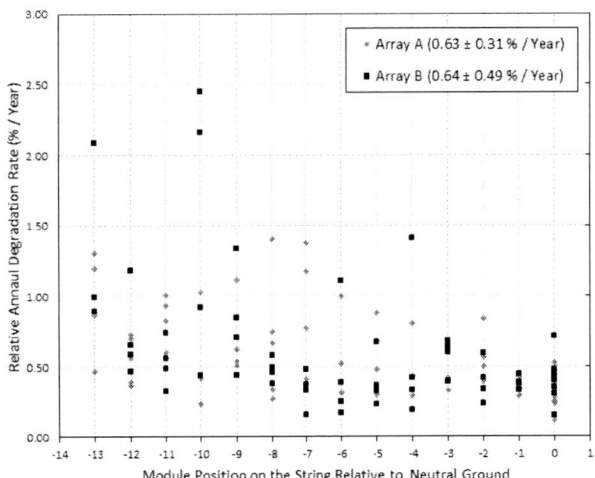

Fig. 5. Degradation rate versus relative module position for positively grounded arrays. Neutral ground is at zero.

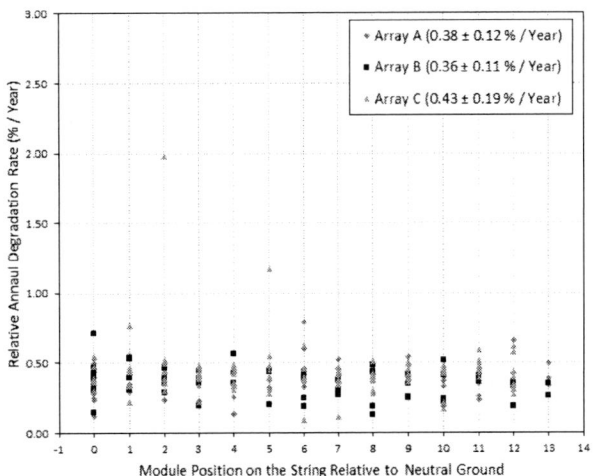

Fig. 6. Degradation rate versus relative module position for negatively grounded arrays. Neutral ground is at zero.

In addition, module parameters were measured in 2015 and compared to starting performance. As shown in Fig. 3, the average relative degradation rate for modules in arrays A, B, and C was approximately 0.6% per annum over the 23 year period. This degradation rate falls within the typical range for crystalline silicon modules which is between 0.5% and 1% per annum [1]. However, the total array performance loss is expected to be greater due to module mismatch and system losses not considered in the current study.

B. Potential Induced Degradation and Moisture Damage

In addition, the location of the module within the arrays also influenced the degradation rate as moisture damage was observed more frequently on modules located in the bottom row of arrays A and B where snow tended to accumulate (Fig. 4). Another important factor to consider was the effect of potential induced degradation (PID) which was first reported in crystalline silicon modules in 2005 [2] [3]. PID is attributed to the voltage potential as a function of position for each cell in a string relative to ground. For the arrays in this study, there were several different string configurations. Fig. 5 and Fig. 6 show the relation between power degradation as a function of module position relative to neutral ground.

Arrays A and B were centrally grounded through Omnion inverters with half the strings positively grounded and the other half negatively grounded. Array C was negatively grounded, first with an Abacus inverter, and then with a

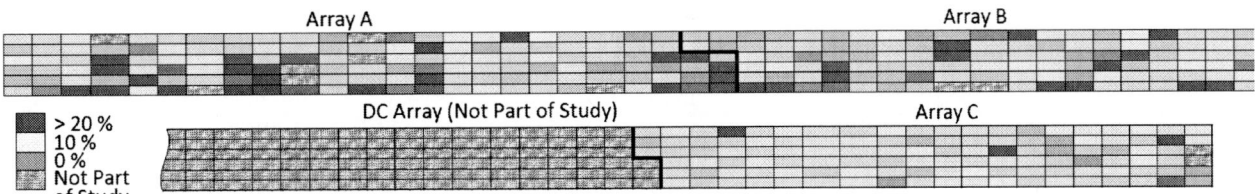

Fig. 4. Module power degradation versus position in arrays A, B, and C. Color code gives cumulative total degradation percentage over the 23 year operating period.

replacement Fronius inverter.

The average degradation rate and standard deviation are given for each array for both grounding conditions. There appears to be little correlation between module positions relative to ground for negatively grounded strings. However, there is relatively higher degradation with positively grounded than negatively grounded strings which may be an indication of PID.

C. Responsivity and Degradation Rate Analysis

In addition to calculating PV module degradation rate by solar simulator efficiency measurements at beginning and end of system operation, the time-dependent PV system degradation rate determined during continuous operation is also a key parameter.

The recorded yield of all three arrays is compared in Fig. 7. Due to system upgrades, the availability of the data logger system in 2003 was relatively low compared to other years. This explains the low yield for 2003 (even though the array was operating normally at this time). Data logger availability and system downtime affected recorded yield in other years as well.

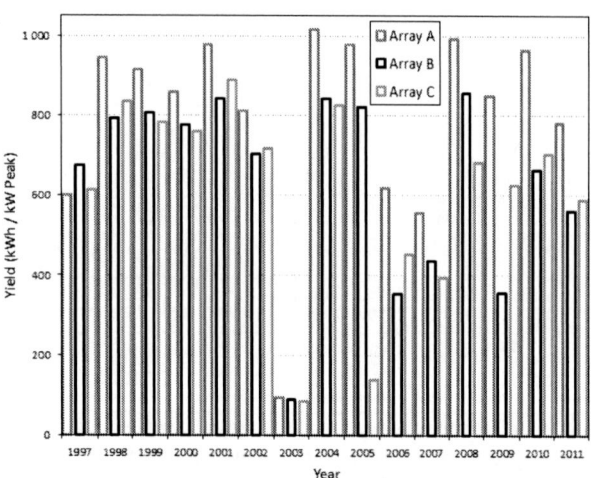

Fig. 7.   Yield of arrays A, B, and C, over system lifetime. Although all three arrays were operated since 1992, regular data monitoring began in 1997.

A recent proposal for calculation of degradation rate (submitted to the IEC for review as IEC 61724-4) proposes a responsivity term from which degradation is determined by a linear fit to the data [4]. The responsivity, which is similar to a temperature corrected performance ratio, is calculated according to (1):

$$Rs = (P_m/P_r) \times (G_{tar}/G_{meas}) \times [1 + \gamma(T_{tar} - T_{mod})] \quad (1)$$

where P_m is the measured power, P_r is the rated power, G_{tar} is the target irradiance, G_{meas} is the measured irradiance, γ is the power temperature coefficient (1/°C), T_{tar} is the target temperature, and T_{mod} is the module temperature. The target irradiance (in this case, 850 ± 50 W/m^2) is chosen since high irradiance conditions give more consistent results. The target temperature is the median module temperature corresponding to the target irradiance data set. For this paper, a Matlab program was used to automate the responsivity analysis. Fig. 8 and Fig. 9 show the mean yearly responsivity from 2004 to 2012 for Array A and B.

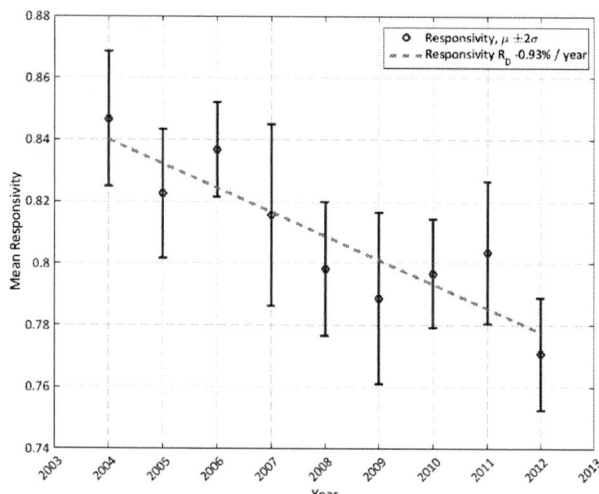

Fig. 8.   Annual mean responsivity for Array A.

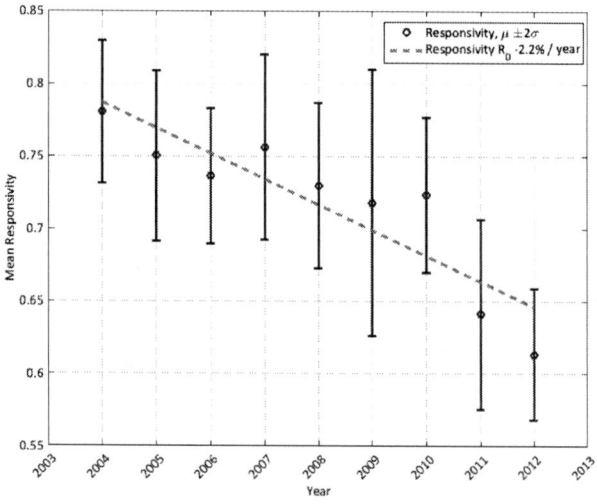

Fig. 9.   Annual mean responsivity for Array B.

Several filters were also applied to narrow the data. In particular, the stability, irradiance, and outlier filters were used to exclude data that might otherwise mask degradation behavior. The stability filter takes the derivative of the instantaneous responsivity with respect to time and removes data fluctuations associated with large negative or positive derivatives. The irradiance filter adjusts the spread of irradiance around the target irradiance to remove a percentage of responsivity data. Lastly, the outlier filter recalculates the mean monthly responsivity by removing a percentage of low and high outliers around the median. For each year of system operation, the degradation rate was calculated by a linear fit to the monthly mean responsivity. The degradation rate and corresponding uncertainty were calculated for each of the three main filters and for different percentages of retained data. After filtering, the degradation rate for each year corresponding to the lowest uncertainty was chosen.

The 2004 to 2012 time period was selected since it corresponds with the installation of a new data logging system for the array using five minute intervals. Array C was excluded from this analysis due to difficulties in accurate measurement of array temperature which affected the calculation of temperature coefficients. The degradation rate is given by the fitted line. For arrays A and B, the measured system degradation rate over the period of operation from 2004 to 2011 was 0.93% per year and 2.2% per year respectively.

III. CONCLUSIONS

Based on solar simulator measurements, the modules examined in this study had an average power loss of 0.6% per year. The system degradation was studied using a solar simulator to examine module performance at the start and end of operation. Furthermore, in addition to this two point measurement, application of the responsivity method provided data on time-dependent degradation. A variety of failure modes such as back-sheet and encapsulant delamination, cell corrosion, and hot-spot formation were also exhibited and will be the subject of future investigation. Multiple failure mechanisms may interact synergistically with complex influences on module stability.

ACKNOWLEDGEMENT

Financial support for this research project was provided by Natural Resources Canada through the Energy Innovation Program. © Her Majesty the Queen in Right of Canada, as represented by the Minister of Natural Resources 2017.

REFERENCES

[1] D. C. Jordan, B. Sekulic, B. Marion and S. Kurtz, "Performance and aging of a 20-year-old silicon PV system," *J. Photovolt.*, vol. 3, no. 5, pp. 744-751, 2015.

[2] S. Pingel, O. Frank, M. Winkler, S. Daryan, T. Geipel, H. Hoehne and J. Berghold, "Potentially induced degradation of solar cells and panels," in *Proceedings of the 35th IEEE PV Specialists Conference*, Honolulu, USA, 2010.

[3] R. Swanson, M. Cudzinovic, D. DeCeuster, V. Desai, J. Jurgens, M. Kaminar, W. Mulligan, L. Rodrigues-Barbarosa, D. Rose, D.

Smith, A. Terao and K. Wilson, "The surface polarization effect in high efficiency silicon solar cells," in *Proceedings of the 15th International Photovoltaic Science and Engineering Conference*, Shanghai, China, 2005.

[4] S. Kurtz, "PNW 82-1061: Photovoltaic system performance – Part 4: degradation rate evaluation method," *International Electrotechnical Commission (IEC)*, Golden, 2016.

Are E-W trackers a better option for future investments in PV sector-A detailed Techno-Commercial Study

Rakesh Bohra, Ramesh Rame Gowda and Mani R. Krishnan

Infosys Ltd., Bangalore, Karnataka , 560100, India

Abstract — **With the growth of PV sector globally, it is need of time to understand the techno-commercial dynamics of investment in E-W trackers. It is evident after declaration in COP21 that world needs huge investment in green energy sector to meet global climate change challenges. Infosys Ltd. has set up a 6.6 MW solar PV plant in Hyderabad Campus thus making it 100 % self sustainable in its green energy demands. Data driven Innovative design ideas and exclusive execution methodologies to mitigate geographical challenges has made this power plant one of its kind across globe. Two different PV technologies in two different orientations including single axis tracking and southward fixed tilt were placed. Before investing further in the PV sector, Infosys did a practical study at its 6.6 MW PV farm to understand the dynamics of extra energy yield on trackers and its value propositions on investment returns.**

Index Terms — **Photovoltaic systems, Maximum power point trackers, solar power generation**

I. INTRODUCTION

Infosys Ltd. is India's second largest IT sector Corporate Company and is involved in setting benchmarks in green energy initiatives. It has set a target to utilize 100 % green energy in its use which accounts approx. 253 million kWh units per annum.

Therefore in this regards we planned our first utility scale on grid solar power plant in Hyderabad 17.4 degree North & 78.6 Degree East. This project was commissioned on 21st December 2015. Since PV industry is on verge to expand exponentially it is the need of time to evaluate the investment strategies in such a way that developers may get favorable returns but not at the cost of project quality. One of the best solution to increase the energy yield of a PV system in field is investing into single axis E-W trackers. This paper will speak in-depth about techno-commercial aspect of investment in E-W tracker.

TABLE I
6.6 MW PROJECT SYNOPSIS

Solar PV Technology	Polycrystalline Silicon	CIS Thin Film
Orientation	Fixed tilt southward & Single axis tracker (E-W)	
Type of Tracker	Hydraulic mechanism and self –powered motorized	
Capacity	3.6 MW	3 MW

The performance of both the technologies viz. polycrystalline silicon and Copper Indium & Selenide (CIS) thin film on E-W trackers and fixed south tilt mounting have been analyzed and the result is published in this paper.

The data to evaluate the performance of this plant has been captured in every one minute of interval.

II. PV MODULES MOUNTING MECHANISMS

Our intention was to understand how the PV dynamics works on trackers and fixed tilt and how it proportionate the energy yield with respect to the investment. We are sure that the simple but innovative methods thus articulated in our execution will be well accepted by the global investors.

In the same field we have installed 3 trackers makes with two PV technologies. Trackers included actuation from hydraulic as well as gear motors on ganged as well as individual row tracker. Modules mounting on these trackers were simple in one case and complex in other due to involvement of extra structure members in the later one.

Fig. 1.   E-W Tracker with Hydraulic and self-powdered motorized Actuation Mechanism.

Our study is concluded over a period of one year of the operation of these trackers and thus is comprised with exclusive data to be shared for investors, researchers and developers.

III. TRACKER SELECTION & EXECUTION METHODOLOGIES

Before finalizing the tracker thorough due diligence was done. We strictly followed the compliance as stated in IEC/TS

978-1-5090-5606-4/17 $31.00 © 2017 IEEE

62727: 2012 and BIS 6756-59 guidelines. It was understood that since the location gets abundant sunlight however there are 60-90 cloudy days as well, we decided to use thin film modules as well as polycrystalline modules both to get the best output in combination. [4][5][6]

A. Execution and Design

The tracker with slew motor was simple and easy to execute since it was not having so many members to be integrated. It comprised with one purling and one simple torque tube. It took 15 days in erecting. While the other tracker with hydraulic actuation was having so many members in structure ,thus making tracker heavier, complex in erection and comparatively difficult to maintain. Many other structural and design parameters were also noticed which distinguished the best tracker out of the others. We will articulate the same in detail during paper submission.

Erection of comparatively complex designed tracker took extra 10 days for same capacity execution with same no. of labors which if quantified coasted us approx. 3000 USD extra. This cost would have been increased for higher capacity Land with thin film technology on tracker was 18 % extra with respect to land for trackers with polycrystalline technology, depending upon shadow, backtracking and pitch distance. Land with trackers will only be 15% extra than on fixed tilt, however it can be well customized as per the requirement by adjusting pitch distance and backtracking mechanisms.

Fig. 2.   (A) & (B), simple designs with cylindrical torque tubes, (c) & (D) a little complex designs with rectangular torque tubes.

B. Operation and Maintenance

Operation and maintenance of these trackers were not at all very difficult the way it is presumed in industry. Tracker without self-power mechanism consumes only 2 kwh/MW/day which was nominal in comparison to the respective generation.

Hardly any big failure was observed in last one year in these trackers which could have adversely affect the system performance. The main thing required for maintaining the tracker was tightening fasteners once in 3 months, changing lubrication oil in hydraulic pistons and greasing the other movable parts. However in the row tracker which was simpler in design and was having motors for actuation, the maintenance cost was even reduced to negligible in comparison to the fixed tilt system. Some of the failures were cracking of piston due to manual mishandling and problems in algorithm programming leading to take tracker in non-tracking mode for time being.

C. Tracker Accuracy Characterization

Record +- 2 % pointing error in the tracker was maintained throughout the operational time for all the trackers. [3]

IV. PERFORMANCE EVALUATION OF DIFFERENT TRACKERS WITH DIFFERENT PV TECHNOLOGIES.

[2]The project was well maintained for the whole year and performance of the system with different permutation combination was analyzed to understand the system feasibility and dynamics. Modules were cleaned twice a month with clean water. We were having CMP22 class pyrometers for GHI and POA along with the same on trackers as well for recording every minute data. The weather system was comprising with temperature sensors, anemometers RH sensors etc. The performance figures thus received has opened up many mysteries regarding trackers thereby giving a clear picture for the investors. We got a clear idea that thin film technology modules if mounted on single axis trackers for geographic locations like Hyderabad will yield best results. Further study on this point is encouraged.

A. Performance Ratio

Performance of the whole PV plant along with trackers was satisfactory. Data was thus collected to analyze the performances thereby concluding the outcome.

Technology	Thin film FT	Thin film SAT	Poly Si_FT	Poly Si_SAT
Average	82%	85%	81%	80%

Performance Ratio of different PV technologies on trackers and fixed tilt is represented as below.

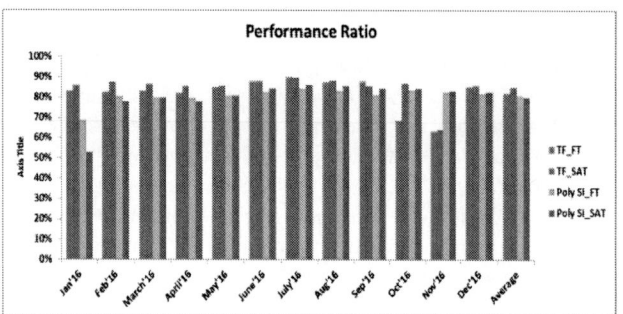

Fig. 3. Performance ratio of different PV technologies on E-W trackers

B. Specific Yield

Specific yield [1] [2] is one of the main parameter to consider while analyzing the performance of a PV plant with different mechanisms of orientations. Data thus analyzed was extraordinary to plot the future investments. It further resolves the magic figure of extra 18-20% over yield on E-W (fixed tilt) tracker (SAT) that Fixed Tilt (FT) which actually is a very varying figure and highly depends on technology and geography. In our case the result was astonishing as represented below.

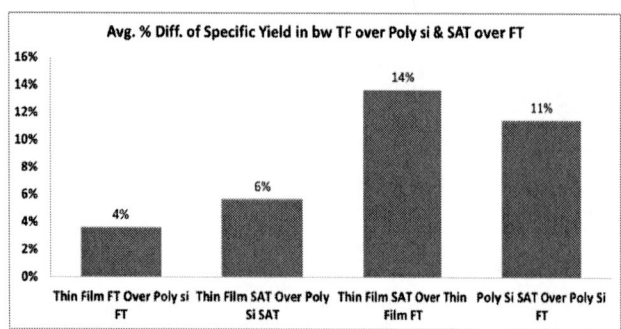

Fig. 4. Specific yield comparison of different PV technologies on E-W tracker & FT (fixed tilt)

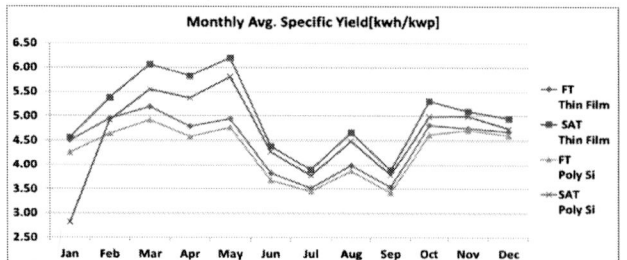

Fig. 5. Specific yield of different PV technologies on E-W trackers& FT (fixed tilt)

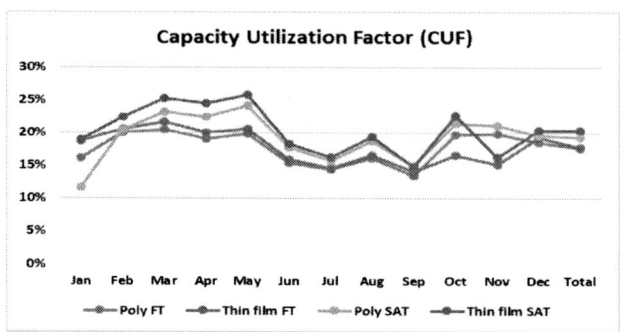

Fig. 6. Capacity Utilization Factor of different PV technologies on E-W (east-west) trackers and FT (fixed tilt)

V. COMMERCIAL FEASIBILITY

A. E-W Tracker Supply and Erection

Trackers are actually energy boosters, but the question is, at how much extra cost. Decreasing module prices today have even made investors to rethink on investments in trackers. E-W Trackers are costing today at the range of 12-14 $ cents per Wp to supply, we can add extra one cent in case to go with self-powered motorized one.0.5 cents /Wp can be added for erecting and installation.

In our case the tracker was costing 16 % extra over fixed tilt and generated 11% extra yield in polycrystalline and 14 % extra yield on thin film. Along with a domestic tracker we used one of the best tracker available in international market which is now costing 18-20 cents per Wp. for small scale project upto 2 MWp. Domestic tracker in India may come in the range of 12-14 $cents for supply, however cost may vary with parameters like project sizing, geographical locations and maintenance. Higher the project size and less challenging the terrain, lower will be the cost. We found that if tracker cost reaches to the level of less than 10 USD cents post significant R&D maintaining all quality standards , there is no way for market to boost while at the same time encouraging the use of high efficiency PV panels with same will also add an advantage.[7]

B. Operations & Maintenance

Operations and Maintenance of E-W trackers today is a hot topic and has been exaggerated unnecessarily. O&M cost for our trackers with hydraulic actuation was 600 USD per MW per year which is almost 0.5% of the domestic tracker cost, however the tracker with simpler design and self-power mechanism, this cost has been further reduced with another 250 USD which becomes almost 0.3 % of tracker cost.

Main problems trackers can face is tightening of fasteners once in 3 months, greasing the moving parts once every quarter and putting lubrications/oil for hydraulic actuation once in 3 months. In case of motorized tracker the problems were found even lesser.

Later in final paper submission we will submit operational statistics in detail for further clarifications that O&M cost is almost negligible while comparing with fixed tilt systems in case the tracker has been selected intellectually and project has been executed properly. However for review, table II and III may help understanding the statistics. We will also put basic guidelines to be kept in mind for quality execution of the trackers which may help in decreasing the probability of faults during execution thereby making trackers easy to maintain.

TABLE II

FAILURE MODES IN E-W TRACKER WITH PREVENTIVE MAINTENANCE (PV) AND BREAKDOWNS (BD)

Self-Powered motorized tracker	Frequency	Ganged Hydraulic tracker	Frequency
lubrication for Gear system (PM)	once in 3 months	lubrication for Gear system (PM)	once in 45 days
SPC Issues (BD)	once in a year	Cylinder failure	once in a year
motor Problem(BD)	once in a year	limit switch corrosion	once in a year
Damper Problem(BD)	once in a year	Fasteners tightening	once in 3 months
Software and controller (BD)	once in a year		
Fastener tightening	once in 3 months		

TABLE III
O&M COST

	Hydraulic actuator	Self-Powered Gear Motorized Tracker	Remark
O&M cost/MW/year	600 USD	*400 USD*	Cost doesn't include Modules cleaning and replacement of material, in our case, nothing was replaced in whole year.

VI. CONCLUSION

The main idea of this project was to deliver a huge set of data for further analytics, to showcase that how the investment dynamics in E-W trackers works in developing countries with extraordinary potential With our research, we also concluded the difference in performances of PV technologies viz. Copper Indium & Selenide (CIS) and Polycrystalline silicon which are widely utilized globally on mass scale. Also Since the market share of single axis trackers is very less globally however it is said to yield 14-18 % more energy provided properly constructed depending on other conditions.

We also concluded that for E-W trackers market to grow in countries like India and even globally, looking into the perspective of decreasing modules prices, our study shows that trackers supply and erection costing less than 11-12 cents/Wp for supply will act as a boosting catalyst for future investments. A cap of 10% extra investment for E-W trackers over fixed tilt is feasible while you get 14 % extra yield with thin-film technology and 11-12% extra energy yield on polycrystalline technology.

We found that E-W trackers with simpler design reduces execution time, maintenance cost and operational failures. E-W Trackers with cylindrical torque tube having male female connectors helps in alignment thereby reducing energy loss due to pointing errors. Trackers with vibration dampers helps

in mitigating issues like micro cracks in PV modules in case of Jerks due to high speed wind and while tracking.

The main result we got was trackers with CIS thin film modules gives 2% extra CUF besides extra land and steel, and 6% extra yield in comparison to Polycrystalline modules, refer Fig 3.

With feasible returns it is also encouraged to go with trackers as it does not clip the power produced during peak time, at the same time comparatively flat curve also helps in stabilizing the grid.

We found that in initial years the maintenance of E-W trackers does not cost much besides there is no failure of main components like batteries or motors and they needs to be replaced.

REFERENCES

[1] Alberto Dolara, Francesco Grimaccia, Sonia Leva, Marco Mussetta, Roberto Faranda, and Moris Gualdoni, "Performance Analysis of a Single-Axis Tracking PV System", IEEE Journal of Photovoltaics, *Vol.2, No.4, October 2012.*

[2] Alami, A. and Batista, R. (2009) Performance Analysis of a Large Scale Photovoltaic Solar Grid Connected System. 34th IEEE Photovoltaic Specialists Conference (PVSC).

[3] Kelly Pickerel, "How does a new single axis tracking method increase solar plant efficiency?",: *Solar power world , June 16 2015.*

[4] BIS Beureau of Indian Standards Doc: ET 28 (6756)

[5] IEC/TS 62548 : 2013 'Photovoltaic (PV) arrays – Design

978-1-5090-5606-4/17 $31.00 © 2017 IEEE

[6] EC/TS 62727 : 2012 'Photovoltaic systems – Specifications for solar trackers' issued by the International Electrotechnical Commission (IEC)

[7] J. Shingleton ,Shingleton Design, LLC ,Auburn, New York One- Axis Trackers – Improved Reliability, Durability, Performance, and Cost Reduction Final Subcontract Technical Status Report 2 May 2006 – 31 August 2007.

Experimental Evaluation of the Performance of Crystalline Si PV Module Degradation after 15-Years of Field Exposure

[1,2]Denio A. Cassini, [+,1]Antonia Sônia A. C. Diniz, [3]Marcelo Machado Viana, [3]Michele C. C. de Oliveira, [3]Vanessa de F. C. Lins, [4]Roberto Zilles, [5]Lawrence L. Kazmerski

[1]GREEN-IPUC, Pontificia Universidade Catolica Minas Gerais –PUC Minas, Belo Horizonte, Brasil
[2]Companhia Energética de Minas Gerais – CEMIG-D, Belo Horizonte, Brasil
[3]Universidade Federal de Minas Gerais – UFMG, Belo Horizonte, Brasil
[4]Universidade de São Paulo – USP, São Paulo, Brasil
[5]Renewable and Sustainable Energy Instutute (RASEI), University of Colorado Boulder and
National Renewable Energy Laboratory, Golden, Colorado, USA
+corresponding author

Abstract — This paper presents the results of the investigation of performance losses of photovoltaic crystalline-silicon modules in Brazil and the identification of durability issues. The PV- modules were located in northern and eastern regions of the state of Minas Gerais, with the oldest module 20-years in operation. These areas are considered subtropical climate zones, with dry winters and rainy summers. The techniques used were: (1) visual inspection procedures to identify the major PV module degradation modes, (2) I-V characterization by solar simulator for the electrical characteristics, (3) IR imaging for hot points identification, and (4) infrared spectroscopy in the IR Region by attenuated-total-reflectance (FTIR-ATR) for encapsulant analysis. A gradual encapsulant discoloration and delamination is attributed to the existing high incidence of ultraviolet radiation and high temperatures. Extensive module encapsulant degradation mechanisms was analyzed and identified. Encapsulation browning was followed by delamination and interconnect corrosion. This paper presents detailed analyses of the operation and reliability of crystalline silicon modules after 15 years in the field. Importantly, these studies provide comparisons between c-Si modules installed at the same region, under the same climate conditions. One set suffers extensive encapsulant degradation—and the other ones, none. The reasons for this are explained.

Index Terms — reliability of PV modules, failure rate, degradation modes and mechanisms, module performance loss.

I. INTRODUCTION

A very important question from PV technology investors is *"how long does the PV system perform well?"*—in order to ensure their investment return. The understanding of how PV modules age is very important for predicting and warranting their performance, for identifying and solving any problems, and for addressing the fundamental considerations for the bankability of photovoltaics (PV) systems–including that of the integrity of the PV module.

The economics of the PV systems can only be competitive if the modules operate reliably with assured longevity for more than 20 years (i.e., the typical warranty period), with power degradation smaller than 20% for this period (typically designated at "less than 1% per year"). The

degradation modes in PV modules dictate the symptoms of failure and degradation mechanisms, and the relative progression of these symptoms. Consequently, a failure mechanism is responsible for the failure mode and could be the origin of one or more failure effects [1].

The degradation modes observed in field-aged modules suggest that the various modes of degradation that are ultimately responsible for performance loss and failure can be of five types: (1) degradation of packaging materials, (2) loss of adhesion, (3) degradation of cell/module interconnects, (4) degradation caused by moisture intrusion, and (5) degradation of the semiconductor device [1]. Jordan and Kurtz [2] recently reviewed the degradation rates from the field-testing studies performed during the last 40 years. They concluded that the degradation rates observed in the PV modules/systems deployed after year 2000 have significantly reduced failure rates over those before the year 2000, indicating a substantial improvement in design, materials, manufacturing, and durability of PV modules [2].

Wohlgemuth reported 20 key results of a study for crystalline-silicon PV modules relating to field failure and warranty return rates based on various reports in the literature. Their analysis shows that less than 0.1% of annual field failure rate on 10-year-old qualified PV modules, 0.005% for PV modules up to 5 years, 0.13% for PV modules manufactured between 1994-2005, and 0.01% annual return rate for PV modules manufactured between 2005-2008. This comprehensive assessment gives some confidence of the reliability of the crystalline Si product—especially that which has been recently introduced into the market [3].

This paper presents the results of the investigation of performance losses of photovoltaic modules and the identification of major degradation mechanisms. These systems have been in operation at least 15 years in Brazil's subtropical climate conditions and hold valuable information for performance road mapping and decision-making for future PV investments, technology bankability, and appropriate technology selection for optimum performance.

II. METHODOLOGY

This study covers an analysis from a PV-module sample chosen from the approximately 1000 stand-alone PV systems that are operated by the Energetic Company of Minas Gerais (CEMIG). These PV systems are installed in the state of Minas Gerais, with the oldest 20 years in operation. They are located at northern and eastern region of Minas Gerais, considered subtropical climate zones, with dry winters and rainy summers. Also, there were some PV systems implemented under several photovoltaic rural electrification demonstration programs, such as US-Brasil program initiated in 1995, PRODEEM in 1997, and Luz Solar in 1999, with the oldest ~20 years in operation.

The investigation of the performance of these PV systems started with an evaluation of the history of PV systems failure from CEMIG database. This was followed by a selection of typical solar-home PV systems based in years of field exposure (CEMIG database) and selection of modules that had been replaced because of performance issues and put into storage. All modules had crystalline-Si solar cells. From the storage, a sample of 20 PV crystalline silicon modules from the oldest systems was chosen, typically the most damaged and exhibiting degradation characteristics. A sample of these PV systems was inspected in the field during technical visits, and the PV modules with the worst degradation modes were taken from the field to be analyzed at in the laboratories of GREEN Solar-IPUC in Belo Horizonte. These modules were inspected following the visual inspection procedures developed by NREL [5] and Freitas Souza [6]. Some degraded modules were removed from the field and brought to the laboratory at GREEN Solar-IPUC to be analyzed and diagnosed.

Following the detailed visual inspection procedure, the modules (both those from the storage and from the field) had their I-V characteristics recorded by at the flash simulator at "Laboratório Fotovoltaico" from University of São Paulo (IEE-USP), and underwent infrared imaging analysis to identify hot spots (using an infrared camera), and chemical analysis of the encapsulant of similar broken modules from the same PV systems. The evaluations also included *Infrared Spectroscopy in the Infrared Region by Attenuated Total Reflectance* (FTIR-ATR) for encapsulant analysis. The modules were also examined for others problems. This included visual inspection for the extent of soiling, the non-uniformity of soiling products (especially for the framed modules), and any indications of cementation on the module glass surfaces.

III. RESULTS

From the visual inspection of the chosen PV module sample, the main degradation modes were identified as gradual encapsulant discoloration and encapsulant delamination—followed by corrosion of the interconnects. This could be caused by the existing high incidence of ultraviolet radiation with high operating temperatures associated with the climate in the northern region of Minas Gerais State, Brazil, and how the encapsulant materials react to them.

The following figures show three of these PV modules (1, 2 and 3) with both types of degradation mechanisms, browning and delamination, and theirs I-V characteristics.

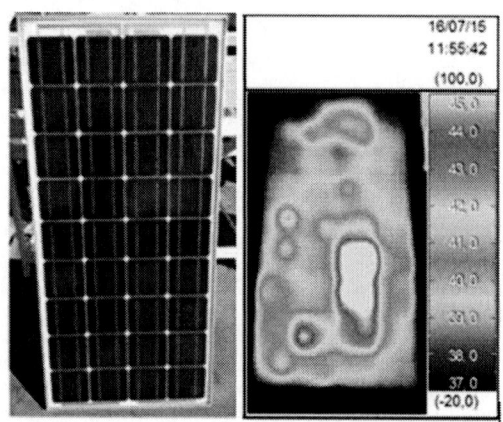

Fig. 1. PV module 1 and its thermography

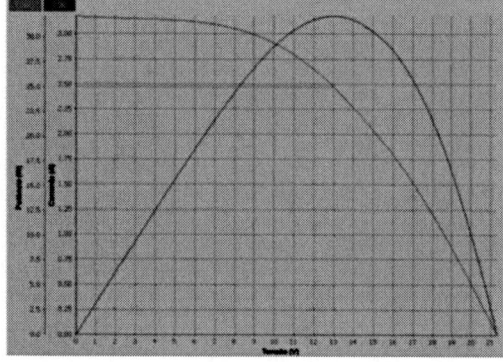

Fig. 2. I-V characteristics module 1

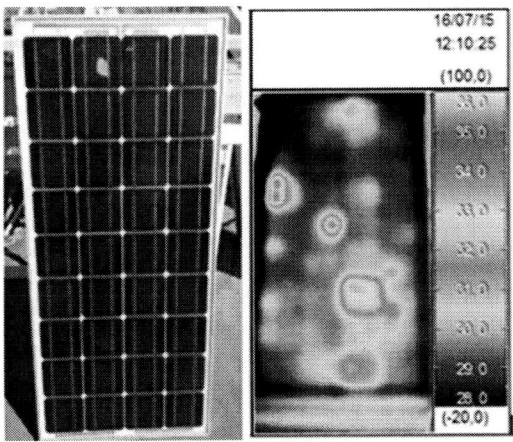

Fig. 3. PV module 2 and its thermography

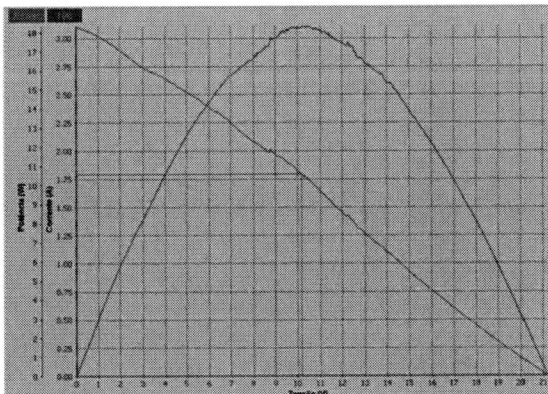

Fig. 4. I-V characteristics module 2

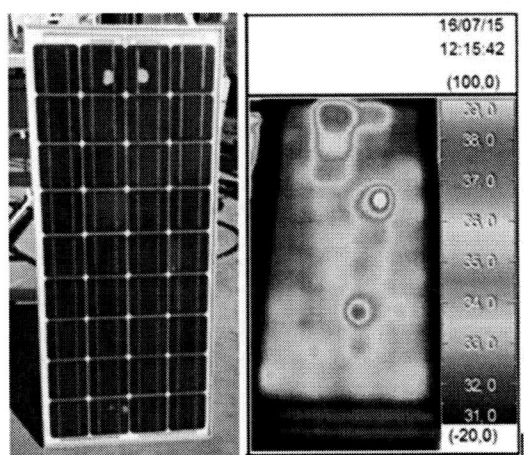

Fig. 5. PV module 3 and its thermography

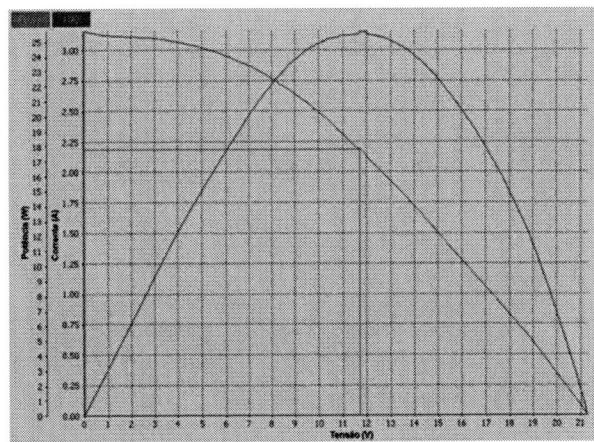

Fig. 6. I-V characteristics module 3

Table 1 summarizes the I-V measurements of the modules 1, 2 and 3, including the data from the data sheet of the reference module. All of them were from the same manufacturer. Table 1 indicates a reduction in all electrical parameters, but the P_{mpp} of the modules was more affected by the degradation of them, with significant increases in series resistance caused by transmission loss in the module surface.

One can observe that the degradation rate in PV modules 1, 2 and 3 were much higher than the 0.5%/year degradation rate reported by literature [1-3], and the 1%/year supported by their warranty. These PV modules were in the field for more than 15 years and were manufactured before 2000. The Ref modules electrical parameters were taken from their reference module's data sheet.

TABLE I
Electrical Parameters from Modules 1,2,3,4

Parameters	Reference Module 1,2,3	Module (1-55/97)	Module (2-55/97)	Module (3-55/97)	Reference Module 4	Module 4 (1-45/96)
Isc (A)	3,5	3,18	3,14	3,19	3,25	3,15
Im (A)	3,17	2,47	1,8	2,18	3,02	2,89
Voc (V)	21,7	21,34	21,08	21,34	18,9	19,06
Vm (V)	17,4	13	10,18	11,71	15	14,36
Pm (W)	55	32,17	18,28	25,56	45,3	41,46
Rs (Ohm)		2,67	6,72	3,95		0,99
Pm variation %		-41,51	-66,76	-53,52		-8,47
Degradation rate % (Anual Average)		2,3	3,7	2,97		0,44

Another similar module (4) but from a different manufacturer was analyzed for comparison (Fig. 7). It was installed in the same region in the same time, near the area of installation of modules 1, 2 and 3. Figure 7 shows its photo and Figure 8 shows its IV characteristics, which indicates no type of degradation.

978-1-5090-5606-4/17 $31.00 © 2017 IEEE

This module also has been in the field for more than 15 years and it was manufacturer also previous to the year 2000.

The comparison of PV module 4 electrical parameters with the data sheet of its reference module is presented in Table 1. As expected, the electrical parameters of the module 4 were slightly reduced, with degradation rate of only 0.6%/year.

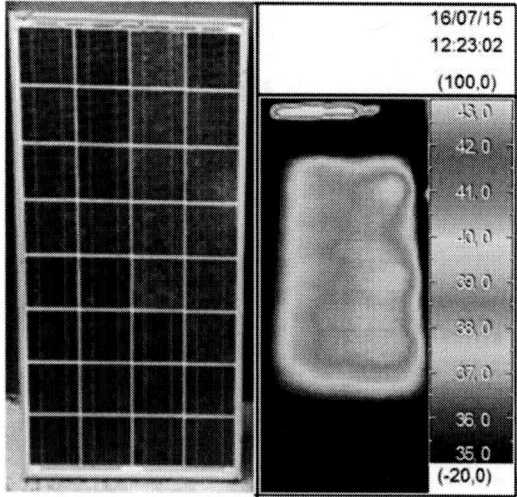

Fig. 7. PV module 4

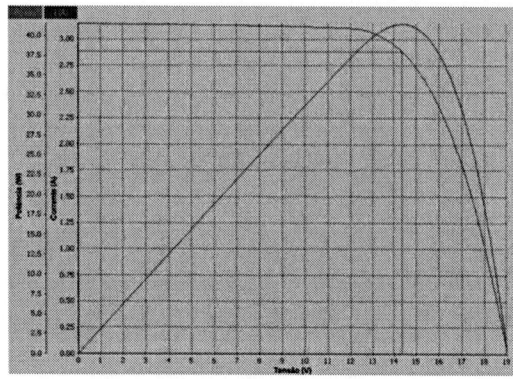

Fig. 8. Current-Voltage characteristics module 4

The exposure to the UV component of solar radiation and high temperature can cause browning of the encapsulant formed by ethylene vinyl acetate copolymer (EVA) and its delamination. The chemical analysis studies of the encapsulant degradation process of the modules 1, 2 and 3 were made. They showed that with exposure to solar radiation EVA undergoes deacetylation of the vinyl-acetate pendant group with the formation of predominantly acetic acid, acetaldehyde and polyenes, which are the discoloring chromophores, or crosslinks

with adjacent chains. Mainly acetic acid and acetaldehyde can catalyze the degradation of EVA beyond they are corrosive of the metallization in crystalline Si PV modules [7,8]. The metal oxidation also occurs between the metal contacts, from the moisture that enters the PV module due to sealing problems in edges related to its incorrect application or rupture by external agents [8]. The glass of similar module to module 4 was analyzed, but no cerium was found in its composition. Thus cerium was not acting to filter the UV as was anticipated. The physical and chemical reasons why this module did not experience discoloration like the other modules is currently being investigation. The initial indications are that the module 4 EVA has a different formulation than the other set from the other manufacturer.

IV. CONCLUSION

This paper presents the results performance loss and degradation mechanism investigations of photovoltaic modules installed in the state of Minas Gerais, Brasil, for periods as long as 20 years. Several degradation modes have been identified from our standard visual inspection methodologies and post-inspection electrical and chemical analyses. We are currently investigating the reasons what the modules installed in the same regions, operating under the same climate conditions, and having the same long-installation times have much-different encapsulation degradation.

ACKNOWLEDGEMENT

The authors would like to thank the Energetic Company of Minas Gerais (CEMIG-D), University of São Paulo (USP) and Pontifical Catholic University of Minas Gerais (PUCMINAS) for their support for the development of this project. Also our team would like to thank FAPEMIG e CNPq for the financial support and guidance. We also gratefully acknowledge the help and inputs of the entire GREEN technical team.

REFERENCES

[1] G. TamizhMani, J. Kuitche, "ABC Report- A Literature Review and Analysis on: Accelerated Lifetime Testing of Photovoltaic Modules", Solar ABCs report, Dec 2012.

[2] D.C. Jordan D.C. and S.R. Kurtz, Photovoltaic Degradation Rates-an Analytical Review. Prog. Photovolt: Res. Appl. 2013; 21:12–29.

[3] J. Wohlgemuth, "Standards for PV Modules and Components - Recent Developments and Challenges". NREL Report number: NREL/CP-5200-56531, 2012.

[4] T. Sarver, A. Al-Qaraghuli, and L.L. Kazmerski, "A comprehensive review of the impact of dust on the use of solar energy: history, investigations, results, literature, and

mitigation approaches," Renewable and Sustainable Energy Reviews 22, 689-744 (2013). N.B. Update of literature and status expected for publication in April 2015.

[5] C. Packard, J. Wohlgemuth, and S. Kurtz, Development of a Visual Inspection Data Collection Tool for Evaluation of Fielded PV Module Condition. National Renewable Energy Laboratory (Technical Report: NREL/TP-5200-56154), 2012.

[6] Francisco Hering Alves de Freitas Souza, *Inspeção e Monitoramento do Desempenho de Sistemas Fotovoltaicos Conectados a Rede Elétrica: Estudo de Caso Real.* 2014. Master thesis – Pontifícia Universidade Catolica de Minas Gerais, Programa de Pós-Graduação em Engenharia Mecânica, Minas Gerais.

[7] F.J. Pern, A.W. Czanderna, K.A. Emery and R.G. Dhere, "Weathering degradation of EVA encapsulant and the effect of its yellowing on solar cell efficiency," Proc. IEEE Photovoltaic Spec. Conf. (IEEE, New York; 1991) pp. 557-561.

[8] Igor Alessandro Silva Carvalho, *Preparação e estudo da fotodegradação de compósitos de matriz polimérica para encapsulamento de módulo fotovoltaico.* 2007. Master thesis, Pós-Graduação em Engenharia de Materiais da REDEMAT, Minas Gerais.

Field investigations of potential-induced degradation (PID) for crystalline silicon PV panels in different climates

Yifeng Chen[1], Peter Hacke[2], Yong Sheng Khoo[3], Kaitlyn VanSant[4], Zigang Wang[1], Wei Luo[3], Jing Chai[3], Chris Deline[2], Yan Wang[3], Armin G. Aberle[3], Pietro P. Altermatt[1], Zhiqiang Feng[1], Sarah Kurtz[2], and Pierre J. Verlinden[1]

[1]State Key Laboratory of PV Science and Technology, Trina Solar, ChangZhou, 213031 China

[2] National Renewable Energy Laboratory (NREL), 15013 Denver West Parkway , Golden, CO, USA

[3] Solar Energy Research Institute of Singapore (SERIS), Singapore 117574

[4] Colorado School of Mines, 1500 Illinois St, Golden, CO, USA.

Abstract — The relationship between field performance and indoor chamber measurement for potential-induced degradation in crystalline silicon PV modules is still unclear. Many acceleration models have been proposed but without strong evidence from field experiments. Our approach focuses on building field testing systems under three different climates: tropical, subtropical, and mountain climates in Singapore, Changzhou (China) and Golden (CO, USA) respectively, in a collaborative testing campaign involving Trina, SERIS and NREL. Each testing system has six groups of modules developed by Trina Solar with different grades of PID resistance from weak to strong. The objective of this 3-year project is to build up the correlation between the outdoors behaviors to chamber tests, and deduce a climate-specific degradation model. A degradation of up to 23% was observed after 18 weeks for the most PID-susceptible module was found in Singapore. Activation energy of 0.94-1.1eV and 0.55-0.63eV were determined for wet and dry conditions, respectively, by fitting the leakage current data in Singapore and in Golden, and an exponential decay of degradation was shown as a function of accumulated charge.

I. Introduction

Potential-induced degradation (PID) is one of the most severe reliability problems for today's crystalline silicon PV industry. Although a significant amount of research has focused on the physical models of PID [1-5], a complete understanding of the PID mechanisms has yet to be established. One of the reasons is that most of the research is based on chamber tests, instead of field observations which require several years. Some field-failed modules can be collected. However, as the fabrication details of such modules are incomplete, it is difficult to extrapolate these observations into lessons that are relevant for current PV module technologies. The relationship between the real field performance and indoor chamber tests are therefore still unclear. The PV industry is seriously concerned with the design and quality control of modules that will exhibit PID resistance in various different climates over the course of the product's warranty.

To address these concerns, we 1) developed PV modules with various grades of PID resistance (6 in total), which cover a large range of PID-resistant characteristics; 2) proposed an international outdoor testing program under three different climates: tropical in Singapore (by SERIS), subtropical in Changzhou (China, by Trina Solar) and mountain, in Golden, CO (USA, at NREL). Our objective is to directly observe PID degradation in the field and develop a correlation with the indoor testing results in various climates conditions for the first time, with this we can try to determine the minimum indoor PID testing necessary to ensure a 25-year lifetime in a given climate for the PV industry.

II. Experimental designs

The 60-cell modules were fabricated with standard full-Al-BSF *p*-type multi-crystalline silicon solar cells. The front side SiN$_x$:H passivation layer was adjusted to yield different PID resistances at the cell level. Meanwhile, different EVA sheets

TABLE I Modules divided in groups (A to F) based on PID test results

Module	Encapsulation	60C./85%R.H./96h at $\pm$1000V	85%R.H./85degC/96h at $\pm$1000V	85%R.H./85degC/600h at $\pm$1000V	85%R.H./85degC/600h w. Cu foil at $\pm$1000V
A type	Backsheet	NG	NG	NG	NG
B type	Backsheet	PASS	PASS	NG	NG
C type	Backsheet	PASS	PASS	PASS	NG
D type	Backsheet	PASS	PASS	PASS	PASS
E type	Backsheet	PASS	PASS	PASS	PASS
F type	Double glass	PASS	PASS	PASS	PASS

with several sets of bulk resistivity were used during module fabrication.

With the combination of the PID-resistant cells designs and module encapsulations, we successfully fabricated 6 different types of modules with PID-resistance ranging from very weak to very strong, as shown in Table I.

Here NG denotes modules that exhibited more than 5% power degradation after the PID chamber test. Therefore, A type modules are considered to be very susceptible to PID when they operate in the field. According to the experimental results of Hoffmann et al. [6], the leakage current density of modules covered with metal foil, like Cu, would be 10 times higher than in ambient relative humidity of 85%. Therefore, one may reasonably suggest that special modules passing the extended PID test (85%R.H./85°C/600h with Cu foil) could demonstrate much stronger PID resistance than the modules that just pass the PID test defined by one of the current IEC standard test conditions (85%R.H./85°C/96h). The power degradation of 6 types of modules as a function of indoor testing time is show in Fig.1.

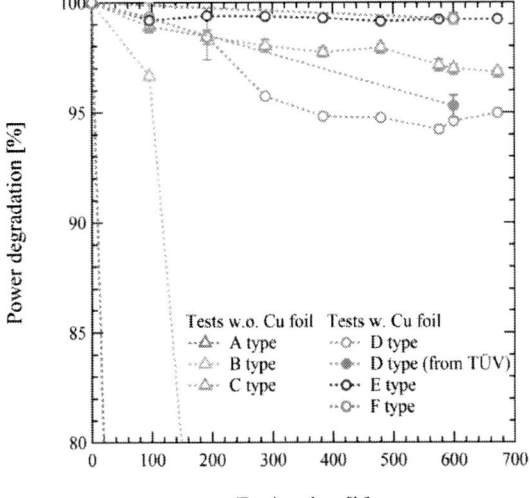

Fig. 1. Normalized module power as a function of indoor PID chamber test (under 85%R.H./85°C/-1000V) time for 6 types of modules, where type A module is completely PID vulnerable. The triangle symbols represent the accelerating measurements without Cu foil, while the circle symbols denote the cases with Cu foil. The solid light blue circles are independently confirmed by TÜV Rheinland.

With careful design, backsheet modules (type D and type E) are also able to pass the most aggressive chamber test, which was independently confirmed by TÜV Rheinland in Shanghai (light blue solid circles). F-type modules are double-glass modules which prevent moisture from penetrating into the module and feature excellent PID-resistant performance. We expect that a module capable of passing such an aggressive indoor test would not suffer from PID degradation over its 25-year lifetime under any real-world conditions.

III. FIELD EXPERIMENTS

For field testing, we chose Changzhou (China), Singapore and Golden, CO (USA) to represent the typical PID tests under subtropical, tropical and mountain climates, respectively. Each testing location has 66 modules categorized as groups A to F. The modules are connected to a fixed resistor to make sure that the modules work near their maximum power point in the daytime. The frames being grounded, the positive poles of the modules are connected to an external voltage source, applying 0V, 400V, 600V, 1000V or 1500V to the modules. The 0V modules are installed in the field without any applied voltage as references to monitor any degradation unrelated to PID.

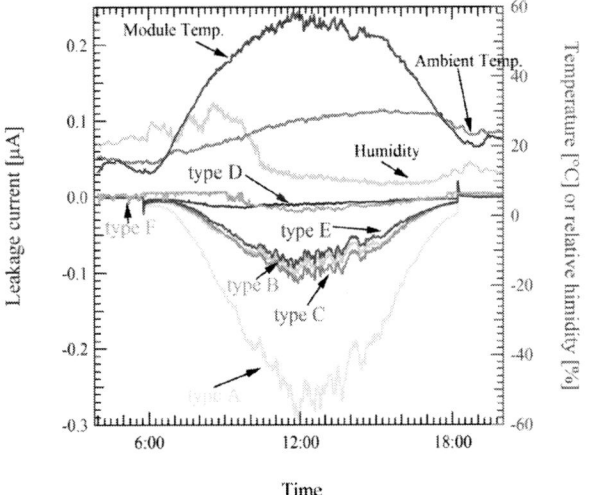

Fig. 2. An example of the data collected in a sunny day in September in the test field of NREL. Blue and green curves are leakage current, while the red and yellow curves represent the temperature and the relative humidity.

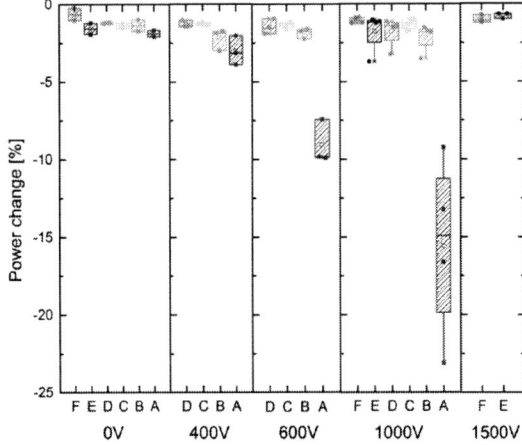

Fig. 3. Power degradation of modules from type A to type F that are installed in the test field of SERIS as a function of the applied voltage varied from 400V to 1500V. As can be seen, the degradation of the PID-vulnerable module (light blue) is approximately proportional to the applied voltage.

978-1-5090-5606-4/17 $31.00 © 2017 IEEE   1923

During the field measurements, all the data including global irradiation, wind speed, RH, module temperatures of each type were monitored. The leakage current data were recorded at only 1000V at three locations. During the night time or low irradiation conditions with irradiation less than 10 W/m², the external voltage source is disconnected and all data were not taken into consideration. All the modules in three locations were installed at the best tilt angles to simulate the real field installation.

Fig. 2 shows an example of the monitored data for a sunny day in September in Golden, the type A modules have the largest leakage current, type B, C and E are quite comparable to each other, and type D and type F (double glass) nearly have no leakage current, which is consistent with the indoor PID results in Fig.1.

Among the three sites, the outdoor test in Singapore is expected to be the most severe, not only the three most important parameters that accelerate PID (high average temperature, high relative humidity and frequent rain-sunshine cycles) are most prevalent in Singapore, but also the modules were installed at 0 degree, that residual rain or morning dew will stay on the module for a longer time only until evaporation. Interestingly, after only 18 weeks of operation, we observed in Singapore: 1)the type A modules, which are PID vulnerable, started to show a degradation of up to 23% (as illustrated in Fig.3) while 2)other groups suffered no degradation or less than 3%. As no significant degradation was observed in the other two test sites, the quick degradation in Singapore indicates that PID in tropical areas could be very aggressive. The outdoor measurements will continue for several years. We intend to regularly report the results and the information acquired during this long-term project with the objective of defining a minimum PID test standard for different climates.

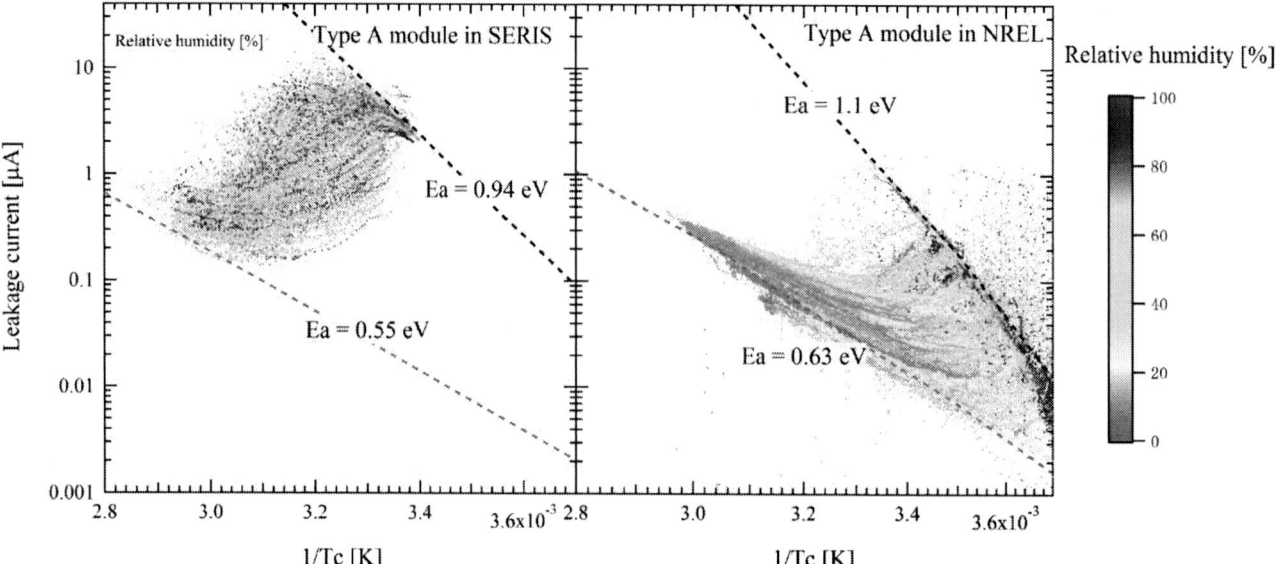

Fig. 4.   The measured leakage current of type A modules (PID vulnerable) as a function of the reverse absolute module temperature in the test field of Singapore (left) and Golden (right) for type A modules. The colors denote the local relative humidity.

Away from the IV parameters, the leakage current could be used as an indicator to study the details of PID [7]. Fig.4 shows the leakage current as a function of the inverse temperature of type A modules in Singapore (left plot) and in Golden (right). The colors in the figure indicate changes in relative humidity.

The leakage current distribution is influenced by the conductivity and hence the current path of charge [7], and greatly impacted by the surface conditions of the modules (wet or dry) [8]. We extracted the activation energy by using activation energy model in the following equation:

$$J_{leakage} = J_0 \exp(-\frac{E_a}{k_B T}) \qquad (1)$$

Here $J_{leakage}$ is the leakage current density, J_0 is the maximum leakage current when the temperature is infinite, E_a the activation energy with unit of eV, k_B the Boltzmann constant and T the temperature. For wet condition (upper part), the activation energy looks to be 0.94 eV and 1.1 eV respectively in Singapore and Golden. For dry condition (upper part), data in Golden show good grouping and yield activation energy of 0.63 eV, however it is hard to define the E_a in dry condition in Singapore as most of the RH are over 40%. A dash line curve with E_a = 0.55 eV is plotted to guide the eyes to see the bottom boundary of the data cloud in Singapore.

As expected, the climate in Singapore is humid and hot. The shape of leakage current as a function of reverse temperature data cloud is quite different between Singapore and Golden, which reflects that the leakage current is quite climate dependent. The activation energy upper and bottom limits in Golden are higher than that of Singapore, which may indicate that the activation energy would depend on the surface condition like soiling, which may change the leakage current from the edges of the module to the solar cells. Generally, the leakage current in Singapore is at least one magnitude higher than that of Golden for type A module.

We try to investigate the relationship between the field power degradation and the leakage current. Fig. 5 plots the modules power loss as a function of the accumulated charge, note that the degradation shown in Fig.5 is biased with reference modules (0V group). Although it seems that the power degradation follows an exponential decay with the cumulative charge, it should be pointed out that the reverse current during the nights to discharge is not taken into account, also it is still unclear if the small degradation (<5%) is due to normal aging.

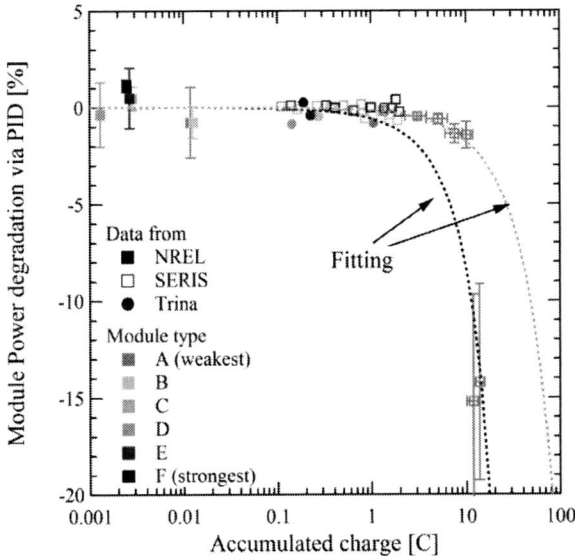

Fig. 5.   The power degradation of modules in outdoor PID tests as a function of accumulated charge. Note that the recovery of the module during low irradiation (and at nights) is not taken into account.

As more and more data are generated, the relationship between the leakage current and the field degradation will become clearer, and they will be presented in following publications.

IV. CONCLUSIONS

Field testing systems were built in tropical, subtropical, and mountain climates in Singapore, Changzhou (China) and Golden (CO, USA) respectively to investigate the relationship between field performance and indoor chamber measurement of PID with 6 types of modules with PID-resistance ranging from very weak to very strong. The strongest one degraded

less than 5% after the most aggressive PID test (85% R.H., 85 degC, 600h, -1000V, and with Cu foil). To the best of our knowledge, this is the best anti-PID result ever reported using EVA encapsulant and optimized SiN_x coating.

A degradation of up to 23% was observed after 18 weeks for the most PID-susceptible module in Singapore. Activation energies of 0.94-1.1eV and 0.55-0.63eV were determined for wet and dry conditions, respectively, by fitting the leakage current data in Singapore and in Golden, and an exponential decay of degradation was shown as a function of accumulated charge.

ACKNOWLEDGEMENT

This work is financially supported by the National High-tech R&D Program of China (863 Program, No. 2015AA050302).

This work was partially supported by the U.S. Department of Energy under Contract No. DE-AC36-08GO28308 with the National Renewable Energy Laboratory. Funding provided by U.S. Department of Energy Office of Energy Efficiency and Renewable Energy Solar Energy Technologies Office.

The U.S. Government retains and the publisher, by accepting the article for publication, acknowledges that the U.S. Government retains a nonexclusive, paid-up, irrevocable, worldwide license to publish or reproduce the published form of this work, or allow others to do so, for U.S. Government purposes.

REFERENCES

[1] R. Swanson, M. Cudzinovic, D. DeCeuster, V. Desai, J. Jürgens, N. Kaminar, W. Mulligan, L. Barbarosa, D. Rose and D. Smith, Proceedings of 15th International Photovoltaic Science and Engineering Conference, Shanghai, China, 2005, 410-411.

[2] S. Pingel, O. Frank, M. Winkler, S. Daryan, T. Geipel, H. Hoehne and J. Berghold, Proceedings of the 35th IEEE Photovoltaic Specialists Conference, Honolulu, HI, United States, 2010, 2817-2822.

[3] P. Hacke, K. Terwilliger, S. Glick, D. Trudell, N. Bosco, S. Johnston and S. Kurtz, Proceedings of the 35th IEEE Photovoltaic Specialists Conference, Honolulu, HI, United States, 2010, 000244-000250.

[4] V. Naumann, D. Lausch, A. Hähnel, J. Bauer, O. Breitenstein, A. Graff, M. Werner, S. Swatek, S. Großer and J. Bagdahn, Solar Energy Materials and Solar Cells, 2014, 120, 383-389.

[5] W. Luo, Y. S. Khoo, P. Hacke, V. Naumann, D. Lausch, S.P. Harvey, J. P. Singh, J. Chai, Y. Wang, A. G. Aberle and S. Ramakrishna, Potential-induced Degradation in Photovoltaic Modules: A Critical Review, Energy & Environmental Science, vol. 10, pp. 43-68, 2016.

[6] S. Hoffmann, M. Koehl, Proceedings of the 28th European Photovoltaic Solar Energy Conference and Exhibition, Paris, France, 2013, 3336-3339.

[7] J. A. del Cueto and T. J. McMahon, Analysis of leakage current in photovoltaic modules under high-voltage bias in the field, Prog. Photovolt: Res. Appl. 2002; 10:15-28

[8] P. Hacke, R. Smith, S. Kurtz, D. Jordan, and J. Wohlgemuth, Modeling current transfer from PV modules based on meteorological data, Proceedings of the 43rd IEEE Photovoltaic Specialists Conference, 2016, 1083-1089

Determining the Power Rate of Change of 353 Plant Inverters Time-series Data Across Multiple Climate Zones, Using a Month-by-Month Data Science Analysis

Alan J. Curran[1], Yang Hu[1], Rojiar Haddadian[1], Jennifer L. Braid[1],
David Meakin[2], Timothy J. Peshek[1], Roger H. French[1]

[1] Case Western Reserve University, Cleveland, OH 44106, United States
[2] SunPower, 77 Rio Robles, San Jose, CA 95134, United States

Abstract—The month-by-month method has been developed in the Solar Durability and Lifetime Extension (SDLE) Research Center with the purpose of improving the power output monitoring of real world photovoltaic (PV) power plants. This method computes the rate of change (ROC) of the system and external variables are ranked based on their correlation with the changing rate of the system to provide insight into the possible degradation mechanisms of PV modules. A data set of non-SunPower, front contacted multi-crystalline modules were analyzed with the month-by-month method. Results show an average linear ROC of $-0.95\%/year$ with an observed correlation between the ROC and the brand of the module, as well as the total age of the system, and the climate zone the system is located in.

I. INTRODUCTION

Solar panels are one of the predominant forms or renewable energy and continue to become more popular as the price decreases and the efficiency increases. The degradation of solar modules can lead to power loss over time, which can be highly problematic if a photovoltaic (PV) system has been designed and built to produce a certain amount of power reliably[1]. Given the potential risk in a power source that may give unexpected diminishing returns over time, it is important to have a strong understanding of how much degradation can be expected when designing a system and when creating modules that are more resistant to degradation.

The DOE SunShot initiative has proposed a set of guidelines for reducing the cost of PV power to 6¢per kWh for utility scale PV[2]. This initiative has set a required power loss rate of 0.2%/year over 30 years of operation, down from a current estimation of about 0.8%/year for existing PV plants[3]. Current standards of monitoring the degradation rates of PV systems are often inaccurate or do not account for all that might be affecting the output of the system, such as maintenance or an inaccurate nameplate power value. Improving the method of quantifying the degradation rate of a system will not only give a more direct pathway to achieving

the SunShot initiative goal, but can also give insight into what factors could be causing these systems to degrade.

Current methods for estimating the degradation rate of a system often involve accelerated testing, which exposes the modules to heat, humidity, and irradiance conditions to induce rapid degradation. The expected real-world performance, i.e. the performance that can be expected from a module exposed to outdoor weather conditions over a long period of time, is then extrapolated to a long term estimation from the observed accelerated degradation. Accelerated testing is important in many cases, including determining the degradation mechanisms and potential stressors, however it cannot be assumed that there will be a strong correlation between accelerated testing and real-world degradation without a proper understanding of the long term performance of real-world PV systems. Some widely used current methods for modeling the long term performance of PV systems usually involve two significant assumptions. The first is the clear sky assumption - that the data should only be considered at close to maximum irradiance without cloud cover[4][5]. The second is the linear degradation assumption - that module degradation will manifest as a linear change over time[5]. In this study, we hope to approach this modeling with the intent of keeping as much data as possible without making assumptions about the final trend of the degradation.

II. DATA SOURCES AND PROCESSING

The data used for this study comes from commercially fielded systems using front contacted multi-crystalline modules. Using industry sources gives us access to large amounts of data over long periods time for many different modules within many different climate zones[6]. The drawback of this is that there many be variable differences between data sets from different sites and the data and sensors many not be as consistent or well monitored as an on-site research

---

978-1-5090-5606-4/17 $31.00 © 2017 IEEE

system. To account for this, the Month-by-Month (MbM) method has been designed to be fault tolerant and usable for many different potential variables, such as different irradiance (Global Horizontal vs. Plane of Array) or temperature readings (module vs. ambient temperature), depending on what was measured. The idea behind this is that the massive amounts of data we obtain using industry sources will make-up for the potential decrease in data quality.

The data consists of long term time series data of 373 inverters. It was taken in 15 minute intervals with variables of power or energy produced during the time interval, as well as instantaneous weather conditions including the ambient temperature, the Global Horizontal Irradiance (GHI), and the wind speed. Metadata for each system is given, including the module brand and model, the age of the system (years since the start of operation), and the location of the system in latitude and longitude which is used to determine the Koppen-Geiger climate zone[7] the system is located in. Ideally, Plane of Array (POA) irradiance would be available instead of GHI as it is a better correlated measurement of the irradiance on the surface of the panel, however GHI is what was measured for these sites and the GHI cannot be accurately converted to POA as the tilt and azimuth angles of the panels are unknown and the data does not always follow the clear sky assumption[8]. GHI still works as a strong predictor of power, however it appears to introduce some seasonality that can be dealt with. To maintain ambiguity, all systems have been given de-identified 7-digit alpha-numeric names.

Analysis of this data was done using R through Rstudio[9]. R is an open source software that specializes in data management and statistical analysis, making it ideal for this application.

The long term scale and relatively small sample interval make these data sets fairly large, on the order of tens or hundreds of Megabytes per system. To keep the analysis time within a reasonable range, the High Performance Computing (HPC) cluster at Case Western Reserve University was utilized[10]. This allowed analysis to be done with R in a much shorter amount of time than a local machine. Previous research within the SDLE group has made use of HPC as well[11].

III. METHODOLOGY

The Month-by-Month (MbM) method is the current Rate of Change (ROC) analysis process used by the SDLE[12]. The ROC can be thought of as the opposite of the commonly used degradation rate (R_d), where a negative ROC indicates a system losing power. This terminology is used to improve clarity of a systems' trend, especially if it shows an increasing power trend. This method isolates the ROC of a PV system from the variability caused by weather conditions. To start, inverter data is segmented into 30 day pseudo-month long sections, and it is assumed that there is no significant degradation occurring within each pseudo-month period. For each month long section, a beta (β) model is created to determine the dependence of the power output on the weather conditions. Once a model is obtained for each month, standard weather conditions are determined over the entire lifetime of the inverter, and each β model is evaluated under these standard conditions. This yields a predicted power output for each month under standard conditions and the slope and intercept of this plot correlate to the ROC of the system.

The next model, the Xi (ξ) model, determines a fit for the predicted power chart. This model can either be linear or piecewise fit, depending on the nature of the predicted power trend. Once the ξ model is determined, the yearly ROC can be calculated from the slope of the model. The final tool is the gamma (γ) model, used to compare the metadata across many inverters and determines the most significant variables affecting the ROC. For example, the gamma model might show that the ROC of a module is most dependent on the brand. Each of the models is explained in more detail below.

A. Data Cleaning

The data cleaning for this method is designed to be as minimal as possible, only removing data errors and nighttime readings. The idea behind this is that there is important system information in more conditions than just the clear sky assumption and we want to capture that. First, all missing and interpolated data is removed from the time series. Then data with an irradiance of less than $50W/m^2$ is removed to remove readings during the night. Next sensor reading errors are removed, including any irradiance over $2000W/m^2$ and power readings significantly higher than the nameplate power. Finally, readings are removed where the power is less than 1% of the maximum power to remove any instances when the system was not producing power during the daytime.

B. β Pseudo-month Predictive Model

Each of the month sections has a power dependence based on the ambient weather conditions. Standard weather data for inverters includes the ambient temperature, the wind speed, and the irradiance (GHI, POA, etc.). The β model is the linear relationship of the power to these variables in the form of equation 1:

$$P_{AC} = \beta_0 + \beta_1 I_{GHI} + \beta_2 T_{Ambient} + \beta_3 W_{windspeed} + \epsilon \quad (1)$$

Where P_{AC} is the AC power, β_n are the variable coefficients, and ϵ is the error of the model. Once a model has been created for each month, the standard weather conditions for the inverter lifetime are determined as the overall average temperature, the overall average wind speed, and the minimum of the maximum irradiance values for each month. These standard conditions are applied to the β model for every month, resulting in a predicted power output for each month. This reduces the inverter data to a single predicted power value for each month. Any long term variation in the predicted monthly power output is determined to be due to the degradation of the system as the weather conditions applied to the β models are constant.

C. ξ Piecewise Regression Model

The ξ model determines the fit of the predicted power trend from the β model. Because each point is derived from a unique β model, each point will have a different precision. The ξ model uses a weighted least squares analysis to increase the significance of the months that have more precise β models and reduce the significance of the months that have less precise β models. This reduces the influence of monthly predicted power values that might have experienced extreme weather conditions, failure, or maintenance that might have affected the *ROC*. The *ROC* of the system is defined by the slope and intercept of the ξ model fit in equation 2:

$$ROC = 12 * \frac{Slope}{Y - intercept} \qquad (2)$$

The fraction of the slope and intercept of the ξ model is multiplied by 12 to convert the result to units of ($\%Change/Year$) This model can also apply a piecewise fit for to plot is non-linearity is observed. Even if there was no break in the power production, a change in the string such as a module or inverter replacement might affect the *ROC*. This would be indicated by a piecewise fit of the resulting predicted power plot. In many cases, events such as failure, replacements, or cleaning can create non-linear trends, some examples of which are show in Figures 3a and 3b.

D. γ Cross-sectional Model

The γ model is an analysis tool that shows the effect of the metadata variables, such as module brand or climate zone, on the system changing rate. An Akaike Information Criterion (AIC) test[13] is used in the γ model to determine which of the variables show greater influence on *ROC*s, indicating a significant effect on the system degradation.

E. Seasonal Decomposition

The time series results from the *MbM* method on this data set show strong seasonality. This is most likely due to the irradiance given in Global Horizontal Irradiance (GHI) instead of the better correlated Plane of Array (POA) irradiance. There is a translation error between the measured GHI and the irradiance that actually reaches solar panel, which determines the power output. In the winter months, the modules have higher predicted power because the GHI measurement pick up less irradiance than what hits the panels. This can be mitigated using seasonal decomposition to remove the seasonal component of the time series, leaving just the trend.

The *stlplus* package in R was used to do this[14]. Each data point represents one month of time and the seasonality is yearly so the *MbM* result is defined as a time series with a frequency of 12 (12 months per year) and the seasonality can be significantly reduced. An example of the seasonal decomposition is shown in Figure 1b.

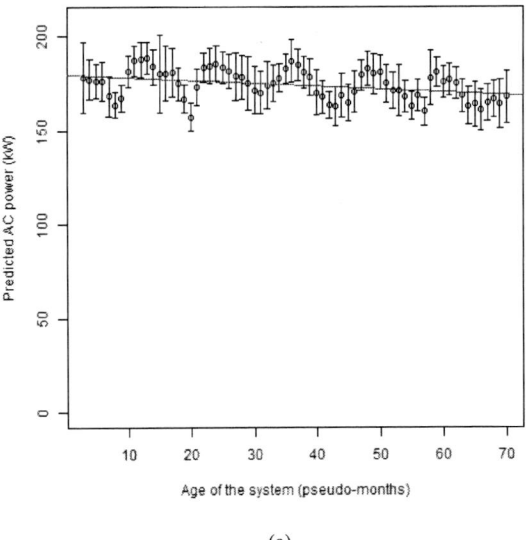

(a)

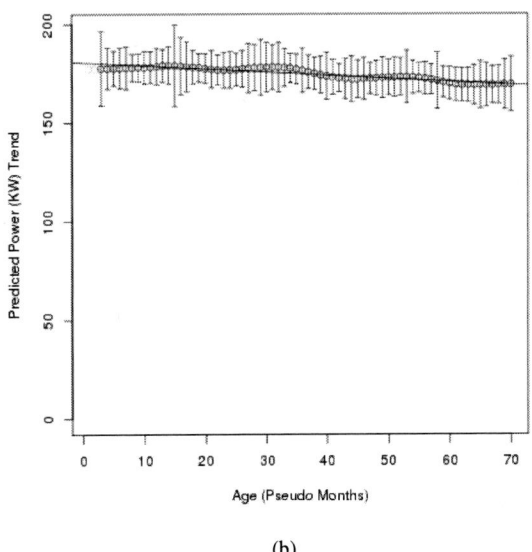

(b)

Fig. 1: (a) Predicted power vs. system age of the "dpa1yun" system with seasonality shown. Each data point represents the power predicted by the β model for each pseudo-month. The error bars are the sigma value for each β model. (b) The "dpa1yun" system after decomposition. The seasonality has been significantly reduced and the trend is strongly linear.

IV. RESULTS

Of the 373 inverters in the data set, 353 were able to be analyzed. Several systems had to be excluded due to large amounts of missing or unusable data, including long term interpolated data that had to be removed. A few systems failed at the decomposition step; a time series with more than one missing season (12 consecutive months in this case) could not be decomposed and were rejected. The a histogram of the ROCs of each system is given in Figure 2. The median ROC of this distribution is $-0.98(\%/year)$ and the mean is $-0.95(\%/year)$.

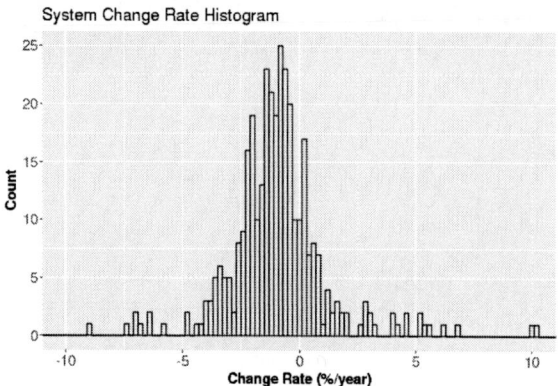

Fig. 2: A histogram of the system ROC distribution. The distribution shows strong normality with a few outliers on both sides.

Given the shape of the distribution and the similarity of the mean and median, this data set shows strong normality. There are several outliers that are seen with extreme ROCs of $\pm10(\%year)$. These are mostly caused by data or system errors not due to degradation. In some cases, the system is not fully monitored or operational at the start of data collection. This would cause a large increase in the performance in the first few months as the power output increases as more modules are connected, leading to a high positive ROC if the increase was significant. Alternately, module failures can cause sharp drops in the power output as fewer modules are available; this is most often observed as a flat line trend followed by a sharp decrease and return to a flat line trend. Examples of systems with events affecting the power are given in Figures 3a and 3b which are two separate inverters from the same site with no applied seasonal decomposition.

In both Figures 3a and 3b, there is a sudden event that changes the power output instead of a gradual change over time that is expected from degradation. Because these systems are located at the same site and the sudden changes in the power trend occur at the same time, it can be assumed that one event caused the change in both of these strings. This can lead to a misinterpretation of the ROC, which is artificially shifted to be more positive or more negative when both systems appear stable when the events are excluded. Given the industry source of this data, the information about plant events during the systems' lifetimes that might have caused these changes are unknown, it could be a change in the string or some sort of sensor error. This also shows that a linear assumption may not always be valid, especially if the system might have experience significant power altering events. This is the case in many of the systems with highly negative or

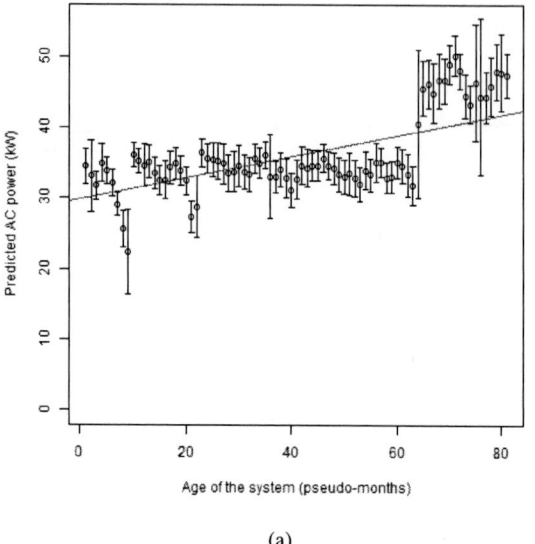

(a)

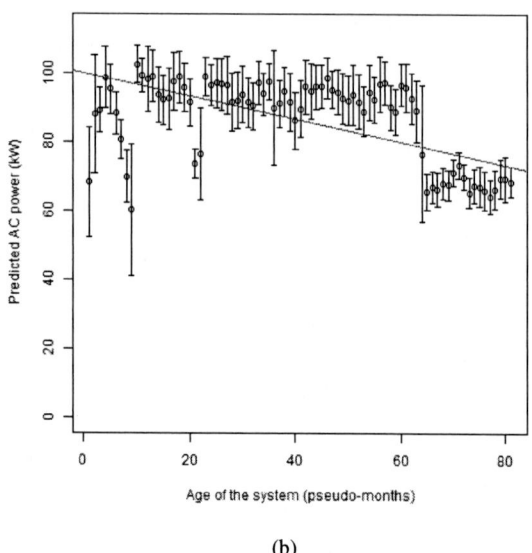

(b)

Fig. 3: (a) System "7voqfp6" showing a highly positive ROC based on a sudden increase in power production. After about 60 months of operation at a fairly consistent output there is a sharp increase in power production, artificially increasing the system ROC. (b) System "qwpkj1y" behaves similarly to the previous plot yet a sudden decrease in power givens a highly negative ROC. This is most likely due to a partial failure of the string. There is also some strong seasonality and the beginning of the time series.

highly positive ROCs. The majority of the systems have a ROC of between about -3 to $1\%/year$.

After the ROC for each system was found a γ cross-sectional model was used to rank the importance of the metadata variables, shown in Equation 3.

$$ROC = \gamma_1 * Module\_Brand + \gamma_2 * System\_Age +$$
$$\gamma_3 * Climate\_Zone + \epsilon \qquad (3)$$

The variables available were the brand and model of the system, the size of the system, the age of the system, and the Koppen-Geiger climate zone the system is located in. The module model variable was removed as there were too many different models and too few modules of each model to be comparable. Most models only had one or two associated systems.

An Akaike Information Criteria was performed using both forwards and backwards regression. The metadata variables in order of significance are the module brand, the system age, and the Koppen-Geiger climate zone. The inverter model could not be included in the γ model as the inverter metadata was unavailable for many of the systems. The significant variables are those that decrease the AIC the most; the greater the decrease the more significant the variable. Insignificant variables increase the AIC when included in the model[13]. The distributions for the brand and climate zone can be seen below in Figures 4a and 4b.

V. DISCUSSION

The results from the MbM method gave an average yearly ROC of $-0.95\%/year$. This is a higher than desired degradation rate, but these systems are fairly old in many cases (up to 15 operating years) and may not be as robust as newer modules. There was a fairly large distribution of ROCs as well, with outliers as far as $\pm 10\%/year$. There are many events that can affect the ROC besides the degradation, such as the addition of more modules or system failures. Given the age of these systems, it is not unusual to find faults such as these that affect the modeled ROC but may not be directly caused by degradation. The normality of the ROC distribution would suggest that errors like these are distributed evenly throughout the data set and do not heavily skew the data set. The results of the γ model suggest that in this data set the module brand has the most significance on the degradation, which is not unexpected. There was also a significant ROC dependence on the age of the system and the climate zone the system was located in. Further investigation will need to be done to see what environmental stressors may have influenced the degradation and ROC of the system.

VI. CONCLUSION

This study used the Month-by-Month (MbM) method developed at the SDLE to determine the yearly power ROC of a sample set of 353 inverters. The linear ROCs of the systems showed a normal distribution with an average of $-0.95\%/year$. This average ROC is higher than set by the DOE SunShot goal, however many of these systems were built close to or more than a decade ago and may not have been designed to withstand degradation as well as modern

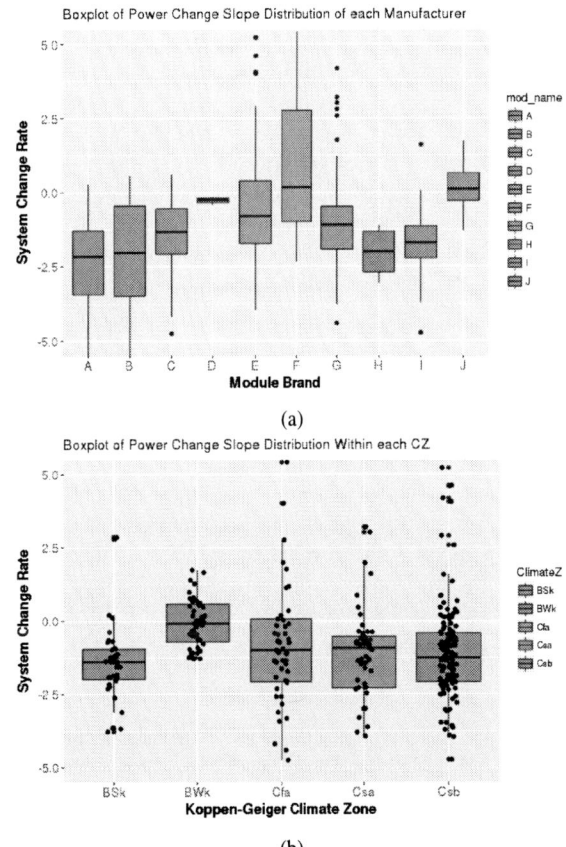

(a)

(b)

Fig. 4: (a) Change rate distribution based on the module brand. Different brands show different performance, both in distribution and in the magnitude of the ROC. (b) Change rate distribution based on the climate zone of the system. There is less difference when compared to the module brand, however there is still a significant difference between some of the climates.

systems. Systems with outlier ROCs were largely caused by what appeared to be anomalies not necessarily due to degradation, such as partial shutdown or the addition of new modules. Instances such as these are natural, given the age of many of these systems and the fact that they come from industry sources. A γ cross-section model showed that the ROC of a system is mostly dependent on the brand of the module, followed by the age and the system, and finally by the climate zone the system is located in.

ACKNOWLEDGMENTS

- This work was supported by the DOE-EERE SunShot award DE-EE-0007140.

- Research was performed at the SDLE Research Center, which was established through funding through the Ohio Third Frontier, Wright Project

Program Award tech 12-004.

- The Global SunFarm Network (GSFN) of the SDLE Research Center was developed in collaboration with Medical Center Company (MCCo) - Cleveland, Replex Plastics, American Electric Power (AEP), Q-Lab Corporation, Underwriters laboratories Taiwan (UL Taiwan), Indian Institute of Technology Gandhinagar (IITGN), Fraunhofer Institute for Solar Energy Systems (ISE), Terraform Power, and SunPower.

- This work made use of the RedCat High Performance Computing Resource in the Core Facility for Advanced Research Computing at Case Western Reserve University.

- Sample and system tracking and data management support was provided by REDCapTM,[15] funded by the Clinical and Translational Science Collaborative (CTSC) grant support (1 UL1 RR024989 from NCRR/NIH).

REFERENCES

[1] D. C. Jordan, T. J. Silverman, J. H. Wohlgemuth, S. R. Kurtz, and K. T. VanSant, "Photovoltaic failure and degradation modes," *Progress in Photovoltaics: Research and Applications*, vol. 25, no. 4, p. 318326, Apr 2017.

[2] "U.s. department of energy sunshot initiative." [Online]. Available: https://www.energy.gov/eere/sunshot/about-sunshot-initiative

[3] D. C. Jordan, S. R. Kurtz, K. VanSant, and J. Newmiller, "Compendium of photovoltaic degradation rates," *Progress in Photovoltaics: Research and Applications*, 2016.

[4] D. C. Jordan and S. R. Kurtz, "Data filtering impact on pv degradation rates and uncertainty," 2012.

[5] "IEC 61724 ed1.0 - photovoltaic system performance monitoring – Guidelines for measurement, data exchange and analysis | IEC webstore | publication abstract, preview, scope." [Online]. Available: http://webstore.iec.ch/webstore/webstore.nsf/Artnum_PK/22954

[6] Y. Hu, M. A. Hosain, T. Jain, Y. R. Gunapati, L. Elkin, G. Zhang, and R. H. French, "Global sunfarm data acquisition network, energy cradle, and time series analysis," in *Energytech, 2013 IEEE*. IEEE, 2013, pp. 1–5.

[7] K. M. Rubel F., "Observed and projected climate shifts 1901 –2100 depicted by world maps of the kppen-geiger climate classification," pp. 135–141, 2010.

[8] J. F. J.S. Stein, W.F. Holmgren and C. Hansen, "Pvlib: Open source photovoltaic performance modeling functions for matlab and python," 2016.

[9] RStudio, "RStudio: Integrated development environment for R," Boston, MA, 2014. [Online]. Available: http://www.rstudio.org/

[10] Y. Hu, V. Y. Gunapati, P. Zhao, D. Gordon, N. R. Wheeler, M. A. Hossain, T. J. Peshek, L. S. Bruckman, G. Q. Zhang, and R. H. French, "A Nonrelational Data Warehouse for the Analysis of Field and Laboratory Data From Multiple Heterogeneous Photovoltaic Test Sites," *IEEE Journal of Photovoltaics*, vol. 7, no. 1, pp. 230–236, Jan. 2017.

[11] T. J. Peshek, J. S. Fada, Y. Hu, Y. Xu, M. A. Elsaeiti, E. Schnabel, M. Khl, and R. H. French, "Insights into metastability of photovoltaic materials at the mesoscale through massive IV analytics," *Journal of Vacuum Science & Technology B*, vol. 34, no. 5, p. 050801, Sep. 2016. [Online]. Available: http://scitation.aip.org/content/avs/journal/jvstb/34/5/10.1116/1.4960628

[12] A. J. C. T. J. P. R. H. F. Yang Hu, Rojiar Haddadian, "Modeling the long-term temporal change of power production for real-world photovoltaic systems," 2017.

[13] T. H. Gareth James, Daniela Witten and R. T. An, "Introduction to statistical learning," 2013.

[14] R. Hafen, *stlplus: Enhanced Seasonal Decomposition of Time Series by Loess*, 2016, r package version 0.5.1. [Online]. Available: https://CRAN.R-project.org/package=stlplus

[15] P. A. Harris, R. Taylor, R. Thielke, J. Payne, N. Gonzalez, and J. G. Conde, "Research electronic data capture (REDCap)A metadata-driven methodology and workflow process for providing translational research informatics support," *Journal of Biomedical Informatics*, vol. 42, no. 2, pp. 377–381, Apr. 2009. [Online]. Available: http://linkinghub.elsevier.com/retrieve/pii/S1532046408001226

Photovoltaic Array Differential Backside Exposure Conditions: Backsheet Degradation and Site Design

Andrew Fairbrother[1], Julien Avenet[1], Yadong Lyu[1], Matthew Boyd[1], Scott Julien[2], Kai-Tak Wan[2], Liang Ji[3], Kenneth Boyce[3], Sebastien Merzlic[4], Amy Lefebvre[4], Greg O'Brien[4], Yu Wang[5], Laura Bruckman[5], Roger French[5], Michael Kempe[6], Brian Dougherty[1], Xiaohong Gu[1]

[1]Engineering Laboratory, National Institute of Standards and Technology (NIST), Gaithersburg, MD 20899 USA

[2]Department of Mechanical & Industrial Engineering, Northeastern University, Boston, MA 02115

[3]Underwriters Laboratories, Northbrook, IL 60062

[4]Fluoropolymers R&D, Arkema, King of Prussia, PA 19406

[5]SDLE Research Center, Materials Science & Engineering, Case Western Reserve University, Cleveland, OH 44106

[6]National Renewable Energy Laboratory, Golden, CO 80401

Abstract — Backsheet exposure conditions and degradation are shown to vary significantly within an array. This is demonstrated by a survey of the NIST ground-mounted photovoltaic array, which is comprised of over 1000 modules with a polyester-based backsheet. Site and array design factors are found to have significant effects on the exposure conditions, and thus degradation of the backsheets. Backside irradiance is found to be the most significant exposure factor which varies within the array.

Index Terms — photovoltaic array, reliability, backsheet, PEN, degradation, exposure conditions, array design.

I. INTRODUCTION

Backsheets play a key role in protecting photovoltaic modules from the effects of weather and providing electrical insulation of the cells. Polymeric backsheets are typically multilayer laminates, consisting of an outer layer intended to withstand weathering, a core layer for electrical insulation, and an inner layer to promote adhesion to the module encapsulant.[1] Because they are composed of polymeric materials they are often sensitive to exposure conditions (e.g. light, temperature, and humidity) over the lifetime of a module and will experience some form of degradation.[2], [3] For decades the industry has been dominated by fluoropolymer-based backsheet outer layers, which withstand weathering remarkably well, however, continued pressure to reduce costs has led to development of non-fluoropolymer-based backsheets. Some backsheet non-fluorinated polymers are based on polyester and polyamide materials. The drawback is that these materials must be modified to withstand weathering, and even with modification, they may still be more susceptible than fluoropolymers. Of these newer backsheet materials polyester-based backsheets are among the most notable, and constitute a growing segment of the market for backsheets.

This work details the backsheet degradation and differential backside exposure conditions present at the NIST

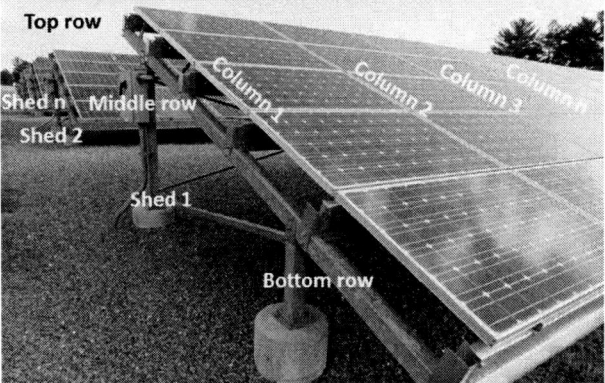

Fig. 1. Layout of the NIST ground-mounted PV array (top) and orientation of modules within the array, including the naming system for module location (bottom).

ground-mounted 271 kW PV array, which has been active since August 2012 and employs modules with a polyethylene naphthalate-based (PEN) backsheet.[4] Site and array design factors such as module spacing, orientation, array length, height, angle, ground cover, play a critical role in backsheet exposure, specifically to albedo light. Backsheet defects and degradation were characterized by non-destructive methods of visual inspection, colorimetry (yellow index), glossimetry, and Fourier transform infrared spectroscopy (FTIR). Temperature and irradiance were measured in various positions of the array and correlated to observed differences in backsheet properties. The results demonstrate that site design factors can have a significant effect on backsheet degradation over the >20 year lifetime of a PV module, and are especially critical when utilizing materials which are more sensitive to environmental exposure.

II. EXPERIMENTAL DETAILS*

The NIST ground-mounted PV array consists of 1152 modules with a total rated capacity of 271 kW. It is located at the main NIST site in Gaithersburg, Maryland (USA), an area with a hot humid continental climate. The modules are oriented at 20° and arranged into five sheds, each with five rows and 48 columns (except in shed 5, which contains only four rows of modules), as shown in Figure 1. The site became active in August 2012, and the dates of inspection were 22 October 2016 and 18 November 2016, corresponding to over four years of field exposure. To characterize backsheet defects and degradation, four methods were used. First, visual inspection of all modules was carried out, and photographs were made of modules with representative types of degradation. Second, color (yellow index) and gloss were measured for approximately 240 modules using two different handheld instruments, one measuring color and 60° gloss, and the other measuring 20°, 60°, and 85° gloss. For this, three points were measured on each module: 1) junction box region; 2) middle region; and 3) region opposite junction box. Thirty to fifty modules were measured in each shed in alternating columns (i.e. odd column numbers) for three rows in each column, designated as the top, middle, and bottom rows. This nomenclature is detailed in Fig. 1. Finally, FTIR was measured on about 40 modules at various positions within the array using a handheld instrument with an attenuated total reflection (ATR) attachment.

III. RESULTS AND DISCUSSION

During visual inspection two main types of backsheet damage or degradation were observed: 1) burn-marks (from hot spots) and 2) busbar bumps and ripples. Figure 2 shows the two main backsheet defects present in the NIST PV array. The burn marks are a result of hot spots in damaged solar cells, and they can be divided into high- and low-temperature burns. In the former, presumably higher temperatures result in rupture of the backsheet. In the latter, lower temperatures lead to discoloration of the backsheet, but it has not (yet) ruptured. Approximately 3% of the modules are affected with burn-marks, and affected modules usually have 1 to 2 marks. The second major defect are bumps which appear along busbar and cell interconnects. The origin of these bumps is not clear, though they may be related to delamination, gas pockets, or

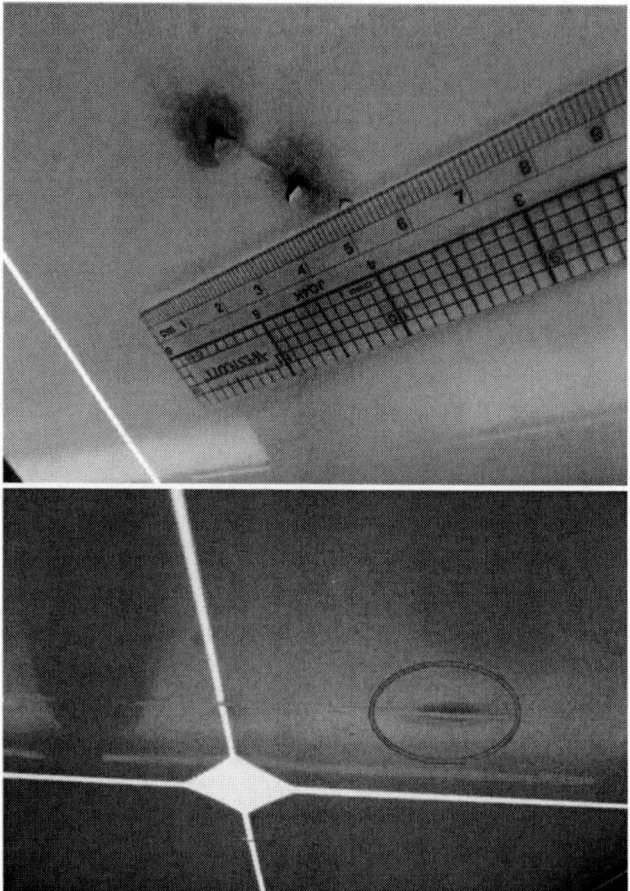

Fig. 2. The main backsheet defects observed in the array: burn-marks (top) and busbar bumps (bottom).

manufacturing defects, such as excess ribbon wiring. About 5% to 10% of the modules have busbar bumps, and affected modules typically have multiple bumps. Additional "defects" that are not necessarily degradation related were also observed, such as the presence of wasp nests, scratches from installation, and tapes or adhesives presumably leftover from installation.

Visual inspection is well suited to detect advanced degradation, but gradual, slowly occurring degradation is also expected. Because of this, non-destructive optical and spectroscopic tests were made. Of these, yellowing and gloss were found to be highly dependent on the module position within the array. Figure 3 shows the yellow index and gloss (60°) for the middle row modules as a function of column and shed. There is a strong edge effect where the 3 to 4 module columns closest to the edge of the array have higher yellowing and lower gloss than the modules in the center of the sheds. Table I shows the average values of these properties for the modules in the center of the sheds, as well as for the most extreme edge module (i.e. highest deviation from the mean). It illustrates that the height of the module and proximity to the array edge significantly affect the properties. Other location

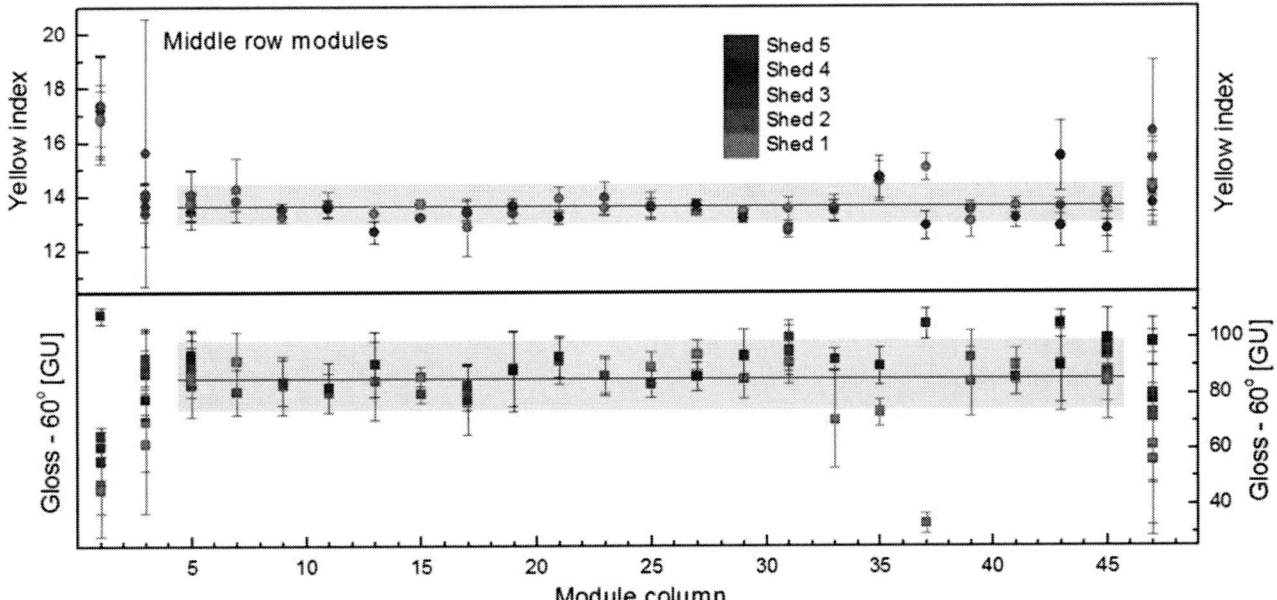

Fig. 3. Yellow index and gloss (60°) of the middle row modules and their dependence on position (column and shed). Each point is the average of three measurements from one module, and the error bar is the standard deviation. The solid line and colored box represent the mean and standard deviation of the center modules (i.e. excluding the edge effect).

related factors include top and bottom modules in the first and fifth sheds. These rows are excluded from the Table, because they exhibit unique yellowing and gloss patterns that are due to the grass border (as opposed to rock for all other modules).

FTIR was used to identify the chemistry of the backsheet material and any degradation products. As previously stated all modules have a PEN-based backsheet, and while not shown here, the peak intensities in the carbonyl region (≈ 1700 cm^{-1}) vary depending on the module position. Specifically, the intensity of the carboxylic acid peak (1698 cm^{-1}) increases with respect to the ester peak (1715 cm^{-1}). In general, the modules with higher yellowing have undergone more significant chemical changes compared to the lower-yellowing backsheets.

Polymer degradation can occur during exposure to light (esp. ultraviolet), heat, and humidity. Due to the trends observed in yellowing and gloss, it is clear that differential exposure conditions exist within the array, which should be quantified. Temperature and irradiance were measured at different modules positions, while absolute humidity levels are not expected to vary at the length scale of the array.

Temperature has been recorded every minute since January 2016, and temperature profiles, including daily average, maximum, minimum, and rates of changes, do not vary significantly for any positions. For four positions measured (different rows and an edge module) the maximum temperature was ≈ 65 °C in 2016, and the median summer daily max was ≈ 52 °C (May 2016 to August 2016). As such, temperature is not expected to have led to the difference in backsheet properties observed.

Irradiance was measured around noon on 16 December 2016, 5 April 2017, and 28 April 2017. Both total and spectral irradiances were found to vary widely in the array, a behavior which has already been recognized in the field of bifacial solar cells.[5] As might be ascertained from Figure 3 and Table 1, irradiance is higher on the edge and top row modules. The reason for this is two-fold: 1) the orientation of light blocking structures (the array itself) which leads to the greatest difference in irradiance; and 2) differences in spectral albedo due to ground cover. For instance, grass – which borders the first and fifth sheds of the array – have very low UV albedo compared to rock, sand, or dirt.[6]

IV. CONCLUSIONS

A survey of the NIST ground-mounted PV array shows dissimilar backsheet degradation patterns. The array consists of over 1000 modules with a PEN-based backsheet, and as such represents one of the first extensive field surveys of this newer class of non-fluoropolymer-based backsheets. Modules mounted higher and near the edge of the array experience significantly more weathering than the lower modules in the

TABLE I
YELLOW INDEX AND GLOSS VALUES (AVERAGE ± STANDARD DEVIATION) AVERAGED FROM THE MODULES IN COLUMNS 5-45, AND VALUES FROM THE EDGE MODULES WITH GREATEST DEVIATION FROM THE MEAN.

Module position	Yellow index		Gloss (60°) [GU]	
	Center	Edge	Center	Edge
Top rack	18.2 ± 0.6	19.4	50 ± 12	24
Middle rack	13.6 ± 0.7	17.4	86 ± 13	46
Bottom rack	12.4 ± 1.1	16.4	91 ± 12	71

978-1-5090-5606-4/17 $31.00 © 2017 IEEE

middle of the array. These findings highlight implications for site and array design, which can significantly influence backsheet exposure over module lifetime.

*Certain commercial equipment, instruments, or materials are identified in this work in order to adequately detail the experimental procedure. Such identification is not intended to imply recommendation or endorsement by the National Institute of Standards and Technology, nor is it intended to imply that the materials and equipment identified are necessarily the best available for this purpose.

ACKNOWLEDGEMENT

This work was funded in part by the U.S. Department of Energy, SunShot project, under award number DE-EE0007143.

REFERENCES

[1] M. Köntges, S. Kurtz, C. Packard, U. Jahn, K. A. Berger, and K. Kato, Performance and reliability of photovoltaic systems: subtask 3.2: Review of failures of photovoltaic modules: IEA PVPS task 13: external final report IEA-PVPS. Sankt Ursen: International Energy Agency, Photovoltaic Power Systems Programme, 2014.

[2] G. Oreski and G. M. Wallner, "Aging mechanisms of polymeric films for PV encapsulation," Solar Energy, vol. 79, no. 6, pp. 612–617, 2005.

[3] W. Gambogi et al., "Weathering and durability of PV backsheets and impact on PV module performance," in Proceedings of SPIE 8825, p. 88250B, 2013.

[4] M. T. Boyd, "High-Speed Monitoring of Multiple Grid-Connected Photovoltaic Array Configurations and Supplementary Weather Station," Journal of Solar Energy Engingeering, vol. 139, no. 3, p. 034502, 2017.

[5] A. Lindsay, M. Chiodetti, P. Dupeyrat, D. Binesti, E. Lutun, and K. Radouane, "Key Elements in the Design of Bifacial PV Power Plants," in Proceedings of the 31st European Photovoltaic Solar Energy Conference and Exhibition, pp. 1764-1769, 2015.

[6] M. Blumthaler and W. Ambach, "Solar UVB-Albedo of Various Surfaces," Photochemistry and Photobiology, vol. 48, no. 1, pp. 85–88, 1988.

Study on Random Failure of Crystalline Silicon Solar Modules in the Field

Xuefang Jiang[a,b], Fumei Wang[a], Ao Wang[a], Hong Yang[a,*], He Wang[a], Jie Ding[c], Junjun Zhang[c], Jingsheng Huang[c]

[a]MOE Key laboratory for Nonequilibrim Synthese and Modulation of Condensed Matter, School of Science, Xi'an Jiaotong University, Xi'an 710049, People's Republic of China

[b]School of Science, Xi'an Polytechnic University, Xi'an 710048, People's Republic of China

[c]China electric power research institute, Nanjing, 210003, People's Republic of China

*Corresponding author: Hong Yang, hongy126@126.com

Abstract — Although the reliability of photovoltaic modules has been greatly improved in recent years, some new failure events have occurred due to changes of module material and process. In this paper, the random failure events of photovoltaic power plant in recent years are analyzed and summarized. The new failure mechanism is studied, such as bypass diode, connector and so on. The failure probability of various failure factors is given out. The field experiences of reliability in this article provide the foundation for further study on reliability growth. It is helpful to improve the reliability of photovoltaic modules.

Index Terms - photovoltaic modules, failure mechanism, bypass diode, connector, failure probability

I. INTRODUCTION

Crystalline silicon solar modules are the critical component of photovoltaic (PV) generation system and often stated as being the most reliable element in PV systems. At present, PV manufacturers give the guarantee that PV modules work for at least 25 years in outdoors. Modules are generally exposed to complex environments for a long time, such as moisture, high temperature, strong UV irradiation, and so on. The degradation of terrestrial PV modules has been an issue of concern to the PV industry for a long time [1]-[4]. Understanding the various degradation and failure modes can make the manufacturers to improve the materials and the processes involved so as to prevent premature failure in the field. Since 1950 many people have made significant contributions in this field [5]-[9]. Previously PV power plants are small scale, in recent years, the number of large scale plants increase rapidly, reliability has been greatly improved. The quality assurance is that power degradation of modules of 10 years does not exceed 10%, 25 years not more than 20%. In addition, the installation process has improved, so the failure case is not the same as the original, there will be some new random failure events in large scale plants.

In this paper, the new random failure events of large scale grid-connected PV power plant (see Fig. 1) are analyzed. The new failure mechanism, the failure probability and the improvement methods of various failure factors are given out. The field experiences of reliability in this article provide the foundation for further study on reliability growth.

Fig. 1.   The scene of photovoltaic power plant.

II. FAILURE FACTORS

A. Bypass Diode

The junction box of PV modules is composed of box cover, box body, diode, cable and connector. The diode can be divided into bypass diode and anti-reverse charge diode. It is well known that bypass diode is one important component in junction box. When hot spot appears and solar cells are mismatch in PV modules, the bypass diode will play a protective effect [10-11].

At present, the 12SQ045 R-6 Schottky diode is dedicated diode for solar junction box, which uses item by item packaging. The diode is cylindrical structure. The advantage of junction box with a cylindrical diode is that the internal space of box body is large, which is convenient for product structure design and injection molding. The disadvantage is the poor heat dissipation that can cause bypass diode overheat, ultimately the junction box will be burned.

Today many manufacturers adopt TO-220 encapsulation which use patch type diode and have large area of heat dissipation, by the air flow to dissipate heat of diode quickly. TO-220 encapsulation has better performance on heat dissipation than cylindrical diode, which can guarantee junction box work safely as hot spot appears in PV modules. But TO-220 encapsulation also occur problems in the use of power plants.

As shown in Fig. 2, in October 22, 2015 the junction box of PV modules was found burnt out in some PV power plant. In which the junction boxes adopted TO-220 encapsulation. By analysis the distance between two pin centers is 2.54 mm, between edges is about 1.8 mm in TO-220 encapsulation. The distance between the two pad edges is even smaller, perhaps only 1mm when it is soldered to the circuit. The potential difference between the two pins is 220 V, when the thyristor is not conductive. Creepage distance should be greater than 2.5 mm as the potential difference is 220 V. Therefore the creepage distance cannot meet the requirements in TO-220 encapsulation, which can lead to the junction box of PV modules burn out.

Fig. 2.   Burn out junction box in power plant.

B. MC4 Connector

Connector is a part of circuit that connects solar cell modules into a system. MC4 connector is one of the types of connectors.

The connector of PV modules is not noticeable small part, however, the quality of connector is substandard, which may cause fire and bring huge losses to the power plant.

The primary cause of the fire is that the increasing resistance of connector leads to temperature increasing in the circuit, further beyond the temperature range of plastic shell and metal device. Therefore the failure factors of MC4 connector are plastic shell and metal device.

One of the reasons for the failure of the metal device is the fact that the MC4 connectors of different companies are inserted each other.

The core components cannot effective contact for a long time, when the MC4 connectors of different manufacturers are inserted. Circulating current cannot be guaranteed, which is the reason that MC4 connectors can't be inserted between the different manufacturers. When there is a gap between the male plug and female plug, the increase of resistance can lead to arc phenomenon. It will result in fire in the dry environment. When the shells and sealed devices cooperate among different manufacturers, the size and tolerance are inconsistent which cause the original IP grade failure. Thus the internal environment of MC4 connectors will be damaged leading to failure. Fig. 3 shows the burned MC4 connectors due to mutual insertion in the power plant.

Therefore reliable quality MC4 connectors should be adopted in the early construction of PV power plant. Only one manufacturer's MC4 connectors can be used in the same power plant.

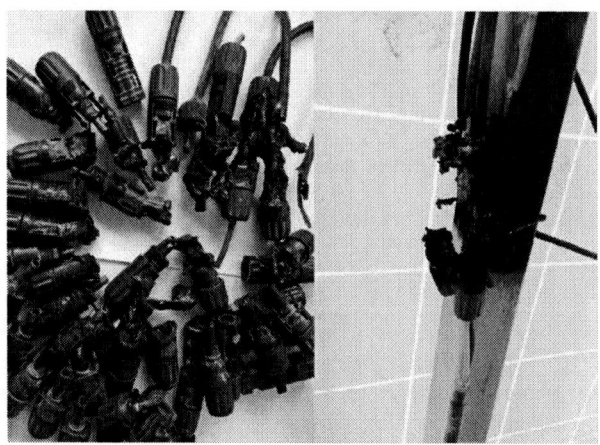

Fig. 3.   Photograph of burned MC4 connectors.

C. Tempered Glass

The tempered glass is an important auxiliary material in the production of PV modules. Tempered glass in the front of module can effectively protect the module from the impact of various objects [12].

In the PV industry tempered glass mainly has ultra-white figured tempered glass and heat reflective coating tempered glass. Currently, the main product of ultra-white figured glass is low iron tempered glass. Tempered glass has the possibility of spontaneous breakage when the temperature changes significantly. It is the main reason of tempered glass failure. Fig. 4 shows that tempered glass of module had been broken in power plant. The causes of spontaneous breakage can simply summarized as the following.

The quality of tempered glass is defective. There is impurity or nickel sulfide crystal in the glass. The surface of the glass has scratches and other defects due to processing or improper operation.

The distribution of stress is uneven in the tempered glass. The temperature gradient in the direction of the thickness of the glass is uneven and asymmetrical when the offset glass is heated or cooled.

The toughened degree has effect. The experiments show that the stress is greater, the toughened degree is higher, and the amount of explosive is greater.

People use improperly. The marginal areas of tempered glass are the most vulnerable to break under force, especially the four corners that is very easy to spontaneous breakage by external impact.

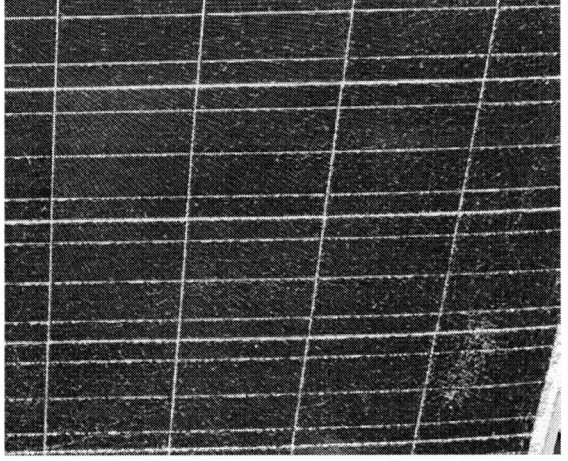

Fig. 4.   Breakage tempered glass in front of solar cells.

The main measures to solve spontaneous breakage of tempered glass are controlling toughened stress, homogenization treatment (HST) and so on.

D. EVA

EVA and glass are the most important parts in glass package technology. Currently，the use of EVA is modified EVA films [13]. EVA is used as raw material. It is heated after adding modifier, extruded molding in the end [14]. Transmittance is more than 90%, degree of crosslinking is about 90%; peel strength between glass and films is more than 40 N/cm.

Because of the structure of EVA, there may be problems such as delamination, bubble and so on in the long-term use, which can lead to module failing. The delamination of EVA on the top surface of the solar cells shows up as whitish patches on the cells, in contrast with the blue color of the antireflective coating of the solar cells (see Fig. 5).

The insufficient bond force between EVA and glass is the direct cause of delamination. The bond force between EVA and

glass is destroyed or EVA cannot fully infiltrate the glass surface, which will lead to the two interface bonding failure.

Bubbles are mainly divided into two categories: one is a weak layer between glass and EVA interfaces, the appearance of it is discontinuous bubbles. After heating and a little pressure, the bubbles will disappear (interface re bonding). The other is caused by small molecular gases, which mainly come from EVA peroxide decomposition (not completely curing), residual soldering flux and water vapor. These kinds of bubbles are generally not easy to extend or disappear.

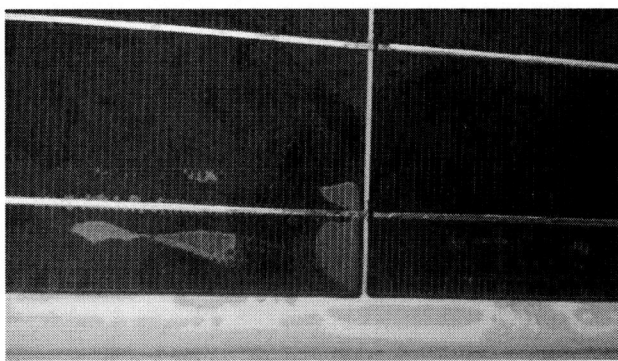

Fig. 5.   Delamination of EVA in the module.

The solution to prevent delamination and bubble emergence is to adjust the process parameters, adjust the lamination time.

III. FAILURE PROBABILITY

Recent the data feedback from PV power plant are analyzed, the failure factors of modules are summarized. There are 436000 PV modules in the power plant, have been running for 3 years or so. The total number of various short-term failures of PV modules is 18. Fig. 6 shows the percentage of various failure factors.

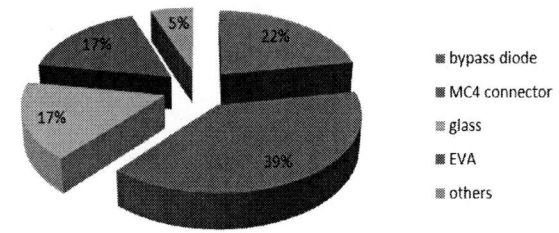

Fig. 6.   Distribution of short-term failure factors of PV modules about 3 years.

IV. CONCLUSION

In view of the random failure events of large scale PV power plant, it is necessary to improve the process to meet the requirements of creepage distance, as TO-220 package is used to solve the heat dissipation problem of bypass diodes. In addition, it should pay more attention to MC4 connectors cannot mutually insert. Moreover the imitation of MC4 connector cannot be used. The tempered glass of modules should clean regularly in order to enhance output power.

The feedback from PV power plant indicate that modules failure caused by various failure factors is a very low percentage, and it will not seriously affect the modules long-term reliability and lifetime. In the future, we will continue to carry out the reliability growth test in order to make the PV modules more reliable.

ACKNOWLEDGEMENT

The authors would like to thank the support of Natural Science Foundation of China (Grant No. 61376067 and 61274050). This study was also supported by the National Key Technology Research and Development Program of the Ministry of Science and Technology of China (Grant No. 2015BAA09B01).

REFERENCES

[1] D.L. King , M.A. Quintana , J.A. Kratochvil , D.E. Ellibee , B.R. Hansen, "Photovoltaic module performance and durability following long-term field exposure," *Progress in Photovoltaics Research & Applications*, vol. 8, pp.241–256，1998.

[2] C.R. Osterwald, A. Anderberg, S. Rummel and L. Ottoson, "Degradation analysis of weathered crystalline-silicon PV modules," in *29th IEEE Photovoltaic Specialists Conference*, 2002, pp.1392-1395.

[3] J.H. Wohlgemuth, D.W. Cunningham, P. Monus, J. Miller, "Long term reliability of photovoltaic modules," *IEEE World Conference on Photovoltaic Energy Conversion*, vol. 2, pp. 2050-2053, 2006.

[4] N. Kim, H. Kang, Kyung-Jun Hwang, C. Han, W. Hong, D. Kim, E. Lyu, H. Kim, "Study on the degradation of different types of backsheets used in PV module under accelerated conditions," *Solar Energy Materials & Solar Cells*, vol.120, pp.543-548, 2014.

[5] E.D. Dunlop, D. Halton, "The performance of crystalline silicon photovoltaic solar modules after 22 years of continuous outdoor exposure," *Progress in Photovoltaics Research & Applications*, vol.14, pp.53-64, 2006.

[6] Zhiyong Xia, J.H. Wohlgemut, D. W. Cunningham, "A lifetime prediction of PV encapsulant and backsheet via time temperature superposition principle," *Conference Record of the IEEE Photovoltaic Specialists Conference,* pp.000523-000526, 2009.

[7] P. Sánchez-Friera , M. Piliougine , J. Peláez , J.Carretero , M.S.D. Cardona, "Analysis of degradation mechanisms of crystalline silicon PV modules after 12 years of operation in Southern Europe, " *Progress in Photovoltaics Research & Applications*, vol.19, pp.658-666, 2011.

[8] He Wang , Ao Wang , Hong Yang , Junjun Zhang , Jingsheng Huang, "Performance degradation of crystalline silicon solar module in various ultraviolet radiation area," in *43th IEEE Photovoltaic Specialists Conference*, 2016, pp.1757-1760.

[9] He Wang, Ao Wang, Hong Yang, Jingsheng Huang, "Study on the thermal stress distribution of crystalline silicon solar cells in BIPV," *Energy Procedia*, vol.88, pp.429-435, 2016.

[10] W. Herrmann , W. Wiesner , W. Vaassen, "Hot spot investigations on PV modules-new concepts for a test standard and consequences for module design with respect to bypass diodes," in *26th IEEE Photovoltaic Specialist Conference*, 1997, pp. 1129-1132.

[11] C. Greacen, D. Green, "The role of bypass diodes in the failure of solar battery charging stations in Thailand," *Solar Energy Materials & Solar Cells*, vol. 70, pp.141－149, 2001.

[12] M.R. Mitchell , R.E. Link , T.M. Shimpi , K.L. Barth , R.A. Enzenroth, "Effect of Hail Impact on Thermally Tempered Glass Substrates Used for Processing CdTe PV Modules," *Journal of Testing & Evaluation*, vol. 36, pp. 207-212, 2007.

[13] F.J.Pren, "Ethylene-vinyl acetate (EVA) encapsulants for photovoltaic modules: Degradation and discoloration mechanisms and formulation modifications for improved photostability," *Macromolecular Materials & Engineering*, vol.252, pp.195-216, 1997.

[14] M.D. Kempe , G.J. Jorgensen , K.M. Terwilliger , T.J. Mcmahon , C.E. Kennedy, "Acetic acid production and glass transition concerns with ethylene-vinyl acetate used in photovoltaic devices," *Solar Energy Materials & Solar Cells*, vol.91, pp. 315－329, 2007.

Potential Induced Degradation (PID) Power Loss Correlation to Leakage and Reverse Bias Currents

Michalis Florides, Georgios Konstantinou, Venizelos Venizelou, George Makrides and
George E. Georghiou

FOSS Research Centre for Sustainable Energy, PV Technology Laboratory, Department of Electrical and
Computer Engineering, University of Cyprus, 75 Kallipoleos Street, 1678, Nicosia, Cyprus.

Abstract — **This paper presents the results of power loss of multi-crystalline solar cells due to Potential Induced Degradation (PID) and its relation to the leakage and reverse bias currents of the cells under test. An outdoor test infrastructure was constructed at the PV Lab, University of Cyprus and a bias voltage of 1000 V DC was applied between the cell's wire terminals and its frame. The resulting leakage current was measured and its dependence on ambient temperature, relative humidity and solar irradiance was derived. In addition, the power loss of the cell was related to leakage and reverse bias currents as the degradation was progressing in order to establish whether these parameters could be used for early PID detection. This was achieved by Flash tests and correlation of the power reduction to the measured leakage and reverse bias currents.**

I. INTRODUCTION

Potential Induced Degradation (PID) has been known since 1978, however, it only recently appeared as a degradation mechanism for PV systems due to the increased system voltage up to 1000 V DC.

As a result of this high voltage, a high electrical potential develops between each module terminals and its grounded frame. In addition, the use of transformer-less inverters which do not allow for grounding of the negative terminal of a PV module string, means that half of the modules are under positive bias and half under negative bias. The modules under negative bias, and especially the ones at the end of the string, are susceptible to PID [1]. It is commonly accepted that sodium ions (Na+) drift under the influence of the electric field that results from the negative bias voltage, which subsequently causes shunting of the cells by the Na+ ions. [2], [3].

Even though the causes of PID are known, there is still no widely accepted detection mechanism, which can be used in the field to flag a warning of PID development in the PV modules under operation. The most common method of detecting PID is to remove the modules and test them with electroluminescence (EL) in the lab, but this is time consuming and costly. Another method reported is the use of a drone copter with a thermal camera to identify PID affected modules [4]. The main aim of this work is to find a way for early and reliable detection of PID under operating conditions in the field.

Leakage current (LC) appears due to the high voltage bias between the module terminals and its aluminium frame. Since this high voltage is the driving mechanism for PID, the change in leakage current will be studied as PID progresses. Indoor tests have shown that leakage current does not change with PID progression [5]. Reverse bias current (I_{rev}), which is related to the shunt resistance (R_{sh}), is another parameter worth studying. This appears when a solar cell is reverse biased. Because of the shunting nature of PID, the reverse bias current increases (i.e. shunt resistance decreases) as PID increases. The decreasing shunt resistance was verified under indoor conditions and it has been demonstrated it can be used as an indication for PID before even a significant power loss occurs [6].

In this paper the progression of leakage and reverse bias currents under outdoor conditions is examined in an attempt to identify methods for early PID detection. Our focus is to develop methods for detecting PID by using electrical quantities, which are cheap to measure and can be integrated in the system installation.

II. EXPERIMENTAL SETUP

A. Infrastructure

For this test, an outdoor experimental setup has been built as shown in Fig. 1 and multi-crystalline silicon cells have been tested. The cells were PID prone with a cell area of 243.4 cm^2, 3 mm thick glass, x-23-105 EVA and TPT Akasol PV1000V back-sheet.

A metallic structure was used to mount the solar cells and the various sensors (pyranometer, ambient temperature (T_{amb}), relative humidity (RH) and surface wetness (SW) sensors) used during the experiment. Electrical boxes housing the leakage current sensing circuit and the data logger (Fig. 1, inset A) were mounted on the structure as well. A high voltage power supply (HV PSU) was used to apply 1000 V DC bias voltage to the solar cells, which were isolated from the metallic structure with rubber holders.

The equipment and the accuracy of the different sensors is summarised in Table I.

Fig 1. PID experimental setup: metallic structure with solar cells, sensors (pyranometer, ambient temperature, relative humidity and surface wetness), leakage current sensing circuit and data logger.

TABLE I
EQUIPMENT AND MEASUREMENT ACCURACY

Equipment	Accuracy
HV PSU (Keithley 2410)	0.05%
Leakage Current measurement	30% @ 10nA 6% @ 100nA 3% @ 1000nA
Pyranometer (Hukseflux SR11)	First Class
RH sensor (Vaisala HMDW110)	4%
T_{amb} sensor (Vaisala HMDW110)	0.2%
Surface Wetness sensor (237 Leaf)	-

B. Electrical connection

The positive and negative wires of the cell were shorted together and aluminium tape was applied at half the perimeter of each cell frame (Fig. 1, inset B). The HV PSU negative terminal (-ve) was connected to the cell frame (aluminium tape) and the positive terminal (+ve) to the shorted wires of the cell.

III. METHODOLOGY

The following methodology was used to study the progression of PID in the samples under test and relate it to leakage and reverse bias currents.

A. Outdoor Test

The HV PSU output voltage was set to -1000 V continuously. All the sensors were connected to the data logger, which was sampling at every second and averaging over a period of 1 minute.

B. PID Progression

The PID progression was monitored by Flash tests of the cell at standard test conditions (STC). The recorded power was compared with the measured leakage current (LC) and reverse bias current (I_{rev}), in order to see if there is any correlation between them as PID was progressing at outdoor conditions. The cell power (P_{cell}) was also calculated as a percentage by dividing the recorded power with the initial power of the cell recorded before the test had started.

C. Leakage Current Estimation

The leakage current (LC) of the cell was estimated based on the data of cell temperature (T_{cell}) and relative humidity (RH). The cell temperature was calculated by estimating the temperature rise caused by the global plane of array irradiance (G_{POA}) and adding it to the ambient temperature (T_{amb}). The leakage current estimation was done based on the Arrhenius equation with an activation energy of 75 kJ/mol [7] and the variation of surface conductivity relative humidity thresholds as a function of cell temperature [8] was considered.

The following equations were used to model the estimated leakage current.

$$LC_{est} = 10^{16} * e^{\frac{-E_a}{R*(T_{cell}+273)}} + 5 * RH_{mod} \quad (1)$$

$$T_{cell} = T_{amb} + G_{POA} * \frac{25}{1000} \quad (2)$$

$$RH_{mod} = 50 - 0.1667 * (T_{cell} - 25) \quad (3)$$

Where:
LC_{est}: Estimated leakage current (nA)
E_a: Activation energy, 75 kJ/mol
R: Gas constant, 8.314 J/mol/K
T_{cell}: Cell temperature (°C)
RH_{mod}: Modelled relative humidity (%)
T_{amb}: Ambient temperature (°C)
G_{POA}: Global plane of array irradiance (W/m²)

For the estimation of the cell temperature a temperature rise of 25 °C at 1000 W/m² was used which is common in Cyprus. The -0.1667 coefficient in equation 3 (RH_{mod}) was calculated based on the data from [8].

D. Reverse Bias Current Measurement

The cell was reverse biased under dark conditions and the reverse bias current (I_{rev}) was measured at various reverse

978-1-5090-5606-4/17 $31.00 © 2017 IEEE

bias voltages at room temperature. The voltage was varied from -0.1 V to -15 V. This is useful for establishing its rate of change with PID progression at outdoor conditions.

IV. RESULTS

Continuous monitoring of the cell has been taking place and the days on the x-axis of the plots correspond to the days since the start of the relevant experiment. The x-axis tick is placed at 12:00 o'clock of each day.

A. Power Loss vs Leakage Current

Two cells were used for this test. Fig. 2, presents the measured leakage current (LC) of the two cells under test (cell 1 and cell 2), ambient temperature (T_{amb}), relative humidity (RH), surface wetness (SW) and G_{POA}. The estimated cell temperature (T_{cell}) is shown as well.

Generally, the two cells experience leakage current (LC) similar in magnitude except in two cases when the relative humidity (RH) exceeds 85% and the surface wetness (SW) increases significantly. This is mainly due to the difference in the amount of 'dirt' transferred by hands during reinstallation of the cells after the indoor tests and the leftovers of the surface cleaning of the cells before the Flash test. The more the 'dirt' left on the surface of the cell the easier the accumulation of moisture and dust and, hence, the more the increase of the leakage current (LC).

In addition, more or less equal magnitudes of leakage current (LC) can be observed during day and night. This is because of the high temperature during the day and the high relative humidity (RH) during the night.

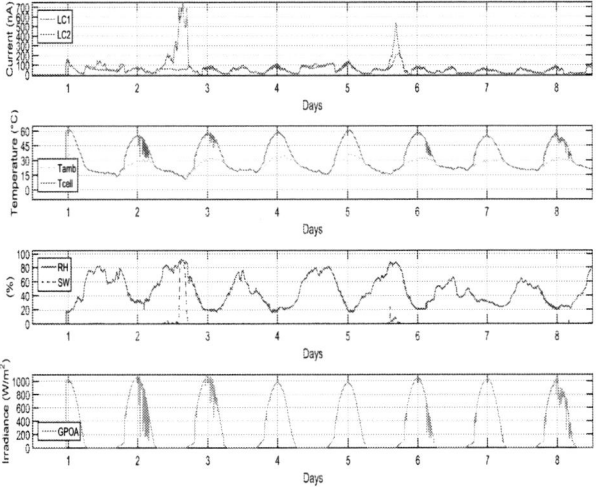

Fig 2. Measured LC1, LC2, T_{amb}, RH, SW and G_{POA} and estimated T_{cell}. Short duration drops of the measured LC to 0 nA on days 3, 6 and 8 are due to interruption for indoor testing.

In order to investigate if the leakage current (LC) changes with PID progression, the measured and estimated leakage currents were compared and plotted alongside the cell power (P_{cell}) as shown in Fig. 3 and Fig. 4.

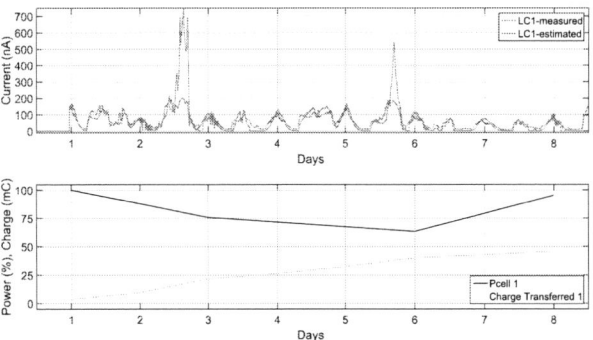

Fig 3. Measured and estimated LC1, measured cell power (P_{cell}) and calculated charge transfer. Short duration drops of the measured LC to 0 nA on days 3, 6 and 8 are due to interruption for indoor testing.

Cell 1 (Fig. 3) was under -1000 V DC bias for 8 days and its power dropped to 63.3% from its initial power (2.57 W) on the sixth day. During the last two days, the weather conditions were such that the cell's power recovered to 95.3%. In the last two days, the cell temperature was high as the previous days of degradation, however, the relative humidity (RH) was generally lower leading to reduced leakage current (LC). This was promoting recovery instead of further degradation. This agrees with the charge transfer rate since during power reduction (days 1-6) its slope was greater than when the power was recovering (days 6-8).

A similar behaviour can be seen for cell 2 (Fig. 4). Its power dropped to 68.6% from its initial power (2.57 W) on the sixth day as well and recovered to 91.7% during the last two days. Similarly, during power reduction (days 1-6) the charge transferred slope was greater than when the power was recovering (days 6-8).

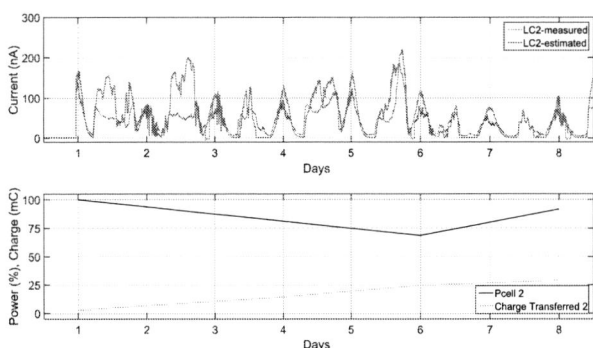

Fig 4. Measured and estimated LC2, measured cell power (P_{cell}) and calculated charge transfer. Short duration drops of the measured LC to 0 nA on days 3, 6 and 8 are due to interruption for indoor testing.

Fig. 5 shows the Flash test results for the two cells for days 1, 6 and 8. For both cells, an increase of the I-V curve slope close to 0 V indicates a reduction of the shunt resistance of the cell. The shunt resistance (R_{sh}) of the cells with PID progression is shown in Table II. There is a dramatic reduction in the shunt resistance on day 6 which correlates with the reduction of the power of the cell. It can also be seen that in order for a significant power loss to occur, the shunt resistance has to drop below 1 Ω.

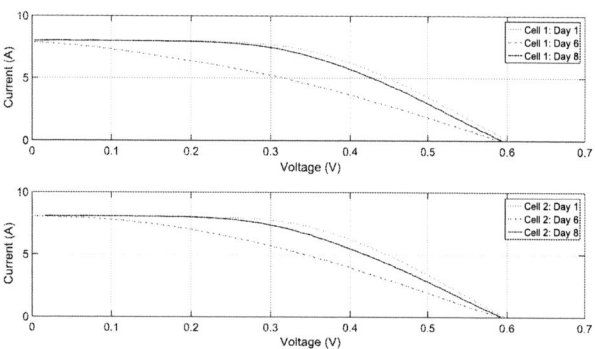

Fig 5. Flash test results at STC for cells 1 and 2 at days 1, 6 and 8.

TABLE II
POWER AND SHUNT RESISTANCE OF THE CELLS WITH PID
PROGRESSION

Day	Cell 1 Power (W)	Cell 1 R_{shunt} (Ω)	Cell 2 Power (W)	Cell 2 R_{shunt} (Ω)
1	2.57	6.34	2.57	7.55
3	1.94	0.92	2.24	2.55
6	1.62	0.16	1.76	0.27
8	2.44	4.51	2.35	5.69

From the data (Fig. 3 and Fig. 4), the estimated (based on T_{cell} and RH) and measured leakage current (LC) are in good agreement, especially during the day when the leakage current is mostly affected by the temperature. At night, the estimated current is not in very good agreement with the measured current since it is more difficult to predict the surface wetness (SW) of the cells and hence surface conductivity due to the unpredictable dirt and dust accumulation on the cell. Nevertheless, it can be deduced that the leakage current does not depend on PID progression (i.e. change in P_{cell}) but only on ambient conditions. Therefore, it is apparent that the leakage current cannot be used for PID detection but more data is required to verify this. More cells are currently being tested and indoor testing is undertaken to ascertain this in a more predictable environment.

B. Power Loss vs Reverse Bias Current

Another cell (cell 3) was used for this test and Fig. 6 shows the results of the reverse bias current (I_{rev}) with PID progression. Days 4 and 7 were omitted to avoid confusion since the reverse bias current decreased due to some recovery of the cell. The current reduction for the omitted days can be seen in Fig. 7 for measurements at a reverse bias voltage of -0.5 V.

It can be observed that as PID progresses the reverse bias current increases due to the reduction of the shunt resistance of the cell. The change in current is more pronounced at low reverse bias voltages.

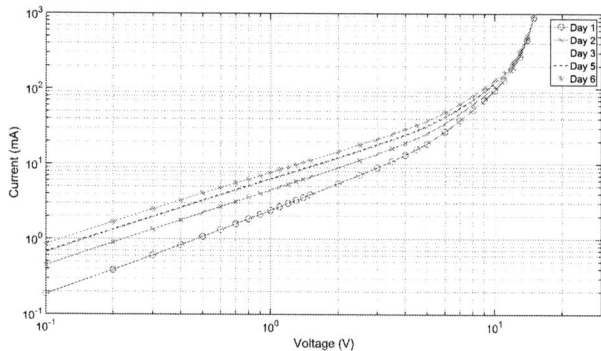

Fig 6. Reverse bias current (I_{rev}) vs reverse bias voltage (V_{rev}) as PID was progressing on a daily basis.

During the test, even though the cell power did not degrade (Fig. 7) the reverse bias current increased by almost one order of magnitude. This behaviour makes reverse bias current a promising electrical quantity for early PID detection in the field. More work is being undertaken to verify this concept.

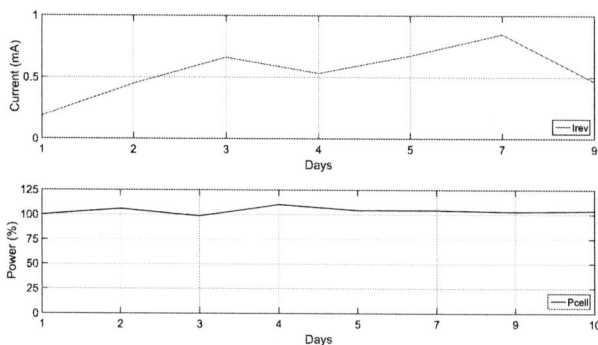

Fig 7. Reverse bias current at -0.5 V reverse bias voltage and cell power progression with PID.

V. Conclusions

Multi-crystalline solar cells have been tested outdoors in order to examine the relation of the leakage and reverse bias currents to PID progression (i.e. change in power loss). The main goal is to use those quantities to detect PID in the field. From the outdoor results, it seems that leakage current (LC) does not vary with PID progression but it only depends on ambient conditions. On the other hand, the reverse bias current (I_{rev}) changed with PID progression by almost an order of magnitude while the power of the cell did not degrade to a large extend. Hence, it is a promising electrical quantity for early PID detection before any significant power loss appears at outdoor conditions.

Acknowledgements

This work was funded through the PVEXPERT project (ΔΙΑΚΡΑΤΙΚΕΣ/ΚΥ-ΙΣΡ/0114/13) which is co-financed by the European Regional Development Fund and the Republic of Cyprus through the Cyprus Research Promotion Foundation.

Simon Koch from PI Berlin is gratefully acknowledged for his help with the PID prone cells.

References

[1] S. Pingel, O. Frank, M. Winkler, T. Geipel, H. Hoehne, and J. Berghold, "Potential Induced Degradation of Solar Cells and Panels," in *35th IEEE Photovoltaic Specialists Conference (PVSC)*, 2010, pp. 002817–002822.

[2] V. Naumann, C. Hagendorf, S. Grosser, M. Werner, and J. Bagdahn, "Micro Structural Root Cause Analysis of Potential Induced Degradation in c-Si Solar Cells," *Energy Procedia*, vol. 27, pp. 1–6, 2012.

[3] S. P. Harvey, J. Aguiar, P. Hacke, S. Johnston, and M. Al-Jassim, "Sodium Accumulation at Potential-Induced Degradation Shunted Areas in Polycrystalline Silicon Modules," in *43rd IEEE Photovoltaic Specialists Conference*, 2016.

[4] T. Kaden, K. Lammers, and H. J. Moller, "Power Loss Prognosis from Thermographic Images of PID Affected Silicon Solar Modules," *Sol. Energy Mater. Sol. Cells*, vol. 142, pp. 24–28, 2015.

[5] M. Schwark *et al.*, "Investigation of Potential Induced Degradation (PID) of Solar Modules from Different Manufacturers," in *Industrial Electronics Society, IECON 2013 - 39th Annual Conference of the IEEE*, 2013, pp. 8090–8097.

[6] M. B. Koentopp, M. Krober, and C. Taubitz, "Toward a PID Test Standard: Understanding and Modeling of Laboratory Tests and Field Progression," *IEEE J. Photovoltaics*, vol. 6, no. 1, pp. 252–257, 2016.

[7] S. Hoffmann and M. Koehl, "Effect of Humidity and Temperature on the Potential-Induced Degradation," *Prog. Photovoltaics Res. Appl.*, vol. 22, no. 2, pp. 173–179, 2014.

[8] H. Nagel, M. Glatthaar, and S. W. Glunz, "Quantitative assessment of the local leakage current in PV modules for degradation prediction," in *31st European PV Solar Energy Conference and Exhibition*, 2015.

Performance Study of Various PV Module Technologies in Desert Conditions

Jim J John, Ammar Elnosh, Anwar Almheiri, Wadhah Alzahmi, Marco Stefancich, Pedro Banda

Research and Development Center, Dubai Electricity and Water Authority (DEWA), Dubai, United Arab Emirates (UAE)

Abstract — **The performance degradation of 31 different types of PV modules continuously measured for more than 1 year in desert condition of Dubai, UAE has been reported. Statistical decomposition based on LOWESS is applied to six modules which had two years of data while linear regression was used for the rest of the modules which had less than two years' data. The average annual degradation rates for BF c-Si (Bifacial crystalline Si) module (1.4%/yr) and mc-Si (1.25%/yr) is much higher compared to the other four mc-Si modules studied based on LOWESS smoothing. It was found that the monocrystalline silicon modules showed almost no degradation, while the multicrystalline silicon modules degradation rate varied from 0.34% to 6.36% per year based on wc-PR (weather-corrected performance ratio). The CdTe modules showed high degradation rate of 2.4% per year, whereas CIGS modules showed less than 1% per year.**

I. Introduction

Desert areas have very high potential for PV application, due to their high global irradiance. These arid regions can be some of the extreme conditions for PV modules. High ambient temperatures, strong winds and highly dusty environment cause not only temporary decreases in transmittance but also non-reversible degradation of the surface, which limits overall yield [1].

Middle-east countries fall under the arid and semi-arid regions as per the Koppen-Geiger climate classification. These countries are becoming increasingly important markets for photovoltaic (PV) installations. The renewable energy market analysis done by IRENA [2] for the GCC countries suggest that these countries will have 30%-84% of their renewable energy targets coming from PV and BIPV technologies (approx. 40GW) by the year 2030. Since the number of PV installations in these regions is set to increase tremendously, performance study of PV modules in the arid, semi-arid regions are of high interest to all segments of the PV industry, from module manufacturers to system integrators and financiers.

The degradation of fielded modules is usually studied with regression fitting to continuous data assuming a linear degradation over the long operational period. Depending on the quality and type of data available, various other methods are usually also employed such as comparing the annual PR [3], periodic indoor evaluations [4], statistical decomposition for small uncertainties and shorter observation time (not less than 2 years) [5], or regular outdoor measurements under similar meteorological conditions [6].

The statistical decomposition method based on the locally weighted scatterplot smoothing (LOWESS) [7] to extract the degradation is used in this study. This method decomposes the time series into seasonal, trend and remainder components by applying a sequence of LOWESS. The trend component is the low-frequency variation in the data which removes the seasonal variations. Since robustness weight is considered during the LOWESS smoothing, outliers are assigned with small weights during regression and cause only small uncertainties to the decomposition.

The weather-corrected PR (wc-PR) introduced by Dierauf et al [8] is also a metric that can be used to study the performance without the influence of ambient conditions. This metric should be constant in a year if the PV modules don't show any significant degradation. Therefore, in this study, we use statistical decomposition to extract the degradation rates from modules that are operating for two years or more. Wc-PR and linear P_{mpp} degradation is assumed for modules less than two years due to non-suitability of seasonal decomposition techniques.

Table 1
Overview of modules installed at DEWA Outdoor Test Facility

Parameter	c-Si	BF c-Si	mc-Si (poly)	CdTe	CIGS
No. of test panels	7	3	19	1	1
Rated Power P_{mpp} (W)	250-290	270-300	235-275	105	165
Current at rated power I_{mpp} (A)	8.14-9.02	8.06-8.57	7.71-9.03	1.6	2.23
Voltage at rated power V_{mpp} (V)	30.01-31.56	30.46-32.16	29.89-31.03	70.97	69.35
Short-circuit current I_{sc} (A)	8.61-9.52	8.85-9.18	8.17-9.48	1.74	2.5
Open-circuit voltage V_{oc} (V)	37.63-38.93	38.52-42.28	37.48-38.83	90.42	86.2
I_{sc} temperature coefficient α (%/K)	0.03-0.05	0.03-0.05	0.03-0.05	0.048	0.006
V_{oc} temperature coefficient β (-%/K)	0.29-0.34	0.28-0.33	0.3-0.34	0.33	0.292
P_{mpp} temperature coefficient γ (-%/K)	0.42-0.46	0.40-0.48	0.41-0.66	0.35	0.392

BF c-Si, Bifacial monocrystalline silicon

II. Methodology

The measurements for this study were performed at the Outdoor solar testing facility of the Dubai Electricity and Water Authority (DEWA) R&D center. This facility is located within the planned 5000MW Mohammad bin Rashid (MBR) Solar Park. The site consists of 31 different PV modules from different manufacturers. The installed PV technologies range from monocrystalline silicon (c-Si) and polycrystalline silicon (mc-Si) to thin film technologies such as Cadmium Telluride (CdTe) and Copper Indium Gallium Selenide (CIGS). All the modules are installed on mounting racks facing south that are oriented at the maximum annual energy yield of 25° for the latitude of Dubai.

Out of all 31 modules, 6 modules were installed in Jan 2015 and the rest were installed in three different periods – March 2015, May 2015 and Dec 2015. The distance between the module and the ground is approximately 1 meter to ensure good ventilation. The specifications of the modules in this study are shown in Table I. Full electrical parameters of the modules are deliberately withheld to keep the specific module make and model anonymous. From the PV modules, we measure current-voltage (IV) curves at time interval of 10 mins, with a Hocherl & Hackl Series ZS Electronic Load. In between IV curve measurements, the modules are kept at maximum power point and P_{mpp} (maximum power) values are recorded every 30 seconds using this instrument. The module temperature is measured at the back of each module with a PT100 temperature sensor. Environmental parameters including plane of array (POA) solar irradiance, ambient temperature, wind speed and wind direction are logged simultaneously with the P_{mpp} data. Indoor I-V curve measurements using solar simulator were performed for all the modules before its installation. All degradation calculation will use the indoor I-V curve parameters as reference instead of the nameplate values, which is less accurate.

The six modules that has been recorded for two years (until Jan 2017), are studied using the statistical decomposition method LOWESS. The LOWESS smoothing is applied to the monthly PR and the maximum normalized STC P_{mpp} (with respect to indoor measured STC) in a month after a cleaning session. Since, the rest of the panels have less than two years' data, a linear degradation model is assumed. The P_{mpp} degradation is calculated using weather-corrected PR (wc-PR) and by comparing the most recent I-V curve parameters with the initial values.

The data is processed in MATLAB and several filtering steps are performed to eliminate measurements that do not fulfil several quality criteria. The filtration process for calculating the wc-PR is as per the literature by Dierauf et al [8] [9]. The monthly PR (uncorrected) values were calculated for all POA values above $50W/m^2$. For the STC values derived from outdoor measurement, the following filters and corrections were used: (i) The data collected after cleaning sessions; (ii) P_{mpp} data chosen closer to the solar noon and for irradiance between 950 and 1050 W/m^2; (iii) more than 15 data points per day; (iv) Stability filter applied for irradiance data (five consecutive data points, 2 minutes, shall not have more than 2% deviation); and (v) P_{mpp} is corrected to $1000W/m^2$ and 25°C.

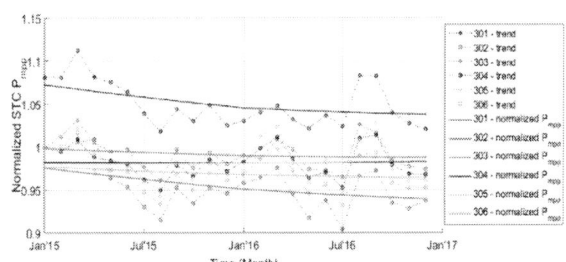

Fig. 1. P_{mpp} normalized to its measured indoor STC value, plotted over Jan'15 – Dec'16. The decomposed trend of the normalized P_{mpp} of 1 BF c-Si (301) and 5 mc-Si modules (302-306) is shown.

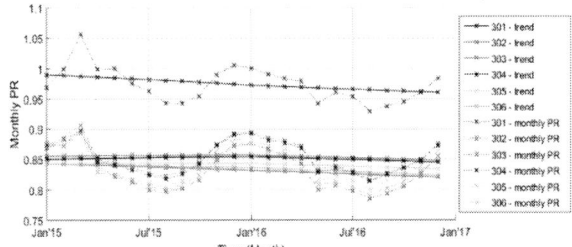

Fig. 2. Monthly PR and its decomposed trend of 1 BF c-Si (301) and 5 mc-Si modules (302-306) is plotted from Jan'15 – Dec'16.

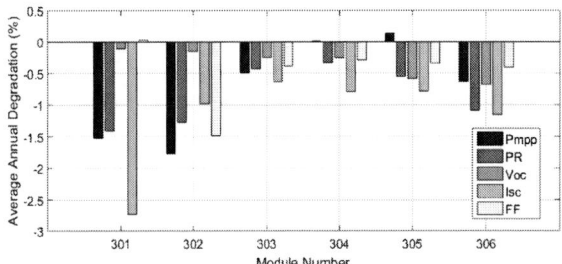

Fig. 3. Average annual degradation rate (%) of PR and the IV curve parameters: I_{sc}, V_{oc} and FF for 1 BF c-Si (301) and 5 mc-Si modules (302-306) is plotted from Jan'15 – Dec'16. Average annual degradation rate calculated using the normalized P_{mpp} is also plotted for comparison.

III. RESULTS AND DISCUSSIONS

A. Degradation Trend using LOWESS for six modules

Fig. 1 shows the P_{mpp} normalized to its measured initial indoor STC value of 1 BF c-Si (301) and 5 mc-Si modules (302-306) from January 2015 to December 2016. Fig. 2 shows the measured monthly PR of the 6 modules for the same duration. After applying the LOWESS decomposition, the trends of normalized P_{mpp} and monthly PR are shown in their respective figures. Modules 301 and 302 show a higher degradation rate than the other four modules. Fig. 3 shows the average annual degradation rates of the decomposed trends of the IV curve parameters (I_{sc}, V_{oc}, and FF) for the 6 modules. The average annual degradation rates calculated from the decomposed trends of the normalized P_{mpp} and monthly PR can also be seen in fig. 3. The average annual degradation rates for modules 301 (1.4%/yr) and 302 (1.25%/yr) is much higher than the manufacturer guarantee of less than 1% per year. The main contributing parameter for degradation in 301 is due to the degradation in I_{sc} which indicates the reduction in optical properties of the module. The test facility is in a high soiling environment and therefore, a soiling loss of 5-10% can be observed in a week. Since the PV modules are cleaned only once in a week, the effect of soiling can be a reason for the degradation in the optical properties. In module 302, the degradation is coming mainly from the decrease in FF (indication of degradation in metal conductors) and then I_{sc} parameter. Module 305 and 306 shows highest decrease (~0.4%/yr) in V_{oc} parameter among the 6 modules indicating deterioration of PV cells.

B. Linear degradation of 25 PV modules

The remaining 25 PV modules had less than 2 years' data and therefore, not suitable for the statistical decomposition study. Hence, a linear regression model is assumed in calculating the degradation rates which is a less accurate method. Table II shows the degradation rates calculated using the wc-PR and the

Table II
The degradation of the PV modules in the DEWA test bed over the period of study

Panel ID	PV Type	Period of Study (Months)	Degradation (in %/ yr)			
			wc-PR	I_{sc}	V_{oc}	FF
307	mc-Si	21	-1.03	-0.71	-0.56	-1.24
308	c-Si	21	0.78	1.34	0.27	-1.10
309	mc-Si	21	-2.34	0.47	-0.26	-3.28
310	mc-Si	21	-0.79	-0.11	-0.67	-1.08
311	mc-Si	21	-0.77	0.52	-0.62	-1.58
312	mc-Si	21	0.17	0.03	-0.11	-0.73
313	BF c-Si	21	-5.58	-3.84	-0.40	-1.38
314	mc-Si	21	-1.90	-2.41	0.04	-0.10
315	CIGS	21	-0.97	-1.01	-0.21	-0.93
316	CdTe	21	-2.40	0.42	-0.33	-2.17
317	mc-Si	21	-0.34	1.58	1.19	-0.05
318	c-Si	19	1.09	-0.13	0.17	1.08
319	mc-Si	19	-1.64	-1.09	-0.13	-0.77
320	mc-Si	19	-0.96	-1.14	0.11	0.02
321	c-Si	19	1.18	-0.69	1.28	1.17
322	mc-Si	19	-2.65	-1.98	-0.01	0.11
323	c-Si	19	0.26	-0.77	1.35	0.91
324	mc-Si	13	-6.36	-4.48	-3.01	-1.95
325	c-Si	13	-1.12	-2.06	-0.75	-1.06
326	mc-Si	13	-2.82	-2.12	-1.69	-1.62
327	c-Si	13	0.66	-3.38	0.17	1.46
328	BF c-Si	13	-3.42	-3.30	-3.53	-2.00
329	mc-Si	13	-3.09	-3.82	-2.51	-1.50
330	c-Si	13	0.27	-2.54	-0.86	-1.02
331	mc-Si	13	-5.12	-4.88	-2.85	-2.55

IV parameters. Multicrystalline silicon modules constitute majority of the modules in the test bed. Out of the 14 modules, 5 modules (310, 311, 312, 317 and 320) showed degradation less than 1% per year. The IV parameter that is causing the degradation is FF followed by V_{oc}, except for module 302, where its degradation is dominated by reduction in I_{sc}. Modules

324 and 331 show the highest degradation with 6.36% and 5.12% per year, respectively. All the IV parameters seem to be contributing to their degradation, dominated by I_{sc} and then V_{oc} and FF. Interestingly, most of the monocrystalline silicon modules show almost no degradation except for module 325. The BF c-Si modules showed significant degradation with 3.42% (328) and 5.58% (313) per year. The main reason for the degradation is due to reduction in I_{sc}. Module 313 showed high FF degradation, whereas module 328 showed high reduction in both V_{oc} and FF. In the thin-film PV module technology category, CdTe (2.4%/yr) showed higher degradation than CIGS (0.97%/yr). The degradation in CdTe is mainly due to FF reduction followed by V_{oc}.

IV. CONCLUSION

In this study, the performance degradation of 31 different types of PV modules continuously measured for more than 1 year in desert condition of Dubai has been reported. Statistical decomposition based on LOWESS is applied to six modules which had two years of data while linear regression was used for the rest of the modules which had less than two years' data. The average annual degradation rates for BF c-SI module (301,1.4%/yr) and mc-Si (302, 1.25%/yr) is much higher compared to the other four modules studied based on LOWESS smoothing. It was found that the monocrystalline silicon modules showed almost no degradation, while the multicrystalline silicon modules degradation rate varied from 0.34% to 6.36% per year based on wc-PR. The CdTe modules showed high degradation rate of 2.4% per year, whereas CIGS modules showed less than 1% per year. The IV parameter responsible for the degradation is also studied.

A more detailed analysis is planned for the next literature which will also include the investigation of reduction in IV parameters using Visual, Infrared and Electroluminescence imaging.

ACKNOWLEDGMENT

The authors would like to acknowledge DEWA for funding the project. Authors would also like to thank TUV Rheinland for supporting DEWA R&D center with the setup and operation of the Outdoor Solar Test Facility.

REFERENCES

[1] J. Herrmann, T. Lorenz, K. Slamova, K. E, M. Koehl and W. K, "Desert applications of PV modules," in *Photovoltaic Specialist Conference (PVSC), 2014 IEEE 40th* , Denver, 2014.

[2] IRENA, "Renewable Energy Market Analysis: The GCC Region," IRENA, Abu Dhabi, 2016.

[3] T. Ishii, T. Takashima and K. Otani, "Long-term performance degradation of various kinds of photovoltaic modules under moderate climatic conditions," *Progr. Photovoltaics, Res. Appl.,* vol. 19, pp. 170-179, 2011.

[4] A. Skoczek, T. Sample and E. D. Dunlop, "The results of performance measurements of field-aged crystalline silicon photovoltaic modules," *Progr. Photovoltaics, Res. Appl.,* vol. 17, pp. 227-240, 2009.

[5] J. Y. Ye, T. Reindl, A. G. Aberle and T. M. Walsh, "Perforamnce degradation of various PV module technologies in tropical singapore," *IEEE Journal of PV,* vol. 4, no. 5, pp. 1288-1294, 2014.

[6] G. Makrides, B. Zinsser, G. E. Georghiou, M. Schubert and J. H. Werner, "Degradation of different photovoltaic technologies under field conditions," in *Proc. 35th IEEE Photovoltaic Spec. Conf.,* Honolulu, 2010.

[7] R. B. Cleveland, C. W. S, J. E. McRae and I. Terpenning, "STL: A seasonal-trend decomposition procedure based on loess," *J. Official Statist.,* vol. 6, pp. 63-73, 1990.

[8] T. Dierauf, A. Growitz, S. Kurtz, J. L. B. Cruz, E. Riley and C. Hansen, "Weather-corrected Perforamnce Ratio," NREL/TP-5200-57991, 2013, Golden CO, 2013.

[9] J. J. John, A. Elnosh, A. B. Took and P. Banda, "Evalaution of various PV module technologies in desert condition using weather-corrected performance ratio," in *NREL PV Module Reliabilty Workshop (Under Review),* Denver, 2017.

High-Speed Measurements of Generated Power and its Relationship to Weather Observations at Yoshinogari Mega Solar Power Plant

Makoto Kasu, Shigeomi Hara, and Takumi Uematsu

Department of Electrical and Electronic Engineering, Saga University, Saga, 840-8502, Japan

Abstract — We constructed a high-speed measurement system of electric power generation and weather observation. To explore the long-term reliability of operating a megawatt solar power plant (MSPP), the power generation was measured every second. The weather observations were monitored every 100ms. On a relatively ideal clear sunny day, we confirmed that the operating current of the MSPP is proportional to the solar radiation. The operating voltage is affected by the panel temperature. Consequently, the generated power can be estimated directly from the solar radiation and panel temperature.

I. INTRODUCTION

The construction of megawatt-class solar panel plants (MSPP) is accelerating. In Japan, MSPPs are supported economically by a feed-in tariff (FIT) system, which should continue for twenty years. However, nobody can expect such long-time reliability from outdoor-operated solar power generation systems. Therefore, long-term measurements [1] and theoretical analysis of the inside solar panel were reported [2].

The purpose of this study is to construct a high-speed system to measure power generation and for weather observation in a commercially operated outdoor MSPP system and explore its long-time reliability until its degradation and failure. In this paper, we show the high-speed measurement results of generated power and weather observation and discuss the relationship between them.

II. HIGH-SPEED MEASUREMENT SYSTEM

The measurement system was constructed at Yoshinogari Mega Solar Power (Kanzaki City, Saga Prefecture, Japan), as shown in Fig. 1. The total generated power was 13 MW. Inside the MSPP, measurements were performed in the first section of a 10-kW solar power system. In the 10-kW system, there were four strings, numbered 1 to 4 in order of height. Each string consisted of 12 x 240 W solar power modules made of polysilicon. The panels have an angle of 10° from the ground surface toward the south. Figure 2 shows a circuit of the measurement of the photovoltaic (PV) system. The DC output of four strings was connected in parallel to a power control system (PCS). The PCS performed maximum power point tracking (MPPT).

As shown in Fig. 2, a string consists of 12 modules at the same height which are connected in series, and a series of the string consisting of 12 modules at four different height are connected in parallel. The measurement system measured the operating current of each strings 1–4 separately and the common operating voltage every second. Separately, the temperature of the backside of each solar panel was measured every 100ms by a thermocouple. In parallel, an integrated meteorological sensor and a pyranometer were situated one meter from the east side of the 10-kW PV system at the same height of the panels. The integrated meteoroidal sensor measures atmospheric temperature, humidity, and solar radiation (2nd-standard pyranometer and Si sensor), wind direction and speed every 100ms. The panels of pyranometer and Si sensor have an angle of the same 10° toward the south from the ground surface. All data are collected at the same data logger at the same time.

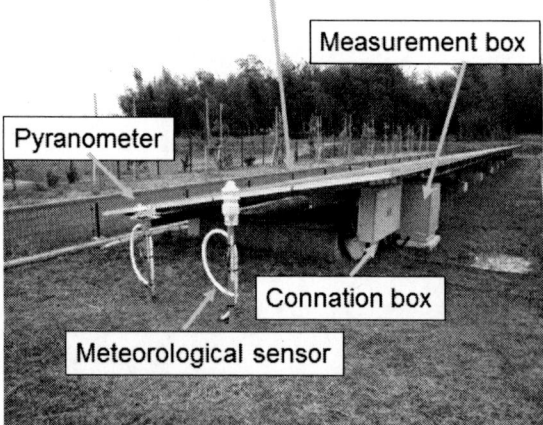

Fig. 1. High-speed measurement system, meteorological sensor, and pyranometer located 1m the east from the solar panel system.

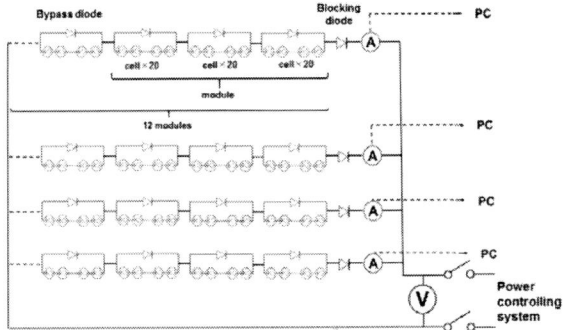

Fig. 2. Circuit of High-speed measurement system in the PV system.

III. EXPERIMENTAL RESULTS

Figure 3 shows the time change of electric power generation and solar radiation on May 30, 2016. Here we chose data on May 30, 2016, because it was an ideal clear sky day. The electric power generation was calculated to be the product of the operating current of strings no. 1–4 and the common operating voltage. Between 7:30 and 8:30, a sudden decrease and recover in the radiation and power occurred at the same time, therefore we understand that their decrease and recover are caused by passing clouds. Between 10:00 and 14:00, we see slightly lower generated power compared with the solar radiation. In order to understand the difference between the generation power and the solar radiation, we measure and ~~analysis~~ analyze operation current and voltage and various weather observations in the next.

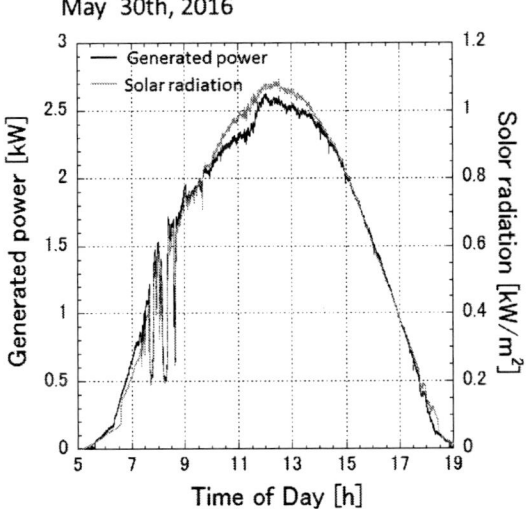

Fig. 3. Time change of generated power and solar radiation.

Figure 4 shows the time change of operating current and solar radiation on the same clear sky day. The operating current was always almost completely proportional to the solar radiation. The current generated in the solar cell is generated from light intensity. Therefore, this linear relation between the current and solar radiation can be well explained in terms of the electron-hole pair generated by the light intensity.

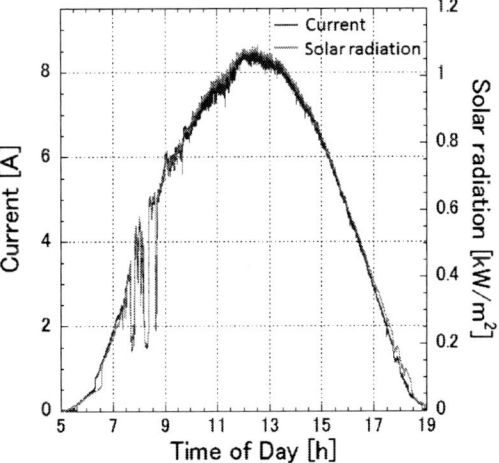

Fig. 4. Time change of operating current and solar radiation.

Figure 5 shows the time change of operating voltage and panel backside temperature on the same clear sky day. In the morning, the operating voltage started at 350 V, but afterwards continued to decrease gradually from 350 to 300 V. In the afternoon, it increased from 300 and returned to the same 350 V as the value at 7:00.

Figure 6 shows the time change of panel temperature and the operation voltage. From 5:00 to 11:30, the panel temperature continued to increase from 18 to 60°C except the periods of clouds passing. However, at 11:30, the panel temperature decreased suddenly from 60 to 45 ~~oC~~ °C and afterwards was kept at a relatively lower 45-50°C. After 15:00 the panel temperature decreased from 45 to 22 °C. We see that the operating voltage decreased by the temperature increase in the panel.

Fig. 6 shows the time change of the panel temperature and the wind speed. At 11:30 the panel temperature decreased from 60 to 50 °C and until 15:00 it was kept at 50 °C. In the same period, we have found the increase of the wind speed. Thus, the wind cooled the panel temperature and consequently the operation voltage increased from 300 to 310 V, and eventually the generation power increased as shown in Fig. 3.

978-1-5090-5606-4/17 $31.00 © 2017 IEEE

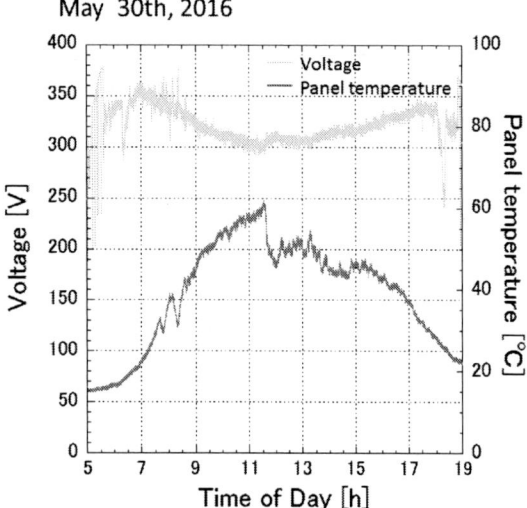

Fig. 5. Time change of operating voltage and panel temperature.

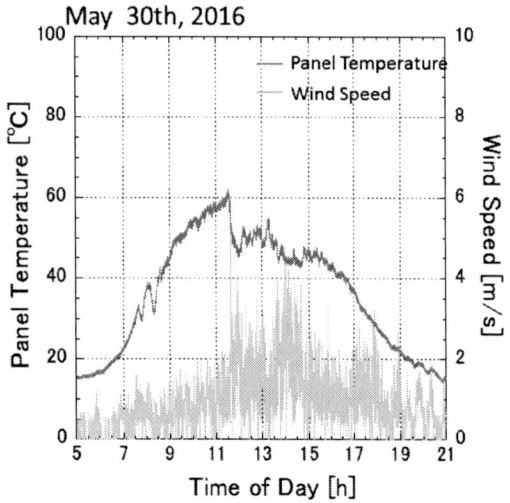

Fig. 6. Time change of the wind speed and operating voltage.

Figure 7 shows the time change of operating current and voltage on the same clear sky day. As explained in Fig. 2, electric power generation was the product of the operating current of strings 1–4 and the common operation voltage. Sometimes, the solar radiation decreased suddenly because of passing clouds. As shown in Fig. 4, the operation current is almost completely proportional to the solar radiation. However, the operation voltage decreased from 350 to 300 V̶ V by an increase in the panel temperature from 15 to 60 °C. The panel temperature is affected by the wind. Therefore, the time changes of the operation current and voltages as shown in Fig. 7 were observed.

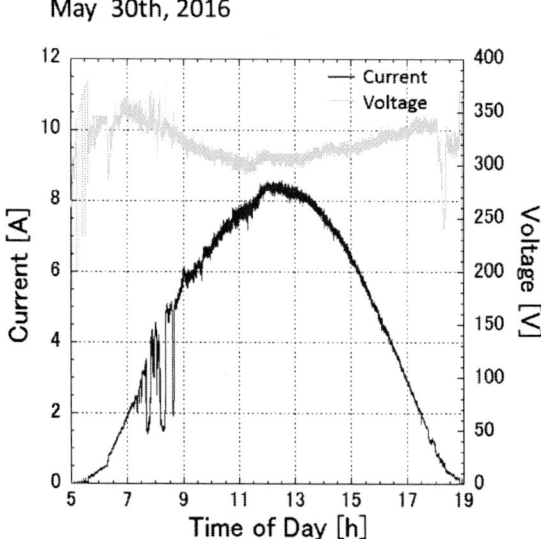

Fig. 7. Time change of operation current and voltage.

IV. CONCLUSIONS

We constructed a high-speed measurement system at an operating megawatt solar power plant and measured the electric power generation every second and solar radiation every 100ms. We have found that the operation current was almost proportional to the solar radiation, the operation voltage was increased slightly by a decrease of the panel temperature, which was caused by an increase of the wind speed. In this way, we showed here that electric power generation, the product of operation current and voltage, could be estimated from the solar radiation, the panel temperature, and the wind speed.

ACKNOWLEDGEMENTS

We deeply appreciate fruitful discussion from Mr. Kensuke Sato, Mr. Kazuhiko Oda, Mr. Kazuhiko Babasaki, Mr. Yasuki Masutomi at NTT Facilities Co. and Drs. Atsushi Masuda, Yasuo Chiba at AIST. This work is supported by New Energy

978-1-5090-5606-4/17 $31.00 © 2017 IEEE

Industrial Technology Development Organization (NEDO) project.

REFERENCES

[1] Y. Ueda, M. Konagai, K. Kurokawa, M. Kudo, K. Hakuta, T. Oozeki, "Five years operation results of different crystalline-Si PV systems and analysis of the degradation rate in the field," *Conference Record of the IEEE Photovoltaic Specialists Conference*, vol. 6745148, pp. 3266-3269, 2013.

[2] S. Hara, M. Kasu, "Estimation method of solar cell temperature using meteorological data in mega solar power plant, *IEEE Journal of Photovoltaics*, vol. 6, pp. 1255-1260, 2016.

Impact of Missing Data on the Estimation of Photovoltaic System Degradation Rate

Andreas Livera, Alexander Phinikarides, George Makrides and George E. Georghiou

FOSS Research Centre for Sustainable Energy, Photovoltaic Technology Laboratory,
Department of Electrical and Computer Engineering, University of Cyprus,
75 Kallipoleos Avenue, P.O. Box 20537 Nicosia, 1678, Cyprus

Abstract — In this paper, the impact of missing data and the robustness of commonly used statistical techniques to calculate the annual degradation rate of crystalline-Silicon (c-Si) photovoltaic (PV) systems was analyzed. In addition, the performance of different imputation techniques was assessed in order to develop an optimal methodology for treating missing data for degradation rate estimation studies. The results obtained clearly demonstrate that the application of the different statistical methods to estimate the annual degradation rate is sensitive to the amount of missing data, since all the statistical methods underestimated the degradation rate consistently with the increasing level of missing data. In addition, the application of the Seasonal Decomposition (Seas) technique yielded robust annual degradation rate estimates since for a level of 40 % of missing data, the absolute percentage error (APE) of the annual degradation rate estimated with all statistical techniques, was less than 7.5 %. Finally, Classical Seasonal Decomposition (CSD) was shown to be the most robust technique for estimating the degradation rate when imputation was applied, while Autoregressive Integrated Moving Average (ARIMA) was the most successful technique in providing robust degradation rate estimates when not applying any imputation.

Index Terms — degradation, imputation, performance ratio, monitoring, photovoltaic systems, missing data, outages.

I. INTRODUCTION

The accurate evaluation of the degradation rate of photovoltaic (PV) systems is crucial in order to reduce investment risks and hence to further increase the bankability of the technology [1]. In particular, the degradation rate estimation of PV systems is influenced by a number of data associated parameters including the resolution, sampling, duration and integrity [2].

With respect to data integrity, missing data caused by outages or data acquisition failures is a common problem of monitoring systems which directly affects the estimation of PV performance metrics and degradation rate analysis. A common approach for dealing with missing data is to filter out (i.e. ignore or delete) missing data and analyze only the remaining data [3, 4]. In this context, another option for handling missing data is to impute the missing data with replacement values using different techniques and models such as Last Observation Carried Forward (LOCF), mean substitution [5, 6] and imputation using regression models [1].

The impact that missing data and imputation techniques pose to the degradation rate estimates of PV systems is an area which is not yet fully explored due to lack of well-defined methodologies. The results obtained from a recent study [7], indicated that degradation rate estimates are both sensitive to the amount of missing data and the statistical technique used to calculate the degradation rate from monthly performance ratio (PR) time-series.

The scope of this paper is to examine the sensitivity of the estimated degradation rate of PV systems to missing data and to develop an optimal methodology for treating missing data through the analysis of data-sets which contain high-resolution measurements from the operation of a typical crystalline-Silicon (c-Si) PV array installed in Cyprus.

II. METHODOLOGY

A. Experimental setup

A large number of c-Si grid-connected PV systems has been monitored at the outdoor PV test facility of the University of Cyprus (UCY) since 2006. The systems are installed side-by-side and are approximately of 1 kWp power capacity each, as shown in Fig. 1.

Fig. 1.   Outdoor PV test facility at the UCY.

The fixed plane PV systems were installed due South, at the optimum annual energy yield angle for Cyprus of 27.5°. The monitoring of the PV systems started at the beginning of June

2006 and both meteorological and PV system measurements were acquired and stored with the use of an advanced measurement platform according to IEC 61724 [8].

The platform comprises of solar irradiance (Kipp & Zonen CM21 pyranometer), temperature (PT 100) and electrical (potential divider, shunt resistor, NRZ) sensors connected to a central data acquisition system that stores data every second. The measurements were also averaged every fifteen-minutes and stored in the database. The monitored meteorological measurements include the total irradiance in the plane of the array (POA), G_I, ambient temperature, T_{am}, module temperature, T_m, wind speed, S_W, and direction, α_W. The electrical parameters measured include DC current of the array, I_A, voltage, V_A, and power, P_A, at the maximum power point (MPP) as well as the AC power to the utility grid, P_{TU}.

Throughout the evaluation period, all the PV arrays and calibrated irradiance sensors were kept clean to minimize the effect of soiling.

B. PV system performance time series

Initially, the recorded fifteen-minute average data-sets of G_I and P_A were used to construct the monthly PV array performance ratio (PR_A) time series. These data-sets, defined as the complete data-sets, were initially corrected for known outage periods due to sensor and human errors, inverter and array downtimes and system disconnections. The observed outage periods were meticulously kept in log files to ensure quality monitoring. Common outage periods such as inverter shut-off due to very low irradiation on cloudy days, energy lost due to shading or other unpredictable events were not corrected in order to avoid introducing any bias in the energy yield estimates.

The PR_A was used as a metric to indicate the overall effect of losses on the array's rated output. The PR_A is defined as the ratio of the array yield, Y_A, and the reference yield, Y_r:

$$PR_A = \frac{Y_A}{Y_r} \qquad (1)$$

where Y_A is defined as the ratio of the monthly array energy output per kWp of installed PV array and Y_r is the ratio of the monthly in-plane irradiation, $H_{I,m}$, and the reference in-plane irradiance, $G_{I,ref}$ (1000 W/m^2).

The monthly PR_A time series of the benchmarked c-Si PV system, over the evaluation period June 2006 – June 2011, is depicted in Fig. 2. The monthly PR_A plot shows the typical seasonal profile of the mono-c-Si PV system, exhibiting higher PR during the winter compared to the summer seasons, due to the higher module operating temperatures of the summer.

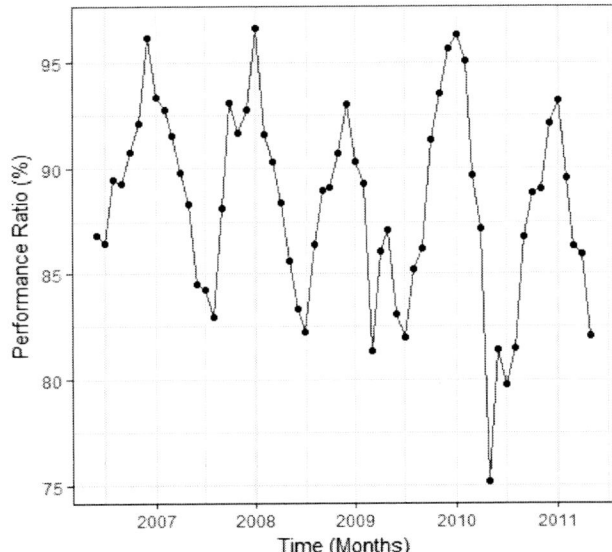

Fig. 2. Monthly PR_A time series of the investigated c-Si PV system, monitored at the outdoor PV test facility of the UCY, over the evaluation period June 2006 - June 2011.

C. Generation of missing data

Unbiased artificially missing PV performance data-sets were created using a random forest approach and were inserted to the complete data-sets of the investigated PV system. This approach uses a random forest trained on the observed values (fifteen-minute average measurements of G_I and P_A) of the completed data-sets to predict the missing values. By this approach, as many data points as desired can be sampled out of the observed values and the respective fields are replaced by "Not Available (NA)" values.

For a given complete data-set (without missing values), random sampling was performed from 1 % to 40 % of the total amount of data points, in order to select data points which were designated as Missing Completely at Random (MCAR) and were subsequently removed from the data-set. Since the missingness mechanism of the data was MCAR, the probability of being missing for the observed values of G_I and P_A was the same.

The resulting data-sets were defined as the incomplete data-sets, which were then analyzed with advanced imputation techniques.

D. Missing data imputation

In order to mitigate the effect of missing data, the incomplete data-sets were treated with missing data imputation techniques and models such as imputation by Linear Interpolation (Linear), weighted Moving Average (MA), Last Observation Carried Forward (LOCF), Kalman Smoothing and State Space Models (Kalman) and Seasonal Decomposition (Seas). Hence, missing (NA) values were replaced with statistical estimates and the resulting data-sets were defined as the imputed data-sets.

E. Degradation rate estimation

For the purpose of this investigation, four commonly used statistical trend extraction methods, Linear Regression (LR), Classical Seasonal Decomposition (CSD), Holt-Winters (HW) exponential smoothing and Autoregressive Integrated Moving Average (ARIMA) models, were applied to the constructed monthly PV array PR time series of the c-Si PV system to calculate the annual degradation rate. The PR_A time series were constructed using the complete, incomplete and imputed data-sets over the first five years of field exposure.

LR is the simplest and most commonly used method for trend extraction and can be expressed as:

$$Y = aX + \beta + e \qquad (2)$$

where α is the slope of the trend, β is the intercept and e denotes the error (the deviation of the observation from the linear relationship). The LR technique tries to fit the above equation by minimizing the sum of squared residuals, by using the ordinary least squares method. Thereby, the annual degradation rate (DR) (in units of %/year) is calculated as follows:

$$DR = a \times 12 \times 100\% \qquad (3)$$

CSD is a widely used method which decomposes a time series object by moving averages into seasonal, trend and residual components. There are two basic decomposition models: the additive (which is useful when the seasonal variation is relatively constant over time) and the multiplicative (which is useful when the seasonal variation increases over time). In this work, the additive model was used for time series decomposition since the amplitude of the seasonal component was mostly stable throughout the years.

In order to calculate the trend component, a 12-month centered moving average was applied to the time series. For a $2 \times k$ moving average, where k is the seasonal period ($k=12$ because of the number of months in a year), the centered moving average at time t was found as:

$$T_t = \frac{1}{2}\left(\frac{1}{k}\sum_{i=t-m}^{t+m-1} Y_i + \frac{1}{k}\sum_{i=t-m+1}^{t+m} Y_i \right) \qquad (4)$$

where Y is the original time series, T_t is the trend at time t ($t>m$), and m is defined as the half-width of a moving average, $m=k/2$. The smoothing effect of the moving average formed the trend of each time series, which by the definition given in the above equation, was half a seasonal period shorter on each end. Afterwards, LR is applied to the trend component to compute the value of annual degradation rate.

HW is a model-based method, in which triple exponential smoothing is applied to the time series [1]. Triple exponential smoothing takes into account seasonal changes as well as trends, through the minimization of the squared one-step ahead prediction error. HW exponential smoothing requires the general additive model in (5), where m_n is the level component, b_n is the slope component and c_{n-S+l} is the relevant seasonal component, S is the seasonal period and are given in

(6), (7) and (8), respectively.

$$\hat{y}_{n+l|n} = m_n + b_n + c_{n-s+l} \quad l=1,2,\dots \qquad (5)$$

$$m_t = a_0\left(y_t - c_{t-s}\right) + \left(1 - a_0\right)\left(m_{t-l} + b_{t-l}\right) \qquad (6)$$

$$b_t = a_1\left(m_t - m_{t-l}\right) + \left(1 - a_1\right)b_{t-l} \qquad (7)$$

$$c_t = a_2 \frac{y_t}{m_t} + \left(1 - a_2\right)c_{t-s} \qquad (8)$$

with α_0, α_1 and α_2 lying between 0 and 1. As before, LR is then applied to the smoothed time series to compute the value of the annual degradation rate.

The multiplicative ARIMA method is a more advanced and complex method than other classical methods since it can effectively deal with seasonal variations, random errors, outliers and level shifts and can therefore be used to specify a model which removes all autocorrelations in the model residuals. The general model for multiplicative ARIMA is given in (9) and is abbreviated as ARIMA(p,d,q)(P,D,Q), where p is the auto-regressive (AR) order, d is the differencing order, q is the moving average (MA) order, P is the seasonal AR order, D is the seasonal differencing order and Q is the seasonal MA order.

$$\phi(T)\phi_s(T^s)\nabla^d \nabla_s^D y_t = \theta(T)\theta_s(T^s)e_t \qquad (9)$$

where T is the delay operator, $\Phi(T)=(1-\varphi_1 T-\dots \varphi_p T^p)$ is an auto-regressive polynomial in T of degree p, $\Phi_s(T^s)$ is an autoregressive polynomial in T^s of degree P_s, $\theta(T)$ is a moving average polynomial in T^s of degree q and $\theta_s(T^s)$ is a moving average polynomial of degree Q_s in T^s. The operators $\nabla^d=(1-T)^d$ and $\nabla^D=(1-T)^D$, respectively, are non-seasonal and seasonal differencing operators which capture non-stationarity. The residuals, e_t, are independent, identically distributed (i.i.d) with mean zero and variance σ^2, therefore resembling white noise [9].

In order to find the optimal ARIMA model, the time series (y_t) was initially checked for stationarity and then transformed using differencing to achieve stationarity, as necessary. The lags p, q, P, Q of the model were determined from the autocorrelation function (ACF) and the partial autocorrelation function (PACF) [10]. The model selection procedure yielded multiple models that fit the time series well. The optimum model is the one with the lowest order (i.e. parsimonious), with the lowest mean-square-error (MSE) and the minimum value of the corrected Akaike Information Criterion (AIC). In order to validate the goodness of fit, the residuals were checked for Gaussian white noise (GWN) properties (i.e. un-correlated, normally distributed) [1].

Subsequently, the annual degradation rate is calculated using the gradient of the obtained linear fit on the optimal ARIMA model and multiplying it by twelve.

F. Standard statistical measure

In order to compare the degradation rate estimates obtained

using the complete, incomplete and imputed data-sets, the absolute percentage error (*APE*) was used as a measure for determining the precision of the different imputation techniques. The *APE* is defined as:

$$APE = \left| \frac{A_t - P_t}{A_t} \right| \times 100\% \qquad (10)$$

where A_t is the actual value and P_t is the predicted value.

III. RESULTS

The degradation rates for the investigated c-Si PV system clearly demonstrated that missing data can greatly affect the value of the annual degradation rate. The results outlined in Fig. 3 – 6, clearly demonstrate that the application of the different statistical methods to estimate the annual degradation rate is sensitive to the amount of missing data, since all the statistical methods underestimated the degradation rate consistently with the increasing level of missing data. In particular, for 40 % of missing data, the *APE* of the annual degradation rate was 38 %, 37 %, 35 % and 34 % when using the HW, LR, CSD and ARIMA techniques to model the trend, respectively, compared to initial estimates using the complete data-sets (without missing data). In addition, for a level of 5 % of missing data which is generally considered as the threshold limit for indicating or not a low quality data acquisition system [11], it is obvious from Fig. 3 that the *APE* of the annual degradation rate using LR to model the trend, was 16 % when no imputation was applied, while when using the Seas imputation technique, the *APE* of the annual degradation rate was 1.5 %. Moreover, the application of the Seas technique yielded robust annual degradation rates with all statistical techniques, since for a level of 40 % of missing data the *APE* was less than 7.5 %.

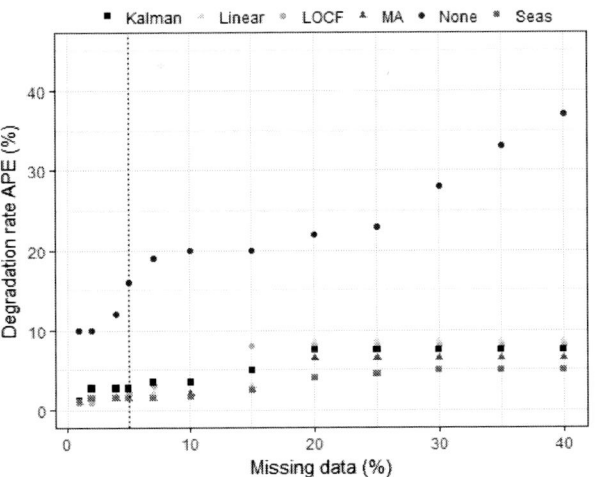

Fig. 3. Annual degradation rate *APE* estimated with LR for 1 – 40 % level of missing data. The 5 % missing data threshold limit of data acquisition systems is also shown as a guideline.

Furthermore, the low *APE* (for all the imputation techniques) of the estimated degradation rates using CSD at a range of 1 - 40 % of missing data, demonstrated that CSD is the most robust technique for missing data, as shown in Fig. 4.

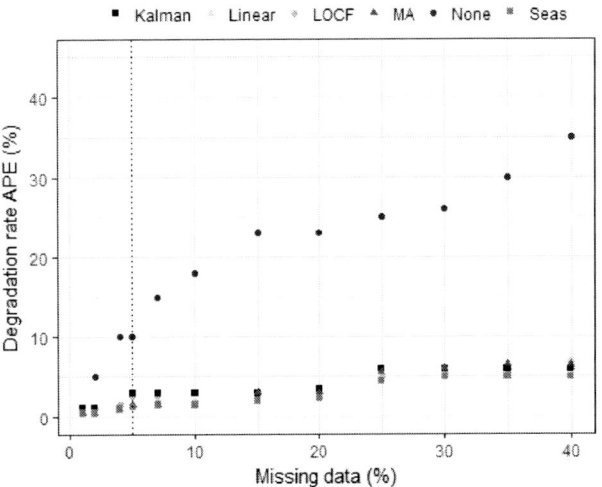

Fig. 4. Annual degradation rate *APE* estimated with CSD for 1 – 40 % level of missing data. The 5 % missing data threshold limit of data acquisition systems is also shown as a guideline.

Additionally, for the same level of missing data, the HW method exhibited higher *APE* dispersion for all the imputation techniques, as shown in Fig. 5, when compared to CSD.

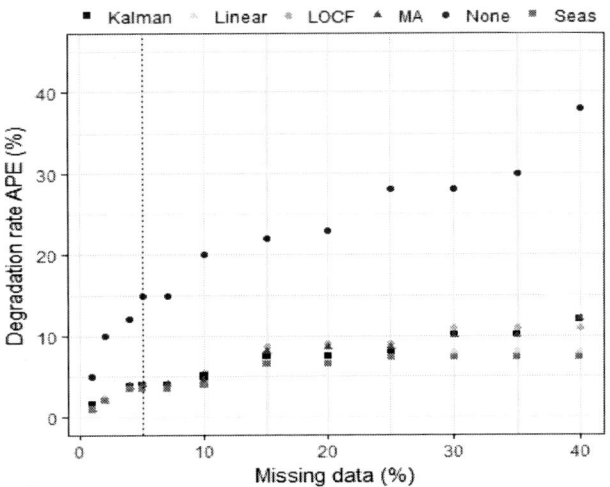

Fig. 5. Annual degradation rate *APE* estimated with HW exponential smoothing model for 1 – 40 % level of missing data. The 5 % missing data threshold limit of data acquisition systems is also shown as a guideline.

Finally, ARIMA was the most successful technique in providing robust degradation rate estimates when not applying any imputation. For a level of 40 % of missing data, the *APE* of the annual degradation rate estimated using the incomplete data-sets, was 34 %.

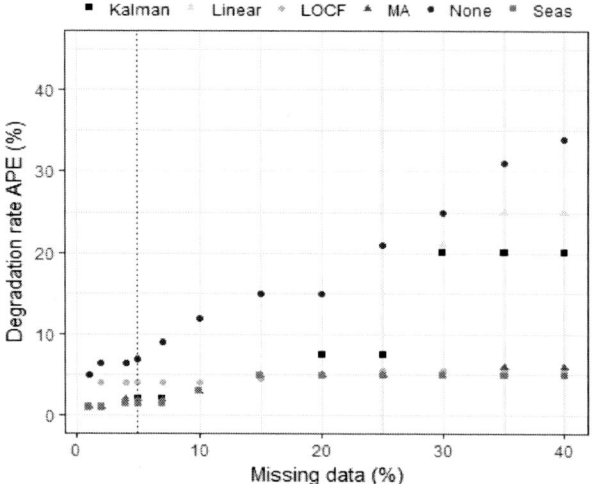

Fig. 6.   Annual degradation rate *APE* estimated with ARIMA model for 1 – 40 % level of missing data. The 5 % missing data threshold limit of data acquisition systems is also shown as a guideline.

IV. CONCLUSION

Missing data can have a significant effect on the estimation of the degradation rate of PV arrays and dealing with such data is an important consideration before commencing to further analysis of performance metrics. Appropriate corrective actions for missing data, using different data imputation techniques, were considered in order to improve the degradation rate estimates and the performance of each imputation technique was assessed at different levels of randomly inserted missing data.

The results clearly demonstrated that for 40 % of missing multivariate instantaneous data points, the *APE* of the annual degradation rate was 38 %, 37 %, 35 % and 34 % when using the HW, LR, CSD and ARIMA models to extract the trend, respectively, compared to initial estimates using the complete data-sets (without missing data).

Furthermore, the application of the different imputation techniques yielded more accurate estimates when compared to estimates without using imputation. The Seasonal Decomposition (Seas) Missing Value Imputation technique proved to be the most reliable data correction method, even at a 40 % level of missing data, exhibiting a 5 % deviation when compared to initial estimates for LR, CSD and ARIMA degradation estimation techniques.

Finally, CSD was shown to be the most robust technique for estimating the degradation rate when imputation was applied while ARIMA was the most successful technique in providing robust degradation rate estimates when not applying any imputation.

ACKNOWLEDGEMENT

This work was funded through the IPERMON project (KOINA/SOLAR-ERA.NET/1214/08) which was co-financed by the European Regional Development Fund and the Republic of Cyprus through the Cyprus Research Promotion Foundation (DESMI 2009-2010).

REFERENCES

[1] A. Phinikarides, N. Kindyni, G. Makrides and G. E. Georghiou, "Review of photovoltaic degradation rate methodologies," *Renew. Sustain. Energy Rev.*, vol. 40, pp. 143-152, 2014.

[2] D. C. Jordan and S. R. Kurtz, "Photovoltaic Degradation Rates-an Analytical Review," *Prog. Photovoltaics Res. Appl.*, vol. 21, no. 1, pp. 12-29, 2013.

[3] A. Karahalios, L. Baglietto, K. J Lee, D. R English, J. B Carlin and J. A Simpson, "The impact of missing data on analyses of a time-dependent exposure in a longitudinal cohort: a simulation study", in *Emerging Themes in Epidemiology*, 2013.

[4] T. McCandless, S. E. Haupt, G. Young, "Replacing missing data for ensemble systems - The Effects of Missing Data", in *7th Artificial Intelligence and Its Applications to the Environmental Sciences Conference*, 2009.

[5] B. Wilmots, Y. Shen, E. Hermans, D. Ruan, "Missing data treatment: Overview of possible solutions", Diepenbeek: Policy Research Centre Mobility and Public Works, 2011.

[6] E.A.P. Gustavo, A. Batista and M. C. Monard, "An Analysis of Four Missing Data Treatment Methods for Supervised Learning", *in Applied Artificial Intelligence*, vol. 17, pp. 519-533, 2003.

[7] A. Phinikarides, G. Makrides, and G. E. Georghiou, "Estimation of the degradation rate of fielded photovoltaic arrays in the presence of measurement outages," in *32nd EU-PVSEC*, 2016, pp. 1754-1757.

[8] IEC 61724:1998, Photovoltaic system performance monitoring - Guidelines for measurement, data exchange and analysis, 1st ed. Geneva, Switzerland: IEC, 1998.

[9] A. Phinikarides, G. Makrides, B. Zinsser, M. Schubert, and G. E. Georghiou, "Analysis of photovoltaic system performance time series: seasonality and performance loss," Renew. Energy, vol. 77, pp. 51-63, 2015.

[10] A. Phinikarides, G. Makrides, N. Kindyni, A. Kyprianou, and G. E. Georghiou, "ARIMA modeling of the performance of different photovoltaic technologies," *in 39th IEEE PVSC*, 2013.

[11] A. Woyte, M. Richter, D. Moser, N. Reich, M. Green, S. Mau, and H. Georg Beyer. 2014. "Analytical Monitoring of Grid-Connected Photovoltaic Systems - Good Practice for Monitoring and Performance Analysis." Report IEA-PVPS T13-03: 2014. IEA PVPS.

978-1-5090-5606-4/17 $31.00 © 2017 IEEE

Field Degradation and Failures of Aged Crystalline Silicon PV modules in Mexico

D. Martínez Escobar*, P. A. Sánchez-Pérez*, Rocío de la luz Santos Magdaleno*, José Ortega Cruz*,
Sai Tatapudi†, Aarón Sánchez Juárez*, GovindaSamy TamizhMani†

*Instituto de Energías Renovables, Privada Xochicalco S/N, Temixco, Morelos 62580, Mexico
†Photovoltaic Reliability Laboratory, Arizona State University, Mesa, Arizona 85212, USA

Abstract—This paper presents the visual defects and performance degradation rates of crystalline silicon photovoltaic (PV) modules exposed in two different climatic conditions (Hot-Humid and Temperate) of Mexico between 10 and 25 years. An extensive evaluation of these modules was performed using different visual, optical and electrical tools including 86-point visual inspection, electroluminescence imaging, infrared imaging, module-level visible-near IR reflectance and current-voltage measurements. The dominant visual defects were determined to be encapsulant discoloration, encapsulant delamination and cell metallization corrosion. The average annual power degradation rate was determined to be between 1.4% and 1.5% for all the modules surveyed.

Index Terms—crystalline silicon, degradation rates, photovoltaic systems, reliability.

I. INTRODUCTION

Photovoltaic (PV) industry has grown from a few MW to 222 GW installed capacity in 2015 worldwide [1]. In the past few years, the PV market has also increased significantly in Mexico. Recently, the Mexican government, through its electrical energy reform program, approved the implementation of several large utility-scale PV power plants. As shown in the current paper, the reliability lessons learned from the old Mexican climate-specific PV systems would be greatly helpful to the sustainability of these growing PV market in Mexico.

Large PV plants are relatively new in Mexico. Before the approval of the electrical energy reform by the Mexican government, only small-scale water pumping systems and stand-alone lighting system were prevalent, and the market size was very limited. After the introduction of the reform program, the market size dramatically increased in Mexico. The installed capacity of PV system in Mexico grew from 131 MW in 2014 to 234 MW in 2015, and in 2020 a 2.1 GW market is predicted for Mexico [2]. This is a great opportunity for investors, manufacturers and installers in the solar market; however, the current market growth alone cannot guarantee the sustainability of the industry in the long term unless the Mexican climate-specific reliability problems of PV modules are identified and mitigated before and after plant installations. As shown in the current paper, the reliability lessons learned from the old Mexican climate-specific PV systems would be greatly helpful to the sustainability of the PV market in Mexico. Although many of the materials used in the PV modules have recently been modified by the manufacturers to meet the price per watt ($/W) demand of the industry, the lessons learned in this investigation with old modules with long term field exposure can still serve as a useful reference and a basis for the selection or elimination of new materials used in the current PV modules or in the future designs.

The purpose of this paper is to present the visual defects and performance degradation rates of crystalline silicon PV modules installed in Mexico between 10 and 25 years ago. To identify the dominant defects and determine the degradation rates, an extensive evaluation of these modules was performed using different visual, optical and electrical tools including 86-point visual inspection, electroluminescence imaging, infrared imaging, module-level visible-near IR, reflectance and current-voltage measurements.

II. METHODOLOGY

The primary goal of this work is to present the field observed defects and degradation rates of field aged crystalline silicon modules in two different climate conditions in Mexico: Hot-Humid and Temperate. We carried out a degradation analysis on forty-seven (47) PV modules, aged between 10 and 25 years, of Siemens company with models M65/75 and SP75. The degradation modes and rates for the PV modules exposed in diverse climatic conditions experienced in various countries have already been reported [3]–[7]. The current study is undertaken from a Mexican climate-specific reliability perspective using several new tools and approaches including the test procedure used in IEC-61215:2005 standard and the visual inspection procedure identified in the NREL's visual inspection sheet. The data utilized for this study was collected by evaluating the PV in the following sequence:

- First, a detailed visual inspection was performed using a checklist following the IEC-61215:2005 standard and the visual inspection sheet developed by NREL. The visual inspection results were classified and grouped by climate condition and manufacturer.
- Second, the electroluminescence imaging (EL) was performed to detect cracked cell issues and metallization corrosion issues.
- Third, the infrared imaging (IR) was performed to detect hotspots and thermal non-uniformity among the cells in each module.
- Fourth, surface reflectance measurement was obtained in different locations of the module using a hand-held spectrophotometer to detect changes in the optical properties.

- Finally, outdoor current-voltage curves were obtained and translated to STC (Standard Testing Conditions) to determine the electrical performance parameters per IEC 61215:2005.

A. Visual inspection

The visual inspection procedure allows detecting the 86-point visual defects or failures including encapsulant browning, burn marks on the back sheet, corrosion of fingers and interconnects, back sheet bubbles, encapsulant delamination, and so on. We inspected all the modules in this work using various outdoor and indoor testing tools. All the climate-specific visual defects were classified and grouped for further analysis. Also, since the impact of soiling in the module performance is beyond the scope of this work, we thoroughly cleaned each module before undertaking all the measurements.

B. EL imaging

Electroluminescence (EL) images were taken in a dark room and under forward bias conditions. For every module, we used the Isc value from the nameplate and the voltage necessary to achieve that current at room temperature. Using this method, we were able to detect the cell cracks, finger detachment and metallic corrosion and related issues of the cells [8].

C. IR imaging

Two main factors that can induce the cell overheating issue are shading and cell mismatch. This issue is typically addressed with a use of bypass diodes in the modules [8]. To see the hotspots susceptibility, we placed all the PV modules in short-circuit configuration and exposed them to a minimum irradiance of $800 \, \mathrm{W \, m^{-2}}$ until the module reached the thermal equilibrium. Next, we took the IR image as close possible to the module and register the module temperature and ambient temperature. The criteria for detecting hotspots was a temperature difference of $20\,°C$ between the average temperature of all cells and the highest temperature cell. All the hotspots obtained were under open-circuit conditions.

D. I-V Curve

The current-voltage curve (I-V curve) is the primary method to determine the electrical performance of a solar cell or PV module. To obtain the degradation rate of the electrical parameters, we used the 1-point approach because the initial I-V data was not available to us [9]. For this work, all the modules were thoroughly cleaned to remove any soiling on the surface and took 10 I-V curves for each module. We used the average values to determine the degradation rate. The average uncertainty in the maximum power measurement was determined to be 1.6

E. Optical characterization

Another useful tool to measure the optical properties of the materials within a module is a hand held reflectance spectrophotometer. The spectrophotometer used in this work can measure the reflectance properties of the materials in a non-destructive way. It can be used to compare the reflectance

of different degraded and non-degraded areas including encapsulant browned areas and encapsulant bleached areas.

III. VISUAL INSPECTION RESULTS

Fig. 1 shows a summary of the major defects observed in all the modules inspected in this study. We noted that encapsulant discoloration (browning) is in all the modules and the second most frequent degradation defect is the delamination of the encapsulant in between the solar cell-encapsulant interphase. Other common defects found were corrosion, chalking of the backsheet and backsheet burn marks. Also, it is important to note that 68% of the modules analyzed presented more than one hotspot.

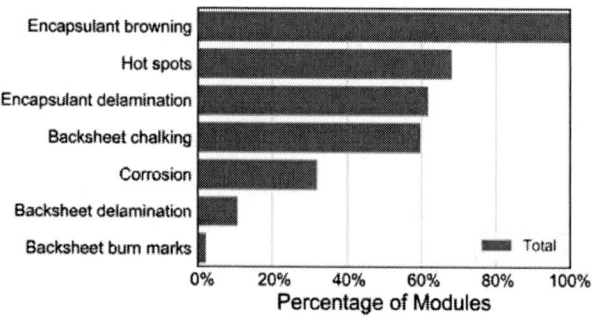

Fig. 1. Normalized visual inspection results of 47 PV modules surveyed (10-25 years of exposure) in Hot-Humid and Temperate climate in Mexico.

A. Encapsulant discoloration and delamination

Encapsulant discoloration happens when the material degrades due to long term exposure to the combined effect of UV radiation and high temperatures. This defect affects the current generation, and hence the power, produced by the solar cells and modules. Depending on this degradation impact, the discoloration goes from yellow in the initial years to brown in later years [10]. It has been reported that long-term degradation is linearly related to the UV exposure and is responsible for 0.2% to 0.6% power loss per year [9], [11].

As shown in Fig. 2, it was found that PV modules (SP75) with 10 to 15 years of exposure in Hot-Humid climate presented light discoloration at the cell center and modules exposed to 25 years (M65/75) in temperate climate presented dark browning in the center of the cells and light browning at the edges of the cells. However, the PV modules exposed for 25 years to hot-humid climate (M65) presented dark brown discoloration in almost the entire cell area. Further material level analysis of the encapsulant is needed to explain this behavior of the degradation mechanism.

Delamination is an easy defect to detect visually. This defect occurs when the polymer of the encapsulant loses adhesion properties to the cell or the glass. Severe optical decoupling losses occur due to encapsulant delamination. Recent studies on the moisture penetration in PV modules deployed in the field show that moisture penetration results into delamination and reduces the true active area of the cells in a PV module

978-1-5090-5606-4/17 $31.00 © 2017 IEEE

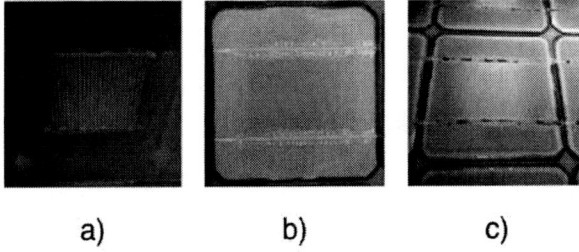

Fig. 2. Fluorescent images showing the discoloration of the EVA in modules with different exposure time. a) 15 years to Hot-Humid climate (SP75), b) 25 years to Temperate climate (M65/75) and c) 25 years to Hot-Humid climate (M65/75). Images taken using UV lamp.

[12] and can induce corrosion inside the module which will lead to increase in series resistance and power loss.

Along the metallization bus bars and fingers, 58% of the M65/75 modules showed encapsulant delamination whereas 47% of the SP75 in Hot-Humid climate. The exact location of the delamination is near the edge as well in the center with an approximate coverage area of 40% of the whole cell. For temperate climate, 100% of the M65/75 modules presented delamination near the edge and in the center but with a lower coverage area about 10%.

We measured the reflectance on various parts of the cells through the glass superstrate using a handheld module-level UV-Vis-NIR spectrophotometer. Fig. 3 shows the exact location of the reflectance measurement and the reflectance values for the three different cells from a module retrieved from a Hot- Humid climate with 25 years of exposure. These three cells have about 40% severe discoloration at the center of the cell and delamination as well. Also, it can be observed that the reflectance is dependent on the wavelength; at lower wavelengths, the reflectance loss is greater (about 12%), and at higher wavelengths, the loss is about 5%.

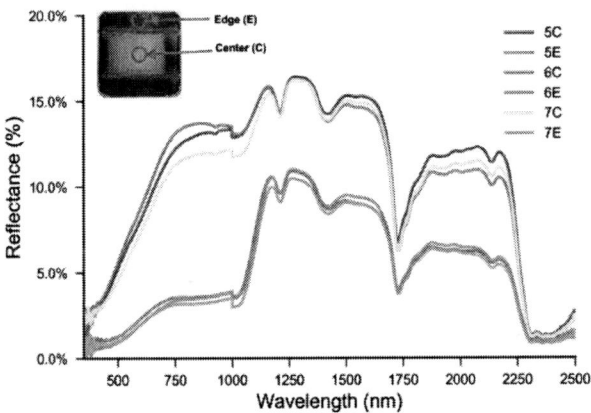

Fig. 3. Surface reflectance measurement of three solar cells (5, 6, 7) on the module surface (C= Center and E = Edge of the solar cell)

B. Corrosion

Corrosion is a significant problem in Hot-Humid climates. It is one of the principal degradation modes that causes series resistance increase. Corrosion may be caused by moisture ingress through the back sheet or laminate edge [13] to the internal structure of the module and reacts with metallic part; it is accelerated inside the module when there is a combination of humidity and high temperatures [14]. For this work, 37% of the PV modules installed in Hot-Humid climates and 11% for temperate climates presented corrosion. From this group, 75% of the PV modules where M65/75 and 25% SP75.

Fig. 4 shows two representative cells from a M65/75 module side by side exposed to temperate climate and Hot-Humid. It can be observed that for Hot-Humid climate there is severe corrosion all over the bus bars and fingers. Also, there is delamination on the edge and delamination on the edge of the cell and in the grid lines.

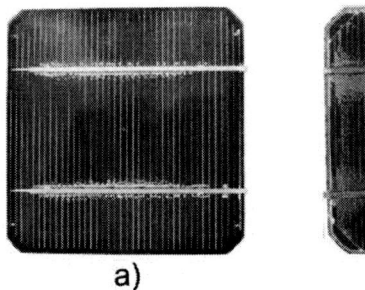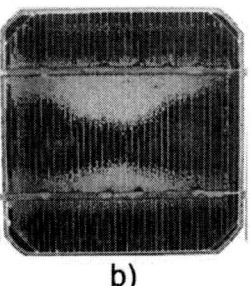

Fig. 4. a) Small delamination in grid lines and fingers in temperate climate and b) delamination and corrosion for Hot-Humid climates.

C. Backsheet delamination and chalking

Delamination and chalking of the backsheet were not a very common degradation mode. 11% of the total module survey presented backsheet delamination with 11% for modules in hot-humid climates and 11% for temperate climates. However, 60% of the modules presented chalking of the backsheet with the same percentage for hot-humid and temperate climates. This can indicate that this degradation mode is related to the quality of the material and not necessarily climate dependent.

D. Hotspot formation and other visual defects found

Hotspot can be caused by thermal expansion/contraction of interconnects or solder bonds, intermetallic compound formation between solder material and interconnect/metallization materials, shadowing, faulty cell or cells in a string, low shunt resistance cells and failure of bypass diode under open-circuit conditions [13]. From the survey, we found that 68% of the modules in Hot-Humid climates had hotspots, from this group 47% are M65/75 and 89% SP75. For temperate climates, 67% of the modules presented hotspots. Fig. 5 shows two of the survey modules with more than one hotspots for Hot-Humid climates and temperate climate.

978-1-5090-5606-4/17 $31.00 © 2017 IEEE

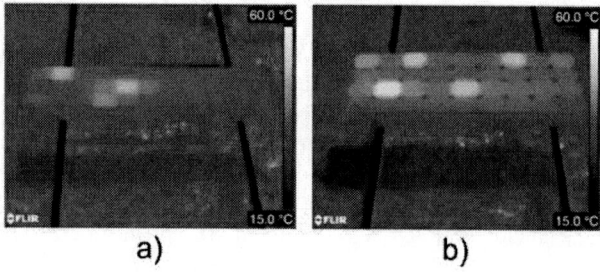

Fig. 5. IR image of two modules with multiple Hotspots. a) Temperate climate with 25 years of exposure (M65/75) and b) hot humid with 10 years of exposure (SP75).

Other important defects found in overall the modules were soiling (light and severe), cracked cells, broken glass, and so on. However, those defects were not identified frequently. Most of them are related to external failures or installations problems which are not related to module design or manufacturing quality.

IV. PERFORMANCE DEGRADATION

The degradation rates have been calculated based on measured data and nameplate data (1-point approach), and the number of years exposed. If any unrealistic negative degradation value is observed, then it is assumed that the nameplate value is underrated for those specific modules. Fig. 6 and Fig. 7 shows the degradation percentage per year of the electrical parameters Isc, Voc, Pmax and FF for all the modules.

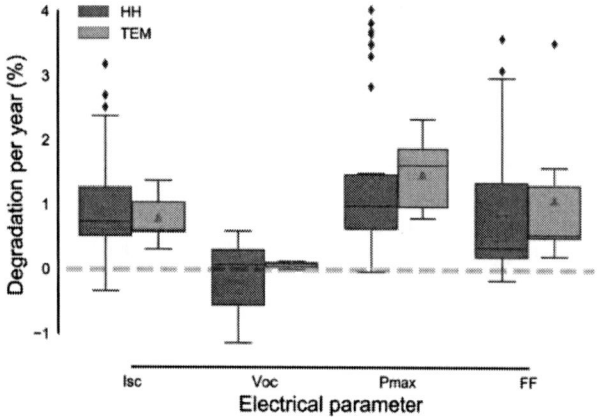

Fig. 6. Electrical degradation rate per year of the modules analyzed and grouped by climate: Hot-Humid (HH) and Temperate (TEM)

From Fig. 6 it can be observed that modules exposed to Hot-Humid (HH) climate present a degradation of Isc between -0.3% and 3.2% with an average of 1.0% per year, Voc ranges between -1.1% and 0.6% with an average of -0.1%/year and Pmax between -0.1% and 4.0% with an average of 1.4%/year. The negative values are probably caused by the underrating of the nameplate values. For Temperate climate Isc degradation

values ranges between 0.3% to 1.4% with an average of 0.8%/year, Voc between - 0.01% to 0.1% with an average of 0.1%/year, Pmax between 0.8% to 2.3% with an average of 1.5%/year.

Also, we can see that for Pmax there is a group of outliers data and it all comes from modules in a specific installation in HH climate. The degradation obtained for this site goes as high as a 2.5%/year degradation rate per year. This last degradation value may be related to the mounting configuration of the PV modules [15]. For this site, the modules were installed on the rooftop with a low air gap. The higher operating temperature of the rooftop modules under humid ambient conditions could be the main cause of the higher degradation rate as compared to the other rack mounted sites.

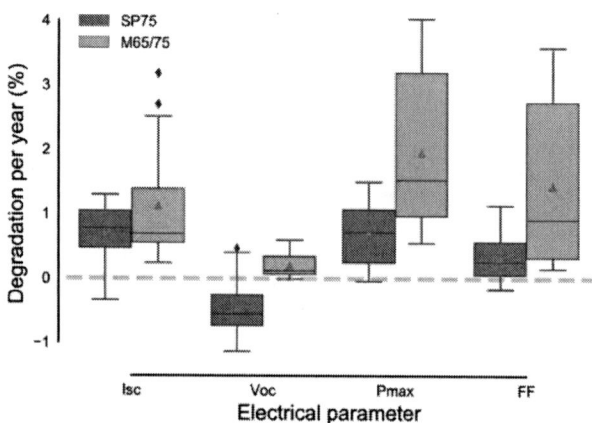

Fig. 7. Electrical degradation rate per year of the modules analyzed and grouped by manufacturer/design: m65/75 (Siemens) and SP75 (Siemens)

Fig. 7 shows that the model M65/75 has higher degradation rates mainly due to FF and Isc losses. We obtained negative degradation rates for Voc for modules SP75 which means that the nameplate used for the 1-point approach was underrated. Also, it is important to note that the degradation rate is extremely sensitive to the module design, not just the technology alone (both are mono-Si technology but two different designs from the same manufacturer).

Fig. 8 shows the degradation data grouped by years of exposure greater than 20 years and less than 20 years. It can be observed that the group of modules with more than 20 years have a wider range of degradation of Pmax that is between 0.6%/year and 4.0% with an average of 2.1%/year. The lower degradation rates for the newer designs (less than 20 years old modules) could be attributed to the design improvements.

In general, all the power losses of the PV modules surveyed are attributed to Isc and FF drops. Isc drops are mostly related to encapsulant discoloration, and FF losses are related to the increase in the series resistance due to the corrosion and/or intermetallic compound formation along with thermo-mechanical fatigue

978-1-5090-5606-4/17 $31.00 © 2017 IEEE

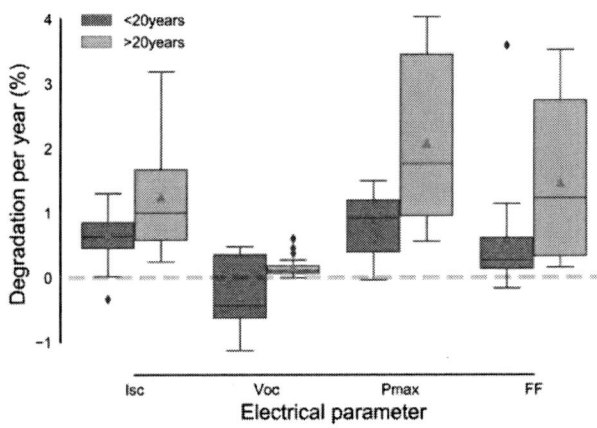

Fig. 8. Electrical degradation rate per year of the modules analyzed and grouped by time of exposure

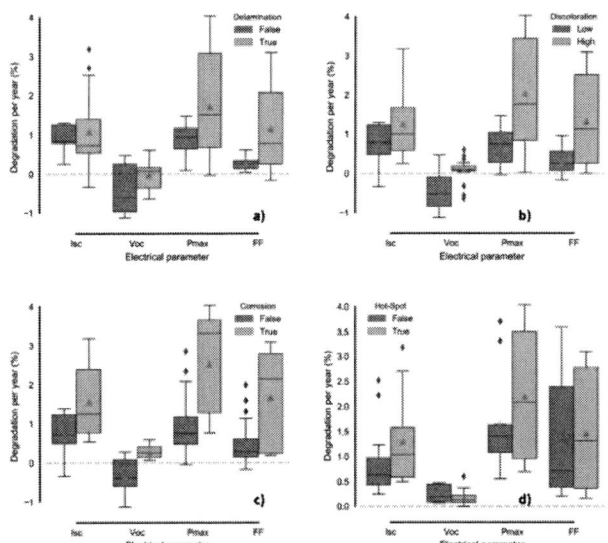

Fig. 9. Electrical degradation rate per year for the modules analyzed grouped by the presence of degradation/defects. a) Grouped by delamination, b) grouped by discoloration, c) grouped by corrosion, d) grouped by hotspot.

If we group the degradation data by visual defect, it can be observed the correlation of the visual degradation and its individual impact on the electrical parameters of the PV modules. Fig. 9 a) shows the electrical degradation rate per year for modules that presented delamination between the encapsulant and the solar cell. For PV modules with delamination, it was obtained an average degradation rate for Pmax of 1.7%/year whereas PV modules without delamination a value of 1.0%/year. This can indicate that the presence of delamination in PV modules is related to optical decoupling and the increase in series resistance, as inferred by the higher FF degradation rate.

From Fig. 9 b), it can be observed that modules with high discoloration has a bigger average degradation of Isc as well as FF than modules with low discoloration (see Fig. 2 to observe the difference of high and low discoloration). We would expect that discoloration will affect more Isc than Voc and FF. We observed that modules with high discoloration also found to be experiencing higher level of corrosion and this would explain the increase of the FF degradation and furthermore its contribution to Isc drop.

Fig. 9 c) shows the results of grouping the data if corrosion exists. It can be seen a similar behavior as the correlation of delamination and discoloration. This result is expected since all the modules surveyed presents discoloration and, in general, the corrosion is a consequence of delamination. For this group of PV modules, we obtained the highest average degradation rate for Pmax of 2.5%/year and PV modules without corrosion a value of 1.0%/year.

Lastly, Fig. 9 d) shows the results of grouping the data by the appearance of hotspots in modules M65/75. From the box plot, it can be seen that modules with hotspots had higher Pmax degradation rates than modules without them. This observation indicates that there is a correlation between hotspots appearance and the loss of Pmax.

V. CONCLUSIONS

There is a considerable increase in demand for solar PV technology in Mexico. To support the Mexican PV industry, reliability data of PV technology for the local climatic conditions is required.

In this work, an extensive evaluation of PV modules was performed using different visual, optical and electrical tools. The visual inspection indicated that out the most common degradation modes are browning, delamination, corrosion and hotspots. Using the 1-point approach, we obtained an average degradation rates for all modules of 1.4% for Hot-Humid climate and 1.5% for Temperate climate. However, the degradation rate is heavily dependent on the module design and mounting type (rack mounted vs roof mounted), not just the climate. For the rack-mounted M65/75 modules aged more than 20 years, the average degradation rate is 1.7%/year for Temperate climate. For rack-mounted aged less than 20 years, the average degradation rate of 1.3%/year for M65/75 in Hot-Humid and 0.7%/year for SP75 modules in Temperate climate. Similarly, for the roof mounted M65/75 modules aged more than 20 years, the average degradation rate is 2.5%/year for Hot-Humid and 1.0%/year for Temperate climate (see Table I for summary).

It was found that PV modules (25 years old) from temperate climates presented two levels of browning: lighter at the edges of the cell and darker at the center of the cell. However, the modules (25 years old) from hot-humid climate showed dark browning throughout the cells. These two level of browning combined with delamination and corrosion seriously affect the long-term durability of the PV modules in Mexico climates.

Table I

SUMMARY OF THE PMAX AVERAGE DEGRADATION RATE PER YEAR GROUPED BY MOUNTING CONFIGURATION, MODEL, CLIMATE AND TIME OF EXPOSURE.

Mounting	Model	Climate	<20years	>20years
Rack	M65/75	HH	1.3%	-
		TEM	-	1.7%
	SP75	HH	0.7%	-
Roof	M65/75	HH	-	2.5%
		TEM	-	1.0%

ACKNOWLEDGMENTS

This research was partially supported by CeMIE-Sol through the project P-29: *"Desarrollo de un laboratorio nacional para la evaluación de la conformidad de módulos y componentes de sistemas e instalaciones fotovoltaicas - LANEFV"*. We would like to thank the Photovoltaic Reliability Laboratory of the Arizona State University (ASU-PRL) for providing access to their facilities. Finally, we would also thank Amilcar Reyes Roldan and Jorge Luis Carnalla Ortiz for the measurements and data collection.

REFERENCES

[1] IEA, "World - Renewable and Waste Energy Statistics (Edition 2016)," 2016.

[2] CRE, "Contratos de interconexión en pequeña y mediana escala," 2015.

[3] P. Rajput, G. Tiwari, O. Sastry, B. Bora, and V. Sharma, "Degradation of mono-crystalline photovoltaic modules after 22years of outdoor exposure in the composite climate of India," *Solar Energy*, vol. 135, pp. 786–795, 2016.

[4] S. Chandel, M. N. Naik, V. Sharma, and R. Chandel, "Degradation analysis of 28 year field exposed mono-c-Si photovoltaic modules of a direct coupled solar water pumping system in western Himalayan region of India," *Renewable Energy*, vol. 78, pp. 193–202, 2015.

[5] S. Chattopadhyay, R. Dubey, V. Kuthanazhi, J. J. John, C. S. Solanki, A. Kottantharayil, B. M. Arora, K. L. Narasimhan, V. Kuber, J. Vasi, A. Kumar, and O. S. Sastry, "Visual Degradation in Field-Aged Crystalline Silicon PV Modules in India and Correlation With Electrical Degradation," *IEEE Journal of Photovoltaics*, vol. 4, pp. 1470–1476, 2014.

[6] J. M. Kuitche, R. Pan, and G. TamizhMani, "Investigation of Dominant Failure Mode(s) for Field-Aged Crystalline Silicon PV Modules Under Desert Climatic Conditions," *IEEE Journal of Photovoltaics*, vol. 4, pp. 814–826, 2014.

[7] D. C. Jordan, B. Sekulic, B. Marion, and S. R. Kurtz, "Performance and Aging of a 20-Year-Old Silicon PV System," *IEEE Journal of Photovoltaics*, vol. 5, pp. 744–751, 2015.

[8] M. Munoz, M. Alonso-García, N. Vela, and F. Chenlo, "Early degradation of silicon PV modules and guaranty conditions," *Solar Energy*, vol. 85, pp. 2264–2274, 2011.

[9] D. C. Jordan and S. R. Kurtz, "Photovoltaic Degradation Rates-an Analytical Review," *Progress in Photovoltaics: Research and Applications*, vol. 21, pp. 12–29, 2011.

[10] D. Berman and D. Faiman, "EVA browning and the time-dependence of I-V curve parameters on PV modules with and without mirror-enhancement in a desert environment," *Solar Energy Materials and Solar Cells*, vol. 45, no. 4, pp. 401–412, 1997.

[11] C. Osterwald, A. Anderberg, S. Rummel, and L. Ottoson, "Degradation analysis of weathered crystalline-silicon PV modules," *in 29th IEEE Photovoltaic Specialists Conference*, pp 1392–1395, 2002.

[12] A. Gxasheka, E. van Dyk, and E. Meyer, "Evaluation of performance parameters of PV modules deployed outdoors," *Renewable Energy*, vol. 30, pp. 611–620, 2005.

[13] J. Kuitche and G. TamizhMani, "Accelerated lifetime testing of photovoltaic modules solar america board for codes and standards," *A report of Solar America Board for Codes and Standards*, 2013.

[14] L. A. Escobar and W. Q. Meeker, "A Review of Accelerated Test Models," *Statistical Science*, vol. 21, 552–577, 2006.

[15] V. Sharma, O. Sastry, A. Kumar, B. Bora, and S. Chandel, "Degradation analysis of a-Si (HIT) hetro-junction intrinsic thin layer silicon and m-C-Si solar photovoltaic technologies under outdoor conditions," *Energy*, vol. 72, pp. 536–546, 2014.

Rapid Shutdown with Panel Level Electronics - A suitable safety measure?

Adam Cordova, Christopher Merz, Gerd Bettenwort, Markus Hopf, Hannes Knopf, Joachim Laschinski

SMA Solar Technology AG, 34266 Niestetal, Germany

Abstract — In the last years, requirements for a rapid shutdown of the PV generator at string level during an emergency came up. In order to comply with the new requirements, a variety of vendors entered the market with technical solutions. Contrary to intentions, an increase of serious emergency situations, even fires [1], had to be observed. The recent NEC 2017 [2] requires, amongst others, a rapid shutdown on module level. This tightens the situation because it leads to even more components in the PV system using Panel Level Electronics (PLE). Hence, this paper will discuss the harsh operating conditions and system failures that can occur within a PV plant and which directly affect the design of PLE.

I. INTRODUCTION

Looking at the PV plants of the last 30 years it becomes quiet clear that the safety philosophy has changed together with technology. In the course of this evolution the system technology developed from low voltage solar fields operated by experts and researchers to the actual 1,000 V module strings installed and automatically operated in the household of normal residential customers. Starting with almost no preventive measures the PV plants now are equipped with many electrical safety precautions: Double electrical insulation, residual current protective device, DC disconnector, etc.. Additionally during the last few years hazard sources came into focus which could cause thermal problems (electrical arcs) as well as minimization of the risk potential for firefighters in an emergency case.

This results in the short term introduction of new regulations for PV plants and BOS components in several countries. Other than intended, this actually led to an increase of serious emergency situations, even fires (see Fig. 1).

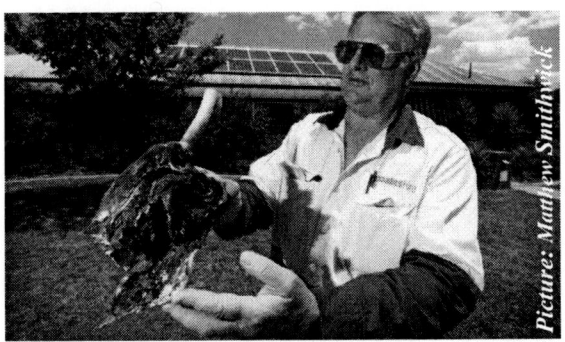

Fig. 1: DC isolators were recalled in Australia in May 2014 [3]

This shows that the integration of single safety measures would not increase the plant's safety necessarily! This paper deals with the improvement of plant's safety by an approach considering the complete system technology including operational experience.

Purpose of this paper is to identify chances and challenges of PLE for safety functions. Different from other studies focused on a specific product this investigation is targeted on the functions PLE could offer for the complete PV plant. Beyond that the work is based on experience of nearly 30 years of PV plant operation in a large range of countries all over the world with different local safety rules. Amongst others the paper addresses the following questions:

- What are the real safety risks of a PV plant for installers, operators, emergency personnel and for the building itself?
- What data exists to demonstrate the status of a PV plants safety presently?
- Is it possible to identify the main causes affecting the safety of a PV plant?
- Which safety guidelines are or will be applied in different countries (in particular NEC 2017) and what is the experience up to now?
- What's the impact on reliability and availability of PV plants with PLE during its expected lifetime?

The results of the analysis above are used to identify the reason of safety risks as well as appropriate technical measures for safe operation. It takes into account not only different technical solutions but also its handling and its impact on operation and maintenance. This basic approach and the broad scope should be of high interest for all stakeholders in PV manufacturing and application. All conclusions derived from this evaluation are generally admitted and are transferable to any PV application. This also supports the work of technical committees concerned with the safety of PV plants in different countries.

II. MAIN FINDINGS

A. Normative Requirements

The requirement for Rapid Shutdown of PV generators on module level is attended by the recent publication of the NEC

2017. Figure 2 shows a drawing of the most important requirements regarding voltage limits and boundaries.

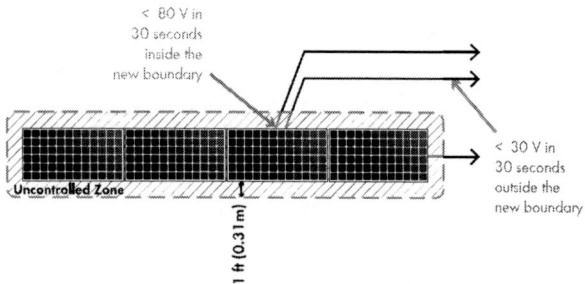

Fig. 2: Voltage limits and boundaries of the NEC 2017, Section 690.12 (B) (2) according to [4]

In summary the voltage within a PV generator is limited to module level (80 V) in case of an emergency shutdown. This requirement automatically leads to the need for electronic circuits and switches in every junction box: Panel Level Electronics (PLE).

In addition to this well-known requirement, the NEC 2017 Section 690.12 (B) offers two alternatives to the use of PLE. First, the PV array shall be listed or field labeled as a rapid shutdown PV array. Thus, it should be installed and used in accordance with the instructions included with this listing or field labeling. Secondly, "PV arrays with no exposed wiring methods, no exposed conductive parts, and installed more than 2.5 m (8 ft) from exposed grounded conductive parts or ground shall not be required to comply with 690.12(B)(2)" [2]. Therefore, the local conditions have to be considered in order to achieve the most suitable technical solution for every PV plant design beyond the use of PLE (see Fig. 3).

Fig. 3: Safety has to start with a clear plant layout

B. Safety of present PV plants

Motivation and main argument for the new normative requirements is the protection of firefighters from being exposed to high string voltages during rescue operation. Doubts about those arguments come up if relevant statistics on safety issues of PV plants are considered. A Fraunhofer ISE study [5] provided statistics relating to PV-related fires and the number of fatalities in Germany over the last 20 years. According to this study only 0.006 % of all PV plants have caused a fire resulting in serious damage. Furthermore, no firefighter has been injured by PV power so far while putting out a fire. These numbers are even more considerable due to the fact, that in Germany not even a string level rapid shutdown is common practice.

Following best practice while extinguishing a fire in an electrical system seems to be very efficient to protect first responders. According to a study from TÜV Rheinland and Fraunhofer ISE [6] there are minimum distances between the fire hose and a 1,500 Vdc electrical system to ensure a safe work of the firefighters (see Table I).

TABLE I: MINIMUM DISTANCES BETWEEN FIRE HOSE AND AN ELECTRICAL DC-SYSTEM WITH 1,500 V ACCORDING TO [6]

Water jet type	Minimum distance
Spray water jet	1 m
Full water jet	5 m

The statistics and examples of best practice mentioned above indicate that conventional PV systems with string technology are safe and can be managed in an emergency.

Compared to the conventional string technology, a rapid shutdown at module level using PLE will automatically lead to a higher number of components in the PV system. Hence, an accurate analysis of the operating conditions and system failures is mandatory to assess the possible risks of PLE concerning the PV plant.

C. System failures and operating conditions

Those operating conditions and system failures have been defined and analyzed in the public funded project "PV-Firebreaker" [7] under the guidance of TÜV Rheinland with collaboration of SMA et al. (see Table II).

TABLE II: OPERATING CONDITIONS & SYSTEM FAILURES ACCORDING TO [7]

Operating conditions
extended environmental conditions
reverse and bypass currents due to shading
capacitive switching currents
asynchronous switching
ageing
System failures
ground faults within the PV generator
short circuits within the PV generator
installation faults (i.e. reverse connection)
lightning discharge

The current national and international standards concerning switching components are lacking some of those PV specific challenges. Hence, new test criteria have been developed by the public funded project "PV-Firebreaker" taking into account those operating conditions and system failures.

The next step would be to improve the existing standards with these experience and new test criteria. This task is now up to the relevant standardization committees.

D. Reliability & Availability

Renewable Energies already plays an important role at least in some countries. According to PV Market Alliance [8] there was an installed worldwide PV capacity of more than 300 GWp by the end of 2016. It is obvious, that PV will become a fundamental part of the future energy supply.

To ensure the availability of electricity, a high reliability of all components is essential comparable to the quality of conventional power stations. With respect to the PV plant this means that every part of the PV system has to fulfill the reliability targets and the usual service life.

Concerning PLE, there are some key factors in the design and application to achieve those reliability and durability targets.

The PLE has to be well designed and technically mature. This results from the harsh operating conditions and system failures that have been mentioned above. It is absolutely necessary, that those PV specific requirements are considered during the design phase to ensure a reliable operation of the PLE.

SMA internal research showed, that the environmental conditions at the PV generator and thus for PLE are comparable to those in automotive applications, which usually set a high standard concerning reliability. Thus, a low component count could help to reduce the failure rate of the device. This can be achieved by a modular design which is tailored to the specific requirements of the PLE. For instance, a pure switching device doesn't necessarily need an optimizing function.

Finally, high durability is achieved if active parts of the PLE stay inactive most of the time. Power optimizers with buck converter topology only have to perform buck converting if the corresponding PV modules are shaded, while other parts of the PV generator see full insolation. At virtually homogeneously insolation, all buck converters switch off. Thus, the power loss is reduced to a minimum.

III. SUMMARY

The experience of 20 years [5] shows that conventional PV systems with string technology are safe and can be managed even in an emergency. The consequent use of best practice allows a safe work of first responders while extinguishing a fire.

Current standards often don't consider the operating conditions and system failures that are specific to PV installations. Panel Level Electronics have to withstand and manage those conditions and failures to allow a safe and reliable operation.

On the basis of the mentioned operating conditions and system failures test criteria have been defined [7]. Those test criteria have to be fixed in future test standards to help the designers of PLE to develop safe and reliable devices.

PLE has to be well designed and technically mature with the lowest component count and low power loss to reach the reliability and durability targets.

REFERENCES

[1] Australian Competition & Consumer Commission: *NSW: Recall of Solar Isolators*, May 20, 2014; https://www.productsafety.gov.au

[2] National Fire Protection Association (NFPA): *National Electrical Code Handbook*; 14th Edition 2017

[3] Thalia McPherson: *"Why weren't we warned?"*, The Border Mail, Feb. 6, 2015; http://www.bordermail.com.au

[4] SMA Solar Technology AG, *Evaluating the Case for Module-Level Shutdown*, 2015

[5] Fraunhofer ISE: *Recent facts about Photovoltaics in Germany*, 2015

[6] TÜV Rheinland, Fraunhofer ISE: *Assessment of the fire risk in photovoltaic systems and elaboration of safety concepts for minimization of risks*, 2015

[7] TÜV Rheinland et. al.: *PV-Firebreaker, final report (FKZ 0325596)*, 2016

[8] Sam Pothecary : *"PVMA figures show 75 GW of solar PV was installed in 2016"*, PV magazine, Jan. 19, 2017, https://www.pv-magazine.com

Investigating a New Operating Point For PV Panels Seeking Maximum Life Span

Bechara Nehme[1, 2], Nacer K. M'Sirdi[2], Tilda Akiki[1]

[1]Department of Electrical and Electronics Engineering, Faculty of Engineering, Holy Spirit University of Kaslik (USEK), B.P. 446 Jounieh, Lebanon

[2]Aix Marseille Université, CNRS, ENSAM, Université de Toulon, LSIS UMR 7296, 13397, Marseille France.

Abstract — **In this paper we try to control the operation mode of PV panels to achieve a higher lifespan. In fact, PV panels are subject to several degradation modes: Potential Induced Degradation, Light Induced Degradation, UltraViolet Degradation, Moisture Induced Degradation, and cell cracks. The common factor that affects all the above degradation modes is temperature. In fact high temperatures lead to higher degradation process. In this paper we present an enhanced thermal model of PV panels. We simulate its behavior as a function of its operation. We define a new operating point to achieve a higher lifespan.**

Index Terms — **degradation modes, temperature effect, thermal model, joule effect.**

I. INTRODUCTION

PV (Photovoltaic) panels share in energy production has increased. Their reliability and efficiency consist of two important issues among investors and photovoltaic specialists.

Our previous works focused on reviewing and modeling degradation modes of PV panels [1, 2]. PV panels are subject to **PID (Potential Induced Degradation)**, **LID (Light Induced Degradation)**, **UVD (UltraViolet Degradation)**, **MID (Moisture Induced Degradation)**, and cell cracks. The above mentioned degradation modes are affected by external environmental conditions like: irradiance, relative humidity, and temperature. The common factor is temperature. Besides, these degradations follow Arrhenius equations; they depend exponentially on temperature.

In literature researchers had developed thermal models of PV panels. A. D. JONES and al. developed a thermal model of PV panels based on energy balance [3]. G.M. Tina and al. added the effect of the operating voltage or the extracted power on the thermal energy balance of a PV panel [4]. Ruhi Bharti and al. conducted experiences that showed the effect of the electrical load on the **NOCT (Nominal Operating Cell Temperature)** [5]. M. Rosa-Clot and al. studied a hybrid thermal-PV panel; the **TEPSI (Thermal Electric Solar Panel Integration)** [6].

In this paper, we build a more precise thermal model of PV panels in which we take into consideration the effect of electric power generated by the system and the joule effect of the series and shunt resistors of the equivalent circuit. The joule effect is directly linked to the operation point of PV panels. Precisely it depends on the square of the current output and the square of the operational voltage. In the present work we suggest an operating point that leads to decreasing the temperature of PV panels thus mitigating their degradation.

In the following section we will recall the equivalent circuit of PV cell. Then we will develop a thermal model of a PV module. At the end simulation and experimental results will be shown.

II. EQUIVALENT CIRCUIT OF A PV PANEL

A PV cell is modeled as a current source (I_{SC}) representing the photo-generated current and depending on the irradiance intensity. Two diodes D1 and D2 are mounted in inverse parallel with the current source and represent the recombination current happening in the neutral region and in the depletion region. A parallel shunt resistor (R_{sh}) models the leakage current. All the above are connected in series with a resistor (R_s) that represents the resistance of the front grid and of the interconnects. Having a panel constituted of n series cells, their series resistances will be added to have an equivalent series resistance of $n \times R_s$.

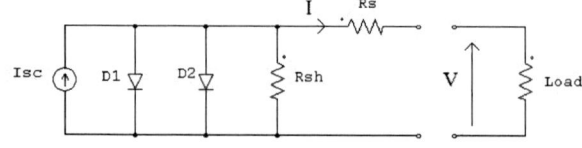

Fig. 1.   Electric equivalent circuit of a PV cell.

The output current of the PV cell is given by the following formula:

$$I = I_{SC} - I_{01} \times \left(\exp\left(\frac{q \times (V_c + R_s \times I)}{K_B \times T_{pv}} \right) - 1 \right)$$

$$-I_{02} \times \left(\exp\left(\frac{q \times (V_c + R_s \times I)}{2 \times K_B \times T_{pv}} \right) - 1 \right) - \frac{V_c + R_s \times I}{R_{sh}}$$

Where:

I_{SC} : is the photo-generated current in A.

I_{01} : the dark saturation current (in neutral region) in A.

I_{02} : the dark saturation current (in depletion region) in A.

R_s : series resistance in Ω.

R_{sh} : shunt resistance in Ω.

V_c : the output voltage of the cell in V.

q : electron charge $1.602 \times 10^{-19} C$.

K_B : Boltzmann constant $1.3806488 \times 10^{-23} m^2.kg.s^{-2}.K^{-1}$.

III. THERMAL MODEL OF A PV PANEL

A PV panel resembles a flat plate that receives solar irradiation and generates electrical power. The incident irradiation is reduced due to reflection. The electrical output power generated by the PV panel depends on its operating voltage. In fact, the P-V characteristic is non linear and presents a mountain shape. The electric output power represents a sink to the thermal system model. The output electric power of a PV panel is given by the following formula:

$$P_{ele} = V \times I \leq 0.18 \times P_{sol} \qquad (1)$$

Where:

P : electric output power in W.

V : operating voltage of the panel in V.

I : current of the panel in A.

P_{sol} : total solar power in W.

The conversion efficiency of conventional PV panels is limited to 18%. The remaining input power will generate heat to the panel. This heat will be dissipated to the exterior via convection and radiation [7]. The conduction is neglected because the panel nearly presents no contact to solids.

The thermal convection flux to ambient air is expressed by the following formula:

$$\dot{Q}_{conv} = h \times (T_{amb} - T_{pv}) \qquad (2)$$

Where:

$\dot{Q}_{conv}$: convective heat flux in J.m^{-2}.

h : heat transfer coefficient in W.m^{-2}.K^{-1}.

T_{amb} : ambient temperature in K.

T_{pv} : PV panel temperature in K.

The heat transfer coefficient depends on many parameters: The wind velocity, the air density, the tilt angle and the dimensions of the panel. The latter parameters are hard or costly to control; this defends our strategy in controlling the electrical operating point of the panel.

The thermal radiation flux is expressed by the following formula:

$$\dot{Q}_{rad} = \varepsilon \times \sigma \times (T_{amb}^4 - T_{pvs}^4) \qquad (3)$$

Where:

$\dot{Q}_{rad}$: radiative heat flux in J.m^{-2}.

ε : emissivity.

σ : Stephan-Boltzmann constant 5.67×10^{-8} W.m^{-2}.K^{-4}.

T_{pvs} : PV panel surface temperature in K.

The front surface emissivity is taken as 0.9, the back surface emissivity is taken as 0.84 [3].

In our model we add the effect of the series resistance and the shunt resistance that heat the PV panel by the Joule effect. The internal heat sources of the panel are due to the Joule effect of the series and shunt resistors. In fact, current passing through the printed Ag fingers and through the Cu bus-bars will lead to a localized heat generation modeled by the series resistor R$_s$. Besides, current passing through a path around the cell will lead to a localized heat generation modeled by the series resistor R$_{sh}$. The joule effect is given by the following formula:

$$P_j = n \times R_s \times I^2 + n \times \frac{\left(\frac{V}{n} + R_s \times I \right)^2}{R_{sh}} \qquad (4)$$

Where:

P_j : Joule's power in W.

n : number of cells.

The PV panel receives two heat sources: The incident irradiance and the internal joule generated heat. The panel generates electrical power to the load and dissipates power by convection and radiation. The thermal energy balance of a PV panel can be illustrated in figure 2.

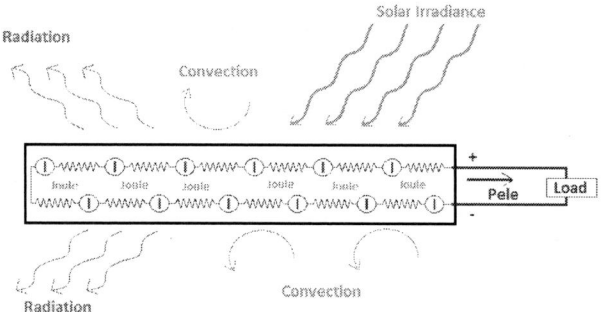

Fig. 2.   Thermal energy balance of a PV panel.

IV. SIMULATION RESULTS

We built a thermal model of a PV panel using Comsol software. The wind velocity is set to $1m.s^{-1}$. The panel is horizontal. The panel dimensions are 1.2m x 0.54m x 0.01m. It is constituted of 36 cells in parallel. Under STC, it can deliver 5.5A and 100W. The radiation heat dissipation happens from the front and back surfaces. The emissivity of the front surface is taken as 0.9 and that of the back surface is taken as 0.84. The operating voltage of the panel was swept from 2V to 21V with a step of 0.1V. The input power irradiance is swept from 100 $W.m^{-2}$ to 1100 $W.m^{-2}$ with a step of 100$W.m^{-2}$. For each operating point, the temperature is calculated. Results are shown in Fig. 4.

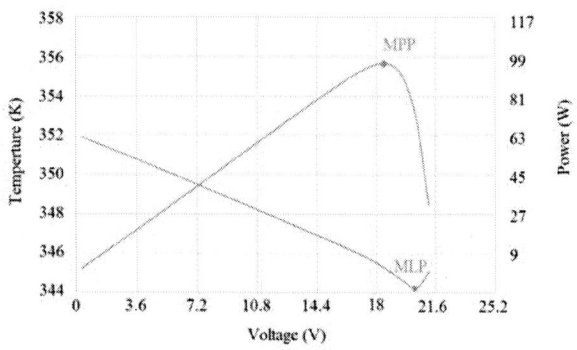

Fig. 3.   Output power in orange and panel's temperature in blue.

We can see that the minimum temperature is reached for a voltage of 20.52V even though the **MPP (Maximum power point)** is located at 18.36V.

We locate a point of operation that will operate the panel at a lower temperature; thus giving longer life span. We call this point the **MLP (Maximum Life Point)**. This point is always at a higher voltage than the MPP. Comparing the MLP with the MPP we can see a reduction of around 19% in power production and a decrease of around 1.5 degrees in temperature.

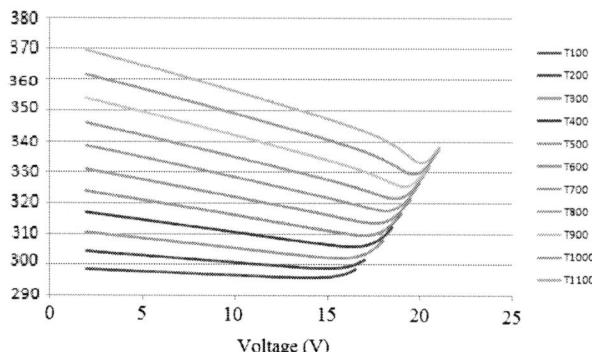

Fig. 4.   Panel temperature for different irradiations and operating points.

We can see from figure 4 that our approach works fine for different illumination levels. From here future perspectives derive to choose when to operate at MPP or at reduced temperature.

Additional information is gathered from the developed model. In fact, a small difference in temperature is noted between the center of the panel and the edge of the panel. Temperature is higher at the center of the panel. The latter is highlighted in figure 5 where we show the temperature repartition at the front surface of the panel.

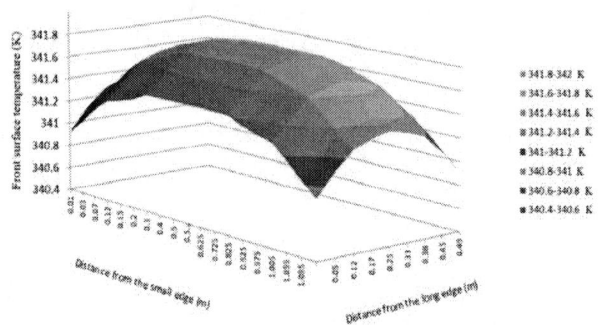

Fig. 5.   Front surface temperature of the Panel.

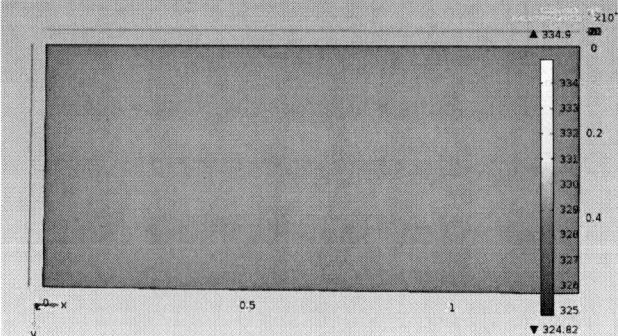

Fig. 6.   Intercell and cell temperature at the front surface.

In addition, cells regions presented higher temperature than intercell regions. This is explained by the fact that in cell regions, a localized heat source is present (R_s and R_{sh}). The results are shown in figure 6.

V. EXPERIMENTAL RESULTS

In order to validate the simulation results we built an experimental apparatus. The indoor apparatus can emulate the external conditions and record the temperature of the panel. We built a PV system composed of a 100W PV panel (Monochrystalline, Solar Innova SI-ESF-M-M90-100W), halogen projectors, resistive load, DC/DC converter, and back surface mounted thermocouples. Two thermocouples are used to measure the temperature of the panel. Their average temperature is noted. The DC/DC converter is used to vary the operating voltage of the panel. A variation in the operating voltage gave the following results (figure 7) that matched our previous simulation results:

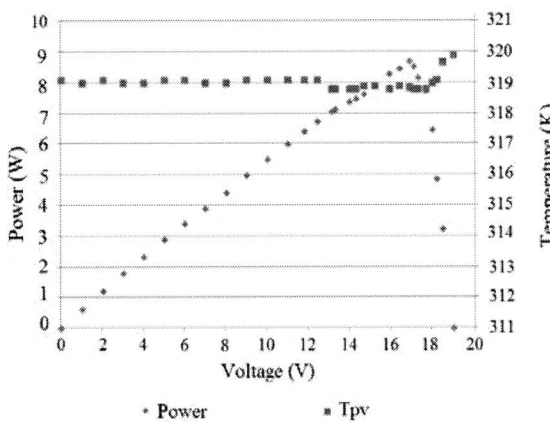

Fig. 7.   Experimental output power in green and panel's temperature in red.

A second experimentation was performed with variable wind speed. Two fans were implemented. The ambient temperature is 27°C, and the relative humidity is 92%. The results show the same thermal behavior. Figures 8, 9, and 10 present the experimentation results with low, medium, and high wind speeds respectively. We can see that the temperature of the back of the cell (points in red) is higher than the temperature of the back of the spaces between the cells (points in green). The latter aligns with the simulation results shown in figure 6.

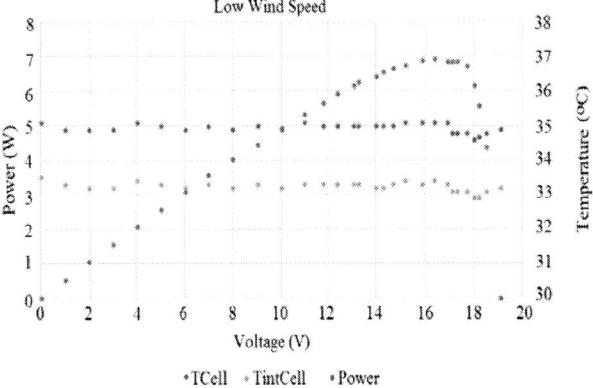

Fig. 8.   Thermal behavior of the PV panel as a function of the operating voltage with low wind speed.

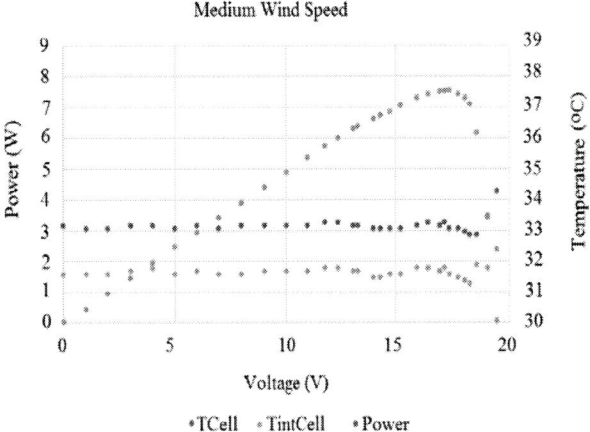

Fig. 9.   Thermal behavior of the PV panel as a function of the operating voltage with medium wind speed.

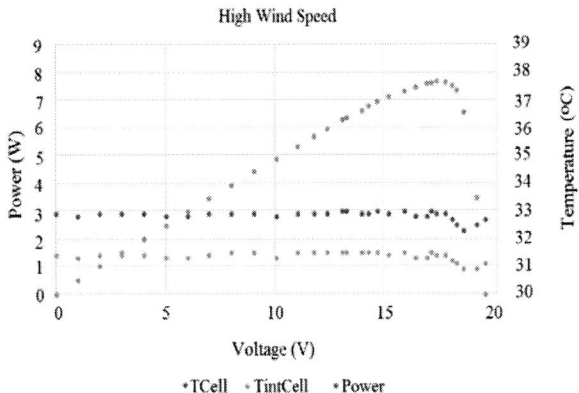

Fig. 10.   Thermal behavior of the PV panel as a function of the operating voltage with high wind speed.

IV. DISCUSSION

In this paper, we presented a new thermal model of PV panels. To the existing models developed in literature we added the effect of internal heat sources emerging from the series and shunt resistors. The model has been validated using an experimental apparatus. The apparatus needs to be enhanced to get higher irradiance, and the equivalent temperature of the lights must meet solar spectrum.

A new point (MLP) of operation of PV panels has been identified where the panel will operate at lower temperature which yields lower degradation rates and thus higher lifespan of PV panels.

IV. CONCLUSION

Degradation of PV panel has become a major concern for investors. In this paper we reached to find an operating point that will lead to increasing the life span of PV panels. In reality, degradation is exponentially affected by temperature. In our study we developed a thermal model for PV panels taking into consideration the joule effect of the series resistor. A point located at the right of the MPP corresponds to our investigation. In future works we will develop an algorithm which objective is to alleviate the degradation process of PV panels without losing high power generation.

IV. ACKNOWLEDGMENT

The authors would like to thank the **AUF (Agence Universitaire de la Francophonie)** and the **HCR (Higher Center of Research)** of USEK University for their financial support.

REFERENCES

[1] B. Nehme, N. K. M'Sirdi, T. Akiki, and A. Naamane, "Contribution to the modeling of ageing effects in pv cells and modules," *Energy Procedia*, vol. 62, pp. 565-575, 2014.

[2] B. Nehme, N. K. M'Sirdi, and T. Akiki, "A geometric approach for pv modules degradation," to be published in *REDEC-2014*.

[3] Jones A. D., and C. P. Underwood, "A thermal model for photovoltaic systems," *Solar Energy EDEC-2014*, vol. 70.4, pp. 349-359, 2001.

[4] Tina GM, and Scrofani S, "Electrical and thermal model for PV module temperature evaluation," *Electrotechnical Conference, 2008. MELECOM 2008. The 14th IEEE Mediterranean.*

[5] R. Bharti, J. Kuitche, and M. G. TamizhMani, "Nominal operating cell temperature (NOCT): effects of module size, loading and solar spectrum," *Photovoltaic specialists conference, 2009. 34th IEEE PVSC2009,* p. 001657.

[6] M. Rosa-Clot, P. Rosa-Clot, and G. M. Tina, "TEPSI: Thermal Electric Solar Panel Integration," *Solar Energy*, vol. 85, pp. 2433-2442, 2011.

[7] F. M. White, *Heat and Mass Transfer*. USA: Addison Wesley, 1991.

Power Generation Evaluation of Large-Scale Photovoltaic Systems Located on Inclined Plane

Naotaka Oka[1], Yasuhito Takahashi[1], Koji Fujiwara[1], Kazuyuki Hidaka[2], and Hiroshi Morita[2]

[1]Doshisha University, 1-3 Tataramiyakodani, Kyotanabe, Kyoto 610-0321, Japan

[2]Kinden Corporation, 3-1-1 Saganakadai, Kizugawa, Kyoto 619-0223, Japan

Abstract — The authors first describe the modelling of a photovoltaic module (PV) based on the experimental data. Then, the models for estimating the irradiance on inclined plane are compared. In addition, discussed is the amount of the power generation of PV systems located on undulating landscape. As a result, the optimal combination for estimating the irradiance on inclined plane is Erbs model for decomposition model and Perez model for transposition model. Moreover, the amount of the power generation is influenced by a difference of short circuit current of the PV module. This is caused by a difference of the plane-of-array irradiance.

Index Terms — decomposition model, large- scale photovoltaic system, plane-of-array irradiance, power generation evaluation, transposition model.

I. INTRODUCTION

The main point of this study is to examine the effect of the irradiance on photovoltaic (PV) array with various geographical conditions. The demand of PV systems in Japan is increasing thanks to Feed-in Tariff Scheme; however, due to its limited national land and a mountainous region, there is not enough flat area to install large-scale PV systems more. Thus, as a solution, PV systems installed along the surface of the mountain have come up.

Generally, PV arrays face the south on a flat area. However, when a PV system is located on undulating landscape, it may not face the south depending on its installation angle. As a result, a difference of the installation angle of PV arrays could decrease the amount of the power generation.

There are numerous previous research on estimating plane-of-array (POA) irradiance [1]–[3]. However, an estimation model for POA irradiance suitable for Japanese climate and geographical conditions has not been deeply discussed. In this paper, the influences of output differences in PV arrays by the POA irradiance are examined. For the examination, current vs. voltage (*I-V*) and power vs. voltage (*P-V*) characteristic of the thin film PV module are modelled. Next, the models for estimating the POA irradiance such as decomposition models and transposition models are compared. Lastly, the amounts of the power generation on the PV systems located at hilly area are examined.

II. MODELLING PHOTOVOLTAIC MODULE

The *I-V* characteristic of a thin film PV module during the light irradiation can be calculated with the following equation:

$$I = I_{ph} - I_o \left\{ \exp \frac{q(V + R_s I)}{nmk(T + 273.15)} - 1 \right\} - \frac{V + R_s I}{R_{sh}}. \quad (1)$$

Here, *I* is the output current of a PV module, *V* is its terminal voltage. The quantities q, k, m, n and T denote the electronic charge (1.6×10^{-19} C), the Boltzmann constant (1.38×10^{-23} J/K), the number of serially-connected cells in a PV module, the diode ideality factor and the PV module temperature, respectively. The *I-V* characteristic of a PV module is calculated by using four parameters: the photoelectric current I_{ph}, the reverse saturation current I_o, the series resistance R_s, and the shunt resistance R_{sh}. These parameters are defined as functions of irradiance and PV module temperature. In addition, the value of n is set as 1.5 in this paper.

Fig. 1 shows the comparison between the measurement and simulation values of *I-V* and *P-V* characteristic of the thin film PV module. The specifications of the thin film PV module used for the study are the short circuit current (I_{sc}) = 2.2 A, the open circuit voltage (V_{oc}) = 112.0 V, and the maximum power (P_m) = 170.0 W. From Fig. 1, the measurement and simulation values agree well. Moreover, a difference of the maximum power point is 0.27 %. The above results enable us to confirm sufficient reproducibility of *I-V* and *P-V* characteristic to evaluate the amount of the power generation on PV systems.

III. COMPARISON OF ESTIMATION MODELS FOR IRRADIANCE ON INCLINED PLANE

Though the best method of measuring POA irradiance is setting the pyranometer on the inclined plane, it is difficult to measure it on all the inclined planes at PV systems located on undulating landscape. Thus, it is necessary to estimate the POA irradiance from the global horizontal irradiance (GHI). To estimate the POA irradiance from the GHI, it needs to be decomposed into the direct normal irradiance (DNI) and the diffuse horizontal irradiance (DHI) at first. Then, three components of POA irradiance: the direct irradiance incident

978-1-5090-5606-4/17 $31.00 © 2017 IEEE

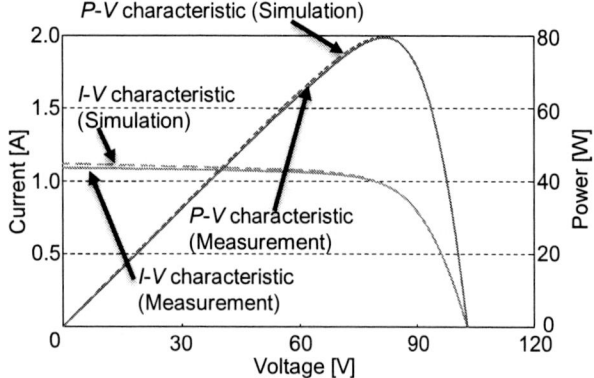

Fig. 1. Comparison of measurement and simulation values of *I-V* and *P-V* characteristic of thin film PV module used in this study.

on POA ($I_{b\beta\gamma}$), the ground reflected diffuse irradiance ($I_{r\beta\gamma}$), and the sky diffuse irradiance on POA ($I_{s\beta\gamma}$) need to be derived from DNI and DHI. Finally, POA irradiance is derived by summing up the three components. Here, the models for deriving POA irradiance are compared to clarify the best model for accurately estimate the POA irradiance.

A. Decomposition model

Currently, Erbs [4], Orgill and Hollands [5], and Reindl [6] models are used as suitable models for the decomposition of the GHI into the DNI and the DHI [7]. In addition, Udagawa and Kimura model [8], which is known for the suitable decomposition model in Japan [9], is also considered for the comparison. Each model is explained below.

In Erbs model, the DHI is derived by the following equations:
interval: $K_t \leq 0.22$

$$I_d = I_g(1.0 - 0.09K_t), \qquad (2)$$

interval: $0.22 < K_t \leq 0.80$

$$I_d = I_g(0.9511 - 0.1604K_t + 4.388K_t^2 - 16.638K_t^3, \qquad (3)$$
$$+ 12.336K_t^4)$$

and interval: $K_t > 0.80$

$$I_d = 0.1651I_g. \qquad (4)$$

In Orgill and Hollands model, the DNI is derived by the following equations:
interval: $K_t < 0.35$

$$K_d = 1.0 - 0.249K_t, \qquad (5)$$

interval: $0.35 \leq K_t \leq 0.75$

$$K_d = 1.577 - 1.84K_t, \qquad (6)$$

and interval: $K_t > 0.75$

$$K_d = 0.177, \qquad (7)$$

$$I_b = \frac{I_g(1 - K_d)}{\sin h}. \qquad (8)$$

In Reindl 2 model, the DHI is derived by the following equations:
interval: $0 \leq K_t \leq 0.30$

$$I_d = I_g(1.020 - 0.249K_t + 0.0123\sin h), \qquad (9)$$

interval: $0.30 < K_t < 0.78$

$$I_d = I_g(1.400 - 1.749K_t + 0.177\sin h), \qquad (10)$$

and interval: $K_t \geq 0.78$

$$I_d = 0.486K_t - 0.182\sin h. \qquad (11)$$

In Udagawa and Kimura model, the DNI is derived by the following equations:

$$K = Q(0.5163 + 0.333\sin h + 0.00803\sin^2 h) \qquad (12)$$

interval: $I_g < K$

$$I_b = I_e(2.277 - 1.258\sin h + 0.2396\sin^2 h)K_t^3, \qquad (13)$$

and interval: $I_g \geq K$

$$I_b = I_e(-0.43 + 1.43K_t). \qquad (14)$$

Here, K_t is the clearness index, K_d is the diffuse fraction on a horizontal surface, $\sin h$ is the sun altitude, I_g is GHI, I_b is DNI, I_d is DHI, and I_e is the extraterrestrial horizontal irradiance.

Table I shows the accuracy comparison of the Erbs, Orgill and Hollands, Reindl 2, and Udagawa and Kimura models based on root mean square error (RMSE) and mean bias error (MBE). The measurement of the DHI was taken in 2002 (the number of sample is 144,060, data on January 9, 11, February 12, March 2, 10, April 21, May 10, June 10, August 3‐ 5, September 24, October 13‐ 16, November 30, December 1‐ 2, 28‐ 30 are excluded by data loss) at the roof top of Doshisha University, Kyotanabe Campus, Yutokukan-Nishi Building (Latitude: 34° 8′ N, Longitude: 135° 8′ E). From Table I, the RMSE and MBE values obtained from the Erbs model are lower than any other models. Therefore, it can be confirmed that the estimated value from the Erbs model has the highest reproducibility to the measurement value.

B. Transposition model

To obtain the POA irradiance, $I_{b\beta\gamma}$, $I_{r\beta\gamma}$, $I_{s\beta\gamma}$ are needed to be derived by the transposition model. Commonly-used transposition models are Isotropic [10], Sandia (King) [11], Hay and Davies [12], and Perez [13] models. Sandia and Isotropic model are identical except that Sandia model derives the value of ground albedo. On this examination, the value of

TABLE I

COMPARISON OF DECOMPOSITION MODELS (ERBS, ORGILL AND HOLLANDS, REINDL 2, AND UDAGAWA AND KIMURA MODELS)

Model	Erbs	Orgill and Hollands	Reindl 2	Udagawa and Kimura
RMSE [W/m²]	131.34	135.06	139.56	169.41
MBE [W/m²]	50.57	60.58	71.38	124.12

TABLE II
COMPARISON OF TRANSPOSITION MODELS (ISOTROPIC, HAY AND DAVIES, AND PEREZ MODELS)

Model	Perez	Hay and Davies	Isotropic
RMSE [W/m^2]	31.60	51.61	49.82
MBE [W/m^2]	-12.42	-33.55	-31.01

ground albedo is assumed as a constant of 0.2 which is typically used value. Thus, Sandia model is not considered in this comparison.

By using the transposition model, $I_{b\beta\gamma}$ is calculated from the DNI with the following equation:

$$\cos\theta = \sin\delta(\sin\varphi\sin\beta - \cos\varphi\sin\beta\cos\gamma)$$
$$+ \cos\delta\cos\omega(\cos\varphi\cos\beta + \sin\varphi\sin\beta\cos\gamma) \qquad (15)$$
$$+ \cos\delta\sin\beta\sin\gamma\sin\omega$$

$$I_{b\beta\gamma} = I_b \times \cos\theta. \qquad (16)$$

Here, θ is the incidence angle of the sun of the considered plane, δ is the declination angle, φ is the latitude of the location, and ω is the hour angle. β and γ denote the tilt and azimuth angle of a PV array, respectively.

Next, $I_{r\beta\gamma}$ is calculated from the GHI with the following equation:

$$I_{r\beta\gamma} = \frac{I_g\rho(1-\cos\beta)}{2}. \qquad (17)$$

Here, ρ is the ground albedo.

As for $I_{s\beta\gamma}$, it depends on the model. For example, in Isotropic model, the distribution of the diffuse irradiance in the sky is assumed to be uniform. Contrary, in the anisotropic models such as Hay and Davies and Perez models, it is assumed to be non-uniform. In Hay and Davies model, the distribution of the diffuse irradiance is divided into two components which are sky isotropic diffuse and circumsolar diffuse. In Perez model, the distribution of the diffuse irradiance is divided into three components which are sky isotropic diffuse, circumsolar diffuse, and horizon brightening diffuse. Each model is explained below.

In Isotropic model, $I_{s\beta\gamma}$ is derived by the following equations:

$$I_{s\beta\gamma} = \frac{I_d(1+\cos\beta)}{2}. \qquad (18)$$

In Hay and Davies model, $I_{s\beta\gamma}$ is derived by the following equations:

$$I_{s\beta\gamma} = I_d\left\{\left(\frac{I_b}{I_e}\right)\cos\theta + \left(1-\frac{I_b}{I_e}\right)\frac{1+\cos\beta}{2}\right\}. \qquad (19)$$

In Perez model, $I_{s\beta\gamma}$ is derived by the following equations:

$$\cos\theta_z = \sin\delta\sin\varphi\sin\beta + \cos\omega\cos\varphi\cos\beta \qquad (20)$$
$$a = \max(0, \cos\theta) \qquad (21)$$
$$b = \max(0.087, \cos\theta_z) \qquad (22)$$

$$\varepsilon = \frac{\dfrac{(I_d + I_b)}{I_d} + \kappa\theta_z{}^3}{1+\kappa\theta_z{}^3} \qquad (23)$$

$$\Delta = \frac{I_d \times m}{I_o} \qquad (24)$$

$$F_1 = \max\left(0, F_{11}(\varepsilon) + F_{12}(\varepsilon)\Delta + F_{13}(\varepsilon)\theta_z\right) \qquad (25)$$

$$F_2 = F_{21}(\varepsilon) + F_{22}(\varepsilon)\Delta + F_{23}(\varepsilon)\theta_z \qquad (26)$$

$$I_{s\beta\gamma} = I_d\left\{(1-F_1)\left(\frac{1+\cos\beta}{2}\right) + F_1\frac{a}{b} + F_2\sin\beta\right\}. \qquad (27)$$

Here, θ_z is the solar zenith angle, κ is a constant equal to 1.041, m is the air mass, ε is the sky's clearness, and Δ is the sky's brightness. F_1 and F_2 are dimensionless coefficients denoting the degree of the circumsolar and the horizon/zenith anisotropy, respectively. F_{11}, F_{12}, F_{13}, F_{21}, F_{22}, F_{23} are coefficients and the each value is expressed on [13].

Table II shows the accuracy comparison of the transposition models using RMSE and MBE. The data were measured at inclined angle of 15° and the measurement was taken at the same place and during the same period as the decomposition model examination. Erbs model is used as the decomposition model. From Table II, Perez model shows the lowest RMSE and MBE values. Thus, it indicates that Perez model is the most suitable transposition model.

From these results, in the following examinations, Erbs model is used as the decomposition model and Perez model is used as the transposition model.

IV. EXAMINATIONS ON PHOTOVOLTAIC SYSTEMS LOCATED ON INCLINED PLANE

To observe the effect of the inclined and azimuth angles on PV arrays, the amount of the power generation were examined. Fig. 2 shows a PV array constitution in which a PV string is composed of seven serially connected PV modules with bypass diode and a PV array with 132 circuits (528 parallel connections). All PV modules are facing same direction. A fuse is connected on each circuit. In this examination, the data set named METPV (MEteorological Test data for PhotoVoltaic systems)-11, the hourly meteorological data set at 837 meteorological stations throughout Japan, are used [14] for the GHI and the air temperature data in Himeji, Japan (Latitude 34° 49′ N, Longitude 134° 40′ E). Here, the temperature of the PV module is derived with the following equation from JIS (Japanese Industrial Standards) method [15]

$$T = 18.4 \times Irr + T_a. \qquad (28)$$

Here, Irr is the irradiance [kW/m^2], and T_a is the air temperature.

First, Table III shows the annual cumulative amount of the irradiance and the power generation of the PV system shown in Fig. 2 with various azimuth angles from 0° to -90° (facing from the south to the east) in steps of 15° at inclined degree of

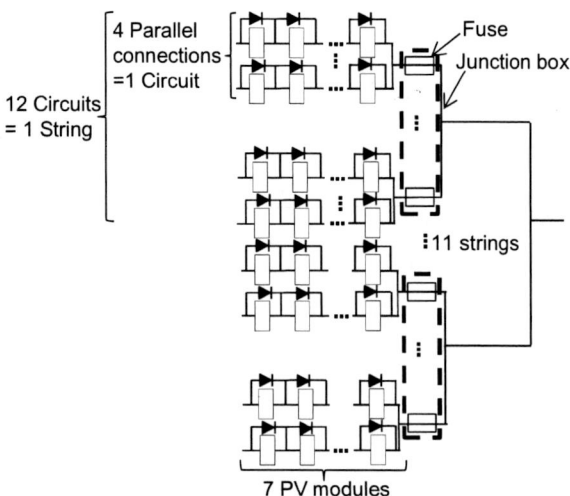

Fig. 2. PV array constitution (a PV string is composed of seven serially-connected PV modules with bypass diode and a PV array with 132 circuits [528 parallel connections]).

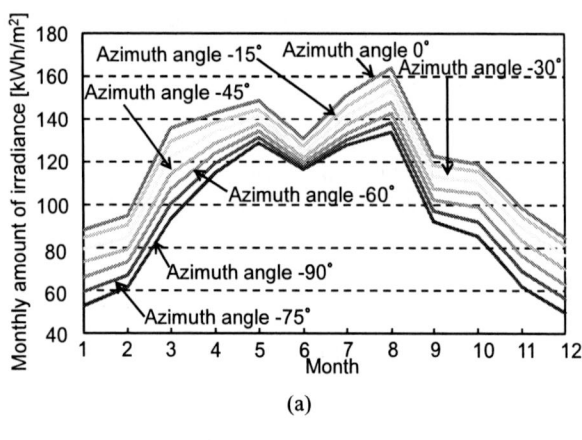

(a)

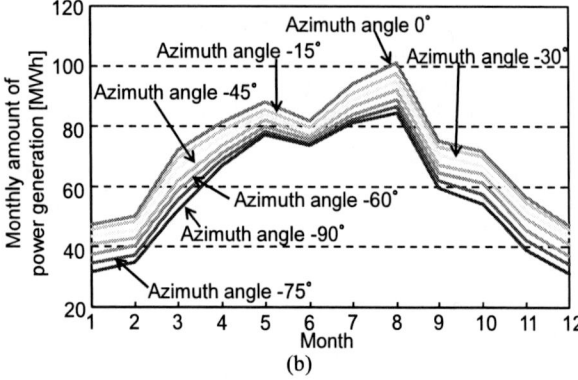

(b)

Fig. 3. Change in monthly cumulative amount with various azimuth angles from 0° to -90° in steps of 15° at inclined degree of 26.6°: (a) irradiance, (b) power generation.

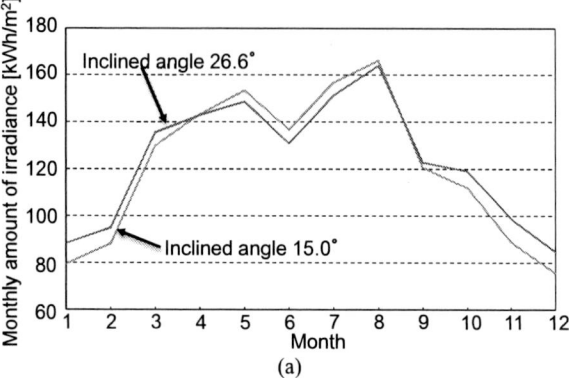

(a)

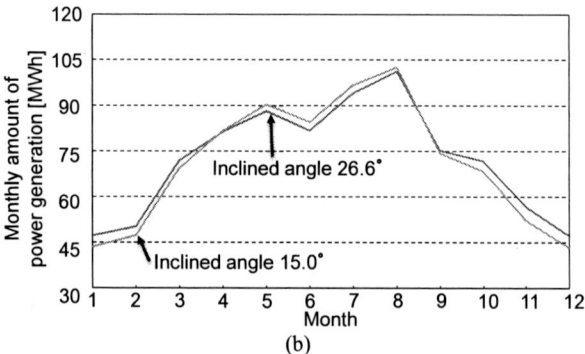

(b)

Fig. 4. Change in monthly cumulative amount when inclined angle is 15° or 26.6° and azimuth angle is 0°: (a) irradiance, (b) power generation.

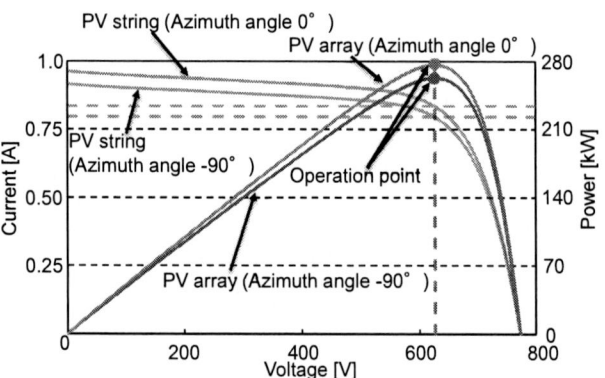

Fig. 5. I-V characteristic of PV strings located on inclined angle of 26.6° and azimuth angle of 0° or −90° and P-V characteristic of PV arrays under the condition of 600 W/m² and 30 °C.

978-1-5090-5606-4/17 $31.00 © 2017 IEEE

TABLE III
CHANGE IN ANNUAL CUMULATIVE AMOUNT OF IRRADIANCE AND POWER GENERATION BY DIFFERENCE OF AZIMUTH ANGLE AT INCLINED ANGLE OF 26.6°

Azimuth angle [deg]	0	-15	-30	-45	-60	-75	-90
Annual cumulative amount of irradiance [kWh/m²]	1481.6	1430.4	1370.8	1306.8	1241.1	1178.5	1119.4
Annual cumulative amount of power generation [MWh]	866.99	839.99	810.37	778.43	744.69	714.06	686.39

TABLE IV
CHANGE IN ANNUAL CUMULATIVE AMOUNT OF IRRADIANCE AND POWER GENERATION BY DIFFERENCE OF INCLINED ANGLE AT AZIMUTH ANGLE OF 0°

Inclined angle [deg]	15	26.6
Annual cumulative amount of irradiance [kWh/m²]	1450.7	1481.6
Annual cumulative amount of power generation [MWh]	854.67	866.99

TABLE V
CONSIDERED COMBINATION OF INCLINED AND AZIMUTH ANGLES OF PV SYSTEM

Pattern	I	II	III	IV
Inclined angle [deg]	15	26.6	15	26.6
Azimuth angle [deg]	0	0	-90	-90

TABLE VI
COMPARISON OF ANNUAL CUMULATIVE AMOUNT OF IRRADIANCE AND POWER GENERATION ON VARIOUS INSTALLATION CONDITIONS

Pattern	I	II	III	IV
Annual cumulative amount of irradiance [kWh/m²]	1450.7	1481.6	1226.8	1119.4
Annual cumulative amount of power generation [MWh]	854.67	866.99	744.44	686.39

26.6°. From Table III, it can be confirmed that as azimuth angle becomes bigger, the annual cumulative amount of the irradiance and the power generation become small. Moreover, the largest annual cumulative amount of the irradiance and the power generation is obtained at azimuth angle of 0°. This is because from (15) and (16), $I_{b\beta\gamma}$ becomes the largest when azimuth angle is 0°.

Fig. 3 shows monthly change in the monthly cumulative amount of the irradiance and the power generation with various azimuth angles from 0° to -90° in steps of 15° at inclined degree of 26.6°. From Fig. 3, it can be confirmed that a difference of the monthly cumulative amount of the irradiance and the power generation by azimuth angle is bigger during the winter than during the summer. This is

caused by the inclination of the Earth's axis. During the winter, the brevity of daytime causes the sunrise time to be late. As a result, the PV array cannot obtain lots of irradiance when it is facing the east. Hence, a difference of the monthly cumulative amount by azimuth angle becomes big.

Next, Table IV shows the annual cumulative amount of the irradiance and the power generation when the inclined angle is 15° or 26.6° and azimuth degree of 0°. From Table IV, it can be confirmed that the annual cumulative amount of the irradiance and the power generation is larger at inclined angle of 26.6°. This is because as the angle of inclined plane becomes larger, $I_{r\beta\gamma}$ increases from (17). However, $I_{s\beta\gamma}$ and $I_{b\beta\gamma}$ vary by location information such as longitude and latitude from (15) and (20). Thus, it can be confirmed that the optimal angle of inclined plane may differ depending on the location of the PV array. For this examination, the better inclined angle is 26.6° from the result.

Fig. 4 shows monthly change in the monthly cumulative amount of the irradiance and the power generation when inclined angle is 15° or 26.6° and azimuth degree of 0°. From Fig. 4, it can be confirmed that monthly cumulative amount of the irradiance and the power generation is bigger at inclined angle of 15° during April to August compared to inclined angle of 26.6°. This is also caused by the inclination of the Earth's axis. On the northern hemisphere, the sun elevation becomes higher around the summer solstice, because the North Pole faces the sun. Therefore, the smaller inclined angle obtains more sunlight.

Lastly, Table V shows various installation conditions of the PV system. Table VI shows the comparison of the annual cumulative amount of the irradiance and the power generation. From Table VI, the amount of the power generation is the biggest at pattern II where it has the biggest amount of the irradiance. In contrast, pattern IV has the smallest amount of the power generation, where it has the smallest amount of the irradiance.

Fig. 5 shows *P-V* characteristic of PV arrays and *I-V* characteristic of PV strings whose inclined plane is 26.6° and azimuth angle is 0° or −90° under the condition of 600 W/m² and 30 °C. From Fig. 5, I_{sc} and P_m of the PV string located on azimuth angle of 0° is bigger than the PV string located on azimuth angle of −90° due to a difference on amount of the irradiance obtained. Though the amount of irradiance also effects on V_{oc} from (28), the influence of it is smaller compared to I_{sc}. Thus, V_{oc} of the PV strings located on azimuth angle of 0° and −90° are nearly the same. Therefore, the PV string located on azimuth angle of −90° produces less power.

V. CONCLUSIONS

Examinations on the effect of a PV array with various installation conditions were considered by simulation. The results of the examinations clearly demonstrate that Erbs model is the most suitable decomposition model and Perez model is the most suitable transposition model. In addition, the results show that I_{sc} of PV arrays are different due to a difference in irradiance; however, V_{oc} of PV arrays are nearly the same on a PV array located on a different inclined and azimuth angle. Therefore, the amount of the power generation decreases by geographical condition where PV arrays are installed. In addition, the annual cumulative amount of the irradiance and the power generation become the biggest when inclined angle is 26.6° and azimuth angle is 0° at the location investigated in this study. Moreover, as a PV array faces to the east, the annual cumulative amount of the irradiance and the power generation become smaller.

REFERENCES

[1] M. Lave, W. Hayes, A. Phol, and C. W. Hansen, "Evaluation of global horizontal irradiance to plane-of-array irradiance models at locations across the United States," *IEEE Journal of Photovoltaics*, vol. 5, no. 2, pp. 597-606, 2015.

[2] K. Passow, M. Lee, and A. F. Panchula, "Evaluation of diffuse decomposition models," in Proc. of *43rd IEEE Photovoltaic Specialists Conference*, 2016, pp.997-1001.

[3] G. M. Tina, C. Ventura, and S. D. Fiore, "Sub-hourly irradiance models on the plane of array for photovoltaic energy forecasting applications," in Proc. of *38th IEEE Photovoltaic Specialists Conference*, 2012, pp.1321-1326.

[4] D. G. Erbs, S. A. Klein, and J. A. Duffie, "Estimation of the diffuse radiation fraction for hourly, daily and monthly-average global radiation," *Sol. Energy*, vol. 28, no. 4, pp. 293-302, 1982.

[5] J. F. Orgill and K. G. T. Hollands, "Correlation equation for hourly diffuse radiation on a horizontal surface," *Sol. Energy*, vol. 19, no. 4, pp. 357-359, 1977.

[6] D. T. Reindl, W. A. Beckman, and J. A. Duffie, "Diffuse fraction correlations," *Sol. Energy*. vol. 45, no. 1, pp. 1-7, 1990.

[7] S. Dervishi and A. Mahdavi, "Computing diffuse fraction of global horizontal solar radiation: A model comparison," *Sol. Energy*. vol. 86, no. 6, pp. 1796-1802, 2012.

[8] M. Udagawa and K. Kimura, "The estimation of direct solar radiation from global radiation," *Transactions of AIJ*, no. 267, pp.83-90, 1978 (in Japanese).

[9] K. Emura and E. Sakamoto, "Evaluation of models to estimating solar radiation on tilted surface," the Osaka City University Bulletin of the Sciences of Living, vol. 40, pp. 71-78, 1992 (in Japanese).

[10] B. Y. H. Liu and R. C. Jordan, "The interrelationship and characteristic distribution of direct, diffuse and total solar radiation," *Sol. Energy*. vol. 4, no. 3, pp. 1-19, 1960.

[11] D. King, Simple Sandia Sky Diffuse Model. (2015, Jan. 29). [Online]. Available: https://pvpmc.sandia.gov/modeling-steps/1-weather-design-inputs/plane-of-array-poa-irradiance/calculating-poa-irradiance/poa-sky-diffuse/simple-sandia-sky-diffuse-model/

[12] J. Hay and J. Davies, "Calculation of the solar radiation incident on an inclined surface," in Proc. of *1st Can. Sol. Radiation Data Workshop*, pp.59-72, 1980.

[13] R. Perez, P. Ineichen, R. Seals, J. Michalsky, and R. Stewart, "Modeling daylight availability and irradiance components from direct and global irradiance," *Sol. Energy*, vol. 44, no. 5, pp. 271-289, 1990.

[14] A. Itagaki, H. Okamura, and M. Yamada, "Preparation of meteorological data set throughout Japan for suitable design of PV systems," in Proc. of *3rd World Conference on Photovoltaic Energy Conversion*, 2003, pp.2074-2077.

[15] Japanese Industrial Standards C8907 "Estimation method of generating electric energy by PV power system" (in Japanese).

Investigating the impact of solar cells partial shading on photovoltaic modules by thermography

David Pera, José A. Silva, Sara Costa and João M. Serra

Instituto Dom Luiz – Faculdade de Ciências Universidade de Lisboa, Ed. C1 – Campo Grande, 1749-016 Lisboa, Portugal

Abstract — In order to understand how partial shading at the cell level affects modules normal operation, a standard monocrystalline silicon PV module, divided in three equivalent cell strings by bypass diodes, was submitted to several shading fractions ranging between 10% and 100% of a solar cell area. The modules were characterized by thermal imaging using a thermal camera and I-V curves were also obtained. A loss in the current produced in the string containing the shaded cell was verified. Such current loss increases with the shaded area, plummeting the electrical power produced by the module by one third for larger shading fractions. For all the shadings tested, a significant and non-uniform increase of temperature of the shaded cell was detected. For shaded fractions of 50% and 70%, the temperature differences between the shaded cell and the rest of the module were as high as 36.6 °C and 56.9 °C respectively.

Index Terms — monocrystalline silicon modules, partial shading, thermography.

I. INTRODUCTION

The global deployment of PV installations has registered a large increase in the last years, in the end of 2016, the global installed capacity surpassed the 300 GW electricity accounted for more than approximately 1.8% of the global electricity consumption [1]. In the next years, it is expected that the PV installations will continue to increase at a high rate, and the importance of the photovoltaic energy for the world electricity production will continue to grow. Such growing importance of the photovoltaic energy, requires an increased surveillance of the PV systems performance and reliability. Also the spreading of urban integrated PV installations, increases the occurrence of partial shading of the modules, mainly due to neighboring constructions or infrastructures. Partial shading of PV modules represents a severe threat to the reliability of PV systems, in addition to the transient energy yield losses induced by the module temporary shading [2], in a partial shaded PV module, the shaded solar cells stop producing electricity and can work on reverse bias, starting to dissipate power via Joule heating thus creating overheated regions on the module (hotspots) that, over certain temperatures, can originate irreversible damages in the PV modules components [3, 4]. This phenomenon is well known as it has been the subject of several experimental studies [5, 6], and different numerical models have been proposed [7, 8].

Bypass diodes can isolate the solar cell affected by shading, and are thus a very effective way to limit partial shadowing induced losses, preventing possible irreversible damages, and are standard constituent of industrial PV modules and systems. Recently more elaborated strategies to avoid the damages induced by partial shading module shading have been proposed [9].

In this article, we present a study that aims to understand how the partial and total shading of one solar cell affects the module normal operation and impacts on its temperature distribution. For this purpose a standard monocrystalline silicon PV module was submitted to different solar cell area shading fractions. The importance of the shading position was also inspected by placing the shadowing in two different positions.

The PV device was characterized by obtaining the I-V curves and their thermal imaging before and during the imposed shading.

II. EXPERIMENTAL

A. Description of the system and field conditions

The studied system is part of a micro generation grid connected system with a total capacity of 3.6 kW$_p$, installed in the *"Campus Solar"* of the Faculty of Sciences of University of Lisbon (Portugal). The PV device analyzed in this study is a *Sanyo HIP-210NKHE*, multicrystalline silicon module with a nominal power of 210 W$_p$. The module is part of a string of three similar modules, all south oriented with 34° of inclination, connected to *a SMA* inverter, *Sunny Boy SB 700* (Fig. 1), and it is composed by 72 cells divided into 3 strings protected with external bypass diodes, distributed according to the diagram on Fig. 2.

978-1-5090-5606-4/17 $31.00 © 2017 IEEE

Fig. 1. Front view of grid connected string of multicrystaline photovoltaic modules in *"Campus Solar"*, Lisbon. The studied PV module is the one on the right.

The electric characteristic curves were acquired using an I-V tracer, *EKO Instruments MP-160*, and the global horizontal irradiance (GHI) was measured with a locally installed pyranometer.

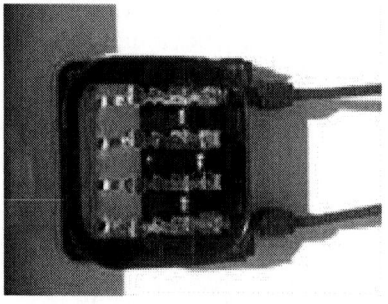

Fig. 2. View of an open junction box of *Sanyo HIP-210NKHE* modules. Inside is possible to view the conduction stripes and three bypass diodes.

The thermal studies were carried out by thermography. The equipment used was a passive cooled micro-bolometer of amorphous silicon camera, *Gobi-2833*, with a spectral sensitivity between 8μm and 14μm. Its sensor has a 25 μm pitch and a 384 x 288 resolution. The camera was placed, facing the module at a distance of 3 meters, avoiding sun radiation blockage, for contextual pictures of the whole system. In these cases the angle between the PV system and the optics was about 34º, the same of the modules inclination. For cells close-ups, the camera was placed the more perpendicular as possible at a distance of one meter. For each thermal acquisition an ambient temperature correction was applied to the thermographic calibration.

The experimental activities were done in June, during cloudless day periods with GHI levels ranging the interval between 982 W.m⁻² and 1014 W.m⁻².

B. Procedure

The partial cell shading effects in the PV system were evaluated in terms of electrical performance impact and in temperature variation of the module and test cell. As described before the tested module was protected with three bypass diodes as shown in the diagram of Fig.3, where the partially shaded cell is represented by the gray filled one, in the right bottom corner.

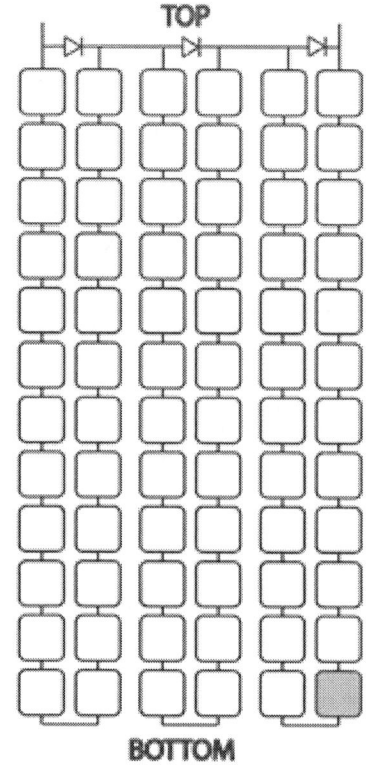

Fig. 3. Wire diagram of the analyzed photovoltaic module, showing the three string composition and bypass diodes location.

To induce the shadings black square cards masks dimensioned to fit 10%, 20%, 50%, 70% and 100% of the cell area were used. For each partial shading, the masks were placed at two different positions: the cell center and close to the right bottom corner. Fig.4 illustrates the partial shading configurations tested.

The experimental procedure consisted in measuring a reference I-V curve before shading, followed by a 15 minutes period of cell shading with the system connected to the grid. A thermogram of both system and cell was also captured before shading. At the end of the exposition period, the system was reconnect to the I-V tracer and new thermograms were acquired.

978-1-5090-5606-4/17 $31.00 © 2017 IEEE

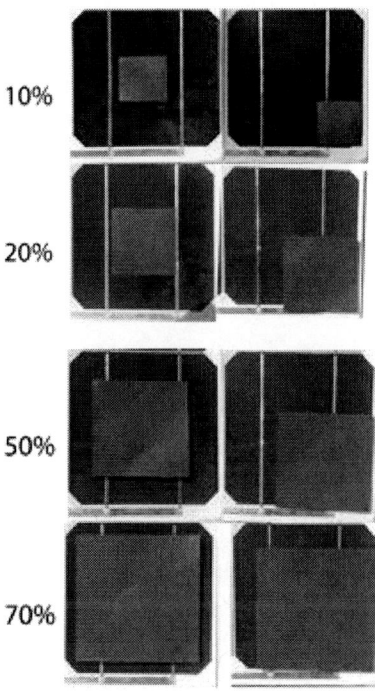

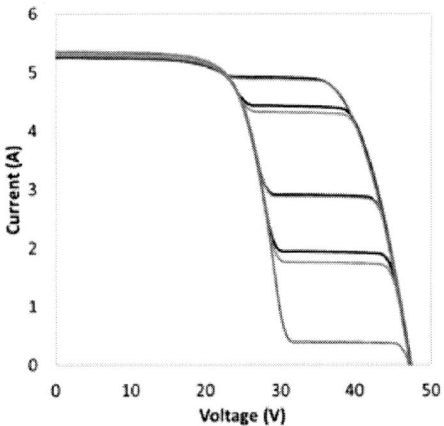

Fig. 5. I-V characteristic curves of the shaded PV module. The black lines correspond to the center shading tests and gray ones to the right-bottom shades. In both cases, the increasing shadow fraction corresponds to descending current steps.

Fig. 4. Tested configurations of cell partial shading. The percentage represents the ratio of the black cards area and the cell total area.

III. DISCUSSION OF RESULTS

As expected when a partial or full shade occurs on a cell its polarization changes and the current photo-generated in the rest of the string elements flows in the opposite direction promoting a voltage drop. Therefore the diode becomes forward biased bypassing the generated current to the next element in the string protecting the module production. The isolation of the failing element by the diode minimizes the drop of the maximum power point of the module, which otherwise would follow the current of the shaded element. As can be seen in Fig. 5, this effect starts to occur with the minimum shade fraction tested, in this case 10% of the cell's area. The larger is shadowed area fraction, the lower is the contribution of the isolated string photo-generated current. Regarding the topology of the shades tested, it was shown that this effect is not sensible to the position of the shade on the cell. In both configurations the photo-current is inversely correlated to the shaded area, as it can be observed in Fig. 5.

In terms of output power loss of the system, a non-linear progression with the increase of the shadow fraction was observed (Fig. 6). The losses stabilized asymptotically at 35% of relative power for the full shaded cell case, representing around of 1/3 of the nominal power. At this point the diode is bypassing all the 24 cell directly linked to the shaded one.

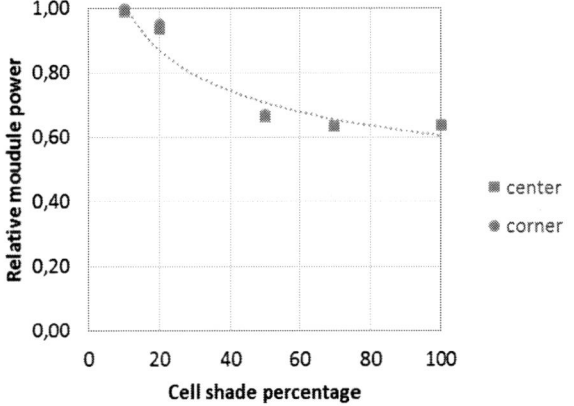

Fig. 6. Relative module power output variation with the shadow fraction at one cell for both centered and cornered configurations.

Regarding heating effects, the reversed current caused by the shading is in part dissipated in the diode, being the remaining dissipated in the shaded cell, which under these conditions behaves like a resistive element. In the thermograms presented in Fig. 7, the Joule heating effect caused by the shading is notorious. The temperatures achieved rise with the shade fraction as more current is dissipated in the shaded cell.

In average the temperature amplitudes relative to the initial homogenous module temperature were: 10.5 °C for 10% and 20%; 36.6 °C for 50% and 56.9 °C for 70%. Temperatures as high as 85 °C and 115 °C were measured for the cases of 50% and 70% of partial shade, respectively. Without significant impact on the temperature magnitude, the different shades positioning promoted variations on the distributions of temperature along the cell surface. For the lower shade fractions (10% and 20%), the shade centered configuration produced more homogeneous temperature distributions.

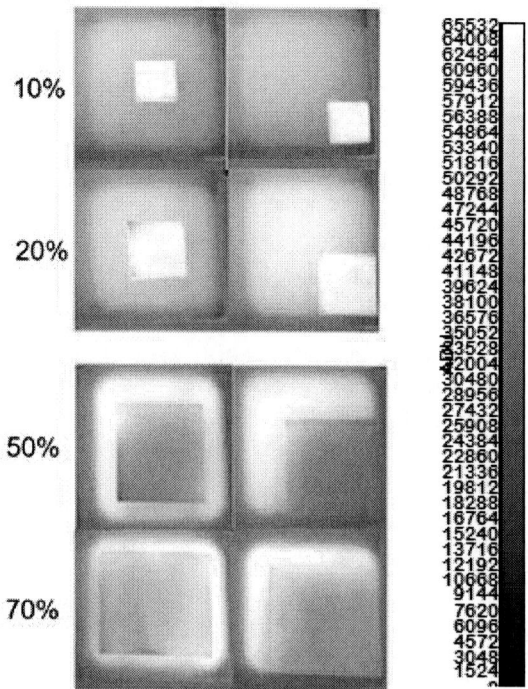

Fig. 7. Thermographic images of the partial shading configurations tested. The color scale represents ADUs (analog-to-digital units), which are proportional to temperature.

Fig. 8. Thermographic image of the photovoltaic system, for a 70% cell area partial shading, placed in the cell's corner. Besides the shaded cell heating, the heating of string was also detected.

For the higher shading fractions, 50% and 70%, the temperature was significantly higher on the contiguous edges. In the case of the corner shadows, the higher temperatures seem to occur next to the opposite corner. Other Joule heating effect was observed along the entire string blocked by the bypass diode for shade areas equal and greater than 50% of the cell area (Fig. 8).

Interestingly despite of the high temperatures registered on the shaded cell, the average module temperature remains stable, hence no significant V_{oc} variations were measured.

IV. CONCLUSIONS

Different partial shadings scenarios on a single PV module's cell where tested. Those differed in terms of cell area percentage and positioning. The results revealed that induced shadowing creates significant losses in the module current and power which increase with the cell shaded area. The experiments also confirmed that such losses are not sensible to the shading positioning on the cell.

For higher shading fractions very significant temperature differences between the shaded cell and the rest of the module where observed. Despite the important temperature gradients detected near the shaded region, the average module temperature remained stable as shown by non-significant variation of V_{oc} values between the positioning configurations for each tested shadowing fraction.

The module temperatures observed in this study, if kept overtime (up to 115ºC, with an ambient temperature ranging 24ºC - 26ºC), may lead to an accelerated degradation of the module elements, to a premature aging of the module and ultimately to its failure.

ACKNOWLEDGEMENT

This work has been funded by FCT (Portugal), through the grant: SFRH/BPD/82540/2011 and the project: UID/GEO/50019/2013.

REFERENCES

[1] International Energy Agency –Power Systems Program, "2015 Snapshot Global Photovoltaic Markets, 2016".

[2] D. Pera, M. C. Brito, J. Maia-Alves, J.M. Serra, C. Silva, A.M. Vallêra, "Optimization Model for High Density Photovoltaic Power Plants by Maximization of the Return on Investment", Proceedings 26th European PVSEC, pp. 4211 – 4214, 2011.

[3] F. Martínez-Moreno, J. Muñoz, and E. Lorenzo, "Experimental model to estimate shading losses on PV arrays", Solar Energy Materials and Solar Cells, vol. 94, pp. 2298-2303, 2010.

[4] S. Wendlandt, A. Drobisch, T. Buseth, S. Krauter and P. Grunow, "Hot spot risk analysis on silicon cell modules", 25th European Photovoltaic Solar Energy Conference and Exhibition, 2010, pp. 4002-4006.

[5] F. Martínez-Moreno, J. Muñoz, and E. Lorenzo, "Experimental model to estimate shading losses on PV arrays", Solar Energy Materials and Solar Cells, vol. 94, pp. 2298-2303, 2010.

[6] S. Silvestre, S. Kichou, A. Chouder, G. Nofuentes, and E. Karatepe, "Analysis of current and voltage indicators in grid connected PV (photovoltaic) systems working in faulty and partial shading conditions", Energy, vol. 86, pp. 42-50, 2015.

[7] Giaffreda, D., Omaña, M., Rossi, D., & Metra, C. (2011, October). Model for thermal behavior of shaded photovoltaic cells under hot-spot condition. In Defect and fault tolerance in VLSI and nanotechnology systems (DFT), 2011 IEEE international symposium on (pp. 252-258). IEEE.

[8] Geisemeyer, I., Fertig, F., Warta, W., Rein, S., & Schubert, M. C. (2014). Prediction of silicon PV module temperature for hot spots and worst case partial shading situations using spatially resolved lock-in thermography. Solar Energy Materials and Solar Cells, 120, 259-269.

[9] S. Daliento, F. Di Napoli, P. Guerriero, and V. d'Alessandro, "A modified bypass circuit for improved hot spot reliability of solar panels subject to partial shading", Solar Energy, vol. 134, pp. 211-218, 2016.

Annual Degradation Rate and its Linearity Analysis using Metered kWh Data

Christopher Raupp and GovindaSamy TamizhMani

Arizona State University, Photovoltaic Reliability Laboratory (ASU-PRL)

ABSTRACT

The degradation rate analysis of 38 commercial crystalline-silicon (c-Si) photovoltaic (PV) power plants fielded for up to 16 years in a hot-dry climate of Arizona has been evaluated. Based on the metered kWh data and modeled weather data of these systems, the linearity and degradation rate analyses of PV modules are statistically determined using three different performance approaches: PR (Performance Ratio), PI (Performance Index), and Raw kWh. The evaluation approaches developed and used in this work are validated using the in-field current-voltage measurements previously performed in four commercial systems.

Index terms: Performance, Degradation, Power Plants, Statistical

I. INTRODUCTION

It is a known fact that the photovoltaic (PV) modules deployed in the field degrade in performance overtime. Some PV technologies such as CIGS, CdTe, and a-Silicon modules are also known to degrade non-linearly with time whereas c-Si has been traditionally assumed to degrade linearly. The rates of degradation can also vary based on environmental conditions such as UV dose, ambient temperature, humidity and etc. It has previously been reported that c-Si modules located in harsher climates that experience hot-dry and hot-humid conditions tend to degrade at more than the industry standard of 1.0%/yr whereas modules located in the milder climates, cold and temperate, degrade at less than 1.0%/yr. In this work, we evaluated 38 commercial c-Si PV systems to statistically determine the hot-dry climate specific degradation rate of the c-Si modules and to assess whether the degradation trend is truly linear or not. Out of these 38 systems, 34 systems are installed on the ASU campuses. All evaluated power plants are located in the Phoenix-Metro area of Arizona, with the majority of sites (20) being fixed rooftop systems and parking structures (11). All other sites were other types of fixed tilt structures for either shading for pedestrians (3), fix tilted ground mount systems (3), or ground mounted 1-axis trackers (1). The evaluation approaches developed and used in this work are validated using the in-field current-voltage measurements previously performed in four commercial systems installed outside of ASU campuses [1]. Using data sets available from a large number of PV systems, as compared to smaller number of systems reported by our group [2], is presented in this paper. A detailed analysis on this study is reported elsewhere [3].

II. METHODOLOGY

Based on the metered kWh data of the thirty eight systems and weather data corresponding to these systems, the linearity and degradation rate analyses of PV modules were statistically determined using three different approaches: PR (Performance Ratio), PI (Performance Index), and Raw kWh. Since these systems did not have their own plane of array (POA) irradiance and module temperatures, they were determined through the selection of most accurate irradiance and temperature models specific to the hot-dry climate of Arizona. There are 12 decomposition models and 10 transposition models published in the literature for modeling the POA irradiance from GHI and DNI. Therefore, a total of 120 possible combinations were evaluated to select the best possible combination specific to this climate. Similarly, at least nine different temperature models have been reported in the literature. Due to the page limitation of this paper, the selection procedure of the Arizona-specific POA irradiance and module temperature models are not presented here but are made available elsewhere [3].

2.1 Degradation Rate Analysis using PR

The performance ratio (PR) is the ratio of measured energy to expected energy (based on nameplate data and measured/modeled irradiance). For all evaluated systems, daily PR values were calculated using measured kWh data and the calculated expected energy based on the system rating and modeled irradiance data. The methodology of calculating performance ratio is similar to that previously reported by Shrestha *et al.* [2], but was slightly modified in order to fit the type of data that was available in this study. Daily PR values were calculated for each system with any period with an irradiance higher than 40 W/m² being analyzed as this was considered to be the limit for daytime operating conditions. When processing the data, any days that were shown to obviously not fit the overall trend of the year-to-year data sets were removed. The median PR of each month for each year was calculated by taking the median of remaining non-removed data points corresponding to that particular month. In order to calculate the degradation rates experienced by the systems, the slope of a line for a particular month vs. number of years was then determined to be the degradation rate per year for that particular month leading to 12 slopes for 12 months of a year. Based on the previous research done by Shrestha *et al.* [2], the months that have the least amount of variability in satellite data are those from April to October.

These months' degradations were then averaged together to produce the true degradation rate experienced by the PV modulus in the system.

2.2 Degradation Rate Analysis using PI

Performance index (PI) is a more accurate representation than PR since it corrects not only for insolation, but also for other losses such as module temperature, soiling, balance of system (BOS), wiring, and etc. In this study, the expected energy generated by the systems' was corrected for irradiance, temperature, module mismatch, inverter efficiencies, and a wiring loss of a nominal 1%. A 3.3% module mismatch loss, based on previous in field measurements from ASUI-PRL, was used. The same procedure as that of the PR procedure was used for calculating degradations rates using the PI approach.

2.3 Degradation Rate Analysis using Raw kWh

Figure 1 shows the flowchart for the newly developed analytical process in this work.

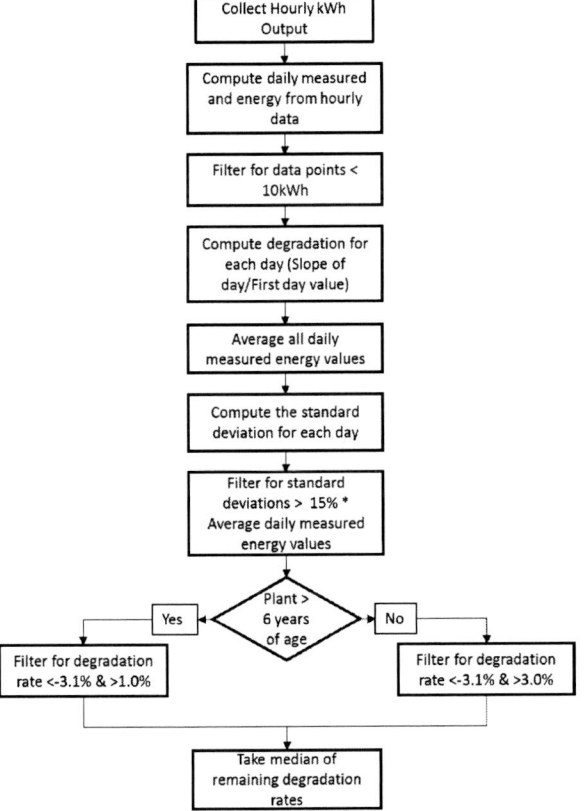

Fig. 1. Flowchart of newly developed kWh degradation method based on statistical analysis

III. RESULTS AND DISCUSSION

3.1 Degradation Rate Determination

In order to provide the most accurate performance and degradation rate assessment of the evaluated systems, it was crucial to decide on the most accurate POA irradiance and thermal models specific to the climatic region. In order to determine the validity of multiple performance metrics, the degradation rate found through in-field I-V measurements reported by Shrestha et al. [2] with measured POA irradiance was used. The system used to validate all performance models was chosen due to the fact the system was only unavailable for 59 days of the total 3,431 days the system was evaluated for. The plot shown in Fig. 2 indicates the availability of the system with any down times resulting in a break in the bar chart (white spaces).

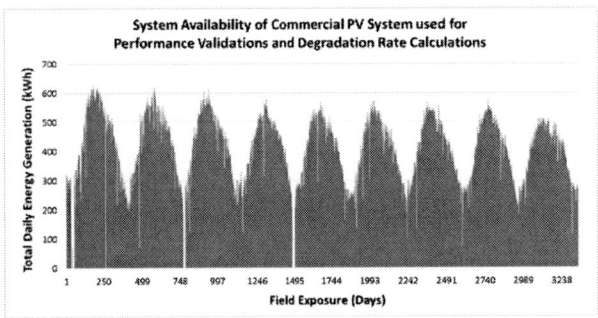

Fig. 2. System availability check of commercial PV system used for performance validations and degradation rate calculations

After having determined the best irradiance and thermal models specific to the climatic region [3], and validating the three performance metrics (PR, PI, and kWh) against in field I-V measurements, the three performance metrics were applied to the 34 systems of ASU campuses and the results are shown in Figures 3, 4 and 5, respectively.

The output of modified raw kWh analysis was validated against four previously measured commercial PV systems. The degradation rates of the four systems were calculated using in field I-V measurements, the most accurate linear method of degradation rate analysis (i.e. "true" degradation). The systems were all located in the same region as the other systems analyzed in this study and contain c-Si PV modules, so the use of them in this validation would be ideal. The results of the measured vs calculated degradation rates from kWh data is shown in Figure 6.

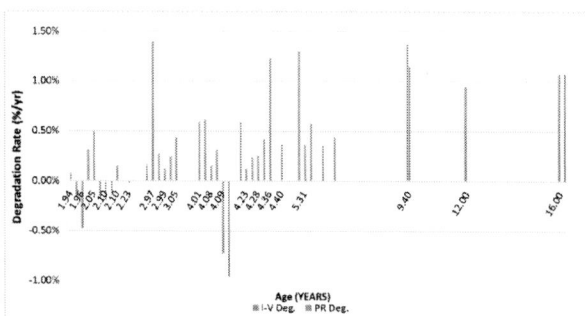

Fig. 3. PR degradation rates of all systems and measured I-V degradation rates of 4 previously measured systems

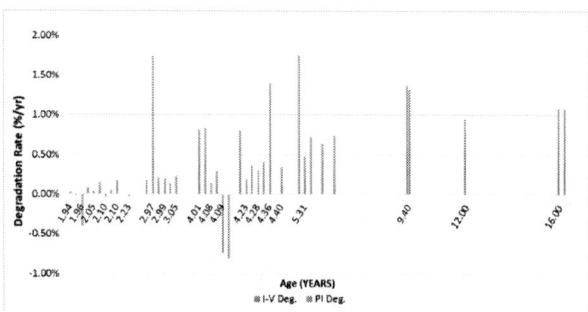

Fig. 4. PI degradation rates of all systems and measured I-V degradation rates of 4 previously measured systems

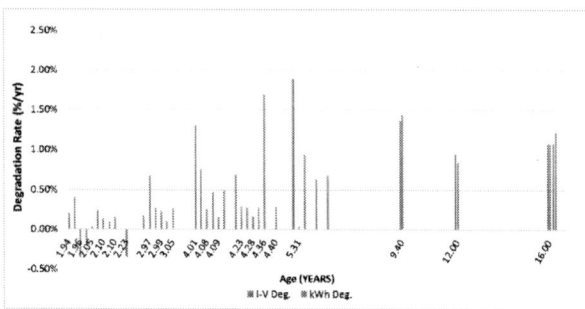

Fig. 5. kWh degradation rates of all systems and measured I-V degradation rates of 4 previously measured systems.

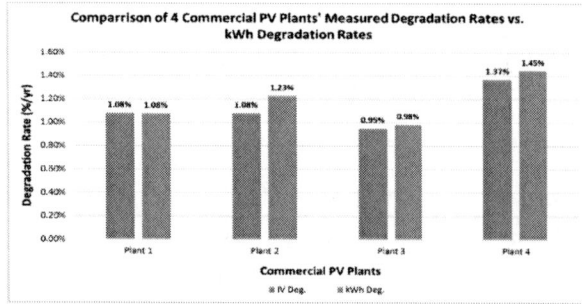

Fig. 6. Comparison between newly developed kWh analysis and in field measured I-V data (9.4-16 years old plants with clipping and shading issues)

In a few systems, negative degradation rates (indicating increased performance over time) have been surprisingly observed. Slightly higher degradation rates for the system-kWh approach as compared to the module-IV method shown in Figure 6 may be attributed to site specific issues over the years including: shading losses, blown fuses, heavy soil deposition, etc. Increases in performance over time, Figure 7, could possibly be attributed to a number of unknown events over the system lifetime including: warranty replacement with new modules (due to safety or performance issues), inverters that were down for long periods of time before either being fixed or replaced, and any other system downtime issues that was corrected at a later date. Systems that were also shown to be clipping for the majority of the year, as seen in Figure 8, were also found to be a problem during analysis since it was not possible to see the "peak" generation of the system and analyze whether that "peak" generation was decreasing or not. Due to these uncertainties in most of these calculated degradation rates shown in Fig. 3 through Fig. 5 (especially for the newer systems), the systems that have unexpected degradation rate behaviors (very high, near zero or negative degradation rates) or had instances of clipping were removed for the determination of the extent of degradation linearity (linear or non-linear degradation) presented in the next section. This study indicates that the degradation rate determination based on PR, PI or kWh method for the newer systems (less than four years old systems shown in Figs. 3, 4 and 5) cannot be fully trusted due to very low number of data points (less than 4 data points) infested with several issues including irradiance modeling from the SolarAnywhere data (from zero degree tilt to POA irradiance), significant insolation variation from SolarAnywhere data in one or more of these four years, unknown events (cutting the shading trees, variation in soiling etc.), and frequent or extended clipping in newer plants. The degradation rate determination using PR and PI method can be significantly improved if a ground mount POA irradiance data is used and the other issues including clipping are appropriately accounted. The kWh method can also be improved by addressing the issues related to year-to-year clipping variation, shading variation etc.

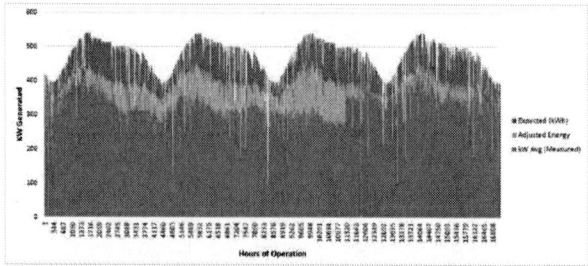

Fig. 7. Hourly kWh generation for one of the ASU systems that shows a high initial degradation rate and then a sudden increase in performance in the later years possibly due to cutting of the shading trees or replacing underperforming modules/strings under warranty.

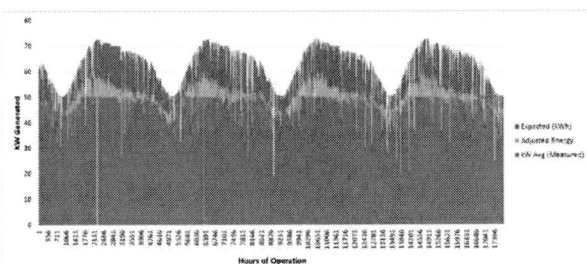

Fig. 8. Hourly kWh generation for one of the ASU systems that shows inverter clipping and thus the unavailability of looking at a degradation trend

A list of all sites with their calculated PI degradation rates is shown below. The list also indicates whether a system was used for the analysis or not.

Table 1. List of all evaluated sites and their calculated degradation rates based on PI method

System Name	Age (Years)	DC System Size (kW)	PI Deg. Rate (%/yr)	Used for Analysis
ASU-U	1.94	230.4	0.03%	Yes
ASU-AC	1.94	252.54	-0.02%	No
ASU-AH	1.96	86.22	-0.41%	No
ASU-X	2.04	250.56	0.08%	Yes
ASU-AF	2.05	310.32	0.04%	No
ASU-AA	2.10	204.6	0.15%	Yes
ASU-AB	2.10	224.4	-0.04%	No
ASU-Z	2.10	89.1	0.05%	Yes
ASU-R	2.10	66	0.18%	No
ASU-T	2.23	254.1	-0.03%	No
ASU-S	2.92	69.3	0.17%	No
ASU-J	2.97	214.86	1.73%	No
ASU-Y	2.97	138.6	0.22%	No
ASU-AE	2.99	246.06	0.21%	Yes
ASU-Q	3.03	65.82	0.15%	No
ASU-AD	3.05	166.38	0.23%	No
ASU-AG	4.01	2132.48	0.82%	No
ASU-L	4.01	705.6	0.83%	No
ASU-G	4.08	144.48	0.13%	Yes
ASU-E	4.09	57.12	0.29%	Yes
ASU-I	4.09	188.16	-0.74%	No
ASU-N	4.12	497.28	-0.80%	No
ASU-K	4.22	168	0.81%	No
ASU-B	4.23	63.84	0.19%	No
ASU-D	4.28	94.08	0.38%	Yes
ASU-H	4.28	64.68	0.30%	Yes
ASU-P	4.28	67.2	0.41%	No
ASU-F	4.36	94.08	1.39%	No
ASU-C	4.40	63.84	0.34%	Yes
ASU-G	5.25	289.8	1.76%	No
ASU-V	5.31	248.64	0.48%	No
ASU-M	5.32	141.12	0.73%	Yes
ASU-A	5.66	150.15	0.64%	Yes
ASU-O	6.00	23.1	0.74%	Yes
CT	9.40	97.2	1.33%	Yes
G	12.00	249.9	Only for kWh analysis	
BRO1	16.00	113.4	Only for kWh analysis	
BRO2	16.00	113.4	Only for kWh analysis	

3.2 Degradation Linearity Analysis

Figures 3 through 5 are presenting interesting data: if the degradation is linear, then the degradation rate should be the same (for example 1%/year), irrespective of the age of the power plant. Since this does not seem to be the case, it could be construed that the degradation of crystalline silicon PV systems is not truly linear. To determine this possibility in detail, distribution of yearly PI for the evaluated PV power plants was analyzed. In order to view whether or not a system is having a linear degradation rate, a minimum of 4 data points was determined to be needed since any systems with less than 4 years of age would potentially show a high linear correlation due to the lack of sufficient data. This imposed data restriction in our analysis resulted in only 10 qualified systems (out of 38 systems) for the degradation trend analysis. Figure 9 shows the results obtained for one of the ten systems with age of 9.4 years (system code CT) using the PI method.

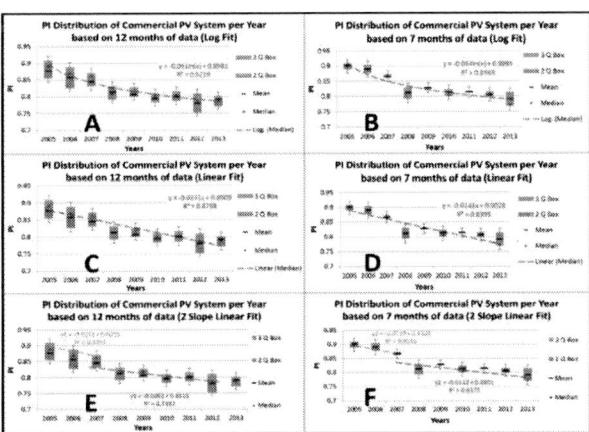

Fig. 9. Degradation trend (linearity) analysis of a 9.4 years old PV system (system code: CT) using the PI method

In Fig. 9, the degradation trend analysis is shown using the median with: **A)** Distribution of PI values per year based on 12 months data using a log fit, **B)** Distribution of PI values per year based on 7 months data using a log fit, **C)** Distribution of PI values per year based on 12 months data using a linear fit, **D)** Distribution of PI values per year based on 7 months data using a linear fit, **E)** Distribution of PI values per year based on 12 months data using a 2 slope linear fit, and **F)** Distribution of PI values per year based on 7 months data using a 2 slope linear fit.

The same trend analysis was conducted on the other 9 remaining systems. Data for three out of these nine systems are shown in Fig. 10 through Fig. 12. The six remaining sites are not shown here due to page limitations, but have similar trends as that of Fig. 10 through 12 and provided elsewhere [3].

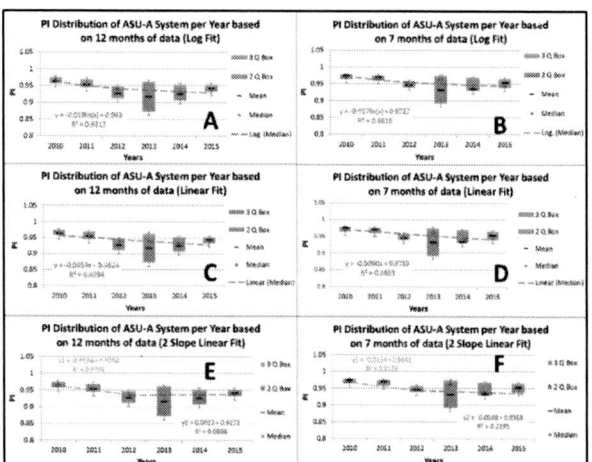

Fig. 10. Degradation trend (linearity) analysis of a 5.6 years old PV system (system code: ASU-A) using the PI method

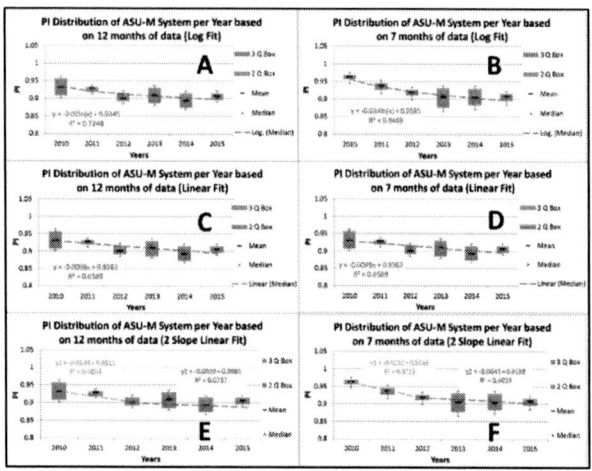

Fig. 11. Degradation trend (linearity) analysis of a 5.3 years old PV system (system code: ASU-M) using the PI method

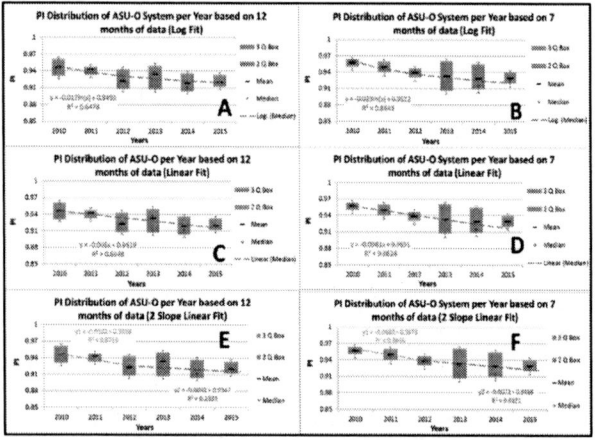

Fig. 12. Degradation trend (linearity) analysis of a 5.3 years old PV system (system code: ASU-O) using the PI method

When looking at the results from these figures (Fig. 9 through 12), it can be seen that the general trend is for the logarithmic degradation to have a higher correlation value than that of a linear trend line, but as the systems increase in age, the difference in correlation values decreases. What this may tend to indicate is that in the initial years, the systems do not degrade linearly (or the data is infested with several issues identified in the previous section), but after a period of time, the degradation rates may seem to then level out into a consistent linear degradation. Fig. 13 shows a comparison of the calculated degradation rate of the "good" ASU systems for PR, PI, and kWh methods as compared to I-V measurements previously found at older PV sites [1, 2] .

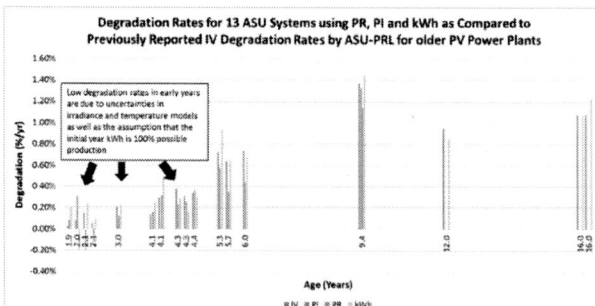

Fig. 13. Degradation rates for 13 evaluated ASU PV systems using PR, PI, and kWh methods as compared to previously reported I-V degradation rates [1, 2]

From Fig. 13, it can be noticed that the calculated degradation rates for systems that have less than 5 years of field exposure are significantly lower than what would typically be expected to be found in hot climates (about 1% per year). There could several reasons for this unexpected deviation and the two major reasons are given below. The first reason could be due to the fact there are uncertainties that arise from the use of modeling irradiance and module temperature values from SolarAnywhere data, instead of having physically measured POA values using ground mounted weather stations. The irradiance values, as taken from SolarAnywhere, have a built in uncertainty of 5% before even being converted to POA irradiance values. By using this data source, it is possible that the year over year (YOY) degradation rate may occasionally have an increase in performance from the first year to the second. The second reason for this occurrence may stem from the fact that the insolation from one year to another may have changed drastically in which one year had more rain than another, or more soil deposition occurred, or etc. These types of performance behaviors do occur as shown by Figure 14 below.

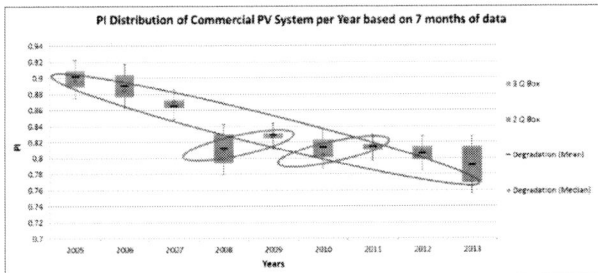

Fig. 14. Overview of degradation trend of a commercial PV system where some year over year degradation rates show a positive slope (red ovals) as compared to the overall negative trend (blue oval)

For the long-term warranty and energy predictions, the linear degradation rate assumption typically used by the industry appears to be practically acceptable; however, this linearity assumption may need to be verified with a large number of older systems (> 10 years of age) rather than just 10 systems, mostly aged between 4 and 6 years. The continuation of this study as the systems increase in age or evaluating other older systems would be beneficial to develop a definitive answer to the overall degradation linearity trend question.

CONCLUSION

The best POA irradiance and module temperature models for developing accurate operating conditions, degradation rates, and degradation trend analysis have been determined for the hot-dry climate of Arizona and is presented elsewhere [3]. Using these optimized models for PR, PI, and kWh performance evaluations, it has been determined that PI is the second best method, after the I-V method, for calculating degradation rates; encouragingly, the kWh degradation rate calculation method also gives similar results as that of the PI method without the need of using complex models for module temperature and POA irradiance determination. While the developed kWh degradation approach does produce better results than those of the PR calculation, but cannot be used to determine any specific system component losses since it is purely a statistical calculation without having any other operating conditions being taken into consideration.

The trend and rate of degradation for crystalline silicon PV systems was found to be slightly nonlinear with logarithmic degradation rates having more correlation than that of the single slope linear degradation rates. When reviewing this trend, it was only observed in systems that were 5 or more years old, since any systems younger then this had too few data points to make conclusions about the trend of the data sets. The overall average degradation rate (using the standard one slope method) of the 13 ASU PV power plants that were analyzed showed an average degradation rate of only 0.31%yr. This value is significantly lower than the

expected value of ~1.0%/yr and should not be used when looking at the overall degradation rate of the hot-dry climate of Arizona since the uncertainties in irradiance and temperature models and assumptions in the kWh method heavily dominate calculated degradation rates for systems less than 5 years of age.

This study shows that the rate and trend of degradation can only be assumed as true when: A) there is a large unclipped data set available in which a system has 5 or more years of field exposure, B) the on-site measured values (POA irradiance, module temperature, weather, etc.) are used for calculating performance and degradation rate analysis instead of modeled data, and C) in-field measured IVs are performed when the system(s) is initially installed and then are measured again at set incremental periods of time.

Overall, the typical linear degradation rate assumption used by the industry appears to be practically acceptable; however, this linearity assumption may need to be verified with a large number of older systems.

REFERENCES

[1] J. Mallineni, B. Knisely, K. Yedidi, S. Tatapudi, J. Kuitche and G. TamizhMani. "Evaluation of 12-Year-Old PV Power Plant in Hot-Dry Desert Climate: Potential Use of Field Failure Metrics for Financial Risk Calculation," *IEEE PVSC*, June 2014.

[2] S. Shrestha, J. Mallineni, K. Yedidi, B. Knisely, S. Tatapudi, J. Kuitche and G. TamizhMani "Determination of Dominant Failure Modes using FMECA on the Field Deployed c-Si Modules under Hot-Dry Desert Climate," IEEE Journal of Photovoltaics, February 2015.

[3] Christopher Raupp, "Climate-Specific Degradation Rate and Linearity Analysis of Photovoltaic Power Plants using Performance Ratio, Performance Index, and Raw kWh Methods" MS Thesis, Arizona State University, 2016 [Free downloading: repository.asu.edu]

Electrical Performance Analysis of a 27 kW Grid-Connected PV System with Soiling and Shading in Morelos Mexico

P. A. Sánchez-Pérez, D. Martínez Escobar, E. O. Ángel Ruiz, R. Santos Magdaleno, José Ortega Cruz,
A. Sánchez Juárez,
Instituto de Energías Renovables, Privada Xochicalco S/N, Temixco, Morelos 62580, Mexico

Abstract—The favorable solar resource in Mexico and the current issues with the oil industry have encouraged the use of renewable energies such as solar photovoltaic systems connected to the grid. This paper presents the performance results of a polycrystalline silicon PV system connected to the grid in Morelos, Mexico. It provides information about the actual energy yield of newer PV plants under local climate conditions, temperature stress, shading and soiling. The PV system generated 35.15 MWh during one year of operation while a simulation of the same system generated 45.81 MWh. The average performance ratio (PR) obtained was 0.75 that is was higher that the simulated value of 0.71 due to overestimation of the solar resource.

Index Terms—photovoltaic system, performance ratio, electrical performance, grid-connected, soiling.

I. INTRODUCTION

The solar photovoltaic (PV) industry has experienced an exponential growth worldwide, and in Mexico, new policies are in development to adopt PV technology to generate megawatts of electricity. In 2015, the total amount of installed PV capacity increased by 204% compared to 2014. This value reflects in a total installed PV capacity of 234 MW by 2015. In Mexico, it is prognosticated that by 2020 the total installed capacity will be up to 2.16 GW [1]. However, there is still more challenges to adopt the PV technology.

Nowadays, PV modules have a longer lifetime cycle due to the increase in the quality and durability of the materials. There are multiple manufacturers that offers at least 20 years of guarantee, and most of them includes a certification that ensures performance and safety. Nevertheless, the real installation conditions may surpass the limits of a certification scope and put the PV installation to over stress the causing a potential fault and loss in the power output.

Countries that are new in adopting utility PV plants, it is a common situation to have an absence of performance climatic specific data. To increase the confidence of the government initiative and improve the availability of data it behooves to people in the market report system performance and analyze extreme cases that trend to appear in Mexico.

In this work, we present the analysis of the current electrical performance under a real stress condition of a 27 kWp PV system located in Morelos, Mexico. The PV installation is partially shading and under local climate and soiling conditions. The information obtained provide useful data about the principal factors that impact the electrical performance of new

Table I
CHARACTERISTICS OF THE SITE AND PV SYSTEM

Category	Parameter	Value
Site	Coordinate	N18°50.36′, W99°11.53′
	Elevation	1213 m.a.s.l
Module	Technology	p-Si
	P_{max}	250
	Isc	8.65
	Voc	37.93
	α	0.01%/°C
	β	-0.31%/°C
	γ	-0.50%°C
Array	P_{STC}	9 kW
	No. modules	36
	Configuration	12 Series x 3 Parallel
Mounting	Type	Fixed
	Tilt	10°
	Azimuth	0° (Facing south)

PV power plants in the local area and will give an insight to how to avoid this stress conditions.

The primary goal of this work is to understand and determine the actual electrical performance of the PV power plant of interest. Then using a simulation, compared the obtained values with the measured values to see the difference and the accuracy of the simulation.

II. CHARACTERISTICS OF THE SITE AND PV SYSTEM

The system is in the facilities of the Agricultural Innovation Center of Morelos in the city of Emiliano Zapata, Morelos, Mexico. It consists of 108 PV modules of 250 W divided into three arrays of 9 kW each directed couple to a 10 kW inverter. Each DC lines from the junction boxes connect to a 10 kW (3-phase) Fronius IG Plus V 10.0-3 Delta inverter with a DC/AC conversion efficiency of 96.2 %. The PV modules are in a rack structure mounted on the roof of the building oriented north-south and tilted 10° due to the building an owner's requirements; see Table I for a summary of the site and characteristics of the system.

A. Climate and system monitoring

This PV system has an in-plane Solar radiation sensor (G_i), one back module temperature (T_m) and ambient temperature (T_{amb}) that records every 5 minutes the average of the time interval. All the sensor merges to a Sensor Box that is near the inverter location. Each inverter has his own DC and AC measuring system that also connects to the Sensor Box via Ethernet cable. The data saves in the internal memory of the Sensor Box, and it can be manually downloaded to do the analysis. For this work, we downloaded the data after one year of recording in 5 minutes intervals from January 2016 to December 2016.

III. METHODOLOGY

We divided this work into three main results: simulated, measured and system losses due to shading and soiling. For the first and second section, we used all the monitoring parameters of the PV system to estimate the year energy output. To evaluate the performance of the PV system we used the following energy metrics defined in the IEC 616214-1:2017 standard:

- total AC energy generated by the PV system (E_{out}),
- final yield (Y_f),
- reference yield (Y_r),
- performance ratio (PR).

A. Energy metrics

The most simple parameter to calculate is AC Energy output of the PV array. It was computed by summing the multiplication of the instantaneous power (P_{out}) and the sampling interval (τ) over the whole year,

$$E_{out} = \sum_k P_{out,k} \tau_k, \tag{1}$$

where the P_{out} is kW and τ is 5 minutes converted into hours.

The energy yields are technical parameters that relate the quantity of energy delivered with the power installed in real operating conditions. They indicate the relative electricity generation to its rated capacity in a given period. Theses parameters allow comparing the performance of the PV systems in different locations. The final yield (Y_f) is the AC energy output of the PV system (E_{out}) per rated or nominal power (P_0) of the installed PV array,

$$Y_f = \frac{E_{out}}{P_0}, \tag{2}$$

where P_0 is in kW.

The reference yield (Y_f) is the ratio of the total in-plane solar radiation (H_i) to the PV array reference irradiance.

$$Y_r = \frac{H_i}{G_{i,ref}}, \tag{3}$$

where H_i is in kWh/m^2 and $G_{i,ref}$ is the irradiance at which P_0 is determined in kW/m^2.

The performance ratio (PR) is a dimensionless relationship between the PV system's final yield Y_f and its reference yield Y_r. It compares the energy power that the system injects to the grid (E_{out}) and the ideal energy yield to determine the overall electrical performance. It is defined as:

$$PR = \frac{Y_f}{Y_r}. \tag{4}$$

The PR is perhaps the most important metric because it determines how effectively the plant converts sunlight into AC energy delivered to the grid about what energy we would expect from the panel nameplate rating. This metric quantifies the overall effect of losses due to system availability, inverter efficiency, wiring, cell mismatch, elevated PV module temperature, reflection from the front surface, soiling, system down-time, shading, and component failures [2]. Besides in the literature there is, at least, two corrections for the PR considering variation in the module temperature.

Once the PR ratio has been determined it can be corrected using the STC (25 °C) temperature:

$$PR_n = \frac{\sum_k P_{out} \tau_k}{\sum_k \frac{(C_{n,k} P_0) H_i}{G_{i,\text{ref}}}} \tag{5}$$

where $C_{n,k}$ is the adjustment factor defined as

$$C_{STC,k} = 1 + \gamma(T_{mod,k} - 25°C) \tag{6}$$

where γ is the maximum power temperature coefficient provided by the manufacturer in units of °C and $T_{m,k}$ is the measured module temperature.

The second correction is by changing eq. (6) with the annual average module temperature $T_{m,avg}$

$$C_{TEM,k} = 1 + \gamma(T_{m,k} - T_{m,avg}). \tag{7}$$

The losses factors that are inherently incorporated in the PR are in general weather dependent. The module temperature (T_m) is the indicative of weather changes. To better assess the performance of the PV system the seasonal variation must be avoided to reduce uncertainty and estimation errors. Usually, for large PV plants, the PR is corrected to STC temperature. However, this correction results in a higher PR if the ambient temperature is above 25 °C, because of that, the PR calculated is not representative of the performance PV system. To improve the PR value and to cut out the seasonal variability, an average module temperature factor can be established to correct the PR. For this work, we present the energy results of the PV system analyzed with the PR, and the STC corrected PR and the annual average temperature PR.

B. Energy Simulation

For this work, we simulated the electrical performance of the PV system prior installation to estimate the first performance metrics. We used the System Advisor Model (SAM) [3] with the information provided in Table I along with the NREL-NSRDB to simulate the energy output of the PV system. We focused in the final electricity generation E_{out}, the plane of array irradiance (H_i) and the performance ratio.

C. Soiling and shading losses

The installation is partially shaded in the morning, almost all the year, being December the largest shade projected by a three that is next to the building. The shade lasts for at least one (1) hour and reduces the power output of two (2) PV arrays. Using the power output information, we evaluated the losses due to shading.

Finally, soiling is an important issue in the PV electrical performance since it can reduce the power output drastically by reducing the current generation. We evaluated the soiling impact in the PV system by randomly picking two modules from the installation and leave the in-situ soiling to deposit for four months. After that time, we measured the individual I-V curve of the soiled module, Then we cleaned it and took another IV curve to determine the power loss.

IV. RESULTS AND DISCUSSION

A. Simulation

The performance of the 27 kW PV system was simulated using SAM from the NREL that include all the climate time series and performance metrics (see Table II for summarized results). Fig. 1 shows the daily average final yield per month with its standard deviation (black line) and its PR. It can be observed that the final yield varies daily from 4.23 h/d (January) the lowest to 5.05 h/d (April) the highest. The performance ratio goes from 0.67 in April to 0.72 in December. We obtained a final energy output of 45,811.64 kWh/year and 0.71 average PR.

The total generated energy output measured was 35,156.15 kWh/year. The average daily solar resource measured for this locations is 5.76 hours; this quantity exceeds the typical average annual daily for the whole country, which is 5 hours. July was the month of the greater solar resource, 6.35 hours and January 2017 had the lowest value 4.91 hours. The average daily yield per month obtained goes from 3.34 kWh/kWp to 4.93 kWh/kWp. These values give an annual final year of 1302.08 kWh/kWp without counting August. The typical average annual yields for a specific country are 1372 kWh/kWp for India [4] and 1361 kWh/kWp for Spain [5]. The average performance ratio obtained from one year of measurement was 0.75 with values within the range of 0.68 to 0.82. January was the month with the lowest PR and June with the highest PR. Although June was not the month with the highest solar resource either the maximum value of energy injected into the grid, it was the month with lower losses, mainly due to climate conditions such as wind speed shading and soiling.

Fig 2 shows the daily average final yield obtained with its performance ratio and using the corrections method of STC and year average module temperature. For STC we used $25.0\,^{\circ}\mathrm{C}$ and for TEM we used the weighted mean (by irradiance) module temperature of $52.7\,^{\circ}\mathrm{C}$.

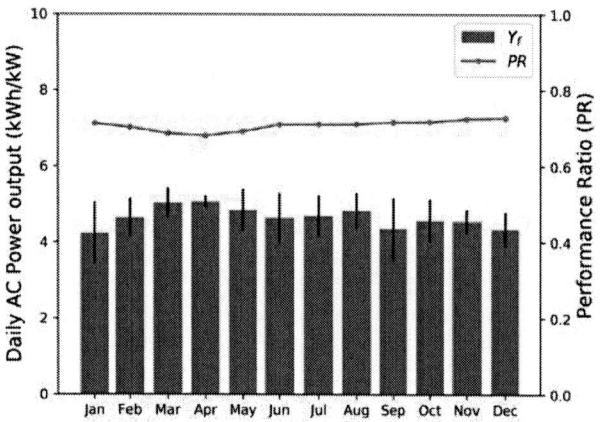

Fig. 1. Simulated average final yield and performance ratio of the 27 kW PV system.

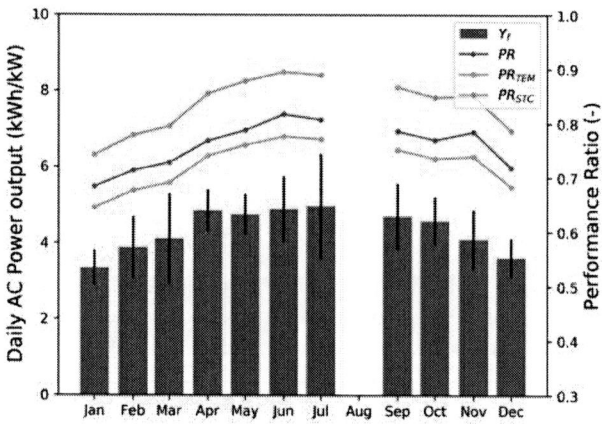

Fig. 2. Measured average final yield, performance ratio and two correction methods a of the 27 kW PV system.

B. Measured results

We recorded the data for one year of operation from January 1st of 2016 to December 31th of 2016. During this time, the system had an unexpected shutdown from July 8th until September 9th due to problems with the electrical service provider. Once the problem was solved, the system started functioning again. While in this period, the climate data monitoring system continued running without any problems.

Table III summarize the results of the PV system performance analysis. It can be seen that the average module temperature obtained was 44.01 °C with a maximum average in August. The ambient temperature fluctuates between 27.36 °C to 32.61 °C.

Fig 3 shows a closeup of the daily average variation of the performance ratio. It can be seen that PR_{STC} is at least 10% higher than PR. Using PR_{STC} to evaluate the PV plant performance could generate inconsistency with the real conditions of the plant. Since the average ambient temperature obtained is higher than STC conditions, PR_{STC} does not give much information about the real performance of the PV system. PR_{TEM} shows lower values since it uses the highest temperature correction. However, this might recreate the real performance condition since it is using the temperature at which the module is going to operate. For climates that have an average temperature greater than STC conditions should use the method two to correct the performance ratio.

978-1-5090-5606-4/17 $31.00 © 2017 IEEE

Table II
DAILY AND MONTHLY ENERGY STATISTICS SIMULATED OF THE 27 KW PV SYSTEM OF ONE YEAR OF DATA.

| Month | Daily temperature | | | | | | Daily irradiation | | | Monthly values | | | |
| | T_m | | | T_{amb} | | | H_i | H_i | H_i | H_i | E_{out} | Y_f | PR |
	mean [°C]	max [°C]	min [°C]	mean [°C]	max [°C]	min [°C]	mean [kWh/m²]	max [kWh/m²]	min [kWh/m²]	sum [kWh/m²]	sum [kWh/m²]	mean [kWh/kW]	mean -
2016-January	43.33	53.11	29.05	26.26	29.93	21.47	5.96	6.98	2.16	184.68	3543.10	4.23	0.71
2016-February	45.48	55.89	33.40	28.49	32.57	24.27	6.62	7.42	4.39	191.86	3631.07	4.64	0.70
2016-March	47.35	53.31	40.37	29.72	34.10	26.62	7.33	7.85	5.34	227.30	4210.19	5.03	0.69
2016-April	49.14	57.89	44.91	31.14	34.45	26.89	7.44	7.59	6.86	223.21	4098.76	5.06	0.68
2016-May	46.28	53.23	29.45	29.11	33.72	22.44	7.01	7.73	3.16	217.37	4051.46	4.84	0.69
2016-June	41.49	48.78	33.48	24.23	26.87	21.83	6.56	7.56	3.71	196.68	3761.01	4.64	0.71
2016-July	41.22	47.51	32.56	24.03	26.17	22.11	6.61	7.54	3.98	205.05	3925.54	4.69	0.71
2016-August	41.08	47.09	33.92	23.44	25.47	21.32	6.79	7.73	5.43	210.55	4038.40	4.82	0.71
2016-September	38.84	44.65	31.36	23.43	24.82	22.03	6.09	7.62	2.79	182.75	3529.49	4.36	0.72
2016-October	41.39	49.35	33.97	23.26	24.69	20.97	6.38	7.22	4.76	197.66	3823.99	4.57	0.72
2016-November	40.10	47.99	33.19	23.03	24.27	21.32	6.28	7.05	5.37	188.42	3684.85	4.55	0.72
2016-December	41.45	49.93	27.80	23.76	26.68	20.06	5.97	6.47	3.80	179.13	3513.78	4.34	0.73
Year	**43.10**	-	-	**25.83**	-	-	**6.59**	-	-	**2404.66**	**45811.64**	**4.65**	**0.71**

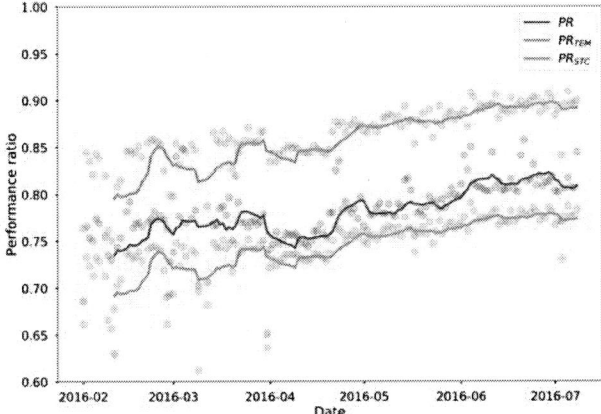

Fig. 3. Performance ratio trend using two correction, performance ratio corrected by mean temperature (eq. 7) and by STC temperature (eq. 6). Lines represent 10 per. moving average for curve smoothing.

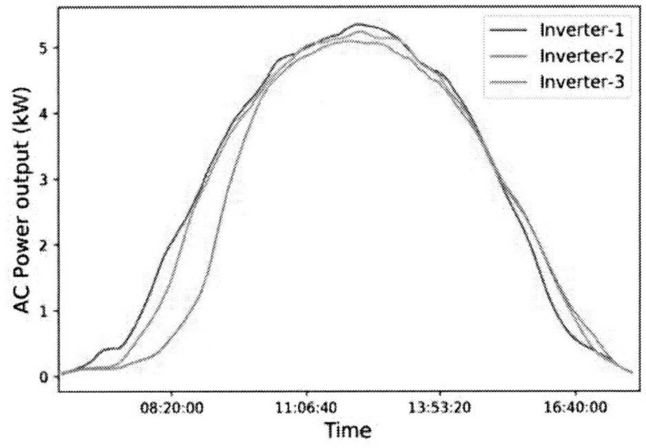

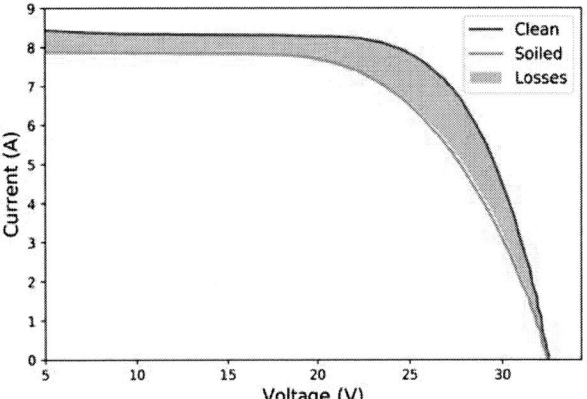

Fig. 4. Hourly average power output of three inverters for the month of December.

C. Shading and losses

As mentioned before, the system is partially shaded by a tree that is located in the neighborhood. The tree projects its shade to the PV array until approximately 10:00 AM. In December, the shadow hits its maximum shading to one array and a portion of the other. Fig 4 shows the impact of the shading in the average power output of December. We estimate that the losses due shading are approximately 50% until 10:00 AM then recovering to its standard power. This losses translate to 9% of the total output energy of December.

Finally, Fig 5 shows the losses in current due to soiling deposition. It can be seen that after fourth months of soiling, the power loss is up to 12% in the max power point. This could indicate that PV installations near the area should have a cleaning schedule to avoid this losses. The next step of this research is to evaluate the soiling deposition rate per month of the site. However, to do this more data is required with rain fall monitoring.

Fig. 5. Average I-V curve of soiled module vs cleaned. All curves were translated to STC conditions using method one of IEC 60891.

Table III
DAILY AND MONTHLY ENERGY STATISTICS MEASURED OF THE 27 KW PV SYSTEM OF ONE YEAR OF DATA.

| | Daily temperature | | | | | | Daily irradiation | | | Monthly values | | | |
| | T_m | | | T_{amb} | | | H_i | H_i | H_i | H_i | E_{out} | Y_f | PR |
Month	mean [°C]	max [°C]	min [°C]	mean [°C]	max [°C]	min [°C]	mean [kWh/m²]	max [kWh/m²]	min [kWh/m²]	sum [kWh/m²]	sum [kWh/m²]	mean [kWh/kW]	mean -
2016-February	42.06	45.64	31.09	28.11	30.39	22.32	5.47	6.19	3.29	158.65	3010.45	3.84	0.71
2016-March	42.15	49.53	23.06	29.03	33.63	19.51	5.70	7.58	1.94	176.69	3428.80	4.10	0.73
2016-April	46.15	49.47	34.07	32.52	34.89	25.97	6.31	7.47	3.41	189.38	3917.51	4.84	0.77
2016-May	45.81	49.64	38.89	32.61	35.36	28.76	6.01	7.31	4.58	186.32	3949.37	4.72	0.79
2016-June	42.91	46.98	33.13	29.25	31.78	25.78	5.99	7.02	3.20	179.74	3955.09	4.88	0.82
2016-July	46.55	50.72	40.06	29.26	31.75	26.11	6.35	7.22	4.53	196.73	1065.41	4.93	0.73
2016-August	47.24	51.54	34.28	29.19	31.17	25.33	6.05	7.18	3.00	187.51	-	-	-
2016-September	44.98	51.03	33.29	28.42	30.39	23.93	6.11	7.19	3.46	183.28	2885.84	4.65	0.75
2016-October	44.67	47.94	37.5	28.93	30.62	26.54	5.94	6.75	3.87	184.10	3822.63	4.57	0.77
2016-November	41.35	46.23	25.67	27.36	30.73	22.04	5.24	6.21	1.46	157.10	3309.28	4.09	0.79
2016-December	42.92	44.33	40.73	28.6	30.08	26.9	5.02	5.55	4.61	155.54	3019.73	3.61	0.72
2017-January	41.28	43.29	38.44	27.55	29.02	23.93	4.91	5.91	3.68	152.25	2792.04	3.34	0.68
Year	**44.01**	-	-	**29.24**	-	-	**5.76**	-	-	**22107.2**	**35156.15**	**4.32**	**0.75**

V. CONCLUSIONS

In this work, we performed an evaluation of a 27 kW PV system installed in Mexico. It was obtained a measured PR of 0.75, and it is comparable with other results from different analysis in similar climates [6], [7]. The final energy output was 35.15 MWh with a daily mean final yield of 4.32 kWh.

The simulation was very close to the values calculated in the PV plant. However, the average PR of the simulated system was lower than estimated PR, because the daily mean irradiation of the measured values were lower than the simulated. Also, the final energy output of the simulated was higher because the over estimation of the parameters of the region and the shutdown of the system for more than one month.

The performance ratio is a good metric to estimate the feasibility of a PV power plant. However, we obtained that using the STC correction overestimate the real value. For this site, the average module temperature correction seems to be a more realistic result and approximation to predict the energy output of a PV power plant.

We found that there is a considerable amount of soiling that is deposit in the PV installation. This might cause a considerable energy loss. The impact in output power of one PV module was 12% of the maximum power.

ACKNOWLEDGMENTS

This research was supported by the program FORDECYT through the project 100603: *"Estudio sobre el uso de la energía solar en aplicaciones residenciales, industriales y comerciales en diferentes estados del país"*.

We would like to thank the partial funding support by CeMIE-Sol through the project P-29: *"Desarrollo de un laboratorio nacional para la evaluación de la conformidad de módulos y componentes de sistemas e instalaciones foto-voltaicas - LANEFV"*.

Finally, we would also thank Amilcar Reyes Roldan and Jorge Luis Carnalla Ortiz for the measurements and data collection.

REFERENCES

[1] IRENA, *Renewable capacity statistics*, 2016.
[2] T. Dierauf, A. Growitz, S. Kurtz, J. L. B. Cruz, E. Riley, and C. Hnasen. "Weather-corrected performance ratio". Contract no. 303, pp 275-3000, 2013.
[3] System Advisor Model Version 2016.3.14 (SAM 2016.3.14). National Renewable Energy Laboratory. Golden, CO., 2017.
[4] B. S. Kumar and K. Sudhakar, "Performance evaluation of 10 MW grid connected solar photovoltaic power plant in India," *Energy Reports*, vol. 1, pp. 184–192, 2015.
[5] M. S. de Cardona and L. López, "Evaluation of a grid-connected photovoltaic system in southern Spain," *Renewable Energy*, vol. 15, no. 1-4, pp. 527–530, 1998.
[6] Y. Ueda, K. Kurokawa, K. Kitamura, M. Yokota, K. Akanuma, and H. Sugihara, "Performance analysis of various system configurations on grid-connected residential PV systems," *Solar Energy Materials and Solar Cells*, vol. 93, pp. 945–949, 2009.
[7] M. Mpholo, T. Nchaba, and M. Monese, "Yield and performance analysis of the first grid-connected solar farm at Moshoeshoe I International Airport Lesotho," *Renewable Energy*, vol. 81, pp. 845–852, 2015.
[8] F. Dimroth, S. Kurtz, G. Sala, A. Bett, F. Dimroth, S. Kurtz, G. Sala, and A. W. Bett, "Preface: 7th International Conference on Concentrating Photovoltaic Systems." AIP, 2011.

Modified STC Correction Procedure for Assessing PV Module Degradation in Field Surveys

Hemant K. Singh, R. Dubey, S. Zachariah, K. L. Narasimhan, B. M. Arora, A. Kottantharayil, J. Vasi

National Centre for Photovoltaic Research and Education (NCPRE)
Department of Electrical Engineering, IIT Bombay, Powai, Mumbai 400076, India

Abstract — In this paper, a new method is explored for translating measured *I-V* characteristics of a module to STC conditions, with emphasis on suitability for PV field survey. The prediction accuracy of our proposed 'Standard Irradiance Desired Temperature' (SIDT) method has been compared to a modified IEC 60891 procedure-1 and the Anderson model-based procedure. It has been shown that the proposed SIDT method results in a normalized root mean square error of less than ±1 % in maximum power, whereas the other two methods result in an error of more than ± 2 %. The SIDT method is also shown to be better for predicting other parameters at STC conditions.

Index Terms — correction procedure, field survey, module degradation, STC, SIDT

I. INTRODUCTION

Degradation in performance of solar modules with age is a well-known phenomenon and its estimation and tracking under various geographical and environmental conditions gives important information [1]. There has been a surge in solar PV based power plant installations in recent years. Tracking the performance of the solar modules would be key to understanding the reliability of the installations. However, this is best done at the site in a field survey. The degradation in performance parameters (like maximum power P_{max}), is calculated by comparing the measured value to the initial value (e.g. the factory-provided value), which would be at Standard Test Conditions (STC) of 1000 W/m^2 irradiance and 25°C module temperature. In an on-site field survey, the choice of irradiance and temperature are not in our control [2], [3], and the parameters measured under field conditions must be translated to STC. There are many translation procedures, for example, IEC 60891 and ASTM 1036 [4], [5]. These works well if the irradiance is above a certain threshold. Also, for good accuracy we require measurements under several irradiance and temperature conditions. Due to limitations of available irradiance and temperature conditions at the site, as well as the time-consuming requirement of multiple measurements, such translation procedures may not always be suitable for a field survey, though they work well for laboratory-based studies. For example, during the 'All-India Surveys of PV Module Reliability' [2], [3] which we have been conducting to assess degradation of performance in the field, several tens of modules are measured at a site within a few hours. In this paper, a new method of translating to STC has been explored. We describe the procedure, and estimate the accuracy of the proposed translation procedure (compared with existing translation procedures) by undertaking a series of experiments.

II. EXPERIMENTAL METHOD

A. The Proposed Standard Irradiance Desired Temperature (SIDT) Correction Procedure

In the proposed SIDT correction (or translation) procedure, the measured *I-V* characteristics at any irradiance (G_m) and temperature (T_m) are first corrected for standard irradiance (1000 W/m^2) using the following equations:

$$V_{SIDT} = V_m \qquad (1)$$

$$I_{SIDT} = I_m \times \frac{1000}{G_m} \qquad (2)$$

where V_m, I_m are the values on the *measured I-V* characteristics at irradiance G_m and temperature T_m; V_{SIDT}, I_{SIDT} are the values on the *irradiance corrected I-V* characteristics at 1000 W/m^2 and temperature T_m.

After irradiance correction of the *I-V* characteristics, important performance parameters like open circuit voltage (V_{oc}), short circuit current (I_{sc}), voltage at maximum power (V_{max}), current at maximum power (I_{max}), fill factor (*FF*), maximum power (P_{max}) are available at standard irradiance and 'desired' temperature (SIDT) condition of 1000 W/m^2 and T_m. After this step, the temperature correction has to be performed. For this, we use the usual temperature coefficients of V_{oc}, I_{sc} and P_{max}, to obtain the values at standard temperature i.e. 25°C, using:

$$X_{STC} = X_{SIDT} + \Phi \times (25 - T_m) \qquad (3)$$

where X_{STC} is the temperature corrected $V_{oc,STC}$, $I_{sc,STC}$, or $P_{max,STC}$, and X_{SIDT} is $V_{oc,SIDT}$, $I_{sc,SIDT}$, or $P_{max,SIDT}$, as obtained earlier, and Φ is the appropriate temperature coefficient α, β or γ (in A/°C, V/°C or W/°C) as reported at STC condition by the manufacturer. (If α, β and γ are given in units of %/°C, then they need to be multiplied by their respective V_{oc}, I_{sc} and P_{max} values as reported for STC conditions.) The main advantage of the SIDT method described here is that since α, β and γ are generally quite irradiance dependent, and often *specified only at STC*, the correct value gets used in the temperature

978-1-5090-5606-4/17 $31.00 © 2017 IEEE

translation step of the procedure. For translating I_{max} and V_{max} to the standard temperature, we use temperature coefficients of 0 and γ respectively, which are reasonable approximations.

B. Modified version of IEC 60891 and Anderson model based methods suited for PV field survey

Since the IEC 60891 correction procedure would require measurements at least at three different irradiance and module temperatures [4], we used modified equations keeping in mind the field constraints of using *I-V* data at a single irradiance and module temperature, as actually encountered at site [2]. The modified version used IEC 60891 correction procedure-1 but keeping internal series resistance factor R_s and curve correction factor k equal to zero, as these values are difficult to assess under field conditions [2]. The modified translation equations based on IEC 60891 correction procedure-1 [2], [4] are:

$$V_{STC} = V_m + \beta.(25 - T_m) \qquad (4)$$

$$I_{STC} = I_m + I_{sc,m}.(\frac{1000}{G_m} - 1) + \alpha.(25 - T_m) \qquad (5)$$

Here V_m, I_m are the values on the measured *I-V* characteristics; V_{STC}, I_{STC} are the corresponding values on the corrected *I-V* characteristics for STC; G_m is the measured irradiance and T_m is measured temperature of the module; $I_{sc,m}$ is the short circuit current estimated from the measured *I-V* characteristics; α, and β are temperature coefficients (in A/°C, V/°C) as reported at STC condition.

Similarly, the modified translation equations based on Anderson model [6]–[9] are:

$$V_{STC} = \frac{V_m}{1 + \beta(T_m - 25)} \qquad (6)$$

$$I_{STC} = \frac{I_m \times 1000}{G_m(1 + \alpha(T_m - 25))} \qquad (7)$$

The symbols have usual meanings as mentioned above. However, in this case the temperature coefficients α, and β are in °C^{-1} unit as per reported STC condition.

C. Measurement of I-V characteristics for evaluation of the correction procedures

We generated experimental data for evaluating the three different correction (or translation) procedures using a multi c-Si module having nominal maximum power of 140 W (with positive tolerance as per module data sheet). The module was heated in outdoor sunlight to about 70°C and then brought back into the lab where *I-V* characteristics were measured using a Spire solar simulator (5600SLP Blue) for different

irradiances (500, 600, 700, 800, 900 and 1000 W/m^2) and temperatures (65, 55, 45, 35, and 25°C) as the module cooled down. The measurement of course included the STC condition of 1000 W/m^2 and 25°C. For checking the accuracy of the translation procedure, the module parameters actually measured at STC were used as the reference.

D. Normalized root mean square error (NRMSE) calculation

For comparing the accuracy of different translation methods, normalized root mean square errors (NRMSE) were calculated using following equation:

$$NRMSE = \frac{1}{X_{STC,m}}\sqrt{\frac{\sum_{i=1}^{n}(X_{STC,i} - X_{STC,m})^2}{n}} \qquad (8)$$

where $X_{STC,i}$ is the predicated parameter (V_{oc}, I_{sc}, V_{max}, I_{max}, FF, or P_{max}) for STC condition using the correction procedure at radiation/temperature i and $X_{STC,m}$ is the same performance parameter actually measured at STC condition.

III. RESULTS AND DISCUSSION

According to the proposed SIDT correction procedure, first the irradiance correction for the measured *I-V* data is performed. Fig. 1 show the *I-V* plot measured for the case of 800 W/m^2 and 45°C temperature, and also the SIDT corrected *I-V* at 45°C. The latter is compared with the actual measured *I-V* at 1000 W/m^2 and 45°C, and these are seen to compare very well.

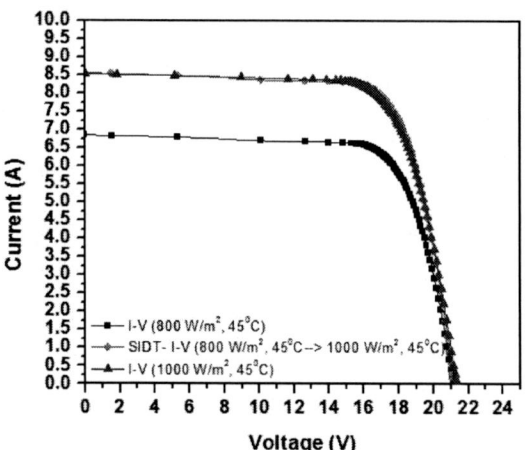

Fig. 1. Irradiance corrected *I-V* plot at T_m=45°C and the corresponding as-measured *I-V* plots for 800 W/m^2 and 1000 W/m^2 irradiance at temperature T_m= 45°C.

After estimating the performance parameters for standard irradiance at temperature T_m, the temperature correction is done to predict the performance parameters for STC. Here temperature corrections were done using the estimated temperature coefficients (α, β and γ) from measured I-V parameters (I_{sc}, V_{oc} and P_{max}) at different temperatures (65, 55, 45, 35, and 25^0C) at standard irradiance (1000 W/m^2) condition instead of factory provided temperature coefficients as the module used for study was one year old. This was done to avoid any uncertainty in temperature coefficient values which might influence the STC correction procedure comparison, as degradation in module may also cause degradation or change in temperature coefficients over time. Table I shows the predicted P_{max} for STC conditions from measured I-V data at various irradiances and module temperatures using the SIDT correction procedure. The bold entry corresponding to the 1000 W/m^2 irradiance and 25°C would be the measured STC value of P_{max}. It can be seen from the table that for a wide range of temperature and irradiance conditions, the maximum deviation of the predicted STC P_{max} from the reference STC value is not more than 2.6 W.

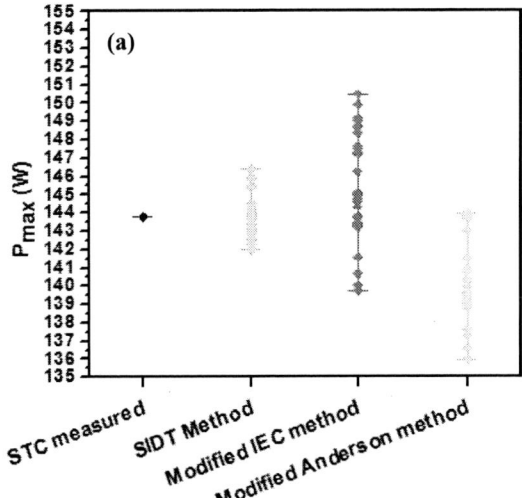

TABLE I
STC CORRECTED P$_{MAX}$ AS ESTIMATED USING PROPOSED SIDT CORRECTION METHOD

P_{max} (W)	Temperature (^{0}C)				
Irradiance (W/m^2)	25	35	45	55	65
500	143.0	142.0	142.9	144.0	145.8
600	143.7	142.5	143.4	144.5	146.4
700	143.9	142.1	142.3	142.5	143.0
800	143.9	143.2	144.3	145.5	145.9
900	143.9	143.2	144.2	145.4	145.9
1000	**143.8**	142.5	142.6	142.9	143.6

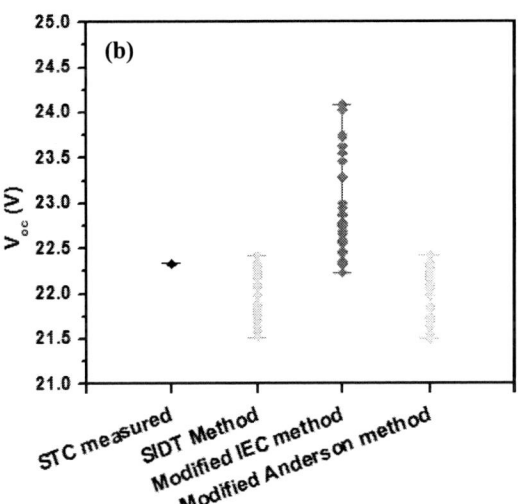

A similar exercise was done for the parameters V_{oc}, I_{sc}, V_{max}, I_{max} and FF, along with P_{max}, for all the measured conditions of irradiance (500-1000 W/m^2) and temperature (25 - 65°C) using the SIDT method as well as the modified IEC 60891 method, and Anderson model based method discussed above. Fig. 2(a)-(d) shows the STC estimated P_{max}, V_{oc}, I_{sc}, and FF comparison respectively for the three different methods. It can be seen from Fig. 2(a) and (d) that SIDT method gives less spread in estimated STC P_{max} and FF values compared to the other two correction procedures. Though, the deviation in predicted STC V_{oc} and I_{sc} values, from actual STC measured values are about the same for the case of SIDT method and Anderson model based method (see Fig. 2(b) and (c)), and much less than the modified IEC method. Similar observation can also be made from Table II which shows the normalized root mean square error (NRMSE) with respect to the STC value for the different performance parameters calculated using these three methods.

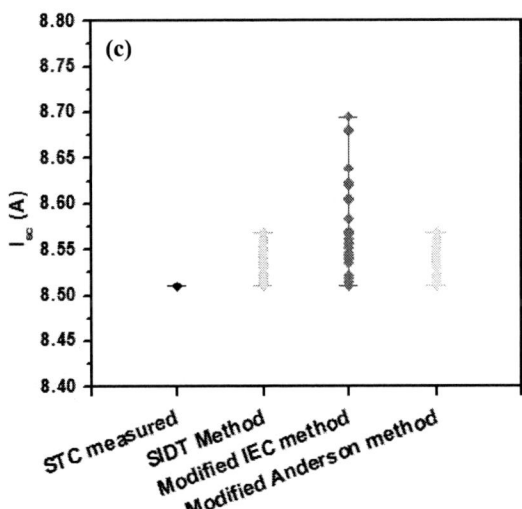

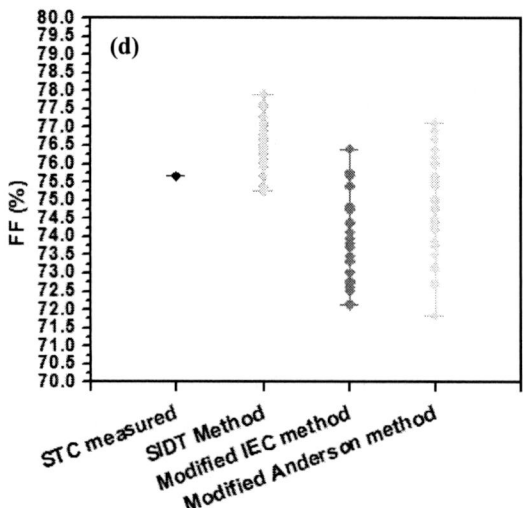

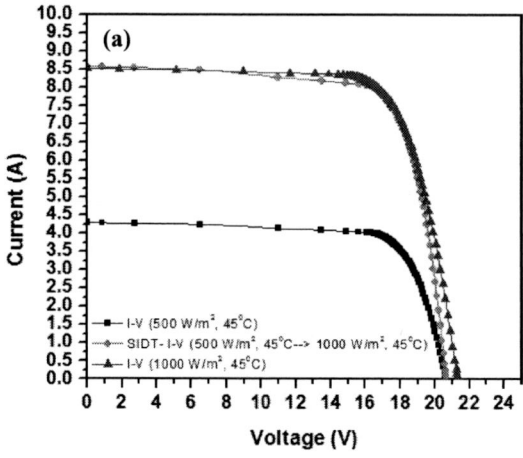

Fig. 2. STC measured and 25 STC corrected values from measured I-V characteristics at different irradiance G_m (500, 600, 700, 800, 900 and 1000 W/m²) and temperature T_m (65, 55, 45, 35, and 25°C) conditions using the SIDT method, the modified IEC 60891 method and Anderson model based method for parameter (a). P_{max}, (b). V_{oc}, (c). I_{sc}, and (d). FF.

TABLE II
NRSME OF DIFFERENT PERFORMANCE PARAMETERS FOR THREE CORRECTION PROCEDURES

NRSME (%)	Correction Procedure		
Parameter	SIDT	Modified IEC 60891	Modified Anderson method
P_{max}	± 0.85	± 2.27	± 2.82
V_{oc}	± 1.88	± 3.47	± 1.96
I_{sc}	± 0.36	± 0.95	± 0.36
V_{max}	± 1.27	± 2.10	± 1.70
I_{max}	± 0.95	± 0.56	± 1.32
FF	± 1.46	± 2.61	± 2.07

It can be seen from Table II that the SIDT correction procedure leads to minimum error for all performance parameters (P_{max}, V_{oc}, I_{sc}, V_{max}, and FF, with the exception of I_{max}) compared to the other two methods. Within the SIDT method, it can be seen that there is a relatively high error in V_{oc}, V_{max} and FF prediction. This is due to an inherent limitation in translating the series resistance from one irradiance to another. The resistance effects are not well captured in the translation method, especially when I-V translation is done from low irradiance and high temperature to SIDT conditions. One example can be seen from Fig. 3(a) and (b) which shows the as measured and SIDT translation based I-V characteristics of the module at given irradiance and temperature conditions.

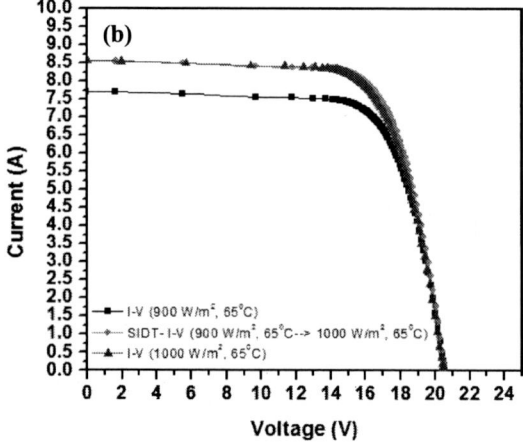

Fig. 3. Irradiance corrected I-V plot and the corresponding as-measured I-V plots. (a). Irradiance 500 W/m² and 1000 W/m² having module temperature T_m=45°C; (b). Irradiance 900 W/m² and 1000 W/m² having module temperature T_m=65°C.

Comparing Fig. 1, Fig. 3(a) and (b), one can see that there is a slight deviation in SIDT irradiance corrected I-V characteristics near P_{max} for the 500 W/m² case having module temperature of 45°C compared to the 800 W/m² case having module temperature of 45°C and 900 W/m² case having module temperature of 65°C. This indicates that the module electrical behavior under low irradiance and high temperature is different from high irradiance high temperature and is not getting translated using the translation procedure to targeted high irradiance level. However, it can be noticed from Table II that the estimated error in these parameters (V_{oc}, V_{max} and FF) prediction using the SIDT method are lesser than the estimated error in prediction of corresponding parameters using the other two methods. Also, the SIDT method gives relatively much better estimate of P_{max} among all the three discussed methods. Therefore, the proposed SIDT method is

978-1-5090-5606-4/17 $31.00 © 2017 IEEE

the most appropriate procedure for translation from a single measured *I-V* curve with inadequate information available for the series resistance, curve correction factor, and irradiance dependence of temperature coefficients. Also, as mentioned earlier, the temperature coefficients can also degrade over time. Micro-cracks, degradation in cell passivation, and many other factors which govern the recombination in cells influence the temperature coefficients. Hence, it is recommended to use the time-zero temperature coefficient values. Most of the module manufactures mention the values of temperature coefficients for their modules measured at factory. These factory specified temperature coefficient values can be considered as time-zero temperature coefficients for the given module, as degradation of these coefficients with time would then be automatically captured in the overall module performance degradation estimate. This is very useful for assessing the degradation of various parameters in the field.

In summary, the proposed SIDT method works well for a wide range of temperature and irradiance conditions. In particular, it performs better than the modified IEC 60891 procedure 1 and the Anderson procedure It is particularly appropriate for field survey situations where a large number of modules need to be assessed during a day in whatever prevalent ambient conditions.

IV. CONCLUSION

A new approach has been presented to calculate useful module performance parameters at STC from the *I-V* data measured at typical field irradiance and module temperature conditions. The proposed Standard Irradiance Desired Temperature (SIDT) correction procedure operates using a two-step procedure, first translating the data to standard irradiance and then using known values of temperature coefficients at standard irradiance to translate to STC. The procedure has been tested in the laboratory for a multi c-Si module using measured *I-V* characteristics under different temperature and irradiance conditions similar to those encountered in actual field survey conditions. It is shown that the proposed SIDT correction procedure is more accurate in predicting the STC values of module performance parameters like maximum power, open circuit voltage, short circuit current, fill factor etc. than the modified IEC 60891 procedure 1 and the Anderson procedure. It therefore provides a better estimation of degradation of these parameters. The SIDT

method is well suited for field surveys as it requires a minimum set of measurements, and is relatively forgiving of lack of knowledge of some of the module details.

ACKNOWLEDGEMENT

This work is supported by National Centre for Photovoltaic Research and Education (NCPRE) funded by Ministry of New and Renewable Energy (MNRE), Government of India through the Project No. 31/09/2015-16/PVSE-R&D dated 15th June 2016. One of the authors (RD) also thank Dr. Dirk Jordan from National Renewable Energy Laboratory for useful discussions.

REFERENCES

[1] D. C. Jordan and S. R. Kurtz, "Photovoltaic Degradation Rates-an Analytical Review," *Prog. Photovoltaics Res. Appl.*, vol. 21, no. 1, pp. 12–29, Jan. 2013.

[2] R. Dubey, S. Chattopadhyay, V. Kuthanazhi, J. John, F. Ansari, S. Rambabu, B. M. Arora, A. Kottantharayil, K. L. Narasimhan, J. Vasi, B. Bora, Y. K. Singh, K. Yadav, M. Banger, R. Singh, and O. S. Sastry, "All-India Survey of Photovoltaic Module Reliability: 2014," National Centre for Photovoltaic Research and Education, Mumbai, India, 2016, available online at http://www.ncpre.iitb.ac.in/uploads/ All_India_Survey_of_Photovoltaic_Module_Reliability_2014. pdf.

[3] R. Dubey et al., "Performance of Field-Aged PV Modules in India: Results from the 2016 All India Survey of PV Module Reliability" to be presented at the *44th IEEE Photovoltaic Specialists Conference*, 2017.

[4] IEC, "International Standard IEC 60891 Edition 2.0," *IEC*, vol. 2009, pp. 1–4, 2009.

[5] ASTM, "ASTM E 1036-15," *ASTM*, pp. 1–8, 2015.

[6] C. M. Whitaker and J. D. Newmiller, "Photovoltaic Module Energy Rating Procedure", Final Subcontract Report, Endecon Engineering San Ramon, California, 1998.

[7] B. Kroposki, W. Marion, D. L. King, W. E. Boyson, and J. A. Kratochvil, "Comparison of Module Performance Characterization Methods," in *Conference Record of the Twenty-Eighth IEEE Photovoltaic Specialists Conference - 2000*, 2000, pp. 1407–1411.

[8] B. Marion, "Validation of a Photovoltaic Module Energy Ratings Procedure at NREL," in *NCPV Program Review Meeting*, 2000, pp. 1–4.

[9] R. M. Smith, D. C. Jordan, and S. R. Kurtz, "Outdoor PV module degradation of current-voltage parameters," in *World renewable energy forum*, 2012, no. May, pp. 1–9.

Degradation Models of Photovoltaic Module Backsheets Exposed to Diverse Real World Condition

Yu Wang*, Sebastien Merzlic†, Andrew Fairbrother‡, Scott Julien§, Lucas Fridman*, Camille Loyer†, Amy L. Lefebvre†, Gregory O'Brien†, Xiaohong Gu‡, Liang Ji¶, Ken Boyce¶, Michael Kempe‖, Kai-tak Wan§, Roger H. French*, Laura S. Bruckman*

* Department of Materials Science and Engineering, SDLE Research Center, Case Western Reserve University,
10900 Euclid Avenue, Cleveland, OH 44106, USA
†Fluoropolymers R&D, Arkema, Inc. King of Prussia, PA 19406, USA
‡Engineering laboratory, National Institute of Standards and Technology, Gaithersburg, MD 20899, USA
§ Mechanical & Industrial Engineering, Northeastern University, Boston, MA 02115, USA
¶Renewable Energy, Underwriters Laboratories, Northbrook, IL 60062, USA
‖Photovoltaics Research, National Renewable Energy Laboratory, Golden, CO, USA

Abstract—Degradation of polymeric backsheets is a result of accumulated synergetic effects of kinds of environmental stresses. In this study, degradation of the optical properties of 16 commercial photovoltaic module backsheets exposed in 5 different climatic zones was investigated. In addition to yellowness index and gloss measurement, Fourier transform infrared spectroscopy was conducted on the outer layer of a PV backsheet to study the chemical changes during exposure. Several simple linear regression models of yellowness index and gloss as function of degradation factors were developed with the R language. Statistical parameters of the models indicates a good fit to data and a high predictive accuracy.

Index Terms—Photo-voltaic modules, backsheet, reliability, prediction model

I. Introduction

The backsheet of a photovoltaic module plays an essential role in protecting the solar cell from outside environmental aging factors and guaranteeing at least 20 years lifetime [1] [2]. Each layer of the backsheet function to be durable, safe and reliable in its service environment[3]. Some commonly observed degradation behaviors of backsheets are yellowing, gloss-loss, delamination and cracking. These effects can subsequently induce the decrease of insulating property of backsheets, bring in moisture, which in turn leads to degradation of encapsulate layer[4][5], corrosion of metals and causes safety hazards and module performance failure consequently[2]. Therefore, it is relevant to study the degradation of photovoltaic (PV) backsheets and full understanding of backsheet performance under real world circumstances is a basic step.

It is well-known that degradation of field-aged PV module backsheets is complex and depends on various factors[2]. Köppen-Geiger climate classification, which was created aim to investigate the relationship between climate and vegetation[6], provides an different way to study the multi environmental stressors in one simple variable and is introduced as a representative of synergetic effects of temperature, humidity and other environmental aging factors.

Prediction of backsheet behavior during different conditions helps manufacturers and customers estimate the lifetime of PV modules without waiting 20-25 years[7]. Simple linear regression models are developed in this project to predict backsheet performance under particular environments with data collected on field-aged modules by Rstudio, a software suitable for big data analysis.

Two types of degradation model are developed: Stress/Response (S/R) model and Stress/Mechanism/Response (S/M/R) model. The S/R model provides a direct connection with net stress that induces degradation of polymeric material and degradation behaviors, such as yellowing and gloss-loss. The stress variable used in S/R model can wither be single factor (climatic zone, material) or multifactor (irradiance, temperature, humidity and so on)[8]. On the other hand, the S/M/R model, which incorporates degradation mechanisms in the semi-supervised generalized structural equation model, takes advantages of both material science and statistics. It is able to show the important variable and identify the pathway of degradation[9].

II. Experiments

A. Module Selection and Characterization

A sample pool consisting of 16 commercial modules that belong to 12 brands was studied in this project. Those photovoltaic modules were retrieved from the field after exposure in 5 climatic zones worldwide for 2 to 28 years. Retrieved modules were cut into pieces and sent to each member of the DOE project for measurement and study. Backsheet materials of studied modules correspond to commonly used polymers in the PV industry: polyvinylidene fluoride (PVDF), polyamide (PA), tetrafluoreothylene hexafluoropropylene and vinylidene fluoride copolymer (THV), polyvinyl fluoride (PVF), polyethylene terephthalate (PET), fluorethylene vinyl ether (FEVE). The explaination of climatic zone is shown in Table I. To quantify degradation, the yellowness index (YI) and gloss values of retrieved modules were measured. Additionally, Fourier transform infrared spectroscopy with attenuated total reflection (FTIR-ATR) was conducted on the outer layer of backsheets to investigate chemical change.

978-1-5090-5606-4/17 $31.00 © 2017 IEEE

TABLE I
DESCRIPTION OF THE KÖPPEN-GEIGER CLIMATE CLASSIFICATION

Climatic zone	Description
Csb	Warm-summer Mediterranean climate
Cfa	Humid subtropical climate
Csa	Hot-summer Mediterranean climate
Dfb	Mild/cool summer subtype climate
Dfa	Cold, without dry season, hot summer

TABLE II
SUMMARY OF S/R MODEL OF YELLOWNESS INDEX AND GLOSS AS
FUNCTION OF OUTER LAYER MATERIAL

Model Response	YI
R^2	0.876
adjusted R^2	0.8697
p-value	$< 2.2e - 16$
Model Response	Gloss
R^2	0.9885
adjusted R^2	0.988
p-value	$< 2.2e - 16$

B. Model Development

Module backsheets information, including material, exposure history, exposure conditions, and degradation data, was compiled in one dataframe. The whole dataframe was imported into Rstudio for data analysis and modeling. Models were developed with 80 % data (training set) and evaluated with the remaining 20 % data (testing set). The model generated was evaluated by statistical parameters (R^2 adjusted R^2, and Akaike information criterior (AIC) value) and the model with the most significant statistical parameters is selected.

III. RESULTS AND DISCUSSION

A. Stress/Response Model: Yellowness Index and Gloss as Function of Outer Layer Material

The optical properties of backsheets are directly related to the polymeric material used in outer layer. All modules used for model development were exposed in same Dfa (Cleveland, Ohio) climate zone for 4 years. There are two types of PVDF used, which are named as PVDF-a and PVDF-b to distinguish between the two. The PVDF-a is blended with acrylic, which is identified with FTIR spectroscopy.

The generated models are shown as (1) and (2):

$$YI = 0.4191 + 0.4649M_1$$
$$+2.3753M_2 + 5.3861M_3 \quad (1)$$
$$+(-1.4224)M_4 + 3.9161M_5$$

$$Gloss = 16.8620 + (-3.3850)M_1$$
$$+54.5190M_2 + (-10.9965)M_3 \quad (2)$$
$$+(-13.1905)M_4 + 40.0545M_5$$

where:

$M_1 = 1$, if outer layer material is PVDF-a;
$M_1 = 0$, if outer layer material is not PVDF-a;
$M_2 = 1$, if outer layer material is PET;
$M_2 = 0$, if outer layer material is not PET;
$M_3 = 1$, if outer layer material is PVF;
$M_3 = 0$, if outer layer material is not PVF;
$M_4 = 1$, if outer layer material is THV;
$M_4 = 0$, if outer layer material is not THV;
$M_5 = 1$, if outer layer material is PVDF-b;
$M_5 = 0$, if outer layer material is not PVDF-b;

The summary of this model is shown in Table II.

The estimated coefficient represents the yellowness index difference between different backsheet materials and the control group (FEVE material) after 4 years exposure in Dfa

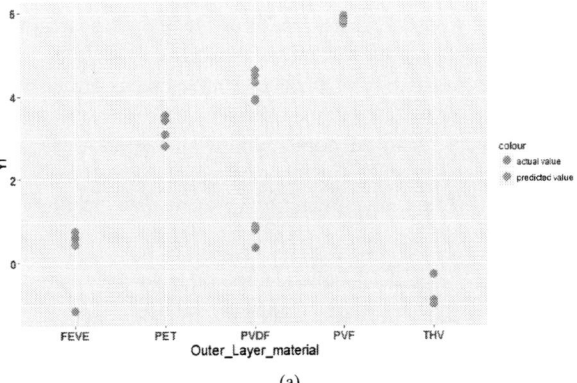

(a)

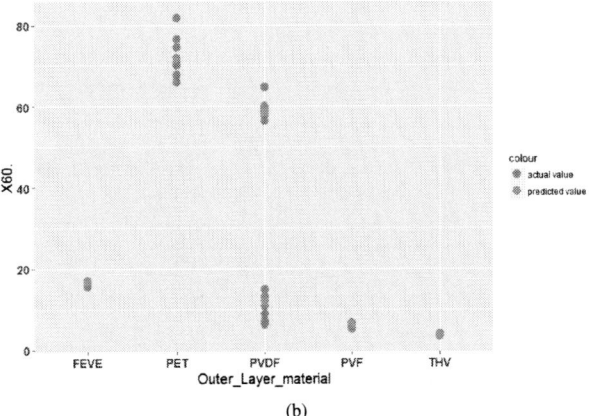

(b)

Fig. 1. Plot of Real Value and Predicted Value generated with Model (1) (a) and Model (2) (b). The predicted values are similar with real data, which represents a nice fitting of models.

climatic zone. It is obvious that the first grade PVDF shows similar yellowness index value with FEVE backsheet. Moreover, two types of PVDF material display large distinction in yellowness index. As mentioned earlier, there are acrylic peaks found in FTIR spectrum of PVDF-b, which may explain this discrepancy.

The R^2 is used to assess the model fitting and is the percentage of the response variable variation that is explained by the model, while the adjusted R^2 is a modified R^2 that adjusted to the number of explanatory terms in a model relative to the number of data points. It eliminates the bias by comparing the sample size to the number of variables in the regression model. Adjusted R^2 and R^2 of models with value near to 1 indicates a good model fit to data [10] and high accuracy of model, which also correspond to the comparison plot of real value and predicted value shown in Fig.1.

B. Stress/Response Model: Yellowness Index and Gloss as Function of Climatic Zone

In addition to the outer layer material, the degradation of backsheets relates to the environmental conditions directly. Modules with PA were exposed to two different climatic zones (Cfa and Csa) for similar exposure years (4 years and 5 years). The yellowness index and gloss at 60° are shown in Fig.2 It is observed that the YI of modules exposed in Csa climate zone has large deviation from mean value, which may due to inhomogeneity of surface. Severe cracks are observed on backsheet of this module. YI is highest near the module frame, especially in the corner, and lowest at center far from the frame and cracks. The stress/response models of YI and gloss as function of climatic zone were developed((3) and (4)).

$$YI = 4.325 + 7.965C_1 + 2.049C_2 \quad (3)$$

$$Gloss = 1.517 + (-1.8095)C_1 + (-8.5440)C_2 \quad (4)$$

where:
$C_1 = 1$, if climatic zone is Cfa;
$C_1 = 0$, if climatic zone is not Cfa;
$C_2 = 1$, if climatic zone is Csa;
$C_2 = 0$, if climatic zone is not Csa;

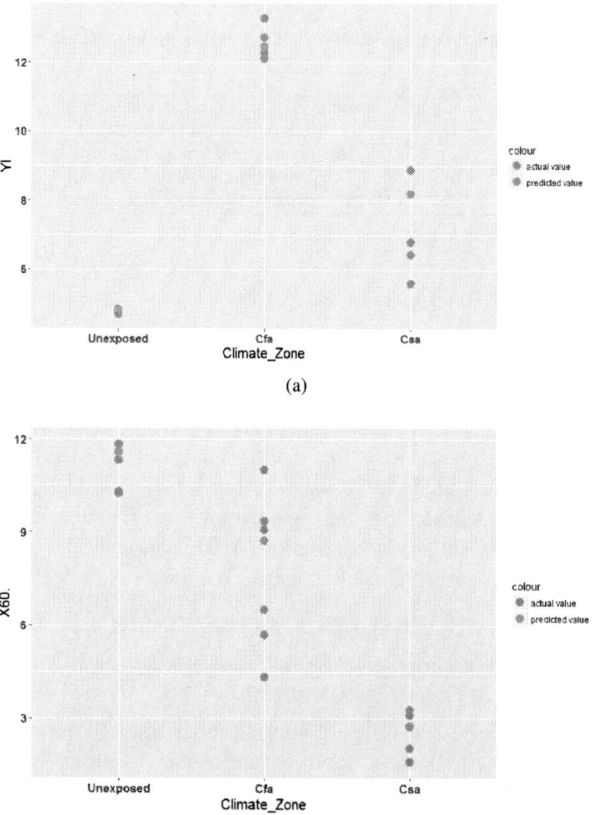

Fig. 2. Plot of Real Value and Predicted Value generated with Model (3) (a) and Model (4) (b). The predicted values are similar with real data, which represents a nice fitting of models.

TABLE III
SUMMARY OF S/R MODEL OF YELLOWNESS INDEX AND GLOSS AS FUNCTION OF CLIMATIC ZONE

Model Response	YI
R^2	0.914
adjusted R^2	0.8567
p-value	0.0252
Model Response	Gloss
R^2	0.9068
adjusted R^2	0.9036
p-value	$< 2.2e - 16$

The estimated parameters (slopes) in model (3) and (4) represent the effect of different weathering factors of climatic zone on PA backsheets. Comparing the yellowness index and gloss model, the two climatic zones of Csa and Cfa induce different responses of yellowing and gloss-loss of polyamide, which may relate to different degradation mechanisms or change of degradation mechanisms due to different degrees of environmental stresses. The comparison plot of real value with predicted value exhibits the well-fitting of data, which is corresponding to the high R^2 and adjusted R^2 in Table III.

As mentioned above, it is observed that the module exposed in Csa showed severe cracks, while another module that exposed in Cfa only had microcracks (Fig.3). The occurrence of cracks may relate to the weathering effect of the two climatic zones. The difference between two climatic zones

is different level of humidity: the Cfa has higher humidity than Csa during summer. The moisture acts as plasticizer for PA material and increases the ductility of PA [11], which may decrease the likelihood of crack formation. While PA exposes in hot and dry environment, it loss the moisture quickly, the material became brittle. We are conducting the measurement of crack quantification, and hoping to connect the crack and delamination of PV backsheet with gloss-loss or yellowing.

C. Stress/Response Model: Yellowness Index and Gloss as Function of Outer layer material and climatic zone

The degradation models consisting of outer layer material and climatic zone as independent variables are developed although the data is not balanced. The AIC value[12] is used to evaluate whether the variable is statistically important to the response. The AIC value of models that adds different variable one by one is listed in Table IV. The best model is selected with the smallest AIC. Therefore, the model is developed with both variables. If the AIC value drops most when one variable is added, then that variable shows the highest effect on response. From the Table IV, it is found that the climatic zone showed larger effect on yellowing of

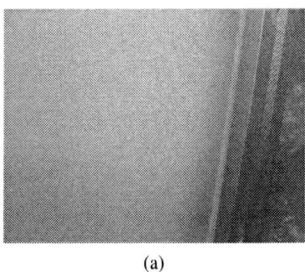

(a)

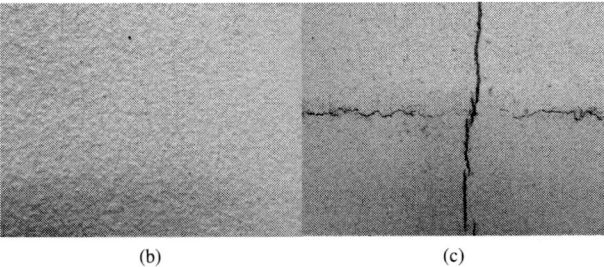

(b)   (c)

Fig. 3. Photos of PV modules with PA backsheet exposed to two climatic zones: unexposed (a), Cfa (b), Csa (c). The backsheet of module exposed in Csa climatic zones show severe cracks.

backsheet than the backsheet outer layer material, while the outer layer material has larger influence in gloss-loss.

TABLE IV
AIC VALUE OF MODELS OF YELLOWNESS INDEX AND GLOSS AS FUNCTION OF TWO INDEPENDENT VARIABLES

Varaible added	Null	Climatic Zone	Outer layer polymer
AIC(YI)	283.17	62.98	45.43
Varaible added	Null	Outer layer polymer	Climatic Zone
AIC(Gloss)	2539.98	1814.93	1808.53

D. Stress/Mechanism/Response Model: Yellowness Index and Gloss as Function of Climatic Zone

It is reported that yellowing of PA material under exposure is related to oxidation of amine end groups. Oxidation products such as ketones and aldehydes may further form carboxylic acid and other carbonyl compounds [13]. The FTIR spectra of PA were normalized to the intensity at 1633cm-1, which was assigned to the carbonyl stretching in the amide group, and was not sensitive to the thermal oxidation.

Intensity value of peak in the wavenumber range of 1700-1735 cm-1, which represents an absorption band of carbonyl compound, is chosen and named with oxidation product as a mechanistic variable. The peak near 3080 cm-1, which was assigned to combination of N-H and C-N bonds after deconvolution[13], decreases a little. It corresponds to the decrease of the peak at 1540 cm-1 that corresponds to C-N stretch and C (O)-N-H bend in amide group. Data of modules with polyamide outer layer material including yellowness index, gloss, and FTIR intensity at 1715 cm-1 and 3080 cm-1 after normalization are distilled and put into dataframe. The

Stress/Mechanism/Response model is created with Rstudio package sgSEM and shown in Fig.4.

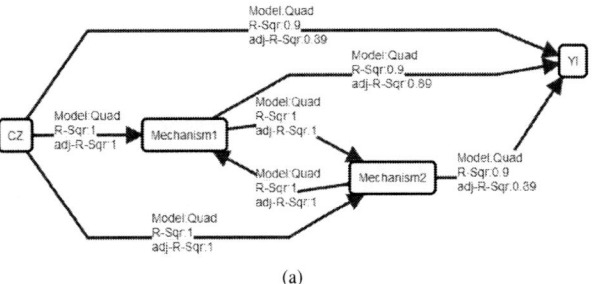

(a)

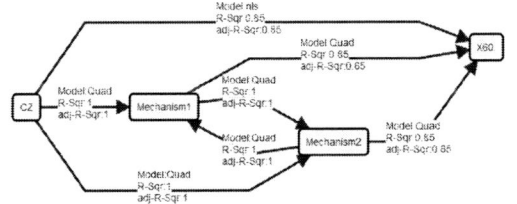

(b)

Fig. 4. The sgSEM plot of Stress/Mechanism/Response model with yellowness index (a) and gloss (b) as function of climatic zones. The mechanism1 represents the increase of FTIR intensity at 1715cm-1, and the mechsnim2 represent the decrease of FTIR at 3080 cm-1. The mechanism1, which is oxidation of polyamide, is found to have higher correlation with climatic zone and degradation of polyamide.

IV. SUMMARY AND CONCLUSION

Degradation by yellowing and gloss-loss of several types commercial PV module backsheets exposed in different climatic zone is studied. Simple linear regression models of yellowing of PV module backsheets, which cross compares the degradation of different types of PV backsheet exposed to distinct climatic zones, are developed based on data collected on retrieved modules. With obtained statistical parameters, it is stated that high prediction accuracy was achieved though generated models under particular condition.

The introduction of Köppen-Geiger climatic classification also benefits the investigation of complex effect of the real-world environment. Incorporated with the degradation mechanism study, the stress/mechanism/model provides a deeper understanding of backsheet degradation, which will be conducive to enhancement of backsheet properties and the elongation of lifetime of photovoltaic module ultimately.

ACKNOWLEDGMENT

This work was funded in part by the U.S. Department of Energy, SunShot project, under award number DE-EE0007143.

*Certain commercial equipment, instruments, or materials are identified in this work in order to adequately detail the experimental procedure. Such identification is not intended to imply recommendation or endorsement by the National Institute of Standards and Technology, nor is it intended to imply that the materials and equipment identified are necessarily the best available for this purpose.

Disclaimer: This report was prepared as an account of work sponsored by an agency of the United States Government. Neither the United States Government nor any agency thereof, nor any of their employees, makes any warranty, express or implied, or assumes any legal liability or responsibility for the accuracy, completeness, or usefulness of any information, apparatus, product, or process disclosed, or represents that its use would not infringe privately owned rights. Reference herein to any specific commercial product, process, or service by trade name, trademark, manufacturer, or otherwise does not necessarily constitute or imply its endorsement, recommendation, or favoring by the United States Government or any agency thereof. The views and opinions of authors expressed herein do not necessarily state or reflect those of the United States Government or any agency thereof.

REFERENCES

[1] Klaus J. Geretschlager, Gernot M. Wallner, Jorg Fischer, Structure and basic properties of photovoltaic module backsheet films, Solar Energy Materials and Solar Cells, vol 144, pp 451-456, 2016.

[2] M. A. Quintana, D. L. King, T. J. McMahon and C. R. Osterwald, Commonly Observed Degradation in Field-Agied Photovoltaic Modules.

[3] Chiao-Chi Lin, Peter J. Krommenhoek, Stephanie S. Watson, Xiaohong Gu, Depth profiling of degradation of multilayer photovoltaic backsheets after accelerated laboratory weathering: cross-sectional Raman imaging, Solar Energy Materials & Solar Cells, vol 144, pp 289-299, 2016.

[4] Philip hulsmann, karl-Anders Weiss, Simulation of water ingress in to PV-modules: IEC-testing versus outdoor exposure, Solar Energy, vol 115, pp 347-353, 2015.

[5] G. J. Jorgensen, K. M. Terwilliger, J. A. DelCueto, S. H. Glick, M.D. Kempe, J. W. Pankow, F. J. Pern, T. J. McMahon, Moisture transport, adhesion, and corrosion protection of PV module packaging magerials, Solar Energy Materials & Solar Cells, vol 90, pp 2739-2775, 2006.

[6] M. C. Peel, B. L. Finlayson and T. A. McMahon, Updatad world map of the Köppen-Geiger climate classification, Hydrology and Earth System Sciences, vol 11, pp 1633-1644, 2007.

[7] Manuel Vazquez and Ignacio Rey-Stolle, Photovoltaic Module Reliability Model Based on Field Degradation Studies, Progress in Photovoltaics: Research and Applications, vol 16, pp 419-433, 2008.

[8] Murray MP, Bruckman LS, French RH; Photodegradation in a stress and response framework: poly(methyl methacrylate) for solar mirrors and lens. J. Photon. Energy. 0001;2(1):022004-022004. doi:10.1117/1.JPE.2.022004.

[9] French, Roger H., et al. "Degradation science: Mesoscopic evolution and temporal analytics of photovoltaic energy materials." Current Opinion in Solid State and Materials Science 19.4 (2015): 212-226.

[10] Laura S. Bruckman, Nicholos R.Wheeler, Junheng Ma, Ethan Wang, Carl K. Wang, Ivan Chou, Jiayang Sun, And Roger H. French, Statistical and Domain Analytics Applied to PV Module Lifetime and Degradation Science, IEEE Access. pp 384-403, (2013).

[11] Jia, Nanying, and V. Kagan. "Mechanical Performance of Polyamides with Influence of Moisture and TemperatureAccurate evaluation and better understanding." Journal: Plastics Failure: Analysis and Prevention (2001): 95-104.

[12] Aho, K.; Derryberry, D.; Peterson, T. (2014), "Model selection for ecologists: the worldviews of AIC and BIC", Ecology, 95: 631636, doi:10.1890/13-1452.1

[13] Rongfu Li, Xingzhou Hu, Study in discoloration mechanism of polyaimide 6 during thermo-oxidative degradation, Polymer Degradation and Stability, vol 62, pp 523-528, 1998.

Addressing Hotspots in the Product Environmental Footprint of CdTe Photovoltaics

Parikhit Sinha[1] and Andreas Wade[2]

[1]First Solar, Tempe, AZ, 85281, USA, [2]First Solar, Mainz, 55116, Germany

Abstract — **A pilot study within the European Commission's "Single Market for Green Products" initiative has been developing standardized methods for measuring, verifying, and communicating a product's environmental performance. These methods provide the basis for identifying and then addressing hotspots in the product environmental footprint of photovoltaics. In the case of CdTe PV, hotspots related to electricity usage in manufacturing, glass content in PV modules, and metal content in modules and balance of systems can be addressed through technology innovation. Transitioning to a larger, lighter (per unit area), more efficient, and still recyclable version of current thin film modules is expected to reduce the product environmental footprint of CdTe PV to a factor of 3.5-4 below that of an average PV module.**

Index Terms — **environmental management, product life cycle management, thin films.**

I. INTRODUCTION

Since 2013, the European Commission has been conducting a pilot study within its "Single Market for Green Products" initiative. The pilot study has been testing rules, verification approaches, and communication vehicles for measuring a product's environmental performance. The product group "photovoltaic electricity generation" has been part of the pilot study since 2014, including development of product environmental footprint (PEF) category rules for conducting life cycle assessment of photovoltaic (PV) modules [1], a screening study implementing these rules on different photovoltaic technologies [2], and supporting studies by PV manufacturers [3] testing those rules on their own products. The objective of this study is to evaluate how hotspots identified in the product environmental footprint of CdTe PV may be addressed with technology innovation.

II. METHODS

Life cycle assessment (LCA) has been conducted with Simapro (V. 8.2.0) software and Ecoinvent (V. 2.2+) unit processes [4]. Life cycle impacts for PV modules were assessed in accordance with ISO 14040/14044, the PEF category rules [1], and the PEF guide [5]. The environmental footprint was quantified using the 15 environmental indicators proposed by the European Union [5] as implemented in the ILCD 2011 Midpoint V. 1.09 impact method (long term emissions excluded [1]) with equal weight assigned to the factors. Carbon footprint was estimated with 100-yr global warming potentials (GWP-100), including a GWP-100 of 25

for methane. In addition to the default 15 indicators, three additional indicators have been evaluated (cumulative energy demand non-renewable, cumulative energy demand renewable, and nuclear waste), with long term emissions excluded [1]. Normalization factors from Benini et al. [6] were the basis to determine the environmental relevance of the different environmental impacts.

The functional unit is 1 kWh of DC electricity generated by a photovoltaic module given an average European irradiation. A product lifetime of 30 years is assumed [1]. Based upon average irradiation conditions of optimally oriented PV modules in Europe and a default 0.70%/yr module degradation rate, an average annual energy yield of 975 kWh/kWp is assumed [1].

The product system of the electricity production with a PV module consists of the three stages of manufacturing, use, and end-of-life (Fig. 1). The manufacturing of PV modules includes the supply chain of raw materials as well as the manufacturing process. The product system also includes the mounting system required for a 3 kWp roof mount PV plant. The inverter and the AC cabling are not part of the product system. Although the predominant application of CdTe PV modules is in large commercial and utility scale power plants, the analysis of 3 kWp roof mount systems was included in this study to ensure comparability with the PEF screening study [2] and the representative product portrayed in that study. The use phase includes electricity production and maintenance.

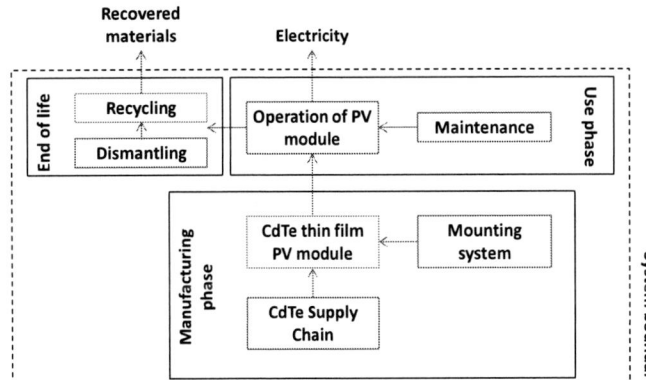

Fig. 1. Product system of electricity produced with a CdTe photovoltaic module, with processes of the foreground (product specific) and background [4] system marked with blue and red color, respectively.

TABLE I
LIFE CYCLE INVENTORY OF CdTe PV MODULE MANUFACTURING FOR SERIES 6 (2017-2018) MODULES

Explanations	Name	Location	Infrastructure-Process	Unit	photovoltaic laminate, CdTe, at plant (2017-2018 estimated)		uncertaintyType	StandardDeviation 95%	GeneralComment (Pedigree Matrix)
	Location				MY	US			
	InfrastructureProcess				1	1			
	Unit				m2	m2			
prod-uct	photovoltaic laminate, CdTe, at plant	MY	1	m2	1				
	photovoltaic laminate, CdTe, at plant	US	1	m2		1			
energy	electricity, medium voltage, at grid	MY	0	kWh	3.34E+1	-	1	1.07	(1,1,1,1,1,3,BU:1.05)
	electricity, medium voltage, at grid	US	0	kWh	-	3.48E+1	1	1.07	(1,1,1,1,1,3,BU:1.05)
	natural gas, burned in boiler modulating >100kW	RER	0	MJ	-	2.08E+1	1	1.07	(1,1,1,1,1,3,BU:1.05)
infra-struct.	Photovoltaic panel factory CdTe	US	1	unit	-	4.00E-6	1	3.04	(2,2,1,1,1,3,BU:3)
	Photovoltaic panel factory CdTe	MY	1	unit	4.00E-6	-	1	3.04	(2,2,1,1,1,3,BU:3)
materials	tap water, at user	RER	0	kg	2.07E+2	1.93E+2	1	1.07	(1,1,1,1,1,3,BU:1.05)
	tempering, flat glass	RER	0	kg	5.34E+0	5.42E+0	1	1.07	(1,1,1,1,1,3,BU:1.05)
	copper, at regional storage	RER	0	kg	3.22E-3	3.27E-3	1	1.07	(1,1,1,1,1,3,BU:1.05)
	silicone product, at plant	RER	0	kg	1.17E-1	1.19E-1	1	1.08	(1,1,1,1,1,3,BU:1.05)
	solar glass, low-iron, at regional storage	RER	0	kg	6.94E+0	7.18E+0	1	1.07	(1,1,1,1,1,3,BU:1.05)
	flat glass, uncoated, at plant	RER	0	kg	5.34E+0	5.42E+0	1	1.07	(1,1,1,1,1,3,BU:1.05)
	glass fibre reinforced plastic, polyamide, injection moulding, at	RER	0	kg	1.08E-1	1.08E-1	1	1.16	(1,4,3,3,1,3,BU:1.05)
	ethylvinylacetate, foil, at plant	RER	0	kg	3.85E-1	3.91E-1	1	1.07	(1,1,1,1,1,3,BU:1.05)
	cadmium telluride, semiconductor-grade, at plant	US	0	kg	2.21E-2	2.29E-2	1	1.07	(1,1,1,1,1,3,BU:1.05)
	nitric acid, 50% in H2O, at plant	RER	0	kg	5.72E-2	5.72E-2	1	1.16	(1,4,3,3,1,3,BU:1.05)
	sulphuric acid, liquid, at plant	RER	0	kg	3.93E-2	3.93E-2	1	1.16	(1,4,3,3,1,3,BU:1.05)
	sodium chloride, powder, at plant	RER	0	kg	4.53E-2	4.53E-2	1	1.16	(1,4,3,3,1,3,BU:1.05)
	hydrogen peroxide, 50% in H2O, at plant	RER	0	kg	1.67E-2	1.67E-2	1	1.16	(1,4,3,3,1,3,BU:1.05)
	isopropanol, at plant	RER	0	kg	2.08E-3	2.08E-3	1	1.16	(1,4,3,3,1,3,BU:1.05)
	sodium hydroxide, 50% in H2O, production mix, at plant	RER	0	kg	4.93E-2	4.93E-2	1	1.16	(1,4,3,3,1,3,BU:1.05)
	chemicals inorganic, at plant	GLO	0	kg	9.26E-3	1.06E-2	1	1.07	(1,1,1,1,1,3,BU:1.05)
	chemicals organic, at plant	GLO	0	kg	2.68E-2	3.35E-2	1	1.16	(1,1,1,1,1,3,BU:1.05)
	nitrogen, liquid, at plant	RER	0	kg	7.32E-2	7.32E-2	1	1.16	(1,4,3,3,1,3,BU:1.05)
	aluminium alloy, AlMg3, at plant	RER	0	kg	1.67E+0	1.69E+0	1	1.07	(1,1,1,1,1,3,BU:1.05)
	chromium steel 18/8, at plant	RER	0	kg	1.11E-2	1.13E-2	1	1.07	(1,1,1,1,1,3,BU:1.05)
	EUR flat pallet	RER	0	p	1.45E-2	1.45E-2	1	1.07	(1,1,1,1,1,3,BU:1.05)
trans-port	transport, lorry >16t, fleet average	RER	0	tkm	1.02E-1	6.95E+0	1	2.00	(1,1,1,1,1,3,BU:2)
	transport, freight, rail	RER	0	tkm	2.11E+0	-	1	2.00	(1,1,1,1,1,3,BU:2)
	transport, transoceanic freight ship	OCE	0	tkm	3.12E+1	-	1	2.00	(1,1,1,1,1,3,BU:2)
disp-osal	disposal, municipal solid waste, 22.9% water, to sanitary landfill	CH	0	kg	2.52E-1	4.75E-1	1	1.16	(1,1,1,1,1,3,BU:1.05)
	treatment, sewage, unpolluted, to wastewater treatment, class 3	CH	0	m3	-	8.63E-2	1	1.07	(1,1,1,1,1,3,BU:1.05)
emissions air	Heat, waste	-	-	MJ	1.20E+2	1.25E+2	1	1.29	(3,4,3,3,1,5,BU:1.05)
	Cadmium	-	-	kg	9.56E-9	9.56E-9	1	5.00	(1,1,1,1,1,3,BU:5)
	Copper	-	-	kg	7.39E-9	7.39E-9	1	5.00	(1,1,1,1,1,3,BU:5)
	Lead	-	-	kg	4.35E-9	4.35E-9	1	5.00	(1,1,1,1,1,3,BU:5)
	Nitric acid	-	-	kg	3.00E-4	3.00E-4	1	5.00	(1,1,1,1,1,3,BU:5)
emissions water	Cadmium, ion	-	-	kg	3.62E-8	3.62E-8	1	3.00	(1,1,1,1,1,3,BU:3)
	Copper	-	-	kg	1.76E-7	1.76E-7	1	3.00	(1,1,1,1,1,3,BU:3)
	Lead	-	-	kg	2.58E-8	2.58E-8	1	3.00	(1,1,1,1,1,3,BU:3)
	Nitrate	-	-	kg	2.59E-2	2.59E-2	1	3.00	(1,1,1,1,1,3,BU:3)
	Zinc	-	-	kg	1.34E-7	1.34E-7	1	3.00	(1,1,1,1,1,3,BU:3)

TABLE II
LIFE CYCLE INVENTORY OF SERIES 6 CdTe PV SYSTEM RECYCLING (MODULES, CABLING, AND MOUNTING STRUCTURES)

	Name	Location	Infrastructure Process	Unit	CdTe PV module takeback + recycling	Avoided burden from recycling, CdTe PV module, mounted construction	Uncertainty Type	Standard Deviation?	General Comment (Pedigree Matrix)
					2017-2018 (estimated)				
	Location				RER	RER			
	InfrastructureProcess				0	0			
	Unit				m2	m2			
product	CdTe PV module takeback + recycling	RER	0	m2	1.00E+0				
	Avoided burden from recycling, CdTe PV module, mounted construction	RER	0	m2		1.00E+0			
energy	Electricity, medium voltage, at grid	DE	0	kWh	4.38E+0		1	1.07	(2,4,1,1,1,3)
auxiliaries	Water, deionised, at plant/CH U	CH	0	kg	5.42E+0		1	1.07	(2,4,1,1,1,3)
	Sulphuric acid, liquid, at plant	RER	0	kg	8.33E-2		1	1.07	(2,4,1,1,1,3)
	Hydrogen peroxide, 50% in H2O, at plant	RER	0	kg	5.71E-1		1	1.07	(2,4,1,1,1,3)
	Sodium hydroxide, 50% in H2O, production mix, at plant	RER	0	kg	1.04E-1		1	1.07	(2,4,1,1,1,3)
transport	Transport, lorry >16t, fleet average	RER	0	tkm	1.25E+1		1	2.00	(2,4,1,1,1,3)
disposal	Treatment, PV cell production effluent, to wastewater treatment, class 3	CH	0	m3	4.79E-3		1	1.23	(2,4,1,1,1,3)
	Disposal, plastics, mixture, 15.3% water, to municipal incineration	CH	0	kg	6.16E-1		1	1.23	(2,4,1,1,1,3)
	Disposal, inert waste, 5% water, to inert material landfill	CH	0	kg	1.28E-1		1	1.23	(2,4,1,1,1,3)
emissions air	Cadmium	-	-	kg	5.89E-9		1	1.28	(2,4,1,1,1,3)
emissions water	Cadmium, ion	-	-	kg	8.92E-8		1	2.09	(2,4,1,1,1,3)
avoided energy	Natural gas, high pressure, at consumer	RER	0	MJ		-1.49E+01	1	2.00	(2,4,1,1,1,3)
	Heavy fuel oil, at regional storage	RER	0	MJ		-9.64E+00	1	2.00	(2,4,1,1,1,3)
avoided materials	copper, at regional storage	RER	0	kg		-3.84E-01	1	2.00	(2,4,1,1,1,3)
	copper, secondary, at refinery	RER	0	kg		3.84E+00	1	2.00	(2,4,1,1,1,3)
	aluminium, primary, at plant	RER	0	kg		-3.05E+00	1	2.00	(2,4,1,1,1,3)
	aluminium, secondary, from old scrap, at plant	RER	0	kg		3.05E+00	1	2.00	(2,4,1,1,1,3)
	pig iron, at plant	GLO	0	kg		-9.70E-01	1	2.00	(2,4,1,1,1,3)
	Cadmium sludge, from zinc electrolysis, at plant	GLO	0	kg		-2.84E-02	1	2.00	(2,4,1,1,1,3)
	Copper telluride cement, from copper production	GLO	0	kg		-3.22E-02	1	2.00	(2,4,1,1,1,3)
	Silica sand	DE	0	kg		-6.28E+00	1	2.00	(2,4,1,1,1,3)
	Soda, powder, at plant	RER	0	kg		-2.49E+00	1	2.00	(2,4,1,1,1,3)
	Limestone, milled, packed, at plant	CH	0	kg		-4.34E+00	1	2.00	(2,4,1,1,1,3)
avoided emissions air	Carbon dioxide	-	-	kg		-2.26E+0	1	2.00	(2,4,1,1,1,3)

The end-of-life covers the dismantling of the modules including transport to a recycling facility and the recycling process itself. Potential benefits for recycling are quantified and allocated 50:50 to the electricity production and the products made of recycled material, respectively [1]. Recycling of aluminum, steel, and copper in the balance of system (BOS) is evaluated, with benefits estimated after subtracting existing recycled content in those metals (32%, 37%, and 44 %, respectively [4]).

Life cycle inventory (LCI) data for Series 6 CdTe PV module manufacturing and end-of-life recycling are shown in Tables I and II. BOS and Series 4 CdTe PV LCI data are from [3]. PV modules are manufactured in the United States (US) and Malaysia (MY) with relative production capacities of 13.8% and 86.2%, respectively. PV module characteristics are shown in Table III, with average module efficiency of 15.5% based on 2015 production of the Series 4 module [7], and 17.0-18.0% based on 2017-2018 production of the Series 6 module [8].

III. RESULTS AND DISCUSSION

The environmental performance of CdTe PV modules is summarized in Table IV and Fig. 2 for a wide variety of impact categories, including those related to ecosystems, human health, and natural resources. Life cycle environmental impacts for Series 6 systems are lower than those for Series 4 systems with the exception of human toxicity, non-cancer effects and ozone depletion, where slight increases for Series 6 are due to the addition of a module frame.

Total life cycle environmental impacts for Series 4 and Series 6 systems are mainly driven by mineral, fossil and renewable resource depletion (34-37%), human toxicity cancer effects (20-22%), human toxicity non-cancer effects (11-13%), and by freshwater ecotoxicity (11-12%) (Fig. 2). Within the impact category mineral, fossil and renewable resource depletion, the extraction of cadmium and tellurium are the main contributors.

Table III. Characteristics of CdTe PV modules

	Unit	First Solar Series 4 (2015)	First Solar Series 6 (2017-2018)
Length |width| thickness	mm	1200|600|6.8	2005|1230|5.4
Area	m²/module	0.72	2.47
Weight	kg/module	12	35
	kg/m² module	16.67	14.17
Front glass	kg/module	5.7	16.6
Back glass	kg/module	5.7 (tempered)	13.0 (heat strengthened)
Encapsulation	-	Laminate material with edge seal	
Frame material	-	None (frameless)	Aluminum
Efficiency	%	15.5	17.0-18.0

Table IV. Characterized environmental impact results (based on European deployment, average annual energy yield of 975 kWh/kWp, and 30 year lifetime).

3kWp installation, roof mounted (total all life stages, recycling benefits included)			
Impact category	Unit per kWh DC electricity	First Solar Series 4	First Solar Series 6
Climate change	kg CO2 eq	1.94E-02	1.66E-02
Ozone depletion	kg CFC-11 eq	8.78E-10	9.47E-10
Human toxicity, non-cancer effects	CTUh	4.95E-09	5.11E-09
Human toxicity, cancer effects	CTUh	5.97E-10	5.16E-10
Particulate matter	kg PM2.5 eq	9.95E-06	7.72E-06
Ionizing radiation HH	kBq U235 eq	9.06E-04	7.83E-04
Photochemical ozone formation	kg NMVOC eq	7.43E-05	5.62E-05
Acidification	molc H+ eq	1.46E-04	1.10E-04
Terrestrial eutrophication	molc N eq	2.76E-04	2.07E-04
Freshwater eutrophication	kg P eq	3.60E-06	3.51E-06
Marine eutrophication	kg N eq	2.54E-05	1.91E-05
Freshwater ecotoxicity	CTUe	7.63E-02	7.50E-02
Land use	kg C deficit	1.19E-02	8.61E-03
Water resource depletion	m3 water eq	7.83E-05	6.07E-05
Mineral, fossil & ren resource depletion	kg Sb eq	3.09E-06	2.58E-06
Cumulative energy demand non renewable	MJ	2.90E-01	2.47E-01
Cumulative energy demand renewable	MJ	3.63E+00	3.62E+00
Nuclear waste	m3 HAA eq	2.12E-11	1.84E-11

Human toxicity cancer effects and human toxicity non-cancer effects are dominated by the installation and mounting system and the supply chains of aluminum, copper and steel therein, including the addition of an aluminum frame for the Series 6 module. The freshwater ecotoxicity impacts are also mainly caused by the installation and mounting system and the supply chain of copper therein, as well as the disposal of plastics from industrial electronic waste, which is associated with the production of the electric installation.

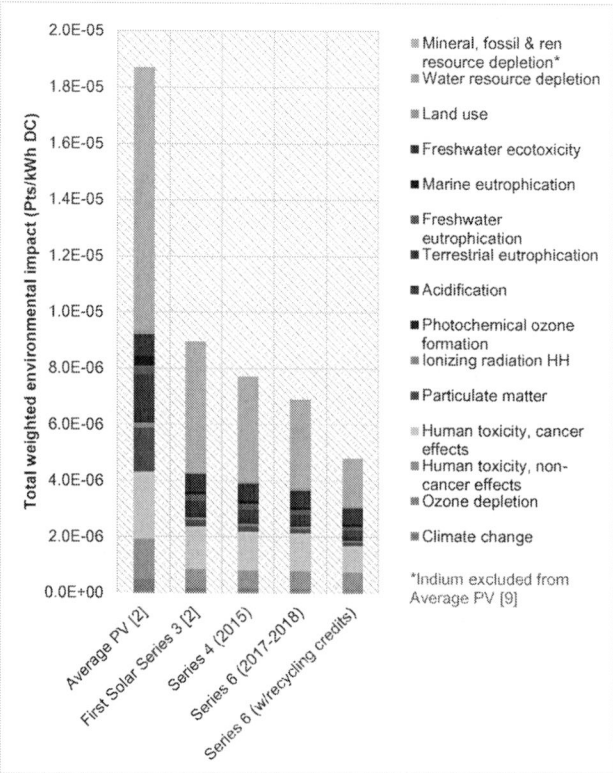

Fig. 2. Total environmental impact results (normalized [6] and equally weighted) of 1 kWh of DC electricity produced with a 3 kWp roof mounted installation (based on European deployment, average annual energy yield of 975 kWh/kWp, and 30 year lifetime), with recycling benefits excluded except where indicated.

Excluding recycling benefits, the total weighted environmental impacts of the First Solar Series 6 module are about 10% lower than the Series 4 module, about 20% lower than the Series 3 module, and about three times lower than the average PV module (with indium contribution to mineral, fossil, and non-renewable resource depletion excluded due to anomalous results [9]) (Fig. 2). When recycling benefits are included, the total weighted environmental impacts of the First Solar Series 6 module are about four times lower than the average PV module (Fig. 2). The latter is a representative (virtual) product composed of market-share weighted averages of different PV technologies in 2012 and based on global supply chain data from 2011 and module efficiencies shown in Table V [2]. From 2011 to 2015, the global supply chain of module production by region has remained reasonably consistent, with an increase of production in China and Taiwan from about 65% in 2011 to about 70% in 2015 [10].

Table V. Assumptions related to representative product

	Average PV [2]		Average PV (2015)	
	Market share	Module efficiency	Market share [10]	Module efficiency [10]
CdTe	6.3%	14.0%	4.0%	15.6%
CIS	3.5%	10.8%	1.7%	13.8%
micromorph-Si	4.5%	10.0%	-	-
multi-c-Si	45.2%	14.7%	69.5%	~16%
mono-c-Si	40.5%	15.1%	23.9%	~17%

Accounting for the technology market shares and increased module efficiencies in 2015, the environmental impacts of the average PV module shown in Fig. 2 may be reduced by about 10%. Accordingly, the total weighted environmental impacts of the First Solar Series 6 module (recycling benefits included) are about 3.5 times lower than the average PV module (2015).

As shown in Fig. 3, the most important contributors (hotspots) to CdTe PV environmental impacts are the supply chain of electricity consumed in various stages of the life cycle of PV electricity, the supply chain of aluminum and steel required in the mounting structure and module frame, the supply chain of copper required mainly in the electric installations of the PV systems, flat glass production, the supply chains of cadmium, and tellurium as abiotic resource hotspots, and transport services provided by transoceanic ships. These hotspots are discussed in turn, with respect to how they may be addressed by Series 6 technology.

A. Supply chain of electricity

In 2011, First Solar developed a sustainability target of reducing its corporate carbon footprint (metric tons CO2-eq per MW produced) by 35% by 2016 relative to a 2008 base year. The single largest contributor to First Solar's corporate carbon footprint is purchased electricity for PV module manufacturing. By end of 2015, the target has been surpassed [11], and a future 2021 target of 45% below base year emissions has been established. The latter includes a targeted 10% reduction in electricity usage per m^2 of module produced for Series 6 compared with Series 4 [12]. Since these reductions are based on equipment design loads and have not been confirmed with operational data, no reductions in electricity usage per module produced are considered in this study.

In addition to a series of energy efficiency projects involving facilities management (e.g., lighting, HVAC controls, compressed air), sub-metering has been installed on manufacturing tools to enable comparative monitoring of manufacturing lines and continuous improvement. In 2014-2015, the electric utility serving First Solar's Kulim, Malaysia facility has also transitioned from a mix of fossil fuel generation to 100% combined cycle natural gas generation, thereby lowering the carbon intensity of the electricity supply [11].

Leading contributors to impact categories (electricity, 3kWp installation, roof mounted with CdTe PV technology, characterized)	Aluminium, copper, and/or steel consumption in mounting system and electrical installation	Copper consumption in PV panel production	Electricity consumption in PV panel production	Electricity consumption in PV panel recycling	Flat and solar glass production	Harvested solar energy	Semiconductor materials in PV panel production	Transoceanic transport of PV panels	Water use in PV panel production
Climate change	X		X		X				
Ozone depletion	X		X						
Human toxicity, non-cancer effects	X								
Human toxicity, cancer effects	X	X							
Particulate matter	X		X		X			X	
Ionizing radiation HH	X		X					X	
Photochemical ozone formation	X		X		X			X	
Acidification	X				X			X	
Terrestrial eutrophication	X				X			X	
Freshwater eutrophication	X		X						
Marine eutrophication	X				X			X	
Freshwater ecotoxicity	X								
Land use	X		X						
Water resource depletion	X		X	X					X
Mineral, fossil & ren resource depletion							X		
Cumulative energy demand non renewable	X		X						
Cumulative energy demand renewable						X			
Nuclear waste	X		X						

Fig. 3. Environmental hotspots for 1 kWh of DC electricity produced with a 3 kWp CdTe PV roof mounted system (based on European deployment, average annual energy yield of 975 kWh/kWp, and 30 year lifetime).

B. Aluminum, Steel, and Copper in BOS

The demand for metals (aluminum and steel) in mounting structures per MWp installed will decrease with the larger, more efficient, framed Series 6 module. This module will be the same length as a 72-cell c-Si PV module (~2 m) but will be wider (~1.2 m), enabling more rapid installation of arrays with fewer mounting materials required per unit area.

C. Flat glass production

The Series 6 module will involve a reduction in thickness of front and back glass (from 3.2 to 2.8 mm and 3.2 to 2.2

mm, respectively) relative to the current Series 4 module, thereby reducing the demand for flat glass. The reduced glass weight will also lower the environmental impact of transport services per unit area of module.

D. Supply chain of cadmium and tellurium

Demand side management strategies have led to about a 50% reduction in the semiconductor intensity (usage per Watt) of CdTe PV modules since 2009 [13]. PV recycling also largely addresses the resource depletion hotspot and provides an additional source of semiconductor material for new PV modules. As of 2016, First Solar PV modules had approximately 8% recycled semiconductor content, which will likely increase as greater volumes of end-of-life modules return from the field [14].

E. Module efficiency

Improvements in module efficiency proportionally lower all of the above hotspots. Series 6 module efficiency is planned for 17-18% in the near term [8] and >19% in the medium term, with research cells already exceeding 22% efficiency [15].

IV. CONCLUSION

Life cycle environmental impact hotspots for CdTe PV technology have been identified through use of the PEF category rules within the European Commission's "Single Market for Green Products" initiative. These hotspots include the supply chains of electricity, aluminum, steel, and copper, flat glass production, transoceanic transport, and the supply chains of cadmium and tellurium as abiotic resource hotspots. Technology innovation for CdTe PV related to Series 6 technology will address hotspots through larger area modules with higher efficiency, reduced glass thickness, and continued recyclability. As a result, the total

weighted environmental impacts of the Series 6 module are about 3.5-4 times lower than an average PV module.

REFERENCES

[1] R. Frischknecht, P. Stolz, G. Heath, A. Wade, L. Francke, I. Theologitis, C. Olson, and J. Tian, "Product Environmental Footprint Category Rules: Production of Photovoltaic Modules Used in Photovoltaic Power Systems for Electricity Generation," Technical Secretariat PEF Pilot PV Electricity Generation, Brussels, 2017.

[2] P. Stolz, R. Frischknecht, F. Wyss, and M. de Wild-Scholten, "PEF screening report of electricity from photovoltaic panels in the context of the EU Product Environmental Footprint Category Rules (PEFCR) Pilots, v. 2.0," treeze Ltd. and SmartGreenScans, Uster, Switzerland, 2016.

[3] First Solar, "First Solar Series 4 PV System Product Environmental Footprint," 2016.

[4] KBOB, eco-bau and IPB, "ecoinvent Datenbestand v2.2+; Grundlage für die KBOB-Empfehlung 2009/1:2014: Ökobilanzdaten im Baubereich, Stand April 2014," Koordinationskonferenz der Bau- und Liegenschaftsorgane der öffentlichen Bauherren c/o BBL Bundesamt für Bauten und Logistik, 2014.

[5] European Commission, "Commission Recommendation of 9 April 2013 on the use of common methods to measure and communicate the life cycle environmental performance of products and organisations." *Official Journal of the European Union*, vol. L 124, pp. 1-210, 2013.

[6] L. Benini, L. Mancini, S. Sala, S. Manfredi, E. M. Schau, and R. Pant, "Normalisation method and data for Environmental Footprints," European Commission Joint Research Centre, Institute for Environment and Sustainability, Ispra, Italy, 2014.

[7] First Solar, "Key Quarterly Financial Data," 2016.

[8] First Solar, "First Solar Series 6 Data Sheet," 2016.

[9] PV Thin and treeze, "Addressing the indium anomaly in the PEFCR for PV electricity generation," 2016.

[10] Fraunhofer ISE, "Photovoltaics Report," 2016.

[11] First Solar, "CDP Carbon Questionnaire Response," 2016.

[12] First Solar, "Energy Plan," 2016.

[13] J. Trivedi, "CdTe thin film technology overview – achievements and opportunities," in *MMTA International Minor Metals Conference*, Toronto, Canada, 2015.

[14] M. Marwede and A. Reller, "Future recycling flows of tellurium from cadmium telluride photovoltaic waste," *Resources, Conservation and Recycling*, vol. 69, pp. 35– 49, 2012.

[15] First Solar, "Analyst Day Meeting, Technology Update," 2016.

Photovoltaic Smart Home System - Dubai Case Study

Ammar Natsheh, Marwa ALJaziri, Maitha Moosa, Gharibah Essa, and Hassa Moosa

Higher Colleges of Technology, Dubai Women's College, Dubai, P.O Box 26266, United Arab Emirates

Abstract — **Based on challenging and development in the world, renewable energy sources and smart technology should play an important role to improve the consumer life and to reduce the unstable climates which are one of the major effects of global warming. This paper presents the operation, appliance, and analysis to ensure the system integrates photovoltaic (PV) clean energy sources to have a smart home. The proposed system provides affordable and reliable clean electricity supply to manage a smart home. The advanced system enables homeowners to control all home devices through an application on a smart panel, phones or other smart devices. The proposed system is expected to save electricity cost of home use and control the home devices in an easy way. The aim is to use the clean and renewable energy to maintain minimum standards of living and comfortable lifestyle.**

I. INTRODUCTION

A solar smart home system is a system of combination between renewable energy sources (photovoltaic) and smart households which provide to coordinated consumer needs. This type of association system is a result to save electricity bill, control households devices and to be environment ego friendly which would increase the convenience, security and entertainment of the clients. The main feature in it are the photovoltaic system which is one of the emerging concepts that integrate from clean and renewable sources that convert direct sunlight to electricity power for residential users. A PV system works with a smart system to control all household systems and devices while saving energy energy and reducing electricity cost. It also provides consumers with the opportunity to control and monitor household appliances and systems by using a particular app "SMART-BUS". The microcontroller "ATMEL" converts the system to a multiplex system that supply a home's electric devices permitting homeowners to communicate with their household devices in an easy way. This protocol requires hardware and software to support different purposes, including controlling lighting, air-conditioning, and sound systems, to provide an easy and smart way of living. The alternative strategic feature by enable the development of this system and implement it in several sectors not only in homes. To embed the important of renewable energy in future from wasting and to implement technology as well, all people should collaborate with each other to share the responsibility to manage the power and save it. This paper will include these main points:

- Operation of the system
- Economy
- Survey

II. OPERATION

In a photovoltaic smart home, it have been used PV as an electric source for all of the home's smart equipment devices operation. The most important component of a solar smart home is a battery storage inverter, which acts as the interface between the PV inverter and the high-voltage battery which is an integral part of the smart home equipment. The sun's time can be exploited by charging the battery, which is consumed after sunset to operate the home appliances. After using an inverter to convert the 12V, 24V and 48V DC voltage to 220V AC voltage, it will go directly to the power supply the main switch DP which control all switches then from the bracket goes to the relay which controls the lights, audio, and RGB controller. All devices are connected to each other as a bus connection from the main Distribution Panel (DP) to the smart devices. For RG45, the connection shall go to a router or modem to provide Wi-Fi capabilities; thus allowing users to control all systems (e.g. lighting, TV, sound system) using IOS or android. Moreover, to illustrate the point, these figures show the use of the application to control the smart home:

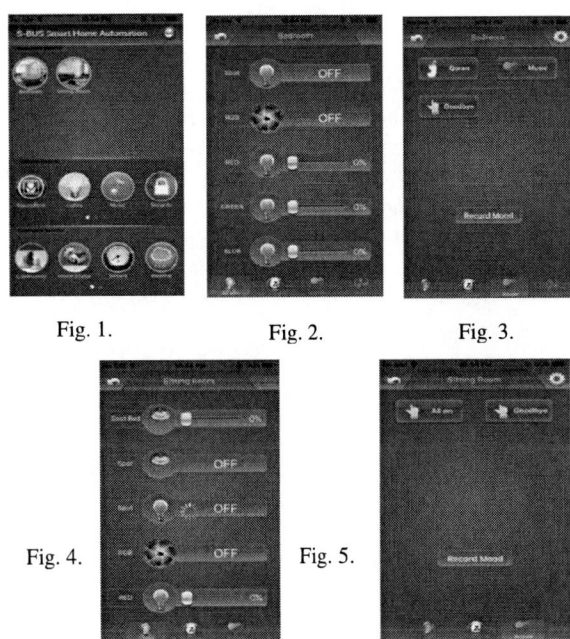

Fig. 1.   Fig. 2.   Fig. 3.

Fig. 4.   Fig. 5.

The application consists of two zones: Bedrooms and Living Rooms (Fig. 1). The system is controlled through a Smart Panel or an application for a smart device. An ON/OFF switch controls the lights' intensity and color (Fig. 2). An ON/OFF switch also controls the audio's volume and sound selection (Fig. 3). An ON/OFF switch controls a living room spotlight and RGB (Fig. 4). Finally, all systems can be switched off by using a master switch GOODBYE (Fig. 5). Fig. 6 shows the operation of the proposed system.

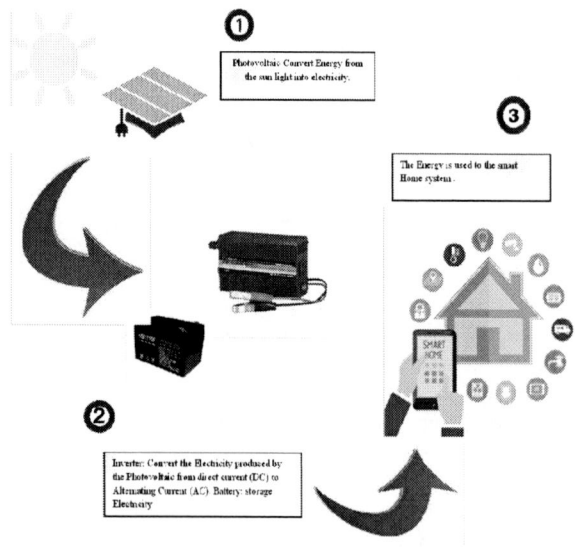

Fig. 6. Operation

III. ECONOMY

Dubai economy appreciates a competitive of environmental and cost that creates an idea to help consumer and society. Here are some questions represents all the economic benefits of introducing smart homes in Dubai.

- How many household in Dubai?

Recently the estimation of Dubai household is around 600,000. Looking farther ahead it is expected to grow slightly every month.

- What is average electricity bill in UAE for a household?

Although the average electricity bill to live in Dubai will vary substantially will base on this calculation
Average consumptions = (3600+2320+7800+8986+2053 (kWh))/5 = 4951.8 kWh

Average cost = 2000×600,000×7.5/100=9000000AED
2000×600,000×9/100 =108000000AED

951.8×600,000×10.5/100 = 60000000AED
Total = 90000000+108000000+60000000=258000000AED
Convert to $ = 258000000×3.6723= 947453400$

- What is the average carbon emission in UAE?

Electricity is one of the main causes of carbon emissions caused by combustion of fuel and natural gas which accounts for 33%, or 64.89 million tons of greenhouse gasses. In 2012 Dubai Electricity and Water Authority's (DEWA) total carbon emissions were 18.26 million metric tons of carbon dioxide (MTCO2). The main emission CO2 comes from the combustion of natural gas to generate power and desalinate water.

- What is the cost to the country to create energy from PV panel?

This calculation shows the cost of using electricity in whole Dubai:
Dubai energy consumption = number of the houses × average consumption
= 4951.8 × 600,000 = 2971080000 kWh
(Dubai consumption)/month= 2971080000/30=99036000
99036000/ (8 hour) =12379500 kWh
The cost of PV for Dubai = 12379500 kWh × 0.5$
= 6189750000$
If the smart home for whole Dubai is
= 6189750000$ × 2 = 12379500000$

- What is the potential benefit of using solar energy?

The reason why most families in Dubai are desperate to use solar energy as a source of energy that cut down electricity bill which reduces by 35% and more while smart system saves around 30 to 60 percentage of bill electricity.

- What is the potential benefit of reducing energy bill to the customer and the increase in their spending power?

Base on this calculation has been figure out that paying a bill for one year will save money for 25 years later, as it will save around 95% of electricity for whole Dubai.

IV. SURVEY

The following graphs show a survey for the photovoltaic smart home system as a case study in Dubai with 27 responses as faculty and students in Dubai women's college based on many issues as shown to give an awareness of the importance of renewable energy in general and photovoltaic in specific with smart systems.

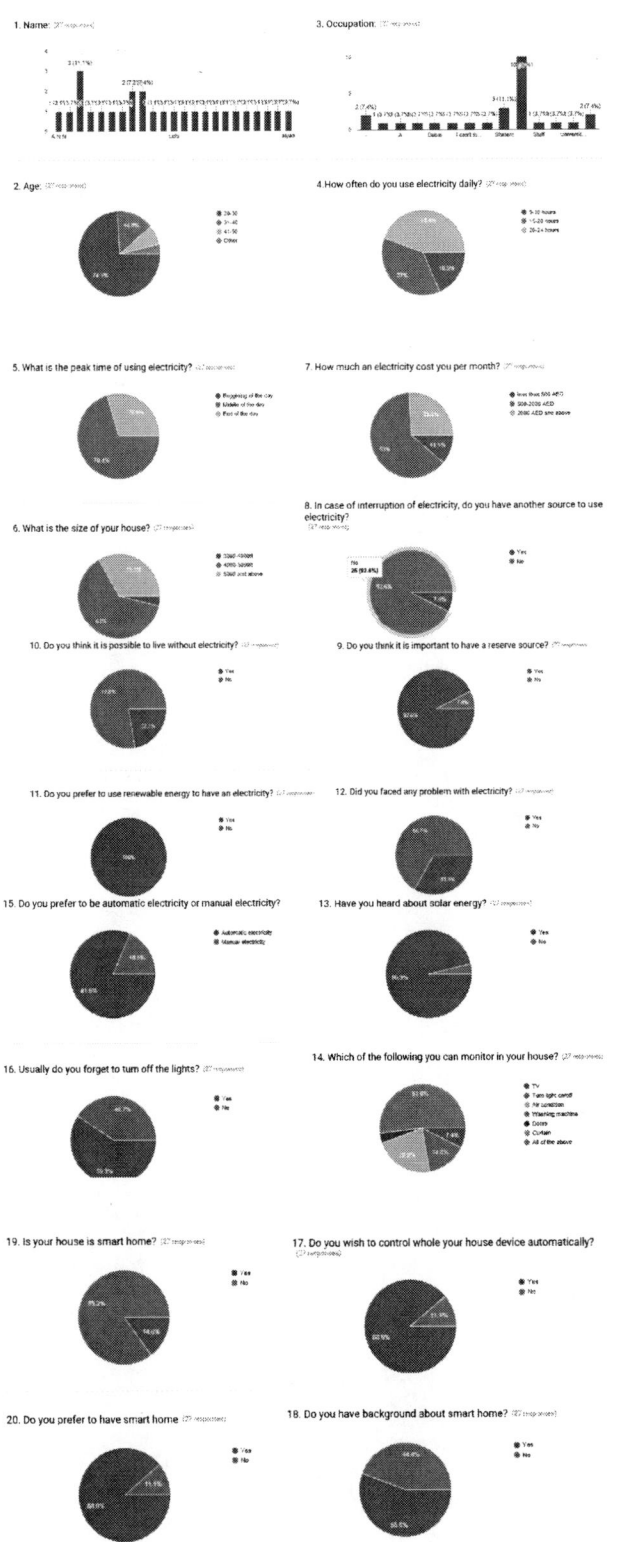

V. CONCLUSION

This project focuses on the photovoltaic Smart Home System. It describes the system's operation, application and benefits. It results in the efficient usage of a photovoltaics smart home system, reducing energy bills and minimizing greenhouse gas emissions. The aim is to save on electricity bills, control the operation of household devices, while maintaining an eco-friendly environment. The results ensure energy reduction, money saving and a comfortable lifestyle.

REFERENCES

[1] N. Radhika, K. Sivalingam, V. Anand, "Network architecture for smart grids," International Conference on Computer Communication and Electrical Technology (ICCCET) (2011), pp. 24-26.

[2] Jinsoo Han, Haeryong Lee, Kwang-Roh Park, "Remote-controllable and energy-saving room architecture based on ZigBee communication," IEEE Transactions on Consumer Electronics, 55 (1) (2009), pp. 264-268.

[3] W.-H.E. Liu, K. Liu, D. Pearson, "Consumer-centric smart grid," IEEE PES Innovative Smart Grid Technologies (ISGT) (2011), pp. 1-6.

Direct Drive Photovoltaic Milk Chilling Experience in Kenya

Robert Foster[1], Brian Jensen[2], Brian Dugdill[1], Wendy Hadley[1], Bruce Knight[1b]
Abdul Faraj[3], Johnson Kyalo Mwove[3],
[1] Winrock International, Arlington, Virginia ([1] USA and [1b] Kenya)
[2] SunDanzer, Inc., Tucson, Arizona (USA)
[3] Egerton University, Department of Dairy and Food Sciences (Njoro, Kenya)

Abstract - The *Photovoltaics for Sustainable Milk for Africa through Refrigeration Technology (PV-SMART)* uses direct drive PV for off-grid milk cooling under the USAID Powering Agriculture Energy Grand Challenge Program (PAEGC). The project is implemented by Winrock International and SunDanzer Refrigeration Inc. refrigeration. The PV chillers are direct drive using a dc compressor that chills ice stored in the walls of the refrigeration unit. This eliminates the need for electro-chemical batteries and successfully chills milk to 10°C in 2 hours and 4°C by morning preventing bacteriological growth. Farmers can receive a premium price for providing higher quality, refrigerated evening milk to dairy processors that would otherwise spoil. These PV farm milk chillers are the first of their kind with 80 pilot units deployed in Kenya.

I. Introduction

The *Photovoltaics for Sustainable Milk for Africa through Refrigeration Technology (PV-SMART)* project aims to tackle off-grid milk cooling under the United States Agency for International Development (USAID) Powering Agriculture Energy Grand Challenge Program (PAEGC). In collaboration with Winrock International (WI), SunDanzer Refrigeration Inc. (SDZR), a leading US solar refrigeration technology company, was awarded a USAID grant to implement the project. PV-SMART began in 2013 to develop the world's first on-farm solar milk chiller. Direct drive PV refrigerators were developed nearly 20 years ago and are commonly used for vaccine refrigerators for remote clinics [1]. This project has scaled up this concept to develop an affordable direct drive with thermal ice storage capable of chilling up to about 40 liters of evening milk. The project has piloted 80 of these innovative milk chillers in Kenya and soon in Rwanda, allowing farmers to sell between 5 to 40 extra liters per day of higher quality evening milk. This results in additional farmer income, ranging from US$60 to US$500 extra income per month, depending on the number of cows per farm. Solar investment payback averages about 6 months for these smallholder farmers.

This 4-year project began in 2013 and is implemented in collaboration with the County Governments and Agriculture Ministries of Baringo and Nakuru, and the Department of Dairy and Food Science and Technology at Egerton University located at Njoro, and various partner dairy cooperatives and dairy processors in Kenya, as well as with Chloride-Exide leading field installations.

There are over 850,000 smallholder dairy farmers in Kenya, about 85 percent of who do not have access to the national electric power grid. Diesel fuel is expensive and logistics difficult to deliver to small rural dairy farmers. Thus, there has not been an economical method available for on-farm milk chilling for the vast majority smallholder dairy farmers in Kenya and other less developed regions globally. The typical Kenyan dairy farmer has about 3 to 5 cows, producing an average of about 8 liters per day of milk per cow (typically ~60% as morning milk and ~40% as evening milk). Dairy cooperatives have an organized morning milk collection system, but normally do not accept evening milk since by morning due to high bacteriological counts growing overnight. Due to the lack of on-farm refrigeration, evening milk is either forced consumed, sold cheaply to nearby neighbors or hawkers, or spoils. Only about 40% of milk produced nationally is processed in Kenya.

Fig. 1.   PV array installation for FMC in Mogotio, Kenya

This failing in upstream milk production causes milk spoilage and lost farm earnings. It also causes poor quality milk and further losses in earnings along the downstream dairy value chain. Of the milk that does arrive, much of it still has a high bacterial count due to lack of refrigeration, resulting in poor quality dairy products. Farmers could receive a premium price for better quality, refrigerated milk; dairy processors could charge a premium for better quality products if milk can be kept cool all the way from cow to consumer; especially during the all-important first four hours after milking that determine quality.

In order to enhance the value of milk from remote producers, *PV-SMART* has developed an affordable solar powered farm milk chiller (FMC) so these producers can deliver cool milk rather than warm to the central collection stations. The farmers also use FMCs on the farm to preserve other produce such as eggs, meat, fruits and vegetables. Besides demonstrating the technology proof of concept, *PV-SMART* is also working with stakeholders like Kenyan SACCOs (Savings and Credit Cooperative Associations) to provide financing for solar technologies like FMCs that can increase on-farm productivity and incomes. Farmers often need access to technology and credit on reasonable terms to finance the initial purchase of solar power systems, which have higher capital costs but lower operating costs when compared to traditional remote generation energy technologies like diesel gen-sets.

PV-SMART has a four-phase implementation strategy for developing, disseminating, and financing FMCs in Kenya:

Year 1—Technology Development: Designed and tested prototype solar farm milk chiller (FMC) by scaling up an off-the shelf photovoltaic refrigerator (PVR) model. Surveys were conducted of smallholder dairy farmer's needs, and besides milk chilling, a wire basket was added for perishable household food items, as well as two 5V USB ports for daytime cell phone, radio, and LED lantern charging.

Year 2—Pilot Phase 1: Piloted the world's first solar FMCs to 39 smallholder dairy farms in Baringo and Nyandarua Counties of Kenya. A baseline control unit was also installed at the Egerton University (EU) Department of Dairy and Food Science for more in-depth testing and milk quality evaluations. Field evaluations and farmer surveys were conducted by EU.

Year 3— Field Testing Pilot Phase 2: Based on Phase 1 field experience, design and testing of a next generation prototype was developed with 40 units deployed to Kenya: (i) design adaptations notably moving from a ground to roof mounted PV system; (ii) established a local dealer network with Chloride-Exide; and (iii) begin financing units to farmer with Kenyan Savings and Credit Cooperative Associations (SACCOs), starting with Skyline SACCO in Mogotio. Demonstrations are also expanding to new regions with NGOs like Mercy Corps and GiZ, including for camel milk chilling.

Year 4— Commercial Rollout Phase 3: Planning is underway to expand the solar FMC technology commercially both inside and outside of Kenya with Chloride-Exide and SACCOS

II. DIRECT DRIVE PHOTOVOLTAIC REFRIGERATION

The SunDanzer direct drive solar farm milk chiller (FMC) uses a vapor compression refrigeration system and directly couples a PV array to a dc compressor. The FMC uses thermal phase change material (PCM - ice storage) instead of electro-chemical battery storage. By storing ice in the walls of the refrigerator, it eliminates the needs of battery storage; ice never wears out and provides sufficient energy storage to cool 40L of milk overnight. The embodied technologies were originally developed in support of NASA's future planetary missions' refrigeration requirements in the late 1990's [2, 3].

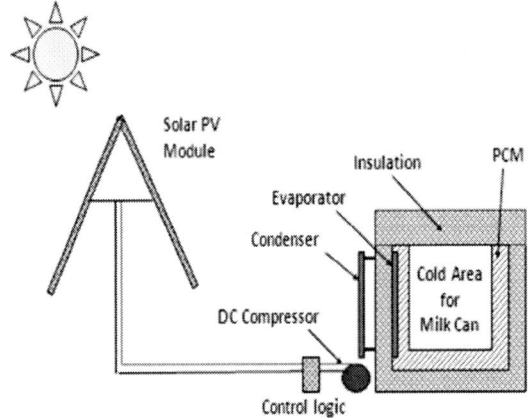

Fig. 2.   PV direct drive milk chiller array, note E-W orientation.

Fig. 3.   PV powered chiller using thermal ice storage and brine bags to chill evening milk.

Fig. 4.   Rooftop PV installation for Mogotio dairy farmer FMC.

In addition to increasing the quality of product for farmers, the solar FMC has two 5 V USB plug ports capable of charging cell phones, radios, and solar lanterns which farmers use for themselves or rent out. Likewise, there are dietary and health benefits for farmer families to store vegetables, fruits, meat, etc. in the solar FMCs, which also reduces the frequency of trips to town to purchase fresh produce.

A. "Fixed Tracking" Array

The solar FMC system also uses an innovative East-West "fixed tracking" array to maximize compressor run time and not daily energy production. Conventional tracking increases both capital and operating costs due to additional hardware and maintenance requirements for moving parts. While a conventional equatorial facing PV array power output is well over the required power requirements for the FMC at solar noon; the East-West array's output wattage supplies the required compressor power for a longer period. Thus, increasing compressor run time providing longer operating hours for farmers to chill more milk. While this approach does not maximize energy usage, it does maximize ice production over the course of the day. PV prices have come down sufficiently that fixed tracking is a viable economic option over tracking without the future maintenance concerns. This type of approach works especially well in equatorial latitudes like Kenya. This simple approach provides reliable performance [4].

Fig. 5.   PV direct drive PV array with "fixed E-W tracking."

With the compressor running most of the daylight hours due to the E-W "fixed tracking" array (the array is not actually moving like a conventional tracker, but is fixed with half the array facing East and the other half facing West to maximize daily compressor run time). Ice is formed and stored into the walls of the PVR. Thus, there is no need for expensive battery storage and replacements. Ice does not wear out. Testing at New Mexico State University for NASA and SDZR on an early prototype PVR with ice storage was successful [2] and led to the development of direct drive vaccine PVRs using ice storage. The proven PVR technology was then increased in size for larger scale milk chilling.

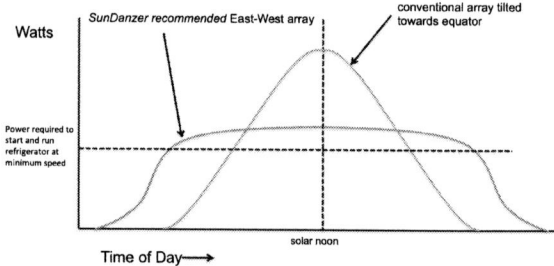

Fig. 6.   E-W "fixed tracking" array designed to maximize daily compressor run time rather than energy capture.

In order to maximize heat transfer, the solar FMC incorporates brine bags, which do not freeze at 0°C, that are placed around the milk cans to increase heat transfer and cool milk quickly. Milk naturally contains antibacterial agents to protect the suckling young from potential infectious diseases; these antibacterial agents also slow bacteriological growth – the cause of milk souring. This effective natural protection is called the lactoperoxidase system, and has both bacteriostatic and bactericidal effects against some milk spoilage microflora for about the first four hours after milking. Bacteriological growth is further retarded when milk temperatures fall below 10°C and is essentially halted at 4°C. The FMC chills 25 liters of milk down to 10°C in a couple of hours, and the milk temperature by morning is about 4°C [4].

Since dairy farmers typically milk cows twice a day, once in the morning (~60%) and again in the evening (~40%), evening milk must be consumed or sold to neighbors or hawkers at a cut price. Even so, much of the evening milk spoils overnight. About 60% of milk in Kenya is not processed mostly due to the lack of on-farm chilling options. Solar FMCs increase farmer incomes from selling milk that would otherwise spoil, and some innovative dairy processors making cheese and yoghurt who need better quality product offer a Quality Milk Payment system incentive to dairy farmers. Solar chilling provides farmers the means to improve their milk quality and overall sales.

III. OPERATIONAL FINDINGS

WI in collaboration with Egerton University Dairy and Food Sciences Department has been monitoring and evaluating (M&E) the performance and benefits of the Solar FMCs installed on the Mogotio and Ngorika Cooperative dairy farms.

PV-SMART team is also monitoring solar irradiance at Ngorika and Egerton University sites. A few selected milk cans have a Hobo data logger installed on them to monitor milk can temperature data. The temperature probe is installed on the can with foam over it so that it measures true milk can temperature only.

The findings are based on milk can temperature data collected, co-op milk sales, surveys with the end-users, milk can temperature data, and field observations by the WI and Egerton University team. All of the original 40 piloted solar FMCs have functioned for 2.5 years with no failures, and no failures in the next 40 Phase 2 units as well. There was one Phase 1 unit which had a refrigerant leak upon delivery due to springing a refrigerant leak over some extremely rough roads traversed for delivery; this was easily repaired by a local refrigeration technician from Chloride-Exide for US$40 and the unit has functioned with no issues ever since. The solar FMC technology couples mature PV technology with mature vapor-compression technology and is very reliable.

Milk temperature: The FMCs work well to chill 25 liters of evening milk to $4^{O}C$ and lower. If some milk is not removed the next morning and left throughout the day, small quantities of milk can freeze, indicating the prototype FMC may have 'spare' cooling capacity for Kenya. The figure below shows daily milk cooling cycle for one of the farmers, milk temperature is repeatedly cooled to about 5°C. Note that the farmer puts the milk can outside in the direct sunlight for drying after cleaning so the can heats up to above ambient peak temperatures during the daytime.

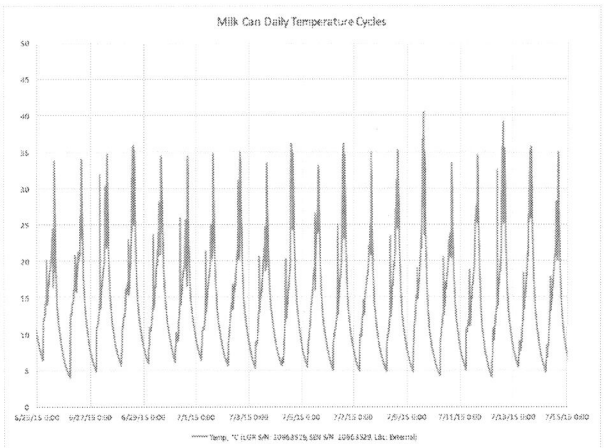

Fig. 7.   Example Daily milk cooling temperature cycles on a dairy farm: milk can temperatures °C. Note after cans are cleaned they are placed outside in the sun to dry and UV disinfect during the day.

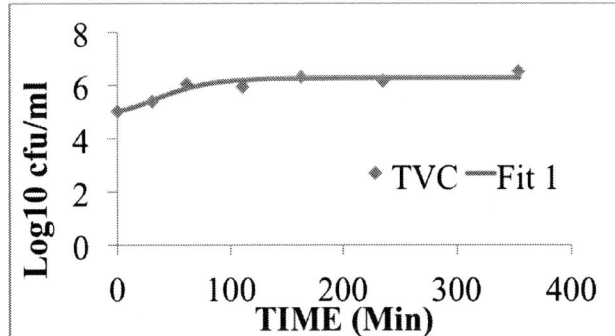

Fig. 8.   Total Viable Count (TVC) of bacteria show there is no bacterial growth once the milk is placed in a solar farm milk chiller.

Cooling Capacity: About half the users did not have enough evening milk to fill the pilot FMC. About ¼ of users utilize full capacity at 25 liters either from their own production or by also combining production with that of their neighbors, thus earning extra income for milk handling/bulking. Another user also earns extra income from charging cell phones from the two USB ports installed on the refrigeration unit, as much as US$1 per day (rate of US$0.10 per phone charge).

Some users are storing/cooling milk in plastic containers in the FMC and already exceeding the original design capacity by chilling up to 40 liters milk in food grade plastic containers, achieving overnight cooling to 'below $10^{O}C$, indicating the potential of the unit to chill more milk adequately.

IV. IMPACTS AND RESULTS

PV-SMART is piloting 80 solar FMCs in Baringo, Kisumu, Nakuru, Nyandarua, and Wajir counties, as well as one control unit at Egerton University. The first 40 Phase 1 Pilot solar FMCs have operated flawlessly in the Nakuru milkshed of Kenya with no equipment failures in the first 2.5 years.

978-1-5090-5606-4/17 $31.00 © 2017 IEEE

Table 1
Change in farmer evening milk usage with solar Farm Milk Chillers (FMC)

Location	Did milk sales increase?	Evening Milk Usage			
		Before FMC Installation		After FMC Installation	
		Forced Home consumption	Local sales	Forced Home cons. / Local sales	Sales to Dairy Processors/Coops
Mogotio **Hot climate**	No (7.7%) Yes (92.3%)	69%	31%	8%	92%
Ngorika **Cool climate**	No (16.7%) Yes (66.7%)	27%	73%	0%	100%

Source: Egerton University

The average dairy farmer chilled about 25 L of evening milk to 4°C; a few farmers chill as much as 40 L every night. Milk quality is maintained after milking and there have been zero rejections of solar chilled milk for any of the participating dairy farmers using solar FMCs, unlike from before. An informal farmer to farmer milk supply network was also organically created by solar FMC owners with excess capacity provided to their neighbors through FMC sharing (rent, barter, or purchase).

Over 92 percent of Mogotio farmers (lower elevation and hotter climate) and 67 percent of Ngorika farmers (higher elevation and cooler climate) reported increased milk sales directly attributable to chilling evening milk using solar FMCs. Other on-farm production factors include a severe drought in 2017 reducing forage and milk production for some farmers.

Fig. 9.   Satisfied Ngorika dairy farmer with solar FMC unit chilling 40 liters of evening milk every day for sale the next morning.

Thus, milk quantity and potential incremental gross earnings gain at current milk prices is excellent for these pilot units, with simple payback ranging anywhere from six months to one year depending on user milk production. From the initial surveys users sell between 5 and 45 liters of extra evening milk each day, indicating gross incremental income gains ranging from US$50 to $650 per month. Kenyan small scale financial credit institutions (SACCOs) have begun financing PVRs during Phase 2 of PV-SMART.

Field surveys found that 83 percent of the Phase 1 pilot FMC farmers felt the solar FMC technology was worth the initial cost and is a worthwhile investment. Financing is key as over 70 percent of small holder farmers prefer a short-term loan mechanism to purchase the technology.

V. CONCLUSIONS

PV direct drive solar milk chillers (FMCs) have been used in Kenya with zero failures in the first 2.5 years of operation. Smallholder dairy farmers have sold between 5 and 40 extra liters of evening milk each day, depending on their dairy herd size. Resulting farmer income gains ranged from US$60 to $500 per month, with expected FMC payback typically in less than a year. This type of solar milk chilling uses no batteries and has no regular maintenance requirements. There are over 5 million smallholder dairy farmers in East Africa who can benefit from this technology, not to mention the millions more of other off-grid smallholder dairy farmers in the rest of Africa, Asia, and Latin America that can also benefit from solar FMCs to improve livelihoods and delivered milk quality.

REFERENCES

[1] R. Foster, M. Ghassemi, A. Cota, <u>Solar Energy</u>, Renewable Energy and the Environment Series, Volume 2, Taylor and Francis Publishing, CRC Press, A. Ghassemi (Ed.), ISBN: 13:9781420075663, Boca Raton, Florida, August, 2009.

[2] R. Foster, R., L. Estrada, D. Bergeron, "Photovoltaic Direct Drive Refrigerator with Ice Storage: Preliminary Monitoring Results," ISES Solar World Congress, Adelaide, Australia, 2001.

[3] M. Ewert, D. Bergeron, R. Foster, L. Estrada, and O. LaFleur, "Photovoltaic Direct-Drive Battery-Free Refrigerator Field Test Results," SOLAR 2002, ASES, ASME, NPSC, Reno, Nevada, June 15-20, 2002.

[4] R. Foster, B. Jensen, B. Dugdill, B. Knight, A. Faraj, J. K. Mwove, and W. Hadley, "Solar Milk Cooling: Smallholder Dairy Farmer Experience in Kenya," ISES Solar World Congress, Daegu, South Korea, November 12, 2015, 11 pp.

Cost Optimization of Decommissioning and Recycling CdTe PV Power Plants

V. Fthenakis[1], Z. Zhang[1] and J.-K Choi[2],

[1]Center for Life Cycle Analysis Columbia University, Mudd 926, 500 W 120th street, New York, NY 10027
[2]Mechanical Engineering/ Renewable and Clean Energy Program, 300 College Park, Dayton, OH 45409

Abstract — This paper evaluates the cost of decommissioning, at the end of their lives, utility-scale ground-mount CdTe PV power plants in the U.S.-SW, based on publicly available industry data and General Algebraic Modeling System (GAMS) optimization, to maximize the net revenue from this activity. The recycling of structure and modules could create a profit of $ 1.58 per module area, the major revenue coming from the copper of the cables and the steel and aluminum of the module mounting and support. Planning for easy recovery of materials from the BOS in addition to those from the modules could further add to the profitability of recycling.

I. INTRODUCTION

With the rapid global deployment of photovoltaic (PV) power systems, the future cumulative amount of end-of-life (EOL) waste PV panels is projected to increase rapidly.[1] Without proper treatment, the wastes could cause environmental impacts from potential leaching of regulated metals commonly found in PV modules such as compounds of lead, cadmium, and selenium. Also, the loss of energy-intensive resources (e.g., solar-grade silicon, glass, aluminum, and copper) and rare metals (e.g., silver, indium, tellurium and gallium) could in the future lead to resource depletion.

Most of the initially installed PV modules globally are currently operating. For CdTe PV, the post-installation annual breakage rate is 0.05% per year in the first ten years of operation and 0.01% per year in operating years 11-25 [2].

The first-generation solar modules are silicon-based, and do not require any rare materials other than Ag. The second-generation solar modules, like thin-film CdTe and CIGS modules, involve elements of limited availability (e.g., Te, Ga, In). Thus, the EOL solar modules could become a significant secondary source of these materials. Also, recycling of EOL PV modules resolves environmental concerns associated with waste management.

This paper describes a cost optimization of decommissioning EOL CdTe PV power plants and recycling the elements from the modules and the metals from the supporting and mounting structures. We used industry and market data and optimized the rate of recycling using the General Algebraic Modeling System (GAMS), extending an earlier module recycling GAMS model described by Choi and Fthenakis (2010) [3].

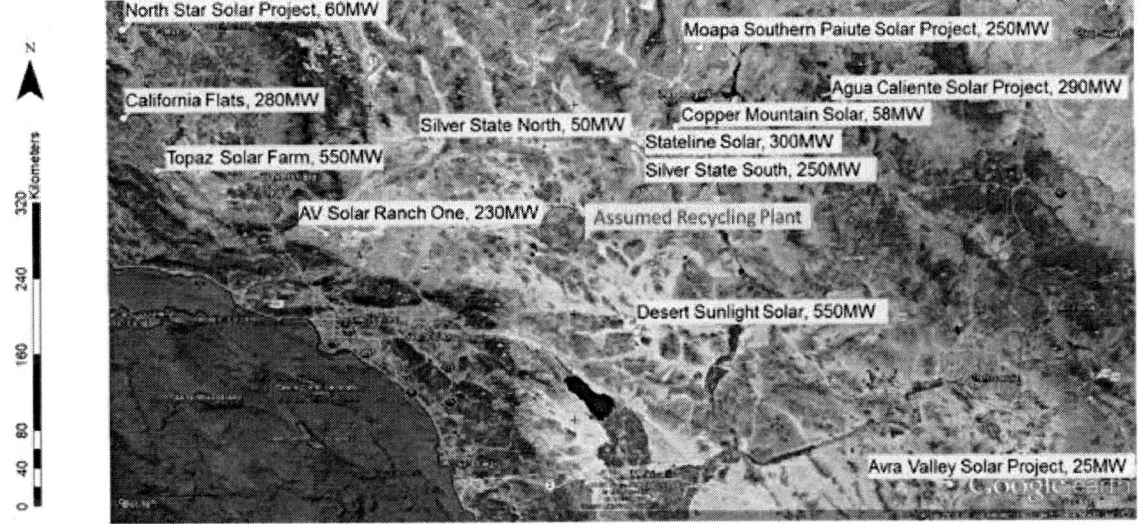

Figure 1. Locations of 12 projects developed by First Solar and the assumed recycling plant location based on capacities and distances

We examined decommissioning and recycling scenarios from large CdTe PV plants in the southwest of the United States. Twelve of First Solar's PV power plants are in CA, AZ and NV, and those represent 2893 MWac, or over half of First Solar's existing projects within the United States.

Fig. 1 shows a potential location of a recycling plant based on its distance to those 12 solar farms and their capacities, to minimize the total transportation cost.

II. METHODOLOGY

A. Recycling Process

First Solar has established the recycling of their modules and manufacturing scrap since the beginning of their operations. Currently, First Solar recycles manufacturing scrap and field returns in their production facilities and sends glass fragments to glass product manufacturers and a "filter cake" containing Cd, Te, Cu to a CdTe supplier, who extracts and purifies tellurium and cadmium in conjunction with their primary CdTe operations. In some cases, plastic is also recovered for use in rubber products; however, it was not included in our model optimization.

The top diagram in Fig. 2 shows the original cadmium flow in the whole industrial chain of First Solar modules. Specifically, the CdTe supplier extracts and purifies tellurium and cadmium byproducts from copper and zinc producers, respectively.

A future option called closed-loop recycling could involve the separation in the recycling center of Cd and Te using selective ion exchange, followed by purification and synthesis of CdTe in the same or a different operation [4]. The bottom diagram in Fig.2 represents this closed loop option.

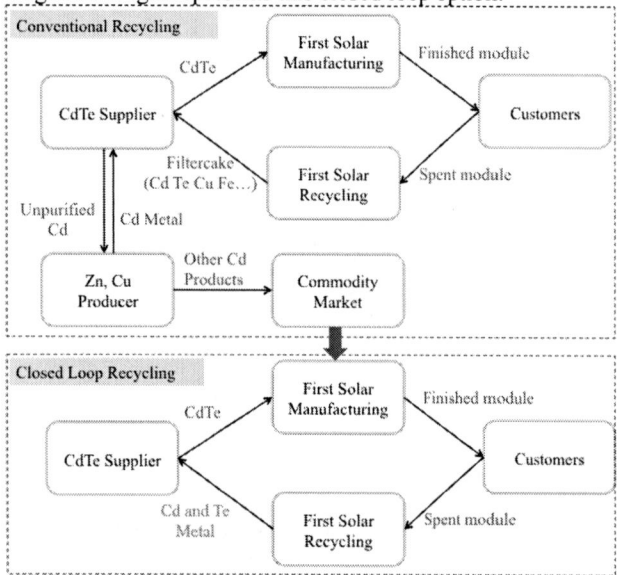

Figure 2. Recycling options

This integrated closed loop option not only simplifies the industrial chain, but could also reduce the total recycling cost.

The mounting and support structure and the cables will first be removed and delivered for metal recovery, while the modules are sent to a recycling plant, where a dry crushing and wet etching process is used as the basic delamination and metal-glass separation process in First Solar. The EOL modules will be loaded onto the auto-loading system, followed by a shredder and a hammer mill. Then the module fragments are immerged into acid solutions that leach the metals out and the later are recovered with selective pH precipitation. In subsequent steps,

the materials are separated from the liquids and are recovered as feedstock in glass product manufacturing.

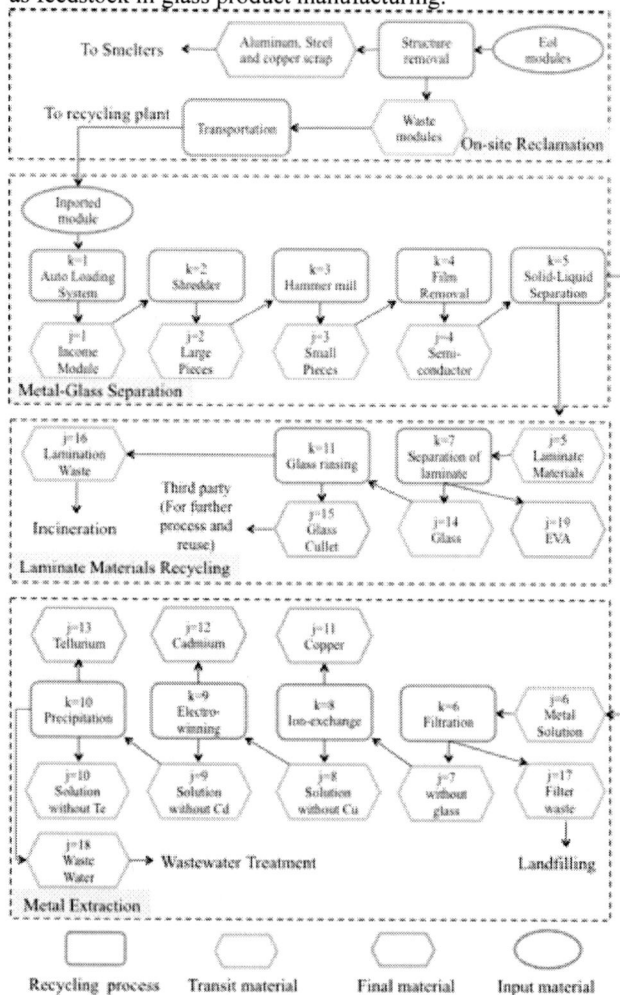

Figure 3. CdTe module recycling process map

In the conventional commercial recycling process, the leachate with the metal compounds is condensed in a thickening tank, and forms filter cake. The filter cake is then transported to a third party for processing.

However, as mentioned, the closed loop recycling would enable a recycler to directly use the leach solution to extract the metals. The leachate, or metal solution, goes through a 3-stage filtration to get rid of glass scrap, and the filter waste (e.g. glass fines) is landfilled. [5] Then the leach solution passes from a series of ion-exchange columns that retain copper and cadmium. The last step is to precipitate tellurium from the effluent solution.

Another recycling route is the recycling of glass. First the encapsulant layer is separated from the glass cullet in a vibrating screen. The separated glass is then discharged and rinsed for further recycling. At the end, all the plastic waste is either landfilled, handled in incineration plants for energy recovery, or used in rubber products. [6][7]

B. GAMS optimization model

We used GAMS to optimize the choice between recycling and storage of decommissioned parts. The objective is to maximize the net revenue from the recycling process described above, thus:

Objective function is to maximize:

$$z = \sum_{j,k,t} V_{jk}\Omega_{jkt} - \sum_{i,t} N_{it}T_i - \sum_{i,k,t} C_k Y_{it} - \sum_{i,t} I_t T_{it}\Lambda_{it} \quad (1)$$

Subject to:

$$\Omega_{jkt} = E_{jk} \sum_i Y_{it} \quad (2)$$

$$\Lambda_{it} = \Lambda_{i(t-1)} + N_{it} - Y_{it} \quad (3)$$

Indexes (shown in Fig. 3):

i Solar farm; $i \in \{1, \dots, 12\}$
j Transit or output material; $j \in \{1, \dots, 22\}$
k Decommissioning and recycling processes; $k \in \{1, \dots, 13\}$
t Monthly time series; $t \in \{1, \dots, 204\}$

Parameters:

V_{jk} Revenue/Cost of material j derived from k [\$/kg]
N_{it} Quantity of modules from i in period t [kg]
T_i Transportation cost of modules from i [\$/kg]
C_k Processing cost on k [\$/kg]
I_t Percentage of inventory cost to the incoming cost [%]
E_{jk} Weight percentage of j derived from k [%]

Variables:

Ω_{jkt} Quantity of j derived from k in period t [kg]
Y_{it} Quantity of modules from j recycled in period t [kg]
Λ_{it} Quantity of modules from i stored as inventory in t [kg]

The first term in (1) stands for the revenue/expense from a recyclable material/waste output. It includes all the on-site structure materials, cables and optional cable trays. Additionally, note that those on-site materials will be separately sent to a nearby smelter and the extra cost will thus be considered inclusively. The second term is a fixed transportation, proportional to the distance from each solar farm to the assumed location for the recycling plant. The third term gives the processing cost for modules step by step, and the last term indicates the corresponding inventory cost on each group of incoming modules. Equations (2) and (3) constrain the material balance and inventory balance respectively. Besides, the limit of processing capacity and the minimum safety storage level also play a part in the optimization problem.

C. Base scenario description

• Plant capacity

The capacity of the recycling plant is assumed to be 60,000 metric tons per year. Considering the operational continuity of recycling plant and the maximum storage time length, the plant is primarily designed for the large number of EOL modules in 2038-2042, during which more than 80% of the solar modules from the projects in Fig. 1 are expected to be decommissioned.

The capacity in the base scenario will guarantee that all the incoming modules could be recycled within the following 2 years.

• Optimized inventory level

The inventory level refers to the quantity of modules stored in warehouse in a certain period for future recycling. If inventories are too high, then money would be tied up in capital; if not sufficient, then stock-outs may paralyze the whole recycling chain. Some operations typically have 0.5~1 % of the year's capacity in stock.

• Solar panel

The specifications of the CdTe PV panels vary among the 12 projects. The reference solar panel in this research is assumed to be First Solar Series 3™ Black PV module; nevertheless they all have the same area (0.72 m²) and weight (12 kg), and have a 25-year performance warranty and a 10-year product warranty.

The approximate composition of these panels is listed in Table I. [8]

TABLE I
MATERIAL COMPOSITION AND RECYCLING RATIO

	Weight [kg]	wt [%]	Recycling rate [%]
Glass	11.527	96.061	90
Copper	0.00132	0.011*	90
Cadmium	0.00708	0.059*	90
Tellurium	0.00900	0.075*	90
Encapsulant	0.314	2.614	-
Other	0.142	1.180	-
Total	12.000	100.000	-

* 5% fluctuation in wt (%)

• On-site reclamation

The on-site reclamation costs include two parts, the aboveground and underground structures removal. The aboveground reclamation cost is proportional to the plant capacity, while the costs of underground structure removal is also related to the percentage of recycled underground cables and the unit digging cost.

For ground-mounted PV systems, several scenarios were explored regarding the recycling of wires on the ground. In the basic scenario, all cables underground would be dug out and recycled. However in reality, the recycling decision on the buried cables usually depends on the local permitting authority. We further examine this in a sensitivity analysis.

• Optional cable tray

In the base case, we assumed all cables were buried underground, while we also explored the scenario when the cables were mounted in above-ground cable trays, which contain steel and aluminum that also need to be recycled when reclaiming the cables.

Table II shows the major parameters of the basic scenario.

TABLE II
MODELING PARAMETERS FOR BASE CASE

Description	Unites
Weight per module	12 kg
Area per module	0.72 m²
Capital cost for inventory	1.96%/month
Inventory safety level	0.5% of year's capacity
Steel weight in cable tray scenario	3.3 tonne/MWac
Aluminum weight in cable tray scenario	2.1 tonne/MWac
CdTe recovery rate	90%
Metal reclamation loss	10%
Transportation cost (22t truck)	$1.2 per km
Distance to smelters	160 km
Structure scrap value	50% of its metal price

III. RESULTS

Table III gives the breakdown of the net revenue on recycling. The importing cost on transportation is a fixed cost, which relates to the weight of incoming modules.

TABLE III
BREAKDOWN OF COST MODEL

Revenue/cost	Involved Variables	Optimized value ($/module area*)	
		With Cable Tray	**Without Cable Tray**
Module revenue	Recycling Quantity	1.668	
Recycling waste disposal cost		-0.038	
Structure scrap revenue	Recovery rate of underground cables	4.634	4.439
On-site cost		-2.761	-3.265
Cost of importing modules	Transportation unit cost	-0.207	
Cost of transporting structure scraps to smelters		-0.067	-0.064
Cost of processing	Recycling quantity	-1.596	
Cost of inventories	Inventory quantity	-0.053	
Total		1.578	0.882

note that module area in this normalization includes all the corresponding mounting and support structure.

Fig. 2a shows the amount of end-of-life modules for recycling and inventory in each period.

In the first 7-8 years, the quantity of retired modules would be much less than those in the following years, and careful planning would be needed to manage recycling capacity. A second designing plan considers for the continuity of the operation, and is compared to the original one.

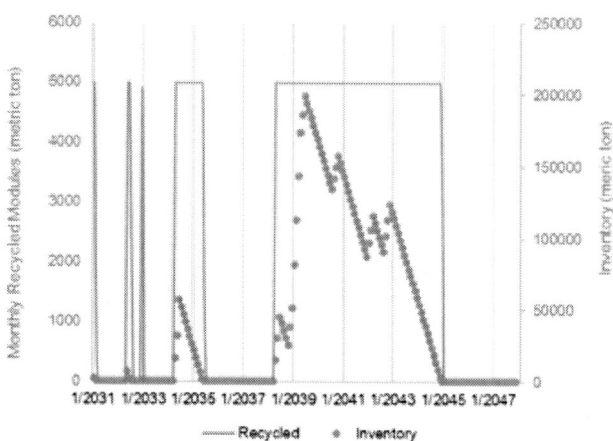

Figure 2a. Quantity of inventory and recycling in base scenario

As shown in Fig. 2b, the recycling plant will be constructed in 3 phases. Starting with a small capacity of 8,000 ton/yr, the plant will expand to 30,000 ton/yr after 3 years, then finally increased to its max capacity of 60,000 ton/yr in 2038, when numerous modules are imported into the plant. The inventory cost will correspondingly increase by 15%.

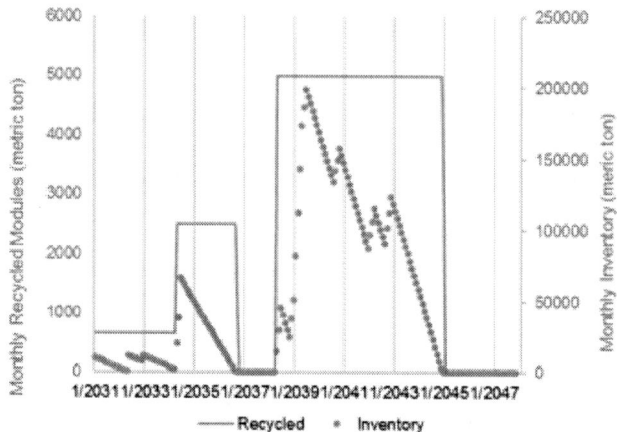

Figure 2b. Quantity of inventory and recycling in 3-phase mode

First Solar has a development plan on designing a smaller and mobile recycling facility[9], which could minimize the transportation costs and could become a solution for the operational continuity problem during the early low recycling volume period.

The on-site reclamation cost could vary with location, especially the digging cost for the buried cable. In the case where most of the wires in the PV system are buried underground, reclamation plans may assume that part or all of copper may be abandoned. However, as discussed below, we found considerable value in digging out and recovering underground cables. From the specifications of the target 12 solar farms, we estimated the potential revenue of digging out the end-of-life cables.

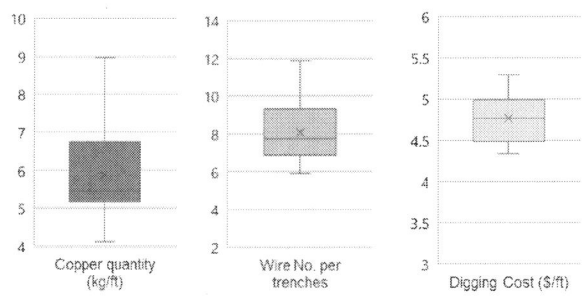

Figure 3. Box plots on the copper quantity, wire number and digging cost of buried cable

As shown in Fig. 3, every foot of underground cable conduits contains about 8 cables accounting for about 6 kg of copper. Consider the copper scrap price is between 50% to 95% of its metal price [10], then in most cases the revenue is profitable. Besides, some local authorities require the removal of underground structure. Thus, in the base scenario, we assumed that 100% of the buried cables will be removed and recycled.

IV. UNCERTAINTY ANALYSIS

A. Plant capacity

As discussed in the previous part, the capacity of the recycling plant not only affects the inventory cost on imported modules, but also affects the continuity of its operation. Besides, it may also be restricted by regulation on the maximum storage year of waste. Two operation modes, referring to the 1-phase and 3-phase modes respectively, were discussed in the base scenario. The sensitivity of the capacity is analyzed accordingly in Table IV.

However, the capacity has a great impact on the max storage time. In the base case, it takes no more than 2 years to recycle a group of imported modules, while with a disposal capacity of 70,000 metric ton/yr it takes only 1 year. Besides, it saves more than 1 million dollars for storage while it also costs more on the initial capital investment. Of course, the choice of capacity should also be based on the potential CdTe solar projects in the future.

Respecting to the continuity of operation, the 3-phase mode (Fig. 2b) performs much better than the 1-phase mode (Fig. 2a), despite the relatively higher inventory cost.

TABLE IV
SENSITIVITY ANALYSIS ON RECYCLING CAPACITY

| Capacity (ton/yr) | Operation Mode | Inventory Cost | | Max. Storage year | Intermittent Interval (months) | Net Revenue ($/module) |
		$/module	Additional cost (over base scenario)			
70000	1-phase	-0.039	-$632,367	1.1	69	1.592
	3-phase	-0.049	-$180,676		18	1.583
65000	1-phase	-0.046	-$316,184	1.5	67	1.585
	3-phase	-0.055	$90,338		18	1.577
60000	1-phase	-0.053	$0 (Base)	2.0	66	1.578
	3-phase	-0.062	$406,522		18	1.569
55000	1-phase	-0.063	$451,691	2.6	64	1.569
	3-phase	-0.070	$767,875		18	1.561
50000	1-phase	-0.073	$903,382	3.3	62	1.558
	3-phase	-0.080	$1,219,565		18	1.551

B. Recovery value

The reselling prices of recycled materials determine the revenue of recycling, which is a most influential factor in the cost structure. Identifying the most sensitive material helps the company make better choices and increase the profit.

In the sensitivity analysis, the total cost is re-calculated when the selling price or disposal fee of a recovered material is increased by 10%. A comparison among different materials is shown in Fig. 4, with the higher recycling revenue associated with copper and steel under both scenarios with/without cable trays. The variation of the revenue caused by the selling price of the recovered materials is considerable. Tellurium has by far the highest recovery value per unit mass, so it will be less influenced by transportation costs. .

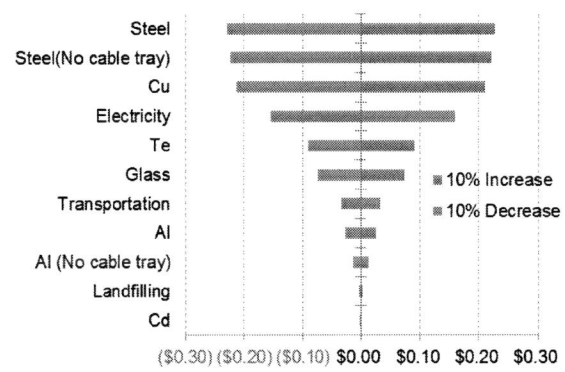

Figure 4. Net revenue sensitivity analysis assuming a 10% fluctuation in each material price or processing cost

The market price varies substantially, especially for metals, so we investigated the sensitivity of net recycling revenue to changes in the prices of the recovered materials. Fig. 5 shows the results of a low price and a high price sensitivity scenario. In the low-price scenario, the market price was set to 50% of its value in the basic scenario, and in the high price scenario it was set to 150%. From these results, we see that the recycling process could be very profitable when the market prices for the recovered materials are relatively high, and aluminum is a major contributor to this profit.

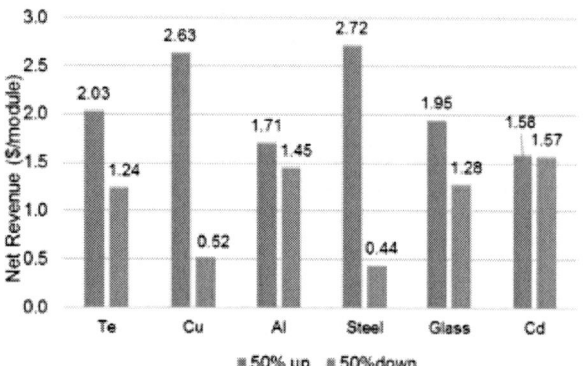

Figure 5. Net revenue sensitivity analyses assuming a +/-50% change in each material price under cable tray scenario

5.3 The recovery rate of buried cables and digging cost

It is helpful to see how the recycling revenue changes with

TABLE V
SENSITIVITY ANALYSIS ON DIGGING COST AND RECOVERY
RATIO ($/module area)

($/MWac) Cost % Recovery	4,000	6,000	8,000	10,000	12,000
100%	1.133	1.008	0.882	0.756	0.630
75%	0.641	0.547	0.453	0.358	0.264
50%	0.149	0.086	0.023	-0.040	-0.103
25%	-0.344	-0.375	-0.407	-0.438	-0.470
0%	-0.836				

the recovery rate of buried cables and the related digging cost, since copper substantially contributes to the profit. Table V provides the net revenue under different recovery rates and cost referring to buried cables.

The recovery of underground cables would be profitable since each foot of trench contains copper wires valuing $24-52, while the digging cost is only around $5/ft assuming depths of 2-4 ft. However, recovery of buried copper may not be accounted for in PV decommissioning plans.

VI. CONCLUSIONS

From the result of the optimization model, the whole recycling process could make a profit of up to $1.58 per module,

the major revenue coming from the copper, steel and aluminum of the module mounting and support.

The cost varies with the fluctuation in the quantity of incoming modules and structures. The selling prices of recovered materials, especially steel, Cu and Al also greatly affect the net revenue from recycling. Based on our preliminary results and discussion the following suggestions for reducing the cost of recycling are made:

Regional transit centers could be set up. A transit center would serve as a buffer between solar plants and a recycling plant, making its operation more stable. Also, with a transit center, the recycling plant could control its inventory level and rate of processing to minimize the total cost.

In the cost model, the revenue could also be time related. As shown by sensitivity analysis the influence of market prices could be significant; thus, it will be useful to simulate the effect of market fluctuations.

ACKNOWLEDGEMENT

We thank Dr. Parikhit Sinha, First Solar, for pointing us to data sources and meticulously answering our inquiries.

REFERENCES

[1] International Renewable Energy Agency (IRENA) and International Energy Agency Photovoltaic Power Systems (IEA-PVPS). "End-of-Life Management: Solar Photovoltaic Panels". T12-06:2016.

[2] P. Sinha and A. Wade. "Assessment of leaching tests for evaluating potential environmental impacts of PV module field breakage." IEEE Journal of Photovoltaics 5.6 (2015): 1710-1714, 2015.

[3] J-K Choi and V. M. Fthenakis, "Economic feasibility of recycling photovoltaic modules." Journal of Industrial Ecology 14.6: 947-964, 2015.

[4] V. M. Fthenakis and W. Wang, "Extraction and separation of Cd and Te from CdTe photovoltaic manufacturing scrap,"Prog. Photovoltaic Res. Appl., 14(4), 363-371, 2006.

[5] P. Sinha, M. Cossette, J. F. Ménard, "End-of-life CdTe PV Recycling with Semiconductor Refining, in 27th EU PVSEC, Frankfurt, Germany, 2012 pp. 4653–4656.

[6] V. M. Fthenakis, P. Duby, W. Wang, et al. "Recycling of CdTe photovoltaic modules: recovery of cadmium and tellurium," in 21st European photovoltaic solar energy conference. Dresden, German, 2006: 4-8.

[7] W.Wang and V.M. Fthenakis. "Kinetics Study on Separation of Cadmium from Tellurium in Acidic Solution Media Using Cation Exchange Resin", Journal of Hazardous Materials, B125, 80-88, 2005.

[8] V. M. Fthenakis, M. Fuhrman, J. Heiser, et al. "Emissions and encapsulation of cadmium in CdTe PV modules during fires". Prog. Photovoltaics: Res. Appl., 13(8): 713-723, 2005.

[9] S. Raju, First Solar's Industry Leading PV Technology and Recycling Program. Solar Power International, Chicago, Illinois, USA.

[10] Department of the Interior, "United States Copper Metal and Scrap Use and Trade Patterns, 1995–2014". Scientific Investigations Report, 2016-5075.

Challenges for decision makers when feed-in tariffs or net metering schemes change to incentives dependent on a high share of self-consumed electricity

Mattias Gustafsson

UNIVERSITY OF GÄVLE, Faculty of Engineering and Sustainable Development, 801 76 GÄVLE, SWEDEN

Abstract — **When there are differences in economic value of self-consumed and exported electricity, profitable PV installations are dependent on accurate predictions of self-consumed electricity. In this study, minute-based data of PV production and electricity use were logged in a single-family house in Sweden. It is shown that when self-consumed electricity is measured, a low time resolution and different electric meter configurations can result in 60% lower registered self-consumed electricity than predicted. When feed-in tariffs or net metering schemes change to incentives dependent on the fraction of self-consumed electricity, the market and electric meter infrastructure must be prepared to avoid market disturbances.**

Index Terms — **government, policy, market incentives, feed-in tariffs, net metering, self-consumption**

I. INTRODUCTION

The global PV market continues to grow, with China, Japan and USA the leading markets. The PV market is still driven by different market incentives, the most common being feed-in tariffs (FiT), direct subsidies or tax breaks and incentivized self-consumption or net metering [1]. FiT is a simple but effective instrument to develop a PV market. The electricity produced to the grid is paid with a pre-defined constant level or decreasing (constant decreasing or with adjustments) level of the FiT. A net metering scheme allows the user to produce electricity to the grid and then credit used electricity, just like if the electric meter is running backwards. The net metering scheme is commonly limited to crediting electricity on a monthly or yearly basis.

In FiT and net metering schemes, the electricity produced by PV systems is encouraged to be exported to the grid and the difference in economic value between the produced excess electricity and the "true" value for the electricity is commonly paid by official bodies or utilities [1]. Retail grid parity has been reached for PV installations in some countries and more countries will follow due to lower costs for PV installations and increased costs for conventional power generation [2]. This means that market incentives such as FiT and net metering schemes will be more limited in the future and

the profitability of the PV system will be dependent on the installation cost, annual production of electricity and value of the electricity produced without subsidies.

When there are differences in economic value between self-consumed and exported electricity, profitable rooftop PV installations are dependent on accurate predictions of the amount of self-consumed electricity within the buildings.

To make a proper prediction of self-consumed electricity it is common to start with a simulation tool to estimate the PV production. Simulation tools most commonly use climate and load data with one-hour resolution. Prediction of PV production for common simulation tools is fairly accurate (within ±8%) [3] even if impact of shadowing and temperature issues can cause larger errors.

When the yearly amount of self-consumed electricity is predicted the internal load needs to be known. For existing buildings not in a construction phase, there is often available consumption data from the local electric grid owner. If the electric meter is a so-called "smart meter" those data are available with different resolution depending on the time frame in which the grid owners collect data from the meter. Common time frames are 1 hour, 30 minutes and 15 minutes. Because the climate data used commonly use one-hour resolution the self-consumption is also commonly predicted with one-hour resolution [4].

An import/export metering arrangement can be made with two individual electric meters where the net value is calculated from monitored values. Another arrangement is to use a bidirectional meter where imported and exported electricity is monitored and stored in one unit, which is the technology investigated here. A bidirectional "smart meter" records the power use momentarily and the recorded values are accumulated to values in the chosen time frame (commonly 1 hour, 30 or 15 minutes). The accumulated values are then collected by the grid owner [5,6].

978-1-5090-5606-4/17 $31.00 © 2017 IEEE

Fig. 1. The investigated single-family house in Gävle, Sweden.

When combining simulations of PV production and the internal use of electricity in buildings, the resolution is important. The solar irradiation during a cloudy day changes quickly. The same applies to electricity use in buildings [4,7,8]. When investigating the self-consumption of electricity, the quick changes in both PV production and use of electricity can overestimate the predicted self-consumption and a short time resolution is recommended [4,7,8]. A study on a typical single-family house in Switzerland shows that the self-consumption was overestimated by 8% when the resolution of used data is low. The investigated house has a yearly electricity use of 4.9 MWh and a PV installation of 5 kW_p [9].

In this investigation, a PV system on a single-family house in Sweden is used as a reference and the difficulties in predicting the share of self-consumed electricity are investigated. A discussion of bidirectional smart meters, the value of produced electricity and governmental challenges concludes the abstract.

II. METHOD

The investigated house, shown in Fig. 1, is a single-family house situated in Gävle, Sweden. It was built in 1983 and has a heated living area of 171 m^2. The annual mean temperature in Gävle is 5.2 °C and there is no air cooling unit installed. The main source of heating is a small ground source heat pump connected to the space heating system and a solar thermal system for domestic hot water. When the heat pump and solar thermal system cannot supply the required heat demand, heat is delivered by an electric immersion heater.

The electricity bought from the electricity retail company is approximately 10,000 kWh/year. The building has a PV installation of 2.6 kW_p with a single phase inverter. As is common in Sweden, the building is connected to all three phases to the grid.

The use of electricity on each phase and production from the PV installation were logged with an interval of one minute with an Egauge data logger [10]. When hourly values were evaluated, they were calculated out of minute-based data. When investigating how the electric meter configuration affects the amount of self-consumed electricity, the PV production was halved and doubled to also investigate a 1.3, 3.9 and 5.2 kW_p PV installation.

This study uses production data from an already installed PV system but similar results should be obtained if monitored irradiation data were used to predict the PV production in a planning process of an installation. In addition to the measured data for the PV production, a prediction of the annual PV production was made in the simulation tool PVsyst [11].

III. CONFIGURATIONS OF BIDIRECTIONAL METERS

The reported values from a bidirectional electric meter depend on the number of phases connected to the meter and the configuration of the meter. This can result in different monitored values of the amount of self-consumed electricity even though two buildings are identical and the PV production and internal loads are equal [12].

Fig. 2 shows a building connected to all three phases and with a momentary PV production of 3 kW on phase one with a single-phase inverter or 1 kW on each phase with a three-phase inverter. The total instantaneous internal load is 3 kW and divided as 1.5 kW on phase one, 1.25 kW on phase two and 0.25 kW on phase three.

Example of internal load and PV production

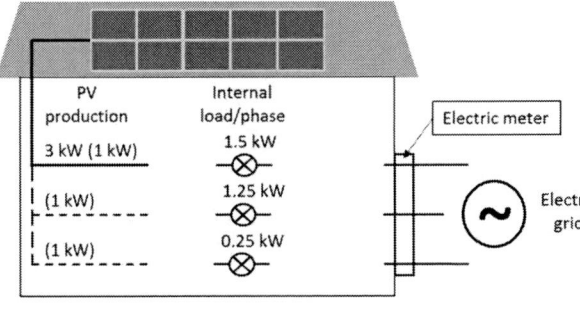

Fig. 2. The instantaneous values used to calculate the recorded values with an internal load of 1.5 kW, 1.25 kW and 0.25 kW on the different phases. The PV system momentarily produces 3 kW on phase one with a single-phase inverter or 1 kW on each phase with a three-phase inverter.

There are two main configurations of a bidirectional meter. The first configuration records each phase individually (imported or exported). The sum of imported and exported electricity is accumulated in different registers. The accumulated result in each register is the monitored value for the used time frame collected by the grid owner. Here this configuration is called individual measurements of phases. The second configuration records the sum of electricity imported in all phases and subtracts the exported electricity. The net value is stored in one register and accumulated to the monitored value. Here this configuration is called sum measurement of phases. The different recorded values according to Fig. 2 are presented in Table 1.

Table 1. The instantaneous recorded values for bidirectional meters configured to record the phases individually or the sum of the phases.

	1-phase inverter		3-phase inverter	
	Imported electricity	Exported electricity	Imported electricity	Exported electricity
Individual measurement of phases	1.5 kW	1.5 kW	0.75 kW	0.75 kW
Sum measurement of phases	0 kW	0 kW	0 kW	0 kW

IV. RESULTS

The measured PV production for the system during 2015 was 2004 kWh. The PV system on the investigated building consists of 10 modules with a peak power of 260 watts each, produced by PV Enterprise. The modules are called PVE-P6H-260W and have a voltage of 34.9 Volts at maximum power point at standard test conditions. The inverter is a Steca Grid 3000 with an operating voltage between 350 to 700 volts. Due to the system voltage when operating at maximum power point and the operating voltage of the inverter, the modules and the inverter are not properly matched. When simulating the PV system in PVsyst, the poor system component match prevents the program from performing a simulation, therefore a Steca Grid 3203 with an operating voltage between 250 to 800 volts is used in the simulation. The annual PV production simulated in PVsyst was 2014 kWh/year which corresponds well with measured production for 2015. The simulation was done with the settings for shadowing, according to module strings.

Measured PV production with hourly data compared with minute-based data for 1 July 2015 is presented in Fig 3. The hourly data of electricity use compared with minute based data for 1 July 2015 is shown in Fig. 4.

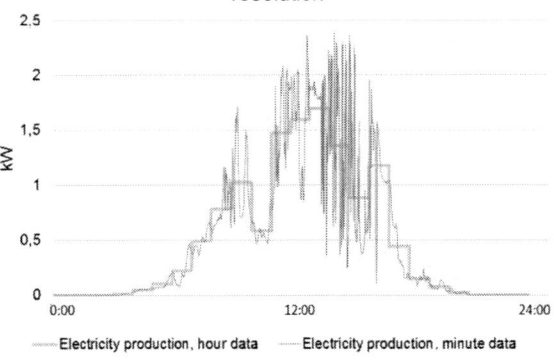

Fig. 3. PV production with hourly data compared with minute-based data for 1 July 2015.

Electricity use with one hour and one minute resolution

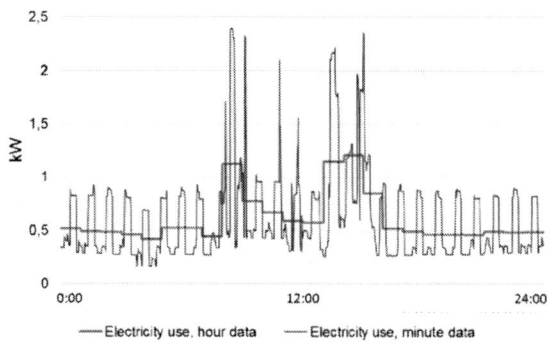

Fig. 4. Electricity use in the building with hourly data compared with minute-based data for 1 July 2015.

Fig. 3 and Fig. 4 show that hourly values smooth out the variations in PV production and electricity use. Similar results have been shown by [7,12].

Produced electricity and self-consumed electricity is compared for hourly and minute-based data and are presented in Fig. 5. The electric meter configuration is the sum of phases. There is a noticeable difference, especially during spring and summer. The amount of self-consumed electricity with hourly resolution is 1207 kWh and with minute-based resolution 1098 kWh. This means that an hourly resolution overestimates the self-consumption by 10% compared to a minute-based resolution for the investigated year.

Difference in amount of self-consumed electricity for one hour and one minute resolution

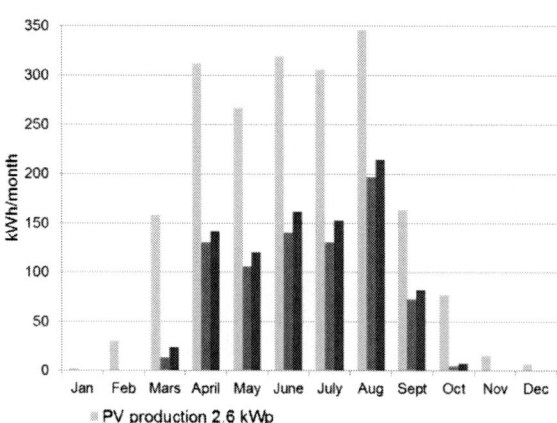

Fig. 5. Produced and self-consumed electricity for hourly and minute-based values each month for 2015.

The PV production and self-consumed electricity for the different PV system sizes, the two types of electric meter configurations and a single-phase inverter (connected to all three phases respectively) or a three-phase inverter is presented in Table 2.

Table 2. Self-consumed electricity for four different PV system sizes with different electric meter configurations and for a single-phase inverter (connected to each phase respectively) or a three-phase inverter. Measured with minute-based data.

PV system size	1.3 kWp	2.6 kWp	3.9 kWp	5.2 kWp
PV production	1002 kWh	2004 kWh	3006 kWh	4008 kWh
Self-consumption, 1- or 3-phase inverter, sum measurement of phases	734 kWh	1098 kWh	1344 kWh	1526 kWh
Self-consumption, 3-phase inverter, individual measurements of phases	655 kWh	964 kWh	1179 kWh	1342 kWh
Self-consumption, 1-phase inverter connected to phase 1, individual measurements of phases	431 kWh	569 kWh	662 kWh	728 kWh
Self-consumption, 1-phase inverter connected to phase 2, individual measurements of phases	362 kWh	483 kWh	559 kWh	613 kWh
Self-consumption, 1-phase inverter connected to phase 3, individual measurements of phases	386 kWh	539 kWh	640 kWh	713 kWh

For the analyzed 1.3 kWp PV installation, the self-consumed amount of electricity varies between 362 kWh and 734 kWh depending on the electric meter configuration and whether a single phase or three-phase inverter is used. Total PV production is 1002 kWh. For the largest analyzed PV installation, 5.2 kWp, the self-consumed amount of electricity varies between 613 kWh and 1526 kWh. The PV production for the largest PV installation is 4008 kWh.

V. DISCUSSION

FiT and net metering schemes are simple and effective instruments to develop a PV market. But as PV system costs decrease and the PV market increases, other or no market incentives are preferable to avoid uncontrolled market development. Since FiT and net metering schemes encourage the user to export electricity to the grid and the economic value is not, or to a small degree, dependent on the share of self-consumed electricity, estimations of profitable PV installations are quite simple and are dependent on the installation cost and annual PV production. When there is a difference in economic value between self-consumed and exported electricity it is more complicated to estimate the profitability of the PV system since other factors are involved.

When estimating the share of self-consumed electricity, hourly values of electricity use and irradiation result in an overestimation of predicted amount of self-consumed electricity. The overestimation in the analyzed building is approximately 10% compared to minute-based values.

When bidirectional "smart meters" are installed in a multi-phase installation, the configuration of the meter can affect the monitored self-consumption of electricity substantially. When the 2.6 kWp PV installation is made with a single-phase inverter connected to phase two, the measured amount of self-consumed electricity is 483 kWh for the investigated year when the electric meter configuration is individual measurements of phases. If the predicted amount of self-consumed electricity is estimated with hourly resolution of climate data and the internal load of electricity, the amount of self-consumed electricity is 1207 kWh if the electric meter configuration is sum measurement of phases. This means that the measured amount of self-consumed electricity is 60% lower than predicted. When there is a difference in economic value between self-consumed electricity and produced excess electricity, this affects the profitability of a PV installation substantially.

In most countries, single-family houses are not connected to all three phases. Other building types (with other activities) can get similar differences in self-consumed electricity due to the electric meter configuration but probably not as prominent as single-family houses. Large installed electrical appliances are commonly connected to all three phases and therefore the internal load is most likely more evenly divided between the phases. In USA where split-phase connections are common for single-family houses, different electric meter configuration can appear if the meter configuration is not regulated nationally or regionally.

VI. CONCLUSIONS

- When the self-consumption is compared for hourly data and minute-based data of PV production and electricity use, hourly data overestimate the self-consumption by 10% for the 2.6 kWp PV installation for the investigated year.

- When the different electric meter configurations are compared, the variations in annual amount of self-consumed electricity can vary by a factor between 2 and 3 depending on the PV system size.

- When the predicted amount of self-consumed electricity is estimated with hourly resolution of PV production and internal load of electricity, the measured amount of self-consumed electricity is 60% lower than predicted for the 2.6 kWp PV installation. This is when a single phase inverter is used, connected to phase two, and the electric meter configuration is individual measurement of phases.

There is a challenge in future incentive systems to avoid abrupt market development disturbances due to non-economical PV installations. When FiT or net metering schemes change to market conditions where there are differences in economic value for self-consumed and exported electricity, the rules and regulations, market and electric meter infrastructure must be prepared to manage the changes to avoid market development disturbances.

ACKNOWLEDGEMENT

The work has been carried out under the auspices of the industrial post-graduate school Reesbe, which is financed by the Knowledge Foundation (KK-stiftelsen).

REFERENCES

[1] IEA-PVPS, "Trends 2015 In Photovoltaic Applications," Report IEA PVPS T1-27:2015. Photovoltaic Power Systems Programme 2015.

[2] Y. Karneyeva and R. Wüstenhagen, "Solar feed-in tariffs in a post-grid parity world: The role of risk, investor diversity and business models," Energy Policy, 2017;106:445-56.

[3] J. Freeman, J. Whitmore, N. Blair and A. P. Dobos, "Validation of multiple tools for flat plate photovoltaic modeling against measured data," 2014 IEEE 40th Photovoltaic Specialist Conference (PVSC), Denver, CO, 2014:1932-1937.

[4] S. Cao and K. Sirén, "Impact of simulation time-resolution on the matching of PV production and household electric demand," Applied Energy, 2014;128:192-208.

[5] Elforsk. "Electric meters for local production of electricity" (Hantering av elmätning vid småskalig produktion och andra udda belastningsfall), Report 09-107, 2009.

[6] Forum of Regulators Working Group, "Evolving Net-metering Model Regulation for rooftop based solar PV projects," Report August 2013.

[7] A. Wright and S. Firth, "The nature of domestic electricity-loads and effects of time averaging on statistics and on-site generation calculations," Applied Energy, 2007;84:389-403.

[8] J. Widén, E Wäckelgård, J Paatero, P Lund, "Impacts of different data averaging times on statistical analysis of distributed domestic photovoltaic systems," Solar Energy 2010;84:492-500.

[9] N. Wyrsch, Y. Reisen and C. Ballif, "Effect of the fluctuation of PV production and electricity demand on the PV electricity self-consumption," 28th PVSEC, Paris, 2013.

[10] eGauge Systems, Available: https://www.egauge.net [2016-02-12].

[11] PVsyst, Available: www.pvsyst.com [2016-02-12].

[12] M. Gustafsson, B. Karlsson and M. Rönnelid, "How the electric meter configuration affect the monitored amount of self-consumed and produced excess electricity from PV systems—Case study in Sweden," Energy and Buildings, 2017;138:60-8.

Procedures to Make Projects about Renewable Energy Generation Connected to the grid in Colombia

J. A. Hernandez[1]*, C. A. Arredondo[2]**, D. J. Rodriguez[1]

[1]Universidad Distrital Francisco José de Caldas, Bogotá – Colombia, *jhernandez@udistrital.edu.co;
[2]Universidad de Medellín, Medellín – Colombia, **caarredondo@udem.edu.co

Abstract — this article addresses the needed formalities in order to present projects about renewable energy generation connected to the grid in Colombia. The definitions given by the Household Utilities Law and the Electricity Law, known as Laws 142 and 143 of 1994, respectively, about the generation activity and the topics to consider established by the second one for the development of grid connected projects that every interested individual must accomplish are discussed. The applicable normativity is presented for generation projects with power rating from 500kW and 20MW, which are considered by the regulation as a minor generators plants as well as the objectives that the Colombian state must accomplish, according to the Law in the Non-Conventional Energy Sources (FNCE for its acronym in Spanish) integration to the National Interconnected System (SIN). Finally, registration characteristics for generation projects, environmental licenses, connection points assignment, and energy commercialization procedures established by the law are given. On May of 2014 the Colombian government issued the Law 1715 [1] promoting and encouraging the use and implementation of Non-Conventional Energy Sources in the country. But, nowadays the Law 1715 has huge gap specifically for the regulation of this sources (FNCE), and any operator or electricity Generation Company with interest in the developing of energy generation from FNCE has to accomplish the current regulation (Laws 142 and 143 of 1994). This paper presents the methodology that any solar PV project connected to the grid must accomplish with the current Colombian regulatory framework.

Keywords: solar projects, Colombia, regulations and laws.

I. INTRODUCTION

Energy policy is established as a tool of modern societies, directed for the use of non-conventional renewable energy and its integration into the grid. Colombia, after the adoption of Kyoto Protocol by Law 629 of December 27th, 2000 has the commitment to promote sustainable development electricity generation projects to reduce the use of fossil fuels, by using new technologies that address climate change issues [2]. In 1992 Colombia suffered an energy crisis as a consequence of the period of drought caused by "El Niño Phenomena", leading to the National Congress to issue the laws 142 and 143 of 1994, also known as the Household Utilities Law and the Electricity Law, respectively [3, 4]. With the enacting of these laws, begins the regulation of power generation activities, among others.

The development of energy generation projects involves considerations in terms of interconnection, environmental issues and others. The Energy and Gas Regulatory Commission (CREG by its acronym in Spanish), in its role to ensure the adequate and reliable service delivery through the efficient use of different energy sources and the fulfillment of its duties as defined in Article 23rd of Law 143 of 1994, the Mining and Energy Planning Unit (UPME by its acronym in Spanish), is the entity in charge and with the responsibility to evaluate the economic and social suitability for the development of non-conventional energy sources and its uses [5], and the Ministry of Environment and Territorial Development (MADS), under Title VII of the Law 99 of 1993 about environmental licenses, are the entities that have regulated the policies for the implementation of generation projects.

A. Goals for the participation of non-conventional energy sources connected to SIN

Through Law 697 of 2001 *"By which the rational and efficient use of energy is encouraged, the use of alternative energy is promoted and other provisions are established"* article 9 gives the directions for solar, wind, geothermal, biomass and small hydropower exploitation not exceeding the equivalent of 10 MW as Non-Conventional Energy Sources (FNCE by its acronym in Spanish) [6]. Accomplishing this this law, among other decrees and resolutions, the Ministry of Mines and Energy issued the Resolution 018-0919 of 2010 which states in its article 7 the goals for the National Interconnected System (SIN) on FNCE [7]. The goals are showed in table 1.

TABLE 1. GOALS OF FNCE PARTICIPATION IN SIN, MINISTRY OF MINES AND ENERGY

Goals of the FNCE participation in the national interconnected system	
2015	3.5%
2020	6.5%

B. Applicable regulations

Generation projects with power between 500 kW and 20 MW to be interconnected to the national grid system in Colombia, must take into account the next considerations:

- All the generation projects are conducted by laws 142 and 143. Those laws establish CREG as the authority to regulate the activities in the energy sector.
- CREG lays down the rules for planning and coordinate the operation of the national interconnected system, according to the article 23, of law 143.
- CREG's Resolution 025 of 1995 establishes the Network Code as fundamental part of the Operation Regulation for the National Interconnected System. Resolution 070 of 1998 establishes the Distribution Regulation.
- Law 143 settled the Mining Energy Planning Unit (UPME) to elaborate and to update the National Energy Plan and the Expansion Plan for the Electric Sector.
- CREG's Resolution 003 of 1994 regulates the electric energy transport in the Regional Transmission System and Local Distribution System.
- Projects with an installed capacity of 20 MW, Resolution 086 of 1996 define those projects as *"Minor Plants"* and regulates their generation activity.
- Ministry of Environment and Territorial Development (MADS), gives the regulations and the environmental licenses that the projects must accomplish (Decree 2041 of 2014) [8].
- CREG's Resolution 106 of 2006, establishes the general procedures to assign the appropriated connection point to the National Interconnected System.
- UPME's Resolution 0520 of 2007 modifies the Resolution 0638 of the same year, advising that any generation project to be connected to National Interconnected System must be registered with this body.
- CREG's Resolution 039 of 2001, establishes the available options to commercialize the energy coming from the *"Minor Plant"*.

II. GRID CONNECTION

The operator or generation pretending or seeking to develop generation projects connected to the grid in Colombia, must manage a series of procedures established by the law, in different government agencies.

Four key issues have been established for the interested operator to be considered during the execution of the project. The issues should be addressed by the understanding that they are not completely independent from each other, and that there

is not a defined sequence, as well as there are additional procedures that should be taken into account. Figure 1 presents the scheme that refers the fundamental aspects for the Generation Project procedures according to Laws 142 and 143 of 1994 [3, 4].

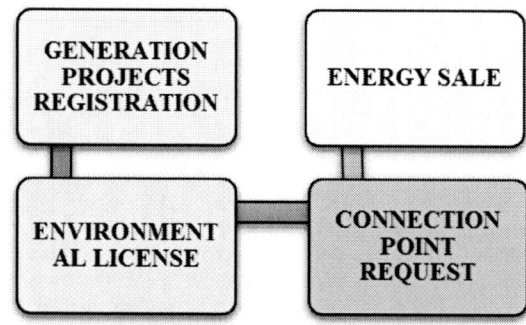

Fig. 1. Conducting generation projects for more than 500 kW and less than 20 MW connected to SIN, authors

According to the legal provisions given by the UPME in Laws 142 and 143 of 1994, and the UPME's resolution 0520 of 2007, this unit establishes that the generation projects must be registered to have information about its planning process [5]: *"The registration of generation projects ease to the UPME the access to the technical and economic information concerning to generation projects, in order to be able to establish different alternatives of expansion in generation in the Reference Expansion Plan for Generation and Transmission".*

This Registration process has three stages: first stage – prefeasibility, second stage – feasibility, and the third stage – final design (Fig 2.). Those stages must be fulfilled sequentially and have to completely accomplish the requirements of the UPME's resolution 0638 of 2007, which states that the generation developer has to make the legal procedure with the environmental authority for the environmental license and with the UPME for the connection point request.

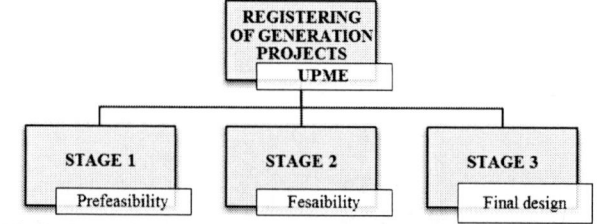

Fig. 2. Stages for the registering of generation projects, authors

For the first stage, the developer has to introduce the project and is necessary to have the prefeasibility study (to begin with the registration process). For the registration on the second stage, it is necessary a certification that has been registered on the first stage and that has been also registered with the environmental agency. On the third stage, a certification showing that complies with the requirements for the previous stage, the final designs, the environmental license, the connection point request approved and other requirements. The process has no cost for the developer.

For the environmental license, the Article 9 of the Decree 2041 of 2014 and must be accomplished. The environmental license is the *"authorization given by the environmental authority for the project execution, work or activity... And carry all the permissions, authorizations and/or concessions for the use, exploitation and/or affection of the renewable energy resources needed for the lifetime period of the project, work or activity"* [8].

For energy generation projects with capacities between 500 kW and 20 MW, the environmental bodies with authority award are defined on the article 9 of the Decree 2041 of 2014 are:
-   Regional Autonomous Corporations (CAR), and sustainable development offices
-   Municipalities, districts, and metropolitan areas (when the population is higher than one million of inhabitants within its urban perimeter according to article 66 of Law 99 of 1993) [10]
-   Environmental authorities created by the Law 768 of 2002.

According with the articles 9 and 18 of the Decree 2041 of 2014, projects surpassing the 10 MW of installed capacity will require environmental license, but those projects with capacity lower than 10 MW should ask for a statement about environmental diagnosis alternatives. Figure 3 shows the environmental license requirements.

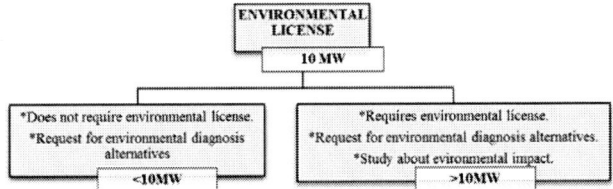

Fig. 3. Environmental license, authors

The environmental diagnosis alternatives "it aims to give information to evaluate and to compare the different options requested by the applying, under which it is possible to develop a project, work or activity" [8]. The applicant has to submit a written application to the Regional Autonomous Corporation, asking to determine if the project requires environmental diagnosis alternatives, attaching a clear description of the project, its objectives, the scope and its localization with blueprints and coordinates.

The environmental impact study "is the basic instrument for decision making concerning the projects, works or activities requiring environmental license and will be entailed in all the cases stated by the law".

The study must be carried out based on reference terms issued by the Environmental and Sustainable Development Ministry, or the environmental authorities (when it has not been issued by the ministry) 15 working days after the submission of the application.

For the connection point, the CREG's Resolution 106 of 2006, defines that the generator developer must present a study with the requisition to the National Transmission or the Network Operator or the UPME [10]. The CREG resolutions 001 and 004 of 1994, stablishes the access to the National Connection System, as well as Regional Transmission System, and the Local Distribution System as follows:
"For the case of generator, minor plants, self-generator or co-generators with the intention to connect their systems directly to the Regional Transmission System, and/or to the Local Distribution System, the procedure for the connection is ruled by whichever applies what is arranged on CREG Resolutions 025 of 1995 and CREG 030 of 1996 and other standards that modify or replace them"

The connection study must include the financial and technical feasibility of the project. The developer of the project should take into account the connection opportunities that periodically are presented by the energy transportation agents to the UPME as a reference to carry out the technical study for the connection according to:
-   Terms for Connection to the National Interconnected System for minor plants
-   Connection to the National Transmission System or the Connection to the Regional Transmission System, and to the Local Distribution System.

For minor generation plants to the National Interconnected System applies the article 4 of the CREG Resolution 086 of 2006, and for the Connection to the National Transmission System applies the stablished by the Annex Code for Connection (CREG Resolution 025 of 1995: *"...which stablishes the minimum technical requirements for the designing, building, mounting, commission, operation, and maintenance that every user should accomplish for its Connection to the National Transmission System"*).

For the energy sale, the CREG's Resolution 039 of 2001, defines the available options for the energy commercialization [11]. Generation plants lower than 10 MW, do not have access to the Central Dispatch and are not able to participate in the Electricity Wholesale Market. Generated energy can be commercialized following the next guidelines:

978-1-5090-5606-4/17 $31.00 © 2017 IEEE

- Generated energy can be sold to a market agent of the regulated market (with no existence of economical link between the buyer and seller).
- Generated energy can be offered to a marketer by means of public notice of the regulated market.
- Generated energy can be sold to generators or marketers for non-regulated users.

On the other hand, minor plants with power capacity between 10 MW and 20 MW have access to Central Dispatch participating in the Electricity Wholesale Market. If the generator is not given access to the Central Dispatch, generated energy can be sold, fulfilling with the previously mentioned terms for plants with effective generation capacity lower than 10 MW and the others contained on the article 3 of CREG's Resolution 039 of 2001[11]

IV. Conclusions

The government of Colombia through Law 697 of 2001, recognized the necessity for the integration of Non-Renewable Energy Sources through participation and incorporation goals into the national grid, this indicates and shows the growing importance in how to diversify energy matrix and the energy sources in the country.

The procedure for renewable energy projects that must be transacted before national entities is similar to the process used for generation projects with plant using conventional energy sources. International experience has shown that projects that use non-conventional energy sources require a different treatment to manage them because of the use new technologies, the deployment of non-conventional energy sources.

The entities that are responsible for designing connection codes and associated rules should start managing working groups in order to identify the needed recommendations for the integration of these new generation units based on non-conventional energy sources to the National Interconnected System in Colombia.

The making of renewable energy projects in Colombia needs clear policies, as well as regulation and procedures to promote and deploy those projects, allowing the growth of the energy sector.

References

[1] LEY 1715 de 2014. Diario Oficial No. 49.150 de 13 de mayo de 2014. Bogotá, Colombia. 2014.
[2] Ley 629 de 2001. Diario Oficial número 44.272 del 27 de diciembre de 2000. Bogotá, Colombia. 2000.
[3] Ley 142 de 1994. Diario Oficial número 41.433 del 11 de julio de 1994. Bogotá, Colombia. 1994.
[4] Ley 142 de 1994. Diario Oficial número 41.433 del 11 de julio de 1994. Bogotá, Colombia. 1994.
[5] UPME (Unidad de Planeación Minero Energética), "Registro Proyectos de Generación. Enero 2014" SIEL, 2014.
[6] Ley 692 de 2001. Diario Oficial número 44.573 del 5 de octubre de 2001. Bogotá, Colombia. 2001.
[7] Resolución 018919 de 2010. Diario Oficial No. 47.728 de 2 de junio de 2010. Bogotá, Colombia. 2010.
[8] Decreto 2041de 2014. Bogotá, Colombia. 2014.
[9] Ley 99 de 1993. Bogotá, Colombia. 1993
[10] Resolución 106 de 2006. Diario Oficial No. 46.495 de 28 de diciembre de 2006. Bogotá, Colombia. 2006.
[11] Resolución CREG 039 de 2001. Bogotá, Colombia 2001.

A Critical Analysis on the Thin Crystalline Silicon PV Module of the Lightweight PV System

Meixi Chen[1], Abhishek Iyer[2], Cheng-Hao Shih[3], Lado Kurdgelashvili[3], and Robert Opila[1]

[1]Department of Material Science and Engineering, University of Delaware, Newark, Delaware, 19716, USA

[2]Department of Electrical and Computer Engineering, University of Delaware, Newark, Delaware, 19716, USA

[3]Department of Energy and Environmental Policy, University of Delaware, Newark, Delaware, 19716, USA

Abstract — This work provides an analysis of the thin crystalline silicon PV module for a Lightweight PV (LPV) system. We first examine the advantages of the LPV system and then discuss the technical and economic challenges of the thin crystalline silicon PV module in assisting LPV system development for future market penetration. A critical analysis is conducted on the efficiency and yield challenge, and the system weight and costs reduction of thin crystalline silicon PV for the future development of the LPV system. It has shown that the key in speeding up thin silicon LPV market diffusion lies in developing cost-effective passivation and light trapping methods, as well as high-yield thin silicon fabrication technologies.

I. INTRODUCTION

Compared with a conventional crystalline silicon PV ballasted system which has a distributed weight of 3-4 pounds per square foot (psf), a Lightweight PV (LPV) system, which uses fewer metal components, can weigh less than 2.5 pounds per square foot [1,2]. This lightweight character can reduce the roof load[1] concern and help the LPV system to penetrate more existing membrane-based commercial rooftop and load-constrained industrial rooftop market [1,3,4].

Beyond the weight reduction advantage, the LPV system also provides several critical deployment advantages such as low Balance-of-System (BOS) hardware costs due to fewer metallic components, lower transportation costs due to the integration of modules with racking hardware, and lower installation costs due to its lightweight character [1]. These advantages can help LPV systems to be deployed on more existing rooftops and speed up future PV market diffusion.

II. THE TECHNICAL AND ECONOMIC CHALLENGES TO THE THIN CRYSTALLINE SILICON PV MODULE

For a conventional solar cell built on 180μm thick silicon substrate, the silicon wafer costs around half of the overall cell [5]. A thinner silicon substrate is a good approach in lowering the solar cell price. This was the biggest motivation to going thin silicon or alternate materials in solar about a decade back when the polysilicon costs were high but today these costs have plummeted down drastically. In this research, we conduct a critical analysis on the technical and economic value of the thin crystalline silicon module for the future LPV system development.

A. Efficiency Limitation

Even though a thinner silicon is known to provide a higher open circuit voltage due to the lower bulk recombination and more efficient carrier extraction [6], challenges exist when the thickness of silicon decreases. Two of the biggest problems include the current reduction caused by strong surface recombination and the light absorption decline for silicon with thickness under 50μm; both resulting in power conversion efficiency losses. Research has been focused on designing light trapping systems and surface passivation structures to compensate the disadvantages of thinner silicon wafers to achieve or maintain a high power conversion efficiency. Models has been developed with the light trapping system and an effective surface passivation, and the efficiency can be maintained in a large thickness range as low as 10μm [7].

Currently, the predominant thin layers used for surface passivation are silicon nitride and silicon oxide. Silicon nitride are generally deposited by PECVD, plasma enhanced chemical vapor deposition, with a temperature around 400 °C [8]. Silicon oxide is thermally grown and it requires a temperature over 1000°C [9]. Recently, PECVD deposited aluminum oxide is found out can provide excellent passivation as well [10]. However, besides the high thermal budget, another problem with these thin films is that the

---

[1] The roof structures for installing PV system must be designed to sustain the additional dead loads (static load) and live loads (dynamic load) of the PV system. For a conventional PV system, the panels and racking will add approximately 3 pounds to each square foot of collector area. In addition to the static load, other live loads such as snow, rain runoff should also be considered for PV installation [2].

plasma used in PECVD induces surface damage where the defects can act as strong recombination centers.

In resolving the problem of plasma damage and high thermal budget, Opila et al. reported the design of organic-inorganic hybrid solar cells using room-temperature solution based chemical passivation, shown in Fig. 1 [11,12]. It has been found that silicon can be passivated immediately when dipping in QHY, quinhydrone solution, without applying plasma or an elevated temperature. The utilization of QHY molecules not only provides excellent passivation, but also greatly lowers the cost and simplifies solar cell fabrication.

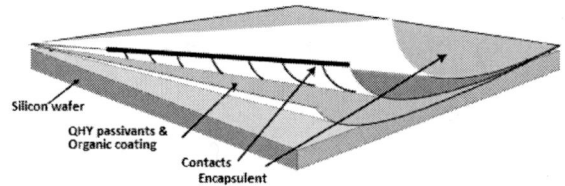

Fig. 1. Organic-inorganic hybrid solar cells using QHY, quinhydrone, as passivants

On the other hand, the problem of light absorption decline can be resolved by applying light trapping structures. Zhang et al. demonstrated that no efficiency loss was observed when decreasing the thickness of silicon from 180μm to 20μm, when applying the metallic-nanoparticle light trapping system [6]. Based on this result, we investigate the cost improvement when silicon substrate thickness is reduced from 180μm, the conventional silicon wafer thickness, to 20μm, where it can still maintain the 18% efficiency. This result gives a 90% decrease in the amount of silicon used in making a solar cell.

B. Yield of The Thin Silicon

The new techniques developed in silicon industry decrease the wafer thickness of about 10–20 μm per year while maintaining a good yield [13]. The typical thickness for single crystalline Si solar cells is currently 180–200μm using multi-wire slurry sawing (MWSS), the most widely used technique nowadays.

When thinning wafers down, the breakage during the silicon manufacture and subsequent cell fabrication steps increases dramatically. S. Choi et al reported that the yield of multi-wire sawing process was as low as 58% and 26% for 140 and 120 μm-thick wafers, respectively, and only 5% for wafers under 100 μm-thick wafers. However, the yield of 100 μm-thick wafer can be significantly enhanced to 83.5% when modifying the MWSS by mirror polishing the brick surface [14].

Nevertheless, 83.5% can hardly meet the industrial requirement. A traditional crystalline factory has a mechanical yield around 98% and an electrical yield of 97%-98%. To go from cell level to module level, this process is about 99% mechanical yield and overall 99% electrical yield.

The key to low manufacturing cost in a crystalline module operation lies in maximizing the first pass mechanical yield[2]. New technologies need to be developed to have industrial disruption in silicon wafer manufacture.

C. Cell Cost Reduction

In M. Woodehouse's study in 2013, the wafer manufacturing cost for 180μm wafer is estimated to be 76 $/m^2 [5]. In order to compare the cost reduction from 180μm wafers to 80μm, 76 $/m^2 is adjusted to 64$/m^2 when assuming the price of polysilicon stay the same, $23/kg, for both cases. Fig. 2 lists the estimated costs, where fabrication costs for 180μm and 80μm wafers include maintenance, depreciation, energy, labor, saw wire, minimum required margin etc., based on Woodhouse's study. It can be seen here the wafering cost per m^2 decreases to 34% when using 80μm wafers, with decreases in the silicon ingot, other materials, energy and labor [5]. This is mainly due to the development of new technology like kerf-less wafering. It is expected that the immersing wafer manufacture techniques will continue to contribute to the cost reduction, and here we assume the wafer manufacture cost can be halved if reducing the thickness from 80μm to 20μm, which gives 11$/m^2.

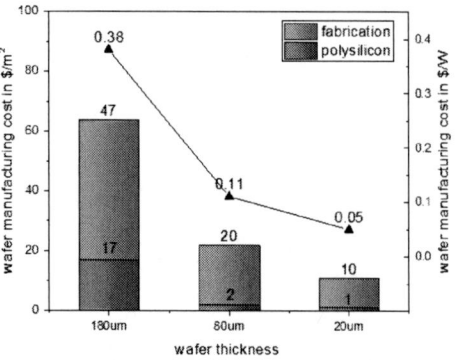

Fig. 2. Wafer manufacturing costs for 180μm, 80μm and 20μm thick polysilicon. Data of 180μm and 80μm wafers are acquired from reference 5. Data for 20μm is hypothesis. The fabrication costs include maintenance, depreciation, energy, labor, saw wire, minimum required margin etc. Cost in $/W are based on calculation with efficiencies of 16.7% for 180μm wafers, and 20% for 80μm and 20μm, details see contents [5].

The wafer manufacturing cost can be further converted into $/W using the equation below, where η is the efficiency.

$$ wafer\ cost\ \$/W = \frac{wafer\ cost\ \$/m^2}{\eta * 1000\ W/m^2} \qquad (1) $$

For a conventional 180μm polysilicon cells, η= 16.7%, and for 80μm cells, η of 20% has been achieved [5]. Thus, the costs for 180μm, 80μm and 20μm wafers are calculated to be

[2] S. Shea, Beamreach Solar, private communication. January 2017

0.38, 0.11 and 0.05 \$/W respectively, where the hypothetical η of 20μm cells is 20% based on analysis in section A. It shows that wafer costs can be potentially reduced to 13% when substituting 180μm silicon with 20μm, if advanced wafering technologies are applied in the future.

Wafer costs take around 50% of the whole cell price for cells with a wafer thickness over 140μm [5]. It is reasonable to assume the cell fabrication costs like depreciation, maintenance, labor, energy and cell materials other than the silicon are the same for 180μm and 20μm wafers, which is also true for 80 μm wafers. Thus, with a deduction of 13% of the wafer cost, the cell price will decrease to 50%+50%*13%=57%. Decreasing the wafer thickness would be a good approach in realizing the SunShot 2030 three cents/kWh goal [15]. And combined with low cost storage could lead to 50% of US electricity supply from solar [16].

D. Module Weight Reduction and System Cost Reduction

Assuming the 20 μm-thick silicon wafer can be achieved with new technologies, a decrease from 180 μm to 20 μm gives a 89% weight decrease in silicon substrate, and 14%*89%=12% decrease of the overall module weight, since the silicon wafer contributes 14% of the entire module weight [17]. Considering a typical solar module weighs 20-50 pounds, and the module with racking system weigh 3-4 pounds/ft², a reduction of 2-6 pounds/module can be expected, and 2.6-3.5 pounds/ft² for the whole system with racking.

The reduction of the module weight attributed by thinner silicon substrate is less significant comparing with other approaches of weight reduction, such as lowering the weight of glasses or mounting system, which takes a bigger portion of around 70% of the system. In addition, the direct labor in installation is unlikely to be affected by the 2-6 lbs/module weight reduction brought by the thinner silicon substrates, neither the shipping costs or packaging costs.

Thus, the key of costs reduction for thin silicon PV system lies in the lower raw material cost and a cost-effective thin silicon fabrication technology.

III. CONCLUSION

In this work, we conduct a critical analysis on the efficiency and yield challenge, as well as the system weight and costs reduction on the thin crystalline silicon cells, summarized in Fig. 3.

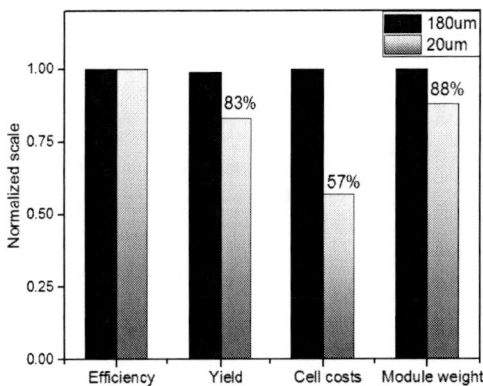

Fig. 3. Comparison between 180μm and 20μm thick wafers. Assuming the 20% efficiency can be maintained by surface passivation and light trapping system, yield is estimated to be around 83% based on studies in [14]. The cell costs dropping to 57% based on analysis in section C.

Light trapping system and surface passivation, especially the room-temperature solution based chemical passivation developed by Opila et al., are good approaches in overcoming the efficiency limitation. Meanwhile, a cost-effective wafer fabrication process with maximized yield is critical for the next generation thin silicon PV manufacture. Thus, the critical aspect in developing thin silicon LPV module lies in investigating cost-effective passivation methods, as well as high-yield thin silicon fabrication technologies. With anticipated sizable reduction in the other parts of the PV system cost (e.g., soft costs, BOS cost, etc.), opportunity to further reduce module costs will also become important. All these improvements can contribute to economic sustainability of PV systems.

ACKNOWLEDGEMENTS

This material is based upon work primarily supported by the National Science Foundation (NSF) and the Department of Energy (DOE) under NSF CA No. EEC-1041895. Any opinions, findings and conclusions or recommendations expressed in this material are those of the author(s) and do not necessarily reflect those of NSF or DOE.

REFERENCES

[1] E. Holton, A. Halbe, A. Garney, J. Whitbeck, K. Sharpe, D. Metacarpa, and P. Haldar, "Cost and Market Analysis of Integrative Lightweight PV Systems for Low-Slope Commercial Rooftops," in *40th IEEE Photovoltaic Specialist Conference*, 2014, pp. 2740-2742

[2] L. Lisell, T. Tetreault, A. Watson, "Solar Ready Buildings Planning Guide," National Renewable Energy Laboratory, 2009.

[3] G. Housser, A. Halbe, K. Sharpe, P. Haldar, and F. Babineau, "Lightweight, Zero-Penetration, Pre-formed Support Molds Adapted for Rigid Thin-Film Solar Modules," in *42nd IEEE Photovoltaic Specialist Conference*, 2015

[4] A. Halbe, K. Sharpe, G. Housser, D. Metacarpa, and P. Haldar, "Demonstration of PV Modules with Lightweight Mounting Systems on Commercial Rooftops," in *42nd IEEE Photovoltaic Specialist Conference*, 2015

[5] A. Goodrich, P. Hacke, Q. Wang, B. Sopori, R. Margolis, T. L. James, and M. Woodhouse, "A wafer-based monocrystalline silicon photovoltaics road map: Utilizing known technology improvement opportunities for further reductions in manufacturing costs," *Solar Energy Materials & Solar Cells,* vol. 114, pp. 110–135, 2013.

[6] Y. Zhang, N. Stokes, B. Jia, S. Fan, and M. Gu, "Towards ultra-thin plasmonic silicon wafer solar cells with minimized efficiency loss," *Scientific Reports*, 4: 4939, 2014.

[7] C. Petti, B. Newman, R. Brainard, and J. Li, "Optimal thickness for crystalline silicon solar cells," Twin Creek Technologies, Inc, http://howcanihelpsandiego.com/wpcontent/uploads/2012/02/Silicon-Solar-Cell-OptimalThickness.pdf

[8] J. Schmidt, M. Kerr, "Highest-quality surface passivation of low-resistivity p-type silicon using stoichiometric PECVD silicon nitride," *Solar Energy Materials and Solar Cells*, vol. 65, pp. 585-591, 2001.

[9] M.J. Kerr, J. Schmidt, A. Cuevas, and J. H. Bultman, "Surface recombination velocity of phosphorus-diffused silicon solar cell emitters passivated with plasma enhanced chemical vapor deposited silicon nitride and thermal silicon oxide," *Journal of Applied Physics*, vol. 89, pp. 3821-3826, 2001.

[10] J. Schmidt, F. Werner, B. Veith, D. Zielke, S. Steingrube, P.P. Altermatt, S. Gatz, T. Dullweber, and R. Brendel, "Advances in the Surface Passivation of Silicon Solar Cells," *Energy Procedia*, vol. 15, pp. 30 – 39, 2012.

[11] N. Kotulak, M. Chen, N. Schreiber, R. L. Opila, "Time and Light-Dependence of Electrical Passivation of c-Si Surfaces with Quinhydrone Constituent Solutions," in *40th IEEE Photovoltaic Specialist Conference*, 2014, pp. 1529 - 1533

[12] N.A. Kotulak, M. Chen, N. Schreiber, K. Jones, R. L. Opila, "Examining the free radical bonding mechanism of benzoquinone– and hydroquinone–methanol passivation of silicon surfaces," *Applied Surface Science*, vol. 354, pp. 469–474, 2015.

[13] A. Bidiville, "Wafer sawing processes: from microscopic phenomena to macroscopic properties," *Ph.D. Dissertation*, University of Neuchatel, May, 2010.

[14] S. Choi, B. Jang, J. Kim, H. Song, T. Baek, and M. Han, "Effects of mirror polishing brick surface on process yield of multi-wire slicing for thin single-crystalline silicon solar cells." *Solar Energy*, vol. 122, pp. 1170-1179, 2015.

[15] J. R. Albertus, D Feldman, R Fu, et al. "Technology advances needed for photovoltaics to achieve widespread grid price parity." *Progress in photovoltaics: research and applications*, 24(9): 1272-1283, 2016.

[16] US DOE SunShot. "The SunShot Initiative's 2030 Goal: 3¢ per Kilowatt Hour for Solar Electricity." 2016.

[17] C. Olson, B. Geerligs, M. Goris, I. Bennett, and J. Clyncke, "Current and future priorities for mass and material in silicon pv module recycling," *EU PVSEC*, Paris, 2013, 6BV.8.2

Photovoltaic Module Manufacturing Costs, Average Prices and Industry Balance 2006-2016

Paula Mints

SPV Market Research, San Jose, California, USA, 95118

Zhengshan J Yu

Arizona State University, Tempe, Arizona 85281 USA

Abstract — **During its over 40-year history as a terrestrial source of electricity the global photovoltaic industry has existed in an energy ecosystem in which demand for grid connected installations is driven by incentives, while demand for off grid installations operates under close to free market conditions. To continue receiving government subsidies and incentives the PV industry has faced significant price pressure culminating in constrained margins that persist.**

Index Terms — **PV, Photovoltaic, Monocrystalline, Multicrystalline, inverters, single axis tracking, PPA, insurance, warranties**

I. INTRODUCTION

In general, photovoltaic industry participants and observers assume that prices are cost based and that periods of low prices are the result of oversupply situations. When prices tick up the assumption is that the industry is in balance. This paper will explore the relationship between PV module raw materials and other components, including backsheets and EVA, overtime, and manufacturer margins as well as the relationship of poor quality and margin constraint on system performance.

The assumption of balance ignores the pressure of incentives and subsidies on industry pricing behavior, pressure from buyers who are also under margin pressure and industry wide acceptance of narrow margins as standard.

Manufacturers in other industry require a gross margin on product sales of 45% at least to cover direct manufacturing costs including variable costs as well as other costs necessary to running a business. Photovoltaic manufacturer margins are in the negative to 16% range.

Pressure on margins has led to lower quality components for module assembly, increasing failure rates, poor production and higher costs for system investors and developers.

Manufacturer inventory is sold at a lower rate than the hoped for ASP of the manufacturer, based on market conditions including a) the level of inventory carried by the demand side and b) the cost of carrying in house inventory. Inventory is a cost to a manufacturer and, depending on market conditions; it may be cost effective and efficient to sell this inventory below the original cost to manufacture it.

When sold, this category is factored into the price point to the first point of sale in the market (first buyer).

Historically, PV industry pricing has not been cost-based. In fact, there have been long stretches of PV industry history during which manufacturers priced technology at or below the cost of production. The current situation of low ASPs for PV technology is an example of aggressive pricing strategy, also

II. PV INDUSTRY GROWTH 2006 TO 2016

From 2006 through 2016 PV industry shipments grew by a compound annual rate of 43%, with commercial installations of all types including multi-megawatt growing at a CAGR of 53%.

Figure 1 presents PV shipments and application growth for the period observed.

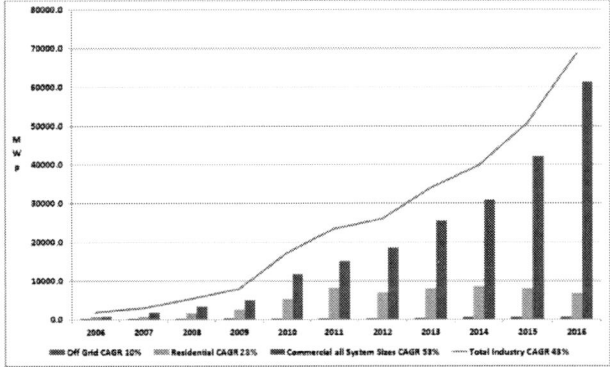

Fig. 1. Photovoltaic Industry Growth 2006 to 2016

From 2011 through 2016 PV industry installations grew by a compound annual rate of 25%.

Figure 2 presents PV shipments, installations, defective modules and inventory from 2011 through 2016.

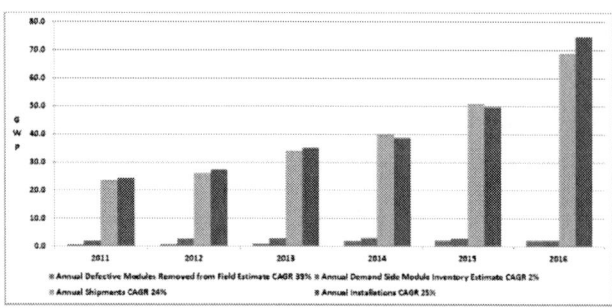

Fig. 2.   PV Shipments and Installations 2006 through 2016

The rate of defective modules removed from the field has been increasing. Current module problems in the field can be traced to 2004 when the EU FiT began driving demand.

In 2009 China had ramped capacity sufficiently and began dominating shipments. The concept of Tier two and three manufacturers can be traced to this period, but, also to the entrance of investors (banks, etc.).

Modules are now failing in the field as early as five years after installation with failures in China and India likely more severe.

This means that the module as a percentage of failure modes in the field is increasing and could be higher than the estimated 6%, to 10% or higher, taking share from inverters.

A few years ago, module failures were almost 100% due to module assembly, now they are primarily because of poor quality inputs including the cell. Low quality backsheets, AR coating and EVA, primarily from China, are found in ALL products.

Module failures in the field increase O&M costs for the developer, lead to poor electricity production and a lower revenue stream and strain warranty obligations and system performance guarantees.

From 2011 through 2016 PV industry defective modules grew by a compound annual rate of 29%.

Figure 2 presents PV defective modules and inventory from 2011 through 2016.

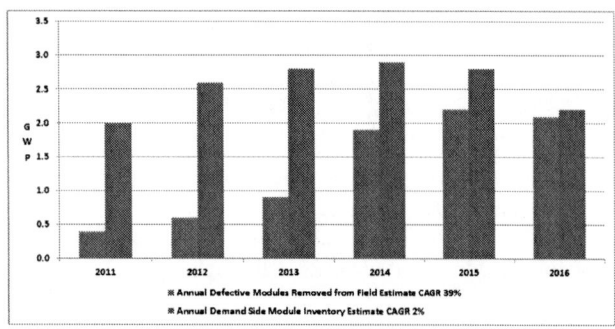

Fig. 3.   PV Industry Defective Modules and Inventory 2006 through 2016

III. POLYSILICON THROUGH MODULE AVERAGE COSTS AND PRICES

Margin pressure has led manufacturers to cut corners in manufacturing.

Polysilicon manufacturers have faced significant price pressure resulting is constrained margin.

Figure 4 presents average polysilicon costs and prices from 2014 through a short term forecast to 2019.

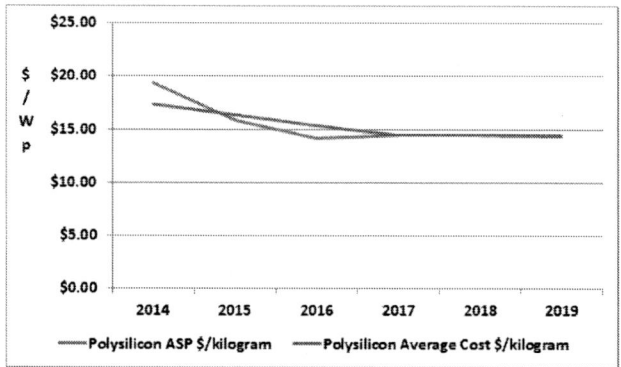

Fig. 4.   Polysilicon Average Costs and Prices 2014 through 2019

Wafer manufacturers have faced significant price pressure resulting is constrained margin.

Figure 5 presents average wafer costs and prices from 2014 through a short term forecast to 2019.

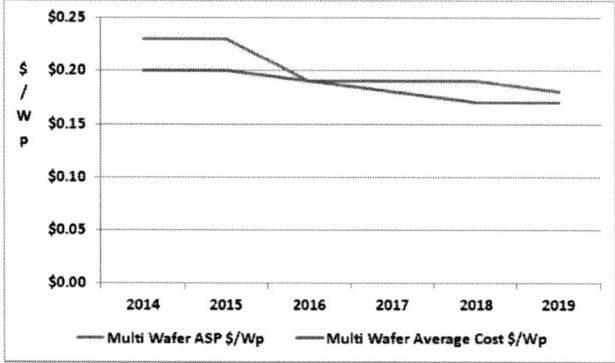

Fig. 5.   Multi Wafer Average Costs and Prices 2014 through 2019

Cell manufacturers have faced significant price pressure resulting is constrained margin.

Figure 6 presents average multicrystalline cell costs and prices from 2014 through a short term forecast to 2019.

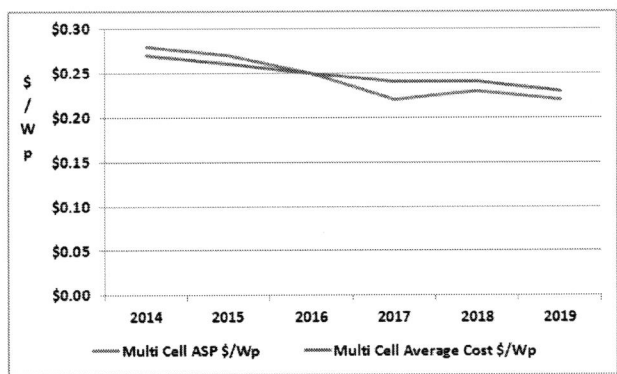

Fig. 6. Photovoltaic Shipments and Average Module Prices 1985 through 1995

Module manufacturers have perhaps faced the most significant pressure and continue to face expectations of lower prices.

Figure 6 presents average module costs and prices from 2014 through a short term forecast to 2019.

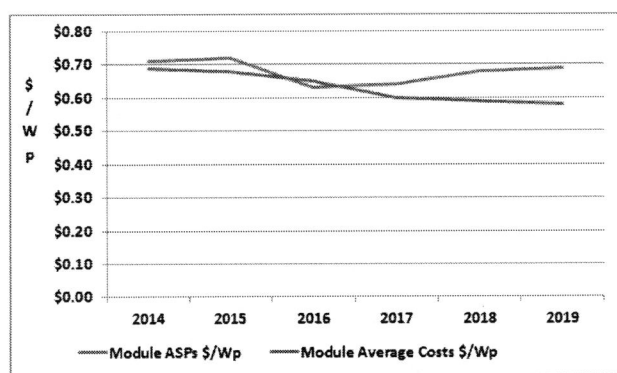

Fig. 7. Average Module Prices 2014 through 2019

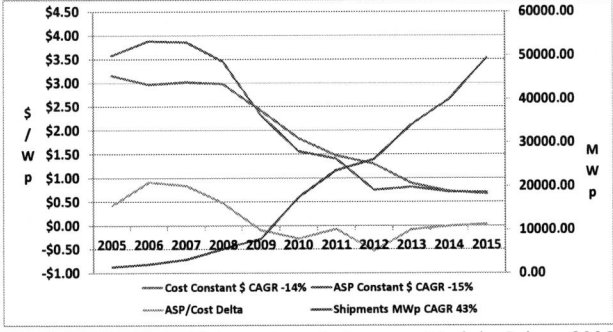

Fig. 4. Photovoltaic Shipments and Average Module Prices 2005 through 2015

VI. CONCLUSION

The goal of this paper is to observe industry pricing and buying behavior through the lens of margin erosion overtime as well as the impact on system performance and developer costs in terms of insurance and O&M.

REFERENCES

[1] P. Mints, SPV Market Research, *Global Markets for Photovoltaic Modules and Systems And Five Year Application Forecast 2014-2019*, August 2016

[2] P. Mints, SPV Market Research, *Photovoltaic Manufacturer Capacity, Shipments, Prices and Revenues, 2015/2016*, April 2016

Solar Cell and Wind Energy Replacement Of Power Plants Globally

Larry Partain,[1,2] Shirley Hansen,[2] Dirk Bennett,[2] Richard Hansen,[2] Allan Newlands,[2] Lewis Fraas[3]

[1]Solar Cell Electricity, Los Altos, CA 94022 USA, [2]BridgePoint Consortium, Los Altos, CA 94022 USA,
[3]JX Crystals, Inc., Issaquah, WA 98027 USA

Abstract — Analysis of atmospheric CO_2 concentrations, global mean surface temperatures, and a "Goldilocks" planet model that is neither too hot nor cold identifies how civilization could continue thriving beyond 2050-2080 (or earlier). It starts with the replacement of one of the largest current sources of atmospheric CO_2, fossil-fueled power plants, using solar cells and wind energy in an economic and timely manner. The reasons this is technically feasible, as soon as 2028, are reviewed, limited primarily now by global public policies. The major remaining technical challenges mainly involve accommodations of the daily and seasonal variability of solar cell and wind energy.

I. INTRODUCTION

Atmospheric carbon dioxide concentrations have increased sharply since the 1870s due to constantly increased burning of fossil fuels, a consequence of the Industrial Revolution [1],[2]. This is responsible for most of the global warming over the past 50 years [3]. Because natural processes cannot quickly remove CO_2, future emissions will influence the climate system for millennia (see Ref. [2], Figs. 5 and 7). In the meantime global warming generates significant impacts harmful to society including threats to the world's thriving global civilization [4]. This human-induced climate change requires urgent action [3]. One of the largest current sources of US atmospheric CO_2 emissions is fossil-fueled power plants, compared to the other two major sources, transportation and industry [5],[6]. The US emissions of CO_2 are exceeded only by those of China [7].

After 40 years of "Swanson's Law" learning curve development, solar cells (photovoltaics) are finally poised to provide major near term positive impacts on climate change [2]. Solar cells' continued exponential cumulative-capacity-growth and exponentially-falling-production cost have recently reached nuclear and fossil-fuel competitive levels without subsidies [8]. Combined with wind they now provide compelling carbon-free-energy alternatives that are economical and available for a billion years or more at magnitudes ample to match anticipated global energy use (Ref. [2], Figs. 1 and 8). Currently in the US there are three times as many jobs in the renewable solar cell and wind energy sector than in coal [9], even though they supply less than 8% of US grid electricity [10]. This is consistent with the projection [11] that renewable energy provides more jobs than the fossil-fuel jobs it replaces.

II. HOW IMPORTANT

Modeling, by the UN Intergovernmental Panel on Climate Change (IPCC) [1] indicates that the rise (or anomaly) in the global mean surface temperature has been caused by greenhouse gas emissions (primarily cumulative CO_2) into the atmosphere since the 1870s, that should be halted if detrimental changes to the planet's climate are to be avoided [2]. A qualitative "Goldilocks Model" (Fig. 1) has been

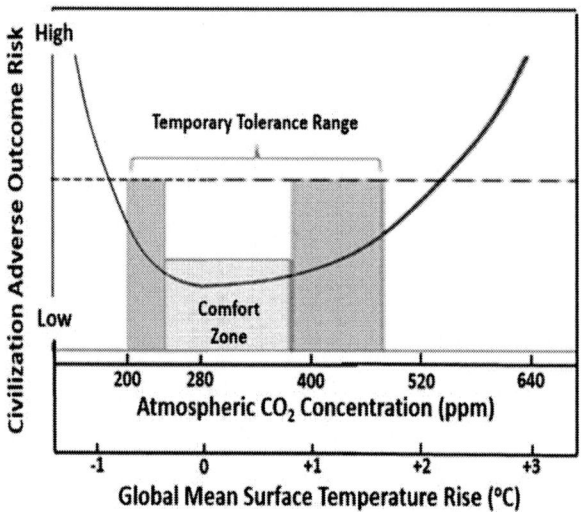

Fig. 1 The "Goldilocks" model of the planet that is not too hot nor too cold for global civilization to continue to thrive.

proposed for the planet with climate and environmental factors that are not too large nor too small to prevent adverse outcomes [12]. Here its qualitative vertical axis is defined to be civilization adverse outcome risk, up to levels that threaten the collapse of global civilization [4]. Its horizontal axis is specifically designated to indicate global mean surface temperature changes (since the 1870s) which do not make the planet too hot nor too cold. A simplifying but major assumption is made here that the range of tolerable temperature change is small enough that a linear relationship with atmospheric CO_2 concentration change is approximately accurate.

978-1-5090-5606-4/17 $31.00 © 2017 IEEE

Going back at least 8000 years (since the end of the last ice age about 12,000 years ago), the atmospheric CO_2 concentration remained around 280 ppm ($\pm$10 ppm) until its rapid, consistent and non-linear rise for approximately the last 140 years starting in the 1870s [2]. This makes the Fig. 1 CO_2 concentration of 280 ppm correspond to a zero change in mean surface temperature. However if one looks back over 400,000 years [2], the atmospheric CO_2 concentration has fallen by 80 ppm (to 200 ppm) four times as the planet descended into glaciation. Since failure of agriculture is a leading explanation for the collapse of multiple early civilizations [4] and agricultural plants cannot grow when ice and snow are the only available soil, the 200 ppm value is taken here as the lower bound for the civilization-adverse-outcome-risk zone in Fig. 1.

The current measured global atmospheric CO_2 concentration is 400 ppm [13], a larger 120 ppm change and increase. The accompanying rise in global mean surface temperatures has been 1°C. These calibrate the Fig. 1 horizontal axis. Using Paris Agreement [14] estimates of 1.5 to 2°C rise as the threshold for a threat to the thriving global civilization, these specify the non-symmetric shape of the Fig. 1 "Goldilocks" curve. The Fig. 2 "Keeling Curve"

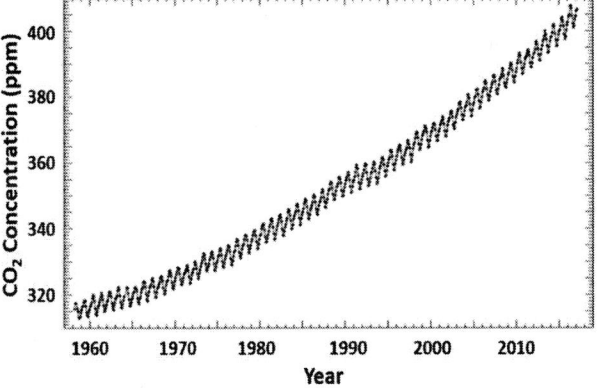

Fig. 2 The "Keeling Curve" global atmospheric CO_2 concentration measured on Mauna Loa in Hawaii since 1958.

measurements of cumulative global atmospheric CO_2 concentration shows its faster-than-linear increase since 1958 [13]. Its mean linear growth rate from 2005 to 2015 was 1.9 ppm/yr. At this linear rate, the time line to the 1.5 to 2°C Paris Agreement threshold, for a direct threat to a continued thriving civilization, would be by approximately 2050 to 2080 should fossil fuels continue to be burned at current rates unless carbon capture and storage can be made to work. This time line shortens if the Fig. 2 growth trend continues its faster-than-linear rise.

It is particularly relevant here to note that the birth of 6 early civilizations all occurred independently and began to flourish 3500-5100 years ago [15],[16] since the end of the last ice age (Table I). These were spread all around the globe

when most had no known initial contacts or interactions. Their common denominator was the presence of *homo sapiens* (around globally for about 200,000 years) [15] and their locations in climatic and terrestrial environments supportive of agriculture development over millennia. This is strong evidence that the 280 ppm atmospheric CO_2 concentrations on earth have spontaneously fostered *homo sapiens* (i.e. human) civilizations that formed and began to flourish. It supports the 280 ppm CO_2 concentration value as the middle "comfort zone" point of the Fig. 1 "Goldilocks" climate-change thriving-civilizations curve. Thus this well compares to the CO_2 concentration versus time data (see Ref. [1], Fig. 7), where the 280 $\pm$ 10 ppm range is indeed the central comfort-zone region.

TABLE I
EARLY CIVILIZATIONS' BIRTH CHRONOLOGY

Egypt	circa 3100 B.C.E.
Mesopotamia	circa 3000 B.C.E.
India	circa 3000 B.C.E.
China	circa 2000 B.C.E.
Central Asia	circa 2000 B.C.E.
Yucatan Mexico	circa 1500 B.C.E.

Over the last half billion years, five climate extreme extinction events have naturally occurred where 75 to 96% of all species were lost [17]. Three involved global climate cooling and two global climate heating [16]. There is no question whether climate change with global mean temperature shifts can pose threats to thriving global civilization. The question is how much and how soon could *homo sapiens* activities themselves initiate (or conversely prevent) global climate changes and temperatures shifts that pose such a global civilization threat. On smaller regional scales *homo sapiens* societies have repeatedly carried out controllable activities that directly led to their societal collapses [18]. For the global scale, such threats arise when worldwide fundamental global threshold limits are breached from contributing factors like over population or deforestation [19].

Global mean seawater surface temperature rise is one climate change example of such a fundamental threshold limit threat. At seawater's current level of 1°C rise (since the 1870's), coral reefs are increasingly bleaching with many proceeding to die. These include large sections of Australia's Great Barrier Reef [20]-[22]. This specific reef is the world's biggest single structure made by living organisms and it can be seen from outer space.

As oceans cover over 70% of the planet's surface they absorb 93% of the heat from global warming [23]. Even though reefs themselves cover less than 1% of ocean floor, they are home to 25% of all marine species, including 4000 fish species [24],[25]. These reefs are spread over large expanses of the world (Fig. 3). A consequence is that 500 million people now rely on reefs for their food, coastal protection and livelihoods [26]. Depending on how quickly reefs are lost due to rising sea water temperatures [22], the societal and political disruptions could cause mass migrations, dwarfing those caused by the Syrian civil war [27], and challenging a reluctant world to accommodate. Particularly note that this specific global coral reef damage and threat is already clearly occurring right now, as opposed to becoming obvious some several decades into the future. Just 1°C sea water change causes coral reefs to bleach [28].

Fig. 3   The distribution of coral reefs (red dots) around the world.

Other thriving-civilization-threat factors made worse by global warming include melting Artic sea ice, ocean acidification, extreme weather related events like droughts, wildfires, storms, flooding, sea level rise, storm surges, agricultural failure, malnutrition, pandemic disease, starvation, mass migrations and increased conflicts [2],[16]. Although avoidable, in combination with other risks (e.g. overpopulation, deforestation), collapse of global civilization is a plausible outcome [4],[18].

III. HOW SOON

In 2015 the renewables' cost of US grid electricity, supplied by utility scale solar cells and by onshore wind turbines, became directly cost competitive with electricity supplied by fossil- and nuclear-fueled power plants [8]. They all lay in the 4 to 15 cents per kW-hr range (Fig. 4). Solar

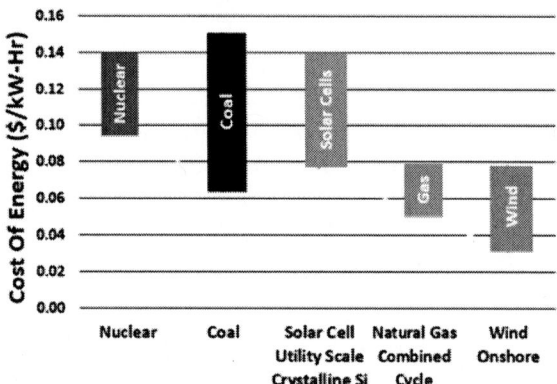

Fig. 4 The levelized cost of grid electricity from nuclear, fossil fuel and renewable solar cell and wind energy resources in 2015.

cells, at 8 to 14 cents per kW-hr, were directly competitive with both nuclear- and coal-fueled power plants. Onshore US wind turbines at 4 to 8 cents per kW-hr were as-low-as or even-lower-than that generated in combined cycle natural gas power plants. However, a continuing challenge in developed

regions like the US and Germany was how best to handle the variable combination of wind and solar cells when the sun does not shine and the wind does not blow, both daily and seasonally (see Fig. 5) [29]. The hourly energy distribution of renewables in California is similar to that of Fig. 4 (compare to Fig. 1 of Ref. [30]).

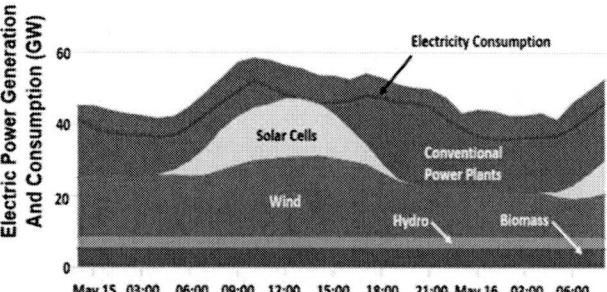

Fig. 5 The electricity generation and consumption of Germany on May 15, 2016.

However it is encouraging that all Dutch electric trains now run only on onshore and offshore wind power as of Jan. 1, 2017 [31]. It is further encouraging that Great Britain had its first 24 hour period without coal power generation since the Industrial Revolution [32].

The 51 GW (GW=10⁹W) of solar cells shipped in 2015, consistent with "Swanson's Law," and their distribution by type are shown in Fig. 6 [33]. Over 90% were from crystalline silicon (c-Si) wafers in both their mono and multi crystalline forms. Less than 10% were from thin film cells of cadmium telluride (CdTe), copper-indium-selenide/copper-indium-gallium-selenide (CIS/CIGS) and amorphous silicon (a-Si). For sustainability note that Si is the second most abundant element in the earth's crust [34].

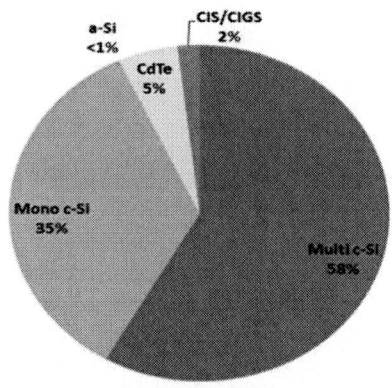

2015 Shipments 50.8-GW

Fig. 6.   The breakdown of the solar cells shipped in 2015.

Most of the mono and multi crystalline silicon cell modules use essentially the planar and abrupt p/n junction with front and rear metal electrodes configuration developed by Chapin *et al* at Bell Labs in the 1950s. However mono crystalline silicon's very highest performance commercial modules, exceeding 20% efficiencies [35], are the modified non-planar configuration with all the metal electrodes on the back of these solar cells so as not to block the incoming sunlight. These are the interdigitated back contact solar cell, designed and developed by Richard Swanson by 2008 [36].

Wind offers a particularly attractive partner for solar cells because it adds affordable power in the early mornings and evenings and winters when sunlight is not available (Fig. 5). Studies [2] indicate that as wind energy is added to grids supplied by renewables, minimal storage (like hydro or batteries) is required to approach the daily and yearly grid reliability expected in developed economies if the grid-and-wind region is large enough. One development scenario may be that developing economies adopt solar cell and wind energy sources for their lower cost, but that with time, their systems grow ever more reliable as these economies grow and produce grids of ever growing size and reliability. In other words, as a grid region grows large enough, there is wind almost always blowing somewhere.

Note the sharp increase in the conventional-power-plant generation of about 20 GW between 13:00 and 18:00 hours in Fig. 5 and the about 10 GW decrease between 6:00 and 10:00 hours on this single day in Germany. Such variability of solar cell and wind power generation and the challenges they present, as they are combined with their steadier non-solar-cell-wind-electricity sources, presents the same characteristics as the famous "duck" shaped curve (see Fig. 2 of Ref. [30]) long anticipated by the California Independent Service Operator that manages 80% of the California grid. To date such variability management has been well handled on both grids but this is an area demanding close attention that could profit from continuing innovations. To reach essentially renewables-only operation likely requires further innovations that could include [2] combinations of demand response, super grids, smart grids, space mirrors [37] and new and improved forms of storage and/or of carbon neutral biofuels.

The cost of solar cells, and of the electricity they produce, has long been decreasing consistently at a fast "Swanson's Law" learning curve rate [2]. This even exceeds the falling cost of onshore wind. Solar cells should soon provide electricity at a lower price than wind as well as that of coal, natural gas and nuclear. Interestingly, the latest 2016 report [38] of completed auctions, started in January with a large contract in India for 6.4 cents per kW-hr. for utility-scale-solar-cells and then fell by August to 2.91 cents per kW-hr. in Chile. It was roughly half the price of competing coal. This matches the lowest value of onshore wind plotted in Fig. 3. If this continues into 2017 it represents a major energy turning point as solar cell electricity cost drop below and likely stay below that from onshore wind.

Extrapolations always involve substantial risks [2] but they have long been done, with some trepidation. An example is the exponential growth Moore's Law guidance for the development and investments into integrated circuit technology for computers. With appropriate caution, one should note that the similar exponential "Swanson's Law" learning curve growth, for cumulative installed solar cell capacity (CC). In 2015, it reached 1% (at 0.2 TW) of the world's 2014 electricity generating capacity, with cost also ever falling exponentially. If CC doubling every 2 years since 2006 continues [2], the solar cell cumulative installed capacity would reach 10% in 2021 and 100% in 2028 of all the world's 2014 electric energy use. This is substantially earlier than others have anticipated. Such a transition is now mainly dependent on global public policies since a lack of competitive cost (Fig. 4) is no longer an impediment.

By 2016 the large markets of both Germany [39] and the State of California [40] supplied 30% and 38% respectively of their electricity consumption from renewable energy sources that were dominated by wind and solar cell installations [29],[40]. The California Energy Commission reported 27% from renewables in 2016 [40] but that did not include large hydro. Adding in California's 11% large hydro contribution for 2016 [41] gives the 38% California total. Note that Germany is the world's fourth largest economy [42]. If California were a country it would be the world's sixth largest economy [43]. For comparisons in 2016 respectively, China and the US supplied 24% and 15% of their electricity from renewables [44],[45] and were the 2nd and 1st largest economies [42].

Another energy turning point occurred earlier in 2015 when renewables spending ($154.1 billion) by the emerging markets countries was for the first time more than that spent ($153.7) by the wealthier 35 member nations of the Organization for Economic Cooperation and Development (OECD) [46]. A direct result is that the largest near term projected renewables markets are now China, Chile, Brazil, Uruguay, South Africa and India. Since 2005, China has led the world in the development of new solar cell energy capacity, with the US in third place behind Europe [2]. If the above trend continues, the US and European growth could soon be overtaken by India or by Brazil in combination with others.

The 2015 global shipments of 51 billion watts of solar cell modules (Fig. 6) [33], combined with a "Swanson's Law" 2014 data point for module cost of $0.65/W [2], gives a rough total estimate of a $33 billion market for the 2015 market size (within a factor of 3). This compares to an independent 2015 solar cell market estimate of over $35 billion [47]. These consistently indicate how significant this solar cell market is becoming.

Numerical integration of the analytical approximation of "Swanson's Law" indicates that the rate of change of cumulative production costs investment, in just solar cell modules alone, could rise into the $1 trillion per year range (in 2012 $) by 2028, using the Moore's-Law-like doubling of cumulative capacity every 2 years (empirically observed since 2006). This relates estimated-future-cumulative-installed capacity to time (see Figs. 1-3 of Ref. [2]). Within a factor of 2-3, this yearly rate of change also estimates the $1 trillion yearly market range by this same 2028 date. If accurate this installed solar cell electricity capacity estimate alone would approach all of the 2014 world electricity consumption [48],[49] at very competitive cost compared to fossil fuel and nuclear alternatives.

The world's depleteable energy resources, including nuclear, can only last a finite length of time. For current consumption rates one estimate is 100 years [2]. This drops to less than 50 years, if the dirtiest and expensive component (coal), is eliminated.

IV. SUMMARY

The "Goldilocks" modeling [12],[16] here suggests there is a scientific basis for arguing that our planet needs to stay not too hot nor too cold for the global civilization as we know it to continue to thrive [4] beyond 2050-2080 (or earlier). The data collected and analyzed so far suggest quite a narrow range for this tolerable global mean surface temperature. According to the Paris Agreement [14] this should be limited to a rise of 1.5 to 2°C (compared to the rise so far of 1°C [2]). This might be achieved, if all the fossil-fuel-burning release of CO_2 into the atmosphere halts, within the next few decades. *Otherwise the global climate grows incrementally worse with each passing year.* A first step of halting is global replacement of one of the largest current generators of atmospheric CO_2, namely fossil-fueled power plants [5]-[7], with renewable solar cell and wind power as soon as 2028. *It largely depends now on global public policies.* This can provide critical additional time to address more challenging but similar replacements in the transportation (including airplanes [2]) and industry (including concrete and steel production) energy sectors.

The future outcomes of human-generated climate change by the turn of this century involves both high risk and high uncertainty [12]. An indication of this risk from one bad but plausible pathway is contained in the summary observation by Gladwell [50] that "…societies, as often as not, aren't murdered. They commit suicide: they slit their wrists and then, in the course of many decades, stand by passively and watch themselves bleed to death". This followed from Diamond's failed-societies analysis [18]. Diamond also gave striking, instructive and sobering comparisons to ongoing contemporary societies responding to change.

Four plausible global warming pathways (Ref. [2], Fig. 4) were defined by the IPCC [1] but none were indicated as more plausible than any others [12]. The worst IPCC pathway (continue business as usual) involved a greater than 4°C global mean temperature rise. Figure 1 and its discussion here indicate such a rise could well indeed present a very high risk to a continued thriving global civilization. In the meantime global mean atmospheric CO_2 concentration (Fig. 2) has risen at ever faster rates since at least 1958 indicating an urgent need for remedial action [3],[13].

Here we recommend a promising start toward a more positive global renewables pathway. It is appealing but nevertheless has its own set of remaining challenges needing focused and priority attention, starting with public policy and then with solar cell and wind variability accommodations.

REFERENCES

[1] "Climate change 2013, Summary for Policymakers," *The Physical Science Basis, Working Group I, Contribution to the Fifth Assessment Report of the Intergovernmental Panel on Climate Change,* Fig. SPM.10. https://www.ipcc.ch/pdf/assessment-report/ar5/wg1/WGIAR5_SPM_brochure_en.pdf. Accessed Apr. 24, 2017.

[2] L. Partain, R. Hansen, S. Hansen, D. Bennett, A. Newlands, L. Fraas, "'Swanson's Law' plan to mitigate global climate change," in *43rd IEEE Photovoltaic Specialist Conference,* pp. 3335-3340, 2016.

[3] *American Geophysical Union Policy Statement on Climate Change,* Adopted 2003, Revised and Reaffirmed 2007, 2012, 2013. https://news.agu.org/press-release/american-geophysical-union-releases-revised-position-statement-on-climate-change/. Accessed July 23, 2016.

[4] P. Ehrlich, A. Ehrlich, "Can a collapse of global civilization be avoided?" *Proc R Soc B* , vol. 280, 20122845, 2013.

[5] "Total U.S. greenhouse gas emissions by economic sector in 2014," *US Environmental Protection Agency.* https://www.epa.gov/ghgemissions/sources-greenhouse-gas-emissions. Accessed Dec. 3, 2016.

[6] B. Plumer, "Power plants are no longer America's biggest climate problem. Transportation is," *vox.com,* Jun. 13, 2016. http://www.vox.com/2016/6/13/11911798/emissions-electricity-versus-transportation. Accessed Apr. 18, 2017.

[7] "Emissions by country 2011, global greenhouse gas emissions data," *US Environmental Protection Agency.* https://www.epa.gov/ghgemissions/global-greenhouse-gas-emissions-data. Accessed Dec. 15, 2016.

[8] "Levelized Cost of Energy Analysis 9.0," *Lazard Power, Energy & Infrastructure Group,* New York, 2015. https://www.lazard.com/media/2390/lazards-levelized-cost-of-energy-analysis-90.pdf. Accessed Nov. 6, 2016.

[9] N. Popovich, "Today's Energy Jobs Are in Solar, Not Coal," *New York Times,* Apr. 25, 2017. https://www.nytimes.com/interactive/2017/04/25/climate/todays-energy-jobs-are-in-solar-not-coal.html?hp&_r=0. Accessed Apr. 26, 2017.

[10] T. Hodge, "Natural gas expected to surpass coal in mix of fuel for US power generation in 2016," *Today in Energy,* US Energy Information Administration, Mar. 16, 2016. https://eia.gov/todayinenergy/detail.php?id=25392. Accessed Apr. 29, 2017.

[11] M. Jacobson, M. Delucchi, G. Bazouin, Z. Bauer, C. Heavey, E. Fisher, S. Morris, D. Piekutowski, T. Vencilla T. Yeskooa, "100% clean and renewable wind, water, and sunlight (WWS) all-sector energy roadmaps for the 50 United States*," Energy Environ. Sci.,* vol. 8, pp. 2093-2117, 2015.

[12] A. McMichael, "Impediments for comprehensive research on climate change and health," *Int. J. Environ. Res. Public Health* 10, pp. 6096-6105, 2013.

[13] Primary Mauna Loa CO_2 Record, Scripps Institute of Oceanography, University of California San Diego. http://scrippsco2.ucsd.edu/data/atmospheric_co2/primary_mlo_co2_record. Accessed Apr. 3, 2017.

[14] "The Paris Agreement, Article 2.1 (a)," *United Nations Framework Convention on Climate Change*, 2015. http://unfccc.int/paris_agreement/items/9485.php. Accessed Apr. 25, 2017.

[15] J. Spielvogel, "The first civilizations," in *Western Civilization,* 9th Ed., Cengage Learning, Stamford, CT, pp. 3-30, 2015.

[16] A. McMichael, *Climate Change and the Health of Nations,* Oxford, New York, pp. 81-82, 2017

[17] V. Richter, "The big five mass extinctions," *Cosmos Magazine.* https://cosmosmagazine.com/palaeontology/big-five-extinctions. Accessed Apr. 30, 2017.

[18] J. Diamond, *Collapse, how societies choose to fail or succeed,* Penguin, New York, 2006.

[19] D. Meadows, D. Meadows, J. Randers, W. Behrens, *The limits to growth,* Universe, New York, 1972

[20] "Great Barrier Reef: Two-thirds damaged in 'unprecedented' damage," *BBC*, April 10, 2017.

[21] Michael Slezak, "The Great Barrier Reef: a catastrophe laid bare," *The Guardian*, June 6, 2016.

[22] D. Normile, "El Niño's warmth devastating reefs worldwide," *Science,* Mar. 29, 2016.

[23] L. Dahlman, "Climate Change: Ocean Heat Content," *NOAA Climate.gov*, Jul. 14, 2015. https://www.climate.gov/news-features/understanding-climate/climate-change-ocean-heat-content. Accessed Apr. 30, 2017.

[24] "Coral reefs", *World Wildlife Foundation Global.* http://wwf.panda.org/about_our_earth/blue_planet/coasts/coral_reefs/. Accessed Apr. 30, 2017.

[25] "Importance of Coral Reefs." *NOAA Ocean Service Education,* Revised Mar. 28, 2008. http://oceanservice.noaa.gov/education/kits/corals/coral07_importance.html. Accessed Apr. 30, 2017.

[26] C. Wilkinson, (ed.) *2004 Status of Coral Reefs of the World: 2004.* Vol. 1. Australian Institute of Marine Science. Townsville, Queensland, Australia. p. 301.

[27] J. Hogg, "Saddled with 2 million Syrian refugees, Turkey shows signs of strain," *World News*, Sept. 15, 2015. http://www.reuters.com/article/us-mideast-crisis-migrants-turkey-analys-idUSKCN0RF1PX20150915. Accessed April 18, 2017.

[28] T. Ainsworth, R. Gates, Corals' microbial sentinels, *Science* vol. 352, pp. 1518-1519, 2017.

[29] J. Shankleman, J. Shankleman, "Germany just got almost all of its power from renewable energy," *Bloomberg News*, May 16, 2016. http://www.bloomberg.com/news/articles/2016-05-16/germany-just-got-almost-all-of-its-power-from-renewable-energy. Accessed July 16, 2016.

[30] L. Partain, L. Fraas, "Displacing California's Coal and Nuclear Generation with Solar PV and Wind by 2022 Using Vechicle-to-Grid Energy Storage," *42nd IEEE Photovoltaic Specialists Conf.,* 2015.

[31] "Dutch electric trains become 100% powered by wind energy," *The Guardian, Agence France-Presse in The Hague,* Jan. 10, 2017. https://www.theguardian.com/world/2017/jan/10/dutch-trains-100-percent-wind-powered-ns. Accessed Jan. 12, 2017.

[32] G. Brown, "British power generation achieves first ever coal-free day," *The Guardian*, Apr. 21, 2017. https://www.theguardian.com/environment/2017/apr/21/britain-set-for-first-coal-free-day-since-the-industrial-revolution. Accessed Apr. 25, 2017.

[33] P. Mints, "Trends in the Solar Cell Market: 2015 PV Shipments," *IEEE Silicon Valley Photovoltaics Society Meeting,* Palo Alto, CA, May 11, 2016. http://sites.ieee.org/scv-photovoltaic/event/2016-05/. Accessed July, 16, 2016.

[34] B. Skinner, "Earth resources," *Proc. Natl. Acad. Sci.* vol. 76, pp. 4212-4217, 1979.

[35] M. Green, K. Emery, Y. Hishikawa, W. Warta, E. Dunlop, "Solar cell efficiency tables (Version 45)," *Prog. Photovoltaics* vol. 23(1), pp. 1-9, 2015.

[36] R. Swanson, "Device physics for back-side contact solar cells," *33rd IEEE Photovoltaic Spec. Conf.*, 2008.

[37] L. Fraas, G. Landis, A. Palisoc, P. Jaffe, "Space Mirror Development for Solar Electric Power And the International Space Station," *43rd IEEE Photovoltaic Specialist Conf.,* Portland, 2016.

[38] "World energy hits a turning point: Solar that's cheaper than wind," *The Japan Times* (2016). http://www.japantimes.co.jp/news/2016/12/18/business/world-energy-hits-turning-point-solar-thats-cheaper-wind/#.WFbDmYWcHIU. Accessed Dec. 18, 2016.

[39] "Germany's energy consumption and power mix in charts, 2016 preliminary results," *Clean Energy Wire Fact Sheet.* https://www.cleanenergywire.org/factsheets/germanys-energy-consumption-and-power-mix-charts. Accessed Apr. 25, 2017.

[40] *California Energy Commission – Tracking Progress,* Last updated Dec. 22, 2016. http://www.energy.ca.gov/renewables/tracking_progress/documents/renewable.pdf. Accessed Apr. 25, 2017.

[41] "What are we doing to green the grid?", *California ISO* http://www.caiso.com/informed/Pages/CleanGrid/default.aspx. Accessed Jan. 29, 2017.

[42] A. Gray, "The world's 10 biggest economies in 2017," *World Economic Forum*, Mar. 9, 2017. https://www.weforum.org/agenda/2017/03/worlds-biggest-economies-in-2017/. Accessed Apr. 24, 2017.

[43] C. Bruno, "Is California Really the 6th Largest Economy in the World?" *RealClear Markets,* Jul. 14, 2016. http://www.realclearmarkets.com/articles/2016/07/14/is_california_really_the_6th_largest_economy_in_the_world_102263.html. Accessed Apr. 24, 2017.

[44] "Power statistics China 2016," *Chinese European Energy News,* Feb. 9, 2017. http://ceenews.info/en/power-statistics-china-2016-huge-growth-of-renewables-amidst-thermal-based-generation/. Accessed May 24, 2017.

[45] "What is the U.S. electricity generation by energy source?" *U.S. Energy Administration Agency*, last updated April 18, 2017. https://www.eia.gov/tools/faqs/faq.php?id=427&t=3. Accessed May 24, 2017.

[46] "Mapping the new global frontiers for Bloomberg New Energy Finance," *Climatescope 2016*, Dec. 22, 2016. http://global-climatescope.org/en/. Accessed Dec. 22, 2016.

[47] "Solar cells market size worth over $100bn by 2024," *Global Market Insights Inc.*, www.gminsights.com. Dec. 6, 2016. https://globenewswire.com/news-release/2016/12/06/895373/0/en/Solar-Cells-Market-size-worth-over-100bn-by-2024-Global-Market-Insights-Inc.html. Accessed Dec. 23, 2016.

[48] " 2014 Snapshot of global PV market," *IEA (OECD), Report IEA PVPS* T1-26, 2015. http://www.ieapvps.org/fileadmin/dam/public/report/technical/PVPS_report_-_A_Snapshot_of_Global_PV_-_1992-2014.pdf. Accessed June 28,2015.

[49] "Country Comparison: Electricity – Installed Generating Capacity," " US Central Intelligence Agency. https://www.cia.gov/library/publications/the-world-factbook/rankorder/2236rank.html. Accessed April 22, 2017.

[50] M. Gladwell, The Vanishing," *The New Yorker*, Jan. 3, 2005. http://www.newyorker.com/magazine/2005/01/03/the-vanishing-2. Accessed May 5, 2017.

Analysis of Light Environment under Solar Panels and Crop Layout

Deng Wang[1], Yaojie Sun[1*], Yandan Lin[1*], and Yuan Gao[2]

[1]Department of Illuminating Engineering and Light Sources, Fudan University, Shanghai, 200433, China
[2]Delft University of Technology, Delft, 2628CD, the Netherlands

Abstract — This paper studies the solar radiation distribution under solar panels in the effective growth period of crops by building the model of photovoltaic power station with Ecotect. The simulation results show that the area beneath the photovoltaic panels receives less solar irradiation. The area covered with no solar panel reveals better irradiation condition. The irradiation also distributes differently at various heights. According to the different sunlight demands, crops can be classified into three major categories, which include six minor classes. Light adaption point (LAP) is defined as an indicator to approximate whether positive crops can flower and bear fruit normally. Based on sunshine hours and photosynthetically active radiation (PAR), we put forward a method of crop layout by combining their different sunlight demands and different light distribution under solar panels. A list of recommended species for this model is also given.

Index Terms — solar radiation distribution, PAR, sunlight hours, crop selection, ecotect.

I. INTRODUCTION

Photovoltaic power generation has been developed rapidly and studied widely in the world in recent years [1]-[3]. However, the construction of large-scale photovoltaic power station in the open will occupy some agricultural land. Under this circumstance, some scholars have proposed complementary solutions of food crops and photovoltaic power generation [4]-[5]. Marrou et al studied the growth rate of crops affected in the specific shade of photovoltaic panels [6]. Klaring et al studied the response of tomato in greenhouse compartments to constrain the intensity of solar radiation [7]. Marco Cossu et al studied the solar radiation distribution inside a greenhouse and the effects of shading from the PV module on crop productivity [8]. There are also many similar researches which relate to plants selection under urban viaducts. Shenglin Wei et al analyzed the light intensity of greening space on roads in Suzhou and summarized the light density distribution characteristic and the accommodated plant types and their light-responsive properties [9]. Lihua Yin presented a method of comparing PAR and light requirement character of the shade plant in order to configure plants [10]. Xueying Wang studied the light condition and diurnal variation under three different viaducts in Shanghai by combining the indexes of photosynthetic photon flux, light compensation and light requirement level [11]. Up to now, no literature clearly points out which crop we can grow for different light distribution areas among photovoltaic arrays. In this paper, the distribution of light resources among solar PV panels is simulated by building a photovoltaic power station model. Then, the characteristics of solar radiation distribution are analyzed. We focused on two factors that have the most important effects on plant growth: PAR and sunshine hours. According to the relationship between crops and sunlight demands, we come forward a method of divisional crop layout for areas among PV arrays. It provides theoretical basis for crop selection.

II. SET UP THE MODEL

A. Location and climate data

We choose Zhangjiakou city, China (41.1°N, 114.7°E) as a sample site. The city's solar energy resources are very rich, geographical sunshine hours there reach 2756 to 3062 hours, total solar radiation per square meter ranges from 1500 to 1700 kwh, belonging to the second type solar radiation area. Use Weather Tool the attached software of Ecotect Analysis 2011 software (© 2009 Autodesk, Inc., U.S.A.) to view climate changes of Zhangjiakou city, particularly the changes in sunshine associated with this study. From meteorological data, on June 22 (near summer solstice), Zhangjiakou's sunrise first is at 4:54 and sunset is at 19:50, sunshine hours reaches 15 h; on December 22 (near winter solstice), the sunrise first is at 7:46 and sunset is at 16:50, sunshine hours reaches 9 h. The months with daily average temperature above 20℃ are June, July and August. In May and October, the daily average temperature is between 10℃ and 20℃. The daily average temperature in April is 9.7℃, close to 10℃. For other months, the daily average temperature is much smaller than 10℃, even below 0℃. Based on the above analysis, for months from April to October, the daily mean temperature is stable at 10℃ and above, that is, the effective period of crop growth in this study. From April 1 to October 31, the common sunrise and sunset period is from 7:00 to 17:00, which provides the basis for the light environment simulation of solar panels.

B. Model establishment and simulation

Computer simulation software can effectively improve field experiment efficiency and get more comprehensive, overall, long-term observation effect in a limited time. The effective growth period of crops is mainly concentrated on the period from April 1 to October 31. Therefore, the model is mainly analyzed around this time period. The main influencing factors of crops growth are photosynthetically active radiation and sunlight hours. These two indicators will be the main factors for simulation. The spatial stereo model of PV arrays is established by using Ecotect. As for solar power station, 260 Wp PV modules are vertically mounted in two rows (2x22) on a bracket unit and 22 PV modules are connected in series as a

978-1-5090-5606-4/17 $31.00 © 2017 IEEE

group string. The group string is 22 m long, 3.32 m wide, 0.1 m thick. Select 1MW photovoltaic power plant, configure two 500 Kw inverters and a 1000 KVA transformer. Each inverter access 96 PV group strings (total 549.12 kWp). From Pvsyst we can calculate that, the optimum inclination of the PV module is 39 degrees. The distance between the north and south center of the PV arrays is 7 meters and the height of the lower part of PV module is one meter from the ground. As the presence of cubic stone piers that play the role of fixing PV panels will reduce the planting area, in the actual project they should be buried under the ground surface. In this model, these stone piers are built below the horizontal level. Since the light distribution in the region between the photovoltaic panels is approximately symmetrical. The light distribution between any two adjacent PV strings can be replaced by the selected area marked with dark blue below. As is shown in Fig. 1, this specific region is 22 m long, 7 m wide.

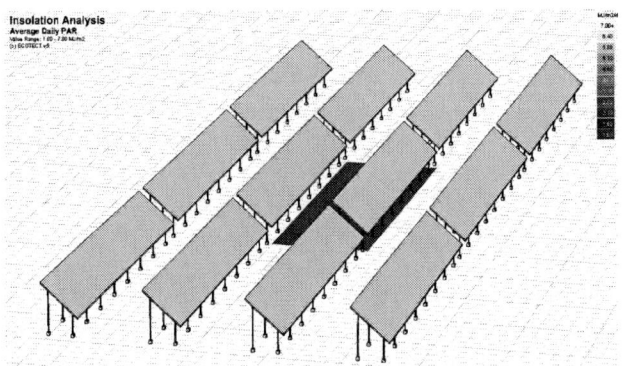

Fig.1. Three-dimensional model of photovoltaic power station

After the establishment of the above model, import the weather data of CHN-Zhangjiakou in the plug-in of Weather Tool and set up corresponding analysis grid. Find out CALCULATIONS column in Analysis Gird panel, choose Insolation Levels and click Perform Calculation. Then, select the Sky Factor & Photosynthetically Active Radiation and Shading, Overshadowing and Sunlight Hours. Time range is set as 7:00 -17:00, date range is the valid period of the crop growth (From April 1 to October 31). Computer simulation analysis can accurately, efficiently and quickly get the temporal and spatial distribution characteristics of PAR and sunlight duration which are most concerned with crop growth.

III. SIMULATION RESULTS

A. Sunshine hours distribution

Due to the lower part of PV modules are one-meter-high from the horizontal plane, in order to obtain light distribution condition in the vertical space in photovoltaic power station we choose 0 m, 0.5 m, 1 m these three different heights to analyze light environment. The simulation results of daily

average sunshine duration distribution at different heights are shown in Fig. 2.

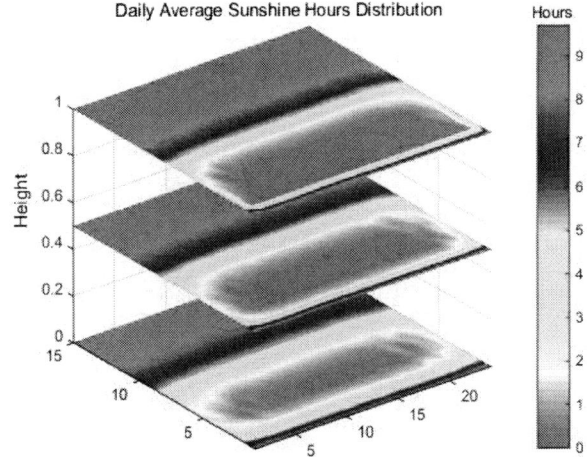

Fig.2. Daily average sunshine hours distribution at different heights

As we can see from Fig. 2, in the horizontal direction, sunshine duration is approximately linearly distributed. The space below solar panels is a weak irradiation area where sunlight hours is even 0 h, and the area that has no shelter above gets better sunshine. In the vertical direction, for strong solar radiation zone, the sunshine time will increase as the height increases, whereas for the weak radiation zone, the higher the height is, the less sunshine hours it will receive.

TABLE 1
EFFECTS OF DIFFERENTS HEIGHTS ON DAILY MEAN SUNSHINE
HOURS

Contrast height	0 m	0.5 m	1 m
Full sunshine hours (h)	10	10	10
Average value (h)	5.08	5.13	5.33
Maximum value (h)	8.99	9.38	9.70
Minimum value (h)	0.15	0	0
<20% of full Sunshine hours	32.9%	38.6%	38.8%
>50% of full Sunshine hours	48.6%	49.1%	51.4%

Table 1 shows the comparison of sunrise hours at different heights. With the altitude rising, high sunshine duration area, low sunshine hours area, average sunshine hours and maximum sunshine hours all have an increasing trend. However, the minimum sunshine hours show an opposite trend.

B. PAR distribution

Fig. 3 shows the simulation result of daily average photosynthetically active radiation distribution at the height of 0 m, 0.5 m, 1 m respectively. In general, the distribution of daily mean PAR is similar to that of average sunshine hours.

978-1-5090-5606-4/17 $31.00 © 2017 IEEE   2049

Through comparative analysis, we will find that for area with 0 hours sunshine, average daily PAR is about 1 MJ/m²·d to 2 MJ/m²·d. It is due to the presence of diffuse light.

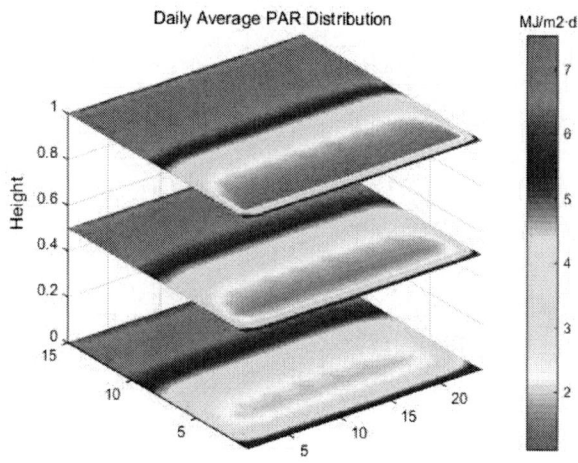

Fig.3. Daily average PAR distribution at different heights

Table 2 shows the comparison of daily mean PAR at different heights. With the altitude rising, high PAR area, low PAR area, average PAR and maximum PAR all have an increasing trend. However, minimum PAR shows the reverse trend.

TABLE 2
EFFECTS OF DIFFERENT HEIGHTS ON DAILY MEAN PAR

Contrast height	0 m	0.5 m	1m
Fullsunlight PAR(MJ/m²·d)	7.84	7.84	7.84
Average PAR (MJ/m²·d)	4.73	4.75	4.83
Maximum PAR(MJ/m²·d)	7.05	7.33	7.52
Minimum PAR (MJ/m²·d)	1.91	1.48	1.10
High PAR area ratio (>5.5 MJ/m²·d)	40.0%	44.3%	47.%
Low PAR area ratio (<3.0 MJ/m²·d)	31.4%	36.4%	37.1%

IV. DISCUSSION AND ANALYSIS

A. Crop classification

The variety of crops that can be planted under PV modules need to have a certain ability to adapt to the light environment of corresponding area. In the effective period of crop growth, light compensation point (LCP), light saturation point (LSP) and so on can be references to crops selection. The level of light compensation point is one of signs that plants can grow under low light intensity, which is also an important critical point for plant growth and development [12]. LCP and LSP generally change with the plant species, different parts of the same plant, growth and development stage, individual and

group, outside temperature conditions [13]. In general, plants with high saturation and high compensation point are positive or positive shade-tolerant plants and those with lower saturation and lower compensation point are negative or negative shade-tolerant palnts. In the daytime, the common light intensity and photosynthesis active radiation have the following basic conversion relationship [14].

$$1 \text{ lx} = 0.0185 \text{ } \mu mol \cdot m^{-2} \cdot s^{-1} \qquad (1)$$

$$1 \text{ lx} = 0.00402 \text{ W/m}^2 \qquad (2)$$

From the above two formulas, we can derive the relationship between the physical light intensity and photosynthetic intensity required by plants [10].

$$1 \text{ MJ/m}^2 \cdot d = 127.79 \text{ } \mu mol \cdot m^{-2} \cdot s^{-1} \qquad (3)$$

Through cluster analysis of photosynthetic parameters of plants [15], according to their different shade tolerance, plants can be divided into three major categories which include six minor classes shown in Table 3. Class III, Class II and Class I are respectively positive crops, neutral crops, negative crops.

B. Light adaption point

Crops are not general plants, most crops such as vegetables and fruits are positive crops, some other complex factors need to be considered. For Class I, it is necessary to ensure that average daily PAR received in the planting area is greater than LCP. Class II crops are neutral crops which can not be exposed to the intense solar radiation.

However, for Class III, LSP can only distinguish which crops can adapt to higher solar radiation. When crops can flower and produce fruit normally, the minimum solar radiation demand at this moment is defined as light adaption point (LAP). LAP is an important indicator to make a distinction among positive crops. For melons and fruits, LAP is about 70% of LSP. For root vegetables and leafy vegetables, LAP is about 50% of LSP. For example, tomato's light saturation point is about 70 klx (1295 $\mu mol \cdot m^{-2} \cdot s^{-1}$), which can grow normally in the light environment of 40 klx (740 $\mu mol \cdot m^{-2} \cdot s^{-1}$) to 50 klx (925 $\mu mol \cdot m^{-2} \cdot s^{-1}$).

For flowering crops, there will be the influence of sunlight hours (critical photoperiod), when in the actual project, we should take attention to this indicator. Otherwise, flowering crops may not bloom. 20% of the total sunshine duration is used as another important reference standard for planting negative plants [16].

C. Lay out according to PAR

Through comprehensive consideration of heights of most crops, it is suitable to select the plane at the height of 0.2 m to configure crops with different light demands. As is shown in Fig. 4, the orange area belongs to low light area, accounting

TABLE 3
CROPS CLASSIFICATION

Category		LSP μmol·m⁻²·s⁻¹	PAR MJ/m²·d	LCP μmol·m⁻²·s⁻¹	PAR MJ/m²·d
I	I—A	<400	<3.13	<22	<0.17
	I—B			>22	>0.17
II	II—A	400~700	3.13~5.48	<22	<0.17
	II—B			>22	>0.17
III	III—A	>700	>5.48	<22	<0.17
	III—B			>22	>0.17

for 33% of the total width. The minimum value of average daily PAR is 1.79 MJ/m²·d (228.7 μmol·m⁻²·s⁻¹), the maximum value is 3.12 MJ/m²·d (398.7 μmol·m⁻²·s⁻¹). Because Class I-A crops that have very low light saturation point and light compensation point are typically shade plants, it is feasible to plant in this area a small number of Class I-A that meet light requirement and this area is mainly suitable for planting Class I-B crops. The bright red and pink area is a strong solar radiation area with average daily PAR fluctuating between 5.5 MJ/m²·d (702.8 μmol·m⁻²·s⁻¹) and 7.21 MJ/m²·d (921.4 μmol·m⁻²·s⁻¹), this area is suitable for planting Class III-A and Class III-B crops. It is distributed on both sides, one side accounts for 39% of the total width and the other accounts for 4%. Other areas where daily average PAR ranges from 3.12 MJ/m²·d to 5.5 MJ/m²·d can grow class II-A and II-B crops. Fig. 6 (a) shows specific plane configuration of crops based on daily average PAR.

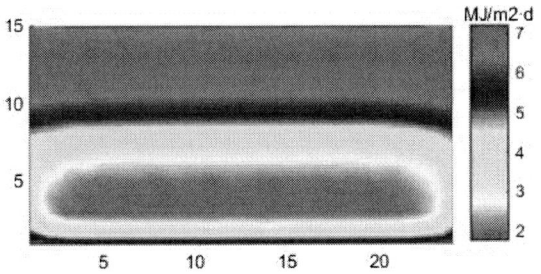

Fig.4. Daily average PAR distribution at the height of 0.2 m

D. Layout according to sunlight hours

Daily average sunshine hours distribution at the height of 0.2 m is shown in Fig. 5. Set 20% and 50% of full sunshine duration as two reference boundaries. From Fig.6 (b) we can see that the region that receives more than 50% of the full sunshine duration has two parts, accounting for half of the total width. This area is suitable for planting Class III-A and Class III-B crops due to the presence of sufficient direct light. The area that has daily average sunlight duration between 20% and 50% of the total sunshine is only 17% of the overall width. This area has a certain degree of shade, suitable for planting crops of Class II-A and Class II-B. The area that receives average sunlight hours less than 20% of the total sunshine

accounts for 33%. This area has almost no direct light, it is better to plant Class I-A and Class I-B crops. The influence of sunlight hours on Class II and III crops is obvious. Compared with planar configuration based on daily mean PAR, we can find that the range of Class III increases by 7%, the range of Class II crops decreases by 7%. There is only a 2% change in planting area of Class I crops.

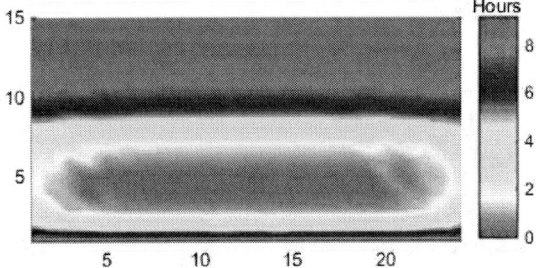

Fig.5. Daily average sunlight hours distribution at the height of 0.2 m

E. Layout according to double indexes

Both PAR and sunshine duration have important effects on crop growth. Here, we can combine with the common influence of these two factors to layout crops. Let Fig.6 (a) and Fig.6 (b) coincide, if two regions of different types overlap, let low solar radiation areas cover high light areaes. Fig.6 (c) shows crops arrangement based on these two factors. As can be seen from the figure below, after the combination of two indicators, the effect of the final crop configuration is similar to that of matching according to PAR, the only difference is that the range of Class II reduces by 2%, corresponding to the range of Class I increases by 2%. Class III accounts for 43% of the overall width, Class II accounts for 22% of the total width and Class I accounts for 35% of the overall width. In the practical application, when laying out crops, it is necessary to combine the specific values of light environmental indicators and light compensation point, light saturation point and light adaption point of crops to match them in the corresponding position. Light compensation point of Class A crops are lower than that of Class B. Compared with crops of Class A, Class B requires more light, in such cases, crops of Class B should be planted on the outside, crops

978-1-5090-5606-4/17 $31.00 © 2017 IEEE

of Class A should be planted on the inside. On two ends of analysis area, the distribution of light environment is not as linear as the middle area. Actually, it is feasible to regard the light distribution of marginal areas as linear for crops which have no strict light requirements.

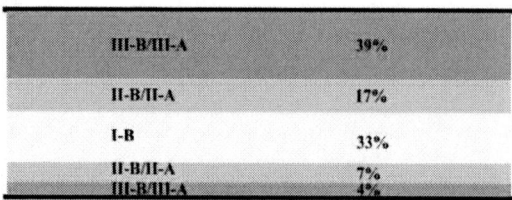

a. Based on PAR

b. Based on sunshine hours

c. Based on double indexes

Fig. 6 Crop layout on the planting plane

Combine with literature research [17]-[19], practice study and Baidu Encyclopedia website, just from the light demand, this paper presents a list of recommended crops for photovoltaic power station in Zhangjiakou city shown in Table 4 and Fig. 7 gives a specific model for crops layout in this model.

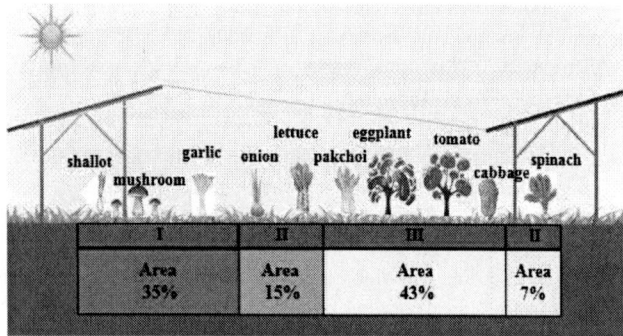

Fig.7. Crop layout model for Zhangjiakou city

From left to right, planted between any two rows of PV strings are respectively shallot, mushroom, garlic, onion, lettuce, pakchoi, eggplant, tomato, cabbage, spinach. The height of crops should not exceed the connection line between the front and rear panels. In this case, the PV array will not be blocked by shadows. In the actual regional planning, it is best to take a part of the weak light area as the walkway. When the PAR value is less than 0.17 MJ/m^2·d, the area can be defined as the dead zone.

V. CONCLUSION

Through the above analysis, solar radiation distribution at any position on different plane under solar panels can be simulated by Ecotect. Among the light correlation factors affecting crop growth, light intensity and sunshine hours have the most significant influence. In this work, PAR and sunlight duration are set as light environment indicators for crop growth matching. Crops are classified into three major categories which include six minor classes. For Class III crops, LAP is another indicator to approximately determine whether positive crops can flower and bear fruit normally. For Class I, only LCP is a critical point. Class II has less strict light requirement. Then, crop layout can be obtained by combining the correlation between their sunlight demands and different solar radiation distribution in different areas. In actual project, local soil properties and climatic conditions also need to be considered. In addition, a list of recommended crops for different areas among photovoltaic power station in Zhangjiakou city is presented. This paper only matches crops from the perspective of macroscopic solar radiation demand and other plant growth factors also need to be studied. There is also a balance between crop farming and power generation remained to be discussed.

ACKNOWLEDGEMENT

The authors thank National Natural Science Foundation of China (61374129) and Shanghai Science and Technology Commission project（14441900400）.

REFERENCES

[1] J. L. Bernal and D. L. Rodolfo, "Economical and environment analysis of grid connected photovoltaic systems in Spain," *Renewable Energy*, vol. 31, pp. 1107-1128, 2006.

[2] L. Y. Seng, G. Lalchand and G. M. S. Lin, "Economical, environmental and technical analysis of building integrated photovoltaic systems in Malaysia," *Energy Policy*, vol. 36, pp. 2130-2142, 2008.

[3] J. L. Yang, H. C. Yu, L. Ge, "Solar photovoltaic power generation technology," *Beijing Publishing House of Electronics Industry,* 2009.

[4] A. Yano, M. Kadowaki, A. Furue, N. Tamaki, T. Tanaka, E. Hiraki, Y. Kato, F. Ishizu and S. Noda, "Shading and electrical

TABLE 4
CROPS RECOMMENDATION LIST FOR ZHANGJIAKOU CITY

Class	Speices(Cultiar)		
III	Cucumber(Xintai mici)	Tomato(Zhong Shu No.4)	Potato (Tai Shan No.1)
	Cabbage(Zhong gan No.11)	Chinese chive (791)	Pakchoi(Shanghai qing)
	Lettuce(Ji nan Lettuce)	Spinach(Yuan ye)	Snapbean(Fengshou No.1)
	Eggplant (Luqie No.1)	Watermalon(Fengshou No.2)	Mushmelon (Qitian No.1)
	Pepper (Qie men pepper)	Radish (Luluobu No.1)	White gourd (Fen pi)
	Onion (Ziluo onion)	Sponge gourd (Ordinary)	Balsampear(Bincheng)
	Leaf lettuce(Boli)	Pumpkin(Yunnan black seeds)	Carrot (Wuchun shen)
	Chinese plate cabbage	Lentil (Green lentil)	Taro (Duozi Taro)
	Cauli flower (French snow ball)	Asparagus bean (Zhijiang No.28)	
II	Garlic(Cang shan)	Celery (America celery)	Leaf beet (Green stem)
	Welsh onion (Zhang qiu)	Leaf lettuce(Boli)	Heading lettuce (Emperor)
	Onion(Ziluo onion)	Carrot (general)	Potato (Tai Shan No.1)
	Ginger (Laiwu ginger)	Taro (Duozi Taro)	Water spinach(White flower)
	Pepper (general)	Chinesecabbage(Lubai No.8)	
I	Oyster mushroon	Garlic (genaral)	Malabar spinach
	Hotbed chives	Crowndaisy chrysanthemum	Some edible mushroom

features of a photovoltaic array mounted inside the roof of an east-west oriented greenhous," *Biosystems Engineering*, vol. 106, No. 4, pp. 367–377, 2010.

[5] C. Dupraz, H. Marrou, G. Talbot, L. Dufour, A. Nogier and Y. Ferard, "Combining solar photovoltaic panels and food crops for optimising land use: Towards new agrivoltaic schemes," *Renewable Energy*, vol. 36, pp. 2725–2732, 2011.

[6] H. Marrou, L. Guilioni, L. Dufour, C. Dupraz and J. Wery, "Microclimate under agrivoltaic systems: Is crop growth rate affected in the partial shade of solar panels?" *Agricultural and Forest Meteorology*, vol. 177, pp. 117–132, 2013.

[7] H. P. Klaring and A. Krumbein, "The effect of constraining the intensity of solar radiation on the photosynthesis, growth, yield and product quality of tomato," *Agro Crop Sci*, vol. 199, pp. 351–359, 2013.

[8] Marco Cossu, Lelia Murgia, Luigi Ledda, A. Paola, Deligios, Antonella Sirigu and Antonio Pazzona, "Solar radiation distribution inside a greenhouse with south-oriented photovoltaic roofs and effects on crop productivity," *Applied Energy*, vol. 133, pp. 89–100, 2014.

[9] Shenglin Wei and Minhong Jiang, "The distribution characteristics of light intensity and plants application for greening space on ground level roads under elevated highways of diverse structures, " *Chinese Landscape Architecture*, Vol.32, pp.78-82, 2016.

[10] Lihua Yin and Min Wan, "Research on regularities of natural light in green space under urban viaducts in Wuhan and suggestions on the greening," *Chinese Landscape Architecture*, pp. 79-83, 2014.

[11] Xueying Wang, Yafen Xin, Liangda Song and Liangjun Da, "Characteristics of light condition under viaduct and suitable

greening distribution in city, "*Chinese Journal of Ecoloogy*, vol. 25, pp. 938-943, 2006.

[12] Chenglin Huang, Songling Fu, Shuyun Liang and Yifan Ji, "Relationships between light and physiological characters of five climbing plants," *Chinese Journal of Applied Ecology*, vol. 15, pp. 1131-1134, 2004.

[13] Yi Wang and Xiaoli Wei, "Advance on the Effects of Different Light Environments on Growth, Physiological Biochemistry and morphostructure of Plant," *Journal of Mountain Agriculture and Biology*, vol. 29, pp. 353-359, 2010.

[14] R. W. Langhans and T. W. Tibbitts, "Plant Growth Chamber Manual" 1997.

[15] Lihua Yin, "Research of Shaded Green Space Landscape Basedon Natural Light Environment under the Urban Viaducts——Case Study on Viaducts in Wuhan City," chap. 5, pp. 141-142, 2012.

[16] Minzhen Lu, Xiaochang Xu, Shucheng Feng and Baichuan Wang, "Selection of vertical greening plants of the columns under the viaduct," *Journal of Plant Resources and Environment*, vol. 6, pp. 63-64, 1997.

[17] Zhenxian Zhang, Xuyuan Zhou and Liping Chen, "The Characteristics of Photosynthesis and Transpiration in Major Vegerable Crops," *Acta Horticulture Sinica*, vol. 24, pp. 155-160, 1997.

[18] Xizhen Ai, Zhenxian Zhang, Xiuhua Yang, Gongjun Shi and Lei Dai, "The Characteristics of Photosynthesis and Transpiration in some Vegetable Crops," *Acta Horticulture Sinica*, vol. 27, pp. 371-373, 2000.

[19] Shuting Dong, "Studies on the Photosynthetic Traits in Various Species of Crops," *Journa of Shangdong Agricultural University*, vol. 18, pp. 89-92, 1987.

Interface effects of alkali treatment on Cu-rich thin film solar cells

Hossam Elanzeery[*], Finn Babbe, Anastasiya Zelenina, Michele Melchiorre and Susanne Siebentritt

Laboratory for Photovoltaics, Physics and Materials Science Research Unit,
University of Luxembourg, Belvaux, L-4422, Luxembourg

Abstract — **CuInSe₂ absorbers grown under Cu-excess have better current collection compared to those grown as Cu-poor. However, cells based on Cu-excess absorbers show lower efficiency due to a worse absorber-buffer interface where the interface recombination is the dominant recombination path. In this paper, potassium fluoride post-deposition treatment is used to improve this interface, as evidenced by an activation energy of the main recombination path close to the band gap energy. The treatment also succeeded in eliminating the 200-meV step observed in admittance measurements; a main characteristic of Cu-excess CuInSe₂ cells.**

I. INTRODUCTION

Copper indium gallium di-selenide Cu(In,Ga)Se$_2$ or CIGS are one of the most promising absorber materials to thin film solar cells due to their high efficiencies reaching 22.6 % for the laboratory scale [1] and 16.5 % for the large module scale [2]. CuInSe$_2$ (CIS) is the ternary compound characterized by its simpler structure compared to CIGS but with lower bandgap and with lower efficiency (15 %) [3]. From the phase diagram [4], it can be observed that CIS can be grown under a wide compositional range either in the copper poor region ([Cu]/[In] < 1), or under Cu-excess ([Cu]/[In] > 1). CIS absorbers grown under Cu-excess -referred to in the text as "Cu-rich"- are characterized by the formation of the stoichiometric phase at [Cu]/[In] = 1 with an additional copper selenide (Cu$_2$Se) secondary phase, which can be etched [5]-[6]. Cu-rich absorbers are characterized by having larger grains, higher mobility, lower defect densities and higher carrier collection efficiencies leading to superior electronic properties compared to Cu-poor ones [7]-[9]. On the other side, solar cells based on Cu-rich absorbers are exhibiting lower efficiencies compared to Cu-poor ones mainly due to a strong decrease in open circuit voltage (V_{OC}) and fill factor (FF). These electrical losses are attributed to recombination at the interface between CIS absorber and cadmium sulphide (CdS) buffer layers [10]. These limitations were overcome by an In-Se post deposition treatment (PDT). This PDT forms a Cu-poor surface on the top of the stoichiometric absorber. These treated absorbers show the same efficiencies as Cu-poor ones due to an improved interface and higher V_{OC}, as in Cu-poor cells [11]-[12]. Alkali treatments are another form of PDT that have the advantage of improving the electrical performance of CIGS solar cells. The alkali treatment improves both the absorber surface and the V_{OC}. Potassium fluoride (KF) is one of those alkali treatments forming a completely Cu-depleted absorber surface and improving the absorber/buffer heterojunction quality [13]-[14]. There are two main types of KF PDT. The first one is an in-situ treatment. In this type; the absorber deposition, KF deposition and annealing steps are performed in the same deposition chamber during the same deposition process. The other type is an ex-situ treatment where the absorber deposition, KF deposition and annealing are performed in separate steps [9]. Both KF PDTs lead to an improvement in the V_{OC}. In case of the first type of KF PDT, both V_{OC} and efficiency are improved as reported in [1,13,15]. However, for the second type of KF PDT, only the V_{OC} was improved in [9,16,17] due to the hygroscopic nature of KF if the KF deposition and annealing steps are not performed in the same chamber. In this contribution, we introduce a third type that is an in-situ KF PDT with an etching step where the deposition and annealing of KF are performed in the same chamber but after getting the absorbers out of its deposition chamber. The details of this treatment are explained in [18,19]. The main focus of this paper is to improve the interface between Cu-rich absorber and buffer layers using KF alkali treatment. The effects of in-situ KF PDT on the interface recombination, apparent doping and admittance steps of Cu-rich CIS absorbers are analyzed and compared.

II. EXPERIMENTAL

A. Synthesis of CuInSe₂ solar cells

Cu, In and Se are deposited on molybdenum coated soda lime glass (SLG) substrates using a Veeco molecular beam epitaxy (MBE) system through a one-stage evaporation process at a substrate temperature of 590 ^{0}C. The Cu/In ratio used was higher than one, forming Cu-rich absorbers. Cu-rich absorbers are then etched using a 10 % aqueous potassium cyanide (KCN) solution for 5 minutes to remove excess Cu$_2$Se secondary phases [6,20]. A KF PDT step is then performed on the Cu-rich etched absorbers. KF is deposited in the same MBE tool used for absorber fabrication using Physical Vapor Deposition (PVD) technique. The absorbers are annealed while KF is evaporated at temperatures between 350-390 ^{0}C for a duration between 4-12 minutes under the presence of Se targeting thicknesses less than 30 nm. The treated CIS absorbers are then rinsed in de-ionized water removing excess fluorides. Those PVD-treated KF absorbers will be referred to in the text as "Cu-rich treated". After that, a CdS buffer layer is deposited using Chemical Bath Deposition (CBD) followed by sputtering of intrinsic-zinc oxide (i-ZnO) and biased-ZnO [21] window layers using an Aja Orion 8 sputtering tool. Finally, nickel-aluminium contact grids are deposited using a Ferrotec electron-beam evaporation tool.

978-1-5090-5606-4/17 $31.00 © 2017 IEEE

B. Characterization

Current-Voltage measurements as a function of Temperature (IVT) were performed using a Keithley 2440 5A Source Meter and the Admittance and Capacitance-Voltage (CV) profiles were obtained using an Agilent E4980A Precision LCR Meter. The IVT and admittance measurements were completed using a CTI-cryogenic closed cycle helium cryostat with a temperature range between 320 – 20 K for both measurements. The sample temperature was determined using a calibrated Si diode mounted on a molybdenum-coated glass substrate identical to the one used for the absorber fabrication. The IVT measurements were performed from -1.5 V to 1.5 V. Temperature dependence of open circuit voltage is extracted from the IVT measurements performed using a cold-mirror halogen lamp in order to identify the activation energy of the dominant recombination path. The calibration of the illumination intensity was based on the short-circuit current density (J_{SC}) obtained from IV measurements where cells are illuminated by a cold mirror halogen lamp with an intensity of 100 mW/cm² calibrated by a silicon reference solar cell. The admittance spectrum was measured under zero bias in a frequency range between 100 - 1M Hz with an AC-voltage level of 30 mV.

III. RESULTS

The KF PDT successfully increased the efficiency from 7.1 % to 9.5 % with an increase in both; the V_{OC} from 356 mV to 404 mV and the FF from 48 % to 58 %. In order to understand this improvement and the effect of the KF PDT on the interface between the absorber and buffer layers, IVT measurements were performed.

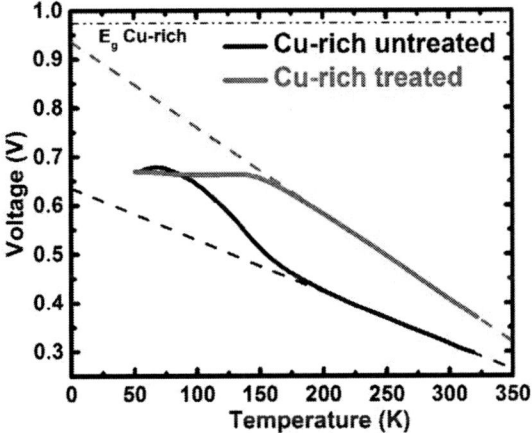

Figure 1: Temperature dependence of voltage for Cu-rich CIS solar cells before and after KF PDT. Dashed double dotted line represents the bandgap of Cu-rich cells as extrapolated from EQE measurements.

The activation energy (E_a) of the dominant recombination path is extracted from the extrapolation of the open circuit voltage dependence on temperature to 0 K [22] as presented in Fig. 1. Based on that, it can be concluded that the improvement in the open circuit voltage is due to an improvement in the dominant recombination path that shows a significantly higher activation energy, close to the energy of the band gap as a result of the treatment. A possible mechanism behind this improvement can be unveiled by the CV measurements at room temperature presented by the Mott-Schottky plot in Fig. 2.

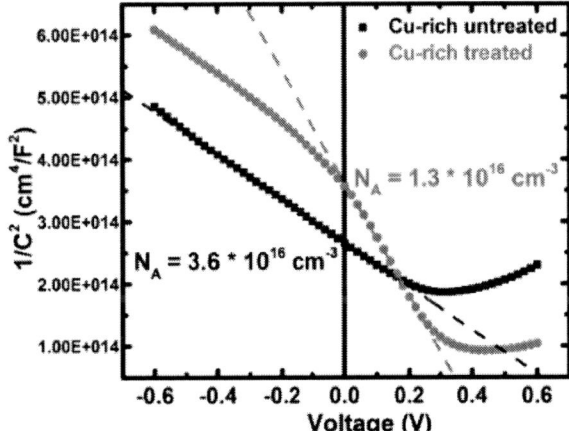

Figure 2: Capacitance-Voltage (CV) measurements for Cu-rich CIS solar cells before and after KF PDT. The apparent doping (N_A) is extracted from the slope of the linear fitting represented by the dashed lines.

From Fig. 2, it can be deduced that the apparent doping decreased from 3.6 x 10¹⁶ to 1.3 x 10¹⁶ cm⁻³. If we assume that this decrease in apparent doping corresponds to a real decrease in the absorber doping, we can discuss a reason for the reduced interface recombination. The decrease in absorber doping leads to a less steep valence band at the interface between the absorber and the buffer layers. Consequently, the less steep valence band helps to minimize recombination of charge carriers by tunneling near the interface leading to higher activation energies observed in Fig. 1. In addition to that, the admittance measurements presented in Fig. 3 (a, b), reveal the disappearance of the 200 meV energy step; deduced from the Arrhenius plot presented in Fig. 3 (c, d), that is a main characteristic for Cu-rich CIS absorbers after the KF PDT. This 200 meV step could either be a defect that was removed by the treatment or a barrier that decreased after improving the interface between the absorber and buffer layers as a consequence of the KF PDT. We observe this step in all Cu-rich cells and it is removed by alkali treatments (KF-PDT with both in-situ and ex-situ treatment). It could therefore be related to an interface defect or barrier.

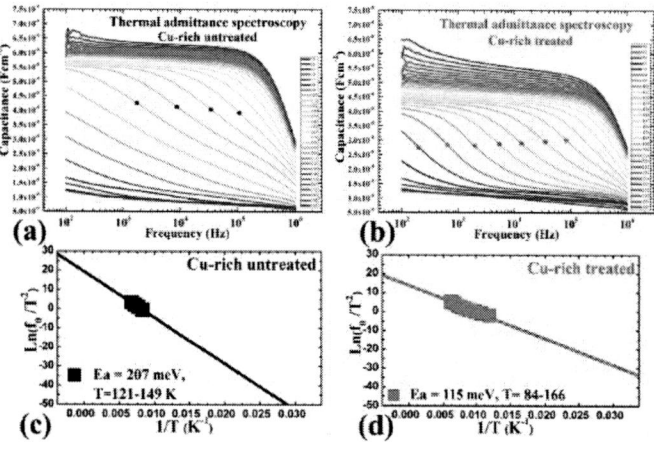

Figure 3: Capacitance-Frequency measurements for Cu-rich CIS solar cells (a) before and (b) after KF PDT. The activation energy of the main capacitance step is deduced from the Arrhenius plot (c) before and (d) after KF PDT.

IV. Conclusion

The KF in-situ treatment with etching step improved the interface between the absorber and buffer layers for cells grown under Cu-excess. This is due to an improvement in the dominant recombination path that shows a higher activation energy as a result of the treatment. The KF PDT was able to modify the interface by decreasing the apparent doping leading to an increase in the efficiency from 7.1 % to 9.5 % with an increase in both; the V_{OC} from 356 mV to 404 mV and the FF from 48 % to 58 %. Finally, the KF PDT succeeded in eliminating the 200 meV admittance step that is a main characteristic of Cu-rich CIS solar cells. Thus, KF treatment is able to improve the surface and the bulk of Cu-rich CuInSe₂, not only of Cu-poor CuInSe₂.

Acknowledgment

This work has been supported by Luxembourg National Research Fund (FNR) in the framework of the CURI-K project, which is gratefully acknowledged.

References

[1] P. Jackson, R. Wuerz, D. Hariskos, E. Lotter, W. Witte and M. Powalla, "Effects of heavy alkali elements in Cu(In,Ga)Se₂ solar cells with efficiencies up to 22.6%", *Physica Status Solidi RRL*, 10, 2016, pp. 583–586.

[2] S. Yang, K. Lin, W. Lee and W. Lo, "Achievement of 16.5% total area efficiency on 1.09m² CIGS modules in TSMC solar production line", *IEEE Photovoltaic Specialist Conference (PVSC)*, 42, 2015, pp. 1-3.

[3] J. AbuShama, R. Noufi, S. Johnston, S. Ward and X. Wu, "Improved performance in CuInSe₂ and surface modified

CuGaSe₂ Solar Cells", *IEEE Photovoltaics Specialists Conference and Exhibition*, 2005, pp. 299-302.

[4] T. Gödecke, T. Haalboom and F. Ernst, "Phase equilibria of Cu-In-Se II. The In-In₂Se₃-Cu₂Se-Cu subsystem", *Z. Metallkd.*, 91, 2000, pp. 635-650.

[5] V. Depredurand, D. Tanaka, Y. Aida, M. Carlberg, N. Fevre and S. Siebentritt, "Current loss due to recombination in "Cu-rich" CuInSe₂ solar cells", *Journal of Applied Physics*, 115, 2014, 044503.

[6] Y. Hashimoto, N. Kohara, T. Negami, M. Nishitani and T. Wada, "Surface characterization of chemically treated Cu(In, Ga)Se₂ Thin Films", *Japan Journal of Applied Physics*, 35, 1996, pp. 4760-4764.

[7] S. Siebentritt, L. Gütay, D. Regesch, Y. Aida and V. Depredurand, "Why do we make Cu(In,Ga)Se₂ solar cells non-stoichiometric?", *Solar Energy Materials and Solar Cells*, 119, 2013, pp. 18–25.

[8] F. Werner, D. Colombara, M. Melchiorre, N. Valle, B. El Adib, C. Spindler and S. Siebentritt, "Doping mechanism in pure CuInSe₂", *Journal of Applied Physics*, 119, 2016, 173103.

[9] H. Elanzeery, F. Babbe, M. Melchiorre, A. Zelenina and S. Siebentritt, "Potassium fluoride ex-situ treatment on both Cu-rich and Cu-poor CuInSe₂ thin film solar cells", *IEEE Journal of Photovoltaics*, 2017.

[10] M. Turcu, O. Pakma and U. Rau "Interdependence of absorber composition and recombination mechanism in Cu(In,Ga)(Se,S)₂ heterojunction solar cells", *Applied Physics Letter*, 80, 2002, 2598.

[11] Y. Aida, V. Depredurand, J. Larsen, H. Arai, D. Tanaka, M. Kurihara and S. Siebentritt, "Cu-rich CuInSe₂ solar cells with a Cu-poor surface", *Progress in Photovoltaics: Research and Applications*, 23, 2015, pp. 754–764.

[12] T. Bertram, V. Depredurand and S. Siebentritt, "In-Se surface treatment of Cu-rich grown CuInSe₂", *40th IEEE Photovoltaic Specialist Conference (PVSC)*, Denver, CO, USA, 2014, pp. 3633-3636.

[13] A. Chirila, P. Reinhard, F. Pianezzi, P. Bloesch, A. Uhl, C. Fella, L. Kranz, D. Keller, C. Gretener, H. Hagendorfer, D. Jaeger, R. Erni, S. Nishiwaki, S. Buecheler and A. Tiwari, "Potassium-induced surface modification of Cu(In,Ga)Se₂ thin films for high-efficiency solar cells", *Nature Materials*, 12, 2013, pp. 1107-1111.

[14] F. Pianezzi, P. Reinhard, A. Chirila, B. Bissig, S. Nishiwaki, S. Buecheler and A. N. Tiwari, "Unveiling the effects of post-deposition treatment with different alkaline elements on the electronic properties of CIGS thin film solar cells", *Phys. Chem. Chem. Phys.*, 16, 2014, 8843.

[15] A. Laemmle, R. Wuerz and M. Powalla, "Efficiency enhancement of Cu(In,Ga)Se₂ thin-film solar cells by a post-deposition treatment with potassium fluoride", *Physica Status Solidi RRL*, 7, 2013, pp. 631–634.

[16] R. Kamada, T. Yagioka, S. Adachi, A. Handa, K. F. Tai, T. Kato and H. Sugimoto, "New World Record Cu(In,Ga)(Se,S)₂ Thin Film Solar Cell Efficiency Beyond 22%", *Proceedings of IEEE Photovoltaic Specialist Conference (PVSC)*, Portland, OR, USA, 2016, pp. 1287-1291.

[17] P. Pistor, D. Greiner, C. Kaufmann, S. Brunken, M. Gorgoi, A. Steigert, W. Calvet, I. Lauermann, R. Klenk, T. Unold and M. Lux-Steiner, "Experimental indication for band gap widening of chalcopyrite solar cell absorbers after potassium fluoride treatment", *Applied Physics Letter*, 105, 2014, 063901.

[18] F. Babbe, H. Elanzeery, M. Melchiorre and S. Siebentritt, "CuInSe₂ absorber layer grown under copper excess with a copper poor surface formed by a KF post deposition treatment", *proceedings IEEE PVSC*, 2017.

[19] F. Babbe, H. Elanzeery, M. Melchiorre, A. Zelenina and S. Siebentritt, "In-situ potassium fluoride treatment of on both Cu rich and Cu poor CuInSe$_2$ thin film solar cells", *submitted*, 2017.

[20] V. Depredurand, T. Bertram and S. Siebentritt, "Influence of the Se environment on Cu-rich CIS devices", *Physica B*, 439, 2014, pp. 101-104.

[21] M. Hala, S. Fujii, A. Redinger, Y. Inoue, G. Rey, M. Thevenin, V. Depredurand, T. P. Weiss, T. Bertram and S. Siebentritt, "Highly conductive ZnO films with high near infrared transparency", *Progress in Photovoltaics*, 23, 2015, pp. 1630-1641.

[22] S. S. Hegedus and W. N. Shafarman, "Thin-Film Solar Cells: Device Measurements and Analysis", *Progress in Photovoltaics*, 12, 2004, pp. 155-176.

Increased V_{OC} and FF in $ZnO_{1-x}S_x$-buffered $CuIn_{1-x}Ga_xSe_2$ Solar Cells by Cadmium Partial Electrolyte Treatment

Andreas Bauer, Dimitrios Hariskos, and Wiltraud Wischmann

Zentrum für Sonnenenergie- und Wasserstoff-Forschung Baden-Württemberg (ZSW)
Meitnerstr. 1, 70563 Stuttgart, Germany

Abstract—We applied a Cd partial electrolyte treatment (Cd-PE) to condition the surface of $CuIn_{1-x}Ga_xSe_2$ (CIGS) layers for solar cells in combination with a $ZnO_{1-x}S_x$ buffer by chemical bath deposition or sputtering. The treatment led to increased V_{OC}, FF, and efficiency. We find that a Cd-PE leads to an improved CIGS/buffer layer heterojunction quality in $ZnO_{1-x}S_x$-buffered devices. We use advanced characterization methods as well as simulations in order to further understand the mechanism.

Index Terms—photovoltaic cells, chalcopyrite, ZnO, ZnS, Cd, simulation.

I. INTRODUCTION

$CuIn_{1-x}Ga_xSe_2$ (CIGS) solar cell efficiencies already climbed as high as 22.6% [1]. Despite this accomplishment, there remains significant potential, some of which can be attributed to the buffer layer. CdS is most commonly used as a buffer layer, even though its optical properties, i.e. its small band gap (2.4 eV), results in a significant photocurrent loss. In order to counter photocurrent losses, buffer layers comprising wider band gaps are subject to research activities. Besides Cd-free buffer layer materials such as e.g. In_2S_3, $Zn_{1-x}Sn_xO$, and $Zn_{1-x}Mg_xO$, $ZnO_{1-x}S_x$ (ZOS) is one of the most often investigated materials.

Although $ZnO_{1-x}S_x$ has a sufficiently high band gap that leads to the desired significant gain in photocurrent, V_{OC} and FF losses occur [2]. As a consequence, net efficiency drops below the reference set by CdS-buffered devices [2]. If either one of these losses were compensated for, $ZnO_{1-x}S_x$-buffered devices would be as efficient as or even outperform CdS-buffered devices.

We assume that the lowered efficiency of $ZnO_{1-x}S_x$-buffered devices compared to CdS-buffered devices originates from a fundamental effect at or near the CIGS surface or eventually in the CIGS layer itself. This assumption is supported by in-house investigations showing that efficiencies close to those of CdS-buffered devices can be achieved for various $ZnO_{1-x}S_x$ compositions as well as deposition techniques.

Messaoud *et al.* have recently shown that a cadmium partial electrolyte treatment (Cd-PE) of CIGS and $Cu_2ZnSn(S,Se)_4$ (CZTS) absorber surfaces can significantly improve the CIGS/CdS as wells as the CZTS/CdS heterojunction quality [3]. The impact of a Cd-PE has already been investigated by Jiang *et al.* [4], concluding that the Cd-PE promotes a stronger band bending at the CIGS surface.

For our experiments, we combine the Cd-PE treatment to CIGS absorbers with different average Ga contents. Additionally, we use common buffer layer materials such as CdS and $ZnO_{1-x}S_x$, where for the latter we use chemical bath deposition (CBD) and sputtering (sp.) as deposition techniques. Complementary to measurement data an diode analysis, we use simulations to attain a more profound insight and to support our theory.

II. EXPERIMENTAL

For the basic device, Mo is sputtered on soda lime glass followed by co-evaporation of CIGS in an in-line multi-stage process. We investigate two average compositions with average GGI ([Ga]/([Ga]+[In]) in %) of 27% (lower) or 33% (higher).

For Cd-PE [5], we use a standard CdS process [6] in which de-ionized water replaces thiourea. The Cd-PE process time is 8 minutes. Afterwards, the samples are cleaned in a de-ionized water ammonia solution to remove eventually present residuals.

Three buffer types are compared: standard CdS as a reference (chemical bath, CBD), and $ZnO_{1-x}S_x$ either by CBD or by sputtering (sp.). For sputtering, a mixed target is used with 50 mol.% ZnS and 50 mol.% ZnO. The deposition is done by in-line process. Typical power densities are ca. 1 W/cm^2 and pure argon is used as the sputter gas. Different highly resistive window layers are applied, depending on the buffer layer and its deposition technique: ZnO for CBD-CdS, $Zn_{0.75}Mg_{0.25}O$ for CBD-$ZnO_{1-x}S_x$, and no highly resistive window layer for sputtered $ZnO_{1-x}S_x$.

All cells are finalized by a highly conductive layer of $ZnO:Al_2O_3$ on which a Ni/Al/Ni-grid is electron-beam evaporated.

We use SCAPS [7] for device simulation. Our models are based on experimental data we collected from EQE (external quantum efficiency), current-voltage curves, capacitance profiling, depth profiling, and compositional analysis. In order to develop a theory that covers all experimental findings, we select a set of suitable samples for more detailed investigations.

III. DISCUSSION

After performing the Cd-PE, the CIGS surface conditions apparently change whereas the results clearly depend on the

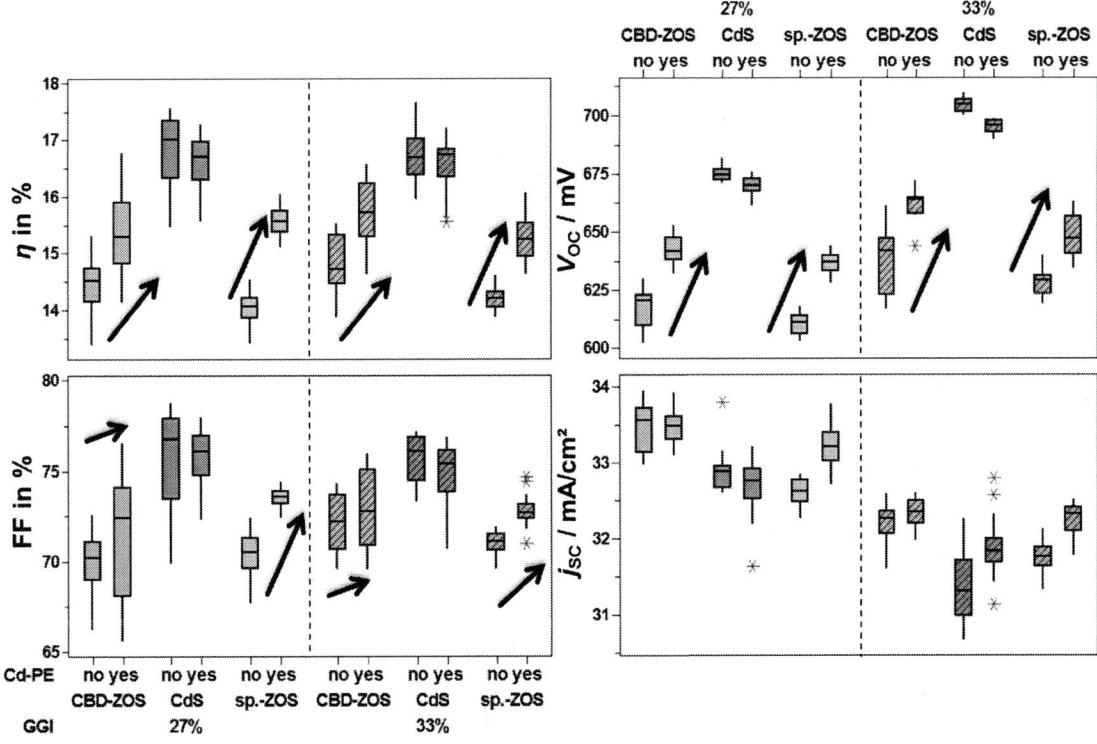

(a) Solar cell parameters.

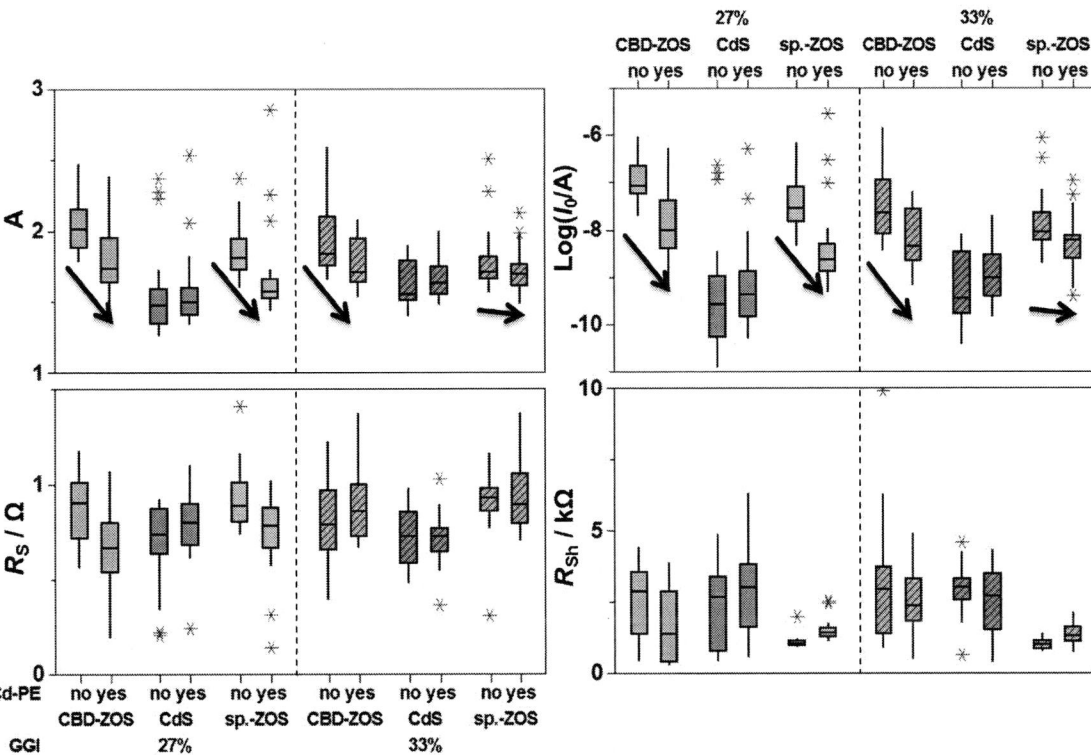

(b) Parameters from diode analysis in the one diode model.

Fig. 1. Result of a Cd-PE on CdS and $ZnO_{1-x}S_x$ buffer layers (CBD, sputtering). For $ZnO_{1-x}S_x$ buffered devices, a Cd-PE improves diode parameters by significantly lowering diode factor A and reverse saturation current I_0. Data after light soak at 298K.

buffer layer material (Fig. 1a). From Fig. 1a it is evident that CdS-buffered devices already achieved their best possible performance level and do not benefit from the Cd-PE. In contrast, the Cd-PE increases V_{OC} and FF for all $ZnO_{1-x}S_x$-buffered devices regardless of which deposition technique was used. This result is valid for both investigated average GGI concentrations in the CIGS layer.

For a closer investigation by junction analysis, we use the one-diode equivalent-circuit model. This model allows us to separate the effect of junction parameters from resistance parameters (Fig. 1b). We can show that the junction quality improves by Cd-PE treatment as the diode ideality factor A and reverse saturation current I_0 decrease. Although the Cd-PE also has a slight impact on the resistance parameters, the gain in V_{OC} and FF can be attributed to the improved junction quality.

In order to check the reproducibility, we applied the Cd-PE treatment to other CIGS batches buffered with CBD-$ZnO_{1-x}S_x$. We found the trend to be reproducible, but not the absolute gains in V_{OC} and FF.

For representative $ZnO_{1-x}S_x$ buffered devices (CBD and sputtered for each GGI level), we recorded quantified depth profiles by GDOES and XRF (glow discharge optical emission spectroscopy, x-ray fluorescence spectroscopy, Fig. 2). Moreover in case of a lower GGI (i.e. 27%) and sputtered $ZnO_{1-x}S_x$, we use their depth profiles as input for building simulation models.

Fig. 2 shows all recorded data either grouped by varying GGI (GGI and CGI ([Cu]/([Ga]+[In] in %), upper row) or by Cd-PE treatment (Cd and Na concentration, lower row). Therefrom we can conclude the chemical effect of a Cd-PE.

The CGI is not affected by the Cd-PE and only slightly varies for lower and higher GGI.

The Cd signal is consequentially very weak as the available overall amount of Cd is expected being very low. However, the Cd signal for the Cd-PE sample indicates a diffusion profile with increased Cd in the front half of the sample. The sodium concentration does not clearly correlate with the Cd-PE and appears to be rather randomly shaped and distributed.

Fig. 3 summarizes the optoelectronic data of all samples shown in Fig. 2. The IV data shows all samples that match the criteria. In case of a lower GGI, the effect of the Cd-PE is more evident than for a higher GGI despite the effect being visible for both GGI levels in Fig. 1a.

The most interesting data is shown by the Mott-Schottky plots (middle row). Independent of the buffer layer deposition method and the average GGI, the Cd-PE shifts all curves to the right i.e. higher voltages. The simulation subsection addresses this finding and its consequences in more detail later on. The buffer layer absorption edge and absorption strength (or layer thickness) remain unchanged as well as almost the whole EQE itself.

We use simulations in order to narrow down the effect of the Cd-PE and combine all available experimental data, especially data from capacitance and depth profiling. For the basic model, we refer to Niemegeers *et al.* [8], Kniese *et al.* [9], and to [10]. Based on this input, we select a suitable pair of devices with sputtered $ZnO_{1-x}S_x$ and lower GGI, with and without Cd-PE (Fig. 4). The basic model is shown in inset of Fig. 4b and the key results are summarized in Tab. I. In general, all simulated FF values are higher than the experimental values, though we consider it as a purely

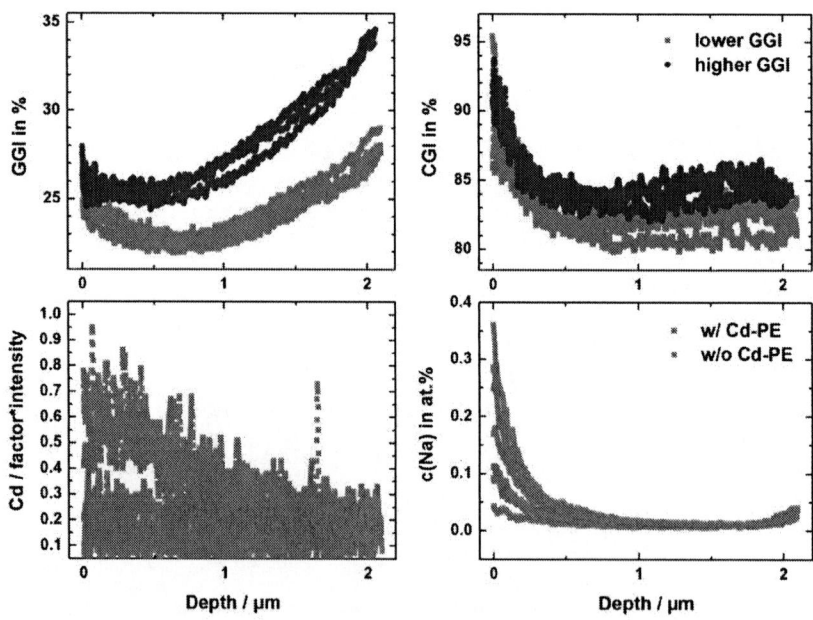

Fig. 2. Depth profile data recorded for selected $ZnO_{1-x}S_x$ buffered devices with lower and higher GGI as well as w/o and w/ Cd-PE treatment. Please note: the Cd signal could not be quantified and edge effects might result in too-high diffusion depth.

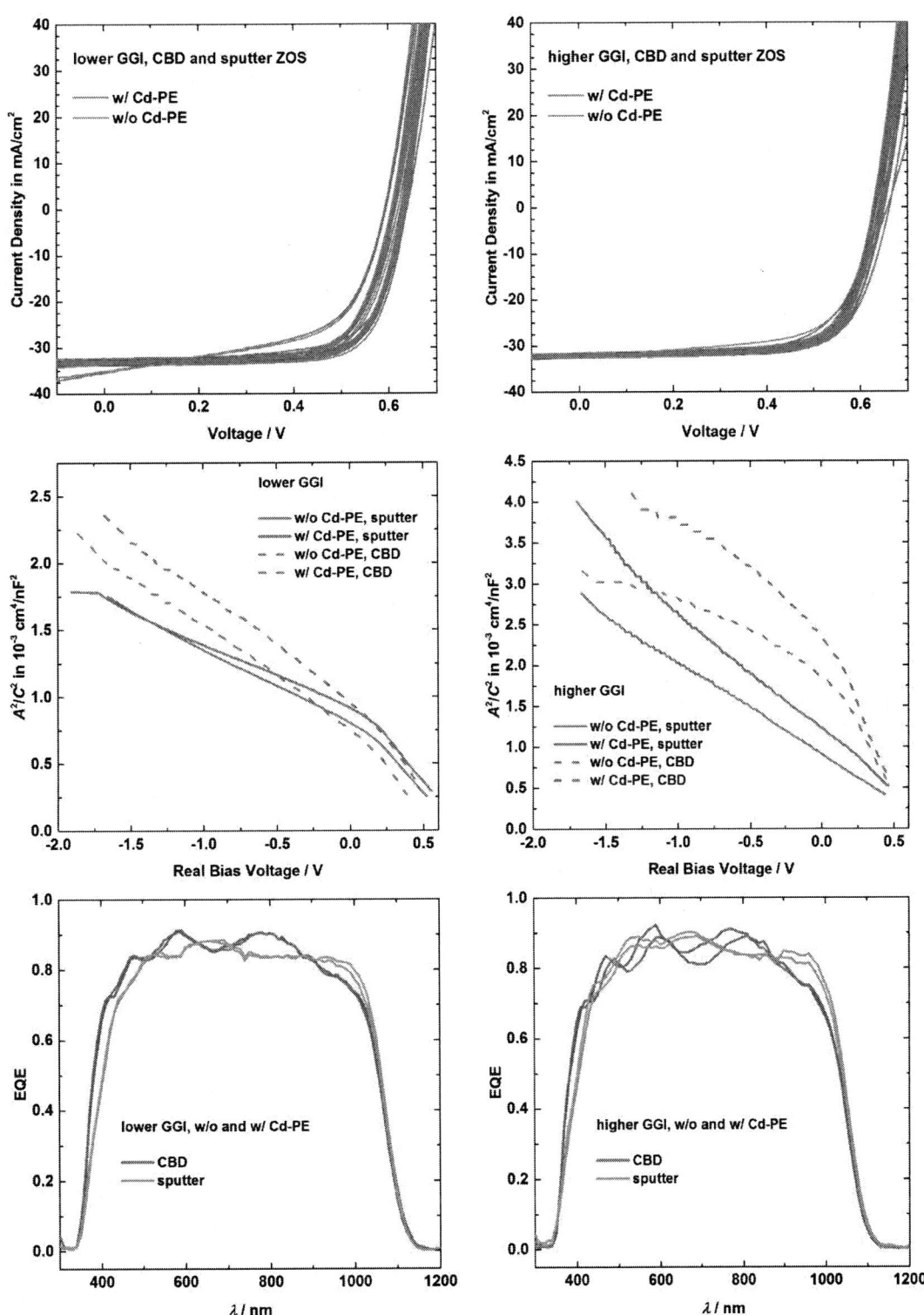

Fig. 3. Optoelectronic data for selected samples with $ZnO_{1-x}S_x$ buffer and both GGIs (lower, higher). IV curves: IV data for all samples matching the data constraints. While IV data and Mott-Schottky data are quite different, the EQE is almost identical for the same GGI.

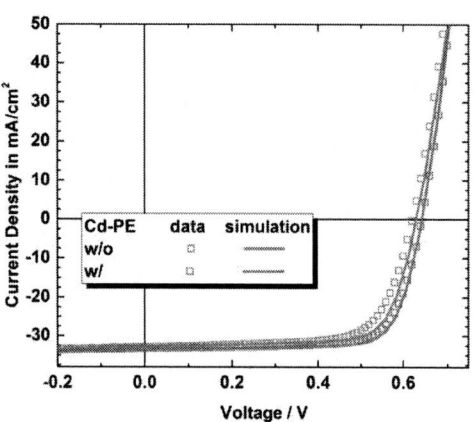

(a) IV characteristics.

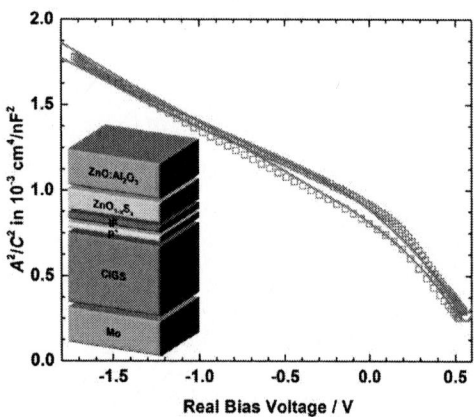

(b) Mott-Schottky plots. Inset: basic layer stack.

Fig. 4. Measurement data and simulated curves.

systematic error that should not compromise our conclusions.

As first step we established a device model for the Cd-PE treated device (reference device/simulation) from which we deduced the model for a non treated device. The deduction is done step by step, i.e. parameter by parameter, to highlight the impact of each parameter. Each time, the simulation of the Mott-Schottky plot is repeated as we take more parameters into account (Tab. II: none, ohmic resistances, traps).

However, after the final step, simulation data and experimental data (both w/o Cd-PE) do not match except for the short circuit current. When taking into account both EQEs (w/o and w/ Cd-PE), it is evident that they are almost identical (Fig. 3). As V_{OC} and FF are still too high compared to the experimental data, a further increased trap density would indeed lower V_{OC} and FF but simultaneously contradict the EQE data. For this reason we had to limit the trap density to a value so that both EQEs match.

TABLE I
SUMMARY OF SIMULATED KEY RESULTS

parameter	layer	Cd-PE no	yes
acceptor density (left) in $1/cm^3$	CIGS	1.9E16	2.4E16
acceptor density (right) in $1/cm^3$	CIGS	6E15	7E15
doping parameter b (beta function)	CIGS	130	160
trap density in $1/cm^3$	CIGS	2.5E13	1.5E13
acceptor traps in $1/cm^3$	p^+	6E17	5E17
R_S / Ω cm^2	device	0.8	0.7
R_{shunt} / Ω cm^2	device	500	1000

Without taking additional traps and ohmic resistance changes into account, an increased p^+ acceptor defect density and altered CIGS doping profile can only explain a small V_{OC} loss (5 mV) and a moderate FF loss (0.7%).

Accordingly, simulation results indicate that in case of Cd-PE treated devices, the trap-assisted recombination must also be lower as well which would imply that trap passivation must have taken place. One possible model e.g. involves Cd_{Cu} donor-type effects close to the conduction band edge which might lower e.g. the net acceptor defect density (= doping) in the strongly doped p^+ layer as well as the defect density in the CIGS layer. Such a passivation extending deep into the CIGS layer might be possible as the Cd signal in Fig. 2 extends beyond ca. 1.0 μm from the surface.

Regarding the p^+ layer, a reduced acceptor (defect) concentration would have the same effect as described in [4] where the Cd-PE promotes a stronger band bending and a higher built-in voltage at the CIGS/buffer heterojunction. The latter is supported by our CV data in every case as shown by Fig. 3.

IV. CONCLUSION

Although the efficiency gap between CdS- and $ZnO_{1-x}S_x$-buffered CIGS solar cells still persists, we were able to reduce this gap by a partial electrolyte treatment with cadmium (Cd-PE) [5]. The Cd-PE procedure can significantly improve the efficiency by simultaneously increasing V_{OC} and FF of $ZnO_{1-x}S_x$-buffered devices. This result can be attributed to an improved heterojunction quality at the CIGS/buffer layer interface. The Cd-PE procedure works for CBD-$ZnO_{1-x}S_x$ as well as for sputtered $ZnO_{1-x}S_x$ buffer layers. However, the success of the Cd-PE procedure can critically depend on the CIGS layer.

ACKNOWLEDGMENT

We would like to thank our many colleagues at ZSW for their support, W. Hempel for GDOES depth profiling, P. von Bismarck for XRF analysis, and T. Friedlmeier for careful reading and corrections. We also thank the German Federal Ministry for Economic Affairs and Energy under the contract numbers 0325715A and 0324179 for funding.

978-1-5090-5606-4/17 $31.00 © 2017 IEEE

TABLE II

DEVICE PARAMETERS FROM EXPERIMENT AND SIMULATION AFTER EACH SIMULATION STEP.

data	experiment	experiment	simulation (reference)	simulation	simulation	simulation
Cd-PE	no	yes	yes	no	no	no
$R_{S,shunt}$				same as reference	from data	from data
trap density				same as reference	same as reference	from data
η in %	14.3	15.8	16.5	16.1	15.7	15.3
V_{OC} / mV	621	643	643	638	637	631
FF in %	69.9	73.5	76.5	75.9	74.2	73.5
j_{SC} in mA/cm^2	32.9	33.3	33.5	33.2	33.2	32.9

REFERENCES

[1] P. Jackson, R. Wuerz, D. Hariskos, E. Lotter, W. Witte, and M. Powalla, "Effects of heavy alkali elements in Cu(In,Ga)Se$_2$ solar cells with efficiencies up to 22.6%," *Physica Status Solidi Rapid Research Letters*, vol. 10, pp. 583–586, 2016.

[2] T. M. Friedlmeier, P. Jackson, A. Bauer, D. Hariskos, O. Kiowski, R. Wuerz, and M. Powalla, "Improved Photocurrent in Cu(In,Ga)Se$_2$ Solar Cells: From 20.8% to 21.7% Efficiency with CdS Buffer and 21.0% Cd-Free," *IEEE Journal of Photovoltaics*, vol. 5, pp. 1487–1491, 2015.

[3] K. Ben Messaoud, M. Buffire, G. Brammertz, H. ElAnzeery, S. Oueslati, J. Hamon, B. J. Kniknie, M. Meuris, M. Amlouk, and J. Poortmans, "Impact of the Cd2+ treatment on the electrical properties of Cu$_2$ZnSnSe$_4$ and Cu(In,Ga)Se$_2$ solar cells," *Progress in Photovoltaics: Research and Applications*, vol. 23, pp. 1608–1620, 2015.

[4] C.-S. Jiang, R. Noufi, K. Ramanathan, H. R. Moutinho, and M. M. Al-Jassim, "Electrical modification in Cu(In,Ga)Se$_2$ thin films by chemical bath deposition process of CdS films," *Journal of Applied Physics*, vol. 97, p. 053701, 2005.

[5] K. Ramanathan, F. Hasoon, S. Smith, D. Young, M. Contreras, P. Johnson, A. Pudov, and J. Sites, "Surface treatment of CuInGaSe$_2$ thin films and its effect on the photovoltaic properties of solar cells," *Journal of Physics and Chemistry of Solids*, vol. 64, pp. 1495 – 1498, 2003, 13th International Conference on Ternary and Multinary Compounds.

[6] D. Hariskos, P. Jackson, W. Hempel, S. Paetel, S. Spiering, R. Menner, W. Wischmann, and M. Powalla, "Method for a High-Rate Solution Deposition of Zn(O,S) Buffer Layer for High-Efficiency Cu(In,Ga)Se$_2$-Based Solar Cells," *IEEE Journal of Photovoltaics*, vol. 6, pp. 1321–1326, 2016.

[7] M. Burgelman, P. Nollet, and S. Degrave, "Modelling polycrystalline semiconductor solar cells," *Thin Solid Films*, vol. 361, pp. 527–532, 2000, symposium on Chalcogenide Semiconductors for Photovoltaics at the 1999 E-MRS Spring Conference, Strasbourg, France, Jun 01-04, 1999.

[8] A. Niemegeers, M. Burgelman, R. Herberholz, U. Rau, D. Hariskos, and H.-W. Schock, "Model for electronic transport in Cu(In,Ga)Se$_2$ solar cells," *Progress in Photovoltaics: Research and Applications*, vol. 6, pp. 407–421, 1998.

[9] R. Kniese, M. Powalla, and U. Rau, "Characterization of the CdS/Cu (In,Ga)Se$_2$ interface by electron beam induced currents," *Thin Solid Films*, vol. 515, pp. 6163–6167, 2007.

[10] A. Bauer, S. Sharbati, and M. Powalla, "Systematic survey of suitable buffer and high resistive window layer materials in CuIn$_{1-x}$Ga$_x$Se$_2$ solar cells by numerical simulations," *Solar Energy Materials and Solar Cells*, vol. 165, pp. 119 – 127, 2017.

Passivating and Carrier-selective Contacts – Basic Requirements and Implementation

S.W. Glunz, M. Bivour, C. Messmer, F. Feldmann, R. Müller, C. Reichel,
A. Richter, F. Schindler, J. Benick, M. Hermle

Fraunhofer Institute for Solar Energy Systems, Heidenhofstr. 2, 79110 Freiburg, Germany

Abstract — Effective passivating and carrier-selective contacts are based on the deposition of thin-film systems on the silicon wafer which results in a spatial decoupling of carrier generation and carrier separation in the device. They provide excellent passivation, high carrier selectivity and low contact resistivity. We have performed dedicated experiments to visualize these properties, thus allowing a better understanding of practical passivating and carrier-selective contact systems. Based on this analysis we recommend a measurement procedure to analyze carrier-selective contacts. The proper design of contact systems as the TOPCon structure and its careful implementation in a solar cell process featuring a simple both-sides contacted cell architecture allows to achieve excellent efficiencies of 25.7% and 21.9% on monocrystalline and multicrystalline silicon, respectively.

I. INTRODUCTION

Traditional crystalline silicon solar cells feature doped junctions like diffused phosphorus "emitters" or alloyed aluminum "back surface fields". Such technologies are well-understood and robust in mass production. To obtain a high degree of carrier selectivity and to screen the highly recombinative metal/silicon interface, it is necessary to increase the concentration of the doping profile significantly. Since this increased doping level within the silicon absorber increases the level of intrinsic charge carrier recombination, i.e. Auger recombination, this concept limits inherently the efficiency potential. To increase the efficiency potential cell structures like the PERL [1] cell have been developed in which the area of the metal contact and the respective local doping is strongly reduced resulting in maximum efficiencies of 25% [2]. To reduce the process complexity and to allow cell efficiencies closer to the theoretical limit of 29.4% [3], it is beneficial to separate the contact systems from the silicon wafer, thus, decoupling the carrier separation from carrier generation. This concept is crucial for the so-called passivating and carrier-selective contacts where a thin-film contact system is deposited on the silicon wafer. The best known example is the a-Si/c-Si heterojunction [4] as successfully used by Panasonic or Kaneka for their record cells. Using an interdigitated back contact structure Kaneka was able to achieve an efficiency of 26.6% [5]. Also the highly efficient interdigitated back contact solar cells fabricated by SunPower [6] are utilizing passivating contacts. The TOPCon [7] and POLO [8] technologies are based on a thin intermediate oxide and a heavily doped nano- or polycrystalline silicon layer and have led to excellent efficiencies of 25% and higher. Recently, also "silicon-free"

thin films as metal oxide layers like MoO_x have also shown a very high potential [9–11].

This contribution will give an introduction to the practical principles for the fabrication of effective passivating and carrier-selective contacts. In order to illustrate the impact of important prerequisites such as interface passivation or carrier separation, we have performed dedicated experiments. This allows to deduce design rules for the ideal contact system. Finally, the integration of passivating and carrier-selective contacts in a solar cell is discussed since the performance at device level is the final proof of a proper design of new contact systems.

II. PREREQUISITES FOR AN IDEAL CONTACT: PASSIVATION AND CARRIER-SELECTIVITY

The "ideal" contact for a high-efficiency solar cell should exhibit the following properties

1. (Self-)Passivation:
   Reduce recombination at contact to a minimum
   → High implied voltage iV_{oc}
   (splitting of the quasi-Fermi levels)
2. Carrier-selectivity:
   Allow one carrier type to pass through the contact while blocking the other one.
   → High external V_{oc}
3. Low contact resistivity ρ_c:
   → High FF (i.e. low ohmic voltage drop at mpp)

As shown in the following chapters these properties (which are not strictly independent) can be instructively visualized in a set of dedicated experiments which are helpful to understand and design an ideal contact.

Frequently, the properties "passivating" and "carrier-selective" are used as synonyms, however, both properties on their own are necessary but not sufficient conditions for an ideal contact. While a dielectric surface passivation layer will provide excellent passivation, it is not carrier-selective since it blocks both carrier types. On the other hand, a carrier-selective contact following the definition of Würfel, Cuevas and Würfel [12] provides a strong difference in conductivity for holes and electrons. A diffusion profile as the classical phosphorus "emitter" features a rectifying and carrier-selective contact characteristic by providing a strong difference in the local conductivity for holes and electrons.

978-1-5090-5606-4/17 $31.00 © 2017 IEEE

However, the lack of surface passivation and the relatively high doping concentration which increases intrinsic Auger recombination [3, 13, 14] will result in a poor external V_{oc}, limited by a low iV_{oc}. This loss can be reduced by increasing the doping only at the metal contact (selective emitter) allowing this classical contact to become more and more "passivating".

However, in our view it makes sense to distinguish these classical approaches from the actual developments summarized as "passivating and carrier-selective contacts". The most striking difference is that the **contact system is spatially separated from the silicon absorber material**. In this paper we will focus on such systems which in all successful implementations consists of

1. an intermediate passivating layer (IPL) at the interface to the crystalline silicon absorber and
2. a carrier separation layer (CSL) between the passivation layer and the metal contact.

In many actual systems the IPL is based on an intrinsic amorphous silicon or an ultra-thin silicon oxide layer, both providing a low interface state density, D_{it}, and thus an excellent chemical passivation. Theoretically it may be possible to avoid this additional layer but it seems that the main functions of the CSL as high conductivity is not compatible with surface passivation. In fact the IPL was so important for the inventors of the HIT-technology that they decided to mention it in the abbreviation of their technology (HIT: Heterojunction with *Intrinsic Thin layer*) [15].

The CSL is in charge of carrier separation i.e. it supplies different conductivities for electrons and holes in or near to the contact region. This functionality can be obtained in different ways:

1. High band gap materials with asymmetric band off-set
2. High doping concentration to achieve different conductivities for holes and electrons
3. High and low work-function material (like metal oxides) well known from conductor–insulator semiconductor (e.g. MIS) systems [16]

Especially in case 2 and 3 which are the most relevant implementations for silicon, a strong bend bending is generated in the silicon absorber leading to an induced junction which will leads to carrier-selectivity while reducing the detrimental effects introduced by heavy doping. Additionally, it supports the passivation quality of the contact by field effect passivation. In case 3 an additional doping profile might be generated from the in-diffusion of doping atoms from the CSL into the absorber. In order to visualize the effect of both, interface passivation and carrier selectivity, we have performed various experiments.

III. INFLUENCE OF INTERFACE PASSIVATION

In this experiment symmetric *n*-type samples with an *n*-TOPCon contact system (see Fig. 1) with different doping concentration were fabricated. The splitting of the Quasi-Fermi levels i.e. the implied V_{oc} was measured using QSSPC [17] before and after a hydrogen passivation step (see Fig. 2).

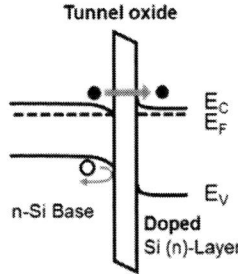

Fig. 1 n-TOPCon structure used for the experiment in Fig.2. The band bending due to the high doping level assists the chemical passivation by the tunnel oxide and allows for carrier selectivity and finally a low contact resistivity.

Before hydrogen passivation (red dots), the chemical passivation of the thin oxide layer is negligible. Therefore, a reduction of the recombination is only achieved by increasing the doping level in the CSL comparable to a classical "back surface field" (BSF). It is obvious that the passivation quality improves with doping concentration but even for a very high doping level an iV_{oc} value of only 670 mV is achieved. After hydrogenation (black squares) the quality of the thin oxide layer is strongly improved. Additionally, the hydrogenation passivates defects within the CSL which additionally improves the electrical quality of the contact system. Finally, very high iV_{oc} values around 730 mV can be achieved independently of the doping concentration. This experiment underlines the importance of an excellent surface passivation.

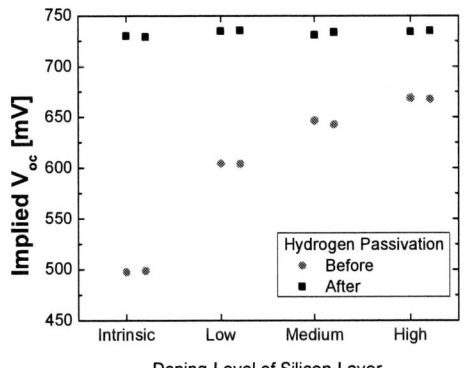

Fig. 2 Implied voltage of a TOPCon system with different doping levels before and after hydrogen passivation demonstrating the influence of the chemical interface passivation.

978-1-5090-5606-4/17 $31.00 © 2017 IEEE

IV. CARRIER SELECTIVITY

In order to visualize the effect of carrier selectivity, we have chosen a system of a thin amorphous silicon passivation layer and a MoO_x layer [11]. Together with an ITO layer on top this structure forms the hole contact on a *n*-type crystalline silicon wafer (see Fig. 3 inset)[1]. We have used three different techniques i.e. ALD [18], sputtering and evaporation to deposit the MoO_x layer. By changing the deposition parameters we are able to change the composition of these layers which has a direct effect on the work function of the MoO_x layer. The work function i.e. the resulting band bending, V_{bb}, was measured using surface photo-voltage (SPV) [11]. As a reference sample a classical heterojunction system with *p*-doped amorphous silicon on the intrinsic a-Si layer was fabricated.

We have measured the implied voltage (iV_{oc}), i.e. the splitting of the quasi-Fermi levels in the silicon absorber and the voltage at the external contacts (V_{oc}). The difference between these two values is an excellent measure for the ability of the contact to separate carriers (carrier selectivity).[2]

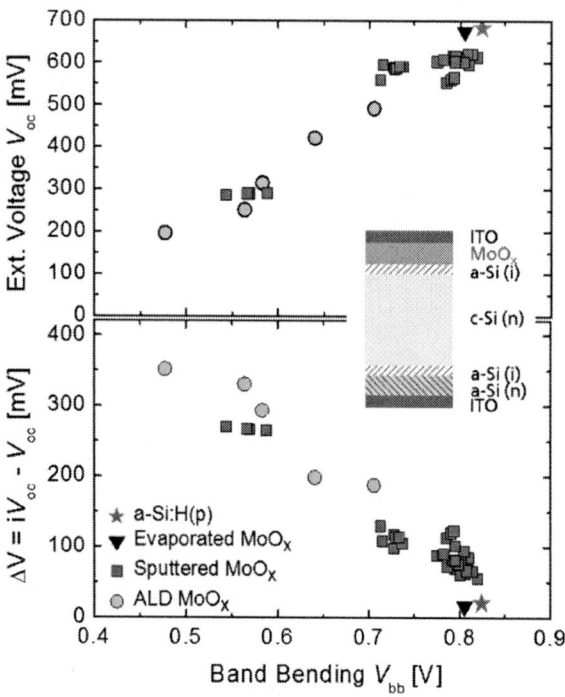

Fig. 3 Influence of the band bending on the carrier selectivity. The process parameters of different deposition techniques (sputtering, evaporation and ALD) have been varied to change the composition of MoO_x layers and thus the resulting band bending.

---

[1] The electron rear contact of all devices is formed by a heterojunction based on *n*-doped amorphous silicon on the intrinsic a-Si layer.

[2] Another metrics for carrier selectivity would be the ratio of V_{oc} and iV_{oc} as suggested in [19].

Fig. 3 shows impressively that only if sufficient band bending and thereby a proper induced p/n-junction is provided by the MoO_x layer, it is possible to achieve good carrier selectivity and thus high external voltages. For low band bending values voltage losses, $\Delta V = iV_{oc} - V_{oc}$, of more than 300 mV are observed. The observed trend is independent of the deposition technology. The best MoO_x systems show excellent external voltages and selectivity comparable to the one of the SHJ reference system. With such layers we were able to fabricate cells with front MoO_x hole contacts resulting in an efficiency of 22.1% [20]. In fact the well-known concept of conductor-insulator-semiconductor (e.g. MIS) systems [16] experiences a renaissance by the use of new material systems.

V. CONTACT RESISTIVITY

If good carrier selectivity is achieved then last but not least a low contact resistivity has to be achieved to allow for a high fill factor of the final device. However, if carrier selectivity is not high enough, i.e the voltage loss $\Delta V = iV_{oc} - V_{oc}$ is not negligible, reasonable measurements of the contact resistivity are often not possible since such samples show non-ohmic characteristics.

Fig. 4 shows the measured contact resistivity ρ_c and recombination current J_0 of a TOPCon structure [21]. For a wide range of annealing temperatures, it is possible to achieve excellent values for both parameters. If the annealing temperature is chosen too high (> 900°C) the thin oxide layer breaks up and the passivation quality is very low while ρ_c is decreased.

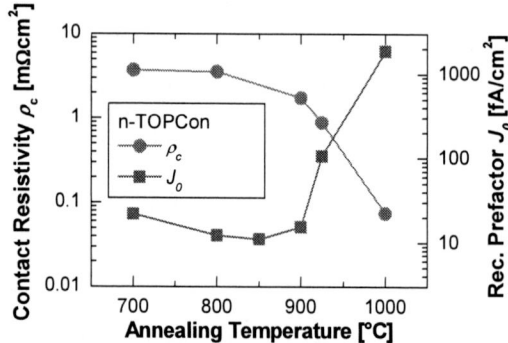

Fig. 4 Contact resistivity and recombination characteristics of a TOPCon structure.

VI. RECOMMENDED MEASUREMENT PROCEDURE

The combination of low J_0 and low ρ_c is a sufficient condition for good contact quality and is used in literature to classify different contact systems [22] and to predict their efficiency potential [23]. Thus, although it would be sufficient to determine J_0 and ρ_c to evaluate the quality of a contact system, we suggest the following procedure for the analysis of novel contact schemes.

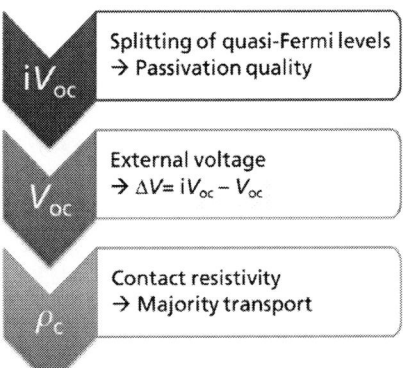

Fig. 5 Measurement procedure for the characterization of passivating and carrier-selective contacts.

The additional step, i.e. the determination of the voltage loss $iV_{oc} - V_{oc}$ is an excellent and direct check of the carrier selectivity. It improves the understanding of contact structures and helps identifying potential limitations due to by contact properties as the workfunction and induced band bending (see Fig. 3). Therefore, is especially useful when screening new contact systems. It can be performed on a structure which is very similar to the final cell structure and uses a currentless measurement method as SunsVoc. Furthermore, the same structure can be used for the determination of iV_{oc} and V_{oc}. Contact structures which exhibit a significant voltage loss often show a non-ohmic behavior and can be excluded from subsequent contact resistivity measurements or investigations at solar cell level. A more detailed discussion for a large variety of contact systems is given in [24].

VII. THE FINAL PROOF: DEVICE PERFORMANCE

If all three properties, i.e. (i) passivation, (ii) selectivity and (iii) low resistivity are achieved, the performance of the contact has to be proven at the device level. Only if the solar cell exhibits parameters superior to the ones using classical concepts as diffused emitters, back surface fields, or local point contacts (PERC), we can really benefit from the new contact system. In order to proof the potential of our TOPCon structure we have it as the rear contact of monocrystalline n-type cells with boron front emitters (see Fig. 6).

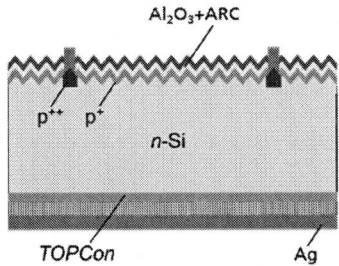

Fig. 6 Sketch of solar cell with diffused front boron emitter and full-area rear passivated contact (TOPCon) on monocrystalline silicon.

With this cell structure we have achieved a calibrated efficiency of 25.7% on small area (4 cm^2) which represents the actual efficiency record for cells with both-sides contacted cell architecture [25]. On large cell area (100 cm^2) with a similar cell structure but featuring a homogeneous emitter we were able to realize efficiencies up to 24.5% [26].

TABLE I

Cell results on monocrystalline silicon
(measured by Fraunhofer ISE Callab)

Area [cm^2]	V_{oc} [mV]	J_{sc} [mA/cm^2]	FF [%]	η [%]
4 (da)	725	42.5	83.3	25.7
100 (ap)	713	41.4	83.1	24.5

In order to proof if the TOPCon structure is also suited for multicrystalline silicon we have prepared cells exhibiting a black silicon texture, a homogeneous boron emitter and a TOPCon rear contact [27] (see Fig. 7).

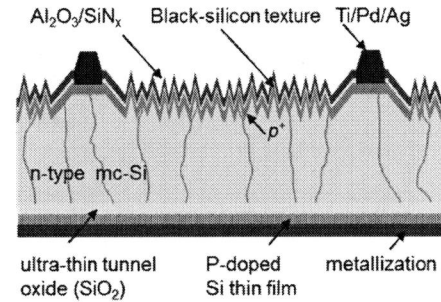

Fig. 7 Sketch of solar cell with diffused front boron emitter and full-area rear passivated contact (TOPCon) on multicrystalline silicon.

n-type high-performance multicrystalline (HP mc) silicon was grown at Fraunhofer ISE and analyzed with photoluminescence imaging [28]. It exhibits high lifetimes with a narrow distribution even after the full cell process including boron diffusion.

Fig. 8 shows the prediction of the efficiency potential of our HP mc-Si material using ELBA (Efficiency Limiting Bulk recombination Analysis) [29] taking into account the cell structure shown in Fig. 7. Not only the high efficiency potential of the good grains is remarkable but especially that even the areas with higher crystal defect density show very performance resulting in excellent area-weighted average values. Our analysis shows that the material should allow conversion efficiencies in the range of 22% [28].

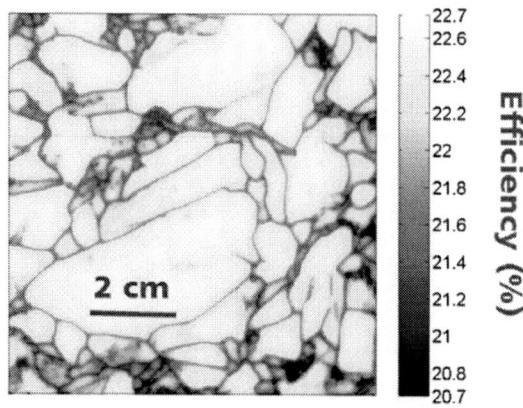

Fig. 8 Prediction of efficiency potential using ELBA of high-performance multicrystalline silicon grown at Fraunhofer ISE.

In fact we were able to achieve an efficiency of 21.9% experimentally (confirmed by Fraunhofer ISE) [27]. This result represents the new world record for multicrystalline silicon and shows impressively that the fabrication of TOPCon structures on multicrystalline silicon is feasible i.e. growing a highly functional thin tunnel oxide on material with different crystal orientation.

TABLE II

Cell results on multicrystalline silicon
(measured by Fraunhofer ISE Callab)

Area [cm^2]	V_{oc} [mV]	J_{sc} [mA/cm^2]	FF [%]	η [%]
4 (ap)	672.6	40.8	79.7	21.9

VIII. CONCLUSION

The measurement of J_o and ρ_c is sufficient to determine the quality of a passivating and carrier-selective contact. However, especially when screening new contact systems, the determination of the voltage loss $\Delta V = iV_{oc} - V_{oc}$ is a very valuable additional measurement step. It is a direct measure for the carrier selectivity and helps to get deeper insight into the physics of the contact. Furthermore, contact systems with non-ohmic characteristics can be excluded from further evaluation.

A well-designed contact system like TOPCon helps to boost the efficiency of solar cells and might be an interesting option as a successor of the PERC/L technology since it allows a simple one-dimensional cell architecture with high efficiency potential [30]. Using TOPCon as the rear contact of both-sides contacted n-type cells with boron emitter we were able to achieve an efficiency of 25.7% and 21.9% on mono-crystalline and multicrystalline silicon, respectively.

REFERENCES

[1] J. Zhao, A. Wang, and M. A. Green, "24% efficient PERL structure silicon solar cells," in *21st IEEE Photovoltaic Specialists Conference Kissimmee*, 1990, pp. 333–335.

[2] M. A. Green, "The path to 25% silicon solar cell efficiency: History of silicon cell evolution," *Prog. Photovolt: Res. Appl.*, vol. 17, no. 3, pp. 183–189, 2009.

[3] A. Richter, S. W. Glunz, F. Werner, J. Schmidt, and A. Cuevas, "Improved quantitative description of Auger recombination in crystalline silicon," *Phys. Rev. B*, vol. 86, no. 16, p. 165202, 2012.

[4] K. Masuko *et al.*, "Achievement of more than 25% conversion efficiency with crystalline silicon heterojunction solar cell," *IEEE J. Photovoltaics*, vol. 4, no. 6, pp. 1433–1435, 2014.

[5] K. Yoshikawa *et al.*, "Silicon heterojunction solar cell with interdigitated back contacts for a photoconversion efficiency over 26%," *Nat. Energy*, vol. 2, p. 17032, 2017.

[6] D. D. Smith *et al.*, "Toward the practical limits of silicon solar cells," *IEEE Journal of Photovoltaics*, vol. 4, no. 6, pp. 1465–1469, 2014.

[7] F. Feldmann, M. Bivour, C. Reichel, M. Hermle, and S. W. Glunz, "Passivated rear contacts for high-efficiency n-type Si solar cells providing high interface passivation quality and excellent transport characteristics," *Sol. Energy Mater. Sol. Cells*, vol. 120, pp. 270–274, 2014.

[8] R. Peibst *et al.*, "A simple model describing the symmetric I–V characteristics of p polycrystalline Si/n monocrystalline Si, and n polycrystalline Si/p monocrystalline Si junctions," *IEEE J. Photovoltaics*, vol. 4, no. 3, pp. 841–850, 2014.

[9] C. Battaglia *et al.*, "Silicon heterojunction solar cell with passivated hole selective MoOx contact," *Appl. Phys. Lett.*, vol. 104, no. 11, p. 113902, 2014.

[10] J. Bullock *et al.*, "Efficient silicon solar cells with dopant-free asymmetric heterocontacts," *Nat. Energy*, vol. 1, no. 3, p. 15031, 2016.

[11] M. Bivour, J. Temmler, H. Steinkemper, and M. Hermle, "Molybdenum and tungsten oxide: High work function wide band gap contact materials for hole selective contacts of silicon solar cells," *Sol. Energy Mater. Sol. Cells*, vol. 142, pp. 34–41, 2015.

[12] U. Würfel, A. Cuevas, and P. Würfel, "Charge Carrier Separation in Solar Cells," pp. 1–9.

[13] A. Cuevas, G. Giroult-Matlakowski, P. A. Basore, C. Dubois, and R. R. King, "Extraction of the surface recombination velocity of passivated phosphorus-doped silicon emitters," in *1st World Conference on Photovoltaic Energy Conversion Hawaii*, 1994, pp. 1446–1449.

[14] P. P. Altermatt *et al.*, "Numerical modeling of highly doped Si:P emitters based on Fermi–Dirac statistics and self-consistent material parameters: P emitters based on

Fermi-Dirac statistics and self-consistent material parameters," *J. Appl. Phys.*, vol. 92, no. 6, p. 3187, 2002.

[15] M. Taguchi *et al.*, "HIT^{TM} cells - high-efficiency crystalline Si cells with novel structure," *Prog. Photovolt: Res. Appl.*, vol. 8, no. 5, pp. 503–513, 2000.

[16] R. Singh, M. A. Green, and K. Rajkanan, "Review of conductor-insulator-semiconductor (CIS) solar cells," *Solar Cells*, vol. 3, no. 2, pp. 95–148, 1981.

[17] R. A. Sinton, A. Cuevas, and M. Stuckings, "Quasi-steady-state photoconductance, a new method for solar cell material and device characterization," in *25th IEEE Photovoltaic Specialists Conference Washington DC*, 1996, pp. 457–460.

[18] M. Bivour, B. Macco, J. Temmler, W. Kessels, and M. Hermle, "Atomic layer deposited molybdenum oxide for the hole-selective contact of silicon solar cells," *Energy Procedia*, vol. 92, pp. 443–449, 2016.

[19] D. Pysch, C. Meinhard, N.-P. Harder, M. Hermle, and S. W. Glunz, "Analysis and optimization approach for the doped amorphous layers of silicon heterojunction solar cells," *J. Appl. Phys.*, vol. 110, no. 9, p. 94516, 2011.

[20] L. Neusel, M. Bivour, and H. Hermle, "Selectivity issues of MoO$_x$ based hole contacts," in *SiliconPV*, 2017.

[21] F. Feldmann *et al.*, "Advanced passivated contacts and their application to high-efficiency solar cells," in *NREL Workshop*, 2015.

[22] A. Cuevas *et al.*, "Skin care for healthy silicon solar cells," in *42nd IEEE Photovoltaic Specialists Conference (PVSC)*, 2015, pp. 1–6.

[23] R. Brendel, M. Rienäcker, and R. Peibst, "A quantitative measure for the carrier selectivity of contacts to solar cells," in *32nd EU PVSEC*, 2016.

[24] M. Bivour, "Silicon heterojunction solar cells: Analysis and basic understanding," PhD thesis, Albert-Ludwigs Universität, Freiburg, 2016.

[25] A. Richter *et al.*, "n-Type Si solar cells with passivating electron contact: Identifying sources for efficiency limitations by wafer thickness and resistivity variation," *Solar Energy Materials and Solar Cells*, 2017.

[26] F. Feldmann, B. Steinhauser, S. Kluska, M. Hermle, and S. W. Glunz, "Evaulation of TOPCon technology on large area solar cells," in *EU PVSEC*, 2017.

[27] J. Benick *et al.*, "High-efficiency n-type HP mc silicon solar cells," *IEEE J. Photovoltaics*, 2017.

[28] F. Schindler *et al.*, "Material Limits of Silicon from State-of-the-Art Photoluminescence Imaging Techniques," *Solar Energy Materials and Solar Cells*, 2017.

[29] B. Michl *et al.*, "Efficiency limiting bulk recombination in multicrystalline silicon solar cells," *Sol. Energy Mater. Sol. Cells*, vol. 98, pp. 441–447, 2012.

[30] S. W. Glunz *et al.*, "The irresistible charm of a simple current flow pattern – 25% with a solar cell featuring a full-area back contact," in *31st EU PVSEC*, 2015, pp. 259–263.

First-principles modeling of alkali metal post deposition treatment effects in CIGS solar cells

Maria Fedina, Hannu-Pekka Komsa, Ville Havu and Martti J. Puska

Aalto university, Espoo, P.O. Box 11000,

Abstract — The efficiencies of Cu(In, Ga)Se$_2$(CIGS) solar cells have increased very fast, thanks the alkali post deposition treatment (PDT). In the present work, we have considered the role of alkali metal atoms in the efficiency enhancement. First, we have investigated the effect of alkali metal atoms in the bulk CuInSe$_2$(CIS) absorber and at the grain boundaries in terms of formation and migration energies and charge transition levels. We found that the copper sublattice is the most preferable for all the alkali metal atoms. A detailed comparison between different alkali metal atoms, with respect to the behavior at grain boundaries and in grain interiors has been done. Moreover, we have studied how alkali metal atoms contribute in the formation of the secondary phases. The secondary phase formation during the PDT process has been suggested on the basis of calculated reaction enthalpies.

I. Introduction

Solar cells based on the CIGS absorber have reached the 22,6% efficiency in the lab, by using the alkali metal atoms PDT treatment.[1] The positive effect of sodium and potassium incorporation in the CIGS absorber layer has been shown in many recently published papers [1, 2, 3]. The last investigation by Jackson et al.[1] showed clear evidence that the addition of heavier alkali (such as Rb and Cs) with light alkali (Na and K) gives the highest efficiency solar cells.

In spite of the remarkable increase in the CIGS solar cell efficiency, the mechanism of the alkali metal effect on absorber layer microstructure is still debated. It is well known, that Na improves the device efficiency by increasing the p-type conductivity of the absorber layer. [2] One explanation is that Na impurities behave as acceptors in CIGS materials. However, previous formation energy calculations of Na impurities in the bulk of CIGS do not confirm this suggestion. The incorporation of Na increases also the open-circuit voltage and the fill factor. [3] K contributes by improving the surface morphology, which allows thinner CdS layers with smaller optical losses. [3] Rb and Cs PDT treatment leads to an improved diode quality. [1] However, in spite of the extensive theoretical and experimental research the microscopic origin of the efficiency improvement is still unknown.

The PDT process results in formation of alkali metal impurities inside the CIGS grain interiors (GIs). Near grain boundaries (GB) and near the surface there exists also Cu-depleted regions and alkali secondary phases.[2, 4] In the present work, we studied two different regions of the CIGS absorber layer, i.e., the bulk GI and the GB region, as well as the p-n junction near the GICS – CdS buffer-layer interface.

Alkali metal atoms in GI regions may act as impurities or as constituents of secondary phases. We have calculated formation energies and migration barriers for Na, K, Rb, and Cs impurities in CuInSe$_2$ (CIS) within the framework of the density-functional theory implemented in the VASP program package. [5] In our presentation, we will discuss the interplay between the alkali metal impurities and the native point defects in CIS. Moreover, the most important parameters of the secondary phases, such as lattice constants, band gaps, and heats of formation have been calculated. Mechanisms for secondary phase formation in CIGS will be discussed.

Due to low free energies, alkali metal atoms accumulate preferably at GBs and near the CIGS surface resulting in formation of secondary phases. We have calculated formation energies of alkali metal atoms near different types of GBs close to the surface and we will discuss the prominent trends between different alkali metal atoms.

II. Computational Details

The calculation were carried out with the density functional theory implemented in the program package VASP.[5] In order to avoid the band gap problem, the hybrid HSE06 exchange-correlation functional was used. The portion of the Hartree-Fock exchange, α, and the range-separation parameter ω were kept as the default values. The projector augmented-wave (PAW) method [6] was employed with a cut-off energy of 450 eV for the plane-wave basis set.

The impurity formation energies for alkali metal atoms were derived from the equation [25]

$$E_{f,i} = E_{tot,i}^{def} - E_{tot}^{bulk} - \sum_j n_j \mu_j + q E_f + E_{corr},$$

where $E_{tot,i}^{defect}$ is the total energy of a supercell with the alkali metal impurity of type i, $E_{tot,i}^{bulk}$ the total energy of the corresponding pristine supercell, μ_j the chemical potential of the atom of type j, q the charge state of the impurity, E_f the Fermi level and E_{corr} is a correction due to artificial electrostatic interactions when using the supercell approximation. Defects were modeled using a 128-atom supercell, which is sufficiently large to prevent significant wave function overlap and also reduces the spurious electrostatic errors in the case of charged defects.[7]

The migration barriers were calculated by using the state-of-the-art CI-NEB method[8]. All of the migration barrier were obtained with the PBE exchange-correlation functional. It gives reasonable results with lower computational cost than the HSE06 functional.

III. RESULTS AND DISCUSSION

The PDT process may results in the formation of alkali metal impurities. Therefore, we investigated properties of the alkali metal impurities by calculating formation energies and migration barriers. The specific chemical conditions can be taken into account by using different sets of chemical potentials. CIGS solar cell devices with the highest efficiencies are usually Cu-poor and grown under the selenium atmosphere. In the present work we have chosen the same chemical conditions as in experiment (point A in Figure 1a)). It corresponds to Se-poor, Cu-rich, In-rich, and Alk-rich conditions.

The alkali metal chemical potentials were obtained from the heat of formation of $AlkInSe_2$. The conditions for Cu, In, Se are taken from the top-right corner (Cu-rich, Se-poor) of the $CuInSe_2$ stability area:

$\Delta\mu_{Se} = 0$,
$\Delta\mu_{In} = (H_f(In_2Se_3) - 3\Delta\mu_{Se})/2$,
$\Delta\mu_{Cu} = H_f(CuInSe_2) - \Delta\mu_{In} - 2\Delta\mu_{Se}$,
$\Delta\mu_{Alk} = H_f(CuInSe_2) - H_f(AlkInSe_2) + \Delta\mu_{Cu}$.

All the different alkali metal atoms prefer to accumulate in the Cu sublattice for the chemical potential set considered and form simple point defects or their complexes. A low formation energy for alkali metal impurities on Cu site can be associated with the weak covalent bond between Cu and Se. It should be pointed out that all alkali metal atoms in the Cu sublattice are always neutral and do not affect the charge carrier concentrations.

Alkali metal atoms on the In site in CIS are donors and may influence the Fermi level position. These defects produce two transition levels in the band gap and may act as electron traps, which can be detrimental to device properties. Due to the -1 and -2 charge states the formation energies of these defects decrease very rapidly with the increasing Fermi energy. Here our results are in contrast with the previous results by Oikkonen et al. and Ghorbani et al. [9, 10] They observed that Na_{In} produces just one transition level or has only one stable charge stage. The difference can be explained by different supercell sizes and electrostatic correction schemes used.

The migration barrier for Li is much higher than those for the other considered alkali metal atoms. The obtained value of 0.62 eV is in a good agreement with the result by Maeda et al. [11] The migration barrier for K is smaller by 0.12 eV than that for Na. A similar trend was observed by Maeda et al., [11]. The migration barrier for Na is 0.28 eV which fits well to the previous theoretical results by Oikkonen et al. [9] and Maeda et al., [11] as well as to the experimental results. [12] Rb has a migration barrier slightly larger than that for K, but still smaller than those for Na and Cs, i.e., 0.28 and 0.37 eV, respectively. For Li and Na the migration barriers are wider than for all the other alkali metal elements. The extremely low migration barrier for K and low formation energy for the K-Cu vacancy complex leads to a low activation energy for K diffusion.

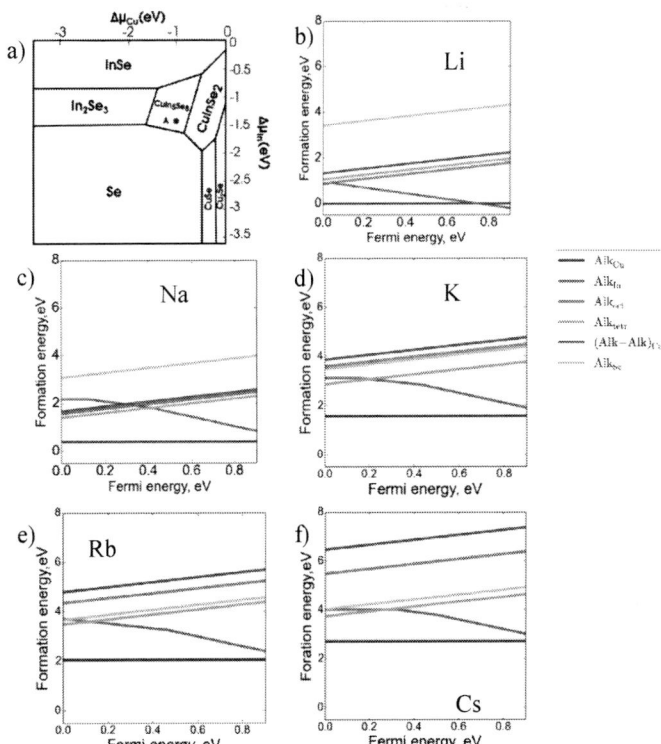

Fig. 1.a- Chemical potential diagrams for $CuInSe_2$ .Formation energies for alkali metal impurities, b,c,d,e,f - the formation energies for Li, Na, K, Rb, and Cs , respectively. The formation energies corresponding to the points A, in chemical potential diagram(a).

The extremely low migration barrier for K and low formation energy for the K-Cu vacancy complex leads to a low activation energy for K diffusion. The activation energies for Li, Na and K are smaller than for other alkali, therefore it will diffuse into the CIGS layer and accumulate on the copper sublattice. Since the concentration of each alkali element does not exceed 0.1 % the diffusion is not limited by the concentration of the Cu-vacancy which is considerably higher in the Cu deficient region. However, the diffusion is limited by the size of the Cu depleted region. In a stoichiometric region diffusion of alkali elements will be blocked due to negligible concentration of Cu vacancies and a high migration barrier for Cu. [9] Due to this fact Li, Na and K will accumulate near the boundary between the copper depleted region and the almost-stoichiometric CIGS. The concentration profile of the alkali elements measured by Jackson et al. [1] supports our findings.

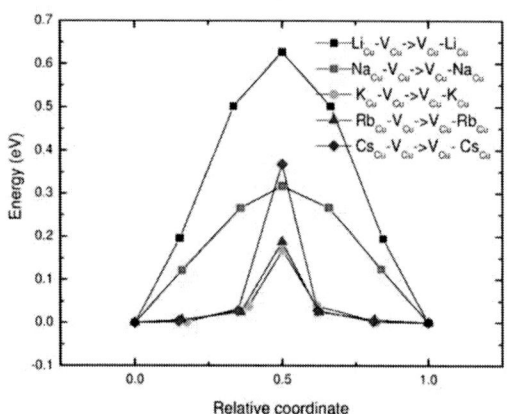

Fig. 2. Migration barriers for Li, Na, K, Rb, and Cs, within the vacancy diffusion mechanism. The PBE functional has been used in the CI-NEB calculation.

IV. CONCLUSIONS

Li, Na, and K can diffuse into the material and accumulate on the copper sublattice thanks to their high mobility. Na in the copper sublattice is an electrically neutral defect and does not affect the electronic properties of the CIGS absorber. However, Li and Na can also replace indium and in this case it is a donor defect.

The most favorable positions for all alkali metals atoms are on the copper sublattice where they are electrically neutral throughout the band gap. Due to high formation energies, concentrations of heavier alkali metal atoms on indium and selenium sites as well as concentrations of interstitial defects are negligible. All alkali metal atoms can diffuse via the vacancy mechanism in the absorber layer, however, Li, Na and K can also use interstitial mechanism. The distribution of alkali metal atoms inside the material may be explained by their different activation energies.

ACKNWOLEDGEMENTS

This work has received funding from the European Union's Horizon 2020 research and innovation programme under grant agreement no. 641004 (Sharc25).

REFERENCES

[1] P. Jackson, R. Wuerz, D. Hariskos, E. Lotter, W. Witte, and M. Powalla, physica status solidi (RRL) Rapid Research Letters , 10, 583 (2016).

[2] P. Reinhard, B. Bissig, F. Pianezzi, Chemistry of Materials 27, 5755 (2015).

[3] F. Pianezzi, P. Reinhard, A. Chirila, Phys. Chem. Chem. Phys. 16, 8843 (2014).

[4] P. Jackson, D. Hariskos, R. Wuerz, O. Kiowski, A. Bauer, T. M. Friedlmeier, and M. Powalla, physica status solidi (RRL) Rapid Research Letters 9, 28 (2015).

[5] G. Kresse and J. Furthm¨uller, Phys. Rev. B 54, 11169(1996).

[6] P.E.Blöhl. Phys.Rev. B 50, 17953 (1994).

[7] C. Freysoldt, J. Neugebauer, and C. G. Van de Walle, physica status solidi (b) 248, 1067 (2011).

[8] G. Henkelman, H, Jönnsson. The Jornal of Chemical physics 2000,113.

[9] L. E. Oikkonen, M. G. Ganchenkova, A. P. Seitsonen, and R. M. Nieminen, Journal of Physics: Condensed Matter 26, 345501 (2014).

[10] E. Ghorbani, J. Kiss, H. Mirhosseini, G. Roma, M. Schmidt, J. Windeln, T. D. Khne, and C. Felser, The Journal of Physical Chemistry C 119, 25197 (2015).

[11] T. Maeda, A. Kawabata, and T. Wada, Japanese Journal of Applied Physics 54, 08KC20 (2015).

[12] A. Laemmle, R.Wuerz, T. Schwarz, O. Cojocaru-Mirdin, P.-P. Choi, and M. Powalla, Journal of Applied Physics 115, 154501 (2014).

Exploring silicon carbide- and silicon oxide-based layer stacks for passivating contacts to silicon solar cells

P. Löper[1], G. Nogay[1], P. Wyss[1], M. Hyvl[2], P. Procel[3], J. Stuckelberger[1], A. Ingenito[1], I. Mack[1], Q. Jeangros[1], M. Ledinsky[2], A. Fejfar[2], C. Allebé[4], J. Horzel[4], M. Despeisse[4], F. Crupi[3], F.-J. Haug[1], C. Ballif[1,4]

[1] École Polytechnique Fédérale de Lausanne (EPFL), Institute of Microengineering (IMT), Photovoltaics and Thin-Film Electronics Laboratory, Rue de la Maladière 71b, 2002 Neuchâtel, Switzerland

[2] Laboratory of Nanostructures and Nanomaterials, Institute of Physics, Academy of Sciences of the Czech Republic, v. v. i., Cukrovarnická 10, 162 00 Prague, Czech Republic

[3] Department of Computer Engineering, Modeling, Electronics and Systems Engineering, University of Calabria, Rende, 87036, Italy

[4] CSEM SA, PV-Center, Jaquet-Droz 1. 2002 Neuchâtel, Switzerland

Abstract — We present the development of passivating contacts for high-efficiency silicon solar cells using silicon oxide (SiO_x) and silicon carbide (SiC_x)-based layers. We discuss a comprehensive optimization of a SiC_x-based passivating hole contact reaching implied open circuit voltages >715 mV. In addition, we introduce a passivating hole contact based on nanocrystalline SiO_x (nc-SiO_x) targeting compatibility with higher process temperatures as well as increased optical transparency for front side application. First planar test devices employing nc-SiO_x-based passivating contacts for both charge carrier types are presented, yielding short-circuit current densities >34 mA/cm^2 and fill factors >78%, showing that efficient current extraction is possible despite the added SiO_x phase and also indicating potential optical advantages of the concept.

I. INTRODUCTION

Passivating contacts prepared as a stack of a thin silicon oxide layer capped with a doped silicon layer, also called poly-Si, POLO or TOPCON contacts or junctions, yield low recombination current densities and contact resistivities and have led to conversion efficiencies of 25.3 % [1]. While best performance is shown with electron-selective contacts doped with phosphorous, remarkable results have been achieved also with hole-selective contacts [2, 3].

Besides ground-breaking efficiencies in the laboratory, passivating contacts are also interesting candidates for future industrial solar cell concepts as their fabrication is compatible with conventional high-temperature processes of industrial c-Si solar cells, such as phosphorous diffusion. Passivating *hole* contacts are thus interesting candidates for an 'upgrade' over PERC-type rear side passivation technology' on p-type wafers. For such a solar cell, the rear passivating hole contact would either be grown prior to front side phosphorous diffusion and then be co-diffused, or after diffusion and annealed in a subsequent step such as firing of the front side Ag paste. Ideally, for co-diffusion, the passivating contact layer stack would be applied by a single-sided technique. Silicon-rich silicon carbide (SiC_x) layers deposited by plasma-enhanced chemical vapour deposition (PECVD) present an attractive material to fulfil these requirements.

In the first part of this paper, we present recent developments towards a hole-selective rear contact based on silicon-carbide [4].

In the second part of this paper, we present developments towards improved thermal stability using boron-doped silicon oxide (SiO_x). In addition to being compatible with a higher thermal budget, silicon oxide layers furthermore present an approach to render Si-based passivating contacts more transparent. We present preliminary surface passivation results achieved with boron-doped SiO_x-based contacts. In addition, we investigate both the SiO_x(p)-based contact introduced here as well as the SiO_x(n)-based contact presented previously [5] in solar cell test devices.

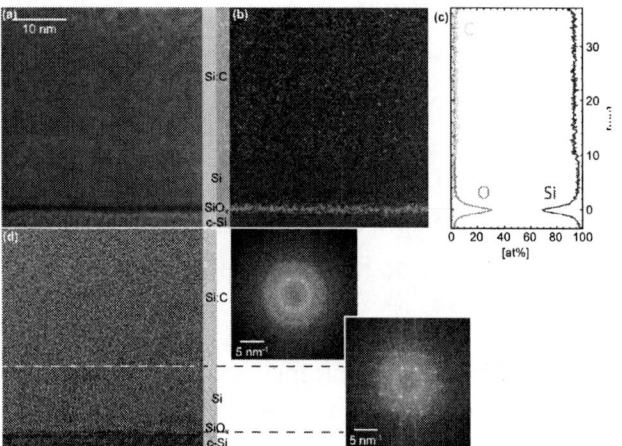

Fig. 1. Transmission electron micrographs of the chem-SiOx/Si(i)/SiCx(p) stack. (a) STEM HAADF image, (b) EDX map, (c) corresponding EDX line scan and (d) HRTEM micrograph along with Fourier transforms calculated at the positions of the Si buffer layer (bottom) and the SiCx(p) layer (top). Reprinted with permission from [2]. Copyright 2016 American Chemical Society.

II. BORON-DOPED SILICON CARBIDE REAR CONTACT

Passivating contacts consisting of a thin chemically prepared silicon oxide (denoted chem-SiOx hereinafter) and boron-doped SiCx layers deposited by PECVD were prepared on shiny-etched 2 Ωcm p-type FZ wafers as described in detail in [4]. In a first set of experiments we explored the annealing conditions and observed that the chem-SiOx layer disappeared after the layer stack had been annealed at 900°C. This motivated us to insert an additional, undoped Si inter layer between the chem-SiOx and the doped SiCx layer to suppress a chemical reaction of the chem-SiOx layer with the SiCx(p) layer. The full structure Si wafer/chem-SiOx/Si/SiCx(p) is shown in transmission electron microscopy (TEM) images in Fig. 1. The layer sequence is well visible in the STEM-HAADF as well as in the HRTEM and the EDX image. The

effect of the Si interlayer thickness on passivation is shown in Fig. 2 for annealing at 800°C for 5 min dwell time. Without Si inter layer, the open-circuit voltage (V_{OC}) is only 685 mV, but it is boosted to 700 mV with the addition of a 2 nm Si inter layer and attains a maximum for a 10 nm inter layer at 707 mV. In a second experiment, we explored doping and annealing conditions further and optimized the selectivity of the contact until reaching 718 mV and a contact resistivity of 17 mΩcm2 [6]. Finally, we implemented the SiCx(p)-based passivating hole rear contact in first test devices, achieving promising efficiencies of up to 21.9% so far [6].

II. MIXED-PHASE SILICON OXIDE/SILICON CONTACTS

To combine a passivating hole rear contact with a phosphorous-diffused front side, a wider process window with higher peak temperatures and longer dwell times than discussed above for the SiCx(p) contact would facilitate device integration. To this end, we use nanocrystalline silicon oxide (nc-SiOx) prepared by PECVD as presented in [5]. The nc-SiOx layer is deposited on the chem-SiOx layer, and capped with a doped nanocrystalline Si (nc-Si) layer. More in detail, we investigate boron-doped nc-SiOx/Si(p) deposited by PECVD from mixtures of carbon dioxide (CO_2), silane (SiH_4) and hydrogen (H_2). In a first step we prepared symmetric lifetime samples on 10 Ωcm p-type FZ wafers and investigate surface passivation with photoconductance decay (PCD) measurements. Fig. 3 depicts the dependence of the saturation current density on the CO_2 flow and the annealing dwell time for a dwell temperature of 900°C. Between 0.25 and 0.75 relative CO_2 flow passivation improves strongly, but decreases slightly for even higher CO_2 flows. Annealing in forming gas (FGA, 4% H_2 in N_2) at 450°C yields a clear improvement. Optimum performance for the stack on a planar surface is reached for annealing 180 min at 900°C followed by a FGA, reaching a saturation current density of 33 fA/cm².

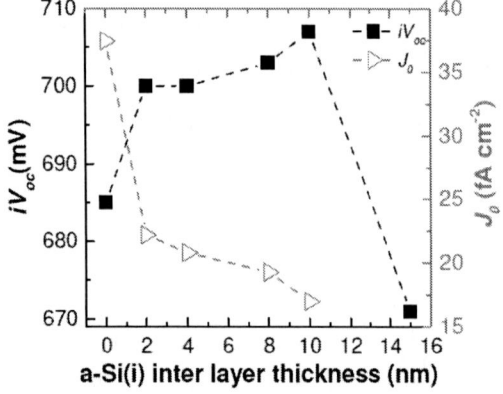

Fig. 2. Impact of the Si inter layer thickness on implied open circuit voltage (iVOC) and emitter saturation current density (J_0). Reprinted with permission from [2]. Copyright 2016 American Chemical Society.

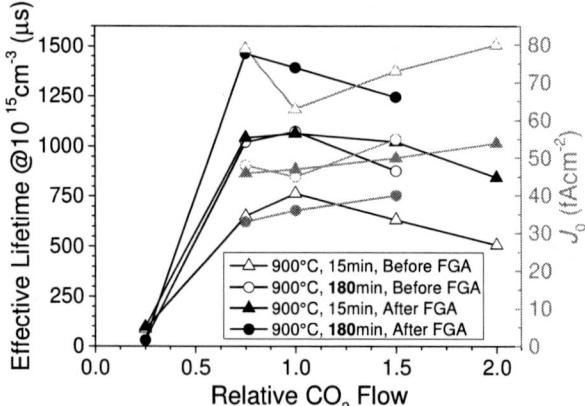

Fig. 3. Effective lifetime and saturation current density of 10 Ωcm p-type FZ wafers passivated with a chem-SiOx/nc-SiOx/Si(p) stack on both sides and annealed at 15 min or 180 min at 900°C.

978-1-5090-5606-4/17 $31.00 © 2017 IEEE

In addition to supporting a higher thermal budget than SiCx-based layers, SiO_x-based layers do also offer the advantage of a wider band gap and a lower refractive index than attainable with SiC_x layers. In a previous study we presented results on $SiO_x/Si(n)$-based contacts, targeting highly-transparent full-area passivating contacts. In order to exploit the optical potential of the $SiO_x(n)$ and $SiO_x(p)$ layers and to characterize them at device level, we implemented them in proof-of-concept solar cells. In a first batch, the active cell area was defined by the area of the ITO electrodes prepared by sputtering through a shadow mask, but the SiO_x-based contacts were applied to the full wafer area. The proof-of-concept solar cells show higher V_{OC} values on n-type wafers than on p-type wafers, likely due to the higher bulk lifetime. The most pronounced effect is observed for the short circuit current (J_{SC}) with the $SiO_x(n)$-based front side showing 2.4 mA/cm2 higher current than with $SiO_x(p)$ front side, independent of whether the layer is applied as front surface field or front emitter. The higher J_{SC} attained with the $SiO_x/Si(n)$-based layers compared to the $SiO_x/Si(p)$-based layers indicates the higher transparency of the former. Remarkable fill factors of up to 80% are achieved, demonstrating efficient charge carrier extraction by the mixed-phase SiO_x/Si-contacts.

IV. CONCLUSION

Layer stacks based on SiC_x and SiO_x were investigated as passivating contacts for silicon solar cells. Excellent passivation was shown with a chem-$SiO_x/Si(i)/SiC_x(p)$ layer stack with promising iV_{OC}>718 mV and efficiencies >21.9% obtained with proof-of-concept devices. To achieve a broader process window, passivating contacts based on boron-doped mixed-phased SiO_x/Si layer stacks were investigated. Good passivation with J_0=33 fAcm^{-2} was shown. To investigate optical properties and charge carrier transport, the SiOx-based contacts were implemented in co-annealed planar proof-of-concept devices. One side of the planar proof-of-concept solar cells was contacted with a p-type Boron-doped, and the other side contacted with an n-type phosphorous-doped SiO_x/Si layer stack. With the $SiO_x(n)$ employed as full area front side contact, J_{SC}>34mAcm^{-2} and FF>78% was obtained, showing that efficient current extraction is possible despite adding the optically transparent SiO_x phase. In a next step we will transfer this technology to the front side of textured solar cells.

REFERENCES

[1] S. W. Glunz , F. Feldmann, A. Richter, M. Bivour, C. Reichel, H. Steinkemper, *et al.*, in *Proceedings of the 22nd European Photovoltaic Solar Energy Conference*, Hamburg, Germany, 2015, pp. 259-63.

[2] F. Feldmann, M. Bivour, C. Reichel, H. Steinkemper, M. Hermle, and S. W. Glunz, *Solar Energy Materials and Solar Cells,* vol. 131, pp. 46-50, 2014.

[3] U. Römer, R. Peibst, T. Ohrdes, B. Lim, J. Krügener, E. Bugiel, *et al.*, *Solar Energy Materials and Solar Cells,* vol. 131, pp. 85-91, 2014.

[4] G. Nogay, J. Stuckelberger, P. Wyss, Q. Jeangros, C. Allebe, X. Niquille, *et al.*, *ACS Appl Mater Interfaces,* vol. 8, pp. 35660–35667, Dec 13 2016.

[5] J. Stuckelberger, G. Nogay, P. Wyss, Q. Jeangros, C. Allebé, F. Debrot, *et al.*, *Solar Energy Materials and Solar Cells,* vol. 158, pp. 2-10, 2016.

[6] G. Nogay, J. Stuckelberger, P. Wyss, E. Rucavado, C. Allebe, T. Koida, *et al.*, *Solar Energy Materials & Solar Cells,* 2017, under review.

Efficient electron contacts for *n*-type silicon solar cells using Magnesium metal, oxide, and fluoride

Yimao Wan[1], Chris Samundsett[1], James Bullock[2], Di Yan[1], Thomas Allen[1], Jun Peng[1], Jie Cui[1], Mark Hettick[2], Ali Javey[2] and Andres Cuevas[1]

[1]Research School of Engineering, Australian National University, Canberra, ACT 0200 Australia

[2]Department of Engineering and Computer Science, University of California, Berkeley, CA 94720, USA

Abstract — **The simplicity and low thermal budget of dopant-free, selective carrier contact materials is opening up new possibilities for low-cost, high performance silicon photovoltaics. This paper describes recent progress in the exploration of magnesium and related materials such as magnesium oxide and fluoride as electron-selective contacts to lightly doped *n*-type c-Si. A direct comparison between the three materials permits to identify significant differences in transport and recombination behaviour. All three are incorporated in *n*-type silicon solar cells to demonstrate their actual performance at the device level. In particular, a simple aluminium electrode, functionalized with a nanoscale Mg-based layer, significantly enhances the performance of n-type c-Si solar cells, to a conversion efficiency of ~20%, equivalent to the performance of the standard p-type silicon solar cells with an alloyed Al full-area hole contact.**

I. Introduction

Crystalline silicon (c-Si) has been dominating worldwide photovoltaic (PV) production for decades, with a global market share of around 93%, making it unequivocally the most important PV technology of all time. The majority of commercialised c-Si PV devices are based on a simple solar cell architecture—a *p*-type c-Si wafer with a front phosphorus diffusion and a full-area aluminium (Al) alloyed rear back surface region. The success of this architecture is largely due to the simple, low-cost formation of a highly doped p^+ region upon alloying, which leads to a low contact resistance for hole transport and a moderate level of recombination suppression at the rear surface. When Al is directly deposited on *n*-type c-Si, however the contact behaves in a rectifying fashion, even if it is not alloyed with the silicon, and results in a high contact resistance, despite the small difference (~0.1–0.2 eV) that exists between the Al work function and the electron affinity of silicon and the consequently low barrier height predicted by the Schottky-Mott rule. This behaviour is widely attributed to the Fermi-level pinning phenomenon, consequence of a high density of bandgap states or defects at the metal/semiconductor interfaces, which leads to a relatively high Schottky barrier height (Φ_B) of ~0.65 eV. This in turn hinders the flow of electrons out of the *n*-type silicon wafer.

The approaches for addressing this problem can be inferred by looking into the dependence of contact resistivity ρ_c, on the Schottky barrier height Φ_B, and the surface doping

concentration of the semiconductor N_d, which is given by $\rho_c \propto \exp\left(\frac{\Phi_B}{\sqrt{N_d}}\right)$. Historically, an Ohmic contact to *n*-type silicon wafers has been achieved by means of heavy phosphorus doping at the surface of the solar cells (i.e., increasing N_d) via thermal diffusion or plasma-assisted deposition of doped silicon layers. Despite its success in producing record-efficiency silicon solar cells, doping usually creates process complexity and requires a high temperature, in excess of 800 °C for the thermal diffusion of dopants or for the re-crystallization of deposited silicon layers [1]. Noxious and air-reactive gases are normally used in silicon heterojunction solar cell technology to introduce dopants in hydrogenated amorphous silicon layers deposited by PECVD [2].

Another obvious approach to reducing ρ_c is to reduce Φ_B. One straightforward technique to reduce Φ_B for electron transport is the utilization of a metal layer with a very low work function, such as calcium or magnesium. The barrier height according to the Schottky-Mott rule would be negative; in practice the measured value is still relatively low of ~0.35 eV on *n*-type c-Si. An alternative, or complementary, technique is the de-pinning of the Fermi-level by inserting an interfacial layer between the outer metal electrode and the inner silicon absorber. The interlayer functions as a passivating layer to reduce the density of states/defects at the metal/silicon interface while being conductive enough to allow significant transport of carriers through it. One or several of the following interlayer properties are desirable for achieving a low contact resistivity: (*i*) low, or negative, conduction band offset to c-Si, (*ii*) low bulk resistivity, which is associated with tunnelling effective mass and bulk carrier concentration, and (*iii*) possible capability of modifying (reducing) the overall work function of the outer metal layer. Based on abovementioned conditions, this work reports the progress in exploring magnesium related layers—metal (Mg) [3], oxide (MgO$_x$) [4] and fluoride (MgF$_x$) [5]—as electron-selective contacts to *n*-type c-Si by investigating (*i*) the contact properties, (*ii*) solar cell performance, and (*iii*) device thermal stability.

II. Results and Discussion

A. Contact resistivity

The electrical contact behaviour was evaluated by measuring the contact resistivity ρ_c using the method devised by Cox and Strack. Figure 1(a) shows a series of

978-1-5090-5606-4/17 $31.00 © 2017 IEEE

representative *I–V* measurements of samples with different contact layers on *n*-type c-Si. Note that the *I–V* measurements of the five groups of samples were performed on the same pad area (~1.96 cm^2) for direct comparison. As can be seen, the sample with Al directly on *n*-type c-Si exhibits somewhat Ohmic contacts for large area pads, but then turns to rectifying behaviour when pad areas become smaller (not shown here, see Ref [MgO$_x$]). A more severe rectifying contact behaviour was observed for sample with Al on a-Si:H passivated c-Si. A high contact resistance, or rectifying behavior, between the Al metal and *n*-type c-Si is presumably attributable to the presence of a large surface potential barrier known to exist at this surface. By contrast, the insertion of nanoscale Mg-related thin film improves dramatically the contact behavior, facillitating an Ohmic contact (i.e., linear *I–V* curve) between the Al electrode and the *n*-type c-Si substrate.

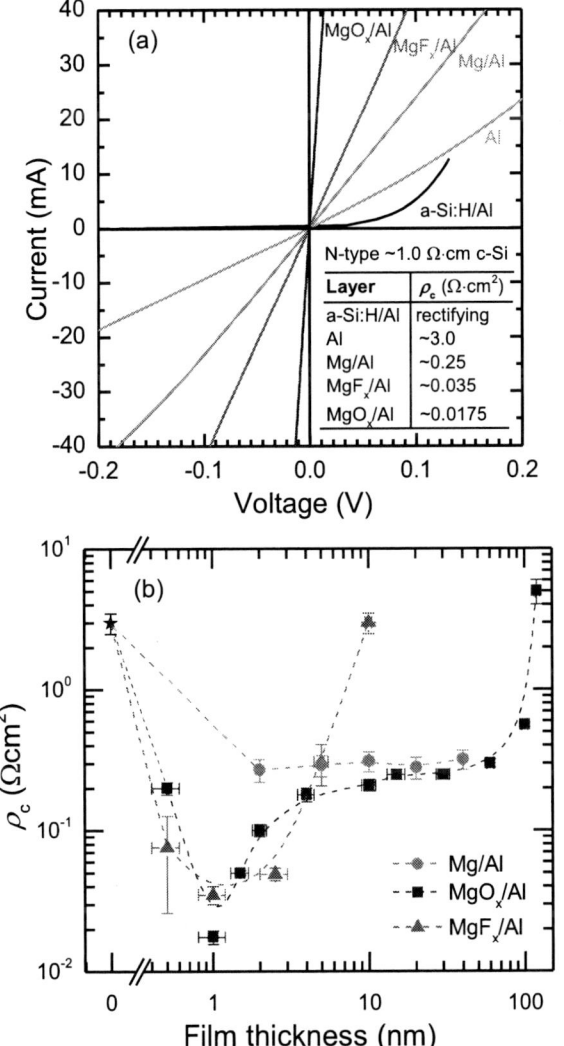

Figure 1: (a) A series of representative *I–V* measurements of samples with different contact layers on *n*-type c-Si. (b) Dependence of contact resistivity ρ_c on film thickness of Mg, MgO$_x$ and MgF$_x$.

The mechanisms underlying the low resistance for

electron transport provided by the three films (Mg, MgO$_x$ and MgF$_x$) seem significantly different. Whilst the higher electron conductivity provided by the Mg/Al contact structure is mainly attributable to the lower work function (~3.7 eV) of Mg metal, the high electron conductivity provided by the MgF$_x$/Al contact structure can be attributed to (*i*) a low work function of ~3.5 eV, as revealed by the XPS measurements, and/or (*ii*) electron tunneling through a reduced barrier width. XPS measurements of MgO$_x$/Al contact revealed a work function of ~4.1 eV, which is comparable to that of Al metal, but interestingly, they also showed a substantial substoichiometry (x=0.75) of the MgO$_x$ film; Hall effect measurements indicated that it has a high electron concentration and a high conductivity, hence a metallic behaviour. In addition, the low resistance for electron transport provided by the MgO$_x$/Al contact structure is also likely assisted by the release, or unpinning, of the Fermi-level, by passivating a significant number of interface states, as evidenced by an increased V_{OC} of the corresponding solar cell (see Figure 2).

Furthermore, the three Mg-related interlayers show different behaviours of ρ_c contingent upon film thickness. Whilst there is little dependence of ρ_c on Mg metal thickness, we can observe a strong impact of MgO$_x$ and MgF$_x$ thickness on ρ_c. It can be seen that ρ_c first decreases dramatically as the MgO$_x$ or MgF$_x$ thickness increases from 0 to 1 nm, and then increases slowly for MgO$_x$ but sharply for MgF$_x$. It is interesting to note that the contact is still Ohmic even when the MgO$_x$ thickness reaches 60 nm, consistent with the high conductivity and high electron concentration in this measured film. In contrast, when the MgF$_x$ thickness exceeds 5 nm, the contact behaves in a rectifying fashion, which implies that the MgF$_x$ bulk film conductivity is significantly lower than that of the MgO$_x$ film.

B. Solar cell performance

The *J–V* photovoltaic characteristic curves under one sun standard illumination are plotted in Figure 2(b) and associated electrical parameters are summarized in Table I. As can be seen, the cell with Al directly on *n*-type Si exhibits a low efficiency mainly due to poor V_{OC} and *FF*. On the other hand, the insertion of a 1 nm thick MgO$_x$ layer enhances substantially all cell parameters, leading to an efficiency of 20%. Further, Al directly on a-Si:H passivated *n*-Si (i.e., without Mg) rear side behaves as a severe Schottky diode that opposes the photovoltaic effect of the front *pn* junction, leading to a typical S-shape *J–V* curve, and resulting in a conversion efficiency of only 2.2%. In contrast, devices with the insertion of a Mg or MgF$_x$ layer between the a-Si:H and the Al show significant improvement in cell performance, exhibiting conversion efficiencies of 19% and 20.1%, respectively.

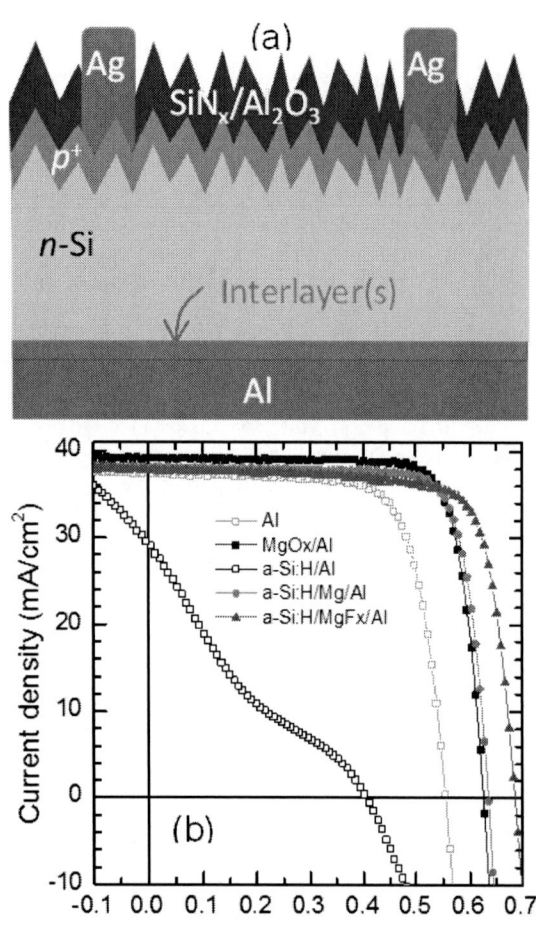

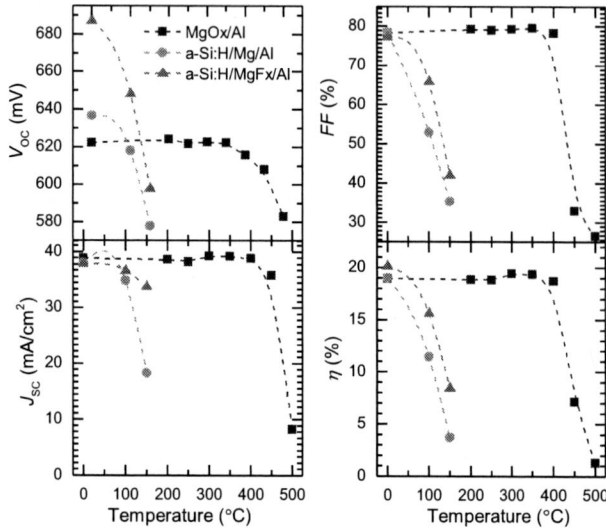

Figure 3: Electrical parameters as function of annealing temperature in air ambient for 10 minutes.

Figure 2: (a) Schematic of *n*-type front-junction silicon solar cells and (b) associated *J-V* curves of several types of solar cells with different full-area rear electron contacts.

C. Device thermal stability

While the Mg-related electron-selective contacts enhance *n*-type c-Si solar cells' performance, their thermal stability remains an important consideration for device longevity and module encapsulation. Cells with the three interlayers were annealed in air for 10 minutes on a hotplate for temperature ≤ 150 °C and in a thermal furnace for temperature ≥ 200 °C. It can be seen that the performance of cells with Mg or MgF$_x$ interlayers deteriorates conspicuously upon annealing even at 100 °C, possibly due to alloying of Al with Mg or MgF$_x$ and further migration into the a-Si:H layer. On the othe hand, the cell with a MgO$_x$ interlayer is essentially stable up to 400 °C, and starts deteriorating only when the temperature increases further. The result implies the MgO$_x$ is preferable in terms of stability. Nevertheless, further optimisation of the passivating qualities of this material is required.

III. CONCLUSION

This paper described recent progress in the exploration of magnesium and related materials such as magnesium oxide and fluoride as electron-selective contacts to lightly doped n-type c-Si. A direct comparison between the three materials permits to identify significant differences in transport and recombination behaviour. All three are incorporated in *n*-type silicon solar cells to demonstrate their actual performance at the device level. In particular, a simple aluminium electrode, functionalized with a nanoscale Mg-based layer, significantly enhances the performance of *n*-type c-Si solar cells, to a conversion efficiency of ~20%, equivalent to the performance of the standard p-type silicon solar cells with an alloyed Al full-area hole contact.

References

[1] S. Glunz, F. Feldmann, A. Richter, M. Bivour, C. Reichel, H. Steinkemper, *et al.*, "The irresistible charm of a simple current flow pattern - approaching 25% with a solar cell featuring a full-area back contact," in *31st European Photovoltaic Solar Energy Conference and Exhibition*, Hamburg, Germany, September, 2015.

[2] K. Masuko, M. Shigematsu, T. Hashiguchi, D. Fujishima, M. Kai, N. Yoshimura, *et al.*, "Achievement of More Than 25% Conversion Efficiency With Crystalline Silicon Heterojunction Solar Cell," *Photovoltaics, IEEE Journal of,* vol. 4, pp. 1433-1435, 2014.

[3] Y. Wan, C. Samundsett, D. Yan, T. Allen, J. Peng, J. Cui, *et al.*, "A magnesium/amorphous silicon passivating contact for n-type crystalline silicon solar cells," *Applied Physics Letters,* vol. 109, p. 113901, 2016.

[4] Y. Wan, C. Samundsett, J. Bullock, M. Hettick, T. Allen, D. Yan, *et al.*, "Conductive and Stable Magnesium Oxide Electron-Selective Contacts for Efficient Silicon Solar Cells," *Advanced Energy Materials,* Accepted.

[5] Y. Wan, C. Samundsett, J. Bullock, T. Allen, M. Hettick, D. Yan, *et al.*, "Magnesium Fluoride Electron-Selective Contacts for Crystalline Silicon Solar Cells," *ACS Applied Materials & Interfaces,* vol. 8, pp. 14671-14677, 2016/06/15 2016.

Graded $(Al_zGa_{1-z})_xIn_{1-x}P$ Window-Emitter Structures for Improved Short-Wavelength Response

Jacob T. Boyer, Daniel L. Lepkowski, Daniel J. Chmielewski, Steven A. Ringel, Tyler J. Grassman

Ohio State University, Columbus, Ohio, 43201, USA

Abstract — $Ga_{0.51}In_{0.49}P$ solar cells with compositionally-graded $(Al_zGa_{1-z})_xIn_{1-x}P$ window-emitter structures, both lattice-matched and pseudomorphically (tensile) strained, were grown via molecular beam epitaxy using a digital alloy based grading approach. Preliminary results suggest that the resulting drift field and elimination of discrete interfaces near the top surface of these devices helps mitigate the impact of surface/interface recombination and sub-optimal window and/or emitter material quality. A significant increase (>10% absolute quantum efficiency) in carrier collection in the UV-blue ($\lambda < 500$ nm) was observed, along with a commensurate J_{SC} increase of ~10% relative versus standard, fixed-composition reference cells.

I. INTRODUCTION

In the development of multijunction solar cells with further increasing numbers of junctions, maximization of potential efficiency requires increasingly larger bandgap top cells and associated window layer materials. Proper partitioning of the solar spectrum into increasingly smaller bins increases the criticality of collecting short wavelength photons in the topmost cell, especially under AM0. Within the standard (i.e. non-nitride) III-V materials regime, achievement of top cells with direct gaps significant larger than the standard lattice-matched $Ga_{0.51}In_{0.49}P$ is only possible with $(Al_zGa_{1-z})_xIn_{1-x}P$ alloys, either through increasing relative Al content (z) or by shifting toward smaller lattice constants (x), both of which introduce materials quality difficulties.

Top cell improvement is also somewhat limited by choice of compatible window layer materials. The widest compatible bandgaps available belong to the $Al_xIn_{1-x}P$ alloy system. Wider direct gaps than that of the GaAs-matched composition (~2.6 eV) are achievable with increased Al fraction, but this comes with smaller lattice constants (and thus lattice mismatch), introducing growth and device design complexities. Additionally, these materials are partially limited by the ~2.4 eV indirect gap across the lattice constant range from $Al_{0.51}In_{0.49}P$ to pure AlP. Even a very thin layer is still a slight parasitic absorber for at least some fraction of the high energy photons.

A potential solution to these issues may lie in the use of a compositionally-graded combined window-emitter structure instead of the conventional abrupt heteroface design. Whereas conventional window/emitter heterostructures can be limited by non-negligible interfacial recombination velocities [1], a continuous (linear) compositionally graded layer would possess no discrete interfaces and should effectively eliminate such interfacial recombination loss currents. Additionally, because of the high doping typically used in window and emitter layers, Auger and enhanced Shockley-Read-Hall (SRH) recombination often significantly reduce the collection efficiency for photocarriers generated in this region, requiring careful device design to minimize the effects. In this case, a graded structure with a continuous increase in bandgap toward the surface should provide a drift field that would quickly sweep minority carriers away from the relatively defective region before they can undergo recombination. This drift field may also serve to help pull carriers away from the front surface, thereby also effectively reducing the window layer surface recombination velocity.

Further, a continuous (linear) grade also affords some ability to extend out to even wider bandgap pseudomorphically-strained (tensile) $Al_xIn_{1-x}P$ terminal ("window") compositions. By keeping the total strained-layer thickness low, and providing no actual interfaces where dislocations can easily form, it should be possible to prevent defect formation and further enhance the drift field strength. Overall, such a structure should significantly improve short- to mid-wavelength performance, allowing for enhanced collection of carriers generated not only in the "emitter" region, but also in the window itself.

The beneficiality of such graded structures within the AlGaInP materials system has been hinted at via time-resolved photoluminescence measurements on lattice-matched (to GaAs) [2] and prototype GaInP solar cells [3]. However, the structures used and comparisons made do not unequivocally resolve the actual improvements, nor do they elucidate the relative impacts of optical versus electronic effects. That is, it is not clear whether the cell benefits from improvement via reduction of interface recombination velocity, Auger and/or SRH loss, or via reduction in parasitic optical absorption, or both.

The work presented herein, based on two new designs using continuously graded structures either throughout the entire emitter, or through the entire emitter and window (see Figure 1) is intended to help provide some explanation regarding the two improvement mechanisms. These designs are also expected, based upon band structure and optical properties, to perform superior to the previously-reported designs due to an expected reduction of both window/emitter recombination and parasitic optical absorption. To test this concept, prototype structures were grown by molecular beam epitaxy (MBE) and fabricated devices yielded highly promising initial results, which are discussed herein. While these effects are strongly coupled together, we discuss the results in light of simple

978-1-5090-5606-4/17 $31.00 © 2017 IEEE

optical absorption calculations to elucidate the relative enhancement due to each mechanism: optical, bulk electronic, or interfacial electronic.

II. APPROACH

Several $Ga_{0.51}In_{0.49}P$ cells, lattice-matched to GaAs, were grown via MBE with different window and emitter designs, as outlined below and presented in Figure 1:

1) **Reference cells:** standard fixed-composition 20 nm $Al_{0.53}In_{0.47}P$ window and 50 nm $Ga_{0.51}In_{0.49}P$ emitter; grown at 0.5 µm/hr (sample **A**) and 1.0 µm/hr (sample **C**).

2) **Half-grade emitter cell (lattice-matched):** emitter graded from $Ga_{0.51}In_{0.49}P$ to $(Al_{0.5}Ga_{0.5})_{0.51}In_{0.49}P$, with a discrete heterostructure 20 nm $Al_{0.53}In_{0.47}P$ window (sample **B**); grown at 0.5 µm/hr.

3) **Full-grade window-emitter cell: (pseudomorphic)** combined window and emitter continuously graded from $Ga_{0.51}In_{0.49}P$ to lattice-matched $Al_{0.53}In_{0.47}P$ to pseudomorphic $Al_{0.75}In_{0.25}P$ (sample **D**); grown at 1.0 µm/hr.

Figure 1. Schematic device structures (not to scale) of the three cell designs considered in this work. Color gradients indicate compositional grading.

The $Ga_{0.51}In_{0.49}P$-based cells were grown on p-type GaAs(100) offcut 6° toward the nearest (111)A. Growth rates of either 0.5 µm/hr ("Half-Grade", **A** and **B**) or 1.0 µm/hr ("Full-Grade," **C** and **D**) were used; equal growth rate standard heterostructure window/emitter reference cells were grown for direct comparison. The slower growth rate was initially used as a remedy for concerns about excessive shutter transients, but following careful observations it was determined that such a precaution was unnecessary. The graded n^+-doped window-emitter structures were grown on nominally identical p-$Ga_{0.51}In_{0.49}P$ base layers with p^+-doped homostructure back surface field (BSF) layers. All samples were capped with an n^+-GaAs contact layer and annealed for 5 min under As_2 flux at 680°C immediately following completion of the growth.

Alloy compositions between the $Ga_{0.51}In_{0.49}P$ and $Al_{0.51}In_{0.49}P$ or $Al_{0.75}In_{0.25}P$ end points were achieved using a digital alloy approach based on variable-timed shuttering sequences (as

opposed to effusion cell temperature ramping). The maximum thickness of any layer (period) was < 5 nm, with most layers < 2 nm in order to maintain as close to bulk-like properties as possible (as opposed to superlattices) [4].

Devices were fabricated using standard optical lithography processes, including a BCl_3-based dry etch within an inductively-couple plasma reactive ion etcher for mesa (2 mm × 2 mm) isolation. Ti/Au and Ni/GeAu metal stacks were used for p- and n-type ohmic contacts, respectively. Front-side grid metal coverage was 8.67% and no antireflection coatings were applied; reflectance was 30-35% across the range of interest. Light current-voltage (LIV) measurements were performed under a single-zone simulated AM0 spectrum. Quantum efficiency (QE) and reflectance measurements were performed using a custom, microscope-based small spot system with lock-in detection.

III. RESULTS & DISCUSSION

The initial designs presented in Fig. 1 were considered in order to elucidate the relative contributions of increased optical transparency from wider bandgap $Al_xIn_{1-x}P$ window materials, the induced drift field due to sloped valence bands, and the elimination of interfaces. Material parameters for the AlGaInP system consisting of first-principles effective masses [5] and band offsets [6], as well as calculated bandgaps using standard tabulated parameters spanning the total range of alloy compositions were used to calculate 1D band diagrams via BandEng [7]. The calculated band diagrams of these structures are given in Figure 2.

The Half-Grade (**B**) emitter valence band is found to match that of the $Al_{0.53}In_{0.47}P$ at a composition of $(Al_{0.5}Ga_{0.5})_{0.51}In_{0.49}P$, which was the endpoint of the grade. At higher Al content the bandgap changes to the much flatter indirect X-band. This design thus ensures a continuously sloped valence band (i.e.

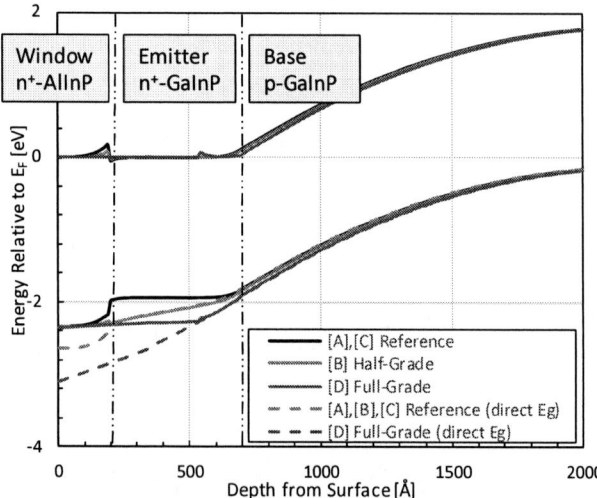

Figure 2. Simulated band diagrams for the structures studied in this work. When the indirect X-band is lower, the direct Γ-band is given in a dashed line for the pseudomorphic $Al_xIn_{1-x}P$, [D]; and LM AlInP, [A],[B],[C] windows.

978-1-5090-5606-4/17 $31.00 © 2017 IEEE

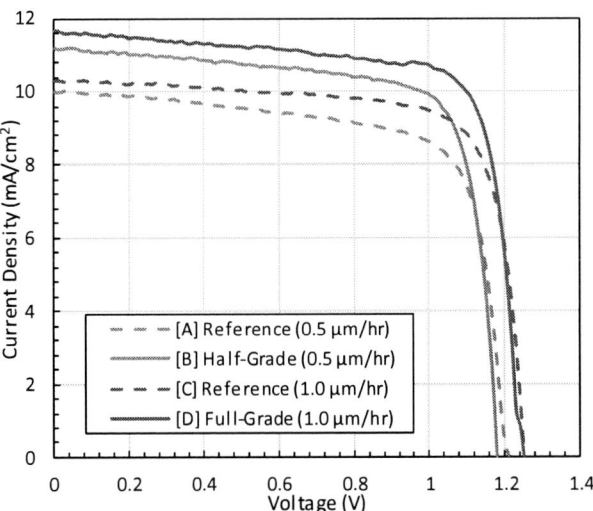

Figure 3. Representative LIV curves collected under simulated AM0 for samples [A]-[D] introduced in Figure 1. Corresponding metrics are tabulated in Table 1.

Table 1. Extracted AM0 LIV metrics corresponding to Figure 2.

Description	Growth Rate [μm/hr]	V_{OC} [V]	J_{SC} [mA/cm^2]	FF [%]
[A] Reference	0.5	1.19	9.99	72.5
[B] Half-Grade	0.5	1.18	11.19	75.6
[C] Reference	1.0	1.25	10.28	76.0
[D] Full-Grade	1.0	1.23	11.67	76.8

constant electric field) across the entire emitter layer, with little to no band offset at the $Al_{0.53}In_{0.47}P/(Al_{0.5}Ga_{0.5})_{0.51}In_{0.49}P$ interface. The Full-Grade **(D)** structure has a much steeper valence band slope up until ~20 nm above the *p/n* junction, with a relatively flat profile extending the rest of the way (~50 nm) to the surface. The direct bandgap, however, continues to increase as the Al-content in the window is increased, so the net optical absorption from this thicker 'window' will be reduced compared to the direct bandgap of the standard $Al_{0.53}In_{0.47}P$ window for the Reference and Half-Grade structures.

Figure 3 presents AM0 LIV curves collected from samples [A]-[D], with extracted metrics—V_{OC}, J_{SC}, and fill factor (FF)—provided in Table 1. Immediately observable are the significant improvements in J_{SC} for both graded structures relative to their associated reference cells, with the Half-Grade structure (sample **B**) showing an improvement of +1.2 mA/cm^2 (+12.0% relative) and the Full-Grade structure (sample **D**) at +1.4 mA/cm^2 (+13.5% relative). An improvement in FF of +3% absolute is observed for the Half-Grade cell over its reference, while the improvement for the Full-Grade device is just under +1%. Note that V_{OC} was found to improve by about 0.05 V for the 1.0 μm/hr versus the 0.5 μm/hr structures, and the FF by a small amount, as well, suggesting a difference in overall

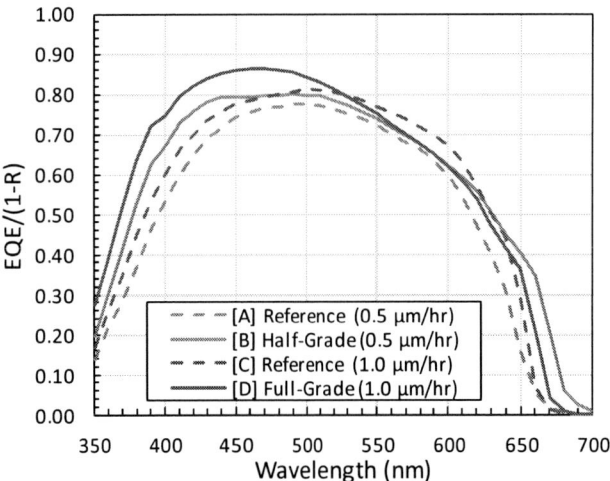

Figure 4. Reflectance-corrected QE curves on all cells [A]-[D]. Within each material growth rate category, the graded structure (solid) is compared to the reference cell (dashed).

material quality. However, the graded window-emitter structure cells showed little to no differences in V_{OC} (no more than 0.02 V) versus the comparable reference devices.

The significant increases in J_{SC} for both graded designs indicates an improvement in both carrier photogeneration and collection. The total magnitude of improvement is similar in each, so extracting the exact contribution from reduced window/emitter interface recombination velocity (IRV), the drift-field sweeping of the carriers out of the emitter, or the generally wider bandgaps is not trivial.

Figure 4 presents reflectance-corrected external quantum efficiency, or EQE/(1-R). While transmission is not accounted for here, at wavelengths shorter than ~600 nm, where full absorption is expected, this value is expected to be equivalent to internal quantum efficiency (IQE). It can be seen here that the Reference (**A**) and Half-Grade (**B**) cells show an overall reduced QE (as anticipated based on J_{SC}), most likely a result of the slow growth rate (0.5 μm/hr). The Full-Grade (**D**) and its associated Reference (**C**), which were grown at 1.0 μm/hr, show higher QE across the entire spectral range measured, indicating an overall higher material quality.

Clear from the QE curves of both graded structures, regardless of growth rate, is that the short wavelength response ($\lambda < 500$ nm) is significantly enhanced over the comparable reference cells, commensurate with the increased J_{SC} values extracted from the LIV measurements. Because all else is nominally equal versus each comparable reference cell, it is evident that the new window-emitter designs are the source of this difference. Even in the case of the Half-Grade structure (**B**), which still possesses a discrete physical heterostructure interface, the higher QE suggests that the field produced by the graded region is indeed helping to sweep carriers away from said interface; this observation is consistent previous reports [2].

978-1-5090-5606-4/17 $31.00 © 2017 IEEE

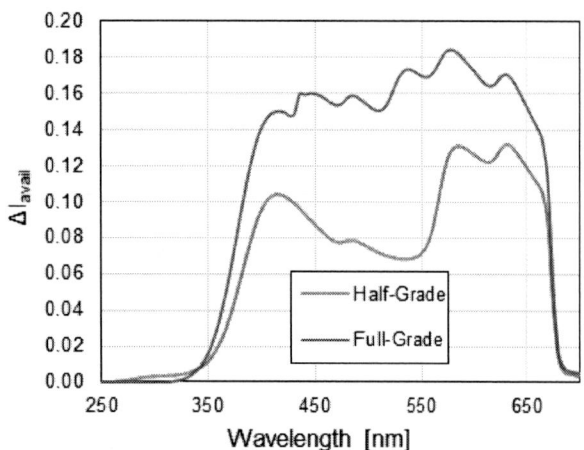

Figure 5. Relative available photocurrents enabled by the increased optical transparencies of the two graded window-emitter structures discussed herein.

Interestingly, the Full-Grade cell **(D)**, which possesses no discrete heterostructure interface, appears to exhibit an even greater relative enhancement in the short wavelength region, including higher signal into the UV, suggesting that the persistent field across the entire window-emitter structure, even within the nearly flat region after the direct-to-indirect gap transition, may indeed be providing enhanced carrier collection all the way to the surface (or at least near it). However, a slightly reduced signal in the 500 – 650 nm region appears to be tempering the total current (and likely voltage) improvement. It is unclear what the source of this effect is, although it is most likely something related to some growth non-ideality, which may be corroborated by the variance in EQE cutoff of 660 – 680 nm.

Figure 5 presents there results from a simplistic optical absorption to help describe the relative improvements offered by the increased transparency of the graded structures. Wavelength-resolved absorption coefficients, $\alpha(\lambda)$, were calculated based on extinction coefficients obtained via spectroscopic ellipsometry. For the sake of simplicity, the absorption behavior of the graded structures was approximated using optical data reported for multiple discrete compositions throughout the graded range of $(Al_zGa_{1-z})_{0.51}In_{0.49}P$ [8] and metamorphic $Al_xIn_{1-x}P$ [9]. The average absorption due to these discrete layers should approximate the net effect of the continuously graded layers, at least to a reasonable first order accuracy. A standard photon absorption model, $I = I_0 \cdot \exp[-\alpha(\lambda) \cdot x]$, was then used to calculate the wavelength-resolved photocurrent available to the cell after passing through the window/emitter structures, relative to the total input photon intensity (I/I_0).

To effectively deconvolute the impact of optical effects (increased transparency) from electrical (induced drift field), the window/emitter IQE was assumed to be zero, thus acting only as a parasitic filter. Figure 5 then plots the difference in available photocurrent, ΔI_{avail}, of the Half-Grade and Full-Grade cells versus their associated Reference cells. This is, of course, a major oversimplification, but it allows the analysis to focus entirely on optical effects.

Comparing the two designs, we find, as should be expected, that the Full-Grade offers better optical transparency overall, with a relative total available photocurrent increase of 17.8%, while the Half-Grade offers an optical increase of 11.0%. The higher overall bandgaps, basically reduced optical opacity, of the Full-Grade thus offers a nearly 2x increase in relative potential photocurrent improvement than the Half-Grade. However, the observed differences are not quite so stark, with the Full-Grade showing a J_{SC} improvement of +13.5% versus the Half-Grade's +12.0%. While this overly simplistic modeling is by no means accurate, it does suggest that the improvements observed are not purely optical and that electronic effects are likely playing an important role.

Finally, with respect to the pseudomorphically-strained Full-Grade structure, it is worth consideration as to whether any dislocations are being formed. This is especially important in consideration of potential compatibility with inverted growth methodology, where any dislocations formed would thread throughout the entire rest of the multijunction structure. Initial electron channeling contrast imaging (ECCI) [10] analysis of both sample **(D)** and an inverted reverse-graded window-emitter test structure revealed the presence of no dislocations, indicating that the graded region is truly pseudomorphic. However, further verification must be performed to be conclusive.

IV. SUMMARY

Various graded window-emitter structures were grown on $Ga_{0.51}In_{0.49}P$ cells via MBE to test the improvement in short wavelength performance. Initial LIV and QE results indicate that a compositionally graded window-emitter structure offers the potential for significantly enhanced short wavelength generation and collection efficiency. Most likely there are several mechanisms occurring at the device level, including a built-in electric field and the elimination of interfaces, which help to overcome material level deficiencies (poor material quality due to growth defects and high doping). This concept shows promise in many multijunction solar cell applications, especially those requiring increasingly wider band gap top cells, especially for deployment under AM0.

ACKNOWLEDGEMENTS

This work was supported by the Air Force Research Laboratory, Space Vehicles Directorate (contract #FA9453-14-C-0373).

REFERENCES

[1] R. R. King, J. H. Ermer, D. E. Joslin, M. Haddad, J. W. Eldredge, N. H. Karam, B. Keyes, R. K. Ahrenkiel, "Double

Heterostructures for Characterization of Bulk Lifetime and Interface Recombination Velocity in III-V Multijunction Solar Cells," in *2nd World Conf. Photovolt. Sol. En. Conv.*, 86 (1998).

[2] Y. Chang, H. Kuo, T. Lu, F. Lai, S. Kuo, L. Laih, L. Laih, S. Wang, "Efficiency improvement of single-junction $In_{0.5}Ga_{0.5}P$ solar cell with compositional grading p-emitter/window capping configuration," *Jap. J. Appl. Phys. 49(12R)*, 122301, 2010.

[3] J. Faucher, Y. Sun, D. Jung, D. Martín, T. Masuda and M. L. Lee, "AlGaInP solar cells with high internal quantum efficiency grown by molecular beam epitaxy," *2016 IEEE 43rd Photovoltaic Specialists Conference (PVSC)*, Portland, OR, 2016, pp. 0035-0039.

[4] O. Kwon, Y. Lin, J. Boeckl, S.A. Ringel, "Growth and properties of digitally-alloyed AlGaInP by solid source molecular beam epitaxy," *J. Electron. Mater. 34(10)*, 1301 (2005).

[5] A. Abdollahi, M. M. Golzan, K. Aghayar, "First-principles investigation of electronic properties of $Al_xIn_{1-x}P$ semiconductor alloy," *J. Mater. Sci. 51*, 7343-7354 (2016).

[6] A. Wadehra, J. W. Nicklas, J. W. Wilkins, "Band offsets of semiconductor heterostructures: A hybrid density functional study," *Appl. Phys. Lett. 97*, 092119 (2010).

[7] M. Grundmann, "Band Engineering BandEng" 1D band diagram simulator, http://my.ece.ucsb.edu/mgrundmann/bandeng.htm.

[8] H. Kato, S. Adachi, H. Nakanishi, K. Ohtsuka., "Optical properties of $(Al_xGa_{1-x})_{0.5}In_{0.5}P$ quaternary alloys," *Jap. J. Appl. Phys. 33(1A)*, 186-192, (1994).

[9] T. J. Kim, S. Y. Hwang, J. S. Byun, D. E. Aspnes, E. H. Lee, J. D. Song, C.-T. Liang, Y.-C. Chang, H. G. Park, J. Choi, J. Y. Kim, Y. R. Kang, J. C. Park, Y. D. Kim, "Dielectric functions and interband transitions of $In_xAl_{1-x}P$ alloys," *Curr. Appl. Phys. 14*, 1273-1276 (2014).

[10] S. D. Carnevale, J. I. Deitz, J. A. Carlin, Y. N. Picard, D. W. McComb, M. De Graef, S. A. Ringel, T. J. Grassman, "Applications of Electron Channeling Contrast Imaging for the Rapid Characterization of Extended Defects in III-V/Si Heterostructures," *IEEE J. Photovolt. 5(2)*, 676 (2015).

Integration of Quantum Dots and Quantum Wells into InGaAs Metamorphic Subcell for Radiation Hard 3-J ELO IMM Photovoltaics

Zachary S. Bittner[1], Hyun Kum[1], Michael A. Slocum[1], George T. Nelson[1], Rao Tatavarti[2]
Andree Wibowo[2], and Seth M. Hubbard[1]

[1]NanoPower Research Laboratory, Rochester Institute of Technology
New York, United States
[2]MicroLink Devices, Niles, Illinois, United States

Abstract—**In this work, a Sentaurus TCAD model of inverted metamorphic triple junction solar cells is demonstrated along with experimental 3J current-voltage and external quantum efficiency measurements for the purpose of predicting end-of-life efficiency improvement from incorporating quantum wells or quantum dots into a metamorphic $In_{0.3}Ga_{0.7}As$ subcell in a 3J solar cell. A radiation-hard "end-of life optimized" design is presented and end-of life efficiency enhancement is predicted from calculated QD and QW absorption coefficients. The resulting QW enhanced solar cell design is predicted to have a 6% relative efficiency enhancement over the optimized baseline design.**

I. Introduction

A major concern for the longevity of satellites in Medium Earth orbit (MEO), or orbital radii from 1.8 to 2.5 times the Earth's radius(R_e) is high energy radiation exposure leading to the formation of Frenkel defects, the main cause of solar cell degradation in space. Two methods of mitigating the effects of radiation induced defects are to increase the mass-specific power and increase beginning-of-life (BoL) capacity. The second method is to increase the radiation tolerance of the solar cell. Inverted metamorphic multijunction (IMM) technology increases the mass-specific power at BoL via removal of the substrate and can enable substrate reuse via epitaxial lift-off (ELO)[1], but may result in a solar cell with lower radiation tolerance due to poor radiation tolerance of the InGaAs metamorphic subcell [2]. In this work, we propose the inclusion of quantum dots (QDs) or quantum wells (QWs) to improve the radiation tolerance of the subcell. Studies on InAs/GaAs QDSCs have shown that the QD contribution to spectral responsivity remained relatively constant out to fluences that resulted in an 80% reduction in carrier collection in the bulk material, and QD photoluminescence emission sustained over 50% of the beginning-of-life (BoL) intensity at an order of magnitude higher fluence as compared to bulk GaAs[3].

Bailey et al.[4] demonstrated that it was possible to add QDs to a GaAs cell without degrading the open circuit voltage, and QDs[5] have been used to engineer the middle cell bandgap in Ge-based upright structures with minimal change in open circuit voltage, avoiding top cell degradation from increased surface roughness or unmitigated strain from the QD layers presented a challenge in maintaining top-cell

current collection[5], and requires a 200 meV shift in middle cell bandgap to match current to the Ge subcell[6]. InAs/GaAs QDs have been demonstrated in the GaAs subcell in inverted metamorphic multijunction (IMM) solar cells with minimal open circuit voltage degradation[7], but in the inverted metamorphic multijunction (IMM) cell architecture, the current limiting subcell at end of life (EoL) is the metamorphic InGaAs subcell [2]. Increasing the In fraction would reduce bandgap and could increase EoL current generation, but does not change the intrinsic radiation damage tolerance of the material.

InAs QD growth on an $In_{0.3}Ga_{0.7}As$ (InGaAs) metamorphic template was previously demonstrated by Slocum et al. [8] , but QDs have yet to be incorporated into an InGaAs subcell of a 3J IMM solar cell. While QDs have been demonstrated to be radiation hard[3], the predicted absorption strength of InAs QDs on InGaAs is around 25 cm^{-1} at the absorption peak. For this reason QWs with In/Ga compositional variation will also be investigated. The goal of this work is to both model and experimentally demonstrate ELO IMM 3J solar cells with strain-balanced InAs QDs or InGaAs QWs in an InGaAs subcell.

II. Characterization of Nanostructures

Test structures were grown in an Aixtron 3x2" close-coupled showerhead metallorganic chemical vapor deposition (MOCVD) reactor with standard metallorganic precursors and AsH$_3$. Characterization of InAs QDs grown on an InGaAs metamorphic template were presented by Slocum et al. [8] Optical parameters from an 8-band k·p simulation were used and will be presented in the device modeling section. Because the predicted absorption coefficient of InAs QDs on InGaAs is so low, QWs are also considered. NextNano++, a commercial Schrodinger-poisson solver that employs an 8-band k·p model, was used to calculate band structure and absorption coefficient of InGaAs quantum wells. Thicknesses and compositions of 7nm $In_{0.4}Ga_{0.6}As$ QWs in 9nm $In_{0.2}Ga_{0.8}As$ barriers were selected via photoluminescence (PL) characterization with conditions that minimize wafer curvature measured with in-situ reflectance monitoring. Figure 1 shows PL along with calculated absorption. QW ground state absorption strength is predicted to be around 5000 cm^{-1},

978-1-5090-5606-4/17 $31.00 © 2017 IEEE

nearly 250 times the predicted absorption strength of InAs QDs on InGaAs.

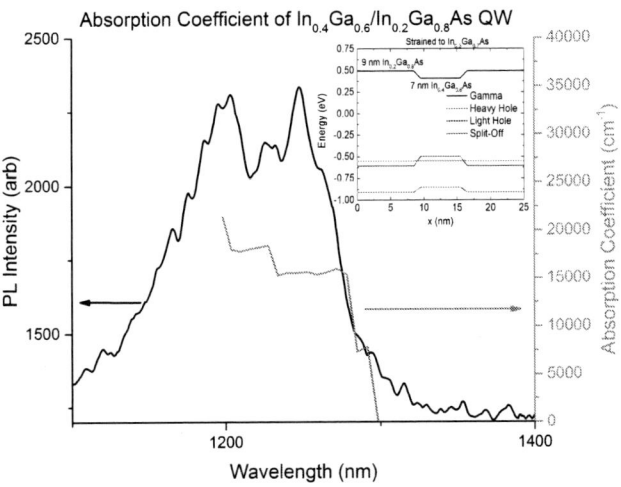

Fig. 1. Experimental photolumeniscence of 7nm $In_{0.4}Ga_{0.6}As$ QW in 9nm $In_{0.2}Ga_{0.8}As$ barrier grown on relaxed $In_{0.3}Ga_{0.7}As$ plotted along with absorption coefficient of QW calculated in NextNano. The inset shows the calculated band structure.

III. DEVICE MODELING

Before modeling the inclusion of QDs and QWs in the InGaAs subcell, A triple junction device model was developed in Sentaurus TCAD, a physics based semiconductor device simulation software package. Calibration of the device model was done by comparing the simulated current-voltage and external quantum efficiency to a triple junction ELO IMM solar cell at BoL and EoL. For the modeled device, ELO templates on GaAs substrates with pre-grown In(Al)GaP top cells were sent from Microlink Devices Inc. to Rochester Institute of Technology where the GaAs subcell was overgrown. Samples were then sent to Microlink Devices for growth of the metamorphic buffer and $In_{0.3}Ga_{0.7}As$ subcell and for ELO and fabrication. Demonstration of high quality overgrowth is important because overgrowth will be employed for the addition of QWs and QDs in the InGaAs subcell.

Devices were grown with a 1.2 μm In(Al)GaP subcell and 3.6 μm GaAs and InGaAs subcell with GaAs tunnel junctions connecting each subsequent cell. Devices were then fabricated via ELO for IV and EQE measurements. It was then irradiated with $2x10^{15}$ 1MeV e^-/cm^2 and electrical characteristics were remeasured. The design is not optimized for EoL conditions but the thick base design assists in fitting material damage coefficients.

Current-Voltage characteristics are shown in Figure 2 along with modeled IVs in Synopsis Sentaurus^{TM} TCAD. Solid lines show BoL results while dotted lines show EoL results. Starting with BoL, the experimental device had an open circuit voltage (V_{OC}) of 2.93 V, a short circuit current density (J_{SC}) of 15.41 mA/cm^2, and an AM0 efficiency of 28% at BoL while the simulated device had a V_{OC} of 2.89 V, a short

circuit current density (J_{SC}) of 15.35 mA/cm^2, and an AM0 efficiency of 28%. Accuracy in subcell characteristics were verified via EQE measurements.

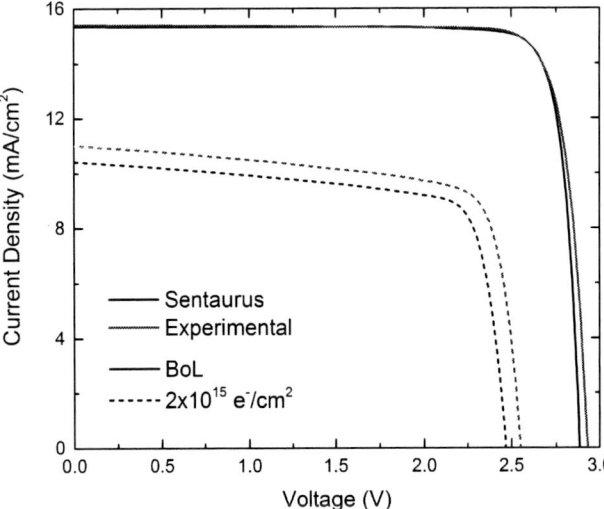

Fig. 2. AM0 IV characteristics of 3J ELO IMM at BoL and EoL after 1MeV electron irradiation.

Figure 3 is a comparison between experimental and simulated EQE at beginning of life. EQE was integrated across the AM0 spectrum to calculate an EQE integrated J_{SC}. The experimental In(Al)GaP, GaAs , and InGaAs subcells integrated EQE are 17.1, 15.4, and 15.6 mA/cm^2 respectively. Simulated subcell integrated EQE are 16.26, 15.2, and 16.68 mA/cm^2 respectively. Disagreement between top cell measured and simulated EQE is due to variation in ARC thickness.

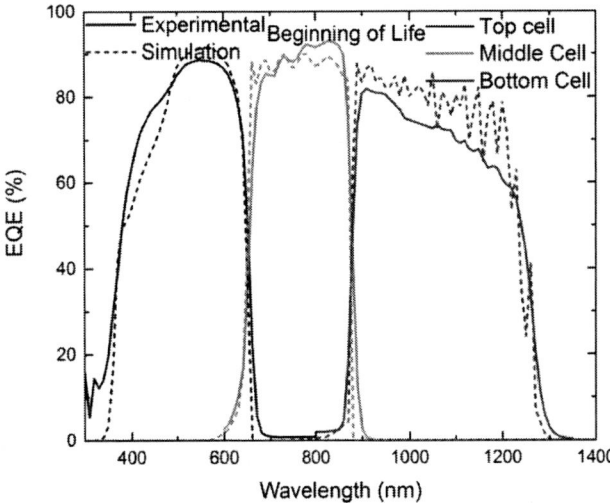

Fig. 3. BoL experimental and simulated EQE characteristics of each subcell in 3J IMM solar cell.

Returning to Figure 2, the experimental device had a V_{OC} of 2.55 V, a J_{SC} of 11.03 mA/cm^2, and an AM0 efficiency of

15.3% at EoL while the simulated device had a V_{OC} of 2.47 V, a short circuit current density (J_{SC}) of 10.42 mA/cm^2, and an AM0 efficiency of 14.1%. EoL EQE measurements and simulations are shown in Figure 4. The experimental subcell integrated EQE at EoL are 15.3, 11.6, and 11.2 mA/cm^2 respectively. Simulated subcell integrated EoL EQE are 14.2, 12.4, and 9.4 mA/cm^2 respectively. Experimental damage coefficients were found to be 3×10^{-7} for In(Al)GaP, 1×10^{-8} for GaAs, and 5×10^{-7} for InGaAs.

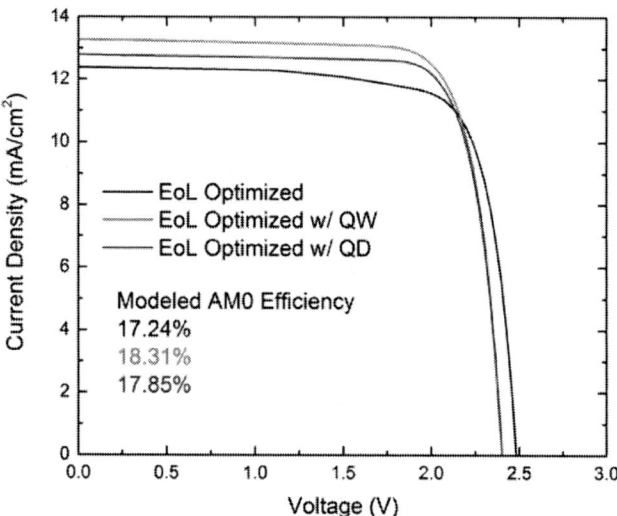

Fig. 5. Simulated JV characteristics of 3J IMM solar cells with a modeled irradiation of $2\times10^{15}\,e^-$/cm^2. The control design is shown in black, a device containing QWs in the InGaAs subcell in red, and a device containing QDs in the InGaAs subcell in blue.

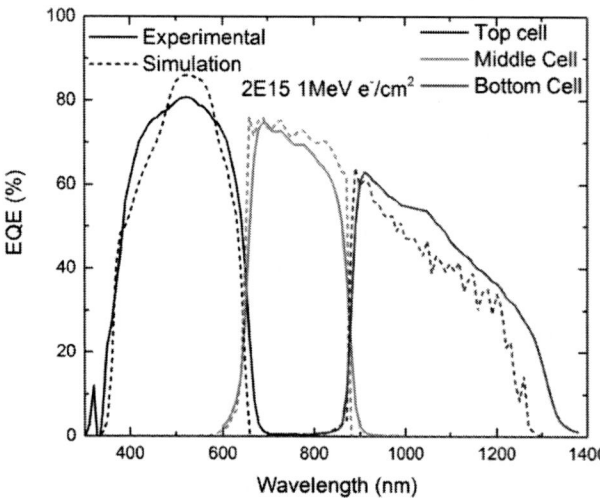

Fig. 4. IV characteristics of control and QDSC 3J devices derived from EL and EQE spectra. Control cell characteristics are shown in black and QDSC characteristics are shown in red. The inset shows a comparison between control and QD-containing GaAs subcells.

InAs and InGaAs 40x multi-quantum well structures were added to an i-region in an InGaAs subcell in the Sentaurus TCAD 3J model with layer and barrier thicknesses consistent with [8] for QDs and the QW simulation presented above along with calculated absorption coefficients. Layer thicknesses were then optimized to closely current-matched EoL conditions. The EoL optimized control device had a 300 nm In(Al)GaP top cell, a 1200 nm GaAs subcell, and a 1200nm InGaAs subcell. The top cell thickness was increased to 350 nm for the EoL optimized devices with QWs and QDs and both designs employed a 100nm InGaAs emitter and a 1100 nm InGaAs base cladding the QD/QW region. Figure 5 shows resulting current-voltage characteristics. A V_{OC} reduction from nearly 2.5 V to 2.4 V from the addition of interfaces and an i-region into the InGaAs subcell, which represents a significant portion of the EoL V_{OC} of a 1 eV subcell, however the reduction of the V_{OC} is compensated for via J_{SC} enhancement. The predicted EoL efficiency for the three devices is 17.24%, 18.31%, and 17.85% for control, QW, and QD respectively. The increase in EoL efficiency for the QW device represents a 6% relative efficiency enhancement.

In the final work, device results for 3J IMM solar cells with QDs and QWs in InGaAs subcells will be presented. Current enhancement will be compared with that predicted from calculated absorption

REFERENCES

[1] R. Tatavarti, A. Wibowo, G. Martin, F. Tuminello, C. Youtsey, G. Hillier, N. Pan, M. W. Wanlass, and M. Romero. InGaP/GaAs/InGaAs inverted metamorphic (IMM) solar cells on 4 #x2033; epitaxial lifted off (ELO) wafers. In *2010 35th IEEE Photovoltaic Specialists Conference*, pages 002125–002128, June 2010.

[2] P. Patel, D. Aiken, A. Boca, B. Cho, D. Chumney, M. B. Clevenger, A. Cornfeld, N. Fatemi, Y. Lin, J. McCarty, F. Newman, P. Sharps, J. Spann, M. Stan, J. Steinfeldt, C. Strautin, and T. Varghese. Experimental Results From Performance Improvement and Radiation Hardening of Inverted Metamorphic Multijunction Solar Cells. *IEEE J. Photovoltaics*, 2(3):377–381, jul 2012.

[3] C. Kerestes, C.D. Cress, B.C. Richards, D.V. Forbes, Yong Lin, Z. Bittner, S.J. Polly, P. Sharps, and S.M. Hubbard. Strain effects on radiation tolerance of triple-junction solar cells with InAs quantum dots in the GaAs junction. *IEEE Journal of Photovoltaics*, 4(1):224–232, January 2014.

[4] Christopher G. Bailey, David V. Forbes, Ryne P. Raffaelle, and Seth M. Hubbard. Near 1 v open circuit voltage InAs/GaAs quantum dot solar cells. *Applied Physics Letters*, 98:163105, 2011.

[5] Christopher Kerestes, Stephen Polly, David Forbes, Christopher Bailey, Adam Podell, John Spann, Pravin Patel, Benjamin Richards, Paul Sharps, and Seth Hubbard. Fabrication and analysis of multijunction solar cells with a quantum dot (In)GaAs junction: Fabrication and analysis of multijunction QD solar cells. *Progress in Photovoltaics: Research and Applications*, 22(11):1172–1179, November 2014.

[6] Seth M Hubbard, Christopher Bailey, Stephen Polly, Cory Cress, John Andersen, David Forbes, and Ryne Raffaelle. Nanostructured photovoltaics for space power. *Journal of Nanophotonics*, 3(1):031880–031880–16, October 2009.

[7] Z. S. Bittner, M. A. Slocum, G. T. Nelson, R. Tatavarti, and S. M. Hubbard. Novel InAs/GaAs QD subcell design for radiation hard 3-J ELO IMM solar cell. In *2016 IEEE 43rd Photovoltaic Specialists Conference (PVSC)*, pages 3421–3424, June 2016.

[8] M. A. Slocum, G. Nelson, S. Hellstrm, B. Smith, A. Wibowo, R. Tatavarti, and S. M. Hubbard. Growth of InAs quantum dots in a metamorphic InGaAs bottom cell of an inverse metamorphic solar cell. In *2016 IEEE 43rd Photovoltaic Specialists Conference (PVSC)*, pages 2111–2114, June 2016.

Proton Irradiation of 3J Solar Cells at Low Temperature

Seonyong Park[1], Jacques C. Bourgoin[1], Olivier Cavani[1], Sandrine Picard[2], Jérôme Bourcois[2],
Victor Khorenko[3], Carsten Baur[4] and Bruno Boizot[1]

[1]Laboratoire des Solides Irradiés, CNRS-UMR 7642, CEA-DRF-IRAMIS, Ecole Polytechnique,
Université Paris-Saclay, Palaiseau Cedex, 91120, France
[2]CSNSM, Université Paris-Sud, CNRS/IN2P3, Université Paris-Saclay, 91405, Orsay, France
[3]AZUR SPACE Solar Power GmbH, Theresienstr.2, Heilbronn, 74072, Germany
[4]European Space Agency, Keplerlaan 1, Noordwijk, 2201AZ, The Netherlands

Abstract — III-V compound based triple junction solar cells and its component cells have been irradiated with 1 MeV protons at various temperatures, ranging from 100 to 300 K. Degradations of the cell performance (short circuit current I_{SC}, open circuit voltage V_{OC} and maximum power P_{MAX}) have been monitored versus fluence. At low temperature, the rate of degradation of I_{SC} of the bottom cell is very rapid, in such a way that this cell becomes the limiting cell even for low fluences. A possible explanation for the origin of such degradation, based on a non-uniform defect distribution, is suggested.

Index Terms — triple junction solar cell, current limiting cell, LILT, space charge by irradiation.

I. INTRODUCTION

Triple junction (TJ) Gallium-Indium-Phosphor/Gallium-Arsenide/Germanium (GaInP/GaAs/Ge) solar cells are now routinely used for space applications and their behavior under proton and electron irradiations has been well studied in the case of room temperature (RT) operation [1]-[5]. Attempts are now made to use these cells for deep space missions, i.e. in conditions of low-intensity and low-temperature, the so-called LILT conditions. Until recently, the degradation under such conditions was estimated from RT irradiation followed by low temperature (LT) measurements, owing to the difficulties to perform LT irradiation with in-situ measurements [6]-[7].

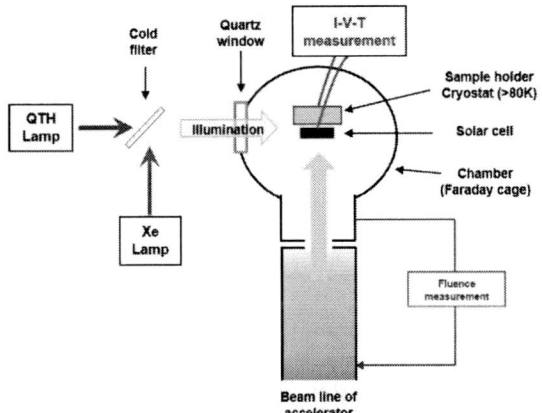

Fig. 1.   Diagram showing the in-situ set up for solar cell irradiation at low temperature

However, in the last years, we have made in-situ LILT studies after LT proton and electron irradiations on TJ cells and associated component cells, i.e. cells having the same structure than TJ cells but in which only one sub-cell: GaInP (Top), GaAs (Middle), or Ge (Bottom) junctions is electrically active. All cells were produced by AZUR SPACE Solar Power [8]-[12]. Here, we report the temperature dependences, of open circuit voltage (V_{OC}), short circuit current (I_{SC}), and maximum power (P_{MAX}) of TJ cells and component cells, after 1 MeV proton irradiations, in the range of 100 to 300 K, with various fluences. In particular, we found that the degradation rate of I_{SC} of the bottom cell is very large at low temperature (compared to the degradation rate at room temperature), which results in current limitation of the TJ cell by the bottom sub-cell even for limited fluences.

II. EXPERIMENTAL CONDITIONS

To realize relevant irradiation conditions of deep space missions at the lab scale, we used a cryostat chamber containing the solar cell positioned on a Cu holder, allowing in-situ electrical measurements under illumination and particle irradiation in the range 80 to 400 K (Fig. 1). The chamber has two windows: one for the beam line of the accelerator and the other one for the illumination. Low temperature is achieved by pumping liquid nitrogen through a pipe which surrounds the Cu holder. The temperature of the cell is directly measured with a calibrated CERNOX® temperature sensor soldered on the cell, during a pre-irradiation stage. For proton irradiation, the chamber is directly connected to the proton beam line of the accelerator (ARAMIS of SCALP facility, CSNSM). The irradiation has been performed at 1 MeV; the fluence varies from 1×10^{10} to 1.6×10^{12} cm^{-2} with a flux of 1×10^{9} cm^{-2}.s^{-1}. The solar simulator, which delivers 3.7 % of AM0 (condition corresponding to the Jupiter environment) is realized by using two different lamps: a Xenon (Xe) lamp and a Quartz Tungsten Halogen (QTH) lamp, separated by a cold filter selecting the high energy part of the Xe lamp and the low energy part of the QTH lamp. The calibration of the simulator is carried out using reference cells (Top, Middle, Bottom component cells provided by AZUR SPACE). The size of the solar cell is 2 cm x 2 cm, and their thickness is either 80 or 140 μm.

978-1-5090-5606-4/17 $31.00 © 2017 IEEE

III. RESULTS

A. Triple Junction Cells

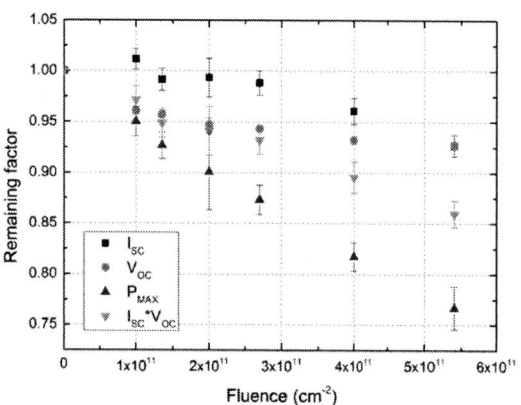

Fig. 2. Remaining factor of I_{SC}, V_{OC}, P_{MAX} of TJ solar cells after 1 MeV proton irradiation at 123 K with various fluences

TJ solar cells were irradiated by 1 MeV protons at 123 K with fluences ranging from 1×10^{11} to 5.4×10^{11} cm^{-2}. Fig. 2 shows the remaining factors of I_{SC}, V_{OC}, P_{MAX}, and the product I_{SC} x V_{OC} measured in-situ after irradiation (The remaining factors are obtained dividing EOL (End-Of-Life) values by BOL (Beginning-Of-Life) ones). We observe that I_{SC} for the TJ cell degrades from a certain fluence (approximatively 3×10^{11} cm^{-2}), while V_{OC} degrades faster at the beginning and then becomes less and less affected by the irradiation. On the other hand, the degradation of P_{MAX} is more significant than that of the product V_{OC} x I_{SC} for large fluences, which indicates a significant degradation of the fill factor.

B. Component Cells

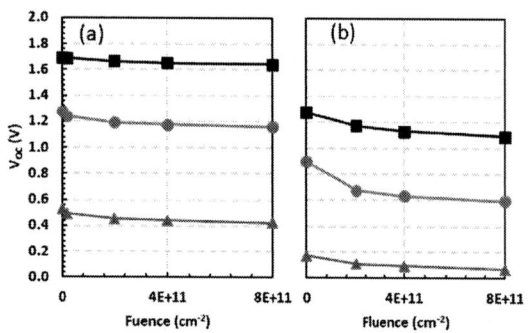

Fig. 3. Variations of V_{OC} of component cells after irradiation with cumulative fluences at (a) 123 K and (b) 300 K: black square – Top, red circle – Middle, blue triangle – Bottom, respectively.

Cumulative irradiations at two different temperatures (123 and 300 K) were carried out for Top, Middle and Bottom component cells (see Fig. 3 and Fig. 4 for V_{OC} and I_{SC},

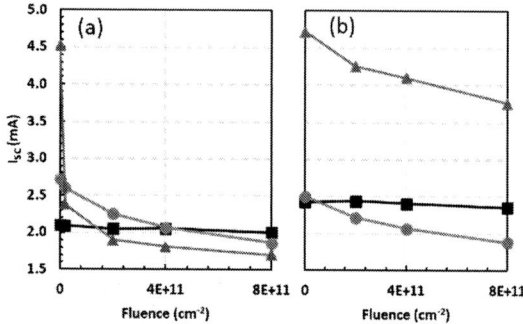

Fig. 4. Degradation of I_{SC} of component cells after 1 MeV proton irradiation with cumulative fluences at 123 and 300 K: black square – Top, red circle – Middle, blue triangle – Bottom, respectively.

respectively). As shown in Fig. 3, at low temperature, V_{OC} for all cells is not significantly degraded even after the irradiation with the highest fluence (8×10^{11} cm^{-2}). However, Fig. 4 shows significant changes of I_{SC}: in particular, the current of the Bottom cell exhibits a sharp drop at low fluence (partly attributed to photon recycling effect (PRE) [13] as we shall

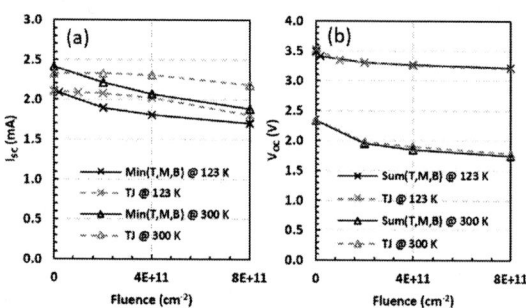

Fig. 5. Degradation of (a) I_{SC} and (b) V_{OC} of TJ cells and simulation with component cells (from Fig. 3 and Fig. 4) after irradiation with accumulative fluences at 123 and 300 K.

discuss elsewhere) which should induce a change of the current limiting cell (from Top to Bottom cell) in a TJ cell.

Fig. 5 presents the results of accumulative irradiations of TJ cells at 123 and 300 K including the expected values derived from the component cells in same conditions: I_{SC} of the TJ cell is obtained by selecting the minimum current value from that of component cells; V_{OC} of the TJ cell is obtained by adding the three V_{OC} values of each component cell.

IV. DISCUSSION

At low temperature, the changes of I_{SC} and V_{OC} of the component cells are in agreement with the corresponding changes in the TJ cells (see Fig. 5). However, P_{MAX} of the TJ cell appears to reflect a significant degradation of the fill factor as illustrated in Fig. 2. This indicates the existence of the excess

978-1-5090-5606-4/17 $31.00 © 2017 IEEE

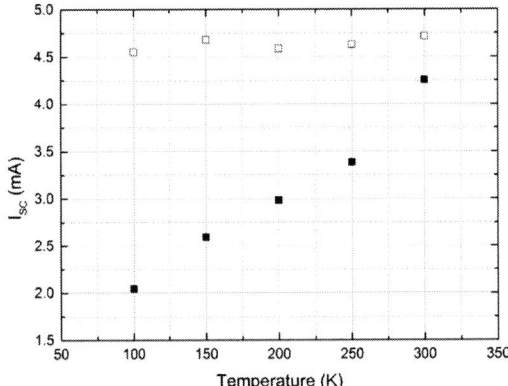

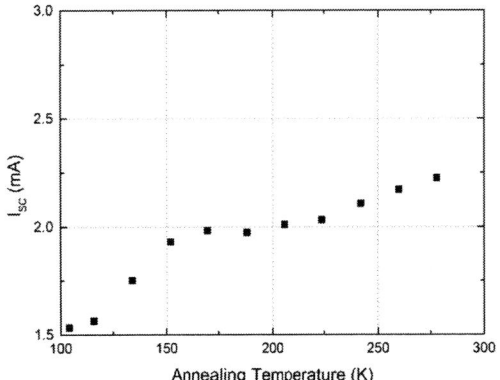

Fig. 6. I_{SC} of BOL (empty square) and EOL (filled square) bottom cells from 100 to 300 K. EOL fluence of 2 x 10^{11} cm^{-2}.

Fig. 7.   Isochronal annealing of I_{SC} of a Bottom cell after irradiation with a fluence of 1.6 x 10^{12} cm^{-2} at 100 K.

recombination current induced by the electric field, as we shall discuss elsewhere.

The degradations of the short circuit current of the Bottom cell at low and room temperatures are several order of magnitude different (Fig. 6). This is more than reasonably expected from the annealing behavior of the created defects (Fig. 7) which indicate a factor of less than 2.

One possible explanation is that the defects created at LT are not uniformly distributed, but decorate the proton tracks. The protons are uniformly distributed but the damage created by these protons is strongly localized along the proton track. The volume associated with a proton track is not available for carrier transport.

Indeed, the defects being charged, they create space charge regions which repel free carriers: then electrical conduction occurs parallel to these tracks (which are also parallel to the electric field). Assuming that all the degradation is a result of the change in I_{SC} produced by N proton tracks, it corresponds to a reduction in the active surface of the cell, S_r, by a quantity equal to $\pi r^2 N$, where r is the radius of the space charge region, and N is the total number of irradiating protons. Thus, r can be evaluated by equation (1):

$$r = \sqrt{\frac{S_r}{\pi \times N}} \qquad (1)$$

According to Fig. 4, we assume that PRE is vanished after the first irradiation with a fluence of 2x10^{10} cm^{-2} ; we thus consider the I_{SC} of this step as an initial value (2.37 mA). Then, irradiation with the fluence of 2x10^{11} cm^{-2} of the Bottom cell results in about 20 % change of I_{SC} (from 2.37 to 1.9 mA). This implies that 20 % of the cell surface is now electrically inactive: 0.8 cm^2 of reduced surface S_r for N = 8x10^{11} (product of the fluence 2x10^{11} cm^{-2} and of the cell size 4 cm^2); hence, r = 5.6x10^{-10} m.

V. CONCLUSION

Using accumulative proton irradiations of TJ cells and component cells at various temperatures, we observed that the behavior of a TJ cell under proton irradiation at low temperature is significantly different from its behavior at 300 K. We found that the Bottom cell appears to be the current limiting cell at 123 K even for low fluences (above few 10^{11} cm^{-2}). Furthermore, the degradation rate of I_{SC} of the Bottom cell is strongly temperature dependent.

We can therefore conclude that the evaluation of the degradation induced by protons at low temperature cannot be derived from results obtained by room temperature irradiation.

VI. ACKNOWLEDGMENTS

Present work is supported by the European Space Agency under contract no. 4000109645. Proton irradiation experiments have been supported by EMIR federation (EMIR project no. 15-5727).

REFERENCES

[1] A. W. Bett, C. Baur, F. Dimroth, W. Guter, M. Meusel, and G. Strobl, "Recent developments in III-V multi-junction space solar cells," in *7th European Space Power Conference*, 2005.

[2] M. Meusel, C. Baur, W. Guter, M. Hermle, F. Dimroth, A. W. Bett, T. Bergunde, R. Dietrich, R. Kern, W. Köstler, M. Nell, W. Zimmermann, G. LaRoche, G. Strobl, S. Taylor, C. Signorini, and G. Hey, "Development status of European multi-junction space solar cells with high radiation hardness," in *20th European Photovoltaic Solar Energy Conference*, 2005, p. 20.

[3] A. W. Bett, F. Dimroth, W. Guter, R. Hoheisel, E. Oliva, S. P. Philipps, J. Schone, G. Siefer, M. Steiner, A. Wekkeli, E. Welser, M. Meusel, W. Kostler, G. Strobl, "Highest Efficiency Multi-Junction Solar Cell for Terrestrial and Space Applications," in *24nd European Photovoltaic Solar Energy Conference and Exhibition*, 2009.

978-1-5090-5606-4/17 $31.00 © 2017 IEEE

[4] M. Imaizumi, T. Takamoto, T. Sumita, T. Ohshima, M. Yamaguchi, S. Matsuda, A. Ohi, and T. Kamiya, "Study of Radiation Response on Single-junction Component Sub-cells in Triple-junction Solar Cells," in *3rd World Conference on Photovoltaic Energy Conversion*, 2003, p. 599.

[5] T. Nakamura, M. Imaizumi, S. Sato, and T. Ohshima, "Change in I-V Characteristics of Subcells in a Multi-Junction Solar Cell due to Radiation Irradiation," in *38th IEEE Photovoltaic Specialist Conference*, 2012, p. 2846.

[6] C. J. Gelderloos, K. B. Miller, R. J. Walters, G. P. Summers, and S. R. Messenger, "Low Intensity Low Temperature Performance of Advanced Solar Cells," in *29th IEEE Photovoltaics Specialist Conference*, 2002, p.804.

[7] P. Stella, G. Davis, R. Mueller, and S. Distefano, "The Environmental Perfoemance at Low Intensity, Low Temperature (LILT) of High Efficiency Triple Juncion Solar Cells," in *2nd International Energy Conversion Engineering Conference*, 2004.

[8] C. Baur, V. Khorenko, G. Siefer, J. C. Bourgoin, M. Casale, R. Campesato, S. Duzellier, and V. Inguimbert, "Development Status of Triple-junction Solar Cells Optimized for Low Intensity Low Temperature Applications," in *39th IEEE Photovoltaic Specialist Conference*, 2013, p. 3237.

[9] V. Khorenko, C. Baur, G. Siefer, M. Casale, R. Campesato, S. Duzellier, and V. Inguimbert, "Performance Analysis of Azur 3G28 Triple-junction Solar Cells Optimized for Operation in Jupiter Environment," in *10th European Space Power Conference*, 2014.

[10] J. C. Bourgoin, B. Boizot, K. Khirouni, and V. Khorenko, "On the Prediction of Solar Cell Degradation in Space," in *10th European Space Power Conference*, 2014.

[11] S. Park, J. C. Bourgoin, O. Cavani, V. Khorrenko, C. Baur, and B. Boizot, "Origin of the Degradation of Triple Junction Solar Cells at Low Temperature," in *11th European Space Power Conference*, 2016.

[12] V. Khorenko, C. Baur, G. Siefer, M. Schachtner, S. Park, B. Boizot, J. C. Bourgoin, M. Casale, and R. Campesato, "BOL and EOL Characterization of Azur 3G LILT Solar Cells for ESA JUICE Mission," in *11th European Space Power Conference*, 2016.

[13] C. Baur, M. Meusel, F. Dimroth, and A. W. Bett, "Investigation of Ge component cells," in *31th IEEE Photovoltaic Specialist Conferences*, 2005, p.675.

Ultra-thin GaAs solar cells: radiation tolerance and space applications

Louise C. Hirst[1], Michael K. Yakes[1], Jeffery. H. Warner[1], Mitchell F. Bennett[2], Kenneth J. Schmieder[1], Stephanie Tomasulo[1], Erin Cleveland[1], Sergey Maximenko[1], James Moore[3], Robert J. Walters[1] and Phillip P. Jenkins[1]

1) U. S. Naval Research Laboratory, 4555 Overlook Ave. SW., Washington DC 20375

2) Sotera Defense Solutions, Inc., Annapolis Junction, Maryland 20701-1067

3) The George Washington University, 2121 I Street NW, Washington, DC 20037, USA

Abstract — **Radiation tolerance is a critical concern for space power applications. In this paper we demonstrate intrinsic radiation tolerance in ultra-thin (<100nm) GaAs devices, without the need for heavy, rigid coverglass. It is shown that in the ultra-thin geometry carrier collection efficiency and corresponding short circuit current do not degrade in the presence of radiation induced defects. By employing scattering structures for light management solar absorption can be compensated in an ultra-thin film to exploit this intrinsic radiation tolerance and enable new mission profiles in challenging space environments.**

I. Introduction

Ultra-thin (<100 nm) III-V photovoltaic devices offer new opportunities for space power. The ultra-thin geometry can provide high specific power in a fully flexible form factor, while also providing unprecedented radiation tolerance without the need for heavy and cumbersome shielding [1]. These features allow for new satellite designs such as arrays of ultra-thin sheets which self-assemble for large area sensing. They also enable new extended mission profiles in hostile, high energy environments such as Earths Van Allen belts and the Galilean moons of Jupiter. In this paper we experimentally demonstrate intrinsic radiation tolerance in ultra-thin GaAs solar cells and study the underlying physical origins of this phenomenon. We also discuss the required design features of the ultra-thin solar cell, in particular the use of scattering structures, necessary to develop this technology for space power applications.

II. Intrinsic Radiation Tolerance

Radiation induced defects in a semiconductor material act as recombination centers for photo-excited carriers. In traditional "thick" devices the presence of defects reduces carrier lifetime, which in turn reduces carrier collection efficiency and device current. Device voltage and fill factor are also degraded. In an ultra-thin solar cell, photo-excited carriers only have a short distance to travel to the device terminals. Provided the transit time of the carriers remains short relative to the recombination lifetime, carrier collection efficiency will not be effected by the presence of defects thus device current will not be degraded by radiation exposure with an ultra-thin device design.

The principal of intrinsic radiation tolerance has been demonstrated by comparing radiation effects in two GaAs solar cells of different thickness: 80 nm and 800 nm. These devices are a p on n design, grown by molecular beam epitaxy (structures given in table 1). The devices where processed with an annealed Ni/Ge/Au (5/30/200 nm) rear contact, an unannealed Cr/Au (10/100 nm) front contact with 6% shadowing and an $H_3PO_4:H_2O_2:H_2O$ (1:8:50) mesa etch. The structures were designed specifically to compare radiation effects and therefore the devices were fabricated to be as similar as possible, except for emitter and base thickness. No special features, such as substrate removal, antireflection coatings or scattering structures, were included to improve absorption in the 80 nm device.

Samples were exposed to 3 MeV protons at the Auburn University Pelletron. Cells from the same wafer were exposed with fluence ranging from 10^{10} to 10^{14} p+/cm². Light and dark current-voltage characteristics were measured for the different exposures using a Keithley 2401 source meter and a class B Oriel solar simulator with Xenon bulb, calibrated for an AM0 spectrum with a GaAs cell. External quantum efficiency measurements were made using a chopped Xenon source, CVI Instruments Digikrom monochromator and SR830 lock-in amplifier.

Layer	Thickness (nm)	Material	Doping (cm⁻³)
cap	25	GaAs	Be $1x10^{19}$
window	20	$Al_{0.3}GaAs$	Be $5x10^{18}$
emitter	xE	GaAs	Be $1x10^{18}$
base	xB	GaAs	Si $5x10^{17}$
bsf	20	$Al_{0.3}GaAs$	Si $5x10^{18}$
buffer	100	GaAs	Si $5x10^{18}$
substrate		GaAs	n+

TABLE 1: Solar cell structures. For the 80 nm cell xE=xB=40 nm. For the 800 nm xE=xB=400 nm.

It was observed that short circuit current (Isc) does not degrade as a function of fluence in the 80 nm cell over the full range tested in this study (up to 10^{14} p$^+$/cm^2 equivalent to ~10^{12} MeV/g). Over the same range Isc of the 800 nm cell degrades to a remaining factor of 0.26. For comparison with a more traditional "thick" device architecture, data adapted from Turowski et al.[2] for a 3.5 μm cell are also shown. Isc of the thickest cell degrades fastest (figure 1a).

As expected, the 80 nm cell has the lowest absolute Isc for low exposures. In accordance with Beer's Law, <10% of incident photons are absorbed in a single pass of the ultra-thin cell structure. The rest are transmitted through the device and absorbed in the substrate, and therefore do not contribute to useful current. Despite this severe current penalty, after 10^{14} p$^+$/cm^2 the absolute Isc of the ultra-thin device exceeds that of the thicker cells (figure 1b).

The origins of this striking effect are highlighted in the external quantum efficiency measurements (figure 1e). At short wavelengths (<450 nm), pre-exposure EQE values for both 80 nm and 800 nm cells are the same. This occurs because photons within this wavelength range are absorbed within the first 80 nm. At longer wavelengths the EQE of the 80 nm cell is suppressed because of transmission. Post-exposure, the 80 nm cell EQE is unchanged indicating that radiation induced defects do not degrade carrier collection efficiency, while the 800 nm cell shows reduced EQE for all wavelengths. Notably, the EQE of the 800 nm cell drops below that of the 80 nm cell for short wavelengths.

Radiation induced defects also degrade device voltage and it was observed that the open circuit voltage of both thin and thick devices degraded at a similar rate as a function of fluence. The resulting max power (relative and absolute) are given in figures 1 c&d respectively. Fluence has been converted into equivalent days in a highly elliptical orbit (HEO) with a 25 μm cover glass, calculated using Non-Ionizing Energy Loss (NIEL) [3]. Despite >90% transmission losses in the 80 nm cell, absolute max power exceeds that of the 800 nm cell after an equivalent of a year in HEO. This highlights the potential for ultra-thin photovoltaics to enable new mission profiles in the most challenging space environments.

III. Ultra-thin Device Design Requirements

While the ultra-thin solar cells used in this study demonstrate unprecedented radiation tolerance, improving the pre-exposure Jsc and corresponding max power will be essential to developing a useful space power technology. Light management architectures can be implemented in ultra-thin cells to improve absorption and achieve a higher pre-exposure Jsc. In this way, the intrinsic radiation tolerance of this geometry can to be exploited for new space power systems. Scattering structures have been implemented extensively for terrestrial PV applications to increase solar energy conversion conversion efficiency. These designs operate by scattering light into optical modes outside the

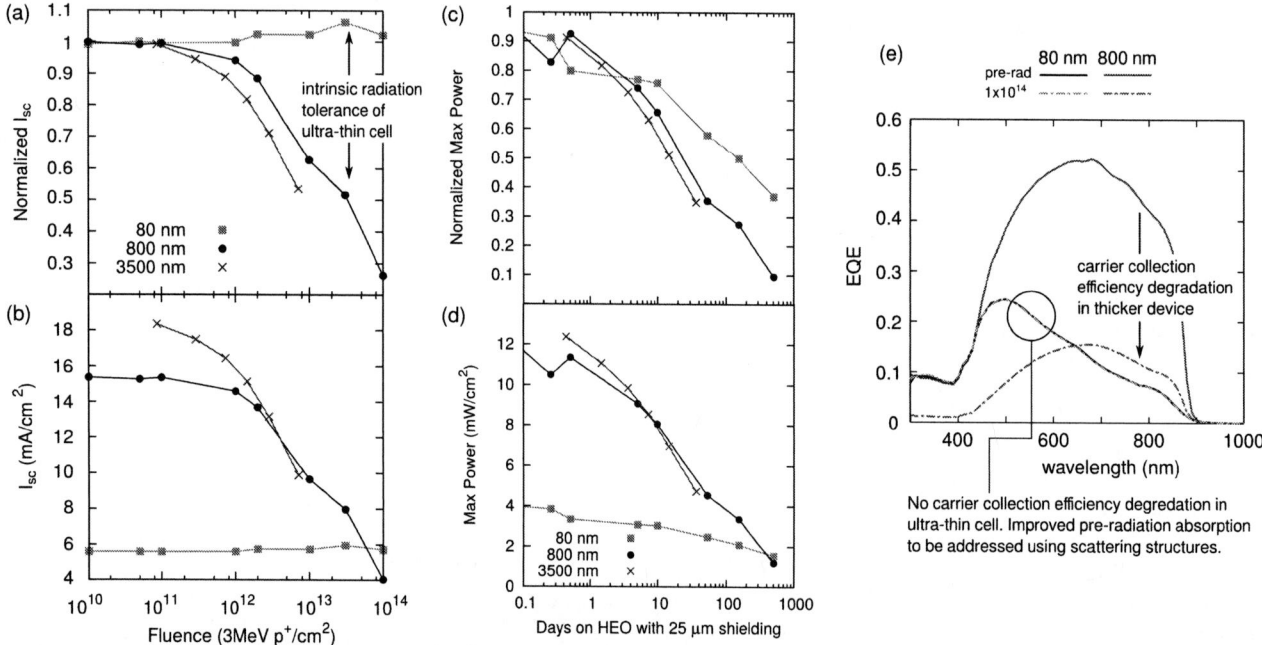

Figure 1. Data showing intrinsic radiation tolerance in an ultra-thin geometry [1]. a&b) Normalized and absolute short circuit current as a function of 3 MeV proton fluence. Current does not degrade in the ultra-thin geometry, however, pre-exposure current must be improved using light management. c&d) Normalized and absolute maximum power output as a function of equivalent time spent in a highly elliptical Earth orbit. c) External quantum efficiency of 80 nm and 800 nm devices, pre and post radiation exposure. Carrier collection efficiency does not degrade in the ultra-thin geometry.

escape cone at the interface between the semiconductor and the air so that light is totally internally reflected within the device and propagates laterally along the thin layer [4,5]. In this way, the optical pathlength is extended to achieve full solar absorption. In recent years, photovoltaic device designs incorporating scattering structures have been developed employing nanophotonic [6,7] and plasmonic schemes [8,9].

Strong solar absorption in ultra-thin layers (<100 nm) of GaAs have been demonstrated. Massiot et al. [10] achieved 80% absorption in a 25 nm GaAs film using a 2D metallic grating. This principle has also been applied to devices by Vandamme et al. [11], who showed a Jsc of 20.7 mA/cm2 for cells with absorber thickness 220 nm, exceeding the pre-exposure Jsc of the 800 nm cell in this study. This work demonstrates the potential to significantly improving absorption in ultra-thin geometries using light management.

IV. Conclusions

In this paper we experimentally demonstrate intrinsic radiation tolerance of ultra-thin GaAs photovoltaics. It is identified that carrier collection efficiency is not degraded in the presence of radiation induced defects in this geometry. Ultra-thin cells processed on their growth substrate have a substantial transmission penalty resulting in a low beginning of life short circuit current and corresponding output power, relative to equivalent thicker devices. However, after exposure to a fluence of 10^{14} p$^+$/cm^2, equivalent to ~500 days in HEO with 25 μm cover glass, the output power of the ultra-thin cell exceeds that of thicker devices, despite the transmission penalty. Substrate removal and integration of scattering structures into the device design are essential for improving the ultra-thin cell beginning of life performance and developing a practical space power technology for high radiation environments.

References

[1] L.C. Hirst, M. K. Yakes, J. H. Warner, M. F. Bennett, K. J. Schmieder, R. J. Walters, P. P. Jenkins,
Intrinsic radiation tolerance of ultra-thin GaAs solar cells , *Appl. Phys. Lett.*, 109, 033908, 2016.

[2] M. Turowski, T. Bald, A. Raman, A. Fedoseyev, J. H. Warner, C. D. Cress, and R. J. Walters, "Simulating the radiation response of GaAs solar cells using a defect-based TCAD model," IEEE Trans. Nucl. Sci. 60, 2477–2485, 2013.

[3] S. R. Messenger, G. P. Summers, E. A. Burke, R. J. Walters, and M. A. Xapsos, "Modeling solar cell degradation in space: A comparison of the nrl displacement damage dose and the jpl equivalent fluence approaches," Prog. Photovolt. 9, 103–121, 2001.

[4] E. Yablonovitch, "Statistical ray optics," J. Opt. Soc. Am. 72, 899–907, 1982.

[5] P. Campbell and M. A. Green, "Light trapping properties of pyramidally textured surfaces," J. Appl. Phys. 62, 243–249, 1987.

[6] Z. Yu, A. Raman, and S. Fan, "Fundamental limit of nanophotonic light trapping in solar cells," Proc. Natl. Acad. Sci. U.S.A. 107, 17491–17496, 2010.

[7] N. P. Hylton, X. F. Li, V. Giannini, K. H. Lee, N. J. Ekins-Daukes, J. Loo, D. Vercruysse, P. V. Dorpe, H. Sodabanlu, M. Sugiyama, and S. A. Maier, "Loss mitigation in plasmonic solar cells: aluminium nanoparticles for broadband photocurrent enhancements in GaAs photodiodes," Sci. Rep. 3, 2874, 2013.

[8] K. Catchpole and A. Polman, "Plasmonic solar cells," Opt. Express 16, 21793–21800, 2008.

[9] K. Nakayama, K. Tanabe, and H. A. Atwater, "Plasmonic nanoparticle enhanced light absorption in GaAs solar cells," Appl. Phys. Lett. 93, 121904, 2008.

[10] I. Massiot, N. Vandamme, N. Bardou, C. Dupuis, A. Lemaitre, J.-F. Guillemoles, and S. Collin, "Metal nanogrid for broadband multiresonant light-harvesting in ultrathin GaAs layers," ACS Photon. 1, 878–884, 2014.

[11] N. Vandamme, H. L. Chen, A. Gaucher, B. Behaghel, A. Lemaitre, A. Cattoni, C. Dupuis, N. Bardou, J. F. Guillemoles, and S. Collin, "Ultrathin GaAs solar cells with a silver back mirror," IEEE J. Photovolt. 5, 565–570, 2015.

978-1-5090-5606-4/17 $31.00 © 2017 IEEE

Large Area Multijunction III-V Space Solar Cells Over 31% Efficiency

X.Q. Liu, C. Fetzer, P. Chiu, M. Haddad, X. Zhang, R. Cravens, D. Law, J. Ermer,
J. Krogen, S. Sharma, J. Hanley

Spectrolab, Inc., a Boeing Company, Sylmar, CA, 91342, USA

Abstract — This paper explores development and progress for large area III/V multijunction space solar cells with efficiencies at or above 31%. XTJ Prime SuperCells with area over 70 cm², average 30.7% AM0 efficiency with a maximum of 31.9%. Post radiation testing indicates an EOL efficiency of 26.7% for 1e15 1 MeV e-/cm² fluences after ECSS annealing. Upright Metamorphic triple junction (UMM3J) devices were developed with breakthroughs in optimizing cell design and uniformity on 150 mm wafers, and demonstrated SuperCell build of average AM0 efficiency at 31.3% for 73.5 cm² cells in area with best efficiency at 31.5%. UMM3J devices were also evaluated under typical GEO and interplanetary environments for low-irradiance and low-temperature (LILT). Finally, next generation space solar cell UMM4J devices were investigated, and high quality 1.9 eV subcells were demonstrated.

I. INTRODUCTION

Lattice matched GaInP/GaInAs/Ge triple junction (3J) solar cells with slightly different types of architectures have been the dominant product for space missions. Currently, Spectrolab is flying XTJ 29.5% SuperCells on orbit on multiple spacecraft from the state-of-the-art 150 mm diameter wafer process line. Here a SuperCell refers to any epitaxial III/V multijunction solar cell we produce from two cells per 150 mm wafer. Recently, we introduced an improved variant of the XTJ family designated the XTJ Prime [1]. This improved variant of lattice matched devices is the only currently qualified cell achieving over 3 Watts per cell in SuperCell sizes. Spectrolab's guiding principle continues to be addressing the space photovoltaic marketplace with increasing value through improved end-of-life cost for power (EOL $/W). This aim is more than increased efficiency alone. It addresses both mission total power needs as well as the total cost to provide that power on orbit.

One of the major opportunities to improve the conversion efficiency in state-of-the-art of a 3J device is redistributing the excess current from Ge subcell to higher band gap and hence capturing incident photons at closer to their incident energy. A pathway to improve the current distribution amongst the subcells is to employ a lattice mismatched subcell design beneath the middle subcell. Previous efforts have demonstrated inverted growth techniques to produce an inverted metamorphic (IMM) design [2]. The additional complexity of substrate removal and attachment as well as additional subcell epitaxial growth greatly increases the cost, complexity, and

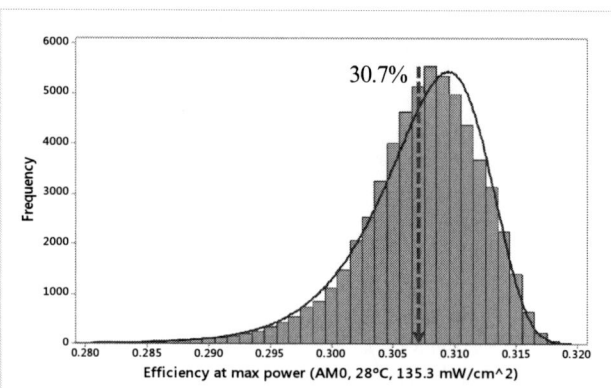

Fig. 1. Histogram of production SuperCell XTJ Prime averaging 30.7% for over 60,000 cells.

flight risk for such devices. In contrast, the upright metamorphic (UMM) architecture accesses the more ideal distribution of subcell band gaps and hence higher efficiencies, without the added fabrication cost of IMM cells [3].

Spectrolab has a significant history of improving the UMM triple junction (UMM3J) solar cells [3, 4] and previously demonstrated 28.8% under AM0 conditions (28°C, 135.3 mW/cm²) and 41.6% efficiency under terrestrial concentration AM1.5D, 484 suns, 25°C conditions for UMM triple junction (UMM3J) design [4-6]. Recently Spectrolab further improved UMM3J large area solar cells intended for space PV application.

Based upon the experience of a long history of development of metamorphic multiple-junction (MM) solar cells, including IMM3J, IMM4J, IMM5J for space, and UMM3J, UMM4J for terrestrial PV, Spectrolab recently moved forward in developing the UMM4J solar cell with a bandgap combination of 1.9/1.4/1.1/0.67eV for space PV. This paper explores the recent advances for UMM4J development at Spectrolab.

II. EXPERIMENTAL METHOD

All XTJ Prime, UMM3J, and UMM4J epitaxial semiconductor layers were grown by OMVPE in a commercial Veeco K475 large-scale reactor. The process used industry standard growth techniques, temperatures, precursor gases and liquid organometallic sources. The reactor was operated at reduced pressure under standard growth conditions for GaInP, GaInAs and similar alloys. Growths were conducted on p-type,

978-1-5090-5606-4/17 $31.00 © 2017 IEEE

epi-ready vicinal (001) Ge of 150 mm diameter. In-situ curvature measurements were made using a Veeco combined RealTemp emissivity corrected pyrometer and Deflectometer (or DRT) head located on a top viewport of the reactor targeted at the center point of the 150 mm outer ring of wafers. The DRT head monitors wafer stress status change in-situ and in-time, which was not available when Spectrolab first developed UMM3J and UMM4J structures.

III. RESULTS & DISCUSSION

A. Production 30.7% XTJ Prime SuperCells

Building upon the heritage of the AIAA S-111 and 112 qualified XTJ solar cell, XTJ Prime makes subtle modifications to the epitaxial layer structure to achieve a higher efficiency. After passing a battery of validation tests to evaluate the changes in performance for design characterization, SuperCell production for the product commenced in 2016. Fig. 1 shows a production histogram of one example of efficiency of SuperCells with areas >70-cm^2 produced to date. The total is for over 60,000 cells and averages 30.7% efficiency (AM0, 28°C, 135.3 mW/cm^2) as tested on a Spectrolab X-25 Solar Simulator with appropriate subcell setup standards. The population totals over 175 kW of space photovoltaic power. The distribution is well fit with a Weibull distribution peaked at 30.9%. The maximum cell attained an efficiency of 31.9%. Approximately 1/3 of the population exceeds 31% efficiency. Contained within the population are multiple areas of devices. We observe no decrease in performance for up to the largest cells, contained within the population, inclusive of our largest size cells producing up to 3.4W per cell.

Spectrolab characterized the XTJ Prime solar cell to enable the JPL method of mission performance prediction[7] to both the AIAA S-111-2005 and ECSS-E-ST-20-08C section 7.5.15 standards.[8-9] More specifically, the ECSS standard incorporates a 48 hour light soak anneal under AM0 illumination at 25°C with an additional 24 hour dark anneal at 60°C. The annealing allows some recovery of the performance to simulate spaceflight conditions. Fig. 2 shows the 1 MeV absolute measured efficiency for without (solid) and with (dashed) annealing for XTJ Prime up to 1e16 e/cm^2 fluence.

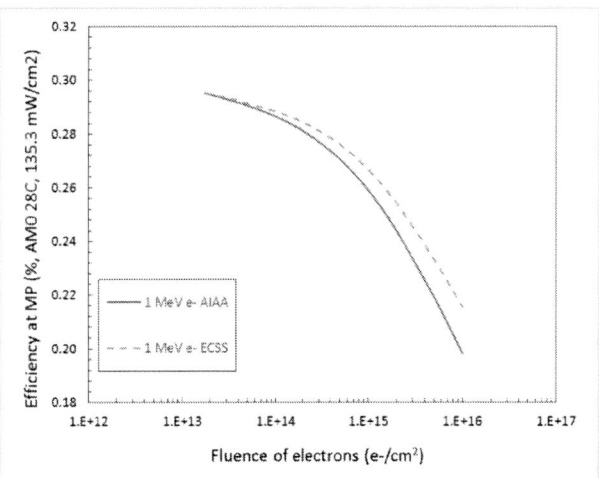

Fig. 2. AM0 efficiency for XTJ Prime under 1 MeV electron radiation for no anneal (solid, AIAA) and with annealing (dashed, ECSS).

At standard GEO mission fluence of 1e15 e/cm^2, XTJ Prime is 26.0% and 26.7% efficient without and with anneal, respectively. At the highest fluence tested of 1e16 e/cm^2, XTJ Prime reaches 21.5% with anneal.

We observed that the electron radiation for all energies and fluences tested recover with the ECSS annealing. Similarly, all proton testing (from 0.1 MeV to 10 MeV in energy) shows similar recovery, albeit at a reduced recovery observed in the electron radiation testing. Furthermore, we observe that recovery for protons that stop within the cell is different than those that fully penetrate the stack. The implications of such differentiated annealing recovery is that all components of the JPL method calculation must be done with a consistent basis, either with or without the annealing. Mixing a mission fluence generated without annealing and simply assuming the ECSS recovery for the 1 MeV electron only will lead to significant over-estimation of remaining power. We expect that these observations are general for all space solar cells.

B. UMM3J Initial Results

With the advanced tool capability and previous experiences built from the UMM structure development, UMM3J SuperCells were developed. Fig. 3 (a) shows the diagram of the UMM3J cell architecture, a band gap combination of 1.85eV/1.3eV/0.67eV is designed and built on 150 mm Ge wafers. One of the biggest challenges is the uniformity across the 150 mm wafers due to the wafer bow caused by the residual strain from the metamorphic structure. Fig. 3(b) shows a picture of UMM3J SuperCell of area 73.5cm^2 under forward bias. The top cell electroluminescence image shows the good uniformity across one of the two SuperCells from a 150 mm wafer.

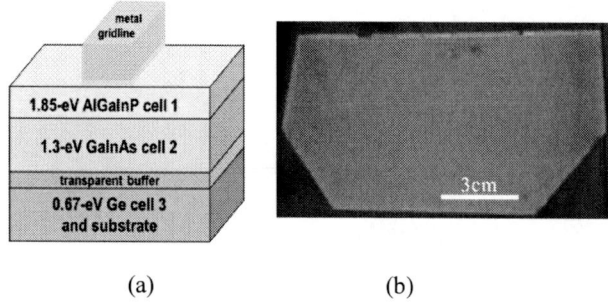

(a)   (b)

Fig. 3. (a) Diagram of the UMM3J design with bandgap combination of 1.85eV/1.3eV/0.65eV; (b) Image of a UMM3J SuperCell under forward bias.

We demonstrated the UMM3J device with a small build of 73.5 cm^2 SuperCells. Fig. 4 shows the QE data of a UMM3J SuperCell. Both the GaInP and InGaAs subcells are showing very high IQE, which is very comparable to lattice matched triple junction IQE, such as XTJ Prime; the EQE current of 1.85 eV top cell is 19.2 mA/cm^2, and 20.2 mA/cm^2 for the 1.30 eV middle cell.

TABLE I

RECENT UMM3J SUPERCELL BUILD CELL PERFORMANCE VERSUS PREVIOUSLY DEMONSTRATED RESULTS [4]

	Units	Previous Demonstration	Current SuperCell Build
Average Voc	V	2.40	2.62
Average Jsc	mA/cm^2	19.6	19.2
Average FF	%	82.6	84.4
Average Eff.	%	n/a	31.3
Best Eff.	%	28.8	31.5
Cell Area	cm^2	4	73.5

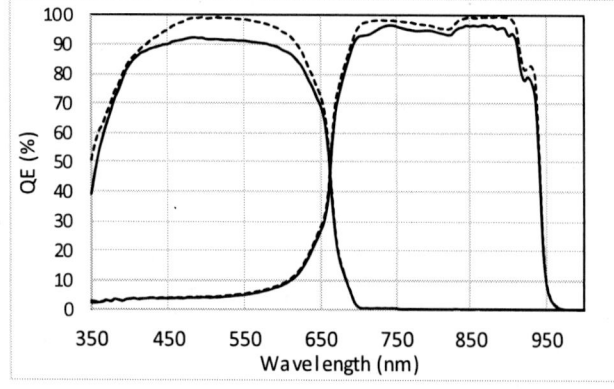

Fig 4. Quantum Efficiency of a UMM3J SuperCell; dash lines are IQE, solid lines are EQE

Fig. 5 shows the illuminated current-voltage (LIV) curve of the best cell at 31.5% efficiency from the build. Compared with Spectrolab's previous demonstration of UMM3J for space [4], significant improvement was achieved as can be seen in table 1. Current build shows the average efficiency of 73.5 cm^2 SuperCells under 1 sun AM0 at 31.3% with the maximum efficiency equal to 31.5% versus previous demonstration of 28.8% efficiency of a 4 cm^2 cell.

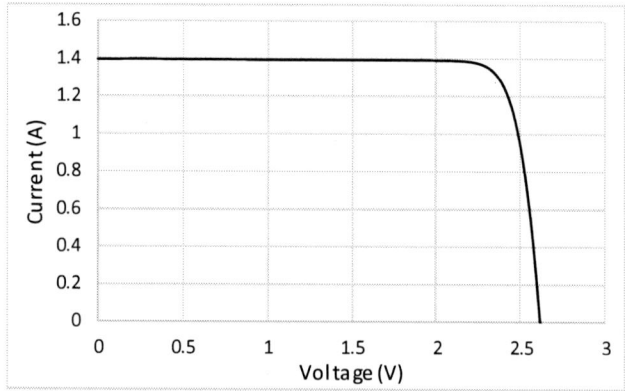

Fig. 5. LIV curve of a 73.5 cm^2 SuperCell under 1 sun AM0 with Spectrolab's X25 simulator; AM0 cell efficiency at 28°C is 31.5%.

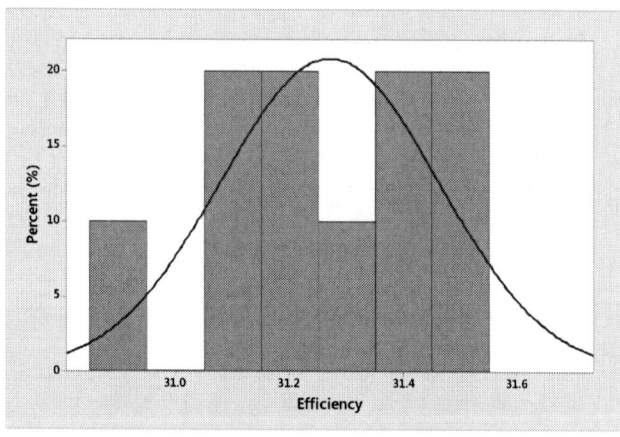

Fig. 6. Histogram of the UMM3J build BOL efficiency profile; the average efficiency is 31.3%

Samples of UMM3J solar cells were also evaluated under Low Intensity Low Temperature (LILT) conditions typical of interplanetary missions to Saturn. Fig. 7 shows an example cell efficiency versus temperature at solar conditions at 5.5 AU. At this distance from the sun, the irradiance is about 50 W/m^2 or about 3.6% of 1 sun AM0 1.0 AU irradiance and the expected solar array operating temperature reaches -140°C. We observe that for both unirradiated and irradiated cells with electrons to 1e15 1 MeV e- fluence, the best UMM3J device is capable of near linear temperature behavior. This result shows the UMM3J device material quality is high enough and suffers no deleterious effects from being mismatched to the substrate. Thus the UMM3J devices produced may be a good candidate for the LILT mission application. For details of the LILT environmental test results, please refer to another paper in this conference [10].

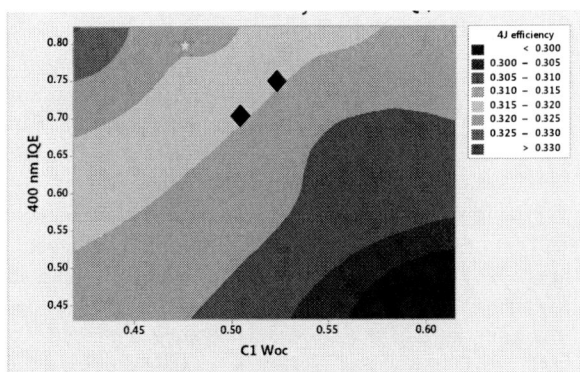

Fig. 8. Modeling of the UMM4J Efficiency profile vs top cell blue response and band gap offset W_{oc}.

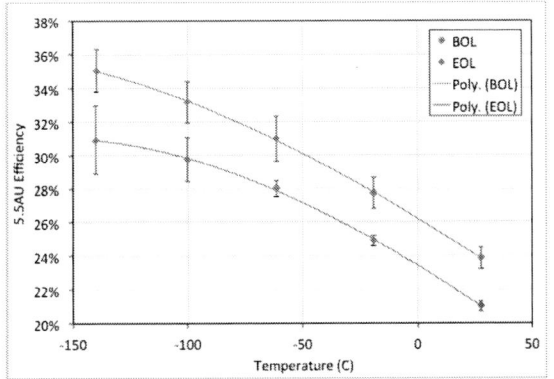

Fig. 7. Cell efficiency versus temperature at 5.5AU. Red color data is EOL efficiency, and the blue color data is BOL control.

C. UMM4J Development

One of the biggest challenges of UMM4J structure is the 1.9 eV AlGaInP top cell, specifically $Al_{0.21}Ga_{0.08}In_{0.71}P$, requiring 21% Al composition and 1.6% lattice mismatch to Ge substrate [11]. As described by Perl et al, AlGaInP subcells degrade in short wavelength response in IQE for increasing Al content [12]. The authors systematically investigated the impact of Al content in the AlGaInP subcell in a lattice matched structure. For optimal growth conditions and device design, the IQE response at 400 nm observed for the lattice matched material drops from 85% for GaInP (0% Al) to as low as 45% for 26% Al. To achieve the design goals of the UMM4J device, this subcell must simultaneously demonstrate a low band gap voltage offset (W_{oc}) and a high short wavelength internal quantum efficiency (as measured by IQE at 400 nm). Fig. 8 shows the contour mapping of the AM0 efficiency results of modelling the combined IQE at 400 nm and band gap voltage offset (W_{oc}) for a UMM4J design. Note that for the lattice matched devices in Fig. 3 of reference 12, a 400 nm response of 70% IQE and W_{oc} of 500 mV are observed for a similar 18% Al subcell of 2.09 eV band gap [12]. Similarly from MBE grown AlGaInP subcells of 73% 400 nm IQE and Woc of 520 mV[13]. These results would be sufficient to reach 31.5% as shown by the diamonds in the UMM4J in Fig. 8.

Presently under development is the top subcell for a UMM device. For a 1.0 cm^2 sample $Al_{0.21}Ga_{0.08}In_{0.71}P$ subcell grown on a metamorphic buffer layer on a 150 mm dia. Ge substrate, we observe a very high internal quantum efficiency, similar to that obtained from the no Al conditions of the XTJ Prime solar cell. As shown in Fig. 9, the IQE of the UMM4J top subcell reaches 77% at 400 nm. The major difference in IQE between the lattice matched case and the metamorphic is due to the absorption from 450 to 600 nm from the AlInP window adjusted to the lattice mismatched composition. Electroluminescence (EL) spectra of the top cell were also measured at room temperature and shows a W_{oc} of 480 mV at a forward bias current of 10 mA/cm^2. Together with the IQE result, the quality of the subcell is sufficient to reach 32% efficiency by the modelling and is shown as a star on Fig. 8. Despite the threading dislocation density residual from the metamorphic mismatch, we observe higher quality AlGaInP subcells compared to the lattice matched work previously presented. This result indicates a strong confidence that future UMM4J devices are posed to reach higher efficiency than present UMM3J or XTJ Prime devices in production.

Fig. 9. Internal quantum efficiency of a 1.9 eV AlGaInP

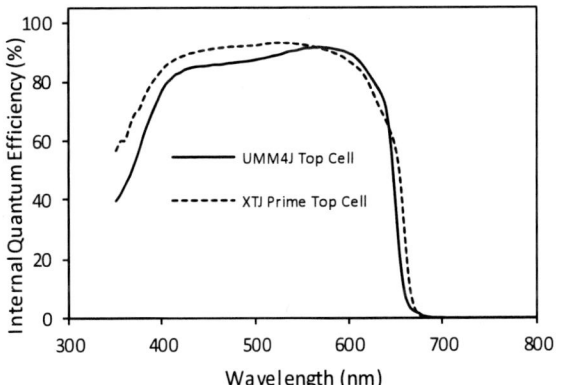

metamorphic subcell (solid) versus that of a lattice matched GaInP subcell from XTJ Prime (dashed) of about the same band gap.

978-1-5090-5606-4/17 $31.00 © 2017 IEEE

IV. CONCLUSIONS

For over 60,000 XTJ Prime solar cells in production average 30.7% efficiency. Radiation testing showed 26.0% and 26.7% at 1e15 e/cm^2 1 MeV electron radiation without and with ECSS annealing respectively. Additionally, we observe that annealing recovery depends on the particle and energy of the radiation. An improved UMM3J solar cell architecture for space were demonstrated. A 73.5cm^2 SuperCell build showed average efficiency of 31.3% under 1 sun AM0 and best cell efficiency of 31.5%. Additional testing under LILT space conditions showed that UMM3J is a viable candidate for deep space missions under LILT environments. For UMM4J, next generation space devices, a high quality 1.9 eV $(Al_{0.21}Ga_{0.79})_{0.xx}In_{0.xx}P$ subcell was demonstrated with 77% IQE response at 400 nm and a band gap voltage offset of 480 mV, sufficient to enable 32% efficient UMM4J devices.

ACKNOWLEDGEMENT

The authors would like to thank Dr. Andreea Boca for providing the LILT testing data; the authors also thank the entire Spectrolab PV team for their contributions and support. In particular this work was enabled by P. Hebert, K. Wan, Dr. H. Yoon, P. Luc, and Dr. R Bardfield and their teams of excellent engineers.

REFERENCES

[1] Spectrolab XTJ Prime datasheet accessed on 1/19/2017 at: http://www.spectrolab.com/DataSheets/cells/XTJ_Prime_Data_Sheet_7-28-2016.pdf.

[2] J. Boisvert, D. Law, R. King, E. Rehder, P. Chiu, D. Bhusari, C. Fetzer, X. Q. Liu, W. Hong, S. Mesropian, R. Woo, K. Edmondson, H. Cotal, D. Krut, S. Singer, S. Wierman, N. Karam, "High Efficiency inverted metamorphic (IMM) solar cells", in *39th IEEE Photovoltaic Specialists Conference*, 2013, p. 2790.

[3] R. King, M. Haddad, T. Isshiki, P. Colter, J. Ermer, H. Yoon, D. E. Joslin, and N. H. Karam, "Metamorphic GaInP/GaInAs/Ge," in *28th IEEE Photovoltaic Specialist Conference*, 2000, p. 982.

[4] C. Fetzer, R. King, P. Colter, K. Edmondson, D. Law, A. Stavrides, H. Yoon, J. Ermer, M. Romero, N. Karam, " High-efficiency metamorphic GaInP/GaInAs/Ge solar cells grown by MOVPE", *J. Crystal Growth*, vol. 261, 2004, p. 341.

[5] R. King, D. Law, K. Edmondson, C. Fetzer, G. Kinsey, H. Yoon, R. Sherrif, N. Karam, "40% efficient metamorphic GaInP/GaInAs/Ge multijunction solar cells", *Applied Physics Letters*, vol. 90, 2007, p. 183516.

[6] R. King, D. Bhusari, D. Larrabee, X. Q. Liu, E. Rehder, K. Edmondson, H. Cotal, R. jones, J. Ermer, C. Fetzer, D. Law, N. Karam, "Solar cell generations over 40% efficiency", *Progress in Photovoltaics*, 2012, p. 801.

[7] B. Anspaugh, "GaAs Solar Cell Radiation Handbook", JPL Publications, Pasadena CA, (1996).

[8] "Qualification and Quality Requirements for Space Solar Cells", AIAA-S-111-2005.

[9] "Photovoltaic Assemblies and Components", Vol 31 July 2008. ECSS-E-ST-20-08C.

[10] A. Boca, J. Grandidier, C. McPheeters, P. Sharps, P. Chiu, X. Q. Liu, and J. Ermer, "Advanced-Architecture High Efficiency Solar cells for Low Irradiance Low Temperature (LILT) Applications", Submitted to the *44th Photovoltaic Specialist Conference*, 2017.

[11] X.Q. Liu, C. M. Fetzer, E. Rehder, H. Cotal, S. Mesropian, D. Law, R. R. King, "Organometallic vapor phase epitaxy growth of upright metamorphic multijunction solar cells" *Journal of Crystal Growth*, vol. 352, 2012, pp. 186-189.

[12] E. E. Perl, J. Simon, J. F. Geisz, W. Olavarria, M. Young, A. Duda, D. J. Friedman, and M. A. Steiner, "Development of a 2.0 eV AlGaInP Solar Cell Grown by OMVPE", *IEEE J. Photovoltaics*, vol. 6, 2016, p. 770.

[13] J. Faucher, Y. Sun, D. Jun, D. Martin, T. Masuda, M. L. Lee, "High efficiency AlGaInP solar cells by molecular beam epitaxy," *Applied. Physics Letters*, vol. 109, 2016, pp. 172105.

Advanced-Architecture High-Efficiency Solar Cells for Low Irradiance Low Temperature (LILT) Applications

Andreea Boca[1], Jonathan Grandidier[1],
Claiborne McPheeters[2], Paul Sharps[2], Philip Chiu[3], Xing-Quan Liu[3] and James Ermer[3]

[1]Jet Propulsion Laboratory, California Institute of Technology, Pasadena CA 91109;
[2]SolAero Technologies Corp., Albuquerque NM 87123; [3]Boeing Spectrolab Inc., Sylmar CA 91342

Abstract — The scientific community and NASA view the outer solar system as a high-priority destination for space exploration and research. The cost and feasibility of future missions to 5-10 AU and beyond would greatly benefit from the availability of high-efficiency solar arrays optimized for low-irradiance low-temperature (LILT) environments. We have therefore set out to develop an advanced-architecture solar cell technology that will produce over 20% more power per unit area at LILT than the current state of practice. This technology will leverage recent advances in bandgap engineering originally developed for 1 AU, such as inverted or upright metamorphic multijunction device architectures. In addition, the new technology will feature a LILT-optimized, radiation-hard design. This paper reports on initial results from JPL's advanced-architecture LILT cell development project, specifically ground-based laboratory measurements of temperature-dependent illuminated current-voltage (LIV) and external quantum efficiency (EQE). Test conditions included irradiances corresponding to Sun distances from 1 AU to 9.5 AU, and junction temperatures between +28 °C and -165 °C. Through such measurements, multiple advanced architectures have been identified which demonstrate superior performance even prior to completing the design optimization for LILT, at both beginning of life (BOL, pre-radiation) and at end of life (EOL, post 1e15 1 MeV e⁻/cm²). For example, we measured 5.5 AU -140 °C average efficiencies of ~35% at BOL and ~31-33% at EOL for inverted metamorphic four-junction cells from SolAero and upright metamorphic triple-junction cells from Spectrolab. Further efficiency improvements are to be expected in the near future as we investigate, identify and eliminate from the device design those mechanisms that limit the cell performance at LILT.

I. INTRODUCTION

The scientific community and NASA view the outer solar system as a high-priority destination for space exploration and research in the current decade [1]. Of particular interest are targets at or beyond the orbit of Jupiter such as Io, the Trojan asteroids, Jupiter-system comets, Enceladus, Titan and Saturn. Solar arrays are low-cost, readily available, and highly reliable, which makes them an attractive alternative to radioisotope thermoelectric generators for such missions to deep-space destinations [2].

Although originally developed for 1 AU Earth-orbiting satellites, solar power has in recent years been successfully employed by missions to more distant destinations. NASA's Juno [3], launched in 2011, is now the farthest solar-powered spacecraft from the Sun, currently in orbit around Jupiter at approximately 5.5 AU. Juno illustrates the current state of

practice (SoP) in solar cell technology intended for low irradiance low temperature (LILT) applications. Specifically, Juno is flying 1 AU-optimized UTJ cells from Spectrolab, which are lattice-matched triple-junction devices; after screening out from the panel build those cells that have poor performance at 5.5 AU +28 °C, the resulting 5.5 AU -140 °C average efficiency is ~30% at beginning of life (BOL), according to ground-based laboratory measurements [4].

The cost and feasibility of future missions to 5-10 AU and beyond would greatly benefit from the availability of high-efficiency solar arrays optimized for LILT environments. And, solar cells are the component which has the single most leveraging effect on the overall power production capability of a solar array. We have therefore set out to develop an advanced-architecture cell technology that will produce over 20% more power per unit area than the current SoP at LILT. This technology will take advantage of recent advances in AM0 bandgap engineering, such as inverted or upright metamorphic multijunction device architectures. In addition, the new technology will feature a LILT-optimized, radiation-hard design, with >36% BOL target efficiency under both Jupiter (5.5 AU -140 °C) and Saturn (9.5 AU -165 °C) conditions.

This paper reports on initial results from the advanced-architecture LILT cell development project led by the NASA Jet Propulsion Laboratory (JPL). Specifically, we show temperature-dependent illuminated current-voltage (LIV) and external quantum efficiency (EQE) data measured on multiple device architectures, under conditions relevant to Sun ranges of up to 9.5 AU.

II. LILT LIV PERFORMANCE STATUS

This section reviews recent characterization data on advanced-architecture cells, obtained in the LILT solar simulator facility at JPL. The device architectures in the study included inverted metamorphic four-junction (IMM4) by SolAero Technologies Corp., and upright metamorphic triple-junction by Boeing-Spectrolab Inc. Both of these cell architectures are currently under development for 1 AU by their respective vendors, and both have yet to be space-qualified per the AIAA-S111 standard [5]. For the initial round of testing presented here, the cell designs were

1 AU-optimized, with no attempt yet made at LILT optimization.

The SolAero IMM4 test articles were coverglassed interconnected cells of active area 27.56 cm² each, quantity 12. The Spectrolab upright metamorphic test articles were bare cells, of active area 3.98 cm² each, quantity 8. For each cell type, half the cells were kept unirradiated at BOL, and the other half were subjected to an end-of-life (EOL) irradiation dose of 1e15 1 MeV e⁻/cm². The LIV test conditions for all cells in this study were: Earth irradiance (1 AU, 1 sun) +28 °C; Jupiter irradiance (5.5 AU, 0.033 suns) between +28 °C and -140 °C; and Saturn irradiance (9.5 AU, 0.011 suns) between +28 °C and -165 °C.

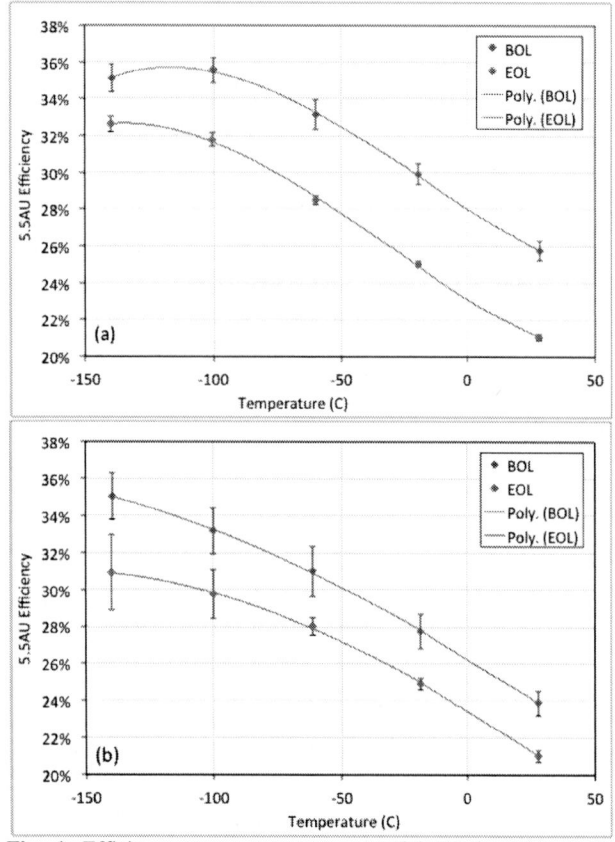

Fig. 1. Efficiency versus temperature at 5.5 AU for (a) inverted metamorphic 4-junction; (b) upright metamorphic 3-junction cells.

The test setup consisted of a light source and a temperature control chamber. The light source was a 2-zone X25 solar simulator, calibrated with top and middle-junction isotype ZTJ [6] standards for testing the SolAero cells, and with XTJ [7] standards for the Spectrolab cells (i.e., for both cell types the spectral calibration was imperfect, given the mismatch between the test articles and the standards). The simulator was capable of producing irradiances for Sun distances between 1 AU (1367 W/m²) and 9.5 AU (15 W/m²) at the test plane. The chamber allowed for cell temperatures from -165 °C to +28 °C, while keeping the test articles in a dry, inert N₂ atmosphere. The test fixture in the chamber consisted of a

vacuum chuck with multiple Kelvin probes, accepting up to eight cells at a time. This enabled thermal contact to the test articles without the need for permanent bonding to a substrate, as well as four-point electrical contacting without the need for permanent soldered connections. The chamber was outfitted with a quartz window for the simulator beam, which had irradiance uniformity of better than ±2% over the test area. The device efficiency was measured by sweeping and averaging over at least three LIV curves per cell at each test condition.

Fig. 1 shows efficiency data measured at 5.5 AU as a function of temperature, with inverted metamorphic four-junction cells in panel (a) and upright metamorphic triple-junction cells in panel (b). For both cell types, the BOL unirradiated cells are shown in blue, whereas the EOL irradiated cells are in red. Markers represent averages, and error bars are standard deviations over the sample population. The lines are third-order polynomial fits to aid the eye. Before plotting in the figure, as-measured efficiency data was divided by a technology-dependent correction factor, which was calculated to be in the range of 1.04-1.05 for both cell types; this was for the purpose of removing the above-mentioned systematic errors in the solar simulator calibration.

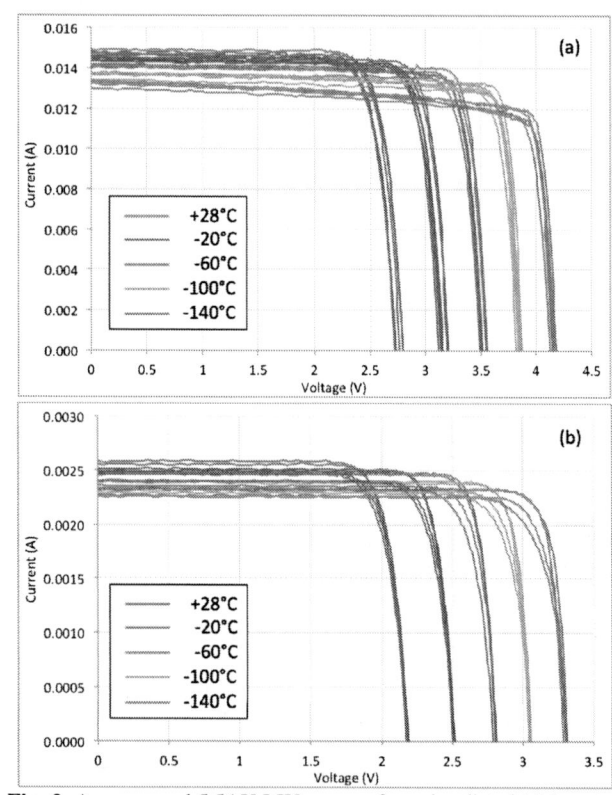

Fig. 2. As-measured 5.5AU LIV sweeps for unirradiated (a) inverted metamorphic 4-junction; (b) upright metamorphic 3-junction cells.

Under 5.5 AU -140 °C conditions relevant to operation at the orbit of Jupiter, the efficiencies as shown in Fig. 1 were: (a) 35.1% ± 0.7% at BOL, and 32.6% ± 0.4% at EOL; (b) 35.0% ± 1.2% at BOL, and 30.9% ± 2.1% at EOL. Therefore, both device architectures are found to have extremely promising performance under Jupiter LILT conditions, especially when considering that these are the initial, 1 AU-optimized designs. Under standard test conditions of 1 AU +28 °C, the efficiencies for the two BOL cell populations were (a) 31.5% ± 0.5%; and (b) 30.3% ± 0.1%.

Fig. 2 shows the 5.5 AU LIV sweeps taken at various temperatures on the BOL cells, with the two architectures in the same order in panels (a) and (b) as for Fig. 1.

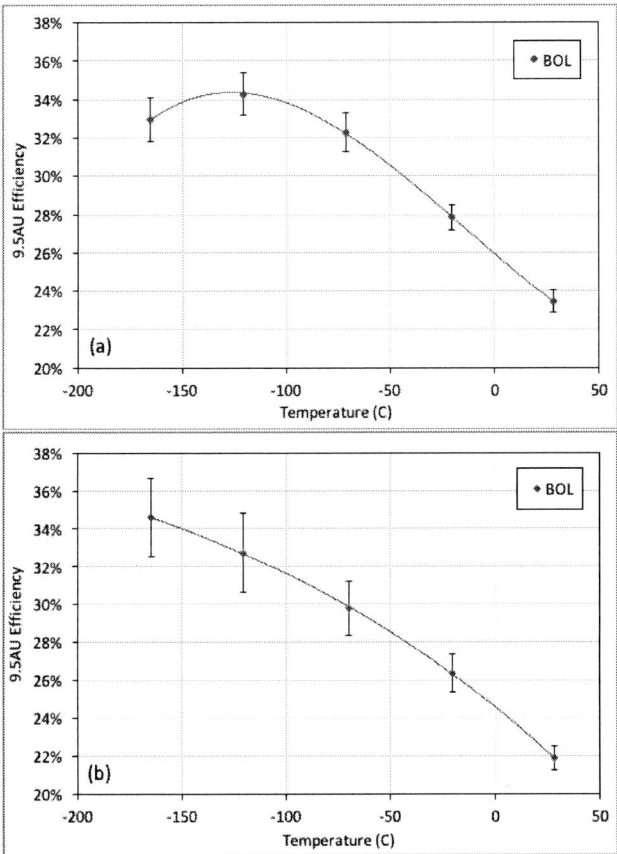

Fig. 3. Efficiency versus temperature at 9.5 AU for (a) inverted metamorphic 4-junction; (b) upright metamorphic 3-junction cells.

Fig. 3 shows efficiency data measured at 9.5 AU on the BOL cells only, as a function of temperature. The plotting conventions and data treatment are analogous to those for Fig. 1. The charged-particle radiation environment at Saturn is relatively benign compared to that at Jupiter [8], meaning that characterization of the cell radiation performance under Saturn LILT conditions is of secondary importance.

Under 9.5 AU -165 °C conditions relevant to operation at the orbit of Saturn, the BOL efficiencies as shown in Fig. 3 were: (a) 33.0% ± 1.1%; (b) 34.6% ± 2.1%. Once again, both device architectures are found to have promising performance under Saturn conditions, even prior to LILT optimization of the device design.

In addition to the advanced architectures shown here, we are also in the process of investigating other promising

alternatives, including IMMX+ from SolAero and several lattice-matched options from Spectrolab.

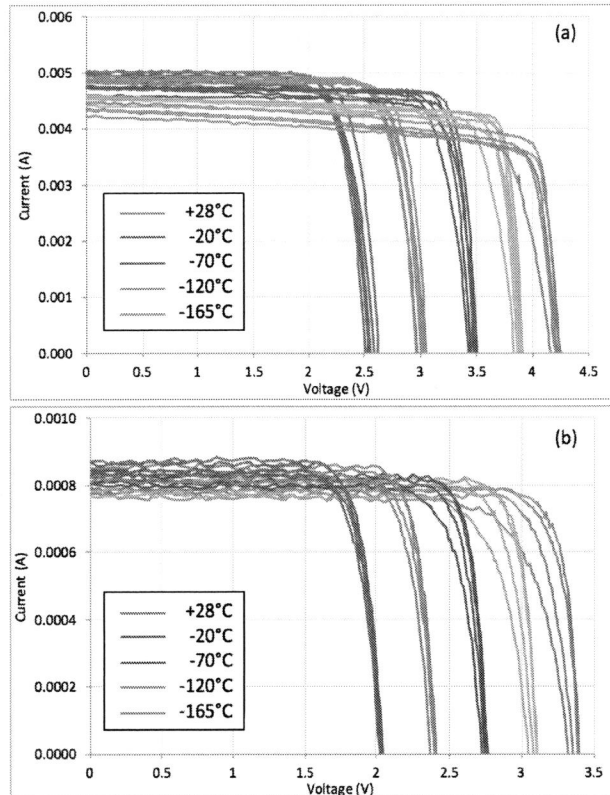

Fig. 4. As-measured 9.5AU LIV sweeps for unirradiated (a) inverted metamorphic 4-junction; (b) upright metamorphic 3-junction cells.

III. EQE AND LILT OPTIMIZATION PLAN

One important consideration in the context of optimizing the cell design for LILT is that of current balance at low temperature. 1 AU-optimized cells tend to be current-balanced at or near +28 °C; however, due to the bandgap shift with temperature, the current-balance condition will not necessarily hold in the -140 to -165 °C range, leading to sub-optimal performance. The situation is further complicated once the cell is exposed to radiation, with the different subcell materials presenting varying degrees of radiation hardness as a function of temperature. This underlines the importance of being able to quantify the low-temperature current generation within each individual subcell, which can be done by way of EQE testing. In addition, thanks to its dependence on minority carrier diffusion length and surface recombination velocity, EQE data can help identify more subtle problems with the quality of specific layers or interfaces, which can also be temperature-dependent.

In response to this need, JPL has recently developed a low-temperature external quantum efficiency test apparatus. The setup is analogous to the LILT LIV lab as described in the previous section, in that it is an external light source coupled with a thermal control chamber. In this case however, the light source consists of a calibrated monochromatic beam of variable wavelength, plus light bias sources that are spectrally relevant to each of the subcells under test. With this setup, we have so far demonstrated the capability to test cells of up to four junctions, at temperatures ranging from +28 °C to -165 °C.

Fig. 3 displays example EQE data sets, that were recently measured on two of the BOL cells in this study. Panel (a) shows one of the IMM4 cells from SolAero, whereas panel (b) shows one of the upright metamorphic triple-junction cells from Spectrolab. The curves represent the EQE data for all the subcells in the multijunction stack as a function of wavelength, with each color corresponding to a different test temperature, in a range relevant to Sun distances from 1 AU to 9.5 AU LILT. The wavelength units are obscured, in order to preserve the competition-sensitive nature of the bandgap combination for each device architecture.

In combination with analytical modeling and additional device and material testing, such data will be used by JPL's advanced cell development project to guide the process of optimizing the device design for LILT.

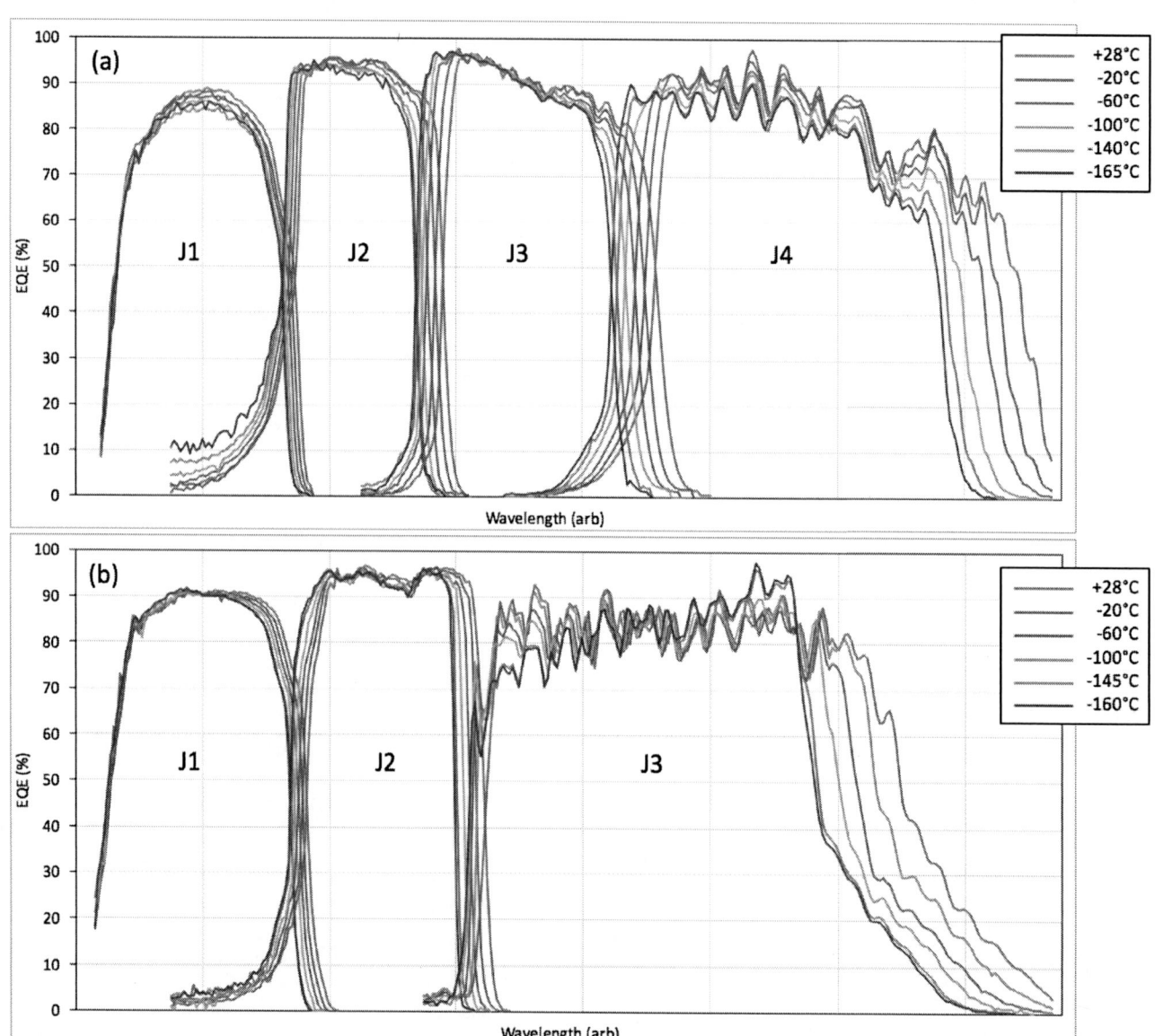

Fig. 3. Example external quantum efficiency data on an unirradiated (a) inverted metamorphic four-junction cell; and (b) upright metamorphic triple-junction cell, at temperatures relevant to sun ranges from 1 AU to 9.5 AU.

IV. CONCLUSIONS

Multiple advanced architectures have been identified which demonstrate high LILT performance at BOL and EOL, under both Jupiter (5.5AU) and Saturn (9.5AU) irradiance conditions. Further improvements are to be expected in the near future as we identify and eliminate from the design those mechanisms that limit cell performance in the LILT environment.

V. ACKNOWLEDGEMENTS

The authors thank Molly Shelton, Robert Kowalczyk and Richard Ewell from JPL for their valuable contributions.

This research was carried out at the Jet Propulsion Laboratory, California Institute of Technology, under a contract with the National Aeronautics and Space Administration.

REFERENCES

[1] "Visions and voyages for planetary science in the decade 2013-2022", National Research Council, 2011.

[2] A. Boca et al., "A data-driven evaluation of the viability of solar arrays at Saturn", *IEEE J. Photovolt.* (accepted), 2017.

[3] S. J. Bolton, "The Juno mission," in *Proc. IAU 269*, 2010.

[4] P. Stella et al., "LILT testing of solar cells for the Juno mission," in *Proc. 21st SPRAT*, 2009.

[5] "Qualification and quality requirements for space solar cells", AIAA S-111A-2014.

[6] "SolAero ZTJ space solar cell", product datasheet, retrieved from www.solaerotech.com on 1/18/2017.

[7] "Spectrolab neXt triple junction (XTJ) solar cells", product datasheet, retrieved from www.spectrolab.com on 1/18/2017.

[8] W. Atwell, "Radiation environments for deep-space missions and exposure estimates," in Proc. AIAA SPACE, 2007, p. 6044.

Copyright 2017. All rights reserved.

Ultra-Lightweight PV module design for Building Integrated Photovoltaics

Ana C. Martins[1], Valentin Chapuis[1], Alessandro Virtuani[1], Christophe Ballif[1, 2]

[1] École Polytechnique Fédérale de Lausanne (EPFL), Institute of Microengineering (IMT),
Photovoltaics and Thin Film Electronics Laboratory (PV-Lab), Rue de la Maladière 71b, 2000 Neuchâtel,
Switzerland

[2] CSEM, PV-center, Jaquet-Droz 1, 2000 Neuchâtel, Switzerland

Abstract — **Most of the existing solutions for Building Integrated PV (BIPV) are based on conventional crystalline-Silicon (c-Si) module architectures (glass-glass or glass-backsheet) exhibiting a relatively high weight (12-20 kg/m²). We are working on the development of robust and reliable lightweight solutions with a weight target of 6 kg/m². Using a composite sandwich architecture and high thermal conductivity materials, we show that it is possible to propose lightweight PV modules compliant with the IEC 61215 thermal cycling test. We further show that we are able to upscale the size of the devices from 2-cells up to 16-cell modules.**

Index Terms — **BIPV, crystalline-silicon module, lightweight, reliability, composite sandwich structure.**

I. INTRODUCTION

Photovoltaic (PV) technology showed an impressive field deployment during the last decades, with estimations showing in 2016 a growth over 2015 of 30% of installed capacity, totaling a cumulative capacity of 295GW [1]. However, in several countries (e.g. Switzerland), the amount of land available for solar fields is extremely limited [2]. Consequently, integration of PV in buildings (Building integrated photovoltaics, BIPV) appears as a high potential solution [3]. Most of the BIPV products currently on the market are based on standard crystalline-Silicon (c-Si) modules architectures exhibiting the drawback of a high weight, due to the presence of one or several glass sheets [4], [5]. Table I shows standard module weights for c-Si photovoltaic modules. The high weight puts constraints on the building envelope and the supporting structure resulting in increased BOS (balance of system) costs and installation limitations in the case of building refurbishment.

The development of lightweight aesthetic PV elements is of high importance for large-scale deployment of BIPV, especially when renovating buildings.

TABLE I
TYPICAL MODULE WEIGHT FOR C-SI PV MODULES

Module layup (c-Si technologies)	Weight, w [kg/m²]
Glass-Backsheet	12 – 16
Glass-Glass	14 – 17
Glass-Glass for BIPV	≥ 20
Proposed lightweight solution	5-7

In this study, we propose an ultra-lightweight PV module based on c-Si technology with a weight of ~6 kg/m². To reach this low weight, the module is built with a glass-free frontsheet and the backsheet is built using a composite sandwich structure, bringing the needed mechanical stiffness to the module [6]–[8]. Due to the use of different stacked materials with different thermal properties, the main challenge is to avoid early failure of the structure during a sub-set of IEC 61215-2:2016 qualification tests.

II. MATERIALS AND METHODS

A. Ultra-Lightweight PV design, processing and testing

PV Module Design

Our ultra-lightweight PV module is based on the use of an innovative composite sandwich structure as a backsheet and a glass-free frontsheet (see Fig. 1). The composite sandwich materials include glass fiber reinforced polymer (GFRP) and a lightweight material with a honeycomb structure [9]. Two type of honeycomb materials are tested having a low or a high thermal conductivity.

Fig. 1. Sketch of our ultra-lightweight PV module design developed for BIPV applications.

Ageing tests methods

Three replicas of 2-cells module were manufactured for each sample design defined in Table II. These modules were subjected to the Thermal Cycling (TC) test, performed according to IEC 61215-2:2016 (but with no current injection) [13]. The condition named *Reference* represent a condition manufactured with a commercially available sandwich structure.

TABLE II
SUMMARY OF MATERIALS AND PROCESSING CONDITIONS USED

Sample	Materials		Processing Conditions
	Adhesive	Core	Process duration
Reference	Thermoset liquid	Low conductivity	Short Proc + 24 h
Sample 1	Adhesive	Low conductivity	Short Proc
Sample 2	Adhesive	High conductivity	Short Proc
Sample 3	Adhesive	High conductivity	Long Proc

B. Differential scanning calorimetry (DSC)

DSC is a widely used technique in PV industry to access the degree of crosslinking of thermosetting adhesives [14]–[16]. In our case, we can use it to evaluate the quality of the sandwich after processing. The tests are performed in a Mettler Toledo DSC1 system operated in single-run mode. The samples consist of 1 mm-thick discs of 5-10mg. Thermograms are recorded under constant nitrogen flow from -20°C to 225°C at a heating rate of 10°C/min, held at 225°C for 1 min and then cooled down to -20°C at a cooling rate of 10°C/min. The enthalpy associated to the crosslinking reaction (ΔH_{cured}) is evaluated from the area of the exothermal peak between 110°C to 200°C. Knowing this parameter allows to calculate the crosslinking degree [17]:

$$X = \frac{\Delta H_{uncured} - \Delta H_{cured}}{\Delta H_{uncured}} \qquad (1)$$

Where X is the degree of crosslinking and $\Delta H_{uncured}$ corresponds to the enthalpy peak of the uncured samples between 110 and 200°C.

C. Four-point bending tests

The resistance of the composite sandwich structure under bending load is investigated by means of four-points bending tests [18], [19]. This test allows us to identity three main parameters: i) bending stiffness, D (resistance of the beam against load), ii) Yield Load, P_y (limit of the elastic region) and the typical failure mode under load (input about the weakest interface and/or component of the structure) [20]. The test is performed on a Walter+Bai AG EC80-MS mechanical testing instrument in displacement control at a rate of 20µm/s, as represented in Fig. 2.

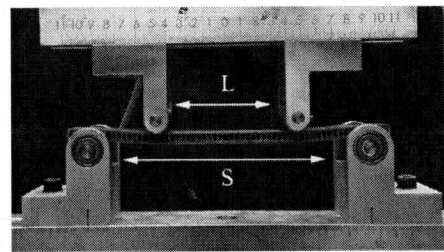

Fig. 2. Setup used to perform four-point bending tests. S represents the outer load span and L the inner load line.

We used an outer load span (S) of 190 mm and an inner load line (L) of 47.5mm. A linear variable displacement transducer (LVDT) is used to measure the load point displacement and the applied load is measure with a 10kN load cell. Coupons of 25mm width and 220mm length are prepared according to conditions presented in Table II. From this test, we obtain load point displacement (δ) versus load (P). For well-bonded cores, the in-plane shear stiffness is sufficiently large so that the overall displacement is only dominated by the bending momentum, leading to Equation 2. The sandwich beam stiffness D can thus be determined from the slope of the bending curve in the elastic regime [20].

$$D = \frac{PL^3}{48\delta} \qquad (2)$$

III. RESULTS AND DISCUSSION

The 2-cells modules (see Table II) are analyzed after 70 and 200 cycles of the TC test, as shown in Fig. 3. After 200 cycles, we obtain a power loss of -1.2%, -2.4% and -1.3% for the *Reference*, *Sample 1* and *Sample 3*, respectively. Thus, according to IEC 61215-2:2016's pass/fail requirements (power drop limited to -5%), all these samples passed successfully the 200 cycles.

From the visual inspection of the 2-cells modules, we see that *Sample 1* and *Sample 2* are delaminated, suggesting that the manufacturing process did not enable a proper crosslinking. In order to evaluate the quality of the sandwich we perform DSC to quantify the degree of crosslinking.

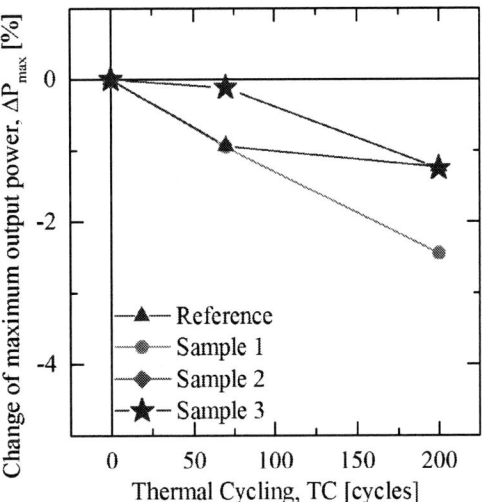

Fig. 3. Results from TC from 2-cells mini-module. Electrical properties are analyzed after 70 and 200 cycles. *Sample 2* failed after few cycles so its electrical performance could not be measured.

A. Analysis of composite sandwich structure quality

Fig. 4 shows the results obtained from the DSC measurements for not processed and processed adhesive (*Sample 1, 2* and *3* with respective processing as shown in Table II). *Sample 1*, processed with a low conductive core, shows a higher degree of crosslinking on the adhesive close to the heating plate than the adhesive far from the heating plate (72% and 1% degree of crosslinking, respectively). The fact that we are using a low conductive core does not allow a good heat transfer through the sandwich, affecting strongly its final quality. These values explain why delamination is observed during TC. The cycles in temperature induce thermal stresses in our modules, consequently, since the quality of our adhesive is very low, the composite sandwich structure will not behave as structural component, but instead, allow our frontsheet to expand when temperature rises and contract when temperature decreases.

By substituting the low conductivity core by a high conductivity core (*Sample 2*) we observe a very small increase of the degree of crosslinking on the top skin but a big decrease on the crosslinking on the bottom skin. The use of a high conductive core allows a good heat transfer from the bottom to the top, so good that most of the heat is lost in the laminator body.

Keeping the high conductivity core but increasing the full processing time, we can increase the stiffness of the sandwich adhesive. For *Sample 3*, the DSC done in both adhesive layers show their high degree of crosslinking and no delamination is observed during TC any more.

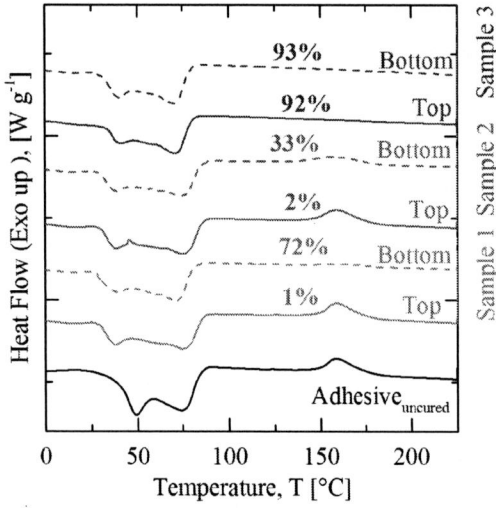

Fig. 4. DSC measurements obtained for an uncured adhesive (no processing), Sample 1, 2 and 3 (top and bottoms refers to close, respectively far from the heating plate).

B. Sandwich bending stiffness

The mechanical properties, bending stiffness and yield load obtained for the different composite sandwich structures are represented in Fig. 5. We can see that the *Reference* condition shows very good results: its bending stiffness is 14.9 N.m^2 and the yield load of about 356.8 N. Its excellent mechanical properties make this sandwich ideal for applications where high stiffness is needed. However, from the manufacturing point of view, this adhesive does not suit the requirements within PV industry, due to the too long manufacturing process. *Sample 1* shows a bending stiffness very low (8.5 N.m^2) and a typical failure mode of skin debond. This failure mode is typical on sandwiches where adhesive does not have enough stiffness and thus cannot transfer the stress properly. *Sample 2* presents even lower bending stiffness and lower yield load due to the very low crosslinking degree as explained in the previous section. From all samples, *Sample 3* is the best candidate, where higher values for bending stiffness are obtained. We found that a bending stiffness of 9.6 N.m^2 is enough to pass the TC test without delamination nor bending of the composite sandwich structure. From all these results, we decide to upscale the solution of *Sample 3*.

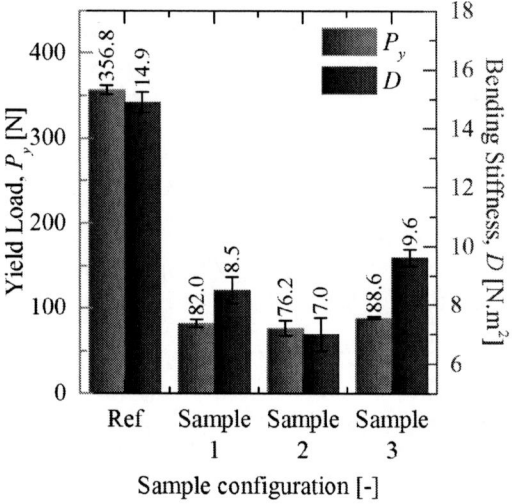

Fig. 5. Bending stiffness of all samples obtained from four-point bending tests.

IV. UPSCALING: FROM MINI- TO MEDIUM-AREA MODULES

In the previous sections, we showed that a high conductivity core and a longer manufacturing process are needed to develop a lightweight module that has good mechanical properties, without bending nor delamination. We upscale this solution to a medium-area module (16-cells) and we test it under TC according to IEC 61215-2:2016 [13] (with current injection). We produced a set of three samples for this test with the size of 810 mm x 810mm. Before introducing the modules in TC we performed an electrical characterization (IV test), the wet leakage current test and the electroluminescence (EL) to analyze the initial state of the solar cells. Visual inspection (VIS) is also performed. Fig. 7 represents module

978-1-5090-5606-4/17 $31.00 © 2017 IEEE   2106

power loss and fill factor loss during TC. We observed a power loss of -2.4% and a loss in fill factor of about -2.2%. These degradation rates are below the -5% pass/fail criteria present at the IEC qualification tests. Moreover, from Electroluminescence (EL) images and VIS we observe that no major change appears after the test (Fig. 7).

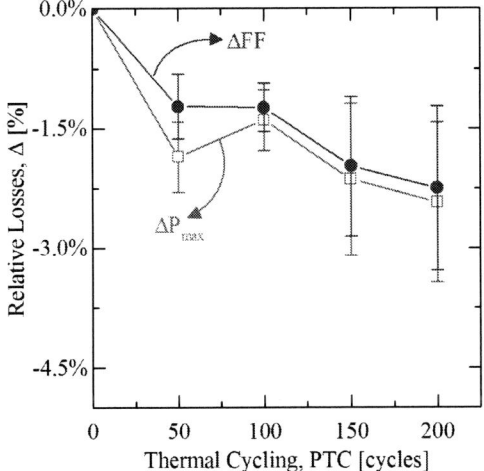

Fig. 6. Relative power and FF losses after TC with current injection of medium-area modules. Electrical properties are analyzed after 50, 100, 150 and 200 cycles.

Initial   **TC – 200 cycles**

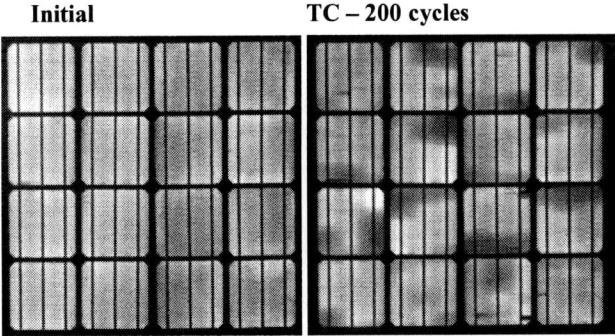

Fig. 7. EL images of one module tested under TC before and after 200 cycles.

The wet leakage test performed with our medium-area modules show that the insulation of our modules under wet operation conditions is very good, showing an insulation resistance higher than the pass/fail criterion of 40MΩ·m². The modules are being subjected to TC testing for an extended number of cycles.

With the aim of designing a BIPV product able to pass IEC certification, we are subjecting our test devices to a set of additional tests such as Damp-Heat, Hail, Mechanical Load tests within other tests. The feedback from these tests are then used to optimize the design and manufacturing process of our devices.

VI. CONCLUSION

In this study, we propose a lightweight PV module with a weight of 6 kg/m² for BIPV (and other) applications. The module is based on a composite backsheet and a glass-free frontsheet. We show that due to the sandwich architecture and the presence of a honeycomb-structured core, the manufacture process and the core thermal conductivity must be adjusted to enable high quality sandwich adhesive after processing. This solution is currently being up-scaled from 2-cells modules to 16-cells modules. The preliminary results obtained after thermal cycling (with power injection) are promising with only -2.4% power loss after 200 cycles and no sign of delamination nor bending. Other modules are currently being tested under different ageing tests to evaluate the influence of humidity, load or hail.

ACKNOWLEDGEMENT

The authors gratefully acknowledge the financial support of the Swiss National Science Foundation (SNSF) through the National Research Programme (NRP) "Energy Turnaround" (NRP 70). Further information on NRP can be found at www.nrp70.ch. We further acknowledge the support of all the staff at PV-Lab and CSEM, in particular Xavier Niquille.

REFERENCES

[1] A. Mathur, "Global solar PV capacity will approach 295 GW in 2016." [Online source]. Available at: http://energy.globaldata.com/media-center/press-releases/power-and-resources/global-solar-pv-capacity-will-approach-295-gw-in-2016-led-by-china-says-globaldata. [Accessed: 02-Jun-2017].

[2] IEA PVPS, "Trends 2016 in photovoltaic applications." Report IEA PVPS T1-30:2016.

[3] L.-E. Perret-Aebi, "New approaches for BIPV elements: from thin film terra-cotta to crystalline white modules," in *13th National Photovoltaic Congress - Swissolar*, Switzerland, 2015.

[4] H. Nussbaumer, M. Klenk, and N. Keller, "Small unit compound module: A new approach for light weight PV modules," in *32nd European Photovoltaic Solar Energy Conference, EUPVSEC*, Munich, 2016, *pp:56-60.*

[5] T. Kajisa, H. Miyauchi, K. Mizuhara, K. Hayashi, T. Tokimitsu, M. Inoue, K. Hara, A. Masuda, "Novel lighter weight crystalline silicon photovoltaic module using acrylic-film as a cover sheet," *Japanese Journal of Applied Physics*, vol. 53, no. 9, pp. 092302, 2014.

[6] D. Zenkert, *An introduction to sandwich construction.* Engineering materials advisory services, 1995.

[7] P.-E. Bourban, *Matériaux composites à matrice organique: constituants, procédés, propriétés*, vol. 15. PPUR presses polytechniques, 2004.

[8] J. M. Davies, *Lightweight Sandwich Construction*. John Wiley & Sons, 2008.

[9] W. S. Burton and A. K. Noor, "Structural analysis of the adhesive bond in a honeycomb core sandwich panel," *Finite Elements in Analysis and Design*, vol. 26, no. 3, pp. 213–227, Jun. 1997.

[10] R. Okada and M. T. Kortschot, "The role of the resin fillet in the delamination of honeycomb sandwich structures," *Composites Science and Technology*, vol. 62, no. 14, pp. 1811–1819, 2002.

[11] M. D. Banea and L. F. M. da Silva, "Adhesively bonded joints in composite materials: An overview," *Proceedings of the IMechE*, vol. 223, no. 1, pp. 1–18, 2009.

[12] M. Kempe, "Overview of scientific issues involved in selection of polymers for PV applications," in *37th IEEE Photovoltaic Specialists Conference*, 2011, pp. 85–90.

[13] IEC-International Electrotechnical Commission, "Terrestrial photovoltaic (PV) modules - Design qualification and type approval - Part 2: Test procedures," 2016.

[14] H.-Y. Li, L.-E. Perret-Aebi, R. Théron, C. Ballif, Y. Luo and R. FM Lange. "Optical transmission as a fast and non-destructive tool for determination of ethylene-co-vinyl acetate curing state in photovoltaic modules." *Progress in Photovoltaics: Research and Applications*, vol. 21, no. 2, pp:187-194, 2013.

[15] W. Stark, M. Jaunich, W. Bohmeyer, and K. Lange, "Investigation of the crosslinking behaviour of ethylene vinyl acetate (EVA) for solar cell encapsulation by rheology and ultrasound," *Polymer Testing*, vol. 31, no. 7, pp. 904–908, 2012.

[16] D. Roşu, C. N. Caşcaval, F. Mustaă, and C. Ciobanu, "Cure kinetics of epoxy resins studied by non-isothermal DSC data," *Thermochimica Acta*, vol. 383, no. 1–2, pp. 119–127, 2002.

[17] H. Li, "Open the Black Box: Understanding the Encapsulation Process of Photovoltaic Modules," 2013.

[18] I. M. Daniel and J. L. Abot, "Fabrication, testing and analysis of composite sandwich beams," *Composites Science and Technology*, vol. 60, no. 12–13, pp. 2455–2463, 2000.

[19] S. Belouettar, A. Abbadi, Z. Azari, R. Belouettar, and P. Freres, "Experimental investigation of static and fatigue behaviour of composites honeycomb materials using four point bending tests," *Composite Structures*, vol. 87, no. 3, pp. 265–273, 2009.

[20] F. Arias, P. JA Kenis, B. Xu, T. Deng, O. JA Schueller, G. M. Whitesides, Y. Sugimura and A. G. Evans, "Fabrication and characterization of microscale sandwich beams," *Journal of Materials Research*, vol. 16, no. 02, pp. 597–605, 2001.

Design it with LSCs; an exploration of applications for Luminescent Solar Concentrator PV technologies.

Wouter Eggink & Angèle Reinders

University of Twente, Faculty of Engineering Technology, Enschede, 7550AE, the Netherlands

Abstract — **This paper shows the results of a design study on possible applications for Luminescent Solar Concentrator PV technologies. The study focused on product integration of LSC PV technologies and was executed by students of Industrial Design Engineering at University of Twente (NL). In total 16 different and highly innovative conceptual designs resulted from this project, which were prototyped at scale to show their feasibility and integration features.**

In this paper we present several concepts, and discuss relevant findings for the integration of LSC PV technologies in future products and buildings. It is shown that the typical material properties of LSCs; low cost, colorful, bendable, and transparent do not only offer a lot of design freedom, but also offer excellent possibilities to incorporate this technology into the overall function and experience of applications.

Keywords — Luminescent Solar Concentrator PV, Product Integrated PV, Building Integrated PV, Design

I. INTRODUCTION

An increased use of PV systems can be achieved by extending the use of existing types of PV systems, or by stimulating the use of PV in new applications. This paper focuses on the latter by exploring novel applications of Luminescent Solar Concentrating PV technologies.

LSCs are light guides, often made from transparent polymer sheet materials which are doped with luminescent dyes [1, 2]. Solar radiation enters the LSC through the large top surface and is absorbed by the luminescent particles which re-emit the radiation at a longer wavelength. Subsequently, a large fraction thereof gets trapped within the light guide by means of total internal reflection. The trapping is interrupted at the interface between the light guide and the attached solar cells, where the radiation gets converted into electricity. Hence, LSCs provide a large solar radiation surface while they employ less semiconducting materials. Hence, LSC-PV modules offer a potentially lower cost per Wp [3], and more design freedom because the polymer light guides can be bent while retaining their efficiency [4]. However, despite these advantages, so far, little experience exists with the application of LSC PV technologies in modules, buildings and products. This while the design features of LSCs allow for interesting form giving such as coloring, and being flexible and transparent. For this reason we liked to explore how new PV applications based on LSC PV technologies could be by the execution of a design project.

This approach fits to the ongoing design-driven research program on product-integrated PV, which started in 2008 [5] and has led to various projects and publications in the past decade, both on the performance of product-integrated PV [6-10] and user experiences with these technologies [11-13], as well as design issues concerning the application of these technologies [14, 15]. The approach also fits to the Human-Technology Relations program of the University of Twente [16] that researches the influence of design features on the adaptation and user acceptance of innovative technologies,

Fig. 1. Examples of conceptual designs with LSC-PV technologies, which have been developed by students of Industrial Design Engineering at University of Twente in 2016. Outdoor street furniture (left), a self-powered table (middle) and a beach shelter with integrated electronics (right).

supported by Future Scenario Development [17-19].

II. PROJECT CONTEXT

As such, a design study was executed on possible applications for LSC-PV technologies within the scope of the master course 'Sources of Innovation', part of the Master in Industrial Design Engineering. This course takes 10 weeks and is especially targeted at developing novel designs for innovative technologies [20]. In the past this approach was also successfully applied to designing with Concentrating PV and other PV technologies [21, 22]. Moreover, the course is supported by a book on designing with sustainable energy technologies [23]. In the fall of 2016 about 40 students executed a design project in groups of two or three. The end results were a feasible concept design worked out to the level of a scaled prototype. The project was supervised by the authors and contained a series of guest lectures by experts on innovation methodologies. In-depth knowledge was provided by Dr. Wilfried van Sark of Utrecht University, who is a long term expert in LSC PV technologies.

II. PV SYSTEMS AND DESIGN

The application of PV systems beyond primary energy production is still limited. Earlier work revealed that the design potential of PV systems is often not fully used [24]. On the other hand, a common method to improve on the application of new technologies is developing demonstrator projects. Also the design potential of innovative technologies like LSCs can be shown by demonstrators.

To explore the design potential of innovative technologies the 'Design & Styling of Future Products' methodology [25] was developed. This method was, among other methods [8], applied throughout the design project. This methodology relies on visual innovation techniques [26], combined with a balanced application of novel styling cues on the one hand and typical, familiar styling cues on the other. The latter ensures that the design results for the innovative technologies are beyond the obvious, but still acceptable as feasible solutions.

II. DESIGN RESULTS

The 16 design results of the student design projects were very diverse, due to the course's emphasis on innovation and creativity. The designed objects were; a tourist shelter, a garden fence, greenhouse panels, safety staircases, parking assistance in the form of trees, outdoor tables, a casing to charge cell phones, a labyrinth for parks, self-powering e-bikes, outdoor furniture, a sundial and a colorful bike parking

with integrated LSC PV technologies. Three of these designs are shown in Figure 1, and in the following we would like to highlight four other examples which represent the broad application possibilities of the LSC technology on several dimensions; scale (big-small), application purpose (professional-leisure), and user experience (active-passive).

A. Example I: AlliSee LSC-Boat

This design makes full use of the design potential of LCS technology with respect to the possibilities for the realization of three dimensional shapes. The project presented a modular design for a small electrically powered dinghy for leisure activities, such as snorkeling and fishing (Figure 2).

Fig. 2.   'Allisee' transparent boat concept by Jullian Claus, Rosan Harmens and Hieu Nguyen.

The transparency of the LSC material adds to the experience by making the underwater world visible to the user. The solar cells that are needed to catch the light that is collected in the LSC hull are located in the reinforced gangway. Segmentation of the LSC area with an additional PV cell grid would be used to minimize the self-absorption of the photons by the material (Figure 3).

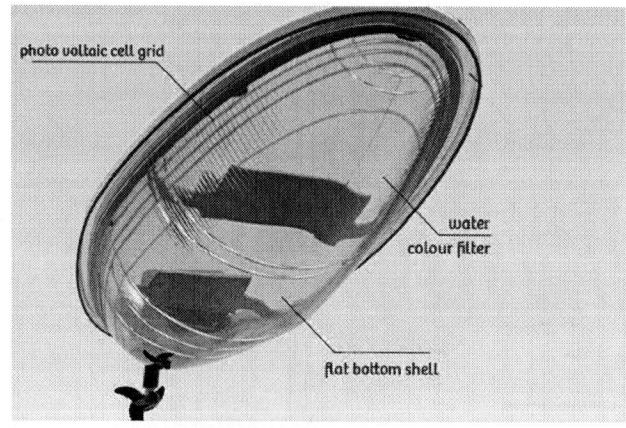

Fig. 3.   LSC transparent sculpted hull design with PV cell grid.

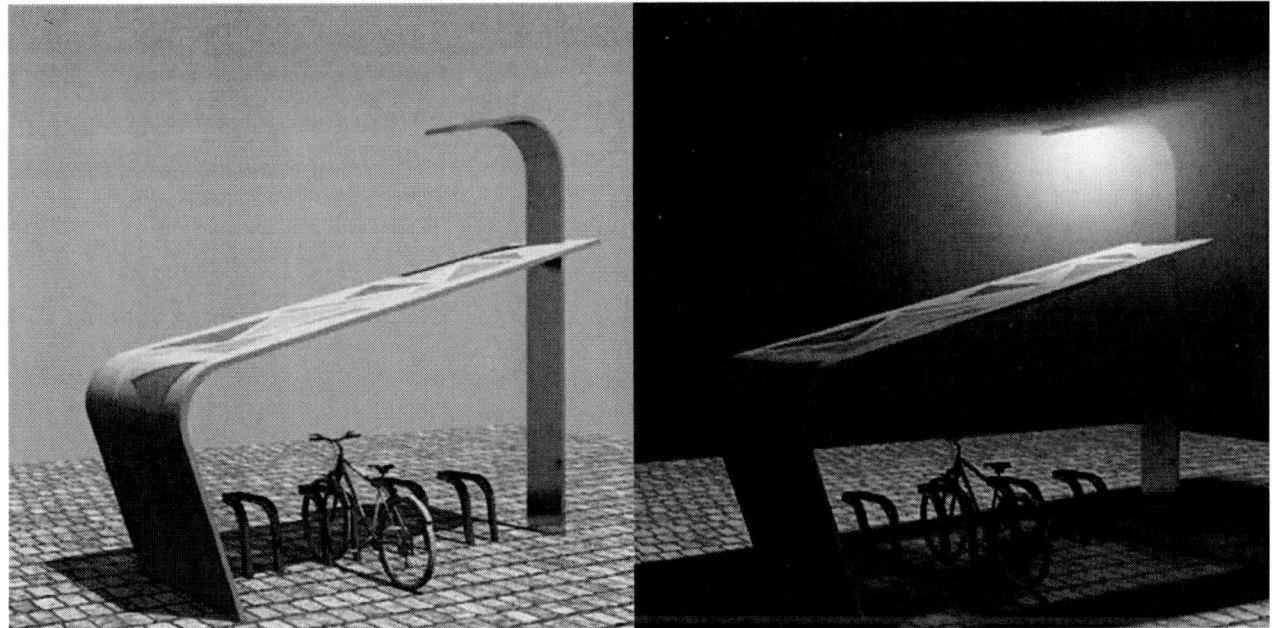

Fig. 4. Colorful effect of the LSC material in the Bike parking concept by Ruben de Noord, Eduard Tudor, and Sem Vossebeld. During daytime (left) and during the night (right)

B. Example II: A bike parking shed with integrated LSCs

In this concept, LSC material is integrated in the roof of a stand-alone bicycle parking. The LSCs generate electrical energy that can be used to charge a lantern for use during the night. The colorful shadow pattern that the LSC material sheds on the floor also adds to the consumers' experience. During the day it draws attention to the sunlight as an energy source and during the night it is supposed to support the feeling of safety (Figure 4). Additionally the LSC PV panels could generate power to charge e-bikes. Provided that the charging station is connected to the grid as well to ensure full power supply during all seasons.

C. Example III: LSC-Sustainable Festival Tent

This sustainable festival tent uses LSC PV modules to generate a part of its own electricity consumption, while at the same time the color of the luminescent dyes is used to create a special atmosphere during day and night (Figure 5).

After dark the LSC panels are lit with the use of led strips that are placed at one edge of the triangles. At the other two edges solar cells are attached (Figure 6). The students calculated that the tent could not provide enough energy when powered by LSC-PV panels only and therefore suggested to incorporate both LSC-PV modules and conventional PV modules in a complete festival set-up (Figure 7).

Fig. 5. Impression of the sustainable festival tent by Rolf van der Toom, Steven Oonk, and Kasper Schriek.

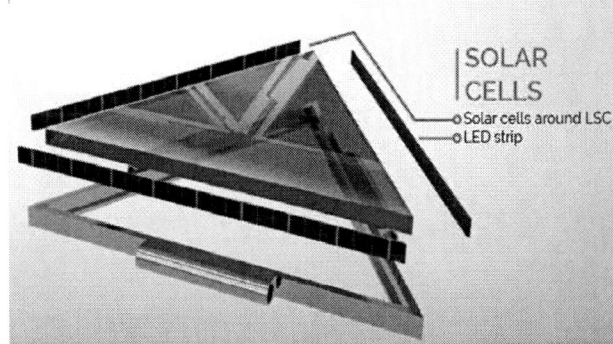

Fig. 6. Impression of the construction of the modular LSC tent segments, with Solar cells and a LED strip around the edges.

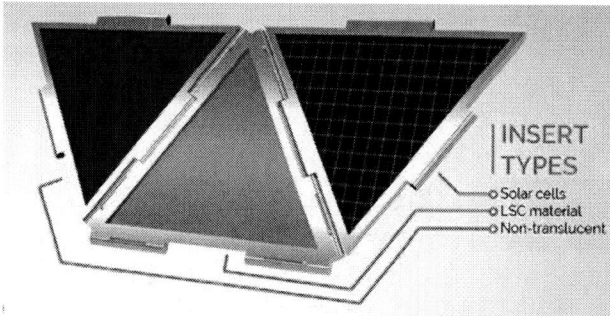

Fig. 7. Combination of LSC-PV and conventional PV modules

D. Example IV: LSC-Safety Strips

This concept consists of a modular system for illuminating outdoor surface edges, in particular stairs and steps (figure 8). The design is self-sufficient. With the use of LSC material in combination with solar cell strips, energy will be converted during daytime and stored by the power controller. In the evening, the energy will be used to activate the LED module inside the casing (figure 9).

Fig. 8. Application example of LSC illuminating safety strips concept by Ashley Hogt.

Fig. 6. Impression of the construction of the modular LSC safety strips with the LSC material (magenta) surrounded by solar cells.

The desired function of the product would be to create a feeling of safety, while walking the stairs when it is dark. To increase the engagement of the user, sensors will be used to respond to the users' movements (figure 10).

Fig. 10. Examples of the interactive use of the LSC safety strips applied on staircases.

II. DISCUSSION

Although the existing LSC-PV applications are primarily building components [27], this design project showed also other possibilities. Despite the assignment 'design a product using luminescent solar concentrator (LSC) technology, that can be applied in the built environment', six out of sixteen groups developed entirely other concepts. Two of these designs were furniture, one a mobile phone case, one boat (example I), and two bicycle concepts. We also saw that the students projects stretched the idea of the 'built environment' beyond limits, with for instance the festival tent concept (example III), a garden fence, and safety strips for stairs.

For all product concepts, the transparency and color of the LSC material played a major role in the appearance of the designs. This is to a certain extend as expected. However besides that, the colorfulness and transparency of the material often also played a major role in the user experience of the designs. The transparency affords the users of the boat to experience the underwater world, the colors add to the nighttime feeling of safety in the bicycle shed and ensures a permanent disco atmosphere in the festival tent. The form freedom that the material offers played in only two of the projects a decisive role in the design, which was most apparent in the LSC boat.

Another observation is that the relatively limited energy yield of LSC PV elements sometimes had to be compensated by the use of additional conventional PV modules. In other cases the energy demand of the chosen applications was kept low to fully be able to use LSC-PV technologies.

II. CONCLUSION

This project showed that the typical material properties of LSCs; low cost, colorful, bendable, and transparent do not only offer a lot of design freedom, but also offer excellent possibilities to incorporate this technology into the overall function and experience of applications.

REFERENCES

[1] Doudart de la Grée, G., A. Papadopoulos, M. Debije, M. Cox, Z. Krumer, A.H.M.E. Reinders, and A. Rosemann. "A new design for luminescent solar concentrating PV roof tiles". In: *Proceedings of 42nd IEEE Photovoltaic Specialists Conference*. 2015. New Orleans, USA: IEEE.

[2] Reinders, A., G. Doudart de la Grée, A. Papadopoulos, A. Rosemann, M.G. Debije, M. Cox, and Z. Krumer. "Leaf Roof – Designing Luminescent Solar Concentrating PV Roof Tiles". In: *Proceedings of 43rd Photovoltaic Specialists Conference (PVSC)*. 2016. Portland: IEEE. pp. 3447-3451.

[3] Van Sark, W.G.J.H.M., K.W.J. Barnham, L.H. Slooff, A.J. Chatten, A. Büchtemann, A. Meyer, . . . D. Vanmaekelbergh, "Luminescent Solar Concentrators – A review of recent results". *Optics Express*, 2008. 16(26): pp. 21773-21792.

[4] Vishwanathan, B., A.H.M.E. Reinders, D.K.G. de Boer, L. Desmet, A.J.M. Ras, F.H. Zahn, and M.G. Debije, "A comparison of performance of flat and bent photovoltaic luminescent solar concentrators". *Solar Energy*, 2015. 112: pp. 120-127.

[5] Reinders, A.H.M.E. "Product-integrated PV applications – How industrial design methods yield innovative PV-powered products". In: *Proceedings of 33rd IEEE Photovoltaic Specialists Conference*. 2008. San Diego.

[6] Apostolou, G. and A.H.M.E. Reinders, "Comparison of the indoor performance of 12 commercial PV products by a simple model". *Energy Science & Engineering*, 2016. 4(1): pp. 69-85.

[7] Durlinger, B., A.H.M.E. Reinders, and M.E. Toxopeus, "A comparative life cycle analysis of low power PV lighting products for rural areas in South East Asia". *Renewable Energy*, 2012. 41: pp. 96-104.

[8] Gorter, T. and A.H.M.E. Reinders, "A comparison of 16 polymer encapsulants for crystalline PV cells applications in PV-powered boats". *Applied Energy*, 2012. 92: pp. 286-297.

[9] Veldhuis, J. and A.H.M.E. Reinders, "Real time irradiance simulation for PV products and building integrated PV in a virtual reality environment". *IEEE Journal of Photovoltaics*, 2012. 2(3): pp. 352-358.

[10] Gorter, T., "Performance evaluation of photovoltaic boats in an early design stage - Numerical simulations with industrial design methods", PhD in *Engineering Technology*, 2014, University of Twente, Enschede.

[11] Apostolou, G. and A.H.M.E. Reinders, "How do users interact with PV-powered products? Investigating 100 lead users and 6 PV products". *Journal of Design Research*, 2016. 14(1): pp. 66-93.

[12] Apostolou, G., "Design Features of Product-Integrated PV: An Evaluation of Various Factors under Indoor Irradiance Conditions", PhD in *Industrial Design Engineering*, 2016, Delft University of Technology, Delft.

[13] Obinna, U., P. Joore, L. Wauben, and A.H.M.E. Reinders, "Insights from stakeholders in five residential smart grid pilot projects in the Netherlands". *Smart Grid and Renewable Energy*, 2016. 7(1-15).

[14] Apostolou, G. and A.H.M.E. Reinders, "Overview of design issues in product-integrated PV". *Journal of Energy Technology*, 2014. 2(3): pp. 229-242.

[15] Reinders, A.H.M.E. and W.G.J.H.M. Van Sark, eds. *Product Integrated PV*. Invited book chapter for Comprehensive Renewable Energy. Vol. 1: Photovoltaic Solar Energy. 2012, Elsevier. 709–732.

[16] Eggink, W. "Where's My Robot? Integrating Human Technology Relations in the Design Curriculum". In: *Proceedings of International Conference on Engineering and Product Design Education; Human Technology Relations*. 2014. Enschede: The Design Society. pp. 87-92.

[17] Eggink, W. and Adri Albert de la Bruheze. "Design Storytelling with Future Scenario Development; envisioning "the museum"". In: *Proceedings of Cumulus 2015*. In print. Milan.

[18] Eggink, W., A. Reinders, and B.v.d. Meulen. "A practical approach to product design for future worlds using scenario-development". In: *Proceedings of 11th Engineering and Product Design Education Conference; Creating a better world*. 2009. Brighton: Institution of Engineering Designers.

[19] Dorrestijn, S., M. Van der Voort, and P.-P. Verbeek, "Future user-product arrangements: Combining product impact and scenarios in design for multi age success". *Technological Forecasting & Social Change*, 2014. 89: pp. 284-292.

[20] Reinders, A., J.d. Borja, and A.d. Boer. "Product-Integrated sustainable energy technologies - Six years of experiences with innovation and sustainability". In: *Proceedings of IASDR 2011, Diversity and Unity*. 2011. Delft: International Association of Design Research Societies.

[21] Eggink, W. and A. Reinders. "The Design and Styling of Technology-based Innovations". In: *Proceedings of IASDR 2013, Consilience and Innovation in Design*. 2013. Tokyo: International Association of Design Research Societies. pp. 001-012.

[22] Reinders, A.H.M.E., M. Wiesenfarth, and R.R. King. "Conceptual product development with integrated concentrating PV systems – CPV in the built environment from a designer's perspective". In: *Proceedings of 39th IEEE Photovoltaic Specialists Conference*. 2013. Tampa (FL), USA: IEEE.

[23] Reinders, A., J.C. Diehl, and H. Brezet, eds. *The Power of Design: Product Innovation in Sustainable Energy Technologies*. 2012, Wiley: West Sussex. 331.

[24] Reinders, A. and W. Eggink. "PV powered Products: The Future is Design and Styling.". In: *Proceedings of European Photovoltaic Solar Energy Conference and Exhibition (EU PVSEC)*. 2015. Hamburg.

[25] Eggink, W., "The Design & Styling of Future Products", In: *The Power of Design: Product Innovation in Sustainable Energy Technologies*, A. Reinders, J.C. Diehl, and H. Brezet (Eds.). 2012, Wiley: West Sussex. pp. 89-99.

[26] Eggink, W. "Disruptive Images: stimulating creative solutions by visualizing the design vision.". In: *Proceedings of 13th Engineering and Product Design Education Conference; Creating a better world*. 2011. London: Institution of Engineering Designers.

[27] Reinders, A., G. Doudart de la Grée, A. Papadopoulos, A. Rosemann, M.G. Debije, M. Cox, and Z. Krumer. "Leaf Roof – Designing Luminescent Solar Concentrating PV Roof Tiles". In: *Proceedings of 43rd Photovoltaic Specialists Conference (PVSC)*. 2016. Portland: IEEE. pp. 3447 - 3451.

Investigating PV-battery 3-terminal Integration Concept as a Self-sustaining Power Solution

Solomon N. Agbo, Oleksandr Astakhov, Uwe Rau and Tsvetelina Merdzhanova

Institute of Energy and Climate Research (IEK-5)- Photovoltaics, Forschungszentrum Julich GmbH, 52428 Julich, Germany

Abstract — **We present here the performance analysis of PV-battery 3-terminal integration concept. In this arrangement, a PV device is integrated monolithically with a battery stack separated only by a common contact layer which makes the use of a control electronic e.g. a DC-DC converter possible. The concept is a versatile power solution for self-sustained operation of small-scale electronics. In this work we investigate this concept by using thin-film multi-junction solar cell to directly charge a Li battery with the aid of a control electronic. Operation of the PV-battery combination with and without charge controller is analyzed under standard AM1.5 illumination.**

Index Terms — **PV energy harvester, self-charging power unit, DC converter, PV-battery integration.**

I. INTRODUCTION

The number of low-power consuming electronics in use globally has seen a rapid increase in recent years. Such electronics as mobile phones, e-book readers, mp3s, sensors to mention but a few out-number the world population. For all these electronics, the battery life is very critical on one hand and the power consumption of the electronics on the other hand. This makes it necessary to focus research on increasing battery energy density and simultaneously minimizing power consumption. In view of reduction of power consumption and with regards to modern trends in development of the so-called internet of things [1] and the wide utilization of sensors and sensor networks, large number of small low power consuming wireless and off-grid devices will be implemented in common human environment. Therefore it is important to evolve a power solution for these low power electronics that will provide self-sustaining operation over years in addition to being compact and environment-friendly. For self-sustaining operation of these low-power electronics, energy requirement should be sourced from its environment [2] using reliable and renewable sources that will not require human intervention [3]. For these reasons, photovoltaic (PV) is a suitable green energy source that can be integrated with a rechargeable battery to provide a self-sustaining power solution for sensors and sensor networks, for example [4]-[6]. Its universal availability can be particularly most suitable for powering distributed electronics in hard-to-reach regions and locations without power grid like deserts, mountains and remote areas.

In this work we present the result of investigation of monolithic integration concept of PV and battery as a compact power solution in the 3-terminal configuration with optional control electronics. Direct coupling of well-matched solar cell and battery may become a very efficient solution especially when the unit is few square cm for extreme low power applications [7]-[8]. However, with control electronics, various possibilities exist for optimization including the ability to use different solar cells with different voltages and efficiencies. Improvements in efficiency of modern and future DC-DC converters will promote use of these units in ultra-low power applications. In view of the given aspects we propose a simple but versatile 3-terminal concept where integrated PV-battery unit can be used with or without power conditioning electronics.

Our proposed monolithically integrated device will be based on state-of-the-arts solid-state lithium ion battery integrated with thin-film silicon solar module hence providing the advantage of being scalable and compact. A special feature of thin-film silicon solar cell based on amorphous silicon is their suitability for application in artificial low light conditions like LED and florescent lights [9]. They have higher efficiencies in these artificial light conditions relative to the efficiency in standard outdoor condition. They are therefore suitable for not just outdoor applications but also indoors.

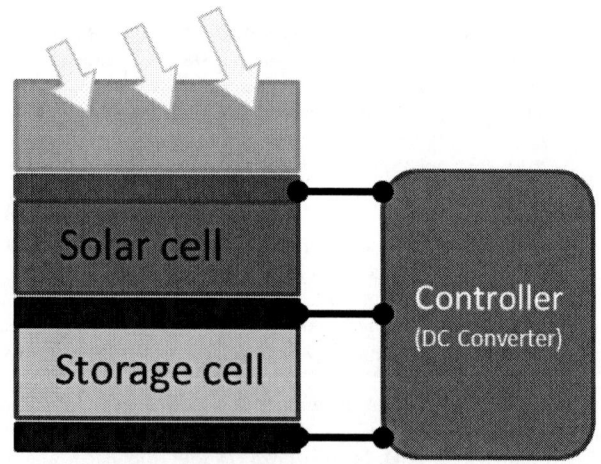

Fig. 1. PV-Battery 3-terminal integration concept showing the solar cell and the storage cell with the extruding terminals to a controller for power conditioning.

The results of this work will be presented in two parts. The first part is the performance of the PV-battery device without control electronics. In the second part the performance profile of the device is shown when aided by a control electronic.

II. Experimental Details

Fig. 1 shows the 3-terminal monolithic PV-battery integration concept consisting of the PV device and the storage cell all on one substrate. The 3-terminals concept as shown makes it possible to use control electronics to adapt the IV characteristics of the solar module to the required charging IV characteristics of the battery in a PV-battery integrated device. Since the actual integration work is still in progress, we have mimicked the PV-battery integrated device by directly connecting the solar cell to the battery. As a prerequisite for optimum charging efficiency in the monolithic integration concept, the PV device IV characteristics are matched to the required charging IV of the battery [7]. We have therefore made the choice of our PV device with this in mind. On illuminating the solar cell, the power gain of the solar cell is used to directly charge the storage battery until complete or full state of charge is reached.

The PV devices used are thin-film silicon multi-junction solar cell and module. The PV devices and their areas have been chosen to match the required charging voltage and current depending on the battery capacity. The module was used for the direct charging of battery without controller while the tandem solar cell was used in the case of charging with the aid of a controller. The tandem solar cell was a $1cm^2$ cell comprising a top thin-film silicon amorphous (a-Si:H) solar cell and a bottom microcrystalline silicon (μc-Si:H) solar cell. The solar module is made from series connection of two triple-junction thin-film silicon solar cell. The triple-junction cell has top and middle cells consisting of a-Si:H absorber layers and bottom cell consist of μc-Si:H. The silicon layers were deposited in a multi-chamber large-area radio frequency plasma-enhanced chemical vapour deposition system with shower-head electrodes.

For more preparation details see [7]-[8]. Aluminum doped zinc oxide in combination with aluminum layer constitute back reflector and back contact. Integrated series connections of both cells are realized through three laser scribing steps in between the several deposition steps [10]-[11]. The module has an active area of 9.6mm² designed to deliver short-circuit current equivalent to the required battery charging current under standard AM 1.5 illumination. We have used two different battery types: a low capacity solid-state electrolyte battery (Type A) and a high capacity non-solid-state electrolyte battery (Type B). Both batteries are commercially-available rechargeable Li batteries. Type A is all-solid thin-film battery "EFL700A39" from STMicroelectronics. It consists of lithium (Li) anode, solid lithium phosphorus oxinitride (LiPON) ceramic electrolyte and lithium cobalt oxide ($LiCoO_2$, LCO) cathode [12]. Its nominal voltage is 3.9 V and can be operated between 3.0 and 4.2 V. The capacity of the battery was measured to be 1.1mAh at $1\mu A$ discharge current. Type B is 130mAh capacity Li battery with lithium cobalt phosphate ($LiCoPO_4$, LCP) cathode and graphite anode in a polymer matrix electrolyte. The battery has a nominal voltage of 3.7V and can be operated between 2.7 and 4.2V.

The PV-battery configuration was tested under AM1.5 illumination (AM1.5 spectrum, 100 $mWcm^{-2}$) with class-A sun-simulator from Wacom. The testing involves first measuring the IV characteristics of the stand-alone solar cell and module and then the charging profile of the PV-battery device under illumination. The photovoltaic parameters of the used cell and module are shown in Table 1. Under AM 1.5 illumination, the PV device delivers the required energy to charge the storage battery cell.

In the first case, we used the multi-junction module to directly charge type A battery without any additional charging control electronics. The current and voltage of the solar module was matched to that of the battery. In the second case, a multi-junction solar cell (tandem) was used to charge type B battery with the aid of a charge controller that also doubles as a dc voltage booster and converter. The multi-junction cell is used here because its delivered current is just within the required limit for the converter hence shades light into the converter operation at such current limit. More so, it is simpler and more efficient than the multi-junction module. The controller is named "SPV1040" control electronic from STMicroelectronics. This is a step-up voltage converter with input voltage in the range 0.5 to 5.5V that can also track the maximum power point of the PV device and deliver the desired photovoltaic parameters for battery charging. We show here only the potential usefulness of this converter. However, its detailed characteristic performance is a subject of a separate contribution under preparation.

TABLE I
PHOTOVOLTAIC PARAMETERS OF PV DEVICES USED IN THIS WORK

	η [%]	FF [%]	I_{sc} [mA]	V_{oc} [V]	I_{mpp} [mA]	V_{mpp} [V]
Tandem cell (1cm²)	12	74.2	12	1.35	10.7	1.12
Triple module (0.96 cm²)	8.1	74.5	0.24	4.34	0.21	3.67

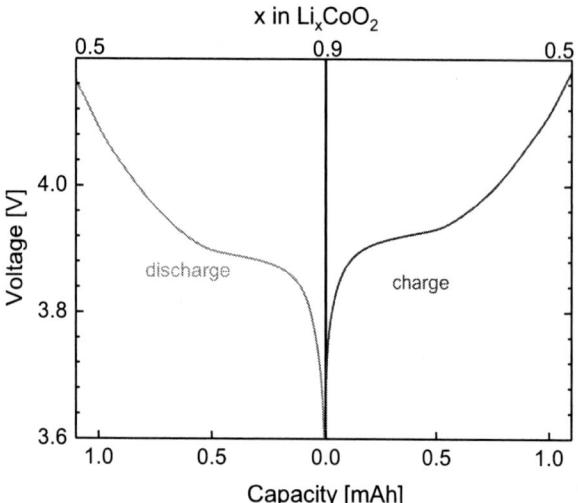

Fig. 2. Discharge-charge curve of battery type A at 0.07 mA constant current showing external measured voltage across the battery verses the state of charge and amount x of lithium in Li_xCoO_2 cathode.

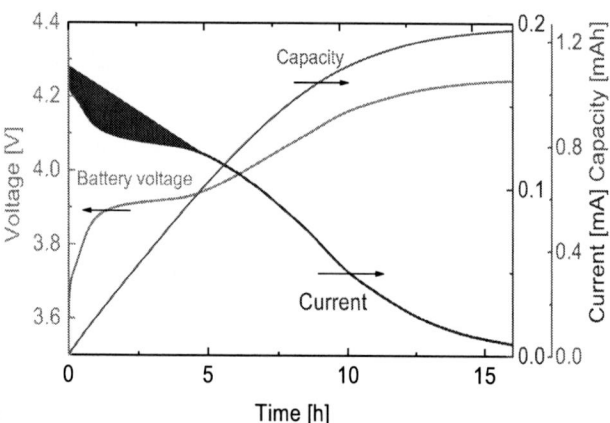

Fig. 3. Charging profile of the test PV-battery combination without control electronics showing the battery voltage, current and capacity as a function of time. The current goes to zero when the battery voltage matches the open-circuit voltage of the PV device. The current can also go to near zero when the battery nears its full capacity hence cannot take in more current.

III. RESULTS AND DISCUSSION

Fig. 2 is the charge and discharge profile of the standalone rechargeable Li ion battery. Here we observe a clear dependence of battery voltage on the capacity as the state of charge builds up. The discharge characteristic shows that the voltage constantly drops from 4.2 V at capacity of 1.1mAh full state of charge (SOC) to 3.9 V at about 50% SOC. There are

no observed distinctive plateau voltages over the entire range of SOC. Conversely, the charge profile shows an initial sharp increase in voltage up to a transition value at 3.9 V at x=0.95. This is in line with earlier report where it is attributed to a single-phase reaction process and the generation of a defect rock salt phase $Li_{0.95}CoO_2$ [13]. Subsequently the voltage increases with SOC to reach the full SOC at 4.2 V.

In Fig. 3 the charge profile of the PV-battery combination under AM1.5 illumination is presented. The battery is charged through the illuminated solar module which delivers a maximum charging current that is close to its short-circuit current I_{sc}. This current drops as the state of charge SOC increases as well as the battery voltage until the voltage is stabilized at around 4.3V and no appreciable current flows through the battery indicating that the battery voltage has matched the open-circuit voltage V_{oc} of the solar module. A rapid increase in the battery capacity is observed from onset of charging reaching values of 1.2 mAh over charging time of 16 hours.

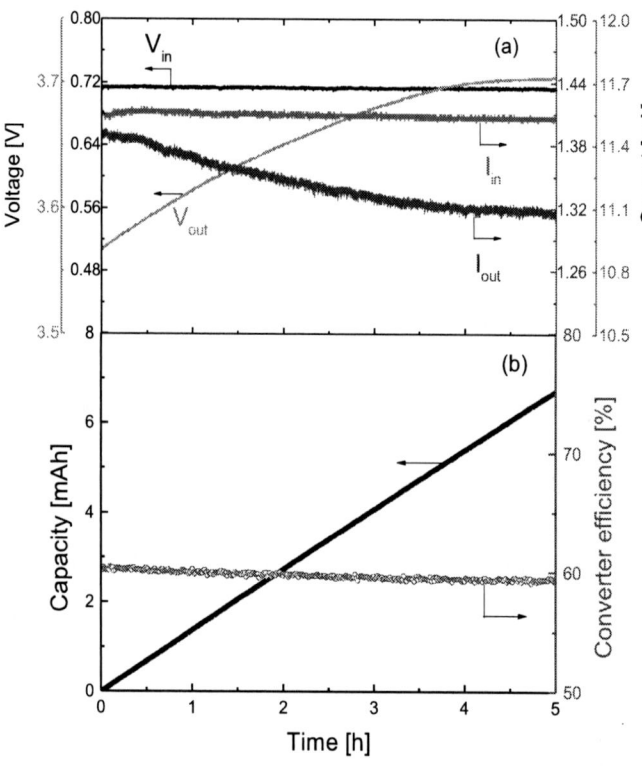

Fig. 4 Charging profile of the test PV-battery combination with a charge controller (dc converter) showing (a) the input and output voltages and currents and (b) the accumulated battery capacity and the dc converter efficiency as a function of charging time. V_{in}, I_{in}, V_{out} and I_{out} represent voltage and current input from the solar cell and voltage and current output from converter respectively.

Fig. 4 shows charging profile of the test PV-battery combination with a charge controller (dc converter) for high capacity non-solid-state electrolyte battery (Type B). The input (from solar cell) and output (to battery) current and voltage as a function of charging time are presented in Fig. 4a. The input voltage of about 0.7V is supplied over the entire charging time of 5 hours and this is stepped up via the converter to charge the battery. The output voltage gradually increases depending on the state of charge of the battery reaching the nominal voltage of 3.7 V. The input voltage is much lower than the voltage at maximum power point because the solar cell here behaves like a current source delivering current value that is nearly constant and nearly same as the short-circuit current. The output current into the battery gradually decreases from the initial value of 1.38 mA as the state of charge increases with time.

In Fig. 4b the accumulated battery capacity and the converter efficiency are presented over the charging time. The gained capacity over this time is 6.7 mAh. The converter efficiency stays at approximately 60%. The converter efficiency is the ratio of the power output from the converter to the input power delivered by the solar cell.

IV. CONCLUSION

PV-battery 3-terminal integration concept was demonstrated using a direct connection of the PV device to the converter and from this to the Li-ion storage battery. We found that this configuration concept results in possibility to use PV devices with lower voltages than the required charging voltage hence introducing more flexibility in the choice of PV and battery types. This flexibility also includes the ability to use other more efficient PV technologies like heterojunction solar cell. Comparison between the case of battery charging with and without controller shows that without a controller the charging process is mainly limited by the photovoltaic parameters of the photovoltaic device. The IV characteristic of the PV device has to be strictly matched to the required charging IV characteristics of the battery. With a converter however, the charging process is not limited by the solar cell device but is mainly affected by the battery and converter properties.

ACKNOWLEDGEMENT

Joachim Kirchhoff, Daniel Weigand and Christoph Zahren are appreciated for technical assistance with the solar cell deposition and measurements. SNA acknowledges Alexander von Humboldt foundation for research fellowship. IEK-1 of Forschungszentrum Jülich GmbH is acknowledged for providing the thin-film battery.

REFERENCES

[1] L. Atzori, A. Iera, and G. Morabito, "The internet of things: a survey," Computer Networks, vol. 54, pp. 2787-2805, 2010.

[2] J. M. Gilbert, and F. Balouchi, "Comparison of energy harvesting systems for wireless sensor networks," International Journal of Automation and Computing, vol. 5, pp. 334–347, 2008.

[3] A. Dewan, S. U Ay, M. N. Karim, and H. Beyenal, "Alternative power sources for remote sensors: A review," Journal of Power Sources, vol. 245, pp. 129- 143, 2014.

[4] C. Alippi, R. Camplani, C. Galperti, M. Roveri, "A robust, adaptive, solar-powered WSN framework for aquatic environmental monitoring," IEEE Sensors Journal vol. 11, pp. 45-55, 2011.

[5] R. Bogue, "Solar-powered sensors: a review of products and applications," Sensor Review, vol. 32, pp. 95- 100, 2012.

[6] H.Yu, Y-Q. Li, Y-H. Shang, B. Su, "Design of a micro photovoltaic system for wireless sensor network," Journal of Functional Materials and Devices, vol. 14, pp. 476-480, 2008.

[7] S. N. Agbo, T. Merdzhanova, S. Yu, H. Tempel, H. Kungl, R.- A. Eichel, U. Rau, and O. Astakhov, "Photoelectrochemical application of thin-film silicon triple-junction solar cell in batteries," Physica Status Solidi. A, vol. 213, pp. 1926-1931, 2016.

[8] S. N. Agbo, T. Merdzhanova, S. Yu, H. Tempel, H. Kungl, R.- A. Eichel, U. Rau, and O. Astakhov, "Development towards cell-to-cell monolithic integration of a thin-film solar cell and lithium-ion accumulator," Journal of Power Sources, vol. 327, pp. 340-344, 2016.

[9] S.N. Agbo, T. Merdzhanova, U. Rau, and O. Astakhov, "Illumination intensity and spectrum-dependent performance of thin-film silicon single and multijunction solar cells," Solar Energy Materials and Solar Cells vol. 159, pp. 427-434, 2017.

[10] US patent: Laser Processing Technique for Fabricating Series-Connected and Tandem Junction Series-Connected Solar Cells Into a Solar Battery, US 4292092 A , 1981.

[11] S. Haas, G. Schöpe, C. Zahren, H. Stiebig, "Analysis of the laser ablation processes for thin-film silicon solar cells," Applied Physics A vol. 92, pp. 755-759, 2008.

[12] EFL700A39 rechargeable solid state lithium thin film battery-Datasheet.
http://www.st.com/content/ccc/resource/technical/document/datasheet. Accessed: 29 November, 2016.

[13] E. Antolini, "LiCoO2: formation, structure, lithium and oxygen nonstoichiometry, electrochemical behaviour and transport properties," Solid State Ionics, vol. 170, pp. 159-171, 2004.

Performance assessment of a BIPV Roofing Tile in outdoor testing

Cristina S. Polo López, Pierluigi Bonomo, Francesco Frontini, Vasco Medici, Lorenzo Nespoli

Institute of Applied Sustainability to the Built Environment (ISAAC), University of Applied Sciences and Arts of Southern Switzerland (SUPSI), Canobbio CH-6952 (Switzerland).

Abstract — In this study, a comparative outdoor test of an innovative BIPV solar tile with a well-established market product is conducted. Two mock-up structures have been built to generate further insights into BIPV efficiency as a function of operative temperatures and back-module ventilation. The two small-scale installations include a complete roof construction package on which the BIPV modules are integrated as a full-roof solution. The test-stands are equipped with a monitoring system aimed to assess the operating conditions both in terms of temperatures and PV performances. Along with the thermal and yield monitoring, the effect of temperature and natural ventilation on the PV production is studied through the identification of a thermal model and the simulation for one of the studied configurations.

Index Terms — BIPV, photovoltaics, solar architecture, active building skin, roof temperatures, outdoor monitoring

I. INTRODUCTION

The role of the building envelope in energy performances of buildings is becoming more and more important since its design accounts for over half of the energy consumption in the residential sector [1], and because it provides also the necessary space for the installation of active energy systems [2]. Building integrated PV (BIPV) can be used in all parts of the building envelope and the roof surfaces are currently the preferred area for installing PV as different survey researches underlined [3]-[4].

Within the European project Construct-PV (FP7-ENERGY-2011-2, http://www.constructpv.eu/), along with other product and process innovation concerning BIPV, an innovative and customizable BIPV solar tile has been developed. The solar roof tiling system has the purpose of achieving a BIPV solution with cost-effective manufacturing and an easy mounting construction system.

In order to assess the performance in real climate conditions, two different demonstration mock-ups have been installed in the summer of 2015 at the SUPSI Campus, in Lugano (Switzerland) and the large-scale demo has been recently built on the rooftop of the NTUA building, in Athens (Greece).

In the section II, a first part (subsections A, B and C) is dedicated to discuss the results, in terms of yield and temperatures, obtained from the outdoor monitoring period. A second part (subsection D) will describe a simulation model developed to better investigate and predict the effect on the power output of operating temperatures, ventilation and thermal resistance of the construction layer on which the PV module is installed. The section III will discuss the emerging results giving also some inputs for further investigation.

II. ON-SITE MEASURING CAMPAIGN AND SIMULATION STUDIES

A. Measurement of Operation Conditions of the BIPV roof

SUPSI mock-up test facility consists in two real small-scale structures where the complete constructive packages of roof were installed. The roofs were tilted at 20°. The Construct-PV solar roof tile (Stand 1, Table I), assembled by the partner Tegola Canadese S.p.a, consists in solar shingles made in laminated PV glass, glued into a rear waterproofing bitumen membrane. The roof tiling consists of six small-size PV tiles containing 10 SWCT monocrystalline cells each, installed by sets of two PV modules connected in series to each MPPT (PV module rated power P_{STC} 48.00 Wp, modules area 0.65 m², temperature coefficient γ_{PMPP} -0.39 %/K). This solar roof solution has been compared with a MegaSlate® in-roof installation (Stand 2, Table I), developed by Meyer Burger AG. The second roof consists of six modules with 30 monocrystalline cells each (PV module rated power P_{STC} 135.00 Wp, module area 0.84 m², temperature coefficient γ_{PMPP} -0.42 %/K) [5].

TABLE I
BIPV ROOFING TILE SYSTEM - CONSTRUCTIVE DETAILS

Stand 1 BIPV Construct-PV Roof	Stand 2 BIPV Megaslate ® Roof
D1: ventilated PV module C5: full-roof integr. PV module	B3: full-roof integrated PV module

In the first stand, the solar tiles were nailed into a plywood board (1,5 cm high) creating a continuous 6 cm thickness ventilation chamber underneath (Stand 1-Construct-PV, modules C5 and D1). In the second stand, a ventilation air-gap of 12 cm (6 cm + 6cm is height of battens and counter-battens) is directly available under the modules since they are mounted onto a wooden counter-batten system (Stand 2, module MegaSlate® B3).

Temperatures were measured with PT100 thermocouple sensors positioned in the different layers of the roofs' packages (Table I, layer 0 to 3) as well as behind the conventional tiles. Other temperature sensors measured the back of module temperatures (T_{bom}). Furthermore, an air flow meter has been placed in the ventilation chamber (Stand 1) to measure the air speed and the volumetric flow in the air gap space. System identification was performed on the collected data and the identified model was used in simulation to investigate the dependence of BIPV efficiency on temperature and natural ventilation in different roof system configurations. The results are presented in the next chapter.

B. Yield Assessment and Performance Monitoring

A performance comparison between the two main BIPV full-roof systems (modules C5 and B3) is presented including a free-standing Construct-PV tile as reference (module D1). The couple of modules C5, nailed directly onto an OSB wood deck, operated at higher temperatures than module B3 mounted through a wooden counter-batten system. Both BIPV systems operated at higher temperatures than rack-mounted D1 modules.

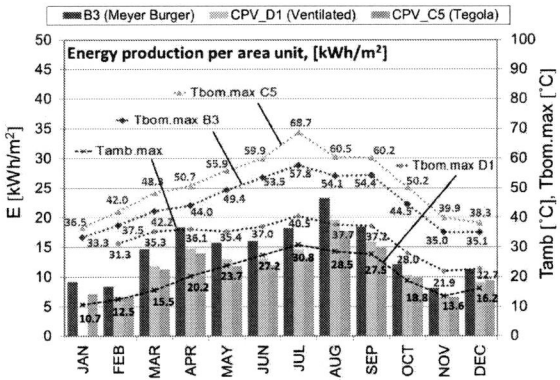

Fig. 1. Average monthly energy production per area unit.

Figure 1 shows the evolution of the yearly average Energy Production per square meter, E_{PV} [kWh/m²], reported to the average outdoor temperature (T_{amb}) and to the back of module temperature (T_{bom}). Results display that the energy output of the new solar tile (C5 in Table 1) is generally lower than the other full-roof system (B3), mainly due to higher operating temperatures, improving in the free-ventilated solution (D1).

Thus the BIPV roof system on counter-battens (B3) performed better than the system installed on the wooden deck (C5). The reasons of this difference have been partially discussed in [5].

In this paper, will focus on the role of the roof construction system for temperature and ventilation control. Other studies have reported comparable results in moderate climate [6]-[7].

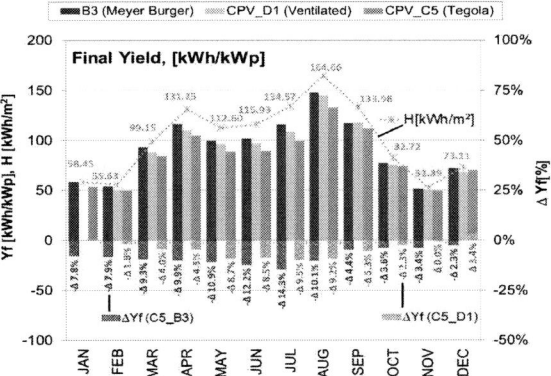

Fig. 2. Average monthly energy yield measurements, Final Yield.

In Figure 2, the average monthly final yield values throughout the year (since January 2016 until December 2016) are reported. The graph also shows the percentage yield variation of the new solar tiles C5 with the free-ventilated D1 ($\Delta Yf_{(C5\_D1)}$) and the other full-roof solution with large BIPV tiles B3 ($\Delta Yf_{(C5\_B3)}$).

C. In-Layer Temperatures

As temperatures are a primary driver for the PV power output, Figure 3 shows the relationship of Performance Ratio (PR) with operating and ambient temperatures.

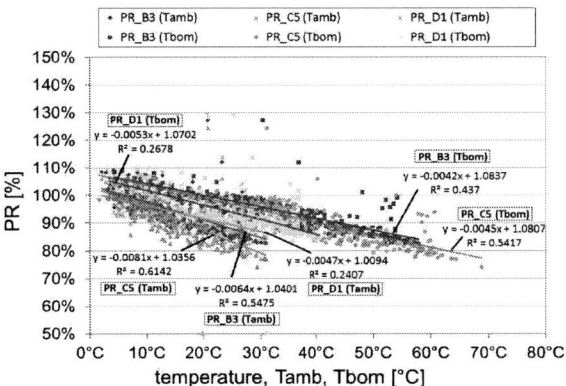

Fig. 3. Trends of performance ratios (PR, %) as function of outdoor ambient temperature and back of modules temperatures.

Figure 3 displays the major differences of PR at high ambient temperatures, as well as a major gradient of the trend lines for C5, meaning a greatest dependency of such tile on climatic conditions for overheating. Along with irradiance, time of day and outdoor temperatures, also further factors such as the wind speed can interact and lead to power output variations. Figure 4 provides some insight over the course of a sunny day where plane of array (*POA*) irradiance, back of module temperature (T_{bom}) and in-layer temperatures (from T_{Lay1}, T_{Lay2} inside the ventilation chamber on the upper and bottom air-gap faces respectively, to T_{Lay3} in the most external layer) are shown. The monitoring period shows great differences, especially at the level of layer 2, under the ventilation chamber. The larger temperature gradient for the B3 system from T_{bom} to T_{Lay2} (temperatures at this level decreases up to even 11.3 °C) is consistent with the convective flow path expected under the modules. In the C5 module the minor gradient, if correlated to the internal air flow in the air-gap, can be interpreted as less-effective ventilation due to the shorter height of the air-gap and the different ridge solution.

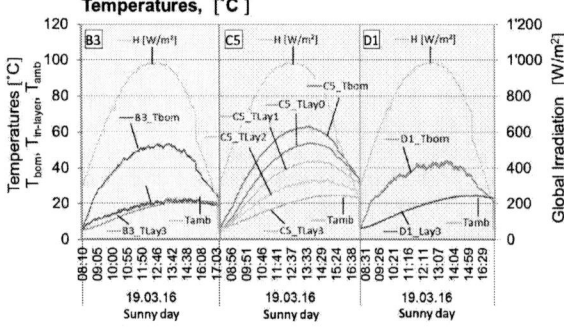

Fig. 4.   Irradiance & Temperatures for March 19th.

D. Effects of operating temperature, ventilation and thermal resistance on energy production

In this section, the **influence of the back of module temperature** (T_{bom}), of the internal ventilation speed (w) and of the thermal resistance of the roof layer under the PV module (R_c) on the energy production, is further investigated, by focusing in detail on the new BIPV roof tiling system C5 mounted on the wooden deck.

In order to compare the effect of the decreased back of module temperature on the energy production, we define

$$P_{ratio} = \frac{\sum_{t=1}^{T} P_t^* - \sum_{t=1}^{T} P_t}{\sum_{t=1}^{T} P_t} \qquad (1)$$

where T is the total number of measurements and P_t^* is a simulated power output from the PV module. P_{ratio} represents the relative change in energy production due to a simulated change in T_{bom}, w or Rc. Since the relation between the electrical PV production and the incident radiation (I) on the plane of array and T_{bom} is almost linear, in order to simulate a

change of energy production due to a change in T_{bom}, it's only needed to fit a function $f(I, T_{bom})$. A plane with the available data has been fitted, using a bi-square loss function, as shown in Figure 5. The resulting adjusted R-squared value is 0.9998, meaning that we can reasonably use the fitted function $f(I, T_{bom})$ as predictor of the power output for new observation of I and T_{bom}. Once $f(I, T_{bom})$ has been fitted, we make T_{bom} change linearly to assess its influence on the energy production. Namely, the I observations are kept unchanged and let T_{bom} change between the observed values and T_a, where T_a is the observed air temperature during the monitored period. That is

$$T_{bom\,n} = T_{bom}(1 - k_t) + T_a k_t \qquad (2)$$

where k_t is a weighting coefficient varying from 0.05 and 1 and T_{bom_n} is the simulated T_{bom}. This means that when k_t is equal to 1, T_{bom_n} is equal to the ambient temperature (perfect thermal exchange).

The results of this study are shown in Figure 6. On the right axis of this panel, the relative change of temperature is also shown, defined as

$$T_{ratio} = \frac{\sum_{t=1}^{T} T_{bom\,t}^* - \sum_{t=1}^{T} T_{bom\,t}}{\sum_{t=1}^{T} T_{bom\,t}} \qquad (3)$$

where $T_{bom\,t}^*$ is the simulated back of module temperature.

The maximum expected improvement in (k_t=1) is slightly below of 8% in terms of power output, according to this model (Figure 6).

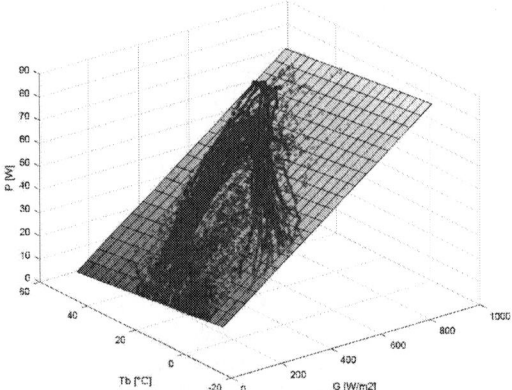

Fig. 5.   Regression function for the electric power of the PV module $f(I, T_{bom})$. The blue dots represent the observed power as a function of of T_{bom} and I, while the grid represents the fitted plane.

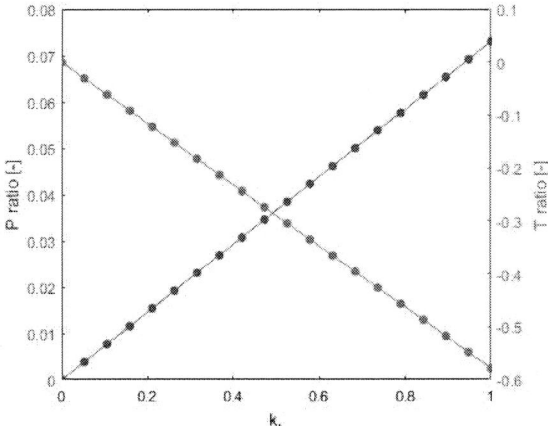

Fig. 6   Effects of operating temperature. Change in P_{ratio} and T_{ratio} as function of kt.

In order to investigate the **effect of the air channel ventilation** rate on the energy production, a relation $T_{bom}=f(w)$ was found. This can be done through the identification of a nonlinear thermal model for the tile, and through the measured T_i, w, I, T_a, where T_i are the temperatures of the different layers. It is known that the heat transfer coefficient due to the buoyancy effect in a tilted air channel can be written as a function of the Rayleigh number [8], which is a function of the temperature difference between the walls and the internal flow. Nevertheless, having measurements for the flow speed inside the air channel under C5 PV module, we can try to identify a heat transfer coefficient in the form of:

$$h = k_1 w^{k_2} \qquad (4)$$

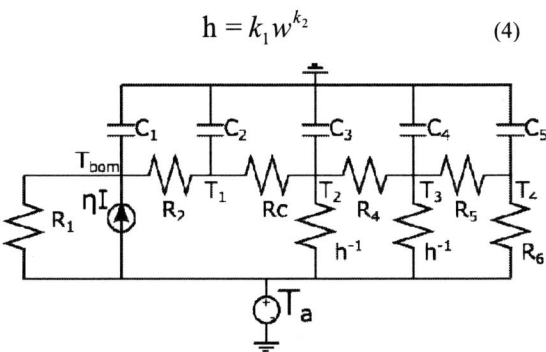

Fig. 7   Identified grey-box model for the thermal dynamics of the BIPV tile C5.

Figure 7 shows the equivalent resistance capacitance (RC) scheme for the identified grey-box model, which represents the thermal dynamics of the BIPV roof system (where C5 PV module is integrated). In this RC scheme, T_1 is the temperature

under the waterproofing membrane layer, T_2 and T_3 are the temperatures on the upper and lower faces of the air channel, respectively. T_4 is the temperature on the lower part of the roof. R_4 represent the linearized radiation coefficient between the air channel walls and η is a coefficient representing the equivalent external surface of the tile.

The influence of the temperature of contiguous layers is modeled as a single resistance, while inside the air duct; both radiation and convection are taken into account. Radiation is modeled by means of a linearized heat transfer coefficient, while convection is modeled using the previously defined non-linear heat transfer coefficient h, dependent on the flow speed inside the channel. Each layer has an associated capacitance, which model its thermal inertia. The air temperature inside the channel is considered to be equal to the external air temperature, which could possibly lead the model to overestimate the cooling effect of the ventilation.

Due to an intrinsic limitation of the data acquisition system, any measurements during nighttime is available so that the parameters of the grey-box model have been estimated by using a rolling horizon identification: instead of trying to minimize the one step-ahead residuals, the squared simulated residuals have been minimized, reinitializing the system states once per day. Since our model has 13 free parameters, in order to better explore the parameter space, a genetic algorithm was used instead of considering a gradient based method.

Once the parameters of the h function have been identified, the model to predict the PV power output at different ventilation speeds can be used. The coefficient k_w is defined such that:

$$w_n = w\, k_w \qquad (5)$$

where w_n is the new wind speed. The results of this study are shown in Figure 8.

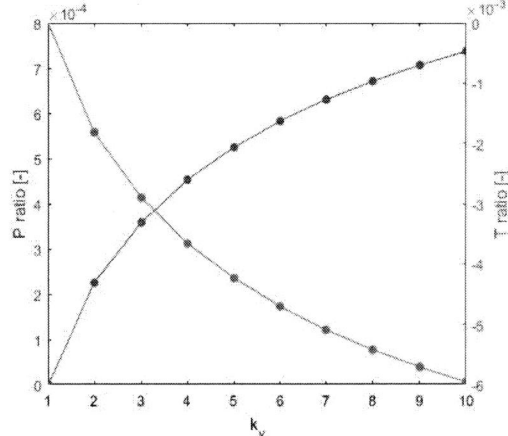

Fig. 8   Effects of ventilation. Change in P_{ratio} and T_{ratio} as function of k_w.

It's can be observed that, due to the change of air flow, the identified thermal model predicts a maximum relative temperature change in the order of 0.6%, which results in a relative change in energy production of about 0.08%. If the boundary conditions are considered as fixed, the only way to change the internal flow rate is by changing the length of the pitch (number of tiles) or other roof construction conditions.

As last, the **effect of changing the insulation layer under the module** (R_c) was investigated. Since this change has an influence on the internal wall temperature, this must be taken into account when simulating the internal air flow speed.

We then replace the measured wind speed with a linear function of the temperature difference between the air walls and I, such that h can be written as:

$$h* = b_1 \left[a_1 I + a_2 (T_{w,1} - T_{w,2}) \right]^{b_2} \qquad (6)$$

where $T_{w,1}$ and $T_{w,2}$ are the simulated wall temperatures of the air channel. The resulting fit has an adjusted *R-squared* value of 0.923. We can finally apply a linear change to the estimated value of the decking board thermal resistance. We let R_c change between 20 and 200 percent of its identified value. We define a coefficient k_{Rc} such that:

$$R_{c\_n} = R_c k_{R_C} \qquad (7)$$

where R_{c_n} is the new resistance.

The results of this analysis are shown in Figure 9. As a first result, we can see how changing R_c affects the P_{ratio} roughly ten times more than changing the air flow speed. However, the maximum relative change in the energy production is about 0.5% and its change is almost symmetric with respect to the reference value of R_c. This preliminary result suggests that the thickness of the insulation layer behind the PV module could be optimized mainly for controlling the heat transfer towards indoor spaces in summer, since its change does not affect significantly the module's power output.

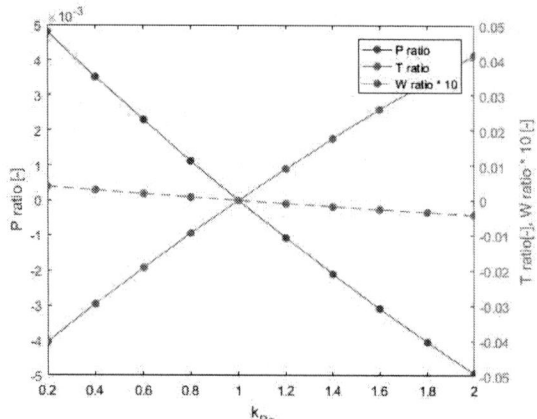

Fig. 9   Effects of thermal resistance. Change in P$_{ratio}$, T$_{ratio}$ and W$_{ratio}$ as function of k$_{Rc}$.

On the other hand, the module lifetime might also be affected by insulation thickness, as higher temperatures accelerate the aging process. This effect needs to be further investigated. It must be stressed that these results are only exploratory and related to the specific installation. The internal air flow velocity due to the stack effect is a nonlinear function of the mean temperatures of the tile internal walls, the solar irradiation, the external air temperature and the tile geometry and used material. This means that the effect of changes in R_c and w on the T_{bom} can be better estimated by means of parametric CFD simulations.

III. CONCLUSIONS

The investigation on how the performance of BIPV solar tiles is affected by temperature and ventilation has been discussed in the paper. Thanks to this study and to the existing literature, a reflection on the following aspects can be done:

- Ventilation strategies related to the roof construction;
- Length of the roof and thickness of ventilation duct;
- Micro-ventilation of tile's joints;
- Thermal resistance of the external decking layer

A. Ventilation strategies related to the technical mounting system and roof construction technology

Usually, standards and regulations provide criteria for dimensioning building skin ventilation chambers. Literature additionally offers inputs that allow understanding benefits of ventilated roofs for indoor spaces. However, texts that explicate effects of the ventilation chamber design for PV cooling to improve energy outputs when installed on the roof are in short supply. As stated in B subsection, some studies focused on different installation techniques, asserted that a counter-batten system contributes more effectively to reduce the T_{bom} rather than a roof with a ventilation chamber. Results obtained in Lugano have confirmed this aspect assessing the effective performances of the two roof systems tested under operation. In our opinion this is a consequence of two main factors discussed ahead: the first is the construction technique, since tiles are placed on a continuous layer (OSB decking) that provides a thermal resistance causing overheating; the second is the size of the air-gap (namely the thickness) that is typically larger in the counter-batten solution.

B. Length of the roof and thickness of ventilation duct

Research has already set that ventilated or micro-ventilated cladding of building skin affect the air flow and the internal comfort of the indoor spaces. Certain studies focused on the optimal thickness in relation to the length of the ventilation channel [9]. The length of a pitched roof and the dimension of the air chamber play a key role, as a consequence of the temperatures in the different roof layers and, namely, the cladding that is T_{bom} in case of a BIPV roof. Regarding the incidence on indoor temperatures, in the case of medium and high irradiation (flow rate between 50 W/m^2 and 100 W/m^2), the previous study shows that air flow temperatures in the ventilation channel increase, by increasing the length of the roof while the effective duct length decreases. This implies a higher indoor temperature of the inhabited spaces. In general, if the length of the roof is very high, the ventilation chamber must have a greater thickness in such a way as to ensure minimum interior environmental comfort (e.g. Swiss technical regulations, SIA 232/1, requires an increase in the height of the ventilation chamber proportional to the length and inversely proportional to the slope of the roof). At this regard, the test stands have some basic construction differences: the counter-batten roof (module B3) has an air gap thickness of 6 cm while the other roof system with decking (module C5) has a total air-gap of 12 cm (4+8cm). Moreover the ridge construction details are slightly different, since the air outlet allows a more direct air flow in the first stand. The length of the pitch are in both 1,80 m so that significant air speed have not been observed. Anyway the monitoring demonstrated that temperatures inside the ventilation chamber in the upper part of the pitch near the ridge, is higher than near the eaves (maximum measured of about 3.8 °C) confirming a stack effect inside the ventilation duct. Even though this paper has not focused in establishing the direct incidence of ventilation on the indoor space comfort, on-site monitoring campaign demonstrated that temperatures inside the ventilation chamber are higher in the case of PV modules mounted directly to the wood deck (Tegola BIPV module C5, T_{Lay2}). This factor would result in a greater diffusion of heat in the successive layers and consequently in the indoor comfort of the building. Apart from the internal comfort and the expected lower power output, also possible effects of the higher operating temperatures on the PV module's durability should be further investigated. In conclusion, the calculations presented in D subsection, confirmed that in our stands there is a low dependence of the T_{bom} in relation to internal ventilation speed, resulting in very little effects in energy production losses. However, the air flow and the temperatures in the air chamber are expected to change considerably by changing the length of the roof, the gap thickness and the construction details of ridge/eaves, so that in a real situation a proper roof ventilation technique is recommended to optimize both the PV module's cooling and the thermal gains for the indoor spaces.

C. Micro-ventilation of tile's joints

A further aspect emerging from literature [10] is that, between the joints generated due to the tile's overlapping, there are interstices that cause the leakage of considerable amounts of air (up to 65% of total air flow in a terracotta tile roof, [10]).

The BIPV module B3 could have benefit due to air spaces between the modules, that calls for further investigations.

D. Thermal resistance of the external decking layer for a roof with the ventilation chamber

Along with the construction technique, also the thermal conductivity of the roof materials influences the temperature distribution in the different layers and consequently the T_{bom} in the case of a BIPV roof. Simulations investigated the effect of changing the thermal resistance of the supporting layer under the BIPV tiles, namely the decking board on which the tile is nailed. A material with a lower thermal resistance in place of the OSB could contribute in improving the PV production even though the observed results from the simulation are not conclusive in this regard. Namely, a higher thermal conductivity corresponds with a greater cooling of the inner surface of PV modules since the heat is more effectively released to the air chamber and thrown away thanks to the stack effect. In next developments the wooden materials for the decking board ($\lambda \approx 0,15$ W/mK) could be replaced by metal ($\lambda \approx 30\text{-}200$ W/mK) such as a corrugated metal sheet.

In conclusion, we can state that a joint approach between energy and construction aspects and investigations on the points discussed above could improve and optimize the energy and cost effectiveness of this solar tile system.

REFERENCES

[1] 2015 Quadrennial Technology Review: *An Assessment of Energy Technologies and Research Opportunities. Chapter 5: Increasing Efficiency of Building Systems and Technologies.* September 2015, U.S. DOE, EERE/DOE, p. 489.

[2] International Energy Agency (IEA) report *Energy Efficiency Indicators Highlights 2016,* OECD/IEA, 2016, p. 144.

[3] E. Delponte, et al., "BIPV in EU28, from Niche to Mass Market: An Assessment of Current Projects and the Potential for Growth through Product Innovation," *EU PVSEC*, 2015.

[4] Cerón I. et al., "State-of-the-art of building integrated photovoltaic products", *Renewable Energy*, 2013, 58, pages 127-133, DOI: 10.1016/j.renene.2013.02.013

[5] C.S. Polo López, et al., "Outdoor Characterization of Innovative BIPV Modules for Roof Application," in *32nd EU PVSEC*, Amsterdam, The Netherlands, 2016, p. 2770 – 2778.

[6] M.T. Muller, J. Rodriguez, and B. Marion, "Performance Comparison of a BIPV Roofing Tile System in Two Mounting Configurations," in *34th IEEE PVSC,* Philadelphia, 2009, p. 8.

[7] M. D'Orazio, et al., "Performance assessment of different roof integrated photovoltaic modules under Mediterranean Climate," *Energy Procedia*, 2013, Volume 42, 20

[8] John Wiley & Sons, *Fundamentals of Heat and Mass Transfer*, Fundamentals of Heat and Mass Transfer, (Paese?) 2006.

[9] M. D'Orazio, "Risultanze di un'indagine sperimentale in funzione del risparmio energetico nei tetti ventilati", *Modulo* n° 265, BE.MA publisher, October 2000, p.919.

[10] E. Di Giuseppe, M. D'Orazio, "Manti permeabili per tetti "traspiranti", *Costruire in Laterizio*, 06/2014, 158, pages 58- 63

Life cycle assessment of transparent organic photovoltaic for window applications

Annick Anctil[1], Eunsang Lee[1], Jack Stephan[2], Anjali Munasinghe[1], Christopher Traverse[3], Richard R. Lunt[3]

Department of Civil and Environmental Engineering, Michigan State University, East Lansing, MI, 48824, USA[1]; Department of Biosystems Engineering, Michigan State University, East Lansing, Michigan, 48824, USA[2,] Chemical Engineering and Materials Science, Michigan State University, East Lansing, Michigan, 48824, USA[3]

Abstract — **The absorption of small molecule material uniquely in the near-infrared region opens new applications for organic solar cells. This work evaluates the environmental and cost benefit of transparent organic solar cells for two different applications using life cycle assessment. The transparent organic solar cells have a positive environmental impact for all scenarios considered.**

I. INTRODUCTION

The installation of solar photovoltaics modules (PV) is motivated in part to reduce the carbon footprint of current electricity production. In transparent solar cells, the near-infrared and ultraviolet parts of the solar spectrum are selectively harvested, while letting visible light pass through. These multifunction devices are specifically enabled by organic and nanostructured materials and help in heat-management (similar to low-e coatings) around buildings. They also have the potential to be extremely affordable by using low-cost materials and by taking advantage of the infrastructure of the built environment to reduce balance-of-systems installation costs. This work investigates the life cycle environmental and cost impact associated with this new technology for building applications in window and skylights for organic photovoltaics (OPV).

II. METHODS

This study is a cradle-to-gate lifecycle analysis as illustrated in Fig. 1 (a). Two different applications are considered: windows and skylights. For the window application, the impact of covering south, west, and east (SEW) windows of the building is compared with adding the north side (SEWN).

A. Energy production model

The potential impact of transparent OPV is evaluated for three locations in the United States where the energy grid, solar insolation, and electricity costs are different as shown in Figure 1(b). Lansing has low solar insolation but high electricity price and environmental impact. By comparison, the two other locations have higher insolation but either low electricity price (Phoenix) or low grid environmental impact (Los Angeles). Energy production as a function of azimuth angle is calculated using RETScreen 4 [1]. The net

environmental benefit (NEB), which includes the benefit from heat absorption ($H_{avoided}$) due to visible transmittance and NIR absorption, is modeled using EnergyPlus 8.0[2]–[4].

$$NEB = H_{avoided} \pm E_{heat\ or\ cooling} + PV_{generation} \qquad (1)$$

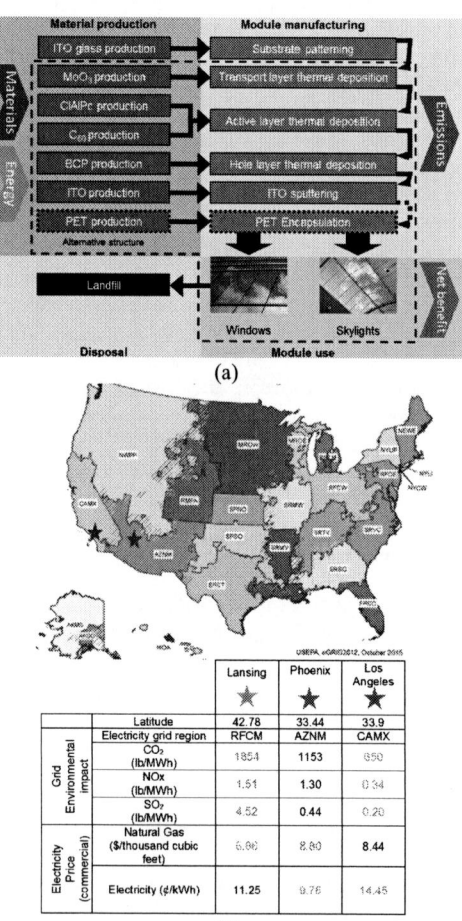

Fig. 1.(a) Scope of this work and b) electric grid subregions [5], associated environmental and cost impact for the 3 locations of interest.

The overall building energy performance is based on the reference-building model from DOE, which complies with ASHRAE Standard 90 [6]. Energy loads are based on previous default values from NREL publication [7]. Two different buildings models are used as illustrated in Figure 2. The medium office is used for the window application while the small office model is used for skylight applications. The medium office building has 33% of window-wall ratio while the small office has 28.5 % and all windows are double pane [3]. The medium office has three floors and a flat roof and the transparent PV is applied to the window on each face of the building. By comparison, the small office has a tilted roof (18.45 Deg.) with a single floor. The skylight is situated on the south side of the roof. The skylight roof ratio follows ASHRAE standard 90.1 that specifies that it must be less than 0.05 [6].

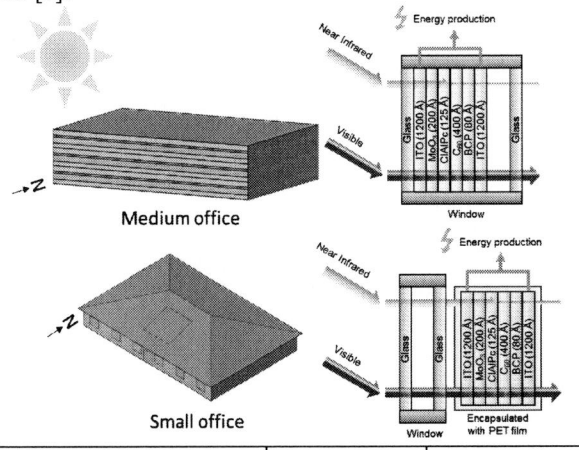

	Medium office	Small office
Total (m2)	4982.2	511.2
Gross wall area (m^2)	1977.7	281.51
Window opening area (m^2)	652.8	59.68
Window-wall ratio (%)	33.0	21.20
Roof area (m^2)	1660.7	598.76
Skylight area (m^2)	-	28.46
Skylight-roof ratio (%)	-	4.75
Window or skylight configuration	3mm clear glass/ 13mm air/ 3mm clear glass	
PV configuration	3mm transparent PV layer/ 13mm air/ 3mm clear glass	
PV application	Window (E, W, S)	Skylight (S)
PV area (m^2)	456.99	28.46

Fig 2. Building architecture and device structures

For the OPV module, two different options are considered as illustrated in Figure 2. The OPV is either directly deposited on the glass inside the double pane window, or deposited on a plastic substrate and encapsulated. The glass option does not allow solar cell replacement until the window reaches end-of-life while the plastic option can be replaced. The plastic solar cells are replaced every five years over the 20 years lifetime of the window. The transparent OPV structure is based on previously published work [8]. In addition to the glass versus plastic alternatives, this work considers best/worst case scenarios for OPV. The worst-case scenario considers 2% efficiency and 50% degradation after 6 years. The best-case scenario considers 6% efficiency and 50% degradation after 10 years.

Optical properties were simulated from 300 nm to 1600 mn based on optical constants measured by variable angle spectroscopic ellipsometry [9] and used to calculate the heat transfer coefficient (U-Factor), solar heat gain constant (SHGC), and visible transmittance (VT) used for the building energy simulation.

B. Life Cycle Assessment

The life cycle inventory consists of material and energy components directly measured during device fabrication. Material consumption is calculated using material density and assuming a 30% material use efficiency as previously measured by the author [10]. The life-cycle greenhouse gases emissions (GHG) and the embodied energy (CED) over the life-cycle of each system is calculated. GHG is assessed using TRACI 2.1 methods and Cumulative Energy Demand Method v1.09 in SimaPro8 [11]. The energy payback time (EPBT) denotes the time needed to compensate for the total cumulative energy-demand (CED) required during the life cycle of the PV system. The annual electricity generation is converted into its equivalent primary-energy, based on the efficiency of electricity conversion at the demand side, using the current average.

IV. RESULTS

A. Energy production

Figure 3 summarizes the building annual energy demand and the electricity production for a 6% plastic solar cell for either window or skylight application. Because of the climate, most of the energy demand in Lansing is from heating while in the other 2 locations, the cooling demand dominates. The dotted line for electricity production represents the SEWN application compared to the SEW. Los Angeles and Phoenix have a similar energy demand, and because of the latitude, electricity production from PV looks similar but the amount produced in Phoenix is higher than in Los Angeles. By comparison, the energy production in Lansing is significantly lower during the winter, but the addition of modules on the north side increases electricity production during the summer.

The energy balance from energy and electricity production does not consider the energy required to fabricate the solar modules. The NEB for cumulative energy demand (CED) is positive for all scenarios considered, as shown in Figure 4. The energy required to fabricate the solar cells is smaller than both the energy demand reduction and the electricity produced. In Lansing, the cooling demand is significantly reduced while the OPV increases the heat demand during the winter. The principal benefit for all window applications is from the energy reduction associated with the air conditioning (AC) cooling system.

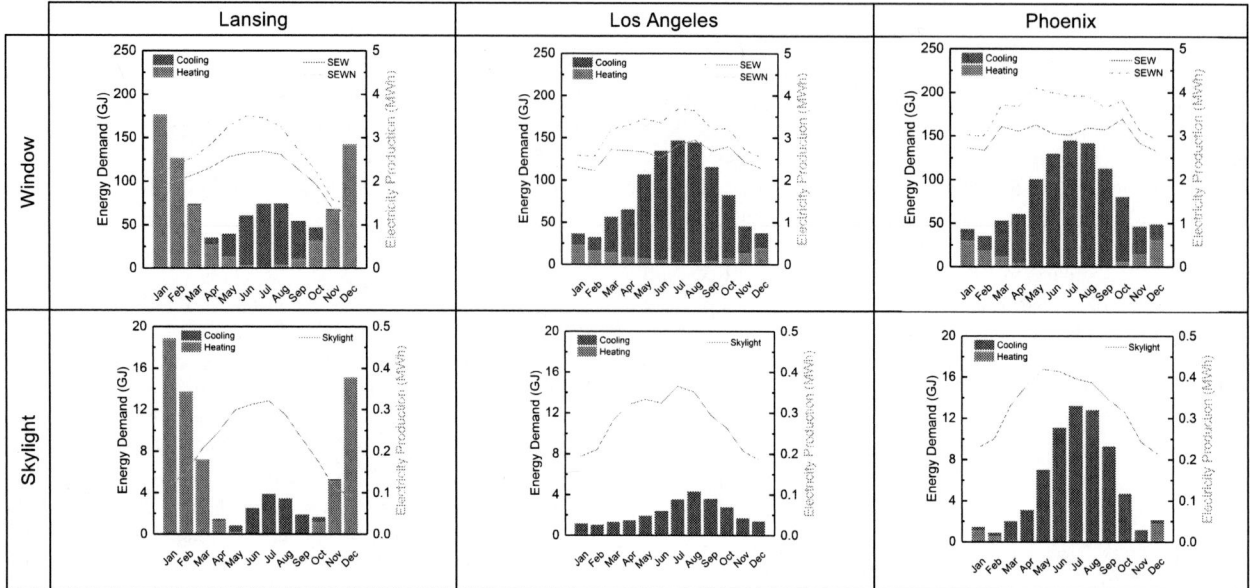

Fig. 3. Annual energy demand and electricity generation for 6% solar cells in window and skylight application

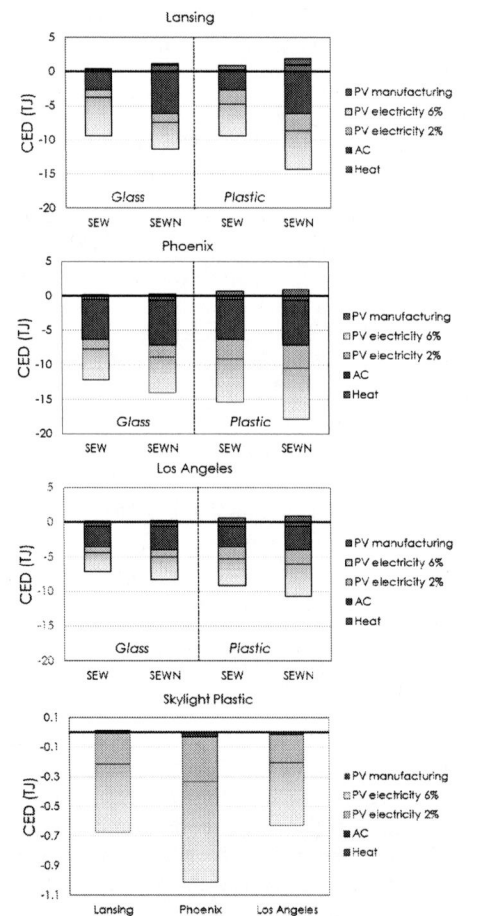

Fig. 4. Net environmental energy balance for window and skylight applications in Lansing, Phoenix, and Los Angeles

One significant difference between the window and skylight application is the relative impact of energy reduction versus electricity production due to the better orientation of the solar photovoltaic modules compared to the vertical window application. In the skylight application, there is negligible energy reduction in the building and most of the energy benefit is associated with the electricity production.

The energy payback time is an additional metric used to evaluate the benefit of solar technology for each scenario. The goal is to calculate how long it takes for the module to produce enough solar energy and save energy to compensate for the energy required to make the solar modules. In Figure 5, for each application, the range of EPBT corresponds to 2 to 6% devices efficiency.

As shown in Figure 5 for the window application, for all scenarios considered, the energy payback time is much lower than the 20 years lifetime period. For the modules integrated into the window, even with PV degradation the maximum payback time is less than 0.7 years. Because the module is replaced four times over the 20 years period for plastic PV, the energy payback time is higher for those solar modules, but still less than three years.

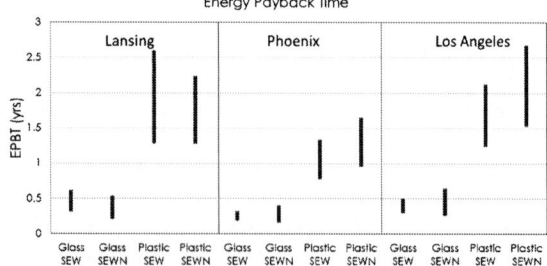

Fig. 5. Energy payback time for glass and plastic solar cells in various applications.

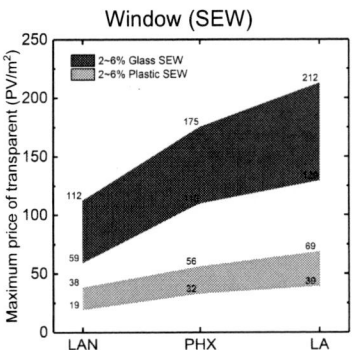 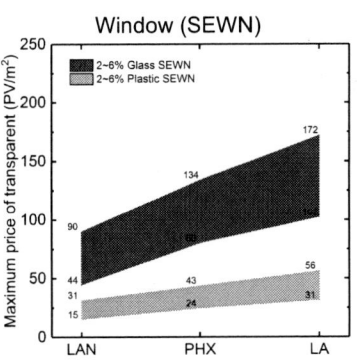 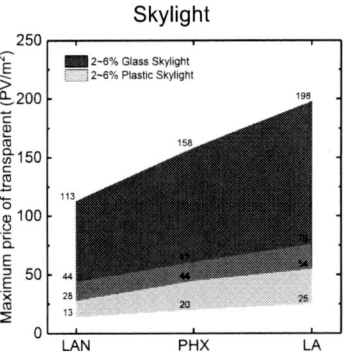

Fig. 6. Maximum price of modules ($/m^2) for window and skylight applications

To evaluate the economic potential of OPV for window and skylight application, the electricity price is used to calculate the energy saving associated with natural gas and electricity consumption. The saving is used to calculate the maximum price for modules for the different cities based on their local energy price and net energy production.

Figure 6 illustrates the price for all scenarios considered. Because of the combined lower net energy benefit of OPV in Lansing, in the worst case scenario, 2% efficient solar cells applied on all sides of the building would need to be less than 15 $/m^2 to be cost-effective By comparison, in Los Angeles where the electricity price is much higher, in the best case scenario where a 6% OPV on glass solar module is installed on SEW sides of the building, the price could be as high as 212 $/m^2. The skylight prices are similar than prices from window application applied on SEW. Finally the energy cost and environmental benefits from three categories: global warming potential, cumulative energy demand and fossil depletion are compared for a 6% plastic solar cell (Figure 7). The reduction is the highest for Los Angeles with more than 30% in all categories. Even though more electricity is produced in Phoenix than in the other 2 locations, the higher electricity price has a greater impact in the other 2 locations. Overall for all criteria, the reduction is more than 10% in all categories considered.

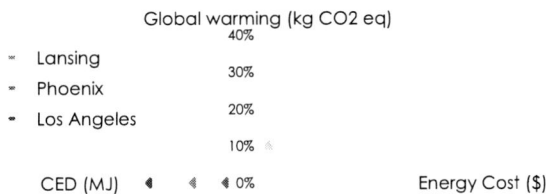

Fig. 7. Environmental and energy cost reduction for each city compared to the baseline model for a 6% plastic solar cells

V. CONCLUSION

The potential of transparent OPV for integration of windows and skylight has been evaluated for three cities with various climate, insolation and energy price which influences the overall benefit of the solar modules. In all cases, the net environmental benefit based on the cumulative energy demand is positive and the energy payback time is less than 3 years.

REFERENCES

[1] Canada, Natural Resources "RETScreen Clean Energy Project Analysis Software." .

[2] P. K. Ng, N. Mithraratne, and H. W. Kua, "Energy analysis of semi-transparent BIPV in Singapore buildings," *Energy Build.*, vol. 66, pp. 274–281, 2013.

[3] Y. T. Chae, J. Kim, H. Park, and B. Shin, "Building energy performance evaluation of building integrated photovoltaic (BIPV) window with semi-transparent solar cells," *Appl. Energy*, vol. 129, pp. 217–227, 2014.

[4] D. B. Crawley, L. K. Lawrie, F. C. Winkelmann, W. F. Buhl, Y. J. Huang, C. O. Pedersen, R. K. Strand, R. J. Liesen, D. E. Fisher, M. J. Witte, and J. Glazer, "EnergyPlus: creating a new-generation building energy simulation program," *Energy Build.*, vol. 33, no. 4, pp. 319–331, Apr. 2001.

[5] EPA, "eGrid." .

[6] "New Construction — Commercial Reference Buildings | Department of Energy." .

[7] M. Deru, K. Field, D. Studer, K. Benne, B. Griffith, P. Torcellini, B. Liu, M. Halverson, D. Winiarski, M. Rosenberg, M. Yazdanian, J. Huang, and D. Crawley, "U.S. Department of Energy commercial reference building models of the national building stock," 2011.

[8] R. R. Lunt and V. Bulovic, "Transparent, near-infrared organic photovoltaic solar cells for window and energy-scavenging applications," *Appl. Phys. Lett.*, vol. 98, no. 11, p. 113305, 2011.

[9] M. Young, C. J. Traverse, R. Pandey, M. C. Barr, and R. R. Lunt, "Angle dependence of transparent photovoltaics in conventional and optically inverted configurations," *Appl. Phys. Lett.*, vol. 103, no. 13, p. 133304, 2013.

[10] A. Anctil, C. W. Babbitt, R. P. Raffaelle, and B. J. Landi, "Cumulative energy demand for small molecule and polymer photovoltaics," *Prog. Photovoltaics Res. Appl.*, vol. 21, no. 7, pp. 1541–1554, Nov. 2013.

[11] PRe, "SimaPro 8." 2014.

AUTHOR INDEX

Aaditya, Gayathri 604
Abbas, A................. 1691, 2457, 3430
Abbas, Ahmed E. 1888
Abbas, Ali 186, 752, 1674
Abbott, Malcolm D.............1322, 2576, 2600
Abdalla, L.B 1245
Abdallah, Amir A...............................3435
Abdallah, Shaimaa A. 219
Abdellaoui, Imane 900
Abdullah, Ahmad 2128
Aberle, Armin.................................2318
Aberle, Armin G. 284, 496, 499, 1922
Ablekim, Tursun 3422
Aboubakr, Benazzouz 487
Abouelkhair, Hussain M........................2324
Abtahi, Amir 638
Abudayyeh, Omar K...............................88
Acebo, Laura 155
Addamane, S. J...................................281
Adewoyin, Adeyinka 2381
Adhikari, Dipendra 2582
Affouda, Chaffra A. 259
Agarwal, Mohit 2330
Agarwal, Sumit................................1777
Agarwal, Vivek............... 2952, 2981, 2986, 3050
Agbo, Solomon N...............................2114
Ager, Joel 3410
Agrawal, Rakesh 1449
Aguiar, Jeff 2702
Aguiar, Jeffrey 2467
Aguirre, Rodolfo 2419
Ahamioje, Joseph A. 2931
Ahanzhamejhad, Ramez H........................170
Ahlswede, E. 3260
Ahlswede, Erik 791
Ahmad, Jawad..................................3096
Ahmed, Benlarabi 487
Ahmed, Nuha.............................658, 2667
Aho, Arto 297, 2520
Aho, T. 1189
Aho, Timo 297
Ahrenkiel, P. 869
Ahrenkiel, Phil 206, 831
Ahrenkiel, Richard K. 3448
Ahrenkiel, S. Phillip 2514
Ahsan, Nazmul 2334
Aierken, Abuduwayiti 226
Aindow, Mark 1522
Aïssa, Brahim 3435
Akaki, Yoji...................................2338
Akari, Shunsuke 2385
Akarm, Muhammad Nadeem 2776
Ake-Sultan, Bernt 2864
Akiki, Tilda 1968
Akimoto, Katsuhiro 33, 160, 900

Akimoto, Naoki.................................. 712
Akiyama, Hidefumi 721, 2781, 3528
Akwari, Chinedum 735, 2446
Al Mahmud, Abdullah 1067
Alahmed, Ahmed............................... 1110
Alam, Giri Wahyu 1498
Alam, Muhammad A........................ 1055, 1259
Alam, Muhammad Ashraful...................... 1904
Alberi, Kirstin................................. 2506
Albin, David 1196, 3305, 3319
Alcubilla, R. 1781
Alcubilla, Ramón 944
Aleman, Monica 2227, 3435
Alexander, Jessica A. 966
Alfadhili, Fadhil K. 730, 815
Al-Fadhili, Fadhil K. 2462
Al-Ghzaiwat, Mutaz 2593
Algora, C. 1210
Algora, Carlos................................ 1204
Alharbi, Fahhad H. 963
Alharthi, Yahya Z. 1018, 1110
Ali, Asad 1228
Ali, Jaffar Moideen Yacob..................... 2318
Ali, Waqar................................... 1228
Alivisatos, A. Paul........................... 1737
Al-Jassim, M. 1196
Al-Jassim, M.M. 1312, 2280, 2785
Al-Jassim, Mowafak.......... 62, 1371, 1381,
1400, 2789, 2887, 3305, 3319
Al-Jassim, Mowafak M. 3147
Aljaziri, Marwa 2011
Alkhayat, Rabee B.............................. 815
Allebé, C. 50, 2073
Allebé, Christophe........................ 3254, 3256
Allen, Thomas 2076
Almheiri, Anwar.............................. 1946
Almonacid, Florencia 2858
Alrashidi, Hameed 2858
Altermatt, Pietro P. 1922, 2220, 3304
Alvarez, Diego Alonso 1339
Alvarez, Genesis 2941
Alvarez, José 2453, 2528
Aly, Shahzada P................................ 963
Alzahmi, Wadhah.............................. 1946
Amdemeskel, Mekbib W......................... 2672
Anctil, Annick................................. 2124
Anderberg, A. 467
Andler, Joseph 1449
Ando, Daisuke 931
Ando, Yasutaka 970
Ando, Yuta 192
Andreani, Lucio 290
Angeles-Ordóñez, G. 142
Annigoni, Eleonora.................... 1395, 2794
Anoma, Marc Abou............................. 1549

AUTHOR INDEX

Anselmo, Andrew..............................74, 2839, 2897
Antony, Aldrin 1755
Anttu, Nicklas...2502
Anyadiegwu, Ifeanacho 970
Anyanwu, Uchechi 319
Araki, Hideaki..2338
Araki, Kenji.............................359, 412, 1479, 1711,
    1714, 1743, 2498, 2548, 2566
Aranguren, G. 643
Archer, Alexander 771
Arehart, A. R. 30, 2414
Arehart, Aaron R. 215, 2446, 3139
Arinze, Ebuka S. 667
Armour, Eric 827
Armour, Eric A. 210, 2506
Armoush, Maher 1058
Arnold, Daniel B. 3002
Arnou, Panagiota 146, 186
Arora, B. M. 396, 1995, 2716
Arora, Brij M. 3478
Arp, Juergen 1411
Arredondo, C. A. 2031
Artegiani, Elisa.....................752, 1669, 2372
Aryal, Krishna...182
Asadirad, Mojtaba..866
Asahi, Shigeo 23
Asgharzadeh, Amir 1537, 1543, 3333
Ashrafee, Tasnuva 735
Aslam, Aasma 2355
Asomoza, René 632
Astakhov, Oleksandr 2114
Aswani, U..1898
Athresh, Eashwer......................................2395, 2399
Atia, Adam A. 3230
Atkins, R. 229
Atlan, Olivier..626
Atwater, Harry A......................512, 521, 558, 572,
    1248, 1589, 1737, 2236
Augarten, Yael..1651
Augusto, André 1589, 2596
Avasthi, Sushobhan........251, 837, 841, 986, 2395, 2399
Avenet, Julien...1933
Avery, J. E. 1863
Awadallah, Osama...3473
Awasthi, Vishnu 2345
Ayala, Orlando..735
Azkona, N. 2740
Azkona, Nekane 2677
Azzolini, Joseph A...608
Baba, Masaaki...1724
Babbe, Finn 151, 2054
Babcock, Sean J...2298
Bachman, Benjamin F. 3381
Badel, N..50
Badosa, Jordi...626

Badr, Ikken 487
Bae, Soohyun 935
Baggu, Murali 2991
Baik, Sungsun 2242
Bailey, C. 845
Bailey, Christopher G. 2298
Bailey, J. 2414
Bailey, Jeff..1686, 3327
Baines, Tom 742, 1445
Baka, Maro 3343
Baker, Rupesh 3172
Bakhshi, Sara 322
Bakker, Klaas 2875
Balaji, Pradeep 2596
Balakrishnan, G. 281
Balasubramaniam, Kavaipatti R. 1704
Baldus-Jeursen, Christopher..............................1908
Ball, Greg 2263
Ballif, C. 50, 2073
Ballif, Christophe 55, 1220, 1395,
    2104, 2794, 3254, 3256, 3435
Baloch, Ahmer A.B..............................963, 1058
Banda, Pedro 1946
Banerje, Rangan 1151
Banerjee, Sanjay K..363
Barahman, Gil 2285
Barakel, Damien 2255
Barnes, T. M. 138
Barnes, Teresa M..3422
Barnett, Allen..315
Barraud, L. 50
Barraud, Loris 3254, 3256
Barrigon, Enrique 2502
Barth, Kurt 424
Bartolo, Robert E. 195
Bartsch, J. 884
Basore, Paul A. 2163
Bastide, Stéphane 3402
Bastola, Ebin 738, 781
Basu, Prabir K..396
Battaglia, A. 1747
Baudrit, Mathieu 2492, 2562
Bauer, Andreas 791, 2058
Bauer, Jan 1376
Bauhuis, G. 1189
Baumann, Thomas 1077
Baumgartner, Franz...1077
Baur, Carsten 541, 2087
Baxter, Jason B..3143
Bearda, Twan 1233
Beauchemin, Ryan D. 102
Becerril-Romero, Ignacio 155
Becker, Jacob J..3366, 3410
Bedair, Salah M. 2195
Belanger, Ted 1427

AUTHOR INDEX

Belletête, Marc 1579
Belluardo, Giorgio 3360, 3482
Bemrrr, Andreas......................................3500
Benamara, Mourad...................................3370
Benatto, Gisele A. Dos Reis 2672, 2682
Benick, J...2064
Benick, Jan 2511
Bennett, Dirk 2042
Bennett, Mitchell.......................................247
Bennett, Mitchell F.............210, 259, 873, 2091
Berardone, Irene 402
Berg, Alexander 1773
Berg, Morgann...3417
Bermel, Peter 1904, 2467
Bernard, Annie.............................2870, 2891
Berry, Joseph J.......................................2176
Bert, J...1733
Bertoni, Mariana.................944, 2610, 2854
Bertoni, Mariana I. 2179, 3309
Besanger, Yvon..3102
Bett, Alexander J......................................1253
Bett, Andreas W. 2511
Bettenwort, Gerd 1965
Beutel, Paul...2511
Beutner, Volker..1855
Bhaduri, Sonali..............................2799, 3478
Bhan, Mohan Krishan 496
Bhandari, Khagendra P................738, 748, 781, 815
Bhatia, A. 1656
Bhatia, Swasti..1755
Bhattacharya, Indranil 3083
Bhattacharya, Sitangshu 2376
Bheemreddy, Venkata 2688
Bialek, Tom 2991
Bidiville, Adrien.......................................1333
Biedenham, Richard E..............................3245
Biegelsen, D. K. 1733
Biiss, M. 2457
Binetti, Simona..1669
Birch, Max T. 2423
Birkmirc, Robert.......................................726
Birkmire, Robert W..................................2637
Bishop, Doug 3275
Bishop, Douglas 726
Bishop, Douglas M. 1441
Bissels, G...1189
Biswas, R...1350
Bittau, F. 3430
Bittau, Francesco 752
Bittner, Zachary...677
Bittner, Zachary S...................18, 202, 2084
Bivour, M...2064
Bivour, Martin..1253
Blakely, Logan..1573
Blanche, Pierre-Alexandre 1147

Bläsi, Benedikt 352
Blasi, David 1531
Blum, Adrienne 2692
Blum, Adrienne L. 2765
Bob, Brion 2258
Bobela, David C. 2506
Bobyl, A.V. 1025, 1811
Boca, Andreea...2099
Boccard, Mathieu 55, 1220, 1317, 1790, 3366
Boeck, Torta...3396
Bohra, Rakesh 1912
Boizot, Bruno 83, 2087
Bolaji, Adewumi 2381
Boley, Allison 2573
Bolke, J. G. 1656
Bonnassieux, Yvan 626
Bonomo, Pierluigi 2118
Book, Felix 1824
Bora, Birinchi 3478
Borgers, Tom 3343
Borgström, Magnus...................................2502
Borgström, Magnus T 1286
Borland, John 2947
Borne, Axel 2864
Borowik, Lukasz.......................................1516
Bosco, Nick.....................................3190, 3200
Bosco, Nick S. 2864
Bosson, Christopher J. 2423
Bostock, Peter 2267
Bothe, Karsten 2692
Bourcois, Jérôme 2087
Bourdin, Vincent 626
Bourgoin, Jacques C. 2087
Bourne, Ben C. 1549
Bousselham, Abdelkader 1058
Bouttcmy, Muriel 2711
Bowden, Stuart 240, 925, 1797, 2719
Bowden, Stuart G. 1589, 2596
Bowen, Leon...1445
Bowers, J.W. 2457, 3430
Bowers, Jake W. 146, 186, 752, 2349
Boyce, Ken 2000
Boyce, Kenneth 1933
Boyd, Matthew..1933
Boyer, Jacob..215
Boyer, Jacob T. 2079, 2554
Boyer-Richard, Soline 2192
Brabec, Christoph J. 1346, 3500
Bradshaw, Geoffrey K. 88, 301, 531
Brady, Brendan 3388
Braga, Daniel Sena 2307
Braid, Jennifer L. 1927, 2697, 3456
Brammertz, G..3260
Brand, A.A. 884
Brates, Nanu 1728

AUTHOR INDEX

Bräuninger, Matthias 3256
Breitenstein, Otwin 1376
Breitwieser, M. 3135
Bremner, S. P. 953
Bremner, Stephen 858, 1215, 1845, 2186, 2569
Bremner, Stephen P. 948
Brendel, Rolf 1366, 3371
Breus, V. 1752
Bright, Jamie M. 1405
Brinnig, Samuel..................................2622
Brito, Pedro P. 2307
Britt, Jeffrey 1455
Brittman, Sarah 2245
Broderick, Robert 3008
Broderick, Robert J. 1435, 1555, 1567, 1573, 3025, 3031
Brolo, Alexander G. 3388
Bruckman, Laura................................1933
Bruckman, Laura S. 2000
Brückner, Sebastian............................2538
Brughera, Céline.................................2492
Brule, Carlton.....................................1728
Brulo, Gregory S. 1469
Bryan, Jonathan..................................1317
Buchanan, Wayne 1196
Büchler, A.................................884, 3135
Buckner, Jessica..................................537
Buerhop, Claudia................................3500
Bukowsky, Colton R. 1737
Bulkin, P. 1781
Bulkin, Pavel 1237
Bullock, James............................59, 2076
Buonassisi, T....................648, 1140, 3295
Buonassisi, Tonio 284, 1264, 1491, 2242, 2532, 2744, 3236, 3290, 3300
Burgers, A.R. 3150
Burgers, Antonius R. 917
Burkhardt, S..1752
Burnham, Laurie 1435
Burroughs, Scott 272, 1469
Busquet, Severine................................1061
Butt, Isaac..182
Cabarrocas, Pere Roca I............. 464, 1237, 2528, 2593
Cachet-Vivier, Christine.......................3402
Caffy, Florent 1516
Calderón-Obaldía, Fausto 626
Calle, Eric 944
Calvo-Barrio, Lorenzo.........................3285
Campa, Andrej....................................1346
Campanelli, Mark................................437
Campbell, Calli M. 3366, 3410
Campesato, Roberta 76, 541, 545
Campos, Cláudio Dias 2307
Camus, Christian 3500
Cañadillas, David........................429, 1116

Canino, A. 1747
Caño, P. 1210
Cao, Huihui...1619
Cao, Wenkai... 696
Cao, Xin..2427
Cao, Yunxue....................392, 1430, 1873, 2918
Cappelluti, F. 1189
Cardona, Dagoberto............................. 670
Cardwell, D. 3511
Cariou, Romain 2511, 2528
Carlin, John A. 215
Carlson, David E. 3442
Carlson, Emily 701
Carneiro, Lucas M. 417
Carolus, Jome 2875
Carpenter, Bernard 537
Carr, Anna J. 1081
Carriere, Jarrett 2833
Carruthers, Steve 3514
Carter, Catrice M.................................3393
Carter, Cedric.....................................2135
Carter, Sam 3514
Casale, Mariacristina 76, 541, 545
Casallas-Moreno, Y. L. 670
Casper, Chadwick 1476
Cassini, Denio A. 1917
Castañeda, Carlos A. Rodríguez 1858
Catthoor, Francky 3343
Cattin, Jean 3435
Cattoni, Andrea 1289
Cavani, Olivier....................................2087
Cédola, A. P. 1189
Cendagorta-Galarza, Manuel................. 429
Cepeda, Kyle 876
Cesar, I. 3150
Cesar, Ilkay 917
Chai, Gaoda 976
Chai, Jing 1922
Chakraborty, Sagnik............................3300
Chamarthi, Phani Kumar......................2952
Chamberlin, Charles 1271
Champliaud, J. 50
Champliaud, Jonathan..........................3435
Champness, C. H. 2388
Chan, Calvin 3417
Chan, Catherine E......................... 2576, 2600
Chan, Mandy 2808
Chan, Maria K.Y. 6, 1256, 2759
Chan, R. 3511
Chandralal, Sreeram 1674
Chandran, Deepak 2986
Chang, Jipeng 1873, 2823
Chang, Sheng-Hao................................1051
Chang, Via-Chung................................1051
Chantana, Jakapan 757, 2385

AUTHOR INDEX

Chaporr, Patrick 2711
Chapuis, Valentin 2104
Chattopadhyay, Kamanio 2811
Chattopadhyay, Shashwata 1850, 1858, 2849, 3478
Chaudhry, Ghulam M. 1018, 1110
Chaujar, Rishu 377
Chaurasia, Saloni 837, 841
Chausseau, Matthieu 2711
Chavali, Raghu Vamsi Krishna 1904
Chavez, Jose J. 2419
Chemisana, Daniel 1339
Chen, Benjamin 2358
Chen, Chien-Hsun 911
Chen, Chun-Chi 1635
Chen, Daniel 2576
Chen, Eric Y. 1598, 3384
Chen, Haiyan 2220
Chen, Hung-Ling 1289
Chen, Junyan 1835, 2732
Chen, Kaifeng 2185
Chen, Kunji 2656
Chen, Lung-Chien 367
Chen, Meixi 326, 999, 2035
Chen, Peng-Wei 2660
Chen, Ran 2576
Chen, Renfang 1241
Chen, Shi-Wei 1627
Chen, Sung-Yu 911
Chen, Tsung-Cheng 329
Chen, Tzu-Yu 1627
Chen, Wanghua 2593
Chen, Weijian 2392
Chen, Y. 2785
Chen, Yang 2502
Chen, Yao- Hui 893, 2664
Chen, Yifeng 1922, 2220
Chen, Yunfei 761, 2427
Chen, Yusi 1835
Chen, Zhi David 1044
Chen, Zihan 2392
Chendo, Michael 2381
Cheng, Y. 14
Cheng, Yan 667
Cheng, Yuh-Jen 1610
Cheng, Zhe 3473
Cheng, Zhongkai 3393
Chenna, Shiva Tarun 1674
Chiang, Cho-Chun 893, 2664
Chiang, Fu-Kuo 198
Chikhalkar, Abhinav 823, 827
Chin, Ken K. 761, 2427
Chinnusamy, Saravanan 980
Chiu, Chun-Yu 1169
Chiu, P. 2094
Chiu, Philip 2099

Chmielewski, Daniel J. 215, 2079, 2554
Cho, Eunhwan 333, 1824, 1838
Cho, Junsik 810
Cho, Yasuo 3323
Choi, Gyu-Seok 2723
Choi, J. -K 2019
Choi, Rae-Won 2723
Choi, Seungkeun 1037
Choi, Sungjin 1758
Chong, Cheemun 2600
Choubisa, Hitarth 1022
Choudhury, K. R. 2312
Chouhan, Arun Singh 986
Choulat, Patrick 2227, 3435
Chow, E.M. 1733
Chowdhury, Ahrar Ahmed 888
Christians, Jeffrey A. 2176
Christmann, G. 50
Chu, Chi-Wei 1051
Chu, Haifeng 1222
Chu, Sheng 1299
Chua, Soo Jin 284
Chuang, Ta-Wei 343, 367, 893, 2664
Chung, Daniel 2707, 3304
Chung, Haejun 1904
Chung, Simon 696, 2186
Ciesla, Alison M. 2576, 2600
Cifuentes, L. 1210
Cifuentes, Luis 1204
Ciocia, Alessandro 3096
Cirino, Daniel A. Merced 3044
Clayton-Warwick, D. 138
Cleveland, Erin 247, 2091
Clinton, Evan A. 305
Cobo-Yepes, Nicolás 2963
Codd, Daniel S. 3245
Cohen, Bat-El 2170
Cole, Wesley J. 2163
Colegrove, E. 1312
Colegrove, Eric 3147, 3319
Coll, Pablo Guimera 2610
Collin, Stéphane 1289, 3147
Collins, Robert W. 807, 2462, 2582, 2646, 3426
Collins, Shamara 802, 1638, 2449, 3413
Comagliotti, Emanuele 3435
Condorelli, G. 1747
Conibeer, Gavin 696, 2186, 2392
Conlon, Benjamin P. 219
Conrad, Brianna 315
Cordeiro, Patricia 2135
Cordova, Adam 1965
Cornagliotti, Emanuele 1804, 2227
Cornaro, Cristina 3482
Cornell, Robert 1275
Correa, J.M. 433

AUTHOR INDEX

Correa-Baena, Juan-Pablo 3300
Cossio, Gabriel 1181
Costa, Sara...1979
Costa, Suellen C. 2307
Côté, Alexandre..................................1908
Cousar, Larry C. 921
Cravens, R. 2094
Crawford, L. 1733
Crupi, F. 2073
Cruz, José Ortega 1959, 1990
Cruz, Leila R. De Oliveira....................2307
Cruz-Campa, Jose Luis 337
Cuevas, Andres...................................2076
Cui, Jie...2076
Cui, Min..1765
Cunningham, Daniel W. 1463
Cunningham, Joseph 3161
Cur, Jie 517
Curran, Alan J. 1927, 2697, 3488
Curvat, L. 50
Cushing, Scott K. 417
Da Fonseca, Jérémy 2492, 2562
Dabney, M. S. 138
D'Abrigeon, Laurent.............................545
Daenen, Michael..................................2875
Dagenais, Mario 195, 1048
Dagyte, Vilgaile 2502
Dahal, Saroj 309, 3123
Dahal, Som..240
Dai, Yushuai 18, 222, 677, 1184
Dalal, V L 1350
Dalal, Vikram 2247
Dalpian, G...1245
Dam-Hansen, Carsten...................2672, 2682
Danel, A. 1747
Dang, Hongmei 2432
Dangate, Milind S. 980
Daniil, Andreana 944
Danzl, F.J.K...3150
Darbali-Zamora, Rachid 2957, 2963
Das, Ujjwal 408, 1473, 1761, 1828, 2667
Das, Ujjwal K. 2637
Datas, Alejandro 2562
Dauskardt, Reinhold....................3190, 3200
Davidsen, Rasmus S. 2672
Davies, J. I. 1210
Davis, Kristopher O. 74, 322, 1804, 3448
Davis, Tracy..537
De Coux, Patricia.................................464
De Melo, O. 2342
De Nicolas, S. Martin.............................50
De Oliveira, Michele C. C.....................1917
De Villers, Bertrand J. Tremolet 1354
De Wolf, Stefaan 55, 3256, 3435
De, F. C. Lins Vanessa 1917

Debnath, M. C. 14
Debnath, Tanmoy 1067
Deboever, Jeremiah.....................1555, 1567
Debrot, F. 50
Debucquoy, Maarten............................1233
Debusschere, Vincent 3102
Deceglie, Michael 2771, 2789
Deceglie, Michael G...........2488, 2804, 3452
Deckerl, D. 1752
Decobert, Jean 2528
Deer, Tanya...1908
Deitz, Julia I. 3139
Delahoy, Alan E. 761, 2427
Delhotal, J. 3224
Deligiannls, D......................................3150
Deline, Chris 116, 1537, 1922, 3184, 3333
Demadrille, Renaud 1516
Demirkan, Korhan 820
Deng, Changhong 1158
Deng, Weiwei 2220
Denk, Patrick 1360
Descoeudres, A. 50
Descoeudres, Antoine 3254
Despeisse, M. 50, 2073
Despeisse, Matthieu...............3254, 3256, 3435
Desrues, Thibaut 2492, 2562
Deutsch, Todd G.....................................47
Devos, Arnaud 464
Dewitt, Daniel 1835
Dey, Anamika 1034
Dhakal, Tara P. 989
Dhere, N. 389, 1701
Di Leo, Paolo 3096
Di Mare, Simone 2372
Di Napoli, Simone 2205
Diaz, Liliana Ruiz 1147
Diercks, David R. 46
Dimitrievska, Mirjana...........................3285
Dimopoulos, Theodoros 178
Dimroth, Frank 2511
Ding, Jie...................................1937, 2823
Dinger, Justin......................................2692, 2765
Diniz, Antonia Sônia A. C. 1917, 2307
Dirriwachter, Antonius B. 3448
Dise, John 132, 1104
Dise, Skip 1427
Dobrich, Anja 2538
Dobroliubov, Aleksandr 2776
Dobson, Kevin 315
Dobson, Kevin D. 658
Dobson, Weston 2692, 2765
Dogan, Yusuf 229
Doi, T. 441
Dominguez, A......................................2342
Dong, Jianfei 2605

AUTHOR INDEX

Doolittle, William A. 305
Dooraghi, Michael.....................1169
Döscher, Henning.........................47
Doty, Matthew F. 1598, 3384
Dougher, Chris............................3245
Dougherty, Brian 1933
Drahi, Etienne 464
Drayton, Jennifer A. 164
Drees, M....................................3511
Dréon, Julie 55
D'Rozario, Julia 18
Drummy, Lawrence F......................966
Du, Chen-Hsun.............................911
Du, Xingzhi................................2558
Du, Zhongming...............198, 767, 1707
Duan, Baosong...................392, 2823
Duan, Wenqi................................346
Dubey, R. 1995
Dubey, Rajiv 1704, 2849, 3478
Dubois, Anne Migan 626
Duenow, J.N. 1312
Duenow, Joel 1196, 3147
Duerinckx, Filip 2227, 3435
Dugan, Roger C............................3055
Dugdill, Brian 2014
Dumbrell, Robert...................420, 3315
Dunham, Scott T. 3119
Dupré, Cécilia.....................2492, 2562
Durand, Olivier 2192
Durose, Ken 742, 1445
Durstock, Michael F......................966
Durygin, Andriy 3473
Dusane, Rajiv O 2330
Dussarrat, Christian 326
Dutt, A.....................................370
Dutt, Ateet 2342
Dutta, P....................................869
Dutta, Pavel 866, 2368
Duttagupta, S.P...........................1898
Eafanti, Joshua...........................3190
Ebe, Falko.................................2996
Ebert, Matthieu............................1531
Ebong, Abasifreke.........................888
Ediger, E...................................2364
Edinger, Stefan............................178
Edoff, Marika..............................796
Edwards, Daniel J.........................3514
Eeles, Alexander.................146, 186
Efthymiou, Venizelos.....................3107
Egbe, Daniel Ayuk Mbi....................1360
Eggink, Wouter............................2109
Ekins-Daukes, Ned........................1339
El Assimi, Taha............................3402
Elangovan, Hemaprabha....................2811
Elanzeery, Hossam...............151, 2054

Eldho, T.I. 1898
El-Henawey, Mohamed.....................2247
Elkhatib, Mohamed...............2141, 2969
Elleuch, Omar.............................359
Ellibee, Donald...........................1543
Ellingson, Randall 2926
Ellingson, Randall J. 1030
Ellingson, Randy J..........738, 748, 781, 815
Ellis, Chase T. 873
Elnosh, Ammar.............................1946
Elsehrawy, Farid 1189
Emery, K.A. 490
Engerer, Nicholas A. 1405
Eriksen, Ryan 2870
Eriksen, Ryan S. 2891
Ermer, J...................................2094
Ermer, James 2099
Ermer, Jim H. 37
Escarra, Matthew D. 37, 3245
Escobar, D. Martínez 1959, 1990
Esfandiari, Parichehr 178
Espinct-Gonzalez, Pilar 558
Espíndola-Rodríguez, Moisés......155, 512, 572, 3265
Espinet-González, Pilar..........521, 1248
Essa, Gharibah 2011
Essig, Stephanie 55, 3254, 3371
Etcheberry, Arnaud 2711
Etgar, Lioz 2170
Eugen, Rene 2864
Evani, Vamsi............802, 1638, 2449
Evans, Garrett Z...........................921
Evstigneev, M............663, 1025, 1811
Evstigneev, Mykhaylo A. 690
Eylers, Katharina 3396
Fada, Justin S. 2697, 3456, 3488
Faes, A....................................50
Fairbrother, Andrew............1933, 2000, 3204
Faleev, Nikolai 1215, 2573
Fan, S......................................3376
Fan, Shanhui 2185, 2732
Fang, Liang 226
Fang, Y....................................1603
Fang, Yi...................................305
Fano, V....................................2740
Fano, Vanesa 2677
Faraj, Abdul 2014
Farnung, Boris............................2267
Farré, Laia Arqués 3285
Farshchi, Rouin 1459, 1686
Faur, Maria................................896
Faur, Orry.................................896
Favre, W...................................1747
Fedina, Maria 2070
Fejfar, A..................................2073
Felder, T..................................2312

AUTHOR INDEX

Feldmann, F. 2064
Feng, Sheng-Kai 343, 367, 893, 2664
Feng, Shien-Ping 1012
Feng, Zhiqiang 1922, 2220
Fenning, David P. 1494, 2245
Ferekides, Chris 802, 1638, 2449, 2467, 3413
Ferekides, Chris S. 1511
Ferekides, Christos 175
Ferguson, Andrew J. 1354
Ferguson, L. 1863
Fernández, Eduardo F. 2858
Fernandez, R. Mis 1691, 2457
Fetzer, C. 2094
Fiducia, Thomas 424
Fields, Brian J. 2618
Filipic, Miha 1233
Filonovich, Sergej 464, 1237
Firth, Peter 1317
Fischer, A. 1603
Fischer, Alec 823
Fischer, Alec M. 305
Fisher, Brent 210, 272, 1469
Fisher, Dallas 989
Fitzgerald, Eugene A. 213
Fleming, Robert A. 1869
Flicker, J. D. 3224
Flicker, Jack 1280
Florides, Michalis 1941
Foldyna, Martin 2528, 2593
Forberich, Karen 1346
Forbes, David V. 3468
Forchhammer, Soren 2682
Forsh, P.A. 1811
Foster, Robert 2014
Fouchier, Marin 3402
Fournel, Frank 2492, 2562
Fraas, L. M. 1863
Fraas, Lewis 2042
France, Ryan M. 47, 232
Fraser, Ray 337
Frederiksen, Kenn H. B. 2682
Freeman, Janine M. 3494
Freiburger, Brennen M. 1869
French, Roger 1933
French, Roger H. 1927, 2000, 2697, 3456, 3488
Freundlich, Alexandre 236, 673, 1452
Fridman, Lucas 2000
Friedman, Daniel 549, 2543
Friedman, Daniel J. 42, 268, 1201
Friend, Mari Paz 429
Fritzsche, M. 1752
Frontini, Francesco 2118
Fthenakis, V. 2019
Fthenakis, V. M. 3230
Fthenakis, Vasilis 3077

Fu, Ran 1259, 1463
Fuhrich, Alexander 3396
Fuhrmann, Bianca 83
Fujiwara, Koji 1973
Fukuda, Tetsuya 931
Funabiki, Shigeyuki 2906
Fung, Tsun H. 2576
Fuyuki, Takashi 2593
Gabetta, Giuseppe 76, 545
Gabor, Andrew M. 74, 2839, 2897
Gaddy, Edward 585
Gahr, Stefan 178
Gai, Boju 549, 2291
Gaiaschi, Sofia 2711
Gallon, Joshua B. 3448
Galtieri, Jason 2975, 3214
Gambogi, W. 2312
Gao, Hui 226
Gao, Peng 1648
Gao, Wei 226
Gao, Y. 869
Gao, Yijun 2392
Gao, Ying 2368
Gao, Yuan 2048, 2605
Gao, Yujie 2870, 2891
García, I. 1210
Garcia, Iván 1204
Garcia, Juan Lopez 402
Garcia-Linares, Pablo 2562
Garg, Vivek 2345
Garner, Sean 2870
Garner, Sean M. 2891
Garnett, Erik C. 2245
Garreau-Iles, L. 2312
Garrillo, Pablo A. Fernández 1516
Garuz, Richard 2255
Gaury, Benoit 1303, 2438
Gdoutos, Eleftherios E. 558
Geelan-Small, Peter 3304
Gehre, Simon 3500
Geissbiihler, J. 50
Geissbuehler, Jonas 3256
Geisz, John 549, 3371
Geisz, John F. 232, 268, 1737, 2195, 3254
Georghiou, George E. 276, 1163, 1941, 1954, 3107
Geraghty, Paul 1342
Gerardi, C. 1747
Gerber, Andreas 1400, 1651
Gerdimenes, Anne 619
Gervasi, Massimo 541
Ghaisas, S.V 389, 1701
Ghimire, Kiran 993
Ghosh, Kunal 716
Gibbs, Jacob M. 730
Gibelli, François 2192

AUTHOR INDEX

Giebink, Noel C. 1469
Giguère, Jean-Benoit 1360
Gilchrist, James B. 966
Gillispie, Kellen 2762
Giordano, Francesco..............................3096
Giraldo, Sergio 3265, 3285
Giussani, A. 845
Giussani, Alessandro..............206, 831, 2514
Givot, Bradley L. 2864
Gladden, Christopher 1476
Glasgow, Nate.....................................1427
Glatthaar, M. 884, 3135
Gloeckler, Markus 1193
Glunz, S. 3135
Glunz, S.W. 2064
Glunz, Stefan W. 1253, 2511
Gokkaya, Huseyin Cem 958
Goldschmidt, Jan Christoph 1253
Golembeski, Andrew A............................3143
Goma, Elias Garcia.................................3462
Gombia, Enos 541
Gona, Michael N. 2349
Gong, Chen...1585
Gong, Jue..2251
Gonzálcz-Díaz, Benjamín 3240
Gonzalez, Maria 259
Gonzalez, S. 3224
Gonzalez, Sigifredo 2147, 3002, 3020
Gonzalez-Díaz, Benjamín 429, 1116
Goodarzi, Mohsen 2707
Gooding, Renee 1280, 1543
Goodnick, S. M. 1603
Goodnick, Stephen 1790
Goodnick, Stephen M. 305, 582, 1797
Gordillo, G. 433, 503
Gordon, Ivan 1233
Gori, Gabriele..76
Gorman, Brian 62, 1371, 1381
Górnez-González, L. A. 2614
Gostein, Michael...........................2808, 2923
Goswamy, Naveen 1908
Gotoh, Kazuhiro...........................1765, 1794
Gottschalg, Ralph............1411, 2827, 3208
Govaerts, Jonathan 3343
Goverde, Hans 3343
Gowda, Ramesh Rame............................1912
Graf, Martin 2511
Graham, Kenneth 1044
Grandidier, Jonathan 2099
Grassman, Tyler J...............182, 215, 2079, 2554, 3139
Greco, Erminio............................76, 541
Grede, Alex J. 1469
Green, Martin............................2213, 2403
Green, Martin A......................................858
Green, Michael 2926

Greenhalgh, R.C. 3430
Gregory, Geoffrey 74
Grévin, Benjamin 1516
Grice, Corey R. 771, 1643, 2473, 3426
Grieco, William J....................................2618
Griffin, Alecia 2870
Griffin, Alecia C. 2891
Grijalva, Santiago..........................1555, 1567
Grini, S. 3269
Große, T. 1752
Großer, Stephan 2232
Grossklaus, Kevin...................................701
Großschädl, Bettina 1329
Grovenor, Chris.....................................424
Grover, Sachit 1193, 2473
Grübel, B. 884
Gu, Fei 1346
Gu, Tian 1473
Gu, Tingyi 1828
Gu, Xiaohong 1933, 2000, 2844, 3195, 3204
Guarracino, Ilaria 1339
Guay, Nathan 1543
Gudla, Sushanth 1389
Guerrero-Lemus, Ricardo 429
Guillemoles, Jean-François..............1289, 2192
Guillevin, N. 3150
Guillevin, Nicolas...................................917
Guina, M. 1189
Guina, Mircea 297, 2520
Guischard, Felix 2836
Gunawan, Oki. 1441, 3275
Gunnarsson, William B............................2443
Guo, D. 1603, 2816
Guo, Hong 1299
Guo, Q. 3
Guo, Qi 226
Guo, Shuwen....................1430, 1873, 2918
Guo, Yongjie 1719
Gupta, Amit Kumar.............2952, 2981, 2986, 3050
Gupta, Mool C. 937
Gupta, Neeti 696
Gupta, Ritesh Kant 1034
Gupta, Shivam 377
Gupta, V. 1733
Gustafsson, Mattias 2025
Guthrey, Harvey 1400, 2887
Gutiérrez, J. R. 2740
Gutiérrez, R. 643
Gutscher, S. 884
Guwaeder, Abdulmunim 1122
Gwak, Jihye 810
Ha, Dongheon 1585
Habermann, D. 1752
Hack, James 999
Hack, James H. 326

AUTHOR INDEX

Hacke, Peter 1371, 1381, 1421, 1922, 2819, 2854, 3305

Hackl, Wolfeanz 178

Haddad, M. 2094

Haddadian, Rojiar 1927

Hadi, Sabina Abdul....................................213, 1741

Hadjipanayi, Maria 276

Hadke, Shreyash 986

Hadley, Wendy..2014

Haegel, Nancy M. 62

Hagendorf, Christian....................................1376, 2232

Hägglund, Carl..796

Hahn, Carina E. 175

Hai, Hoang Tri 931

Haight, Richard 1441

Hajimiri, Ali 521, 558

Hajizadeh, Amin 3092

Halbwax, Mathieu..3402

Hall, Allen 1511

Hallam, Brett J. 2576

Halliday, Douglas P.......................................2423

Hamadani, Behrang H.............................263, 437, 508

Hameiri, Ziv 66, 420, 3290, 3315

Hamon, Gwénaëlle..2528

Hamui, L. 2614

Hamzaoui, Saad 900

Hamzavy, Babak T. 2618

Han, Sang M. 88

Han, Xinyue 1719

Han, Youngsik 2242

Hanada, Toru 940

Handwerker, Carol A. 1449

Haney, Paul 1303

Haney, Paul M...2438

Hanley, J. 2094

Hanna, Amir 1055

Hannappel, Thomas.........................2524, 2538, 2538

Hanriot, Sergio De Morais 2307

Hansen, Clifford............... 1127, 1537, 3184, 3333, 3348

Hansen, Clifford W...........................110, 1543, 1549

Hansen, Ole 2672

Hansen, Richard..2042

Hansen, Shirley 2042

Hao, Xia 160

Hao, Xiaojing...............................858, 2213, 2403

Haohui, L..1140

Haohui, Liu...2744

Haque, K A S M Ehteshamul 346

Haque, M. D. 552

Hara, Shigeomi.......................................1950, 3339

Hara, Tomoya..2548

Harari, Joseph...3402

Hardikar, Kedar 2688

Häring, Adrian 2263

Hariskos, Dimitrios..2058

Harmand, Jean-Christophe...................................1289

Harris, Christian 319

Harris, James.................................1835, 2732

Harris, Tom 2991

Harvey, Steven 2887

Harvey, Steven P. 1371, 1381, 2702, 3305, 3319

Haschke, Jan 3435

Haslinger, Michael 1804

Hassan, Ibrahim A. I. 2858

Hatch, S. 14

Hatton, Peter D. 2423

Hauch, Jens...3500

Haug, F.-J. 2073

Hausgen, Paul E. 102

Hausmann, J. 1752

Havu, Ville 2070

Haysom, Joan E. 1094

He, Junwen 1469, 1737

He, Qiuxiang 3304

He, Wenshuang..392

Hea, Wenshuang...2823

Heben, Michael...2926

Heben, Michael J. 170, 730, 748, 815, 1030, 2462

Hegedus, Steven................408, 1473, 1761, 1828, 2667

Hegedus, Steven S.. 658

Heidmann, Berit 3396

Heilbrunner, Herwig..1360

Heilscher, Gerd..2996

Heinz, F. D. 3135

Heinze, Matthias...2263

Heller, Dominic..1077

Henes, Dan 1094

Hentz, Sandrine 966

Hermle, M. 2064

Hermle, Martin.......................................1253, 2511

Hernandez, J. A. 2031

Hernández, Johan..1143

Hernandez, Joseph 3473

Hernandez-Alvidrez, Javier.................................2153

Hernández-Gutiérrez, C. A. 670

Hernández-Rodríguez, Cecilio 429

Herrera, Daniel J...219

Herrmann, W.. 107

Herz, Magnus...3360

Heta, Y..2312

Hetterich, Michael 1682, 2216

Hettick, Mark 59, 823, 2076

Heurlin, Magnus...1286

Hickey, Benjamin 1459

Hidaka, Kazuyuki 1973

Higa, M. 441

Hilfiker, M. 2364

Hill, Alex 1893

Himwas, Chalermchai...1289

Hindi, Basel...1058

AUTHOR INDEX

Hinken, David 2692
Hinojosa, M. 1210
Hinzer, Karin 1094
Hirai, Masakazu 1769
Hirata, Yoichi 613
Hirose, Kotaro 3323
Hirstl, Louise C. 2091
Hishikawa, Y. 441, 1003
Hishikawa, Yoshihiro 480, 2781
Ho, Jian Wei 496
Ho, Wen-Jeng 343, 367, 893, 2664
Hoang, Bao 96
Ho-Baillie, Anita.......................858, 1845, 2569
Hobbs, William B. 2618
Hoerteis, Matthias 914
Hoex, Bram 517
Hoff, Thomas 132, 1104
Hofmann, Johannes 2407
Hoheisel, Raymond 247, 3514
Höhn, Oliver 352
Holman, Z. C. 3376
Holman, Zachary 1790, 3366
Holman, Zachary C. 1220, 1228, 1317,
     1322, 1820, 3250
Holmgren, William F. 110, 1127
Holzmann, Daniel..914
Hong, Chung-Yu 294
Hong, Keunkee 399
Honsberg, Christiana.................................827, 3088
Honsberg, Christiana B. 240, 305, 582,
     681, 1215, 1841, 2573
Hopf, Markus 1965
Horenstein, Mark..2870
Horenstein, N Mark 2891
Horner, Greg S. 3448
Horowitz, Kelsey A.W. 1259, 1463
Horzel, J...50, 2073
Hoshii, Takuya 2334
Hosokawa, Kazuya 613
Hossain, Istiaque 2247
Hossain, Mohammad A. 3456
Hossain, Mohammad I..963
Howard, John M. 2443
Hsi, Edward...1275
Hsu, Chia-Jhe 1623
Hsu, Chih An 1638, 2449, 3413
Hsu, Lung-Hsing 1610
Hsu, Shun-Chieh 1606, 1623
Hsu, Shu-Tsung 445, 448, 476
Hsu, Wei-Lun 1048
Hsu, Yu -Chen...888
Hu, Chehao 229
Hu, Cheng-Shun 329
Hu, Hailin 1858
Hu, Juejun 1473

Hu, Lilei...3129
Hu, Long 2392
Hu, Yang 1927
Hu, Yicong 2392
Huang, Jialiang..2213
Huang, Jingsheng 1937, 2823
Huang, Jing-Shun....................512, 521, 558, 572, 1248
Huang, Shujuan 2392
Huang, Vi-Wen 1631
Huang, Weijing...................................1873, 2918
Huang, Wei-Ming 1627, 1631
Huang, Wen-Hsi 385
Huang, Ying-Yuan 1807
Huang, Yi-Wen 1627
Huang, Yu-Ming 1606
Huang, Yu-Ting 1012
Huang, Z. 3260
Huayamave, Victor...2839
Hubbard, S. M. 552, 845, 2755
Hubbard, Seth 677
Hubbard, Seth M............................18, 202, 206, 222,
     831, 1184, 2084, 2298, 2514
Huber, Christian 2216
Hudson, A.I. 2755
Huey, Bryan D. 1522
Huffaker, D.L. 2755
Huffaker, Diana 202
Huhn, Vito..1651
Huld, Thomas 2167
Hunault, Philippe 2711
Hung, Yung-Jr 1606
Huo, Yijie 1835
Husein, Sebastian 944
Huss, Alexandra M..164
Hussain, Babar 451, 2355
Hussain, Muhammad M. 1055
Hutchings, Douglas..1869
Hutchings, Douglas A..921
Hutter, Oliver S. 1445
Hwang, James...333
Hyvl, M. 2073
Iandolo, Beniamino 2672
Ianno, N.J. 2364
Ichikawa, Yukimi 1769
Idlbi, Basem 2996
Ikki, Osamu 2159
Ilic, Ognjen 1737
Imai, Jun 2906
Imaizumi, Mitsuru 567, 3506
Imtiaz, Syed N...1067
Ingenhoven, Philip...3482
Ingenito, A. 2073
Inns, Daniel 3113
Isabella, Olindo 2605
Isbilir, Kenan 2827

AUTHOR INDEX

Isherwood, Patrick J. M. 2349
Ishii, Tomoaki 455
Ishino, Yuya..757
Ishizuka, Shogo 33
Islam, Kazi..37
Islam, Muhammad Monirul..33, 900
Islam, Raisul...1835
Isoaho, Riku...2520
Iwasaki, Kazuya..2338
Iwata, Naotaka..2642
Iwuoha, Emmanuel..1360
Iyer, Abhishek..326, 999, 2035
Iyer, Parameswar K...1034
Izquierdo-Roca, Victor 3265, 3285
Jackson, Christine...215
Jackson, Philip 2205, 2453
Jacob, David...1549
Jacobson, Arne...1271
Jadkar, S.R...1701
Jaeckel, Bengt...1411
Jae-Yun, Fa-Jun Ma,..1845
Jagdish, A K..2811
Jäger-Waldau, Arnulf...2167
Jahn, Ulrike..3360
Jain, Aditi..333
Jain, Nikhil 42, 46, 232, 578, 2195, 3371
Janoch, Rob...74, 2839, 2897
Jansen, Mark J..1081
Jany, Christophe..2492, 2562
Janz, Stefan 83, 2407
Jaramillo, Adolfo 1143
Jared, Bradley...1473
Jarmar, T..30
Jasti, Naga Prathibha 986
Jaswal, Rohit..3172
Javed, Mehwish Azher..1317
Javey, Ali 59, 823, 2076
Jeangros, Q..2073
Jenkins, P. P. 845
Jenkins, Phillip P.................. 247, 373, 1838, 2091, 3514
Jensen, Brian..2014
Jensen, M. A...3295
Jensen, Mallory A. 1491, 3290, 3300
Jensen, Soren..1196, 3147
Jeong, Woo-Lim..777, 1665
Jhaveri, Janam..1773
Ji, Liang 1933, 2000
Ji, Yaping..37
Ji, Yaping Vera..3245
Jia, Jieyang 1835, 2732
Jian, Ding-Rung..1627, 1631
Jiang, C. S. 1312, 2789
Jiang, C.-S..2280, 2785
Jiang, Chun-Sheng...62, 1371
Jiang, Lian L. 589

Jiang, Lian Lian 120
Jiang, Xuefang 1937
Jiang, Yu 3220
Jimeno, J. C. 643, 2740
Jimeno, Juan Carlos.. 2677
Jin, C. 1781
Jin, Yu 3119
John, Jim J 1946
John, Joachim 1804, 2227
John, Suru Vivian 1360
Johnson, A. D. 1210
Johnson, E.V. 1781
Johnson, Erik V. 2593
Johnson, J. L. 1656, 1661
Johnson, Jay 2135, 2141, 2153, 2969, 3002, 3008
Johnston, S. 2785
Johnston, Steve......................62, 202, 459, 1371,
    1381, 1400, 2213, 2819, 2887, 3305, 3452
Jones, C. Birk...........................2618, 3008, 3155, 3488
Jones, David 1342
Joonwichien, Supawan 904
Joshi, Madhuwanti S.................2952, 2981, 2986, 3050
Joshi, Pranav 2247
Jošt, Marko 1346
Jovanovic, Raka.. 963
Juang, B.C. 2755
Juang, Bor-Chau 202
Juárez, A. Sánchez 1990
Juárez, Aarón Sánchez... 1959
Juhl, Mattias K. 420, 3315
Juhl, Mattias Klaus 66
Julien, Scott... 1933, 2000
Junci, Wang 496
Junda, Maxwell 771
Junda, Maxwell M...................2462, 2582, 3426, 3468
Jung, Jae Hak.. 487
Jung, Jiirgen 2864
Jung, Sang Hoon... 2723
Jung, Sang Hyun 244
Juso, Hiroyuki 3506
Kabalan, Amal.. 2358
Kaczynski, Ryan 1455
Kaizu, Toshiyuki 23
Kaizuka, Izumi 2159
Kakosimos, Konstantinos E...................................... 1888
Kalainatharr, Sivaperuman....................................... 2334
Kalb, J. 1733
Kale, Abhijit... 1801
Kale, Abhijit S. 1777
Kallickal, Johnson 1543, 3348
Kalt, Heinz 1682, 2216
Kamata, N. 552
Kamevama, Satoshi 2642
Kamino, Brett 3256
Kamins, Ted... 1835

AUTHOR INDEX

Kaminski, P.M. 3430
Kaminski, Piotr 1674
Kaminski-Cachopo, Anne 2562
Kamioka, Takefumi 2498, 2548, 2566, 2642
Kanemitsu, Yoshihiko...........................721, 2781
Kanevce, A. 1312
Kanevce, Ana 3147
Kang, Ho Kwan...244
Kang, Min Gu...........................356, 1758, 2723
Kang, Yoonmook 935
Kankiewicz, Adam 132, 1104, 1132, 1427
Kannan, C. V...2716
Kao, Ming-Hsuan...1627
Kaplan, Stephen...600, 1071
Kaplar, R. 3224
Karas, Joseph 925
Karki, Shankar... 182, 735, 807, 2298, 2446, 2646, 3139
Karmarkar, M...1661
Karpowich, Lindsey...914
Karthik, Shravan 3172
Kashkoush, Ismail...322
Kaslin, Remo 1077
Kasry, Amal...2858
Kasu, Makoto 1950, 3339
Kato, Takekazu 1175
Kato, Takuya...160
Katsube, Ryoji..2361
Kaule, Felix..2622
Kausika, Bala Bhavya...........................3014, 3167
Kavaipatti, Balasubramaniam 2799
Kawatsu, Tomoyuki.........................381, 2588
Kazmcrski, Lawrence L. 2799
Kazmerski, L.L. 1245
Kazmerski, Lawrence L. 1917, 2307
Kazumi, Kenji 2361
Keeler, Gordon..1473
Keller, Nico...1077
Kelly, George..1275, 2263
Kelly, Matthew..3514
Kelzenberg, Michael D. 512, 521, 558, 572, 1248
Kempe, M.D. 138
Kempe, Michael 1933, 2000
Kempe, Michael D..3208
Kephart, Jason 785
Kephart, Jason M...3417
Kern, Gregory 2147
Kern, Gregory A. 3020
Kesavan, Arul Varman..1614
Kessels, Wilhelmus M.M. 1817
Kessler, Emily..206
Khadimallah, A...869
Khalili, A...1189
Khan, Imran 802, 1638, 3413
Khan, Imran S...2449
Khan, Mohammad R. 1055

Khan, Taj M. 451
Khanna, Raghav.. 2926
Kharait, Rounak A... 2833
Kharel, Khim 236, 673, 1452
Khatavkar, Sanchit 2716
Khatiwada, D. 869
Khatiwada, Devendra 866
Khatri, Ishwor 192
Khatri, Trijul 377
Khomcnko, Denis V.. 690
Khoo, Yong Sheng... 1922
Khor, Alan 3172
Khoram, Parisa 2245
Khorenko, Victor............................... 83, 2087
Khoury, R. 1781
Kiefer, Fabian 1366
Killam, Alex 2719
Killinger, Sven 126, 1405
Kim, Boram 2201, 2524, 2538
Kim, Chang Zoo 244
Kim, D. 1189
Kim, Dae Young 626
Kim, Dong Seop 399
Kim, Dong-Ho................................... 2631, 2634
Kim, Donghwan 935, 1758
Kim, Hae-Sun 777, 1665
Kim, Hyo Jin 849
Kim, In-Young 777
Kim, Jae Hyun...........................363, 2844, 3195, 3204
Kim, Jin-Hyeok 777
Kim, Jisun 399
Kim, Ka-Hyun 1758
Kim, Kangho... 244
Kim, Kihwan... 810
Kim, Kyoung- Tae 1037
Kim, Min-Soo... 487
Kim, Moon 2759
Kim, Sangpyeong... 240
Kim, Soo Min 2723
Kim, Woo Kyoung 487
Kim, Yeongho 827
Kim, Yong Bae 2723
Kim, Yong Whan.. 849
Kim, Youngjo... 244
Kimbal, Gregory M... 110
Kimura, Daiki 854
Kindole, Dickson 970
Kindvall, Anna.. 785
King, Bruce H. 3155, 3488
King, Richard 827
King, Richard R. 301, 823, 1215, 1841
Kingma, Aldo 541
Kini, Roshan 2926
Kinoshita, Kosuke 1504, 2588
Kirk, A. 3511

AUTHOR INDEX

Kita, Takashi 23
Kleider, Jean-Paul 2528
Klein, Talysa R. 2482, 3371, 3439
Kleinschmidt, Peter..................................2538
Klemm, Hagen W. 3396
Klenk, Markus 1077
Klie, Robert F. 2759
Klimm, Elisabeth 2836
Klise, Katherine A. 3161
Klisel, Geoffrey T. 3494
Kluska, S. 884, 3135
Knight, Bruce 2014
Knopf, Hannes..1965
Ko, Changhee 326
Kobayashi, Jonathan 1061
Koehl, Michael 3488
Koepgel, Ringo 2622
Kogler, Willi 791
Kohlstädt, Markus 1253
Koike, Junichi 931
Koirala, Prakash 2462, 3426
Kojima, Nobuaki.................359, 2498, 2566
Kojima, Takuto 1504, 2588
Komsa, Hannu-Pekka 2070
Konagai, Makoto.................1769, 2627
König, M. 1752
Konstantinou, Georgios............................1941
Kontges, Marc 1366
Kopecek, Radovan 1222
Koschny, T. 1350
Kostylyov, V.P. 1025, 1811
Kostylyov, Vitaliy P. 690
Kotipalli, Ratan 2209
Kottantharayil, A. 1995
Kottantharayil, Anil...........396, 716, 1850, 2799, 3478
Kottokkaran, Ranjith 2247
Kotulak, Nicole.............................999, 1838
Kotulak, Nicole A. 247
Koyama, Koichi.............................1765, 1787
Kozodoy, Peter 1476
Krabb, Peter 178
Krantz, Patrick W. 730
Krasikov, D...2816
Krc, Janez...1346
Krein, Philip T. 3214
Krich, Jacob J. 1294
Krishnan, Mani R. 1912
Krishnan, Sheeja 76
Krishnaswami, Hariharan...............2931, 2936
Krogen, J. 2094
Krügener, Jan...1494
Krut, Dimitri D. 37
Ku, Chen-Hao...329
Kubiniec, Alex 132, 1132, 1427
Kuciauskas, Darius 1679

Kudriavtsev, Yu..670
Kuitche, Joseph 1877, 1883
Kulish, Mykola R......................................690
Kum, Hyun 18, 222, 677, 2084
Kumar, Rajesh..3478
Kumar, Shailendra 2345
Kumar, Sukanya Santhosh........................980
Kumar, Vijay..2716
Kumari, Khushboo 251
Kuo, Hao-Chung 1610, 1627
Kuo, Po-Tsun 1006
Kuo, Ting-Wei 329
Kurdgelashvili, Lado................................2035
Kurihara, Risa 2159
Kurimoto, Yuji 931
Kurokawa, Yasuyoshi 1765
Kurstjens, Rufi 83
Kurtz, Sarah 1275, 1922, 2263, 3190
Kusaki, Kazuki 23
Kuthanazhi, Vivek....................................3478
Kwon, Jung-Dae 2631, 2634
Kwon, Sang Jik 195
Kyureghian, H. 2364
La Centra, Ricci 2870, 2891
Lachaurne, Raphaël 2528, 3402
Lachowicz, A. 50
Lackner, David 2511
Lacroix, Jean-Sébastien 1579
Lafleur-Lambert, Antoine 1360
Lafont, Ombline 2453
Lagumavarapu, Ramesh B. 202
Lai, B. 3295
Lai, Barry..............1494, 2170, 2179, 2245, 3300, 3309
Lai, Yi 1009
Laine, Hannu S. 1491, 1494, 3236
Lakshmanan, Ramakrishnan 2870, 2891
Landgraf, D...1752
Lang, Mario 2216
Lapierre, Ray R. 1294
Larrey, Vincent 2492
Larsen, Ross E. 1354
Larson, Bryon W. 1354
Lasalvia, Vincenzo............881, 1491, 1801, 2242, 3439
Laschinski, Joachim 1965
Lassise, Maxwell.....................................3410
Latham, Joseph 1086
Lau, Derwin 3220
Lau, Kei May 578
Lave, Matthew 1435, 3008, 3025, 3031, 3184, 3348
Lavrova, Olga 1280, 2618, 3488, 3494
Law, D. 2094
Lazarou, Constantinos 276
Le Corre, Alain 2192
Le Donne, Alessia 1669
Le Gall, Sylvain 3402

AUTHOR INDEX

Le Guen, Vincent 70
Le Rouzo, Judikaël 2255
Lebreton, Fabien 464, 1237
Leclerc, Christophe 558, 572
Lecouvey, Christophe 2492, 2562
Ledinek, Dorothea 796
Ledinsky, M. 2073
Lee, Angela 417
Lee, Benjamin G. 881, 1737, 1801, 1832, 2482, 3439
Lee, Calvin 1342
Lee, Dong-Seon 777, 1665
Lee, Eunjoo 399
Lee, Eunsang 2124
Lee, Hae-Seok 935
Lee, Hyeonseok 1012
Lee, Jaejin 244
Lee, Jeong In 356, 1758
Lee, Ji-Hoon 2631, 2634
Lee, Jihwan 1181
Lee, Jinwoo 1455
Lee, Jongwon 681, 1215
Lee, Kan-Hua 359, 412, 1479, 1711,
        1714, 1743, 2498, 2566
Lee, Kyumin 1526
Lee, Kyu-Tae 1469
Lee, M. L. 3376
Lee, Minjoo 2291
Lee, Minjoo L. 42
Lee, Mitch 600
Lee, Mitchell 595
Lee, Seunghun 1253
Lee, Soonil 363
Lee, Yeonbae 1204
Lee, Yun Seog 1441
Lefebvre, Amy 1933
Lefebvre, Amy L. 2000
Lehman, Peter 1271
Lehr, J. 3224
Leilaeioun, M. 3376
Leilaeioun, Mehdi 1322, 1790
Leite, Marina S. 1508, 1585, 2443
Lekx, David 1094
Lemaître, Aristide 1289
Lemus, Ricardo Guerrero 1116, 3240
Lennon, Alison 3220
Lennon, Kyle 3384
Lennon, Kyle R. 1598
Leone, Stephen R. 417
Leonhardt, M. 1752
Leow, Shin Woei 3275
Lepkowski, Daniel 215
Lepkowski, Daniel L. 2079, 2554
Lester, Luke F. 219
Leto, Riccardo 1728
Leu, S. 1752

Levcenco, Sergiu 3396
Levi, D.H. 467, 490
Levi, Dean 483
Levrat, J. 50
Levrat, Jacques 3435
Levy, David H. 3442
Li, Chu Tu 1094
Li, Duanhui 1473
Li, Guan-Yi 343, 367, 893, 2664
Li, Jian 771, 1643
Li, Jian V. 2473, 2728, 2749
Li, Joel B. 3300
Li, Kexue 424
Li, L. 3295
Li, Lan 1473
Li, Li 1175
Li, Lu 1619
Li, Mengjie 3315
Li, Qiang 578
Li, Rui 1094
Li, Siming 3143
Li, Wenjie 3275
Li, Xiaoping 1193
Li, Xinyi 255
Li, Xueying 2170
Li, Y. 869
Li, Yongkuan 2368
Li, Yunjun 907
Li, Yunpeng 2220
Li, Zhanhang 226
Li, Zhuohui 1598, 3384
Liang, B.L. 2755
Liang, Jianbo 2548, 2551
Liao, Anqi 948
Liao, Yuanxun 696
Liao, Yuaxun 2186
Libby, Cara S. 2618
Licht, Abigail 701
Lichty, Marlene L. 2298
Lie, Stener 3275
Lim, Bianca 2318
Lin, Albert 294, 1631
Lin, Albert S. 1627
Lin, Cheng-Shian 1006
Lin, Chien-Chung 1606, 1610, 1623, 1627
Lin, Ching-Fuh 1006, 1009
Lin, Fen 284
Lin, Ming-Yi 1051
Lin, Shang-Pang 1006
Lin, Yan 1100
Lin, Yandan 2048
Lin, Yan-Zhang 1623
Lin, Yida 667
Lin, Yu-Hsuan 911
Lin, Yung-Sheng 329

AUTHOR INDEX

Lin, Zong-Xian 367
Lincoln, Jason 2897
Lincoln, Jason L. 2839
Lincot, Daniel 2453
Linton, John 337
Lipovšek, Benjamin 1346
Lipski, Michael V. 1469
Lisbona, Emilio Fernandez 545
Lisco, F. 1691, 2457
Litjens, Geert 3014
Liu, A. Y. 1485
Liu, Chenxi 3172
Liu, Fang Fang 1648
Liu, Fangyang 2213
Liu, H. 1189
Liu, H.Y. 14
Liu, Haitao 472
Liu, Han-Wen 2660
Liu, Haohui 284, 2532
Liu, Hsiang-Yu 2637
Liu, Huiyun 3370
Liu, Jheng-Jie 343, 893, 2664
Liu, Kanglin 1100
Liu, Mengxia 3129
Liu, Qihang 1245
Liu, Ruimin 2220
Liu, Simon H. 93
Liu, X.Q. 2094
Liu, Xiangxin 198, 767, 1707
Liu, Xinbing 1728
Liu, Xing-Quan 2099
Liu, Zhe. 284
Liu, Zhen 2532
Liu, Zhengjun 1494
Liu, Zhengxin 1241
Livera, Andreas 276, 1954
Liyanage, Geethika K. 170, 730, 815, 2462
Llin, Lourdes Ferre 1339
Lloyd, Alexis 2870
Lloyd, Michael A. 726, 3143
Lnr, Yiming 2558
Loach, Andrew J. 2697
Lodha, Saurabh 716
Lokanath, Sumanth 1275
Loke, Samuel P. 512, 521, 558
Loke, W.K. 1210
Lombardero, I. 1210
Lombez, Laurent 70, 1289, 2192
Lonergan, Mark 802
Long, Yean-San 448, 476
Looney, Erin E. 1491, 3236, 3290, 3300
Löper, P. 2073
Löper, Philipp 55
Lopez, Cristina S. Polo 2118
López, G. 1781

Lopez, Roberto 2728, 2749
López-González, J.M. 1781
López-López, M. 670
Lopez-Marino, Simón 155
Lorentzen, Justin 3514
Lorenzo, Antonio T. 1127
Loser, Ulrich 2272
Lossen, Jan 1222
Lotshaw, W.T. 2755
Lou, Chaogang 1619
Loubar, Anais 2711
Loyer, Camille 2000
Lu, Ching-Ying 1835
Lu, Hongbo 255
Lu, J.P. 1733
Lu, Jiawen 2656
Lu, Kyle B. 3448
Lu, Zhou 1728
Lubenow, Tomas 3333
Lujan, R. 1733
Luka, Tabea 2232
Lumb, Matthew P 210, 247, 259, 272, 873, 2506
Luna, Miguel A. 632
Lunacek, Monte 3008
Lunt, Richard R. 2124
Luo, Shiqiang 976
Luo, Wei 1922
Luo, Yanqi 2170, 2245
Luria, Justin L. 1522
Luther, Joseph M. 2176
Lynn, Kelvin G. 3422
Lyons, Alan 2285
Lyu, Yadong 1933, 2844, 3195, 3204
Lyu, Zheng 1835, 2732
Ma, D. 229
Ma, Fa-Jun 2569
Ma, Xiaokun 1469
Macalpine, Sara 1537
Macco, Bart 1817
Macdonald, D. 1485, 3295
Macdonald, Daniel 2707, 3300
Mack, C. 490
Mack, I. 2073
Mack, Shawn 259, 873
Mackie, Neil 820
Maclaren, Scott 1511
Macmaster, Steven W. 2864
Madani, Keeya 940, 1824
Madsen, C. K. 229
Maeda, P.Y. 1733
Magaña, Ernesto 1494
Magdaleno, R. Santos 1990
Magdaleno, Rocío De La Luz Santos 1959
Magnin, Vincent 3402
Magnone, Lydie 1415

AUTHOR INDEX

Mahadik, N. A. 845
Mahapatra, Chiranjibi 2849
Maia, Cristiana Brasil 2307
Maidaniuk, Yurii 3370
Mailoal, Jonathan 1264
Major, Jonathan D. 742, 1445
Makita, Kikuo 861, 1724
Makoutz, Emily A. 3381
Makrides, George 1163, 1941, 1954, 3107
Malhotra, Raghav 3172
Malik, Roger 1193
Maliya, Heini 226
Malkov, Andrei V. 146, 186
Mallick, Tapas K. 2858
Manda, Surya 761
Mandelis, Andreas 3129
Manganiello, Patrizio 3343
Mangelinck-Noël, Nathalie 1498
Mani, Monto 604
Maniscalco, B. 2457
Maniscalco, Biancamaria 2827
Mann, Colin 1248
Mann, Colin J. 93, 512
Mansfield, Lorelle 1400, 2473
Mansoori, A. 281
Mantel, Claire 2682
Manzoor, Salman 1228, 1322
Marie, Benoit 1498
Marion, Bill 1134, 1537, 1543, 3333, 3348
Markevich, V. P. 1485
Markides, Christos N. 1339
Maros, Aymeric 1215
Marsh, Brett M. 417
Marsillac, Sylvain.................. 182, 735, 807, 2298, 2446, 2646, 3139
Marsillac, Sylvain X. 2582
Marti, Shilpa 2936
Martín, I. 1781
Martin, Mickaël 2492
Martinez, Aaron D. 2536, 3406
Martinez-Morales, Alfredo A. 2881
Martínez-Pérez, Alejandro................3285
Martín-Martín, D. 3376
Martins, Ana C. 2104
Martinson, Alex B.F. 6, 1256
Masada, Isao 1504
Mascarenhas, Angelo 2506
Maser, Jörg 3309
Maser, Jörg M. 2179
Maskell, Douglas L. 120, 589
Mastroianni, Simone 1253
Masuda, Atsushi 1268
Masuda, Shota 1794
Masutomi, Yasuki 3339
Matei, I. 1733

Mather, Barry 1561
Mathew, Leo 363
Mathew, X. 142
Mathews, N. R. 142
Matsubara, Koji 381, 1333
Matsui, Takuya 381, 1333
Matsumoto, Yasuhiro....................... 632
Matsumura, Hideki 1765, 1787
Matsuo, K. 3
Matthew, Leo 2506
Maximenko, S. I. 845
Maximenko, Sergey 2091
Maximenko, Sergey I. 873
May, Matthias M. 2538
Mayberry, Ryan 914
Mazumder, Malay 2870
Mazumder, Malay K. 2891
Mazur, Yuriy I. 3370
Mccandless, Brian 1196, 3319
Mccandless, Brian E....................... 726
Mcclary, Scott A. 1449
Mcclung, Larry 2833
Mcclure, E. L. 845
Mcclure, Elisabeth L. 2298
Mcclure, Harumi 2947
Mccndless, Brian E. 3143
Mccomb, David W. 966, 3139
Mcdanal, A.J. 525
Mcdanold, Byron K. 2864
Mcfavilen, Heather 305, 582
Mcintosh, Keith R. 1322
Mcintyre, Maxwell 1040
Mcintyre, Michael 1086
Mckenna, Russell.......................... 126
Mcmahon, William E. 268, 3381, 3406
Mcmeans, Philip A. 921
Mcpheeters, Claiborne............... 42, 525, 2099
Meakin, David 1927
Medic, V. 2364
Medici, Vasco 2118
Meeker, Michael A. 873
Mehlich, H................................. 1752
Mehta, Hitesh K. 3038
Meier, Florian 2996
Meissner, Dieter 178
Meitl, Matt 272
Meitl, Matthew 873
Melamed, Celeste L. 3406
Melchiorre, Michele 151, 2054
Meleco, A. J. 14
Mellor, Alexander......................... 1339
Melnikov, Alexander 3129
Melvin, Andrew 1
Méndez, Juan A. 3240
Meng, Fanying 1241

AUTHOR INDEX

Meng, Hsin-Fei 1635
Meng, Xiaodong 2854
Menossi, Daniele 752, 1669, 2372
Menozzi, Roberto 2205
Men-Pérez, E. 370
Meot, Jacques 2593
Merdzhanova, Tsvetelina 2114
Merghcim, Julia 3500
Merkle, Agnes 3371
Merz, Christopher 1965
Merzlic, Sebastien 1933, 2000
Messer, Alexander 512
Messer, Alexander J. 521, 558
Messerschmidt, Michael 2682
Messmer, C. 2064
Metzger, W.K. 1312
Metzger, Wyatt K. 1196, 3147, 3305, 3319
Meuris, M. 3260
Mewe, Agnes A. 917
Meyers, Bennet 3354
Mi, Z. 2388
Mi, Zetian 1299
Mia, Md Dalim 2749
Michaelson, Lynne 925
Micheli, Leonardo 2301, 2789, 2804, 2858, 2881
Michl, Bernhard 1329
Mihailetchi, Valentin D. 1222
Mihaylov, Blagovest 1411
Mikofski, Mark 3354
Mikofski, Mark M. 110
Milakovich, Timothy 213
Miller, Bill 1473
Miller, David C. 2789, 2864, 3195, 3208
Miller, Elisa M. 2536
Milleville, Christopher C. 1598, 3384
Mil'shtein, S. 2411
Min, Jung-Hong 777
Minemoto, Takashi 455, 757, 2385
Minkin, L. 1863
Mints, Paula 2039
Miryala, Tejaswini 2646
Mishima, T. D. 14
Mishra, Himani 2376
Misra, Sudhajit 175, 802, 2467
Mitchell, Bernhard 2707, 3304
Mittag, Max 1531
Miyajima, Sakutaro 480
Miyashita, Naoya 854, 2334
Mizuno, Hidenori 1724
Moffett, C.E. 2736
Mohammed, Khaja H. 921
Mohapatra, Soumya Ranjan 3050
Mohr, Christian 83
Monnard, Raphäel 3256
Montenegro, Davis 3055

Montes, Carlos 429
Montgomery, Kyle H. 531
Montiel-Chicharro, Daniel 3208
Moon, Soo-Jin 3256
Moore, A. 2816
Moore, Andrew 1522
Moore, James 1838, 2091
Moore, James E. 259, 272, 373, 2506
Moore, Jay 2947
Moosa, Hassa 2011
Moosa, Maitha 2011
Moradi, Hadis 638, 2941
Moraitis, Panagiotis 3167
Morales, Christophe 2492, 2562
Morales, Cristian 2870, 2891
Morales-Acevedo, A. 670
Morel, Don 802, 1638, 3413
Morgado-Dias, F. 3178
Moriarty, T. 490
Moriarty, Tom 483
Moriki, Akinori 2906
Morin, Jean-Francois 1360
Morishige, Ashley E. 1494, 3236, 3290, 3300
Morita, Hiroshi 1973
Morral, Anna Fontcuberta I. 944
Morris, Jeromie 2996
Morrison, Matthew 229
Mortazavi, Soheyl 2875
Moseley, J. 1312
Moseley, John 62, 1196, 1381, 2887, 3123, 3147
Moser, David 3360, 3482
Moustafa, A. 1747
Moutinho, H.R. 1312, 2280, 2785
Moutinho, Helio 62, 2789, 3305
M'sirdi, Nacer K. 1968
Muaddi, Saad 1110
Mueller, Thomas 496, 2318
Mukherjee, Shaibal 2345
Mulder, P. 1189
Muller, Bjorn 126, 2267
Muller, M. 2280
Muller, Matthew 2294, 2301, 2789, 2804, 2858, 2881
Müller, R. 2064
Munasinghe, Anjali 2124
Munday, Jeremy N. 1585
Mundt, Laura 1253
Mundus, Markus 1253
Munkhammar, Joakim 3067
Muñoz, D. 1747
Munoz, Krystal 925
Munshi, Amit 1674
Munshi, Amita 980
Mur, Pierre 2562
Muralidharan, Pradyumna 1790, 1797
Muramatsu, Kazuo 2642

AUTHOR INDEX

Murphy, J. D. 1485
Murugesan, Arumugam 2172
Muskovin, Eric 537
Mutitu, James.....................................315
Mwove, Johnson Kyalo......................2014
Myers, Matt.......................................525
Nærland, Tine Uberg........................2610
Nagaoka, Akira 1679
Nagarajan, Adarsh............................2991
Nage, M..3150
Nagel, H. 3135
Nägelein, Andreas 2538
Nair, P. R..2716
Nair, Pradeep R...................1015, 1022, 1755
Nakada, Tokio 192
Nakamur, Tetsuya 567
Nakamura, Kyotaro...............1504, 1794, 2498, 2566, 2588, 2642
Nakamura, Shigeyuki 2338
Nakamura, Tetsuya....................562, 3506
Nakano, Yoshiaki.............. 854, 2201, 2524, 2538, 3528
Nakata, Tatsuya 854
Nakatsuka, Shigeru 2385
Nam, Wooseok...................................2242
Nanda, A. 229
Nandal, Vikas 1015
Nanduri, Sai Naga Raghuram.................1018
Naqavi, Ali..............512, 521, 558, 572, 1248
Narasimhan, K.L...........396, 1850, 1995, 3478
Nardone, Marco...........................309, 3123
Naseem, Hameed A...............................921
Natsheh, Ammar 2011
Naumann, Volker..............................1376
Nawara, Witek 3462
Nawaz, Syed F. 1067
Nayfeh, Ammar....................213, 1741
Naylor, Mark....................................914
Nayshevsky, Illya 2285
Ndione, Paul 1253
Needell, David R...............................1737
Neely, J. 3224
Neely, Jason 2141
Neergat, Manoj.................................1704
Nehme, Bechara 1968
Nelson, George T.............202, 206, 222, 1184, 2084
Nemeth, William 1777, 1801, 1817, 1832, 2242, 2702, 3439
Nespoli, Lorenzo...............................2118
Nett, Zach 1737
Neuschitzer, Markus.............................155
Neuwirth, Markus...............................1682
Newlands, Allan 2042
Ng, Annie 958
Ngan, Lauren.....................................600
Nguven, Dac-Trung..............................2192

Nguyen, H. T. 3295
Nguyen, Tinh 3204
Nickel, Benedikt................................3388
Nicolay, S. 50
Nicolay, Sylvain 3256
Niemi, T. 1189
Niesen, Bjoern...................................3256
Nietzold, Tara....................944, 2179, 3309
Nii, Kohdai...................................... 85
Niki, Shigeru 33
Nilsson, Ulf H................................... 2864
Nishikawa, Naoyuki 1385
Nishio, M....................................... 3
Nishioka, Kensuke............................ 480, 1479
Noack, Max 2247
Nobre, André M. 3172
Nobuhara, Shohei................................1175
Nocerino, John.................................. 93
Noda, Naoto 326
Noda, Yoshimasa 970
Nofuentes, Gustavo............................. 2858
Nogay, G.. 2073
Noh, Shinyoung 858
Nonnenmacher, H. J.............................. 1752
Norman, Andrew................................ 1381, 2887
Norman, Andrew G.............................. 2536, 3406
Norwood, Robert A. 1147
Nose, Yoshitaro....................1679, 2361, 2385
Nowakowski, Marilyn L........................ 3524
Nsofor, Ugochukwu 1828
Nukala, Tejeswar................................ 3061
Nunomura, Shota 381
Nurdin, Muhammad 3102
Nussbaumer, Hartmut 1077
Nuzzo, Ralph G................................ 1469, 1737
Nyirjesy, Gabrielle 667
Oberbeck, Lars 3370
O'Brien, Greg 1933
O'Brien, Gregory................................ 2000
Ocaña, Luis 429
O'Carroll, Deirdre M. 3393
Ochoa, M.. 1210
Odden, Jan Ove 2651, 2776
Oehler, Fabrice 1289
Ogawa, Tomoki 2548
Ogura, Atsushi...................1504, 2588, 2642
Ogutman, Kortan 1804, 3448
Oh, Jaewon 1858, 1877, 1883, 2912
Oh, Seung Kyu 866
Oh, Soo-Young................................... 487
Ohdaira, Keisuke.............................. 1385, 1787
Ohigashi, Takashi............................... 2159
Ohshima, H. 441
Ohshima, Takeshi............................ 562, 567
Ohshita, Yoshio 1504, 1794, 2498, 2566, 2588, 2642

AUTHOR INDEX

Ohta, Taisuke 3417
Ok, Young-Woo 333, 1807, 1838
Oka, Naotaka...1973
Okada, Yoshitaka........................ 10, 85, 854, 2334
Okafor, Jonathan O. 219
Okano, Y. 3
Okel, Lars A.G. 1081
Oliva, Florian 3265, 3285
Olopade, Muteeu..2381
Olvera, María De La Luz 632
O'Neill, Mark 525
Oney, Michael F. T 2176
Onno, Arthur..3370
Onunkwo, Ifeoma..2135
Oo, W.M. Hlaing...1661
Opila, Robert..999, 2035
Opila, Robert L. 315, 326
Oreski, Gemot 178
Orlovskaya, Nina A..2324
Ortega, E. 643
Ortega, Pablo...944
Ortiz, Brenden R. 3406
Ortiz-Rivera, Eduardo I. 2957, 2963
Ory, Daniel 70
Oshima, Ryuji...861
Ososanya, Esther..2432
Osowski, M. 3511
Osterwald, C.R. 467, 490
Ota, Yasuyuki..1479
Otaegi, A. 2740
Otaegi, Aloña 2677
Otnes, Gaute 1286, 2502
Ottoson, L. 467, 490
Ouyang, Zi...2403, 3220
Oviedo, Felipe..2744
Ozanne, A. -S...1747
Paap, Scott 1473
Packard, Corinne E. 46
Page, Matthew.......................................1777, 2242
Paggi, Marco 402
Palekis, Vasilios.......... 175, 802, 1511, 1638, 2467, 3413
Palekis, Vasilis..2449
Palitzsch, Wolfram...2272
Palmer, Evan 496
Palmiotti, Elizabeth...1400
Palmquist, Nathan 667
Pan, Hui...1100
Pan, N. 3511
Pan, Zhen 226
Panchal, A. K...3061
Panchal, Ashish K. 3038
Pandey, Rahul 377
Paolone, Mario 1415
Parashar, Parag 1627, 1631
Paraskeva, Vasiliki..276

Parenti, Robert C. 3520
Parikh, Anuja V. 3123
Parikh, Harsh 2682
Park, Chinho..487
Park, Ji-Sang 6, 1256
Park, Joo Hyung 810
Park, Kyung Ho 244
Park, S. 2388
Park, Seonyong 2087
Park, Somin 1044
Park, Sungeun 935
Park, Won-Kyu 244
Partain, Larry...2042
Passow, Kendra 595, 600, 1071
Paszuk, Agnieszka 2524, 2538
Patra, Payal..761
Patterson, Robert J 2392
Paudel, Naba R. 2443
Paul, Douglas 1339
Paul, Nicolas..70
Paul, P. K. 30
Paul, Pran K. 2446, 3139
Paul, Sanjoy 2473, 2749
Paulauskas, Tadas 2759
Paull, P. K. 2414
Paull, Sanjoy 2728
Paulsen, Andrew..3514
Pavgi, Ashwini 1877, 1883
Paviet-Salomon, B. 50
Paviet-Salomon, Bertrand 3256
Pavilonis, Michael...1476
Pavlov, Marko 626
Pavlovsky, Igor 907
Pawar, Vaibhav 2986
Payne, David 315
Payne, David N.R. 2576
Peaker, A. R. 1485
Peale, Robert E. 2324
Peharz, Gerhard 178, 1329
Peibst, Robby 1366, 3371
Pellegrino, Sergio.....................512, 521, 558, 572
Peña, J.L. 1691, 2457
Pena, Juan Luis 1669, 2372
Peña, Ramón 632
Peng, Jun 2076
Peng, Shou 761, 2427
Penning, David P 2170
Peppanen, Jouni 3025
Pera, David 1979
Peraca, Nicolás Márquez 263
Perez, Richard 132, 1104
Pérez-Rodríguez, Alejandro 3265, 3285
Perez-Wurfl, Ivan..315
Perkins, C. 2280
Perkins, Craig... 2702, 2789

AUTHOR INDEX

Perkins, Craig L. 2294
Perl, E. 3376
Perl, Emmett E. 42, 1201
Perna, Allison 2467
Pesala, Bala 2858
Peschel, Gina 3396
Peshek, Timothy J. 1927, 2697, 3456
Peters, I. M. 648, 1140
Peters, Ian Marius 284, 1264, 2532, 2744
Peters, Marius 3236
Petersen, Michael 2682
Peterson, Chris 512
Peterson, Josh 1169
Petoukhoff, Christopher E. 3393
Pfiester, Nicole 701
Phillips, Adam 748
Phillips, Adam B. 170, 730, 815, 1030, 2462
Phillips, Laurie J. 1445
Phillips, Nancy H. 2864
Phinikarides, Alexander 1954
Picard, Sandrine 2087
Piccinelli, Fabio 1669, 2372
Pickel, Tobias 3500
Pierro, Marco 3482
Pieters, Bart E. 1651
Pihan, Etienne 1498
Pillai, Supriya 2403
Pistor, Paul 155, 3285
Piszczor, Michael 525
Pitalúa, Nun 632
Platzer-Björkman, C. 3269
Plessing, Lukas 178
Pleus, Albert 1835
Plochowietz, A. 1733
Podraza, Nikolas 2646
Podraza, Nikolas J. 2462, 2582, 2771, 3426, 3468
Poindexter, Jeremy 3300
Poissant, Yves 1908
Pokharel, Nikhil 831, 2514
Polojärvi, Ville 297
Poncho, Corpuz 2947
Poortmans, J. 3260
Poortmans, Jef 1233
Pop, Sergiu C. 921, 1869
Poplavskyy, Dmitry 1459, 1686
Porter, Ilana J. 417
Potamialis, C. 2457
Pötz, Sandra 178
Pouladi, S. 869
Pouladi, Sara 866
Poulsen, Peter B. 2672, 2682
Powalla, Michael 791
Previtali, Jonathan 1275
Price, Jared S. 1469
Prietl, Christine 1329

Printraza, Nikolas J. 993
Procel, P. 2073
Ptak, Aaron J. 46, 62, 2275
Puska, Martti J. 2070
Puthanveettil, Suresh E. 76
Qazi, Farah 1317
Qian, Gary 667
Qian, Shen 958
Qin, Xuefei 1594
Qiu, Botong 667
Qudsia, Syeda 1317
Quinto, Carlos 429
Quiroz, Jimmy E. 1280
Rada, Jacob 1271
Radhakrishnan, Hariharsudan
   Sivaramakrishnan 1233
Raghavan, Srinivasan 837, 841, 986, 2395, 2399
Ragunathan, Gautham 1181
Rahman, Mosaddequr 1067
Rahn, Christopher D. 1469
Raiker, Gautam A. 3073
Raj, Samuel 284, 496, 499
Rajan, Grace 182, 735, 807, 2298, 2446, 2646
Rajbhandari, Pravakar P. 989
Rajput, Amit Singh 499
Raju, T. Bhim 1034
Raker, David 2926
Rale, Pierre 1289, 3147
Ramakumar, Rama 1122
Ramamurthy, Praveen C. 1614, 2811
Rambabu, Sugguna 3478
Ramic, Zekija 2776
Ramírez, A. 503
Ramirez, A.A. 433
Ramírez, E. A. 433, 503
Ramos, Helena Geirinhas 3178
Ramos, Javier 2255
Ramprasad, Sumukh 496
Ramu, Govind 1275, 2263
Rancoita, P.G. 541
Rand, James 925
Ranjan, Rajeev 2395, 2399
Ranjan, Upasna 2811
Ranjbar, S. 3260
Ransome, Steve 652
Rao, Arun D. 1614
Rao, B.V. 1898
Rao, Rajesh 363, 2506
Rao, Roshan R 604
Raorane, Neha 1755
Raote, Yojak 1022
Rashkin, L. 3224
Rastogi, A.C. 3279
Rathi, M. 869
Rathi, Monika 866, 2368

AUTHOR INDEX

Rathore, Sudharm 2902
Rau, Uwe.......................................1651, 2114
Raupp, Christopher 1984
Ravindra, M. 76
Ravindra, Pramod 2395, 2399
Raychaudhuri, S.....................................1733
Razooqi, Mohammed A. 2462
Recart, Federico 2677
Reddy, Anurag 3528
Reddy, K.S. 2858
Reddy, Rekha 3524
Reed, S. 869
Reedy, Robert C......................................881
Reese, M. O. 138
Regalado-Pérez, E. 142
Reichel, C. 2064
Reichert, Andreas 2407
Reinders, Angèle 2109
Reindl, T. 1140
Reise, Christian 2267
Rejon, V. 1691, 2457
Ren, Zekun 284, 2532, 2744
Ren, Zhiwei..958
Reno, Matthew J. 1555, 1567, 1573,
    1579, 2975, 3025, 3031, 3055
Renteria, E. J..281
Repins, Ingrid.............................2728, 3452
Reusser, Jean 2255
Reyes-Banda, M.G. 142
Rey-Stolle, I. 1210
Rey-Stolle, Ignacio 1204
Rhodes, Christopher 1476
Riaz, Hiba 1741
Ribeyron, P. -J.......................................1747
Ricardo, Julian Do Nascimento 3077
Rich, Geoffrey 600
Richards, J...3224
Richardson, Walter 1116
Richter, A....................................1752, 2064
Richter, Mauricio...................................3360
Riedel, Nicholas...........................2672, 2682
Rienacker, Michael 3371
Riesen, Yannick......................................3435
Rigdon, Terry B. 3448
Riggs, Brian 37
Riggs, Brian C..3245
Riley, Daniel 1537, 3155, 3184, 3348
Riley, Daniel M..........................1543, 1549
Rimmaudo, I. 1691, 2457
Rimmaudo, Ivan 1669, 2372
Rincon-Charris, Amilcar A.....................2963
Ringel, Steven A. 215, 2079, 2446, 2554
Ringleb, Franziska 3396
Rivera, Eduardo I. Ortiz 3044
Riverola, Alberto....................................1339

Robert, Sofie 1804
Roberts, Jesse 3083
Robertson, John 37
Robertson, Kyle W. 1294
Robinson, Charles D. 3155
Rochat, Raphael 326
Rocheleau, Richard E.............................1061
Rockett, A. 30
Rockett, Angus.................182, 1400, 2446
Rockett, Angus A. 1511
Rodrigues, Sandy...................................3178
Rodriguez, D. J. 2031
Rodríguez, Diego J. 1143
Rodríguez, Pedro 2677
Rodríguez-Gallegos, Carlos D. 2318
Roest, Stefan 3462
Roeth, A. J. 14
Rogers, John A. 1469
Rohatgi, Ajeet 333, 940, 1807, 1824, 1838
Roland, Paul J..1030
Roller, John 508
Romanin, Vince 37, 3245
Romeo, Alessandro.................752, 1669, 2372
Ronoh, Geoffrey Kibiegon 970
Rooijakkers, Tom T.H. 1081
Ropp, Michael...............................2147, 3020
Rosales-Ascnsio, Enrique......................3240
Rose, Volker 2179, 3300
Ross, N. 3269
Rotoli, P. 1747
Rounsaville, Brian 940, 1807, 1824
Routhier, Alexander F. 3088
Rowell, David...3524
Roy, Sam 2358
Roy, Tatiana A............................521, 558
Royer, Fabien..558
Rozza, Davide 541
Rubbard, Seth M. 3468
Ruffini, Leia 2453
Ruiz, Carmen M. 2255
Ruiz, E. O. Ángel 1990
Rummel, S. 467
Rupp, B. 1733
Ruppalt, Laura B.....................................873
Russell, Annie 1094
Russell, Richard...........................2227, 3435
Russell, Thomas C.R. 2236
Ruth, Daniel...2301
Ryou, J. 869
Ryou, Jae-Hyun 866, 2368
Saavedra, Michael 1473
Sablon, Kimberly 1181
Sabnis, Sanjeev 2849
Sacchetto, Davide 3256
Sachenko, A.V...............663, 1025, 1811

AUTHOR INDEX

Sachenko, Anatoliy V................................690
Sáenz, M.J...643
Saetre, Tor Oskar 685
Sahayaraj, S..3260
Sahli, Florent......................................3256
Sahraei, N. 648
Sai, Hitoshi..................................381, 1333
Saifullah, Muhammad 810
Sainsbury, Cassidy......................2692, 2765
Saito, K. 3
Saito, Tomohiro.....................................931
Saive, Rebecca.........................1589, 2236
Sakamoto, Katsuyoshi 85
Sakamoto, Norihiko.............................1268
Sakurai, Takeaki....................33, 160, 900
Salamo, Gregory J. 3370
Salavei, Andrei 2372
Salazar, J. 370
Salazar-Duque, John E. 2963
Salo, Kristian 1494
Salome, Pedro.......................................796
Salpakari, Jyri.....................................3236
Salvetat, Thierry 2492
Samoilenko, Yegor 1697
Sampath, W.S........................980, 2736
Sampath, Walajabad 424, 785, 1674
Sampath, Walajabad S. 3417
Sample, Tony.......................................1275
Sampson, Matthew D. 6, 1256
Samuelson, Lars 2502
Samundsett, C.......................................3295
Samundsett, Chris.................................2076
Sánchez, Yudania 155
Sánchez-Pérez, P. A. 1959, 1990
Sanchiz, Joaquín 429
Sandeep, K..396
Sang, Baosheng....................................1455
Sang, Shiyu 472, 1430, 2918
Sangjeong, Myeong 356
Sankaran, M..76
Sankin, I. 2816
Santana, G. 370, 2342, 2614
Santana-Rodríguez, G. 670
Santbergen, Rudi 2605
Santhanam, Parthiban...........................2185
Santos, M. B. 14
Santoyo-Salazar, J.................................370
Saraf, Akash 761
Saraswat, Krishna.................................1835
Sargent, Edward H. 3129
Sarmah, Nabin 2858
Sarvari, Hojjatollah 1044, 2432
Sarwar, Jawad......................................1888
Sasaki, A. 1003
Sastry, O. S. 3478

Sato, Daisuke 1743
Sato, Shin-Ichiro.................................... 562
Sato, S-I. 552
Satzinger, Valentin 178
Saucedo, Edgardo.................155, 3265, 3285
Savin, Hele 944, 1494, 3236
Sawallich, S. 3150
Sayed, Islam E.H. 2195
Sayyah, Arash 2891
Scaccabarozzi, Andrea............................ 1289
Scarpulla, M.A. 1656, 1661
Scarpulla, Michael A. 175, 802, 1679, 2467
Schäfer, Nicolas 2216
Schaller, Richard D. 6
Scheiman, David 1838, 3514
Schelhasl, Laura T. 2176
Scheltens, Frank J. 966
Schenller, E.J. 1701
Schermer, J. 1189
Schindler, F. 2064
Schitthelm, F. 1752
Schlemmer, James................................. 1104
Schmid, Martina 3396
Schmidt, Jan 3371
Schmidt, Thomas................................... 3396
Schmieder, Kenneth J. ...210, 259, 272, 873, 2091, 2506
Schnabe, Thomas................................... 2216
Schnabel, Erdmut................................... 3488
Schnabel, Manuel ... 1817, 2482, 2543, 3254, 3371, 3439
Schnabel, T. 3260
Schnabel, Thomas 791
Schneider, Kevin 1476
Schneller, Ej 389
Schneller, Eric J. 2839, 2897, 3448
Schoenfeld, Winston 2839
Schoenfeld, Winston V. 322, 1804
Schoenfelder, Stephan............................ 2622
Schoenwald, David 2969
Scholl, Jonathan A................................. 1549
Schoop, Urs 1455
Schorch, M. 1752
Schriemer, Henry P................................ 1094
Schubert, M. C. 3135
Schubert, Martin C. 1329
Schulte, Kevin 62
Schulte, Kevin L. 46, 232, 2275
Schulte-Huxel, Henning 1366, 2543, 3371
Schulz, Gerd.. 914
Schulze, Patricia S.C. 1253
Schwabe, Hartmut................................. 2622
Schweiger, M. 107
Sclj, Josefine 619
Scofield, A.C. 2755
Scolari, Enrica 1415
Sculati-Meillaud, Fanny.......................... 2794

AUTHOR INDEX

Seif, Johannes P. 3435
Seigneur, Hubert 2839, 2897
Sellami, Nazmi 2858
Sellers, Andrew 2926
Sellers, Diane G. 3384
Sellers, I. R. 14
Selvamanickam, V. 869
Selvamanickam, Venkat 866, 2368
Semichaevsky, Andrey 319
Sen, Fatih G. 2759
Senaud, L.-L. 50
Sengar, Brajendra S. 2345
Sengupta, Manajit 116, 1169
Senthilarasu, S. 2858
Sepeher, Mohsen M. 1094
Sera, Dezso 1421, 2682
Serra, João M. 1979
Sethia, Saurabh 2902
Seydel, Elisabeth 1682
Shafarman, William N. 26
Shah, S. 1350
Shahirinia, Amir 3092
Shanmugam, Vinodh 2318
Sharma, Ashok K. 396
Sharma, Romika 3300
Sharma, S. 2094
Sharps, Paul 42, 525, 2099
She, Hui 1863
Shen, Chang-Hong 1627
Shen, Zeqing 3393
Shephard, Les E. 1116
Shervin, Kaveh 1452
Shervin, Shahab 866
Shetty, Nishit 876
Shi, Jianwei 1820
Shi, Jiatiwei 1322
Shi, Xuanyi 3220
Shi, Zhan 1037
Shibata, Hajime 33, 1268
Shieh, Jia-Min 1627
Shigekawa, Naoteru 2548, 2551
Shih, Cheng-Hao 2035
Shih, I. 2388
Shih, Ishiang 1299
Shima, D. M. 281
Shimura, H. 1003
Shin, Hyun-Beom 244
Shin, Myunghun 2631
Shin, Seunghyun 935
Shin, Woo Jung 385
Shinde, O.S. 389, 1701
Shirasawa, Katsuhiko 904, 931
Shkrebtii, Anatoli I. 690
Shoji, Yasushi 10
Shore, Andrew 437

Shrestha, Niraj 1030
Shrestha, Santosh 696, 2186
Shu, Chia-Jhe 1606
Shu, Jinn-Kong 1606
Shubhrant, Abhishek 2902
Si, Fai Tong 2605
Siddiki, Mahbube K. 1018, 1110
Sidhu, Navjot Kaur 3279
Siebentritt, Susanne 151, 2054, 2205, 2478
Siepchen, Bastian 761
Sikchang, Hyo 356
Silva, Francois 464, 1237
Silva, José A. 1979
Silvaggio, Amber C. 2554
Silverman, Timothy 1259, 1893
Silverman, Timothy J. 1400, 2771, 3452
Simon, John 42, 46, 62, 1201, 2275
Simon, Kirby 876
Simpson, L. 2280
Simpson, Lin 2294
Simpson, Lin J. 1893, 2789
Sinapis, Kostas 1081, 1090
Singh, Aparna 2902
Singh, Ashish 1034
Singh, Ashish K. 1704
Singh, Hemant K. 1995, 3478
Singh, Rajeev 2762
Singh, Rhythm 1151
Singh, Rubina 1855
Singh, Sukvhinder 2227
Singlr, Vijay P. 2432
Sinha, Archana 3478
Sinha, Parikhit 2005
Sinisuka, Ngapuli I 3102
Sink, Joseph 3333
Sinton, Ronald 2707
Sinton, Ronald A. 2692, 2765
Sio, H. C. 3295
Sio, Hang Cheong 3300
Sites, James R. 164, 1308
Slocum, Michael 677
Slocum, Michael A. 18, 202, 206, 222, 831, 1184, 2084, 2514, 3468
Slooff, Lenneke H. 1081
Smaglik, Nathan 831, 2514
Smestad, Greg P. 2858
Smith, Benjamin 1134
Smith, Brittany L. 18, 1184
Smith, David J. 2573
Smith, Mathew 2941
So, Won-Shup 487
Soares, Gabriela De Amorim 2875
Sodabanlu, Hassanet 854
Söderström, T. 1752
Sofia, Sarah E. 1264

AUTHOR INDEX

Sogabe, Tomah 85, 712
Sokolovskyi, I.O. 663, 1025, 1811
Sokolovskyi, Igor O. 690
Solanki, Chetan S. 3478
Solanki, Chetan Singh 1850
Soltanmohammad, Sina............................26
Soman, Anishkumar 1828
Søndergaard, Sissel Tind 2651
Song, Dengyuan......................................1430
Song, Hee-Eun 356, 1758
Song, Myungkwan 2631, 2634
Song, Tao...1308
Song, Zhaoning........170, 730, 748, 815, 1030
Sonp, Hee-Eun.......................................2723
Sood, Neeru 2858
Sossan, Fabrizio 1415
Soudachanh, A. L....................................281
Sozzi, Giovanna.....................................2205
Spandana, B..396
Spataru, Sergiu...............1421, 2682, 2819
Spaulding, David 820, 1686
Spertino, Filippo....................................3096
Spiering, Stefanie 791
Spinelli, P...3150
Spooner, Ted 1275, 2263
Sreekumar, Nimisha 1755
Sridharan, Akirt..999
Sriramagiri, Gowri 658, 1196
Srivatsan, R. 120, 589
Stark, Cameron......................................1855
Starkl, Hannes...178
Steeman, Rob...337
Steenhoff, Volker 3388
Stefancich, Marco 1946
Steijvers, Henk 2875
Stein, Joshua...................1537, 3333, 3348
Stein, Joshua S...........1543, 3155, 3161, 3184, 3488
Steiner, Myles A. 42, 47, 232, 1201, 2195, 3254
Steinfedt, Jeff...525
Stender, C..3511
Stender, Christopher L...........................3524
Stephan, Jack 2124
Stevens, Margaret 701
Steward, Malia 1037
Stewart, J. 3224
Stika, K. 2312
Stiles, Phil 2833
Stoddard, Nathan 2610
Stokes, Adam 1381, 2887
Stolt, L..30
Stone, Kevin H.......................................2176
Stradins, Paul881, 1491, 1777, 1801, 1817, 1832, 2242, 24
Stradins, Pauls.............................2482, 2702
Strandberg, Rune 706, 2651
Stride, John A..2392

Stuart, Thomas 2926
Stuckelberger, J. 2073
Stuckelberger, Michael 2610, 2854, 3309
Stuckelberger, Michael E........................2179
Stueve, Bill 2808, 2923
Sturm, James C. 1773
Stutz, Elias Z. 944
Su, Bojie 392, 1430, 1873, 2918
Su, Chengfeng.............................392, 1873
Subbiah, Jegadesan 1342
Subedi, Indra 2771, 3468
Subedi, Kamala Khanal 781
Sudbury, Benjamin A. 1322
Suga, Mitsunobu 567
Sugaya, Takeyoshi 562, 861, 1724
Sugimoto, Hiroki 160
Sugiyama, Masakazu 854, 2201, 2524, 2538, 3528
Sugiyama, Mutsumi 192
Sugiyama, Ryo 712
Suhana, Hadi 3102
Sumita, Taishi 3506
Sun, C. 1485
Sun, Ce 2759
Sun, Chang 3300
Sun, Chenguang......................................1241
Sun, Kaiwen 2213
Sun, Qiang 1648
Sun, Qiming 3129
Sun, S. 869
Sun, Sicong..2368
Sun, Wen-Cheng 2227
Sun, Xiaolin...2656
Sun, Xingshu......................1055, 1259, 1904
Sun, Yaojie 2048
Sun, Yubo...2467
Sun, Yukun 2291
Sun, Zeming 937
Supplie, Oliver 2524, 2538
Surya, Charles 958
Sutou, Yuji 931
Sutterlueti, Juergen 652
Suzuki, Ryota 1504, 2588
Swain, Santosh K....................................3422
Swartz, Craig H. 2473, 2749
Sweatt, William 1473
Syu, Hong-Jhang 1009
Szabo, Sandor 2167
Szlufcik, Jozef....................1233, 2227, 3435
Tabet, Nouar.......................963, 1058, 3435
Tacconi, Mauro 541
Tachibana, Shoji.....................................1504
Tadese, Alemu 1104
Tadesse, Alemu 132, 1132, 1427
Tae, Christian 1835
Taekjeong, Kyung 356

AUTHOR INDEX

Takahashi, Akiko 2906
Takahashi, Isao 1765, 1794
Takahashi, Takuji 455
Takahashi, Yasuhito 1973
Takamoto, Tatsuya 3506
Takato, Hidetaka 381, 904, 1724, 3323
Takenouchi, T. 441
Tamaki, Ryo 10
Tamboli, Adele 3254, 3371
Tamboli, Adele C. 578, 2482, 2488,
      2536, 2543, 3381, 3406
Tamizhmani, Govindasamy 1389, 1850,
      1858, 1877, 1883, 1959, 1984, 2789, 2912
Tan, Jin 1158
Tan, Joel M. R. 3275
Tan, K.H. 1210
Tan, Xuehai 761
Tanahashi, Katsuto 3323
Tanahashi, Tadanori...1268
Tanaka, Aki 2642
Tanaka, T. 3
Tanaka, Takahiro 2947
Tang, Chiu C. 2423
Tang, Houjun 1100
Tang, Mingchu 3370
Tang, Tao 2558
Tanke-Pedretti, Anna 1473
Tao, Meng 385, 608
Tao, Yuguo 1824
Tappan, Ian A. 2864
Tassone, Christopher J..2176
Tatapudi, Sai 1850, 1858, 1877, 1959, 2912
Tatapudi, Sai Ravi Vasista 2789
Tatavarti, Rao 1184, 2084
Tatavarti, Sudersena Rao....................................1181
Tate, John Keith 333
Tayagaki, T. 3
Tayyib, Muhammad 2776
Tchemycheva, Maria 1289
Tedeschi, Giampiero 2372
Teena, Percis 3113
Tennyson, Elizabeth M.............................1508, 2443
Terheiden, Barbara...1824
Terukov, E.I. 1025, 1811
Teubner, Thomas 3396
Teymouri, Arastoo 2403
Thanh, Nguyen Cong 1765
Thankalekshmi, Ratheesh R. 3279
Theelen, Mirjam...2875
Theigi, San 881
Theingi, San 1832
Theocharides, Spyros................................1163, 3107
Therrien, Francis 1579
Thibeault, Brian 315
Thimsen, Elijah..876

Thompson, Christopher......................................1196
Thompson, Corey S..1869
Thon, Susanna M...667
Thorseth, Anders...................................... 2672, 2682
Thorsteinsson, Sune 2672, 2682
Thway, Maung 284, 2744
Tidwell, Steven...1086
Timò, Gianluca 290
Tirumalai, Tejas 2923
Tischler, Joseph G...873
Titus, Jochen 820
To, Alexander 517
To, B. 2280
Toberer, Eric S. 2536, 3406
Todorov, Teodor 1441
Togay, M. 2457
Togay, Mustafa 146, 186
Tomasi, A. 50
Tomasi, Andrea 3435
Tomasulo, Stephanie 2091
Tonic, Marko 1346
Toor, Fatima 346, 1537, 1543, 3184, 3333, 3348
Toprasertpong, Kasidit................................ 2201, 2524
Torralba, Encarnacion...3402
Tous, Loïc 2227, 3435
Tracy, Jared.. 3190, 3200
Traverse, Christopher..2124
Trout, T. John...2312
Trupke, Thorsten 66, 420, 2707, 3304, 3315
Tsafarakis, Odysseas 1090
Tsai, Cheng- Ying 3366
Tsai, Jia-Lin 1606
Tsai, Jia-Ling 294
Tseng, Zong-Liang 367
Tsutsumi, S. 3
Tu, Wei-Chen 1051
Tucher, Nico.................................... 352, 1253, 2511
Tukiainen, Antti 297, 2520
Tuminello, F. 3511
Tummala, Abhishiktha 2912
Turek, Marko 2232
Turner, John A. 47
Tuteja, Mohit 1511
Tyagi, Astha 716
Tyler, Kevin 301
Tyson, Tom 925
Tzolov, Marian 1040
Ubukata, Akinori..861
Ueda, Kohsuke 3506
Ueda, T. 1003
Uematsu, Takumi 1950
Ulbricht, Christoph 1360
Ulicná, Sona 146, 186
Uma, B. R. 76
Umishio, Hiroshi 381

AUTHOR INDEX

Unold, Thomas 3396
Unsur, Veysel......................................888
Upadhyaya, Ajay D.....................940, 1807
Upadhyaya, Vijay D.............................333
Upadhyaya, Vijaykumar 940, 1807, 1824
Uprety, Prakash 3468
Urbano, J. Antonio..............................632
Uruena, Angel...........................2227, 3435
Usami, Noritaka 1765, 1794
Utsunomiya, Satoshi 904
Vadiee, E...1603
Vadiee, Ehsan........................305, 827, 1841
Vagidov, Nizami Z. 531
Vähänissi, Ville..................................1494
Vaida, Mihai E. 417
Vaidya, Nina 512, 521, 558, 572, 1248
Vaisman, M. 3376
Vaisman, Michelle......................578, 3381
Vaissiére, Nicolas 2528
Valderrama, Nicolas 1893
Valdivia, Christopher E......................1094
Van Aken, Bas B. 3462
Van Alsburg, Jane..............................1455
Van De Loo, Bas W.H.........................1817
Van Der Heide, Arvid.........................3343
Van Hest, Maikel F.A.M..........2482, 3371, 3439
Van Sark, Wilfried..............................3014
Van Sark, Wilfried G.J.H.M. 1090, 3167
Vandamme, Nicolas 2453
Vandervelde, Thomas E. 701
Vanka, S...2388
Vanka, Srinivas.................................1299
Vansant, Kaitlyn 1922, 3452
Vargas, Carlos...................................3290
Vasi, J...1995
Vasi, Juzer 1850, 3478
Vasileska, D. 1603, 2816
Vasileska, Dragica...................1790, 1797
Vasilyev, Leonid A. 3448
Vasudevan, Saravanan 2172
Vauche, Laura..........................2492, 2562
Vedde, Jan 2682
Veettil, Binesh Puthen 2392
Vehse, Martin 3388
Veinberg-Vidal, Elias.................2492, 2562
Veith-Wolf, Boris 1366
Velappan, Krishnakumar......................761
Venizelou, Venizelos 276, 1163, 1941, 3107
Verbitskiy, V.N..................................1811
Verlinden, Pierre J....................1922, 2220
Vermang, B..3260
Vermang, Bart....................................2209
Verschac, Rodrigo..............................1175
Vetter, E..1752
Viana, Marcelo Machado 1917, 2307

Vignola, Frank..................................1169
Vijh, Aarohi 3520
Vilcot, Jean-Pierre............................3402
Vincent, Nina 1893
Vines, L. 3269
Vinogradova, Tatiana 512
Vinogradova, Tatiana G. 521, 558, 572
Virtuani, Alessandro 1395, 2104, 2794
Vlasiuk, V.M. 1025
Vlasyuk, V.M. 1811
Vleugels, J. 3260
Voarino, Philippe....................2492, 2562
Vogt, Malte Ruben 1366
Von Gastrow, Guillaume 944
Voroshazi, Eszter...............................3343
Voss, Henrik......................................2682
Waddle, John M. 309, 3123
Wade, Andreas 2005
Wagner, Sigurd 1773
Waiis, J.M. 2457
Waldhauser, Wolfgang 1329
Walker, Don.........................93, 512, 1248
Walls, J.M. 1691, 3430
Walls, John 1674
Walls, John M......................146, 186, 752, 2349
Walls, John Michael...........................2827
Walls, Michael 424
Walter, Arnaud 3256
Walters, Joseph 2839, 2897
Walters, R. J. 845
Walters, Robert 3514
Walters, Robert J. 210, 247, 259, 272, 373, 873, 1838, 2091, 2506
Waltmger, A......................................1752
Walukiewicz, W......................................3
Walukiewicz, Wladek.........................1204
Wan, Kai-Tak 1933, 2000
Wan, Ronghua 226
Wan, Yimao 59, 2076
Wang, Ao 1937, 2823
Wang, Baomin 1469
Wang, Changlei 993
Wang, Da-Wei 3220
Wang, Deng..2048
Wang, Feng 1044
Wang, Fumei.............................392, 1937, 2823
Wang, Haotian 1342
Wang, He. 226, 392, 1430, 1648, 1873, 1937, 2823, 2918
Wang, Hongfeng 1215
Wang, Laidong 385
Wang, Mu 3370
Wang, Q. 1733
Wang, Rui 1100
Wang, Shenghao 160
Wang, Shizhen 976

AUTHOR INDEX

Wang, Sisi 2600
Wang, Teng-Yu 2660
Wang, Xiaohui 2432
Wang, Y. 1733
Wang, Y. D. 1733
Wang, Yan 1922
Wang, Yichen 1299
Wang, Yiwang 1100
Wang, Yongqian 2220
Wang, Yu 1933, 2000
Wang, Yu-Cian 2498, 2566
Wang, Zigang 1922
Ward, J. Scott 3254
Warmann, Emily 1248
Warmann, Emily C. 512, 521, 558, 572
Warner, Jeffery. H. 2091
Warren, Emily 3371
Warren, Emily L. 578, 2482, 2488, 2543, 3381
Washington, Lori 3520
Washio, Hidetoshi 3506
Watanabe, Kentaroh............................854, 3528
Watanabe, Yasuyuki 613
Waters, Martin 2923
Watson, S. 648
Watthage, Suneth C. 170, 730, 748, 815, 1030
Watts, John L.R. 2762
Weeber, Arthur 2875
Weick, Clément 2492, 2562
Weigand, William 1790
Weiss, Charlotte 83, 2407
Weiss, Dirk 1264
Weiss, Karl-Anders 2836
Wen, Ching-Chang 329
Wen, Xiaoming 696
Wenham, Stuart R. 2576, 2600
Werner, Florian 2205, 2478
Werner, Jérémie 55, 3256
West, Bradley M. 2179, 3309
Western, N. J. 953
Western, Ned J. 948
Wheeler, Tobias 1476
Whipple, Steven 88
Whiteside, V. R. 14
Wibowo, A. 3511
Wibowo, Andre 1181, 1184, 2084
Wicaksono, S. 1210
Widén, Joakim 3067
Wieghold, Sarah 3300
Wienands, Karl 1253
Wiese, Martin 1531
Wille-Haussmann, Bernhard 126
Williams, J. 1603
Williams, Joshua J. 305, 582
Williams, R. 490
Wilson, Gregory 3236

Wilson, Marshall 322
Wilt, David M. 88, 102, 301, 531
Wilt, Sam 301
Wilterdink, Harrison 2692, 2765
Winkler, Kristina 1253
Winkler, Thilo 3500
Wirsching, Sven 3500
Wirth, Harry 1531
Wischmann, Wiltraud 2058
Wissen, A. 1752
Witte, Wolfram 2205
Witteck, Robert 1366
Wohlgemuth, John 1275
Wojtowicz, Anna 164
Wolden, C. A. 138
Wolden, Colin A. 1697
Wolf, Martin 2692
Wolffersdorff, Paul 595
Wong, Johnson 499, 3113
Wong, Johnson Kai Chi 496
Wong, Lydia H. 3275
Woodhouse, Michael 1259
Woods, Jason 1893
Woods-Robinson, Rachel 3410
Worrell, Ernst 3014
Wright, Lewis D. 146, 186
Wu, Gordon 96
Wu, J. 1189
Wu, Jiang 3370
Wu, Kuen-Yi 911
Wu, Po-Ching 1623
Wu, Ruei-Ying 1635
Wu, Shang-Hsuan 1051
Wu, Teng-Chun 448, 476
Wu, Yonggang 1594
Wu, Yuh-Renn 294
Wu, Zhuopeng 1241
Würfel, Uli 1253
Wyrsch, Nicolas 3435
Wyss, P. 2073
Xia, Hongze 2392
Xia, Zihuan 1594
Xiao, C. 1312, 2785
Xiao, Chuanxiao 62, 1371
Xiao, T. Patrick 2185
Xiao, Zhi Bin 1648
Xie, Yu 116
Xiong, Gang 1193, 2473
Xiong, Zhen 2220, 3304
Xu, Jun 2656
Xu, Ling 2656
Xu, Lu 1737
Xu, Menglei 1233
Xu, Qi 37, 3245
Xu, Qianfeng 2285

AUTHOR INDEX

Xu, Tao 2251
Xu, Xiaojie 3410
Xu, Zhaoran 59
Xue, Muyu 1835, 2732
Yablonovitch, Eli.................................2185
Yachi, Toshiaki.......................................613
Yadav, Karan Shishir 2902
Yadav, Tarun S. 396
Yakes, Michael K. 873, 2091
Yamada, Noboru...................1724, 1743
Yamada, Nobuyuki.................................2906
Yamagami, Takeru 192
Yamagoe, K. 441
Yamaguchi, Hiroshi.................................3506
Yamaguchi, Koichi..................................712
Yamaguchi, Masafumi........359, 412, 1479,
1711, 1714, 1743, 2498, 2548, 2566
Yamaguchi, Seira.................................1385
Yamamichi, Masaaki 1275, 2263
Yamaya, Haruki......................................2159
Yan, Chang..2213
Yan, Di..2076
Yan, Yanfa.............. 771, 993, 1643, 2443, 2473, 3426
Yancey, Billy...2128
Yanchilin, Anton 585
Yang, Fan...2656
Yang, Guangtao 2605
Yang, Hao-Yu....................343, 367, 893, 2664
Yang, Hong............. 392, 1430, 1873, 1937, 2823, 2918
Yang, Jianfeng..2392
Yang, Mohshi..907
Yang, Peter..1100
Yang, Shuying 2697, 3456
Yang, X. 2785
Yang, Yang 2220
Yang, Yi Tong 1648
Yang, Yun-Chie.............................893, 2664
Yang, Zhihao 74
Yao, Li You...1648
Yao, Y. 869, 1752
Yao, Yangyi 1048
Yao, Yao 866, 2368
Yarnaquchi, Koichi....................................85
Yates, Peter..1445
Yaung, K. Nay 3376
Yaung, Kevin Nay 284, 2744
Ye, Feng 2220
Ye, J. 2785
Ye, Qilin...948
Yeh, Chun-Ming.......................................911
Yellowhair, Julius 2870
Yellowhair, Julius E................................2891
Yi, Chuqi..2569
Yilmaz, S...3430
Yoo, Chang Youn 399

Yoon, Howard W......................................437
Yoon, Jongseung 549, 2291
Yoon, S. F. 1210
Yoon, Woojun 373, 1838
Yoshiba, Shuhei 1769
Yoshino, Kenji 1679
Yoshita, M. 1003
Yoshita, Masahiro 2781
You, Bang-Jin 367
You, Liang-Chian 1635
Young, David 46
Young, David L...........1817, 1832, 2275, 3254
Young, James L. 47
Young, Steven 582
Youssef, Amanda........................1491, 2242, 3300
Youtsey, Christopher 3524
Yu, Edward T. 363, 1181
Yu, Jia 2453
Yu, K. M. 3
Yu, Kin Man 1204
Yu, Li-Chieh 3204
Yu, Linwei 2656
Yu, Ming 1193
Yu, Peichen 294, 1610, 1635
Yu, Pei-Chen 1606
Yu, Sun 1522
Yu, Zhengshan J..............1228, 1317, 1322, 2039, 3250
Yuan, Bo 315
Yuan, Lin 2392
Yue, Yao 93
Yun, Jae Ho 810
Zachariah, S. 1995
Zachariah, Sachin 2799, 2849, 3478
Zahler, James..1463
Zahler, James M......................................3245
Zakaria, Naimi...487
Zamora, Rachid Darbali 3044
Zang, Kai...1835
Zapalac, G. 2414
Zapalac, Geordie 820, 3327
Zauner, Andy..1237
Zech, Tobias..1531
Zelenina, Anastasiya 2054, 2478
Zeman, Miro 2605
Zeng, Guoping 907
Zeng, Xulu 1286
Zeyu, L. 1781
Zhai, Yonghui 472
Zhan, Tien-Chien 294
Zhang, Bao 226
Zhang, C. 1603
Zhang, Chaomin...................240, 827, 1215, 1841, 2573
Zhang, Guoqi 2605
Zhang, Hua 3304
Zhang, Huan 2558

AUTHOR INDEX

Zhang, Jili 1100
Zhang, Jing 3384
Zhang, Junjun 1937, 2823
Zhang, Lei 408, 1761, 1828, 2667
Zhang, Liang 2247
Zhang, Liping 1241
Zhang, Nian 2432
Zhang, Qiming 226
Zhang, Wei 255, 1193
Zhang, Weijie 820
Zhang, X. 2094
Zhang, Xiaochen 1567
Zhang, Xue 392, 1430, 1873, 2918
Zhang, Yang 195
Zhang, Yi 696, 2186
Zhang, Yong-Hang 3366, 3410
Zhang, Yufeng 198, 767, 1707
Zhang, Z. 2019
Zhang, Zhilong 2392
Zhang, Zongyi 1594
Zhangl, Xiaochen 1555
Zhao, Dewei 993
Zhao, Hui 392, 1430, 1873, 2918
Zhao, J. 1752
Zhao, Jing 1845
Zhao, Pan 1430, 1873, 2918
Zhao, Xin-Hao 3366
Zhao, Yuan 3366
Zhao, Yuetao 1044
Zhe, Liu 2744
Zheng, N. 869
Zhigunov, D.M. 1811
Zhongbiao, Ye 1044
Zhou, Guomin 472
Zhou, Hang 976, 2558
Zhou, Jian 1594
Zhou, Xiao W. 2419
Zhu, Jiang 3208
Zhu, Lin 721, 3528
Zhu, Yan 66, 3290
Zhu, Ziyao 198
Zide, Joshua M. O. 3384
Zielnik, Allen 3208
Zilles, Roberto 1917
Zilouchian, Ali 638, 2941
Zimmerman, Jeramy D. 3381
Zin, Ngwe 322
Zinaddinov, M. 2411
Zoppi, Guillaume. 742
Zubia, David 2419
Zunger, Alex 1245

IEEE
445 Hoes Lane
Piscataway, NJ 08854-4141

ISBN 978-1-5090-5606-4